SOME PHYSICAL PROPERTIES

Air (dry, at 20° C and 1 atm)
  Density   1.29 kg/m³
  Specific heat at constant pressure 1.00×10^3 J/kg·K
                                                0.240 cal/gm·K

  Ratio of specific heats (γ)   1.40
  Speed of sound   331 m/s
                                                1090 ft/s

Water (20° C and 1 atm)
  Density 1.00×10^3 kg/m³
                                                1.00 gm/cm³

  Speed of sound   1460 m/s
                                                4790 ft/s

  Index of refraction ($\lambda = 5890$Å)   1.33
  Specific heat at constant pressure   4180 J/kg·K
                                                1.00 cal/gm·K

  Heat of fusion (0° C) 3.33×10^5 J/kg
                                                79.7 cal/gm

  Heat of vaporization (100° C) 2.26×10^6 J/kg
                                                539 cal/gm

The Earth
  Mass 5.98×10^{24} kg
  Mean radius 6.37×10^6 m
                                                3960 mi

  Mean earth-sun distance 1.49×10^8 km
                                                9.29×10^7 mi

  Mean earth-moon distance 3.80×10^5 km
                                                2.39×10^5 mi

  Standard gravity   9.81 m/s²
                                                32.2 ft/s²

  Standard atmosphere 1.01×10^5 Pa
                                                14.7 lb/in²
                                                760 mm-Hg
                                                29.9 in-Hg

physics

PARTS I and II, COMBINED

BOOKS BY HALLIDAY (D.)
Introductory Nuclear Physics. Second Edition

BOOKS BY HALLIDAY (D.) AND RESNICK (R.)
Physics, Parts I and II Combined, Third Edition
Physics, Part II, Third Edition
Fundamentals of Physics

BOOKS BY RESNICK (R.) AND HALLIDAY (D.)
Physics, Part I, Third Edition

BOOKS BY R. RESNICK
Introduction to Special Relativity
Available in Paper Edition

Basic Concepts in Relativity and
Early Quantum Theory
Available in Cloth and Paper Editions

BY ROBERT EISBERG AND ROBERT RESNICK
Quantum Physics of Atoms, Molecules,
Solids, Nuclei, and Particles

physics

PARTS I and II, COMBINED
THIRD EDITION

PROFESSOR OF PHYSICS **DAVID HALLIDAY**
UNIVERSITY OF PITTSBURGH

PROFESSOR OF PHYSICS **ROBERT RESNICK**
RENSSELAER POLYTECHNIC INSTITUTE

JOHN WILEY & SONS
NEW YORK   SANTA BARBARA   CHICHESTER   BRISBANE   TORONTO

SUPPLEMENTAL MATERIAL
Student Study Aid

Available for student use with *Physics* as well as with *Fundamentals of Physics* is the *Student Study Guide,* 3rd ed. by Williams, Brownstein and Gray. ISBN 03668-4

This study guide is available from John Wiley and Sons.

Library of Congress Cataloging in Publication (Revised)

Halliday, David-
  Physics.

  Published in 1960 and 1962 under title: Physics for students of science and engineering.
  Includes bibliographical references and index.
  1. Physics   I. Resnick, Robert, joint author
II. Title.
QC21.2.R47   1977   530   77-1295
ISBN 0-471-34530-X
10 9 8 7 6 5 4

preface to
the third edition
of part one

Physics is available in a single volume or in two separate parts; Part I includes mechanics, sound and heat, and Part II includes electromagnetism, optics and quantum physics. The first edition was published in 1960 (*Physics for Students of Science and Engineering*) and the second in 1966 (*Physics*).

The text is intended for students studying calculus concurrently, such as students of science and engineering. The emphasis is on building a strong foundation in the principles of classical physics and on solving problems. Attention is given, however, to practical application, to the most modern theories, and to historical and philosophic issues throughout the book. This is accomplished by inclusion of special sections and thought questions, and by the entire manner of presentation of the material. There is a large set of worked-out examples, interspersed throughout the book, and an extensive collection of problems at the end of each chapter. Much care has been given to pedagogic devices that have proved effective for learning.

It has been eleven years since the publication of the second edition of *Physics*. During that time the book has continued to be well received throughout the world. We have had abundant correspondence with users over those years and concluded that a new edition is now appropriate.

In accordance with the increasing use of metric units in the United States and their general use throughout the world, we have greatly increased the emphasis on the metric system, using the Système Internationale (SI) units and nomenclature throughout. Where it seems to be sensible, in this transition period for the United States, we retain some features of the British Engineering system. To help the student making the transition to the SI to get a physical feeling for its units, we have stressed equivalencies between the two systems, especially in problems and worked-out examples, by frequently presenting the same data in both systems.

The entire book was carefully reviewed for pedagogic improvement, based chiefly on the experience of users and on the most recent scientific literature. As a result, we have rewritten selected areas significantly for improvements in presentation, accuracy, or physics. We have included new worked-out examples for topics or areas needing them. We have modernized all references, added new ones, and have improved many figures for greater clarity. The tables and the appendices have been expanded and updated to give newer data and more information than before. And we have added a supplementary topic on special relativity.

Major improvements have been made in the questions and problems. Overall in Part I there has been a net increase over the second edition of 35% in their number, with 430 out of the total of 1567 being new. The set of questions, now numbering 611 compared with 413 before, covers a wider range of ideas, puts somewhat more stress on current and applied topics, and contains a large increase in up-to-date useful references to the popular scientific literature. We encourage students and teachers to make use of them. As with the questions, most of the previous problems have been retained, though some have been revised for greater clarity. But 225 new tested problems have been added to Part I to improve the coverage of the material and the spread of level for the student and to give the teacher a fresher choice.

To assist students and teachers in organizing and evaluating this large number of problems, 956 now compared with 746 before, we have done several things. First, we have grouped problems within each chapter by section number; namely the first section needed to be covered in order to be able to work out the problem. Then, within each set of section problems, we have arranged the problems in the approximate order of increasing difficulty. Naturally, neither the assignment by section nor by difficulty is absolute, given different ways of solving some problems and different pedagogic values and tastes. Finally, we have coded the illustrations to the problems and have put the answers to the odd-numbered problems right at the end of these problems rather than at the end of the book.

Lastly, we have restyled the physical layout of the book to give it a less crowded appearance than formerly, making it easier now for the student to read the material, to make notations and to differentiate between the various components of each chapter (text, figures, examples, tables, quotes, references, questions, problems, and so forth).

We are grateful to John Wiley and Sons and to Donald Deneck, physics editor, for outstanding cooperation. We acknowledge the valuable assistance of Dr. Edward Derringh with the problem sets and of Mrs. Carolyn Clemente with the wide range of secretarial services required.

We hope that this third edition of *Physics* will contribute to the improvement of physics education.

January 1977
Troy, New York 12181

ROBERT RESNICK
Department of Physics
Rensselaer Polytechnic Institute

Hanover, New Hampshire 03755

DAVID HALLIDAY
3 Clement Road

preface to the third edition of part two

Physics is available in a single volume or in two separate parts; Part One includes mechanics, sound and heat, and Part Two includes electromagnetism, optics and quantum physics. The first edition was published in 1960 (*Physics for Students of Science and Engineering*) and the second in 1966 (*Physics*).

The text is intended for students studying calculus concurrently, such as students of science and engineering. The emphasis is on building a strong foundation in the principles of classical physics and on solving problems. Attention is given, however, to practical application, to the most modern theories, and to historical and philosophic issues throughout the book. This is accomplished by inclusion of special sections and thought questions, and by the entire manner of presentation of the material. There is a large set of worked-out examples, interspersed throughout the book, and an extensive collection of problems at the end of each chapter. Much care has been given to pedagogic devices that have proved effective for learning.

It has been eleven years since the publication of the second edition of *Physics.* During that time the book has continued to be well received throughout the world. We have had abundant correspondence with users over those years and concluded that a new edition is now appropriate.

In accordance with the increasing use of metric units in the United States and their general use throughout the world, we have greatly increased the emphasis on the metric system, using the Système Internationale (SI) units and nomenclature throughout. Where it seems to be sensible, in this transition period for the United States, we retain some features of the British Engineering system.

The entire book was carefully reviewed for pedagogic improvement, based chiefly on the experiences of users—, students and teachers,—and on the most recent scientific literature. As a result, we have rewritten selected areas significantly for improvements in presentation, accuracy, or physics. We have included new worked-out examples for topics or areas needing them. We have modernized all references, added new ones, and have improved many figures for greater clarity. The tables and the appendices have been expanded and updated to give newer data and more information than before. And we have added a supplementary topic on special relativity, in which the applications of this theory, scattered throughout Parts One and Two, are brought together as a cohesive whole.

Some subjects, not included in the second edition, are treated significantly in this third edition of Part Two. These include semiconductors, mutual inductance, earth magnetism, radio astronomy, virtual objects, and optical instruments. The long chapter on electromagnetic oscillations of the second edition has been divided into two chapters here, with extensive rewriting for greater clarity, and an entirely new chapter on alternating currents, for which there has been much demand, has been added.

As in Part One, we have made major improvements in the questions and the problems. In Parts One and Two combined, the number of questions has increased by 57%, from 778 in the second edition to 1219 in the present edition. For the problems the increase is 29%, from 1441 to 1864. Both problems and questions have been carefully edited, and most of the new ones have been classroom tested.

To assist students and teachers in organizing and evaluating the large number of problems, we have done several things. First, we have grouped problems within each chapter by section number, namely the first section needed to be covered in order to be able to work out the problem. Then, within each set of section problems, we have arranged the problems in the approximate order of increasing difficulty. Naturally, neither the assignment by section nor by difficulty is absolute, given different ways of solving some problems and different pedagogic values and tastes. We have coded the illustrations to the problems and have put the answers to the odd-numbered problems right at the end of these problems rather than at the end of the book. Finally, we have blended the supplementary problems, which appeared at the end of Part Two in the second edition, with the problems at the end of each chapter.

We have restyled the physical layout of the book to give it a less crowded appearance than formerly, making it easier now for the student to read the material, to make notations, and to differentiate between the various components of each chapter (text, figures, examples, tables, quotes, references, questions, problems, and so forth). We continue the practice of using somewhat reduced print for material which, in the context of a chapter, is of an advanced, specialized, or historical character.

We are grateful to John Wiley and Sons and to Donald Deneck, physics editor, for outstanding cooperation. We acknowledge the valuable assistance of Dr. Edward Derringh with the problem sets and of Mrs. Carolyn Clemente with the wide range of secretarial services required. We are indebted to the many teachers and students who have sent us constructive criticisms of the 1966 edition and particularly to Robert P. Bauman, Kenneth Brownstein, Robert Karplus, and Brian A. McInnes, who have advised or assisted us in many ways. We hope that this third edition of *Physics* will contribute to the improvement of physics education.

Hanover, New Hampshire 03755   DAVID HALLIDAY
                                       3 Clement Road

January 1978   ROBERT RESNICK
Troy, New York 12181   Department of Physics
                                       Rensselaer Polytechnic Institute

contents

18
FLUID DYNAMICS
385

19
WAVES IN ELASTIC MEDIA
404

20
SOUND WAVES
433

21
TEMPERATURE
457

22
HEAT AND THE FIRST LAW OF THERMODYNAMICS
475

23
KINETIC THEORY OF GASES—I
497

24
KINETIC THEORY OF GASES—II
522

physics

PARTS I and II, COMBINED

physics

PART I

1

measurement

The building blocks of physics are the physical quantities that we use to express the laws of physics. Among these are length, mass, time, force, velocity, density, resistivity, temperature, luminous intensity, magnetic field strength, and many more. Many of these words, such as length and force, are part of our everyday vocabulary. You might say for example: "I will go to any *length* to satisfy you as long as you do not *force* me to do so." In physics, however, we must define words that we associate with physical quantities, such as force and length, clearly and precisely and we must not confuse them with their everyday meanings. In this example the precise scientific definitions of length and force have no connection at all with the uses of these words in the quoted sentence.

We say that we have defined a physical quantity such as mass, for example, when we have laid down a set of procedures, a recipe if you will, for measuring that quantity and assigning a unit, such as the kilogram, to it. That is, we set up a standard. The procedures are quite arbitrary. We can define the kilogram in any way we want. The important thing is to define it in a useful and practical way, and to obtain international acceptance of the definition.

There are so many physical quantities that it becomes a problem as to how to organize them. They are not independent of each other. For a simple example, a speed is the ratio of a length to a time. What we do is select from all possible physical quantities a certain small number that we choose to call basic, all others being derived from them. We then assign standards to each of these basic quantities and to no others. If, for example, we select length as a basic quantity, we choose a standard called the meter (see Section 1-3) and we define it in terms of precise laboratory operations.

1-1
THE PHYSICAL QUANTITIES, STANDARDS, AND UNITS

Several questions arise: (a) How many basic quantities should be selected? (b) Which quantities should they be? (c) Who is going to do the selecting?

The answers to the first two questions are that we select the smallest number of physical quantities that will lead to a complete description of physics in the simplest terms. Many choices are possible. In one system, for example, force is a basic quantity. In the system that we select (see Section 1-2) it is a derived quantity.

The answer to the third question depends on international agreement. The International Bureau of Weights and Measures, located near Paris and established in 1875, is the fountainhead for these matters. It maintains contact with standardizing laboratories throughout the world, our own National Bureau of Standards being an example.* Periodically the General Conference on Weights and Measures, an international body, meets and makes its resolutions and recommendations. Its first meeting was in 1889 and its fourteenth meeting was in 1971.

Once you have set up a basic standard, for length say, you must also set up procedures that allow you to measure the length of any object by comparing it with the standard. This means that the standard must be accessible. Also you want within acceptable limits to get the same answer every time you compare the standard with a given object. This means that the standard must be invariable. These two requirements are often incompatible. If you choose length as a basic quantity, define its standard as the distance between a person's nose and the fingertips of the outstretched arm, and assign the yard as the unit, you have a standard that is certainly accessible but it is not invariable. The demands of science and technology steer us just the other way. We achieve accessibility by creating more readily available secondary, tertiary, etc., standards and we strongly stress invariability.

We often make comparisons with a basic standard in very indirect ways. In the case of length consider the problems of measuring: (a) the distance to the Great Nebula in Andromeda, (b) your height, and (c) the distances between nuclei in the molecule NH_3. It is clear that the comparison techniques vary greatly. For example, you cannot use a ruler directly in either the first or third problem.

1-2
THE INTERNATIONAL SYSTEM OF UNITS**

The 14th General Conference on Weights and Measures (1971), building on the work of preceding conferences and international committees, selected as base units the seven quantities displayed in Table 1-1. This is the basis of the International System of Units, abbreviated SI from the French "Le Système International d'Unites."

Throughout the book we will give many examples of SI derived units, such as velocity, force, electric resistance, and so on that follow from Table 1-1. For example, the SI unit of force, called the *newton* (abbreviation N), is defined in terms of the SI base units as

$$1 \text{ N} = 1 \text{ m·kg/s}^2$$

as we will make clear in Chapter 5.

* See "Standards of Measurement," Allen V. Astin, *Scientific American*, June 1968, and "A Short History of Measurement Standards at the National Physical Laboratory," H. Barrell, *Contemporary Physics 9*, 205 (1968).
** See NBS Special Publication 330, 1972, "The International System of Units (SI)." Write to the U.S. Government Printing Office, Washington, D.C. 20402.

Table 1-1
SI base units

Quantity	Name	Symbol
Length	meter[a]	m
Mass	kilogram	kg
Time	second	s
Electric current	ampere	A
Thermodynamic temperature	kelvin	K
Amount of substance	mole	mol
Luminous intensity	candela	cd

[a] The officially recommended spelling is "metre." However, many SI supporters in this country prefer "meter," which we adopt. We will also use "liter" in preference to the recommended "litre."

Often if we express physical properties such as the radius of the earth or the time interval between two nuclear events in SI units (base or derived), we end up with very large or very small numbers. For convenience, the 14th General Conference on Weights and Measures, again building on previous work, recommended the prefixes shown in Table 1-2. Thus we can write the mean radius of the earth ($=6.37 \times 10^6$ m) as 6.38 Mm and a time interval of the size often encountered in nuclear physics, 2.35×10^{-9} s say, as 2.35 ns. Prefixes for factors greater than unity have Greek roots; those for factors less than unity have Latin roots (except that femto and atto, recently added, have Danish roots).

Table 1-2
SI prefixes

Factor	Prefix	Symbol	Factor	Prefix	Symbol
10^1	deka	da	10^{-1}	deci	d
10^2	hecto	h	10^{-2}	centi	c
10^3	kilo	k	10^{-3}	milli	m
10^6	mega	M	10^{-6}	micro	μ
10^9	giga	G	10^{-9}	nano	n
10^{12}	tera	T	10^{-12}	pico	p
10^{15}	peta	P	10^{-15}	femto	f
10^{18}	exa	E	10^{-18}	atto	a

To fortify Table 1-1 we need seven sets of operational procedures that tell us how to produce in the laboratory the seven SI base units. We will explore those for length, mass, and time in the next three sections.

Two other major systems of units compete with the International System (SI). One is the Gaussian system, in terms of which much of the literature of physics is expressed. We will not use this system in this book. Appendix G gives conversion factors to SI units.

The second is the British system, still in daily use in this country, Britain, and elsewhere. The basic units, in mechanics, are length (the foot), force (the pound), and time (the second). Again Appendix G gives conversion factors to SI units. We will use SI units in this book except that in mechanics we will sometimes use the British system, especially in the early chapters. The British system is being phased out in Britain in favor of the officially adopted International System. In fact, as of 1970, the countries Ceylon (later renamed Sri Lanka), Gambia, Guyana, Jamaica, Liberia, Malawi, Nigeria, Sierra Leone, and the United States had in common the fact that they had not by that date adopted the metric system (which later emerged as SI), or officially indicated that they intended to do so.*

The first international standard of length was a bar of a platinum-iridium alloy called the standard meter, and was kept at the International Bureau of Weights and Measures. The distance between two fine lines engraved on gold plugs near the ends of the bar, when the bar was held at a temperature of 0°C and supported mechanically in a prescribed way, was defined to be one meter. Historically, the meter was intended to be one ten-millionth of the distance from the north pole to the equator along the meridian line through Paris. However, accurate measurements taken after the standard meter bar was constructed showed that it differs slightly (about 0.023%) from its intended value.

Because the standard meter is not very accessible, accurate master copies of it were made and sent to standardizing laboratories throughout the world. These secondary standards were used to calibrate other, still more accessible, measuring rods. Thus until recently every measuring rod or device derived its authority from the standard meter through a complicated chain of comparisons using microscopes and dividing engines. Since 1959 this statement had also been true for the yard, whose legal definition in this country was adopted in that year to be

$$1 \text{ yard} = 0.9144 \text{ meter (exactly)}$$

which is equivalent to

$$1 \text{ in.} = 2.54 \text{ cm (exactly)}$$

There are several objections to the meter bar as the primary standard of length: It is potentially destructible, by fire or war for example, and it is not very accessible. These are not idle threats. When the British Houses of Parliament burned in 1834 the British standard yard and standard pound were destroyed. The International Bureau of Weights and Measures was established by France as a neutral international zone and was, fortunately, so respected by the Nazis during World War II.

Most important, the accuracy with which the necessary intercomparisons of length can be made by the technique of comparing fine scratches using a microscope is no longer satisfactory for modern science and technology. Evidence of this is suggested by the trifling midcourse corrections required on space missions. If, among other things, we did not know the distance to the moon in meters as a function of time with some precision, these missions would be much more difficult.

The suggestion that the length of a light wave be used as a length standard was first made in 1828 by J. Babinet. The later development of the interferometer (see Chapter 45) provided scientists with a precision optical device in which a light wave can be used as a length comparison probe. Visible light has a wavelength of about 0.5 μm (see Table 1-2) and length measurements of bars of even many centimeters long can be made to a small fraction of a wavelength. An accuracy of 1 part in 10^9 in the intercomparison of lengths using light waves is possible.

---

* See "Conversion to the Metric System," Lord Ritchie-Calder, *Scientific American*, July 1970. The journal *Metric News* (Swani Publishing Company, P.O. Box 248, Roscoe, Illinois 61073) gives up-to-date information about "metrification" problems, as does the *Metric System Guide*—Bulletin (J. J. Keller Associates, 145 W. Wisconsin Avenue, Neenah, Wisconsin 54956).

** See "The Metre," H. Barrell, *Contemporary Physics 3*, 415 (1962).

In 1960 the 11th General Conference on Weights and Measures adopted an atomic standard for the meter. The wavelength in vacuum of a particular orange-red radiation, identified by the spectroscopic notation $2p_{10} - 5d_5$ and emitted by atoms of a particular isotope of krypton, Kr^{86}, in electrical discharge was chosen (see Fig. 1-1). Specifically, one meter is now defined to be 1,650,763.73 wavelengths of this light. This number of wavelengths was arrived at by carefully measuring the length of the standard meter bar in terms of these light waves. This comparison was done so that the new standard, based on the wavelength of light, would be as consistent as possible with the old standard based on the meter bar. The new standard permits length comparisons to a factor of ten better than is possible with the meter bar.

figure 1-1
A Kr^{86} light source shown removed from the container in which it is housed. In operation the lamp is cooled with liquid nitrogen. (Courtesy the National Physical Laboratories, Teddington, England. Crown copyright reserved.)

The choice of an atomic standard offers advantages other than increased precision in length measurements. The Kr^{86} atoms are available everywhere, are identical, and emit light of the same wavelength. The particular wavelength chosen is uniquely characteristic of Kr^{86} and is sharply defined. The isotope can readily be obtained in pure form.

Given the atomic length standard as basic we still need convenient secondary standards calibrated against it for practical use. Often, as in measuring intramolecular or interstellar distances, we cannot make a direct comparison to a standard. We must use indirect methods to relate the distance in question to the primary standard of length. For example, we know the distances to nearby stars because their positions against the background of much more distant stars shift as the earth moves around its orbit. If we measure this angular shift (parallax), and if we know the diameter of the earth's orbit in meters, we can calculate the distance to the nearby star.

Table 1-3 shows some measured lengths. Note that they vary by a factor of about 10^{37}.

Table 1-3
Some measured lengths

Length	Meters
Distance to the nearest galaxy (in Andromeda)	2×10^{22}
Radius of our galaxy	6×10^{19}
Distance to the nearest star (Alpha Centauri)	4.3×10^{16}
Mean orbit radius for our most distant planet (Pluto)	5.9×10^{12}
Radius of the sun	6.9×10^{8}
Radius of the earth	6.4×10^{6}
Height of Mt. Everest	8.9×10^{3}
Height of a typical person	1.8×10^{0}
Thickness of a page in this book	1×10^{-4}
Size of a poliomyelitis virus	1.2×10^{-8}
Radius of a hydrogen atom	5.0×10^{-11}
Effective radius of a proton	1.2×10^{-15}

The SI standard of mass is a platinum-iridium cylinder kept at the International Bureau of Weights and Measures and assigned, by international agreement, a mass of one kilogram. Secondary standards are sent to standardizing laboratories in other countries and the masses of other bodies can be found by an equal-arm balance technique to a precision of two parts in 10^{8}.

The U.S. copy of the international standard of mass, known as Prototype Kilogram No. 20, is housed in a vault at the National Bureau of Standards (see Fig. 1-2). It is removed no more than once a year for checking the values of tertiary standards. Since 1889 Prototype No. 20 has been taken to France twice for recomparison with the master kilogram. When it is removed from the vault two people are always present, one to carry the kilogram in a pair of forceps, the second to catch the kilogram if the first person should fall.

Table 1-4 shows some measured masses. Note that they vary by a factor of about 10^{70}. Most masses have been measured in terms of the standard kilogram by indirect methods. For example, we can measure the mass of the earth (see Section 16.3) by measuring in the laboratory the gravitational force of attraction between two lead spheres. Their masses must be known by direct comparison with the standard kilogram, using, say, an equal-arm balance.

On the atomic scale we have a second standard of mass, not an SI unit. It is the mass of the C^{12} atom which, by international agreement, has been assigned an atomic mass of 12 unified atomic mass units (abbreviation u), exactly and by definition. We can find the masses of other atoms to considerable accuracy by using a mass spectrometer. Table 1-5 shows some selected atomic masses, including the probable errors of measurement. We need a second standard of mass because present laboratory techniques permit us to compare atomic masses to each other with greater precision than we can compare them to the standard kilogram. The relationship is approximately

$$1 \text{ u} = 1.660 \times 10^{-27} \text{ kg.}$$

1-4
THE STANDARD
OF MASS

figure 1-2
This is national standard kilogram No. 20 which is kept at the United States National Bureau of Standards. It is an accurate copy of the International standard kept at the International Bureau of Weights and Measures near Paris. The standard kilogram is the platinum cylinder housed under the double bell-jar.

Table 1-4
Some measured masses

Object	Kilograms
Our galaxy	2.2×10^{41}
The sun	2.0×10^{30}
The earth	6.0×10^{24}
The moon	7.4×10^{22}
The waters of the oceans	1.4×10^{21}
An ocean liner	7.2×10^{7}
An elephant	4.5×10^{3}
A person	5.9×10^{1}
A grape	3.0×10^{-3}
A speck of dust	6.7×10^{-10}
A tobacco mosaic virus	2.3×10^{-13}
A penicillin molecule	5.0×10^{-17}
A uranium atom	4.0×10^{-26}
A proton	1.7×10^{-27}
An electron	9.1×10^{-31}

Table 1-5
Some measured atomic masses

Isotope	Mass in Atomic mass units
H^{1}	$1.00782522 \pm 0.00000002$
C^{12}	12.00000000 (exactly)
Cu^{64}	63.9297568 ± 0.0000035
Ag^{102}	101.911576 ± 0.000024
Cs^{137}	136.907074 ± 0.000005
Pt^{190}	189.959965 ± 0.000026
Pu^{238}	238.049582 ± 0.000011

The measurement of time has two aspects. For civil and for some scientific purposes we want to know the time of day so that we can order events in sequence. In most scientific work we want to know how long an event lasts (the time interval). Thus any time standard must be able to answer the questions "At what time does it occur?" and "How long does it last?" Table 1-6 shows the range of time intervals that can be measured. They vary by a factor of about 10^{40}.

We can use any phenomenon that repeats itself as a measure of time. The measurement consists of counting the repetitions. We could use an oscillating pendulum, a mass spring system, or a quartz crystal, for example. Of the many repetitive phenomena in nature the rotation of

1-5
*STANDARD OF TIME**

Table 1-6
Some measured time intervals

Time Interval	Seconds
Age of the earth	1.3×10^{17}
Age of the pyramid of Cheops	1.2×10^{11}
Human life expectancy (USA)	2×10^{9}
Time of earth's orbit around the sun (1 year)	3.1×10^{7}
Time of earth's rotation about its axis (1 day)	8.6×10^{4}
Period of a typical satellite	5.1×10^{3}
Half-life of the free neutron	7.0×10^{2}
Time between normal heartbeats	8.0×10^{-1}
Period of concert-A tuning fork	2.3×10^{-3}
Half-life of the muon	2.2×10^{-6}
Period of oscillation of 3-cm microwaves	1.0×10^{-10}
Typical period of rotation of a molecule	1×10^{-12}
Half-life of the neutral pion	2.2×10^{-16}
Period of oscillation of a 1-MeV gamma ray (calculated)	4×10^{-21}
Time for a fast elementary particle to pass through a medium-sized nucleus (calculated)	2×10^{-23}

* See "Accurate Measurement of Time," Louis Essen, *Physics Today*, 1960.

the earth on its axis, which determines the length of the day, has been used as a time standard for centuries. It is still the basis of our civil time standard, one (mean solar) second being defined to be 1/86,400 of a (mean solar) day. Time defined in terms of the rotation of the earth is called universal time (UT).

Universal time must be measured by astronomical observations extended over several weeks. Thus we need a good terrestrial clock, calibrated by the astronomical observations. Quartz crystal clocks based on the electrically sustained periodic vibrations of a quartz crystal serve well as secondary time standards. The best of these have kept time for a year with a maximum error of 0.02 s.

One of the most common uses of a time standard is to measure frequencies. In the radio range frequency comparisons to a quartz clock can be made electronically to a precision of at least 1 part in 10^{10} and, indeed, we often need such precision. However this precision is about 100 times greater than that with which a quartz clock itself can be calibrated by astronomical observations. To meet the need for a better time standard, atomic clocks have been developed in several countries, using periodic atomic vibrations as a standard.

A particular type of atomic clock, based on a characteristic frequency associated with the Cs^{133} isotope, has been in continuous operation at the National Physical Laboratory in England since 1955. Figure 1-3 shows a similar clock at the National Bureau of Standards in this country.

figure 1-3
Atomic cesium beam frequency standard at the Boulder laboratories of the National Bureau of Standards.

In 1967 the second based on the cesium clock was adopted as an international standard by the Thirteenth General Conference on Weights and Measures. The second was defined as 9,192,631,770 periods of the particular Cs^{133} transition selected. This action increased the accuracy of time measurements to 1 part in 10^{12}, an improvement over the accuracy associated with astronomical methods of about 10^3. If two cesium clocks are operated at this precision, and if there are no other sources of error, the clocks will differ by no more than one second after running for 6000 years. Even better potential atomic clocks are being studied.

Figure 1-4 shows, by comparison with the cesium clock, variations in the rate of rotation of the earth over nearly a three-year period. Note that the earth's rotation rate is high in summer and low in winter (northern hemisphere) and decreases steadily from year to year. You may ask how we can be sure that the rotating earth and not the cesium clock is at fault. There are two answers. (1) The relative simplicity of the atom compared to the earth leads us to account for any difference between these two timekeepers to the earth. Tidal friction between the water and the land, for example, causes a slowing down of the earth's rotation. Also the seasonal motion of the winds introduces a seasonal variation in the rotation. Other variations may be associated with the melting and refreezing of polar icecaps. (2) The solar system contains other timekeepers such as the orbiting planets and the orbiting moons of the planets. The rotation of the earth shows variations with respect to these, too, which are similar to but less accurately observable than the variations shown in Fig. 1-4.

The time standard can be made available at remote locations by radio transmission.* WWV in Colorado and WWVH in Hawaii, operated by the National Bureau of Standards, are examples of such stations. They broadcast on frequencies of 2.5, 5, 10, 15, 20, and 25×10^6 Hz stabilized to 1 part in 10^{11} by comparison with a cesium clock. One hertz (abbreviation Hz) is 1 cycle/s. At 5-min intervals WWV alternately broadcasts an accurate 440 Hz tone (concert A) and a 600 Hz tone. Ten times per hour it broadcasts time signals using a binary digit coding system. Two other stations, WWVB and WWVL, both at Fort Collins, Colorado, provide standards of even higher accuracy for special purposes.

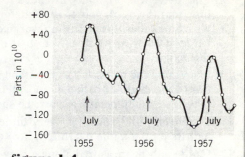

figure 1-4
Variation in the rate of rotation of the earth as revealed by comparison with a cesium clock. (Adapted from L. Essen, *Physics Today*, July 1960.)

questions

1. How would you criticize this statement: "Once you have picked a standard by the very meaning of 'standard' it is invariable"?

2. Many capable investigators, on the evidence, believe in the reality of extrasensory perception. Assuming that ESP is indeed a fact of nature, what physical quantity or quantities would you seek to define to describe this phenomenon quantitatively?

3. According to a point of view adopted by some physicists and philosophers, if we cannot describe procedures for determining a physical quantity, we say that the quantity is undetectable and should be given up as having no physical reality. Not all scientists accept this view. What in your opinion are the merits and drawbacks of this point of view?

4. Do you think that a definition of a physical quantity for which no method of measurement is given has meaning?

5. List characteristics other than accessibility and invariability that you would consider desirable for a physical standard.

6. Can you imagine a system of base units (Table 1-1) in which time was not included?

7. Of the seven base units listed in Table 1-1 only one, the kilogram, has a prefix (see Table 1-2). Would it be wise to redefine the mass of that platinum-iridium cylinder at the International Bureau of Weights and Measures as one gram, rather than one kilogram?

* See "NBS Time and Frequency Dissemination Services; Special Publication 432," National Bureau of Standards, January, 1976 (write to the U.S. Government Printing Office, Washington, D.C. 20402).

8. Can we define temperature as a derived quantity, in terms of length, mass, and time? Think of a pendulum.

9. The meter was originally intended to be one ten-millionth of the meridian line from the north pole to the equator, passing through Paris. In Section 1-3 we learned that this definition was in disagreement with the standard meter bar by 0.023%. Does this mean that the standard meter bar is inaccurate to this extent?

10. In defining the meter bar as the standard of length why specify its temperature? Can length be called a fundamental quantity if another physical quantity, such as temperature, must be specified in choosing a standard?

11. If someone told you that every dimension of every object had shrunk to half its former value overnight, how could you refute this statement?

12. Can length be measured along a curved line? If so, how?

13. Can you suggest a way to measure (a) the radius of the earth, (b) the distance between the sun and the earth, (c) the radius of the sun?

14. Can you suggest a way to measure (a) the thickness of a sheet of paper, (b) the thickness of a soap bubble film, (c) the diameter of an atom?

15. Why do we find it useful to have two standards of mass, the kilogram and the C^{12} atom?

16. How does one obtain the relation between the mass of the standard kilogram and the mass of the C^{12} atom?

17. Is the current standard kilogram of mass accessible, invariable, reproducible, indestructible? Does it have simplicity for comparison purposes? Would an atomic standard be better in any respect? Why don't we use an atomic standard, as we do for length and time?

18. Suggest practical ways by which one could determine the mass of the various objects listed in Table 1-4.

19. Suggest objects whose mass would fall in the wide range in Table 1-4 between that of an ocean liner and all the water in the oceans and estimate their mass.

20. Name several repetitive phenomena occurring in nature which could serve as reasonable time standards.

21. You could define "one second" to be 1.20 pulse beats of the current president of the American Physical Society. Galileo used a similar definition in some of his work. Putting aside considerations of invariability, why is a definition based on the atomic clock better?

22. What criteria should a good clock satisfy?

23. The time it takes the moon to return to a given position as seen against the background of the fixed stars is called a sidereal month. The time interval between identical phases of the moon is called a lunar month. The lunar month is longer than a sidereal month. Why?

24. From what you know about pendula, cite the drawbacks to using the period of a pendulum as a time standard.

25. Can you think of a way to define a length standard in terms of a time standard or vice versa? Think about a pendulum clock. If so, can length and time both be considered as basic quantities?

26. Critics of the metric system often cloud the issue by saying things such as "Instead of buying one pound of butter you will have to ask for 0.452 kg of butter." The implication is that life would be more complicated. How would you refute this?

SECTION 1-2

1. Use the prefixes in Table 1-2 and express (a) 10^6 phones; (b) 10^{-6} phones; (c) 10^1 cards; (d) 10^9 los; (e) 10^{12} bulls; (f) 10^{-1} mates; (g) 10^{-2} pedes; (h) 10^{-9} Nannettes; (i) 10^{-12} boos; (j) 10^{-18} boys; (k) 2×10^2 withits; (l) 2×10^3 mockingbirds. Now that you have the idea, invent a few more similar expres-

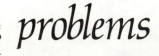

problems

sions. (See, in this connection, p. 61 of *A Random Walk in Science*, edited by R. L. Weber, Crane, Russak, and Co., Inc., New York, 1974).

SECTION 1-3

2. What is your height in meters?

3. Calculate the number of kilometers in 20 miles using only the following conversion factors: 1 mile = 5280 ft, 1 ft = 12 in., 1 in. = 2.54 cm, 1 meter = 100 cm, and 1 km = 1000 meters.   *Answer:* 32.2 km.

4. A rocket attained a height of 300 km. What is this distance in miles?

5. (*a*) In track meets both 100 yards and 100 meters are used as distances for dashes. Which is longer? By how many meters is it longer? By how many feet? (*b*) Track and field records are kept for the mile and the so-called metric mile (1500 meters). Compare these distances.
   Answer: (*a*) 100 meters exceeds 100 yards by 8.56 meters or 28.1 feet. (*b*) One mile exceeds one metric mile by 109 m or 358 ft.

6. Astronomical distances are so large compared to terrestrial ones that much larger units of length are used for easy comprehension of the relative distances of astronomical objects. An *astronomical unit* (AU) is equal to the average distance from the earth to the sun, about 92.9×10^6 miles. A *parsec* is the distance at which one astronomical unit would subtend an angle of 1". A *light-year* is the distance that light, traveling through a vacuum with a speed of 186,000 miles/s, would cover in one year. (*a*) Express the distance from earth to sun in parsecs and in light years. (*b*) Express a light year and a parsec in miles.

7. Master machinists would like to have master gauges (1 in. long, for example) good to 0.0000001 in. Show that the platinum-iridium meter is not measurable to this accuracy but that the Kr^{86} meter is. Use data given in this chapter.
   Answer: Pt-Ir meter bar good to 10^{-7} meter; Kr^{86} standard good to 10^{-9} meter; 10^{-7} in. = 2.5×10^{-9} meter; 10^{-7} meter $> 10^{-9}$ meter.

8. Give the relation between (*a*) a square inch and a square centimeter; (*b*) a square mile and a square kilometer; (*c*) a cubic meter and a cubic centimeter; (*d*) a square foot and a square yard.

9. Assume that the average distance of the sun from the earth is 400 times the average distance of the moon from the earth. Now consider a total eclipse of the sun and state conclusions that can be drawn about (*a*) the relation between the sun's diameter and the moon's diameter; (*b*) the relative volumes of the sun and the moon. (*c*) Find the angle intercepted at the eye by a dime that just eclipses the full moon and from this experimental result and the given distance between the moon and the earth ($= 3.80 \times 10^5$ km) estimate the diameter of the moon.
   Answer: (*a*) $d_{sun}/d_{moon} = 400$. (*b*) $V_{sun}/V_{moon} = 6.4 \times 10^7$. (*c*) 3.5×10^3 km.

SECTION 1-4

10. Using appropriate conversions and data in the chapter, determine the number of hydrogen atoms (isotope number 1) required to obtain one kilogram of mass.

11. If you remember Avogadro's number, you can think of the mass of the earth as being 10 moles of kilograms. What does this statement mean, and how accurate is it? The actual mass of the earth is 5.98×10^{24} kg.
    Answer: Error = 0.67%.

12. (*a*) Assuming that the density (mass/volume) of water is exactly one gram per cubic centimeter, express the density of water in kilograms per liter. (*b*) Suppose that it takes exactly 10 hours to drain a container of 1.00 liter of water. What is the average mass flow rate, in kilograms per second, of water from the container?

13. A convenient substitution for the number of seconds in a year is $\pi \times 10^7$. To within what percentage error is this correct?   *Answer:* −0.44%.

14. (a) A unit of time sometimes used in microscopic physics is the *shake*. One shake equals 10^{-8} s. Are there more shakes in a second than there are seconds in a year? (b) Mankind has existed for about 10^6 years, whereas the universe is about 10^{10} years old. If the age of the universe is taken to be one day, for how many seconds has mankind existed?

15. The maximum speeds of various animals are given roughly as follows in miles per hour: (a) snail, 3×10^{-2}; (b) spider, 1.2; (c) squirrel, 12; (d) man, 28; (e) rabbit, 35; (f) fox, 42; (g) lion, 50; and (h) cheetah, 70. Convert these data to meters per second.
    Answer: (a) 0.013. (b) 0.54. (c) 5.4. (d) 13. (e) 16. (f) 19. (g) 22. (h) 31 m/s.

16. From Fig. 1-2 calculate by what length of time the earth's rotation period in midsummer differs from that in the following spring.

17. Five clocks are being tested in a laboratory. Exactly at noon, as determined by the WWV time signal, on the successive days of a week the clocks read as follows:

Clock	Sun.	Mon.	Tues.	Wed.	Thurs.	Fri.	Sat.
A	12:36:40	12:36:56	12:37:12	12:37:27	12:37:44	12:37:59	12:38:14
B	11:59:59	12:00:02	11:59:57	12:00:07	12:00:02	11:59:56	12:00:03
C	15:50:45	15:51:43	15:52:41	15:53:39	15:54:37	15:55:35	15:56:33
D	12:03:59	12:02:52	12:01:45	12:00:38	11:59:31	11:58:24	11:57:17
E	12:03:59	12:02:49	12:01:54	12:01:52	12:01:32	12:01:22	12:01:12

How would you arrange these five clocks in the order of their relative value as good timekeepers? Justify your choice.
Answer: C, D, A, B, E (best to worst). The important criterion is the constancy of the daily variation, not its magnitude.

18. Assuming that the length of the day uniformly increases by 0.001 s in a century, calculate the cumulative effect on the measure of time over twenty centuries. Such a slowing down of the earth's rotation is indicated by observations of the occurrences of solar eclipses during this period.

19. Express the speed of light, 3×10^8 m/s, in (a) feet/nanosecond and (b) in millimeters/picosecond.   *Answer:* (a) 0.98 ft/ns. (b) 0.3 mm/ps.

20. An astronomical unit (AU) is the average distance of the earth from the sun, approximately 149,000,000 km. The speed of light is about 3.0×10^8 m/s. Express the speed of light in terms of astronomical units per minute.

21. A certain spaceship has a speed of 18,600 mi/h. What is its speed in light-years per century? A light-year is the distance light travels in one year with a speed of 186,000 mi/s.   *Answer:* 2.8×10^{-3} light-years/century.

22. (a) The radius of the proton is about 10^{-15} m; the radius of the observable universe is about 10^{28} cm. Identify a physically meaningful distance which is approximately halfway between these two extremes on a logarithmic scale. (b) The mean life of a neutral pion (an elementary particle) is about 2×10^{-16} s. The age of the universe is about 4×10^9 years. Identify a physically meaningful time interval that is approximately halfway between these two extremes on a logarithmic scale.

2
vectors

A change of position of a particle is called a *displacement*. If a particle moves from position A to position B (Fig. 2-1a), we can represent its displacement by drawing a line from A to B; the direction of displacement can be shown by putting an arrowhead at B indicating that the displacement was *from A to B*. The path of the particle need not necessarily be a straight line from A to B; the arrow represents only the net effect of the motion, not the actual motion.

In Fig. 2-1b, for example, we plot an actual path followed by a particle from A to B. The path is not the same as the displacement AB. If we were to take snapshots of the particle when it was at A and, later, when

figure 2-1

Displacement vectors. *(a)* Vectors AB and A'B' are identical since they have the same length and point in the same direction. *(b)* The actual *path* of the particle in moving from A to B may be the curve shown; the *displacement* remains the vector AB. At some intermediate point P the displacement from A is the vector AP. *(c)* After displacement AB the particle undergoes another displacement BC. The net effect of the two displacements is represented by the vector AC.

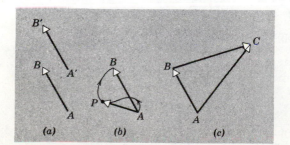

(a)   (b)   (c)

it was at some intermediate position P, we could obtain the displacement vector AP, representing the net effect of the motion during this interval, even though we would not know the actual path taken between these points. Furthermore, a displacement such as $A'B'$ (Fig. 2-1a), which is parallel to AB, similarly directed, and equal in length to AB, represents the same *change* in position as AB. We make no distinction between these two displacements. A displacement is therefore characterized by a *length* and a *direction*.

In a similar way, we can represent a subsequent displacement from B to C (Fig. 2-1c). The net effect of the two displacements will be the same as a displacement from A to C. We speak then of AC as the *sum* or *resultant* of the displacements AB and BC. Notice that this sum is not an algebraic sum and that a number alone cannot uniquely specify it.

Quantities that behave like displacements are called *vectors*.* Vectors, then, are quantities that have both magnitude and direction and combine according to certain rules of addition. These rules are stated below. The displacement vector is a convenient prototype. Some other physical quantities which are vectors are force, velocity, acceleration, the electric field, and the magnetic field. Many of the laws of physics can be expressed in compact form using vectors; derivations involving these laws are often greatly simplified if we do this.

Quantities that can be completely specified by a number and unit and that therefore have magnitude only are called *scalars*. Some physical quantities which are scalars are mass, length, time, density, energy, and temperature. Scalars can be manipulated by the rules of ordinary algebra.

To represent a vector on a diagram we draw an arrow. We choose the length of the arrow proportional to the magnitude of the vector (that is, we choose a scale), and we choose the direction of the arrow to be the direction of the vector, with the arrowhead giving the sense of the direction. For example, a displacement of 40 ft north of east on a scale of 1.0 in. per 10 ft would be represented by an arrow 4.0 in. long, drawn at an angle of 45° above a line pointing east with the arrowhead at the top right extreme. A vector such as this is represented conveniently in printing by a boldface symbol such as **d**. In handwriting it is convenient to put an arrow above the symbol to denote a vector quantity, such as $\vec{d}$.

Often we shall be interested only in the magnitude of the vector and not in its direction. The magnitude of **d** may be written as $|\mathbf{d}|$, called the absolute value of **d**; more frequently we represent the magnitude alone by the italic letter symbol d. The boldface symbol is meant to signify both properties of the vector, magnitude and direction.

Consider now Fig. 2-2 in which we have redrawn and relabeled the vectors of Fig. 2-1c. The relation among these displacements (vectors) can be written as

$$\mathbf{a} + \mathbf{b} = \mathbf{r}. \tag{2-1}$$

The rules to be followed in performing this (vector) addition geometrically are these: On a diagram drawn to scale lay out the displacement

2-2
ADDITION OF VECTORS, GEOMETRICAL METHOD

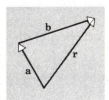

figure 2-2
The vector sum $\mathbf{a} + \mathbf{b} = \mathbf{r}$. Compare with Fig. 2-1c.

---

* The word *vector* means *carrier* in Latin, which suggests a displacement. You might want to review what your analytic geometry and calculus text says about vectors. A good reference that explores the matter in depth is *About Vectors*, by Banesh Hoffman, Prentice-Hall, Englewood Cliffs, N.J., 1966.

vector **a**; then draw **b** with its tail at the head of **a**, and draw a line from the tail of **a** to the head of **b** to construct the vector sum **r**. This is a displacement equivalent in length and direction to the successive displacements **a** and **b**. This procedure can be generalized to obtain the sum of any number of successive displacements.

Since vectors are new quantities, we must expect new rules for their manipulation. The symbol "+" in Eq. 2-1 simply has a different meaning from its meaning in arithmetic or scalar algebra. It tells us to carry out a different set of operations.

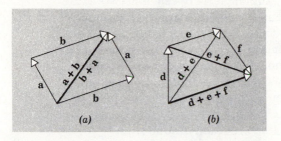

figure 2-3
(a) The commutative law for vector sums, which states that **a** + **b** = **b** + **a**. *(b)* The associative law, which states that **d** + (**e** + **f**) = (**d** + **e**) + **f**.

Using Fig. 2-3 we can prove two important properties of vector addition:

$$\mathbf{a} + \mathbf{b} = \mathbf{b} + \mathbf{a}, \qquad \text{(commutative law)} \qquad (2\text{-}2)$$

and

$$\mathbf{d} + (\mathbf{e} + \mathbf{f}) = (\mathbf{d} + \mathbf{e}) + \mathbf{f}. \qquad \text{(associative law)} \qquad (2\text{-}3)$$

These laws assert that it makes no difference in what order or in what grouping we add vectors; the sum is the same. In this respect, vector addition and scalar addition follow the same rules.

The operation of subtraction can be included in our vector algebra by defining the negative of a vector to be another vector of equal magnitude but opposite direction. Then

$$\mathbf{a} - \mathbf{b} = \mathbf{a} + (-\mathbf{b}) \qquad (2\text{-}4)$$

as shown in Fig. 2-4.

Remember that, although we have used displacements to illustrate these operations, the rules apply to *all* vector quantities.

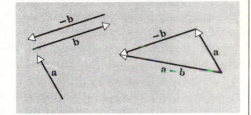

figure 2-4
The vector difference **a** − **b** = **a** + (−**b**).

The geometrical method of adding vectors is not very useful for vectors in three dimensions; often it is even inconvenient for the two-dimensional case. Another way of adding vectors is the analytical method, involving the resolution of a vector into components with respect to a particular coordinate system.

Figure 2-5*a* shows a vector **a** whose tail has been placed at the origin of a rectangular coordinate system. If we drop perpendicular lines from the head of **a** to the axes, the quantities a_x and a_y so formed are called the *components* of the vector **a**. The process is called resolving a vector into its components. Figure 2-5 shows a two-dimensional case for convenience; the extension of our conclusions to three dimensions will be clear.

A vector may have many sets of components. For example, if we rotate the *x*-axis and *y*-axis in Fig. 2-5*a* by 10° counterclockwise, the components of **a** would be different. Furthermore, we may use a nonrec-

2-3
RESOLUTION AND ADDITION OF VECTORS, ANALYTIC METHOD

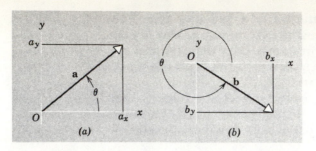

tangular coordinate system, that is, the angle between the two axes need not be 90°. Thus the components of a vector are only uniquely specified if we specify the particular coordinate system being used. The vector need not be drawn with its tail at the origin of the coordinate system to find its components—although we have done so for convenience; the vector may be moved anywhere in the coordinate space and, as long as its angles with the coordinate directions are maintained, its components will be unchanged.

The components a_x and a_y in Fig. 2-5a are readily found from

$$a_x = a \cos \theta \qquad \text{and} \qquad a_y = a \sin \theta, \qquad (2\text{-}5)$$

where θ is the angle that the vector **a** makes with the positive x-axis, measured counterclockwise from this axis. Note that, depending on the angle θ, a_x and a_y can be positive or negative. For example, in Fig. 2-5b, b_y is negative and b_x is positive. The components of a vector behave like scalar quantities because, in any particular coordinate system, only a number with an algebraic sign is needed to specify them.

Once a vector is resolved into its components, the components themselves can be used to specify the vector. Instead of the two numbers a (magnitude of the vector) and θ (direction of the vector relative to the x-axis), we now have the two numbers a_x and a_y. We can pass back and forth between the description of a vector in terms of its components a_x, a_y and the equivalent description in terms of magnitude and direction a and θ. To obtain a and θ from a_x and a_y, we note from Fig. 2-5a that

$$a = \sqrt{a_x^2 + a_y^2} \qquad (2\text{-}6a)$$

and

$$\tan \theta = a_y/a_x. \qquad (2\text{-}6b)$$

The quadrant in which θ lies is determined from the signs of a_x and a_y.

When resolving a vector into components it is sometimes useful to introduce a vector of unit length in a given direction. Thus vector **a** in Fig. 2-6a may be written, for example, as

$$\mathbf{a} = \mathbf{u}_a a, \qquad (2\text{-}7)$$

where $\mathbf{u}_a$ is a *unit vector* in the direction of **a**. Often it is convenient to

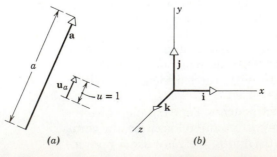

figure 2-6
(a) The vector **a** may be written as $\mathbf{u}_a a$ in which $\mathbf{u}_a$ is a unit vector in the direction of **a**. (b) The unit vector **i**, **j**, and **k**, used to specify the positive x-, y-, and z-directions respectively.

draw unit vectors along the particular coordinate axes chosen. In the rectangular coordinate system the special symbols **i**, **j**, and **k** are usually used for unit vectors in the positive x-, y-, and z-directions, respectively; see Fig. 2-6b. Note that **i**, **j**, and **k** need not be located at the origin. Like all vectors, they can be translated anywhere in the coordinate space as long as their directions with respect to the coordinate axes are not changed.

The vectors **a** and **b** of Fig. 2-5 may be written in terms of their components and the unit vectors as

$$\mathbf{a} = \mathbf{i}a_x + \mathbf{j}a_y \qquad (2\text{-}8a)$$

and

$$\mathbf{b} = \mathbf{i}b_x + \mathbf{j}b_y; \qquad (2\text{-}8b)$$

see Fig. 2-7. The vector relation Eq. 2-8a is equivalent to the scalar relations of Eq. 2-6; each equation relates the vector (**a**, or a and θ) to its components (a_x and a_y). Sometimes we will call quantities such as $\mathbf{i}a_x$ and $\mathbf{j}a_y$ in Eq. 2-8a the *vector components* of **a**; they are drawn as vectors in Fig. 2-7a. The word *component* alone will continue to refer to the scalar quantities a_x and a_y.

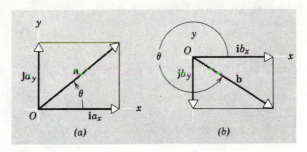

(a)   (b)

figure 2-7
Two examples of the resolution of a vector into its vector components in a particular coordinate system; compare with Figure 2-5.

We now consider the addition of vectors by the analytical method. Let **r** be the sum of the two vectors **a** and **b** lying in the x-y plane, so that

$$\mathbf{r} = \mathbf{a} + \mathbf{b}. \qquad (2\text{-}9)$$

In a given coordinate system, two vectors such as **r** and **a** + **b** can only be equal if their corresponding components are equal, or

$$r_x = a_x + b_x \qquad (2\text{-}10a)$$

and

$$r_y = a_y + b_y. \qquad (2\text{-}10b)$$

These two algebraic equations, taken together, are equivalent to the single vector relation Eq. 2-9. From Eqs. 2-6 we may find r and the angle θ that **r** makes with the x-axis; that is,

$$r = \sqrt{r_x^2 + r_y^2}$$

and

$$\tan \theta = r_y/r_x.$$

Thus we have the following analytic rule for adding vectors: Resolve each vector into its components in a given coordinate system; the algebraic sum of the individual components along a particular axis is the component of the sum vector along that same axis; the sum vector can be reconstructed once its components are known. This method for adding vectors may be generalized to many vectors and to three dimensions (see Problems 13 and 18).

The advantage of the method of breaking up vectors into components, rather than adding directly with the use of suitable trigonometric relations, is that we always deal with right triangles and thus simplify the calculations.

In adding vectors by the analytical method, the choice of coordinate axes determines how simple the process will be. Sometimes the components of the vectors with respect to a particular set of axes are known to begin with, so that the choice of axes is obvious. Other times a judicious choice of axes can greatly simplify the job of resolution of the vectors into components. For example, the axes can be oriented so that at least one of the vectors lies parallel to an axis.

EXAMPLE 1

An airplane travels 130 miles (= 209 km) on a straight course making an angle of 22.5° east of due north. How far north and how far east did the plane travel from its starting point?

We choose the positive x-direction to be east and the positive y-direction to be north. Next (Fig. 2-8) we draw a displacement vector from the origin (starting point), making an angle of 22.5° with the y-axis (north) inclined along the positive x-direction (east). The length of the vector is chosen to represent a magnitude of 130 miles. If we call this vector **d**, then d_x gives the distance traveled east of the starting point and d_y gives the distance traveled north of the starting point. We have

$$\theta = 90.0° - 22.5° = 67.5°,$$

so that (see Eqs. 2-5)

$$d_x = d \cos \theta = (130 \text{ miles}) \cos 67.5° = 50.0 \text{ miles} (= 80.5 \text{ km}),$$

and

$$d_y = d \sin \theta = (130 \text{ miles}) \sin 67.5° = 120 \text{ miles} (= 193 \text{ km}).$$

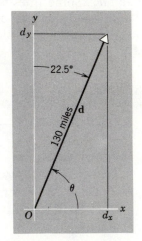

figure 2-8
Example 1

EXAMPLE 2

An automobile travels due east on a level road for 30 km. It then turns due north at an intersection and travels 40 km before stopping. Find the resultant displacement of the car.

We choose a coordinate system fixed with respect to the earth, with the positive x-direction pointing east and the positive y-direction pointing north. The two successive displacements, **a** and **b**, are then drawn as shown in Fig. 2-9. The resultant displacement **r** is obtained from $\mathbf{r} = \mathbf{a} + \mathbf{b}.$ Since **b** has no x-component and **a** has no y-component, we obtain (see Eqs. 2-10)

$$r_x = a_x + b_x = 30 \text{ km} + 0 = 30 \text{ km},$$
$$r_y = a_y + b_y = 0 + 40 \text{ km} = 40 \text{ km}.$$

The magnitude and direction of **r** are then (see Eqs. 2-6)

$$r = \sqrt{r_x^2 + r_y^2} = \sqrt{(30 \text{ km})^2 + (40 \text{ km})^2} = 50 \text{ km},$$

$$\tan \theta = r_y/r_x = \frac{40 \text{ km}}{30 \text{ km}} = 1.33, \qquad \theta = \tan^{-1} (1.33) = 53°.$$

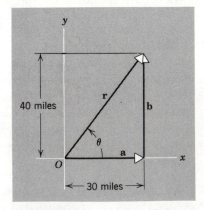

figure 2-9
Example 2

The resultant vector displacement **r** has a magnitude of 50 km and makes an angle of 53° north of east.

EXAMPLE 3

Three coplanar vectors are expressed, with respect to a certain rectangular coordinate system, as

$$\mathbf{a} = 4\mathbf{i} - \mathbf{j},$$

$$\mathbf{b} = -3\mathbf{i} + 2\mathbf{j},$$

and

$$\mathbf{c} = -3\mathbf{j},$$

in which the components are given in arbitrary units. Find the vector **r** which is the sum of these vectors.

From Eqs. 2-10 we have

$$r_x = a_x + b_x + c_x = 4 - 3 + 0 = 1,$$

and

$$r_y = a_y + b_y + c_y = -1 + 2 - 3 = -2.$$

Thus

$$\mathbf{r} = \mathbf{i}r_x + \mathbf{j}r_y$$

$$= \mathbf{i} - 2\mathbf{j}.$$

Figure 2-10 shows the four vectors. From Eqs. 2-6 we can calculate that the magnitude of **r** is $\sqrt{5}$ and that the angle that **r** makes with the positive *x*-axis, measured counterclockwise from that axis, is

$$\tan^{-1}(-2/1) = 297°.$$

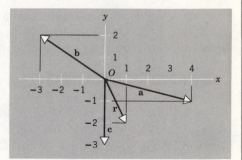

figure 2-10
Three vectors, **a, b,** and **c,** and their vector *sum* **r.**

2-4
MULTIPLICATION OF VECTORS*

We have assumed in the previous discussion that the vectors being added together are of like kind; that is, displacement vectors are added to displacement vectors, or velocity vectors are added to velocity vectors. Just as it would be meaningless to add together scalar quantities of different kinds, such as mass and temperature, so it would be meaningless to add together vector quantities of different kinds, such as displacement and electric field.

However, like scalars, vectors of different kinds can be multiplied by one another to generate quantities of new physical dimensions. Because vectors have direction as well as magnitude, vector multiplication cannot follow exactly the same rules as the algebraic rules of scalar multiplication. We must establish new rules of multiplication for vectors.

We find it useful to define three kinds of multiplication operations for vectors: (1) multiplication of a vector by a scalar, (2) multiplication of two vectors in such a way as to yield a scalar, and (3) multiplication of two vectors in such a way as to yield another vector. There are still other possibilities, but we shall not consider them here.

The multiplication of a vector by a scalar has a simple meaning: The product of a scalar k and a vector **a,** written $k\mathbf{a}$, is defined to be a new vector whose magnitude is k times the magnitude of **a.** The new vector has the same direction as **a** if k is positive and the opposite direction if k is negative. To divide a vector by a scalar we simply multiply the vector by the reciprocal of the scalar.

When we multiply a vector quantity by another vector quantity, we must distinguish between the *scalar* (or *dot*) *product* and the *vector*

---

* The material of this section will be used later in the text. The scalar product is used first in Chapter 7 and the vector product in Chapter 11. The instructor who wishes to postpone this section can do so. Its presentation here gives a unified treatment of vector algebra and serves as a convenient reference for later work.

(or *cross*) *product*. The *scalar product* of two vectors **a** and **b**, written as **a** · **b**, is defined to be

$$\mathbf{a} \cdot \mathbf{b} = ab \cos \phi, \qquad (2\text{-}11)$$

where a is the magnitude of vector **a**, b is the magnitude of vector **b**, and $\cos \phi$ is the cosine of the (smaller) angle ϕ between the two vectors* (see Fig. 2-11).

Since a and b are scalars and $\cos \phi$ is a pure number, the scalar product of two vectors is a scalar. The scalar product of two vectors can be regarded as the product of the magnitude of one vector and the component of the other vector in the direction of the first. Because of the notation **a** · **b** is also called the dot product of **a** and **b** and is spoken as "a dot **b**."

We could have defined **a** · **b** to be any operation we want, for example, to be $a^{1/3}b^{1/4} \tan (\phi/2)$, but this would turn out to be of no use to us in physics. With our definition of the scalar product, a number of important physical quantities can be described as the scalar product of two vectors. Some of them are mechanical work, gravitational potential energy, electrical potential, electric power, and electromagnetic energy density. When such quantities are discussed later, their connection with the scalar product of vectors will be pointed out.

The *vector product* of two vectors **a** and **b** is written as **a** × **b** and is another vector **c**, where **c** = **a** × **b**. The *magnitude* of **c** is defined by

$$c = ab \sin \phi, \qquad (2\text{-}12)$$

where ϕ is the (smaller) angle* between **a** and **b**.

The *direction* of **c**, the vector product of **a** and **b**, is defined to be perpendicular to the plane formed by **a** and **b**. To specify the sense of the vector **c** we must refer to Fig. 2-12. Imagine rotating a right-handed screw whose axis is perpendicular to the plane formed by **a** and **b** so as to turn it *from* **a** *to* **b** through the angle ϕ between them. Then the direction of advance of the screw gives the direction of the vector product **a** × **b** (Fig. 2-12*a*). Another convenient way to obtain the direction of a vector product is the following. Imagine an axis perpendicular to the plane of **a** and **b** through their origin. Now wrap the fingers of the *right hand* around this axis and push the vector **a** into the vector **b** through the smaller angle between them with the fingertips, keeping the thumb erect; the direction of the erect thumb then gives the direction of the vector product **a** × **b** (Fig. 2-12*b*).† Because of the notation, **a** × **b** is also called the cross product of **a** and **b** and is spoken as "a cross **b**."

Notice that **b** × **a** is not the same vector as **a** × **b**, so that the order of factors in a vector product is important. This is not true for scalars because the order of factors in algebra or arithmetic does not affect the resulting product. Actually, **a** × **b** = −**b** × **a** (Fig. 2-12*c*). This can be deduced from the fact that the magnitude $ab \sin \phi$ equals the magnitude

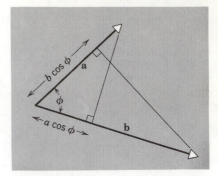

figure 2-11
The scalar product **a** · **b** (= $ab \cos \phi$) is the product of the magnitude of either vector (a, say) by the component of the other vector in the direction of the first vector ($b \cos \phi$, say).

---

* There are two different angles between a pair of vectors, depending on the sense of rotation. We always choose the *smaller* of the two in vector multiplication. In Eq. 2-11 it does not matter because $\cos(2\pi - \phi) = \cos \phi$. But in Eq. 2-12 it does matter because $\sin(2\pi - \phi) = -\sin \phi$.

† The procedures described in Fig. 2-12 are a convention. Two vectors such as **a** and **b** form a plane and there are *two* directions that point away from any plane. We choose the right hand or right-handed screw convention; choosing the left hand or a left-handed screw would have led to the opposite choice for the direction of **a** × **b**.

ba sin ϕ, but the direction of $\mathbf{a} \times \mathbf{b}$ is opposite to that of $\mathbf{b} \times \mathbf{a}$; this is so because the right-handed screw advances in one direction when rotated from $\mathbf{a}$ to $\mathbf{b}$ through ϕ but advances in the opposite direction when rotated from $\mathbf{b}$ to $\mathbf{a}$, through ϕ. You can obtain the same result by applying the right-hand rule.

If ϕ is 90°, $\mathbf{a}$, $\mathbf{b}$, and $\mathbf{c}$ $(= \mathbf{a} \times \mathbf{b})$ are all at right angles to one another and give the directions of a three-dimensional right-handed coordinate system.

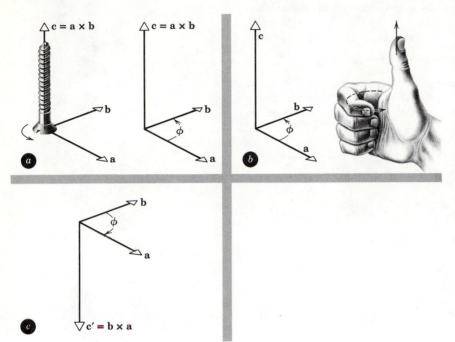

figure 2-12
The vector product. (*a*) In $\mathbf{c} = \mathbf{a} \times \mathbf{b}$, the direction of $\mathbf{c}$ is that in which a right-handed screw advances when turned from $\mathbf{a}$ to $\mathbf{b}$ through the smaller angle. (*b*) The direction of $\mathbf{c}$ can also be obtained from the "right-hand rule": If the right hand is held so that the curled fingers follow the rotation of $\mathbf{a}$ into $\mathbf{b}$, the extended right thumb will point in the direction of $\mathbf{c}$. (*c*) The vector product changes sign when the order of the factors is reversed: $\mathbf{a} \times \mathbf{b} = -\mathbf{b} \times \mathbf{a}$. Apply the right-hand rule or the rule for the advance of a right-handed screw to show that $\mathbf{c}$ and $\mathbf{c}'$ have opposite directions.

The reason for defining the vector product in this way is that it proves to be useful in physics. We often encounter physical quantities that are vectors whose product, defined as above, is a vector quantity having important physical meaning. Some examples of physical quantities that are vector products are torque, angular momentum, the force on a moving charge in a magnetic field, and the flow of electromagnetic energy. When such quantities are discussed later, their connection with the vector product of two vectors will be pointed out.

The scalar product is the simplest product of two vectors. The order of multiplication does not affect the product. The vector product is the next simplest case. Here the order of multiplication does affect the product, but only by a factor of minus one, which implies a direction reversal. Other products of vectors are useful but more involved. For example, a tensor can be generated by multiplying each of the three components of one vector by the three components of another vector. Hence a tensor (of the second rank) has nine numbers associated with it, a vector three, and a scalar only one. Some physical quantities that can be represented by tensors are mechanical and electrical stress, moments and products of inertia, and strain. Still more complex physical quantities are possible. In this book, however, we are concerned only with scalars and vectors.

EXAMPLE 4

A certain vector **a** in the *x*-*y* plane is 250° counterclockwise from the positive *x*-axis and has a magnitude 7.4 units. Vector **b** has magnitude 5.0 units and is directed parallel to the *z*-axis. Calculate (*a*) the scalar product **a** · **b** and (*b*) the vector product **a** × **b**.

(*a*) Because **a** and **b** are perpendicular to one another, the angle ϕ between them is 90° and $\cos \phi = \cos 90° = 0$. Therefore, from Eq. 2-11, the scalar product is

$$\mathbf{a} \cdot \mathbf{b} = ab \cos \phi = ab \cos 90° = (7.4)(5.0)\, 0 = 0,$$

consistent with the fact that neither vector has a component in the direction of the other.

(*b*) The *magnitude* of the vector product is, from Eq. 2-12,

$$|\mathbf{a} \times \mathbf{b}| = ab \sin \phi = (7.4)(5.0) \sin 90° = 37.$$

The *direction* of the vector product is perpendicular to the plane formed by **a** and **b**. Therefore, as shown in Fig. 2-13, it lies in the *x*-*y* plane (perpendicular to **b**) at an angle of $250° - 90° = 160°$ from the $+x$-axis (perpendicular to **a**) in accordance with the right-hand rule.

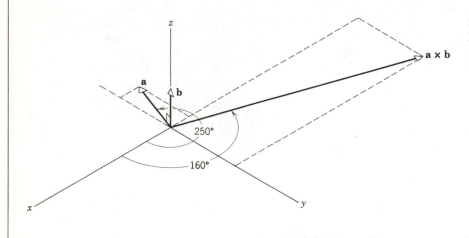

figure 2-13
Example 4

2-5
VECTORS AND THE LAWS OF PHYSICS

Vectors turn out to be very useful in physics. It will be helpful to look a little more deeply into why this is true. Suppose that we have three vectors **a**, **b**, and **r**, which have components $a_x, a_y, a_z; b_x, b_y, b_z;$ and $r_x, r_y, r_z,$ respectively, in a particular coordinate system *xyz*. Let us suppose further that the three vectors are related so that

$$\mathbf{r} = \mathbf{a} + \mathbf{b}. \qquad (2\text{-}13)$$

By a simple extension of Eqs. 2-10 this means that

$$r_x = a_x + b_x; \qquad r_y = a_y + b_y; \qquad \text{and} \qquad r_z = a_z + b_z \qquad (2\text{-}14)$$

Now consider another coordinate system *x'y'z'* which has these properties: (1) its origin does not coincide with the origin of the first, or *xyz*, system and (2) its three axes are not parallel to the corresponding axes in the first system. In other words, the second set of coordinates has been both *translated* and *rotated* with respect to the first.

The components of the vectors **a, b,** and **r** in the new system would all prove, in general, to be different; we may represent them by $a_{x'}, a_{y'}, a_{z'}; b_{x'}, b_{y'}, b_{z'};$ and $r_{x'}, r_{y'}, r_{z'},$ respectively. These new components would be found, however, to be related (see Problem 39) in that

$$r_{x'} = a_{x'} + b_{x'}; \qquad r_{y'} = a_{y'} + b_{y'}; \qquad \text{and} \qquad r_{z'} = a_{z'} + b_{z'}. \qquad (2\text{-}15)$$

That is, in the new system we would find once again (see Eq. 2-13) that

$$\mathbf{r} = \mathbf{a} + \mathbf{b}.$$

In more formal language: Relations among vectors, of which Eq. 2-13 is only one example, are invariant (that is, are unchanged) with respect to translation or rotation of the coordinates. Now it is a fact of experience that the experiments on which the laws of physics are based and indeed the laws of physics themselves are similarly unchanged in form when we rotate or translate the coordinate system. Thus the language of vectors is an ideal one in which to express physical laws. If we can express a law in vector form, the invariance of the law for translation and rotation of the coordinate system is assured by this purely geometrical property of vectors.

It was thought until about 1956 that all laws of physics were invariant under another kind of transformation of coordinates, the substitution of a right-handed coordinate system for a left-handed one (see Fig. 2-14). In that year, however, some experiments involving the decay of certain elementary particles were studied in which the result of the experiment *did* turn out to depend on the "handedness" of the coordinate system used to express the results. In other words, the experiment and its image in a mirror would yield different results!* This surprising result led to a re-examination of the whole question of the symmetry of physical laws; these studies remain among the most challenging in modern physics.

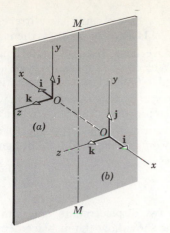

figure 2-14
Showing (a) a left-handed and (b) a right-handed coordinate system. Notice that (a) and (b) are related in that each may be viewed as the image of the other in mirror *MM*. The "handedness" of a coordinate system cannot be changed by rotating it. Note that in (b), $\mathbf{i} \times \mathbf{j} = \mathbf{k}$, whereas in (a), $\mathbf{i} \times \mathbf{j} = -\mathbf{k}$.

questions

1. Three astronauts leave Cape Canaveral, go to the moon and back, and splash down in the Pacific Ocean. An Admiral bids them goodby at the Cape and then sails to the Pacific Ocean in an aircraft carrier where he picks them up. For their respective journeys do the astronauts or the Admiral have the larger displacement?

2. Can two vectors of different magnitude be combined to give a zero resultant? Can three vectors?

3. Can a vector have zero magnitude if one of its components is not zero?

4. Does it make any sense to call a quantity a vector when its magnitude is zero?

5. If three vectors add up to zero, they must all be in the same plane. Make this plausible.

6. Does a unit vector have units?

7. Name several scalar quantities. Is the value of a scalar quantity dependent on the coordinate system chosen?

8. We can order events in time. For example, event b may precede event c but follow event a, giving us a time order of events a, b, c. Hence there is a sense of time, distinguishing past, present, and future. Is time a vector therefore? If not, why not?

9. Do the commutative and associative laws apply to vector subtraction?

10. Can a scalar product be a negative quantity?

11. (a) If $\mathbf{a} \cdot \mathbf{b} = 0$, does it follow that $\mathbf{a}$ and $\mathbf{b}$ are perpendicular to one another? (b) If $\mathbf{a} \cdot \mathbf{b} = \mathbf{a} \cdot \mathbf{c}$, does it necessarily follow that $\mathbf{b}$ equals $\mathbf{c}$?

12. If $\mathbf{a} \times \mathbf{b} = 0$, must $\mathbf{a}$ and $\mathbf{b}$ be parallel to each other? Is the converse true?

13. (a) Show that if all of the components of a vector are reversed in direction, then the vector itself is reversed in direction. (b) Show that if the compo-

---

* C. N. Yang and T. D. Lee were awarded the Nobel prize in 1957 for their theoretical prediction that this would be the case. See "The Overthrow of Parity" by Phillip Morrison, *Scientific American*, April 1957, for a very readable review of this matter.

nents of a vector product are all reversed, then the vector product is not changed. (c) Is a vector product, then, a vector?

14. Thus far we have discussed addition, subtraction, and multiplication of vectors. Why do you suppose that we do not discuss the division of vectors? Is it possible to define such an operation?

15. Must you specify a coordinate system when you (a) add two vectors, (b) form their scalar product, (c) form their vector product, (d) find their components?

16. It is conventional to use the right hand in rules for vector algebra. What changes would be required if a left-hand convention were adopted instead?

problems

SECTION 2-2

1. Consider two displacements, one of magnitude 3 m and another of magnitude 4 m. Show how the displacement vectors may be combined to get a resultant displacement of magnitude (a) 7 m, (b) 1 m, and (c) 5 m.
   Answer: The displacements should be: (a) parallel, (b) antiparallel, (c) perpendicular.

2. What are the properties of two vectors **a** and **b** such that
   (a) $\mathbf{a} + \mathbf{b} = \mathbf{c}$ and $a + b = c$,

   (b) $\mathbf{a} + \mathbf{b} = \mathbf{a} - \mathbf{b}$,

   (c) $\mathbf{a} + \mathbf{b} = \mathbf{c}$ and $a^2 + b^2 = c^2$.

3. Two vectors **a** and **b** are added. Show that the magnitude of the resultant cannot be greater than $a + b$ or smaller than $|a - b|$, where the vertical bars signify absolute value.

4. A car is driven east for a distance of 50 km, then north for 30 km, and then in a direction 30° east of north for 25 km. Draw the vector diagram and determine the total displacement of the car from its starting point.

5. A golfer takes three putts to get his ball into the hole once he is on the green. The first putt displaces the ball 12 ft north, the second 6.0 ft southeast, and the third 3.0 ft southwest. What displacement was needed to get the ball into the hole on the first putt?   *Answer:* 6.0 ft, 20.5° E of N.

6. Vector **a** has a magnitude of 5.0 units and is directed east. Vector **b** is directed 45° west of north and has a magnitude of 4.0 units. Construct vector diagrams for calculating (a) (**a** + **b**) and (b) (**b** − **a**). Estimate the magnitudes and directions of (**a** + **b**) and (**b** − **a**) from your diagrams.

SECTION 2-3

7. Find the sum of the vector displacements **c** and **d** whose components in kilometers along three perpendicular directions are

   $$c_x = 5.0, \ c_y = 0, \ c_z = -2.0; \ d_x = -3.0, \ d_y = 4.0, \ d_z = 6.0.$$
   Answer: $r_x = 2.0$ km; $r_y = r_z = 4.0$ km.

8. (a) A man leaves his front door, walks 1000 ft east, 2000 ft north, and then takes a penny from his pocket and drops it from a cliff 500 ft high. Set up a coordinate system and write down an expression, using unit vectors, for the displacement of the penny. (b) The man then returns to his front door, following a different path on the return trip. What is his resultant displacement for the round trip?

9. Two vectors are given by $\mathbf{a} = 4\mathbf{i} - 3\mathbf{j} + \mathbf{k}$ and $\mathbf{b} = -\mathbf{i} + \mathbf{j} + 4\mathbf{k}$. Find (a) **a** + **b**, (b) **a** − **b**, and (c) a vector **c** such that $\mathbf{a} - \mathbf{b} + \mathbf{c} = 0$.
   Answer: (a) $3\mathbf{i} - 2\mathbf{j} + 5\mathbf{k}$. (b) $5\mathbf{i} - 4\mathbf{j} - 3\mathbf{k}$. (c) Negative of (b).

10. A room has the dimensions 10 ft × 12 ft × 14 ft. A fly starting at one corner ends up at a diametrically opposite corner. (a) What is the magnitude of its displacement? (b) Could the length of its path be less than this distance? Greater than this distance? Equal to this distance? (c) Choose a suitable

coordinate system and find the components of the displacement vector in this frame. (d) If the fly walks rather than flies, what is the length of the shortest path it can take?

11. Given two vectors $\mathbf{a} = 4\mathbf{i} - 3\mathbf{j}$ and $\mathbf{b} = 6\mathbf{i} + 8\mathbf{j}$, find the magnitude and direction of $\mathbf{a}$, of $\mathbf{b}$, of $\mathbf{a} + \mathbf{b}$, of $\mathbf{b} - \mathbf{a}$, and of $\mathbf{a} - \mathbf{b}$.
Answer: The magnitudes are 5, 10, 11, 11, and 11. The angles with the positive x-axis are 323°, 53°, 27°, 80° and 260°.

12. Two vectors of lengths a and b make an angle θ with each other when placed tail to tail. Prove, by taking components along two perpendicular axes, that the length of their sum is

$$r = \sqrt{a^2 + b^2 + 2ab \cos \theta}.$$

13. Generalize the analytical method of resolution and addition to the case of *three or more vectors*, in two dimensions.

14. Two vectors $\mathbf{a}$ and $\mathbf{b}$ have equal magnitudes, say 10 units. They are oriented as shown in Fig. 2-15 and their vector sum is $\mathbf{r}$. Find (a) the x- and y-components of $\mathbf{r}$; (b) the magnitude of $\mathbf{r}$; and (c) the angle $\mathbf{r}$ makes with the x-axis.

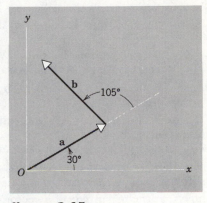

figure 2-15
Problems 14 and 25

15. A particle undergoes three successive displacements in a plane, as follows: 4.0 m southwest, 5.0 m east, 6.0 m in a direction 60° north of east. Choose the y-axis pointing north and the x-axis pointing east and find (a) the components of each displacement, (b) the components of the resultant displacement, (c) the magnitude and direction of the resultant displacement, and (d) the displacement that would be required to bring the particle back to the starting point.
Answer: (a) $a_x = -2.8$ m, $a_y = -2.8$ m;
$b_x = +5.0$ m, $b_y = 0$;
$c_x = +3.0$ m, $c_y = +5.2$ m.
(b) $d_x = +5.2$ m, $d_y = +2.4$ m.
(c) 5.7 m, 25° north of east.
(d) 5.7 m, 25° south of west.

16. Use a scale of 2 m to the inch and add the displacements of Problem 15 graphically. Determine *from your graph* the magnitude and direction of the resultant.

17. A person flies from Washington to Manila. (a) Describe the displacement vector. (b) What is its magnitude if the latitude and longitude of the two cities are 39° N, 77° W, and 15° N, 121° E?   *Answer:* (b) 11,230 km.

18. Generalize the analytical method of resolving and adding two vectors to *three dimensions*.

19. Let N be an integer greater than one; then

$$\cos 0 + \cos \frac{2\pi}{N} + \cos \frac{4\pi}{N} + \cdots + \cos (N-1) \frac{2\pi}{N} = 0;$$

that is,

$$\sum_{n=0}^{n=N-1} \cos \frac{2\pi n}{N} = 0.$$

Also

$$\sum_{n=0}^{n=N-1} \sin \frac{2\pi n}{N} = 0.$$

Prove these two statements by considering the sum of N vectors of equal length, each vector making an angle of $2\pi/N$ with that preceding.

20. A vector **d** has a magnitude 2.5 m and points north. What are the magnitudes and directions of the vectors

    (a) −**d**, (b) **d**/2.0, (c) −2.5**d** and (d) 4.0**d**?

21. In the coordinate system of Fig. 2.6b show that

    $$\mathbf{i} \cdot \mathbf{i} = \mathbf{j} \cdot \mathbf{j} = \mathbf{k} \cdot \mathbf{k} = 1$$

    and

    $$\mathbf{i} \cdot \mathbf{j} = \mathbf{j} \cdot \mathbf{k} = \mathbf{k} \cdot \mathbf{i} = 0.$$

22. In the right-handed coordinate system of Fig. 2-6b show that

    $$\mathbf{i} \times \mathbf{i} = \mathbf{j} \times \mathbf{j} = \mathbf{k} \times \mathbf{k} = 0$$
    $$\mathbf{i} \times \mathbf{j} = \mathbf{k}; \; \mathbf{k} \times \mathbf{i} = \mathbf{j}; \; \mathbf{j} \times \mathbf{k} = \mathbf{i}.$$

23. Show for any vector **a** that (a) $\mathbf{a} \cdot \mathbf{a} = a^2$ and that (b) $\mathbf{a} \times \mathbf{a} = 0$.

24. Use the standard (right-hand) *xyz* system of coordinates. Given vector **a** in the +x-direction, vector **b** in the +y-direction, and the scalar quantity *d*: (a) What is the direction of **a** × **b**? (b) What is the direction of **b** × **a**? (c) What is the direction of **b**/*d*? (d) What is **a** · **b**?

25. For the two vectors in Problem 14, find (a) **a** · **b**, and (b) **a** × **b**.
    Answer: (a) −26. (b) 97**k**.

26. A vector **a** of magnitude ten units and another vector **b** of magnitude six units point in directions differing by 60°. Find (a) the scalar product of the two vectors and (b) the vector product of the two vectors.

27. Show that the area of the triangle contained between the vectors **a** and **b** is $\frac{1}{2}|\mathbf{a} \times \mathbf{b}|$, where the vertical bars signify absolute value (see Fig. 2-16).

28. Show that the magnitude of a vector product gives numerically the area of the parallelogram formed with the two component vectors as sides (see Fig. 2-16). Does this suggest how an element of area oriented in space could be represented by a vector?

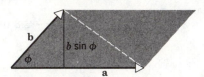

figure 2-16
Problems 27 and 28

29. Show that $\mathbf{a} \cdot (\mathbf{b} \times \mathbf{c})$ is equal in magnitude to the volume of the parallelepiped formed on the three vectors **a**, **b**, and **c**.

30. Prove that two vectors must have equal magnitudes if their sum is perpendicular to their difference.

31. *Scalar product in unit vector notation.* Let two vectors be represented in terms of their coordinates as

    $$\mathbf{a} = \mathbf{i}a_x + \mathbf{j}a_y + \mathbf{k}a_z$$

    and

    $$\mathbf{b} = \mathbf{i}b_x + \mathbf{j}b_y + \mathbf{k}b_z$$

    Show analytically that
    $$\mathbf{a} \cdot \mathbf{b} = a_x b_x + a_y b_y + a_z b_z.$$

    (*Hint:* See Problem 21.)

32. Use the definition of scalar product $\mathbf{a} \cdot \mathbf{b} = ab \cos \phi$ and the fact that $\mathbf{a} \cdot \mathbf{b} = a_x b_x + a_y b_y + a_z b_z$ (see Problem 31) to obtain the angle between the two vectors given by $\mathbf{a} = 3\mathbf{i} + 3\mathbf{j} - 3\mathbf{k}$ and $\mathbf{b} = 2\mathbf{i} + \mathbf{j} + 3\mathbf{k}$.

33. *Vector product in unit vector notation.* Show analytically that $\mathbf{a} \times \mathbf{b} = \mathbf{i}(a_y b_z - a_z b_y) + \mathbf{j}(a_z b_x - a_x b_z) + \mathbf{k}(a_x b_y - a_y b_x)$. (*Hint:* See Problem 22.)

34. Three vectors are given by $\mathbf{a} = 3\mathbf{i} + 3\mathbf{j} - 2\mathbf{k}$, $\mathbf{b} = -\mathbf{i} - 4\mathbf{j} + 2\mathbf{k}$, and $\mathbf{c} = 2\mathbf{i} + 2\mathbf{j} + \mathbf{k}$. Find (a) $\mathbf{a} \cdot (\mathbf{b} \times \mathbf{c})$, (b) $\mathbf{a} \cdot (\mathbf{b} + \mathbf{c})$ and (c) $\mathbf{a} \times (\mathbf{b} + \mathbf{c})$.

35. Let **b** and **c** be the intersecting face diagonals of a cube of edge *a*, as shown in Fig. 2-17. (a) Find the components of the vector **d**, where $\mathbf{d} = \mathbf{b} \times \mathbf{c}$. (b) Find the values of $\mathbf{b} \cdot \mathbf{c}$, of $\mathbf{d} \cdot \mathbf{c}$, and of $\mathbf{d} \cdot \mathbf{b}$. (c) Find the angle between the body diagonal **e**, as shown in Fig. 2-17, and the face diagonal **b**.
    Answer: (a) $d_x = d_z = a^2$, $d_y = -a^2$. (b) $\mathbf{b} \cdot \mathbf{c} = a^2$, $\mathbf{d} \cdot \mathbf{c} = \mathbf{d} \cdot \mathbf{b} = 0$. (c) 35°.

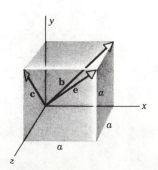

figure 2-17
Problem 35

36. Suppose **a**, **b**, and **c** are any three noncoplanar vectors. They are not necessarily mutually at right angles. (a) show that

$$\mathbf{a} \cdot (\mathbf{b} \times \mathbf{c}) = \mathbf{b} \cdot (\mathbf{c} \times \mathbf{a}) = \mathbf{c} \cdot (\mathbf{a} \times \mathbf{b}).$$

(b) Let

$$\mathbf{A} = \frac{\mathbf{b} \times \mathbf{c}}{v}, \quad \mathbf{B} = \frac{\mathbf{c} \times \mathbf{a}}{v}, \quad \mathbf{C} = \frac{\mathbf{a} \times \mathbf{b}}{v},$$

where $v = \mathbf{a} \cdot (\mathbf{b} \times \mathbf{c})$. Evaluate the dot product of each of $\mathbf{a}$, $\mathbf{b}$, $\mathbf{c}$ with each of $\mathbf{A}$, $\mathbf{B}$, $\mathbf{C}$. (c) If $\mathbf{a}$, $\mathbf{b}$, $\mathbf{c}$ have dimensions of length, what are the dimensions of $\mathbf{A}$, $\mathbf{B}$, $\mathbf{C}$?

37. Two vectors $\mathbf{a}$ and $\mathbf{b}$ have components, in arbitrary units $a_x = 3.2$, $a_y = 1.6$; $b_x = 0.50$, $b_y = 4.5$. (a) Find the angle between $\mathbf{a}$ and $\mathbf{b}$. (b) Find the components of a vector $\mathbf{c}$ which is perpendicular to $\mathbf{a}$, is in the x-y plane, and has a magnitude of 5.0 units.
   Answer: (a) 57°. (b) $c_x = \pm 2.2$ units; $c_y = \mp 4.5$ units.

38. (a) We have seen that the commutative law does *not* apply to vector products, that is, $\mathbf{a} \times \mathbf{b}$ does not equal $\mathbf{b} \times \mathbf{a}$. Show that the commutative law *does* apply to scalar products, that is, $\mathbf{a} \cdot \mathbf{b} = \mathbf{b} \cdot \mathbf{a}$. (b) Show that the distributive law applies to both scalar products and vector products, that is, show that

$$\mathbf{a} \cdot (\mathbf{b} + \mathbf{c}) = \mathbf{a} \cdot \mathbf{b} + \mathbf{a} \cdot \mathbf{c} \text{ and that } \mathbf{a} \times (\mathbf{b} + \mathbf{c}) = \mathbf{a} \times \mathbf{b} + \mathbf{a} \times \mathbf{c}.$$

(c) Does the associative law apply to vector products, that is, does $\mathbf{a} \times (\mathbf{b} \times \mathbf{c})$ equal $(\mathbf{a} \times \mathbf{b}) \times \mathbf{c}$? Does it make any sense to talk about an associative law for scalar products?

SECTION 2-5

39. *Invariance of vector addition under rotation of the coordinate system.* Figure 2-18 shows two vectors $\mathbf{a}$ and $\mathbf{b}$ and two systems of coordinates which differ in that the x and x' axes and the y and y' axes each make an angle ϕ with each other. Prove analytically that $\mathbf{a} + \mathbf{b}$ has the same magnitude and direction no matter which system is used to carry out the analysis.

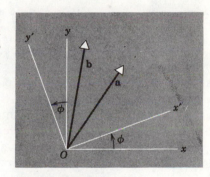

figure 2-18
Problem 39

3
motion in one dimension

Mechanics, the oldest of the physical sciences, is the study of the motion of objects. The calculation of the path of a baseball or of a space probe sent from Earth to Mars is among its problems. So too is the analysis of tracks formed in bubble chambers, representing the collisions, decay, and interactions of elementary particles (see Fig. 10-11 and Appendix F).

When we describe motion we are dealing with that part of mechanics called *kinematics*. When we relate motion to the forces associated with it and to the properties of the moving objects, we are dealing with *dynamics*. In this chapter we shall define some kinematical quantities and study them in detail for the special case of motion in one dimension. In Chapter 4 we discuss some cases of two- and three-dimensional motion. Chapter 5 deals with the more general case of dynamics.

An object can rotate as it moves. For example, a baseball may be spinning while it is moving as a whole in some trajectory. Also, a body may vibrate as it moves, as, for example, a falling water droplet. These complications can be avoided by considering the motion of an idealized body called a *particle*. Mathematically, a particle is treated as a point, an object without extent, so that rotational and vibrational considerations are not involved.

Actually, there is no such thing in nature as an object without extent. The concept of "particle" is nevertheless very useful because real objects often behave to a very good approximation as though they were particles. A body need not be "small" in the usual sense of the word in order to be treated as a particle. For example, if we consider the distance

3-1
MECHANICS

3-2
PARTICLE KINEMATICS

from the earth to the sun, with respect to this distance the earth and the sun can usually be considered to be particles. We can find out a great deal about the motion of the sun and planets, without appreciable error, by treating these bodies as particles. Baseballs, molecules, protons, and electrons can often be treated as particles. Even if a body is too large to be considered a particle for a particular problem, it can always be thought of as made up of a number of particles, and the results of particle motion may be useful in analyzing the problem. As a simplification, therefore, we confine our present treatment to the motion of a particle.

Bodies that have only motion of translation behave like particles. An observer will call motion *translational* if the axes of a reference frame which is imagined rigidly attached to the object, say x', y', and z', always remain parallel to the axes of his own reference frame, say x, y, and z. In Fig. 3-1, for example, we show the translational motion of an object moving from positions A to B to C. Notice that the path taken is not necessarily a straight line. Notice too that throughout the motion every point of the body undergoes the same displacements as every other point. We can assume the body to be a particle because in describing the motion of one point on the body we have described the motion of the body as a whole.

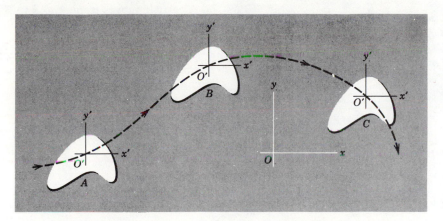

figure 3-1
Translational motion of an object. Translation can occur in three dimensions, but only two are shown for simplicity.

3-3
AVERAGE VELOCITY

The displacement, the velocity, and the acceleration of a particle are vectors. Because this chapter deals with motion in one dimension only, we really do not need the full power of the vector method to deal with it. Nevertheless we find it useful to begin by considering motion in two dimensions (the extension to three is not difficult). From this vantage point we then specialize to the particular case of one-dimensional motion. This procedure allows us to keep in mind the essential vector character of all motion.

The velocity of a particle is the rate at which its position changes with time. The position of a particle in a particular reference frame is given by a position vector drawn from the origin of that frame to the particle. At time t_1, let a particle be at point A in Fig. 3-2a, its position in the x-y plane being described by position vector $\mathbf{r}_1$. At a later time t_2 let the particle be at point B, described by position vector $\mathbf{r}_2$. The displacement vector describing the *change* in position of the particle as it moves from A to B is $\Delta\mathbf{r}$ ($= \mathbf{r}_2 - \mathbf{r}_1$) and the elapsed time for the motion between these points is Δt ($= t_2 - t_1$). The *average velocity* for the particle during this interval is defined by

$$\bar{\mathbf{v}} = \frac{\Delta \mathbf{r}}{\Delta t} = \frac{\text{displacement (a vector)}}{\text{elapsed time (a scalar)}} \qquad (3\text{-}1)$$

A bar above a symbol indicates an average value for the quantity in question.

The quantity $\bar{\mathbf{v}}$ is a vector, for it is obtained by dividing the vector $\Delta \mathbf{r}$ by the scalar Δt. Velocity, therefore, involves both direction and magnitude. Its direction is the direction of $\Delta \mathbf{r}$ and its magnitude is $|\Delta \mathbf{r}/\Delta t|$. The magnitude is expressed in distance units divided by time units, as, for example, meters per second or miles per hour.

The velocity defined by Eq. 3-1 is called an *average* velocity because the measurement of the net displacement and the elapsed time does not tell us anything at all about the motion between A and B. The path may have been curved or straight; the motion may have been steady or erratic. The average velocity involves simply the total displacement and the total elapsed time. For example, suppose a man leaves his house and goes on an automobile trip, returning to his house in a time Δt (five hours, say) after he left it. His average velocity for the trip is zero because his displacement for this particular time interval Δt is zero.

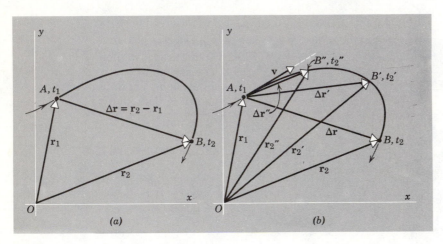

(a)   (b)

figure 3-2
(a) A particle moves from A to B in time Δt $(= t_2 - t_1)$ undergoing a displacement $\Delta \mathbf{r}$ $(= \mathbf{r}_2 - \mathbf{r}_1)$. *The average velocity $\bar{\mathbf{v}}$ between A and B is in the direction of $\Delta \mathbf{r}$. (b)* As B moves closer to A the average velocity approaches the *instantaneous velocity* $\mathbf{v}$ at A; $\mathbf{v}$ is tangent to the path at A.

If we were to measure the time of arrival of the particle at each of many points along the actual path between A and B in Fig. 3-2a, we could describe the motion in more detail. If the average velocity turned out to be the same (in magnitude and direction) between any two points along the path, we would conclude that the particle moved with *constant velocity*, that is, along a straight line (constant direction) at a uniform rate (constant magnitude).

3-4 INSTANTANEOUS VELOCITY

Suppose that a particle is moving in such a way that its average velocity, measured for a number of different time intervals, does *not* turn out to be constant. This particle is said to move with variable velocity. Then we must seek to determine a velocity of the particle at any given instant of time, called the *instantaneous velocity*.

Velocity can vary by a change in magnitude, by a change in direction, or both. For the motion portrayed in Fig. 3-2a, the average velocity during the time interval $t_2 - t_1$ may differ both in magnitude and direction from the average velocity obtained during another time interval $t_2' - t_1$. In Fig. 3-2b we illustrate this by choosing the point B to be successively closer to point A. Points B' and B'' show two intermediate positions of the particle corresponding to the times t_2' and t_2'' and described by position vectors $\mathbf{r}_2'$ and $\mathbf{r}_2''$, respectively. The vector displacements $\Delta\mathbf{r}$, $\Delta\mathbf{r}'$, and $\Delta\mathbf{r}''$ differ in direction and become successively smaller. Likewise, the corresponding time intervals Δt $(= t_2 - t_1)$, $\Delta t'$ $(= t_2' - t_1)$, and $\Delta t''$ $(= t_2'' - t_1)$ become successively smaller.

As we continue this process, letting B approach A, we find that the ratio of displacement to elapsed time approaches a definite limiting value. Although the displacement in this process becomes extremely small, the time interval by which we divide it becomes small also and the ratio is not necessarily a small quantity. Similarly, while growing smaller, the displacement vector approaches a limiting direction, that of the tangent to the path of the particle at A. This limiting value of $\Delta\mathbf{r}/\Delta t$ is called the *instantaneous* velocity at the point A, or the velocity of the particle at the instant t_1.

If $\Delta\mathbf{r}$ is the displacement in a small interval of time Δt, following the time t, the velocity at the time t is the limiting value approached by $\Delta\mathbf{r}/\Delta t$ as both $\Delta\mathbf{r}$ and Δt approach zero. That is, if we let $\mathbf{v}$ represent the instantaneous velocity,

$$\mathbf{v} = \lim_{\Delta t \to 0} \frac{\Delta\mathbf{r}}{\Delta t}.$$

The direction of $\mathbf{v}$ is the limiting direction that $\Delta\mathbf{r}$ takes as B approaches A or as Δt approaches zero. As we have seen, this limiting direction is that of the tangent to the path of the particle at point A.

In the notation of the calculus, the limiting value of $\Delta\mathbf{r}/\Delta t$ as Δt approaches zero is written $d\mathbf{r}/dt$ and is called the *derivative* of $\mathbf{r}$ with respect to t. We have then

$$\mathbf{v} = \lim_{\Delta t \to 0} \frac{\Delta\mathbf{r}}{\Delta t} = \frac{d\mathbf{r}}{dt}. \tag{3-2}$$

The magnitude v of the instantaneous velocity is called the *speed* and is simply the absolute value of $\mathbf{v}$. That is,

$$v = |\mathbf{v}| = |d\mathbf{r}/dt|. \tag{3-3}$$

Speed, being the magnitude of a vector, is intrinsically positive.

Just as a particle is a physical concept making use of the mathematical concept of a point, so here velocity is a physical concept using the mathematical concept of differentiation. In fact, the calculus was invented in order to have a proper mathematical tool for treating fundamental mechanical problems.

In the next section we shall examine the concept of instantaneous velocity in detail for the special case of motion in one dimension, sometimes called rectilinear motion.

3-5
ONE-DIMENSIONAL MOTION—VARIABLE VELOCITY

Here again we approach one-dimensional motion by first considering two-dimensional motion and then considering the special case in which only one dimension is involved.

Figure 3-3 shows a particle moving along a path in the x-y plane. At time t its position with respect to the origin is described by position vector $\mathbf{r}$ (see Fig. 3-3a) and it has a velocity $\mathbf{v}$ (see Fig. 3-3b) tangent to its path as shown. We can write (see Eq. 2-8)

$$\mathbf{r} = \mathbf{i}x + \mathbf{j}y, \tag{3-4}$$

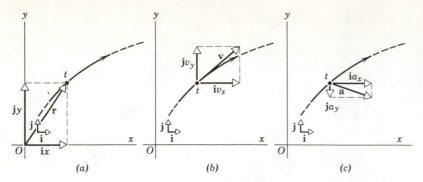

figure 3-3
A particle at time t has (a) a position described by $\mathbf{r}$, (b) an instantaneous velocity $\mathbf{v}$, and (c) an instantaneous acceleration $\mathbf{a}$. The vector components $\mathbf{i}x$ and $\mathbf{j}y$ of Eq. 3-4, $\mathbf{i}v_x$ and $\mathbf{j}v_y$ of Eq. 3-5, and $\mathbf{i}a_x$ and $\mathbf{j}a_y$ of Eq. 3-10 are also shown, as are the unit vectors $\mathbf{i}$ and $\mathbf{j}$.

where $\mathbf{i}$ and $\mathbf{j}$ are unit vectors in the positive x- and y-directions, respectively, and x and y are the (scalar) components of vector $\mathbf{r}$. Because $\mathbf{i}$ and $\mathbf{j}$ are constant vectors, we have, on combining Eqs. 3-2 and 3-4,

$$\mathbf{v} = \frac{d\mathbf{r}}{dt} = \mathbf{i}\frac{dx}{dt} + \mathbf{j}\frac{dy}{dt},$$

which we can express as

$$\mathbf{v} = \mathbf{i}v_x + \mathbf{j}v_y \qquad \text{(two-dimensional motion)}, \qquad (3\text{-}5)$$

where $v_x \ (= dx/dt)$ and $v_y \ (= dy/dt)$ are the (scalar) components of the vector $\mathbf{v}$.

We now consider motion in one dimension only, chosen for convenience to be the x-axis. We must then have $v_y = 0$ so that Eq. 3-5 reduces to

$$\mathbf{v} = \mathbf{i}v_x \qquad \text{(one-dimensional motion)}. \qquad (3\text{-}6)$$

Since $\mathbf{i}$ points in the positive x-direction, v_x will be positive (and equal to $+v$) when $\mathbf{v}$ points in that direction, and negative (and equal to $-v$) when it points in the other direction. Since, in one-dimensional motion, there are only two choices as to the direction of $\mathbf{v}$, the full power of the vector method is not needed, as we have pointed out; we can work with the (scalar) velocity component v_x alone.

The limiting process. As an illustration of the limiting process in one dimension, consider the following table of data taken for motion along the x-axis. The first four columns are experimental data. The symbols refer to Fig. 3-4 in which the particle is moving from left to right, that is, in the positive x-direction. The particle was at position x_1 (100 cm from the origin) at time t_1 (1.00 s). It was at position x_2 at time t_2. As we consider different values for x_2, and different corresponding times t_2, we find

EXAMPLE 1

x_1, cm	t_1, s	x_2, cm	t_2, s	$x_2 - x_1$ $= \Delta x$, cm	$t_2 - t_1$ $= \Delta t$, s	$\Delta x/\Delta t$, cm/s
100.0	1.00	200.0	11.00	100.0	10.00	+10.0
100.0	1.00	180.0	9.60	80.0	8.60	+9.3
100.0	1.00	160.0	7.90	60.0	6.90	+8.7
100.0	1.00	140.0	5.90	40.0	4.90	+8.2
100.0	1.00	120.0	3.56	20.0	2.56	+7.8
100.0	1.00	110.0	2.33	10.0	1.33	+7.5
100.0	1.00	105.0	1.69	5.0	0.69	+7.3
100.0	1.00	103.0	1.42	3.0	0.42	+7.1
100.0	1.00	101.0	1.14	1.0	0.14	+7.1

Equation 3-2, which holds for the general case of motion in three dimensions, is

$$\mathbf{v} = \lim_{\Delta t \to 0} \frac{\Delta \mathbf{r}}{\Delta t} = \frac{d\mathbf{r}}{dt}.$$

For one-dimensional motion along the x-axis we have a similar relation, scalar in character, in which each vector quantity is replaced by its corresponding component or

$$v_x = \lim_{\Delta t \to 0} \frac{\Delta x}{\Delta t} = \frac{dx}{dt}. \tag{3-7}$$

It is clear from the table that as we select values of x_2 closer to x_1, Δt approaches zero and the ratio $\Delta x / \Delta t$ approaches the apparent limiting value $+7.1$ cm/s. At time t_1, therefore, $v_x = +7.1$ cm/s, as closely as we are able to determine from the data. Since v_x is positive, the velocity $\mathbf{v}$ ($= \mathbf{i}v_x$; see Eq. 3-6) points to the right in Fig. 3-4. This is tangent to the path in the direction of motion, as it must be.

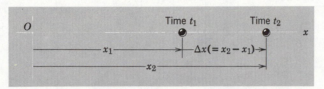

figure 3-4
A particle is moving to the right along the x-axis.

EXAMPLE 2

Figure 3-5a shows six successive "snapshots" of a particle moving along the x-axis with variable velocity. At $t = 0$ it is at position $x = +1.00$ m to the right of the origin; at $t = 2.5$ s it has come to rest at $x = +5.00$ m; at $t = 4.0$ s it has returned to $x = +1.40$ m. Figure 3-5b is a plot of position x versus time t for this motion. The average velocity for the entire 4.0-s interval is the net displacement or change of position ($+0.40$ m) divided by the elapsed time (4.0 s) or $\overline{v_x} = +0.10$ m/s. (We call $\overline{v_x}$ average velocity and v_x velocity, for one-dimensional motion, even though velocity is a vector and not a scalar. This conforms to common usage and should cause no misunderstandings. These quantities are not speeds for they may be negative, whereas speed is intrinsically positive.) The average velocity vector $\overline{\mathbf{v}}$ points in the positive x-direction (that is, to the right in Fig. 3-5a) because the net displacement points in this direction. The quantity $\overline{v_x}$ can be obtained directly from the slope of the dashed line af in Fig. 3-5b, where by slope we mean the ratio of the net displacement gf to the elapsed time ga. (The slope is not the tangent of the angle fag measured on the graph with a protractor. This angle is arbitrary because it depends on the scales we choose for x and t.)

The velocity v_x at any instant is found from the slope of the curve of Fig. 3-5b at that instant. Equation 3-7 is in fact the relation by which the slope of the curve is defined in the calculus. In our example the slope at b, which is the value of v_x at b, is $+1.7$ m/s; the slope at d is zero and the slope at f is -6.2 m/s. When we determine the slope dx/dt at each instant t, we can make a plot of v_x versus t, as in Fig. 3-5c. Note that for the interval $0 < t < 2.5$ s, v_x is positive so that the velocity vector $\mathbf{v}$ points to the right in Fig. 3-5a; for the interval 2.5 s $< t < 4.0$ s v_x is negative so that $\mathbf{v}$ points to the left in Fig. 3-5a.

3-6
ACCELERATION

Often the velocity of a moving body changes either in magnitude, in direction, or both as the motion proceeds. The body is then said to have an acceleration. *The acceleration of a particle is the rate of change of its velocity with time.* Suppose that at the instant t_1 a particle, as in Fig. 3-6, is at point A and is moving in the x-y plane with the instantaneous velocity $\mathbf{v}_1$, and at a later instant t_2 it is at point B and moving

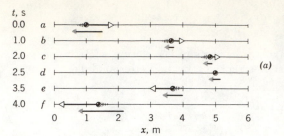

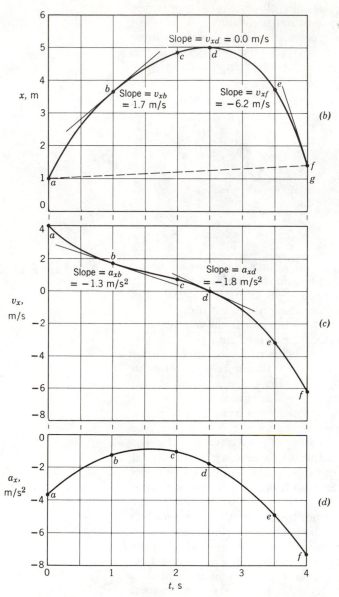

(a)

figure 3-5

(a) Six consecutive "snapshots" of a particle moving along the x-axis. The vector joined to the particle is its instantaneous velocity; that below the particle is its instantaneous acceleration.

(b)

(b) A plot of x versus t for the motion of the particle.

(c)

(c) A plot of v_x versus t.

(d)

(d) A plot of a_x versus t.

with the instantaneous velocity $\mathbf{v}_2$. The *average acceleration* $\bar{\mathbf{a}}$ during the motion from A to B is defined to be the *change of velocity* divided by the time interval, or

$$\bar{\mathbf{a}} = \frac{\mathbf{v}_2 - \mathbf{v}_1}{t_2 - t_1} = \frac{\Delta \mathbf{v}}{\Delta t}. \tag{3-8}$$

The quantity $\bar{\mathbf{a}}$ is a vector, for it is obtained by dividing a vector $\Delta \mathbf{v}$ by a scalar Δt. Acceleration is therefore characterized by magnitude and direction. Its direction is the direction of $\Delta \mathbf{v}$ and its magnitude is

$|\Delta\mathbf{v}/\Delta t|$. The magnitude of the acceleration is expressed in velocity units divided by time units, as for example meters per second per second (written m/s² and read "meters per second squared"), cm/sec², and ft/sec².

We call $\bar{\mathbf{a}}$ of Eq. 3-8 the *average* acceleration because nothing has been said about the time variation of velocity during the interval Δt. We know only the net change in velocity and the total elapsed time. If the change in velocity (a vector) divided by the corresponding elapsed time, $\Delta\mathbf{v}/\Delta t$, were to remain constant, regardless of the time intervals over which we measured the acceleration, we would have *constant* acceleration. Constant acceleration, therefore, implies that the *change* in velocity is uniform with time in direction and magnitude. If there is *no change* in velocity, that is, if the velocity were to remain constant both in magnitude and direction, then $\Delta\mathbf{v}$ would be zero for all time intervals and the acceleration would be zero.

If a particle is moving in such a way that its average acceleration, measured for a number of different time intervals, does not turn out to be constant, the particle is said to have a variable acceleration. The acceleration can vary in magnitude, or in direction, or both. In such cases we seek to determine the acceleration of the particle at any given time, called the instantaneous acceleration.

The *instantaneous acceleration* is defined by

$$\mathbf{a} = \lim_{\Delta t \to 0} \frac{\Delta\mathbf{v}}{\Delta t} = \frac{d\mathbf{v}}{dt}. \qquad (3\text{-}9)$$

That is, the acceleration of a particle at time t is the limiting value of $\Delta\mathbf{v}/\Delta t$ at time t as both $\Delta\mathbf{v}$ and Δt approach zero. The direction of the instantaneous acceleration $\mathbf{a}$ is the limiting direction of the vector change in velocity $\Delta\mathbf{v}$. The magnitude a of the instantaneous acceleration is simply $a = |\mathbf{a}| = |d\mathbf{v}/dt|$. When the acceleration is constant the instantaneous acceleration equals the average acceleration. You should note that the relation of $\mathbf{a}$ to $\mathbf{v}$, in Eq. 3-9, is the same as that of $\mathbf{v}$ to $\mathbf{r}$, in Eq. 3-2.

Two special cases illustrate that acceleration can arise from a change in either the magnitude or the direction of the velocity. In one case we have motion along a straight line with uniformly changing speed (as in Section 3-8). Here the velocity does not change in direction but its magnitude changes uniformly with time. This is a case of constant acceleration. In the second case we have motion in a circle at constant speed (Section 4-4). Here the velocity vector changes continuously in direction but its magnitude remains constant. This, too, is accelerated motion, though the direction of the acceleration vector is not constant. Later we will encounter other important cases of accelerated motion.

From Eqs. 3-5 and 3-9 we can write, for motion in two dimensions as in Fig. 3-3,

$$\mathbf{a} = \frac{d\mathbf{v}}{dt} = \mathbf{i}\frac{dv_x}{dt} + \mathbf{j}\frac{dv_y}{dt}$$

or

$$\mathbf{a} = \mathbf{i}a_x + \mathbf{j}a_y, \qquad (3\text{-}10)$$

where $a_x (= dv_x/dt)$ and $a_y (= dv_y/dt)$ are the (scalar) components of the acceleration vector $\mathbf{a}$ (see Fig. 3-3c).

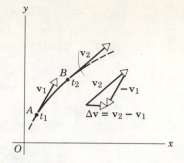

figure 3-6
A particle has velocity $\mathbf{v}_1$ at point A and moves to point B, where its velocity is $\mathbf{v}_2$. The triangle shows the (vector) change in velocity $\Delta\mathbf{v} (= \mathbf{v}_2 - \mathbf{v}_1)$ experienced by the particle as it moves from A to B.

3-7
ONE-DIMENSIONAL MOTION—VARIABLE ACCELERATION

We again restrict ourselves to motion in one dimension only, chosen for convenience to be the x-axis. Since v_y for such motion does not change with time (and is, in fact, zero), a_y, which is dv_y/dt, must also be zero so that

$$\mathbf{a} = \mathbf{i}a_x. \tag{3-11}$$

Since $\mathbf{i}$ points in the positive x-direction, a_x will be positive when $\mathbf{a}$ points in this direction and negative when it points in the other direction.

The motion of Fig. 3-5a is one of variable acceleration along the x-axis. To find the acceleration* a_x at each instant we must determine dv_x/dt at each instant. This is simply the slope of the curve of v_x versus t at that instant. The slope of Fig. 3-5c at point b is -1.3 m/s² and that at point d is -1.8 m/s², as shown in the figure. The result of calculating the slope for all points is shown in Fig. 3-5d. Notice that a_x is negative at all instants, which means that the acceleration vector $\mathbf{a}$ points in the negative x-direction. This means that v_x is always decreasing with time, as is clearly seen from Fig. 3-5c. The motion is one in which the acceleration vector has a constant direction but varies in magnitude (see Fig. 3-5a).

EXAMPLE 3

Let us now further restrict our considerations to motion which not only occurs in one dimension (the x-axis) but for which $a_x =$ a constant. For such constant acceleration the average acceleration for any time interval is equal to the (constant) instantaneous acceleration a_x. Let $t_1 = 0$ and let t_2 be any arbitrary time t. Let v_{x0} be the value of v_x at $t = 0$ and let v_x be its value at the arbitrary time t. With this notation we find a_x (see Eq. 3-8) from

3-8
ONE-DIMENSIONAL MOTION—CONSTANT ACCELERATION

$$a_x = \frac{\Delta v}{\Delta t} = \frac{v_x - v_{x0}}{t - 0}$$

or

$$v_x = v_{x0} + a_x t. \tag{3-12}$$

This equation states that the velocity v_x at time t is the sum of its value v_{x0} at time $t = 0$ plus the change in velocity during time t, which is $a_x t$.

Figure 3-7c shows a graph of v_x versus t for constant acceleration; it is a graph of Eq. 3-12. Notice that the slope of the velocity curve is constant, as it must be because the acceleration a_x ($= dv_x/dt$) has been assumed to be constant, as Fig. 3-7d shows.

When the velocity v_x changes uniformly with time, its average value over any time interval equals one-half the sum of the values of v_x at the beginning and at the end of the interval. That is, the average velocity $\overline{v_x}$ between $t = 0$ and $t = t$ is

$$\overline{v_x} = \tfrac{1}{2}(v_{x0} + v_x). \tag{3-13}$$

This relation would not be true if the acceleration were not constant, for then the curve of v_x versus t would not be a straight line.

---

* As for velocity, we commonly call a_x for one-dimensional motion the acceleration even though acceleration is a vector and a_x is correctly an acceleration component. For one-dimensional motion there is only one component if the axis is chosen along the line of the motion.

If the position of the particle at $t = 0$ is x_0, the position x at $t = t$ can be found from

$$x = x_0 + \overline{v_x}t$$

which can be combined with Eq. 3-13 to yield

$$x = x_0 + \tfrac{1}{2}(v_{x0} + v_x)t. \tag{3-14}$$

The displacement due to the motion in time t is $x - x_0$. Often the origin is chosen so that $x_0 = 0$.

Notice that aside from initial conditions of the motion, that is, the values of x and v_x at $t = 0$ (taken here as $x = x_0$ and $v_x = v_{x0}$), there are four parameters of the motion. These are x, the displacement; v_x, the velocity; a_x, the acceleration; and t, the elapsed time. If we know only that the acceleration is constant, but not necessarily its value, from any two of these parameters we can obtain the other two. For example, if a_x and t are known, Eq. 3-12 gives v_x, and having obtained v_x, we find x from Eq. 3-14.

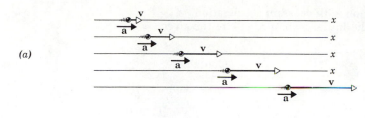

(a)

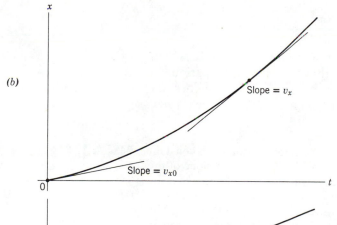

(b)

(c)

(d)

figure 3-7
(a) Five successive "snapshots" of rectilinear motion with constant acceleration. The arrows on the spheres represent **v**; those below represent **a**.

(b) The displacement increases quadratically according to $x = v_{x0}t + \tfrac{1}{2}a_x t^2$. Its slope increases uniformly and at each instant has the value v_x, the velocity.

(c) The velocity v_x increases uniformly according to $v_x = v_{x0} + a_x t$. Its slope is constant and at each instant has the value a_x, the acceleration.

(d) The acceleration a_x has a constant value; its slope is zero. Figure 3-5 shows similar plots for one-dimensional motion in which the acceleration is *not* constant.

In most problems in uniformly accelerated motion, two parameters are known and a third is sought. It is convenient, therefore, to obtain relations between any three of the four parameters. Equation 3-12 contains v_x, a_x, and t, but *not* x; Eq. 3-14 contains, x, v_x, and t but *not* a_x. To complete our system of equations we need two more relations, one containing x, a_x, and t but *not* v_x and another containing x, v_x, and a_x but *not* t. These are easily obtained by combining Eqs. 3-12 and 3-14.

Thus, if we substitute into Eq. 3-14 the value of v_x from Eq. 3-12, we thereby eliminate v_x and obtain

$$x = x_0 + v_{x0}t + \tfrac{1}{2}a_x t^2. \tag{3-15}$$

When Eq. 3-12 is solved for t and this value for t is substituted into Eq. 3-14, we obtain

$$v_x{}^2 = v_{x0}{}^2 + 2a_x(x - x_0). \tag{3-16}$$

Equations 3-12, 3-14, 3-15, and 3-16 (see Table 3-1) are the complete set of equations for motion along a straight line with constant acceleration.

Table 3-1
Kinematic equations for straight line motion with constant acceleration
(The position x_0 and the velocity v_{x0} at the initial instant $t = 0$ are the given initial conditions)

Equation Number	Equation	Contains			
		x	v_x	a_x	t
3-12	$v_x = v_{x0} + a_x t$	×	✓	✓	✓
3-14	$x = x_0 + \tfrac{1}{2}(v_{x0} + v_x)t$	✓	✓	×	✓
3-15	$x = x_0 + v_{x0}t + \tfrac{1}{2}a_x t^2$	✓	×	✓	✓
3-16	$v_x{}^2 = v_{x0}{}^2 + 2a_x(x - x_0)$	✓	✓	✓	×

A special case of motion with constant acceleration is one in which the acceleration is zero, that is, $a_x = 0$. In this case the four equations in Table 3-1 reduce to the expected results $v_x = v_{x0}$ (the velocity does not change) and $x = x_0 + v_{x0}t$ (the displacement changes linearly with time).

EXAMPLE 4

The curve of Fig. 3-7b is a displacement-time graph for motion with constant acceleration; that is, it is a graph of Eq. 3-15 in which $x_0 = 0$. The slope of the tangent to the curve at time t equals the velocity v_x at that time. Notice that the slope increases continuously with time from v_{x0} at $t = 0$. The *rate of increase* of this slope should give the acceleration a_x, which is constant in this case. The curve of Fig. 3-7b is a parabola since Eq. 3-15 is the equation for a parabola having slope v_{x0} at $t = 0$. We obtain, on successive differentiation of Eq. 3-15,

$$x = x_0 + v_{x0}t + \tfrac{1}{2}a_x t^2$$

$$dx/dt = v_{x0} + a_x t \quad \text{or} \quad v_x = v_{x0} + a_x t,$$

which gives the velocity v_x at time t (compare Eq. 3-12), and

$$dv_x/dt = a_x,$$

the constant acceleration. The displacement-time graph for uniformly accelerated rectilinear motion will therefore always be parabolic.

3-9
CONSISTENCY OF UNITS AND DIMENSIONS

You should not feel compelled to memorize relations such as those of Table 3-1. The important thing is to be able to follow the line of reasoning used to obtain the results. These relations will be recalled automatically after you have used them repeatedly to solve problems, partly as a result of the familiarity acquired, but chiefly as a result of the better understanding obtained through application.

We can use any convenient *units* of time and distance in these equations. If we choose to express time in seconds and distance in meters, for self-consistency we must express velocity in m/s and acceleration in m/s². If we are given data in which the units of one quantity, as velocity, are not consistent with the units of another quantity, as acceleration, then before using the data in our equations we should transform both quantities to units that are consistent with one another. Having chosen the units of our fundamental quantities, we automatically determine the units of our derived quantities consistent with them. In carrying out any calculation, always remember to attach the proper units to the final result, for the result is meaningless without this label.

Suppose we wish to find the speed of a particle which has a uniform acceleration of 5.00 cm/s² for an interval of 0.50 h if the particle has a speed of 10.0 ft/s at the beginning of this interval. We decide to choose the foot as our length unit and the second as our time unit. Then

EXAMPLE 5

$$a_x = 5.00 \text{ cm/s}^2 = 5.00 \text{ cm/s}^2 \times \left(\frac{1 \text{ in.}}{2.54 \text{ cm}}\right) \times \left(\frac{1 \text{ ft}}{12 \text{ in.}}\right) = \frac{5.00}{30.5} \text{ ft/s}^2 = 0.164 \text{ ft/s}^2.$$

The time interval

$$\Delta t = t - t_0 = 0.50 \text{ h} \times \left(\frac{60 \text{ min}}{1 \text{ h}}\right) \times \left(\frac{60 \text{ s}}{1 \text{ min}}\right) = 1800 \text{ s}.$$

Note that the conversion factors in large parentheses are equal to unity. Taking the initial time $t_0 = 0$, as in Eq. 3-12, we then have

$$v_x = v_{x0} + a_x t = 10.0 \text{ ft/s} + (0.164 \text{ ft/s}^2)(1800 \text{ s}) = 305 \text{ ft/s}.$$

One way to spot an erroneous equation is to check the *dimensions* of all its terms. The dimensions of any physical quantity can always be expressed as some combination of the fundamental quantities, such as mass, length, and time, from which they are derived. The dimensions of velocity are length (L) divided by time (T); the dimensions of acceleration are length divided by time squared, etc. *In any legitimate physical equation the dimensions of all the terms must be the same.* This means, for example, that we cannot equate a term whose total dimension is a velocity to one whose total dimension is an acceleration. The dimensional labels attached to various quantities may be treated just like algebraic quantities and may be combined, canceled, and so on, just as if they were factors in the equation. For example, to check Eq. 3.15, $x = x_0 + v_{x0}t + \frac{1}{2}a_x t^2$, dimensionally, we note that x and x_0 have the dimension of a length. Therefore the two remaining terms must also have the dimension of a length. The dimension of the term $v_{x0}t$ is

$$\frac{\text{length}}{\text{time}} \times \text{time} = \text{length} \qquad \text{or} \qquad \frac{L}{T} \times T = L,$$

and that of $\frac{1}{2}a_x t^2$ is

$$\frac{\text{length}}{\text{time}^2} \times \text{time}^2 = \text{length} \qquad \text{or} \qquad \frac{L}{T^2} \times T^2 = L.$$

The equation is therefore dimensionally correct. You should check the dimensions of all the equations you use.

The speed of an automobile traveling due east is uniformly reduced from 45.0 miles per hour to 30.0 miles per hour in a distance of 264 ft.

EXAMPLE 6

(a) What is the magnitude and direction of the constant acceleration?

We choose, arbitrarily, the direction from west to east to be the positive x-direction. We are given x and v_x and we seek a_x. The time is not involved. Equation 3.16 is therefore appropriate (see Table 3-1). We have $v_x = +30.0$ mi/h, $v_{x0} = +45.0$ mi/h, $x - x_0 = +264$ ft $= 0.0500$ mi. From Eq. 3-16, $v_x^2 = v_{x0}^2 + 2a_x(x - x_0)$, we obtain

$$a_x = \frac{v_x^2 - v_{x0}^2}{2(x - x_0)},$$

or $\quad a_x = \dfrac{(30.0 \text{ mi/h})^2 - (45.0 \text{ mi/h})^2}{2(0.0500 \text{ mi})} = -1.13 \times 10^4 \text{ mi/h}^2 = -4.58 \text{ ft/s}^2$.

The direction of the acceleration **a** is due west, that is, in the negative x-direction because a_x is negative. The car is slowing down as it moves eastward, as it must do if it is being accelerated toward the west. When the speed of a body is decreasing, we often say that it is decelerating.

(b) How much time has elapsed during this deceleration?

If we use only the original data, Table 3-1 shows that Eq. 3-14 is appropriate. From Eq. 3-14, $x = x_0 + \frac{1}{2}(v_{x0} + v_x)t$, we obtain

$$t = \frac{2(x - x_0)}{v_{x0} + v_x},$$

or

$$t = \frac{(2)(0.0500 \text{ mi})}{(45.0 + 30.0) \text{ mi/h}} = \frac{1}{750} \text{ h} = 4.80 \text{ s}.$$

If we use the derived data of part (a), Eq. 3-12 is appropriate. This gives us a check. From Eq. 3-12, $v_x = v_{x0} + a_x t$, we have

$$t = \frac{v_x - v_{x0}}{a_x}$$

or $\quad t = \dfrac{(30.0 - 45.0) \text{ mi/h}}{-1.13 \times 10^4 \text{ mi/h}^2} = 1.33 \times 10^{-3} \text{ h} = 4.80 \text{ s}.$

(c) If one assumes that the car continues to decelerate at the same rate, how much time would elapse in bringing it to rest from 45.0 mi/h?

Equation 3-12 is useful here. We have $v_{x0} = 45.0$ mi/h, $a_x = -1.13 \times 10^4$ mi/h^2, and the final velocity $v_x = 0$. Then from Eq. 3-12, $v_x = v_{x0} + a_x t$, we obtain

$$t = \frac{v_x - v_{x0}}{a_x},$$

or $\quad t = \dfrac{(0 - 45.0) \text{ mi/h}}{-1.13 \times 10^4 \text{ mi/h}^2} = 4.00 \times 10^{-3} \text{ h} = 14.4 \text{ s}.$

(d) What total distance is required to bring the car to rest from 45.0 mi/h? Equation 3-15 is appropriate here. We have $v_{x0} = 45.0$ mi/h, $a_x = -1.13 \times 10^4$ mi/h^2, $t = 4.00 \times 10^{-3}$ h. From Eq. 3-15, $x = x_0 + v_{x0}t + \frac{1}{2}a_x t^2$, we obtain

$x - x_0 = v_{x0}t + \frac{1}{2}a_x t^2$

$\qquad = (45.0 \text{ mi/h})(4.00 \times 10^{-3} \text{ h}) + \frac{1}{2}(-1.13 \times 10^4 \text{ m/h}^2)(4.00 \times 10^{-3} \text{ h})^2$

$\qquad = 0.0900 \text{ mi} = 475 \text{ ft}.$

EXAMPLE 7

43

FREELY FALLING BODIES   SEC. 3-10

The nucleus of a helium atom (alpha-particle) travels along the inside of a straight hollow tube 2.0 m long which forms part of a particle accelerator. (a) If one assumes uniform acceleration, how long is the particle in the tube if it enters at a speed of 1.0×10^4 m/s and leaves at 5.0×10^6 m/s? (b) What is its acceleration during this interval?

(a) We choose an x-axis parallel to the tube, its positive direction being that in which the particle is moving and its origin at the tube entrance. We are given x and v_x and we seek t. The acceleration a_x is not involved. Hence we use Eq. 3-14, $x = x_0 + \frac{1}{2}(v_{x0} + v_x) t$ with $x_0 = 0$ or

$$t = \frac{2x}{v_{x0} + v_x},$$

$$t = \frac{(2)(2.0 \text{ m})}{(500 + 1) \times 10^4 \text{ m/s}} = 8.0 \times 10^{-7} \text{ s},$$

or 0.80 microseconds (= 0.80 μs).

(b) The acceleration follows from Eq. 3-12, $v_x = v_{x0} + a_x t$, or

$$a_x = \frac{v_x - v_{x0}}{t} = \frac{(500 - 1) \times 10^4 \text{ m/s}}{8.0 \times 10^{-7} \text{ s}} = +6.3 \times 10^{12} \text{ m/s}^2,$$

or 6 trillion meters per second per second! Although this acceleration is enormous by standards of the previous example, it occurs over an extremely short time. The acceleration **a** is in the positive x-direction, that is, in the direction in which the particle is moving, because a_x is positive.

3-10
FREELY FALLING BODIES

The most common example of motion with (nearly) constant acceleration is that of a body falling toward the earth. In the absence of air resistance we find that all bodies, regardless of their size, weight, or composition, fall with the same acceleration at the same point of the earth's surface, and if the distance covered is not too great, the acceleration remains constant throughout the fall. This ideal motion, in which air resistance and the small change in acceleration with altitude are neglected, is called "free fall."

The acceleration of a freely falling body is called the acceleration due to gravity and is denoted by the symbol **g.** Near the earth's surface its magnitude* is approximately 32 ft/s², 9.8 m/s², or 980 cm/s², and it is directed down toward the center of the earth. The variation of the exact value with latitude and altitude will be discussed later (Chapter 16).

The nature of the motion of a falling object was long ago a subject of interest in natural philosophy. Aristotle had asserted that "the downward movement . . . of any body endowed with weight is quicker in proportion to its size." It was not until many centuries later when Galileo Galilei (1564–1642) appealed to experiment to discover the truth, and then publicly proclaimed it, that Aristotle's authority on the matter was seriously challenged. In the later years of his life, Galileo wrote the treatise entitled *Dialogues Concerning Two New Sciences* in which he detailed his studies of motion.

Aristotle's belief that a heavier object will fall faster is a commonly held view. It appears to receive support from a well-known lecture demonstration in which a ball and a sheet of paper are dropped at the same instant, the ball reaching the floor much sooner than the paper. However, when the lecturer first crumples the paper tightly and then repeats the demonstration, both ball and

* See "Absolute value of g at the National Bureau of Standards" by D. R. Tate, *J. Res. NBS 70C*, April–June, 1966.

paper strike the floor at essentially the same time. In the former case, it is the effect of greater resistance of the air which makes the paper fall more slowly than the ball. In the latter case, the effect of air resistance on the paper is reduced and is about the same for both bodies, so that they fall at about the same rate. Of course, a direct test can be made by dropping bodies in vacuum. Even in easily obtainable partial vacuums we can show that a feather and a ball of lead thousands of times heavier drop at rates that are practically indistinguishable.

In Galileo's time, however, there was no effective way to obtain a partial vacuum, nor did equipment exist to time freely falling bodies with sufficient precision to obtain reliable numerical data. Nevertheless, Galileo proved his result by showing first that the character of the motion of a ball rolling down an incline was the same as that of a ball in free fall.* The incline merely served to reduce the effective acceleration of gravity and to slow the motion thereby. Time intervals measured, for example, by the volume of water discharged from a tank could then be used to test the speed and acceleration of this motion.** Galileo showed that if the acceleration along the incline is constant, the acceleration due to gravity must also be constant; for the acceleration along the incline is simply a component of the vertical acceleration of gravity, and along an incline of constant slope the ratio of the two accelerations remains fixed.

He found from his experiments that the distances covered in consecutive time intervals were proportional to the odd numbers 1, 3, 5, 7, . . ., etc. Total distances for consecutive intervals thus were proportional to $1 + 3$, $1 + 3 + 5$, $1 + 3 + 5 + 7$, and so on, that is, to the squares of the integers 1, 2, 3, 4, etc. But if the distance covered is proportional to the square of the elapsed time, velocity acquired is proportional to the elapsed time, a result which is true only if motion is uniformly accelerated. He found that the same results held regardless of the mass of the ball used.

3-11
EQUATIONS OF MOTION IN FREE FALL

We shall select a reference frame rigidly attached to the earth. The y-axis will be taken as positive vertically upward. Then the acceleration due to gravity **g** will be a vector pointing vertically down (toward the center of the earth) in the negative y-direction. (This choice is arbitrary. In other problems it may be convenient to choose down as positive.) Our equations for constant acceleration are applicable here. We simply replace x by y and set $y_0 = 0$ in Eqs. 3-12, 3-14, 3-15, and 3-16, obtaining

$$v_y = v_{y0} + a_y t,$$
$$y = \tfrac{1}{2}(v_{y0} + v_y)t,$$
$$y = v_{y0}t + \tfrac{1}{2}a_y t^2,$$
$$v_y^2 = v_{y0}^2 + 2a_y y,$$

$$(3\text{-}17)$$

and, for problems in free fall, we set $a_y = -g$. Notice that we have chosen the initial position as the origin, that is, we have chosen $y_0 = 0$ at $t = 0$. Note also that g is the magnitude of the acceleration due to gravity.

* See "Galileo's Discovery of the Law of Free Fall' by Stillman Drake, *Scientific American*, May, 1973.
** See "The Role of Music in Galileo's Experiments" by Stillman Drake, *Scientific American*, June, 1975.

A body is dropped from rest and falls freely. Determine the position and speed of the body after 1.0, 2.0, 3.0, and 4.0 s have elapsed.

EXAMPLE 8

We choose the starting point as the origin. We know the initial speed and the acceleration and we are given the time. To find the position we use

$$y = v_{y0}t - \tfrac{1}{2}gt^2.$$

Then, $v_{y0} = 0$ and $g = 32$ ft/s², and with $t = 1.0$ s we obtain

$$y = 0 - \tfrac{1}{2}(32 \text{ ft/s}^2)(1.0 \text{ s})^2 = -16 \text{ ft}.$$

To find the speed with $t = 1.0$ s, we use

$$v_y = v_{y0} - gt$$

and obtain $v_y = 0 - (32 \text{ ft/s}^2)(1.0 \text{ s}) = -32 \text{ ft/s}.$

After 1.0 s of falling from rest, the body is 16 ft (= 4.9 m) below its starting point and has a velocity directed downward whose magnitude is 32 ft/s (= 9.8 m/s); the negative signs for y and v_y show that the associated vectors each point in the negative y-direction, that is, downward.

Show that the values of y, v_y, and a_y obtained at times $t = 2.0$, 3.0, and 4.0 s are those shown in Fig. 3-8 and determine the metric equivalents.

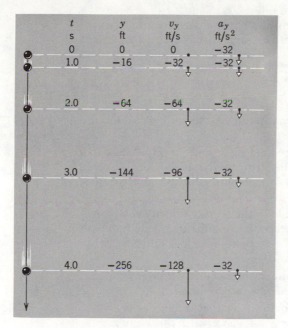

t	y	v_y	a_y
s	ft	ft/s	ft/s²
0	0	0	−32
1.0	−16	−32	−32
2.0	−64	−64	−32
3.0	−144	−96	−32
4.0	−256	−128	−32

figure 3-8
A body in free fall; showing y, v_y, and a_y at particular times t.

A ball is thrown vertically upward from the ground with a speed of 80 ft/s (= 24.4 m/s).

EXAMPLE 9

(a) How long does it take to reach its highest point?

At its highest point, $v_y = 0$, and we have $v_{y0} = +80$ ft/s. To obtain the time t we use $v_y = v_{y0} - gt$, or

$$t = \frac{v_{y0} - v_y}{g}$$

$$t = \frac{(80 - 0) \text{ ft/s}}{32 \text{ ft/s}^2} = 2.5 \text{ s}.$$

(b) How high does the ball rise? Using only the original data, we choose the relation $v_y^2 = v_{y0}^2 - 2gy$, or

$$y = \frac{v_{y0}{}^2 - v_y{}^2}{2g},$$

$$= \frac{(80 \text{ ft/s})^2 - 0}{2 \times 32 \text{ ft/s}^2} = +100 \text{ ft } (= 30.5 \text{ m}).$$

(c) At what times will the ball be 96 ft (= 29 m) above the ground? Using $y = v_{y0}t - \frac{1}{2}gt^2$, we have

$$\tfrac{1}{2}gt^2 - v_{y0}t + y = 0,$$

$$\tfrac{1}{2}(32 \text{ ft/s}^2)t^2 - (80 \text{ ft/s})t + 96 \text{ ft} = 0,$$

or

$$t^2 - 5.0t + 6.0 = 0,$$

which yields $t = 2.0$ s and $t = 3.0$ s.

At $t = 2.0$ s, the ball is moving upward with a speed of 16 ft/s (= 4.9 m/s), for

$$v_y = v_{y0} - gt = 80 \text{ ft/s} - (32 \text{ ft/s}^2)(2.0 \text{ s}) = +16 \text{ ft/s}.$$

At $t = 3.0$ s, the ball is moving downward with the same speed, for

$$v_y = v_{y0} - gt = 80 \text{ ft/s} - (32 \text{ ft/s}^2)(3.0 \text{ s}) = -16 \text{ ft/s}.$$

Notice that in this 1.0-s interval the velocity changed by -32 ft/s ($=-9.8$ m/s), corresponding to an acceleration of -32 ft/s² ($= -9.8$ m/s²).

You should be able to convince yourself that in the absence of air resistance the ball will take as long to rise as to fall the same distance, and that it will have the same speed going down at each point as it had going up.

---

questions

1. Can you think of physical phenomena involving the earth in which the earth cannot be treated as a particle?

2. Each second a rabbit moves half the remaining distance from his nose to a head of lettuce. Does he ever get to the lettuce? What is the limiting value of his average velocity? Draw graphs showing his velocity and position as time increases.

3. Average speed can mean the magnitude of the average velocity vector. Another meaning given to it is that average speed is the total length of path traveled divided by the elapsed time. Are these meanings different? If so, give an example.

4. When the velocity is constant, does the average velocity over any time interval differ from the instantaneous velocity at any instant?

5. Is the average velocity of a particle moving along the x-axis $\frac{1}{2}(v_{x0} + v_x)$ when the acceleration is not uniform? Prove your answer with the use of graphs.

6. Does the speedometer on an automobile register speed as we defined it?

7. (a) Can a body have zero velocity and still be accelerating? (b) Can a body have a constant speed and still have a varying velocity? (c) Can a body have a constant velocity and still have a varying speed?

8. Can an object have an eastward velocity while experiencing a westward acceleration?

9. Can the direction of the velocity of a body change when its acceleration is constant?

10. Can a body be increasing in speed as its acceleration decreases? Explain.

11. Of the following situations, which one is impossible? (a) A body having velocity east and acceleration east; (b) a body having velocity east and acceleration west; (c) a body having zero velocity but acceleration not zero; (d) a body having constant acceleration and variable velocity; (e) a body having constant velocity and variable acceleration.

12. If a particle is released from rest ($v_{y0} = 0$) at $y_0 = 0$ at the time $t = 0$, Eq. 3-17 for constant acceleration says that it is at position y at two different times,

namely, $+\sqrt{2y/a_y}$ and $-\sqrt{2y/a_y}$. What is the meaning of the negative root of this quadratic equation?

13. What happens to our kinematic equations under the operation of time reversal, that is, replacing t by $-t$? Explain.

14. Consider a ball thrown vertically up. Taking air resistance into account, would you expect the time during which the ball rises to be longer or shorter than the time during which it falls?

15. (a) A body is thrown upwards with a certain speed on a world where the acceleration due to gravity is double that on earth. How high does it rise compared to the height it rises on earth? (b) If the initial speed were doubled, what change would that make?

16. Can there be motion in two dimensions with acceleration in only one dimension?

17. A person standing on the edge of a cliff at some height above the ground below throws one ball straight up with initial speed u and then throws another ball straight down with the same initial speed. Which ball, if either, has the larger speed when it hits the ground? Neglect air resistance.

18. A tube in the shape of a rectangle with rounded corners is placed in a vertical plane, as shown in Fig. 3-9. You introduce two ball bearings at the upper right-hand corner. One travels by path AB and the other by path CD. Which will arrive first at the lower left-hand corner?

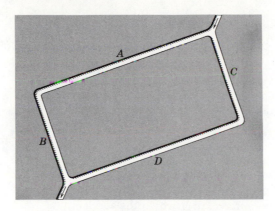

figure 3-9
Question 18.

19. We expect a truly general relation to be valid regardless of the choice of coordinate system. By demanding that general equations be dimensionally consistent we insure that the equations are valid regardless of the choice of units. Is there any need then for units or coordinate systems?

20. From what you know about angular measure, what *dimensions* would you assign to an angle? Can a quantity have units without having dimensions?

21. If m is a light stone and M is a heavy one, according to Aristotle M should fall faster than m. Galileo attempted to show that Aristotle's belief was logically inconsistent by the following argument. Tie m and M together to form a double stone. Then, in falling, m should retard M, because it tends to fall more slowly, and the combination would fall faster than m but more slowly than M; but according to Aristotle the double body $(M + m)$ is heavier than M and hence should fall faster than M.

   If you accept Galileo's reasoning as correct, can you conclude that M and m must fall at the same rate? What need is there for experiment in that case?

   If you believe Galileo's reasoning is incorrect, explain why.

SECTION 3-3

1. How far does a car, moving at 55 mi/h (88 km/h), travel forward during the one second of time that the driver takes to look at an accident on the side on the road?   *Answer:* 81 ft (24 m).

2. The legal speed limit on a thruway is changed from 65 mi/h (105 km/h) to

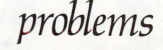

problems

55 mi/h (88.5 km/h) to conserve fuel. How much time is thereby added to the trip from the Buffalo entrance to the New York City exit of the New York Thruway for someone traveling at the legal speed limit over this 435-mile (700 km) stretch of highway?

3. Compare your average speed in the following two cases. (a) You walk 240 ft at a speed of 4.0 ft/s and then run 240 ft at a speed of 10 ft/s along a straight track. (b) You walk for 1.0 min at a speed of 4.0 ft/s and then run for 1.0 min at 10 ft/s along a straight track.   *Answer:* (a) 5.7 ft/s. (b) 7.0 ft/s.

4. A train moving at an essentially constant speed of 60 km/h moves east for 40 min, then in a direction 45° east of north for 20 min, and finally west for 50 min. What is the average velocity of the train during this run?

5. Two trains, each having a speed 40 km/h are headed for each other on the same straight track. A bird that can fly 60 km/h flies off one train when they are 80 km apart and heads directly for the other train. On reaching the other train it flies directly back to the first train, and so forth. (a) How many trips can the bird make from one train to the other before they crash? Explain. (b) What is the total distance the bird travels?
Answer: (a) an infinite number. (b) 60 km.

SECTION 3-6

6. A particle moving along the positive x-axis has the following positions at various times:

x(meters)	0.080	0.050	0.040	0.050	0.080	0.13	0.20
t(seconds)	0.0	1.0	2.0	3.0	4.0	5.0	6.0

(a) Plot displacement (not position) versus time. (b) Find the average velocity of the particle in the intervals 0.0 to 1.0 s, 0.0 to 2.0 s, 0.0 to 3.0 s, 0.0 to 4.0 s. (c) Find the slope of the curve drawn in part a at the points $t = 0.0$, 1.0, 2.0, 3.0, 4.0, and 5.0 s. (d) Plot the slope (units?) versus time. (e) From the curve of part (d) determine the acceleration of the particle at times $t = 2.0$, 3.0, and 4.0 s.

SECTION 3-7

7. The graph of x versus t (see Fig. 3-10a) is for a particle in straight line motion. (a) State for each interval whether the velocity v_x is $+$, $-$, or 0, and whether the acceleration a_x is $+$, $-$, or 0. The intervals are *OA, AB, BC,* and *CD*. (b) From the curve is there any *interval* over which the acceleration is obviously *not constant*? (Ignore the behavior at the end points of the intervals.)

Answer: (a)

	v_x	a_x	(b) No.
OA	+	0	
AB	+	−	
BC	0	0	
CD	−	+	

8. Answer the previous questions for the motion described by the graph of Fig. 3-10b.

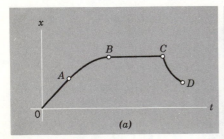

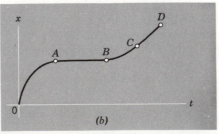

figure 3-10a
Problem 7

figure 3-10b
Problem 8

9. An electron, starting from rest, has an acceleration that increases linearly with time, that is, $a = kt$, the change in acceleration being $k = (1.5 \text{ m/s}^2)/\text{s}$. (a) Plot a versus t during the first 10-s interval. (b) From the curve of part (a) plot the corresponding v versus t curve and estimate the electron's velocity 5.0 s after its motion starts. (c) From the v versus t curve of part (b) plot the corresponding x versus t curve and estimate how far the electron moved during the first 5.0 s of its motion. *Answer: (b) 19 m/s. (c) 31 m.*

10. The position of a particle moving along the x-axis depends on the time according to the relation

$$x = \frac{v_{x0}}{k}(1 - e^{-kt})$$

in which v_{x0} and k are constants. (a) Plot a curve of x versus t. Notice that $x = 0$ at $t = 0$ and that $x = v_{x0}/k$ at $t = \infty$; that is, the total distance through which the particle moves is v_{x0}/k. (b) Show that the velocity v_x is given by

$$v_x = v_{x0}e^{-kt}$$

so that the velocity decreases exponentially with time from its initial value of v_{x0}, coming to rest only in infinite time. (c) Show that the acceleration a_x is given by

$$a_x = -kv_x$$

so that the acceleration is directed opposite to the velocity and has a magnitude proportional to the speed. (d) This particular motion is one with variable acceleration. Give a plausible physical argument explaining how it can take an infinite time to bring to rest a particle that travels a finite distance.

11. A particle moves along the x-axis with a displacement versus time as shown in Fig. 3-11. Sketch roughly curves of velocity versus time and acceleration versus time for this motion.

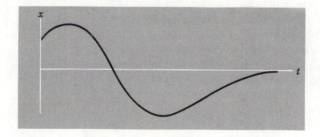

figure 3-11
Problem 11

SECTION 3-8

12. A jumbo jet needs to reach a speed of 225 mi/h (360 km/h) on the runway for takeoff. Assuming a constant acceleration and a runway 1.1 miles (1.8 km) long, what minimum acceleration from rest is required?

13. An automobile increases its speed uniformly from 25 to 55 km/h in one-half minute. A bicycle rider uniformly speeds up to 30 km/h from rest in one-half minute. Compare the accelerations.
Answer: Both accelerations are equal to 0.28 m/s².

14. A rocket-driven sled running on a straight level track is used to investigate the physiological effects of large accelerations on humans. One such sled can attain a speed of 1600 km/h in 1.8 s starting from rest. (a) Assume the acceleration is constant and compare it to g. (b) What is the distance traveled in this time?

15. A rocketship in free space moves with constant acceleration equal to 9.8 m/s². (a) If it starts from rest, how long will it take to acquire a speed one-tenth that of light? (b) How far will it travel in so doing?
Answer: (a) 36 days. *(b)* 4.6×10^{10} km.

16. An arrow while being shot from a bow was accelerated over a distance of 2.0 ft. If its speed at the moment it left the bow was 200 ft/s, what was the average acceleration imparted by the bow? Justify any assumptions you need to make.

17. A subway train accelerates from rest at one station at a rate of 1.20 m/s² for half of the distance to the next station, then decelerates at this same rate for the final half. If the stations are 1100 m apart, find (a) the time of travel between stations and (b) the maximum speed of the train.
    Answer: (a) 60.6 s. (b) 36.4 m/s (= 81.4 mi/h).

18. Suppose that you were called upon to give some advice to a lawyer concerning the physics involved in one of her cases. The question is whether a driver was exceeding a 30 mi/h speed limit before he made an emergency stop, brakes locked and wheels sliding. The length of skid marks on the road was 19.2 ft. The police officer made the assumption that the maximum deceleration of the car would not exceed the acceleration of a freely falling body and arrested the driver for speeding. Was he speeding? Explain.

19. Two trains, one traveling at 60 mi/h and the other at 80 mi/h, are headed toward one another along a straight level track. When they are 2.0 miles apart, both engineers simultaneously see the other's train and apply their brakes. If the brakes decelerate each train at the rate of 3.0 ft/s², determine whether there is a collision.
    Answer: No.

20. A train started from rest and moved with constant acceleration. At one time it was traveling 30 ft/s and 160 ft farther on it was traveling 50 ft/s. Calculate (a) the acceleration, (b) the time required to travel the 160 ft mentioned, (c) the time required to attain the speed of 30 ft/s, (d) the distance moved from rest to the time the train had a speed of 30 ft/s.

21. An electron with initial velocity $v_{x0} = 1.0 \times 10^4$ m/s enters a region of width 1.0 cm where it is electrically accelerated (Fig. 3-12). It emerges with a velocity $v_x = 4.0 \times 10^6$ m/s. What was its acceleration, assumed constant? (Such a process occurs in the electron gun in a cathode-ray tube, used in television receivers and oscilloscopes.)   *Answer:* 8.0×10^{14} m/s².

22. A meson is shot with speed 5.00×10^6 m/s into a region where an electric field produces an acceleration on the meson of magnitude 1.25×10^{14} m/s² directed opposite to the initial velocity. (a) How far does the meson travel before coming to rest? (b) How long does the meson remain at rest?

23. A car moving with constant acceleration covers the distance between two points 180 ft apart in 6.0 s. Its speed as it passes the second point is 45 ft/s. (a) What is its speed at the first point? (b) What is its acceleration? (c) At what prior distance from the first point was the car at rest?
    Answer: (a) 15 ft/s. (b) 5.0 ft/s². (c) 23 ft.

24. The speed of an automobile traveling east is uniformly reduced from 45 mi/h to 30 mi/h in a distance of 264 ft. (a) What is the magnitude and direction of the constant acceleration? (b) How much time has elapsed during this deceleration? (c) If the car continues to decelerate at the same rate, how much time would elapse in bringing it to rest from 45 mi/h? (d) What distance is required to bring the car to rest from 45 mi/h? See Question 8.

25. At the instant the traffic light turns green, an automobile starts with a constant acceleration a_x of 6.0 ft/s². At the same instant a truck, traveling with a constant speed of 30 ft/s, overtakes and passes the automobile. (a) How far beyond the starting point will the automobile overtake the truck? (b) How fast will the car be traveling at that instant? (It is instructive to plot a qualitative graph of x versus t for each vehicle.)
    Answer: (a) 300 ft. (b) 60 ft/s.

26. An automobile traveling 35 mi/h (56 km/h) is 110 ft (35 m) from a barrier when the driver slams on the brakes. Four seconds later the car hits the barrier. (a) What was the automobile's deceleration before impact? (b) How fast was the car traveling at impact?

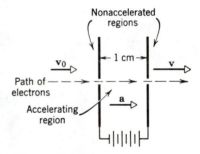

figure 3-12
Problem 21

27. The engineer of a train moving at a speed v_1 sights a freight train a distance d ahead of him on the same track moving in the same direction with a slower speed v_2. He puts on the brakes and gives his train a constant deceleration a. Show that

$$\text{if } d > \frac{(v_1 - v_2)^2}{2a}, \text{ there will be no collision;}$$

$$\text{if } d < \frac{(v_1 - v_2)^2}{2a}, \text{ there will be a collision.}$$

(It is instructive to plot a qualitative graph of x versus t for each train.)

28. A driver's handbook states that an automobile with good brakes and going 50 mi/h can stop in a distance of 186 ft. The corresponding distance for 30 mi/h is 80 ft. Assume that the driver reaction time, during which the acceleration is zero, and the acceleration after he applies the brakes are both the same for the two speeds. Calculate (a) the driver reaction time and (b) the acceleration.

SECTION 3-9

29. The position of a particle along the x-axis depends on the time according to the equation

$$x = at^2 - bt^3,$$

where x is in meters and t in seconds. (a) What dimensions and units must a and b have? For the following, let their numerical values be 3.0 and 1.0, respectively. (b) At what time does the particle reach its maximum positive x-position? (c) What total length of path does the particle cover in the first 4.0 s? (d) What is its displacement during the first 4.0 s? (e) What is the particle's velocity at the end of each of the first four seconds? (f) What is the particle's acceleration at the end of each of the first four seconds? (g) What is the average velocity for the time interval $t = 2.0$ to $t = 4.0$ seconds? *Answer:* (a) a: LT^{-2}, m/s²; b: LT^{-3}, m/s³. (b) $t = 2$ s. (c) 24 m. (d) −16 m. (e) 3.0, 0.0, −9.0, −24.0 m/s. (f) 0.0, −6.0, −12.0, −18.0 m/s². (g) −10 m/s.

SECTION 3-11

30. (a) With what speed must a ball be thrown vertically upward in order to rise to a height of 50 ft? (b) How long will it be in the air?

31. A tennis ball is dropped onto the floor from a height of 4.0 ft. It rebounds to a height of 3.0 ft. If the ball was in contact with the floor for 0.010 s, what was its average acceleration during contact? *Answer:* 3000 ft/s²

32. While thinking of Isaac Newton, a person standing on a bridge overlooking a highway inadvertently drops an apple over the railing just as the front end of a truck passes directly below the railing. If the vehicle is moving at 55 km/h (34 mi/h) and is 12 m (39 ft) long, how far above the truck must the railing be if the apple just misses hitting the rear end of the truck?

33. A lead ball is dropped into a lake from a diving board 16 ft above the water. It hits the water with a certain velocity and then sinks to the bottom with this same constant velocity. It reaches the bottom 5.0 s after it is dropped. (a) How deep is the lake? (b) What is the average velocity of the ball? (c) Suppose all the water is drained from the lake. The ball is thrown from the diving board so that it again reaches the bottom in 5.0 s. What is the initial velocity of the ball? *Answer:* (a) 128 ft. (b) 29 ft/s. (c) 51 ft/s upward.

34. A rocket is fired vertically and ascends with a constant vertical acceleration of 64 ft/s² for 1.0 min. Its fuel is then all used and it continues as a free particle. (a) What is the maximum altitude reached? (b) What is the total time elapsed from take-off until the rocket strikes the earth?

35. A balloon is ascending at the rate of 12 m/s at a height 80 m above the ground when a package is dropped. How long does it take the package to reach the ground? *Answer:* 5.4 s.

36. A stone is dropped into the water from a bridge 144 ft (44 m) above the water. Another stone is thrown vertically down 1.0 s after the first is dropped. Both stones strike the water at the same time. (a) What was the initial speed of the second stone? (b) Plot speed versus time on a graph for each stone, taking zero time as the instant the first stone was released.

37. An open elevator is ascending with a constant speed v (32 ft/s). A ball is thrown straight up by a boy on the elevator when it is a height h (100 ft) above the ground. The initial speed of the ball with respect to the elevator is V_0 (64 ft/s). (a) What is the maximum height attained by the ball? (b) How long does it take for the ball to return to the elevator?
Answer: (a) 244 ft. (b) 4.0 s.

38. An arrow is shot straight up in the air with an initial speed of 250 ft/s. If on striking the ground it imbeds itself 6.0 in. into the ground, find (a) the acceleration (assumed constant) required to stop the arrow and (b) the time required for it to come to rest. Neglect air resistance during the arrow's flight.

39. A parachutist after bailing out falls 50 m without friction. When the parachute opens, he decelerates downward 2.0 m/s². He reaches the ground with a speed 3.0 m/s. (a) How long is the parachutist in the air? (b) At what height did he bail out?   *Answer:* (a) 17 s. (b) 290 m.

40. A shell is fired directly up from a gun; a rocket, propelled by burning fuel, takes off vertically from a launching area. Plot qualitatively (numbers not required) possible graphs of a_y versus t, of v_y versus t, and of y versus t for each. Take $t = 0$ at the instant the shell leaves the gun barrel or the rocket leaves the ground. Continue the plots until the rocket and the shell fall back to earth; neglect air resistance; assume that up is positive and down is negative.

41. If a body travels half its total path in the last second of its fall from rest, find (a) the time and (b) height of its fall. (c) Explain the physically unacceptable solution of the quadratic time equation.
Answer: (a) 3.4 s. (b) 57 m.

42. Two bodies begin a free fall from rest from the same height 1.0 s apart. How long after the first body begins to fall will the two bodies be 10 m apart?

43. A steel ball bearing is dropped from the roof of a building (the initial velocity of the ball is zero). An observer standing in front of a window 4.0 ft high notes that the ball takes $\frac{1}{8}$ s to fall from the top to the bottom of the window. The ball bearing continues to fall, makes a completely elastic collision with a horizontal sidewalk, and reappears at the bottom of the window 2.0 s after passing it on the way down. How tall is the building? (The ball will have the same speed at a point going up as it had going down after a completely elastic collision.)   *Answer:* 68 ft.

44. Water drips from the nozzle of a shower onto the floor 81 in. below. The drops fall at regular intervals of time, the first drop striking the floor at the instant the fourth drop begins to fall. Find the location of the individual drops when a drop strikes the floor.

45. An elevator ascends with an upward acceleration of 4.0 ft/s². At the instant its upward speed is 8.0 ft/s, a loose bolt drops from the ceiling of the elevator 9.0 ft from the floor. Calculate (a) the time of flight of the bolt from ceiling to floor and (b) the distance it has fallen relative to the elevator shaft.   *Answer:* (a) 0.71 s. (b) 2.3 ft.

46. A dog sees a flowerpot sail up and then back past a window 5.0 ft (1.5 m) high. If the total time the pot is in sight is 1.0 s, find the height above the window that the pot rises.

4
motion in a plane

In this chapter we return to a consideration of motion in two dimensions taken, for convenience, to be the *x-y* plane. Figure 4-1 shows a particle at time *t* moving along a curved path in this plane. Its *position*, or displacement from the origin, is measured by the vector **r;** its *velocity* is indicated by the vector **v** which, as we have seen in Section 3-4, must be tangent to the path of the particle. The *acceleration* is indicated by the vector **a;** the direction of **a,** as we shall see more explicitly later, does not bear any unique relationship to the path of the particle but depends rather on the rate at which the velocity **v** changes with time as the particle moves along its path.

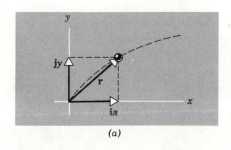
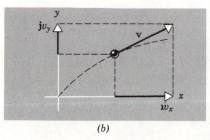
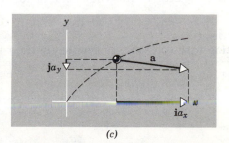

(a)   *(b)*   *(c)*

figure 4-1
A particle moves along a curved path in the *x-y* plane. *(a)* Its position **r**, *(b)* its velocity **v**, and *(c)* its acceleration **a** are shown at time *t*, along with the vector components of those vectors. Note that x, y, v_x, v_y, and a_x are positive but that a_y is negative. Compare to Fig. 3-3.

The vectors **r, v,** and **a** are interrelated (see Eqs. 3-4, 3-5, and 3-10) and can be expressed in terms of their components, using unit vector notation, as

$$\mathbf{r} = \mathbf{i}x + \mathbf{j}y, \tag{4-1}$$

$$\mathbf{v} = \frac{d\mathbf{r}}{dt} = \mathbf{i}v_x + \mathbf{j}v_y, \tag{4-2}$$

53

and
$$\mathbf{a} = \frac{d\mathbf{v}}{dt} = \mathbf{i}a_x + \mathbf{j}a_y. \tag{4-3}$$

These equations can easily be extended to three dimensions by adding to them the terms $\mathbf{k}z$, $\mathbf{k}v_z$, and $\mathbf{k}a_z$, respectively in which $\mathbf{k}$ is a unit vector in the z-direction.

In Chapter 3 we considered the special case in which the particle moved in one dimension only, say along the x-axis, where the vectors $\mathbf{r}$, $\mathbf{v}$, and $\mathbf{a}$ were directed along this axis, either in the positive x-direction or the negative x-direction. The components y, v_y, and a_y were zero and we described the motion in terms of equations relating the scalar quantities x, v_x, and a_x. Or, when the particle moved along the y-axis, the components x, v_x, and a_x were zero and the motion was described in terms of equations relating the scalar quantities y, v_y, and a_y. In this chapter we consider motion in the x-y plane so that, in general, both sets of components have nonzero values.

Let us consider first the special case of motion in a plane with constant acceleration. Here, as the particle moves, the acceleration $\mathbf{a}$ does not vary either in magnitude or in direction. Hence the components of $\mathbf{a}$ also will not vary, that is, $a_x = $ constant and $a_y = $ constant. We then have a situation which can be described as the sum of two component motions occurring simultaneously with constant acceleration along each of two perpendicular directions. The particle will move, in general, along a curved path in the plane. This may be so even if one component of the acceleration, say a_x, is zero, for then the corresponding component of the velocity, say v_x, may have a constant, nonzero value. An example of this latter situation is the motion of a projectile which follows a curved path in a vertical plane and, neglecting the effects of air resistance, is subject to a constant acceleration $\mathbf{g}$ directed down along the y-axis only.

We can obtain the general equations for plane motion with constant $\mathbf{a}$ simply by setting

$$a_x = \text{constant} \quad \text{and} \quad a_y = \text{constant}.$$

The equations for constant acceleration, summarized in Table 3-1, then apply to both the x- and y-components of the position vector $\mathbf{r}$, the velocity vector $\mathbf{v}$, and the acceleration vector $\mathbf{a}$ (see Table 4-1).

4-2
MOTION IN A PLANE WITH CONSTANT ACCELERATION

Table 4-1
Motion with constant acceleration in the x-y plane

Equation No.	x-Motion Equations	Equation No.	y-Motion Equations
4-4a	$v_x = v_{x0} + a_x t$	4-4a'	$v_y = v_{y0} + a_y t$
4-4b	$x = x_0 + \frac{1}{2}(v_{x0} + v_x)t$	4-4b'	$y = y_0 + \frac{1}{2}(v_{y0} + v_y)t$
4-4c	$x = x_0 + v_{x0}t + \frac{1}{2}a_x t^2$	4-4c'	$y = y_0 + v_{y0}t + \frac{1}{2}a_y t^2$
4-4d	$v_x^2 = v_{x0}^2 + 2a_x(x - x_0)$	4-4d'	$v_y^2 = v_{y0}^2 + 2a_y(y - y_0)$

The two sets of equations in Table 4-1 are related in that the time parameter t is the same for each, since t represents the time at which the particle, moving in a curved path in the x-y plane, occupied a position described by the position components x and y.

The equations of motion in Table 4-1 may also be expressed in vector form. For example, substituting Eqs. 4-4a, 4a' into Eq. 4-2 yields

$$\mathbf{v} = \mathbf{i}v_x + \mathbf{j}v_y$$

$$= \mathbf{i}(v_{x0} + a_x t) + \mathbf{j}(v_{y0} + a_y t)$$

$$= (\mathbf{i}v_{x0} + \mathbf{j}v_{y0}) + (\mathbf{i}a_x + \mathbf{j}a_y)t.$$

The first quantity in parentheses is the initial velocity vector $\mathbf{v}_0$ (see Eq. 4-2) and the second is the (constant) acceleration vector $\mathbf{a}$ (see Eq. 4-3). Thus the vector relation

$$\mathbf{v} = \mathbf{v}_0 + \mathbf{a}t \qquad (4\text{-}5a)$$

is equivalent to the two scalar relations Eqs. 4-4a, a' in Table 4-1. It shows simply and compactly that the velocity $\mathbf{v}$ at time t is the sum of the initial velocity $\mathbf{v}_0$ which the particle would have in the absence of acceleration plus the (vector) change in velocity, $\mathbf{a}t$, acquired during the time t under the constant acceleration $\mathbf{a}$. Similarly, the scalar equations 4-4c, c' are equivalent to the single vector equation

$$\mathbf{r} = \mathbf{r}_0 + \mathbf{v}_0 t + \tfrac{1}{2}\mathbf{a}t^2, \qquad (4\text{-}5b)$$

which is also easily interpreted. The proof of this and other relations is left to Problem 3.

4-3
PROJECTILE MOTION

An example of curved motion with constant acceleration is projectile motion. This is the two-dimensional motion of a particle thrown obliquely into the air. The ideal motion of a baseball or a golf ball is an example of projectile motion.* We assume that we can neglect the effect of the air on this motion.

The motion of a projectile is one of constant acceleration g, directed downward, and thus should be described by the equations in Table 4-1. There is no horizontal component of acceleration. If we choose a coordinate system with the positive y-axis vertically upward, we may put $a_y = -g$ and $a_x = 0$ in these equations.

Let us further choose the origin of our coordinate system to be the point at which the projectile begins its flight (see Fig. 4-2). Hence the origin will be the point at which the ball leaves the thrower's hand or the fuel in the rocket burns out, for example. In Table 4-1 this choice of origin implies that $x_0 = y_0 = 0$. The velocity at $t = 0$, the instant the projectile begins its flight, is $\mathbf{v}_0$, which makes an angle θ_0 with the positive x-direction. The x- and y-components of $\mathbf{v}_0$ (see Fig. 4-2) are then

$$v_{x0} = v_0 \cos \theta_0 \qquad \text{and} \qquad v_{y0} = v_0 \sin \theta_0.$$

Because there is no horizontal component of acceleration, the horizontal component of the velocity will be constant. In Eq. 4-4a we set $a_x = 0$ and $v_{x0} = v_0 \cos \theta_0$, so that

$$v_x = v_0 \cos \theta_0. \qquad (4\text{-}6a)$$

The horizontal velocity component retains its initial value throughout the flight.

The vertical component of the velocity will change with time in accordance with vertical motion with constant downward acceleration.

---

* See Galileo Galilei, *Dialogues Concerning Two New Sciences*, the "Fourth Day," for a fascinating discussion of Galileo's research on projectiles.

In Eq. 4-4a' we set

$$a_y = -g \quad \text{and} \quad v_{y0} = v_0 \sin \theta_0,$$

so that

$$v_y = v_0 \sin \theta_0 - gt. \tag{4-6a'}$$

The vertical velocity component is that of free fall. Indeed, if we view the motion of Fig. 4-2 from a reference frame that moves to the right with a speed v_{x0}, the motion will be that of an object thrown vertically upward with an initial speed $v_0 \sin \theta_0$.

The magnitude of the resultant velocity vector at any instant is

$$v = \sqrt{v_x{}^2 + v_y{}^2}. \tag{4-7}$$

The angle θ that the velocity vector makes with the horizontal at that instant is given by

$$\tan \theta = \frac{v_y}{v_x}.$$

The velocity vector is tangent to the path of the particle at every point, as shown in Fig. 4-2.

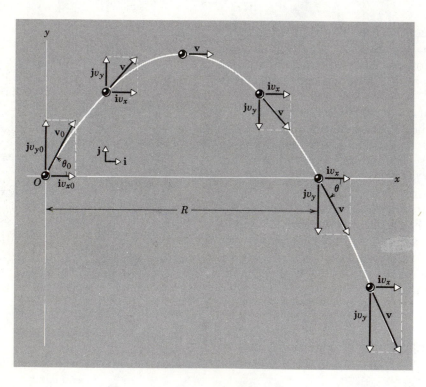

figure 4-2
The trajectory of a projectile, showing the initial velocity $\mathbf{v}_0$ and its vector components and also the velocity $\mathbf{v}$ and its vector components at five later times. Note that $v_x = v_{x0}$ throughout the flight. The distance R is called the range.

The x-coordinate of the particle's position at any time, obtained from Eq. 4-4c with $x_0 = 0$, $a_x = 0$, and $v_{x0} = v_0 \cos \theta_0$, is

$$x = (v_0 \cos \theta_0)t. \tag{4-6c}$$

The y-coordinate, obtained from Eq. 4-4c' with $y_0 = 0$, $a_y = -g$, and $v_{y0} = v_0 \sin \theta_0$, is

$$y = (v_0 \sin \theta_0)t - \tfrac{1}{2}gt^2. \tag{4-6c'}$$

Equations 4-6c, c' give us x and y as functions of the common parameter t, the time of flight. By combining and eliminating t from them, we obtain

$$y = (\tan \theta_0)x - \frac{g}{2(v_0 \cos \theta_0)^2}\, x^2, \qquad (4\text{-}8)$$

which relates y to x and is the equation of the trajectory of the projectile. Since v_0, θ_0, and g are constants, this equation has the form

$$y = bx - cx^2,$$

the equation of a parabola. Hence the trajectory of a projectile is parabolic.*

A plane is flying at a constant horizontal velocity of 500 km/h at an elevation of 5.0 km toward a point directly above its target. At what angle of sight ϕ should a survival package be released to strike the target (Fig. 4-3)?

EXAMPLE 1

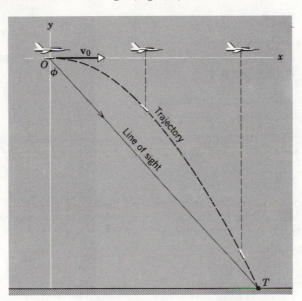

figure 4-3
Example 1. A survival package is released from an airplane with horizontal velocity $\mathbf{v_0}$.

We choose a reference frame fixed with respect to the earth, its origin O being the release point. The motion of the package at the moment of release is the same as that of the plane. Hence the initial package velocity $\mathbf{v_0}$ is horizontal and its magnitude is 500 km/h. The angle of projection θ_0 is zero.

We find the time of fall from Eq. 4-6c'. With $\theta_0 = 0$ and $y = 5.0$ km this gives

$$t = \sqrt{\frac{2y}{g}} = \sqrt{-\frac{(2)(-5.0 \times 10^3 \text{ m})}{(9.8 \text{ m/s}^2)}} = 31.9 \text{ s}$$

Note that the time of fall does not depend on the speed of the plane for a horizontal projection. (See, however, Problem 11.)

The horizontal distance traveled by the package in this time is given by Eq. 4-6c, $x = (v_0 \cos \theta_0)t$, or

$$x = (500 \text{ km/h}) \times 10^3 \text{m/km} \times (1 \text{ h/3600 s}) \times (31.9 \text{ s}) = 4430 \text{ m}.$$

so that the angle of sight (Fig. 4-3) should be

$$\phi = \tan^{-1} \frac{x}{|y|} = \tan^{-1} \frac{4430 \text{ m}}{5000 \text{ m}} = 42°.$$

Does the motion of the package appear to be parabolic when viewed from a reference frame fixed with respect to the plane?

---

* See "Galileo's Discovery of the Parabolic Trajectory" by Stillman Drake and James MacLachlan in *Scientific American*, March 1975.

A soccer player kicks a ball at an angle of 37° from the horizontal with an initial speed of 50 ft/s. (A right triangle, one of whose angles is 37°, has sides in the ratio 3:4:5, or 6:8:10.) Assuming that the ball moves in a vertical plane:

EXAMPLE 2

(a) Find the time t_1 at which the ball reaches the highest point of its trajectory.

At the highest point, the vertical component of velocity v_y is zero. Solving Eq. 4-6a' for t, we obtain

$$t = \frac{v_0 \sin \theta_0 - v_y}{g}.$$

With

$$v_y = 0, \qquad v_0 = 50 \text{ ft/s}, \qquad \theta_0 = 37°, \qquad g = 32 \text{ ft/s}^2,$$

we have

$$t_1 = \frac{[50(\frac{6}{10}) - 0] \text{ ft/sec}}{32 \text{ ft/sec}^2} = \frac{15}{16} \text{ s}.$$

(b) How high does the ball go?

The maximum height is reached at $t = 15/16$ s. By using Eq. 4-6c',

$$y = (v_0 \sin \theta_0)t - \tfrac{1}{2}gt^2,$$

we have

$$y_{\max} = (50 \text{ ft/s})(\tfrac{6}{10})(\tfrac{15}{16} \text{ s}) - \tfrac{1}{2}(32 \text{ ft/s}^2)(\tfrac{15}{16})^2 \text{ s}^2 = 14 \text{ ft}.$$

(c) What is the range of the ball and how long is it in the air?

The horizontal distance from the starting point at which the ball returns to its original elevation (ground level) is the *range* R. We set $y = 0$ in Eq. 4-6c' and find the time t_2 required to transverse this range. We obtain

$$t_2 = \frac{2v_0 \sin \theta_0}{g} = \frac{2(50 \text{ ft/s})(\tfrac{6}{10})}{32 \text{ ft/s}^2} = \frac{15}{8} \text{ sec}.$$

Notice that $t_2 = 2t_1$. This corresponds to the fact that the same time is required for the ball to go up (reach its maximum height from ground) as is required for the ball to come down (reach the ground from its maximum height).

The range R can then be obtained by inserting this value t_2 for t in Eq. 4-6c. We obtain, from $x = (v_0 \cos \theta_0)t$,

$$R = (v_0 \cos \theta_0)t_2 = (50 \text{ ft/s})(\tfrac{8}{10})(\tfrac{15}{8} \text{ s}) = 75 \text{ ft}.$$

(d) What is the velocity of the ball as it strikes the ground? From Eq. 4-6a we obtain

$$v_x = v_0 \cos \theta_0 = (50 \text{ ft/s})(\tfrac{8}{10}) = 40 \text{ ft/s}.$$

From Eq. 4-6a' we obtain for $t = t_2 = \tfrac{15}{8}$ s,

$$v_y = v_0 \sin \theta_0 - gt = (50 \text{ ft/s})(\tfrac{6}{10}) - (32 \text{ ft/s}^2)(\tfrac{15}{8} \text{ s}) = -30 \text{ ft/s}.$$

Hence, from Eq. 4-7,

$$v = \sqrt{v_x^2 + v_y^2} = \sqrt{(40 \text{ ft/s})^2 + (-30 \text{ ft/s})^2} = 50 \text{ ft/s},$$

and

$$\tan \theta = v_y/v_x = -\tfrac{30}{40},$$

so that $\theta = -37°$, or 37° *clockwise* from the x-axis. Notice that $\theta = -\theta_0$, as we expect from symmetry (Fig. 4-2).

In a favorite lecture demonstration an air gun is sighted at an elevated target which is released in free fall by a trip mechanism as the "bullet" leaves the muzzle. No matter what the initial speed of the bullet, it always hits the falling target.

EXAMPLE 3

The simplest way to understand this is the following. If there were no acceleration due to gravity, the target would not fall and the bullet would move

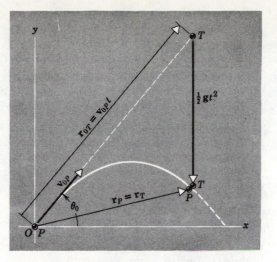

figure 4-4

Example 3. In the motion of a projectile, its displacement from the origin at any time t can be thought of as the sum of two vectors: $\mathbf{v}_{0P}t$, directed along $\mathbf{v}_{0P}$, and $\frac{1}{2}\mathbf{g}t^2$, directed down.

along the line of sight directly into the target (Fig. 4-4). The effect of gravity is to cause each body to accelerate down at the same rate from the position it would otherwise have had. Therefore, in the time t, the bullet will fall a distance $\frac{1}{2}gt^2$ from the position it would have had along the line of sight and the target will fall the same distance from its starting point. When the bullet reaches the line of fall of the target, it will be the same distance below the target's initial position as the target is and hence the collision. If the bullet moves faster than shown in the figure (v_0 larger), it will have a greater range and will cross the line of fall at a higher point; but since it gets there sooner, the target will fall a correspondingly smaller distance in the same time and collide with it. A similar argument holds for slower speeds.

For an equivalent analysis, let us use Eq. 4-5b.

$$\mathbf{r} = \mathbf{r}_0 + \mathbf{v}_0 t + \tfrac{1}{2}\mathbf{a}t^2$$

to describe the positions of the projectile and the target at any time t. For the projectile P, $\mathbf{r}_0 = 0$ and $\mathbf{a} = \mathbf{g}$, and we have

$$\mathbf{r}_P = \mathbf{v}_{0P}t + \tfrac{1}{2}\mathbf{g}t^2.$$

For the target T, $\mathbf{r}_0 = \mathbf{r}_{0T}$, $\mathbf{v}_0 = 0$, and $\mathbf{a} = \mathbf{g}$, leading to

$$\mathbf{r}_T = \mathbf{r}_{0T} + \tfrac{1}{2}\mathbf{g}t^2.$$

If there is a collision, we must have $\mathbf{r}_P = \mathbf{r}_T$. Inspection shows that this will always occur at a time t given by $\mathbf{r}_{0T} = \mathbf{v}_{0P}t$, that is, in the time t ($= r_{0T}/v_{0P}$) required for the projectile to travel to the target position along the line of sight, assuming that its initial velocity remains unchanged.

4-4
UNIFORM CIRCULAR MOTION

In Section 3-6 we saw that acceleration arises from a change in velocity. In the simple case of free fall the velocity changed in magnitude only, but not in direction. For a particle moving in a circle with constant speed, called *uniform circular motion*, the velocity vector changes continuously in direction but not in magnitude. We seek now to obtain the acceleration in uniform circular motion.

The situation is shown in Fig. 4-5a. Let P be the position of the particle at the time t and P' its position at the time $t + \Delta t$. The velocity at P is $\mathbf{v}$, a vector tangent to the curve at P. The velocity at P' is $\mathbf{v}'$, a vector tangent to the curve at P'. Vectors $\mathbf{v}$ and $\mathbf{v}'$ are equal in magnitude, the speed being constant, but their directions are different. The length of path traversed during Δt is the arc length PP', which is equal to $v\,\Delta t$, v being the constant speed.

Now redraw the vectors $\mathbf{v}$ and $\mathbf{v}'$, as in Fig. 4-5b, so that they originate

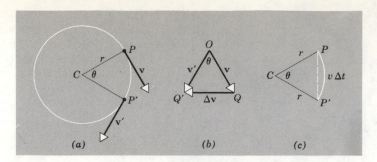

figure 4-5
Uniform circular motion. The
particle travels around a circle at
constant speed. Its velocity at two
points P and P' is shown. Its
change in velocity in going from
P to P' is Δv.

at a common point. We are free to do this as long as the magnitude and direction of each vector are the same as in Fig. 4-5a. This diagram (Fig. 4-5b) enables us to see clearly the *change in velocity* as the particle moved from P to P'. This change, $v' - v = \Delta v$, is the vector which must be added to v to get v'. Notice that it points inward, approximately toward the center of the circle.

Now the triangle OQQ' formed by v, v', and Δv is similar to the triangle CPP' (Fig. 4-5c) formed by the chord PP' and the radii CP and CP'. This is so because both are isosceles triangles having the same vertex angle; the angle θ between v and v' is the same as the angle PCP' because v is perpendiculat to CP and v' is perpendicular to CP'. We can therefore write

$$\frac{\Delta v}{v} = \frac{v\,\Delta t}{r}, \qquad \text{approximately,}$$

the chord PP' being taken equal to the arc length PP'. This relation becomes more nearly exact as Δt is diminished, since the chord and the arc then approach each other. Notice also that Δv approaches closer and closer to a direction perpendicular to v and v' as Δt is diminished and therefore approaches closer and closer to a direction pointing to the exact center of the circle. It follows from this relation that

$$\frac{\Delta v}{\Delta t} = \frac{v^2}{r}, \qquad \text{approximately,}$$

and in the limit when $\Delta t \to 0$ this expression becomes exact. We therefore obtain

$$a = \lim_{\Delta t \to 0} \frac{\Delta v}{\Delta t} = \frac{v^2}{r} \qquad (4\text{-}9)$$

as the magnitude of the acceleration. The direction of a is instantaneously along a radius inward toward the center of the circle.

Figure 4-6 shows the instantaneous relation between v and a at various points of the motion. The magnitude of v is constant, but its direction changes continuously. This gives rise to an acceleration a which is also constant in magnitude (but not zero) but continuously changing in direction. The velocity v is always tangent to the circle in the direction of motion; the acceleration a is always directed radially inward. Because of this, a is called a radial, or *centripetal*, acceleration. Centripetal means "seeking a center."

Both in free fall and in projectile motion a is constant in direction and magnitude and we can use the equations developed for constant acceleration (see Table 4-1). We cannot use these equations for uniform circular motion because a varies in direction and is therefore not constant.

The units of centripetal acceleration are the same as those of an acceleration resulting from a change in the magnitude of a velocity.

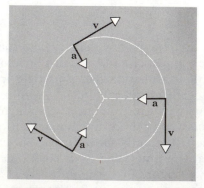

figure 4-6
In uniform circular motion the acceleration a is always directed toward the center of the circle and hence is perpendicular to v.

Dimensionally, we have

$$\frac{v^2}{r} = \left(\frac{\text{length}}{\text{time}}\right)^2 \bigg/ \text{length} = \frac{\text{length}}{\text{time}^2} \quad \text{or} \quad \frac{L}{T^2},$$

which are the dimensions of acceleration. The units therefore may be ft/s² and m/s², among others.

The acceleration resulting from a change in direction of a velocity is just as real and just as much an acceleration in every sense as that arising from a change in magnitude of a velocity. By definition, acceleration is the time rate of change of velocity, and velocity, being a vector, can change in direction as well as magnitude. If a physical quantity is a vector, its directional aspects cannot be ignored, for their effects will prove to be every bit as important and real as those produced by changes in magnitude.

It is worth emphasizing at this point that there need not be any motion in the direction of an acceleration and that there is no fixed relation in general between the directions of **a** and **v.** In Fig. 4-7 we give examples in which the angle between **v** and **a** varies from 0 to 180°. Only in one case, $\theta = 0°$, is the motion in the direction of **a.**

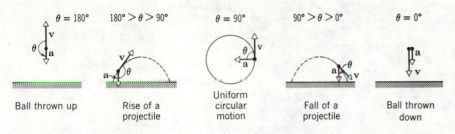

figure 4-7
Showing the relation between **v** and **a** for various motions.

The moon revolves about the earth, making a complete revolution in 27.3 days. Assume that the orbit is circular and has a radius of 239,000 miles. What is the magnitude of the acceleration of the moon toward the earth?

We have $r = 239,000$ mi $= 3.85 \times 10^8$ m. The time for one complete revolution, called the period, is $T = 27.3$ d $= 2.36 \times 10^6$ s. The speed of the moon (assumed constant) is therefore

$$v = 2\pi r/T = 1020 \text{ m/s}.$$

The centripetal acceleration is

$$a = \frac{v^2}{r} = \frac{(1020 \text{ m/s})^2}{3.85 \times 10^8 \text{ m}} = 0.00273 \text{ m/s}^2, \quad \text{or only } 2.78 \times 10^{-4} \text{ g}.$$

EXAMPLE 4

Calculate the speed of an earth satellite, assuming that it is traveling at an altitude h of 140 miles above the surface of the earth where $g = 30$ ft/s². The radius R of the earth is 3960 mi.

Like any free object near the earth's surface the satellite has an acceleration g toward the earth's center. It is this acceleration that causes it to follow the circular path. Hence the centripetal acceleration is g, and from Eq. 4-9, $a = v^2/r$, we have

$$g = v^2/(R + h),$$

or

$$v = \sqrt{(R + h)g} = \sqrt{(3960 \text{ mi} + 140 \text{ mi})(5280 \text{ ft/mi})(30 \text{ ft/s}^2)}$$
$$= 2.55 \times 10^4 \text{ ft/s} = 17,400 \text{ mi/h}.$$

EXAMPLE 5

Let us now derive Eq. 4-9 using vector methods. Figure 4-8a shows a particle in uniform circular motion about the origin O of a reference frame. For this motion the polar coordinates r, θ are more useful than the rectangular coordinates x, y because r remains constant throughout the motion and θ increases in a simple linear way with time; the behavior of x and y during such motion is more complex. The two sets of coordinates are related by

$$r = \sqrt{x^2 + y^2} \qquad \text{and} \qquad \theta = \tan^{-1} y/x \qquad (4\text{-}10a)$$

or by the reciprocal relations

$$x = r \cos \theta \qquad \text{and} \qquad y = r \sin \theta. \qquad (4\text{-}10b)$$

In rectangular coordinate systems we used the unit vectors $\mathbf{i}$ and $\mathbf{j}$ to describe motion in the x-y plane. Here we find it more convenient to introduce two new unit vectors $\mathbf{u}_r$ and $\mathbf{u}_\theta$. These, like $\mathbf{i}$ and $\mathbf{j}$, have unit length and are dimensionless; they designate direction only.

The unit vector $\mathbf{u}_r$ at any point is in the direction of increasing $\mathbf{r}$ at that point; it is directed radially outward from the origin. The unit vector $\mathbf{u}_\theta$ at any point is in the direction of increasing θ at that point; it is always tangent to a circle through the point in a counterclockwise direction. As Fig. 4-8a shows, $\mathbf{u}_r$ and $\mathbf{u}_\theta$ are at right angles to each other. The unit vectors $\mathbf{u}_r$ and $\mathbf{u}_\theta$ differ from the unit vectors $\mathbf{i}$ and $\mathbf{j}$ in that the directions of $\mathbf{u}_r$ and $\mathbf{u}_\theta$ vary from point to point in the plane; the unit vectors $\mathbf{u}_r$ and $\mathbf{u}_\theta$ are thus *not* constant vectors.

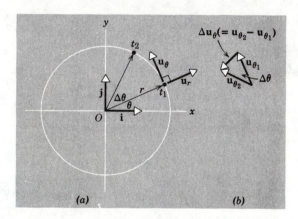

(a)   (b)

figure 4-8

(a) A particle moving counterclockwise in a circle of radius r. (b) The unit vectors $\mathbf{u}_{\theta 1}$ and $\mathbf{u}_{\theta 2}$ at times t_1 and t_2 respectively, and the change $\Delta \mathbf{u}_\theta$ ($= \mathbf{u}_{\theta 2} - \mathbf{u}_{\theta 1}$).

In terms of $\mathbf{u}_r$ and $\mathbf{u}_\theta$ the motion of a particle moving counterclockwise at uniform speed v in a circle about the origin in Fig. 4-8a can be described by the vector equation

$$\mathbf{v} = \mathbf{u}_\theta v. \qquad (4\text{-}11)$$

This relation tells us, correctly, that the direction of $\mathbf{v}$ (which is the same as the direction of $\mathbf{u}_\theta$) is tangent to the circle and that the magnitude of $\mathbf{v}$ is the constant quantity v (because the magnitude of $\mathbf{u}_\theta$ is unity).

To find the acceleration we combine Eqs. 4-3 and 4-11, yielding

$$\mathbf{a} = \frac{d\mathbf{v}}{dt} = \frac{d\mathbf{u}_\theta}{dt} v. \qquad (4\text{-}12)$$

Note that v in Eq. 4-11 is a constant, but $\mathbf{u}_\theta$ is not since its direction changes as the particle moves. To evaluate $d\mathbf{u}_\theta/dt$, consider Fig. 4-8b which shows the unit vectors $\mathbf{u}_{\theta 1}$ and $\mathbf{u}_{\theta 2}$ corresponding to an elapsed time Δt ($= t_2 - t_1$) for the moving particle. The vector $\Delta \mathbf{u}_\theta$ ($= \mathbf{u}_{\theta 2} - \mathbf{u}_{\theta 1}$) points radially inward toward the origin in the limiting case as $\Delta t \to 0$. In other words, $d\mathbf{u}_\theta$ at any point has the direction of $-\mathbf{u}_r$. The angle between $\mathbf{u}_{\theta 2}$ and $\mathbf{u}_{\theta 1}$ in the figure is $\Delta \theta$, which is the angle swept out by a radial line from the origin to the particle in time Δt. The magnitude of $\Delta \mathbf{u}_\theta$ is simply $\Delta \theta$; bear in mind that the vectors $\mathbf{u}_{\theta 1}$ and $\mathbf{u}_{\theta 2}$ in Fig. 4-8b have the magnitude unity. Thus

$$\frac{d\mathbf{u}_\theta}{dt} = -\mathbf{u}_r \lim_{\Delta t \to 0} \frac{\Delta \theta}{dt} = -\mathbf{u}_r \frac{d\theta}{dt}$$

and, from Eq. 4-12,

$$\mathbf{a} = \frac{d\mathbf{u}_\theta}{dt} v = -\mathbf{u}_r \frac{d\theta}{dt} v. \qquad (4\text{-}13)$$

Now, $d\theta/dt$ is the uniform angular rotation rate of the particle and is given by

$$\frac{d\theta}{dt} = \frac{2\pi \text{ radians}}{\text{time for one revolution}} = \frac{2\pi}{2\pi r/v} = \frac{v}{r}.$$

Putting this into Eq. 4-13 leads us finally to

$$\mathbf{a} = -\mathbf{u}_r \frac{v^2}{r} \qquad (4\text{-}14)$$

which tells us that the acceleration in uniform circular motion has a magnitude v^2/r (see Eq. 4-9) and points radially inward (note the factor $-\mathbf{u}_r$). The vector relation Eq. 4-14 thus tells us *both* the magnitude *and* the direction of the centripetal acceleration $\mathbf{a}$. Note that, as expected, $\mathbf{a}$ has a constant magnitude but changes continually in direction because $\mathbf{u}_r$ changes continually in direction.

4-5 TANGENTIAL ACCELERATION IN CIRCULAR MOTION

We now consider the more general case of circular motion in which the speed v of the moving particle is *not* constant. We shall use vector methods in polar coordinates.

As before, the velocity is given by Eq. 4-11, or

$$\mathbf{v} = \mathbf{u}_\theta v$$

except that, in this case, not only $\mathbf{u}_\theta$ but also v varies with time. Recalling the formula for the derivative of a product, we obtain for the acceleration

$$\mathbf{a} = \frac{d\mathbf{v}}{dt} = \mathbf{u}_\theta \frac{dv}{dt} + v \frac{d\mathbf{u}_\theta}{dt} \qquad (4\text{-}15)$$

In Eq. 4-12 the first term in this equation was not present because, v being there assumed to be constant, dv/dt was zero. The last term in Eq. 4-15 reduces, as we saw in the last section, to $-\mathbf{u}_r(v^2/r)$. We can now write Eq. 4-15 as

$$\mathbf{a} = \mathbf{u}_\theta a_T - \mathbf{u}_r a_R, \qquad (4\text{-}16)$$

in which $a_T = dv/dt$ and $a_R = v^2/r$. The first term, $\mathbf{u}_\theta a_T$, is the vector component of $\mathbf{a}$ that is tangent to the path of the particle and arises from a change in the *magnitude* of the velocity in circular motion (see Fig. 4-9). This term and a_T are called the *tangential acceleration*. The second term $-\mathbf{u}_r a_R$ is the vector component of $\mathbf{a}$ directed radially in toward the center of the circle and arises from a change in the *direction* of the velocity in circular motion (see Fig. 4-9). This term and a_R are called the *centripetal acceleration*.

The magnitude of the instantaneous acceleration is

$$a = \sqrt{a_T{}^2 + a_R{}^2} \qquad (4\text{-}17)$$

If the speed is constant, then $a_T = dv/dt = 0$ and Eq. 4-16 reduces to Eq. 4-14.

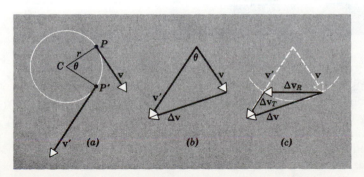

figure 4-9
(a) In nonuniform circular motion the speed is variable. *(b)* The change in velocity $\Delta \mathbf{v}$ in going from P to P' is made up of two parts: *(c)* $\Delta \mathbf{v}_R$ caused by the change in direction of $\mathbf{v}$, and $\Delta \mathbf{v}_T$ caused by the change in magnitude of $\mathbf{v}$. In the limit as $\Delta t \to 0$, $\Delta \mathbf{v}_R$ points toward the center C of the circle and $\Delta \mathbf{v}_T$ is tangent to the circular path.

When the speed v is not constant, a_T is not zero and a_R varies from point to point. If the speed changes at a rate that is not constant, then a_T will also vary from point to point.

If the motion is not circular, the formulas for a_T ($= dv/dt$) and for a_R ($= v^2/r$) can still be applied if instead of using for r the magnitude of the radius vector from the origin, we substitute the radius of curvature of the path at the instantaneous position of the particle. Then a_T gives the component of acceleration tangent to the curve at that position, and a_R gives the component of acceleration normal to the curve at that position. Figure 4-10 shows the track left in a liquid-hydrogen-filled bubble chamber by an energetic electron that spirals inward. The electron loses energy as it traverses the liquid in the chamber so that its speed v is being reduced steadily. Thus there is at every point a tangential acceleration a_T given by dv/dt. The centripetal acceleration a_R at any point is given by v^2/r, where r is the radius of curvature of the track at the point in question; both v and r become smaller as the particle loses energy. The force causing the electron to spiral is produced by a magnetic field present in the bubble chamber and at right angles to the plane of Fig. 4-10 (see Chapter 33).

figure 4-10
A track left in a 10-in. liquid-hydrogen-filled bubble chamber by an energetic spiralling electron. (Courtesy Lawrence Radiation Laboratory.) This picture is one of a number in a collection prepared for easy stereoscopic viewing and published, with explanatory material, as *Introduction to the Detection of Nuclear Particles in a Bubble Chamber*, The Ealing Press, Cambridge, Massachusetts (1964). When viewed stereoscopically the electron is seen to be moving toward the reader as it moves in along the spiral. Its velocity vector at any point, thus, does not lie in the plane of the figure, but tilts up out of it; its motion is thus three-dimensional, rather than two-dimensional as we assumed for other examples in this chapter.

4-6
RELATIVE VELOCITY AND ACCELERATION

In earlier sections we considered the addition of velocities in a particular reference frame. Let us now consider the relation between the velocity of an object as determined by one observer S (= reference frame S) and the velocity of the same object as determined by another observer S' (= reference frame S') who is moving with respect to the first.

Consider observer S fixed to the earth, so that his reference frame is the earth. The other observer S' is moving on the earth—for example, a passenger sitting on a moving train—so that his reference frame is the train. They each follow the motion of the same object, say an automobile on a road or a man walking through the train. Each observer will record a displacement, a velocity, and an acceleration for this object measured *relative to his reference frame.* How will these measurements compare? In this section we consider only the case in which the second frame is in motion with respect to the first with a *constant* velocity **u**.

In Fig. 4-11 the reference frame S represented by the x- and y-axes can be thought of as fixed to the earth. The shaded region indicates another reference frame S', represented by x'- and y'-axes, which moves along the x-axis with a constant velocity **u**, as measured in the S-system; it can be thought of as drawn on the floor of a railroad flatcar.

Initially, a particle (say a ball on the flatcar) is at a position called A in the S-frame and called A' in the S'-frame. At a time t later the flatcar and its S' reference frame have moved a distance ut to the right and the particle has moved to B. The displacement of the particle from its initial position *in the S-frame* is the vector **r** from A to B. The displacement of the particle from its initial position *in the S'-frame* is the vector **r'** from A' to B. These are different vectors because the reference point A' of the moving frame has been displaced a distance ut along the x-axis during the motion. From the figure we see that **r** is the vector sum of **r'** and **u**t:

$$\mathbf{r} = \mathbf{r}' + \mathbf{u}t. \qquad (4\text{-}18)$$

Differentiating Eq. 4-18 leads to

$$\frac{d\mathbf{r}}{dt} = \frac{d\mathbf{r}'}{dt} + \mathbf{u}.$$

But $d\mathbf{r}/dt = \mathbf{v}$, the instantaneous velocity of the particle measured in the S-frame, and $d\mathbf{r}'/dt = \mathbf{v}'$, the instantaneous velocity of the same particle measured in the S' frame, so that

$$\mathbf{v} = \mathbf{v}' + \mathbf{u}. \qquad (4\text{-}19)$$

Hence the velocity of the particle relative to the S-frame, **v**, is the vector sum of the velocity of the particle relative to the S'-frame, **v'**, and the velocity **u** of the S'-frame relative to the S-frame.

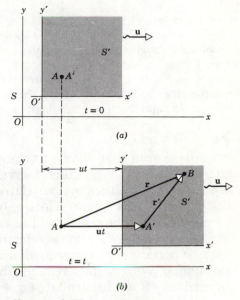

figure 4-11
Two reference frames, S ($= x, y$) and S' ($= x', y'$); S' moves to the right, relative to S, with speed u.

EXAMPLE 6

(*a*) The compass of an airplane indicates that it is heading due east. Ground information indicates a wind blowing due north. Show on a diagram the velocity of the plane with respect to the ground.

The object is the airplane. The earth is one reference frame (S) and the air is the other reference frame (S') moving with respect to the first. Then

   u is the velocity of the air with respect to the ground.
   v' is the velocity of the plane with respect to the air
   v is the velocity of the plane with respect to the ground.

In this case **u** points north and **v'** points east. Then the relation $\mathbf{v} = \mathbf{v}' + \mathbf{u}$ determines the velocity of the plane with respect to the ground, as shown in Fig. 4-12*a.*

The angle α is the angle N of E of the plane's course with respect to the ground and is given by

$$\tan \alpha = u/v'.$$

The airplane's speed with respect to the ground is given by

$$v = \sqrt{(v')^2 + u^2}.$$

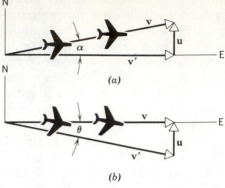

For example, if the air-speed indicator shows that the plane is moving relative to the air at a speed of 200 mi/h, and if the speed of the wind with respect to the ground is 40.0 mi/h, then

$$v = \sqrt{(200)^2 + (40.0)^2} \text{ mi/h} = 204 \text{ mi/h}$$

is the ground speed of the plane and

$$\alpha = \tan^{-1} \frac{40.0}{200} = 11° \, 20'$$

gives the course of the plane N of E.

(b) Now draw the vector diagram showing the direction the pilot must steer the plane through the air for the plane to travel due east with respect to the ground.

He would naturally head partly into the wind. His speed relative to the earth will therefore be less than before. The vector diagram is shown in Fig. 4-12b. You should calculate θ and v, using the previous data for **u** and **v**′.

figure 4-12
Example 6

We have seen that different velocities are assigned to a particle by different observers when the observers are in relative motion. These velocities always differ by the relative velocity of the two observers, which here is a constant velocity. It follows that when the particle velocity changes, the change will be the same for both observers. Hence they each measure the same acceleration for the particle. *The acceleration of a particle is the same in all reference frames moving relative to one another with constant velocity*; that is, **a** = **a**′. This result follows in a formal way if we differentiate Eq. 4-19. Thus $d\mathbf{v}/dt = d\mathbf{v}'/dt + d\mathbf{u}/dt$; but $d\mathbf{u}/dt = 0$ when **u** is constant, so that **a** = **a**′.

questions

1. In projectile motion when air resistance is negligible, is it ever necessary to consider three-dimensional motion rather than two-dimensional?

2. In broad jumping does it matter how high you jump? What factors determine the span of the jump?

3. Why doesn't the electron in the beam from an electron gun fall as much because of gravity as a water molecule in the stream from a hose? Assume horizontal motion initially in each case.

4. At what point in the path of a projectile does it have its minimum speed? its maximum?

5. Suppose you could vary the angle of incline θ of a plane surface that is fixed at a hinge line to a horizontal table top. How should you choose θ so that the balls dropped vertically and rebounding elastically from the incline have the maximum range?

6. What advantage is there, if any, in measuring angles in radians rather than in degrees?

7. An aviator, pulling out of a dive, follows the arc of a circle. He was said to have "experienced 3g's" in pulling out of the dive. Explain what this statement means.

8. Describe qualitatively the acceleration acting on a bead which, sliding along a frictionless wire, moves inward with constant speed along a spiral.

9. Could the acceleration of a projectile be represented in terms of a radial and a tangential component at each point of the motion? If so, is there any advantage to this representation?

10. Over a short distance a circular arc is a good approximation to a parabola. What then is the radius r of the circular arc approximating the motion of a projectile, of initial speed v_0 and angle θ_0, near the top of its path?

11. A boy sitting in a railroad car moving at constant velocity throws a ball

straight up into the air. Will the ball fall behind him? In front of him? Into his hand? What happens if the car accelerates forward or goes around a curve while the ball is in the air?

12. A man on the observation platform of a train moving with constant velocity drops a coin while leaning over the rail. Describe the path of the coin as seen by (a) the man on the train, (b) a person standing on the ground near the track, and (c) a person in a second train moving in the opposite direction to the first train on a parallel track.

13. A bus with a vertical windshield moves along in a rainstorm at speed v_b. The raindrops fall vertically with a terminal speed v_r. At what angle do the raindrops strike the windshield?

14. Drops are falling vertically in a steady rain. In order to go through the rain from one place to another in such a way as to encounter the least number of raindrops, should you move with the greatest possible speed, the least possible speed, or some intermediate speed?

15. What is wrong with this picture (Fig. 4-13)? The sailor is running with the wind.

16. An elevator is descending at a constant speed. A passenger takes a coin from his pocket and drops it to the floor. What accelerations would (a) the passenger and (b) a person at rest with respect to the elevator shaft observe for the falling coin?

figure 4-13
Question 15

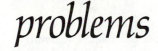

SECTION 4-1

1. Prove that for a vector **a** defined by

$$\mathbf{a} = \mathbf{i}a_x + \mathbf{j}a_y + \mathbf{k}a_z$$

the scalar components are given by

$$a_x = \mathbf{i} \cdot \mathbf{a}, \, a_y = \mathbf{j} \cdot \mathbf{a}, \text{ and } a_z = \mathbf{k} \cdot \mathbf{a}.$$

SECTION 4-2

2. A particle moves so that its position as a function of time is

$$\mathbf{r}(t) = \mathbf{i} + 4t^2\mathbf{j} + t\mathbf{k}.$$

(a) Write expressions for its velocity and acceleration as functions of time. (b) What is the shape of the particle's trajectory?

3. Show (a) that Eqs. 4-4b, b' can be expressed in vector form as

$$\mathbf{r} = \mathbf{r}_0 + \tfrac{1}{2}(\mathbf{v}_0 + \mathbf{v})t,$$

and (b) Eqs. 4-4 c, c' as

$$\mathbf{r} = \mathbf{r}_0 + \mathbf{v}_0 t + \tfrac{1}{2}\mathbf{a}t^2.$$

Also, show (c) that Eqs. 4-4d, d' can be combined to give

$$\mathbf{v} \cdot \mathbf{v} = \mathbf{v}_0 \cdot \mathbf{v}_0 + 2\mathbf{a} \cdot (\mathbf{r} - \mathbf{r}_0).$$

SECTION 4-3

4. Consider a projectile at the top of its trajectory. (a) What is its speed in terms of v_0 and θ_0? (b) What is its acceleration? (c) How is the direction of its acceleration related to that of its velocity? (See Question 10.)

5. A ball rolls off the edge of a horizontal table top 4.0 ft high. If it strikes the floor at a point 5.0 ft horizontally away from the edge of the table, what was its speed at the instant it left the table?   *Answer:* 10 ft/s.

6. A rifle with a muzzle velocity of 1500 ft/s shoots a bullet at a target 150 ft away. How high above the target must the rifle be aimed so that the bullet will hit the target?

7. (a) Show that the range of a projectile having an initial speed v_0 and angle

of projection θ_0 is $R = (v_0^2/g) \sin 2\theta_0$. Then show that a projection angle of 45° gives the maximum range (Fig. 4-14). (b) Show that the maximum height reached by the projectile is $y_{max} = (v_0 \sin \theta_0)^2/2g$. (c) Find the angle of projection at which the range and the maximum height of a projectile are equal. *Answer: (c) 76°.*

8. A projectile is fired horizontally from a gun located 144 ft (44 m) above a horizontal plain with a muzzle speed of 800 ft/s (240 m/s). (a) How long does the projectile remain in the air? (b) At what horizontal distance does it strike the ground? (c) What is the magnitude of the vertical component of its velocity as it strikes the ground?

9. A ball is thrown from the ground into the air. At a height of 9.1 m the velocity is observed to be $\mathbf{v} = 7.6\mathbf{i} + 6.1\mathbf{j}$ in m/s (x-axis horizontal, y-axis vertical). (a) To what maximum height will the ball rise? (b) What will be the total horizontal distance traveled by the ball? (c) What is the velocity of the ball (magnitude and direction) the instant before it hits the ground? *Answer: (a) 11 m. (b) 23 m. (c) 17 m/s, 63° below the horizontal.*

10. Electrons, nuclei, atoms, and molecules, like all forms of matter, will fall under the influence of gravity. Consider separately a beam of electrons, of nuclei, of atoms, and of molecules traveling a horizontal distance of 1.0 m. Let the average speed be for an electron 3.0×10^7 m/s, for a thermal neutron 2.2×10^3 m/s, for a neon atom 5.8×10^2 m/s, and for an oxygen molecule 4.6×10^2 m/s. Let the beams move through vacuum with initial horizontal velocities and find by how much their paths deviate from a straight line (vertical displacement in 1.0 m) due to gravity. How do these results compare to that for a beam of golf balls (use reasonable data)? What is the controlling factor here?

11. A dive bomber, diving at an angle of 53° with the vertical, releases a bomb at an altitude of 730 m. The bomb hits the ground 5.0 s after being released. (a) What is the speed of the bomber? (b) How far did the bomb travel horizontally during its flight? (c) What were the horizontal and vertical components of its velocity just before striking the ground? *Answer: (a) 200 m/s. (b) 810 m. (c) $v_h = 160$ m/s, $v_v = 170$ m/s.*

12. A football is kicked off with an initial speed of 64 ft/s at a projection angle of 45°. A receiver on the goal line 60 yd away in the direction of the kick starts running to meet the ball at that instant. What must be his minimum speed if he is to catch the ball before it hits the ground? (See, in this connection, "Catching a Baseball" by Seville Chapman in *American Journal of Physics*, October 1968.)

13. In a cathode-ray tube a beam of electrons is projected horizontally with a speed of 1.0×10^9 cm/s into the region between a pair of horizontal plates 2.0 cm long. An electric field between the plates exerts a constant downward acceleration on the electrons of magnitude 1.0×10^{17} cm/s². Find (a) the vertical displacement of the beam in passing through the plates and (b) the velocity of the beam (direction and magnitude) as it emerges from the plates. *Answer: (a) 2.0 mm. (b) $v_x = 1.0 \times 10^9$ cm/s, $v_y = 0.2 \times 10^9$ cm/s down.*

14. A batter hits a pitched ball at a height of 4.0 ft above the ground so that its angle of projection is 45° and its initial speed is 110 ft/s. The ball is hit fair down the left field line where a 24-ft high fence is located 320 ft from home plate. Will the ball clear the fence?

15. Galileo, in his *Two New Sciences*, states that "for elevations (angles of projection) which exceed or fall short of 45° by equal amounts, the ranges are equal. . . ." Prove this statement. See Fig. 4-14.

16. A ball rolls off the top of a stairway with a horizontal velocity of magnitude 5.0 ft/s. The steps are 8.0 in. high and 8.0 in. wide. Which step will the ball hit first?

17. (a) Show that if the acceleration due to gravity changes by an amount dg, the range of a projectile (see Problem 7) of given initial speed v_0 and angle

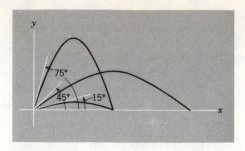

figure 4-14
Problems 7, and 15

of projection θ_0 changes by dR where $dR/R = -dg/g$. (b) If the acceleration due to gravity changes by a small amount Δg (say by going from one place to another), the range for a given projectile system will change as well. Let the change in range by ΔR. If Δg, ΔR are small enough, we may write $\Delta R/R = -\Delta g/g$. In 1936, Jesse Owens established a world's running broad jump record of 8.09 m at the Olympic Games at Berlin ($g = 9.8128$ m/s²). By how much would his record have differed if he had competed instead in 1956 at Melbourne ($g = 9.7999$ m/s²)? (In this connection see "Bad Physics in Athletic Measurements," by P. Kirkpatrick, *American Journal of Physics*, February 1944.)
Answer: His record would have been longer by about 1 cm.

18. A juggler manages to keep five balls in motion, throwing each sequentially up a distance of 3.0 m. (a) Determine the time interval between successive throws. (b) Give the positions of the other balls at the instant when one reaches his hand. (Neglect the time taken to transfer balls from one hand to the other.)

19. A cannon is arranged to fire projectiles, with initial speed v_0, directly up the face of a hill of elevation angle α, as shown in Fig. 4-15. At what angle from the horizontal should the cannon be aimed to obtain the maximum possible range R up the face of the hill? *Answer:* $\pi/4 + \alpha/2$.

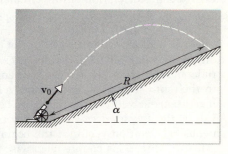

figure 4-15
Problem 19

20. The kicker on a football team can give the ball an initial speed of 25 m/s. Within what angular range must he kick the ball if he is to just score a field goal from a point 50 m in front of the goalposts whose horizontal bar is 3.44 m above the ground?

21. A radar observer on the ground is "watching" an approaching projectile. At a certain instant he has the following information: the projectile has reached maximum altitude and is moving horizontally with a speed v; the straight-line distance to the projectile is l; the line of sight to the projectile is an angle θ above the horizontal. (a) Find the distance D between the observer and the point of impact of the projectile. D is to be expressed in terms of the observed quantities v, l, and θ and the known value of g. Assume a flat earth; assume also that the observer lies in the plane of the projectile's trajectory. (b) Does the projectile pass over his head or strike the ground before reaching him?
Answer: (a) $D = v\sqrt{(2l/g)\sin\theta} - l\cos\theta$. (b) The projectile will pass over the observer's head if D is positive and will fall short if D is negative.

22. Projectiles are hurled at a horizontal distance R from the edge of a cliff of height h in such a way as to land a horizontal distance x from the bottom of the cliff. If you want x to be as small as possible, how would you adjust θ_0 and v_0, assuming that v_0 can be varied from zero to some finite maximum value and that θ_0 can be varied continuously? Only one collision with the ground is allowed (see Fig. 4-16).

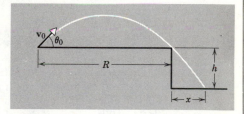

figure 4-16
Problem 22

SECTION 4-4

23. Certain neutron stars (extremely dense stars) are believed to be rotating at about 1 rev/s. If such a star has a radius of 20 km, what is the acceleration of an object on the equator of the star? *Answer:* 8×10^5 m/s².

24. A magnetic field will deflect a charged particle perpendicular to its direction of motion. An electron experiences a radial acceleration of 3.0×10^{14} m/s² in one such field. What is its speed if the radius of its curved path is 0.15 m?

25. In Bohr's model of the hydrogen atom an electron revolves around a proton in a circular orbit of radius 5.28×10^{-11} m with a speed of 2.18×10^6 m/s. What is the acceleration of the electron in the hydrogen atom?
Answer: 9.00×10^{22} m/s².

26. A particle rests on the top of a hemisphere of radius R. Find the smallest horizontal velocity that must be imparted to the particle if it is to leave the hemisphere without sliding down it.

27. What is the acceleration of an object (a) on the equator and (b) at latitude

60°, due to rotation of the earth? (c) By what factor would the speed of the earth's rotation have to increase for a body on the equator to require an acceleration of g to keep it on the earth?
Answer: (a) 3.4×10^{-2} m/s². (b) 1.7×10^{-2} m/s². (c) 17.

28. A boy whirls a stone in a horizontal circle 6.0 ft (1.8 m) above the ground by means of a string 4.0 ft (1.2 m) long. The string breaks, and the stone flies off horizontally, striking the ground 30 ft (9.1 m) away. What was the centripetal acceleration during circular motion?

29. A particle P travels with constant speed counterclockwise on a circle of radius 3.0 m and completes 1.0 rev in 20 s (Fig. 4-17). The particle passes through O at $t = 0$. Starting from the origin O, find (a) the magnitude and direction of the position vectors 5.0 s, 7.5 s, and 10 s later; (b) the magnitude and direction of the displacement in the 5.0-s interval from the fifth to the tenth second; (c) the average velocity vector in this interval; (d) the instantaneous velocity vector at the beginning and at the end of this interval; (e) the average acceleration vector in this interval; and (f) the instantaneous acceleration vector at the beginning and at the end of this interval. (Measure directions counterclockwise from the x-axis in Fig. 4-17.)
Answer: (a) 4.2 m, 45°; 5.5 m, 68°; 6.0 m, 90°. (b) 4.2 m, 135°. (c) 0.85 m/s, 135°. (d) 0.94 m/s, 90°; 0.94 m/s, 180°. (e) 0.27 m/s², 225°. (f) 0.30 m/s², 180°; 0.30 m/s², 270°.

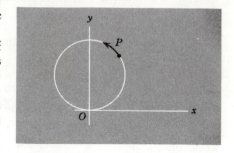

figure 4-17
Problem 29

30. (a) Write an expression for the position vector **r** for a particle describing uniform circular motion, using rectangular coordinates and the unit vectors **i** and **j**. (b) From (a) derive vector expressions for the velocity **v** and the acceleration **a**. (c) Prove that the acceleration is directed toward the center of the circular motion.

31. (a) Express the unit vectors $\mathbf{u}_r$ and $\mathbf{u}_\theta$ in terms of **i**, **j**, and the angle θ in Fig. 4-8. (b) Write an expression, using the unit vectors $\mathbf{u}_\theta$ and $\mathbf{u}_r$, for the position vector **r** for a particle describing uniform circular motion and from it derive Eq. 4-11, $\mathbf{v} = \mathbf{u}_\theta v$.

32. A particle in uniform circular motion about the origin O has a speed v. (a) Show that the time Δt required for it to pass through an angular displacement $\Delta \theta$ is given by

$$\Delta t = \frac{2\pi r}{v} \Delta\theta/360°,$$

where $\Delta\theta$ is in degrees and r is the radius of the circle. (b) Refer to Fig. 4-18, and by taking x and y components of the velocities at points 1 and 2 show that $\overline{a_x} = 0$ and $\overline{a_y} = -0.9 \ v^2/r$, for a pair of points symmetric about the y-axis with $\Delta\theta = 90°$. (c) Show that if $\Delta\theta = 30°$, $\overline{a_x} = 0$ and $\overline{a_y} = -0.99 \ v^2/r$. (d) Argue that $\overline{a_y} \to -v^2/r$ as $\Delta\theta \to 0$ and that circular symmetry requires this answer for each point on the circle.

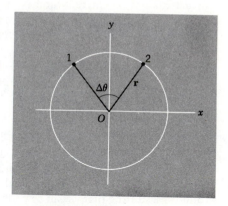

figure 4-18
Problem 32

SECTION 4-5
33. A particle moves in a plane according to

$$x = R \sin \omega t + \omega R t,$$
$$y = R \cos \omega t + R,$$

where ω and R are constants. This curve, called a *cycloid*, is the path traced out by a point on the rim of a wheel which rolls without slipping along the x-axis. (a) Sketch the path. (b) Calculate the instantaneous velocity and acceleration when the particle is at its maximum and minimum value of y.
Answer: (b) At minimum y: $v_x = v_y = a_x = 0$; $a_y = +\omega^2 R$. At maximum y: $v_x = 2\omega R$; $v_y = a_x = 0$; $a_y = -\omega^2 R$.

SECTION 4-6
34. Snow is falling vertically at a constant speed of 8.0 m/s. (a) At what angle from the vertical and (b) with what speed do the snowflakes appear to be falling as viewed by the driver of a car traveling on a straight road with a speed of 50 km/h?

35. A train travels due south at 88.2 ft/s (relative to ground) in a rain that is blown toward the south by the wind. The path of each raindrop makes the angle 21.6° with the vertical, as measured by an observer stationary on the earth. An observer seated in the train, however, sees perfectly vertical tracks of rain on the windowpane. Determine the speed of each raindrop relative to the earth. *Answer:* 240 ft/s.

36. A helicopter is flying in a straight line over a level field at a constant speed of 4.9 m/s and at a constant altitude of 4.9 m. A package is ejected horizontally from the helicopter with an initial velocity of 12 m/s relative to the helicopter, and in a direction opposite to the helicopter's motion. (*a*) Find the initial velocity of the package relative to the ground. (*b*) What is the horizontal distance between the helicopter and the package at the instant the package strikes the ground? (*c*) What angle does the velocity vector of the package make with the ground at the instant before impact?

37. Find the speeds of two objects if, when they move uniformly toward each other, they get 4.0 m closer each second, and, when they move uniformly in the same direction with the original speeds, they get 4.0 m closer each 10 seconds. *Answer:* 2.2 m/s, 1.8 m/s.

38. A man can row a boat 4.0 mi/h in still water. (*a*) If he is crossing a river where the current is 2.0 mi/h, in what direction will his boat be headed if he wants to reach a point directly opposite from his starting point? (*b*) If the river is 4.0 mi wide, how long will it take him to cross the river? (*c*) How long will it take him to row 2.0 mi *down* the river and then back to his starting point? (*d*) How long will it take him to row 2.0 mi *up* the river and then back to his starting point? (*e*) In what direction should he head the boat if he wants to cross in the smallest possible time?

39. An airplane has a speed of 135 mi/h in still air. It is flying straight north so that it is at all times directly above a north-south highway. A ground observer tells the pilot by radio that a 70 mi/h wind is blowing, but neglects to tell him the wind direction. The pilot observes that in spite of the wind he can travel 135 miles along the highway in one hour. In other words, his ground speed is the same as if there were no wind. (*a*) What is the direction of the wind? (*b*) What is the heading of the plane, that is, the angle between its axis and the highway?

Answer: (*a*) From 75° E of S. (*b*) 30° E of N. Substituting W for E gives a second solution.

40. A pilot is supposed to fly due east from *A* to *B* and then back again to *A* due west. The velocity of the plane in air is **v′** and the velocity of the air with respect to the ground is **u.** The distance between *A* and *B* is *l* and the plane's air speed **v′** is constant. (*a*) If $u = 0$ (still air), show that the time for the round trip is $t_0 = 2l/v'$. (*b*) Suppose that the air velocity is due east (or west). Show that the time for a round trip is then

$$t_E = \frac{t_0}{1 - u^2/(v')^2}.$$

(*c*) Suppose that the air velocity is due north (or south). Show that the time for a round trip is then

$$t_N = \frac{t_0}{\sqrt{1 - u^2/(v')^2}}.$$

(*d*) In parts (*b*) and (*c*) one must assume that $u < v'$. Why?

41. A person walks up a stalled escalator in 90 s. When standing on the same escalator, now moving, he is carried up in 60 s. How much time would it take him to walk up the moving escalator? *Answer:* 36 s.

42. A man wants to cross a river 500 m wide. His rowing speed (relative to the water) is 3000 m/h. The river flows with a speed of 2000 m/h. If the man's walking speed on shore is 5000 m/h, (*a*) find the path (combined rowing and walking) he should take to get to the point directly opposite his starting point in the shortest time. (*b*) How long does it take?

5
particle dynamics—I

In Chapters 3 and 4, we studied the motion of a particle, with emphasis on motion along a straight line or in a plane. We did not ask what "caused" the motion; we simply described it in terms of the vectors **r,** **v,** and **a.** Our discussion was thus largely geometrical. In this chapter and the next we discuss the causes of motion, an aspect of mechanics called *dynamics.* As before, bodies will be treated as though they were single particles. Later in the book we shall treat groups of particles and rigid bodies as well.

The motion of a given particle is determined by the nature and the arrangement of the other bodies that form its *environment.* Table 5-1 shows some "particles" and possible environments for them.

In what follows, we limit ourselves to the very important special case of gross objects moving at speeds that are small compared to c, the speed of light; this is the realm of *classical mechanics.* Specifically, we shall not inquire here into such questions as the motion of an electron in a uranium atom or the collision of two protons whose speeds are, say, $0.90c$. The first inquiry would involve us with the quantum theory and the second with the theory of relativity. We leave consideration of these theories, of which classical mechanics is a special case (see Section 6-4), to later.

The central problem of classical mechanics is this: (1) We are given a particle whose characteristics (mass, charge, magnetic dipole moment, etc.) we know. (2) We place this particle, with a known initial velocity, in an environment of which we have a complete description. (3) Problem: what is the subsequent motion of the particle?

This problem was solved, at least for a large variety of environments, by Isaac Newton (1642–1727) when he put forward his laws of motion and formulated his law of universal gravitation. The program for solving this problem, in terms of our present understanding of classical mechanics,* is: (1) We introduce the concept of *force* **F** and define it in terms of the acceleration **a** experienced by a particular standard body. (2) We develop a procedure for assigning a *mass m* to a body so that we may understand the fact that different particles of the same kind experience different accelerations in the same environment. (3) Finally, we try to find ways of calculating the forces that act on particles from the properties of the particle and of its environment; that is, we look for *force laws*. Force, which is at root a technique for relating the environment to the motion of the particle, appears both in the laws of motion (which tell us what acceleration a given body will experience under the action of a given force) and in the force laws (which tell us how to calculate the force that will act on a given body in a given environment). The laws of motion and the force laws, taken together, constitute the laws of mechanics, as the sketch suggests.

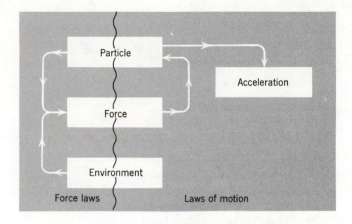

The program of mechanics cannot be tested piecemeal. We must view it as a unit and we shall judge it to be successful if we can say "yes" to these two questions. (1) Does the program yield results that agree with experiment? (2) Are the force laws simple in form? It is the crowning glory of Newtonian mechanics that we can indeed answer each of these questions in the affirmative.

In this section we have used the terms *force* and *mass* rather unprecisely, having identified force with the influence of the environment, and mass with the resistance of a body to be accelerated when a force acts on it, a property often called inertia. In later sections we shall refine these primitive ideas about force and mass.

5-2
NEWTON'S FIRST LAW

For centuries the problem of motion and its causes was a central theme of natural philosophy, an early name for what we now call physics. It was not until the time of Galileo and Newton, however, that dramatic progress was made. Isaac Newton, born in England in the year of Gali-

---

* See "Presentation of Newtonian Mechanics" by Norman Austern, *American Journal of Physics*, September 1961, "On the Classical Laws of Motion" by Leonard Eisenbud, *American Journal of Physics*, March 1958, and "The Laws of Classical Motion: What's F? What's m? What's a?" by Robert Weinstock, *American Journal of Physics*, October 1961, for expositions of the laws of classical mechanics as we now view them.

Table 5-1

	System	The Particle	The Environment
1.	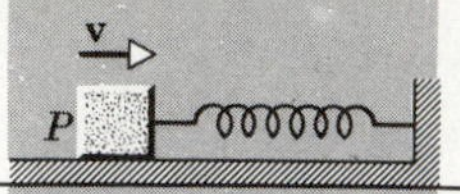	A block	The spring; the rough surface
2.	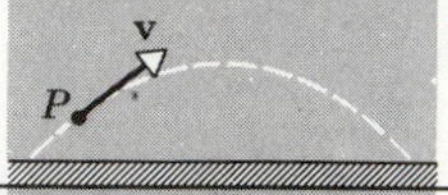	A golf ball	The earth
3.	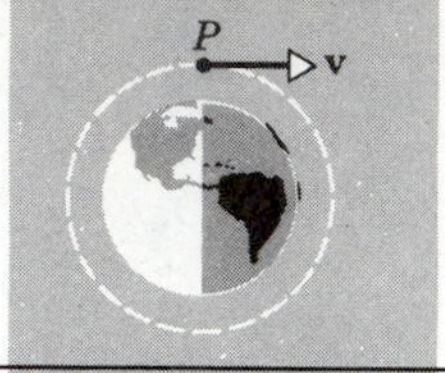	A satellite	The earth
4.	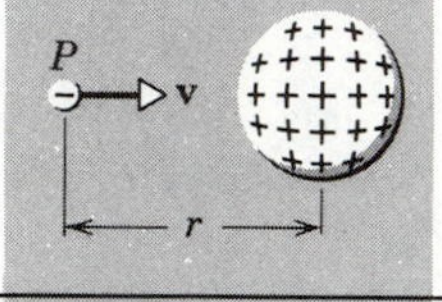	An electron	A large uniformly charged sphere
5.	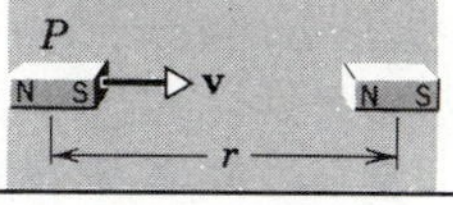	A bar magnet	A second bar magnet

leo's death, is the principal architect of classical mechanics.* He carried to full fruition the ideas of Galileo and others who preceded him. His three laws of motion were first presented (in 1686) in his *Philosophiae Naturalis Principia Mathematica*, usually called the *Principia*.

Before Galileo's time most philosophers thought that some influence or "force" was needed to keep a body moving. They thought that a body was in its "natural state" when it was at rest. For a body to move in a straight line at constant speed, for example, they believed that some external agent had to continually propel it; otherwise it would "naturally" stop moving.

If we wanted to test these ideas experimentally, we would first have to find a way to free a body from all influences of its environment or from all forces. This is hard to do, but in certain cases we can make the forces very small. If we study the motions as we make the forces smaller and smaller, we shall have some idea of what the motion would be like if the external forces were truly zero.

Let us place our test body, say a block, on a rigid horizontal plane. If we let the block slide along this plane, we notice that it gradually slows down and stops. This observation was used, in fact, to support the idea that motion stopped when the external force, in this case the hand initially pushing the block, was removed. We can argue against this idea, however, reasoning as follows: Let us repeat our experiment, now using a smoother block and a smoother plane and providing a lubricant. We notice that the velocity decreases more slowly than before. Let us use still smoother blocks and surfaces and better lubricants. We find that the block decreases in velocity at a slower and slower rate and travels

---

* Newton also invented the (fluxional) calculus, conceived the idea of universal gravitation and formulated its law, and discovered the composition of white light. He was a skillful experimenter and a mathematician of first rank as well as what today would be called theoretical physicist.

farther each time before coming to rest.* We can now extrapolate and say that if all friction could be eliminated, the body would continue indefinitely in a straight line with constant speed. Some external force is necessary to *change* the velocity of a body but no external force is necessary to *maintain* the velocity of a body. Our hand, for example, exerts a force on the block when it sets it in motion. The rough plane exerts a force on it when it slows it down. Both of these forces produce a change in the velocity, that is, they produce an acceleration.

This principle was adopted by Newton as the first of his three laws of motion. Newton stated his first law in these words: *"Every body persists in its state of rest or of uniform motion in a straight line unless it is compelled to change that state by forces impressed on it."*

Newton's first law is really a statement about reference frames. For, in general, the acceleration of a body depends on the reference frame relative to which it is measured. The first law tells us that, if there are no nearby objects (and by this we mean that there are no forces because every force must be associated with an object in the environment), then it is possible to find a family of reference frames in which a particle has no acceleration. The fact that bodies stay at rest or retain their uniform linear motion in the absence of applied forces is often described by assigning a property to matter called inertia. Newton's first law is often called the law of inertia and the reference frames to which it applies are called inertial frames. Such frames are assumed to be fixed with respect to the distant stars.

In nearly all cases in this book we will apply the laws of classical mechanics from the point of view of an observer in an inertial frame. It is possible to solve problems in mechanics using a noninertial frame, such as a frame rotating with respect to the fixed stars, but to do so we have to introduce forces that cannot be associated with objects in the environment. We will discuss this in Chapters 6, 11, and 16. A reference frame attached to the earth can be considered to be an inertial frame for most practical purposes. We shall see in Chapter 16 how good an approximation this is.

Notice that there is no distinction in the first law between a body at rest and one moving with a constant velocity. Both motions are "natural" in the absence of forces. That this is so becomes clear when a body at rest in one inertial frame is viewed from a second inertial frame, that is, a frame moving with constant velocity with respect to the first. An observer in the first frame finds the body to be at rest; an observer in the second frame finds the same body to be moving with uniform velocity. Both observers find the body to have no acceleration, that is, no change in velocity, and both may conclude from the first law that no force acts on the body.

Notice, too, that by implication there is no distinction in the first law between the absence of all forces and the presence of forces whose resultant is zero. For example, if the push of our hand on the book exactly counteracts the force of friction on it, the book will move with uniform velocity. Hence another way of stating the first law is: *If no net force acts on a body, its acceleration* **a** *is zero.*

If there *is* an interaction between the body and objects present in the

* You may have experimented in the laboratory with a dry ice puck. This is a device which can be made to move over a smooth horizontal surface, floating on a layer of CO_2 gas. The friction between the puck and the surface is very low indeed and it is hard to measure any reduction in speed for path lengths of practical dimensions.

environment, the effect may be to change the "natural" state of the body's motion. To investigate this we must now examine carefully the concept of force.

Let us refine our concept of force by defining it operationally. In our everyday language force is associated with a push or a pull, perhaps exerted by our muscles. In physics, however, we need a more precise definition. We define force here in terms of the acceleration that a given standard body experiences when placed in a suitable environment.

As a standard body we find it convenient to use (or rather to imagine that we use!) the standard kilogram (see Fig. 1-2). This body has been selected as our standard of mass and has been assigned, by definition, a mass m_0 of exactly 1 kg. Later we will describe how masses are assigned to other bodies.

As for an environment we place the standard body on a horizontal table having negligible friction and we attach a spring to it. We hold the other end of the spring in our hand, as in Fig. 5-1a. Now we pull the spring horizontally to the right so that by trial and error the standard body experiences a measured uniform acceleration of 1.0 m/s². We then declare, as a matter of definition, that the spring (which is the significant body in the environment) is exerting a constant force whose magnitude we will call "1.00 newton," or in SI notation: 1.00 N, on the standard body. We note that, in imparting this force, the spring is kept stretched an amount Δl beyond its normal unextended length, as Fig. 5-1b shows.

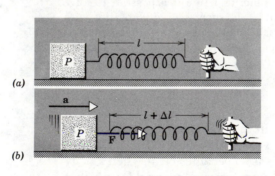

figure 5-1
(a) A "particle" P (the standard kilogram) at rest on a horizontal frictionless surface. (b) The body is accelerated by pulling the spring to the right.

We can repeat the experiment, either stretching the spring more or using a stiffer spring, so that we measure an acceleration of 2.00 m/s² for the standard body. We now declare that the spring is exerting a force of 2.00 N on the standard body. In general, if we observe this particular standard body to have an acceleration *a* in a particular environment, we then say that the environment is exerting a force *F* on the standard body, where *F* (in newtons) is numerically equal to *a* (in m/s²).

Now let us see whether force, as we have defined it, is a *vector* quantity. In Fig. 5-2b we assigned a magnitude to the force *F*, and it is a simple matter to assign a direction to it as well, namely, the direction of the acceleration that the force produces. However, to be a vector it is not enough for a quantity to have magnitude and direction; it must also obey the laws of vector addition described in Chapter 2. We can learn only from experiment whether forces, as we defined them, do indeed obey these laws.

Let us arrange to exert a 4.00-N force along the *x*-axis and a 3.00-N force along the *y*-axis and let us apply these forces simultaneously to the standard body placed, as before, on a horizontal, frictionless surface. What will be the acceleration of the standard body? We would find by experiment that it was 5.00 m/s², directed along a line that makes an angle of 37° with the x-axis. In other words, we would say that the standard body was experiencing a force of 5.00 N in this same direction. This same result can be obtained by adding the 4.00-N and 3.00-N forces vectorially according to the parallelogram method. Experiments of this kind show conclusively that forces are vectors; they have magnitude; they have direction; they add according to the parallelogram law.

The result of experiments of this general type is often stated as follows: *When several forces act on a body, each produces its own acceleration independently. The resulting acceleration is the vector sum of the several independent accelerations.*

5-4
MASS; NEWTON'S SECOND LAW

In Section 5-3 we considered only the accelerations given to one particular object, the standard kilogram. We were able thereby to define forces quantitatively. What effect would these forces have on other objects? Because our standard body was chosen arbitrarily in the first place, we know that for any given object the acceleration will be directly proportional to the force applied. The significant question remaining then is: What effect will the *same force* have on *different objects?* Everyday experience gives us a qualitative answer. The same force will produce different accelerations on different bodies. A baseball will be accelerated more by a given force than will an automobile. In order to obtain a quantitative answer to this question we need a method to measure mass, the property of a body which determines its resistance to a change in its motion.

Let us attach a spring to our standard body (the standard kilogram, to which we have arbitrarily assigned a mass m_0 = one kg, exactly) and arrange to give it an acceleration a_0 of, say 2.00 m/s², using the method of Fig. 5-1b. Let us measure carefully the extension Δl of the spring associated with the force that the spring is exerting on the block.

Now we remove the standard kilogram and substitute an arbitrary body, whose mass we label m_1. We apply the same force (the one that accelerated the standard kilogram 2.00 m/s²) to the arbitrary body (by stretching the spring by the same amount) and we measure an acceleration a_1 of, say, 0.50 m/s².

We *define* the ratio of the masses of the two bodies to be the inverse ratio of the accelerations given to these bodies by the same force, or

$$m_1/m_0 = a_0/a_1 \qquad \text{(same force } \mathbf{F} \text{ acting)}.$$

In this example we have, numerically,

$$m_1 = m_0(a_0/a_1) = 1.00 \text{ kg } [(2.00 \text{ m/s}^2)/(0.50 \text{ m/s}^2)]$$
$$= 4.00 \text{ kg}.$$

The second body, which has only one-fourth the acceleration of the first body when the same force acts on it, has, by definition, four times the mass of the first body. Hence mass may be regarded as a quantitative measure of inertia.

If we repeat the preceding experiment with a different common force acting, we find the ratio of the accelerations, a_0'/a_1', to be the same as in the previous experiment, or

$$m_1/m_0 = a_0/a_1 = a_0'/a_1'.$$

The ratio of the masses of two bodies is thus independent of the common force used.

Furthermore, experiment shows that we can consistently assign masses to any body by this procedure. For example, let us compare a second arbitrary body with the standard body, and thus determine its mass, say m_2. We can now compare the two arbitrary bodies, m_2 and m_1, directly, obtaining accelerations a_2'' and a_1'' when the same force is applied. The mass ratio, defined as usual from

$$m_2/m_1 = a_1''/a_2'', \qquad \text{(same force acting)}$$

turns out to have the same value that we obtain by using the masses m_2 and m_1 previously determined by direct comparison with the standard.

We can show, in still another experiment of this type, that if objects of mass m_1 and m_2 are fastened together, they behave mechanically as a single object of mass $(m_1 + m_2)$. In other words, masses add like (and are) scalar quantities.

We can now summarize all the experiments and definitions described above in one equation, the fundamental equation of classical mechanics,

$$\mathbf{F} = m\mathbf{a}. \tag{5-1}$$

In this equation $\mathbf{F}$ is the (vector) *sum* of *all* the forces acting *on* the body, m is the mass of the body, and $\mathbf{a}$ is its (vector) acceleration. Equation 5-1 may be taken as a statement of Newton's second law. If we write it in the form $\mathbf{a} = \mathbf{F}/m$, we can easily see that the acceleration of the body is directly proportional to the resultant force acting on it and parallel in direction to this force and that the acceleration, for a given force, is inversely proportional to the mass of the body.

Notice that the first law of motion is contained in the second law as a special case, for if $\mathbf{F} = 0$, then $\mathbf{a} = 0$. In other words, if the resultant force on a body is zero, the acceleration of the body is zero. Therefore in the absence of applied forces a body will move with constant velocity or be at rest (zero velocity), which is what the first law of motion says. Therefore of Newton's three laws of motion only two are independent, the second and the third (Section 5-5). The division of translational particle dynamics that includes only systems for which the resultant force $\mathbf{F}$ is zero is called *statics*.

Equation 5-1 is a vector equation. We can write this single vector equation as three scalar equations,

$$F_x = ma_x, \qquad F_y = ma_y, \qquad \text{and} \quad F_z = ma_z, \tag{5-2}$$

relating the x, y, and z components of the resultant force (F_x, F_y, and F_z) to the x, y, and z components of acceleration (a_x, a_y, and a_z) for the mass m. It should be emphasized that F_x is the sum of the x-components of *all* the forces, F_y is the sum of the y-components of *all* the forces, and F_z is the *sum* of the z-components of *all* the forces acting on m.

5-5
NEWTON'S THIRD LAW OF MOTION

Forces acting on a body originate in other bodies that make up its environment. Any single force is only one aspect of a mutual interaction between *two* bodies. We find by experiment that when one body exerts a force on a second body, the second body always exerts a force on the first. Furthermore, we find that these forces are equal in magnitude but opposite in direction. A single isolated force is therefore an impossibility.

If one of the two forces involved in the interaction between two bodies is called an "action" force, the other is called the "reaction" force. *Either* force may be considered the "action" and the other the "reaction." Cause and effect is *not* implied here, but a mutual simultaneous interaction *is* implied.

This property of forces was first stated by Newton in his third law of motion: *"To every action there is always opposed an equal reaction; or, the mutual actions of two bodies upon each other are always equal, and directed to contrary parts."*

In other words, if body A exerts a force on body B, body B exerts an equal but oppositely directed force on body A; and furthermore the forces lie along the line joining the bodies. Notice that the action and reaction forces, which always occur in pairs, act on *different* bodies. If they were to act on the same body, we could never have accelerated motion because the resultant force on every body would always be zero.

Imagine a boy kicking open a door. The force exerted by the boy B on the door D accelerates the door (it flies open); at the same time, the door D exerts an equal but opposite force on the boy B, which decelerates the boy (his foot loses forward velocity). The boy will be painfully aware of the "reaction" force to his "action," particularly if his foot is bare.

The following examples illustrate the application of the third law and clarify its meaning.

EXAMPLE 1

Consider a man pulling horizontally on a rope attached to a block on a horizontal table as in Fig. 5-2. The man pulls on the rope with a force $\mathbf{F}_{MR}$. The rope exerts a reaction force $\mathbf{F}_{RM}$ on the man. According to Newton's third law, $\mathbf{F}_{MR} = -\mathbf{F}_{RM}$. Also, the rope exerts a force $\mathbf{F}_{RB}$ on the block, and the block exerts a reaction force $\mathbf{F}_{BR}$ on the rope. Again according to the third law, $\mathbf{F}_{RB} = -\mathbf{F}_{BR}$.

Suppose that the rope has a mass m_R. Then, in order to start the block and rope moving from rest, we must have an acceleration, say $\mathbf{a}$. The only forces acting *on the rope* are $\mathbf{F}_{MR}$ and $\mathbf{F}_{BR}$, so that the resultant force on it is $\mathbf{F}_{MR} + \mathbf{F}_{BR}$, and this must be different from zero if the rope is to accelerate. In fact, from the

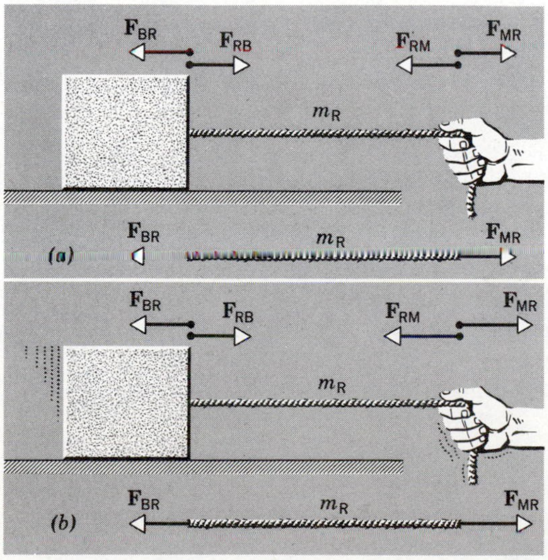

figure 5-2
Example 1. A man pulls on a rope attached to a block. *(a)* The forces exerted on the rope by the block and by the man are equal and opposite. Thus the resultant horizontal force on the rope is zero, as is shown in the free-body diagram. The rope does not accelerate. *(b)* The force exerted on the rope by the man exceeds that exerted by the block. The net horizontal force has magnitude $F_{MR} - F_{BR}$ and points to the right. Thus the rope is accelerated to the right. The block is also acted upon by a frictional force not shown here.

second law we have

$$\mathbf{F}_{MR} + \mathbf{F}_{BR} = m_R \mathbf{a}$$

Since the forces and the acceleration are along the same line, we can drop the vector notation and write the relation between the *magnitudes* of the vectors, namely

$$F_{MR} - F_{BR} = m_R a.$$

We see therefore that in general $\mathbf{F}_{MR}$ does not have the same magnitude as $\mathbf{F}_{BR}$ (Fig. 5-2*b*). These two forces act on the *same* body and are *not* action and reaction pairs.

According to Newton's third law the magnitude of $\mathbf{F}_{MR}$ always equals the magnitude of $\mathbf{F}_{RM}$, and the magnitude of $\mathbf{F}_{RB}$ always equals the magnitude of $\mathbf{F}_{BR}$. However, only if the acceleration $\mathbf{a}$ of the system is zero will we have the pair of forces $\mathbf{F}_{MR}$ and $\mathbf{F}_{RM}$ equal in magnitude to the pair of forces $\mathbf{F}_{RB}$ and $\mathbf{F}_{BR}$ (Fig. 5-2*a*). In this special case only, we could imagine that the rope merely transmits the force exerted by the man to the block without change. This same result holds in principle if $m_R = 0$. In practice, we never find a massless rope. However, we can often neglect the mass of a rope, in which case the rope is assumed to transmit a force unchanged. The force exerted at any point in the rope is called the *tension* at that point. We may measure the tension at any point in the rope by cutting a suitable length from it and inserting a spring scale; the tension is the reading of the scale. The tension is the same at all points in the rope only if the rope is unaccelerated or assumed to be massless.

EXAMPLE 2

Consider a spring attached to the ceiling and at the other end holding a block at rest (Fig. 5-3*a*). Since no body is accelerating, all the forces on any body will add vectorially to zero. For example, the forces on the suspended block are $\mathbf{T}$, the tension in the stretched spring, pulling vertically up on the mass, and $\mathbf{W}$, the pull of the earth acting vertically down on the body, called its weight. These are drawn in Fig. 5-3*b*, where we show only the block for clarity. There are no other forces on the block.

In Newton's second law, $\mathbf{F}$ represents the *sum* of *all* the forces acting *on* a body, so that for the block

$$\mathbf{F} = \mathbf{T} + \mathbf{W}.$$

The block is at rest so that its acceleration is zero, or $\mathbf{a} = 0$. Hence, from the relation $\mathbf{F} = m\mathbf{a}$, we obtain $\mathbf{T} + \mathbf{W} = 0$, or

$$\mathbf{T} = -\mathbf{W}.$$

The forces act along the same line, so that their magnitudes are equal, or

$$T = W.$$

Therefore the tension in the spring is an exact measure of the weight of the block. We shall use this result later in presenting a static procedure for measuring forces.

It is instructive to examine the forces exerted on the spring; they are shown in Fig. 5-3*c*. $\mathbf{T}'$ is the pull of the block on the spring and is the reaction force of the action force $\mathbf{T}$. $\mathbf{T}'$ therefore has the same magnitude as $\mathbf{T}$, which is W. $\mathbf{P}$ is the upward pull of the ceiling on the spring, and $\mathbf{w}$ is the weight of the spring, that is, the pull of the earth on it. Since the spring is at rest and all forces act along the same line, we have

$$\mathbf{P} + \mathbf{T}' + \mathbf{w} = 0,$$

or

$$P = W + w.$$

The ceiling therefore pulls up on the spring with a force whose magnitude is the sum of the weights of the block and spring.

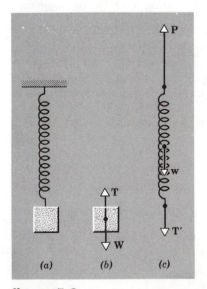

figure 5-3

Example 2. *(a)* A block is suspended by a spring. *(b)* A free-body diagram showing all the vertical forces exerted on the block. *(c)* A similar diagram for the vertical forces on the spring.

From the third law of motion, the force exerted by the spring on the ceiling, **P**′, must be equal in magnitude to **P,** which is the reaction force to the action force **P**′. **P**′ therefore has a magnitude $W + w$.

In general, the spring exerts different forces on the bodies attached at its different ends, for $P' \neq T$. In the special case in which the weight of the spring is negligible, $w = 0$ and $P' = W = T$. Therefore a weightless spring (or cord) may be considered to transmit a force from one end to the other without change.

It is instructive to classify all the forces in this problem according to action and reaction pairs. The reaction to **W,** a force exerted by the earth on the block, must be a force exerted by the block on the earth. Similarly, the reaction to **w** is a force exerted by the spring on the earth. Because the earth is so massive, we do not expect these forces to impart a noticeable acceleration to the earth. Since the earth is not shown in our diagrams, these forces have not been shown. The forces **T** and **T**′ are action-reaction pairs, as are **P** and **P**′. Notice that although $\mathbf{T} = -\mathbf{W}$ in our problem, these forces are *not* an action-reaction pair because they act on the *same* body.

5-6
SYSTEMS OF MECHANICAL UNITS

Unit force is defined as a force that causes a unit of acceleration when applied to a unit mass. In SI terms unit force is the force that will accelerate a one-kg mass at one m/s^2; we have seen that this unit is called the newton (abbreviation, N). In the cgs (centimeter, gram, second) system unit force is the force that will accelerate a one-g mass at one cm/s^2; this unit is called the *dyne.* Since 1 kg $= 10^3$ g and 1 $m/s^2 = 10^2\ cm/s^2$, it follows that 1 N $= 10^5$ dynes.

In each of our systems of units we have chosen mass, length, and time as our fundamental quantities. Standards were adopted for these fundamental quantities and units defined in terms of these standards. Force appears as a derived quantity, determined from the relation $\mathbf{F} = m\mathbf{a}.$

In the BE (British engineering) system of units, however, *force,* length, and time are chosen as the fundamental quantities and mass is a derived quantity. In this system, mass is determined from the relation $m = F/a$. The standard and unit of force in this system is the *pound.* Actually, the pound of force was originally defined to be the pull of the earth on a certain standard body at a certain place on the earth. We can get this force in an operational way by hanging the standard body from a spring at the particular point where the earth's pull on it is defined to be one lb of force. If the body is at rest, the earth's pull on the body, its weight W, is balanced by the tension in the spring. Therefore $T = W =$ one lb, in this instance. We can now use this spring (or any other one thus calibrated) to exert a force of one lb on any other body; to do this we simply attach the spring to another body and stretch it the same amount as the pound force had stretched it. The standard body can be compared to the kilogram and it is found to have the mass 0.45359237 kg. The acceleration due to gravity at the certain place on the earth is found to be 32.1740 ft/s^2. The pound of force can therefore be defined from $F = ma$ as the force that accelerates a mass of 0.45359237 kg at the rate of 32.1740 ft/s^2.

This procedure enables us to compare the pound-force with the newton. Using the fact that 32.1740 ft/s^2 equals 9.8066 m/s^2, we find that

$$1\ \text{lb} = (0.45359237\ \text{kg})(32.1740\ ft/s^2)$$
$$= (0.45359237\ \text{kg})(9.8066\ m/s^2)$$
$$\cong 4.45\ \text{N}.$$

The unit of mass in the British engineering system can now be derived. It is defined as the mass of a body whose acceleration is 1 ft/s^2

when the force on it is 1 lb; this mass is called the slug. Thus, in this system

$$F[\text{lb}] = m[\text{slugs}] \times a[\text{ft/s}^2].$$

Legally the pound is a unit of mass but in engineering practice the pound is treated as a unit of force or weight. This has given rise to the terms pound-mass and pound-force. The pound-mass is a body of mass 0.45359237 kg; no standard block of metal is preserved as the pound-mass, but, like the yard, it is defined in terms of the SI standard. The pound-force is the force that gives a standard pound an acceleration equal to the standard acceleration of gravity, 32.1740 ft/s². As we shall see later, the acceleration of gravity varies with distance from the center of the earth, and this "standard acceleration" is, therefore, the value at a particular distance from the center of the earth. (Any point at sea level and 45°N latitude is a good approximation.)

In this book only forces will be measured in pounds. Thus the corresponding unit of mass is the slug. The units of force, mass, and acceleration in the three systems are summarized in Table 5-2.

Table 5-2
Units in $F = ma$

Systems of Units	Force	Mass	Acceleration
SI	newton (N)	kilogram (kg)	m/s²
Cgs	dyne	gram (g)	cm/s²
BE	pound (lb)	slug	ft/s²

The *dimensions* of force are the same as those of mass times acceleration. In a system in which mass, length, and time are the fundamental qualities, the dimensions of force are, therefore, mass × length/time², or MLT^{-2}. We shall arbitrarily adopt mass, length, and time as our fundamental mechanical quantities.

The three laws of motion that we have described are only part of the program of mechanics that we outlined in Section 5-1. It remains to investigate the *force laws*, which are the procedures by which we calculate the force acting on a given body in terms of the properties of the body and its environment. Newton's second law

$$\mathbf{F} = m\mathbf{a} \tag{5-3}$$

is essentially not a law of nature but a definition of force. We need to identify various functions of the type:

$\mathbf{F} =$ a function of the properties of the particle and of the environment $\qquad$ (5-4)

so that we can, in effect, eliminate $\mathbf{F}$ between Eqs. 5-3 and 5-4, thus obtaining an equation that will let us calculate the acceleration of a particle in terms of the properties of the particle and its environment. We see here clearly that force is a concept that connects the acceleration of the particle on the one hand with the properties of the particle and its environment on the other. We indicated earlier that one criterion for declaring the program of mechanics to be successful would be the

5-7
THE FORCE LAWS

discovery that *simple* laws of the type of Eq. 5-4 do indeed exist. This turns out to be the case, and this fact constitutes the essential reason that we "believe" the laws of classical mechanics. If the force laws had turned out to be very complicated, we would not be left with the feeling that we had gained much insight into the workings of nature.

The number of possible environments for an accelerated particle is so great that a detailed discussion of all the force laws is not feasible in this chapter. We shall, however, indicate in Table 5-3 the force laws that apply to the five particle-plus-environment situations of Table 5-1. At appropriate places throughout the text we will discuss these and other force laws in detail; several of the laws in Table 5-3 are approximations or special cases.

Table 5-3
The force laws for the systems of table 5-1

System	Force Law
1. A block propelled by a stretched spring over a rough horizontal surface	(a) Spring force: $F = -kx$, where x is the extension of the spring and k is a constant that describes the spring; **F** points to the right; see Chapter 15 (b) Friction force: $F = \mu mg$, where μ is the coefficient of friction and mg is the weight of the block; **F** points to the left; see Chapter 6
2. A golf ball in flight	$F = mg$; **F** points down (see Section 5-8)
3. An artificial satellite	$F = GmM/r^2$, where G is the *gravitational constant*, M the mass of the earth, and r the orbit radius; **F** points toward the center of the earth; see Chapter 16. This is *Newton's law of universal gravitation*
4. An electron near a positively charged sphere	$F = (1/4\pi\epsilon_0)eQ/r^2$, where ϵ_0 is a constant, e is the electron charge, Q is the charge on the sphere, and r is the distance from the electron to the center of the sphere; **F** points to the right; see Chapter 26. This is *Coulomb's law of electrostatics*
5. Two bar magnets	$F = (3\mu_0/2\pi)\mu^2/r^4$, where μ_0 is a constant, μ is the magnetic dipole moment of each magnet, and r is the center-to-center separation of the magnets; we assume that $r \gg l$, where l is the length of each magnet; **F** points to the right

5-8 WEIGHT AND MASS

The *weight* of a body is the gravitational force exerted on it by the earth. Weight, being a force, is a vector quantity. The direction of this vector is the direction of the gravitational force, that is, toward the center of the earth. The magnitude of the weight is expressed in force units, such as pounds or newtons.

When a body of mass m is allowed to fall freely, its acceleration is that of gravity **g** and the force acting on it is its weight **W**. Newton's second law, **F** = m**a,** when applied to a freely falling body, gives us **W** = m**g.** Both **W** and **g** are vectors directed toward the center of the earth. We can therefore write

$$W = mg, \tag{5-5}$$

where W and g are the magnitudes of the weight and acceleration vectors. To keep an object from falling we have to exert on it an upward force equal in magnitude to W, so as to make the net force zero. In Fig. 5-3a the tension in the spring supplies this force.

We stated previously that g is found experimentally to have the same value for all objects *at the same place*. From this it follows that the ratio of the weights of two objects must be equal to the ratio of their masses. Therefore a chemical balance, which actually is an instrument for comparing two downward forces, can be used in practice to compare masses. If a sample of salt in one pan of a balance is pulling down on that pan with the same force as is a standard one gram-mass on the other pan, we know* that the mass of salt is equal to one gram. We are likely to say that the salt "weighs" one gram, although a gram is a unit of mass, not weight. However, it is always important to distinguish carefully between weight and mass.

We have seen that the weight of a body, the downward pull of the earth on that body, is a vector quantity. The mass of a body is a scalar quantity. The quantitative relation between weight and mass is given by $\mathbf{W} = m\mathbf{g}$. Because $\mathbf{g}$ varies from point to point on the earth, $\mathbf{W}$, the weight of a body of mass m, is different in different localities. Thus, the weight of a one kg-mass in a locality where g is 9.80 m/s² is 9.80 N; in a locality where g is 9.78 m/s², the same one kg-mass weighs 9.78 N. If these weights were determined by measuring the amount of stretch required in a spring to balance them, the difference in weight of the same one kg-mass at the two different localities would be evident in the slightly different stretch of the spring at these two localities. Hence, unlike the mass of a body, which is an intrinsic property of the body, the weight of a body depends on its location relative to the center of the earth. Spring scales read differently, balances the same, at different parts of the earth.

We shall generalize the concept of weight in Chapter 16 in which we discuss universal gravitation. There we shall see that the weight of a body is zero in regions of space where the gravitational effects are nil, although the inertial effects, and hence the mass of the body, remain unchanged from those on earth. In a space ship free from the influence of gravity it is a simple matter to lift a large block of lead ($\mathbf{W} = 0$), but the astronaut would still stub his toe if he were to kick the block ($m \neq 0$).

It takes the same force to accelerate a body in gravity-free space as it does to accelerate it along a horizontal frictionless surface on earth, for its mass is the same in each place. But it takes a greater force to hold the body up against the pull of the earth on the earth's surface than it does high up in space, for its weight is different in each place.

Often, instead of being given the mass, we are given the weight of a body on which forces are exerted. The acceleration $\mathbf{a}$ produced by the force $\mathbf{F}$ acting on a body whose weight has a magnitude W can be obtained by combining Eq. 5-3 and Eq. 5-5. Thus from $\mathbf{F} = m\mathbf{a}$ and $W = mg$ we obtain

$$m = W/g, \quad \text{so that} \quad \mathbf{F} = (W/g)\mathbf{a}. \qquad (5\text{-}6)$$

The quantity W/g plays the role of m in the equation $F = ma$ and is, in fact, the mass of a body whose weight has the magnitude W. For example, a man whose weight is 160 lb at a point where $g = 32.0$ ft/s² has a mass $m = W/g = (160 \text{ lb})/(32.0 \text{ ft/s}^2) = 5.00$ slugs. Notice that his weight at another point where $g = 32.2$ ft/s² is $W = mg = (5.00 \text{ slugs})(32.2 \text{ ft/s}^2) = 161$ lb.

---

* Corrections for buoyancy, owing to the different volumes of air displaced by the salt and the standard, must be made. We discuss these in Chapter 17.

In Section 5-3 we defined force by measuring the acceleration imparted to a standard body by pulling on it with a stretched spring. That may be called a dynamic method for measuring force. Although convenient for the purposes of definition, it is not a particularly practical procedure for the measurement of forces. Another method for measuring forces is based on measuring the change in shape or size of a body (a spring, say) on which the force is applied when the body is unaccelerated. This may be called the static method of measuring forces.

The idea of the static method is to use the fact that when a body, under the action of several forces, has zero acceleration, the vector sum of all the forces acting on the body must be zero. This is, of course, just the content of the first law of motion. A single force acting on a body would produce an acceleration; this acceleration can be made zero if we apply another force to the body equal in magnitude but oppositely directed. In practice we seek to keep the body at rest. If now we choose some force as our unit force, we are in a position to measure forces. The pull of the earth on a standard body at a particular point can be taken as the unit force, for example.

The instrument most commonly used to measure forces in this way is the spring balance. It consists of a coiled spring having a pointer at one end that moves over a scale. A force exerted on the balance changes the length of the spring. If a body weighing 1.00 N is hung from the spring, the spring stretches until the pull of the spring on the body is equal in magnitude but opposite in direction to its weight. A mark can be made on the scale next to the pointer and labeled "1.00-N force." Similarly, 2.00-N, 3.00-N, etc., weights may be hung from the spring and corresponding marks can be made on the scale next to the pointer in each case. In this way the spring is calibrated. We assume that the force exerted on the spring is always the same when the pointer stands at the same position. The calibrated balance can now be used to measure any suitable unknown force, not merely the pull of the earth on some body.

The third law is tacitly used in our static procedure because we assume that the force exerted by the spring on the body is the same in magnitude as the force exerted by the body on the spring. This latter force is the force we wish to measure. The first law is used too, because we assume $\mathbf{F}$ is zero when $\mathbf{a}$ is zero. It is worth noting again here that if the acceleration were not zero, the body of weight W would not stretch the spring to the same length as it did with $\mathbf{a} = 0$. In fact, if the spring and attached body of weight W were to fall freely under gravity so that $\mathbf{a} = \mathbf{g}$, the spring would not stretch at all and its tension would be zero.

5-9
A STATIC PROCEDURE FOR MEASURING FORCES

It will be helpful to write down some procedures for solving problems in classical mechanics and to illustrate them by several examples. Newton's second law states that the vector sum of all the forces acting on a body is equal to its mass times its acceleration. The first step in problem solving is therefore: (1) Identify the body to whose motion the problem refers. Lack of clarity on the point as to what has been or should be picked as "the body" is a major source of mistakes. (2) Having selected "the body," we next turn our attention to the objects in "the environment" because these objects (inclined planes, springs, cords, the earth, etc.) exert forces on the body. We must be clear as to the nature of these forces. (3) The next step is to select a suitable (inertial) reference frame. We should position the origin and orient the coordinate axes so

5-10
SOME APPLICATIONS OF NEWTON'S LAWS OF MOTION

as to simplify the task of our next step as much as possible. (4) We now make a separate diagram of the body alone, showing the reference frame and *all* of the forces acting *on* the body. This is called a *free-body* diagram. (5) Finally we apply Newton's second law, in the form of Eq. 5-2, to each component of force and acceleration.

The following examples illustrate the method of analysis used in applying Newton's laws of motion. Each body is treated as if it were a particle of definite mass, so that the forces acting on it may be assumed to act at a point. Strings and pulleys are considered to have negligible mass. Although some of the situations picked for analysis may seem simple and artificial, they are the prototypes for many interesting real situations; but, more important, the method of analysis – which is the chief thing to understand – is applicable to all the modern and sophisticated situations of classical mechanics, even sending a spaceship to Mars.

Fig. 5-4*a* shows a weight W hung by strings. Consider the knot at the junction of the three strings to be "the body." The body remains at rest under the action of the three forces shown in Fig. 5-4*b*. Suppose we are given the magnitude of one of these forces. How can we find the magnitude of the other forces?

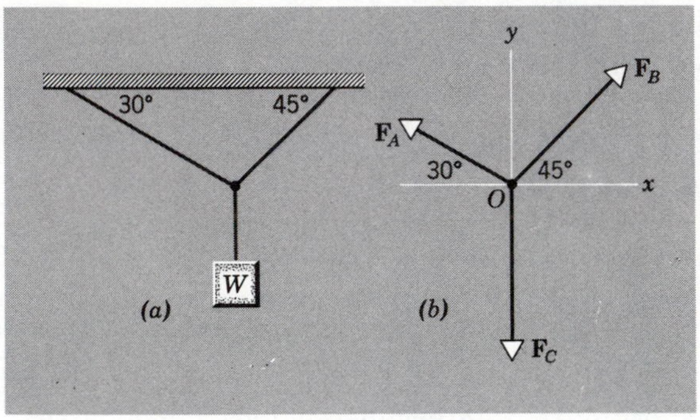

EXAMPLE 3

(a)

(b)

figure 5-4
Example 3. *(a)* A weight is suspended by strings. *(b)* A free-body diagram showing all the forces acting on the knot. The strings are assumed to be weightless.

$\mathbf{F}_A$, $\mathbf{F}_B$, and $\mathbf{F}_C$ are *all* the forces acting *on* the body. Since the body is unaccelerated (actually at rest), $\mathbf{F}_A + \mathbf{F}_B + \mathbf{F}_C = 0$. Choosing the x- and y-axes as shown, we can write this vector equation as three scalar equations:

$$F_{Ax} + F_{Bx} = 0,$$

$$F_{Ay} + F_{By} + F_{Cy} = 0,$$

using Eq. 5-2. The third scalar equation for the z-axis is simply

$$F_{Az} = F_{Bz} = F_{Cz} = 0.$$

That is, the vectors all lie in the x-y plane so that they have no z-components.

From the figure we see that

$$F_{Ax} = -F_A \cos 30° = -0.866F_A,$$

$$F_{Ay} = F_A \sin 30° = 0.500F_A,$$

and

$$F_{Bx} = F_B \cos 45° = 0.707F_B,$$

$$F_{By} = F_B \sin 45° = 0.707F_B.$$

Also,

$$F_{Cy} = -F_C = -W,$$

because the string C merely serves to transmit the force on one end to the junction at its other end. Substituting these results into our original equations, we obtain

$$-0.866F_A + 0.707F_B = 0,$$

$$0.500F_A + 0.707F_B - W = 0.$$

If we are given the magnitude of any one of these three forces, we can solve these equations for the other two. For example, if $W = 100$ N, we obtain $F_A = 73.3$ N and $F_B = 89.6$ N.

We wish to analyze the motion of a block on a smooth incline.

(a) *Static case.* Figure 5-5a shows a block of mass m kept at rest on a smooth plane, inclined at an angle θ with the horizontal, by means of a string attached to the vertical wall. The forces acting *on* the block are shown in Fig. 5-5b. $\mathbf{F}_1$ is the force exerted *on* the block by the string; $m\mathbf{g}$ is the force exerted *on* the block by the earth, that is, its weight; and $\mathbf{F}_2$ is the force exerted *on* the block by the inclined surface. $\mathbf{F}_2$, called the normal force, is normal to the surface of contact because there is no frictional force between the surfaces.* If there were a frictional force, $\mathbf{F}_2$ would have a component parallel to the incline. Because we wish to analyze the motion of *the block*, we choose ALL the forces acting ON the *block.* You will note that the block will exert forces on other bodies in its environment (the string, the earth, the surface of the incline) in accordance with the action-reaction principle; these forces, however, are not needed to determine the motion of the block because they do not act on the block.

Suppose θ and m are given. How do we find F_1 and F_2? Since the block is unaccelerated, we obtain

$$\mathbf{F}_1 + \mathbf{F}_2 + m\mathbf{g} = 0.$$

It is convenient to choose the x-axis of our reference frame to be along the incline and the y-axis to be normal to the incline (Fig. 5-5b). With this choice of coordinates, only one force, $m\mathbf{g}$, must be resolved into components in solving the problem. The two scalar equations obtained by resolving $m\mathbf{g}$ along the x- and y-axes are

$$F_1 - mg \sin \theta = 0, \quad \text{and} \quad F_2 - mg \cos \theta = 0,$$

from which F_1 and F_2 can be obtained if θ and m are given.

(b) *Dynamic case.* Now suppose that we cut the string. Then the force $\mathbf{F}_1$, the pull of the string on the block, will be removed. The resultant force on the block will no longer be zero, and the block will accelerate. What is its acceleration? From Eq. 5-2 we have $F_x = ma_x$ and $F_y = ma_y$. Using these relations we obtain

$$F_2 - mg \cos \theta = ma_y = 0,$$

and

$$-mg \sin \theta = ma_x,$$

which yield

$$a_y = 0, \quad a_x = -g \sin \theta.$$

The acceleration is directed down the incline with a magnitude of $g \sin \theta$.

EXAMPLE 4

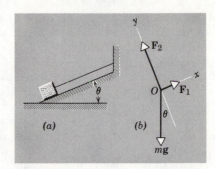

figure 5-5
Example 4. (a) A block is held on a smooth inclined plane by a string. (b) A free-body diagram showing all the forces acting on the block.

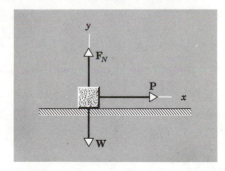

figure 5-6
Example 5. A block is being pulled along a smooth table. The forces acting on the block are shown.

EXAMPLE 5

Consider a block of mass m pulled along a smooth horizontal surface by a horizontal force $\mathbf{P}$, as shown in Fig. 5-6. $\mathbf{F}_N$ is the normal force exerted on the block by the frictionless surface and $\mathbf{W}$ is the weight of the block.

(a) If the block has a mass of 2.0 kg, what is the normal force?

From the second law of motion with $a_y = 0$, we obtain

$$F_y = ma_y \quad \text{or} \quad F_N - W = 0.$$

Hence, $F_N = W = mg = (2.0 \text{ kg})(9.8 \text{ m/s}^2) = 20$ N.

* The normal force is an example of a constraining force, one which limits the freedom of movement a body might otherwise have. It is an elastic force arising from small deformations of the bodies in contact, which are never perfectly rigid as we often tacitly assume.

(b) What force P is required to give the block a horizontal velocity of 4.0 m/s in 2.0 s starting from rest?

The acceleration a_x follows from

$$a_x = \frac{v_x - v_{x0}}{t} = \frac{4.0 \text{ m/s} - 0}{2.0 \text{ s}} = 2.0 \text{ m/s}^2.$$

From the second law, $F_x = ma_x$ or $P = ma_x$. The force P is then

$$P = ma_x = (2.0 \text{ kg})(2.0 \text{ m/s}^2) = 4.0 \text{ N}.$$

Figure 5-7a shows a block of mass m_1 on a smooth horizontal surface pulled by a massless string which is attached to a block of mass m_2 hanging over a pulley. We assume that the pulley has no mass and is frictionless and that it merely serves to change the direction of the tension in the string at that point. The magnitude of the tension is the same throughout a massless string (see Example 2). Find the acceleration of the system and the tension in the string.

Suppose we choose the block of mass m_1 as the body whose motion we investigate. The forces on this block, taken to be a particle, are shown in Fig. 5-7b. **T**, the tension in the string, pulls on the block to the right; m_1g is the downward pull of the earth on the block and $\mathbf{F}_N$ is the vertical force exerted on the block by the smooth table. The block will accelerate in the x-direction only, so that $a_{1y} = 0$. We, therefore, can write

$$F_N - m_1g = 0 = m_1a_{1y}, \tag{5-7}$$

and

$$T = m_1a_{1x}.$$

From these equations we conclude that $F_N = m_1g$. We do not know T, so we cannot solve for a_{1x}.

To determine T we must consider the motion of the block m_2. The forces acting on m_2 are shown in Fig. 5-7c. Because the string and block are accelerating, we cannot conclude that T equals m_2g. In fact, if T were to equal m_2g, the resultant force on m_2 would be zero, a condition holding only if the system is not accelerated.

The equation of motion for the suspended block is

$$m_2g - T = m_2a_{2y}. \tag{5-8}$$

The direction of the tension in the string changes at the pulley and, because the string has a fixed length, it is clear that

$$a_{2y} = a_{1x},$$

so that we can represent the acceleration of the system as simply a. We then obtain from Eqs. 5-7 and 5-8

$$m_2g - T = m_2a, \tag{5-9}$$

and

$$T = m_1a.$$

These yield

$$m_2g = (m_1 + m_2)a, \tag{5-10}$$

or

$$a = \frac{m_2}{m_1 + m_2}g,$$

and

$$T = \frac{m_1m_2}{m_1 + m_2}g, \tag{5-11}$$

which gives us the acceleration of the system a and the tension in the string T.

Notice that the tension in the string is always less than m_2g. This is clear from Eq. 5-11, which can be written

$$T = m_2g \frac{m_1}{m_1 + m_2}.$$

EXAMPLE 6

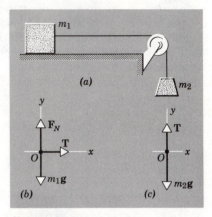

figure 5-7

Example 6. (a) Two masses are connected by a string; m_1 lies on a smooth table, m_2 hangs freely. (b) A free-body diagram showing all the forces acting on m_1. (c) A similar diagram for m_2.

Notice also that a is always less than g, the acceleration due to gravity. Only when m_1 equals zero, which means that there is no block at all on the table, do we obtain $a = g$ (from Eq. 5-10). In this case $T = 0$ (from Eq. 5-9).

We can interpret Eq. 5-10 in a simple way. The net unbalanced force on the system of mass $m_1 + m_2$ is represented by m_2g. Hence, from $F = ma$, we obtain Eq. 5-10 directly.

To make the example specific, suppose $m_1 = 2.0$ kg and $m_2 = 1.0$ kg. Then

$$a = \frac{m_2}{m_1 + m_2} g = \tfrac{1}{3}g = 3.3 \text{ m/s}^2,$$

and

$$T = \frac{m_1 m_2}{m_1 + m_2} g = (\tfrac{2}{3})(9.8) \text{ kg m/s}^2 = 6.5 \text{ N.}$$

EXAMPLE 7

Consider two unequal masses connected by a string which passes over a frictionless and massless pulley, as shown in Fig. 5-8a. Let m_2 be greater than m_1. Find the tension in the string and the acceleration of the masses.

We consider an *upward* acceleration *positive*. If the acceleration of m_1 is a, the acceleration of m_2 must be $-a$. The forces acting on m_1 and on m_2 are shown in Fig. 5-8b in which T represents the tension in the string.

The equation of motion for m_1 is

$$T - m_1g = m_1a$$

and for m_2 is

$$T - m_2g = -m_2a.$$

Combining these equations, we obtain

$$a = \frac{m_2 - m_1}{m_2 + m_1} g, \tag{5-12}$$

and

$$T = \frac{2m_1 m_2}{m_1 + m_2} g.$$

For example, if $m_2 = 2.0$ slugs and $m_1 = 1.0$ slug,

$$a = (32/3.0) \text{ ft/s}^2 = g/3,$$

$$T = (\tfrac{4}{3})(32) \text{ slug ft/s}^2 = 43 \text{ lb.}$$

Notice that the magnitude of T is always intermediate between the weight of the mass m_1 (32 lb in our example) and the weight of the mass m_2 (64 lb in our example). This is to be expected, since T must exceed m_1g to give m_1 an upward acceleration, and m_2g must exceed T to give m_2 a downward acceleration. In the special case when $m_1 = m_2$, we obtain $a = 0$ and $T = m_1g = m_2g$, which is the static result to be expected.

Figure 5-8c shows the forces acting on the massless pulley. If we treat the pulley as a particle, all the forces can be taken to act through its center. P is the

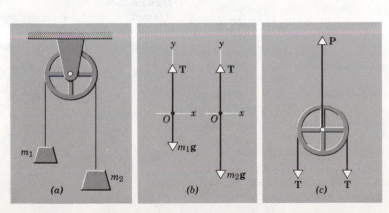

figure 5-8

Example 7. (a) Two unequal masses are suspended by a string from a pulley (Atwood's machine). (b) Free-body diagrams for m_1 and m_2. (c) Free-body diagram for the pulley, assumed massless.

upward pull of the support on the pulley and T is the downward pull of each segment of the string on the pulley. Since the pulley has no translational motion,

$$P = T + T = 2T.$$

If we were to drop our assumption of a massless pulley and assign a mass m to it, we would then be required to include a downward force mg on the support. Also, as we shall see later, the rotational motion of the pulley results in a different tension in each segment of the string. Friction in the bearings also affects the rotational motion of the pulley and the tension in the strings.

EXAMPLE 8

Consider an elevator moving vertically with an acceleration **a**. We wish to find the force exerted by a passenger on the floor of the elevator.

Acceleration will be taken *positive upward* and *negative downward*. Thus positive acceleration in this case means that the elevator is either moving upward with increasing speed or is moving downward with decreasing speed. Negative acceleration means that the elevator is moving upward with decreasing speed or downward with increasing speed.

From Newton's third law the force exerted by the passenger on the floor will always be equal in magnitude but opposite in direction to the force exerted by the floor on the passenger. We can therefore calculate either the action force or the reaction force. When the forces acting on the passenger are used, we solve for the latter force. When the forces acting on the floor are used, we solve for the former force.

The situation is shown in Fig. 5-9: The passenger's true weight is **W** and the force exerted on him by the floor, called **P,** is his *apparent* weight in the accelerating elevator. The resultant force acting on him is $\mathbf{P} + \mathbf{W}$. Forces will be taken as positive when directed upward. From the second law of motion we have

$$F = ma,$$

or

$$P - W = ma, \qquad (5\text{-}13)$$

where m is the mass of the passenger and a is his (and the elevator's) acceleration.

Suppose, for example, that the passenger weighs 160 lb and the acceleration is 2.0 ft/s² upward. We have

$$m = \frac{W}{g} = \frac{160 \text{ lb}}{32 \text{ ft/s}^2} = 5.0 \text{ slugs,}$$

and from Eq. 5-13,

$$P - 160 \text{ lb} = (5.0 \text{ slugs})(2.0 \text{ ft/s}^2)$$

or

$$P = \text{apparent weight} = 170 \text{ lb.}$$

If we were to measure this force directly by having the passenger stand on a spring scale fixed to the elevator floor (or suspended from the ceiling), we would find the scale reading to be 170 lb for a man whose weight is 160 lb. The passenger feels himself pressing down on the floor with greater force (the floor is pressing upward on him with greater force) than when he and the elevator are at rest. Everyone experiences this feeling when an elevator starts upward from rest.

If the acceleration were taken as 2.0 ft/s² downward, then $a = -2.0$ ft/s² and $P = 150$ lb for the passenger. The passenger who weighs 160 lb feels himself pressing down on the floor with less force than when he and the elevator are at rest.

If the elevator cable were to break and the elevator were to fall freely with an acceleration $a = -g$, then P would equal $W + (W/g)(-g) = 0$. Then the passenger and floor would exert no forces on each other. The passenger's apparent weight, as indicated by the spring scale on the floor, would be zero. Such a situation is

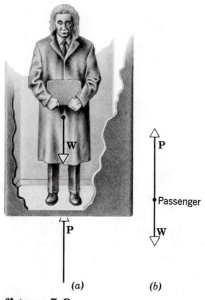

(a)   (b)

figure 5-9
Example 8. *(a)* A passenger stands on the floor of an elevator. *(b)* A free-body diagram for the passenger.

popularly referred to as "weightlessness." The passenger's weight (the pull of gravity on him) has not changed, of course, but the force he exerts on the floor and the reaction force of the floor on him are zero.

questions

1. What is your mass in slugs? Your weight in newtons?
2. Why do you fall forward when a moving train decelerates to a stop and fall backward when a train accelerates from rest? What would happen if the train rounded a curve at constant speed?
3. A block of mass m is supported by a cord C from the ceiling, and another cord D is attached to the bottom of the block (Fig. 5-10). Explain this: If you give a sudden jerk to D, it will break, but if you pull on D steadily, C will break.
4. A horse is urged to pull a wagon. The horse refuses to try, citing Newton's third law as his defense: " 'The pull of the horse on the wagon is equal but opposite to the pull of the wagon on the horse.' If I can never exert a greater force on the wagon than it exerts on me, how can I ever start the wagon moving?" asks the horse. How would you reply?
5. Comment on whether the following pairs of forces are examples of action-reaction: (a) the earth attracts a brick; the brick attracts the earth; (b) a propellered airplane pulls air in toward the plane; the air pushes the plane forward; (c) a horse pulls forward on a cart, accelerating it; the cart pulls backward on the horse; (d) a horse pulls forward on a cart without moving it; the cart pulls back on the horse; (e) a horse pulls forward on a cart without moving it; the earth exerts an equal and opposite force on the cart.
6. Criticize the statement, often made, that the mass of a body is a measure of the "quantity of matter" in it.
7. Using force, length, and time as fundamental quantities, what are the dimensions of mass?
8. Is the definition of mass that we have given limited to objects initially at rest?
9. Comment on the following statements about mass and weight taken from examination papers. (a) Mass and weight are the same physical quantities expressed in different units; (b) mass is a property of one object alone whereas weight results from the interaction of two objects; (c) the weight of an object is proportional to its mass; (d) the mass of a body varies with changes in its local weight.
10. A horizontal force acts on a mass which is free to move. Can it produce an acceleration if the force is less than the weight of that mass?
11. Does the acceleration of a freely falling body depend upon the weight of the body?
12. A bird alights on a stretched telegraph wire. Does this change the tension in the wire? If so, by an amount less than, equal to, or greater than the weight of the bird?
13. In Fig. 5-11, we show four forces which are about equal in magnitude. What combination of three forces, acting together on the same body, might keep that body in translational equilibrium?
14. Why do raindrops fall with constant speed during the later stages of their descent?
15. In a tug of war, three men pull on a rope to the left at A and three men pull to the right at B with forces of equal magnitude. Now a weight of 5.0 lb is hung vertically from the center of the rope. (a) Can the men get the rope AB to be horizontal? (b) If not, explain. If so, determine the magnitude of the forces required at A and B to do this.
16. Both the following statements are true; explain them. Two teams having a tug of war must always pull equally hard on one another. The team that pushes harder against the ground wins.

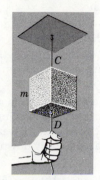

figure 5-10
Question 3

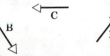

figure 5-11
Question 13

17. A massless rope is strung over a frictionless pulley. A monkey holds onto one end of the rope and a mirror, having the same weight as the monkey, is attached to the other end of the rope at the monkey's level. Can the monkey get away from his image seen in the mirror (a) by climbing up the rope, (b) by climbing down the rope, (c) by releasing the rope?

18. Two objects of equal mass rest on opposite pans of a trip scale. Does the scale remain balanced when it is accelerated up or down in an elevator?

19. You stand on the large platform of a spring scale and note your weight. You then take a step on this platform and notice that the scale reads less than your weight at the beginning of the step and more than your weight at the end of the step. Explain.

20. A weight is hung by a cord from the ceiling of an elevator. From the following conditions, choose the one in which the tension in the cord will be greatest . . . least? (a) elevator at rest; (b) elevator rising with uniform speed; (c) elevator descending with decreasing speed; (d) elevator descending with increasing speed.

21. A woman stands on a spring scale in an elevator. In which case below will the scale record the minimum reading . . . the maximum reading? (a) elevator stationary; (b) elevator cable breaks, free fall; (c) elevator accelerating upward; (d) elevator accelerating downward; (e) elevator moving at constant velocity.

22. Under what circumstances would your weight be zero? Does your answer depend on the choice of a reference system?

problems

SECTION 5-4

1. Two blocks, mass m_1 and m_2, are connected by a light spring on a horizontal frictionless table. Find the ratio of their accelerations a_1 and a_2 after they pulled apart and then released.   *Answer: $a_1/a_2 = m_2/m_1$.*

SECTION 5-5

2. (a) Two 10-lb weights are attached to a spring scale as shown in Fig. 5-12(a). What is the reading of the scale? (b) A single 10-lb weight is attached to a spring scale which itself is attached to a wall, as shown in Fig. 5-12(b). What is the reading of the scale?

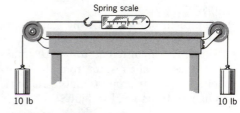

figure 5-12(a)
Problem 2(a)

3. Two blocks are in contact on a frictionless table. A horizontal force is applied to one block, as shown in Fig. 5-13. (a) If $m_1 = 2.0$ kg, $m_2 = 1.0$ kg, and $F = 3.0$ N, find the force of contact between the two blocks. (b) Show that if the same force F is applied to m_2 rather than to m_1, the force of contact between the blocks is 2.0 N, which is not the same value derived in (a). Explain.   *Answer: (a) 1.0 N.*

SECTION 5-8

4. A space traveler whose mass is 75 kg leaves the earth. Compute his weight (a) on the earth, (b) on Mars, where $g = 3.8$ m/s², and (c) in interplanetary space. (d) What is his mass at each of these locations?

SECTION 5-10

5. A car moving initially at a speed of 50 mi/h (80 km/h) and weighing 3000 lb (13,000 N) is brought to a stop in a distance of 200 ft (61 m). Find (a) the braking force, and (b) the time required to stop. Assuming the same braking force, find (c) the distance, and (d) the time required to stop if the car was going 25 mi/h (40 km/h) initially.
Answer: (a) 1300 lb (5400 N). (b) 5.5 s (5.5 s). (c) 50 ft (15 m). (d) 2.7 s (2.7 s).

6. A body of mass m is acted on by two forces $\mathbf{F}_1$ and $\mathbf{F}_2$, as shown in Fig. 5-14. If $m = 5.0$ kg, $F_1 = 3.0$ N, and $F_2 = 4.0$ N, find the vector acceleration of the body.

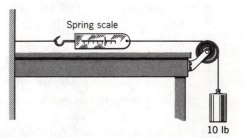

figure 5-12(b)
Problem 2(b)

7. An electron is projected horizontally at a speed of 1.2×10^7 m/s into an

electric field which exerts a constant vertical force of 4.5×10^{-16} N on it. The mass of the electron is 9.1×10^{-31} kg. Determine the vertical distance the electron is deflected during the time it has moved forward 3.0 cm horizontally. *Answer: 1.5 mm.*

8. A body of mass 2.0 slugs is acted on by the downward force of gravity and a horizontal force of 130 lb. Find (a) its acceleration and (b) its velocity as functions of time, assuming it starts from rest.

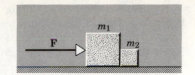

figure 5-13
Problem 3

9. An electron travels in a straight line from the cathode of a vacuum tube to its anode, which is exactly 1.0 cm away. It starts with zero speed and reaches the anode with a speed of 6.0×10^6 m/s. (a) Assume constant acceleration and compute the force on the electron. Take the electron's mass to be 9.1×10^{-31} kg. This force is electrical in origin. (b) Compare it with the gravitational force on the electron, which we neglected when we assumed straight line motion. *Answer: (a) 1.6×10^{-15} N. (b) 8.9×10^{-30} N.*

10. A man of mass 80 kg (weight $mg = 176$ lb) jumps down to a concrete patio from a window ledge only 0.50 m (1.6 ft) above the ground. He neglects to bend his knees on landing, so that his motion is arrested in a distance of about 2.0 cm (0.79 in). (a) What is the average acceleration of the man from the time his feet first touch the patio to the time he is brought fully to rest? (b) With what average force does this jump jar his bone structure?

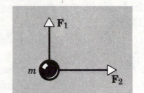

figure 5-14
Problem 6

11. Let the only forces acting on two bodies be their mutual interactions. If both bodies start from rest, show that the distances traveled by each are inversely proportional to the respective masses of the bodies.

12. Determine the frictional force of the air on a body of mass 0.25 kg falling with an acceleration of 9.2 m/s².

13. A charged sphere of mass 3.0×10^{-4} kg is suspended from a string. An electric force acts horizontally on the sphere so that the string makes an angle of 37° with the vertical when at rest. Find (a) the magnitude of the electric force and (b) the tension in the string. *Answer: (a) 2.2×10^{-3} N. (b) 3.7×10^{-3} N.*

14. A block of mass M is pulled along a horizontal frictionless surface by a rope of mass m, as shown in Fig. 5-15. A horizontal force $\mathbf{P}$ is applied to one end of the rope. (a) Show that the rope *must* sag, even if only by an imperceptible amount. Then, assuming that the sag is negligible, find (b) the acceleration of rope and block, (c) the force that the rope exerts on the block M, and (d) the tension in the rope at its midpoint.

figure 5-15
Problem 14

figure 5-16
Problem 15

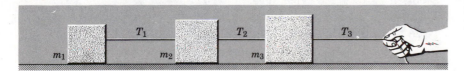

15. Three blocks are connected, as shown in Fig. 5-16, on a horizontal frictionless table and pulled to the right with a force $T_3 = 60$ N. If $m_1 = 10$ kg, $m_2 = 20$ kg, and $m_3 = 30$ kg, find the tensions T_1 and T_2. Draw an analogy to bodies being pulled in tandem, such as an engine pulling a train of coupled cars. *Answer: $T_1 = 10$ N, $T_2 = 30$ N.*

16. A rocket and its payload have a total mass of 50,000 kg (weight $mg = 110,250$ lb). How large is the thrust of the rocket engine when (a) the rocket is "hovering" over the launch pad, just after ignition, and (b) when the rocket is accelerating upward at 20 m/s² (66 ft/s²)?

17. How could a 100-lb object be lowered from a roof using a cord with a breaking strength of 87 lb without breaking the cord? *Answer: Lower object with an acceleration ≥ 4.2 ft/s².*

18. A block is released from rest at the top of a frictionless inclined plane 16 m long. It reaches the bottom 4.0 s later. A second block is projected up the plane from the bottom at the instant the first block is released in such a way that it returns to the bottom simultaneously with the first block. (a) Find the acceleration of each block on the incline. (b) What is the initial velocity of the second block? (c) How far up the incline does it travel? (d) What angle does the plane make with the horizontal?

19. A block of mass $m_1 = 3.0$ slugs on a smooth inclined plane of angle 30° is connected by a cord over a small frictionless pulley to a second block of mass $m_2 = 2.0$ slugs hanging vertically (Fig. 5-17). (a) What is the acceleration of each body? (b) What is the tension in the cord?
    Answer: (a) 3.2 ft/s². (b) 58 lb.

20. A block is projected up a frictionless inclined plane with a speed v_0. The angle of incline is θ. (a) How far up the plane does it go? (b) How long does it take to get there? (c) What is its speed when it gets back to the bottom? Find numerical answers for $\theta = 30°$ and $v_0 = 8.0$ ft/s.

21. An elevator weighing 6000 lb is pulled upward by a cable with an acceleration of 4.0 ft/s². (a) What is the tension in the cable? (b) What is the tension when the elevator is accelerating downward at 4.0 ft/s², but is still moving upward?
    Answer: (a) 6800 lb. (b) 5300 lb.

22. A lamp hangs vertically from a cord in a descending elevator. The elevator has a deceleration of 8.0 ft/s² (2.4 m/s²) before coming to a stop. (a) If the tension in the cord is 20 lb (89 N), what is the mass of the lamp? (b) What is the tension in the cord when the elevator ascends with an acceleration of 8.0 ft/s² (2.4 m/s²)?

23. An 80-kg man is parachuting and experiencing a downward acceleration of 2.5 m/s². The mass of the parachute is 5.0 kg. (a) What is the value of the upward force exerted on the parachute by the air? (b) What is the value of the downward force exerted by the man on the parachute?
    Answer: (a) 620 N. (b) 580 N.

24. A research balloon of total mass M is descending vertically with downward acceleration a. How much ballast must be thrown from the car to give the balloon an *upward* acceleration a?

25. An elevator consists of the elevator cage (A), the counterweight (B), the driving mechanism (C), and the cable and pulleys as shown in Fig. 5-18. The mass of the cage is 1100 kg and the mass of the counterweight is 1000 kg. Neglect friction and the mass of the cable and pulleys. The elevator accelerates upward at 2.0 m/s² and the counterweight accelerates downward at the same rate. (a) What is the value of the tension T_1? (b) T_2? (c) What force is exerted on the cable by the driving mechanism?
    Answer: (a) 1.3×10^4 N. (b) 0.78×10^4 N. (c) 5.2×10^3 N, toward the counterweight.

26. A 100-kg man lowers himself to the ground from a height of 10 m by means of a rope passed over a frictionless pulley and attached to a 70-kg sandbag. (a) With what speed does the man hit the ground? (b) Is there anything he could do to reduce the speed with which he hits the ground?

27. Someone exerts a force **F** directly up on the axle of the pulley shown in Fig. 5-19. Consider the pulley and string to be massless and the bearing frictionless. Two bodies, m_1 of mass 1.0 kg and m_2 of mass 2.0 kg, are attached, as shown, to the opposite ends of the string which passes over the pulley. The body m_2 is in contact with the horizontal floor. (a) Draw a free body diagram for the pulley and for each of the masses. (b) What is the largest value the force F may have so that m_2 will remain at rest on the floor? (c) What is the tension in the string if the upward force **F** is 100 N? (d) With the tension determined in part (c), what is the acceleration of m_1?
    Answer: (b) 39 N. (c) 50 N. (d) 40 m/s², upward.

28. A 10-kg monkey is climbing a massless rope attached to a 15-kg mass over a (frictionless!) tree limb. (a) Explain quantitatively how the monkey can

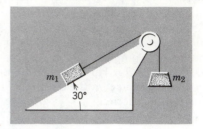

figure 5-17
Problem 19

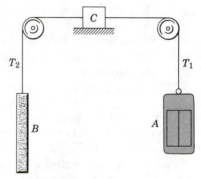

figure 5-18
Problem 25

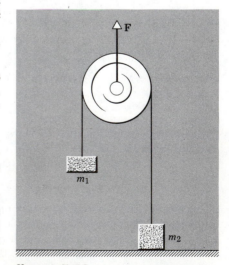

figure 5-19
Problem 27

climb up the rope so that he can raise the 15-kg mass off the ground. If, after the mass has been raised off the ground, the monkey stops climbing and holds on to the rope, what will now be (b) his acceleration and (c) the tension in the rope?

29. A plumb bob hanging from the ceiling of a railroad car acts as an accelerometer. (a) Derive the general expression relating the horizontal acceleration *a* of the car to the angle θ made by the bob with the vertical. (b) Find *a* when $\theta = 20°$. (c) Find θ when $a = 5.0$ ft/s². *Answer:* (a) $a = g \tan \theta$. (b) 12 ft/s². (c) 8.9°.

30. A uniform flexible chain of length *l*, with weight per unit length λ, passes over a small, frictionless, massless pulley. It is released from a rest position with a length of chain *x* hanging from one side and a length $l - x$ from the other side. Find the acceleration *a* as a function of *x*.

31. Two particles, each of mass *m*, are connected by a light string of length 2*l*, as shown in Fig. 5-20. A continuous force **F** is applied at the midpoint of the string ($x = 0$) at right angles to the initial position of the string. Show that the acceleration of *m* in the direction at right angles to **F** is given by

$$a_r = \frac{F}{2m} \frac{x}{\sqrt{l^2 - x^2}}$$

in which *x* is the perpendicular distance of one of the particles from the line of action of **F**. Discuss the situation when $x = l$.

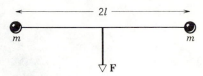

figure 5-20
Problem 31

32. A chain consisting of five links, each of mass 0.10 kg, is lifted vertically with a constant acceleration of 2.5 m/s², as shown in Fig. 5-21. Find (a) the forces acting between adjacent links, (b) the force **F** exerted on the top link by the agent lifting the chain, and (c) the *net* force acting on each link.

33. *Terminal velocity.* The resistance of the air to the motion of bodies in free fall depends on many factors, such as the size of the body and its shape, the density and temperature of the air, and the velocity of the body through the air. A useful assumption, only approximately true, is that the resisting force f_R is proportional to the velocity and oppositely directed; that is, $\mathbf{f}_R = -k\mathbf{v}$, where *k* is a constant whose value in any particular case is determined by factors other than velocity.

Consider free fall of an object from rest through the air.

(a) Show that Newton's second law gives

$$mg - kv = ma \quad \text{or} \quad mg - k\frac{dy}{dt} = m\frac{d^2y}{dt^2}.$$

(b) Show that the body ceases to accelerate when it reaches a velocity $v_T = mg/k$, called the *terminal velocity*.

(c) Prove, by substituting it in the equation of motion of part (a), that the velocity varies with time as

$$v = v_T(1 - e^{-kt/m})$$

and plot *v* versus *t*.

(d) Sketch qualitatively curves of *y* versus *t* and *a* versus *t* for this motion, noting that the initial acceleration is *g* and the final acceleration is zero.

figure 5-21
Problem 32

34. A right triangular wedge of mass *M* and angle θ, supporting a cubical block of mass *m* on its side, rests on a horizontal table, as shown in Fig. 5-22. (a) What horizontal acceleration *a* must *M* have relative to the table to keep *m* stationary relative to the wedge, assuming frictionless contacts? (b) What horizontal force **F** must be applied to the system to achieve this result, assuming a frictionless table top? (c) Suppose no force is supplied to *M* and both surfaces are frictionless. Describe the resulting motion.

35. A block, mass *m*, slides down a frictionless incline making an angle θ with an elevator floor. Find its acceleration relative to the incline in the following cases. (a) Elevator descends at constant speed *v*. (b) Elevator ascends at constant speed *v*. (c) Elevator descends with acceleration *a*. (d) Elevator

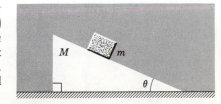

figure 5-22
Problem 34

descends with deceleration *a*. (*e*) Elevator cable breaks. (*f*) In part (*c*) above, what is the force exerted on the block by the incline?

Answer: (*a*) $g \sin \theta$ down the incline. (*b*) $g \sin \theta$ down the incline. (*c*) $(g - a)$ $\sin \theta$ down the incline. (*d*) $(g + a) \sin \theta$ down the incline. (*e*) Zero. (*f*) $m(g - a) \cos \theta$.

6
particle dynamics—II

In Chapter 5 we considered particle dynamics for bodies subject to a force that was constant in both magnitude and direction. The forces that we dealt with were exerted by the earth or by taut cords, that is, they were either gravitational or elastic. In this chapter we consider another kind of force, that resulting from friction.

We shall also discuss the dynamics of uniform circular motion, in which the force, although constant in magnitude, changes in direction with time. In Chapter 10 we shall consider problems in which the force, although constant in direction, changes in magnitude with time, as when one body exerts a transient force on another during a collision. Finally, in Chapter 15, we shall consider problems in which the force changes in *both* magnitude *and* direction with time, such as the force exerted by a spring on an oscillating mass suspended from it.

6-1
INTRODUCTION

If we project a block of mass m with initial velocity $\mathbf{v}_0$ along a long horizontal table, it eventually comes to rest. This means that, while it is moving, it experiences an average acceleration $\bar{\mathbf{a}}$ that points in the direction opposite to its motion. If (in an inertial frame) we see that a body is being accelerated, we always associate a force, defined from Newton's second law, with the motion. In this case we declare that the table exerts a force of friction, whose average value is $m\bar{\mathbf{a}}$, on the sliding block.

Actually, whenever the surface of one body slides over that of an-

6-2
*FRICTIONAL FORCES**

* See ''The Friction of Solids'' by E. H. Freitag, in *Contemporary Physics*, Vol. 2, 1961, p. 198, for a good general reference; see also the article ''Friction'' in Britannica 3.

other, each body exerts a frictional force on the other. The frictional force on each body is in a direction opposite to its motion relative to the other body. Frictional forces automatically oppose the motion and never aid it. Even when there is no relative motion, frictional forces may exist between surfaces.

Although we have ignored its effects up to now, friction is very important in our daily lives. Left to act alone it brings every rotating shaft to a halt. In an automobile, about 20% of the engine power is used to counteract frictional forces. Friction causes wear and seizing of moving parts and many engineering man-hours are devoted to reducing it. On the other hand, without friction we could not walk; we could not hold a pencil and if we could it would not write; wheeled transport as we know it would not be possible.

We want to know how to express frictional forces in terms of the properties of the body and its environment; that is, we want to know the force law for frictional forces. In what follows we consider the sliding (not rolling) of one dry (unlubricated) surface over another. As we shall see later, friction, viewed at the microscopic level, is a very complicated phenomenon* and the force laws for dry, sliding friction are empirical in character and approximate in their predictions. They do not have the elegant simplicity and accuracy that we find for the gravitational force law (Chapter 16) or for the electrostatic force law (Chapter 26). It is remarkable, however, considering the enormous diversity of surfaces one encounters, that many aspects of frictional behavior can be understood qualitatively on the basis of a few simple mechanisms.

Consider a block at rest on a horizontal table as in Fig. 6-1. Attach a spring to it to measure the force required to set the block in motion. We find that the block will not move even though we apply a small force. We say that our applied force is balanced by an opposite frictional force exerted on the block by the table, acting along the surface of contact. As we increase the applied force we find some definite force at which the block just begins to move. Once motion has started, this same force produces accelerated motion. By reducing the force once motion has started, we find that it is possible to keep the block in uniform motion without acceleration; this force may be small, but it is never zero.

The frictional forces acting between surfaces at rest with respect to each other are called forces of *static friction*. The maximum force of static friction will be the same as the smallest force necessary to start motion. Once motion is started, the frictional forces acting between the surfaces usually decrease so that a smaller force is necessary to maintain uniform motion. The forces acting between surfaces in relative motion are called forces of *kinetic friction*.

The maximum force of static friction between any pair of dry unlubricated surfaces follows these two empirical laws. (1) It is approximately independent of the area of contact, over wide limits and (2) it is proportional to the normal force. The normal force, sometimes called the loading force, is the one which either body exerts on the other at right angles to their mutual interface. It arises from the elastic deformation of the bodies in contact, such bodies never really being entirely rigid. For a block resting on a horizontal table or sliding along it, the normal force is equal in magnitude to the weight of the block. Because

---

* See, for example, "Stick and Slip" by Ernest Rabinowicz, in *Scientific American*, May 1956.

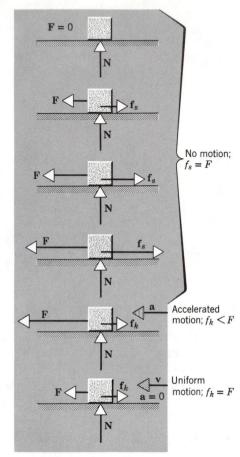

figure 6-1
A block being put into motion as applied force **F** overcomes frictional forces. In the first four drawings the applied force is gradually increased from zero to magnitude $\mu_s N$. No motion occurs until this point because the frictional force always just balances the applied force. The instant F becomes greater than $\mu_s N$, the block goes into motion, as is shown in the fifth drawing. In general, $\mu_k N < \mu_s N$; this leaves an unbalanced force to the left and the block accelerates. In the last drawing F has been reduced to equal $\mu_k N$. The net force is zero, and the block continues with constant velocity.

the block has no vertical acceleration, the table must be exerting a force on the block that is directed upward and is equal in magnitude to the downward pull of the earth on the block, that is, equal to the block's weight.

The ratio of the magnitude of the maximum force of static friction to the magnitude of the normal force is called the *coefficient of static friction* for the surfaces involved. If f_s represents the magnitude of the force of static friction, we can write

$$f_s \leq \mu_s N, \tag{6-1}$$

where μ_s is the coefficient of static friction and N is the magnitude of the normal force. The equality sign holds only when f_s has its maximum value.

The force of kinetic friction f_k between dry, unlubricated surfaces follows the same two laws as those of static friction. (1) It is approximately independent of the area of contact over wide limits and (2) it is proportional to the normal force. The force of kinetic friction is also reasonably independent of the relative speed with which the surfaces move over each other.

The two laws of friction above were first discovered experimentally by Leonardo da Vinci (1452–1519). Leonardo's statement of the two laws was remarkable, coming as it did about two centuries before the concept of force was developed by Newton. Leonardo's formulation was: (1) "Friction made by the same weight will be of equal resistance at the beginning of the movement though the contact may be of different breadths or lengths" and (2) "Friction produces double the amount of effort if the weight be doubled." The French scientist, Charles A. Coulomb, (1736–1806) did many experiments on friction and pointed out the difference between static and kinetic friction.

The ratio of the magnitude of the force of kinetic friction to the magnitude of this normal force is called the *coefficient of kinetic friction*. If f_k represents the magnitude of the force of kinetic friction,

$$f_k = \mu_k N, \tag{6-2}$$

where μ_k is the coefficient of kinetic friction.

Both μ_s and μ_k are dimensionless constants, each being the ratio of (the magnitudes of) two forces. Usually, for a given pair of surfaces $\mu_s > \mu_k$. The actual values of μ_s and μ_k depend on the nature of both the surfaces in contact. Both μ_s and μ_k can exceed unity, although commonly they are less than one. Notice that Eqs. 6-1 and 6-2 are relations between the *magnitudes only* of the normal and frictional forces. These forces are always directed perpendicularly to one another.

On the atomic scale even the most finely polished surface is far from plane. Figure 6-2, for example, shows an actual profile, highly magnified, of a steel surface that would be considered to be highly polished. One can readily believe that when two bodies are placed in contact, the actual microscopic area of contact is much less than apparent macroscopic area of contact; in a particular case these areas can be easily in the ratio of 1 to 10^4.

The actual (microscopic) area of contact is proportional to the normal force, because the contact points deform plastically under the great stresses that develop at these points. Many contact points actually become "cold-welded" together. This phenomenon, *surface adhesion*, occurs because at the contact points the molecules on opposite sides of the surface are so close together that they exert strong intermolecular forces on each other.

When one body (a metal, say) is pulled across another, the frictional resistance is associated with the rupturing of these thousands of tiny welds,

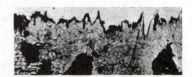

figure 6-2

A highly magnified view of a section of a finely polished steel surface. The section was cut at an angle so that vertical distances are exaggerated by a factor of ten with respect to horizontal distances. The surface irregularities are several thousand atomic diameters high. From *Friction and Lubrication of Solids*, by F. P. Bowden and D. Tabor, Clarendon Press, 1950.

which continually reform as new chance contacts are made (see Fig. 6-3). Radio-active tracer experiments have shown that, in the rupturing process, small fragments of one metallic surface may be sheared off and adhere to the other surface. If the relative speed of the two surfaces is great enough, there may be local melting at certain contact areas even though the surface as a whole may feel only moderately warm.

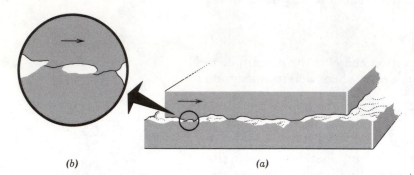

(b)   (a)

figure 6-3
Sliding friction. *(a)* The upper body is sliding to the right over the lower body in this enlarged diagram. *(b)* A further enlarged view showing two spots where surface adhesion has occurred. Force is required to break these welds apart and maintain the motion.

The coefficient of friction depends on many variables, such as the nature of the materials, surface finish, surface films, temperature, and extent of contamination. For example, if two carefully cleaned metal surfaces are placed in a highly evacuated chamber so that surface oxide films do not form, the coefficient of friction rises to enormous values and the surfaces actually become firmly "welded" together. The admission of a small amount of air to the chamber so that oxide films may form on the opposing surfaces reduces the coefficient of friction to its "normal" value.

With these complications it is not surprising that there is no exact theory of dry friction and that the laws of friction are empirical. The surface adhesion theory of friction for metals leads to a ready understanding of the two laws of friction mentioned above however. (1) The microscopic contact area, which determines the frictional force f_k, is proportional to the normal force N and thus f_k is proportional to N, as Eq. 6-2 shows. (2) The fact that the frictional force is independent of the apparent area of contact means, for example, that the force required to drag a metal "brick" along a metal table is the same no matter which face of the brick is in contact with the table. We can understand this only if the microscopic area of contact is the same for all positions of the brick, and this is indeed the case. With the largest face down, there are a relatively large number of relatively small area contacts supporting the load; with the smallest face down there are fewer contacts (because the apparent contact area is smaller), but the area of individual contact is larger by just the same factor because of the higher pressure exerted by the up-ended brick on this smaller number of contacts supporting the same load.

The frictional force that opposes one body *rolling* over another is much less than that for a sliding motion and this, indeed, is the advantage of the wheel over the sledge. This reduced friction is due in large part to the fact that, in rolling, the microscopic contact welds are "peeled" apart rather than "sheared" apart as in sliding friction. This will reduce the frictional force by a large factor.

Frictional resistance in dry, sliding, friction can be considerably reduced by lubrication. A mural in a grotto in Egypt dating back to 1900 B.C. shows a large stone statue being pulled on a sledge while a man in front of the sledge pours lubricating oil in its path. A still more effective technique is to introduce a layer of gas between the sliding surfaces; the dry ice puck and the gas-supported bearing are two examples. Friction can be reduced still further by suspending a rotating object in an evacuated space by means of magnetic forces. J. W. Beams, for example, has spun a 30-lb rotor of this type at 1000 rev/s; when the drive was cut off, the rotor lost speed at the rate of only 1 rev/s in a day.*

* See "Ultrahigh-Speed Rotation," Jesse W. Beams in *Scientific American,* April 1961.

Examples of the application of the empirical force law for friction follow. The coefficients of friction given are assumed to be constant. Actually μ_k can be regarded as a good average value that is not greatly different from the value at any particular speed in the range.

A block is at rest on an inclined plane making an angle θ with the horizontal, as in Fig. 6-4a. As the angle of incline is raised, it is found that slipping just begins at an angle of inclination θ_s. What is the coefficient of static friction between block and incline?

EXAMPLE 1

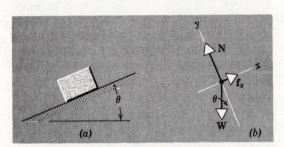

figure 6-4
Example 1. (a) A block at rest on a rough inclined plane. (b) A free-body force diagram for the block.

The forces acting on the block, considered to be a particle, are shown in Fig. 6-4b. **W** is the weight of the block, **N** the normal force exerted on the block by the inclined surface, and $\mathbf{f}_s$ the tangential force of friction exerted by the inclined surface on the block. Notice that the resultant force exerted by the inclined surface on the block, $\mathbf{N} + \mathbf{f}_s$, is no longer perpendicular to the surface of contact, as was true for smooth surfaces ($\mathbf{f}_s = 0$). The block is at rest, so that

$$\mathbf{N} + \mathbf{f}_s + \mathbf{W} = 0.$$

Resolving our forces into x- and y-components, along the plane and the normal to the plane, respectively, we obtain

$$N - W \cos \theta = 0,$$

$$f_s - W \sin \theta = 0.$$

(6-3)

However, $f_s \leq \mu_s N$. If we increase the angle of incline slowly until slipping just begins, then for that angle, $\theta = \theta_s$ and we can use $f_s = \mu_s N$. Substituting this into Eqs. 6-3, we obtain

$$N = W \cos \theta_s$$

and

$$\mu_s N = W \sin \theta_s,$$

so that

$$\mu_s = \tan \theta_s.$$

Hence measurement of the angle of inclination at which slipping just starts provides a simple experimental method for determining the coefficient of static friction between two surfaces.

You can use similar arguments to show that the angle of inclination θ_k required to maintain a *constant speed* for the block as it slides down the plane, once it has been started by tapping, is given by

$$\mu_k = \tan \theta_k,$$

where $\theta_k < \theta_s$. With the aid of a ruler you can now determine μ_s and μ_k for a coin sliding down your textbook.

Consider an automobile moving along a straight horizontal road with a speed v_0. If the coefficient of static friction between the tires and the road is μ_s, what is the shortest distance in which the automobile can be stopped?

The forces acting on the automobile, considered to be a particle, are shown

EXAMPLE 2

in Fig. 6-5. The car is assumed to be moving in the positive x-direction. If we assume that f_s is a constant force, we have uniformly decelerated motion.

From the relation (see Eq. 3-16)

$$v^2 = v_0^2 + 2ax,$$

with the final speed $v = 0$, we obtain

$$x = -v_0^2/2a,$$

where the minus sign means that **a** points in the negative x-direction.

To determine a, apply the second law of motion to the x-component of the motion:

$$-f_s = ma = (W/g)a \quad \text{or} \quad a = -g(f_s/W).$$

From the y components we obtain

$$N - W = 0 \quad \text{or} \quad N = W,$$

so that

$$\mu_s = f_s/N = f_s/W$$

and

$$a = -\mu_s g.$$

Then the distance of stopping is

$$x = -v_0^2/2a = v_0^2/2g\mu_s. \tag{6-4}$$

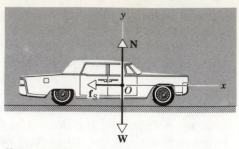

figure 6-5
Example 2. The forces on a decelerating automobile.

The greater the initial speed, the longer the distance required to come to a stop; in fact, this distance varies as the square of the initial velocity. Also, the greater the coefficient of static friction between the surfaces, the less the distance required to come to a stop.

We have used the coefficient of static friction in this problem, rather than the coefficient of sliding friction, because we assume there is no sliding between the tires and the road. We have neglected rolling friction. Furthermore, we have assumed that the maximum force of static friction ($f_s = \mu_s N$) operates because the problem seeks the shortest distance for stopping. With a smaller static frictional force the distance for stopping would obviously be greater. The correct braking technique required here is to keep the car just on the verge of skidding. If the surface is smooth and the brakes are fully applied, sliding may occur. In this case μ_k replaces μ_s, and the distance required to stop is seen to increase from Eq. 6-4.

The assumption that the car is a particle is valid if the wheels are locked (skidding). When the wheels rotate, internal forces (and torques) in the brake drums must be considered to understand work and energy ideas (see Questions 3, 4, and 5 of Chapter 8), though the result (Eq. 6-4) is correct. The rotation of the wheels is explicitly considered in Chapter 13.

As a specific example, if $v_0 = 60$ mi/h = 88 ft/s = 97 km/h, and $\mu_s = 0.60$ (a typical value), we obtain

$$x = \frac{v_0^2}{2\mu_s g} = \frac{(88 \text{ ft/s})^2}{2(0.60)(32 \text{ ft/s}^2)} = 200 \text{ ft} = 61 \text{ m}.$$

Notice that the mass of the car does not appear in Eq. 6-4. How can you explain the practice of "weighing down" a car in order to increase safety in driving on icy roads? (Hint: See Prob. 6-2.)

How do the forces of friction modify the results of the examples of Section 5-10?

6-3
THE DYNAMICS OF UNIFORM CIRCULAR MOTION

In Section 4-4 we pointed out that if a body is moving at uniform speed v in a circle of radius r, it experiences a centripetal acceleration **a** whose magnitude is v^2/r. The direction of **a** is always radially inward toward the center of rotation. Thus **a** is a variable vector because, even though its magnitude remains constant, its direction changes continuously as the motion progresses.

Recall that there need not be any motion in the direction of an acceleration. In general, there is no fixed relation between the directions of the acceleration **a** and the velocity **v** of a particle, as Fig. 4-7 shows. As it happens, for a particle in uniform circular motion the acceleration **a** and velocity **v** are always at right angles to each other.

Every accelerated body must have a force **F** acting on it, defined by Newton's second law (**F** = m**a**). Thus (assuming that we are in an inertial frame), if we see a body undergoing uniform circular motion, we can be certain that a net force **F**, given in magnitude by

$$F = ma = mv^2/r$$

must be acting on the body; the body is *not* in equilibrium. The direction of **F** at any instant must be the direction of **a** at that instant, namely, radially inward. We must always be able to account for this force by pointing to a particular object in the environment that is exerting the force on the circulating, accelerating body.

If the body in uniform circular motion is a disc on the end of a string moving in a circle on a frictionless horizontal table as in Fig. 6-6, the force **F** on the disc is provided by the tension **T** in the string. This force **T** is the net force acting on the disc. It accelerates the disc by constantly changing the direction of its velocity so that the disc moves in a circle. **T** is always directed toward the pin at the center and its magnitude is mv^2/R. If the string were to be cut where it joins the disc, there would be no net force exerted on the disc. The disc would then move with constant speed in a straight line along the direction of the tangent to the circle at the point at which the string was cut. Hence, to keep the disc moving in a circle, a force must be supplied to it pulling it *inward* toward the center.

Forces responsible for uniform circular motion are called *centripetal* forces because they are directed "toward the center" of the circular motion. To label a force as "centripetal," however, simply means that it always points radially inward; the name tells us nothing about the nature of the force or about the body that is exerting it. Thus, for the revolving disc of Fig. 6-6, the centripetal force is an elastic force provided by the string; for the moon revolving around the earth the centripetal force is the gravitational pull of the earth on the moon; for an electron circulating about an atomic nucleus the centripetal force is electrostatic. A centripetal force is not a new kind of force but simply a way of describing the behavior with time of forces that are attributable to specific bodies in the environment. Thus a force can be centripetal *and* elastic, centripetal *and* gravitational, or centripetal *and* electrostatic, among other possibilities.

Let us consider some examples of forces that act centripetally.

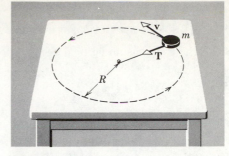

figure 6-6
A disk *m* moves with constant speed in a circular path on a horizontal frictionless surface. The only horizontal force acting on *m* is the centripetal force **T** with which the string pulls on the body.

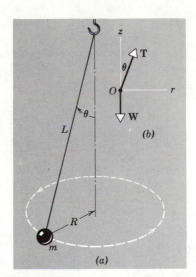

figure 6-7
Example 3. *(a)* A mass *m* suspended from a string of length *L* swings so as to describe a circle. The string describes a right circular cone of semiangle θ. *(b)* A free-body force diagram for *m*.

EXAMPLE 3

The Conical Pendulum. Figure 6-7a shows a small body of mass *m* revolving in a horizontal circle with constant speed *v* at the end of a string of length *L*. As the body swings around, the string sweeps over the surface of a cone. This device is called a *conical pendulum*. Find the time required for one complete revolution of the body.

If the string makes an angle θ with the vertical, the radius of the circular path is $R = L \sin \theta$. The forces acting on the body of mass *m* are **W**, its weight, and **T**, the pull of the string, as shown in Fig. 6-7b. It is clear that **T** + **W** ≠ 0. Hence, the resultant force acting on the body is nonzero, which is as it should be because a force is required to keep the body moving in a circle with constant speed.

We can resolve **T** at any instant into a radial and a vertical component

$$T_r = T \sin \theta \qquad \text{and} \qquad T_z = T \cos \theta.$$

Since the body has no vertical acceleration,

$$T_z - W = 0.$$

But

$$T_z = T \cos \theta \qquad \text{and} \qquad W = mg,$$

so that

$$T \cos \theta = mg.$$

The radial acceleration is v^2/R. This acceleration is supplied by T_r, the radial component of **T**, which is the centripetal force acting on m. Hence

$$T_r = T \sin \theta = mv^2/R.$$

Dividing this equation by the preceding one, we obtain

$$\tan \theta = v^2/Rg, \qquad \text{or} \qquad v^2 = Rg \tan \theta,$$

which gives the constant speed of the bob. If we let τ represent the time for one complete revolution of the body, then

$$v = \frac{2\pi R}{\tau} = \sqrt{Rg \tan \theta}$$

or

$$\tau = \frac{2\pi R}{v} = \frac{2\pi R}{\sqrt{Rg \tan \theta}} = 2\pi \sqrt{R/(g \tan \theta)}.$$

But $R = L \sin \theta$, so that

$$\tau = 2\pi \sqrt{(L \cos \theta)/g}.$$

This equation gives the relation between τ, L, and θ. Notice that τ, called the *period* of motion, does not depend on m.

If $L = 1.0$ m and $\theta = 30°$, what is the period of the motion? We have

$$\tau = 2\pi \sqrt{\frac{(1.0 \text{ m})(0.866)}{9.8 \text{ m/s}^2}} = 1.9 \text{ s}.$$

The Rotor. In many amusement parks* we find a device called the rotor. The rotor is a hollow cylindrical room which can be set rotating about the central vertical axis of the cylinder. A person enters the rotor, closes the door, and stands up against the wall. The rotor gradually increases its rotational speed from rest until, at a predetermined speed, the floor below the person is opened downward, revealing a deep pit. The person does not fall but remains "pinned up" against the wall of the rotor. Find the coefficient of friction necessary to prevent falling.

The forces acting on the person are shown in Fig. 6-8. **W** is the person's weight, $\mathbf{f}_s$ is the force of static friction between person and rotor wall, and **P** is the centripetal force exerted by the wall on the person necessary to keep him moving in a circle. Let the radius of the rotor be R and the final speed of the passenger be v. Since the person does not move vertically, but experiences a radial acceleration v^2/R at any instant, we have

$$f_s - W = 0$$

and

$$P(= ma) = (W/g)(v^2/R).$$

If μ_s is the coefficient of static friction between person and wall necessary to

* See "Physics and the Amusement Park" by John L. Roeder in *The Physics Teacher*, September 1975.

EXAMPLE 4

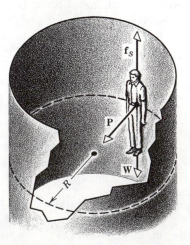

figure 6-8
Example 4. The forces on a person in a "rotor" of radius R.

prevent slipping, then $f_s = \mu_s P$ and

$$f_s = W = \mu_s P$$

or

$$\mu_s = \frac{W}{P} = \frac{gR}{v^2}.$$

This equation gives the minimum coefficient of friction necessary to prevent slipping for a rotor of radius R when a particle on its wall has a speed v. Notice that the result does not depend on the person's weight.

As a practical matter the coefficient of friction between the textile material of clothing and a typical rotor wall (canvas) is about 0.40. For a typical rotor the radius is 2.0 m, so that v must be about 7.0 m/s or 25 km/h or more.

EXAMPLE 5

Let the block in Fig. 6-9a represent an automobile or railway car moving at constant speed v on a *level* road-bed around a curve having a radius of curvature R. In addition to two vertical forces, namely, the force of gravity **W** and a normal force **N**, a horizontal centripetal force **P** acts on the car. In the case of the automobile this centripetal force is supplied by a sidewise frictional force exerted by the road on the tires; in the case of the railway car the centripetal force is supplied by the rails exerting a sidewise force on the inner rims of the car's wheels. Neither of these sidewise forces can be safely relied upon to be large enough at all times and both cause unnecessary wear. Hence, the roadbed is *banked* on curves, as shown in Fig. 6-9b. In this case, the normal force **N** has not only a vertical component, as before, but also a horizontal component which supplies the centripetal force necessary for uniform circular motion; no additional sidewise forces are needed, therefore, with a properly banked roadbed.

The correct angle θ of banking can be obtained as follows. There is no vertical acceleration, so that

$$N \cos \theta = W.$$

The centripetal force is $N \sin \theta$, so that $N \sin \theta = mv^2/R$. Dividing the latter equation by the former and setting $W = mg$, we obtain

$$\tan \theta = v^2/Rg$$

Notice that the proper angle of banking depends upon the speed of the car and

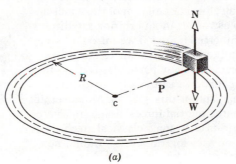

(a)

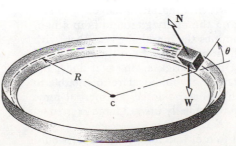

(b)

figure 6-9
Examples.
(a) a level roadbed.
(b) a banked roadbed.

the curvature of the road. For a given curvature, the road is banked at an angle corresponding to an expected average speed. Often curves are marked by signs giving the proper speed for which the road was banked.

Check the banking formula for the limiting cases $v = 0$; $R \rightarrow \infty$; v large; and R small. Also note the similarity between Fig. 6-7 of Example 3 and Fig. 6-9b of this example.

All forces in nature can be classified under four headings, each with a different relative strength: (1) gravitational forces, which are relatively very weak, (2) electromagnetic forces, which are of intermediate strength, (3) nuclear forces which bind neutrons and protons in the nucleus and are the strongest of all, and (4) the weak interaction force, which is involved in the β-decay of nuclei and in the interactions of many elementary particles (see Appendix F).

These forces are "real" in the sense that we can associate them with specific objects in the environment. Such forces as the tension in a rope, the force of friction, the force that we exert on a wall by pushing on it, or the force exerted by a compressed spring are electromagnetic forces; all are macroscopic manifestations of the (electromagnetic) attractions and repulsions between atoms.

In our treatment of classical mechanics so far we have assumed that our measurements and observations were made from an inertial frame. This, we recall, is a reference frame that is either at rest or is moving at constant velocity with respect to the average positions of the fixed stars; it is the set of reference frames defined by Newton's first law, namely, that set of frames in which a body will not be accelerated ($\mathbf{a} = 0$) if there are no identifiable force-producing bodies in its environment ($\mathbf{F} = 0$). The choice of a reference frame is always ours to make, so that if we choose to select only inertial frames, we do not restrict in any way our ability to apply classical mechanics to natural phenomena.

Nevertheless we can, if we find it convenient, apply classical mechanics from the point of view of an observer in a *noninertial frame*. Such a frame might be one that is attached to a falling body or one that is rotating (and therefore accelerating) with respect to the fixed stars. We sometimes choose a noninertial reference frame when we consider, for example, the separation of liquids of different density in a spinning centrifuge, the global circulation of the winds on the rotating earth, or the experiences of an astronaut in an orbiting satellite.

We can apply classical mechanics in noninertial frames if we introduce non-Newtonian forces called *inertial forces*. Unlike the forces that we have examined so far, we cannot associate inertial forces with any particular body in the environment of the particle on which they act and we cannot classify them into any of the categories listed in the first paragraph of this section. Moreover, if we view the particle from an inertial frame, the inertial forces disappear. These forces are, then, simply a technique that permits us to apply classical mechanics in the normal way to events if we insist on viewing the events from a noninertial reference frame.

Consider a rotating merry-go-round on which a marble is lodged against a raised rim at the outer edge. An observer on the merry-go-round is in a noninertial frame. As he kneels down and examines the marble he sees that, with respect to him, it is not moving; if he pulls it away a bit from the rim toward the center of rotation, he observes that it moves back again, as if under the influence of a force directed radially outward. He would declare the marble to be in equilibrium under the action of this outward force (an inertial force called, in this case, a *centrifugal force*) and the radially inward force exerted by the rim.

An observer on the ground (an inertial frame) watching the marble would describe it differently. He would declare the marble to be in uniform circular

6-4
CLASSIFICATION OF FORCES; INERTIAL FORCES

motion, accelerated radially inward with $a = v^2/R$. The inward force $\mathbf{F}$ exerted by the rim on the marble accounts for this acceleration from Newton's second law, or $F = ma = mv^2/R$. The marble is definitely *not* in equilibrium from the point of view of this observer or of an observer in any inertial frame. Only if the rim were *not* exerting this inward force would the marble move with uniform speed in a straight line and be in equilibrium. This observer would find no trace of a force directed radially outward on the marble (the inertial force) and, indeed, there is no room for such a force in his analysis of the motion.

It is clear from this simple example that the radially outward inertial force (or centrifugal force) noted by the observer on the rotating merry-go-round must have a magnitude mv^2/R. Thus the magnitude of the inertial force depends on the speed of the particle *as seen from another reference frame*, namely, the ground; the speed of the particle in its own (rotating) reference frame is zero.

This example illustrates why inertial forces are non-Newtonian, namely, Newton's third law of motion does not apply to them. That is, there is no reaction force to the inertial (action) force. In the rotating frame, if the rim were *not* present we would have an inertial (centrifugal) force acting on the marble without any reaction force of the marble on another body. When the rim *is* present we have two forces acting on the *same* body, the centripetal force due to the rim and the inertial (centrifugal) force each acting on the marble. The marble is viewed as being in equilibrium under the influence of two forces acting on it but, as we have seen, we can have one force without the other. In an inertial frame, on the other hand, the (action) force of the rim is the only force on the marble and the marble exerts a (reaction) force on the rim, equal in magnitude but oppositely directed. If one wished to use the terms *centripetal* and *centrifugal* here, he would have an action-reaction pair acting on *different* bodies, consistent with Newton's third law. But in the accelerated frame, the forces called by these names act on the same body and are not an action-reaction pair.

In more general terms we might say that the expression $\mathbf{F} = m\mathbf{a}$ used in an inertial frame is changed to $\mathbf{F} - m\mathbf{a} = 0$ in a noninertial frame and that the interpretation given to the term $-m\mathbf{a}$ in the latter case is that it is an (inertial) force existing only in the accelerated frame which permits one to regard the object acted upon as always being in equilibrium. In this sense it is sometimes simpler to use a noninertial frame to describe motion, such as circular motion, the object being regarded as at rest in such a frame.

In mechanical problems, then, we have two choices: (1) select an *inertial frame* as a reference frame and consider only "real" forces, that is, forces that we can associate with definite bodies in the environment or (2) select a *noninertial frame* as a reference frame and consider not only the "real" forces but suitably defined inertial forces. Although we usually choose the first alternative, we sometimes choose the second; both are completely equivalent and the choice is a matter of convenience. We shall discuss noninertial frames and inertial forces further in Chapters 11 and 16.

6-5 CLASSICAL MECHANICS, RELATIVISTIC MECHANICS, AND QUANTUM MECHANICS

In these first chapters we have laid the groundwork of classical mechanics. We have presented the laws of motion and have given several examples of the force laws. In later chapters we shall discuss other kinds of forces and shall continue to develop the structure of the theory. Here we want to point out where classical mechanics stands in the framework of modern physics.

Physics is not a static body of doctrine but a developing science. Historically there have been long periods of deep concern with a certain class of problem, culminating, often rather suddenly and in unexpected ways, in a "breakthrough" in the form of a new, more comprehensive theory.* This occurred about 1690 (Newtonian mechanics), about 1870 (Maxwell's theory of electromagnetism), 1905 (Einstein's theory of relativity), and about 1925 (quantum

* See "The Structure of Scientific Revolutions" by Thomas Kuhn, The University of Chicago Press, 1970.

mechanics). Some physicists believe that our present concern for problems in the area of elementary particles (see Appendix F) will lead us eventually to another major "breakthrough."

As physics has evolved, many things have changed, such as the problems to be solved and the tools we use to investigate them. But through it all the general method of inquiry or process of solution remains basically the same. Thus earlier theories of physics are found to have limited ranges of validity and to be special cases of more comprehensive theories, which in turn are found to have limitations, and so on. However, independent of any particular area or problem in physics, we always demand that theory meet the test of experiment, we search for quantities that are invariant, we are guided by a belief in the simplicity and symmetry of nature, and we seek and use analogies and models. Major unifying concepts arise which are valid in all domains of physics, such as the conservation laws. All this is important to understand for its own sake, independent of mastery of any particular special topic, and is exemplified throughout the book. If, in addition to mastering classical mechanics, the student comes to understand this process, he will find it much easier to understand and master such theories as relativity theory and quantum theory, wherein the same method of inquiry applies but whose areas of application, unlike those of classical mechanics, are not a familiar part of his daily life experience.

Classical mechanics, like all theories in physics, is based on observations of things that happen in nature. It will help to point out how limited are our normal experiences of natural phenomena. This is particularly true during our formative years which is the period when we develop our intuitive notions (often false!) of what is "common sense" in natural events and what is not.

For example, the highest speed that can be used to transmit signals from one point to another is the speed of light ($c = 186{,}000$ mi/s $= 3.00 \times 10^8$ m/s) and this seems to set an upper limit to the speeds of material objects. However, gross objects, even the fastest of them, such as jet planes or earth satellites, have speeds v that are very much less than c. For an earth satellite moving at 17,000 mi/h, v/c is only 0.00025. Classical mechanics was built up over several centuries on a body of observations of relatively slow-moving objects such as planets, balls rolling down inclined planes, and falling bodies. Our experience with moving objects has indeed been limited, until the last few decades, to a tiny fraction of the range of possible speeds.

During these last decades it has become possible to make measurements on small particles, of potentially high speed, such as electrons, protons, and other fundamental particles. A proton accelerated in the 30-billion electron volt accelerator at the Brookhaven National Laboratories has, for example, $v/c = 0.98$. Are we to expect that the laws of classical mechanics, which work so beautifully when $v/c \ll 1$, will also describe correctly the collisions, decays, and interactions of these elementary particles moving at such high speeds? This is the grossest kind of extrapolation and indeed we find by experiment that it simply does not work; classical mechanics gives answers that do not agree with experiment if the speeds of the objects involved are appreciable compared to the speed of light. This does not make us think less of classical mechanics, which serves so well in the region of low speed, precisely the very important region of our daily experiences. We are led, however, to view classical mechanics as a special case of a more general theory which would hold for all speeds up to the speed of light.

Einstein, in 1905, first proposed this more general theory, the *special theory of relativity*.* We shall discuss it again later but will state here its fundamental postulate. This is that the speed of light c is the *same* for all observers in inertial frames, no matter what the motion of the light source may be. In other words, if a light source is moving directly toward you at a speed v, you would measure the same value for c, if you observed a light pulse passing you, no matter what the value of v; you would also obtain speed c for the light pulse if the source were rushing away from you at speed v. If this basic assumption seems to violate "common sense," we must realize that our intuitive feelings are based on

* For a summary of special relativity, see Supplementary Topic V.

"common sense at low speeds." We have no direct experience in our daily activities about what really happens in nature at high speeds. Furthermore, all of Einstein's predictions (1) agree with experiment and (2) reduce to the predictions of classical mechanics at low speeds.

We list here just one of the predictions of the theory of relativity that is at variance with classical mechanics. If two observers watch an object moving parallel to the common $x - x'$-axis in Fig. 4-11, they will find, from Eq. 4-19,

$$v = v' + u, \tag{6-5}$$

where v' is the speed as measured by observer S', v is that measured by observer S, and u is the relative speed of separation of the two reference frames. Note that there is nothing in Eq. 6-5 to prevent v from exceeding c if v' and u are large enough.

The theory of relativity predicts that Eq. 6-5 is a special case of a more general formula, namely,

$$v = \frac{v' + u}{1 + v'u/c^2}. \tag{6-6}$$

Note that for $v' \ll c$ and $u \ll c$ Eq. 6-6 does indeed reduce to Eq. 6-5. Also, if $v' < c$ and $u < c$, then v cannot exceed c. If $v' = u = 0.8\,c$, for example, Eq. 6-6 yields $v = 0.975\,c$; Eq. 6-5, on the other hand, yields $v = 1.6\,c$, which is contrary to experience.

For gross objects, Eqs. 6-5 and 6-6 give the same results within experimental error, so that we naturally use the simpler, Eq. 6-5. If two satellites moving in opposite directions have speeds $v' = u = 17,000$ mi/h, the denominator in Eq. 6-6 has the value 1.0000000007, so that the speed v of one satellite as seen from the other differs very slightly from the value $v' + u$ predicted by Eq. 6-5. It would take speeds almost 3000 *times* as great as above, nearly 50 million mi/h, generally achievable only in the subatomic domain, to obtain a difference as great as one-half of one percent in the two formulas.

We point out a second way in which our daily experiences are limited, namely, that all the objects that we normally deal with have masses that greatly exceed, for example, the electron mass ($m = 9.11 \times 10^{-31}$ kg). This turns out to have an interesting consequence, closely related to the very concept of "particle" on which classical mechanics is based. We have not hesitated to assign a mass m, a position x, and a velocity v_x to a particle, assumed to be moving along the x-axis.* If we are asked within what accuracy Δx and Δv_x we could measure the position x and the velocity v_x respectively, we would be inclined to say that, although there might be limits in practice there are none in principle and, with sufficient attention to methods of measurement, we can specify x and v_x as closely as we wish. Experiment seems to confirm this view for large objects like golf balls.

When we deal with objects of very small mass, however, such as electrons, we learn that the very procedures of measurement introduce fundamental uncertainties and that, in fact, the more precise our knowledge of x becomes the less precise is our knowledge of v_x and conversely. We can express this in terms of the famous Heisenberg uncertainty relation, which we write as

$$\Delta x \cong \frac{h}{m\,\Delta v_x} \tag{6-7}$$

in which h (Planck's constant) is a fundamental constant of nature and has the value $h = 6.63 \times 10^{-34}$ kg m²/s. Equation 6-7 shows clearly that if Δv_x is very small (which means that we know v_x very precisely), then Δx must be relatively large (which means that we do not know x very precisely). Thus it does not seem possible to measure *both* the position *and* the velocity of a particle to any given precision at the same time. If we cannot do this, then our whole concept of a particle as a mass point following a trajectory, which is a basic concept of classical mechanics, is open to question.

---

* We assume $v_x \ll c$ so that considerations of relativity do not enter this new discussion.

Just as for relativity theory, these considerations of quantum mechanics simply do not make any difference for the gross objects of our daily experience. Consider a ball bearing with a speed of 10^3 m/s and a mass of 1.0 g ($= 10^{-3}$ kg). Let us assume that we know the speed to be accurate to 0.1%, which means that $\Delta v_x = 0.001 \times 10^3 = 1$ m/s. The uncertainty in the position of the ball bearing is now given by Eq. 6-7 as

$$\Delta x \cong \frac{6.63 \times 10^{-34} \text{ kg m}^2/\text{s}}{(10^{-3} \text{ kg})(1 \text{ m/s})} \cong 7 \times 10^{-31} \text{ m}$$

This is such a small distance (being 10^{-15} times smaller than an atomic nucleus!) that we could not possibly detect any limitation on the measurement of x set by Eq. 6-7.

Consider, however, not a ball bearing but an electron ($m = 9.11 \times 10^{-31}$ kg) whose speed is measured to be 2×10^6 m/s, which is about the speed of an electron in a hydrogen atom. If we assume that we know this speed to be accurate to, say, 1%, then $\Delta v_x = 0.01 \times 2 \times 10^6$ m/s $= 2 \times 10^4$ m/s. The uncertainty in position predicted by Eq. 6-7 is then

$$\Delta x \cong \frac{6.63 \times 10^{-34} \text{ kg m}^2/\text{s}}{(9.11 \times 10^{-31} \text{ kg})(2 \times 10^{-4} \text{ m/s})} = 3 \times 10^{-8} \text{ m}.$$

Since the radius of a hydrogen atom is about 5×10^{-11} m we see that the uncertainty with which we can locate the electron in the hydrogen atom, assuming that we have measured its speed as accurately as we claim, is 600 times the radius of the atom! The concept of "particle" does not mean much under these circumstances. This simply means that we cannot use classical mechanics to describe the motions of electrons in atoms; we need quantum mechanics.

The situation is very much like that of relativity theory. Ideas that we find acceptable in a certain region of experience (ball bearings) fall down when we apply them to a region outside our direct normal experience (electrons in atoms). Once again the solution is the same: Classical mechanics turns out to be an important special case of a more general theory. In this case the general theory is that of quantum mechanics developed about 1925 to 1926 by Heisenberg, Schrödinger, Born, and others. Once again, quantum mechanics does not detract from the merit of classical mechanics, which continues to give results that agree admirably with experiment for particles of relatively large mass.

The situation most remote from our daily experience deals with particles that have *both* small mass *and* high speed. Here we must use a still more general theory, *relativistic quantum mechanics*, which combines both relativity theory and quantum mechanics; such a theory was first developed by Dirac in 1927.

In the rest of our treatment of mechanics we return to the familiar special case of our daily experience, that of relatively massive and relatively slow-moving objects (classical mechanics). From time to time we will point out parenthetically how the predictions of classical mechanics must be modified when we depart from this region of experience.

questions

1. There is a limit beyond which further polishing of a surface *increases* rather than decreases frictional resistance. Can you explain this?

2. Is it unreasonable to expect a coefficient of friction to exceed unity?

3. How could a person who is at rest on completely frictionless ice covering a pond reach shore? Could he do this by walking, rolling, swinging his arms, or kicking his feet? How could a person be placed in such a position in the first place?

4. Explain how the range of your car's headlights limits the safe driving speed at night.

5. Your car skids across the center line on an icy highway. Should you turn the front wheels in the direction of skid or in the opposite direction (a) when

you want to avoid a collision with an oncoming car, (b) when no other car is near but you want to regain control of the steering?

6. If you want to stop the car in the shortest distance on an icy road, should you (a) push hard on the brakes to lock the wheels, (b) push just hard enough to prevent slipping, or (c) "pump" the brakes?

7. Discuss how the choice of angle for maximum range of a projectile would be affected by the resistance of the air to motion of the projectile through it.

8. Why are the train roadbeds and highways banked on curves?

9. How does the earth's rotation affect the apparent weight of a body at the equator?

10. Explain why a plumb bob will not hang exactly in the direction of the earth's gravitational attraction at most latitudes.

11. Suppose you need to measure whether a table top in a train is truly horizontal. If you use a spirit level, can you determine this when the train is moving down or up a grade? When the train is moving along a curve? (Hint: there are two horizontal components.)

12. In the conical pendulum of Example 3, what happens to the period τ and the speed v when $\theta = 90°$? Why is this angle not achievable physically? Discuss the case for $\theta = 0°$.

13. A coin is put on a photograph turntable. The motor is started, but before the final speed of rotation is reached, the coin flies off. Explain.

14. Suppose that a body that is acted upon by exactly two forces is accelerated. Does it then follow that (a) the body cannot move with constant speed? (b) the velocity can never be zero? (c) the sum of the two forces cannot be zero? (d) the two forces must act in the same line?

15. A car is riding on a country road that resembles a roller coaster track. If the car travels with uniform speed, compare the force it exerts on a horizontal section of the road to the force it exerts on the road at the top of a hill and at the bottom of a hill. Explain.

16. A passenger in the front seat of a car finds himself sliding toward the door as the driver makes a sudden left turn. Describe the forces on the passenger and on the car at this instant if (a) the motion is viewed from a reference frame attached to the earth and (b) if attached to the car.

17. Astronauts in the orbiting Skylab spacecraft want to keep a daily record of their weight. Can you think how they might do it, considering that they are 'weightless'?

18. What conclusion might a physicist draw if, while standing in an elevator, he observes that unequal masses hung over a pulley remain balanced, that is, there is no tendency for the pulley to turn?

19. Explain how the question "What is the linear velocity of a point on the equator?" requires an assumption about the reference frame used. Show how the answer changes as you change reference frames.

20. What is the distinction between inertial reference frames and those differing only by a translation or rotation of the axes?

SECTION 6-2

1. A hockey puck weighing 0.25 lb (1.1 N) slides on the ice for 50 ft (15 m) before it stops. (a) If its initial speed was 20 ft/s (6.1 m/s), what is the force of friction between puck and ice? (b) What is the coefficient of kinetic friction? *Answer:* (a) 0.031 lb (0.14 N). (b) 0.12 (0.13).

problems

2. Suppose that only the rear wheels of an automobile can accelerate it, and that half the total weight of the automobile is supported by those wheels. (a) What is the maximum acceleration attainable if the coefficient of static friction between tires and road is μ_s? (b) Take $\mu_s = 0.35$ and get a numerical value for this acceleration.

3. Frictional heat generated by the moving ski is the chief factor promoting sliding in skiing. The ski sticks at the start, but once in motion will melt the snow beneath it. Waxing the ski makes it water repellent and reduces friction with the film of water. A magazine reports that a new type of plastic ski is even more water repellent and that on a gentle 700-ft slope in the Alps, a skier reduced his time from 61 to 42 s with new skis. (a) Determine the average accelerations for each pair of skis. (b) Assuming a 3°-slope compute the coefficient of kinetic friction for each case.
   Answer: (a) 0.38 ft/s²; 0.79 ft/s². (b) 0.041; 0.028.

4. A fireman weighing 160 lb (710 N) slides down a vertical pole with an average acceleration of 10 ft/s² (3 m/s²). What is the average vertical force he exerts on the pole?

5. A man drags a 150-lb crate across a floor by pulling on a rope inclined 15° above the horizontal. (a) If the coefficient of static friction is 0.50, what tension in the rope is required to start the crate moving? (b) If $\mu_k = 0.35$, what is the initial acceleration of the crate?
   Answer: (a) 68 lb. (b) 4.2 ft/s².

6. A cube of weight W rests on a rough inclined plane which makes an angle θ with the horizontal. (a) What is the minimum force necessary to start the cube moving *down* the plane? (b) What is the minimum force necessary to start the cube moving *up* the plane? (c) What is the minimum *horizontal* (transverse to the slope) force necessary to start the cube moving down the plane?

7. The handle of a floor mop of mass m makes an angle θ with the vertical direction. Let μ_k be the coefficient of kinetic friction between mop and floor, and μ_s be the coefficient of static friction between mop and floor. Neglect the mass of the handle. (a) Find the magnitude of the force F directed along the handle required to slide the mop with uniform velocity across the floor. (b) Show that if θ is smaller than a certain angle θ_0, the mop cannot be made to slide across the floor no matter how great a force is directed along the handle. (c) What is the angle θ_0?
   Answer: (a) $\mu_k mg/(\sin\theta - \mu_k \cos\theta)$. (c) $\theta_0 = \tan^{-1}\mu_s$.

8. A piece of ice slides down a 45°-incline in twice the time it takes to slide down a frictionless 45°-incline. What is the coefficient of kinetic friction between the ice and the incline?

9. A block slides down an inclined plane of slope angle φ with constant velocity. It is then projected up the same plane with an initial speed v_0. (a) How far up the incline will it move before coming to rest? (b) Will it slide down again?   *Answer:* (a) $v_0^2/4g \sin\varphi$. (b) No.

10. A student wants to determine the coefficients of static friction and kinetic friction between a box and a plank. He places the box on the plank and gradually raises the plank. When the angle of inclination with the horizontal reaches 30°, the box starts to slip and slides 4.0 m down the plank in 4.0 s. What are the coefficients of friction?

11. A horizontal force F of 12 lb pushes a block weighing 5.0 lb against a vertical wall (Fig. 6-10). The coefficient of static friction between the wall and the block is 0.60 and the coefficient of kinetic friction is 0.40. Assume the block is not moving initially. (a) Will the block start moving? (b) What is the force exerted on the block by the wall?
   Answer: (a) No. (b) A 12-lb force to the left and a 5.0-lb force up.

12. A 10-lb block of steel is at rest on a horizontal table. The coefficient of static friction between block and table is 0.50. (a) What is the magnitude of the horizontal force that will just start the block moving? (b) What is the magnitude of a force acting upward 60° from the horizontal that will just start the block moving? (c) If the force acts down at 60° from the horizontal, how large can it be without causing the block to move?

13. Block B in Fig. 6-11 weighs 160 lb (710 N). The coefficient of static friction

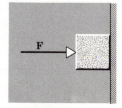

figure 6-10
Problem 11

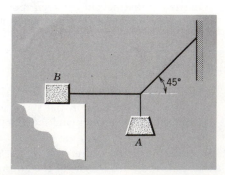

figure 6-11
Problem 13

between block and table is 0.25. Find the maximum weight of block A for which the system will be in equilibrium. *Answer:* 40 lb (180 N).

14. Two masses, $m_1 = 1.65$ kg and $m_2 = 3.30$ kg, attached by a massless rod parallel to the incline on which both slide, as shown in Fig. 6-12, travel down along the plane with m_1 trailing m_2. The angle of incline is $\theta = 30°$. The coefficient of kinetic friction between m_1 and the incline is $\mu_1 = 0.226$; between m_2 and the incline the corresponding coefficient is $\mu_2 = 0.113$. Compute (a) the tension in the rod linking m_1 and m_2 and (b) the common acceleration of the two masses. (c) Would the answers to (a) and (b) be changed if m_2 trails m_1?

15. A 4.0-kg block is put on top of a 5.0-kg block. In order to cause the top block to slip on the bottom one, held fixed, a horizontal force of 12 N must be applied to the top block. The assembly of blocks is now placed on a horizontal, frictionless table (Fig. 6-13). Find (a) the maximum horizontal force F which can be applied to the lower block so that the blocks will move together, and (b) the resulting acceleration of the blocks. *Answer:* (a) 27 N. (b) 3.0 m/s².

16. A railroad flatcar is loaded with crates having a coefficient of static friction 0.25 with the floor. If the train is moving at 30 mi/h (48 km/h), in how short a distance can the train be stopped without letting the crates slide?

17. A 40-kg slab rests on a frictionless floor. A 10-kg block rests on top of the slab (Fig. 6-14). The static coefficient of friction between the block and the slab is 0.60 while the kinetic coefficient is 0.40. The 10-kg block is acted upon by a horizontal force of 100 N. What are the resulting accelerations of (a) the block, and (b) the slab? *Answer:* (a) 6.1 m/s². (b) 0.98 m/s².

18. In Fig. 6-15, A is a 10-lb (44-N) block and B is a 5.0-lb (22-N) block. (a) Determine the minimum weight (block C) which must be placed on A to keep it from sliding, if μ_s between A and the table is 0.20. (b) The block C is suddenly lifted off A. What is the acceleration of block A, if μ_k between A and the table is 0.20?

19. An 8.0-lb block and a 16-lb block connected together by a string slide down a 30° inclined plane. The coefficient of kinetic friction between the 8.0-lb block and the plane is 0.10; between the 16-lb block and the plane it is 0.20. Find (a) the acceleration of the blocks and (b) the tension in the string, assuming that the 8.0-lb block leads. (c) Describe the motion if the blocks are reversed. *Answer:* (a) 11 ft/s². (b) 0.46 lb. (c) Blocks move independently, unless they subsequently collide.

20. Body B weighs 100 lb and body A weighs 32 lb (Fig. 6-16). Given $\mu_s = 0.56$ and $\mu_k = 0.25$, (a) find the acceleration of the system if B is initially at rest and (b) find the acceleration if B is moving initially.

21. A block of mass m slides in an inclined right-angled trough as in Fig. 6-17. If the coefficient of kinetic friction between the block and the material composing the trough is μ_k, find the acceleration of the block. *Answer:* $g(\sin \theta - \sqrt{2}\,\mu_k \cos \theta)$.

SECTION 6-3

22. In the Bohr model of the hydrogen atom, the electron revolves in a circular orbit around the nucleus. If the radius is 5.3×10^{-11} meters and the electron makes 6.6×10^{15} rev/s, find (a) the acceleration (magnitude and direction) of the electron and (b) the centripetal force acting on the electron. (This force is due to the attraction between the positively charged nucleus and the negatively charged electron.) The mass of the electron is 9.1×10^{-31} kg.

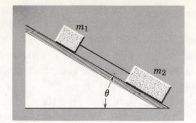

figure 6-12
Problem 14

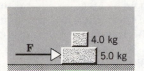

figure 6-13
Problem 15

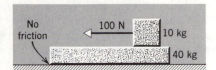

figure 6-14
Problem 17

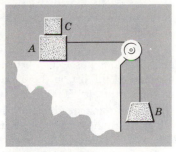

figure 6-15
Problem 18

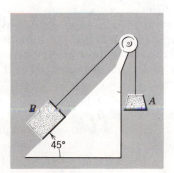

figure 6-16
Problem 20

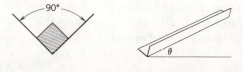

figure 6-17
Problem 21

23. A mass m on a frictionless table is attached to a hanging mass M by a cord through a hole in the table (Fig. 6-18). Find the condition (v and r) with which m must spin for M to stay at rest.   *Answer: $v^2/r = Mg/m$.*

24. Show that the periods of two conical pendula of different lengths which are hung from a ceiling and rotate with their bobs an equal distance below the ceiling are equal.

25. A small coin is placed on a flat, horizontal turntable. The turntable is observed to make three revolutions in 3.14 s. (*a*) What is the speed of the coin when it rides without slipping at a distance 5.0 cm from the center of the turntable? (*b*) What is the acceleration (magnitude and direction) of the coin in part (*a*)? (*c*) What is the frictional force acting on the coin in part (*a*) if the coin has a mass of 2.0 g? (*d*) What is the coefficient of static friction between the coin and the turntable if the coin is observed to slide off the turntable when it is more than 10 cm from the center of the turntable?
Answer: (a) 30 cm/s. (b) 180 cm/s², radially inward. (c) 3.6×10^{-3} N. (d) 0.37.

26. A block of mass m at the end of a string is whirled around in a vertical circle of radius R. Find the critical speed below which the string would become slack at the highest point?

27. A circular curve of highway is designed for traffic moving at 40 mi/h. (*a*) If the radius of the curve is 400 ft, what is the correct angle of banking of the road? (*b*) If the curve is not banked, what is the minimum coefficient of friction between tires and road that would keep traffic from skidding at this speed?   *Answer: (a) 15°. (b) 0.27.*

28. A driver's manual states that a driver traveling at 30 mi/h (48 km/h) and desiring to stop as quickly as possible travels 33 ft (10 m) before his foot reaches the brake. He travels an additional 68 ft (21 m) before coming to rest. (*a*) What coefficient of friction is assumed in these calculations? (*b*) What is the minimum radius for turning a corner at 30 mi/h (48 km/h) without skidding?

29. A 5000-lb airplane loops at a speed of 200 mi/h. Find (*a*) the radius of the largest circular loop possible, (*b*) the net force on the plane at the bottom of this loop, and (*c*) the lift on the plane at the bottom of this loop.
Answer: (a) 2700 ft. (b) 5000 lb. (c) 10,000 lb.

30. A 150-lb student on a steadily rotating Ferris wheel has an apparent weight of 125 lb at his highest point. (*a*) What is his apparent weight at the lowest point? (*b*) What would be his apparent weight at the highest point if the speed of the Ferris wheel were doubled?

31. Assume that the standard kilogram would weigh exactly 9.80 N at sea level on the earth's equator if the earth did not rotate. Then take into account the fact that the earth does rotate so that this mass moves in a circle of radius 6.40×10^6 m (earth's radius) at a constant speed of 465 m/s. (*a*) Determine the centripetal force needed to keep the standard moving in its circular path. (*b*) Determine the force exerted by the standard kilogram on a spring balance from which it is suspended at the equator (its weight).
Answer: (a) 0.0338 N. (b) 9.77 N.

32. An old streetcar rounds a corner on unbanked tracks. (*a*) If the radius of the tracks is 30 ft and the car's speed is 10 mi/h, what angle with the vertical will be made by the loosely hanging hand straps? (*b*) Is there a force acting on these straps? If so, is it a centripetal or centrifugal force? Do your answers depend on what reference frame you choose?

33. A particle of mass $M = 0.305$ kg moves counterclockwise in a horizontal circle of radius $r = 2.63$ m with uniform speed $v = 0.754$ m/s as in Fig. 6-19. Determine at the instant $\theta = 322°$ (measured counterclockwise from the positive x-direction) the following quantities: (*a*) the x-component of the velocity; (*b*) the y-component of the acceleration; (*c*) the total force on the particle; (*d*) the component of the total force on the particle in the direction of its velocity.
Answer: (a) 0.464 m/s. (b) 0.133 m/s². (c) 6.59×10^{-2} N. (d) Zero.

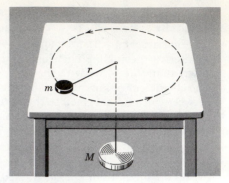

figure 6-18
Problem 23

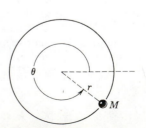

figure 6-19
Problem 33

34. A 1.0-kg ball is attached to a rigid vertical rod by means of two massless strings each 1.0 m long. The strings are attached to the rod at points 1.0 m apart. The system is rotating about the axis of the rod, both strings being taut and forming an equilateral triangle with the rod, as shown in Fig. 6-20. The tension in the upper string is 25 N. (a) Draw the free-body diagram for the ball. (b) What is the tension in the lower string? (c) What is the net force on the ball at the instant shown in the figure? (d) What is the speed of the ball?

35. An airplane is flying in a horizontal circle at a speed of 300 mi/h (480 km/h). If the wings of the plane are tilted 45° to the vertical, what is the radius of the circle the plane is flying?   *Answer: 1.1 mi (1.8 km).*

36. Because of the rotation of the earth, a plumb bob may not hang exactly along the direction of the earth's gravitational pull (its weight) but deviate slightly from that direction. Calculate the deviation (a) at 40° latitude, (b) at the poles, and (c) at the equator.

37. Imagine that the disc of Fig. 6-6 is attached to a spring rather than a string. The unstretched length of the spring is l_o and the tension in the spring increases in direct proportion to its elongation, the tension per unit elongation being k. If the disc revolves with a frequency ν (revolutions per unit time), show that (a) the radius R of the uniform circular motion is $kl_o/(k - 4\pi^2 m\nu^2)$ and (b) the tension T in the spring is $4\pi^2 mkl_o\nu^2/(k - 4\pi^2 m\nu^2)$.

38. A very small cube of mass m is placed on the inside of a funnel (Fig. 6-21) rotating about a vertical axis at a constant rate of ν rev/s. The wall of the funnel makes an angle θ with the horizontal. If the coefficient of static friction between the cube and the funnel is μ and the center of the cube is a distance r from the axis of rotation, what are (a) the largest and (b) the smallest values of ν for which the cube will not move with respect to the funnel?

figure 6-20
Problem 34

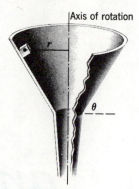

figure 6-21
Problem 38

7
work and energy

A fundamental problem of particle dynamics is to find how a particle will move when we know the forces that act on it. By "how a particle will move" we mean how its position varies with time. If the motion is one-dimensional, the problem is to find x as a function of time, $x(t)$. In the previous two chapters we solved this problem for the special case of a constant force. The method used is this. We find the resultant force $\mathbf{F}$ acting on the particle from the appropriate force law. We then substitute $\mathbf{F}$ and the particle mass m into Newton's second law of motion. This gives us the acceleration $\mathbf{a}$ of the particle; or

$$\mathbf{a} = \mathbf{F}/m.$$

If the force $\mathbf{F}$ and the mass m are constant, the acceleration $\mathbf{a}$ must be constant. Let us choose the x-axis to be along the direction of this constant acceleration. We can then find the speed of the particle from Eq. 3-12,

$$v = v_0 + at,$$

and the position of the particle from Eq. 3-15 (with $x_0 = 0$), or

$$x = v_0 t + \tfrac{1}{2}at^2;$$

note that, for simplicity and convenience, we have dropped the subscript x in these equations. The last equation gives us directly what we usually want to know, namely $x(t)$, the position of the particle as a function of time.

The problem is more difficult, however, when the force acting on a particle is *not constant*. In such a case we still obtain the acceleration of the particle, as before, from Newton's second law of motion. However, in order to get the speed or position of the particle, we can no longer use the formulas previously developed for constant acceleration because the acceleration now is *not* constant. To solve such problems, we use the mathematical process of integration, which we consider in this chapter.

We confine our attention to forces that vary with the position of the particle in its environment. This type of force is common in physics. Some examples are the gravitational forces between bodies, such as the sun and earth or earth and moon, and the force exerted by a stretched spring on a body to which it is attached. The procedure used to determine the motion of a particle subject to such a force leads us to the concepts of work and kinetic energy and to the development of the work-energy theorem, which is the central feature of this chapter. In Chapter 8 we consider a broader view of energy, embodied in the law of conservation of energy, a concept which has played a major role in the development of physics.

Consider a particle acted on by a force. In the simplest case the force **F** is constant and the motion takes place in a straight line in the direction of the force. In such a situation we define the work done by the force on the particle as the product of the magnitude of the force F and the distance d through which the particle moves. We write this as

$$W = Fd.$$

However, the constant force acting on a particle may not act in the direction in which the particle moves. In this case we define the work done by the force on the particle as the product of the component of the force along the line of motion by the distance d the body moves along that line. In Fig. 7-1 a constant force **F** makes an angle ϕ with the x-axis and acts on a particle whose displacement along the x-axis is **d.** If W represents the work done by **F** during this displacement, then according to our definition

$$W = (F \cos \phi)d. \tag{7-1}$$

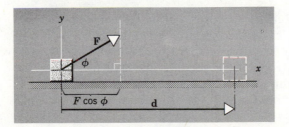

figure 7-1
A force **F** makes the block undergo a displacement **d**. The component of **F** that does the work has magnitude $F \cos \phi$; the work done is $Fd \cos \phi \ (= \mathbf{F} \cdot \mathbf{d})$.

Of course, other forces must act on a particle that moves in this way (its weight and the frictional force exerted by the plane, to name two). A particle acted on by only a single force may have a displacement in a direction other than that of this single force, as in projectile motion. But it cannot move in a straight line unless the line has the same direction as that of the single force applied to it. *Equation 7-1 refers only to the work done on the particle by the particular force* **F.** *The work done on the particle by the other forces must be calculated separately. The total work done on the particle is the sum of the works done by the separate forces.*

When ϕ is zero, the work done by **F** is simply Fd, in agreement with our previous equation. Thus, when a horizontal force draws a body horizontally, or when a vertical force lifts a body vertically, the work done by the force is the product of the magnitude of the force by the distance moved. When ϕ is 90°, the force has no component in the direction of motion. That force then does no work on the body. For instance, the vertical force holding a body a fixed distance off the ground does no work on the body, even if the body is moved horizontally over the ground. Also, the centripetal force acting on a body in motion does no work on that body because the force is always at right angles to the direction in which the body is moving. Of course, a force does no work on a body that does not move, for its displacement is then zero. In Fig. 7-2 we illustrate common examples in which a force applied to a body does no work on that body.

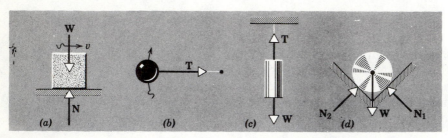

figure 7-2
Work is not always done by a force that is applied to a body. *(a)* The block is moving to the right at constant speed v over a frictionless surface. Work is not done by either the weight **W** or the normal force **N**. *(b)* The ball moves in a circle under the influence of a centripetal force **T**. There is a centripetal acceleration **a** but no work is done by **T**. In both *(a)* and *(b)* the forces being considered (**W**, **N**, and **T**) are at right angles to the displacement so that $W = \mathbf{F} \cdot \mathbf{d} = Fd \cos \phi = Fd \cos 90° = 0$. *(c)* A cylinder hangs from a cord. No work is done either by **T**, the tension in the cord, or by **W** the weight of the cylinder. *(d)* A cylinder rests in a groove; no work is done by **W**, **N**$_1$ or **N**$_2$. In both *(c)* and *(d)* the work done by the individual forces is zero because the displacement is zero.

Notice that we can write Eq. 7-1 either as $(F \cos \phi)d$ or $F(d \cos \phi)$. This suggests that the work can be calculated in two different ways: Either we multiply the magnitude of the displacement by the component of the force in the direction of the displacement or we multiply the magnitude of the force by the component of the displacement in the direction of the force. These two methods always give the same result.

Work is a *scalar*, although the two quantities involved in its definition, force and displacement, are vectors. In Section 2-4 we defined the *scalar product* of two vectors as the scalar quantity that we find when we multiply the magnitude of one vector by the component of a second vector along the direction of the first. We promised in that section that we would soon run across physical quantities that behave like scalar products. Equation 7-1 shows that work is such a quantity. In the terminology of vector algebra we can write this equation as

$$W = \mathbf{F} \cdot \mathbf{d}, \qquad (7\text{-}2)$$

where the dot indicates a scalar (or dot) product. Equation 7-2 for **F** and **d** corresponds to Eq. 2-11 for **a** and **b**.

Work can be either positive or negative. If the particle on which a force acts has a component of motion opposite to the direction of the force, the work done by that force is negative. This corresponds to an obtuse angle between the force and displacement vectors. For example, when a person lowers an object to the floor, the work done on the object by the upward force of his hand holding the object is negative. In this case ϕ is 180°, for **F** points up and **d** points down.

Work as we have defined it (Eq. 7-2) proves to be a very useful concept in physics. Our special definition of the word "work" does not correspond to the colloquial usage of the term. This may be confusing. A person holding a heavy weight at rest in the air may say that he is doing hard work—and he may work hard in the physiological sense—but from the point of view of physics we say that he is not doing any work. We say this because the applied force causes no displacement. The word *work* is used only in the strict sense of Eq. 7-2. In many scientific fields words are borrowed from our everyday language and are used to name a very specific concept. The words "basic" and "cell," for example, mean quite different things in chemistry and biology than in everyday language.

The *unit* of work is the work done by a unit force in moving a body a unit distance in the direction of the force. In SI units the unit of work is 1 *newton-meter*, called 1 *joule* (abbreviation J). In the British engineering system the unit of work is the *foot-pound*. In cgs systems the unit of work is 1 *dyne-centimeter*, called 1 *erg*. Using the relations between the newton, the dyne, and the pound, and the meter, the centimeter, and foot, we obtain 1 joule $= 10^7$ ergs $= 0.7376$ ft · lb.

EXAMPLE 1

A block of mass 10.0 kg is to be raised from the bottom to the top of an incline 5.00 m long and 3.00 m off the ground at the top. Assuming frictionless surfaces, how much work must be done by a force parallel to the incline pushing the block up at *constant speed* at a place where $g = 9.80$ m/s².

The situation is shown in Fig. 7-3a. The forces acting on the block are shown in Fig. 7-3b. We must first find P, the magnitude of the force pushing the block up the incline. Because the motion is not accelerated, the resultant force parallel to the plane must be zero. Thus

$$P - mg \sin \theta = 0,$$

or

$$P = mg \sin \theta = (10.0 \text{ kg})(9.80 \text{ m/s}^2)(\tfrac{3}{5}) = 58.8 \text{ N}.$$

Then the work done by **P**, from Eq. 7-1 with $\phi = 0°$, is

$$W = \mathbf{P} \cdot \mathbf{d} = Pd \cos 0° = Pd = (58.8 \text{ N})(5.00 \text{ m}) = 294 \text{ J}.$$

If a man were to raise the block vertically without using the incline, the work he would do would be the vertical force mg times the vertical distance or

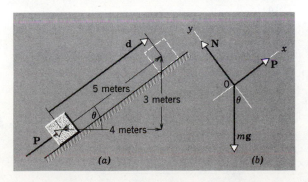

(a)   (b)

figure 7-3
Example 1. *(a)* A force **P** displaces a block a distance **d** up an inclined plane which makes an angle θ with the horizontal. *(b)* A free-body force diagram for the block.

$$(98.0 \text{ N})(3.00 \text{ m}) = 294 \text{ J},$$

the same as before. The only difference is that with the incline he could apply a smaller force ($P = 58.8$ N) to raise the block than is required without the incline ($mg = 98.0$ N); on the other hand, he had to push the block a greater distance (5.00 m) up the incline than he had to raise the block directly (3.00 m).

A boy pulls a 10-lb sled 30 ft along a horizontal surface at a *constant speed.* What work does he do on the sled if the coefficient of kinetic friction is 0.20 and his pull makes an angle of 45° with the horizontal?

EXAMPLE 2

figure 7-4
Example 2. *(a)* A boy displaces a sled an amount **d** by pulling with a force **P** on a rope that makes an angle ϕ with the horizontal. *(b)* A free-body diagram for the sled.

The situation is shown in Fig. 7-4a and the forces acting on the sled are shown in Fig. 7-4b. **P** is the boy's pull, **w** the sled's weight, **f** the frictional force, and **N** the normal force exerted by the surface on the sled. The work done by the boy on the sled is

$$W = \mathbf{P} \cdot \mathbf{d} = Pd \cos \phi.$$

To evaluate this we first must determine P, whose value has not been given. To obtain P we refer to the force diagram.

The sled is unaccelerated, so that from the second law of motion we obtain

$$P \cos \phi - f = 0,$$

and

$$P \sin \phi + N - w = 0.$$

We know that f and N are related by

$$f = \mu_k N.$$

These three equations contain three unknown quantities, P, f, and N. To find P we eliminate f and N from these equations and solve the remaining equation for P. You should verify that

$$P = \mu_k w / (\cos \phi + \mu_k \sin \phi).$$

With $\mu_k = 0.20$, $w = 10$ lb, and $\phi = 45°$ we obtain

$$P = (0.20)(10 \text{ lb})/(0.707 + 0.141) = 2.4 \text{ lb}.$$

Then with $d = 30$ ft, the work done by the boy on the sled is

$$W = Pd \cos \phi = (2.4 \text{ lb})(30 \text{ ft})(0.707) = 51 \text{ ft} \cdot \text{lb}.$$

The vertical component of the boy's pull **P** does no work on the sled. Notice, however, that it reduces the normal force between the sled and the surface ($N = w - P \sin \phi$) and thereby reduces the magnitude of the force of friction ($f = \mu_k N$).

Would the boy do more work, less work, or the same amount of work on the sled if he pulled horizontally instead of at 45° from the horizontal? Do any of the other forces acting on the sled do work on it?

Let us now consider the work done by a force that is not constant. We consider first a force that varies in magnitude only. Let the force be given as a function of position $F(x)$ and assume that the force acts in the x-direction. Suppose a body is moved along the x-direction by this force. What is the work done by this variable force in moving the body from x_1 to x_2?

In Fig. 7-5 we plot F versus x. Let us divide the total displacement into a large number of small equal intervals Δx (Fig. 7-5a). Consider the small displacement Δx from x_1 to $x_1 + \Delta x$. During this small displacement the force F has a nearly constant value and the work it does, ΔW, is approximately

$$\Delta W = F \, \Delta x, \tag{7-3}$$

where F is the value of the force at x_1. Likewise, during the small displacement from $x_1 + \Delta x$ to $x_1 + 2\Delta x$, the force F has a nearly constant value and the work it does is approximately $\Delta W = F \, \Delta x$, where F is the value of the force at $x_1 + \Delta x$. The total work done by F in displacing the body from x_1 to x_2, W_{12}, is approximately the sum of a large number of terms like that of Eq. 7-3, in which F has a different value for each term. Hence

$$W_{12} = \sum_{x_1}^{x_2} F \, \Delta x, \tag{7-4}$$

where the Greek letter sigma (Σ) stands for sum over all intervals from x_1 to x_2.

To make a better approximation we can divide the total displacement from x_1 to x_2 into a larger number of equal intervals, as in Fig. 7-5b, so that Δx is smaller and the value of F at the beginning of each interval is more typical of its values within the interval. It is clear that we can obtain better and better approximations by taking Δx smaller and smaller so as to have a larger and larger number of intervals. We can obtain an exact result for the work done by F if we let Δx go to zero and the number of intervals go to infinity. Hence the exact result is

$$W_{12} = \lim_{\Delta x \to 0} \sum_{x_1}^{x_2} F \, \Delta x. \tag{7-5}$$

The relation

$$\lim_{\Delta x \to 0} \sum_{x_1}^{x_2} F \, \Delta x = \int_{x_1}^{x_2} F \, dx,$$

as you may have learned in your calculus course, *defines the integral of F with respect to x from x_1 to x_2.* Numerically, this quantity is exactly equal to the area between the force curve and the x-axis between the limits x_1 and x_2 (Fig. 7-5c). Hence, graphically an integral can be interpreted as an area. The symbol $\int$ is a distorted S (for *sum*) and symbolizes the integration process. We can write the total work done by F in displacing a body from x_1 to x_2 as

$$W_{12} = \int_{x_1}^{x_2} F(x) \, dx. \tag{7-6}$$

As an example, consider a spring attached to a wall. Let the (horizontal) axis of the spring be chosen as an x-axis, and let the origin, $x = 0$, coincide with the endpoint of the spring in its normal, unstretched state. We assume that the positive x-direction points away from the

7-3
WORK DONE BY A VARIABLE FORCE—ONE-DIMENSIONAL CASE

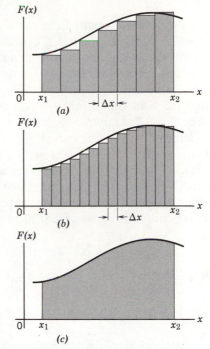

figure 7-5

Computing $\int_{x_1}^{x_2} F(x) \, dx$ amounts to finding the area under the curve $F(x)$ between the limits x_1 and x_2. This can be done approximately as in the top drawing *(a)* by dividing the area into a few strips, each of width Δx. The areas of the rectangles are then summed to give a rough value of the area. In the middle drawing *(b)* the strips are narrower and the value for the area becomes more exact as the errors at the tops of the rectangles become smaller. In the bottom drawing *(c)* the strips are only infinitesimal in width. The measurement of area is exact, since the errors at the tops of the rectangles go to zero as the strip width dx goes to zero.

wall. In what follows we imagine that we stretch the spring so slowly that it is essentially in equilibrium at all times ($\mathbf{a} = 0$).

If we stretch the spring so that its endpoint moves to a position x, the spring will exert a force on the agent doing the stretching given to a good approximation by

$$F = -kx, \qquad (7\text{-}7)$$

where k is a constant called the *force constant* of the spring. Equation 7-7 is the *force law* for springs. The direction of the force is always opposite to the displacement of the endpoint from the origin. When the spring is stretched, $x > 0$ and F is negative; when the spring is compressed, $x < 0$ and F is positive. The force exerted by the spring is a *restoring force* in that it always points toward the origin. Real springs will obey Eq. 7-7, known as *Hooke's law*, if we do not stretch them beyond a limited range. We can think of k as the magnitude of the force per unit elongation. Thus very stiff springs have large values of k.

To stretch a spring we must exert a force F' on it equal but opposite to the force F exerted by the spring on us. The applied force* is therefore $F' = kx$ and the work done by the applied force in stretching the spring so that its endpoint moves from x_1 to x_2 is†

$$W_{12} = \int_{x_1}^{x_2} F'(x)\ dx = \int_{x_1}^{x_2} (kx)\ dx = \tfrac{1}{2}kx_2{}^2 - \tfrac{1}{2}kx_1{}^2.$$

If we let $x_1 = 0$ and $x_2 = x$, we obtain

$$W = \int_0^x (kx)\ dx = \tfrac{1}{2}kx^2. \qquad (7\text{-}8)$$

This is the work done in stretching a spring so that its endpoint moves from its unstretched position to x. Note that the work to *compress* a spring by x is the same as that to stretch it by x because the displacement x is squared in Eq. 7-8; either sign for x gives a positive value for W.

We can also evaluate this integral by computing the area under the force-displacement curve and the x-axis from $x = 0$ to $x = x$. This is drawn as the white area in Fig. 7-6. The area is a triangle of base x and altitude kx. The white area is therefore

$$\tfrac{1}{2}(x)(kx) = \tfrac{1}{2}kx^2.$$

in agreement with Eq. 7-8.

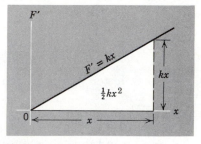

figure 7-6
The force exerted in stretching a spring is $F' = kx$. The area under the force curve is the work done in stretching the spring a distance x and can be found by integrating or by using the formula for the area of a triangle.

The force $\mathbf{F}$ acting on a particle may vary in direction as well as in magnitude, and the particle may move along a curved path. To compute the work in this general case we divide the path into a large number of small displacements $\Delta\mathbf{r}$, each pointing along the path in the direction of motion. Figure 7-7 shows two selected displacements for a particular situation; it also shows the value of $\mathbf{F}$ and the angle ϕ between $\mathbf{F}$ and $\Delta\mathbf{r}$ at each location. We can find the amount of

7-4
WORK DONE BY A VARIABLE FORCE— TWO-DIMENSIONAL CASE

---

* If the applied force were different from $F' = kx$, we would have a net unbalanced force acting on the spring and its motion would be accelerated. To compute the work done we would have to specify exactly what the applied force is at each point. No matter what the force turned out to be, the work done would always be the same for the same displacement x_1 to x_2, providing the spring has the same speed initially and finally. However, it is much easier to use the simple force $F' = kx$ in calculating the work done. Such an applied force leads to unaccelerated motion. It is in order to be able to use this simple force that we specified unaccelerated motion in the first place.

† The student just becoming familiar with calculus should consult the list of integrals in Appendix I.

work done on the particle during a displacement $\Delta \mathbf{r}$ from

$$dW = \mathbf{F} \cdot \Delta \mathbf{r} = F \cos \phi \, \Delta r \qquad (7\text{-}9)$$

The work done by the variable force $\mathbf{F}$ on the particle as the particle moves, say, from a to b in Fig. 7-7 is found very closely by adding up (summing) the elements of work done over each of the line segments that make it up. As the line segments $\Delta \mathbf{r}$ become smaller they may be replaced by differentials $d\mathbf{r}$ and the sum over the line segments may be replaced by an intergral, as in Eq. 7-6. The work is then found from

$$W_{ab} = \int_a^b \mathbf{F} \cdot d\mathbf{r} = \int_a^b F \cos \phi \, dr. \qquad (7\text{-}10a)$$

We cannot evaluate this integral until we are able to say how F and ϕ in Eq. 7-10a vary from point to point along the path; both are functions of the x- and y-coordinates of the particle in Fig. 7-7.

We can obtain another equivalent expression for Eq. 7-10a by expressing $\mathbf{F}$ and $d\mathbf{r}$ in terms of their components. Thus $\mathbf{F} = \mathbf{i} F_x + \mathbf{j} F_y$ and $d\mathbf{r} = \mathbf{i} \, dx + \mathbf{j} \, dy$, so that $\mathbf{F} \cdot d\mathbf{r} = F_x \, dx + F_y \, dy$. In this evaluation recall (see Problem 21, Chapter 2) that $\mathbf{i} \cdot \mathbf{i} = \mathbf{j} \cdot \mathbf{j} = 1$ and $\mathbf{i} \cdot \mathbf{j} = \mathbf{j} \cdot \mathbf{i} = 0$. Substituting this result into Eq. 7-10a, we obtain

$$W_{ab} = \int_a^b (F_x \, dx + F_y \, dy) \qquad (7\text{-}10b)$$

Integrals such as those in Eqs. 7-10a and 7-10b are called *line integrals*.

As an example of a variable force consider a particle of mass m suspended from a weightless cord of length l. This is called a simple pendulum. Let us displace the particle along a circular path of radius l from $\phi = 0$ to $\phi = \phi_0$ by applying a force that is always *horizontal*. We can apply such a force by pulling horizontally on the particle with an attached string, for example. The particle will then have been displaced a vertical distance h. Figure 7-8a shows the situation and Fig. 7-8b shows the forces acting on the particle in the arbitrary position ϕ. The applied force is $\mathbf{F}$, $\mathbf{T}$ is the tension in the cord, and mg the weight of the particle.

Again we assume that there is no acceleration (the reason is the same as before), so that in practice the motion must be very slow. The force $\mathbf{F}$ is always horizontal, but the displacement $d\mathbf{r}$ is along the arc. The direction of $d\mathbf{r}$ depends on the value of ϕ and is tangent to the circle at each point. $\mathbf{F}$ will vary in magnitude in such a way as to balance the horizontal component of the tension. Notice that the angle between $\mathbf{F}$ and $d\mathbf{r}$ is equal to the angular displacement ϕ in this case.

The work done as the mass m moves from $\phi = 0$ to $\phi = \phi_0$ under the action of the force $\mathbf{F}$ is

$$W = \int_{\phi=0}^{\phi=\phi_0} \mathbf{F} \cdot d\mathbf{r} = \int_{\phi=0}^{\phi=\phi_0} F \cos \phi \, dr \qquad (7\text{-}10a)$$

or

$$W = \int_{x=0, y=0}^{r=(l-h)\tan \phi_0, \, y=h} (F_x \, dx + F_y \, dy). \qquad (7\text{-}10b)$$

Let us evaluate Eq. 7-10b.

Note that, from Newton's first law (see Fig. 7-8b)

$$F_x = T \sin \phi \qquad \text{and} \qquad mg = T \cos \phi.$$

Eliminating T between these relations gives us

$$F_x = mg \tan \phi.$$

We also note in Fig. 7-8b that $F_y = 0$. Substituting these values for F_x and F_y into Eq. 7-10b yields

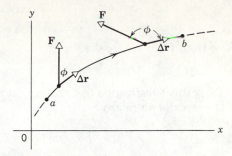

figure 7-7
How F and ϕ might change along a path. As $\Delta \mathbf{r} \to 0$ we may replace it by the differential $d\mathbf{r}$, which always points in the direction of the velocity of the moving object, since $\mathbf{v} = d\mathbf{r}/dt$, and hence is tangent to the path at all points.

EXAMPLE 3

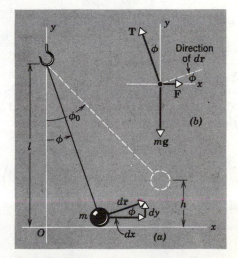

figure 7-8
Example 3. (a) A simple pendulum. A mass point m is suspended on a string of length l. Its maximum displacement is ϕ_0. (b) A free-body force diagram for the mass subjected to an applied horizontal force.

$$W = \int_{x=0,y=0}^{x=(l-h)\tan\phi_0, y=h} mg \tan\phi \, dx.$$

Now from Fig. 7-8a we see that

$$\tan\phi = dy/dx \qquad \text{or} \qquad \tan\phi \, dx = dy.$$

Making this substitution and noting that the integral depends only on the variable y, we obtain finally

$$W = \int_{y=0}^{y=h} (mg) \, dy = mg \int_0^h dy = mgh.$$

You should now try to compute the work done in displacing the particle along the arc with constant speed by applying a force that is always directed *along the arc*. Here it will be simpler to work with Eq. 7-10a, using the tantential force and taking $dr = l \, d\phi$. The result will be the same as before, $W = mgh$. Notice that both these results are the same as the work that would be done in raising a mass m vertically through a height h.

What work has been done on the particle by the tension $\mathbf{T}$ in the string?

7-5
KINETIC ENERGY AND THE WORK-ENERGY THEOREM

In our previous examples of work done by forces, we dealt with *unaccelerated* objects. In such cases the *resultant force* acting on the object is *zero*. Let us suppose now that the *resultant force* acting on an object is *not zero*, so that the object is *accelerated*. The conditions are the same in all respects to those that exist when a single unbalanced force acts on the object.

The simplest situation to consider is that of a *constant resultant force* $\mathbf{F}$. Such a force, acting on a particle of mass m, will produce a constant acceleration $\mathbf{a}$. Let us choose the x-axis to be in the common direction of $\mathbf{F}$ and $\mathbf{a}$. What is the work done by this force on the particle in causing a displacement x? We have (for constant acceleration) the relations

$$a = \frac{v - v_0}{t}$$

and

$$x = \frac{v + v_0}{2} \cdot t,$$

which are Eqs. 3-12 and 3-14 respectively (in which we have dropped the subscript x, for convenience, and chosen $x_0 = 0$ in the last equation). Here v_0 is the particle's speed at $t = 0$ and v its speed at the time t. Then the work done is

$$W = Fx = max$$

$$= m\left(\frac{v - v_0}{t}\right)\left(\frac{v + v_0}{2}\right)t = \tfrac{1}{2}mv^2 - \tfrac{1}{2}mv_0^2. \qquad (7\text{-}11)$$

We call one-half the product of the mass of a body and the square of its speed the kinetic energy of the body. If we represent kinetic energy by the symbol K, then

$$K = \tfrac{1}{2}mv^2. \qquad (7\text{-}12)$$

We may then state Eq. 7-11 in this way: *The work done by the resultant force acting on a particle is equal to the change in the kinetic energy of the particle.*

Although we have proved this result for a constant force only, it holds whether the resultant force is constant or variable. Let the resultant force vary in magnitude (but not in direction), for example. Take the displacement to be in the direction of the force. Let this direction be the x-axis. The work done by the resultant force in displacing the particle from x_0 to x is

$$W = \int \mathbf{F} \cdot d\mathbf{r} = \int_{x_0}^{x} F \, dx.$$

But from Newton's second law we have $F = ma$, and we can write the acceleration a as

$$a = \frac{dv}{dt} = \frac{dv}{dx} \cdot \frac{dx}{dt} = \frac{dv}{dx} v = v \frac{dv}{dx}.$$

Hence

$$W = \int_{x_0}^{x} F \, dx = \int_{x_0}^{x} mv \frac{dv}{dx} dx = \int_{v_0}^{v} mv \, dv = \tfrac{1}{2}mv^2 - \tfrac{1}{2}mv_0^2. \quad (7\text{-}13)$$

A more general case is that in which the force varies both in direction and magnitude and the motion is along a curved path, as in Fig. 7-7. (See Problem 8.) Once again we find that the work done on a particle by the resultant force is equal to the change in the kinetic energy of the particle.

The work done *on* a particle by the *resultant* force is *always* equal to the change in the kinetic energy of the particle:

$$W \text{ (of the } resultant \text{ force)} = K - K_0 = \Delta K. \quad (7\text{-}14)$$

Equation 7-14 is known as the *work-energy theorem* for a particle.

Notice that when the speed of the particle is constant, there is no change in kinetic energy and the work done by the resultant force is zero. With uniform circular motion, for example, the speed of the particle is constant and the centripetal force does no work on the particle. A force at right angles to the direction of motion merely changes the *direction* of the velocity and not its magnitude. Only when the *resultant* force has a component along the direction of motion does it change the speed of the particle or its kinetic energy. Work is done on a particle by that component of the resultant force along the line of motion. This agrees with our definition of work in terms of a scalar product, for in $\mathbf{F} \cdot d\mathbf{r}$ only the component of $\mathbf{F}$ along $d\mathbf{r}$ contributes to the product.

If the kinetic energy of a particle decreases, the work done on it by the resultant force is negative. The displacement and the component of the resultant force along the line of motion are oppositely directed. The work done *on* the particle by the force is the negative of the work done *by* the particle on whatever produced the force. This is a consequence of Newton's third law of motion. Hence Eq. 7-14 can be interpreted to say that the kinetic energy of a particle *decreases* by an amount just equal to the amount of work which the particle *does*. A body is said to have energy stored in it because of its motion; as it does work it slows down and loses some of this energy. Therefore, *the kinetic energy of a body in motion is equal to the work it can do in being brought to rest.* This result holds whether the applied forces are constant or variable.

The units of kinetic energy and of work are the same. Kinetic energy, like work, is a scalar quantity. The kinetic energy of a group of particles is simply the (scalar) sum of the kinetic energies of the individual particles in the group.

A neutron, one of the constituents of a nucleus, is found to pass two points 6.0 m apart in a time interval of 1.8×10^{-4} s. Assuming its speed was constant, find its kinetic energy. The mass of a neutron is 1.7×10^{-27} kg.

EXAMPLE 4

We find the speed from

$$v = \frac{d}{t} = \frac{6.0 \text{ m}}{1.8 \times 10^{-4} \text{ s}} = 3.3 \times 10^4 \text{ m/s}.$$

The kinetic energy is

$$K = \tfrac{1}{2}mv^2 = (\tfrac{1}{2})(1.7 \times 10^{-27} \text{ kg})(3.3 \times 10^4 \text{ m/s})^2 = 9.3 \times 10^{-19} \text{ J}.$$

For purposes of nuclear physics the joule is a very large energy unit. A unit more commonly used is the *electron volt* (eV), which is equal to 1.60×10^{-19} J. The kinetic energy of the neutron in our example can then be expressed as

$$K = (9.3 \times 10^{-19} \text{ J})\left(\frac{1 \text{ eV}}{1.60 \times 10^{-19} \text{ J}}\right) = 5.8 \text{ eV}.$$

Assume the force of gravity to be constant for small distances above the surface of the earth. A body is dropped from rest at a height h above the earth's surface. What will its kinetic energy be just before it strikes the ground?

EXAMPLE 5

The gain in kinetic energy is equal to the work done by the resultant force, which here is the force of gravity. This force is constant and directed along the line of motion. so that the work done by gravity is

$$W = \mathbf{F} \cdot \mathbf{d} = mgh.$$

Initially the body has a speed $v_0 = 0$ and finally a speed v. The gain in kinetic energy of the body is

$$\tfrac{1}{2}mv^2 - \tfrac{1}{2}mv_0^2 = \tfrac{1}{2}mv^2 - 0.$$

Equating these two equivalent terms we obtain

$$K = \tfrac{1}{2}mv^2 = mgh$$

as the kinetic energy of the body just before it strikes the ground.

The speed of the body is then

$$v = \sqrt{2gh}.$$

You should show that in falling from a height h_1 to a height h_2 a body will increase its kinetic energy from $\tfrac{1}{2}mv_1^2$ to $\tfrac{1}{2}mv_2^2$, where

$$\tfrac{1}{2}mv_2^2 - \tfrac{1}{2}mv_1^2 = mg(h_1 - h_2).$$

In this example we are dealing with a constant force and a constant acceleration. The methods developed in previous chapters should be useful here too. Can you show how the results obtained by energy considerations could be obtained directly from the laws of motion for uniformly accelerated bodies?

A block weighing 8.0 lb (= 35.6 N) slides on a horizontal frictionless table with a speed of 4.0 ft/s (= 1.22 m/s). It is brought to rest in compressing a spring in its path. By how much is the spring compressed if its force constant is 0.25 lb/ft (= 1.35 N/m)?

EXAMPLE 6

The kinetic energy of the block is

$$K = \tfrac{1}{2}mv^2 = \tfrac{1}{2}(w/g)v^2.$$

This kinetic energy is equal to the work W that the block can do before it is brought to rest. The work done in compressing the spring a distance x beyond its unstretched length is

$$W = \tfrac{1}{2}kx^2,$$

so that

$$\tfrac{1}{2}kx^2 = \tfrac{1}{2}(w/g)v^2$$

or

$$x = \sqrt{\frac{w}{gk}}\,v = \sqrt{\frac{8.0}{(32)(0.25)}}\,4.0 \text{ ft} = 4.0 \text{ ft } (= 1.22 \text{ m}).$$

7-6
SIGNIFICANCE OF THE WORK-ENERGY THEOREM

The work-energy theorem does *not* represent a new, independent law of classical mechanics. We have simply *defined* work and kinetic energy and *derived* the relation between them directly from Newton's second law. The work-energy theorem is useful, however, for solving problems in which the work done by the resultant force is easily computed and in which we are interested in finding the particle's speed at certain positions. Of greater significance, perhaps, is the fact that the work-energy theorem is the starting point for a sweeping generalization in physics. It has been emphasized that the work-energy theorem is valid when W is interpreted as the work done by the *resultant* force acting on the particle. However, it is helpful in many problems to compute separately the work done by certain types of force and give special names to the work done by each type. This leads to the concepts of different types of energy and the principle of the conservation of energy, which is the subject of the next chapter.

7-7
POWER

Let us now consider the time involved in doing work. The same amount of work is done in raising a given body through a given height whether it takes one second or one year to do so. However, the *rate at which work is done* is often more interesting to us than the total work performed.

We define *power* as the time rate at which work is done. The average power delivered by an agent is the total work done by the agent divided by the total time interval, or

$$\overline{P} = W/t.$$

The instantaneous power delivered by an agent is

$$P = dW/dt. \qquad (7\text{-}15)$$

If the power is constant in time, then $P = \overline{P}$ and

$$W = Pt.$$

In the International System of units, the unit of power is 1 joule/sec, which is called 1 *watt* (abbreviation W). This unit of power is named in honor of James Watt who made major improvements to the steam engines of his day that pointed the way toward today's more efficient engines. In the British engineering system, the unit of power is 1 ft · lb/sec. Because this unit is quite small for practical purposes, a larger unit, called the *horsepower* (abbreviation hp), has been adopted. Actually Watt himself suggested as a unit of power the power delivered by a horse as an engine. One horsepower was chosen to equal 550 ft · lb/sec. One horsepower is equal to about 746 watts or about three-fourths of a kilowatt. A horse would not last very long at that rate.

Work can also be expressed in units of power × time. This is the origin of the term *kilowatt-hour*, for example. One kilowatt-hour is the work done in 1 hour by an agent working at a constant rate of 1 kW.

EXAMPLE 7

An automobile uses 100 hp and moves at a uniform speed of 60 mi/h (= 88 ft/s). What is the forward thrust exerted by the engine on the car?

$$P = \frac{W}{t} = \frac{\mathbf{F} \cdot \mathbf{d}}{t} = \mathbf{F} \cdot \mathbf{v}.$$

The forward thrust **F** is in the direction of motion given by **v**, so that

$$P = Fv,$$

and

$$F = \frac{P}{v} = \left(\frac{100 \text{ hp}}{88 \text{ ft/s}}\right)\left(\frac{550 \text{ ft} \cdot \text{lb/sec}}{1 \text{ hp}}\right) = 630 \text{ lb}.$$

Why doesn't the car accelerate?

questions

1. Can you think of other words like "work" whose colloquial meanings are often different from their scientific meanings?

2. Suppose that three constant forces act on a particle as it moves from one position to another. Is the work done on the particle by the resultant of these three forces equal to the sum of the work done by each of the three forces acting separately?

3. The inclined plane (see Example 1) is a simple machine which enables us to do work with the application of a smaller force than is otherwise necessary. The same statement applies to a wedge, a lever, a screw, a gear wheel, and a pulley. Do such machines save us work?

4. In a tug of war one team is slowly giving way to the other. What work is being done and by whom?

5. The work done by frictional forces is always negative. Can you explain this?

6. A man exerts a constant force on a fixed wall and does no mechanical work on it. Why does he feel tired doing this?

7. You lift a bowling ball from the floor and put it on a table. Two forces act on the ball: its weight $-m\mathbf{g}$, and your upward force $+m\mathbf{g}$. These two forces cancel each other so that it would seem that no work is done. On the other hand you know that you have done some work. What is wrong?

8. You cut a spring in half. What is the relationship of the spring constant k for the original spring to that for either of the half-springs?

9. Springs A and B are identical except that A is stiffer than B, that is, $k_A > k_B$. On which spring is more work expended if (a) they are stretched by the same amount, (b) they are stretched by the same force?

10. Does kinetic energy depend on the direction of the motion involved? Can it be negative?

11. The work done by the resultant force is always equal to the change in kinetic energy. Can it happen that the work done by one of the component forces alone will be greater than the change in kinetic energy? If so, give examples.

12. You throw a ball vertically in the air and catch it. What does the work-energy theorem say qualitatively about the free flight during this round trip? Answer the question first neglecting air resistance and second, taking it into account.

13. When two children play catch on a train, does the kinetic energy of the ball depend on the speed of the train? Does the reference frame chosen affect your answer? If so, would you call kinetic energy a scalar quantity? (See Problem 21.)

14. Does the work done by the resultant force acting on a particle depend on the (inertial) reference frame of the observer? Does the change in kinetic energy so depend?

15. A man rowing a boat upstream is at rest with respect to the shore. (a) Is he

doing any work? (b) If he stops rowing and moves down with the stream, is any work being done on him?

16. Does the power needed to raise a box onto a platform depend on how fast it is raised?

17. You lift some books from a library shelf to a higher shelf in time t. Does the work that you do depend on (a) the mass of the books, (b) the weight of the books, (c) the height of the upper shelf above the floor, (d) the time t, and (e) whether you lift the books sideways or directly upward?

18. We hear a lot about the "energy crisis." Would it be more accurate to speak of a "power crisis"?

problems

SECTION 7-2

1. A man pushes a 60-lb (270-N) block 30 ft (9.1 m) along a level floor at constant speed with a force directed 45° below the horizontal. If the coefficient of kinetic friction is 0.20, how much work does the man do on the block?
   Answer: 450 ft · lb (610 J).

2. A block of mass $m = 3.57$ kg is drawn at constant speed a distance $d = 4.06$ m along a horizontal floor by a rope exerting a constant force of magnitude $F = 7.68$ N making an angle $\theta = 15.0°$ above the horizontal. Compute (a) the total work done on the block, (b) the work done by the rope on the block, (c) the work done by friction on the block, and (d) the coefficient of kinetic friction between block and floor.

3. A 100-lb block of ice slides down an incline 5.0 ft long and 3.0 ft high. A man pushes up on the ice parallel to the incline so that it slides down at constant speed. The coefficient of friction between the ice and the incline is 0.10. Find (a) the force exerted by the man, (b) the work done by the man on the block, (c) the work done by gravity on the block, (d) the work done by the surface of the incline on the block, (e) the work done by the resultant force on the block, and (f) the change in kinetic energy of the block.
   Answer: (a) 52 lb. (b) −260 ft · lb. (c) 300 ft · lb. (d) −40 ft · lb. (e) Zero. (f) Zero.

4. A crate weighing 500 lb (2200 N) is suspended from a rope 40 ft (12 m) long. The crate is then pushed aside 4.0 ft (1.2 m) from the vertical and held there. (a) What force directed along the arc is needed to keep the crate in this position? (b) Is work being done in holding it there? (c) Was work done in moving it aside? If so, how much? (d) Does the tension in the rope perform any work on the crate?

5. A cord is used to lower vertically a block of mass M a distance d at a constant downward acceleration of g/4. Find the work done by the cord on the block.
   Answer: −3Mgd/4.

SECTION 7-3

6. (a) Estimate the work done by the force shown on the graph (Fig. 7-9) in displacing a particle from $x = 1$ m to $x = 3$ m. Refine your method to see how

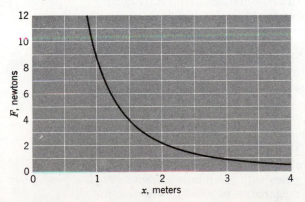

figure 7-9
Problem 6

close you can come to the exact answer of 6 J. (b) The curve is given analytically by $F = a/x^2$ where $a = 9$ N · m². Show how to get the work done by the rules of integration.

7. A single force acts on a body in rectilinear motion. A plot of velocity versus time for the body is shown in Fig. 7-10. Find the sign (positive or negative) of the work done *by* the force *on* the body in each of the intervals *AB*, *BC*, *CD*, and *DE*.

Answer: *AB*, *BC*, *CD*, *DE*.
    +   0   −   +

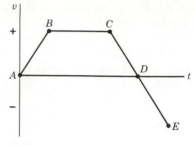

figure 7-10
Problem 7

SECTION 7-4

8. When the force **F** varies both in direction and magnitude and the motion is along a curved path, the work done by **F** is obtained from $dW = \mathbf{F} \cdot d\mathbf{r}$, the subsequent integration being taken along the curved path. Notice that both F and ϕ, the angle between **F** and $d\mathbf{r}$, may vary from point to point (see Fig. 7-7). Show that for two- or three-dimensional motion

$$W = \tfrac{1}{2}mv^2 - \tfrac{1}{2}mv_0^2,$$

where v is the final speed and v_0 the initial speed.

SECTION 7-5

9. From what height would an automobile have to fall to gain the kinetic energy equivalent to what it would have when going 60 mi/h (97 km/h)?
Answer: 120 ft (37 m).

10. A running man has half the kinetic energy of a boy half his mass. The man speeds up by 1.0 m/s and then has the same kinetic energy as the boy. What were the original speeds of (a) man and (b) boy?

11. If a 2.9×10^5 kg (weight $mg = 6.4 \times 10^5$ lb) Saturn V rocket with an Apollo spacecraft attached must achieve an escape velocity of 11.2 km/s (25,000 mi/h) near the surface of the earth, how much energy must the fuel contain? Would the system actually need as much, or would it need more? Why?   Answer: 1.8×10^{13} J (1.3×10^{13} ft · lb).

12. A proton starting from rest is accelerated in a cyclotron to a final speed of 3.0×10^7 m/s (about one-tenth the speed of light). How much work, in electron volts, is done on the proton by the electrical force of the cyclotron that accelerates it? 1 eV = 1.6×10^{-19} J.

13. A 30-g bullet initially traveling 500 m/s penetrates 12 cm into a wooden block. What average force does it exert on the block?
Answer: 3.1×10^4 N.

14. An outfielder throws a baseball with an initial speed of 60 ft/s. An infielder catches the ball at the same level when its speed is reduced to 40 ft/s. What work was done in overcoming the resistance of the air? The weight of a baseball is 9.0 oz.

15. A proton (nucleus of the hydrogen atom) is being accelerated in a linear accelerator. In each stage of such an accelerator the proton is accelerated along a straight line at 3.6×10^{15} m/s². If a proton enters such a stage moving initially with a speed of 2.4×10^7 m/s, and the stage is 3.5 cm long, compute (a) its speed at the end of the stage and (b) the gain in kinetic energy resulting from the acceleration. Take the mass of the proton to be 1.67×10^{-27} kg and express the energy in electron volts; 1 eV = 1.6×10^{-19} J.
Answer: (a) 2.9×10^7 m/s. (b) 1.3×10^6 eV.

16. Show *from considerations of work and kinetic energy* that, assuming the driver jams on the brakes, the stopping distance for a car of mass m moving with speed v along a level road is $v^2/2\mu_k g$, where μ_k is the coefficient of kinetic friction between tires and road. (See Example 2, Chapter 6, and Questions 3, 4, 5 of Chapter 8.)

17. A 5.0-kg block moves in a straight line on a horizontal frictionless surface under the influence of a force that varies with position as shown in Fig. 7-11. (a) How much work is done by the force as the block moves from the

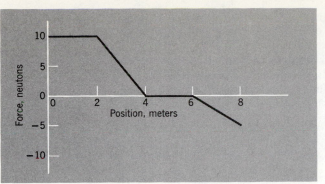

figure 7-11
Problem 17

origin to $x = 8.0$ m? (b) If the particle's speed passing through the origin
was 4.0 m/s, with what speed does it pass the point $x = 8.0$ m?

Answer: (a) 25 J. (b) 5.1 m/s.

18. A helicopter is used to lift a 160-lb (710-N) astronaut 50 ft (15 m) vertically
from the ocean by means of a cable. The acceleration of the astronaut is
$g/10$. (a) How much work is done by the helicopter on the astronaut? (b)
How much work is done by the gravitational force on the astronaut? (c)
With what speed does the astronaut reach the helicopter?

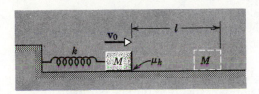

figure 7-12
Problem 19

19. The block of mass M shown in Fig. 7-12 initially has a velocity v_0 to the
right and its position is such that the spring exerts no force on it, that is, the
spring is not stretched or compressed. The block moves to the right a distance l before stopping in the dotted position shown. The spring constant
is k and the coefficient of kinetic friction between block and table is μ_k. As
the block moves the distance l, (a) what is the work done on it by the friction force? (b) What is the work done on it by the spring force? (c) Are there
other forces acting on the block, and, if so, what work do they do? (d) What
is the total work done on the block? (e) Use the work-energy theorem to find
the value of l in terms of M, v_0, μ_k, g. and k.

Answer: (a) $-\mu_k Mgl$. (b) $-kl^2/2$. (c) Gravity and the vertical thrust of the
table, which do no work. (d) $-(\mu_k Mgl + kl^2/2)$.
(e) $(\sqrt{\mu_k^2 M^2 g^2 + v_0^2 kM} - \mu_k Mg)/k$.

20. (a) A mass of 0.675 kg on a frictionless table is attached to a string which
passes through a hole in the table at the center of the horizontal circle in
which the mass moves with constant speed. If the radius of the circle is
0.500 m and the speed is 10.0 m/s, compute the tension in the string. (b) It is
found that drawing an additional 0.200 m of the string down through the
hole, thereby reducing the radius of the circle to 0.300 m, has the effect of
multiplying the original tension in the string by 4.63. Compute the total
work done by the string on the revolving mass during the reduction of the
radius.

21. *Work and Kinetic Energy in Moving Reference Frames.* Consider two observers, one whose frame is attached to the ground and another whose frame
is attached, say, to a train moving with uniform velocity **u** with respect to
the ground. Each observes that a particle, initially at rest with respect to
the train, is accelerated by a constant force applied to it for time t in the
forward direction.

(a). Show that for each observer the work done by the force is equal to the gain in kinetic energy of the particle, but that one observer measures these quantities to be $\frac{1}{2}ma^2t^2$, whereas the other observer measures them to be $\frac{1}{2}ma^2t^2 + maut$. Here a is the common acceleration of the particle of mass m.

(b). Explain the differences in work done by the same force in terms of the different distances through which the observers measure the force to act during the time t. Explain the different final kinetic energies measured by each observer in terms of the work the particle could do in being brought to rest relative to each observer's frame.

SECTION 7-7

22. If a 128-lb (570 N) woman runs up a flight of stairs having a rise of 14 ft (4.3 m) in 3.5 s, what average power must she supply?

23. 1200 m³ of water passes each second over a waterfall 100 m high. Assuming that three-fourths of the kinetic energy gained by the water in falling is converted to electrical energy by a hydroelectric generator, what is the power output of the generator?   *Answer:* 8.8×10^5 kW.

24. The loaded cab of an elevator has a mass m of 3.0×10^3 kg and moves 200 m up the shaft in 20 s. At what average rate does the cable do work on the cab?

25. A horse pulls a cart horizontally with a force of 40 lb at an angle of 30° above the horizontal and moves along at a speed of 6.0 mi/h. (a) How much work does the horse do in 10 minutes? (b) What is the power output of the horse? *Answer:* (a) 1.8×10^5 ft · lb. (b) 0.55 hp.

26. What power is developed by a grinding machine whose wheel has a radius of 8.0 in. and runs at 2.5 rev/s when the tool to be sharpened is held against the wheel with a force of 40 lb? The coefficient of friction between the tool and the wheel is 0.32.

27. A satellite rocket weighing 100,000 lb acquires a speed of 4000 mi/h in 1.0 min after launching. (a) What is its kinetic energy at the end of the first minute? (b) What is the average power expended during this time, neglecting frictional and gravitational forces?
Answer: (a) 5.4×10^{10} ft · lb. (b) 1.6×10^6 hp.

28. A net force of 5.0 N acts on a 15 kg-body initially at rest. Compute (a) the work done by the force in the first, second, and third second and (b) the instantaneous power exerted by the force at the end of the third second.

29. A force acts on a 3.0-kg particle in such a way that the position of the particle as a function of time is given by $x = 3t - 4t^2 + t^3$, where x is in meters and t is in seconds. (a) Find the work done by the force during the first 4.0 s. (b) At what instantaneous rate is the force doing work on the particle at the instant $t = 1.0$ s?   *Answer:* (a) 530 J. (b) 12 W.

30. The force required to tow a boat at constant velocity is proportional to the speed. If it takes 10 hp (7500 W) to tow a certain boat at a speed of 2.5 mi/h (4.0 km/h), how much power does it take to tow it at a speed of 7.5 mi/h (12 km/h)?

31. A body of mass m accelerates uniformly from rest to a speed v_f in time t_f. (a) Show that the work done on the body as a function of time t, in terms of v_f and t_f, is

$$\tfrac{1}{2}m \frac{v_f^2}{t_f^2} t^2.$$

(b) As a function of time t, what is the instantaneous power delivered to the body? (c) What is the instantaneous power at the end of 5.0 s delivered to a 3200-lb body which accelerates to 60 mi/h in 10 s?
Answer: (b) $mv_f^2 t/t_f^2$. (c) 70 hp.

32. A truck can move up a road having a grade of 1.0-ft rise every 50 ft with a speed of 15 mi/h. The resisting force is equal to 1/25 the weight of the truck. How fast will the same truck move down the hill with the same horsepower?

33. A 1.5×10^6 W railroad locomotive accelerates a train from a speed of 10 m/s to 25 m/s at full power in 6.0 minutes. (a) Neglecting friction, calculate the mass of the train. (b) Find the speed of the train as a function of time during the interval. (c) Find the force accelerating the train as a function of time during the interval. (d) Find the distance moved by the train during the interval.

Answer: (a) 2.1×10^6 kg. (b) $\sqrt{100 + 1.4\, t}$ m/s.
   (c) $(1.5 \times 10^6)/\sqrt{100 + 1.4\, t}$ N. (d) 6.9 km.

8
the
conservation
of energy

In Chapter 7 we derived the work-energy theorem from Newton's second law of motion. This theorem says that the work W done by the resultant force $\mathbf{F}$ acting on a particle as it moves from one point to another is equal to the change ΔK in the kinetic energy of the particle, or

$$W = \Delta K. \tag{8-1}$$

Often several forces act on a particle, the resultant force $\mathbf{F}$ being their vector sum, that is, $\mathbf{F} = \mathbf{F}_1 + \mathbf{F}_2 + \cdots \mathbf{F}_n$, in which we assume that n forces act. The work done by the resultant force $\mathbf{F}$ is the algebraic sum of the work done by these individual forces, or $W = W_1 + W_2 + \cdots W_n$. Thus we can write the work-energy theorem (Eq. 8-1) as

$$W_1 + W_2 + \cdots + W_n = \Delta K. \tag{8-2}$$

In this chapter we shall consider systems in which a particle is acted upon by various kinds of forces and we shall compute W_1, W_2, and so on, for these forces; this will lead us to define different kinds of energy such as potential energy and heat energy. The process culminates in the formulation of one of the great principles of science, the *conservation of energy principle*.

Let us first distinguish between two types of forces, *conservative* and *nonconservative*. We shall consider an example of each type and we discuss each example from several different, but related, points of view.

Imagine a spring fastened at one end to a rigid wall as in Fig. 8-1. Let us slide a block of mass m with velocity $\mathbf{v}$ directly toward the spring; we assume that the horizontal plane is frictionless and that the spring is ideal, that is, that it obeys Hooke's law (Eq. 7-7)

$$F = -kx, \tag{8-3}$$

where F is the force exerted by the spring when its free end is displaced through a distance x; we assume further that the mass of the spring is so small compared with that of the block that we can neglect the kinetic energy of the spring. Thus, in the system (mass + spring), all the kinetic energy is concentrated in the mass.

After the block touches the spring, the speed and hence the kinetic energy of the block decrease until finally the block is brought to rest by the action of the spring force, as in Fig. 8-1b. The block now reverses its motion as the compressed spring expands. It gains speed and kinetic energy and, when it comes once again to its position of initial contact with the spring, we find that it has the same speed and kinetic energy as it had originally; only the direction of motion has changed. The block loses kinetic energy during one part of its motion but gains it all back during the other part of its motion as it returns to its starting point (Fig. 8-1c).

We have interpreted the kinetic energy of a body as its ability to do work by virtue of its motion. It is clear that at the completion of a round trip the ability of the block in Fig. 8-1 to do work remains the same; it has been *conserved*. The elastic force exerted by an ideal spring, and other forces that act in this same way, are called *conservative*. The force of gravity is also conservative; if we throw a ball vertically upward, it will (if we assume air resistance to be negligible) return to our hand with the same kinetic energy that it had when it left our hand.

If, however, a particle on which one or more forces act returns to its initial position with either more or less kinetic energy than it had initially, then in a round trip its ability to do work has been changed. In this case the ability to do work has *not* been conserved and at least one of the forces acting is labeled *nonconservative*.

To illustrate a nonconservative force let us assume that the surfaces of the block and the plane in Fig. 8-1 are not frictionless but rather that a force of friction **f** is exerted by the plane on the block. The frictional force opposes the motion of the block no matter which way the block is moving and we find that the block returns to its starting point with *less* kinetic energy than it had initially. Since we showed in our first experiment that the spring force was conservative, we must attribute this new result to the action of the friction force.* We say that this force, and other forces that act in this same way, are *nonconservative*. The induction force in a betatron (Section 35-6) is also a nonconservative force. Instead of dissipating kinetic energy, however, it generates it, so that an electron moving in the circular betatron orbit will return to its initial position with *more* kinetic energy than it had there originally. In a round trip the electron gains kinetic energy, as it must do if the betatron is to be effective.

We can define conservative force from another point of view, that of the work done by the force on the particle. In our first example above, the work done by the elastic spring force on the block while the spring was being compressed was negative, because the force exerted on the block by the spring (to the left in Fig. 8-1a) was directed opposite to the

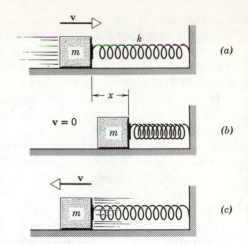

figure 8-1
(a) A block of mass m is projected with speed v against a spring. (b) The block is brought to rest by the action of the spring force. (c) The block has regained its initial speed v as it returns to its starting point.

---

*Actually two other forces act on the block in Fig. 8-1, its weight **W** and the normal force **N** exerted by the plane. Because these act at right angles to the motion, they cannot change the kinetic energy of the block and hence do not enter into this discussion.

displacement of the block (to the right in Fig. 8-1a). While the spring was being extended the work that the spring force did on the block was positive (force and displacement in the same direction). In our first example the net work done on the block by the spring force during a complete cycle, or round trip, is zero.

In our second example we considered the effect of the frictional force. The work done on the block by this force was negative for each portion of the cycle because the frictional force always opposed the motion. Hence the work done by friction in a round trip cannot be zero. In general, then: *A force is conservative if the work done by the force on a particle that moves through any round trip is zero. A force is non-conservative if the work done by the force on a particle that moves through any round trip is not zero.*

The work-energy theorem shows that this second way of defining conservative and nonconservative forces is fully equivalent to our first definition. If there is no change in the kinetic energy of a particle moving through any round trip then $\Delta K = 0$ and, from Eq. 8-1, $W = 0$ and the resultant force acting must be conservative. Similarly, if $\Delta K \neq 0$, then from Eq. 8-1, $W \neq 0$ and at least one of the forces acting must be nonconservative.

We can look into this matter in a little more detail. When friction is present in the system of Fig. 8-1, four forces act on the block, the resultant force being

$$F = F_s + W + N + f$$

in which the forces are the spring force F_s, the weight of the block W, the normal force exerted on the block by the plane N, and the frictional force f. We can write Eq. 8-2, the work-energy theorem, as

$$W_s + W_W + W_N + W_f = \Delta K,$$

where the terms on the left are the work done on the block by the four forces above. We have seen that for a round trip $W_s = 0$. Similarly, $W_W = W_N = 0$ because the corresponding forces are at right angles to the displacement of the block. Thus the change in kinetic energy is due entirely to W_f, the work done by the frictional force.

We can consider the difference between conservative and nonconservative forces in still a third way. Suppose a particle goes from a to b along path 1 and back from b to a along path 2 as in Fig. 8-2a. Several forces may act on the particle during this round trip; we consider each force separately. If the force being considered is conservative, the work done on the particle by that particular force for the round trip is zero, or

$$W_{ab,1} + W_{ba,2} = 0,$$

which we can write as

$$W_{ab,1} = -W_{ba,2}.$$

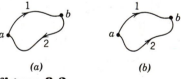

(a)   *(b)*

figure 8-2

That is, the work in going from a to b along path 1 is the negative of the work in going from b to a along path 2. However, if we cause the particle to go from a to b along path 2, as shown in Fig. 8-2b, we merely reverse the direction of the previous motion along 2, so that

$$W_{ab,2} = -W_{ba,2}.$$

Hence

$$W_{ab,1} = W_{ab,2},$$

which tells us that the work done on the particle by a conservative force in going from a to b is the same for either path.

Paths 1 and 2 can be any paths at all as long as they go from a to b;

and a and b can be chosen to be any two points at all. We always find the same result if the force is conservative. Hence, we have another equivalent definition of conservative and nonconservative forces: *A force is conservative if the work done by it on a particle that moves between two points depends only on these points and not on the path followed. A force is nonconservative if the work done by that force on a particle that moves between two points depends on the path taken between those points.*

To illustrate this third (equivalent) definition of conservative forces, let us consider a second kind of conservative force, that due to gravity. Suppose that we take a stone of mass m in our hand and raise it to a height h above the ground, going from a to b by several different paths as in Fig. 8-3. We already know that in a round trip the total work done by a conservative force is zero and that the gravitational force is conservative. The work done *on* the stone by gravity along the return path bca is simply mgh. Hence, because gravity is a conservative force, the work done by gravity *on* the stone along any of the paths from a to b must be $-mgh$, for only if this is true can the total work done by gravity in a round trip be zero. This means that gravity does negative work on the stone as it moves from a to b, or, to put it another way, work must be done *against* gravity along any of the paths ab. You can compute directly the result that the work done by gravity along any path from a to b equals $-mgh$. For any of these paths can be decomposed into infinitesimal displacements which are alternately horizontal and vertical; no work is done by gravity in horizontal displacements, and the net vertical displacement is the same in all cases. Hence the work done by gravity on the stone moving from a to b depends only on the positions of a and b and not at all on the path taken.

For a nonconservative force, such as friction, the work done is *not* independent of the path taken between two fixed points. We need only point out that as we push a block over a (rough) table between any two points a and b by various paths, the distance traversed varies and so does the work done by the frictional force. It depends on the path.

The definitions of conservative force which we have given are equivalent to one another. Which one we use depends only on convenience. The round-trip approach shows clearly that kinetic energy is conserved when conservative forces act. If we wish to develop the idea of potential energy, however, the path independence statement is preferable.

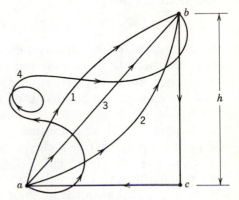

figure 8-3
A stone is raised from a to b via various paths 1, 2, 3, and 4.

8-3
POTENTIAL ENERGY

In this section we shall focus attention not on the moving block of Fig. 8-1 but rather on the isolated system (block + spring). Instead of saying that the block is moving we prefer, from this point of view, to say that the configuration of the system is changing. We measure both the position of the block and the configuration of the system at any instant by the same parameter x, namely, the displacement of the free end of the spring from its normal position, corresponding to an unstretched spring. The kinetic energy of the system is the same as that of the block because we have assumed the spring to be massless.

We have seen that the kinetic energy of the system of Fig. 8-1 decreases during the first half of the motion, becomes zero, and then increases during the second half of the motion. If there is no friction, the kinetic energy of the system when it has regained its initial configuration returns to its initial value.

Under these circumstances (conservative forces acting) it makes sense to introduce the concept of *energy of configuration*, or *potential energy U*, and to say that if K for the system changes by ΔK as the configuration of the system changes (that is, as the block moves in the system of Fig. 8-1), then U for the system must change by an equal but opposite amount so that the sum of the two changes is zero, or

$$\Delta K + \Delta U = 0. \qquad (8\text{-}4a)$$

Alternatively, we can say that any change in kinetic energy K of the system is compensated for by an equal but opposite change in the potential energy U of the system so that their sum remains constant throughout the motion, or

$$K + U = \text{a constant.} \qquad (8\text{-}4b)$$

The potential energy of a system represents a form of stored energy which can be fully recovered and converted into kinetic energy. We cannot associate a potential energy with a nonconservative force such as the force of friction because the kinetic energy of a system in which such forces act does *not* return to its initial value when the system returns to its initial configuration.

Equations 8-4 apply to a closed system of interacting objects, such as the (mass + spring) system of Fig. 8-1. In this example, because we have taken the spring to be effectively massless, the kinetic energy may be associated with the moving mass alone. The block slows down (or speeds up) because a force is exerted on it *by the spring*; it is appropriate, then, to associate the potential energy of the system with this force, that is to say, with the spring. Thus in this simple case we say that kinetic energy, localized in the mass, decreases during the first part of the motion whereas potential energy, localized in the spring, increases during this same time.*

Equations 8-4 are essentially bookkeeping statements about energy. They, and the concept of potential energy, have no real meaning, however, until we have shown how to calculate U as a function of the configuration of the system within which the conservative forces act; in the example of Fig. 8-1 this means that we must be able to calculate $U(x)$, where x is the spring displacement.

To refine our concept of potential energy U let us consider the work-energy theorem, $W = \Delta K$, in which W is the work done by the resultant force on a particle as it moves from a to b. For simplicity let us assume that only a single force **F** acts on the particle; this is effectively true in the system of Fig. 8-1. If **F** is conservative, we can combine the work-energy theorem (Eq. 8-1) with Eq. 8-4a, obtaining

$$W = \Delta K = -\Delta U. \qquad (8\text{-}5a)$$

The work W done by a conservative force depends only on the starting and the end points of the motion and not on the path followed between them. Such a force can depend only on the position of a particle; it does not depend on the velocity of the particle or on the time, for example.

For motion in one dimension, Eq. 8-5a becomes

---

* Just as we assumed the spring to be effectively massless we also assume the block to be rigid, that is, effectively "springless." In a more general system, kinetic and potential energy could each be present in various parts of the system, in varying proportions as the system configuration changes.

$$\Delta U = -W = -\int_{x_0}^{x} F(x) \, dx, \tag{8-5b}$$

the particle moving from x_0 to x. Equation 8-5b shows how to calculate the change in potential energy ΔU when a particle, acted on by a conservative force $F(x)$, moves from point a, described by x_0, to point b, described by x. The equation shows that we can only calculate ΔU if the force $\mathbf{F}$ depends only on the position of the particle (that is, on the system configuration), which is equivalent to saying that potential energy has meaning only for conservative forces.

Now that we know that the potential energy U depends on the position of the particle only, we can write Eq. 8-4b as

$$\tfrac{1}{2}mv^2 + U(x) = E \qquad \text{(one-dimension)} \tag{8-6a}$$

in which E, which remains constant as the particle moves, is called the *total mechanical energy*. Suppose that the particle moves from point a (where its position is x_0 and its speed is v_0) to point b (where its position is x and its speed is v); the total mechanical energy E must be the same for each system configuration when the force is conservative, or, from Eq. 8-6a,

$$\tfrac{1}{2}mv^2 + U(x) = \tfrac{1}{2}mv_0^2 + U(x_0). \tag{8-6b}$$

The quantity on the right depends only on the initial position x_0 and the initial speed v_0, which have definite values; it is, therefore, *constant during the motion*. This is the constant total mechanical energy E. Notice that force and acceleration do not appear in this equation, only position and speed. Equations 8-6 are often called the *law of conservation of mechanical energy* for conservative forces.

In many problems we find that although some of the individual forces are not conservative, they are so small that we can neglect them. In such cases we can use Eqs. 8-6 as a good approximation. For example, air resistance may be present but may have so little effect on the motion that we can ignore it.

Notice that, instead of starting with Newton's laws, we can simplify problem solving when conservative forces alone are involved by starting with Eqs. 8-6. This relation is derived from Newton's laws, of course, but it is one step closer to the solution (the so-called first integral of the motion). We often solve problems without analyzing the forces or writing down Newton's laws by looking instead for something in the motion that is constant; here the mechanical energy is constant and we can write down Eqs. 8-6 as the first step.

For one-dimensional motion we can also write the relation between force and potential energy (Eq. 8-5b) as

$$F(x) = -\frac{dU(x)}{dx}. \tag{8-7}$$

To show this, substitute this expression for $F(x)$ into Eq. 8-5b and observe that you get an identity. Equation 8-7 gives us another way of looking at potential energy. *The potential energy is a function of position whose negative derivative gives the force.*

You may have noticed that we wrote down the quantity $U(x)$ in Eqs. 8-6 although we are only able to calculate *changes* in U (from Eq. 8-5b) and not U itself. Let us imagine that a particle moves from a to b along the x-axis and that a single conservative force $F(x)$ acts on it. To assign a value to U_b, the potential energy at point b, let us write

$$\Delta U = U_b - U_a,$$

or (see Eq. 8-5b),

$$U_b = \Delta U + U_a = -\int_{x_a}^{x_b} F(x)\,dx + U_a. \tag{8-8}$$

We cannot assign a value to U_b until we have assigned one to U_a. If point b is any arbitrary position x, so that $U_b = U(x)$, we give meaning to $U(x)$ by choosing point a to be some convenient reference position, described by $x_a = x_0$, and by arbitrarily assigning a value to the potential energy $U_a = U(x_0)$ when the body is at that point. Thus Eq. 8-8 becomes

$$U(x) = -\int_{x_0}^{x} F(x)\,dx + U(x_0). \tag{8-9}$$

The potential energy when the body is at the reference position, that is, $U(x_0)$, is usually given the arbitrary value zero.

It is often convenient to choose the reference position x_0 to be that at which the force acting on the particle is zero. Thus the force exerted by a spring is zero when the spring has its normal unstretched length; we usually say that the potential energy is also zero for this condition. Also, the attraction of the earth on a body decreases as the body moves away from the earth, becoming zero at an infinite distance. We usually take infinity as our reference position and assign the value zero to the potential energy associated with the gravitational force at that position (see Chapter 16). So far, however, we have been more concerned with the gravitational pull on bodies such as baseballs, etc., which, in comparison to the earth's radius, never move very far from the earth's surface. Here the gravitational force ($= mg$) is essentially constant and we find it convenient to take the zero of potential energy to be, not at infinity, but at the surface of the earth.

The effect of changing the coordinate of the standard reference position x_0, or of the arbitrary value assigned to $U(x_0)$, is simply to change the value of $U(x)$ by an added constant. The presence of an arbitrary added constant in the potential energy expression (Eq. 8-9) makes no difference to the equations that we have written so far. This simply adds the same constant term to each side of Eq. 8-6b, for example, leaving that equation unchanged. Furthermore, changing $U(x)$ by an added constant does not change the force calculated from Eq. 8-7 because the derivative of a constant is zero. All this simply means that the choice of a reference point for potential energy is immaterial because we are always concerned with *differences* in potential energy, rather than with any absolute value of potential energy at a given point.

There is a certain arbitrariness in specifying kinetic energy also. In order to determine speed, and hence kinetic energy, we must specify a reference frame. The speed of a man sitting on a train is zero if we take the train as a reference frame, but it is not zero to an observer on the ground who sees the man move by with uniform velocity. The value of the kinetic energy depends on the reference frame used by the observer. Hence the important thing about mechanical energy E, which is the sum of the kinetic and the potential energies, is *not* its actual value during a given motion (this depends on the observer) but the fact that this value *does not change* during the motion for any particular observer when the forces are conservative.

Let us now calculate the potential energy in one-dimensional motion for two examples of conservative forces, the force of gravity for motions near the earth's surface and the elastic restoring force of an (ideal) stretched spring.

For the force of gravity we take the one-dimensional motion to be vertical, along the y-axis. We take the positive direction of the y-axis to be upward; the force of gravity is then in the negative y-direction, or downward. We have $F(y) = -mg$, a constant. The potential energy at position y is found from Eq. 8-9, or

$$U(y) = -\int_0^y F(y)\, dy + U(0) = -\int_0^y (-mg)\, dy + U(0) = mgy + U(0).$$

The potential energy can be taken as zero where $y = 0$, so that $U(0) = 0$, and

$$U(y) = mgy. \qquad (8\text{-}10)$$

The gravitational potential energy is then mgy. The relation $F(y) = -dU/dy$ (Eq. 8-7) is satisfied, for $-d(mgy)/dy = -mg$. We choose $y = 0$ to be at the surface of the earth for convenience, so that the gravitational potential energy is zero at the earth's surface and increases linearly with altitude y.

If we compare points y and $y = 0$, the conservation of kinetic plus potential energy, Eq. 8-6b, gives us the relation

$$\tfrac{1}{2}mv^2 + mgy = \tfrac{1}{2}mv_0^2.$$

This is mathematically equivalent to the well-known result (see Eq. 3-17),

$$v^2 = v_0^2 - 2gy.$$

If our particle moves from a height h_1 to a height h_2, we can use Eq. 8-6b to obtain

$$\tfrac{1}{2}mv_1^2 + mgh_1 = \tfrac{1}{2}mv_2^2 + mgh_2.$$

This result is equivalent to that of Example 5, Chapter 7. The total mechanical energy E is constant and is conserved during the motion, even though the kinetic energy and the potential energy vary as the configuration of the system (particle + earth) changes.

A second example of a conservative force is that exerted by an elastic spring on a body of mass m attached to it moving on a horizontal frictionless surface. If we take $x_0 = 0$ as the position of the end of the spring when unextended, the force exerted *on* the mass when the spring is stretched a distance x from its unextended length is $F = -kx$. The potential energy is obtained from Eq. 8-9,

$$U(x) = -\int_0^x F(x)\, dx + U(0) = -\int_0^x (-kx)\, dx + U(0).$$

If we choose $U(0) = 0$, the potential energy, as well as the force, is zero when the spring is unextended, and

$$U(x) = -\int_0^x (-kx)\, dx = \tfrac{1}{2}kx^2.$$

The result is the same whether we stretch or compress the spring, that is, whether x is plus or minus.

The relation $F(x) = -dU/dx$ (Eq. 8-7) is satisfied, for $-d(\tfrac{1}{2}kx^2)/dx = -kx$. The elastic potential energy of the spring is then

$$U(x) = \tfrac{1}{2}kx^2. \tag{8-11}$$

The body of mass m will undergo a motion in which the total energy E is conserved (Fig. 8-4). From Eq. 8-6b we have

$$\tfrac{1}{2}mv^2 + \tfrac{1}{2}kx^2 = \tfrac{1}{2}mv_0^2.$$

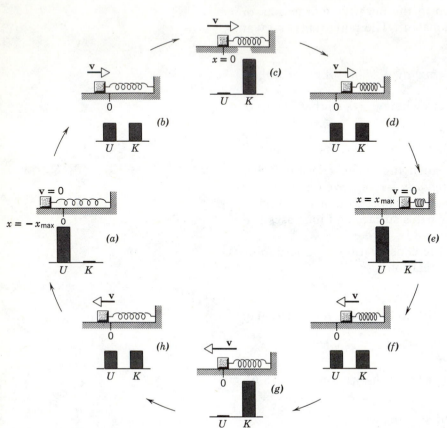

figure 8-4
A mass attached to a spring slides back and forth on a frictionless surface. The system is called a harmonic oscillator. The motion of the mass through one cycle is illustrated. Starting at the left (9 o'clock) the mass is in its extreme left position and momentarily at rest: $K = 0$. The spring is extended to its maximum length: $U = U_{max}$. (K and U are illustrated in the bar graphs below each sketch.) An eighth-cycle later (next drawing) the mass has gained kinetic energy, but the spring is no longer so elongated; K and U have here the same value, $K = U = U_{max}/2$. At the top this spring is neither elongated nor compressed and the speed is a maximum: $U = 0$, $K = K_{max} = U_{max}$. The cycle continues, with the total energy $E = K + U$ always the same: $E = K_{max} = U_{max}$. The harmonic oscillator will be analyzed more closely in Chapter 15.

Here v_0 is the speed of the particle for $x = 0$. Physically we achieve such a result by stretching the spring with an applied force to some position, x_m, and then releasing the spring. Notice that at $x = 0$ the energy of the system (particle + spring) is all kinetic. At $x = x_m$ (the maximum value of x), v must be zero, so that here the system energy is all potential. At $x = x_m$, we have

$$\tfrac{1}{2}kx_m^2 = \tfrac{1}{2}mv_0^2$$

or

$$x_m = \sqrt{m/k}\; v_0.$$

For positions between x_1 and x_2, Eq. 8-6b gives

$$\tfrac{1}{2}kx_1^2 + \tfrac{1}{2}mv_1^2 = \tfrac{1}{2}kx_2^2 + \tfrac{1}{2}mv_2^2.$$

We have seen that *the kinetic energy of a body is the work that a body can do by virtue of its motion.* We express the kinetic energy by the formula $K = \tfrac{1}{2}mv^2$. We cannot give a similar universal formula by which potential energy can be expressed. *The potential energy of a system of bodies is the work that the system of bodies can do by virtue of the relative position of its parts, that is, by virtue of its configuration.* In each case we must determine how much work the system can do in passing from one configuration to another and then take this as the difference in potential energy of the system between these two configurations.

The potential energy of the spring depends on the relative position of the parts of the spring. Work can be obtained by allowing the spring to return from its extended to its unextended length, during which time it exerts a force through a distance. If a mass is attached to the spring, as in our example, the mass will be accelerated by this force and the potential energy will be converted to kinetic energy. In the gravitational case an object occupies a position with respect to the earth. The potential energy is a property of the object and the earth, considered as a system of bodies. It is the relative position of the parts of this system that determines its potential energy. The potential energy is greater when the parts are far apart than when they are close together. The loss of potential energy is equal to the work done in this process. This work is converted into kinetic energy of the bodies. In our example we ignored the kinetic energy acquired by the earth itself as an object fell toward it. In principle, this object exerts a force on the earth and causes it to acquire an acceleration, relative to some inertial frame. The resulting change in speed, however, is extremely small, and in spite of the enormous mass of the earth, its additional kinetic energy is negligible compared to that acquired by the falling object. This will be proved in a later chapter. In other cases, such as in planetary motion where the masses of the objects in our system may be comparable, we cannot ignore any part of the system. In general, *potential energy* is not assigned to either body separately but *is considered a joint property of the system.*

EXAMPLE 1

What is the change in gravitational potential energy when a 1600-lb (= 7117 N) elevator moves from street level to the top of the Empire State Building, 1250 ft (= 381 m) above street level?

The gravitational potential energy of the system (elevator + earth) is $U = mgy$. Then

$$\Delta U = U_2 - U_1 = mg(y_2 - y_1).$$

But

$$mg = W = 1600 \text{ lb} \quad \text{and} \quad y_2 - y_1 = 1250 \text{ ft,}$$

so that

$$\Delta U = 1600 \times 1250 \text{ ft} \cdot \text{lb} = 2.00 \times 10^6 \text{ ft} \cdot \text{lb} = 2.71 \times 10^6 \text{ J.}$$

EXAMPLE 2

As an example of the simplicity and usefulness of the energy method of solving dynamical problems, consider the problem illustrated in Fig. 8-5. A block of mass m slides down a curved frictionless surface. The force exerted by the surface on the block is always normal to the surface and to the direction of the

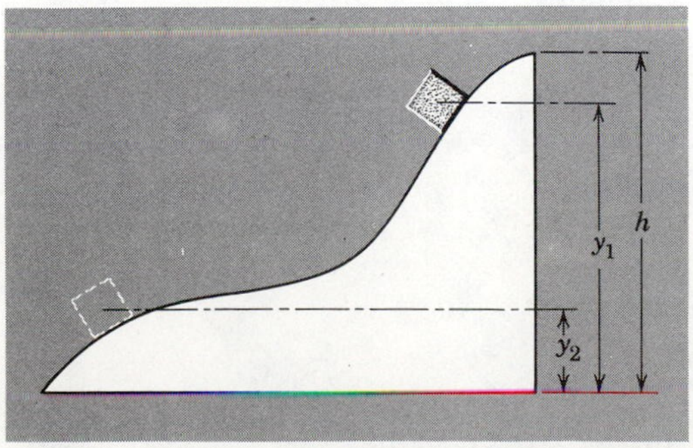

figure 8.5
Example 2. A block sliding down a frictionless curved surface.

motion of the block, so that this force does no work. Only the gravitational force does work on the block and that force is conservative. The mechanical energy E is, therefore, conserved and we can write at once

$$\tfrac{1}{2}mv_1{}^2 + mgy_1 = \tfrac{1}{2}mv_2{}^2 + mgy_2.$$

This gives

$$v_2{}^2 = v_1{}^2 + 2g(y_1 - y_2).$$

The speed at the bottom of the curved surface depends only on the initial speed and the change in vertical height but does not depend at all on the shape of the surface. In fact, if the block is initially at rest at $y_1 = h$, and if we set $y_2 = 0$, we obtain

$$v_2 = \sqrt{2gh}.$$

At this point you should recall the independence of path feature of work done by conservative forces and should be able to justify applying the ideas developed for one-dimensional motion to this two-dimensional example.

In this problem the value of the force depends on the slope of the surface at each point. Hence, the acceleration is not constant but is a function of position. To obtain the speed by starting with Newton's laws we would first have to determine the acceleration at each point and then integrate the acceleration over the path. We avoid all this labor by starting at once from the fact that the mechanical energy is constant throughout the motion.

The spring in a spring gun has a force constant of 4.0 lb/in. (= 7.0 N/cm). It is compressed 2.0 in. (= 5.1 cm) from its natural length, and a ball weighing 0.030 lb (= 0.133 N) is put into the barrel against it. Assuming no friction and a horizontal gun barrel, with what speed will the ball leave the gun when released?

The force is conservative so that mechanical energy is conserved in the process. The initial mechanical energy is the elastic potential energy of the spring, $\tfrac{1}{2}kx^2$, and the final mechanical energy is the kinetic energy of the ball, $\tfrac{1}{2}mv^2$. Hence,

$$\tfrac{1}{2}kx^2 = \tfrac{1}{2}mv^2$$

or

$$v = \sqrt{\frac{k}{m}}\, x = \sqrt{\frac{48\ \text{lb/ft}}{(0.030\ \text{lb})/(32\ \text{ft/s}^2)}}\ (\tfrac{1}{6}\ \text{ft}) = 38\ \text{ft/s}\ (= 11.6\ \text{m/s}).$$

EXAMPLE 3

Equation 8-6a gives the relation between coordinate and speed for one-dimensional motion when the force depends on position only. The force and the acceleration have been eliminated in arriving at this equation. To complete the solution of the dynamical problem we must eliminate the speed and determine position as a function of time.

We can do this in a formal way, as follows. From Eq. 8-6a we have

$$\tfrac{1}{2}mv^2 + U(x) = E.$$

Solving for v, we obtain

$$v = \frac{dx}{dt} = \sqrt{\frac{2}{m}\,[E - U(x)]}, \qquad (8\text{-}12)$$

or

$$\frac{dx}{\sqrt{\dfrac{2}{m}\,[E - U(x)]}} = dt.$$

Then the function $x(t)$ may be found by solving for x the equation

8-5
THE COMPLETE SOLUTION OF THE PROBLEM FOR ONE-DIMENSIONAL FORCES DEPENDING ON POSITION ONLY

$$\int_{x_0}^{x} \frac{dx}{\sqrt{\frac{2}{m}\left[E - U(x)\right]}} = \int_{t_0}^{t} dt = t - t_0. \qquad (8\text{-}13)$$

Here the particle is taken to be at x_0 at the time t_0 and E is the constant total energy. In applying this equation, the sign of the square root taken corresponds to whether **v** points in the positive or in the negative x-direction. When **v** changes direction during the motion it may be necessary to carry out the integration separately for each part of the motion.

Even when this integral cannot be evaluated or when the resulting equation cannot be solved to give an explicit solution for $x(t)$, the equation of energy conservation gives us useful information about the solution. For example, for a given total energy E, Eq. 8-12 tells us that the particle is restricted to those regions on the x-axis where $E > U(x)$. We cannot have an imaginary speed or a negative kinetic energy physically, so that $E - U(x)$ must be zero or greater. Furthermore, we can obtain a good qualitative description of the types of motion possible by plotting $U(x)$ versus x. This description depends on the fact that the speed is proportional to the square root of the difference between E and U.

For example, consider the potential energy function shown in Fig. 8-6. This could be thought of as an actual profile of a frictionless roller coaster, but in general it can represent the potential energy of a nongravitational system. Since we must have $E \geqq U(x)$ for real motion, the lowest total energy possible is E_0. At this value of the total energy, $E_0 = U$ and the kinetic energy must be zero. The particle must be at rest at the point x_0. At a slightly higher energy E_1 the particle can move between x_1 and x_2 only. As it moves from x_0 its speed decreases on approaching either x_1 or x_2. At x_1 or x_2 the particle stops and reverses its direction. These points x_1 and x_2 are, therefore, called *turning points* of the motion. At a total energy E_2 there are four turning points, and the particle can oscillate in either one of the two potential valleys. At the total energy E_3 there is only one turning point of the motion, at x_3. If the particle is initially moving in the negative x-direction, it will stop at x_3 and then move in the positive x-direction. It will speed up as U decreases and slow down as U increases. At energies above E_4 there are no turning points, and the particle will not reverse direction. Its speed will change according to the value of the potential at each point.

At a point where $U(x)$ has a minimum value, such as at $x = x_0$, the slope of the curve is zero so that the force is zero, that is, $F(x_0) = -(dU/dx)_{x=x_0} = 0$. A particle at rest at this point will remain at rest. Furthermore, if the particle is displaced slightly in either direction, the force, $F(x) = -dU/dx$, will tend to return it, and it will oscillate about the equilibrium point. This equilibrium point is, therefore, called a point of *stable equilibrium*.

At a point where $U(x)$ has a maximum value, such as at $x = x_4$, the slope of the curve is zero so that the force is again zero, that is, $F(x_4) = -(dU/dx)_{x=x_4} = 0$. A particle at rest at this point will remain at rest. However, if the particle is dis-

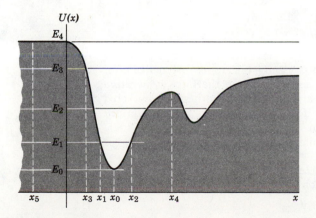

figure 8-6
A potential energy curve.

placed even the slightest distance from this point, the force, $F(x) = -dU/dx$, will tend to push it farther away from the equilibrium position. Such an equilibrium point is, therefore, called a point of *unstable equilibrium*.

In an interval in which $U(x)$ is constant, such as near $x = x_5$, the slope of the curve is zero so that the force is zero, that is, $F(x_5) = -(dU/dx)_{x=x_5} = 0$. Such an interval is called one of neutral equilibrium, since a particle can be displaced slightly without experiencing either a repelling or a restoring force.

From this it is clear that if we know the potential energy function for the region of x in which the body moves, we know a great deal about the motion of the body.

EXAMPLE 4

The potential energy function for the force between two atoms in a diatomic molecule can be expressed approximately as follows:

$$U(x) = \frac{a}{x^{12}} - \frac{b}{x^6}$$

where a and b are positive constants and x is the distance between atoms.

(a) At what values of x is $U(x)$ equal to zero? At what value of x is $U(x)$ a minimum?

In Fig. 8-7a we show $U(x)$ versus x. The values of x at which $U(x)$ equals zero are found from

$$\frac{a}{x^{12}} - \frac{b}{x^6} = 0.$$

Hence

$$x^6 = \frac{a}{b} \qquad x = \sqrt[6]{\frac{a}{b}}.$$

$U(x)$ also becomes zero as $x \to \infty$ [see figure or put $x = \infty$ into equation for $U(x)$], so that $x = \infty$ is also a solution.

The value of x at which $U(x)$ is a minimum is found from

$$\frac{d}{dx}U(x) = 0.$$

That is,

$$\frac{-12a}{x^{13}} + \frac{6b}{x^7} = 0$$

or

$$x^6 = \frac{2a}{b} \qquad x = \sqrt[6]{\frac{2a}{b}}.$$

(b) Determine the force between the atoms.
From Eq. 8-7

$$F(x) = -\frac{d}{dx}U(x),$$

$$F = \frac{-d}{dx}\left(\frac{a}{x^{12}} - \frac{b}{x^6}\right) = \frac{12a}{x^{13}} - \frac{6b}{x^7}.$$

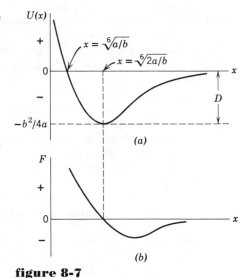

figure 8-7
Example 4. (a) The potential energy and (b) the force between two atoms in a diatomic molecule as a function of the distance x between atoms.

We plot the force as a function of the separation between the atoms in Fig. 8-7b. When the force is positive (from $x = 0$ to $x = \sqrt[6]{2a/b}$), the atoms are repelled from one another (force directed toward increasing x). When the force is negative (from $x = \sqrt[6]{2a/b}$ to $x = \infty$), the atoms are attracted to one another (force directed toward decreasing x). At $x = \sqrt[6]{2a/b}$ the force is zero; this is the equilibrium point and is a point of stable equilibrium.

(c) Assume that one of the atoms remains at rest and that the other moves along x. Describe the possible motions.

From the analysis of this section it is clear that the atom oscillates about the equilibrium separation at $x = \sqrt[6]{2a/b}$, much as a particle sliding up and down the frictionless hills of the potential valley.

(d) The energy needed to break up the molecule into separate atoms is called the dissociation energy. What is the dissociation energy of the molecule?

If one atom has enough kinetic energy to get over the potential hill, it will no longer be bound to the other atom. Hence, the dissociation energy D equals the change in potential energy from the minimum value at $x = \sqrt[6]{2a/b}$ to the value at $x = \infty$. This is simply

$$U(x = \infty) - U\left(x = \sqrt[6]{\frac{2a}{b}}\right) = 0 - \left(\frac{a}{4a^2/b^2} - \frac{b}{2a/b}\right) = \frac{b^2}{4a}.$$

If the kinetic energy at the equilibrium position is equal to or greater than this value, the molecule will dissociate.

So far we have discussed potential energy and energy conservation for one-dimensional systems in which the force was directed along the line of motion. We can easily generalize the discussion to three-dimensional motion.

If the work done by the force **F** depends only on the end points of the motion and is independent of the path taken between these points, the force is conservative. We define the potential energy U by analogy with the one-dimensional system and find that it is a function of three space coordinates, that is, $U = U(x,y,z)$. Again we obtain an expression for the conservation of mechanical energy.

The generalization of Eq. 8-5b to motion in three dimensions is

$$\Delta U = -\int_{x_0}^{x} F_x \, dx - \int_{y_0}^{y} F_y \, dy - \int_{z_0}^{z} F_z \, dz \qquad (8\text{-}5c)$$

or, more compactly in vector notation,

$$\Delta U = -\int_{\mathbf{r}_0}^{\mathbf{r}} \mathbf{F}(\mathbf{r}) \cdot d\mathbf{r} \qquad (8\text{-}5d)$$

in which ΔU is the change in potential energy for the system as the particle moves from the point (x_0, y_0, z_0), described by the position vector $\mathbf{r}_0$, to the point (x, y, z), described by the position vector $\mathbf{r}$. F_x, F_y, and F_z are the components of the conservative force $\mathbf{F}(\mathbf{r}) = \mathbf{F}(x, y, z)$.

The generalization of Eq. 8-6b to three-dimensional motion is

$$\tfrac{1}{2}mv^2 + U(x,y,z) = \tfrac{1}{2}mv_0^2 + U(x_0, y_0, z_0) \qquad (8\text{-}6c)$$

which can be written in vector notation as

$$\tfrac{1}{2}m\mathbf{v} \cdot \mathbf{v} + U(\mathbf{r}) = \tfrac{1}{2}m\mathbf{v}_0 \cdot \mathbf{v}_0 + U(\mathbf{r}_0) \qquad (8\text{-}6d)$$

in which $\mathbf{v} \cdot \mathbf{v} = v_x^2 + v_y^2 + v_z^2 = v^2$ and $\mathbf{v}_0 \cdot \mathbf{v}_0 = v_{0x}^2 + v_{0y}^2 + v_{0z}^2 = v_0^2$. Likewise Eq. 8-6a becomes

$$\tfrac{1}{2}mv^2 + U(x,y,z) = E$$

in three dimensions, E being the constant total mechanical energy.

Finally, the generalization of Eq. 8-7 to three dimensions is

$$\mathbf{F}(\mathbf{r}) = -\mathbf{i}\frac{\partial U}{\partial x} - \mathbf{j}\frac{\partial U}{\partial y} - \mathbf{k}\frac{\partial U}{\partial z}.$$

If we substitute this expression for **F** into Eq. 8-5d, we again obtain an identity. In vector language the conservative force **F** is said to be the negative of the *gradient* of the potential energy $U(x,y,z)$.

You can show that all these expressions reduce to the correct one-dimensional equations for motion along the x-axis.

8-6
TWO- AND THREE-DIMENSIONAL CONSERVATIVE SYSTEMS

Consider the single pendulum, Section 7-4, Fig. 7-8a. The motion of the system is in the x-y plane, that is, it is a two-dimensional motion. The tension in the cord is always at right angles to the motion of the suspended particle, so that this force does no work on the particle. If the pendulum is displaced through some angle and is then released, only the gravitational force of attraction exerted on the particle by the earth does work on it. Since this force is conservative, we can use the equation of energy conservation in two dimensions,

$$\tfrac{1}{2}mv^2 + U(x,y) = E.$$

But $U(x,y)$ equals mgy, where y is taken as zero at the lowest point of the arc ($\phi = 0°$). Then,

$$\tfrac{1}{2}mv^2 + mgy = E.$$

The particle is pulled through an angle ϕ_0 before being released. The potential energy there is mgh, corresponding to a height $y = h$ above the reference point. At the release point ($\phi = \phi_0$) the speed and the kinetic energy are zero so that the potential energy equals the total mechanical energy at that point.

Hence,

$$E = mgh$$

and

$$\tfrac{1}{2}mv^2 + mgy = mgh,$$

or

$$\tfrac{1}{2}mv^2 = mg(h - y).$$

The maximum speed occurs at $y = 0$, where $v = \sqrt{2gh}$.
The minimum speed occurs at $y = h$, where $v = 0$.
At $y = 0$ the energy is entirely kinetic, the potential energy being zero.
At $y = h$ the energy is entirely potential, the kinetic energy being zero.
At intermediate positions the energy is partly kinetic and partly potential.

Notice that $U \le E$ at all points of the motion; the pendulum cannot rise higher than $y = h$, its initial release point.

EXAMPLE 5

So far we have considered only the action of a single conservative force on a particle. Starting from the work-energy theorem, or

$$W_1 + W_2 + \cdots + W_n = \Delta K \qquad (8\text{-}2)$$

we saw that, if only one force, say $\mathbf{F}_1$, was acting and if it was conservative, then we could represent the work W_1 that it did on the particle as a decrease in potential energy ΔU_1 of the system (see Eq. 8-5a), or

$$W_1 = -\Delta U_1.$$

Combining this with Eq. 8-2 yielded

$$\Delta K + \Delta U_1 = 0.$$

If several conservative forces such as gravity, an elastic spring force, an electrostatic force, etc., are acting, we can easily extend these two equations to

$$\Sigma\, W_c = -\Sigma\, \Delta U \qquad (8\text{-}14a)$$

and

$$\Delta K + \Sigma\, \Delta U = 0 \qquad (8\text{-}14b)$$

in which $\Sigma\, W_c$ is the sum of the work done by the various (conservative) forces and the ΔU's are the changes in the potential energy of the system associated with these forces. The quantity on the left of Eq. 8-14b is simply ΔE, the change in the total mechanical energy, for the case in

8-7
NONCONSERVATIVE FORCES

which several conservative forces are acting on a particle. We can write this equation then as

$$\Delta E = 0 \qquad \text{(conservative forces)}, \qquad (8\text{-}15)$$

which tells us that, as the system configuration changes the total mechanical energy E for the system remains constant.

Let us now suppose that, in addition to the several conservative forces, a single nonconservative force due to friction acts on the particle. We can then write Eq. 8-2 as

$$W_f + \Sigma\, W_c = \Delta K,$$

where $\Sigma\, W_c$ is again the sum of the work done by the conservative forces and W_f is the work done by friction. We can recast this (see Eq. 8-14a) as

$$\Delta K + \Sigma\, \Delta U = W_f. \qquad (8\text{-}16)$$

Equation 8-16 shows that, if a frictional force acts, the total mechanical energy is *not* constant, but changes by the amount of work done by the frictional force. We can write Eq. 8-16 as

$$\Delta E = E - E_0 = W_f. \qquad (8\text{-}17)$$

Since W_f, the work done by friction *on* the particle, is always negative, it follows from Eq. 8-17 that the final mechanical energy $E\ (= K + \Sigma\, U)$ is less than the initial mechanical energy $E_0\ (= K_0 + \Sigma\, U_0)$.

Friction is an example of a dissipative force, one which does negative work on a body and tends to diminish the total mechanical energy of the system. If we had used another nonconservative force, then W_f in Eqs. 8-16 and 8-17 would be replaced by a term W_{nc}, showing again that the total mechanical energy E of the system is *not* constant, but changes by the amount of work done by the nonconservative force. Hence, *only when there are no nonconservative forces, or when we neglect the work they do, can we assume conservation of mechanical energy.*

What happened to the "lost" mechanical energy in the case of friction? It is transformed into internal energy U_{int}, resulting in a temperature rise. The internal energy developed is exactly equal to the mechanical energy dissipated. We shall have much more to say about internal energy in later chapters.

Just as the work done by a conservative force *on* an object is the negative of the potential energy gain, so the work done by a frictional force *on* an object is the negative of the internal energy gained. In other words, the internal energy produced is equal to the work done *by* the object. Then we can replace W_f in Eq. 8-17 by $-U_{int}$, in which U_{int} is the internal energy produced, or

$$\Delta E + U_{int} = 0. \qquad (8\text{-}18)$$

This asserts that there is no change in the sum of the mechanical and internal energy of the system when only conservative and frictional forces act on the system. Writing this equation as $U_{int} = -\Delta E$ we see that the loss of mechanical energy equals the gain in internal energy.

An object with an initial velocity v_0 of 14 m/s falls from a height of 240 m and buries itself in 0.20 m of sand. The mass of the body is 1.0 kg. Find the average resistive force exerted by the sand on the body. Neglect air resistance and solve the problem by considerations of work and energy.

EXAMPLE 6

The kinetic energy of the body just as it enters the sand is

$$K = \tfrac{1}{2}mv_0^2 + mgh,$$

where m is the mass of the body and h is the height of fall.

Also, from the work-energy principle, we have (approximately)

$$K = \overline{F}\, s,$$

where $\overline{F}$ is the average resistive force and s is the distance of penetration into the ground.

Equating and solving for $\overline{F}$ gives

$$\overline{F} = \frac{mv_0^2}{2\,s} + \frac{mgh}{s}$$

$$= \frac{(1.0 \text{ kg})(14 \text{ m/s})^2}{2(0.20 \text{ m})} + \frac{(1.0 \text{ kg})(9.8 \text{ m/s}^2)(240 \text{ m})}{(0.20 \text{ m})}$$

$$= 12{,}250 \text{ N}.$$

For what equations in this chapter are the first two equations of this example special cases?

What error do we make by neglecting (in comparison to h) the extra distance of fall s before the object is brought to rest? Show that this is equivalent to neglecting mg in comparison to $\overline{F}$ in arriving at the resultant force to be used in the work-energy theorem. Such terms are not always negligible in practice (see Problem 19, for example).

EXAMPLE 7

A 44-N block is thrust up a 30° inclined plane with an initial speed of 5.0 m/s. It is found to travel 1.5 m along the plane, stop, and slide back to the bottom. Compute the force of friction $\mathbf{f}$ (assumed to have a constant magnitude) acting on the block and find the speed v of the block when it returns to the bottom of the inclined plane.

Consider first the upward motion. At the top, where this motion ends,

$$E = K + U = 0 + (44 \text{ N})(1.5 \text{ m})(\sin 30°) = 33 \text{ J}.$$

At the bottom, where this motion begins,

$$E_0 = K_0 + U_0 = \tfrac{1}{2}\left(\frac{44 \text{ N}}{9.8 \text{ m/s}^2}\right)(5.0 \text{ m/s})^2 + 0 = 57 \text{ J}.$$

But

$$W_f = -fs = -f(1.5 \text{ m})$$

and

$$E - E_0 = W_f,$$

so that

$$33 \text{ J} - 57 \text{ J} = -f(1.5 \text{ m})$$

and

$$f = 16 \text{ N}.$$

Now consider the downward motion. The block returns to the bottom of the inclined plane with a speed v. Then, at the bottom, where this motion ends,

$$E = K + U = \tfrac{1}{2}\left(\frac{44 \text{ N}}{9.8 \text{ m/s}^2}\right)v^2 + 0 = \left(\frac{22}{9.8} \text{ N s}^2/\text{m}\right)v^2.$$

At the top, where this motion begins,

$$E_0 = K_0 + U_0 = 0 + (44 \text{ N})(1.5 \text{ m})(\sin 30°) = 33 \text{ J}.$$

But

$$W_f = -(16 \text{ N})(1.5 \text{ m}) = -24 \text{ J}$$

and

$$E - E_0 = W_f,$$

so that

$$\left(\frac{22}{9.8} \text{ N s}^2/\text{m}\right)v^2 - 33 \text{ J} = -24 \text{ J}$$

and

$$v = 2.0 \text{ m/s}.$$

We can extend the discussion of the previous section by considering not only conservative forces and the force of friction but also other, nonfrictional, nonconservative forces. We can regroup the work-energy theorem (Eq. 8-2)

8-8
THE CONSERVATION OF ENERGY

$$W_1 + W_2 + \cdots + W_n = \Delta K$$

as

$$\Sigma W_c + W_f + \Sigma W_{nc} = \Delta K \qquad (8\text{-}19)$$

in which ΣW_c is the total work done on the particle by conservative forces, W_f is the work done by friction, and ΣW_{nc} is the total work done by nonconservative forces other than friction. We have seen that each conservative force can be associated with a potential energy and that friction is associated with internal energy, or

$$\Sigma W_c = -\Sigma \Delta U$$

and

$$W_f = -U_{int,}$$

so that Eq. 8-19 becomes

$$\Sigma W_{nc} = \Delta K + \Sigma \Delta U + U_{int}.$$

Now whatever the W_{nc} are, it has always been possible to find new forms of energy which corresponds to this work. We can then represent ΣW_{nc} by another change of energy term on the right-hand side of the equation, with the result that we can always write the work-energy theorem as

$$0 = \Delta K + \Sigma \Delta U + U_{int} + (\text{change in other forms of energy}).$$

In other words, the total energy—kinetic plus potential plus internal plus all other forms—does not change. *Energy may be transformed from one kind to another, but it cannot be created or destroyed; the total energy is constant.*

This statement is a generalization from our experience, so far not contradicted by observation of nature. It is called the *principle of the conservation of energy.* Often in the history of physics this principle seemed to fail. But its apparent failure stimulated the search for the reasons. Experimentalists searched for physical phenomena besides motion that accompany the forces of interaction between bodies. Such phenomena have always been found. With work done against friction, internal energy is produced; in other interactions energy in the form of sound, light, electricity, etc., may be produced. Hence the concept of energy was generalized to include forms other than kinetic and potential energy of directly observable bodies. This procedure, which relates the mechanics of bodies observed to be in motion to phenomena which are not mechanical or in which motion is not directly detected, has linked mechanics to all other areas of physics. The energy concept now permeates all of physical science and has become one of the unifying ideas of physics.*

* See for example, "Concept of Energy in Mechanics," by R. B. Lindsay in *The Scientific Monthly*, October 1957.

In subsequent chapters we shall study various transformations of energy — from mechanical to internal, mechanical to electrical, nuclear to internal, etc. It is during such transformations that we measure the energy changes in terms of work, for it is during these transformations that forces arise and do work.

Although the principle of the conservation of kinetic plus potential energy is often useful, we see that it is a restricted case of the more general principle of the conservation of energy. Kinetic and potential energy are conserved only when conservative forces act. Total energy is *always* conserved.

8-9
MASS AND ENERGY

One of the great conservation laws of science had been the law of conservation of matter. From a philosophical point of view an early statement of this general principle was given by the Roman poet Lucretius, a contemporary of Julius Caesar, in his celebrated work *De Rerum Natura*. Lucretius wrote "Things cannot be born from nothing, cannot when begotten be brought back to nothing." It was a long time before this concept was established as a firm scientific principle. The principal experimental contribution was made by Antoine Lavoisier (1743–1794), regarded by many as the father of modern chemistry. He wrote in 1789 "We must lay it down as an incontestable axiom, that in all the operations of art and nature, nothing is created; an equal quantity of matter exists both before and after the experiment . . . and nothing takes place beyond changes and modifications in the combinations of these elements."

This principle, subsequently called the conservation of mass, proved extremely fruitful in chemistry and physics. Serious doubts as to the general validity of this principle were raised by Albert Einstein in his papers introducing the theory of relativity. Subsequent experiments on fast-moving electrons and on nuclear matter confirmed his conclusions.

Einstein's findings suggested that, if certain physical laws were to be retained, the mass of a particle had to be redefined as

$$m = \frac{m_0}{\sqrt{1 - v^2/c^2}}. \tag{8-20}$$

Here m_0 is the mass of the particle when at rest with respect to the observer, called the *rest mass*; m is the mass of the particle measured as it moves at a speed v relative to the observer; and c is the speed of light, having a constant value of approximately 3×10^8 m/s. Experimental checks of this equation can be made, for example, by deflecting high-speed electrons in magnetic fields and measuring the radii of curvature of their path. The paths are circular and the magnetic force a centripetal one ($F = mv^2/r$, F and v being known). At ordinary speeds the difference between m and m_0 is too small to be detectable. Electrons, however, can be emitted from radioactive nuclei with speeds greater than nine-tenths that of light. In such cases the results (Fig. 8-8 shows early data) confirm Eq. 8-20.

It is convenient to let the ratio v/c be represented by β. Then Eq. 8-20 becomes

$$m = m_0(1 - \beta^2)^{-1/2}.$$

To find the kinetic energy of a body, we compute the work done by the resultant force in setting the body in motion. In Section 7-5 we obtained

$$K = \int_0^v \mathbf{F} \cdot d\mathbf{r} = \tfrac{1}{2}m_0 v^2$$

for kinetic energy, when we assumed a constant mass m_0. Suppose now instead we take into account the variation of mass with speed and use $m = m_0(1 - \beta^2)^{-1/2}$ in our previous equation. We find (Problem 29, Chapter 9) that the kinetic energy is no longer given by $\tfrac{1}{2}m_0 v^2$ but instead is

$$K = mc^2 - m_0 c^2 = (m - m_0)c^2 = \Delta m c^2. \qquad (8\text{-}21)$$

The kinetic energy of a particle is, therefore, the product of c^2 and the *increase in mass* Δm resulting from the motion.

Now, at small speeds we expect the relativistic result to agree with the classical result. By the binomial theorem we can expand $(1 - \beta^2)^{-1/2}$ as

$$(1 - \beta^2)^{-1/2} = 1 + \tfrac{1}{2}\beta^2 + \tfrac{3}{8}\beta^4 + \tfrac{5}{16}\beta^6 + \cdots.$$

At small speeds $\beta = v/c \ll 1$ so that all terms beyond β^2 are negligible. Then

$$K = (m - m_0)c^2 = m_0 c^2[(1 - \beta^2)^{-1/2} - 1]$$
$$= m_0 c^2(1 + \tfrac{1}{2}\beta^2 + \cdots - 1) \cong \tfrac{1}{2}m_0 c^2 \beta^2 = \tfrac{1}{2}m_0 v^2,$$

which is the classical result. Notice also that when K equals zero, $m = m_0$ as expected.

The basic idea that energy is equivalent to mass can be extended to include energies other than kinetic. For example, when we compress a spring and give it elastic potential energy U, its mass increases from m_0 to $m_0 + U/c^2$. When we add heat energy in amount Q to an object, its mass increases by an amount Δm, where Δm is Q/c^2. We arrive at a principle of *equivalence of mass and energy*: For every unit of energy E of *any* kind supplied to a material object, the mass of the object increases by an amount

$$\Delta m = E/c^2.$$

This is the famous Einstein formula

$$E = \Delta m c^2. \qquad (8\text{-}22)$$

In fact, since rest mass itself is just one form of energy, we can now assert that a body at rest has an energy $m_0 c^2$ by virtue of its rest mass.

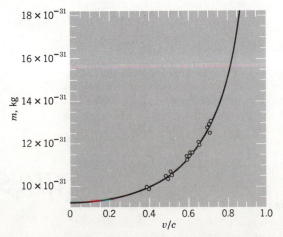

figure 8-8
The way an electron's mass increases as its speed relative to the observer increases. The solid line is a plot of $m = m_0(1 - v^2/c^2)^{-1/2}$, and the circles are adapted from experimental values obtained by Bucherer and Neumann in 1914. The curve tends toward infinity as $v \rightarrow c$.

This is called its *rest energy*. If we now consider a closed system, the principle of the conservation of energy, as generalized by Einstein, becomes

$$\Sigma\,(m_0c^2 + \mathscr{E}) = \text{constant}$$

or

$$\Delta(\Sigma\,m_0c^2 + \Sigma\,\mathscr{E}) = 0,$$

where $\Sigma\,m_0c^2$ is the total rest energy and $\Sigma\,\mathscr{E}$ is the total energy of *all other* kinds. As Einstein wrote, "Pre-relativity physics contains two conservation laws of fundamental importance, namely the law of conservation of energy and the law of conservation of mass; these two appear there as completely independent of each other. Through relativity theory they melt together into *one* principle."

Because the factor c^2 is so large, we would not expect to be able to detect changes in mass in ordinary mechanical experiments. A change in mass of 1 g would require an energy of 9×10^{13} joules. But when the mass of a particle is quite small to begin with and high energies can be imparted to it, the relative change in mass may be readily noticeable. This is true in nuclear phenomena, and it is in this realm that classical mechanics breaks down and relativistic mechanics receives its most striking verification.

A beautiful example of exchange of energy between rest mass and other forms is given by the phenomenon of pair annihilation or pair production. In this phenomenon an electron and a positron, elementary material particles differing only in the sign of their electric charge, can combine and literally disappear. In their place we find high-energy radiation, called γ-radiation, whose radiant energy is exactly equal to the rest mass plus kinetic energies of the disappearing particles. The process is reversible, so that a materialization of rest mass from radiant energy can occur when a high enough energy γ-ray, under proper conditions, disappears; in its place appears a positron-electron pair whose total energy (rest mass + kinetic) is equal to the radiant energy lost.

EXAMPLE 8

Consider a quantitative example. On the atomic mass scale the unit of mass is 1.66×10^{-27} kg approximately. On this scale the rest mass of the proton (the nucleus of a hydrogen atom) is 1.00731 and the rest mass of the neutron (a neutral particle, one of the constituents of all nuclei except hydrogen) is 1.00867. A deuteron (the nucleus of heavy hydrogen) is known to consist of a neutron and a proton; the rest mass of the deuteron is found to be 2.01360. The rest mass of the deuteron is *less than* the combined rest masses of neutron and proton by 0.00238 atomic mass units. The discrepancy is equivalent in energy to

$$E = \Delta mc^2 = (0.00238 \times 1.66 \times 10^{-27}\text{ kg})(3.00 \times 10^8\text{ m/s})^2$$

$$= 3.57 \times 10^{-13}\text{ joules} = 2.22 \times 10^6\text{ eV}.$$

When a neutron and a proton combine to form a deuteron, this exact amount of energy is given off in the form of γ-radiation. Similarly, it is found that the same amount of energy must be *added* to the deuteron to break it up into a proton and a neutron. This energy is therefore called the *binding energy* of the deuteron.

---

questions

1. Mountain roads rarely go straight up the slope but wind up gradually. Explain why.

2. Is any work being done on a car moving with constant speed along a straight level road?

3. An automobile of mass m and speed v is moving along a highway. The

driver jams on the brakes and the car skids to a halt. In what form does the lost kinetic energy of the car appear?

4. In the above question, assume that the driver "pumps" the brakes in such a way that there is no skidding or sliding. In this case, in what form does the lost kinetic energy of the car appear?

5. An automobile accelerates from rest to a speed v, under conditions such that no slipping of the driving wheels occurs. Where does the kinetic energy of the car come from? In particular, is it true that it is provided by the work done on the car by the (static) frictional force exerted by the road on the car?

6. If it takes no work to hold up a heavy object, why is it tiring?

7. What happens to the potential energy an elevator loses in coming down from the top of a building to a stop at the ground floor?

8. In Example 2 (see Fig. 8-5) we asserted that the speed at the bottom does not depend at all on the *shape* of the surface. Would this still be true if friction were present?

9. Give physical examples of unstable equilibrium. Of neutral equilibrium. Of stable equilibrium.

10. Explain, using work and energy ideas, how a child pumps a swing up to large amplitudes from a rest position. (See "How to Make a Swing Go" by R. V. Hesheth, *Physics Education*, July 1975.)

11. A swinging pendulum eventually comes to rest. Is this a violation of the law of conservation of energy?

12. A scientific article ("The Energetic Cost of Moving About" by V. A. Tucker, *American Scientist* July–August 1975) asserts that walking and running are extremely inefficient forms of locomotion and that much greater efficiency is achieved by birds, fish, and bicyclists. Can you suggest an explanation?

13. Two disks are connected by a stiff spring. Can one press the upper disk down enough so that when it is released it will spring back and raise the lower disk off the table (see Fig. 8-9)? Can mechanical energy be conserved in such a case?

14. In the case of work done against friction, the amount of heat energy generated is independent of the velocity (or inertial reference frame) of the observer. That is, different observers would assign the same quantity of mechanical energy transformed into heat energy due to friction. How can this be explained, considering that such observers measure different quantities of total work done and different changes in kinetic energy in general (see Problem 21, Chapter 7)?

15. Must *all* nonconservative forces be dissipative, as friction is? Could $\Sigma\, W_{nc}$ be greater than zero, in principle?

16. An object is dropped and observed to bounce to one and one-half times its original height. What conclusion can you draw from this observation?

17. The driver of an automobile traveling at speed v suddenly sees a brick wall at a distance d directly in front of him. To avoid crashing, is it better for him to slam on the brakes or to turn the car sharply away from the wall? (Hint: Consider the force required in each case.)

18. A spring is kept compressed by tying its ends together tightly. It is then placed in acid and dissolves. What happened to its stored potential energy?

figure 8-9
Question 13

SECTION 8-3

1. A body moving along the x-axis is subject to a force repelling it from the origin, given by $F = kx$. (a) Find the potential energy function $U(x)$ for the motion and write down the conservation of energy condition. (b) Describe the motion of the system and show that this is the kind of motion we would expect near a point of unstable equilibrium.   *Answer:* $(a) - kx^2/2$.

2. If the magnitude of the force of attraction between a particle of mass m_1 and one of mass m_2 is given by

problems

$$F = k\frac{m_1m_2}{x^2}$$

where k is a constant and x is the distance between the particles, find (a) the potential energy function and (b) the work required to increase the separation of the masses from $x = x_1$ to $x = x_1 + d$.

3. A chain is held on a frictionless table with one-fifth of its length hanging over the edge. If the chain has a length l and a mass m, how much work is required to pull the hanging part back on the table?   *Answer: mgl/50.*

SECTION 8-4

4. A 2.0-g (weight $mg = 0.071$-oz) penny is pushed down on a vertical spring, compressing the spring by 1.0 cm (0.39 in.). The force constant of the spring is 40 N/m (2.7 lb/ft). How far above this original position will the penny fly if it is released?

5. A 200-lb man jumps out a window into a fire net 30 ft below. The net stretches 6.0 ft before bringing him to rest and tossing him back into the air. What is the potential energy of the stretched net if no energy is dissipated by nonconservative forces?   *Answer: 7200 ft · lb.*

6. A 2.0-kg (0.14-slug) block is dropped from a height of 0.40 m (1.3 ft) onto a spring of force constant $k = 1960$ N/m (134 lb/ft). Find the maximum distance the spring will be compressed (neglect friction).

7. Show that for the same initial speed v_0, the speed v of a projectile will be the same at all points at the same elevation, regardless of the angle of projection.

8. A certain peculiar spring is found *not* to conform to Hooke's law. The force (in newtons) it exerts when stretched a distance x (in meters) is found to have magnitude $52.8x + 38.4x^2$ in the direction opposing the stretch. (a) Compute the total work required to stretch the spring from $x = 0.50$ to $x = 1.00$ m. (b) With one end of the spring fixed, a particle of mass 2.17 kg is attached to the other end of the spring when it is extended by an amount $x = 1.00$ m. If the particle is then released from rest, compute its *speed* at the instant the spring has returned to the configuration in which the extension is $x = 0.50$ m. (c) Is the force exerted by the spring conservative or nonconservative? Explain.

9. It is claimed that large trees can evaporate as much as 1 ton (910 kg mass) of water per day. (a) Assuming the average rise of water to be 30 ft (9.1 m) from the ground, how much energy (in kW · h) must be supplied to do this? (b) What is the average power if the evaporation is assumed to occur during 12 hours of the day?   *Answer: (a) 2.3×10^{-2} kW · h. (b) 1.9 W.*

10. An object is attached to a vertical spring and slowly lowered to its equilibrium position. This stretches the spring by an amount d. If the same object is attached to the same vertical spring but permitted to fall instead, through what maximum distance does it stretch the spring?

11. A body falls from rest from a height h. Determine the kinetic energy and the potential energy of the body as a function (a) of time and (b) of height. Graph the expressions and show that their sum, the total energy, is constant in each case.

SECTION 8-5

12. A particle moves along a line in a region in which its potential energy varies as in Fig. 8-10. (a) Sketch, with the same scale on the abscissa, the force $F(x)$ acting on the particle. Indicate on the graph the approximate numerical scale for $F(x)$. (b) If the particle has a constant total energy of 4.0 joules, sketch the graph of its kinetic energy. Indicate the numerical scale on the $K(x)$ axis.

13. An α-particle (helium atom nucleus) in a large nucleus is bound by a potential like that shown in Fig. 8-11. (a) Construct a function of x, which has

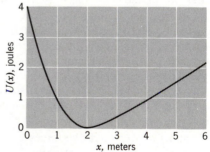

figure 8-10
Problem 12

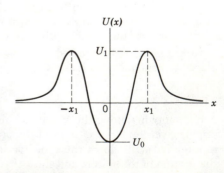

figure 8-11
Problem 13

this general shape, with a minimum value U_0 at $x=0$ and a maximum value U_1 at $x = x_1$ and $x = -x_1$. (b) Determine the force between the α-particle and the nucleus as a function of x. (c) Describe the possible motions.

SECTION 8-6

14. The string in Fig. 8-12 has a length $l = 4.0$ ft. When the ball is released, it will swing down the dotted arc. How fast will it be going when it reaches the lowest point in its swing?

15. A frictionless roller coaster of mass m starts at point A with speed v_0, as in Fig. 8-13. Assume that the roller coaster can be considered as a particle and that it always remains on the track. (a) What will be the speed of the roller coaster at points B and C? (b) What constant deceleration is required to stop it at E if the brakes are applied at point D?
   Answer: (a) $v_B = v_0$; $v_C = \sqrt{v_0{}^2 + gh}$. (b) $(v_0{}^2 + 2gh)/2L$.

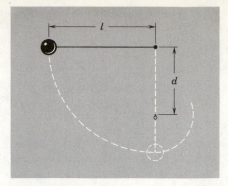

figure 8-12
Problems 14, 27, 30

figure 8-13
Problem 15

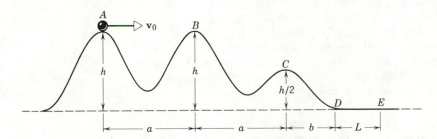

16. What force corresponds to a potential energy $U = -ax^2 + bxy + z$?

17. The potential energy corresponding to a certain two-dimensional force field is given by $U(x,y) = \frac{1}{2}k(x^2 + y^2)$. (a) Derive F_x and F_y and describe the vector force at each point in terms of its coordinates x and y. (b) Derive F_r and F_θ and describe the vector force at each point in terms of the polar coordinates r and θ of the point. (c) Can you think of a physical model of such a force?
   Answer: (a) $F_x = -kx$; $F_y = -ky$; **F** points toward the origin. (b) $F_r = -kr$, $F_\theta = 0$.

18. The so-called Yukawa potential

$$U(r) = -\frac{r_0}{r}\, U_0 e^{-r/r_0}$$

gives a fairly accurate description of the interaction between nucleons (i.e., neutrons and protons, the constituents of the nucleus). The constant r_0 is about 1.5×10^{-15} meter and the constant U_0 is about 50 MeV. (a) Find the corresponding expression for the force of attraction. (b) To show the short range of this force, compute the ratio of the force at $r = 2r_0$, $4r_0$, and $10r_0$ to the force at $r = r_0$.

19. An ideal massless spring S can be compressed 1.0 m by a force of 100 N. This same spring is placed at the bottom of a frictionless inclined plane which makes an angle of $\theta = 30°$ with the horizontal (see Fig. 8-14). A 10-kg mass M is released from the top of the incline and is brought to rest momentarily after compressing the spring 2.0 m. (a) Through what distance does the mass slide before coming to rest? (b) What is the speed of the mass just before it reaches the spring?   *Answer:* (a) 4.1 m. (b) 4.5 m/s.

20. The magnitude of the force of attraction between the positively charged nucleus and the negatively charged electron in the hydrogen atom is given by

$$F = k\frac{e^2}{r^2}$$

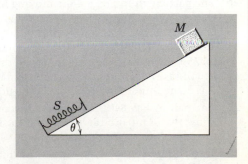

figure 8-14
Problem 19

where e is the charge of the electron, k is a constant, and r is the separation between electron and nucleus. Assume that the nucleus is fixed. The electron, initially moving in a circle of radius R_1 about the nucleus, jumps suddenly into a circular orbit of smaller radius R_2. (a) Calculate the change in kinetic energy of the electron, using Newton's second law. (b) Using the relation between force and potential energy, calculate the change in potential energy of the atom. (c) Show by how much the mechanical energy of the atom has changed in this process. (This energy is given off in the form of radiation.)

21. A light rigid rod of length l has a mass m attached to its end, forming a simple pendulum. It is inverted and then released. What are (a) the speed v at its lowest point and (b) the tension T in the suspension at that instant? (c) The same pendulum is next put in a horizontal position and released from rest. At what angle from the vertical will the tension in the suspension equal the weight in magnitude?   *Answer: (a) $2\sqrt{gl}$. (b) 5 mg. (c) 71°.*

22. A simple pendulum of length l, the mass of whose bob is m, is observed to have a speed v_0 when the cord makes an angle θ_0 with the vertical $(0 < \theta_0 < \pi/2)$, as in Fig. 8-15. In terms of g and the foregoing given quantities, determine (a) the speed v_1 of the bob when it is at its lowest position; (b) the least value v_2 that v_0 could have if the cord is to achieve a horizontal position during the motion; (c) the speed v_3 such that if $v_0 > v_3$, the pendulum will not oscillate but rather will continue to move around in a vertical circle.

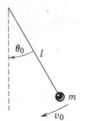

figure 8-15
Problem 22

23. A simple pendulum is made by tying a 2.0-kg stone to a string 4.0 m long. The stone is projected perpendicular to the string, away from the ground, with the string at an angle of 60° with the vertical. It is observed to have a speed of 8.0 m/s when it passes its lowest point. (a) What was the speed of the stone at the moment of release? (b) What is the largest angle with the vertical that the string will reach during the stone's motion? (c) Using the lowest point of the swing as the zero of gravitational potential energy, what is the total mechanical energy of the system?
Answer: (a) 5.0 m/s. (b) 80°. (c) 64 J.

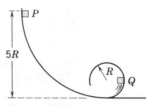

figure 8-16
Problem 24

24. A small block of mass m slides along the frictionless loop-the-loop track shown in Fig. 8-16. (a) If it starts from rest at P, what is the resultant force acting on it at Q? (b) At what height above the bottom of the loop should the block be released so that the force exerted on it by the track at the top of the loop is equal to its weight?

25. A point mass m starts from rest and slides down the surface of a frictionless solid sphere of radius r as in Fig. 8-17. Measure angles from the vertical and potential energy from the top. Find (a) the change in potential energy of the mass with angle; (b) the kinetic energy as a function of angle; (c) the radial and tangential accelerations as a function of angle; (d) the angle at which the mass flies off the sphere.
Answer: (a) $-mgr(1 - \cos\theta)$. (b) $mgr(1 - \cos\theta)$. (c) $2g(1 - \cos\theta)$; $g\sin\theta$. (d) $\cos^{-1}2/3$.

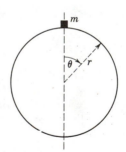

figure 8-17
Problem 25

26. The particle m in Fig. 8-18 is moving in a vertical circle of radius R inside a track. There is no friction. When m is at its lowest position, its speed is v_0. (a) What is the minimum value v_m of v_0 for which m will go completely around the circle without losing contact with the track? (b) Suppose v_0 is $0.775v_m$. The particle will move up the track to some point at P at which it will lose contact with the track and travel along a path shown roughly by the dashed line. Find the angular position θ of point P.

27. The nail in Fig. 8-12 is located a distance d below the point of suspension. Show that d must be at least $0.6l$ if the ball is to swing completely around in a circle centered on the nail.

28. Two children are playing a game in which they attempt to hit a small box on the floor using a spring-loaded marble gun placed horizontally on a frictionless table (Fig. 8-19). The first child compresses the spring 1.0 cm and the

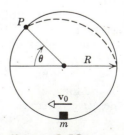

figure 8-18
Problem 26

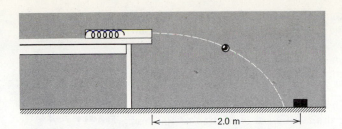

figure 8-19
Problem 28

159   PROBLEMS   CHAP. 8

marble falls 20 cm short of the target, which is 2.0 m horizontally from the edge of the table. How far should the second child compress the spring so that the same marble falls into the box?

29. An escalator joins one floor with another one 25 ft (7.6 m) above. The escalator is 40 ft (12 m) long and moves along its length at 2.0 ft/s (0.61 m/s). (a) What power must its motor deliver if it is required to carry a maximum of 100 persons per minute, of average mass 5.0 slugs (73 kg)? (b) A 160-lb (710-N) man walks up the escalator in 10 s. How much work does the motor do on him? (c) If this man turned around at the middle and walked down the escalator so as to stay at the same level in space, would the motor do work on him? If so, what power does it deliver for this purpose? (d) Is there any (other?) way the man could walk along the escalator without consuming power from the motor?
Answer: (a) 6700 ft · lb/s (9100 W). (b) 2000 ft · lb (2700 J).

30. Suppose that the string of Fig. 8-12 is very elastic, made of rubber, say, and that the string is unextended at length l when the ball is released. (a) Explain why you would expect the ball to reach a low point greater than a distance l below the point of suspension. (b) Show, using dynamic and energy considerations, that if Δl is small compared to l, the string will stretch by an amount $\Delta l = 3\,mg/k$, where k is the assumed force constant of the string. Notice that the larger k is, the smaller Δl is, and the better the approximation $\Delta l \ll l$. (c) Show, under these circumstances, that the speed of the ball at the bottom is $v = \sqrt{2g(l - 3mg/2k)}$, *less* than it would be for an inelastic string $(k = \infty)$. Give a physical explanation for this result using energy considerations.

SECTION 8-7

31. Two snow-covered peaks at elevations of 3500 m and 3400 m are separated by a valley. A ski-run extends from the top of the higher peak to the top of the lower one, with a total length of 3000 m. (a) A skier starts from rest on the higher peak. With what speed will he arrive at the top of the lower peak if he goes as fast as possible, never trying to slow down? Neglect friction. (b) Make a rough estimate of how large a coefficient of friction with the snow could be tolerated without preventing him from reaching the lower peak.   *Answer:* (a) 44 m/s. (b) Approximately 1/10.

32. A projectile of mass 9.4 kg is fired straight up with an initial speed of 470 m/s. How much higher would it have gone if the air resistance did not dissipate the energy of 6.8×10^5 J that it does?

33. Show that when friction is present in an otherwise conservative mechanical system, the rate at which mechanical energy is dissipated equals the frictional force times the speed at that instant, or

$$\frac{d}{dt}(K + U) = -fv$$

34. A boy is seated on the top of a hemispherical mound of ice (Fig. 8-20). He is given a very small push and starts sliding down the ice. (a) Show that he leaves the ice at a point whose height is $2R/3$ if the ice is frictionless. (b) If there is friction between the ice and the boy, would he fly off at a greater or lesser height than in (a)?

figure 8-20
Problem 34

35. A 1.0-kg (weight $mg = 2.2$ lb) block collides with a horizontal weightless spring of force constant 2.0 N/m (0.14 lb/ft) (Fig. 8-21). The block compresses the spring 4.0 m (13 ft) from the rest position. Assuming that the coefficient of kinetic friction between block and horizontal surface is 0.25, what was the speed of the block at the instant of collision?
Answer: 7.2 m/s (23 ft/s).

36. A body of mass m starts from rest down a plane of length l inclined at an angle θ with the horizontal. (a) Take the coefficient of friction to be μ and find the body's speed at the bottom. (b) How far, d, will it slide horizontally on a similar surface after reaching the bottom of the incline? Solve by using energy methods and solve again using Newton's laws directly.

37. A 4.0-kg block starts up a 30° incline with 128 J of kinetic energy. How far will it slide up the plane if the coefficient of friction is 0.30?
Answer: 4.3 m.

38. A 40-lb body is pushed up a frictionless 30° inclined plane 10 ft long by a horizontal force **F**. (a) If the speed at the bottom is 2.0 ft/s and at the top is 10 ft/s, how much work is done by **F**? (b) What is the magnitude of the force **F**? (c) Suppose the plane is not frictionless, and that $\mu_k = 0.15$. How far up the plane goes the body go?

39. A particle slides along a track with elevated ends and a flat central part, as shown in Fig. 8-22. The flat part has a length $l = 2.0$ m. The curved portions of the track are frictionless. For the flat part the coefficient of kinetic friction is $\mu_k = 0.20$. The particle is released at point A which is at a height $h = 1.0$ m above the flat part of the track. Where does the particle finally come to rest?   *Answer:* In the center of the flat part.

40. A very light rigid rod whose length is l has a ball of mass m attached to one end (Fig. 8-23). The other end is pivoted frictionlessly in such a way that the ball moves in a vertical circle. The system is launched from the horizontal position A with downward initial velocity $\mathbf{v}_0$. The ball just reaches point D and then stops. (a) Derive an expression for v_0 in terms of l, m, and g. (b) What is the tension in the rod when the ball is at B? (c) A little sand is placed on the pivot, after which the ball just reaches C when launched from A with the same speed as before. How much work is done by friction during this motion? (d) How much total work is done by friction before the ball finally comes to rest at B after oscillating back and forth several times?

41. The cable of a 4000-lb elevator in Fig. 8-24 snaps when the elevator is at rest at the first floor so that the bottom is a distance $d = 12$ ft above a cushioning spring whose spring constant is $k = 10,000$ lb/ft. A safety device clamps the guide rails so that a constant friction force of 1000 lb opposes the motion of the elevator. (a) Find the speed of the elevator just before it hits the spring. (b) Find the distance s that the spring is compressed. (c) Find the distance that the elevator will "bounce" back up the shaft. (d) Using the conservation of energy principle, find approximately the total distance that the elevator will move before coming to rest. Why is the answer not exact?   *Answer:* (a) 24 ft/s. (b) 3.0 ft. (c) 9.0 ft. (d) 49 ft.

SECTION 8-9

42. A vacuum diode consists of a cylindrical anode enclosing a cylindrical cathode. An electron with a potential energy relative to the anode of 4.8×10^{-16} J leaves the surface of the cathode with zero initial speed. Assume that the electron does not collide with any air molecules and that the gravitational force is negligible. (a) What kinetic energy would the electron have when it strikes the anode? (b) Take 9.1×10^{-31} kg as the mass of the electron and find its final speed. (c) Were you justified in using classical relations for kinetic energy and mass rather than the relativistic ones?

43. What is the speed of an electron with a kinetic energy of (a) 1.0×10^5 eV $(1.2 \times 10^{-14}$ ft · lb), (b) 1.0×10^6 eV $(1.2 \times 10^{-13}$ ft · lb)?
Answer: (a) 1.6×10^8 m/s (103,000 mi/s). (b) 2.8×10^8 m/s (175,000 mi/s).

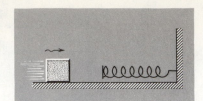

figure 8-21
Problem 35

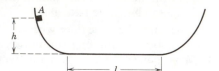

figure 8-22
Problem 39

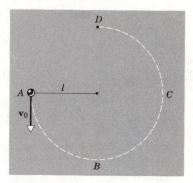

figure 8-23
Problem 40

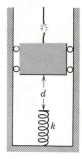

figure 8-24
Problem 41

44. The United States consumed about 1.6×10^{12} kW · h of electrical energy in 1970. How many kilograms of matter would have to be completely destroyed to yield this energy?

45. A nuclear reactor generating plant supplies 60 MW (8.0×10^4 hp) of useful power steadily for a year. (a) How much energy, in joules (ft · lb), did it supply? (b) Assuming that, in addition, 90 MW (12×10^4 hp) of power is wasted in heat production, determine the mass (weight) converted to energy in a year at this plant.
Answer: (a) 1.9×10^{15} J (1.4×10^{15} ft · lb). (b) 52 g (1.9 oz).

46. How much matter would have to be converted into energy in order to accelerate a 1.0-kiloton spaceship from rest to a speed of $(1/10)c$?

47. An electron (rest mass 9.1×10^{-31} kg) is moving with a speed $0.99\, c$. (a) What is its total energy? (b) Find the ratio of the Newtonian kinetic energy to the relativistic kinetic energy in this case?   *Answer:* (a) 5.8×10^{-13} J. (b) 0.08.

48. (a) The rest mass of a body is 0.010 kg. What is its mass when it moves at a speed of 3.0×10^7 m/s relative to the observer? At 2.7×10^8 m/s? (b) Compare the classical and relativistic kinetic energies for these cases. (c) What if the observer, or measuring apparatus, is riding on the body?

49. Equation (8-21), $K = (m - m_0)c^2$, is the usual relativistic equation for kinetic energy. (a) Show that, by using Eq. (8-20), $m = m_0(1 - \beta^2)^{-1/2}$, we can also express the relativistic kinetic energy as $K = \dfrac{m}{m + m_0}\, mv^2$. (b) Contrast the way these two expressions reduce to the classical result as $m \to m_0$ or $v/c \to 0$. (See "Parallels between Relativistic and Classical Dynamics for Introductory Courses" by Donald E. Fahnline, *American Journal of Physics*, June 1975.)

50. It is believed that the sun obtains its energy by a fusion process in which four hydrogen atoms are transformed into a helium atom with the emission of energy in various forms of radiation. If a hydrogen atom has a rest mass of 1.0081 atomic mass units (see Example 7) and a helium atom has a rest mass of 4.0039 atomic mass units, calculate the energy released in each fusion process.

9
conservation of linear momentum

So far we have treated objects as though they were particles, having mass but no size. In translational motion each point on a body experiences the same displacement as any other point as time goes on, so that the motion of one particle represents the motion of the whole body. But even when a body rotates or vibrates as it moves, there is one point on the body, called the *center of mass*, that moves in the same way that a single particle subject to the same external forces would move. Figure 9-1 shows the simple parabolic motion of the center of mass of an Indian club thrown from one performer to another; no other point in the club moves in such a simple way. Note that if the club were moving in pure translation (see Fig. 3-1), then *every* point in it would experience the same displacements as does the center of mass in Fig. 9-1. For this reason the motion of the center of mass of a body is called the translational motion of the body.

When the system with which we deal is not a rigid body, a center of mass (whose motion can also be described in a relatively simple way) can be assigned, even though the particles that make up the system may be changing their positions with respect to each other in a relatively complicated way as the motion proceeds. In this section we define the center of mass and show how to calculate its position. In the next section we discuss the properties that make it useful in describing the motion of extended objects or systems of particles.

Consider first the simple case of a system of two particles m_1 and m_2 at distances x_1 and x_2 respectively, from some origin O. We define a point C, the center of mass of the system, as a distance x_{cm} from the origin O, where x_{cm} is defined by

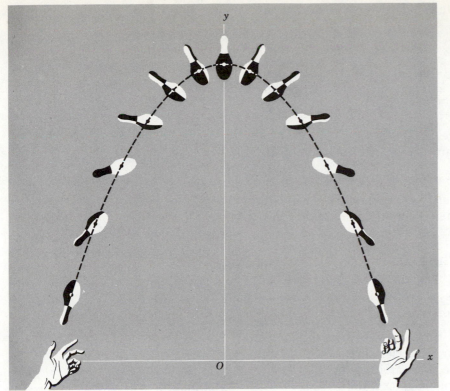

figure 9-1
An Indian club is thrown from one performer to another. Even though it rotates and spins around its axis, as shown, there is one point on its axis, the *center of mass*, that follows a simple parabolic path.

$$x_{cm} = \frac{m_1 x_1 + m_2 x_2}{m_1 + m_2}. \tag{9-1}$$

This point (Fig. 9-2) has the property that the product of the total mass of the system $M(= m_1 + m_2)$ times the distance of this point from the origin is equal to the sum of the products of the mass of each particle by its distance from the origin; that is,

$$(m_1 + m_2)x_{cm} = Mx_{cm} = m_1 x_1 + m_2 x_2.$$

In Eq. 9-1, x_{cm} can be regarded as the *mass-weighted mean* of x_1 and x_2.

An analogy might help to fix this idea. Suppose, for example, that we are given two boxes of nails. In one box we have n_1 nails all having the same length l_1; in the other box we have n_2 nails all having the same length l_2. We are asked to get the mean length of the nails. If $n_1 = n_2$, the mean length is simply $(l_1 + l_2)/2$. But if $n_1 \neq n_2$, we must allow for the fact that there are more nails of one length than another by a "weighting" factor for each length. For l_1 this factor is $n_1/(n_1 + n_2)$ and for l_2 this factor is $n_2/(n_1 + n_2)$, the fraction of the total number of nails in each box. Then the weighted-mean length is

$$\bar{l} = \left(\frac{n_1}{n_1 + n_2}\right)l_1 + \left(\frac{n_2}{n_1 + n_2}\right)l_2$$

or

$$\bar{l} = \frac{n_1 l_1 + n_2 l_2}{n_1 + n_2}.$$

The center of mass, defined in Eq. 9-1, is then a weighted-mean displacement where the "weighting" factor for each particle is the fraction of the total mass that each particle has.

If we have n particles, $m_1, m_2, \cdots, m_n$ along a straight line, by defini-

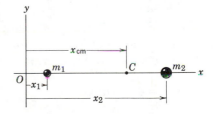

figure 9-2
The center of mass of the two masses m_1 and m_2 lies on the line joining m_1 and m_2 at C, a distance x_{cm} from the origin.

tion the center of mass of these particles relative to some origin is

$$x_{cm} = \frac{m_1 x_1 + m_2 x_2 + \cdots + m_n x_n}{m_1 + m_2 + \cdots + m_n} = \frac{\Sigma\, m_i x_i}{\Sigma\, m_i}, \qquad (9\text{-}2)$$

where $x_1, x_2, \cdots, x_n$ are the distances of the masses from the origin from which x_{cm} is measured. The symbol Σ represents a summation operation, in this case over all n particles. The sum

$$\Sigma\, m_i = M$$

is the total mass of the system. We can then rewrite Eq. 9-2 in the form

$$M x_{cm} = \Sigma\, m_i x_i. \qquad (9\text{-}2a)$$

Suppose now that we have three particles *not* in a straight line; they will lie in a *plane*, as in Fig. 9-3. The center of mass C is defined and located by the coordinates x_{cm} and y_{cm}, where

$$x_{cm} = \frac{m_1 x_1 + m_2 x_2 + m_3 x_3}{m_1 + m_2 + m_3}, \qquad (9\text{-}3)$$

$$y_{cm} = \frac{m_1 y_1 + m_2 y_2 + m_3 y_3}{m_1 + m_2 + m_3},$$

in which x_1, y_1 are the coordinates of the particle of mass m_1; x_2, y_2 are those of m_2; and x_3, y_3 are those of m_3. The coordinates x_{cm}, y_{cm} of the center of mass are measured from the same arbitrary origin.

For a large number of particles lying in a plane, the center of mass is at x_{cm}, y_{cm}, where

$$x_{cm} = \frac{\Sigma\, m_i x_i}{\Sigma\, m_i} = \frac{1}{M} \Sigma\, m_i x_i \quad \text{and} \quad y_{cm} = \frac{\Sigma\, m_i y_i}{\Sigma\, m_i} = \frac{1}{M} \Sigma\, m_i y_i \qquad (9\text{-}4)$$

in which $M\ (= \Sigma\, m_i)$ is the total mass of the system.

For a large number of particles not necessarily confined to a plane but *distributed in space*, the center of mass is at x_{cm}, y_{cm}, z_{cm}, where

$$x_{cm} = \frac{1}{M} \Sigma\, m_i x_i, \quad y_{cm} = \frac{1}{M} \Sigma\, m_i y_i, \quad z_{cm} = \frac{1}{M} \Sigma\, m_i z_i. \qquad (9\text{-}5a)$$

In vector notation each particle in the system can be described by a position vector $\mathbf{r}_i$ in a particular reference frame and the center of mass can be located by a position vector $\mathbf{r}_{cm}$. These vectors are related to x_i, y_i, z_i, and x_{cm}, y_{cm}, z_{cm} in Eq. 9-5a by

$$\mathbf{r}_i = \mathbf{i} x_i + \mathbf{j} y_i + \mathbf{k} z_i$$

and

$$\mathbf{r}_{cm} = \mathbf{i} x_{cm} + \mathbf{j} y_{cm} + \mathbf{k} z_{cm}.$$

Thus the three scalar equations of Eq. 9-5a can be replaced by a single vector equation

$$\mathbf{r}_{cm} = \frac{1}{M} \Sigma\, m_i \mathbf{r}_i \qquad (9\text{-}5b)$$

in which the sum is a vector sum. You can prove that Eq. 9-5b is true by substituting the expressions given for $\mathbf{r}_i$ and $\mathbf{r}_{cm}$ just above into Eq. 9-5b. Note the economy of expression permitted by the use of vectors. Equation 9-5b shows that, if the origin of our reference frame is at the center of mass (which means that $\mathbf{r}_{cm} = 0$), then $\Sigma\, m_i \mathbf{r}_i = 0$ for the system.

Equations 9-5 are the most general case for a collection of particles. Equations 9-1 through 9-4 are special instances of this one. The location

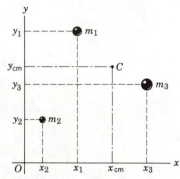

figure 9-3
The center of mass of the three masses m_1, m_2, and m_3 lies at point C, with coordinates x_{cm}, y_{cm}. C lies in the same plane as that of the triangle formed by the three masses.

of the center of mass is independent of the reference frame used to locate it (see Problem 1). *The center of mass of a system of particles depends only on the masses of the particles and the positions of the particles relative to one another.*

A rigid body, such as a meter stick, can be thought of as a system of closely packed particles. Hence it also has a center of mass. The number of particles (atoms, for example) in the body is so large and their spacing so small, however, that we can treat such a body as though it has a continuous distribution of mass. To obtain the expression for the center of mass of a continuous body, let us begin by subdividing the body into n small elements of mass Δm_i located approximately at the points x_i, y_i, z_i. The coordinates of the center of mass are then given approximately by

$$x_{cm} = \frac{\Sigma \, \Delta m_i x_i}{\Sigma \, \Delta m_i}, \quad y_{cm} = \frac{\Sigma \, \Delta m_i y_i}{\Sigma \, \Delta m_i}, \quad z_{cm} = \frac{\Sigma \, \Delta m_i z_i}{\Sigma \, \Delta m_i}.$$

Now let the elements of mass be further subdivided so that the number of elements n tends to infinity. The points x_i, y_i, z_i will locate the mass elements more precisely as n is increased and will locate them exactly as n becomes infinite. The continuous body is then subdivided into an infinite number of infinitesimal mass elements. We can now give the coordinates of the center of mass precisely as

$$x_{cm} = \lim_{\Delta m_i \to 0} \frac{\Sigma \, \Delta m_i x_i}{\Sigma \, \Delta m_i} = \frac{\int x \, dm}{\int dm} = \frac{1}{M} \int x \, dm,$$

$$y_{cm} = \lim_{\Delta m_i \to 0} \frac{\Sigma \, \Delta m_i y_i}{\Sigma \, \Delta m_i} = \frac{\int y \, dm}{\int dm} = \frac{1}{M} \int y \, dm, \qquad (9\text{-}6a)$$

$$z_{cm} = \lim_{\Delta m_i \to 0} \frac{\Sigma \, \Delta m_i z_i}{\Sigma \, \Delta m_i} = \frac{\int z \, dm}{\int dm} = \frac{1}{M} \int z \, dm.$$

In these expressions dm is the differential element of mass at the point x, y, z, and $\int dm$ equals M, where M is the total mass of the body. For a continuous body the summation of Eq. 9-5a is replaced by the integration of Eq. 9-6a.

The vector expression that is equivalent to the three scalar expressions of Eq. 9-6a is

$$\mathbf{r}_{cm} = \frac{1}{M} \int \mathbf{r} \, dm \qquad (9\text{-}6b)$$

As before, the summation of Eq. 9-5b has been replaced by an integration. Once again we see that if the origin of our reference frame is at the center of mass (that is, if $\mathbf{r}_{cm} = 0$), then $\int \mathbf{r} \, dm = 0$ for the body. This integral, and the corresponding sum $\Sigma \, m_i \mathbf{r}_i$ of Eq. 9-5b, is called the *first moment of mass for the system.*

Often we deal with homogeneous objects having a point, a line, or a plane of symmetry. Then the center of mass will lie at the point, on the line, or in the plane of symmetry. For example, the center of mass of a homogeneous sphere (which has a point of symmetry) will be at the center of the sphere, the center of mass of a cone (which has a line of symmetry) will be on the axis of the cone, etc. We can understand that this is so because, from symmetry, the first moment of mass ($\int \mathbf{r} \, dm$) is zero at the center of a sphere, somewhere along the axis of a cone, etc. If follows from Eq. 9-6b that $\mathbf{r}_{cm} = 0$ for such points which means that the center of mass is located at these points.

EXAMPLE 1

Locate the center of mass of three particles of mass $m_1 = 1.0$ kg, $m_2 = 2.0$ kg, and $m_3 = 3.0$ kg at the corners of an equilateral triangle 1.0 m on a side.

Choose the x-axis along one side of the triangle as shown in Fig. 9-4. Note that m_3 is then $\sqrt{3}/2$ m along the y-axis. Then,

$$x_{cm} = \frac{\Sigma \, m_i x_i}{\Sigma \, m_i} = \frac{(1.0 \text{ kg})(0) + (2.0 \text{ kg})(1.0 \text{ m}) + (3.0 \text{ kg})(\frac{1}{2} \text{ m})}{(1.0 + 2.0 + 3.0) \text{ kg}} = \frac{7}{12} \text{ m},$$

$$y_{cm} = \frac{\Sigma \, m_i y_i}{\Sigma \, m_i} = \frac{(1.0 \text{ kg})(0) + (2.0 \text{ kg})(0) + (3.0 \text{ kg})(\sqrt{3}/2 \text{ m})}{(1.0 + 2.0 + 3.0) \text{ kg}} = \frac{\sqrt{3}}{4} \text{ m}.$$

The center of mass C is shown in the figure. Why is it not at the geometric center of the triangle?

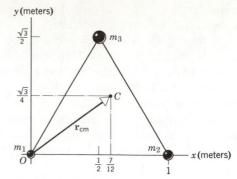

figure 9-4
Example 1. Finding the center of mass C of three unequal masses forming an equilateral triangle.

EXAMPLE 2

Find the center of mass of the triangular plate of Fig. 9-5.

If we can divide a body into parts such that the center of mass of each part is known, we can usually find the center of mass of the body simply. The triangular plate may be divided into narrow strips parallel to one side. The center of mass of each strip lies on the line which joins the middle of that side to the opposite vertex. But we can divide up the triangle in three different ways, using this process for each of three sides. Hence the center of mass lies at the intersection of the three lines which join the middle of each side with the opposite vertices. This is the only point that is common to the three lines.

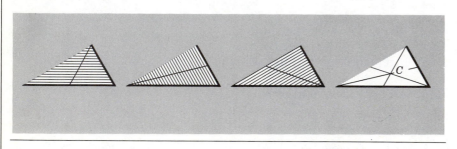

figure 9-5
Example 2. Finding the center of mass C of a triangular plate.

Now we can discuss the physical importance of the center-of-mass concept. Consider the motion of a group of particles whose masses are $m_1, m_2, \ldots, m_n$ and whose total mass is M. For the time being we will assume that mass neither enters nor leaves the system so that the total mass M of the system remains constant with time. In Section 9-7 we shall consider systems in which M is not constant; a familiar example is a rocket, which expels hot gases as its fuel burns, thus reducing its mass.

From Eq. 9-5b we have, for our fixed system of particles,

$$M\mathbf{r}_{cm} = m_1\mathbf{r}_1 + m_2\mathbf{r}_2 + \cdots + m_n\mathbf{r}_n,$$

where $\mathbf{r}_{cm}$ is the position vector identifying the center of mass in a particular reference frame. Differentiating this equation with respect to time, we obtain

$$M\frac{d\mathbf{r}_{cm}}{dt} = m_1\frac{d\mathbf{r}_1}{dt} + m_2\frac{d\mathbf{r}_2}{dt} + \cdots + m_n\frac{d\mathbf{r}_n}{dt} \qquad (9\text{-}7)$$

9-2
MOTION OF THE CENTER OF MASS

or

$$Mv_{cm} = m_1 v_1 + m_2 v_2 + \cdots + m_n v_n,$$

where v_1 is the velocity of the first particle, etc., and $dr_{cm}/dt \, (= v_{cm})$ is the velocity of the center of mass.

Differentiating Eq. 9-7 with respect to time, we obtain

$$M\frac{dv_{cm}}{dt} = m_1\frac{dv_1}{dt} + m_2\frac{dv_2}{dt} + \cdots + m_n\frac{dv_n}{dt} \qquad (9\text{-}8)$$

$$= m_1 a_1 + m_2 a_2 + \cdots + m_n a_n,$$

where a_1 is the acceleration of the first particle, etc., and $dv_{cm}/dt \, (= a_{cm})$ is the acceleration of the center of mass of the system. Now, from Newton's second law, the force F_1 acting on the first particle is given by $F_1 = m_1 a_1$. Likewise, $F_2 = m_2 a_2$, etc. We can then write Eq. 9-8 as

$$Ma_{cm} = F_1 + F_2 + \cdots + F_n. \qquad (9\text{-}9)$$

Hence *the total mass of the group of particles times the acceleration of its center of mass is equal to the vector sum of all the forces acting on the group of particles.*

Among all these forces will be *internal* forces exerted by the particles on each other. However, from Newton's third law, these internal forces will occur in equal and opposite pairs, so that they contribute nothing to the sum. Hence the internal forces can be removed from the problem. The right-hand sum in Eq. 9-9 represents the sum of only the *external* forces acting on all the particles. We can then rewrite Eq. 9-9 as simply

$$Ma_{cm} = F_{ext}. \qquad (9\text{-}10)$$

This states that *the center of mass of a system of particles moves as though all the mass of the system were concentrated at the center of mass and all the external forces were applied at that point.*

Notice that we obtain this simple result without specifying the nature of the system of particles. The system can be a rigid body in which the particles are in fixed positions with respect to one another, or it can be a collection of particles in which there may be all kinds of internal motion. Whatever the system is, and however its individual parts may be moving, its center of mass moves according to Eq. 9-10.

Hence, instead of treating bodies as single particles as we have done in previous chapters, we can treat them as collections of particles. Then we can obtain the translational motion of the body, that is, the motion of its center of mass, by assuming that all the mass of the body is concentrated at its center of mass and all the external forces are applied at that point.* This, in fact, is the procedure that we followed implicitly in all our force diagrams and problem solving.

Aside from justifying and making more concrete our previous procedure, we have now found how to describe the translational motion of a system of particles and how to describe the translational motion of a body which may be rotating as well. In this chapter and the next we apply this result to the linear motion of a system of particles. In later chapters we shall see how it simplifies the analysis of rotational motion.

---

* When the external force is gravity, it acts through the *center of gravity* of the body. In every case we have considered, the center of gravity coincides with the center of mass, which is a more general concept. We will discuss the conditions under which these points are different for a body in Chapter 14.

EXAMPLE 3

Consider three particles of different masses acted on by external forces, as shown in Fig. 9-6. Find the acceleration of the center of mass of the system.

First we find the coordinates of the center of mass. From Eq. 9-3,

$$x_{cm} = \frac{(8.0 \times 4) + (4.0 \times -2) + (4.0 \times 1)}{16} \, m = 1.8 \, m,$$

$$y_{cm} = \frac{(8.0 \times 1) + (4.0 \times 2) + (4.0 \times -3)}{16} \, m = 0.25 \, m.$$

These are shown as C in Fig. 9-6.

To obtain the acceleration of the center of mass, we first determine the resultant external force acting on the system consisting of the three particles. The x-component of this force is

$$F_x = 14 \, N - 6.0 \, N = 8.0 \, N,$$

and the y-component is

$$F_y = 16 \, N.$$

Hence the resultant external force has a magnitude

$$F = \sqrt{(8.0)^2 + (16)^2} \, N = 18 \, N,$$

and makes an angle θ with the x-axis given by

$$\tan \theta = \frac{16 \, N}{8.0 \, N} = 2.0 \quad \text{or} \quad \theta = 63°.$$

Then, from Eq. 9-10, the acceleration of the center of mass is

$$a_{cm} = \frac{F}{M} = \frac{18 \, N}{16 \, kg} = 1.1 \, m/s^2,$$

making an angle of 63° with the x-axis.

Although the three particles will change their relative positions as time goes on, the center of mass will move, as shown, with this constant acceleration.

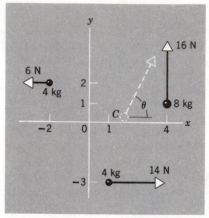

figure 9-6
Example 3. Finding the motion of the center of mass of three masses, each subjected to a different force. The forces all lie in the plane defined by the particles. The distances indicated along the axes are in meters.

The *momentum* of a single particle is a vector **p** defined as the product of its mass m and its velocity **v**. That is,

$$\mathbf{p} = m\mathbf{v}. \tag{9-11}$$

Momentum, being the product of a scalar by a vector, is itself a vector. Because it is proportional to **v**, the momentum **p** of a particular particle depends on the reference frame of the observer; we must always specify this frame.

Newton, in his famous *Principia*, expressed the second law of motion in terms of momentum (which he called "quantity of motion"). Expressed in modern terminology Newton's second law reads: *The rate of change of momentum of a body is proportional to the resultant force acting on the body and is in the direction of that force.* In symbolic form this becomes

$$\mathbf{F} = \frac{d\mathbf{p}}{dt}. \tag{9-12}$$

If our system is a single particle of (constant) mass m, this formulation of the second law is equivalent to the form $\mathbf{F} = m\mathbf{a}$, which we have used up to now. That is, if m is a constant, then

$$\mathbf{F} = \frac{d\mathbf{p}}{dt} = \frac{d}{dt}(m\mathbf{v}) = m\frac{d\mathbf{v}}{dt} = m\mathbf{a}.$$

9-3
LINEAR MOMENTUM OF A PARTICLE

The relations $\mathbf{F} = m\mathbf{a}$ and $\mathbf{F} = d\mathbf{p}/dt$ for single particles are completely equivalent in classical mechanics.

In relativity theory (See Supplementary Topic V) the second law for a single particle in the form $\mathbf{F} = m\mathbf{a}$ is not valid. However, it turns out that Newton's second law in the form $\mathbf{F} = d\mathbf{p}/dt$ *is* still a valid law *if* the momentum $\mathbf{p}$ for a single particle is defined not as $m_0\mathbf{v}$ but as

$$\mathbf{p} = \frac{m_0\mathbf{v}}{\sqrt{1 - v^2/c^2}}. \tag{9-13}$$

This result suggested a new definition of mass (compare Eqs. 9-11 and 9-13)

$$m = \frac{m_0}{\sqrt{1 - v^2/c^2}},$$

so that the momentum could still be written as $\mathbf{p} = m\mathbf{v}$; see Section 8-9. In this equation v is the speed of the particle, c is the speed of light, and m_0 is the "rest mass" of the body (its mass when $v = 0$). From this definition we must expect the mass of a particle to increase with its speed. Elementary particles such as electrons, protons, etc., may acquire enormous speeds, comparable to the speed of light. This concept can be put to a direct test in such cases because the increase in mass over the rest mass for such particles is large enough to measure accurately. Results of all such experiments indicate that this effect is real and given exactly by the equation above. (See for example, Fig. 8-8.)

9-4
LINEAR MOMENTUM OF A SYSTEM OF PARTICLES

Suppose that instead of a single particle we have a system of n particles, with masses m_1, m_2, etc. We shall continue to assume, as we did in Section 9-2, that no mass enters or leaves the system, so that the mass $M (= \Sigma\, m_i)$ of the system remains constant with time. The particles may interact with each other and external forces may act on them as well. Each particle will have a velocity and a momentum. Particle 1 of mass m_1 and velocity $\mathbf{v}_1$ will have a momentum $\mathbf{p}_1 = m_1\mathbf{v}_1$, for example. The system as a whole will have a *total momentum* $\mathbf{P}$ in a particular reference frame, which is defined to be simply the vector sum of the momenta of the individual particles in that same frame, or

$$\mathbf{P} = \mathbf{p}_1 + \mathbf{p}_2 + \cdots + \mathbf{p}_n \tag{9-14}$$
$$= m_1\mathbf{v}_1 + m_2\mathbf{v}_2 + \cdots + m_n\mathbf{v}_n.$$

If we compare this relation with Eq. 9-7, we see at once that

$$\mathbf{P} = M\mathbf{v}_{cm}, \tag{9-15}$$

which is an equivalent definition for the momentum of a system of particles. In words, Eq. 9-15 states: *The total momentum of a system of particles is equal to the product of the total mass of the system and the velocity of its center of mass.*

We have seen (Eq. 9-10) that Newton's second law for a system of particles can be written as

$$\mathbf{F}_{ext} = M\mathbf{a}_{cm} \tag{9-10}$$

in which $\mathbf{F}_{ext}$ is the vector sum of all the *external* forces acting on the system; we recall that the internal forces acting between particles cancel in pairs because of Newton's third law (see Fig. 9-7). If we differentiate Eq. 9-15 with respect to time we obtain, for an assumed constant mass M,

$$\frac{d\mathbf{P}}{dt} = M\frac{d\mathbf{v}_{cm}}{dt} = M\mathbf{a}_{cm}. \tag{9-16}$$

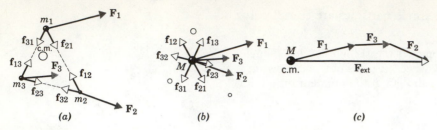

figure 9-7

Relationship between the forces acting on a system of three masses m_1, m_2, and m_3. (a) All the forces acting on each mass are shown here, as well as the location of the center of mass. On m_1 act forces $\mathbf{f}_{21}$ and $\mathbf{f}_{31}$ exerted by m_2 and m_3 respectively, as well as $\mathbf{F}_1$, a force from some external agent. Similar sets of forces act on m_2 and m_3. However, according to Newton's third law, internal forces $\mathbf{f}_{31}$ and $\mathbf{f}_{13}$ must be equal and opposite and must both lie along the line of centers of m_1 and m_3. Similar statements hold for the other two pairs of action-reaction forces. (b) If we are interested only in the motion of the system as a whole, we may consider all the forces to act on a mass $M = m_1 + m_2 + m_3$, located at the center of mass. Owing to the equality of the action-reaction pairs of internal forces as just stated, they cancel each other identically, leaving only the three external forces $\mathbf{F}_1$, $\mathbf{F}_2$, and $\mathbf{F}_3$. We add these graphically in (c) to yield a net force $\mathbf{F}_{ext}$ acting on the center of mass of the system.

Comparison of Eqs. 9-10 and 9-16 allows us to write Newton's second law for a system of particles in the form

$$\mathbf{F}_{ext} = \frac{d\mathbf{P}}{dt}. \tag{9-17}$$

This equation is the generalization of the single-particle equation $\mathbf{F} = d\mathbf{p}/dt$ (Eq. 9-12) to a system of many particles, no mass entering or leaving the system. Equation 9-17 reduces to Eq. 9-12 for the special case of a single particle, there being only external forces on a one-particle system.

Suppose that the sum of the external forces acting on a system is zero. Then, from Eq. 9-17,

$$\frac{d\mathbf{P}}{dt} = 0 \quad \text{or} \quad \mathbf{P} = \text{a constant}.$$

When the resultant external force acting on a system is zero, the total vector momentum of the system remains constant. This simple but quite general result is called *the principle of the conservation of linear momentum.* We shall see that it is applicable to many important physical situations.

The conservation of linear momentum principle is the second of the great conservation principles that we have met so far, the first being the conservation of energy principle. Later we shall meet several others, among them the conservation of electric charge and of angular momentum. Conservation principles are of theoretical and practical importance in physics because they are simple and universal. They are all cast in the form: While the system is changing there is one aspect of the system that remains unchanged. Different observers, each in his own reference frame, would all agree, if they watched the same changing system, that the conservation laws applied to the system. For the con-

9-5
CONSERVATION OF LINEAR MOMENTUM

servation of linear momentum, for example, observers in different reference frames would assign different values of **P** to the linear momentum of the system, but each would agree (assuming $\mathbf{F}_{ext} = 0$) that his own value of **P** remained unchanged as the particles that make up the system move about.

The total momentum of a system can only be changed by external forces acting on the system. The internal forces, being equal and opposite, produce equal and opposite changes in momentum which cancel each other. For a system of particles

$$\mathbf{p}_1 + \mathbf{p}_2 + \cdots + \mathbf{p}_n = \mathbf{P},$$

so that when the total momentum **P** is constant we have

$$\mathbf{p}_1 + \mathbf{p}_2 + \cdots + \mathbf{p}_n = \text{a constant} = \mathbf{P}_0. \qquad (9\text{-}18)$$

The momenta of the individual particles may change, but their sum remains constant if there is no external force.

Momentum is a vector quantity. Equation 9-18 is therefore equivalent to three scalar equations, one for each coordinate direction. Hence the conservation of linear momentum supplies us with three conditions on the motion of a system to which it applies. The conservation of energy on the other hand supplies us with only one condition on the motion of a system to which it applies, because energy is a scalar.

The law of the conservation of linear momentum holds true even in atomic and nuclear physics, although Newtonian mechanics does not. Hence this conservation law must be more fundamental than the Newtonian principles. In our derivation of this principle we must have made more rigid assumptions than we needed to. This is true even in the framework of classical mechanics. Recall the key role played by Newton's third law in this deduction of momentum conservation. This law was used to justify the assumption that the sum of the internal forces acting on all the particles is zero. However, it is somewhat artificial to regard the internal forces in a piece of matter as resulting from pairs of equal and opposite forces between the various pairs of atoms. These internal forces are actually many-body forces, depending on not only the relative separation and orientation of two atoms but also on the positions and orientations of neighboring atoms. If it were possible to prove our assumption without using Newton's third law, the law of conservation of linear momentum would not depend on the validity of the third law of motion. Actually we can prove this assumption on the basis of a much less stringent requirement than that the third law should hold. The proof lies outside the scope of this text.[*]

Consider first a problem in which an external force acts on a system of particles. Recall our previous discussion of projectile motion (Chapter 4). Now let us imagine that our projectile is a fireworks shell that explodes while in flight. The path of the shell is shown in Fig. 9-8. We assume that the air resistance is negligible. The system is the shell, the earth is our reference frame, and the external force is that of gravity. At the point x_1 the shell explodes and shell fragments are blown in all directions. What can we say about the motion of this system thereafter?

The forces of the explosion are all *internal forces*; they are forces exerted by part of the system on other parts of the system. These forces may change the momenta of all the individual fragments from the values they had when they made up the shell, but they cannot change the *total* vector momentum of the system. Only an external force can change the total momentum. The external

9-6
SOME APPLICATIONS OF THE MOMENTUM PRINCIPLE

EXAMPLE 4

---

[*] See "On Newton's Third Law and the Conservation of Momentum" by E. Gerjuoy, *American Journal of Physics*, November 1949.

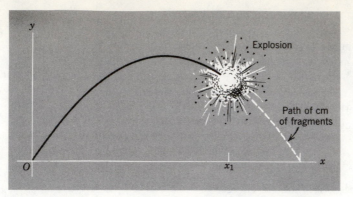

figure 9-8
Example 4. A projectile, following the usual parabolic trajectory, bursts at x_1. The center of mass of the fragments continues along the same parabolic path.

force, however, is simply that due to gravity. Because a system of particles as a whole moves as though all its mass were concentrated at the center of mass with the external force applied there, the center of mass of the fragments will continue to move in the parabolic trajectory that the unexploded shell would have followed. The change in the total momentum of the system attributable to gravity is the same whether the shell explodes or not.

What can you say about the *mechanical energy* of the system before and after the explosion?

Consider now two blocks A and B, of masses m_A and m_B, coupled by a spring and resting on a horizontal frictionless table. Let us pull the blocks apart and stretch the spring, as in Fig. 9-9, and then release the blocks. Describe the subsequent motion.

EXAMPLE 5

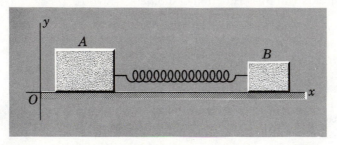

figure 9-9
Example 5. Two blocks A and B, resting on a frictionless surface, are connected by a spring. If they are held apart and then released, the sum of their momenta remains zero.

If the system consists of the two blocks and spring, then after we have released the blocks there is no net *external* force acting on the system. We can therefore apply the conservation of linear momentum to the motion. The momentum of the system before the blocks were released was zero in the reference frame shown attached to the table, so the momentum must remain zero thereafter. The total momentum can be zero even though the blocks move because momentum is a *vector* quantity. One block will have positive momentum (A moves in the $+x$ direction) and the other block will have negative momentum (B moves in the $-x$ direction). From the conservation of momentum we have

initial momentum = final momentum.

$$0 = m_B \mathbf{v}_B + m_A \mathbf{v}_A.$$

Therefore

$$m_B \mathbf{v}_B = -m_A \mathbf{v}_A$$

or

$$\mathbf{v}_A = -\frac{m_B}{m_A} \mathbf{v}_B.$$

For example, if m_A is 2 kg and m_B is 1 kg, then $\mathbf{v}_A$ will always be one-half $\mathbf{v}_B$ in magnitude and oppositely directed as the blocks move.

The kinetic energy of block A is $\frac{1}{2}m_A v_A^2$ and can be written as $(m_A v_A)^2/2m_A$ and that of block B is $\frac{1}{2}m_B v_B^2$ and can be written as $(m_B v_B)^2/2m_B$. But

$$\frac{K_A}{K_B} = \frac{2m_B(m_A v_A)^2}{2m_A(m_B v_B)^2} = \frac{m_B}{m_A},$$

in which $m_A v_A$ equals $m_B v_B$ because of momentum conservation. The kinetic energies of the blocks at any instant are inversely proportional to their respective masses. Because mechanical energy is conserved also, the blocks will continue to oscillate back and forth, the energy being partly kinetic and partly potential. What is the motion of the center of mass of this system?

If mechanical energy is not conserved, as would be true if friction were present, the motion will die out as the energy is dissipated. Can we apply the conservation of linear momentum in this case? Explain.

EXAMPLE 6

As an example of recoil, consider radioactive decay. An α-particle (the nucleus of a helium atom) is emitted from a uranium-238 nucleus, originally at rest, with a speed of 1.4×10^7 m/s and a kinetic energy of 4.1 MeV (million electron volts). Find the recoil speed of the residual nucleus (thorium-234).

We think of the system (thorium $+ \alpha$-particle) as initially bound and forming the uranium nucleus. The system then fragments into two separate parts. The momentum of the system before fragmentation is zero. In the absence of external forces, the momentum after fragmentation is also zero. Hence,

initial momentum = final momentum,

$$0 = M_\alpha \mathbf{v}_\alpha + M_{Th}\mathbf{v}_{Th},$$

$$\mathbf{v}_{Th} = -\frac{M_\alpha}{M_{Th}}\mathbf{v}_\alpha.$$

The ratio of the α-particle mass to the thorium nucleus mass, M_α/M_{Th}, is 4/234 and $v_\alpha = 1.4 \times 10^7$ m/s. Hence,

$$v_{Th} = -(4/234)(1.4 \times 10^7 \text{ m/s}) = -2.4 \times 10^5 \text{ m/s}.$$

The minus sign indicates that the residual thorium nucleus recoils in a direction exactly opposite to the motion of the α-particle, so as to give a resultant vector momentum of zero.

How can we compute the kinetic energy of the recoiling nucleus (see previous example)? Where does the energy of the fragments come from?

EXAMPLE 7

Consider now the apparently simple example of a ball thrown up from the earth by a person and then caught by him on its return. To simplify matters we can consider the person to be part of the earth since he does not lose contact with it. We also assume that air resistance is negligible.

The system being considered consists of the earth and the ball. The gravitational forces between the parts of the system are now internal forces. Let us choose a reference frame in which the system (earth + ball) is at rest. When the ball is thrown up, the earth must recoil as seen by an observer in this reference frame. The momentum of the system (earth + ball) is zero initially and no external forces act. Therefore, momentum is conserved and the total momentum remains zero throughout the motion. The upward momentum acquired by the ball is balanced by an equal and opposite downward momentum of the earth. We have

initial momentum = final momentum,

$$0 = m_B \mathbf{v}_B + m_E \mathbf{v}_E,$$

$$m_B \mathbf{v}_B = -m_E \mathbf{v}_E.$$

Here m_B and m_E are the masses of ball and earth respectively and $\mathbf{v}_B$ and $\mathbf{v}_E$ are the velocities of the ball and the earth in our selected reference frame. Owing to the enormous mass of the earth in comparison with the ball, the recoil speed of the earth is negligibly small.

As the ball and earth separate, the internal force of gravitational attraction pulls them together until they cease separating and begin to approach one another. As the ball falls toward the earth, the earth falls toward the ball with an equal but oppositely directed momentum. As the ball is caught, its momentum is neutralized by (and it neutralizes) the momentum of the earth. Both objects lose their relative motion, the total momentum is still zero, and the original situation before throwing is restored.

You will recall that when we discussed the conservation of energy in the presence of gravitational potential, we neglected to consider the motion of the earth itself. We took the surface of the earth as our zero level of gravitational potential energy. The reference position did not matter, because we were concerned only with *changes* in potential energy. However, in computing changes in kinetic energy, we assumed that the earth remained stationary, as in the case of the ball thrown up from the earth.

In principle, we cannot ignore the change in the kinetic energy of the earth itself. For example, when the ball falls toward the earth, the earth is slightly accelerated toward the ball. We neglected this fact before because we assumed that the change in kinetic energy of the earth is negligible. This result is not obvious, because although the earth's speed will certainly be small, its mass is enormous and the kinetic energy acquired might be significant. To settle the point we compute the ratio of the kinetic energy of the earth to that of the ball. Using $m_E v_E = m_B v_B$ from momentum conservation, we have

$$\frac{K_E}{K_B} = \frac{\frac{1}{2}m_E v_E^2}{\frac{1}{2}m_B v_B^2} = \frac{\frac{1}{2}(m_E v_E)^2}{\frac{1}{2}(m_B v_B)^2} \cdot \frac{m_B}{m_E} = \frac{m_B}{m_E}.$$

Since the mass of the ball m_B is negligibly small compared to the mass of the earth m_E, the kinetic energy acquired by the earth, K_E, is negligibly small compared to that of the ball, K_B. For example, if $m_B = 6$ kg (a rather massive ball), then, since $m_E = 6 \times 10^{24}$ kg, $K_E/K_B = 10^{-24}$!

Notice that this problem is identical in principle to Example 5. The differences are only those of detail; in one the potential energy is elastic and in the other the potential energy is gravitational; in one the masses are pictured as of the same order of magnitude, and in the other they are of very different orders of magnitude.

9-7
SYSTEMS OF VARIABLE MASS

So far we have dealt only with systems in which the total system mass M remained constant with time. Now we consider systems in which mass enters or leaves the system while we are observing it, dM/dt being positive in the former case and negative in the latter.

Figure 9-10a shows a system of mass M whose center of mass is moving with velocity $\mathbf{v}$ as seen from a particular reference frame. An external force $\mathbf{F}_{\text{ext}}$ acts on the system. At a time Δt later the configuration has changed to that shown in Fig. 9-10b. A mass ΔM has been ejected from the system, its center of mass moving with velocity $\mathbf{u}$ as seen by our observer. The system mass is reduced to $M - \Delta M$ and the velocity $\mathbf{v}$ of the center of mass of the system is changed to $\mathbf{v} + \Delta \mathbf{v}$.

The student may imagine the system of Fig. 9-10 to represent a rocket. It ejects hot gas from its orifice at a fairly high speed, decreasing its own mass and increasing its own speed. In a rocket the loss of mass is continuous during the burning process. The external force $\mathbf{F}_{\text{ext}}$ is *not* the thrust of the rocket but is the force of gravity on the rocket and the resisting force of the atmosphere.

To analyze the situation let us, for the time being, define the system to be one of constant mass. This means that in Fig. 9-10b, we shall include in our system not only the mass $M - \Delta M$ of the body but also the ejected mass ΔM, the total mass of the system being the M of Fig. 9-10a. Doing so permits us to apply

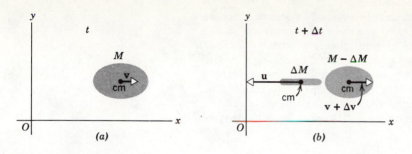

the results that we have derived so far for constant mass systems. We shall see that this approach leads us to the form of Newton's second law for systems in which the mass is not constant.

From Eq. 9-17

$$\mathbf{F}_{\text{ext}} = \frac{d\mathbf{P}}{dt} \qquad (9\text{-}17)$$

we can write, as an approximate result for the finite time interval Δt,

$$\mathbf{F}_{\text{ext}} \cong \frac{\Delta \mathbf{P}}{\Delta t} = \frac{\mathbf{P}_f - \mathbf{P}_i}{\Delta t}$$

in which $\mathbf{P}_f$ is the (final) system momentum in Fig. 9-10b and $\mathbf{P}_i$ is the (initial) system momentum for Fig. 9-10a. But $\mathbf{P}_f = (M - \Delta M)(\mathbf{v} + \Delta \mathbf{v}) + \Delta M \mathbf{u}$ and $\mathbf{P}_i = M\mathbf{v}$. This leads to

$$\mathbf{F}_{\text{ext}} \cong \frac{[(M - \Delta M)(\mathbf{v} + \Delta \mathbf{v}) + \Delta M \mathbf{u}] - [M\mathbf{v}]}{\Delta t}$$

$$= M\frac{\Delta \mathbf{v}}{\Delta t} + [\mathbf{u} - (\mathbf{v} + \Delta \mathbf{v})]\frac{\Delta M}{\Delta t}. \qquad (9\text{-}19)$$

Now, if we let Δt approach zero, the configuration of Fig. 9-10b approaches that of Fig. 9-10a; that is, $\Delta \mathbf{v}/\Delta t$ approaches $d\mathbf{v}/dt$, the acceleration of the body in Fig. 9-10a. The quantity ΔM is the mass ejected in Δt: this leads to a decrease in the mass M of the original body. Since dM/dt, the change in mass of the body with time, is intrinsically negative in this case, the positive quantity $\Delta M/\Delta t$ is replaced by $-dM/dt$ as Δt approaches zero. Finally, $\Delta \mathbf{v}$ goes to zero as Δt approaches zero. Making these changes in Eq. 9-19 leads to

$$\mathbf{F}_{\text{ext}} = M\frac{d\mathbf{v}}{dt} + \mathbf{v}\frac{dM}{dt} - \mathbf{u}\frac{dM}{dt} \qquad (9\text{-}20a)$$

or

$$\mathbf{F}_{\text{ext}} = \frac{d}{dt}(M\mathbf{v}) - \mathbf{u}\frac{dM}{dt}, \qquad (9\text{-}20b)$$

which is Newton's second law, defining the external forces on a body (like that of Fig. 9-10a) whose mass is changing.

Note that these equations reduce to the familiar forms $\mathbf{F}_{\text{ext}} = M\mathbf{a}$ and $\mathbf{F}_{\text{ext}} = (d/dt)(M\mathbf{v})$ respectively for the special case of a body of constant mass ($dM/dt = 0$). It is important to note that we *cannot* derive a general expression for Newton's second law for variable mass systems by treating the mass in $\mathbf{F}_{\text{ext}} = d\mathbf{P}/dt = d(M\mathbf{v})/dt$ as a variable. For this leads to

$$\mathbf{F}_{\text{ext}} = d(M\mathbf{v})/dt = M\,d\mathbf{v}/dt + \mathbf{v}\,dM/dt,$$

which is only a special case of the more general Eq. 9-20, namely, the case in which either (a) $dM/dt = 0$, a system of constant mass, or (b) $\mathbf{u} = 0$, a special choice of reference frame. We *can* use $\mathbf{F}_{\text{ext}} = d\mathbf{P}/dt$ to analyze variable mass systems *only* if we apply it to an *entire system of constant mass* having parts among which there is an interchange of mass. This indeed is what we have done in deriving Eqs. 9-20. The importance of the momentum formulation $\mathbf{F}_{\text{ext}} = d\mathbf{P}/dt$ in classical physics lies in the fact that it highlights momentum conservation and gives us a simple, physical way to treat complicated systems. Since the choice of what we will take as the system is ours to make, we can always choose a system of constant mass by defining our system broadly enough.

However, it is often convenient, as in rocket problems, to choose a system whose mass varies with time. In such cases we apply Newton's second law of Eqs. 9-20 in a form that is sometimes more convenient and interpretable more physically. The quantity $\mathbf{u} - (\mathbf{v} + \Delta\mathbf{v})$ in Eq. 9-19 is just $\mathbf{v}_{rel}$, the relative velocity of the ejected mass with respect to the main body. Therefore Eq. 9-20a may be written as

$$M\frac{d\mathbf{v}}{dt} = \mathbf{F}_{ext} + (\mathbf{u} - \mathbf{v})\frac{dM}{dt} \qquad (9\text{-}21a)$$

or

$$M\frac{d\mathbf{v}}{dt} = \mathbf{F}_{ext} + \mathbf{v}_{rel}\frac{dM}{dt}. \qquad (9\text{-}21b)$$

The last term in Eq. 9-21b, $[\mathbf{v}_{rel}(dM/dt)]$, is the rate at which momentum is being transferred into (or out of) the system by the mass that the system has ejected (or collected). It can be interpreted as the force exerted *on* the system by the mass that leaves it (or joins it). For a rocket, this term is called the *thrust* and it is the rocket designer's aim to make it as large as possible. Inspection of Eq. 9-21b shows that this requires that the rocket eject as much mass per unit time as possible and that the speed of the ejected mass relative to the rocket be as high as possible. We can rewrite Eq. 9-21b as

$$M\frac{d\mathbf{v}}{dt} = \mathbf{F}_{ext} + \mathbf{F}_{reaction}$$

in which $\mathbf{F}_{reaction}$ ($= \mathbf{v}_{rel}\,dM/dt$) is the reaction force exerted on the system by the mass that leaves it.

EXAMPLE 8

A machine gun is mounted on a car that can roll with negligible friction on a horizontal surface as in Fig. 9-11a. The mass of the system (car + gun) at a particular instant is M. At that same instant the gun is firing bullets of mass m whose velocity, in the reference frame shown, is $\mathbf{u}$. The velocity of the car in this same frame is $\mathbf{v}$ and the velocity of the bullets *with respect to the car* is $\mathbf{u} - \mathbf{v}$. The number of bullets fired per unit time is n. What is the acceleration of the car?

We select the car and gun as our system. Because its mass M is variable, we apply Newton's second law in the form given in Eq. 9-21b. Since no net external force acts on the system, we have $\mathbf{F}_{ext} = 0$ in that equation, yielding

$$M\frac{d\mathbf{v}}{dt} = \mathbf{v}_{rel}\frac{dM}{dt}.$$

Now $d\mathbf{v}/dt$ is $\mathbf{a}$, the acceleration of the system; $\mathbf{v}_{rel}$ is $\mathbf{u} - \mathbf{v}$, pointing to the left in Fig. 9-11a, and dM/dt is $-mn$. Inserting these in the equation above yields

$$\mathbf{a} = \frac{d\mathbf{v}}{dt} = -\frac{\mathbf{v}_{rel}(mn)}{M}.$$

This shows that $\mathbf{a}$ points in the direction opposite to $\mathbf{v}_{rel}$, that is, $\mathbf{a}$ points to the right in Fig. 9-11a. If $v_{rel} = 500$ m/s, $m = 10$ g, $n = 10$/s, and $M = 200$ kg at some instant, then at that instant

$$a = \frac{(500 \text{ m/s})(10^{-2} \text{ kg})(10/\text{sec})}{200 \text{ kg}} = 0.25 \text{ m/s}^2.$$

The magnitude of the average "thrust" of the ejected bullets on the system (car + gun) at this instant is given by

$$F = v_{rel}nm = (500 \text{ m/s})(10/\text{s})(10^{-2} \text{ kg})$$

$$= 50 \text{ N}.$$

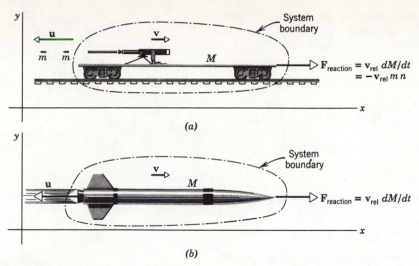

figure 9-11
(a) Example 8. A machine gun is fixed to a car that rolls with negligible friction. The gun fires bullets of mass m at a rate (number per unit time) n, the velocity of the bullets with respect to the gun being $\mathbf{u} - \mathbf{v}$. At the instant shown some bullets have already left the system. The velocities indicated for the car and the bullets are those that would be measured by an observer in a reference frame fixed to the rails as shown. The reaction force on the system is $\mathbf{F} = -mn\mathbf{v}_{rel} = (dM/dt)\mathbf{v}_{rel}$. (b) A rocket moves through space with negligible external forces. Gas particles are ejected from the exhaust, the particles having a velocity $\mathbf{u} - \mathbf{v}$ with respect to the rocket. The rate at which mass is expelled at the exhaust is $-dM/dt$. The reaction force on the rocket is $\mathbf{F} = (dM/dt)\mathbf{v}_{rel}$. The velocities indicated for the rocket and exhaust gases are relative to the ground.

In Figure 9-11b we show the analogous situation for a rocket. It is instructive to view this problem from the point of view of Newton's third law and the momentum principle. Choose a fixed-mass system (rocket + gas) and attach a reference frame to its center of mass. The rocket forces a jet of hot gases from its exhaust; this is the action force. The jet of hot gases exerts a force on the rocket, propelling it forward. This is the reaction force. These forces are internal forces in the system (rocket + gas). In the absence of external forces the total momentum of the system is constant (the center of mass, initially at rest, remains at rest). The individual parts of the system (rocket and gases) may change their momentum, however; with respect to the center of mass frame, the hot gases acquire momentum in the backward direction and the rocket acquires an equal amount of momentum in the forward direction.

You can analyze the system (bullets + car and gun) in a similar way.

EXAMPLE 9

A rocket weighs 30,000 lb when fueled up on the launching pad. It is fired vertically upward and, at burnout, weighs 10,000 lb. Gases are exhausted at the rate of 10 slugs/s with a velocity of 5000 ft/s, relative to the rocket (exhaust velocity), both quantities being assumed to be constant while the fuel is burning.

(a) What is the thrust? The thrust $\mathbf{F}$ is the last term in Eq. 9-21b, or

$$F = v_{rel}\frac{dM}{dt} = (5000 \text{ ft/s})(10 \text{ slugs/s}) = 50,000 \text{ lb.}$$

Note that initially, when the fuel tanks are full, the net upward force acting on the rocket (neglecting air resistance) is the thrust (50,000 lb) minus the initial weight (30,000 lb) or 20,000 lb. Just before burnout the net upward force is 50,000 lb minus 10,000 lb or 40,000 lb.

(b) If we could neglect *all external forces*, including gravity and air resistance, what would be the speed of the rocket at burnout?

If we put $\mathbf{F}_{ext} = 0$ in Eq. 9-21b, we have

$$M\frac{d\mathbf{v}}{dt} = \mathbf{v}_{rel}\frac{dM}{dt} \quad \text{or} \quad d\mathbf{v} = \mathbf{v}_{rel}\frac{dM}{M}.$$

Integrating this expression (see Appendix I) from the instant the velocity is $\mathbf{v}_0$ and the mass of the rocket is M_0 to the instant when the velocity is $\mathbf{v}$ and the mass of the rocket is M, we obtain

$$\int_{\mathbf{v}_0}^{\mathbf{v}} d\mathbf{v} = \mathbf{v}_{rel}\int_{M_0}^{M}\frac{dM}{M},$$

the exhaust velocity being assumed constant during this time. This yields

$$\mathbf{v} - \mathbf{v}_0 = -\mathbf{v}_{rel}\ln(M_0/M) = -\mathbf{v}_{rel}\ln\left(1 + \frac{M_0 - M}{M}\right).$$

Hence the change in velocity of the rocket in any interval of time depends only on the exhaust velocity (being opposite in direction from it) and on the fraction of mass exhausted during that time interval.

In our example, $v_0 = 0$ and $M_0/M = (30{,}000/10{,}000) = 3.0$, so that the speed of the rocket at burnout is

$$v = v_{rel}\ln(M_0/M) = (5000 \text{ ft/s}) \ln 3.0 = 3800 \text{ mi/h}.$$

If the external forces of gravity and air resistance were taken into account, the final speed would be smaller.*

Assuming that the rocket starts from rest $(v_0 = 0)$ with an initial mass M_0 and reaches a final velocity v_f at burnout when its mass is M_f, we can write the rocket equation above as

$$\frac{M_f}{M_0} = e^{-v_f/v_{rel}}$$

in which v_{rel} is the exhaust velocity.

The classical rocket (or variable mass) equations imply that the speed of the rocket can increase to any value provided only that the rocket expels enough propellant so that the final remaining mass is sufficiently small. However, we know from relativistic mechanics that a rocket cannot be accelerated to a speed equal to or greater than the speed of light. Once the rocket's speed approaches the relativistic range the classical equations are no longer applicable. One must take into account the variation of inertial mass of a particle with speed and the relativistic velocity formula. The resulting equations apply to a relativistic rocket.†

EXAMPLE 10

Sand drops from a stationary hopper at a rate dM/dt onto a conveyor belt moving with velocity $\mathbf{v}$ in the reference frame of the laboratory, as in Fig. 9-12. What force is required to keep the belt moving at a speed v?

This is a clear-cut example of a force associated with change of mass alone, the velocity being constant. We take as our system the belt of varying mass so that Eq. 9-21b applies. We must put $d\mathbf{v}/dt = 0$ in that equation because the velocity of the belt is constant. Furthermore, to an observer at rest on the belt, the falling sand (and the hopper) would appear to have a horizontal motion with speed v in a direction opposite to that shown for the belt in the laboratory. Therefore $\mathbf{v}_{rel} = -\mathbf{v}$ in Eqs. 9-21. More formally, $\mathbf{v}_{rel} = \mathbf{u} - \mathbf{v}$; but $\mathbf{u} = 0$, so that $\mathbf{v}_{rel} = -\mathbf{v}$. Making these substitutions yields

---

* For an exact solution of the classical rocket problem see "Variable-Mass Dynamics" by J. L. Meriam, *Journal of Engineering Education*, December 1960.

† See "The Equation of Motion for Relativistic Particles and Systems with a Variable Rest Mass," by Kalman B. Pomeranz, *American Journal of Physics*, December 1964.

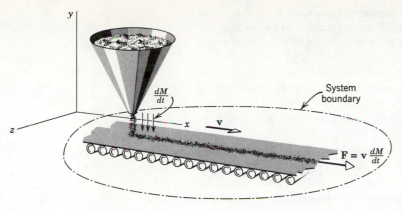

figure 9-12
Example 10. Sand drops from a hopper at a rate dM/dt onto a conveyer belt moving with velocity **v** in the reference frame of the laboratory. The force **F** required to keep the belt moving at constant velocity is **v** dM/dt. The hopper is at rest in the reference frame shown.

$$0 = \mathbf{F}_{ext} - \mathbf{v}\frac{dM}{dt}$$

or

$$\mathbf{F}_{ext} = \mathbf{v}\frac{dM}{dt}.$$

In this example, dM/dt is positive because the system is gaining mass with time. Hence, as expected, the necessary external force must point in the direction in which the belt moves. Note that, in the absence of friction, the mass of the belt itself does not enter the problem.

The power supplied by the external force is

$$P = \mathbf{F} \cdot \mathbf{v} = \mathbf{v} \cdot \mathbf{F} = \mathbf{v} \cdot (\mathbf{v}\, dM/dt) = v^2(dM/dt).$$

Since $v = $ a constant, we can write this as

$$P = \frac{d(Mv^2)}{dt} = 2\frac{d}{dt}\left(\frac{1}{2}Mv^2\right) = 2\frac{dK}{dt}.$$

This tells us that the power required to keep the belt moving is *twice* the rate at which the kinetic energy of the system is increasing; note that we need not consider the kinetic energy of the belt itself because — its speed being constant — its kinetic energy does not change. It is clear that mechanical energy is not conserved in this case. Where is the other half of the power going? In which of the previous examples did we have conservation of momentum without conservation of mechanical energy?

The student should be able to solve Example 10 alternatively by choosing a fixed-mass system and applying the momentum principle.*

---

questions

1. Must there necessarily be any mass at the center of mass of a system?
2. Does the center of mass of a solid body necessarily lie within the body? If not, give examples.
3. How is the center of mass concept related to the concept of geographic center of the country? To the population center of the country? What can you conclude from the fact that the geographic center differs from the population center?
4. An amateur sculptor decides to portray a bird (Fig. 9-13). Luckily the final model is actually able to stand upright. The model is formed of a single sheet of metal of uniform thickness. Of the points shown, which is most likely to be the center of mass?

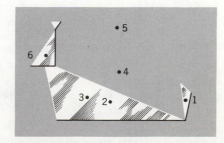

figure 9-13
Question 4

* See "Force, Momentum Change, and Motion" by Martin S. Tiersten, *American Journal of Physics*, January 1969, for an excellent general reference on systems of fixed and variable mass.

5. The location of the center of mass of a group of particles with respect to those particles does not depend on the reference frame used to describe the system. Is that so? Can you choose a reference frame whose origin is actually at the center of mass?

6. If only an external force can change the state of motion of the center of mass of a body, how does it happen that the internal force of the brakes can bring a car to rest?

7. Can a body have energy without having momentum? Explain. Can a body have momentum without having energy? Explain.

8. A light and a heavy body have equal kinetic energies of translation. Which one has the larger momentum?

9. A bird is in a wire cage hanging from a spring balance. Is the reading of the balance when the bird is flying about greater than, less than, or the same as that when the bird sits in the cage?

10. Can a sailboat be propelled by air blown at the sails from a fan attached to the boat?

11. A canoeist in a still pond can reach shore by jerking sharply on the rope attached to the bow of the canoe. How do you explain this? (yes she can!— its true).

12. How might a person standing at rest on a frictionless horizontal surface get altogether off of it?

13. A man stands still on a large sheet of slick ice; in his hand he holds a lighted firecracker. He throws the firecracker into the air. Describe briefly, but as exactly as you can, the motion of the center of mass of the firecracker and the motion of the center of mass of the system consisting of man and firecracker. It will be most convenient to describe each motion during each of the following periods: (a) after he throws the firecracker, but before it explodes; (b) between the explosion and the first piece of firecracker hitting the ice; (c) between the first fragment hitting the ice and the last fragment landing; (d) during the time when all fragments have landed but none has reached the edge of the ice.

14. As stated in the text one cannot use the equation $\mathbf{F}_{ext} = d(M\mathbf{v})/dt$ for a system of variable mass. To show this (a) put the equation in the equivalent form $\left(\mathbf{F}_{ext} - M\dfrac{d\mathbf{v}}{dt}\right)/(dM/dt) = \mathbf{v}$ and (b) show that one side of this equation has the same value in all inertial frames, whereas the other side does not. Hence the equation cannot be generally valid. (c) Show that Eq. 9-20 leads to no such contradiction.

15. You throw an ice cube with velocity $\mathbf{v}$ into a hot gravity-free, evacuated space. The cube gradually melts to liquid water and then boils to water vapor. (a) Is it a system of particles all the time? (b) If so, is it the same system of particles? (c) Does the motion of the center of mass undergo any abrupt changes? (d) Does the total linear momentum change? (e) Would your answers change if the space were not gravity free?

16. In 1920 a prominent newspaper editorialized as follows about the pioneering rocket experiments of Robert H. Goddard, dismissing the notion that a rocket could operate in a vacuum: "That Professor Goddard, with his 'chair' in Clark College and the countenancing of the Smithsonian Institution, does not know the relation of action to reaction, and of the need to have something better than a vacuum against which to react—to say that would be absurd. Of course, he seems only to lack the knowledge ladled out daily in high schools." What is wrong with this argument?

17. The final velocity of the final stage of a multistage rocket is much greater than the final velocity of a single-stage rocket of the same total weight and fuel supply. Explain this fact.

18. As a rocket expels burned fuel the location of the center of mass of the rocket (in a frame attached to the rocket) changes. Must one take this into account in an exact solution of the rocket problem?

19. Explain clearly the distinction between the origin of the varying mass of a classical system and that of a relativistic system.

20. Can you think of variable mass systems other than the examples given in the text?

problems

SECTION 9-1

1. Show that the ratio of the distances of two particles from their center of mass is the inverse ratio of their masses.

2. Experiments using the diffraction of electrons show that the distance between the centers of the carbon (C) and oxygen (O) atoms in the carbon monoxide gas molecule is 1.130×10^{-10} m. Locate the center of mass of a CO molecule relative to the carbon atom.

3. The mass of the moon is about 0.013 times the mass of the earth, and the distance from the center of the moon to the center of the earth is about 60 times the radius of the earth. How far is the center of mass of the earth-moon system from the center of the earth? Take the earth's radius to be 6400 km.   *Answer: 4900 km.*

4. The masses and coordinates of four particles are as follows: 5.0 kg, $x = y = 0.0$ cm; 3.0 kg, $x = y = 8.0$ cm; 2.0 kg, $x = 3.0$ cm, $y = 0.0$ cm; 6.0 kg, $x = -2.0$ cm, $y = -6.0$ cm. Find the coordinates of the center of mass of this collection of particles.

5. In the ammonia (NH_3) molecule, the three hydrogen (H) atoms form an equilateral triangle, the distance between centers of the atoms being 1.628×10^{-10} m, so that the center of the triangle is 9.39×10^{-11} m from each hydrogen atom. The nitrogen (N) atom is at the apex of a pyramid, the three hydrogens constituting the base (see Fig. 9-14). The hydrogen-nitrogen distance is 1.014×10^{-10} m. Locate the center of mass relative to the nitrogen atom.
   Answer: 6.74×10^{-12} m toward the plane of the hydrogens, along the axis of symmetry.

6. Find the center of mass of a homogenous semicircular plate. Let a be the radius of the circle.

figure 9-14
Problem 5

SECTION 9-2

7. Two blocks of masses 1.0 kg (weight 2.2 lb) and 3.0 kg (weight 6.6 lb) connected by a spring rest on a frictionless surface. If the two are given velocities such that the first travels at 1.7 m/s (5.6 ft/s) toward the center of mass which remains at rest, what is the velocity of the second?
   Answer: 0.57 m/s (1.9 ft/s), toward center of mass.

8. Two particles P and Q are initially at rest 1.0 m apart. P has a mass of 0.10 kg and Q a mass of 0.30 kg. P and Q attract each other with a constant force of 1.0×10^{-2} N. No external forces act on the system. (a) Describe the motion of the center of mass. (b) At what distance from P's original position do the particles collide?

9. A man of mass m clings to a rope ladder suspended below a balloon of mass M. The balloon is stationary with respect to the ground. (a) If the man begins to climb the ladder at a speed v (with respect to the ladder), in what direction and with what speed (with respect to the earth) will the balloon move? (b) What is the state of motion after the man stops climbing?

   Answer: (a) down, $\dfrac{m}{m + M} v$. (b) Balloon again stationary.

10. A cannon and a supply of cannon balls are inside a sealed railroad car as in Fig. 9-15. The cannon fires to the right; the car recoils to the left. The cannon balls remain in the car after hitting the far wall. Show that no matter

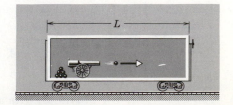

figure 9-15
Problem 10

how the cannon balls are fired the railroad car cannot travel more than its length L, assuming it starts from rest.

11. A dog, weighing 10 lb, is standing on a flatboat so that he is 20 ft from the shore. He walks 8.0 ft on the boat toward shore and then halts. The boat weighs 40 lb, and one can assume there is no friction between it and the water. How far is he from the shore at the end of this time? (Hint: The center of mass of boat + dog does not move. Why?) The shoreline is also to the left in Fig. 9-16. *Answer: 13.6 ft.*

figure 9-16
Problem 11

12. A ball of mass m and radius R is placed inside a larger hollow sphere with the same mass and inside radius $2R$. The combination is at rest on a frictionless surface in the position shown in Fig. 9-17. The smaller ball is released, rolls around the inside of the hollow sphere, and finally comes to rest at the bottom. How far will the larger sphere have moved during this process?

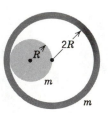

figure 9-17
Problem 12

13. Ricardo, mass 80 kg, and Carmelita are enjoying Lake Merced at dusk in a 30-kg canoe. When the canoe is at rest in the placid water they change seats, which are 3.0 m apart and symmetrically located with respect to the canoe's center. Ricardo notices that the canoe moved 0.40 m relative to a submerged log, and calculates Carmelita's mass, which she has declined to tell him. What is it? *Answer: 58 kg.*

14. An 80-kg man is standing at the rear of a 400-kg iceboat that is moving at 4.0 m/s across ice that may be considered to be frictionless. He decides to walk to the front of the 18 m-long boat and does so at a speed of 2.0 m/s with respect to the boat. How far did the boat move across the ice while he was walking?

SECTION 9-3

15. How fast must an 1800-lb (mass = 816 kg) Volkswagen travel (a) to have the same momentum as a 5850-lb (mass = 2650 kg) Cadillac going 10 mi/h (16 km/h)? (b) To have the same kinetic energy? (c) Make the same calculations using a 10-ton (mass = 9080 kg) truck instead of a Cadillac?
Answer: (a) 33 mi/h (52 km/h). (b) 18 mi/h (29 km/h). (c) 110 mi/h (180 km/h); 33 mi/h (52 km/h).

16. A 50-g ball is thrown into the air with an initial speed of 15 m/s at an angle of 45°. (a) What are the values of the kinetic energy of the ball initially and

just before it hits the ground? (b) Find the corresponding values of the momentum (magnitude and direction). (c) Show that the change in momentum is just equal to the weight of the ball multiplied by the time of flight.

17. A 5.0-kg object with a speed of 30 m/s strikes a steel plate at an angle of 45° and rebounds at the same speed and angle (Fig. 9-18). What is the change (magnitude and direction) of the linear momentum of the object?
Answer: 210 kg · m/s, perpendicular to the plate.

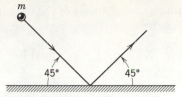

figure 9-18
Problem 17

18. Two bodies, each made up of weights from a set, are connected by a light cord which passes over a light, frictionless pulley with a diameter of 5.0 cm. The two bodies are at the same level. Each originally has a mass of 500 g. (a) Locate their center of mass. (b) Twenty grams are transferred from one body to the other, but the bodies are prevented from moving. Locate the center of mass. (c) The two bodies are now released. Describe the motion of the center of mass and determine its acceleration.

SECTION 9-4

19. A 200-lb man standing on a surface of negligible friction kicks forward a 0.10-lb stone lying at his feet so that it acquires a speed of 10 ft/s. What velocity does the man acquire as a result?
Answer: 5.0×10^{-3} ft/s, backward.

20. A pellet gun fires ten 2.0-g pellets per second with a speed of 500 m/s. The pellets are stopped by a rigid wall. (a) What is the momentum of each pellet? (b) What is the kinetic energy of each pellet? (c) What is the average force exerted by the pellets on the wall?

21. A machine gun fires 50-g bullets at a speed of 1000 m/s. The gunner, holding the machine gun in his hands, can exert an average force of 180 N against the gun. Determine the maximum number of bullets he can fire per minute.
Answer: 220 bullets per minute.

22. A very flexible uniform chain of mass M and length L is suspended from one end so that it hangs vertically, the lower end just touching the surface of a table. The upper end is suddenly released so that the chain falls onto the table and coils up in a small heap, each link coming to rest the instant it strikes the table. Find the force exerted by the table on the chain at any instant, in terms of the weight of chain already on the table at that moment.

SECTION 9-5

23. A body of mass 8.0 kg is traveling at 2.0 m/s under the influence of no external force. At a certain instant an internal explosion occurs, splitting the body into two chunks of 4.0 kg mass each; 16 J of translational kinetic energy are imparted to the two-chunk system by the explosion. Neither chunk leaves the line of the original motion. Determine the speed and direction of motion of each of the chunks after the explosion.
Answer: One chunk comes to rest. The other moves ahead with a speed of 4.0 m/s.

24. The last stage of a rocket is traveling at a speed of 25,000 ft/s (7600 m/s). This last stage is made up of two parts which are clamped together, namely, a rocket case with a mass of 20 slugs (290 kg) and a payload capsule with a mass of 10 slugs (150 kg). When the clamp is released, a compressed spring causes the two parts to separate with a relative speed of 3000 ft/s (910 m/s). (a) What are the speeds of the two parts after they have separated? Assume that all velocities are along the same line. (b) Find the total kinetic energy of the two parts before and after they separate and account for the difference, if any.

25. A radioactive nucleus, initially at rest, decays by emitting an electron and a neutrino at right angles to one another. The momentum of the electron is 1.2×10^{-22} kg·m/s and that of the neutrino is 6.4×10^{-23} kg·m/s. (a) Find the

direction and magnitude of the momentum of the recoiling nucleus. (b) The mass of the residual nucleus is 5.8×10^{-26} kg. What is its kinetic energy of recoil?

Answer: (a) 1.4×10^{-22} kg · m/s, 150° from the electron track and 120° from the neutrino track. (b) 1.0 eV.

26. Each minute, a special game-warden's machine gun fires 220 10-g rubber bullets with a muzzle velocity of 1200 m/s. How many bullets must be fired at an 85-kg animal charging toward the warden at 4.0 m/s in order to stop the animal in its tracks? (Assume the bullets travel horizontally and drop to the ground after striking the target.)

27. A vessel at rest explodes, breaking into three pieces. Two pieces, having equal mass, fly off perpendicular to one another with the same speed of 30 m/s. The third piece has three times the mass of each other piece. What is the direction and magnitude of its velocity immediately after the explosion?
Answer: 14 m/s, 135° from either other piece.

28. A shell is fired from a gun with a muzzle velocity of 1500 ft/s, at an angle of 60° with the horizontal. The shell explodes into two fragments of equal mass 50 s after leaving the gun. One fragment, whose speed immediately after the explosion is zero, falls vertically. How far from the gun does the other fragment land, assuming level terrain?

29. A block of mass m rests on a wedge of mass M which, in turn, rests on a horizontal table, as shown in Fig. 9-19. All surfaces are frictionless. If the system starts at rest with point P of the block a distance h above the table, find the velocity of the wedge the instant point P touches the table.

$$\text{Answer: } \sqrt{\frac{2m^2gh\,\cos^2\alpha}{(M+m)(M+m\,\sin^2\alpha)}}.$$

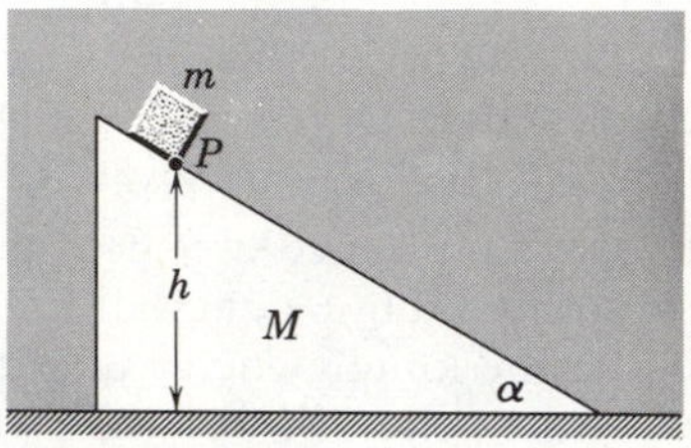

figure 9-19
Problem 29

SECTION 9-7

30. (a) Show that the rocket speed is equal to the exhaust speed when the ratio M_0/M is e (about 2.7). Specify the coordinate system in which this result holds. (b) Show also that the rocket speed is twice the exhaust speed when M_0/M is e^2 (about 7.4).

31. A rocket is moving away from the solar system at a speed of 6.0×10^3 m/s. It fires its rocket engine, which ejects exhaust with a relative velocity of 3.0×10^3 m/s. The mass of the rocket at this time is 4.0×10^4 kg, and it experiences an acceleration of 2.0 m/s². (a) What is the velocity of the exhaust relative to the solar system? (b) At what rate was exhaust ejected during the firing?   *Answer:* (a) 3.0×10^3 m/s. (b) 27 kg/s.

32. A widely used rocket fuel is kerosene and liquid oxygen, capable of giving an exhaust velocity v_{rel} of 8000 ft/s (about 1.5 mi/s). (a) Neglect gravity and the weight of fuel tanks, pumps, etc., and find how many pounds of this fuel one needs for each pound of payload in order to get a rocket, starting from rest, to reach a velocity of 7.5 mi/s (the velocity of escape from the earth is 7.0 mi/s). (b) In the Mariner probe to Mars the initial weight was about 200,000 lb and the payload about 500 lb, a "fuel" to payload ratio of 400 to 1. Starting a rocket from rest, what final velocity is achievable under these circumstances? (c) The actual final rocket velocity was about 15 mi/s, much greater than the value found in (b). Explain this, considering the following factors: the external forces and weight neglected in (a) must be taken into account; the rocket uses a number of stages; the initial rocket velocity is that of the earth's surface, in a reference frame attached to the sun.

33. A 6000-kg rocket is set for vertical firing. If the exhaust speed is 1000 m/s, how much gas must be ejected each second to supply the thrust needed (a) to overcome the weight of the rocket, and (b) to give the rocket an initial upward acceleration of 20 m/s²?   *Answer:* (a) 59 kg/s. (b) 180 kg/s.

34. Consider a particle acted on by a force having the same direction as its velocity. (a) Using the relativistic relation $F = d(mv)/dt$ for a single particle, show that

$$F \, ds = mv \, dv + v^2 \, dm,$$

where ds is an infinitesimal displacement. (b) Using the relativistic relation $v^2 = (1 - m_0^2/m^2)c^2$, show that

$$mv \, dv = \frac{m_0^2 c^2}{m^2} \, dm.$$

(c) Substitute the relations for $mv \, dv$ and v^2 into result (a) and show that

$$W = \int F \, ds = (m - m_0)c^2.$$

35. A railroad flatcar of weight W can roll without friction along a straight horizontal track as shown. Initially a man of weight w is standing on the car which is moving to the right with speed v_0. What is the change in velocity of the car if the man runs to the left (Fig. 9-20) so that his speed relative to the car is v_{rel} just before he jumps off at the left end? *Answer:* $wv_{rel}/(W + w)$.

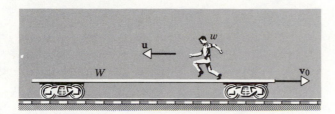

figure 9-20
Problem 35

36. Assume that the car in Problem 35 is initially at rest. It holds n men each of weight w. If each man in succession runs with a relative velocity v_{rel} and jumps off the end, do they impart to the car a greater velocity than if they all run and jump at the same time?

37. A toboggan weighing 12 lb and carrying 80 lb of sand slides from rest down an icy slope 300 ft long, inclined 30° below the horizontal. The toboggan has a hole in the bottom, so that the sand leaks out at the rate of 5.0 lb/s. How long does it take the toboggan to reach the bottom of the slope? *Answer:* 6.1 s.

38. Two long barges are floating in the same direction in still water, one with a speed of 10 km/h and the other with a speed of 20 km/h. While they are passing each other, coal is shoveled from the slower to the faster one at a rate of 1000 kg/min. How much additional force must be provided by the driving engines of each barge if neither is to change speed? Assume that the shoveling is always perfectly sideways and that the frictional forces between the barges and the water do not depend on the weight of the barges.

39. A jet airplane is traveling 180 m/s (600 ft/s). The engine takes in 68 m³ (2400 ft³) of air making a mass of 70 kg (4.8 slugs) each second. The air is used to burn 2.9 kg (0.20 slugs) of fuel each second. The energy is used to compress the products of combustion and to eject them at the rear of the plane at 490 m/s (1600 ft/s) relative to the plane. Find (a) the thrust of the jet engine and (b) the delivered power (horsepower). *Answer:* (a) 2.3×10^4 N (5100 lb). (b) 4.1×10^6 W (5600 hp).

40. A freight car, open at the top, weighing 10 tons, is coasting along a level track with negligible friction at 2.0 ft/s when it begins to rain hard. The raindrops fall vertically with respect to the ground. What is the speed of the car when it has collected 0.50 ton of rain? What assumptions, if any, must you make to get your answer?

41. A flexible inextensible string of length l is threaded into a smooth tube, into which it snugly fits. The tube contains a right-angled bend, and is positioned in the vertical plane so that one arm is vertical and the other horizontal. Initially, at $t = 0$, a length y_0 of the string is hanging down in the vertical

arm. The string is released and slides through the tube, so that at any subsequent time t later, it is moving with a speed dy/dt, where $y(t)$ is the length
of the string that is then hanging vertically. (a) Show that in terms of the
variable mass problem $\mathbf{v}_{\mathrm{rel}} = 0$, so that the equation of motion has the form
$m\, d\mathbf{v}/dt = \mathbf{F}_{\mathrm{ext}}$. ($b$) Show that the specific equation of motion is $(d^2y/dt^2) = gy$.
(c) Show that conservation of mechanical energy leads to $(dy/dt)^2 - gy^2 = a$
constant, and that this is consistent with (b). (d) Show that $y = (y_0/2)(e^{\sqrt{g/l}\,t} +
e^{-\sqrt{g/l}\,t})$ is a solution to the equation of motion [by substitution into (b)] and
discuss the solution.

10
collisions

We learn much about atomic, nuclear, and elementary particles experimentally by observing collisions between them. On a larger scale we can interpret such things as the properties of gases in terms of particle collisions. In this chapter we apply the principles of conservation of energy and conservation of momentum to the collisions of particles.

In a collision a relatively large force acts on each colliding particle for a relatively short time. The basic idea of a "collision" is that the motion of the colliding particles (or of at least one of them) changes rather abruptly and that we can make a relatively clean separation of times that are "before the collision" and those that are "after the collision."

When a bat strikes a baseball for example, the beginning and the end of the collision can be determined fairly precisely. The bat is in contact with the ball for an interval that is quite short in comparison to the time during which we are watching the ball. During the collision the bat exerts a large force on the ball (Fig. 10-1). This force varies with time in a complex way that we can measure only with difficulty. Both the ball and the bat are deformed during the collision.* Forces that act for a time that is short compared to the time of observation of the system are called *impulsive* forces.

When an alpha particle (He⁴) "collides" with a nucleus of gold (Au¹⁹⁷), the force acting between them may be the well-known repulsive electrostatic force associated with the charges on the particles. The particles may not "touch," but we still may speak of a "collision" because a relatively strong force, acting for a time that is short in comparison to the

10-1
WHAT IS A COLLISION?

figure 10-1
A high-speed flash photograph of a bat striking a baseball. Notice the deformation of the ball, indicating the enormous magnitude of the impulsive force at this instance. (Courtesy Harold E. Edgerton, Massachusetts Institute of Technology, Cambridge, Mass.)

---

* See "Batting the Ball" by P. Kirkpatrick, *American Journal of Physics*, August 1963.

time that the alpha particle is under observation, has a marked effect on the motion of the alpha particle.

When a proton (H^1 or p) with energy of, say, 25 MeV, "collides" with a nucleus of, say, a silver isotope (perhaps Ag^{107}), the particles may actually "touch," the predominant force then acting between them being, not the electrostatic repulsive force, but the strong, short-range, attractive nuclear force (see page 106). The proton may enter the silver nucleus, forming a compound structure. A short time later—the "collision interval" may be 10^{-18} s—this compound structure may break up into *two different* particles, according to a scheme such as

$$p + Ag^{107} \rightarrow \alpha + Pd^{104},$$

in which α (= He^4) is an alpha particle. Thus we may broaden the concept of collision to include events (usually called *reactions*) in which the identities of the interacting particles change during the event. The conservation principles are applicable to all these examples.

We may, if we wish, broaden our definition of "collision" even further to include the spontaneous decay of a single particle into two or more other particles. An example is the decay of the elementary particle called the *sigma particle* into two other particles, the *pion* and the *neutron* (see Appendix I) or

$$\Sigma^- \rightarrow \pi^- + n.$$

Although two bodies do not come in contact in this process (unless we consider it in reverse), it has many features in common with collisions: (1) there is a clean distinction between "before the event" and "after the event," and (2) the laws of conservation of momentum and energy permit us to learn much about such processes by studying the "before" and "after" situations, even though we may know little about the force laws that operate during the "event" itself.

In studying collisions in this chapter our aim will be this: given the initial motions of the colliding particles, what can we learn about their final motions from the principles of conservation of momentum and energy, assuming that we know nothing about the forces acting during the collision?

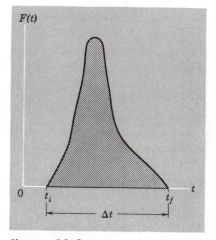

figure 10-2
How an impulsive force $F(t)$ might vary with time during a collision starting at time t_i and ending at t_f.

10-2
IMPULSE AND MOMENTUM

Let us assume that Fig. 10-2 shows the magnitude of the force exerted on a body during a collision. We assume that the force has a constant direction. The collision begins at time t_i and ends at time t_f, the force being zero before and after collision. From Eq. 9-12 we can write the change in momentum $d\mathbf{p}$ of a body in a time dt during which a force $\mathbf{F}$ acts on it as

$$d\mathbf{p} = \mathbf{F}\,dt. \qquad (10\text{-}1)$$

We can find the change in momentum of the body during a collision by integrating over the time of collision, that is,

$$\mathbf{p}_f - \mathbf{p}_i = \int_{\mathbf{p}_i}^{\mathbf{p}_f} d\mathbf{p} = \int_{t_i}^{t_f} \mathbf{F}\,dt \qquad (10\text{-}2)$$

in which the subscripts i (= *initial*) and f (= *final*) refer to the times before and after the collision, respectively. The integral of a force over the time interval during which the force acts is called the *impulse* **J** of the force. Hence the change in momentum of a body acted on by an impulsive force is equal to the impulse. Both impulse and momentum are vectors and both have the same units and dimensions.

The impulsive force represented in Fig. 10-2 is assumed to have a constant direction. The impulse of this force, $\int_{t_i}^{t_f} \mathbf{F}\, dt$, is represented in magnitude by the area under the force-time curve.*

Consider now a collision between two particles, such as those of masses m_1 and m_2, shown in Fig. 10-3. During the brief collision these particles exert large forces on one another. At any instant $\mathbf{F}_1$ is the force exerted on particle 1 by particle 2 and $\mathbf{F}_2$ is the force exerted on particle 2 by particle 1. By Newton's third law these forces at any instant are equal in magnitude but oppositely directed.

The change in momentum of particle 1 resulting from the collision is

$$\Delta \mathbf{p}_1 = \int_{t_i}^{t_f} \mathbf{F}_1\, dt = \overline{\mathbf{F}_1}\, \Delta t$$

in which $\overline{\mathbf{F}_1}$ is the average value of the force $\mathbf{F}_1$ during the time interval of the collision $\Delta t = t_f - t_i$.

The change in momentum of particle 2 resulting from the collision is

$$\Delta \mathbf{p}_2 = \int_{t_i}^{t_f} \mathbf{F}_2\, dt = \overline{\mathbf{F}_2}\, \Delta t$$

in which $\overline{\mathbf{F}_2}$ is the average value of the force $\mathbf{F}_2$ during the time interval of the collision $\Delta t = t_f - t_i$.

If no other forces act on the particles, then $\Delta \mathbf{p}_1$ and $\Delta \mathbf{p}_2$ give the total change in momentum for each particle. But we have seen that at each instant $\mathbf{F}_1 = -\mathbf{F}_2$, so that $\overline{\mathbf{F}_1} = -\overline{\mathbf{F}_2}$, and therefore

$$\Delta \mathbf{p}_1 = -\Delta \mathbf{p}_2.$$

If we consider the two particles as an isolated system, the total momentum of the system is

$$\mathbf{P} = \mathbf{p}_1 + \mathbf{p}_2,$$

and the total *change* in momentum of the system as a result of the collision is zero, that is,

$$\Delta \mathbf{P} = \Delta \mathbf{p}_1 + \Delta \mathbf{p}_2 = 0.$$

Hence, *if there are no external forces* the total momentum of the system is not changed by the collision. The impulsive forces acting during the collision are internal forces which have no effect on the total momentum of the system.

We have defined a collision as an interaction which occurs in a time Δt that is negligible compared to the time during which we are observing the system. We can also characterize a collision as an event in which the external forces that may act on the system are negligible compared to the impulsive collision forces. When a bat strikes a baseball, a golf club strikes a golf ball, or one billiard ball strikes another,

10-3
CONSERVATION OF MOMENTUM DURING COLLISIONS

figure 10-3
Two "particles" m_1 and m_2, in collision, experience equal and opposite forces along their line of centers, according to Newton's third law; $\mathbf{F}_2(t) = -\mathbf{F}_1(t)$.

---

* The impulse **J**, defined from Eq. 10-2, does not depend critically on the precise values of t_i and t_f as long as these times are far enough apart to include the crosshatched area of Fig. 10-2. For reasons that will appear later we usually choose t_i and t_f with a separation that is *just large enough* to make a clean distinction between the "collision" and the "before and after intervals."

external forces act on the system. Gravity or friction exerts forces on these bodies, for example; these external forces may not be the same on each colliding body nor are they necessarily canceled by other external forces. Even so it is quite safe to neglect these external forces during the collision and to assume momentum conservation provided, as is almost always true, that the external forces are negligible compared to the impulsive forces of collision. As a result the change in momentum of a particle during a collision arising from an external force is negligible compared to the change in momentum of that particle arising from the impulsive collisional force (Fig. 10-4).

For example, when a bat strikes a baseball, the collision lasts only a small fraction of a second. Because the change in momentum is large and the time of collision is small, it follows from

$$\Delta \mathbf{p} = \overline{\mathbf{F}} \, \Delta t$$

that the average impulsive force $\overline{\mathbf{F}}$ is relatively large. Compared to this force, the external force of gravity is negligible. *During the collision* we can safely ignore this external force in determining the change in motion of the ball; the shorter the duration of the collision the more likely this is to be true.

In practice, therefore, *we can apply the principle of momentum conservation during collisions if the time of collision is small enough.* We can then say that the momentum of a system of particles just before the particles collide is equal to the momentum of the system just after the particles collide.

We can always calculate the motions of bodies after collision from their motions before collision if we know the forces that act during the collision, and if we can solve the equations of motion. Often we do not know these forces. However, the principle of conservation of momentum must hold during the collision. We already know that the principle of conservation of total energy holds. Although we may not know the details of the interaction, we can use these principles in many cases to predict the results of the collision.

Collisions are usually classified according to whether or not *kinetic energy* is conserved in the collision. When kinetic energy is conserved, the collision is said to be *elastic*. Otherwise, the collision is said to be *inelastic*. Collisions between atomic, nuclear, and fundamental particles are *sometimes* (but not always) elastic. These are, in fact, the only truly elastic collisions known. Collisions between gross bodies are always inelastic to some extent. We can often treat such collisions as approximately elastic, however, as, for example, collisions between ivory or glass balls. When two bodies stick together after collision, the collision is said to be *completely inelastic*. For example, the collision between a bullet and a block of wood into which it is fired is completely inelastic when the bullet remains embedded in the block. The term completely inelastic does *not* mean that all the initial kinetic energy is lost; as we shall see, it means rather that the loss is as great as is consistent with momentum conservation.

Even if the forces of collision are not known, we can find the motions of the particles after collision from the motions before collision, provided the collision is completely inelastic, or, if the collision is elastic, provided the collision is a one-dimensional one. For a one-dimensional collision the relative motion after collision is along the

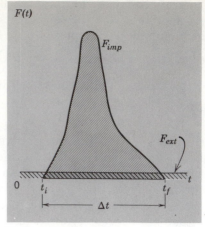

figure 10-4
During a collision, the impulsive force F_{imp} is generally much greater than any external forces F_{ext}, which may act on the system.

10-4
COLLISIONS IN ONE DIMENSION

same line as the relative motion before collision. We restrict ourselves to one-dimensional motion for the present.

Consider first an *elastic* one-dimensional collision. We can imagine two smooth nonrotating spheres moving initially along the line joining their centers, then colliding head-on and moving along the same straight line without rotation after collision (see Fig. 10-5). These bodies exert forces on each other during the collision that are along the initial line of motion, so that the final motion is also along this same line.

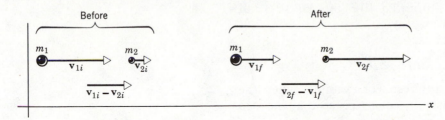

figure 10-5
Two spheres before and after an elastic collision. The velocity, $\mathbf{v}_{1i} - \mathbf{v}_{2i}$, of m_1 relative to m_2 before collision is equal to the velocity, $\mathbf{v}_{2f} - \mathbf{v}_{1f}$, of m_2 relative to m_1 after collision.

The masses of the spheres are m_1 and m_2, the (scalar) velocity components being v_{1i} and v_{2i} before collision and v_{1f} and v_{2f} after collision.* We take the positive direction of the momentum and velocity to be to the right. We assume, unless we specify otherwise, that the speeds of the colliding particles are low enough so that we need not use the relativistic expressions for momentum and kinetic energy. Then from conservation of momentum we obtain

$$m_1 v_{1i} + m_2 v_{2i} = m_1 v_{1f} + m_2 v_{2f}.$$

Because we are considering an elastic collision the kinetic energy is conserved by definition and we obtain

$$\tfrac{1}{2} m_1 v_{1i}^2 + \tfrac{1}{2} m_2 v_{2i}^2 = \tfrac{1}{2} m_1 v_{1f}^2 + \tfrac{1}{2} m_2 v_{2f}^2.$$

It is clear at once that, if we know the masses and the initial velocities, we can calculate the two final velocities v_{1f} and v_{2f} from these two equations.

The momentum equation can be written as

$$m_1(v_{1i} - v_{1f}) = m_2(v_{2f} - v_{2i}), \qquad (10\text{-}3)$$

and the energy equation can be written as

$$m_1(v_{1i}^2 - v_{1f}^2) = m_2(v_{2f}^2 - v_{2i}^2). \qquad (10\text{-}4)$$

Dividing Eq. 10-4 by Eq. 10-3, and assuming $v_{2f} \neq v_{2i}$ and $v_{1f} \neq v_{1i}$ (see Question 7), we obtain

$$v_{1i} + v_{1f} = v_{2f} + v_{2i}$$

and, after rearrangement,

$$v_{1i} - v_{2i} = v_{2f} - v_{1f} \qquad (10\text{-}5)$$

This tells us that in an elastic one-dimensional collision, the relative velocity of approach before collision is equal to the relative velocity of separation after collision.

---

* The notation used is easy to interpret and to remember and reveals much information in a simple compact way. The number subscripts, such as 1 and 2, specify the particle and the letter subscripts, i and f, indicate initial value (before the collision) and final value (after the collision), respectively.

To find the velocity components v_{1f} and v_{2f} after collision from the velocity components v_{1i} and v_{2i} before collision, we can use any two of the three previous numbered equations. Thus from Eq. 10-5

$$v_{2f} = v_{1i} + v_{1f} - v_{2i}.$$

Inserting this into Eq. 10-3 and solving for v_{1f}, we find that

$$v_{1f} = \left(\frac{m_1 - m_2}{m_1 + m_2}\right)v_{1i} + \left(\frac{2m_2}{m_1 + m_2}\right)v_{2i}.$$

Likewise, inserting $v_{1f} = v_{2f} + v_{2i} - v_{1i}$ (from Eq. 10-5) into Eq. 10-3 and solving for v_{2f}, we obtain

$$v_{2f} = \left(\frac{2m_1}{m_1 + m_2}\right)v_{1i} + \left(\frac{m_2 - m_1}{m_1 + m_2}\right)v_{2i}.$$

There are several cases of special interest. For example, when the colliding particles have the same mass, m_1 equals m_2 so that the two previous equations become simply

$$v_{1f} = v_{2i} \quad \text{and} \quad v_{2f} = v_{1i}.$$

That is, in a one-dimensional elastic collision of two particles of equal mass, the particles simply exchange velocities during collision.

Another case of interest is that in which one particle m_2 is initially at rest. Then v_{2i} equals zero and

$$v_{1f} = \left(\frac{m_1 - m_2}{m_1 + m_2}\right)v_{1i}, \qquad v_{2f} = \left(\frac{2m_1}{m_1 + m_2}\right)v_{1i}.$$

Of course, if $m_1 = m_2$ also, then $v_{1f} = 0$ and $v_{2f} = v_{1i}$ as we expect. The first particle is "stopped cold" and the second one "takes off" with the velocity the first one originally had. If, however, m_2 is very much greater than m_1, we obtain

$$v_{1f} \cong -v_{1i} \quad \text{and} \quad v_{2f} \cong 0.$$

That is, when a light particle collides with a very much more massive particle at rest, the velocity of the light particle is approximately reversed and the massive particle remains approximately at rest. For example, suppose that we drop a ball vertically onto a horizontal surface attached to the earth. This is in effect a collision between the ball and the earth. If the collision is elastic, the ball will rebound with a reversed velocity and will reach the same height from which it fell.

If, finally, m_2 is very much smaller than m_1, we obtain

$$v_{1f} \cong v_{1i} \qquad v_{2f} \cong 2v_{1i}.$$

This means that the velocity of the massive incident particle is virtually unchanged by the collision with the light stationary particle, but that the light particle rebounds with approximately twice the velocity of the incident particle. The motion of a bowling ball is hardly affected by collision with an inflated beach ball of the same size, but the beach ball bounces away quickly.

Neutrons produced in a reactor from the fission of uranium atoms move very fast and must be slowed down if they are to produce more fissions. Assuming that they make elastic collisions with the nuclei at rest, what material should be picked to moderate (that is, to slow down) the neutrons in the reactor? We can answer this from the considerations just discussed. If the stationary targets were massive nuclei, like lead, the neutrons would simply bounce back with practically the same speed they had initially. If the stationary targets were very much lighter than the neutrons, like electrons, the neutrons would move on with practically the same velocity they had initially. However, if the stationary targets are particles of nearly the same mass, the neutrons will be brought almost to rest in a (head-on) collision with them. Hence, hydrogen, whose nucleus (proton) has nearly the same mass as a neutron, should be most effective. Other considerations affect the choice of a moderator for neutrons, but momentum and energy considerations alone limit the choice to the lighter elements.

If a collision is *inelastic* then, by definition, the kinetic energy is not conserved. The final kinetic energy may be less than the initial value, the difference being ultimately converted to heat energy or to potential energy of deformation in the collision, for example; or the final kinetic energy may exceed the initial value, as when potential energy is released in the collision. In any case, the conservation of momentum still holds, as does the conservation of *total* energy.

Let us consider finally a *conpletely inelastic* collision. The two particles stick together after collision, so that there will be a final common velocity $\mathbf{v}_f$. It is not necessary to restrict the discussion to one-dimensional motion. Using only the conservation of momentum principle, we find

$$m_1\mathbf{v}_{1i} + m_2\mathbf{v}_{2i} = (m_1 + m_2)\mathbf{v}_f. \qquad (10\text{-}6)$$

This determines $\mathbf{v}_f$ when $\mathbf{v}_{1i}$ and $\mathbf{v}_{2i}$ are known.

EXAMPLE 1

A baseball weighing 0.35 lb is struck by a bat while it is in horizontal flight with a speed of 90 ft/s. After leaving the bat the ball travels with a speed of 110 ft/s in a direction opposite to its original motion. Determine the impulse of the collision.

We cannot calculate the impulse from the definition $\mathbf{J} = \int \mathbf{F}\,dt$ because we do not know the force exerted on the ball as a function of time. However, we have seen (Eq. 10-2) that the change in momentum of a particle acted on by an impulsive force is equal to the impulse. Hence

$$\mathbf{J} = \text{change in momentum} = \mathbf{p}_f - \mathbf{p}_i$$

$$= m\mathbf{v}_f - m\mathbf{v}_i = \left(\frac{W}{g}\right)(\mathbf{v}_f - \mathbf{v}_i).$$

Assuming arbitrarily that the direction of $\mathbf{v}_i$ is positive, the impulse is then

$$J = \left(\frac{0.35\text{ lb}}{32\text{ ft/s}^2}\right)(-110\text{ ft/s} - 90\text{ ft/s}) = -2.2\text{ lb·s}$$

The minus sign shows that the direction of the impulse acting on the ball is opposite that of the original velocity of the ball.

We cannot determine the force of the collision from the data we are given. Actually, any force whose impulse is -2.2 lb·s will produce the same change in momentum. For example, if the bat and ball were in contact for 0.0010 s, the average force during this time would be

$$\bar{F} = \frac{\Delta p}{\Delta t} = \frac{-2.2\text{ lb·s}}{0.0010\text{ s}} = -2200\text{ lb}.$$

For a shorter contact time the average force would be greater. The actual force would have a maximum value greater than this average value.

How far would gravity cause the baseball to fall during its collision time?

(a) By what fraction is the kinetic energy of a neutron (mass m_1) decreased in a head-on elastic collision with an atomic nucleus (mass m_2) initially at rest?

EXAMPLE 2

The initial kinetic energy of the neutron K_i is $\frac{1}{2}m_1v_{1i}^2$. Its final kinetic energy K_f is $\frac{1}{2}m_1v_{1f}^2$. The fractional decrease in kinetic energy is

$$\frac{K_i - K_f}{K_i} = \frac{v_{1i}^2 - v_{1f}^2}{v_{1i}^2} = 1 - \frac{v_{1f}^2}{v_{1i}^2}.$$

But, for such a collision,

$$v_{1f} = \left(\frac{m_1 - m_2}{m_1 + m_2}\right)v_{1i},$$

so that

$$\frac{K_i - K_f}{K_i} = 1 - \left(\frac{m_1 - m_2}{m_1 + m_2}\right)^2 = \frac{4m_1m_2}{(m_1 + m_2)^2}.$$

(b) Find the fractional decrease in the kinetic energy of a neutron when it collides in this way with a lead nucleus, a carbon nucleus, and a hydrogen nucleus. The ratio of nuclear mass to neutron mass ($= m_2/m_1$) is 206 for lead, 12 for carbon, and 1 for hydrogen.

For lead, $m_2 = 206m_1$,

$$\frac{K_i - K_f}{K_i} = \frac{4 \times 206}{(207)^2} = 0.02 \quad \text{or} \quad 2\%.$$

For carbon, $m_2 = 12m_1$,

$$\frac{K_i - K_f}{K_i} = \frac{4 \times 12}{(13)^2} = 0.28 \quad \text{or} \quad 28\%.$$

For hydrogen, $m_2 = m_1$,

$$\frac{K_i - K_f}{K_i} = \frac{4 \times 1}{(2)^2} = 1 \quad \text{or} \quad 100\%.$$

These results explain why paraffin, which is rich in hydrogen, is far more effective in slowing down neutrons than is lead.

The Ballistic Pendulum. The ballistic pendulum is used to measure bullet speeds. The pendulum is a large wooden block of mass M hanging vertically by two cords. A bullet of mass m, traveling with a horizontal speed v_i, strikes the pendulum and remains embedded in it (Fig. 10-6). If the collision time (the time required for the bullet to come to rest with respect to the block) is very small compared to the time of swing of the pendulum, the supporting cords remain approximately vertical during the collision. Therefore, no external horizontal force acts on the system (bullet + pendulum) during collision, and the horizontal component of momentum is conserved. The speed of the system after collision v_f is much less than that of the bullet before collision. This final speed can be easily determined, so that the original speed of the bullet can be calculated from momentum conservation.

EXAMPLE 3

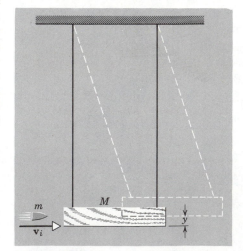

The initial momentum of the system is that of the bullet mv_i, and the momentum of the system just after collision is $(m + M)v_f$, so that

$$mv_i = (m + M)v_f.$$

After the collision is over, the pendulum and bullet swing up to a maximum height y, where the kinetic energy left after impact is converted into gravitational potential energy. Then, using the conservation of mechanical energy *for this part of the motion*, we obtain

$$\tfrac{1}{2}(m + M)v_f^2 = (m + M)gy.$$

Solving these two equations for v_i, we obtain

$$v_i = \frac{m + M}{m}\sqrt{2gy}.$$

Hence, we can find the initial speed of the bullet by measuring m, M, and y.

figure 10-6
Example 3. A ballistic pendulum consisting of a large wooden block of mass M suspended by cords. When a bullet of mass m and velocity $\mathbf{v}_i$ is fired into it, the block swings, rising a maximum distance y.

The kinetic energy of the bullet initially is $\frac{1}{2}mv_i^2$ and the kinetic energy of the system (bullet + pendulum) just after collision is $\frac{1}{2}(m + M)v_f^2$. The ratio is

$$\frac{\frac{1}{2}(m + M)v_f^2}{\frac{1}{2}mv_i^2} = \frac{m}{m + M}.$$

For example, if the bullet has a mass $m = 5$ g and the block has a mass $M = 2000$ g, only about one-fourth of 1% of the original kinetic energy remains; over 99% is converted to other forms of energy, such as heat energy.

The velocity of the center of mass of two particles is not changed by their collision, for the collision, whether elastic or inelastic, does not change the total momentum of the system of two particles, only the distribution of momentum between the two particles. The momentum of the system can be written (Eq. 9-15) as $\mathbf{P} = (m_1 + m_2)\mathbf{v}_{cm}$. If no external forces act on the system, then $\mathbf{P}$ is constant before and after the collision, and the center of mass moves with uniform velocity throughout.

If we choose a reference frame attached to the center of mass, then in this center-of-mass reference frame, $\mathbf{v}_{cm} = 0$ and $\mathbf{P} = 0$. There is a great simplicity and symmetry in describing collisions with respect to the center of mass, and it is customary to do so in nuclear physics. For whether collisions are elastic or inelastic, momentum is conserved, and in the center of mass reference frame the total momentum is zero. These results hold in two and three dimensions as well as in one because momentum is a vector quantity.

As an example, consider a head-on elastic collision between two particles m_1 and m_2. Let m_2 equal $3m_1$ and let m_2 be at rest, so that v_{2i} equals zero in the laboratory reference frame. The total momentum of the two particles is just that of the incident particle m_1v_{1i}, so that

$$m_1v_{1i} = (m_1 + m_2)v_{cm}$$

or

$$v_{cm} = \left(\frac{m_1}{m_1 + m_2}\right)v_{1i} = \frac{1}{4}v_{1i}.$$

After the collision, m_1 has a velocity $v_{1f} = -\frac{1}{2}v_{1i}$ and m_2 has a velocity $v_{2f} = \frac{1}{2}v_{1i}$. The total momentum of the two particles $(m_1v_{1f} + m_2v_{2f})$ is the same as before the collision, and the motion of the center of mass is unchanged (check this). In Fig. 10-7a we show a series of "snapshots" of the collision taken at equal time

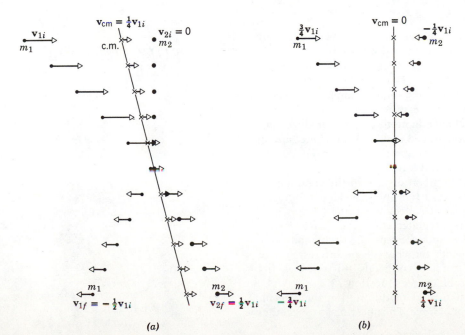

(a) (b)

figure 10-7
(a) An elastic collision in the laboratory reference frame. (b) The same elastic collision in the center-of-mass reference frame.

intervals as seen in the laboratory reference frame. In Fig. 10-7b we show the same situation as seen in the center-of-mass reference frame, where v_{cm} is zero. Notice the symmetry of the particles' motions when described in this way. The particle coming from the left has a speed $\frac{3}{4}v_{1i}$ with respect to the center of mass (where v_{1i} is the speed of m_1 in the laboratory frame) and recedes with this same speed. The particle coming from the right has a speed $\frac{1}{4}v_{1i}$ with respect to the center of mass and recedes with this same speed.

If the collision is completely inelastic, the motion after collision is simply that of two particles moving along together at the center of mass. In Figs. 10-8a and 10-8b we show how the collision of Fig. 10-7, now assumed to be completely inelastic, would be described in the laboratory and the center-of-mass reference frames, respectively. These figures are exactly like the previous ones until the collision occurs; after the collision, however, the motion of the center of mass describes that of the entire system.

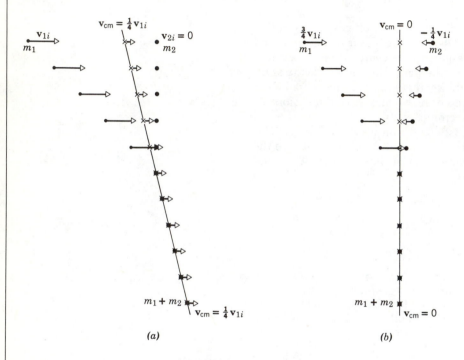

(a)   (b)

figure 10-8
(a) A completely inelastic collision in the laboratory reference frame. *(b)* The same completely inelastic collision in the center-of-mass reference frame. In each case the motion before collision is the same as that of Fig. 10-7.

The distinction between kinetic energy and momentum and the relationship of these concepts to force were not clearly understood until late in the eighteenth century. Scientists argued whether kinetic energy or momentum was the "true" measure of the effect of a force on a body. Descartes argued that when bodies interact, all that can happen is the transfer of momentum from one body to another, for the total momentum of the universe remains constant; hence the only "true" measure of a force is the change in momentum it produces in a given time. Leibnitz attacked this view and argued that the "true" measure of a force is the change it produces in kinetic energy (called by him *vis viva* or living force, taken to be twice what we now call kinetic energy).

In his treatise on mechanics (1743), D'Alembert dismissed the argument as being pointless and arising from a confusion of terminology. The cumulative effect of a force can be measured by its integrated effect over *time*, $\int F\, dt$, which produces a change in momentum, *or* by its integrated effect over *space*, $\int F\, dx$, which produces a change in kinetic energy. Both concepts are useful and valid, although different. Which one we use depends on what we are interested in or what is more convenient. As our present study of collisions illustrates, we frequently use both concepts in the same problem (see Question 23).

A more modern view is to look for quantities of the motion that are invariant, rather than focusing on the concept of force. The question as to whether

10-5
THE "TRUE" MEASURE OF A FORCE

the energy or the momentum is the "real" quantity of motion becomes point-less for there is no unique "quantity of motion." Instead, *both* energy *and* momentum may be regarded as invariant quantities of the motion in that for an isolated system the total of each of these quantities, summed up over all parts of the system, remains constant with time. There may be an exchange of energy, and of momentum, between different parts of an isolated system, but the total of each quantity is conserved.

10-6
COLLISIONS IN TWO AND THREE DIMENSIONS

In two or three dimensions (except for a completely inelastic collision) the conservation laws alone cannot tell us the motion of particles after a collision if we know the motion before the collision. For example, for a two-dimensional elastic collision, which is the simplest case, we have four unknowns, namely the two components of velocity for each of two particles after collision; but we have only three known relations between them, one for the conservation of kinetic energy and a con-servation of momentum relation for each of the two dimensions. Hence we need more information than just the initial conditions. When we do not know the actual forces of interaction, as is often the case, the additional information must be obtained from experiment. It is sim-plest to specify the angle of recoil of one of the colliding particles.

Let us consider what happens when one particle is projected at a target particle which is at rest. This case is not as restrictive as it may seem, for we can always pick our reference frame to be one in which the target particle is at rest before the collision. Much experimental work in nuclear physics involves projecting nuclear particles at a target which is stationary in the laboratory reference frame. In such collisions, because of momentum conservation, the motion is in a plane deter-mined by the lines of recoil of the colliding particles. The initial mo-tion need not be along the line joining the centers of the two particles. The force of interaction may be electromagnetic (in which we include "contact" forces; see page 106), gravitational, or nuclear. The par-ticles need not "touch"; strong forces, which act at relatively close dis-tances of approach and for a time short compared to the observation time, deflect the particles from their initial courses.

A typical situation is shown in Fig. 10-9. The distance b between the initial line of motion and a line parallel to it through the center of the

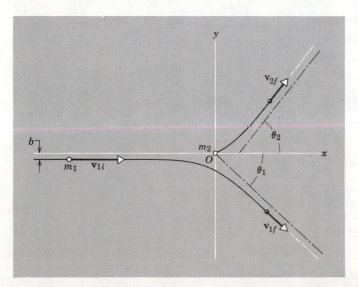

figure 10-9
Two particles, m_1 and m_2, undergoing a collision. The open circles indicate their positions before collision, the shaded ones after collision. Initially m_2 is at rest. The impact parameter b is the distance by which the collision misses being head-on.

target particle is called the *impact parameter*. This is a measure of the directness of the collision, $b = 0$, corresponding to a head-on collision. The direction of motion of the incident particle m_1 after collision makes an angle θ_1 with the initial direction, and the target projectile m_2, initially at rest, moves in a direction after collision making an angle θ_2 with the initial direction of the incident projectile. Applying the conservation of momentum, which is a vector relation, we obtain two scalar equations; for the x-component of motion we have

$$m_1 v_{1i} = m_1 v_{1f} \cos \theta_1 + m_2 v_{2f} \cos \theta_2,$$

and for the y-component

$$0 = m_1 v_{1f} \sin \theta_1 - m_2 v_{2f} \sin \theta_2.$$

Let us now assume that the collision is *elastic*. Here the conservation of kinetic energy applies and we obtain a third equation,

$$\tfrac{1}{2} m_1 v_{1i}^2 = \tfrac{1}{2} m_1 v_{1f}^2 + \tfrac{1}{2} m_2 v_{2f}^2.$$

If we know the initial conditions (m_1, m_2, and v_{1i}), we are left with four unknowns (v_{1f}, v_{2f}, θ_1, and θ_2) but only three equations relating them. We can determine the motion after collision only if we specify a value for one of these quantities, such as θ_1.

Two skaters collide and embrace, as Fig. 10-10 suggests. One, whose mass m_1 is 70 kg ($W_1 = m_1 g = 150$ lb), is initially moving east at a speed v_1 of 6.0 km/h (3.7 mi/h). The other, whose mass m_2 is 50 kg ($W_2 = m_2 g = 110$ lb), is initially moving north at a speed v_2 of 8.0 km/h (5.0 mi/h). (a) What is the final velocity of the couple? (b) What fraction of the initial kinetic energy of the skaters is lost because of the collision?

EXAMPLE 4

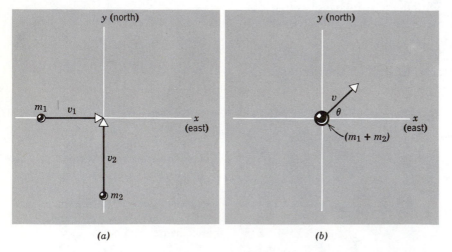

(a)   (b)

figure 10-10
Example 4. *(a)* initial situation. *(b)* final situation.

(a) Figure 10-10 shows the initial and final situations. Because $\mathbf{P} = \mathbf{P}_f$ (no external forces act) we can write for the x-component of momentum

$$m_1 v_1 = (m_1 + m_2) v \cos \theta,$$

and for the y-component of momentum

$$m_2 v_2 = (m_1 + m_2) v \sin \theta.$$

Dividing the second equation by the first, we get

$$\tan \theta = \frac{m_2 v_2}{m_1 v_1} = \frac{(50 \text{ kg})(8.0 \text{ km/h})}{(70 \text{ kg})(6.0 \text{ km/h})}$$

$$= 0.95 \quad \text{or} \quad \theta = 43°,$$

which gives the direction of the final velocity.

Then, from the y-component equation we have

$$v = \frac{m_2 v_2}{(m_1 + m_2) \sin \theta}$$

$$= \frac{(50 \text{ kg})(8.0 \text{ km/h})}{(70 \text{ kg} + 50 \text{ kg}) \sin 43°}$$

$$= 4.9 \text{ km/h} \ (3.0 \text{ mi/h}),$$

which gives the magnitude of the final velocity.

(b) The initial kinetic energy of the skaters is

$$K_i = \tfrac{1}{2} m_1 v_1{}^2 + \tfrac{1}{2} m_2 v_2{}^2$$

$$= \tfrac{1}{2} (70 \text{ kg})(6.0 \text{ km/h})^2 + \tfrac{1}{2} (50 \text{ kg})(8.0 \text{ km/h})^2$$

$$= \left(2860 \frac{\text{kg km}^2}{\text{h}^2}\right)\left(\frac{1 \text{ h}}{3600 \text{ s}}\right)^2\left(\frac{10^3 \text{ m}}{1 \text{ km}}\right)^2$$

$$= 220 \text{ J}.$$

The final kinetic energy of the couple is

$$K_f = \tfrac{1}{2} (m_1 + m_2) v^2$$

$$= \tfrac{1}{2} (70 \text{ kg} + 50 \text{ kg})(4.9 \text{ km/h})^2$$

$$= \left(1440 \frac{\text{kg km}^2}{\text{h}^2}\right)\left(\frac{1 \text{ h}}{3600 \text{ s}}\right)^2\left(\frac{10^3 \text{ m}}{1 \text{ km}}\right)^2$$

$$= 110 \text{ J}.$$

Hence,

$$\frac{K_i - K_f}{K_i} = \frac{220 \text{ J} - 110 \text{ J}}{220 \text{ J}} = \frac{1}{2}$$

so that 50% of the initial kinetic energy is lost in the collision.

EXAMPLE 5

A gas molecule having a speed of 300 m/s collides elastically with another molecule of the same mass which is initially at rest. After the collision the first molecule moves at an angle of 30° to its initial direction. Find the speed of each molecule after collision and the angle made with the incident direction by the recoiling target molecule.

This example corresponds exactly to the situation discussed before the last example, with $m_1 = m_2$, $v_{1i} = 300$ m/s, and $\theta_1 = 30°$. Setting m_1 equal to m_2, we have the relations

$$v_{1i} = v_{1f} \cos \theta_1 + v_{2f} \cos \theta_2,$$

$$v_{1f} \sin \theta_1 = v_{2f} \sin \theta_2,$$

and

$$v_{1i}{}^2 = v_{1f}{}^2 + v_{2f}{}^2.$$

We must solve for v_{1f}, v_{2f}, and θ_2. To do this we square the first equation (rewriting it as $v_{1i} - v_{1f} \cos \theta_1 = v_{2f} \cos \theta_2$), and add this to the square of the second equation (noting that $\sin^2 \theta + \cos^2 \theta = 1$); we obtain

$$v_{1i}{}^2 + v_{1f}{}^2 - 2v_{1i}v_{1f} \cos \theta_1 = v_{2f}{}^2.$$

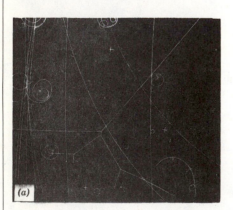

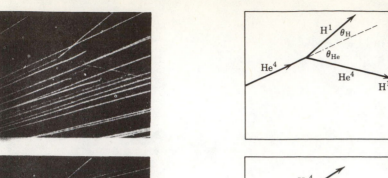

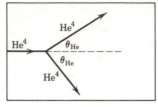

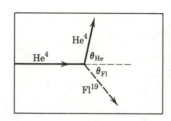

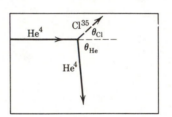

(a)

(b)

(c)

(d)

figure 10-11
Photographs of trajectories of particles undergoing collisions in a cloud chamber, a device that makes these paths visible. The chamber contains saturated water vapor. If the vapor is slightly compressed and then allowed to expand quickly, the water vapor will condense in droplets along the trajectory. The incident particle in all four cases is a helium nucleus (He⁴, or α). In *(a)* the target is a hydrogen nucleus (H¹ or *p*). The other tracks are similar, except that in *(b)* the target is another He⁴ nucleus, whereas in *(c)* and *(d)* the targets are fluorine and chlorine nuclei respectively. In general, the particles do not lie in the plane of the photograph. Stereoscopic photos are required for a complete analysis.

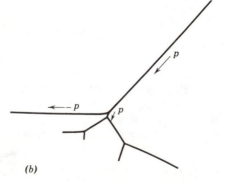

(a)

(b)

figure 10-12
(a) Four proton-proton collisions in a 10″-diameter bubble chamber. The original high-energy proton entered from the upper right. The spiral tracks are low energy electrons. The other tracks passing through the chamber are mesons of various kinds. Stereoscopic viewing shows that the angle between the outgoing tracks in each case is 90°. This is not apparent in the figure because the tracks do not lie in the plane of the figure. *(b)* A schematic representation of the proton tracks in *(a)*. (Photo courtesy Laurence Radiation Laboratory.)

Combining this with the third equation, we obtain

$$2v_{1f}^2 = 2v_{1i}v_{1f} \cos \theta_1$$

or (because $v_{1f} \neq 0$)

$$v_{1f} = v_{1i} \cos \theta_1 = (300 \text{ m/s})(\cos 30°)$$

or

$$v_{1f} = 260 \text{ m/s}.$$

From the third equation

$$v_{2f}^2 = v_{1i}^2 - v_{1f}^2 = (300 \text{ m/s})^2 - (260 \text{ m/s})^2,$$

or

$$v_{2f} = 150 \text{ m/s}.$$

Finally, from the second equation

$$\sin \theta_2 = (v_{1f}/v_{2f}) \sin \theta_1$$

$$= (260/150)(\sin 30°) = 0.866$$

or

$$\theta_2 = 60°.$$

The two molecules move apart at right angles ($\theta_1 + \theta_2 = 90°$ in Fig. 10-9).

You should be able to show that in an elastic collision between particles of equal mass, one of which is initially at rest, the recoiling particles always move off at right angles to one another.

In Fig. 10-11, we show photographs of four elastic nuclear collisions that take place in a Wilson cloud chamber.* The tracks of the particles are made visible by the trail of droplets left in their wake. In each case the incident particle is an α-particle (He^4) and the target nucleus is essentially at rest before collision. Notice that as the target mass increases, the angle between the recoiling particles increases (see Problem 42). In case (b) where the target is also an α-particle, stereoscopic photos show that the recoiling particles move off at right angles; the angle is not quite a right angle in the figure because the particles do not lie in the plane of the figure.

Figure 10-12 shows a series of four successive elastic collisions between protons caused when a high energy proton enters a bubble chamber† filled with liquid hydrogen, which supplies the target protons. The tracks of the particles are made visible in this case by the trail of bubbles left in their wake. Since the interacting particles are of equal mass and the collisions are elastic, the particles recoil at right angles to each other; this is apparent when the tracks of Fig. 10-12a are viewed stereoscopically.

10-7 CROSS SECTION

Although we have introduced the concept of the impact parameter b to describe collisions (see Fig. 10-9), it must be clear that, when we deal with particles of atomic or subatomic dimensions, we cannot define the track of the incident particle or the location of the target particle pre-

---

*In 1927, the English physicist, C. T. R. Wilson, received the Nobel prize for inventing the cloud chamber; his investigations started along an entirely different line, namely, an attempt to produce in the laboratory a certain atmospheric phenomenon observed on Ben Nevis, a mountain in Scotland.

†In 1960, the American physicist, Donald Glaser, received the Nobel prize for inventing the bubble chamber; it is said that the concept occurred to him while watching bubbles form in a glass of beer.

cisely enough. In practice, as when we bombard a thin target foil with a beam of α-particles from a cyclotron, we must deal in a statistical way with a large number of independent collisions between the α-particles and the nuclei in the target; the impact parameters for individual collisions cannot be determined.

The situation is much the same as if we were firing a machine gun at random (in the dark, say) at the side of a distant barn of area A on which someone had hung a number of small dinner plates, each of area σ, in random (but not overlapping) positions. If the number of plates is q and if the rate at which bullets strike the barn is R_0, what is the rate R at which plates are broken? It is, on the basis of the random character of the events,

$$R = R_0(\sigma q/A), \qquad (10\text{-}7a)$$

where σq is the total area of all the plates. We could, in fact, use this relation to measure σ, the geometrical area of a single plate. Solving for σ yields

$$\sigma = RA/R_0 q \qquad (10\text{-}7b)$$

which permits us to find σ from measured values of R, A, R_0, and q. We may call σ the *cross section* for the event consisting of the impact of a bullet on a plate.

Let us now consider a more restricted class of event, namely the impact of a bullet on a plate causing the plate to break into (say) exactly five pieces. The rate R_5 at which events of this kind occur is much less than the rate R at which the events described above occur. We may assign a *cross section* σ_5 to these restricted events and may measure it, by analogy with Eq. 10-7, from

$$\sigma_5 = R_5 A/R_0 q. \qquad (10\text{-}8)$$

We can consider other ways of breaking plates such as breaking into thirteen pieces, breaking so that one fragment has an area equal to half the plate or more, breaking so that one fragment flies vertically upward, and so on. Each of these events can be assigned its own cross section σ_x by measuring the rate R_x at which the events occur. *None of these cross sections necessarily has anything to do with the geometrical area of the plate; all are measures of the probability of occurrence of the events to which they are assigned.* Cross sections are important because they are identified with single events and are independent of the details of particular experimental setups. In Eq. 10-8, for example, we would find the same value for σ_5 no matter how large the barn (A), how many plates (q), or how rapid the rate of fire of the machine gun (R_0); the measured value of R_5 would always be such as to yield the same measured value for σ_5.

Similarly, in nuclear physics we often bombard targets with nuclear projectiles, measure the rate at which events of a selected type occur, and assign a cross section to those events. For example, let us bombard a thin gold foil (Au^{197}) with deuterons (H^2, or d) whose energy is, say, 30 Mev. Many events can occur, among them (1) elastic scattering of the deuteron into the forward hemisphere, (2) elastic scattering of the deuteron into the backward hemisphere, (3) inelastic scattering of the deuteron between the angles of 30° and 60° with the direction of the incident beam, (4) the nuclear reaction $d + Au^{197} \rightarrow p + Au^{198}$, and (5) the nuclear reaction $d + Au^{197} \rightarrow n + Hg^{198}$, in which n represents a neutron. Each of these events (and many others that could be written down) has its own cross section σ_x which allows us to calculate the rate R_x at which these events occur if we know the details of the experimental arrangement. The ultimate goal of all the experiments is to understand the nature of nuclear forces.

Let the area of the foil exposed to the beam be A and the thickness of the foil be x. If there are n target particles per unit volume in the foil, the total number of available target particles is nAx. If the *effective area* (i.e., the cross section) for the event we are concerned with is σ_x, the total *effective area* of all the nuclei is $nAx\sigma_x$. If R_0 is the rate at which projectiles strike the target and R_x is the rate at which the events in which we are interested occur, we have, because of the random nature of the events (see Eq. 10-7a),

$$\frac{R_x}{R_0} = \frac{(nAx\sigma_x)}{A}$$

or

$$R_x = R_0 nx\sigma_x. \tag{10-9}$$

Thus we can measure σ_x for the event by measuring R_x, R_0, n, and x and substituting into Eq. 10-9. Cross sections are commonly expressed in *barns* or submultiplies thereof; one barn $= 10^{-28}$ m^2.

Cross sections almost always depend on the energy of the incident particle, often exhibiting sharp peaks as the energy is varied. This simply means that at certain rather precisely defined characteristic energies the reaction is much more likely to "go" than at others. In much the same way a steel plate will vibrate with large amplitude at a series of rather precisely defined characteristic frequencies. We shall study such "resonances" in Chapter 20 and elsewhere.

Figure 10-13 shows one of the thousands of cross section curves that have been measured for various atomic and nuclear process. Let a highly collimated beam of neutrons of kinetic energy K fall on a thin foil of cadmium. The process to which the cross section σ in Fig. 10-13 refers is any process (absorption, elastic scattering, or inelastic scattering) that results in the removal of a neutron from the collimated beam. The numbers above the various peaks show the particular isotope responsible for

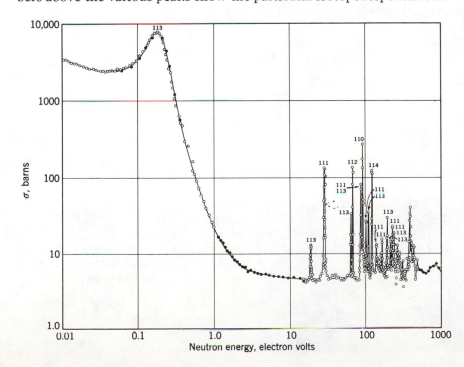

figure 10-13

The cross section as a function of energy for processes that remove neutrons from a collimated beam as they pass through cadmium. Note that both scales are logarithmic.

that peak; this can be learned from other experiments using foils made of the separated isotopes. The strong peak labeled "113" that occurs at 0.17 electron volts is caused by the reaction

$$Cd^{113} + n \rightarrow Cd^{114} + \gamma$$

in which γ represents a gamma ray. This reaction, which has a peak cross section of 7600 barns, is responsible for the very large absorbing power of cadmium for slow neutrons. Note that both scales in Fig. 10-13 are logarithmic.

EXAMPLE 6

(a) About 1910, Geiger and Marsden, working under Ernest Rutherford at the University of Manchester, performed a series of classic experiments that established the fact that atoms consisted of a small nucleus surrounded by a cloud of electrons rather than a sphere of distributed positive and negative charges, as Thomson had suggested earlier.

This experiment was in essence that shown very schematically in Fig. 10-14. Here α-particles from a polonium source are allowed to strike a gold foil 4.0×10^{-7} m thick. It is found that although most of the α-particles pass through the foil (forward scattering), about 1 in 6.17×10^4 are scattered backward, that is, are deflected through an angle greater than 90°. The number of gold atoms per unit volume in the foil is $5.9 \times 10^{28}/m^3$. What is the scattering cross section in barns for backward scattering (1 barn = 10^{-28} m²)?

From

$$n x \sigma = \text{fraction scattered backward}$$

we have

$$(5.9 \times 10^{28}/m^3)(4.0 \times 10^{-7} \text{ m})\sigma = 1/(6.17 \times 10^4)$$

or

$$\sigma = 6.9 \times 10^{-28} \text{ m}^2 = 6.9 \text{ barns}.$$

This is the cross section for backward scattering.

(b) Rutherford reasoned that the backward scattering could not be caused by electrons in the atom; the α-particles are so much more massive than the electrons that they would hardly be deflected at all by them, let alone be scattered backward. He then suggested the nuclear model of the atom, attributing the scattering to collisions between α-particles and the massive positive core of the atom, the nucleus.

Assuming that the cross section for backward scattering is approximately equal to the area offered by a gold nucleus for direct collisions, estimate the *effective* size of a gold nucleus.

If the *effective* radius of the gold nucleus is taken to be r, we have

$$\sigma = \pi r^2,$$

$$r^2 = \sigma/\pi = 6.9 \times 10^{-28} \text{ m}^2/\pi,$$

or

$$r = 1.5 \times 10^{-14} \text{ m}.$$

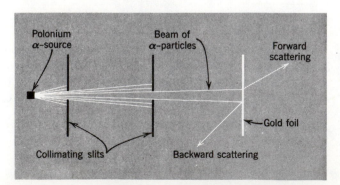

figure 10-14
Example 6. α-particles stream from a polonium source and a beam is formed by collimating slits. Some of the α-particles are scattered backward by the gold foil target; the rest pass through the foil.

This is the approximate radius of a gold *nucleus* which compares with the value of about 1.5×10^{-10} m for the gold *atom*. Hence the massive nucleus is concentrated in a very small region of the atom (about 1 part in 10^{12} by volume).

10-8 REACTIONS AND DECAY PROCESSES

We stated in Section 10-1 that reactions and radioactive decay processes, for atoms, nuclei, and elementary particles, can be treated by the same methods used in collision studies, namely: We can apply the principles of conservation of linear momentum and energy to the (well-defined) periods "before the event" and "after the event." For these processes we must use the conservation of *total* energy because kinetic energy is *not* conserved. In this section we only consider examples in which the speeds of the particles are negligible with respect to the speed of light. This means that we may use the classical expressions for momentum and energy and need not use the relativistic expressions.

EXAMPLE 7

Nuclear Reactions. A thin film containing a fluorine (F^{19}) compound is bombarded by a beam of protons (p) which has been accelerated to an energy of 1.85 MeV (million electron volts; 1 MeV = 1.60×10^{-13} J) in a Van de Graaff accelerator. Some of the protons interact with the fluorine nuclei to produce the following nuclear reaction:

$$F^{19} + p \rightarrow O^{16} + \alpha$$

It is observed that the α-particles (which are helium nuclei) that emerge at *right angles* to the incident proton beam (see Fig. 10-15) have speeds of 1.95×10^7 m/s. What can you learn about the reaction by applying the laws of conservation of linear momentum and of total energy? The masses involved are, to a precision good enough for our purposes,

$$m_p = 1.01 \text{ u} \qquad m_O = 16.0 \text{ u}$$

$$m_F = 19.0 \text{ u} \qquad m_\alpha = 4.00 \text{ u},$$

in which 1 u (*unified atomic mass unit*) = 1.66×10^{-27} kg.

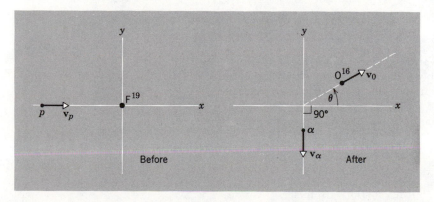

figure 10-15
Example 7. The nuclear reaction $p + F^{19} \rightarrow \alpha + O^{16}$, showing the situation before and after the event, in the laboratory reference frame.

The x- and y-components of linear momentum are conserved, which means that they have the same values before and after the reaction. In the laboratory reference frame of Fig. 10-15, then

$$m_p v_p = m_O v_O \cos \theta \qquad \text{(x-component)} \qquad (10\text{-}10)$$

and

$$0 = m_\alpha v_\alpha - m_O v_O \sin \theta \qquad \text{(y-component)} \qquad (10\text{-}11)$$

For the conservation of total energy we write

$$Q + \tfrac{1}{2}m_p v_p^2 = \tfrac{1}{2}m_0 v_0^2 + \tfrac{1}{2}m_\alpha v_\alpha^2 \qquad (10\text{-}12)$$

in which it is clear that Q is the amount by which the kinetic energy of the system after the reaction exceeds the kinetic energy of the system before the reaction. Note that we have assumed that the particles are moving slowly enough so that we may use the classical expression for kinetic energy $(\tfrac{1}{2}mv^2)$ rather than the relativistic one $[m_0 c^2(1/\sqrt{1 - v^2/c^2} - 1)]$. If Q is positive, kinetic energy must be generated by the reaction.

The energy represented by Q can only come from differences in the rest energies of the particles before or after the reaction, according to Einstein's well-known relation $E = \Delta mc^2$ (see Section 8-9). Thus (if Q is positive), we expect that the rest mass of the system after the reaction would be slightly less than its rest mass before the reaction and that Q would indeed be given by the Einstein relation

$$Q = \Delta mc^2$$

$$= [(m_p + m_F) - (m_\alpha + m_0)]c^2. \qquad (10\text{-}13)$$

Note that Eqs. 10-12 and 10-13 are independent relations for Q, being connected through Einstein's mass-energy relation.

The three conservation equations contain just three unknowns, v_0, θ, and Q. To find Q from them let us first eliminate θ between the first two equations by squaring and adding (recalling that $\cos^2 \theta + \sin^2 \theta = 1$). We obtain

$$m_p^2 v_p^2 + m_\alpha^2 v_\alpha^2 = m_0^2 v_0^2.$$

We can now eliminate v_0 between this relation and Eq. 10-12. You can show that, after a little rearrangement, we obtain

$$Q = K_\alpha(1 + m_\alpha/m_0) - K_p(1 - m_p/m_0). \qquad (10\text{-}14)$$

From the data given we know that $K_p(= \tfrac{1}{2}m_p v_p^2) = 1.85$ MeV and

$$K_\alpha = \tfrac{1}{2}m_\alpha v_\alpha^2$$

$$= \tfrac{1}{2}(4.00 \text{ u})(1.66 \times 10^{-27} \text{ kg/u})(1.95 \times 10^7 \text{ m/s})^2$$

$$= (1.26 \times 10^{-12} \text{ J})(1 \text{ MeV}/1.60 \times 10^{-13} \text{ J})$$

$$= 7.88 \text{ MeV}.$$

We may now calculate Q from Eq. 10-14 as

$$Q = (7.88 \text{ MeV})(1 + 4.00/16.0) - (1.85 \text{ MeV})(1 - 1.01/16.0) = 8.13 \text{ MeV}.$$

Thus, by using the principles of conservation of linear momentum and total energy, we can calculate Q for the reaction without making any observations on the recoiling O^{16} nucleus. If we want to know v_0 and θ for this nucleus we can easily calculate them from Eqs. 10-10 and 10-11.

The result $Q = 8.13$ MeV is an important bit of information about the reaction. From Eq. 10-13, which is a relation for Q independent of Eq. 10-14, we can now calculate that the decrease in rest mass during the reaction is given by

$$\Delta m = Q/c^2$$

$$= (8.13 \text{ MeV} \times 1.60 \times 10^{-13} \text{ J/MeV})/(3.00 \times 10^8 \text{ m/s})^2$$

$$= (1.44 \times 10^{-29} \text{ kg})(1 \text{ u}/1.66 \times 10^{-27} \text{ kg})$$

$$= 0.00873 \text{ u}.$$

We can verify this result by calculating Δm $[= (m_p + m_F) - (m_\alpha + m_0)]$ from very precise measurements of the four separate masses made in a mass spectrometer (see Problem 47). The excellent agreement that we get shows once again the essential validity of Einstein's mass-energy relationship.

1. Explain how conservation of momentum applies to a handball bouncing off a wall.

2. How can you reconcile the sailing of a sailboat into the wind with the conservation of momentum principle?

3. Is it true that the acceleration of a baseball after it has been hit does not depend on who hit it?

4. Many features on cars, such as collapsible steering wheels and padded dashboards, are meant to transfer momentum more safely for passengers. Explain, using the impulse concept.

5. C. R. Daish (see "At Impact, Clubhead Travels 100 Mph," *Museum*, December 1973) states that, for professional golfers, the initial speed of the ball of the clubhead is about 140 mi/h. He also says: (a) ". . . if the Empire State Building could be swung at the ball at the same speed as the clubhead, the initial ball velocity would only be increased by about 2% . . ."; and (b) that, once the golfer has started his downswing, camera clicking, sneezing, etc., can have no effect on the motion of the ball. Can you give qualitative arguments to support these two statements?

6. The blades of a turbine are usually curved rather than flat in shape so that the fluid striking them follows a path resembling a u-turn. Convince yourself about the fluid's motion and explain the advantage of the curved shape over the flat one.

7. It is obvious from inspection of Eqs. 10-3 and 10-4 that a valid solution to the problem of finding the final velocities of two particles in a one-dimensional elastic collision is $v_{1f} = v_{1i}$ and $v_{2f} = v_{2i}$. What does this mean physically?

8. Consider a one-dimensional elastic collision between a given incoming body A and a body B initially at rest. How would you choose the mass of B, in comparison to the mass of A, in order that B should recoil with (a) the greatest speed, (b) the greatest momentum, and (c) the greatest kinetic energy?

9. A football player, momentarily at rest on the field, is about to catch a football when he is tackled by a running player on the other team. This is certainly a collision (inelastic!) and momentum must be conserved. In the reference frame of the football field there is momentum before the collision but there seems to be none after the collision. Is linear momentum really conserved?

10. Steel is more elastic than rubber. Explain what this means.

11. Two clay balls of equal mass and speed strike each other head-on, stick together, and come to rest. Kinetic energy is certainly not conserved. What happened to it? Is momentum conserved?

12. Discuss the possibility that, if only we could take into account internal motions of atoms and such in bodies, *all* collisions are elastic.

13. In commenting on the fact that kinetic energy is not conserved in a totally inelastic collision, a student observed that kinetic energy clearly is not conserved in an explosion and that a totally inelastic collision is merely the reverse of an explosion. Is this a useful or valid observation?

14. A sand glass is being weighed on a sensitive balance, first when sand is dropping in a steady stream from the upper to the lower part and then again after the upper part is empty. Are the two weights the same or not? Explain your answer.

15. Give a plausible explanation for the breaking of wooden boards or of bricks by a karate punch (see "Karate Strikes" by Jearl D. Walker, *American Journal of Physics*, October 1975).

16. In which one of the following cases is the linear momentum of the italicized objects most nearly conserved? (a) a *ball* falling freely in vacuum; (b) an *automobile* making a turn at constant speed; (c) a *rubber ball* as it bounces from the floor; (d) *two balls* as they collide at right angles; (e) a *bullet* and the *gun* from which it is fired by a man holding the gun.

17. If (only) two particles collide, are we ever forced to resort to a three-dimensional description to describe the event? Explain.

18. In a two body collision *in the center-of-mass reference frame* the momenta of the particles are equal and opposite to one another both before and after the collision. Is the line of relative motion necessarily the same after collision as before? Under what conditions would the magnitudes of the velocities of the bodies increase? decrease? remain the same as a result of the collision?

19. When dealing with atoms, nuclei, or elementary particles, what does it mean to say that two such bodies "touch" during a collision?

20. When the forces of interaction between two particles have an infinite range, such as the mutual gravitational attraction between two bodies, can the cross section for "collision" be finite? Is it at all useful to regard this interaction as a collision?

21. Why does the computation of the radius of the gold nucleus in Example 5 give only an approximate answer?

22. Could we determine in principle the cross section for a collision by using only one bombarding particle and one target particle? In practice?

23. We have seen that the conservation of momentum may apply whether kinetic energy is conserved or not. What about the reverse, that is, does the conservation of kinetic energy imply the conservation of momentum in classical physics? (See "Connection between Conservation of Energy and Conservation of Momentum" by Carl G. Adler, *Am. J. Phys.*, May 1976.)

problems

SECTION 10-2

1. A ball of mass m and speed v strikes a wall perpendicularly and rebounds with undiminished speed. If the time of collision is t, what is the average force exerted by the ball on the wall?   *Answer:* $2mv/t$.

2. A stream of water impinges on a stationary "dished" turbine blade, as shown in Fig. 10-16. The speed of the water is u, both before and after it strikes the curved surface of the blade, and the mass of water striking the blade per unit time is constant at the value μ. Find the force exerted by the water on the blade.

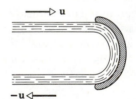

figure 10-16
Problem 2

3. A 150-g (0.01 slug) ball is moving at a speed of 40 m/s (130 ft/s) when it is struck by a bat that reverses its direction and gives it a speed of 60 m/s (200 ft/s). What average force was exerted by the bat if it was in contact with the ball for 5.0 ms?   *Answer:* 3000 N (660 lb).

4. A 1.0-kg ball drops vertically onto the floor with a speed of 25 m/s. It rebounds with an initial speed of 10 m/s. (a) What impulse acts on the ball during contact? (b) If the ball is in contact for 0.020 s, what is the average force exerted on the floor?

5. A cue strikes a billiard ball, exerting an average force of 50 N over a time of 10 ms. If the ball has mass 0.20 kg, what speed does it have after impact? *Answer:* 2.5 m/s.

6. A croquet ball (mass 0.50 kg) is struck by a mallet, receiving the impulse shown in the graph (Fig. 10-17). What is the ball's velocity just after the force has become zero?

7. The force on a 10-kg (0.69 slug) object increases uniformly from zero to 50 N (11 lb) in 4.0 s. What is the object's final speed if it started from rest? *Answer:* 10 m/s (32 ft/s).

8. A golfer hits a golf ball imparting to it an initial velocity of magnitude 5.0×10^3 cm/s directed 30° above the horizontal. Assuming that the mass of the ball is 25 g and the club and ball are in contact for 0.010 s find (a) the impulse imparted to the ball; (b) the impulse imparted to the club; (c) the average force exerted on the ball by the club; (d) the work done on the ball.

9. A stream of water from a hose is sprayed on a wall. If the speed of the water is 5.0 m/s (16 ft/s) and the hose sprays 300 cm³/s (0.011 ft³/s), what is the

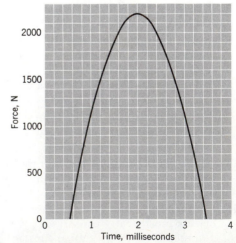

figure 10-17
Problem 6

average force exerted on the wall by the stream of water? Assume that the water does not spatter back appreciably. The density of water is 1.0 g/cm³ (1.9 slug/ft³). *Answer:* 1.5 N (0.33 lb).

10. Two spacecraft are separated by exploding a small charge placed between them. If the masses of the crafts are 1200 kg and 1800 kg and the impulse of the force of the explosion is 600 N·s, what is the relative speed of recession of the two craft?

SECTION 10-4

11. A bullet of mass 10 g strikes a ballistic pendulum of mass 2.0 kg. The center of mass of the pendulum rises a vertical distance of 12 cm. Assuming the bullet remains embedded in the pendulum, calculate its initial speed. *Answer:* 310 m/s.

12. A 6.0-kg box sled is traveling across the ice at a speed of 9.0 m/s when a 12-kg package is dropped into it vertically. Describe the subsequent motion of the sled.

13. (a) Show that in a one-dimensional elastic collision the speed of the center of mass of two particles, m_1 moving with initial speed v_{1i} and m_2 moving with initial speed v_{2i} is

$$v_{cm} = \left(\frac{m_1}{m_1 + m_2}\right)v_{1i} + \left(\frac{m_2}{m_1 + m_2}\right)v_{2i}.$$

(b) Use the expressions obtained for v_{1f} and v_{2f}, the particles' speeds after collision, to derive the same result for v_{cm} *after* the collision.

14. A body of 2.0 kg mass makes an elastic collision with another body at rest and continues to move in the original direction but with one-fourth of its original speed. What is the mass of the struck body?

15. In a breech-loading automatic firearm of early vintage the reloading mechanism at the rear of the bore is activated when the breech-block, which recoils after the bullet is fired, compresses a spring by a predetermined amount d. (a) Show that the speed v of the bullet of mass m must be at least $d\sqrt{kM/m}$ on firing, for automatic loading, where k is the force constant of the spring and M is the mass of the breech-block. (b) In what sense, if any, can this process be regarded as a collision?

16. A steel ball weighing 1.0 lb is fastened to a cord 27 in. long and is released when the cord is horizontal. At the bottom of its path the ball strikes a 5.0-lb steel block initially at rest on a frictionless surface (Fig. 10-18). The collision is elastic. Find (a) the speed of the ball and (b) the speed of the block just after the collision.

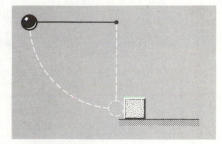

figure 10-18
Problem 16

17. A bullet weighing 1.0×10^{-2} lb (mass = 4.5×10^{-3} kg) is fired horizontally into a 4.0-lb (mass = 1.8 kg) wooden block at rest on a horizontal surface. The coefficient of kinetic friction between block and surface is 0.20. The bullet comes to rest in the block which moves 6.0 ft (1.8 m). Find the speed of the bullet. *Answer:* 3500 ft/s (1100 m/s).

18. Two pendulums each of length l are initially situated as in Fig. 10-19. The first pendulum is released and strikes the second. Assume that the collision is completely inelastic and neglect the mass of the strings and any frictional effects. How high does the center of mass rise after the collision?

19. Two particles, one having twice the mass of the other, are held together with a compressed spring between them. The energy stored in the spring is 60 J. How much kinetic energy does each particle have after they are released? *Answer:* 20 J for the heavy particle; 40 J for the light particle.

20. A railroad freight car weighing 32 tons and traveling 5.0 ft/s overtakes one weighing 24 tons traveling 3.0 ft/s in the same direction. (a) Find the speed of the cars after collision and the loss of kinetic energy during collision if the cars couple together. (b) If the collision is elastic, the freight cars do not couple but separate after collision. What are their speeds?

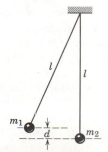

figure 10-19
Problem 18

21. An electron collides elastically with a hydrogen atom initially at rest. The initial and final motions are along the same straight line. What fraction of the electron's initial kinetic energy is transferred to the hydrogen atom? The mass of the hydrogen atom is 1840 times the mass of the electron. *Answer: 0.22%.*

22. A block of mass $m_1 = 100$ kg is at rest on a very long frictionless table, one end of which is terminated in a wall. Another block of mass m_2 is placed between the first block and the wall and set in motion to the left with constant speed v_{2i}, as in Fig. 10-20. Assuming that all collisions are completely elastic, find the value of m_2 for which both blocks move with the same velocity after m_2 has collided once with m_1 and once with the wall. The wall has infinite mass effectively.

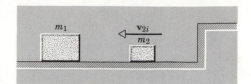

figure 10-20
Problem 22

23. An electron, mass m, collides head-on with an atom, mass M, initially at rest. As a result of the collision a characteristic amount of energy E is stored internally in the atom. What is the minimum initial speed v_0 that the electron must have? (Hint: Conservation principles lead to a quadratic equation for the final electron velocity v and a quadratic equation for the final atom velocity V. The minimum value v_0 follows from the requirement that the radical in the solutions for v and V be real.)

$$\text{Answer: } v_0 = \left(2E\frac{M+m}{Mm}\right)^{1/2}.$$

24. A ball of mass m is projected with speed v_i into the barrel of a spring-gun of mass M initially at rest on a frictionless surface; see Fig. 10-21. The mass m sticks in the barrel at the point of maximum compression of the spring. No energy is lost in friction. What fraction of the initial kinetic energy of the ball is stored in the spring?

figure 10-21
Problem 24

25. A box is put on a scale that is adjusted to read zero when the box is empty. A stream of pebbles is then poured into the box from a height h above its bottom at a rate of μ (pebbles per second). Each pebble has a mass m. If the collisions between the pebbles and the box are completely inelastic, find the scale reading at time t after the pebbles begin to fill the box. Determine a numerical answer when $\mu = 100$ s^{-1}, $h = 25$ ft, $mg = 0.010$ lb, and $t = 10$ s. *Answer: 11 lb.*

26. A scale is adjusted to read zero. Particles fall from a height of 9.0 ft (2.7 m) colliding with the balance pan of the scale; the collisions are elastic, that is, the particles rebound upward with the same speed. If each particle has a mass of 1/128 slug (110 g) and collisions occur at the rate of 32 s^{-1}, what is the scale reading in pounds (kg)?

27. Mass m_1 collides head-on with m_2, initially at rest, in a completely inelastic collision. (a) What is the kinetic energy of the system before collision? (b) What is the kinetic energy of the system after collision? (c) What fraction of the original kinetic energy was converted into heat energy? (d) Let $\mathbf{v}_{cm}$ be the velocity of the center of mass of the system. View the collision from a primed reference frame moving with the center of mass so that $v_{1i}' = v_{1i} - v_{cm}$, $v_{2i}' = -v_{cm}$. Repeat parts (a), (b), and (c), as seen by an observer in this reference frame. Is the mechanical energy converted to heat energy the same in each case? Explain.
    Answer: (a) $m_1v_{1i}^2/2$. (b) $m_1^2v_{1i}^2/2(m_1 + m_2)$. (c) $m_2/(m_1 + m_2)$.
    (d) $m_1m_2v_{1i}^2/2(m_1 + m_2)$; zero; one; yes.

28. An elevator is moving up at 6.0 ft/s in a shaft. At the instant the elevator is 60 ft from the top, a ball is dropped from the top of the shaft. The ball rebounds elastically from the elevator roof. (a) To what height can it rise relative to the top of the shaft? (b) Do the same problem assuming the elevator is moving down at 6.0 ft/s.

29. A block of mass $m_1 = 2.0$ kg slides along a frictionless table with a speed of 10 m/s. Directly in front of it, and moving in the same direction, is a block of mass $m_2 = 5.0$ kg moving at 3.0 m/s. A massless spring with a spring constant $k = 1120$ N/m is attached to the backside of m_2, as shown in Fig. 10-22.

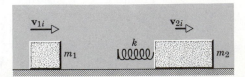

figure 10-22
Problem 29

When the blocks collide, what is the maximum compression of the spring? Assume that the spring does not bend and always obeys Hooke's law.
Answer: 0.25 m.

30. The two masses on the right of Fig. 10-23 are slightly separated and initially at rest; the left mass is incident with speed v_0. Assuming head-on elastic collisions, (*a*) if $M \leq m$, show that there are two collisions and find all final velocities; (*b*) if $M > m$, show that there are three collisions and find all final velocities.

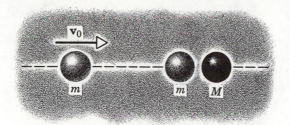

figure 10-23
Problem 30

31. Consider a situation such as that in the previous problem (10-30) but in which the collisions now may be either all elastic, all inelastic, or some elastic and some inelastic; also, the masses are now m, m', and M. Show that to transfer the maximum kinetic energy from m to M the intermediate body should have a mass $m' = \sqrt{mM}$, that is, the geometric mean of the adjacent masses. (It is interesting to note that this same relation exists between masses of successive layers of air in the exponential horn in acoustics. See "Energy Transfer in One-Dimensional Collisions of Many Objects" by John B. Hart and Robert B. Herrmann, *American Journal of Physics*, January, 1968.)

SECTION 10-6

32. Two vehicles A and B are traveling west and south, respectively, toward the same intersection where they collide and lock together. Before the collision A (total weight, 900 lb) is moving with a speed of 40 mi/h and B (total weight, 1200 lb) has a speed of 60 mi/h. Find the magnitude and direction of the velocity of the (interlocked) vehicles immediately after the collision.

33. Two balls A and B, having different but unknown masses, collide. A is initially at rest when B has a speed v. After collision B has a speed $v/2$ and moves at right angles to its original motion. (*a*) Find the direction in which ball A moves after collision. (*b*) Can you determine the speed of A from the information given? Explain.
Answer: (a) 117° from the final direction of B. (b) No.

34. A billiard ball moving at a speed of 2.2 m/s strikes an identical stationary ball a glancing blow. After the collision one ball is found to be moving at a speed of 1.1 m/s in a direction making a 60° angle with the original line of motion. (*a*) Find the velocity of the other ball. (*b*) Can the collision be inelastic, given these data?

35. An α particle collides with an oxygen nucleus, initially at rest. The α-particle is scattered at an angle of 64° from its initial direction of motion and the oxygen nucleus recoils at an angle of 51° on the other side of this initial direction. What is the ratio, α-particle to nucleus, of the final speeds of these particles? The mass of the oxygen nucleus is four times that of the α-particle. *Answer: 3.46.*

36. A deuteron is a nuclear particle made up of one proton and one neutron. Its mass is about 3.4×10^{-24} g. A deuteron, accelerated by a cyclotron to a speed of 10^9 cm/s, collides with another deuteron at rest. (*a*) If the two particles stick together head-on to form a helium nucleus, find the speed of the resulting nucleus. (*b*) The helium nucleus then breaks up into a neutron with

a mass of about 1.7×10^{-24} g and a helium isotope of mass 5.1×10^{-24} g. If the neutron is given off at right angles to the direction of the original velocity with a speed of 5.0×10^8 cm/s, find the magnitude and direction of the velocity of the helium isotope.

37. A certain nucleus, at rest, disintegrates into three particles. Two of them are detected, with masses and velocities as shown in Fig. 10-24. (*a*) What is the momentum of the third particle, which is known to have a mass of 12×10^{-27} kg? (*b*) How much energy was involved in the disintegration process.   *Answer:* (*a*) $(-1.0 \mathbf{i} + 0.64 \mathbf{j}) \times 10^{-19}$ N·s. (*b*) 1.1×10^{-12} J.

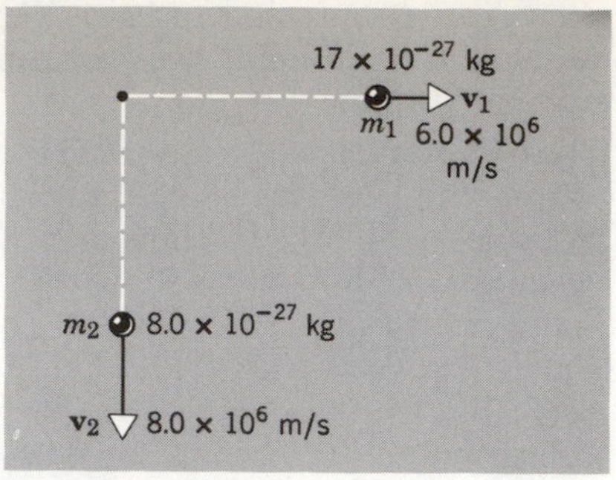

38. In 1932 Chadwick, in England, demonstrated the existence and properties of the neutron (one of the fundamental particles making up the atom) with the device shown in Fig. 10-25. In an evacuated chamber, a sample of radioactive polonium decays to yield α-rays (helium nuclei). These nuclei impinge on a block of beryllium inducing a process whereby neutrons are emitted. (In the reaction He and Be combine to form radioactive carbon, which decays to stable carbon + neutrons.) The neutrons strike a film of paraffin (CH_4), releasing hydrogen nuclei which are detected in an ionization chamber. In other words, an elastic collision occurs in which the momentum of the neutron is partially transferred to the hydrogen nucleus.

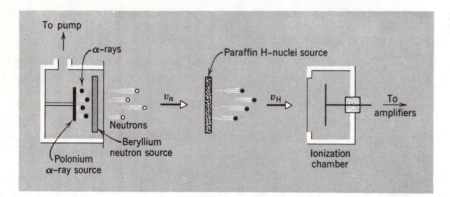

figure 10-24
Problem 37

figure 10-25
Problem 38

(*a*) Find an expression for the maximum speed V_H that the hydrogen nucleus (mass m_H) can achieve. Let the incoming neutrons have mass m_n and speed v_n. (Hint: Will more energy be transferred in a head-on collision or in a glancing collision?)

(*b*) One of Chadwick's goals was to find the mass of his new particle. Inspection of expression (*a*) which contains this parameter, however, shows that *two* unknowns are present, v_n and m_n (v_H is known; it can be measured with the ionization chamber). To eliminate the unknown v_n, he substituted a paracyanogen (CN) block for the paraffin. The neutrons then underwent elastic collisions with nitrogen nuclei instead of hydrogen nuclei. Of course, expression (*a*) still holds if v_N is written for v_H and m_N for m_H. Therefore if v_H and v_N are measured in separate experiments, v_N can be eliminated between the two expressions for hydrogen and nitrogen to yield a value for m_n. Chadwick's values were

$$v_H = 3.3 \times 10^9 \text{ cm/s},$$

$$v_N = 0.47 \times 10^9 \text{ cm/s}.$$

What is his value for m_n? How does this compare with the established value $m_n = 1.00867$ u? (Take $m_H = 1.0$ u, $m_N = 14$ u.)

39. A ball with an initial speed of 10 m/s collides elastically with two identical balls whose centers are on a line perpendicular to the initial velocity and which are initially in contact with each other (Fig. 10-26). The first ball is aimed directly at the contact point and all the balls are frictionless. Find the velocities of all three balls after the collision. (Hint: The directions of the

two originally stationary balls can be found by considering the direction of the impulse they receive during the collision.)

Answer: $\mathbf{v}_2$ and $\mathbf{v}_3$ will be at 30° to $\mathbf{v}_0$ and will have a magnitude of 6.9 m/s. $\mathbf{v}_1$ will be in the opposite direction to $\mathbf{v}_0$ and will have magnitude 2.0 m/s.

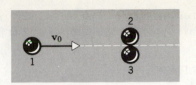

figure 10-26
Problem 39

40. After a totally inelastic collision, two objects of the same mass and initial speed are found to move away together at half their initial speed. Find the angle between the initial velocities of the objects.

41. Show that a slow neutron that is scattered through 90° in an elastic collision with a deuteron, initially at rest in a tank of heavy water, loses two-thirds of its initial kinetic energy to the struck deuteron.

42. Show that, in the case of an elastic collision between a particle of mass m_1 with a particle of mass m_2 initially at rest, (a) the maximum angle θ_m through which m_1 can be deflected by the collision is given by $\cos^2\theta_m = 1 - m_2^2/m_1^2$, so that $0 \leq \theta_m \leq \pi/2$, when $m_1 > m_2$; (b) $\theta_1 + \theta_2 = \pi/2$, when $m_1 = m_2$; (c) θ_1 can take on all values between 0 and π, when $m_1 < m_2$.

SECTION 10-7

43. A sphere of radius r_1 impinges on a sphere of radius r_2. What is the cross section for a contact collision?   *Answer:* $\pi(r_1 + r_2)^2$.

44. A beam of slow neutrons strikes an aluminum foil 1.0×10^{-5} m thick. Some neutrons are captured by the aluminum that becomes radioactive and decays by emitting an electron (β^-) forming silicon:

$$n + Al^{27} \rightarrow Al^{28} \rightarrow Si^{28} + \beta^-.$$

Suppose the neutron flux is 3.0×10^{16} $m^{-2} \cdot s^{-1}$ and the neutron capture cross section is 0.23 b. How many transmutations per unit area will occur each second?

45. A beam of fast neutrons impinges on a 5.0-mg sample of Cu^{65}, a stable isotope of copper. A possibility exists that the copper nucleus may capture a neutron to form Cu^{66}, which is radioactive and decays to Zn^{66}, which is again stable. If a study of the electron emission of the copper sample implies that 4.6×10^{11} neutron captures occur each second, what is the neutron capture cross section in barns for this process? The intensity of the neutron beam is 1.1×10^{18} neutrons $m^{-2} \cdot s^{-1}$.   *Answer:* 90 barns.

46. In a thick foil there are a great many layers of target particles so that the number of projectile particles reaching a layer will depend on how many have been scattered out by previous layers. Let the number of particles reaching a layer at a depth s be N and the number lost by scattering from that layer by $-dN$; then show that

$$-\frac{dN}{N} = n\sigma ds$$

and

$$N = N_0 e^{-n\sigma s}$$

where N_0 is the number of particles incident on the face of the foil $(s = 0)$ of unit area and n is the number of scatterers per unit volume.

SECTION 10-8

47. The precise masses in the reaction

$$p + F^{19} \rightarrow \alpha + O^{16}$$

have been determined by mass spectrometer measurements and are

$$m_p = 1.00783 \text{ u} \qquad m_\alpha = 4.00260 \text{ u}$$
$$m_F = 18.99840 \text{ u} \qquad m_O = 15.99491 \text{ u}$$

Calculate the Q of the reaction from these data and compare with the Q calculated in Example 7 from reaction studies.   *Answer:* 8.14 MeV.

48. An elementary particle called Σ^-, at rest in a certain reference frame, decays spontaneously into two other particles according to

$$\Sigma^- \rightarrow \pi^- + n.$$

The masses are

$$m_\Sigma = 2340.5\ m_e$$
$$m_\pi = 273.2\ m_e$$
$$m_n = 1838.65\ m_e$$

where m_e is the electron mass. (a) How much kinetic energy is generated in this process? (b) Which of the decay products (π^- and n) gets the larger share of this kinetic energy? Of the momentum?

49. The Q of the reaction in which a U^{236} nucleus at rest splits into just two fragments of masses 132 u and 98 u is 192 MeV. (a) How much energy was lost through radiation? (b) What is the speed of each fragment? (c) What is the kinetic energy of each fragment?

   Answer: (a) 5400 MeV. (b) $v_{132} = 1.09 \times 10^7$ m/s; $v_{98} = 1.47 \times 10^7$ m/s.
   (c) $K_{132} = 81.7$ MeV; $K_{98} = 110$ MeV.

11 rotational kinematics

So far we have dealt mostly with the translational motion of single particles or of rigid bodies, that is, of bodies whose parts all have a fixed relationship to each other. No real body is truly rigid, but many bodies, such as molecules, steel beams, and planets, are rigid enough so that, in many problems, we can ignore the fact that they warp, bend, or vibrate. As Fig. 3-1 suggests, we say that a rigid body moves in pure *translation* if each particle of the body undergoes the same displacement as every other particle in any given time interval.

In this chapter, however, we are interested in *rotation* rather than translation. For the time being we again restrict ourselves to single particles and to rigid bodies, which means that we shall not consider such rotational motions as those of the solar system or of water in a spinning beaker. *We* shall also *deal only with rotation about axes that remain fixed in the reference frame in which we observe the rotation.*

Figure 11-1 shows the rotational motion of a rigid body about a fixed axis, in this case the *z*-axis of our reference frame. Let *P* represent a particle in the rigid body, arbitrarily selected and described by the position vector **r**. We then say that: *A rigid body moves in pure rotation if every particle of the body* (such as *P* in Fig. 11-1) *moves in a circle, the centers of which are on a straight line called the axis of rotation* (the *z*-axis in Fig. 11-1). If we draw a perpendicular from any point in the body to the axis, each such line will sweep through the same angle in any given time interval as another such line. Thus we can describe the pure rotation of a rigid body by considering the motion of any one of the particles (such as *P*) that make it up. (We must rule out, however, particles that are on the axis of rotation. Why?)

The general motion of a rigid body is a combination of translation

11-1 ROTATIONAL MOTION

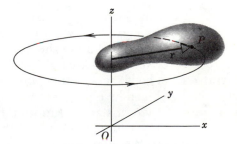

figure 11-1
A rigid body rotating about the z-axis. Each point in the body, such as *P*, describes a circle about this axis.

and rotation however, rather than one of pure rotation. We can locate a rigid body that is moving in pure translation by giving the three coordinates x, y, z of any point in it (its center of mass, say) in a particular reference frame. For a body that rotates as it moves translationally we need, in the most general case, three more coordinates, such as angles, to specify the orientation of the body with respect to the reference frame. Figure 11-2 (see also Fig. 9-1) shows a special case of rigid body motion combining translation and rotation. The figure is an extension of Fig. 3-1 in which the body now rotates as it moves translationally. To locate this body we must not only locate point O in the body in the xy reference frame but we must also say how the $x'y'$ reference frame, which is fixed in the body, is oriented with respect to the xy frame.

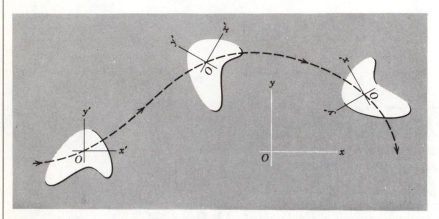

figure 11-2

A rigid body moving in combined translational and rotational motion as seen from reference frame x, y. Notice that the reference frame fixed on the body (x', y') changes its orientation with respect to x, y as the motion proceeds. Compare with Figs. 3-1 and 9-1. This figure represents a special case in that the translational motion occurs in two dimensions only (the xy plane) and the rotational motion occurs about an axis that maintains a fixed direction (the z'-axis).

As we saw in Chapter 9 we can describe the translational motion of any system of particles — whether rigid or not — whether rotating or not — by imagining that all of the mass M of the body is concentrated at the center of mass and that $\mathbf{F}_{ext}$, the resultant of the external forces acting on the body, acts at this point. The acceleration of the center of mass is then given by Eq. 9-10 or $\mathbf{F}_{ext} = M\mathbf{a}_{cm}$. It is very helpful to be able to represent the translational motion of a rigid body by the motion of a single point — its center of mass; all that is left is to determine its rotational motion. We shall discuss such combined translational and rotational motions in the next chapter. This will be simpler to do after we have studied pure rotation about a fixed axis.

We now return, therefore, to the pure rotation of a rigid body about a fixed axis (Fig. 11-1). First, we must *describe* the rotational motion. We call this description rotational kinematics; we must define the variables of angular motion and relate them to each other, just as in particle kinematics (see Chapter 4) we defined the variables of translational motion and related them to each other. The next part of our program is to relate the rotational motion of a body to the properties of the body and of its environment. This is rotational dynamics. In this chapter we study the kinematics of rotation. We develop the dynamics of rotation in the next chapter.

In Fig. 11-1 let us pass a plane through P at right angles to the axis of rotation. This plane, which cuts through the rotating body, contains the circle in which particle P moves. Figure 11-3 shows this plane, as we look downward on it from above, along the z-axis in Fig. 11-1.

We can tell exactly where the entire rotating body is in our reference frame if we know the location of any single particle (P) of the body in this frame. Thus, for the kinematics of this problem, we need only consider the (two-dimensional) motion of a particle in a circle.

The angle θ in Fig. 11-3 is the angular position of particle P with respect to the reference position. We arbitrarily choose the positive sense of rotation in Fig. 11-3 to be counterclockwise, so that θ increases for counterclockwise rotation and decreases for clockwise rotation.

It is convenient to measure θ in radians* rather than in degrees. By definition θ is given in radians by the relation

$$\theta = s/r,$$

in which s is the arc length shown in Fig. 11-3.

Let the body of Fig. 11-3 be rotating counterclockwise. At time t_1 the angular position of P is θ_1 and at a later time t_2 its angular position is θ_2. This is shown in Fig. 11-4, which gives the positions of P and of the position vector $\mathbf{r}$ at these times; the outline of the body itself has been omitted in that figure for simplicity. The *angular displacement* of P will be $\theta_2 - \theta_1 = \Delta\theta$ during the time interval $t_2 - t_1 = \Delta t$. We define the *average angular speed* $\overline{\omega}$ of particle P in this time interval as

$$\overline{\omega} = \frac{\theta_2 - \theta_1}{t_2 - t_1} = \frac{\Delta\theta}{\Delta t}.$$

We define the *instantaneous angular speed* ω as the limit approached by this ratio as Δt approaches zero:

$$\omega = \lim_{\Delta t \to 0} \frac{\Delta\theta}{\Delta t} = \frac{d\theta}{dt}. \tag{11-1}$$

For a rigid body all radial lines fixed in it perpendicular to the axis of rotation rotate through the same angle in the same time, so that the angular speed ω about this axis is the same for each particle in the body. Thus ω is characteristic of the body as a whole. Angular speed has the dimensions of an inverse time (T^{-1}); its units are commonly taken to be radians/second (rad/s) or revolutions/second (rev/s).

If the angular speed of P is not constant, then the particle has an angular acceleration. Let ω_1 and ω_2 be the instantaneous angular speeds at the times t_1 and t_2 respectively; then the *average angular acceleration* $\overline{\alpha}$ of the particle P is defined as

$$\overline{\alpha} = \frac{\omega_2 - \omega_1}{t_2 - t_1} = \frac{\Delta\omega}{\Delta t}.$$

The *instantaneous angular acceleration* is the limit of this ratio as Δt

11-2
ROTATIONAL KINEMATICS— THE VARIABLES

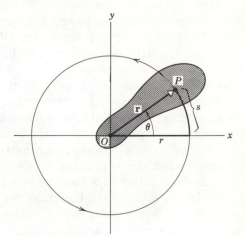

figure 11-3
A cross sectional view of the rigid body of Fig. 11-1, showing point P and vector $\mathbf{r}$ of that figure. Point P, which is fixed in the rotating body, rotates counterclockwise about the origin in a circle of radius r.

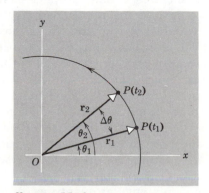

figure 11-4
The reference line r (= OP), fixed in the body of Figs. 11-1 and 11-3, is displaced through angle $\Delta\theta$ (= $\theta_2 - \theta_1$) in time Δt (= $t_2 - t_1$).

---

* The radian is a purely geometrical unit having no physical dimension because it is the ratio of two lengths. Since the circumference of a circle of radius r is $2\pi r$, there are 2π rad in a complete circle, that is, $\theta = 2\pi r/r = 2\pi$. Therefore 2π rad $= 360°$, π rad $= 180°$, and 1 rad $\cong 57.3°$.

approaches zero, or

$$\alpha = \lim_{\Delta t \to 0} \frac{\Delta \omega}{\Delta t} = \frac{d\omega}{dt}. \qquad (11\text{-}2)$$

Because ω is the same for all particles in the rigid body, it follows from Eq. 11-2 that α must be the same for each particle and thus α, like ω, is a characteristic of the body as a whole. Angular acceleration has the dimensions of an inverse time squared (T^{-2}); its units are commonly taken to be radians/second^2 (rad/s^2) or revolutions/second^2 (rev/s^2).

The rotation of a particle (or a rigid body) *about a fixed axis* has a formal correspondence to the translational motion of a particle (or a rigid body) *along a fixed direction*. The kinematical variables are θ, ω, and α in the first case and x, v, and a in the second. These quantities correspond in pairs: θ to x, ω to v, and α to a. Note that the angular quantities differ dimensionally from the corresponding linear quantities by a length factor. Note, too, that all six quantities may be treated as scalars in this special case. For example, a particle at any instant can be moving in one direction or the other along its straight-line motion, corresponding to a positive or a negative value for v; similarly a particle at any instant can be rotating in one direction or another about its fixed axis, corresponding to a positive or a negative value for ω.

When, in translational motion, we remove the restriction that the motion be along a straight line and consider the general case of motion in three dimensions along a curved path, the linear variables x, v, and a reveal themselves as the scalar components of the kinematic vectors **r**, **v**, and **a**. In Section 11-4, we shall see to what extent the rotational kinematic variables reveal themselves as vectors when we remove the restriction of a fixed axis of rotation.

11-3
ROTATION WITH CONSTANT ANGULAR ACCELERATION

For translational motion of a particle or a rigid body along a fixed direction, such as the x-axis, we have seen (in Chapter 3) that the simplest type of motion is that in which the acceleration a is zero. The next simplest type corresponds to $a = $ a constant (other than zero); for this motion we derived the equations of Table 3-1, which connect the kinematic variables x, v, a, and t in all possible combinations.

For the rotational motion of a particle or a rigid body around a fixed axis the simplest type of motion is that in which the angular acceleration α is zero (such as uniform circular motion). The next simplest type of motion, in which $\alpha = $ a constant (other than zero), corresponds exactly to linear motion with $a = $ a constant (other than zero). As before, we can derive four equations linking the four kinematic variables θ, ω, α, and t in all possible combinations. You can either derive these angular equations by the methods used to derive the linear equations (see Example 2) or you may write them down at once by substituting corresponding angular quantities for the linear quantities in the linear equations.

We list both sets of equations in Table 11-1, having chosen $x_0 = 0$ and $\theta_0 = 0$ in these relations for simplicity. Here ω_0 is the angular speed at the time $t = 0$. You should check these equations dimensionally before verifying them. Both sets of equations hold not only for particles but also for rigid bodies.

For the angular quantities, we arbitrarily select one of the two possible directions of rotation about the fixed axis as the direction in which θ is increasing. From Eq. 11-1 ($\omega = d\theta/dt$) we see that if θ is increasing with

Table 11-1
Motion with constant linear or angular acceleration

	Translational Motion (Fixed Direction)	Rotational Motion (Fixed Axis)	
(3-12)	$v = v_0 + at$	$\omega = \omega_0 + \alpha t$	(11-3)
(3-14)	$x = \dfrac{v_0 + v}{2} t$	$\theta = \dfrac{\omega_0 + \omega}{2} t$	(11-4)
(3-15)	$x = v_0 t + \frac{1}{2} a t^2$	$\theta = \omega_0 t + \frac{1}{2}\alpha t^2$	(11-5)
(3-16)	$v^2 = v_0^2 + 2ax$	$\omega^2 = \omega_0^2 + 2\alpha\theta$	(11-6)

time, ω is positive. Similarly, from Eq. 11-2 ($\alpha = d\omega/dt$), we see that if ω is increasing with time, α is positive. There are corresponding sign conventions for the linear quantities.

EXAMPLE 1

A grindstone has a constant angular acceleration α of 3.0 rad/s². Starting from rest a line, such as OP in Fig. 11-5, is horizontal. Find (a) the angular displacement of the line OP (and hence of the grindstone) and (b) the angular speed of the grindstone 2.0 s later.

(a) α and t are given; we wish to find θ. Hence, we use Eq. 11-5,

$$\theta = \omega_0 t + \tfrac{1}{2}\alpha t^2.$$

At $t = 0$, we have $\omega = \omega_0 = 0$ and $\alpha = 3.0$ rad/s². Therefore, after 2.0 s,

$$\theta = (0)(2.0 \text{ s}) + \tfrac{1}{2}(3.0 \text{ rad/s}^2)(2.0 \text{ s})^2 = 6.0 \text{ rad} = 0.96 \text{ rev}.$$

(b) α and t are given; we wish to find ω. Hence, we use Eq. 11-3

$$\omega = \omega_0 + \alpha t,$$

and

$$\omega = 0 + (3.0 \text{ rad/s}^2)(2.0 \text{ s}) = 6.0 \text{ rad/s}.$$

Using Eq. 11-6 as a check, we have

$$\omega^2 = \omega_0^2 + 2\alpha\theta,$$

$$\omega^2 = 0 + (2)(3.0 \text{ rad/s}^2)(6.0 \text{ rad}) = 36 \text{ rad}^2/\text{s}^2,$$

$$\omega = 6.0 \text{ rad/s}.$$

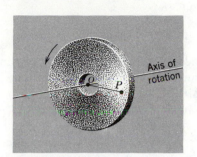

figure 11-5
Example 1. The line OP is attached to a grindstone rotating as shown about an axis through O that is fixed in the reference frame of the observer.

EXAMPLE 2

Derive the equation $\omega = \omega_0 + \alpha t$ for constant angular acceleration.

(a) Starting from the definition of angular acceleration,

$$\alpha = \frac{d\omega}{dt},$$

we have

$$\alpha \, dt = d\omega$$

or

$$\int \alpha \, dt = \int d\omega.$$

But α is a constant, so that

$$\alpha \int dt = \int d\omega.$$

If at $t = 0$ we call the angular speed ω_0, then

$$\alpha \int_0^t dt = \int_{\omega_0}^\omega d\omega$$

or

$$\alpha t = \omega - \omega_0$$

and

$$\omega = \omega_0 + \alpha t.$$

(b) We can also derive the result by making use of the fact that the average acceleration equals the instantaneous acceleration when the acceleration is constant. The average acceleration is

$$\bar{\alpha} = \frac{\omega - \omega_0}{t - t_0}.$$

For constant acceleration we have $\alpha = \bar{\alpha}$. Letting $t_0 = 0$, we obtain

$$\alpha = \frac{\omega - \omega_0}{t}$$

or

$$\omega = \omega_0 + \alpha t.$$

Compare this derivation with that of the corresponding linear relation $v = v_0 + at$ in Section 3-8.

The linear displacement, velocity, and acceleration are vectors. The corresponding angular quantities *may* be vectors also, for in addition to a magnitude we must also specify a direction for them, namely, the direction of the axis of rotation in space. Because we considered rotation only about a fixed axis, we were able to treat θ, ω, and α as scalar quantities. If the direction of the axis changes, however, we can no longer avoid the question "are rotational quantities vectors?" We can find out only by seeing whether or not they obey the laws of vector addition.

Let us discuss first the angular displacement θ. The magnitude of the angular displacement of a body is the angle through which the body turns. Angular displacements, however, are *not* vectors because they do *not* add like vectors. For example, give two successive rotations θ_1 and θ_2 to a book which initially lies in a horizontal plane. Let rotation θ_1 be a 90° clockwise turn about a vertical axis through the center of the book as we view it from above. Let θ_2 be a 90° clockwise turn about a north-south axis through the center of the book as we view it looking north. In one case, apply operation θ_1 first and then θ_2. In the other case, apply operation θ_2 first and then θ_1. You should try this for yourself. Now, if angular displacements are vector quantities, they must add like vectors. In particular, they must obey the law of vector addition $\boldsymbol{\theta}_1 + \boldsymbol{\theta}_2 = \boldsymbol{\theta}_2 + \boldsymbol{\theta}_1$, which tells us that the order in which we add vectors does not affect their sum. This law fails for finite angular displacements (see exercise above and also Fig. 11-6a). Hence finite angular displacements are *not* vector quantities.

Suppose that instead of 90° rotations we had made 3° rotations. The result of $\boldsymbol{\theta}_1 + \boldsymbol{\theta}_2$ would still differ from the result of $\boldsymbol{\theta}_2 + \boldsymbol{\theta}_1$, but the difference would be much smaller. In fact, as the two angular displacements are made smaller, the difference between the two sums rapidly disappears (Fig. 11-6b,c). If the angular displacements are made infinitesimal, the order of addition no longer affects the result. Hence *infinitesimal angular displacements are vectors.*

Quantities defined in terms of infinitesimal angular displacements may themselves be vectors. For example, the angular velocity is $\boldsymbol{\omega} = d\boldsymbol{\theta}/dt$. Since $d\boldsymbol{\theta}$ is a vector and dt a scalar, the quotient is a vector.

11-4
ROTATIONAL QUANTITIES AS VECTORS

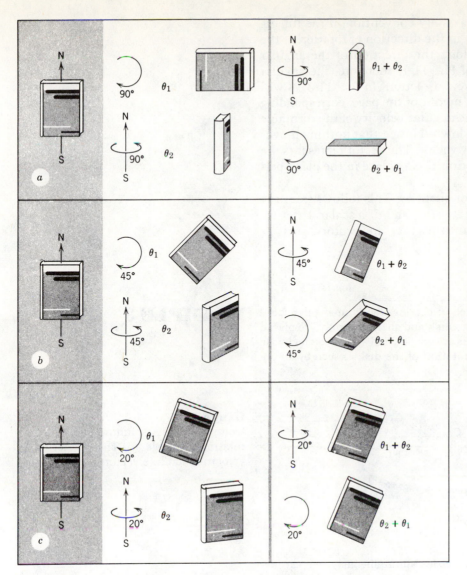

figure 11-6
(a) A book rotated θ_1 (90° as shown about an axis at right angles to the page) and then θ_2 (90° as shown about a north-south axis) has a different final orientation than if rotated first through θ_2 and then θ_1. This property is called the noncommutivity of finite angles under addition: $\theta_1 + \theta_2 \neq \theta_2 + \theta_1$. (b) The middle group is the same except that the angular displacements are smaller, being 45°. Although the final orientations still differ, they are much nearer each other. (c) The lower group repeats the experiment for 20° displacements. We see here that $\theta_1 + \theta_2 \cong \theta_2 + \theta_1$. As θ_1, $\theta_2 \to 0$, the final positions approach each other. Finite angles under addition tend to commute as the angles become very small. Infinitesimal angles *do* commute under addition, making it possible to treat them as vectors.

Therefore the angular velocity is a vector. In Fig. 11-7a, for example, we represent the angular velocity $\boldsymbol{\omega}$ of the rotating rigid body by an arrow drawn along the axis of rotation; in Fig. 11-7b we represent the rotation of a particle (such as P in Fig. 11-7a) about a fixed axis in just the same way. The length of the arrow is made proportional to the magnitude of the angular velocity. The sense of the rotation determines the direction

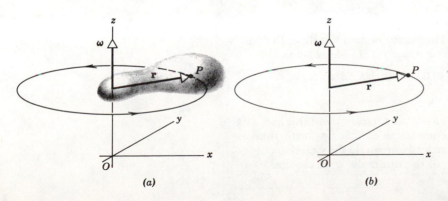

(a)   (b)

figure 11-7
The angular velocity ω of (a) a rotating rigid body and (b) a rotating particle, about a fixed axis.

in which the arrow points along the axis. By convention, if the fingers of the *right hand* curl around the axis in the direction of rotation of the body, the extended thumb points along the direction of the angular velocity vector. For the rigid body of Fig. 11-1, therefore, the angular velocity vector will be in the positive z-direction. In Fig. 11-3, $\boldsymbol{\omega}$ will be perpendicular to the page pointing up out of the page, corresponding to the counter-clockwise rotation. The angular velocity of the turntable of a phonograph is a vector pointing down. Notice that nothing moves in the direction of the angular velocity vector. The vector represents the angular velocity of the rotational motion taking place in the plane perpendicular to it.

Angular acceleration is also a vector quantity. This follows from the definition $\boldsymbol{\alpha} = d\boldsymbol{\omega}/dt$, in which $d\boldsymbol{\omega}$ is a vector and dt a scalar. Later we shall encounter other rotational quantities that are vectors, such as torque and angular momentum.

A disk spins on a horizontal shaft mounted in bearings, with an angular speed ω_1 of 100 rad/s as in Fig. 11-8a. The entire disk and shaft assembly are placed on a turntable rotating about a vertical axis at $\omega_2 = 30.0$ rad/s, counterclockwise as we view it from above. Describe the rotation of the disk as seen by an observer in the room.

EXAMPLE 3

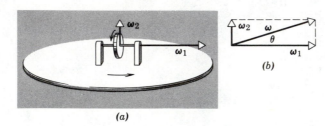

(a)

figure 11-8
Example 3. *(a)* A spinning disc on a rotating turntable. *(b)* The angular velocities add like vectors.

The disk is subject to two angular velocities simultaneously; we can describe its resultant motion by the vector sum of these vectors. The angular velocity $\boldsymbol{\omega}_1$ associated with the shaft rotation has a magnitude of 100 rad/s and occurs about an axis that is not fixed but, as seen by an observer in the room, rotates in a horizontal plane at 30 rad/s. The angular velocity $\boldsymbol{\omega}_2$ associated with the turntable is fixed vertically and has a magnitude of 30 rad/s.

The resultant angular velocity of the disk $\boldsymbol{\omega}$ is the vector sum of $\boldsymbol{\omega}_1$ and $\boldsymbol{\omega}_2$. The magnitude of $\boldsymbol{\omega}$ is

$$\omega = \sqrt{\omega_1{}^2 + \omega_2{}^2} = \sqrt{(100 \text{ rad/s})^2 + (30.0 \text{ rad/s})^2}$$

$$= 104 \text{ rad/s}.$$

The direction of $\boldsymbol{\omega}$ is not fixed in our observer's reference frame but rotates at the same angular rate as the turntable. The vector $\boldsymbol{\omega}$ does not lie in the horizontal plane but points above it by an angle θ (see Fig. 11-8b), where

$$\theta = \tan^{-1} \omega_2/\omega_1 = \tan^{-1} (30.0 \text{ rad/s})/(100 \text{ rad/s})$$

$$= \tan^{-1} 0.300 = 16.7°$$

We can describe the motion of the disk as a simple rotation about this new axis (whose direction in our observer's reference frame is changing with time as described above) at an angular rate of 104 rad/s. How would the situation change if the direction of rotation of the disk, or of the turntable, were changed?

In Sections 4-4 and 4-5 we discussed the linear velocity and accelera-tion of a particle moving in a circle. When a rigid body rotates about a fixed axis, every particle in the body moves in a circle. Hence we can describe the motion of such a particle either in linear variables or in angular variables. The relation between the linear and angular variables enables us to pass back and forth from one description to another and is very useful.

Consider a particle at P in the rigid body, a distance r from the axis through O. This particle moves in a circle of radius r as the body rotates, as in Fig. 11-9a. The reference position is Ox. The particle moves through a distance s along the arc when the body rotates through an angle θ, such that

$$s = \theta r \qquad (11\text{-}7)$$

where θ is in radians.

11-5
RELATION BETWEEN LINEAR AND ANGULAR KINEMATICS FOR A PARTICLE IN CIRCULAR MOTION — SCALAR FORM

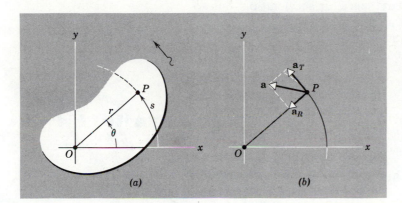

figure 11-9
(a) A rigid body rotates about a fixed axis through O perpendicular to the page. The point P sweeps out an arc s which subtends an angle θ. (b) The acceleration **a** of point P has components $\mathbf{a}_T$ (tangential) where $a_T = \alpha r$ and $\mathbf{a}_R$ (radial) where $a_R = v^2/r = \omega^2 r$ (ω = angular speed).

Differentiating both sides of this equation with respect to the time, and noting that r is constant, we obtain

$$\frac{ds}{dt} = \frac{d\theta}{dt} r.$$

But ds/dt is the linear speed of the particle at P and $d\theta/dt$ is the angular speed ω of the rotating body so that

$$v = \omega r. \qquad (11\text{-}8)$$

This is a relation between the *magnitudes* of the linear velocity and the angular velocity; the linear speed of a particle in circular motion is the product of the angular speed and the distance r of the particle from the axis of rotation.

Differentiating Eq. 11-8 with respect to the time, we have

$$\frac{dv}{dt} = \frac{d\omega}{dt} r.$$

But dv/dt is the magnitude of the *tangential* component of acceleration of the particle (see Section 4-5) and $d\omega/dt$ is the magnitude of the an-gular acceleration of the rotating body, so that

$$a_T = \alpha r. \qquad (11\text{-}9)$$

Hence the magnitude of the tangential component of the linear acceler-ation of a particle in circular motion is the product of the magnitude of the angular acceleration and the distance r of the particle from the axis of rotation.

We have seen that the *radial* component of acceleration is v^2/r for a particle moving in a circle. This can be expressed in terms of angular speed by use of Eq. 11-8. We have

$$a_R = \frac{v^2}{r} = \omega^2 r. \qquad (11\text{-}10)$$

The resultant acceleration of point P is shown in Fig. 11-9b.

Equations 11-7 through 11-10 enable us to describe the motion of one point on a rigid body rotating about a fixed axis *either* in angular variables *or* in linear variables. We might ask why we need the angular variables when we are already familiar with the equivalent linear variables. The answer is that the angular description offers a distinct advantage over the linear description when various points on the same rotating body must be considered. Different points on the same rotating body do not have the same linear displacement, speed, or acceleration, but *all* points on a rigid body rotating about a fixed axis do have the same *angular* displacement, speed, or acceleration at any instant. By the use of angular variables we can describe the motion of the body as a whole in a simple way.

EXAMPLE 4

If the radius of the grindstone of Example 1 is 0.50 m, calculate (*a*) the linear or tangential speed of a particle on the rim, (*b*) the tangential acceleration of a particle on the rim, and (*c*) the centripetal acceleration of a particle on the rim, at the end of 2.0 s.

We have $\alpha = 3.0$ rad/s², $\omega = 6.0$ rad/s after 2.0 s, and $r = 0.50$ m. Then,

(*a*)
$$\begin{aligned} v &= \omega r \\ &= (6.0 \text{ rad/s})(0.50 \text{ m}) \\ &= 3.0 \text{ m/s} \quad \text{(linear speed)}; \end{aligned}$$

(*b*)
$$\begin{aligned} a_T &= \alpha r \\ &= (3.0 \text{ rad/s}^2)(0.50 \text{ m}) \\ &= 1.5 \text{ m/s}^2 \quad \text{(tangential acceleration)}; \end{aligned}$$

(*c*)
$$\begin{aligned} a_R &= v^2/r = \omega^2 r \\ &= (6.0 \text{ rad/s})^2(0.50 \text{ m}) \\ &= 18 \text{ m/s}^2 \quad \text{(centripetal acceleration)}. \end{aligned}$$

(*d*) Are the results the same for a particle halfway in from the rim, that is, at $r = 0.25$ m?

The *angular* variables are the same for this point as for a point on the rim. That is, once again

$$\alpha = 3.0 \text{ rad/s}^2, \qquad \omega = 6.0 \text{ rad/s}.$$

But now $r = 0.25$ m, so that for this particle

$$v = 1.5 \text{ m/s}, \qquad a_T = 0.75 \text{ m/s}^2, \qquad a_R = 9.0 \text{ m/s}^2.$$

Notice that the relations deduced in the previous section are relations between *scalar* quantities, both the linear and angular variables being expressed in scalar form. Let us now use vector methods, making an analysis essentially like that of Section 4-5 except that we now introduce the angular variables. This will illustrate, for a familiar special case, the more general approach and prepare the way for situations in which vector methods are essential. We continue to restrict ourselves to rotation about a fixed axis.

11-6
RELATION BETWEEN LINEAR AND ANGULAR KINEMATICS FOR A PARTICLE IN CIRCULAR MOTION—VECTOR FORM

Figure 11-10a shows a particle P, rotating about a fixed axis through the origin, at times t and $t + \Delta t$. The particle moves in a circle of constant radius r; beyond this there are no restrictions on its motion and in general ω and α may have values that vary as the particle moves. We can express the restriction to a constant radius by

$$\mathbf{r} = \mathbf{u}_r r, \tag{11-11}$$

in which $\mathbf{u}_r$ is a unit vector in the direction of $\mathbf{r}$.

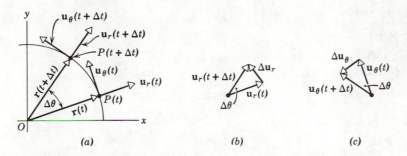

(a)   (b)   (c)

figure 11-10
(a) The particle P rotates through an angle $\Delta\theta$ in time Δt. The unit vectors, in polar coordinates, are shown at each point. (b) The change in $\mathbf{u}_r$; note that $\Delta\mathbf{u}_r$, as $\Delta\theta \to 0$, points in the direction of $\mathbf{u}_\theta$. (c) The change in $\mathbf{u}_\theta$; note that $\Delta\mathbf{u}_\theta$, as $\Delta\theta \to 0$, points in the direction of $-\mathbf{u}_r$.

Differentiating Eq. 11-11, remembering that r (but not $\mathbf{r}$ or $\mathbf{u}_r$, since their directions change) is a constant, we have

$$\frac{d\mathbf{r}}{dt} = \frac{d\mathbf{u}_r}{dt} r. \tag{11-12}$$

Now $d\mathbf{r}/dt$ is $\mathbf{v}$, the linear velocity of the particle. To evaluate $d\mathbf{u}_r/dt$, consider Fig. 11-10b, which shows the unit vector $\mathbf{u}_r$ for two different positions of P, corresponding to a rotation through a (small) angle $\Delta\theta$. Using the definition of angular measure in radians we obtain the *magnitude* of the (vector) change $\Delta\mathbf{u}_r$ in $\mathbf{u}_r$ from

$$\Delta u_r = (1)\,\Delta\theta,$$

in which the factor (1) reminds us that the two unit vectors in Fig. 11-10b have unit length. The above equation will be correct if $\Delta\theta$ is small enough so that we can neglect the difference between the chord and the arc in the small triangle in Fig. 11-10b. The change in $\mathbf{u}_r$ is a vector, $\Delta\mathbf{u}_r$, whose magnitude is given by the above equation; its *direction*, again assuming that $\Delta\theta$ is small enough, is given by the unit vector $\mathbf{u}_\theta$. This follows because, if $\Delta\mathbf{u}_r$ in Fig. 11-10b is translated to point P in Fig. 11-10a, we see that, as $\Delta\theta \to 0$, it points in the direction of $\mathbf{u}_\theta$. Thus we find

$$\Delta\mathbf{u}_r \cong \mathbf{u}_\theta\,\Delta\theta.$$

Dividing by Δt and allowing Δt to approach zero, we have

$$\frac{d\mathbf{u}_r}{dt} = \mathbf{u}_\theta \frac{d\theta}{dt} = \mathbf{u}_\theta\omega. \tag{11-13}$$

Substituting these results into Eq. 11-12 yields, then,

$$\mathbf{v} = \mathbf{u}_\theta\omega r. \tag{11-14a}$$

The scalar relationship that corresponds to this is

$$v = \omega r \tag{11-14b}$$

and is one of the relationships, obtained before, connecting the linear speed v of a particle in circular motion with its angular speed ω.

To find the relation between linear and angular acceleration we differentiate Eq. 11-14a, remembering that r is a constant although $\mathbf{u}_\theta$ and ω vary. We have

$$\frac{d\mathbf{v}}{dt} = \mathbf{u}_\theta \frac{d\omega}{dt} r + \omega \frac{d\mathbf{u}_\theta}{dt} r. \tag{11-15}$$

Now $d\mathbf{v}/dt = \mathbf{a}$, the linear acceleration of the particle and $d\omega/dt = \alpha$, its angular acceleration. From Fig. 11-10c, guided by the derivation of Eq. 11-13, you should be able to prove that

$$\frac{d\mathbf{u}_\theta}{dt} = -\mathbf{u}_r\omega. \tag{11-16}$$

The minus sign comes in because when we translate $\Delta\mathbf{u}_\theta$ in Fig. 11-10c to point P, we see that, as $\Delta\theta \to 0$, it points radially inward, in the direction opposite to $\mathbf{u}_r$.

Making these substitutions into Eq. 11-15 yields

$$\mathbf{a} = \mathbf{u}_\theta\alpha r - \mathbf{u}_r\omega^2 r. \tag{11-17}$$

Thus, as we know from Section 4-5, $\mathbf{a}$ has a radial (or centripetal) component $\mathbf{a}_R$ and a tangential component $\mathbf{a}_T$. Their magnitudes, from Eq. 11-17, are

$$a_T = \alpha r \tag{11-18a}$$

and (using Eq. 11-14b)

$$a_R = \omega^2 r = v^2/r. \tag{11-18b}$$

The last is the familiar result derived in Section 4-4. In Supplementary Topic I we derive the relations between the linear and angular kinematic variables for a particle free to move in a plane but not restricted to circular motion. Equations 11-14a and 11-17 will prove to be special cases of the more general relationships derived there.

Equations 11-14a and 11-17 are relations between the linear kinematic variables in vector form and the angular kinematic variables in scalar form. We should be able to derive relationships in which *each* set of variables is expressed in vector form. Let us do so now. This will be especially useful in cases where the axis of rotation is not fixed.

Figure 11-11 shows the vectors $\mathbf{r}$, $\mathbf{v}$, $\mathbf{a}_T$, $\mathbf{a}_R$, $\boldsymbol{\omega}$, and $\boldsymbol{\alpha}$ for the rotating particle of Fig. 11-7b. The angular quantities are on the axis of rotation, pointing in the direction given by the right-hand rule of page 22. We declare — and shall prove — that the relationships we seek are

$$\mathbf{v} = \boldsymbol{\omega} \times \mathbf{r} \tag{11-19}$$

and

$$\mathbf{a} = \mathbf{a}_T + \mathbf{a}_R, \tag{11-20a}$$

in which

$$\mathbf{a}_T = \boldsymbol{\alpha} \times \mathbf{r} \quad \text{and} \quad \mathbf{a}_R = \boldsymbol{\omega} \times \mathbf{v}. \tag{11-20b}$$

In Section 2-4 (which you may wish to reread) we defined the vector product of two vectors. If $\mathbf{c} = \mathbf{a} \times \mathbf{b}$, then the *magnitude of* $\mathbf{c}$ is $ab \sin \phi$, where ϕ is the angle between $\mathbf{a}$ and $\mathbf{b}$. In applying this part of the definition to Eqs. 11-19 and 11-20 we note (see Fig. 11-11) that $\boldsymbol{\omega}$ and $\mathbf{r}$, $\boldsymbol{\omega}$ and $\mathbf{v}$, and $\boldsymbol{\alpha}$ and $\mathbf{r}$ are each mutually perpendicular; thus the angle ϕ for each of these three pairs of vectors is 90°. In Eq. 11-19 we have, for magnitudes

$$v = \omega r \sin 90° = \omega r,$$

which is exactly Eq. 11-14b. In Eqs. 11-20b we have, for magnitudes

$$a_R = \omega v = \omega(\omega r) = \omega^2 r$$

and

$$a_T = \alpha r.$$

These relations agree with Eqs. 11-18b and a exactly.

It remains to be seen whether *directions* are correctly given by Eqs. 11-19 and 11-20b. For the vector product $\mathbf{c} = \mathbf{a} \times \mathbf{b}$, Fig. 2-12 shows that the direction of $\mathbf{c}$ is found by sweeping $\mathbf{a}$ into $\mathbf{b}$ through the (smaller) angle between them with the fingers of the right hand; the extended right thumb then points in the direc-

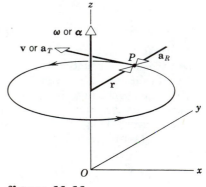

figure 11-11
The directions of the vectors $\mathbf{r}$, $\mathbf{v}$, $\mathbf{a}_T$, $\mathbf{a}_R$, ω and α for a particle rotating in a circle about the z-axis.

tion of **c**. You can readily check that, in Fig. 11-11, the directions of the vectors **v**, $\mathbf{a}_T$ and $\mathbf{a}_R$ are indeed correctly given by Eqs. 11-19 and 11-20*b*.

questions

1. In Section 11-1 we stated that, in general, six variables are required to locate a rigid body with respect to a particular reference frame. How many variables are required to locate the body of Fig. 11-2 with respect to the *x-y* frame shown in that figure? If this number is not six, account for the difference.

2. An irregular body is free to rotate about its center of mass which is placed at the origin of a reference frame. How would you specify its orientation?

3. Could the angular quantities θ, ω, and α be expressed in terms of degrees instead of radians in the kinematical equations?

4. Explain why the radian measure of angle is equally satisfactory for all systems of units. Is the same true for degrees?

5. If a car's speedometer is set to read at a speed proportional to the rotational speed of its rear wheels, is it necessary to correct the reading when snow tires replace regular ones?

6. How could you express simply the relationship between the angular velocities of a pair of gears which are coupled?

7. A wheel is rotating about an axis through its center perpendicular to the plane of the wheel. Consider a point on the rim. When the wheel rotates with *constant angular velocity*, does the point have a radial acceleration? A tangential acceleration? When the wheel rotates with *constant angular acceleration*, does the point have a radial acceleration? A tangential acceleration? Do the magnitudes of these accelerations change?

8. Suppose you were asked to determine the equivalent distance traveled by a phonograph needle in playing, say, a 12-in., $33\frac{1}{3}$ rpm record. What information do you need? Discuss from the points of view of reference frames (*a*) fixed in the room, (*b*) fixed on the rotating record, and (*c*) fixed on the record arm.

9. (*a*) Describe the vector that would represent the angular velocity of the earth rotating about its axis. (*b*) Describe the vector that would represent the angular velocity of the earth rotating about the sun.

10. It is convenient to picture rotational vectors as lying along the axis of rotation. Is there any reason why they could not be pictured as merely parallel to the axis, but located anywhere? Recall that we are free to slide a displacement vector along its own direction or translate it sideways without changing its value.

11. In a centrifuge particles will tend to separate from the fluid in which they are suspended if their density (mass/volume) differs from that of the fluid. Discuss the dynamical principles upon which the operation of a centrifuge depends. View the situation from both an inertial (laboratory) frame and a noninertial (rotating) frame.

12.* A marksman stands at the center of a merry-go-round firing at a target fixed to a post on its outer perimeter. How, if at all, must the man take into account the (constant) angular velocity of the merry-go-round if he is to hit the target? What if the positions of marksman and target were reversed?

13.* A man on a merry-go-round rotating at constant angular velocity ω releases a cake of ice that he had been holding fixed to the merry-go-round at a radial distance r_0 from the center. Describe the motion of the ice in the reference frame of (*a*) a ground observer and (*b*) the man on the merry-go-round. Neglect frictional forces but describe all other forces.

14.* A man on a rotating merry-go-round kicks a cake of ice outward along a radial line. What is its subsequent motion as seen by an observer (*a*) on the

* See Supplementary Topic I.

merry-go-round and (b) on the ground? Assume that frictional forces may be neglected.

SECTION 11-2

problems

1. What is the angular speed of (a) the second hand of a watch; (b) of the minute hand?
   Answer: (a) 0.10 rad/s. (b) 1.7×10^{-3} rad/s.

2. A phonograph record on a turntable rotates at 33 rev/min. What is the linear speed of a point on the record at the needle at (a) the beginning and (b) the end of the recording? The distances of the needle from the turntable axis are 5.9 and 2.9 in., respectively, at these two positions.

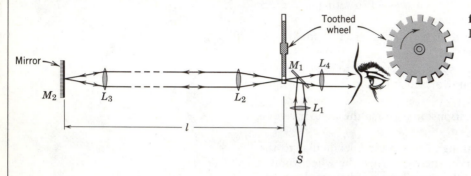

figure 11-12
Problem 3

3. One method of measuring the speed of light makes use of a rotating toothed wheel. A beam of light passes through a slot at the outside edge of the wheel, as in Fig. 11-12, travels to a distant mirror, and returns to the wheel just in time to pass through the next slot in the wheel. One such toothed wheel has a radius of 5.0 cm and 500 teeth at its edge. Measurements taken when the mirror was 500 m from the wheel indicated a speed of light of 3.0×10^5 km/s. (a) What was the (constant) angular speed of the wheel? (b) What was the linear speed of a point on its edge?
   Answer: (a) 3.8×10^3 rad/s. (b) 190 m/s.

4. If an airplane propeller of radius 5.0 ft (1.5 m) rotates at 2000 rev/min and the airplane is propelled at a ground speed of 300 mi/h (480 km/h), what is the speed of a point on the tip of the propeller, as seen by (a) the pilot and (b) an observer on the ground?

5. The angular position of a point on the rim of a rotating wheel is described by $\theta = 4.0t - 3.0t^2 + t^3$, where θ is in radians and t in seconds. What is the average acceleration for the time interval which begins at $t = 2.0$ s and ends at $t = 4.0$ s?
   Answer: 12 rad/s².

6. The angle turned through by the flywheel of a generator during a time interval t is given by

$$\theta = at + bt^3 - ct^4,$$

   where a, b, and c are constants. What is the expression for its angular acceleration?

7. A wheel rotates with an angular acceleration α given by

$$\alpha = 4at^3 - 3bt^2,$$

   where t is the time and a and b are constants. If the wheel has an initial angular speed ω_0, write the equations for (a) the angular speed and (b) the angle turned through as functions of time.
   Answer: (a) $\omega_0 + at^4 - bt^3$. (b) $\theta_0 + \omega_0 t + at^5/5 - bt^4/4$.

8. A planet P revolves around the sun S in a circular orbit, with the sun at the center, which is coplanar with and concentric to, the circular orbit of the earth E around the sun. P and E revolve in the same direction. The times required for the revolution of P and E around the sun are T_P and T_E. Let T_S

be the time required for P to make one revolution around the sun relative to E: show that $1/T_S = 1/T_E - 1/T_P$. Assume $T_P > T_E$.

9. A solar day is the time interval between two successive appearances of the sun overhead at a given longitude, that is, the time for one complete rotation of the earth relative to the sun. A sidereal day is the time for one complete rotation of the earth relative to the fixed stars, that is, the time interval between two successive overhead observations of a fixed direction in the heavens called the vernal equinox. (a) Show that there is exactly one less (mean) solar day in a year than there are (mean) sidereal days in a year. (b) If the (mean) solar day is exactly 24 hours, how long is a (mean) sidereal day?　　　　　　　　　　　　　　　　　　　　　　*Answer:* (b) 23 h 56 min.

SECTION 11-3

10. The angular speed of an automobile engine is increased from 1200 rev/min to 3000 rev/min in 12 s. (a) What is its angular acceleration, assuming it to be uniform? (b) How many revolutions does the engine make during this time?

11. A phonograph turntable rotating at 78 rev/min slows down and stops in 30 s after the motor is turned off. (a) Find its (uniform) angular acceleration. (b) How many revolutions did it make in this time?
Answer: (a) −0.27 rad/s². (b) 20.

12. A heavy flywheel rotating on its axis is slowing down because of friction in its bearings. At the end of the first minute its angular velocity is 0.90 of its angular velocity ω_0 at the start. Assuming constant frictional forces, find its angular velocity at the end of the second minute.

13. While waiting to board a helicopter, you notice that the rotor's motion changed from 300 rev/min to 225 rev/min in one minute. (a) Find the average angular acceleration during the interval. (b) Assuming that this acceleration remains constant, calculate how long it will take for the rotor to stop. (c) How many revolutions will the rotor make after your second observation?　　　　　　　　　　　*Answer:* (a) −0.13 rad/s². (b) 4.0 min. (c) 340.

14. A wheel has a constant angular acceleration of 3.0 rad/s². In a 4.0-s interval, it turns through an angle of 120 rad. Assuming the wheel started from rest, how long had it been in motion at the start of this 4.0-s interval?

15. A uniform disk rotates about a fixed axis starting from rest and accelerating with constant angular acceleration. At one time it is rotating at 10 rev/s. After completing 60 more complete revolutions its angular speed is 15 rev/s. Calculate (a) the angular acceleration, (b) the time required to complete the 60 revolutions mentioned, (c) the time required to attain the 10 rev/s angular speed, and (d) the number of revolutions from rest until the time the disk attained the 10 rev/s angular speed.
Answer: (a) 1.04 rev/s². (b) 4.8 s. (c) 9.6 s. (d) 48.

16. A flywheel completes 40 revolutions as it slows from an angular speed of 1.5 rad/s to a complete stop. Assuming uniform acceleration, (a) what is the time required for it to come to rest? (b) What is the angular acceleration? (c) How much time is required for it to complete the first one-half of the 40 revolutions?

17. An automobile traveling 60 mi/h (97 km/h) has wheels of 30 in. (76 cm) diameter. (a) What is the angular speed of the wheels about the axle? (b) If the car is brought to a stop uniformly in 30 turns, what is the angular acceleration? (c) How far does the car advance during this braking period?
Answer: (a) 70 rad/s (71 rad/s). (b) −13 rad/s² (−13 rad/s²). (c) 240 ft (72 m).

18. A body moves in the x–y plane such that $x = R \cos \omega t$ and $y = R \sin \omega t$. Here x and y are the coordinates of the body, t is the time, and R and ω are constants. (a) Eliminate t between these equations to find the equation of the curve in which the body moves. What is this curve? What is the meaning of the constant ω? (b) Differentiate the equations for x and y with respect to the time to find the x and y components of the velocity of the body, v_x and

v_y. Combine v_x and v_y to find the magnitude and direction of **v**. Describe the motion of the body. (c) Differentiate v_x and v_y with respect to the time to obtain the magnitude and direction of the resultant acceleration.

19. Wheel A of radius $r_A = 10$ cm is coupled by a belt B to wheel C of radius $r_C = 25$ cm, as shown in Fig. 11-13. Wheel A increases its angular speed from rest at a uniform rate of $\pi/2$ rad/s². Determine the time for wheel C to reach a rotational speed of 100 rev/min, assuming the belt does not slip.
Answer: 17 s.

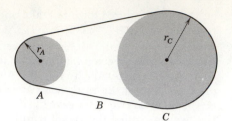

figure 11-13
Problem 19

SECTION 11-5

20. (a) What is the angular speed about the polar axis of a point on the earth's surface at a latitude of 45°N? (b) What is the linear speed? (c) How do these compare with the similar values for a point at the equator?

21. The earth's orbit about the sun is almost a circle. (a) What is the angular velocity of the earth (regarded as a particle) about the sun and (b) its average linear speed in its orbit? (c) What is the acceleration of the earth with respect to the sun?
Answer: (a) 2.0×10^{-7} rad/s. (b) 3.0×10^4 m/s. (c) 6.0×10^{-3} m/s².

22. What is the angular speed of a car rounding a circular turn of radius 360 ft (110 m) at 30 mi/h (48 km/h)?

23. What is the acceleration of a point on the rim of a 12-in. (30 cm) diameter record rotating at 33.3 rev/min?   *Answer:* 6.1 ft/s² (1.8 m/s²).

24. What is the ratio of the acceleration, associated with the earth's rotation, of a point on the equator, to the acceleration of the earth itself, associated with its motion around the sun? Assume a circular orbit.

25. The flywheel of a steam engine runs with a constant angular speed of 150 rev/min. When steam is shut off, the friction of the bearings and of the air brings the wheel to rest in 2.2 h. (a) What is the average angular acceleration of the wheel? (b) How many rotations will the wheel make before coming to rest? (c) What is the tangential linear acceleration of a particle distant 50 cm from the axis of rotation when the flywheel is turning at 75 rev/min? (d) What is the magnitude of the total linear acceleration of the particle in part (c)?
Answer: (a) -2.0×10^{-3} rad/s². (b) 10^4 rev. (c) -1.0 mm/s². (d) 31 m/s².

26. A rigid body, starting at rest, rotates about a fixed axis with constant angular acceleration α. Consider a particle a distance r from the axis. Express (a) the radial acceleration and (b) the tangential acceleration of this particle in the body in terms of α, r and the time t. (c) If the resultant acceleration of the particle at some instant makes an angle of 60° with the tangential acceleration, what total angle has the body turned through to that instant?

SECTION 11-6

27. Derive Eq. 11-20 by differentiation of Eq. 11-19.

28.* An insect of mass 8.0×10^{-2} g walks out with a constant speed of 1.6 cm/s along a radial line marked on a phonograph turntable rotating at a constant angular velocity of $33\frac{1}{3}$ rev/min. Find (a) the velocity and (b) the acceleration of the insect as seen by the ground observer when the insect is 12 cm from the axis of rotation. (c) What must the minimum coefficient of friction be to allow the insect to get all the way to the edge of the turntable (radius = 16 cm) without slipping?

29.* A virus particle, mass 1.0×10^{-7} g, in solution in a centrifuge is, at a particular moment, at a distance of 6.5 cm from the axis of rotation and moving radially outward at a relatively constant speed of 2.0 mm/s. The centrifuge is rotating at 55,000 rev/min. Discuss the motion quantitatively, giving the magnitude of all forces and accelerations as viewed from a reference frame (a) rotating with the centrifuge and (b) fixed in the laboratory.

* See Supplementary Topic I.

12
rotational dynamics I

In Chapter 11 we considered the kinematics of rotation. In this chapter, following the pattern of our study of translational motion, we study the causes of rotation, a subject called *rotational dynamics*. Rotating systems are made up of particles and we have already learned how to apply the laws of classical mechanics to the motion of particles. For this reason rotational dynamics should contain no features that are fundamentally new. In the same way rotational kinematics contained no basic new features, the rotational parameters θ, ω, and α being related to corresponding translational parameters x, v, and a for the particles that make up the rotating system. As in Chapter 11, however, it is very useful to recast the concepts of translational motion into a new form, especially chosen for its convenience in describing rotating systems.

We restricted our kinematical studies in Chapter 11 to a single but important special case, the rotation of a rigid body about an axis that is fixed in the reference frame in which we make our measurements. In studying rotational dynamics we start from a more fundamental point of view, that of a single particle viewed from an inertial reference frame. Later we shall generalize to systems of many particles, including the special case of a rigid body rotating about a fixed axis. In Chapter 13 we shall discuss the rotation of rigid bodies about axes that are *not* fixed in an inertial reference frame.

In translational motion we associate a *force* with the *linear acceleration* of a body. In rotational motion, what quantity shall we associate with the *angular acceleration* of a body? It cannot be simply force because, as experiment with a heavy revolving door teaches us, a given

12-1
INTRODUCTION

12-2
TORQUE ACTING ON A PARTICLE

231

force (vector) can produce various angular accelerations of the door depending on where the force is applied and how it is directed; a force applied to the hinge line cannot produce any angular acceleration, whereas a force of given magnitude applied at right angles to the door at its outer edge produces a maximum acceleration.

We shall call the rotational analogue of force *torque* and shall now define it for the special case of a single particle observed from an inertial reference frame. Later we shall extend the torque concept to systems of particles (including rigid bodies) and shall show that torque is intimately associated with angular acceleration.

If a force **F** acts on a single particle at a point P whose position with respect to the origin O of the inertial reference frame is given by the displacement vector **r** (Fig. 12-1), the *torque* τ acting on the particle *with respect to the origin O* is defined as

$$\tau = \mathbf{r} \times \mathbf{F}. \tag{12-1}$$

Torque is a vector quantity. Its magnitude is given by

$$\tau = rF \sin \theta, \tag{12-2a}$$

where θ is the angle between **r** and **F**; its direction is normal to the plane formed by **r** and **F**. The sense is given by the right-hand rule for the vector product of two vectors, namely, one swings **r** into **F** through the smaller angle between them with the curled fingers of the right hand; the direction of the extended thumb then gives the direction of τ.

Torque has the same dimensions as force times distance, or in terms of our assumed fundamental dimensions, M, L, T, it has the dimensions ML^2T^{-2}. These are the same as the dimensions of work. However, torque and work are very different physical quantities. Torque is a vector and work is a scalar, for example. The unit of torque may be the newton-meter (N · m) or pound-foot (lb · ft), among other possibilities.

Notice (Eq. 12-1) that the torque produced by a force depends not only on the magnitude and on the direction of the force but also on the point of application of the force relative to the origin, that is, on the vector **r**. In particular, when particle P is at the origin, so that the line of action of **F** passes through the origin, **r** is zero and the torque τ about the origin is zero.

We can also write the magnitude of τ (Eq. 12-2a) either as

$$\tau = (r \sin \theta) F = Fr_\perp, \tag{12-2b}$$

or as

$$\tau = r(F \sin \theta) = rF_\perp, \tag{12-2c}$$

in which, as Fig. 12-2a shows, $r_\perp (= r \sin \theta)$ is the component of **r** at right angles to the line of action of **F**, and $F_\perp (= F \sin \theta)$ is the component of **F** at right angles to **r**. Torque is often called the *moment of force* and $r_\perp$ in Eq. 12-2b is called the *moment arm*. Equation 12-2c shows that only the component of **F** perpendicular to **r** contributes to the torque. In particular, when θ equals 0 or 180°, there is no perpendicular component $(F_\perp = F \sin \theta = 0)$; then the line of action of the force passes through the origin and the moment arm $r_\perp$ about the origin is also zero. In this case both Eq. 12-2b and Eq. 12-2c show that the torque τ is zero.

If, as in Fig. 12-2b, we reverse the direction of **F**, the magnitude of τ remains unchanged but the direction of τ is reversed. Similarly, if, as in Fig. 12-2c, we reverse **r**, thereby changing the point of application of **F**, the magnitude of τ remains unchanged but the direction of τ is again reversed.

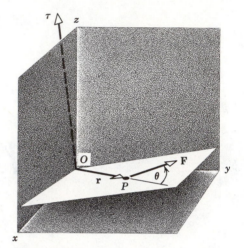

figure 12-1
A force is applied to a particle P, displaced **r** relative to the origin. The force vector makes an angle θ with the radius vector **r**. The torque τ about O is shown. Its direction is perpendicular to the plane formed by **r** and **F** with the sense given by the right-hand rule.

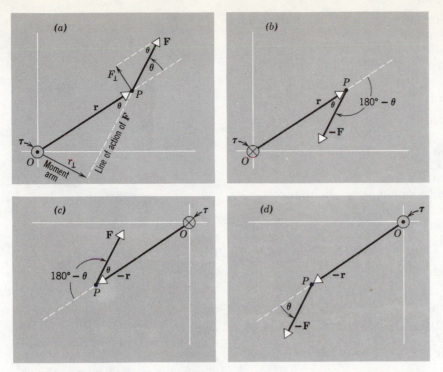

figure 12-2
The plane shown is that defined by **r** and **F** in Fig. 12-1. *(a)* The magnitude of τ is given by $Fr_\perp$ (Eq. 12-2b) or by $rF_\perp$ (Eq. 12-2c). *(b)* Reversing **F** reverses the direction of τ but leaves its magnitude unchanged. *(c)* Reversing **r** also reverses the direction of τ but leaves its magnitude unchanged. *(d)* Reversing **F** and **r** leaves both the direction and magnitude of τ unchanged. The directions of τ are represented by $\odot$ (perpendicularly out of the figure, the symbol representing the tip of an arrow) and by $\otimes$ (perpendicularly into the figure, the symbol representing the tail of an arrow).

If, as in Fig. 12-2d, we reverse *both* **r** and **F**, then both the magnitude and the direction of τ remain unchanged. These results follow formally from the facts that: (1) $\sin \theta = \sin (180° - \theta)$, so that Eq. 12-2a for the magnitude of τ is unaffected; (2) reversing the direction of *one* vector in a vector product (either **r** *or* **F**) reverses the direction of the product; and (3) reversing the directions of *both* vectors in a vector product (both **r** *and* **F**) leaves the direction of the product unchanged. You should verify the directions of τ shown in Fig. 12-2, using the right-hand rule.

12-3
ANGULAR MOMENTUM OF A PARTICLE

We have found *linear momentum* to be useful in dealing with the translational motion of single particles or of systems of particles (including rigid bodies). For example, linear momentum is conserved in collisions. For a single particle the linear momentum is $\mathbf{p} = m\mathbf{v}$ (Eq. 9-11); for a system of particles it is $\mathbf{P} = M\mathbf{v}_{cm}$ (Eq. 9-15) in which M is the total system mass and $\mathbf{v}_{cm}$ is the velocity of the center of mass. In rotational motion, what is the analog of linear momentum? We call it *angular momentum* and we define it below for the special case of a single particle. Later we shall broaden the definition to include systems of particles and shall show that angular momentum, as we define it, is as useful a concept in rotational motion as linear momentum is in translational motion.

Consider a particle of mass m and linear momentum **p** at a position **r** relative to the origin O of an inertial reference frame (Fig. 12-3). We define the *angular momentum* **l** of the particle *with respect to the origin O* to be

$$\mathbf{l} = \mathbf{r} \times \mathbf{p}. \qquad (12-3)$$

Note that we must specify the origin O in order to define the position vector **r** in the definition of angular momentum.

Angular momentum is a vector. Its magnitude is given by

$$l = rp \sin \theta, \qquad (12-4a)$$

where θ is the angle between $\mathbf{r}$ and $\mathbf{p}$; its direction is normal to the plane formed by $\mathbf{r}$ and $\mathbf{p}$. The sense is given by the right-hand rule, namely, one swings $\mathbf{r}$ into $\mathbf{p}$, through the smaller angle between them, with the curled fingers of the right hand; the extended right thumb then points in the direction of $\mathbf{l}$.

We can also write the magnitude of $\mathbf{l}$ either as

$$l = (r \sin \theta) p = p r_{\perp}, \qquad (12\text{-}4b)$$

or as

$$l = r(p \sin \theta) = r p_{\perp}, \qquad (12\text{-}4c)$$

in which $r_{\perp} (= r \sin \theta)$ is the component of $\mathbf{r}$ at right angles to the line of action of $\mathbf{p}$ and $p_{\perp} (= p \sin \theta)$ is the component of $\mathbf{p}$ at right angles to $\mathbf{r}$. Angular momentum is often called *moment of* (linear) *momentum* and $r_{\perp}$ in Eq. 12-4b is often called the *moment arm*. Equation 12-4c shows that only the component of $\mathbf{p}$ perpendicular to $\mathbf{r}$ contributes to the angular momentum. When the angle θ between $\mathbf{r}$ and $\mathbf{p}$ is 0 or 180°, there is no perpendicular component $(p_{\perp} = p \sin \theta = 0)$; then the line of action of $\mathbf{p}$ passes through the origin and $r_{\perp}$ is also zero. In this case both Eqs. 12-4b and 12-4c show that the angular momentum l is zero.

We now derive an important relation between torque and angular momentum. We have seen that $\mathbf{F} = d(m\mathbf{v})/dt = d\mathbf{p}/dt$ for a particle. Let us take the vector product of $\mathbf{r}$ with both sides of this equation, obtaining

$$\mathbf{r} \times \mathbf{F} = \mathbf{r} \times \frac{d\mathbf{p}}{dt}.$$

But $\mathbf{r} \times \mathbf{F}$ is the torque, or moment of a force, about O. We can then write

$$\boldsymbol{\tau} = \mathbf{r} \times \frac{d\mathbf{p}}{dt}. \qquad (12\text{-}5)$$

Next we differentiate Eq. 12-3 and obtain

$$\frac{d\mathbf{l}}{dt} = \frac{d}{dt}(\mathbf{r} \times \mathbf{p}).$$

Now the derivative of a vector product is taken in the same way as the derivative of an ordinary product, except that we must not change the order of the terms. We have

$$\frac{d\mathbf{l}}{dt} = \frac{d\mathbf{r}}{dt} \times \mathbf{p} + \mathbf{r} \times \frac{d\mathbf{p}}{dt}.$$

But $d\mathbf{r}$ is the vector displacement of the particle in the time dt so that $d\mathbf{r}/dt$ is the instantaneous velocity $\mathbf{v}$ of the particle. Also, $\mathbf{p}$ equals $m\mathbf{v}$, so that we can rewrite the equation as

$$\frac{d\mathbf{l}}{dt} = (\mathbf{v} \times m\mathbf{v}) + \mathbf{r} \times \frac{d\mathbf{p}}{dt}.$$

Now $\mathbf{v} \times m\mathbf{v} = 0$, because the vector product of two parallel vectors is zero. Therefore,

$$\frac{d\mathbf{l}}{dt} = \mathbf{r} \times \frac{d\mathbf{p}}{dt}. \qquad (12\text{-}6)$$

Inspection of Eqs. 12-5 and 12-6 shows that

$$\boldsymbol{\tau} = d\mathbf{l}/dt, \qquad (12\text{-}7)$$

which states that *the time rate of change of the angular momentum of*

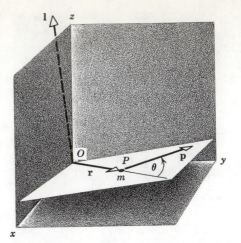

figure 12-3

A particle of mass m is at point P displaced $\mathbf{r}$ relative to the origin, and has linear momentum $\mathbf{p}$. The vector $\mathbf{p}$ makes an angle θ with the radius vector $\mathbf{r}$. The angular momentum $\mathbf{l}$ of the particle with respect to origin O is shown. Its direction is perpendicular to the plane formed by $\mathbf{r}$ and $\mathbf{p}$ with the sense given by the right-hand rule.

a particle is equal to the torque acting on it. This result is the rotational analog of Eq. 9-12, which stated that the time rate of change of the linear momentum of a particle is equal to the force acting on it, that is, that $\mathbf{F} = d\mathbf{p}/dt$.

Equation 12-7, like all vector equations, is equivalent to three scalar equations, namely,

$$\tau_x = (dl_x/dt)_x, \quad \tau_y = (dl_y/dt)_y, \quad \tau_z = (dl_z/dt)_z. \tag{12-8}$$

Hence, the x-component of the applied torque is given by the x-component of the change with time of the angular momentum. Similar results hold for the y- and z-directions.

A particle of mass m is released from rest at point a in Fig. 12-4, falling parallel to the (vertical) y-axis. (a) Find the torque acting on m at any time t, with respect to origin O. (b) Find the angular momentum of m at any time t, with respect to this same origin. (c) Show that the relation $\tau = dl/dt$ (Eq. 12-7) yields a correct result when applied to this familiar problem.

(a) The torque is given by Eq. 12-1 or $\boldsymbol{\tau} = \mathbf{r} \times \mathbf{F}$, its magnitude being given by

$$\tau = rF \sin \theta.$$

In this example $r \sin \theta = b$ and $F = mg$ so that

$$\tau = mgb = \text{a constant}.$$

Note that the torque is simply the product of the force (mg) times the moment arm (b). The right-hand rule shows that $\boldsymbol{\tau}$ is directed perpendicularly into the figure.

(b) The angular momentum is given by Eq. 12-3 or $\mathbf{l} = \mathbf{r} \times \mathbf{p}$, its magnitude being given by

$$l = rp \sin \theta.$$

In this example $r \sin \theta = b$ and $p = mv = m(gt)$ so that

$$l = mgbt.$$

The right-hand rule shows that $\mathbf{l}$ is directed perpendicularly into the figure, which means that $\mathbf{l}$ and $\boldsymbol{\tau}$ are parallel vectors. The vector $\mathbf{l}$ changes with time in magnitude only, its direction always remaining the same in this case.

(c) Since $d\mathbf{l}$, the change in $\mathbf{l}$, and $\boldsymbol{\tau}$ are parallel, we can replace the vector relation $\boldsymbol{\tau} = d\mathbf{l}/dt$ by the scalar relation

$$\tau = dl/dt.$$

Using the expressions for τ and l from (a) and (b) above we have

$$mgb = \frac{d}{dt}(mgbt) = mgb,$$

which is an identity. Thus the relation $\boldsymbol{\tau} = d\mathbf{l}/dt$ yields correct results in this simple case. Indeed, if we cancel the constant b out of the first two terms above and if we substitute for gt the equivalent quantity v, we have

$$mg = \frac{d}{dt}(mv).$$

Since $mg = F$ and $mv = p$, this is the familiar result $F = dp/dt$. Thus, as we indicated earlier, relations such as $\boldsymbol{\tau} = d\mathbf{l}/dt$, though often vastly useful, are not new basic postulates of classical mechanics but are rather the reformulation of the Newtonian laws for rotational motion.

Note that the values of τ and l depend on our choice of origin, that is, on b. In particular, if $b = 0$, then $\tau = 0$ and $l = 0$.

EXAMPLE 1

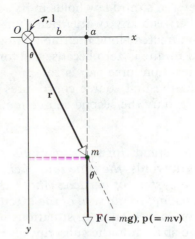

figure 12-4
Example 1. A particle of mass m drops vertically from point a. The torque and the angular momentum about O are directed perpendicularly into the figure, as shown by the symbol $\otimes$ at O.

So far we have talked only about single particles. Let us now consider a system of many particles. To calculate the total angular momentum **L** of a system of particles about a given point, we must add vectorially the angular momenta of all the individual particles of the system about this same point. For a system containing n particles we have, then,

$$\mathbf{L} = \mathbf{l}_1 + \mathbf{l}_2 + \cdots + \mathbf{l}_n = \sum_{i=1}^{i=n} \mathbf{l}_i,$$

in which the (vector) sum is taken over all particles in the system.

As time goes on, the total angular momentum **L** of the system about a fixed reference point (which we choose, as in our basic definition of **l** in Eq. 12-3, to be the origin of an inertial reference frame) may change. This change, $d\mathbf{L}/dt$, can arise from two sources: (1) torques exerted on the particles of the system by internal forces between the particles and (2) torques exerted on the particles of the system by external forces.

If Newton's third law holds in its so-called strong form, that is, if the forces between any two particles not only are equal and opposite but are also directed along the line joining the two particles, then the total internal torque is zero because the torque resulting from each internal action-reaction force pair is zero.

Hence the first source contributes nothing. For our reference point, therefore, only the second source remains, and we can write

$$\tau_{ext} = d\mathbf{L}/dt, \tag{12-9}$$

where τ_{ext} stands for the *sum* of *all* the *external* torques acting on the system. In words, *the time rate of change of the total angular momentum of a system of particles about the origin of an inertial reference frame is equal to the sum of the* external *torques acting on the system.* Later, for convenience, in situations in which no confusion is likely to arise, we shall drop the subscript on τ_{ext}.

Equation 12-9 is the generalization of Eq. 12-7 to many particles. When we have only one particle, there are no internal forces or torques. This relation (Eq. 12-9) holds whether the particles that make up the system are in motion relative to each other or whether they have fixed spatial relationships, as in a rigid body.

Equation 12-9 is the rotational analog of Eq. 9-17

$$\mathbf{F}_{ext} = d\mathbf{P}/dt \tag{9-17}$$

which tells us that for a system of particles (rigid body or not) the resultant external *force* acting on the system equals the time rate of change of the *linear momentum* of the system.

As we have derived it, Eq. 12-9 holds when τ and **L** are measured with respect to the origin of an inertial reference frame. We may well ask whether it still holds if we measure these two vectors with respect to an arbitrary point (a particular particle, say) in the moving system. In general, such a point would move in a complicated way as the body or system of particles translated, tumbled, and changed its configuration and Eq. 12-9 would *not* apply to such a reference point. However, if the reference point is chosen to be the center of mass of the system, even though this point is not fixed in our reference frame, then Eq. 12-9 *does* hold.* This is another remarkable property of the center of mass. Thus we can separate the general motion of a system of particles into the

* See Problem 10 of this chapter and K. R. Symon, *Mechanics*, 3d ed., Addison-Wesley Publishing Co., 1972, Section 4.2.

translational motion of its center of mass (Eq. 9-17) and rotational motion about its center of mass (Eq. 12-9).

We shall now confine our attention to an important special case of a system of particles, *a rigid body*. In a rigid body the particles in the system always maintain the same positions with respect to one another. In studying the rotation of a rigid body we shall consider first the special case, often encountered, in which the axis of rotation is fixed* in an inertial reference frame. Later we shall investigate more general systems and motions.

Let us now imagine a rigid body rotating with angular speed ω about an axis that is fixed in a particular inertial frame, as in Fig. 11-1. Each particle in such a rotating body has a certain amount of kinetic energy. A particle of mass m at a distance r from the axis of rotation moves in a circle of radius r with an angular speed ω about this axis and has a linear speed $v = \omega r$. Its kinetic energy therefore is $\frac{1}{2}mv^2 = \frac{1}{2}mr^2\omega^2$. The total kinetic energy of the body is the sum of the kinetic energies of its particles.

If the body is rigid, as we assume in this section, ω is the same for all particles. The radius r may be different for different particles. Hence the total kinetic energy K of the rotating body can be written as

$$K = \frac{1}{2}(m_1r_1^2 + m_2r_2^2 + \cdots)\omega^2 = \frac{1}{2}(\Sigma\, m_ir_i^2)\omega^2.$$

The term $\Sigma\, m_ir_i^2$ is the sum of the products of the masses of the particles by the squares of their respective distances from the axis of rotation. If we denote this quantity by I, then

$$I = \Sigma\, m_ir_i^2 \qquad (12\text{-}10)$$

is called the *rotational inertia*, or moment of inertia,† of the body with respect to the particular axis of rotation.

Note that *the rotational inertia of a body depends on the particular axis about which it is rotating* as well as on the shape of the body and the manner in which its mass is distributed. Rotational inertia has the dimensions ML^2 and is usually expressed in $kg \cdot m^2$ or $slug \cdot ft^2$.

In terms of rotational inertia we can now write the kinetic energy of the rotating rigid body as

$$K = \frac{1}{2}I\omega^2. \qquad (12\text{-}11)$$

This is analogous to the expression for the kinetic energy of translation of a body, $K = \frac{1}{2}Mv^2$. We have already seen that the angular speed ω is analogous to the linear speed v. Now we see that the rotational inertia I is analogous to the mass, or the translational inertia M. Although the

12-5
KINETIC ENERGY OF ROTATION AND ROTATIONAL INERTIA

---

* As stated in Section 12-4, we can separate the general motion of a system of particles into translational motion of its center of mass and rotational motion about its center of mass. Hence the considerations of this chapter apply also to rotations about an axis that is *not* fixed in an inertial reference frame, provided (1) the axis passes through the center of mass and (2) the moving axis always has the same direction in space, that is, the axis at one instant is parallel to the axis at any other instant. Although we shall often refer to a "fixed axis" in what follows we shall always mean to include this special case of a moving axis.

† The term *moment of inertia* is widely used for this second moment of mass even though there are first, third, and other moments of mass. We choose to emphasize the term *rotational inertia*, however, chiefly because I (the rotational inertia) plays the same role, we shall see, in rotational motion as M (the mass, or the translational inertia) plays in translational motion.

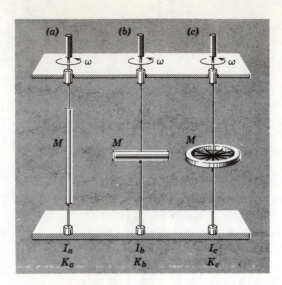

figure 12-5
An experiment to show that
$I_a < I_b < I_c$. The three lead bodies
have the same mass M but the mass
is distributed differently about the
axis of rotation.

mass of a body does not depend on its location, the rotational inertia of a body does depend on the axis about which it is rotating.

We should understand that the rotational kinetic energy given by Eq. 12-11 is simply the sum of the ordinary translational kinetic energy of all the parts of the body and not a new kind of energy. Rotational kinetic energy is simply a convenient way of expressing the kinetic energy for a rotating rigid body.

Equations 12-10 and 12-11 show that the rotational energy of a body, for a given angular speed ω, depends not only on the mass of the body but also on the way that mass is distributed around the axis of rotation. The experiment suggested in Fig. 12-5 makes this convincing. The figure shows three identical aluminum shafts, to each of which is attached a body of mass M, made of lead. In (a) the mass is very close to the shaft so that the quantities r_i in Eq. 12-10 for the particles that make up the body are relatively small, in (b) the particles are, on the average, farther from the shaft and in (c), in which the body is a flywheel, they are still farther, corresponding to still larger values of r_i.

Now let us twist each handle until each shaft, starting from rest, is spinning at the same measured angular speed ω. We know from experience that we shall need to do relatively little work on shaft (a), somewhat more work on shaft (b), and still more on shaft (c). In fact, if we were not certain which body was attached to which shaft we could label the shafts with confidence using this technique. Since the work done on each shaft is equal to the kinetic energy $\frac{1}{2}I\omega^2$ imparted to each shaft, the experimental result, that $K_a < K_b < K_c$ when each shaft has the same angular speed ω, leads to the conclusion that $I_a < I_b < I_c$. This is just what we expect from the defining equation for I (Eq. 12-10). We shall see in Section 12-6 that just as the mass M, which we may call the translational inertia, is a measure of the resistance a body offers to a change in its translational motion, so I, the rotational inertia, is a measure of the resistance a body offers to a change in its rotational motion about a given axis.

Consider a body consisting of two spherical masses of 5.0 kg each connected by a light rigid rod 1.0 m long (Fig. 12-6). Treat the spheres as point particles and neglect the mass of the rod. Determine the rotational inertia (or moment of

EXAMPLE 2

inertia) of the body (a) about an axis normal to it through its center at C, and (b) about an axis normal to it through one sphere.

(a) If the axis is normal to the page through C, we have

$$I_C = \Sigma \, m_i r_i^2 = m_a r_a^2 + m_b r_b^2$$

$$= (5.0 \text{ kg})(0.50 \text{ m})^2 + (5.0 \text{ kg})(0.50 \text{ m})^2 = 2.5 \text{ kg} \cdot \text{m}^2.$$

(b) If the axis is normal to the page through A or B, we have

$$I_A = m_a r_a^2 + m_b r_b^2 = (5.0 \text{ kg})(0 \text{ m})^2 + (5.0 \text{ kg})(1.0 \text{ m})^2 = 5.0 \text{ kg} \cdot \text{m}^2,$$

$$I_B = m_a r_a^2 + m_b r_b^2 = (5.0 \text{ kg})(1.0 \text{ m})^2 + (5.0 \text{ kg})(0 \text{ m})^2 = 5.0 \text{ kg} \cdot \text{m}^2.$$

Hence the rotational inertia of this rigid dumbell model is twice as great about an axis through an end as it is about an axis through the center.

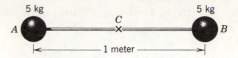

figure 12-6
Example 2. Calculating the rotational inertia of a dumbell.

For a body that is not composed of discrete point masses but is instead a continuous distribution of matter, the summation in $I = \Sigma \, m_i r_i^2$ becomes an integration. We imagine the body to be subdivided into infinitesimal elements, each of mass dm. We let r be the distance from such an element to the axis of rotation. Then the rotational inertia is obtained from

$$I = \int r^2 \, dm, \tag{12-12}$$

where the integral is taken over the whole body. The procedure by which the summation Σ of a discrete distribution is replaced by the integral $\int$ for a continuous distribution is the same as that discussed for the center of mass in Section 9-1.

For bodies of irregular shape the integrals may be hard to evaluate. For bodies of simple geometrical shape the integrals are relatively easy when an axis of symmetry is chosen as the axis of rotation.

Let us illustrate the procedure for an annular cylinder (or ring) about the cylinder axis (Fig. 12-7). The most convenient mass element is an infinitesimally thin cylinder shell of radius r, thickness dr, and length L. If the density of the material, that is, the mass per unit volume, is called ρ, then

$$dm = \rho \, dV,$$

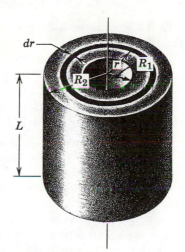

figure 12-7
Calculating the rotational inertia of an annular cylinder.

where dV is the volume of the cylindrical shell of mass dm. We have

$$dV = (2\pi r \, dr)L,$$

so that

$$dm = 2\pi L \rho r \, dr.$$

Then the rotational inertia about the cylinder axis is

$$I = \int r^2 \, dm = 2\pi L \int_{R_1}^{R_2} \rho r^3 \, dr.$$

Here R_1 is the radius of the inner cylindrical wall and R_2 is the radius of the outer cylindrical wall.

If this body did not have a uniform constant density, we would have to know how ρ depends on r before we could carry out the integration. Let us assume for simplicity that the density is uniform. Then

$$I = 2\pi L \rho \int_{R_1}^{R_2} r^3 \, dr = 2\pi L \rho \, \frac{R_2{}^4 - R_1{}^4}{4} = \rho \pi (R_2{}^2 - R_1{}^2) L \, \frac{R_2{}^2 + R_1{}^2}{2}.$$

The mass M of the annular cylinder is the product of its density ρ by its volume $\pi(R_2{}^2 - R_1{}^2)L$, or

$$M = \rho\pi(R_2{}^2 - R_1{}^2)L.$$

The rotational inertia of the *annular cylinder* (or ring) of mass M, inner radius R_1 and outer radius R_2, is therefore

$$I = \tfrac{1}{2}M(R_1{}^2 + R_2{}^2)$$

about the cylinder axis.

If the inner radius is zero, R_1 equals zero, and we have a *solid cylinder* (or disk). Then

$$I = \tfrac{1}{2}MR^2$$

about the cylinder axis, where R is the radius of the solid cylinder of mass M.

A *hoop* can be thought of as a very thin-walled hollow cylinder. In that case

$$R_1 \cong R_2 \cong R,$$

and

$$I = MR^2$$

is the rotational inertia of a hoop of mass M and radius R about the cylinder axis.

This result for the thin hoop is obvious since every mass point in the hoop is the same distance R from the central axis. For the solid cylinder (or disk) having the *same mass* as the hoop, the rotational inertia (or moment of inertia) would naturally be less than that of the hoop, because most of the cylinder (or disk) is less than a distance R from the axis.

The rotational inertias about certain axes of some common solids (of uniform density) are listed in Table 12-1. Each of these results can be derived by integration in a manner similar to that of our illustration. The total mass of the body is denoted by M in each equation.

There is a simple and very useful relation between the rotational inertia I of a body about any axis and its rotational inertia I_{cm} with respect to a parallel axis *through the center of mass*. If M is the total mass of the body and h the distance between the two axes, the relation is

$$I = I_{cm} + Mh^2. \tag{12-13}$$

The proof of this relation (parallel-axis theorem) follows. Let C be the center of mass of the arbitrarily shaped body whose cross section is shown in Fig. 12-8.

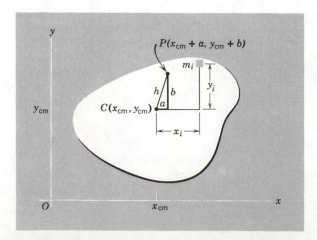

figure 12-8
Derivation of the parallel-axis theorem. Knowing the rotational inertia about an axis through C, we can find its value about a parallel axis through P.

Table 12-1

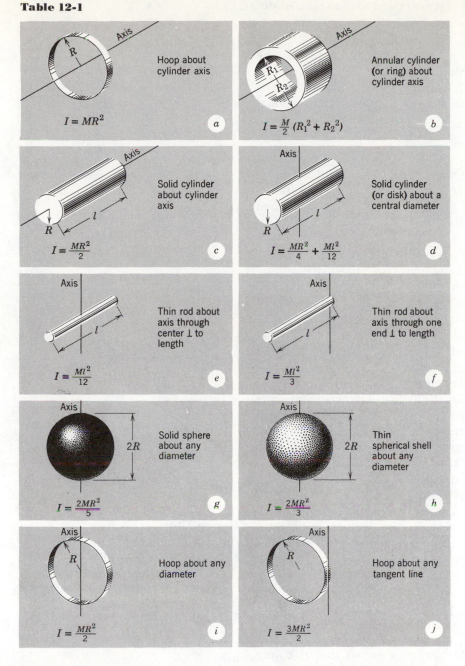

Hoop about cylinder axis

$$I = MR^2$$

a

Annular cylinder (or ring) about cylinder axis

$$I = \frac{M}{2}(R_1^2 + R_2^2)$$

b

Solid cylinder about cylinder axis

$$I = \frac{MR^2}{2}$$

c

Solid cylinder (or disk) about a central diameter

$$I = \frac{MR^2}{4} + \frac{Ml^2}{12}$$

d

Thin rod about axis through center ⊥ to length

$$I = \frac{Ml^2}{12}$$

e

Thin rod about axis through one end ⊥ to length

$$I = \frac{Ml^2}{3}$$

f

Solid sphere about any diameter

$$I = \frac{2MR^2}{5}$$

g

Thin spherical shell about any diameter

$$I = \frac{2MR^2}{3}$$

h

Hoop about any diameter

$$I = \frac{MR^2}{2}$$

i

Hoop about any tangent line

$$I = \frac{3MR^2}{2}$$

j

The center of mass has coordinates x_{cm} and y_{cm}. We choose the x-y plane to include C, so that z_{cm} equals zero. Consider an axis through C at right angles to the plane of the paper and another axis parallel to it through P at $(x_{cm} + a)$ and $(y_{cm} + b)$. The distance between the axes is $h = \sqrt{a^2 + b^2}$. Then the square of the distance of a particle from the axis through C is $x_i^2 + y_i^2$, where x_i and y_i measure the coordinates of a mass element m_i relative to the axis through C. The square of its distance from an axis through P is $(x_i - a)^2 + (y_i - b)^2$. Hence the rotational inertia about an axis through P is

$$I = \Sigma\, m_i[(x_i - a)^2 + (y_i - b)^2]$$
$$= \Sigma\, m_i(x_i^2 + y_i^2) - 2a\, \Sigma\, m_i x_i - 2b\, \Sigma\, m_i y_i + (a^2 + b^2)\, \Sigma\, m_i.$$

From the definition of center of mass,

$$\Sigma\, m_i x_i = \Sigma\, m_i y_i = 0,$$

so that the two middle terms are zero. The first term is simply the rotational inertia about an axis through the center of mass I_{cm} and the last term is Mh^2. Hence it follows that $I = I_{cm} + Mh^2$.

With the aid of this formula several of the results of Table 12-1 can be deduced from previous results. For example, (f) follows from (e), and (j) follows from (i) with the aid of Eq. 12-13. The formula will prove to be especially useful in problems that combine rotational and translational motion.

In this section we continue to study the special case of a rigid body confined to rotate about an axis that is fixed* in an inertial reference frame. First we shall review the concept of torque as applied to such a rigid body; then we shall show how the torque is related to the angular acceleration of the body about this axis.

Suppose that we apply a torque τ to one of the particles in a rigid body. Since all the particles of a truly rigid body maintain a fixed spatial relationship to all the other particles that make up the body, the torque may be said to act on the rigid body as a whole. In general, the vector τ will not lie along the axis around which the body is free to rotate. We are not concerned in this section with the actual torques applied to the body but only with the components of these torques that lie along the axis.† Only these components can cause the body to rotate about this axis. Torque components perpendicular to the axis tend to turn the axis from its fixed position. We have specifically assumed, however, that the axis maintains a fixed direction. The body may, for example, be attached to a shaft that is held in a fixed position by bearings at each end; if an applied torque has a component at right angles to the shaft, tending to turn it, the bearings will automatically apply an equal and opposite counter-torque to the shaft, canceling the effect of this component.

In Fig. 12-9 (compare Fig. 11-3) we show a section through a rigid body that is free to rotate about the z-axis of an inertial reference frame. A force $\mathbf{F}$, taken for convenience to be in the x-y plane of the section, acts on a particle at point P in the body, the position of P with respect to the rotational axis (the z-axis) being defined by the vector $\mathbf{r}$. The torque acting on the particle at P may be said to act on the rigid body as a whole and is given by Eq. 12-1, or

$$\tau = \mathbf{r} \times \mathbf{F}.$$

Because we have chosen $\mathbf{r}$ and $\mathbf{F}$ to lie in the x-y plane, the torque τ will point along the z-axis. The right-hand rule shows that it points perpendicularly *out of* the plane of Fig. 12-9. If $\mathbf{r}$ and $\mathbf{F}$ did *not* lie in the plane of the figure, τ would not be parallel to the z-axis and we would concern ourselves here only with the component of τ along this axis. The magnitude of τ is given by Eq. 12-2 or

$$\tau = rF \sin \theta$$

which, as we have seen, can also be written as $\tau = rF_\perp$ or $\tau = Fr_\perp$.

12-6
ROTATIONAL DYNAMICS OF A RIGID BODY

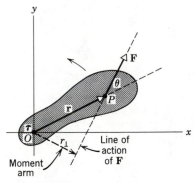

figure 12-9

A force $\mathbf{F}$ acts on the particle P in a rigid body, exerting a torque $\tau = \mathbf{r} \times \mathbf{F}$ on the body, with respect to an axis through O at right angles to the plane of the figure. The moment arm $r_\perp$ is also shown, as is the torque τ, which is a vector emerging perpendicularly from the page.

---

* See the footnote on page 237.

† As for any other vector, we can speak of the vector component of a torque in any given direction, such as a given axis. For torque—and for other angular quantities—we also often speak of the component *around* a given direction or axis. The meaning is the same.

EXAMPLE 3

A wagon wheel is free to rotate about a horizontal axis through O. A force of 10 lb is applied to a spoke at the point P, 1.0 ft from the center. OP makes an angle of 30° with the horizontal (x-axis) and the force is in the plane of the wheel making an angle of 45° with the horizontal (x-axis). What is the torque on the wheel?

The angle between the displacement vector $\mathbf{r}$ from O to P and the applied force $\mathbf{F}$ (Fig. 12-10) is θ, where

$$\theta = 45° - 30° = 15°.$$

Then the magnitude of the torque is

$$\tau = rF \sin \theta$$

$$= (1.0 \text{ ft})(10 \text{ lb})(\sin 15°) = 2.6 \text{ lb} \cdot \text{ft}.$$

It is clear that we can obtain this same result from $\tau = rF_\perp$ or $\tau = Fr_\perp$ as well (see Eqs. 12-2). The torque ($\boldsymbol{\tau} = \mathbf{r} \times \mathbf{F}$) is a vector pointing out $\odot$ along the axis through O having a magnitude 2.6 lb · ft.

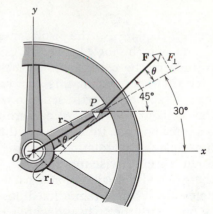

figure 12-10
Example 3.

We now investigate the relationship between the torque applied to the rigid body of Fig. 12-9 and the angular acceleration of this body. Let us observe the rigid body for an infinitesimal time dt, during which it will rotate through an infinitesimal angle $d\theta$. We have seen earlier that we can describe the rotation of a rigid body about a fixed axis by examining the motion of any single point fixed in the body, such as P in Fig. 12-9. For convenience, then, we ignore the body itself in Fig. 12-11 and focus our attention on the representative point P and on the vector $\mathbf{r}$ which locates point P with respect to the axis of rotation.

During the time dt, the point P will move an infinitesimal distance ds along a circular path of radius r as the body rotates through an infinitesimal angle $d\theta$, where

$$ds = r \, d\theta.$$

The work dW done by this force during this infinitesimal rotation is

$$dW = \mathbf{F} \cdot d\mathbf{s} = F \cos \phi \, ds = (F \cos \phi)(r \, d\theta),$$

where $F \cos \phi$ is the component of $\mathbf{F}$ in the direction of $d\mathbf{s}$.

The term $(F \cos \phi)r$, however, is the magnitude of the instantaneous torque exerted by $\mathbf{F}$ on the rigid body about the axis perpendicular to the page through O, so that

$$dW = \tau \, d\theta. \tag{12-14}$$

This differential expression for the work done in rotation (about a fixed axis) is equivalent to the expression $dW = F \, dx$ for the work done in translation (along a straight line).

To obtain the rate at which work is done in rotational motion (about a fixed axis), we divide both sides of Eq. 12-14 by the infinitesimal time interval dt during which the body is displaced through $d\theta$, obtaining

$$\frac{dW}{dt} = \tau \frac{d\theta}{dt}$$

or

$$P = \tau\omega,$$

giving the instantaneous power P. This last expression is the rotational analog of $P = Fv$ for translational motion (along a straight line).

If now a number of forces $\mathbf{F}_1$, $\mathbf{F}_2$, etc., are applied to the body in the

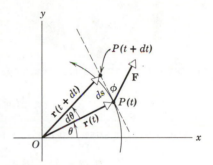

figure 12-11
In time dt point P in the rigid body of Fig. 12-9 moves a distance ds along the arc of a circle of radius r. The rigid body (not shown) and the vector $\mathbf{r}$ that locates point P in it each rotate through an angle $d\theta$ during this interval.

plane normal to its axis of rotation, the work done by these forces on the body in a rotation $d\theta$ will be

$$dW = F_1 \cos \phi_1 r_1 \, d\theta + F_2 \cos \phi_2 r_2 \, d\theta + \cdots ,$$

$$= (\tau_1 + \tau_2 + \cdots) \, d\theta = \tau \, d\theta,$$

where $r_1 \, d\theta$ equals ds_1, the displacement of the point at which $\mathbf{F}_1$ is applied, and ϕ_1 is the angle between $\mathbf{F}_1$ and $d\mathbf{s}_1$, etc., and where τ is now the magnitude of the *component* of the *resultant* torque along the axis through O. In computing this sum each torque is considered positive or negative according to the sense in which it alone would tend to rotate the body about its axis. We can arbitrarily call the torque associated with a force positive if the effect of the force, acting alone, is to produce a counterclockwise rotation; then the torque is negative if the effect is to produce a clockwise rotation.

There is no internal motion of particles within a truly rigid body. The particles always maintain a fixed position relative to one another and move only with the body as a whole. Hence there can be no dissipation of energy within a truly rigid body. We can therefore equate the rate at which work is being done on the body to the rate at which its kinetic energy is increasing. The rate at which work is being done on the rigid body is

$$\frac{dW}{dt} = \tau \frac{d\theta}{dt} = \tau\omega. \tag{12-15}$$

The rate at which the kinetic energy of the rigid body is increasing is

$$\frac{d}{dt} \left(\tfrac{1}{2} I \omega^2 \right).$$

But I is constant because the body is rigid and the axis is fixed. Hence

$$\frac{d}{dt} \left(\tfrac{1}{2} I \omega^2 \right) = \tfrac{1}{2} I \frac{d}{dt} (\omega^2) = I\omega \frac{d\omega}{dt} = I\omega\alpha. \tag{12-16}$$

Equating the right-hand members of Eqs. 12-15 and 12-16, we obtain

$$\tau\omega = I\alpha\omega,$$

or

$$\tau = I\alpha. \tag{12-17}$$

Equation 12-17 refers to the rotational motion of a rigid body about a fixed axis. The torque $\boldsymbol{\tau}$, the angular velocity $\boldsymbol{\omega}$, and the angular acceleration $\boldsymbol{\alpha}$ are all constrained to point along this axis, in one direction or the other. The equivalent translational case is that in which the force $\mathbf{F}$ acting on a body, its velocity $\mathbf{v}$, and its acceleration $\mathbf{a}$ all point along a given straight line, in one direction or the other.

The above six quantities are vectors, but when they are directed along a fixed line, they can have only two directions. By taking one of these directions as $+$ and the other as $-$, we can treat these vectors algebraically and deal with their magnitudes only. Thus, in deriving Eq. 12-17 ($\tau = I\alpha$), we have simply transformed Newton's second law ($F = Ma$), written in scalar form suitable to describe rectilinear motion, into rotational terms. This suggests that just as we associate a force with the linear acceleration of a body, so we may associate a torque with the angular acceleration of a body about a given axis. The rotational inertia I is a measure of the resistance a body offers to having its rotational motion changed by a given torque just as the translational inertia, or mass, M is a measure of the resistance a body offers to having its translational motion changed by a given force.

In Table 12-2 we compare the translational motion of a rigid body along a straight line with the rotational motion of a rigid body about a fixed axis.

Table 12-2

Rectilinear Motion		Rotation about a Fixed Axis	
Displacement	x	Angular displacement	θ
Velocity	$v = \dfrac{dx}{dt}$	Angular velocity	$\omega = \dfrac{d\theta}{dt}$
Acceleration	$a = \dfrac{dv}{dt}$	Angular acceleration	$\alpha = \dfrac{d\omega}{dt}$
Mass (translational inertia)	M	Rotational inertia	I
Force	$F = Ma$	Torque	$\tau = I\alpha$
Work	$W = \int F\,dx$	Work	$W = \int \tau\,d\theta$
Kinetic energy	$\frac{1}{2}Mv^2$	Kinetic energy	$\frac{1}{2}I\omega^2$
Power	$P = Fv$	Power	$P = \tau\omega$
Linear momentum	Mv	Angular momentum	$I\omega$

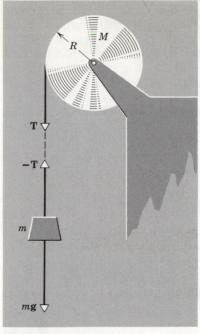

figure 12-12
Example 4. A steady downward force **T** produces rotation of the disk. Example 5. Here **T** is supplied by the falling mass m.

The rotation of a rigid body about a fixed axis (to which $\tau = I\alpha$ applies) is not the most general kind of rotary motion in that the body may not be rigid and the axis may not be fixed in an inertial reference frame. In this general case Eq. 12-9, or $\tau_{ext} = d\mathbf{L}/dt$, applies. As we have already pointed out, this is equivalent to Newton's second law for the general translational motion of a system of particles, namely, Eq. 9-17, or $\mathbf{F}_{ext} = d\mathbf{P}/dt$.

In the rest of this chapter we confine ourselves to the rotations of rigid bodies about fixed axes. In Chapter 13 we shall consider some more general kinds of rotary motion.

EXAMPLE 4

A uniform disk of radius R and mass M is mounted on an axle supported in fixed frictionless bearings, as in Fig. 12-12. A light cord is wrapped around the rim of the wheel and a steady downward pull **T** is exerted on the cord. Find the angular acceleration of the wheel and the tangential acceleration of a point on the rim.

The torque about the central axis is $\tau = TR$, and the rotational inertia of the disk about its central axis is $I = \frac{1}{2}MR^2$. From

$$\tau = I\alpha,$$

we have

$$TR = (\tfrac{1}{2}MR^2)\alpha,$$

or

$$\alpha = \frac{2T}{MR}.$$

If the mass of the disk is taken to be $M = 2.50$ kg, its radius $R = 0.20$ m, and the force $T = 5.0$ N, then

$$\alpha = \frac{(2)(5.0 \text{ N})}{(2.50 \text{ kg})(0.20 \text{ m})} = 20 \text{ rad/s}^2.$$

The tangential acceleration of a point on the rim is given by

$$a = R\alpha = (20 \text{ rad/s}^2)(0.20 \text{ m}) = 4.0 \text{ m/s}^2.$$

Suppose that we hang a body of mass m from the cord in the previous problem. Find the angular acceleration of the disk and the tangential acceleration of a point on the rim in this case.

EXAMPLE 5

Now, let T be the tension in the cord. Since the suspended body will accelerate downward, the magnitude of the downward pull of gravity on it, mg, must exceed the magnitude of the upward pull of the cord on it, T. The acceleration a of the suspended body is the same as the tangential acceleration of a point on the rim of the disk. From Newton's second law

$$mg - T = ma.$$

The resultant torque on the disk is TR and its rotational inertia is $\frac{1}{2}MR^2$, so that from

$$\tau = I\alpha$$

we obtain

$$TR = \frac{1}{2}MR^2\alpha.$$

Using the relation $a = R\alpha$, we can write this last equation as

$$2T = Ma.$$

Solving the first and last equations simultaneously leads to

$$a = \left(\frac{2m}{M + 2m}\right)g,$$

and

$$T = \left(\frac{Mm}{M + 2m}\right)g.$$

If now we let the disk have a mass $M = 2.50$ kg and a radius $R = 0.20$ m as before, and we let the suspended body weigh 5.0 N, we obtain

$$a = \frac{2mg}{M + 2m} = \frac{(2)(5.0 \text{ N})}{(2.50 \text{ kg}) + 2(5/9.8) \text{ kg}} = 2.85 \text{ m/s}^2,$$

$$\alpha = \frac{a}{R} = \frac{(2.85 \text{ m/s}^2)}{0.20 \text{ m}} = 14.3 \text{ rad/s}^2.$$

Notice that the accelerations are less for a suspended 5.0 N body than they were for a steady 5.0 N pull on the string (Example 4). This corresponds to the fact that the tension in the string supplying the torque is now less than 5.0 N, namely

$$T = \frac{Mmg}{M + 2m} = \frac{(2.50 \text{ kg})(5.0 \text{ N})}{(2.50 + 1.0) \text{ kg}} = 3.6 \text{ N}.$$

The tension in the string must be less than the weight of the suspended body if the body is to accelerate downward.

Assuming that the disk of Example 5 starts from rest, compute the work done by the applied torque on the disk in 2.0 s. Compute also the increase in rotational kinetic energy of the disk.

EXAMPLE 6

Since the applied torque is constant, the resulting angular acceleration is constant. The total angular displacement in constant angular acceleration is obtained from Eq. 11-5,

$$\theta = \omega_0 t + \frac{1}{2}\alpha t^2,$$

in which

$$\omega_0 = 0, \qquad \alpha = 14.3 \text{ rad/s}^2, \qquad t = 2.0 \text{ s},$$

so that

$$\theta = 0 + (\tfrac{1}{2})(14.3 \text{ rad/s}^2)(2.0 \text{ s})^2 = 28.6 \text{ rad}.$$

For constant torque the work done in a finite angular displacement is

$$W = \tau(\theta_2 - \theta_1),$$

in which

$$\tau = TR = (3.6 \text{ N})(0.20 \text{ m}) = 0.72 \text{ N} \cdot \text{m},$$

and

$$\theta_2 - \theta_1 = \theta = 28.6 \text{ rad}.$$

Therefore

$$W = (0.72 \text{ N} \cdot \text{m})(28.6 \text{ rad}) = 20.5 \text{ J}.$$

This work must result in an increase in rotational kinetic energy of the disk. Starting from rest the disk acquires an angular speed ω. The rotational energy is $\frac{1}{2}I\omega^2 = \frac{1}{2}(\frac{1}{2}MR^2)\omega^2$. To obtain ω we use Eq. 11-3,

$$\omega = \omega_0 + \alpha t,$$

in which

$$\omega_0 = 0, \qquad t = 2.0 \text{ s}, \qquad \alpha = 14.3 \text{ rad/s}^2,$$

so that

$$\omega = 0 + (14.3 \text{ rad/s}^2)(2.0 \text{ s}) = 28.6 \text{ rad/s}.$$

Then

$$\tfrac{1}{2}I\omega^2 = (\tfrac{1}{4})(2.50 \text{ kg})(0.20 \text{ m})^2(28.6 \text{ rad/s})^2 = 20.5 \text{ J},$$

as before. Hence the increase in kinetic energy of the disk is equal to the work done by the resultant force on the disk, as it must be.

EXAMPLE 7

Show that the conservation of mechanical energy holds for the system of Example 5.

The resultant force acting on the system is the force of gravity on the suspended body. This is a conservative force. Viewing the system as a whole, we see that the suspended body loses potential energy U as it descends,

$$U = mgy,$$

where y is the vertical distance through which the block descends. At the same time the suspended body gains kinetic energy of translation and the disk gains kinetic energy of rotation. The total gain in kinetic energy is

$$\tfrac{1}{2}mv^2 + \tfrac{1}{2}I\omega^2,$$

where v is the linear speed of the suspended mass. We must show then that

$$mgy = \tfrac{1}{2}mv^2 + \tfrac{1}{2}I\omega^2.$$

For the linear motion starting from rest we have $v^2 = 2ay$. From Example 5, we obtained $a = 2\,mg/(M + 2m)$. Hence

$$mgy = \frac{mgv^2}{2a} = \tfrac{1}{2}\,mv^2\left(\frac{g}{a}\right) = \tfrac{1}{2}\,mv^2\left(\frac{M + 2m}{2m}\right) = \tfrac{1}{4}\,(M + 2m)v^2.$$

We also know that $\omega = v/R$ and $I = \tfrac{1}{2}MR^2$. Substituting these relations into the right-hand side of the conservation equation, we obtain

$$\tfrac{1}{2}mv^2 + \tfrac{1}{2}I\omega^2 = \tfrac{1}{2}mv^2 + \tfrac{1}{2}(\tfrac{1}{2}MR^2)(v^2/R^2) = \tfrac{1}{4}(M + 2m)v^2.$$

The mechanical energy is therefore conserved.

EXAMPLE 8

Derive the relation $L = I\omega$, shown in Table 12-2, for the angular momentum of a rigid body confined to rotate about a fixed axis.

Starting from the scalar relation $\tau = I\alpha$ and the definition of α ($= d\omega/dt$), we may write

$$\tau = I\alpha = I(d\omega/dt) = d(I\omega)/dt,$$

in which the last step is justified by the fact that I is a constant for a given rigid body and a specified (fixed) axis of rotation.

Next we use the vector relation $\tau_{\text{ext}} = d\mathbf{L}/dt$ (Eq. 12-9) and write the corresponding relation for the *scalar components*, τ and dL, of τ_{ext} and $d\mathbf{L}$ along the fixed axis of rotation, obtaining

$$\tau = dL/dt.$$

Simply by comparing the two equations above we obtain the relation sought, namely

$$L = I\omega. \qquad (12\text{-}18)$$

Like Eq. 12-17 ($\tau = I\alpha$), this is a scalar relation holding for the rotation of a rigid body about a fixed axis. L is the component along that axis of the vector angular momentum $\mathbf{L}$ of the rigid body and I must refer to that same axis.

Equation 12-18 is the rotational analog of the expression $P = Mv$ for the *linear* momentum of a rigid body of mass M in pure translational motion with linear speed v. It gives the *angular* momentum about a fixed axis of a rigid body having rotational inertia I and angular speed ω about that same axis.

Up until now we have considered only bodies rotating about some fixed axis. If a body is rolling, however, it is rotating about an axis and also moving translationally. Therefore it would seem that the motion of rolling bodies must be treated as a combination of translational and rotational motion. It is also possible, however, to treat a rolling body as though its motion is one of pure rotation. We wish to illustrate the equivalence of the two approaches.

Consider, for example, a cylinder rolling along a level surface, as in Fig. 12-13. At any instant the bottom of the cylinder is at rest on the surface, since it does not slide. The axis normal to the diagram through the point of contact P is called the *instantaneous axis of rotation*. At that instant the linear velocity of every particle of the cylinder is directed at right angles to the line joining the particle and P and its magnitude is proportional to this distance. This is the same as saying that the cylinder is rotating about a fixed axis through P with a certain angular speed ω, *at that instant*. Hence, at a given instant the motion of the body is equivalent to a pure rotation. The total kinetic energy can, therefore, be written as

$$K = \tfrac{1}{2}I_P\omega^2, \qquad (12\text{-}19)$$

where I_P is the rotational inertia about the axis through P.

Let us now use the parallel axis theorem, which tells us that

$$I_P = I_{\text{cm}} + MR^2,$$

where I_{cm} is the rotational inertia of the cylinder of mass M and radius R about a parallel axis through the center of mass. Equation 12-19 now becomes

$$K = \tfrac{1}{2}I_{\text{cm}}\omega^2 + \tfrac{1}{2}MR^2\omega^2. \qquad (12\text{-}20)$$

The quantity $R\omega$ is the speed with which the center of mass of the cylinder is moving with respect to the fixed point P. Let $R\omega = v_{\text{cm}}$. Equation 12-20 then becomes

$$K = \tfrac{1}{2}I_{\text{cm}}\omega^2 + \tfrac{1}{2}Mv_{\text{cm}}^2. \qquad (12\text{-}21)$$

Now notice that the speed of the center of mass with respect to P is the same as the speed of P with respect to the center of mass. Hence, the angular speed ω of the center of mass about P as seen by someone at P is the same as the angular speed of a particle at P about C as seen

12-7
THE COMBINED TRANSLATIONAL AND ROTATIONAL MOTION OF A RIGID BODY

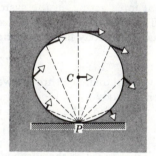

figure 12-13
A rolling body may at any instant be thought of as rotating about a perpendicular axis through its point of contact P.

by someone at C (moving along with the cylinder). This is equivalent to saying that any reference line in the cylinder turns through the same angle in a given time whether it is observed from a reference frame fixed with respect to the surface on which the cylinder is rolling or from a frame moving translationally with respect to this fixed frame. We can therefore interpret Eq. 12-21, which was derived on the basis of a pure rotational motion, in another way; that is, the first term, $\frac{1}{2}I_{cm}\omega^2$, is the kinetic energy the cylinder would have if it were merely rotating about an axis through its center of mass, without translational motion; and the second term, $\frac{1}{2}Mv_{cm}^2$, is the kinetic energy the cylinder would have if it were moving translationally with the speed of its center of mass, without rotating. Notice that there is now no reference at all to the instantaneous axis of rotation. In fact, Eq. 12-21 applies to any body that is moving and rotating about an axis perpendicular to its motion whether or not it is rolling on a surface.

The combined effects of translation of the center of mass and rotation about an axis through the center of mass are equivalent to a pure rotation with the same angular speed about an axis through the point of contact of a rolling body.

To illustrate this result simply, let us consider the instantaneous speed of various points on the rolling cylinder. If the speed of the center of mass (as seen by an observer fixed with respect to the surface) is v_{cm}, the instantaneous angular speed about an axis through P is $\omega = v_{cm}/R$. A point Q at the top of the cylinder will therefore have a speed $\omega 2R = 2v_{cm}$ at that instant. The point of contact P is instantaneously at rest. Hence, from the point of view of pure rotation about P, the situation is as shown in Fig. 12-14.

Now let us regard the rolling as a combination of translation of the center of mass and rotation about the cylinder axis through C. If we consider translation only, all points on the cylinder have the same speed v_{cm}, the speed of the center of mass. This is shown in Fig. 12-15a. If we consider the rotation only, the center is at rest, whereas the point Q at the top has a speed ωR in the x-direction and the point P at the bottom of the cylinder has a speed ωR in the $-x$-direction. This is shown in Fig. 12-15b. Now let us combine these two results. Recalling that $\omega = v_{cm}/R$, we obtain

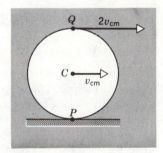

figure 12-14
Since Q and C have the same angular velocity about P, therefore Q, being twice as far from P, moves with twice the linear velocity of C.

for the point Q $\qquad v = v_{cm} + \omega R = v_{cm} + \dfrac{v_{cm}}{R}R = 2v_{cm}$,

for the point C $\qquad v = v_{cm} + 0 = v_{cm}$,

for the point P $\qquad v = v_{cm} - \omega R = v_{cm} - \dfrac{v_{cm}}{R}R = 0$.

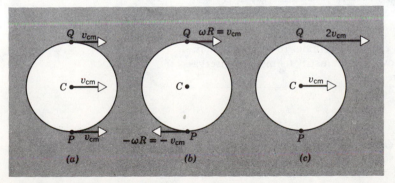

figure 12-15
(a) For pure translation, all points move with the same velocity. (b) For pure rotation about C, opposite points move with opposite velocities. (c) Combined rotation and translation is obtained by adding together corresponding vectors in (a) and (b).

This result, shown in Fig. 12-15c is exactly the same as that obtained from the purely rotational point of view, Fig. 12-14.

Consider a solid cylinder of mass M and radius R rolling down an inclined plane without slipping. Find the speed of its center of mass when the cylinder reaches the bottom.

EXAMPLE 9

The situation is illustrated in Fig. 12-16. We can use the conservation of energy to solve this problem. The cylinder is initially at rest. In rolling down the incline the cylinder loses potential energy of an amount Mgh, where h is the height of the incline. It gains kinetic energy equal to

$$\tfrac{1}{2}I_{cm}\omega^2 + \tfrac{1}{2}Mv^2,$$

where v is the linear speed of the center of mass and ω is the angular speed about the center of mass at the bottom.

We have then the relation

$$Mgh = \tfrac{1}{2}I_{cm}\omega^2 + \tfrac{1}{2}Mv^2,$$

in which

$$I_{cm} = \tfrac{1}{2}MR^2 \quad \text{and} \quad \omega = \frac{v}{R}.$$

Hence

$$Mgh = \tfrac{1}{2}(\tfrac{1}{2}MR^2)\left(\frac{v}{R}\right)^2 + \tfrac{1}{2}Mv^2 = (\tfrac{1}{4} + \tfrac{1}{2})Mv^2,$$

$$v^2 = \tfrac{4}{3}gh \quad \text{or} \quad v = \sqrt{\tfrac{4}{3}gh}.$$

The speed of the center of mass would have been $v = \sqrt{2gh}$ if the cylinder had *slid* down a *frictionless* incline. The speed of the rolling cylinder is, therefore, less than the speed of the sliding cylinder, because for the rolling cylinder, part of the lost potential energy has been transformed into rotational kinetic energy, leaving less available for the translational part of the kinetic energy. Although the rolling cylinder arrives later at the bottom of the incline than an identical sliding cylinder started at the same time down a frictionless, but otherwise identical, incline, both arrive at the bottom with the same amount of energy; the rolling cylinder happens to be rotating as it moves, whereas the sliding one does not rotate as it moves.

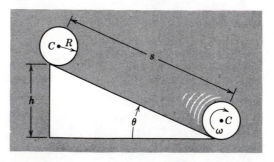

figure 12-16
Example 9. A cylinder rolling down an incline.

Notice that static friction is needed to cause the cylinder to rotate. Remembering that friction is a dissipative force, how can you justify using the conservation of mechanical energy in this problem?

The previous result was derived by use of energy methods. Solve the same problem using only dynamical methods.

EXAMPLE 10

The force diagram is shown in Fig. 12-17. Mg is the weight of the cylinder

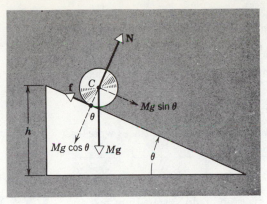

figure 12-17
Example 10. Dynamic solution of
the motion of a cylinder rolling
down an incline.

acting vertically down through the center of mass,* N is the normal force exerted by the incline on the cylinder, and f is the force of static friction acting along the incline at the point of contact.

The *translational* motion of a body is obtained by assuming that all the external forces act at its center of mass. Using Newton's second law, we obtain

$$N - Mg \cos \theta = 0 \qquad \text{for motion normal to the incline,}$$

and

$$Mg \sin \theta - f = Ma \qquad \text{for motion along the incline.}$$

The *rotational* motion about the center of mass follows from

$$\tau = I_{cm}\alpha.$$

Neither N nor Mg can cause rotation about C because their lines of action pass through C, and they have zero moment arms. The force of friction has a moment arm R about C, so that

$$fR = I_{cm}\alpha.$$

But

$$I_{cm} = \tfrac{1}{2}MR^2 \qquad \text{and} \qquad \alpha = \frac{a}{R}$$

so that

$$f = I_{cm}\alpha/R = Ma/2.$$

Substituting this into the second translational equation, we find

$$a = \tfrac{2}{3}g \sin \theta.$$

That is, the acceleration of the center of mass for the rolling cylinder ($\tfrac{2}{3}g \sin \theta$) is less than the acceleration of the center of mass for the cylinder sliding down the incline ($g \sin \theta$).

This result holds at any instant, regardless of the position of the cylinder along the incline. The center of mass moves with constant linear acceleration. To obtain the speed of the center of mass, starting from rest, we use the relation

$$v^2 = 2as,$$

so that

$$v^2 = 2(\tfrac{2}{3}g \sin \theta)s = \tfrac{4}{3}g\frac{h}{s}s = \tfrac{4}{3}gh$$

or

$$v = \sqrt{\tfrac{4}{3}gh}.$$

* In drawing the vector diagram for this problem we tacitly assume that the total weight of the body can be thought of as acting at the center of mass. We saw in Section 9-2 that this is justified for analyzing the translational motion. However, later in the problem we use this result in analyzing the rotational motion as well. We shall justify this procedure in Section 14-3, where it is shown that the *weight* of a body can be considered to act at its center of mass for both translational *and* rotational motion.

This result is the same as that obtained before by the energy method. The energy method is certainly simpler and more direct. However, if we are interested in knowing what the forces are, such as **N** and **f**, we must use a dynamical method.

This method determines the minimum force of static friction needed for rolling:

$$f = Ma/2 = (M/2)(\tfrac{2}{3}g \sin \theta) = \tfrac{1}{3}Mg \sin \theta.$$

What would happen if the force of static friction between the surfaces were less than this value?

EXAMPLE 11

A sphere and a cylinder, having the same mass and radius, start from rest and roll down the same incline. Which body gets to the bottom first?

For a sphere I_{cm} equals $\tfrac{2}{5}MR^2$. Using the dynamical method we obtain

$$Mg \sin \theta - f = Ma, \qquad \text{translation of cm,}$$

$$fR = I_{\mathrm{cm}}\alpha = (\tfrac{2}{5}MR^2)(a/R), \qquad \text{rotation about cm,}$$

or

$$f = \tfrac{2}{5}Ma \quad \text{and} \quad a = \tfrac{5}{7}g \sin \theta, \quad \text{sphere.}$$

For the cylinder (Example 10)

$$a = \tfrac{2}{3}g \sin \theta, \qquad \text{cylinder.}$$

Hence the acceleration of the center of mass of the sphere is at all times greater than the acceleration of the center of mass of the cylinder. Since both bodies start from rest at the same instant, the sphere will reach the bottom first.

Which body has the greater rotational energy at the bottom? Which body has the greater translational energy at the bottom?

Note carefully that neither the mass nor the radius of the rolling object enters the previous results. How then would we expect the behavior of cylinders of different mass and radii to compare? How would we expect the behavior of spheres of different mass and radii to compare? How would the behavior of a cylinder and sphere having different masses and radii compare?

EXAMPLE 12

A uniform solid cylinder of radius r and mass m is given an initial angular velocity ω_0 and then dropped on a flat horizontal surface. The coefficient of kinetic friction between the surface and the cylinder is μ_k. Initially the cylinder slips but after a time t pure rolling begins. (a) What is the velocity V the center of mass at the time t? (b) What is the value of t?

(a) Figure 12-18 shows the forces that act on the cylinder.

The acceleration a of the center of mass is constant, since all the forces are constant, so that for the translational motion we can write

$$F = ma = m\left(\frac{V_f - V_i}{t - 0}\right).$$

Here, $V_i = 0$ and $V_f = V$, the velocity at t when pure rolling begins. Also, the resultant force F is $\mu_k mg$, so that

$$\mu_k mg = mV/t. \qquad (12\text{-}22)$$

The angular acceleration α about an axis through the center of mass is also constant (why?), so that for the rotational motion we can write

$$\tau = I\alpha = I\left(\frac{\omega_f - \omega_i}{t - 0}\right).$$

Here, $\omega_f = \omega = V/r$, the angular velocity at time t, and $\omega_i = \omega_0$. Also, the magni-

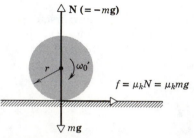

figure 12-18
Example 12

tude of the resultant torque τ is $\mu_k \, mg \, r$. The torque causes an angular deceleration so that

$$\mu_k mg \; r = (\tfrac{1}{2}mr^2)\left(\frac{\omega_0 - V/r}{t}\right). \qquad (12\text{-}23)$$

If we eliminate t from our two equations (e.g., divide Eq. 12-23 by Eq. 12-22) and solve for V (please do the algebra), we obtain

$$V = \tfrac{1}{3}\omega_0 r.$$

Note that V does not depend on the value of m, g, or μ_k. What, however, if any of these quantities were zero?

(b) By eliminating V from Eqs. 12-22 and 12-23 we can solve for t (please do the algebra) and find

$$t = \frac{\omega_0 r}{3\mu_k g}.$$

It is worth noting that in this problem neither mechanical energy, linear momentum, nor angular momentum are conserved, but the *changes* in momentum and angular momentum are directly related because the force of friction is responsible for both.

---

questions

1. What are the dimensions of angular momentum? Can you find any significance in the fact that they are the same as those of energy multiplied by time?

2. Is the vector product of two vectors necessarily an axial vector?*

3. Can the mass of a body be considered as concentrated at its center of mass for purposes of computing its rotational inertia?

4. About what axis would a uniform cube have its minimum rotational inertia?

5. If two circular disks of the same weight and thickness are made from metals having different densities, which disk, if either, will have the larger rotational inertia about its central axis?

6. The rotational inertia of a body of rather complicated shape is to be determined. The shape makes a mathematical calculation from $\int r^2 \, dm$ exceedingly difficult. Suggest ways in which the rotational inertia could be measured experimentally.

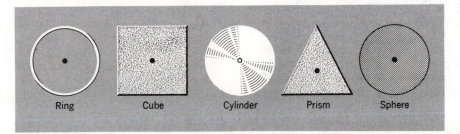

figure 12-19
Question 7

Ring Cube Cylinder Prism Sphere

7. Five solids are shown in cross section (Fig. 12-19). The cross sections have equal heights and maximum widths. The axes of rotation are perpendicular to the sections through the points shown. The solids have equal masses. Which one has the largest rotational inertia about a perpendicular axis through the center of mass? Which the smallest?

* See Supplementary Topic II.

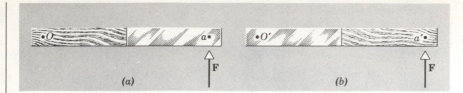

figure 12-20
Question 8

8. In Fig. 12-20a a meter stick, half of which is wood—the other half steel—is pivoted at the wooden end at O and a force is applied to the steel end at a. In Fig. 12-20b the stick is pivoted at the steel end at O' and the same force is applied at the wooden end at a'. Does one get the same angular acceleration in each case? Explain.

9. A person can distinguish between a raw egg and a hard-boiled one by spinning each one on the table. Explain how. Also, if you stop a spinning raw egg with your fingers and release it very quickly, it will resume spinning. Why?

10. Torque has the same dimensions as work or energy. Is torque, therefore, work or energy?

11. Comment on each of these assertions about skiing. (a) In downhill racing one wants skis that do not turn easily. (b) In slalom racing, one wants skis that turn easily. (c) Therefore, the rotational inertia of downhill skis should be larger than that of slalom skis. (See "The Physics of Ski Turns" by J. I. Shonie and D. L. Nordick in *The Physics Teacher*, December 1972.)

12. Considering that there is low friction between skis and snow and that the skier's center of mass is about over the center of the skis, how does a skier exert torques to turn or to stop a turn? (See "The Physics of Ski Turns" by J. I. Shonie and D. L. Nordick in *The Physics Teacher*, December 1972.)

13. Do the expressions for a and T in Example 5 give reasonable results for the special cases in which $g = 0$, $M = 0$, $M \to \infty$, $m = 0$, and $m \to \infty$?

14. The total momentum of a system of particles does not depend on the motions of the particles relative to the center of mass of the system. Can a similar statement be made about the total kinetic energy of a system of particles?

15. A cylindrical drum, pushed along by a board from an initial position shown in Fig. 12-21, rolls forward on the ground a distance $l/2$, equal to half the length of the board. There is no slipping at any contact. Where is the board then? How far has the man walked?

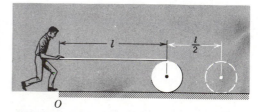

figure 12-21
Question 15

16. For storing wind energy or solar energy, flywheels have been suggested. The amount of energy that can be stored in a flywheel depends on the density and tensile strength of the material making up the flywheel and for a given weight one wants the lowest density strong material available. (See "Flywheels" by R. F. Post and S. F. Post, *Scientific American*, December 1973.) Can you make this plausible?

17. A solid wooden sphere rolls down two different inclined planes of the same height but different angles of inclines. Will it reach the bottom with the same speed in each case? Will it take longer to roll down one incline than the other? If so, which one and why?

18. Two heavy disks are connected by a short rod of much smaller radius. The system is placed on an inclined plane so that the disks hang over the sides and the system rolls down on the rod without slipping (Fig. 12-22). Near the bottom of the incline the disks touch the horizontal table top and the system takes off with greatly increased translational speed. Explain carefully.

19. When a logger cuts down a tree he makes a cut on the side facing the direction in which he wants it to fall. Explain why. Would it be safe to stand directly behind the tree on the opposite side of the fall?

20. Consider a straight stick standing on end on (frictionless) ice. What would be the path of its center of mass if it falls?

figure 12-22
Question 18

21. A yo-yo is resting on a horizontal table and is free to roll (Fig. 12-23). If the string is pulled by a horizontal force such as $\mathbf{F}_1$, which way will the yo-yo roll? What happens when the force $\mathbf{F}_2$ is applied (its line of action passes through the point of contact of the yo-yo and table)? If the string is pulled vertically with the force $\mathbf{F}_3$, what happens?

22. You are looking at the wheel of an automobile traveling at constant speed. Some one says to you: "The top of the wheel is moving twice as fast as the axle but the bottom is not moving at all." Can you accept this statement? Discuss.

23. State Newton's three laws of motion in words suitable for rotating bodies.

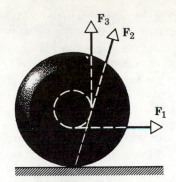

figure 12-23
Question 21

problems

SECTION 12-2

1. (a) Given that $\mathbf{r} = \mathbf{i}x + \mathbf{j}y + \mathbf{k}z$ and $\mathbf{F} = \mathbf{i}F_x + \mathbf{j}F_y + \mathbf{k}F_z$, find the torque $\boldsymbol{\tau} = \mathbf{r} \times \mathbf{F}$. (b) Show that if $\mathbf{r}$ and $\mathbf{F}$ lie in a given plane, then $\boldsymbol{\tau}$ has no component in that plane.

$Answer$: (a) $\mathbf{i}(yF_z - zF_y) + \mathbf{j}(zF_x - xF_z) + \mathbf{k}(xF_y - yF_x)$.

2. Show that the angular momentum about any point of a single particle moving with constant velocity remains constant throughout the motion.

SECTION 12-3

3. A particle P with mass 2.0 kg has position $\mathbf{r}$ and velocity $\mathbf{v}$ as shown in Fig. 12-24. It is acted on by the force $\mathbf{F}$. All three vectors lie in a common plane. Presume that $r = 3.0$ m, $v = 4.0$ m/s, and $F = 2.0$ N. Compute (a) the angular momentum of the particle and (b) the torque acting on the particle. What are the directions of these two vectors?

$Answer$: (a) 12 kg · m²/s; out of page. (b) 3.0 N · m; out of page.

4. If we are given r, p, and θ, we can calculate the angular momentum of a particle from Eq. 12-4a. Sometimes, however, we are given the components (x, y, z) of $\mathbf{r}$ and (p_x, p_y, p_z) of $\mathbf{p}$ instead. (a) Show that the components of l along the x-, y-, and z-axes are then given by

$$l_x = yp_z - zp_y,$$
$$l_y = zp_x - xp_z,$$
$$l_z = xp_y - yp_x.$$

(b) Show that if the particle moves only in the x-y plane, the resultant angular momentum vector has only a z-component.

5. (a) In Example 1, express $\mathbf{F}$ and $\mathbf{r}$ in terms of unit vectors and compute $\boldsymbol{\tau}$. Do the same in Example 3. (b) In example 1, express $\mathbf{p}$ and $\mathbf{r}$ in unit vectors and compute $\mathbf{l}$.

$Answer$: (a) $\boldsymbol{\tau} = +\mathbf{k}mgb$; 2.6 $\mathbf{k}$, lb · ft. (b) $\mathbf{l} = +\mathbf{k}mgbt$.

SECTION 12-4

6. In Fig. 12-25 are shown the lines of action and the moment arms of two forces about the origin O. Imagine these forces to be acting on a rigid body pivoted at O, all vectors shown being in the plane of the figure, and find the magnitude and the direction of the resultant torque on the body.

7. Two particles, each of mass m and speed v, travel in opposite directions along parallel lines separated by a distance d. Show that the vector angular momentum of this system of particles is the same no matter what point is taken as the origin.

8. Three particles, each of mass m, are fastened to each other and to a rotation axis by three light strings each with length l as shown in Fig. 12-26. The combination rotates around the rotational axis with angular velocity ω in such a way that the particles remain in a straight line. (a) Calculate the rotational inertia of the combination about O. (b) What is the angular mo-

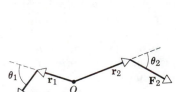

figure 12-24
Problem 3

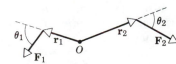

figure 12-25
Problem 6

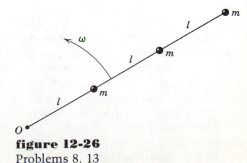

figure 12-26
Problems 8, 13

mentum of the middle particle? (c) What is the total angular momentum of the three particles? Express your answers in terms of m, l, and ω.

9. Starting from Newton's third law, prove that the resultant internal torque on a system of particles is zero.

10. *Relation between the Resultant External Torque and the Angular Momentum of a System of Particles about the Center of Mass of the System.* Let $\mathbf{r}_{cm}$ be the position vector of the center of mass C of a system of particles with respect to the origin O of an inertial reference frame, and let $\mathbf{r}_i'$ be the position vector of the ith particle, of mass m_i, with respect to the center of mass C. Hence $\mathbf{r}_i = \mathbf{r}_{cm} + \mathbf{r}_i'$ (see Fig. 12-27). Now define the total angular momentum of the system of particles relative to the center of mass C to be $\mathbf{L}' = \sum_i \mathbf{r}_i' \times \mathbf{p}_i'$, where $\mathbf{p}_i' = m_i\, d\mathbf{r}_i'/dt$.

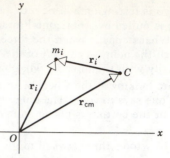

figure 12-27
Problem 10

(a) Show that $\mathbf{p}_i' = m_i d\mathbf{r}_i/dt - m_i d\mathbf{r}_{cm}/dt = \mathbf{p}_i - m_i\mathbf{v}_{cm}$. (b) Show next that $d\mathbf{L}'/dt = \sum_i \mathbf{r}_i' \times d\mathbf{p}_i'/dt$. (c) Combine the results of (a) and (b) and, using the definition of center of mass and Newton's third law, show that $\tau'_{ext} = d\mathbf{L}'/dt$, where τ'_{ext} is the sum of all the external torques acting on the system about its center of mass.

SECTION 12-5

11. Assume the earth to be a sphere of uniform density. (a) What is its rotational kinetic energy? Take the radius of the earth to be 6.4×10^3 km and the mass of the earth to be 6.0×10^{24} kg. (b) Suppose this energy could be harnessed for our use. For how long could the earth supply 1.0 kW of power to each of the 4.2×10^9 persons on earth?
Answer: (a) 2.6×10^{29} J. (b) 2.0×10^9 yr.

12. The oxygen molecule has a total mass of 5.30×10^{-26} kg and a rotational inertia of 1.94×10^{-46} kg·m² about an axis through the center perpendicular to the line joining the atoms. Suppose that such a molecule in a gas has a mean speed of 500 m/s and that its rotational kinetic energy is two-thirds of its translational kinetic energy. Find its average angular velocity.

13. Presume that the strings in Problem 8 are all replaced with uniform rods, each of mass M. (a) What is the total rotational inertia of the system about O? (b) What is the rotational kinetic energy of the system?
Answer: (a) $14\,ml^2 + 9\,Ml^2$. (b) $(7\,m + 9\,M/2)l^2\omega^2$.

14. (a) Show that a solid cylinder of mass M and radius R is equivalent to a thin hoop of mass M and radius $R/\sqrt{2}$, for rotation about a central axis. (b) The radial distance from a given axis at which the mass of a body could be concentrated without altering the rotational inertia of the body about that axis is called the *radius of gyration*. Let k represent radius of gyration and show that

$$k = \sqrt{I/M}.$$

This gives the radius of the "equivalent hoop" in the general case.

15. A thin rod of length l and mass m is suspended freely from one end. It is pulled aside and swung about a horizontal axis, passing through its lowest position with an angular speed ω. How high does its center of mass rise above its lowest position? Neglect friction and air resistance.
Answer: $l^2\omega^2/6\,g$.

16. (a) Prove that the rotational inertia of a thin rod of length l about an axis through its center perpendicular to its length is $I = \frac{1}{12}Ml^2$. (See Table 12-1.) (b) Use the parallel-axis theorem to show that $I = \frac{1}{3}Ml^2$ when the axis of rotation is through one end perpendicular to the length of the rod.

17. (a) Show that the sum of the rotational inertias of a plane laminar body about any two perpendicular axes in the plane of the body is equal to the rotational inertia of the body about an axis through their point of intersection perpendicular to the plane. (b) Apply this to a circular disk to find its rotational inertia about a diameter as axis.   *Answer:* (b) $MR^2/4$.

18. Show that the rotational inertia of a rectangular plate of sides a and b about an axis perpendicular to the plate through its center is $\frac{1}{12}M(a^2 + b^2)$.

19. A meter stick is held vertically with one end on the floor and is then allowed to fall. Find the speed of the other end when it hits the floor, assuming that the end on the floor does not slip. *Answer: 5.4 m/s.*

20. A tall chimney cracks near its base and falls over. Express (a) the radial and (b) the tangential linear acceleration of the top of the chimney as a function of the angle θ made by the chimney with the vertical. (c) Can the resultant linear acceleration exceed g? (d) The chimney cracks up during the fall. Explain how this can happen. (See 'More on the Falling Chimney' by Albert A. Bartlett, in *The Physics Teacher*, September, 1976.)

SECTION 12-6

21. An automobile engine develops 100 hp $(7.5 \times 10^4$ W) when rotating at a speed of 1800 rev/min. What torque does it deliver?
Answer: 290 ft · lb (400 N · m).

22. Calculate (a) the torque, (b) the energy, and (c) the average power required to accelerate the earth from rest to its present angular speed about its axis in one day.

23. A pulley having a rotational inertia of 1.0×10^4 g · cm² and a radius of 10 cm is acted upon by a force, applied tangentially at its rim, that varies in time as $F = 0.50\,t + 0.30\,t^2$, where F is in newtons and t is in seconds. If the pulley was initially at rest, find its angular velocity after 3.0 seconds.
Answer: 5.0×10^2 rad/s.

24. A wheel of mass M and radius of gyration k (see Problem 14) spins on a fixed horizontal axle passing through its hub. The hub rubs the axle of radius a at only the topmost point, the coefficient of kinetic friction being μ_k. The wheel is given an initial angular velocity ω_0. Assume uniform deceleration and find (a) the elapsed time and (b) the number of revolutions before the wheel comes to a stop.

25. A uniform steel rod of length 1.20 m and mass 6.40 kg has attached to each end a small ball of mass 1.06 kg. The rod is constrained to rotate in a horizontal plane about a vertical axis through its midpoint. At a certain instant it is observed to be making 39.0 rev/s. Because of axle friction it comes to rest 32.0 s later. Compute, assuming a constant frictional torque, (a) the angular acceleration, (b) the retarding torque exerted by axle friction, (c) the total work done by the axle friction, and (d) the number of revolutions executed during the 32.0 s. (e) Suppose, however, that the frictional torque is known not to be constant. Which, if any, of the quantities (a), (b), (c), or (d) can still be computed without requiring any additional information? If such exists, give its value.
Answer: (a) -7.67 rad/s². (b) -11.7 N · m. (c) 4.58×10^4 J. (d) 624 rev. (e) The total work done; 4.58×10^4 J.

26. The angular momentum of a flywheel having a rotational inertia of 0.125 kg · m² $(9.22 \times 10^{-2}$ slug · ft²) decreases from 3.0 (2.21) to 2.0 (1.48) kg · m²/s (slug · ft²/s) in a period of 1.5 s. (a) What is the average torque acting on the flywheel during this period? (b) Assuming a uniform angular acceleration, through how many revolutions will the flywheel have turned? (c) How much work was done? (d) What was the average power supplied by the flywheel?

27. In an Atwood's machine (Fig. 5-9) one block has a mass of 500 g and the other a mass of 460 g. The pulley, which is mounted in horizontal friction-less bearings, has a radius of 5.0 cm. When released from rest the heavier block is observed to fall 75 cm in 5.0 s. What is the rotational inertia of the pulley?
Answer: 1.4×10^{-2} kg · m².

28. A uniform spherical shell rotates about a vertical axis on frictionless bearings (Fig. 12-28) A light cord passes around the equator of the shell, over a

pulley, and is attached to a small object that is otherwise free to fall under the influence of gravity. What is the speed of the object after it has fallen a distance h from rest?

29. A 6.0-lb block is put on a plane inclined 30° to the horizontal and is attached by a cord parallel to the plane over a pulley at the top to a hanging block weighing 18 lb. The pulley weighs 2.0 lb and has a radius of 0.33 ft. The coefficient of kinetic friction between block and plane is 0.10. Find (a) the acceleration of the hanging block and (b) the tension in the cord on each side of the pulley. Assume the pulley to be a uniform disk.
Answer: (a) 19 ft/s². (b) T_{18} = 7.6 lb; T_6 = 7.0 lb.

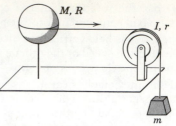

figure 12-28
Problem 28

SECTION 12-7

30. A hoop of radius 10 ft (3.0 m) weighs 320 lb (mass = 150 kg). It rolls along a horizontal floor so that its center of mass has a speed of 0.50 ft/s (0.15 m/s). How much work has to be done to stop it?

31. An automobile has a total mass of 1700 kg. It accelerates from rest to 40 km/h in 10 s. Each wheel has a mass of 32 kg and a radius of gyration (see Problem 14) of 0.30 m. Find, for the end of the 10-s interval, (a) the rotational kinetic energy of each wheel about its axle, (b) the total kinetic energy of each wheel; (c) the total kinetic energy of the automobile.
Answer: (a) 990 J. (b) 3000 J. (c) 1.1×10^5 J.

32. Show that a cylinder will slip on an inclined plane of inclination angle θ if the coefficient of static friction between plane and cylinder is less than $\frac{1}{3} \tan \theta$.

33. A 10-ft-long ladder rests against a wall and makes an angle of 60° with the horizontal floor. If it starts to slip, where is the instantaneous axis of rotation?
Answer: 5.0 ft horizontally from the wall and $5\sqrt{3}$ ft vertically above the ground.

34. A sphere rolls up an inclined plane of inclination angle 30°. At the bottom of the incline the center of mass of the sphere has a translational speed of 16 ft/s. (a) How far does the sphere travel up the plane? (b) How long does it take to return to the bottom?

35. A body of radius R and mass m is rolling horizontally without slipping with speed v. It then rolls up a hill to a maximum height h. If $h = 3\ v^2/4\ g$, (a) what is the body's rotational inertia? (b) What might the body be?
Answer: (a) $\frac{1}{2}mR^2$. (b) A solid circular cylinder.

36. A small sphere rolls without slipping on the inside of a large hemisphere whose axis of symmetry is vertical. It starts at the top from rest. (a) What is its kinetic energy at the bottom? What fraction is rotational? What translational? (b) What normal force does the small sphere exert on the hemisphere at the bottom? Take the radius of the small sphere to be r, that of the hemisphere to be R, and let m be the mass of the sphere.

37. A uniform disk, of mass M and radius R, lies on one side initially at rest on a frictionless horizontal surface. A constant force $\mathbf{F}$ is then applied tangentially at its perimeter by means of a string wrapped around its edge. Describe the subsequent (rotational and translational) motion of the disk.
Answer: $\alpha = 2\ F/MR$; $a = F/M$.

38. A tape of negligible mass is wrapped around a cylinder of mass M, radius R. The tape is pulled vertically upward at a speed that just prevents the center of mass from falling as the cylinder unwinds the tape. (a) What is the tension in the tape? (b) How much work has been done on the cylinder once it has reached an angular velocity ω? (c) What is the length of tape unwound in this time?

39. A cylinder of length L and radius R has a weight W. Two cords are wrapped around the cylinder, one near each end, and the cord ends are attached to hooks on the ceiling. The cylinder is held horizontally with the two cords exactly vertical and is then released (Fig. 12-29). Find (a) the tension in each

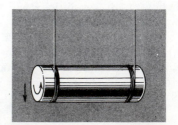

figure 12-29
Problem 39

cord as they unwind and (b) the linear acceleration of the cylinder as it falls. *Answer:* (a) W/6. (b) 2 g/3.

40. A homogeneous sphere starts from rest at the upper end of the track shown in Fig. 12-30 and rolls without slipping until it rolls off the right-hand end. If $H = 204$ ft and $h = 64$ ft and the track is horizontal at the right-hand end, determine the distance to the right of point A at which the ball strikes the horizontal base line.

41. A length l of flexible tape is tightly wound. It is then allowed to unwind as it rolls down a steep incline that makes an angle θ with the horizontal, the upper end of the tape being tacked down (Fig. 12-31). Show that the tape unwinds completely in a time $T = \sqrt{3} \, l/g \sin \theta$.

42. A small solid marble of mass m and radius r rolls without slipping along the loop-the-loop track shown in Fig. 12-32, having been released from rest somewhere on the straight section of track. (a) From what minimum height above the bottom of the track must the marble be released in order that it just stay on the track at the top of the loop? (The radius of the loop-the-loop is R; assume $R \gg r$.) (b) If the marble is released from height $6 R$ above the bottom of the track, what is the horizontal component of the force acting on it at point Q?

43. A yo-yo is made from two uniform disks of radius R and combined mass m. The short shaft connecting the disks has a very small radius r. A string of length $(L + R)$ is wrapped around the shaft several times by an expert yo-yo operator, who releases it with zero speed. Assume the string is vertical at all times. (a) What is the tension in the string during the descent and subsequent ascent of the yo-yo? (b) How long does it take the yo-yo to return to the operator's hands?

Answer: (a) $mgR^2/(R^2 + 2r^2)$, ascent and descent. (b) $\dfrac{2}{r}\sqrt{L(2\,r^2 + R^2)/g}$.

44. A uniform solid cylinder of radius R is given an angular velocity ω_0 about its axis and is then dropped vertically onto a flat horizontal table. The table is not frictionless, so the cylinder begins to move as it slips. What is the velocity of the center of mass of the cylinder when pure rolling sets it?

45. A solid cylinder of weight 50 lb (mass = 23 kg) and radius 3.0 in (7.6 cm) has a light thin tape wound around it. The tape passes over a light, smooth fixed pulley to a 10-lb (mass = 4.5 kg) body, hanging vertically (Fig. 12-33). If the plane on which the cylinder moves is inclined 30° to the horizontal, find (a) the linear acceleration of the cylinder down the incline and (b) the tension in the tape, assuming no slipping.
Answer: (a) 1.4 ft/s² (0.47 m/s²). (b) 11 lb (48 N).

46. A billiard ball is struck by a cue as in Fig. 12-34. The line of action of the applied impulse is horizontal and passes through the center of the ball. The initial velocity $\mathbf{v}_0$ of the ball, its radius R, its mass M, and the coefficient of friction μ between the ball and the table are all known. How far will the ball move before it ceases to slip on the table?

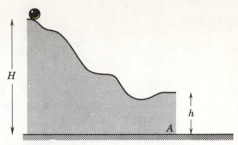

figure 12-30
Problem 40

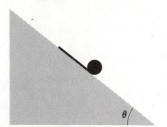

figure 12-31
Problem 41

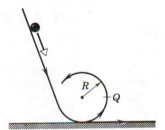

figure 12-32
Problem 42

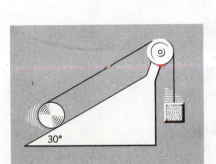

figure 12-33
Problem 45

figure 12-34
Problem 46

13

rotational dynamics II and the conservation of angular momentum

In Chapter 12 we discussed the dynamics of the rotational motion of a rigid body about an axis that was fixed in an inertial reference frame. We saw that the scalar relation $\tau = I\alpha$ (Eq. 12-17), in which only torque components along the axis of rotation were considered, was sufficient to solve dynamical problems in this special case.

In this chapter we shall first consider the rotation of a rigid body about an axis that is *not* fixed in an inertial reference frame. To solve dynamical problems in this more general case we shall use the general (vector) relation for rotational motion, namely,

$$\tau = d\mathbf{L}/dt \qquad (12\text{-}9)$$

in which we have dropped the subscript on τ_{ext} for convenience.

Later we shall consider once more the rotation of particles and rigid bodies about fixed axes. This time, however, we specifically examine the action of torques which have components at right angles to the axis. Our point of departure here will not be Eq. 12-17 $(\tau = I\alpha)$ but again the more general Eq. 12-9 $(\tau = d\mathbf{L}/dt)$.

Finally we shall consider systems on which no external torques act and shall introduce the important principle of *conservation of angular momentum*.

Figure 13-1*a* shows a top spinning about its axis of symmetry, the point of the top being fixed at the origin *O* of an inertial reference frame. We know from experience that the axis of such a rapidly spinning top will move around the vertical axis, sweeping out a cone. This motion is called *precession*. Let us see if we can predict this motion from the prin-

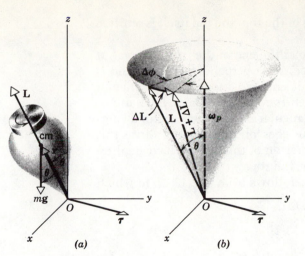

ciples of classical mechanics and, in particular, if we can calculate ω_p, the angular speed of the precessional motion.

At the instant shown in Fig. 13-1a the top has an angular velocity $\boldsymbol{\omega}$ about its own axis. It also has an angular momentum **L** about this same *axis*,* the axis making an angle θ with the vertical.

Two forces act on the top, an upward force on the pivot at O and the pull of gravity, or weight, which acts downward at the center of mass. The upward force passes through O and thus can exert no torque about that point because its moment arm is zero. The weight mg, however, exerts a torque about O given by

$$\boldsymbol{\tau} = \mathbf{r} \times \mathbf{F} = \mathbf{r} \times m\mathbf{g},$$

where **r** locates the center of mass with respect to the pivot. This equation requires that $\boldsymbol{\tau}$ be perpendicular to the plane formed by **r** and mg; application of the right-hand rule shows that its direction is as shown in Fig. 13-1a. Note that $\boldsymbol{\tau}$, as well as **L** and **r**, rotates about the axis at angular speed ω_p as the top precesses.

When a torque acts on a rigid body it changes the angular momentum of that body according to the fundamental relation (Eq. 12-9)

$$\boldsymbol{\tau} = d\mathbf{L}/dt. \qquad \text{(Eq. 12-9)}$$

Being a vector, **L** can change in magnitude, in direction, or in both. Equation 12-9 shows that the *change* in **L** (that is, $d\mathbf{L}$) must point in the direction of $\boldsymbol{\tau}$. Figure 13-1a shows us that $\boldsymbol{\tau}$ is *at right angles* to **L**; thus the change in **L** brought about by the action of the torque must also be at right angles to **L**.

To examine the matter quantitatively let us observe the top for a time Δt. During this interval a change in **L** of

$$\Delta \mathbf{L} = \boldsymbol{\tau} \, \Delta t \qquad (13\text{-}1)$$

is predicted by Eq. 12-9 (if Δt is small enough). This change $\Delta \mathbf{L}$, which, like $\boldsymbol{\tau}$, is at right angles to **L**, is displayed in Fig. 13-1b where we see the

---

* The vector $\boldsymbol{\omega}$ always points along the (fixed) axis of rotation of a spinning body but, in general, the vector **L** does not (see Section 13-3). For bodies with symmetry about the rotational axis, however, both $\boldsymbol{\omega}$ and **L** point along this axis, assuming that the axis is fixed. We can assume that $\boldsymbol{\omega}$ and **L** are coaxial for the spinning top of Fig. 13-1a if $\omega \gg \omega_p$, that is, if the precession rate is relatively slow so that the axis, although not fixed, changes direction only slowly.

cone swept out by the precessing axis of the top; the top itself is omitted here for clarity.

The angular momentum of the top at the end of the time interval Δt is the vector sum of $\mathbf{L}$ and $\Delta\mathbf{L}$. Since $\Delta\mathbf{L}$ is perpendicular to $\mathbf{L}$ and is assumed to be very small in magnitude compared to it, the new angular momentum vector has the same *magnitude* as the old one but a different *direction*. Hence the head of the angular momentum vector swings around in a horizontal circle as time goes on (Fig. 13-1b). Since this vector always lies along the axis of rotation of the top, we have qualitatively accounted for the precession of the top.

The angular speed of precession ω_p follows from Fig. 13-1b in which

$$\omega_p = \Delta\phi/\Delta t.$$

But, since $\Delta L \ll L$, we have (see Eq. 13-1),

$$\Delta\phi \cong \Delta L/L \sin\theta = \tau \, \Delta t/L \sin\theta$$

or

$$\omega_p = \Delta\phi/\Delta t = \tau/L \sin\theta. \qquad (13\text{-}2a)$$

Since (Fig. 13-1a)

$$\tau = rmg \sin(180° - \theta) = rmg \sin\theta$$

we have finally

$$\omega_p = mgr/L. \qquad (13\text{-}2b)$$

Notice that the precessional angular velocity is independent of θ and varies inversely as the magnitude of the angular momentum. If the angular momentum is large, the precessional angular velocity will be small.

We can express Eq. 13-2b in vector form. We start by rewriting Eq. 13-2a as

$$\tau = \omega_p L \sin\theta.$$

Now $\boldsymbol{\omega}_p$ is a vector pointing vertically upward in Fig. 13-1b, and θ in that figure is the angle between $\boldsymbol{\omega}_p$ and $\mathbf{L}$. We recognize the right side of the above equation as the magnitude of the vector product $\boldsymbol{\omega}_p \times \mathbf{L}$ and we see that this equation gives the magnitude of $\boldsymbol{\tau}$ in the vector relation

$$\boldsymbol{\tau} = \boldsymbol{\omega}_p \times \mathbf{L}. \qquad (13\text{-}3)$$

This is the general vector expression relating the precessional angular velocity to $\boldsymbol{\tau}$ and to $\mathbf{L}$; you should show that Eq. 13-2b may be derived readily from it. Application of right-hand rule to Fig. 13-1b shows that the order of factors on the right side of Eq. 13-3 is correct, that is, $\boldsymbol{\omega}_p \times \mathbf{L}$ gives the correct direction as well as the correct magnitude for $\boldsymbol{\tau}$.

EXAMPLE 1

A student holds a rim-loaded bicycle wheel, rotating at a relatively high angular speed ω, with its shaft horizontal as in Fig. 13-2a. Her physics instructor now asks her to turn the shaft rapidly (for a time Δt) so that the shaft points at a small angle $\Delta\theta$ above the horizontal as in Fig. 13-2b. The instructor also asks the student to keep the shaft in a vertical plane at all times. What torques must the student exert on the shaft if she is to follow these instructions?

The student will be well aware from the strain in her wrists that she must exert a torque on the shaft simply to hold it in a horizontal position. This torque, which is needed to counteract the turning effect of the force of gravity that acts at the center of mass, is directed along a horizontal axis and emerges perpendicularly out of the plane of Fig. 13-2. The student must supply this torque whether or not the wheel is rotating.

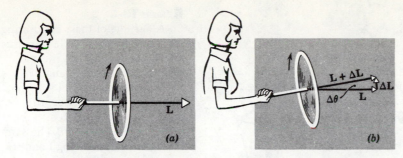

figure 13-2
(a) A student holds a heavy, rim-loaded, rapidly spinning bicycle wheel by the shaft and (b) tilts the shaft upward from the horizontal through a small angle.

If now the student turns the shaft of the spinning wheel upward, she will find that the wheel will swerve around to her right, perhaps rather violently, so that she will have failed to keep the shaft in a vertical plane. If she is to keep the shaft in this plane while she is tilting it upward, she must exert a torque on the shaft (about an almost vertical axis) tending to turn it to her left to counteract this effect. Let us see why this is so.

In tilting the shaft one changes its angular momentum **L**, in time Δt, by an amount $\Delta \mathbf{L}$, as Fig. 13-2b shows. During this interval, then, the student must exert an average torque on the wheel given from Eq. 12-9 as

$$\bar{\tau} = \Delta \mathbf{L}/\Delta t;$$

the magnitude of $\bar{\tau}$ is given by

$$\bar{\tau} = \Delta L/\Delta t = L \sin \Delta\theta/\Delta t.$$

This average torque $\bar{\tau}$ has the same direction as $\Delta \mathbf{L}$, that is, it is approximately vertically upwards if the angle $\Delta\theta$ in Fig. 13-2b is not too large. We can see that such a torque would tend to turn the shaft to the left if the wheel was not rotating. This torque *must* be supplied by the student as she is tilting the shaft of the spinning wheel upward; if she fails to do so, the shaft will not remain in a vertical plane.

You should experiment with such a spinning wheel yourself, working out the relationships between the vectors **L**, $\Delta \mathbf{L}$, and **τ**. If one is not available you can experiment with a toy gyroscope, although this fails to give the kinesthetic appreciation of $\tau = d\mathbf{L}/dt$ that is provided by a rim-loaded, rapidly spinning wheel.

There is an analogy between the experiment of Fig. 13-2 and another experiment in which the student is asked to swing a heavy weight (attached to a stout cord) around in a horizontal circle at constant speed. In this latter experiment the student, during a time Δt, must change the direction of the *linear momentum* **P** of the weight, leaving its magnitude unchanged. To do so, she must supply a *force* that points at right angles to **P** (in the direction of $\Delta \mathbf{P}$), that is, radially inward. In the experiment of Fig. 13-2 the student must, during a time Δt, change the direction of the *angular momentum* **L** of the wheel, leaving its magnitude unchanged. To do so she must supply a *torque* that points at right angles to **L** (in the direction of $\Delta \mathbf{L}$), that is, vertically upward.*

13-3 ANGULAR MOMENTUM AND ANGULAR VELOCITY

In this section it is our purpose to examine the relationship between the angular momentum and the angular velocity for particles and rigid bodies rotating about an axis fixed in an inertial reference frame.

First we consider a single particle of mass m moving with speed v in a circle about the z-axis of an inertial reference frame as in Fig. 13-3. Its angular velocity **ω** points upward and can be taken to lie along the z-axis. Its angular momentum **l** with respect to the origin O of the reference frame is given by Eq. 12-3, or

$$\mathbf{l} = \mathbf{r} \times \mathbf{p},$$

* This analogy is explored by A. E. Benfield, *American Journal of Physics*, September 1958. See also Problem 5.

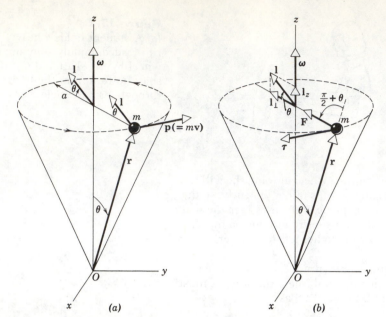

figure 13-3

(a) A particle of mass m rotating with speed v in a circle of radius a about the z-axis of an inertial reference frame. The angular momentum about O, $\mathbf{l}$ $(= \mathbf{r} \times \mathbf{p})$, is shown; for convenience, this vector is also shown translated to the center of the circle. (b) The same configuration, showing $\mathbf{l}$ and its components and also the centripetal force $\mathbf{F}$ and the torque τ about O.

where $\mathbf{r}$ and $\mathbf{p}$ $(= m\mathbf{v})$ are shown in the figure. The vector $\mathbf{l}$ is perpendicular to the plane formed by $\mathbf{r}$ and $\mathbf{p}$, which means that $\mathbf{l}$ is *not* parallel to $\boldsymbol{\omega}$. Note that $\mathbf{l}$ has a (vector) component $\mathbf{l}_z$ which is parallel to $\boldsymbol{\omega}$, but it has another (vector) component $\mathbf{l}_\perp$ which is perpendicular to $\boldsymbol{\omega}$. Note too, that if we choose our origin to lie in the plane of the circulating particle, then $\mathbf{l}$ *is* parallel to $\boldsymbol{\omega}$.

The perhaps unexpected result that $\mathbf{l}$ and $\boldsymbol{\omega}$ are not parallel in this simple case may cause some concern. However, this result is quite in accord with the general relationship $\boldsymbol{\tau} = d\mathbf{l}/dt$ for a torque acting on a single particle. The vector $\mathbf{l}$ is changing with time as the particle motion proceeds, the change being entirely in direction and not in magnitude, just as it was for the precessing top in the preceding section. Since the right side of the preceding relationship $(= d\mathbf{l}/dt)$ has a nonzero value, the left side $(= \boldsymbol{\tau})$ must also have a nonzero value; that is, a torque must act on the particle with respect to origin O.

There is indeed such a torque. For if the particle moves in a circle, a centripetal force $\mathbf{F}$ must act on it, as in Fig. 13-3b. We may imagine that $\mathbf{F}$ is provided by the tension in a light cord that ties the rotating particle to the z-axis. The torque about O is provided by $\mathbf{F}$ and is given by Eq. 12-1.

$$\boldsymbol{\tau} = \mathbf{r} \times \mathbf{F}.$$

The torque $\boldsymbol{\tau}$ is tangent to the circle (perpendicular to the plane formed by $\mathbf{r}$ and $\mathbf{F}$) and in the direction shown in Fig. 13-3b, as you may verify from the right-hand rule.

Show that the moving particle of Fig. 13-3 satisfies the relation $\boldsymbol{\tau} = d\mathbf{l}/dt$ quantitatively.

   The proof is along the same lines as that of Section 13-2 for the spinning top because, from a vector point of view, the two problems are identical. In each case we have the precession of an angular momentum vector ($\mathbf{L}$ for the top and $\mathbf{l}$ for the particle of Fig. 13-3) about a vertical axis, at a rate which we called ω_p for the top and which we call ω for the particle. In each case we have a torque which is always at right angles to the plane formed by $\mathbf{L}$ (or $\mathbf{l}$) and $\boldsymbol{\omega}_p$ (or $\boldsymbol{\omega}$).

   Thus, since the two problems are formally identical, it suffices to inquire

EXAMPLE 2

whether the rotating particle of Fig. 13-3 obeys the vector equation for precession ($\tau = \omega_p \times L$; Eq. 13-3). This equation was derived for the precessing top directly from—and is directly equivalent to—the relation $\tau = dL/dt$ (Eq. 12-9). We can write the relation $\tau = \omega_p \times L$ in terms of magnitudes as

$$\tau = \omega l \sin (90° - \theta) = \omega l \cos \theta, \qquad (13\text{-}3)$$

in which we have substituted ω for ω_p and l for L and have noted in Fig. 13-3a that the angle between ω and l is $90° - \theta$. For τ and l, again using the notation of Fig. 13-3a, we can write

$$\tau = Fr \sin (90° + \theta) = [m\omega^2(r \sin \theta)](r)(\cos \theta)$$

and

$$l = rp \sin 90° = r(mv) = (r)(m)[\omega(r \sin \theta)],$$

in which $r \sin \theta$ is the radius a of the circle in which the particle moves, $(90° + \theta)$ is the angle between r and F, and $90°$ is the angle between r and p. Substituting these two expressions into Eq. 13-3 yields

$$m\omega^2 r^2 \sin \theta \cos \theta = \omega(m\omega r^2 \sin \theta) \cos \theta,$$

which is an identity. In terms of *magnitudes* we have proved our point. Refer to Fig. 13-3 and make certain that the *direction of* τ is that of dl/dt (Eq. 12-7), or alternatively of $\omega \times l$ (Eq. 13-2b).

Let us now investigate the relationship between l_z and ω for the particle of Fig. 13-3. From Example 2 we have

$$l = mr^2 \omega \sin \theta.$$

From Fig. 13-3b we see that

$$l_z = l \sin \theta = m\omega r^2 \sin^2 \theta.$$

Now $r \sin \theta = a$, the radius of the circle in which the particle moves. This leads to

$$l_z = ma^2 \omega, \qquad (13\text{-}4)$$

in which ma^2 is the rotational inertia I of the particle with respect to the z-axis. Thus

$$l_z = I\omega, \qquad (13\text{-}5)$$

which is to be compared with Eq. 12-18 ($L = I\omega$) for the rotation of a rigid body about a fixed axis. Note that the vector relation $l = I\omega$ is *not* correct in this case because l and ω do *not* point in the same direction. However, l_z and ω *do*, so that we could write Eq. 13-5 in vector form as $l_z = I\omega$.

Now let us add another particle of mass m to the system of Fig. 13-3. In particular, let us add this particle in the same orbit, moving with the same speed, but always at a diametrically opposite point, on the other side of the axis of rotation. The angular momentum l_2 with respect to O for this second particle will have the same magnitude as that of l_1 for the first particle and it will make the same angle $(90° - \theta)$ with the z-axis, but it will have a different orientation around that axis. As Fig. 13-4a shows, l_2 will lie in a plane formed by ω and by l_1 but will be on the opposite side of the z-axis from l_1. The vectors l_1 and l_2 include an angle of $180° - 2\theta$.

The total angular momentum L of the system of two particles is the vector sum of the angular momenta of the separate particles, that is, $L = l_1 + l_2$. The resultant vector L, as Fig. 13-4b shows, points along the z-axis (in the direction of ω) and is constant in magnitude. Note that

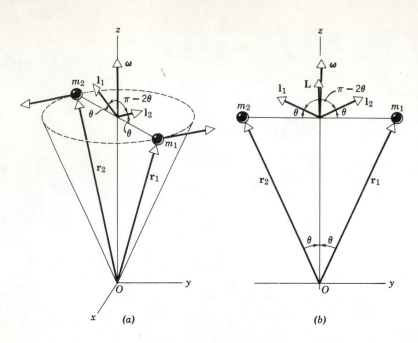

figure 13-4
(a) Two particles of mass m rotating as in Fig. 13-3 but maintaining diametrically opposite positions. (b) A cross section through the two particles, showing that the total angular momentum $\mathbf{L}$ $(= \mathbf{l}_1 + \mathbf{l}_2)$ for the two-particle system points along the axis of rotation, in the same direction as $\boldsymbol{\omega}$.

this statement is true no matter where the origin O is located along the axis of rotation.

The fact that $\mathbf{L} =$ a constant (in both magnitude and direction), for this two-particle system, means that $d\mathbf{L}/dt = 0$, which in turn (Eq. 12-9) means that $\tau = 0$ for this system. Convince yourself (Fig. 13-3b will be helpful) that this is the case, the torques for the two particles about O being equal in magnitude but oppositely directed so that the torque acting on the two-particle system is zero.

The fact that $\boldsymbol{\omega}$ and $\mathbf{L}$ point in the same direction in this problem but did not for the case of a single particle can be traced to the fact that, in the two-particle system, the particles have the same mass and are in diametrically opposite positions at the same distance from the rotation axis.

We can now extend our system to a rigid body, made up of many particles. If the body is symmetric about the axis of rotation, by which we mean that for every mass element in the body there must be an identical mass element diametrically opposite the first element and at the same distance from the axis of rotation, then the body can be regarded as made up of sets of particle pairs of the kind we have been discussing. Since $\mathbf{L}$ and $\boldsymbol{\omega}$ are parallel for all such pairs, they are also parallel for rigid bodies that possess this kind of symmetry. Note that, in Table 12-1, all the systems except f and j meet this criterion.

For such symmetrical rigid bodies $\mathbf{L}$ and $\boldsymbol{\omega}$ are parallel and we can write Eq. 12-18 $(L = I\omega)$ in vector form as

$$\mathbf{L} = I\boldsymbol{\omega}. \tag{13-6}$$

Do not forget, however, that if $\mathbf{L}$ stands for the *total* angular momentum, then Eq. 13-6 applies *only* to bodies that have symmetry* about

---

*We have oversimplified the symmetry requirement. Every rigid body, no matter how irregular its shape, has three perpendicular axes through its center of mass, about each of which $\mathbf{L}$ and $\boldsymbol{\omega}$ have the same direction, being related by $\mathbf{L} = I\boldsymbol{\omega}$. These axes are called the *principal axes*. The axis of a figure of revolution is always a principal axis, as are axes at right angles to it through the center of mass. In general, however, $\mathbf{L}$ and $\boldsymbol{\omega}$ point in different directions for axes that are not principal axes. See Arnold Sommerfeld, *Mechanics*, Chapter IV, Academic Press, New York, (1964 paperback edition).

the (fixed) rotational axis. If **L** stands for the vector component of angular momentum along the rotational axis (that is, for L_z), then Eq. 13-6 is equivalent to Eq. 12-8 and holds for *any* rigid body, symmetrical or not, that is rotating about a fixed axis.

EXAMPLE 3

In Example 5, Chapter 12, find the acceleration of the falling mass by direct application of Eq. 12-9 ($\tau = d\mathbf{L}/dt$).

The system of Fig. 12-12, consisting of the wheel M and the mass m is acted on by two external forces, the downward pull of gravity $m\mathbf{g}$ acting on mass m and the upward force exerted by the bearings of the shaft of the cylinder, which we take as our origin. The tension in the cord is an internal force and does not act from the outside on the system (wheel + weight). Only the first of these external forces exerts a torque about the origin and its magnitude is $(mg)R$.

The angular momentum of the system about the origin at any instant is

$$L = I\omega + (mv)R,$$

in which $I\omega$ is the angular momentum of the (symmetrical) disk and $(mv)R$ is the angular momentum (= linear momentum × moment arm) of the falling mass about the origin. Both these contributions to L point in the same direction, namely, perpendicularly out of the plane of Fig. 12-12.

Applying $\tau = d\mathbf{L}/dt$ (in scalar form) yields

$$(mg)R = \frac{d}{dt}(I\omega + mvR)$$

$$= I(d\omega/dt) + mR(dv/dt)$$

$$= I\alpha + mRa.$$

Since $a = \alpha R$ and $I = \frac{1}{2}MR^2$, this reduces to

$$mgR = (\tfrac{1}{2}MR^2)(a/R) + mRa$$

or

$$a = \frac{2\,mg}{M + 2m}.$$

EXAMPLE 4

A simple example of an unsymmetrical rotating rigid body is a dumbbell-type rod whose bar makes an angle θ with the fixed axis of rotation passing through its center of mass. The rod rotates at a constant angular velocity ω about this axis, the vector $\boldsymbol{\omega}$ thus pointing along this axis, as shown in Fig. 13-5. Experience tells us that such a system is "unbalanced" or "lop-sided," and if it were not securely fastened to the vertical shaft near C, it would break away from the shaft at high angular velocities. It would tend to move until the angle θ becomes 90°, in which limiting position the system would then be symmetrical about the shaft.

(*a*) Show qualitatively that in the unsymmetrical case, shown in Fig. 13-5, **L** and $\boldsymbol{\omega}$ are not parallel.

Each particle of mass m has an angular momentum with respect to C given by $\mathbf{r} \times \mathbf{p}$ for that particle. At the instant shown the upper particle is moving into the page at right angles to it, and the lower particle is moving out of the page at right angles to it. The momentum vectors of the two masses are therefore equal but opposite, and so are their position vectors with respect to C. Hence, by application of the right-hand rule in $\mathbf{r} \times \mathbf{p}$, we find that l is the same for each particle and that their sum, the total angular momentum vector **L** of the dumbbell, is as shown in the figure, at right angles to the bar in the plane of the page. Hence **L** and $\boldsymbol{\omega}$ are *not* parallel at this instant. It is clear that as the dumbbell itself rotates, the angular momentum vector, while constant in magnitude, rotates around the fixed axis of rotation.

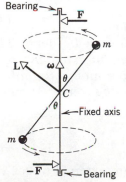

figure 13-5
Example 4.

(b) The fact that $\mathbf{L}$ and $\boldsymbol{\omega}$ do not point in the same direction is perfectly consistent with the fundamental relation $\boldsymbol{\tau} = d\mathbf{L}/dt$. We have seen twice before (see Section 13-2 and Example 1) that an angular momentum vector of constant magnitude that rotates around a fixed axis must have associated with it a torque $\boldsymbol{\tau}$ that is at right angles to the plane formed by $\mathbf{L}$ and $\boldsymbol{\omega}$. At the instant shown in Fig. 13-5 this plane is the plane of the figure. Is there such a torque in this problem and if so, where does it come from?

There is indeed such a torque and it arises from the unbalanced sideways forces exerted by the bearings on the shaft and transmitted by the shaft to the dumbbell bar. At the instant shown in the figure the upper end of the dumbbell would tend to move outward to the right. The shaft would be pulled to the right against the upper bearing, which in turn exerts a force $\mathbf{F}$ on the shaft that points to the left. Similarly, the lower end of the dumbbell tends to move outwards to the left. The shaft would be pulled to the left against the lower bearing, which in turn exerts a force $-\mathbf{F}$ on the shaft that points to the right. The torque $\boldsymbol{\tau}$ about C as a result of these forces points perpendicularly out of the page, at right angles to the plane formed by $\mathbf{L}$ and $\boldsymbol{\omega}$, and in the right direction to account for the rotary motion of $\mathbf{L}$ (you should check this).

The forces $\mathbf{F}$ and $-\mathbf{F}$ lie in the plane of Fig. 13-5 at the instant shown. As the dumbbell rotates, these forces, and therefore the torque $\boldsymbol{\tau}$, rotate with it, so that $\boldsymbol{\tau}$ always remains at right angles to the plane formed by $\boldsymbol{\omega}$ and $\mathbf{L}$ (compare with Fig. 13-1) The rotating forces $\mathbf{F}$ and $-\mathbf{F}$ cause a "wobble" in the upper and lower bearings. The bearings and their supports must be made strong enough to provide these forces. For a symmetrical rotating body there is no bearing wobble and the shaft rotates smoothly.

Bearing wobble and internal strains can cause serious practical problems when objects, such as turbine rotors, are made to rotate at high speeds. Although designed to be symmetrical, such rotors, because of small errors of blade placement, etc., may be slightly unsymmetrical. They may be restored to symmetry by the addition or removal of metal at appropriate places; this is done by spinning the wheel in a special device such that bearing wobble can be measured quantitatively and the appropriate corrective measure computed and indicated automatically. We are all familiar with lead weights placed at strategic points on automobile tire rims to reduce wobble at high speeds due to unbalance.

13-4
CONSERVATION OF ANGULAR MOMENTUM

In Chapter 12, we found that the time rate of change of the total angular momentum of a system of particles about a point fixed in an inertial reference frame (or about the center of mass) is equal to the sum of the *external* torques acting on the system, that is,

$$\boldsymbol{\tau}_{\text{ext}} = d\mathbf{L}/dt. \tag{12-9}$$

Suppose now that $\boldsymbol{\tau}_{\text{ext}} = 0$; then $d\mathbf{L}/dt = 0$ so that $\mathbf{L} = $ a constant.

When the resultant external torque acting on a system is zero, the total vector angular momentum of the system remains constant. This is the *principle of the conservation of angular momentum.*

For a system of n particles, the total angular momentum $\mathbf{L}$ about some point is

$$\mathbf{L} = \mathbf{l}_1 + \mathbf{l}_2 + \cdots + \mathbf{l}_n.$$

When the resultant external torque on the system is zero, we have

$$\mathbf{L} = \text{a constant} = \mathbf{L}_0, \tag{13-7}$$

where $\mathbf{L}_0$ is the constant total angular momentum vector. The angular momenta of the individual particles may change, but their vector sum $\mathbf{L}_0$ remains constant in the absence of a net external torque.

Angular momentum is a vector quantity so that Eq. 13-7 is equivalent to three scalar equations, one for each coordinate direction through the reference point. The conservation of angular momentum therefore supplies us with three conditions on the motion of a system to which it applies.

For a system consisting of a rigid body rotating about an axis (the z-axis, say) that is fixed in an inertial reference frame, we have

$$\mathbf{L}_z = I\boldsymbol{\omega}, \tag{13-6}$$

where $\mathbf{L}_z$ is the component of the angular momentum along the rotation axis and I is the rotational inertia for this same axis. It is possible for the rotational inertia I of a rotating body to change by rearrangement of its parts. If no net external torque acts, then $\mathbf{L}_z$ must remain constant and, if I does change, there must be a compensating change in ω. The principle of conservation of angular momentum in this case is expressed as

$$I\omega = I_0\omega_0 = \text{a constant}. \tag{13-8}$$

Equation 13-8 holds not only for rotation about a fixed axis but also about an axis through the center of mass of the system that moves so that it always remains parallel to itself (see footnote, p. 237).

Acrobats, divers, ballet dancers, ice skaters, and others often use this principle. Because I depends on the square of the distance of the parts of the body from the axis of rotation, a large variation is possible by extending or pulling in the limbs. Consider the diver* in Fig. 13-6. Let us assume that as he leaves the diving board he has a certain angular speed ω_0 about a horizontal axis through the center of mass, such that he would rotate through half a turn before he strikes water. If he wishes to make a one and one-half turn somersault instead, in the same time, he must triple his angular speed. Now there are no external forces acting

figure 13-6
A diver leaves the diving board with arms and legs outstretched and with some initial angular velocity. Since no torques are exerted on him about his center of mass, L ($= I\omega$) is constant while he is in the air. When he pulls his arms and legs in, since L decreases, ω increases. When he again extends his limbs, his angular velocity drops back to its initial value. Notice the parabolic motion of his center of mass, common to all two-dimensional motion under the influence of gravity.

* See ''The Mechanics of Swimming and Diving'' by R. L. Page in *The Physics Teacher*, February 1976, for an interesting biomechanical analysis.

on him except gravity, and gravity exerts no torque about his center of mass. His angular momentum therefore remains constant, and $I_0\omega_0 = I\omega$. Since $\omega = 3\omega_0$, the diver must change his rotational inertia about the horizontal axis through the center of mass from the initial value I_0 to a value I, such that I equals $\frac{1}{3}I_0$. This he does by pulling in his arms and legs toward the center of his body. The greater his initial angular speed and the more he can reduce his rotational inertia, the greater the number of revolutions he can make in a given time.

We should notice that the rotational kinetic energy of the diver is not constant. In fact, in our example, since

$$I\omega = I_0\omega_0$$

and

$$I < I_0,$$

it follows that

$$\tfrac{1}{2}I\omega^2 > \tfrac{1}{2}I_0\omega_0^2,$$

and the diver's rotational kinetic energy *increases*. This increase in energy is supplied by the diver, who does work when he pulls the parts of his body together.

In a similar way the ice skater and ballet dancer can increase or decrease the angular speed of a spin about a vertical axis. A cat manages to land on its feet after a fall by using the same principles, the tail serving as a useful, but unessential, extra appendage.

EXAMPLE 5

A small object of mass m is attached to a light string which passes through a hollow tube. The tube is held by one hand and the string by the other. The object is set into rotation in a circle of radius r_1 with a speed v_1. The string is then pulled down, shortening the radius of the path to r_2 (Fig. 13-7). Find the new linear speed v_2 and the new angular speed ω_2 of the object in terms of the initial values v_1 and ω_1 and the two radii.

The downward pull on the string is transmitted as a radial force on the object. Such a force exerts a zero torque on the object about the center of rotation. Since no torque acts on the object about its axis of rotation, its angular momentum in that direction is constant. Hence

initial angular momentum = final angular momentum,

$$mv_1r_1 = mv_2r_2,$$

and

$$v_2 = v_1\left(\frac{r_1}{r_2}\right).$$

Since $r_1 > r_2$, the object speeds up on being pulled in.

In terms of angular speed, since v_1 equals ω_1r_1 and v_2 equals ω_2r_2,

$$mr_1^2\omega_1 = mr_2^2\omega_2$$

and

$$\omega_2 = \left(\frac{r_1}{r_2}\right)^2 \omega_1,$$

so that there is an even greater increase in angular speed over the initial value (see Problem 31). What effect does the force of gravity (the object's weight) have on this analysis?

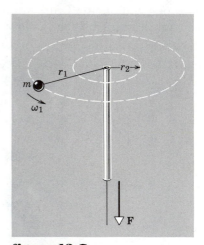

figure 13-7
Example 5. A mass at the end of a cord moves in a circle of radius r_1 with angular speed ω_1. The cord passes down through a tube. **F** supplies the centripetal force.

EXAMPLE 6

A student sits on a stool that is free to rotate about a vertical axis. He holds his arms extended horizontally with an 8.0-lb (= 35.6 N) weight in each hand. The instructor sets him rotating with an angular speed of 0.50 rev/s. Assume that friction is negligible and exerts no torque about the vertical axis of rotation.

Assume that the rotational inertia of the student remains constant at 4.0 slug · ft² (= 5.4 kg · m²) as he pulls his hands to his sides and that the change in rotational inertia is due only to pulling the weights in. Take the original distance of the weights from the axis of rotation to be 3.0 ft (= 0.91 m) and their final distance 0.50 ft (= 0.15 m). Find the final angular speed of the student.

The only external force is gravity acting through the center of mass, and that exerts no torque about the axis of rotation. Hence the angular momentum is conserved about this axis and

$$\text{initial angular momentum} = \text{final angular momentum,}$$

$$I_0\omega_0 = I\omega.$$

We have

$$I = I_{\text{student}} + I_{\text{weights}},$$

$$I_0 = 4.0 + 2\left(\frac{8.0}{32}\right)(3.0)^2 = 8.5 \text{ slug} \cdot \text{ft}^2 \ (= 11.6 \text{ kg} \cdot \text{m}^2),$$

$$I = 4.0 + 2\left(\frac{8.0}{32}\right)\left(\frac{1}{2}\right)^2 = 4.1 \text{ slug} \cdot \text{ft}^2 \ (= 5.6 \text{ kg} \cdot \text{m}^2),$$

$$\omega_0 = 0.50 \text{ rev/s} = \pi \text{ rad/s}.$$

Therefore

$$\omega = \frac{I_0}{I}\omega_0 = \frac{8.5}{4.1}\pi \text{ rad/s} = 2.1\pi \text{ rad/s} \cong 1.0 \text{ rev/s}.$$

The final angular speed is approximately doubled.

If we had allowed for the decrease in I caused by the arms being pulled in, the final angular speed would have been much greater.

What change would friction make? Is kinetic energy conserved as the student pulls in his arms and then puts them out again, assuming there is no friction? Explain.

EXAMPLE 7

A classroom demonstration that illustrates the vector nature of the law of conservation of angular momentum is worth considering.

A student stands on a platform that can rotate only about a vertical axis. In his hand he holds the axle of a rim-loaded bicycle wheel with its axis vertical. The wheel is spinning about this vertical axis with an angular speed ω_0, but the student and platform are at rest. The student tries to change the direction of rotation of the wheel. What happens?

Let us choose as the system the student plus platform plus wheel. The initial total angular momentum of this system is $I_0\omega_0$, arising from the spinning wheel, I_0 being the rotational inertia of the wheel about its axis and ω_0 pointing vertically upward. Figure 13-8a shows the initial condition.

The student next turns the axis of the wheel through an angle θ from the vertical (to do this he must supply a torque; see Example 1. This torque, however, is internal to the system as we have defined it). Since there is no external component of torque on the system about the vertical axis, the vertical component of angular momentum of the system must be conserved. The wheel, however, is now spinning about an axis making an angle θ with the vertical so that it contributes a vertical component of angular momentum of only $I_0\omega_0 \cos \theta$ to the system. Hence the student and platform must supply the additional angular momentum about the vertical axes, and they begin to rotate about a vertical axis. This extra vertical angular momentum $I_p\omega_p$ when added to $I_0\omega_0 \cos \theta$ must equal the initial vertical angular momentum of the system $I_0\omega_0$. That is,

$$I_p\omega_p = I_0\omega_0(1 - \cos \theta).$$

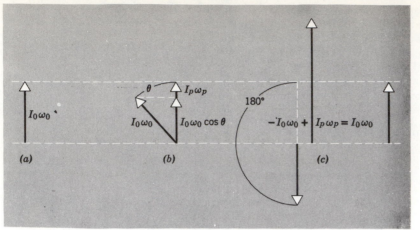

figure 13-8
Example 7. (a) The initial angular momentum of the system is shown. In (b), the wheel has been tilted an angle θ. Since no external torque in the vertical direction has been exerted on the system, the angular momentum in that direction must be conserved. The deficit, $(1 - \cos \theta)I_0\omega_0$, is made up by rotation of the student and platform. In (c), the wheel has been tilted 180°. The deficit is now $2I_0\omega_0$, which is now, as before, made up by the student and platform.

This is shown in Fig. 13-8b. I_p is the rotational inertia of student and platform with respect to the vertical axis, and ω_p is their angular speed about this axis.

If the student turns the wheel through an angle $\theta = 180°$, the student and platform acquire a vertical angular momentum of $2I_0\omega_0$. The total vertical angular momentum of the system is still being conserved at the initial value $I_0\omega_0$, as shown in Fig. 13-8c.

Consider now the angular momentum of the wheel alone. As the student turns the axis of the wheel through an angle θ he exerts a torque on it which lasts for the time Δt that it takes to reorient the shaft. The vertical component of the reaction to this "torque-impulse" acts on the student and accounts for the vertical angular momentum acquired by him and the platform.

The wheel, held with its shaft fixed at an angle θ with the vertical axis, precesses about this axis just like the top of Fig. 13-1. As for the top, a horizontal torque, which always remains at right angles to the plane defined by the vertical axis and the axis of the wheel, must be provided, in this case by the student.

The precise analysis of the motion of this system depends only on the application of the equation $\boldsymbol{\tau} = d\mathbf{L}/dt$ and the vector nature of the quantities involved. You might want to work this out in more detail, as an exercise.

13-5
SOME OTHER ASPECTS OF THE CONSERVATION OF ANGULAR MOMENTUM

The conservation of angular momentum principle holds in atomic and nuclear physics as well as in celestial and macroscopic regions. Because Newtonian mechanics does not hold in the atomic and nuclear domain, this conservation law must be more fundamental than Newtonian principles. In our derivation of this principle we must have made more rigid assumptions than we needed to. This is true even in the framework of classical mechanics. Note the key role played by Newton's third law in our deduction of this conservation principle. This law was used to justify the assumption that the sum of the internal torques was zero. It was necessary to assert not only that the action and reaction forces were equal and opposite (the "weak" form of the third law) but also that these forces were directed along the line joining the two particles (the "strong" form of the third law). The strong form is known to be violated in some electromagnetic interactions. However, the assumption that the sum of the internal torques in a system of particles is zero can be proven on the basis of a much less stringent requirement than that the third law should hold.*

The law of conservation of angular momentum, as we have formulated it, holds for a system of bodies whenever the bodies can be treated as particles, that is, whenever effects due to the rotation of the individual bodies can be neglected. When the individual bodies have rotation, the conservation of angular momentum principle is still valid, providing we include the angular momentum asso-

* See E. Gerjuoy, American Journal of Physics, Vol. 17, 477 (1949).

ciated with this rotation. However, the bodies then are no longer simple particles whose motion can be described by particle dynamics.

In atomic and nuclear physics we find that the "elementary particles" such as electrons, protons, mesons, neutrons, etc. (see Appendix I) have angular momentum associated with an intrinsic spinning motion, as well as with orbital motion about some external point. When we use the law of conservation of total angular momentum we must include this *spin* angular momentum in the total. A fundamental aspect of atomic, molecular, and nuclear systems is that their angular momenta can take on only definite discrete values, rather than a continuum of values. Angular momentum is said to be *quantized*. Hence, angular momentum plays a central role in the description of the behavior of such systems (see Problems 9 and 10). These ideas will be developed in later chapters.

If we were to regard the sun, planets, and satellites as particles having no intrinsic spinning motion, the angular momentum of the solar system would turn out not to be constant. But these bodies do have intrinsic rotations; in fact, tidal forces convert some of the intrinsic spinning angular momentum

Table 13-1
Summary of equations for rotary motion

Eq. No.	Equation	Remarks
	I. Defining Equations	
12-1	$\tau = \mathbf{r} \times \mathbf{F}$	Torque on a particle about a point O, due to a resultant force $\mathbf{F}$
	$\tau_{\text{ext}} = \Sigma\tau_i = \Sigma(\mathbf{r}_i \times \mathbf{F}_i)$	Resultant external torque on a system of particles about a point O
12-3	$\mathbf{l} = \mathbf{r} \times \mathbf{p}$	Angular momentum of a particle about a point O
	$\mathbf{L} = \Sigma\mathbf{l}_i = \Sigma(\mathbf{r}_i \times \mathbf{p}_i)$	Resultant angular momentum of a system of particles about a point O
	II. General Relations	
12-7	$\tau = d\mathbf{l}/dt$	The law of motion for a single particle acted on by a torque. It is the rotational analog of $\mathbf{F} = d\mathbf{p}/dt$ (Eq. 9-12). Equation 12-7 holds only if τ and $\mathbf{l}$ are measured with respect to any point O fixed in an inertial reference frame
12-9	$\tau_{\text{ext}} = d\mathbf{L}/dt$	The law of motion for a system of particles acted on by a resultant external torque τ_{ext}. It is the rotational analog of $\mathbf{F} = d\mathbf{P}/dt$ (Eq. 9-7). Equation 12-9 holds only if τ_{ext} and $\mathbf{L}$ are measured with respect to (a) any point O fixed in an inertial reference frame *or* (b) the center of mass of the system
	III. Special Case of a Rigid Body Rotating about an Axis Fixed in an Inertial Reference Frame (see footnote, p. 237)	
12-17	$\tau = I\alpha$	α is constrained to lie along the axis; I must also refer to this axis and τ must be the scalar component of τ_{ext} directed along this same axis. It is the rotational analog of $F = Ma$ for rectilinear motion
12-18	$L = I\omega$	ω is constrained to lie along the axis; I must also refer to this axis and L must be the scalar component along this axis of the total angular momentum. If the rotation axis has special symmetry (that is, if it is a principal axis; see footnote, p. 266), then $\mathbf{L}$ and ω are both axially directed. It is the rotational analog of $P = Mv$ for rectilinear motion

into orbital angular momentum of the planets and satellites. When we use the law of conservation of the angular momentum, we must include this spin angular momentum in the total. The conservation of angular momentum plays a key role in the evaluation of theories of the origin of the solar system, the contraction of giant stars, and other problems in astronomy. We will consider some astronomical applications in Chapter 16.

The basis for this rather simple way of analyzing the total angular momentum of atomic or astronomical systems is a theorem (see Problem 15) that the *total* angular momentum **L** of any system with respect to the origin of an inertial reference frame may be computed by adding the angular momentum with respect to its center of mass (*spin* angular momentum) to the angular momentum arising from the motion of the center of mass with respect to the origin (*orbital* angular momentum).

The conservation laws of total energy, of linear momentum, and of angular momentum are fundamental to physics, being valid in all modern physical theories. We shall have occasion to use them many times in later chapters.

The subject of the rotary motions of particles and rigid bodies is reasonably complicated, so much so in fact that a completely general treatment is beyond our scope here. It seems advisable to collect in one place all equations dealing with rotational dynamics and to comment on the conditions under which they can be used. We do this in Table 13-1 (see previous page.)

13-6
ROTATIONAL DYNAMICS —A REVIEW

questions

1. We have encountered many vector quantities so far, including position, displacement, velocity, acceleration, force, momentum, and angular momentum. Which of these are defined independent of the choice of the origin in the reference frame?

2. (a) In Example 1, why would merely turning the shaft up send the wheel to the student's right? (b) If the student is anchored to the floor of a large spaceship that is drifting in a region free from gravity, in what way, if any, would this affect her performance of the experiment?

3. If the top of Fig. 13-1 were not spinning, it would tip over. If its spin angular momentum is large compared to the change caused by the applied torque, the top precesses. What happens in between when the top spins slowly?

4. A Tippy-Top, having a section of a spherical surface of large radius on one end and a stem for spinning it on the opposite end, will rest on its spherical surface with no spin but flips over when spun, so as to stand on its stem. Explain. (See "The Tippy-Top" by George D. Freier, *The Physics Teacher*, January 1967.) If you can't find a Tippy-Top, use a hard-boiled egg; the "stand-on-end" behavior of the spinning egg is most easily followed if you put an ink mark on the "pointed" end of the egg.

5. A famous physicist (R. W. Wood), who was fond of practical jokes, mounted a rapidly spinning flywheel in a suitcase which he gave to a porter with instructions to follow him. What happens when the porter is led quickly around a corner? Explain in terms of $\tau = d\mathbf{L}/dt$.

6. A single-engine airplane must be "trimmed" to fly level. (Trimming consists of raising one aileron and lowering the opposite one.) Why is this necessary? Is this necessary on a twin-engine plane under normal circumstances?

7. The propeller of an aircraft rotates clockwise as seen from the rear. When the pilot pulls upward out of a steep dive, he finds it necessary to apply left rudder at the bottom of the dive if he is to maintain his heading. Explain.

8. Why does a long bar help a tightrope walker to keep her balance?

9. You are walking along a narrow rail and you start to lose your balance. If you start falling to the right, which way do you turn your body to regain balance? Explain.

10. Describe, in terms of $\tau = d\mathbf{L}/dt$, the rotational dynamics of the wheels on a fast train going around a curve.

11. Can you suggest a simple theory to explain the stability of a moving bicycle? (See "The Stability of the Bicycle" by David E. H. Jones, *Physics Today*, April 1970.)

12. Explain, in terms of angular momentum and rotational inertia, exactly how one "pumps up" a swing. (See "Pumping on a Swing" by P. L. Tea and H. Falk, *American Journal of Physics*, December 1968; "The Child's Swing" by B. F. Gore, *American Journal of Physics*, March 1970; "On Initiating the Motion in a Swing" by J. T. McMullan, *American Journal of Physics*, May 1972 and "How Children Swing" by S. M. Curry, *American Journal of Physics*, October 1976.)

13. In order to get a billiard ball to roll without sliding from the start the cue must hit the ball not at the center (that is, a height above the table equal to the ball's radius R) but exactly at a height $\frac{2}{5} R$ *above* the center. Explain. (See Arnold Sommerfeld, *Mechanics*, Volume I of Lectures on Theoretical Physics, Academic Press, New York (1964 paperback edition), pp. 158-161, for a supplement on the mechanics of billiards. See also "Some Pitfalls in Demonstrating Conservation of Momentum" by H. L. Armstrong, *American Journal of Physics*, January 1968.)

14. There are points on a bat where, if the ball is hit there, your hands will sting and the bat might break. Explain. (See "Batting the Ball" by P. Kirkpatrick, *American Journal of Physics*, August 1963.)

15. Assume that a uniform rod rests in a vertical position on a surface of negligible friction. The rod is then given a horizontal blow at its lower end. Describe the motion of the center of mass of the rod and of its upper endpoint.

16. A cylinder rotates with angular speed ω about an axis through one end, as in Fig. 13-9. Choose an appropriate origin and show qualitatively the vectors $\mathbf{L}$ and $\boldsymbol{\omega}$. Are these vectors parallel? Do symmetry considerations enter here?

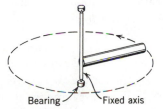

figure 13-9
Question 16

17. Consider the motion of a football tumbling through the air. Is the angular momentum with respect to the center of mass of the football conserved in flight? Does the magnitude or the direction of the angular velocity change with respect either to axes fixed in space or in the body?

18. In Chapter 1 the melting of the polar icecaps was cited as a possible cause of the variation in the earth's time of rotation. Explain.

19. Many great rivers flow toward the equator. What effect does the sediment they carry to the sea have on the rotation of the earth?

20. A man turns on a rotating table with an angular speed ω. He is holding two equal masses at arm's length. Without moving his arms, he drops the two masses. What change, if any, is there in his angular speed? Is the angular momentum conserved? Explain.

21. In Example 5, if the string is released suddenly back to where the object can travel in a circle of radius r_1, will the object return to its original speed? What happens if one repeatedly pulls down on and suddenly releases the string? Explain the behavior in terms of work-energy and torque-angular momentum considerations.

22. A circular turntable rotates at constant angular velocity about a vertical axis. There is no friction and no driving torque. A circular pan rests on the turntable and rotates with it: see Fig. 13-10. The bottom of the pan is covered with a layer of ice of uniform thickness, which is, of course, also rotating with the pan. The ice melts but none of the water escapes from the pan. Is the angular velocity now greater than, the same as, or less than the original velocity? Give reasons for your answer.

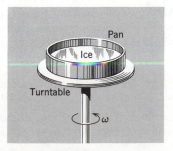

figure 13-10
Question 22

problems

SECTION 13-1

1. The time integral of a torque is called the *angular impulse*. Starting from $\tau = d\mathbf{L}/dt$, show that the resultant angular impulse equals the change in angular momentum. This is the rotational analog of the linear impulse-momentum theorem.

SECTION 13-2

2. A top is spinning at 30 Hz (cycles/s) about an axis making an angle of 30° with the vertical. Its mass is 0.50 kg (3.4×10^{-2} slug) and its rotational inertia is 5.0×10^{-4} kg · m² (3.7×10^{-4} slug · ft²). The center of mass is 4.0 cm (1.6 in.) from the pivot point. If the spin is clockwise as seen from above, what is the magnitude and direction of the angular velocity of precession?

3. A gyroscope consists of a rotating disc of 0.05-m radius suitably mounted at the center of a 0.12-m axle so that it can spin and precess freely. Find the rate of precession (in rev/min) if the axle is supported at one end and is horizontal, and the gyroscope's spin rate is 1000 rev/min.
Answer: 43 rev/min.

4. The gyroscope of Problem 3 is modified by attaching a small weight to the distant end of the axle. Find the new rate of precession (in rev/min) as a function of the ratio $r = $ (mass of added weight)/(mass of gyroscope disc).

SECTION 13-3

5. Start from Eq. 11-20b, $\mathbf{a}_R = \boldsymbol{\omega} \times \mathbf{v}$, for a particle in circular motion and show that the force required for uniform circular motion is $\mathbf{F} = \boldsymbol{\omega} \times \mathbf{p}$. Compare this to Eq. 13-2b, $\tau = \boldsymbol{\omega}_p \times \mathbf{L}$, and explain how the precessing spinning top can be regarded as a rotational analog to uniform circular motion.

6. Two wheels, *A* and *B*, are arranged by a belt system as in Fig. 13-11. The radius of *B* is three times the radius of *A*. What would be the ratio of the rotational inertias I_A/I_B if (a) both wheels have the same angular momenta? (b) both wheels have the same rotational kinetic energy?

7. Show that $\mathbf{L} = I\boldsymbol{\omega}$ for the two-particle system of Fig. 13-4.

8. Figure 13-12 shows a symmetrical rigid body rotating about a fixed axis. The origin of coordinates is fixed for convenience at the center of mass. Prove, by summing over the contributions made to the angular momentum by all of the mass elements m_i into which the body is divided, that $\mathbf{L} = I\boldsymbol{\omega}$, where $\mathbf{L}$ is the *total* angular momentum.

9. (a) Assume that the electron moves in a circular orbit about the proton in a hydrogen atom. If the centripetal force on the electron is supplied by an electrical force $e^2/4\pi\epsilon_0 r^2$, where e is the magnitude of the charge of an electron and of a proton, r is the orbit radius, and ϵ_0 is a constant, show that the radius of the orbit is

$$r = \frac{e^2}{4\pi\epsilon_0 m v^2},$$

where m is the mass of the electron and v is its speed.

   (b) Assume now that the angular momentum of the electron about the nucleus can only have values that are integral multiples n of $h/2\pi$, where h is a constant called *Planck's constant*. Show that the only electronic orbits possible are those with a radius

$$r = \frac{nh}{2\pi m v}.$$

   (c) Combine these results to eliminate v and show that the only orbits consistent with both requirements have radii

$$r = \frac{n^2 \epsilon_0 h^2}{\pi m e^2}.$$

Hence the allowed radii are proportional to the square of the integers $n = 1, 2, 3$, etc. When $n = 1$, r is smallest and has the value 0.528×10^{-10} meter.

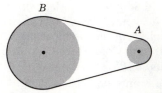

figure 13-11
Problem 6

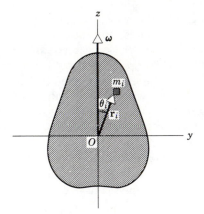

figure 13-12
Problem 8

10. In 1913, Niels Bohr postulated that any mechanical rotating system with rotational inertia I can have an angular momentum whose values can take on only integral multiples of a particular number $h/2\pi = 1.054 \times 10^{-34}$ J · s. In other words,

$$L = I\omega = n(h/2\pi),$$

where n is any positive integer or zero. We say that L is quantized because it is no longer allowed to have any value whatsoever. (a) Show that this postulate restricts the kinetic energy the rotating system can have to a set of discrete values, that is, that the energy is quantized. (b) Consider the so-called *rigid rotator*, consisting of a mass m constrained to rotate in a circle of radius R. With what angular speeds could the mass rotate if the postulate were correct? What values of kinetic energy may it assume? (c) Draw an energy-level diagram of some sort indicating how the spacing between the energy levels varies as n increases. It might look something like Fig. 13-13. Certain low-energy diatomic molecules behave like a rigid rotator.

figure 13-13
Problem 10

11. Using data in the appendix, find (a) the angular momentum of the earth's spin about its own axis, (b) the angular momentum of the earth's orbital motion about the sun.
    Answer: (a) 7.1×10^{33} kg · m²/s. (b) 2.7×10^{40} kg · m²/s.

12. A stick has a mass 0.30 slug (4.4 kg) and a length 4.0 ft (1.2 m). It is initially at rest on a frictionless horizontal plane and is struck perpendicularly by a horizontal impulsive force of impulse 3.0 lb · s (13 N · s) at a distance $l = 1.5$ feet (0.46 m) from the center. Determine the subsequent motion.

13. The moon revolves about the earth in such a way that we always see the same face of the moon. (a) How are the spin and orbital parts of the angular momentum of the moon related? (b) By how much would its spin angular momentum have to change if we were to be able to see all the moon's surface during the course of a month?
    Answer: (a) $L_{\text{spin}}/L_{\text{orbital}} = \frac{2}{5}(R_m/R_{e-m})^2$ in which R_m is the lunar radius and R_{e-m} is the earth-moon distance. (b) Increase or decrease by $\frac{1}{2}$ of present value.

14. A cylinder rolls down an inclined plane of angle θ. Show, by direct application of Eq. 12-9 ($\tau = dL/dt$), that the acceleration of its center of mass is $\frac{2}{3}g \sin \theta$. Compare this method with that used in Example 10 of Chapter 12.

15. *Relation between the Total Angular Momentum of a System of Particles and the Orbital and Spin Angular Momenta.* The total angular momentum of a system of particles relative to the origin O of an inertial reference frame is given by $\mathbf{L} = \Sigma \mathbf{r}_i \times \mathbf{p}_i$, where $\mathbf{r}_i$ and $\mathbf{p}_i$ are measured with respect to O.
    (a) Use the relations $\mathbf{r}_i = \mathbf{r}_{cm} + \mathbf{r}_i'$ and $\mathbf{p}_i = m_i\mathbf{v}_{cm} + \mathbf{p}_i'$ of Problem 10 of Chapter 12 to express $\mathbf{L}$ in terms of the positions $\mathbf{r}_i'$ and momenta $\mathbf{p}_i'$ relative to the center of mass C. (b) Use the definition of center of mass and the definition of angular momentum $\mathbf{L}'$ with respect to the center of mass (problem 10 of Chapter 12) to obtain $\mathbf{L} = \mathbf{L}' + \mathbf{r}_{cm} \times M\mathbf{v}_{cm}$. (c) Show how this result can be interpreted as regarding the total angular momentum to be the sum of spin angular momentum (angular momentum relative to the center of mass) and orbital angular momentum (angular momentum of the motion of the center of mass C with respect to O if all the system's mass were concentrated at C).

16. A thin rectangular sheet, of length a and width b, rotates about one of its diagonals with constant angular speed ω, the axis being fixed in an inertial reference frame. Find the direction and the magnitude of the angular momentum $\mathbf{L}$ with respect to an origin at the center of mass.

17. The axis of the cylinder in Fig. 13-14 is fixed. The cylinder is initially at rest. The block of mass M is initially moving to the right without friction with speed v_1. It passes over the cylinder to the dotted position. When it first makes contact with the cylinder, it slips on the cylinder, but the friction is large enough so that slipping ceases before M loses contact with the

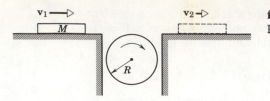

figure 13-14
Problem 17

cylinder. The cylinder has a radius R and a rotational inertia I. Find the final speed v_2 in terms of v_1, M, I, and R. This can be done most easily by using the relation between impulse and change in momentum.
Answer: $v_1/(1 + I/MR^2)$.

18. A stick, length l, lies on a frictionless horizontal table. It has a mass M and is free to move in any way on the table. A hockey puck m, moving as shown in Fig. 13-15, with speed v collides elastically with the stick. (a) What quantities are conserved in the collision? (b) What must be the mass m of the puck so that it remains at rest immediately after the collision?

19. At what point below the suspension at one end of a uniform rod of length $2\,L$, hanging vertically, should you strike it to start its oscillatory motion without imparting an initial horizontal reaction force at the point of suspension? *Answer:* 4 $L/3$.

20. Two cylinders having radii R_1 and R_2 and rotational inertias I_1 and I_2, respectively, are supported by fixed axes perpendicular to the plane of Fig. 13-16. The large cylinder is initially rotating with angular velocity ω_0. The small cylinder is moved to the right until it touches the large cylinder and is caused to rotate by the frictional force between the two. Eventually, slipping ceases, and the two cylinders rotate at constant rates in opposite directions. (a) Find the final angular velocity ω_2 of the small cylinder in terms of I_1, I_2, R_1, R_2, and ω_0. (b) Is total angular momentum conserved in this case?

21. A billiard ball, initially at rest, is given a sharp impulse by a cue. The cue is held horizontally a distance h above the centerline as in Fig. 13-17. The ball leaves the cue with a speed v_0 and, because of its "forward english," eventually acquires a final speed of $\frac{9}{7}v_0$. Show that

$$h = \tfrac{4}{5}R,$$

where R is the radius of the ball.

22. In Problem 21, imagine **F** to be applied below the centerline. (a) Show that it is impossible, with this "reverse english," to reduce the forward speed to zero, without rolling having set in, unless $h = R$. (b) Show that it is impossible to give the ball a backward velocity unless **F** has a downward, vertical component.

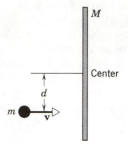

figure 13-15
Problem 18

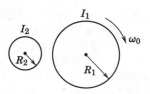

figure 13-16
Problem 20

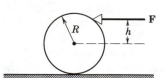

figure 13-17
Problem 21

SECTION 13-4

23. A man stands on a frictionless rotating platform which is rotating with an angular speed of 1.0 rev/s (Hz); his arms are outstretched and he holds a weight in each hand. With his hands in this position the total rotational inertia of the man, the weights, and the platform is 6.0 kg · m². If by drawing in the weights the man decreases the rotational inertia to 2.0 kg · m², (a) what is the resulting angular speed of the platform? (b) By how much is the kinetic energy increased? *Answer:* (a) 3.0 Hz. (b) By a factor of 3.

24. Two skaters, each of mass 50 kg, approach each other along parallel paths separated by 3.0 m. They have equal and opposite velocities of 10 m/s. The first skater carries a long light pole, 3.0 m long, and the second skater grabs the end of it as he passes. (Assume frictionless ice.) (a) Describe quantitatively the motion of the skaters after they are connected by the pole. (b) By pulling on the pole, the skaters reduce their distance apart to 1.0 m. What is their motion then? (c) Compare the kinetic energy of the system in parts (a) and (b). Where does the change come from?

25. A wheel is rotating with an angular speed of 800 rev/min on a shaft whose rotational inertia is negligible. A second wheel, initially at rest and with twice the rotational inertia of the first, is suddenly coupled to the same shaft. (a) What is the angular speed of the resultant combination of the shaft and two wheels? (b) Account for any changes in rotational kinetic energy experienced by this system.
   Answer: (a) 267 rev/min. (b) The system loses two-thirds of its kinetic energy.

26. With center and spokes of negligible mass, a certain bicycle wheel has a thin rim of radius 1.14 ft and weight 8.36 lb; it can turn on its axle with negligible friction. A man holds the wheel above his head with the axis vertical while he stands on a turntable free to rotate without friction; the wheel rotates clockwise, as seen from above, with an angular speed 57.7 rad/s, and the turntable is initially at rest. The rotational inertia of wheel-plus-man-plus-turntable about the common axis of rotation is 1.54 slug · ft². (a) The man's hand suddenly stops the rotation of the wheel (relative to the turntable). Determine the resulting angular velocity (magnitude and direction) of the system. (b) The experiment is repeated with noticeable friction introduced into the axle of the wheel, which, starting from the same initial angular speed (57.7 rad/s) gradually comes to rest (relative to the turntable) while the man holds the wheel as described above. (The turntable is still free to rotate without friction.) Describe what happens to the system, giving as much quantitative information as the data permit.

27. The rotor of an electric motor has a rotational inertia $I_m = 2 \times 10^{-3}$ kg · m² about its central axis. The motor is mounted parallel to the axis of a space probe having a rotational inertia $I_p = 12$ kg · m² about its axis. Calculate the number of revolutions required to turn the probe through 30° about its axis.
   Answer: 500 rev.

28. In a lecture demonstration, a toy train track is mounted on a large wheel that is free to turn with negligible friction about a vertical axis. A toy train of mass m is placed on the track and, with the system initially at rest, the electrical power is turned on. The train reaches a steady speed v with respect to the track. What is the angular velocity ω of the wheel, if its mass is M and its radius R? (Neglect the mass of the spokes of the wheel.)

29. A girl (mass M) stands on the edge of a frictionless merry-go-round (mass $10\,M$, radius R, rotational inertia I) that is not moving. She throws a rock (mass m) in a horizontal direction that is tangent to the outer edge of the merry-go-round. The speed of the rock, relative to the ground, is v. What are (a) the angular speed of the merry-go-round and (b) the linear speed of the girl after the rock is thrown?
   Answer: (a) $mvR/(I + MR^2)$. (b) $vmR^2/(I + mR^2)$.

30. In a playground there is a small merry-go-round of radius 4.0 ft (1.2 m) and mass 12.0 slugs (180 kg). The radius of gyration (see Problem 14 of Chapter 12) is 3.0 ft (0.91 m). A child of mass 3.0 slugs (44 kg) runs at a speed of 10 ft/s (3.0 m/s) tangent to the rim of the merry-go-round when it is at rest and then jumps on. Neglect friction between the bearings and the shaft of the merry-go-round and find the angular velocity of the merry-go-round and child.

31. A uniform flat disk of mass M and radius R rotates about a horizontal axis through its center with angular speed ω_0. (a) What is its kinetic energy? Its angular momentum? (b) A chip of mass m breaks off the edge of the disk at an instant such that the chip rises vertically above the point at which it broke off (Fig. 13-18). How high above the point does it rise before starting to fall? (c) What is the final angular speed of the broken disk? The final angular momentum and energy?
   Answer: (a) $MR^2\omega_0^2/4$; $MR^2\omega_0/2$. (b) $R^2\omega_0^2/2$ g. (c) ω_0; $(M/2 - m)R^2\omega_0$; $(M/2 - m)R^2\omega_0^2/2$.

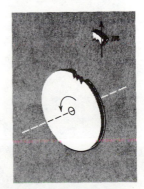

figure 13-18
Problem 31

32. A cockroach, mass m, runs counterclockwise around the rim of a lazy Susan (a circular dish mounted on a vertical axle) of radius R and rotational

inertia I with frictionless bearings. The cockroach's speed (relative to the earth) is v, whereas the lazy Susan turns clockwise with angular speed ω_0. The cockroach finds a bread crumb on the rim and of course, stops. (a) What is the angular speed of the lazy Susan after the cockroach stops? (b) Is energy conserved?

33. In Example 5 compare the kinetic energies of the object in two different orbits. Use the work-energy theorem to explain the difference quantitatively.

34. A particle is projected horizontally along the interior of a smooth hemispherical bowl of radius r which is kept at rest (Fig. 13-19). We wish to find the initial speed v_0 required for the particle to just reach the top of the bowl. Find v_0 as a function of θ_0, the initial angular position of the particle. (*Hint:* Use conservation principles.)

35. On a large horizontal frictionless circular track, radius R, lie two small masses m and M, free to slide on the track. Between the two masses is squeezed a spring which, however, is not fastened to m and M. The two masses are held together by a string. (a) If the string breaks, the compressed spring (assumed massless) shoots off the two masses in opposite directions; the spring itself is left behind. The balls collide when they again meet on the track (Fig. 13-20). Where does this collision take place? (You might find it convenient to express the answer in terms of the angle M travels through.) (b) If the potential energy initially stored in the spring was U_0, what is the time it takes after the string breaks for the collision to take place? (c) Assuming the collision to be perfectly elastic and head-on, where will the balls again collide after the first collision?

　　Answer: (a) $2\pi m/(m + M)$ rad. (b) $[2\pi^2 mMR^2/(m + M)U_0]^{1/2}$. (c) At the point of origin.

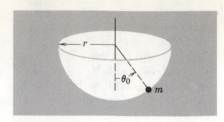

figure 13-19
Problem 34

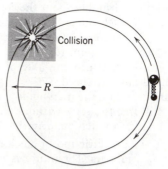

figure 13-20
Problem 35

14
equilibrium of rigid bodies

The towers supporting a suspension bridge must be strong enough so that they do not collapse under the weight of the bridge and its traffic load; the landing gear of an aircraft must not collapse if the pilot makes a poor landing; the tines of a fork must not bend when we cut a tough steak. In all such problems the engineer is concerned that these presumed rigid structures do indeed remain rigid under the forces, and the associated torques, that act on them.

In such problems the engineer must ask two questions. (1) What forces and torques act on the presumed rigid body? (2) Considering its design and the materials used, will the body remain rigid under the action of these forces and torques? In this chapter we are concerned only with the first of these questions; the engineering student will deal at length with the second question in later courses.

We note that the presumed rigid bodies of the preceding section (that is, the bridge towers, the landing gear, and the fork) are in *mechanical equilibrium*. A rigid body is in mechanical equilibrium if, as viewed from an inertial reference frame, (1) the linear acceleration $\mathbf{a}_{cm}$ of its center of mass is zero and (2) its angular acceleration $\boldsymbol{\alpha}$ about any axis fixed in this reference frame is zero.

This definition does not require the body to be at rest with respect to the observer but only to be unaccelerated. Its center of mass, for example, may be moving with constant velocity $\mathbf{v}_{cm}$ and the body may be rotating about a fixed axis with constant angular velocity $\boldsymbol{\omega}$. If the body is actually at rest (so that $\mathbf{v}_{cm} = 0$ and $\boldsymbol{\omega} = 0$), we often speak of *static equilibrium*, the central subject of this chapter. However, as we shall see, the restrictions imposed on the forces and torques are the same whether or not the equilibrium is static. Furthermore, we can transform any case of (nonstatic) equilibrium to one of static equilibrium by choosing an appropriate new reference frame.

The translational motion of a rigid body of mass M is governed by Eq. 9-10, or

$$\mathbf{F}_{ext} = M\mathbf{a}_{cm},$$

in which $\mathbf{F}_{ext}$ is the vector sum of all the external forces acting on the body. Because $\mathbf{a}_{cm}$ must be zero for equilibrium, the first condition of equilibrium (static or otherwise) is: *The vector sum of all the external forces acting on a body in equilibrium must be zero.*

We can write condition (1) as

$$\mathbf{F} = \mathbf{F}_1 + \mathbf{F}_2 + \cdots = 0, \tag{14-1}$$

in which we have dropped the subscript on $\mathbf{F}_{ext}$ for convenience. This vector equation leads to three scalar equations.

$$F_x = F_{1x} + F_{2x} + \cdots = 0,$$
$$F_y = F_{1y} + F_{2y} + \cdots = 0, \tag{14-2}$$
$$F_z = F_{1z} + F_{2z} + \cdots = 0,$$

which state that the sum of the components of the forces along each of any three mutually perpendicular directions is zero.

The second requirement for equilibrium is that $\boldsymbol{\alpha} = 0$ for any axis. Since the angular acceleration of a rigid body is associated with torque — recall that $\tau = I\alpha$ for a fixed axis — we can state this second condition of equilibrium (static or otherwise) as: *The vector sum of all the external torques acting on a body in equilibrium must be zero.*

We can write condition (2) as

$$\boldsymbol{\tau} = \boldsymbol{\tau}_1 + \boldsymbol{\tau}_2 + \cdots = 0. \tag{14-3}$$

This vector equation leads to three scalar equations

$$\tau_x = \tau_{1x} + \tau_{2x} + \cdots = 0,$$
$$\tau_y = \tau_{1y} + \tau_{2y} + \cdots = 0, \tag{14-4}$$
$$\tau_z = \tau_{1z} + \tau_{2z} + \cdots = 0,$$

which state that, at equilibrium, the sum of the components of the torques acting on a body, along each of any three mutually perpendicular directions, is zero.

The resultant torque τ in Eq. 14-3, which must be zero for mechanical equilibrium, is defined with respect to a particular origin O. The quantities τ_x, τ_y, and τ_z in Eq. 14-4 are the scalar components of τ and refer to any set of three mutually perpendicular axes whose origin is at O, no matter how these axes are oriented in space. This follows because, if a vector is zero, its scalar components must be zero no matter how we orient the axes of the reference frame. You may wonder whether the choice of an origin is essential. The answer — as we shall show below — is that it is not, because (for a body in translational equilibrium), if $\tau = 0$ for any single origin O it is also zero for any other origin in the reference frame. The substance of this paragraph then is that condition 2 is satisfied for a body in translational equilibrium if we can show either that (a) $\tau = 0$ with respect to *any* one point (Eq. 14-3) or that (b) the torque components along *any* three mutually perpendicular axes are zero (Eq. 14-4).

Let us now assume that we have a rigid body in translational equilibrium, so that $\mathbf{F} = \mathbf{F}_1 + \mathbf{F}_2 + \cdots = 0$ (Eq. 14-1). We now wish to show that the torque about *any* point (such as P in Fig. 14-1) will be zero if the torque about one particular point (such as O in Fig. 14-1) is zero. The figure shows three of the n forces, $\mathbf{F}_1$, $\mathbf{F}_2 \cdots \mathbf{F}_n$, applied at various points on a rigid body and pointing in various directions. The points of application with respect to O are identified by displacement vectors, of which $\mathbf{r}_1$ is an example. The arbitrary point P is identified by displacement vector $\mathbf{r}_P$; the vector $\mathbf{r}_1 - \mathbf{r}_P$ locates the point of application of $\mathbf{F}_1$ with respect to point P.

We can write for the resultant torque about O (see Eq. 12-1)

$$\tau_O = \mathbf{r}_1 \times \mathbf{F}_1 + \mathbf{r}_2 \times \mathbf{F}_2 + \cdots + \mathbf{r}_n \times \mathbf{F}_n$$

and for the torque about P,

$$\tau_P = (\mathbf{r}_1 - \mathbf{r}_P) \times \mathbf{F}_1 + (\mathbf{r}_2 - \mathbf{r}_P) \times \mathbf{F}_2 + \cdots + (\mathbf{r}_n - \mathbf{r}_P) \times \mathbf{F}_n.$$

We can expand the latter equation as

$$\tau_P = [\mathbf{r}_1 \times \mathbf{F}_1 + \mathbf{r}_2 \times \mathbf{F}_2 + \cdots + \mathbf{r}_n \times \mathbf{F}_n] - [\mathbf{r}_P \times (\mathbf{F}_1 + \mathbf{F}_2 + \cdots + \mathbf{F}_n)].$$

Now if, as we have assumed, the first condition of equilibrium is satisfied, then $\mathbf{F}_1 + \mathbf{F}_2 + \cdots + \mathbf{F}_n = 0$ and the second term above in the square brackets vanishes. The first term in the square brackets is simply τ_O (see above) so that, under these conditions

$$\tau_P = \tau_O.$$

Thus, for a body in translational equilibrium, if $\tau_O = 0$, then $\tau_P = 0$, where P is an arbitrary point.

Hence we have *six independent conditions* on our forces for a body to be in equilibrium. These conditions are the six algebraic relations of Eqs. 14-2 and 14-4. These six conditions are a condition on each of the six degrees of freedom of a rigid body, three translational and three rotational.

Often we deal with problems in which all the forces lie in a plane. Then we have only three conditions on our forces: The sum of their components must be zero for each of any two perpendicular directions in the plane, and the sum of their torques about any one axis perpendicular to the plane must be zero. These conditions correspond to the three degrees of freedom for motion in a plane, two of translation and one of rotation.

We shall limit ourselves henceforth mostly to planar problems in order to simplify the calculations. This does not impose any fundamental restriction on the general principles. Also, as a matter of convenience, we shall consider only the case of static equilibrium, in which bodies are actually at rest in our chosen inertial reference frame.

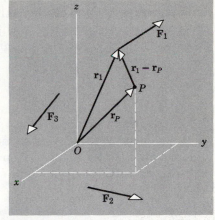

figure 14-1
We display three of the n forces, $\mathbf{F}_1$, $\mathbf{F}_2$, $\mathbf{F}_3$ ... $\mathbf{F}_n$, that act on a rigid body, not shown. In the text we show that if $\tau = 0$ for point O it also vanishes for any point such as P, assuming that the body is in translational equilibrium.

14-3
CENTER OF GRAVITY

One of the forces encountered in rigid-body motions is the force of gravity. Actually, for an extended body, this is not just one force but the resultant of a great many forces. Each particle in the body is acted on by a gravitational force. If the body of mass M is imagined to be divided into a large number of particles, say n, the gravitational force exerted by the earth on the ith particle of mass m_i is $m_i\mathbf{g}$. This force is directed down toward the earth. If the acceleration due to gravity $\mathbf{g}$ is the same everywhere in a region, we say that a uniform gravitational field exists there; that is, $\mathbf{g}$ has the same magnitude and direction everywhere in that region. For a rigid body in a uniform gravitational field, **g** must be the same for each particle in the body and the weight forces on the particles must be parallel to one another. If we assume that the earth's gravitational field is uniform, we can show that all the individual weight forces acting on a body can be replaced by a single force $M\mathbf{g}$ acting down at the center of mass of the body. This is equivalent to showing that the individual weight forces, acting downward, can be counteracted in their accelerating effects by a single force **F** $(= -M\mathbf{g})$ acting upward, *provided this force* **F** *is applied at the center of mass of the body.*

Figure 14-2 shows two typical particles or mass elements m_1 and m_2, selected from the n such elements into which the rigid body has been divided. An upward force **F** $(= -M\mathbf{g})$ is applied at a certain point O. It remains to show that the body is in mechanical equilibrium *if* (and only if) point O is the center of mass. Condition 1 for equilibrium (Eq. 14-1) has already been satisfied by our choice of the magnitude and direction of **F**. That is,

$$\mathbf{F} + m_1\mathbf{g} + m_2\mathbf{g} + \cdots + m_n\mathbf{g} = 0,$$

or

$$\mathbf{F} = -(m_1 + m_2 + \cdots + m_n)\mathbf{g} = -M\mathbf{g},$$

which corresponds to our assumption.

It remains to prove that $\tau = 0$ for any single point in the body, such as O. This is the second condition for equilibrium. By choosing O as our origin we insure that the torque of **F** about this point is zero, because the moment arm of **F** is zero for this point. The torque about O due to the gravitational pull on the mass elements is

$$\boldsymbol{\tau} = \mathbf{r}_1 \times m_1\mathbf{g} + \mathbf{r}_2 \times m_2\mathbf{g} + \cdots + \mathbf{r}_n \times m_n\mathbf{g}$$

which (because m_1, m_2, etc., are scalars) we can write as

$$\boldsymbol{\tau} = m_1\mathbf{r}_1 \times \mathbf{g} + m_2\mathbf{r}_2 \times \mathbf{g} + \cdots + m_n\mathbf{r}_n \times \mathbf{g}.$$

Because **g** is the same in each term, we can factor it out, obtaining

$$\boldsymbol{\tau} = (m_1\mathbf{r}_1 + m_2\mathbf{r}_2 + \cdots + m_n\mathbf{r}_n) \times \mathbf{g}$$

$$= \left(\sum_1^n m_i\mathbf{r}_i \right) \times \mathbf{g},$$

in which the sum is taken over all the mass elements that make up the body.

Now *if* point O is the center of mass of the body, the sum above is zero. This follows from the definition of the center of mass (see Eq. 9-5b and the discussion following it). We conclude then that *if* point O is the center of mass, then $\tau = 0$ and the second condition for mechanical equilibrium is satisfied.*

Thus the gravitational forces acting on the individual mass elements that make up a rigid body are equivalent in their translational and rotational effects to a single force equal to $M\mathbf{g}$, the total weight of the body, acting at the center of mass. The point of application of the equivalent resultant gravitational force is often called the *center of gravity.*

The fact that the center of gravity and the center of mass coincided came about only because we assumed that the earth's gravitational field **g** was the same for all parts of the rigid body. Actually this assumption is not strictly true,

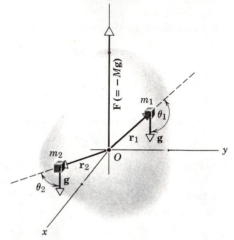

figure 14-2
An irregular body is divided into n mass elements of which two typical elements m_1 and m_2 are shown. In the text we prove that, if the gravitational field is uniform, the body can be held in translational and rotational equilibrium by a single force $\mathbf{F} = (-M\mathbf{g})$ directed upward and applied at the center of mass of the body.

---

* A body can be suspended in rotational equilibrium from *any* point; see Fig. 14-4. However, the center of mass is the *only* point that gives $= 0$ for all rotational orientations of the body.

for the magnitude of **g** changes with distance from the center of the earth and furthermore the direction of **g** is radially in toward the center of the earth from any point (Chapter 16). To see the effect this has, let us consider a very long uniform stick inclined to the vertical in the earth's gravitational field, as in Fig. 14-3. The center of gravity of a body is the point at which the equivalent resultant gravitational force on it acts. This point must be the same as the point at which a single oppositely directed force is applied for the body to be kept in equilibrium. If the field were uniform, a single upward force of magnitude Mg at the center of mass would keep the stick in translational and rotational equilibrium. But the field is not uniform, and the value of g at m_1 is less than the value of g at m_n. The point at which a single force must be applied to keep the body in equilibrium is therefore at a point P some distance below the center of mass. Furthermore, if the orientation of the body is changed, the position of the point P, required for application of an equilibrium force, changes. Hence center of gravity really has little usefulness in such a case. Not only does it not coincide with the center of mass, but its position changes with respect to the body as the body is moved.

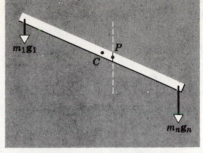

figure 14-3
The center of mass C and center of gravity P in reality do not coincide, since the earth's gravitational field is not uniform.

Because almost all problems in mechanics involve objects having dimensions small compared to the distances over which **g** changes appreciably, we can assume that **g** is uniform over the body. The center of mass and the center of gravity can then be taken as the same point. In fact, we can use this coincidence to determine experimentally the center of mass in irregularly shaped objects. For example, let us locate the center of mass of a thin plate of irregular shape, as in Fig. 14-4. We suspend the body by a cord from some point A on its edge. When the body is at rest, the center of gravity must lie directly under the point of support somewhere on the line Aa, for only then can the torque resulting from the cord and the weight add to zero. We next suspend the body from another point B on its edge. Again, the center of gravity must lie somewhere on Bb. The only point common to the lines Aa and Bb is O, the point of intersection, so that this point must be the center of gravity. If now we suspend the body from any other point on its edge, as C, the vertical line Cc will pass through O. Since we have assumed a uniform field, the center of gravity coincides with the center of mass, which is therefore located at O.

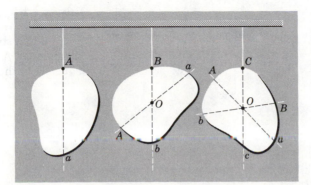

figure 14-4
Since the center of mass O always hangs directly below the point of suspension, hanging a plate from two different points determines O.

In applying the conditions for equilibrium (zero resultant force and zero resultant torque about any axis), we can clarify and simplify the procedure by proceeding as follows.

First, we draw an imaginary boundary around the system under consideration. This assures that we see clearly just what body or system of

14-4
EXAMPLES OF EQUILIBRIUM

bodies it is to which we are applying the laws of equilibrium. This process is called *isolating the system.*

Second, we draw vectors representing the magnitude, direction, and point of application of all *external* forces. An external force is one that acts from outside the boundary which was drawn earlier. Examples of external forces often encountered are gravitational forces and forces transmitted by strings, wires, rods, and beams which cross the boundary. A question sometimes arises about the direction of a force. In this case make an imaginary cut through the member transmitting the force at the point where it crosses the boundary. If the ends of this cut tend to pull apart, the force acts outward. If you are in doubt, choose the direction arbitrarily. A negative value for a force in the solution means that the force acts in the direction opposite to that assumed. Note that only external forces acting on the system need be considered; all internal forces cancel one another in pairs.

Third, we choose a convenient coordinate system along whose axes we resolve the external forces before applying the first condition of equilibrium (Eq. 14-2). The object here is to simplify the calculations. The preferable coordinate system is usually obvious.

Fourth, we choose a convenient coordinate system along whose axes we resolve the external torques before applying the second condition of equilibrium (Eq. 14-4). The object again is to simplify calculations and we may use different coordinate systems in applying the two conditions for static equilibrium if this proves to be convenient. Suppose that an axis passes through the point at which two forces concur and is at right angles to the plane formed by these forces; these forces will automatically have no torque component along (or about) this axis. The torque components resulting from all external forces must be zero about any axis for equilibrium. Internal torques will cancel in pairs and need not be considered.

(a) A uniform steel meter bar rests on two scales at its ends (Fig. 14-5). The bar weighs 4.0 lb. Find the readings on the scales.

EXAMPLE 1

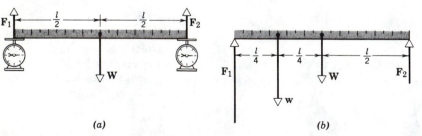

(a)   (b)

figure 14-5
(a) Example 1a. A uniform steel bar rests on two spring scales. (b) Example 1b. A weight is suspended a quarter of the way from one end.

Our system is the bar. The forces acting on the bar are **W**, the gravitational force acting down at the center of gravity, and $\mathbf{F}_1$ and $\mathbf{F}_2$, the forces exerted upward on the bar at its ends by the scales. These are shown in Fig. 14-5a. By Newton's third law, the force exerted by a scale on the bar is equal and opposite to that exerted by the bar on the scale. Therefore, to obtain the readings on the scales, we must determine the magnitudes of $\mathbf{F}_1$ and $\mathbf{F}_2$.

For translational equilibrium (Eq. 14-1) the condition is

$$\mathbf{F}_1 + \mathbf{F}_2 + \mathbf{W} = 0.$$

All forces act vertically, so that if we choose the y-axis to be vertical, no other axes need be considered. Then we get the scalar equation

$$F_1 + F_2 - 4.0 \text{ lb} = 0.$$

For rotational equilibrium, the component of the resultant torque on the bar must be zero about *any* axis. We have seen that it is enough to show that the torque components are zero for any set of three mutually perpendicular axes. These components are certainly zero for any two perpendicular axes that lie in the plane of Fig. 14-5*a* (Why?). It remains to require that the resultant torque is zero about any one axis at right angles to the plane of the figure. Let us choose an axis through the center of gravity. Then, taking clockwise rotation as positive and counterclockwise rotation as negative, the condition for rotational equilibrium (Eq. 14-4) is

$$F_1\left(\frac{l}{2}\right) - F_2\left(\frac{l}{2}\right) + W(0) = 0,$$

or

$$F_1 - F_2 = 0.$$

Combining the two equations, we obtain

$$F_1 + F_2 = 2F_1 = 2F_2 = 4.0 \text{ lb},$$

$$F_1 = F_2 = 2.0 \text{ lb}.$$

Each scale reads 2.0 lb, as we might have expected.

If we had chosen an axis through one end of the bar, we would have obtained the same result. For example, taking torques about an axis through the right end, we obtain

$$F_1(l) - W\left(\frac{l}{2}\right) + F_2(0) = 0,$$

or

$$F_1 = \frac{W}{2} = \frac{4.0 \text{ lb}}{2} = 2.0 \text{ lb}.$$

Combining this with $F_1 + F_2 = 4.0$ lb, we obtain $F_2 = 2.0$ lb, as before.

(*b*) Suppose that a 6.0-lb block is placed at the 25-cm mark on the meter bar. What do the scales read now?

The external forces acting on the bar are shown in Fig. 14-5*b*, where **w** is the force exerted on the bar by the block. The first condition for equilibrium is

$$F_1 + F_2 - W - w = 0.$$

With $W = 4.0$ lb and $w = 6.0$ lb, we obtain

$$F_1 + F_2 = 10 \text{ lb}.$$

If we take an axis through the left end of the bar, the second condition for equilibrium is

$$w\left(\frac{l}{4}\right) + W\left(\frac{l}{2}\right) - F_2(l) = 0.$$

With $W = 4.0$ lb and $w = 6.0$ lb, we obtain

$$F_2 = 3.5 \text{ lb}.$$

Putting this result into the first equation, we obtain

$$F_1 + 3.5 \text{ lb} = 10 \text{ lb},$$

$$F_1 = 6.5 \text{ lb}.$$

The left-hand scale reads 6.5 lb and the right-hand scale reads 3.5 lb at equilibrium.

Why do we obtain only two conditions on the forces in this problem rather than the three conditions expected for problems in which all forces lie in the same plane?

(a) A 60-ft ladder weighing 100 lb rests against a wall at a point 48 ft above the ground. The center of gravity of the ladder is one-third the way up. A 160-lb man climbs halfway up the ladder. Assuming that the wall (but not the ground) is frictionless, find the forces exerted by the system on the ground and on the wall.

EXAMPLE 2

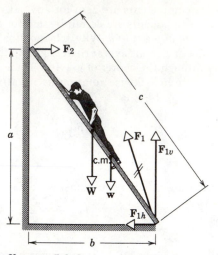

The forces acting on the ladder are shown in Fig. 14-6. **W** is the weight of the man standing on the ladder and **w** is the weight of the ladder itself. A force $\mathbf{F}_1$ is exerted by the ground on the ladder. $\mathbf{F}_{1v}$ is the vertical component and $\mathbf{F}_{1h}$ is the horizontal component of this force (due to friction). The wall, being frictionless, can exert only a force normal to its surface, called $\mathbf{F}_2$. We are given the following data:

$$W = 160 \text{ lb}, \qquad w = 100 \text{ lb},$$
$$a = 48 \text{ ft}, \qquad c = 60 \text{ ft}.$$

From the geometry we conclude that $b = 36$ ft. The line of action of **W** intersects the ground at a distance $b/2$ from the wall and the line of action of **w** intersects the ground at a distance $2b/3$ from the wall.

We choose the x-axis to be along the ground and the y-axis along the wall. Then, the conditions on the forces for translational equilibrium (Eq. 14-2) are

$$F_2 - F_{1h} = 0,$$

$$F_{1v} - W - w = 0.$$

For rotational equilibrium (Eq. 14-4) choose an axis through the point of contact with the ground and obtain

figure 14-6
Example 2.

$$F_2(a) - W\left(\frac{b}{2}\right) - w\left(\frac{b}{3}\right) = 0.$$

Using the data given, we obtain

$$F_2(48 \text{ ft}) - (160 \text{ lb})(18 \text{ ft}) - (100 \text{ lb})(12 \text{ ft}) = 0,$$

$$F_2 = 85 \text{ lb},$$

$$F_{1h} = F_2 = 85 \text{ lb},$$

$$F_{1v} = 160 \text{ lb} + 100 \text{ lb} = 260 \text{ lb}.$$

By Newton's third law the forces exerted by the ground and the wall on the ladder are equal but opposite to the forces exerted by the ladder on the ground and the wall, respectively. Therefore, the normal force on the wall is 85 lb, and the force on the ground has components of 260 lb down and 85 lb to the right.

(b) If the coefficient of static friction between the ground and the ladder is $\mu_s = 0.40$, how high up the ladder can the man go before it starts to slip?

Let x be the fraction of the total length of the ladder the man can climb before slipping begins. Then our equilibrium conditions are

$$F_2 - F_{1h} = 0,$$

$$F_{1v} - W - w = 0,$$

and

$$F_2 a - Wbx - w\left(\frac{b}{3}\right) = 0.$$

Now we obtain

$$F_2(48 \text{ ft}) = (160 \text{ lb})(36 \text{ ft})x + (100 \text{ lb})(12 \text{ ft}),$$

$$F_2 = (120x + 25) \text{ lb}.$$

Hence

$$F_{1h} = (120x + 25) \text{ lb},$$

and, as before,

$$F_{1v} = 260 \text{ lb}.$$

The maximum force of static friction is given by

$$F_{1h} = \mu_s F_{1v} = (0.40)(260\ lb) = 104\ lb.$$

Therefore,

$$F_{1h} = (120x + 25)\ lb = 104\ lb$$

and

$$x = \tfrac{79}{120},$$

so that the man can climb up the ladder

$$60x\ ft = 39.5\ ft$$

before slipping begins.

In this example the ladder is treated as a one-dimensional object, with only one point of contact at the wall and ground. You should reflect on how this limits consideration of the less artificial case of two contact points at each end.

The reason for assuming that the wall is frictionless is discussed later in this section. Can you guess what it is?

EXAMPLE 3

A uniform beam is hinged at the wall. A wire connected to the wall a distance d above the hinge is attached to the other end of the beam. The beam makes an angle of 30° with the horizontal when a weight w is hung from a string fastened to the end of the beam. If the beam has a weight W and a length l, find the tension in the wire and the forces exerted by the hinge on the beam.

The situation is depicted in Fig. 14-7, in which all the forces acting on the beam are shown. The wire pulling on the beam makes some angle α with the horizontal so that the tension $\mathbf{T}$ in the wire has horizontal and vertical components $\mathbf{T}_h$ and $\mathbf{T}_v$, respectively, as shown. The force $\mathbf{F}$ exerted by the hinge on the beam also has horizontal and vertical components $\mathbf{F}_h$ and $\mathbf{F}_v$, respectively. $\mathbf{W}$ is the weight of the beam, acting at its center of gravity, and $\mathbf{w}$ is the tension in the string that transmits the weight of the suspended body to the beam.

Choosing our axes to be horizontal and vertical, we obtain for translational equilibrium

$$F_v + T_v - W - w = 0,$$

and

$$F_h - T_h = 0.$$

Choosing an axis through the point of intersection of $\mathbf{T}$ and $\mathbf{w}$ (Why?), we obtain for rotational equilibrium

$$F_v(l\ \cos 30°) - F_h(l\ \sin 30°) - \frac{W(l\ \cos 30°)}{2} = 0.$$

Our unknowns are T_h, T_v, F_h, and F_v. Let us assign the following values to the other quantities:

$$W = 60\ N, \qquad w = 40\ N, \qquad l = 3.0\ m, \qquad d = 2.0\ m.$$

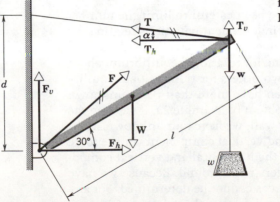

figure 14-7
Example 3.

Therefore

(1)
$$F_v + T_v = 100 \text{ N},$$

(2)
$$F_h = T_h,$$

and

$$F_v(3)(0.866) = F_h(1.5) + (60)(1.5)(0.866),$$

or

(3)
$$F_v = F_h(5.0/8.66) + 30 \text{ N}.$$

Recall that we have four unknowns, namely F_v, F_h, T_v, and T_h. We need another relation between these quantities if we are to solve the problem. This relation follows from the fact that $\mathbf{T}_v$ and $\mathbf{T}_h$ must add to give a resultant vector $\mathbf{T}$ directed along the wire. The wire cannot supply or support a force transverse to its orientation. (Notice that this is not true for the beam, however.) Hence our fourth relation is

$$T_v = T_h \tan \alpha,$$

where $\tan \alpha = (d - l \sin 30°)/l \cos 30° = 1.0/5.2$, so that

(4)
$$T_v = T_h/5.2.$$

Combining (1) and (4) we obtain

$$F_v = 100 \text{ N} - T_h/5.2.$$

Combining (2) and (3), we obtain

$$F_v = T_h(5.0/8.66) + 30 \text{ N}.$$

Solving these equations simultaneously, we obtain

$$T_h = 91.0 \text{ N},$$

and

$$F_v = 82.5 \text{ N}.$$

From (2) we obtain

$$F_h = 91.0 \text{ N}.$$

From (1) we obtain

$$T_v = 17.5 \text{ N}.$$

The tension in the wire will then be

$$T = \sqrt{T_h^2 + T_v^2} = 92.7 \text{ N},$$

and the hinge will exert a horizontal force of 91.0 N and a vertical force of 82.5 N on the beam.

In the preceding examples we have been careful to limit the number of unknown forces to the number of independent equations relating the forces. When all the forces act in a plane, we can have only three independent equations of equilibrium, one for rotational equilibrium about any axis normal to the plane and two others for translational equilibrium in the plane. However, we often have more than three unknown forces. For example, in the ladder problem of Example 2a, if we drop the artificial assumption of a frictionless wall, we have four unknown scalar quantities, namely, the horizontal and vertical components of the force acting on the ladder at the wall and the horizontal and vertical components of the force acting on the ladder at the ground. Because we have only three scalar equations, these forces cannot be determined. For any value assigned to one unknown force, the other three forces can be determined. But if we have no basis for assigning any particular value to an unknown force, there are an infinite number of solutions mathematically possible. We must therefore find another independent relation between the unknown forces if we hope to solve the problem uniquely.

Another simple example of such underdetermined structures is the automobile. In this case we wish to determine the forces exerted by the ground on each of the four tires when the car is at rest on a horizontal surface. If we assume that these forces are normal to the ground, we have four unknown scalar quantities. All other forces, such as the weight of the car and passengers, act normal to the ground. Therefore, we have only three independent equations giving the equilibrium conditions, one for translational equilibrium in the single direction of all the forces and two for rotational equilibrium about the two axes perpendicular to each other in a horizontal plane. Again the solution of the problem is indeterminate, mathematically. A four-legged table with all its legs in contact with the floor is a similar example.

Of course, since there is actually a unique solution to any real physical problem, we must find a physical basis for the additional independent relation between the forces that enable us to solve the problem. The difficulty is removed when we realize that structures are never perfectly rigid, as we have tacitly assumed throughout. Actually our structures will be somewhat deformed. For example, the automobile tires and the ground will be deformed, as will the ladder and wall. The laws of elasticity and the elastic properties of the structure determine the nature of the deformation and will provide the necessary additional relation between the four forces. A complete analysis therefore requires not only the laws of rigid body mechanics but also the laws of elasticity. In courses of civil and mechanical engineering, many such problems are encountered and analyzed in this way. We shall not consider the matter further here.

14-5
STABLE, UNSTABLE, AND NEUTRAL EQUILIBRIUM OF RIGID BODIES IN A GRAVITATIONAL FIELD

In Chapter 8 we saw that the gravitational force is a conservative force. For conservative forces we can define a potential energy function $U(x,y,z)$, where U is related to **F** by

$$F_x = -\partial U/\partial x, \qquad F_y = -\partial U/\partial y, \qquad F_z = -\partial U/\partial z.$$

At points where $\partial U/\partial x$ is zero, a particle subject to this conservative force will be in translational equilibrium in the x-direction, for then F_x equals zero. Likewise, at points where $\partial U/\partial y$ or $\partial U/\partial z$ are zero, a particle will be in translational equilibrium in the y- and z-directions, respectively. The derivative of U at a point will be zero when U has an extreme value (maximum or minimum) at that point or when U is constant with respect to the variable coordinate.

When U is a minimum, the particle is in *stable* equilibrium; any displacement from this position will result in a restoring force tending to return the particle to the equilibrium position. Another way of stating this is to say that if a body is in stable equilibrium, work must be done on it by an external agent to change its position. This results in an increase in its potential energy.

When U is a maximum, the particle is in *unstable* equilibrium; any displacement from this position will result in a force tending to push the particle farther from the equilibrium position. In this case no work must be done on the particle by an external agent to change its position; the work done in displacing the body is supplied internally by the conservative force, resulting in a decrease in potential energy.

When U is constant, the particle is in *neutral* equilibrium. In this case a par-

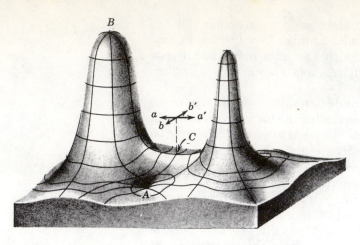

figure 14-8

A gravitational potential surface, which may be thought of as a real surface. A particle placed at *A*, *B* or *C* remains at rest; a plane tangent to any of these points is horizontal. We say that a particle here is in equilibrium. At *A*, a particle, if slightly displaced, tends to return to *A*. *A* represents a point of *stable equilibrium*. At *B*, a particle, if slightly displaced, tends to increase its displacement. Thus *B* represents a point of *unstable equilibrium*. At *C*, the particle, if slightly displaced in the direction *aa'*, will tend to return to *C*, but if it is displaced in direction *bb'*, it will tend to increase its displacement. *C* is called a *saddle point* since a saddle has somewhat this shape. Neutral equilibrium, experienced by a particle anywhere on a plane horizontal surface is not illustrated.

ticle can be displaced slightly without experiencing either a repelling or restoring force.

Notice that a particle can be in equilibrium with respect to one coordinate without necessarily being in equilibrium with respect to another coordinate, as for example a freely falling ball. Furthermore, a particle may be in stable equilibrium with respect to one coordinate and in unstable equilibrium with respect to another coordinate, as for example a particle at a saddle point (Fig. 14-8).

All these remarks apply to particles, that is, to translational motion. Suppose now we treat a rigid body. We must consider rotational equilibrium as well as translational equilibrium. The problem of a rigid body in a gravitational field is particularly simple, however, because *all the gravitational forces on the particles of the rigid body can be considered to act at one point, both for translational and rotational purposes.* We can replace this entire rigid body, for purposes of equilibrium under gravitational forces, by a single particle having the equivalent mass at the center of gravity.

For example, consider a cube at rest on one side on a horizontal table. The center of gravity is shown at the center of the central cross section of the cube in Fig. 14-9a. Let us supply a force to the cube so as to rotate it without its slipping about an axis along an edge. Notice that the center of gravity is raised and that work is done on the cube, which increases its potential energy. If the force is removed, the cube tends to return to its original position, its increased potential energy being converted into kinetic energy as it falls back. This initial position is, therefore, one of *stable* equilibrium. In terms of a particle of equivalent mass at the center of gravity, this process is described by the dotted line which indicates the path taken by the center of gravity during this motion. The particle is seen to have a minimum potential energy in the position of stable equilibrium, as required. We can conclude that the rigid body will be in stable equilibrium if the application of any force can raise the center of gravity of the body but not lower it.

If the cube is rotated until it balances on an edge, as in Fig. 14-9b, then once again the cube is in equilibrium. This equilibrium position is seen to be un-

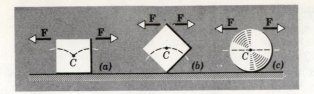

figure 14-9
Equilibrium of an extended body. *(a)* A cube resting on one side is in *stable equilibrium* since its center of gravity *C* is raised if the cube is tipped by a horizontal force **F**. *(b)* A cube resting on one edge is in *unstable equilibrium* since *C* falls if the cube is tipped by **F**. *(c)* A circular cylinder is in *neutral equilibrium* since *C* neither rises nor falls when **F** is applied. Compare these criteria for equilibrium with those given in Fig. 14-8. How are the criteria in the two figures related?

stable. The application of even the slightest horizontal force will cause the cube to fall away from this position with a decrease of potential energy. The particle of equivalent mass at the center of gravity follows the dotted path shown. At the position of unstable equilibrium this particle has a maximum potential energy, as required. We can conclude that the rigid body will be in unstable equilibrium if the application of any horizontal force tends to lower the center of gravity of the body.

The neutral equilibrium of a rigid body is illustrated by a cylinder on a horizontal table (Fig. 14-9c). If the cylinder is subjected to any horizontal force, the center of gravity is neither raised nor lowered but moves along the horizontal dotted line. The potential energy of the cylinder is constant during the displacement, as is that of the particle of equivalent mass at the center of gravity. The system has no tendency to move in any direction when the applied force is removed. A rigid body will be in neutral equilibrium if the application of any horizontal force neither raises nor lowers the center of gravity of the body.

Under what circumstances would a *suspended* rigid body be in stable equilibrium? When would a *suspended* rigid body be in unstable equilibrium, and when would it be in neutral equilibrium?

questions

1. Are Eqs. 14-1 and 14-3 both necessary and sufficient conditions for mechanical equilibrium? For static equilibrium?

2. A wheel rotating at constant angular velocity ω about a fixed axis is in mechanical equilibrium because no net external force or torque acts on it. However, the particles that make up the wheel undergo a centripetal acceleration **a** directed toward the axis. Since $\mathbf{a} \neq 0$ how can the wheel be said to be in equilibrium?

3. Give several examples of a body which is not in equilibrium, even though the resultant of all the forces acting on it is zero.

4. If a body is not in translational equilibrium, will the torque about any point be zero if the torque about some particular point is zero?

5. Which is more likely to break in use, a hammock stretched tightly between two trees or one that sags quite a bit? Prove your answer.

6. A ladder is at rest with its upper end against a wall and the lower end on the ground. Is it more likely to slip when a man stands on it at the bottom or at the top? Explain.

7. In Example 2, if the wall were rough, would the empirical laws of friction supply us with the extra condition needed to determine the extra (vertical) force exerted by the wall on the ladder?

8. In Example 3, why isn't it necessary to consider friction at the hinge?

9. A picture hangs from a wall by two wires. What orientation should the wires have to be under minimum tension? Explain how equilibrium is possible

with any number of orientations and tensions, even though the picture has a definite mass.

10. Show how to use a spring balance to weigh objects well beyond the maximum reading of the balance.

11. Do the center of mass and the center of gravity coincide for a building? For a lake? Under what conditions does the difference between the center of mass and the center of gravity of a body become significant?

12. If a rigid body is thrown into the air without spinning, it does not spin during its flight, provided air resistance can be neglected. What does this simple result imply about the location of the center of gravity?

13. Explain, using forces and torques, how a tree can maintain equilibrium in a high wind.

14. Is there such a thing as a truly rigid body?

15. You are sitting in the driver's seat of a parked automobile. You are told that the forces exerted upward by the ground on each of the four tires are different. Discuss the factors that enter into a consideration of whether this statement is true or false.

16. A uniform block, in the shape of a rectangular parallelepiped of sides in the ratio 1:2:3, lies on a horizontal surface. In which position, that is, on which of its three different faces, can it be said to be most stable, if any?

17. A virus particle in a rotating liquid-filled centrifuge tube is in uniform circular motion (that is, in *accelerated* motion) as viewed by an observer in the laboratory. An observer rotating with the centrifuge, however, would declare the particle to be *unaccelerated*. Explain how the virus particle can be in equilibrium for this second observer but not for the first.

18. In Chapter 5 we *defined* force in terms of acceleration from the relation $\mathbf{F} = m\mathbf{a}$. For a body in equilibrium, however, there are no accelerations. How, then, can we give meaning to the forces acting on such a body?

problems

SECTION 14-2

1. Prove that when only three forces act on a body in equilibrium, they must be coplanar and their lines of action must meet at a point or at infinity.

2. A uniform sphere of weight w and radius r is being held by a rope attached to a frictionless wall a distance L above the center of the sphere, as in Fig. 14-10. Find (a) the tension in the rope and (b) the force exerted on the sphere by the wall.

3. A uniform sphere of weight w lies at rest wedged between two inclined planes of inclination angles θ_1 and θ_2 (Fig. 14-11). (a) Assume that no friction is involved and determine the forces (directions and magnitude) that the planes exert on the sphere. (b) What change would it make, in principle, if friction were taken into account?
Answer: (a) $F_1 = w \sin \theta_2 / \sin (\theta_2 - \theta_1)$; $F_2 = w \sin \theta_1 / \sin (\theta_2 - \theta_1)$; normal to the planes.

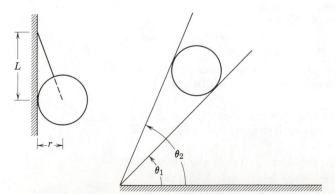

figure 14-10
Problem 2

figure 14-11
Problem 3

4. Two identical uniform smooth spheres, each of weight W, rest as shown in Fig. 14-12 at the bottom of a fixed, rectangular container. Find, in terms of W, the forces acting on the spheres by (a) the container surfaces and (b) by one another, if the line of centers of the spheres makes an angle of 45° with the horizontal.

5. A flexible chain of weight W hangs between two fixed points, A and B, at the same level, as shown in Fig. 14-13. Find (a) the vector force exerted by the chain on each end point and (b) the tension in the chain at the lowest point.

Answer: (a) $\dfrac{W}{2 \sin \theta}$, tangent to chain. (b) $\dfrac{1}{2} W \cot \theta$.

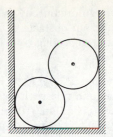

figure 14-12
Problem 4

SECTION 14-3

6. A nonuniform bar of weight W is suspended at rest in a horizontal position by two light cords as shown in Fig. 14-14; the angle one cord makes with the vertical is $\theta = 36.9°$; the other makes the angle $\phi = 53.1°$ with the vertical. If the length l of the bar is 6.1 m, compute the distance x from the left-hand end to the center of gravity.

7. A circular section of radius r is cut out of a uniform disk of radius R, the center of the hole being $R/2$ from the center of the original disk. Locate the center of gravity of the resulting body.

Answer: Along a line from the center of the hole through the center of the disk, beyond the latter point by a distance $Rr^2/2(R^2 - r^2)$.

figure 14-13
Problem 5

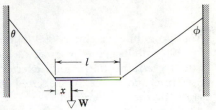

figure 14-14
Problem 6

SECTION 14-4

8. A beam is carried by three men, one man at one end and the other two supporting the beam between them on a crosspiece so placed that the load is equally divided among the three men. Find where the crosspiece is placed. Neglect the mass of the crosspiece.

9. In Fig. 14-15 a man is trying to get his car out of the mud on the shoulder of a road. He ties one end of a rope tightly around the front bumper and the other end tightly around a telephone pole 60 ft away. He then pushes sideways on the rope at its midpoint with a force of 125 lb, displacing the center of the rope 1.0 ft from its previous position and the car almost moves. What force does the rope exert on the car? (The rope stretches somewhat under the tension.)   *Answer:* 1900 lb.

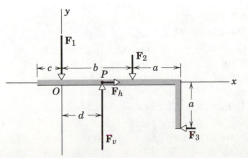

figure 14-15
Problem 9

10. Forces $\mathbf{F}_1$, $\mathbf{F}_2$, and $\mathbf{F}_3$ act on the structure of Fig. 14-16 as shown. We wish to put the structure in equilibrium by applying a force, at a point such as P, whose vector components are $\mathbf{F}_h$ and $\mathbf{F}_v$. We are given that $a = 2.0$ m, $b = 3.0$ m, $c = 1.0$ m, $F_1 = 20$ N, $F_2 = 10$ N, and $F_3 = 5.0$ N. Find (a) F_h, (b) F_v, and (c) d.

figure 14-16
Problem 10

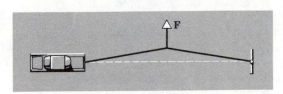

11. What force F applied horizontally at the axle of the wheel is necessary to raise the wheel over an obstacle of height h? Take r as the radius of the wheel and W as its weight (Fig. 14-17).

Answer: $W\sqrt{h(2r-h)}/(r-h)$.

12. A trap door in a ceiling is 3.0 ft (0.91 m) square, weighs 25 lb (mass = 11 kg), and is hinged along one side with a catch at the opposite side. If the center of gravity of the door is 4.0 in. (10 cm) from the door's center and closer to the hinged side, what forces must (a) the catch and (b) the hinge sustain?

13. A meter stick balances on a knife edge at the 50.0-cm mark. When two nickels are stacked over the 12.0-cm mark, the loaded stick is found to balance at the 45.5-cm mark. A nickel has a mass of 5.0 g. What is the mass of the meter stick? Try this technique and check your answer experimentally.   Answer: 74.4 g.

14. A balance is made up of a rigid rod free to rotate about a point not at the center of the rod. It is balanced by unequal weights placed in the pans at each end of the rod. When an unknown mass m is placed in the left-hand pan, it is balanced by a mass m_1 placed in the right-hand pan, and similarly when the mass m is placed in the right-hand pan, it is balanced by a mass m_2 in the left-hand pan. Show that

$$m = \sqrt{m_1 m_2}.$$

15. An automobile weighing 3000 lb (mass = 1360 kg) has a wheel base of 120 in. (305 cm). Its center of gravity is located 70.0 in. (178 cm) behind the front axle. Determine (a) the force exerted on each of the front wheels (assumed the same) and (b) the force exerted on each of the back wheels (assumed the same) by the level ground.

Answer: (a) 625 lb (2780 N). (b) 875 lb (3890 N).

16. A crate in the form of a 4.0-ft cube contains a piece of machinery whose design is such that the center of gravity of the crate and its contents is located 1.0 ft above its geometrical center. (a) If the crate is to be slid down a ramp without tipping over, what is the maximum angle which the ramp may make with the horizontal? (b) What is the maximum value for the coefficient of static friction between the crate and the ramp in this case such that the crate will just begin to slide?

17. A door 7.0 ft (2.1 m) high and 3.0 ft (0.91 m) wide weighs 60 lb (mass = 27 kg). A hinge 1.0 ft (0.30 m) from the top and another 1.0 ft (0.30 m) from the bottom each support half the door's weight. Assume that the center of gravity is at the geometrical center of the door and determine the horizontal and vertical force components exerted by each hinge on the door. Answer: 30 lb (130 N) vertical; 18 lb (80 N) horizontal, oppositely directed.

18. Four bricks, each of length l, are put on top of one another (see Fig. 14-18) in such a way that part of each extends beyond the one beneath. Show that the largest equilibrium extensions are (a) top brick overhanging the one below by $l/2$, (b) second brick from top overhanging the one below by $l/4$, and (c) third brick from top overhanging the bottom one by $l/6$.

19. The system shown in Fig. 14-19 is in equilibrium. The mass hanging from the end of the strut S weighs 500 lb (mass = 230 kg), and the strut itself weighs 100 lb (mass = 45 kg). Find (a) the tension T in the cable and (b) the force exerted on the strut by the pivot P.

Answer: (a) 1500 lb (6800 N). (b) F_h = 1300 lb (5900 N); F_v = 1350 lb (6100 N).

20. A 100-lb plank, of length l = 20 ft, rests on the ground and on a frictionless roller (not shown) at the top of a wall of height h = 10 ft (see Fig. 14-20). The center of gravity of the plank is at its center. The plank remains in equilibrium for any value of $\theta \geqslant 70°$, but slips if $\theta < 70°$. (a) Draw a diagram showing all forces acting on the plank. (b) Find the coefficient of static friction between the plank and the ground.

21. A thin horizontal bar AB of negligible weight and length l is pinned to a vertical wall at A and supported at B by a thin wire BC that makes an angle θ

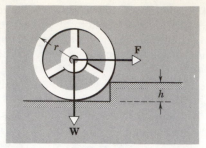

figure 14-17
Problem 11

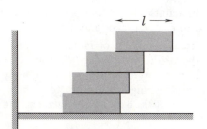

figure 14-18
Problem 18

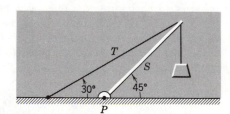

figure 14-19
Problem 19

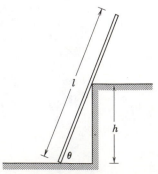

figure 14-20
Problem 20

with the horizontal. A weight W can be moved anywhere along the bar as defined by the distance x from the wall (Fig. 14-21). (a) Find the tension T in the thin wire as a function of x. Find (b) the horizontal and (c) the vertical components of the force exerted on the bar by the pin at A.
Answer: (a) $Wx/(l \sin \theta)$. (b) $Wx/(l \tan \theta)$. (c) $W(1 - x/l)$.

22. A homogeneous sphere of radius r and weight **W** slides along the floor under the action of a constant horizontal force **P** applied to a string, as shown in Fig. 14-22. (a) Show that if μ is the coefficient of kinetic friction between sphere and floor, the height h is given by $h = r(1 - \mu W/P)$. (b) Show that the sphere is not in translational equilibrium under these circumstances. Is there any point about which the sphere is in rotational equilibrium? (c) Can one get the sphere to be in both rotational *and* translational equilibrium by a different choice of h? By a different direction for **P**? Explain.

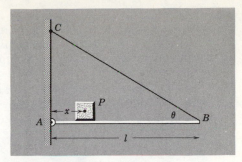

figure 14-21
Problem 21

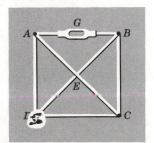

figure 14-22
Problem 22

23. In the stepladder shown in Fig. 14-23 AC and CE are 8.0 ft long and hinged at C. BD is a tie rod 2.5 ft long, halfway up. A man weighing 192 lb climbs 6.0 ft along the ladder. Assuming that the floor is frictionless and neglecting the weight of the ladder, find (a) the tension in the tie rod and (b) the forces exerted on the ladder by the floor. (*Hint:* It will help to isolate parts of the ladder in applying the equilibrium conditions.)
Answer: (a) 47 lb. (b) $F_A = 120$ lb; $F_E = 72$ lb.

24. A cubical box is filled with sand and weighs 200 lb (890 N). It is desired to "roll" the box by pushing horizontally on one of the upper edges. (a) What minimum force is required? (b) What minimum coefficient of static friction is required? (c) Is there a more efficient way to roll the box? If so, find the lowest possible force that would be required to be applied directly to the box.

25. By means of a turnbuckle G, a tension force **T** is produced in bar AB of the square frame $ABCD$ in Fig. 14-24. Determine the forces produced in the other bars. The diagonals AC and BD pass each other freely at E. Symmetry considerations can lead to considerable simplification in this and similar problems.
Answer: Bars AD, BC, and DC are in tension (force T); diagonals AC and BD are in compression (force $\sqrt{2}\ T$).

26. A well-known problem is the following (see, for example, *Scientific American*, November 1964, p. 128): Uniform bricks are placed one upon another in such a manner as to have the maximum offset. This is accomplished by having the center of gravity of the top brick directly above the edge of the brick below it, the center of gravity of the two top bricks combined directly above the edge of the third brick from the top, etc. (a) Justify this criterion for maximum offset. (b) Show that, if the process is continued downward, one can obtain as large an offset as he wants. (Martin Gardner, in the article referred to above, states: "With 52 playing cards, the first placed so that its end is flush with a table edge, the maximum overhang is a little more than $2\frac{1}{4}$ card lengths. . . .") (c) Suppose now, instead, one piles up uniform bricks so that the end of one brick is offset from the one below it by a constant fraction, $1/n$, of a brick length l. How many bricks, N, can one use in this process before the pile will fall over? Check the plausibility of your answer for $n = 1$, $n = 2$, $n = \infty$.

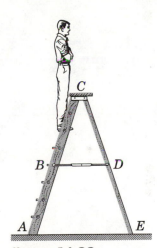

figure 14-23
Problem 23

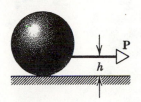

figure 14-24
Problem 25

SECTION 14-5

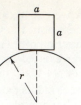

figure 14-25
Problem 28

27. A bowl having a radius of curvature r rests on a rough horizontal table. Show that the bowl will be in stable equilibrium about the center point at its bottom only if the center of mass of the material piled up in the bowl is not as high as r above the center of the bowl.

28. A cube of uniform density and edge a is balanced on a cylindrical surface of radius r as shown in Fig. 14-25. Show that the criterion for stable equilibrium of the cube, assuming that friction is sufficient to prevent slipping, is $r > a/2$.

15
oscillations

Any motion that repeats itself in equal intervals of time is called *periodic motion*. As we shall see, the displacement of a particle in periodic motion can always be expressed in terms of sines and cosines. Because the term *harmonic* is applied to expressions containing these functions, periodic motion is often called *harmonic motion*.

If a particle in periodic motion moves back and forth over the same path, we call the motion *oscillatory* or *vibratory*. The world is full of oscillatory motions. Some examples are the oscillations of the balance wheel of a watch, a violin string, a mass attached to a spring, atoms in molecules or in a solid lattice, and air molecules as a sound wave passes by.

Many oscillating bodies do not move back and forth between precisely fixed limits because frictional forces dissipate the energy of motion. Thus a violin string eventually stops vibrating and a pendulum stops swinging. We call such motions *damped* harmonic motions. Although we cannot eliminate friction from the periodic motions of gross objects, we can often cancel out its damping effect by feeding energy into the oscillating system so as to compensate for the energy dissipated by friction. The main spring of a watch and the falling weight in a pendulum clock supply external energy in this way, so that the oscillating system, that is, the balance wheel or the pendulum, moves as if it were undamped.

Not only mechanical systems can oscillate. Radio waves, microwaves, and visible light are oscillating magnetic and electric field vectors. Thus a tuned circuit in a radio and a closed metal cavity in which microwave energy is introduced can oscillate electromagnetically. The analogy is close, being based on the fact that *mechanical and electro-*

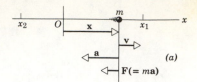

magnetic oscillations are described by the same basic mathematical equations. We will make the most of this analogy in later chapters.

The *period T* of a harmonic motion is the time required to complete one round trip of the motion, that is, one complete oscillation or *cycle*. The *frequency* of the motion ν is the number of oscillations (or cycles) per unit of time. The frequency is therefore the reciprocal of the period, or

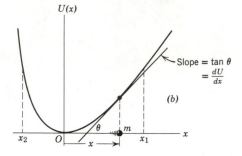

$$\nu = 1/T. \qquad (15\text{-}1)$$

The SI unit of frequency is the cycle per second, or *hertz* (Hz).* The position at which no net force acts on the oscillating particle is called its *equilibrium* position. The *displacement* (linear or angular) is the distance (linear or angular) of the oscillating particle from its equilibrium position at any instant.

Let us focus attention on a particle oscillating back and forth along a straight line between fixed limits. Its displacement **x** changes periodically in both magnitude and direction. Its velocity **v** and acceleration **a** also vary periodically in magnitude and direction and, in view of the relation **F** = m**a,** so does the force **F** acting on the particle.

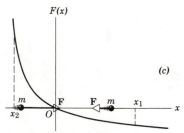

Forces associated with harmonic motion are the most general kinds of forces that we have discussed so far. In the earlier chapters we dealt only with constant forces (and accelerations). Later, when we considered forces that are not constant but instead vary with time, we examined a force (and thus an acceleration) that varied in direction although its magnitude was constant (the centripetal force of Section 6-3), and a force (and thus an acceleration) which varied in magnitude although its direction was constant (the impulsive force of Section 10-1). Here, in harmonic motion, the force, and the acceleration, vary both in direction *and* magnitude.

In terms of energy, we can say that a particle undergoing harmonic motion passes back and forth through a point (its equilibrium position) at which its potential energy is a minimum. A swinging pendulum is a good example, its potential energy being a minimum at the bottom of the swing, that is, at the equilibrium position. Figure 15-1a shows the generalized case of a particle oscillating between the limits x_1 and x_2, O being the equilibrium position. Figure 15-1b shows the corresponding potential energy curve, which has a minimum value at that position. The force acting on the particle at any position is derivable from the potential energy function; it is given by Eq. 8-7,

$$F = -dU/dx, \qquad (8\text{-}7)$$

and is displayed in Fig. 15-1c. The force is zero at the equilibrium position O, points to the right (that is, has a positive value) when the particle is to the left of O, and points to the left (that is, has a negative value) when the particle is to the right of O. The force is *a restoring force* because it always acts to accelerate the particle in the direction of its equilibrium position. Hence in harmonic motion the position of equilibrium is always one of *stable* equilibrium.

The total mechanical energy E for an oscillating particle is the sum of its kinetic energy and its potential energy, or

$$E = K + U \qquad (15\text{-}2)$$

figure 15-1
(a) A particle of mass m oscillates harmonically between points x_1 and x_2, O being the equilibrium position. *(b)* The potential energy of the particle as a function of position. The force acting on the particle at position x is given by $F = -dU/dx$. *(c)* The force acting on the particle as a function of position x; note that the force is directed toward the equilibrium position.

---

* This frequency unit is named after Heinrich Hertz (1857–1894) whose research provided the experimental confirmation of the electromagnetic waves predicted theoretically by James Clerk Maxwell (1831–1879).

in which E remains constant if no nonconservative forces, such as the force of friction, are acting. Figure 15-2 shows E for the motion of Fig. 15-1. Note how Eq. 15-2 is satisfied for the particle in the typical position shown. The particle cannot move outside the limits x_1 and x_2 because, in these regions, U exceeds E. This, as Eq. 15-2 shows, would require the kinetic energy to be negative, an impossibility.

For a given environment, that is, for a given function $U(x)$, an oscillating particle can have various total energies, depending on how we set it into motion initially. Thus the total energy may be E', rather than E, in which case the limits of oscillation would be x_1' and x_2', as Fig. 15-2 shows, rather than x_1 and x_2.

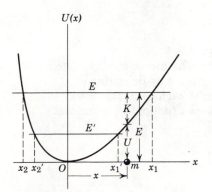

figure 15-2
The total mechanical energy E for the motion of Fig. 15-1 is shown. If the total mechanical energy of the oscillating particle is reduced to E', the limits of oscillation are reduced to x_1' and x_2' respectively.

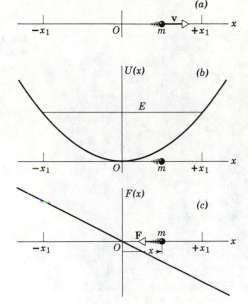

figure 15-3
(a) A particle of mass m oscillates with simple harmonic motion between points $+x_1$ and $-x_1$, O being the equilibrium position. (b) The potential energy $U(x)$ and the total mechanical energy E. (c) The force acting on the particle. Compare this figure carefully with Fig. 15-1, which illustrates the general case of harmonic motion.

15-2
THE SIMPLE HARMONIC OSCILLATOR

Let us consider an oscillating particle (Fig. 15-3a) moving back and forth about an equilibrium position through a potential that varies as

$$U(x) = \tfrac{1}{2}kx^2 \qquad (15\text{-}3)$$

in which k is a constant; see Fig. 15-3b. The force acting on the particle is given by Eq. 8-7, or

$$F(x) = -dU/dx = -d(\tfrac{1}{2}kx^2)/dx = -kx; \qquad (15\text{-}4)$$

see Fig. 15-3c. Such an oscillating particle is called a *simple harmonic oscillator* and its motion is called *simple harmonic motion*. In such motion, as Eq. 15-3 shows, the potential energy curve varies as the square of the displacement, and, as Eq. 15-4 shows, the force acting on the particle is proportional to the displacement but is opposite to it in direction. In simple harmonic motion the limits of oscillation are equally spaced about the equilibrium position. This is not true for the more general motion of Fig. 15-1 which, although harmonic, is not simple harmonic. The magnitude of the maximum displacement, that is, the quantity x_1 in Fig. 15-3, always taken as positive, is called the *amplitude* of the simple harmonic motion.

You will have recognized Eq. 15-3 [$U(x) = \tfrac{1}{2}kx^2$] as the expression for the potential energy of an "ideal" spring, compressed or extended by a distance x; see Section 8-4. In this same section an ideal spring was defined as one in which the force exerted by the stretched or compressed spring is given by $F(x) = -kx$ (see Eq. 15-4), k being called the *force constant*.

Hence, a body of mass m attached to an ideal spring of force constant k and free to move over a frictionless horizontal surface is an example of a simple harmonic oscillator (see Fig. 15-4). Note that there is a position (the equilibrium position; see Fig. 15-4b) in which the spring exerts no force on the body. If the body is displaced to the right (as in Fig. 15-4a), the force exerted by the spring on the body points to the left and is given by $F = -kx$. If the body is displaced to the left (as in Fig. 15-4c), the force points to the right and is also given by $F = -kx$. In each case the force is a *restoring* force. The motion of the oscillating mass is *simple harmonic motion*.

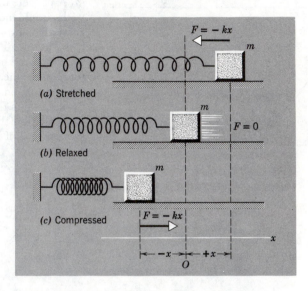

(a) Stretched

(b) Relaxed

(c) Compressed

figure 15-4
A simple harmonic oscillator. The force exerted by the spring is shown in each case. The block slides on a frictionless horizontal table.

Let us apply Newton's second law, $F = ma$, to the motion of Fig. 15-4. For F we substitute $-kx$ (from Eq. 15-4) and for the acceleration a we put in d^2x/dt^2 ($= dv/dt$). This gives us

$$-kx = m\frac{d^2x}{dt^2}$$

or

$$\frac{d^2x}{dt^2} + \frac{k}{m}x = 0. \qquad (15\text{-}5)$$

This equation involves derivatives and is, therefore, called a *differential equation*. To solve this equation means to determine how the displacement x of the particle must depend on the time t in order that the equation be satisfied. When we know how x depends on time, we know the motion of the particle; thus, Eq. 15-5 is called the *equation of motion* of a simple harmonic oscillator. We shall solve this equation and describe the motion in detail in the next section.

The simple harmonic oscillator problem is important for two reasons: First, most problems involving mechanical vibrations reduce to that of the simple harmonic oscillator at small amplitudes of vibration, or to a combination of such vibrations. This is equivalent to saying that if we consider a small enough portion of the restoring force curve of Fig. 15-1c (around the origin), it becomes arbitrarily close to a straight line which, as Fig. 15-3c shows, is characteristic of simple harmonic motion. Or, in other words, the potential energy curve of Fig. 15-1b for general oscillatory motion reduces to that of Fig. 15-3b for simple harmonic oscillation when the amplitude of vibration is made sufficiently small about the equilibrium position O.

Second, as we have indicated, equations like Eq. 15-5 occur in many physical problems in acoustics, in optics, in mechanics, in electrical circuits, and even in atomic physics. The simple harmonic oscillator exhibits features common to many physical systems.

Equation 15-4 ($F = -kx$) is an empirical relation known as *Hooke's law*. It is a special case of a more general relation, dealing with the deformation of elastic bodies, discovered by Robert Hooke (1635–1703).* It is obeyed by springs and other elastic bodies provided the deformation is not too great. If the solid is deformed beyond a certain point, called its *elastic limit*, it will not even return to its original shape when the applied force is removed (Fig. 15-5). It turns out that Hooke's law holds almost up to the elastic limit for many common materials. The range of applied forces over which Hooke's law is valid is called the "proportional region." Beyond the elastic limit, the force can no longer be specified by a potential energy function, because the force then depends on many factors including the speed of deformation and the previous history of the solid.

Notice that the restoring force and potential energy function of the simple harmonic oscillator are the same as that of a solid deformed in one dimension in the proportional region. If the deformed solid is released, it will vibrate, just as the simple harmonic oscillator does. Therefore, as long as the amplitude of the vibration is small enough, that is, as long as the deformation remains in the proportional region, mechanical vibrations behave exactly like simple harmonic oscillators. It is easy to generalize this discussion to show that any problem involving mechanical vibrations of small amplitude in three dimensions reduces to a combination of simple harmonic oscillators.

The vibrating string or membrane, sound vibrations, the vibrations of atoms in solids, and electrical or acoustical oscillations in a cavity can be described in a form which is mathematically identical to a system of harmonic oscillators. The analogy enables us to solve problems in one area by using the techniques developed in other areas.

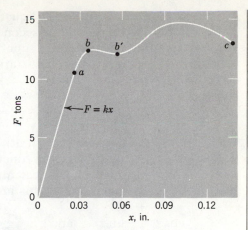

figure 15-5
Typical graph of applied force F versus resulting elongation of an aluminum bar under tension. The sample was a foot long and a square inch in cross section. Notice that we may write $F = kx$ only for the portion Oa, since beyond this point the slope is no longer constant but varies in a complicated way with x. At some point such as b (the *elastic limit*) the sample does not return to its original length when the applied force is removed. Between b and b' the elongation increases, even though the force is held constant; the material flows like a viscous fluid. At c, the sample can be stretched no farther; any increase in elongation results in the sample's breaking in two. The applied force is equal in magnitude to the restoring force so that no minus sign appears in the relation $F = kx$.

15-3
SIMPLE HARMONIC MOTION

Let us now solve the equation of motion of the simple harmonic oscillator,

$$\frac{d^2x}{dt^2} + \frac{k}{m}x = 0. \tag{15-6}$$

Recall that any system of mass m upon which a force $F = -kx$ acts will be governed by this equation. In the case of a spring, the proportionality constant k is the force constant of the spring, which is a measure of its stiffness. In other oscillating systems the proportionality constant k may be related to other physical features of the system, as we shall see later. We can use the oscillating mass-spring system as our prototype.

---

* Hooke first expressed his law in 1676 as a Latin cryptogram *ceiiinossssttuv*. Two years later he deciphered this as *ut tensio sic vis*, which we may translate as: *the* (*ex*)*tension is proportional to the force*.

Equation 15-6 is a differential equation. It gives a relation between a function of the time $x(t)$ and its second time derivative d^2x/dt^2. To find the position of the particle as a function of the time, we must find a function $x(t)$ which satisfies this relation.

We can rewrite Eq. 15-6 as

$$\frac{d^2x}{dt^2} = -\left(\frac{k}{m}\right)x. \tag{15-7}$$

Equation 15-7 then requires that $x(t)$ be some function whose second derivative is the negative of the function itself, except for a constant factor k/m. We know from the calculus, however, that the sine function or the cosine function has this property.* For example,

$$\frac{d}{dt} \cos t = -\sin t \quad \text{and} \quad \frac{d^2}{dt^2} \cos t = -\frac{d}{dt} \sin t = -\cos t.$$

This property is not affected if we multiply the cosine function by a constant A. We can allow for the fact that the sine function will do as well, and for the fact that Eq. 15-7 contains a constant factor, by writing as a tentative solution of Eq. 15-7,

$$x = A \cos (\omega t + \phi). \tag{15-8}$$

Here since

$$\cos (\omega t + \phi) = \cos \phi \cos \omega t - \sin \phi \sin \omega t = a \cos \omega t + b \sin \omega t,$$

the constant ϕ allows for any combination of sine and cosine solutions. Hence, with the (as yet) unknown constants A, ω, and ϕ, we have written as general a solution to Eq. 15-7 as we can. In order to determine these constants such that Eq. 15-8 is actually the solution of Eq. 15-7, we differentiate Eq. 15-8 twice with respect to the time. We have

$$\frac{dx}{dt} = -\omega A \sin (\omega t + \phi)$$

and

$$\frac{d^2x}{dt^2} = -\omega^2 A \cos (\omega t + \phi).$$

Putting this into Eq. 15-7, we obtain

$$-\omega^2 A \cos (\omega t + \phi) = -\frac{k}{m} A \cos (\omega t + \phi).$$

Therefore, if we choose the constant ω such that

$$\omega^2 = \frac{k}{m}, \tag{15-9}$$

then

$$x = A \cos (\omega t + \phi)$$

is in fact a solution of the equation of a simple harmonic oscillator.

The constants A and ϕ are still undetermined and, therefore, still completely arbitrary. This means that any choice of A and ϕ whatsoever will satisfy Eq. 15-7, so that a large variety of motions is possible for the oscillator. Actually, this is characteristic of a differential equation of motion, for such an equation does not describe just one single motion but a group or family of possible motions which have some features in common but differ in other ways. In this case ω is common to all the allowed motions, but A and ϕ may differ among them. We

---

* Harmonic motion is not only periodic but also bounded. Only the sine and cosine functions (or combinations of them) have both these properties.

shall see later that A and ϕ are determined for a particular harmonic motion by how the motion starts.

Let us find the *physical* significance of the constant ω. If we increase the time t in Eq. 15-8 by $2\pi/\omega$, the function becomes

$$x = A \cos [\omega(t + 2\pi/\omega) + \phi],$$
$$= A \cos (\omega t + 2\pi + \phi),$$
$$= A \cos (\omega t + \phi).$$

That is, the function merely repeats itself after a time $2\pi/\omega$. Therefore, $2\pi/\omega$ is the *period* of the motion T. Since $\omega^2 = k/m$, we have

$$T = \frac{2\pi}{\omega} = 2\pi \sqrt{\frac{m}{k}}. \tag{15-10}$$

Hence, all motions given by Eq. 15-7 have the same period of oscillation, and this is determined only by the mass m of the oscillating particle and the force constant k of the spring. The *frequency* ν of the oscillator is the number of complete vibrations per unit time and is given by

$$\nu = \frac{1}{T} = \frac{\omega}{2\pi} = \frac{1}{2\pi} \sqrt{\frac{k}{m}}. \tag{15-11}$$

Hence,
$$\omega = 2\pi\nu = \frac{2\pi}{T}. \tag{15-12}$$

The quantity ω is called the *angular frequency*; it differs from the frequency ν by a factor 2π. It has the dimension of reciprocal time (the same as angular speed) and its unit is the radian/second. In Section 15-6 we shall give a geometric meaning to this angular frequency.

The constant A has a simple physical meaning. The cosine function takes on values from -1 to 1. The *displacement* x from the central equilibrium position $x = 0$, therefore, has a maximum value of A; see Eq. 15-8. We call A $(= x_{max})$ the *amplitude* of the motion. Because A is not fixed by our differential equation, motions of various amplitudes are possible, but all have the same frequency and period. *The frequency of a simple harmonic motion is independent of the amplitude of the motion.*

The quantity $(\omega t + \phi)$ is called the *phase* of the motion. The constant ϕ is called the *phase constant*. Two motions may have the same amplitude and frequency but differ in phase. If $\phi = -\pi/2$, for example,

$$x = A \cos (\omega t + \phi) = A \cos (\omega t - 90°)$$
$$= A \sin \omega t$$

so that the displacement is zero at the time $t = 0$. When $\phi = 0$, the displacement $x = A \cos \omega t$ is a maximum at the time $t = 0$. Other initial displacements correspond to other phase constants.

The amplitude A and the phase constant ϕ of the oscillation are determined by the initial position and speed of the particle. These two initial conditions will specify A and ϕ exactly.* Once the motion has started, however, the particle will continue to oscillate with a constant amplitude and phase constant at a fixed frequency, unless other forces disturb the system.

---

* A phase constant may be increased by any integral multiple of 2π, or of $360°$, and it will still describe the motion equally well.

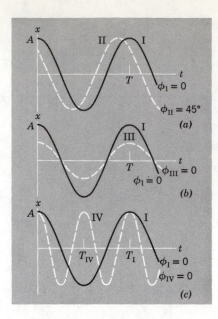

figure 15-6
Several solutions of the simple harmonic oscillator equation. (a) Both solutions have the same amplitude and period but differ in phase by 45°. (b) Both have the same period and phase constant but differ in amplitude by a factor of 2. (c) Both have the same phase constant and amplitude but differ in period by a factor of 2.

In Fig. 15-6 we plot the displacement x versus the time t for several simple harmonic motions described by Eq. 15-8. Three comparisons are made. In Fig. 15-6a, I and II have the same amplitude and frequency but differ in phase by $\phi = \pi/4$ or 45°. In Fig. 15-6b, I and III have the same frequency and phase constant but differ in amplitude by a factor of 2. In Fig. 15-6c, I and IV have the same amplitude and phase constant but differ in frequency by a factor $\frac{1}{2}$ or in period by a factor 2. Study these curves carefully to become familiar with the terminology used in simple harmonic motion.

Another distinctive feature of simple harmonic motion is the relation between the displacement, the velocity, and the acceleration of the oscillating particle. Let us compare these quantities for curve I of Fig. 15-6, which is typical. In Fig. 15-7 we plot separately the displacement x versus the time t, the velocity $v = dx/dt$ versus the time t, and the acceleration $a = dv/dt = d^2x/dt^2$ versus the time t. The equations of these curves are

$$x = A \cos (\omega t + \phi),$$

$$v = \frac{dx}{dt} = -\omega A \sin (\omega t + \phi), \qquad (15\text{-}13)$$

$$a = \frac{dv}{dt} = -\omega^2 A \cos (\omega t + \phi).$$

For the case plotted we have taken $\phi = 0$. The units and scale of displacement, velocity, and acceleration are omitted for simplicity of comparison. Notice that (see Eq. 15-13) the maximum displacement is A, the maximum speed is ωA, and the maximum acceleration is $\omega^2 A$.

When the displacement is a maximum in either direction, the speed is zero because the velocity must now change its direction. The acceleration at this instant, like the restoring force, has a maximum value but is directed opposite to the displacement. When the displacement is zero, the speed of the particle is a maximum and the acceleration is zero, corresponding to a zero restoring force. The speed increases as the particle moves toward the equilibrium position and then decreases as it moves out to the maximum displacement, just as for a pendulum bob.

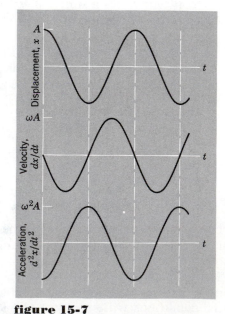

figure 15-7
The relations between displacement, velocity, and acceleration in simple harmonic motion. The phase constant ϕ is zero in this particular case since the displacement is maximum at $t = 0$; see Eq. 15-8.

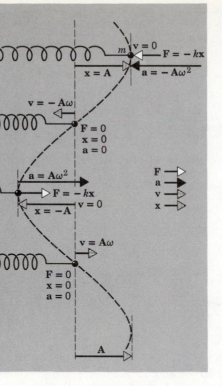

figure 15-8
The force acting on, and the acceleration, velocity and displacement of a mass m undergoing simple harmonic motion. Compare carefully with Fig. 15-7.

In Fig. 15-8 we show the instantaneous values of $\mathbf{x}$, $\mathbf{v}$, and $\mathbf{a}$ at four instants in the motion of a particle oscillating at the end of a spring.

Equation 15-2 tells us that for harmonic motion, including simple harmonic motion, in which no dissipative forces act, the total mechanical energy E $(= K + U)$ is conserved (remains constant). We can now study this in more detail for the special case of simple harmonic motion, for which the displacement is given by

15-4
ENERGY CONSIDERATIONS IN SIMPLE HARMONIC MOTION

$$x = A \cos (\omega t + \phi). \tag{15-8}$$

The potential energy U at any instant is given by

$$U = \tfrac{1}{2}kx^2$$

$$= \tfrac{1}{2}kA^2 \cos^2 (\omega t + \phi). \tag{15-14}$$

The potential energy has a maximum value of $\tfrac{1}{2}kA^2$. During the motion the potential energy varies between zero and this maximum value, as the curves in Fig. 15-9a and 15-9b show.

The kinetic energy K at any instant is $\tfrac{1}{2}mv^2$. Using the relations

$$v = dx/dt = -\omega A \sin (\omega t + \phi)$$

and

$$\omega^2 = k/m,$$

we obtain

$$K = \tfrac{1}{2}mv^2,$$

$$= \tfrac{1}{2}m\omega^2A^2 \sin^2 (\omega t + \phi),$$

$$= \tfrac{1}{2}kA^2 \sin^2 (\omega t + \phi). \tag{15-15}$$

The kinetic energy, therefore, has a maximum value of $\tfrac{1}{2}kA^2$ or $\tfrac{1}{2}m(\omega A)^2$, in agreement with the maximum speed ωA noted earlier. During the

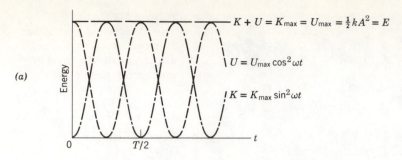

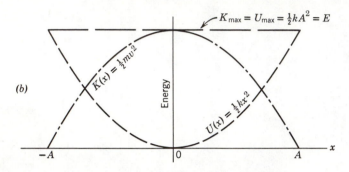

figure 15-9

Energies of a simple harmonic oscillator. *(a)* Potential energy $(-\,-\,-\,-)$, kinetic energy $(-\,\cdot\,-)$, and total energy $(\mathrm{-\!-}\;\mathrm{-\!-})$ plotted as a function of time. *(b)* Potential, kinetic, and total energy plotted as a function of displacement from the equilibrium position. Compare with Fig. 8-4.

motion the kinetic energy varies between zero and this maximum value, as shown by the curves in Fig. 15-9*a* and 15-9*b*.

The total mechanical energy is the sum of the kinetic and the potential energy. Using Eqs. 15-14 and 15-15 we obtain

$$E = K + U = \tfrac{1}{2}kA^2 \sin^2 (\omega t + \phi) + \tfrac{1}{2}kA^2 \cos^2 (\omega t + \phi) = \tfrac{1}{2}kA^2. \qquad (15\text{-}16)$$

We see that the total mechanical energy is constant, as we expect, and has the value $\tfrac{1}{2}kA^2$. At the maximum displacement the kinetic energy is zero, but the potential energy has the value $\tfrac{1}{2}kA^2$. At the equilibrium position the potential energy is zero, but the kinetic energy has the value $\tfrac{1}{2}kA^2$. At other positions the kinetic and potential energies each contribute energy whose sum is always $\tfrac{1}{2}kA^2$. This constant total energy is shown in Fig. 15-9*a* and 15-9*b*. *The total energy of a particle executing simple harmonic motion is proportional to the square of the amplitude of the motion.* It is clear from Fig. 15-9*a* that the *average* kinetic energy for the motion during one period is exactly equal to the *average* potential energy and that each of these average quantities is $\tfrac{1}{4}kA^2$.

Equation 15-16 can be written quite generally as

$$K + U = \tfrac{1}{2}mv^2 + \tfrac{1}{2}kx^2 = \tfrac{1}{2}kA^2. \qquad (15\text{-}17)$$

From this relation we obtain $v^2 = (k/m)(A^2 - x^2)$ or

$$v = \frac{dx}{dt} = \pm \sqrt{\frac{k}{m}(A^2 - x^2)}. \qquad (15\text{-}18)$$

This relation shows clearly that the speed is a maximum at the equilibrium position $x = 0$ and zero at the maximum displacement $x = A$. In fact, we can start from the conservation of energy principle, Eq. 15-17 (in which $\tfrac{1}{2}kA^2 = E$), and by integration of Eq. 15-18 obtain the displacement as a function of time. The result is identical with Eq. 15-8 which we deduced from the differential equation of the motion, Eq. 15-6. (See Problem 29.)

The effect of dissipative forces will be discussed in Section 15-9.

The horizontal spring of Fig. 15-4 is found to be stretched 3.0 in. from its equilibrium position when a force of 0.75 lb acts on it. Then a 1.5-lb body is attached to the end of the spring and is pulled 4.0 in. along a horizontal frictionless table from the equilibrium position. The body is then released and executes simple harmonic motion.

(a) What is the force constant of the spring?

A force of 0.75 lb *on* the spring produces a displacement of 0.25 ft. Hence,

$$k = F/x = 0.75 \text{ lb}/0.25 \text{ ft} = 3.0 \text{ lb/ft}.$$

Why didn't we use $k = -F/x$ here?

(b) What is the force exerted by the spring on the 1.5-lb body just before it is released?

The spring is stretched 4.0 in. or $\frac{1}{3}$ ft. Hence, the force exerted *by* the spring is

$$F = -kx = -(3.0 \text{ lb/ft})(\tfrac{1}{3} \text{ ft}) = -1.0 \text{ lb}.$$

The minus sign indicates that the force is directed opposite to the displacement.

(c) What is the period of oscillation after release?

$$T = 2\pi \sqrt{\frac{m}{k}} = 2\pi \sqrt{\frac{1.5/32}{3.0}} \text{ s} = \frac{\pi}{4} \text{ s} = 0.79 \text{ s}.$$

This corresponds to a frequency $\nu(= 1/T)$ of 1.3 Hz and to an angular frequency $\omega(= 2\pi\nu)$ of 8.0 rad/s.

(d) What is the amplitude of the motion?

The maximum displacement corresponds to zero kinetic energy and a maximum potential energy. This is the initial condition before release, so that the amplitude is the initial displacement of 4.0 in. Hence, $A = \frac{1}{3}$ ft.

(e) What is the maximum speed of the vibrating body?

From Eq. 15-13, $v_{max} = \omega A = (2\pi/T)A$,

$$v_{max} = \left(\frac{2\pi}{\pi/4} \text{ s}^{-1}\right)(\tfrac{1}{3} \text{ ft}) = 2.7 \text{ ft/s}.$$

The maximum speed occurs at the equilibrium position, where $x = 0$. This value is achieved twice in each period, the velocity being -2.7 ft/s when the body passes through $x = 0$ after release and $+2.7$ ft/s when the body passes through $x = 0$ on the return trip.

(f) What is the maximum acceleration of the body?

From Eq. 15-13, $a_{max} = \omega^2 A = (k/m)A$,

$$a_{max} = \left(\frac{3.0}{1.5/32}\right)\left(\frac{1}{3}\right) \text{ ft/s}^2 = 21 \text{ ft/s}^2.$$

The maximum acceleration occurs at the ends of the path where $x = \pm A$ and $v = 0$. Hence, $a = -21$ ft/s² at $x = +A$ and $a = +21$ ft/s² at $x = -A$, the acceleration and displacement being oppositely directed.

(g) Compute the velocity, the acceleration, and the kinetic and potential energies of the body when it has moved in halfway from its initial position toward the center of motion.

At this point,
$$x = \frac{A}{2} = \tfrac{1}{6} \text{ ft,}$$

so that from Eq. 15-18,

$$v = -\frac{2\pi}{T} \sqrt{A^2 - x^2}$$

$$= -\frac{2\pi}{\pi/4} \sqrt{\left(\tfrac{1}{3}\right)^2 - \left(\tfrac{1}{6}\right)^2} \text{ ft/s} = -\frac{4}{\sqrt{3}} \text{ ft/s} = -2.3 \text{ ft/s,}$$

$$a = -\frac{k}{m}x = \frac{-3.0}{1.5/32}\left(\frac{1}{6}\right) \text{ ft/s}^2 = -11 \text{ ft/s}^2,$$

$$K = \tfrac{1}{2}mv^2 = \left(\frac{1}{2}\right)\left(\frac{1.5}{32}\right)\left(\frac{4}{\sqrt{3}}\right)^2 \text{ ft} \cdot \text{lb} = \tfrac{1}{8} \text{ ft} \cdot \text{lb},$$

$$U = \tfrac{1}{2}kx^2 = \left(\frac{1}{2}\right)(3)\left(\frac{1}{6}\right)^2 \text{ ft} \cdot \text{lb} = \tfrac{1}{24} \text{ ft} \cdot \text{lb}.$$

(*h*) Compute the total energy of the oscillating system.

Since the total energy is conserved, we can compute it at any stage of the motion. Using previous results, we obtain

$$E = K + U = \left(\tfrac{1}{8} + \tfrac{1}{24}\right) \text{ ft} \cdot \text{lb} = \tfrac{1}{6} \text{ ft} \cdot \text{lb}, \qquad \text{(particle at } x = A/2\text{)}$$

$$E = U_{\max} = \tfrac{1}{2}kx_{\max}^2 = \left(\frac{1}{2}\right)(3)\left(\frac{1}{3}\right)^2 \text{ ft} \cdot \text{lb} = \tfrac{1}{6} \text{ ft} \cdot \text{lb}, \qquad \text{(particle at } x = A\text{)}$$

$$E = K_{\max} = \tfrac{1}{2}mv_{\max}^2 = \left(\frac{1}{2}\right)\left(\frac{1.5}{32}\right)\left(\frac{8}{3}\right)^2 \text{ ft} \cdot \text{lb} = \tfrac{1}{6} \text{ ft} \cdot \text{lb}. \qquad \text{(particle at } x = 0\text{)}$$

(*i*) What is the displacement of the body as a function of time?

In general, we have

$$x = A \cos(\omega t + \phi).$$

We have already found that $A = \tfrac{1}{3}$ ft. We must now determine ω and ϕ. We obtain

$$\omega = \frac{2\pi}{T} = \frac{2\pi}{\pi/4} = 8 \text{ rad/s},$$

so that, with our particular units,

$$x = \tfrac{1}{3} \cos(8t + \phi).$$

At the time $t = 0$, $x = \tfrac{1}{3}$ ft, so that at that instant

$$x = \tfrac{1}{3} \cos\phi = \tfrac{1}{3}$$

or

$$\phi = 0 \text{ rad}.$$

Therefore, with $A = \tfrac{1}{3}$ ft, $\omega = 8$ rad/s, and $\phi = 0$ rad, we obtain

$$x = \tfrac{1}{3} \cos 8t.$$

This describes the motion of the body, where x is in feet, t is in seconds, and the angle $8t$ is in radians.

A few physical systems that move with simple harmonic motion are considered here. We will discuss others from time to time throughout the text.

The Simple Pendulum. A simple pendulum is an idealized body consisting of a point mass, suspended by a light inextensible cord. When pulled to one side of its equilibrium position and released, the pendulum swings in a vertical plane under the influence of gravity. The motion is periodic and oscillatory. We wish to determine the period of the motion.

Figure 15-10 shows a pendulum of length *l*, particle mass *m*, making an angle θ with the vertical. The forces acting on *m* are *m***g**, the gravita-

15-5
APPLICATIONS OF SIMPLE HARMONIC MOTION*

---

* See "A Repertoire of S.H.M." by Eli Maor, *The Physics Teacher*, October 1972, for a full discussion of 16 physical systems that exhibit simple harmonic motion.

tional force, and **T**, the tension in the cord. Choose axes tangent to the circle of motion and along the radius. Resolve mg into a radial component of magnitude $mg \cos \theta$, and a tangential component of magnitude $mg \sin \theta$. The radial components of the forces supply the necessary centripetal acceleration to keep the particle moving on a circular arc. The tangential component is the restoring force acting on m tending to return it to the equilibrium position. Hence, the restoring force is

$$F = -mg \sin \theta.$$

Notice that the restoring force is not proportional to the angular displacement θ but to $\sin \theta$ instead. The resulting motion is, therefore, not simple harmonic. However, *if the angle θ is small*, $\sin \theta$ is very nearly equal to θ in radians.* The displacement along the arc is $x = l\theta$, and for small angles this is nearly straight-line motion. Hence, assuming

$$\sin \theta \cong \theta,$$

we obtain

$$F = -mg\theta = -mg\frac{x}{l} = -\left(\frac{mg}{l}\right)x.$$

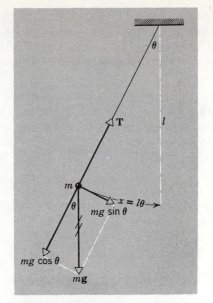

figure 15-10
The forces acting on a simple pendulum are the tension **T** in the string and the weight mg of mass m. The magnitudes of the radial and tangential components of mg are labeled.

For *small displacements*, therefore, the restoring force is proportional to the displacement and is oppositely directed. This is exactly the criterion for simple harmonic motion. The constant mg/l represents the constant k in $F = -kx$. Check the dimensions of k and mg/l. The period of a simple pendulum when its amplitude is small is

$$T = 2\pi\sqrt{\frac{m}{k}} = 2\pi\sqrt{\frac{m}{mg/l}} \quad \text{or} \quad T = 2\pi\sqrt{\frac{l}{g}}. \quad (15\text{-}19)$$

Notice that the period is independent of the mass of the suspended particle.

When the amplitude of the oscillation is not small, the general equation for the period can be shown to be

$$T = 2\pi\sqrt{\frac{l}{g}}\left(1 + \frac{1}{2^2}\cdot\sin^2\frac{\theta_m}{2} + \frac{1}{2^2}\cdot\frac{3^2}{4^2}\cdot\sin^4\frac{\theta_m}{2} + \cdots\right). \quad (15\text{-}20)$$

Here θ_m is the maximum angular displacement and the succeeding terms become smaller and smaller. The period can then be computed to any desired degree of accuracy by taking enough terms in the infinite series. When $\theta_m = 15°$, corresponding to a total to-and-fro angular displacement of $30°$, the true period differs from that given by Eq. 15-19 by less than 0.5%.

Because the period of a simple pendulum is practically independent of the amplitude, the pendulum is useful as a timekeeper. As damping forces reduce the amplitude of swing, the period remains very nearly unchanged. In a pendulum clock energy is supplied automatically by an escapement mechanism to compensate for frictional loss. The pendulum clock with escapement was invented by Christian Huygens (1629–1695).

The simple pendulum also provides a convenient method for measuring the value of g, the acceleration due to gravity. We need not perform a free-fall experiment here, but instead we merely measure l and T.

* For example,.

θ	$\sin \theta$	Difference, %
0° = 0.00000 rad	0.00000	0.00
2° = 0.03491 rad	0.03490	0.03
5° = 0.08727 rad	0.08716	0.24
10° = 0.17453 rad	0.17365	0.50
15° = 0.26180 rad	0.25882	1.14

The Torsional Pendulum. In Fig. 15-11 we show a disk suspended by a wire attached to the center of mass of the disk. The wire is securely fixed to a solid support and to the disk. At the equilibrium position of the disk a radial line is drawn from its center to P, as shown. If the disk is rotated in a horizontal plane to the radial position Q, the wire will be twisted. The twisted wire will exert a torque on the disk tending to return it to the position P. This is a restoring torque. For small twists the restoring torque is found to be proportional to the amount of twist, or the angular displacement (Hooke's law), so that

$$\tau = -\kappa\theta. \tag{15-21}$$

Here κ is a constant that depends on the properties of the wire and is called the *torsional* constant. The minus sign shows that the torque is directed opposite to the angular displacement θ. Equation 15-21 is the condition for *angular simple harmonic motion.*

The equation of motion for such a system is

$$\tau = I\alpha = I\frac{d\omega}{dt} = I\frac{d^2\theta}{dt^2},$$

so that, on using Eq. 15-21, we obtain

$$-\kappa\theta = I\frac{d^2\theta}{dt^2}$$

or

$$\frac{d^2\theta}{dt^2} = -\left(\frac{\kappa}{I}\right)\theta. \tag{15-22}$$

Notice the similarity between Eq. 15-22 for simple angular harmonic motion and Eq. 15-7 for simple linear harmonic motion. In fact, the equations are mathematically identical. We have simply substituted angular displacement θ for linear displacement x, rotational inertia I for mass m, and torsional constant κ for force constant k. By substituting these correspondences, we find the solution of Eq. 15-22, therefore, to be a simple harmonic oscillation in the angle coordinate θ, namely

$$\theta = \theta_m \cos(\omega t + \phi). \tag{15-23}$$

Here, θ_m is the maximum angular displacement, that is, the amplitude of the angular oscillation. In Fig. 15-11 the disk oscillates about the equilibrium position $\theta = 0$ (line OP), the total angular range being $2\theta_m$ (from OQ to OR).

The period of the oscillation by analogy with Eq. 15-10 is

$$T = 2\pi\sqrt{\frac{I}{\kappa}}. \tag{15-24}$$

If κ is known and T is measured, the rotational inertia I about the axis of rotation of any oscillating rigid body can be determined. If I is known and T is measured, the torsional constant κ of any sample of wire can be determined.

Many laboratory instruments involve torsional oscillations, notably the galvanometer. The Cavendish balance is a torsional pendulum (Chapter 16). The balance wheel of a watch is another example of angular harmonic motion, the restoring torque here being supplied by a spiral hairspring.

figure 15-11
The torsional pendulum. The line drawn from the center to P oscillates between Q and R, sweeping out an angle $2\theta_m$ where θ_m is the (angular) amplitude of the motion.

EXAMPLE 2

A thin rod of mass 0.10 kg and length 0.10 m is suspended by a wire which passes through its center and is perpendicular to its length. The wire is twisted and the rod set oscillating. The period is found to be 2.0 s. When a flat body in the shape of an equilateral triangle is suspended similarly through its center of mass, the period is found to be 6.0 s. Find the rotational inertia of the triangle about this axis.

The rotational inertia of the rod is $Ml^2/12$ (see Table 12-1). Hence

$$I_{rod} = \frac{(0.10 \text{ kg})(0.10 \text{ m})^2}{12} = 8.3 \times 10^{-5} \text{ kg} \cdot \text{m}^2.$$

From Eq. 15-24,

$$\frac{T_{rod}}{T_{triangle}} = \left(\frac{I_{rod}}{I_{triangle}}\right)^{1/2} \quad \text{or} \quad I_{triangle} = I_{rod}\left(\frac{T_t}{T_r}\right)^2,$$

so that

$$I_{triangle} = (8.3 \times 10^{-5} \text{ kg} \cdot \text{m}^2)\left(\frac{6.0 \text{ s}}{2.0 \text{ s}}\right)^2 = 7.5 \times 10^{-4} \text{ kg} \cdot \text{m}^2.$$

Does the amplitude of the oscillation affect the period in these cases?

The Physical Pendulum. Any rigid body mounted so that it can swing in a vertical plane about some axis passing through it is called a physical pendulum. This is a generalization of the simple pendulum in which a weightless cord holds a single particle. Actually all real pendulums are physical pendulums.

For convenience we choose our pendulum to be a laminar body, such as may be cut out from a sheet of plywood with a jigsaw, and we choose the axis of oscillation to be at right angles to the plane of this body. We lose nothing essential by this restriction.

In Fig. 15-12 a body of irregular shape is pivoted about a horizontal frictionless axis through P and displaced from the equilibrium position by an angle θ. The equilibrium position is that in which the center of mass of the body, C, lies vertically below P. The distance from pivot to center of mass is d, the rotational inertia of the body about an axis through the pivot is I, and the mass of the body is M. The restoring torque for an angular displacement θ is

$$\tau = -Mgd \sin \theta$$

and is due to the tangential component of the force of gravity. Since τ is proportional to $\sin \theta$, rather than θ, the condition for simple angular harmonic motion does not, in general, hold here. For small angular displacements, however, the relation $\sin \theta \cong \theta$ is, as before, an excellent approximation, so that for *small amplitudes*,

$$\tau = -Mg \, d \, \theta$$

or

$$\tau = -\kappa\theta,$$

where

$$\kappa = Mgd.$$

But

$$\tau = I\frac{d^2\theta}{dt^2} = I\alpha,$$

so that

$$\frac{d^2\theta}{dt^2} = \frac{\tau}{I} = -\frac{\kappa}{I}\theta.$$

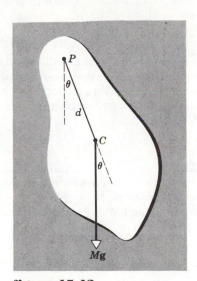

figure 15-12
A laminar physical pendulum, with center of mass C, is pivoted at P and displaced an angle θ from its equilibrium position (when C hangs directly below P). Its weight Mg supplies a restoring torque.

Hence, the period of a physical pendulum oscillating with small amplitude is

$$T = 2\pi\sqrt{\frac{I}{\kappa}} = 2\pi\sqrt{\frac{I \cdot}{Mgd}}. \qquad (15\text{-}25)$$

At larger amplitudes the physical pendulum still has a harmonic motion, but not a simple harmonic one.

Notice that this treatment applies to a laminar object of *any* shape and that the pivot can be located *anywhere*. As a special case consider a point mass m suspended at the end of a weightless string of length l. Here

$$I = ml^2, \qquad M = m, \qquad d = l,$$

so that

$$T = 2\pi\sqrt{\frac{I}{Mgd}} = 2\pi\sqrt{\frac{l}{g}},$$

which is the period of a simple pendulum with small amplitude. The physical pendulum is often used for accurate determinations of g.

Equation 15-25 can be solved for the rotational inertia I, giving

$$I = \frac{T^2 Mgd}{4\pi^2}. \qquad (15\text{-}26)$$

The quantities on the right are all directly measurable. The center of mass can be determined by suspension as was shown in Fig. 14-4. Hence, the rotational inertia about an axis of rotation (other than through the center of mass) of a body of any shape can be determined by suspending the body as a physical pendulum from that axis.

EXAMPLE 3

Find the length of a simple pendulum whose period is equal to that of a particular physical pendulum.

Equating the period of a simple pendulum to that of a physical pendulum, we obtain

$$T = 2\pi\sqrt{\frac{l}{g}} = 2\pi\sqrt{\frac{I}{Mgd}}$$

or

$$l = \frac{I}{Md}. \qquad (15\text{-}27)$$

Hence, as far as its period of oscillation is concerned, the mass of a physical pendulum may be considered to be concentrated at a point whose distance from the pivot is $l = I/Md$. This point is called the *center of oscillation* of the physical pendulum. Notice that it depends on the location of the pivot for any given body.

EXAMPLE 4

A disk is pivoted at its rim (Fig. 15-13). Find its period for small oscillations and the length of the equivalent simple pendulum.

The rotational inertia of a disk about an axis through its center is $\frac{1}{2}Mr^2$, where r is the radius and M is the mass of the disk. The rotational inertia about the pivot at the rim is

$$I = \tfrac{1}{2}Mr^2 + Mr^2 = \tfrac{3}{2}Mr^2.$$

The period then, with $d = r$, is

$$T = 2\pi\sqrt{\frac{I}{Mgr}} = 2\pi\sqrt{\frac{3}{2}\frac{Mr^2}{Mgr}} = 2\pi\sqrt{\frac{3}{2}\frac{r}{g}},$$

independent of the mass of the disk.

The simple pendulum having the same period has a length

$$l = \frac{I}{Mr} = \tfrac{3}{2}r$$

or three-fourths the diameter of the disk. The center of oscillation of the disk pivoted at P is, therefore, at O, a distance $\tfrac{3}{2}r$ below the point of support. Is any particular mass required of the equivalent simple pendulum?

If we pivot the disk at a point midway between the rim and the center, as at O, we find that $I = \tfrac{3}{4}Mr^2$ and $d = \tfrac{1}{2}r$. The period T is

$$T = 2\pi\sqrt{\tfrac{3}{2}r/g}$$

just as before. This illustrates a general property of the center of oscillation O and the point of support P, namely, if the pendulum is pivoted about a new axis through O, its period is unchanged and P becomes the new center of oscillation.

If the disk were pivoted at the center, what would be its period of oscillation?

EXAMPLE 5

The center of oscillation of a physical pendulum has another interesting property. If an impulsive force (assumed horizontal and in the plane of oscillation) acts at the center of oscillation, no reaction is felt at the point of support. Prove this for an impulsive force $\mathbf{F}$ acting toward the left at point O in Fig. 15-13. Assume the pendulum to be initially at rest.

This is a case of combined translation and rotation (see Sec. 12-7). The translation effect, acting alone, would make P in Fig. 15-13 move to the left with an acceleration

$$\mathbf{a}_{\text{left}} = F/M.$$

The rotational effect, acting alone, would produce a clockwise angular acceleration about C of

$$\alpha = \tau/I$$
$$= (F)(\tfrac{1}{2}r)/(\tfrac{1}{2}Mr^2)$$
$$= F/Mr.$$

Because of this angular acceleration P would move to the right with an acceleration

$$\mathbf{a}_{\text{right}} = \alpha r$$
$$= (F/Mr)(r) = F/M.$$

Thus $\mathbf{a}_{\text{left}} = -\mathbf{a}_{\text{right}}$ and there is no movement at point P.

When viewed from this point of view the center of oscillation is often called the *center of percussion*. Baseball players know that unless the ball hits the bat at just the right spot (center of percussion) the impact will sting their hands. The "sting" has a different direction depending on whether the ball strikes on one side or the other of this spot.

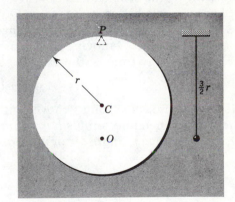

figure 15-13
Example 4. A physical pendulum consisting of a disk pivoted at the edge *(P)*, along with a simple pendulum having the same period. O is the center of oscillation.

EXAMPLE 6

The period of a disk of radius 10.2 cm executing small oscillations about a pivot at its rim is measured to be 0.784 s. Find the value of g, the acceleration due to gravity at that location.

From $T = 2\pi\sqrt{\tfrac{3}{2}r/g}$, we obtain

$$g = \frac{6\pi^2 r}{T^2}.$$

With $T = 0.784$ s and $r = 0.102$ m, we obtain

$$g = \frac{6\pi^2 \cdot 0.102}{(0.784)^2}\ \text{m/s}^2 = 9.82\ \text{m/s}^2.$$

Let us consider the relation between simple harmonic motion along a straight line and uniform circular motion. This relation is useful in describing many features of simple harmonic motion. It also gives a simple geometric meaning to the angular frequency ω and the phase constant ϕ. Uniform circular motion is also an example of a combination of simple harmonic motions, a phenomenon we deal with rather often in wave motion.

In Fig. 15-14 Q is the point moving around a circle of radius A with a constant angular speed of ω, expressed, say, in radians/second. P is the perpendicular projection of Q on the horizontal diameter, along the x-axis. Let us call Q the *reference point* and the circle on which it moves the *reference circle*. As the reference point revolves, the projected point P moves back and forth along the horizontal diameter. The x-component of Q's displacement is always the same as the displacement of P; the x-component of the velocity of Q is always the same as the velocity of P; and the x-component of the acceleration of Q is always the same as the acceleration of P.

Let the angle between the radius OQ and the x-axis at the time $t = 0$ be called ϕ. At any later time t, the angle between OQ and the x-axis is $(\omega t + \phi)$, the point Q moving with constant angular speed. The x-coordinate of Q at any time is, therefore,

$$x = A\cos{(\omega t + \phi)}. \tag{15-28}$$

Hence, the projected point P moves with simple harmonic motion along the x-axis. Therefore, *simple harmonic motion can be described as the projection along a diameter of uniform circular motion.*

15-6
RELATION BETWEEN SIMPLE HARMONIC MOTION AND UNIFORM CIRCULAR MOTION

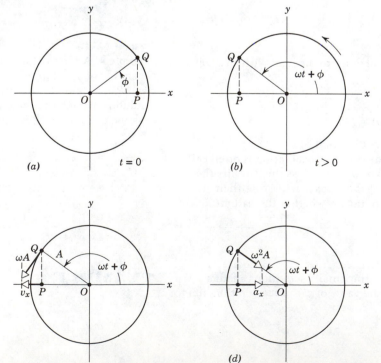

(a) $t = 0$

(b) $t > 0$

(d)

figure 15-14
The relation of simple harmonic motion to uniform circular motion. Q moves in uniform circular motion and P in simple harmonic motion. Q has angular speed ω, P angular frequency ω. (a, b) The x-component of Q's displacement is always equal to P's displacement. (c) The x-component of Q's velocity is always equal to P's velocity. (d) The x-component of Q's acceleration is always equal to P's acceleration.

The angular frequency ω of simple harmonic motion is the same as the angular speed of the reference point. The frequency of the simple harmonic motion is the same as the number of revolutions per unit time of the reference point. Hence, $\nu = \omega/2\pi$ or $\omega = 2\pi\nu$. The time for a complete revolution of the reference point is the same as the period T of the simple harmonic motion. Hence, $T = 2\pi/\omega$ or $\omega = 2\pi/T$. The phase of the simple harmonic motion, $\omega t + \phi$, is the angle that OQ makes with the x-axis at any time t (Fig. 15-14b,c,d). The angle that OQ makes with the x-axis at the time $t = 0$ (Fig. 15-14a) is ϕ, the phase constant or initial phase of the motion. The amplitude of the simple harmonic motion is the same as the radius of the reference circle.

The tangential velocity of the reference point Q has a magnitude of ωA. Hence, the x-component of this velocity (Fig. 15-14c) is

$$v_x = -\omega A \sin (\omega t + \phi).$$

This relation gives a negative v_x when Q and P are moving to the left and a positive v_x when Q and P are moving to the right. Notice that v_x is zero at the end points of the simple harmonic motion, where $\omega t + \phi$ is zero and π, as required.

The acceleration of point Q in uniform circular motion is directed radially inward and has a magnitude of $\omega^2 A$. The acceleration of the projected point P is the x-component of the acceleration of the reference point Q (Fig. 15-14d). Hence,

$$a_x = -\omega^2 A \cos (\omega t + \phi)$$

gives the acceleration of the point executing simple harmonic motion. Notice that a_x is zero at the midpoints of the simple harmonic motion, where $\omega t + \phi = \pi/2$ or $3\pi/2$, as required.

These results are all identical with those of simple harmonic motion along the x-axis; see Eqs. 15-13.

If we had taken the perpendicular projection of the reference point onto the y-axis, instead, we would have obtained for the motion of the y-projected point

$$y = A \sin (\omega t + \phi). \tag{15-29}$$

This is again a simple harmonic motion. It differs only in phase from Eq. 15-28, for if we replace ϕ by $\phi - \pi/2$, then $\cos (\omega t + \phi)$ becomes $\sin (\omega t + \phi)$. It is clear that the projection of uniform circular motion along *any* diameter gives a simple harmonic motion.

Conversely, uniform circular motion can be described as a combination of two simple harmonic motions. It is that combination of two simple harmonic motions, occuring along perpendicular lines, which have the same amplitude and frequency but differ in phase by 90°. When one component is at the maximum displacement, the other component is at the equilibrium point. If we combine these components (Eqs. 15-28 and 15-29), we obtain at once the relation

$$r = \sqrt{x^2 + y^2} = A.$$

By writing the relations for v_y and a_y (you should do this) and combining corresponding quantities, we obtain also the relations

$$v = \sqrt{v_x{}^2 + v_y{}^2} = \omega A,$$

$$a = \sqrt{a_x{}^2 + a_y{}^2} = \omega^2 A.$$

These relations correspond respectively to the magnitudes of the displacement, the velocity, and the acceleration in uniform circular motion.

It will be possible for us to analyze many complicated motions as combinations of individual simple harmonic motions. Circular motion is a particularly simple combination. In the next section we shall consider other combinations of simple harmonic motions.

In example 1 we considered a body executing a horizontal simple harmonic motion. The equation of that motion (units?) was

$$x = \tfrac{1}{3} \cos 8t.$$

This motion can also be represented as the projection of uniform circular motion along a horizontal diameter.

(a) Give the properties of the corresponding uniform circular motion.

The x-component of the circular motion is given by

$$x = A \cos (\omega t + \phi).$$

Therefore, the reference circle must have a radius $A = \tfrac{1}{3}$ ft, the initial phase or phase constant must be $\phi = 0$, and the angular velocity must be $\omega = 8$ rad/s, in order to obtain the equation $x = \tfrac{1}{3} \cos 8t$ for the horizontal projection.

(b) From the motion of the reference point determine the time required for the body to come halfway in toward the center of motion from its initial position.

As the body moves halfway in, the reference point moves through an angle of $\omega t = 60°$ (Fig. 15-15). The angular velocity is constant at 8 rad/s so that the time required to move through 60° is

$$t = \frac{60°}{\omega} = \frac{\pi/3 \text{ rad}}{8 \text{ rad/s}} = \frac{\pi}{24} \text{ s} = 0.13 \text{ s}.$$

The time may also be computed directly from the equation of motion. Thus,

$$x = \tfrac{1}{3} \cos 8t \quad \text{and} \quad x = \frac{A}{2} = \tfrac{1}{6}.$$

Hence

$$\tfrac{1}{6} = \tfrac{1}{3} \cos 8t \quad \text{or} \quad 8t = \cos^{-1}(\tfrac{1}{2}) = \frac{\pi}{3}.$$

Therefore,

$$t = \frac{\pi}{24} \text{ s} = 0.13 \text{ s}.$$

EXAMPLE 7

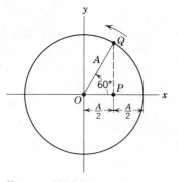

figure 15-15
Example 7. The particles Q and P of Fig. 15-14 are shown for $\omega t = 60°$. Since ω is known, t may be found.

Often two linear simple harmonic motions *at right angles* are combined. The resulting motion is the sum of two independent oscillations. Consider first the case in which the frequencies of the vibrations are the same, such as

$$\begin{aligned} x &= A_x \cos (\omega t + \phi_x), \\ y &= A_y \cos (\omega t + \phi_y). \end{aligned} \qquad (15\text{-}30)$$

The x- and y-motions have different amplitudes and different phase constants, however.

If the phase constants are the same so that $\phi_x = \phi_y = \phi$, the resulting motion is a straight line. This can be shown analytically, for when we eliminate t from the equations

$$x = A_x \cos (\omega t + \phi) \qquad y = A_y \cos (\omega t + \phi),$$

we obtain

$$y = (A_y/A_x)x.$$

15-7
COMBINATIONS OF HARMONIC MOTIONS

This is the equation of a straight line, whose slope is A_y/A_x. In Fig. 15-16a and b we show the resultant motion for two cases, $A_y/A_x = 1$ and $A_y/A_x = 2$. In these cases both the x- and y-displacements reach a maximum at the same time and reach a minimum at the same time. They are in phase.

If the phase constants are different, the resulting motion will not be a straight line. For example, if the phase constants differ by $\pi/2$, the maximum x-displacement occurs when the y-displacement is zero and vice versa. When the amplitudes are equal, the resulting motion is circular; when the amplitudes are unequal, the resulting motion is elliptical. Two cases, $A_y/A_x = 1$ and $A_y/A_x = 2$, are shown in Fig. 15-16c and d, for $\phi_x = \phi_y + \pi/2$. The cases $A_y/A_x = 1$ and $A_y/A_x = 2$, for $\phi_x = \phi_y - \pi/4$, are shown in Fig. 15-16e and f.

All possible combinations of two simple harmonic motions at right angles having the same frequency correspond to *elliptical* paths, the circle and straight line being special cases of an ellipse. This can be shown analytically by combining Eqs. 15-30 and eliminating the time; you can show that the resulting equation is that of an ellipse. The shape of the ellipse depends only on the ratio of the amplitudes, A_y/A_x, and the *difference* in phase between the two oscillations, $\phi_x - \phi_y$. The actual motion can be either clockwise or counterclockwise, depending on which component leads in phase.

A simple way to produce such patterns is by means of an oscillo-

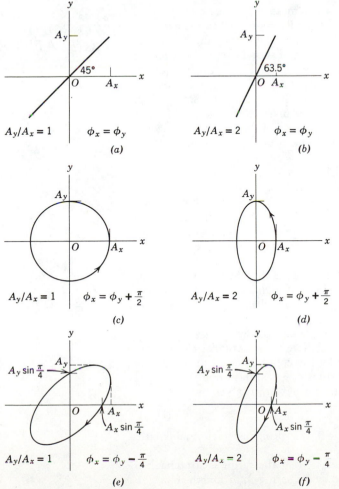

figure 15-16
Simple harmonic motions in two dimensions. *(a)* The amplitudes of x and y (namely A_x and A_y) are the same, as are their phase constants. *(b)* y's amplitude is twice x's but their phase constants are the same. *(c)* Their amplitudes are equal, but x leads y in phase by 90°. *(d)* Same as *(c)* except that y's amplitude is twice x's. *(e)* Equal amplitudes, but x lags y in phase by 45°. *(f)* Same as *(e)* except that y's amplitude is twice x's.

scope. In this, electrons are deflected by each of two electric fields at right angles to one another. The field strengths alternate sinusoidally with the same frequency, but their phases and amplitudes can be varied. In this way the electrons can be made to trace out the various patterns discussed above on a fluorescent screen. We can also produce these patterns mechanically by means of a pendulum swinging with small amplitude but not confined to one vertical plane. Such combinations of two simple harmonic motions at right angles having the same frequency are particularly important in the study of polarized light and alternating current circuits.

Combinations of simple harmonic motions of the same frequency in the *same direction*, but with different amplitudes and phases, are of special interest in the study of diffraction and interference of light, sound, and electromagnetic radiation. This will be discussed later in the text.

If two oscillations of *different frequencies* are combined at right angles, the resulting motion is more complicated. The motion is not even periodic unless the two component frequencies ω_1 and ω_2 are the ratio of two integers (see Problem 49). Oscillations of different frequencies in the same direction may also be combined. The treatment of this motion is particularly important in the case of sound vibrations and will be discussed in Chapter 20.

15-8 TWO-BODY OSCILLATIONS

The simple harmonic oscillator of Fig. 15-4 is a mass m coupled by a spring of force constant k to a solid wall. The wall is rigidly connected to the earth, so that this system is really a two-body system, connected by a spring, one of the bodies being effectively of infinite mass. This solid support remains at rest in an inertial reference frame so that the change in length of the spring is equal to the displacement of the mass m; the other end of the spring does not move. In this case we defined the potential energy $U(x)$ of the oscillating system of Fig. 15-4 to be a function of the displacement x of the mass m alone (see Figs. 15-3, 9). This again is equivalent to assuming that one end of the spring is connected to an infinite mass so that the extension of the spring is determined by the motion of mass m alone.

Often in nature we find two-body oscillating systems in which we *cannot* take the mass of one of the bodies to be infinite and we must consider the motions of both bodies in an appropriate inertial reference frame. Examples are diatomic molecules such as H_2, CO, HCl, etc., which can oscillate along their axis of symmetry. The coupling between the atoms that make up these molecules is electromagnetic, but we may imagine them, for our purpose, to be connected by a tiny, massless spring.

The surprising thing about two-body oscillators is that, by redefining terms slightly and by introducing a new concept (that of *reduced mass*), we can describe the oscillations by exactly the same equations that we have already derived for the (effectively) one-body system of Fig. 15-4. Let us prove this.

Figure 15-17a shows two bodies m_1 and m_2 connected by a (massless) spring of force constant k; the system is free to oscillate on a frictionless horizontal surface. We locate the ends of the spring by the coordinates $x_1(t)$ and $x_2(t)$, as shown. The length of the spring at any instant is $x_1 - x_2$. If its normal, unstressed length is l, then the *change* in length of the spring, $x(t)$, is given by

$$x = (x_1 - x_2) - l. \tag{15-31}$$

If x is positive, the spring is stretched, if $x = 0$, the spring has its normal length, and if x is negative, it is compressed.

In Fig. 15-17a we assume, for concreteness, that the spring is stretched, so that $x > 0$. We show also the force $\mathbf{F}$ exerted by the spring on m_2 and the force $-\mathbf{F}$

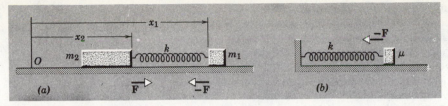

figure 15-17
(a) Two bodies of masses m_1 and m_2 connected by a (massless) spring whose unstressed length is l. (b) A single body of mass μ (the reduced mass) connected by an identical spring to a rigid wall.

exerted on m_1. These two forces are equal and opposite, as the figure shows, and have the common magnitude $F = kx$.

If we apply Newton's second law, $F = ma$, to masses m_1 and m_2, we obtain

$$m_1 \frac{d^2x_1}{dt^2} = -kx$$

and

$$m_2 \frac{d^2x_2}{dt^2} = +kx.$$

Let us now multiply the first equation by m_2 and the second equation by m_1 and subtract. We obtain

$$m_1 m_2 \frac{d^2x_1}{dt^2} - m_1 m_2 \frac{d^2x_2}{dt^2} = -m_2 kx - m_1 kx,$$

which we can write as

$$\frac{m_1 m_2}{m_1 + m_2} \frac{d^2}{dt^2}(x_1 - x_2) = -kx. \qquad (15\text{-}32)$$

Let us call the quantity $m_1 m_2/(m_1 + m_2)$, which has the dimensions of mass, the *reduced mass* of the system and give it the symbol μ; that is,

$$\mu = \frac{m_1 m_2}{m_1 + m_2}. \qquad (15\text{-}33)$$

Because l is a constant, $d^2(x_1 - x_2)/dt^2 = d^2x/dt^2$ (see Eq. 15-31) and Eq. 15-32 now can be written as

$$\frac{d^2x}{dt^2} + \frac{k}{\mu}x = 0. \qquad (15\text{-}34)$$

This is identical in form to Eq. 15-5 which we developed for the single-body oscillation of Fig. 15-4. The differences are that (1) x in Eq. 15-34 is the *relative* displacement of the two blocks from their equilibrium positions (see Eq. 15-31) rather than the displacement of a single block from its equilibrium position, and (2) μ is the *reduced mass* of the pair of blocks rather than the mass of a single block.

Note from Eq. 15-33, which we can write either as

$$\mu = m_1 \frac{m_2}{m_1 + m_2} = m_2 \frac{m_1}{m_1 + m_2}$$

or as

$$\frac{1}{\mu} = \frac{1}{m_1} + \frac{1}{m_2},$$

that (for finite masses) μ is always *smaller* than m_1 or m_2; hence the name *reduced* mass. Equation 15-34 leads, by way of the derivation that follows Eq. 15-6, to

$$v = \frac{1}{2\pi} \sqrt{\frac{k}{\mu}} \qquad \text{or} \qquad T = 2\pi \sqrt{\frac{\mu}{k}} \qquad (15\text{-}35)$$

for the frequency and period of oscillation of the system of Fig. 15-17a. It is clear that this system has the same frequency and period as a single block of mass μ, connected by a similar spring to a rigid wall, as in Fig. 15-17b. Hence, the two-body oscillation of Fig. 15-17a is equivalent to the one-body oscillation of Fig. 15-17b. One particle moves relative to the other particle as though the other particle were fixed and the mass of the moving one were reduced to μ. The reduced mass concept is applied widely in physics.

We can solve Eq. 15-34, as in Section 15-3, to yield these relations:

$$x = A \cos (\omega t + \phi),$$

$$v = dx/dt = -\omega A \sin (\omega t + \phi),$$

and

$$a = dv/dt = -\omega^2 A \cos (\omega t + \phi).$$

They are identical with Eqs. 15-13 except that here x, v, and a are the *relative* displacement, velocity, and acceleration, respectively, of the two blocks. Thus

$$x = (x_1 - x_2) - l,$$

$$v = dx/dt = v_1 - v_2, \tag{15-36}$$

and

$$a = dv/dt = a_1 - a_2,$$

in which the subscripts refer to the two blocks.

The potential energy of a two-body, simple harmonic oscillator is given by $U(x) = \frac{1}{2}kx^2$ which shows clearly, because x depends on the positions of both blocks (see Eq. 15-36), that the potential energy is a characteristic of the system as a whole.

Many actual two-body oscillators, although harmonic, are not simple harmonic; their potential energy curves, like that of Fig. 8-7a which refers to a diatomic molecule, are not parabolic. Even such oscillators, however, behave like simple harmonic oscillators for small enough amplitudes of oscillation about the equilibrium position. Note, too, that x in Fig. 8-7a has a different meaning than we have assigned to it in this chapter; it is the actual separation, rather than (see Eq. 15-36) the difference between the actual separation and the equilibrium separation. Thus in Fig. 8-7a the stable equilibrium position corresponds, not to $x = 0$ as in Fig. 15-2, but to $x = \sqrt[6]{2a/b}$. This change is only a change in the origin of the x-axis of the potential energy curve and has no fundamental significance.

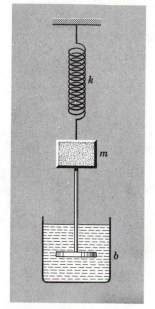

figure 15-18
A damped harmonic oscillator. A disk is attached to the mass and immersed in a fluid which exerts a damping force $-b \, dx/dt$. The elastic restoring force is $-kx$.

15-9
DAMPED HARMONIC MOTION

Up to this point we have assumed that no frictional forces act on the oscillator. If this assumption held strictly, a pendulum or a weight on a spring would oscillate indefinitely. Actually, the amplitude of the oscillation gradually decreases to zero as a result of friction. The motion is said to be damped by friction and is called *damped harmonic motion*. Often the friction arises from air resistance or internal forces. The magnitude of the frictional force usually depends on the speed. In most cases of interest the frictional force is proportional to the velocity of the body but directed opposite to it. An example of a damped oscillator is shown in Fig. 15-18.

The equation of motion of the damped simple harmonic oscillator is given by the second law of motion, $F = ma$, in which F is the sum of the restoring force $-kx$ and the damping force $-b \, dx/dt$. Here b is a positive constant. We obtain

$$F = ma,$$

or

$$-kx - b\frac{dx}{dt} = m\frac{d^2x}{dt^2}$$

or

$$m\frac{d^2x}{dt^2} + b\frac{dx}{dt} + kx = 0. \tag{15-37}$$

If b is small, the solution of this differential equation (given without proof)* is

$$x = Ae^{-bt/2m} \cos{(\omega' t + \phi)}, \qquad (15\text{-}38)$$

where

$$\omega' = 2\pi\nu' = \sqrt{\frac{k}{m} - \left(\frac{b}{2m}\right)^2}. \qquad (15\text{-}39)$$

In Fig. 15-19 we plot the displacement x as a function of the time t for oscillatory motion with small damping.

We can interpret the solution as follows. First, the frequency is smaller and the period is longer when friction is present. Friction slows down the motion, as might be expected. If no friction were present, b would equal zero and ω' would equal $\sqrt{k/m}$ or ω, which is the angular frequency of undamped motion. When friction is present, ω' is less than ω, as shown by Eq. 15-39. Second, the amplitude of the motion gradually decreases to zero. The time interval τ during which the amplitude drops to $1/e$ of its initial value is called the *mean lifetime* of the oscillation. The amplitude factor is $Ae^{-bt/2m}$, so that $\tau = 2m/b$. Once again, if there were no friction present, b would equal zero and the amplitude would have the constant value A as time went on; the lifetime would be infinite.

If the force of friction is great enough, b becomes so large that Eq. 15-38 is no longer a valid solution of the equation of motion.* Then the motion will not be periodic at all. The body merely returns to its equilibrium position when released from its initial displacement A.

In damped harmonic motion the energy of the oscillator is gradually dissipated by friction and falls to zero in time.

Thus far we have discussed only the natural oscillations of a body, that is, the oscillations that occur when the body is displaced and then released. For a mass attached to a spring the natural frequency is

$$\omega = 2\pi\nu = \sqrt{\frac{k}{m}}$$

in the absence of friction and

$$\omega' = 2\pi\nu' = \sqrt{\frac{k}{m} - \left(\frac{b}{2m}\right)^2},$$

in the presence of a small frictional force bv.

A different situation arises, however, when the body is subject to an oscillatory external force. As examples, a bridge vibrates under the influence of marching soldiers, the housing of a motor vibrates owing to periodic impulses from an irregularity in the shaft, and a tuning fork vibrates when exposed to the periodic force of a sound wave. The oscillations that result are called *forced oscillations*. These forced oscillations have the frequency of the *external force* and not the natural frequency of the body. However, the response of the body depends on the relation between the forced and the natural frequency. A succession of small impulses applied at the proper frequency can produce an oscillation of large amplitude. A child using a swing learns that by pumping at proper time intervals he can make the swing move with a large amplitude. The problem of forced oscillations is a very general one. Its solution is useful in acoustic systems, alternating current circuits, and atomic physics as well as in mechanics.

The equation of motion of a forced oscillator follows from the second law of motion. In addition to the restoring force $-kx$ and the damping force $-b\,dx/dt$, we have also the applied oscillating external force. For simplicity let this external force be given by $F_m \cos{\omega'' t}$. Here F_m is the maximum value of the external force and ω'' $(= 2\pi\nu'')$ is its angular frequency. We can imagine such a force applied directly to the suspended mass of Fig. 15-18, if we wish, for concreteness.

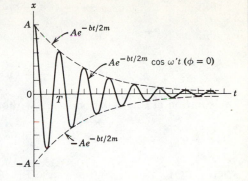

figure 15-19
Damped harmonic motion plotted versus time. The motion is oscillatory with everdecreasing amplitude. The amplitude $(- - -)$ is seen to start with value A and decay exponentially to zero as $t \to \infty$.

15-10
FORCED OSCILLATIONS AND RESONANCE

* See, for example, K. R. Symon, *Mechanics*, third edition, Addison-Wesley Publishing Company, 1971, Section 2.9.

From
$$F = ma,$$

we obtain
$$-kx - b\frac{dx}{dt} + F_m \cos \omega''t = m\frac{d^2x}{dt^2}$$

or
$$m\frac{d^2x}{dt^2} + b\frac{dx}{dt} + kx = F_m \cos \omega''t. \tag{15-40}$$

The solution of this equation (given without proof)* is

$$x = \frac{F_m}{G} \sin (\omega''t - \phi), \tag{15-41}$$

where
$$G = \sqrt{m^2(\omega''^2 - \omega^2)^2 + b^2\omega''^2}, \tag{15-42}$$

and
$$\phi = \cos^{-1}\frac{b\omega''}{G}. \tag{15-43}$$

Let us consider the resulting motion in a qualitative way.

Notice (Eq. 15-41) that the system vibrates with the frequency of the driving force, ω'', rather than with its natural frequency ω, and that the motion is un-damped harmonic motion.

The simplest case is that in which there is no damping, which means that $b = 0$ in Eq. 15-42. The factor G, which has the value $|m(\omega''^2 - \omega^2)|$ for $b = 0$, is large when the frequency of the driving force ω'' is very different from the nat-ural undamped frequency of the system ω. This means that the amplitude of the resultant motion, F_m/G, is small. As the driving frequency approaches the natu-ral frequency, that is, as $\omega'' \to \omega$, we see that $G \to 0$ and the amplitude $F_m/G \to \infty$. Actually some damping is always present so that the amplitude of oscillation, although it may become large, remains finite in practice.

For actual, damped oscillators (for which $b \neq 0$ in Eq. 15-42), there is a char-acteristic value of the driving frequency ω'' at which the amplitude of oscilla-tion is a maximum. This condition is called *resonance*† and the value of ω'' at which resonance occurs is called the *resonant frequency*. The smaller the damping in a given system the closer is the resonant frequency to the natural undamped frequency ω. Frequently the damping is small enough so that the resonant frequency can be taken to equal the natural undamped frequency ω with small error. Likewise, for small damping, the natural undamped frequency $\omega\ (= \sqrt{k/m})$ can be taken to equal the natural damped frequency ω' (see Eq. 15-39) with small error.

In Fig. 15-20 we have drawn five curves giving the amplitude of the forced vibrations as a function of the ratio of the driving frequency ω'' to the undamped natural frequency ω. Each of the five curves corresponds to a different value of the damping constant b. Curve (a) shows the amplitude when $b = 0$, that is, when there is no damping. In this case, as we have seen, the amplitude becomes infinite at $\omega'' = \omega$ because energy is being fed into the system continuously by the applied force and none of it is dissipated. In practice, some friction is al-ways present, so the amplitude reaches a large, but finite, value. Of course, when the amplitude gets so large that Hooke's law no longer holds and the elastic limit is exceeded, the system is no longer governed by Eq. 15-40. Often the system breaks, as in the Tacoma Bridge disaster (Fig. 15-21). Curves (b) and (c) give the amplitude of forced vibration for two cases of increasing damping.

The displacement caused by a constant force F_m applied to a system with a force constant k is simply F_m/k. Notice (Fig. 15-20) that the amplitude of the

---

* *Ibid.*, Section 2.10.

† Resonance, defined here to occur at the frequency at which the forced oscillations have their maximum amplitude, may be defined in other ways as, for example, at the frequency at which maximum power is transferred from the driving unit to the oscillating system or at which the speed of the oscillating mass is a maximum. The definitions are not equivalent; we will discuss the matter further when we deal with forced electrical oscil-lations; see Problem 55.

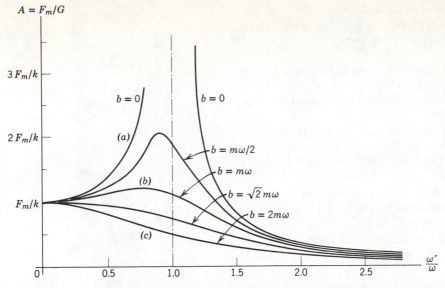

figure 15-20
The amplitude of a driven damped simple harmonic oscillator is plotted versus the ratio of the driving frequency ω'' to the undamped natural frequency ω. Curves for five different degrees of damping are shown; curve *(a)* shows no damping and curve *(c)* high damping. We notice that the resonant peak moves nearer and nearer the vertical line at $\omega''/\omega = 1$ as b becomes smaller and smaller.

figure 15-21
On July 1, 1940, the Tacoma Narrows Bridge at Puget Sound, Washington, was completed and opened to traffic. Just four months later a mild gale set the bridge oscillating until the main span broke up, ripping loose from the cables and crashing into the water below. The wind produced a fluctuating resultant force in resonance with a natural frequency of the structure. This caused a steady increase in amplitude until the bridge was destroyed. Many other bridges were later redesigned to make them aerodynamically stable.

forced vibrations is rather large compared to this static displacement. A column of soldiers marching in step across a bridge can set it vibrating with a destructively large amplitude if the frequency of their steps happens to be some natural frequency of the bridge. This is the reason why soldiers break step when crossing a bridge. Resonance considerations are very important in many electrical, acoustic, and atomic devices, as we shall see later.

questions

1. Give some examples of motions that are approximately simple harmonic. Why are motions that are exactly simple harmonic rare?

2. A typical screen-door spring is tension-stressed in its normal state, that is, adjacent turns cling to each other and resist separation. Does such a spring obey Hooke's law?

3. Is Hooke's law obeyed, even approximately, by a diving board? A trampoline? A coiled spring made of lead wire?

4. A spring has a force constant k, and a mass m is suspended from it. The spring is cut in half and the same mass is suspended from one of the halves. How are the frequencies of oscillation, before and after the spring is cut, related?

5. An unstressed spring has a force constant k. It is stretched by a weight hung from it to an equilibrium length well within the elastic limit. Does the spring have the same force constant k for displacements from this new equilibrium position?

6. Suppose we have a block of unknown mass and a spring of unknown force constant. Show how we can predict the period of oscillation of this block-spring system simply by measuring the extension of the spring produced by attaching the block to it.

7. Any real spring has mass. If this mass is taken into account, explain qualitatively how this will change our expressions for the period of oscillation of a spring-and-mass system (see Problem 31).

8. Can one have an oscillator which even for smaller amplitudes is not simple harmonic? That is, can one have a nonlinear restoring force in an oscillator even at arbitrarily small amplitudes?

9. How are each of the following properties of a simple harmonic oscillator affected by doubling the amplitude: period, force constant, total mechanical energy, maximum velocity, maximum acceleration?

10. What changes could you make in a harmonic oscillator that would double the maximum speed of the oscillating mass?

11. We think of energy exchange for a mass-spring system as a transfer between U and K, their sum E remaining constant; see Fig. 15-9. Suppose a mass is oscillating between two stretched springs, as in Fig. 15-23. A student says: "Consider the mass instantaneously at rest at one end of its limit of oscillation. Here $K = 0$. However, when the mass starts to move toward its equilibrium position K increases. Also because $U = \frac{1}{2}kx^2$ (Eq. 8-11) both springs increase their potential energy because the sign of x (compression or extension) does not matter. Therefore K and U both increase. How can their sum $(= E)$ be constant?"

What is wrong with this argument?

12. A person stands on a bathroom-type scale which rests on a platform suspended by a large spring. The whole system executes simple harmonic motion in a vertical direction. Describe the variation in scale reading during a period of motion.

13. Could we ever construct a simple pendulum?

14. Could standards of mass, length, and time be based on properties of a pendulum? Explain.

15. Show that as the amplitude θ_m in Eq. 15-20 approaches 180° the period approaches infinity. Is this reasonable?

16. Predict by qualitative arguments whether a pendulum oscillating with large amplitude will have a period longer or shorter than the period for oscillations with small amplitude. (Consider extreme cases.)

17. What happens to the frequency of a swing as its oscillations die down from large amplitude to small?

18. How is the period of a pendulum affected when its point of suspension is (a) moved horizontally with acceleration a; (b) moved vertically upward with acceleration a; (c) moved vertically downward with acceleration $a < g$. Which case, if any, applies to a pendulum mounted on a cart rolling down an inclined plane?

19. Why was an axis through the center of mass excluded in using Eq. 15-26 to determine I? Does this equation apply to such an axis? How can you determine I for such an axis using physical pendulum methods?

20. A hollow sphere is filled with water through a small hole in it. It is hung by a long thread and, as the water slowly flows out of the hole at the bottom,

one finds that the period of oscillation first increases and then decreases. Explain.

21. (a) The effect of the mass, m, of the cord attached to the bob, of mass M, of a pendulum is to increase the period over that for a simple pendulum in which $m = 0$. Make this plausible. (b) Although the effect of the mass of the cord on the pendulum is to increase its period, a cord of length l swinging without anything on the end ($M = 0$) has a period less than that of a simple pendulum of length l. Make that plausible. (See "Effect of the Mass of the Cord on the Period of a Simple Pendulum," by H. L. Armstrong, *American Journal of Physics*, June, 1976.)

22. Two pendula, each consisting of a disk attached to a light bar, are identical except for the coupling between disk and bar. In one the bar is rigidly mounted to the disk; in the other ball-bearings are used so that the disk would be free to spin about the end of the bar, for example. Both pendula are hung, pulled aside to the same height, and released. Which has the greater period? Explain.

23. Will the frequency of oscillation of a torsional pendulum change if it is taken to the moon? a simple pendulum? a mass-spring oscillator? a physical pendulum?

24. How can a pendulum be used so as to trace out a sinusoidal curve?

25. What component simple harmonic motions would give a figure 8 as the resultant motion?

26. Is there a connection between the F vs. x relation at the molecular level and the macroscopic relation between F and x in a spring?

27. (a) Under what circumstances would the reduced mass of a two-body system be equal to the mass of one body? Explain. (b) What is the reduced mass if the bodies have equal mass? (c) Do cases (a) and (b) give the extreme values of the reduced mass?

28. Why are damping devices often used on machinery? Give an example.

29. Give some examples of common phenomena in which resonance plays an important role.

30. The lunar ocean tide is much more important than the solar ocean tide (see Question 18 of Chapter 16, for example). The opposite is true for tides in the earth's atmosphere, however. Explain this, using resonance ideas, given the fact that the atmosphere has a natural period of oscillation of nearly 12 hours.

problems

SECTION 15-3

1. A 4.0-kg block extends a spring 16 cm from its unstretched position. The block is removed and a 0.50-kg body is hung from the same spring. If the spring is then stretched and released, what is its period of oscillation?
   Answer: 0.28 s.

2. A 2.0-kg mass hangs from a spring. A 300-g body hung below the mass stretches the spring 2.0 cm farther. If the 300 g body is removed and the mass is set into oscillation, find the period of motion.

3. The scale of a spring balance reading from 0 to 32 lb is 4.0 in. long. A package suspended from the balance is found to oscillate vertically with a frequency of 2.0 Hz. How much does the package weigh?
   Answer: 19 lb.

4. An automobile can be considered to be mounted on a spring as far as vertical oscillations are concerned. The springs of a certain car are adjusted so that the vibrations have a frequency of 3.0 Hz. (a) What is the spring's force constant if the car weighs 3200 lb? (b) What will the vibration frequency be if five passengers, averaging 160 lb each, ride in the car?

5. (a) Show that the general relations for the period and frequency of any simple harmonic motion are

$$T = 2\pi\sqrt{-\frac{x}{a}} \qquad \text{and} \qquad \nu = \frac{1}{2\pi}\sqrt{-\frac{a}{x}}.$$

(b) Show that the general relations for the period and frequency of any simple angular harmonic motion are

$$T = 2\pi\sqrt{-\frac{\theta}{\alpha}} \qquad \text{and} \qquad \nu = \frac{1}{2\pi}\sqrt{-\frac{\alpha}{\theta}}.$$

6. The endpoint of a spring vibrates with a period of 2.0 s when a mass m is attached to it. When this mass is increased by 2.0 kg the period is found to be 3.0 s. Find the value of m.

7. A particle executes linear harmonic motion about the point $x = 0$. At $t = 0$ it has displacement $x = 0.37$ cm and zero velocity. The frequency of the motion is 0.25 Hz. Determine (a) the period, (b) the angular frequency, (c) the amplitude, (d) the displacement at time t, (e) the velocity at time t, (f) the maximum speed, (g) the maximum acceleration, (h) the displacement at $t = 3.0$ s, and (i) the speed at $t = 3.0$ s.
   Answer: (a) 4.0 s. (b) $\pi/2$ rad/s. (c) 0.37 cm. (d) 0.37 cos $(\pi t/2)$, in centimeters. (e) -0.58 sin $(\pi t/2)$, in centimeters per second. (f) 0.58 cm/s. (g) 0.91 cm/s². (h) Zero. (i) 0.58 cm/s.

8. A small body of mass 0.10 kg ($W = mg = 0.22$ lb) is undergoing simple harmonic motion of amplitude 1.0 m (3.3 ft) and period 0.20 s. (a) What is the maximum value of the force acting on it? (b) If the oscillations are produced by a spring, what is the force constant of the spring?

9. The vibration frequencies of atoms in solids at normal temperatures are of the order 10^{13} Hz. Imagine the atoms to be connected to one another by "springs." Suppose that a single silver atom vibrates with this frequency and that all the other atoms are at rest. Compute the force constant of a single spring. One mole of silver has a mass of 108 g and contains 6.02 $\times 10^{23}$ atoms. Assume that the atom interacts only with its nearest neighbor.
   Answer: 710 N/m.

10. A block is on a piston which is moving vertically with a simple harmonic motion of period 1.0 s. (a) At what amplitude of motion will the block and piston separate? (b) If the piston has an amplitude of 5.0 cm, what is the maximum frequency for which the block and piston will be in contact continuously?

11. A block is on a horizontal surface which is moving horizontally with a simple harmonic motion of frequency 2.0 Hz. The coefficient of static friction between block and plane is 0.50. How great can the amplitude be if the block does not slip along the surface?   *Answer:* 3.1 cm.

12. The end of one of the prongs of a tuning fork that executes simple harmonic motion of frequency 1000 Hz has an amplitude of 0.40 mm. Find (a) the maximum acceleration and maximum speed of the end of the prong and (b) the acceleration and the speed of the end of the prong when it has a displacement 0.20 mm. (c) Express the end's position as a function of time if it is at equilibrium when $t = 0$.

13. A body oscillates with simple harmonic motion according to the equation

$$x = 6.0 \cos (3\pi t + \pi/3)$$

where x is in meters, t is in seconds, and the numbers in the parentheses are in radians. What is (a) the displacement, (b) the velocity, (c) the acceleration, and (d) the phase at the time $t = 2.0$ s. Find also (e) the frequency ν, and (f) the period of the motion.
   Answer: (a) 3.0 m. (b) -49 m/s. (c) -270 m/s². (d) 20 rad. (e) 1.5 Hz. (f) 0.67 s.

14. A loudspeaker produces a musical sound by the oscillation of a diaphragm. If the amplitude of oscillation is limited to 1.0×10^{-3} mm, what frequencies will result in the acceleration of the diaphragm exceeding g?

15. Two particles execute simple harmonic motion of the same amplitude and frequency along the same straight line. They pass one another when going in opposite directions each time their displacement is half their amplitude. What is the phase difference between them? *Answer:* 120°.

16. Two particles oscillate in simple harmonic motion along a common straight line segment of length A. Each particle has a period of 1.5 s but they differ in phase by 30°. (a) How far apart are they (in terms of A) 0.50 s after the lagging particle leaves one end of the path? (b) Are they moving in the same direction, toward each other, or away from each other at this time?

17. A massless spring of force constant 7.0 N/m is cut into halves. (a) What is the force constant of each half? (b) The two halves, suspended separately, support a block of mass M (see Fig. 15-22). If the system vibrates at a frequency of 3.0 Hz, what is the value of the mass M?
Answer: (a) 14 N/m. (b) 79 g.

18. A uniform spring whose unstressed length is l has a force constant k. The spring is cut into two pieces of unstressed lengths l_1 and l_2, where $l_1 = nl_2$ and n is an integer. What are the corresponding force constants k_1 and k_2 in terms of n and k? Check your result for $n = 1$ and $n = \infty$.

19. Two springs are attached to a mass m and to fixed supports as shown in Fig. 15-23. Show that the frequency of oscillation in this case is

$$\nu = \frac{1}{2\pi}\sqrt{\frac{k_1 + k_2}{m}}.$$

(The electrical analog of this system is a series combination of two capacitors.)

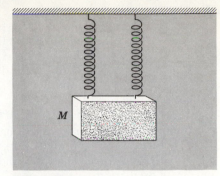

figure 15-22
Problem 17

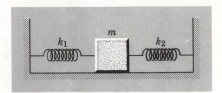

figure 15-23
Problem 19

figure 15-24
Problem 20

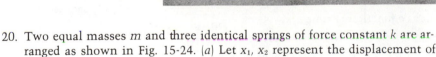

20. Two equal masses m and three identical springs of force constant k are arranged as shown in Fig. 15-24. (a) Let x_1, x_2 represent the displacement of each mass from its equilibrium position and show that

$$m\frac{d^2x_1}{dt^2} = k(x_2 - 2x_1)$$

and

$$m\frac{d^2x_2}{dt^2} = k(x_1 - 2x_2).$$

(b) Find the frequencies of vibration for the system by assuming a solution of the form $x_1 = A_1 \sin \omega t$ and $x_2 = A_2 \sin \omega t$.

21. Two springs are joined and connected to a mass m as shown in Fig. 15-25. The surfaces are frictionless. If the springs separately have force constants k_1 and k_2, show that the frequency of oscillation of m is

$$\nu = \frac{1}{2\pi}\sqrt{\frac{k_1k_2}{(k_1 + k_2)m}}.$$

(The electrical analog of this system is a parallel connection of two capacitors.)

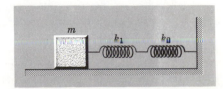

figure 15-25
Problem 21

22. The force of interaction between two atoms in certain diatomic molecules can be represented by $F = -a/r^2 + b/r^3$, in which a and b are positive constants and r is the separation distance of the atoms. Make a graph of F vs. r. Then (a) show that the separation at equilibrium is b/a; (b) show that for small oscillations about this equilibrium separation the force constant is a^4/b^3; (c) find the period of this motion.

23. A massless spring of force constant 19 N/m (1.3 lb/ft) hangs vertically. A body of mass 0.20 kg (W = mg = 0.44 lb) is attached to its free end and then released. Assume that the spring was unstretched before the body was released. Find (a) how far below the initial position the body descends, (b) the frequency, and (c) the amplitude of the resulting motion, assumed to be simple harmonic.
    Answer: (a) 0.21 m (0.68 ft). (b) 1.6 Hz (1.5 Hz). (c) 0.11 m (0.34 ft).

24. An oscillating mass-spring system has a mechanical energy of 1.0 J (0.74 ft · lb), amplitude of 0.10 m (0.33 ft), and maximum speed of 1.0 m/s (3.3 ft/s). Find (a) the force constant of the spring, (b) the mass, and (c) the frequency of oscillation.

25. (a) When the displacement is one-half the amplitude A, what fraction of the total energy is kinetic and what fraction is potential in simple harmonic motion? (b) At what displacement is the energy half kinetic and half potential?
    Answer: (a) $\frac{3}{4} : \frac{1}{4}$. (b) $A/\sqrt{2}$.

26. (a) Prove that in simple harmonic motion the average potential energy equals the average kinetic energy when the average is taken with respect to time over one period of the motion, and that each average equals $\frac{1}{4}kA^2$. (See Fig. 15-9a.) (b) Prove that when the average is taken with respect to position over one cycle, the average potential energy equals $\frac{1}{6}kA^2$ and the average kinetic energy equals $\frac{1}{3}kA^2$. (See Fig. 15-9b.) (c) Explain physically why the two results above (a and b) are different.

27. *Vertical Spring in a Uniform Gravitational Field.* Consider a massless spring of force constant k in a uniform gravitational field. Attach a mass m to the spring. (a) Show that if $x = 0$ marks the slack position of the spring, the static equilibrium position is given by $x = mg/k$ (see Fig. 15-26). (b) Show that the equation of motion of the mass-spring system is

$$m \frac{d^2x}{dt^2} + kx = mg$$

and that the solution for the displacement as a function of time is $x = A \cos (\omega t + \phi) + mg/k$, where $\omega = \sqrt{k/m}$ as before. (c) Show, therefore, that the system has the same ω, v, a, ν, and T in a uniform gravitational field as in the absence of such a field, with the one change that the equilibrium position has been displaced by mg/k. (d) Now consider the energy of the system, $\frac{1}{2}mv^2 + \frac{1}{2}kx^2 + mg(h - x) = $ constant, and show that time differentiation leads to the equation of motion of part (b). (e) Show that when the mass falls from $x = 0$ to the static equilibrium position, $x = mg/k$, the loss in gravitational potential energy goes half into a gain in elastic potential energy and half into a gain in kinetic energy. (f) Finally, consider the system in motion at the static equilibrium position. Compute separately the change in gravitational potential energy and in elastic potential energy when the mass moves *up* through a displacement A, and when the mass moves *down* through a displacement A. Show that the *total* change in potential energy is the same in each case, namely $\frac{1}{2}kA^2$.

   In view of the results (c) and (f), one can simply ignore the uniform gravitational field in the analysis merely by shifting the reference position from $x = 0$ to $x_0 = x - mg/k = 0$. The new potential energy curve [$U(x_0) = \frac{1}{2}kx_0^2 + $ constant] has the same parabolic shape as the potential energy curve in the absence of a gravitational field [$U(x) = \frac{1}{2}kx^2$].

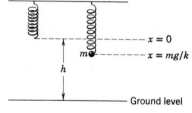

figure 15-26
Problem 27

28. An 8.0-lb block is suspended from a spring with a force constant of 3.0 lb/in. A bullet weighing 0.10 lb is fired into the block from below with a speed of 500 ft/s and comes to rest in the block. (a) Find the amplitude of the resulting simple harmonic motion. (b) What fraction of the original kinetic energy of the bullet is stored in the harmonic oscillator? Is energy lost in this process? Explain your answer.

29. Start from Eq. 15-17 for the conservation of energy (with $\frac{1}{2}kA^2 = E$) and

obtain the displacement as a function of the time by integration of Eq. 15-18. Compare with Eq. 15-8.

30. Attach a solid cylinder to a horizontal massless spring so that it can *roll without slipping* along a horizontal surface, as in Fig. 15-27. The force constant k of the spring is 3.0 N/m. If the system is released from rest at a position in which the spring is stretched by 0.25 m, find (a) the translational kinetic energy and (b) the rotational kinetic energy of the cylinder as it passes through the equilibrium position. (c) Show that under these conditions the center of mass of the cylinder executes simple harmonic motion with a period

$$T = 2\pi \sqrt{3M/2k},$$

where M is the mass of the cylinder.

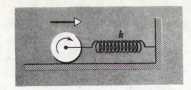

figure 15-27
Problem 30

31. If the mass of a spring m_s is not negligible but is small compared to the mass m of the object suspended from it, the period of motion is $T = 2\pi \sqrt{(m + m_s/3)/k}$. Derive this result. (*Hint:* The condition $m_s \ll m$ is equivalent to the assumption that the spring stretches proportionally along its length.) (See H. L. Armstrong, *American Journal of Physics*, 37, 447 (1969) for a complete solution of the general case.)

SECTION 15-5

32. What is the length of a simple pendulum whose period is 1.00 s at a point where $g = 32.2$ ft/s²?

33. A simple pendulum of length 1.00 m (3.28 ft) makes 100 complete oscillations in 204 s at a certain location. What is the acceleration due to gravity at this point?   *Answer:* 9.49 m/s² (31.1 ft/s²).

34. A solid sphere of mass 2.0 kg ($W = mg = 4.4$ lb) and diameter 0.30 m (0.98 ft) is suspended on a wire. Find the period of angular oscillation for small displacements if the torque constant of the wire is 6.0×10^{-3} N · m/rad (4.4×10^{-3} lb · ft/rad).

35. A circular hoop of radius 2.0 ft and weight 8.0 lb is suspended on a horizontal nail. (a) What is its frequency of oscillation for small displacements from equilibrium? (b) What is the length of the equivalent simple pendulum?   *Answer:* (a) 0.45 Hz. (b) 4.0 ft.

36. Determine the largest amplitude of a simple pendulum such that Eq. 15-19 for the period is correct to within 1.0%.

37. A long uniform rod of length l and mass m is free to rotate in a horizontal plane about a vertical axis through its center. A spring with force constant k is connected horizontally between the end of the rod and a fixed wall as shown in Fig. 15-28. What is the period of the small oscillations that result when the rod is pushed slightly to one side and released?
Answer: $2\pi \sqrt{m/3k}$.

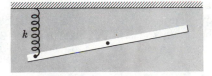

figure 15-28
Problem 37

38. The balance wheel of a watch vibrates with an angular amplitude of π radians and a period of 0.50 s. Find (a) the maximum angular speed of the wheel, (b) the angular speed of the wheel when its displacement is $\pi/2$ rads, and (c) the angular acceleration of the wheel when its displacement is $\pi/4$ radians.

39. (a) What is the frequency of a simple pendulum 2.0 m long? (b) Assuming small amplitudes, what would its frequency be in an elevator accelerating upward at a rate of 2.0 m/s². (c) What would its frequency be in free fall?
Answer: (a) 0.35 Hz. (b) 0.39 Hz. (c) Zero.

40. A simple pendulum of length l and mass m is suspended in a car that is traveling with a constant speed v around a circle of radius R. If the pendulum undergoes small oscillations in a radial direction about its equilibrium position, what will its frequency of oscillation be?

41. Prove, for the generalized physical pendulum of Fig. 15-12, that the centers of oscillation and percussion coincide. See Examples 4 and 5 for a special case.

42. A pendulum is formed by pivoting a long thin rod of length l and mass m about a point on the rod which is a distance d above the center of the rod. (a) Find the small amplitude period of this pendulum in terms of d, l, m, and g. (b) Show that the period has a *minimum* value when $d = l/\sqrt{12} = 0.289l$.

43. A disk 1.0 m in diameter is cut from a metal sheet. The disk is made to swing as a pendulum by drilling a small hole in it and mounting it on a nail driven into a wall. Let l be the distance from the nail to the center of the plate. (a) For what value or values of l will the period be 1.7 s? (b) Suppose you want the period to be as small as possible. What value of l would you use?   *Answer:* (a) 0.30 m; 0.42 m. (b) 0.35 m.

44. (a) Show that the maximum tension in the string of a simple pendulum, when the amplitude θ_m is small, is $mg(1 + \theta_m{}^2)$. (b) At what position of the pendulum is the tension a maximum?

SECTION 15-7

45. Electrons in an oscilloscope are deflected by two mutually perpendicular electric fields in such a way that at any time t the displacement is given by

$$x = A \cos \omega t, \qquad y = A \cos (\omega t + \phi_y).$$

(a) Describe the path of the electrons and determine its equation when $\phi_y = 0°$. (b) When $\phi_y = 30°$. (c) When $\phi_y = 90°$.
Answer: (a) Straight line, $y = \pm x$. (b) Ellipse, $y^2 - \sqrt{3}xy + x^2 = A^2/4$. (c) Circle, $x^2 + y^2 = A^2$.

46. Sketch the path of a particle which moves in the x-y plane according to the equations $x = A \cos (\omega t - \pi/2)$, $y = 2A \cos (\omega t)$, in which x and y are in meters and t is in seconds.

47. The figure shown in Fig. 15-29 is the result of combining the two simple harmonic motions $x = A_x \cos \omega_x t$ and $y = A_y \cos (\omega_y t + \phi_y)$. (a) What is the value of A_x/A_y? (b) What is the value of ω_x/ω_y? (c) What is the value of ϕ_y? *Answer:* (a) 1.0, (b) 0.50, (c) $\pm\frac{1}{2}\pi$.

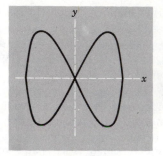

figure 15-29
Problem 47

48. A particle, mass m, moves in a fixed plane along the trajectory $\mathbf{r} = \mathbf{i}\, A \cos \omega t + \mathbf{j}\, A \cos 3\,\omega t$. (a) Sketch the trajectory of the particle. (b) Find the particle's angular momentum as a function of time. (c) Find the force acting on the particle. Also find (d) its potential energy and (e) its total energy as functions of time. (f) Is the motion periodic? If so, what is the period?

49. *Lissajous Figures.* When oscillations at right angles are combined, the frequencies for the motion of the particle in the x- and y-directions need not be equal, so that in the general case Eqs. 15-30 become

$$x = A_x \cos (\omega_x t + \phi_x) \qquad \text{and} \qquad y = A_y \cos (\omega_y t + \phi_y).$$

The path of the particle is no longer an ellipse but is called a *Lissajous curve*, after Jules Antoine Lissajous who first demonstrated such curves in 1857. (a) If ω_x/ω_y is a rational number, so that the angular frequencies ω_x and ω_y are "commensurable," then the curve is closed and the motion repeats itself at regular intervals of time. Assume $A_x = A_y$ and $\phi_x = \phi_y$ and draw the Lissajous curve for $\omega_x/\omega_y = \frac{1}{2}, \frac{1}{3}$, and $\frac{2}{3}$. (b) Let ω_x/ω_y be a rational number, either $\frac{1}{2}, \frac{1}{3}$, or $\frac{2}{3}$ say, and show that the shape of the Lissajous curve depends upon the phase difference $\phi_x - \phi_y$. Draw curves for $\phi_x - \phi_y = 0$, $\pi/4$, and $\pi/2$ rad. (c) If ω_x/ω_y is not a rational number, then the curve is "open." Convince yourself that after a long time the curve will have passed through every point lying in the rectangle bounded by $x = \pm A_x$ and $y = \pm A_y$, the particle never passing twice through a given point with the same velocity. For definiteness, assume $\phi_x = 0$ throughout.

SECTION 15-8

50. (a) What is the reduced mass of each of the following diatomic molecules: O_2, HCl, and CO? Express your answers in unified atomic mass units, the mass of a hydrogen atom being approximately 1.00 u. (b) An HCl molecule is known to vibrate at a fundamental frequency of $\nu = 8.7 \times 10^{13}$ Hz.

What is the effective "force constant" k for the coupling forces between the atoms? In terms of your experience with ordinary springs, would you say that this "molecular spring" is relatively stiff or not?

51. (a) Show that when $m_2 \to \infty$ in Eq. 15-33, $\mu \to m_1$. (b) Show that the effect of a noninfinite wall $(m_2 < \infty)$ on the oscillations of a mass m_1 at the end of a spring attached to the wall is to reduce the period, or increase the frequency, of oscillation compared to (a). (c) Show that when $m_2 = m_1$ the effect is as though the spring were cut in half, each mass oscillating independently about the center of mass at the middle.

52. The spring in Fig. 15-17a has a force constant $k = 250$ N/m (17 lb/ft). Let $m_1 = 1.0$ kg $(W_1 = m_1g = 2.2$ lb) and $m_2 = 3.0$ kg $(W_2 = m_2g = 6.6$ lb). (a) What is the oscillation frequency of the two-body system? (b) What is the ratio K_1/K_2 of the kinetic energies of the bodies?

53. Show that the kinetic energy of the two-body oscillator of Fig. 15-17a is given by $K = \frac{1}{2}\mu v^2$, where μ is the reduced mass and v $(= v_1 - v_2)$ is the relative velocity. It may help to note that linear momentum is conserved while the system oscillates.

SECTION 15-9

54. For the system shown in Fig. 15-18, the block has a mass of 1.5 kg and the spring constant $k = 8.0$ N/m. Suppose the block is pulled down a distance of 12 cm and released. If the friction force is given by $-b\, dx/dt$, where $b = 0.23$ kg/s, find the number of oscillations made by the block during the time interval required for the amplitude to fall to one-third of its initial value.

SECTION 15-10

55. Starting from Eq. 15-41, find the velocity v $(= dx/dt)$, in forced oscillatory motion. Show that the velocity amplitude is $v_m = F_m/[(m\omega'' - k/\omega'')^2 + b^2]^{1/2}$.

The equations of Section 15-10 are identical in form with those representing an electrical circuit containing a resistance R, an inductance L, and a capacitance C in series with an alternating emf $V = V_m \cos \omega''t$. Hence, b, m, k, and F_m are analogous to R, L, $1/C$, and V_m, respectively, and x and v are analogous to electric charge q and current i, respectively. In the electrical case the current amplitude i_m, analogous to the velocity amplitude v_m above, is used to describe the quality of the resonance.

16
gravitation

From at least the time of the Greeks two problems were the subjects of searching inquiry: (1) the tendency of objects such as stones to fall to earth when released, and (2) the motions of the planets, including the sun and the moon, which were classified with the planets in those times. In early days these problems were thought of as completely separate. It is one of Newton's achievements that, building on the work of his predecessors, he saw them clearly as aspects of a single problem and subject to the same laws.

In 1665 the 23-year-old Newton was driven from Cambridge to Lincolnshire when the college was dismissed because of the plague. About 50 years later he wrote, ". . . in the same year (1665) I began to think of gravity extending to the orb of the Moon . . . and having thereby compared the force requisite to keep the Moon in her orb with the force of gravity at the surface of the earth, and found them to answer pretty nearly."

Newton's young friend William Stukeley wrote of having tea with Newton under some apple trees when Newton said that the setting was the same as when he got the idea of gravitation. "It was occasion'd by the fall of an apple,† as he sat in a contemplative mood . . . and thus by degrees he began to apply this property of gravitation to the motion of the earth and the heavenly bodys . . ." (See Fig. 16-1.)

We can compute the acceleration of the moon toward the earth from its period of revolution and the radius of its orbit. We obtain 0.0089 ft/s²

16-1
*HISTORICAL INTRODUCTION**

---

* See "A Background to Newtonian Gravitation" by V. V. Raman in *The Physics Teacher*, November 1972.
† There is little basis for the belief that the apple hit Newton on the head!

(see Example 4, Chapter 4). This value is about 3600 times smaller than g, the acceleration due to gravity at the surface of the earth. Newton, guided as he says by Kepler's third law (see below and see Problem 25), sought to account for this difference by assuming that the acceleration of a falling body is inversely proportional to the square of its distance from the earth.

The question of what we mean by "distance from the earth" immediately arises. Newton eventually came to regard every particle of the earth as contributing to the gravitational attraction it had on other bodies. He made the daring assumption that the mass of the earth could be treated as if it were all concentrated at its center. (See Section 16-6.)

We can treat the earth as a particle with respect to the sun, for example. It is not obvious, however, that we can treat the earth as a particle with respect to an apple located only a few feet above its surface. If we make this assumption, however, a falling body near the earth's surface is a distance of one earth radius from the effective center of attraction of the earth, or 4000 mi. The moon is about 240,000 mi away. The inverse square of the ratio of these distances is $(4000/240,000)^2 = 1/3600$, in agreement with the ratio of the accelerations of the moon and the apple. In Newton's words, quoted above, it does indeed "answer pretty nearly."

Newton did not publish his conclusions in full until 1678, some 22 years after he had conceived the basic ideas. He did so then in his master-work, the *Principia*. Quite apart from the apple-earth problem which we mentioned above, there was a real uncertainty about the radius of the earth, a needed parameter in the calculations. Finally, there was Newton's general reluctance to publish anything; he was a shy and introspective man and abhorred controversy. Bertrand Russell wrote of him: "If he had encountered the sort of opposition with which Galileo had to contend, it is probable that he would never have published a line." Edmund Halley, of Halley's comet fame, virtually forced Newton to publish the *Principia*. The mathematician Augustus DeMorgan wrote of Halley: ". . . but for him, in all human probability, that work would not have been thought of, nor when thought of written, nor when written printed."

In the *Principia* Newton went beyond the apple-earth and the moon-earth problems and extended his law of gravitation to *all* bodies, in a way that we will discuss in the next section.

There are three overlapping realms in which we can discuss gravitation. (1) The gravitational attraction between two bowling balls, for example, although measurable by sensitive techniques, is too weak to fall within our ordinary sense perceptions. (2) The attraction of ourselves and objects around us by the earth is a controlling feature of our lives from which we can escape only by extreme measures. The designers of our space program have the gravitational force constantly in mind as a central and controlling factor. (3) On a cosmic scale, that is, in the realm of the solar system and of the formation and interaction of stars and galaxies, gravitation is by far the dominant force.

The earliest serious attempts to explain the kinematics of the solar system were made by the Greeks. Ptolemy (Claudius Ptolemaus, second century A.D.) developed a geocentric (Ptolemaic) scheme for the solar system in which, as the name implies, the earth remains stationary at the center whereas the planets, including the sun and the moon, revolve around it. This should not be a surprising deduction. The earth seems to us to be a substantial body. Shakespeare referred to it as ". . .

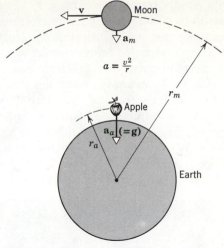

figure 16-1
Both the moon and the apple are accelerated toward the center of the earth. The difference in their motions arises because the moon has a tangential velocity v whereas the apple does not.

this goodly frame, the earth. . . ." Even today, in teaching navigational astronomy, we use a geocentric reference frame and in ordinary conversation we use terms such as "sunrise," which imply such a frame.

Simple circular orbits cannot account for the complicated motions of the planets so that Ptolemy had to use the concept of epicycles, in which a planet moves around a circle whose center moves around another circle centered on the earth. (See Fig. 16-2b.) He also had to resort to several other geometrical arrangements, each of which preserved the supposed sanctity of the circle as a central feature of planetary motions. We now know that it is not a circle that is fundamental but an ellipse, with the sun at one focus (see below).

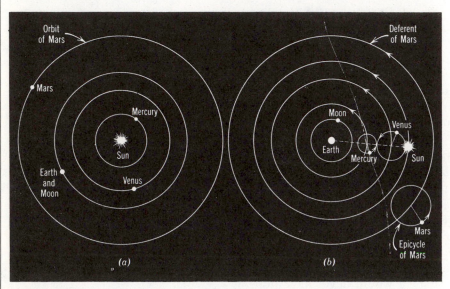

figure 16-2
(a) The Copernican view of the solar system. The sun is at the center and the planets move around it. (b) The Ptolemaic view of the solar system. The earth is at the center and the planets move around it. Both investigators introduced geometrical complexities to explain the complex motion of the planets. In (b), for example, Mars travels about a (circular) epicycle whose center travels about a (circular) deferent. The arrangements of Copernicus, essentially equally complex, are not shown. The basic difference is whether or not the sun or the earth is to be at the center of the planetary motions. (See *The Crime of Galileo*, by Giorgio De Santillana, Chicago: University of Chicago Press, 1955. See also *The Copernican Revolution*, by Thomas S. Kuhn, Cambridge, Mass.: Harvard University Press, 1957.)

In the sixteenth century Copernicus (1473–1543) proposed a heliocentric (Copernican) scheme, in which the sun was at the center of the solar system, the earth moving about it as one of its planets: see Fig. 16-2a. It is often thought that the Copernicus scheme is so much simpler than that of Ptolemy that it should have been adopted at once. This is not true. Copernicus still believed in the sanctity of circles and his use of epicycles and other arrangements was about as great as that of Ptolemy; these are not shown in Fig. 16-2a. Copernicus however, by putting the sun at the center of things, gave a much simpler description and more natural explanation of certain features of planetary motion. Above all he laid the indispensable groundwork from which our modern view of the solar system developed.

The growing controversy over the two theories stimulated astronomers to obtain more accurate observational data. Such data were compiled by Tycho Brahe* (1546–1601), who was the last great astronomer to make observations without the use of a telescope.† His data on planetary motions were analyzed and interpreted for about twenty years by Johannes Kepler (1571–1630), who had been Brahe's assistant. Kepler found important regularities in the motion of the planets. These regularities are known as *Kepler's three laws of planetary motion.*

1. All planets move in elliptical orbits having the sun as one focus (the law of orbits).

2. A line joining any planet to the sun sweeps out equal areas in equal times (the law of areas).

3. The square of the period of any planet about the sun is proportional to the cube of the planet's mean distance from the sun (the law of periods).

Kepler's laws lent strong support to the Copernican theory. They showed the great simplicity with which planetary motions could be described when the sun was taken as the reference body. However, these laws were empirical; they simply described the observed motion of the planets without any theoretical interpretation. Kepler had no concept of force as a cause of such regularities.** In fact, the concept of force was not yet clearly formulated. It was, therefore, a great triumph for Newton's ideas that he could *derive* Kepler's laws from his laws of motion and his law of gravitation. Newton's law of gravitation in this case required each planet to be attracted toward the sun with a force proportional to the mass of the planet and inversely proportional to the square of its distance from the sun.

In this way Newton was able to account for the motion of the planets in the solar system and of bodies falling near the surface of the earth with one common concept. He thereby synthesized into one theory the previously separate sciences of terrestrial mechanics and celestial mechanics. The real scientific significance of Copernicus' work lies in the fact that the heliocentric theory opened the way for this synthesis.‡ Subsequently, on the assumption that the earth rotates and revolves about the sun, it became possible to explain such diverse phenomena as the daily and the annual apparent motion of the stars, the flattening of the earth from a spherical shape, the behavior of the tradewinds, and many other things that could not have been tied together so simply in a geocentric theory.

It is instructive to review the development of our understanding of the motions of the bodies in the solar system in terms of the program of classical mechanics that we outlined in Chapter 5; see page 73. Historically, there were four "breakthroughs."

---

* See "Copernicus and Tycho" by Owen Gingerich, in *Scientific American*, December 1973.

† The first scientifically useful telescope was built in 1609 by Galileo. With it he discovered the inner moons of Jupiter and the phases of Venus. Galileo was a strong advocate of the Copernican theory and used his observations to argue in its behalf. Newton, incidentally, invented a telescope, the reflecting type.

** See 'How Did Kepler Discover His First Two Laws' by Curtis Wilson in *Scientific American*, March 1972.

‡ Newton's work certainly built on or was influenced by the work of others. Among them we must include Galileo, Kepler, Halley, and Hooke.

1. Copernicus pointed out that the sun and not the earth is the central body of the solar system. In today's language he gave us a reference frame (the sun) much more suitable than the one previously used (the earth) for describing the motions of the solar system. Among other advantages, the Copernican frame, fixed with respect to the sun but not rotating with it, is essentially an *inertial* reference frame; the reference frame fixed to the revolving earth on which we live cannot be so considered for problems involving planetary motions.

2. Brahe made accurate measurements of the motions of the planets as viewed from the earth. He provided the necessary observational data that made further progress possible.

3. Kepler, studying Brahe's data, deduced from it the three simple empirical laws of planetary motion that we have discussed above. Adopting Copernicus' reference frame, he displayed the kinematic information about planetary motions in simple form.

4. Newton discovered the laws of motion for mechanical systems in general as well as the particular force law that applies to the motions of the planets, namely the law of universal gravitation.

Thus, over a span of about 200 years, we see emerging (1) the appropriate reference frame, (2) accurate kinematical information, (3) the empirical laws of planetary motion, and (4) the general laws of classical mechanics and the force law appropriate to planetary motion.

16-2 THE LAW OF UNIVERSAL GRAVITATION

The force between any two particles having masses m_1 and m_2 separated by a distance r is an attraction acting along the line joining the particles and has the magnitude

$$F = G \frac{m_1 m_2}{r^2},$$ (16-1)

where G is a universal constant having the same value for all pairs of particles.

This is Newton's law of universal gravitation. It is important to stress at once many features of this law in order that we understand it clearly.

First, the gravitational forces between two particles are an action-reaction pair. The first particle exerts a force on the second particle that is directed toward the first particle along the line joining the two. Likewise, the second particle exerts a force on the first particle that is directed toward the second particle along the line joining the two. These forces are equal in magnitude but oppositely directed.

The universal constant G must not be confused with the **g** which is the acceleration of a body arising from the earth's gravitational pull on it. The constant G has the dimensions L^3/MT^2 and is a scalar; **g** has the dimensions L/T^2, is a vector, and is neither universal nor constant.

Notice that Newton's law of universal gravitation is not a defining equation for any of the physical quantities (force, mass, or length) contained in it. According to our program for classical mechanics in Chapter 5, force is defined from Newton's second law, $\mathbf{F} = m\mathbf{a}$. The essence of this law, however, is the assumption that the force on a particle, so defined, can be related in a simple way to measurable properties of the particle and of its environment, that is, the existence of simple force laws is assumed. The law of universal gravitation is such a simple law. The constant G must be found from experiment. Once G is determined for a given pair of bodies, we can use that value in the law of gravitation to determine the gravitational forces between any other pair of bodies.

Notice also that Eq. 16-1 expresses the force between mass *particles*. If we want to determine the force between extended bodies, as for example the earth and the moon, we must regard each body as decomposed into particles. Then the interaction between all particles must be computed. Integral calculus makes such a calculation possible. Newton's motive in developing the calculus arose in part from a desire to solve such problems. In general, it is incorrect to assume that all the mass of a body can be concentrated at its center of mass for gravitational purposes. This assumption is correct for uniform spheres, however, a result that we shall use often and shall prove in Section 16-6.

Implicit in the law of universal gravitation is the idea that the gravitational force between two particles is independent of the presence of other bodies or the properties of the intervening space. The correctness of this idea depends on the correctness of the deductions using it and has so far been borne out. This fact has been used by some to rule out the possible existence of "gravity screens."

We can express the law of universal gravitation in vector form. Let the displacement vector $\mathbf{r}_{12}$ point *from* the particle of mass m_1 *to* the particle of mass m_2, as Fig. 16-3a shows. The gravitational force $\mathbf{F}_{21}$, exerted *on* m_2 by m_1, is given in direction and magnitude by the vector relation

$$\mathbf{F}_{21} = -G\frac{m_1 m_2}{r_{12}{}^3}\mathbf{r}_{12} \qquad (16\text{-}2a)$$

in which r_{12} is the magnitude of $\mathbf{r}_{12}$. The minus sign in Eq. 16-2a shows that $\mathbf{F}_{21}$ points in a direction opposite to $\mathbf{r}_{12}$; that is, the gravitational force is *attractive*, m_2 feeling a force directed toward m_1 (see Fig. 16-3). That Eq. 16-2a is indeed an inverse square law can be seen by writing it as $\mathbf{F}_{21} = -(Gm_1m_2/r_{12}{}^2)(\mathbf{r}_{12}/r_{12})$; here the displacement vector divided by its own magnitude, $\mathbf{r}_{12}/r_{12}$, is simply a unit vector $\mathbf{u}_r$ in the direction of the displacement. If we express the relation in scalar form by equating the magnitudes of each side, a factor r_{12} in the numerator cancels one of the factors of $r_{12}{}^3$ in the denominator and the inverse square relation of Eq. 16-1 results.

The force exerted *on* m_1 by m_2 is clearly

$$\mathbf{F}_{12} = -G\frac{m_2 m_1}{r_{21}{}^3}\mathbf{r}_{21}. \qquad (16\text{-}2b)$$

Note, in Eqs. 16-2, that $\mathbf{r}_{21} = -\mathbf{r}_{12}$ (see Fig. 16-3a,b) so that, as we expect, $\mathbf{F}_{12} = -\mathbf{F}_{21}$ (see Fig. 16-3c); that is, the gravitational forces acting on the two bodies form an action-reaction pair.

To determine the value of G it is necessary to measure the force of attraction between two known masses. The first accurate measurement was made by Lord Cavendish in 1798. Significant improvements were made by Poynting and Boys in the nineteenth century. The present accepted value of G is*

$$G = 6.6720 \times 10^{-11}\ \text{N} \cdot \text{m}^2/\text{kg}^2,$$

accurate to about 0.0006×10^{-11} N · m²/kg². In the British engineering system this value is 3.436×10^{-8} lb · ft²/slug².

The constant G can be determined by the maximum deflection

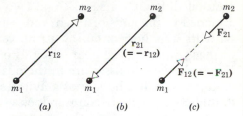

figure 16-3
The force exerted on m_2 (by m_1), $\mathbf{F}_{21}$, is directed opposite to the displacement, $\mathbf{r}_{12}$, of m_2 from m_1. The force exerted on m_1 (by m_2), $\mathbf{F}_{12}$, is directed opposite to the displacement, $\mathbf{r}_{21}$, of m_1 from m_2. $\mathbf{F}_{21} = -\mathbf{F}_{12}$, the forces being an action-reaction pair.

16-3
THE CONSTANT OF UNIVERSAL GRAVITATION, G

* See "A New Determination of the Constant of Gravitation" by A. H. Cook, in *Contemporary Physics*, May 1968, for a good review of the principles and methods used.

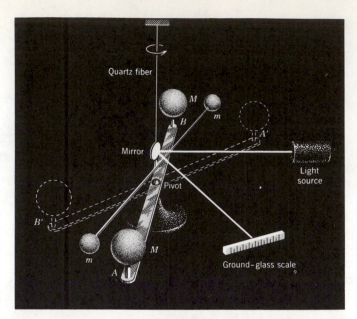

figure 16-4
The Cavendish balance, used for experimental verification of Newton's law of universal gravitation. Masses m,m are suspended from a fiber. Masses M,M can rotate on a stationary support. An image of the lamp filament is reflected by the mirror attached to m,m onto the scale so that any rotation of m,m can be measured.

method illustrated in Fig. 16-4. Two small balls, each of mass m, are attached to the ends of a light rod. This rigid "dumbbell" is suspended, with its axis horizontal, by a fine vertical fiber. Two large balls each of mass M are placed near the ends of the dumbbell on opposite sides. When the large masses are in the positions A and B, the small masses are attracted, by the law of gravitation, and a torque is exerted on the dumbbell rotating it counterclockwise, as viewed from above. When the large masses are in the positions A' and B', the dumbbell rotates clockwise. The fiber opposes these torques as it is twisted. The angle θ through which the fiber is twisted when the balls are moved from one position to the other is measured by observing the deflection of a beam of light reflected from the small mirror attached to it. If the masses and their distances of separation and the torsional constant of the fiber are known, we can calculate G from the measured angle of twist. The force of attraction is very small so that the fiber must have an extremely small torsion constant if we are to obtain a detectable twist. In Example 1 at the end of this section some data are given from which G can be calculated.

The masses in the Cavendish balance of Fig. 16-4 are, of course, not particles but extended objects. Since they are uniform spheres, however, they act gravitationally as though all their mass were concentrated at their centers (Section 16-6).

Because G is so small, the gravitational forces between bodies on the earth's surface are extremely small and can be neglected for ordinary purposes. For example, two spherical objects each having a mass of 100 kg (about 220-lb weight) and separated by 1.0 m at their centers attract each other with a force

$$F = \frac{Gm_1m_2}{r^2} = \frac{(6.67 \times 10^{-11}) \times (100) \times (100)}{(1.0)^2}\,\mathrm{N}$$

$$= 6.7 \times 10^{-7}\,\mathrm{N}$$

or about 1.5×10^{-7} lb! The Cavendish experiment must be a very delicate one indeed. Even so, it is often performed as an experiment in an introductory physics laboratory.

The large gravitational force which the earth exerts on all bodies near its surface is due to the extremely large mass of the earth. In fact, we can determine the mass of the earth from the law of universal gravitation and the value of G calculated from the Cavendish experiment. For this reason Cavendish is said to have been the first person to "weigh" the earth. Consider the earth, mass M_e, and an object on its surface of mass m. The force of attraction is given both by

$$F = mg$$

and

$$F = \frac{GmM_e}{R_e^2}.$$

Here R_e is the radius of the earth, which is the separation of the two bodies, and g is the acceleration due to gravity at the earth's surface. Combining these equations we obtain

$$M_e = \frac{g\,R_e^2}{G} = \frac{(9.80 \text{ m/s}^2)(6.37 \times 10^6 \text{ m})^2}{6.67 \times 10^{-11} \text{ N} \cdot \text{m}^2/\text{kg}^2} = 5.97 \times 10^{24} \text{ kg}$$

or

$$6.6 \times 10^{21} \text{ tons} \quad \text{"weight."}$$

If we were to divide the total mass of the earth by its total volume, we would obtain the average density of the earth. This turns out to be 5.5 g/cm³, or about 5.5 times the density of water. The average density of the rock on the earth's surface is much less than this value. We conclude that the interior of the earth contains material of density greater than 5.5 g/cm³. From the Cavendish experiment we have obtained information about the nature of the earth's core. (See Question 7 and Problem 16.)

EXAMPLE 1

Let the small spheres of Fig. 16-4 each have a mass of 10.0 g and let the light rod be 50.0 cm long. The period of torsional oscillation of this system is found to be 769 s. Then two large fixed spheres each of mass 10.0 kg are placed near each suspended sphere so as to produce the maximum torsion. The angular deflection of the suspended rod is then 3.96×10^{-3} rad and the distance between centers of the large and small spheres is 10.0 cm. Calculate the universal constant of gravitation G from these data.

The period of torsional oscillation is given by Eq. 15-24,

$$T = 2\pi \sqrt{\frac{I}{\kappa}}.$$

For the rigid dumbbell, if we neglect the contribution of the light rod,

$$I = \Sigma\, mr^2 = (10.0 \text{ g})(25.0 \text{ cm})^2 + (10.0 \text{ g})(25.0 \text{ cm})^2$$

or

$$I = 1.25 \times 10^{-3} \text{ kg} \cdot \text{m}^2.$$

With $T = 769$ s, we can obtain the torsional constant κ as

$$\kappa = \frac{4\pi^2 I}{T^2} = \frac{(4\pi^2)(1.25 \times 10^{-3} \text{ kg} \cdot \text{m}^2)}{(769 \text{ s})^2} = 8.34 \times 10^{-8} \text{ kg} \cdot \text{m}^2/\text{s}^2.$$

The relation between the applied torque and the angle of twist is $\tau = \kappa\theta$. We now know κ and the value of θ at maximum deflection.

The torque arises from the gravitational forces exerted by the large spheres on the small ones. This torque will be a maximum for a given separation when the line joining the centers of these spheres is at right angles to the rod. The force on *each* small sphere is

$$F = \frac{GMm}{r^2},$$

and the moment arm for each force is half the length of the rod ($l/2$). Then,

$$\text{torque} = \text{force} \times \text{moment arm}$$

or

$$\tau = 2\,\frac{GMm}{r^2}\,\frac{l}{2}.$$

Combining this with

$$\tau = \kappa\theta,$$

we obtain

$$G = \frac{\kappa\theta r^2}{Mml} = \frac{(8.34 \times 10^{-8}\ \text{kg} \cdot \text{m}^2/\text{s}^2)(3.96 \times 10^{-3}\ \text{rad})(0.100\ \text{m})^2}{(10.0\ \text{kg})(0.0100\ \text{kg})(0.500\ \text{m})}$$

$$= 6.63 \times 10^{-11}\ \text{N} \cdot \text{m}^2/\text{kg}^2.$$

Notice that this result is about 1% lower than the accepted value. What have we neglected in this calculation that might account for this difference?

The gravitational force on a body is proportional to its mass, as Eq. 16-1 shows. This proportionality between gravitational force and mass is the reason we ordinarily consider the theory of gravitation to be a branch of mechanics, whereas theories of other kinds of force (electromagnetic, nuclear, etc.) may not be.

16-4
INERTIAL AND
GRAVITATIONAL MASS

An important consequence of this proportionality is that we can measure a mass by measuring the gravitational force on it. This can be done by using a spring balance or by comparing the gravitational force on one mass with that on a standard mass, as in a balance; in other words, we can determine the mass of a body by weighing it. This gives us a more practical and more convenient method of measuring mass than is given by our original definition of mass (Section 5-4).

The question arises whether these two methods really measure the same property. The word *mass* has been used in two quite different experimental situations. For example, if we try to push a block that is at rest on a horizontal frictionless surface, we notice that it requires some effort to move it. The block seems to be inert and tends to stay at rest, or if it is moving, it tends to keep moving. Gravity does not enter here at all. It would take the same effort to accelerate the block in gravity-free space. It is the mass of the block which makes it necessary to exert a force to change its motion. This is the mass occurring in $\mathbf{F} = m\mathbf{a}$ in our original experiments in dynamics. We call this mass m the *inertial mass*. Now there is a different situation which involves the mass of the block. For example, it requires effort just to hold the block up in the air at rest above the earth. If we do not support it, the block will fall to the earth with accelerated motion. The force required to hold up the block is equal in magnitude to the force of gravitational attraction between it and the earth. Here inertia plays no role whatever; the property of material bodies, that they are attracted to other objects such as the earth, does play a role. The force is given by

$$F = G\,\frac{m'M_e}{R_e^2},$$

where m' is the *gravitational mass* of the block. *Are the gravitational mass m' and the inertial mass m of the block really the same?* Let us look more carefully at this.

Consider two particles A and B of gravitational masses m_A' and m_B' acted on by a third particle C of gravitational mass m_C'. Let the third particle be an equal distance r from the other two. Then, the gravitational force exerted on A by C is

$$F_{AC} = G\,\frac{m_A'm_C'}{r^2},$$

and the gravitational force exerted on B by C is

$$F_{BC} = G\frac{m_B{}'m_C{}'}{r^2}.$$

The ratio of the gravitational forces on A and B is the ratio of their gravitational masses; that is,

$$\frac{F_{AC}}{F_{BC}} = \frac{m_A{}'}{m_B{}'}.$$

Now suppose that the third body C is the earth. Then F_{AC} and F_{BC} are what we have called the *weights* of bodies A and B. Hence,

$$\frac{W_A}{W_B} = \frac{m_A{}'}{m_B{}'}.$$

Therefore, the law of universal gravitation contains within it the result that the weights of various bodies, at the same place on the earth, are exactly proportional to their *gravitational masses*.

Now suppose we measure the inertial masses m_A and m_B of the particles A and B by dynamical experiments, perhaps using a spring as in Section 5-4. Having done this, we then let these particles fall to the earth from a given place and measure their accelerations. We find experimentally that objects of different *inertial masses* all fall with the same acceleration g arising from the earth's gravitational pull. But the earth's gravitational pulls on these bodies are their weights, so that using the second law of motion we obtain

$$W_A = m_A g,$$

$$W_B = m_B g,$$

or

$$\frac{W_A}{W_B} = \frac{m_A}{m_B}.$$

In other words, the weights of bodies at the same place on the earth are exactly proportional to their *inertial masses* as well. Hence, inertial mass and gravitational mass are at least proportional to one another. In fact, they appear to be identical.

Newton devised an experiment to test directly the apparent equivalence of inertial and gravitational mass. If we go back (Section 15-5) and look up the derivation of the period of a simple pendulum, we find that the period (for small angles) was given by

$$T = 2\pi\sqrt{\frac{ml}{m'g}},$$

where m in the numerator refers to the inertial mass of the pendulum bob and m' in the denominator is the gravitational mass of the pendulum bob, such that $m'g$ gives the gravitational pull on the bob. Only if we assume that m equals m', as we did there implicitly, do we obtain the expression

$$T = 2\pi\sqrt{\frac{l}{g}}$$

for the period. Newton made a pendulum bob in the form of a thin shell. Into this hollow bob he put different substances, being careful always to have the same *weight* of substance as determined by a balance. Hence, in all cases the force on the pendulum was the same at the same angle. Because the external shape of the bob was always the same, even the air resistance on the moving pendulum was the same. As one substance replaced another inside the bob, any difference in acceleration could only be due to a difference in the *inertial* mass. Such a difference would show up by a change in the period of the pendulum. But in all cases Newton found the period of the pendulum to be the same, always given by $T = 2\pi\sqrt{l/g}$. Hence, he concluded that $m = m'$ and that inertial and gravitational masses are equivalent.

In 1909, Eötvös devised an apparatus which could detect a difference of 5 parts in 10⁹ in gravitational force. He found that equal inertial masses always experienced equal gravitational forces within the accuracy of his apparatus. A refined version of the Eötvös experiment was reported in 1964 by R. H. Dicke and his collaborators, who improved the accuracy of the original experiment by a factor of several hundred.*

In classical physics the equivalence of gravitational and inertial mass was looked upon as a remarkable accident having no deep significance. But in modern physics this equivalence is regarded as a clue leading to a deeper understanding of gravitation (see Section 16-13). This was, in fact, an important clue leading to the development of the general theory of relativity.

Up to this point we have taken the acceleration due to gravity g as a constant. From Newton's law of gravitation, however, it is apparent that g will vary with altitude, that is, with distance from the center of the earth. We have already pointed this out specifically in the moon-apple discussion. Let us compute the change in g that occurs as we proceed outward from the earth's surface. From Eq. 16-1,

16-5
VARIATIONS IN ACCELERATION DUE TO GRAVITY

$$F = G\frac{m_1 m_2}{r^2},$$

we obtain, on differentiating with respect to r,

$$dF = -2\frac{G m_1 m_2}{r^3}\,dr.$$

Combining these two equations, we obtain

$$\frac{dF}{F} = -2\frac{dr}{r}.$$

Therefore, the fractional change in F is twice the fractional change in r. The minus sign indicates that the force decreases as the separation distance increases. If we let m_1 be the earth's mass and m_2 the object's mass, the gravitational force on the object attributable to the earth is

$$F = m_2 g$$

directed toward the earth. If we differentiate this expression, we obtain

$$dF = m_2 dg,$$

and on dividing this equation by the previous one we find that

$$\frac{dF}{F} = \frac{dg}{g} = -2\frac{dr}{r}. \tag{16-3}$$

For example, in going up 16 km from the earth's surface, r changes from about 6400 km to 6416 km, a relative increase of 1/400. Therefore,† g must change by about −1/200 over this distance, or from about 980 cm/s² to about 975 cm/s². Hence, g is really very nearly constant near the earth's surface at a given latitude. At higher altitudes, such as those for a typical satellite orbit or for the moon's orbit, g drops appreciably, as Table 16-1 shows.

---

* See "The Eötvös Experiment," by R. H. Dicke, *Scientific American*, December 1961, for an elegant review of this subject.

† Equation 16-3 is a differential expression and is exact. The corresponding expression obtained when dr is replaced by a finite change Δr is a good approximation, provided that $\Delta r/r$ is very small.

Table 16-1
Variation of g with altitude at 45° latitude

Altitude, meters	g, meters/second²	Altitude, meters	g, meters/second²
0	9.806	32,000	9.71
1,000	9.803	100,000	9.60
4,000	9.794	500,000	8.53
8,000	9.782	1,000,000[1]	7.41
16,000	9.757	380,000,000[2]	0.00271

[1] Typical satellite orbit altitude (= 620 mi).
[2] Radius of moon's orbit (= 240,000 mi).

Measurements of **g** are an essential source of information about the shape of the earth. To define the problem more closely we usually consider not the earth itself but an imaginary closed surface called the *geoid*. Over the oceans the geoid is defined to coincide with mean sea level, whereas over the continents it is defined as a continuation of this level; in principle the position of the geoid can be found by digging small sea-level canals across the continents and noting the mean water level. The geoid is a surface of constant gravitational potential; at any point the direction of a plumb line is at right angles to it.

The ancient Greeks believed the earth to be round and one of them, Eratosthenes (c 276–194 B.C.), measured the radius of the earth on the assumption that it is a sphere. He obtained a value of 7400 km, which is to be compared with the modern value of 6371 km. Later it was learned by measurement that, to a good second approximation, the geoid is not a sphere but is an ellipsoid of revolution, flattened along the earth's rotational axis and bulging at the equator. The equatorial radius, in fact, exceeds the polar radius by 21 km. This flattening is caused by centrifugal effects in the rotating, plastic earth. The geoidic surface is not exactly ellipsoidal, lying outside the ellipsoid of closest fit under mountain masses and inside it over the oceans.

The fact that the equator is farther from the center of the earth than are the poles means that there should be a steady increase in the measured value of g as one goes from the equator (latitude 0°) to either pole (latitude 90°). This is shown in Table 16-2. As Example 2 shows, however, about half of this variation can be accounted for by another effect, namely, the change in the effective value of g caused by the earth's rotation. If the earth were rotating fast enough, for example, objects on its surface at the equator would seem to be weightless, which means that the effective value of $g = (W/m)$ would be zero. For all rotational speeds less than this critical value, g has a definite nonzero value which is, however, less than the value it would have at the same point on a nonrotating earth.

Table 16-2
Variation of g with latitude at sea level

Latitude	g, meters/second²	Latitude	g, meters/second²
0°	9.78039	50°	9.81071
10°	9.78195	60°	9.81918
20°	9.78641	70°	9.82608
30°	9.79329	80°	9.83059
40°	9.80171	90°	9.83217

In 1959, it was observed that the orbit of the Vanguard artificial earth satellite, calculated using values of g based on an ellipsoidal geoid, did not agree exactly with the observed orbit. It was concluded that the geoid is best approximated not by an ellipsoid of revolution but by a slightly pear-shaped figure, the small end of the "pear" being in the northern hemisphere and extending about 15 m above the reference ellipsoid. The motion of a satellite is governed at all times by the value of **g** at its position. Thus an artificial earth satellite forms a useful probe to explore the values of **g** near the surface of the earth and from this to deduce information about the shape of the geoid.*

Effect on g of the rotation of the earth. Figure 16-5 is a schematic view of the earth looking down on the north pole. In it we show an enlarged view of a body of mass m hanging from a spring balance at the equator. The forces on this body are the upward pull of the spring balance w, which is the apparent weight of the body, and the downward pull of the earth's gravitational attraction $F = GmM_e/R_e^2$. This body is not in equilibrium because it experiences a centripetal acceleration a_R as it rotates with the earth. There must, therefore, be a net force acting on the body toward the center of the earth. Consequently, the force F of gravitational attraction (the true weight of the body) must exceed the upward pull of the balance w (the apparent weight of the body).

From the second law of motion we obtain

$$F - w = ma_R,$$

$$\frac{GM_e m}{R_e^2} - mg = ma_R,$$

$$g = \frac{GM_e}{R_e^2} - a_R \qquad \text{at the equator.}$$

At the poles $a_R = 0$ so that

$$g = \frac{GM_e}{R_e^2} \qquad \text{at the poles.}$$

This is the value of g we would obtain anywhere (assuming a spherical earth) were the rotation of the earth to be neglected.

Actually the centripetal acceleration is not directed in toward the center of the earth other than at the equator. It is directed perpendicularly in toward the earth's axis of rotation at any given latitude. The detailed analysis is, therefore, really a two-dimensional one. However, the extreme case is at the equator. There

$$a_R = \omega^2 R_e = \left(\frac{2\pi}{T}\right)^2 R_e = \frac{4\pi^2 R_e}{T^2},$$

in which ω is the angular speed of the earth's rotation, T is the period, and R_e is the radius of the earth. Using the values

$$R_e = 6.37 \times 10^6 \text{ m,}$$

$$T = 8.64 \times 10^4 \text{ s,}$$

we obtain

$$a_R = 0.0336 \text{ m/s}^2.$$

Referring to Table 16-2, we see that this effect is enough to account for more than half the difference between the observed values of g at low and high latitudes.

EXAMPLE 2

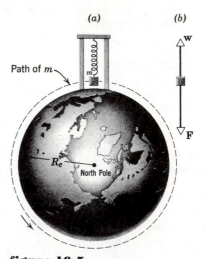

(a)   *(b)*

Path of m

R_e

North Pole

w

F

figure 16-5
Example 2. Effect of the earth's rotation on the weight of a body as measured by a spring balance.

---

* See "Satellite Orbits and Their Geophysical Implications," by D. G. King-Hele, *Contemporary Physics*, April 1961, "Refining the Shape of the Earth" by D. G. King-Hele and G. E. Cook, *Nature*, 246, 86 (1973), and "The Shape of the Earth" by D. G. King-Hele, *Science*, June 25, 1976.

We have already used the fact that a large sphere attracts particles outside it, just as though the mass of the sphere were concentrated at its center. Let us now prove this result.

Consider a uniformly dense spherical shell whose thickness t is small compared to its radius r (Fig. 16-6). We seek the gravitational force it exerts on an external particle P of mass m.

We assume that each particle of the shell exerts on P a force which is proportional to the mass of the small part, inversely proportional to the square of the distance between that part of the shell and P, and directed along the line joining them. We must then obtain the resultant force on P attributable to all parts of the spherical shell.

The small part of the shell at A attracts m with a force $\mathbf{F}_1$. A small part of equal mass at B, equally far from m but diametrically opposite A, attracts m with a force $\mathbf{F}_2$. The resultant of these two forces on m is $\mathbf{F}_1 + \mathbf{F}_2$. Notice, however, that the vertical components of these two forces cancel one another and that the horizontal components, $F_1 \cos \alpha$ and $F_2 \cos \alpha$, are equal. By dividing the spherical shell into pairs of particles like these, we can see at once that all transverse forces on m cancel in pairs. A small mass in the upper hemisphere exerts a force having an upward component on m that will annul the downward component of force exerted on m by an equal symmetrically located mass in the lower hemisphere of the shell. To find the resultant force on m arising from the shell, we need consider only horizontal components.

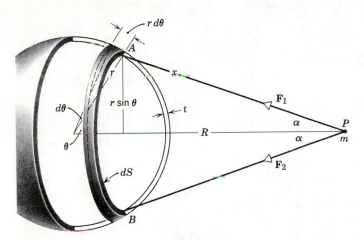

figure 16-6
Gravitational attraction of a section dS of a spherical shell of matter on a particle of mass m.

Let us take as our element of mass of the shell a circular strip labeled dS in the figure. Its length is $2\pi(r \sin \theta)$, its width is $r\, d\theta$, and its thickness is t. Hence, it has a volume

$$dV = 2\pi t r^2 \sin \theta\, d\theta.$$

Let us call the density ρ, so that the mass within the strip is

$$dM = \rho\, dV = 2\pi t \rho r^2 \sin \theta\, d\theta.$$

The force exerted by dM on the particle of mass m at P is horizontal and has the value

$$dF = G \frac{m\, dM}{x^2} \cos \alpha$$

$$= 2\pi G t \rho m r^2 \frac{\sin \theta\, d\theta}{x^2} \cos \alpha. \qquad (16\text{-}4)$$

The variables x, α, and θ are related. From the figure we see that

$$\cos \alpha = \frac{R - r \cos \theta}{x}. \qquad (16\text{-}5)$$

Since, by the law of cosines,

$$x^2 = R^2 + r^2 - 2Rr \cos \theta, \qquad (16\text{-}6)$$

we have

$$r \cos \theta = \frac{R^2 + r^2 - x^2}{2R}. \qquad (16\text{-}7)$$

On differentiating Eq. 16-6, we obtain

$$2x \, dx = 2Rr \sin \theta \, d\theta$$

or

$$\sin \theta \, d\theta = \frac{x}{Rr} dx. \qquad (16\text{-}8)$$

We now put Eq. 16-7 into Eq. 16-5 and then put Eqs. 16-5 and 16-8 into Eq. 16-4. As a result we eliminate θ and α and obtain

$$dF = \frac{\pi G t \rho m r}{R^2} \left(\frac{R^2 - r^2}{x^2} + 1 \right) dx.$$

This is the force exerted by the circular strip dS on the particle m.

We must now consider every element of mass in the shell and sum up over all the circular strips in the entire shell. This operation is an integration over the shell with respect to the variable x. But x ranges from a minimum value of $R - r$ to a maximum value $R + r$.

Since

$$\int_{R-r}^{R+r} \left(\frac{R^2 - r^2}{x^2} + 1 \right) dx = 4r,$$

we obtain the resultant force

$$F = \int_{R-r}^{R+r} dF = G \frac{(4\pi r^2 \rho t)m}{R^2} = G \frac{Mm}{R^2}, \qquad (16\text{-}9)$$

where

$$M = (4\pi r^2 t \rho)$$

is the total mass of the shell. This is exactly the same result we would obtain for the force between *particles* of mass M and m separated a distance R. We have proved, therefore, that *a uniformly dense spherical shell attracts an external mass point as if all its mass were concentrated at its center.*

A solid sphere can be regarded as composed of a large number of concentric shells. If each spherical shell has a uniform density, even though different shells may have different densities, the same result applies to the solid sphere. Hence, a body such as the earth, the moon, or the sun, to the extent that they are such spheres, may be regarded gravitationally as point particles to bodies outside them.

Notice that our proof applies only to spheres and then only when the density is constant over the sphere or a function of radius alone.

An interesting result of some significance is the force exerted by a spherical shell on a particle *inside* it. This force is *zero*. To prove this we refer to Fig. 16-7, where m is shown inside the shell. Notice that R is now smaller than r. The limits of our integration over x are now $r - R$ to $R + r$. But

$$\int_{r-R}^{R+r} \left(\frac{R^2 - r^2}{x^2} + 1 \right) dx = 0,$$

so that $F = 0$.

This last result, although not obvious, is plausible because the mass elements of the shell on opposite sides of m now exert forces of opposite directions on m inside. The total annulment depends on the fact that the force varies precisely as an inverse square of the separation distance of two particles. (See Problem 18.) Important consequences of this result will be discussed in the chapters on electricity. There we shall see that the electrical force between charged par-

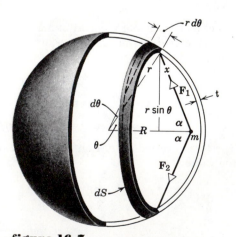

figure 16-7

Gravitational attraction of a section dS of a spherical shell of matter on a particle of mass m. Here the particle is inside the shell.

ticles also depends inversely on the square of the distance between them. A consequence of interest in gravitation is that the gravitational force exerted by the earth on a particle decreases as the particle goes deeper into the earth, *assuming a constant density for the earth*, for the portions of matter in shells external to the position of the particle exert no force on it, the force becoming zero at the center of the earth. Hence, g would be a maximum at the earth's surface and decrease both outward and inward from that point *if the earth had constant density*. Can you imagine a spherically symmetric distribution of the earth's mass* which would not give this result? (See Problem 16.)

EXAMPLE 3

Suppose a tunnel could be dug through the earth from one side to the other along a diameter, as shown in Fig. 16-8.

(a) Show that the motion of a particle dropped into the tunnel is simple harmonic motion. Neglect all frictional forces and assume that the earth has a uniform density.

The gravitational attraction of the earth for the particle at a distance r from the center of the earth arises entirely from that portion of matter of the earth in shells internal to the position of the particle. The external shells exert no force on the particle. Let us assume that the earth's density is uniform at the value ρ. Then the mass inside a sphere of radius r is

$$M' = \rho V' = \rho \frac{4\pi r^3}{3}.$$

This mass can be treated as though it were concentrated at the center of the earth for gravitational purposes. Hence, the force on the particle of mass m is

$$F = \frac{-GM'm}{r^2}.$$

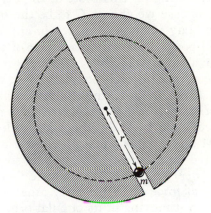

figure 16-8
Example 3. Particle moving in a tunnel through the earth.

The minus sign is used to indicate that the force is attractive and directed toward the center of the earth.

Substituting for M', we obtain

$$F = -G\frac{(\rho 4\pi r^3)m}{3r^2} = -\left(G\rho\frac{4\pi m}{3}\right)r = -kr.$$

Here $G\rho 4\pi m/3$ is a constant, which we have called k. The force is, therefore, proportional to the displacement r but oppositely directed. This is exactly the criterion for simple harmonic motion.

(b) If mail were delivered through this chute, how much time would elapse between deposit at one end and delivery at the other end?

The period of this simple harmonic motion is

$$T = 2\pi\sqrt{\frac{m}{k}} = 2\pi\sqrt{\frac{3m}{G\rho 4\pi m}} = \sqrt{\frac{3\pi}{G\rho}}.$$

Let us take $\rho = 5.51 \times 10^3$ kg/m³ and $G = 6.67 \times 10^{-11}$ N · m²/kg². This gives

$$T = \sqrt{\frac{3\pi}{G\rho}} = \sqrt{\frac{3\pi}{(6.67 \times 10^{-11})(5.51 \times 10^3)}} \text{ s} = 5050 \text{ s} = 84.2 \text{ min.}$$

The time for delivery is one-half period, or about 42 min. Notice that this time is independent of the mass of the mail.

The earth does not really have a uniform density. Suppose ρ were some function of r, rather than a constant. What effect would this have on our problem?

---

* See, in this connection, "Comment on 'The Radial Variation of g in a Spherically Symmetric Mass with Nonuniform Density'" by K. Sundaralingam, in *American Journal of Physics*, September 1974.

The motions of bodies in the solar system can be deduced from the laws of motion and the law of universal gravitation. As Kepler pointed out (see page 337), all planets move in elliptical orbits, the sun being at one focus. We can learn a lot about planetary motion by considering the special case of circular orbits. We shall neglect the forces between planets, considering only the interaction between the sun and a given planet. These considerations apply equally well to the motion of a satellite (natural or artificial) about a planet.

Consider two spherical bodies of masses M and m moving in circular orbits under the influence of each other's gravitational attraction. The center of mass of this system of two bodies lies along the line joining them at a point C such that $mr = MR$ (Fig. 16-9). If there are no external forces acting on this system, the center of mass has no acceleration. In this case we choose C to be the origin of our reference frame. The large body of mass M moves in an orbit of constant radius R and the small body of mass m in an orbit of constant radius r, both having the same angular velocity ω. In order for this to happen, the gravitational force acting on each body must provide the necessary centripetal acceleration. Because these gravitational forces are simply an action-reaction pair, the centripetal forces must be equal but oppositely directed. That is, $m\omega^2 r$ (the magnitude of the centripetal force exerted by M on m) must equal $M\omega^2 R$ (the magnitude of the centripetal force exerted by m on M). That this is so follows at once, for $mr = MR$ so that $m\omega^2 r = M\omega^2 R$. The specific requirement, then, is that the gravitational force on either body must equal the centripetal force needed to keep it moving in its circular orbit, that is,

$$\frac{GMm}{(R + r)^2} = m\omega^2 r. \qquad (16\text{-}10)$$

If one body has a much greater mass than the other, as in the case of the sun and a planet, its distance from the center of mass is much smaller than that of the other body. Let us therefore assume that R is negligible compared to r. Equation 16-10 then becomes

$$GM_s = \omega^2 r^3,$$

where M_s is the mass of the sun. If we express the angular velocity in terms of the period of the revolution, $\omega = 2\pi/T$, we obtain

$$GM_s = \frac{4\pi^2 r^3}{T^2}. \qquad (16\text{-}11)$$

This is a basic equation of planetary motion; it holds also for elliptical orbits if we define r to be the semi-major axis of the ellipse. Let us consider some of its consequences.

One immediate consequence of Eq. 16-11 is that it predicts Kepler's third law of planetary motion in the special case of circular orbits. For we can express Eq. 16-11 as

$$T^2 = \frac{4\pi^2}{GM_s} r^3.$$

Notice that the mass of the planet is not involved in this expression. Here, $4\pi^2/GM_s$ is a constant, the same for all planets.

When the period T and radius r of revolution are known for any planet, Eq. 16-11 can be used to determine the mass of the sun. For example, the earth's period is

$$T = 365 \text{ days} = 3.15 \times 10^7 \text{ s},$$

16-7
THE MOTIONS OF PLANETS AND SATELLITES

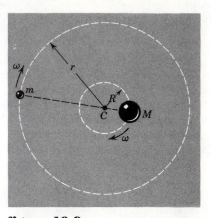

figure 16-9
Two bodies moving in circular orbits under the influence of each other's gravitational attraction. They both have the same angular velocity ω.

and its orbital radius is

$$r = 93 \times 10^6 \text{ mi} = 1.5 \times 10^{11} \text{ m}.$$

Hence,

$$M_s = \frac{4\pi^2 r^3}{GT^2} = \frac{(4\pi^2)(1.5 \times 10^{11} \text{ m})^3}{(6.67 \times 10^{-11} \text{ N} \cdot \text{m}^2/\text{kg}^2)(3.15 \times 10^7 \text{ s})^2} \cong 2.0 \times 10^{30} \text{ kg.}$$

The mass of the sun is thus about 300,000 times the mass of the earth. The error made in neglecting R compared to r is seen to be trivial, for

$$R = \frac{m}{M} r = \frac{1}{300,000} r \cong 300 \text{ mi}; \qquad \frac{R}{r} 100\% \cong \frac{1}{3000} \text{ of } 1\%.$$

In a similar manner we can determine the mass of the earth from the period and radius of the moon's orbit about the earth. (See Problem 22.)

If we know the mass of the sun M_s and the period of revolution T of any planet about it, we can determine the radius of the planet's orbit r from Eq. 16-11. Since the period is easily obtained from astronomical observations, this method of determining a planet's distance from the sun is fairly reliable.

Equation 16-11 holds also for the motion of artificial satellites about the earth; we need only substitute the mass of the earth M_e for M_s in that equation.

Kepler's second law of planetary motion (see page 337) must, of course, hold for circular orbits. In such orbits both ω and r are constant so that equal areas are swept out in equal times by the line joining a planet and the sun. For the exact elliptical orbits, however, or for any orbit in general, both r and ω will vary. Let us consider this case.

Figure 16-10 shows a particle revolving about C along some arbitrary path. The area swept out by the radius vector in a very short time interval Δt is shown shaded in the figure. This area, neglecting the small triangular region at the end, is one-half the base times the altitude or approximately $\frac{1}{2}(r\omega \, \Delta t) \cdot r$. This expression becomes more exact in the limit as $\Delta t \to 0$, the small triangle going to zero more rapidly than the large one. The rate at which area is being swept out instantaneously is therefore

$$\lim_{\Delta t \to 0} \frac{\frac{1}{2}(r\omega \, \Delta t)(r)}{\Delta t} = \frac{1}{2}\omega r^2.$$

But $m\omega r^2$ is simply the angular momentum of the particle about C. Hence, Kepler's second law, which requires that the rate of sweeping

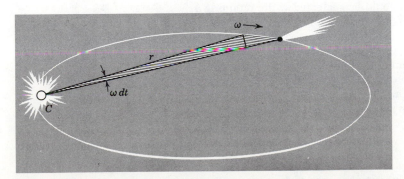

figure 16-10
A comet* moving along an elliptical path with the sun C at the focus of the ellipse. In time dt the comet (or a planet) sweeps out an angle $d\theta = \omega dt$.

* See "The Nature of Comets" by Fred L. Whipple, in *Scientific American*, February 1974, for a fascinating discussion of the properties and possible origin of comets.

out of area $\frac{1}{2}\omega r^2$ be constant, is entirely equivalent to the statement that *the angular momentum of any planet about the sun remains constant*. The angular momentum of the particle about C cannot be changed by a force on it directed toward C. Kepler's second law would, therefore, be valid for any *central force*, that is, any force directed toward the sun. The exact nature of this force—how it depends on distance of separation or other properties of the bodies—is not revealed by this law.

It is Kepler's first law that requires the gravitational force to depend exactly on the inverse square of the distance between two bodies, that is, on $1/r^2$. Only such a force, it turns out, can yield planetary orbits which are elliptical with the sun at one focus.

A planet revolves around the sun in a elliptical orbit of eccentricity e. Find the ratio of the time spent by the planet between the ends of the minor axis close to the sun to the period of revolution.

By Kepler's first law, the sun is at one focus of the ellipse. (In Fig. 16-11, the ellipse shown has a much larger eccentricity than does the orbit of any planet in the solar system.) The major axis (length $2a$) and the minor axis (length $2b$) intersect at the center C of the ellipse and the distance CF from the center of the ellipse to the focus F is ea by the definition of eccentricity. Notice that for a circular orbit the eccentricity would be zero.

Let the period of revolution be T, and the time required for the planet to travel from B to D, on that part of the ellipse which is close to the sun, be t. Then, if A = area of the ellipse and A' = shaded area, we have, by the conservation of angular momentum (or by the equivalent statement that the rate of sweeping out of area is constant),

$$\frac{A}{T} = \frac{A'}{t}.$$

But, $A' = \frac{1}{2}A - A''$, where A'' = area of triangle BDF. Therefore,

$$\frac{t}{T} = \frac{A'}{A} = \frac{\frac{1}{2}A - A''}{A} = \frac{1}{2} - \frac{A''}{A} = \frac{1}{2} - \frac{\frac{1}{2}(2b)(ae)}{\pi ab}$$

and

$$\frac{t}{T} = \frac{1}{2} - \frac{e}{\pi}.$$

Notice that this reduces to $\frac{1}{2}$ for a circular orbit $(e = 0)$. Why is the ratio less than one-half for elliptical orbits?

EXAMPLE 4

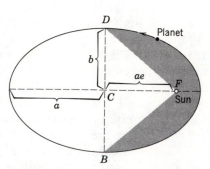

figure 16-11
Example 4. A planet revolves around the sun in an elliptical orbit.

Newton's laws of motion and his law of universal gravitation are in almost complete agreement with astronomical observations.* In our calculation we considered the motion of a planet about the sun as a "two-body" problem. However, we saw that the motion of the sun could be neglected while retaining a high degree of accuracy because of the large ratio of solar mass to planetary mass. This reduced the problem to that of motion of one body about a center of force. If we had required greater accuracy, we would have had to include the sun's motion in our problem (see Problem 28). In fact, for an exact treatment we would have to take into account the effect of other planets and satellites on the motion of sun and planet. This "many-body" problem is quite formidable, but it can be solved by approximation methods to a high degree of accuracy. The results of such calculations are in excellent agreement with astronomical observations.

---

*The major axis of the elliptical orbit of Mercury rotates slightly more than that predicted by Newtonian mechanics when the perturbing influence of other planets is included. This effect is accounted for by the general theory of relativity.

A basic fact of gravitation is that two masses exert forces on one another. We can think of this as a direct interaction between the two mass particles, if we wish. This point of view is called *action-at-a-distance*, the particles interacting even though they are not in contact. Another point of view is the *field* concept, which regards a mass particle as modifying the space around it in some way and setting up a *gravitational field*. This field then acts on any other mass particle in it, exerting the force of gravitational attraction on it. The field, therefore, plays an intermediate role in our thinking about the forces between mass particles. According to this view we have two separate parts to our problem. First, we must determine the field established by a given distribution of mass particles; and secondly, we must calculate the force that this field exerts on another mass particle placed in it.

For example, consider the earth as an isolated mass. If a body is now brought in the vicinity of the earth, a force is exerted on it. This force has a definite direction and magnitude at each point in space. The direction is radially in toward the center of the earth and the magnitude is *mg*. We can, therefore, associate with each point near the earth a vector **g** which is the acceleration that a body would experience if it were released at this point. We call **g** the *gravitational field strength* at the point in question. Since

$$\mathbf{g} = \frac{\mathbf{F}}{m}, \tag{16-12}$$

we may define gravitational field strength at any point as the gravitational force per unit mass at that point.* We calculate the force from the field simply by multiplying **g** by the mass *m* of the particle placed at any point.

The gravitational field is an example of a *vector field*, each point in this field having a vector associated with it. There are also scalar fields, such as the temperature field in a heat-conducting solid. The gravitational field arising from a fixed distribution of matter is also an example of a *stationary field*, because the value of the field at a given point does not change with time.

The field concept is particularly useful for understanding electromagnetic forces between moving electric charges. It has distinct advantages, both conceptually and in practice, over the action-at-a-distance concept. The field concept was not used in Newton's day. It was developed much later by Faraday for electromagnetism and only then applied to gravitation. Subsequently, this point of view was adopted for gravitation in the general theory of relativity. The chief purpose of introducing it here is to give the student an early familiarity with a concept that proves to be important in the development of physical theory.

---

* In Eq. 16-12 *g* is defined as the gravitational force per unit mass; at a point *P* a distance *R* from the center of a spherically symmetric mass *M*, it is given by GM/R^2. This *g* differs from the *g* whose magnitude is displayed in Tables 16-1 and 16-2 in that, as Example 2 shows, the centripetal acceleration of a body moving around the earth is already taken into account so that what is described in these tables is an *effective g*. For example, the effective *g* in an orbiting earth satellite is zero, as we have all seen on television transmissions from such satellites. This is because GM/R^2 in Example 2 is exactly equal to a_R in that example. However, the gravitational field at the site of the orbiting satellite, which is given just by GM/R^2, is *not* zero.

In Chapter 15 we derived the formula for the period of a simple pendulum, $T = 2\pi \sqrt{l/g}$. Keeping in mind that the earth's gravitational field is not uniform over large distances, as was assumed for small distances, what is the longest period a simple pendulum could have in the vicinity of the earth's surface?

The formula $T = 2\pi \sqrt{l/g}$, although not applicable when g varies over the pendulum's path, suggests that we increase the length of the pendulum. Let us make the length infinite. The pendulum bob would then travel along the arc of a circle of infinite radius, that is, along a straight line, as shown in Fig. 16-12. The direction of the earth's gravitational field is everywhere radially in toward the center of the earth, so that its direction changes along the arc. Let us assume that the bob of mass m has an amplitude that is small compared to the radius of the earth. Then the bob is always a distance R_e, the earth's radius, from the center of the earth, to a good approximation. Then the force F on m is

$$F = \frac{GM_e m}{R_e^2} = mg,$$

where M_e is the mass of the earth. This force is directed toward the earth's center as shown. The component of this vector force along x, the line of motion of the bob, is

$$F_x = F \cos \theta = -F\frac{x}{R_e} = -\frac{GM_e m}{R_e^3} x,$$

where the minus sign indicates that the force is directed opposite to the displacement from $x = 0$. We can write this as

$$F_x = -kx,$$

where $k = GM_e m/R_e^3$, a constant.

The formula for the period of a simple harmonic oscillator is $T = 2\pi \sqrt{m/k}$. Hence,

$$T = 2\pi \sqrt{\frac{m}{k}} = 2\pi \sqrt{\frac{m}{GM_e m/R_e^3}} = 2\pi \sqrt{\frac{R_e}{GM_e/R_e^2}} = 2\pi \sqrt{\frac{R_e}{g}},$$

because g at the earth's surface equals GM_e/R_e^2. Putting in $R_e = 6.37 \times 10^6$ m and $g = 9.80$ m/s², we obtain $T = 84.3$ min as the longest period of a simple pendulum in the vicinity of the earth's surface. (See Question 37.)

EXAMPLE 5

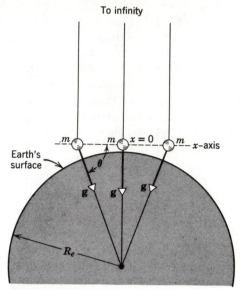

figure 16-12
Example 5. A simple pendulum suspended at infinity.

In Chapter 8 we discussed the gravitational potential energy of a particle (mass m) and the earth (mass M). We considered only the special case in which the particle remains close to the earth so that we could assume the gravitational force acting on the particle to be constant for all positions of the particle. In this section we remove that restriction and consider particle-earth separations that may be appreciably greater than the earth's radius.

Equation 8-5b, which we may write as,

$$\Delta U = U_b - U_a = -W_{ab}, \qquad (8\text{-}5b)$$

defines the change ΔU in the potential energy of any system, in which a conservative force (gravity, say) acts, as the system changes from configuration a to configuration b. W_{ab} is the work done by that conservative force as the system changes.

The potential energy of the system in any arbitrary configuration b is (see Eq. 8-5b)

$$U_b = -W_{ab} + U_a. \qquad (16\text{-}13)$$

To give a value to U_b we must (arbitrarily) choose configuration a to be

16-9
GRAVITATIONAL POTENTIAL ENERGY

some agreed-upon reference configuration and we must assign to U_a some (arbitrarily) agreed-upon value, usually zero.

In Chapter 8 we chose as a reference configuration for the earth-particle system that in which the particle is resting on the surface of the earth and we assigned to this configuration the potential energy $U_a = 0$. When the particle is at a height y above the surface of the earth, the potential energy $U (= U_b)$ is given from Eq. 16-13 as

$$U = -W_{ab} + 0 = -(-mg)(y) = mgy.$$

The conservative force in question, gravity, points down and has the value $(-mg)$; the displacement of the particle $(+y)$ points up from the reference level; hence the difference in sign for these quantities.

For the more general case, in which the restriction $y \ll R$ (in which R is the radius of the earth) is not imposed, we find it convenient to select a different reference configuration, namely that in which the particle and the earth are infinitely far apart. We assign the value zero to the potential energy of the system in this configuration. Thus the zero-potential-energy configuration is also the zero-force configuration. We made a similar choice when we defined the zero-energy configuration of a spring to be its normal unstressed state, for which the restoring force is zero.

When the particle of mass m is a distance r from the center of the earth, the system potential energy is given by Eq. 16-13 as

$$U(r) = -W_{\infty r} + 0 \qquad (16\text{-}14)$$

in which $W_{\infty r}$ is the work done by the conservative force (gravity) on the particle as the particle moves in from infinity to a distance r from the center of the earth. For simplicity we assume for the present that the particle moves toward the earth along a radial line. The gravitational force $F(r)$ acting on the particle (assuming $r \geqslant R$) will then be $-GMm/r^2$, the minus sign indicating an attractive force, that is, a force that pulls the particle toward the earth. We may then find $U(r)$ from Eq. 16-14 as

$$U(r) = -W_{\infty r}$$

$$= -\int_{\infty}^{r} F(r)dr$$

$$= -\int_{\infty}^{r} \left(-\frac{GMm}{r^2}\right)dr = -\frac{GMm}{r}\bigg|_{\infty}^{r}$$

$$= -\frac{GMm}{r}. \qquad (16\text{-}15)$$

The minus sign indicates that the potential energy is negative at any finite distance; that is, the potential energy is zero at infinity and decreases as the separation distance decreases. This corresponds to the fact that the gravitational force exerted on the particle by the earth is attractive. As the particle moves in from infinity, the work $W_{\infty r}$ done by this force on the particle is positive, which means, from Eq. 16-14, that $U(r)$ is negative.

Equation 16-15 holds no matter what path is followed by the particle in moving in from infinity to radius r. We can show this by breaking up any arbitrary path into infinitesimal steplike portions, which are drawn alternately along the radius and perpendicular to it (Fig. 16-13). No work is done along perpendicular segments, such as AB, because along them the force is perpendicular to the displacement. But the work done along the radial parts of the path, such as BC, adds up to the work done in going directly along a radial path, such as AE. The work done in moving the particle between two points in a gravitational field is, therefore, independent of the actual path connecting these points. Hence, the gravitational force is a *conservative* force.

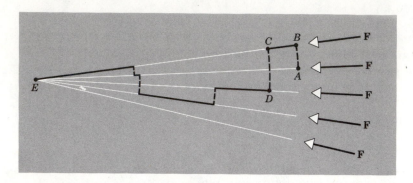

figure 16-13
Work done in taking a mass from A to E is independent of the path.

Equation 16-15 shows that the potential energy of the particles M and m is a characteristic of the system $M + m$. The potential energy is a property of the *system* of bodies, rather than of either body alone. The potential energy changes whether M or m is displaced; each is in the gravitational field of the other. Nor does it make any sense to assign part of the potential energy to M and part of it to m. Often, however, we *do* speak of the potential energy of a body m (planet or stone, say) in the gravitational field of a much more massive body M (sun or earth, respectively). The justification for speaking as though the potential energy belongs to the planet or to the stone alone is this: When the potential energy of a system of two bodies changes into kinetic energy, the lighter body gets most of the kinetic energy. The sun is so much more massive than a planet that the sun receives hardly any of the kinetic energy; and the same is true for the earth in the earth-stone system.

We can derive the gravitational force from the potential energy. The relation for spherically symmetric potential energy functions is $F = -dU/dr$; see Eq. 8-7. This relation is the converse of Eq. 16-15. From it we obtain

$$F = -\frac{dU}{dr} = -\frac{d}{dr}\left(-\frac{GMm}{r}\right) = -\frac{GMm}{r^2}. \qquad (16\text{-}16)$$

The minus sign here shows that the force is an attractive one, directed inward along a radius opposite to the radial displacement vector.

We can, if we wish, associate a scalar field with gravitation. We first define, quite generally, the *gravitational potential V* as the *gravitational potential energy per unit mass* of a body in a gravitational field. Then, for the spherically symmetrical body of mass M,

$$V = \frac{U(r)}{m} = -\frac{GM}{r}. \qquad (16\text{-}17)$$

Associated with every point in the space around a mass M we then have a number, the gravitational potential. This gives us a *scalar field*, potential being a scalar quantity. To determine the force exerted by this field on a mass particle m placed in it, we simply compute $-dV/dr$ at the point in question and multiply by m. The force has a magnitude $-m\,dV/dr$ and a direction radially in toward the center of force M.

Escape velocity. We can readily find the gravitational potential energy of a particle of mass m at the surface of the earth as (Eq. 16-15) $U(R) = -GM_e m/R_e$. The amount of work required to move a body from the surface of the earth to infinity is $GM_e m/R_e$, or about 6.0×10^7 J/kg. If we could give a projectile more than this energy at the surface of the earth, then, neglecting the resistance of the earth's atmosphere, it would escape from the earth never to return. As it proceeds outward its kinetic energy decreases and its potential energy increases, but its speed is never reduced to zero. The critical initial speed, called the escape speed v_0, such that the projectile does not return, is given by

$$\tfrac{1}{2}mv_0{}^2 = \frac{GM_e m}{R_e}$$

or

$$v_0 = \sqrt{2\frac{GM_e}{R_e}} = 7.0 \text{ mi/s (25,000 mi/h)} = 11.2 \text{ km/s.}$$

Should a projectile be given this initial speed, it would escape from the earth. For initial speeds less than this the projectile will return. Its kinetic energy becomes zero at some finite distance from the earth and the projectile falls back to earth.*

The lighter molecules in the earth's upper atmosphere can attain high enough speeds by thermal agitation to escape into outer space. Hydrogen gas, which must have been present in the earth's atmosphere a long time ago, has now disappeared from it. Helium gas escapes at a steady rate, much of it resupplied by radioactive decay from the earth's crust. The escape speed for the sun is much too great to allow hydrogen to escape from its atmosphere. On the other hand, the speed of escape on the moon is so small that it can hardly keep any atmosphere at all. (See Problem 30.)

EXAMPLE 6

16-10
POTENTIAL ENERGY FOR MANY-PARTICLE SYSTEMS

If two particles are separated by a distance r, their potential energy is given from Eq. 16-14 as

$$U(r) = -W_{\infty r} \qquad (16\text{-}14)$$

in which $W_{\infty r}$ is the work done *by the gravitational force* as the particles move from an infinite separation to separation r. We now give another interpretation to $U(r)$.

Let us balance out the gravitational force by an *external force* applied by some external agent and let us arrange it so that, at all times, this external force is equal and opposite to the gravitational force for each particle. The work done by the *external* force as the particles move from an infinite separation to separation r is not $W_{\infty r}$ but $-W_{\infty r}$; this follows because the displacements are the same but the forces are equal and opposite. Thus we may interpret Eq. 16-14 as follows: *The potential energy of a system of particles is equal to the work that must be done by an external agent to assemble the system, starting from the standard reference configuration*

Thus, if you lift a stone of mass m a distance of y above the earth's surface, you are the external agent (separating earth and stone) and the work you do in "assembling the system" is $+mgy$, which is also the potential energy. Similarly, the work done *by an external agent* as a body of mass m moves in from infinity to a distance r from the earth

---

* We have ignored the forces exerted on the projectile by bodies other than the earth. At sufficiently great distances from the earth we must take into account the gravitational forces arising from the moon, the planets, the sun etc., so that we can no longer use the simple "two-body" result. A projectile can escape from the earth by being "captured" by another astronomical body, for example, in such "many-body" cases.

is *negative* because the agent must exert a restraining force on the body; this is in agreement with Eq. 16-14.

These considerations hold for systems that contain more than two particles. Consider three bodies of masses m_1, m_2, and m_3. Let them initially be infinitely far from one another. The problem is to compute the work done by an external agent to bring them into the positions shown in Fig. 16-14. Let us first bring m_2 in toward m_1 from an infinite separation to the separation $\mathbf{r}_{12}$. The work done against the gravitational force exerted by m_1 on m_2 is $-Gm_1m_2/r_{12}$. Now let us bring m_3 in from infinity to the separation $\mathbf{r}_{13}$ from m_1 and $\mathbf{r}_{23}$ from m_2. The work done against the gravitational force exerted by m_1 on m_3 is $-Gm_1m_3/r_{13}$ and that against the gravitation force exerted by m_2 on m_3 is $-Gm_2m_3/r_{23}$. The total work done in assembling this system is the total potential energy of the system

$$-\left(\frac{Gm_1m_2}{r_{12}} + \frac{Gm_1m_3}{r_{13}} + \frac{Gm_2m_3}{r_{23}}\right).$$

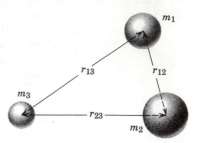

figure 16-14
Three masses m_1, m_2, and m_3 brought together from infinity.

Notice that no vector operations are needed in this procedure.

No matter how we assemble the system, that is, regardless of which bodies are moved or which paths are taken, we always find this same amount of work required to bring the bodies into the configuration of Fig. 16-13 from an initial infinite separation. The potential energy must, therefore, be associated with the system rather than with any one or two bodies. If we wanted to separate the system into three isolated masses once again, we would have to supply an amount of energy

$$+\left(\frac{Gm_1m_2}{r_{12}} + \frac{Gm_1m_3}{r_{13}} + \frac{Gm_2m_3}{r_{23}}\right).$$

This energy may be regarded as a sort of binding energy holding the mass particles together in the configuration shown.

Just as we can associate elastic potential energy with the compressed or stretched configuration of a spring holding a mass particle, so we can associate gravitational potential energy with the configuration of a system of mass particles held together by gravitational forces. Similarly, if we want to think of the elastic potential energy of a particle as being stored in the spring, so we can think of the gravitational potential energy as being stored in the gravitational field of the system of particles. A change in either configuration results in a change of potential energy.

These concepts occur again when we meet forces of electric or magnetic origin, or, in fact, of nuclear origin. Their application is rather broad in physics. The advantage of the energy method over the dynamical method is derived from the fact that the energy method uses scalar quantities and scalar operations rather than vector quantities and vector operations. When the actual forces are not known, as is often the case in nuclear physics, the energy method is essential.

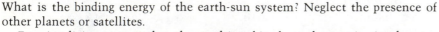

EXAMPLE 7

What is the binding energy of the earth-sun system? Neglect the presence of other planets or satellites.

For simplicity assume that the earth's orbit about the sun is circular at a radius r_{es}. The work done against the gravitational force to bring the earth and sun from an infinite separation to a separation r_{es} is

$$-G\frac{M_sM_e}{r_{es}} \cong -5.0 \times 10^{33}\text{ J},$$

where we take $M_s \cong 330{,}000 M_e$, $M_e = 6.0 \times 10^{24}$ kg, $r_{es} = 150 \times 10^9$ m. The minus sign indicates that the force is attractive, so that work is done by the gravitational force. It would take an equivalent amount of work by an outside agent to separate these bodies completely from rest. Because the kinetic energy of the earth in its orbit is half the magnitude of the potential energy of the earth-sun system, only half of this work is needed to break up the system, so that the effective binding energy, assuming that the earth-sun-system is at rest after breakup, is about 2.5×10^{33} J.

What effect does the presence of the moon and other planets have on the energy binding the earth to the solar system?

Consider again the motion of a body of mass m (planet or satellite, say) about a massive body of mass M (sun or earth, say). We shall consider M to be at rest in an inertial reference frame with the body m moving about it in a circular orbit. The potential energy of the system is

$$U(r) = -\frac{GMm}{r},$$

where r is the radius of the circular orbit. The kinetic energy of the system is

$$K = \tfrac{1}{2} m \omega^2 r^2$$

the sun being at rest. From the equation preceding Eq. 16-11 we obtain

$$\omega^2 r^2 = \frac{GM}{r},$$

so that

$$K = \frac{1}{2} \frac{GMm}{r}.$$

The total energy is

$$E = K + U = \frac{1}{2} \frac{GMm}{r} - \frac{GMm}{r} = -\frac{GMm}{2r}. \qquad (16\text{-}18)$$

This energy is constant and is negative. Now the kinetic energy can never be negative, but from Eq. 16-18 we see that it must go to zero as the separation goes to infinity. The potential energy is always negative, except for its zero value at infinite separation. The meaning of the total negative energy, then, is that the system is a closed one, the planet m always being bound to the attracting solar center M and never escaping from it (Fig. 16-15).

Even when we consider elliptical orbits, in which r and ω vary, the total energy is negative. It is also constant, corresponding to the fact that gravitational forces are conservative. Hence, both the total energy and the total angular momentum are constant in planetary motion. These quantities are often called *constants of the motion*. We obtain the actual orbit of a planet with respect to the sun by starting with these conservation relations and eliminating the time variable by use of the laws of dynamics and gravitation. The result is that planetary orbits are elliptical.

In the earlier theories of the atom, as in the Bohr theory of the hydrogen atom, these identical mechanical relations are used in describing the motion of an electron about an attracting nuclear center. These same relations are used for open orbits (total energy positive) as in the experiments of Rutherford on the scattering of charged nuclear particles. Central forces, and particularly inverse square forces, are often encountered in physical systems.

16-11
ENERGY CONSIDERATIONS IN THE MOTIONS OF PLANETS AND SATELLITES

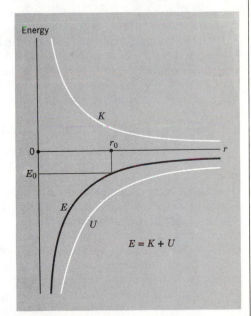

figure 16-15
Kinetic energy K, potential energy U, and total energy $E = U + K$ of a body in circular planetary motion. A planet with total energy $E_0 < 0$ will remain in an orbit of radius r_0. The farther the planet is from the sun, the greater (that is, less negative) its (constant) total energy E. To escape from the center of force and still have kinetic energy at infinity, it would need positive total energy.

In describing the experiments which were fundamental to our definitions of force and mass, we had to assume some reference frame relative to which accelerations could be measured. If the reference frame itself were erratically accelerated, we would not observe any regularity in our measured accelerations. As a matter of fact, our laboratory experiments are performed in a reference frame which is fixed to the earth. We have already discussed the effect that the rotation of the earth about its own axis has on our measurements. What effect does the motion of the earth as a whole about the sun, or some other cosmic body, have?

The acceleration of the earth with respect to the sun is $\omega^2 r$ or about 6×10^{-3} m/s². It would seem at first that this acceleration, small as it is, might prove disturbing in experiments involving small forces. That this is not the case, however, follows from the universality of the law of gravitation. Not only the earth but also the masses we use in our laboratory apparatus are accelerated toward the sun at practically the same rate.

Let us compute the error made in neglecting the earth's orbital acceleration. The acceleration of the earth toward the sun is k/r^2, where r is the distance from the center of the sun and the center of the earth and k is GM_s. Consider now a body on that side of the earth most distant from the sun. We can imagine that we are weighing it on a spring scale, for example. Then, the part of its acceleration toward the sun which is due to the gravitational attraction of the sun itself is

$$\frac{k}{(r + r_0)^2} = \frac{k}{r^2}\left(1 - \frac{2r_0}{r} + \cdots + \text{much smaller terms}\right),$$

where r_0 is the radius of the earth. The *difference* between the acceleration of the earth due to the sun's attraction (that is, k/r^2) and the acceleration of the apparatus due to the sun's attraction (the expression above) would be less than $(k/r^2)(2r_0/r)$. But $2r_0/r$ is about 10^{-4}. The difference, then, would be less than 10^{-4} of the earth's acceleration, or less than 10^{-6} m/s². The relative acceleration of the body and the earth due to the sun's attraction is about one-ten-millionth as strong as the gravitational acceleration of the body due to the earth. The moon has a similar effect of comparable magnitude on the body. Hence, only if we were measuring to one part in a million, would we need to consider seriously the accelerating nature of a reference frame attached to the earth. For almost all practical purposes the earth is good enough as an inertial reference frame.

16-12 THE EARTH AS AN INERTIAL REFERENCE FRAME

16-13 THE PRINCIPLE OF EQUIVALENCE

Consider two reference frames: (1) a nonaccelerating (inertial) reference frame S in which there is a uniform gravitational field and (2) a reference frame S' which is accelerating uniformly with respect to an inertial frame but in which there is no gravitational field. In his general theory of relativity, Albert Einstein showed that two such frames are exactly equivalent physically. That is, experiments carried out under the same conditions in these two frames should give the same results. This is the *principle of equivalence*.

Suppose that a spaceship is at rest in an inertial reference frame S in which there is a uniform gravitational field, say at the surface of the earth. Inside the spaceship objects such as an apple, which are released, will fall with an acceleration, say **g**, in the gravitational field; objects which are at rest—such as an astronaut sitting on the floor or a package on a spring scale attached to the ceiling—will experience a force, exerted by the floor or the spring respectively, opposing their weight.

Now suppose that the rockets are turned on and that the spaceship proceeds to a region of outer space where there is no gravitational field. Let the acceleration of the spaceship, our new frame S', be $\mathbf{a} = -\mathbf{g}$ with respect to the inertial reference frame S; that is, the ship is accelerating away from the earth beyond the region where the earth's field (or any other gravitational field) is appreciable. The conditions in the spaceship will now be similar to those in a spaceship at rest on the surface of the earth. Inside the ship, if the astronaut releases an apple, it will accelerate downward relative to the spaceship with an acceleration **g.** In fact, since all bodies that are free of any forces move with uniform velocity relative to the inertial frame *S*, all such bodies appear to fall with the *same* acceleration **g** with respect to the spaceship, *S'*. Furthermore, objects which are at rest relative to the spaceship—such as an astronaut sitting on the floor or a package on a spring scale attached to the ceiling—will experience forces indistinguishable from the forces which balanced their weight in the case when the spaceship was at rest in a gravitational field in *S*.

Indeed, if the astronaut did not know that rockets were accelerating his ship from *S*, he would be justified in concluding that he was in a gravitational field—a field whose pull accelerated the falling apple in *S'* and whose pull required that a balancing force be applied to the package (the tension in the spring) and to the spaceman (the normal force of the floor) to keep them at rest in *S'*. The astronaut simply could not tell the difference, from observations in his own frame, between a situation in which his ship was accelerating relative to an inertial frame in a region having no gravitational field and a situation in which the spaceship was unaccelerated in an inertial frame in which a uniform gravitational field existed. The two situations are exactly equivalent.

Einstein pointed out that, from the principle of equivalence, it follows that one cannot speak of the absolute acceleration of a reference frame, only a relative one, just as it followed from the special theory of relativity that one cannot speak of the absolute velocity of a reference frame, only a relative one. It also follows from the principle of equivalence that inertial mass and gravitational mass are equal. For all bodies which are free of any forces will move with uniform velocity relative to an inertial reference frame no matter what their inertial masses are, and they should, therefore, all have the same acceleration relative to an accelerated reference frame. Hence, from the principle of equivalence of *S* and *S'*, all bodies should fall with the same acceleration in a homogeneous gravitational field.

From the discussion so far we see that a uniform gravitational field can be imitated by a "field of acceleration." Indeed, a uniform gravitational field can be "transformed away" by transforming to a reference frame accelerating in the direction of the field with an acceleration equal in magnitude to that due to the field. In this new frame a particle whose motion was originally subject to a gravitational field is now a free particle. For example, in an artificial earth satellite an apple released by an astronaut will not fall relative to the satellite and the astronaut himself will be free of the forces which countered the pull of gravity before launching, so that he feels weightless. In general, however, gravitational fields, such as that of the earth, are not uniform through all space, so that one cannot replace the gravitational fields throughout space simply by transforming to a single reference frame accelerating with respect to the source of the field. One would need a different accelerated frame at each point in space to imitate the entire gravitational field.

questions

1. Modern observational astronomy and navigation procedures make use of the geocentric (or Ptolemaic) point of view (by using the rotating "celestial sphere"). Is this wrong? If not, then what criterion determines the system (the Copernican or Ptolemaic) we use? When would we use the heliocentric (or Copernican) system?

2. If the force of gravity acts on all bodies in proportion to their masses, why doesn't a heavy body fall correspondingly faster than a light body?

3. How does the weight of a body vary en route from the earth to the moon? Would its mass change?

4. At the earth's surface a freely suspended object is given a horizontal blow by a hammer. The object is taken to the moon, suspended freely, and given an equal horizontal blow with the same hammer. How is the horizontal speed resulting on the moon related to the horizontal speed on the earth?

5. Would we have more sugar to the pound at the pole or the equator? What about sugar to the kilogram?

6. What approximately is the *gravitational* force of attraction between a normal woman and a typical man 10 m away? When they are dancing? Compare with typical body weights.

7. Does the concentration of the earth's mass near its center change the variation of g with height compared with a homogeneous sphere? How?

8. Because the earth bulges near the equator, the source of the Mississippi River, although high above sea level, is nearer to the center of the earth than is its mouth. How can the river flow "uphill"?

9. The earth is an oblate spheroid because of the "flattening" effect of the earth's rotation. Is a degree of latitude larger or smaller near either pole than near the equator? Why?

10. Why can we learn more about the shape of the earth by studying the motion of an artificial satellite than by studying the motion of the moon?

11. How can one determine the mass of the moon?

12. One clock is based on an oscillating spring, the other on a pendulum. Both are taken to Mars. Will they keep the same time there that they kept on Earth? Will they agree with each other? Explain. Mars has a mass 0.1 that of Earth and a radius half as great.

13. From Kepler's second law and observations of the sun's motion as seen from the earth, we can conclude that the earth is closer to the sun during winter in the Northern hemisphere than during summer. Why isn't it colder in summer than in winter?

14. Does the law of universal gravitation require the planets of our solar system to have the actual orbits observed? Would planets of another star, similar to our Sun, have the same orbits? Suggest factors that might have determined the special orbits observed.

15. How is the orbital speed of a planet related to its (assumed circular) orbital radius?

16. The gravitational force exerted by the sun on the moon is about twice as great as the gravitational force exerted by the earth on the moon. Why then doesn't the moon escape from the earth (during a solar eclipse, for example)?

17. Explain why the following reasoning is wrong. "The sun attracts all bodies on the earth. At midnight, when the sun is directly below, it pulls on an object in the same direction as the pull of the earth on that object; at noon, when the sun is directly above, it pulls on an object in a direction opposite to the pull of the earth. Hence, all objects should be heavier at midnight (or night) than they are at noon (or day)."

18. The gravitational attraction of the sun and the moon on the earth produces tides. The sun's tidal effect is about half as great as that of the moon's. The direct pull of the sun on the earth, however, is about 175 times that of the moon. Why is it then that the moon causes the larger tides?

19. If lunar tides slow down the rotation of the earth (owing to friction), the angular momentum of the earth decreases. What happens to the motion of the moon as a consequence of the conservation of angular momentum? Does the sun (and solar tides play a role here? (See "Tides and the Earth-Moon System" by Peter Goldreich, *Scientific American*, April 1972 and "Tides of the British Seas" by Frank Sandon, in *Physics Education*, June 1975.)

20. Would you expect the total energy of the solar system to be constant? The total angular momentum? Explain your answers.

21. Discuss how the period of a simple pendulum changes if it is in an assembly that a rocket will propel from earth to a stable satellite orbit about the earth.

22. Does a rocket really need the escape speed of 25,000 mi/h initially to escape from the earth?

23. Objects at rest on the earth's surface move in circular paths with a period of 24 h. Are they "in orbit" in the sense that an earth satellite is in orbit?

Why not? What would the length of the "day" have to be to put such objects in true orbit?

24. An artificial satellite of the earth releases a package. Neglecting effects of air resistance, determine whether the package will strike the earth and if so, where—at a point ahead, directly below, or behind the satellite at the instant of impact, or directly under the satellite at release time?

25. Neglecting air friction and technical difficulties, can a satellite be put into an orbit by being fired from a huge cannon at the earth's surface? Explain.

26. Can a satellite move in a stable orbit in a plane not passing through the earth's center? Explain.

27. As measured by an observer on earth would there be any difference in the periods of two satellites, each in a circular orbit near the earth in an equatorial plane, but one moving eastward and the other westward?

28. After Sputnik I was put into orbit we were told that it would not return to earth but would burn up in its *descent*. Considering the fact that it did not burn up in its *ascent*, how is this possible?

29. In which case do astronauts experience greater acceleration, when being launched into orbit or on reentry and return to earth?

30. Show that a satellite may speed down: that is, show that if frictional forces cause a satellite to lose total energy, it will move into an orbit closer to the earth and may have increased kinetic energy.

31. An artificial satellite is in a circular orbit about the earth. How will its orbit change if one of its rockets is momentarily fired (a) toward the earth, (b) away from the earth, (c) in a forward direction, (d) in a backward direction, (e) at right angles to the plane of the orbit?

32. Inside a spaceship what difficulties would you encounter in walking? In jumping? In drinking?

33. We have all seen TV transmissions from orbiting satellites and watched objects floating around in effective zero gravity. Suppose an astronaut, bracing himself against the satellite frame, kicks a floating bowling ball. Will he stub his toe? Explain.

34. If a planet of given density were made larger, its force of attraction for an object on its surface would increase because of the planet's greater mass but would decrease because of the greater distance from the object to the center of the planet. Which effect predominates?

35. Consider a hollow spherical shell. How does the gravitational potential inside compare with that on the surface? What is the gravitational field strength inside?

36. A stone is dropped along the center of a deep vertical mine shaft. Assume no air resistance but consider the earth's rotation. Will the stone continue along the center of the shaft? If not, describe its motion.

37. Use qualitative arguments to explain why the following four periods are equal (all are 84 min, assuming a uniform earth density): (a) time of revolution of a satellite just above the earth's surface; (b) period of oscillation of mail in a tunnel through the earth; (c) period of simple pendulum having a length equal to the earth's radius in a uniform field 9.8 N/kg; (d) period of an infinite simple pendulum in the earth's real gravitational field.

38. The "action-at-a-distance" view of the gravitational force implies that the action is instantaneous. Actually, present physical theory assumes that gravitation propagates with a finite speed and this is taken into account in the modification of classical physics represented by general relativity theory. (See "Gravitational Waves—a Progress Report" by Jonothan L. Logan, in *Physics Today*, March 1973 for a discussion of the ideas and attempts at experimental verification.) What would happen to classical deductions if it were assumed there that the action were not instantaneous? (See also "Infinite Speed of Propagation of Gravitation in Newtonian Physics" by I. J. Good, *American Journal of Physics*, July 1975.)

39. Can one regard gravity as a "fictitious" force arising from the acceleration of one's reference frame relative to an inertial reference frame, rather than a "real" force?

problems

SECTION 16-2

1. How far from the earth must a body be along a line toward the sun so that the sun's gravitational pull balances the earth's? The sun is 9.3×10^7 mi away and its mass is $3.24 \times 10^5 M_e$.   *Answer:* 1.6×10^5 mi.

SECTION 16-3

2. What is the percentage change in the acceleration of the earth toward the sun from a total eclipse of the sun to the point where the moon is on a side of the earth directly opposite the sun?

SECTION 16-5

3. At what altitude above the earth's surface would the acceleration of gravity be 4.9 m/s²? The mass of the earth is 6.0×10^{24} kg and its mean radius is 6.4×10^6 m.   *Answer:* 2.6×10^6 m.

4. (a) What is the period of a "seconds pendulum" (period $= 2$ s on earth) on the surface of the moon? The moon's mass is 7.35×10^{22} kg and its radius is 1,720 km. (b) Why should a "seconds pendulum" have a period of two seconds rather than one second?

5. Certain neutron stars (extremely dense stars) are believed to be rotating at about one revolution per second. If such a star has a radius of 20 km, what must be its mass so that objects on its surface will be attracted to the star and not "thrown off" by the rapid rotation?   *Answer:* 4.7×10^{24} kg.

6. The fact that g varies from place to place over the earth's surface drew attention when Jean Richer in 1672 took a pendulum clock from Paris to Cayenne, French Guiana, and found that it·lost 2.5 min/day. If $g = 9.81$ m/s² in Paris, what is g in Cayenne?

7. If a pendulum has a period of exactly one second at the Equator, what would be its period at the South Pole?   *Answer:* 0.9974 s.

8. Masses m, assumed equal, hang from strings of different lengths on a balance at the surface of the earth, as shown in Fig. 16-16. If the strings have negligible mass and differ in length by h, (a) show that the error in weighing, associated with the fact that W' is closer to the earth than W, is $W' - W = 8\pi G\rho mh/3$ in which ρ is the mean density of the earth (5.5 g/cm³). (b) Find the difference in length which will give an error of one part in a million.

9. A scientist is making a precise measurement of g at a certain point in the Indian Ocean (on the equator) by timing the swings of a pendulum of accurately known construction. To provide a stable base the measurements are conducted in a submerged submarine. It is observed that a slightly different result for g is obtained when the submarine is moving eastward through the point than when it is moving westward, the speed in each case being 16 km/h. Account for this difference and calculate the fractional error $\Delta g/g$ in g.   *Answer:* 6.6×10^{-5}.

10. A body is suspended on a spring balance in a ship sailing along the equator with a speed v. (a) Show that the scale reading will be very close to $W_0(1 \pm 2\omega v/g)$, where ω is the angular speed of the earth and W_0 is the scale reading when the ship is at rest. (b) Explain the plus or minus.

SECTION 16-6

11. Two concentric shells of uniform density of mass M_1 and M_2 are situated as shown in Fig. 16-17. Find the force on a particle of mass m when the particle is located at (a) $r = a$, (b) $r = b$, and (c) $r = c$. The distance r is measured from the center of the shells.
Answer: (a) $G(M_1 + M_2)m/a^2$. (b) GM_1m/b^2. (c) Zero.

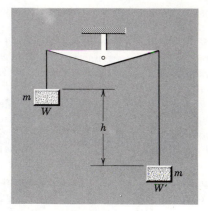

figure 16-16
Problem 8

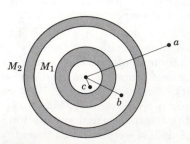

figure 16-17
Problem 11

12. With what speed would mail pass through the center of the earth if it were delivered by the chute of Example 3?

13. The sun, mass 2.0×10^{30} kg, is revolving about the center of the Milky Way galaxy, which is 2.4×10^{20} m away. It completes one revolution every 2.5×10^8 yr. Estimate the number of stars in the Milky Way, assuming a circular orbit.　　　　　　　　　　　　　　*Answer:* 6.5×10^{10}.

14. Consider an inertial reference frame whose origin is fixed at the center of mass of the system earth + falling object. (*a*) Show that the acceleration toward the center of mass of *either* body is independent of the mass of that body. (*b*) Show that the mutual, or relative, acceleration of the two bodies depends on the sum of the masses of the two bodies. Comment on the meaning, then, of the statement that a body falls toward the earth with an acceleration that is independent of its mass.

15. The following problem is from the 1946 "Olympic" examination of Moscow State University (see Fig. 16-18): A spherical hollow is made in a lead sphere of radius R, such that its surface touches the outside surface of the lead sphere and passes through its center. The mass of the sphere before hollowing was M. With what force, according to the law of universal gravitation, will the lead sphere attract a small sphere of mass m, which lies at a distance d from the center of the lead sphere on the straight line connecting the centers of the spheres and of the hollow?

$$Answer:\ \frac{GmM}{d^2}\left[1 - \frac{1}{8(1 - R/2d)^2}\right].$$

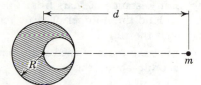

figure 16-18
Problem 15

16. The variation of g in the earth's interior is given in the accompanying table. The earth's radius is 6400 km.

Depth, km	g, m/s²	Depth, km	g, m/s²
0	9.82	1400	9.88
33	9.85	1600	9.86
100	9.89	1800	9.85
200	9.92	2000	9.86
300	9.95	2200	9.90
413	9.98	2400	9.98
600	10.01	2600	10.09
800	9.99	2800	10.26
1000	9.95	2900	10.37
1200	9.91	4000	8.00

Within the earth's central core (below 2900 km) the values of g diminish monotonically (not linearly) from 10.37 m/s² to zero. The actual variation of g below 4000 km is uncertain. (*a*) Plot qualitatively g versus r (where r *is the distance from the earth's center*) from 0 to 6400 km. (*b*) Explain carefully how the earth's density must vary as we proceed from its surface to its center in order to account for this variation of g. (*c*) Take $\rho = 1$ at the surface (its average value is actually 3.0 g/cm³), and plot qualitatively ρ versus r. Assume throughout that ρ and g are spherically symmetrical.

17. (*a*) Show that in a chute through the earth along a chord line, rather than along a diameter, the motion of an object will be simple harmonic; assume a uniform earth density. (*b*) Find the period. (*c*) Will the object attain the same maximum speed along a chord as it does along a diameter?
Answer: (*b*) 84 min. (*c*) No.

18. Consider a mass particle at a point P anywhere inside a spherical shell of matter. Assume the shell is of uniform thickness and density. Construct a narrow double cone with apex at P intercepting areas dA_1 and dA_2 on the shell (Fig. 16-19). (*a*) Show that the resultant gravitational force exerted on the particle at P by the intercepted mass elements is zero. (*b*) Show then

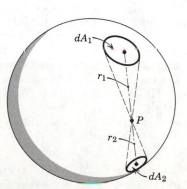

figure 16-19
Problem 18

that the resultant gravitational force of the entire shell on an internal particle is zero everywhere. (This method was devised by Newton.)

SECTION 16-7

19. (a) Can a satellite be sent out to a distance where it will revolve about the earth with an angular velocity equal to that at which the earth rotates, so that it remains always above the same point on the earth? (b) What would be the radius of the orbit of such a so-called synchronous earth satellite? *Answer:* (a) Yes. The plane of the orbit must be equatorial. (b) 4.2×10^4 km.

20. (a) With what horizontal speed must a satellite be projected at 100 mi (160 km) above the surface of the earth so that it will have a circular orbit about the earth? Take the earth's radius as 4000 mi (6400 km). (b) What will be the period of revolution?

21. The mean distance of Mars from the sun is 1.52 times that of Earth from the sun. Find the number of years required for Mars to make one revolution about the sun. *Answer:* 1.87 yr.

22. Determine the mass of the earth from the period T and the radius r of the moon's orbit about the earth: $T = 27.3$ days and $r = 2.39 \times 10^5$ mi (3.85×10^5 km).

23. (a) Satellite A is in a circular earth orbit with radius R and satellite B is in a circular earth orbit with radius $4R$. Calculate the ratio of the periods of revolution, T_A/T_B. (b) A pendulum and a mass-spring system oscillate with approximately the same frequency on the earth's surface. How do their frequencies compare if they are mounted first in satellite A and then in satellite B?
Answer: (a) $T_A/T_B = \frac{1}{8}$. (b) The pendulum frequency is zero; the mass-spring frequency is unchanged.

24. Consider an artificial satellite in a circular orbit about the earth. State how the following properties of the satellite vary with the radius r of its orbit: (a) period; (b) kinetic energy; (c) angular momentum; (d) speed.

25. Show how, guided by Kepler's third law (p. 337), Newton could deduce that the force holding the moon in its orbit, assumed circular, must vary as the inverse square of the distance from the center of the earth.

26. If a satellite in an elliptical orbit about the earth has a perigee (closest distance of approach) of 300 km above the surface of the earth, and an apogee (furthest distance of approach) of 2000 km above the surface of the earth, then what is the ratio of the orbital speed at perigee to that at apogee?

27. Three identical bodies of mass M are located at the vertices of an equilateral triangle with side L. At what speed must they move if they all revolve under the influence of one another's gravity in a circular orbit circumscribing the triangle while still preserving the equilateral triangle?
Answer: $\sqrt{GM/L}$.

28. (a) Show that the two-body problem of Section 16-7 can be simplified to a one-body problem by use of the reduced mass concept of Section 15-8. That is, show that if we use $\mu = mM/(m + M)$ instead of m, we may solve for the motion of m relative to M exactly as though M were the origin of our inertial reference frame. (b) Show that the assumption made in Section 16-7 that R is negligibly small compared to r is equivalent to assuming that the reduced mass μ is equal to m. (c) Compare μ for the earth-sun system with the earth's mass; compare μ for the moon-earth system with the moon's mass. (d) If we were to use the reduced mass μ of the two-body system instead of m, how would this affect the equations of Section 16-7?

SECTION 16-9

29. Mars has a mean diameter of 6900 km, Earth one of 1.3×10^4 km. The mass of Mars is $0.11M_e$. (a) How does the mean density of Mars compare with that of Earth? (b) What is the value of g on Mars? (c) What is the escape velocity on Mars? *Answer:* (a) $\rho_M = 0.73 \rho_E$. (b) 3.7 m/s². (c) 5.0 km/s.

30. (a) Show that to escape from the atmosphere of a planet a necessary condition for a molecule is that it have a speed such that $v^2 > 2GM/r$, where M is the mass of the planet and r is the distance of the molecule from the center of the planet. (b) Determine the escape speed from the earth for an atmospheric particle 1000 km above the earth's surface. (c) Do the same for Mars.

31. It is conjectured that a "burned-out" star could collapse to a "gravitational radius," defined as the radius for which the work needed to remove an object of mass m_0 from the star's surface to infinity equals the rest energy m_0c^2 of the object. Show that the gravitational radius of the sun is GM_s/c^2 and determine its value in terms of the sun's present radius. (For a review of this phenomenon see "Black Holes: New Horizons in Gravitational Theory" by Philip C. Peters, in *American Scientist*, Sept.–Oct. 1974.)
Answer: $2 \times 10^{-6} R_s$.

32. Show that the velocity of escape from the sun at the earth's distance from the sun is $\sqrt{2}$ times the speed of the earth in its orbit, assumed to be a circle.

33. A projectile is fired vertically from the earth's surface with an initial speed of 10 km/s. Neglecting atmospheric friction, how far above the surface of the earth would it go? Take the earth's radius as 6400 km.
Answer: 2.6×10^4 km.

34. A rocket is accelerated to a speed of $v = 2\sqrt{gR_e}$ near the earth's surface and then coasts upward. (a) Show that it will escape from the earth. (b) Show that very far from the earth its speed is $V = \sqrt{2gR_e}$.

35. Physicists have speculated about the possible existence of bodies with negative mass; for such hypothetical bodies it is postulated that m in the formulas of physics should be replaced by $-m$. Suppose that two particles, of mass $+m$ and $-m$ respectively, are placed a distance d apart. Show (a) the force acting on each and (b) the acceleration of each. Describe the expected motion, assuming that both particles are initially at rest, and show that this motion does not violate the laws of conservation of linear momentum or of mechanical energy. Such negative-mass particles have not yet been found.
Answer: (a) The force, from Newton's law of gravitation, is repulsive. (b) The accelerations, from Newton's second law, point in the same direction, from the negative to the positive mass.

36. A sphere of matter, mass M, radius a, has a concentric cavity, radius b, as shown in cross section in Fig. 16-20. (a) Sketch the gravitational force F exerted by the sphere on a particle of mass m, located a distance r from the center of the sphere, as a function of r in the range $0 \le r \le \infty$. Consider points $r = 0$, b, a, and ∞ in particular. (b) Sketch the corresponding curve for the potential energy $U(r)$ of the system. (c) From these graphs, how would you obtain graphs of the gravitational field strength and the gravitational potential due to the sphere?

37. Two particles of mass m and M are initially at rest an infinite distance apart. Show that at any instant their relative velocity of approach attributable to gravitational attraction is $\sqrt{2G(M + m)/d}$, where d is their separation at that instant.

figure 16-20
Problem 36

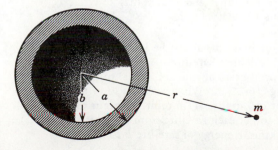

38. How long will it take a comet, moving in a parabolic path, to move from its point of closest approach to the sun through an angle of 90°, measured at the sun? Let the distance of closest approach to the sun be equal to the radius of the earth's orbit, assumed circular. (*Hint:* See Example 4 and Problem 32.)

SECTION 16-10

39. In a double star, two stars of mass 3×10^{30} kg each rotate about their common center of mass, 10^{11} m away. (*a*) What is their common angular speed? (*b*) Suppose that a meteorite passes through this center of mass moving at right angles to the orbital plane of the stars. What must its speed be if it is to escape from the gravitational field of the double star?
Answer: (*a*) 2×10^{-7} rad/s. (*b*) 9×10^4 m/s.

40. An 800-kg mass and a 600-kg mass are separated by 0.25 m. (*a*) What is the gravitational field strength due to these masses at a point 0.20 m from the 800-kg mass and 0.15 m from the 600-kg mass? (*b*) What is the gravitational potential at this point due to these same masses?

41. Masses of 200 and 800 g are 12 cm apart. (*a*) Find the gravitational force on an object of unit mass situated at a point on the line joining the masses 4.0 cm from the 200-g mass. (*b*) Find the gravitational potential energy per unit mass at that point. (*c*) How much work is needed to move this object to a point 4.0 cm from the 800-g mass along the line of centers?
Answers: (*a*) Zero. (*b*) -10×10^{-5} erg/g. (*c*) -5.0×10^{-6} erg.

42. For interstellar travel, a spaceship must overcome the sun's gravitational field as well as that of the earth. (*a*) What is the total amount of energy required for a 1.0-kiloton (equivalent to 9.1×10^5 kg) spaceship to free itself from the combined earth-sun gravitational field starting from an orbit 300 mi (480 km) above the earth's surface? Neglect all other bodies in the solar system. (*b*) What fraction of this energy is used to overcome the sun's field?

43. (*a*) Write an expression for the potential energy of a body of mass m in the gravitational field of the earth and moon. Let M_e be the earth's mass, M_m the moon's mass, R the distance from the earth's center, and r the distance from the moon's center. (*b*) At what point between the earth and moon will the total gravitational field strength attributable to the earth and moon be zero? (*c*) What will be the gravitational potential and the gravitational field strength on the earth's surface? (*d*) Answer for the moon's surface.
Answer: (*a*) $-Gm(M_e/R + M_m/r)$. (*b*) 3.4×10^8 m from earth. (*c*) -6.3×10^7 J/kg; 9.8 m/s². (*d*) -3.9×10^6 J/kg; 1.6 m/s².

SECTION 16-11

44. Consider two satellites A and B of equal mass m, moving in the same circular orbit of radius r around the earth E but in opposite senses of rotation and therefore on a collision course (see Fig. 16-21). (*a*) In terms of G, M_e, m, and r, find the total mechanical energy $E_A + E_B$ of the two-satellite-plus-earth system before collision. (*b*) If the collision is completely inelastic so that wreckage remains as one piece of tangled material (mass $= 2m$), find the total mechanical energy immediately after collision. (*c*) Describe the subsequent motion of the wreckage.

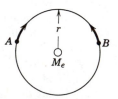

figure 16-21
Problem 44

45. (*a*) Does it take more energy to get a satellite up to 1000 mi above the earth than to put it in orbit once it is there? (*b*) What about 2000 mi? (*c*) What about 3000 mi? Take the earth's radius to be 4000 miles.
Answer: (*a*) No. (*b*) The same. (*c*) Yes.

46. Two earth satellites, A and B, each of mass m, are to be launched into (nearly) circular orbits about the earth's center. Satellite A is to orbit at an altitude of 4000 mi. Satellite B is to orbit at an altitude of 12,000 mi. The radius of the earth R_e is 4000 mi (Fig. 16-22). (*a*) What is the ratio of the potential energy of satellite B to that of satellite A, in orbit? (Explain the result in terms of the work required to get each satellite from its orbit to infinity.) (*b*) What is the ratio of the kinetic energy of satellite B to that of

satellite A, in orbit? (c) Which satellite has the greater total energy if each has a mass of 1.0 slug? By how much?

47. A pair of stars rotates about a common center of mass. One of the stars has a mass M which is twice as large as the mass m of the other, that is, M = 2m. Their centers are a distance d apart, d being large compared to the size of either star. (a) Derive an expression for the period of rotation of the stars about their common center of mass in terms of d, m, and G. (b) Compare the angular momenta of the two stars about their common center of mass by calculating the ratio L_m/L_M. (c) Compare the kinetic energies of the two stars by calculating the ratio K_m/K_M.
Answer: (a) $2\pi d^{3/2}/\sqrt{3Gm}$. (b) 2. (c) 2.

48. A satellite travels initially in an approximately circular orbit 640 km above the surface of the earth; its mass is 220 kg. (a) Determine its speed. (b) Determine its period. (c) For various reasons the satellite loses mechanical energy at the (average) rate of 1.4×10^5 J per complete orbital revolution. Adopting the reasonable approximation that the trajectory is a "circle of slowly diminishing radius," determine the distance from the surface of the earth, the speed, and the period of the satellite at the end of its 1500th orbital revolution. (d) What is the magnitude of the average retarding force? (e) Is angular momentum conserved?

49. A particle of mass m is subject to an attractive central force of magnitude k/r^2, k being a constant. If at the instant when the particle is at an extreme position in its closed orbit, at a distance a from the center of force, its speed is $\sqrt{k/2ma}$, find (a) the other extreme position, and (b) the speed of the particle at this position.
Answer: (b) $3\sqrt{k/2ma}$.

SECTION 16-12
50. *Foucault Pendulum.* A pendulum whose upper end is attached so as to allow the pendulum to swing freely in any direction can be used to repeat an experiment first shown publicly by Foucault in Paris in 1851. If the pendulum is set oscillating, the plane of oscillation slowly rotates with respect to a line drawn on the floor, even though the tension in the wire supporting the bob and the gravitational pull of the earth on the bob lie in a vertical plane. (a) Show that this is a result of the fact that the earth is not an inertial reference frame. (b) Show that for a Foucault pendulum at a latitude angle θ, the period of rotation of the plane is $(24/\sin\theta)$ h. (c) Explain in simple terms the result at $\theta = 90°$ (the poles) and $\theta = 0°$ (the equator).

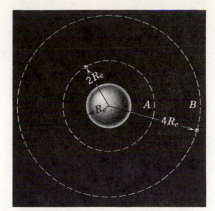

figure 16-22
Problem 46

17
fluid statics

It is customary to classify matter, viewed macroscopically, into solids and fluids. A *fluid* is a substance that can *flow*. Hence the term fluid includes liquids and gases. Such classifications are not always clearcut. Some fluids, such as glass or pitch, flow so slowly that they behave like solids for the time intervals that we usually work with them. Plasma, which is highly ionized gas, does not fit easily into any of these categories; it is often called a "fourth state of matter" to distinguish it from the solid, the liquid, and the gaseous state. Even the distinction between a liquid and a gas is not clearcut because, by changing the pressure and temperature properly, it is possible to change a liquid (water, say) into a gas (steam, say) without the appearance of a meniscus and without boiling; the density and viscosity change in a continuous manner throughout the process.* In this text, however, we will define a fluid as it is ordinarily understood, and we will be interested only in those properties of fluids connected with their ability to flow. Therefore, the same basic laws control the static and dynamic behavior of both liquids and gases in spite of the differences between them that we observe at ordinary pressures.

For solids, which have a definite size and shape, we formulated the mechanics of rigid bodies, modified by the laws of elasticity for bodies that cannot be considered perfectly rigid. Since fluids change their shape readily and, in the case of gases, have a volume equal to that of the container in which they are confined, we must develop new techniques for solving problems in fluid mechanics. Our applications of

---

* Pressures higher than the so-called critical point pressure must be employed to do this; for H_2O the critical point pressure is 218 atmospheres.

mechanics to continuous media, both solids and fluids, are based on Newton's laws of motion combined with the appropriate force laws. For fluids, as for solids, however, we find it convenient to develop special formulations of these basic laws.

17-2
PRESSURE AND DENSITY

There is a difference in the way a surface force acts on a fluid and on a solid. For a solid there are no restrictions on the direction of such a force, but for a fluid at rest the surface force must always be directed at right angles to the surface. For a fluid at rest cannot sustain a tangential force; the fluid layers would simply slide over one another when subjected to such a force. Indeed, it is the inability of fluids to resist such tangential forces (or shearing stresses) that gives them their characteristic ability to change their shape or to flow.

It is convenient, therefore, to describe the force acting on a fluid by specifying the *pressure p*, which is defined as the magnitude of the *normal* force per unit surface area. Pressure is transmitted to solid boundaries or across arbitrary sections of fluid *at right angles* to these boundaries or sections at every point. Pressure is a scalar quantity. The SI unit of pressure is the *pascal* (abbreviation Pa, 1 Pa = 1 N/m²). This unit is named after the French scientist Blaise Pascal (1623–1662) (see Section 17-4). Other units are bar (1 bar = 10^5 Pa), lb/in.², atmosphere (1 atm = 14.7 lb/in.² = 101,325 Pa), and mm-Hg (760 mm-Hg = 1 atm).

A fluid under pressure exerts a force on any surface in contact with it. Consider a closed surface containing a fluid (Fig. 17-1). An element of the surface can be represented by a vector **S** whose magnitude gives the area of the element and whose direction is taken to be the outward normal to the surface of the element. Then the force **F** exerted by the fluid against this surface element is

$$\mathbf{F} = p\mathbf{S}.$$

Since **F** and **S** have the same direction, the pressure p can be written as

$$p = \frac{F}{S}.$$

We assume that the element of area S is small enough so that the pressure p, defined as above, is independent of the size of the element S. The pressure may actually vary from point to point on the surface.

The density ρ of a homogeneous fluid (its mass divided by its volume) may depend on many factors, such as its temperature and the pressure to which it is subjected. For liquids the density varies very little over wide ranges in pressure and temperature, and we can safely treat it as a constant for our present purposes; see entries under *Water*

figure 17-1
An element of surface S can be represented by a vector **S**, equal to its area in magnitude and normal to it in direction.

Table 17-1
Densities of some materials and objects in kg/meter³

Interstellar space	$10^{-18} - 10^{-21}$
Best laboratory vacuum	$\sim 10^{-17}$
Hydrogen: at 0°C and 1.0 atm	9.0×10^{-2}
Air: at 0°C and 1.0 atm	1.3
at 100°C and 1.0 atm	0.95
at 0°C and 50 atm	6.5
Styrofoam	$\sim 1 \times 10^{2}$
Ice	0.92×10^{3}
Water: at 0°C and 1.0 atm	1.000×10^{3}
at 100°C and 1.0 atm	0.958×10^{3}
at 0°C and 50 atm	1.002×10^{3}
Aluminum	2.7×10^{3}
Mercury	1.36×10^{4}
Platinum	2.14×10^{4}
The earth: average density	5.52×10^{3}
density of core	9.5×10^{3}
density of crust	2.8×10^{3}
The sun: average density	1.4×10^{3}
density at center	$\sim 1.6 \times 10^{5}$
White dwarf stars (central densities)	$10^{8} - 10^{15}$
A uranium nucleus	$\sim 10^{17}$

in Table 17-1. The density of a gas, however, is very sensitive to changes in temperature and pressure; see entries under *Air* in Table 17-1. This table shows the range of densities that occur in nature. Note that the variation is by a factor of about 10^{38}.

17-3
THE VARIATION OF PRESSURE IN A FLUID AT REST

If a fluid is in equilibrium, every portion of the fluid is in equilibrium. Let us consider a small element of fluid volume submerged within the body of the fluid. Let this element have the shape of a thin disk and be a distance y above some reference level, as shown in Fig. 17-2a. The thickness of the disk is dy and each face has an area A. The mass of this element is $\rho A\, dy$ and its weight is $\rho g A\, dy$. The forces exerted on the element by the surrounding fluid are perpendicular to its surface at each point (Fig. 17-2b).

The resultant horizontal force is zero, for the element has no horizontal acceleration. The horizontal forces are due only to the pressure

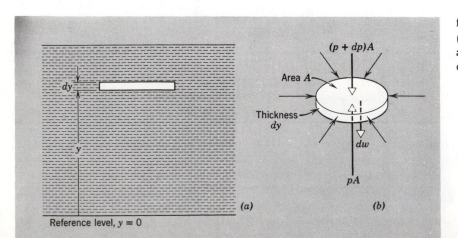

(a)

(b)

figure 17-2
(a) A small volume element of fluid at rest. (b) The forces on the element.

of the fluid, and by symmetry the pressure must be the same at all points within a horizontal plane at y.

The fluid element is also unaccelerated in the vertical direction, so that the resultant vertical force on it must be zero. However, the vertical forces are due not only to the pressure of the fluid on its faces but also to the weight of the element. If we let p be the pressure on the lower face and $p + dp$ the pressure on its upper face, the upward force is pA (exerted on the lower face) and the downward force is $(p + dp)A$ (exerted on the upper face) plus the weight of the element dw. Hence, for vertical equilibrium

$$pA = (p + dp)A + dw$$
$$= (p + dp)A + \rho gA \, dy,$$

and
$$\frac{dp}{dy} = -\rho g. \tag{17-1}$$

This equation tells us how the pressure varies with elevation above some reference level in a fluid in static equilibrium. As the elevation increases (dy positive), the pressure decreases (dp negative). The cause of this pressure variation is the weight per unit cross-sectional area of the layers of fluid lying between the points whose pressure difference is being measured.

The quantity ρg is often called the *weight density* of the fluid; it is the weight per unit volume of the fluid. For water, for example, the weight density is 62.4 lb/ft³ (= 9800 N/m³).

If p_1 is the pressure at elevation y_1, and p_2 the pressure at elevation y_2 above some reference level, integration of Eq. 17-1 gives

$$\int_{p_1}^{p_2} dp = -\int_{y_1}^{y_2} \rho g \, dy$$

or
$$p_2 - p_1 = -\int_{y_1}^{y_2} \rho g \, dy. \tag{17-2}$$

For liquids ρ is practically constant because liquids are nearly incompressible, and differences in level are rarely so great that any change in g need be considered. Hence, taking ρ and g as constants, we obtain

$$p_2 - p_1 = -\rho g(y_2 - y_1) \tag{17-3}$$

for a homogeneous liquid.

If a liquid has a free surface, this is the natural level from which to measure distances. To change our reference level to the top surface, we take y_2 to be the elevation of the surface, at which point the pressure p_2 acting on the fluid is usually that exerted by the earth's atmosphere p_0. We take y_1 to be at any level and we represent the pressure there as p. Then,

$$p_0 - p = -\rho g(y_2 - y_1).$$

But $y_2 - y_1$ is the depth h below the surface at which the pressure is p (see Fig. 17-3), so that

$$p = p_0 + \rho gh. \tag{17-4}$$

This shows clearly that the pressure is the same at all points at the same depth.

For gases ρ is comparatively small and the difference in pressure at two points is usually negligible (see Eq. 17-3). Thus, in a vessel con-

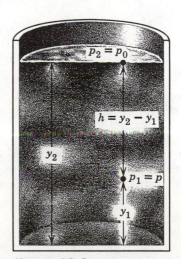

figure 17-3
A liquid whose top surface is open to the atmosphere.

taining a gas the pressure can be taken as the same everywhere. However, this is not the case if $y_2 - y_1$ is very great. The pressure of the air varies greatly as we ascend to great heights in the atmosphere. In fact, in such cases the density ρ varies with altitude and ρ must be known as a function of y before we can integrate Eq. 17-2.

EXAMPLE 1

We can get a reasonable idea of the variation of pressure with altitude in the earth's atmosphere if we assume that the density ρ is proportional to the pressure. This would be very nearly true if the temperature of the air remained the same at all altitudes. Using this assumption, and also assuming that the variation of g with altitude is negligible, find the pressure p at an altitude y above sea level.

From Eq. 17-1 we have

$$\frac{dp}{dy} = -\rho g.$$

Since ρ is proportional to p, we have

$$\frac{\rho}{\rho_0} = \frac{p}{p_0},$$

where ρ_0 and p_0 are the known values of density and pressure at sea level. Then,

$$\frac{dp}{dy} = -g\rho_0 \frac{p}{p_0},$$

so that

$$\frac{dp}{p} = -\frac{g\rho_0}{p_0} dy.$$

Integrating this from the value p_0 at the point $y = 0$ (sea level) to the value p at the point y (above sea level), we obtain

$$\ln \frac{p}{p_0} = -\frac{g\rho_0}{p_0} y$$

or

$$p = p_0 e^{-g(\rho_0/p_0)y}.$$

However,

$$g = 9.80 \text{ m/s}^2, \qquad \rho_0 = 1.20 \text{ kg/m}^3 \text{ (at 20°C)},$$

$$p_0 = 1.01 \times 10^5 \text{ N/m}^2 = 1.01 \times 10^5 \text{ Pa},$$

so that

$$g \frac{\rho_0}{p_0} = 1.16 \times 10^{-4} \text{ m}^{-1} = 0.116 \text{ km}^{-1}.$$

Hence,

$$p = p_0 e^{-ay}.$$

where $a = 0.116 \text{ km}^{-1}$.

We have seen that because liquids are almost incompressible the lower layers are not noticeably compressed by the weight of the upper layers superimposed on them and the density ρ is practically constant at all levels. For gases at uniform temperature the density ρ of any layer is proportional to the pressure p at that layer. The variation of pressure with distance above the bottom of the fluid for a gas is different from that for a liquid. Figure 17-4 shows the pressure distribution in water and in air.

Equation 17-3 gives the relation between the pressures at any two points in a fluid, regardless of the shape of the containing vessel. For no matter what the shape of the containing vessel, two points in the fluid can be connected by a path made up of vertical and horizontal steps. For example, consider points A and B in the homogeneous liquid con-

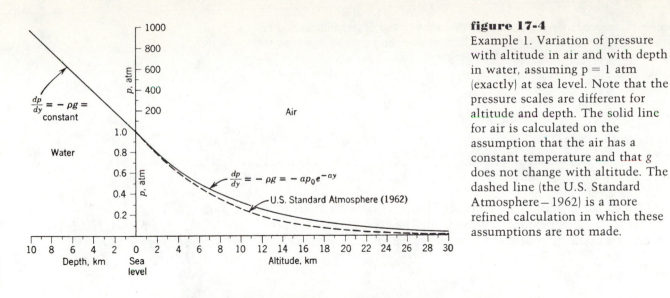

figure 17-4
Example 1. Variation of pressure with altitude in air and with depth in water, assuming p = 1 atm (exactly) at sea level. Note that the pressure scales are different for altitude and depth. The solid line for air is calculated on the assumption that the air has a constant temperature and that g does not change with altitude. The dashed line (the U.S. Standard Atmosphere – 1962) is a more refined calculation in which these assumptions are not made.

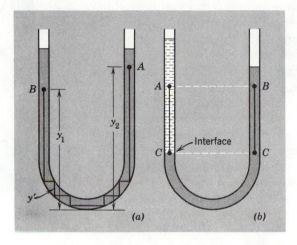

figure 17-5
(a) The difference in pressure between two points A and B in a homogeneous liquid depends only on their difference in elevation $y_2 - y_1$. (b) Two points A and B at the same elevation can be at different pressures if the densities there differ.

tained in the U-tube of Fig. 17-5a. Along the zigzag path from A to B there is a difference in pressure $\rho g y'$ for each vertical segment of length y', whereas along each horizontal segment there is no change in pressure. Hence, the difference in pressure $p_B - p_A$ is ρg times the algebraic sum of the vertical segments from A to B, or $\rho g(y_2 - y_1)$.

If the U-tube contains different imiscible liquids, say a dense liquid in the right tube and a less dense one in the left tube, as shown in Fig. 17-5b, the pressure can be different at the same level on different sides. In the figure the liquid surface is higher in the left tube than in the right. The pressure at A will be greater than at B. The pressure at C is the same on both sides, but the pressure falls less from C to A than from C to B, for a column of liquid of unit cross-sectional area connecting A and C will weigh less than a corresponding column connecting B and C.

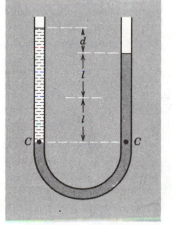

figure 17-6
Example 2.

EXAMPLE 2

A U-tube is partly filled with water. Another liquid, which does not mix with water, is poured into one side until it stands a distance d above the water level on the other side, which has meanwhile risen a distance l (Fig. 17-6). Find the density of the liquid relative to that of water.

In Fig. 17-6 points C are at the same pressure. Hence, the pressure drop from C to each surface is the same, for each surface is at atmospheric pressure.

The pressure drop on the water side is $\rho_w g 2l$; the $2l$ comes from the fact that the water column has risen a distance l on one side and fallen a distance l on the other side, from its initial position. The pressure drop on the other side is $\rho g(d + 2l)$, where ρ is the density of the unknown liquid. Hence,

$$\rho_w g 2l = \rho g(d + 2l)$$

and

$$\frac{\rho}{\rho_w} = \frac{2l}{(2l + d)}.$$

The ratio of the density of a substance to the density of water is called the *relative density* (or the *specific gravity*) of that substance.

Figure 17-7 shows a liquid in a cylinder that is fitted with a piston to which we may apply an external pressure p_0. The pressure p at any arbitrary point P a distance h below the upper surface of the liquid is given by Eq. 17-4, or

$$p = p_0 + \rho g h.$$

Let us increase the external pressure by an arbitrary amount Δp_0 (which need not be small compared to p_0). Since liquids are virtually incompressible, the density ρ in the preceding equations remains essentially constant during the process. The equation shows that, to this extent, the change in pressure Δp at the arbitrary point P is equal to Δp_0. This result was stated by Blaise Pascal (see p. 371) and is called *Pascal's principle*. It is usually given as follows: Pressure applied to an enclosed fluid is transmitted undiminished to every portion of the fluid and the walls of the containing vessel. This result is a necessary consequence of the laws of fluid mechanics, rather than an independent principle.

Although we often assume liquids to be incompressible, they are, in fact, slightly compressible. This means that a change of pressure applied to one portion of a liquid propagates through the liquid as a wave at the speed of sound in that liquid. Once the disturbance has died out and equilibrium is established, it is found that Pascal's principle is valid. The principle holds for gases with slight complications of interpretation caused by the large volume changes that may occur when the pressure on a confined gas is changed.

Archimedes' principle is also a necessary consequence of the laws of fluid statics. When a body is wholly or partly immersed in a fluid (either a liquid or a gas) at rest, the fluid exerts pressure on every part of the body's surface in contact with the fluid. The pressure is greater on the parts immersed more deeply. The resultant of all the forces is an upward force called the *buoyancy* of the immersed body. We can determine the magnitude and direction of this resultant force quite simply as follows.

The pressure on each part of the surface of the body certainly does not depend on the material the body is made of. Let us suppose, then, that the body, or as much of it as is immersed, is replaced by fluid like the surroundings. This fluid will experience the pressures that acted on the immersed body (Fig. 17-8) and will be at rest. Hence, the resultant upward force on it will equal its weight and will act vertically upward through its center of gravity. From this follows *Archimedes' principle*, namely, that a body wholly or partly immersed in a fluid is buoyed up with a force equal to the weight of the fluid displaced by the body. We have seen that the force acts vertically up through the center of gravity of the fluid before its displacement. The corresponding point in the immersed body is called its *center of buoyancy*.

17-4
PASCAL'S PRINCIPLE AND ARCHIMEDES' PRINCIPLE

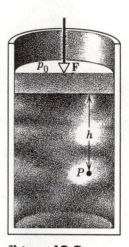

figure 17-7
A fluid in a cylinder fitted with a movable piston. The pressure at any point P is due not only to the weight of the fluid above the level of P but also to the force exerted by the piston.

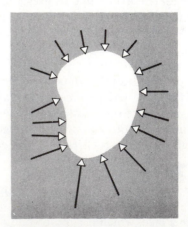

figure 17-8
Illustrating Archimedes' principle. The fluid exerts a resultant upward force on the immersed body.

What fraction of the total volume of an iceberg is exposed? The density of ice is $\rho_i = 0.92$ gram/cm^3 and that of sea water is $\rho_w = 1.03$ gram/cm^3. The weight of the iceberg is

$$W_i = \rho_i V_i g,$$

where V_i is the volume of the iceberg; the weight of the volume V_w of sea water displaced is the buoyant force

$$B = \rho_w V_w g.$$

But B equals W_i, for the iceberg is in equilibrium, so that

$$\rho_w V_w g = \rho_i V_i g,$$

and

$$\frac{V_w}{V_i} = \frac{\rho_i}{\rho_w} = \frac{0.92}{1.03} = 89\%.$$

The volume of water displaced V_w is the volume of the submerged portion of the iceberg, so that 11% of the iceberg is exposed.

EXAMPLE 3

Evangelista Torricelli (1608–1647) devised a method for measuring the pressure of the atmosphere by his invention of the mercury barometer in 1643.* The mercury barometer is a long glass tube that has been filled with mercury and then inverted in a dish of mercury, as in Fig. 17-9. The space above the mercury column contains only mercury vapor, whose pressure is so small at ordinary temperatures that it can be neglected. It is easily shown (see Eq. 17-3) that the atmospheric pressure p_0 is

$$p_0 = \rho g h.$$

Most pressure gauges use atmospheric pressure as a reference level and measure the difference between the actual pressure and atmospheric pressure, called the *gauge pressure*. The actual pressure at a point in a fluid is called the *absolute pressure*. Gauge pressure is given either above or below atmospheric pressure.

The pressure of the atmosphere at any point is numerically equal to the weight of a column of air of unit cross-sectional area extending from that point to the top of the atmosphere. The atmospheric pressure at a point, therefore, decreases with altitude. There are variations in atmospheric pressure from day to day because the atmosphere is not static. The mercury column in the barometer will have a height of about 76 cm at sea level, varying with the atmospheric pressure. A pressure equivalent to that exerted by exactly 76 cm of mercury at 0°C under standard gravity, $g = 32.174$ ft/s^2 = 980.665 cm/s^2, is called *one atmosphere* (1 atm). The density of mercury at this temperature is 13.5950 gram/cm^3. Hence, one atmosphere is equivalent to

$$1 \text{ atm} = (13.5950 \text{ gram/cm}^3)(980.665 \text{ cm/s}^2)(76.00 \text{ cm})$$

$$= 1.013 \times 10^5 \text{ N/m}^2 \ (\equiv 1.013 \times 10^5 \text{ Pa})$$

$$= 2116 \text{ lb/ft}^2$$

$$= 14.70 \text{ lb/in.}^2$$

Often pressures are specified by giving the height of mercury column,

17-5
MEASUREMENT OF PRESSURE

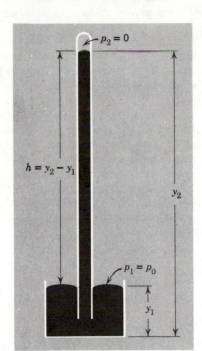

figure 17-9
The Torricelli barometer.

* See *The History of the Barometer*, by W. E. K. Middleton, The Johns Hopkins Press (1964) for a fascinating account of the development of the concept of atmospheric pressure and of devices to measure it.

at 0°C under standard gravity, which exerts the same pressure. This is the origin of the expression "centimeters of mercury (cm-Hg)" or "inches of mercury (in-Hg)" pressure. Pressure is the ratio of force to area, however, and not a length.

Torricelli described his experiments with the mercury barometer in letters in 1644 to his friend Michelangelo Ricci in Rome. In them he says that the aim of his investigation was "not simply to produce a vacuum, but to make an instrument which shows the mutations of the air, now heavier and dense, and now lighter and thin." On hearing of the Italian experiments, Blaise Pascal, in France, reasoned that if the mercury column was held up simply by the pressure of the air, the column ought to be shorter at a high altitude. He tried it on a church steeple in Paris, but desiring more decisive results, he wrote to his brother-in-law to try the experiment on the Puy de Dôme, a high mountain in Auvergne. There was a difference of 3 inches in the height of the mercury, "which ravished us with admiration and astonishment." Pascal himself constructed a barometer using red wine and a glass tube 46 feet long.

The chief significance of these experiments at the time was the realization it brought that an evacuated space could be created. Aristotle believed that a vacuum could not exist, and as late a writer as Descartes held the same view. For 2000 years philosophers spoke of the "horror" that nature had for empty space—the *horror vacui*. Because of this nature was said to prevent the formation of a vacuum by laying hold of anything nearby and with it instantly filling up any vacuated space. Hence, the mercury or wine should fill up the inverted tube because "nature abhorred a vacuum." The experiments of Torricelli and Pascal showed that there were limitations to nature's ability to prevent a vacuum. They created a sensation at the time. The goal of producing a vacuum became more of a practical reality through the development of pumps by Otto von Guericke in Germany around 1650 and by Robert Boyle in England around 1660. Even though these pumps were relatively crude, they did provide a tool for experimentation. With a pump and a glass jar, an experimental space could be provided in which to study how the properties of heat, light, sound, and later electricity and magnetism, are affected by an increasingly rarefied atmosphere. Although even today we cannot completely remove every trace of gas from a closed vessel, these seventeenth-century experimenters freed science from the bugaboo of *horror vacui* and spurred efforts to create highly evacuated systems.

Interestingly, within several decades in the seventeenth century no fewer than six important instruments were developed. They are the barometer, air pump, pendulum clock, telescope, microscope, and thermometer. All excited great wonder and curiosity.

The open-tube manometer (Fig. 17-10) measures gauge pressure. It consists of a U-shaped tube containing a liquid, one end of the tube being open to the atmosphere and the other end being connected to the system (tank) whose pressure p we want to measure. From Eq. 17-4 we obtain

$$p - p_0 = \rho g h.$$

Thus the gauge pressure, $p - p_0$, is proportional to the difference in height of the liquid columns in the U-tube. If the vessel contains gas under high pressure, a dense liquid like mercury is used in the tube; water can be used when low gas pressures are involved.

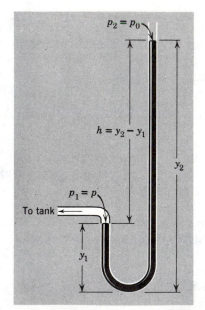

figure 17-10
The open-tube manometer, as used to measure the pressure in a tank.

EXAMPLE 4

An open-tube mercury manometer (Fig. 17-10) is connected to a gas tank. The mercury is 39.0 cm higher on the right side than on the left when a barometer nearby reads 75.0 cm-Hg. What is the absolute pressure of the gas? Express the answer in cm-Hg, atm, Pa, and lb/in.²

The gas pressure is the pressure at the top of the left mercury column. This is the same as the pressure at the same horizontal level in the right column. The pressure at this level is the atmospheric pressure (75.0 cm-Hg) plus the pressure exerted by the extra 39.0-cm column of Hg, or (assuming standard values of mercury density and gravity) a total of 114 cm-Hg. Therefore, the absolute pressure of the gas is

$$114 \text{ cm-Hg} = \tfrac{114}{76} \text{ atm} = 1.50 \text{ atm} = 1.52 \times 10^5 \text{ Pa.} = (1.50)(14.7) \text{ lb/in.}^2$$

$$= 22.1 \text{ lb/in.}^2.$$

What is the gauge pressure of the gas?

questions

1. Make an estimate of the average density of your body. Explain a way in which you could get an accurate value using ideas in this chapter.

2. Persons confined to bed are less likely to develop sores on their bodies if they use a water bed rather than an ordinary mattress. Explain.

3. (a) Two bodies (for example, balls) have the same shape and size but one is denser than the other. Assuming the air resistance to be the same on each, show that when they are released simultaneously from the same height the heavier body will reach the ground first. (b) Two bodies (for example, raindrops) have the same shape and density but one is larger than the other. Assuming the air resistance to be proportional to the body's speed through the air, which body will fall faster?

4. Water is poured to the same level in each of the three vessels shown, all of the same base area (Fig. 17-11). If the pressure is the same at the bottom of each vessel, the force experienced by the base of each vessel is the same. Why then do the three vessels have different weights when put on a scale? This apparently contradictory result is commonly known as the "hydrostatic paradox."

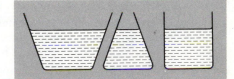
figure 17-11
Question 4.

5. Can a mountain climber climb high enough so that the atmospheric pressure is reduced to one-half of its sea-level value?

6. (a) An ice cube is floating in a glass of water. When the ice melts, will the water level rise? Explain. (b) If the ice cube contains a piece of lead, the water level will fall when the ice melts. Explain.

7. When a slice of lemon is first put into a cup of tea it sinks to the bottom. Later it is found to be floating. What is a likely explanation?

8. Does Archimedes' law hold in a vessel in free fall? In a satellite moving in a circular orbit? Explain.

9. A spherical bob made of cork floats half submerged in a pot of tea at rest on the earth. Will the cork float or sink aboard a spaceship coasting in free space? On the surface of Jupiter?

10. Two hollow bodies of equal weight and volume and having the same shape, except that one has an opening at the bottom and the other is sealed, are immersed to the same depth in water. Is less work required to immerse one than the other? If so, which one and why?

11. A ball floats on the surface of water in a container exposed to the atmosphere. Will the ball remain immersed at its former depth or will it sink or rise somewhat if (a) the container is covered and the air is removed or (b) the container is covered and the air is compressed?

12. Explain why an inflated balloon will rise to a definite height once it starts to rise, whereas a submarine will always sink to the bottom of the ocean once it starts to sink, if no changes are made.

13. Explain how a submarine rises, falls, and maintains a fixed depth. Do fish use the same principles? (See "The Buoyancy of Marine Animals" by Eric Denton in *Scientific American*, July 1960 and "Submarine Physics" by G. P. Harnwell in *American Journal of Physics*, March 1948.)

14. A soft plastic bag weighs the same when empty as when filled with air at atmospheric pressure. Why? Would the weights be the same if measured in a vacuum?

15. A leaky tramp steamer that is barely able to float in the North Sea steams up the Thames estuary toward the London docks. It sinks before it arrives. Why?

16. Is it true that a floating object will only be in stable equilibrium if its center of buoyancy lies above its center of gravity? Illustrate with examples.

17. Why didn't American Indians put seats in their canoes?

18. Very often a sinking ship will turn over as it becomes immersed in water. Why?

19. According to Example 3, 89% of an iceberg is submerged. Yet occasionally icebergs turn over, with possibly disastrous results to nearby shipping. How can this happen considering that so much of their mass is below sea level?

20. A barge filled with scrap iron is in a canal lock. If the iron is thrown overboard, what happens to the water level in the lock?

21. A bucket of water is suspended from a spring balance. Does the balance reading change when a piece of iron suspended from a string is immersed in the water? When a piece of cork is put in the water?

22. Logs dropped upright into a pond do not remain upright, but float "flat" in the water. Explain.

23. Explain why a uniform wooden stick which will float horizontally if it is not loaded will float vertically if enough weight is added to one end. (See Problem 30.)

24. A solid cylinder is placed in a container in contact with the base. When liquid is poured into the container, none of it goes beneath the solid, which remains closely in contact with the base. Is there a buoyant force on the solid? Explain.

25. Estimate with some care the buoyant force exerted by the atmosphere on you.

26. Although there are practical difficulties it is possible in principle to float an ocean liner in a few barrels of water. How would you go about doing this?

27. Is the water at the bottom of the Marianas Trench (11,000 m deep) substantially (a) less or (b) more buoyant than the water on the surface?

28. Mountain climbers use aneroid barometers to estimate their altitude. How can they be useful, considering that the atmospheric pressure at a given location is not constant?

29. What is wrong with this statement? "The atmospheric pressure is 753 mm-Hg when the atmosphere supports, in a barometer, a mercury column 753 mm long."

30. In a barometer, how important is it that the inner diameter of the barometer tube be uniform? That the barometer tube be absolutely vertical?

31. An open-tube manometer has one tube twice the diameter of the other. Explain how this would affect the operation of the manometer. Does it matter which end is connected to the chamber whose pressure is to be measured?

32. Explain how a physician can determine your blood pressure.

33. Liquid containers tend to leak when taken aloft in an airplane. Why? Does it matter whether or not they are right-side up? Does it matter whether or not they are initially completely full?

34. Assuming that a mercury barometer at standard atmospheric pressure on the earth's surface reads 76.0 cm height of mercury column, estimate what would be the height of the mercury column in an artificial earth satellite in orbit about the earth.

35. An open bucket of water is on a smooth plane inclined at an angle α to the

horizontal. Find the equilibrium inclination to the horizontal of the free surface of the water when (a) the bucket is held at rest, $a = 0$ and $v = 0$; (b) the bucket is allowed to slide down at constant speed, $a = 0$, $v = $ constant; (c) the bucket slides down without restraint, $a = $ constant. If the plane is curved so that $a \neq$ constant, what will happen?

36. If a U-tube containing water is rotated about a vertical axis through the center of one limb, the water level will fall in one limb and rise in the other compared to the rest position. Explain carefully. (See Problem 18.)

37. Explain how it can be that pressure is a scalar quantity when forces, which are vectors, can be produced by the action of pressures.

38. A thin-walled pipe will burst more easily if, when there is a pressure differential between inside and outside, the excess pressure is on the outside. Explain.

39. We have considered liquids under compression. Can liquids be put under tension? If so, will they tear under sufficient tension as do solids? (See "The Tensile Strength of Liquids" by Robert E. Apfel in *Scientific American*, December 1972.)

problems

SECTION 17-2

1. Find the pressure increase in the fluid in a syringe when a nurse applies a force of 42 N to the syringe's piston of radius 1.1 cm.
   Answer: 1.1×10^5 Pa.

2. An airtight box having a lid with an area of 12 in.² is partially evacuated. If a force of 108 lb is required to pull the lid off the box, and the outside atmospheric pressure is 15 lb/in.², what was the pressure in the box?

3. In 1654 Otto von Guericke, burgomeister of Magdeburg and inventor of the air pump, gave a demonstration before the Imperial Diet in which two teams of eight horses could not pull apart two evacuated brass hemispheres. (a) Show that the force F required to pull apart the hemispheres is $F = \pi R^2 P$ where R is the (outside) radius of the hemispheres and P is the difference in pressure outside and inside the sphere (Fig. 17-12). (b) Taking R equal to 1.0 ft and the inside pressure as 0.10 atm, what force would the team of horses have had to exert to pull apart the hemispheres? (c) Why were two teams of horses used? Would not one team prove the point just as well?
   Answer: (b) 6000 lb.

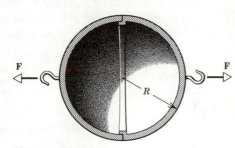

figure 17-12
Problem 3

SECTION 17-3

4. Find the total pressure, in lb/in.² (Pa), 500 ft (150 m) below the surface of the ocean. The relative density of sea water is 1.03 and the atmospheric pressure at sea level is 14.7 lb/in.² (1.0×10^5 Pa).

5. Estimate the hydrostatic difference in blood pressure in a person of height 1.83 m (6.00 ft), between the brain and the foot, assuming that the density of blood is 1.06×10^3 kg/m³ (2.06 slug/ft³).
   Answer: 1.90×10^4 Pa (2.75 lb/in.²).

6. The human lungs can operate against a pressure differential of less than one-twentieth a standard atmosphere. If a diver uses a snorkel (long tube) for breathing, how far below water level can he swim?

7. Find the pressure in the atmosphere 16 km (10 mi) above sea level.
   Answer: 1.6×10^4 Pa (2.3 lb/in.²).

8. The height at which the pressure in the atmosphere is just $1/e$ that at sea level is called the *scale height* of the atmosphere at sea level. (a) Show that the scale height H at sea level is also the height of an atmosphere that has the same density everywhere as at sea level and that will exert the same pressure at sea level as the actual infinite atmosphere does. (b) Show that the scale height at sea level is 8.6 km.

9. What would be the height of the atmosphere if the air density (a) were constant and (b) decreased linearly to zero with height? Assume a sea-level density of 1.3 kg/m³.   *Answer:* (a) 8.0×10^3 m. (b) 16×10^3 m.

10. A swimming pool has the dimensions 80 ft × 30 ft × 8.0 ft. (a) When it is filled with water, what is the force (due to the water alone) on the bottom? On the ends? On the sides? (b) If you are concerned with whether or not the concrete walls will collapse, is it appropriate to take the atmospheric pressure into account?

11. A simple U-tube contains mercury. When 13.6 cm of water is poured into the right arm, how high does the mercury rise in the left arm from its initial level?   *Answer:* 0.50 cm.

12. Water stands at a depth D behind the vertical upstream face of a dam, as shown in Fig. 17-13. Let W be the width of the dam. (a) Find the resultant horizontal force exerted on the dam by the gauge pressure of the water and (b) the net torque due to the gauge pressure of the water exerted about a line through O parallel to the width of the dam. (c) What is the line of action of the equivalent resultant force?

13. Three liquids that will not mix are poured into a cylindrical container. The amounts and densities of the liquids are 0.50 liter, 2.6 g/cm³; 0.25 liter, 1.0 g/cm³; and 0.40 liter, 0.80 g/cm³. What is the total force acting on the bottom of the container? (Ignore the contribution due to the atmosphere.) *Answer:* 18 N.

14. Two identical cylindrical vessels with their bases at the same level each contain a liquid of density ρ. The area of either base is A, but in one vessel the liquid height is h_1 and in the other h_2. Find the work done by gravity in equalizing the levels when the two vessels are connected.

15. (a) Consider a container of fluid subject to a *vertical upward* acceleration a. Show that the pressure variation with depth in the fluid is given by

$$p = \rho h \, (g + a),$$

where h is the depth and ρ is the density. (b) Show also that if the fluid as a whole undergoes a *vertical downward* acceleration a, the pressure at a depth h is given by

$$p = \rho h (g - a).$$

(c) What is the state of affairs in free fall?

16. (a) Consider the horizontal acceleration of a mass of liquid in an open tank. Acceleration of this kind causes the liquid surface to drop at the front of the tank and to rise at the rear. Show that the liquid surface slopes at an angle θ with the horizontal, where $\tan \theta = a/g$, a being the horizontal acceleration. (b) How does the pressure vary with h, the *vertical* depth below the surface?

17. The surface of contact of two fluids of different densities that are at rest and do not mix is horizontal. Prove this general result (a) from the fact that the potential energy of a system must be a minimum in stable equilibrium; (b) from the fact that at any two points in a horizontal plane in either fluid the pressures are equal.

18. (a) A fluid mass is rotating at constant angular velocity ω about the central vertical axis of a cylindrical container. Show that the variation of pressure in the radial direction is given by

$$\frac{dp}{dr} = \rho \omega^2 r.$$

(b) Take $p = p_c$ at the axis of rotation $(r = 0)$ and show that the pressure p at any point r is

$$p = p_c + \tfrac{1}{2}\rho\omega^2 r^2.$$

(c) Show that the liquid surface is of paraboloidal form (Fig. 17-14); that is, a

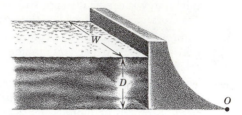

figure 17-13
Problem 12

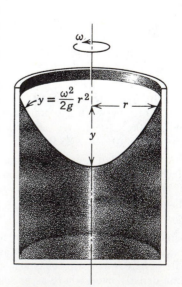

figure 17-14
Problem 18

vertical cross section of the surface is the curve $y = \omega^2 r^2/2g$. (d) Show that the variation of pressure with depth is $dp = \rho g\, dh$.

SECTION 17-4

19. A piston of small cross-sectional area a is used in the hydraulic press to exert a small force f on the enclosed liquid. A connecting pipe leads to a larger piston of cross-sectional area A (Fig. 17-15). (a) What force F will the larger piston sustain? (b) If the small piston has a diameter of 1.5 in. and the large piston one of 21 in., what weight on the small piston will support 2.0 tons on the large piston? *Answer:* (a) fA/a. (b) 20 lb.

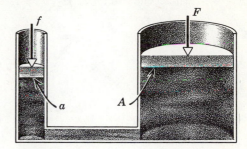

figure 17-15
Problem 19

20. A cubical object of dimensions L (2.0 ft) on a side and weight W (1000 lb) in a vacuum is suspended by a rope in an open tank of water of density ρ (2.0 slug/ft³) as in Fig. 17-16. (a) Find the total downward force exerted by the water and the atmosphere on the top of the object of area A (4.0 ft²). (b) Find the total force on the bottom of the object. (c) Find the tension in the rope.

21. (a) What is the minimum area of a block of ice 1.0 ft. (0.3 m) thick floating on water that will hold up an automobile weighing 2500 lb (mass = 1100 kg)? (b) Does it matter where the car is placed on the block of ice? *Answer:* (a) 500 ft² (46 m²). (b) Yes.

22. Three boys each of weight W (80 lb) make a log raft by lashing together logs of diameter D (1.0 ft) and length L (6.0 ft). How many logs will be needed to keep them afloat? Take the relative density of wood to be 0.80.

23. A block of wood floats in water with two-thirds of its volume submerged. In oil it has 0.90 of its volume submerged. Find the density of (a) the wood and (b) the oil. *Answer:* (a) 6.7×10^2 kg/m³. (b) 7.4×10^2 kg/m³.

24. A block of wood has a mass of 3.67 kg and a relative density of 0.60. It is to be loaded with lead so that it will float in water with 0.90 of its volume immersed. What weight of lead is needed (a) if the lead is on top of the wood? (b) if the lead is attached below the wood? The density of lead is 1.13×10^4 kg/m³.

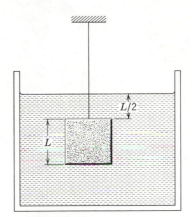

figure 17-16
Problem 20

25. Assume the density of brass weights to be 8.0 g/cm³ and that of air to be 0.0012 g/cm³. What percent error arises from neglecting the buoyancy of air in weighing an object of mass m and density ρ on a beam balance? *Answer:* $0.12(1/\rho - 1/8)$, with ρ in g/cm³.

26. A hollow spherical iron shell floats almost completely submerged in water. If the outer diameter is 2.00 ft and the relative density of iron is 7.80, find the inner diameter.

27. An iron casting containing a number of cavities weighs 60 lb (mass = 27 kg) in air and 40 lb (mass = 18 kg) in water. What is the volume of the cavities in the casting? Assume the relative density of iron to be 7.8. *Answer:* 0.20 ft³ (5.5×10^{-3} m³).

28. A cube floating on mercury has one-fourth of its volume submerged. If enough water is added to cover the cube, (a) what fraction of its volume will remain immersed in mercury? (b) Does the answer depend on the shape of the body?

29. A U-tube is filled with a single homogeneous liquid. The liquid is temporarily depressed in one side by a piston. The piston is removed and the level of the liquid in each side oscillates. Show that the period of oscillation is $\pi\sqrt{2L/g}$ where L is the total length of the liquid in the tube.

30. A cylindrical wooden log is loaded with lead at one end so that it floats upright in water as in Fig. 17-17. The length of the submerged portion is $l = 8.0$ ft. The log is set into vertical oscillation. (a) Show that the oscillation is simple harmonic. (b) Find the period of the oscillation. Neglect the fact that the water has a damping effect on the motion.

31. A long uniform wooden bar with square cross section floats on water either with two opposite surfaces parallel to the water or with all four surfaces at 45° with the water. Which of these positions is assumed for densities of 0.20, 0.50, and 0.80 g/cm³?

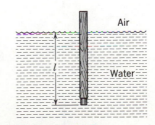

figure 17-17
Problem 30

32. The tension in a string holding a solid block below the surface of a liquid (of density greater than the solid) is T_0 when the containing vessel (Fig. 17-18) is at rest. Show that the tension T, when the vessel has an upward vertical acceleration a, is given by $T_0(1 + a/g)$.

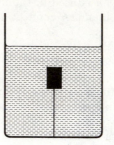

figure 17-18
Problem 32

18
fluid dynamics

One way of describing the motion of a fluid is to divide the fluid into infinitesimal volume elements, which we may call fluid particles, and to follow the motion of each of these particles. This is a formidable task. We would give coordinates x, y, z to each such fluid particle and would specify these as functions of the time t. The coordinates x, y, z at the time t of the fluid particle which was at x_0, y_0, z_0 at the time t_0 would be determined by functions $x(x_0,y_0,z_0,t_0,t)$, $y(x_0,y_0,z_0,t_0,t)$, $z(x_0,y_0,z_0,t_0,t)$, which then describe the motion of the fluid. This procedure is a direct generalization of the concepts of particle mechanics and was first developed by Joseph Louis Lagrange (1736–1813).

There is a treatment, developed by Leonhard Euler (1707–1783), which is more convenient for most purposes. In it we give up the attempt to specify the history of each fluid particle and instead specify the density and the velocity of the fluid at each point in space at each instant of time. This is the method we shall follow here. We describe the motion of the fluid by specifying the density $\rho(x,y,z,t)$ and the velocity $\mathbf{v}(x,y,z,t)$ at the point (x,y,z) at the time t. We thus focus our attention on what is happening at a particular point in space at a particular time, rather than on what is happening to a particular fluid particle. Any quantity used in describing the state of the fluid, for example the pressure p, will have a definite value at each point in space and at each instant of time. Although this description of fluid motion focuses attention on a point in space rather than on a fluid particle, we cannot avoid following the fluid particles themselves, at least for short time intervals dt. For it is the particles, after all, and not the space points, to which the laws of mechanics apply. In order to understand the nature of the simplifications we shall make, let us consider first some general characteristics of fluid flow.

18-1
GENERAL CONCEPTS OF FLUID FLOW

1. Fluid flow can be *steady* or *nonsteady*. When the fluid velocity **v** at any given point is constant in time, the fluid motion is said to be steady. That is, at any given point in a steady flow the velocity of each passing fluid particle is always the same. At some other point a particle may travel with a different velocity, but every other particle which passes this second point behaves there just as this particle did when it passed this point. These conditions can be achieved at low flow speeds; a gently flowing stream is an example. In nonsteady flow, as in a tidal bore, the velocities **v** *are* a function of the time. In the case of turbulent flow, such as rapids or a waterfall, the velocities vary erratically from point to point as well as from time to time.

2. Fluid flow can be *rotational* or *irrotational*. If the element of fluid at each point has no net angular velocity about that point, the fluid flow is irrotational. We can imagine a small paddle wheel immersed in the moving fluid (Fig. 18-1). If the wheel moves without rotating, the motion is irrotational; otherwise it is rotational. Rotational flow includes vortex motion, such as whirlpools.

3. Fluid flow can be *compressible* or *incompressible*. Liquids can usually be considered as flowing incompressibly. But even a highly compressible gas may sometimes undergo unimportant changes in density. Its flow is then practically incompressible. In flight at speeds much lower than the speed of sound in air (described by subsonic aerodynamics), the motion of the air relative to the wings is one of nearly incompressible flow. In such cases the density ρ is a constant, independent of x, y, z, and t, and the mathematical treatment of fluid flow is thereby greatly simplified.

4. Finally, fluid flow can be *viscous* or *nonviscous*. Viscosity in fluid motion is the analog of friction in the motion of solids. In many cases, such as in lubrication problems, it is extremely important. Sometimes, however, it is negligible. Viscosity introduces tangential forces between layers of fluid in relative motion and results in dissipation of mechanical energy.

We shall confine our discussion of fluid dynamics for the most part to *steady, irrotational, incompressible, nonviscous* flow. The mathematical simplifications resulting should be obvious. We run the danger, however, of making so many simplifying assumptions that we are no longer talking about a meaningfully real fluid.* Furthermore, it is sometimes difficult to decide whether a given property of a fluid—its viscosity, say—can be neglected in a particular situation. In spite of all this, the restricted analysis that we are going to give has wide application in practice, as we shall see.

figure 18-1
We place a small free-floating paddle wheel in a flowing liquid. If it rotates, we call the flow *rotational*; if not, we call the flow *irrotational*.

18-2
STREAMLINES

In steady flow the velocity **v** at a given point is constant in time. Consider the point P (Fig. 18-2) within the fluid. Since **v** at P does not change in time, every particle arriving at P will pass on with the same speed in the same direction. The same is true about the points Q and R. Hence, if we trace out the path of the particle, as is done in the figure, that curve will be the path of every particle arriving at P. This curve is called a *streamline*. A streamline is parallel to the velocity of the fluid particles at every point. No two streamlines can cross one another, for if

* Richard Feynman has pointed out that John von Neumann called this idealized fluid "dry water."

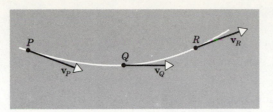

figure 18-2
A particle passing through points
P, Q, and R traces out a streamline,
assuming steady flow. Any other
particle passing through P must be
traveling along the same streamline
in steady flow.

they did, an oncoming fluid particle could go either one way or the
other, and the flow could not be steady. In steady flow the pattern of
streamlines in a flow is stationary with time.*

In principle we can draw a streamline through every point in the
fluid. Let us assume steady flow and select a finite number of streamlines to form a bundle, like the streamline pattern of Fig. 18-3. This
tubular region is called a *tube of flow*. The boundary of such a tube consists of streamlines and is always parallel to the velocity of the fluid
particles. Hence, no fluid can cross the boundaries of a tube of flow and
the tube behaves somewhat like a pipe of the same shape. The fluid that
enters at one end must leave at the other.

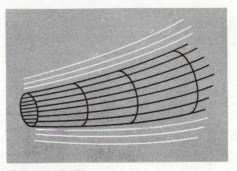

figure 18-3
A tube of flow made up of a bundle
of streamlines.

In Fig. 18-4 we have drawn a thin tube of flow. The velocity of the fluid
inside, although parallel to the tube at any point, may have different
magnitudes at different points. Let the speed be v_1 for fluid particles at
P and v_2 for fluid particles at Q. Let A_1 and A_2 be the cross-sectional
areas of the tubes perpendicular to the streamlines at the points P and Q,
respectively. In the time interval Δt a fluid element travels approximately the distance $v\,\Delta t$. Then the mass of fluid Δm_1 crossing A_1 in the
time interval Δt is approximately

$$\Delta m_1 = \rho_1 A_1 v_1 \,\Delta t$$

or the *mass flux* $\Delta m_1 / \Delta t$ is approximately $\rho_1 A_1 v_1$. We must take Δt small
enough so that in this time interval neither v nor A varies appreciably
over the distance the fluid travels. In the limit as $\Delta t \to 0$, we obtain the
precise definitions:

$$\text{mass flux at } P = \rho_1 A_1 v_1,$$

and

$$\text{mass flux at } Q = \rho_2 A_2 v_2,$$

18-3
THE EQUATION OF
CONTINUITY

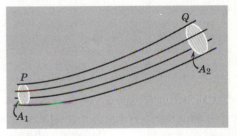

figure 18-4
A tube of flow used in proving the
equation of continuity.

where ρ_1 and ρ_2 are the fluid densities at P and Q respectively. Because
no fluid can leave through the walls of the tube and there are no
"sources" or "sinks" wherein fluid can be created or destroyed in the
tube, the mass crossing each section of the tube per unit time must be
the same. In particular, the mass flux at P must equal that at Q:

$$\rho_1 A_1 v_1 = \rho_2 A_2 v_2,$$

or

$$\rho A v = \text{constant.} \tag{18-1}$$

* The family of streamlines in a fluid is so drawn that, at any point in the fluid, the
direction of the instantaneous velocity **v** for the fluid particle at that point is tangent to
the streamline at that point. In nonsteady flow the pattern of streamlines in the fluid
changes as time goes on and the path of an individual fluid particle through the fluid
does not coincide with a streamline of a given instant. The streamline and the line of
motion of the particle touch each other at the point, locating the particle at the instant
in question. The path or line of motion and the streamline coincide only for steady flow.

This result (Eq. 18-1) expresses the law of conservation of mass in fluid dynamics.

Would you expect Eq. 18-1 to hold when the flow is (a) nonsteady, (b) rotational, (c) compressible, or (d) viscous?

In the more general case in which sources or sinks are present and in which the density varies with time as well as position, mass must still be conserved and we can write (without proof) an *equation of continuity* that expresses this fact. It is

$$\frac{\partial(\rho v_x)}{\partial x} + \frac{\partial(\rho v_y)}{\partial y} + \frac{\partial(\rho v_z)}{\partial z} + \frac{\partial \rho}{\partial t} = S \tag{18-2}$$

in which v_x, v_y, and v_z are the velocity components of the fluid; like the density ρ they vary both with position and time.*

Let us consider a small volume element in such a fluid. It can be shown that:

1. The sum of the first three terms of Eq. 18-2 gives the net outflow, per unit volume, of mass from the volume element.
2. The fourth term gives the rate, per unit volume, at which mass is accumulating within the volume element.
3. The last term, S, gives the rate, per unit volume, at which mass is being introduced into volume element from a "source" (if S is positive) or is disappearing from the volume element into a "sink" (if S is negative).

It is clear that, with these interpretations of its terms, Eq. 18-2 is a statement of the conservation of mass for fluid flow. Is this equation dimensionally correct?

If $S = 0$ in Eq. 18-2, there are no sources or sinks. If the sum of the first three terms is negative, there is a net *inflow* of mass to the volume element. Thus the mass contained in the element must increase with time as fluid "piles up." This is in agreement with Eq. 18-2 because, for the conditions stated, $\partial \rho / \partial t$ must be positive, which means that the density of the fluid (and thus the mass of the fluid) in the volume element is increasing as time goes on.

If the fluid is incompressible, as we shall henceforth assume, then $\rho_1 = \rho_2$ and Eq. 18-1 takes on the simpler form

$$A_1 v_1 = A_2 v_2.$$

or
$$A v = \text{constant}, \tag{18-3}$$

The product Av gives the *volume flux* or flow rate, as it is often called. Its SI units are m³/s. Notice that it predicts that in steady incompressible flow the speed of flow varies inversely with the cross-sectional area, being larger in narrower parts of the tube. The fact that the product Av remains constant along a tube of flow allows us to interpret the streamline picture somewhat. In a narrow part of the tube the streamlines must crowd closer together than in a wide part. Hence, as the distance between streamlines decreases, the fluid speed must increase. Therefore, we conclude that widely spaced streamlines indicate regions of low speed and closely spaced streamlines indicate regions of high speed.

We can obtain another interesting result by applying Newton's second law of motion to the flow of fluid between P and Q (Fig. 18-4). A fluid particle at P with speed v_1 must be decelerated in the forward direction in acquiring the smaller forward speed v_2 at Q. Hence the fluid is decelerated in going from P to Q. The deceleration can come about

---

* Because these four quantities are functions of more than one variable we have written the derivatives in Eq. 18-2 as partial derivatives.

from a difference in pressure acting on the fluid particle flowing from P to Q or from the action of gravity. In a horizontal tube of flow the gravitational force does not change. Hence we can conclude that in steady horizontal flow the pressure is greatest where the speed is least.

Were you ever in a crowd when it started to push its way through a small opened door? Outside in the back of the crowd the cross-sectional area was large, the pressure was great, but the speed of advance rather small. Through the door of small cross section the pressure was relieved and the speed of advance gratifyingly increased. This particular "human fluid" is compressible and viscous and the flow is sometimes turbulent and rotational.

18-4
BERNOULLI'S
EQUATION*

Bernoulli's equation is a fundamental relation in fluid mechanics. Like all equations in fluid mechanics it is not a new principle but is derivable from the basic laws of Newtonian mechanics. We will find it convenient to derive it from the work-energy theorem (see Section 7-4), for it is essentially a statement of the work-energy theorem for fluid flow.

Consider the nonviscous, steady, incompressible flow of a fluid through the pipeline or tube of flow in Fig. 18-5. The portion of pipe shown in the figure has a uniform cross section A_1 at the left. It is horizontal there at an elevation y_1 above some reference level. It gradually widens and rises and at the right has a uniform cross section A_2. It is horizontal there at an elevation y_2. Let us concentrate on the portion of fluid represented by both cross-shading and horizontal shading and call this fluid the "system." Consider then the motion of the system from the position shown in (a) to that in (b). At all points in the narrow part of the pipe the pressure is p_1 and the speed v_1; at all points in the wide portion the pressure is p_2 and the speed v_2.

The work-energy theorem (see Eq. 7-14) states: *The work done by the*

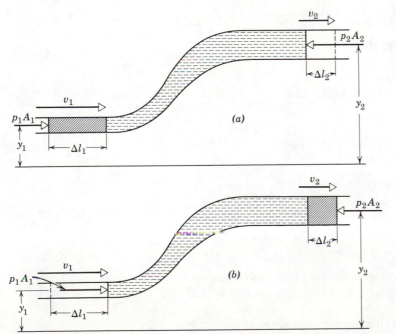

figure 18-5
A portion of fluid (cross-shading and horizontal shading) moves through a section of pipeline from the position shown in (a) to that shown in (b).

* There are eight Bernoullis listed in the Encyclopedia Brittanica (11th ed.). We refer here to Daniel Bernoulli (1700–1782), perhaps the most renowned member of this famous family.

resultant force acting on a system is equal to the change in kinetic energy of the system. In Fig. 18-5 the forces that do work on the system, assuming that we can neglect viscous forces, are the pressure forces p_1A_1 and p_2A_2 that act on the left- and right-hand ends of the system, respectively, and the force of gravity. As fluid flows through the pipe the net effect, as a comparison of Figs. 18-5a and b shows, is to raise an amount of fluid represented by the cross-shaded area in Fig. 18-5a to the position shown in Fig. 18-5b. The amount of fluid represented by the horizontal shading is unchanged by the flow.

We can find the work W done on the system by the resultant force as follows:

1. The work done on the system by the pressure force p_1A_1 is $p_1A_1 \, \Delta l_1$.

2. The work done on the system by the pressure force p_2A_2 is $-p_2A_2 \, \Delta l_2$. Note that it is negative, which means that positive work is done *by* the system.

3. The work done on the system by gravity is associated with lifting the cross-shaded fluid from height y_1 to height y_2 and is $-mg(y_2 - y_1)$ in which m is the mass of fluid in either cross-shaded area. It too is negative because work is done by the system *against* the gravitational force.

The work W done *on* the system by the *resultant* force is found by adding these three terms, or

$$W = p_1A_1 \, \Delta l_1 - p_2A_2 \, \Delta l_2 - mg(y_2 - y_1).$$

Now $A_1 \, \Delta l_1 (= A_2 \, \Delta l_2)$ is the volume of the cross-shaded fluid element, which we can write as m/ρ, in which ρ is the (constant) fluid density. Recall that the two fluid elements have the same mass, so that in setting $A_1 \, \Delta l_1 = A_2 \, \Delta l_2$ we have assumed the fluid to be incompressible. With this assumption we have

$$W = (p_1 - p_2)(m/\rho) - mg(y_2 - y_1). \qquad (18\text{-}4a)$$

The change in kinetic energy of the fluid element is

$$\Delta K = \tfrac{1}{2}mv_2{}^2 - \tfrac{1}{2}mv_1{}^2. \qquad (18\text{-}4b)$$

From the work-energy theorem (Eq. 7-14) we then have

$$W = \Delta K$$

or

$$(p_1 - p_2)(m/\rho) - mg(y_2 - y_1) = \tfrac{1}{2}mv_2{}^2 - \tfrac{1}{2}mv_1{}^2, \qquad (18\text{-}5a)$$

which can be rearranged to read

$$p_1 + \tfrac{1}{2}\rho v_1{}^2 + \rho g y_1 = p_2 + \tfrac{1}{2}\rho v_2{}^2 + \rho g y_2. \qquad (18\text{-}5b)$$

Since the subscripts 1 and 2 refer to *any* two locations along the pipeline, we can drop the subscripts and write

$$p + \tfrac{1}{2}\rho v^2 + \rho g y = \text{constant}. \qquad (18\text{-}6)$$

Equation 18-6 is called *Bernoulli's equation* for steady, nonviscous, incompressible flow. It was first presented by Daniel Bernoulli in his *Hydrodynamica* in 1738.

Bernoulli's equation is strictly applicable only to steady flow, the quantities involved being evaluated along a streamline. In our figure the streamline used is along the axis of the pipeline. If the flow is irrotational, however, it can be shown (see Problem 25 for a special case) that the constant in Eq. 18-6 is the same for *all* streamlines.

In a nonviscous incompressible fluid we cannot change the temperature of the fluid by mechanical means. Hence, Bernoulli's equation, as we stated it, refers to isothermal (constant temperature) processes. It is possible, however, to change the temperature of a nonviscous *compressible* fluid by mechanical means. We can generalize this equation to include a compressible fluid by adding to the left of Eq. 18-6 a term u, which represents the *internal energy* per unit volume of the fluid. This term (and the pressure p) will have a value that depends on the temperature.

If the flow is *viscous*, forces of a frictional nature act on the fluid so that some of the work done that appeared as a change in kinetic energy in the nonviscous case appears now as heat energy in the fluid. We must then write Eq. 18-5a as

$$(p_1 - p_2)(m/\rho) - mg(y_2 - y_1) = \tfrac{1}{2}mv_2{}^2 - \tfrac{1}{2}mv_1{}^2 + Q$$

where Q represents the heat energy generated in the viscous flow from point 1 to point 2. In practice, Bernoulli's equation can be modified accordingly by use of empirical corrections for conversion of mechanical energy to heat energy. However, if the pipe is smooth and the diameter is large compared to the length, and if the fluid flows slowly and has a small viscosity, the heat energy generated is negligible.

Just as the statics of a particle is a special case of particle dynamics, so fluid statics is a special case of fluid dynamics. It should come as no surprise, therefore, that the law of pressure change with height in a fluid at rest is included in Bernoulli's equation as a special case. For let the fluid be at rest; then $v_1 = v_2 = 0$ and Eq. 18-5b becomes

$$p_1 + \rho g y_1 = p_2 + \rho g y_2$$

or

$$p_2 - p_1 = -\rho g(y_2 - y_1),$$

which is the same as Eq. 17-3.

In Eq. 18-6 all terms have the dimension of a pressure (check this). The pressure $p + \rho g h$, which would be present even if there were no flow ($v = 0$), is called the *static pressure*; the term $\tfrac{1}{2}\rho v^2$ is called the *dynamic pressure*.

Bernoulli's equation can be used to determine fluid speeds by means of pressure measurements. The principle generally used in such measuring devices is the following: The equation of continuity requires that the speed of the fluid at a constriction increase; Bernoulli's equation then shows that the pressure must fall there. That is, for a horizontal pipe $\tfrac{1}{2}\rho v^2 + p$ equals a constant; if v increases and the fluid is incompressible, p must decrease. This result was also deduced from dynamic considerations in Section 18-3.

18-5
APPLICATIONS OF BERNOULLI'S EQUATION AND THE EQUATION OF CONTINUITY

1
The Venturi Meter

This (Fig. 18-6) is a gauge put in a flow pipe to measure the flow speed of a liquid. A liquid of density ρ flows through a pipe of cross-sectional area A. At the throat the area is reduced to a and a manometer tube is attached, as shown. Let the manometer liquid, such as mercury, have a density ρ'. By applying Bernoulli's equation and the equation of continuity at points 1 and 2, you can show that the speed of flow at A is

$$v = a\sqrt{\frac{2(\rho' - \rho)gh}{\rho(A^2 - a^2)}}.$$

If we want the volume flux or flow rate R, which is the volume of liquid transported past any point per second, we simply compute

$$R = vA.$$

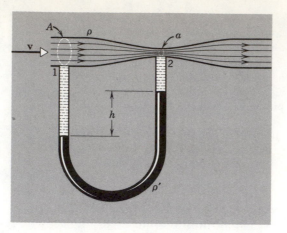

2
The Pitot Tube

This device (Fig. 18-7) is used to measure the flow speed of a gas. Consider the gas, say air, flowing past the openings at a. These openings are parallel to the direction of flow and are set far enough back so that the velocity and pressure outside the openings have the free-stream values. The pressure in the left arm of the manometer, which is connected to these openings, is then the static pressure in the gas stream, p_a. The opening of the right arm of the manometer is at right angles to the stream. The velocity is reduced to zero at b and the gas is stagnant at that point. The pressure at b is the full *ram pressure*, p_b. Applying Bernoulli's equation to points a and b, we obtain

$$p_a + \tfrac{1}{2}\rho v^2 = p_b,$$

where, as shown in the figure, p_b is greater than p_a. If h is the difference in height of the liquid in the manometer arms and ρ' is the density of the manometer liquid, then

$$p_a + \rho' g h = p_b.$$

Comparing these two equations, we find

$$\tfrac{1}{2}\rho v^2 = \rho' g h$$

or

$$v = \sqrt{\frac{2gh\rho'}{\rho}},$$

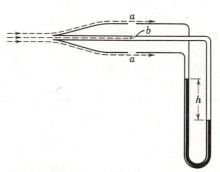

figure 18-7
Cross-sectional diagram of a Pitot tube.

which gives the gas speed. This device can be calibrated to read v directly and is then known as an air-speed indicator.

3
Dynamic Lift

Dynamic lift is the force that acts on a body, such as an airplane wing, a hydrofoil, or a helicopter rotor, by virtue of its motion through a fluid. We must distinguish it from *static lift*, which is the buoyant force that acts on a balloon or an iceberg in accord with Archimedes' principle (Section 17-4).

Figure 18-8 shows the streamlines about an airfoil (or wing cross section) attached to an aircraft.* Let us choose the aircraft as our frame of reference, as in a wind tunnel experiment, and let us assume that the air is moving past the wing from right to left.

---

* See "Bernoulli and Newton in Fluid Mechanics," Norman F. Smith, *The Physics Teacher*, November 1972.

See also *The Flettner Ship*, an article by Albert Einstein in his book *Essays in Science*, Philosophical Library, New York. The Flettner ship, like a sailboat, derives its motive power from the wind. Instead of a sail it has a large cylinder that is caused to rotate about a vertical axis by a small motor. The resulting dynamic "lift" (in this case, horizontal) propels the vessel.

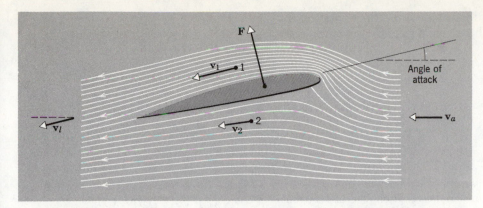

figure 18-8
Streamlines about an airfoil. The velocity of the approaching air $\mathbf{v}_a$ is horizontal. That of the leaving air $\mathbf{v}_l$ has a downward component. Thus, because the airfoil has forced the air down, the air, from Newton's third law, must have forced the airfoil up. This is represented by the "lift" $\mathbf{F}$.

The *angle of attack* of the wing causes air to be deflected downward. From Newton's third law the reaction of this downward force of the wing on the air is an upward force $\mathbf{F}$, the lift, exerted by the air on the wing.

The pattern of streamlines is consistent. Above the wing (point 1) the streamlines are closer together than they are below the wing (point 2). Thus $\mathbf{v}_1 > \mathbf{v}_2$ and, from Bernoulli's principle, $p_1 < p_2$, which must be true if there is to be a lift.

As our final example let us compute the thrust on a rocket produced by the escape of its exhaust gases. Consider a chamber (Fig. 18-9) of cross-sectional area A filled with a gas of density ρ at a pressure p. Let there be a small orifice of cross-sectional area A_0 at the bottom of the chamber. We wish to find the speed v_0 with which the gas escapes through the orifice.

Let us write Bernoulli's equation (Eq. 18-5b) as

$$p_1 - p_2 = \rho g(y_2 - y_1) + \tfrac{1}{2}\rho(v_2^2 - v_1^2).$$

For a gas the density is so small that we can neglect the variation in pressure with height in a chamber (see Section 17-3). Hence, if p represents the pressure p_1 in the chamber and p_0 represents the atmospheric pressure p_2 just outside the orifice, we have

$$p - p_0 = \tfrac{1}{2}\rho(v_0^2 - v^2)$$

or

$$v_0^2 = \frac{2(p - p_0)}{\rho} + v^2, \qquad (18\text{-}7)$$

where v is the speed of the flowing gas inside the chamber and v_0 is the *speed of efflux* of the gas through the orifice. Although a gas is compressible and the flow may become turbulent, we can treat the flow as steady and incompressible for pressure and efflux speeds that are not too high.

Now let us assume continuity of mass flow (in a rocket engine this is achieved when the mass of escaping gas equals the mass of gas created by burning the fuel), so that (for an assumed constant density)

$$Av = A_0 v_0.$$

If the orifice is very small so that $A_0 \ll A$, then $v_0 \gg v$, and we can neglect v^2 compared to v_0^2 in Eq. 18-7. Hence, the speed of efflux is

$$v_0 = \sqrt{\frac{2(p - p_0)}{\rho}}. \qquad (18\text{-}8)$$

4
Thrust on a Rocket

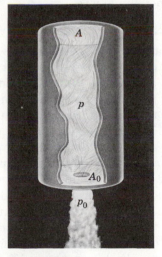

figure 18-9
Fluid streaming out of a chamber.

If our chamber is the exhaust chamber of a rocket, the thrust on the rocket (Section 9-7) is $v_0\, dM/dt$. But the mass of gas flowing out in time dt is $dM = \rho A_0 v_0\, dt$, so that

$$v_0 \frac{dM}{dt} = v_0 \rho A_0 v_0 = \rho A_0 v_0^2,$$

and from Eq. 18-8 the thrust is

$$2A_0(p - p_0). \qquad (18\text{-}9)$$

In Newtonian particle mechanics the derivation of the laws of conservation of linear momentum and angular momentum makes explicit use of Newton's third law of motion. The internal forces and torques in a mechanical system cancel one another because of this third law, leaving only the external forces and torques to contribute to the momenta. In the case of a fluid the internal forces are represented by the pressure within the fluid. But the very concept of pressure itself contains Newton's third law implicitly. The force produced by pressure exerted in one direction across any surface element is equal and opposite to the force exerted in the opposite direction across the same surface element. Also, each of these two forces is applied at the same place, namely at the surface element. Both forces must have the same line of action. Hence, in the equations for the time rate of change of linear momentum or of angular momentum of a fluid, the internal pressures will cancel out. We can conclude then that the time rate of change of the total linear momentum in a volume V of moving fluid is equal to the total *external force* acting on it. Likewise, the time rate of change of the total angular momentum in a volume V of moving fluid is equal to the total *external torque* acting on it. The conservation laws of linear and angular momentum follow.

18-6
CONSERVATION OF MOMENTUM IN FLUID MECHANICS

In the chapter on gravitation we saw how to summarize the physical state of affairs near masses by use of a field. Each point in the field can be regarded as having a vector associated with it, namely **g,** the gravitational force per unit mass at that point. Or, alternately, we can associate a scalar quantity with each point in space, namely the gravitational potential V. We can then draw a surface, called an equipotential surface, through all points that have the same potential. We draw several such surfaces, the potential on one differing by a constant amount from that on the next one, and so on. The gravitational force at any point is then directed along a line passing through this point perpendicular to these surfaces, and its magnitude is determined from the rate of change of potential with distance in this direction, as indicated by the spacing and orientation of the equipotential surfaces. By drawing in lines of force we can picture vividly how space is affected by the presence of mass.

Likewise, in fluid dynamics we can summarize the physical state of affairs within a moving fluid by means of a field of flow. In general, the field of flow is a *vector* field. We associate a vector quantity with each point in space, namely the flow velocity **v** at that point. For a steady flow the field of flow is stationary. Of course, even in this case a particular fluid particle may still have a variable velocity as it moves from point to point in the field. The field gives the properties of the space from which we deduce the behavior of particles in that space. If the flow is irrotational, as well as steady, we call it *potential flow*. Then the flow velocity **v** can be related to a velocity potential ψ, just as in gravitation **g** can be related to the gravitational potential V. If we draw in surfaces of equal velocity potential, as we drew in surfaces of equal gravitational

18-7
FIELDS OF FLOW

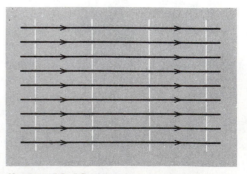

figure 18-10
Streamlines (horizontal) and surfaces of equal velocity potential (vertical) for a homogensous field of flow.

potential, we can deduce **v** from the equipotential flow surfaces just as **g** is deduced from the equipotential gravitational surfaces. Hence, a field for potential flow is analogous to a conservative force field.

A flowing fluid mass can always be divided into tubes of flow. When the flow is steady, the tubes remain unchanged in shape and the fluid that is at one instant in a tube remains inside this tube thereafter. We have seen that the flow velocity inside a tube of flow is parallel to the tube and has a magnitude inversely proportional to the area of the cross section (Eq. 18-1). Let us assign such cross sections to the tubes that the constant of proportionality is the same for all of them; if possible we take this constant to be unity. That is, the volume flux is the same for all tubes, namely unit flux. Then the magnitude of the flow velocity can be determined from the areas of the cross sections of the tubes of flow. There is another procedure equivalent to this which consists of setting up a unit area perpendicular to the direction of flow and drawing through it just as many streamlines as the number of units of magnitude of the velocity at that point.

Let us consider some examples of fields of flow. For drawing purposes we consider only *two-dimensional* examples. In these the flow velocity is the same at all points on a line perpendicular to the plane at any point.

In Fig. 18-10 we have drawn a *homogeneous field of flow*. Here all the streamlines are parallel and the flow velocity **v** is the same at all points. We have seen that there are two equivalent ways of deriving the relative magnitudes of the flow velocities from such fields of flow: (*a*) from the widths of the tubes of flow and (*b*) from the distances between lines of equal velocity potential. The latter method applies to steady irrotational flow only. For such flows we draw in the lines of equal velocity potential as dashed lines.

In Fig. 18-11 we show the field for a *uniform rotation* (see Problem 18, Chapter 17). Here *v* is proportional to *r*. In Fig. 18-12 we draw the field of flow of a *vortex*. In this case *v* is proportional to $1/r$ (see Problem 29). Notice that both

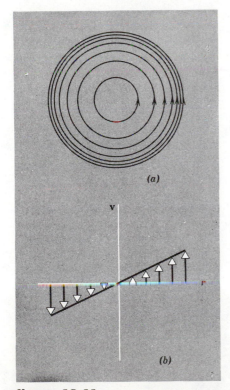

figure 18-11
(*a*) Uniform rotational field of flow.
(*b*) Variation of fluid velocity from the center.

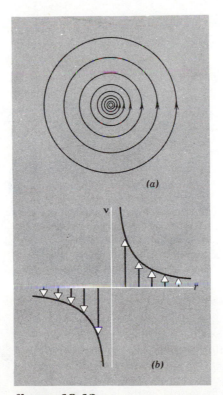

figure 18-12
(*a*) Vortical field of flow. (*b*) Variation of fluid velocity from the center.

uniform rotation and vortex motion are represented by circular streamlines but are entirely different kinds of flow. Obviously, the *shapes* of the streamlines give only limited information; their spacing is needed too.

Figure 18-13 represents the field of flow for a *source*. All streamlines are directed radially outward. The source is a line through the center perpendicular to the paper emitting a mass per unit time Q. The field of flow around a linear *sink* is the same as the source except for the sign of the flow, which is directed radially inward.

For a linear source and linear sink which have the same strengths, Q and $-Q$, and are slightly separated, we obtain the combined field called *linear dipole flow*, shown in Fig. 18-14.

As we shall see later the electrostatic field, the magnetic field, and the field of flow for an electric current are also vector fields. In this connection, the homogeneous field (Fig. 18-10) corresponds to the electric field of a plane capacitor, the source field or sink field (Fig. 18-13) correspond to the electric field of a cylindrical capacitor or straight wire of positive or negative charge respectively, and the linear dipole field (Fig. 18-14) corresponds to the electric field of two oppositely charged wires. In all these the field of flow is potential flow and the electric fields are conservative.

The homogeneous field of Fig. 18-10 also represents the magnetic field inside a solenoid. The vortex field of Fig. 18-12 represents the magnetic field around a straight current-carrying wire. This last is an example of a field that is rotational (about the vortex axis).

Because of these analogies between fluid and electromagnetic fields, we can often determine a field of flow, which is difficult to calculate by present mathematical methods, by experimental measurements on appropriate electrical devices.

As we have seen throughout this chapter, the basic field ideas and conservation principles find application in many areas of physics. We shall encounter them many times again.

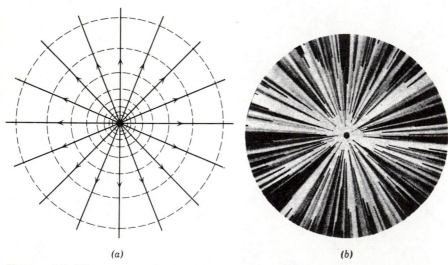

(a) *(b)*

figure 18-13

(a) Flow from a linear source. *(b)* Fluid flow map of the same. The map in this figure is made by allowing water to flow between a horizontal layer of plate glass and a horizontal layer of plaster. In *(b)* the water comes up through a hole in the center of the plaster and flows out toward the edges. The direction of the flow is made visible by sprinkling the plaster with potassium permanganate crystals which dissolve and color the water a deep purple. (The fluid flow map was made and photographed by Professor A. D. Moore at the University of Michigan, and is taken from *Introduction to Electric Fields*, by W. E. Rogers, McGraw-Hill Book Co., 1954.)

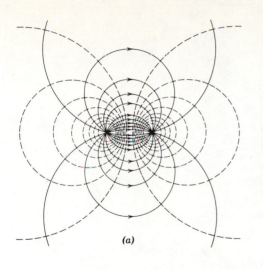

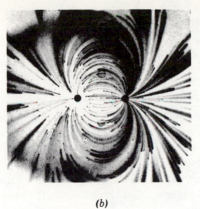

(a)   (b)

figure 18-14
(a) Linear dipole flow. The source is on the left, the sink on the right. (b) A fluid flow map of the same. (The fluid flow map was made and photographed by Professor A. D. Moore at the University of Michigan, and is taken from *Introduction to Electric Fields*, by W. E. Rogers, McGraw-Hill Book Co., 1954.)

questions

1. Briefly describe what is meant by each of the following and illustrate with an example: (*a*) steady fluid flow; (*b*) nonsteady fluid flow; (*c*) rotational fluid flow; (*d*) irrotational fluid flow; (*e*) compressible fluid flow; (*f*) incompressible fluid flow; (*g*) viscous fluid flow; (*h*) nonviscous fluid flow.

2. Can you assign a coefficient of static friction between two surfaces, one of which is a fluid surface?

3. It is found that liquid will flow faster and more smoothly from a sealed can when two holes are punctured in the can than when one hole is made. Explain.

4. List all the assumptions made in deriving Bernoulli's equation (Eq. 18-6).

5. Describe the forces acting on an element of fluid as it flows through a pipe of nonuniform cross section.

6. In a lecture demonstration a ping-pong ball is kept in midair by a vertical jet of air. Is the equilibrium stable, unstable, or neutral? Explain.

7. The height of the liquid in the standpipes indicates that the pressure drops along the channel, even though the channel has a uniform cross section and the flowing liquid is incompressible (Fig. 18-15). Explain.

8. The taller the chimney the better the draft taking the smoke out of the fireplace. Explain. Why doesn't the smoke pour into the room containing the fireplace?

9. (*a*) Explain how a pitcher can make a baseball curve to his right or left? Can we justify applying Bernoulli's equation to such a spinning baseball? (See the Smith reference on p. 392 for an explanation.) (*b*) Why is it easier to throw a curve with a tennis ball than with a baseball?

10. Not only a ball with a rough surface but also a smooth ball can be made to curve when thrown, but these balls will curve in *opposite* directions. Why? (See "Effect of Spin and Speed on the Curve of a Baseball; and the Magnus Effect for Smooth Spheres" by Lyman J. Briggs, in *American Journal of Physics*, November 1959.)

11. Two rowboats moving parallel to one another in the same direction are pulled toward one another. Two automobiles moving parallel are also pulled together. Explain such phenomena on the basis of Bernoulli's equation.

12. In building "skyscrapers," what forces produced by the movement of air must be counteracted? How is this done? (See "The Wind Bracing of Buildings" by Carl W. Condit in *Scientific American*, February 1974.)

13. Can the action of a parachute in retarding free fall be explained by Bernoulli's equation?

14. Liquid is flowing inside a horizontal pipe which has a constriction along its length. Vertical tube manometers are attached at both the wide portion and the narrow portion of the pipe. If a stopcock at the exit end is closed, will the liquid in the manometer tubes rise or fall? Explain.

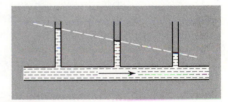

figure 18-15
Question 7

15. A stream of water from a faucet becomes narrower as it falls. Explain.

16. Can you explain why water flows in a continuous stream down a vertical pipe, whereas it breaks into drops when falling freely?

17. How does the flush toilet work? Really. (See *Flushed with Pride: The Story of Thomas Crapper*, by W. Reyburn, Englewood Cliffs, N.J.: Prentice-Hall, 1969.)

18. Can you explain why an object falling from a great height reaches a steady terminal speed?

19. Bernoulli's equation (Eq. 18-6) is a statement of energy conservation for fluid motion. In connection with the Venturi meter (p. 391) can you see a formal relationship to energy changes occurring in a roller coaster when it dips down into a valley and climbs up the other side?

20. Sometimes people remove letters from envelopes by cutting a sliver from a narrow end, holding it firmly and blowing toward it. Does Bernoulli's equation play a role in this enterprise? Explain.

21. On takeoff would it be better for an airplane to move into the wind or with the wind? On landing . . . ?

22. Does the difference in pressure between the lower and upper surfaces of an airplane wing depend on the altitude of the moving plane? Explain.

23. The accumulation of ice on an airplane wing may change its shape in such a way that its lift is greatly reduced. Explain.

24. How is an airplane able to fly upside down?

25. An aeronautical engineer claims that he can design a helicopter that will make a "soft" landing without causing a "down draft." Explain whether or not you think this is possible and why.

26. "The characteristic banana-like shape of most returning boomerangs has hardly anything to do with their ability to return. . . . The essential thing is the cross section of the arms, which should be more convex on one side than on the other, like the wing profile of an airplane." (From, "The Aerodynamics of Boomerangs" by Felix Hess, in *Scientific American*, November 1968.) Explain.

27. What powers the flight of soaring birds? (See "The Soaring Flight of Birds" by C. D. Cone, Jr. in *Scientific American*, April 1962.)

28. Why does the factor "2" appear in Eq. 18-9, rather than "1"? One might naively expect that the thrust would simply be the pressure difference times the area, that is, $A_0(p - p_0)$.

29. The destructive effect of a tornado (twister) is greater near the center of the disturbance than near the edge. Explain.

30. When a stopper is pulled from a filled basin, the water drains out while circulating like a small whirlpool. The angular velocity of a fluid element about a vertical axis through the orifice appears to be greatest near the orifice. Explain.

31. Is it true that in bathtubs in the northern hemisphere the water drains out with a counterclockwise rotation and in those in the southern hemisphere with a clockwise rotation? If so, explain and predict what would happen at the equator. (See "Bath-Tub Vortex" by Ascher H. Shapiro in *Nature*, December 15, 1962.)

32. The longer the board and the shallower the water, the farther will a surf board skim across the water. Explain. (See "The Surf Skimmer" by R. D. Edge, in *American Journal of Physics*, July 1968.)

33. When poured from a teapot water has a tendency to run along the underside of the spout. Explain. (See "The Teapot Effect . . . a Problem" by Markus Reiner, in *Physics Today*, September 1956.)

34. Prairie dogs live in large colonies in complex interconnected burrow systems. They face the problem of maintaining a sufficient air supply to their burrows to avoid suffocation. They avoid this by building conical earth mounds about some of their many burrow openings. In terms of Bernoulli's

equation (Eq. 18-6) how does this air conditioning scheme work? Note that because of viscous forces the wind speed over the prairie is less close to ground level than it is even a few inches higher up. (See *New Scientist*, p. 191, 27 January 1972.)

35. Use the criterion of the paddle wheel (Fig. 18-1) to determine which flow fields (Figs. 18-10 through 18-14) are rotational.

36. In steady flow the velocity vector **v** at any point is constant. Can there then be accelerated motion of the fluid particles? Discuss.

SECTION 18-3

1. A garden hose having an internal diameter of 0.75 in. is connected to a lawn sprinkler that consists of an enclosure with 24 holes, each 0.050 in. in diameter. If the water in the hose has a speed of 3.0 ft/s, at what speed does it leave the sprinkler holes? *Answer:* 28 ft/s.

2. How much work is done by pressure in forcing 50 ft³ (1.4 m³) of water through a 0.50-in. (13-mm) pipe if the difference in pressure at the two ends of the pipe is 15 lb/in.² (1.0×10^5 Pa)?

3. Water flows continuously from the outlet of a faucet of internal diameter d at an initial speed v_0. Determine the diameter of the stream in terms of the distance h below the outlet. (Neglect air resistance and assume droplets are not formed.)

   Answer: $d\left[\dfrac{v_0}{\sqrt{v_0^2 + 2gh}}\right]^{1/2}$.

4. Water is pumped steadily out of a flooded basement at a speed of 5.0 m/s through a uniform hose of radius 1.0 cm. The hose passes out through a window 3.0 m above the water line. How much power is supplied by the pump?

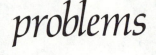

problems

SECTION 18-4

5. A hollow tube has a disc DD attached to its end. When air is blown through the tube, the disc attracts the card CC. Let the area of the card be A and let v be the average airspeed between CC and DD (Fig. 18-16), calculate the resultant upward force on CC. Neglect the card's weight.
   Answer: $\frac{1}{2}\rho v^2 A$, where ρ is the density of air.

6. In a horizontal oil pipeline of constant cross-sectional area the pressure decrease between two points 1000 ft apart is 5.0 lb/in.². What is the energy loss per cubic foot of oil per unit distance?

7. Figure 18-17 shows liquid discharging from an orifice in a large tank at a distance h below the water level. (a) Apply Bernoulli's equation to a streamline connecting points 1, 2, and 3, and show that the speed of efflux is

$$v = \sqrt{2gh}.$$

This is known as Torricelli's law. (b) If the orifice were curved directly up-

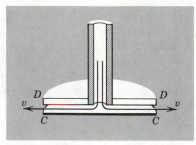

figure 18-16
Problem 5

figure 18-17
Problem 7

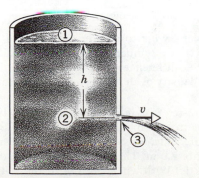

ward, how high would the liquid stream rise? (c) How would viscosity or turbulence affect the analysis?   *Answer:* (b) It would rise to height h.

8. A tank is filled with water to a height H. A hole is punched in one of the walls at a depth h below the water surface (Fig. 18-18). (a) Show that the dis-. tance x from the foot of the wall at which the stream strikes the floor is given by $x = 2\sqrt{h(H - h)}$. (b) Could a hole be punched at another depth so that this second stream would have the same range? If so, at what depth?

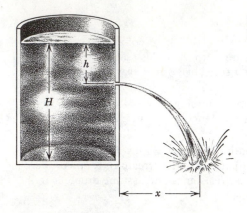

figure 18-18
Problem 8

9. The upper surface of water in a standpipe is a height H above level ground. (a) At what depth h should a small hole be put to make the emerging horizontal water stream strike the ground at the maximum distance from the base of the standpipe? (b) What is this maximum distance?
Answer: (a) H/2. (b) H.

10. (a) Consider the stagnant air at the front edge of a wing and the air rushing over the wing surface at a speed v. Assume pressure at the leading edge to be approximately atmospheric and find the greatest value possible for v in streamline flow; assume air is incompressible and use Bernoulli's equation. Take the density of air to be 1.2×10^{-3} g/cm³. (b) How does this compare with the speed of sound of 770 mi/h? Can you explain the difference? Why should there be any connection between these quantities?

11. If a person blows air with a speed of 15 m/s across the top of one side of a U-tube containing water, what will be the difference between the water levels on the two sides? Assume the density of air is 1.2 kg/m³.
Answer: 1.4 cm.

12. A siphon is a device for removing liquid from a container that cannot be tipped. It operates as shown in Fig. 18-19. The tube must initially be filled, but once this has been done the liquid will flow until its level drops below the tube opening at A. The liquid has density ρ and negligible viscosity. (a) With what speed does the liquid emerge from the tube at C? (b) What is the pressure in the liquid at the topmost point B? (c) What is the greatest possible height h_1 that a siphon may lift water?

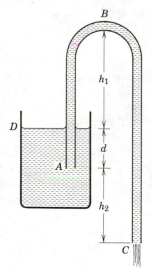

figure 18-19
Problem 12

SECTION 18-5

13. A Pitot tube is mounted on an airplane wing to determine the speed of the plane relative to the air, which is at a temperature of 0°C. The tube contains alcohol and indicates a level difference of 26 cm. What is the plane's speed relative to the air? The density of alcohol is 0.81×10^3 kg/m³.
Answer: 200 km/h.

14. Models of torpedoes are sometimes tested in a horizontal pipe of flowing water, much as a wind tunnel is used to test model airplanes. Consider a circular pipe of internal diameter 10 in. and a torpedo model, aligned along the axis of the pipe, with a diameter of 2.0 in. The torpedo is to be tested with water flowing past it at 8.0 ft/s. (a) With what speed must the water

flow in the unconstricted part of the pipe? (b) What will the pressure difference be between the constricted and unconstricted parts of the pipe?

15. Water is moving with a speed of 5.0 m/s through a pipe with a cross-sectional area of 4.0 cm². The water gradually descends 10 m as the pipe increases in area to 8.0 cm². (a) What is the speed of flow at the lower level? (b) If the pressure at the upper level is 1.50×10^5 Pa, what is the pressure at the lower level?   Answer: (a) 2.5 m/s. (b) 2.6×10^5 Pa.

16. Suppose that two tanks, 1 and 2, each with a large opening at the top, contain different liquids. A small hole is made in the side of each tank at the same depth h below the liquid surface, but the hole in tank 1 has half the cross-sectional area of the hole in tank 2. (a) What is the ratio ρ_1/ρ_2 of the densities of the fluids if it is observed that the mass flux is the same for the two holes? (b) What is the ratio of the flow rates (volume flux) from the two tanks? (c) To what height above the hole in the second tank should fluid be added or drained to equalize the flow rates?

17. A small plane has a wing area (each wing) of 100 ft² (9.3 m²). At a certain air speed, air flows over the upper wing surface at 160 ft/s (49 m/s) and over the lower wing surface at 130 ft/s (40 m/s). What is the weight of the plane? Assume that the plane travels at constant velocity and that the lift effects associated with the fuselage and tail assembly are small. Discuss the lift if the plane, flying at the same air speed, is (a) in level flight, (b) climbing at 15°, and (c) descending at 15°. Take the density of air to be 2.33×10^{-3} slug/ft³ (1.2 kg/m³).
Answer: 2000 lb (8900 N). Lift is the same in all three cases.

18. If the speed of air flow past the lower surface of a wing is 350 ft/s, what speed of flow over the upper surface will give a lift of 20.0 lb/in.²? Take the density of air to be 2.33×10^{-3} slug/ft³.

19. Consider a uniform U-tube with a diaphragm at the bottom and filled with a liquid to different heights in each arm (see Fig. 18-20). Now imagine that the diaphragm is punctured so that the liquid flows from left to right. (a) Show that application of Bernoulli's principle to points 1 and 3 leads to a contradiction. (b) Explain why Bernoulli's principle is not applicable here. (*Hint:* Is the flow steady?)

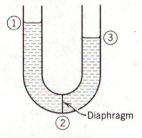

figure 18-20
Problem 19

20. Calculate the speed of efflux of a liquid from an opening in a tank, taking into account the velocity of the top surface of the liquid, as follows. (a) Show, from Bernoulli's equation, that

$$v_0^2 = v^2 + 2gh$$

where v is the speed of the top surface. (b) Then consider the flow as one big tube of flow and obtain v/v_0 from the equation of continuity, so that

$$v_0 = \sqrt{2gh/[1 - (A_0/A)^2]}$$

where A is the tube cross section at the top and A_0 is the tube cross section at the opening. (c) Then show that if the hole is small compared to the area of the surface,

$$v_0 \cong \sqrt{2gh} \, [1 + \tfrac{1}{2}(A_0/A)^2].$$

21. By applying Bernoulli's equation and the equation of continuity to points 1 and 2 of Fig. 18-6, show that the speed of flow at the entrance is

$$v = a \sqrt{\frac{2(\rho' - \rho)gh}{\rho(A^2 - a^2)}}.$$

22. A Venturi meter has a pipe diameter of 10 in. and a throat diameter of 5.0 in. If the water pressure in the pipe is 8.0 lb/in.² and in the throat is 6.0 lb/in.², determine the rate of flow of water in ft³/s (volume flux).

23. Consider the Venturi tube of Fig. 18-6 without the manometer. Let A equal $5a$. Suppose the pressure at A is 2.0 atm. (a) Compute the values of v at A and v' at a that would make the pressure p' at a equal to zero. (b) Compute the corresponding volume flow rate if the diameter at A is 5.0 cm. The

phenomenon at a when p' falls to nearly zero is known as *cavitation*. The water vaporizes into small bubbles.

Answer: (a) 20 m/s. (b) 8.0×10^{-3} m³/s.

SECTION 18-6

24. (a) Consider a stream of fluid of density ρ with speed v_1 passing *abruptly* from a cylindrical pipe of cross-sectional area a_1 into a wider cylindrical pipe of cross-sectional area a_2 (see Fig. 18-21). The jet will mix with the surrounding fluid and, after the mixing, will flow on almost uniformly with an average speed v_2. Without referring to the details of the mixing, use momentum ideas to show that the increase in pressure due to the mixing is approximately

$$p_2 - p_1 = \rho v_2 (v_1 - v_2).$$

(b) Show from Bernoulli's principle that in a *gradually* widening pipe we would get

$$p_2 - p_1 = \tfrac{1}{2}\rho(v_1{}^2 - v_2{}^2).$$

(c) Find the loss of pressure due to the abrupt enlargement of the pipe. Can you draw an analogy with elastic and inelastic collisions in particle mechanics?

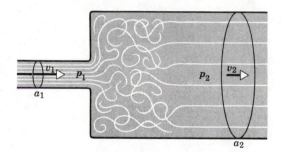

figure 18-21
Problem 24

SECTION 18-7

25. Show that the constant in Bernoulli's equation (Eq. 18-6) is the same for *all* streamlines in the case of the steady, irrotational flow of Fig. 18-10.

26. A force field is conservative if $\oint \mathbf{F} \cdot d\mathbf{s} = 0$. The circle on the integration sign means that the integration is to be taken along a closed curve (a round trip) in the field. A flow is a potential flow (hence irrotational) if $\oint \mathbf{v} \cdot d\mathbf{s} = 0$ for every closed path in the field.

Using this criterion, show that the fields of Figs. (a) 18-10 and (b) 18-13 are fields of potential flow.

27. The so-called Poiseuille field of flow is shown in Fig. 18-22. The spacing of the streamlines indicates that although the motion is rectilinear, there is a velocity gradient in the transverse direction. Show that such a flow is rotational.

28. In flows that are sharply curved centrifugal effects are appreciable. Consider an element of fluid which is moving with speed v along a streamline of a curved flow in a horizontal plane (Fig. 18-23).

(a) Show that $dp/dr = \rho v^2/r$, so that the pressure increases by an amount $\rho v^2/r$ per unit distance perpendicular to the streamline as we go from the concave to the convex side of the streamline.

(b) Then use Bernoulli's equation and this result to show that vr equals a constant, so that speeds increase toward the center of curvature. Hence, streamlines that are uniformly spaced in a straight pipe will be crowded toward the inner wall of a curved passage and widely spaced toward the outer wall. This problem should be compared to Problem 18 of Chapter 17

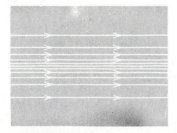

figure 18-22
Problem 27

in which the curved motion is produced by rotating a container. There the speed varied directly with r, but here it varies inversely.

(c) Show that this flow is irrotational.

29. Before Newton proposed his theory of gravitation, a model of planetary motion proposed by René Descartes was widely accepted. In Descartes' model the planets were caught in and dragged along by a whirlpool of ether particles centered around the sun. Newton showed that this vortex scheme contradicted observations, for: (a) The speed of an ether particle in the vortex varies inversely as its distance from the sun. (b) The period of revolution of such a particle varies directly as the square of its distance from the sun. (c) This result contradicts Kepler's third law. Prove (a), (b), and (c).

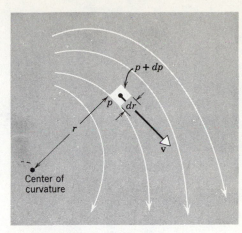

figure 18-23
Problem 28

19
waves in elastic media

Wave motion appears in almost every branch of physics. We are all familiar with water waves. There are also sound waves, as well as light waves, radio waves, and other electromagnetic waves. One formulation of the mechanics of atoms and subatomic particles is called wave mechanics. Clearly the properties and behavior of waves are very important in physics.

In this chapter and the next we confine our attention to waves in deformable or elastic media. These waves, among which ordinary sound waves in air are one example, might be called *mechanical waves*. They originate in the displacement of some portion of an elastic medium from its normal position, causing it to oscillate about an equilibrium position. Because of the elastic properties of the medium, the disturbance is transmitted from one layer to the next. This disturbance, or wave, consequently progresses through the medium. Note that the medium itself does not move as a whole along with the wave motion; the various parts of the medium oscillate only in limited paths. For example, in water waves small floating objects like corks show that the actual motion of various parts of the water is slightly up and down and back and forth. Yet the water waves move steadily along the water. As they reach floating objects they set them in motion, thus transferring energy to them.* Energy can be transmitted over considerable distances by wave motion. The energy in the waves is the kinetic and potential energy of the matter, but the transmission of the energy comes about by its being passed along from one part of the matter to the next, not by any long-range motion of the matter itself. Mechanical waves are charac-

19-1
MECHANICAL WAVES

* See "Ocean Waves," by Willard Bascom, *Scientific American*, August 1959.

terized by the transport of energy through matter by the motion of a disturbance in that matter without any corresponding bulk motion of the matter itself.

It is necessary to have a material medium to transmit mechanical waves. We do not need such a medium, however, to transmit electromagnetic waves, light passing freely, for example, through the near vacuum of space from the stars. The properties of the medium that determine the speed of a wave through that medium, as we will see in Section 19-5, are its inertia and its elasticity. All material media, including, say, air, water, and steel, possess these properties and can transmit mechanical waves. It is the elasticity that gives rise to the restoring forces on any part of the medium displaced from its equilibrium position; it is the inertia that tells us how this displaced portion of the medium will respond to these restoring forces. Together these two factors determine the wave speed.

19-2 TYPES OF WAVES

In listing water waves, light waves, and sound waves as examples of wave motion, we are classifying waves according to their broad physical properties. Waves can be classified in other ways.

We can distinguish different kinds of mechanical waves by considering how the motions of the particles of matter are related to the direction of propagation of the waves themselves. If the motions of the matter particles conveying the wave are perpendicular to the direction of propagation of the wave itself, we then have a *transverse* wave. For example, when a vertical string under tension is set oscillating back and forth at one end, a transverse wave travels down the string; the disturbance moves along the string but the string particles vibrate at right angles to the direction of propagation of the disturbance (Fig. 19-1a).

Light waves are not mechanical waves. The disturbance that travels along is not a motion of matter but an electromagnetic field (Chapter 41). But because the electric and magnetic fields are perpendicular to the direction of propagation, light waves are also transverse waves.

If, however, the motion of the particles conveying a mechanical wave is back and forth along the direction of propagation, we then have a *longitudinal wave*. For example, when a vertical spring under tension is set oscillating up and down at one end, a longitudinal wave travels along the spring; the coils vibrate back and forth in the direction in which the disturbance travels along the spring (Fig. 19-1b). Sound waves in a gas are longitudinal waves. We shall discuss them in greater detail in Chapter 20.

Some waves are neither purely longitudinal nor purely transverse. For example, in waves on the surface of water the particles of water move both up and down and back and forth, tracing out elliptical paths as the water waves move by.

Waves can also be classified as one-, two-, and three-dimensional waves, according to the number of dimensions in which they propagate energy. Waves moving along the string or the spring of Fig. 19-1 are one-dimensional. Surface waves or ripples on water, caused by dropping a pebble into a quiet pond, are two-dimensional. Sound waves and light waves which emanate radially from a small source are three-dimensional.

Waves may be classified further according to the behavior of a particle of the matter conveying the wave during the course of time the wave propagates. For example, we can produce a *pulse* traveling down a

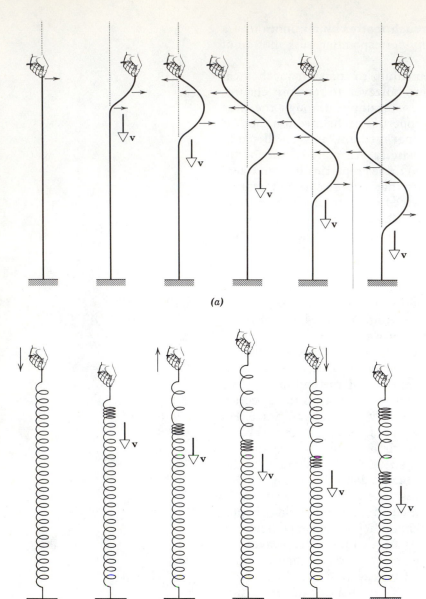

(a)

(b)

figure 19-1
(a) In a transverse wave the particles of the medium (stretched string) vibrate at right angles to the direction in which the wave itself is propagated. *(b)* In a longitudinal wave the particles of the medium (stretched spring) vibrate in the same direction as that in which the wave itself is propagated.

stretched string by applying a single sidewise movement at its end. Each particle remains at rest until the pulse reaches it, then it moves during a short time, and then it again remains at rest. If we continue to move the end of the string back and forth (Fig. 19-1a), we produce a *train of waves* traveling along the string. If our motion is periodic, we produce a *periodic train of waves* in which each particle of the string has a periodic motion. The simplest special case of a periodic wave is a *simple harmonic wave* which gives each particle a simple harmonic motion.

Consider a three-dimensional pulse. We can draw a surface through all points undergoing a similar disturbance at a given instant. As time goes on, this surface moves along showing how the pulse propagates. We can draw similar surfaces for subsequent pulses. For a periodic wave we can generalize the idea by drawing in surfaces, all of whose points are in the same phase of motion. These surfaces are called *wavefronts*. If the medium is homogeneous and isotropic, the direction of propagation is always at right angles to the wavefront. A line normal to the wavefronts, indicating the direction of motion of the waves, is called a *ray*.

Wavefronts can have many shapes. If the disturbances are propagated in a single direction, the waves are called *plane waves*. At a given instant conditions are the same everywhere on any plane perpendicular to the direction of propagation. The wavefronts are plane and the rays are parallel straight lines (Fig. 19-2a). Another simple case is that of *spherical waves*. Here the disturbance is propagated out in all directions from a point source of waves. The wavefronts are spheres and the rays are radial lines leaving the point source in all directions (Fig. 19-2b). Far from the source the spherical wavefronts have very small curvature, and over a limited region they can often be regarded as plane. Of course, there are many other possible shapes for wavefronts.

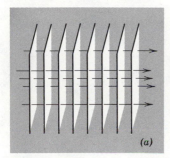 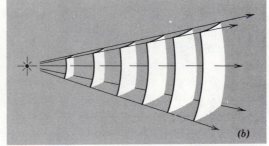

figure 19-2
(a) A plane wave. The planes represent wavefronts spaced a wavelength apart, and the arrows represent rays. *(b)* A spherical wave. The rays are radial and the wavefronts, spaced a wavelength apart, from spherical shells. Far out from the source, however, small portions of the wavefronts become nearly plane.

We shall refer to all these wave types as we progress through the wave phenomena of physics. In this chapter we often use the transverse wave in a string to illustrate the general properties of waves. In the next chapter we shall see the consequences of these properties for sound, a longitudinal mechanical wave. Later in the text we will discuss the properties of nonmechanical waves such as light waves.

19-3
TRAVELING WAVES

Let us consider a long string stretched in the x-direction along which a transverse wave is traveling. At some instant of time, say $t = 0$, the shape of the string can be represented by

$$y = f(x) \qquad t = 0, \tag{19-1}$$

where y is the transverse displacement of the string at the position x. In Fig. 19-3a we show a possible waveform (a pulse) on the string at $t = 0$. Experiment shows that as time goes on such a wave travels along the string without changing its form, provided internal frictional losses are small enough. At some time t later the wave has traveled a distance vt to the right, where v is magnitude of the wave velocity, assumed constant. The equation of the curve at the time t is therefore

$$y = f(x - vt) \qquad t = t. \tag{19-2}$$

This gives us the same waveform about the point $x = vt$ at time t as we

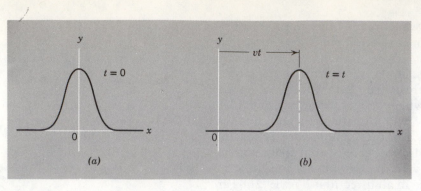

figure 19-3
(a) The shape of a stretched string
(in this case a pulse) at $t = 0$.
(b) At a later time t the pulse has
traveled to the right a distance
$x = vt$.

had about $x = 0$ at the time $t = 0$ (Fig. 19-3b). Equation 19-2 is the general equation representing a wave of *any* shape *traveling to the right*. To describe a particular shape we must specify exactly what the function f is.*

Let us look more carefully at this equation. If we wish to follow a particular part (or phase) of the wave as times goes on, then in the equation we look at a particular value of y (say, the top of the pulse just described). Mathematically this means we look at how x changes with t when $(x - vt)$ has some particular fixed value. We see at once that as t increases x must increase in order to keep $(x - vt)$ fixed. Hence, Eq. 19-2 does in fact represent a wave traveling to the right (increasing x as time goes on). If we wished to represent a wave *traveling to the left*, we would write

$$y = f(x + vt), \qquad (19\text{-}3)$$

for here the position x of some fixed phase $(x + vt)$ of the wave decreases as time goes on. The velocity of a particular phase of the wave is easily obtained. For a particular phase of a wave traveling to the right we require that

$$x - vt = \text{constant}.$$

Then differentiation with respect to time gives

$$\frac{dx}{dt} - v = 0 \qquad \text{or} \qquad \frac{dx}{dt} = v, \qquad (19\text{-}4)$$

so that v is really the *phase velocity* of the wave. For a wave traveling to the left we obtain $-v$, in the same way, as its phase velocity.†

The general equation of a wave can be interpreted further. Note that for any fixed value of the time t the equation gives y as a function of x. This defines a curve, and this curve represents the actual shape of the string at this chosen time. It gives us a snapshot of the wave at this time. Suppose, on the other hand, we wish to focus our attention on one point of the string, that is, a fixed value of x. Then the equation gives us y as a function of the time t. This describes how the transverse position of this point on the string changes with time.

* When we say that "y is a function of $(x - vt)$," we mean that the variables x and t occur only in the combination $x - vt$. For example, $\sin k(x - vt)$, $\log (x - vt)$, and $(x - vt)^3$ are functions of $x - vt$, but $x^2 - vt^2$ is not.
† In disturbances that can be represented as a group of waves, the energy may be transported with a velocity different from the phase velocity of any individual wave. This group velocity will be considered in Chapter 41 in connection with electromagnetic waves. Until then whenever we use the term wave velocity we mean the phase velocity of the wave.

The argument just presented holds for longitudinal waves as well as for transverse waves. The analogous longitudinal example is that of a long straight tube of gas whose axis is taken as the x-axis, and the wave or pulse is a pressure change traveling along the tube. Then the same reasoning leads us to an equation, having the form of Eqs. 19-2 and 19-3, which gives the pressure variations with time at all points of the tube. (See Section 20-3.)

Let us now consider a particular waveform, whose importance will soon become clear. Suppose that at the time $t = 0$ we have a wavetrain along the string given by

$$y = y_m \sin \frac{2\pi}{\lambda} x. \qquad (19\text{-}5)$$

The wave shape is a sine curve (Fig. 19-4). The maximum displacement y_m is the *amplitude* of the sine curve. The value of the transverse displacement y is the same at x as it is at $x + \lambda$, $x + 2\lambda$, etc. The symbol λ is called the *wavelength* of the wavetrain and represents the distance between two adjacent points in the wave having the same phase. As times goes on let the wave travel to the right with a phase velocity v. Hence, the equation of the wave at the time t is

$$y = y_m \sin \frac{2\pi}{\lambda} (x - vt). \qquad (19\text{-}6)$$

Notice that this has the form required for a traveling wave (Eq. 19-2).

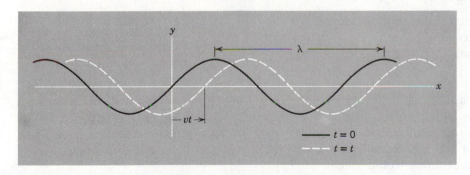

$t = 0$
$t = t$

figure 19-4
At $t = 0$, the string has a shape $y = y_m \sin 2\pi x/\lambda$ (solid line). At a later time t the sine wave has moved to the right a distance $x = vt$, and the string has a shape given by $y = y_m \sin 2\pi(x - vt)/\lambda$.

The *period* T is the time required for the wave to travel a distance of one wavelength λ, so that

$$\lambda = vT. \qquad (19\text{-}7)$$

Putting this relation into the equation of the wave, we obtain

$$y = y_m \sin 2\pi \left(\frac{x}{\lambda} - \frac{t}{T} \right). \qquad (19\text{-}8)$$

From this form it is clear that y, at any given time, has the same value at $x + \lambda$, $x + 2\lambda$, etc., as it does at x, and that y, at any given position, has the same value at the time $t + T$, $t + 2T$, etc., as it does at the time t.

To reduce Eq. 19-8 to a more compact form, we define two quantities, the *wave number* k and the *angular frequency* ω (see Eq. 15-12). They are given by

$$k = \frac{2\pi}{\lambda} \qquad \text{and} \qquad \omega = \frac{2\pi}{T}. \qquad (19\text{-}9)$$

In terms of these quantities, the equation of a sine wave traveling to the right (positive x-direction) is

$$y = y_m \sin (kx - \omega t). \qquad (19\text{-}10a)$$

For a sine wave traveling to the left (negative x-direction), we have

$$y = y_m \sin (kx + \omega t). \qquad (19\text{-}10b)$$

Comparing Eqs. 19-7 and 19-9, we see that the phase velocity v of the wave is given by

$$v = \frac{\lambda}{T} = \frac{\omega}{k}. \qquad (19\text{-}11)$$

In the traveling waves of Eqs. 19-10a and 19-10b we have assumed that the displacement y is zero at the position $x = 0$ at the time $t = 0$. This, of course, need not be the case. The general expression for a sinusoidal wavetrain traveling to the right is

$$y = y_m \sin (kx - \omega t - \phi),$$

where ϕ is called the phase constant. For example, if $\phi = -90°$, the displacement y at $x = 0$ and $t = 0$ is y_m. This particular example is

$$y = y_m \cos (kx - \omega t),$$

for the cosine function is displaced by 90° from the sine function.

If we fix our attention on a given point of the string, say $x = \pi/k$, the displacement y at that point can be written* as

$$y = y_m \sin (\omega t + \phi).$$

This is similar to Eq. 15-29 for simple harmonic motion. Hence, any particular element of the string undergoes simple harmonic motion about its equilibrium position as this wavetrain travels along the string.

19-4 THE SUPERPOSITION PRINCIPLE

It is an experimental fact that for many kinds of waves *two or more waves can traverse the same space independently of one another.* The fact that waves act independently of one another means that the displacement of any particle at a given time is simply the sum of the displacements that the individual waves alone would give it. This process of vector addition of the displacements of a particle is called *superposition.* For example, radio waves of many frequencies pass through a radio antenna; the electric currents set up in the antenna by the superposed action of all these waves are very complex. Nevertheless, we can still tune to a particular station, the signal that we receive from it being in principle the same as that which we would receive if all other stations were to stop broadcasting. Likewise, in sound we can listen to notes played by individual instruments in an orchestra, even though the sound wave reaching our ears from the full orchestra is very complex.

For waves in deformable media the superposition principle holds whenever the mathematical relation between the deformation and the restoring force is one of simple proportionality. Such a relation is expressed mathematically by a linear equation. For electromagnetic waves the superposition principle holds because the mathematical relations between the electric and magnetic fields are linear.

The superposition principle seems so obvious that it is worthwhile to point out that it does not always hold. Superposition fails when the equations governing

---

* Using the fact that $\sin (\pi - \theta) = \sin \theta$.

wave motion are not linear. Physically this happens when the wave disturbance is relatively large and the ordinary linear laws of mechanical action no longer hold. For example, beyond the elastic limit Hooke's law no longer holds and the linear relation $F = -kx$ can no longer be used.

As for sound, violent explosions create shock waves. Although shock waves are longitudinal elastic waves in air, they behave differently from ordinary sound waves. The equation governing their propagation is quadratic, and superposition does not hold. With two very loud notes the ear hears something more than just the two individual notes. Those familiar with high-fidelity apparatus will know that "intermodulation distortion" between two tones arises when the system fails to combine the tones linearly, and that this distortion is more apparent when the amplitude of the tones is high. A more obvious physical example is water waves. Ripples cannot travel independently across breakers as they can across gentle swells.

The importance of the superposition principle physically is that, where it holds, it makes it possible to analyze a complicated wave motion as a combination of simple waves. In fact, as was shown by the French mathematician J. Fourier (1768–1830), all that we need to build up the most general form of periodic wave are simple harmonic waves. Fourier showed that any periodic motion of a particle can be represented as a combination of simple harmonic motions. For example, if $y(t)$ represents the motion of a source of waves having a period T, we can analyze $y(t)$ as follows:

$$y(t) = A_0 + A_1 \sin \omega t + A_2 \sin 2\omega t + A_3 \sin 3\omega t + \cdots$$
$$+ B_1 \cos \omega t + B_2 \cos 2\omega t + B_3 \cos 3\omega t + \cdots$$

where $\omega = 2\pi/T$. This expression is called a Fourier series. The A's and B's are constants which have definite values for any particular periodic motion $y(t)$. (See Fig. 19-5, for example.) If the motion is not periodic, as

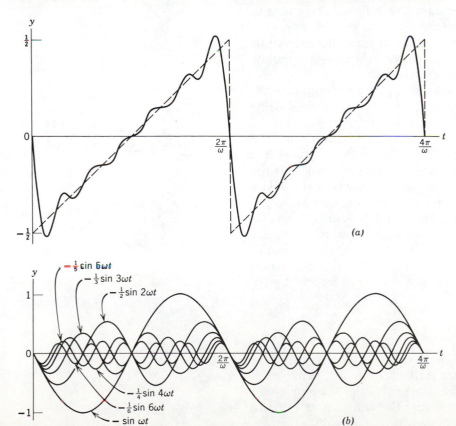

(a)

(b)

figure 19-5

(a) The dashed line is a sawtooth wave commonly encountered in electronics. It can be written $y(t) = (\omega/2\pi)t - \frac{1}{2}$ for $0 < t < 2\pi/\omega$, as $y(t) = (\omega/2\pi)t - \frac{3}{2}$ for $2\pi/\omega < t < 4\pi/\omega$, etc. The Fourier series for this function is $y(t) = (1/\pi)[-\sin \omega t - \frac{1}{2} \sin 2\omega t - \frac{1}{3} \sin 3\omega t - \ldots]$. The solid line is the sum of the first six terms of this series and can be seen to approximate the sawtooth quite closely, except for overshooting near the discontinuities. As more terms of the series are included, the approximation becomes better and better. (b) Here are shown the first six terms of the Fourier series which, when added together, yield the solid curve in (a).

a pulse, the sum is replaced by an integral—the so-called Fourier integral. Hence, any motion of a source of waves can be represented in terms of simple harmonic motions. Because the motion of the source creates the waves, it should come as no surprise that the waves themselves can be analyzed as combinations of simple harmonic waves. Herein lies the importance of simple harmonic motion and simple harmonic waves.

When the elasticity of the medium is such that (for mechanical waves) Hooke's law is not exactly obeyed, then a wave pulse produced at the end of a stretched string may change its shape as it travels along the string. Although each of the component harmonic waves travels without changing its shape, the speed of each component is now different for each frequency (or wavelength). This phenomena is called *dispersion* and the medium is said to be *dispersive* for the wave type in question. As a result the pulse shape can change and the pulse speed may depend on the details of its initial shape. Examples of nondispersive situations are mechanical waves propagated along an ideal (perfectly flexible) stretched string and electromagnetic waves (including light) propagated through a vacuum. Examples of dispersive situations are ocean waves and light waves propagated through a transparent medium such as glass.

Another way in which the wave pulse may change its shape is by loss of mechanical energy to the medium or its surroundings; for example, by air resistance, viscosity, or internal friction. Then the amplitude of the wave decreases with time and the wave is said to be *attenuated*.

For the moment, we will assume that the medium is nondispersive and that there is no dissipation of energy as the wave travels through the medium.

19-5
WAVE SPEED

Given the characteristics of the medium it should be possible to calculate the wave speed from the basic principles of Newtonian mechanics. In this section we continue to focus our attention on transverse waves in a stretched string and in Supplementary Topic III we show how to calculate the speed of such waves in the most general way. Here we consider two other approaches—a treatment based on dimensional analysis and a somewhat less general mechanical analysis in which we compute the speed of a transverse pulse along a stretched string.

We stated in Section 19-1 that the wave speed for a medium depends on the elasticity of the medium and on its inertia. For a stretched string the elasticity is measured by the tension F in the string; the greater the tension the greater will be the elastic restoring force on an element of the string that is pulled sideways. The inertia characteristic is measured by μ, the mass per unit length of the string. Assuming then, that the wave speed v depends only on F and μ, we can use dimensional analysis to find how v depends on these quantities. In terms of mass M, length L, and time T, the dimensions of F are MLT^{-2} and the dimensions of μ are ML^{-1}. The only way these dimensions can be combined to get a velocity (which has the dimensions LT^{-1}) is to take the square root of F/μ. That is, F/μ has the dimensions L^2T^{-2} and $\sqrt{F/\mu}$ has the dimensions LT^{-1} of a velocity. Dimensional analysis cannot account for any dimensionless quantities, so that the result

$$v = \sqrt{\frac{F}{\mu}} \qquad (19\text{-}12)$$

may or may not be complete. The most we can say is that the wave speed is equal to a dimensionless constant times $\sqrt{F/\mu}$. The value of the constant can be obtained from a mechanical analysis of the problem or from experiment. These methods show that the constant is equal to unity and that Eq. 19-12 is correct as it stands.

Now let us *derive* the velocity of a pulse in a stretched string by a mechanical analysis. In Fig. 19-6 we show a wave pulse proceeding from right to left in the string with a speed v. We can imagine the entire string to be moved from left to right with this same speed so that the wave pulse remains fixed in space, whereas the particles composing the string successively pass through the pulse. This simply means that, instead of taking our reference frame to be the walls between which the string is stretched, we choose a reference frame which is in uniform motion with respect to that one. Because Newton's laws involve only accelerations, which are the same in both frames, we can use them in either frame. We just happen to choose a more convenient frame.

We consider a small section of the pulse of length Δl to form an arc of a circle of radius R, as shown in the diagram. If μ is the mass per unit length of the string, the so-called linear density, then $\mu \, \Delta l$, is the mass of this element. The tension F in the string is a tangential pull at each end of this small segment of the string. The horizontal components cancel and the vertical components are each equal to $F \sin \theta$. Hence, the total vertical force is $2F \sin \theta$. Because θ is small, we can take $\sin \theta \cong \theta$ and

$$2F \sin \theta = 2F\theta = 2F\frac{(\Delta l/2)}{R} = F\frac{\Delta l}{R}.$$

This gives the force supplying the centripetal acceleration of the string particles directed toward O. Now the centripetal force acting on a mass $\mu \, \Delta l$ moving in a circle of radius R with speed v is $\mu \, \Delta l \, v^2/R$; see Section 6-3. Notice that the tangential velocity v of this mass element along the top of the arc is horizontal and is the same as the pulse speed. Combining the equivalent expressions just given we obtain

$$F\frac{\Delta l}{R} = \frac{\mu \, \Delta l \, v^2}{R}$$

or

$$v = \sqrt{\frac{F}{\mu}}.$$

If the amplitude of the pulse were very large compared to the length of the string, we would have been unable to use the approximation $\sin \theta \cong \theta$. Furthermore, the tension F in the string would be changed by the presence of the pulse, whereas we assumed F to be unchanged from the original tension in the stretched string. Therefore, our result, like superposition, holds only for relatively small transverse displacements of the string—which case, however, is widely applicable in practice. Notice also that the wave speed is independent of the shape of the wave, for no particular assumption about the actual shape of the pulse was used in the proof.

The frequency of a wave is naturally determined by the frequency of the source. The speed with which the wave travels through a medium is determined by the properties of the medium, as previously illustrated. Once the frequency ν and speed v of the wave are determined, the wavelength λ is fixed. In fact, from Eq. 19-7 and the relation, $\nu = 1/T$, we have

$$\lambda = \frac{v}{\nu}. \tag{19-13}$$

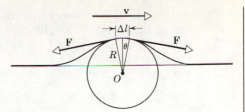

figure 19-6
Derivation of wave speed by considering the forces on a section of string of length Δl.

A transverse sinusoidal wave is generated at one end of a long horizontal string by a bar which moves the end up and down through a distance of 0.50 cm. The motion is continuous and is repeated regularly 120 times per second.

(*a*) If the string has a linear density of 0.25 kg/m and is kept under a tension of 90 N, find the speed, amplitude, frequency, and wavelength of the wave motion.

EXAMPLE 1

The end moves 0.25 cm away from the equilibrium position, first above it, then below it; therefore, the amplitude y_m is 0.25 cm.

The entire motion is repeated 120 times each second so that the frequency is 120 vibrations per second, or 120 Hz.

The wave speed is given by $v=\sqrt{F/\mu}$. But $F=90$ N and $\mu=0.25$ kg/m, so that

$$v=\sqrt{\frac{90\text{ N}}{0.25\text{ kg/m}}}=19\text{ m/s}.$$

The wavelength is given by $\lambda = v/\nu$, so that

$$\lambda=\frac{19\text{ m/s}}{120\text{ vib/s}}=16\text{ cm}.$$

(*b*) Assuming the wave moves in the $+x$-direction and that, at $t=0$, the end of the string described by $x=0$ is in its equilibrium position $y=0$, write the equation of the wave.

The general expression for a transverse sinusoidal wave moving in the $+x$-direction is

$$y=y_m \sin(kx-\omega t-\phi).$$

Requiring that $y=0$ for the conditions $x=0$ and $t=0$ yields

$$0=y_m \sin(-\phi),$$

which means that the phase constant ϕ may be taken to be zero. You should show that integral multiples of π yield the same final results. Hence for this wave

$$y=y_m \sin(kx-\omega t),$$

and with the values just found,

$$y_m=0.25\text{ cm},$$

$$\lambda=16\text{ cm} \quad\text{or}\quad k=\frac{2\pi}{\lambda}=\frac{2\pi}{16\text{ cm}}=0.39\text{ cm}^{-1},$$

$$v=19\text{ m/s}=1900\text{ cm/s} \quad\text{or}\quad \omega=vk=(1900\text{ cm/s})(0.39\text{ cm}^{-1})=740\text{s}^{-1}=740\text{ Hz},$$

we obtain as the equation for the wave

$$y=0.25\sin(0.39x-740t)$$

where x and y are in centimeters and t is in seconds.

EXAMPLE 2

As this wave passes along the string, each particle of the string moves up and down at right angles to the direction of the wave motion. Find the velocity and acceleration of a particle 2.0 ft from the end.

The general form of this wave is

$$y=y_m \sin(kx-\omega t)=y_m \sin k(x-vt).$$

The v in this equation is the constant horizontal velocity of the wavetrain. What we are after now is the velocity of a particle in the string through which this wave moves; this particle velocity is neither horizontal nor constant. In fact, each particle moves vertically, that is, in the y-direction. In order to determine the particle velocity, which we shall designate by the symbol u, let us fix our attention on a particle at a particular position x—that is, x is now a constant in this equation—and ask how the particle displacement y changes with time. With x constant we obtain

$$u=\frac{\partial y}{\partial t}=-y_m\omega\cos(kx-\omega t),$$

in which the *partial derivative* $\partial y/\partial t$ reminds us that although in general y is a function of both x and t, we here assume that x remains constant so that t becomes the only variable. The acceleration a of the particle at this (constant) value of x is

$$a = \frac{\partial^2 y}{\partial t^2} = \frac{\partial u}{\partial t} = -y_m \omega^2 \sin{(kx - \omega t)} = -\omega^2 y.$$

This shows that for each particle through which this transverse sinusoidal wave passes we have precisely SHM (simple harmonic motion), for the acceleration a is proportional to the displacement y, but oppositely directed.

For a particle at $x = 62$ cm with the wave of Example 1, in which

$$y_m = 0.25 \text{ cm}, \qquad k = 0.39 \text{ cm}^{-1}, \qquad \omega = 740 \text{ s}^{-1},$$

we obtain

$$u = -y_m v \cos{(kx - \omega t)}$$

or

$$u = -0.25 (740) \cos{[(0.39)(62) - (740)t]} = -185 \cos{(24 - 740\,t)}$$

and

$$a = -\omega^2 y$$

or

$$a = -(740)^2\, 0.25 \sin{[(0.39)(62) - (740)t]} = -13.7 \times 10^4 \sin{(24 - 740\,t)}$$

where t is expressed in seconds u in cm/s and a in cm/s². Can you describe the motion of this particle at the time $t = 4$ s?

In Fig. 19-7 we draw an element of the stretched string at some position x and at a particular time t. The *transverse* component of the tension in the string exerted by the element to the left of x on the element of the right of x is

$$F_{\text{trans}} = -F\frac{\partial y}{\partial x}.$$

F is the tension in the string; $\partial y / \partial x$ gives the tangent of the angle made by the direction of F with the horizontal at the time t in question and, because we assume small displacements, this can be taken equal also to the sine of the angle. The transverse force is in the direction of increasing y; in the figure the slope is negative, so the transverse force is positive. The transverse velocity of the particle at x is $\partial y / \partial t$, which may be positive or negative. The power being expended by the force at x, or the energy passing through the position x per unit time in the positive x-direction (see Section 7-6), is

$$P = F_{\text{trans}}\, u = \left(-F\frac{\partial y}{\partial x}\right)\frac{\partial y}{\partial t}.$$

Suppose that the wave on the string is the simple sine wave

$$y = y_m \sin{(kx - \omega t)}.$$

Then the magnitude of the slope at x is

$$\frac{\partial y}{\partial x} = ky_m \cos{(kx - \omega t)}, \qquad [t = \text{constant}]$$

and the transverse force is

$$-F\frac{\partial y}{\partial x} = -Fky_m \cos{(kx - \omega t)}.$$

The transverse velocity of a particle of the string at x is

$$u = \frac{\partial y}{\partial t} = -\omega y_m \cos{(kx - \omega t)}, \qquad [x = \text{constant}].$$

Hence, the power transmitted through x is

$$P = (-Fky_m)(-\omega y_m) \cos^2{(kx - \omega t)},$$

$$= y_m^2 k\omega F \cos^2{(kx - \omega t)}.$$

19-6
POWER AND INTENSITY IN WAVE MOTION

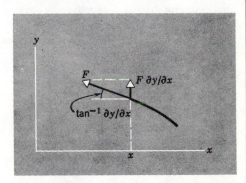

figure 19-7
The transverse component of the tension in the string at each point x is $F\,(\partial y/\partial x)$.

Notice that the power or rate of flow of energy is not constant. The power is not constant because the power input oscillates. As the energy is passed along the string, it is stored in each element of string as a combination of kinetic energy of motion and potential energy of deformation. The situation is much like that in an alternating current circuit; there energy is stored both in the inductor and in the capacitor and the power input also oscillates. For a string the power is absorbed by internal friction and viscous effects and appears as heat energy; in the circuit the power is expended in the resistor and appears as heat energy. The power input to the string or the circuit is often taken to be the *average* over one period of motion. The average power delivered is

$$\bar{P} = \frac{1}{T} \int_{t}^{t+T} P \, dt,$$

where T is the period. Using the fact that the average value of $\sin^2 \theta$ or $\cos^2 \theta$ over one cycle is $\frac{1}{2}$, we obtain for the string

$$\bar{P} = \frac{1}{2} y_m^2 k \omega F = 2\pi^2 y_m^2 \nu^2 \frac{F}{v},$$

a result which does not depend on x or t. For the string, however, $v = \sqrt{F/\mu}$, so that

$$\bar{P} = 2\pi^2 y_m^2 \nu^2 \mu v.$$

The fact that the rate of transfer of energy depends on the square of the wave amplitude and square of the wave frequency is true in general, holding for all types of waves.

Confirm that, if we had picked a wave traveling in the negative x-direction, we would have obtained the negative of this result. That is, the wave delivers power in the direction of wave propagation.

In a three-dimensional wave, such as a light wave or a sound wave from a point source, it is more significant to speak of the *intensity* of the wave. Intensity is defined as the power transmitted across a unit area normal to the direction in which the wave is traveling. Just as with power in the wave in a string, the intensity of a space wave is always proportional to the square of the amplitude.

As a wave progresses through space, its energy may be absorbed. For example, in a viscous medium, such as syrup or lead, mechanical waves would rapidly decay in amplitude and disappear, owing to absorption of energy by internal friction. In most cases of interest to us, however, absorption will be negligible. Throughout this chapter we have assumed that there is no loss of energy in a given wave, no matter how far it travels.

Spherical waves travel from a source of waves whose power output, assumed constant, is P; see Fig. 19-8. Find how the wave intensity depends on the distance from the source. We assume that the medium is isotropic and that the

EXAMPLE 3

figure 19-8
Example 3.

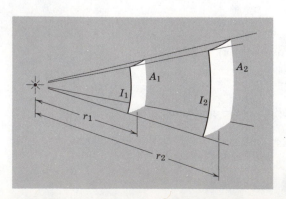

source radiates uniformly in all directions, that is, that its emission is spherically symmetrical.

The intensity of a three-dimensional wave is the power transmitted across a unit area normal to the direction of propagation. As the wavefront expands from a distance r_1 from the source at the center to a distance r_2, its surface area increases from $4\pi r_1^2$ to $4\pi r_2^2$. If there is no absorption of energy, the total energy transported per second by the wave remains constant at the value P, so that

$$P = 4\pi r_1^2 I_1 = 4\pi r_2^2 I_2,$$

where I_1 and I_2 are the wave intensities at r_1 and r_2 respectively. Hence,

$$\frac{I_1}{I_2} = \frac{r_2^2}{r_1^2}$$

and the wave intensity varies inversely as the square of its distance from the source. Since the intensity is proportional to the square of the amplitude, the amplitude of the wave must vary inversely as the distance from the source.

Interference refers to the physical effects of superimposing two or more wavetrains. Let us consider two waves of equal frequency and amplitude traveling with the same speed in the same direction $(+x)$ but with a phase difference ϕ between them. The equations of the two waves will be

$$y_1 = y_m \sin (kx - \omega t - \phi) \tag{19-14}$$

and

$$y_2 = y_m \sin (kx - \omega t). \tag{19-15}$$

We can rewrite the first equation in two equivalent forms

$$y_1 = y_m \sin \left[k\left(x - \frac{\phi}{k}\right) - \omega t \right] \tag{19-14a}$$

or

$$y_1 = y_m \sin \left[kx - \omega\left(t + \frac{\phi}{\omega}\right) \right]. \tag{19-14b}$$

Equations 19-14a and 19-15 suggest that if we take a "snapshot" of the two waves at any time t, we will find them displaced from one another along the x-axis by the constant distance ϕ/k. Equations 19-14b and 19-15 suggest that if we station ourselves at any position x, the two waves will give rise to two simple harmonic motions having a constant time difference ϕ/ω. This gives some insight into the meaning of the phase difference ϕ.

Now let us find the resultant wave, which, on the assumption that superposition occurs, is the sum of Eqs. 19-14 and 19-15 or

$$y = y_1 + y_2 = y_m[\sin (kx - \omega t - \phi) + \sin (kx - \omega t)].$$

From the trigonometric equation for the sum of the sines of two angles

$$\sin B + \sin C = 2 \sin \tfrac{1}{2}(B + C) \cos \tfrac{1}{2}(C - B), \tag{19-16}$$

we obtain

$$y = y_m \left[2 \sin\left(kx - \omega t - \frac{\phi}{2}\right)\cos \frac{\phi}{2} \right],$$

$$= \left(2y_m \cos \frac{\phi}{2}\right)\sin\left(kx - \omega t - \frac{\phi}{2}\right). \tag{19-17}$$

This resultant wave corresponds to a new wave having the same frequency but with an amplitude $2y_m \cos (\phi/2)$. If ϕ is *very small* (compared to 180°), the resultant amplitude will be nearly $2y_m$. That is, when

19-7
INTERFERENCE OF WAVES

ϕ is very small, $\cos(\phi/2) \cong \cos 0° = 1$. When ϕ is *zero*, the two waves have the same phase everywhere. The crest of one corresponds to the crest of the other and likewise for the troughs. The waves are then said to interfere constructively. The resultant amplitude is just twice that of either wave alone. If ϕ is near 180°, on the other hand, the resultant amplitude will be nearly zero. That is, when $\phi \cong 180°$, $\cos(\phi/2) \cong \cos 90° = 0$. When ϕ is *exactly* 180°, the crest of one wave corresponds exactly to the trough of the other. The waves are then said to interfere destructively. The resultant amplitude is zero.

In Fig. 19-9*a* we show the superposition of two wavetrains almost in phase (ϕ small) and in Fig. 19-9*b* the superposition of two wavetrains almost 180° out of phase ($\phi \cong 180°$). Notice that in these figures the algebraic sum of the ordinates of the thin (component) curves at any value of x equals the ordinate of the thick (resultant) curve. The sum of two waves can, therefore, have different values, depending on their phase relations.

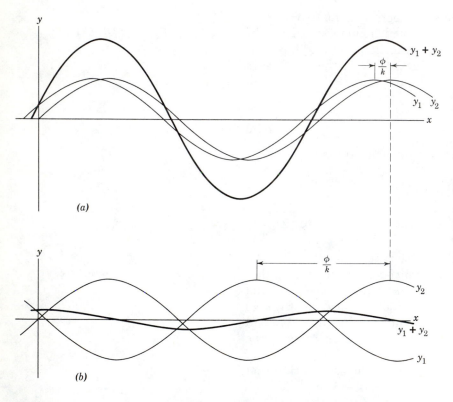

(a)

(b)

figure 19-9

(a) The superposition of two waves of equal frequency and amplitude that are almost in phase results in a wave of almost twice the amplitude of either component. *(b)* The superposition of two waves of equal frequency and amplitude and almost 180° out of phase results in a wave whose amplitude is nearly zero. Note that in both the resultant frequency is unchanged. (The drawings correspond to the instant $t = 0$.)

The resultant wave will be a sine wave, even when the amplitudes of the component sine waves are unequal. Figure 19-10, for example, illustrates the addition of two sine waves of the same frequency and velocity but different amplitudes. The resultant amplitude depends on the phase difference, which is taken as zero in this figure. The result for other phase differences could be obtained by shifting one of the component waves sideways with respect to the other and would give a smaller resultant amplitude. The smallest resultant amplitude would be the difference in the amplitudes of the components, obtained when the phases differ by 180°. However, the resultant is always a sine wave. The addition of any number of sine waves having the same frequency and velocity gives a similar result. The resultant waveform will always have a constant amplitude because the component waves (and their

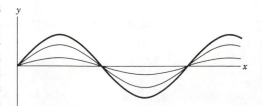

figure 19-10

The addition of two waves of the same frequency and phase but differing amplitudes (light lines) yields a third wave of the same frequency and phase (heavy line).

resultant) all move with the same velocity and maintain the same relative position. The actual state of affairs can be pictured by having all the waves in Figs. 19-9 and 19-10 move toward the right with the same speed.

In practice, interference effects are obtained from wavetrains which originate in the same source (or in sources having a fixed phase relationship to one another) but which follow different paths to the point of interference. The phase difference ϕ between the waves arriving at a point can be calculated by finding the difference between the paths traversed by them from the source to the point of interference. The path difference is ϕ/k or $(\phi/2\pi)\lambda$. When the path difference is 0, λ, 2λ, 3λ, etc., so that $\phi = 0$, 2π, 4π, etc., the two waves interfere constructively. For path differences of $\frac{1}{2}\lambda$, $\frac{3}{2}\lambda$, $\frac{5}{2}\lambda$, etc., ϕ is π, 3π, 5π, etc., and the waves interfere destructively. We shall return to these matters later in more detail.

19-8
COMPLEX WAVES

The waves we have considered thus far have been of the simple harmonic type, in which the displacements at any time are represented by a sine curve. We have seen that superposition of any number of such waves having the same frequency and velocity, but arbitrary amplitudes and phases, still gives rise to a resultant wave of this simple type. If, however, we superimpose waves that have *different frequencies*, the resulting wave is *complex*. In a complex wave the motion of a particle is no longer simple harmonic motion, and the wave shape is no longer a sine curve. In this section we consider only the qualitative aspects of complex waves. The analytical treatment of such waves will be given when we encounter physical situations described by them. We will look at the results of adding graphically two or more waves traveling with the same speed in the same direction but having various relative frequencies, amplitudes, and phases.

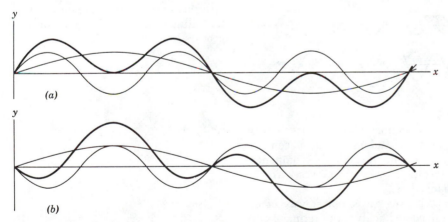

figure 19-11
The addition of two waves with a frequency ratio 3:1 (light lines) yields a wave whose shape (heavy line) depends on the phase relationship of the components. Compare *(a)* and *(b)*.

In Figs. 19-11*a* and 19-11*b* we add two waves having the same amplitude but having frequencies in the ratio 3 to 1; the phase relation is changed from *a* to *b* and we see how changing the phase relation may produce a resultant of very different form. If these represent sound waves, our eardrums will vibrate in a way represented by the resultant in each case, but we will hear and interpret these as the two original frequencies, regardless of their phase relation. If the resultant waves represent visible light, our eyes will receive the same sensation of a mixture of two colors, regardless of the phase relation of the components.

In Fig. 19-12 three waves of different frequencies and amplitudes are added. The resultant complex wave is quite different from a simple periodic wave and, in this respect resembles waveforms normally generated by musical instru-

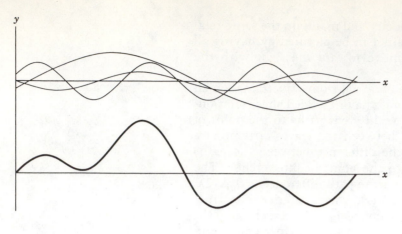

figure 19-12
The addition of three waves (top)
of differing frequencies yields a
complex waveform (bottom).

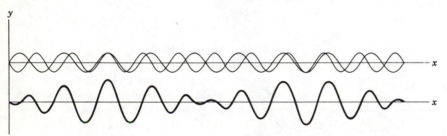

figure 19-13
The addition (heavy line) of two
waves of widely differing frequency
(light lines).

ments. In Fig. 19-13 a wave of very high frequency is added to one of very low frequency. Each component frequency is clearly discernible in the resultant. In Fig. 19-14 two waves of nearly the same frequency are added. The resultant wave consists of groups which, in the case of sound, produce the familiar phenomenon of beats (Section 20-6).

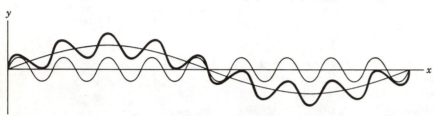

figure 19-14
The addition (bottom) of two waves
with nearly the same frequency
(top), illustrating the phenomenon
of beats (see Chapter 20).

In all of these figures the resultant wave is obtained under the assumption that the principle of superposition holds, by simply adding the displacements caused by the individual waves at every point. Because all the component waves travel with the same velocity, the resultant waveform moves with this same velocity and the wave shape is unchanged.

The cathode-ray oscilloscope (Chapter 27) gives the simplest way of observing how complex waves can be synthesized and analyzed in terms of simple harmonic waves.

19-9 STANDING WAVES

In a one-dimensional body of finite size, such as a taut string held by two clamps a distance *l* apart, traveling waves in the string are reflected from the boundaries of the body, that is, from the clamps. Each such reflection gives rise to a wave traveling in the string in the opposite direction. The reflected waves add to the incident waves according to the principle of superposition.

Consider two wavetrains of the same frequency, speed, and amplitude which are traveling in *opposite directions* along a string. Two such waves may be represented by the equations

$$y_1 = y_m \sin (kx - \omega t),$$

$$y_2 = y_m \sin (kx + \omega t).$$

Hence, the resultant may be written as

$$y = y_1 + y_2 = y_m \sin (kx - \omega t) + y_m \sin (kx + \omega t) \qquad (19\text{-}18a)$$

or, making use of the trigonometric relation of Eq. 19-16, as

$$y = 2y_m \sin kx \cos \omega t. \qquad (19\text{-}18b)$$

Equation 19-18b is the equation of a *standing* wave.* Notice that a particle at any particular point x executes simple harmonic motion as time goes on, and that all particles vibrate with the same frequency. In a traveling wave each particle of the string vibrates with the same amplitude. *Characteristic of a standing wave, however, is the fact that the amplitude is not the same for different particles but varies with the location x of the particle.*† In fact, the amplitude, $2y_m \sin kx$, has a *maximum* value of $2y_m$ at positions where

$$kx = \frac{\pi}{2}, \frac{3\pi}{2}, \frac{5\pi}{2}, \text{ etc.}$$

or

$$x = \frac{\lambda}{4}, \frac{3\lambda}{4}, \frac{5\lambda}{4}, \text{ etc.}$$

These points are called *antinodes* and are spaced one-half wavelength apart. The amplitude has a *minimum* value of zero at positions where

$$kx = \pi, 2\pi, 3\pi, \text{ etc.}$$

or

$$x = \frac{\lambda}{2}, \lambda, \frac{3\lambda}{2}, 2\lambda, \text{ etc.}$$

These points are called *nodes* and are spaced one-half wavelength apart. The separation between a node and an adjacent antinode is one-quarter wavelength.

It is clear that energy is not transported along the string to the right or to the left, for energy cannot flow past the nodal points in the string which are permanently at rest. Hence, the energy remains "standing" in the string, although it alternates between vibrational kinetic energy and elastic potential energy. We call the motion a wave motion because we can think of it as a superposition of waves traveling in opposite directions (Eq. 19-18a). We can equally well regard the motion as an oscillation of the string as a whole (Eq. 19-18b), each particle oscillating with SHM of angular frequency ω and with an amplitude that depends on its location. Each small part of the string has inertia and elasticity, and the string as a whole can be thought of as a collection of coupled oscillators. Hence, the vibrating string is the same in principle‡ as a spring-mass system, except that a spring-mass system has only one natural frequency, and a vibrating string has a large number of natural frequencies (Section 19-10).

---

* Standing waves may also be produced in finite bodies of two or three dimensions; see Chapters 20 and 38 for examples.

† The combining waves moving in opposite directions along the string will still produce standing waves even if their amplitudes are unequal. We consider only the equal-amplitude case here; see Problem 29, however.

‡ For a general discussion see "On the Teaching of 'Standing Waves,'" J. Rekveld, *American Journal of Physics*, March 1958.

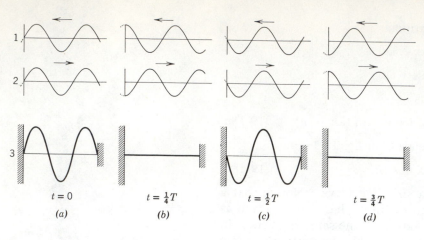

figure 19-15
Standing waves as the superposition of left- and right-going waves; 1 and 2 are the components, 3 the resultant.

In Fig. 19-15, in (a), (b), (c), and (d), we show a standing wave pattern separately at intervals of one-quarter of a period in the lower figures, 3. The traveling waves, one moving in the positive x-direction and the other moving in the negative x-direction, whose superposition can be considered to give rise to the standing wave, are shown for the same quarter-period intervals in the upper figures 2 and 1. Standing waves can also be produced with electromagnetic waves and with sound waves.

In Fig. 19-16 we show how the energy associated with the oscillating string shifts back and forth between kinetic energy of motion K and potential energy of deformation U during one cycle. Compare this with Fig. 8-4, which shows the same thing for a mass-spring oscillator. Oscillating strings often vibrate so rapidly that the eye perceives only a blur whose shape is that of the envelope of the motion; see Fig. 19-17.

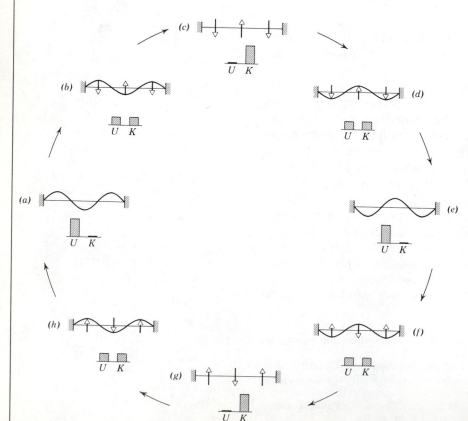

figure 19-16
A standing wave in a stretched string, showing one cycle of oscillation. At (a) the string is momentarily at rest and the energy of the system is all potential energy of elastic deformation associated with the transverse displacement of the string. (b) An eighth-cycle later the displacement is reduced and the string is in motion. The two arrows show the velocities of the string particles at the positions shown. K and U have the same value. (c) The string is not displaced, but its particles have their maximum speeds; the energy is all kinetic. The motion continues until the initial condition (a) is reached after which the cycle continues to repeat itself.

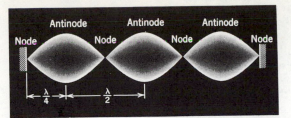

figure 19-17
The envelope of a standing wave, corresponding to a time exposure of the motion, and showing the patterns of nodes and antinodes.

The superposition of an incident wave and a reflected wave, being the sum of two waves traveling in opposite directions, will give rise to a standing wave. We shall now consider the process of reflection of a wave more closely. Suppose a pulse travels down a stretched string which is fixed at one end, as shown in Fig. 19-18a. When the pulse arrives at that end, it exerts an upward force on the support. The support is rigid, however, and does not move. By Newton's third law the support exerts an equal but oppositely directed force on the string. This reaction force generates a pulse at the support, which travels back along the string in a direction opposite to that of the incident pulse. We say that the incident pulse has been *reflected* at the fixed endpoint of the string. Notice that the reflected pulse returns with its transverse displacement reversed. If a wavetrain is incident on the fixed endpoint, a reflected wavetrain is generated at that point in the same way. The displacement of any point along the string is the sum of the displacements caused by the incident and reflected wave. Since the endpoint is fixed, these two waves must always interfere destructively at that point so as to give zero displacement there. Hence, the reflected wave is always 180° out of phase with the incident wave at a fixed boundary. We say that *on reflection from a fixed end a wave undergoes a phase change of 180°.*

Let us now consider the reflection of a pulse at a free end of a stretched string, that is, at an end that is free to move transversely. This can be achieved by attaching the end to a very light ring free to slide without friction along a transverse rod, or (see later) to a long and very much lighter string. When the pulse arrives at the free end, it exerts a force on the element of string there. This element is accelerated and its inertia carries it past the equilibrium point; it "overshoots" and exerts a reaction force on the string. This generates a pulse which travels back along the string in a direction opposite to that of the incident pulse. Once again we get reflection, but now at a free end. The free end will obviously suffer the maximum displacement of the particles on the string; an incident and a reflected wavetrain must interfere constructively at that point if we are to have a maximum there. Hence, the reflected wave is always in phase with the incident wave at that point (see Fig. 19-18b). We say that *at a free end a wave is reflected without change of phase.*

Hence, when we have a standing wave in a string, there will be a node at a fixed end (Fig. 19-18a) and an antinode at a free end (Fig. 19-18b). These ideas will be applied to sound waves and electromagnetic waves in subsequent chapters.

In the treatment just given we have assumed that there is total reflection at the boundary. In general, at a boundary there is partial reflection and partial transmission. For example, suppose that instead of being attached to a rigid wall the string is attached to another string. At the boundary joining the strings the incident wave will be partly reflected and partly transmitted. The amplitude

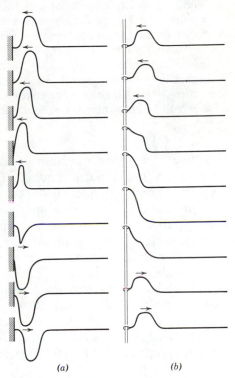

(a)   (b)

figure 19-18
(a) Reflection of a pulse at the fixed end of a string. The drawings are spaced uniformly in time. The phase is changed by 180° on reflection. *(b)* Reflection of a pulse at an end free to move in a transverse direction. (The string is attached to a ring which slides vertically without friction.) The phase is unchanged on reflection.

of the reflected wave will be less than that of the incident wave because a transmitted wave continues along the second string and carries away some of the incident energy. If the second string has a greater linear density than the first, the wave reflected back into the first will still suffer a phase shift of 180° on reflection. But because its amplitude is less than the incident wave, the boundary point will not be a node and will move. Thus a net energy transfer occurs along the first string into the second. If the second string has a smaller linear density than the first, partial reflection occurs without change of phase, but once again energy is transmitted to the second string. In practice the best way to realize a "free end" for a string is to attach it to a long and very much lighter string. The energy transmitted is negligible, and the second string serves to maintain the tension in the first one.

It is of interest to note that the transmitted wave travels with a different speed than the incident and reflected waves. The wave speed is determined by the relation $v = \sqrt{F/\mu}$; the tension is the same in both strings, but their densities are different. Hence, the wave travels more slowly in the denser string. The frequency of the transmitted wave is the same as that of the incident and reflected waves. Waves having the same frequency but traveling with different speeds have different wavelengths. Hence, from the relation $\lambda = v/\nu$ we conclude that in the denser string, where v is less, the wavelength is shorter. This phenomenon of change of wavelength as a wave passes from one medium to another will be encountered frequently in our study of light waves.

19-10 RESONANCE

In general, whenever a system capable of oscillating is acted on by a periodic series of impulses having a frequency equal or nearly equal to one of the natural frequencies of oscillation of the system, the system is set into oscillation with a relatively large amplitude. This phenomenon is called *resonance* (see Section 15-10) and the system is said to resonate with the applied impulses.

Consider a string fixed at both ends. Oscillations or standing waves can be established in the string. The only requirement we have to satisfy is that the endpoints be nodes. There may be any number of nodes in between or none at all, so that the wavelength associated with the standing waves can take on many different values. The distance between adjacent nodes is $\lambda/2$, so that in a string of length l there must be exactly an integral number n of half wavelengths, $\lambda/2$. That is,

$$\frac{n\lambda}{2} = l$$

or

$$\lambda = \frac{2l}{n}, \qquad n = 1, 2, 3, \cdots.$$

But $\lambda = v/\nu$ and $v = \sqrt{F/\mu}$, so that the natural frequencies of oscillation of the system are

$$\nu = \frac{n}{2l} \sqrt{\frac{F}{\mu}}, \qquad n = 1, 2, 3, \cdots. \qquad (19-19)$$

If the string is set vibrating and left to itself, the oscillations gradually die out. The motion is damped by dissipation of energy through the elastic supports at the ends and by the resistance of the air to the motion. We can pump energy into the system by applying a driving force. If the driving frequency is near that of any natural frequency of the string, the string will vibrate at that frequency with a large amplitude. Because the string has a large number of natural frequencies, resonance can occur at many different frequencies. A mass-spring system, by contrast, has only one resonant frequency. The difference is associated with the fact that in the mass-spring system the inertia characteristic is concentrated ("lumped") in one part of the system—the mass—and the elastic characteristic is concentrated in a separate part of the system—the spring. We say that this system has *lumped elements*.

A stretched string, on the other hand, is said to have *distributed elements* because every element of the string has both inertia and elastic characteristics. In the mass-spring system, there is only one way to exchange energy between kinetic and potential forms as the system oscillates; energy in kinetic form must be associated with the moving mass and energy in potential form must be associated with the deformed spring. In the stretched string, however, masslike (inertia) and springlike (elasticity) elements are uniformly distributed along the string. There are many possible ways, rather than a single way, of exchanging energy between kinetic and potential forms as the system oscillates, corresponding to the sequence of allowed values for *n* in Eq. 19-19.

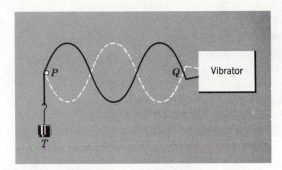

figure 19-19
Standing waves in a driven string when the natural and driving frequencies are very nearly equal.

Resonance in a string is often demonstrated by attaching a string to a fixed end, by means of a weight attached to it over a pulley, and connecting the other end to a vibrator, as shown in Fig. 19-19. The transverse oscillations of the vibrator set up a traveling wave in the string which is reflected back from the fixed end. The frequency of the waves is that of the vibrator, and the wavelength is determined by $\lambda = v/\nu$. The fixed end P is a node, but the end Q vibrates and is not. If we now vary the tension in the string by changing the hanging weight, for example, we can change the wavelength. Changing the tension changes the wave velocity, and the wavelength changes in proportion to the velocity, the frequency being constant. Whenever the wavelength becomes nearly equal to $2l/n$, where l is the length of the string, we obtain standing waves of great amplitude. The string now vibrates in one of its natural modes and resonates with the vibrator. The vibrator does work on the string to maintain these oscillations against the losses due to damping. The amplitude builds up only to the point at which the vibrator expends all its energy input against damping losses. The point Q is almost a node because the amplitude of the vibrator is small compared to that of the string.

Hence, with damping, the resonant frequency is almost, but not quite, a natural frequency of the string. One endpoint is a node, the other almost a node. In between there are points that are almost nodes, points at which the amplitude is very small. These points cannot be true nodes, for energy must flow along the string past them from the vibrator. This situation is analogous to the resonance condition for a damped harmonic oscillator with driving force, discussed in Section 15-10. There, too, the resonant frequency was almost the same as the natural frequency of the system, and the amplitude was large but not infinite.

If no damping were present, the resonant frequency would be exactly a natural frequency. Then the amplitude would build up to infinity as the energy is pumped in. In practice, the system would cease to obey Hooke's law, or the small-oscillations condition, as the amplitude becomes large and the system would break. This happens even with damping, when the damping is small or the driving force is large (as in the Tacoma Bridge disaster, Fig. 15-21).

If the frequency of the vibrator is much different from a natural frequency of the system, as given by Eq. 19-19, the wave reflected at P on returning to Q may be much out of phase with the vibrator, and it can do work on the vibrator. That is, the string can give up some energy to the vibrator as well as receive energy from it. The "standing" wave pattern is not fixed in form but wiggles about. On the average the amplitude is small and not much different from that of the vibrator. This situation is analogous to the erratic motion of a swing being pushed periodically with a frequency other than its natural one. The displacement of the swing is rather small.

Hence, the string absorbs peak energy from the vibrator at resonance. Tuning a radio is an analogous process. By tuning a dial the natural frequency of an alternating current in the receiving circuit is made equal to the frequency of the waves broadcast by the station desired. The circuit resonates with the transmitted signals and absorbs peak energy from the signal. We shall encounter resonance conditions again in sound, in electromagnetism, in optics, and in atomic and nuclear physics. In these areas, as in mechanics, the system will absorb peak energy from the source at resonance and relatively little energy off resonance.

EXAMPLE 4

In a demonstration with the apparatus just described, the vibrator has a frequency $\nu = 20$ Hz and the string has a linear density $\mu = 1.56 \times 10^{-4}$ slug/ft $(= 7.47 \times 10^{-3}$ kg/m) and a length $l = 24$ ft $(= 7.3$ m). The tension F is varied by pulling down on the end of the string over the pulley. If the demonstrator wants to show resonance, starting with one loop and then with two, three, and four loops, what force must he exert on the string?

At resonance,

$$\nu = \frac{n}{2l} \sqrt{\frac{F}{\mu}}.$$

Hence, the tension F is given by

$$F = \frac{4l^2 \nu^2 \mu}{n^2}.$$

For one loop, $n = 1$, so that

$$F_1 = 4l^2 \nu^2 \mu = 4(24 \text{ ft})^2 (20 \text{ s}^{-1})^2 (1.56 \times 10^{-4} \text{ slug/ft}) = 144 \text{ lb} (= 640 \text{ N}).$$

For two loops, $n = 2$, and

$$F_2 = \frac{4l^2 \nu^2 \mu}{4} = \frac{F_1}{4} = 36 \text{ lb} (= 160 \text{ N}).$$

Likewise, for three and four loops

$$F_3 = \frac{F_1}{(3)^2} = 16 \text{ lb} (= 71 \text{ N}),$$

$$F_4 = \frac{F_1}{(4)^2} = 9 \text{ lb} (= 40 \text{ N}).$$

Hence, the demonstrator gradually relaxes the tension to obtain resonance with an increasing number of loops. Although the resonant frequency is always the same under these circumstances, the speed of propagation and the wavelength at resonance decrease proportionately.

Taking damping into account, are the tensions given exactly correct?

If the tension were kept fixed, giving a definite wave speed, would we obtain more than one resonance condition by varying the frequency of the vibrator?

questions

1. How could you prove experimentally that energy is associated with a wave?

2. Energy can be transferred by particles as well as by waves. How can we experimentally distinguish between these methods of energy transfer?

3. Can a wave motion be generated in which the particles of the medium vibrate with angular simple harmonic motion? If so, explain how and describe the wave.

4. Are torsional waves transverse or longitudinal? Can they be considered as a superposition of two waves, which are either transverse or longitudinal?

5. How can one create plane waves? Spherical waves?

6. The following functions in which A is a constant are of the form $f(x \pm vt)$:

$$y = A(x - vt), \qquad y = A(x + vt)^2,$$
$$y = A\sqrt{x - vt}, \qquad y = A\ln(x + vt).$$

Explain why these functions are not useful in wave motion.

7. Can one produce on a string a wave form which has a discontinuity in slope at a point, that is, it has a sharp corner? Explain.

8. How do the amplitude and the intensity of surface water waves vary with the distance from the source?

9. The inverse square law does not apply exactly to the decrease in intensity of sounds with distance. Why not?

10. When two waves interfere, does one alter the progress of the other?

11. When waves interfere, is there a loss of energy? Explain your answer.

12. Why don't we observe interference effects between the light beams emitted from two flashlights or between the sound waves emitted by two violins.

13. As Fig. 19-15 shows, twice during a cycle the configuration of standing waves in a stretched string is a straight line, exactly what it would be if the string were not vibrating at all. Discuss from the point of view of energy conservation.

14. If two waves differ only in amplitude and are propagated in opposite directions through a medium, will they produce standing waves? Is energy transported? Are there any nodes? (See Problem 29.)

15. The partial reflection of wave energy by discontinuities in the path of transmission is usually wasteful and can be minimized by insertion of "impedance matching" devices between the sections of the path bordering on the discontinuity. For example, a megaphone helps match the air column of mouth and throat to the air outside the mouth. Give other examples and explain qualitatively how such devices minimize reflection losses (see Problem 29).

16. Consider the standing waves in a string to be a superposition of traveling waves and explain, using superposition ideas, why there are no true nodes in the resonating string of Fig. 19-19, even at the "fixed" end. (*Hint:* Consider damping effects.)

17. Standing waves in a string are demonstrated by an arrangement such as that of Fig. 19-19. The string is illuminated by a fluorescent light and the vibrator is driven by the same electric outlet that powers the light. The string exhibits a curious color variation in the transverse direction. Explain.

18. In the discussion of transverse waves in a string we have dealt only with displacements in a single plane, the x-y plane. If all displacements lie in one plane, the wave is said to be *plane polarized.* Can there be displacements in a plane other than the single plane dealt with? If so, can two differently plane-polarized waves be combined? What appearance would such a combined wave have?

19. A wave transmits energy. Does it transfer momentum? Can it transfer angular momentum? (See Question 18.) (See "Energy and Momentum Transport in String Waves" by D. W. Juenker, *American Journal of Physics*, January 1976.)

problems

SECTION 19-3

1. The speed of electromagnetic waves in vacuum is 3.0×10^8 m/s. (a) Wavelengths in the visible part of the spectrum (light) range from about 4.0×10^{-7} m in the violet to about 7.0×10^{-7} m in the red. What is the range of frequencies of light waves? (b) The range of frequencies for shortwave radio (for example, FM radio and VHF television) is 1.5 MHz (megahertz; see Table 2, Chapter 1) to 300 MHz. What is the corresponding wavelength range? (c) X-rays are also electromagnetic. Their wavelength range extends from about 5.0 nm (nanometer; see Table 2, Chapter 1) to about 1.0×10^{-2} nm. What is the frequency range for X-rays?
   Answer: (a) 400 THz (THz = terahertz; see Table 2, Chapter 1) to 800 THz.
   (b) 1.0 m to 200 m. (c) 6.0×10^4 THz to 3.0×10^7 THz.

2. Show that $y = y_m \sin (kx - \omega t)$ may be written in the alternative forms

$$y = y_m \sin k (x - vt), \qquad y = y_m \sin 2\pi \left(\frac{x}{\lambda} - \nu t\right),$$

$$y = y_m \sin \omega \left(\frac{x}{v} - t\right), \qquad y = y_m \sin 2\pi \left(\frac{x}{\lambda} - \frac{t}{T}\right).$$

3. The equation of a transverse wave traveling along a very long string is given by $y = 6.0 \sin (0.020\pi x + 4.0\pi t)$, where x and y are expressed in cm and t in seconds. Calculate (a) the amplitude, (b) the wavelength, (c) the frequency, (d) the speed, (e) the direction of propagation of the wave, and (f) the maximum transverse speed of a particle in the string.
   Answer: (a) 6.0 cm. (b) 100 cm. (c) 2.0 Hz. (d) 200 cm/s. (e) negative x-direction. (f) 75 cm/s.

4. A sinusoidal wave travels along a string. If the time for a particular point to move from maximum displacement to zero displacement is 0.17 s, what are (a) the period, and (b) frequency? (c) If the wavelength is 1.4 m what is the speed of the wave?

5. A wave of frequency 500 Hz has a velocity of 350 m/s. (a) How far apart are two points 60° out of phase? (b) What is the phase difference between two displacements at a certain point at times 10^{-3} s apart?
   Answer: (a) 12 cm. (b) 180°.

6. Write the equation for a wave traveling in the negative direction along the x-axis and having an amplitude 0.010 m, a frequency 550 Hz, and a speed 330 m/s.

7. (a) A continuous sinusoidal longitudinal wave is sent along a coiled spring from a vibrating source attached to it. The frequency of the source is 25 Hz, and the distance between successive rarefactions in the spring is 24 cm. Find the wave speed. (b) If the maximum longitudinal displacement of a particle in the spring is 0.30 cm and the wave moves in the $-x$-direction, write the equation for the wave. Let the source be at $x = 0$ and the displacement $x = 0$ when $t = 0$ be zero.
   Answer: (a) 600 cm/s. (b) $y = 0.30 \sin (0.26x + 160t)$, with x and y in cm and t in seconds.

SECTION 19-5

8. What is the speed of a transverse wave in a rope of length 2.0 m (6.6 ft) and mass 0.060 kg (0.0041 slug) under a tension of 500 N (110 lb)?

9. The linear density of a vibrating string is 1.3×10^{-4} kg/m. A transverse wave is propagating on the string and is described by the equation $y = 0.021 \sin (x + 30t)$, where x and y are measured in meters and t in seconds. What is the tension in the string?   *Answer:* 0.12 N.

10. A continuous sinusoidal wave is traveling on a string with velocity 80 cm/s. The displacement of the particles of the string at $x = 10$ cm is found to vary with time according to the equation $y = 5.0 \sin (1.0 - 4.0t)$ in cm. The linear density of the string is 4.0 g/cm. (a) What is the frequency of the wave? (b) What is the wavelength of the wave? (c) Write the general equation giving the transverse displacement of the particles of the string as a function of position and time. (d) Calculate the tension in the string.

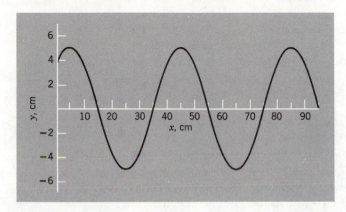

figure 19-20
Problem 11

11. A simple harmonic transverse wave is propagating along a string toward the left (or $-x$) direction. Figure 19-20 shows a plot of the displacement as a function of position at time $t = 0$. The string tension is 3.6 N and its linear density is 25 g/m. Calculate (a) the amplitude, (b) the wavelength, (c) the wave speed, (d) the period, and (e) the maximum speed of a particle in the string. (f) Write an equation describing the traveling wave.
Answer: (a) 5.0 cm. (b) 40 cm. (c) 12 m/s. (d) 0.033 s. (e) 9.4 m/s. (f) 5.0 sin $(0.16x + 190t + 0.93)$, with x and y in cm and t in seconds.

12. Prove that the slope of a string at any point x is numerically equal to the ratio of the particle speed to the wave speed at that point.

13. A uniform circular hoop of string is rotating clockwise in the absence of gravity (see Fig. 19-21). The tangential speed is v_0. Find the speed of waves traveling on this string. (*Remark:* The answer is independent of the radius of the circle and the mass per unit length of the string!)   *Answer:* v_0.

14. A uniform rope of mass m and length L hangs from a ceiling. (a) Show that the speed of a transverse wave in the rope is a function of y, the distance from the lower end, and is given by $v = \sqrt{gy}$. (b) Show that the time it takes a transverse wave to travel the length of the rope is given by $t = 2\sqrt{L/g}$. (c) Does the actual mass of the rope affect the results of (a) and (b)?

figure 19-21
Problem 13

SECTION 19-6

15. Spherical waves are emitted from a 1.0-watt source in an isotropic non-absorbing medium. What is the wave intensity 1.0 m from the source?
Answer: 0.080 W/m².

16. (a) Show that the intensity I (the energy crossing unit area per unit time) is the product of the energy per unit volume u and the speed of propagation v of a wave disturbance. (b) Radio waves travel at a speed of 3.0×10^8 m/s $(9.8 \times 10^8$ ft/s). Find the energy density in a radio wave 480 km (300 mi) from a 50,000-W (67-hp) source, assuming the waves to be spherical and the propagation to be isotropic.

17. A line source emits a cylindrical expanding wave. Assuming the medium absorbs no energy, find how (a) the intensity and (b) the amplitude of the wave depend on the distance from the source.
    Answer: (a) Proportional to r^{-1}. (b) Proportional to $r^{-1/2}$.

18. (a) From Example 2 show that the maximum speed of a particle in a string through which a sinusoidal wave is passing is $u = y_m \omega$. (b) In Example 2 we saw that the particles in the string oscillate with simple harmonic motion. The mechanical energy of each particle is the sum of its potential and kinetic energies and is always equal to the maximum value of its kinetic energy. Consider an element of string of mass $\mu \Delta x$ and show that the energy per unit length of the string is given by

$$E_1 = 2\pi^2 \mu \nu^2 y_m^2.$$

    (c) Show finally that the average power or average rate of transfer of energy is the product of the energy per unit length and the wave speed. (d) Do these results hold only for a sinusoidal wave?

19. A wave travels out uniformly in all directions from a point source. (a) Justify the following expression for the displacement y of the medium at any distance r from the source:

$$y = \frac{Y}{r} \sin k(r - vt).$$

    Consider the speed, direction of propagation, periodicity, and intensity of the wave. (b) What are the dimensions of the constant Y?
    Answer: (b) L^2.

SECTION 19-7

20. Determine the amplitude of the resultant motion when two sinusoidal motions having the same frequency and traveling in the same direction are combined, if their amplitudes are 3.0 cm and 4.0 cm and they differ in phase by $\pi/2$ rad.

21. A source S and a detector D of high-frequency waves are a distance d apart on the ground. The direct wave from S is found to be in phase at D with the wave from S that is reflected from a horizontal layer at an altitude H (Fig. 19-22). The incident and reflected rays make the same angle with the reflecting layer. When the layer rises a distance h, no signal is detected at D. Neglect absorption in the atmosphere and find the relation between d, h, H, and the wavelength λ of the waves.

    Answer: $\lambda = 2\sqrt{4(H + h)^2 + d^2} - 2\sqrt{4H^2 + d^2}$.

22. Four component sine waves have frequencies in the ratio 1, 2, 3, and 4 and amplitudes in the ratio 1, $\frac{1}{2}$, $\frac{1}{3}$, and $\frac{1}{4}$, respectively. When $t = 0$, at $x = 0$, the first and third components are 180° out of phase with the second and fourth components. Plot the resultant waveform and discuss its nature.

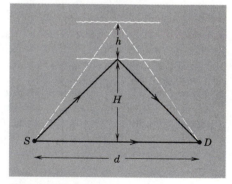

figure 19-22
Problem 21

figure 19-23
Problem 23

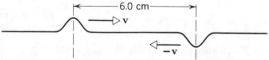
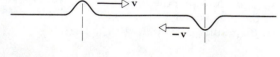

23. Two pulses are traveling along a string in opposite directions, as shown in Fig. 19-23. (a) If the wave velocity is 2.0 m/s and the pulses are 6.0 cm apart, sketch the patterns after 5.0, 10, 15, 20, 25 ms. (b) What has happened to the energy at $t = 15$ ms?
    Answer: (b) Even though the displacement is zero at this instant, the transverse velocities are not. The energy is all kinetic.

24. Three component sinusoidal waves have the same period, but their amplitudes are in the ratio 1, $\frac{1}{2}$, and $\frac{1}{3}$ and their phase angles are 0, $\pi/2$, and π respectively. Plot the resultant waveform and discuss its nature.

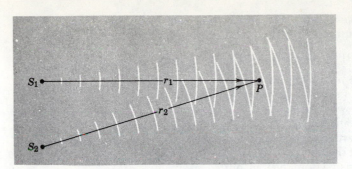

figure 19-24
Problem 25

25. Consider two point sources S_1 and S_2 in Fig. 19-24 which emit waves of the same frequency and amplitude. The waves start in the same phase, and this phase relation at the sources is maintained throughout time. Consider points P at which r_1 is nearly equal to r_2. (a) Show that the superposition of these two waves gives a wave whose amplitude varies with the position P approximately according to

$$\frac{2Y}{r} \cos \frac{k}{2} (r_1 - r_2),$$

in which $r = (r_1 + r_2)/2$. (b) Then show that total annulment occurs when $r_1 - r_2 = (n + \frac{1}{2})\lambda$, n being any integer, and that total re-enforcement occurs when $r_1 - r_2 = n\lambda$.

The locus of points whose difference in distance from two fixed points is a constant is a hyperbola, the fixed points being the foci. Hence each value of n gives a hyperbolic line of constructive interference and a hyperbolic line of destructive interference. At points at which r_1 and r_2 are not approximately equal (as near the sources), the amplitudes of the waves from S_1 and S_2 differ and the annulments are only partial.

SECTION 19-9

26. The equation of a transverse wave traveling in a string is given by

$$y = 10 \cos (0.0079x - 13t - 0.89),$$

in which x and y are expressed in centimeters and t in seconds. Write down the equation of a wave which, when added to the given one, would produce standing waves on the rope.

27. A string vibrates according to the equation

$$y = 0.5 \sin \frac{\pi x}{3} \cos 40\pi t,$$

where x and y are in centimeters and t is in seconds. (a) What are the amplitude and velocity of the component waves whose superposition can give rise to this vibration? (b) What is the distance between nodes? (c) What is the velocity of a particle of the string at the position $x = 1.5$ cm when $t = \frac{9}{8}$ s? *Answer:* (a) 0.25 cm, 120 cm/s. (b) 3.0 cm. (c) Zero.

28. Two transverse sinusoidal waves travel in opposite directions along a string. Each has an amplitude of 0.30 cm and a wavelength of 6.0 cm. The speed of a transverse wave in the string is 1.5 m/s. Plot the shape of the string at each of the following times: $t = 0$ (arbitrary), $t = 5.0$, $t = 10.0$, $t = 15.0$, $t = 20.0$ ms.

29. If an incident traveling wave is only partially reflected from a boundary, the resulting superposition of two waves having different amplitudes and traveling in opposite directions gives a standing wave pattern of waves whose envelope is shown in Fig. 19-25. The standing wave ratio (SWR) is defined as $(A_i + A_r)/(A_i - A_r) = A_{max}/A_{min}$, and the percent reflection is defined as the ratio of the average power in the reflected wave to the average power in the incident wave, times 100. (a) Show that for 100% reflection SWR $= \infty$ and that for no reflection SWR $= 1$. (b) Show that a measurement

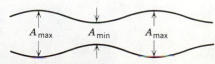

figure 19-25
Problem 29

of the SWR just before the boundary reveals the percent reflection occurring at the boundary according to the formula

$$\% \text{ reflection} = [(SWR - 1)^2/(SWR + 1)^2] \times 100.$$

30. Two strings of linear density μ_1 and μ_2 are knotted together at $x = 0$ and stretched to a tension F. A wave $y = A \sin k_1(x - v_1 t)$ in the string of density μ_1 reaches the junction between the two strings, at which it is partly transmitted into the string of density μ_2 and partly reflected. Call these waves $B \sin k_2(x - v_2 t)$ and $C \sin k_1(x + v_1 t)$, respectively. (a) Assuming that $k_2 v_2 = k_1 v_1 = \omega$ and that the displacement of the knot arising from the incident and reflected waves is the same as that arising from the transmitted wave, show that $A = B + C$. (b) If it is assumed that both strings near the knot have the same slope (why?), i.e., that dy/dx in string $1 = dy/dx$ in string 2, show that

$$C = A \frac{(k_2 - k_1)}{(k_2 + k_1)}$$

$$= A \frac{v_1 - v_2}{v_1 + v_2}.$$

Under what conditions is C negative?

31. Consider a standing wave that is the sum of two waves traveling in opposite directions but otherwise identical. Show that the energy in each loop of the standing wave is $2\pi^2 y_m^2 \nu v$.

SECTION 19-10

32. In a laboratory experiment on standing waves a string 3.0 ft (0.9 m) long is attached to the prong of an electrically driven tuning fork which vibrates perpendicular to the length of the string at a frequency of 60 vib/s (60 Hz). The weight of the string is 0.096 lb (mass $= 0.044$ kg). (a) What tension must the string be under (weights are attached to the other end) if it is to vibrate in four loops? (b) What would happen if the tuning fork is turned so as to vibrate parallel to the length of the string?

33. Vibrations from a 600-cycle/s tuning fork set up standing waves in a string clamped at both ends. The wave speed for the string is 400 m/s. The standing wave has four loops and an amplitude of 2.0 mm. (a) What is the length of the string? (b) Write an equation for the displacement of the string as a function of position and time.
    Answer: (a) 1.3 m. (b) $2.0 \times 10^{-3} \sin 9.4x \cos 3800t$, where x and y are in meters and t in seconds.

34. An aluminum wire of length $l_1 = 60.0$ cm and of cross-sectional area 1.00×10^{-2} cm^2 is connected to a steel wire of the same cross-sectional area. The compound wire, loaded with a block m of mass 10.0 kg, is arranged as shown in Fig. 19-26 so that the distance l_2 from the joint to the supporting pulley is 86.6 cm. Transverse waves are set up in the wire by using an external source of variable frequency. (a) Find the lowest frequency of excitation for which standing waves are observed such that the joint in the wire is a node. (b) What is the total number of nodes observed at this frequency, excluding the two at the ends of the wire? The density of aluminum is 2.60 g/cm^3, and that of steel is 7.80 g/cm^3.

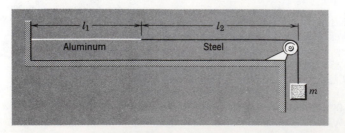

figure 19-26
Problem 34

20
sound waves

Sound waves are longitudinal mechanical waves. They can be propagated in solids, liquids, and gases. The material particles transmitting such a wave oscillate in the direction of propagation of the wave itself. There is a large range of frequencies within which longitudinal mechanical waves can be generated, sound waves being confined to the frequency range which can stimulate the human ear and brain to the sensation of hearing. This range is from about 20 cycles/sec (or 20 Hz) to about 20,000 Hz and is called the *audible* range. A longitudinal mechanical wave whose frequency is below the audible range is called an *infrasonic* wave, and one whose frequency is above the audible range is called an *ultrasonic* wave.

Infrasonic waves of interest are usually generated by large sources, earthquake waves being an example.* The high frequencies associated with ultrasonic waves† may be produced by elastic vibrations of a quartz crystal induced by resonance with an applied alternating electric field (piezoelectric effect). It is possible to produce ultrasonic frequencies as high as 6×10^8 Hz in this way; the corresponding wavelength in air is about 5×10^{-5} cm, the same as the length of visible light waves.

Audible waves originate in vibrating strings (violin, human vocal cords), vibrating air columns (organ, clarinet), and vibrating plates and membranes (xylophone, loudspeaker, drum). All of these vibrating elements alternately compress the surrounding air on a forward movement and rarefy it on a backward movement. The air transmits these dis-

20-1
AUDIBLE, ULTRASONIC, AND INFRASONIC WAVES

---

* See "Long Earthquake Waves," by Jack Oliver, *Scientific American*, March 1959.
† See "Applications of Ultrasonics" by Margaret F. Cracknell and Arthur P. Cracknell, *Contemporary Physics*, January 1976.

turbances outward from the source as a wave. Upon entering the ear, these waves produce the sensation of sound. Waveforms which are approximately periodic or consist of a small number of approximately periodic components give rise to a pleasant sensation (if the intensity is not too high), as, for example, musical sounds.* Sound whose waveform is nonperiodic is heard as noise. Noise can be represented as a superposition of periodic waves, but the number of components is very large.

In this chapter we deal with the properties of longitudinal mechanical waves, using sound waves as the prototype.

Sound waves, if unimpeded, will spread out in all directions from a source. It is simpler to deal with one-dimensional propagation, however, than with three-dimensional propagation, so that we consider first the transmission of longitudinal waves in a tube.

Figure 20-1 shows a piston at one end of a long tube filled with a compressible medium. The vertical lines divide the compressional (fluid) medium into thin "slices," each of which contains the same mass of fluid. Where the lines are relatively close together the fluid pressure and density are greater than they are in the normal undisturbed fluid, and conversely. We shall treat the fluid as a continuous medium and ignore for the time being the fact that it is made up of molecules that are in continual random motion.

If we push the piston of Fig. 20-1 forward, the fluid in front of it is compressed, the fluid pressure and density rising above their normal undisturbed values. The compressed fluid moves forward, compressing the fluid layers next to it, and a compressional pulse travels down the tube. If we then withdraw the piston, the fluid in front of it expands, its pressure and density falling below their normal undisturbed values; a pulse of rarefaction travels down the tube. These pulses are similar to transverse pulses traveling along a string, except that the oscillating fluid elements are displaced along the direction of propagation (longitudinal) instead of at right angles to this direction (transverse). If the piston oscillates back and forth, a continuous train of compressions and rarefactions will travel along the tube (Fig. 20-1). As for transverse waves in a string (see Section 19-5) we should be able, using Newton's laws of motion, to express the speed of propagation of this longitudinal wave in terms of an elastic and an inertial property of the medium. We now do so.

For the moment, let us assume that the tube is very long so that we can ignore reflections from the far end. As for the string of Fig. 19-6, we will consider not an extended wave but a single (compressional) pulse that we might generate by giving the piston in Fig. 20-1 a short, rapid, inward stroke.

Figure 20-2 shows such a pulse (labeled "compressional zone") traveling at speed v along the tube from left to right. For simplicity we have assumed this pulse to have sharply defined leading and trailing edges and to have a uniform fluid pressure and density in its interior. When we analyzed the motion of a transverse pulse in a string, we found it convenient to choose a reference frame in which the pulse remained stationary; we will do this here also. In Fig. 20-2, then, the compressional zone remains stationary in our reference frame while the fluid moves through it from right to left with speed v, as shown.

Let us follow the motion of the element of fluid contained between the ver-

20-2
PROPAGATION AND SPEED OF LONGITUDINAL WAVES

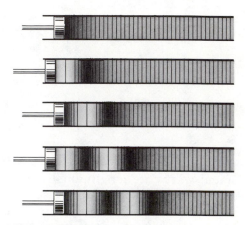

figure 20-1
Sound waves generated in a tube by an oscillating piston. The vertical lines divide the compressible medium in the tube into layers of equal mass.

---

* A good general reference on the scientific properties of musical sound is 'The Acoustical Foundations of Music' by John Backus, W. W. Norton & Co., Inc., New York, 1969.

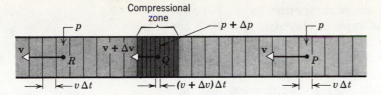

figure 20-2
A compressional pulse travels along a gas-filled tube. In a reference frame in which the undisturbed gas is at rest the pulse moves from left to right with speed v. We view the pulse, however, from a reference frame in which the pulse is stationary; in such a frame the gas outside the pulse streams through the tube from right to left with speed v, as shown. Note that Δv is negative.

tical lines at P in Fig. 20-2. This element moves forward at speed v until it strikes the compressional zone. While it is entering this zone it encounters a difference of pressure Δp between its leading and its trailing edges. The element is compressed and *decelerated*, moving with a lower speed $v + \Delta v$ within the zone, the quantity Δv being negative. The element eventually emerges from the left face of the zone where it expands to its original volume and the pressure differential Δp acts to *accelerate* it to its original speed v. The figure shows the element at point R, having passed through the compressional zone and moving again with speed v, as at P.

Let us apply Newton's laws to the fluid element while it is entering the compressional zone. The resultant force acting during entry points to the right in Fig. 20-2 and has magnitude

$$F = (p + \Delta p)A - pA = \Delta p A$$

in which A is the cross-sectional area of the tube.

The length of the element outside the compressional zone (at P, say) is $v \Delta t$, where Δt is the time required for the element to move past any given point. The volume of the element is thus $vA \Delta t$ and its mass is $\rho_0 vA \Delta t$, where ρ_0 is the density of the fluid outside the compressional zone. The deceleration a experienced by the element as it enters the zone is $-\Delta v/\Delta t$; because Δv is inherently negative, a is positive, which means that, like the force $\Delta p A$ in Fig. 20-2, it points to the right. Thus Newton's second law

$$F = ma$$

yields

$$\Delta p A = (\rho_0 vA \ \Delta t) \frac{-\Delta v}{\Delta t},$$

which we may write as

$$\rho_0 v^2 = \frac{-\Delta p}{\Delta v/v}.$$

Now the fluid that would occupy a volume $V = Av \ \Delta t$ at P is compressed by an amount $A(\Delta v) \ \Delta t = \Delta V$ on entering the compressional zone. Hence,

$$\frac{\Delta V}{V} = \frac{A \ \Delta v \ \Delta t}{Av \ \Delta t} = \frac{\Delta v}{v}$$

and we obtain

$$\rho_0 v^2 = \frac{-\Delta p}{\Delta V/V}.$$

The ratio of the change in pressure on a body, Δp, to the fractional change in volume resulting, $-\Delta V/V$, is called the *bulk modulus of elasticity* B of the body. That is, $B = -V \ \Delta p/\Delta V$. B is positive because an increase in pressure causes a decrease in volume. In terms of B, the speed of the longitudinal pulse in the medium of Fig. 20-2 is

$$v = \sqrt{B/\rho_0}. \tag{20-1}$$

A more extended analysis than given above shows that Eq. 20-1 applies not only to rectangular pulses of the type displayed in Fig. 20-2 but also to pulses of any shape and to extended wave trains. Notice that the speed of the wave is determined by the properties of the medium through which it propagates, and that an elastic property B and an inertial property ρ_0 are involved. Table 20-1 gives the speed of longitudinal (sound) waves in various media.

Table 20-1
Speed of sound

Medium	Tempera-ture, °C	Speed	
		m/s	ft/s
Air	0	331.3	1,087
Hydrogen	0	1,286	4,220
Oxygen	0	317.2	1,041
Water	15	1,450	4,760
Lead	20	1,230	4,030
Aluminum	20	5,100	16,700
Copper	20	3,560	11,700
Iron	20	5,130	16,800
Extreme values			
Granite		6,000	19,700
Vulcanized rubber	0	54	177

If the medium is a gas, such as air, it is possible to express B in terms of the undisturbed gas pressure p_0. For a sound wave in a gas we obtain

$$v = \sqrt{\gamma p_0 / \rho_0,}$$

where γ is a constant called the ratio of specific heats for the gas (Chapter 23).

If the medium is a solid, for a thin rod the bulk modulus is replaced by a stretch modulus (called Young's modulus). If the solid is extended, we must allow for the fact that, unlike a fluid, a solid offers elastic resistance to tangential or shearing forces and the speed of longitudinal waves will depend on the shear modulus as well as the bulk modulus.

20-3 TRAVELING LONGITUDINAL WAVES

Consider again the continuous train of compressions and rarefactions traveling down the tube of Fig. 20-1. As the wave advances along the tube, each small volume element of fluid oscillates about its equilibrium position. The displacement is to the right or left along the x-direction of propagation of the wave. For convenience let us represent the displacement of any such volume element (or layers of elements that move in the same way) from its equilibrium position at x by the letter y. It is to be understood that the displacement y is *along the direction of propagation* for a longitudinal wave, whereas for a transverse wave the displacement y is *at right angles to the direction of propagation*. Then the equation of a longitudinal wave traveling to the right may be written as

$$y = f(x - vt).$$

For the particular case of a simple harmonic oscillation we may have

$$y = y_m \cos \frac{2\pi}{\lambda} (x - vt).$$

In this equation v is the speed of the longitudinal wave, y_m is its amplitude, and λ is its wavelength; y gives the displacement of a particle at time t from its equilibrium position at x. As before, we may write this more compactly as

$$y = y_m \cos (kx - \omega t). \qquad (20\text{-}2)$$

It is usually more convenient to deal with pressure variations in a sound wave than with the actual displacements of the particles conveying the wave. Let us therefore write the equation of the wave in terms of the pressure variation rather than in terms of the displacement.

From the relation

$$B = -\frac{\Delta p}{\Delta V / V},$$

we have

$$\Delta p = -B \frac{\Delta V}{V}.$$

Just as we let y represent the displacement from the equilibrium position x, so we now let p represent the *change* from the undisturbed pressure p_0. Then p replaces Δp, and

$$p = -B \frac{\Delta V}{V}.$$

If a layer of fluid at pressure p_0 has a thickness Δx and cross-sectional area A, its volume is $V = A \, \Delta x$. When the pressure changes, its volume will change by $A \, \Delta y$, where Δy is the amount by which the thickness of the layer changes during compression or rarefaction. Hence,

$$p = -B \frac{\Delta V}{V} = -B \frac{A \, \Delta y}{A \, \Delta x}.$$

As we let $\Delta x \to 0$ so as to shrink the fluid layer to infinitesimal thickness, we obtain

$$p = -B \frac{\partial y}{\partial x}. \qquad (20\text{-}3)$$

We have used partial derivative notation because (see Eq. 20-2) y is a function of both x and t and we take the latter quantity as constant in this discussion. If the particle displacement is simple harmonic, then, from Eq. 20-2, we obtain

$$\frac{\partial y}{\partial x} = -k y_m \sin (kx - \omega t),$$

and from Eq. 20-3

$$p = B k y_m \sin (kx - \omega t). \qquad (20\text{-}4)$$

Hence, the pressure variation at each position x is also simple harmonic.

Because $v = \sqrt{B/\rho_0}$, we can write Eq. 20-4 more conveniently as

$$p = [k \rho_0 v^2 y_m] \sin (kx - \omega t).$$

Recall that p represents the change from standard pressure p_0. The term in brackets represents the maximum change in pressure and is called the *pressure amplitude*. If we denote this by P, then

$$p = P \sin (kx - \omega t), \qquad (20\text{-}5)$$

where

$$P = k \rho_0 v^2 y_m. \qquad (20\text{-}6)$$

Hence, a sound wave may be considered either as a displacement wave or as a pressure wave. If the former is written as a cosine function, the latter will be a sine function and vice versa. The displacement wave is thus 90° out of phase with the pressure wave. That is, when the displacement from equilibrium at a point is a maximum or a minimum, the excess pressure there is zero; when the displacement at a point is zero, the excess or deficiency of pressure there is a maximum. Equation 20-6 gives the relation between the pressure amplitude (maximum variation of pressure from equilibrium) and the displacement amplitude (maximum variation of position from equilibrium). You should check the dimensions of each side of Eq. 20-6 for consistency. What units may the pressure amplitude have?

The intensity of a wave is proportional to the square of the displacement amplitude of the wave; see Section 19-6. We have just shown that for sound waves the pressure amplitude is proportional to the displacement amplitude. Hence, the intensity of a sound wave is proportional to the square of the pressure amplitude. In fact, when the intensity is expressed in terms of the pressure amplitude, the frequency does not appear explicitly in the expression (see Problem 14). Hence, by measuring pressure changes, the intensities of sounds having *different* frequencies can be compared directly. For this reason instruments that measure pressure changes are preferred to those that measure displacement amplitude. As we shall see in Example 1, the displacement amplitudes would be difficult to measure in any case.

EXAMPLE 1

(*a*) The maximum pressure variation P that the ear can tolerate in loud sounds is about 28 N/m² (= 28 Pa). Normal atmospheric pressure is about 100,000 Pa. Find the corresponding maximum displacement for a sound wave in air having a frequency of 1000 Hz.

From Eq. 20-6 we have

$$y_m = \frac{P}{k\rho_0 v^2}.$$

From Table 20-1, $v = 331$ m/s so that

$$k = \frac{2\pi}{\lambda} = \frac{2\pi v}{v} = \frac{2\pi \times 10^3}{331} \text{ m}^{-1} = 19 \text{ m}^{-1}.$$

The density of air ρ_0 is 1.22 kg/m³. Hence, for $P = 28$ Pa we obtain

$$y_m = \frac{28}{(19)(1.22)(331)^2} \text{ m} = 1.1 \times 10^{-5} \text{ m}.$$

The displacement amplitudes for the *loudest* sounds are about 10^{-5} m, a very small value indeed.

(*b*) In the faintest sound that can be heard at 1000 Hz the pressure amplitude is about 2.0×10^{-5} Pa. Find the corresponding displacement amplitude.

From $y_m = P/k\rho_0 v^2$, using these values for k, v, and ρ_0, we obtain, with $P = 2.0 \times 10^{-5}$ N/m²,

$$y_m \cong 8 \times 10^{-12} \text{ m} \cong 10^{-11} \text{ m}.$$

This is smaller than the radius of an atom, which is about 10^{-10} m! How can it be that the ear responds to such a small displacement?

In our analysis we have ignored the molecular structure of matter and treated the fluid as a continuous medium. In gases, however, the spaces between molecules are large compared to the diameters of the molecules. The molecules move about at random. The oscillations produced by a sound wave passing through are superimposed on this random thermal motion. An impulse given to one molecule is passed on to another molecule only after the first one has moved through the empty space between them and collided with the second. From this brief discussion, would you ever expect the speed of sound to exceed the average molecular speed in a fluid?

Longitudinal waves traveling along a gas-filled tube are reflected at the ends of the tube, just as transverse waves in a stretched string are reflected at its ends. Interference between the waves traveling in opposite directions gives rise to standing longitudinal waves.

If the end of the tube is closed, the reflected wave is 180° out of phase with the incident wave. This result is a necessary consequence of the fact that the displacement of the small volume elements at a closed end must always be zero. Hence, a closed end is a displacement *node*. If the end of the tube is open, the fluid elements there are free to move. However, the nature of the reflection there depends on whether the tube is wide or narrow compared to the wavelength. If the tube is narrow compared to the wavelength, as in most musical instruments, the reflected wave has nearly the same phase as the incident wave. Then the open end is almost a displacement *antinode*. The exact antinode is usually somewhere near the opening, but the effective length of the air columns of a wind instrument, for example, is not as definite as the length of a string fixed at both ends.

Standing longitudinal waves in a gas column can be dramatically demonstrated by means of the apparatus shown in Fig. 20-3. A source of longitudinal waves, such as the speaker of an audio oscillator at S, sets up vibrations in a flexible diaphragm at one end of the tube. Gas fills the tube from the inlet and passes slowly out through regularly spaced small openings along the top. The escaping gas is lit, giving a series of flames. The frequency of the audio oscillator is varied and when a frequency is found at which the gas column is in resonance, the amplitude of the standing longitudinal waves becomes rather large; then we can see a wavelike variation in the height and width of the gas flames along the tube. The interval between nodes or antinodes is clearly visible. By continuing to vary the frequency we can pass from one resonance condition to another. The natural modes of oscillation of the gas column are determined by the effective length of the column and the wave speed. The wavelength λ at resonance can be taken to be twice the distance between adjacent nodes (or antinodes), and knowing the frequency ν of the source at resonance, we can determine the wave speed in the gas under these conditions from $v = \nu\lambda$. In practice there are more flexible and accurate ways to measure the speed of sound in gases. (See Problem 21 and Example 2.)

In Fig. 20-3 the nodes and antinodes, N and A, refer to the particle *displacements* in the standing wave. At a displacement node, the pressure variations (above and below the average) are a maximum. Hence,

20-4
STANDING LONGITUDINAL WAVES

figure 20-3
Flames show the presence of standing waves in a tube filled with illuminating gas. *A* and *N* refer to *displacement* antinodes and nodes, respectively.

a displacement node corresponds to a pressure antinode. At a displacement antinode the pressure remains constant with time. Hence, a displacement antinode corresponds to a pressure node.

This can be understood physically by realizing that two small volume elements of gas on opposite sides of a displacement node are vibrating in *opposite phase*. Hence, when they approach each other, the pressure at this node is increasing, and when they recede from each other, the pressure at this node is decreasing. Two small elements of gas which are on opposite sides of a displacement antinode vibrate *in phase* and therefore give rise to no pressure variations at the antinode.

If a string fixed at both ends is bowed, transverse vibrations travel along the string; these disturbances are reflected at the fixed ends, and a standing wave pattern is formed. The natural modes of vibration of the string are excited and these vibrations give rise to longitudinal waves in the surrounding air which transmits them to our ears as a musical sound.

We have seen (Section 19-10) that a string of length l, fixed at both ends, can resonate at frequencies given by

$$\nu_n = \frac{n}{2l} v = \frac{n}{2l} \sqrt{\frac{F}{\mu}}, \qquad n = 1, 2, 3, \ldots . \qquad (20\text{-}7)$$

Here v is the speed of the transverse waves in the string whose super-position can be thought of as giving rise to the vibrations; the speed $v \ (= \sqrt{F/\mu})$ is the same for all frequencies. At any one of these frequencies the string will contain a whole number n of loops between its ends, and the condition that the ends be nodes is met (Fig. 20-4).

20-5
VIBRATING SYSTEMS AND SOURCES OF SOUND

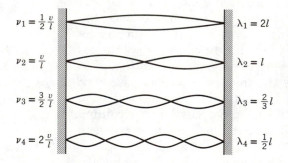

figure 20-4
The first four modes of vibration of a string fixed at both ends. Note that $\nu_n \lambda_n = v = \sqrt{F/\mu}$.

The lowest frequency, $\sqrt{F/\mu}/2l$, is called the *fundamental* frequency ν_1 and the others are called *overtones*. Overtones whose frequencies are integral multiples of the fundamental are said to form a harmonic series. The fundamental is the first harmonic. The frequency $2\nu_1$ is the first overtone or the second harmonic, the frequency $3\nu_1$ is the second overtone or the third harmonic, and so on.

If the string is initially distorted so that its shape is the same as *any one* of the possible harmonics, it will vibrate at the frequency of that particular harmonic, when released. The initial conditions usually arise from striking or bowing the string, however, and in such cases not only the fundamental but many of the overtones are present in the resulting vibration. We have a superposition of several natural modes of oscillation. The actual displacement is the sum of the several harmonics with various amplitudes; see Fig. 19-12. The impulses that are sent through the air to the ear and brain give rise to one net effect which is characteristic of the particular stringed instrument. The quality of the sound of a particular note (fundamental frequency) played by an instrument is determined by the number of overtones present and their respective intensities. Figure 20-5 shows the sound spectra and corresponding waveforms for the violin and piano.*

An organ pipe is a simple example of sound originating in a vibrating air column. If both ends of a pipe are open and a stream of air is directed against an edge, standing longitudinal waves can be set up in the tube. The air column will then resonate at its natural frequencies of vibration, given by

$$\nu_n = \frac{n}{2l} v, \qquad n = 1, 2, 3, \ldots .$$

Here v is the speed of the longitudinal waves in the column whose superposition can be thought of as giving rise to the vibrations, and n is the number of half wavelengths in the length l of the column. As with the bowed string, the fundamental and overtones are excited at the same time.

In an open pipe the fundamental frequency corresponds (approximately) to a displacement antinode at each end and a displacement node in the middle, as shown in Fig. 20-6a. The succeeding drawings of Fig. 20-6a show three of the overtones, the second, third, and fourth harmonics. Hence, in an open pipe the fundamental frequency is $v/2l$ and *all* harmonics are present.

In a closed pipe the closed end is a displacement node. Figure 20-6b shows the modes of vibration of a closed pipe. The fundamental frequency is $v/4l$ (approximately), which is one-half that of an open pipe of the same length. The only overtones present are those that give a

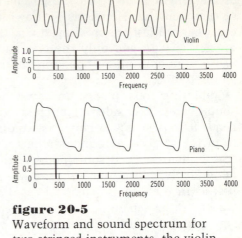

figure 20-5

Waveform and sound spectrum for two stringed instruments, the violin and the piano. The fundamental frequency in both cases is 440 cycles/sec (concert A). In each diagram we show only four cycles of the wave. The sound spectrum shows the relative amplitude of the various harmonic components of the wave. Notice the presence of loud higher harmonics (especially the fifth) in the violin spectrum.

figure 20-6

(a) The first four modes of an open organ pipe. The distance from the center line of the pipe to the light lines drawn inside the pipe shows the displacement amplitude at each place. N and A mark the locations of the displacement nodes and antinodes. Note that *both* ends of the pipe are open. (b) The first four modes of vibration of a closed organ pipe. Notice that the even-numbered harmonics are absent and the upper end of the pipe is closed.

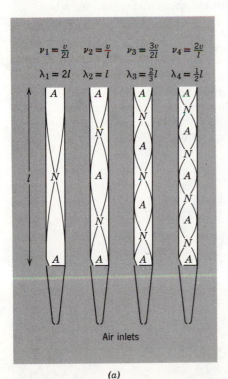

(a)

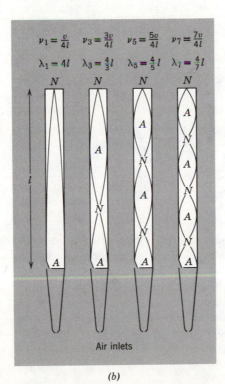

(b)

---

* See "The Physics of the Piano" by E. Donnell Blackham in *Scientific American*, December 1965 and "The Physics of the Violin" by Carleen M. Hutchins in *Scientific American*, November 1962.

displacement node at the closed end and an antinode (approximately) at the open end. Hence, as is shown in Fig. 20-6b, the second, fourth, etc., harmonics are missing. In a closed pipe the fundamental frequency is $v/4l$, and only the *odd* harmonics are present. The quality of the sounds from an open pipe is therefore different from that from a closed pipe.

Vibrating rods, plates, and stretched membranes also give rise to sound waves. Consider a stretched flexible membrane, such as a drumhead. If it is struck a blow, a two-dimensional pulse travels outward from the struck point and is reflected again and again at the boundary of the membrane. If some point of the membrane is forced to vibrate periodically, continuous trains of waves travel out along the membrane. Just as in the one-dimensional case of the string, so here too standing waves can be set up in the two-dimensional membrane. Each of these standing waves has a certain frequency natural to (or characteristic of) the membrane. Again the lowest frequency is called the fundamental and the others are overtones. Generally, a number of overtones are present along with the fundamental when the membrane is vibrating. These vibrations may excite sound waves of the same frequency.

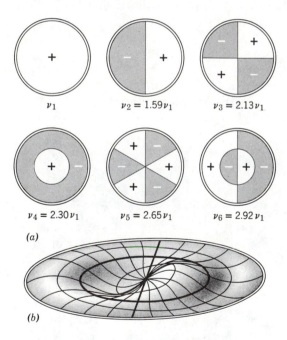

v_1 $v_2 = 1.59v_1$ $v_3 = 2.13v_1$

$v_4 = 2.30v_1$ $v_5 = 2.65v_1$ $v_6 = 2.92v_1$

(a)

(b)

figure 20-7
(a) The first six modes of vibration of a circular drumhead clamped around its periphery. The lines represent nodes, the circumference being a node in every case. The + and − signs represent opposite displacements; at an instant when the + areas are raised, the − areas will be depressed. Note that the frequency of each mode is not an integral multiple of the fundamental v_1 as is the case for strings and tubes. (b) A sketch of a drum-head vibrating in mode v_6. The displacement shown here is exaggerated for clarity.

The nodes of a vibrating membrane are lines rather than points (as in a vibrating string) or planes (as in a pipe). Since the boundary of the membrane is fixed, it must be a nodal line. For a circular membrane fixed at its edge, possible modes of vibration together with their nodal lines are shown in Fig. 20-7. The natural frequency of each mode is given in terms of the fundamental v_1. Notice that the frequencies of the overtones are *not* harmonics, that is, they are not integral multiples of v_1. Vibrating rods also have a nonharmonic set of natural frequencies. Rods and plates have limited use as musical instruments for this reason.

In general, we find that all elastic bodies will vibrate freely with a definite set of frequencies for a given set of boundary or end conditions. These frequencies are called proper frequencies, characteristic frequencies, or *eigenfrequencies**

* *Eigen* — from the German — meaning *own, individual, characteristic.*

of the system. In general, the eigenfrequencies do *not* form a harmonic series, although some of them may be related as the ratio of whole numbers. In all these cases we have standing waves, and certain regions of the bodies stay at rest all the time. These nodes are curves in two-dimensional bodies and surfaces in three-dimensional bodies.

Recall that for a vibrating string the equation describing a standing wave (see Eq. 19-18b) is of the type

$$y = 2y_m \cos 2\pi\nu t \, \sin \frac{2\pi x}{\lambda}.$$

This holds for a string fixed at both ends ($y = 0$ at $x = 0$ and $x = n\lambda/2$). The picture of the string at any time is determined by the equation

$$y = C \sin \frac{2\pi x}{\lambda} = C \sin \frac{n\pi x}{l} \qquad (t = \text{constant}),$$

where C is a constant "scale factor," whose value varies with time; l is the length of the string, and n is an integer specifying the mode of vibration (the harmonic). This function $\sin 2\pi x/\lambda$ fixes the position of the nodes and is called the proper function, characteristic function, or *eigenfunction* of the string.

Likewise, the nodes of *any* vibrating elastic body are fixed by certain functions of position which are called the eigenfunctions of the problem. In general, these functions are *not* sinusoidal functions but are functions that become zero for certain values of the coordinates. The determination of these functions and the corresponding values of the eigenfrequencies is an important problem in atomic, nuclear, and solid-state physics. They characterize the behavior of such systems. It is in quantum mechanics that the procedure has been successfully worked out for microscopic systems. The results, however, bear a striking analogy to the results of classical vibration and wave theory, as applied to macroscopic systems.

EXAMPLE 2

Figure 20-8 shows a simple apparatus that can be used to measure the speed of sound in air by resonance methods. A vibrating tuning fork of frequency ν is held near the open end of a tube. The tube is partly filled with water. The length of the air column can be varied by changing the water level. It is found that the sound intensity is a maximum when the water level is gradually lowered from the top of the tube a distance a. Thereafter, the intensity reaches a maximum again at distances d, $2d$, $3d$, etc., below the level at a. Find the speed of sound in air.

The sound instensity reaches a maximum when the air column resonates with the tuning fork. The air column acts like a tube closed at one end. The standing wave pattern consists of a node at the water surface and an antinode near the open end. Since the frequency of the source is fixed and the speed of sound in the air column has a definite value, resonance occurs at one specific wavelength,

$$\lambda = \frac{v}{\nu}.$$

The distance d between successive resonance positions is therefore the distance between adjacent nodes. (See Fig. 20-8.) Hence,

$$d = \frac{\lambda}{2} \qquad \text{or} \qquad \lambda = 2d.$$

Combining equations we find

$$2d = \frac{v}{\nu} \qquad \text{or} \qquad v = 2d\nu.$$

In an experiment with a fork of frequency $\nu = 1080$ cycles/s, d is found to be 15.3 cm. Hence,

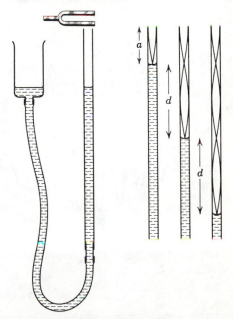

figure 20-8
Example 2. Measuring the speed of sound in air. The water level in the tube can be adjusted by raising or lowering the reservoir on the left which is connected to the tube by a rubber hose.

$$\lambda = 2d = 30.6 \text{ cm}$$

and

$$v = \nu\lambda = (1080)(0.306) \text{ m/s} = 330 \text{ m/s}.$$

What significance does the distance a have? Could gases other than air be used conveniently in this apparatus?

When two wavetrains of the *same frequency* travel along the same line in *opposite directions*, standing waves are formed in accord with the principle of superposition. We may characterize these waves by drawing a plot of the amplitude of oscillation as a function of distance, as in Fig. 20-4. This illustrates a type of interference that we can call *interference in space*.

The same principle of superposition leads us to another type of interference, which we can call *interference in time*. It occurs when two wavetrains of slightly *different frequency* travel in the *same direction*. With sound such a condition exists when, for example, two adjacent piano keys are struck simultaneously.

Consider some one point in space through which the waves are passing. In Fig. 20-9a we plot the displacements produced at such a point by the two waves separately as a function of time. For simplicity we have assumed that the two waves have equal amplitude, although this is not necessary. The resultant vibration at that point as a function of time is the sum of the individual vibrations and is plotted in Fig. 20-9b. We see that the *amplitude* of the resultant wave at the given point is not constant but *varies with time*. In the case of sound the varying amplitude gives rise to variations in loudness which are called *beats*. Two strings may be tuned to the same frequency by tightening one of them while sounding both until the beats disappear.

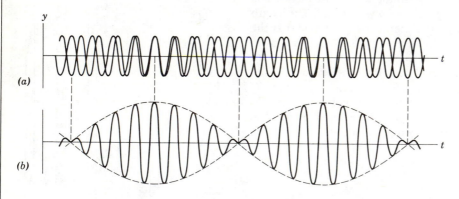

(a)

(b)

figure 20-9
The beat phenomenon. Two waves of slightly different frequencies, shown in (a), combine in (b) to give a wave whose amplitude (dashed line) varies periodically with time. Compare with Fig. 19-14, which shows the same phenomenon displayed as a function of distance.

Let us represent the displacement at the point produced by one wave as

$$y_1 = y_m \cos 2\pi\nu_1 t,$$

and the displacement at the point produced by the other wave of equal amplitude as

$$y_2 = y_m \cos 2\pi\nu_2 t.$$

By the superposition principle, the resultant displacement is

$$y = y_1 + y_2 = y_m(\cos 2\pi\nu_1 t + \cos 2\pi\nu_2 t),$$

and since

$$\cos a + \cos b = 2 \cos \frac{a - b}{2} \cos \frac{a + b}{2},$$

this can be written as

$$y = \left[2y_m \cos 2\pi \left(\frac{\nu_1 - \nu_2}{2} \right) t \right] \cos 2\pi \left(\frac{\nu_1 + \nu_2}{2} \right) t. \qquad (20\text{-}8)$$

The resulting vibration may then be considered to have a frequency

$$\bar{\nu} = \frac{\nu_1 + \nu_2}{2},$$

which is the average frequency of the two waves, and an amplitude given by the expression in brackets. Hence, the amplitude itself varies with time with a frequency

$$\nu_{\text{amp}} = \frac{\nu_1 - \nu_2}{2}.$$

If ν_1 and ν_2 are nearly equal, this term is small and the amplitude fluctuates slowly. This phenomenon is a form of amplitude modulation which has a counterpart (side bands) in AM radio receivers.

A beat, that is, a maximum of amplitude, will occur whenever

$$\cos 2\pi \left(\frac{\nu_1 - \nu_2}{2} \right) t$$

equals 1 or −1. Since *each* of these values occurs once in each cycle (see Fig. 19-14), the number of beats per second is *twice* the frequency ν_{amp} or $\nu_1 - \nu_2$. Hence, the number of beats per second equals the difference of the frequencies of the component waves. Beats between two tones can be detected by the ear up to a frequency of about seven per second. At higher frequencies individual beats cannot be distinguished in the sound produced.

20-7 THE DOPPLER EFFECT

When a listener is in motion toward a stationary source of sound, the pitch (frequency) of the sound heard is higher than when he is at rest. If the listener is in motion away from the stationary source, he hears a lower pitch than when he is at rest. We obtain similar results when the source is in motion toward or away from a stationary listener. The pitch of the whistle of the locomotive is higher when the source is approaching the hearer than when it has passed and is receding.

Christian Johann Doppler (1803–1853), an Austrian, in a paper of 1842, called attention to the fact that the color of a luminous body must be changed by relative motion of the body and the observer. This *Doppler effect*, as it is called, applies to waves in general. Doppler himself mentions the application of his principle to sound waves. An experimental test was carried out in Holland in 1845 by Buys Ballot, ", , , using a locomotive drawing an open car with several trumpeters."

We now consider the application of the Doppler effect to sound waves, treating only the special case in which the source and observer move along the line joining them. Let us adopt a reference frame at rest in the medium through which the sound travels. Figure 20-10 shows a source of sound S at rest in this frame and an observer O (note the ear) moving *toward* the source at a speed v_o. The circles represent wavefronts, spaced one wavelength apart, traveling through the medium. If the observer were at rest in the medium, he would receive vt/λ waves in time t, where v is the speed of sound in the medium and λ is the wave-

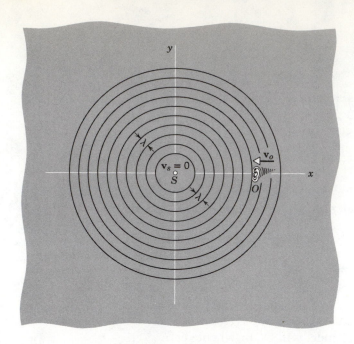

length. Because of his motion toward the source, however, he receives $v_0 t/\lambda$ *additional* waves in this same time t. The frequency ν' that he hears is the number of waves received per unit time or

$$\nu' = \frac{vt/\lambda + v_o t/\lambda}{t} = \frac{v + v_o}{\lambda} = \frac{v + v_o}{v/\nu}.$$

That is,

$$\nu' = \nu \frac{v + v_o}{v} = \nu\left(1 + \frac{v_o}{v}\right). \tag{20-9a}$$

The frequency ν' heard by the observer is the ordinary frequency ν heard at rest plus the increase $\nu(v_o/v)$ arising from the motion of the observer. When the observer is in motion *away* from the stationary source, there is a *decrease* in frequency $\nu(v_o/v)$ corresponding to the waves that do not reach the observer each unit of time because of his receding motion. Then

$$\nu' = \nu\left(\frac{v - v_o}{v}\right) = \nu\left(1 - \frac{v_o}{v}\right). \tag{20-9b}$$

Hence, the general relation holding when the *source is at rest* with respect to the medium but the *observer is moving* through it is

$$\nu' = \nu\left(\frac{v \pm v_o}{v}\right), \tag{20-9}$$

where the *plus* sign holds for motion *toward* the source and the *minus* sign holds for motion *away* from the source. Notice that the cause of the change here is the fact that the observer intercepts more or fewer waves each second because of his motion through the medium.

When the *source* is in motion *toward* a stationary observer, the effect is a shortening of the wavelength (see Fig. 20-11), for the source is following after the approaching waves and the crests therefore come closer together. If the frequency of the source is ν and its speed is v_s, then during each vibration it travels a distance v_s/ν and each wavelength is shortened by this amount. Hence, the wavelength of the sound arriving

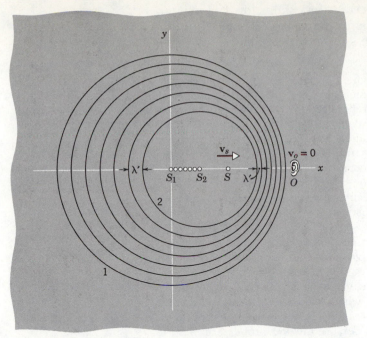

figure 20-11
The Doppler effect due to motion of the source. The observer is at rest. Wavefront 1 was emitted by the source when it was at S_1, wavefront 2 was emitted when it was at S_2, etc. At the instant the "snapshot" was taken, the source was at S.

at the observer is not $\lambda = v/\nu$ but $\lambda' = v/\nu - v_s/\nu$. Therefore, the frequency of the sound heard by the observer is *increased*, being

$$\nu' = \frac{V}{\lambda'} = \frac{V}{(v - v_s)/\nu} = \nu\left(\frac{v}{v - v_s}\right). \qquad (20\text{-}10a)$$

If the source moves *away* from the observer, the wavelength emitted is v_s/ν greater than λ, so that the observer hears a *decreased* frequency, namely

$$\nu' = \frac{V}{(v + v_s)/\nu} = \nu\left(\frac{v}{v + v_s}\right). \qquad (20\text{-}10b)$$

Hence, the general relation holding when the *observer is at rest* with respect to the medium but the *source is moving* through it is

$$\nu' = \nu\left(\frac{v}{v \mp v_s}\right), \qquad (20\text{-}10)$$

where the *minus* sign holds for motion *toward* the observer and the *plus* sign holds for motion *away* from the observer. Notice that the cause of the change here is the fact that the motion of the source through the medium shortens or increases the wavelength transmitted through the medium.

If both source *and* observer move through the transmitting medium, the student should be able to show that the observer hears a frequency

$$\nu' = \nu\left(\frac{v \pm v_0}{v \mp v_s}\right), \qquad (20\text{-}11)$$

where the upper signs (+ numerator, − denominator) correspond to the source and observer moving along the line joining the two in the direction *toward* the other, and the lower signs in the direction *away* from the other. Notice that Eq. 20-11 reduces to Eq. 20-9 when $v_s = 0$ and to Eq. 20-10 when $v_0 = 0$, as it must.

If a vibrating tuning fork on its resonating box is moved rapidly toward a wall, the observer will hear two notes of different frequency. One is the note heard directly from the receding fork and is lowered in pitch by the motion. The other note is due to the waves reflected from the wall, and this is raised in pitch. The superposition of these two wave trains produces beats.

The Doppler effect is important in light. The speed of light is so great that only astronomical or atomic sources, which have high velocities compared to terrestrial macroscopic sources, show pronounced Doppler effects. The astronomical effect consists of a shift in the wavelength observed from light emitted by elements on moving astronomical bodies compared to the wavelength observed from these same elements on earth. (See Chapter 42). An easily observed consequence of the Doppler effect is the broadening (or spread in frequency) of the radiation emitted from hot gases. This broadening results from the fact that the emitting atoms or molecules move in all directions and with varying speeds relative to the observing instruments, so that a spread of frequencies is detected.

There are differences, however, in the Doppler effect formula for light and for sound. In sound it is not just the relative motion of source and observer that determines the frequency change. In fact, as we have seen, even when the relative motion is the same (v_o in Eq. 20-9a equals v_s in Eq. 20-10a), we obtain different quantitative results, depending on whether the source or the observer is moving. This difference occurs because v_o and v_s are relative to the medium in which the sound wave is propagated and because this medium determines the wave speed. Light, however, does not require a material medium for its transmission, and the speed of light relative to the source or the observer is always the same value c, regardless of the motion of these bodies relative to each other. This is a basic postulate of the special theory of relativity (See Supplementary Topic V). Hence, for light only the relative motion of source and observer can lead to physical changes, there being no material medium to use as a reference frame. Although the Doppler formula for light (Chapter 42) differs from that for sound, the effects are qualitatively the same. We can apply Eq. 20-10 to light as a good approximation if v_s is taken to mean the *relative* velocity of source and observer and if v_s is very small compared to the velocity of light.

EXAMPLE 3

Show that Eqs. 20-9 and 20-10 become practically identical when the speed of the sources and the observer are small compared to the speed of sound in the medium.

Let $v_0 = v_s = u$. That is, let u represent the speed of observer *or* source. Then Eq. 20-9 becomes

$$v' = v\left(1 \pm \frac{u}{v}\right).$$

We must show then that Eq. 20-10,

$$v' = v\left(\frac{v}{v \mp u}\right),$$

reduces to the previous form when $u/v \ll 1$.

We can rewrite Eq. 20-10 as

$$v' = v\left(\frac{1}{1 \mp u/v}\right).$$

Now by the binomial expansion

$$\left(\frac{1}{1 \mp u/v}\right) = \left(1 \mp \frac{u}{v}\right)^{-1} = 1 \pm \frac{u}{v} + \left(\frac{u}{v}\right)^2 \pm \cdots.$$

But if u/v is sufficiently small compared to unity that we may neglect $(u/v)^2$ and higher powers, then

$$\left(\frac{1}{1 \mp u/v}\right) \cong 1 \pm \frac{u}{v},$$

and Eq. 20-10 becomes
$$\nu' \cong \nu\left(1 \pm \frac{u}{v}\right),$$

the same as Eq. 20-9.

As a numerical example take $u = 73.0$ mi/h (= 117.5 km/h). The speed of sound in air is about 730 mi/h (= 1175 km/h). Then if the source has a speed $v_s = u = 73.0$ mi/h toward the stationary observer, the frequency heard by the observer is Eq. 20-10,

$$\nu' = \nu\left(\frac{v}{v - v_s}\right) = \nu\left(\frac{730}{730 - 73.0}\right)$$

or
$$\frac{\nu'}{\nu} = 1.11.$$

If the observer has a speed $v_o = u = 73.0$ mi/h toward the stationary source, the frequency heard by the observer is Eq. 20-9,

$$\nu' = \nu\left(\frac{v + v_o}{v}\right) = \nu\left(\frac{730 + 73.0}{730}\right)$$

or
$$\frac{\nu'}{\nu} = 1.10.$$

Hence, when $u/v = 73.0/730 = 1/10$, the percentage difference in the frequency heard between that for the moving observer and that for the moving source, the relative motion being the same, is only 1%.

When v_0 or v_s becomes comparable in magnitude to v, the formulas just given for the Doppler effect usually must be modified. One modification is required because the linear relation between restoring force and displacement assumed

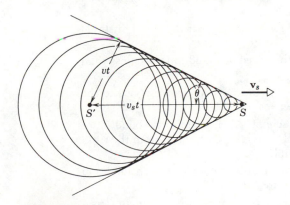

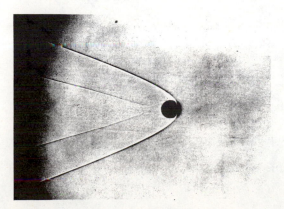

figure 20-12

Top, a group of wavefronts associated with a projectile moving with supersonic speed. The wavefronts are spherical and their envelope is a cone. The student should see the relation between this figure and the previous one. Bottom, a spark photograph of a projectile undergoing this motion. (U.S. Navy Photograph.)

up until now may no longer hold in the medium. The speed of wave propagation is then no longer the normal phase velocity, and the wave shapes change in time. Components of the motion at right angles to the line joining source and observer also contribute to the Doppler effect at these high speeds. When v_0 or v_s exceeds v, the Doppler formula does not apply; for example, if $v_s > v$, the source will get ahead of the wave in one direction; if $v_0 > v$ and the observer moves away from the source, the wave will never catch up with the observer.

There are many instances in which the source moves through a medium at a speed greater than the phase velocity of the wave in that medium. In such cases the wavefront takes the shape of a cone with the moving body at its apex. Some examples are the bow wave from a speedboat on the water and the "shock wave" from an airplane or projectile moving through the air at a speed greater than the velocity of sound in that medium (supersonic speeds). The Cerenkov radiation consists of light waves emitted by charged particles which move through a medium with a speed greater than the phase velocity of light in that medium.*

In Fig. 20-12 we show the present positions of the spherical waves which originated at various positions of the source during its motion. The radius of each sphere at this time is the product of the wave speed v and the time t which has elapsed since the source was at its center. The envelope of these waves is a cone whose surface makes an angle θ with the direction of motion of the source. From the figure we obtain the result

$$\sin \theta = \frac{v}{v_s}.$$

For water waves the cone reduces to a pair of intersecting lines. In aerodynamics the ratio v_s/v is called the *Mach number*.

questions

1. List some sources of infrasonic waves. Of ultrasonic waves.
2. Ultrasound can be used to reveal internal structures of the body. It can, for example, distinguish between liquid and soft human tissues far better than can X-rays. Discuss. (See Problem 4.)
3. What experimental evidence is there for assuming that the speed of sound is the same for all wavelengths?
4. Give a qualitative explanation why the speed of sound in lead is less than that in copper.
5. What quantity, if any, for transverse waves in a string corresponds to the pressure amplitude for longitudinal waves in a tube?
6. A bell is rung for a short time in a school. After a while its sound is inaudible. Trace the sound waves and the energy they transfer from the time of emission until they become inaudible.
7. How can we experimentally locate the positions of nodes and antinodes in a string? In an air column? On a vibrating surface?
8. What physical properties of a sound wave corresponds to the human sensation of pitch, of loudness, and of tone quality?
9. What is the difference between a violin note and the same note sung by a human voice that enables us to distinguish between them?
10. Bells frequently sound much less pleasant than pianos or violins. Why?
11. Does your singing really sound better in a shower? If so, are there physical reasons for this?
12. Discuss the factors that determine the range of frequencies in your voice and the quality of your voice.
13. Explain the origin of the sound in ordinary whistling.

* See "Cerenkov Radiation: its Origin, Properties and Applications," by J. V. Jelley in *Contemporary Physics*, October 1961.

14. What is the common purpose of the valves of a cornet and the slide of a trombone?*

15. The bugle has no valves. How then can we sound different notes on it? To what notes is the bugler limited? Why?*

16. The pitch of the wind instruments rises and that of the string instruments falls as an orchestra "warms up." Explain.*

17. Would a plucked violin string oscillate for a longer or shorter time if the violin had no sounding board? Explain.*

18. Explain how bowing a violin string gets it to vibrate.*

19. Explain the audible tone produced by drawing a wet finger around the rim of a wine glass.

20. Explain how a stringed instrument is "tuned."*

21. A tube can act like an acoustic filter, discriminating against the passage through it of sound of frequencies different from the natural frequencies of the tube. The muffler of an automobile is an example. (a) Explain how such a filter works. (b) How can we determine the cut-off frequency, below which frequency sound is not transmitted?

22. Two ships with steam whistles of the same pitch sound off in the harbor. Would you expect this to produce an interference pattern with regions of high and low intensity?

23. Can sound waves from a *single* tuning fork interfere? How can you explain that the fork is much less audible in certain directions than in others?

24. Two identical tuning forks emit notes of the same frequency. Explain how you might hear beats between them.

25. Suppose that, in the Doppler effect for sound, the source and receiver are at rest in some reference frame but the transmitting medium is moving with respect to this frame. Will there be a change in wavelength, or in frequency, received?

26. Is there a Doppler effect for sound when the observer or the source moves at right angles to the line joining them? How then can we determine the Doppler effect when the motion has a component at right angles to this line?

27. A satellite emits radio waves of constant frequency. These waves are picked up on the ground and made to beat against some standard frequency. The beat frequency is then sent through a loudspeaker and one "hears" the satellite signals. Describe how the sound changes as the satellite approaches, passes overhead, and recedes from the detector on the ground.

28. Discuss factors that improve the acoustics in music halls.†

29. A lightning flash dissipates an enormous amount of energy and is essentially instantaneous. How is that energy transformed into the sound waves of thunder and why is that sound often a spread-out sequence of noises?‡

30. Transverse waves in a string can be polarized (see, for example, Question 18 of Chapter 19). Can sound waves be polarized?

31. Bats can examine the characteristics of objects—such as size, shape, distance, direction, motion—by sensing the way the high-frequency sounds

* See the following articles for discussions of the physics of musical instruments: "Acoustics of the Flute" by John W. Coltman, in *Physics Today*, November 1968. "The Physics of Wood Winds" by Arthur H. Benade, in *Scientific American*, October 1960. "The Physics of Brasses" by Arthur H. Benade, in *Scientific American*, July 1973. "The Physics of the Piano" by E. Dornell Blackham, in *Scientific American*, December 1965. "The Physics of Violins" by Carleen M. Hutchins, in *Scientific American*, November 1962. "The Electronic Music Synthesizer and the Physics of Music" by W. M. Hartman, in *American Journal of Physics*, September 1975.

† See "The Development of Architectural Acoustics" by Robert S. Shankland in *American Scientist*, March–April 1972.

‡ See "Thunder" by Arthur A. Few, in *Scientific American*, July 1975.

they emit are reflected off the objects back to the bat. Discuss qualitatively each of these features. (See "Information Content of Bat Sonar Echoes" by J. A. Simmons, D. J. Howell, and N. Suga in *American Scientist*. March–April 1975.

problems

SECTION 20-2

1. The lowest pitch detectable as sound by the average human ear is about 20 Hz and the highest is about 20,000 Hz. What is the wavelength of each in air?   *Answer:* 17 m; 1.7 cm.

2. Bats emit ultrasonic waves. The shortest wavelength emitted in air by a bat is about 0.13 in. (3.3 mm). What is the highest frequency a bat can emit?

3. A sound wave has a frequency of 440 Hz. What is the wavelength of this sound (a) in air and (b) in water?   *Answer:* (a) 75 cm. (b) 3.3 m.

4. Intense ultrasound of frequency 10 MHz is used to modify or destroy tumors in soft tissue. (a) What is the wavelength in air of such a sound wave? (b) If the speed of sound in tissue is 1500 m/s, what is the wavelength of this wave in tissue?

5. (a) A conical loudspeaker has a diameter of 6.0 in. At what frequency will the wavelength of the sound it emits in air be equal to its diameter? Be ten times its diameter? Be one-tenth its diameter? (b) Make the same calculations for a speaker of diameter 12 in. (*Note:* If the wavelength is large compared to the diameter of the speaker, the sound waves spread out almost uniformly in all directions from the speaker, but when the wavelength is small compared to the diameter of the speaker, the wave energy is propagated mostly in the forward direction.)
Answer: (a) 2.2; 0.22; 22 kHz. (b) 1.1; 0.11; 11 kHz.

6. (a) A rule for finding your distance from a lightning flash is to count seconds from the time you see the flash until you hear the thunder and then divide the count by five. The result is supposed to give the distance in miles. Explain this rule and determine the percent error in it at standard conditions. (b) Can you devise a similar rule for the distance in kilometers?

7. The speed of sound in a certain metal is V. One end of a long pipe of that metal of length l is struck a hard blow. A listener at the other end hears two sounds, one from the wave that has traveled along the pipe and the other from the wave that has traveled through the air. (a) If v is the speed of sound in air, what time interval t elapses between the two sounds? (b) Suppose $t = 1.0$ s and the metal is iron. Find the length l.
Answer: (a) $l(V - v)/Vv$. (b) 350 m.

8. Two spectators at a soccer game in a large stadium see, and a moment later hear, the ball being kicked on the playing field. If the time delay for one spectator is 0.90 s and for the other 0.60 s, and lines through each spectator and the player kicking the ball meet at an angle of 90°, (a) how far is each spectator from the player? (b) How far are the spectators from each other?

9. A stone is dropped into a well. The sound of the splash is heard 3.0 s later. What is the depth of the well?   *Answer:* 41 m.

SECTION 20-3

10. The pressure in a traveling sound wave is given by the equation

$$p = 1.5 \sin \pi(x - 330t).$$

where x is in meters, t in seconds, and p in pascals. Find (a) the pressure amplitude, (b) the frequency, (c) the wavelength, and (d) the speed of the wave.

11. Two waves give rise to pressure variations at a certain point in space given by

$$p_1 = P \sin 2\pi\nu t,$$

$$p_2 = P \sin 2\pi(\nu t - \phi).$$

What is the pressure amplitude of the resultant wave at this point when $\phi = 0$, $\phi = \frac{1}{4}$, $\phi = \frac{1}{6}$, $\phi = \frac{1}{8}$? All ϕ's are measured in radians.
Answer: 2.00 P; 1.41 P; 1.73 P; 1.85 P.

12. In Fig. 20-13 we show an acoustic interferometer, used to demonstrate the interference of sound waves. S is a diaphragm that vibrates under the influence of an electromagnet. D is a sound detector, such as the ear or a microphone. Path SBD can be varied in length, but path SAD is fixed.

The interferometer contains air, and it is found that the sound intensity has a minimum value of 100 units at one position of B and continuously climbs to a maximum value of 900 units at a second position 1.65 cm from the first. Find (a) the frequency of the sound emitted from the source, and (b) the relative amplitudes of the waves arriving at the detector for either of the two positions of B. (c) How can it happen that these waves have different amplitudes, considering that they originate at the same source?

13. A spherical sound source is placed at P_1 near a reflecting wall AB and a microphone is located at point P_2, as shown in Fig. 20-14. The frequency of the sound source P_1 is variable. Find two different frequencies for which the sound intensity, as observed at P_2, will be a maximum. The speed of sound in air is 1100 ft/s. Assume the paths of the interfering waves to be parallel.
Answer: 31 Hz; 94 Hz.

14. Show that the intensity of a sound wave (a) when expressed in terms of the pressure amplitude P, is given by

$$I = \frac{p^2}{2\rho_0 v},$$

where v is the speed of the wave and ρ_0 is the standard density of air, and, (b) when expressed in terms of the displacement amplitude y_m, is given by

$$I = 2\pi^2 \rho_0 v y_m^2 \nu^2,$$

where ν is the frequency of the wave. (c) If two sound waves, one in air and one in water, are equal in intensity, what is the ratio of the pressure amplitude of the wave in water to that of the wave in air? (d) If the pressure amplitudes are equal instead, what is the ratio of the intensities of the waves?

15. A sound wave of frequency 1000 Hz propagating through air has a pressure amplitude of 10 Pa. What are the (a) wavelength, (b) particle displacement amplitude, and (c) maximum particle speed?
Answer: (a) 33 cm. (b) 40 μm. (c) 2.5 cm/s.

16. A note of frequency 300 Hz has an intensity of 1.0 μW/m². What is the amplitude of the air vibrations caused by this sound?

17. A certain loudspeaker produces a sound with a frequency of 2000 Hz and an intensity of 1.2 × 10⁻⁷ hp/ft² (9.6 × 10⁻⁴ W/m²) at a distance of 20 ft (6.1 m). Presume that there are no reflections and that the loudspeaker emits the same in all directions. (a) What would be the intensity at 100 ft (30 m)? (b) What is the displacement amplitude at 20 ft (6.1 m)? (c) What is the pressure amplitude at 20 ft (6.1 m)?
Answer: (a) 4.8 × 10⁻⁹ hp/ft² (4.0 × 10⁻⁵ W/m²). (b) 5.7 × 10⁻⁷ ft (1.7 × 10⁻⁷ m). (c) 1.3 × 10⁻⁴ lb/in.² (0.88 Pa).

18. Two sources of sound are separated by a distance of 10 m. They both emit sound at the same amplitude and frequency, 300 Hz, but they are 180° out of phase. At what points along the line between them will the sound intensity be at a relative minimum due to destructive interference?

19. The violin section in some symphony orchestras is divided into two parts,

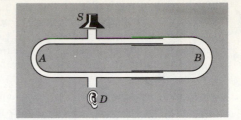

figure 20-13
Problem 12

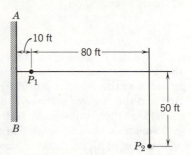

figure 20-14
Problem 13

one placed on each side of the conductor. Consider two violinists 8.0 m apart, symmetrically placed with respect to the conductor and each 5.0 m from him. If the power output from each is 1.0×10^{-4} W, what are (a) the intensity of each playing alone as heard by the conductor, and (b) the combined intensity of both playing together (the same note) as heard by the conductor? *Answer:* (a) 3.2×10^{-7} W/m². (b) 4.6×10^{-7} W/m².

20. Two loudspeakers, S_1 and S_2, each emit sound of frequency 200 Hz uniformly in all directions (three-dimensional waves). S_1 has an acoustic output of 1.2×10^{-3} W and S_2 one of 1.8×10^{-3} W. The loudspeakers are 7.0 m apart. Consider a point P which is 4.0 m from S_1 and 3.0 m from S_2. (a) How are the phases of the two waves arriving at P related? What is the intensity of sound at P (b) if S_2 is turned off (S_1 on), (c) if S_1 is turned off (S_2 on), and (d) with both S_1 and S_2 on?

SECTION 20-5

21. In Fig. 20-15 a rod R is clamped at its center and a disk D at its end projects into a glass tube, which has cork filings spread over its interior. A plunger P is provided at the other end of the tube. The rod is set into longitudinal vibration and the plunger is moved until the filings form a pattern of nodes and antinodes (the filings form well-defined ridges at the pressure antinodes). If we know the frequency ν of the longitudinal vibrations in the rod, a measurement of the average distance d between successive antinodes determines the speed of sound v in the gas in the tube. Show that

$$v = 2\nu d.$$

This is *Kundt's method* for determining the speed of sound in various gases.

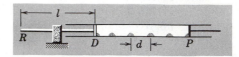

figure 20-15
Problem 21

22. If a violin string is tuned to a certain note, by how much must the tension in the string be increased if it is to emit a note of double the original frequency (that is, a note one octave higher in pitch)?

23. An open organ pipe has a fundamental frequency of 300 Hz. The first overtone of a closed organ pipe has the same frequency as the first overtone of the open pipe. How long is each pipe? *Answer:* 55 cm; 41 cm.

24. A 3.0-m skip rope is used essentially in its fundamental mode of oscillation. If the rope has a mass of 1.0 kg and the children are pulling back with a force of 10 N, what is the frequency of oscillation?

25. A certain violin string is 50 cm long between its fixed ends and has a mass of 2.0 g. The string sounds an A note (440 Hz) when played without fingering. Where must one put one's finger to play a C (528 Hz)?
Answer: 8.3 cm from one end.

26. The strings of a cello have a length L. (a) By what length l must they be shortened by fingering to change the pitch by a frequency ratio r? (b) Find l, if $L = 0.80$ m and $r = 6/5$, $5/4$, $4/3$, and $3/2$.

27. The water level in a vertical glass tube 1.0 m long can be adjusted to any position in the tube. A tuning fork vibrating at 660 Hz is held just over the open top end of the tube. At what positions of the water level will there be resonance? *Answer:* Water filled to a height of 7/8, 5/8, 3/8, or 1/8 m.

28. S in Fig. 20-16 is a small loudspeaker driven by an audio oscillator and amplifier, adjustable in frequency from 1000 to 2000 Hz only. D is a piece of cylindrical sheetmetal pipe 18.0 in. long. (a) If the speed of sound in air is 1130 ft/s at the existing temperature, at what frequencies will resonance occur when the frequency emitted by the speaker is varied from 1000 to 2000 Hz? (b) Sketch the displacement nodes for each. Neglect end effects.

29. A well with vertical sides and water at the bottom resonates at 7.0 Hz and at no lower frequency. The air in the well has a density of 1.1 kg/m³ (2.1×10^{-3} slug/ft³), a pressure of 9.5×10^4 Pa (13.8 lb/in.²), and a ratio of specific heats of 7/5. How deep is the well? *Answer:* 12 m (41 ft).

figure 20-16
Problem 28

30. The period of a pulsating variable star may be estimated by considering the star to be executing radial longitudinal pulsations in the fundamental

standing wave mode; that is, the radius varies periodically with the time, with a displacement antinode at the surface. (a) Would you expect the center of the star to be a displacement node or antinode? (b) By analogy with the open organ pipe, show that the period of pulsation T is given by

$$T = 4R/v_s,$$

where R is the equilibrium radius of the star and v_s is the average sound speed. (c) Typical white dwarf stars have pressures of 10^{22} Pa, densities of 10^{10} kg/m³, ratio of specific heats of 4/3, and radius 0.009 solar radii. What is the approximate pulsation period of a white dwarf? (See "Pulsating Stars" by John R. Percy, in *Scientific American*, June 1975.)

31. A tube 1.0 m (3.3 ft) long is closed at one end. A stretched wire is placed near the open end. The wire is 0.30 m (0.98 ft) long and has a mass of 0.010 kg (6.9×10^{-4} slug). It is fixed at both ends and vibrates in its fundamental mode. It sets the air column in the tube into vibration at its fundamental frequency by resonance. Find (a) the frequency of oscillation of the air column and (b) the tension in the wire.
   Answer: (a) 83 Hz (82 Hz). (b) 82 N (18 lb).

32. A 31.6-cm violin string with linear density 0.65 g/m is placed near a loudspeaker that is fed by an audio-oscillator of variable frequency. It is found that the string is set in oscillation only at the frequencies 880 and 1320 Hz as the frequency of the oscillator is varied continuously over the range 500 to 1500 Hz. What is the tension in the string?

SECTION 20-6

33. Two identical piano wires have a fundamental frequency of 600 Hz when kept under the same tension. What fractional increase in the tension of one wire will lead to the occurrence of six beats per second when both wires vibrate simultaneously?
   Answer: 2.0%.

34. A tuning fork of unknown frequency makes three beats per second with a standard fork of frequency 384 Hz. The beat frequency decreases when a small piece of wax is put on a prong of the first fork. What is the frequency of this fork?

SECTION 20-7

35. A bullet is fired with a speed of 2200 ft/s. Find the angle made by the shock wave with the line of motion of the bullet.
   Answer: 30°.

36. Calculate the speed of the projectile illustrated in the photograph in Fig. 20-12. Assume the speed of sound in the medium through which the projectile is traveling to be 380 m/s.

37. The speed of light in water is about three-fourths the speed of light in vacuum. A beam of high-speed electrons from a betatron emits Cerenkov radiation in water, the wavefront being a cone of angle 120°. Find the speed of the electrons in the water?
   Answer: 2.6×10^8 m/s.

38. A jet plane passes overhead at a height of 5000 m and a speed of Mach 1.5 (that is, 1.5 times the speed of sound). (a) Find the angle made by the shock wave with the line of motion of the jet. (b) How long after the jet has passed directly overhead will the shock wave reach the ground?

39. A whistle of frequency 540 Hz rotates in a circle of radius 2.00 ft an an angular speed of 15.0 rad/s. What is (a) the lowest and (b) the highest frequency heard by a listener a long distance away at rest with respect to the center of the circle?
   Answer: (a) 525 Hz. (b) 555 Hz.

40. (a) Could you go through a red light fast enough to have it appear green? (b) If so, would you get a ticket for speeding? Take $\lambda = 620$ nm (= 620 nanometer = 620×10^{-9} m; see Table 1-2) for red light, $\lambda = 540$ nm for green light, and $c = 3.0 \times 10^8$ m/s as the speed of light.

41. A bat is flittering about in a cave, navigating very effectively by the use of ultrasonic bleeps (short emissions lasting a millisecond or less and repeated several times a second). Assume that the sound emission frequency of the

bat is 39,000 Hz. During one fast swoop directly toward a flat wall surface, the bat is moving at 1/40 of the speed of sound in air. What frequency does he hear reflected off the wall?   *Answer:* 41,000 Hz.

42. A source of sound waves of frequency 1080 Hz moves to the right with a speed of 108 ft/s relative to the ground. To its right is a reflecting surface moving to the left with a speed of 216 ft/s relative to the ground. Take the speed of sound in air to be 1080 ft/s and find (*a*) the wavelength of the sound emitted in air by the source, (*b*) the number of waves per second arriving at the reflecting surface, (*c*) the speed of the reflected waves, (*d*) the wavelength of the reflected waves.

43. A siren emitting a sound of frequency 1000 Hz moves away from you toward a cliff at a speed of 10 m/s. (*a*) What is the frequency of the sound you hear coming directly from the siren? (*b*) What is the frequency of the sound you hear reflected off the cliff? (*c*) Could you hear the beat frequency? Take the speed of sound in air as 330 m/s.
Answer: (*a*) 970 Hz. (*b*) 1030 Hz. (*c*) No. It is too high.

44. Microwaves, which travel with the speed of light, are reflected from a distant airplane approaching the wave source. It is found that when the reflected waves are beat against the waves radiating from the source the beat frequency is 990 Hz. If the microwaves are 0.10 m in wavelength, what is the approach speed of the airplane?

45. Radar measurements in pursuing situations are relatively inaccurate compared to rest situations. (*a*) Consider a radar unit at rest and show that the difference dv between the frequency reflected off a car moving at a speed V and the transmitted frequency v is given approximately by $dv/v = 2V/c$. (*b*) Now consider the radar unit to be in a pursuing vehicle moving at a speed v and show that $dv/v = 2(v - V)/c$. Discuss various cases and justify the first sentence: in particular, consider (*c*) the case in which the pursuer (police) is moving at the same speed as the speeder. What Doppler shift is observed in this case?   *Answer:* None.

46. Equation 20-11 for the Doppler effect assumes a reference frame at rest in the medium through which the sound travels. Suppose instead that the reference frame is fixed to the earth and that the medium moves with speed v_m from source to observer. How must you modify Eq. 20-11 in this (more general) case?

47. A girl is sitting near the open window of a train that is moving at a velocity of 10.0 m/s to the east. The girl's uncle stands near the tracks and watches the train move away. The locomotive whistle vibrates at 500 Hz. The air is still. (*a*) What frequency does the uncle hear? (*b*) What frequency does the girl hear? A wind begins to blow from the east at 10 m/s. (*c*) What frequency does the uncle now hear? (*d*) What frequency does the girl now hear?
Answer: (*a*) 485 Hz. (*b*) 500 Hz. (*c*) 486 Hz. (*d*) 500 Hz.

48. A woman standing on the ground beside a highway blows her whistle (pitch 800 Hz) to alert three colleagues waiting 200 m away, one each to the north, east, and south. A fourth confederate is driving west at 40 m/s. A steady wind at 4.0 m/s is blowing from south to north. What is the frequency of the sound heard by each of the waiting women and the driving woman?

21
temperature

In analyzing physical situations we usually focus our attention on some portion of matter which we separate, in our minds, from the environment external to it. We call such a portion the *system*. Everything outside the system which has a direct bearing on its behavior we call the *environment*. We then seek to determine the behavior of the system by finding how it interacts with its environment. For example, a ball can be the system and the environment can be the air and the earth. In free fall we seek to find how the air and the earth affect the motion of the ball. Or the gas in a container can be the system, and a movable piston and a Bunsen burner can be the environment. We seek to find how the behavior of the gas is affected by the action of the piston and burner. In all such cases we must choose suitable observable quantities to describe the behavior of the system. We classify these quantities, which are gross properties of the system measured by laboratory operations, as *macroscopic*. For processes in which heat is involved the laws relating the appropriate macroscopic quantities (which include pressure, volume, temperature, internal energy, and entropy, among others) form the basis for the science of *thermodynamics*. Many of the macroscopic quantities (pressure, volume, and temperature, for example) are directly associated with our sense perceptions. We can also adopt a *microscopic* point of view. Here we consider quantities that describe the atoms and molecules that make up the system, their speeds, energies, masses, angular momenta, behavior during collisions, etc. These quantities, or mathematical formulations based on them, form the basis for the science of *statistical mechanics*. The microscopic properties are not directly associated with our sense perceptions.

For any system the macroscopic and the microscopic quantities must

21-1
MACROSCOPIC AND MICROSCOPIC DESCRIPTIONS

be related because they are simply different ways of describing the same situation. In particular, we should be able to express the former in terms of the latter. The pressure of a gas, viewed macroscopically, is measured operationally using a manometer (Fig. 17-10). Viewed microscopically it is related to the average rate per unit area at which the molecules of the gas deliver momentum to the manometer fluid as they strike its surface. In Section 23-4 we will make this microscopic definition of pressure quantitative. Similarly (see Section 23-5), the temperature of a gas may be related to the average kinetic energy of translation of the molecules.

If the macroscopic quantities can be expressed in terms of the microscopic quantities, we should be able to express the laws of thermodynamics quantitatively in the language of statistical mechanics. We can indeed do this. In the words of R. C. Tolman:

The explanation of the complete science of thermodynamics in terms of the more abstract science of statistical mechanics is one of the greatest achievements of physics. In addition, the more fundamental character of statistical mechanical considerations makes it possible to supplement the ordinary principles of thermodynamics to an important extent.

We begin our examination of heat phenomena in this chapter with a study of temperature. As we progress we shall try to gain a deeper understanding of these phenomena by interweaving the microscopic and the macroscopic description — statistical mechanics and thermodynamics. The interweaving of the microscopic and the macroscopic points of view is characteristic of present-day physics.

21-2
THERMAL EQUILIBRIUM — THE ZEROTH LAW OF THERMODYNAMICS

The sense of touch is the simplest way to distinguish hot bodies from cold bodies. By touch we can arrange bodies in the order of their hotness, deciding that A is hotter than B, B than C, etc. We speak of this as our *temperature* sense. This is a very subjective procedure for determining the temperature of a body and certainly not very useful for purposes of science. A simple experiment, suggested in 1690 by John Locke, shows the unreliability of this method. Let a person immerse his hands, one in hot water, the other in cold. Then let him put both hands in water of intermediate hotness. This will seem cooler to the first hand and warmer to the second hand. Our judgment of temperature can be rather misleading. Further, the range of our temperature sense is limited. What we need is an objective, numerical, measure of temperature.

To begin with, we should try to understand the meaning of temperature. Let an object A which feels cold to the hand and an identical object B which feels hot be placed in contact with each other. After a sufficient length of time, A and B give rise to the same temperature sensation. Then A and B are said to be in *thermal equilibrium* with each other. We can generalize the expression "two bodies are in thermal equilibrium" to mean that the two bodies are in states such that, if the two *were* connected, the combined systems would be in thermal equilibrium. The logical and operational test for thermal equilibrium is to use a third or test body, such as a thermometer. This is summarized in a postulate often called *the zeroth law of thermodynamics: If A and B are each in thermal equilibrium with a third body C (the "thermometer"), then A and B are in thermal equilibrium with each other.*

This discussion expresses the idea that the temperature of a system is a property which eventually attains the same value as that of other systems when all these systems are put in contact. This concept agrees with the everyday idea of temperature as the measure of the hotness or coldness of a system, because as far as our temperature sense can be trusted, the hotness of all objects becomes the same after they have been in contact long enough. The idea contained in the zeroth law, although simple, is not obvious. For example, Jones and Smith each know Green, but they may or may not know each other. Two pieces of iron attract a magnet but they may or may not attract each other.

A more formal, but perhaps more fundamental phrasing of the zeroth law is: *There exists a scalar quantity called* temperature, *which is a property of all thermodynamic systems (in equilibrium states), such that temperature equality is a necessary and sufficient condition for thermal equilibrium.* This statement* justifies our use of temperature as a thermodynamic variable; the formulation given above is the corollary of this new statement. Speaking loosely, the essence of the zeroth law is: *there exists a useful quantity called "temperature."*

21-3 MEASURING TEMPERATURE

There are many measurable physical properties that vary as our physiological perception of temperature varies. Among these are the volume of a liquid, the length of a rod, the electrical resistance of a wire, the pressure of a gas kept at constant volume, the volume of a gas kept at constant pressure, and the color of a lamp filament. Any of these properties can be used in the construction of a thermometer—that is, in the setting up of a particular "private" temperature scale. Such a temperature scale is established by choosing a particular thermometric substance and a particular thermometric property of this substance. We then define the temperature scale by an *assumed* continuous monotonic relation between the chosen thermometric property of our substance and the temperature as measured on our (private) scale. For example, the thermometric substance may be a liquid in a glass capillary tube and the thermometric property can be the length of the liquid column; or the thermometric substance may be a gas kept in a container at constant volume and the thermometric property can be the pressure of the gas; and so forth. *We must realize that each choice of thermometric substance and property—along with the assumed relation between property and temperature—leads to an individual temperature scale whose measurements need not necessarily agree with measurements made on any other independently defined temperature scale.*

This apparent chaos in the definition of temperature is removed by universal agreement, within the scientific community, on the use of a particular thermometric substance, a particular thermometric property, and a particular functional relation between measurements of that property and a universally accepted temperature scale. A private temperature scale defined in any other way can then always be calibrated against the universal scale. We describe such a universal scale in Section 21-5 and an equivalent one in Section 25-6.

Suppose that we have chosen a thermometric substance. Let us represent by X the thermometric property that we wish to use in setting up a temperature scale. We arbitrarily choose the following linear function of the property X as the temperature T which the appropriate thermometer, and any system in thermal equilibrium with it, has:

$$T(X) = aX. \qquad (21\text{-}1)$$

---

* See J. S. Thomsen, "A Restatement of the Zeroth Law of Thermodynamics," *American Journal of Physics*, 30, 294, 1962.

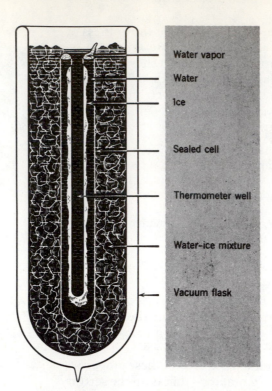

Water vapor

Water

Ice

Sealed cell

Thermometer well

Water-ice mixture

Vacuum flask

figure 21-1
The National Bureau of Standards triple-point cell. It contains pure water and is sealed after all air has been removed. It is then immersed in a water-ice bath. The system is at the triple point when ice, water, and vapor are all present, and in equilibrium, inside the cell. The thermometer to be calibrated is immersed in the central well.

In this expression a is a constant which we must still evaluate. By choosing this linear form for $T(X)$ we have fixed it so that *equal temperature differences*, or temperature intervals, *correspond to equal changes in X*. This means, for example, that every time the mercury column in the mercury-in-glass thermometer changes in length by one unit, the temperature changes by a definite fixed amount, no matter what the starting temperature. It also follows that two temperatures, measured with the same thermometer, are in the same ratio as their corresponding X's, that is,

$$\frac{T(X_1)}{T(X_2)} = \frac{X_1}{X_2}.$$

To determine the constant a, and hence to calibrate the thermometer, we specify a *standard fixed point* at which all thermometers must give the same reading for temperature T. This fixed point is chosen to be that at which ice, liquid water, and water vapor coexist in equilibrium and is called the *triple point of water*. This state can be achieved only at a definite pressure and is unique (Fig. 21-1). The water vapor pressure at the triple point is 4.58 mm-Hg. The temperature at this standard fixed point was arbitrarily set at 273.16 degrees Kelvin and was abbreviated 273.16° K. Later,* the name *kelvin* (symbol K) replaced degree Kelvin (symbol °K) and the unit of thermodynamic temperature was defined as follows: *The kelvin, unit of thermodynamic temperature, is the fraction 1/273.16 of the thermodynamic temperature of the triple point of water.*†

If we indicate values at the triple point by the subscript tr, then, for any thermometer,

$$\frac{T(X)}{T(X_{tr})} = \frac{X}{X_{tr}},$$

---

* Adopted in 1967 at the Tenth General Conference on Weights and Measures.
† The triple point of water was chosen over the freezing point (previously used) because the former is more reproducible, by a factor of ten, than the latter.

where, for *all* thermometers,

$$T(X_{tr}) = 273.16 \text{ K},$$

so that

$$T(X) = 273.16 \text{ K} \frac{X}{X_{tr}}. \qquad (21\text{-}2)$$

Hence, when the thermometric property has the value X, the temperature T, on the particular private scale selected, is given in K by $T(X)$, when the value of X and X_{tr} are inserted on the right-hand side of this equation.

We can now apply Eq. 21-2 to several thermometers. For a liquid-in glass thermometer X is L, the length of the liquid column, and Eq. 21-2 yields

$$T(L) = 273.16 \text{ K} \frac{L}{L_{tr}}.$$

For a gas at constant pressure, X is V, the volume of the gas, and

$$T(V) = 273.16 \text{ K} \frac{V}{V_{tr}} \qquad \text{(constant } P\text{)}.$$

For a gas at constant volume, X is P, the gas pressure, and

$$T(P) = 273.16 \text{ K} \frac{P}{P_{tr}} \qquad \text{(constant } V\text{)}.$$

For a platinum resistance thermometer, X is R, the electrical resistance, and

$$T(R) = 273.16 \text{ K} \frac{R}{R_{tr}},$$

and likewise for other thermometric substances and thermometric properties.

EXAMPLE 1

A certain platinum resistance thermometer has a resistance R of 90.35 ohms when its bulb is placed in a triple-point cell like that of Fig. 21-1. What temperature is defined by Eq. 21-2 if the bulb is placed in an environment such that its resistance is 96.28 ohms?

From Eq. 21-2,

$$T(X) = 273.16 \text{ K} \frac{X}{X_{tr}}$$

$$= (273.16 \text{ K})\left(\frac{96.28}{90.35}\right) = 280.6 \text{ K}.$$

Note that this temperature is on a private scale, defined by applying Eq. 21-2 to a particular device, the platinum resistance thermometer.

The question now arises whether the value we obtain for the temperature of a system depends on the choice of the thermometer we use to measure it. We have insured by definition that all the different kinds of thermometers will agree at the standard fixed point, but what happens at other points? We can imagine a series of tests in which the temperature of a given system is measured simultaneously with many different thermometers. Results of such tests show that the thermometers all read differently. Even when different thermometers of the same kind are used, such as constant-volume gas thermometers using different gases, we obtain different temperature readings for a given system in a given state.

Hence, to obtain a definite temperature scale, we must select one particular kind of thermometer as the standard. The choice will be made, not on the basis of experimental convenience, but by inquiring whether the temperature scale defined by a particular thermometer proves to be a useful quantity in the formulation of the laws of physics. The smallest variation in readings is found among different constant-volume gas thermometers, which suggests that we choose a gas as the standard thermometric substance. It turns out that as the amount of gas used in such a thermometer, and therefore its pressure, is reduced, the variation in readings between gas thermometers using different kinds of gas is reduced also. Hence, there seems to be something fundamental about the behavior of a constant-volume thermometer containing a gas at low pressure.

21-4
THE CONSTANT VOLUME GAS THERMOMETER

If the volume of a gas is kept constant, its pressure depends on the temperature and increases steadily with rising temperature. The constant-volume gas thermometer uses the pressure at constant volume as the thermometric property.

The thermometer is shown diagrammatically in Fig. 21-2. It consists of a bulb of glass, porcelain, quartz, platinum or platinum-iridium (depending on the temperature range over which it is to be used), connected by a capillary tube to a mercury manometer. The bulb containing some gas is put into the bath or environment whose temperature is to be measured; by raising or lowering the mercury reservoir the mercury in the left branch of the U-tube can be made to coincide with a fixed reference mark, thus keeping the confined gas at a constant volume. Then we read the height of the mercury in the right branch. The pressure of the confined gas is the difference of the heights of the mercury columns (times ρg) plus the atmospheric pressure, as indicated by the barometer. In practice the apparatus is very elaborate and we must make many corrections, for example, (1) to allow for the small volume change owing to slight contraction or expansion of the bulb and (2) to allow for the fact that not all the confined gas (such as that in the capillary) has been immersed in the bath. Assume that all corrections have been made, and let P be the corrected value of the pressure at the temperature of the bath. Then the temperature is given provisionally (see below) by

$$T(P) = 273.16 \text{ K} \frac{P}{P_{tr}} \quad \text{(constant } V\text{)}. \quad (21\text{-}3)$$

The constant-volume thermometer, used as described below, is the thermometer which serves to establish the temperature scale used universally in scientific work today.

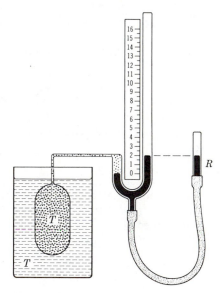

figure 21-2
A representation of a constant-volume gas thermometer. As long as the mercury in the left manometer tube remains at the same position on the scale (zero) the volume of the confined gas will be constant. The meniscus can always be brought to the zero position by raising or lowering reservoir R.

21-5
IDEAL GAS TEMPERATURE SCALE

Let a certain amount of gas be put into the bulb of a constant-volume gas thermometer so that when the bulb is surrounded by water at the triple point the pressure P_{tr} is equal to a definite value, say 80 cm-Hg. Now surround the bulb with steam condensing at 1-atm pressure and, with the volume kept constant at its previous value, measure the gas

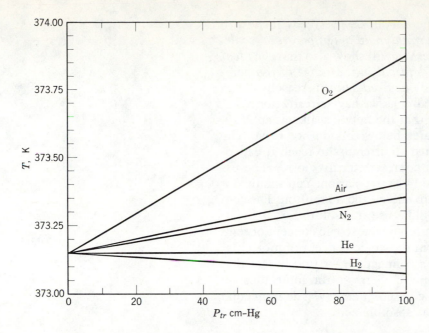

figure 21-3
The readings of a constant-volume gas thermometer for the temperature T of condensing steam as a function of P_{tr}, when different gases are used. As the amount of gas in the thermometer is reduced its pressure P_{tr} at the triple point decreases. Note that at a particular P_{tr} the values of T given by different gas thermometers differ. The discrepancy is small but measurable, being about 0.2% in the most extreme cases shown (O_2 and H_2 at 100 cm-Hg; note that the entire vertical axis covers only 1.00 K). Helium gives nearly the same T at all pressures (the curve is almost horizontal) so that its behavior is the most similar to that of an ideal gas over the entire range shown.

pressure P_s, the pressure at the steam point, in this case, P_{s80}. Then calculate the temperature provisionally from $T(P_{s80}) = 273.16$ K $(P_{s80}/80$ cm-Hg). Next remove some of the gas so that P_{tr} has a smaller value, say 40 cm-Hg. Then measure the new value of P_s and calculate another provisional temperature from $T(P_{s40}) = 273.16$ K $(P_{s40}/40$ cm-Hg). Continue this same procedure, reducing the amount of gas in the bulb again, and at this new lower value of P_{tr} calculating the temperature at the steam point $T(P_s)$. If we plot the values $T(P_s)$ against P_{tr} and have enough data, we can extrapolate the resulting curve to the intersection with the axis where $P_{tr} = 0$.

In Fig. 21-3, we plot curves obtained from such a procedure for constant-volume thermometers of some different gases. These curves show that the temperature readings of a constant-volume gas thermometer depend on the gas used at ordinary values of the reference pressure. However, as the reference pressure is decreased, the temperature readings of constant-volume gas thermometers using different gases approach the same value. Therefore, *the extrapolated value of the temperature depends only on the general properties of gases and not on any particular gas.* We therefore define an *ideal gas temperature scale* by the relation

$$T = 273.16 \text{ K } \lim_{P_{tr} \to 0} \left(\frac{P}{P_{tr}} \right) \qquad \text{(constant } V\text{).} \qquad (21\text{-}4)$$

Our standard thermometer is therefore chosen to be a constant-volume gas thermometer using a temperature scale defined by Eq. 21-4.

Although our temperature scale is independent of the properties of any one particular gas, it does depend on the properties of gases in general (that is, on the properties of a so-called ideal gas). Therefore, to measure a temperature, a gas must be used at that temperature. The lowest temperature that can be measured with any gas thermometer is about 1 K. To obtain this temperature we must use low-pressure helium, for helium becomes a liquid at a temperature lower than any other gas. Therefore we cannot give experimental meaning to temperatures below about 1 K, by means of a gas thermometer.

We would like to define a temperature scale in a way that is *inde-*

pendent of the properties of any particular substance. We will show in Section 25-6 that the *absolute thermodynamic temperature scale,* called the Kelvin scale, is such a scale. We will show also that *the ideal gas scale and the Kelvin scale are identical in the range of temperatures in which a gas thermometer may be used.* For this reason we can write "K" after an ideal gas temperature, as we have already done.

We will also show in Section 25-6 that the Kelvin scale has an *absolute zero* of 0 K and that temperatures below this do not exist. The absolute zero of temperature has defied all attempts to reach it experimentally, although it is possible to come arbitrarily close.* The existence of the absolute zero is inferred by extrapolation. You should not think of absolute zero as a state of zero energy and no motion. The conception that all molecular action would cease at absolute zero is incorrect. This notion assumes that the purely macroscopic concept of temperature is strictly connected to the microscopic concept of molecular motion. When we try to make such a connection, we find in fact that as we approach absolute zero the kinetic energy of the molecules approaches a finite value, the so-called zero-point energy. The molecular energy is a minimum, but not zero, at absolute zero.

Table 21-1
Some temperatures‡ (K)

Carbon thermonuclear reaction	5×10^8
Helium thermonuclear reaction	10^8
Solar interior	10^7
Solar corona	10^6
Shock wave in air at Mach 20	2.5×10^4
Luminous nebulae	10^4
Solar surface	6×10^3
Tungsten melts	3.6×10^3
Lead melts	6.0×10^2
Water freezes	2.7×10^2
Oxygen boils (1 atm)	9.0×10^1
Hydrogen boils (1 atm)	2.0×10^1
Helium (He^4) boils at 1 atm	4.2
He^3 boils at attainable low pressure	3.0×10^{-1}
Adiabatic demagnetization of paramagnetic salts	10^{-3}
Adiabatic demagnetization of nuclei	10^{-6}

‡ See *Scientific American*, September 1954, special issue on heat.

In Table 21-1 we list the temperatures, on the Kelvin scale, of various bodies and processes.

Two temperature scales in common use are the Celsius† and the Fahrenheit scales. These are defined in terms of the Kelvin scale, which is the fundamental temperature scale in science.

The Celsius temperature scale uses the unit "degree Celsius" (sym-

21-6
THE CELSIUS AND FAHRENHEIT SCALES

* It is possible to prepare systems that have *negative Kelvin temperatures.* Surprisingly enough, such temperatures are *not* reached by passing through 0 K but by proceeding through infinite temperatures. That is, negative temperatures are not 'colder' than absolute zero but instead are 'hotter' than infinite temperatures. See *Science by Degrees,* by Castle, Emmerich, Heikes, Miller, and Rayne, published by Walker and Company, New York, 1965. The absolute zero remains experimentally unattainable.

† This scale, based on a scale invented by a Swede named Celsius in 1742, was called the "centigrade" scale until 1948, when the Ninth General Conference on Weights and Measures decided that the name should be changed.

bol °C) equal to the unit "kelvin." If we let T_C represent the Celsius temperature, then

$$T_C = T - 273.15° \qquad (21\text{-}5)$$

relates the Celsius temperature $T_C(°C)$ and the Kelvin temperature $T(K)$. We see that the triple point of water (= 273.16 K by definition) corresponds to 0.01° C. By experiment the temperature at which ice and air-saturated water are in equilibrium at atmospheric pressure—the so-called ice point—proves to be 0.00° C and the temperature at which steam and liquid water are in equilibrium at 1-atm pressure—the so-called steam point—proves to be 100.00° C.

The Fahrenheit scale, still in use in some English-speaking countries (England itself adopted the Celsius scale for commercial and civil use in 1964) is not used in scientific work. The relationship between the Fahrenheit and Celsius scales is defined to be

$$T_F = 32 + \tfrac{9}{5}T_C.$$

From this relation we can conclude that the ice point (0.00° C) equals 32.0° F, that the steam point (100.0° C) equals 212.0° F, and that one Fahrenheit degree is exactly $\tfrac{5}{9}$ as large as one Celsius degree. In Fig. 21-4 we compare the Kelvin, Celsius, and Fahrenheit scales.

Let us now summarize the ideas of the last few sections. The standard fixed point in thermometry is the triple point of water which is arbitrarily assigned a value of 273.16 K. The constant-volume gas thermometer is the standard thermometer. The extrapolated gas scale is used to define the ideal gas temperature from $T = 273.16 \text{ K} \lim_{P_{tr} \to 0} (P/P_{tr})$. This scale is identical with the (absolute thermodynamic) Kelvin scale in the range in which a gas thermometer can be used.

By using the standard thermometer in this way, we can experimentally determine other reference points for temperature measurements, called fixed points. We list the basic fixed points adopted for experimental reference in Table 21-2. The temperatures can be expressed on

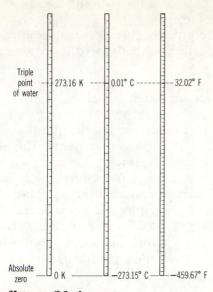

figure 21-4
The Kelvin, Celsius, and Fahrenheit temperature scales.

21-7
THE INTERNATIONAL PRACTICAL TEMPERATURE SCALE

Table 21-2
Fixed points on the international practical temperature scale[a]

Substance	State	Temperature	
		K	°C
Hydrogen	Triple point	13.81	−259.34
Hydrogen	Boiling point[b]	17.042	−256.108
Hydrogen	Boiling point	20.28	−252.87
Neon	Boiling point	27.102	−246.048
Oxygen	Triple point	54.361	−218.789
Oxygen	Boiling point	90.188	−182.962
Water[c]	Triple point	273.16	0.01
Water[c]	Boiling point	375.15	100
Zinc	Freezing point	692.73	419.58
Silver	Freezing point	1235.08	961.93
Gold	Freezing point	1337.58	1064.43

[a] The so-called IPTS-68, adopted in 1968 by the International Committee on Weights and Measures.
[b] This boiling point is for a pressure of 25/76 atm. All other boiling points (and all freezing points) are for a pressure of 1 atm.
[c] The water used should have the isotopic composition of sea water.

the Celsius scale, with the use of Eq. 21-5, once the Kelvin temperature is determined.

Determining ideal gas temperatures is painstaking. It would not make sense to use this procedure to determine temperatures for all work.

Hence, an International Practical Temperature Scale (IPTS) was adopted in 1927 (revised in 1948 and again in 1968) to provide a scale that can be used easily for practical purposes, such as for calibration of industrial or scientific instruments. This scale consists of a set of recipes for providing in practice the best possible approximations to the Kelvin scale. A set of fixed points, the basic points in Table 21-2, is adopted, and a set of instruments is specified to be used in interpolating between these fixed points and in extrapolating beyond the highest fixed point. The IPTS-68 departs from the Kelvin scale at temperatures between the fixed points, but the difference is usually negligible. The IPTS-68 has become the legal standard in nearly all countries.

Common effects of temperature changes are changes in size and changes of state of materials. Let us consider changes of size which occur without changes of state. Consider a simple model of a crystalline solid. The atoms are held together in a regular array by forces of electrical origin. The forces between atoms are like those that would be exerted by a set of springs connecting the atoms, so that we can visualize the solid body as a microscopic bedspring (Fig. 21-5). These "springs" are quite stiff (Problem 9, Chapter 15), and there are about 10^{22} of them per cubic centimeter. At any temperature the atoms of the solid are vibrating. The amplitude of vibration is about 10^{-9} cm, about one-tenth of an atomic diameter, and the frequency about 10^{13} Hz.

When the temperature is increased the average distance between atoms increases, which leads to an expansion of the whole solid body. The change in *any* linear dimension of the solid, such as its length, width, or thickness, is called a linear expansion. If the length of this linear dimension is l, the change in length, arising from a change in temperature ΔT, is Δl. We find from experiment that, if ΔT is small enough, this change in length Δl is proportional to the temperature change ΔT and to the original length l. Hence, we can write

$$\Delta l = \alpha l \, \Delta T, \qquad (21\text{-}6)$$

where α, called the *coefficient of linear expansion*, has different values for different materials. Rewriting this formula we obtain

$$\alpha = \frac{1}{l} \frac{\Delta l}{\Delta T},$$

so that α has the meaning of a fractional change in length per degree temperature change.

Strictly speaking, the value of α depends on the actual temperature and the reference temperature chosen to determine l (see Problem 13). However, its variation is usually negligible compared to the accuracy with which engineering measurements need to be made. We can safely take it as a constant for a given material, independent of the temperature. In Table 21-3 we list the experimental values for the average coefficient of linear expansion of several common solids. For all the substances listed, the change in size consists of an expansion as the

21-8
TEMPERATURE EXPANSION

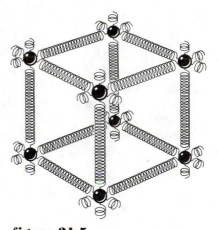

figure 21-5
A solid behaves in many ways as if it is a microscopic "bedspring" in which the molecules are held together by elastic forces.

Table 21-3
Some values* of $\bar{\alpha}$

Substance	$\bar{\alpha}$ (per C°)	Substance	$\bar{\alpha}$ (per C°)
Aluminum	23×10^{-6}	Hard rubber	80×10^{-6}
Brass	19×10^{-6}	Ice	51×10^{-6}
Copper	17×10^{-6}	Invar	0.7×10^{-6}
Glass (ordinary)	9×10^{-6}	Lead	29×10^{-6}
Glass (pyrex)	3.2×10^{-6}	Steel	11×10^{-6}

* For the range 0° C to 100° C; except −10° C to 0° C for ice.

temperature rises, for $\bar{\alpha}$ is positive. The order of magnitude of the expansion is about 1 millimeter per meter length per 100 Celsius degrees.**

EXAMPLE 2

A steel metric scale is to be ruled so that the millimeter intervals are accurate to within about 5×10^{-5} mm at a certain temperature. What is the maximum temperature variation allowable during the ruling?

From Eq. 21-6,

$$\Delta l = \alpha l \ \Delta T,$$

we have

$$5 \times 10^{-5} \text{ mm} = (11 \times 10^{-6}/C°)(1.0 \text{ mm}) \ \Delta T$$

in which we have used $\bar{\alpha}$ for steel, taken from Table 21-3. This yields $\Delta T \cong$ 5 C°. The temperature maintained during the ruling process must be maintained when the scale is being used and it must be held constant to within about 5 C°.

Note (see Table 21-3) that if the alloy invar is used instead of steel, then for the same required tolerance one can permit a temperature variation of about 75 C°; or for the same temperature variation ($\Delta T = 5$ C°) the tolerance achieved would be more than an order of magnitude better.

On the microscopic level thermal expansion of a solid suggests an increase in the average separation between the atoms in the solid. The potential energy curve for two adjacent atoms in a crystalline solid as a function of their internuclear separation is an asymmetric curve like that of Fig. 21-6. As the atoms move close together, their separation decreasing from the equilibrium value r_0, strong repulsive forces come into play and the potential curve rises steeply ($F = -dU/dr$); as the atoms move farther apart, their separation increasing from the equilibrium value, somewhat weaker attractive forces take over and the potential curve rises more slowly. At a given vibrational energy the separation of the atoms will change periodically from a minimum to a maximum value, the average separation being greater than the equilibrium separation because of the asymmetric nature of the potential energy curve. At still higher vibrational energy the average separation will be even greater. The effect is enhanced by the fact that in taking a time average of the motion one must allow for the longer time spent at extreme separations (lower vibrational speeds). Because the vibrational energy increases as the temperature rises, the average separation between atoms increases with temperature and the solid as a whole expands.

Note that if the potential energy curve were symmetric about the equilibrium separation, then no matter how large the amplitude of the vibration becomes the average separation would correspond to the equilibrium separation. Hence, thermal expansion is a direct consequence of the deviation from

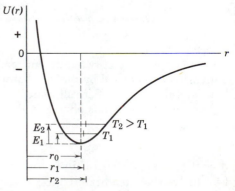

figure 21-6
Potential energy curve for two adjacent atoms in a crystalline solid as a function of internuclear separation. The equilibrium separation is r_0. Because the curve is asymmetric the average separation (r_1, r_2) increases as the temperature (T_1, T_2), and hence the vibrational energy (E_1, E_2), increases.

** One Celsius degree (1 C°) is a temperature *interval* (ΔT_C) of one unit measured on a Celsius scale. One degree Celsius (1° C) is a specific temperature reading (T_C) on that scale.

symmetry (i.e., the assymetry) of the potential energy curve characteristic of solids.

Some crystalline solids, in certain temperature regions, may contract as the temperature rises. The above analysis remains valid if one assumes that only compressional (i.e., longitudinal) modes of vibration exist or that these modes predominate. However, solids may vibrate in shear-like (i.e., transverse) modes as well and these modes of vibration will allow the solid to contract as the temperature rises, the average separation of the planes of atoms decreasing. For certain types of crystalline structure and in certain temperature regions these transverse modes of vibration may predominate over the longitudinal ones, giving a net negative coefficient of thermal expansion.

It should be emphasized that the microscopic models presented here are oversimplifications of a complex phenomenon which can be treated with greater insight with the use of thermodynamics and quantum theory.

For many solids, called *isotropic*, the percent change in length for a given temperature change is the same for all lines in the solid. The expansion is quite analogous to a photographic enlargement, except that a solid is three-dimensional. Thus, if you have a flat plate with a hole punched in it, $\Delta l/l$ $(= \alpha \Delta T)$ for a given ΔT is the same for the length, the thickness, the face diagonal, the body diagonal, and the hole diameter. Every line, whether straight or curved, lengthens in the ratio α per degree temperature rise. If you scratch your name on the plate, the line representing your name has the same fractional change in length as any other line. The analogy to a photographic enlargement is shown in Fig. 21-7.

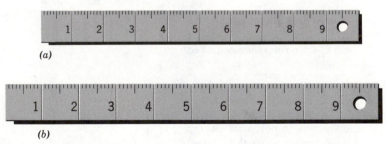

(a)

(b)

figure 21-7
The same steel rule at two different temperatures. On expansion every dimension is increased by the same proportion: the scale, the numbers, the hole, and the thickness are all increased by the same factor. (The expansion shown, from (a) to (b), is obviously exaggerated, for it would correspond to an imaginary temperature rise of about 100,000 C°!)

With these ideas in mind, you should be able to show (see Problems 14 and 16) that to a high degree of accuracy the fractional change in area A per degree temperature change for an isotropic solid is 2α, that is,

$$\Delta A = 2\alpha A \ \Delta T,$$

and the fractional change in volume V per degree temperature change for an isotropic solid is 3α, that is,

$$\Delta V = 3\alpha V \ \Delta T.$$

Because the shape of a fluid is not definite, only the change in volume with temperature is significant. Gases respond strongly to temperature or pressure changes, whereas the change in volume of liquids with changes in temperature or pressure is very much smaller. If we let β represent the coefficient of volume expansion for a liquid, that is,

$$\beta = \frac{1}{V}\frac{\Delta V}{\Delta T},$$

we find that β is relatively independent of the temperature. Liquids typically expand with increasing temperature, their volume expansion being generally about ten times greater than that of solids.

However, the most common liquid, water, does not behave like other liquids. In Fig. 21-8 we show the expansion curve for water. Notice that above 4° C water expands as the temperature rises, although not linearly. As the temperature is lowered from 4 to 0° C, however, water expands instead of contracting. Such an expansion with decreasing temperature is not observed in any other common liquid; it is observed in rubberlike substances and in certain crystalline solids over limited temperature intervals. The density of water is a maximum at 4° C, where its value* is 1000 kg/m³ or 1.000 g/cm³. At all other temperatures its density is less. This behavior of water is the reason why lakes freeze first at their upper surface.

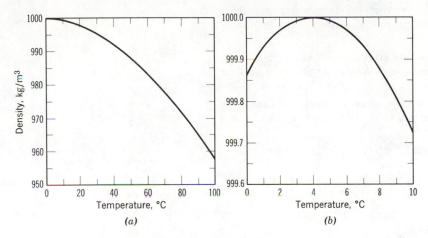

figure 21-8
(a) The variation with temperature of density of water under atmospheric pressure. (b) The variation between 0 and 10° C in more detail.

questions

1. Is temperature a microscopic or macroscopic concept?
2. Are there physical quantities other than temperature that tend to equalize if two different systems are joined?
3. Give a reasonable explanation for this: a piece of ice and a thermometer are suspended in an insulated evacuated enclosure so that they are not in contact and yet the thermometer reading decreases for a time.
4. Can a temperature be assigned to a vacuum?
5. Does our "temperature sense" have a built-in sense of direction; that is, does hotter necessarily mean higher temperature, or is this just an arbitrary convention? Celsius, by the way, originally chose the steam point as 0° C and the ice point as 100° C.
6. Figure 21-1 shows an apparatus by which the triple point of water is realized. How would you modify this apparatus to realize the freezing point of water?
7. How would you suggest measuring the temperature of (a) the sun, (b) the

*It is to this value of *unit* maximum density of water that the relative sizes of the kilogram and meter were originally supposed to correspond. Accurate measurements show, however, that the international standards of mass and length do not correspond exactly to this value. The maximum density of water is actually 999.973 kg/m³ at 3.98° C.

earth's upper atmosphere, (c) an insect, (d) the moon, (e) the ocean floor, and (f) liquid helium?

8. Is one gas any better than another for purposes of a standard constant-volume gas thermometer? What properties are desirable in a gas for such purposes?

9. State some objections to using water-in-glass as a thermometer. Is mercury-in-glass an improvement?

10. Can you explain why the column of mercury first descends and then rises when a mercury-in-glass thermometer is put in a flame?

11. What do the Celsius and Fahrenheit temperature conventions have in common?

12. Considering the Celsius, Fahrenheit, and Kelvin scales, does any one stand out as "Nature's scale"? Discuss.

13. What are the dimensions of α, the coefficient of linear expansion? Does the value of α depend on the unit of length used? When F° are used instead of C° as a unit of temperature change, does the numerical value of α change? If so, how?

14. A metal ball can pass through a metal ring. When the ball is heated, however, it gets stuck in the ring. What would happen if the ring, rather than the ball, were heated?

15. A bimetallic strip, consisting of two different metal strips riveted together, is used as a control element in the common thermostat. Explain how it works.

16. Two strips, one of iron and one of zinc, are riveted together side by side to form a straight bar which curves when heated. Why is it that the iron is on the inside of the curve?

17. Explain how the period of a pendulum clock can be kept constant with temperature by attaching tubes of mercury to the bottom of the pendulum. (See Problem 32.)

18. Explain why some rubberlike substances contract with rising temperature. (See Question 25, Chapter 25.)

19. Explain why the apparent expansion of a liquid in a bulb does not give the true expansion of the liquid.

20. Why do liquids typically have much larger volume coefficients of expansion than solids?

21. Does the change in volume of a body when its temperature is raised depend on whether the body has cavities inside, other things being equal? Consider a solid sphere and a hollow sphere, for example.

22. What difficulties would arise if you defined temperature in terms of the density of water?

23. Explain why lakes freeze first at the surface.

24. What is it that causes water pipes to burst in the winter?

25. What can you conclude about how the melting point of ice depends on pressure from the fact that ice floats on water?

SECTION 21-3

problems

1. A *resistance thermometer* is a thermometer in which the thermometric property is electrical resistance. We are free to define temperatures measured by such a thermometer in Kelvins to be directly proportional to the resistance R, measured in ohms. A certain resistance thermometer is found to have a resistance R of 90.35 ohms when its bulb is placed in water at the triple point temperature (273.16 K). What temperature is indicated by the thermometer if the bulb is placed in an environment such that its resistance is 96.28 ohms?   *Answer:* 291.1 K.

2. It is an everyday observation that hot and cold objects cool down or warm up to the temperature of their surroundings. If the temperature ΔT between an object and its surroundings is not too great, the rate of cooling or warming is approximately proportional to the temperature difference between the object and its surroundings; that is,

$$\frac{d \, \Delta T}{dt} = -K \, \Delta T,$$

where K is a constant. The minus sign appears because ΔT decreases with time if ΔT is positive and vice versa. This is known as *Newton's law of cooling*. (a) On what factors does K depend? What are its dimensions? (b) If at some instant $t = 0$ the temperature difference is ΔT_0, show that it is

$$\Delta T = \Delta T_0 e^{-Kt}$$

at a time t later.

3. A mercury-in-glass thermometer is placed in boiling water for a few minutes and then removed. The temperature readings at various times after removal are as follows:

t, s	T, °C	t, s	T, °C	t, s	T, °C	t, s	T, °C
0	98.4	25	65.1	100	50.3	700	26.5
5	76.1	30	63.9	150	43.7	1000	26.1
10	71.1	40	61.6	200	38.8	1400	26.0
15	67.7	50	59.4	300	32.7	2000	26.0
20	66.4	70	55.4	500	27.8	3000	26.0

Plot K as a function of time, assuming Newton's law of cooling to apply [see Problem 2]. To what extent are you justified in assuming that Newton's law of cooling applies here?

SECTION 21-5

4. If the ideal gas temperature at the steam point is 373.15 K, what is the limiting value of the ratio of the pressures of a gas at the steam point and at the triple point of water when the gas is kept at constant volume?

5. Let p_{tr} be the pressure in the bulb of a constant-volume gas thermometer when the bulb is at the triple-point temperature of 273.16 K and p the pressure when the bulb is at room temperature. Given three constant-volume gas thermometers: for No. 1 the gas is oxygen and $p_{tr} = 20$ cm-Hg; for No. 2 the gas is also oxygen but $p_{tr} = 40$ cm-Hg; for No. 3 the gas is hydrogen and $p_{tr} = 30$ cm-Hg. The measured values of p for the three thermometers are p_1, p_2, and p_3. (a) An approximate value of the room temperature T can be obtained with each of the thermometers using

$$T_1 = 273.16 \text{ K} \frac{p_1}{20 \text{ cm-Hg}}; \quad T_2 = 273.16 \text{ K} \frac{p_2}{40 \text{ cm-Hg}};$$

$$T_3 = 273.16 \text{ K} \frac{p_3}{30 \text{ cm-Hg}}.$$

Mark each of the following statements "true" or "false": (1) With the method described, all three thermometers will give the same value of T. (2) The two oxygen thermometers will agree with each other but not with the hydrogen thermometer. (3) Each of the three will give a different value of T. (b) In the event that there is disagreement among the three thermometers, explain how you would change the method of using them to cause all three to give the same value of T.
Answer: (a) (1) False; (2) false; (3) true. (b) Take the limiting value as $p_{tr} \to 0$.

SECTION 21-6

6. (a) The temperature of the surface of the sun is about 6000 K. Express this on the Fahrenheit scale. (b) Express normal human body temperature, 98.6° F, on the Celsius scale. (c) In the continental United States, the highest officially recorded temperature is 134° F at Death Valley, California, and the lowest is −70° F at Rogers Pass, Montana. Express these extremes on the Celsius scale. (d) Express the normal boiling point of oxygen, −183°C, on the Fahrenheit scale. (e) At what Celsius temperature would you find a room to be uncomfortably warm?

7. At what temperature do the following pairs of scales give the same reading? (a) Fahrenheit and Celsius. (b) Fahrenheit and Kelvin. (c) Celsius and Kelvin. *Answer:* (a) −40°. (b) 575°. (c) Not possible.

SECTION 21-7

8. In the interval between 0 and 660° C, a platinum resistance thermometer of definite specifications is used for interpolating temperatures on the International Practical Temperature Scale. The temperature T_C is given by a formula for the variation of resistance with temperature:

$$R = R_0(1 + AT_c + BT_c^2).$$

R_0, A, and B are constants determined by measurements at the ice point, the steam point, and the sulphur point. (a) If R equals 10.000 ohms at the ice point, 13.946 ohms at the steam point, and 24.817 ohms at the sulphur point, find R_0, A, and B. (b) Plot R versus T_C in the temperature range from 0 to 660° C.

SECTION 21-8

9. The Pyrex glass mirror in the telescope at the Mount Palomar Observatory has a diameter of 200 in. The temperature ranges from −10° to 50° C on Mount Palomar. Determine the maximum change in the diameter of the mirror. *Answer:* 0.038 in.

10. A circular hole in an aluminum plate is 1.000 in. (2.540 cm) in diameter at 0° C. What is its diameter when the temperature of the plate is raised to 100° C?

11. Steel railroad tracks are laid when the temperature is 0° C. A standard section of rail is then 12.0 m long. What gap should be left between rail sections so that there is no compression when the temperature gets as high as 42° C? *Answer:* 0.55 cm.

12. A steel rod is 3.000 cm in diameter at 25° C. A brass ring has an interior diameter of 2.992 cm at 25° C. At what common temperature will the ring just slide onto the rod?

13. Show that if α is treated as a variable, dependent on the temperature T, then

$$L = L_0 \left[1 + \int_{T_0}^{T} \alpha(T)\, dT \right]$$

where L_0 is the length at a reference temperature T_0.

14. The area A of a rectangular plate is ab. Its coefficient of linear expansion is α. After a temperature rise ΔT, side a is longer by Δa and side b is longer by Δb. Show that if we neglect the small quantity $\Delta a\, \Delta b/ab$ (see Fig. 21-9), then $\Delta A = 2\alpha A\, \Delta T$.

15. A glass window is exactly 20 cm (7.9 in.) by 30 cm (11.8 in.) at 10° C. By how much has its area increased when its temperature is 40° C? *Answer:* 0.32 cm² (0.050 in.²).

16. Prove that, if we neglect extremely small quantities, the change in volume of a solid on expansion through a temperature rise ΔT is given by $\Delta V = 3\alpha V\, \Delta T$ where α is the coefficient of linear expansion.

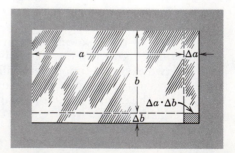

figure 21-9
Problem 14

17. Find the change in volume of an aluminum sphere of 10.0-cm (3.94-in.) radius when it is heated from 0° to 100° C.   *Answer:* 29 cm³ (1.8 in.³).

18. When the temperature of a "copper" penny is raised by 100 C°, its diameter increases by 0.18%. To two significant figures give the percent increase in the (a) area of a face, (b) thickness, (c) volume, and (d) mass of the penny. (e) What is the coefficient of linear expansion?

19. Density is mass per unit volume. If the volume V is temperature dependent, so is the density ρ. Show that the change in density $\Delta\rho$ with change in temperature ΔT is given by

$$\Delta\rho = -\beta\rho\,\Delta T$$

where β is the volume coefficient of expansion. Explain the minus sign.

20. Show that when the temperature of a liquid in a barometer changes by ΔT, and the pressure is constant, the height h changes by $\Delta h = \beta h\,\Delta T$ where β is the coefficient of volume expansion.

21. (a) Show that if the lengths of two rods of different solids are inversely proportional to their respective coefficients of linear expansion at some initial temperature, the difference in length between them will be constant at all temperatures. (b) What should be the lengths of a steel and a brass rod at 0° C so that at all temperatures their difference in length is 0.30 m? *Answer:* (b) Steel, 71 cm; brass, 41 cm.

22. Consider a mercury-in-glass thermometer. Assume that the cross-section of the capillary is constant at A_0, and that V_0 is the volume of the bulb of mercury at 0.00° C. If the mercury just fills the bulb at 0.00° C, show that the length of the mercury column in the capillary at a temperature $t°$ C is

$$l = \frac{V_0}{A_0}(\beta - 3\alpha)t,$$

that is, proportional to the temperature, where β is the volume coefficient of expansion of mercury and α is the linear coefficient of expansion of glass.

23. Imagine an aluminum cup of 0.1 liter capacity filled with mercury at 12° C. How much mercury, if any, will spill out of the cup if the temperature is raised to 18° C? (The coefficient of volume expansion of mercury is $1.8 \times 10^{-4}/C°$.)   *Answer:* 70 mm³.

24. A clock pendulum made of Invar has a period of 0.500 s at 20° C. If the clock is used in a climate where the temperature averages 30° C, what correction (approximately) is necessary at the end of 30 days to the time given by the clock?

25. (a) Prove that the change in rotational inertia I with temperature of a solid object is given by $\Delta I = 2\alpha I\,\Delta T$. (b) Prove that the change in period t of a physical pendulum with temperature is given by $\Delta t = \frac{1}{2}\alpha t\,\Delta T$.

26. Consider a uniform solid brass cylinder of mass $M = 0.50$ kg and radius $R = 0.030$ m. The cylinder is placed in frictionless bearings and set to rotate about its cylinder axis with an angular velocity $\omega = 60$ rad/s. (a) What is the angular momentum of the cylinder and how much work is required to reach this rate of rotation, starting from rest? After the cylinder has reached the state of rotation just described we heat it, without mechanical contact, from room temperature (20° C) to 100° C. Take the mean coefficient of linear expansion of brass to be $\bar\alpha = 2.0 \times 10^{-5}/C°$. Find the fractional changes, if any, in (b) the angular velocity, (c) the angular momentum, and (d) the kinetic energy of rotation of the cylinder.

27. A 1.0-m long vertical glass tube is half-filled with a liquid at 20° C. How much will the height of the liquid column change when the tube is heated to 30° C? Take $\bar\alpha_{glass} = 1.0 \times 10^{-5}/C°$ and $\bar\beta_{liquid} = 4 \times 10^{-5}/C°$. *Answer:* Increases by 0.10 mm.

28. A solid aluminim cylinder is suspended by a flexible steel belt attached to opposite walls at the same level, as shown in Fig. 21-10. It is required that the axis C of the cylinder not be moved by thermal expansions and contrac-

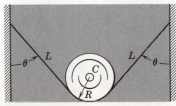

figure 21-10
Problem 28

tions of the cylinder and belt. The angle $\theta = 50°$ and remains practically unaffected by temperature changes. Find the radius R of the cylinder when $T = 290$ K if $L = 2.5$ m at this temperature. (Neglect the weight of the belt.)

29. Two vertical glass tubes filled with a liquid are connected at their lower ends by a horizontal capillary tube. One tube is surrounded by a bath containing ice and water in equilibrium (0.0° C), the other by a hot-water bath (t). The difference in height of the liquids in the two columns is Δh, and h_0 is the height of the column at 0.0° C. (a) Show how this apparatus (Fig. 21-11), first used in 1816 by Dulong and Petit, can be used to measure the true coefficient of volume expansion β of a liquid (rather than the differential expansion between glass and liquid). (b) Determine β if $t = 16.0°$ C, $h_0 = 126$ cm, and $\Delta h = 1.50$ cm.   *Answer:* (b) $7.44 \times 10^{-4}/C°$.

30. An aluminum cube 20 cm on an edge floats on mercury. How much further will the block sink down when the temperature rises from 270 K to 320 K? (The coefficient of volume expansion of mercury is $1.8 \times 10^{-4}/C°$.)

31. The distance between the towers of the main span of the Golden Gate Bridge at San Francisco is 4200 ft. The sag of the cable halfway between the towers at 50° F is 470 ft. Take $\alpha = 6.5 \times 10^{-6}/F°$ for the cable and compute (a) the change in length of the cable and (b) the change in sag for a temperature change from −20 to 110° F. Assume no bending or separation of the towers and a parabolic shape for the cable.   *Answer:* (a) 3.7 ft. (b) 6.5 ft.

32. A glass tube nearly filled with mercury is attached in tandem to the bottom of an iron pendulum rod 100 cm long. How high must the mercury be in the glass tube so that the center of mass of this pendulum will not rise or fall with changes in temperature? (The cross-sectional area of the tube is equal to that of the iron rod. Neglect the mass of the glass. Iron has a density of 7.87×10^3 kg/m³ and a linear coefficient of expansion equal to $12 \times 10^{-6}/C°$. The coefficient of volume expansion of mercury is $18 \times 10^{-5}/C°$.)

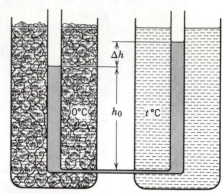

figure 21-11
Problem 29

22
heat and the first law of thermodynamics

When two systems at different temperatures are placed together, the final temperature reached by both systems is somewhere between the two starting temperatures. This is a common observation. Humans have long sought for a deeper understanding of such phenomena. Up to the beginning of the nineteenth century, they were explained by postulating that a material substance, *caloric*, existed in every body. It was believed that a body at high temperature contained more caloric than one at a low temperature. When the two bodies were put together, the body rich in caloric lost some to the other until both bodies reached the same temperature. The caloric theory was able to describe many processes, such as heat conduction or the mixing of substances in a calorimeter, in a satisfactory way. However, the concept of heat as a *substance*, whose total amount remained constant, eventually could not stand the test of experiment. Nevertheless, we still describe many common temperature changes as the transfer of "something" from one body at a higher temperature to one at the lower, and this "something" we call heat. A useful but nonoperational definition, is: *heat is that which is transferred between a system and its surroundings as a result of temperature differences only.*

Eventually it became generally understood that heat is a form of energy rather than a substance. The first conclusive evidence that heat could not be a substance was given by Benjamin Thompson (1753–1814), an American who later became Count Rumford of Bavaria. In a paper read before the Royal Society* in 1798 he wrote:

22-1
HEAT, A FORM OF ENERGY

---

* Rumford, an American, was instrumental in founding the Royal Institution in London. On the other hand, the Smithsonian Institution in Washington was founded on the basis of a £100,000 bequest from the estate of an Englishman, James Smithson (1765–1829).

I . . . am persuaded, that a habit of keeping the eyes open to everything that is going on in the ordinary course of the business of life has oftener led, as it were by accident, or in the playful excursions of the imagination . . . to useful doubts and sensible schemes for investigation and improvement, than all the more intense meditations of philosophers, in the hours expressly set apart for study. It was by accident that I was led to make the Experiments of which I am about to give an account.

Rumford made this discovery while supervising the boring of cannon for the Bavarian government. To prevent overheating, the bore of the cannon was kept full of water. The water was replenished as it boiled away during the boring process. It was accepted that caloric had to be supplied to water to boil it. The continuous production of caloric was explained by assuming that when a substance was more finely sub-divided, as in boring, its capacity for retaining caloric became smaller, and that the caloric released in this way was what caused the water to boil. Rumford observed in specific experiments, however, that the water boiled away even when his boring tools became so dull that they were no longer cutting or subdividing matter.

He wrote after ruling out by experiment all possible caloric interpretations,

. . . in reasoning on this subject, we must not forget to consider that most re-markable circumstance, that the source of Heat generated by friction, in these Experiments, appeared evidently to be *inexhaustible* . . . it appears to me to be extremely difficult, it not quite impossible, to form any distinct idea of any thing capable of being excited and communicated in the manner the Heat was excited and communicated in these Experiments, except it be *MOTION*.

Here we have the germ of the idea that the mechanical work ex-pended in the boring process was responsible for the creation of heat. The idea was not clearly put until much later, by others. Instead of the continuous disappearance of mechanical energy and the continuous cre-ation of heat, neither obeying any conservation principle, the whole process is now viewed as a transformation of energy from one form to another, the total energy being conserved.

Although the concept of energy and its conservation seems self-evident today, it was a novel idea as late as the 1850s and had eluded such men as Galileo and Newton. Throughout the subsequent history of physics this conservation idea led to new discoveries. Its early his-tory was remarkable in many ways. Several thinkers arrived at this great concept at about the same time; at first, all of them either met with a cold reception or were ignored. The principle of the conservation of energy was established independently by Julius Mayer (1814–1878) in Germany, James Joule (1818–1889) in England, Hermann von Helm-holtz (1821–1894) in Germany, and L. A. Colding (1815–1888) in Den-mark.*

It was Joule who showed by experiment that, when a given quantity of mechanical energy is converted to heat, the same quantity of heat

---

* After the posthumous publication of his *Reflections on the Motive Power of Fire* (in 1872, 40 years after his death) it became clear that Sadi Carnot (1796–1832) had arrived at the conservation of energy principle before all the others. It should give some food for thought to realize that the five men who first understood the conservation of energy prin-ciple were all young and all had major professional interests outside the field of physics: Mayer (medicine; age 28), Helmholtz (physiology; age 32), Colding (engineering; age 27), Joule (industrial management—he inherited his father's brewery; age 25), and Carnot (en-gineering; age 34). Rumford, age 45, was an old man by comparison.

is always developed. Thus, the equivalence of heat and mechanical work as two forms of energy was definitely established.

Helmholtz first expressed clearly the idea that not only heat and mechanical energy but all forms of energy are equivalent, and that a given amount of one form cannot disappear without an equal amount appearing in some of the other forms.

22-2
QUANTITY OF HEAT AND SPECIFIC HEAT

The unit of heat Q used to be defined* quantitatively in terms of a specified change produced in a body during a specified process. Thus, if the temperature of one kilogram of water is raised from 14.5 to 15.5° C by heating, we say that one *kilocalorie* (kcal) of heat has been added to the system. The *calorie* (= 10^{-3} kcal) is also used as a heat unit. (Incidentally, the "calorie" used to measure the energy content of foods is actually a kilocalorie.) In the engineering system the unit of heat is the *British thermal unit* (Btu), which is defined as the heat necessary to raise the temperature of one pound of water from 63 to 64° F.

The reference temperatures are stated because, near room temperature, there is a slight variation in the heat needed for a one-degree temperature rise with the temperature interval chosen. We will neglect this variation for most practical purposes. The heat units are related as follows:

$$1.000 \text{ kcal} = 1000 \text{ cal} = 3.968 \text{ Btu}.$$

Substances differ from one another in the quantity of heat needed to produce a given rise of temperature in a given mass. The ratio of the amount of heat energy ΔQ supplied to a body to its corresponding temperature rise ΔT is called the *heat capacity C* of the body; that is,

$$C = \text{heat capacity} = \frac{\Delta Q}{\Delta T}.$$

The word "capacity" may be misleading because it suggests the essentially meaningless statement "the amount of heat a body can hold," whereas what is meant is simply the energy that must be added as heat in order to raise the temperature of the body one degree.

The heat capacity per unit mass of a body, called *specific heat*, is characteristic of the material of which the body is composed:

$$c = \frac{\text{heat capacity}}{\text{mass}} = \frac{\Delta Q}{m \, \Delta T}. \tag{22-1}$$

We properly speak, on the one hand, of the heat capacity of a penny but, on the other, of the specific heat of copper.

Neither the heat capacity of a body nor the specific heat of a material is constant but depends on the location of the temperature interval. The previous equations give only average values for these quantities in the temperature range of ΔT. In the limit, as $\Delta T \to 0$, we can speak of the specific heat at a particular temperature T.

The heat that must be given to a body of mass m, whose material has a specific heat capacity c, to increase its temperature from T_i to T_f, is, assuming $\Delta T \ll T_f - T_i$,

$$Q = \Sigma \, \Delta Q = \sum_{T_i}^{T_f} mc \, \Delta T. \tag{22-2}$$

---

* We shall see in Section 22-5 how the calorie is now defined.

In the differential limit this becomes

$$Q = m \int_{T_i}^{T_f} c \, dT \qquad (22\text{-}3)$$

where c is a function of the temperature. At ordinary temperatures and over ordinary temperature intervals, specific heats can be considered to be constants. Figure 22-1 shows the variation in the specific heat of water with temperature. Information of this sort is obtained by using an electrical heating coil to supply heat at a rate that can be accurately determined. We see from the graph that the specific heat of water varies less than 1% from its value of 1.000 cal/g·C° at 15° C.

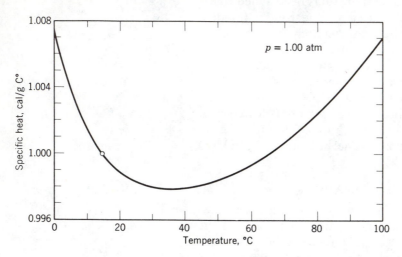

figure 22-1
The variation with temperature of the specific heat of water at a pressure of 1.00 atm. The circle, located at 15° C, suggests the definition of the calorie.

Equation 22-1 does not define specific heat uniquely. We must also specify the conditions under which the heat ΔQ is added to the specimen. We have implied that the condition is that the specimen remain at normal (constant) atmospheric pressure while we add the heat. This is a common condition, but there are many other possibilities, each leading, in general, to a different value for c. To obtain a unique value for c we must specify the conditions, such as specific heat at constant pressure c_p, specific heat at constant volume c_v, etc.

Table 22-1
Values for c_p for some solids
(at room temperature and for $p = 1.0$ atm)

Substance	Specific heat cal/g C°	Specific heat J/g C°	Molecular weight g/mol	Molar heat capacity cal/mol C°	Molar heat capacity J/mol C°
Aluminum	0.215	0.900	27.0	5.82	24.4
Carbon	0.121	0.507	12.0	1.46	6.11
Copper	0.0923	0.386	63.5	5.85	24.5
Lead	0.0305	0.128	207	6.32	26.5
Silver	0.0564	0.236	108	6.09	25.5
Tungsten	0.0321	0.134	184	5.92	24.8

Table 22-1 (second and third columns) shows the specific heats at constant pressure of some solid elements; we will discuss the specific heats of gases later. You should realize from the way the calorie and the Btu are defined that 1 cal/g·C° = 1 kcal/kg C° = 1 Btu/lb F°, exactly. Note that the specific heat of water, equal to 1.00 cal/g·C°, is large compared to that of most substances.

EXAMPLE 1

A 75-gram block of copper, taken from a furnace, is dropped into a 300-gram glass beaker containing 200 grams of water. The temperature of the water rises from 12 to 27° C. What was the temperature of the furnace?

This is an example of two systems originally at different temperatures reaching thermal equilibrium after contact. No mechanical energy is involved, only heat exchange. Hence,

$$\text{heat from copper} = \text{heat to (beaker + water)},$$

$$m_C c_C (T_C - T_e) = (m_G c_G + m_W c_W)(T_e - T_W).$$

The subscript C stands for copper, G for glass, and W for water. The initial copper temperature is T_C, the initial beaker water temperature is T_W, and T_e is the final equilibrium temperature. Substituting the given values, with $c_C = 0.093$ cal/g·C°, $c_G = 0.12$ cal/g·C°, and $c_W = 1.0$ cal/g·C°, we obtain

$$(75 \text{ g})(0.093 \text{ cal/g·C°})(T_C - 27° \text{ C}) = [(300 \text{ g})(0.12 \text{ cal/g·C°})$$
$$+ (200 \text{ g})(1.0 \text{ cal/g·C°})](27° \text{ C} - 12° \text{ C})$$

or, solving for T_C, $\qquad\qquad T_C = 530°$ C

What approximations, both experimental and theoretical, were used implicitly to arrive at this answer?

22-3
MOLAR HEAT CAPACITIES OF SOLIDS

From the second column of Table 22-1 we conclude that the specific heats of solids vary widely from one material to another. However quite a different story emerges if we compare samples of materials that contain the same number of molecules rather than samples that have the same mass. We can do this by expressing specific heats (called when so expressed *molar heat capacities*) in cal/mol·C° rather than in cal/g·C°.* In 1819 Dulong and Petit pointed out that the molar heat capacities of all substances, with few exceptions (see carbon in Table 22-1), have values close to 6 cal/mol C°. The molar heat capacity, listed in the fifth and sixth columns of Table 22-1, is found by multiplying the specific heat (second and third columns) by the molecular weight (fourth column). We see that the amount of heat required *per molecule* to raise the temperature of a solid by a given amount seems to be about the same for almost all materials. This is striking evidence for the molecular theory of matter.

Actually molar heat capacities vary with temperature, approaching zero as $T \to 0$ K and approaching the Dulong-Petit value as $T \to \infty$. Since the number of molecules rather than the kind of molecule seems to be important in determining the heat required to increase the temperature of a body by a given amount, we are led to expect that the molar heat capacities of different substances will vary with temperature in much the same way. Figure 22-2 shows that, indeed, the molar heat capacities of various substances can be made to fall on the same curve by a simple, empirical adjustment in the temperature scale. The horizontal scale in Fig. 22-2 is the dimensionless ratio T/T_D, where T is the Kelvin temperature and T_D is a characteristic temperature, called the *Debye temperature*, that has a particular constant value for each material. For lead, T_D has the empirical value of 88 K and for carbon, $T_D = 1860$ K. From these data you can show that a scale value of $T/T_D = 0.600$ corresponds to $T = 53$ K for lead but to $T = 1120$ K for carbon. Alternatively, room temperature (~ 300 K) corresponds to $T/T_D = 3.4$ for lead and to $T/T_D = 0.16$ for carbon. Thus

---

* A mole (abbr. *mol*) of any substance is the amount of the substance that contains a specified number of elementary entities, namely, 6.02252×10^{23}, called Avogadro's number. This number is the result of the defining relation that one mol of carbon atoms (actually, of the isotope C^{12}) shall have a mass of 12 g, exactly. The *gram molecular weight* M of a substance is the number of grams per mole of that substance. Thus the gram molecular weight of ordinary oxygen molecules is 32.0 g/mol. Although the mole is an amount of substance, we cannot translate it into mass, as grams, until we specify what the elementary entity is; it may be atoms, molecules, ions, electrons, other particles, or specified groups of such particles.

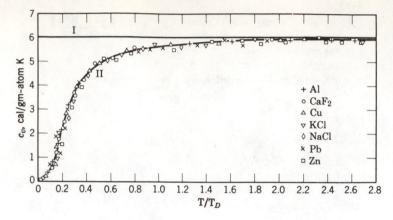

figure 22-2
The molar heat capacities (c_v) showing a few selected points only. Line I represents the Dulong and Petit rule and curve II represents a theory due to Debye.

we see from Fig. 22-2 that in the early days, when only room temperature specific heats were available, lead would conform to the Dulong and Petit rule but carbon would seem to be an exception.

The straight line I in Fig. 22-2 is the Dulong and Petit value of 1819; it agrees with experiment at high temperature but fails at low temperatures. It corresponds to the assumption that every atom in a solid vibrates independently like a classical oscillator. Curve II is due to Debye (1912). In the Debye theory, a characteristic temperature T_D, which is directly related to a vibrational frequency characteristic of the material, can be obtained independent of specific heat experiments. One then uses quantum principles to analyze the coupled vibrations of the atoms in a solid and obtains a specific heat formula which, in terms of the dimensionless ratio T/T_D, is the same for all substances. The excellent agreement of this formula (curve II) with experiment is a triumph of quantum physics.*

The materials displayed in Fig. 22-2 are "normal" in that they do not melt, boil, change their crystal structure, etc., in the temperature range indicated. Specific heat measurements, which tells us how a solid absorbs energy as its temperature is raised, are a sensitive probe to detect such molecular, atomic, or electronic rearrangements. Figure 22-3, for example, shows the specific heat of tantalum near 4.39 K. Below this transition temperature tantalum loses all its electric resistance—it becomes superconducting. Above this temperature it has the resistance expected of a normal metal.

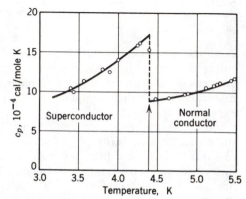

figure 22-3
The specific heat of tantalum near its superconducting transition temperature.

22-4
HEAT CONDUCTION

The transfer of energy arising from the temperature difference between adjacent parts of a body is called *heat conduction*. Consider a slab of material of cross-sectional area A and thickness Δx, whose faces are kept at different temperatures. We measure the heat ΔQ that flows perpendicular to the faces in a time Δt. Experiment shows that ΔQ is proportional to Δt and to the cross-sectional area A for a given temperature difference ΔT, and that ΔQ is proportional to $\Delta T/\Delta x$ for a given Δt and A, providing both ΔT and Δx are small. That is,

$$\frac{\Delta Q}{\Delta t} \propto A \frac{\Delta T}{\Delta x} \qquad \text{approximately.} \qquad (22\text{-}4a)$$

---

* The data reported in Fig. 22-2 are values of c_v but those in Table 22-1 are c_p. The former is easier to calculate theoretically because the thermal expansion need not be taken into account, but (for solids) the latter is much easier to measure. The two are related by the simple thermodynamic formula

$$c_p = c_v + T\beta^2/\kappa\rho$$

in which β is the thermal coefficient of volume expansion, $\kappa \ (= -\Delta V/V\Delta p)$ is the (isothermal) compressibility, and ρ is the density. At room temperature the difference between c_p and c_v for typical solids is about 5%.

In the limit of a slab of infinitesimal thickness dx, across which there is a temperature difference dT, we obtain the fundamental law of heat conduction, in which the heat flow H is given by

$$H = -kA \frac{dT}{dx}. \qquad (22\text{-}4b)$$

Here H (measured, say, in cal/s; see Eq. 22-4a) is the time rate of heat transfer across the area A, dT/dx is called the *temperature gradient*, and k is a constant of proportionality called the *thermal conductivity*. We choose the direction of heat flow to be the direction in which x increases; since heat flows in the direction of decreasing T, we introduce a minus sign in Eq. 22-4 (that is, we wish H to be positive when dT/dx is negative).

A substance with a large thermal conductivity k is a good heat conductor; one with a small thermal conductivity k is a poor heat conductor, or a good thermal insulator. The value of k depends on the temperature, increasing slightly with increasing temperature, but k can be taken to be practically constant throughout a substance if the temperature difference between its parts is not too great. In Table 22-2 we list values of k for various substances; we see that metals as a group are better heat conductors than nonmetals, and that gases are poor heat conductors.

Let us apply Eq. 22-4b to a rod of length L and constant cross-sectional area A in which a steady state has been reached (Fig. 22-4). In a

Table 22-2
Thermal conductivities
(Gases at 0° C; others at about room temperature)

	kcal/s·m·C°	J/s·m·C°
Metals		
Aluminum	4.9×10^{-2}	20×10^{1}
Brass	2.6×10^{-2}	11×10^{1}
Copper	9.2×10^{-2}	39×10^{1}
Lead	8.3×10^{-3}	35
Silver	9.9×10^{-2}	41×10^{1}
Steel	1.1×10^{-2}	46
Gases		
Air	5.7×10^{-6}	2.4×10^{-2}
Hydrogen	3.3×10^{-5}	1.4×10^{-1}
Oxygen	5.6×10^{-6}	2.3×10^{-2}
Others		
Asbestos	2×10^{-5}	8×10^{-2}
Concrete	2×10^{-4}	8×10^{-1}
Cork	4×10^{-5}	17×10^{-2}
Glass	2×10^{-4}	8×10^{-1}
Ice	4×10^{-4}	17×10^{-1}
Wood	2×10^{-5}	8×10^{-2}

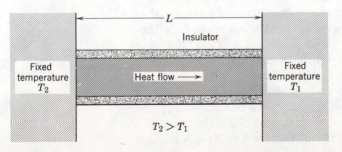

figure 22-4
Conduction of heat through an insulated conducting bar.

steady state the temperature at each point is constant in time. Hence, H is the same at all cross-sections. (Why?) But $H = -kA(dT/dx)$, so that, for a constant k and A, the temperature gradient dT/dx is the same at all cross-sections. Hence, T decreases linearly along the rod so that $-dT/dx = (T_2 - T_1)/L$. Therefore, the time rate of transfer of heat energy is

$$H = kA \frac{T_2 - T_1}{L}. \qquad (22\text{-}5)$$

The phenomenon of heat conduction also shows that the concepts of heat and temperature are distinctly different. Different rods, having the same temperature difference between their ends, may transfer entirely different quantities of heat in the same time.

EXAMPLE 2

Consider a compound slab, consisting of two materials having different thicknesses, L_1 and L_2, and different thermal conductivities, k_1 and k_2. If the temperatures of the outer surfaces are T_2 and T_1, find the rate of heat transfer through the compound slab (Fig. 22-5) in a steady state.

Let T_x be the temperature at the interface between the two materials. Then

$$H_2 = \frac{k_2 A (T_2 - T_x)}{L_2}$$

and

$$H_1 = \frac{k_1 A (T_x - T_1)}{L_1}.$$

In a steady state $H_2 = H_1 = H$, so that

$$\frac{k_2 A (T_2 - T_x)}{L_2} = \frac{k_1 A (T_x - T_1)}{L_1}.$$

Let H be the rate of heat transfer (the same for all sections). Then, solving for T_x and substituting into either of these equations, we obtain

$$H = \frac{A (T_2 - T_1)}{(L_1/k_1) + (L_2/k_2)}.$$

The extension to any number of sections in series is obviously

$$H = \frac{A (T_2 - T_1)}{\Sigma (L_i/k_i)}.$$

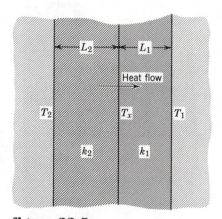

figure 22-5
Example 2. Conduction of heat through two layers of matter with different thermal conductivities.

22-5
THE MECHANICAL EQUIVALENT OF HEAT

We have seen earlier that the foot-pound (Section 7-2) was developed as a unit of work and the Btu (together with the calorie; see Section 22-2) as a unit of heat. Work and heat were thought of as separate concepts until Rumford, in 1798, suggested (Section 22-1) that heat had a mechanical aspect, thus proposing a connection between them. This connection was firmly established in the middle of the nineteenth century as the principle of conservation of energy. This principle asserts that heat and work are each forms of energy and that there should be a definite relationship, called the *mechanical equivalent of heat*, between them. It was Joule, in 1850, who first found by experiment how many foot-pounds of work are equivalent to 1 Btu of heat.

Joule used an apparatus in which falling weights rotated a set of paddles in an insulated water container (Fig. 22-6). In one cycle of operation the falling weights do a known amount of work on the water, of mass m, and we note that the temperature rises by ΔT. Now we *could* have produced this same rise in temperature by transfering heat energy Q to the system, given by

$$Q = mc \, \Delta T.$$

Thus, we measure W, observe ΔT, and calculate Q. The results are

$$1 \text{ Btu } (= 252.0 \text{ cal}) = 777.9 \text{ ft} \cdot \text{lb},$$

that is, 777.9 ft · lb of mechanical work will, when converted entirely into heat energy, generate 1 Btu; that is, it will raise the temperature of one pound of water from 63° F to 64° F. We can write this relation in other units as*

$$1 \text{ cal} = 4.186 \text{ J}.$$

It is appropriate that the SI unit of energy is the joule $(= 1 \text{ N} \cdot \text{m} = 1 \text{ kg} \cdot \text{m}^2/\text{s}^2)$. In modern laboratory practice the calorie is not much used or needed. It is, however, deeply embedded in the literature of science. To permit the continued use of this familiar unit—but to recognize the practical importance of the joule—a new calorie, the *thermochemical calorie*, is defined as

$$1 \text{ calorie (thermochemical)} = 4.184 \text{ joule (exactly)}.$$

In ordinary laboratory practice this calorie does not differ significantly from that defined earlier.

Joule also made other experiments (stirring mercury, forcing water through narrow tubes, rubbing together iron rings in a mercury bath, etc.). His conclusions are noteworthy for (1) the skill and ingenuity that he showed, (2) the accuracy of his final results, which differ only by about 1% from present values, and (3) the influence that they had in convincing scientists of the correctness of the concept that heat, like work, is a form of energy.

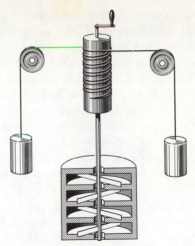

figure 22-6
Joule's arrangement for measuring the mechanical equivalent of heat. The falling weights turn paddles which stir the water in the container, thus raising its temperature.

22-6
HEAT AND WORK

We have seen that *heat is energy that flows from one body to another because of a temperature difference between them.* The idea that heat is something *in* a body, as the caloric theory assumed, contradicts many experimental facts. It is only as it flows, because of a temperature difference, that the energy is called heat energy. If heat were a substance, or a definite kind of energy that kept its identity while contained in a system, it would not be possible to remove heat indefinitely from a system which does not change. Yet Rumford showed that this was possible. In fact, by continually performing mechanical work in Joule's apparatus, we can obtain an indefinite amount of heat out of the water, by connecting it to a cooler system, for example, without changing the condition of the water.

In the same way work is not something of which a system contains a definite amount. We can put an indefinite amount of work into a system, as Joule's apparatus again illustrates. Work, like heat, involves a transfer of energy. In mechanics, work is involved in energy transfers

---

* Henry A. Rowland, in 1879, carried out a painstaking determination of the mechanical equivalent of heat which, to this day, remains a model of careful experimentation. His result differs from the accepted value today by only 1 part in 2000. Rowland graduated from Rensselaer Polytechnic Institute in 1870 and in 1876 became the first Professor of Physics at the then newly established Johns Hopkins University, where he conducted this experiment. See "The Education of an American Scientist, Henry A. Rowland," by Samuel Rezneck, *American Journal of Physics*, February, 1960 and 'Rowland's Physics' by John D. Miller, *Physics Today*, July, 1976.

in which temperature played no role. If heat energy is transmitted by temperature differences, we can distinguish heat and work by defining *work as energy that is transmitted from one system to another in such a way that a difference of temperature is not directly involved.* This definition is consistent with our previous use of the term. That is, in the expression $dW = F\,dx$, the force F can arise from electrical, magnetic, gravitational, and other sources. The term *work* includes all these energy transfer processes, but it specifically excludes energy transfer arising from temperature differences.

Consider another simple example, that of rubbing two surfaces together. There is no limit to the amount of heat that can be removed from this system or to the amount of work that can be put into it, so that there is no definite meaning to phrases such as "the heat in the system" or "the work in the system." The quantities Q and W are not characteristic of the (equilibrium) *state* of the system but rather of the *thermodynamic process* by which the system moves from one equilibrium state to another, by interacting with its environment. It is only during such a process that we can give meaning to heat and work; we can then identify Q with the heat transferred to or from the system and W with the work done on or by the system. The study of such processes and of the changes in energy involved in the performance of work and the flow of heat is the subject matter of *thermodynamics.*

In Fig. 22-7 we consider a general thermodynamic process. We must first state definitely what the system is and what the environment is. In the figure we draw a closed surface surrounding the system to define it. In (*a*) the system is in its *initial state,* in equilibrium with the environment external to it. In (*b*) the system interacts with its environment through some specific *thermodynamic process.* During this process, energy in the form of heat and/or work may go into or out of the system. Arrows representing the flow of Q or W must pierce the surface enclosing the system. In (*c*) the system has reached its *final state,* again in equilibrium with the environment external to it.

Figure 22-8 shows a falling weight which turns a generator, which in turn sends an electric current through a resistor immersed in a water container. Let us choose the system to be the generator and the attached electric circuit, the water, and its container. Then the environment is the weight and the earth, which pulls on the weight. The process consists of letting the weight fall a distance h in the earth's gravitational field. During this process the environment (by means of the cord) does work W on the system. There are no temperature differences between the system and its environment and hence $Q = 0$ for this process.

Our choice of a system in thermodynamic problems is arbitrary. Let us now choose the system to be only the water and its container in Fig. 22-8. The environment now is the generator and attached circuit as well as the weight and the earth. For this choice of system there now *is* a temperature difference between the environment (resistor) and the system (water), and heat Q will flow into the system during the process. No forces act through the system boundary to produce displacements, however, and hence $W = 0$ for this process. This example shows that we must first state definitely what the system is and what the environment is before we can decide whether the change in the state of the system is due to the flow of heat or to the performance of work or both. There will be a transfer of heat between system and environment only when a

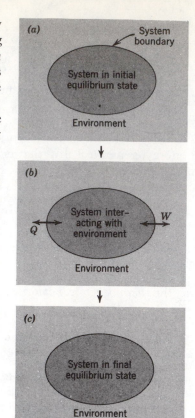

figure 22-7
(*a*) A system in an initial state, in equilibrium with its surroundings. (*b*) A thermodynamic process during which the system may exchange heat Q or work W with its environment. (*c*) A final equilibrium state reached as the result of the process.

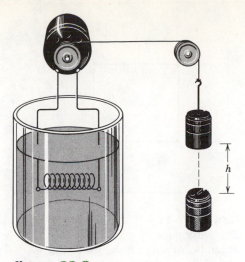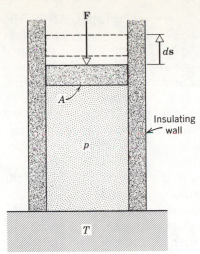

figure 22-8
Heat and work. A weight, in falling,
does work on an electric generator
which sends current through a
resistor which heats the water in
which it is immersed.

Heat reservoir of controllable
temperature T

figure 22-9
Work is done by the gas at pressure
p as it expands against the piston.
Heat may enter or leave the system
from the heat reservoir on which
the cylinder rests.

485 *HEAT AND WORK* SEC. 22-6

temperature difference exists across the system boundary; if no temperature difference exists, the energy transfer involves work.

Let us now compute Q and W for a specific thermodynamic process. Consider a gas in a cylindrical container with a movable piston. Let the gas be the system. Initially it is in equilibrium with the environment external to it (which is the heat reservoir and the piston, shown in Fig. 22-9) and has a pressure p_i and a volume V_i. We can think of the containing walls as the system boundary. Heat can flow into the system or out of it through the bottom of the cylinder and work can be done on the system or by the system by compressing or expanding the gas, respectively, with the piston. Consider a process whereby the system interacts with its environment and reaches a final equilibrium state characterized by a pressure p_f and a volume V_f.

In Fig. 22-9 we show the gas expanding against the piston. The work done by the gas in displacing the piston through an infinitesimal distance ds is

$$dW = \mathbf{F} \cdot d\mathbf{s} = pA \; ds = p \; dV$$

where dV is the differential change in the volume of the gas. In general, the pressure will not be constant during a displacement. To obtain the total work W done on the piston by the gas in a large displacement, we must know how p varies with the displacement. Then we compute the integral

$$W = \int dW = \int_{V_i}^{V_f} p \; dV$$

over the range in volume. This integral can be graphically evaluated as the area under the curve in a $p-V$ diagram, as shown for a special case in Fig. 22-10.

There are many different ways in which the system can be taken

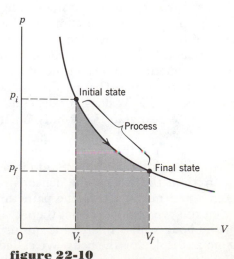

figure 22-10
The work done by a gas is equal to the area under a p-V curve.

from the initial state i to the final state f. For example (Fig. 22-11), the pressure may be kept constant from i to a and then the volume kept constant from a to f. Then the work done by the expanding gas is equal to the area under the line ia. Another possibility is the path ibf, in which case the work done by the gas is the area under the line bf. The continuous curve from i to f is another possible path in which the work done by the gas is still different from the previous two paths. We can see, therefore, that *the work done by a system depends not only on the initial and final states but also on the intermediate states, that is, on the path of the process.*

A similar result follows if we compute the flow of heat during the process. State i is characterized by a temperature T_i and state f by a temperature T_f. The heat flowing into the system, say, depends on how the system is heated. We can heat it at a constant pressure p_i, for example, until we reach the temperature T_f, and then change the pressure at constant temperature to the final value p_f. Or we can first lower the pressure to p_f and then heat it at that pressure to the final temperature T_f. Or we can follow many other paths. Each path gives a different result for the heat flowing into the system. Hence, *the heat lost or gained by a system depends not only on the initial and final states but also on the intermediate states, that is, on the path of the process.* This is an experimental fact. As J. C. Slater has written:

". . . It would be pleasant to be able to say, in a given state of the system, that the system has so and so much heat energy. Starting from the absolute zero of temperature, where we could say that the heat energy was zero, we could heat the body up to the state we were interested in, find $\int dQ$ from absolute zero up to this state, and call that the heat energy. But the stubborn fact remains that we would get different answers if we heated it up in different ways. . . . There is nothing to do about it."

Both heat and work "depend on the path" taken; neither one is independent of the path, and neither one can be conserved alone.

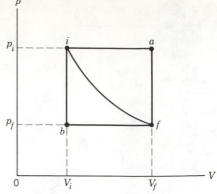

figure 22-11
The work done by a system depends not only on the initial state (i) and the final state (f) but on the intermediate path as well.

22-7
THE FIRST LAW OF THERMODYNAMICS

We can now tie all these ideas together. Let a system change from an initial equilibrium state i to a final equilibrium state f in a definite way, the heat absorbed by the system being Q and the work done by the system being W. Then we compute the $Q - W$. Now we start over and change the system from the same state i to the same state f, but this time in another way by a different path. We do this over and over again, using different paths each time. We find that in every case the quantity $Q - W$ *is the same.* That is, although Q and W separately depend on the path taken, $Q - W$ does *not* depend at all on how we took the system from state i to state f but only on the initial and final (equilibrium) states.

You will recall from mechanics that when an object is moved from an initial point i to a final point f in a gravitational field in the absence of friction, the work done depends only on the positions of the two points and not at all on the path through which the body is moved. From this we concluded that there is a function of the space coordinates of the body whose final value minus its initial value equals the work done in displacing the body. We called it the potential energy function. Now in thermodynamics we find that when a system has its state changed from state i to state f, the quantity $Q - W$ depends *only* on the initial and final coordinates and *not at all* on the path taken between these end points. We conclude that there is a function of the thermodynamic coordinates whose final value minus its initial value equals the change $Q - W$ in the process. We call this function the *internal energy function.*

Now Q is the energy added to the system by the transfer of heat and W is the energy given up by the system in performing work, so that $Q - W$ represents, by definition, *the internal energy change of the system.* Let us represent the internal energy function by the letter U. Then the internal energy of the system in state f, U_f, minus the internal energy of the system in state i, U_i, is simply *the change in internal energy of the system,* and this quantity *has a definite value independent of how the system went from state i to state f.* We have

$$U_f - U_i = \Delta U$$

and

$$\Delta U = Q - W. \tag{22-6}$$

Just as for potential energy, so for internal energy too it is the change that matters. If some arbitrary value is chosen for the internal energy in some standard reference state, its value in any other state can be given a definite value. Equation 22-6 is known as the *first law of thermodynamics.* In applying Eq. 22-6 we must remember that Q is considered positive when heat *enters* the system and W is positive when work is done *by* the system.

If our system undergoes only an infinitesimal change in state, only an infinitesimal amount of heat dQ is absorbed and only an infinitesimal amount of work dW is done, so that the internal energy change dU is also infinitesimal. In such a case, the first law is written in *differential* form as*

$$dU = dQ - dW. \tag{22-7}$$

We may express the first law in words by saying: *Every thermodynamic system in an equilibrium state possesses a state variable called the internal energy U whose change dU in a differential process is given by Eq. 22-7.* Recall that the essential content of the zeroth law of thermodynamics (p. 459) is, speaking loosely: *there exists a useful thermodynamic quantity called "temperature."* The essential content of the first law is: *there exists a useful thermodynamic quantity called "internal energy";* the law also provides, in Eq. 22-6, a recipe for measuring changes in internal energy quantitatively.

The first law of thermodynamics is thought to apply to every process in nature that proceeds between equilibrium states. Note that the *process* may or may not involve equilibrium states. We may apply the first law to the explosion of a firecracker in an insulated steel drum, for example. Because of its generality, the information that the first law gives is far from complete, although exact and correct. There are some very general questions which it cannot answer. For example, although it tells us that energy is conserved in every process, it does not tell us whether any particular process can actually occur. An entirely different generalization, called the second law of thermodynamics, gives us this information, and much of the subject matter of thermodynamics depends on this second law (Chapter 25).

---

* W and Q are not actual functions of the state of a system, that is, they do not depend on the values of the system's coordinates. Hence, dW and dQ are not exact differentials as the term is used in mathematics. All they mean here is a very small quantity. More advanced books write them as đQ and đW to indicate their inexact nature. However, dU is an exact differential, for U is an exact function of the system's coordinates.

We have seen that when a gas expands the work it does on its environment is

$$W = \int p \, dV,$$

where p is the pressure exerted on or by the gas and dV is the differential change in volume of the gas. Consider a special case in which the pressure remains constant while the volume changes by a finite amount, say from V_i to V_f. Then

$$W = \int_{V_i}^{V_f} p \, dV = p \int_{V_i}^{V_f} dV = p(V_f - V_i) \qquad \text{(constant pressure)}.$$

A process taking place at constant pressure is called an *isobaric* process. For example, water is heated in the boiler of a steam engine up to its boiling point and is vaporized to steam; then the steam is superheated, all processes proceeding at a constant pressure.

In Fig. 22-12 we show an isobaric process. The system is H_2O in a cylindrical container. A frictionless airtight piston is loaded with sand to produce the desired pressure on the H_2O and to maintain it automatically. Heat can be transferred from the environment to the system by a Bunsen burner. If the process continues long enough, the water boils and some is converted to steam; we assume that this occurs. The system may expand, very slowly (quasi-statically) but the pressure it exerts on the piston is automatically always the same, for this pressure must be equal to the constant pressure which the piston exerts on the system. If we wedged the piston so that it could not move, or if we added or took away some sand during the heating process, the process would not be isobaric.

Let us consider the boiling process. We know that substances will change their phase from liquid to vapor at a definite combination of values of pressure and temperature. Water will vaporize at 100° C and atmospheric pressure, for example. For a system to undergo a change of phase heat must be added to it, or taken from it, quite apart from the heat necessary to bring its temperature to the required value. Consider the change of phase of a mass m of liquid to a vapor occurring at constant temperature and pressure. Let V_l be the volume of liquid and V_v the volume of vapor. The work done by this substance in expanding from V_l to V_v at constant pressure is

$$W = p(V_v - V_l).$$

Let L represent the heat of vaporization, that is, the heat needed per unit mass to change a substance from liquid to vapor at constant temperature and pressure. Then the heat absorbed by the mass m during the change of state is

$$Q = mL.$$

From the first law of thermodynamics, we have

$$\Delta U = Q - W$$

so that

$$\Delta U = mL - p(V_v - V_l)$$

for this process.

At atmospheric pressure 1.00 g of water, having a volume of 1.00 cm³, becomes 1671 cm³ of steam when boiled. The heat of vaporization of water is 539 cal/g at 1 atm. Hence, if $m = 1.00$ g,

$$Q = mL = 539 \text{ cal.}$$

22-8
SOME APPLICATIONS OF THE FIRST LAW OF THERMODYNAMICS

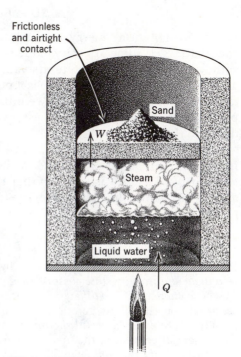

Frictionless and airtight contact

Sand

W

Steam

Liquid water

Q

figure 22-12
Water boiling at constant pressure (isobarically). The pressure is kept constant by the weight of the sand, the piston, and the external atmospheric pressure.

EXAMPLE 3

This quantity, which represents heat *added* to the system from the environment, is positive.

$$W = p(V_v - V_l) = (1.013 \times 10^5 \text{ N/m}^2)[(1671 - 1) \times 10^{-6} \text{ m}^3]$$

$$= 169.5 \text{ J.}$$

This quantity, which represents work done *by* the system on the environment, is positive.

Since 1 cal equals 4.186 J, $W = 41$ cal. Then,

$$\Delta U = U_v - U_l = mL - p(V_v - V_l) = (539 - 41) \text{ cal}$$

$$= 498 \text{ cal.}$$

This quantity is positive; the internal energy of the system *increases* during this process. Hence, of the 539 cal needed to boil 1 g of water (at 100° C and 1 atm), 41 cal go into external work of expansion and 498 cal go into internal energy added to the system. This energy represents the internal work done in overcoming the strong attraction of H_2O molecules for one another in the liquid state.

How would you expect the 80 cal that are needed to melt 1 g of ice to water (at 0° C and 1 atm) to be shared by the external work and the internal energy?

A process that takes place in such a way that no heat flows into or out of the system is called an *adiabatic process*. Experimentally such processes are achieved either by sealing the system off from its surroundings with heat insulating material or by performing the process quickly. Because the flow of heat is somewhat slow, any process can be made practically adiabatic if it is performed quickly enough.

For an adiabatic process Q equals zero, so that from the first law we obtain

$$\Delta U = U_f - U_i = -W.$$

Hence, the internal energy of a system increases exactly by the amount of work done *on* the system in an adiabatic process. If work is done *by* the system in an adiabatic process, the internal energy of the system decreases by exactly the amount of external work it performs. An increase of internal energy usually raises the system's temperature and conversely, a decrease of internal energy usually lowers the system's temperature. A gas that expands adiabatically does external work and its internal energy decreases; such a process is used to attain low temperatures. The increase of temperature during an adiabatic compression of air is well known from the heating of a bicycle pump.

In Fig. 22-13 we show a simple adiabatic process. The system is a gas inside a cylinder made of heat-insulating material. Heat cannot enter the system from its environment or leave the system to the environment. Again we have a pile of sand on a frictionless airtight piston. The only interaction permitted between system and environment is through the performance of work. Such a process can occur when sand is added or removed from the piston, so that the gas can be compressed or can expand against the piston.

Among the many engineering examples of adiabatic processes are the expansion of steam in the cylinder of a steam engine, the expansion of hot gases in an internal combustion engine, and the compression of air in a Diesel engine or in an air compressor. These processes all occur rapidly enough so that only a very small amount of heat can enter or leave the system through its walls during that short time. The compres-

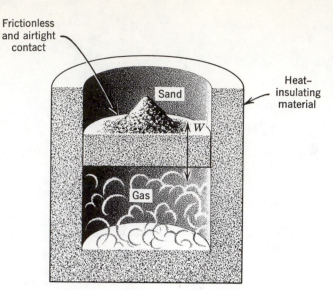

figure 22-13
In an adiabatic process there is no flow of heat to or from the system. Here the walls are insulated and, as sand is removed or added, the volume of the gas changes adiabatically.

sions and rarefactions of a sound wave in a gas are adiabatic (Example 6, Chapter 23).

The most important reason for studying adiabatic processes, however, is that ideal engines use processes that are exactly adiabatic. These ideal engines determine the theoretical limits to the operation and capabilities of real engines. We shall look further into this in Chapter 25.

A process of much theoretical interest is that of *free expansion*. This is an adiabatic process in which no work is performed on or by the system. Something like this can be achieved by connecting one vessel which contains a gas to another evacuated vessel with a stopcock connection, the whole system being enclosed with thermal insulation (Fig. 22-14). If the stopcock is suddenly opened, the gas rushes into the vacuum and expands freely. Because of the heat insulation this process is adiabatic, and because the walls of the vessels are rigid no external work is done on the system. Hence, in the first law we have $Q = 0$ and $W = 0$, so that $U_i = U_f$ for this process. The initial and final internal energies are equal in free expansion.

In free expansion, after we open the stopcock we have no further control over the process. At intermediate states the pressure, volume, and temperature do not have unique values characteristic of the system as a whole, that is, the system passes through nonequilibrium states so that we cannot plot the course of the process by a curve on a p-V diagram. We can plot the initial and final states as points on such plots because they are well-defined, equilibrium states. The free expansion is a good example of an irreversible process; see Section 25-2.

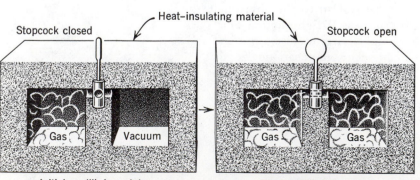

Initial equilibrium state   Final equilibrium state

figure 22-14
Free expansion. There is no change of internal energy U since there is no flow of heat Q and no external work W is done.

1. Give examples to distinguish clearly between temperature and heat.
2. (a) Show how heat conduction and calorimetry could be explained by the caloric theory. (b) List some heat phenomena that cannot be explained by the caloric theory.
3. Give an example of a process in which no heat is transferred to or from the system but the temperature of the system changes.
4. Can heat be considered a form of stored (or potential) energy? Would such an interpretation contradict the concept of heat as energy in process of transfer because of a temperature difference?
5. Apply Eq. 22-1 to boiling water.
6. It is difficult to "boil" eggs in water at the top of a high mountain because water boils there at a relatively low temperature. What is a simple, practical way of overcoming this difficulty?
7. Will a three-minute egg cook any faster if the water is boiling furiously than if it is simmering quietly?
8. Can heat be added to a substance without causing the temperature of the substance to rise? If so, does this contradict the concept of heat as energy in process of transfer because of a temperature difference?
9. Why must heat energy be supplied to melt ice—the temperature doesn't change, after all?
10. (a) Can ice be heated to a temperature above 0° C without its melting? Explain. (b) Can water be cooled to a temperature below 0° C without its freezing? Explain. (See "The Undercooling of Liquids" by David Turnbull in *Scientific American*, January 1965.)
11. Does putting sand on it help you to drive on an icy road? Does your answer depend on the temperature? Explain.
12. Explain the fact that the presence of a large body of water nearby, such as a sea or ocean, tends to moderate the temperature extremes of the climate on adjacent land.
13. Theory shows that the coefficient of linear expansion α (see Sec. 21-8) is proportional to the heat capacity C_v. Show that this is to be expected. (*Hint:* Heat capacity measures the rate of change of the vibrational energy with temperature.)
14. If someone told you that a conventional electric fan not only does not cool the air but heats it slightly, how would you reply?
15. Both heat conduction and wave propagation involve the transfer of energy. Is there any difference in principle between these two phenomena?
16. When a hot body warms a cool one are their temperature changes equal in magnitude? Give examples. Can one then say that temperature passes from one to the other?
17. What connection is there between an object's feeling hot or cold and its heat capacity? Between this and its thermal conductivity?
18. A block of wood and a block of metal are at the *same* temperature. When the blocks feel cold the metal feels colder than the wood; when the blocks feel hot the metal feels hotter than the wood. Explain. At what temperature will the blocks feel equally cold or hot?
19. Explain why your finger sticks to a metal ice tray just taken from the refrigerator.
20. On a winter day the temperature of the inside surface of a wall is much lower than room temperature and that of the outside surface is much higher than the outdoor temperature. Explain.

The physiological mechanisms which maintain man's internal temperature operate in a limited range of external temperature. Explain how this range can be extended at each extreme by the use of clothes. (See "Heat, Cold, and Clothing" by James B. Kelley in *Scientific American*, February 1956.)

22. What requirements for thermal conductivity, specific heat capacity, and coefficient of expansion would you want a material to be used as a cooking utensil to satisfy?

23. Consider that heat can be transferred by convection and radiation, as well as by conduction, and explain why a thermos bottle is double-walled, evacuated, and silvered.

24. The system heating the cabin of a rocket ship seems to fail when the rocket ship is far out in free space. Give one possible explanation.

25. In what way is steady-state heat flow analogous to the flow of an incompressible fluid?

26. Is the mechanical equivalent of heat, J, a physical quantity or merely a conversion factor for converting energy from heat units to mechanical units and vice versa?

27. Defend this statement: "In Joule's experiment on the mechanical equivalent of heat, described in Section 22-5, no heat is involved."

28. Is the temperature of an isolated system (no interaction with the environment) conserved?

29. Is heat the same as internal energy? If not, give an example in which a system's internal energy changes without a flow of heat across the system's boundary.

30. Can one distinguish between whether the internal energy of a body was acquired by heat transfer or acquired by performance of work?

31. If the pressure and volume of a system are given, is the temperature always uniquely determined?

32. Does a gas do any work when it expands adiabatically? If so, what is the source of the energy needed to do this work?

33. A quantity of gas occupies an initial volume V_0 at a pressure p_0 and a temperature T_0. It expands to a volume V (a) at constant temperature and (b) at constant pressure. In which case does the gas do more work?

34. Discuss the process of the freezing of water from the point of view of the first law of thermodynamics. Remember that ice occupies a greater volume than an equal mass of water.

35. A thermos bottle contains coffee. The thermos bottle is vigorously shaken. Consider the coffee as the system. (a) Does its temperature rise? (b) Has heat been added to it? (c) Has work been done on it? (d) Has its internal energy changed?

36. We have seen that "energy conservation" is a universal law of nature. At the same time national leaders urge "energy conservation" upon us (driving slower, etc.). Explain the two quite different meanings of these words.

SECTION 22-2

problems

1. Suppose the specific heat of a substance is found to vary with temperature as

$$c = A + BT^2,$$

where A and B are constants and T is Celsius temperature. Compare the mean specific heat of the substance in a temperature range $T = 0$ to $T = T_0$ to the specific heat at the midpoint $T_0/2$.

Answer: Mean specific heat exceeds that at the midpoint by $BT_0^2/12$.

2. By means of a heating coil energy is transferred at a constant rate to a substance in a thermally insulated container. The temperature of the substance is measured as a function of time. (a) Show how we can deduce the way in which the heat capacity of the body depends on the temperature from this information. (b) Suppose that in a certain temperature range it is found that the temperature T is proportional to t^3, where t is the time. How does the heat capacity depend on T in this range?

3. Calculate the specific heat of a metal from the following data. A container

made of the metal weighs 8.0 lb (mass = 3.6 kg) and contains 30 lb (mass = 14 kg) of water. A 4.0 lb (mass = 1.8 kg) piece of the metal initially at a temperature of 350° F (180° C) is dropped into the water. The container and water initially have a temperature of 60° F (16° C) and the final temperature of the entire system is 65° F (18° C).
Answer: 0.14 Btu/lb · F° (0.099 cal/g · C°).

4. Two 50-g ice cubes are dropped into 200 g of water in a glass. If the water was initially at a temperature of 25° C, and if the ice came directly from a freezer operating at a temperature of −15° C, what will be the final temperature of the drink? The specific heat of ice is approximately 0.50 cal/g · C° in this temperature range and the heat required to melt ice to water is approximately 80 cal/g.

5. A thermometer of mass 0.0550 kg and of specific heat 0.200 cal/g · C° reads 15.0° C. It is then completely immersed in 0.300 kg of water and it comes to the same final temperature as the water. If the thermometer reads 44.4° C and is accurate, what was the temperature of the water before insertion of the thermometer, neglecting other heat losses? *Answer:* 45.5° C.

6. A copper ring has a diameter of exactly 1.00000 in. at its temperature of 0° C. An aluminum sphere has a diameter of exactly 1.00200 in. at its temperature of 100° C. The sphere is placed on top of the ring (Fig. 22-15), and the two are allowed to come to thermal equilibrium, no heat being lost to the surroundings. The sphere just passes through the ring at the equilibrium temperature. What is the ratio of the mass of the sphere to the mass of the ring?

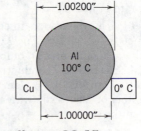

figure 22-15
Problem 6

SECTION 22-3

7. Show that the number of atomic mass units per gram of a substance is equal to the number of particles per mole (Avogadro's number).

SECTION 22-4

8. Consider the rod shown in Fig. 22-4. Suppose $L = 25$ cm, $A = 1.0$ cm², and the material is copper. If $T_2 = 125°$ C, $T_1 = 0°$ C, and a steady state is reached, find (a) the temperature gradient, (b) the rate of heat transfer, and (c) the temperature at a point in the rod 10 cm from the high-temperature end.

9. A cylindrical copper rod of length 1.2 m and cross-sectional area 4.8 cm² is insulated to prevent heat loss through its surface. The ends are maintained at a temperature difference of 100 C° by having one end in a water-ice mixture and the other in boiling water and steam. (a) Find the rate at which heat is transferred along the rod. (b) Find the rate at which ice melts at one end. *Answer:* (a) 3.7 cal/s. (b) 0.046 g/s.

10. (a) Calculate the rate at which body heat flows out through the clothing of a skier, given the following data. The body surface area is 1.8 m² and the clothing is 1.0 cm thick; skin surface temperature is 33° C, whereas the outer surface of the clothing is at −5° C; the thermal conductivity of the clothing is 0.04 W/m · K. (b) How would the answer change if, after a fall, the skier's clothes become soaked with water?

11. Two identical square rods of metal are welded end-to-end as shown in Fig. 22-16a. Assume that 10 cal of heat flows through the rods in 2 min. How long would it take for 10 cal to flow through the rods if they are welded as shown in Fig. 22-16b? *Answer:* 0.5 min.

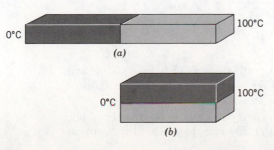

figure 22-16
Problem 11

12. Show that in a compound slab the temperature gradient in each portion is inversely proportional to the thermal conductivity.

13. Assume that the thermal conductivity of copper is twice that of aluminum and four times that of brass. Three metal rods, made of copper, aluminum, and brass, respectively, are each 6.0 in. long and 1.0 in. in diameter. These rods are placed end-to-end, with the aluminum between the other two. The free ends of the copper and brass rods are maintained at 100 and 0° C, respectively. Find the equilibrium temperatures of the copper-aluminum junction and the aluminum-brass junction.
Answer: Cu-Al, 86° C; Al-Brass, 57° C.

14. (a) What is the rate of heat loss in W/m² through a glass window 3.0 mm thick if the outside temperature is −20° F and the inside temperature is +72° F? (b) If a storm window is installed having the same thickness of glass but with an air gap of 7.5 cm between the two windows, what will be the corresponding rate of heat loss?

15. A tank of water has been outdoors in cold weather until a 5.0-cm thick slab of ice has formed on its surface (Fig. 22-17). The air above the ice is at −10° C. Calculate the rate of formation of ice (in cm/h) on the bottom surface of the ice slab. Take the thermal conductivity, density and heat of fusion of ice to be 0.0040 cal/s · cm · C°, 0.92 g/cm³ and 80 cal/g, respectively. Assume that no heat enters or leaves the water through the walls of the tank.
Answer: 0.39 cm/h.

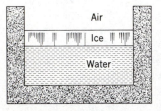

figure 22-17
Problem 15

16. Assuming k is constant, show that the radial rate of flow of heat in a substance between two concentric spheres is given by

$$H = \frac{(T_1 - T_2)4\pi k r_1 r_2}{r_2 - r_1}$$

where the inner sphere has a radius r_1 and temperature T_1, and the outer sphere has a radius r_2 and temperature T_2.

17. Energy released by radioactivity within the earth is conducted outward as heat through the oceans. For purposes of approximate calculation, assume the average temperature gradient within the solid earth beneath the ocean to be 0.07 C°/m and the average thermal conductivity to be 2×10^{-4} kcal/m · s · C°, and (a) determine the rate of heat transfer per square meter. Assume that this is approximately the rate for the entire surface of the earth, and (b) determine how much heat is thereby transferred through the earth's surface each day.
Answer: (a) 1.4×10^{-5} kcal/m² · s. (b) 6.2×10^{14} kcal/day.

18. Assuming k is constant, show that the radial rate of flow of heat in a substance between two coaxial cylinders is given by

$$H = \frac{(T_1 - T_2)2\pi L k}{\ln (r_2/r_1)}$$

where the inner cylinder has a radius r_1 and temperature T_1, and the outer cylinder has a radius r_2 and temperature T_2, each cylinder having a length L.

19. A long tungsten heater wire is rated at 3.0 kW/m and is 5.0×10^{-4} m in diameter. It is embedded along the axis of a ceramic cylinder of diameter 0.12 m. When operating at the rated power, the wire is at 1500° C; the outside of the cylinder is at 20° C. Find the thermal conductivity of the ceramic; use the result given in Problem 18.   *Answer:* 1.8 J/m · s · C°.

SECTION 22-5

20. In a Joule experiment, a mass of 6.00 kg falls through a height of 50.0 m and rotates a paddle wheel that stirs 0.600 kg of water. The water is initially at 15.0° C. By how much does its temperature rise?

21. An energetic athlete dissipates all the energy in a diet of 4000 kcal per day. If he were to release this energy at a steady rate, how would this output

compare with the energy output of a 100-W bulb? (*Note:* The calorie of nutrition is really a kilocalorie, as we have defined it.)
Answer: 1.9 times as great.

22. Power is supplied at the rate of 0.40 hp for 2.0 min in drilling a hole in a 1.0-lb copper block. (*a*) How much heat is generated? (*b*) What is the rise in temperature of the copper if only 75% of the power warms the copper? (*c*) What happens to the other 25%?

23. (*a*) Compute the possible increase in temperature for water going over Niagara Falls, 162 ft high. (*b*) What factors would tend to prevent this possible rise?   *Answer:* (*a*) 0.12 C°.

24. A 2.0-g (1.4 × 10⁻⁴-slug) bullet moving at a speed of 200 m/s (660 ft/s) becomes embedded in a 2.0 kg (0.14-slug) wooden block suspended as a pendulum bob (a ballistic pendulum). Calculate the rise in temperature of the bullet, assuming that all the absorbed energy raises the bullet's temperature.

25. A block of ice at 0° C whose mass is initially 50.0 kg slides along a horizontal surface, starting at a speed of 5.38 m/s and finally coming to rest after traveling 28.3 m. Compute the mass of ice melted as a result of the friction between the block and the surface.   *Answer:* 2.16 g.

26. The specific heat of silver, measured at atmospheric pressure, is found to vary with temperature between 50 and 100 K by the empirical equation

$$c_p = 0.076T - 0.00026T^2 - 0.15,$$

where c_p is in cal/mol · K and T is the Kelvin temperature. Calculate the quantity of heat required to raise 216 g of silver from 50 to 100 K.

27. Count Rumford weighed a metal object at low temperature and then at high temperature to see whether its "caloric content" increased. He concluded that (for gold) the "caloric" did not weigh more than 10⁻⁶ the weight of the sample. (*a*) Should the mass of a sample increase when heated, according to modern theories? (*b*) If so, by what order of magnitude? (*c*) Was Rumford safe in rejecting the caloric theory on this basis, in retrospect?
Answer: (*a*) Yes. (*b*) About 10⁻¹⁴ kg. (*c*) No.

28. Take the average specific heat of copper to be 0.090 cal/g · C° in the temperature range 0 to 1000° C. If 1.00 kg of copper is heated from 0 to 1000° C, by how much does its mass increase?

29. A "flow calorimeter" is used to measure the specific heat of a liquid. Heat is added at a known rate to a stream of the liquid as it passes through the calorimeter at a known rate. Then a measurement of the resulting temperature difference between the inflow and the outflow points of the liquid stream enables us to compute the specific heat of the liquid.

   A liquid of density 0.85 g/cm³ flows through a calorimeter at the rate of 8.0 cm³/s. Heat is added by means of a 250-W electric heating coil, and a temperature difference of 15 C° is established in steady-state conditions between the inflow and outflow points. Find the specific heat of the liquid.
Answer: 2500 J/kg · C°.

30. A chef, upon awaking one morning to find his stove out of order, decides to boil the water for his wife's coffee by shaking it in a thermos flask. Suppose that he uses ½ liter of tap water at 59° F, and that the water falls 1.0 ft each shake, the chef making 30 shakes each minute. Neglecting any loss of heat, how long must he shake the flask before the water boils?

SECTION 22-7

31. Determine the value of *J*, the mechanical equivalent of heat, from the following data: 2000 cal (7.936 Btu) of heat are supplied to a system; the system does 3350 J (2471 ft · lb) of external work during that time; the increase of internal energy during the process is 5030 J (3710 ft · lb).
Answer: 4.190 J/cal (779 ft · lb/Btu).

SECTION 22-8

32. A thermodynamic system is taken from an initial state *A* to another *B* and back again to *A*, via state *C*, as shown by the path *A-B-C-A* in the *p-V* dia-

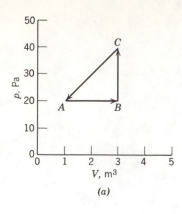
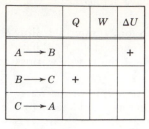

(a)

(b)

figure 22-18
Problem 32

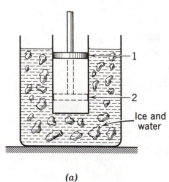

(a)

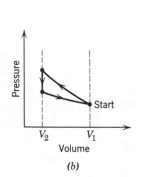

(b)

figure 22-19
Problem 33

gram of Fig. 22-18a. (a) Complete the table in Fig. 22-18b by filling in appropriate + or − indications for the signs of the thermodynamic quantities associated with each process. (b) Calculate the numerical value of the work done by the system for the complete cycle A-B-C-A.

33. Figure 22-19a shows a cylinder containing gas and closed by a movable piston. The cylinder is submerged in an ice-water mixture. The piston is *quickly* pushed down from position (1) to position (2). The piston is held at position (2) until the gas is again at 0° C and then is *slowly* raised back to position (1). Figure 22-19b is a p-V diagram for the process. If 100 g of ice are melted during the cycle, how much work has been done *on* the gas?
Answer: 8000 cal.

34. When a system is taken from state i to state f along the path iaf in Fig. 22-20 it is found that Q = 50 cal and W = 20 cal. Along the path ibf, Q = 36 cal. (a) What is W along the path ibf? (b) If W = −13 cal for the curved return path fi, what is Q for this path? (c) Take U_i = 10 cal. What is U_f? (d) If U_b = 22 cal, what is Q for the process ib? For the process bf?

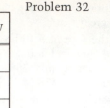

figure 22-20
Problem 34

35. An iron ball is dropped onto a concrete floor from a height of 10 m. On the first rebound it rises to a height of 0.50 m. Assume that all the macroscopic mechanical energy lost in the collision with the floor goes into the ball. The specific heat of iron is 0.12 cal/g · C°. During the collision (a) has heat been added to the ball? (b) Has work been done on it? (c) Has its internal energy changed? If so, by how much? (d) How much has the temperature of the ball risen after the first collision?
Answer: (a) No. (b) Yes. (c) Yes, by +93 J/kg. (d) 0.20 C°.

36. A cylinder has a well-fitted 2.0-kg metal piston whose cross-sectional area is 2.0 cm² (Fig. 22-21). The cylinder contains water and steam at 100° C. The piston is observed to fall slowly at a rate of 0.30 cm/s because heat flows out of the cylinder through the cylinder walls. As this happens, some steam condenses in the chamber. The density of the steam inside the chamber is 6.0×10^{-4} g/cm³. (a) Calculate the rate of condensation of steam. (b) What is the rate of change of internal energy of the steam and water inside the chamber? (c) At what rate is heat leaving the chamber?

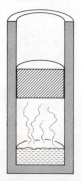

figure 22-21
Problem 36

23
kinetic theory of gases—I

Thermodynamics deals only with macroscopic variables, such as pressure, temperature, and volume. Its basic laws, expressed in terms of such quantities, say nothing at all about the fact that matter is made up of atoms. *Statistical mechanics*, however, which deals with the same areas of science that thermodynamics does, presupposes the existence of atoms. Its basic laws are the laws of mechanics, which are applied to the atoms that make up the system.

No existing electronic computer could solve the problem of applying the laws of mechanics individually to every atom in a gas, say. If there were one, the results of such calculations would be too voluminous to be useful. Fortunately, the detailed life histories of individual atoms in a gas are not important if we want to calculate only the macroscopic behavior of the gas. We apply the laws of mechanics *statistically*, then, and we find that we are able to express all the thermodynamic variables as certain averages of atomic properties. For example, the pressure exerted by a gas on the wall of the containing vessel is the average rate per unit area at which the atoms of the gas transfer momentum to the wall as they collide with it. The number of atoms in a macroscopic system is usually so large that such averages are very sharply defined quantities indeed.

We can apply the laws of mechanics statistically to assemblies of atoms at two different levels. At the level called *kinetic theory* we proceed in a rather physical way, using relatively simple mathematical averaging techniques. In this chapter we will use these methods to enlarge our understanding of pressure, temperature, specific heat, and internal energy at the atomic level. Kinetic theory was developed by Robert Boyle (1627–1691), Daniel Bernoulli (1700–1782), James Joule

23-1
INTRODUCTION

(1818–1889), A. Kronig (1822–1879), Rudolph Clausius (1822–1888), and Clerk Maxwell (1831–1879), among others.* In this book we apply the kinetic theory to gases only, because the interactions between atoms in gases are much weaker than in liquids and solids; this greatly simplifies the mathematical difficulties.

At another level, we can apply the laws of mechanics statistically using techniques that are more formal and abstract than those of kinetic theory. This approach, developed by J. Willard Gibbs (1839–1903) and by Ludwig Boltzmann (1844–1906) among others, is called *statistical mechanics*, a term that includes kinetic theory as a sub-branch. Using these methods one can derive the laws of thermodynamics, thus establishing that science as a branch of mechanics. The fullest flowering of statistical mechanics (*quantum statistics*) involves the statistical application of the laws of quantum mechanics—rather than those of classical mechanics—to many-atom systems.

Let a mass nM of a gas be confined in a container of volume V; M is the molecular weight (grams/mole) and n is the number of moles. The density ρ of the gas is nM/V and it is clear that we can reduce ρ either by removing some gas from the container (reducing n) or by putting the gas in a larger container (increasing V). We find from experiment that, at low enough densities, all gases, no matter what their chemical composition, tend to show a certain simple relationship among the thermodynamic variables p, V, and T. This suggests the concept of an *ideal gas*, one that would have the same simple behavior under all conditions. In this section we give a macroscopic or thermodynamic definition of an ideal gas. In Section 23-3 we will define an ideal gas microscopically, from the standpoint of kinetic theory, and we will see what we can learn by comparing these two approaches.

Given a mass nM of any gas in a state of thermal equilibrium we can measure its pressure p, its temperature T, and its volume V. For low enough values of the density experiment show that (1) for a given mass of gas held at a constant temperature, the pressure is inversely proportional to the volume (Boyle's law) and (2) for a given mass of gas held at a constant pressure, the volume is directly proportional to the temperature (law of Charles and Gay-Lussac). We can summarize these two experimental results by the relation

$$\frac{pV}{T} = \text{a constant} \quad \text{(for a fixed mass of gas).} \quad (23\text{-}1)$$

The volume occupied by a gas at a given pressure and temperature is proportional to its mass. Thus the constant in Eq. 23-1 must also be proportional to the mass of the gas. In Section 22-2 (see Fig. 22-2) we saw the great simplification that occurs in studies of the specific heats of solids if we compare samples of solids that contain the same number of molecules rather than samples which have the same mass in grams. We did this by using the mole as our mass unit. Let us also do that here.

We therefore write the constant in Eq. 23-1 as nR, where n is the number of moles of the gas and R is a constant that must be determined for each gas by experiment. Our expectation that simplicity will emerge if we compare gases on a molar basis is justified because experiment

23-2
IDEAL GAS—A MACROSCOPIC DESCRIPTION

---

* See "John James Waterston and the Kinetic Theory of Gases," by S. G. Brush, in *American Scientist*, June 1961, for an interesting aspect of the history of kinetic theory.

shows that, at low enough densities, *R has the same value for all gases*, namely

$$R = 8.314 \text{ J/mol·K} = 1.986 \text{ cal/mol·K}.$$

R is called the *universal gas constant*. We then write Eq. 23-1 as

$$pV = nRT \qquad (23\text{-}2)$$

and we define an ideal gas as one that obeys this relation *under all conditions*. There is no such thing as a truly ideal gas, but it remains a useful and simple concept connected with reality by the fact that all real gases approach the ideal gas abstraction in their behavior if the density is low enough. Equation 23-2 is called the *equation of state* of an ideal gas.

If we could fill the bulb of an (ideal) constant-volume gas thermometer with an ideal gas, we see from Eq. 23-2 that we could define temperature in terms of its pressure readings, that is,

$$T = 273.16 \text{ K} \frac{p}{p_{tr}} \qquad \text{(ideal gas)}.$$

Here p_{tr} is the gas pressure at the triple point, at which the temperature T_{tr} is 273.16 K by definition. In practice we must fill our thermometer with a real gas and measure the temperature by extrapolating to zero density using Eq. 21-4,

$$T = 273.16 \text{ K} \lim_{p_{tr} \to 0} \frac{p}{p_{tr}} \qquad \text{(real gas)}.$$

If we had an ideal gas available (which we do not), the extrapolation would be unnecessary.

EXAMPLE 1

A cylinder contains oxygen at a temperature of 20° C and a pressure of 15 atm in a volume of 100 liters. A piston is lowered into the cylinder decreasing the volume occupied by the gas to 80 liters and raising the temperature to 25° C. Assuming oxygen to behave like an ideal gas under these conditions, what then is the gas pressure?

From Eq. 23-1, since the mass of gas remains unchanged, we may write

$$\frac{p_i V_i}{T_i} = \frac{p_f V_f}{T_f}.$$

Our initial conditions are

$$p_i = 15 \text{ atm}, \qquad T_i = 293 \text{ K}, \qquad V_i = 100 \text{ liters}.$$

Our final conditions are

$$p_f = ?, \qquad T_f = 298 \text{ K}, \qquad V_f = 80 \text{ liters}.$$

Hence,

$$p_f = \left(\frac{T_f}{V_f}\right)\left(\frac{p_i V_i}{T_i}\right) = \left(\frac{298 \text{ K}}{80 \text{ liters}}\right)\left(\frac{15 \text{ atm} \times 100 \text{ liters}}{293 \text{ K}}\right) = 19 \text{ atm}.$$

EXAMPLE 2

Calculate the work per mole done by an ideal gas which expands isothermally, that is, at constant temperature, from an initial volume V_i to a final volume V_f.

The work done may be represented as

$$W = \int_{V_i}^{V_f} p \ dV.$$

From the ideal gas law we have

$$p = \frac{nRT}{V},$$

so that W/n, the work per mole, is

$$\frac{W}{n} = \int_{V_i}^{V_f} \frac{RT}{V} \, dV.$$

The temperature is constant so that

$$\frac{W}{n} = RT \int_{V_i}^{V_f} \frac{dV}{V} = RT \ln \frac{V_f}{V_i}$$

is the work per mole done by an ideal gas in an isothermal expansion at temperature T from an initial volume V_i to a final volume V_f.

Notice that when the gas expands, so that $V_f > V_i$, the work done by the gas is positive; when the gas is compressed, so that $V_f < V_i$, the work done by the gas is negative. This is consistent with the sign convention adopted for W in the first law of thermodynamics. The work done is shown as the shaded area in Fig. 23-1. The solid line is an isotherm, that is, a curve giving the relation of p to V at a constant temperature.

In practice, how can we keep an expanding or contracting gas at constant temperature?

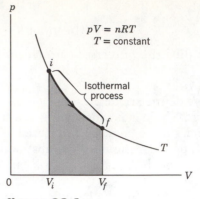

figure 23-1
Example 2. The shaded area represents the work done by n moles of gas in expanding from V_i to V_f with the temperature held constant.

From the microscopic point of view we define an ideal gas by making the following assumptions; it will then be our task to apply the laws of classical mechanics statistically to the gas atoms and to show that our microscopic definition is consistent with the macroscopic definition of the preceding section.

1. *A gas consists of particles, called molecules.* Depending on the gas, each molecule will consist of one atom or a group of atoms. If the gas is an element or a compound and is in a stable state, we consider all its molecules to be identical.

2. *The molecules are in random motion and obey Newton's laws of motion.* The molecules move in all directions and with various speeds. In computing the properties of the motion, we assume that Newtonian mechanics works at the microscopic level. As for all our assumptions, this one will stand or fall depending on whether or not the experimental results it predicts are correct.

3. *The total number of molecules is large.* The direction and speed of motion of any one molecule may change abruptly on collision with the wall or another molecule. Any particular molecule will follow a zigzag path because of these collisions. However, because there are so many molecules we assume that the resulting large number of collisions maintains the over-all distribution of molecular velocities and the randomness of the motion.

4. *The volume of the molecules is a negligibly small fraction of the volume occupied by the gas.* Even though there are many molecules, they are extremely small. We know that the volume occupied by a gas can be changed through a large range of values with little difficulty, and that when a gas condenses the volume occupied by the liquid may be thousands of times smaller than that of the gas. Hence, our assumption is plausible. Later we shall investigate the actual size of molecules and see whether we need to modify this assumption.

23-3
AN IDEAL GAS—MICROSCOPIC DEFINITION

5. *No appreciable forces act on the molecules except during a collision.* To the extent that this is true a molecule moves with uniform velocity between collisions. Because we have assumed the molecules to be so small, the average distance between molecules is large compared to the size of a molecule. Hence, we assume that the range of molecular forces is comparable to the molecular size.

6. *Collisions are elastic and are of negligible duration.* Collisions between molecules and with the walls of the container conserve momentum and (we assume) kinetic energy. Because the collision time is negligible compared to the time spent by a molecule between collisions, the kinetic energy which is converted to potential energy during the collision is available again as kinetic energy after such a brief time that we can ignore this exchange entirely.

Let us now calculate the pressure of an ideal gas from kinetic theory. To simplify matters, we consider a gas in a cubical vessel whose walls are perfectly elastic. Let each edge be of length l. Call the faces normal to the x-axis (Fig. 23-2) A_1 and A_2, each of area l^2. Consider a molecule which has a velocity $\mathbf{v}$. We can resolve $\mathbf{v}$ into components v_x, v_y, and v_z in the directions of the edges. If this particle collides with A_1, it will rebound with its x-component of velocity reversed. There will be no effect on v_y or v_z, so that the change in the particle's momentum will be

final momentum − initial momentum $= -mv_x - (mv_x) = -2mv_x$,

normal to A_1. Hence, the momentum imparted to A_1 will be $2mv_x$, since the total momentum is conserved.

Suppose that this particle reaches A_2 without striking any other particle on the way. The time required to cross the cube will be l/v_x. At A_2 it will again have its x-component of velocity reversed and will return to A_1. Assuming no collisions in between, the round trip will take a time $2l/v_x$. Hence, the number of collisions per unit time this particle makes with A_1 is $v_x/2l$, so that the rate at which it transfers momentum to A_1 is

$$2mv_x \frac{v_x}{2l} = \frac{mv_x^2}{l}.$$

To obtain the total force on A_1, that is, the rate at which momentum is imparted to A_1 by all the gas molecules, we must sum up mv_x^2/l for all the particles. Then, to find the pressure, we divide this force by the area of A_1, namely l^2.

If m is the mass of each molecule, we have

$$p = \frac{m}{l^3}\left(v_{x1}^2 + v_{x2}^2 + \cdots\right),$$

where v_{x1} is the x-component of the velocity of particle 1, v_{x2} is that of particle 2, etc. If N is the total number of particles in the container and n_v is the number per unit volume, then $N/l^3 = n_v$ or $l^3 = N/n_v$. Hence,

$$p = mn_v\left(\frac{v_{x1}^2 + v_{x2}^2 + \cdots}{N}\right).$$

But mn_v is simply the mass per unit volume, that is, the density ρ. The quantity $(v_{x1}^2 + v_{x2}^2 + \cdots)/N$ is the average value of v_x^2 for all the particles in the container. Let us call this $\overline{v_x^2}$. Then

23-4

KINETIC CALCULATION OF THE PRESSURE

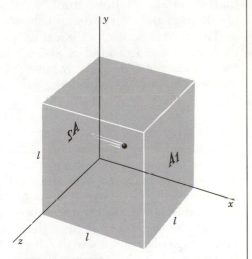

figure 23-2
A cubical box of side l, containing an ideal gas. A molecule is shown moving toward A_1.

$$p = \rho\overline{v_x^2}.$$

For any particle $v^2 = v_x^2 + v_y^2 + v_z^2$. Because we have many particles and because they are moving entirely at random, the average values of v_x^2, v_y^2, and v_z^2 are equal and the value of each is exactly one-third the average value of v^2. There is no preference among the molecules for motion along any one of the three axes. Hence, $\overline{v_x^2} = \tfrac{1}{3}\overline{v^2}$, so that

$$p = \rho\overline{v_x^2} = \tfrac{1}{3}\rho\overline{v^2}. \tag{23-3}$$

Although we derived this result by neglecting collisions between particles, the result is true even when we consider collisions. Because of the exchange of velocities in an elastic collision between identical particles, there will always be some one molecule that will collide with A_2 with momentum mv_x corresponding to the one that left A_1 with this same momentum. Also, the time spent during collisions is negligible compared to the time spent between collisions. Hence, our neglect of collisions is merely a convenient device for calculation. Similarly, we could have chosen a container of any shape—the cube merely simplifies the calculation. Although we have calculated the pressure exerted only on the side A_1, it follows from Pascal's law that the pressure is the same on all sides and everywhere in the interior.*

The square root of $\overline{v^2}$ is called the *root-mean-square* speed of the molecules and is a kind of average molecular speed.† Using Eq. 23-3, we can calculate this root-mean-square speed from measured values of the pressure and density of the gas. Thus,

$$v_{\rm rms} = \sqrt{\overline{v^2}} = \sqrt{\frac{3p}{\rho}}. \tag{23-4a}$$

In Eq. 23-3 we relate a macroscopic quantity (the pressure p) to an average value of a microscopic quantity (that is, to $\overline{v^2}$ or $v_{\rm rms}^2$). However, averages can be taken over short times or over long times, over small regions of space or large regions of space. The average computed in a small region for a short time might depend on the time or region chosen, so that the values obtained in this way may fluctuate. This could happen in a gas of very low density, for example. We can ignore fluctuations, however, when the number of particles in the system is large enough.

EXAMPLE 3

Calculate the root-mean-square speed of hydrogen molecules at 0.00° C and 1.00-atm pressure, assuming hydrogen to be an ideal gas. Under these conditions hydrogen has a density ρ of 8.99×10^{-2} kg/m³. Then, since $p = 1.00$ atm = 1.01×10^5 N/m²,

$$v_{\rm rms} = \sqrt{\frac{3p}{\rho}} = 1840 \text{ m/s}.$$

This is of the order of a mile per second, or 3600 mi/h.

Table 23-1 gives the results of similar calculations for some gases at 0° C. These molecular speeds are roughly of the same order as the speed

* We neglect the weight of the gas, a negligible effect unless the gas is of very large extent, as in the atmosphere. (See Section 17-3 and Problem 26.)

† We will consider this further in Section 24-2 in which we discuss the molecular distribution of speeds.

Table 23-1

Gas	Molecular weight,* g/mol	v_{rms} (at 0° C), m/s	Translational kinetic energy per mole (at 0° C), $\frac{1}{2}Mv_{rms}^2$, J/mol
H_2	2.02	1838	3370
He	4.0	1311	3430
H_2O	18	615	3400
Ne	20.1	584	3420
N_2	28	493	3390
CO	28	493	3390
Air	28.8	485	3280
O_2	32	461	3400
CO_2	44	393	3400

* The molecular weight and the mole are defined on page 479. We will discuss the last column in this table in the next section.

of sound at the same temperature. For example, in air at 0° C, $v_{rms} = 485$ m/s and the speed of sound is 331 m/s and in hydrogen $v_{rms} = 1838$ m/s and sound travels at 1286 m/s. These results are to be expected in terms of our model of a gas; see Prob. 34. We visualize the propagation of sound waves as a directional motion of the molecules as a whole superimposed on their random motion. Hence, the energy of the sound wave is carried as kinetic energy from one molecule to the next one with which it collides. The molecules themselves, in spite of their high speeds, do not move very far during a period of the sound vibration; they are confined to a rather small space by the effects of a large number of collisions.† However, the energy of the sound wave is communicated from one molecule to the next with that high speed, even though we do not expect the speed of sound to be *exactly* equal to v_{rms}, a point that we will clarify in Example 6.

EXAMPLE 4

Assuming that the speed of sound in a gas is the same as the root-mean-square speed of the molecules, show how the speed of sound for an ideal gas would depend on the temperature. (Actually this assumption is only crudely correct. Compare Eq. 23-4a and Eq. 23-15.)

The density of a gas is

$$\rho = \frac{nM}{V}$$

in which M is the molecular weight (grams/mole) and n is the number of moles. Combining this with the ideal gas law

$$pV = nRT$$

yields

$$\frac{p}{\rho} = \frac{RT}{M}.$$

We obtain from Eq. 23-4a

$$v_{rms} = \sqrt{\frac{3p}{\rho}} = \sqrt{\frac{3RT}{M}}, \tag{23-4b}$$

so that the speed of sound v_1 at a temperature T_1 is related to the speed of sound

† This explains why there is a time lag between opening an ammonia bottle at one end of a room and smelling it at the other end. Although molecular speeds are high, the large number of collisions restrains the advance of the ammonia molecules. They diffuse through the air at speeds that are very much less than molecular speeds.

v_2 in the same gas at a temperature T_2 by

$$\frac{v_1}{v_2} = \sqrt{\frac{T_1}{T_2}}.$$

For example, if the speed of sound at 273 K is 332 m/s in air, its speed in air at 300 K will be

$$\sqrt{\tfrac{300}{273}} \times 332 \text{ m/s} = 348 \text{ m/s}.$$

Would our result change if the speed of sound were proportional to, rather than equal to, the root-mean-square speed of the molecules of a gas?

23-5 KINETIC INTERPRETATION OF TEMPERATURE

If we multiply each side of Eq. 23-3 by the volume V, we obtain

$$pV = \tfrac{1}{3}\rho V \overline{v^2},$$

where ρV is simply the total mass of gas, ρ being the density. We can also write the mass of gas as nM, in which n is the number of moles and M is the molecular weight. Making this substitution yields

$$pV = \tfrac{1}{3}nM\overline{v^2}.$$

The quantity $\tfrac{1}{3}nM\overline{v^2}$ is two-thirds the total kinetic energy of translation of the molecules, that is, $\tfrac{2}{3}(\tfrac{1}{2}nM\overline{v^2})$*. We can then write

$$pV = \tfrac{2}{3}(\tfrac{1}{2}nM\overline{v^2}).$$

The equation of state of an ideal gas is

$$pV = nRT.$$

Combining these two expressions, we obtain

$$\tfrac{1}{2}M\overline{v^2} = \tfrac{3}{2}RT. \tag{23-5}$$

That is, *the total translational kinetic energy per mole of the molecules of an ideal gas is proportional to the temperature.* We may say that this result, Eq. 23-5, is necessary to fit the kinetic theory to the equation of state of an ideal gas, or we may consider Eq. 23-5 as a definition of gas temperature on a kinetic theory or microscopic basis. In either case, we gain some insight into the meaning of temperature for gases.

The temperature of a gas is related to the total translational kinetic energy measured with respect to the center of mass of the gas. The kinetic energy associated with the motion of the center of mass of the gas has no bearing on the gas temperature. In Section 23-3 we assumed random motion as part of our statistical definition of an ideal gas and in Section 23-4 we calculated $\overline{v^2}$ on this basis. For a random distribution of molecular velocities with direction the center of mass would be at rest, so that we must use a reference frame in which the center of mass of the gas is at rest. For all other frames the molecules will each have velocities greater by **u** (the velocity of the center of mass in that frame) than in the center of mass frame; hence, the motions will no longer be random and we will obtain different values for $\overline{v^2}$. The temperature of a gas in a container does not increase when we put the container on a moving train!

Let us now divide each side of Eq. 23-5 by Avogadro's number, N_0, which (see page 479, footnote) is the number of molecules per mole of a gas. Thus $M/N_0 (= m)$ is the mass of a single molecule and we have

$$\tfrac{1}{2}(M/N_0)\overline{v^2} = \tfrac{1}{2}m\overline{v^2} = \tfrac{3}{2}(R/N_0)T.$$

*If N is the total number of molecules and m is the mass of each molecule, then $\tfrac{1}{2}mv_1^2 + \tfrac{1}{2}mv_2^2 + \cdots = \tfrac{1}{2}mN\left[\dfrac{v_1^2 + v_2^2 + \cdots}{N}\right] = \tfrac{1}{2}mN\overline{v^2}$ in which $mN (= nM)$ is the total mass of the gas.

Now $\frac{1}{2}m\overline{v^2}$ is the average translational kinetic energy per molecule. The ratio R/N_0—which we call k, the *Boltzmann constant*—plays the role of the gas constant per molecule. We have

$$\tfrac{1}{2}m\overline{v^2} = \tfrac{3}{2}kT \qquad (23\text{-}6)$$

in which*

$$k = \frac{R}{N_0} = \frac{8.314 \text{ J/mol·K}}{6.023 \times 10^{23} \text{ molecules/mol}} = 1.380 \times 10^{-23} \text{ J/molecule·K}$$

We shall return to Boltzmann's constant in Chapter 24.

In the last column of Table 23-1 we list calculated values of $\frac{1}{2}Mv_{\text{rms}}^2$. As Eq. 23-5 predicts, this quantity (the translational kinetic energy per mole) has (closely) the same value for all gases at the same temperatures, 0° C in this case. From Eq. 23-6 we conclude that at the same temperature T the ratio of the root-mean-square speeds of molecules of two different gases is equal to the square root of the inverse ratio of their masses. That is, from

$$T = \frac{2}{3k}\frac{m_1\overline{v_1^2}}{2} = \frac{2}{3k}\frac{m_2\overline{v_2^2}}{2}$$

we obtain

$$\sqrt{\frac{\overline{v_1^2}}{\overline{v_2^2}}} = \frac{v_{1\text{rms}}}{v_{2\text{rms}}} = \sqrt{\frac{m_2}{m_1}}. \qquad (23\text{-}7)$$

We can apply Eq. 23-7 to the diffusion of two different gases in a container with porous walls placed in an evacuated space. The lighter gas, whose molecules move more rapidly on the average, will escape faster than the heavier one. The ratio of the numbers of molecules of the two gases which find their way through the porous walls for a short time interval is equal to the square root of the inverse ratio of their masses, $\sqrt{m_2/m_1}$. This diffusion process is one method of separating (fissionable) U^{235} (0.7% abundance) from a normal sample of uranium containing mostly (nonfissionable) U^{238} (99.3% abundance). To quote from the Smyth report,†

As long ago as 1896 Lord Rayleigh showed that a mixture of gases of different atomic weight could be partly separated by allowing some of it to diffuse through a porous barrier into an evacuated space. Because of their higher average speed the molecules of the light gas diffuse through the barrier faster so that the gas which has passed through the barrier (i.e., the "diffusate") is enriched in the lighter constituent and the residual gas which has not passed through the barrier is impoverished in the lighter constituent. The gas most highly enriched in the lighter constituent is the so-called "instantaneous diffusate"; it is the part that diffuses before the impoverishment of the residue has become appreciable. . . . On the assumption that the diffusion rates are inversely proportional to the square roots of the molecular weights‡ the separation factor for the instantaneous diffusate, called the "ideal separation factor" α, is given by

$$\alpha = \sqrt{M_2/M_1},$$

where M_1 is the molecular weight of the lighter gas and M_2 that of the heavier. Applying this formula to the case of uranium will illustrate the magnitude of the separation problem. Since uranium itself is not a gas, some gaseous com-

---

* See footnote, p. 479.

† *A General Account of the Development of Methods of Using Atomic Energy for Military Purposes* . . . , H. D. Smyth, U.S. Government Printing Office, 1945.

‡ Note that the ratio m_2/m_1 of the masses of the two molecules of different gases is the same as the ratio M_2/M_1 of their molecular weights, because the molecular weights refer to the same number of molecules. Compare Eq. 23-7.

pound of uranium must be used. The only one obviously suitable is uranium hexafluoride, UF_6. . . . Since fluorine has only one isotope, the two important uranium hexafluorides are $U^{235}F_6$ and $U^{238}F_6$; their molecular weights are 349 [g/mol] and 352 [g/mol]. Thus if a small fraction of a quantity of uranium hexafluoride is allowed to diffuse through a porous barrier, the diffusate will be enriched in $U^{235}F_6$ by a factor

$$\alpha = \sqrt{\tfrac{352}{349}} = 1.0043 \ldots$$

To separate the uranium isotopes, many successive diffusion stages (i.e., a cascade) must be used. . . . Studies by Cohen and others show that the best flow arrangement for the successive stages is that in which half the gas pumped into each stage diffused through the barrier, the other (impoverished) half being returned to the feed of the next lower stage. . . . If one desires to produce 99 per cent pure $U^{235}F_6$, and if one uses a cascade in which each stage has a reasonable overall enrichment factor, then it turns out that roughly 4000 stages are required. . . . Most of the material that eventually emerges from the cascade has been recycled many times. Calculation shows that for an actual uranium-separation plant it may be necessary to force through the barriers of the first stage 100,000 times the volume of gas that comes out the top of the cascade (i.e., as desired product $U^{235}F_6$).

Forces between molecules are of electromagnetic origin. All molecules contain electric charges in motion. These molecules are electrically neutral in the sense that the negative charge of the electrons is equal and opposite to the charge of the nuclei. This does not mean, however, that molecules do not interact electrically. For example, when two molecules approach each other, the charges on each are disturbed and depart slightly from their usual positions in such a way that the average distance between opposite charges in the two molecules is a little smaller than that between like charges. Hence, an attractive intermolecular force results. This internal rearrangement takes place only when molecules are fairly close together, so that these forces act only over short distances; they are short-range forces. If the molecules come very close together, so that their outer charges begin to overlap, the intermolecular force becomes repulsive. The molecules repel each other because there is no way for a molecule to rearrange itself internally to prevent repulsion of the adjacent external electrons. It is this repulsion on contact that accounts for the billiard-ball character of molecular collisions in gases. If it were not for this repulsion, molecules would move right through each other instead of rebounding on collision.

Let us assume that molecules are approximately spherically symmetrical. Then we can describe intermolecular forces graphically by plotting the mutual potential energy of two molecules, U, as a function of distance r between their centers. The force F acting on each molecule is related to the potential energy U by $F = -dU/dr$. In Fig. 23-3a we plot a typical $U(r)$. Here we can imagine one molecule to be fixed at O. Then the other molecule will be repelled from O when the slope of U is negative and will be attracted to O when the slope is positive. At r_0 no force acts between the molecules; the slope is zero there. In Fig. 23-3b we plot the mutual force $F(r)$ corresponding to this potential energy function. The line E in Fig. 23-3a represents the total mechanical energy of the colliding molecules. The intersection of $U(r)$ with this line is a "turning point" of the motion (see Section 8-5). The separation of the centers of two molecules at the turning point is the distance of closest approach. The separation distance at which the mutual potential energy is zero may be taken as the approximate distance of closest approach in a collision and hence as the diameter of the molecule. For simple molecules the diameter is about 2.5×10^{-10} m. The forces between molecules practically cease at about 10^{-9} m or 4 diameters apart, so that molecular forces are very short-range ones. The distance r_0 at which the potential is a minimum (the equilibrium point) is about 3.5×10^{-10} m for simple molecules. Of course, different molecules have different sizes and internal arrangement of charges so that intermolecular forces vary from one molecule to

23-6 INTERMOLECULAR FORCES

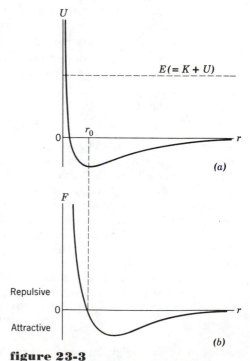

figure 23-3

(a) The mutual potential energy of two molecules versus their separation. E shows their total mechanical energy ($= K + U$). (b) The mutual force, $-dU/dr$, corresponding to this potential energy. U is a minimum at r_0, at which separation $F = 0$.

another. However, they always show the qualitative behavior indicated in the figures.*

In the solid state molecules vibrate about the equilibrium position r_0, their total energy E being negative, that is, lying below the horizontal axis in Fig. 23-3a. The molecules do not have enough energy to escape from the potential valley (that is, from the attractive binding force). The centers of vibration O are more or less fixed in a solid. In a liquid the molecules have greater vibrational energy about centers which are free to move but which remain about the same distance from one another. These molecules have their greatest kinetic energy in the gaseous state. In a gas the average distance between the molecules is considerably greater than the effective range of intermolecular forces, and the molecules move in straight lines between collisions. Clerk Maxwell discusses the relation between the kinetic theory model of a gas and the intermolecular forces as follows: "Instead of saying that the particles are hard, spherical, and elastic, we may if we please say that the particles are centers of force, of which the action is insensible except at a certain small distance, when it suddenly appears as a repulsive force of very great intensity. It is evident that either assumption will lead to the same results."

It is interesting to compare the measured intermolecular forces with the gravitational force of attraction between molecules. If we choose a separation distance of 4×10^{-10} m, for example, the force between two helium atoms is about 6×10^{-13} N. The gravitational force at that separation is about 7×10^{-42} N, smaller than the intermolecular force by a factor of 10^{29}! This is a typical result and shows that gravitational forces are negligible in comparison with intermolecular forces. Although the intermolecular forces appear to be small by ordinary standards, we must remember that the mass of a molecule is so small (about 10^{-26} kg) that these forces can impart instantaneous accelerations of the order of 10^{15} m/s² ($10^{14}g$). These accelerations may last for only a very short time, of course, because one molecule can very quickly move out of the range of influence of the other.

23-7
SPECIFIC HEATS OF AN IDEAL GAS

We picture the molecules in an ideal gas as hard elastic spheres; that is, we assume that there are no forces between the molecules except during collisions and that the molecules are not deformed by collisions. If this is so, there is no internal potential energy and the internal energy of an ideal gas is entirely kinetic. We have already found that the average translational kinetic energy per molecule is $\frac{3}{2}kT$, so that the internal energy U of an ideal gas containing N molecules is†

$$U = \tfrac{3}{2}NkT = \tfrac{3}{2}nRT. \qquad (23\text{-}8)$$

This prediction of kinetic theory says that *the internal energy of an ideal gas is proportional to the Kelvin temperature and depends only on the temperature*, being independent of pressure and volume. With this result we can now obtain information about the specific heats of an ideal gas.

The specific heat of a substance is the heat required per unit mass per unit temperature change. A convenient unit of mass is the mole. The corresponding specific heat is called the molar heat capacity and is represented by C. Only two varieties of molar heat capacity are important for gases, namely, that at constant volume, C_v, and that at constant pressure, C_p.

---

* See "The Force between Molecules," by B. V. Derjaguin, *Scientific American*, July 1960, for a discussion of the measurement of molecular attractions between macroscopic bodies.
† We will see in Section 23-8 that this result applies only to monatomic gases, for which rotational and vibrational energies are not possible. Only in this case can we equate U to the *translational* kinetic energy.

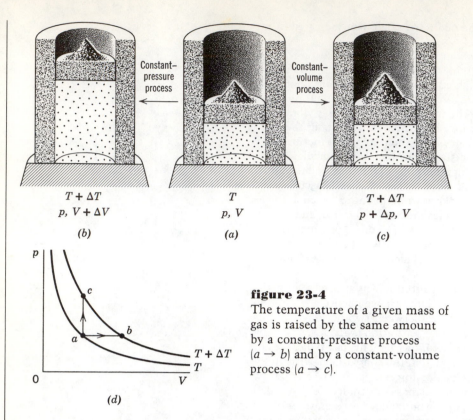

$T + \Delta T$
$p, V + \Delta V$

(b)

T
p, V

(a)

$T + \Delta T$
$p + \Delta p, V$

(c)

(d)

figure 23-4
The temperature of a given mass of gas is raised by the same amount by a constant-pressure process $(a \rightarrow b)$ and by a constant-volume process $(a \rightarrow c)$.

Let us confine a certain number of moles of an ideal gas in a piston-cylinder arrangement as in Fig. 23-4a. The cylinder rests on a heat reservoir whose temperature can be raised or lowered at will, so that we may add heat to the system or remove it, as we wish. The gas has a pressure p such that its upward force on the (frictionless) piston just balances the weight of the piston and its sand load. The state of the system is represented by point a in the p-V diagram of Fig. 23-4d; this diagram shows two isothermal lines, all points on one corresponding to a temperature T and all points on the other to a (higher) temperature $T + \Delta T$.

Now let us raise the temperature of the system by ΔT, by slowly increasing the reservoir temperature. As we do this let us add sand to the piston so that the volume V does not change. This *constant-volume process* carries the system from the initial state of Fig. 23-4a to the final state of Fig. 23-4c. Equivalently, it goes from point a to point c in Fig. 23-4d. Let us apply the first law of thermodynamics

$$\Delta U = Q - W$$

to this process. By definition of C_v we have $Q = nC_v \, \Delta T$. Also, W $(= p \, \Delta V) = 0$ because $\Delta V = 0$. Thus

$$\Delta U = nC_v \, \Delta T. \qquad (23\text{-}9)$$

Let us restore the system to its original state and again raise its temperature by ΔT, this time leaving the sand load undisturbed so that the pressure p does not change. This *constant-pressure process* carries the system from the initial state of Fig. 23-4a to the final state of Fig. 23-4b. Equivalently, it goes from point a to point b in Fig. 23-4d. Let us apply the first law to *this* process. By definition of C_p we have $Q = nC_p \, \Delta T$. Also, $W = p \, \Delta V$. Now for an ideal gas, U depends only on the temperature. Since processes $a \to b$ and $a \to c$ in Fig. 23-4 involve the same change ΔT in temperature, they must also involve the same change ΔU in internal energy, namely, that given by Eq. 23-9. Thus for the constant-pressure process the first law yields

$$nC_p \, \Delta T = nC_v \, \Delta T + p \, \Delta V.$$

Let us apply the equation of state $pV = nRT$ to the constant-pressure process $a \to b$. For p constant we have, by taking differences,

$$p \, \Delta V = nR \, \Delta T.$$

Combining these equations yields

$$C_p - C_v = R. \qquad (23\text{-}10)$$

This shows that the molar heat capacity of an ideal gas at constant pressure is always larger than that at constant volume by an amount equal to the universal gas constant R ($= 8.31$ J/mol·K or 1.99 cal/mol·K). Although Eq. 23-10 is exact only for an ideal gas, it is nearly true for real gases at moderate pressure (see Table 23-2). Notice that in obtaining this result we did not use the specific relation $U = \frac{3}{2}nRT$, but only the fact that U depends on temperature alone.

If we can compute C_v, then Eq. 23-10 will give us C_p and vice versa. We *can* obtain C_v by combining Eq. 23-9 with the kinetic theory result for the internal energy of an ideal gas, $U = \frac{3}{2}nRT$ (Eq. 23-8). Thus, in the limit of differential changes,

$$C_v = \frac{dU}{n \, dT} = \frac{d}{n \, dT} \left[\tfrac{3}{2}nRT \right] = \tfrac{3}{2}R. \qquad (23\text{-}11)$$

This result (about 3 cal/mol·K) turns out to be rather good for monatomic gases. It is, however, in serious disagreement with values obtained for diatomic and polyatomic gases (see Table 23-2). This suggests that Eq. 23-8 is not generally correct (see footnote on p. 507). Since that relation followed directly from the kinetic theory model, we conclude that we must change the model if kinetic theory is to survive as a useful approximation to the behavior of real gases.

Show that for an ideal gas undergoing an adiabatic process $pV^\gamma =$ a constant, where $\gamma = C_p/C_v$.

EXAMPLE 5

Let us apply the first law of thermodynamics

$$Q = \Delta U + W.$$

For an adiabatic process $Q = 0$. For W we put $p \, \Delta V$. Since the gas is assumed to be ideal, U depends only on temperature and, from Eq. 23-9, $\Delta U = nC_v \, \Delta T$. With these substitutions we have

$$0 = nC_v \, \Delta T + p \, \Delta V$$

or

$$\Delta T = -\frac{p \, \Delta V}{nC_v}.$$

For an ideal gas $pV = nRT$, so that, if p, V, and T are allowed to take on small variations,

$$p \, \Delta V + V \, \Delta p = nR \, \Delta T$$

or

$$\Delta T = \frac{p \, \Delta V + V \, \Delta p}{nR}.$$

Equating these two expressions and using Eq. 23-10 ($C_p - C_v = R$), we obtain, after some rearrangement,

$$p \, \Delta V C_p + V \, \Delta p C_v = 0$$

Dividing by pVC_v and recalling that, by definition, $C_p/C_v = \gamma$, we get

$$\frac{\Delta p}{p} + \gamma \frac{\Delta V}{V} = 0.$$

In the limiting case of differential changes this reduces to

$$\frac{dp}{p} + \gamma \frac{dV}{V} = 0,$$

which (assuming γ to be constant) we can integrate as

$$\ln p + \gamma \ln V = \text{a constant}$$

or

$$pV^\gamma = \text{a constant.} \tag{23-12}$$

The value of the constant is proportional to the quantity of gas. In Fig. 23-5 we compare the isothermal and adiabatic behaviors of an ideal gas.

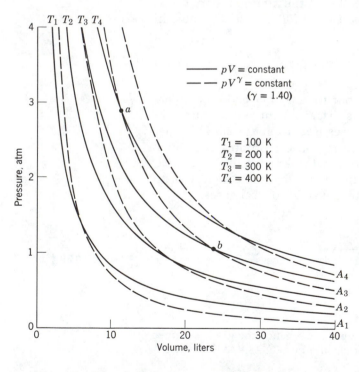

figure 23-5
T_1, T_2, T_3 and T_4 show how the pressure of one mole of an ideal gas changes as its volume is changed, the temperature being held constant (isothermal process). A_1, A_2, A_3 and A_4 show how the pressure of an ideal gas changes as its volume is changed, no heat being allowed to flow to or from the gas (adiabatic process). An adiabatic *increase* in volume (for example going from a to b along A_3) is always accompanied by a *decrease* in temperature, since at a, $T = 400$ K, whereas at b, $T = 300$ K.

In the figure:
— pV = constant
— — pV^γ = constant ($\gamma = 1.40$)

$T_1 = 100$ K
$T_2 = 200$ K
$T_3 = 300$ K
$T_4 = 400$ K

EXAMPLE 6

The compressions and rarefactions in a sound wave are practically adiabatic at audio frequencies. Show that in such a case the speed of sound in an ideal gas is given by

$$v = \sqrt{\frac{\gamma p}{\rho}}.$$

In Chapter 20 we showed the speed of sound to be $v = \sqrt{B/\rho}$, where ρ is the gas density and B is the bulk modulus of the gas, $B = -V(\Delta p/\Delta V)$. However, B will depend on the conditions that prevail as the pressure is changed. If we assume the temperature to remain constant we have, in the limit of differential changes,

$$B_{\text{isothermal}} = -V\left(\frac{dp}{dV}\right)_{\text{isothermal}} \qquad (23\text{-}13)$$

In an isothermal process for an ideal gas we have

$$pV = \text{a constant}$$

or, by differentiation with respect to V,

$$p + V\left(\frac{dp}{dV}\right)_{\text{isothermal}} = 0.$$

Combined with Eq. 23-13 this yields

$$B_{\text{isothermal}} = p.$$

In a sound wave, however, the conditions are not isothermal but closely adiabatic. The appropriate bulk modulus is then

$$B_{\text{adiabatic}} = -V\left(\frac{dp}{dV}\right)_{\text{adiabatic}} \qquad (23\text{-}14)$$

In an adiabatic process for an ideal gas we have

$$pV^{\gamma} = \text{a constant}$$

or, by differentiating with respect to V,

$$p\gamma V^{\gamma-1} + V^{\gamma}\left(\frac{dp}{dV}\right)_{\text{adiabatic}} = 0.$$

This, combined with Eq. 23-14, yields

$$B_{\text{adiabatic}} = \gamma p$$

and, for the speed of sound,

$$v = \sqrt{\frac{B}{\rho}} = \sqrt{\frac{\gamma p}{\rho}}. \qquad (23\text{-}15)$$

To understand why the compressions and rarefactions are adiabatic rather than isothermal, recall that compression of a gas causes a temperature rise and rarefaction a temperature fall unless heat energy is removed or added. Hence, in a gas through which sound propagates, the compressed regions are warmer than the rarefied ones. In principle, heat will be conducted from compression to rarefaction. The rate of heat conduction per unit area, however, depends (see Section 22-4) on the thermal conductivity of the gas and on the distance between compression and adjacent rarefaction, which is half a wavelength. The wavelength of audible sound is much too large for any significant rate of heat flow even in gases that are the best heat conductors. Hence, the conditions are essentially adiabatic in sound propagation and not isothermal. Actually the condition for breakdown of the adiabatic approximation is that the wavelength of the wave be comparable with the mean free path of molecules in the gas, an extreme situation (see Section 24-1).

Newton derived a formula for the speed of sound in 1710, when the only gas law formulated was Boyle's law. He assumed isothermal rather than adiabatic conditions and obtained $v = \sqrt{p/\rho}$ rather than the (correct) value of $\sqrt{\gamma\, p/\rho}$. Newton was able to get good agreement with experimental values by making (then) reasonable corrections to his basic model.* His error and the correct model were pointed out by Laplace in 1816, more than a century later. We must remember that, at that date, the concept of energy was not yet understood and the science of thermodynamics did not exist.

Does this result modify the result obtained in Example 4? Can you now explain why the speed of sound in a gas is not the same as the root-mean-square speed of the gas molecules?

* See "Newton's Derivation of the Velocity of Sound" by Haven Whiteside, *American Journal of Physics*, May 1964.

A modification of the kinetic theory model designed to explain the specific heats of gases was first suggested by Clausius in 1857. Recall that in our model we assumed a molecule to behave like a hard elastic sphere and we treated its kinetic energy as purely translational. The specific heat prediction was satisfactory for monatomic molecules. Further, because of the success of this simple model in other respects in predicting the correct behavior of gases of all kinds over wide temperature ranges, we feel confident that it is the average kinetic energy of translation which determines what we measure as the temperature of a gas.

In the case of specific heats, however, we are concerned with all possible ways of absorbing energy and we must ask whether or not a molecule can store energy internally, that is, in a form other than kinetic energy of translation. This would certainly be so if we pictured a molecule, not as a rigid particle, but as an object with internal structure. For then a molecule could rotate and vibrate as well as move with translational motion. In collisions, the rotational and vibrational modes of motion could be excited, and this would contribute to the internal energy of the gas. Here then is a model which enables us to modify the kinetic theory formula for the internal energy of a gas.

Let us now find the total energy of a system containing a large number of such molecules, where each molecule is thought of as an object having internal structure. The energy will consist of kinetic energy of translation, with terms like $\frac{1}{2}mv_x^2$; of kinetic energy of rotation, with terms like $\frac{1}{2}I\omega_x^2$; of kinetic energy of vibration of the atoms in a molecule, with terms like $\frac{1}{2}\mu v^2$ (where μ is the reduced mass), and of potential energy of vibration of the atoms in a molecule, with terms like $\frac{1}{2}kx^2$. Although other kinds of energy contributions exist, such as magnetic, for gases we can describe the total energy quite accurately by terms such as these. Although these terms have different origins, they all have the same mathematical form, namely, a positive constant times the square of a quantity which can take on negative or positive values. We can show from statistical mechanics that *when the number of particles is large and Newtonian mechanics holds, all these terms have the same average value, and this average value depends only on the temperature.* In other words, the available energy depends only on the temperature and distributes itself in equal shares to each of the independent ways in which the molecules can absorb energy. This theorem, stated here without proof, is called the *equipartition of energy* and was deduced by Clerk Maxwell. Each such independent mode of energy absorption is called a *degree of freedom*.

From Eq. 23-8 we know that the kinetic energy of translation per mole of gaseous molecules is $\frac{3}{2}RT$. The kinetic energy of translation per mole is the sum of three terms, however, namely $\frac{1}{2}M\overline{v_x^2}$, $\frac{1}{2}M\overline{v_y^2}$, and $\frac{1}{2}M\overline{v_z^2}$. The theorem of equipartition requires that each such term contribute the same amount to the total energy per mole, or $\frac{1}{2}RT$ per degree of freedom.

For *monatomic gases* the molecules have only translational motion (no internal structure in kinetic theory), so that $U = \frac{3}{2}nRT$. It follows from Eq. 23-11 that $C_v = \frac{3}{2}R \cong 3$ cal/mol K. Then from Eq. 23-10, $C_p = \frac{5}{2}R$, and the ratio of specific heat is

$$\gamma = \frac{C_p}{C_v} = \frac{5}{3} = 1.67.$$

For a *diatomic gas* we can think of each molecule as having a dumb-

23-8
EQUIPARTITION OF ENERGY

bell shape (two spheres joined by a rigid rod). Such a molecule can rotate about any one of three mutually perpendicular axes. However, the rotational inertia about an axis along the rigid rod should be negligible compared to that about axes perpendicular to the rod, so that the rotational energy should consist of only two terms,* such as $\frac{1}{2}I\omega_y^2$ and $\frac{1}{2}I\omega_z^2$. Each rotational degree of freedom is required by equipartition to contribute the same energy as each translational degree, so that for a diatomic gas having both rotational and translational motion,

$$U = 3n(\tfrac{1}{2}RT) + 2n(\tfrac{1}{2}RT) = \tfrac{5}{2}nRT,$$

or

$$C_v = \frac{dU}{n\,dT} = \tfrac{5}{2}R \cong 5 \text{ cal/mol·K}$$

and

$$C_p = C_v + R = \tfrac{7}{2}R,$$

or

$$\gamma = \frac{C_p}{C_v} = \frac{7}{5} = 1.40.$$

For *polyatomic gases*, each molecule contains three or more spheres (atoms) joined together by rods in our model, so that the molecule is capable of rotating energetically about each of three mutually perpendicular axes. Hence, for a polyatomic gas having both rotational and translational motion,

$$U = 3n(\tfrac{1}{2}RT) + 3n(\tfrac{1}{2}RT) = 3nRT,$$

or

$$C_v = \frac{dU}{n\,dT} = 3R = 6 \text{ cal/mol·K},$$

and

$$C_p = 4R,$$

or

$$\gamma = \frac{C_p}{C_v} = 1.33.$$

Let us now turn to experiment to test these ideas. In Table 23-2 we list the experimentally determined molar heat capacities for common gases at 20° C and 1.0 atm. Notice that for monatomic and diatomic

Table 23-2

Type of Gas	Gas	C_p, cal/mol·K	C_v, cal/mol·K	$C_p - C_v$	$\gamma = C_p/C_v$
Monatomic	He	4.97	2.98	1.99	1.67
	A	4.97	2.98	1.99	1.67
Diatomic	H_2	6.87	4.88	1.99	1.41
	O_2	7.03	5.03	2.00	1.40
	N_2	6.95	4.96	1.99	1.40
	Cl_2	8.29	6.15	2.14	1.35
Polyatomic	CO_2	8.83	6.80	2.03	1.30
	SO_2	9.65	7.50	2.15	1.29
	NH_3	8.80	6.65	2.15	1.31
	C_2H_6	12.35	10.30	2.05	1.20

* We have already ruled out the possibility that a monatomic molecule could rotate. Actually it could spin about any one of three mutually perpendicular axes if it had any extent, such as a finite sphere. Implicitly, therefore, we have adopted a point mass as our model of the atom. Hence, in a diatomic molecule we are rid of one rotational degree of freedom, for point masses joined by a rigid line have no rotational energy about an axis along that line.

gases the values of C_v, C_p, and γ are close to the ideal gas predictions. In some diatomic gases, like chlorine, and in most polyatomic gases the specific heats are larger than the predicted values. Even γ shows no simple regularity for polyatomic gases. This suggests that our model is not yet close enough to reality.

We have not yet considered energy contributions from the vibrations of the atoms in diatomic and polyatomic molecules. That is, we can modify the dumbbell model and join the spheres instead by springs. This new model will greatly improve our results in some cases. Instead of having a theoretical model for all gases, however, we now require an empirical model which differs from gas to gas. We can obtain a reasonably good picture of molecular behavior this way and the empirical model is therefore useful; however it ceases to be fundamental.

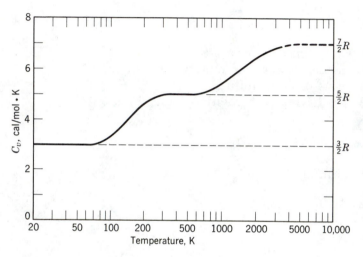

figure 23-6
Variation of the molar heat C_v of hydrogen with temperature. Note that T is drawn on a logarithmic scale. Hydrogen dissociates before 3200 K is reached. The dashed curve is for a diatomic molecule that does not dissociate before 10,000 K is reached.

To see this more clearly, let us consider Fig. 23-6, which shows the variation of the molar heat capacity of hydrogen with temperature. The value of 5 cal/mol·K, which is predicted for diatomic molecules by our model, is characteristic of hydrogen only in the temperature range from about 250 to 750 K. Above 750 K, C_v increases steadily toward 7 cal/mol·K and below 250 K, C_v decreases steadily to 3 cal/mol K. Other gases show similar variations of molar heat with temperature.

Here is a possible explanation. At low temperatures apparently (see Example 7) the hydrogen molecule has translational energy only and, for some reason, cannot rotate. As the temperature rises rotation becomes possible so that at "ordinary" temperatures a hydrogen molecule acts like our dumbbell model. At high temperatures the collisions between molecules cause the atoms in the molecule to vibrate and the molecule ceases to behave as a rigid body. Different gases, because of their different molecular structure, may show these effects at different temperatures. Thus a chlorine molecule appears to vibrate at room temperature.

Although this description is essentially correct, and we have obtained much insight into the behavior of molecules, this behavior contradicts classical kinetic theory. For kinetic theory is based on Newtonian mechanics applied to a large collection of particles, and the equipartition of energy is a necessary consequence of this classical statistical mechanics. But *if equipartition of energy holds, then, no matter what happens to the total internal energy as the temperature* changes, *each part of the energy*—translational, rotational, and vibrational—*must share equally in the change.* There is no classical mechanism for changing one mode of mechanical energy at a time in such a system. Kinetic theory requires that the specific heats of gases be independent of the temperature.

Hence, we have come to the limit of validity of classical mechanics when we try to explain the structure of the atom (or molecule). Just as Newtonian principles break down at very high speeds (near the speed of light), so here in the region of very small dimensions they also break down. Relativity theory modifies Newtonian ideas to account for the behavior of physical systems in the region of high speeds. It is quantum physics that modifies Newtonian ideas to account for the behavior of physical systems in the region of small dimensions. Both relativity theory and quantum mechanics are generalizations of classical theory in the sense that they give the (correct) Newtonian results in the regions in which Newtonian physics has accurately described experimental observations. In the following two chapters we shall confine our attention to the very fruitful application of thermodynamics and the kinetic theory to "classical" systems.

EXAMPLE 7

According to quantum theory the internal energy of an atom (or molecule) is "quantized"; that is, the atom cannot have any of a continuous set of internal energies but *only certain discrete ones*. After being raised from its lowest energy state to some higher one the atom can give up this energy by emitting radiation whose energy equals the difference in energy between the upper and lower internal energy states of the atom.

When two atoms collide, some of their translational kinetic energy may be converted into internal energy of one or both of the atoms. In such a case the collision is inelastic, for translational kinetic energy is not conserved. In a gas, the average translational kinetic energy of an atom is $\frac{3}{2}kT$. When the temperature is raised to a value where $\frac{3}{2}kT$ is about equal to some allowed internal excitation energy of the atom, then an appreciable number of the atoms can absorb enough energy through inelastic collisions to be raised to this higher internal energy state. We can detect this because, after an interval, radiation corresponding to the absorbed energy will be emitted.

(a) Compute the average translational kinetic energy per molecule in a gas at room temperature.

We have, for $T = 300$ K,

$$\frac{3}{2}kT = \frac{3}{2}(1.38 \times 10^{-23} \text{ J/molecule·K})(300 \text{ K})$$

$$= 6.21 \times 10^{-21} \text{ J/molecule}$$

$$= 3.88 \times 10^{-2} \text{ eV/molecule}.$$

This is about $\frac{1}{25}$ eV per molecule. Some molecules will have larger energies and some smaller energies than this average value.

(b) The first allowed (internal) excited state of a hydrogen atom is 10.2 eV above its lowest (ground) state. What temperature is needed to excite a "large" number of hydrogen atoms to emit radiation of this energy?

We require

$$\frac{3}{2}kT = 10.2 \text{ eV}$$

and we have from above

$$\frac{3}{2}k(300 \text{ K}) = \frac{1}{25} \text{ eV}.$$

Hence

$$T = 300 \text{ K} \times 10.2/(\tfrac{1}{25}) \simeq 7.5 \times 10^4 \text{ K}.$$

Actually, because many molecules have energies much greater than the average value, appreciable excitation may occur at somewhat lower temperatures.

We can now appreciate why the kinetic theory assumption, that molecules can be regarded as having no internal structure and collide elastically with one another, holds true at ordinary temperatures. Only at temperatures high enough to give the molecules an average translational kinetic energy comparable to the energy difference between the lowest and the first allowed excited state of the molecule will the internal structure of the molecule change and the collisions become inelastic. Indeed, in retrospect one may say that early evidence that the internal energy of an atom is quantized existed in experiments with gas collisions and that the seeds of quantum theory lay in the kinetic theory of gases.*

questions

1. In discussing the fact that it is impossible to apply the laws of mechanics individually to atoms in a macroscopic system, Mayer and Mayer state: "The very complexity of the problem [that is, the fact that the number of atoms is large] is the secret of its solution." Discuss this sentence.

2. Is there any such thing as a truly continuous body of matter?

3. In kinetic theory we assume that there are a large number of molecules in a gas. Real gases behave like an ideal gas at low densities. Are these statements contradictory? If not, what conclusion can you draw from them?

4. We have assumed that the walls of the container are elastic for molecular collisions. Actually, the walls may be inelastic. In practice this makes no difference as long as the walls are at the same temperature as the gas. Explain.

5. In large-scale inelastic collisions mechanical energy is lost through internal friction resulting in a rise of temperature owing to increased internal molecular agitation. Is there a loss of mechanical energy to heat in an inelastic collision between molecules?

6. What justification is there in neglecting the change in gravitational potential energy of molecules in a gas?

7. We have assumed that the force exerted by molecules on the wall of a container is steady in time. How is this justified?

8. The average velocity of the molecules in a gas must be zero if the gas as a whole and the container are not in translational motion. Explain how it can be that the average *speed* is not zero.

9. Consider a hot, stationary golf ball sitting on a tee and a cold golf ball just moving off the tee after being hit. Can the numerical value of the kinetic energy of the molecules' motion relative to the tee be the same in each case? If so, what is the difference between the two cases?

10. By considering quantities which must be conserved in an elastic collision, show that in general molecules of a gas cannot have the same speeds after a collision as they had before. Is it possible, then, for a gas to consist of molecules which all have the same speed?

11. Justify the fact that the pressure of a gas depends on the *square* of the speed of its particles by explaining the dependence of pressure on the collision frequency and the momentum transfer of the particles.

12. Why does the boiling temperature of a liquid increase with pressure?

13. Pails of hot and cold water are set out in freezing weather. Explain (a) if the pails have lids, the cold water will freeze first but (b) if the pails do not have lids, it is possible for the hot water to freeze first. (Hint: If equal masses of water are taken at two starting temperatures, more rapid evaporation from the hotter one may diminish its mass enough to compensate for the greater temperature range it must cover to reach freezing. See "The Freezing of Hot and Cold Water" by G. S. Kell, *American Journal of Physics*, May 1969.)

* See "On Teaching Quantum Phenomena" by Sir N. F. Mott in *Contemporary Physics*, August 1964.

14. How is the speed of sound related to gas variables in the kinetic theory model?

15. Far above the earth's surface the gas kinetic temperature (see Eq. 23-5) is reported to be the order of 1000 K. However, a person placed in such an environment would freeze to death rather than vaporize. Explain.

16. Why must the time allowed for diffusion separation be relatively short?

17. Suppose we want to obtain U^{238} instead of U^{235} as the end product of a diffusion process. Would we use the same process? If not, explain how the separation process would have to be modified.

18. Considering the diffusion of gases into each other (see footnote on page 503), can you draw an analogy to a large jostling crowd with many "collisions" on a large inclined plane with a slope of a few degrees?

19. Can you describe a centrifugal device for gaseous separation? Is a centrifuge better than a diffusion chamber for separation of gases?

20. Would you expect real molecules to be spherically symmetrical? If not, how would the potential energy function of Fig. 23-3 change?

21. Explain how we might keep a gas at a constant temperature during a thermodynamic process.

22. Explain why the temperature of a gas drops in an adiabatic expansion.

23. If hot air rises, why is it cooler at the top of a mountain than near sea level?

24. Comment on this statement: "There are two ways to carry out an adiabatic process. One is to do it quickly and the other is to do it in an insulated box."

25. A sealed rubber balloon contains a very light gas. The balloon is released and it rises high into the atmosphere. Describe and explain the thermal and mechanical behavior of the balloon.

26. Explain why the specific heat at constant pressure is greater than the specific heat at constant volume.

27. It is more common to excite radiation from gaseous atoms by use of electrical discharge than by thermal methods. Why?

28. *Extensive* quantities have values that depend on what the system's boundaries are, whereas *intensive* quantities are independent of the choice of boundaries. That is, extensive quantities are necessarily defined for a whole system, whereas intensive quantities apply uniformly to any small part of the system. Of the following quantities, determine which are extensive and which are intensive: pressure, volume, temperature, density, mass, internal energy.

SECTION 23-2

problems

1. At 0° C and 1.000-atm pressure the densities of air, oxygen, and nitrogen are, respectively, 1.293 kg/m³, 1.429 kg/m³, and 1.251 kg/m³. Calculate the percentage of nitrogen in the air from these data, assuming only these two gases to be present. *Answer:* 76.4%, by mass.

2. (a) What is the volume occupied by one mole of an ideal gas at standard conditions, that is, pressure of one atmosphere and temperature of 0° C? (b) Show that the number of molecules per cubic centimeter (Loschmidt number) at standard conditions is 2.687×10^{19}.

3. The best vacuum that can be attained in the laboratory corresponds to a pressure of about 10^{-14} atm, or about 10^{-10} mm-Hg. How many molecules are there per cubic centimeter in such a "vacuum" at room temperature? *Answer:* 2.7×10^5.

4. An air bubble of 20 cm³ volume is at the bottom of a lake 40 m deep where the temperature is 4° C. The bubble rises to the surface which is at a temperature of 20° C. Take the temperature of the bubble to be the same as that of the surrounding water and find its volume just before it reaches the surface?

5. Oxygen gas having a volume of 1.0 liter at 40° C and a pressure of 76 cm-Hg expands until its volume is 1.5 liters and its pressure is 80 cm-Hg. Find (a) the mass in moles of oxygen in the system and (b) its final temperature. *Answer: (a) 0.039 mol. (b) 220° C.*

6. An automobile tire has a volume of 1000 in.^{3} and contains air at a gauge pressure of 24 lb/in.^{2} when the temperature is 0° C. What is the gauge pressure of the air in the tires when its temperature rises to 27° C and its volume increases to 1020 in.^{3}?

7. Compute the number of molecules in a gas contained in a volume of 1.00 cm^3 at a pressure of 1.00×10^{-3} atm and a temperature of 200 K. *Answer: 3.67×10^{16}.*

8. If the water molecules in 1.0 g of water were distributed uniformly over the surface of the earth, how many such molecules would there be on 1.0 cm^2 of the earth's surface?

9. Calculate the work done in compressing 1.00 mol of oxygen from a volume of 22.4 l at 0° C and 1.00-atm pressure to 16.8 l at the same temperature. *Answer: 648 J.*

10. Suppose that, as happened historically, we are given Boyle's law

$$pV = a \text{ constant} \quad \text{(constant } T\text{)}$$

and Charles' law

$$V/T = a \text{ constant} \quad \text{(constant } p\text{)}$$

separately. Show how these two laws may be combined to yield

$$pV/T = a \text{ constant.}$$

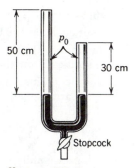

11. A mercury-filled manometer with two unequal arms is sealed off with the same pressure p_0 in the two arms as in Fig. 23-7. The cross-sectional area of the manometer arms is 1.0 cm^2. With the temperature constant, an additional 10 cm^3 of mercury is admitted through the stopcock at the bottom; the level on the left increases 6.0 cm and that on the right increases 4.0 cm. Find the pressure p_0. *Answer: 1.5×10^5 Pa.*

figure 23-7
Problem 11

12. Air that occupies 5.0 ft^3 (0.14 m^3) at 15 lb/in.^{2} (1.034×10^5 Pa) gauge pressure is expanded isothermally to atmospheric pressure and then cooled at constant pressure until it reaches its initial volume. Compute the work done by the gas.

SECTION 23-4

13. The mass of the H$_2$ molecule is 3.3×10^{-24} g (2.3×10^{-28} slug). If 10^{23} hydrogen molecules per second strike 2.0 cm^2 (0.31 in.^{2}) of wall at an angle of 45° with the normal when moving with a speed of 10^5 cm/s (3.3×10^3 ft/s), what pressure do they exert on the wall? *Answer: 2300 Pa (0.35 lb/in.^{2}).*

SECTION 23-5

14. At 273 K. and 1.00×10^{-2} atm the density of a gas is 1.24×10^{-5} g/cm^3. (a) Find v_{rms} for the gas molecules. (b) Find the molecular weight of the gas and identify it.

15. (a) Compute the root-mean-square speed of an argon atom at room temperature (20° C). (b) At what temperature will the root-mean-square speed be half that value? Twice that value? *Answer: (a) 430 m/s. (b) 73 K; 1200 K.*

16. In a gas of uranium hexafluoride there are isotopes U^{235}F$_6$ and U^{238}F$_6$ having molecular weights 349 and 352, respectively. (a) What is the ratio of the rms speeds of these two molecular isotopes? (b) How could this fact be used to separate the isotopes?

17. (a) Determine the average value of the kinetic energy of the particles of an ideal gas at 0.0° C and 100° C. (b) What is the kinetic energy per mole of an ideal gas at these temperatures? *Answer: (a) 5.65×10^{-21} J; 7.72×10^{-21} J. (b) 3400 J; 4650 J.*

18. At what temperature is the average translational kinetic energy of a mole-

cule equal to the kinetic energy of an electron accelerated from rest through a potential difference of one volt (that is, an energy of 1.0 eV)?

19. Oxygen gas at 273 K and 1.00-atm pressure is confined to a cubical container 10 cm on a side. Compare the change in gravitational potential energy of an oxygen molecule falling the height of the box with its mean translational kinetic energy.

    Answer: Ratio of the mean translational kinetic energy to the change in gravitational potential energy is 1.1×10^5.

20. Find the root-mean-square speeds of (a) helium and (b) argon molecules at 40° C from that of oxygen molecules (460 m/s at 0.00° C). The molecular weight of oxygen is 32 g/mol, of argon 40, of helium 4.0.

21. (a) Compute the temperature at which the root-mean-square speed is equal to the speed of escape from the surface of the earth for molecular hydrogen. For molecular oxygen. (b) Do the same for the moon, assuming gravity on its surface to be 0.16 g. (c) The temperature high in the earth's upper atmosphere is about 1000 K. Would you expect to find much hydrogen there? Much oxygen?

    Answer: (a) 1.0×10^4 K; 1.6×10^5 K. (b) 440 K; 7000 K.

22. (a) Consider an ideal gas at 273 K and 1.0-atm pressure. Imagine that the molecules are for the most part evenly spaced at the centers of identical cubes. Using Avogadro's number and taking the diameter of a molecule to be 3.0×10^{-8} cm, find the length of an edge of such a cube and compare this length to the diameter of a molecule. (b) Now consider a mole of water having a volume of 18 cm³. Again imagine the molecules to be evenly spaced at the centers of identical cubes. Find the length of an edge of such a cube and compare this length to the diameter of a molecule.

23. Plot and physically interpret (a) the variation of gas density with temperature for an isobaric (constant-pressure) process and (b) the variation of gas density with pressure for an isothermal process.

24. Water standing in the open at 27° C evaporates due to the escape of some of the surface molecules. The heat of vaporization (540 cal/g) may be found approximately from ϵn, where ϵ is the average energy of the escaping molecules and n is the number of molecules per gram. (a) Find ϵ. (b) How many times greater is ϵ than the average kinetic energy of H_2O molecules, assuming that the kinetic energy is related to temperature in the same way as it is for gases?

25. Consider a given mass of an ideal gas. Compare curves representing constant-pressure, constant-volume, and isothermal processes on (a) a p-V diagram, (b) a p-T diagram, and (c) a V-T diagram. (d) How do these curves depend on the mass of gas chosen?

26. (a) Show that the variation in pressure in the earth's atmosphere, assumed to be isothermal, is given by $p = p_0 e^{-Mgy/RT}$ where M is the molecular weight of the gas. (See Example 1, Chapter 17.) (b) Show also that $n_v = n_{v_0} e^{-Mgy/RT}$ where n_v is the number of molecules per unit volume.

SECTION 23-7

27. (a) What is the internal energy of one mole of an ideal gas at 273 K? (b) Does it depend on volume or pressure? Does it depend on the nature of the gas?

    Answer: (a) 3400 J.

28. One mole of an ideal gas expands adiabatically from an initial temperature T_1 to a final temperature T_2. Prove that the work done by the gas is $C_v(T_1 - T_2)$, where C_v is the molar heat capacity.

29. One mole of an ideal gas undergoes an isothermal expansion. Find the heat flow into the gas in terms of the initial and final volumes and the temperature.

    Answer: $RT \ln V_f/V_i$.

30. The mass of a gas molecule can be computed from the specific heat at constant volume. Take $C_v = 0.075$ kcal/kg · K for argon and calculate (a) the mass of an argon atom and (b) the atomic weight of argon.

31. Take the mass of a helium atom to be 6.66×10^{-27} kg. Compute the specific heat at constant volume for helium gas. *Answer:* 3.11×10^3 J/kg · K.

32. Air at $0.00°$ C and 1.00-atm pressure has a density of 1.291×10^{-3} g/cm³ and the speed of sound in air is 332 m/s at that temperature. Compute the ratio of specific heats of air.

33. Show that the speed of sound in an ideal gas is independent of the pressure and density.

34. The speed of sound in different gases at the same temperature depends on the molecular weight of the gas. Show that $v_1/v_2 = \sqrt{M_2/M_1}$ (constant T) where v_1 is the speed of sound in the gas of molecular weight M_1 and v_2 is the speed of sound in the gas of molecular weight M_2.

35. Show that the speed of sound in air increases about 0.61 m/s for each Celsius degree rise in temperature near $0°$ C.

36. From the knowledge that c_v, the specific heat at constant volume, for a gas in a container is $5R$, what can you conclude about the ratio of the speed of sound in that gas to the root-mean-square speed of its molecules at a temperature T?

37. The following data are the result of accurate experimental measurements: 1.000 mol of a gas occupies a volume of 2.541×10^{-2} m³ at a pressure of 9.480×10^4 Pa when its temperature is 290.0 K. The same mass of gas requires 125.0 cal to raise its temperature from 290.0 to 315.0 K while its volume is held constant. The ratio (c_p/c_v) of its specific heats is 1.430. (a) Use these data to compute the mechanical equivalent of heat J. (b) Account for the fact that your value of J differs from the accepted three-figure value—namely, 4.19 J/cal. *Answer:* (a) 3.86 J/cal.

38. A mass of gas occupies a volume of 4.0 liters at a pressure of 1.0 atm and a temperature of 300 K. It is compressed adiabatically to a volume of 1.0 liter. Determine (a) the final pressure and (b) the final temperature, assuming it to be an ideal gas for which $\gamma = 1.5$.

39. (a) A liter of gas with $\gamma = 1.3$ is at 273 K and 1.0-atm pressure. It is suddenly compressed to half its original volume. Find its final pressure and temperature. (b) The gas is now cooled back to $0°$ C at constant pressure. What is its final volume? *Answer:* (a) 2.5 atm; 340 K. (b) 0.40 liter.

40. A reversible heat engine carries 1.00 mol of an ideal monatomic gas around the cycle shown in Fig. 23-8. Process 1-2 takes place at constant volume, process 2-3 is adiabatic, and process 3-1 takes place at a constant pressure. (a) Compute the heat Q, the change in internal energy ΔU, and the work done W, for each of the three processes and for the cycle as a whole. (b) If the initial pressure at point 1 is 1.00 atm, find the pressure and the volume at points 2 and 3.

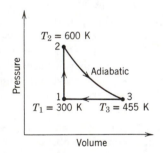

figure 23-8
Problem 40

41. A quantity of ideal gas occupies an initial volume V_0 at a pressure p_0 and a temperature T_0. It expands to a volume V_1 (a) at constant pressure, (b) at constant temperature, (c) adiabatically. Graph each case on a P-V diagram. In which case is Q greatest? Least? In which case is W greatest? Least? In which case is ΔU greatest? Least?

Answer:	greatest	least
Q	a	c
W	a	c
ΔU	a	c

42. A thin tube, sealed at both ends, is 1.0 m long. It lies horizontally, the middle 10 cm containing mercury and the two equal ends containing air at standard atmospheric pressure. If the tube is now turned to a vertical position, by what amount will the mercury be displaced? Assume that the process is (a) isothermal and (b) adiabatic. Which assumption is more reasonable?

SECTION 23-8

43. One mole of oxygen is heated at a constant pressure starting at $0.00°$ C. How much heat energy must be added to the gas to double its volume? *Answer:* 8040 J.

44. An ideal diatomic gas (4.0 moles) at high temperature experiences a temperature increase of 60 K under constant pressure conditions. (a) How much heat was added to the gas? (b) By how much did the internal energy of the gas increase? (c) How much work was done by the gas? (d) By how much did the internal translational kinetic energy of the gas increase?

45. Ten grams of oxygen are heated at constant atmospheric pressure from 27.0 to 127° C. (a) How much heat is transferred to the oxygen? (b) What fraction of the heat is used to raise the internal energy of the oxygen?
    Answer: (a) 920 J. (b) 71%.

46. Calculate the mechanical equivalent of heat from the value of R and the values of C_v and γ for oxygen from Table 23-2.

47. *Avogadro's law* states that under the same condition of temperature and pressure equal volumes of gas contain equal numbers of molecules. Derive this law from kinetic theory using Eq. 23-3 and the equipartition of energy assumption.

48. A room of volume V is filled with a diatomic ideal gas (air) at temperature T_1 and pressure p_0. The air is heated to a higher temperature T_2, the pressure remaining constant at p_0 because the walls of the room are not air-tight. Show that the internal energy content of the air remaining in the room is the same at T_1 and T_2, and that the energy supplied by the furnace to heat the air has all gone to heat the air *outside* the room. If we add no energy to the air, why bother to light the furnace? (Ignore the furnace energy used to raise the temperature of the walls, and consider only the energy used to raise the air temperature.)

49. The atomic weight of iodine is 127. A standing wave in a tube filled with iodine gas at 400 K has nodes that are 6.77 cm apart when the frequency is 1000 Hz. Is iodine gas monatomic or diatomic?   *Answer:* Diatomic.

50. How would you explain the observed value of $C_v = 7.50$ cal/mol · K for gaseous SO_2 at 15.0° C and 1.00 atm?

51. *Dalton's law* states that when mixtures of gases having no chemical interaction are present together in a vessel, the pressure exerted by each constituent at a given temperature is the same as it would exert if it alone filled the whole vessel, and that the total pressure is equal to the sum of the partial pressures of each gas. Derive this law from kinetic theory, using Eq. 23-3.

52. A hydrogen atom, in its lowest (ground) state and moving with 13-eV kinetic energy, collides head-on with another hydrogen atom which is *at rest* in its ground state. (a) Use the conservation laws of energy and momentum to show that this collision must be elastic. The first allowed excited state is about 10.2 eV above the ground state. (b) Show that the minimum initial kinetic energy that the incident atom needs to raise one of the atoms to the first excited state is *twice* the energy difference between ground state and first excited state.

53. (a) A monatomic ideal gas initially at 17° C is suddenly compressed to one-tenth its original volume. What is its temperature after compression? (b) Make the same calculation for a diatomic gas.
    Answer: (a) 1350 K. (b) 730 K.

24
kinetic theory
of gases—II

Between successive collisions a molecule in a gas moves with constant speed along a straight line. The average distance between such successive collisions is called the *mean free path* (Fig. 24-1). If molecules were points, they would not collide at all and the mean free path would be infinite. Molecules, however, are not points and hence collisions occur. If they were so numerous that they completely filled the space available to them, leaving no room for translational motion, the mean free path would be zero. Thus the mean free path is related to the size of the molecules and to their number per unit volume.

Consider the molecules of a gas to be spheres of diameter d. The cross section for a collision is then πd^2. That is, a collision will take place when the centers of two molecules approach within a distance d of one another. An equivalent description of collisions made by any one molecule is to regard that molecule as having a diameter $2d$ and all other molecules as point particles (see Fig. 24-2).

Imagine a typical molecule of equivalent diameter $2d$ moving with speed v through a gas of equivalent point particles and let us assume, for the time being, that the molecule and the point particles exert no forces on each other. In time t our molecule will sweep out a cylinder of cross-sectional area πd^2 and length vt. If n_v is the number of molecules per unit volume, the cylinder will contain $(\pi d^2 vt)n_v$ particles (see Fig. 24-3). Since our molecule and the point particles *do* exert forces on each other, this will be the number of collisions experienced by the molecule in time t. The cylinder of Fig. 24-3 will, in fact, be a broken one, changing direction with every collision.

The mean free path $\bar{l}$ is the average distance between successive collisions. Hence, $\bar{l}$ is the total distance vt covered in time t divided by the

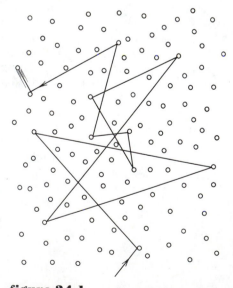

figure 24-1
A molecule traveling through a gas, colliding with other molecules in its path. Of course, all the other molecules are moving in a similar fashion.

number of collisions that take place in this time, or

$$\bar{l} = \frac{vt}{\pi \, d^2 n_v vt} = \frac{1}{\pi n_v d^2}.$$

This equation is based on the picture of a molecule hitting stationary targets. Actually the molecule hits moving targets. The collision frequency is increased as a result (see below) and the mean free path is reduced to

$$\bar{l} = \frac{1}{\sqrt{2} \, \pi \, n_v d^2}. \qquad (24\text{-}1)$$

When the target molecules are moving, the two v's in the first equation above are not the same. The one in the numerator ($= \bar{v}$) is the mean molecular speed measured with respect to the container. The one in the denominator ($= \bar{v}_{\text{rel}}$) is the mean *relative* speed with respect to other molecules; it is this relative speed that determines the collision rate.

We can see qualitatively that $\bar{v}_{\text{rel}} > \bar{v}$. Thus two molecules of speed v moving toward each other have a relative speed of $2v$ ($> v$); two molecules with speed v moving at right angles on a collision course have a relative speed of $\sqrt{2} \, v$ (also $> v$); two molecules moving with speed v in the same direction have a relative speed of zero ($< v$). Thus molecules arriving from *all of the forward hemisphere* and *part of the backward hemisphere* have $\bar{v}_{\text{rel}} > \bar{v}$. The molecules arriving from the rest of the backward hemisphere have $\bar{v}_{\text{rel}} < \bar{v}$ but, since their numbers are smaller, they do not determine the nature of the average over both hemispheres, which yields $\bar{v}_{\text{rel}} > \bar{v}$. A quantitative calculation, taking into account the actual speed distribution of the molecules, gives $\bar{v}_{\text{rel}} = \sqrt{2} \, \bar{v}$.

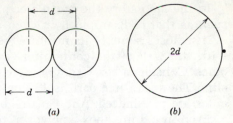

figure 24-2
(a) If a collision occurs when two molecules come within a distance d of each other, the process can be treated equivalently (b) by thinking of one molecule as having an effective diameter $2d$ and the other as being a point mass.

Let us calculate the magnitude of the mean free path and the collision frequency for air molecules at 0° C and 1-atm pressure.

We take 2×10^{-8} cm as an effective molecular diameter d. For the conditions stated, the average speed of air molecules is about 1×10^5 cm/s and there are about 3×10^{19} molecules/cm³. The mean free path is then

$$\bar{l} = \frac{1}{\pi \, \sqrt{2} \, n_v d^2} = \frac{1}{\pi \, \sqrt{2} \, (3 \times 10^{19}/\text{cm}^3)(2 \times 10^{-8} \text{ cm})^2}$$

$$= 2 \times 10^{-5} \text{ cm.}$$

This is about a thousand molecular diameters.

The corresponding collision frequency is

$$\frac{v}{\bar{l}} = (1 \times 10^5 \text{ cm/s})/(2 \times 10^{-5} \text{ cm})$$

$$= 5 \times 10^9/\text{s.}$$

Thus, on the average, *each molecule* makes five billion collisions per second!

EXAMPLE 1

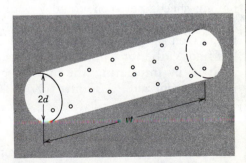

figure 24-3
A molecule of equivalent diameter $2d$ traveling with speed v sweeps out a cylinder of base πd^2 and length vt in a time t. It suffers a collision with every other molecule whose center lies within this cylinder.

In the earth's atmosphere we have seen that the mean free path of air molecules at sea level (760 mm-Hg) is 2×10^{-5} cm. At 100 km above the earth (10^{-3} mm-Hg) the mean free path is 2 mm. At 300 km (10^{-6} mm-Hg) it is 15 cm, and yet there are about 10^8 molecules/cm³ in this region. This emphasizes that molecules are indeed small. At great enough heights the mean free path concept fails because the upward-directed molecules follow ballistic paths and may escape from the atmosphere.

In the laboratory the mean free path concept is useful in situations such as that of Example 1. In even modest laboratory vacuums, however, it loses some of its meaning because nearly all the collisions are with the wall of the containing vessel rather than with other molecules. Consider a box 10 cm on edge containing air at 10^{-7} mm-Hg pressure. The mean free path is 150 cm, so that collisions between molecules are rare indeed. And yet this box contains about 10^{12} molecules!

Even in a finite "box," however, there are some conditions in which particles can travel great distances without striking the walls. In a typical proton synchrotron, used to accelerate protons to the billion-electron-volt range of energies, the protons are constrained by a magnetic field to move in a circular path and may travel *several hundred thousand miles* during the acceleration process. Mean free path considerations are important if the accelerating protons are to have essentially no collisions with residual air molecules. In this case the effective cross section of the proton is so much smaller than that of the air molecules that if we have a vacuum of about 10^{-6} mm-Hg, there is essentially no beam loss by proton scattering from gas molecules inside the vacuum chamber.

24-2
DISTRIBUTION OF MOLECULAR SPEEDS

In Chapter 23 we discussed the root-mean-square speed of the molecules of a gas. However, the speeds of individual molecules vary over a wide range of magnitude; there is a characteristic distribution of molecular speeds for a given gas which depends, as we will see below, on the temperature. If all the molecules of a gas had the same speed v, this situation would not persist for very long because the molecular speeds would be changed by collisions. However, we do not expect many molecules to have speeds $\ll v_{\text{rms}}$ (that is, near zero) or $\gg v_{\text{rms}}$ because such extreme speeds would require an unlikely sequence of preferential collisions.

Clerk Maxwell first solved the problem of the most probable distribution of speeds in a large number of molecules of a gas. His molecular speed distribution law, for a sample of gas containing N molecules, is*

$$N(v) = 4\pi N (m/2\pi kT)^{3/2} v^2 e^{-mv^2/2kT}. \qquad (24\text{-}2)$$

In this equation $N(v)\,dv$ is the number of molecules in the gas sample having speeds between v and $v + dv$. T is the absolute temperature, k is Boltzmann's constant, and m is the mass of a molecule. Note that for a given gas the speed distribution depends only on the temperature. We find N, the total number of molecules in the sample, by adding up (that is, by integrating) the number present in each differential speed interval from zero to infinity, or

$$N = \int_0^\infty N(v)\,dv. \qquad (24\text{-}3)$$

The unit of $N(v)$ is, say, molecules/(cm/s).

In Fig. 24-4 we plot the Maxwell distribution of speeds for molecules of oxygen at two different temperatures. The number of molecules having a speed between v_1 and v_2 equals the area under the curve between the vertical lines at v_1 and v_2. As Eq. 24-3 shows, the area under the speed distribution curve, which is the integral in that equation, is equal to the total number of molecules in the sample. At any

* A derivation of Eq. 24-2 appears in Supplementary Topic IV.

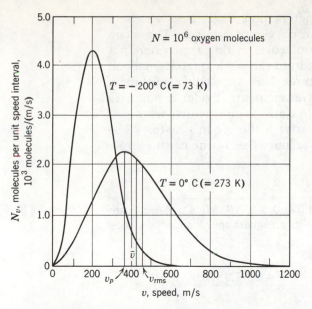

figure 24-4
The Maxwellian distribution of speeds of 10^6 oxygen molecules at two different temperatures. The number of molecules within a certain range of speeds (say, 300 to 600 m/s) is the area under this section of the curve. The complete area under either curve is the total number of molecules (equals 10^6); this area must be the same for each temperature if, as in this case, the curves refer to a given number of molecules. The pressure is lower than atmospheric because oxygen is a liquid at 1.0 atm and 73 K.

temperature the number of molecules in a given speed interval* Δv increases as the speed increases up to a maximum (the most probable speed v_p) and then decreases asymptotically toward zero. The distribution curve is not symmetrical about the most probable speed because the lowest speed must be zero, whereas there is no classical limit to the upper speed a molecule can attain. In this case the *average speed* $\bar{v}$ is somewhat larger than the most probable value. The root-mean-square value, v_{rms}, being the square root of the average of the *squares* of the speeds, is still larger.

As the temperature increases, the root-mean-square speed v_{rms} (and $\bar{v}$ and v_p as well) increases, in accord with our microscopic interpretation of temperature. The range of typical speeds is now greater, so that the distribution broadens. Since the area under the distribution curve (which is the total number of molecules in the sample) remains the same, the distribution must also flatten as the temperature rises. Hence the number of molecules which have speeds greater than some given speed increases as the temperature increases (see Fig. 24-4). This explains many phenomena, such as the increase in the rates of chemical reactions with rising temperature.

The distribution of speeds of molecules in a liquid also resembles the curves of Fig. 24-4. This explains why some molecules in a liquid (the fast ones) can escape through the surface (evaporate) at temperatures well below the normal boiling point. Only these molecules can overcome the attraction of the molecules in the surface and escape by evaporation. The average kinetic energy of the remaining molecules drops correspondingly, leaving the liquid at a lower temperature. This explains why evaporation is a cooling process.

From Eq. 24-2 we see that the distribution of molecular speeds depends on the mass of the molecule as well as on the temperature. The smaller the mass, the larger the proportion of high-speed molecules at

---

* We cannot simply plot the "number of particles having speed v" against v, because there are a finite number of particles and an infinite number of possible speeds. Hence, the probability that a particle has a precisely stated speed, such as $279.343267 \cdots$ m/s, is exactly zero. However, we can divide the range of speeds into intervals and the probability that a particle has a speed somewhere in a given interval (such as 279 m/s to 280 m/s) has a definite nonzero value.

any given temperature. Hence, hydrogen is more likely to escape from the atmosphere at high altitudes than oxygen or nitrogen. The moon has a tenuous atmosphere. For the molecules in this atmosphere not to have a great probability of escaping from the weak gravitational pull of the moon, even at the low temperatures there, we would expect them to be molecules or atoms of the heavier elements. Evidence points to the heavy inert gases, such as krypton and xenon, which were produced largely by radioactive decay early in the moon's history. The atmospheric pressure on the moon is about 10^{-13} of the earth's atmospheric pressure.

EXAMPLE 2

The speeds of ten particles in m/s are 0, 1.0, 2.0, 3.0, 3.0, 3.0, 4.0, 4.0, 5.0, and 6.0. Find (a) the average speed, (b) the root-mean-square speed, and (c) the most probable speed of these particles.

(a) The average speed is

$$\bar{v} = \frac{0 + 1.0 + 2.0 + 3.0 + 3.0 + 3.0 + 4.0 + 4.0 + 5.0 + 6.0}{10} = 3.1 \text{ m/s.}$$

(b) The mean-square speed is

$$\bar{v^2} = \frac{0 + (1.0)^2 + (2.0)^2 + (3.0)^2 + (3.0)^2 + (3.0)^2 + (4.0)^2 + (4.0)^2 + (5.0)^2 + (6.0)^2}{10}$$

$$= 12.5 \text{ m}^2/\text{s}^2$$

and the root-mean-square speed is

$$v_{\text{rms}} = \sqrt{12.5 \text{ m}^2/\text{s}^2} = 3.5 \text{ m/s.}$$

(c) Of the ten particles three have speeds of 3.0 m/s, two have speeds of 4.0 m/s, and the other five each have a different speed. Hence, the most probable speed of a particle v_p is

$$v_p = 3.0 \text{ m/s.}$$

EXAMPLE 3

Use Eq. 24-2 to determine the average speed $\bar{v}$, the root-mean-square speed v_{rms}, and the most probable speed v_p of the molecules in a gas in terms of the gas parameters.

The quantity $N(v)\,dv$ is the number of particles in the sample having a speed between v and $v + dv$, $N(v)$ being given by Eq. 24-2. We find the average speed $\bar{v}$ in the usual way: we multiply the number of particles in each speed interval by a speed v characteristic of that interval; we sum these products over all speed intervals and we divide by the total number of particles. Replacing the summation by an integration, we obtain

$$\bar{v} = \frac{\int_0^\infty N(v)v\,dv}{N}$$

Substituting Eq. 24-2 for $N(v)$ and integrating* we obtain

$$\bar{v} = \sqrt{\frac{8}{\pi}\frac{kT}{m}} = 1.59\sqrt{\frac{kT}{m}} \quad \text{(average speed).}$$

The mean-square speed is given by

---

* Let $\lambda = m/2kT$. From tables of integrals,

$$\int_0^\infty v^2 e^{-\lambda v^2}\,dv = \frac{1}{4}\sqrt{\frac{\pi}{\lambda^3}}; \quad \int_0^\infty v^3 e^{-\lambda v^2}\,dv = \frac{1}{2\lambda^2}; \quad \int_0^\infty v^4 e^{-\lambda v^2}\,dv = \frac{3}{8}\sqrt{\frac{\pi}{\lambda^5}}.$$

$$\overline{v^2} = \frac{\int_0^\infty N(v)v^2\,dv}{N}$$

which yields

$$v_{rms} = \sqrt{\overline{v^2}} = \sqrt{\frac{3kT}{m}} = 1.73\sqrt{\frac{kT}{m}} \quad \text{(root-mean-square speed)}.$$

The most probable speed v_p is the speed at which $N(v)$ has its maximum value. It is given by requiring that

$$\frac{dN(v)}{dv} = 0.$$

By substituting Eq. 24-2 for $N(v)$ we obtain, as you should show,

$$\overline{v}_p = \sqrt{\frac{2kT}{m}} = 1.41\sqrt{\frac{kT}{m}} \quad \text{(most probable speed)}.$$

In Fig. 24-4 we show v_p, $\overline{v}$, and v_{rms} at 0° C for a molecular speed distribution in oxygen.

24·3
EXPERIMENTAL CONFIRMATION OF THE MAXWELLIAN DISTRIBUTION

Maxwell derived his distribution law for molecular speeds (Eq. 24-2) in 1859. At that early date it was not possible to check this law by direct measurement and, indeed, it was not until 1920 that Stern made the first serious attempt to do so. Techniques improved rapidly in the hands of various workers but it was not until 1955 that a high-precision experimental verification of the law (for gas molecules) was provided, by Miller and Kusch of Columbia University.

Their apparatus is shown in Fig. 24-5. The walls of oven O were heated, in one set of experiments, to a uniform temperature of 870 ± 4K, some thallium having been placed in the oven. At this temperature thallium vapor, at a pressure of 3.2×10^{-3} mm-Hg, fills the oven. Some molecules of thallium vapor escape from slit S into the highly evacuated space outside the oven, falling on the rotating cylinder R. This cylinder, of length l, has a number of helical grooves cut into it, only one of them being shown in Fig. 24-5. For a given angular speed ω of the cylinder, only molecules of a sharply defined speed v can pass along the grooves without striking the walls. The speed v can be found from:

$$\text{time of travel along the groove} = \frac{l}{v} = \frac{\phi}{\omega},$$

or
$$v = l\omega/\phi \tag{24-4}$$

in which ϕ (see Fig. 24-5) is the angular displacement between the entrance and

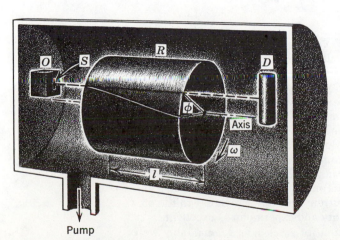

Pump

figure 24-5
The apparatus used by Miller and Kusch to verify the Maxwell speed distribution law. The mechanism for rotating the cylinder is not shown. The whole apparatus is highly evacuated to reduce collisions with the residual gas molecules of the thallium molecules in the beam emerging from slit S.

the exit of a helical groove. Thus the rotating cylinder is a *velocity selector*, the speed selected being proportional to the (controllable) angular speed ω, as Eq. 24-4 shows. One observes the beam intensity recorded by detector D as a function of the selected speed v. Figure 24-6 shows the remarkable agreement between theory (the solid line) and experiment (the triangles and circles) for thallium vapor.

The distribution of speeds in the *beam* (as distinguished from the distribution of speeds in the *oven*) is not proportional to $v^2 e^{-mv^2/2kT}$, as in Eq. 24-2, but to $v^3 e^{-mv^2/2kT}$. Consider a group of molecules in the oven whose speeds lie within a certain small range v_1 to $v_1 + \Delta v$, where v_1 is less than the most probable speed v_p. We can always find another equal speed interval Δv, extending from v_2 to $v_2 + \Delta v$, where v_2, which will be greater than v_p, is chosen so that the two speed intervals contain the same number of molecules. However, more molecules in the latter interval than in the former will escape from slit S to form the beam, because molecules in the latter interval "bombard" the slit with a greater frequency, by precisely the factor v_2/v_1. Thus, other things being equal, fast molecules are favored in escaping from the oven, just in proportion to their speeds, and the molecules in the beam have a "v^3" rather than a "v^2" distribution. This effect is allowed for in computing the theoretical curve of Fig. 24-6.

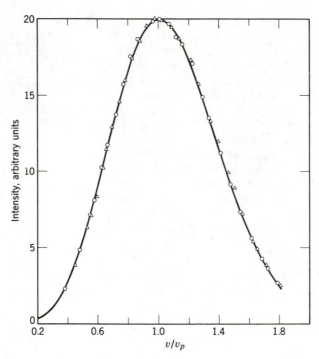

figure 24-6
The solid line shows Maxwell's molecular speed distribution. The circles (O) are experimental points for thallium atoms emerging from an oven at 870 K; the triangles (△) correspond to 944 K. The horizontal scale is a plot of v/v_p where v_p is the most probable speed. When speeds are plotted in this way the distributions for different temperatures should fall on the same curve. At 870 K, $v_p = 376$ m/s and at 944 K, it is 395 m/s. From R. C. Miller and P. Kusch, *Physical Review*, 99, 1314 (1955).

Rainwater and Havens (1946), also of Columbia University, provided a convincing experimental check of the Maxwell speed distribution law by using a "gas" of neutrons. The neutrons were produced (as fast neutrons) in continuous series of short bursts in a cyclotron and allowed to fall on a block of paraffin. By repeated collisions with the nuclei of the block, the neutrons rapidly slowed down and came into thermal equilibrium with the block, behaving like a "neutron gas" in a container. The container, however, is a leaky one because neutrons diffuse out through the walls of the block and move across the laboratory. It is possible, by electronic means, to measure the time between the production of the neutrons in the cyclotron and their arrival at a distant detector after escaping from the paraffin block. Thus one can measure the speed distribution in a collimated beam of escaping neutrons and can compare it to the prediction of Maxwell; the agreement of theory and experiment is excellent.

Although the Maxwell speed distribution for gases agrees remarkably well with observations under ordinary conditions, it fails at high densities, where the basic assumptions of the classical kinetic theory fail. In these regions we must use speed distributions founded on the principles of quantum physics, the Fermi-Dirac and the Bose-Einstein distributions. These quantum distributions closely agree with the Maxwell distribution in the classical region (low density) and agree with experiment where the classical distribution fails. Hence, there are limits to the applicability of the Maxwell distribution, as in fact there are to any theory.

24-4 BROWNIAN MOTION

The prominence given to atomic and molecular theory during the last quarter of the nineteenth century was deplored by many able scientists. In spite of the many quantitative agreements between kinetic theory and the behavior of gases, no proof of the separate existence of atoms and molecules had been obtained, nor had any observation been made that could really demonstrate the continuous motions of the molecules. Ernst Mach (1838–1916) saw no point to "thinking of the world as a mosaic, since we cannot examine its individual pieces of stone." It had been established rather early in the development of kinetic theory that an atom should be about 10^{-7} cm or 10^{-8} cm in diameter. No one actually expected to see an atom or detect the effect of a single atom.

The leader of the opposition to the atomic theory was Wilhelm Ostwald, rightly regarded as "the father of physical chemistry." He was a champion of the principle of the conservation of energy and regarded energy as the ultimate reality. Ostwald argued that with a thermodynamical treatment of a process we know all that is essential about the process and that further mechanical assumptions about the mechanism of the reactions are unproved hypotheses. He abandoned the atomic and molecular theories and fought to free science "from hypothetical conception which lead to no immediate experimentally verifiable conclusions." Other prominent scientists were reluctant to admit the atom as an established scientific fact.

Ludwig Boltzmann felt compelled to protest this attitude in an article in 1897, stressing the indispensability of atomism in natural science. The progress of science is often guided by the analogies of nature's processes which occur in the minds of investigators. Kinetic theory was such a mechanical analogy. As with most analogies it suggests experiments to test the validity of our mental pictures and leads to further investigations and clearer knowledge.

As is always true in such controversies in science, the decision rests with experiment. The earliest and most direct experimental evidence for the reality of atoms was the proof of the atomic kinetic theory provided by the quantitative studies of Brownian motion. These observations convinced both Mach and Ostwald of the validity of the kinetic theory and the atomic description of matter on which it rests. The atomic theory gained unquestioned acceptance in later years when a wide variety of experiments all led to the same values of the fundamental atomic constants.

Brownian motion is named after the English botanist Robert Brown who discovered in 1827 that pollen suspended in water shows a continuous random motion when viewed under a microscope. At first these motions were considered a form of life, but it was soon found that small inorganic particles behave similarly. There was no quantitative explanation of this phenomenon until the development of kinetic theory. Then, in 1905, Albert Einstein developed a theory of Brownian motion.* In his *Autobiographical Notes*, Einstein writes, "My major aim in this was to find facts which would guarantee as much as possible the existence of atoms of definite size. In the midst of this I discovered that, according to atomistic theory, there would have to be a movement of suspended microscopic particles open to observation, without knowing that observations concerning the Brownian motion were already long familiar."

The basic assumption made by Einstein was that particles suspended in a liquid or a gas share in the thermal motions of the medium and that on the average the translational kinetic energy of each particle is $\frac{3}{2}kT$, in accordance with the principle of equipartition of energy. In this view the Brownian motions result from impacts by molecules of the fluid, and the suspended particles acquire the same mean kinetic energy as the molecules of the fluid.

The suspended particles are extremely large compared to the molecules of the fluid and are being continually bombarded on all sides by them. If the particles are sufficiently large and the number of molecules is sufficiently great, equal numbers of molecules strike the particles on all sides at each instant. For smaller particles and fewer molecules the number of molecules striking various sides of the particle at any instant, being merely a matter of chance, may not be equal; that is, fluctuations occur. Hence the particle at each instant suffers an unbalanced force causing it to move this way or that. The particles therefore act just like very large molecules in the fluid, and their motions should be qualitatively the same as the motions of the fluid molecules. If Avogadro's number were infinite, there would be no statistical unbalance (fluctuations) and no Brownian motion. If Avogadro's number were very small, the Brownian motion would be very large. Hence we should be able to deduce the value of Avogadro's number from observations of the Brownian motion. Deeply ingrained in this picture is the idea of molecular motion and the smallness of molecules. The Brownian motion therefore offers a striking experimental test of the kinetic theory hypotheses.

The suspended particles are under the influence of gravity and would settle to the bottom of the fluid were it not for the molecular bombardment opposing this tendency. Since the suspended particles behave like gas molecules we are not surprised to learn that, as for molecules in the atmosphere, their density drops off exponentially with respect to height in the fluid; they form a "miniature atmosphere"; see Example 1, Chapter 17; Problem 26, Chapter 23; and Problem 21, this chapter. Jean Perrin, a French physical chemist, confirmed this prediction in 1908 by determining the numbers of small particles of gum resin suspended at different heights in a liquid drop (Fig. 24-7, left). From his data he deduced a value of Avogadro's number $N_0 = 6 \times 10^{23}$ particles/mol. Perrin also made measurements of the displacements of Brownian particles during many equal time intervals and found that they have the statistical distribution required by kinetic theory and the root-mean-square displacement predicted by Einstein (Fig. 24-7, right).

Among the many subsequent experiments was that of Kappler, in 1931, who observed the Brownian motion of a rather gross object, a small mirror (area 0.77 mm²), mounted on a fine torsion fiber with light reflected from the mirror

---

*Einstein's theory appeared as an article in the same volume of the *Annalen der Physik* which contained his famous paper on the theory of relativity and also his paper on the theory of the photoelectric effect. It was for his work on the photoelectric effect that he won the Nobel prize in 1921.

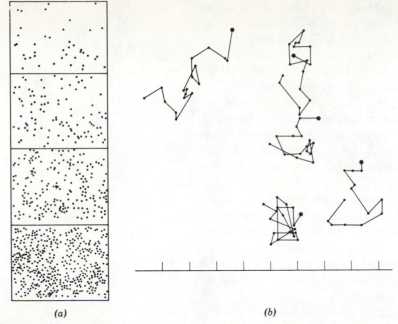

(a)   (b)

figure 24-7
(a) A gum resin suspension contained in a glass vessel viewed in a
microscope by Perrin in 1909. At first the distribution of particles was
uniform, but in time they settled to the steady state distribution shown.
The particles have a diameter of 0.6×10^{-3} cm and the horizontal lines are
10×10^{-3} cm apart. (b) Sketch by V. Henri in 1908 from his cinematographic
study of Brownian movement. Henri used a microscope with a motion-
picture camera which ran 20 frames/s, each exposure being $\frac{1}{320}$ s. The zigzag
lines show the position of five rubber particles as recorded by successive
frames. The lines do not represent the actual paths of the particles for
between exposures the particles may have traveled a similar erratic path.
The scale at the bottom is divided into micrometers (abbr. μm; value 10^{-6} m).

to a moving photographic film. The mirror is mounted in a chamber with gas
at low pressure (10^{-2} mm-Hg); the record on the moving film yields the function $\theta(t)$ (angular displacement as a function of time). This shows clearly the
rotational Brownian motion of the mirror which consists of a series of angular
displacements produced by unbalanced impacts from the molecules. As the
gas pressure is lowered, there is a gradual decrease in the motion. From the
photographic record we can find the angular displacement θ and the angular
velocity ω. The equipartition of energy principle requires that

$$\tfrac{1}{2}I\overline{\omega^2} = \tfrac{1}{2}\kappa\overline{\theta^2} = \tfrac{1}{2}kT,$$

for $\tfrac{1}{2}I\overline{\omega^2}$ is the average rotational kinetic energy of the system and $\tfrac{1}{2}\kappa\overline{\theta^2}$ is the
average potential energy of the system. Here I is the rotational inertia of the
system and κ the torsion constant of the fiber. From his observations Kappler
could calculate Boltzmann's constant k and from the relation $N_0 = R/k$ he
could obtain Avogadro's number. His values were $k = 1.36 \times 10^{-23}$ J/molecule
K $\pm 3\%$ (the accepted value today of 1.380×10^{-23} J/molecule·K being within the
limits of error) and $N_0 = 6.1 \times 10^{23}$ particles/mole.

In the preceding chapter we discussed the behavior of an ideal gas. On
the macroscopic scale its fundamental relationship is the equation of
state

$$pV = nRT.$$

24-5
THE VAN DER WAALS
EQUATION OF STATE

From this equation and the principles of thermodynamics we can show that the internal energy U of a gas depends only on the temperature. Real gases obey these relations fairly well at low densities, but their behavior may become markedly different as the density increases. We cannot neglect these deviations from ideal behavior in accurate scientific work. For example, to establish the Kelvin thermodynamic scale in the laboratory we must know how to make the necessary corrections to the scale of a constant-volume gas thermometer. We must therefore know the behavior of real gases rather accurately. Even more important, perhaps, is the fact that the behavior of real gases gives us information on the nature of intermolecular forces and the structure of molecules.

Kinetic theory provides the microscopic description of the behavior of an ideal gas. We have already suggested how the assumptions of kinetic theory could become invalid if applied to a real gas. Under some conditions we may not be justified in neglecting the facts that the molecules occupy a fraction of the volume available to the gas and that the range of molecular forces is greater than the size of the molecule. At high densities we cannot ignore these effects.

J. D. van der Waals (1837–1923) deduced a modified equation of state which takes these factors into account in a simple way. Let us imagine the molecules to be hard spheres of diameter d. The diameter of such a sphere would correspond to the distance between the centers of molecules at which strong collision forces come into play. During its motion the center of a molecule cannot approach within a distance $d/2$ from a wall or a distance d from the center of another molecule. Hence the actual volume available to a molecule is smaller than the volume of the containing vessel. Just how much smaller depends on how many molecules there are. Let us represent the volume per mole, V/n, by v. Then the "free volume" per mole would be less than this by the "covolume" b. Hence we modify the equation of state from the ideal relation $pv = RT$ to

$$p(v - b) = RT$$

to allow for this. Because of the reduced volume, the number of impacts on the wall increases, thereby increasing the pressure; this relationship was first derived by Clausius.

We can also allow for the effect of attractive forces between molecules in a simple way. Imagine a plane passed through a gas and consider, at any instant, the intermolecular forces which act across it. Each molecule on the left, say, will attract and be attracted by some small number n of those on the right. Now compare this situation with another similar in every way except that the number of molecules per unit volume is doubled. Here any particular molecule on the left will interact on the average with $2n$ of those on the right, for the range of the molecular force is the same, and twice as many molecules now fall into this range. Since there also are twice as many molecules on the left as before which attract in this way, it is clear that the number of attractive pairs across the plane has increased fourfold. Therefore, the effect of these forces varies as the *square* of the number of particles per unit volume or inversely as the square of the volume per mole, that is, as $(1/v)^2$. Because of these intermolecular force bonds, the gas should, for a given external pressure, occupy a volume less than the volume it would occupy as an ideal gas, in which there are no such attractive forces. Or, equivalently, the gas acts as though it is subject to a pressure in excess of the externally applied pressure. This excess pressure is proportional to $(1/v)^2$, or equal to a/v^2 where a is a constant. Hence, we obtain the *van der Waals equation of state* of a gas,

$$\left(p + \frac{a}{v^2}\right)(v - b) = RT. \qquad (24\text{-}5)$$

The values of a and b are to be found from experiment, and in this respect the equation is empirical. We must realize that these corrections to the ideal gas equation of state are of the simplest kind, and that failure of the van der Waals equation in any particular case is evidence that our assumptions are oversimplified for that case. No one simple formula is known which applies to all gases under all conditions.*

We have seen that real gases do not follow the ideal gas law exactly. Our discussion suggests also that for real gases the internal energy U depends on the volume as well as on the temperature. For if there are (long range) attractive forces between molecules, the potential energy increases as the average distance between molecules increases. Hence, we would expect the internal energy of most real gases to increase slightly with the volume at ordinary temperatures, and this is found to be the case. Of course, collisions can be regarded as arising from repulsive forces. If the molecules move rapidly so as to make many collisions, the potential energy of the (short range) repulsive forces may be more important than that of the attractive forces and the internal energy could decrease as the volume increases. This is true for hydrogen and helium at ordinary temperatures. In either case, however, the internal energy U is not a function of temperature alone but depends also on the volume. The dependence of the internal energy of a gas on the volume can be deduced readily from the observed results of the free expansion experiment, discussed in Chapter 22.

EXAMPLE 4

On a pressure-volume diagram compare the behavior of an ideal gas at constant temperature to that of a van der Waals gas.

In Fig. 24-8a we draw the isotherms (curves of constant T) according to the law $pv = RT$. Figure 24-8b shows the isotherms according to the law

$$(p + a/v^2)(v - b) = RT.$$

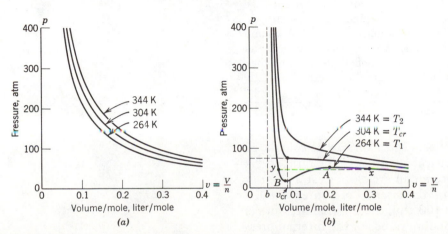

(a)   (b)

figure 24-8
(a) Isotherms for an ideal gas. (b) Isotherms for a van der Waals gas. We have assumed $a = 3.59$ liter^{2}atm/mole^2 and $b = 0.0427$ liter/mole in Eq. 24.4. These values give the best fit of this equation to $p\,V\,T$ data for the real gas CO_2. $T_{cr}\ (= 304$ K) is the critical temperature.

---

* For an interesting discussion of these and related matters see 'Liquids — The Awkward In-between' by J. G. Powles, in *Contemporary Physics*, September 1974.

The ideal gas isotherms are each one branch of a rectangular hyperbola, $pv = $ constant. For the van der Waals gas the pressure varies with volume as

$$p = \frac{RT}{(v - b)} - \frac{a}{v^2}. \tag{24-6}$$

As the volume per mole v decreases from large values, the pressure rises, but the a/v^2 term, which diminishes the pressure, climbs rapidly so that for sufficiently low T the pressure passes through a maximum at A. As v is further decreased, the $RT/(v - b)$ term climbs more rapidly so that the pressure goes through a minimum at B and then rises rapidly without bound as v tends to the value b. At neighboring higher temperatures, the maxima and minima are less pronounced and are closer to the inflection point that lies between them. At the so-called critical temperature $(T = T_{cr})$, they coincide in a horizontal inflection point called the critical point. For temperatures sufficiently higher than the critical temperature T_{cr} the van der Waals isotherms have no inflection point and approach the rectangular-hyperbola behavior of the ideal-gas isotherms. For carbon dioxide the critical temperature is 304 K and the pressure at the critical point is 72.9 atm.

We can obtain the pressure p_{cr}, the molar volume v_{cr}, and the temperature T_{cr} of the critical point quite generally from the conditions that the tangent to the isotherm is horizontal, $dp/dv = 0$ when $T = $ constant, and that the point is an inflection point, $d^2p/dv^2 = 0$ when $T = $ constant. We obtain

$$\frac{dp}{dv} = -\frac{RT}{(v - b)^2} + \frac{2a}{v^3} = 0 \quad (T = \text{constant})$$

and

$$\frac{d^2p}{dv^2} = \frac{2RT}{(v - b)^3} - \frac{6a}{v^4} = 0 \quad (T = \text{constant}).$$

This gives us

$$v_{cr} = 3b$$

and

$$T_{cr} = \frac{8a}{27bR}.$$

Putting these in Eq. 24-6, we obtain

$$p_{cr} = \frac{a}{27b^2}.$$

The isotherms suggest the actual experimental behavior of liquids and gases. The maxima and minima of the isotherms below the critical temperature are not usually observed experimentally. At some point x the gas begins to condense. As the volume is decreased, the pressure remains constant (dotted line) until at y all the gas has been transformed into liquid. Beyond y, as we decrease the volume, we are compressing a liquid, with the consequent sharp rise in pressure needed to make even small volume changes. Actually the portions xA and By of the isotherms can be obtained experimentally by using very pure gases and liquids. We call these supersaturated vapors and supercooled liquids,* and they are in metastable states. The portion AB cannot be reproduced experimentally and is unstable.

The constants a and b in van der Waals equation can be calculated from the experimental values of the critical quantities. The term a/v^2 is called an *internal pressure*. Some values for air are of interest. For air at $0°$ C and external pressure p of 1.00 atm, the internal pressure is 0.0028 atm; at $0°$ C and external pressure p of 100 atm, the internal pressure is 26 atm. For air at $-75°$ C the corresponding values of the internal pressure are 0.0056 atm and 84.5 atm. When a gas expands under pressure

---

* See "The Undercooling of Liquids" by David Turnbull in *Scientific American*, January 1965.

and does work against outside compressing forces, it must also do work against these internal forces. For air at −75° C and 100 atm, the work done against internal forces is nearly as great as that done against external forces. There is an important distinction between internal and external work, however. In the case of external work, energy is transferred from the body to an outside body; in the case of internal work, there is merely a transfer from one kind of energy to another within the body, as from potential to kinetic. The constant b varies from gas to gas, but is usually of the order of 30 cm³/mol. Hence the covolume is about 0.15% of the free volume available to a gas at standard conditions.

Although the van der Waals formula is a good qualitative guide, the quantitative experimental data cannot be matched everywhere with constant values for a and b. The reason is that the model on which the formula is based is still an oversimplification. Instead of assuming that the molecules always have a well-defined diameter, for example, we must use the actual intermolecular force (Fig. 23-3). In this way a more accurate correction to the ideal gas law can be made. Van der Waals knew this would be necessary for accurate quantitative work.

questions

1. Consider the case in which the mean free path is greater than the longest straight line in a vessel. Is this a perfect vacuum for a molecule in this vessel?

2. List effective ways of increasing the number of molecular collisions per unit time in a gas.

3. Give a qualitative explanation of the connection between the mean free path of ammonia molecules in air and the time it takes to smell the ammonia when a bottle is opened across the room.

4. Consider Archimedes' principle applied to a gas. Isn't it true that once we accept a kinetic theory model of a gas, we need a new explanation for this principle? For example, suppose the mean free path of a gas molecule is comparable to the depth of the body immersed in the gas, or greater; what is the origin of the buoyant force then? (See "Archimedes' Principle in Gases" by Alan J. Walton in *Contemporary Physics*, March 1969.)

5. The two opposite walls of a container of gas are kept at different temperatures. Describe the mechanism of heat conduction through the gas.

6. A gas can transmit only those sound waves whose wavelength is long compared with the mean free path. Can you explain this? Where might this limitation arise?

7. If molecules are not spherical, what meaning can we give to d in Eq. 24-1 for the mean free path? In which gases would the molecules act the most nearly as rigid spheres?

8. Suppose we dispense with the hypothesis of elastic collisions in kinetic theory and consider the molecules as centers of force acting at a distance. Does the concept of mean free path have any meaning under these circumstances?

9. Since the actual force between molecules depends on the distance between them, forces can cause deflections even when molecules are far from "contact" with one another. Furthermore, the deflection caused should depend on how long a time these forces act and hence on the relative speed of the molecules. (a) Would you then expect the measured mean free path to depend on temperature, even though the density remains constant? (b) If so, would you expect $\bar{l}$ to increase or decrease with temperature? (c) How does this dependence enter into Eq. 24-1?

10. Justify qualitatively the statement that, in a mixture of molecules of different kinds in complete equilibrium, each kind of molecule has the same

Maxwellian distribution in speed that it would have if the other kinds were not present.

11. What observation is good evidence that not all molecules of a body are moving with the same speed at a given temperature?

12. The Maxwellian distribution of speeds among molecules in a gas is shown in Fig. 24-4. How would you expect the Maxwellian distribution of *velocities* to look? What would the average velocity be?

13. The fraction of molecules within a given range Δv of the root-mean-square speed decreases as the temperature of a gas rises. Explain why.

14. (a) Do half the molecules in a gas in thermal equilibrium have speeds greater than v_p? Than v? Than v_{rms}?
   (b) Which speed, v_p, $\bar{v}$, or v_{rms}, corresponds to a molecule having average kinetic energy?

15. The slit system in Fig. 24-5 selects only those molecules moving in the $+x$-direction. Does this destroy the validity of the experiment as a measure of the distribution of speeds of molecules moving in all directions?

16. Why did Rainwater and Havens, in their investigation of the speed distribution of neutrons (page 528), select paraffin as a material to bring fast neutrons rather quickly into thermal equilibrium?

17. List examples of the Brownian motion in physical phenomena.

18. Would Brownian motion occur in gravity-free space?

19. A golf ball is suspended from the ceiling by a long thread. Explain in detail why its Brownian motion is not readily apparent.

20. We have defined n_v to be the number of molecules per unit volume in a gas. If we define n_v for a very small volume in a gas, say one equal to ten times the volume of an atom, then n_v fluctuates with time through the range of values zero to some maximum value. How then can we justify a statement that n_v has a definite value at every point in the gas?

21. Show that as the volume per mole of a gas increases, the van der Waals equation tends to the equation of state of an ideal gas.

22. The covolume b in van der Waals equation is often taken to be four times the actual volume of the gas molecules themselves. What factors would have to be taken into account to obtain such a result?

23. Keeping in mind that internal energy of a body consists of kinetic energy and potential energy of its particles, how would you distinguish between the internal energy of a body and its temperature?

problems

SECTION 24-1

1. The mean free path of nitrogen molecules at $0°$ C and 1 atm is 0.80×10^{-5} cm. At this temperature and pressure there are 2.7×10^{19} molecules/cm³. What is the molecular diameter?   *Answer:* 3.2×10^{-8} cm.

2. In a certain particle accelerator the protons travel around a circular path of diameter 75 ft in a chamber of 10^{-6} mm-Hg pressure and 273 K temperature. (a) Estimate the number of gas molecules per cubic centimeter at this pressure. (b) What is the mean free path of the gas molecules under these conditions if the molecular diameter is 2.0×10^{-8} cm?

3. At what frequency would the wavelength of sound be of the order of the mean free path in oxygen at 1-atm pressure and $0°$ C? Take the diameter of the oxygen molecule to be 3.00×10^{-8} cm.   *Answer:* 3.5×10^9 Hz.

4. What is the mean free path for 15 spherical jelly beans in a bag that is vigorously shaken? Take the volume of the bag to be 1.0 l and the diameter of a jelly bean to be 1.0 cm.

5. At 2500 km above the earth's surface the density is about one molecule/cm³. (a) What mean free path is predicted by Eq. 24-1 and (b) what is its significance under these conditions?
   Answer: (a) 7×10^9 km. (b) The answer to (a) has little significance because,

at this altitude, nearly all molecules would follow collisionless ballistic paths in the earth's gravitational field, and many would escape from the atmosphere.

6. The mean free path $\bar{l}$ of the molecules of a gas may be determined from measurements (e.g., from measurement of the viscosity of the gas). At 20° C and 75 cm-Hg pressure such measurements yield values of $\bar{l}_A$ (argon) = 9.9×10^{-6} cm and $\bar{l}_{N_2}$ (nitrogen) = 27.5×10^{-6} cm. (a) Find the ratio of the effective cross-section diameters of argon and nitrogen. (b) What would the value be of the mean free path of argon at 20° C and 15 cm-Hg? (c) What would the value be of the mean free path of argon at −40° C and 75 cm-Hg?

7. A molecule of hydrogen (diameter 1.0×10^{-8} cm) escapes from a furnace $(T = 4000$ K) with the root-mean-square speed into a chamber containing atoms of cold argon (diameter 3.0×10^{-8} cm) at a density of 4.0×10^{19} atoms/cm³. (a) What is the speed of the hydrogen molecule? (b) If the molecule and an argon atom collide, what is the closest distance between their centers, considering each as spherical? (c) What is the initial number of collisions per unit time experienced by the hydrogen molecule?
Answer: (a) 7.1 km/s. (b) 2.0×10^{-8} cm. (c) 5.0×10^{10} collisions/s.

8. The mean free path of a molecule is $\bar{l}$. Prove that the probability that a molecule will go at least a distance x before having its next collision is $e^{-x/\bar{l}}$.

9. For a gas in which all molecules travel with the same speed $\bar{v}$, show that $\bar{v}_{rel} = \frac{4}{3}v$ rather than $\sqrt{2}\,\bar{v}$ (which is the result obtained when we consider the actual distribution of molecular speeds). See p. 523.

SECTION 24-2

10. It is found that the most probable speed of molecules in a gas at an equilibrium temperature T_2 is the same as the root-mean-square speed of the molecules in this gas when its equilibrium temperature is T_1. Find T_2/T_1.

11. You are given the following group of particles (N_i represents the number of particles which have a speed v_i).

N_i	v_i(cm/s)
2	1.00
4	2.00
6	3.00
8	4.00
2	5.00

(a) Compute the average speed $\bar{v}$. (b) Compute the root-mean-square speed v_{rms}. (c) Among the five speeds shown, which is the most probable speed v_p for the entire group? Answer: (a) 3.2 cm/s. (b) 3.4 cm/s. (c) 4.0 cm/s.

12. Consider the distribution of speeds shown in Fig. 24-9. (a) List v_{rms}, $\bar{v}$, and v_p in the order of increasing speed. (b) How does this compare with the Maxwellian distribution?

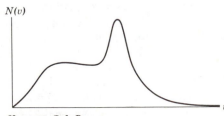

figure 24-9
Problem 12

13. A gas consists of N particles. (a) Show that $v_{rms} \geq \bar{v}$ regardless of the form of the distribution of speeds. (b) When does the equality hold?
Answer: (b) When all the speeds are the same.

14. A hypothetical gas of N particles has the speed distribution shown in Fig. 24-10. ($N_v = 0$ for $v > 2v_0$.) (a) Evaluate a in terms of N and v_0. (b) Find the number of particles with speeds between $1.5v_0$ and $2.0v_0$. (c) Find the average speed of the particles.

15. A container of volume 1000 cm³ contains argon at a pressure of 3.0×10^5 Pa and a temperature of 300 K. The atomic weight of argon is 40. (a) How many argon atoms are in the container? (b) What is the average speed of these atoms? (c) How many atoms strike an area of 1.0×10^{-3} cm² on one of the container walls in one second? (d) If this area is a hole, and all the atoms striking the hole leave the container, how long will it take for the number of atoms in the container to fall to $1/e$ of its initial value?
Answer: (a) 7.2×10^{22}. (b) 400 m/s. (c) 7.2×10^{20}. (d) 100 s.

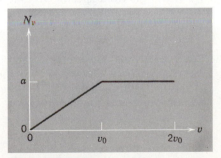

figure 24-10
Problem 14

SECTION 24-3

16. In the apparatus of Miller and Kusch (Fig. 24-5) the length l of the rotating cylinder is 20.4 cm and the angle ϕ is $(2\pi/74.7)$ rad. What rotational speed corresponds to a selected speed v of 200 m/s?

SECTION 24-4

17. Calculate the root-mean-square speed of smoke particles of mass 5.0×10^{-14} g in air at $0°$ C and 1.0-atm pressure.   *Answer:* 1.5 cm/s.

18. Particles of mass 6.2×10^{-14} g are suspended in a liquid at $27°$ C and are observed to have a root-mean-square speed of 1.4 cm/s. Calculate Avogadro's number from the equipartition theorem and these data.

19. The average speed of hydrogen molecules at $0°$ C is 1694 m/s. Compute the average speed of colloidal particles of "molecular weight" 3.2×10^6 g/mol. *Answer:* 1.3 m/s.

20. Very small solid particles, called grains, exist in interstellar space. They are continually bombarded by hydrogen atoms of the surrounding interstellar gas. As a result of these collisions, the grains execute Brownian movement in both translation and rotation. Assume the grains are uniform spheres of diameter 4.0×10^{-6} cm and density 1.0 g/cm^3, and that the temperature of the gas is 100 K. Find (a) the root-mean-square speed of the grains between collisions and (b) the approximate rate (rev/s) at which the grains are spinning.

21. Colloidal particles in solution are buoyed up by the liquid in which they are suspended. Let ρ' be the density of liquid and ρ the density of the particles. If V is the volume of a particle, show that the number of particles per unit volume in the liquid varies with height as

$$n_v = n_{v0} \exp\left[-\frac{N_0}{RT} V(\rho - \rho')gh\right].$$

This equation was tested by Perrin in his Brownian motion studies.

SECTION 24-5

22. The constant a in van der Waals equation is (a) 0.37 N · m^4/mol^2 for CO_2 and (b) 0.025 N · m^4/mol^2 for hydrogen. Compute the internal pressures for these gases for values of v/v_0 (where $v_0 = 22.4$ l/mol) of 1, 0.01, and 0.001.

23. (a) The constant b in van der Waals equation is 43 cm^3/mol for CO_2. Using the value for a in the previous problem, compute the pressure at $0°$ C for a specific volume of 0.55 l/mol, assuming van der Waals equation to be strictly true. (b) What is the pressure under these same conditions, assuming CO_2 behaves as an ideal gas?
Answer: (a) 3.3×10^6 Pa. (b) 4.1×10^6 Pa.

24. Van der Waals b for oxygen is 32 cm^3/mol. Assume b is four times the actual volume of a mole of "billiard-ball" O_2 molecules and compute the diameter of an O_2 molecule.

25. Calculate the work done in an isothermal expansion of one mole of a van der Waals gas from specific volume v_i to v_f.

Answer: $RT \ln \dfrac{v_f - b}{v_i - b} + a(1/v_f - 1/v_i)$.

26. The constants a and b in the van der Waals equation are different for different substances. Show, however, that if we take $v_{cr}, p_{cr},$ and T_{cr} as the units of specific volume, pressure, and temperature, the van der Waals equation becomes identical for all substances.

25

entropy and the second law of thermodynamics

The first law of thermodynamics states that energy is conserved. However, we can think of many thermodynamic processes which conserve energy but which actually never occur. For example, when a hot body and a cold body are put into contact, it simply does not happen that the hot body gets hotter and the cold body colder. Or again, a pond does not suddenly freeze on a hot summer day by giving up heat to its environment. *And yet neither of these processes violates the first law of thermodynamics.* Similarly, the first law does not restrict our ability to convert work into heat or heat into work, except that energy must be conserved in the process. And yet in practice, although we can convert a given quantity of work completely into heat, we have never been able to find a scheme that converts a given amount of heat completely into work. The second law of thermodynamics deals with this question of whether processes, assumed to be consistent with the first law, do or do not occur in nature. Although the ideas contained in the second law may seem subtle or abstract, in application they prove to be extremely practical.

Consider a typical system in thermodynamic equilibrium, say a mass M of a (real) gas confined in a cylinder-piston arrangement of volume V, the gas having a pressure p and a temperature T. In an equilibrium state these thermodynamic variables remain constant with time. Suppose that the cylinder, whose walls are an (ideal) heat insulator but whose base is an (ideal) heat conductor is placed on a large heat reservoir maintained at this same temperature T, as in Fig. 22-9. Now let us change the system to another equilibrium state in which the temperature T is the

25-1
INTRODUCTION

25-2
REVERSIBLE AND IRREVERSIBLE PROCESSES

539

same but the volume V is reduced by one-half. Of the many ways in which we could do this we discuss two extreme cases.

I. We depress the piston very rapidly; we then wait for equilibrium with the reservoir to be re-established. During this process the gas is turbulent and its pressure and temperature are not well defined; we cannot plot the process as a continuous line on a p-V diagram because we would not know what value of pressure (or temperature) to associate with a given volume. The system passes from one equilibrium state i to another f through a series of nonequilibrium states (Fig. 25-1a).

II. We depress the piston (assumed to be frictionless) exceedingly slowly—perhaps by adding sand to the top of the piston—so that the pressure, volume, and temperature of the gas are, at all times, well-defined quantities. We first drop a few grains of sand on the piston. This will reduce the volume of the system a little and the temperature will tend to rise; the system will depart from equilibrium, but only slightly. A small amount of heat will be transferred to the reservoir and in a short time the system will reach a new equilibrium state, its temperature again being that of the reservoir. Then we drop a few more grains of sand on the piston, reducing the volume further. Again we wait for a new equilibrium state to be established, and so forth. By many repetitions of this procedure we finally reduce the volume by one-half. During this entire process the system is never in a state differing much from an equilibrium state. If we imagine carrying out this procedure with still smaller successive increases in pressure, the intermediate states will depart from equilibrium even less. By indefinitely increasing the number of changes and correspondingly decreasing the size of each change, we arrive at an ideal process in which the system passes through a continuous succession of equilibrium states, which we can plot as a continuous line on a p-V diagram (Fig. 25-1b). During this process a certain amount of heat Q is transferred from the system to the reservoir.

Processes of type I are called *irreversible* and those of type II are called *reversible. A reversible process is one that, by a differential change in the environment, can be made to retrace its path.* Thus as we cause the piston to move slowly downward, in II, the external pressure on the piston exceeds the pressure exerted on it by the gas by only a differential amount dp. If at any instant we reduce the external pressure ever so slightly (by removing a few sand grains), so that it is *less than* the internal gas pressure by dp, the gas will expand instead of contracting and the system will retrace the equilibrium states through which it has just passed.* In practice all processes are irreversible, but we can approach reversibility arbitrarily closely by making appropriate experimental refinements. The strictly reversible process is a simple and useful abstraction that bears a similar relation to real processes that the ideal gas abstraction does to real gases.

The process described in II is not only reversible but *isothermal*, be-

---

* Not all processes carried out very slowly are reversible. For example, if the piston in our example exerted a frictional force against the cylinder walls, it would not reverse its motion if we made only a differential change dp in the external pressure. We would have to make a change Δp that might be an appreciable fraction of p. Thus our criterion for reversibility, which involves a response of the system to a *differential* change in the environment, is not met. The word *quasi-static* is used to describe processes that are carried out slowly enough so that the system passes through a continuous sequence of equilibrium states; a quasi-static process may or may not be reversible. See "Thermodynamics of an Irreversible Quasi-Static Process" by John S. Thomsen, *American Journal of Physics,* **28,** 119, 1960.

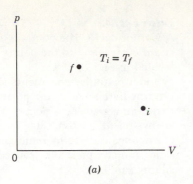 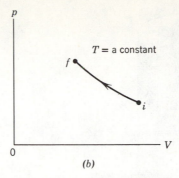

figure 25-1
We cause a real gas to go from an initial equilibrium state i described by p_i, V_i, T_i to a final equilibrium state f described by p_f, V_f $(= \frac{1}{2} V_i)$, and T_f $(= T_i)$. We carry out the process *(a)* irreversibly, and *(b)* reversibly.

cause we have assumed that the temperature of the gas differs at all times by only a differential amount dT from the (constant) temperature of the reservoir on which the cylinder rests.

We could also reduce the volume *adiabatically* by removing the cylinder from the heat reservoir and putting it on a nonconducting stand. In an adiabatic process no heat is allowed to enter or to leave the system. An adiabatic process can be either reversible or irreversible — the definition does not exclude either. In a reversible adiabatic process we move the piston exceedingly slowly — perhaps using the sand-loading technique; in an irreversible adiabatic process we shove the piston down quickly.

The temperature of the gas will rise during an adiabatic compression because, from the first law, with $Q = 0$, the work W done in pushing down the piston must appear as an increase ΔU in the internal energy of the system. W will have different values for different rates of pushing down the piston, being given by $\int p\, dV$ — that is, by the area under a curve on a p-V diagram — only for reversible processes (for which p has a well-defined value). Thus ΔU and the corresponding temperature change ΔT will not be the same for reversible and irreversible adiabatic processes.

Suppose that we have a system (a real gas, say) in an equilibrium state in a cylinder-piston arrangement. By using our ability to make changes in the environment of the system we can carry out, at our pleasure, a wide variety of processes. We can let the gas expand or we can compress it; we can add or subtract energy in the form of heat; we can do these things and others irreversibly or reversibly. We can also choose to carry out a sequence of processes such that the system returns to its original equilibrium state; we call this a *cycle*. If the processes involved are all reversible, we call it a *reversible cycle*.

Figure 25-2 shows a reversible cycle on a p-V diagram. Along the curve *abc* we allow the system to expand, and the area under this curve represents the work done by the system during the expansion. Along the curve *cda*, which returns the system to its original state, we compress the system, and the area under this curve represents the work we must do on the system during the compression. Hence, the *net* work done by the system is represented by the area enclosed by the curve and is positive. If we decided to traverse the cycle in the opposite sense, that is, expanding along *adc* and compressing along *cba*, the net work done by the system would be the negative of that of the previous case.

An important reversible cycle is the *Carnot cycle*, introduced by Sadi Carnot in 1824. We shall see later that this cycle will determine the

25-3
THE CARNOT CYCLE

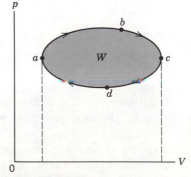

figure 25-2
A p-V diagram of a gas undergoing a reversible cycle. The shaded area W represents the net work done by the gas in the cycle.

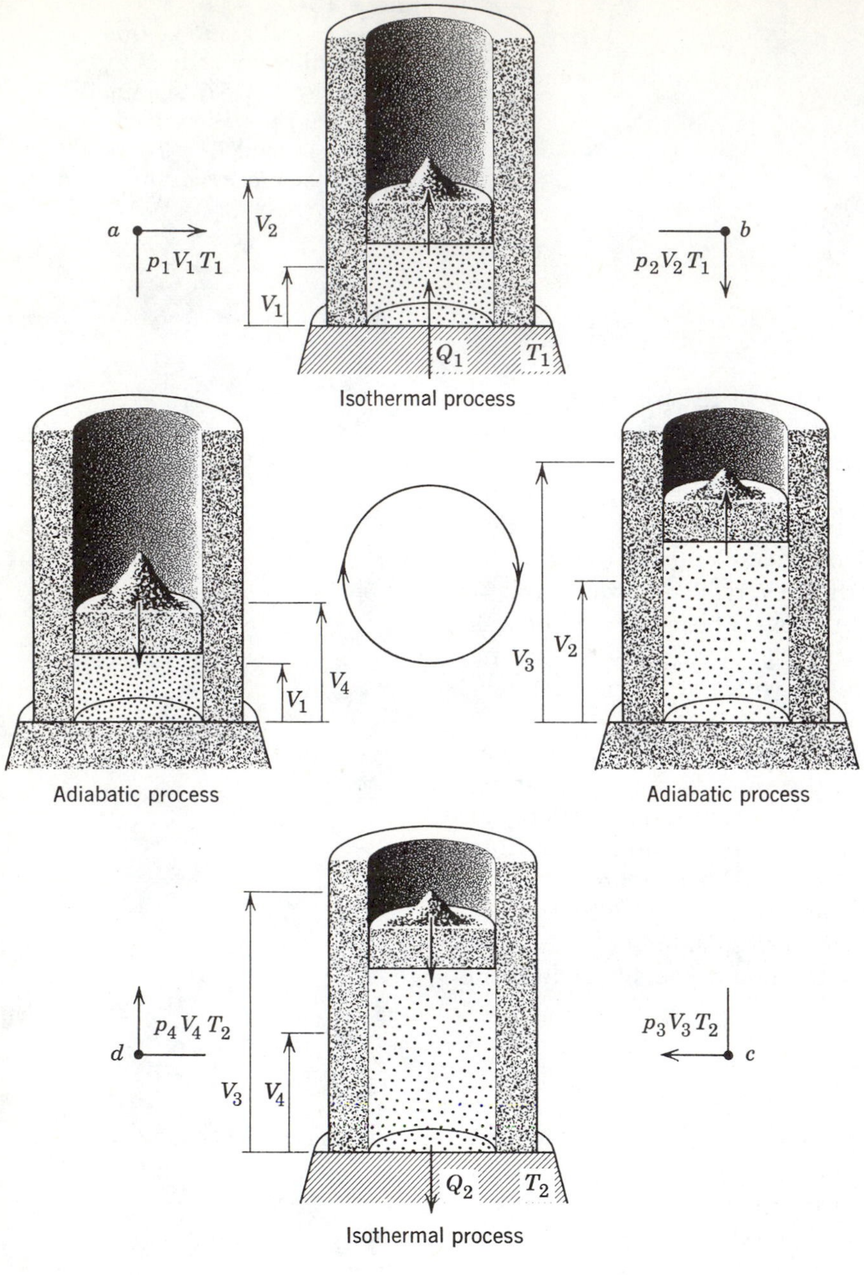

a •→
$p_1 V_1 T_1$

V_2
V_1
Q_1 T_1
Isothermal process

←
$p_2 V_2 T_1$ ↓
b

Adiabatic process

V_4
V_1

V_3
V_2

Adiabatic process

d •
$p_4 V_4 T_2$ ↑

V_3 V_4

Q_2 T_2
Isothermal process

$p_3 V_3 T_2$ ↓
←
c

figure 25-3

A Carnot cycle. The points a, b, c, and d correspond to the points so labeled in Fig. 25-4. The cylinder-piston arrangements show intermediate steps in the processes that connect adjacent points. The arrows on the pistons suggest expansions (caused by removing sand) and compressions (caused by adding sand).

limit of our ability to convert heat into work. The system consists of a "working substance," such as a gas, and the cycle is made up of two isothermal and two adiabatic reversible processes. The working substance, which we can think of as an ideal gas for concreteness, is contained in a cylinder with a heat-conducting base and nonconducting walls and piston. We also provide, as part of the environment, a heat reservoir in the form of a body of large heat capacity at a temperature T_1, another reservoir of large heat capacity at a temperature T_2, and a nonconducting stand. We carry out the Carnot cycle in four steps, as shown in Fig. 25-3. The cycle is shown on the p-V diagram of Fig. 25-4.

Step 1. The gas is in an initial equilibrium state represented by p_1, V_1, T_1 (a, Fig. 25-4). We put the cylinder on the heat reservoir at temperature T_1, and allow the gas to expand slowly to p_2, V_2, T_1

(b, Fig. 25-4). During the process heat energy Q_1 is absorbed by the gas by conduction through the base. The expansion is isothermal at T_1 and the gas does work in raising the piston and its load.

Step 2. We put the cylinder on the nonconducting stand and allow the gas to expand slowly further (by reducing the piston load) to p_3, V_3, T_2 (c, Fig. 25-4). The expansion is adiabatic because no heat can enter or leave the system. The gas does work in raising the piston and its temperature falls to T_2.

Step 3. We put the cylinder on the (colder) heat reservoir T_2 and compress the gas slowly to p_4, V_4, T_2 (d, Fig. 25-4). During the process heat energy Q_2 is transferred from the gas to the reservoir by conduction through the base. The compression is isothermal at T_2 and work is done *on* the gas by the piston and its load.

Step 4. We put the cylinder on the nonconducting stand and compress the gas slowly to the initial condition p_1, V_1, T_1. The compression is adiabatic because no heat can enter or leave the system. Work is done on the gas and its temperature rises to T_1.

The net work W done by the system during the cycle is represented by the area enclosed by path *abcd* of Fig. 25-4. The net amount of heat energy received by the system in the cycle is $Q_1 - Q_2$, where Q_1 is the heat absorbed in Step 1 and Q_2 is that given up in Step 3. The initial and final states are the same so that there is no net change in the internal energy U of the system. Hence, from the first law of thermodynamics,

$$W = Q_1 - Q_2 \qquad (25\text{-}1)$$

for the cycle, in which Q_1 and Q_2 are taken as positive quantities. The result of the cycle is that heat has been converted into work by the system. Any required amount of work can be obtained by simply repeating the cycle. Hence, the system acts like a *heat engine*.

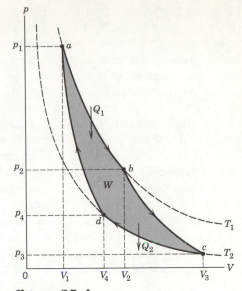

figure 25-4
The Carnot cycle illustrated in the previous figure, plotted on a p-V diagram for an ideal gas as the working substance.

We have used an ideal gas as an example of a working substance. The working substance can be anything at all, although the p-V diagrams for other substances would be different. Common heat engines use steam or a mixture of fuel and air, or fuel and oxygen as their working substance. Heat may be obtained from the combustion of a fuel such as gasoline or coal, or from the annihilation of mass in nuclear fission processes in nuclear reactors. Heat may be discharged at the exhaust or to a condenser. Although real heat engines do not operate on a reversible cycle, the Carnot cycle, which is reversible, gives useful information about the behavior of any heat engine. It is especially important because, as we shall see, it sets an upper limit to the performance of real engines and thereby gives us a goal to work toward.

The efficiency e of a heat engine is the ratio of the net work done by the engine during one cycle to the heat taken in from the high temperature source in one cycle.* Hence,

$$e = \frac{W}{Q_1} = \frac{Q_1 - Q_2}{Q_1} = 1 - \frac{Q_2}{Q_1}. \qquad (25\text{-}2)$$

* The definition reflects the economic importance of engines. Work W is the desirable output; the heat Q_1, is the input that must be paid for in the form, say, of a fuel bill. An efficient engine has a large ratio of W to Q_1.

Equation 25-2 shows that the efficiency of a heat engine is less than one (100%) so long as the heat Q_2 delivered to the exhaust is not zero. Experience shows that every heat engine rejects some heat during the exhaust stroke. This represents the heat absorbed by the engine that is not converted to work in the process.

We may choose to carry out the Carnot cycle by starting at any point, such as a in Fig. 25-4, and traversing each process in a direction opposite to that of the arrowheads in that figure. Then an amount of heat Q_2 is *removed* from the lower temperature reservoir at T_2, and an amount of heat Q_1 is *delivered* to the higher temperature reservoir at T_1; work must be done *on* the system by an outside agency. In this reversed cycle work must be done *on* the system which extracts heat from the lower temperature reservoir. Any amount of heat can be removed from this reservoir by simply repeating the reverse cycle. Hence, the system acts like a *refrigerator*, transferring heat from a body at a lower temperature (the freezing compartment) to one at a higher temperature (the room) by means of work supplied to it (the electric power input).

EXAMPLE 1

Show that the efficiency of a Carnot engine using an ideal gas as the working substance is $e = (T_1 - T_2)/T_1$.

Along the isothermal path ab, the temperature, and hence the internal energy of the ideal gas, remains constant. From the first law, the heat Q_1 absorbed by the gas in its expansion must be equal to the work W_1 done in this expansion. From Example 2, Chapter 23, we have,

$$Q_1 = W_1 = nRT_1 \ln (V_2/V_1).$$

Likewise, in the isothermal compression along the path cd, we have

$$Q_2 = W_2 = nRT_2 \ln (V_3/V_4).$$

On dividing the first equation by the second, we obtain

$$\frac{Q_1}{Q_2} = \frac{T_1 \ln (V_2/V_1)}{T_2 \ln (V_3/V_4)}.$$

From the equation describing an isothermal process for an ideal gas we obtain for the paths ab and cd

$$p_1V_1 = p_2V_2,$$
$$p_3V_3 = p_4V_4.$$

From the equation describing an adiabatic process for an ideal gas we have for paths bc and da

$$p_2V_2^\gamma = p_3V_3^\gamma,$$
$$p_4V_4^\gamma = p_1V_1^\gamma.$$

Multiplying these four equations together and canceling the factor $p_1p_2p_3p_4$ appearing on both sides, we obtain

$$V_1V_2^\gamma V_3V_4^\gamma = V_2V_3^\gamma V_4V_1^\gamma,$$

from which

$$(V_2V_4)^{\gamma-1} = (V_3V_1)^{\gamma-1}$$

and

$$V_2/V_1 = V_3/V_4.$$

Using this result in our expression for Q_1/Q_2, we see that

$$Q_1/Q_2 = T_1/T_2, \tag{25-3}$$

so that

$$e = 1 - Q_2/Q_1 = 1 - T_2/T_1$$

or

$$e = \frac{Q_1 - Q_2}{Q_1} = \frac{T_1 - T_2}{T_1}.$$

The temperatures T_1 and T_2 are those measured on the ideal gas scale described in Chapter 21.

The first heat engines constructed were very inefficient devices. Only a small fraction of the heat absorbed at the high-temperature source could be converted to useful work. Even as engineering design improved, a sizable fraction of the absorbed heat was still discharged at the lower-temperature exhaust of the engine, remaining unconverted to mechanical energy. It remained a hope to devise an engine that could take heat from an abundant reservoir, like the ocean, and convert it completely into useful work. Then it would not be necessary to provide a source of heat at a higher temperature than the outside environment by burning fuels (Fig. 25-5). Likewise, we might hope to be able to devise a refrigerator that simply transfers heat from a cold body to a hot body, without requiring the expense of outside work (Fig. 25-6). *Neither of these hopeful ambitions violates the first law of thermodynamics.* The heat engine would simply convert heat energy completely into mechanical energy, the total energy being conserved in the process. In the refrig-

25-4
THE SECOND LAW OF THERMODYNAMICS

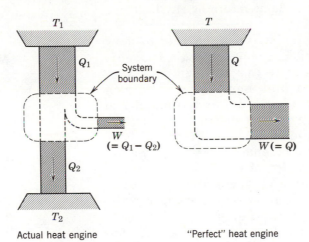

Actual heat engine   "Perfect" heat engine

figure 25-5
In an actual heat engine, some of the heat Q_1 taken in by the engine is converted into work W, but the rest is rejected as heat Q_2. In a "perfect" heat engine all the heat input would be converted into work output.

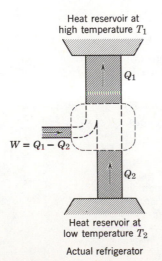

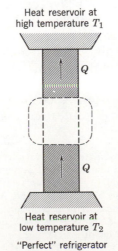

Actual refrigerator   "Perfect" refrigerator

figure 25-6
In an actual refrigerator, work W is needed to transfer heat from a low-temperature to a high-temperature reservoir. In a "perfect" refrigerator, heat would flow from the low-temperature to the high-temperature reservoir without any work being done on the engine.

erator, the heat energy would simply be transferred from cold body to hot body without any loss of energy in the process. Nevertheless *neither of these ambitions has ever been achieved*, and there is reason to believe they never will be.

The *second law of thermodynamics*, which is a generalization of experience, is an assertion that such devices do not exist. There have been many statements of the second law, each emphasizing another facet of the law, but all can be shown to be equivalent to one another. Clausius stated it as follows: *It is impossible for any cyclical machine to produce no other effect than to convey heat continuously from one body to another at a higher temperature.* This statement rules out our ambitious refrigerator, for it implies that to convey heat continuously from a cold to a hot object it is necessary to supply work by an outside agent. We know from experience that when two bodies are in contact, heat energy flows from the hot body to the cold body. The second law rules out the possibility of heat energy flowing from cold to hot body in such a case and so determines the direction of transfer of heat. The direction can be reversed only by an expenditure of work.

Kelvin (with Planck) stated the second law in words equivalent to these: *A transformation whose only final result is to transform into work heat extracted from a source which is at the same temperature throughout is impossible.** This statement rules out our ambitious heat engine, for it implies that we cannot produce mechanical work by extracting heat from a single reservoir without returning any heat to a reservoir at a lower temperature.

To show that the two statements are equivalent we need to show that, if either statement is false, the other statement must be false also. Suppose Clausius' statement were false so that we could have a refrigerator operating without needing a work input. We could use an ordinary engine to remove heat from a hot body, to do work and to return part of the heat to a cold body. But by connecting our "perfect" refrigerator into the system, this heat would be returned to the hot body without expenditure of work and would become available again for use by the heat engine. Hence, the combination of an ordinary engine and the "perfect" refrigerator would constitute a heat engine which violates the Kelvin-Planck statement. Or we can reverse the argument. If the Kelvin-Planck statement were incorrect, we could have a heat engine which simply takes heat from a source and converts it completely into work. By connecting this "perfect" heat engine to an ordinary refrigerator, we could extract heat from the hot body, convert it completely to work, use this work to run the ordinary refrigerator, extract heat from the cold body, and deliver it plus the work converted to heat by the refrigerator to the hot body. The net result is a transfer of heat from cold to hot body without expenditure of work and this violates Clausius' statement.

The second law tells us that many processes are irreversible. For example, Clausius' statement specifically rules out a simple reversal

* This statement needs to be supplemented if we extend thermodynamics to the region of negative Kelvin temperatures. All other formulations of the second law, and indeed, all other laws of thermodynamics apply to negative temperatures without revision. See an article, "Thermodynamics and Statistical Mechanics at Negative Absolute Temperatures" by N. F. Ramsey, in *Temperature, Its Measurement and Control in Science and Industry*, Vol. 3, Part 1, Reinhold Publishing Co., New York, 1962, or "Negative Temperatures and Negative Dissipation" by Stefan Machlup, in *American Journal of Physics*, November 1975.

of the process of heat transfer from hot body to cold body. Not only will some processes not run backward by themselves, but no combination of processes can undo the effect of an irreversible process without causing another corresponding change elsewhere. In later sections we shall develop these ideas more fully and formulate the second law quantitatively.

Carnot first wrote scientifically on the theory of heat engines. In 1824 he published *Reflections on the Motive Power of Heat.* By then the steam engine was commonly used in industry. Carnot wrote:

25-5
THE EFFICIENCY OF ENGINES

In spite of labor of all sorts expended on the steam engine, and in spite of the perfection to which it has been brought, its theory is very little advanced. . . .

The production of motion in the steam engine is always accompanied by a circumstance which we should particularly notice. This circumstance is the passage of caloric from one body where the temperature is more or less elevated to another where it is lower. . . .

The motive power of heat is independent of the agents employed to develop it; its quantity is determined solely by the temperature of the bodies between which, in the final result, the transfer of the caloric occurs.

Hence, Carnot directed attention to the facts that the difference in temperature was the real source of "motive power," that the transfer of heat played a significant role, and that the choice of working substance was of no theoretical importance.

Carnot's achievement was remarkable when we recall that the mechanical equivalence of heat and the conservation of energy principle were not known in 1824. In his later papers, published posthumously in 1872, it became clear that Carnot had foreseen the principle of the conservation of energy and had made an accurate determination of the mechanical equivalent of heat. He had planned a program of research which included all the important developments in the field made by other investigators during the following several decades. However, he died during a cholera epidemic in 1832 at the age of 36, leaving it to others to extend his work. It was William Thomson (later Lord Kelvin) who modified Carnot's reasoning to bring it into accord with the mechanical theory of heat, and who, together with Clausius, successfully developed the science of thermodynamics.

Carnot developed the concept of a reversible engine and the reversible cycle named after him. He stated a theorem of great practical importance: *The efficiency of all reversible engines operating between the same two temperatures is the same, and no irreversible engine working between the same two temperatures can have a greater efficiency than this.* Clausius and Kelvin showed that this theorem was a necessary consequence of the second law of thermodynamics. Notice that nothing is said about the working substance, so that the efficiency of a reversible engine is independent of the working substance and depends only on the temperatures. Furthermore, a reversible engine operates at the maximum efficiency possible for any engine working between the same two temperature limits. The proof of this theorem follows.

Let us call the two reversible engines H and H'. They operate between the temperatures T_1 and T_2 where $T_1 > T_2$. They may differ, say, in their working substance or in their initial pressures and lengths of stroke. We choose H to run forward and H' to run backward (as a refrigerator). The forward-running engine

H takes in heat energy Q_1 at T_1 and gives out heat energy Q_2 at T_2. The backward-running engine (refrigerator) H' takes in heat Q_2' at T_2 and gives out heat Q_1' at T_1. We now connect the engines mechanically and adjust the stroke lengths so that the work done per cycle by H is just sufficient to operate H' (Fig. 25-7). Suppose the efficiency e of H were greater than the efficiency e' of H'. Then

$$e > e', \quad \text{(assumption)}$$

or

$$\frac{Q_1 - Q_2}{Q_1} > \frac{Q_1' - Q_2'}{Q_1'}.$$

Since the work per cycle done by one engine equals the work per cycle done on the other engine,

$$W = W',$$

or

$$Q_1 - Q_2 = Q_1' - Q_2'.$$

Comparing these relations, we see that (since $Q_1 - Q_2 > 0$)

$$\frac{1}{Q_1} > \frac{1}{Q_1'}$$

or

$$Q_1 < Q_1'.$$

Hence (from the work equality),

$$Q_2 < Q_2'.$$

Thus, the hot source gains heat $Q_1' - Q_1$ (positive) and the cool source loses heat $Q_2' - Q_2$ (positive). But no work is done in the process by the combined system $H + H'$ so that we have transferred heat from a body at one temperature to a body at a higher temperature without performing work—in direct contradiction to Clausius' statement of the second law. Hence, we conclude that e cannot be greater than e'. Likewise, by reversing the engines we can use the same reasoning to prove that e' cannot be greater than e. Hence,

$$e = e',$$

proving the first part of Carnot's theorem.

Now suppose that H is an *irreversible* engine. Then by the exact same procedure we can prove that e_{ir} cannot be greater than e'. But H cannot be reversed, so we cannot prove that e' cannot be greater than e_{ir}. Therefore, e_{ir} is either equal to or less than e'. Since $e' = e = e_{reversible}$, we have

$$e_{irreversible} \leq e_{reversible},$$

thus proving the second part of Carnot's theorem.

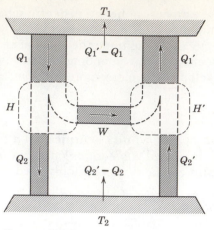

figure 25-7
Proof of Carnot's theorem.

EXAMPLE 2

A steam engine takes steam from the boiler at 200° C (225 lb/in.² pressure) and exhausts directly into the air (14 lb/in.² pressure) at 100° C. What is its maximum possible efficiency?

Using the result of Example 1 (which applies to this case by virtue of Carnot's theorem, which we have just proved) we have

$$e = \frac{T_1 - T_2}{T_1} = \frac{473 \text{ K} - 373 \text{ K}}{473 \text{ K}} \times 100\% = 21.1\%.$$

Actual efficiencies of about 15% are usually realized. Energy is lost by friction, turbulence, and heat conduction. Lower exhaust temperatures on more complicated steam engines may raise the maximum possible efficiency to 35% and the actual efficiency to 20%. The efficiency of an ordinary automobile engine is about 22% and that of a large Diesel oil engine about 40%.

The efficiency of a reversible engine is independent of the working substance and depends only on the two temperatures between which the engine works. Since $e = 1 - Q_2/Q_1$, then Q_2/Q_1 can depend only on the temperatures. This led Kelvin to suggest a new scale of temperature. If we let θ_1 and θ_2 represent these two temperatures, his defining equation is

$$\theta_1/\theta_2 = Q_1/Q_2.$$

That is, two temperatures on this scale are to each other as the heats absorbed and rejected, respectively, by a Carnot engine operating between these temperatures. Such a temperature scale is called the *thermodynamic* (or *Kelvin*) *temperature scale.*

To complete the definition of the thermodynamic scale, we assign the arbitrary value of 273.16 to the temperature of the triple point of water. Hence, $\theta_{tr} = 273.16$ K. Then for a Carnot engine operating between reservoirs at the temperatures θ and θ_{tr}, we have

$$\frac{\theta}{\theta_{tr}} = \frac{Q}{Q_{tr}}$$

or
$$\theta = 273.16 \text{ K} \frac{Q}{Q_{tr}}. \tag{25-4}$$

If we compare this with the corresponding equation for the ideal gas temperature T, namely

$$T = 273.16 \text{ K} \lim_{p_{tr} \to 0} \frac{p}{p_{tr}}, \tag{25-5}$$

we see that on the thermodynamic scale Q plays the role of a thermometric property. However, Q does not depend on the characteristics of any substance because a Carnot engine is independent of the nature of the working substance. Therefore, we obtain a scale of temperature which is free of the objection we can raise to the ideal gas scale of Chapter 21, and in fact we arrive at a fundamental definition of temperature.

The definition of thermodynamic temperature enables us to rewrite the equation for the efficiency of a reversible engine as

$$e = \frac{Q_1 - Q_2}{Q_1} = \frac{\theta_1 - \theta_2}{\theta_1}. \tag{25-6}$$

But we have shown (Example 1) that the efficiency of a Carnot engine using an ideal gas as working substance is

$$e = \frac{Q_1 - Q_2}{Q_1} = \frac{T_1 - T_2}{T_1} \tag{25-7}$$

where T is the temperature given by the constant-volume thermometer containing the ideal gas. Hence, $Q_1/Q_2 = T_1/T_2$ and $Q_1/Q_2 = \theta_1/\theta_2$. Since $\theta_{tr} = T_{tr} = 273.16$ and $\theta/\theta_{tr} = T/T_{tr}$, it follows that $\theta = T$. Hence, *if an ideal gas were available for use in a constant-volume thermometer, the thermometer would yield the thermodynamic* (or *Kelvin*) *temperature.* We have seen that, although an ideal gas is not available, measurements made using the limiting process of Eq. 25-5 with real gases correspond to ideal gas behavior. We shall treat the ideal gas scale and the thermodynamic scale as identical and we shall use the designation K interchangeably for each, as in fact we have already done.

In practice, we cannot have a gas below 1 K. One of the methods used in measuring temperature below 1 K employs the thermodynamic scale directly. The ratio of two thermodynamic temperatures is the ratio of two heats transferred during two isothermal processes bounded by the same two adiabatics (Fig. 25-8). The location of the adiabatic boundaries (on the p-V diagram) can be found experimentally, and the heats transferred during two nearly reversible isothermal processes can be measured with great precision.

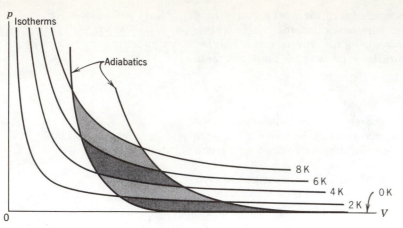

figure 25-8
A series of Carnot cycles tending toward absolute-zero temperature, as used in establishing the thermodynamic scale of temperature. The difference in slope between isothermals and adiabatics has here been exaggerated for clarity.

From the equations

$$T = 273.16 \text{ K} \frac{Q}{Q_{tr}} \quad \text{or} \quad \frac{T}{T_{tr}} = \frac{Q}{Q_{tr}},$$

it is clear that the heat Q transferred in an isothermal process between two given adiabatics decreases as the temperature T decreases. Conversely, the smaller Q is the lower the corresponding temperature T is. Now the smallest possible value of Q is zero and the corresponding T is absolute zero. That is, *if a system undergoes a reversible isothermal process with no transfer of heat, the temperature at which this process takes place is the absolute zero*. Hence, at absolute zero, an isothermal and an adiabatic process are identical (Fig. 25-8).

This definition of absolute zero applies to all substances and is independent of the properties of any one of them. Notice that no reference is made to molecules or molecular energy and that we have obtained a purely macroscopic definition of absolute zero.

The efficiency of a Carnot engine is

$$e = 1 - \frac{T_2}{T_1},$$

which is the maximum possible efficiency any engine can have operating between temperatures T_1 and T_2. To obtain 100% efficiency, T_2 must be zero. Only when the low-temperature reservoir is at absolute zero will all the heat absorbed at the high-temperature reservoir be converted to work.

The fundamental feature of all cooling processes is that the lower the temperature, the more difficult it is to go still lower. This experience has led to the formulation of the *third law of thermodynamics*, which can be stated in one form as follows: *It is impossible by any procedure, no matter how idealized, to reduce any system to the absolute zero of temperature in a finite number of operations*. Hence, because we cannot obtain a reservoir at absolute zero, a heat engine with 100% efficiency is a practical impossibility.

The zeroth law of thermodynamics is related to the concept of *temperature T* and the first law is related to the concept of *internal energy U*. In this and the following sections we show that the second law of thermodynamics is related to a thermodynamic variable called *entropy, S*, and that we can express the second law quantitatively in terms of this variable. We start by considering a Carnot cycle. For such a cycle we have seen (Eq. 25-3) that

$$\frac{Q_1}{T_1} = \frac{Q_2}{T_2},$$

25-7
ENTROPY—REVERSIBLE PROCESSES

in which the Q's were taken as positive quantities, that is, we dealt with the magnitudes, or absolute values, only of the Q's. *If we now interpret them again as algebraic quantities, Q being positive when heat enters the system and negative when heat leaves the system,* we can write this relation as

$$\frac{Q_1}{T_1} + \frac{Q_2}{T_2} = 0.$$

This equation states that the sum of the algebraic quantities Q/T is zero for a Carnot cycle.

As a next step, we assert that *any* reversible cycle is equivalent, to as close an approximation as we wish, to an assembly of Carnot cycles. Figure 25-9a shows an arbitrary reversible cycle superimposed on a family of isotherms. We can approximate the actual cycle by connecting the isotherms by suitably chosen adiabatic lines (Fig. 25-9b), thus forming an assembly of Carnot cycles. You should convince yourself that traversing the individual Carnot cycles in Fig. 25-9b is exactly equivalent, in terms of heat transferred and work done, to traversing the jagged sequence of isotherms and adiabatic lines that approximates the actual cycle. This is so because adjacent Carnot cycles have a common isotherm and the two traversals, in opposite directions, cancel each other in the region of overlap as far as heat transfer and work done are concerned. By making the temperature interval between the isotherms in Fig. 25-9b small enough we can approximate the actual cycle as closely as we wish by an alternating sequence of isotherms and adiabatic lines.

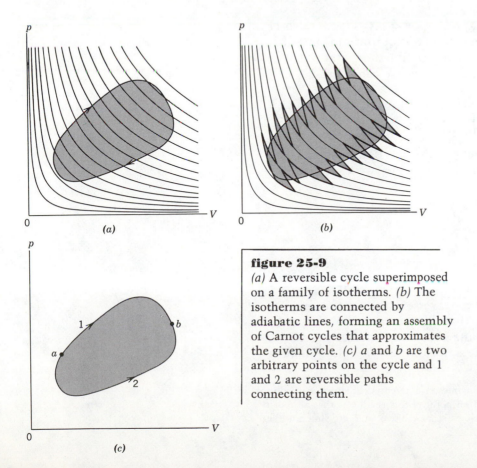

(a)

(b)

p

1

a

b

2

V

0

(c)

figure 25-9
(a) A reversible cycle superimposed on a family of isotherms. (b) The isotherms are connected by adiabatic lines, forming an assembly of Carnot cycles that approximates the given cycle. (c) a and b are two arbitrary points on the cycle and 1 and 2 are reversible paths connecting them.

We can write, then, for the isothermal-adiabatic sequence of lines in Fig. 25-9b,

$$\sum \frac{Q}{T} = 0,$$

or, in the limit of infinitesimal temperature differences between the isotherms of Fig. 25-9b,*

$$\oint \frac{đQ}{T} = 0, \tag{25-8}$$

in which $\oint$ indicates that the integral is evaluated for a complete traversal of the cycle, starting (and ending) at any arbitrary point of the cycle.

If the integral of a quantity around any closed path is zero, that quantity is called a state variable, that is, it has a value that is characteristic only of the state of the system, regardless of how that state was arrived at. We call the variable in this case the *entropy* S and we have, from Eq. 25-8,

$$dS = \frac{đQ}{T} \quad \text{and} \quad \oint dS = 0. \tag{25-9}$$

Common units for entropy are J/K or cal/K.

Gravitational potential energy U_g, internal energy U, pressure p, and temperature T are other state variables and equations of the form $\oint dX = 0$ hold for each of them, where for X we substitute the appropriate symbol. Heat Q and work W are *not* state variables and we know that, in general, $\oint đQ \neq 0$ and $\oint đW \neq 0$, as the student can easily show for the special case of a Carnot cycle.

The property of a state variable expressed by $\oint dX = 0$ can also be expressed by saying that $\int dX$ between any two equilibrium states has the same value for all (reversible) paths connecting those states. Let us prove this for the state variable called entropy. We can write Eq. 25-9 (see Fig. 25-9c) as

$$_1\!\int_a^b dS + {}_2\!\int_b^a dS = 0 \tag{25-10}$$

where a and b are arbitrary points and 1 and 2 describe the paths connecting these points. Since the cycle is reversible, we can write Eq. 25-10 as

$$_1\!\int_a^b dS - {}_2\!\int_a^b dS = 0$$

or

$$_1\!\int_a^b dS = {}_2\!\int_a^b dS \tag{25-11}$$

In Eq. 25-11 we have simply decided to traverse path 2 in the opposite direction, that is, from a to b rather than from b to a. We do this by changing the order of the limits in the second integral of Eq. 25-10, which requires that we also change the sign of the integral, thus yielding

---

* See footnote on page 487. $đQ$ represents an inexact differential because Q is not a function of the state of the system. The central point of this section is that although $đQ$ is an inexact differential $đQ/T$ ($= dS$) *is* exact, so that S, like p, V, T, etc. (but not like Q or W), is a state variable.

Eq. 25-11. This latter equation tells us that the quantity $\int_a^b dS$ between any two equilibrium states of the system, such as a and b, is independent of the path connecting those states, for 1 and 2 are quite arbitrary paths. Recall our almost identical discussion in Section 8-2, where we introduced the concept of a conservative force.

The change in entropy between a and b in Fig. 25-9c is, then

$$S_b - S_a = \int_a^b dS = \int_a^b \frac{dQ}{T} \qquad \text{(reversible process)}, \qquad (25\text{-}12)$$

where the integral is evaluated over *any reversible path* connecting these two states.

25-8 ENTROPY— IRREVERSIBLE PROCESSES

In Section 25-7 we spoke only of reversible processes. However, entropy, like all state variables, depends only on the state of the system and we must be able to calculate the change in entropy for irreversible processes, provided only that they begin and end in equilibrium states. Let us consider two examples.

1. *Free Expansion.* As in Section 22-7 (see Fig. 22-14) let a gas double its volume by expanding into an evacuated enclosure. Since no work is done against the vacuum, $W = 0$ and, since the gas is enclosed by nonconducting walls, $Q = 0$. From the first law, then $\Delta U = 0$ or

$$U_i = U_f \qquad (25\text{-}13)$$

where i and f refer to the initial and final (equilibrium) states. If the gas is an ideal gas, then U depends on temperature alone and not on the pressure or the volume so that Eq. 25-13 implies $T_i = T_f$.

The free expansion is certainly irreversible because we lose control of the environment once we turn the stopcock in Fig. 22-14. There is, however, an entropy difference $S_f - S_i$ between the initial and final equilibrium states, but we cannot calculate it from Eq. 25-12 because that relation applies only to reversible paths; if we tried to use that equation, we would have the immediate difficulty that $Q = 0$ for the free expansion and—further—we would not know how to assign meaningful values of T to the intermediate, nonequilibrium states.

How, then, do we calculate the difference $S_f - S_i$ between these two states? We do so by finding a *reversible* path (*any* reversible path) that connects the states i and f and we calculate the entropy change for that path. In the free expansion a convenient reversible path (assuming an ideal gas) is an isothermal expansion from V_i to $V_f (= 2V_i)$. This corresponds to the isothermal expansion carried out between the points a and b of the Carnot cycle of Fig. 25-4. It represents quite a different set of operations from the free expansion and has in common with it *only* the fact it connects the same set of equilibrium states, i and f. From Eq. 25-12 and Example 1 we have

$$S_f - S_i = \int_i^f \frac{dQ}{T} = nR \ln (V_f/V_i)$$

$$= nR \ln 2.$$

This is positive so that the entropy of the system *increases* in this irreversible, adiabatic process.

2. *Heat Conduction.* For another example consider two bodies that are similar in every respect except that one is at a temperature T_1 and the other at temperature T_2, where $T_1 > T_2$. If we put both objects in contact inside a box with nonconducting walls, they will eventually reach a common temperature T_m, somewhere between T_1 and T_2. Like the free expansion, the process is irreversible because we lose control of the environment once we put the two bodies in the box. Like the free expansion this process is also (irreversibly) adiabatic because no heat enters or leaves the system during the process.

To calculate the entropy change for the system during this process we must again find a *reversible* process connecting the same initial and final states and calculate the system entropy change by applying Eq. 25-12 to that process. We can do so if we imagine that we have at our disposal a heat reservoir of large heat capacity whose temperature T is at our control, by turning a knob, say. We first adjust the reservoir temperature to T_1 and put the first (hotter) object in contact with the reservoir. We then *slowly* (reversibly) lower the reservoir temperature from T_1 to T_m, extracting heat from the hot body as we do so. The hot body *loses* entropy in this process, the amount being approximately

$$\Delta S_1 = -\frac{Q}{T_{1,m}}$$

where $T_{1,m}$ is the average of T_1 and T_m and Q is the heat extracted.

We then adjust our reservoir temperature to T_2 and place it in contact with the second (cooler) object. We then *slowly* (reversibly) raise the reservoir temperature from T_2 to T_m, adding heat to the cool body as we do so. The cool body *gains* entropy in this process, the amount being approximately

$$\Delta S_2 = +\frac{Q}{T_{2,m}},$$

where $T_{2,m}$ is the average of T_2 and T_m and Q is the heat added. Note that the two Q's are identical.

The two bodies are now at the same temperature T_m and the system, which consists of these two bodies, is now in its final equilibrium state. The change in entropy for the complete system is

$$S_f - S_i = \Delta S_1 + \Delta S_2$$

$$= -\frac{Q}{T_{1,m}} + \frac{Q}{T_{2,m}}.$$

Since $T_{1,m} > T_{2,m}$ we have $S_f > S_i$. Again, as for the free expansion, the entropy of the system has *increased* in this irreversible, adiabatic process.

In each of these examples we must distinguish carefully between the actual (irreversible) process (free expansion or heat conduction) and the reversible process that we introduce just so that we can calculate the entropy change in the actual process. We can choose *any* reversible process, as long as it connects the same initial and final state as the actual process; all such reversible processes will yield the same entropy change because this depends only on the initial and final states and not on the process connecting them—be it reversible or irreversible.

We are now ready to formulate the second law of thermodynamics in terms of entropy. Since this law is a generalization from experience we cannot *prove* it but can write it down and show that our statement is in agreement with experiment and is equivalent to other formulations of the second law that we have given earlier. In this spirit we assert that the second law is: *A natural process that starts in one equilibrium state and ends in another will go in the direction that causes the entropy of the system plus environment to increase.*

Following our pattern for the zeroth law and the first law of thermodynamics (see page 487) the essence of the second law, speaking loosely, is: *There exists a useful thermodynamic variable called entropy.* The second law also tells us how to use this variable to predict whether a particular process will occur in nature.

The two experiments of Section 25-8 (free expansion and heat conduction) are consistent with the second law. The entropy of the system *increased* in each of these irreversible processes. Note that the entropy of the environment in these two cases remains unchanged because, both being carried out in adiabatic enclosures, there was no interchange of heat with the environment. Thus, as required by our statement of the second law, the entropy of the system plus environment increased for each of these (natural) processes.

In the form that we have written it the second law applies only to irreversible processes because only such processes have a "natural direction." Indeed (see Section 25-1) the understanding of the natural directions of such processes is the main concern of the second law. Reversible processes can go equally well in either direction, however, and *for reversible processes the entropy of the system plus environment remains unchanged.* This is so because if heat dQ is transferred from the environment to the system, the entropy of the environment *decreases* by dQ/T whereas that of the system *increases* by dQ/T, the net change for the system plus environment being zero. The fact that the process is reversible means that the environment and the system can differ in temperature by only a differential amount dT when the heat transfer takes place; this is in sharp contrast to our (irreversible) heat conduction problem of the previous section, in which the temperature difference of the two bodies placed in contact was large.

Another class of processes of particular interest are adiabatic processes (reversible or irreversible); they involve no transfer of heat with the environment so that the only entropy change possible is that of the system. From our statement of the second law and from our remarks about reversible processes in the paragraph above, we conclude that

$$S_f = S_i \qquad \text{(reversible adiabatic process)}$$

and

$$S_f > S_i \qquad \text{(irreversible adiabatic process)},$$

where S_f and S_i are the final and initial entropies of the system.

Our statement of the second law is consistent with the Clausius statement (page 546) which declares that there is no such thing as a "perfect" refrigerator (see Fig. 25-6). If there were, the entropy of the lower temperature reservoir would *decrease* by Q/T_2; that of the upper temperature reservoir would *increase* by Q/T_1; that of the system would remain unchanged because the system traverses a cycle, returning to its starting point. Thus the net change in the entropy of the system plus environment is a *decrease*, because $T_2 < T_1$. This violates the statement of the second law that we have just given and, if we wish to retain the statement, we must conclude (with Clausius) that there is no such thing as a "perfect" refrigerator.

Our statement of the second law is also consistent with the Kelvin-Planck statement (page 546) which declares that there is no such thing as a "perfect" heat engine (see Fig. 25-5). If there were, the entropy of the reservoir at temperature T would *decrease* by Q/T; that of the system would remain unchanged because the system traverses a cycle, returning to its starting point. Thus the net change of entropy of the system plus environment is a *decrease*. This violates the statement of the second law that we have just given and, if we wish to retain the statement, we must conclude (with Kelvin) that there is no such thing as a "perfect" heat engine.

EXAMPLE 3

Compute the entropy change of a system consisting of 1.00 kg of ice at 0° C which melts (reversibly) to water at that same temperature. The latent heat of melting is 79.6 cal/g.

The requirement that we melt the ice *reversibly* means that we must put it in contact with a heat reservoir whose temperature exceeds 0° C by only a differential amount; if we lower the reservoir temperature until it is a differential amount below 0° C, the melted ice will begin to freeze. Since the process is reversible, we can use Eq. 25-12 to compute the entropy change of the system. The temperature remains constant at 273 K. Therefore,

$$S_{water} - S_{ice} = \int_0^Q \frac{dQ}{T} = \frac{1}{T} \int_0^Q dQ = \frac{Q}{T}.$$

But

$$Q = 10^3 \text{ g} \times 79.6 \text{ cal/g} = 7.96 \times 10^4 \text{ cal}$$

or

$$S_{water} - S_{ice} = \frac{7.96 \times 10^4}{273} \text{ cal/K} = 292 \text{ cal/K}$$
$$= 1220 \text{ J/K}.$$

In this example of reversible melting the entropy change of the *system plus environment* is zero, as it must be for all reversible processes. The entropy change calculated above is the increase in entropy of the *system*; there is an exactly equal decrease in entropy of the environment (−1220 J/K) associated with the heat that leaves the reservoir (environment), at 273 K, to melt the ice.

In practice, melting is likely to be irreversible, as when we put an ice cube in a glass of water at room temperature. This process has only one natural direction—the ice will melt. The entropy of the system plus environment will *increase* in this process as required by the second law. The (irreversible) heat conduction example of the previous section should make this understandable.

EXAMPLE 4

Calculate the entropy change that an ideal gas undergoes in a reversible isothermal expansion from a volume V_i to a volume V_f.

From the first law

$$dU = dQ - p\, dV.$$

But $dU = 0$, since U depends only on temperature for an ideal gas and the temperature is constant. Hence,

$$dQ = p\, dV$$

and

$$dS = \frac{dQ}{T} = \frac{p\, dV}{T}.$$

But

$$pV = nRT,$$

so that

$$dS = nR \frac{dV}{V}$$

and

$$S_f - S_i = \int_{V_i}^{V_f} nR \frac{dV}{V} = nR \ln \frac{V_f}{V_i}. \qquad (25\text{-}14)$$

Since $V_f > V_i$, $S_f > S_i$ and the *entropy of the gas increases*.

In order to carry out this process we must have a reservoir at temperature T which is in contact with the system and supplies the heat to the gas. Hence, the *entropy of the reservoir decreases* by $\int dQ/T[= nR \ln (V_f/V_i)]$, so that in this process the entropy of system plus environment does not change. As in the previous example, this is characteristic of a reversible process.

25-10
ENTROPY AND DISORDER

Entropy is associated with disorder and the second law statement that in natural processes the entropy of the (system + environment) tends to increase is equivalent to saying that the disorder of the (system + environment) tends to increase.

There are two approaches to this point of view and we discuss each in turn. The first approach is qualitative and provides an intuitive sense of the equivalence of entropy and disorder. The second is quite formal and provides the solid quantitative base for this equivalence.

From the qualitative point of view let us consider three examples, the first two of which we have discussed in Section 25-8. All are "natural processes" in that there is no doubt whatever as to the direction in which, left to themselves, they will go. Let us now see *qualitatively* in what sense the final (equilibrium) state is more disordered than the initial state.

1. *Free Expansion.* In a free expansion (Section 22-7) the gas molecules confined to one-half of a box are permitted to fill the entire box. By any reasonable definition of the word disorder the system has become more disordered, in the same sense that disorder increases if the litter on one vacant lot is spread over two lots. More precisely, the disorder has increased because we have lost some of our ability to classify molecules. The statement: "The molecules are in the box" is weaker from this point of view than the statement: "The molecules are in the left half of the box."

2. *Heat Conduction.* In this example two bodies of different temperatures T_1 and T_2 come to a uniform intermediate temperature T when they are placed in contact. Here again the system has become more disordered in this natural process because we have lost some of our ability to classify molecules. The statement: "All molecules in the system correspond, by way of Eq. 23-6, to temperature T" is weaker from this point of view than the statement: "All molecules in body A correspond to temperature T_1 and all molecules in body B correspond to temperature T_2." It is clear that some order has been lost in this process.

3. *A Stirred Coffee Cup.* Suppose that you stir a cup of coffee and then remove the spoon. In the initial state there is an ordered motion of the swirling coffee. In the final equilibrium state there is random molecular motion. Surely disorder is increased in this natural and irreversible process.

Let us now discuss the *quantitative* relationship between entropy and disorder. In statistical mechanics we give a precise meaning to disorder and we express its connection with entropy by the relation

$$S = k \ln w. \qquad (25\text{-}15)$$

Here, k is Boltzmann's constant, S is the entropy of the system, and w, which we may call the *disorder parameter*, is the probability that the system will exist in the state it is in relative to all the possible states it could be in. This equation connects a thermodynamic or macroscopic quantity, the entropy, with a statistical or microscopic quantity, the probability.

Let us illustrate by computing the change in entropy of an ideal gas in an isothermal expansion. Here the number of molecules and the temperature do not change, but the volume does. The probability that a given molecule may be found in a region having a volume V is proportional to V; that is, the greater V is, the greater the chance of finding it in V. Hence, the probability of finding a *single* molecule in V is

$$w_1 = cV$$

where c is a constant. The probability of finding N molecules simultaneously in the volume V is the N-fold product of w_1. That is, the probability of a state consisting of N molecules in a volume V is

$$w = w_1{}^N = (cV)^N. \qquad (25\text{-}16)$$

For example, if the probability of finding a single molecule in V is $\frac{1}{2}$ (that is, there is a 50% chance of its being in V and a 50% chance of its being outside V), the probability of finding two molecules in V is $\frac{1}{4}$. There are four equally probable states here (both in; both out; one in, the other out; one out, the other in), and only one of them is a state with both molecules in V.

If we now combine Eq. 25-15 and Eq. 25-16, we obtain

$$S = kN \,(\ln c + \ln V).$$

Hence, the difference in entropy between a state of volume V_f and a state of volume V_i (temperature and number of molecules remaining constant) is

$$S_f - S_i = kN \,(\ln c + \ln V_f) - kN \,(\ln c + \ln V_i)$$

$$= kN \ln \frac{V_f}{V_i} = \frac{RN}{N_0} \ln \frac{V_f}{V_i} = nR \ln \frac{V_f}{V_i}$$

in exact agreement with the strictly thermodynamic result of Eq. 25-14.

It is on the basis of Eq. 25-16 that we stated above that disorder increases during a free expansion; that equation yields $(cV)^N$ for the disorder parameter before expansion and $(c2V)^N$ for that parameter when the volume is doubled by the expansion.

One must use care not to identify intuitive qualitative ideas of "disorder" as mixed-up-ness with the quantitative meaning we have given the term here. There is a *correlation*, of course, between the qualitative idea of "disorder" and entropy defined either on the macroscopic or microscopic level, but the *identity* exists only for the precise meaning we have given to disorder.*

The statistical definition of entropy, Eq. 25-15, connects the thermodynamic and the statistical mechanical pictures and enables us to put the second law of thermodynamics on a statistical basis. The direction in which natural processes take place (toward higher entropy) is determined by the laws of probability (toward a more probable state). The equilibrium state is the state of maximum entropy thermodynamically and the most probable state statistically. We have seen, however, that fluctuations may occur about an equilibrium distribution (for example, the Brownian motion). From this point of view, then, it is not absolutely certain that the entropy increases in every spontaneous process. The entropy may sometimes decrease. If we waited long enough, even the most improbable states might occur: the water in a pond suddenly freezing on a hot summer day or a local vacuum occurring suddenly in a room. Although such occurrences are possible, the probability of their happening, when computed, turns out to be incredibly small. Hence, the second law of thermodynamics shows us the most probable course of events, not the only possible ones. But its area of application is so broad and the chance of nature's contradicting it so small that it occupies the distinction of being one of the most useful and general laws in all sciences.

---

* For specific examples, see "Entropy and Disorder" by P. G. Wright, in *Contemporary Physics*, November 1970.

questions

1. What requirements should a system meet in order to be in thermodynamic equilibrium?

2. Are any of these phenomena reversible? (*a*) breaking an empty soda bottle; (*b*) mixing a cocktail; (*c*) melting an ice cube in a glass of iced tea; (*d*) burning a log of firewood; (*e*) puncturing an automobile tire; (*f*) finishing the "Unfinished Symphony"; (*g*) writing this book.

3. Give some examples of irreversible processes in nature.

4. In the irreversible process of Fig. 25-1*a* can we calculate the work done in terms of an area on a *p-V* diagram? Is any work done?

5. Can a given amount of mechanical energy be converted completely into heat energy? If so, give an example.

6. Can you suggest a reversible process whereby heat can be added to a system? Would adding heat by means of a Bunsen burner be a reversible process?

7. Give a qualitative explanation of how frictional forces between moving surfaces produce heat energy. Does the reverse process (heat energy producing relative motion of these surfaces) occur? Can you give a plausible explanation?

8. A block returns to its initial position after dissipating mechanical energy to heat through friction. Is this process thermodynamically reversible?

9. To carry out a Carnot cycle we need not start at point *a* in Fig. 25-4, but may equally well start at points *b*, *c*, or *d*, or indeed any intermediate point. Explain.

10. If a Carnot engine is independent of the working substance, then perhaps real engines should be similarly independent, to a certain extent. Why then, for real engines, are we so concerned to find suitable fuels such as coal, gasoline, or fissionable material? Why not use stones as a fuel?

11. Couldn't we just as well define the efficiency of an engine as $e = W/Q_2$ rather than as $e = W/Q_1$? Why don't we?

12. Under what conditions would an ideal heat engine be 100% efficient?

13. What factors reduce the efficiency of a heat engine from its ideal value?

14. In order to increase the efficiency of a Carnot engine most effectively, would you increase T_1, keeping T_2 constant, or would you decrease T_2, keeping T_1 constant?

15. Can a kitchen be cooled by leaving the door of an electric refrigerator open? Explain.

16. Is a heat engine operating between the warm surface water of a tropical ocean and the cooler water beneath the surface a possible concept? Is the idea practical? (See "Solar Sea Power" by Clarence Zener, *Physics Today*, January 1973.)

17. Is there a change in entropy in purely mechanical motions?

18. Two samples of a gas initially at the same temperature and pressure are compressed from a volume *V* to a volume (*V*/2), one isothermally, the other adiabatically. In which sample is the final pressure greater? Does the entropy of the gas change in either process?

19. Suppose we had chosen to represent the state of a system by its entropy and its absolute temperature rather than by its pressure and volume. (*a*) What would a Carnot cycle look like on a *T-S* diagram? (*b*) What physical significance, if any, can be attached to the area under a curve on a *T-S* diagram?

20. Consider a box containing a very small number of molecules, say five. It must sometimes happen by chance that all of these molecules find themselves in the left half of the box, the right half being completely empty. This is just the reverse of a free expansion, a process that we have declared to be *irreversible*. What is your explanation?

21. Show that the total entropy increases when work is converted into heat by friction between sliding surfaces. Describe the increase in disorder.

22. Comment on the statement "A heat engine converts disordered mechanical motion into organized mechanical motion."

23. When we put cards together in a deck or put bricks together to build a house, for example, we increase the order in the physical world. Does this violate the second law of thermodynamics? Explain.

24. The process of human birth seems to involve an increase in order. Does this process then violate the rule governing the entropy of a system? (See "Thermodynamics of Evolution" by Prigogine, Nicolis, and Babloyantz in *Physics Today*, November 1972.)

25. A rubber band feels warmer than its surroundings immediately after it is quickly stretched; it becomes noticeably cooler when it is allowed to contract rapidly; and a rubber band supporting a load contracts on being heated. Explain these observations using the fact that the molecules of rubber consist of intertwined and cross-linked long chains of atoms in roughly random orientation.

26. Explain the statement "Cosmic rays continually *decrease the entropy of the earth* on which they fall." Does this contradict the second law of thermodynamics?

27. Heat energy flows from the sun (surface temperature 6000 K) to the earth (surface temperature 300 K). Show that the entropy of the earth-sun system increases during this process.

28. Is it true that the heat energy of the universe is steadily growing less available? Is so, why?

29. Can one use terrestrial thermodynamics, which is known to apply to bounded and isolated bodies, for the whole universe? If so, is the universe bounded and from what is the universe isolated?

30. The first, second and third laws of thermodynamics may be paraphrased respectively as follows: (*1*) You can't win; (*2*) You can't even break even; (*3*) You can't get out of the game. Explain in what sense these are permissible restatements.

31. Discuss the following comment of Panofsky and Phillips: "From the standpoint of formal physics there is only one concept which is asymmetric in the time, namely entropy. But this makes it reasonable to assume that the second law of thermodynamics can be used to ascertain the sense of time independently in any frame of reference; that is, we shall take the positive direction of time to be that of statistically increasing disorder, or increasing entropy. . . ." (See, in this connection, "The Arrow of Time" by David Layzer, in *Scientific American*, December 1975.)

SECTION 25-3

1. An ideal gas heat engine operates in a Carnot cycle between 227 and 127° C. It absorbs 6.0×10^4 cal at the higher temperature. (*a*) How much work per cycle is this engine capable of performing? (*b*) What is the efficiency of the engine?   *Answer:* (*a*) 1.2×10^4 cal. (*b*) 20%.

2. In a Carnot cycle, the isothermal expansion of an ideal gas takes place at 400 K and the isothermal compression at 300 K. During the expansion 500 cal of heat energy are transferred to the gas. Determine (*a*) the work performed by the gas during the isothermal expansion, (*b*) the heat rejected from the gas during the isothermal compression, (*c*) the work done on the gas during the isothermal compression.

3. If the Carnot cycle is run backward, we have an ideal refrigerator. A quantity of heat Q_2 is taken in at the lower temperature T_2 and a quantity of heat Q_1 is given out at the higher temperature T_1. The difference is the work W that must be supplied to run the refrigerator. (*a*) Show that

$$W = Q_2 \frac{T_1 - T_2}{T_2}.$$

problems

(b) The *coefficient of performance K* of a refrigerator is defined as the ratio of the heat extracted from the cold source to the work needed to run the cycle. Show that ideally

$$K = \frac{T_2}{T_1 - T_2}.$$

In actual refrigerators K has a value of 5 or 6.

(c) In a mechanical refrigerator the low-temperature coils are at a temperature of $-13°$ C, and the compressed gas in the condenser has a temperature of $27°$ C. What is the theoretical coefficient of performance?
Answer: (c) 6.5.

4. How is the efficiency of a reversible ideal heat engine related to the coefficient of performance of the reversible refrigerator obtained by running the engine backward?

5. (a) A Carnot engine operates between a hot reservoir at 320 K and a cold reservoir at 260 K. If it absorbs 500 joules of heat at the hot reservoir, how much work does it deliver? (b) If the same engine, working in reverse, functions as a refrigerator between the same two reservoirs, how much work must be supplied to remove 1000 J of heat from the cold reservoir?
Answer: (a) 94 J. (b) 230 J.

6. (a) How much work must be done to extract 1.0 J of heat from a reservoir at $7°$ C and transfer it to one at $27°$ C by means of a refrigerator using a Carnot cycle? (b) From one at $-73°$ C to one at $27°$ C? (c) From one at $-173°$ C to one at $27°$ C? (d) From one at $-223°$ C to one at $27°$ C?

7. (a) Plot accurately a Carnot cycle on a p-V diagram for 1.0 mol of an ideal gas. Let point a correspond to $p = 1.0$ atm, $T = 300$ K, and let b correspond to $p = 0.50$ atm, $T = 300$ K; take the low temperature reservoir to be at 100 K. Let $\gamma = 1.5$. (b) Compute graphically the work done in this cycle. (c) Compute the work analytically.   *Answer:* (c) 1150 J.

8. In a two-stage Carnot heat engine a quantity of heat Q_1 is absorbed at a temperature T_1, work W_1 is done, and a quantity of heat Q_2 is expelled at a lower temperature T_2 by the first stage. The second stage absorbs the heat expelled by the first, does work W_2, and expels a quantity of heat Q_3 at a lower temperature T_3. Prove that the efficiency of the combination engine is $(T_1 - T_3)/T_1$.

SECTION 25-5

9. A combination mercury-steam turbine takes saturated mercury vapor from a boiler at $876°$ F and exhausts it to heat a steam boiler at $460°$ F. The steam turbine receives steam at this temperature and exhausts it to a condenser at $100°$ F. What is the maximum efficiency of the combination?
Answer: 58%.

10. Apparatus that liquefies helium is in a room at 300 K. If the helium in the apparatus is at 5 K, what is the minimum ratio of heat energy delivered to the room to the heat energy removed from the helium?

11. Suppose a deep shaft were drilled in the earth's crust near one of the poles where the surface temperature is $-40°$ C to a depth where the temperature is $800°$ C. (a) What is the theoretical limit to the efficiency of an engine operating between these temperatures? (b) If all of the heat released into the low temperature reservoir were used to melt ice that was initially at $-40°$ C, at what rate could water at $0°$ C be produced by a power plant having an output of 100 MW? The specific heat of ice is 0.50 cal/g · C°; its heat of fusion is 80 cal/g.   *Answer:* (a) 78%. (b) 239 kg/s.

12. The motor in a refrigerator has a power output of 200 W. If the freezing compartment is at 270 K and the outside air is at 300 K, assuming ideal efficiency, what is the maximum amount of heat that can be extracted from the freezing compartment in 10 min?

13. In a heat pump, heat Q_2 is extracted from the outside atmosphere at T_2 and a larger quantity of heat Q_1 is delivered to the inside of the house at T_1,

with the performance of work W. (a) Draw a schematic diagram of a heat pump. (b) How does it differ in principle from a refrigerator? In practical use? (c) How are Q_1, Q_2, and W related to one another? (d) Can a heat pump be reversed for use in summer? Explain. (e) What advantages does such a pump have over other heating devices?

14. In a heat pump, heat from the outdoors at $-5°$ C is transferred to a room at $17°$ C, energy being supplied by an electric motor. How many joules of heat will be delivered to the room for each joule of electric energy consumed, ideally?

15. A gasoline internal combustion engine can be approximated by the cycle shown in Figure 25-10. Assume an ideal gas and use a compression ratio of $4 : 1$ ($V_4 = 4V_1$). Assume $p_2 = 3p_1$. (a) Determine the pressure and temperature of each of the vertex points of the p-V diagram in terms of p_1, T_1 and the ratio of specific heats of the gas. (b) What is the efficiency of this cycle?
  Answer: (a) $T_2 = 3T_1$
  $T_3 = 3(4)^{1-\gamma}T_1$
  $T_4 = (4)^{1-\gamma}T_1$
  $p_2 = 3p_1$
  $p_3 = (3)(4)^{-\gamma}p_1$
  $p_4 = (4)^{-\gamma}p_1$.
  (b) $1 - (4)^{1-\gamma}$.

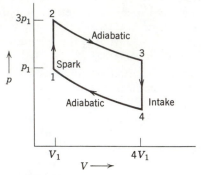

figure 25-10
Problem 15

SECTION 25-6

16. Using the equation of state of an ideal gas and the equation describing an adiabatic process for an ideal gas, show that the slope, dp/dV, on a p-V diagram of an adiabatic can be written as $-\gamma p/V$ and of an isothermal can be written as $-p/V$. From these results prove that adiabatics are steeper curves than isothermals.

17. Suppose that we were to take as our measure of temperature $-1/T$ rather than T. The unit of this new measure might be the Nivlek (Kelvin spelled backwards) degree (°N). Write a sequence of temperatures in °N extending from positive to negative values of T. (See footnote, page 464.)

SECTION 25-7

18. A mole of a monatomic ideal gas is taken from an initial state of pressure p and volume V to a final state of pressure $2p$ and volume $2V$ by two different processes. (I) It expands isothermally until its volume is doubled, and then its pressure is increased at constant volume to the final state. (II) It is compressed isothermally until its pressure is doubled, and then its volume is increased at constant pressure to the final state.

   Show the path of each process on a p-V diagram. For each process calculate in terms of p and V (a) the heat absorbed by the gas in each part of the process; (b) the work done by the gas in each part of the process; (c) the change in internal energy of the gas $U_f - U_i$; (d) the change in entropy of the gas $S_f - S_i$.

SECTION 25-8

19. Heat can be removed from water at $0°$ C and atmospheric pressure without causing the water to freeze, if done with little disturbance of the water. Suppose the water is cooled to $-5.0°$ C before ice begins to form. What is the change in entropy per unit mass occurring during the sudden freezing that then takes place?   *Answer:* -0.30 cal/g $\cdot$ K.

20. In a specific heat experiment, 200 g of aluminum ($c_p = 0.215$ cal/g $\cdot$ C°) at $100°$ C is mixed with 50 g of water at $20°$ C. Find the difference in entropy of the system at the end from its value before mixing.

21. An 8.00 g ice cube at $-10.0°$ C is dropped into a thermos flask containing 100 cm³ of water at $20.0°$ C. What is the change in entropy of the system when a final equilibrium state is reached? The specific heat of ice is 0.52 cal/g $\cdot$ C°.   *Answer:* $+0.15$ cal/K.

22. A 10-g ice cube at $-10°$ C is placed in a lake whose temperature is $+15°$ C. Calculate the change in entropy of the system as the ice cube comes to thermal equilibrium with the lake.

SECTION 25-9

23. (a) Show that when a substance of mass m having a constant specific heat c is heated from T_1 to T_2 the entropy change is

$$S_2 - S_1 = mc \ln \frac{T_2}{T_1}.$$

(b) Does the entropy of the substance decrease on cooling? (c) If so, does the total entropy of the universe decrease in such a process?
Answer: (b) Yes. (c) No.

24. Four moles of an ideal gas are caused to expand from a volume V_1 to a volume V_2 $(= 2V_1)$. (a) If the expansion is isothermal at the temperature $T = 400$ K, find the work done by the expanding gas. (b) Find the change in entropy, if any. (c) If the expansion were reversibly adiabatic instead of isothermal, would the change in entropy be positive, negative, or zero?

25. A brass rod is in contact thermally with a heat reservoir at $127°$ C at one end and a heat reservoir at $27°$ C at the other end. (a) Compute the total change in the entropy arising from the process of conduction of 1200 cal of heat through the rod. (b) Does the entropy of the rod change in the process?   *Answer:* (a) $+1.0$ cal/K. (b) No.

26. One mole of hydrogen gas and 1.0 mole of nitrogen gas are in adjacent containers at the same pressure p and temperature T. The pressure and temperature are such that both gases behave virtually ideally. (a) If the rms speed of the H_2 molecules is 1850 m/s at temperature T, what will the rms speed be of the N_2 molecules? (b) For which gas will a larger percentage or fraction of the molecules have speeds within ± 50 m/s of the rms speed? (c) If the containers are connected so that the H_2 and N_2 mix, will the change in entropy be positive, negative, or zero?

27. An object of constant heat capacity C is heated from an initial temperature T_i to a final temperature T_f, by being placed in contact with a heat reservoir at T_f. Represent the process on a graph of C/T versus T and (a) show graphically that the total change in entropy ΔS (object plus reservoir) is positive, and (b) show how the use of heat reservoirs at intermediate temperatures would allow the process to be carried out in a way that makes ΔS as small as desired.

28. (a) A body of finite mass is originally at temperature T_2, higher than that of a heat reservoir at a temperature T_1. An engine operates in infinitesimal cycles between the body and the reservoir until it lowers the temperature of the body from T_2 to T_1. Prove that the maximum work obtainable from the engine is $W_{max} = Q - T_1(S_2 - S_1)$, where $S_1 - S_2$ is the entropy change in the body and Q is the heat extracted from the body by the engine. (b) A body of finite mass is originally at temperature T_1, the same as that of a heat reservoir. A refrigerator operates in infinitesimal cycles between the body and reservoir until it lowers the temperature of the body from T_1 to T_0. Prove that the minimum amount of work which must be supplied to the refrigerator is $W_{min} = T_1(S_1 - S_0) - Q$, where $S_0 - S_1$ is the entropy change in the body and Q is the heat extracted from the body by the refrigerator.

SECTION 25-10

29. In general, the probability w_{12} of a complex event, which consists of two unrelated simple events, is equal to the product of their respective probabilities w_1, w_2. The entropy S_{12} of a complex system which consists of two simple systems is just the sum of their respective entropies, S_1, S_2. Show that Eq. 25-15, which relates probability and entropy, is consistent with the additive property of entropy and the multiplicative property of probability for a complex system.

30. (a) Compute the change in entropy of a deck of cards caused by taking the 52 cards of a particular randomly dealt bridge hand and stacking them into a pile, the cards arranged in a specific, predetermined order. (b) Compare, in order of magnitude, this entropy change with thermodynamic entropy changes.

physics

PART
TWO

26
charge and matter

The science of electricity has its roots in the observation, known to Thales of Miletus in 600 B.C., that a rubbed piece of amber will attract bits of straw. The study of magnetism goes back to the observation that naturally occurring "stones" (that is, magnetite) will attract iron. These two sciences developed quite separately until 1820, when Hans Christian Oersted (1777–1851) observed a connection between them, namely, that an *electric* current in a wire can affect a *magnetic* compass needle (Section 33-1).

The new science of electromagnetism was developed further by many workers, of whom one of the most important was Michael Faraday (1791–1867). It fell to James Clerk Maxwell (1831–1879) to put the laws of electromagnetism in essentially the form in which we know them today. These laws, called *Maxwell's equations,* are displayed in Table 40-2, which you may want to examine at this time. These laws play the same role in electromagnetism that Newton's laws of motion and of gravitation do in mechanics.

Although Maxwell's synthesis of electromagnetism rests heavily on the work of his predecessors, his own contribution was central and vital. Maxwell deduced that light is electromagnetic in nature and that its speed can be found by making purely electric and magnetic measurements. Thus the science of optics was intimately connected with those of electricity and of magnetism. The scope of Maxwell's equations is remarkable, including as it does the fundamental principles of all large-scale electromagnetic and optical devices such as motors, radio, television, microwave radar, microscopes, and telescopes.

The development of classical electromagnetism did not end with Maxwell. The English physicist Oliver Heaviside (1850–1925) and es-

26-1
ELECTROMAGNETISM— A PREVIEW

pecially the Dutch physicist H. A. Lorentz (1853–1928) contributed substantially to the clarification of Maxwell's theory. Heinrich Hertz (1857–1894)* took a great step forward when, more than twenty years after Maxwell set up his theory, he produced in the laboratory electromagnetic "Maxwellian waves" of a kind that we would now call short radio waves. It remained for Marconi and others to exploit the practical application of the electromagnetic waves of Maxwell and Hertz.

Present interest in electromagnetism takes two forms. At the level of engineering applications Maxwell's equations are used constantly and universally in the solution of a wide variety of practical problems. At the level of the foundations of the theory there is a continuing effort to extend its scope in such a way that electromagnetism is revealed as a special case of a more general theory. Such a theory would also include (say) the theories of gravitation and of quantum physics. This grand synthesis has not yet been achieved.

The rest of this chapter deals with electric charge and its relationship to matter. We can show that there are *two kinds* of charge by rubbing a glass rod with silk and hanging it from a long thread as in Fig. 26-1. If a second rod is rubbed with silk and held near the rubbed end of the first rod, the rods will repel each other. On the other hand, a rod of plastic (Lucite, say) rubbed with fur will *attract* the glass rod. Two plastic rods rubbed with fur will repel each other. We explain these facts by saying that rubbing a rod gives it an *electric charge* and that the charges on the two rods exert forces on each other. Clearly the charges on the glass and on the plastic must be different in nature.

Benjamin Franklin (1706–1790), who, among his other achievements, was the first American physicist‡, named the kind of electricity that appears on the glass *positive* and the kind that appears on the plastic (sealing wax or shell-lac in Franklin's day) *negative;* these names have remained to this day. We can sum up these experiments by saying that *like charges repel and unlike charges attract.*

Electric effects are not limited to glass rubbed with silk or to plastic rubbed with fur. Any substance rubbed with any other under suitable conditions will become charged to some extent; by comparing the unknown charge with a glass rod which had been rubbed with silk or a plastic rod which had been rubbed with fur, it can be labeled as either positive or negative.

The modern view of bulk matter is that, in its normal or neutral state, it contains equal amounts of positive and negative electricity. If two bodies like glass and silk are rubbed together, a small amount of charge is transferred from one to the other, upsetting the electric neutrality of each. In this case the glass would become positive, the silk negative.

26-2
ELECTRIC CHARGE†

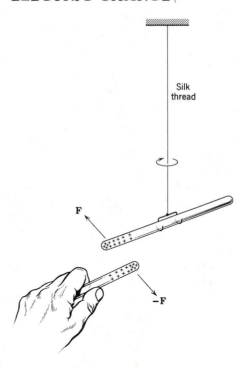

Silk thread

F

−F

figure 26-1
Two positively charged glass rods repel each other.

---

* "Heinrich Hertz," by P. and E. Morrison, *Scientific American,* December 1957.
† To learn about practical applications of static electric charges, as in fly-ash precipitators, paint sprayers, electrostatic copying machines, etc., see "Modern Electrostatics" by A. W. Bright, *Physics Education,* **9,** 381 [1974], and "Electrostatics" by A. D. Moore, *Scientific American,* March 1972.
‡ The science historian I. Bernard Cohen of Harvard University says of Franklin in his book *Franklin and Newton:* "To say . . . that had Franklin 'Not been famous as a publisher and a statesman, he might never have been heard of as a scientist,' is absolutely wrong. Just the opposite is more nearly the case; his international fame and public renown as a scientist was in no small measure responsible for his success in international statesmanship." See also "The Lightning Discharge" by Richard E. Orville, *The Physics Teacher,* January 1976 for a description of Franklin's famous kite experiment and a review of modern concepts about the nature of lightning.

A metal rod held in the hand and rubbed with fur will not seem to develop a charge. It is possible to charge such a rod, however, if it is furnished with a glass or plastic handle and if the metal is not touched with the hands while rubbing it. The explanation is that metals, the human body, and the earth are *conductors* of electricity and that glass, plastics, etc., are *insulators* (also called *dielectrics*).

In conductors electric charges are free to move through the material, whereas in insulators they are not. Although there are no perfect insulators, the insulating ability of fused quartz is about 10^{25} times as great as that of copper, so that for many practical purposes some materials behave as if they were perfect insulators.

In metals a fairly subtle experiment called the Hall effect (see Section 33-5) shows that only negative charge is free to move. Positive charge is as immobile as it is in glass or in any other dielectric. The actual charge carriers in metals are the *free electrons*. When isolated atoms are combined to form a metallic solid, the outer electrons of the atom do not remain attached to individual atoms but become free to move throughout the volume of the solid. For some conductors, such as electrolytes, both positive and negative charges can move.

A class of materials called *semiconductors* is intermediate between conductors and insulators in its ability to conduct electricity. Among the elements, silicon and germanium are well-known examples. Semiconductors have many practical applications, among which is their use in the construction of transistors. The way a semiconductor works cannot be described adequately without some understanding of the basic principles of quantum physics. Figure 26-2, however, suggests the principal features of the distinction between conductors, semiconductors, and insulators.

In solids, electrons have energies that are restricted to certain levels, the levels being confined to certain bands. The intervals between bands are forbidden, in the sense that electrons in the solid may not possess such energies. Electrons are assigned two to a level and they may not increase their energy (which means that they may not move freely through the solid) unless there are empty levels at higher energies into which they can readily move.

Figure 26-2a shows a conductor, such as copper. Band 1 is only partially filled so that electrons can easily move to the higher empty levels and thus travel through the solid. Figure 26-2b shows a (intrinsic) semiconductor such as silicon. Here band 1 is completely filled but band 2 is so close energetically that electrons can easily "jump" (absorbing energy from, say, thermal fluctuations) into the unfilled levels of that band. Figure 22-6c shows an insulator, such as sodium chloride. Here again band 1 is filled, but band 2 is too far above band 1 energetically to permit any appreciable number of the band-1 electrons to jump the energy gap.

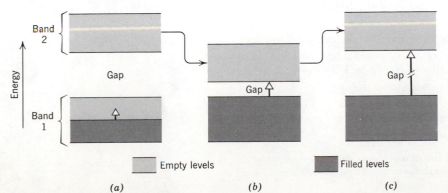

figure 26-2
Suggesting (*a*) a conductor, (*b*) an intrinsic semiconductor, and (*c*) an insulator. In (*b*) the gap is relatively small but in (*c*) it is relatively large. In intrinsic semiconductors the electrical conductivity can often be greatly increased by adding very small amounts of other elements such as arsenic or boron, a process called "doping."

Charles Augustin Coulomb (1736–1806) measured electrical attractions and repulsions quantitatively and deduced the law that governs them. His apparatus, shown in Fig. 26-3, resembles the hanging rod of Fig. 26-1, except that the charges in Fig. 26-3 are confined to small spheres a and b.

If a and b are charged, the electric force on a will tend to twist the suspension fiber. Coulomb canceled out this twisting effect by turning the suspension head through the angle θ needed to keep the two charges at the particular distance apart in which he was interested. The angle θ is then a relative measure of the electric force acting on charge a. The device of Fig. 26-3 is called a *torsion balance*; a similar arrangement was used later by Cavendish to measure gravitational attractions (Section 16-3).

Coulomb's first experimental results can be represented by

$$F \propto \frac{1}{r^2}.$$

Here F is the magnitude of the interaction force that acts on each of the two charges a and b; r is their distance apart. These forces, as Newton's third law requires, act along the line joining the charges but point in opposite directions. Note that the magnitude of the force on each charge is the same, even though the charges may be different.

The force between charges depends also on the magnitude of the charges. Specifically, it is proportional to their product. Although Coulomb did not prove this rigorously, he implied it and thus we arrive at

$$F \propto \frac{q_1 q_2}{r^2}, \tag{26-1}$$

where q_1 and q_2 are relative measures of the charges on spheres a and b. Equation 26-1, which is called *Coulomb's law*, holds only for charged objects whose sizes are much smaller than the distance between them. We often say that it holds only for *point charges*.

Coulomb's law resembles the inverse square law of gravitation which was already more than 100 years old at the time of Coulomb's experiments; q plays the role of m in that law. In gravity, however, the forces are always attractive; this corresponds to the fact that there are two kinds of electricity but (apparently) only one kind of mass.

Our belief in Coulomb's law does not rest quantitatively on Coulomb's experiments. Torsion balance measurements are difficult to make to an accuracy of better than a few percent. Such measurements could not, for example, convince us that the exponent in Eq. 26-1 is exactly 2 and not, say, 2.01. In Section 28-7 we show that Coulomb's law can also be deduced from an indirect experiment (1971) which shows that the exponent in Eq. 26-1 lies between the limits $2 \pm 3 \times 10^{-16}$.

Although we have established the physical concept of electric charge, we have not yet defined a unit in which it may be measured. It is possible to do so operationally by putting equal charges q on the spheres of a torsion balance and by measuring the magnitude F of the force that acts on each when the charges are a measured distance r apart. One could then define q to have a unit value if a unit force acts on each charge when the charges are separated by a unit distance and one can give a name to the unit of charge so defined.*

---

* This scheme is the basis for the definition of the unit of charge called the *statcoulomb*. However, in this book we do not use this unit or the systems of units of which it is a part; see Appendix L, however.

26-4
COULOMB'S LAW

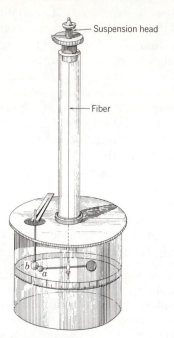

Suspension head

Fiber

b a

figure 26-3
Coulomb's torsion balance, from his 1785 memoir to the French Academy of Sciences.

For practical reasons having to do with the accuracy of measurements, the SI unit of charge is not defined using a torsion balance but is derived from the unit of electric current. If the ends of a long wire are connected to the terminals of a battery, it is common knowledge that an *electric current i* is set up in the wire. We visualize this current as a flow of charge. The SI unit of current is the *ampere* (abbr. A). In Section 34-4 we describe the operational procedures in terms of which the ampere is defined.

The SI unit of charge is the *coulomb* (abbr. C). *A coulomb is defined as the amount of charge that flows through any cross section of a wire in 1 second if there is a steady current of 1 ampere in the wire.* In symbols

$$q = it, \tag{26-2}$$

where q is in coulombs if i is in amperes and t is in seconds. Thus, if a wire is connected to an insulated metal sphere, a charge of 10^{-6} C can be put on the sphere if a current of 1.0 A exists in the wire for 10^{-6} s.

A copper penny has a mass of 3.1 g. Being electrically neutral, it contains equal amounts of positive and negative electricity. What is the magnitude q of these charges? A copper atom has a positive nuclear charge of 4.6×10^{-18} C and a negative electronic charge of equal magnitude.

The number N of copper atoms in a penny is found from the ratio

$$\frac{N}{N_0} = \frac{m}{M},$$

where N_0 is the Avogadro number, m the mass of the coin, and M the atomic weight of copper. This yields, solving for N,

$$N = \frac{(6.0 \times 10^{23} \text{ atoms/mole})(3.1 \text{ g})}{64 \text{ g/mole}} = 2.9 \times 10^{22} \text{ atoms.}$$

The charge q is

$$q = (4.6 \times 10^{-18} \text{ C/atom})(2.9 \times 10^{22} \text{ atoms}) = 1.3 \times 10^5 \text{ C.}$$

In a 100-watt, 110-volt light bulb the current is 0.91 ampere. Verify that it would take 40 h for a charge of this amount to pass through this bulb.

EXAMPLE 1

Equation 26-1 can be written as an equality by inserting a constant of proportionality. Instead of writing this simply as, say, k, it is usually written in a more complex way as $1/4\pi\epsilon_0$ or

$$F = \frac{1}{4\pi\epsilon_0} \frac{q_1 q_2}{r^2}. \tag{26-3}$$

Certain equations that are derived from Eq. 26-3, but are used more often than it is, will be simpler in form if we do this.

In SI units we can measure q_1, q_2, r, and F in Eq. 26-3 in ways that do not depend on Coulomb's law. Numbers with units can be assigned to them. There is no choice about the so-called *permittivity constant* ϵ_0; it must have that value which makes the right-hand side of Eq. 26-3 equal to the left-hand side. This (measured) value turns out to be*

$$\epsilon_0 = 8.854187818 \times 10^{-12} \text{ C}^2/\text{N} \cdot \text{m}^2.$$

---

* For practical reasons this value is not actually measured by direct application of Eq. 26-3 but in an equivalent although more circuitous way.

In this book the value 8.9×10^{-12} C²/N · m² will be accurate enough for most problems. For direct application of Coulomb's law or in any problem in which the quantity $1/4\pi\epsilon_0$ occurs we may use, with sufficient accuracy for this book,

$$1/4\pi\epsilon_0 = 9.0 \times 10^9 \text{ N·m}^2/\text{C}^2.$$

EXAMPLE 2

Let the total positive and the total negative charges in a copper penny be separated to a distance such that their force of attraction is 1.0 lb (= 4.5 N). How far apart must they be?

We have (Eq. 26-3)

$$F = \frac{1}{4\pi\epsilon_0} \frac{q_1 q_2}{r^2}.$$

Putting $q_1 q_2 = q^2$ (see Example 1) and solving for r yields

$$r = q \sqrt{\frac{1/4\pi\epsilon_0}{F}} = 1.3 \times 10^5 \text{ C} \sqrt{\frac{9.0 \times 10^9 \text{ N · m}^2/\text{C}^2}{4.5 \text{ N}}}$$

$$= 5.8 \times 10^9 \text{ m} = 3.6 \times 10^6 \text{ mi}.$$

This is 910 earth radii and it suggests that it is not possible to upset the electrical neutrality of gross objects by any very large amount. What would be the force between the two charges if they were placed 1.0 m apart?

If more than two charges are present, Eq. 26-3 holds for every pair of charges. Let the charges be q_1, q_2, q_3, etc.; we calculate the force exerted on any one (say q_1) by all the others from the vector equation

$$\mathbf{F}_1 = \mathbf{F}_{12} + \mathbf{F}_{13} + \mathbf{F}_{14} + \cdots, \qquad (26\text{-}4)$$

where $\mathbf{F}_{12}$, for example, is the force exerted on q_1 by q_2.

EXAMPLE 3

Figure 26-4 shows three fixed charges q_1, q_2, and q_3. What force acts on q_1? Assume that $q_1 = -1.0 \times 10^{-6}$ C, $q_2 = +3.0 \times 10^{-6}$ C, $q_3 = -2.0 \times 10^{-6}$ C, $r_{12} = 15$ cm, $r_{13} = 10$ cm, and $\theta = 30°$.

From Eq. 26-3, ignoring the signs of the charges, since we are interested only in the magnitudes of the forces,

$$F_{12} = \frac{1}{4\pi\epsilon_0} \frac{q_1 q_2}{r_{12}^2}$$

$$= \frac{(9.0 \times 10^9 \text{ N · m}^2/\text{C}^2)(1.0 \times 10^{-6} \text{ C})(3.0 \times 10^{-6} \text{ C})}{(1.5 \times 10^{-1} \text{ m})^2}$$

$$= 1.2 \text{ N}$$

and

$$F_{13} = \frac{(9.0 \times 10^9 \text{ N · m}^2/\text{C}^2)(1.0 \times 10^{-6} \text{ C})(2.0 \times 10^{-6} \text{ C})}{(1.0 \times 10^{-1} \text{ m})^2}$$

$$= 1.8 \text{ N}.$$

The directions of $\mathbf{F}_{12}$ and $\mathbf{F}_{13}$ are as shown in the figure.

The components of the resultant force $\mathbf{F}_1$ acting on q_1 (see Eq. 26-4) are

$$F_{1x} = F_{12x} + F_{13x} = F_{12} + F_{13} \sin\theta$$

$$= 1.2 \text{ N} + (1.8 \text{ N})(\sin 30°) = 2.1 \text{ N}$$

and

$$F_{1y} = F_{12y} + F_{13y} = 0 - F_{13} \cos\theta$$

$$= -(1.8 \text{ N})(\cos 30°) = -1.6 \text{ N}.$$

Find the magnitude of $\mathbf{F}_1$ and the angle it makes with the x-axis.

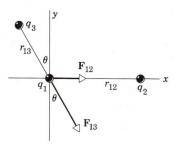

figure 26-4
Example 3. Showing the forces exerted on q_1 by q_2 and q_3.

In Franklin's day electric charge was thought of as a continuous fluid, an idea that was useful for many purposes. The atomic theory of matter, however, has shown that fluids themselves, such as water and air, are not continuous but are made up of atoms. Experiment shows that the "electric fluid" is not continuous either but that it is made up of integral multiples of a certain minimum electric charge. This fundamental charge, to which we give the symbol e, has the magnitude $1.6021892 \times 10^{-19}$ C. Any physically existing charge q, no matter what its origin, can be written as ne where n is a positive or a negative integer.

When a physical property such as charge exists in discrete "packets" rather than in continuous amounts, the property is said to be *quantized*. Quantization is basic to modern quantum physics. The existence of atoms and of particles such as the electron and the proton indicates that *mass* is quantized also. Later you will learn that several other properties prove to be quantized when suitably examined on the atomic scale; among them are energy and angular momentum.

The quantum of charge e is so small that the "graininess" of electricity does not show up in large-scale experiments, just as we do not realize that the air we breathe is made up of atoms. In an ordinary 110-volt, 100-watt light bulb, for example, 6×10^{18} elementary charges enter and leave the bulb every second.

There exists today no theory that predicts the quantization of charge (or the quantization of mass, that is, the existence of fundamental particles such as protons, electrons, muons, and so ons). Even assuming quantization, the classical theory of electromagnetism and Newtonian mechanics are incomplete in that they do not correctly describe the behavior of charge and matter on the atomic scale. The classical theory of electromagnetism, for example, describes correctly what happens when a bar magnet is thrust through a closed copper loop; it fails, however, if we wish to explain the magnetism of the bar in terms of the atoms that make it up. The more detailed theories of quantum physics are needed for this and similar problems.

Matter as we ordinarily experience it can be regarded as composed of three kinds of particles, the proton, the neutron, and the electron. Table 26-1 shows their masses and charges. Note that the masses of the neutron and the proton are approximately equal but that the electron is less massive by a factor of about 1840.

26-5
CHARGE IS QUANTIZED

26-6
CHARGE AND MATTER

Table 26-1
Some properties of three particles

Particle	Symbol	Charge	Mass
Proton	p	$+e$	$1.6726485 \times 10^{-27}$ kg
Neutron	n	0	$1.6749543 \times 10^{-27}$ kg
Electron	e^-	$-e$	9.109534×10^{-31} kg

Atoms are made up of a dense, positively charged *nucleus*, surrounded by a cloud of electrons; see Fig. 26-5. The nucleus varies in radius from about 1×10^{-15} m for hydrogen to about 7×10^{-15} m for the heaviest atoms. The outer diameter of the electron cloud, that is, the diameter of the atom, lies in the range 1–3×10^{-10} m, about 10^5 times larger than the nuclear diameter.

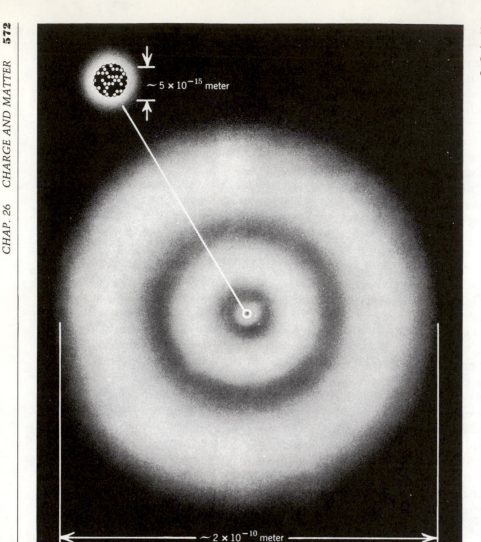

figure 26-5
An atom, suggesting the electron
cloud and, above, an enlarged view
of the nucleus.

EXAMPLE 4

The distance r between the electron and the proton in the hydrogen atom is about 5.3×10^{-11} m. What are the magnitudes of (a) the electrical force and (b) the gravitational force between these two particles?

From Coulomb's law,

$$F_e = \frac{1}{4\pi\epsilon_0} \frac{q_1 q_2}{r^2}$$

$$= \frac{(9.0 \times 10^9 \text{ N} \cdot \text{m}^2/\text{C}^2)(1.6 \times 10^{-19} \text{ C})^2}{(5.3 \times 10^{-11} \text{ m})^2}$$

$$= 8.1 \times 10^{-8} \text{ N}.$$

The gravitational force is given by Eq. 16-1, or

$$F_g = G \frac{m_1 m_2}{r^2}$$

$$= \frac{(6.7 \times 10^{-11} \text{ N} \cdot \text{m}^2/\text{kg}^2)(9.1 \times 10^{-31} \text{ kg})(1.7 \times 10^{-27} \text{ kg})}{(5.3 \times 10^{-11} \text{ m})^2}$$

$$= 3.7 \times 10^{-47} \text{ N}.$$

Thus the electrical force is about 10^{39} times stronger than the gravitational force.

The significance of Coulomb's law goes far beyond the description of the forces acting between charged spheres. This law, when incorporated into the structure of quantum physics, correctly describes (a) the electric forces that bind the electrons of an atom to its nucleus, (b) the forces that bind atoms together to form molecules, and (c) the forces that bind atoms or molecules together to form solids or liquids. Thus most of the forces of our daily experience that are not gravitational in nature are electrical. A force transmitted by a steel cable is basically an electrical force because, if we pass an imaginary plane through the cable at right angles to it, it is only the attractive electrical interatomic forces acting between atoms on opposite sides of the plane that keep the cable from parting. We ourselves are an assembly of nuclei and electrons bound together in a stable configuration by Coulomb forces.

In the atomic *nucleus* we encounter a new force which is neither gravitational nor electrical in nature. This strong attractive force, which binds together the protons and neutrons that make up the nucleus, is called simply *the nuclear force* or *the strong interaction*. If this force were not present, the nucleus would fly apart at once because of the strong Coulomb repulsion force that acts between its protons. The nature of the nuclear force is only partially understood today and forms the central problem of present day researches in nuclear physics.

EXAMPLE 5

What repulsive Coulomb force exists between two protons in a nucleus of iron? Assume a separation of 4.0×10^{-15} m.

From Coulomb's law,

$$F = \frac{1}{4\pi\epsilon_0} \frac{q_1 q_2}{r^2}$$

$$= \frac{(9.0 \times 10^9 \text{ N} \cdot \text{m}^2/\text{C}^2)(1.6 \times 10^{-19} \text{ C})^2}{(4.0 \times 10^{-15} \text{ m})^2}$$

$$= 14 \text{ N}.$$

This enormous repulsive force (3.2 lb acting on a single proton) must be more than compensated for by the strong attractive nuclear forces. This example, combined with Example 4, shows that nuclear binding forces are much stronger than atomic binding forces. Atomic binding forces are, in turn, much stronger than gravitational forces for the same particles separated by the same distance.

The repulsive Coulomb forces acting between the protons in a nucleus make the nucleus less stable than it otherwise would be. The spontaneous emission of alpha particles from heavy nuclei and the phenomenon of nuclear fission are evidences of this instability.

The fact that heavy nuclei contain significantly more neutrons than protons is still another effect of the Coulomb forces. Consider Fig. 26-6 in which a particular atomic species is represented by a circle, the coordinates being Z, the number of protons in the nucleus (that is, the *atomic number*), and N, the number of neutrons in the nucleus (that is, the *neutron number*). Stable nuclei are represented by filled circles and radioactive nuclei, that is, nuclei which disintegrate spontaneously, emitting electrons or α-particles, by open circles. Note that all elements (xenon, for example, for which $Z = 54$; see arrow) exist in a number of different forms, called *isotopes*.

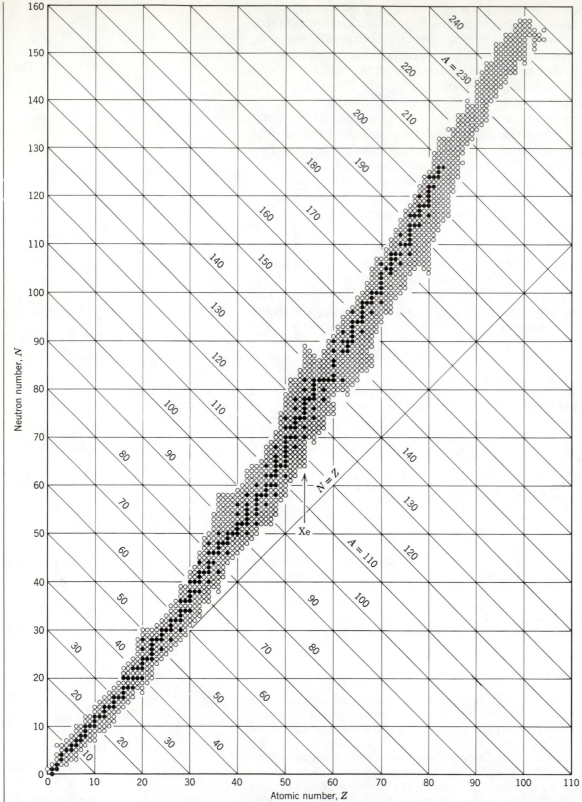

Neutron number, N

Atomic number, Z

figure 26-6
The filled circles represent stable nuclei and the open ones represent radioactive nuclei. Note, for example, that xenon (Z = 54) has 26 isotopes, 9 of them stable and 17 radioactive. Each xenon isotope has 54 protons (and 54 extranuclear electrons for neutral atoms). The number of neutrons ranges from N = 64 to N = 89 and the mass number A (= N + Z) ranges from 118 to 143. No other element has as many isotopes.

Figure 26-6 shows that light nuclei, for which the Coulomb forces are relatively unimportant,* lie on or close to the line labeled "$N = Z$" and thus have about equal numbers of neutrons and protons. The heavier nuclei have a pronounced neutron excess, U^{238} having 92 protons and $238 - 92$ or 146 neutrons.† In the absence of Coulomb forces we would assume, extending the $N = Z$ rule, that the most stable nucleus with 238 particles would have 119 protons and 119 neutrons. However, such a nucleus, if assembled, would fly apart at once because of Coulomb repulsion. Relative stability is found only if 27 of the protons are replaced by neutrons, thus diluting the total Coulomb repulsion. Even in U^{238} Coulomb repulsion is still very important because (a) this nucleus is radioactive and emits α-particles and (b) it may break up into two large fragments (fission) if it absorbs a neutron; both processes result in separation of the nuclear charge and are Coulomb repulsion effects. Figure 26-6 shows that all nuclei with $Z \gtrsim 83$ are unstable.

We have pointed out that matter, as we ordinarily experience it, is made up of electrons, neutrons, and protons. Nature exhibits much more variety than this, however. There are very many more particles than these. Appendix F, which lists some properties of some of these particles, shows that, like the more familiar particles of Table 26-1, their charges are quantized, the quantum of charge again being e. An understanding of the nature of these particles and of their relationships to each other is one of the most significant research goals of modern physics.

When a glass rod is rubbed with silk, a positive charge appears on the rod. Measurement shows that a negative charge of equal magnitude appears on the silk. This suggests that rubbing does not create charge but merely transfers it from one object to another, disturbing slightly the electrical neutrality of each. This hypothesis of the *conservation of charge* has stood up under close experimental scrutiny both for large-scale events and at the atomic and nuclear level; no exceptions have ever been found.

An interesting example of charge conservation comes about when an electron (charge $= -e$) and a positron (charge $= +e$) are brought close to each other. The two particles may simply disappear, converting all their rest mass into energy according to the well-known $E = mc^2$ relationship; this annihilation process was described in Section 8-9. The energy appears in the form of two oppositely directed gamma rays, which are similar in character to X-rays; thus:

$$e^- + e^+ \rightarrow \gamma + \gamma. \tag{26-5}$$

The net charge is zero both before and after the event so that charge is conserved. Rest mass is *not* conserved, being turned completely into energy.

Another example of charge conservation is found in radioactive decay, of which the following process is typical:

$$U^{238} \rightarrow Th^{234} + He^4. \tag{26-6}$$

The radioactive "parent" nucleus, U^{238}, contains 92 protons (that is, its

26-7
CHARGE IS CONSERVED

---

* Coulomb forces are important in relation to the strong nuclear attractive forces only for large nuclei, because Coulomb repulsion occurs between *every pair* of protons in the nucleus but the attractive nuclear force does not. In U^{238}, for example, every proton exerts a force of repulsion on each of the other 91 protons. However, each proton (and neutron) exerts a nuclear attraction on only a small number of other neutrons and protons that happen to be near it. As we proceed to larger nuclei, the amount of energy associated with the repulsive Coulomb forces increases much faster than that associated with the attractive nuclear forces.

† The superscript in this notation is the *mass number* $A (= N + Z)$. This is the total number of particles in the nucleus. See the sloping lines in Figure 26-6.

atomic number $Z = 92$). It disintegrates spontaneously by emitting an α-particle (He⁴; $Z = 2$) transmuting itself into the nucleus Th²³⁴, with $Z = 90$. Thus the amount of charge present before disintegration $(+92e)$ is the same as that present after the disintegration.

An additional example of charge conservation is found in nuclear reactions, of which the bombardment of Ca⁴⁴ with cyclotron-accelerated protons is typical. In a particular collision a neutron may emerge from the nucleus, leaving Sc⁴⁴ as a "residual" nucleus:

$$Ca^{44} + p \rightarrow Sc^{44} + n.$$

The sum of the atomic numbers before the reaction $(20 + 1)$ is exactly equal to the sum of the atomic numbers after the reaction $(21 + 0)$. Again charge is conserved.

A final example of charge conservation is the decay of the K-meson (see Appendix F) which, in one mode, goes as

$$K^0 \rightarrow \pi^+ + \pi^-.$$

The resultant charge is zero both before and after this decay process.

questions

1. You are given two metal spheres mounted on portable insulating supports. Find a way to give them equal and opposite charges. You may use a glass rod rubbed with silk but may not touch it to the spheres. Do the spheres have to be of equal size for your method to work?

2. In Question 1, find a way to give the spheres equal charges of the same sign. Again, do the spheres need to be of equal size for your method to work?

3. A charge rod attracts bits of dry cork dust which, after touching the rod, often jump violently away from it. Explain.

4. In Section 26-2 can there not be four kinds of charge, that is, on glass, silk, plastic, and fur? What is the argument against this?

5. If you rub a coin briskly between your fingers, it will not seem to become charged by friction. Why?

6. If you walk briskly down the carpeted corridor of a hotel, you often experience a spark upon touching a door knob. (a) What causes this? (b) How might it be prevented?

7. Why do electrostatic experiments not work well on humid days?

8. An insulated rod is said to carry an electric charge. How could you verify this and determine the sign of the charge?

9. If a charged glass rod is held near one end of an insulated uncharged metal rod as in Fig. 26-7, electrons are drawn to one end, as shown. Why does the flow of electrons cease? There is an almost inexhaustible supply of them in the metal rod.

10. In Fig. 26-7 does any net electrical force act on the metal rod? Explain.

11. A person standing on an insulated stool touches a charged, insulated conductor. Is the conductor discharged completely?

12. (a) A positively charged glass rod attracts a suspended object. Can we conclude that the object is negatively charged? (b) A positively charged glass rod repels a suspended object. Can we conclude that the object is positively charged?

13. Is the Coulomb force that one charge exerts on another changed if other charges are brought nearby?

14. You are given a collection of small charged spheres, the sign and magnitude of the charge and the mass of the sphere being at your disposal. Do you think

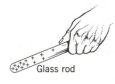

Metal

Glass rod

Insulating support

figure 26-7
Questions 9, 10

that a *stable* equilibrium position is possible, involving only electrostatic forces? Test several arrangements. A rigorous answer is not required.

15. Suppose that someone told you that in Eq. 26-3 the product of the charges (q_1q_2) should be replaced by their algebraic sum $(q_1 + q_2)$. What experimental facts refute this statement? What if the square root of their product $\sqrt{q_1q_2}$ were proposed?

16. The quantum of charge is 1.60×10^{-19} C. Is there a corresponding single quantum of mass?

17. A nucleus U^{238} splits into two identical parts. Are the nuclei so produced likely to be stable or radioactive?

18. In the decay mode

$$\Xi^0 \rightarrow \Lambda + \pi^0$$

what is the charge of the Λ particle? See Appendix F.

19. Verify the fact that the decay schemes for the elementary particles in Appendix F are consistent with charge conservation.

20. What does it mean to say that a physical quantity is (a) quantized or (b) conserved? Give some examples.

problems

SECTION 26-4

1. The electrostatic force between two like ions that are separated by a distance of 5.0×10^{-10} m is 3.7×10^9 N. (a) What is the charge on each ion? (b) How many electrons are missing from each ion?
   Answer: (a) 3.2×10^{-19} C (b) Two.

2. Two fixed charges, $+1.0 \times 10^{-6}$ C and -3.0×10^6 C, are 10 cm apart. (a) Where may a third charge be located so that no force acts on it? (b) Is the equilibrium of this third charge stable or unstable?

3. Each of two small spheres is charged positively, the combined charge being 5.0×10^{-5} C. If each sphere is repelled from the other by a force of 1.0 N when the spheres are 2.0 m apart, how is the total charge distributed between the spheres?   *Answer:* 1.2×10^{-5} C and 3.8×10^{-5} C.

4. Two equal positive point charges are held a fixed distance $2a$ apart. A point test charge is located in a plane which is normal to the line joining these charges and midway between them. (a) Calculate the radius r of the circle of symmetry in this plane for which the force on the test charge has a maximum value. (b) What is the direction of this force, assuming a positive test charge.

5. A certain charge Q is to be divided into two parts, q and $Q - q$. What is the relationship of Q to q if the two parts, placed a given distance apart, are to have a maximum Coulomb repulsion?   *Answer:* $q = \frac{1}{2}Q$.

6. How far apart must two protons be if the electrical repulsive force acting on either one is equal to its weight at the earth's surface. The mass of a proton is 1.7×10^{-27} kg.

7. Two equal positive charges, Q, are fixed at a distance $2a$ apart. The force on a small positive test charge q midway between them is zero. If the test charge is displaced a short distance either (a) toward one of the charges or (b) at right angles to the line joining the charges, find the direction of the force on it. Is the equilibrium stable or unstable in each case?
   Answer: (a) Toward the original position; stable (b) Away from the original position; unstable.

8. Two *free* point charges $+q$ and $+4q$ are a distance l apart. A third charge is so placed that the entire system is in equilibrium. Find the location, magnitude, and sign of the third charge. Is the equilibrium stable?

9. Two similar conducting balls of mass m are hung from silk threads of length l and carry similar charges q as in Fig. 26-8. Assume that θ is so small that $\tan \theta$ can be replaced by its approximate equal, $\sin \theta$. To this approximation (a) show that

$$x = \left(\frac{q^2 l}{2\pi\epsilon_0 mg}\right)^{1/3}$$

where x is the separation between the balls. (b) If $l = 120$ cm, $m = 10$ g, and $x = 5.0$ cm, what is q? *Answer: (b) $\pm 2.4 \times 10^{-8}$ C.*

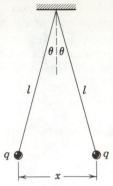

figure 26-8
Problems 9, 10, 11

10. Assume that each ball in Problem 9 is losing charge at the rate of 1.0×10^{-9} C/s. At what instantaneous relative speed $(= dx/dt)$ do the balls approach each other initially?

11. If the balls of Fig. 26-8 are conducting, (a) what happens to them after one is discharged? (b) Find the new equilibrium separation.
Answer: (a) They touch and repel. (b) 3.1 cm.

12. The charges and coordinates of two charged particles held fixed in the x-y plane are: $q_1 = +3.0 \times 10^{-6}$ C; $x = 3.5$ cm, $y = 0.50$ cm, and $q_2 = -4.0 \times 10^{-6}$ C; $x = -2.0$ cm, $y = 1.5$ cm. (a) Find the magnitude and direction of the force on q_2. (b) Where you locate a third charge $q_3 = +4.0 \times 10^{-6}$ C such that the total force on q_2 is zero?

13. Two identical conducting spheres, having charges of opposite sign, attract each other with a force of 0.108 N when separated by 0.500 m. The spheres are connected by a conducting wire, which is then removed, and thereafter repel each other with a force of 0.0360 N. What were the initial charges on the spheres?
Answer: $\pm 1.0 \times 10^{-6}$ C; $\mp 3.0 \times 10^{-6}$ C.

14. Two engineering students (John at 200 lb and Mary at 100 lb) are 100 ft apart. Let each have an 0.01% imbalance in their amount of positive and negative charge, one student being positive and the other negative. Estimate roughly the electrostatic force of attraction between them. (Hint: Replace the students by equivalent spheres of water.)

15. Two equally charged particles, 3.2×10^{-3} m apart, are released from rest. The acceleration of the first particle is observed to be 7.0 m/s² and that of the second to be 9.0 m/s². If the mass of the first particle is 6.3×10^{-7} kg, what are (a) the mass of the second particle and (b) the common charge?
Answer: (a) 4.9×10^{-7} kg. (b) 7.1×10^{-11} C.

16. (a) How many electrons would have to be removed from a penny to leave it with a charge of $+1.0 \times 10^{-7}$ C (b) What fraction of the electrons in the penny does this correspond to?

17. (a) What equal positive charges would have to be placed on the earth and on the moon to neutralize their gravitational attraction? (b) Do you need to know the lunar distance to solve this problem? (c) How many tons of hydrogen would be needed to provide the positive charge calculated in a?
Answer: (a) 5.7×10^{13} C. (b) No. (c) 630 tons.

18. Estimate *roughly* the number of coulombs of positive charge in a glass of water. Assume the volume of the water to be 250 cm³.

19. Protons in the cosmic rays strike the earth's upper atmosphere at a rate, averaged over the earth's surface, of 0.15 protons/cm²·s. What total current does the earth receive from beyond its atmosphere in the form of incident cosmic ray protons? The earth's radius is 6.4×10^6 m. *Answer: 0.12 A.*

20. Three charged particles lie on a straight line and are separated by a distance d as shown in Fig. 26-9. Charges q_1 and q_2 are held fixed. If q_3 is free to move but in fact remains stationary, then how are q_1 and q_2 related?

figure 26-9
Problem 20

21. Three small balls, each of mass 10 g, are suspended separately from a common point by silk threads, each 1.0 m. long. The balls are identically charged and hang at the corners of an equilateral triangle 0.1 m long on a side. What is the charge on each ball? *Answer: 6.0×10^{-8} C.*

22. Three point charges of $+4.0 \times 10^{-6}$ C are fixed at the corners of an equilateral triangle of side 10 cm. What force (magnitude and direction) acts on a typical charge?

23. A charge Q is fixed at each of two opposite corners of a square. A charge q is placed at each of the other two corners. (a) If the resultant electrical force on Q is zero, how are Q and q related? (b) Could q be chosen to make the resultant force on *every* charge zero?
Answer: (a) $Q = -2\sqrt{2}\, q$. (b) No.

24. In Fig. 26-10 what is the resultant force on the charge in the lower left corner of the square? Assume that $q = 1.0 \times 10^{-7}$ C and $a = 5.0$ cm. The changes are fixed in position.

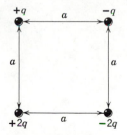

figure 26-10
Problem 24

25. A cube of edge a carries a point charge q at each corner. (a) Show that the magnitude of the resultant force on any one of the charges is

$$F = \frac{0.262q^2}{\epsilon_0 a^2}.$$

(b) What is the direction of **F** relative to the cube edges?
Answer: (b) Along a body diagonal, directed away from the cube.

26. Figure 26-11 shows a long insulating, massless rod of length l, pivoted at its center and balanced with a weight W at a distance x from the left end. At the left and right ends of the rod are attached positive charges q and $2q$, respectively. A distance h directly beneath each of these charges is a fixed positive charge Q. (a) Find the distance x for the position of the weight when the rod is balanced. (b) What value should h have so that the rod exerts no vertical force on the bearing when balanced? Neglect the interaction between charges at the opposite ends of the rod.

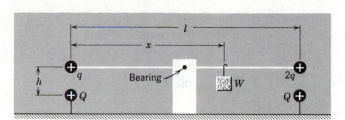

figure 26-11
Problem 26

SECTION 26-7

27. An electron is projected with an initial speed of 3.24×10^5 m/s directly toward a proton that is essentially at rest. If the electron is initially a great distance from the proton, at what distance from the proton is its speed instantaneously equal to twice its initial value? (Hint: Use the work-energy theorem.)
Answer: 1.6×10^{-9} m.

28. In the radioactive decay of U^{238} (see Eq. 26-6) the center of the emerging α-particle is, at a certain instant, 9.0×10^{-15} m from the center of the residual nucleus Th^{234}. At this instant (a) what is the force on the α-particle and (b) what is its acceleration?

27
the electric field

With every point in space near the earth we can associate a *gravitational field* vector **g** (see Eq. 16-12). This is the gravitational acceleration that a test body, placed at that point and released, would experience. If m is the mass of the body and **F** the gravitational force acting on it, **g** is given by

$$\mathbf{g} = \mathbf{F}/m. \qquad (27\text{-}1)$$

This is an example of a *vector field*. For points near the surface of the earth the field is often taken as *uniform*; that is, **g** is the same for all points.

The flow of water in a river provides another example of a vector field, called a *flow field* (see Section 18-7). Every point in the water has associated with it a vector quantity, the velocity **v** with which the water flows past the point. If **g** and **v** do not change with time, the corresponding fields are described as *stationary*. In the case of the river note that even though the water is moving, the vector **v** at any point does not change with time for steady-flow conditions.

If we place a test charge in the space near a charged rod, an electrostatic force will act on the charge. We speak of an *electric field* in this space. In a similar way we speak of a *magnetic field* in the space around a bar magnet. In the classical theory of electromagnetism the electric and magnetic fields are central concepts. In this chapter we deal with electric fields associated with charges viewed from a reference frame in which they are at rest, that is, with *electrostatics*.

Before Faraday's time, the force acting between charged particles was thought of as a direct and instantaneous interaction between the two particles. This *action-at-a-distance* view was also held for magnetic

580

and for gravitational forces. Today we prefer to think in terms of electric fields as follows:

1. Charge q_1 in Fig. 27-1 sets up an electric field in the space around itself. This field is suggested by the shading in the figure; later we shall show how to represent electric fields more concretely.
2. The field acts on charge q_2; this shows up in the force **F** that q_2 experiences.

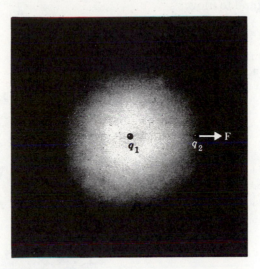

figure 27-1
Charge q_1 sets up a field that exerts a force **F** on charge q_2.

The field plays an intermediary role in the forces between charges. There are two separate problems: (a) calculating the fields that are set up by given distributions of charge and (b) calculating the forces that given fields will exert on charges placed in them. We think in terms of

$$\text{charge} \rightleftharpoons \text{field} \rightleftharpoons \text{charge}$$

and not, as in the action-at-a-distance point of view, in terms of

$$\text{charge} \rightleftharpoons \text{charge}.$$

In Fig. 27-1 we can also imagine that q_2 sets up a field and that this field acts on q_1, producing a force $-\mathbf{F}$ on it in accord with Newton's third law. The situation is completely symmetrical, each charge being immersed in a field associated with the other charge.

If the only problem in electromagnetism was that of the forces between stationary charges, the field and the action-at-a-distance points of view would be perfectly equivalent. Suppose, however, that q_1 in Fig. 27-1 suddenly accelerates to the right. How quickly does the charge q_2 learn that q_1 has moved and that the force which it (q_2) experiences must increase? Electromagnetic theory predicts that q_2 learns about q_1's motion by a *field disturbance* that emanates from q_1, traveling with the speed of light. The action-at-a-distance point of view requires that information about q_1's acceleration be communicated *instantaneously* to q_2; this is not in accord with experiment.* Accelerating electrons in the antenna of a radio transmitter influence electrons in a distant receiving antenna only after a time l/c where l is the separation of the antennas and c is the speed of light.

* By introducing other considerations it *is* possible to develop a consistent program of electromagnetism from the action-at-a-distance point of view. This is not commonly done however and we will not do so in this book.

To define the electric field operationally, we place a small test charge q_0 (assumed positive for convenience) at the point in space that is to be examined, and we measure the electrical force **F** (if any) that acts on this body. The *electric field* **E** at the point is defined as*

$$\mathbf{E} = \mathbf{F}/q_0. \qquad (27\text{-}2)$$

Here **E** is a vector because **F** is one, q_0 being a scalar. The direction of **E** is the direction of **F**, that is, it is the direction in which a resting positive charge placed at the point would tend to move.

The definition of gravitational field **g** is much like that of electric field, except that the mass of the test body rather than its charge is the property of interest. Although the units of **g** are usually written as m/s², they could also be written as N/kg (Eq. 27-1); those for **E** are N/C (Eq. 27-2). Thus both **g** and **E** are expressed as a force divided by a property (mass or charge) of the test body.

What is the magnitude of the electric field **E** such that an electron, placed in the field, would experience an electrical force equal to its weight?

EXAMPLE 1

From Eq. 27-2, replacing q_0 by e and F by mg, where m is the electron mass, we have

$$E = \frac{F}{q_0} = \frac{mg}{e}$$

$$= \frac{(9.1 \times 10^{-31}\ \text{kg})(9.8\ \text{m/s}^2)}{1.6 \times 10^{-19}\ \text{C}}$$

$$= 5.6 \times 10^{-11}\ \text{N/C}.$$

This is a very weak electric field. Which way will **E** have to point if the electric force is to cancel the gravitational force?

In applying Eq. 27-2 we must use a test charge as small as possible. A large test charge might disturb the primary charges that are responsible for the field, thus changing the very quantity that we are trying to measure. Equation 27-2 should, strictly, be replaced by

$$\mathbf{E} = \lim_{q_0 \to 0} \frac{\mathbf{F}}{q_0}. \qquad (27\text{-}3)$$

This equation instructs us to use a smaller and smaller test charge q_0, evaluating the ratio $\mathbf{F}/q_0$ at every step. The electric field **E** is then the limit of this ratio as the size of the test charge approaches zero.†

The concept of the electric field as a vector was not appreciated by Michael Faraday, who always thought in terms of *lines of force*. The lines of force still form a convenient way of visualizing electric-field patterns. We shall use them for this purpose but we shall not employ them quantitatively.

The relationship between the (imaginary) lines of force and the electric field vector is this:

---

* This definition of **E**, though conceptually sound and quite appropriate to our present purpose, is rarely carried out in practice because of experimental difficulties. **E** is normally found by calculation from more readily measurable quantities such as the electric potential; see Section 29-7.

† Of course, q_0 can never be less than the electronic charge e

1. The tangent to a line of force at any point gives the *direction* of **E** at that point.
2. The lines of force are drawn so that the number of lines per unit cross-sectional area (perpendicular to the lines) is proportional to the *magnitude* of **E**. Where the lines are close together E is large and where they are far apart E is small.

It is not obvious that it is possible to draw a continuous set of lines to meet these requirements. Indeed, it turns out that if Coulomb's law were not true, it would *not* be possible to do so; see Problem 7.

Figure 27-2 shows the lines of force for a uniform sheet of positive charge. We assume that the sheet is infinitely large, which, for a sheet of finite dimensions, is equivalent to considering only those points whose distance from the sheet is small compared to the distance to the nearest edge of the sheet. A positive test charge, released in front of such a sheet, would move away from the sheet along a perpendicular line. Thus the electric field vector at any point near the sheet must be at right angles to the sheet. The lines of force are uniformly spaced, which means that **E** has the same magnitude for all points near the sheet.

Figure 27-3 shows the lines of force for a negatively charged sphere. From symmetry, the lines must lie along radii. They point inward because a free positive charge would be accelerated in this direction. The electric field E is not constant but decreases with increasing distance from the charge. This is evident in the lines of force, which are farther apart at greater distances. From symmetry, E is the same for all points that lie a given distance from the center of the charge.

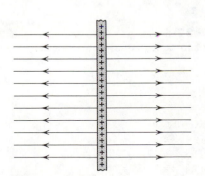

figure 27-2
Lines of force for a section of an infinitely large sheet of positive charge.

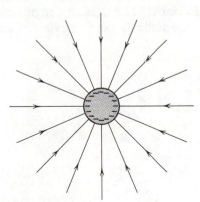

figure 27-3
Lines of force for a negatively charged sphere.

In Fig. 27-3 how does E vary with the distance r from the center of the charged sphere?

    Suppose that N lines terminate on the sphere. Draw an imaginary concentric sphere of radius r; the number of lines per unit cross-sectional area at every point on the sphere is $N/4\pi r^2$. Since E is proportional to this, we can write that

$$E \propto 1/r^2.$$

We derive an exact relationship in Section 27-4. How does E vary with distance from an infinitely long uniform cylinder of charge?

EXAMPLE 2

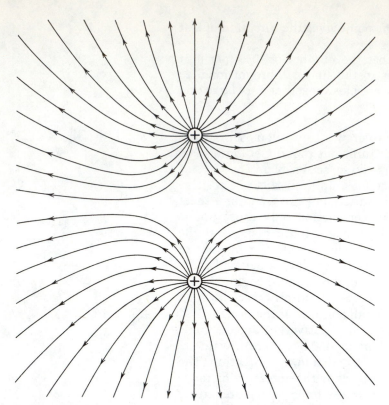

figure 27-4
Lines of force for two equal
positive charges.

Figures 27-4 and 27-5 show the lines of force for two equal like charges and for two equal unlike charges, respectively. Michael Faraday, as we have said, used lines of force a great deal in his thinking. They were more real for him than they are for most scientists and engineers today. It is possible to sympathize with Faraday's point of view. Can

figure 27-5
Lines of force for equal but opposite
charges.

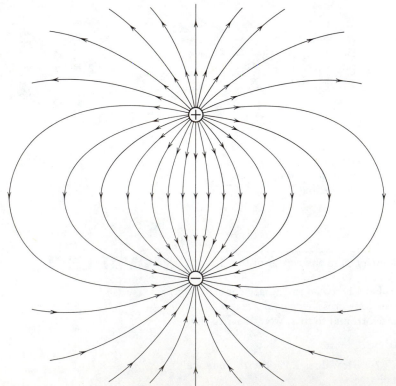

we not almost "see" the charges being pushed apart in Fig. 27-4 and pulled together in Fig 27-5 by the lines of force? Compare Fig. 27-5 with Fig. 18-14, which represents a flow field. Figure 27-6 shows a representation of lines of force around charged conductors, using grass seeds suspended in an insulating liquid.

Lines of force give a vivid picture of the way **E** varies through a given region of space. However, the equations of electromagnetism (see Table 40-2) are written in terms of the electric field **E** and other field vectors and not in terms of the lines of force. The electric field **E** varies in a perfectly continuous way as any path in the field is traversed; see Fig. 27-7.

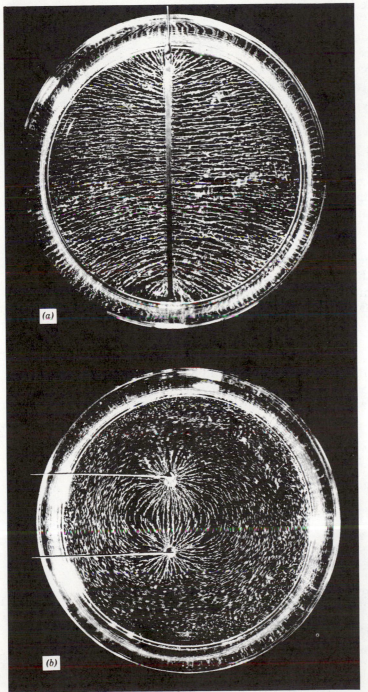

(a)

(b)

figure 27-6
Photographs of the patterns of electric lines of force around (a) a charged plate (compare Fig. 27-2), and (b) two rods with equal and opposite charges (compare Fig. 27-5). The patterns were made by suspending grass seed in an insulating liquid. (Courtesy Educational Services Incorporated, Watertown, Mass.)

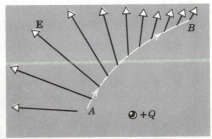

figure 27-7
E varies continuously as we move along any path *AB* in the field set up by point charge +Q. In general, the path *AB* and the field vectors **E** will not lie in the plane of the figure.

In this section we deal with the charge-field interaction by showing how we may calculate **E** for various points near given charge distributions, starting with the simple case of a point charge q.

Let a test charge q_0 be placed a distance r from a point charge q. The magnitude of the force acting on q_0 is given by Coulomb's law, or

$$F = \frac{1}{4\pi\epsilon_0} \frac{qq_0}{r^2}.$$

The electric field at the site of the test charge is given by Eq. 27-2, or

$$E = \frac{F}{q_0} = \frac{1}{4\pi\epsilon_0} \frac{q}{r^2}. \tag{27-4}$$

The direction of **E** is on a radial line from q, pointing outward if q is positive and inward if q is negative.

To find **E** for a group of point charges: (*a*) Calculate $\mathbf{E}_n$ due to each charge at the given point *as if it were the only charge present.* (*b*) Add these separately calculated fields vectorially to find the resultant field **E** at the point. In equation form,

$$\mathbf{E} = \mathbf{E}_1 + \mathbf{E}_2 + \mathbf{E}_3 + \cdots = \Sigma\mathbf{E}_n \qquad n = 1, 2, 3, \ldots. \tag{27-5}$$

The sum is a vector sum, taken over all the charges. Equation 27-5 (like Eq. 26-4) is an example of the *principle of superposition* which states, in this context, that at a given point the electric fields due to separate charge distributions simply add up (vectorially) or superimpose independently. The principle of superposition is very important in physics. It applies, for example, to gravitational and magnetic situations as well.*

If the charge distribution is a continuous one, the field it sets up at any point P can be computed by dividing the charge into infinitesimal elements dq. The field $d\mathbf{E}$ due to each element at the point in question is then calculated, treating the elements as point charges. The magnitude of $d\mathbf{E}$ (see Eq. 27-4) is given by

$$dE = \frac{1}{4\pi\epsilon_0} \frac{dq}{r^2}, \tag{27-6}$$

where r is the distance from the charge element dq to the point P. The resultant field at P is then found from the superposition principle by adding (that is, integrating) the field contributions due to all the charge elements, or,

$$\mathbf{E} = \int d\mathbf{E}. \tag{27-7}$$

The integration, like the sum in Eq. 27-5, is a vector operation; in Example 5 we will see how such an integral is handled in a simple case.

An electric dipole. Figure 27-8 shows a positive and a negative charge of equal magnitude q placed a distance $2a$ apart, a configuration called an electric dipole. The pattern of lines of force is that of Fig. 27-5, which also shows an electric dipole. What is the field **E** due to these charges at point P, a distance r along the perpendicular bisector of the line joining the charges? Assume $r \gg a$.

EXAMPLE 3

---

* Formally, the principle of superposition holds in physics only to the extent that the differential equation defining the situation is linear. To the extent that the amplitudes of mechanical or electromagnetic oscillations become relatively large the principle tends to fail. We do not discuss such cases in this book. In particular, the principle holds absolutely in electrostatics.

Equation 27-5 gives the vector equation

$$\mathbf{E} = \mathbf{E}_1 + \mathbf{E}_2,$$

where, from Eq. 27-4,*

$$E_1 = E_2 = \frac{1}{4\pi\epsilon_0} \frac{q}{a^2 + r^2}.$$

The vector sum of $\mathbf{E}_1$ and $\mathbf{E}_2$ points vertically downward and has the magnitude

$$E = 2E_1 \cos \theta.$$

From the figure we see that

$$\cos \theta = \frac{a}{\sqrt{a^2 + r^2}}.$$

Substituting the expressions for E_1 and for $\cos \theta$ into that for E yields

$$E = \frac{2}{4\pi\epsilon_0} \frac{q}{(a^2 + r^2)} \frac{a}{\sqrt{a^2 + r^2}} = \frac{1}{4\pi\epsilon_0} \frac{2aq}{(a^2 + r^2)^{3/2}}.$$

If $r \gg a$, we can neglect a in the denominator; this equation then reduces to

$$E \cong \frac{1}{4\pi\epsilon_0} \frac{(2a)(q)}{r^3}. \tag{27-8a}$$

The essential properties of the charge distribution in Fig. 27-8, the magnitude of the charge q and the separation $2a$ between the charges, enter Eq. 27-8a only as a product. This means that, if we measure E at various distances from the electric dipole (assuming $r \gg a$), we can never deduce q and $2a$ separately but only the product $2aq$; if q were doubled and a simultaneously cut in half, the electric field *at large distances from the dipole* would not change.

The product $2aq$ is called the *electric dipole moment p*. Thus we can rewrite this equation for E, *for distant points along the perpendicular bisector*, as

$$E = \frac{1}{4\pi\epsilon_0} \frac{p}{r^3}. \tag{27-8b}$$

The result for distant points *along the dipole axis* (see Problem 25) and the general result for any distant point (see Problem 28) also contain the quantities $2a$ and q only as the product $2aq$ $(= p)$. The variation of E with r in the general result for distant points is also as $1/r^3$, as in Eq. 27-8b.

The dipole of Fig. 27-8 is two equal and opposite charges placed close to each other so that their separate fields at distant points almost, but not quite, cancel. On this point of view it is easy to understand that $E(r)$ for a dipole varies at large distances as $1/r^3$ (Eq. 27-8b), whereas for a point charge $E(r)$ drops off more slowly, namely as $1/r^2$ (Eq. 27-4).

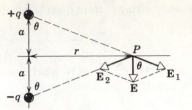

figure 27-8
Example 3.

EXAMPLE 4

Figure 27-9 shows a charge q_1 $(= +1.0 \times 10^{-6}$ C$)$ 10 cm from a charge q_2 $(= +2.0 \times 10^{-6}$ C$)$. At what point on the line joining the two charges is the electric field zero?

The point must lie between the charges because only here do the forces exerted by q_1 and q_2 on a test charge (no matter whether it is positive or negative) oppose each other. If $\mathbf{E}_1$ is the electric field due to q_1 and $\mathbf{E}_2$ that due to q_2, we must have

$$E_1 = E_2$$

or (see Eq. 27-4)

$$\frac{1}{4\pi\epsilon_0} \frac{q_1}{x^2} = \frac{1}{4\pi\epsilon_0} \frac{q_2}{(l - x)^2}.$$

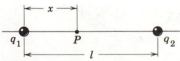

figure 27-9
Example 4.

*Note that the r's in Eq. 27-4 and in this equation have different meanings.

where x is the distance from q_1 and l equals 10 cm. Solving for x yields

$$x = \frac{l}{1 + \sqrt{q_2/q_1}} = \frac{10 \text{ cm}}{1 + \sqrt{2}} = 4.1 \text{ cm}.$$

Supply the missing steps. On what basis was the second root of the resulting quadratic equation discarded?

EXAMPLE 5

Ring of charge. Figure 27-10 shows a ring of charge q and radius a. Calculate **E** for points on the axis of the ring a distance x from its center.

Consider a differential element of the ring of length ds, located at the top of the ring in Fig. 27-10. It contains an element of charge given by

$$dq = q\,\frac{ds}{2\pi a}$$

where $2\pi a$ is the circumference of the ring. This element sets up a differential electric field $d\mathbf{E}$ at point P.

The resultant field $\mathbf{E}$ at P is found by integrating the effects of all the elements that make up the ring. From symmetry this resultant field must lie along the ring axis. Thus only the component of $d\mathbf{E}$ parallel to this axis contributes to the final result. The component perpendicular to the axis is canceled out by an equal but opposite component established by the charge element on the opposite side of the ring.

Thus the general vector integral (Eq. 27-7)

$$\mathbf{E} = \int d\mathbf{E}$$

becomes a scalar integral $E = \int dE \cos \theta.$

The quantity dE follows from Eq. 27-6, or

$$dE = \frac{1}{4\pi\epsilon_0}\frac{dq}{r^2} = \frac{1}{4\pi\epsilon_0}\left(\frac{q\,ds}{2\pi a}\right)\frac{1}{a^2 + x^2}.$$

From Fig. 27-10 we have $\cos \theta = \dfrac{x}{\sqrt{a^2 + x^2}}.$

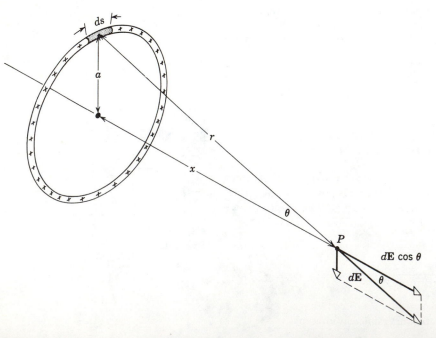

figure 27-10
Example 5

Noting that, for a given point P, x has the same value for all charge elements and is not a variable and that s is the variable of integration, we obtain

$$E = \int dE \cos \theta = \int \frac{1}{4\pi\epsilon_0} \frac{q \, ds}{(2\pi a)(a^2 + x^2)} \frac{x}{\sqrt{a^2 + x^2}}$$

$$= \frac{1}{4\pi\epsilon_0} \frac{qx}{(2\pi a)(a^2 + x^2)^{3/2}} \int ds.$$

The integral is simply the circumference of the ring $(= 2\pi a)$, so that

$$E = \frac{1}{4\pi\epsilon_0} \frac{qx}{(a^2 + x^2)^{3/2}}.$$

Does this expression for E reduce to an expected result for $x = 0$? For $x \gg a$ we can neglect a in the denominator of this equation, yielding

$$E \cong \frac{1}{4\pi\epsilon_0} \frac{q}{x^2}.$$

This is an expected result (compare Eq. 27-4) because at great enough distances the ring behaves like a point charge q.

EXAMPLE 6

Infinite line of charge. Figure 27-11 shows a section of an infinite line of charge whose linear charge density (that is, the charge per unit length, measured in C/m) has the constant value λ. Calculate the field **E** a distance y from the line.

The magnitude of the field contribution dE due to charge element $dq \, (= \lambda \, dx)$ is given, using Eq. 27-6, by

$$dE = \frac{1}{4\pi\epsilon_0} \frac{dq}{r^2} = \frac{1}{4\pi\epsilon_0} \frac{\lambda \, dx}{y^2 + x^2}.$$

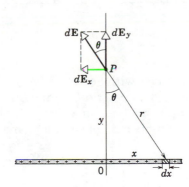

figure 27-11
Example 6. A section of an infinite line of charge.

The vector $d\mathbf{E}$, as Fig. 27-11 shows, has the components

$$dE_x = -dE \sin \theta \quad \text{and} \quad dE_y = dE \cos \theta.$$

The minus sign in front of dE_x indicates that $d\mathbf{E}_x$ points in the negative x direction. The x and y components of the resultant vector **E** at point P are given by

$$E_x = \int dE_x = -\int_{x=-\infty}^{x=+\infty} \sin \theta \, dE \quad \text{and} \quad E_y = \int dE_y = \int_{x=-\infty}^{x=+\infty} \cos \theta \, dE.$$

E_x must be zero because every charge element on the right has a corresponding element on the left such that their field contributions in the x direction cancel. Thus **E** points entirely in the y direction. Because the contributions to E_y from the right- and left-hand halves of the rod are equal, we can write

$$E = E_y = 2 \int_{x=0}^{x=+\infty} \cos \theta \, dE.$$

Note that we have changed the lower limit of integration and have introduced a compensating factor of two.

Substituting the expression for dE into this equation gives

$$E = \frac{\lambda}{2\pi\epsilon_0} \int_{x=0}^{x=\infty} \cos\theta \, \frac{dx}{y^2 + x^2}.$$

From Fig. 27-11 we see that the quantities θ and x are not independent. We must eliminate one of them, say x. The relation between x and θ is (see figure)

$$x = y \tan\theta.$$

Differentiating, we obtain $dx = y \sec^2\theta \, d\theta.$

Substituting these two expressions leads finally to

$$E = \frac{\lambda}{2\pi\epsilon_0 y} \int_{\theta=0}^{\theta=\pi/2} \cos\theta \, d\theta.$$

You should check this step carefully, noting that the limits must now be on θ and not on x. For example, as $x \to +\infty$, $\theta \to \pi/2$, as Fig. 27-11 shows. This equation integrates readily to

$$E = \frac{\lambda}{2\pi\epsilon_0 y} (\sin\theta) \Big|_0^{\pi/2} = \frac{\lambda}{2\pi\epsilon_0 y}.$$

You may wonder about the usefulness of solving a problem involving an infinite rod of charge when any actual rod must have a finite length (see Problem 23). However, for points close enough to finite rods and not near their ends, the equation that we have just derived yields results that are so close to the correct values that the difference can be ignored in many practical situations. It is usually unnecessary to solve exactly every geometry encountered in practical problems. Indeed, if idealizations or approximations are not made, the vast majority of significant problems of all kinds in physics and engineering cannot be solved at all.

27-5
A POINT CHARGE IN AN ELECTRIC FIELD

Here and in the following section, in contrast with Section 27-4, we investigate the other half of the charge-field interaction, namely, if we are given a field **E**, what forces and torques will act on a charge configuration placed in it? We start with the simple case of a point charge in a uniform electric field.

An electric field will exert a force on a charged particle given by (Eq. 27-2)

$$\mathbf{F} = \mathbf{E}q.$$

This force will produce an acceleration

$$\mathbf{a} = \mathbf{F}/m,$$

where m is the mass of the particle. We will consider two examples of the acceleration of a charged particle in a uniform electric field. Such a field can be produced by connecting the terminals of a battery to two parallel metal plates which are otherwise insulated from each other. If the spacing between the plates is small compared with the dimensions of the plates, the field between them will be fairly uniform except near the edges. Note that in calculating the motion of a particle in a field set up by external charges the field due to the particle itself (that is, its *self-field*) is ignored. In the same way, the earth's gravitational field can exert no net force on the earth itself but only on a second object, say a stone, placed in that field.

A particle of mass m and charge q is placed at rest in a uniform electric field (Fig. 27-12) and released. Describe its motion.

EXAMPLE 7

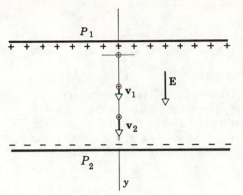

figure 27-12
A charge is released from rest in a uniform electric field set up between two oppositely charged metal plates P_1 and P_2.

The motion resembles that of a body falling in the earth's gravitational field. The (constant) acceleration is given by

$$a = \frac{F}{m} = \frac{qE}{m}.$$

The equations for uniformly accelerated motion (Table 3-1) then apply. With $v_0 = 0$, they are

$$v = at = \frac{qEt}{m},$$

$$y = \tfrac{1}{2}at^2 = \frac{qEt^2}{2m},$$

and

$$v^2 = 2ay = \frac{2qEy}{m}.$$

The kinetic energy attained after moving a distance y is found from

$$K = \tfrac{1}{2}mv^2 = \tfrac{1}{2}m\left(\frac{2qEy}{m}\right) = qEy.$$

This result also follows directly from the work-energy theorem because a constant force qE acts over a distance y.

Deflecting an electron beam. Figure 27-13 shows an electron of mass m and charge e projected with speed v_0 at right angles to a uniform field **E**. Describe its motion.

EXAMPLE 8

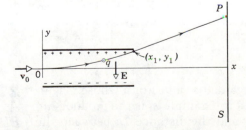

figure 27-13
Example 8. An electron is projected into a uniform electric field.

The motion is like that of a projectile fired horizontally in the earth's gravitational field. The considerations of Section 4-3 apply, the horizontal (x) and vertical (y) motions being given by

$$x = v_0 t$$

and

$$y = \tfrac{1}{2}at^2 = \frac{eE}{2m}t^2.$$

Eliminating t yields

$$y = \frac{eE}{2mv_0^2}x^2 \qquad (27\text{-}9)$$

for the equation of the trajectory.

When the electron emerges from the plates in Fig. 27-13, it travels (neglecting gravity) in a straight line tangent to the parabola of Eq. 27-9 at the exit point. We can let it fall on a fluorescent screen S placed some distance beyond the plates. Together with other electrons following the same path, it will then make itself visible as a small luminous spot; this is the principle of the electrostatic *cathode-ray oscilloscope.*

The electric field between the plates of a cathode-ray oscilloscope is 1.2×10^4 N/C. What deflection will an electron experience if it enters at right angles to the field with a kinetic energy of 2000 eV $(= 3.2 \times 10^{-16}$ J), a typical value? The deflecting assembly is 1.5 cm long.

Recalling that $K_0 = \tfrac{1}{2}mv_0^2$, we can rewrite Eq. 27-9 as

$$y = \frac{eEx^2}{4K_0}.$$

If x_1 is the horizontal position of the far edge of the plate, y_1 will be the corresponding deflection (see Fig. 27-13), or

$$y_1 = \frac{eEx_1^2}{4K_0}$$

$$= \frac{(1.6 \times 10^{-19}\ \text{C})(1.2 \times 10^4\ \text{N/C})(1.5 \times 10^{-2}\ \text{m})^2}{(4)(3.2 \times 10^{-16}\ \text{J})}$$

$$= 3.4 \times 10^{-4}\ \text{m} = 0.34\ \text{mm}.$$

The deflection measured, not at the deflecting plates but at the fluorescent screen, is much larger.

EXAMPLE 9

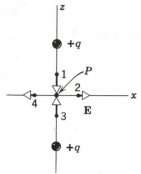

figure 27-14
Example 10. The electric field at four points near a point P which is centered between two equal positive charges.

A positive point test charge q_0 is placed halfway between two equal fixed positive charges q. What force acts on it at or near this point P?

From symmetry the force *at* the point is zero so that the particle is in equilibrium; the nature of the equilibrium remains to be found. Figure 27-14 (compare Fig. 27-4) shows the $\mathbf{E}$ vectors for four points near P. If the test charge is moved along the z axis, a *restoring* force is brought into play; however, the equilibrium is unstable for motion along the x (and y) axes. Thus we have the three-dimensional equivalent of *saddle point equilibrium*; see Fig. 14-8. What is the nature of the equilibrium for a negative test charge?

EXAMPLE 10

An electric dipole moment can be regarded as a vector $\mathbf{p}$ whose magnitude p, for a dipole like that described in Example 3, is the product $2aq$ of the magnitude of either charge q and the distance $2a$ between the charges. The *direction* of $\mathbf{p}$ for such a dipole is from the *negative* to the *positive* charge. The vector nature of the electric dipole moment permits us to cast many expressions involving electric dipoles into concise form, as we shall see.

27-6
A DIPOLE IN AN ELECTRIC FIELD

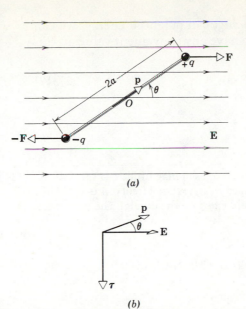

(a)

(b)

Figure 27-15a shows an electric dipole formed by placing two charges $+q$ and $-q$ a fixed distance $2a$ apart. The arrangement is placed in a uniform *external* electric field $\mathbf{E}$, its dipole moment $\mathbf{p}$ making an angle θ with this field. Two equal and opposite forces $\mathbf{F}$ and $-\mathbf{F}$ act as shown, where

$$F = qE.$$

The net force is clearly zero, but there is a net torque about an axis through O (see Eq. 12-2) given by

$$\tau = 2F(a \sin \theta) = 2aF \sin \theta.$$

Combining these two equations and recalling that $p = (2a)(q)$, we obtain

$$\tau = 2aqE \sin \theta = pE \sin \theta. \qquad (27\text{-}10)$$

Thus an electric dipole placed in an external electric field $\mathbf{E}$ experiences a torque tending to align it with the field. Equation 27-10 can be written in vector form as

$$\boldsymbol{\tau} = \mathbf{p} \times \mathbf{E}, \qquad (27\text{-}11)$$

the appropriate vectors being shown in Fig. 27-15b.

Work (positive or negative) must be done by an external agent to change the orientation of an electric dipole in an external field. This work is stored as potential energy U in the system consisting of the dipole and the arrangement used to set up the external field. If θ in Fig. 27-15a has the initial value θ_0, the work required to turn the dipole axis to an angle θ is given (see Table 12-2) from

$$W = \int dW = \int_{\theta_0}^{\theta} \tau \, d\theta = U,$$

where τ is the torque exerted by the agent that does the work. Combining this equation with Eq. 27-10 yields

$$U = \int_{\theta_0}^{\theta} pE \sin \theta \, d\theta = pE \int_{\theta_0}^{\theta} \sin \theta \, d\theta$$

$$= pE \left(-\cos \theta \right) \Big|_{\theta_0}^{\theta}.$$

Since we are interested only in *changes* in potential energy, we can choose the reference orientation θ_0 to have any convenient value, in this case 90°. This gives

$$U = -pE \cos \theta \qquad (27\text{-}12)$$

or, in vector terms,

$$U = -\mathbf{p} \cdot \mathbf{E}. \qquad (27\text{-}13)$$

An electric dipole consists of two opposite charges of magnitude $q = 1.0 \times 10^{-6}$ C separated by $d = 2.0$ cm. The dipole is placed in an external field of 1.0×10^5 N/C.

EXAMPLE 11

(a) What maximum torque does the field exert on the dipole? The maximum torque is found by putting $\theta = 90°$ in Eq. 27-10 or

$$\tau = pE \sin \theta = qdE \sin \theta$$

$$= (1.0 \times 10^{-6} \text{ C})(0.020 \text{ m})(1.0 \times 10^5 \text{ N/C})(\sin 90°)$$

$$= 2.0 \times 10^{-3} \text{ N} \cdot \text{m}.$$

(b) How much work must an external agent do to turn the dipole end for end, starting from a position of alignment $(\theta = 0)$? The work is the difference in potential energy U between the positions $\theta = 180°$ and $\theta = 0$. From Eq. 27–12,

$$W = U_{180°} - U_{0°} = (-pE \cos 180°) - (-pE \cos 0)$$

$$= 2pE = 2qdE$$

$$= (2)(1.0 \times 10^{-6} \text{ C})(0.020 \text{ m})(1.0 \times 10^5 \text{ N/C})$$

$$= 4.0 \times 10^{-3} \text{ J}.$$

---

questions

1. Name as many scalar fields and vector fields as you can.

2. (a) In the gravitational attraction between the earth and a stone, can we say that the earth lies in the gravitational field of the stone? (b) How is the gravitational field due to the stone related to that due to the earth?

3. A positively charged ball hangs from a long silk thread. We wish to measure E at a point in the same horizontal plane as that of the hanging charge. To do so, we put a positive test charge q_0 at the point and measure F/q_0. Will F/q_0 be less than, equal to, or greater than E at the point in question?

4. Taking into account the quantization of electric charge (the single electron providing the basic charge unit), how can we justify the procedure suggested by Eq. 27-3?

5. In exploring electric fields with a test charge we have often assumed, for convenience, that the test charge was positive. Does this really make any difference in determining the field? Illustrate in a simple case of your own devising.

6. Electric lines of force never cross. Why?

7. In Fig. 27-4 why do the lines of force around the edge of the figure appear, when extended backward, to radiate uniformly from the center of the figure?

8. Figure 27-2 shows that E has the same value for all points in front of an infinite uniformly charged sheet. Is this reasonable? One might think that the field should be stronger near the sheet because the charges are so much closer.

9. If a point charge q of mass m is released from rest in a nonuniform field, will it follow a line of force?

10. A point charge is moving in an electric field at right angles to the lines of force. Does any force **F** act on it?

11. In Fig. 27-7 path AB is not a line of force. How can you tell?

12. In Fig. 27-6, why should grass seeds line up with electric lines of force? Grass seeds normally carry no electric charge. See "Demonstration of the Electric Fields of Current-Carrying Conductors" by O. Jefimenko, *American Journal of Physics.* January 1962.

13. Two point charges of unknown magnitude and sign are a distance d apart. The electric field is zero at one point between them, on the line joining them. What can you conclude about the charges?

14. Compare the way E varies with r for (a) a point charge (Eq. 27-4), (b) a dipole (Eq. 27-8a), and (c) a quadrupole (Problem 33).

15. Charges $+Q$ and $-Q$ are fixed a distance L apart and a long straight line is drawn through them. What is the direction of **E** on this line for points (a) between the charges, (b) outside the charges in the direction of $+Q$, and (c) outside the charges in the direction of $-Q$?

16. Two point charges of unknown sign and magnitude are fixed a distance L apart. Can we have $\mathbf{E} = 0$ for off-axis points (excluding ∞)? Explain.

17. In what way does Eq. 27-8b fail to represent the lines of force of Fig. 27-5 if we relax the requirement that $r \gg a$?

18. If two dipoles of moments $\mathbf{p}_1$ and $\mathbf{p}_2$ are superimposed, is the dipole moment of the resulting configuration given by $\mathbf{p}_1 + \mathbf{p}_2$?

19. In Fig. 27-5 the force on the lower charge points up and is finite. The crowding of the lines of force, however, suggests that E is infinitely great at the site of this (point) charge. A charge immersed in an infinitely great field should have an infinitely great force acting on it. What is the solution to this dilemma?

20. An electric dipole is placed in a *nonuniform* electric field. Is there a net force on it?

21. An electric dipole is placed at rest in a uniform external electric field, as in Fig. 27-15a, and released. Discuss its motion.

22. An electric dipole has its dipole moment $\mathbf{p}$ aligned with a uniform external electric field **E**. (a) Is the equilibrium stable or unstable? (b) Discuss the nature of the equilibrium if $\mathbf{p}$ and **E** point in opposite directions.

problems

SECTION 27-2

1. What is the magnitude of a point charge chosen so that the electric field 50 cm away has the magnitude 2.0 N/C?   *Answer:* 5.6×10^{-11} C.

2. What is the magnitude and direction of an electric field that will balance the weight of (a) an electron and (b) an alpha particle?

3. An electric field **E** with an average magnitude of about 150 N/C points downward in the earth's atmosphere. We wish to "float" a sulfur sphere weighing 1.0 lb in this field by charging it. (a) What charge (sign and magnitude) must be used? (b) Why is the experiment not practical? Give a qualitative reason supported by a very rough numerical calculation to prove your point.
   Answer: (a) -0.030 C. (b) Sphere would blow up because of mutual Coulomb repulsion.

4. At some instant the velocity components of an electron moving between two charged parallel plates are $v_x = 1.5 \times 10^5$ m/s and $v_y = 0.30 \times 10^4$ m/s. If the electric field between the plates is given by $\mathbf{E} = \mathbf{j}\, 1.2 \times 10^4$ N/C, (a) what is the acceleration of the electron? (b) When the x-coordinate of the electron has changed by 2.0 cm what will be the velocity of the electron?

5. A particle having a charge of -2.0×10^{-9} C is acted on by a downward electric force of 3.0×10^{-6} N in a uniform electric field. (*a*) What is the strength of the electric field? (*b*) What is the magnitude and direction of the electric force exerted on a proton placed in this field? (*c*) What is the gravitational force on the proton? (*d*) What is the ratio of the electric to the gravitational forces in this case?

Answer: (*a*) 1.5×10^3 N/C. (*b*) 2.4×10^{-16} N, up. (*c*) 1.6×10^{-26} N. (*d*) 1.5×10^{10}.

6. A uniform vertical field **E** is established in the space between two large parallel plates. In this field one suspends a small conducting sphere of mass *m* from a string of length *l*. Find the period of this pendulum when the sphere is given a charge $+q$ if the lower plate (*a*) is charged positively; (*b*) is charged negatively.

SECTION 27-3

7. Assume that the exponent in Coulomb's law is not "two" but *n*. Show that for n ≠ 2 it is impossible to construct lines that will have the properties listed for lines of force in Section 27-3. For simplicity, treat an isolated point charge.

8. Sketch qualitatively the lines of force associated with a thin, circular, uniformly charged disk of radius *R*. (Hint: Consider as limiting cases points very close to the surface and points very far from it.) Show the lines only in a plane containing the axis of the disk.

9. Sketch qualitatively the lines of force between two concentric conducting spherical shells, charge $+ q_1$ being placed on the inner sphere and $-q_2$ on the outer. Consider the cases $q_1 > q_2$, $q_1 = q_2$, $q_1 < q_2$.

10. (*a*) Sketch qualitatively the lines of force associated with three long parallel lines of charge, in a perpendicular plane. Assume that the intersections of the lines of charge with such a plane form an equilateral triangle and that each line of charge has the same linear charge density λ (C/*m*). (*b*) Discuss the nature of the equilibrium of a test charge placed on the central axis of the charge assembly.

11. In Fig. 27-4 consider two neighboring lines of force leaving the upper charge at small angles with a straight line connecting the charges. If the angle between their tangents for points near the charge is θ, it becomes $\theta/\sqrt{2}$ at great distances. Verify this statement and explain it. (Hint: Consider how the lines must behave both close to either charge and far from the charges.)

SECTION 27-4

12. Three charges are arranged in an equilateral triangle as in Fig. 27-16. What is the direction of the force on $+q$?

13. Two equal and opposite charges of magnitude 2.0×10^{-7} C are 15 cm apart. (*a*) What are the magnitude and direction of **E** at a point midway between the charges? (*b*) What force (magnitude and direction) would act on an electron placed there?

Answer: (*a*) 6.4×10^5 N/C, toward the negative charge. (*b*) 1.0×10^{-13} N, toward the positive charge.

14. Two point charges of magnitude $+2.0 \times 10^{-7}$ C and $+8.5 \times 10^{-8}$ C are 12 cm apart. (*a*) What electric field does each produce at the site of the other? (*b*) What force acts on each?

15. Two point charges of unknown magnitude and sign are placed a distance *d* apart. (*a*) If it is possible to have **E** $= 0$ at any point *not* between the charges but on the line joining them, what are the necessary conditions and where is the point located? (*b*) Is it possible, for any arrangement of two point charges, to find *two* points (neither at infinity) at which **E** $= 0$; if so, under what conditions?

Answer: (*a*) Charges must be opposite in sign, the nearer charge being smaller in magnitude than the farther charge. (*b*) No.

16. (*a*) In Fig. 27-17 locate the point (or points) at which the electric field is zero. (*b*) Sketch qualitatively the lines of force. Take $a = 50$ cm.

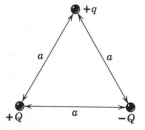

figure 27-16
Problem 12

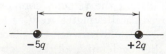

figure 27-17
Problem 16

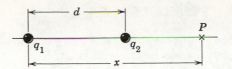

figure 27-18
Problem 17

17. Two point charges are a distance d apart (Fig. 27-18). Plot $E(x)$, assuming $x = 0$ at the left-hand charge. Consider both positive and negative values of x. Plot E as positive if E points to the right and negative if E points to the left. Assume $q_1 = +1.0 \times 10^{-6}$ C, $q_2 = +3.0 \times 10^{-6}$ C, and $d = 10$ cm.

18. What is E in magnitude and direction at the center of the square of Fig. 27-19? Assume that $q = 1.0 \times 10^{-8}$ C and $a = 5.0$ cm.

19. In Fig. 27-8 assume that both charges are positive. (a) Show that E at point P in that figure, assuming $r \gg a$, is given by

$$E = \frac{1}{4\pi\epsilon_0} \frac{2q}{r^2}.$$

(b) What is the direction of E? (c) Is it reasonable that E should vary as r^{-2} here and as r^{-3} for the dipole of Fig. 27-8?
Answer: (b) At right angles to the axis and away from it

20. Charges $+q$ and $-2q$ are fixed a distance d apart as in Fig. 27-20. (a) Find E at points A, B, and C. (b) Sketch roughly the lines of force.

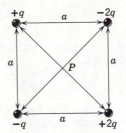

figure 27-19
Problem 18

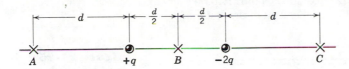

figure 27-20
Problem 20

21. Calculate E (direction and magnitude) at point P in Fig. 27-21.
Answer: $E = q/\pi\epsilon_0 a^2$, along bisector, away from triangle.

22. A thin glass rod is bent into a semicircle of radius R. A charge $+Q$ is uniformly distributed along the upper half and a charge $-Q$ is uniformly distributed along the lower half, as shown in Fig. 27-22. Find the electric field E at P, the center of the semicircle.

23. A thin nonconducting rod of finite length l carries a total charge q, spread uniformly along it. Show that E at point P on the perpendicular bisector in Fig. 27-23 is given by

$$E = \frac{q}{2\pi\epsilon_0 y} \frac{1}{\sqrt{l^2 + 4y^2}}.$$

Show that as $l \to \infty$ this result approaches that of Example 6.

24. An electron is constrained to move along the axis of the ring of charge in Example 5. Show that the electron can perform oscillations whose frequency is given by

$$\omega = \sqrt{\frac{eq}{4\pi\epsilon_0 m a^3}}.$$

This formula holds only for small oscillations, that is, for $x \ll a$ in Fig. 27-10. (Hint: Show that the motion is simple harmonic and use Eq. 15-11.)

25. *Axial field due to an electric dipole.* In Fig. 27-8, consider a point a distance r from the center of the dipole along its axis. (a) Show that, at large values of r, the electric field is

$$E = \frac{1}{2\pi\epsilon_0} \frac{p}{r^3},$$

which is twice the value given for the conditions of Example 3. (b) What is the direction of E?
Answer: (b) Parallel to p.

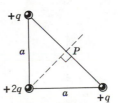

figure 27-21
Problem 21

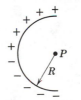

figure 27-22
Problem 22

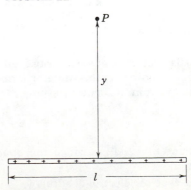

figure 27-23
Problem 23

26. For the ring of charge in Example 5, show that the maximum value of E occurs at $x = a/\sqrt{2}$.

27. Consider the ring of charge of Example 5. Suppose that the charge q is not distributed uniformly over the ring but that charge q_1 is distributed uniformly over half the circumference and charge q_2 is distributed uniformly over the other half. Let $q_1 + q_2 = q$. (a) Find the component of the electric field at any point on the axis directed *along* the axis and compare with the uniform case of Example 5. (b) Find the component of the electric field at any point on the axis *perpendicular* to the axis and compare with the uniform case of Example 5.

Answer: (a) $E = \dfrac{1}{4\pi\epsilon_0} \dfrac{qx}{(a^2 + x^2)^{3/2}}$; (b) $E = \dfrac{1}{2\pi^2\epsilon_0} \dfrac{(q_1 - q_2)a}{(a^2 + x^2)^{3/2}}$.

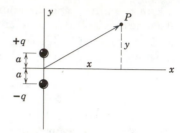

28. *Field due to an electric dipole.* Show that the components of **E** due to a dipole are given, at distant points, by

$$E_x = \frac{1}{4\pi\epsilon_0} \frac{3pxy}{(x^2 + y^2)^{5/2}}$$

$$E_y = \frac{1}{4\pi\epsilon_0} \frac{p(2y^2 - x^2)}{(x^2 + y^2)^{5/2}},$$

where x and y are coordinates of a point in Fig. 27-24. Show that this general result includes the special results of Eq. 27-8b and of Problem 25.

figure 27-24
Problem 28.

29. A "semi-infinite" insulating rod (Fig. 27-25) carries a constant charge per unit length of λ. Show that the electric field at the point P makes an angle of 45° with the rod and that this result is independent of the distance R.

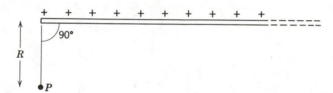

figure 27-25
Problem 29.

30. A nonconducting hemispherical cup of inner radius a has a total charge q spread uniformly over its inner surface. Find the electric field at the center of curvature.

31. A thin nonconducting rod is bent to form the arc of a circle of radius a and subtends an angle θ_0 at the center of the circle. A total charge q is spread uniformly along its length. Find the electric field at the center of the circle in terms of a, q, and θ_0.

Answer: $E = \dfrac{q}{2\pi\epsilon_0\theta_0 a^2} \sin(\theta_0/2)$.

32. A thin circular disk of radius a is charged uniformly so as to have a charge per unit area of σ. Find the electric field on the axis of the disk at a distance r from the disk.

33. *Electric quadrupole.* Figure 27-26 shows a typical electric quadrupole. It consists of two dipoles whose effects at external points do not quite cancel. Show that the value of E on the axis of the quadrupole for points distant r from its center (assume $r \gg a$) is given by

$$E = \frac{3Q}{4\pi\epsilon_0 r^4}$$

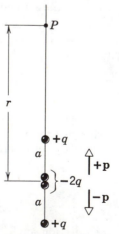

figure 27-26
Problem 33.

where Q $(= 2qa^2)$ is called the *quadrupole moment* of the charge distribution.

34. One type of "electric quadrupole" is formed by four charges located at the vertices of a square of side $2a$. Point P lies a distance R from the center of the quadrupole on a line parallel to two sides of the square as shown in

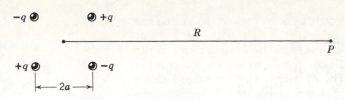

figure 27-27
Problem 34.

Fig. 27-27. For $R \gg a$, show that the electric field at P is approximately given by

$$E = \frac{3(2qa^2)}{2\pi\epsilon_0 R^4}.$$

(Hint: Treat the quadrupole as two dipoles.)

SECTION 27-5

35. A uniform electric field exists in a region between two oppositely charged plates. An electron is released from rest at the surface of the negatively charged plate and strikes the surface of the opposite plate, 2.0 cm away, in a time 1.5×10^{-8} s. (a) What is the speed of the electron as it strikes the second plate? (b) What is the magnitude of the electric field $\mathbf{E}$?
    Answer: (a) 2.7×10^6 m/s. (b) 1.0×10^3 N/C.

36. An electron moving with a speed of 5.0×10^8 cm/s is shot parallel to an electric field of strength 1.0×10^3 N/C arranged so as to retard its motion. (a) How far will the electron travel in the field before coming (momentarily) to rest and (b) how much time will elapse? (c) If the electric field ends abruptly after 0.8 cm, what fraction of its initial kinetic energy will the electron lose in traversing it?

37. (a) What is the acceleration of an electron in a uniform electric field of 1.0×10^6 N/C? (b) How long would it take for the electron, starting from rest, to attain one-tenth the speed of light? Assume that Newtonian mechanics holds.
    Answer: (a) 1.8×10^{17} m/s². (b) 1.7×10^{-10} s.

38. An electron is projected as in Fig. 27-28 at a speed of 6.0×10^6 m/s and at an angle θ of 45°; $E = 2.0 \times 10^3$ N/C (directed upward), $d = 2.0$ cm, and $l = 10.0$ cm. (a) Will the electron strike either of the plates? (b) If it strikes a plate, where does it do so?

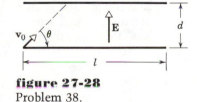

figure 27-28
Problem 38.

39. *Millikan's oil drop experiment.* R. A. Millikan set up an apparatus (Fig. 27-29) in which a tiny, charged oil drop, placed in an electric field $\mathbf{E}$, could be

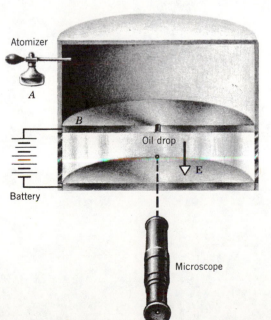

figure 27-29, Problem 39
Millikan's oil drop apparatus.
Charged oil drops from the
atomizer fall through the hole in
the central plate.

"balanced" by adjusting E until the electric force on the drop was equal and opposite to its weight. If the radius of the drop is 1.64×10^{-4} cm and E at balance is 1.92×10^{5} N/C, (a) what charge is on the drop in terms of the electronic charge e? (b) Why did Millikan not try to balance electrons in his apparatus instead of oil drops? The density of the oil is 0.851 g/cm³. (Millikan first measured the electronic charge in this way. He measured the drop radius by observing the limiting speed that the drops attained when they fell in air with the electric field turned off. He charged the oil drops by irradiating them with bursts of X-rays.) See *The Electron* by Robert Millikan, 2d ed., University of Chicago Press, 1924.

Answer: (a) 5.0 e. (b) Cannot see electrons; also E at balance would be inconveniently small.

40. In a particular early run (1911), Millikan observed that the following measured charges, among others, appeared at different times on a single drop:

$$6.563 \times 10^{-19} \text{ C} \qquad 13.13 \times 10^{-19} \text{ C} \qquad 19.71 \times 10^{-19} \text{ C}$$
$$8.204 \times 10^{-19} \text{ C} \qquad 16.48 \times 10^{-19} \text{ C} \qquad 22.89 \times 10^{-19} \text{ C}$$
$$11.50 \ \times 10^{-19} \text{ C} \qquad 18.08 \times 10^{-19} \text{ C} \qquad 26.13 \times 10^{-19} \text{ C}$$

What value for the elementary charge e can be deduced from these data?

SECTION 27-6

41. *Dipole in a nonuniform field.* (a) Derive an expression for dE/dz at a point midway between two equal positive charges, where z is the distance from one of the charges, measured along the line joining them. (b) Would there be a force on a small dipole placed at this point, its axis being aligned with the z axis? Recall that $\mathbf{E} = 0$ at this point.

Answer: (a) $dE/dz = -8q/\pi\epsilon_0 d^3$, where d is the distance between the charges. (b) Yes.

42. Find the frequency of oscillation of an electric dipole, of moment p and rotational inertia I, for small amplitudes of oscillation about its equilibrium position in a uniform electric field E.

43. A charge $q = 3.0 \times 10^{-6}$ C is 30 cm from a small dipole along its perpendicular bisector. The magnitude of the force on the charge is 5.0×10^{-6} N. Show on a diagram (a) the direction of the force on the charge, (b) the direction of the force on the dipole, and (c) determine the magnitude of the force on the dipole.

Answer: (a) Opposite to **p**. (b, c) 5.0×10^{-6} N, parallel to **p**.

28
gauss's law

In the preceding chapter we saw how we could use Coulomb's law to calculate **E** at various points if we knew enough about the distribution of charges that set up the field. This method always works. It is straightforward but, except in the simplest cases, laborious. Given a versatile enough computor, however, we can always find the answer to any such problem, no matter how complicated.

We can express Coulomb's law in another form, called *Gauss's law*. If we use this formulation, the calculations are not laborious. However, the number of problems that we can solve by the Gauss law formulation is small. Those that we can solve we solve with grace and elegance but, by and large, the Gauss law formulation is more useful for the insights that it gives rather than for practical problem solving.

Before we discuss Gauss's law we must develop a new concept, that of the flux of a vector field.

Flux (symbol Φ) is a property of all vector fields. We are concerned in this chapter with the flux Φ_E of the electric field **E**. By way of introduction however, let us discuss semiquantitatively the more familiar flux of fluid flow Φ_v (see Chapter 18). The word *flux* is derived from the Latin word *fluere* (to flow).

Figure 28-1 shows a stationary, uniform field of fluid flow (water, say) characterized by a constant flow vector **v**, the constant velocity of the fluid at any given point. Figure 28-1*a* suggests, in cross section, a hypothetical plane surface, a circle of radius R and area $\mathbf{A}_a$, immersed in the flow field at right angles to **v**. The mass flux $\Phi_{v,a}$ (kg/s) through

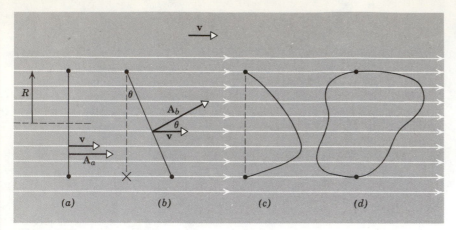

(a)   (b)   (c)   (d)

figure 28-1

figure 28-1
Showing four hypothetical surfaces immersed in a stationary, uniform flow field of an incompressible fluid (water, say) characterized by a constant field vector **v**, the velocity of the fluid at any given point. The horizontal lines are stream lines. R, in all four cases, is the radius of a circle at right angles to the stream lines.

this surface is given by

$$\Phi_{v,a} = \rho \, v \, A_a \qquad (28\text{-}1a)$$

in which ρ is the fluid density (kg/m³). Check that the dimensions are correct. We may also write this equation in vector notation as

$$\Phi_{v,a} = \rho \mathbf{v} \cdot \mathbf{A}_a. \qquad (28\text{-}1b)$$

Note that flux is a scalar.

Figure 28-1b suggests a plane surface whose projected area $(A_b \cos \theta)$ is equal to A_a. It seems clear that the mass flux Φ_v (kg/s) through surface b must be the same as that through surface a. To gain some insight we can write

$$\Phi_{v,b} = \Phi_{v,a} = \rho v A_a = \rho v (A_b \cos \theta)$$

$$= \rho \mathbf{v} \cdot \mathbf{A}_b. \qquad (28\text{-}2)$$

Figure 28-1c suggests a curved hypothetical surface whose projected area is said, without proof, to equal A_a. Once more it seems clear that $\Phi_{v,c} = \Phi_{v,a}$.

Figure 28-1d suggests a *closed* surface, the preceding three having been open. We assert that the flux $\Phi_{v,d}$ for this closed surface in this flow field is zero and justify it by noting that the amount of fluid (kg/s) that enters the left portion of the surface per unit time also leaves through the right portion. In this case the fluid (assumed incompressible) neither builds up nor disappears within the surface. In the language of Chapter 18 we say that there happen to be no sources or sinks of fluid within the surface. Every stream line that enters on the left leaves on the right.

After these preliminaries we now turn our attention from Φ_v to Φ_E, the flux of the electric field. It may seem that in the latter case nothing is flowing. In a formal sense, however, Eqs. 28-1b and 28-2 do not concern themselves with flow either but deal with the (constant in this case) field vector **v**. If, in Fig. 28-1 we replace **v** by **E** and regard the stream lines as lines of force, all the discussion of this section remains true.

Finally, in what follows we deal only with closed surfaces immersed in the **E** field. This is because we are concerned here with Gauss's law, which is expressed only in terms of closed surfaces.

In the flow of incompressible fluids it is not true in general that, as in the special case of Fig. 28-1d, $\Phi_v = 0$ for *all* closed surfaces. There may be sources or sinks of fluid within the surface, as suggested in Fig. 18-14. In such cases $\Phi_v \neq 0$.

In the same way it is not true that $\Phi_E = 0$ for every closed surface. There *are* sources (positive charges; here $\Phi_E > 0$) and sinks (negative charges; here $\Phi_E < 0$) of **E** that may be located within the hypothetical closed surface immersed in the **E** field.

28-3
FLUX OF THE ELECTRIC FIELD

For closed surfaces in an electric field we shall see below that Φ_E is positive if the lines of force point outward everywhere and negative if they point inward. Figure 28-2 shows two equal and opposite charges and their lines of force. Curves S_1, S_2, S_3, and S_4 are the intersections with the plane of the figure of four hypothetical closed surfaces. From the statement just given, Φ_E is positive for surface S_1 and negative for S_2. Φ_E for surface S_3 (compare Fig. 28-1d) is zero. We shall discuss Φ_E for surface S_4 in Section 28-4. The flux of the electric field is important because Gauss's law, one of the four basic equations of electromagnetism (see Table 40-2), is expressed in terms of it.

To define Φ_E precisely, consider Fig. 28-3, which shows an arbitrary closed surface immersed in a nonuniform electric field. Let the surface be divided into elementary squares ΔS, each of which is small enough so that it may be considered to be plane. Such an element of area can be represented as a vector $\Delta \mathbf{S}$, whose magnitude is the area ΔS; the direction of $\Delta \mathbf{S}$ is taken as the *outward-drawn normal* to the surface.

At every square in Fig. 28-3 we can also construct an electric field vector **E**. Since the squares have been taken to be arbitrarily small, **E** may be taken as constant for all points in a given square.

The vectors **E** and $\Delta \mathbf{S}$ that characterize each square make an angle θ with each other. Figure 28-3b shows an enlarged view of the three

figure 28-2
Two equal and opposite charges. The dashed lines represent the intersection with the plane of the figure of hypothetical closed surfaces.

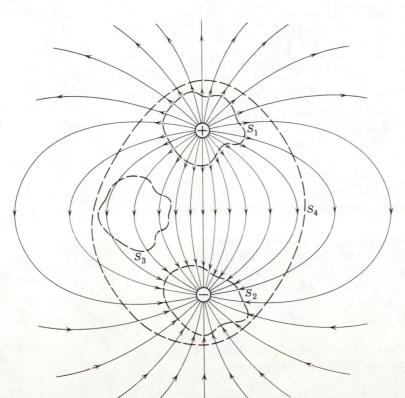

figure 28-3
(*a*) A hypothetical surface immersed in a nonuniform electric field.
(*b*) Three elements of area on this surface, shown enlarged.

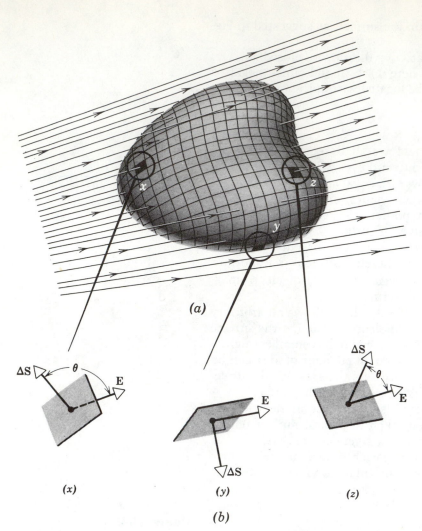

(*a*)

(*x*)   (*y*)   (*z*)

(*b*)

squares on the surface of Fig. 28-3*a* marked *x*, *y*, and *z*. Note that at *x*, $\theta > 90°$ (**E** points in); at *y*, $\theta = 90°$ (**E** is parallel to the surface); and at *z*, $\theta < 90°$ (**E** points out).

A semiquantitative definition of flux is

$$\Phi_E \cong \Sigma \mathbf{E} \cdot \Delta \mathbf{S}, \qquad (28\text{-}3)$$

which instructs us to add up the scalar quantity $\mathbf{E} \cdot \Delta \mathbf{S}$ for all elements of area into which the surface has been divided. For points such as *x* in Fig. 28-3 the contribution to the flux is negative; at *y* it is zero and at *z* it is positive. Thus if **E** is everywhere outward, $\theta < 90°$, $\mathbf{E} \cdot \Delta \mathbf{S}$ will be positive, and Φ_E for the entire surface will be positive; see Fig. 28-2, surface S_1. If **E** is everywhere inward, $\theta > 90°$, $\mathbf{E} \cdot \Delta \mathbf{S}$ will be negative, and Φ_E for the surface will be negative; see Fig. 28-2, surface S_2. From Eq. 28-3 we see that the appropriate SI unit for Φ_E is the newton meter²/coulomb (N · m²/C).

The exact definition of electric flux is found in the differential limit of Eq. 28-3. Replacing the sum over the surface by an integral over the surface yields

$$\Phi_E = \oint \mathbf{E} \cdot d\mathbf{S}. \qquad (28\text{-}4)$$

This *surface integral* indicates that the surface in question is to be divided into infinitesimal elements of area $d\mathbf{S}$ and that the scalar quantity $\mathbf{E} \cdot d\mathbf{S}$ is to be evaluated for each element and the sum taken for the entire surface. The circle on the integral sign indicates that the surface of integration is a closed surface.*

Figure 28-4 shows a hypothetical closed cylinder of radius R immersed in a uniform electric field $\mathbf{E}$, the cylinder axis being parallel to the field. What is Φ_E for this closed surface?

EXAMPLE 1

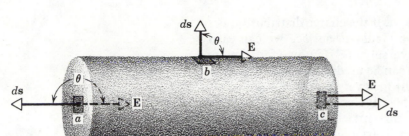

figure 28-4
Example 1. A closed cylindrical surface immersed in a uniform field $\mathbf{E}$ parallel to its axis.

The flux Φ_E can be written as the sum of three terms, an integral over (a) the left cylinder cap, (b) the cylindrical surface, and (c) the right cap. Thus

$$\Phi_E = \oint \mathbf{E} \cdot d\mathbf{S}$$
$$= \int_{(a)} \mathbf{E} \cdot d\mathbf{S} + \int_{(b)} \mathbf{E} \cdot d\mathbf{S} + \int_{(c)} \mathbf{E} \cdot d\mathbf{S}.$$

For the left cap, the angle θ for all points is 180°, $\mathbf{E}$ has a constant value, and the vectors $d\mathbf{S}$ are all parallel. Thus

$$\int_{(a)} \mathbf{E} \cdot d\mathbf{S} = \int E \cos 180° \, dS$$
$$= -E \int dS = -ES,$$

where $S \, (= \pi R^2)$ is the cap area. Similarly, for the right cap,

$$\int_{(c)} \mathbf{E} \cdot d\mathbf{S} = +ES,$$

the angle θ for all points being zero here. Finally, for the cylinder wall,

$$\int_{(b)} \mathbf{E} \cdot d\mathbf{S} = 0,$$

because $\theta = 90°$, hence $\mathbf{E} \cdot d\mathbf{S} = 0$ for all points on the cylindrical surface. Thus

$$\Phi_E = -ES + 0 + ES = 0.$$

As we shall see in Section 28-4 we expect this because there are no sources or sinks of $\mathbf{E}$, that is, charges, within the closed surface of Fig. 28-4. Lines of (constant) $\mathbf{E}$ enter at the left and emerge at the right, just as in Fig. 28-1d.

---

* Similarly, a circle on a *line* integral sign indicates a closed *path*. It will be clear from the context and from the differential element $(d\mathbf{S}$ in this case) whether we are dealing with a surface integral or a line integral.

Gauss's law, which applies to any closed hypothetical surface (called a *Gaussian surface*), gives a connection between Φ_E for the surface and the net charge q enclosed by the surface. It is

$$\epsilon_0 \Phi_E = q \qquad (28\text{-}5)$$

or, using Eq. 28-4,

$$\epsilon_0 \oint \mathbf{E} \cdot d\mathbf{S} = q. \qquad (28\text{-}6)$$

The fact that Φ_E proves to be zero in Example 1 is predicted by Gauss's law because no charge is enclosed by the Gaussian surface in Fig. 28-4 ($q = 0$).

Note that q in Eq. 28-5 (or in Eq. 28-6) is the *net* charge, taking its algebraic sign into account. If a surface encloses equal and opposite charges, the flux Φ_E is zero. Charge *outside* the surface makes no contribution to the value of q, nor does the exact location of the inside charges affect this value.

Gauss's law can be used to evaluate $\mathbf{E}$ if the charge distribution is so symmetric that by proper choice of a Gaussian surface we can easily evaluate the integral in Eq. 28-6. Conversely, if $\mathbf{E}$ is known for all points on a given closed surface, Gauss's law can be used to compute the charge inside. If $\mathbf{E}$ has an outward component for every point on a closed surface, Φ_E, as Eq. 28-4 shows, will be positive and, from Eq. 28-6, there must be a net positive charge within the surface (see Fig. 28-2, surface S_1). If $\mathbf{E}$ has an inward component for every point on a closed surface, there must be a net negative charge within the surface (see Fig. 28-2, surface S_2). Surface S_3 in Fig. 28-2 encloses no charge, so that Gauss's law predicts that $\Phi_E = 0$. This is consistent with the fact that lines of $\mathbf{E}$ pass directly through surface S_3, the contribution to the integral on one side canceling that on the other. For surface S_4 in Fig. 28-2, $\Phi_E = 0$ because the algebraic sum of the charges within the surface is zero. Put another way, as for surface S_3, as many lines of force leave the surface as enter it.

28-4
GAUSS'S LAW†

28-5
GAUSS'S LAW AND COULOMB'S LAW

Coulomb's law can be deduced from Gauss's law and symmetry considerations. To do so, let us apply Gauss's law to an isolated point charge q as in Fig. 28-5. Although Gauss's law holds for any surface whatever, information can most readily be extracted for a spherical surface of radius r centered on the charge. The advantage of this surface is that, from symmetry, $\mathbf{E}$ must be normal to it and must have the same (as yet unknown) magnitude for all points on the surface.

In Fig. 28-5 both $\mathbf{E}$ and $d\mathbf{S}$ at any point on the Gaussian surface are directed radially outward. The angle between them is zero and the quantity $\mathbf{E} \cdot d\mathbf{S}$ becomes simply $E \, dS$. Gauss's law (Eq. 28-6) thus reduces to

$$\epsilon_0 \oint \mathbf{E} \cdot d\mathbf{S} = \epsilon_0 \oint E \, dS = q.$$

Because E is constant for all points on the sphere, it can be factored from inside the integral sign, leaving

$$\epsilon_0 E \oint dS = q,$$

where the integral is simply the area of the sphere.* This equation gives

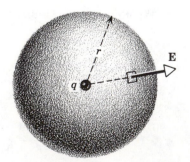

figure 28-5
A spherical Gaussian surface of radius r surrounding a point charge q.

---

* The usefulness of Gauss's law depends on our ability to find a surface over which, from symmetry, both E and θ (see Fig. 28-3) have constant values. Then $E \cos \theta$ can be factored out of the integral and E can be found simply, as in this example.

† See "Gauss," Ian Stewart, *Scientific American*, July 1977 for a fascinating account of the life of this remarkable man.

$$\epsilon_0 E(4\pi r^2) = q$$

or
$$E = \frac{1}{4\pi\epsilon_0}\frac{q}{r^2}. \qquad (28\text{-}7)$$

Equation 28-7 gives the magnitude of the electric field **E** at any point a distance r from an isolated point charge q; compare Eq. 27-4. The direction of **E** is already known from symmetry.

Let us put a second point charge q_0 at the point at which **E** is calculated. The magnitude of the force that acts on it (see Eq. 27-2) is

$$F = Eq_0.$$

Combining with Eq. 28-7 gives

$$F = \frac{1}{4\pi\epsilon_0}\frac{qq_0}{r^2},$$

which is precisely Coulomb's law. Thus we have deduced Coulomb's law from Gauss's law and considerations of symmetry.

Gauss's law is one of the fundamental equations of electromagnetic theory and is displayed in Table 40-2 as one of Maxwell's equations. Coulomb's law is not listed in that table because, as we have just proved, it can be deduced from Gauss's law and from simple assumptions about the symmetry of **E** due to a point charge.

It is interesting to note that writing the proportionality constant in Coulomb's law as $1/4\pi\epsilon_0$ (see Eq. 26-3) permits a particularly simple form for Gauss's law (Eq. 28-5). If we had written the Coulomb law constant simply as k, Gauss's law would have to be written as $(1/4\pi k)\Phi_E = q$. We prefer to leave the factor 4π in Coulomb's law so that it will not appear in Gauss's law or in other much used relations that will be derived later.

28-6 AN INSULATED CONDUCTOR

Gauss's law can be used to make an important prediction, namely: *An excess charge, placed on an insulated conductor, resides entirely on its outer surface.* This hypothesis was shown to be true by experiment (see Section 28-7) before either Gauss's law or Coulomb's law was advanced. Indeed, the experimental proof of the hypothesis is the experimental foundation upon which both laws rest: We have already pointed out that Coulomb's torsion balance experiments, although direct and convincing, are not capable of great accuracy. In showing that the italicized hypothesis is predicted by Gauss's law, we are simply reversing the historical situation.

Figure 28-6 is a cross section of an insulated conductor of arbitrary shape carrying an excess charge q. The dashed lines show a Gaussian surface that lies a small distance below the actual surface of the conductor. Although the Gaussian surface can be as close to the actual surface as we wish, it is important to keep in mind that the Gaussian surface is *inside* the conductor.

When an excess charge is placed at random on an insulated conductor, it will set up electric fields inside the conductor. These fields act on the charge carriers of the conductor (electrons) and cause them to move, that is, they set up internal currents. These currents redistribute the excess charge in such a way that the internal electric fields are automatically reduced in magnitude. Eventually the electric fields inside the conductor become zero everywhere, the currents automatically stop, and electrostatic conditions prevail. This redistribution of charge normally takes place in a time that is negligible for most purposes. What

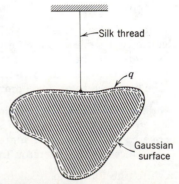

Silk thread

q

Gaussian surface

figure 28-6
An insulated metallic conductor.

can be said about the distribution of the excess charge when such electrostatic conditions have been achieved?

If, at electrostatic equilibrium, **E** is zero everywhere inside the conductor, it must be zero for every point on the Gaussian surface because this surface lies inside the conductor. This means that the flux Φ_E for this surface must be zero. Gauss's law then predicts (see Eq. 28-5) that there must be no net charge inside the Gaussian surface. If the excess charge q is not *inside* this surface, it can only be *outside* it, that is, *it must be on the actual surface of the conductor.*

28-7
EXPERIMENTAL PROOF OF GAUSS'S AND COULOMB'S LAWS

Let us turn to the experiments that prove that the hypothesis of Section 28-6 is true. For a simple test, charge a metal ball and lower it with a silk thread deep into a metal can as in Fig. 28-7. Touch the ball to the inside of the can; when the ball is removed from the can, all its charge will have vanished. When the metal ball touches the can, the ball and can together form an "insulated conductor" to which the hypothesis of Section 28-6 applies. That the charge moves entirely to the outside surface of the can can be shown by touching a small insulated metal object to the can; only on the *outside* of the can will it be possible to pick up a charge.

Benjamin Franklin seems to have been the first to notice that there can be no charge inside an insulated metal can. In 1755 he wrote to a friend:

> I electrified a silver pint cann, on an electric stand, and then lowered into it a cork-ball, of about an inch diameter, hanging by a silk string, till the cork touched the bottom of the cann. The cork was not attracted to the inside of the cann as it would have been to the outside, and though it touched the bottom, yet when drawn out, it was not found to be electrified by that touch, as it would have been by touching the outside. The fact is singular. You require the reason; I do not know it. . . .

About ten years later Franklin recommended this "singular fact" to the attention of his friend Joseph Priestley (1733–1804). In 1767 (about twenty years before Coulomb's experiments) Priestley checked Franklin's observation and, with remarkable insight, realized that the inverse square law of force followed from it. Thus the indirect approach is not only more accurate than the direct approach of Section 26-4 but was carried out earlier.

Priestley, reasoning by analogy with gravitation, said that the fact that no electric force acted on Franklin's cork ball when it was surrounded by a deep metal can is similar to the fact (see Section 16-6) that no gravitational force acts on a mass inside a spherical shell of matter; if gravitation obeys an inverse square law, perhaps the electrical force does also. In Priestley's words:

> May we not infer from this [that is, Franklin's experiment] that the attraction of electricity is subject to the same laws with that of gravitation and is therefore according to the squares of the distances; since it is easily demonstrated that were the earth in the form of a shell, a body in the inside of it would not be attracted to one side more than another?

Michael Faraday also carried out experiments designed to show that excess charge resides on the outside surface of a conductor. In particular, he built a large metal-covered box which he mounted on insulating supports and charged with a powerful electrostatic generator. In Faraday's words:

> I went into the cube and lived in it, and using lighted candles, electrometers, and all other tests of electrical states, I could not find the least influence upon them . . . though all the time the outside of the cube was very powerfully charged, and large sparks and brushes were darting off from every part of its outer surface.

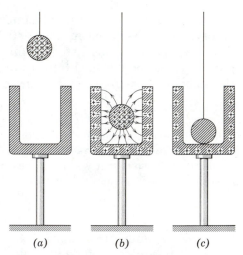

figure 28-7
The *entire* charge on the ball is transferred to the *outside* of the can. This statement and the discussion of the first paragraph of Section 28-7 are strictly correct only if the can is provided with a conducting lid which can be closed after the ball is inserted; otherwise we cannot define "outside surface."

table 28-1
Test of Coulomb's inverse square law[a]

Experimenters	Date	n
Benjamin Franklin[c]	1755	—
Joseph Priestley[c]	1767	". . . according to the squares of the distance . . ."
John Robison[b]	1769	⩽ 0.06
Henry Cavendish[b,c]	1773	⩽ 0.02
Charles A. Coulomb	1785	a few percent at most
James Clerk Maxwell[c]	1873	⩽ 5×10^{-5}
Samuel J. Plimpton and Willard E. Lawton[c,d]	1936	⩽ 2×10^{-9}
Edwin R. Williams, James E. Faller, and Henry A. Hill[c,e]	1971	⩽ 2×10^{-16}

[a] Values of n (see Eq. 28-8) are subject to a probable error, not shown. All results are consistent with $n = 0$.
[b] Robison's and Cavendish's results were not made public until after Coulomb had published his results.
[c] These are "Gauss's law" experiments, in the spirit of Fig. 28-7. The others are direct tests of Coulomb's law.
[d] Work done at Worcester Polytechnic Institute.
[e] Work done at Wesleyan University.

For many reasons it is important to know whether or not the exponent in Coulomb's law is exactly "2"; experiments based on Gauss's law can help to determine this. Let us write Coulomb's law as

$$F = \frac{1}{4\pi\epsilon_0} \frac{q_1 q_2}{r^{2+n}} \tag{28-8}$$

in which $n = 0$ yields an exact inverse square law.* Table 28-1 shows the progress made in determining how close n in Eq. 28-8 approaches zero.

Figure 28-8 is an idealized sketch of the apparatus of Plimpton and

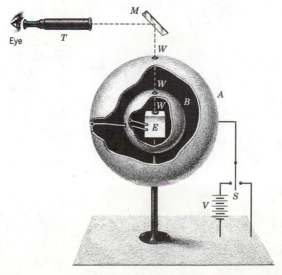

figure 28-8
A representation of the apparatus of Plimpton and Lawton.

* If $n \neq 0$ Maxwell's equations (Table 40-2), including Gauss's law, must be modified and the photon, the particle aspect of light, must have a nonzero mass, contrary to the usual expectation. See "The Mass of the Photon" by Alfred Scharff Goldhaber and Michael Martin Nieto, *Scientific American*, January 1976 for a readable account, including magnetic as well as electric field considerations.

Lawton; see Table 28-1. It consists in principle of two concentric metal shells, A and B, the former being 5 ft in diameter. The inner shell contains a sensitive electrometer E connected so that it will indicate whether any charge moves between shells A and B. If the shells are connected electrically, any charge placed on the shell assembly should reside entirely on shell A (see Section 28-6) if Gauss's law—and thus Coulomb's law—are correct as stated.

By throwing switch S to the left, a substantial charge can be placed on the sphere assembly. If any of this charge moves to shell B, it will have to pass through the electrometer and will cause a deflection, which can be observed optically using telescope T, mirror M, and windows W.

However, when the switch S is thrown alternately from left to right, thus connecting the shell assembly either to the battery or to the ground, no effect is observed on the galvanometer. Knowing the sensitivity of their electrometer, Plimpton and Lawton calculated that n in Eq. 28-8 has the value shown in Table 28-1.

28-8
GAUSS'S LAW—SOME APPLICATIONS

Gauss's law can be used to calculate $\mathbf{E}$ if the symmetry of the charge distribution is high. One example of this, the calculation of $\mathbf{E}$ for a point charge, has already been discussed (Eq. 28-7). Here we present other examples.

EXAMPLE 2

Spherically symmetric charge distribution. Figure 28-9 shows a spherical distribution of charge of radius R. The *charge density ρ* (that is, the charge per unit volume, measured in C/m³) at any point depends only on the distance of the point from the center and not on the direction, a condition called *spherical symmetry*. Find an expression for E for points (a) outside and (b) inside the charge distribution. Note that the object in Fig. 28-9 cannot be a conductor or, as we have seen, the excess charge would reside on its surface.

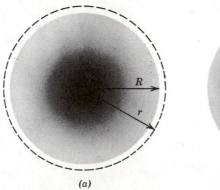
(a)

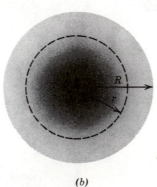
(b)

figure 28-9
Example 2. A spherically symmetric charge distribution, showing two Gaussian surfaces. The density of charge, as the shading suggests, varies with distance from the center but not with direction.

Applying Gauss's law to a spherical Gaussian surface of radius r in Fig. 28-9a (see Section 28-5) leads exactly to Eq. 28-7, or

$$E = \frac{1}{4\pi\epsilon_0} \frac{q}{r^2}, \qquad (28-7)$$

where q is the total charge. Thus for points outside a spherically symmetric distribution of charge, the electric field has the value that it would have if the charge were concentrated at its center. This reminds us that a spherically symmetric distribution of mass m behaves gravitationally, for outside points, as if the mass were concentrated at its center. At the root of this similarity lies the fact that both Coulomb's law and the law of gravitation are inverse square laws. The gravitational case was proved in detail in Section 16-6; the proof using Gauss's law in the electrostatic case is certainly much simpler.

Figure 28-9b shows a spherical Gaussian surface of radius r drawn *inside* the charge distribution. Gauss's law (Eq. 28-6) gives

$$\epsilon_0 \oint \mathbf{E} \cdot d\mathbf{S} = \epsilon_0 E(4\pi r^2) = q'$$

or

$$E = \frac{1}{4\pi\epsilon_0} \frac{q'}{r^2},$$

in which q' is that part of q contained within the sphere of radius r. The part of q that lies outside this sphere makes no contribution to $\mathbf{E}$ at radius r. This corresponds, in the gravitational case (Section 16-6), to the fact that a spherical shell of matter exerts no gravitational force on a body inside it.

A special case of a spherically symmetric charge distribution is a uniform sphere of charge. For such a sphere, which would be suggested by uniform shading in Fig. 28-9, the charge density ρ would have a constant value for all points within a sphere of radius R and would be zero for all points outside this sphere. For points inside such a uniform sphere of charge we can put

$$q' = q \frac{\frac{4}{3}\pi r^3}{\frac{4}{3}\pi R^3}$$

or

$$q' = q \left(\frac{r}{R}\right)^3$$

where $\frac{4}{3}\pi R^3$ is the volume of the spherical charge distribution. The expression for E then becomes

$$E = \frac{1}{4\pi\epsilon_0} \frac{qr}{R^3}. \tag{28-9}$$

This equation becomes zero, as it should, for $r = 0$. Note that Eqs. 28-7 and 28-9 give the same result, as they must, for points on the surface of the charge distribution (that is, if $r = R$). Note that Eq. 28-9 does not apply to the charge distribution of Fig. 28-9b because the charge density, suggested by the shading, is *not* constant in that case.

EXAMPLE 3

The Thomson atom model. At one time the positive charge in the atom was thought to be distributed uniformly throughout a sphere with a radius of about 1.0×10^{-10} m, that is, throughout the entire atom. Calculate the electric field at the surface of a gold atom ($Z = 79$) on this (erroneous) assumption. Neglect the effect of the electrons.

The positive charge of the atom is Ze or $(79)(1.6 \times 10^{-19}$ C$)$. Equation 28-7 yields, for E at the surface,

$$E = \frac{1}{4\pi\epsilon_0} \frac{q}{r^2}$$

$$= \frac{(9.0 \times 10^9 \text{ N·m}^2/\text{C}^2)(79)(1.6 \times 10^{-19} \text{ C})}{(1.0 \times 10^{-10} \text{ m})^2}$$

$$= 1.1 \times 10^{13} \text{ N/C}.$$

Figure 28-10 is a plot of E as a function of distance from the center of the atom,

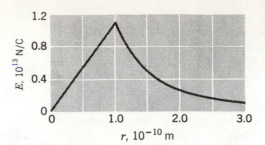

using Eqs. 28-7 and 28-9. We see that E has its maximum value on the surface and decreases linearly to zero at the center (see Eq. 28-9). Outside the sphere E decreases as the inverse square of the distance (see Eq. 28-7).

EXAMPLE 4

The Rutherford, or nuclear, atom model. We shall see in Section 28-9 that the positive charge of the atom is *not* spread uniformly throughout the atom (see Example 3) but is concentrated in a small region (the *nucleus*) at the center of the atom. For gold the radius of the nucleus is about 6.9×10^{-15} m. What is the electric field at the nuclear surface? Again neglect effects associated with the atomic electrons.

The problem is the same as that of Example 3, except that the radius is much smaller. This will make the electric field at the surface larger, in proportion to the ratio of the squares of the radii. Thus

$$E = (1.1 \times 10^{13} \text{ N/C}) \frac{(1.0 \times 10^{-10} \text{ m})^2}{(6.9 \times 10^{-15} \text{ m})^2}$$

$$= 2.3 \times 10^{21} \text{ N/C}.$$

This is an enormous electric field, much stronger than could be produced and maintained in the laboratory. It is about 10^8 times as large as the field calculated in Example 3.

EXAMPLE 5

An infinite line of charge.* Figure 28-11 shows a section of an infinite line of charge, the *linear charge density* λ (that is, the charge per unit length, measured in C/m) being constant for all points on the line. Find an expression for E at a distance r from the line.

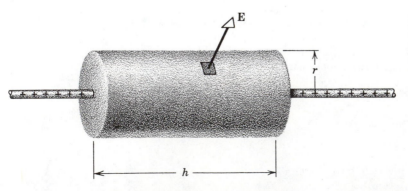

figure 28-11
Example 5. An infinite line of charge, showing a closed, coaxial cylindrical Gaussian surface.

---

* There are in practice, of course, no such things as infinite lines or sheets of charge. It is a convenient abstraction however, as, for example, was the ideal gas concept developed in Chapter 23. In all cases discussed in this book the assumption of infinite lines (or sheets) of charge allows us to ignore end (or edge) effects.

From symmetry, **E** due to a uniform linear charge can only be radially directed. As a Gaussian surface we choose a circular cylinder of radius r and length h, closed at each end by plane caps normal to the axis. E is constant over the cylindrical surface and the flux of **E** through this surface is $E(2\pi rh)$ where $2\pi rh$ is the area of the surface. There is no flux through the circular caps because **E** here lies in the surface at every point.

The charge enclosed by the Gaussian surface of Fig. 28-11 is λh. Gauss's law (Eq. 28-6),

$$\epsilon_0 \oint \mathbf{E} \cdot d\mathbf{S} = q,$$

then becomes

$$\epsilon_0 E(2\pi rh) = \lambda h,$$

whence

$$E = \frac{\lambda}{2\pi \epsilon_0 r}. \qquad (28\text{-}10)$$

The direction of **E** is radially outward for a line of positive charge.

Note how much simpler is the solution using Gauss's law than that using integration methods, as in Example 6, Chapter 27. Note too that the solution using Gauss's law is possible only if we choose our Gaussian surface to take full advantage of the cylindrical symmetry of the electric field set up by a long line of charge. We are free to choose any surface, such as a cube or a sphere, for a Gaussian surface. Even though Gauss's law holds for all such surfaces, they are not all useful for the problem at hand; only the cylindrical surface of Fig. 28-11 is appropriate in this case.

Gauss's law has the property that it provides a useful technique for calculation only in problems that have a certain degree of symmetry, but in these problems the solutions are strikingly simple.

An infinite sheet of charge. Figure 28-12 shows a portion of a thin, *nonconducting*, infinite sheet of charge, the *surface charge density* σ (that is, the charge per unit area, measured in C/m²) being constant. What is **E** at a distance r in front of the sheet?

EXAMPLE 6

figure 28-12
Example 6. An infinite nonconducting sheet of charge pierced by a cylindrical Gaussian surface.

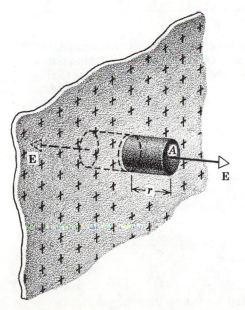

A convenient Gaussian surface is a closed cylinder of cross-sectional area A and height $2r$, arranged to pierce the plane as shown. From symmetry, **E** points at right angles to the end caps and away from the plane. Since **E** does not pierce the cylindrical surface, there is no contribution to the flux from this source. Thus Gauss's law,

$$\epsilon_0 \oint \mathbf{E} \cdot d\mathbf{S} = q$$

becomes

$$\epsilon_0(EA + EA) = \sigma A$$

where σA is the enclosed charge. This gives

$$E = \frac{\sigma}{2\epsilon_0}. \qquad (28\text{-}11)$$

Note that E is the same for all points on each side of the plane; compare Fig. 27-2. Although an infinite sheet of charge cannot exist physically, this derivation is still useful in that Eq. 28-11 yields substantially correct results for real (not infinite) charge sheets if we consider only points not near the edges whose distance from the sheet is small compared to the dimensions of the sheet.

EXAMPLE 7

A charged conductor. Figure 28-13 shows a *conductor* carrying on its surface a charge whose surface charge density at any point is σ; in general σ will vary from point to point. What is **E** for points a short distance above the surface?

The direction of **E** for points close to the surface is at right angles to the surface, pointing away from the surface if the charge is positive. If **E** were *not* normal to the surface, it would have a component lying in the surface. Such a component would act on the charge carriers in the conductor and set up surface currents. Since there are no such currents under the assumed electrostatic conditions, **E** must be normal to the surface.

The magnitude of **E** can be found from Gauss's law using a small flat closed cylinder of cross section A as a Gaussian surface. Since **E** equals zero everywhere inside the conductor (see Section 28-6), the only contribution to Φ_E is through the plane cap of area A that lies outside the conductor. Gauss's law

$$\epsilon_0 \oint \mathbf{E} \cdot d\mathbf{S} = q$$

becomes

$$\epsilon_0(EA) = \sigma A$$

where σA is the net charge within the Gaussian surface. This yields

$$E = \frac{\sigma}{\epsilon_0}. \qquad (28\text{-}12)$$

Comparison with Eq. 28-11 shows that the electric field is twice as great near a *conductor* carrying a charge whose surface charge density is σ as that near a *nonconducting sheet* with the same surface charge density. Compare the Gaussian surfaces in Figs. 28-12 and 28-13 carefully. In Fig. 28-12 lines of force leave the surface through *each* end cap, an electric field existing on *both* sides of the sheet. In Fig. 28-13 the lines of force leave only through the *outside* end cap, the inner end cap being inside the conductor where no electrical field exists. If we assume the same surface charge density and cross-sectional area A for the two Gaussian surfaces, the enclosed charge $(= \sigma A)$ will be the same. Since, from Gauss's law, the flux Φ_E must then be the same in each case, it follows that $E = (\Phi_E/A)$ must be twice as large in Fig. 28-13 as in Fig. 28-12. It is helpful to note that in Fig. 28-12 half the flux emerges from one side of the surface and half from the other, whereas in Fig. 28-13 all the flux emerges from the outside surface.

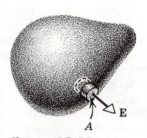

figure 28-13
Example 7. A charged insulated conductor, showing a Gaussian surface.

Ernest Rutherford (1871–1937) was first led, in 1911, to assume that the atomic nucleus existed when he tried to interpret some experiments carried out at the University of Manchester by his collaborators H. Geiger and E. Marsden.* The results of Examples 3 and 4 played an important part in Rutherford's analysis of these experiments.

28-9
THE NUCLEAR MODEL OF THE ATOM

*See "The Birth of the Nuclear Atom," E. N. da C. Andrade, *Scientific American,* November 1956. See also Example 6, Chapter 10.

These workers, at Rutherford's suggestion, allowed a beam of α-particles* to strike and be deflected by a thin film of a heavy element such as gold. They counted the number of particles deflected through various angles φ. Figure 28-14 shows the experimental setup schematically. Figure 28-15 shows the paths taken by typical α-particles as they scatter from atoms in a gold foil; the angles φ through which the α-particles are deflected range from 0 to 180° as the character of the collision varies from "grazing" to "head-on."

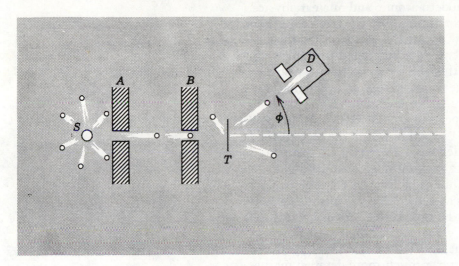

figure 28-14
Experimental arrangement for studying the scattering of α-particles. Particles from radioactive source *S* are allowed to fall on a thin metal "target" *T*; α-particles scattered by the target through an (adjustable) angle φ are counted by detector *D*.

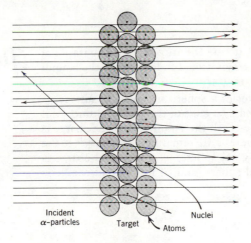

Incident α-particles   Target   Atoms   Nuclei

figure 28-15
The deflection of the incident α-particles depends on the nature of the nuclear collision. (From "The Birth of the Nuclear Atom," E. N. Andrade, *Scientific American*, November 1956.)

The electrons in the gold atom, being so light, have almost no effect on the motion of an oncoming α-particle; the electrons are themselves strongly deflected, just as a swarm of insects would be by a stone hurled through them. Any deflection of the α-particle must be caused by the repulsive action of the positive charge of the gold atom, which was known to possess most of the mass of the atom.

At the time of these experiments most physicists believed in the so-called "plum pudding" model of the atom that had been suggested by J. J. Thomson (1856–1940). In this view (see Example 3) the positive

---

* α-particles are helium nuclei that are emitted spontaneously by some radioactive materials such as radium. They move with speeds of the order of one-thirtieth that of light when so emitted.

charge of the atom was thought to be spread out through the whole atom, that is, through a spherical volume of radius about 10^{-10} m. The electrons were thought to vibrate about fixed centers inside this sphere.

Rutherford showed that this model of the atom was not consistent with the α-scattering experiments and proposed instead the nuclear model of the atom that we now accept. Here the positive charge is confined to a very much smaller sphere whose radius is about 10^{-14} m (the *nucleus*). The electrons move around this nucleus and occupy a roughly spherical volume of radius about 10^{-10} m. This brilliant deduction by Rutherford laid the foundation for modern atomic and nuclear physics.

The feature of the α-scattering experiments that attracted Rutherford's attention at once was that a few α-particles are deflected through very large angles, up to 180°. To scientists accustomed to thinking in terms of the "plum pudding" model, this was a very surprising result. In Rutherford's words: "It was quite the most incredible event that ever happened to me in my life. It was almost as incredible as if you had fired a 15-inch shell at a piece of tissue paper and it came back and hit you."

The α-particle must pass through a region in which the electric field is very high indeed in order to be deflected so strongly.* Example 3 shows that, in Thomson's model, the maximum electric field is 1.1×10^{13} N/C. Compare this with the value calculated in Example 4 for a point on the surface of a gold nucleus (2.3×10^{21} N/C). Thus the deflecting force acting on an α-particle can be up to 10^8 times as great if the positive charge of the atom is compressed into a small enough region (the nucleus) at the center of the atoms. Rutherford made his hypothesis about the existence of nuclei only after a much more detailed mathematical analysis than that given here.

questions

1. By analogy with Φ_E, how would you define the flux Φ_g of a gravitational field? What is the flux of the earth's gravitational field through the boundaries of a room, assumed to contain no matter?

2. A point charge is placed at the center of a spherical Gaussian surface. Is Φ_E changed (a) if the surface is replaced by a cube of the same volume, (b) if the sphere is replaced by a cube of one-tenth the volume, (c) if the charge is moved off-center in the original sphere, still remaining inside, (d) if the charge is moved just outside the original sphere, (e) if a second charge is placed near, and outside, the original sphere, and (f) if a second charge is placed inside the Gaussian surface?

3. In Gauss's law,

$$\epsilon_0 \oint \mathbf{E} \cdot d\mathbf{S} = q,$$

is $\mathbf{E}$ the electric field attributable to the charge q?

4. Show that Eq. 18-3 illustrates what might be called *Gauss's law for incompressible fluids*, or

$$\Phi_v = \oint \mathbf{v} \cdot d\mathbf{S} = 0.$$

5. A surface encloses an electric dipole. What can you say about Φ_E for this surface?

6. Suppose that a Gaussian surface encloses no net charge. Does Gauss's law require that $\mathbf{E}$ equal zero for all points on the surface? Is the converse of this statement true, that is, if $\mathbf{E}$ equals zero everywhere on the surface, does Gauss's law require that there be no net charge inside?

* The chance that a big deflection can result from the combined effects of many small deflections can be shown to be very small.

7. Is Gauss's law useful in calculating the field due to three equal charges located at the corners of an equilateral triangle? Explain.

8. The use of line, surface, and volume densities of charge to calculate the charge contained in an element of a charged object implies a continuous distribution of charge, whereas, in fact, charge on the microscopic scale is discontinuous. How, then, is this procedure justified?

9. Is **E** necessarily zero inside a charged rubber balloon if the balloon is (a) spherical or (b) sausage-shaped? For each shape assume the charge to be distributed uniformly over the surface.

10. A spherical rubber balloon carries a charge that is uniformly distributed over its surface. How does E vary for points (a) inside the balloon, (b) at the surface of the balloon, and (c) outside the balloon, as the balloon is blown up?

11. We have seen that there are sources (positive charges) and sinks (negative charges) for the **E** field. Are there sources and/or sinks for (a) the **v** field in fluid flow and (b) for the gravitational **g** field?

12. Is it precisely true that Gauss's law states that the total number of lines of force crossing any closed surface in the outward direction is proportional to the net positive charge enclosed within the surface?

13. In Section 28-5 we have seen that Coulomb's law can be derived from Gauss's law. Does this *necessarily* mean that Gauss's law can be derived from Coulomb's law?

14. A large, insulated, hollow conductor carries a charge $+q$. A small metal ball carrying a charge $-q$ is lowered by a thread through a small opening in the top of the conductor, allowed to touch the inner surface, and then withdrawn. What is then the charge on (a) the conductor and (b) the ball?

15. Can we deduce from the argument of Section 28-6 that the electrons in the wires of a house wiring system move along the surfaces of those wires. If not, why not?

16. Does Gauss's law, as applied in Section 28-6, require that all the conduction electrons in an insulated conductor reside on the surface?

17. In Section 28-6, we assumed that **E** equals zero everywhere inside a conductor. However, there are certainly very large electric fields inside the conductor, at points close to the electrons or to the nuclei. Does this invalidate the proof of Section 28-6?

18. It is sometimes said that excess charge resides entirely on the outer surface of a conductor because like charges repel and try to get as far away as possible from one another. Comment on this plausibility argument.

19. Would Gauss's law hold if the exponent in Coulomb's law were not exactly two?

20. As you penetrate a uniform sphere of charge, E should decrease because less charge lies inside a sphere drawn through the observation point. On the other hand, E should increase because you are closer to the center of this charge. Which effect predominates and why?

21. Given a spherically symmetric charge distribution (not of uniform density of charge), is E necessarily a maximum at the surface? Comment on various possibilities.

22. Explain in your own words the factor of 2 that distinguishes Eq. 28-11 from Eq. 28-12.

23. Does Eq. 28-7 hold true for Fig. 28-9a if (a) there is a concentric spherical cavity in the body, (b) a point charge Q is at the center of this cavity, and (c) the charge Q is inside the cavity but not at its center?

24. An atom is normally *electrically neutral*. Why then should an α-particle be deflected by the atom under any circumstances?

25. If an α-particle, fired at a gold nucleus, is deflected through 135°, can you conclude (a) that any force has acted on the α-particle or (b) that any net work has been done on it?

26. Explain in your own words why the α-scattering experiments of Rutherford and his colleagues (see Example 4) render the Thomson atom model (see Example 3) untenable.

problems

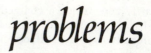

SECTION 28-1

1. Calculate Φ_E through a hemisphere of radius R. The field of **E** is uniform and is parallel to the axis of the hemisphere.   *Answer:* $\pi R^2 E$.

2. A butterfly net is in a uniform electric field as shown in Fig. 28-16. The rim, a circle of radius a, is aligned perpendicular to the field. Find the electric flux through the netting.

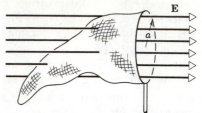

figure 28-16
Problem 2

SECTION 28-3

3. In Example 1 compute Φ_E for the cylinder if it is turned so that its axis is perpendicular to the electric field. Calculate the flux directly without using Gauss's law.   *Answer:* Zero.

4. A point charge of 1.0×10^{-6} C is at the center of a cubical Gaussian surface 0.50 m on edge. What is Φ_E for the surface?

SECTION 28-4

5. Charge on an originally uncharged insulated conductor is separated by holding a positively charged rod nearby, as in Fig. 28-17. What can you learn from Gauss's law about the flux for the five Gaussian surfaces shown? The induced negative charge on the conductor is equal to the positive charge on the rod.

   Answer: $+q$ = charge on rod. $\Phi_{S_1} = q/\epsilon_0$. $\Phi_{S_2} = -q/\epsilon_0$. $\Phi_{S_3} = q/\epsilon_0$. $\Phi_{S_4} = 0$. $\Phi_{S_5} = q/\epsilon_0$.

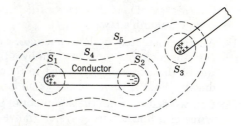

figure 28-17
Problem 5

6. A uniformly charged conducting sphere of 1.0 m diameter has a surface charge density of 8.0 C/m². What is the total electric flux leaving the surface of the sphere?

7. The intensity of the earth's electric field near its surface is ~ 130 N/C, pointing down. What is the earth's charge, assuming that this field is caused by it?   *Answer:* -6×10^5 C.

8. A point charge q is placed at one corner of a cube of edge a. What is the flux through each of the cube faces? (Hint: Use Gauss's law and symmetry arguments.)

9. "Gauss's law for gravitation" is

$$\frac{1}{4\pi G} \Phi_g = \frac{1}{4\pi G} \oint \mathbf{g} \cdot d\mathbf{S} = -m,$$

where m is the enclosed mass and G is the universal gravitation constant (Section 16-3). Derive Newton's law of gravitation from this.

10. The electric field components in Fig. 28-18 are $E_x = bx^{1/2}$, $E_y = E_z = 0$, in which $b = 800$ N/C $\cdot$ m$^{1/2}$. Calculate (a) the flux Φ_E through the cube and (b) the charge within the cube. Assume that $a = 10$ cm.

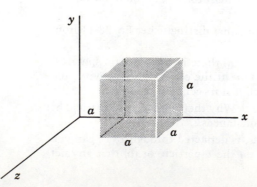

figure 28-18
Problem 10

SECTION 28-6

11. Equation 28-12 ($E = \sigma/\epsilon_0$) gives the electric field at points near a charged conducting surface. Show that this equation leads to a believable result when applied to a conducting sphere of radius r, carrying a charge q.
    Answer: It leads to $E = q/4\pi\epsilon_0 r^2$.

12. An insulated conductor carries a net charge of $+10 \times 10^{-6}$ C. Inside the conductor is a hollow cavity inside of which is a point charge Q of $+3.0 \times 10^{-6}$ C. What is the charge (a) on the cavity wall and (b) on the outer surface of the conductor?

13. Figure 28-19 shows a point charge of 1.0×10^{-7} C at the center of a spherical cavity of radius 3.0 cm in a piece of metal. Use Gauss's law to find the electric field (a) at point a, halfway from the center to the surface and (b) at point b.
    Answer: (a) 4.0×10^6 N/C. (b) Zero.

14. Figure 28-20 shows a spherical nonconducting shell of charge of uniform density ρ (C/m^3). Plot E for distances r from the center of the shell ranging from zero to 30 cm. Assume that $\rho = 1.0 \times 10^{-6}$ C/m^3, $a = 10$ cm, and $b = 20$ cm.

15. An uncharged, spherical, thin, metallic shell has a point charge q at its center. Give expressions for the electric field (a) inside the shell and (b) outside the shell, using Gauss's law. (c) Has the shell any effect on the field due to q? (d) Has the presence of q any effect on the shell? (e) If a second point charge is held outside the shell, does this outside charge experience a force? (f) Does the inside charge experience a force? (g) Is there a contradiction with Newton's third law here?
    Answer: (a) $E = q/4\pi\epsilon_0 r^2$, radially outward. (b) Same as (a). (c) No. (d) Yes, charges are induced on the surfaces. (e) Yes. (f) No. (g) No.

16. A thin, metallic, spherical shell of radius a carries a charge q_a. Concentric with it is another thin, metallic, spherical shell of radius b ($b > a$) carrying a charge q_b. Use Gauss's law to find the electric field at radial points r where (a) $r < a$; (b) $a < r < b$; (c) $r > b$. (d) Discuss the criterion one would use to determine how the charges are distributed on the inner and outer surfaces of each shell.

17. A *nonconducting* sphere of radius a is placed at the center of a spherical *conducting* shell of inner radius b and outer radius c, as in Fig. 28-21. A charge $+Q$ is distributed uniformly through the inner sphere (charge density ρ, C/m^3). The outer shell carries $-Q$. Find $E(r)$, (a) within the sphere ($r < a$), (b) between the sphere and the shell ($a < r < b$), (c) inside the shell ($b < r < c$), and (d) outside the shell ($r > c$). (e) What charges appear on the inner and outer surfaces of the shell?
    Answer: (a) $E = (Q/4\pi\epsilon_0 a^3)r$. (b) $E = Q/4\pi\epsilon_0 r^2$. (c) Zero. (d) Zero. (e) Inner, $-Q$; Outer, zero.

18. An irregularly shaped conductor has an irregularly shaped cavity inside. A charge $+Q$ is placed on the conductor but there is no charge inside the cavity. Show that (a) $E = 0$ inside the cavity and (b) there is no charge on the cavity walls.

19. Two concentric conducting spherical shells have radii $R_1 = 0.145$ m and $R_2 = 0.207$ m. The inner sphere bears a charge -6.00×10^{-8} C. An electron escapes from the inner sphere with negligible speed. Assuming that the region between the spheres is a vacuum, compute the speed with which the electron strikes the outer sphere.
    Answer: 2.0×10^7 m/s.

20. The spherical region $a < r < b$ carries a charge per unit volume of $\rho = A/r$, where A is constant. At the center ($r = 0$) of the enclosed cavity is a point charge Q. What should the value of A be so that the electric field in the region $a < r < b$ has constant magnitude?

21. A solid nonconducting sphere carries a uniform charge per unit volume ρ. Let $\mathbf{r}$ be the vector from the center of the sphere to a general point P within the sphere. (a) Show that the electric field at P is given by $\mathbf{E} = \rho\mathbf{r}/3\epsilon_0$. (b) A spherical cavity is created in the above sphere, as shown in Fig. 28-22.

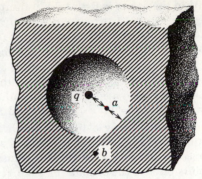

figure 28-19
Problem 13

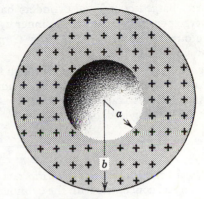

figure 28-20
Problem 14

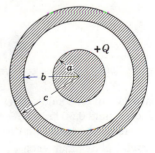

figure 28-21
Problem 17

figure 28-22
Problem 21

Using superposition concepts show that the electric field at all points within the cavity is $\mathbf{E} = \rho\mathbf{a}/3\epsilon_0$ (uniform field), where $\mathbf{a}$ is the vector connecting the center of the sphere with the center of the cavity. Note that both these results are independent of the radii of the sphere and the cavity.

22. Figure 28-23 shows a section through a long, thin-walled metal tube of radius R, carrying a charge per unit length λ on its surface. Derive expressions for E for various distances r from the tube axis, considering both (a) $r > R$ and (b) $r < R$. Plot your results for the range $r = 0$ to $r = 5$ cm, assuming that $\lambda = 2.0 \times 10^{-8}$ C/m and $R = 3.0$ cm.

23. A long conducting cylinder (length l) carrying a total charge $+q$ is surrounded by a conducting cylindrical shell of total charge $-2q$, as shown in cross section in Fig. 28-24. Use Gauss's law to find (a) the electric field at points outside the conducting shell, (b) the distribution of the charge on the conducting shell, and (c) the electric field in the region between the cylinders.
    Answer: (a) $E = q/2\pi\epsilon_0 lr$, radially inward. (b) $-q$ on both inner and outer surfaces. (c) $E = q/2\pi\epsilon_0 lr$, radially outward.

24. Two charged concentric cylinders have radii of 3.0 cm and 6.0 cm. The charge per unit length on the inner cylinder is 5.0×10^{-6} C/m and that on the outer cylinder is -7.0×10^{-6} C/m. Find the electric field at (a) $r = 4.0$ cm and (b) at $r = 8.0$ cm.

25. Figure 28-25 shows a section through two long concentric cylinders of radii a and b. The cylinders carry equal and opposite charges per unit length λ. Using Gauss's law prove (a) that $E = 0$ for $r > b$ and for $r < a$ and (b) that between the cylinders E is given by

$$E = \frac{1}{2\pi\epsilon_0} \frac{\lambda}{r}.$$

26. In Problem 25 a positive electron revolves in a circular path of radius r, between and concentric with the cylinders. What must be its kinetic energy K? Assume $a = 2.0$ cm, $b = 3.0$ cm, and $\lambda = 3.0 \times 10^{-8}$ C/m.

27. Charge is distributed uniformly throughout an infinitely long cylinder of radius R. (a) Show that E at a distance r from the cylinder axis ($r < R$) is given by

$$E = \frac{\rho r}{2\epsilon_0},$$

where ρ is the density of charge (C/m³). (b) What result do you expect for $r > R$?   Answer: (b) $\rho R^2/2\epsilon_0 r$.

28. A metal plate 8.0 cm on a side carries a total charge of 6.0×10^{-6} C. (a) Estimate the electric field 0.50 cm above the surface of the plate near the plate's center. (b) Estimate the field at a distance of .3.0 m.

29. Two large metal plates face each other as in Fig. 28-26 and carry charges with surface charge density $+\sigma$ and $-\sigma$, respectively, on their inner surfaces. What is $\mathbf{E}$ at points (a) to the left of the sheets, (b) between them, and (c) to the right of the sheets. Consider only points not near the edges whose distance from the sheets is small compared to the dimensions of the sheet.
                Answer: (a) Zero. (b) $E = \sigma/\epsilon_0$, to the left. (c) Zero.

30. Two large metal plates of area 1.0 m² face each other. They are 5.0 cm apart and carry equal and opposite charges on their inner surfaces. If E between the plates is 55 N/C, what is the charge on the plates? Neglect edge effects.

31. Two large nonconducting sheets of positive charge face each other as in Fig. 28-27. What is $\mathbf{E}$ at points (a) to the left of the sheets, (b) between them, and (c) to the right of the sheets? Assume the same surface charge density σ for each sheet. Consider only points not near the edges whose distance from the sheets is small compared to the dimensions of the sheet. (Hint: $\mathbf{E}$ at any point is the vector sum of the separate electric fields set up by each sheet.)
                Answer: (a) $E = \sigma/\epsilon_0$, to the left. (b) $E = 0$. (c) $E = \sigma/\epsilon_0$, to the right.

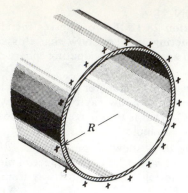

figure 28-23
Problem 22

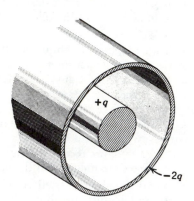

figure 28-24
Problem 23

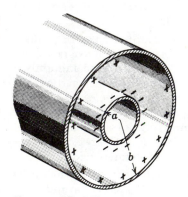

figure 28-25
Problem 25

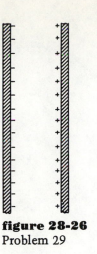

figure 28-26
Problem 29

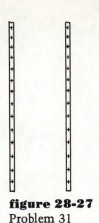

figure 28-27
Problem 31

32. A nonconducting plane slab of thickness d has a uniform volume charge density ρ. Find the magnitude of the electric field at all points in space both (a) inside and (b) outside the slab.

33. A 100-eV electron is fired directly toward a large metal plate that has a surface charge density of -2.0×10^{-6} C/m^2. From what distance must the electron be fired if it is to just fail to strike the plate?   *Answer:* 0.44 mm.

34. A small sphere whose mass m is 1.0×10^{-3} g carries a charge q of 2.0×10^{-8} C. It hangs from a silk thread which makes an angle of 30° with a large, charged nonconducting sheet as in Fig. 28-28. Calculate the surface charge density σ for the sheet.

35. Show that stable equilibrium under the action of electrostatic forces alone is impossible. (Hint: Assume that at a certain point P in an $\mathbf{E}$ field a charge $+q$ would be in stable equilibrium if it were placed there—which it is not. Draw a spherical Gaussian surface about P, imagine how $\mathbf{E}$ must point on this surface, and apply Gauss's law.)

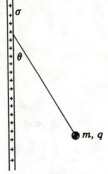

figure 28-28
Problem 34

SECTION 28-9

36. A gold foil used in a Rutherford scattering experiment is 3×10^{-5} cm thick. (a) What fraction of its surface area is "blocked out" by gold nuclei, assuming a nuclear radius of 6.9×10^{-15} m? Assume that no nucleus is screened by any other. (b) What fraction of the volume of the foil is occupied by the nuclei? (c) What fills all the rest of the space in the foil?

37. An α-particle, approaching the surface of a nucleus of gold, is a distance equal to one nuclear radius $(6.9 \times 10^{-15}$ m) away from that surface. What are (a) the force on the α-particle and (b) its acceleration at that point? The mass of the α-particle, which may be treated as a point, is 6.7×10^{-27} kg.
     Answer: (a) 190 N. (b) 2.9×10^{28} m/s^2.

29
electric potential

The electric field around a charged rod can be described not only by a (vector) electric field **E** but also by a scalar quantity, the *electric potential V*. These quantities are intimately related, and often it is only a matter of convenience which is used in a given problem.

To find the *electric potential difference* between two points A and B in an electric field, we move a test charge q_0 from A to B, always keeping it in equilibrium, and we measure the work W_{AB} that must be done by the agent moving the charge. The electric potential difference* is defined from

$$V_B - V_A = \frac{W_{AB}}{q_0}. \qquad (29\text{-}1)$$

The work W_{AB} may be (*a*) positive, (*b*) negative, or (*c*) zero. In these cases the electric potential at B will be (*a*) higher, (*b*) lower, or (*c*) the same as the electric potential at A.

The SI unit of potential difference that follows from Eq. 29-1 is the joule/coulomb. This combination occurs so often that a special unit, the *volt* (abbr. V), is used to represent it; that is,

1 volt = 1 joule/coulomb.

Usually point A is chosen to be at a large (strictly an infinite) distance from all charges, and the electric potential V_A at this infinite distance is arbitrarily taken as zero. This allows us to define the *electric potential*

29-1
ELECTRIC POTENTIAL

---

* This definition of potential difference, though conceptually sound and suitable for our present purpose, is rarely carried out in practice because of technical difficulties. Equivalent and more technically feasible methods are usually adopted.

622

at a point. Putting $V_A = 0$ in Eq. 29-1 and dropping the subscripts leads to

$$V = \frac{W}{q_0}, \qquad (29\text{-}2)$$

where W is the work that an external agent must do to move the test charge q_0 from infinity to the point in question. Keep in mind that *potential differences* are of fundamental concern and that Eq. 29-2 depends on the arbitrary assignment of the value zero to the potential V_A at the reference position (infinity); this reference potential could equally well have been chosen as any other value, say -100 V. Similarly, any other agreed-upon point could be chosen as a reference position. In many problems the earth is taken as a reference of potential and assigned the value zero.

Bearing in mind the assumptions made about the reference position, we see from Eq. 29-2 that V near an isolated positive charge is positive because positive work must be done by an outside agent to push a (positive) test charge in from infinity. Similarly, the potential near an isolated negative charge is negative because an outside agent must exert a restraining force on (that is, must do negative work on) a (positive) test charge as it comes in from infinity. Electric potential as defined in Eq. 29-2 is a scalar because W and q_0 in that equation are scalars.

Both W_{AB} and $V_B - V_A$ in Eq. 29-1 are independent of the path followed in moving the test charge from point A to point B. If this were not so, point B would not have a unique electric potential (with respect to point A as a defined reference position) and the concept of potential would not be useful.

Let us now show that the potential difference is path independent for the field due to the point charge of Fig. 29-1. Then we shall show that path independence holds in *all* electrostatic situations. Figure 29-1 shows two points A and B set up in the field of a point charge q, assumed positive.

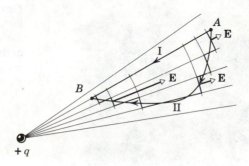

figure 29-1
A test charge q_0 is moved from A to B in the field of charge q along either of two paths. The open arrows show **E** at three points on path II.

Let us imagine a positive test charge q_0 being moved by an external agent from point A to point B by two different paths. Path I is a simple radial line. Path II is a completely arbitrary path between the two points.

We may approximate path II by a broken path made up of alternating elements of circular arc and of radius. Since these elements can be arbitrarily small, the broken path can be made arbitrarily close to the actual path. On path II the external agent does work *only along the radial segments* because along the arcs the force **F** and the displacement $d\mathbf{l}$ are at right angles, $\mathbf{F} \cdot d\mathbf{l}$ being zero in such cases. The sum of the work done on the radial segments that make up path II is the same as the work done on path I because each path has the same array of radial segments. Since path II is arbitrary, we have proved that the work done is the same for *all* paths connecting A and B.

This proof holds only for the field due to a point charge. However, any charge distribution (discrete or continuous) can be considered as made up of an assembly of point charges or differential charge elements. Hence, from the principle of superposition, we conclude that path independence for the electrostatic potential holds for *all* electrostatic charge configurations.

The locus of points, all of which have the same electric potential, is called an *equipotential surface*. A family of equipotential surfaces, each surface corresponding to a different value of the potential, can be used to give a general description of the electric field in a certain region of space. We have seen earlier (Section 27-3) that electric lines of force can also be used for this purpose; in later sections (see, for example, Fig. 29-15) we explore the intimate connection between these two ways of describing the electric field.

No work is required to move a test charge between any two points on an equipotential surface. This follows from Eq. 29-1,

$$V_B - V_A = \frac{W_{AB}}{q_0},$$

because W_{AB} must be zero if $V_A = V_B$. This is true, because of the path independence of potential difference, even if the path connecting A and B does not lie entirely in the equipotential surface.

Figure 29-2 shows an arbitrary family of equipotential surfaces. The work to move a charge along paths I and II is zero because both these paths begin and end on the same equipotential surface. The work to move a charge along paths I' and II' is not zero but is the same for each path because the initial and the final potentials are identical; paths I' and II' connect the same pair of equipotential surfaces.

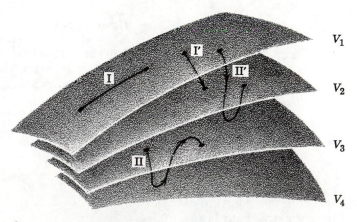

V_1

V_2

V_3

V_4

figure 29-2
Portions of four equipotential surfaces. The heavy lines show four paths along which a test charge is moved.

From symmetry, the equipotential surfaces for a spherical charge are a family of concentric spheres. For a uniform field they are a family of planes at right angles to the field. In all cases (including these two examples) the equipotential surfaces are at right angles to the lines of force and thus to **E** (see Fig. 29-15). If **E** were *not* at right angles to the equipotential surface, it would have a component lying in that surface. Then work would have to be done in moving a test charge about on the surface. Work cannot be done if the surface is an equipotential, so **E** must be at right angles to the surface.

There is a strong analogy between electrostatic forces and gravitational forces, based on the fact that their fundamental laws are inverse square laws (see Eqs. 26-3 and 16-1):

$$F_E = \frac{1}{4\pi\epsilon_0}\frac{q_1 q_2}{r^2} \quad \text{and} \quad F_g = G\,\frac{m_1 m_2}{r^2},$$

and also on the fact that the forces are proportional to the magnitude of the test body (charge in one case, mass in the other). Thus we can define the *gravitational potential* V_g (compare Eq. 29-2) from

$$V_g = \frac{W}{m},$$

where W is the work required to move a test body of mass m from infinity to the point in question. Gravitational equipotential surfaces can also be constructed; they prove to be everywhere at right angles to the gravitational field vector **g**. For a uniform gravitational field, such as that near the surface of the earth, these surfaces are horizontal planes. This correlates with the facts that (*a*) no net work is required to move a stone of mass m between two points with the same elevation and (*b*) the same net work is required to move a stone along any path starting on a given horizontal surface and ending on another. Gravitational forces, like coulombic forces, are conservative; see Section 8-2.

29-2
POTENTIAL AND THE ELECTRIC FIELD

Let A and B in Fig. 29-3 be two points in a uniform electric field **E**, set up by an arrangement of charges not shown, and let A be a distance d in the field direction from B. Assume that a positive test charge q_0 is moved, by an external agent and without acceleration, from A to B along the straight line connecting them.

The electric force on the charge is $q_0\mathbf{E}$ and points down. To move the charge in the way we have described we must counteract this force by applying an external force **F** of the same magnitude but directed up. The work W done by the agent that supplies this force is

$$W_{AB} = Fd = q_0 Ed. \tag{29-3}$$

Substituting this into Eq. 29-1 yields

$$V_B - V_A = \frac{W_{AB}}{q_0} = Ed. \tag{29-4}$$

This equation shows the connection between potential difference and field strength for a simple special case. Note from this equation that another SI unit for **E** is the volt/meter (V/m). You may wish to prove that a volt/meter is identical with a newton/coulomb (N/C); this latter unit was the one first presented for **E** in Section 27-2.

In Fig. 29-3 B has a higher potential than A. This is reasonable because an external agent would have to do positive work to push a positive test charge from A to B. Figure 29-3 could be used as it stands to illustrate the act of lifting a stone from A to B in the uniform gravitational field near the earth's surface.

What is the connection between V and **E** in the more general case in which the field is *not* uniform and in which the test body is moved along a path that is *not* straight, as in Fig. 29-4? The electric field exerts a force $q_0\mathbf{E}$ on the test charge, as shown. To keep the test charge from accelerating, an external agent must apply a force **F** chosen to be exactly equal to $-q_0\mathbf{E}$ for all positions of the test body.

If the external agent causes the test body to move through a displacement $d\mathbf{l}$ along the path from A to B, the element of work done by the

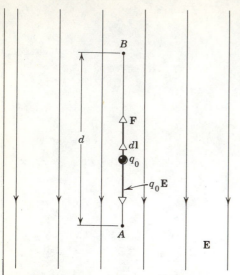

figure 29-3

A positive test charge q_0 is moved from A to B in a uniform electric field $\mathbf{E}$ by an external agent that exerts a force $\mathbf{F}$ on it.

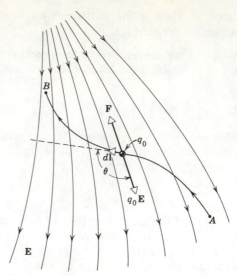

figure 29-4

A positive test charge q_0 is moved from A to B in a nonuniform electric field by an external agent that exerts a force $\mathbf{F}$ on it.

external agent is $\mathbf{F} \cdot d\mathbf{l}$. To find the total work W_{AB} done by the external agent in moving the test charge from A to B, we add up (that is, integrate) the work contributions for all the infinitesimal segments into which the path is divided. This leads to

$$W_{AB} = \int_A^B \mathbf{F} \cdot d\mathbf{l} = -q_0 \int_A^B \mathbf{E} \cdot d\mathbf{l}.$$

Such an integral is called a *line integral*. Note that we have substituted $-q_0\mathbf{E}$ for its equal, $\mathbf{F}$.

Substituting this expression for W_{AB} into Eq. 29-1 leads to

$$V_B - V_A = \frac{W_{AB}}{q_0} = -\int_A^B \mathbf{E} \cdot d\mathbf{l}. \tag{29-5}$$

If point A is taken to be infinitely distant and the potential V_A at infinity is taken to be zero, this equation gives the potential V at point B, or, dropping the subscript B,

$$V = -\int_\infty^B \mathbf{E} \cdot d\mathbf{l}. \tag{29-6}$$

These two equations allow us to calculate the potential difference between any two points (or the potential at any point) if $\mathbf{E}$ is known at various points in the field.

In Fig. 29-3 calculate $V_B - V_A$ using Eq. 29-5. Compare the result with that obtained by direct analysis of this special case (Eq. 29-4).

EXAMPLE 1

In moving the test charge the element of path $d\mathbf{l}$ always points in the direction of motion; this is upward in Fig. 29-3. The electric field $\mathbf{E}$ in this figure points down so that the angle θ between $\mathbf{E}$ and $d\mathbf{l}$ is 180°.

Equation 29-5 then becomes

$$V_B - V_A = -\int_A^B \mathbf{E} \cdot d\mathbf{l} = -\int_A^B E \cos 180°\, dl = \int_A^B E\, dl.$$

E is constant for all parts of the path in this problem and can thus be taken outside the integral sign, giving

$$V_B - V_A = E \int_A^B dl = Ed,$$

which agrees with Eq. 29-4, as it must.

EXAMPLE 2

In Fig. 29-5 let a test charge q_0 be moved without acceleration from A to B over the path shown. Compute the potential difference between A and B.

For the path AC we have $\theta = 135°$ and, from Eq. 29-5,

$$V_C - V_A = -\int_A^C \mathbf{E} \cdot d\mathbf{l} = -\int_A^C E \cos 135° \, dl = \frac{E}{\sqrt{2}} \int_A^C dl.$$

The integral is the length of the line AC which is $\sqrt{2}d$. Thus

$$V_C - V_A = \frac{E}{\sqrt{2}} (\sqrt{2}d) = Ed.$$

Points B and C have the same potential because no work is done in moving a charge between them, $\mathbf{E}$ and $d\mathbf{l}$ being at right angles for all points on the line CB. In other words, B and C lie on the same equipotential surface at right angles to the lines of force. Thus

$$V_B - V_A = V_C - V_A = Ed.$$

This is the same value derived for a direct path connecting A and B, a result to be expected because the potential difference between two points is path independent.

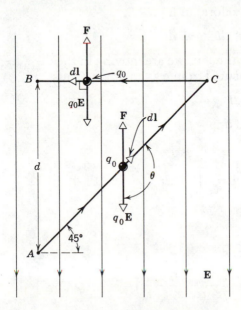

figure 29-5
Example 2. A test charge q_0 is moved along path ACB in a uniform electric field by an external agent.

Figure 29-6 shows two points A and B near an isolated positive point charge q. For simplicity we assume that A, B, and q lie on a straight line. Let us compute the potential difference between points A and B, assuming that a test charge q_0 is moved without acceleration along a radial line from A to B.

29-3
POTENTIAL DUE TO A POINT CHARGE

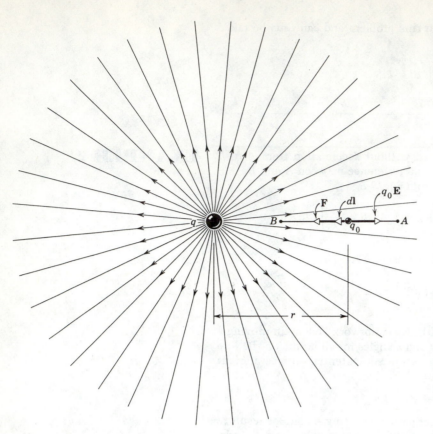

figure 29-6
A test charge q_0 is moved by an external agent from A to B in the field set up by a positive point charge q.

In Fig. 29-6 **E** points to the right and $d\mathbf{l}$, which is always in the direction of motion, points to the left. Therefore, in Eq. 29-5,

$$\mathbf{E} \cdot d\mathbf{l} = E \cos 180° \, dl = -E \, dl.$$

However, as we move a distance dl to the left, we are moving in the direction of decreasing r because r is measured from q as an origin. Thus

$$dl = -dr.$$

Combining yields $\qquad \mathbf{E} \cdot d\mathbf{l} = E \, dr.$

Substituting this into Eq. 29-5 gives

$$V_B - V_A = -\int_A^B \mathbf{E} \cdot d\mathbf{l} = -\int_{r_A}^{r_B} E \, dr.$$

Combining with Eq. 27-4,

$$E = \frac{1}{4\pi\epsilon_0} \frac{q}{r^2}$$

leads to $\qquad V_B - V_A = -\frac{q}{4\pi\epsilon_0} \int_{r_A}^{r_B} \frac{dr}{r^2} = \frac{q}{4\pi\epsilon_0} \left(\frac{1}{r_B} - \frac{1}{r_A} \right).$ $\qquad$ (29-7)

Choosing reference position A to be at infinity (that is, letting $r_A \to \infty$), choosing $V_A = 0$ at this position, and dropping the subscript B leads to

$$V = \frac{1}{4\pi\epsilon_0} \frac{q}{r}.$$ $\qquad$ (29-8)

This equation shows clearly that equipotential surfaces for an isolated point charge are spheres concentric with the point charge (see Fig. 29-15a). A study of the derivation will show that this relation also holds for points external to spherically symmetric charge distributions.

What must the magnitude of an isolated positive point charge be for the electric potential at 10 cm from the charge to be $+100$ V?

EXAMPLE 3

Solving Eq. 29-8 for q yields

$$q = V4\pi\epsilon_0 r = (100 \text{ V})(4\pi)(8.9 \times 10^{-12} \text{ C}^2/\text{N}\cdot\text{m}^2)(0.10 \text{ m})$$
$$= 1.1 \times 10^{-9} \text{ C}.$$

This charge is comparable to charges that can be produced by friction.

What is the electric potential at the surface of a gold nucleus? The radius is 6.6×10^{-15} m and the atomic number $Z = 79$.

EXAMPLE 4

The nucleus, assumed spherically symmetrical, behaves electrically for external points as if it were a point charge. Thus we can use Eq. 29-8, or, recalling that the proton charge is 1.6×10^{-19} C,

$$V = \frac{1}{4\pi\epsilon_0}\frac{q}{r} = \frac{(9.0 \times 10^9 \text{ N}\cdot\text{m}^2/\text{C}^2)(79)(1.6 \times 10^{-19} \text{ C})}{6.6 \times 10^{-15} \text{ m}}$$

$$= 1.7 \times 10^7 \text{ V}.$$

The potential at any point due to a group of point charges is found by (a) calculating the potential V_n due to each charge, as if the other charges were not present and (b) adding the quantities so obtained, or (see Eq. 29-8)

29-4
A GROUP OF POINT CHARGES

$$V = \sum_n V_n = \frac{1}{4\pi\epsilon_0} \sum_n \frac{q_n}{r_n}, \qquad (29\text{-}9)$$

where q_n is the value of the nth charge and r_n is the distance of this charge from the point in question. The sum used to calculate V is an *algebraic sum* and not a vector sum like the one used to calculate **E** for a group of point charges (see Eq. 27-5). Herein lies an important computational advantage of potential over electric field.

If the charge distribution is continuous, rather than being a collection of points, the sum in Eq. 29-9 must be replaced by an integral, or

$$V = \int dV = \frac{1}{4\pi\epsilon_0} \int \frac{dq}{r}, \qquad (29\text{-}10)$$

where dq is a differential element of the charge distribution, r is its distance from the point at which V is to be calculated, and dV is the potential it establishes at that point.

What is the potential at the center of the square of Fig. 29-7? Assume that $q_1 = +1.0 \times 10^{-8}$ C, $q_2 = -2.0 \times 10^{-8}$ C, $q_3 = +3.0 \times 10^{-8}$ C, $q_4 = +2.0 \times 10^{-8}$ C, and $a = 1.0$ m.

EXAMPLE 5

The distance r of each charge from P is $a/\sqrt{2}$ or 0.71 m. From Eq. 29-9

$$V = \sum_n V_n = \frac{1}{4\pi\epsilon_0}\frac{q_1 + q_2 + q_3 + q_4}{r}$$

$$= \frac{(9.0 \times 10^9 \text{ N}\cdot\text{m}^2/\text{C}^2)(1.0 - 2.0 + 3.0 + 2.0) \times 10^{-8} \text{ C}}{0.71 \text{ m}}$$

$$= 500 \text{ V}.$$

Is the potential constant within the square? Does any point inside have a negative potential? Sketch roughly the intersection of the plane of Fig. 29-7 with the equipotential surface corresponding to zero volts.

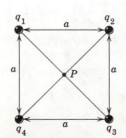

figure 29-7
Example 5

A charged disk. Find the electric potential for points on the axis of a uniformly charged circular disk whose surface charge density is σ (see Fig. 29-8).

Consider a charge element dq consisting of a flat circular strip of radius y and width dy. We have

$$dq = \sigma(2\pi y)(dy),$$

where $(2\pi y)(dy)$ is the area of the strip. All parts of this charge element are the same distance $r'\ (= \sqrt{y^2 + r^2})$ from axial point P so that their contribution dV to the electric potential at P is given by Eq. 29-8, or

$$dV = \frac{1}{4\pi\epsilon_0}\frac{dq}{r'} = \frac{1}{4\pi\epsilon_0}\frac{\sigma 2\pi y\, dy}{\sqrt{y^2 + r^2}}.$$

The potential V is found by integrating over all the strips into which the disk can be divided (Eq. 29-10), or

$$V = \int dV = \frac{\sigma}{2\epsilon_0}\int_0^a (y^2 + r^2)^{-1/2} y\, dy$$

$$= \frac{\sigma}{2\epsilon_0}(\sqrt{a^2 + r^2} - r).$$

This general result is valid for all values of r. In the special case of $r \gg a$ the quantity $\sqrt{a^2 + r^2}$ can be approximated as

$$\sqrt{a^2 + r^2} = r\left(1 + \frac{a^2}{r^2}\right)^{1/2} = r\left(1 + \frac{1}{2}\frac{a^2}{r^2} + \cdots\right) \cong r + \frac{a^2}{2r},$$

in which the quantity in parentheses in the second member of this equation has been expanded by the binomial theorem (see Appendix I). This equation means that V becomes

$$V \cong \frac{\sigma}{2\epsilon_0}\left(r + \frac{a^2}{2r} - r\right) = \frac{\sigma\pi a^2}{4\pi\epsilon_0 r} = \frac{1}{4\pi\epsilon_0}\frac{q}{r},$$

where $q\ (= \sigma\pi a^2)$ is the total charge on the disk. This limiting result is expected because the disk behaves like a point charge for $r \gg a$.

EXAMPLE 6

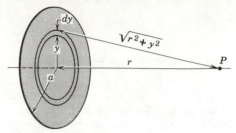

figure 29-8
Example 6. A point P on the axis of a uniformly charged circular disk of radius a.

Two equal charges of opposite sign, $\pm q$, separated by a distance $2a$, constitute an electric dipole; see Example 3, Chapter 27. The electric dipole moment $\mathbf{p}$ has the magnitude $2aq$ and points from the negative charge to the positive charge. Here we derive an expression for the electric potential V at any point of space due to a dipole, provided only that the point is not too close to the dipole $(r \gg a)$.

A point P is specified by giving the quantities r and θ in Fig. 29-9. From symmetry, it is clear that the potential will not change as point P rotates about the z axis, r and θ being fixed. Thus we need only find $V(r,\theta)$ for any plane containing this axis; the plane of Fig. 29-9 is such a plane. Applying Eq. 29-9 gives,

$$V = \sum_n V_n = V_1 + V_2 = \frac{1}{4\pi\epsilon_0}\left(\frac{q}{r_1} - \frac{q}{r_2}\right) = \frac{q}{4\pi\epsilon_0}\frac{r_2 - r_1}{r_1 r_2},$$

which is an exact relationship.

We now limit consideration to points such that $r \gg 2a$. These approximate relations then follow from Fig. 29-9:

$$r_2 - r_1 \cong 2a\cos\theta \quad \text{and} \quad r_1 r_2 \cong r^2,$$

and the potential reduces to

$$V = \frac{q}{4\pi\epsilon_0}\frac{2a\cos\theta}{r^2} = \frac{1}{4\pi\epsilon_0}\frac{p\cos\theta}{r^2}, \tag{29-11}$$

29-5
POTENTIAL DUE TO A DIPOLE

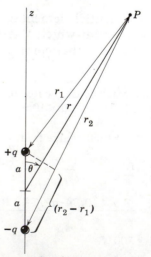

figure 29-9
A point P in the field of an electric dipole.

in which $p \ (= 2aq)$ is the dipole moment. Note that V vanishes everywhere in the equatorial plane ($\theta = 90°$). This reflects the fact that it takes no work to bring a test charge in from infinity along the perpendicular bisector of the dipole. For a given radius, V has its greatest positive value for $\theta = 0$ and its greatest negative value for $\theta = 180°$. Note that the potential does not depend separately on q and $2a$ but only on their product p.

It is convenient to call *any* assembly of charges, for which V at distant points is given by Eq. 29-11, an *electric dipole*. Two point charges separated by a small distance behave this way, as we have just proved. However, other charge configurations also obey Eq. 29-11. Suppose that by measurement at points outside an imaginary box (Fig. 29-10) we find a pattern of lines of force that can be described quantitatively by Eq. 29-11. We then declare that the object inside the box is an *electric dipole*, that its axis is the line zz', and that its dipole moment **p** points vertically upward.

Many molecules have electric dipole moments. That for H_2O in its vapor state is 6.1×10^{-30} C·m. Figure 29-11 is a representation of this molecule, showing the three nuclei and the surrounding electron cloud. The dipole moment **p** is represented by the arrow on the axis of symmetry of the molecule. In this molecule the effective center of positive charge does not coincide with the effective center of negative charge. It is precisely because of this separation that the dipole moment exists.

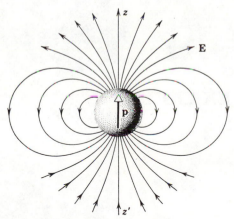

figure 29-10
If an object inside the spherical box sets up the electric field shown (described quantitatively by Eq. 29-11), it is an electric dipole.

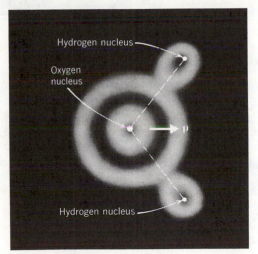

figure 29-11
A schematic representation of a water molecule, showing the three nuclei, the electron cloud, and the orientation of the dipole moment.

Atoms, and many molecules, do not have permanent dipole moments. However, dipole moments may be induced by placing any atom or molecule in an external electric field. The action of the field (Fig. 29-12) is to separate the centers of positive and of negative charge. We say that the atom becomes *polarized* and acquires an *induced electric dipole moment*. Induced dipole moments disappear when the electric field is removed.

Electric dipoles are important in situations other than atomic and molecular ones. Radio and radar antennas are often in the form of a metal wire or rod in which electrons surge back and forth periodically. At a certain time one end of the wire or rod will be negative and the other end positive. Half a cycle later the

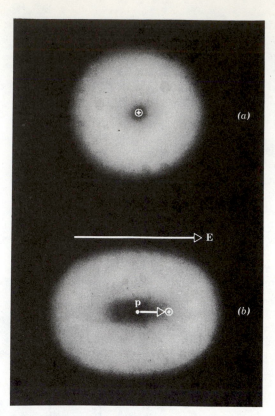

(a)

E

p

(b)

figure 29-12
(a) An atom, showing the nucleus and the electron cloud. The center of negative charge coincides with the center of positive charge, that is, with the nucleus. (b) If an external field **E** is applied, the electron cloud is distorted so that the center of negative charge, marked by the dot, and the center of positive charge no longer coincide. An electric dipole appears. The distortion is greatly exaggerated.

polarity of the ends is exactly reversed. This is an *oscillating* electric dipole. It is so named because its dipole moment changes in a periodic way with time.

An electric quadrupole. An electric quadrupole, of which Fig. 27-26 is an example, consists of two electric dipoles so arranged that they almost, but not quite, cancel each other in their electric effects at distant points. Calculate $V(r)$ for points on the axis of this quadrupole.

Applying Eq. 29-9 to Fig. 27-26 yields,

$$V = \sum_n V_n = \frac{1}{4\pi\epsilon_0} \left(\frac{q}{r-a} - \frac{2q}{r} + \frac{q}{r+a} \right)$$

$$= \frac{q}{4\pi\epsilon_0} \frac{2a^2}{(r-a)(r)(r+a)}.$$

Assuming that $r \gg a$ allows us to put $a = 0$ in the sum and difference terms in the denominator, yielding

$$V = \frac{1}{4\pi\epsilon_0} \frac{Q}{r^3},$$

where $Q \; (= 2qa^2)$ is the *electric quadrupole moment* of the charge assembly of Fig. 27-26. Note that V varies (a) as $1/r$ for a point charge (see Eq. 29-8), (b) as $1/r^2$ for a dipole (see Eq. 29-11), and (c) as $1/r^3$ for a quadrupole.

Note too that (a) a dipole is two equal and opposite charges that do not quite coincide in space so that their electric effects at distant points do not quite cancel, and (b) a quadrupole is two equal and opposite dipoles that do not quite coincide in space so that their electric effects at distant points again do not quite cancel. This pattern can be extended to define higher orders of charge distribution such as *octupoles*.

EXAMPLE 7

The potential at points at distances from an arbitrary charge distribution (continuous or discrete) that are large compared with the size of the distribution can always be written as the sum of separate potential distributions due to (a) a single charge—sometimes, in this context, called a *monopole*—(b) a dipole, (c) a quadrupole, etc. This process is called an *expansion in multipoles* and is a very useful technique in problem solving.

If we raise a stone from the earth's surface, the work that we do against the earth's gravitational attraction is stored as *potential energy* in the system earth + stone. If we release the stone, the stored potential energy changes steadily into kinetic energy as the stone drops. After the stone comes to rest on the earth, this kinetic energy, equal in magnitude just before the time of contact to the originally stored potential energy, is transformed into thermal energy in the system earth + stone.

A similar situation exists in electrostatics. Consider two charges q_1 and q_2 a distance r apart, as in Fig. 29-13. If we increase the separation between them, an external agent must do work that will be positive if the charges are opposite in sign and negative otherwise. The energy represented by this work can be thought of as stored in the system $q_1 + q_2$ as *electric potential energy*. This energy, like all varieties of potential energy, can be transformed into other forms. If q_1 and q_2, for example, are charges of opposite sign and we release them, they will accelerate toward each other, transforming the stored potential energy into kinetic energy of the accelerating masses. The analogy to the earth + stone system is exact, save for the fact that electric forces may be either attractive or repulsive whereas gravitational forces are always attractive.

We define the electric potential energy of a system of point charges as the work required to assemble this system of charges by bringing them in from an infinite distance. We assume that the charges are all at rest when they are infinitely separated, that is, they have no initial kinetic energy.

In Fig. 29-13 let us imagine q_2 removed to infinity and at rest. The *electric potential* at the original site of q_2, caused by q_1, is given by Eq. 29-8, or

$$V = \frac{1}{4\pi\epsilon_0}\frac{q_1}{r}.$$

If q_2 is moved in from infinity to the original distance r, the work required is, from the definition of electric potential, that is, from Eq. 29-2,

$$W = Vq_2. \tag{29-12}$$

Combining these two equations and recalling that this work W is pre-

29-6
*ELECTRIC POTENTIAL ENERGY**

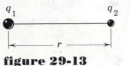

figure 29-13

---

*In mechanics the concept of *potential energy* (of compressed springs, bodies in the earth's gravitational field, etc.) is perhaps more commonly used than that of *potential* (gravitational, say). In electricity and magnetism, on the other hand, *potential* is perhaps more commonly used than *potential energy*. A shorthand distinction is that potential is potential energy *per unit charge.* With potential in units of volts and potential energy in units of joules, the volt is equivalent, therefore, to a joule per coulomb.

cisely the *electric potential energy U* of the system $q_1 + q_2$ yields

$$U \ (= W) = \frac{1}{4\pi\epsilon_0} \frac{q_1 q_2}{r_{12}}. \tag{29-13}$$

The subscript of r emphasizes that the distance involved is that between the point charges q_1 and q_2.

For systems containing more than two charges the procedure is to compute the potential energy for every pair of charges separately and to add the results algebraically. This procedure rests on a physical picture in which (*a*) charge q_1 is brought into position, (*b*) q_2 is brought from infinity to its position near q_1, (*c*) q_3 is brought from infinity to its position near q_1 and q_2, etc.

The potential energy of continuous charge distributions (an ellipsoid of charge, for example) can be found by dividing the distribution into differential elements dq, treating each such element as a point charge, and using the procedures of the preceding paragraph, with the summation process replaced by an integration. We have not considered such problems in this text.

EXAMPLE 8

Two protons in a nucleus of U^{238} are 6.0×10^{-15} m apart. What is their mutual electric potential energy?

From Eq. 29-13

$$U = \frac{1}{4\pi\epsilon_0} \frac{q_1 q_2}{r} = \frac{(9.0 \times 10^9 \ \text{N} \cdot \text{m}^2/\text{C}^2)(1.6 \times 10^{-19} \ \text{C})^2}{6.0 \times 10^{-15} \ \text{m}}$$

$$= 3.8 \times 10^{-14} \ \text{J} = 2.4 \times 10^5 \ \text{eV}.$$

EXAMPLE 9

Three charges are held fixed as in Fig. 29-14. What is their mutual electric potential energy? Assume that $q = 1.0 \times 10^{-7}$ C and $a = 10$ cm.

The total energy of the configuration is the sum of the energies of each pair of particles. From Eq. 29-13,

$$U = U_{12} + U_{13} + U_{23}$$

$$= \frac{1}{4\pi\epsilon_0} \left[\frac{(+q)(-4q)}{a} + \frac{(+q)(+2q)}{a} + \frac{(-4q)(+2q)}{a} \right]$$

$$= -\frac{10}{4\pi\epsilon_0} \frac{q^2}{a}$$

$$= -\frac{(9.0 + 10^9 \ \text{N} \cdot \text{m}^2/\text{C}^2)(10)(1.0 \times 10^{-7} \ \text{C})^2}{0.10 \ \text{m}} = -9.0 \times 10^{-3} \ \text{J}.$$

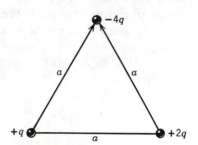

figure 29-14
Example 9. Three charges are fixed, as shown, by external forces.

The fact that the total energy is negative means that negative work would have to be done to assemble this structure, starting with the three charges separated and at rest at infinity. Expressed otherwise, 9.0×10^{-3} J of work must be done to dismantle this structure, removing the charges to an infinite separation from one another.

When, as is common practice, infinity is taken as the zero of electric potential, a positive potential energy, in simple problems like these, (as in Example 8) corresponds to repulsive electric forces and a negative potential energy (as in this example) to attractive electric forces. If the protons in Example 8 were not held in place by attractive (nonelectrical) nuclear forces, they would move away from each other. If the three particles in this example were released from their fixed positions, in which they are held by external forces, they would move toward each other.

We have stated that V and $\mathbf{E}$ are equivalent descriptions of electric fields, and have determined (Eq. 29-6) how to calculate V from $\mathbf{E}$. Let us now consider how to calculate $\mathbf{E}$ if we know V throughout a certain region.

This problem has already been solved graphically. If $\mathbf{E}$ is known at every point in space, the lines of force can be drawn; then a family of equipotentials can be sketched in by drawing surfaces at right angles. These equipotentials describe the behavior of V. Conversely, if V is given as a function of position, a set of equipotential surfaces can be drawn. The lines of force can then be found by drawing lines at right angles, thus describing the behavior of $\mathbf{E}$. It is the mathematical equivalent of this second graphical process that we seek here. Figure 29-15 shows three examples of lines of force and of the corresponding equipotential surfaces.

29-7
CALCULATION OF E FROM V

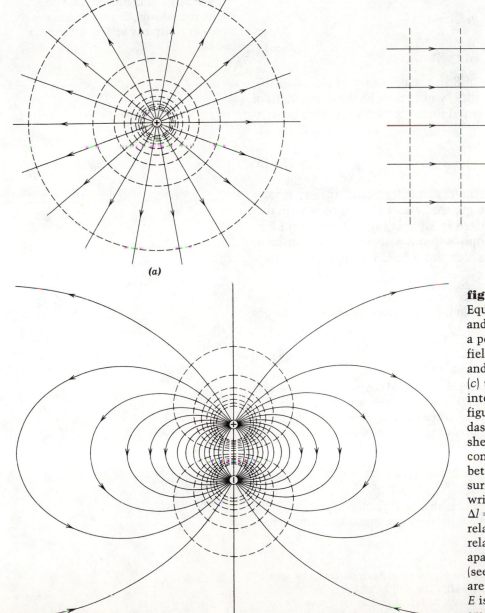

(a)

(b)

(c)

figure 29-15
Equipotential surfaces (dashed lines) and lines of force (solid lines) for (a) a point charge, (b) a uniform electric field set up by charges not shown, and (c) an electric dipole. In (a) and (c) the dashed lines represent intersections with the plane of the figure of closed surfaces; in (b) the dashed lines represent infinite sheets. In all figures there is a constant difference of potential ΔV between adjacent equipotential surfaces. Thus from Eq. 29-14, written for the case of $\theta = 180°$ as $\Delta l = -\Delta V/E$, the surfaces will be relatively close together where E is relatively large and relatively far apart where E is small. Similarly (see Section 27-3) the lines of force are relatively close together where E is large and far apart where E is small. See discussion and figures of Section 18-7 for other examples.

Figure 29-16 shows the intersection with the plane of the figure of a family of equipotential surfaces. The figure shows that **E** at a typical point P is at right angles to the equipotential surface through P, as it must be.

Let us move a test charge q_0 from P along the path marked $\Delta \mathbf{l}$ to the equipotential surface marked $V + \Delta V$. The work that must be done by the agent exerting the force **F** (see Eq. 29-1) is $q_0 \Delta V$.

From another point of view we can calculate the work from*

$$\Delta W = \mathbf{F} \cdot \Delta \mathbf{l},$$

where **F** is the force that must be exerted on the charge to overcome exactly the electrical force $q_0 \mathbf{E}$. Since **F** and $q_0 \mathbf{E}$ have opposite signs and are equal in magnitude,

$$\Delta W = -q_0 \mathbf{E} \cdot \Delta \mathbf{l} = -q_0 E \cos(\pi - \theta)\, \Delta l = q_0 E \cos \theta\, \Delta l.$$

These two expressions for the work must be equal, which gives

$$q_0\, \Delta V = q_0 E \cos \theta\, \Delta l$$

or

$$E \cos \theta = \frac{\Delta V}{\Delta l}. \qquad (29\text{-}14)$$

Now $E \cos \theta$ is the component of **E** in the direction $-\mathbf{l}$ in Fig. 29-16; the quantity $-E \cos \theta$, which we call E_l, would then be the component of **E** in the $+\mathbf{l}$ direction. In the differential limit Eq. 29-14 can then be written as

$$E_l = -\frac{dV}{dl}. \qquad (29\text{-}15)$$

In words, this equation says: If we travel through an electric field along a straight line and measure V as we go, the rate of change of V with distance that we observe, when changed in sign, is the component of **E** in that direction. The minus sign implies that **E** points in the direction of decreasing V, as in Fig. 29-16. It is clear from Eq. 29-15 that appropriate units for **E** are volt/meter (V/m).

There will be one direction $\mathbf{l}$ for which the quantity $-dV/dl$ is a maximum. From Eq. 29-15, E_l will also be a maximum for this direction and will in fact be E itself. Thus

$$E = -\left(\frac{dV}{dl}\right)_{\max}. \qquad (29\text{-}16)$$

The maximum value of dV/dl at a given point is called the *potential gradient* at that point. The direction $\mathbf{l}$ for which dV/dl has its maximum value is always at right angles to the equipotential surface, corresponding to $\theta = 0$ in Fig. 29-16.

If we take the direction $\mathbf{l}$ to be, in turn, the directions of the x, y, and z axes, we can find the three components of **E** at any point, from Eq. 29-15.

$$E_x = -\frac{\partial V}{\partial x}; \qquad E_y = -\frac{\partial V}{\partial y}; \qquad E_z = -\frac{\partial V}{\partial z}. \qquad (29\text{-}17)$$

Thus if V is known for all points of space, that is, if the function $V(x, y, z)$ is known, the components of **E**, and thus **E** itself, can be found by taking derivatives.†

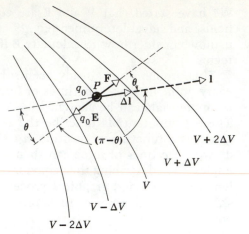

figure 29-16
A test charge q_0 is moved from one equipotential surface to another along an arbitrarily selected direction marked **l**.

---

* We assume that the equipotentials are so close together that **F** is constant for all parts of the path $\Delta \mathbf{l}$. In the limit of a differential path ($d\mathbf{l}$) there will be no difficulty.

† The symbol $\partial V/\partial x$ denotes a *partial derivative*. In taking this derivative of the function $V(x, y, z)$ the quantity x is to be viewed as a variable and y and z are to be regarded as constants. Similar considerations hold for $\partial V/\partial y$ and $\partial V/\partial z$.

Calculate $E(r)$ for a point charge q, using Eq. 29-16 and assuming that $V(r)$ is given as (see Eq. 29-8)

EXAMPLE 10

$$V = \frac{1}{4\pi\epsilon_0} \frac{q}{r}.$$

From symmetry, **E** must be directed radially outward for a (positive) point charge. Consider a point P in the field a distance r from the charge. It is clear that $-dV/dl$ at P has its greatest value if the direction l is identified with that of r. Thus, from Eq. 29-16,

$$E = -\frac{dV}{dr} = -\frac{d}{dr}\left(\frac{1}{4\pi\epsilon_0} \frac{q}{r}\right)$$

$$= -\frac{q}{4\pi\epsilon_0} \frac{d}{dr}\left(\frac{1}{r}\right) = \frac{1}{4\pi\epsilon_0} \frac{q}{r^2}.$$

This result agrees exactly with Eq. 27-4, as it must.

E *for a dipole.* Figure 29-17 shows a (distant) point P in the field of a dipole located at the origin of an xy-axis system. V is given by Eq. 29-11, or

EXAMPLE 11

$$V = \frac{1}{4\pi\epsilon_0} \frac{p\cos\theta}{r^2}.$$

Calculate **E** as a function of position.

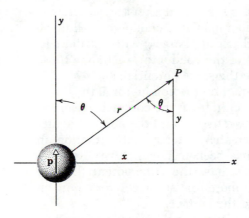

figure 29-17
Showing a point P in the field of an electric dipole **p**.

From symmetry, **E**, for points in the plane of Fig. 29-17, lies in this plane. Thus it can be expressed in terms of its components E_x and E_y, E_z being zero. Let us first express the potential function in rectangular coordinates rather than polar coordinates, making use of

$$r = (x^2 + y^2)^{1/2} \quad \text{and} \quad \cos\theta = \frac{y}{(x^2 + y^2)^{1/2}}.$$

The result is

$$V = \frac{p}{4\pi\epsilon_0} \frac{y}{(x^2 + y^2)^{3/2}}.$$

We find E_y from Eq. 29-17, recalling that x is to be treated as a constant in this calculation:

$$E_y = -\frac{\partial V}{\partial y} = -\frac{p}{4\pi\epsilon_0} \frac{(x^2 + y^2)^{3/2} - y\frac{3}{2}(x^2 + y^2)^{1/2}(2y)}{(x^2 + y^2)^3}$$

$$= -\frac{p}{4\pi\epsilon_0} \frac{x^2 - 2y^2}{(x^2 + y^2)^{5/2}}.$$

Note that putting $x = 0$ describes distant points along the dipole axis (that is, the y axis), and the expression for E_y reduces to

$$E_y = \frac{2p}{4\pi\epsilon_0} \frac{1}{y^3}.$$

This result agrees exactly with that found in Chapter 27 (see Problem 25), for, from symmetry, E_x equals zero on the dipole axis.

Putting $y = 0$ in the expression for E_y describes distant points in the median plane of the dipole and yields

$$E_y = -\frac{p}{4\pi\epsilon_0} \frac{1}{x^3},$$

which agrees exactly with the result found in Chapter 27 (see Example 3), for, again from symmetry, E_x equals zero in the median plane. The minus sign in this equation indicates that **E** points in the negative y direction (see Fig. 29-10).

The component E_x is also found from Eq. 29-17, recalling that y is to be taken as a constant during this calculation:

$$E_x = -\frac{\partial V}{\partial x} = -\frac{py}{4\pi\epsilon_0} (-\tfrac{3}{2})(x^2 + y^2)^{-5/2}(2x)$$

$$= \frac{3p}{4\pi\epsilon_0} \frac{xy}{(x^2 + y^2)^{5/2}}.$$

As expected, E_x vanishes both on the dipole axis ($x = 0$) and in the median plane ($y = 0$); see Fig. 29-10.

29-8
AN INSULATED CONDUCTOR

We proved in Section 28-6, using Gauss's law, that after a steady state is reached an excess charge q placed on an insulated conductor will move to its outer surface. We now assert that this charge q will distribute itself on this surface so that all points of the conductor, including those on the surface and those inside, have the same potential.

Consider any two points A and B in or on the conductor. If they were not at the same potential, the charge carriers in the conductor near the point of lower potential would tend to move toward the point of higher potential. We have assumed, however, that a steady-state situation, in which such currents do not exist, has been reached; thus all points, both on the surface and inside it, must have the same potential. Since the surface of the conductor is an equipotential surface, **E** for points on the surface must be at right angles to the surface.

We saw in Section 28-6 that a charge placed on an insulated conductor will spread over the surface until **E** equals zero for all points inside. We now have an alternative way of saying the same thing; the charge will move until all points of the conductor (surface points and interior points) are brought to the same potential, for if V is constant in the conductor, then **E** is zero everywhere in the conductor ($E_l = -dV/dl$).

Figure 29-18a is a plot of potential against radial distance for an isolated spherical conducting shell of 1.0-m radius carrying a positive charge of 1.0×10^{-6} C. For points outside the shell $V(r)$ can be calculated from Eq. 29-8 because the charge q behaves, for such points, as if it were concentrated at the center of the sphere. Equation 29-8 is correct right up to the surface of the shell. Now let us push the test charge through the surface, assuming that there is a small hole, and into the interior. No extra work is needed because no electrical forces act on the test charge once it is inside the shell. Thus the potential everywhere inside is the same as that on the surface, as Fig. 29-18a shows.

Figure 29-18b shows the electric field for this same sphere. Note that E equals zero inside. The lower of these curves can be derived from the

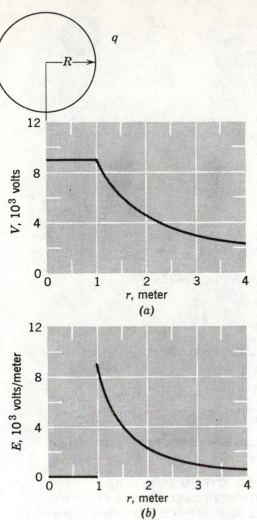

upper by differentiation with respect to r, using Eq. 29-16; the upper can be derived from the lower by integration with respect to r, using Eq. 29-6.

Figure 29-18 holds without change if the conductor is a solid sphere rather than a spherical shell as we have assumed. It is constructive to compare Fig. 29-18b (conducting shell or sphere) with Fig. 28-10, which holds for a *nonconducting* sphere. Try to understand the difference between these two figures, bearing in mind that in the first the charge lies on the surface, whereas in the second it was assumed to have been spread uniformly throughout the volume of the sphere.

Finally we note that, as a general rule, the charge density tends to be high on isolated conducting surfaces whose radii of curvature are small, and conversely. For example, the charge density tends to be relatively high on sharp points and relatively low on plane regions on a conducting surface. The electric field E at points immediately above a charged surface is proportional to the charge density σ so that E may also reach very high values near sharp points. Glow discharges from sharp points during thunderstorms are a familiar example. The lightning rod acts in this way to neutralize charged clouds and thus prevent lightning strokes.

We can examine the qualitative relationship between σ and the curvature of the surface in a particular case by considering two spheres of different radii con-

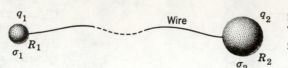

figure 29-19
Two spheres connected by a long
fine wire.

nected by a very long fine wire; see Fig. 29-19. Suppose that this entire assembly is raised to some arbitrary potential V. The (equal) potentials of the two spheres are, from Eq. 29-8,*

$$V = \frac{1}{4\pi\epsilon_0} \frac{q_1}{R_1} = \frac{1}{4\pi\epsilon_0} \frac{q_2}{R_2},$$

which yields

$$\frac{q_1}{q_2} = \frac{R_1}{R_2}, \qquad (29\text{-}18)$$

where q_1 is the charge on the sphere of radius R_1 and q_2 is the charge on the sphere of radius R_2.

The *surface charge densities* for each sphere are given by

$$\sigma_1 = \frac{q_1}{4\pi R_1{}^2} \quad \text{and} \quad \sigma_2 = \frac{q_2}{4\pi R_2{}^2}.$$

Dividing gives

$$\frac{\sigma_1}{\sigma_2} = \frac{q_1}{q_2} \frac{R_2{}^2}{R_1{}^2}.$$

Combining with Eq. 29-18 yields

$$\frac{\sigma_1}{\sigma_2} = \frac{R_2}{R_1},$$

which is consistent with our qualitative statement above. Note that the larger sphere has the larger total charge but the smaller charge density.

The fact that σ, and thus $\mathbf{E}$, can become very large near sharp points is important in the design of high-voltage equipment. *Corona discharge* can result from such points if the conducting object is raised to high potential and surrounded by air. Normally air is thought of as a nonconductor. However, it contains a small number of ions produced, for example, by the cosmic rays. A positively charged conductor will attract negative ions from the surrounding air and thus will slowly neutralize itself.

If the charged conductor has sharp points, the value of E in the air near the points can be very high. If the value is high enough, the ions, as they are drawn toward the conductor, will receive such large accelerations that, by collision with air molecules, they will produce vast additional numbers of ions. The air is thus made much more conducting, and the discharge of the conductor by this corona discharge may be very rapid indeed. The air surrounding sharp conducting points may even glow visibly because of light emitted from the air molecules during these collisions.

The electrostatic generator was conceived by Lord Kelvin in 1890 and put into useful practice in essentially its modern form by R. J. Van de Graaff in 1931. It is a device for producing electric potential differences of the order of several millions of volts. Its chief application in physics is the use of this potential difference to accelerate charged particles to

29-9
*THE ELECTROSTATIC
GENERATOR*

* Equation 29-8 holds only for an isolated point charge or spherically symmetric charge distribution. The spheres must be assumed to be so far apart that the charge on either one has a negligible effect on the distribution of charge on the other.

high energies. Beams of energetic particles made in this way can be used in many different "atom-smashing" experiments. The technique is to let a charged particle "fall" through a potential difference V, gaining kinetic energy as it does so.

Let a particle of (positive) charge q move in a vacuum under the influence of an electric field from one position A to another position B whose electric potential is lower by V. The electric potential energy of the system is reduced by qV because this is the work that an external agent would have to do to restore the system to its original condition. This decrease in potential energy appears as kinetic energy of the particle, or

$$K = qV. \qquad (29\text{-}19)$$

K is in joules if q is in coulombs and V in volts. If the particle is an electron or a proton, q will be the quantum of charge e.

If we adopt the quantum of charge e as a unit in place of the coulomb, we arrive at another unit for energy, the *electron volt* (abbr. eV), which we have used before and is used extensively in atomic and nuclear physics. By substituting into Eq. 29-19,

$$1 \text{ electron volt} = (1 \text{ quantum of charge})(1 \text{ volt})$$

$$= (1.60 \times 10^{-19} \text{ coulomb})(1.00 \text{ volt})$$

$$= 1.60 \times 10^{-19} \text{ joule.}$$

The electron volt can be used interchangeably with any other energy unit. Thus a 10-g object moving at 1000 cm/s can be said to have a kinetic energy of 3.1×10^{18} eV. Most physicists would prefer to express this result as 0.50 J, the electron volt being inconveniently small. In atomic, nuclear, and elementary particle physics, however, the electron volt (eV) and its multiples the keV (= 10^3 eV), MeV (= 10^6 eV), and the GeV (= 10^9 eV) are the usual units of choice.

The electrostatic or Van de Graaff generator. Figure 29-20, which illustrates the basic operating principle of the electrostatic generator, shows a small sphere of radius r placed inside a large spherical shell of radius R. The two spheres carry charges q and Q, respectively. Calculate their potential difference.

The potential of the large sphere is caused in part by its own charge and in part because it lies in the field set up by the charge q on the small sphere. From Eq. 29-8,

$$V_R = \frac{1}{4\pi\epsilon_0}\left(\frac{Q}{R} + \frac{q}{R}\right).$$

The potential of the small sphere is caused in part by its own charge and in part because it is inside the large sphere; see Fig. 29-18a. From Eq. 29-8,

$$V_r = \frac{1}{4\pi\epsilon_0}\left(\frac{q}{r} + \frac{Q}{R}\right).$$

The potential difference is

$$V_r - V_R = \frac{q}{4\pi\epsilon_0}\left(\frac{1}{r} - \frac{1}{R}\right).$$

Thus, assuming q is positive, the inner sphere will always be higher in potential than the outer sphere. If the spheres are connected by a fine wire, the charge q will flow *entirely* to the outer sphere, regardless of the charge Q that may already be present.

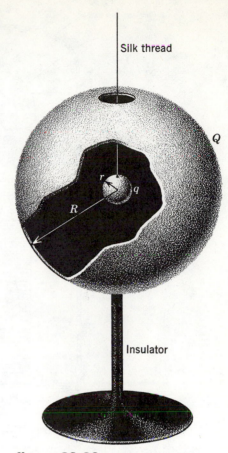

figure 29-20
Example 12. A small charged sphere of radius r is suspended inside a charged spherical shell whose outer surface has radius R.

EXAMPLE 12

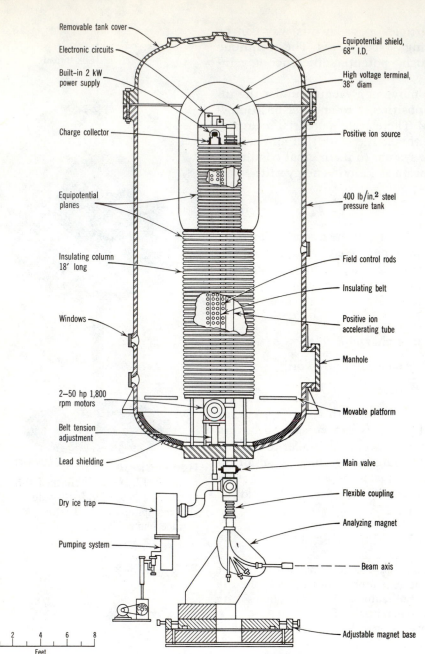

Removable tank cover

Electronic circuits

Built-in 2 kW
power supply

Charge collector

Equipotential
planes

Insulating column
18' long

Windows

2–50 hp 1,800
rpm motors

Belt tension
adjustment

Lead shielding

Dry ice trap

Pumping system

Equipotential shield,
68" I.D.

High voltage terminal,
38" diam

Positive ion source

400 lb/in.² steel
pressure tank

Field control rods

Insulating belt

Positive ion
accelerating tube

Manhole

Movable platform

Main valve

Flexible coupling

Analyzing magnet

Beam axis

Adjustable magnet base

0 2 4 6 8
Feet

figure 29-21
An electrostatic generator at MIT
capable of producing 9-MeV protons.
The proton beam is accelerated
vertically downward, being deflected
into a horizontal plane by the
analyzing magnet shown at the
bottom. (Courtesy of J. G. Trump.)

From another point of view, we note that since the spheres when
electrically connected form a single conductor at electrostatic equilibrium there can be only a single potential. This means that $V_r - V_R = 0$,
which can occur only if $q = 0$.

In actual electrostatic generators charge is carried into the shell on
rapidly moving belts made of insulating material. Charge is "sprayed"
onto the belts outside the shell by corona discharge from a series of
sharp metallic points connected to a source of moderately high potential
difference. Charge is removed from the belts inside the shell by a similar
series of points connected to the shell. Electrostatic generators can be
built commercially to accelerate protons to energies up to 10 MeV,
using a single acceleration. Figure 29-21 shows a schematic diagram of
an electrostatic generator at MIT that can produce 9-MeV protons.

Generators can be built in which the accelerated particles are subject to two or three successive accelerations.

questions

1. Are we free to call the potential of the earth +100 volts instead of zero? What effect would such an assumption have on measured values of (a) potentials and (b) potential differences?

2. What would happen to a person on an insulated stand if his potential was increased by 10,000 volts?

3. Do electrons tend to go to regions of high potential or of low potential?

4. Suppose that the earth has a net charge that is *not* zero. Is it still possible to adopt the earth as a standard reference point of potential and to assign the potential $V = 0$ to it?

5. Does the potential of a positively charged insulated conductor have to be positive? Give an example to prove your point.

6. Can two different equipotential surfaces intersect?

7. An electrical worker is accidentally electrocuted and a newspaper account reported: "He accidentally touched a high voltage cable and 20,000 volts of electricity surged through his body." Criticize this statement.

8. Does the amount of work per unit charge required to transfer electric charge from one point to another in an electrostatic field depend on the amount of charge transferred?

9. Advice to mountaineers caught in lightning and thunderstorms is (a) get rapidly off peaks and ridges and (b) put both feet together and crouch in the open, only the feet touching the ground. What is the basis for this good advice?

10. If **E** equals zero at a given point, must V equal zero for that point? Give some examples to prove your answer.

11. If you know **E** at a given point, can you calculate V at that point? If not, what further information do you need?

12. In Fig. 29-2 is the electric field **E** greater at the left or the right side of the figure?

13. In Fig. 29-6 is it necessary to assume that A, B, and q lie on a straight line in order to prove that Eq. 29-8 is true?

14. Is the uniformly charged, nonconducting disk of Example 6 a surface of constant potential? Explain.

15. Why can an isolated atom not have a permanent electric dipole moment?

16. If V equals a constant throughout a given region of space, what can you say about **E** in that region?

17. In Section 16-6 we saw that the gravitational field strength is zero inside a spherical shell of matter. The electrical field strength is zero not only inside an isolated charged spherical conductor but inside an isolated conductor of *any* shape. Is the gravitational field strength inside, say, a cubical shell of matter zero? If not, in what respect is the analogy not complete?

18. How can you insure that the electric potential in a given region of space will have a constant value?

19. A charge is placed on an insulated conductor in the form of a perfect cube. What will be the relative charge density at various points on the cube (surfaces, edges, corners); what will happen to the charge if the cube is in air?

20. We have seen (Fig. 29-18a) that the potential inside a thin conducting spherical shell is the same as that on its surface. (a) What if the sphere is solid? (b) What if it is solid and irregularly shaped? (c) What if it is solid, is irregularly shaped, and has an irregularly shaped cavity inside? In particular, what is V within the hollow cavity? (d) Same as (c) but suppose a point charge is suspended within the cavity?

21. A closed metal box in the form of a pyramid is placed on an insulating support and charged to a potential $+V$. Is the average potential inside the pyramid (a) greater than, (b) equal to, or (c) less than V?

22. An isolated conducting spherical shell carries a negative charge. What will happen if a positively charged metal object is placed in contact with the shell interior? Assume that the positive charge is (a) less than, (b) equal to, and (c) greater than the negative charge in magnitude.

23. An uncharged metal sphere suspended by a silk thread is placed in a uniform external electric field **E**. What is the magnitude of the electric field for points inside the sphere? Is your answer changed if the sphere carries a charge?

problems

SECTION 29-1

1. In a typical lightning flash the potential difference between discharge points is about 10^9 V and the quantity of charge transferred is about 30 C. How much ice would it melt at 0°C if all the energy released could be used for this purpose?   *Answer:* 99 tons.

2. A charge q is distributed uniformly throughout a nonconducting spherical volume of radius R. (a) Show that the potential a distance r from the center, where $r < R$, is given by

$$V = \frac{q(3R^2 - r^2)}{8\pi\epsilon_0 R^3}.$$

(b) Is it reasonable that, according to this expression, V is not zero at the center of the sphere?

3. A gold nucleus contains a positive charge equal to that of 79 protons. An α-particle $(Z = 2)$ has a kinetic energy K at points far from this charge and is traveling directly toward the charge. The particle just touches the surface of the charge (assumed spherical) and is reversed in direction. (a) Calculate K, assuming a nuclear radius of 5.0×10^{-15} m. (b) The actual α-particle energy used in the experiments of Rutherford and his collaborators (see Section 28-9) was 5.0 MeV. What do you conclude?
   Answer: (a) 45 MeV. (b) In these important experiments the α-particles approached, but did not "touch" the gold nuclei.

4. (a) Through what potential difference must an electron fall, according to Newtonian mechanics, to acquire a speed v equal to the speed c of light? (b) Newtonian mechanics fails as $v \to c$. Therefore, using the correct relativistic expression for the kinetic energy

$$K = mc^2 \left[\frac{1}{\sqrt{1 - (v/c)^2}} - 1 \right]$$

in place of the Newtonian expression $K = \frac{1}{2}mv^2$, determine the actual electron speed acquired in falling through the potential difference computed in (a). Express this speed as an appropriate fraction of the speed of light.

SECTION 29-2

5. An infinite charged sheet has a surface charge density σ of 1.0×10^{-7} C/m². How far apart are the equipotential surfaces whose potentials differ by 5.0 V?
   Answer: 0.89 mm.

6. Two large parallel conducting plates are 10 cm apart and carry equal but opposite charges on their facing surfaces. An electron placed midway between the two plates experiences a force of 1.6×10^{-15} N. What is the potential difference between the plates?

7. In the Millikan oil drop experiment (see Fig. 27-29) an electric field of 1.92×10^5 N/C is maintained at balance across two plates separated by 1.50 cm. Find the potential difference between the plates.   *Answer:* 2900 V.

SECTION 29-3

8. Consider a point charge with $q = 1.5 \times 10^{-8}$ C. (a) What is the radius of an equipotential surface having a potential of 30 V? (b) Are surfaces whose potentials differ by a constant amount (1.0 V, say) evenly spaced in radius?

SECTION 29-4

9. A point charge has $q = +1.0 \times 10^{-6}$ C. Consider point A which is 2.0 m distant and point B which is 1.0 m distant in a direction diametrically opposite, as in Fig. 29-22a. (a) What is the potential difference $V_A - V_B$? (b) Repeat if points A and B are located as in Fig. 29-22b.
Answer: (a) -4500 V. (b) Same as (a) because potential is a scalar quantity.

10. In Fig. 29-23, locate the points (a) where $V = 0$ and (b) where $\mathbf{E} = 0$. Consider only points on the axis and choose $d = 1.0$ m.

11. In Fig. 29-23 (see Problem 10) sketch qualitatively (a) the lines of force and (b) the intersections of the equipotential surfaces with the plane of the figure. (Hint: Consider the behavior close to each point charge and at considerable distances from the pair of charges.)

12. (a) In Fig. 29-24 derive an expression for $V_A - V_B$. (b) Does your result reduce to the expected answer when $d = 0$? When $q = 0$?

13. For the charge configuration of Fig. 29-25, show that $V(r)$ for points on the vertical axis, assuming $r \gg a$, is given by

$$V = \frac{1}{4\pi\epsilon_0}\left(\frac{q}{r} + \frac{2qa}{r^2}\right).$$

Is this an expected result? (Hint: The charge configuration can be viewed as the sum of an isolated charge and a dipole.)

SECTION 29-5

14. Calculate the dipole moment of a water molecule under the assumption that all ten electrons in the molecule circulate symmetrically about the oxygen atom, that the OH distance is 0.96×10^{-8} cm, and that the angle between the two OH bonds is 104°. Compare with the value quoted on p. 631; see Fig. 29-11.

SECTION 29-6

15. A particle of mass m, charge $q > 0$, and initial kinetic energy K is projected (from "infinity") toward a heavy nucleus of charge Q, assumed to have a fixed position in our reference frame. (a) If the aim is "perfect," how close to the center of the nucleus is the particle when it comes instantaneously to rest? (b) With a particular imperfect aim the particle's closest approach to the nucleus is twice the distance determined in part (a). Determine the speed of the particle at this closest distance of approach.
Answer: (a) $qQ/4\pi\epsilon_0 K$. (b) $\sqrt{K/m}$.

16. The charges and coordinates of two charges located on the x-y plane are: $q_1 = +3.0 \times 10^{-6}$ C; $x = +3.5$ cm, $y = +0.50$ cm, and $q_2 = -4.0 \times 10^{-6}$ C; $x = -2.0$ cm, $y = +1.5$ cm. (a) Find the electric potential at the origin. (b) How much work must be done to locate these charges at their given positions, starting from infinity?

17. Derive an expression for the work required to put the four charges together as indicated in Fig. 29-26. *Answer:* $-0.21 q^2/\epsilon_0 a$.

18. A particle of charge Q is kept in a fixed position at a point P and a second particle of mass m, having the same charge Q, is initially held at rest a distance r_1 from P. The second particle is then released and is repelled from the first one. Determine its speed at the instant it is a distance r_2 from P. Let $Q = 3.1 \times 10^{-6}$ C, $m = 2.0 \times 10^{-5}$ kg, $r_1 = 9.0 \times 10^{-4}$ m, and $r_2 = 25 \times 10^{-4}$ m.

19. What is the electric potential energy of the charge configuration of Fig. 29-7? Use the numerical values of Example 5. *Answer:* -6.4×10^{-7} J.

20. In the rectangle shown in Fig. 29-27, the sides have lengths 5.0 cm and 15.0 cm, $q_1 = -5.0 \times 10^{-6}$ C and $q_2 = +2.0 \times 10^{-6}$ C. (a) What is the electric potential at corner B? At corner A? (b) How much work is involved in moving

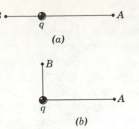

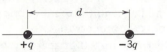

figure 29-22
Problem 9

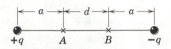

figure 29-23
Problems 10, 11

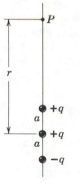

figure 29-24
Problem 12

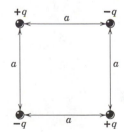

figure 29-25
Problem 13

figure 29-26
Problem 17

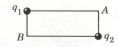

figure 29-27
Problem 20

a third charge $q_3 = +3.0 \times 10^{-6}$ C *from B to A* along a diagonal of the rectangle? (c) In this process, is external work converted into electrostatic potential energy or vice versa? Explain.

21. Two charges q ($= +2.0 \times 10^{-6}$ C) are fixed in space a distance d (2.0 cm) apart, as shown in Fig. 29-28. (a) What is the electric potential at point C? (b) You bring a third charge q ($+2.0 \times 10^{-6}$ C) very slowly from infinity to C. How much work must you do? (c) What is the potential energy U of the configuration when the third charge is in place?
    Answer: (a) 2.5×10^6 V. (b) 5.1 J. (c) 6.9 J.

22. Three charges of $+0.1$ C each are placed on the corner of an equilateral triangle, 1.0 m on a side. If energy is supplied at the rate of 1.0 kW, how many days would be required to move one of the charges onto the midpoint of the line joining the other two?

23. Two electrons are 2.0 m apart. Another electron is shot from infinity and comes to rest midway between the two. What must its initial velocity be?
    Answer: 32 m/s.

24. Calculate (a) the electric potential established by the nucleus of a hydrogen atom at the mean distance of the circulating electron ($r = 5.3 \times 10^{-11}$ m), (b) the electric potential energy of the atom when the electron is at this radius, and (c) the kinetic energy of the electron, assuming it to be moving in a circular orbit of this radius centered on the nucleus. (d) How much energy is required to ionize the hydrogen atom? Express all energies in electron volts.

25. A particle of (positive) charge Q is assumed to have a fixed position at P. A second particle of mass m and (negative) charge $-q$ moves at constant speed in a circle of radius r_1, centered at P. Derive an expression for the work W that must be done by an external agent on the second particle in order to increase the radius of the circle of motion, centered at P, to r_2.

    Answer: $W = \dfrac{qQ}{8\pi\epsilon_0}\left[\dfrac{1}{r_1} - \dfrac{1}{r_2}\right].$

26. Devise an arrangement of three point charges, separated by finite distances, that has zero electric potential energy.

27. Figure 29-29 shows an idealized representation of a U^{238} nucleus ($Z = 92$) on the verge of fission. Calculate (a) the repulsive force acting on each fragment and (b) the mutual electric potential energy of the two fragments. Assume that the fragments are equal in size and charge, spherical, and just touching. The radius of the initially spherical U^{238} nucleus is 8.0×10^{-15} m. Assume that the material out of which nuclei are made has a constant density. *Answer:* (a) 3.0×10^3 N. (b) 3.8×10^{-11} J, or 240 MeV.

SECTION 29-7

28. The electric potential varies along the x-axis as shown in the graph of Fig. 29-30. For each of the intervals shown (ignore the behavior at the end points of the intervals) determine the x-component of the electric field and plot E_x vs. x.

figure 29-28
Problem 21

figure 29-29
Problem 27

figure 29-30
Problem 28

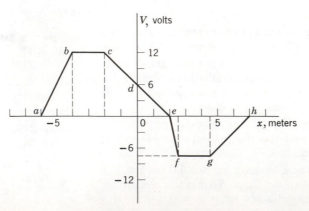

29. (a) Show that the electric potential at a point on the axis of a ring of charge of radius a, computed directly from Eq. 29-10, is given by

$$V = \frac{1}{4\pi\epsilon_0} \frac{q}{\sqrt{x^2 + a^2}}.$$

(b) From this result derive an expression for E at axial points; compare with the direct calculation of E in Example 5, Chapter 27.

30. What is the potential gradient, in V/m, at the surface of a gold nucleus? See Problem 3.

31. In Example 6 the potential at an axial point for a charged disk was shown to be

$$V = \frac{\sigma}{2\epsilon_0} (\sqrt{a^2 + r^2} - r).$$

(a) From this result show that E for axial points is given by

$$E = \frac{\sigma}{2\epsilon_0} \left(1 - \frac{r}{\sqrt{a^2 + r^2}}\right).$$

(b) Does this expression for E reduce to an expected result for $r \gg a$ and for $r = 0$?

32. (a) Starting from Eq. 29-11, find the magnitude E_r of the radial component of the electric field due to a dipole. (b) For what value of θ is E_r zero?

33. A charge per unit length λ is distributed uniformly along a straight-line segment of length L. (a) Determine the electrostatic potential (chosen to be zero at infinity) at a point P a distance y from one end of the charged segment and in line with it (see Fig. 29-31). (b) Use the result of (a) to compute the component of the electric field at P in the y-direction (along the line). (c) Determine the component of the electric field at P in a direction perpendicular to the straight line.

Answer: (a) $\dfrac{\lambda}{4\pi\epsilon_0} \ln \dfrac{L+y}{y}$.

(b) $\dfrac{\lambda}{4\pi\epsilon_0} \dfrac{L}{y(L+y)}$.

(c) Zero.

34. On a thin rod of length L lying along the x-axis with one end at the origin $(x = 0)$, as in Fig. 29-32, there is distributed a charge per unit length given by $\lambda = kx$, where k is a constant. (a) Taking the electrostatic potential at infinity to be zero, find V at the point P on the y-axis. (b) Determine the vertical component, E_y, of the electric field intensity at P from the result of part (a) and also by direct calculation. (c) Why cannot E_x, the horizontal component of the electric field at P, be found using the result of part (a)?

SECTION 29-8
35. What is the charge density σ on the surface of a conducting sphere of radius 0.15 m whose potential is 200 V?   *Answer:* 1.2×10^{-8} C/m².

36. A charged sphere of radius 1.5 m contains a total charge of 3.0×10^{-6} C. (a) What is the electric field at the sphere's surface? (b) What is the electric potential at the sphere's surface? (c) At what distance from the sphere's surface has the electric potential decreased by 5000 V?

figure 29-31
Problem 33

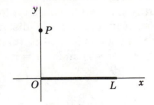

figure 29-32
Problem 34

figure 29-33
Problem 37

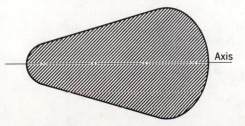

Axis

37. The metal object in Fig. 29-33 is a figure of revolution about the horizontal axis. If it is charged negatively, sketch roughly a few equipotentials and lines of force. Use physical reasoning rather than mathematical analysis.

38. If the earth had a net charge equivalent to 1 electron/m² of surface area (a very artificial assumption), (a) what would the earth's potential be? (b) What would the electric field due to the earth be just outside its surface?

39. A charge of 10^{-8} C can be produced by simple rubbing. To what potential would such a charge raise an insulated conducting sphere of 10-cm radius? *Answer:* 900 V.

40. For the spheres of Fig. 29-19, what is the ratio of electric fields at the surface?

41. Consider a thin, isolated, conducting, spherical shell which is uniformly charged to a constant charge density σ (C/m²). How much work does it take to move a small positive test charge q_0 (a) from the surface of the shell to the interior, through a small hole, (b) from one point on the surface to another, regardless of path, (c) from point to point inside the shell, and (d) from any point P outside the shell over any path, whether or not it pierces the shell, back to P? (e) For the conditions given, does it matter whether or not the shell is conducting?   *Answer:* (a) Zero. (b) Zero. (c) Zero. (d) Zero. (e) No.

42. A thin, spherical, conducting shell of radius R is mounted on an insulated support and charged to a potential $-V$. An electron is fired from point P a distance r from the center of the shell ($r \gg R$) with an initial speed v_0, directed radially inward. What is the value of v_0 chosen so that the electron will just reach the shell?

43. Can a conducting sphere 10 cm in radius hold a charge of 4×10^{-6} C in air without breakdown? The dielectric strength (minimum field required to produce breakdown) of air at 1 atm is 3×10^6 V/m.   *Answer:* No.

44. Two thin, insulated, concentric conducting spheres of radii R_1 and R_2 carry charges q_1 and q_2. Derive expressions for $E(r)$ and $V(r)$, where r is the distance from the center of the spheres. Plot $E(r)$ and $V(r)$ from $r=0$ to $r=4.0$ m for $R_1 = 0.50$ m, $R_2 = 1.0$ m, $q_1 = +2.0 \times 10^{-6}$ C, and $q_2 = +1.0 \times 10^{-6}$ C. Compare with Fig. 29-18.

45. The space between two concentric spheres of radii r_1 and r_2 is filled with a nonconducting material having a uniform charge density ρ. Find the electric potential V as a function of the distance r from the center of the spheres, considering the regions (a) $r > r_2$, (b) $r_2 > r > r_1$, and (c) $r < r_1$. (d) Do these solutions agree at $r = r_2$ and at $r = r_1$?

Answer: (a) $\dfrac{\rho}{3\epsilon_0} \dfrac{(r_2^3 - r_1^3)}{r}$. (b) $\dfrac{\rho}{3\epsilon_0} \left(\dfrac{3}{2} r_2^2 - \dfrac{r^2}{2} - \dfrac{r_1^3}{r} \right)$. (c) $\dfrac{\rho}{2\epsilon_0} (r_2^2 - r_1^2)$. (d) Yes.

46. Two metal spheres are 3.0 cm in radius and carry charges of $+1.0 \times 10^{-8}$ C and -3.0×10^{-8} C, respectively, assumed to be uniformly distributed. If their centers are 2.0 m apart, calculate (a) the potential of the point halfway between their centers and (b) the potential of each sphere.

47. Two identical conducting spheres of radius $r = 0.15$ m are separated by a distance $a = 10.0$ m. What is the charge on each sphere if the potential of one is $+1500$ V and if the other is -1500 V?   *Answer:* $\pm 2.5 \times 10^{-8}$ C.

48. Two conducting spheres, one of radius 6.0 cm and the other of radius 12.0 cm, each have a charge of 3×10^{-8} C and are very far apart. If the spheres are connected by a conducting wire, find (a) the direction of motion and the magnitude of the charge transferred and (b) the final charge on and potential of each sphere.

49. In Fig. 29-19 let $R_1 = 1.0$ cm and $R_2 = 2.0$ cm. Before the spheres are connected by the fine wire, a charge of 2.0×10^{-7} C is placed on the smaller sphere, the larger sphere being uncharged. Calculate (a) the charge, (b) the surface charge density, and (c) the potential for each sphere after they are connected.
Answer: (a) $q_1 = 0.67 \times 10^{-7}$ C, $q_2 = 1.33 \times 10^{-7}$ C. (b) $\sigma_1 = 2.1 \times 10^{-4}$ C/m², $\sigma_2 = 1.1 \times 10^{-4}$ C/m². (c) $V_1 = V_2 = 6.0 \times 10^4$ V.

50. A spherical drop of water carrying a charge of 3×10^{-11} C has a potential of 500 V at its surface. (*a*) What is the radius of the drop? (*b*) If two such drops of the same charge and radius combine to form a single spherical drop, what is the potential at the surface of the new drop so formed?

SECTION 29-9

51. (*a*) How much charge is required to raise an isolated metallic sphere of 1.0-m radius to a potential of 1.0×10^6 V? Repeat for a sphere of 1.0-cm radius. (*b*) Why use a large sphere in an electrostatic generator since the same potential can be achieved for a smaller charge with a small sphere?
Answer: (*a*) 1.1×10^{-4} C; 1.1×10^{-6} C. (*b*) Because of the larger E field at the surface of the smaller sphere (verify), charge will leak off more rapidly.

52. An alpha particle is accelerated through a potential difference of one million volts in an electrostatic generator. (*a*) What kinetic energy does it acquire? (*b*) What kinetic energy would a proton acquire under these same circumstances? (*c*) Which particle would acquire the greater speed, starting from rest?

53. Let the potential difference between the shell of an electrostatic generator and the point at which charges are sprayed onto the moving belt be 3.0×10^6 V. If the belt transfers charge to the shell at the rate of 3.0×10^{-3} C/s, what power must be provided to drive the belt, considering only electrical forces?
Answer: 9.0 kW.

54. The high-voltage electrode of an electrostatic generator is a charged spherical metal shell having a potential V (+9.0×10^6 V). (*a*) Electrical breakdown occurs in the gas in this machine at a field E (1.0×10^8 V/m). In order to prevent such breakdown, what restriction must be made on the radius r of the shell? (*b*) A long moving rubber belt transfers charge to the shell at 3.0×10^{-4} C/s, the potential of the shell remaining constant because of leakage. What minimum power is required to transfer the charge? (*c*) The belt is of width w (0.50 m) and travels at speed v (30 m/s). What is the surface charge density on the belt?

30
capacitors and dielectrics

Figure 30-1 shows a generalized *capacitor*, consisting of two insulated conductors, *a* and *b*, of arbitrary shape (later, regardless of their geometry, we will call them *plates*). We assume that they are totally isolated from objects in their surroundings and carry equal and opposite charges $+q$ and $-q$, respectively. Every line of force that originates on *a* terminates on *b*. We further assume, for the time being, that the conductors *a* and *b* exist in a vacuum.

The capacitor of Fig. 30-1 is characterized by *q*, the magnitude of the *charge on either conductor*, and by *V*, the *potential difference between the conductors*. Note (*a*) that *q* is *not* the net charge on the capacitor, which is zero, and (*b*) that *V* is *not* the potential of either conductor, referred perhaps to $V \rightarrow 0$ at ∞, but the potential difference between them.

It is not hard to put equal and opposite charges on conductors such as those of Fig. 30-1. We need not charge them separately. All that we have to do is to connect the conductors momentarily to opposite poles of a battery. Equal and opposite charges ($\pm q$) will automatically appear.

For the moment we state without proof that *q* and *V* for a capacitor are proportional to each other, or

$$q = CV \tag{30-1}$$

in which *C*, the constant of proportionality, is called the *capacitance* of the capacitor. We state, again for the moment without proof, that *C* depends on the shapes and relative positions of the conductors. We will show in Section 30-2, for three important special cases, that *C* does indeed depend on these variables. *C* also depends on the medium in

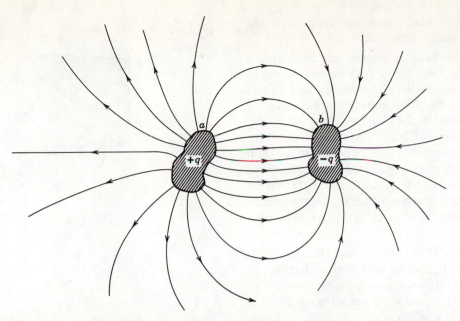

figure 30-1
Two insulated conductors, totally
isolated from their surroundings and
carrying equal and opposite charges,
form a capacitor.

651 CAPACITANCE SEC. 30-1

which the conductors are immersed but for the present (see Section 30-4 however) we assume this to be a vacuum.

The SI unit of capacitance that follows from Eq. 30-1 is the coulomb/volt. A special unit, the *farad* (abbr. F), is used to represent it. It is named in honor of Michael Faraday who, among other contributions, developed the concept of capacitance. Thus

$$1 \text{ farad} = 1 \text{ coulomb/volt}.$$

The submultiples of the farad, the *microfarad* $(1 \ \mu\text{F} = 10^{-6} \text{ F})$ and the *picofarad* $(1 \text{ pF} = 10^{-12} \text{ F})$, are more convenient units in practice.

An analogy can be made between a capacitor carrying a charge q and a rigid container of volume V containing n moles of an ideal gas. The gas pressure p is directly proportional to n for a fixed temperature, according to the ideal gas law (Eq. 23-2)

$$n = \left(\frac{V}{RT}\right) p.$$

For the capacitor (Eq. 30-1)

$$q = (C)V.$$

Comparison shows that the capacitance C of the capacitor, assuming a fixed temperature for the gas, is analogous to the volume V of the container.

Note that any amount of charge can be put on the capacitor, and any mass of gas can be put in the container, up to certain limits. These correspond to electrical breakdown ("arcing over") for the capacitor and to rupture of the walls for the container.

Capacitors are very useful devices, of great interest to engineers and physicists. For example:

1. In this book we stress the importance of *fields* to the understanding of natural phenomena. A capacitor can be used to establish desired electric field configurations for various purposes. In Section 27-5 we described the deflection of an electron beam in a uniform field set up by a capacitor, although we did not use this term in that section. In later sections we discuss the behavior of dielectric materials when placed in an electric field (provided conveniently by a capacitor) and

we shall see how the laws of electromagnetism can be generalized to take the presence of dielectric bodies into account.

2. A second concept stressed in this book is *energy*. By analyzing a charged capacitor we show that electric energy may be considered to be stored in the electric field between the plates and indeed in any electric field, however generated. Because capacitors can confine strong electric fields to small volumes, they can serve as useful devices for storing energy. In electron synchrotrons, which are cyclotron-like devices for accelerating electrons, energy accumulated and stored in a large bank of capacitors over a relatively long period of time is made available intermittently to accelerate the electrons by discharging the capacitor in a much shorter time. Many researches and devices in plasma physics also make use of bursts of energy stored in this way.

3. The electronic age could not exist without capacitors. They are used, in conjunction with other devices, to reduce voltage fluctuations in electronic power supplies, to transmit pulsed signals, to generate or detect electromagnetic oscillations at radio frequencies, to provide electronic time delays, and in many other ways. In most of these applications the potential difference between the plates will not be constant, as we assume in this chapter, but will vary with time, often in a sinusoidal or a pulsed fashion. In later chapters we consider some aspects of the capacitor used as a circuit element.

Figure 30-2 shows a *parallel-plate* capacitor in which the conductors of Fig. 30-1 take the form of two parallel plates of area A separated by a distance d. If we connect each plate momentarily to the terminals of a battery, a charge $+q$ will automatically appear on one plate and a charge $-q$ on the other. If d is small compared with the plate dimensions, the electric field **E** between the plates will be uniform, which means that the lines of force will be parallel and evenly spaced. The laws of electromagnetism (see Problem 14, Chapter 35) require that there be some "fringing" of the lines at the edges of the plates, but for small enough d we can neglect it for our present purpose.

We can calculate the capacitance of this device using Gauss's law, another illustration of the usefulness of this law in situations of simple geometry. Figure 30-2 shows (dashed lines) a Gaussian surface of height h closed by plane caps of area A that are the shape and size of the capacitor plates. The flux of **E** is zero for the part of the Gaussian surface that lies inside the top capacitor plate because the electric field inside a conductor carrying a static charge is zero. The flux of **E** through the vertical wall of the Gaussian surface is zero because, to the extent that we can neglect the fringing of the lines of force, **E** lies in the wall.

This leaves only the face of the Gaussian surface that lies between

30-2
CALCULATING CAPACITANCE

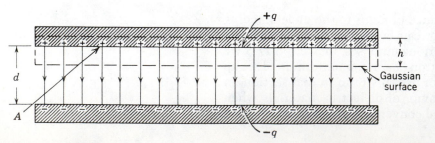

figure 30-2
A parallel-plate capacitor with conductors (plates) of area A. The dashed line represents a Gaussian surface whose height is h and whose top and bottom caps are the same shape and size as the capacitor plates.

the plates. Here **E** is constant and the flux Φ_E is simply EA. Gauss's law (Eq. 28-5) gives

$$\epsilon_0 \Phi_E = \epsilon_0 EA = q. \tag{30-2}$$

The work required to carry a test charge q_0 from one plate to the other can be expressed either as $q_0 V$ (see Eq. 29-1) or as the product of a force $q_0 E$ times a distance d or $q_0 Ed$. These expressions must be equal, or

$$V = Ed. \tag{30-3}$$

Formally, Eq. 30-3 is a special case of the general relation (Eq. 29-5; see also Example 1, Chapter 29)

$$V = -\int \mathbf{E} \cdot d\mathbf{l},$$

where V is the difference in potential between the plates. The integral may be taken over any path that starts on one plate and ends on the other because each plate is an equipotential surface and the electrostatic force is path independent. Although the simplest path between the plates is a perpendicular straight line, Eq. 30-3 follows no matter what path of integration we choose.

If we substitute Eqs. 30-2 and 30-3 into the relation $C = q/V$, we obtain

$$C = \frac{q}{V} = \frac{\epsilon_0 EA}{Ed} = \frac{\epsilon_0 A}{d}. \tag{30-4}$$

Equation 30-4 holds only for capacitors of the parallel-plate type; different formulas hold for capacitors of different geometry. This equation shows, for a particular case, that C does indeed depend on the geometry of the conductors (plates) as we pointed out in Section 30-1. Both A and d are geometrical factors.

In Section 26-4 we stated that ϵ_0, which we first met in connection with Coulomb's law, was not measured in terms of that law because of experimental difficulties. Equation 30-4 suggests that we might determine ϵ_0 by building a capacitor of accurately known plate area and plate spacing and measuring its capacitance experimentally by measuring q and V in the relation $C = q/V$. Thus we can solve Eq. 30-4 for ϵ_0 and find a numerical value in terms of the measured quantities A, d, and C; ϵ_0 has indeed been measured in this way.

EXAMPLE 1

The parallel plates of an air-filled capacitor are everywhere 1.0 mm apart. What must the plate area be if the capacitance is to be 1.0 F?

From Eq. 30-4

$$A = \frac{dC}{\epsilon_0} = \frac{(1.0 \times 10^{-3}\ \text{m})(1.0\ \text{F})}{8.9 \times 10^{-12}\ \text{C}^2/\text{N} \cdot \text{m}^2} = 1.1 \times 10^8\ \text{m}^2.$$

This is the area of a square sheet more than 6 miles on edge; the farad is indeed a large unit.

EXAMPLE 2

A cylindrical capacitor. A cylindrical capacitor consists of two coaxial cylinders (Fig. 30-3) of radii a and b and length l. What is the capacitance of this device? Assume that the capacitor is very long (that is, that $l \gg b$) so that we can ignore the fringing of the lines of force at the ends for the purpose of calculating the capacitance.

As a Gaussian surface construct a coaxial cylinder of radius r and length l, closed by plane caps. Gauss's law (Eq. 28-6)

$$\epsilon_0 \oint \mathbf{E} \cdot d\mathbf{S} = q$$

gives

$$\epsilon_0 E(2\pi r)(l) = q,$$

the flux being entirely through the cylindrical surface and not through the end caps. Solving for E yields

$$E = \frac{q}{2\pi\epsilon_0 r l}.$$

The potential difference between the plates is given by Eq. 29-5 [note that $\mathbf{E}$ and $d\mathbf{l}$ ($= d\mathbf{r}$) point in opposite directions] or

$$V = -\int_a^b \mathbf{E} \cdot d\mathbf{l} = \int_a^b E \, dr = \int_a^b \frac{q}{2\pi\epsilon_0 l} \frac{dr}{r} = \frac{q}{2\pi\epsilon_0 l} \ln \frac{b}{a}.$$

Finally, the capacitance is given by

$$C = \frac{q}{V} = \frac{2\pi\epsilon_0 l}{\ln (b/a)}.$$

Like the relation for the parallel-plate capacitor (Eq. 30-4), this relation also depends only on geometrical factors, l, b, and a.

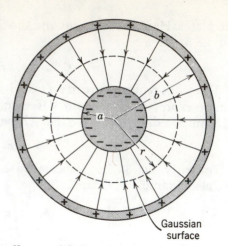

figure 30-3
Example 2. A cross section of a cylindrical capacitor. The dashed circle is a cross section of a closed cylindrical Gaussian surface of radius r and length l.

EXAMPLE 3

The Capacitance of an isolated sphere. In Section 29-8 we showed that the potential of an isolated conducting sphere of radius R carrying a charge q is given by

$$V = \frac{1}{4\pi\epsilon_0} \frac{q}{R}. \tag{30-5}$$

We can regard this sphere as one plate of a capacitor, the other plate being a conducting sphere of infinite radius, with V chosen to be zero on the sphere at infinity.

The capacitance of the sphere of radius R is then given, from Eq. 30-5, by

$$C = \frac{q}{V} = 4\pi\epsilon_0 R. \tag{30-6}$$

Again, the only relevant geometric factor, the sphere radius R, appears.

EXAMPLE 4

Capacitors in parallel. Figure 30-4 shows three capacitors connected in parallel. .What single capacitance C is equivalent to this combination? "Equivalent" means that if the parallel combination and the single capacitor were each in a box with wires a and b connected to terminals, it would not be possible to distinguish the two by electrical measurements external to the box.

The potential difference across each capacitor in a parallel arrangement (Fig. 30-4) *is the same.* This follows because all of the upper plates are connected together and to terminal a, whereas all of the lower plates are connected together and to terminal b. Applying the relation $q = CV$ to each capacitor yields

$$q_1 = C_1 V; \qquad q_2 = C_2 V; \qquad \text{and} \qquad q_3 = C_3 V.$$

The total charge q on the combination is

$$q = q_1 + q_2 + q_3$$
$$= (C_1 + C_2 + C_3)V.$$

The equivalent capacitance C is

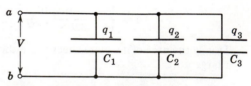

figure 30-4
Example 4. Three capacitors in parallel. The symbol for a capacitor (—||—) is chosen for its simplicity. Although it suggests a parallel-plate capacitor, it is meant to suggest a capacitor of *any* geometry.

$$C = \frac{q}{V} = C_1 + C_2 + C_3.$$

We can easily extend this result to any number of parallel-connected capacitors.

Capacitors in series. Figure 30-5 shows three capacitors connected in series. What single capacitance C is "equivalent" (see Example 4) to this combination?

For capacitors connected in a series arrangement (Fig. 30-5) *the magnitude q of the charge on each plate must be the same.* This is true because the net charge on the part of the circuit enclosed by the dashed line in Fig. 30-5 must be zero; that is, the charge present on these plates initially is zero and connecting a battery between a and b will only produce a charge separation, the *net* charge on these plates still being zero. Assuming that neither C_1 nor C_2 "sparks over," there is no way for charge to enter or leave the volume suggested by the dashed line.

Applying the relation $q = CV$ to each capacitor yields

$$V_1 = q/C_1; \qquad V_2 = q/C_2; \qquad \text{and} \qquad V_3 = q/C_3.$$

The potential difference for the series combination is

$$V = V_1 + V_2 + V_3$$

$$= q\left(\frac{1}{C_1} + \frac{1}{C_2} + \frac{1}{C_3}\right).$$

The equivalent capacitance

$$C = \frac{q}{V} = \frac{1}{\dfrac{1}{C_1} + \dfrac{1}{C_2} + \dfrac{1}{C_3}},$$

or

$$\frac{1}{C} = \frac{1}{C_1} + \frac{1}{C_2} + \frac{1}{C_3}.$$

The equivalent series capacitance is always less than the smallest capacitance in the chain.

EXAMPLE 5

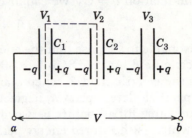

figure 30-5
Example 5. Three capacitors in series.

30-3
ENERGY STORAGE IN AN ELECTRIC FIELD

In Section 29-6 we saw that all charge configurations have a certain *electric potential energy U*, equal to the work W (which may be positive or negative) that must be done to assemble them from their individual components, originally assumed to be infinitely far apart and at rest. This potential energy reminds us of the potential energy stored in a compressed spring or the gravitational potential energy stored in, say, the earth-moon system.

For a simple example, work must be done to separate two equal and opposite charges. This energy is stored in the system and can be recovered if the charges are allowed to come together again. Similarly, a charged capacitor has stored in it an electrical potential energy U equal to the work W required to charge it. This energy can be recovered if the capacitor is allowed to discharge. We can visualize the work of charging by imagining that an external agent pulls electrons from the positive plate and pushes them onto the negative plate, thus bringing about the charge separation; normally the work of charging is done by a battery, at the expense of its store of chemical energy.

Suppose that at a time t a charge $q'(t)$ has been transferred from one plate to the other. The potential difference $V(t)$ between the plates at that moment will be $q'(t)/C$. If an extra increment of charge dq' is transferred, the small amount of additional work needed will be

$$dW = V\,dq = \left(\frac{q'}{C}\right)dq'.$$

If this process is continued until a total charge q has been transferred, the total work will be found from

$$W = \int dW = \int_0^q \frac{q'}{C}\,dq' = \frac{1}{2}\frac{q^2}{C}. \qquad (30\text{-}7)$$

From the relation $q = CV$ we can also write this as

$$W\;(= U) = \tfrac{1}{2}CV^2. \qquad (30\text{-}8)$$

It is reasonable to suppose that the energy stored in a capacitor resides in the electric field. As q or V in Eqs. 30-7 and 30-8 increase, for example, so does the electric field E; when q and V are zero, so is E.

In a parallel-plate capacitor, neglecting fringing, the electric field has the same value for all points between the plates. Thus the *energy density u*, which is the stored energy per unit volume, should also be uniform; u (see Eq. 30-8) is given by

$$u = \frac{U}{Ad} = \frac{\tfrac{1}{2}CV^2}{Ad},$$

where Ad is the volume between the plates. Substituting the relation $C = \epsilon_0 A/d$ (Eq. 30-4) leads to

$$u = \frac{\epsilon_0}{2}\left(\frac{V}{d}\right)^2.$$

However, V/d is the electric field E, so that

$$u = \tfrac{1}{2}\epsilon_0 E^2. \qquad (30\text{-}9)$$

Although we derived this equation for the special case of a parallel-plate capacitor, it is true in general. *If an electric field $\mathbf{E}$ exists at any point in space (a vacuum), we can think of that point as the site of stored energy in amount, per unit volume, of $\tfrac{1}{2}\epsilon_0 E^2$.*

EXAMPLE 6

A capacitor C_1 is charged to a potential difference V_0. This charging battery is then removed and the capacitor is connected as in Fig. 30-6 to an uncharged capacitor C_2.

(a) What is the final potential difference V across the combination?

The original charge q_0 is now shared by the two capacitors. Thus

$$q_0 = q_1 + q_2.$$

Applying the relation $q = CV$ to each of these terms yields

$$C_1 V_0 = C_1 V + C_2 V$$

or

$$V = V_0 \frac{C_1}{C_1 + C_2}.$$

This suggests a way to measure an unknown capacitance (C_2, say) in terms of a known one.

(b) What is the stored energy before and after the switch in Fig. 30-6 is thrown?

The initial stored energy is

$$U_0 = \tfrac{1}{2}C_1 V_0^2.$$

The final stored energy is

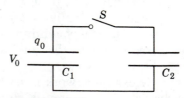

figure 30-6
Example 6. C_1 is charged to a potential difference V_0 and then connected to C_2 by closing switch S.

$$U = \tfrac{1}{2}C_1V^2 + \tfrac{1}{2}C_2V^2 = \tfrac{1}{2}(C_1 + C_2)\left(\frac{V_0C_1}{C_1 + C_2}\right)^2 = \left(\frac{C_1}{C_1 + C_2}\right)U_0.$$

Thus U is less than U_0! For $C_1 = C_2$, in fact, $U = \tfrac{1}{2}U_0$.

This is *not* a violation of the principle of the conservation of energy. The example tacitly assumes an ideal (rather than an actual laboratory) circuit in that the resistance (Chapter 31) and the inductance (Chapter 36) of the connecting wires have both been assumed to be zero. In an actual laboratory circuit the "missing" energy would appear as thermal energy in the wires and/or as energy radiated away from the circuit as electromagnetic radiation (Chapter 41). For a good discussion see "On Conservation of Energy in Electric Circuits" by Camillo Cuvaj, *American Journal of Physics*, 1968.

An isolated conducting sphere of radius R, in a vacuum, carries a charge q. (*a*) Compute the total electrostatic energy stored in the surrounding space. At any radius r from the center of the sphere (assuming $r > R$) E is given by

$$E = \frac{1}{4\pi\epsilon_0}\frac{q}{r^2}.$$

EXAMPLE 7

The energy density at any radius r is found from Eq. 30-9, or

$$u = \tfrac{1}{2}\epsilon_0 E^2 = \frac{q^2}{32\pi^2\epsilon_0 r^4}.$$

The energy dU that lies in a spherical shell between the radii r and $r + dr$ is

$$dU = (4\pi r^2)(dr)u = \frac{q^2}{8\pi\epsilon_0}\frac{dr}{r^2},$$

where $(4\pi r^2)(dr)$ is the volume of the spherical shell. The total energy U is found by integration, or

$$U = \int dU = \frac{q^2}{8\pi\epsilon_0}\int_R^\infty \frac{dr}{r^2} = \frac{q^2}{8\pi\epsilon_0 R}.$$

Note that this relation follows at once from Eq. 30-7 ($U = q^2/2C$), where C (see Example 3) is the capacitance ($= 4\pi\epsilon_0 R$) of an isolated sphere of radius R.

(*b*) What is the radius R_0 of a spherical surface such that half the stored energy lies within it?

In the equation just given we put

$$\tfrac{1}{2}U = \frac{q^2}{8\pi\epsilon_0}\int_R^{R_0} \frac{dr}{r^2}$$

or

$$\frac{q^2}{16\pi\epsilon_0 R} = \frac{q^2}{8\pi\epsilon_0}\left(\frac{1}{R} - \frac{1}{R_0}\right),$$

which yields, after some rearrangement,

$$R_0 = 2R.$$

Thus, most of the energy is stored in space rather close to the sphere.

Equation 30-4 holds only for a parallel-plate capacitor with its plates in a vacuum. Michael Faraday, in 1837, first investigated the effect of filling the space between the plates with a dielectric (see Table 30-1 for a sampling of dielectrics used today). In Faraday's words:

30-4
PARALLEL-PLATE CAPACITOR WITH DIELECTRIC

The question may be stated thus: suppose A an electrified plate of metal suspended in air, and B and C two exactly similar plates, placed parallel to and on each side of A at equal distances and insulated; A will then induce equally to-

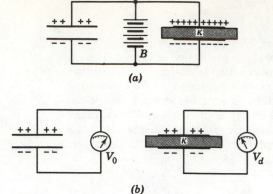

figure 30-7
(a) Battery B supplies the same potential difference to each capacitor; the one on the right has the higher charge. (b) Both capacitors carry the same charge; the one on the right has the lower potential difference, as indicated by the meter readings.

ward B and C [that is, equal charges will appear on these plates]. If in this position of the plates some other dielectric than air, as shell-lac, be introduced between A and C, will the induction between them remain the same? Will the relation of C and B to A be unaltered, notwithstanding the difference of the dielectrics interposed between them?

Faraday answered this question by constructing two identical capacitors, in one of which he placed a dielectric, the other containing air at normal pressure. When both capacitors were charged to the *same potential difference*, Faraday found by experiment that *the charge on the one containing the dielectric was greater than that on the other*; see Fig. 30-7a.

Since q is larger, for the same V, if a dielectric is present, it follows from the relation $C = q/V$ that the capacitance of a capacitor increases if a dielectric is placed between the plates. The ratio of the capacitance with the dielectric* to that without is called the *dielectric constant κ* of the material; see Table 30-1.

Instead of maintaining the two capacitors at the same potential difference, we can place the *same charge* on them, as in Fig. 30-7b. Experiment then shows that the potential difference V_d between the plates of the right-hand capacitor is smaller than that for the left-hand capacitor by the factor $1/\kappa$, or

$$V_d = V_0/\kappa.$$

We are led once again to conclude, from the relation $C = q/V$, that the effect of the dielectric is to increase the capacitance by a factor κ.

For a parallel-plate capacitor we can write, as an experimental result,

$$C = \frac{\kappa \epsilon_0 A}{d}. \tag{30-10}$$

Equation 30-4 is a special case of this relation found by putting $\kappa = 1$, corresponding to a vacuum between the plates. Experiment shows that the capacitance of *all* types of capacitor is increased by the factor κ if the space between the plates is filled with a dielectric. Thus the capacitance of any capacitor can be written as

$$C = \kappa \epsilon_0 L,$$

where L depends on the geometry and has the dimensions of a length. For a parallel-plate capacitor (see Eq. 30-4) L is A/d; for a cylindrical capacitor (see Example 2) it is $2\pi l/\ln (b/a)$.

---

* Assumed to fill completely the space between the plates.

Table 30-1
Properties of some dielectrics*

Material	Dielectric Constant	Dielectric Strength** (kV/mm)
Vacuum	1.00000	∞
Air	1.00054	0.8
Water	78	—
Paper	3.5	14
Ruby mica	5.4	160
Porcelain	6.5	4
Fused quartz	3.8	8
Pyrex glass	4.5	13
Bakelite	4.8	12
Polyethylene	2.3	50
Amber	2.7	90
Polystyrene	2.6	25
Teflon	2.1	60
Neoprene	6.9	12
Transformer oil	4.5	12
Titanium dioxide	100	6

* These properties are at approximately room temperature and for conditions such that the electric field **E** in the dielectric does not vary with time.
** This is the maximum potential gradient that may exist in the dielectric without the occurrence of electrical breakdown. Dielectrics are often placed between conducting plates to permit a higher potential difference to be applied between them than would be possible with air as the dielectric.

EXAMPLE 8

A parallel-plate capacitor has plates with area A and separation d. A battery charges the plates to a potential difference V_0. The battery is then disconnected, and a dielectric slab of thickness d is introduced. Calculate the stored energy both before and after the slab is introduced and account for any difference.

The energy U_0 before introducing the slab is

$$U_0 = \tfrac{1}{2} C_0 V_0^2.$$

After the slab is in place, we have

$$C = \kappa C_0 \quad \text{and} \quad V = V_0/\kappa$$

and thus

$$U = \tfrac{1}{2} C V^2 = \tfrac{1}{2} \kappa C_0 \left(\frac{V_0}{\kappa}\right)^2 = \frac{1}{\kappa} U_0.$$

The energy after the slab is introduced is *less* by a factor $1/\kappa$. The "missing" energy would be apparent to the person who inserted the slab. He would feel a "tug" on the slab and would have to restrain it if he wished to insert the slab without acceleration. This means that he would have to do negative work on it, or, alternatively, that the capacitor + slab system would do positive work on him. This positive work is

$$W = U_0 - U = \tfrac{1}{2} C_0 V_0^2 \left(1 - \frac{1}{\kappa}\right).$$

As expected, $W = 0$ for the case of $\kappa = 1$.

The following section will give some detailed insight into how the "tug" referred to above arises, in terms of the attraction between what we will call "free" charges on the capacitor plates and "induced" charges on the dielectric.

Note from the relation $U = \frac{1}{2}CV^2$ (see Eq. 30-8) that we can derive the *energy density u*, for a parallel-plate capacitor in which a dielectric slab is present, from

$$u = \frac{U}{(Ad)} = \left(\frac{1}{Ad}\right)\left(\tfrac{1}{2}CV^2\right) = \left(\frac{1}{Ad}\right)\left(\tfrac{1}{2}\right)\left(\frac{\epsilon_0\kappa A}{d}\right)(V^2).$$

But $E = V/d$ so that we have

$$u = \tfrac{1}{2}\epsilon_0\kappa E^2.$$

As for Eq. 30-9, this relation, although derived for a parallel-plate capacitor, holds in general; that is, at any point P in a dielectric of constant κ. As we expect, for $\kappa = 1$, this new relation reduces to Eq. 30-9.

30-5 DIELECTRICS— AN ATOMIC VIEW

We now seek to understand, in atomic terms, what happens when we place a dielectric in an electric field. There are two possibilities. The molecules of some dielectrics, like water (see Fig. 29-11), have permanent electric dipole moments. In such materials (called *polar*) the electric dipole moments **p** tend to align themselves with an external electric field, as in Fig. 30-8b; see also section 27-6. Because the molecules are in constant thermal agitation, the degree of alignment will not be complete but will increase as the applied electric field is increased or as the temperature is decreased.

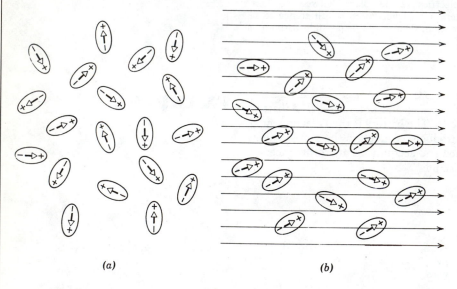

(a)   (b)

figure 30-8
(a) Molecules with a permanent electric dipole moment, showing their random orientation in the absence of an external electric field. (b) An electric field is applied, producing partial alignment of the dipoles. Thermal agitation prevents complete alignment.

Whether or not the molecules have permanent electric dipole moments, they acquire them by *induction* when placed in an electric field. In Section 29-5 we saw that the external electric field tends to separate the negative and the positive charge in the atom or molecule. This *induced electric dipole moment* is present only when the electric field is present. It is proportional to the electric field (for normal field strengths) and is created already lined up with the electric field as Fig. 29-12 suggests.

Let us use a parallel-plate capacitor, carrying a fixed charge q and not connected to a battery (see Fig. 30-7b), to provide a uniform external electric field $\mathbf{E}_0$ into which we place a dielectric slab. The over-all effect of alignment and induction is to separate the center of positive charge of the entire slab slightly from the center of negative charge. The slab, as

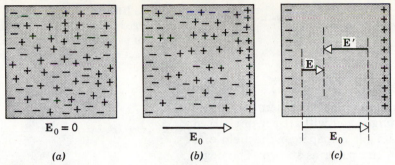

figure 30-9
(a) A dielectric slab, showing the random distribution of plus and minus charges. (b) An external field E_0, established by putting the slab between the plates of a parallel-plate capacitor (not shown), separates the center of plus charge in the slab slightly from the center of minus charge, resulting in the appearance of surface charges. No *net* charge exists in any volume element located in the *interior* of the slab. (c) The surface charges set up a field E' which opposes the external field E_0 associated with the charges on the capacitor plates. The resultant field E ($= E_0 + E'$) in the dielectric is thus less than E_0.

a whole, although remaining electrically neutral, becomes *polarized*, as Fig. 30-9b suggests. The net effect is a pile-up of positive charge on the right face of the slab and of negative charge on the left face; within the slab no excess charge appears in any given volume element. Since the slab as a whole remains neutral, the positive *induced surface charge* must be equal in magnitude to the negative induced surface charge. Note that in this process electrons in the dielectric are displaced from their equilibrium positions by distances that are considerably less than an atomic diameter. There is no transfer of charge over macroscopic distances such as occurs when a current is set up in a conductor.

Figure 30-9c shows that the induced surface charges will always appear in such a way that the electric field set up *by them* (E') opposes the external electric field E_0. The *resultant* field in the dielectric E is the vector sum of E_0 and E'. It points in the same direction as E_0 but is smaller. *If we place a dielectric in an electric field, induced surface charges appear which tend to weaken the original field within the dielectric.*

This weakening of the electric field reveals itself in Fig. 30-7b as a reduction in potential difference between the plates of a charged isolated capacitor when a dielectric is introduced between the plates. The relation $V = Ed$ for a parallel-plate capacitor (see Eq. 30-3) holds whether or not dielectric is present and shows that the reduction in V described in Fig. 30-7b is directly connected to the reduction in E described in Fig. 30-9. More specifically, if a dielectric slab is introduced into an isolated charged parallel-plate capacitor, then

$$\frac{E_0}{E} = \frac{V_0}{V_d} = \kappa \qquad (30\text{-}11)$$

where the symbols on the left refer to Fig. 30-9 and the symbols V_0 and V_d refer to Fig. 30-7b.*

---

* Equation 30-11 does not hold if the battery remains connected while the dielectric slab is introduced. In this case V (hence E) could not change. Instead, the charge q on the capacitor plates would increase by a factor κ, as Fig. 30-7a suggests.

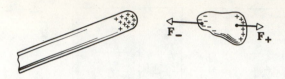

figure 30-10
A charged rod attracts an uncharged
piece of paper because unbalanced
forces act on the induced surface
charges.

Induced surface charge is the explanation of the most elementary fact
of static electricity, namely, that a charged rod will attract uncharged
bits of paper, etc. Figure 30-10 shows a bit of paper in the field of a
charged rod. Surface charges appear on the paper as shown. The negatively charged end of the paper will be pulled toward the rod and the
positively charged end will be repelled. These two forces do not have the
same magnitude because the negative end, being closer to the rod, is in
a stronger field and experiences a stronger force. The net effect is an
attraction. A dielectric body in a *uniform* electric field will not experience a net force.

In Example 8 we pointed out that, if we insert a dielectric slab into a
parallel-plate capacitor carrying a fixed charge q, a force will act on the
slab drawing it into the capacitor. This force is provided by the electrostatic attraction between the charges $\pm q$ on the capacitor plates and the
induced surfaces charges $\mp q'$ on the dielectric slab. When the slab is
only part way into the capacitor neither q nor q' will be uniformly distributed. Sketch qualitatively a possible distribution for q and q' when
the slab is, say, halfway into the capacitor.

30-6
DIELECTRICS AND GAUSS'S LAW

So far our use of Gauss's law has been confined to situations in which
no dielectric was present. Now let us apply this law to a parallel-plate
capacitor filled with a dielectric of dielectric constant κ.

Figure 30-11 shows the capacitor both with and without the dielectric. We assume that the charge q on the plates is the same in each case.
Gaussian surfaces have been drawn after the fashion of Fig. 30-2.

If no dielectric is present (Fig. 30-11a), Gauss's law (see Eq. 30-2)
gives

$$\epsilon_0 \oint \mathbf{E} \cdot d\mathbf{S} = \epsilon_0 E_0 A = q$$

or
$$E_0 = \frac{q}{\epsilon_0 A}. \tag{30-12}$$

If the dielectric is present (Fig. 30-11b), Gauss's law gives

$$\epsilon_0 \oint \mathbf{E} \cdot d\mathbf{S} = \epsilon_0 E A = q - q'$$

or
$$E = \frac{q}{\epsilon_0 A} - \frac{q'}{\epsilon_0 A}, \tag{30-13}$$

in which $-q'$, the *induced surface charge*, must be distinguished from
q, the so-called *free charge* on the plates. These two charges, both of
which lie within the Gaussian surface, are opposite in sign; $q - q'$ is
the *net* charge within the Gaussian surface.

Equation 30-11 shows that in Fig. 30-11

$$E = \frac{E_0}{\kappa}.$$

Combining this with Eq. 39-12, we have

$$E = \frac{E_0}{\kappa} = \frac{q}{\kappa \epsilon_0 A}. \tag{30-14}$$

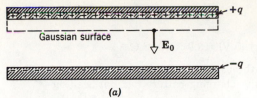

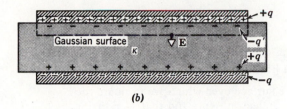

figure 30-11
A parallel-plate capacitor (*a*) without and (*b*) with a dielectric. The charge *q* on the plates is assumed to be the same in each case.

663 DIELECTRICS AND GAUSS'S LAW   SEC. 30-6

Inserting this in Eq. 30-13 yields

$$\frac{q}{\kappa\epsilon_0 A} = \frac{q}{\epsilon_0 A} - \frac{q'}{\epsilon_0 A} \tag{30-15a}$$

or

$$q' = q\left(1 - \frac{1}{\kappa}\right). \tag{30-15b}$$

This shows correctly that the induced surface charge q' is always less in magnitude than the free charge q and is equal to zero if no dielectric is present, that is, if $\kappa = 1$.

Now we write Gauss's law for the case of Fig. 30-11*b* in the form

$$\epsilon_0 \oint \mathbf{E} \cdot d\mathbf{S} = q - q', \tag{30-16}$$

$q - q'$ again being the net charge within the Gaussian surface. Substituting from Eq. 30-15*b* for q' leads, after some rearrangement, to

$$\epsilon_0 \oint \kappa \mathbf{E} \cdot d\mathbf{S} = q. \tag{30-17}$$

This important relation, although derived for a parallel-plate capacitor, is true generally and is the form in which Gauss's law is usually written when dielectrics are present. Note the following:

1. The flux integral now contains a factor κ.

2. The charge q contained within the Gaussian surface is taken to be the *free charge only*. Induced surface charge is deliberately ignored on the right side of this equation, having been taken into account by the introduction of κ on the left side. Equations 30-16 and 30-17 are completely equivalent formulations.

Figure 30-12 shows a dielectric slab of thickness *b* and dielectric constant κ placed between the plates of a parallel-plate capacitor of plate area *A* and separation *d*. A potential difference V_0 is applied with no dielectric present. The battery is then disconnected and the dielectric slab inserted. Assume that $A = 100$ cm², $d = 1.0$ cm, $b = 0.50$ cm, $\kappa = 7.0$, and $V_0 = 100$ V and (*a*) calculate the capacitance C_0 before the slab is inserted.

EXAMPLE 9

From Eq. 30-4, C_0 is:

$$C_0 = \frac{\epsilon_0 A}{d} = \frac{(8.9 \times 10^{-12} \text{ C}^2/\text{N} \cdot \text{m}^2)(10^{-2} \text{ m}^2)}{10^{-2} \text{ m}} = 8.9 \times 10^{-12} \text{ F} = 8.9 \text{ pF}.$$

(b) Calculate the free charge q.
From Eq. 30-1,

$$q = C_0 V_0 = (8.9 \times 10^{-12} \text{ F})(100 \text{ V}) = 8.9 \times 10^{-10} \text{ C}.$$

Because of the technique used to charge the capacitor, the free charge remains unchanged as the slab is introduced. If the charging battery had *not* been disconnected, this would not be the case.

(c) Calculate the electric field in the gap.
Applying Gauss's law in the form given in Eq. 30-17 to the Gaussian surface of Fig. 30-12 (upper plate) yields

$$\epsilon_0 \oint \kappa \mathbf{E} \cdot d\mathbf{S} = \epsilon_0 E_0 A = q,$$

or $$E_0 = \frac{q}{\epsilon_0 A} = \frac{8.9 \times 10^{-10} \text{ C}}{(8.9 \times 10^{-12} \text{ C}^2/\text{N} \cdot \text{m}^2)(10^{-2} \text{ m}^2)} = 1.0 \times 10^4 \text{ V/m}.$$

Note that we put $\kappa = 1$ here because the surface over which we evaluate the flux integral does not pass through any dielectric. Note too that E_0 remains unchanged when the slab is introduced; this derivation takes no specific account of the presence of the dielectric.

(d) Calculate the electric field in the dielectric.

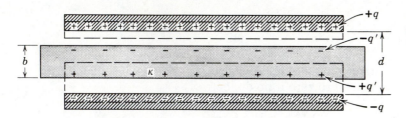

figure 30-12
Example 9. A parallel-plate capacitor containing a dielectric slab.

Applying Eq. 30-17 to the Gaussian surface of Fig. 30-12 (lower plate) yields

$$\epsilon_0 \oint \kappa \mathbf{E} \cdot d\mathbf{S} = \epsilon_0 \kappa E A = q.$$

Note that κ appears here because the surface cuts through the dielectric and that only the free charge q appears on the right. Thus we have

$$E = \frac{q}{\kappa \epsilon_0 A} = \frac{E_0}{\kappa} = \frac{1.0 \times 10^4 \text{ V/m}}{7.0} = 0.14 \times 10^4 \text{ V/m}.$$

(e) Calculate the potential difference between the plates.
Applying Eq. 29-5 to a straight perpendicular path from the lower plate (L) to the upper one (U) yields

$$V = -\int_L^U \mathbf{E} \cdot d\mathbf{l} = -\int_L^U E \cos 180° \; dl = \int_L^U E \; dl = E_0(d - b) + Eb.$$

Numerically

$$V = (1.0 \times 10^4 \text{ V/m})(5 \times 10^{-3} \text{ m}) + (0.14 \times 10^4 \text{ V/m})(5 \times 10^{-3} \text{ m}) = 57 \text{ V}.$$

This contrasts with the original applied potential difference of 100 V; compare Fig. 30-7b.

(f) Calculate the capacitance with the slab in place.
From Eq. 30-1,

$$C = \frac{q}{V} = \frac{8.9 \times 10^{-10} \text{ C}}{57 \text{ V}} = 16 \text{ pF}.$$

When the dielectric slab is introduced, the potential difference drops from 100 to 57 V and the capacitance rises from 8.9 to 16 pF, a factor of 1.8. If the dielectric slab had filled the capacitor, the capacitance would have risen by a factor of $\kappa \; (= 7.0)$ to 62 pF.

For all situations that we encounter in this book our discussion of the behavior of dielectrics in an electric field is adequate. However, the problems that we treat are simple ones, such as that of a rectangular slab placed at right angles to a uniform external electric field. For more difficult problems, such as that of finding **E** at the center of a dielectric ellipsoid placed in a (possibly nonuniform) external electric field, it greatly simplifies the labor and leads to deeper insight if we introduce a new formalism. We do so largely so that students who take a second course in electromagnetism will have some familiarity with the concepts.

Let us rewrite Eq. 30-15a, which applies to a parallel-plate capacitor containing a dielectric, as

$$\frac{q}{A} = \epsilon_0 \left(\frac{q}{\kappa \epsilon_0 A} \right) + \frac{q'}{A}. \qquad (30\text{-}18)$$

The quantity in parentheses (see Eq. 30-14) is simply the electric field E in the dielectric. The last term in Eq. 30-18 is the *induced surface charge per unit area*. We call it the *electric polarization P*, or

$$P = \frac{q'}{A}. \qquad (30\text{-}19)$$

The name is suitable because the induced surface charge q' (also called the polarization charge) appears when the dielectric is polarized.

The electric polarization P can be defined in an equivalent way by multiplying the numerator and denominator in Eq. 30-19 by d, the thickness of the dielectric slab in Fig. 30-11,

$$P = \frac{q'd}{Ad}. \qquad (30\text{-}20)$$

The numerator is the product $q'd$ of the magnitude of the (equal and opposite) polarization charges by their separation. It is thus the induced electric dipole moment of the dielectric slab. Since the denominator Ad is the volume of the slab, we see that the electric polarization can also be defined as the induced electric dipole moment per unit volume in the dielectric. This definition suggests that since the electric dipole moment is a vector the electric polarization is also a vector, its magnitude being P. The direction of **P** is from the negative induced charge to the positive induced charge, as for any dipole. In Fig. 30-13, which shows a capacitor with a dielectric slab filling half the space between the plates, **P** points down.

We can now rewrite Eq. 30-18 as

$$\frac{q}{A} = \epsilon_0 E + P. \qquad (30\text{-}21)$$

The quantity on the right occurs so often in electrostatic problems that we give it the special name *electric displacement D*, or

$$D = \epsilon_0 E + P \qquad (30\text{-}22a)$$

in which

$$D = \frac{q}{A}. \qquad (30\text{-}22b)$$

The name has historical significance only.

Since **E** and **P** are vectors, **D** must also be one, so that in the more general case we have

$$\mathbf{D} = \epsilon_0 \mathbf{E} + \mathbf{P}. \qquad (30\text{-}23)$$

In Fig. 30-13 all three vectors point down and each has a constant magnitude for every point in the dielectric (and also at every point in the air gap) so that the vector nature of Eq. 30-23 is not very important in this case. In more compli-

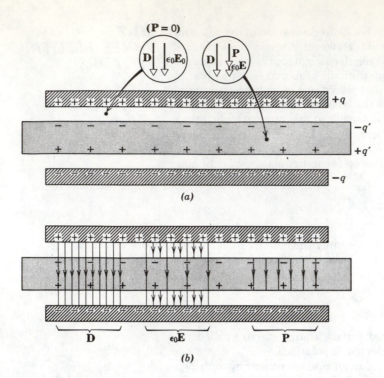

figure 30-13
(a) Showing **D**, ϵ_0**E**, and **P** in the dielectric (*upper right*) and in the gap (*upper left*) for a parallel-plate capacitor. (b) Showing samples of the lines associated with **D** (free charge), ϵ_0**E** (all charges), and **P** (polarization charge).

cated problems, however, **E**, **P**, and **D** may vary in magnitude and direction from point to point.

From their definitions we see the following:

1. **D** (see Eq. 30-22b) is connected with the *free charge* only. We can represent the vector field of **D** by *lines of* **D**, just as we represent the field of **E** by lines of force. Figure 30-13b shows that the lines of **D** begin and end on the free charges.

2. **P** (see Eq. 30-19) is connected with the *polarization charge* only. It is also possible to represent this vector field by lines. Figure 30-13b shows that the lines of **P** begin and end on the polarization charges.

3. **E** is connected with *all* charges that are actually present, whether free or polarization. The lines of **E** reflect the presence of both kinds of charge, as Fig. 30-13b shows. Note (Eqs. 30-19 and 30-22b) that the units for **P** and **D** (C/m²) differ from those of **E** (N/C).

The electric field vector **E**, which is what determines the force that acts on a suitably placed test charge, remains of fundamental interest. **D** and **P** are auxiliary vectors useful as aids in the solution of problems more complex than that of Fig. 30-13.

The vectors **D** and **P** can both be expressed in terms of **E** alone. A convenient starting point is the identity

$$\frac{q}{A} = \kappa\epsilon_0 \left(\frac{q}{\kappa\epsilon_0 A} \right).$$

Comparison with Eqs. 30-14 and 30-22b shows that this, extended to vector form, can be written as

$$\mathbf{D} = \kappa\epsilon_0\mathbf{E}. \qquad (30\text{-}24)$$

We can also write the polarization (see Eqs. 30-19 and 30-15b) as

$$P = \frac{q'}{A} = \frac{q}{A}\left(1 - \frac{1}{\kappa}\right).$$

Since $q/A = D$, we can rewrite this, using Eq. 30-24 and casting the result into vector form, as

$$\mathbf{P} = \epsilon_0(\kappa - 1)\mathbf{E}. \tag{30-25}$$

This shows clearly that in a vacuum $(\kappa = 1)$ the polarization vector $\mathbf{P}$ is zero.* Equations 30-24 and 30-25 show that for isotropic materials, to which a single dielectric constant κ can be assigned, $\mathbf{D}$ and $\mathbf{P}$ both point in the direction of $\mathbf{E}$ at any given point.

The definition of $\mathbf{D}$ given by Eq. 30-24 allows us to write Eq. 30-17, that is, Gauss's law in the presence of a dielectric, simply as

$$\oint \mathbf{D} \cdot d\mathbf{S} = q, \tag{30-26}$$

where, as before, q represents the free charge only, the induced surface charges being excluded.

EXAMPLE 10

In Figure 30-13, using data from Example 9, calculate E, D, and P: (a) in the dielectric and (b) in the air gap.

(a) The electric field in the dielectric is calculated in Example 9 to be 1.43×10^3 V/m. From Eq. 30-24,

$$D = \kappa\epsilon_0 E$$
$$= (7.0)(8.9 \times 10^{-12} \text{ C}^2/\text{N} \cdot \text{m}^2)(1.43 \times 10^3 \text{ V/m})$$
$$= 8.9 \times 10^{-8} \text{ C/m}^2$$

and, from Eq. 30-25,

$$P = \epsilon_0(\kappa - 1)E$$
$$= (8.9 \times 10^{-12} \text{ C}^2/\text{N} \cdot \text{m}^2)(7.0 - 1)(1.43 \times 10^3 \text{ V/m})$$
$$= 7.5 \times 10^{-8} \text{ C/m}^2.$$

(b) The electric field E_0 in the air gap is calculated in Example 9 to be 1.00×10^4 V/m. From Eq. 30-24,

$$D_0 = \kappa\epsilon_0 E_0$$
$$= (1)(8.9 \times 10^{-12} \text{ C}^2/\text{N} \cdot \text{m}^2)(1.00 \times 10^4 \text{ V/m})$$
$$= 8.9 \times 10^{-8} \text{ C/m}^2$$

and, from Eq. 30-25, recalling, as above, that $\kappa = 1$ in the air gap,

$$P_0 = \epsilon_0(\kappa - 1)E_0 = 0.$$

Note that $\mathbf{P}$ vanishes outside the dielectric, $\mathbf{D}$ has the same value in the dielectric and in the gap, and $\mathbf{E}$ has different values in the dielectric and in the gap. Verify that Eq. 30-23 $(\mathbf{D} = \epsilon_0\mathbf{E} + \mathbf{P})$ is correct both in the gap and in the dielectric.

It can be shown from Maxwell's equations that no matter how complex the problem the component of $\mathbf{D}$ *normal* to the surface of the dielectric has the same value on each side of the surface. In this problem $\mathbf{D}$ itself is normal to the surface, there being no component but the normal one. It can also be shown that the component of $\mathbf{E}$ *tangential* to the dielectric surface has the same value on each side of the surface. This *boundary condition*, like the one for $\mathbf{D}$, is trivial

---

*Certain waxes, when polarized in their molten state, retain a permanent polarization after solidifying, even though the external polarizing field is removed. *Electrets*, manufactured in this way, are the electrostatic analog of permanent magnets in that they possess a gross permanent electric dipole moment. Materials from which electrets can be constructed are called *ferroelectric*. Electrets do *not* obey Eq. 30-25 because they have a nonvanishing value of $\mathbf{P}$ even though $\mathbf{E} = 0$.

in this problem, both tangential components being zero. In more complex problems these boundary conditions on **D** and **E** are very important. Table 30-2 summarizes the properties of the electric vectors, **E**, **D**, and **P**.

Table 30-2
Three electric vectors

Name	Symbol	Associated with	Boundary Condition
Electric field	**E**	All charges	Tangential component continuous
Electric displacement	**D**	Free charges only	Normal component continuous
Polarization (electric dipole moment per unit volume)	**P**	Polarization charges only	Vanishes in a vacuum

Defining equation for **E**	$F = qE$	Eq. 27-2
General relation among the three vectors	$D = \epsilon_0 E + P$	Eq. 30-23
Gauss's law when dielectric media are present	$\oint D \cdot dS = q$	Eq. 30-26
	(q = free charge only)	
Empirical relations for certain dielectric materials*	$D = \kappa\epsilon_0 E$	Eq. 30-24
	$P = (\kappa - 1)\epsilon_0 E$	Eq. 30-25

* Generally true, with κ independent of **E**, except for certain materials called *ferroelectrics*; see footnote on page 667.

questions

1. A capacitor is connected across a battery. (*a*) Why does each plate receive a charge of exactly the same magnitude? (*b*) Is this true even if the plates are of different sizes?

2. Can there be a potential difference between two adjacent conductors that carry the same positive charge?

3. The relation $\sigma \propto 1/R$, in which R is the radius of curvature (see Section 29-8), suggests that the charge placed on an isolated conductor concentrates on points and avoids flat surfaces, where $R = \infty$. How do we reconcile this with Fig. 30-2 in which the charge is definitely on the flat surface of either plate?

4. A sheet of aluminum foil of negligible thickness is placed between the plates of a capacitor as in Fig. 30-14. What effect has it on the capacitance if (*a*) the foil is electrically insulated and (*b*) the foil is connected to the upper plate?

5. You are given two capacitors, C_1 and C_2, in which $C_1 \gg C_2$. Can C_1 always hold more charge than C_2? Explain.

6. In Fig. 30-1 suppose that a and b are nonconductors, the charge being distributed arbitrarily over their surfaces. (*a*) Would Eq. 30-1 ($q = CV$) hold, with C independent of the charge arrangements? (*b*) How would you define V in this case?

7. In connection with Eq. 30-1 ($q = CV$) we said that C is a constant. Yet we pointed out (see Eq. 30-4) that it depends on the geometry (and also, as we

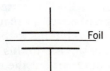

figure 30-14
Question 4.

saw later, on the medium). If C is indeed a constant, with respect to what variables does it remain constant?

8. For a finite A, does Eq. 30-4 ($C = \epsilon_0 A/d$) hold as $d \to \infty$? If not, why not?

9. Suppose that in Example 4 the three capacitors shown are identical parallel-plate capacitors with the same (square) plates of area A and the same plate separation d. Develop an argument, based only on Eq. 30-4 ($C = \epsilon_0 A/d$) that the equivalent capacitance is three times the individual capacitance, as Example 4 predicts.

10. You are given a parallel-plate capacitor with square plates of area A and separation d, in a vacuum. What is the qualitative effect of each of the following on its capacitance? (a) Reduce d. (b) Put a slab of copper between the plates, touching neither plate. (c) Double the area of both plates. (d) Double the area of one plate only. (e) Slide the plates parallel to each other so that the area of overlap is, say, 50%. (f) Double the potential difference between the plates. (g) Tilt one plate so that the separation remains d at one end but is $\frac{1}{2}d$ at the other.

11. Discuss similarities and differences when (a) a dielectric slab and (b) a conducting slab are inserted between the plates of a parallel-plate capacitor. Assume the slab thicknesses to be one-half the plate separation.

12. An oil-filled parallel-plate capacitor has been designed to have a capacitance C and to operate safely at or below a certain maximum potential difference V_m without arcing over. However, the designer did not do a good job and the capacitor occasionally arcs over. What can be done to redesign the capacitor, keeping C and V_m unchanged and using the same dielectric?

13. Would you expect the dielectric constant, for substances containing permanent molecular electric dipoles, to vary with temperature?

14. An isolated conducting sphere is given a positive charge. Does its mass increase, decrease, or remain the same?

15. A dielectric slab is inserted in one end of a charged parallel-plate capacitor (the plates being horizontal and the charging battery having been disconnected) and then released. Describe what happens. Neglect friction.

16. A capacitor is charged by using a battery, which is then disconnected. A dielectric slab is then slipped between the plates. Describe qualitatively what happens to the charge, the capacitance, the potential difference, the electric field, and the stored energy.

17. While a capacitor remains connected to a battery, a dielectric slab is slipped between the plates. Describe qualitatively what happens to the charge, the capacitance, the potential difference, the electric field, and the stored energy. Is work required to insert the slab?

18. Two identical capacitors are connected as shown in Fig. 30-15. A dielectric slab is slipped between the plates of one capacitor, the battery remaining connected. Describe qualitatively what happens to the charge, the capacitance, the potential difference, the electric field, and the stored energy for each capacitor.

19. Show that the dielectric constant of a conductor can be taken to be infinitely great.

20. For a given potential difference does a capacitor store more or less charge with a dielectric than it does without a dielectric (vacuum)? Explain in terms of the microscopic picture of the situation.

21. In this chapter we have assumed electrostatic conditions, that is, the potential difference V between the capacitor plates remains constant. Suppose however that, as it often does in practice, V varies sinusoidally with time with an angular frequency ω. Would you expect the dielectric constant κ to vary with ω?

22. In connection with Section 30-7 describe in your own words the differences between $\mathbf{D}$, $\mathbf{E}$, and $\mathbf{P}$ in Eq. 30-23.

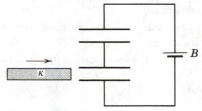

figure 30-15
Question 18.

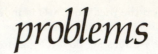

SECTION 30-2

1. A 100-pF capacitor is charged to a potential difference of 50 V, the charging battery then being disconnected. The capacitor is then connected in parallel with a second (initially uncharged) capacitor. If the measured potential difference drops to 35 V, what is the capacitance of this second capacitor? *Answer:* 43 pF.

2. A potential difference of 300 V is applied to a 2.0-μF capacitor and an 8.0-μF capacitor connected in series. (*a*) What are the charge and the potential difference for each capacitor? (*b*) The charged capacitors are reconnected with their positive plates together and their negative plates together, no external voltage being applied. What are the charge and the potential difference for each? (*c*) The charged capacitors in (*a*) are reconnected with plates of *opposite* sign together. What are the charge and the potential difference for each?

3. If we solve Eq. 30-4 for ϵ_0, we see that its SI units are farad/meter. Show that these units are equivalent to those obtained earlier for ϵ_0, namely coulomb^2/newton $\cdot$ meter^2.

4. A parallel-plate capacitor has circular plates of 8.0-cm radius and 1.0-mm separation. What charge will appear on the plates if a potential difference of 100 V is applied?

5. Figure 30-16 shows two capacitors in series, the rigid center section of length b being movable vertically. Show that the equivalent capacitance of the series combination is independent of the position of the center section and is given by

$$C = \frac{\epsilon_0 A}{a - b}.$$

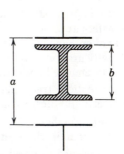

figure 30-16
Problem 5

6. In Fig. 30-17 a variable air capacitor of the type used in tuning radios is shown. Alternate plates are connected together, one group being fixed in position, the other group being capable of rotation. Consider a pile of n plates of alternate polarity, each having an area A and separated from adjacent plates by a distance d. Show that this capacitor has a maximum capacitance of

$$C = \frac{(n - 1)\epsilon_0 A}{d}.$$

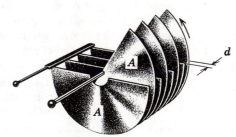

figure 30-17
Problem 6

7. A 6.0-μF capacitor is connected in series with a 4.0-μF capacitor and a potential difference of 200 V is applied across the pair. (*a*) What is the charge on each capacitor? (*b*) What is the potential difference across each capacitor? *Answer:* (*a*) 4.8×10^{-4} C. (*b*) $V_4 = 120$ V; $V_6 = 80$ V.

8. Repeat the previous problem for the same two capacitors connected in parallel.

9. How many 1.0-μF capacitors would need to be connected in parallel in order to store a charge of 1.0 C with a potential of 300 V across the capacitors? *Answer:* 3300.

figure 30-18
Problem 10

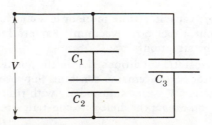

10. In Fig. 30-18 find the equivalent capacitance of the combination. Assume that $C_1 = 10\ \mu$F, $C_2 = 5\ \mu$F, and $C_3 = 4\ \mu$F.

11. In Fig. 30-19 find the equivalent capacitance of the combination. Assume that $C_1 = 10\ \mu$F, $C_2 = 5\ \mu$F, and $C_3 = 4\ \mu$F. *Answer:* 3.2 μF.

12. In Fig. 30-19 suppose that capacitor C_3 breaks down electrically, becoming equivalent to a conducting path. What *changes* in (a) the charge and (b) the potential difference occur for capacitor C_1? Assume $V = 100$ V.

13. A slab of copper of thickness b is thrust into a parallel-plate capacitor as shown in Fig. 30-20; it is exactly halfway between the plates. What is the capacitance (a) before and (b) after the slab is introduced?
    Answer: (a) $\epsilon_0 A/d$. (b) $\epsilon_0 A/(d - b)$.

14. When switch S is thrown to the left in Fig. 30-21, the plates of the capacitor C_1 acquire a potential difference V_0. C_2 and C_3 are initially uncharged. The switch is now thrown to the right. What are the final charges q_1, q_2, q_3 on the corresponding capacitors?

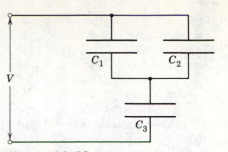

figure 30-19
Problems 11, 12, 31

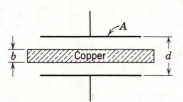

figure 30-20
Problem 13

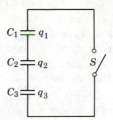

figure 30-21
Problem 14

15. Charges q_1, q_2, q_3 are placed on capacitors of capacitance C_1, C_2, C_3, respectively, arranged in series as shown in Fig. 30-22. Switch S is then closed. What are the final charges q_1', q_2', q_3' on the capacitors?

$$\text{Answer: (a) } q_1' = \frac{(C_1 C_2 + C_1 C_3)q_1 - C_1 C_3 q_2 - C_1 C_2 q_3}{C_1 C_2 + C_1 C_3 + C_2 C_3}.$$

$$\text{(b) } q_2' = \frac{(C_1 C_2 + C_2 C_3)q_2 - C_1 C_2 q_3 - C_2 C_3 q_1}{C_1 C_2 + C_1 C_3 + C_2 C_3}.$$

$$\text{(c) } q_3' = \frac{(C_1 C_3 + C_2 C_3)q_3 - C_1 C_3 q_2 - C_2 C_3 q_1}{C_1 C_2 + C_1 C_3 + C_2 C_3}.$$

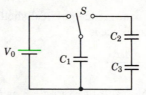

figure 30-22
Problem 15

16. You have material available to construct two parallel-plate capacitors having a combined plate area A. How would you distribute this area between the two capacitors to obtain maximum total capacitance if you intend to connect them (a) in parallel and (b) in series?

17. Capacitors C_1 (1.0 μF) and C_2 (3.0 μF) are each charged to a potential V (100 V) but with opposite polarity, so that points a and c are on the side of the respective positive plates of C_1 and C_2, and points b and d are on the side of the respective negative plates (see Fig. 30-23). Switches S_1 and S_2 are now closed. (a) What is the potential difference between points e and f? (b) What is the charge on C_1? (c) What is the charge on C_2?
    Answer: (a) 50 V. (b) 0.50×10^{-4} C. (c) 1.5×10^{-4} C.

18. Find the equivalent capacitance between points x and y in Fig. 30-24. Assume that $C_2 = 10$ μF and that the other capacitors are all 4.0 μF. (Hint: Apply a potential difference V between x and y and write down all the relationships that involve the charges and potential differences for the separate capacitors.)

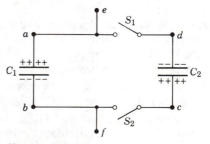

figure 30-23
Problem 17

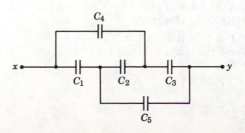

figure 30-24
Problem 18

19. In Fig. 30-25 the battery B supplies 12 V. (a) Find the charge on each capacitor when switch S_1 is closed and (b) when (later) switch S_2 is also closed. Take $C_1 = 1$ μF, $C_2 = 2$ μF, $C_3 = 3$ μF, and $C_4 = 4$ μF.
Answer: (a) $q_1 = 9.0\mu$C; $q_2 = 16\mu$C;
$q_3 = 9.0\mu$C; $q_4 = 16\mu$C.
(b) $q_1 = 8.4\mu$C; $q_2 = 17\mu$C;
$q_3 = 11\mu$C; $q_4 = 14\mu$C.

20. If you have available several 2.0-μF capacitors, each capable of withstanding 200 V without breakdown, how would you assemble a combination having an equivalent capacitance of (a) 0.40 μF or of (b) 1.2 μF, each capable of withstanding 1000 V?

21. Calculate the capacitance of the earth, viewed as a spherical conductor of radius 6400 km.   Answer: 710 μF.

22. A spherical capacitor consists of two concentric spherical shells of radii a and b, with $b > a$. (a) Show that its capacitance is

$$C = 4\pi\epsilon_0 \frac{ab}{b-a}.$$

(b) Does this reduce (with $a = R$) to the result of Example 3 as $b \to \infty$?

23. Suppose that the two spherical shells of a spherical capacitor have their radii approximately equal. Under these conditions the device approximates a parallel-plate capacitor with $b - a = d$. Show that the formula in Problem 22 does indeed reduce to Eq. 30-4 in this case.

24. Two metallic spheres, radii a and b, are connected by a thin wire. Their separation is large compared with their dimensions. A charge Q is put onto this system and then the wire is disconnected. (a) How much charge resides on each sphere? (b) Apply the definition of capacitance to show that the capacitance of this system is $C = 4\pi\epsilon_0 (a + b)$.

25. A capacitor has square plates, each of side a, making an angle of θ with each other as shown in Fig. 30-26. Show that for small θ the capacitance is given by

$$C = \frac{\epsilon_0 a^2}{d}\left(1 - \frac{a\theta}{2d}\right).$$

(Hint: The capacitor may be divided into differential strips which are effectively in parallel.)

SECTION 30-3

26. A parallel-plate air capacitor having area A (40 cm²) and spacing d (1.0 mm) is charged to a potential V (600 V). Find (a) the capacitance, (b) the magnitude of the charge on each plate, (c) the stored energy, (d) the electric field between the plates and (e) the energy density between the plates.

27. What would be the capacitance required to store energy U (10 kW·h) at a potential difference V (1000 V)?   Answer: 72F.

28. Two capacitors (2.0 μF and 4.0 μF) are connected in parallel across a 300-V potential difference. Calculate the total stored energy in the system.

29. A parallel-connected bank of 2000 5.0-μF capacitors is used to store electric energy. What does it cost to charge this bank to 50,000 V, assuming a rate of 10¢/kW·h?   Answer: 35¢.

30. A parallel-plate air capacitor has a capacitance of 100 pF. (a) What is the stored energy if the applied potential difference is 50 V? (b) Can you calculate the energy density for points between the plates?

31. In Fig. 30-19 find (a) the charge, (b) the potential difference, and (c) the stored energy for each capacitor. Assume the numerical values of Problem 11, with $V = 100$ V.
Answer: (a) $q_1 = 2.1 \times 10^{-4}$ C; $q_2 = 1.1 \times 10^{-4}$ C; $q_3 = 3.2 \times 10^{-4}$ C.
(b) $V_1 = V_2 = 21$ V; $V_3 = 79$ V.
(c) $U_1 = 2.2 \times 10^{-3}$ J; $U_2 = 1.1 \times 10^{-3}$ J; $U_3 = 1.3 \times 10^{-2}$ J.

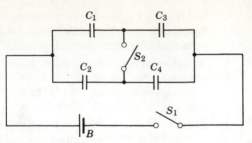

figure 30-25
Problem 19

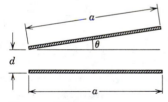

figure 30-26
Problem 25

32. For the capacitors of Problem 2, compute the energy stored for the three different connections of parts (a), (b), and (c). Compare your answers and explain any differences.

33. A parallel-plate capacitor has plates of area A and separation d, and is charged to a potential difference V. The charging battery is then disconnected and the plates are pulled apart until their separation is $2d$. Derive expressions in terms of A, d, and V for (a) the new potential difference, (b) the initial and the final stored energy, and (c) the work required to separate the plates.

   Answer: (a) $V_f = 2V$. (b) $U_i = \frac{1}{2}\frac{\epsilon_0 A V^2}{d}$; $U_f = 2U_i$. (c) $W = \frac{1}{2}\frac{\epsilon_0 A V^2}{d}$.

34. In terms of the original capacitance C, find the work done in inserting a copper slab of thickness $d/2$ in Problem 13 if (a) the potential difference is held constant and (b) if the charge is held constant.

35. An isolated metal sphere whose diameter is 10 cm has a potential of 8000 V. What is the energy density at the surface of the sphere?   Answer: 0.11 J/m³.

36. (a) If the potential difference across a cylindrical capacitor is doubled, the energy stored in the capacitor is changed by what factor? (b) If the radii of the inner and outer cylinders are each doubled, keeping the charge constant, how does the stored energy change?

37. A cylindrical capacitor has radii a and b as in Fig. 30-3. Show that half the stored electric potential energy lies within a cylinder whose radius is

$$r = \sqrt{ab}.$$

38. Show that the plates of a parallel-plate capacitor attract each other with a force given by

$$F = \frac{q^2}{2\epsilon_0 A}.$$

   Prove this by calculating the work necessary to increase the plate separation from x to $x + dx$.

39. Using the result of Problem 38 show that the force per unit area (the electrostatic stress) acting on either capacitor plate is given by $\frac{1}{2}\epsilon_0 E^2$. Actually, this result is true in general, for a conductor of *any* shape with an electric field $\mathbf{E}$ at its surface.

40. A soap bubble of radius R_0 is slowly given a charge q. Because of mutual repulsion of the surface charges, the radius increases slightly to R. The air pressure inside the bubble drops, because of the expansion, to $p(V_0/V)$ where p is the atmospheric pressure, V_0 is the initial volume, and V the final volume. Show that

$$q^2 = 32 \pi^2 \epsilon_0 \, p \, R \, (R^3 - R_0^3).$$

   (Hint: Imagine the bubble to expand a further amount dR. Consider the energy changes associated with (a) the decrease in the stored electric field energy, (b) work ($= p\, dV$) done in pushing back the atmosphere, and (c) work done by the gas in the bubble. Apply conservation of energy. Neglect surface tension.)

SECTION 4

41. A certain substance has a dielectric constant of 2.8 and a dielectric strength of 18×10^6 V/m. If it is used as the dielectric material in a parallel-plate capacitor, what minimum area may the plates of the capacitor have in order that the capacitance be 7.0×10^{-2} μF and that the capacitor be able to withstand a potential difference of 4000 V?   Answer: 0.63 m².

42. A parallel-plate capacitor is filled with two dielectrics as in Fig. 30-27. Show that the capacitance is given by

$$C = \frac{\epsilon_0 A}{d}\left(\frac{\kappa_1 + \kappa_2}{2}\right).$$

   Check this formula for all the limiting cases that you can think of. (Hint: Can you justify regarding this arrangement as two capacitors in parallel?)

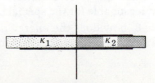

figure 30-27
Problem 42

43. A parallel-plate capacitor is filled with two dielectrics as in Fig. 30-28. Show that the capacitance is given by

$$C = \frac{2\epsilon_0 A}{d}\left(\frac{\kappa_1 \kappa_2}{\kappa_1 + \kappa_2}\right).$$

Check this formula for all the limiting cases that you can think of. (Hint: Can you justify regarding this arrangement as two capacitors in series?)

44. What is capacitance of the capacitor in Fig. 30-29? The plate area is A.

45. For making a capacitor you have available two plates of copper, a sheet of mica (thickness = 0.10 mm, $\kappa = 6$), a sheet of glass (thickness = 2.0 mm, $\kappa = 7$), and a slab of paraffin (thickness = 1.0 cm, $\kappa = 2$). To obtain the largest capacitance, which sheet (or sheets) should you place between the copper plates?   *Answer:* The mica sheet.

SECTION 30-6

46. A parallel-plate capacitor has a capacitance of 100 pF, a plate area of 100 cm², and a mica dielectric ($\kappa = 5.4$). At 50 V potential difference, calculate (a) E in the mica, (b) the magnitude of the free charge on the plates, and (c) the magnitude of the induced surface charge.

47. Two parallel plates of area 100 cm² are each given equal but opposite charges of 8.9×10^{-7} C. Within the dielectric material filling the space between the plates the electric field is 1.4×10^6 V/m. (a) Find the dielectric constant of the material. (b) Determine the magnitude of the charge induced on each dielectric surface.   *Answer:* (a) 7.1. (b) 7.7×10^{-7} C.

48. In Example 9, suppose that the 100-V battery remains connected during the time that the dielectric slab is being introduced. Calculate (a) the charge on the capacitor plates, (b) the electric field in the gap, (c) the electric field in the slab, and (d) the capacitance. For all of these quantities give the numerical values before and after the slab is introduced. Contrast your results with those of Example 9 by constructing a tabular listing.

49. A parallel-plate capacitor has plates of area 0.12 m² and a separation of 1.2 cm. A battery charges the plates to a potential difference of 120 V and is then disconnected. A dielectric slab of thickness 0.4 cm and dielectric constant 4.8 is then placed symmetrically between the plates. (a) Find the capacitance before the slab is inserted. (b) What is the capacitance with the slab in place? (c) What is the free charge q before and after the slab is inserted? (d) Determine the electric field in the space between the plates and dielectric. (e) What is the electric field in the dielectric? (f) With the slab in place what is the potential difference across the plates? (g) How much external work is involved in the process of inserting the slab?
     Answer: (a) 89 pF. (b) 120 pF. (c) 1.1×10^{-8} C; 1.1×10^{-8} C. (d) 10^4 V/m. (e) 2.1×10^3 V/m. (f) 88 V. (g) 1.7×10^{-7} J.

50. In the capacitor of Example 9 the dielectric slab fills half the space between the plates. (a) What percent of the energy is stored in the air gaps? (b) What percent is stored in the slab?

51. A dielectric slab of thickness b is inserted between the plates of a parallel-plate capacitor of plate separation d. Show that the capacitance is given by

$$C = \frac{\kappa \epsilon_0 A}{\kappa d - b(\kappa - 1)}.$$

(Hint: Derive the formula following the pattern of Example 9.) Does this formula predict the correct numerical result of Example 9? Does the formula seem reasonable for the special cases of $b = 0$, $\kappa = 1$, and $b = d$?

52. In Example 8, how does the *energy density* u between the plates compare, before and after the dielectric slab is introduced?

figure 30-28
Problem 43

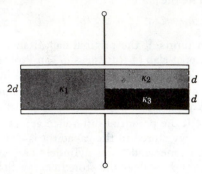

figure 30-29
Problem 44

31
current
and resistance

The free electrons in an isolated metallic conductor, such as a length of copper wire, are in random motion like the molecules of a gas confined to a container. They have no net directed motion along the wire. If we pass a hypothetical plane through the wire, the rate at which electrons pass through it from right to left is the same as the rate at which they pass through from left to right; the *net* rate is zero.*

If the ends of the wire are connected to a battery, an electric field will be set up at every point within the wire. If the potential difference maintained by the battery is 10 V and if the wire (assumed uniform) is 5 m long, the strength of this field at every point will be 2 V/m. This field **E** will act on the electrons and will give them a resultant motion in the direction of −**E**. We say that an *electric current i* is established; if a net charge q passes through any cross section of the conductor in time t, the current (assumed constant) is

$$i = q/t. \qquad (31\text{-}1)$$

The appropriate SI units are amperes (abbr. A) for i, coulombs for q, and seconds for t. Recall (Section 26-4) that Eq. 31-1 is the defining equation for the coulomb and that we have not yet given an operational definition of the ampere; we do so in Section 34-4.

31-1
CURRENT AND
CURRENT DENSITY

*Actually, because the number of electrons is finite, there will be small statistical fluctuations in these rates and a conductor will contain a small, rapidly fluctuating current, even though, averaged over a long enough period of time, the net current i is zero. This is one aspect of the readily measurable *electrical noise* which is so familiar to those who know something about electronics.

If the rate of flow of charge with time is not constant, the current varies with time and is given by the differential limit of Eq. 31-1, or

$$i = dq/dt. \qquad (31\text{-}2)$$

In the rest of this chapter we consider only constant currents.*

The current i is the same for all cross sections of a conductor, even though the cross-sectional area may be different at different points. In the same way the rate at which water (assumed incompressible) flows past any cross section of a pipe is the same even if the cross section varies. The water flows faster where the pipe is smaller and slower where it is larger, so that the volume rate, measured perhaps in liters/second, remains unchanged. This constancy of the electric current follows because charge must be conserved; it does not pile up steadily or drain away steadily from any point in the conductor under the assumed steady-state conditions. In the language of Section 18-3 there are no "sources" or "sinks" of charge.

The existence of an electric field inside a conductor does not contradict Section 28-6, in which we asserted that $\mathbf{E}$ equals zero inside a conductor. In that section, which dealt with a state in which all net motion of charge had stopped (electrostatics), we assumed that the conductor was insulated and that no potential difference was deliberately maintained between any two points on it, as by a battery. In this chapter, which deals with charges in motion, we relax this restriction.

The electric field exerts a force $(= -e\mathbf{E})$ on the electrons in a conductor but this force does not produce a *net* acceleration because the electrons keep colliding with the atoms (strictly, ions, Cu^+ in copper) that make up the conductor. This array of ions, coupled together by strong spring-like forces of electromagnetic origin, is called the *lattice* (see Fig. 21-5). The over-all effect of these collisions is to transfer kinetic energy from the accelerating electrons into vibrational energy of the lattice. The electrons acquire a constant average *drift speed* v_d in the direction $-\mathbf{E}$. There is a close analogy to a ball bearing falling in a uniform gravitational field $\mathbf{g}$ at a constant terminal speed through a viscous oil. The gravitational force $(m\mathbf{g})$ acting on the ball as it falls does not go into increasing the ball's kinetic energy (which is constant) but is transferred to the fluid by molecular collisions, producing a small rise in temperature.

Although in metals the charge carriers are electrons, in electrolytes or in gaseous conductors (plasmas) they may also be positive or negative ions, or both. We need a convention for labeling the directions of currents because charges of opposite sign move in opposite directions in a given field. A positive charge moving in one direction is equivalent in nearly all external effects to a negative charge moving in the opposite direction. Hence, for simplicity and algebraic consistency, *we assume that all charge carriers are positive and we draw the current arrows in the direction that such charges would move.* If the charge carriers are negative, they simply move opposite to the direction of the current arrow (see Fig. 31-1). When we encounter a case (as in the *Hall effect;* see Section 33-5) in which the sign of the charge carriers makes a difference in the external effects, we will disregard the convention and take the actual situation into account.

Current i is a characteristic of a particular conductor. It is a macroscopic quantity, like the mass of an object, the volume of an object, or the length of a rod. A related microscopic quantity is the current density

---

* A much less formal example of current flow is lightning. See "The Lightning Discharge" by Richard E. Orville, *The Physics Teacher,* January 1976.

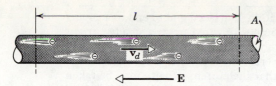

figure 31-1
Electrons drift in a direction opposite
to the electric field in a conductor.

$\mathbf{j}$. It is a vector and is characteristic of a point inside a conductor rather
than of the conductor as a whole. If the current is distributed uniformly
across a conductor of cross-sectional area A, as in Fig. 31-1, the magnitude of the current density for all points on that cross section is

$$j = i/A. \qquad (31\text{-}3)$$

The vector $\mathbf{j}$ at any point is oriented in the direction that a positive
charge carrier would move at that point. An electron at that point moves
in the direction $-\mathbf{j}$. In Fig. 31-1 $\mathbf{j}$ is a constant vector and points to the
left; the electrons drift to the right.

The general relationship between $\mathbf{j}$ and i is that, for a particular surface (which need not be plane) in a conductor, i is the flux of the vector
$\mathbf{j}$ over that surface, or

$$i = \int \mathbf{j} \cdot d\mathbf{S}, \qquad (31\text{-}4)$$

where $d\mathbf{S}$ is an element of surface area and the integral is taken over
the surface in question. Equation 31-3 (written as $i = jA$) is a special
case of this relationship in which the surface of integration is a plane
cross section of the conductor and in which $\mathbf{j}$ is constant over this surface and at right angles to it. However, we may apply Eq. 31-4 to any
surface through which we wish to know the current. Equation 31-4
shows clearly that i is a scalar because the integrand $\mathbf{j} \cdot d\mathbf{S}$ is a scalar.

The arrow often associated with the current in a wire does not indicate that current is a vector but merely shows the *sense* of charge flow. Positive charge carriers either move in a certain direction along the wire or in the opposite direction, these two possibilities being represented by $+$ or $-$ in algebraic equations.
Note that (a) the current in a wire remains unchanged if the wire is bent, tied
into a knot, or otherwise distorted, and (b) the arrows representing the sense of
currents do not in any way obey the laws of vector addition. Thus currents can
not be vectors.

We can compute the drift speed v_d of charge carriers in a conductor
from the current density j. Figure 31-1 shows the conduction electrons
in a wire moving to the right at an assumed constant drift speed v_d.
The number of conduction electrons in the wire is nAl where n is the
number of conduction electrons per unit volume and Al is the volume of
the wire. A charge of magnitude

$$q = (nAl)e$$

passes out of the wire, through its right end, in a time t given by

$$t = \frac{l}{v_d}.$$

The current i is given by

$$i = \frac{q}{t} = \frac{nAle}{l/v_d} = nAev_d.$$

Solving for v_d and recalling that $j = i/A$ (Eq. 31-3) yields

$$v_d = \frac{i}{nAe} = \frac{j}{ne}. \qquad (31\text{-}5)$$

EXAMPLE 1

An aluminum wire whose diameter is 0.10 in. is welded end to end to a copper wire with a diameter of 0.064 in. The composite wire carries a steady current of 10 A. What is the current density in each wire?

The current is distributed uniformly over the cross section of each conductor, except near the junction, which means that the current density may be taken as constant for all points within each wire. The cross-sectional area of the aluminum wire is 0.0079 in.² Thus, from Eq. 31-3,

$$j_{Al} = \frac{i}{A} = \frac{10 \text{ A}}{0.0079 \text{ in.}^2} = 1300 \text{ A/in.}^2.$$

The cross-sectional area of the copper wire is 0.0032 in.² Thus

$$j_{Cu} = \frac{i}{A} = \frac{10 \text{ A}}{0.0032 \text{ in.}^2} = 3100 \text{ A/in.}^2.$$

The fact that the wires are of different materials does not enter into consideration here.

EXAMPLE 2

What is v_d for the copper wire in Example 1?

We can write the current density for the copper wire (3100 A/in.²) as 480 A/cm². To compute n we start from the fact that there is one free electron per atom in copper. The number of atoms per unit volume is dN_0/M where d is the density, N_0 is the Avogadro number, and M is the atomic weight. The number of free electrons per unit volume is then

$$n = \frac{dN_0}{M} = \frac{(9.0 \text{ g/cm}^3)(6.0 \times 10^{23} \text{ atoms/mol})(1 \text{ electron/atom})}{64 \text{ g/mol}}$$

$$= 8.4 \times 10^{22} \text{ electrons/cm}^3.$$

Finally, v_d is, from Eq. 31-5,

$$v_d = \frac{j}{ne} = \frac{480 \text{ A/cm}^2}{(8.4 \times 10^{22} \text{ electrons/cm}^3)(1.6 \times 10^{-19} \text{ C/electron})}$$

$$= 3.6 \times 10^{-2} \text{ cm/s.}$$

It takes 28 seconds for the electrons in this wire to drift 1.0 cm. Would you have guessed that v_d was so low? The drift speed of electrons must not be confused with the speed at which changes in the electric field configuration travel along wires, a speed which approaches that of light. When we apply a pressure to one end of a long water-filled garden hose, a *pressure wave* travels rapidly along the hose. The speed at which *water* moves through the hose is much lower, however.

31-2
RESISTANCE, RESISTIVITY, AND CONDUCTIVITY

If we apply the same potential difference between the ends of geometrically similar rods of copper and of wood, very different currents result. The characteristic of the conductor that enters here is its *resistance*. We define the resistance of a conductor (often called a *resistor*; symbol ⌁⌁⌁⌁⌁) between two points by applying a potential difference V between those points, measuring the current i, and dividing:

$$R = V/i. \qquad (31-6)$$

If V is in volts and i in amperes, the resistance R will be in *ohms* (abbr. Ω).

The flow of charge through a conductor is often compared with the

Table 31-1
Properties of metals as conductors

Metal	Resistivity (at 20°C) 10^{-8} $\Omega \cdot m$	Temperature Coefficient of Resistivity, α, per C° ($\times 10^{-5}$)†
Silver	1.6	380
Copper	1.7	390
Aluminum	2.8	390
Tungsten	5.6	450
Nickel	6.8	600
Iron	10	500
Steel	18	300
Manganin	44	1.0
Carbon*	3500	−50

* Carbon, not strictly a metal, is included for comparison.
† This quantity, defined from

$$\alpha = \frac{1}{\rho} \frac{d\rho}{dT} \tag{31-7}$$

is the fractional change in resistivity $(d\rho/\rho)$ per unit change in temperature. It varies with temperature, the values here referring to 20°C. For copper ($\alpha = 3.9 \times 10^{-3}/C°$) the resistivity increases by 0.39 percent for a temperature increase of 1°C near 20°C. Note that α for carbon is negative, which means that the resistivity *decreases* with increasing temperature.

flow of water through a pipe, which occurs because there is a difference in pressure between the ends of the pipe, established perhaps by a pump. This pressure difference can be compared with the potential difference established between the ends of a resistor by a battery. The flow of water (liters/second, say) is compared with the current (coulombs/second, or amperes). The rate of flow of water for a given pressure difference is determined by the nature of the pipe. Is it long or short? Is it narrow or wide? Is it empty or filled, perhaps with gravel? These characteristics of the pipe are analogous to the resistance of a conductor.

Primary standards of resistance, kept at the National Bureau of Standards, are spools of wire whose resistances have been accurately measured. Because resistance varies with temperature, these standards, when used, are placed in an oil bath at a controlled temperature. They are made of a special alloy, called *manganin*, for which the change of resistance with temperature is very small. They are carefully annealed to eliminate strains, which also affect the resistance. These primary standard resistors are used chiefly to calibrate secondary standards for other laboratories.

Operationally, the primary resistance standards are not measured by using Eq. 31-6 but are measured in an indirect way which involves magnetic fields. Equation 31-6 is, in fact, used to measure *V* by setting up an accurately known current *i* (using a *current balance*; see Section 34-4) in an accurately known resistance *R*. This operational procedure for potential difference is the one normally used in place of the conceptual definition introduced in Section 29-1, in which one measures the work per unit charge required to move a test charge between two points.

Related to resistance is the *resistivity* ρ, which is a characteristic of a material rather than of a particular specimen of a material; it is defined, for isotropic materials,* from

$$\rho = \frac{E}{j}. \tag{31-8a}$$

The resistivity of copper is $1.7 \times 10^{-8} \ \Omega \cdot m$; that of fused quartz is about $10^{16} \ \Omega \cdot m$. Few physical properties are measurable over such a range of values; Table 31-1 lists some electrical properties for common metals.

Often we prefer to speak of the *conductivity* (σ) of a material rather than its resistivity. These are reciprocal quantities, related by

$$\sigma = 1/\rho. \tag{31-8b}$$

The SI units of σ are $(\Omega \cdot m)^{-1}$.

Consider a cylindrical conductor, of cross-sectional area A and length l, carrying a steady current i. Let us apply a potential difference V between its ends. If the cylinder cross sections at each end are equipotential surfaces, the electric field and the current density will be constant for all points in the cylinder and will have the values

$$E = \frac{V}{l} \quad \text{and} \quad j = \frac{i}{A}.$$

We may then write the resistivity ρ as

$$\rho = \frac{E}{j} = \frac{V/l}{i/A}.$$

But V/i is the resistance R which leads to

$$R = \rho \frac{l}{A}. \tag{31-9}$$

V, i, and R are *macroscopic* quantities, applying to a particular body or extended region. The corresponding *microscopic* quantities are $\mathbf{E}$, $\mathbf{j}$, and ρ; they have values at every point in a body. The macroscopic quantities are related to each other by Eq. 31-6 $(V = iR)$ and the microscopic quantities by Eq. 31-8a, which can be written in vector form as $\mathbf{E} = \mathbf{j}\rho$.

The macroscopic quantities can be found by integrating over the microscopic quantities, using relations already given, namely

$$i = \int \mathbf{j} \cdot d\mathbf{S} \tag{31-4}$$

and

$$V_{ab} = -\int_a^b \mathbf{E} \cdot d\mathbf{l}. \tag{29-5}$$

The integral in Eq. 31-4 is a surface integral, carried out over any cross section of the conductor. The integral in Eq. 29-5 is a line integral carried out along an arbitrary line drawn along the conductor, connecting any two equipotential surfaces, identified by a and b. For a long wire connected to a battery equipotential surface a might be chosen as a cross section of the wire near the positive battery terminal and b might be a cross section near the negative terminal.

We can express the resistance of a conductor between a and b in microscopic terms by dividing the two equations, or

$$R = \frac{V_{ab}}{i} = \frac{-\int_a^b \mathbf{E} \cdot d\mathbf{l}}{\int \mathbf{j} \cdot d\mathbf{S}}.$$

---

* These are materials whose properties (electrical in this case) do not vary with direction in the material.

If the conductor is a long cylinder of cross section A and length l, and if points a and b are its ends, the foregoing equation for R (see Eq. 31-8a) reduces to

$$R = \frac{El}{jA} = \rho \frac{l}{A},$$

which is Eq. 31-9.

The macroscopic quantities V, i, and R are of primary interest when we are making electrical measurements on real conducting objects. They are the quantities that one reads on meters. The microscopic quantities $\mathbf{E}$, $\mathbf{j}$, and ρ are of primary importance when we are concerned with the fundamental behavior of matter (rather than of specimens of matter), as we usually are in the research area of *solid state physics*. Thus Section 31-4 deals appropriately with an atomic view of the *resistivity* of a metal and not of the *resistance* of a metallic specimen. The microscopic quantities are also important when we are interested in the interior behavior of irregularly shaped conducting objects.

EXAMPLE 3

A rectangular carbon block has dimensions $1.0\,\text{cm} \times 1.0\,\text{cm} \times 50\,\text{cm}$. (a) What is the resistance measured between the two square ends and (b) between two opposing rectangular faces? The resistivity of carbon at 20°C is $3.5 \times 10^{-5}\,\Omega \cdot \text{m}$.

(a) The area of a square end is $1.0\,\text{cm}^2$ or $1.0 \times 10^{-4}\,\text{m}^2$. Equation 31-9 gives for the resistance between the square ends

$$R = \rho \frac{l}{A} = \frac{(3.5 \times 10^{-5}\,\Omega \cdot \text{m})(0.50\,\text{m})}{1.0 \times 10^{-4}\,\text{m}^2} = 0.18\,\Omega.$$

(b) For the resistance between opposing rectangular faces (area $= 5.0 \times 10^{-3}$ m^2), we have

$$R = \rho \frac{l}{A} = \frac{(3.5 \times 10^{-5}\,\Omega \cdot \text{m})(10^{-2}\,\text{m})}{5.0 \times 10^{-3}\,\text{m}^2} = 7.0 \times 10^{-5}\,\Omega.$$

Thus a given conductor can have any number of resistances, depending on how the potential difference is applied to it. The ratio of resistances for these two cases is 2600. We assume in each that the potential difference is applied to the block in such a way that the surfaces between which the resistance is desired are equipotential. Otherwise Eq. 31-9 would not be valid.

Figure 31-2 shows (solid curve) how the resistivity of copper varies with temperature. Sometimes, for practical use, such data are expressed in equation form. If we are interested in only a limited range of temperatures extending, say, from 0 to 500°C, we can fit a straight line to the curve of Fig. 31-2, making it pass through two arbitrarily selected points; see the dashed line. We choose the point labeled T_0, ρ_0 in the figure as a reference point, T_0 being 0° C in this case and ρ_0 being $1.56 \times 10^{-8}\,\Omega \cdot \text{m}$. The resistivity ρ at any temperature T can be found from the empirical equation of the dashed straight line in Fig. 31-2, which is

$$\rho = \rho_0 [1 + \bar{\alpha}(T - T_0)]. \tag{31-10}$$

This relation shows correctly that $\rho \to \rho_0$ as $T \to T_0$.

If we solve Eq. 31-10 for $\bar{\alpha}$, we obtain

$$\bar{\alpha} = \frac{1}{\rho_0} \frac{\rho - \rho_0}{T - T_0}.$$

Comparison with Eq. 31-7 shows that $\bar{\alpha}$ is a *mean temperature coefficient of resistivity* for a selected pair of temperatures rather than the temperature coefficient of resistivity at a particular temperature, which is the definition of α. For most practical purposes Eq. 31-10 gives results that are within the acceptable range of accuracy.

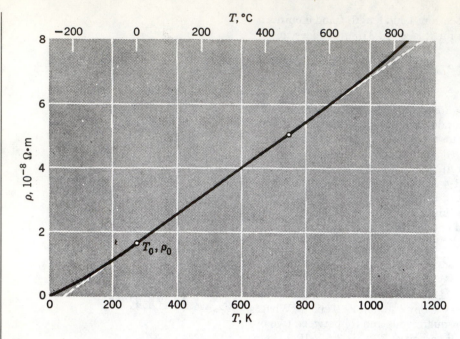

figure 31-2
The resistivity of copper as a function of temperature. The dashed line is an approximation chosen to fit the curve at the two circled points. The point marked T_0, ρ_0 is chosen as a reference point.

The curve of Fig. 31-2 does not go to zero at the absolute zero of temperature, even though it appears to do so, the residual resistivity at this temperature being $0.02 \times 10^{-8}\ \Omega \cdot m$. For many substances the resistance *does* become zero at some low temperature. Figure 31-3 shows the resistance of a specimen of mercury for temperatures below 6 K. In the space of about 0.05 K the resistance drops abruptly to an immeasurably low value. This phenomenon, called *superconductivity*,* was discovered by Kamerlingh Onnes in the Netherlands in 1911. The resistance of materials in the superconducting state seems to be truly zero; currents, once established in closed superconducting circuits, persist for weeks without diminution, even though there is no battery in the circuit. If the temperature is raised slightly above the superconducting point, or if a large enough magnetic field is applied, such currents drop rapidly to zero.

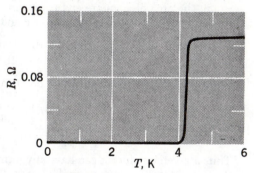

figure 31-3
The resistivity of mercury disappears below about 4 K.

Let us apply a variable potential difference V between the ends of a 100-foot coil of #18 copper wire. For each applied potential difference, let us measure the current i and plot it against V as in Fig. 31-4. The straight line that results means that *the resistance of this conductor is the same no matter what applied voltage we use to measure it*. This important result, which holds for metallic conductors, is known as *Ohm's law*. We assume that the temperature of the conductor is essentially constant throughout the measurements.

Many conductors do not obey Ohm's law. Figure 31-5, for example, shows a V-i plot for a type 2A3 vacuum tube. The plot is not straight and the resistance depends on the voltage used to measure it. Also, the current for this device is almost vanishingly small if the polarity of the applied potential difference is reversed. For metallic conductors the

31-3
OHM'S LAW

* See (a) "Superconductivity" by B. T. Matthias, *Scientific American*, November 1957, (b) "The Search for High-Temperature Superconductors" by B. T. Matthias, *Physics Today*, August 1971; (c) "Large-Scale Applications of Superconductivity" by Brian B. Schwartz and Simon Foner, *Physics Today*, July 1977.

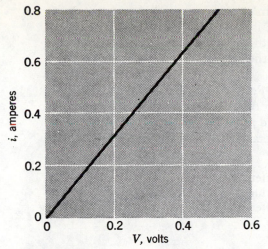

figure 31-4
The current in a particular copper conductor as a function of potential difference. This conductor obeys Ohm's law.

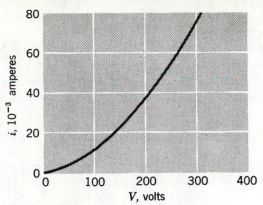

figure 31-5
The current in a type 2A3 vacuum tube as a function of potential difference. This conductor does *not* obey Ohm's law.

current reverses direction when the potential difference is reversed, but its magnitude does not change.

Figure 31-6 shows a typical *V-i* plot for another nonohmic device, a *thermistor*. This is a semiconductor (see Section 26-3) with a large and negative temperature coefficient of resistivity α (see Table 31-1) that varies greatly with temperature. We note that two different currents through the thermistor can correspond to the same potential difference between its ends. Thermistors are often used to measure the rate of energy flow in microwave beams by allowing the microwave beam to fall on the thermistor and heat it. The relatively small temperature rise so produced results in a relatively large change in resistance, which serves as a measure of the microwave power. Modern electronics, and therefore much of the character of our present technological civilization, depends in a fundamental way on the fact that many conductors, such as transistors and vacuum tubes, do *not* obey Ohm's law.

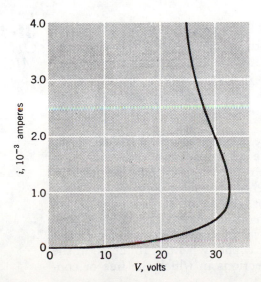

figure 31-6
A plot of current as a function of potential difference in a Western Electric 1-B thermistor. Again, this conductor does not obey Ohm's law. The shape of the curve can be accounted for in terms of the large negative temperature coefficient of resistivity of the material of which the device is made.

We stress that the relationship $V = iR$ is *not* a statement of Ohm's law. A conductor obeys this law only if its V-i curve is linear, that is, if R is independent of V and i. The relationship $R = V/i$ remains as the general definition of the resistance of a conductor whether or not the conductor obeys Ohm's law.

The microscopic equivalent of the relationship $V = iR$ is Eq. 31-8a, or $\mathbf{E} = \mathbf{j}\rho$. A conducting *material* is said to obey Ohm's law if a plot of E versus j is linear, that is, if the resistivity ρ is independent of E and j. Ohm's law is a specific property of certain materials and is not a general law of electromagnetism, for example, like Gauss's law.

A close analogy exists between the flow of charge because of a potential difference and the flow of heat because of a temperature difference. Consider a thin electrically conducting slab of thickness Δx and area A. Let a potential difference ΔV be maintained between opposing faces. The current i is given by Eqs. 31-6 ($i = V/R$) and 31-9 ($R = \rho l/A$), or

$$i = \frac{V_a - V_b}{R} = \frac{(V_a - V_b)A}{\rho l} = -\frac{(V_b - V_a)A}{\rho \Delta x}.$$

In the limiting case of a slab of thickness dx this becomes

$$i = -\frac{1}{\rho} A \frac{dV}{dx}$$

or

$$\frac{dq}{dt} = -\sigma A \frac{dV}{dx}, \qquad (31\text{-}11)$$

where $\sigma \,(= 1/\rho)$, as we have seen (Eq. 31-8b), is the *conductivity* of the material. Since positive charge flows in the direction of decreasing V, we introduce a minus sign into Eq. 31-11, that is, dq/dt is positive when dV/dx is negative.

The analogous heat flow equation (see Section 22-4) is

$$\frac{dQ}{dt} = -kA \frac{dT}{dx}, \qquad (31\text{-}12)$$

which shows that k, the thermal conductivity, corresponds to σ and dT/dx, the temperature gradient, corresponds to dV/dx, the potential gradient. There is more than a formal mathematical analogy between Eqs. 31-11 and 31-12. Both heat energy and charge are carried by the free electrons in a metal; empirically, a good electrical conductor (silver, say) is almost always also a good heat conductor and conversely.

31-4
OHM'S LAW—A MICROSCOPIC VIEW

As we have said earlier, Ohm's law is not a fundamental law of electromagnetism because it depends on the properties of the conducting medium. The law is very simple in form, and it is curious that many conductors obey it so well, whereas other conductors do not obey it at all (see Figs. 31-4, 31-5, and 31-6). Let us see if we can understand why metals obey Ohm's law, which we shall write (see Eq. 31-8a) in the microscopic form $\mathbf{E} = \rho\mathbf{j}$.

In a metal the valence electrons are not attached to individual atoms but are free to move about within the lattice and are called *conduction electrons*. In copper there is one such electron per atom, the other 28 remaining bound to the copper nuclei to form ionic cores.

Although the speed distribution of conduction electrons can be described correctly only in terms of quantum physics, the classical *free-electron model* will suit our purpose. It suffices to consider only a suitably defined average speed $\bar{v}$; for copper $\bar{v} = 1.6 \times 10^8$ cm/s. In the absence of an electric field, the directions in which the free or conduction) electrons move are completely random, like those of the molecules of a gas confined to a container.

The electrons collide constantly with the ionic cores of the conductor, that is, they interact with the lattice, often suffering sudden changes in speed and direction. These collisions remind us of the collisions of gas molecules confined to a container. As in the case of molecular collisions, we can describe electron-lattice collisions by a *mean free path* λ, where λ is the average distance that an electron travels between collisions.*

In an ideal metallic crystal at 0 K electron-lattice collisions would not occur, according to the predictions of quantum physics, that is, $\lambda \to \infty$ as $T \to 0$ K for ideal crystals. Collisions take place in actual crystals because (a) the ionic cores at any temperature T are vibrating about their equilibrium positions in a random way, (b) impurities, that is, foreign atoms, may be present, and (c) the crystal may contain lattice imperfections, such as missing atoms and displaced atoms. On this view it is not surprising that the resistivity of a metal can be increased by (a) raising its temperature, (b) adding small amounts of impurities, and (c) straining it severely, as by drawing it through a die, to increase the number of lattice imperfections.

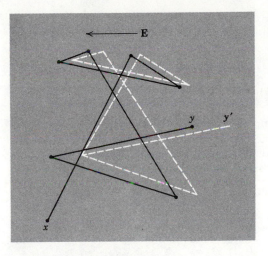

figure 31-7
The solid lines show an electron moving from x to y, making six collisions. The dashed curves show what the electron path *might* have been in the presence of an electric field **E**. Note the steady drift in the direction of $-\mathbf{E}$.

When we apply an electric field to a metal, the electrons modify their random motion in such a way that they drift slowly, in the opposite direction to that of the field, with an average drift speed v_d. This drift speed is very much less (by a factor of something like 10^{10}; see Example 2) than the effective average speed $\bar{v}$ mentioned above. Figure 31-7 suggests the relationship between these two speeds. The solid lines suggest a possible random path followed by an electron in the absence of an applied field; the electron proceeds from x to y, making six collisions on the way. The dashed curves show how this same event *might* have occurred if an electric field **E** had been applied. Note that the electron drifts steadily to the right, ending at y' rather than at y. In preparing Fig. 31-7, it has been assumed that the drift speed v_d is $0.02\bar{v}$; actually, it is more like $10^{-10}\bar{v}$, so that the "drift" exhibited in the figure is greatly exaggerated.

We can calculate the drift speed v_d in terms of the applied electric field E and of $\bar{v}$ and λ. When a field is applied to an electron in the metal

* It can be shown that collisions between electrons occur only rarely and have little effect on the resistivity.

it will experience a force eE which will impart to it an acceleration a given by Newton's second law,

$$a = \frac{eE}{m}.$$

Consider an electron that has just collided with an ion core. The collision, in general, will momentarily destroy the tendency to drift and the electron will have a truly random direction after the collision. During the time interval to the next collision the electron's velocity will have changed, on the average, by $a(\lambda/\bar{v})$ or $a\tau$ where τ is the mean time between collisions. We call this the drift speed v_d, or

$$v_d = a\tau = \frac{eE\tau}{m} \tag{31-13}$$

The electron's motion through the conductor is analogous to the constant terminal rate of fall of a stone in water. The gravitational force F_g on the stone is opposed by a viscous resisting force that is proportional to the velocity, or

$$F_g = mg = bv,$$

where b is a viscous coefficient (see Section 15-9). Thus the constant terminal speed of the stone is

$$v = \left(\frac{1}{b}\right) F_g.$$

We can rewrite Eq. 31-13 as

$$v_d = \left(\frac{\tau}{m}\right) F_E,$$

where F_E ($= eE$) is the electrical force. Comparison of these equations shows that the equivalent "viscous coefficient" for the motion of an electron in a particular conductor is m/τ. If τ is short, the conductor exhibits a greater "viscous effect" on the electron motion, and the drift speed v_d is proportionally lower.

We may express v_d in terms of the current density (Eq. 31-5) and combine with Eq. 31-13 to obtain

$$v_d = \frac{j}{ne} = \frac{eE\tau}{m}.$$

Combining this with Eq. 31-8a ($\rho = E/j$) leads finally to*

$$\rho = \frac{m}{ne^2\tau}. \tag{31-14}$$

Equation 31-14 can be taken as a statement that metals obey Ohm's law *if* we can show that τ does not depend on the applied electric field E. In this case ρ will not depend on E, which (see Section 31-3) is the criterion that a material obey Ohm's law. The quantity τ depends on the speed distribution of the conduction electrons. We have seen that this distribution is affected only very slightly by the application of even a relatively large electric field, since $\bar{v}$ is of the order of 10^8 cm/s and v_d (see Example 2) only of the order of 10^{-2} cm/s, a ratio of 10^{10}. We may be sure that whatever the value of τ is (for copper at 20°C, say) in

---

* See "Drift Speed and Collision Time" by Donald E. Tilley, *American Journal of Physics*, June 1976, for a full discussion.

the absence of a field it remains essentially unchanged when the field is applied. Thus the right side of Eq. 31-14 is independent of E (which means that ρ is independent of E) and the material obeys Ohm's law.

EXAMPLE 4

What are (a) the mean time τ between collisions and (b) the mean free path for free electrons in copper?

(a) From Eq. 31-14 (see also Example 2), we have

$$\tau = \frac{m}{ne^2\rho} = \frac{(9.1 \times 10^{-31}\ \text{kg})}{(8.4 \times 10^{28}/\text{m}^3)(1.6 \times 10^{-19}\ \text{C})^2(1.7 \times 10^{-8}\ \Omega \cdot \text{m})}$$

$$= 2.5 \times 10^{-14}\ \text{s}.$$

(b) The mean free path is

$$\lambda = \tau \bar{v} = (2.5 \times 10^{-14}\ \text{s})(1.6 \times 10^8\ \text{cm/s}) = 4.0 \times 10^{-6}\ \text{cm}.$$

This is about 220 ionic diameters.*

31-5
ENERGY TRANSFERS IN AN ELECTRIC CIRCUIT

Figure 31-8 shows a circuit consisting of a battery B connected to a "black box". A steady current i exists in the connecting wires and a steady potential difference V_{ab} exists between the terminals a and b. The box might contain a resistor, a motor, or a storage battery, among other things.

Terminal a, connected to the positive battery terminal, is at a higher potential than terminal b. If a charge dq moves through the box from a to b, this charge will decrease its electric potential energy by $dq\,V_{ab}$ (see Section 29-6). The conservation-of-energy principle tells us that this energy is transferred in the box from electric potential energy to some other form. What that other form will be depends on what is in the box. In a time dt the energy dU transferred inside the box is then

$$dU = dq\,V_{ab} = i\,dt\,V_{ab}.$$

We find the *rate* of energy transfer P by dividing by the time, or

$$P = \frac{dU}{dt} = iV_{ab}. \tag{31-15}$$

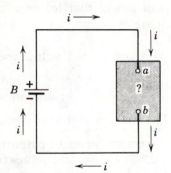

figure 31-8
A battery B sets up a current in a circuit containing a "black box", that is, a box whose contents are not known to us.

If the device in the box is a motor, the energy appears largely as mechanical work done by the motor; if the device is a storage battery that is being charged, the energy appears largely as stored chemical energy in this second battery.

If the device is a resistor, we assert that the energy appears as thermal energy in the resistor. To see this, consider a stone of mass m that falls through a height h. It decreases its gravitational potential energy by mgh. If the stone falls in a vacuum or—for practical purposes—in air, this energy is transformed into kinetic energy of the stone. If the stone falls into the depths of the ocean, however, its speed eventually becomes constant, which means that the kinetic energy no longer increases. The potential energy that is steadily being made available as the stone falls then appears as thermal energy in the stone and the surrounding water. It is the viscous, friction-like drag of the water on the surface of the stone that stops the stone from accelerating, and it is at this surface that thermal energy appears.

* See footnote on p. 686.

The course of the electrons through the resistor is much like that of the stone through water. The electrons travel with a constant drift speed v_d and thus do not gain kinetic energy. The electric potential energy that they lose is transferred to the resistor as thermal energy. On a microscopic scale we can understand this in that collisions between the electrons and the lattice (see Fig. 21-5) increase the amplitude of the thermal vibrations of the lattice; on a macroscopic scale this corresponds to a temperature increase. There can be a flow of heat out of the resistor subsequently, if the environment is at a lower temperature than the resistor.

For a resistor we can combine Eqs. 31-6 ($R = V/i$) and 31-15 and obtain either

$$P = i^2 R \qquad (31\text{-}16)$$

or

$$P = \frac{V^2}{R}. \qquad (31\text{-}17)$$

Note that Eq. 31-15 applies to electrical energy transfer of *all* kinds; Eqs. 31-16 and 31-17 apply only to the transfer of electrical energy to thermal energy in a resistor. Equations 31-16 and 31-17 are known as *Joule's law*. This law is a particular way of writing the conservation-of-energy principle for the special case in which electrical energy is transferred into thermal energy (Joule energy).

The unit of power that follows from Eq. 31-15 is the volt-ampere. We can write it as

$$1 \text{ volt} \cdot \text{ampere} = 1 \text{ volt} \cdot \text{ampere} \left(\frac{1 \text{ joule}}{1 \text{ volt} \times 1 \text{ coulomb}} \right)$$
$$\left(\frac{1 \text{ coulomb}}{1 \text{ ampere} \times 1 \text{ second}} \right)$$
$$= 1 \text{ joule/second}.$$

The first conversion factor in parentheses comes from the definition of the volt (Eq. 29-1); the second comes from the definition of the coulomb. The joule/second is such a common unit that it is given a special name of its own, the *watt* (abbr. W); see Section 7-7. Power is not an exclusively electrical concept, of course, and we can express in watts the power ($= \mathbf{F} \cdot \mathbf{v}$) expended by an agent that exerts a force $\mathbf{F}$ while it moves with a velocity $\mathbf{v}$.

EXAMPLE 5

You are given a 20-ft length of heating wire made of the special alloy Nichrome; it has a resistance of 24 Ω. Can you obtain more heat by winding one coil or by cutting the wire in two and winding two separate coils? In each case the coils are to be connected individually across a 110-V line.

The power P for the single coil is given by Eq. 31-17:

$$P = \frac{V^2}{R} = \frac{(110 \text{ V})^2}{24 \text{ }\Omega} = 500 \text{ W}.$$

The power for a coil of half the length is given by

$$P' = \frac{(110 \text{ V})^2}{12 \text{ }\Omega} = 1000 \text{ W}.$$

There are two "half-coils," so that the total power obtained by cutting the wire in half is 2000 W, or four times that for the single coil. This would seem to suggest that we could buy a 500-W heating coil, cut it in half, and rewind it to obtain 2000 W. Why is this not a practical idea?

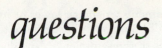

questions

1. Name other physical quantities that, like current, are scalars having a sense represented by an arrow in a diagram.

2. What conclusions can you draw by applying Eq. 31-4 to a closed surface through which a number of wires pass in random directions, carrying steady currents of different magnitudes? Does Gauss's law hold?

3. A potential difference V is applied to a copper wire of diameter d and length l. What is the effect on the electron drift speed of (a) doubling V, (b) doubling l, and (c) doubling d?

4. If the drift speeds of the electrons in a conductor under ordinary circumstances are so slow (see Example 2), why do the lights in a room turn on so quickly after the switch is closed?

5. Can you think of a way to measure the drift speed for electrons by timing their travel along a conductor?

6. Let a battery be connected to a copper cube at two corners defining a body diagonal. Pass a hypothetical plane completely through the cube, tilted at any angle. (a) Is the current i through the plane independent of the position and orientation of the plane? (b) Is there any position and orientation of the plane for which $\mathbf{j}$ is a constant in magnitude, direction, or both? (c) Does Eq. 31-4 hold for all orientations of the plane? (d) Does Eq. 31-4 hold for a closed surface of arbitrary shape, which may or may not lie entirely within the cube? If not, why; if so, what does it predict?

7. In our convention for the direction of current arrows (a) would it have been more convenient, or even possible, to have assumed all charge carriers to be negative? (b) Would it have been more convenient, or even possible, to have labeled the electron as positive, the proton as negative, etc.?

8. Explain in your own words why we can have $\mathbf{E} \neq 0$ inside a conductor in this chapter but that we took $\mathbf{E} = 0$ for granted in Chapter 28 (see, for example, Section 28-6).

9. A potential difference V is applied to a circular cylinder of carbon by clamping it between circular copper electrodes, as in Fig. 31-9. Discuss the difficulty of calculating the resistance of the carbon cylinder, using the relation $R = \rho L/A$.

10. How would you measure the resistance of a pretzel-shaped conductor? Give specific details to clarify the concept.

11. Discuss the difficulties of testing whether the filament of a light bulb obeys Ohm's law.

12. You are given a circular cylinder of aluminum, 1.00 cm in radius and 2.00 cm high. What practical laboratory arrangements would you make if you wanted to measure its resistance between parallel faces (assumed equipotentials) using Eq. 31-9 $(R = \rho l/A)$.

13. You are given a cube of aluminum and access to two battery terminals. How would you connect the terminals to the cube to insure (a) a maximum and (b) a minimum resistance?

14. Does the relation $V = iR$ apply to resistors that do not obey Ohm's law?

15. The temperature coefficient of resistance of a thermistor is negative and varies greatly with temperature. Account qualitatively for the shape of the curve of i versus V for the thermistor of Fig. 31-6.

16. Why are the dashed white lines in Fig. 31-7 curved slightly?

17. A current i enters the top of a copper sphere of radius R and leaves at a diametrically opposite point. Are all parts of the sphere equally effective in dissipating thermal energy?

18. What special characteristics must (a) heating wire and (b) fuse wire have?

19. Equation 31-16 $(P = i^2R)$ seems to suggest that the rate of increase of thermal energy in a resistor is reduced if the resistance is made less; Eq. 31-17 $(P = V^2/R)$ seems to suggest just the opposite. How do you reconcile this apparent paradox?

Copper

Carbon

Copper

figure 31-9
Question 9

20. Is the filament resistance lower or higher in a 500-W light bulb than in a 100-W bulb? Both bulbs are designed to operate on 100 V.

21. Five wires of the same length and diameter are connected in turn between two points maintained at constant potential difference. Will Joule energy be developed at the faster rate in the wire of (a) the smallest or (b) the largest resistance?

22. The windings of a motor (connected to a load) have a resistance of 1.0 Ω. If we apply a potential difference of 100 V to the motor, does it follow that the current through the motor will be 110 V/1.0 Ω = 110 A?

23. A cow and a man are standing in a meadow when lightning strikes the ground nearby. Why is the cow more likely to be killed than the man? The responsible phenomenon is called "step voltage".

problems

SECTION 31-1

1. A current of 5 A exists in a 10-Ω resistor for 4 min. (a) How many coulombs and (b) how many electrons pass through any cross section of the resistor in this time?   *Answer:* (a) 1.2×10^3 C. (b) 7.5×10^{21} electrons.

2. A current is established in a gas discharge tube when a sufficiently high potential difference is applied across the two electrodes in the tube. The gas ionizes; electrons move toward the positive terminal and positive ions toward the negative terminal. What are the magnitude and sense of the current in a hydrogen discharge tube in which 3.1×10^{18} electrons and 1.1×10^{18} protons move past a cross-sectional area of the tube each second?

3. A steady beam of alpha particles ($q = 2e$) traveling with constant kinetic energy 20 MeV carries a current 0.25×10^{-6} A. (a) If the beam is directed perpendicular to a plane surface, how many alpha particles strike the surface in 3.0 s? (b) At any instant, how many alpha particles are there in a given 20-cm length of the beam? (c) Through what potential difference was it necessary to accelerate each alpha particle from rest to bring it to an energy of 20 MeV?   *Answer:* (a) 2.3×10^{12}. (b) 5.0×10^3. (c) 10^7 V.

4. We have 2.0×10^8 doubly charged positive ions per cubic centimeter, all moving north with a speed of 1.0×10^7 cm/s. (a) What is the current density **j**, in magnitude and direction? (b) Can you calculate the total current i in this ionic beam? If not, why?

5. A small but measurable current of 1.0×10^{-10} A exists in a copper wire whose diameter is 0.10 in. Calculate the electron drift speed.
Answer: 1.5×10^{-15} m/s.

6. The belt of an electrostatic generator is 50 cm wide and travels at 30 m/s. The belt carries charge into the sphere at a rate corresponding to 1.0×10^{-4} A. Compute the surface charge density on the belt.

7. A current i enters one corner of a square sheet of copper and leaves at the opposite corner. Sketch arrows for various points within the square to represent the relative values of **j**. Intuitive guesses rather than detailed mathematical analyses are called for.

8. You are given a conducting sphere of 10-cm radius. One wire carries a current of 1.0000020 A into it. Another wire carries a current of 1.0000000 A out of it. How long would it take for the sphere to increase in potential by 1000 V?

SECTION 31-2

9. A steel trolley-car rail has a cross-sectional area of 7.1 in.² What is the resistance of 10 miles of single track? The resistivity of the steel is 6.0×10^{-7} Ω·m.   *Answer:* 2.1 Ω.

10. A square aluminum rod is 1.0 m long and 5.0 mm on edge. (a) What is the resistance between its ends? (b) What must be the diameter of a circular 1.0 m copper rod if its resistance is to be the same?

11. A copper wire and an iron wire of the same length have the same potential

difference applied to them. (a) What must be the ratio of their radii if the current is to be the same? (b) Can the current density be made the same by suitable choices of the radii? *Answer:* (a) 2.4, iron being larger. (b) No.

12. A wire with a resistance of 6.0 Ω is drawn out through a die so that its new length is three times its original length. Find the resistance of the longer wire, assuming that the resistivity and density of the material are not changed during the drawing process.

13. A wire of Nichrome (a nickel-chromium alloy commonly used in heating elements) is 1.0 m long and 1.0 mm² in cross-sectional area. It carries a current of 4.0 A when a 2.0-V potential difference is applied between its ends. What is the conductivity σ of Nichrome?   *Answer:* 2.0×10^6 S.

14. A rod of a certain metal is 1.00 m long and 0.550 cm in diameter. The resistance between its ends (at 20°C) is 2.87×10^{-3} Ω. A round disk is formed of this same material, 2.00 cm in diameter and 1.00 mm thick. (a) What is the resistance between the opposing round faces, assuming equipotential surfaces? (b) What is the material?

15. Two conductors are made of the same material and have the same length. Conductor A is a solid wire of diameter 1.0 mm. Conductor B is a hollow tube of outside diameter 2.0 mm and inside diameter 1.0 mm. What is the resistance ratio, R_A/R_B, measured between their ends?   *Answer:* 3.0

16. Nine copper wires of length l and diameter d are connected in parallel to form a single composite conductor of resistance R. What must be the diameter D of a single copper wire of length l if it is to have the same resistance?

17. A copper wire and an iron wire of equal length l and diameter d are joined and a potential difference V is applied between the ends of the composite wire. Calculate (a) the potential difference across each wire. Assume that $l = 10$ m, $d = 2.0$ mm, and $V = 100$ V. (b) Also calculate the current density in each wire, and (c) the electric field in each wire.
Answer: (a) 15 V (copper); 85 V (iron). (b) 8.5×10^7 A/m². (c) 1.5 V/m (copper); 8.5 V/m (iron).

18. The resistance of an iron wire is 5.9 times that of a copper wire of the same dimensions. What must be the diameter of an iron wire if it is to have the same resistance as a copper wire 0.12 cm in diameter, both wires being the same length?

19. Circularly cylindrical aluminum and copper rods have the same length and are designed to have the same resistance. The resistivity of copper is 0.61 times that of aluminum but its density is 3.3 times that of aluminum. What is the ratio of the mass of the aluminum rod to that of the copper rod?
Answer: 0.50.

20. Conductors A and B, having equal lengths of 40 m and a cross-sectional area of 0.10 m², are connected in series. A potential of 60 V is applied across the terminal points of the connected wires. The resistances of the wires are 40 and 20 Ω respectively. Determine: (a) the resistivities of the two wires; (b) the magnitude of the electric field in each wire; (c) the current density in each wire; (d) the potential difference applied to each conductor.

21. A resistor is in the shape of a truncated right circular cone (Fig. 31-10). The end radii are a and b, the altitude is l. If the taper is small, we may assume that the current density is uniform across any cross section. (a) Calculate the resistance of this object. (b) Show that your answer reduces to $\rho(l/A)$ for the special case of zero taper $(a = b)$.

Answer: (a) $R = \rho \dfrac{l}{\pi a b}$.

22. (a) At what temperature would the resistance of a copper conductor be double its resistance at 0°C? (b) Does this same temperature hold for all copper conductors, regardless of shape or size?

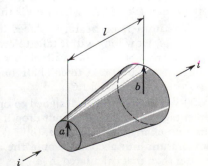

figure 31-10
Problem 21

23. The copper windings of a motor have a resistance of 50 Ω at 20°C when the motor is idle. After running for several hours the resistance rises to 58 Ω. What is the temperature of the windings? *Answer:* 61°C.

24. When a metal rod is heated, not only its resistance but also its length and its cross-sectional area change. The relation $R = \rho l/A$ suggests that all three factors should be taken into account in measuring ρ at various temperatures. (a) If the temperature changes by 1.0°C, what percent changes in R, l, and A occur for a copper conductor? (b) What conclusion do you draw? The coefficient of linear expansion is 1.7×10^{-5}/C°.

SECTION 31-3

25. List in tabular form similarities and differences between the flow of charge along a conductor, the flow of water through a horizontal pipe, and the conduction of heat through a slab. Consider such quantities as what causes the flow, what opposes it, what particles (if any) participate, and the units in which the flow may be measured.

26. (a) Using data from Fig. 31-5, plot the resistance of the vacuum tube as a function of applied potential difference. (b) Repeat for the thermistor of Fig. 31-6.

SECTION 31-4

27. Explain why the momentum which conduction electrons transfer to the ions in a metal conductor does not give rise to a resultant force on the conductor.

Answer: Because of Newton's third law there is no resultant force on the conductor (electrons + ions).

SECTION 31-5

28. Thermal energy is developed in a resistor at a rate of 100 W when the current is 3.0 A. What is the resistance in ohms?

29. An x-ray tube takes a current of 7.0 mA and operates at a potential difference of 80 kV. What power in watts is dissipated? *Answer:* 560 W.

30. A potential difference of 1.0 V is applied to a 100-ft length of #18 copper wire (diameter = 0.040 in.). Calculate (a) the current, (b) the current density, (c) the electric field, and (d) the rate at which thermal energy is developed in the wire.

31. The National Board of Fire Underwriters has fixed safe current-carrying capacities for various sizes and types of wire. For #10 rubber-coated copper wire (wire diameter = 0.10 in.) the maximum safe current is 25 A. At this current, find (a) the current density, (b) the electric field, (c) the potential difference for 1000 ft of wire, and (d) the rate at which thermal energy ($= i^2R$) is developed for 1000 ft of wire.

Answer: (a) 4.9×10^6 A/m². (b) 8.3×10^{-2} V/m. (c) 25 V. (d) 630 W.

32. A beam of 16-MeV deuterons from a cyclotron falls on a copper block. The beam is equivalent to a current of 15×10^{-6} A. (a) At what rate do deuterons strike the block? (b) At what rate is thermal energy produced in the block?

33. A 500-W immersion heater is placed in a pot containing 2.0 liters of water at 20°C. (a) How long will it take to bring the water to boiling temperature, assuming that 80% of the available energy is absorbed by the water? (b) How much longer will it take to boil half the water away?

Answer: (a) 28 min. (b) 1.6 h.

34. A 500-W heating unit is designed to operate from a 115-V line. (a) By what percentage will its heat output drop if the line voltage drops to 110 V? Assume no change in resistance. (b) Taking the variation of resistance with temperature into account, would the actual heat output drop be larger or smaller than that calculated in (a)?

35. A 1250-W radiant heater is constructed to operate at 115 V. (a) What will be the current in the heater? (b) What is the resistance of the heating coil? (c) How many kilocalories are generated in one hour by the heater?

Answer: (a) 11 A. (b) 11 Ω. (c) 1100 kcal.

36. A Nichrome heater dissipates 500 W when the applied potential difference is 110 V and the wire temperature is 800°C. How much power would it dissipate if the wire temperature were held to 200°C by immersion in a bath of cooling oil? The applied potential difference remains the same; $\bar{\alpha}$ for Nichrome is about $4 \times 10^{-4}/C°$.

37. (a) Derive the formulas $P = j^2\rho$ and $P = E^2/\rho$ where P = power per unit volume in a resistor. (b) A cylindrical resistor of radius 0.50 cm and length 2.0 cm has a resistivity of 3.5×10^{-5} $\Omega \cdot$m. What are the current density and potential difference when the power dissipation is 1.0 W?
   Answer: (b) $j = 1.3 \times 10^5$ A/m²; $V = 0.094$ V.

38. It is desired to make a long cylindrical conductor whose temperature coefficient of resistivity at 20°C will be close to zero. (a) If such a conductor is made by assembling alternate disks of iron and carbon, what is the ratio of the thickness of a carbon disk to that of an iron disk? Assume that the temperature remains essentially the same in each disk. (b) What is the ratio of thermal energy generation in a carbon disk to that in an iron disk? See Table 31-1.

39. An electron linear accelerator produces a pulsed beam of electrons. The pulse current is 0.50 A and the pulse duration 0.10 μs. (a) How many electrons are accelerated per pulse? (b) What is the average current for a machine operating at 500 pulses/s? (c) If the electrons are accelerated to an energy of 50 MeV, what are the average and peak power outputs of the accelerator?
   Answer: (a) 3.1×10^{11}. (b) 25 μA. (c) 1200 W (average); 2.5×10^7 W (peak).

32
electromotive
force and circuits

There exist in nature certain devices such as batteries and electric generators which are able to maintain a potential difference between two points to which they are attached. We call such devices seats of *electromotive force* (symbol ε; abbr. emf). In this chapter we do not discuss their internal construction or detailed mode of action but confine ourselves to describing their gross electrical characteristics and to exploring their usefulness in electric circuits.

Figure 32-1a shows a seat of emf B, represented by a battery*, connected to a resistor R. The seat of emf maintains its upper terminal positive and its lower terminal negative, as shown by the $+$ and $-$ signs. In the circuit external to B positive charge carriers would be driven in the direction shown by the arrows marked i. In other words, a clockwise current would be set up.

An emf is represented by an arrow which is placed next to the seat and points in the direction in which the seat, acting alone, would cause a positive charge carrier to move in the external circuit. We draw a small circle on the tail of an emf arrow so that we will not confuse it with a current arrow.

A seat of emf must be able to do work on charge carriers that enter it. In the circuit of Fig. 32-1a, for example, the seat acts to move positive charges from a point of low potential (the negative terminal) through the seat to a point of high potential (the positive terminal). This reminds us of a pump, which can cause water to move from a place of low gravitational potential to a place of high potential.

In Fig. 32-1a a charge dq passes through *any* cross section of the circuit in time dt. In particular, this charge enters the seat of emf ε at its low-potential end and leaves at its high-potential end. The seat must do

* Batteries are far from being the only source of emf. Among others are generators (see Chapter 39); devices activated by temperature differences (thermocouples, etc.); devices activated by light (see "The Photovoltaic Generation of Electricity" by Bruce Chalmers, *Scientific American*, October 1976); the human heart (see "The Electrocardiograph — Teaching Physics to Premeds" by Pierre Lafrance, *The Physics Teacher*, November 1972); certain fish (see "Seeing the World Through a New Sense: Electroreception in Fish" by T. H. Bullock, *American Scientist*, May–June 1973).

an amount of work dW on the (positive) charge carriers to force them to go to the point of higher potential. The emf ε of the seat is defined from

$$\varepsilon = dW/dq. \qquad (32\text{-}1)$$

The unit of emf is the joule/coulomb (see Eq. 29-1) which is the *volt*. We might be inclined to say that a battery has an emf of 1 volt if it maintains a difference of potential of 1 volt between its terminals. This is true only under certain conditions, which we describe in Section 32-4. Note also from Eq. 32-1 that the electromotive force is not actually a force, that is, we cannot measure it in newtons. The name is involved with the early history of the subject.

If a seat of emf does work on a charge carrier, energy must be transferred within the seat. In a battery, for example, chemical energy is transferred into electrical energy. Thus we can describe a seat of emf as a device in which chemical, mechanical, or some other form of energy is changed (reversibly) into electrical energy. The chemical energy provided by the battery in Fig. 32-1a is stored in the electric and the magnetic* fields that surround the circuit. This stored energy does not increase because it is being drained away, by transfer to Joule (thermal) energy in the resistor, at the same rate at which it is supplied. The electric and magnetic fields play an intermediary role in the energy transfer process, acting as a storage reservoir.

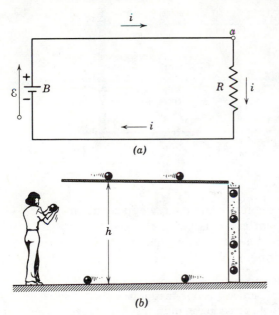

(a)

(b)

figure 32-1
(a) A simple electric circuit and (b) its gravitational analog.

Figure 32-1b shows a gravitational analog of Fig. 32-1a. In the top figure the seat of emf B does work on the charge carriers. This energy, stored in passage as electromagnetic field energy, appears eventually as thermal energy in resistor R. In the lower figure the person, in lifting the bowling balls from the floor to the shelf, does work on them. This energy is stored, in passage, as gravitational field energy. The balls roll slowly and uniformly along the shelf, dropping from the right end into a cylinder full of viscous oil. They sink to the bottom at an essentially

---

* A current in a wire is surrounded by a magnetic field, and this field, like the electric field, can also be viewed as a site of stored energy (see Section 36-4).

constant speed, are removed by a trapdoor mechanism not shown, and roll back along the floor to the left. The energy put into the system by the person appears eventually as thermal energy in the viscous fluid, resulting in a temperature rise. The energy supplied by the person comes from internal (chemical) energy. The circulation of charges in Fig. 32-1a will stop eventually if battery B is not recharged; the circulation of bowling balls in Fig. 32-1b will stop eventually if the person does not replenish the store of internal energy by eating.

Figure 32-2 shows a circuit containing two (ideal) batteries, A and B, a resistor R, and an (ideal) electric motor employed in lifting a weight. The batteries are connected so that they tend to send charges around the circuit in opposite directions; the actual direction of the current is determined by B, which supplies the larger potential difference. The energy transfers in this circuit are

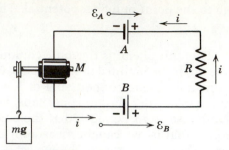

figure 32-2
Two batteries, a resistor, and a motor, connected in a single-loop circuit. It is given that $\varepsilon_B > \varepsilon_A$.

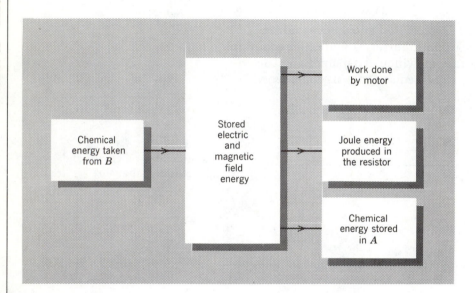

The chemical energy in B is steadily depleted, the energy appearing in the three forms shown on the right. Battery A is being charged while battery B is being discharged. Again, the electric and magnetic fields that surround the circuit act as an intermediary.

It is part of the definition of an emf that the energy transfer process be *reversible*, at least in principle. Recall that a reversible process is one that passes through equilibrium states; its course can be reversed by making an infinitesimal change in the environment of the system; see Section 25-2. A battery, for example, can either be on charge or discharge; a generator can be driven mechanically, producing electrical energy, or it can be operated backward as a motor. The (reversible) energy transfers here are

$$\text{electrical} \rightleftharpoons \text{chemical}$$

and

$$\text{electrical} \rightleftharpoons \text{mechanical.}$$

The energy transfer from electrical energy to thermal energy is not reversible. We can easily raise the temperature of a conductor by supplying electric energy to it, but it is *not* possible to set up a current in a closed copper loop by raising its temperature uniformly. Because of this lack of reversibility, we do not associate an emf with the Joule effect, that is, with energy transfers associated with Eqs. 31-16 and 31-17.

In a time dt an amount of energy given by $i^2R\,dt$ will appear in the resistor of Fig. 32-1a as Joule thermal energy. During this same time a charge $dq\,(= i\,dt)$ will have moved through the seat of emf, and the seat will have done work on this charge (see Eq. 32-1) given by

$$dW = \varepsilon\,dq = \varepsilon i\,dt.$$

From the conservation of energy principle, the work done by the seat must equal the thermal energy, or

$$\varepsilon i\,dt = i^2R\,dt.$$

Solving for i, we obtain

$$i = \varepsilon/R. \tag{32-2}$$

We can also derive Eq. 32-2 by considering that, if electric potential is to have any meaning, a given point can have only one value of potential at any given time. If we start at any point in the circuit of Fig. 32-1a and, in imagination, go around the circuit in either direction, adding up algebraically the changes in potential that we encounter, we must arrive at the same potential when we return to our starting point. In other words, *the algebraic sum of the changes in potential encountered in a complete traversal of the circuit must be zero.*

In Fig. 32-1a let us start at point a, whose potential is V_a,* and traverse the circuit clockwise. In going through the resistor, there is a change in potential of $-iR$. The minus sign shows that the top of the resistor is higher in potential than the bottom, which must be true, because positive charge carriers move of their own accord from high to low potential. As we traverse the battery from bottom to top, there is an *increase* of potential $+\varepsilon$ because the battery does (positive) work on the charge carriers, that is, it moves them from a point of low potential to one of high potential. Adding the algebraic sum of the changes in potential to the initial potential V_a must yield the identical value V_a, or

$$V_a - iR + \varepsilon = V_a.$$

We write this as

$$-iR + \varepsilon = 0,$$

which is independent of the value of V_a and which asserts explicitly that the algebraic sum of the potential changes for a complete circuit traversal is zero. This relation leads directly to Eq. 32-2.

These two ways to find the current in single-loop circuits, based on the conservation of energy and on the concept of potential, are completely equivalent because potential differences are defined in terms of work and energy (see Section 29-1). The statement that the sum of the changes in potential encountered in making a complete loop is zero is called *Kirchhoff's second rule*; for brevity we call it the *loop theorem.* Always bear in mind that this theorem is simply a particular way of stating the law of conservation of energy for electric circuits.

To prepare for the study of more complex circuits, let us examine the rules for finding potential differences; these rules follow from the previous discussion. They are not meant to be memorized but rather to be

---

* The actual value of V_a depends on assumptions made in the definition of potential (as described in Section 29-1). The numerical value of V_a is not important because, as in most electric circuit situations, we are concerned here with *differences* of potential. Point a in Fig. 32-1a (or any other single point in that figure) could be connected to ground (symbol $\doteq$) and assigned the potential $V_a = 0$, following a common practice.

so thoroughly understood that it becomes trivial to re-derive them on each application.

1. If a resistor is traversed in the direction of the current, the change in potential is $-iR$; in the opposite direction it is $+iR$.

2. If a seat of emf is traversed in the direction of the emf, the change in potential is $+\varepsilon$; in the opposite direction it is $-\varepsilon$.

Figure 32-3a shows a circuit which emphasizes that all seats of emf have an intrinsic internal resistance r. This resistance cannot be removed— although we would usually like to do so— because it is an inherent part of the device. The figure shows the internal resistance r and the emf separately, although, actually, they occupy the same region of space.

If we apply the loop theorem, starting at b and going around clockwise, we obtain

$$V_b + \varepsilon - ir - iR = V_b$$

or

$$+\varepsilon - ir - iR = 0.$$

Compare these equations with Fig. 32-3b, which shows the changes in potential graphically. In writing these equations, note that we traversed r and R in the direction of the current and ε in the direction of the emf. The same equation follows if we start at any other point in the circuit or if we traverse the circuit in a counterclockwise direction. Solving for i gives

$$i = \frac{\varepsilon}{R + r}. \tag{32-3}$$

32-3
OTHER SINGLE-LOOP CIRCUITS

figure 32-3
(a) A single-loop circuit. The rectangular block is a seat of emf with internal resistance r. (b) The same circuit is drawn for convenience as a straight line. Directly below are shown the changes in potential that one encounters in traversing the circuit clockwise from point b. Note that the two points marked "b" at the top of this figure are the same point.

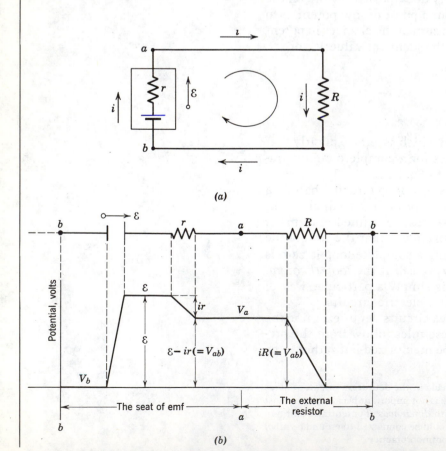

(a)

(b)

Resistors in series. Resistors in series are connected so that there is only one conducting path through them, as in Fig. 32-4. What is the equivalent resistance R of this series combination? The equivalent resistance is the single resistance R which, substituted for the series combination between the terminals ab, will leave the current i unchanged.

Applying the loop theorem (going clockwise from a) yields

$$-iR_1 - iR_2 - iR_3 + \varepsilon = 0$$

or

$$i = \frac{\varepsilon}{R_1 + R_2 + R_3}.$$

For the equivalent resistance R

$$i = \frac{\varepsilon}{R}$$

or

$$R = R_1 + R_2 + R_3. \tag{32-4}$$

The extension to more than three resistors is clear.

We often want to compute the potential difference between two points in a circuit. In Fig. 32-3a, for example, what is the relationship between the potential difference V_{ab} ($= V_a - V_b$) between points b and a and the fixed circuit parameters ε, r, and R? To find this relationship, let us start from point b and traverse the circuit to point a, passing through resistor R against the current. If V_b and V_a are the potentials at b and a, respectively, we have

$$V_b + iR = V_a$$

because we experience an increase in potential in traversing a resistor against the current arrow. We rewrite this relation as

$$V_{ab} = V_a - V_b = +iR,$$

which tells us that V_{ab}, the potential difference between points a and b, has the magnitude iR and that point a is more positive than point b. Combining this last equation with Eq. 32-3 yields

$$V_{ab} = \varepsilon \frac{R}{R + r}. \tag{32-5}$$

To sum up: To find the potential difference between any two points in a circuit start at one point and traverse the circuit to the other, following any path,* and add up algebraically the potential changes encountered. This algebraic sum will be the potential difference. This procedure is similar to that for finding the current in a closed loop, except that here the potential differences are added up over part of a loop and not over the whole loop.

The potential difference between any two points can have only one value; thus we must obtain the same answer for all paths that connect these points. If we consider two points on the side of a hill, the measured difference in gravitational potential (that is, in altitude) between them is the same no matter what path is followed in going from one to the other. In Fig. 32-3a let us also calculate V_{ab}, using a path passing through the seat of emf. We have

EXAMPLE 1

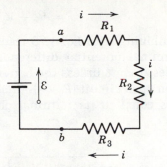

figure 32-4
Example 1. Three resistors are connected in series between terminals a and b.

32-4
POTENTIAL DIFFERENCES

---

* Recall (see Section 29-1) that path independence is a central feature of the potential concept.

$$V_b + \varepsilon - ir = V_a$$

or (see also Fig. 32-3b)

$$V_{ab} = V_a - V_b = +\varepsilon - ir.$$

Again, combining with Eq. 32-3 leads to Eq. 32-5.

The terminal potential difference of the battery V_{ab}, as Eq. 32-5 shows, is less than ε unless the battery has no internal resistance $(r=0)$ or if it is on open circuit $(R=\infty)$; then V_{ab} is equal to ε. Thus the emf of a device is equal to its terminal potential difference *when on open circuit*.

EXAMPLE 2

In Fig. 32-5a let ε_1 and ε_2 be 2.0 V and 4.0 V, respectively; let the resistances r_1, r_2, and R be 1.0 Ω, 2.0 Ω, and 5.0 Ω, respectively. What is the current?

Emfs ε_1 and ε_2 oppose each other, but because ε_2 is larger it controls the direction of the current. Thus i will be counterclockwise. The loop theorem, going clockwise from a, yields

$$-\varepsilon_2 + ir_2 + iR + ir_1 + \varepsilon_1 = 0.$$

Check that the same result is obtained by going around counterclockwise. Also, compare this equation carefully with Fig. 32-5b, which shows the potential changes graphically.

Solving for i yields

$$i = \frac{\varepsilon_2 - \varepsilon_1}{R + r_1 + r_2} = \frac{4.0 \text{ V} - 2.0 \text{ V}}{5.0 \text{ Ω} + 1.0 \text{ Ω} + 2.0 \text{ Ω}}$$

$$= 0.25 \text{ A}.$$

It is not necessary to know in advance what the actual direction of the current is. To show this, let us assume that the current in Fig. 32-5a is clockwise, an assumption that we know is incorrect. The loop theorem then yields (going clockwise from a)

$$-\varepsilon_2 - ir_2 - iR - ir_1 + \varepsilon_1 = 0$$

or

$$i = \frac{\varepsilon_1 - \varepsilon_2}{R + r_1 + r_2}.$$

Substituting numerical values (see above) yields −0.25 A for the current. The minus sign tells us that the current is in the opposite direction from the one we have assumed.

In more complex circuit problems involving many loops and branches it is often impossible to know in advance the correct directions for the currents in all parts of the circuit. We can assume directions for the currents at random. Those currents for which positive numerical values are obtained will have the correct directions; those for which negative values are obtained will be exactly opposite to the assumed directions. In all cases the numerical values will be correct.

EXAMPLE 3

What is the potential difference (a) between points b and a in Fig. 32-5a, and (b) between points a and c?

(a) For points a and b we start at b and traverse the circuit to a, obtaining

$$V_{ab} (= V_a - V_b) = -ir_2 + \varepsilon_2 = -(0.25 \text{ A})(2.0 \text{ Ω}) + 4.0 \text{ V}$$

$$= +3.5 \text{ V}.$$

Thus a is more positive than b and the potential difference (3.5 V) is *less than* the emf (4.0 V); see Fig. 32-5b.

(b) For points c and a, we start at c and traverse the circuit to a, obtaining

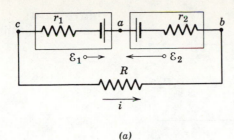

(a)

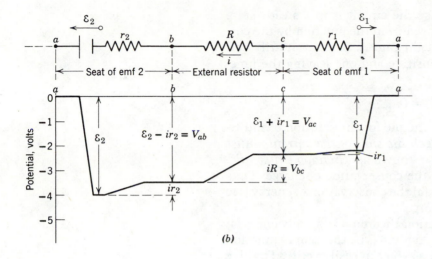

(b)

figure 32-5

701   MULTILOOP CIRCUITS   SEC. 32-5

Examples 2, 3. (a) A single-loop circuit. (b) The same circuit is shown schematically as a straight line, the potential differences encountered in traversing the circuit clockwise from point a being displayed directly below. In the lower figure the potential of point a was assumed to be zero for convenience. Note that the two points marked "a" in this figure are the same point.

$$V_{ac} (= V_a - V_c) = +\varepsilon_1 + ir_1 = +2.0 \text{ V} + (0.25 \text{ A})(1.0 \text{ }\Omega)$$

$$= +2.25 \text{ V}.$$

This tells us that a is at a higher potential than c. The terminal potential difference of ε_1 (2.25 V) is *larger than* the emf (2.0 V); see Fig. 32-5b. Charge is being forced through ε_1 in a direction opposite to the one in which it would send charge if it were acting by itself; if ε_1 is a storage battery, it is being charged at the expense of ε_2.

Let us test the first result by proceeding from b to a along a different path, namely, through R, r_1, and ε_1. We have

$$V_{ab} = iR + ir_1 + \varepsilon_1 = (0.25 \text{ A})(5.0 \text{ }\Omega) + (0.25 \text{ A})(1.0 \text{ }\Omega) + 2.0 \text{ V} = +3.5 \text{ V},$$

which is the same as the earlier result.

32-5
MULTILOOP CIRCUITS

Figure 32-6 shows a circuit containing two loops. For simplicity, we have neglected the internal resistances of the batteries. There are two *junctions*, b and d, and three *branches* connecting these junctions. The branches are the left branch *bad*, the right branch *bcd*, and the central branch *bd*. If the emfs and the resistances are given, what are the currents in the various branches?

We label the currents in the branches as i_1, i_2, and i_3, as shown. Current i_1 has the same value for any cross section of the left branch from b to d. Similarly, i_2 has the same value everywhere in the right branch and i_3 in the central branch. The directions of the currents have been chosen arbitrarily. The careful reader will note that i_3 must point in a direction opposite to the one we have shown. We have deliberately drawn it in wrong to show how the formal mathematical procedures will always indicate this to us.

The three currents i_1, i_2, and i_3 carry charge either toward junction d or away from it. Charge does not accumulate at junction d, nor does it

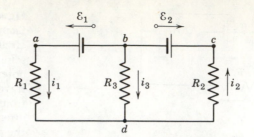

figure 32-6
A multiloop circuit.

drain away from this junction because the circuit is in a steady-state condition. Thus charge must be removed from the junction by the currents at the same rate that it is brought into it. If we arbitrarily call a current approaching the junction positive and one leaving the junction negative, then

$$i_1 + i_3 - i_2 = 0.$$

This equation suggests a general principle for the solution of multi-loop circuits: *At any junction the algebraic sum of the currents must be zero.* This *junction theorem* is also known as *Kirchhoff's first rule.* Note that it is simply a statement of the conservation of charge. Thus our basic tools for solving circuits are (*a*) the conservation of energy (see p. 697) and (*b*) the conservation of charge.

For the circuit of Fig. 32-6, the junction theorem yields only one relationship among the three unknowns. Applying the theorem at junction *b* leads to exactly the same equation, as you can easily verify. To solve for the three unknowns, we need two more independent equations; they can be found from the loop theorem.

In single-loop circuits there is only one conducting loop around which to apply the loop theorem, and the current is the same in all parts of this loop. In multiloop circuits there is more than one loop, and the current in general will not be the same in all parts of any given loop.

If we traverse the left loop of Fig. 32-6 in a counterclockwise direction, the loop theorem gives

$$\mathcal{E}_1 - i_1 R_1 + i_3 R_3 = 0. \qquad (32\text{-}6)$$

The right loop gives

$$-i_3 R_3 - i_2 R_2 - \mathcal{E}_2 = 0. \qquad (32\text{-}7)$$

These two equations, together with the relation derived earlier with the junction theorem, are the three simultaneous equations needed to solve for the unknowns i_1, i_2, and i_3. Doing so yields

$$i_1 = \frac{\mathcal{E}_1(R_2 + R_3) - \mathcal{E}_2 R_3}{R_1 R_2 + R_2 R_3 + R_1 R_3}, \qquad (32\text{-}8a)$$

$$i_2 = \frac{\mathcal{E}_1 R_3 - \mathcal{E}_2(R_1 + R_3)}{R_1 R_2 + R_2 R_3 + R_1 R_3}, \qquad (32\text{-}8b)$$

and

$$i_3 = \frac{-\mathcal{E}_1 R_2 - \mathcal{E}_2 R_1}{R_1 R_2 + R_2 R_3 + R_1 R_3}. \qquad (32\text{-}8c)$$

Supply the missing steps. Equation 32-8*c* shows that no matter what numerical values are given to the emfs and to the resistances, the current i_3 will always have a negative value. This means that it will always point up in Fig. 32-6 rather than down, as we deliberately assumed. The currents i_1 and i_2 may be in either direction, depending on the particular numerical values given.

Verify that Eqs. 32-8 reduce to sensible conclusions in special cases. For $R_3 = \infty$, for example, we find

$$i_1 = i_2 = \frac{\mathcal{E}_1 - \mathcal{E}_2}{R_1 + R_2} \quad \text{and} \quad i_3 = 0.$$

What do these equations reduce to for $R_2 = \infty$?

The loop theorem can be applied to a large loop consisting of the entire circuit *abcda* of Fig. 32-6. This fact might suggest that there are more equations than we need, for there are only three unknowns and we already have three equations written in terms of them. However, the loop theorem yields for this loop

$$-i_1 R_1 - i_2 R_2 - \mathcal{E}_2 + \mathcal{E}_1 = 0,$$

which is nothing more than the sum of Eqs. 32-6 and 32-7. Thus this large loop does not yield another *independent* equation. It will never be found in solving multiloop circuits that there are more independent equations than variables.

Resistors in parallel. Figure 32-7 shows three resistors connected across the same seat of emf. Resistances across which the identical potential difference is applied are said to be in parallel. What is the equivalent resistance R of this parallel combination? The equivalent resistance is that single resistance which, substituted for the parallel combination between terminals *ab*, would leave the current *i* unchanged.

EXAMPLE 4

figure 32-7
Example 4. Three resistors are connected in parallel between terminals *a* and *b*.

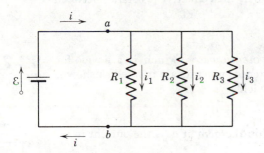

The currents in the three branches are

$$i_1 = \frac{V}{R_1}, \quad i_2 = \frac{V}{R_2}, \quad \text{and} \quad i_3 = \frac{V}{R_3},$$

where V is the potential difference that appears between points *a* and *b*. The total current *i* is found by applying the junction theorem to junction *a*, or

$$i = i_1 + i_2 + i_3 = V\left(\frac{1}{R_1} + \frac{1}{R_2} + \frac{1}{R_3}\right).$$

If the equivalent resistance is used instead of the parallel combination, we have

$$i = \frac{V}{R}.$$

Combining these two equations gives

$$\frac{1}{R} = \frac{1}{R_1} + \frac{1}{R_2} + \frac{1}{R_3}. \tag{32-9}$$

This formula can easily be extended to more than three resistances. Note that the equivalent resistance of a parallel combination is less than any of the resistances that make it up.

A meter to measure currents is called an *ammeter* (or a *milliammeter, microammeter*, etc., depending on the size of the current to be measured). To determine the current in a wire, it is necessary to break or cut the wire and to insert the ammeter, so that the current to be measured passes through the meter (see Fig. 32-8).*

It is essential that the resistance R_A of the ammeter be *small* compared to other resistances in the circuit. Otherwise the presence of the meter will in itself change the current to be measured. An ideal ammeter would have zero resistance. In the circuit of Fig. 32-8 the required condition, assuming that the voltmeter is not connected, is

$$R_A \ll r + R_1 + R_2.$$

A meter to measure potential differences is called a *voltmeter* (or a *millivoltmeter* or *microvoltmeter*, etc.). To find the potential difference between two points in a circuit, it is necessary to connect one of the voltmeter terminals to each of the circuit points, without breaking the circuit (see Fig. 32-8).†

It is essential that the resistance of the voltmeter R_V be *large* compared to any circuit resistance across which the voltmeter is connected. Otherwise the meter will itself constitute an important circuit element and will alter the circuit current and the potential difference to be measured. An ideal voltmeter would have an infinite resistance. In Fig. 32-8 the required condition is

$$R_V \gg R_1.$$

In measuring potential difference in electronic circuits, where the effective circuit resistance may be of the order of 10^6 ohms or higher, it becomes necessary to use an electronic voltmeter, which is an electron tube or transistor device designed specifically to have an extremely high effective resistance between its input terminals.

32-6
MEASURING CURRENTS AND POTENTIAL DIFFERENCES

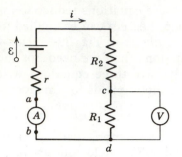

figure 32-8
An ammeter (A) is connected to read the current in a circuit, and a voltmeter (V) is connected to read the potential difference across resistor R_1.

32-7
THE POTENTIOMETER

Figure 32-9 shows the rudiments of a *potentiometer*, which is a device for measuring an unknown emf ε_x. The currents and emfs are marked as shown. Applying the loop theorem to loop *abcd* yields

$$-\varepsilon_x - ir + (i_0 - i)R = 0,$$

where $i_0 - i$, by application of the junction theorem at *a*, is the current in resistor R. Solving for i yields

$$i = \frac{i_0 R - \varepsilon_x}{R + r},$$

in which R is a variable resistor. This relation shows that if R is adjusted to have the value R_x where

$$i_0 R_x = \varepsilon_x, \tag{32-10}$$

the current i in the branch *abcd* becomes zero. To *balance* the potentiometer in this way, R must be adjusted manually until the sensitive current meter G reads zero.

The emf can be obtained from Eq. 32-10 if the current i_0 is known. However, it is standard practice to replace ε_x by a known standard emf ε_s, and once again to adjust R to the zero-current condition. This yields, assuming the current i_0 remains unchanged,

$$i_0 R_s = \varepsilon_s.$$

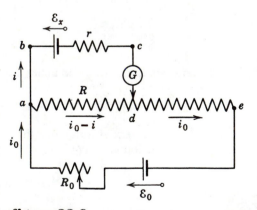

figure 32-9
Elements of a potentiometer.
Point *d* represents a sliding contact.

* The meter must be connected so that the direction of current through it (assuming positive charge carriers) is *into* the meter terminal marked +. Otherwise the meter will deflect in a direction opposite to that intended.
† The voltmeter terminal marked + must be connected to the point of higher potential. Otherwise the meter will deflect in a direction opposite to that intended.

Combining the last two equations yields

$$\mathcal{E}_x = \mathcal{E}_s \frac{R_x}{R_s},$$ (32-11)

which allows us to compare emfs with precision. Note that the internal resistance r of the emf plays no role. In practice, potentiometers are conveniently packaged units, containing a *standard cell* which, after calibration at the National Bureau of Standards or elsewhere, serves as a convenient known standard seat of emf $\mathcal{E}_s$. Switching arrangements for replacing the unknown emf by the standard and arrangements for ascertaining that the current i_0 remains constant are also incorporated.

32-8
RC *CIRCUITS*

The preceding sections dealt with circuits in which the circuit elements were resistors and in which the currents did not vary with time. Here we introduce the capacitor as a circuit element, which will lead us to the concept of time-varying currents. In Fig. 32-10 let switch S be thrown to position a. What current is set up in the single-loop circuit so formed? Let us apply conservation of energy principles.

In time dt a charge $dq\ (= i\ dt)$ moves through any cross section of the circuit. The work done by the seat of emf $(= \mathcal{E}\ dq$; see Eq. 32-1) must equal the energy that appears as thermal energy in the resistor during time $dt\ (= i^2R\ dt)$ plus the increase in the amount of energy U that is stored in the capacitor $[= dU = d(q^2/2C)$; see Eq. 30-7]. In equation form

$$\mathcal{E}\ dq = i^2R\ dt + d\left(\frac{q^2}{2C}\right)$$

or

$$\mathcal{E}\ dq = i^2R\ dt + \frac{q}{C}\ dq.$$

Dividing by dt yields

$$\mathcal{E}\ \frac{dq}{dt} = i^2R + \frac{q}{C}\ \frac{dq}{dt}.$$

But dq/dt is simply i, so that this equation becomes

$$\mathcal{E} = iR + \frac{q}{C}.$$ (32-12)

This equation also follows from the loop theorem, as it must, since the loop theorem was derived from the conservation of energy principle. Starting from point x and traversing the circuit clockwise, we experience an increase in potential in going through the seat of emf and decreases in potential in traversing the resistor and the capacitor, or

$$\mathcal{E} - iR - \frac{q}{C} = 0,$$

which is identical to Eq. 32-12.

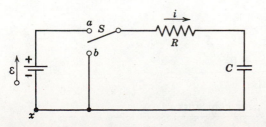

figure 32-10
An *RC* circuit.

We cannot immediately solve Eq. 32-12 because it contains two variables, q and i, which, however, are related by

$$i = \frac{dq}{dt}.$$
(32-13)

Substituting for i into Eq. 32-12 gives

$$\varepsilon = R \frac{dq}{dt} + \frac{q}{C}.$$
(32-14)

Our task now is to find the function $q(t)$ that satisfies this *differential equation*. Although this particular equation is not hard to solve, we choose to avoid mathematical complexity by simply presenting the solution, which is

$$q = C\varepsilon(1 - e^{-t/RC}).$$
(32-15)

We can easily test whether this function $q(t)$ is really a solution of Eq. 32-14 by substituting it into that equation and seeing whether an identity results. Differentiating Eq. 32-15 with respect to time yields

$$\frac{dq}{dt} \, (= i) = \frac{\varepsilon}{R} \, e^{-t/RC}.$$
(32-16)

Substituting q (Eq. 32-15) and dq/dt (Eq. 32-16) into Eq. 32-14 yields an identity, as you should verify. Thus Eq. 32-15 is a solution of Eq. 32-14.

Figure 32-11 shows some plots of Eqs. 32-15 and 32-16 for a particular case. Study of these plots and of the corresponding equations shows that (a) at $t = 0$, $q = 0$, and $i = \varepsilon/R$, and (b) as $t \to \infty$, $q \to C\varepsilon$, and $i \to 0$; that is, the current is initially ε/R and finally zero, and the charge on the capacitor plates is initially zero and finally $C\varepsilon$.

The quantity RC in Eqs. 32-15 and 32-16 has the dimensions of time (since the exponent must be dimensionless) and is called the *capacitive time constant* of the circuit. It is the time at which the charge on the capacitor has increased to within a factor of $(1 - e^{-1})(\cong 63\%)$ of its equilibrium value. To show this, we put $t = RC$ in Eq. 32-15 to obtain

$$q = C\varepsilon(1 - e^{-1}) = 0.63C\varepsilon.$$

Since $C\varepsilon$ is the equilibrium charge on the capacitor, corresponding to $t \to \infty$, the foregoing statement follows.

EXAMPLE 5

After how many time constants will the energy stored in the capacitor in Fig. 32-10 reach one-half its equilibrium value?

The energy is given by Eq. 30-7, or

$$U = \frac{1}{2C} \, q^2,$$

the equilibrium energy U_∞ being $(1/2C)(C\varepsilon)^2$. From Eq. 32-15, we can write the energy as

$$U = \frac{1}{2C} \, (C\varepsilon)^2(1 - e^{-t/RC})^2$$

or

$$U = U_\infty(1 - e^{-t/RC})^2.$$

Putting $U = \frac{1}{2}U_\infty$ yields

$$\tfrac{1}{2} = (1 - e^{-t/RC})^2$$

and solving this relation for t yields finally

$$t = 1.22 \, RC = 1.22 \text{ time constants.}$$

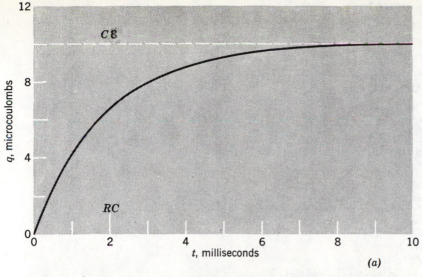

(a)

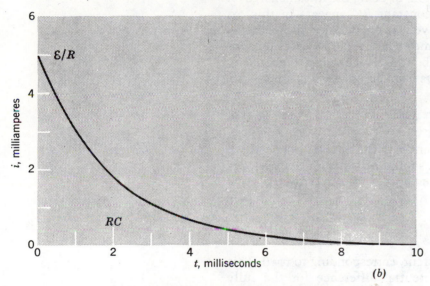

(b)

figure 32-11

If, in Fig. 32-10, we assume that $R = 2000\ \Omega$, $C = 1.0\ \mu F$, and $\varepsilon = 10$ V, then (a) shows the variation of q with t during the charging process and (b) the variations of i with t. The time constant is $RC = 2.0 \times 10^{-3}$ s.

Figure 32-11 shows that if a resistance is included in the circuit, the rate of increase of the charge of a capacitor toward its final equilibrium value is *delayed* in a way measured by the time constant RC. With no resistor present ($RC = 0$), the charge would rise immediately to its equilibrium value. Although we have shown that this time delay follows from an application of the loop theorem to RC circuits, it is important to develop a physical understanding of the causes of the delay.

When switch S in Fig. 32-10 is closed on a, the resistor experiences instantaneously an applied potential difference of ε, and an initial current of ε/R is set up. Initially, the capacitor experiences no potential difference because its initial charge is zero, the potential difference always being given by q/C. The flow of charge through the resistor starts to charge the capacitor, which has several effects. First, the existence of a capacitor charge means that there must now be a potential difference ($= q/C$) across the capacitor; this, in turn, means that the potential difference across the resistor must decrease by this amount, since the sum of the two potential differences must always equal ε. This decrease in the potential difference across R means that the charging current is reduced. Thus the charge of the capacitor builds up and the charging current decreases until the capacitor is fully charged. At this point the full emf ε is applied to the capacitor, there being no potential drop ($iR = 0$) across the resistor. This is precisely the reverse of the initial situation. Review the derivations of Eqs. 32-15 and 32-16 and study Fig. 32-11 with the qualitative arguments of this paragraph in mind.

Assume now that the switch S in Fig. 32-10 has been in position a for a time t such that $t \gg RC$. The capacitor is then fully charged for all practical purposes. The switch S is then thrown to position b. How do the charge of the capacitor and the current vary with time?

With the switch S closed on b, there is no emf in the circuit and Eq. 32-12 for the circuit, with $\varepsilon = 0$, becomes simply

$$iR + \frac{q}{C} = 0. \tag{32-17}$$

Putting $i = dq/dt$ allows us to write, as the differential equation of the circuit (compare Eq. 32-14),

$$R \frac{dq}{dt} + \frac{q}{C} = 0. \tag{32-18a}$$

The solution is

$$q = q_0 e^{-t/RC}, \tag{32-18b}$$

as you may readily verify by substitution, q_0 being the initial charge on the capacitor. The capacitive time constant RC appears in this expression for capacitor discharge as well as in that for the charging process (Eq. 32-15). We see that at a time such that $t = RC$ the capacitor charge is reduced to $q_0 e^{-1}$, which is about 37% of the initial charge q_0.

The current during discharge follows from differentiating Eq. 32-18b, or

$$i = \frac{dq}{dt} = -\frac{q_0}{RC} e^{-t/RC}. \tag{32-19}$$

The negative sign shows that the current is in the direction opposite to that shown in Fig. 32-10. This is as it should be, since the capacitor is discharging rather than charging. Since $q_0 = C\varepsilon$, we can write Eq. 32-19 as

$$i = -\frac{\varepsilon}{R} e^{-t/RC},$$

in which ε/R appears as the initial current, corresponding to $t = 0$. This is reasonable because the initial potential difference for the fully charged capacitor is ε.

The behavior of the RC circuit in Fig. 32-10 during charge and discharge can be studied with a cathode-ray oscilloscope. This familiar laboratory device can display on its fluorescent screen plots of the variation of potential with time. Figure 32-12 shows the circuit of Fig. 32-10 with connections made to display (a) the potential difference V_C across the capacitor and (b) the potential difference V_R across the resistor as functions of time. V_C and V_R are given by

$$V_C = \left(\frac{1}{C}\right) q$$

and

$$V_R = (R)i,$$

the former being proportional to the charge and the latter to the current.

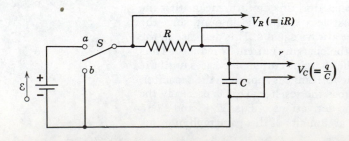

figure 32-12
The circuit of Fig. 32-10 with connections made to display the potential variations across the resistor and the capacitor on a cathode-ray oscilloscope.

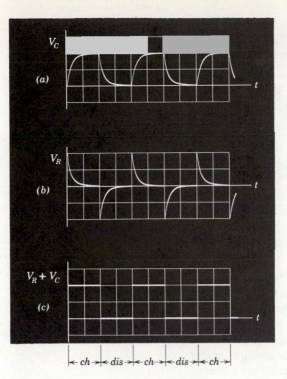

Figure 32-13 shows oscillograph plots of V_C and V_R that result when, in effect, switch S in Fig. 32-10 is thrown regularly back and forth between positions a and b, being left in each position for a time equal to several time constants. Intervals during which the charge is building up are labeled ch and those during which it is decaying are labeled dis.

The "charge" intervals in plot a (see Eq. 32-15) are represented by

$$V = \left(\frac{1}{C}\right) q = \varepsilon(1 - e^{-t/RC})$$

and the "discharge" intervals (see Eq. 32-18b) by

$$V = \left(\frac{1}{C}\right) q = \varepsilon e^{-t/RC}.$$

Note that the current, as indicated by plot b, is in opposite directions during the charge and discharge intervals, in agreement with Eqs. 32-16 and 32-19.

In plot c in Fig. 32-13 the oscillograph has been connected to show the algebraic sum of plots a and b. According to the loop theorem this sum should equal ε during the charge intervals and should be zero during the discharge intervals, when the battery is no longer in the circuit; that is,

$$V_R + V_C = \varepsilon \text{ during charge (see Eq. 32-12),}$$

$$V_R + V_C = 0 \text{ during discharge (see Eq. 32-17).}$$

Plot c is in exact agreement with this expectation.

questions

1. Does the direction of the emf provided by a battery depend on the direction of current flow through the battery?
2. In Fig. 32-2 discuss what changes would occur if we increased the mass m by such an amount that the "motor" reversed direction and became a "generator", that is, a seat of emf.
3. Discuss in detail the statement that the energy method and the loop theorem method for solving circuits are perfectly equivalent.

4. It is possible to generate a 10,000-volt potential difference by rubbing a pocket comb with wool. Why is this large voltage not dangerous when the much lower voltage provided by an ordinary electric outlet is very dangerous?

5. Devise a method for measuring the emf and the internal resistance of a battery.

6. The current passing through a battery of emf ε and internal resistance r is decreased by some external means. Does the potential difference between the terminals of the battery necessarily decrease or increase? Explain.

7. In calculating V_{ab} in Fig. 32-3a, is it permissible to follow a path from a to b that does not lie in the conducting circuit?

8. A 25-watt, 110-volt bulb glows at normal brightness when connected across a bank of batteries. A 500-watt, 110-volt bulb glows only dimly when connected across the same bank. Explain.

9. Under what circumstances can the terminal potential difference of a battery exceed its emf?

10. What is the difference between an emf and a potential difference?

11. Compare and contrast the formulae for the equivalent values of (a) capacitors and (b) resistors, in series and in parallel.

12. On what physical laws do (a) the loop theorem and (b) the junction theorem depend?

13. In Fig. 32-6 convince yourself that i_3 is drawn in the wrong direction. It was drawn so deliberately to make a point. What point?

14. Explain in your own words why the resistance of an ammeter should be very small whereas that of a voltmeter should be very large.

15. In a potentiometer, the internal resistance r of the emf plays no role. Why? See Fig. 32-9.

16. Does the time required for the charge on a capacitor in an RC circuit to build up to a given fraction of its equilibrium value depend on the value of the applied emf?

17. Devise a method whereby an RC circuit can be used to measure very high resistances.

18. In Fig. 32-10 suppose that switch S is closed on a. Explain why, in view of the fact that the negative terminal of the battery is not connected to resistance R, the initial current in R, at $t = 0$, should be ε/R, as Eq. 32-16 predicts.

19. In Fig. 32-10 suppose that switch S is closed on a. Why does the charge on capacitor C not rise instantaneously to $q = C\varepsilon$? After all, the positive battery terminal is connected to one plate of C and the negative terminal to the other.

20. Show that the product RC in Eqs. 32-15 and 32-16 has the dimensions of time, that is, that 1 second = 1 ohm $\times$ 1 farad.

21. Can you construct a water-flow analogy to capacitor discharge, using, for example, two burettes to simulate the capacitor plates and a long capillary tube to simulate the resistor? See "A Water Flow Analogy to Capacitor Discharge" by Ernest L. Madsen, *The Physics Teacher*, April 1976, for a full discussion.

problems

SECTION 32-1

1. A 5.0-A current is set up in an external circuit by a 6.0-V storage battery for 6.0 min. By how much is the chemical energy of the battery reduced?
   Answer: 1.1×10^4 J.

2. A certain car battery (12 V) carries an initial charge Q (120 A·h). Assuming that the potential across the terminals stays constant until the battery is completely discharged, for how many hours can it deliver power P (100 W)?

SECTION 32-2

3. The current in a simple series circuit is 5.0 A. When an additional resistance of 2.0 Ω is inserted, the current drops to 4.0 A. What was the resistance of the original circuit? *Answer:* 8.0 Ω.

4. A battery of emf ε (2.0 V) and internal resistance r (1.0 Ω) is driving a motor. The motor is lifting a weight W (2.0 N) at constant speed v (0.50 m/s). Assuming no power losses, find (*a*) the current i in the circuit and (*b*) the potential difference V across the terminals of the motor.

SECTION 32-3

5. Thermal energy is to be generated in a 0.10-Ω resistor at the rate of 10 W by connecting it to a battery whose emf is 1.5 V. (*a*) What is the internal resistance of the battery? (*b*) What potential difference exists across the resistor? *Answer:* (*a*) 0.050 Ω. (*b*) 1.0 V.

6. A wire of resistance 5.0 Ω is connected to a battery whose emf ε is 2.0 V and whose internal resistance is 1.0 Ω. In 2.0 min (*a*) how much energy is transferred from chemical to electric form? (*b*) How much energy appears in the wire as thermal energy? (*c*) Account for the difference between (*a*) and (*b*).

7. (*a*) In the circuit of Fig. 32-3*a* show that the power delivered to R as thermal energy is a maximum when R is equal to the internal resistance r of the battery. (*b*) Show that this maximum power is $P = \varepsilon^2/4r$.

8. (*a*) In Fig. 32-14 what value must R have if the current in the circuit is to be 0.0010 A? Take $\varepsilon_1 = 2.0$ V, $\varepsilon_2 = 3.0$ V, and $r_1 = r_2 = 3.0$ Ω. (*b*) What is the rate of thermal energy transfer in R?

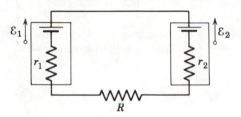

figure 32-14
Problem 8

9. In Fig. 32-3*a* put $\varepsilon = 2.0$ V and $r = 100$ Ω. Plot (*a*) the current and (*b*) the potential difference across R, as functions of R over the range 0 to 500 Ω. Make both plots on the same graph. (*c*) Make a third plot by multiplying together, for each value of R, the two curves plotted. What is the physical significance of this plot?

10. You are given a number of 10-Ω resistors, each capable of dissipating only 1.0 W. What is the minimum number of such resistors that you need to combine in series or parallel combinations to make a 10-Ω resistor capable of dissipating 5.0 W?

SECTION 32-4

11. In Fig. 32-5(*a*) calculate the potential difference between *a* and *c* by considering a path that contains R and ε_2.
Answer: 2.25 V, as expected. See Example 3.

12. In Example 2 an ammeter whose resistance is 0.050 Ω is inserted in the circuit. What percent change in the current results because of the presence of the meter?

13. The section of circuit AB (see Fig. 32-15) absorbs power $P = 50.0$ W and a current $i = 1.0$ A passes through it in the indicated direction. (*a*) What is the potential difference between A and B? (*b*) If the element C does not have internal resistance, what is its emf? (*c*) What is its polarity?
Answer: (*a*) 50 V. (*b*) 48 V. (*c*) B is the negative terminal.

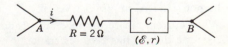

figure 32-15
Problem 13

14. Two batteries having the same emf ε but different internal resistances r_1 and r_2 are connected in series to an external resistance R. Find the value of R that makes the potential difference zero between the terminals of the first battery.

SECTION 32-5

15. Two light bulbs, one of resistance R_1 and the other of resistance R_2 ($<R_1$), are connected (a) in parallel and (b) in series. Which bulb is brighter? *Answer: (a) R_2. (b) R_1.*

16. Two batteries of emf ε and internal resistance r are connected in parallel across a resistor R, as in Fig. 32-18b. (a) For what value of R is the thermal energy delivered to the resistor a maximum? (b) What is the maximum energy dissipation rate?

17. In Fig. 32-6 calculate the potential difference between points c and d by as many paths as possible. Assume that $\varepsilon_1 = 4.0$ V, $\varepsilon_2 = 1.0$ V, $R_1 = R_2 = 10\ \Omega$, and $R_3 = 5\ \Omega$. *Answer: $V_d - V_c = +0.25$ V, by all paths.*

18. Two resistors, R_1 and R_2, may be connected either in series or parallel across a (resistanceless) battery with emf ε. We desire the thermal energy transfer rate for the parallel combination to be five times that for the series combination. If $R_1 = 100\ \Omega$, what is R_2?

19. (a) In Fig. 32-16 what is the equivalent resistance of the network shown? (b) What are the currents in each resistor? Put $R_1 = 100\ \Omega$, $R_2 = R_3 = 50\ \Omega$, $R_4 = 75\ \Omega$, and ε $= 6.0$ V. *Answer: (a) $120\ \Omega$; note that R_2, R_3, and R_4 are in parallel. (b) $i_1 = 50$ mA; $i_2 = i_3 = 20$ mA; $i_4 = 13$ mA.*

20. A copper wire of radius a (0.25 mm) has an aluminum jacket of outside radius b (0.38 mm). (a) If there is a current i (2.0 A) in the wire, find the current in each material. (b) What is the wire length if potential difference V (12 V) maintains the current?

21. By using only two resistance coils—singly, in series, or in parallel—you are able to obtain resistances of 3, 4, 12, and 16 Ω. What are the separate resistances of the coils? *Answer: 4.0 Ω and 12 Ω.*

22. Four 100-W heating coils are to be connected in all possible series-parallel combinations and plugged into a 100-V line. What different rates of Joule (thermal) energy dissipation are possible?

23. (a) In Fig. 32-17 what power appears as thermal energy in R_1? In R_2? In R_3? (b) What power is supplied by ε_1? By ε_2? (c) Discuss the energy balance in this circuit. Assume that $\varepsilon_1 = 3.0$ V, $\varepsilon_2 = 1.0$ V, $R_1 = 5.0\ \Omega$, $R_2 = 2.0\ \Omega$, and $R_3 = 4.0\ \Omega$. *Answer: (a) 0.35 W; 0.050 W; 0.71 W. (b) 1.3 W; −0.16 W.*

24. You are given two batteries of emf ε and internal resistance r. They may be connected either in series or in parallel and are used to establish a current in a resistor R, as in Fig. 32-18. (a) Derive expressions for the current in R for both methods of connection. (b) Which connection yields the larger current if $R > r$ and if $R < r$?

25. In Fig. 32-19 find the current in each resistor and the potential difference between a and b. Put $\varepsilon_1 = 6.0$ V, $\varepsilon_2 = 5.0$ V, $\varepsilon_3 = 4.0$ V, $R_1 = 100\ \Omega$, and $R_2 = 50\ \Omega$. *Answer: $i_1 = 50$ mA, to the right; $i_2 = 60$ mA, down; $V_{ab} = 9.0$ V.*

26. What is the equivalent resistance between the terminal points x and y of the circuits shown in (a) Fig. 32-20a, (b) Fig. 32-20b, and (c) Fig. 32-20c? Assume that the resistance of each resistor is 10 Ω. Do you detect any similarities between Figs. 32-20a and c?

27. (a) Find the three currents in Fig. 32-21. (b) Find V_{ab}. Assume that $R_1 = 1.0\ \Omega$, $R_2 = 2.0\ \Omega$, $\varepsilon_1 = 2.0$ V, and $\varepsilon_2 = \varepsilon_3 = 4.0$ V. *Answer: (a) Left branch, 0.67 A, down; center branch, 0.33 A, up; right branch, 0.33 A, up. (b) 3.3 V.*

28. What current, in terms of ε and R, does the meter M in Fig. 32-22 read? Assume that M has zero resistance.

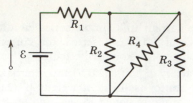

figure 32-16
Problem 19

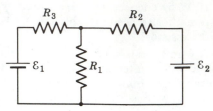

figure 32-17
Problem 23

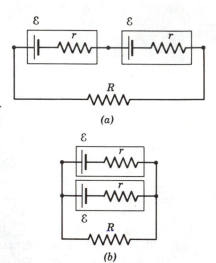

figure 32-18
Problems 16, 24

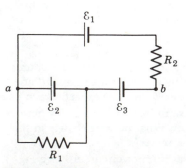

figure 32-19
Problem 25

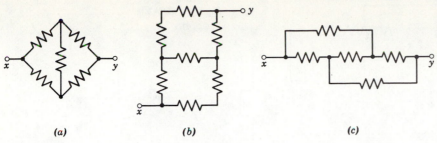

(a) (b) (c)

figure 32-20
Problem 26

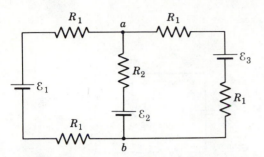

figure 32-21
Problem 27

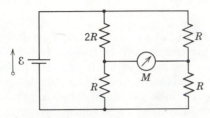

figure 32-22
Problem 28

29. (a) Two batteries of emf and internal resistance ε_1, r_1 and ε_2, r_2, are connected in parallel. Show that the effective emf of this parallel combination is

$$\varepsilon = r\left(\frac{\varepsilon_1}{r_1} + \frac{\varepsilon_2}{r_2}\right)$$

where r is defined from

$$\frac{1}{r} = \frac{1}{r_1} + \frac{1}{r_2}.$$

See "Batteries Connected in Parallel" by J. S. Wallingford and H. W. Jones, *American Journal of Physics*, **36**, 639, (1968) for an extension of this problem to an indefinite number of batteries.

30. N identical batteries of emf ε and internal resistance r may be connected all in series or all in parallel. Show that each arrangement will give the same current in an external resistor R, if $R = r$.

31. *The Wheatstone bridge.* In Fig. 32-23 R_s is to be adjusted in value until points a and b are brought to exactly the same potential. (One tests for this condition by momentarily connecting a sensitive meter between a and b; if these points are at the same potential, the meter will not deflect.) Show that when this adjustment is made, the following relation holds:

$$R_x = R_s \frac{R_2}{R_1}.$$

Unknown resistors (R_x) can be measured in terms of standards (R_s) using this device, which is called a Wheatstone bridge.

32. If points a and b in Fig. 32-23 are connected by a wire of resistance r, show that the current in the wire is

$$i = \frac{\varepsilon(R_s - R_x)}{(R + 2r)(R_s + R_x) + 2R_sR_x},$$

where ε is the emf of the battery. Assume that R_1 and R_2 are equal $(R_1 = R_2 = R)$ and that R_0 equals zero. Is this formula consistent with the result of Problem 31?

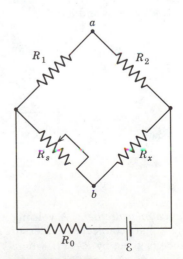

figure 32-23
Problem 31

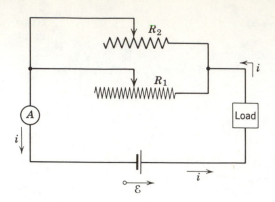

figure 32-24
Problem 33

33. For manual control of the current in a circuit, you can use a parallel combination of variable resistors of the sliding contact type, as in Fig. 32-24, with $R_1 = 20R_2$. (a) What procedure is used to adjust the current to the desired value? (b) Why is the parallel combination better than a single-variable resistor? (c) Can you extend this technique to three resistors in parallel?
Answer: (a) Put R_1 roughly in the middle of its range; adjust current roughly with R_2; make fine adjustment with R_1. (b) Relatively large percentage changes in R_1 cause only small percentage changes in the resistance of the parallel combination, thus permitting fine adjustment. The ratio is 1:21.

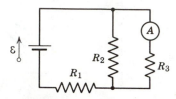

figure 32-25
Problem 34

34. In Fig. 32-25 imagine an ammeter inserted in the branch containing R_3. (a) What will it read, assuming $\varepsilon = 5.0$ V, $R_1 = 2.0\ \Omega$, $R_2 = 4.0\ \Omega$, and $R_3 = 6.0\ \Omega$? (b) The ammeter and the source of emf are now physically interchanged. Show that the ammeter reading remains unchanged.

35. Twelve resistors, each of resistance R ohms, form a cube (see Fig. 32-26). (a) Find R_{AB}, the equivalent resistance of an edge. (b) Find R_{BC}, the equivalent resistance of a face diagonal. (c) Find R_{AC}, the equivalent resistance of a body diagonal. *Answer:* (a) $\frac{7}{12} R$. (b) $\frac{3}{4} R$. (c) $\frac{5}{6} R$.

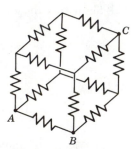

figure 32-26
Problem 35

SECTION 32-6

36. A voltmeter and an ammeter are used to determine two unknown resistances R_1 and R_2, one determination by each of the two methods shown in Fig. 32-27. The voltmeter resistance is 307 Ω and the ammeter resistance is 3.62 Ω; in method (a) the ammeter reads 0.317 A and the voltmeter reads 28.1 V, whereas in method (b) the ammeter reads 0.356 A and the voltmeter 23.7 V. Compute R_1 and R_2.

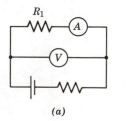 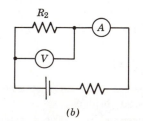

(a)   (b)

figure 32-27
Problem 36

37. *Resistance measurement.* A voltmeter (resistance R_V) and an ammeter (resistance R_A) are connected to measure a resistance R, as in Fig. 32-28a. The resistance is given by $R = V/i$, where V is the voltmeter reading and i is the current *in the resistor* R. Some of the current registered by the ammeter (i') goes through the voltmeter so that the ratio of the meter readings ($= V/i'$)

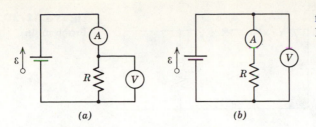

figure 32-28
Problems 37, 38

(a) (b)

gives only an *apparent* resistance reading R'. Show that R and R' are related by

$$\frac{1}{R} = \frac{1}{R'} - \frac{1}{R_V}.$$

Note that if $R_V \gg R$, then $R \cong R'$.

38. *Resistance measurement.* If meters are used to measure resistance, they may also be connected as they are in Fig. 32-28b. Again the ratio of the meter readings gives only an apparent resistance R'. Show that R' is related to R by

$$R = R' - R_A,$$

in which R_A is the ammeter resistance. Note that if $R_A \ll R$, then $R \cong R'$.

39. In Fig. 32-8 assume that $\varepsilon = 5.0$ V, $r = 2.0$ Ω, $R_1 = 5.0$ Ω, and $R_2 = 4.0$ Ω. If $R_A = 0.10$ Ω, what percent error is made in reading the current? Assume that the voltmeter is not present.   *Answer: 0.9%.*

40. In Fig. 32-8 assume that $\varepsilon = 5.0$ V, $r = 20$ Ω, $R_1 = 50$ Ω, and $R_2 = 40$ Ω. If $R_V = 1000$ Ω, what percent error is made in reading the potential differences across R_1? Ignore the presence of the ammeter.

SECTION 32-8

41. How many time constants must elapse before a capacitor in an RC circuit is charged to within 1.0 percent of its equilibrium charge?   *Answer: 4.6.*

42. A 10,000-Ω resistor and a capacitor are connected in series and a 10-V potential is suddenly applied. If the potential across the capacitor rises to 5.0 V in 1.0 μs, what is the capacitance of the capacitor?

43. The potential difference between the plates of a leaky capacitor C (2.0 μF) drops from V_0 to V ($\frac{1}{4}V_0$) in time t (2.0 s). What is the equivalent resistance between the capacitor plates?   *Answer: 7.2×10^5 Ω.*

44. An RC circuit is discharged by closing a switch at time $t = 0$. The initial potential difference across the capacitor is 100 V. If the potential difference has decreased to 1.0 V after 10 s, (a) what will the potential difference be 20 s after $t = 0$? (b) What is the time constant of the circuit?

45. Prove that when switch S in Fig. 32-10 is thrown from a to b all the energy stored in the capacitor is transformed into thermal energy in the resistor. Assume that the capacitor is fully charged before the switch is thrown.

46. A capacitor with capacitance $C = 1.0$ μF and initial stored energy $U_0 = 0.50$ J is discharged through a resistance $R = 1.0 \times 10^6$ Ω. (a) What is the initial charge on the capacitor? (b) What is the current through the resistor when the discharge starts? (c) Determine V_C, the voltage across the capacitor, and V_R, the voltage across the resistor, as a function of time. (d) Express the rate of generation of thermal energy in the resistor as a function of time.

47. A 3.0×10^6-Ω resistor and a 1.0-μF capacitor are connected in a single-loop circuit with a seat of emf with $\varepsilon = 4.0$ V. At 1.0 s after the connection is made, what are the rates at which (a) the charge of the capacitor is increasing, (b) energy is being stored in the capacitor, (c) thermal energy is appearing in the resistor, and (d) energy is being delivered by the seat of emf? *Answer: (a) 9.6×10^{-7} C/s. (b) 1.1×10^{-6} W. (c) 2.8×10^{-6} W. (d) 3.8×10^{-6} W.*

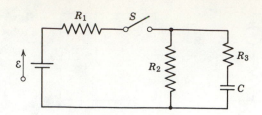

figure 32-29
Problem 48

48. In the circuit of Fig. 32-29 let i_1, i_2, and i_3 be the currents through resistors R_1, R_2, and R_3, respectively, and let V_1, V_2, V_3, and V_C be the corresponding potential differences across the resistors and across the capacitor C. (a) Plot qualitatively, as a function of time after switch S is closed, the currents and voltages listed above. (b) After being closed for a large number of time constants, the switch S is now opened. Plot qualitatively, as a function of time after the switch is opened, the currents and voltages listed above.

49. In the circuit of Fig. 32-30, $\varepsilon = 1200$ V, $C = 6.50$ μF, $R_1 = R_2 = R_3 = 7.30 \times 10^5$ Ω. With C completely uncharged, the switch S is suddenly closed ($t = 0$). (a) Determine the currents through each resistor for $t = 0$ and $t = \infty$. (b) Draw qualitatively a graph of the potential drop V_2 across R_2 from $t = 0$ to $t = \infty$. (c) What are the numerical values of V_2 at $t = 0$ and $t = \infty$? (d) Give the physical meaning of "$t = \infty$" and state a rough, but significant, numerical lower bound, in seconds, for "$t = \infty$" in this case.

   Answer: (a)At $t=0$, $i_1=1.1$ mA, $i_2=i_3=0.55$ mA; at $t = \infty$, $i_1 = i_2 = 0.82$ mA, $i_3 = 0$. (c) At $t = 0$, $V_2 = 400$ V; at $t = \infty$, $V_2 = 600$ V. (d) The time constant is 7.1 s.

figure 32-30
Problem 49

33
the magnetic field

The science of magnetism grew from the observation that certain "stones" (magnetite) would attract bits of iron. The word *magnetism* comes from the district of Magnesia in Asia Minor, which is one of the places at which the stones were found. Figure 33-1 shows a modern permanent magnet, the lineal descendant of these natural magnets. Another "natural magnet" is the earth itself, whose orienting action on a magnetic compass needle has been known since ancient times.

In 1820 Oersted discovered that a current in a wire can also produce magnetic effects, namely, that it can change the orientation of a compass needle.* We pointed out in Section 26-1 how this important discovery linked the then separate sciences of magnetism and electricity. We can intensify the magnetic effect of a current in a wire by forming the wire into a coil of many turns and by providing an iron core. Figure 33-2 shows how this is done in a large electromagnet of a type commonly used for research involving magnetism.

We define the space around a magnet or a current-carrying conductor as the site of a *magnetic field*, just as we defined the space near a charged rod as the site of an electric field. The basic magnetic field vector **B**, which we define in the following section, is called the *magnetic induc-*

33-1
THE MAGNETIC FIELD

---

*In 1878 H. A. Rowland, at the Johns Hopkins University, discovered that a moving charged object (in his case, a rapidly rotating charged disk) also causes magnetic effects. Actually, it is far from obvious that a moving charge is equivalent to a current in a wire. See. "Rowland's Physics" by John D. Miller, *Physics Today*, July 1976 for a scientific biography of this outstanding American physicist, including a detailed discussion of this experiment.

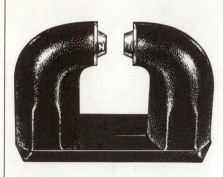

figure 33-1
A permanent magnet. Lines of
magnetic induction leave the north
pole face, marked N, and enter the
south pole face on the other side
of the air gap.

figure 33-2
A research-type electromagnet
showing iron frame F, pole faces P,
and coils C. The pole faces are 12 in.
in diameter. (Courtesy Varian
Associates.)

tion.* We can represent it by *lines of induction*, just as we represent
the electric field by lines of force. As for the electric field (see Section
27-3), the magnetic field vector is related to its lines of induction in this
way:

1. The tangent to a line of induction at any point gives the *direction* of
   B at that point.

2. The lines of induction are drawn so that the number of lines per unit
   cross-sectional area (perpendicular to the lines) is proportional to the
   magnitude of **B.** Where the lines are close together B is large and
   where they are far apart B is small.

As for the electric field, the magnetic field vector **B** is of fundamental
importance, the lines of induction simply giving a graphic representation of the way **B** varies throughout a certain region of space. The *flux*
Φ_B for a magnetic field can be defined in exact analogy with the flux Φ_E
for the electric field, namely $\Phi_B = \int \mathbf{B} \cdot d\mathbf{S}$, in which the integral is taken
over the surface (closed or open) for which Φ_B is defined.

In this chapter we are not concerned with the *causes* of the magnetic
field; we will explore them in the next chapter. Our related concerns
here are (*a*) to *define* operationally the magnitude and direction of **B** at
any point P near, say, a magnet, a current-carrying conductor, or a
moving charge and (*b*) to determine the *effect* of a magnetic field on
objects, such as moving charges, subject to its influence.

As we did for the electric field, let us consider a charge q_0 as a test
body. Let us place the test body *at rest* at a point P near, say, a perma-

33-2
THE DEFINITION† OF **B**

---

* *Magnetic field strength* would be a more suitable name for **B,** but it has been usurped for
historical reasons by another vector connected with the magnetic field (see Section 37-8).
Often in this book we shall call **B,** for brevity, the "magnetic field" in analogy to calling **E**
the "electric field".

† The definition and method of measurement of **B** given in this section, though conceptually sound and suitable for our present purpose, are not carried out in practice because
of experimental difficulties. The following section will show how **B** may be more conveniently measured in the laboratory.

nent magnet such as that of Fig. 33-1. We would then find by experiment that *no force* (attributable *only* to the presence or absence of the magnet) acts on q_0. However, if we fire test body q_0 through point P with a velocity **v**, we find that a *sideways force* **F** acts on it if the magnet is in place; by a sideways force we mean a force at right angles to **v**. We shall define **B** at point P in terms of **F**, **v**, and q_0.

If we vary the direction of **v** through point P, keeping the magnitude of **v** unchanged, we find that although **F** will always remain at right angles to **v**, its magnitude F will change. For a particular orientation of **v** (and also for the opposite orientation $-$**v**) the force **F** becomes zero. We define this orientation as the *direction* of **B**, the specification of the *sense* of **B** (that is, the way it points along this line) being left to the more complete definition of **B** that we give below.

Having found the direction of **B**, we are now able to orient **v** so that the test charge moves at right angles to **B**. We will find that the force **F** is now a maximum, and we define the *magnitude* of **B** from the measured magnitude of this maximum force $F_\perp$, or

$$B = \frac{F_\perp}{q_0 v}. \qquad (33\text{-}1)$$

Let us regard this definition of **B** (in which we have specified its magnitude and direction, but not its sense) as preliminary to the complete vector definition that we now give: *If we fire a positive test charge q_0 with velocity **v** through a point P and if a force **F** acts on the moving charge, a magnetic field **B** is present at point P, where **B** is the vector that satisfies the relation*

$$\mathbf{F} = q_0 \mathbf{v} \times \mathbf{B}, \qquad (33\text{-}2)$$

v, q_0, and **F** being measured quantities. The magnitude of the magnetic deflecting force **F**, according to the rules for vector products, is given by *

$$F = q_0 v B \sin\theta, \qquad (33\text{-}3)$$

in which θ is the angle between **v** and **B**.

Figure 33-3 shows the relations among the vectors. We see from Eq. 33-2 that **F**, being at right angles to the plane formed by **v** and **B**, will always be at right angles to **v** (and also to **B**) and thus will always be a sideways force. Equation 33-2 is consistent with the observed facts that (a) the magnetic force vanishes as $v \to 0$, (b) the magnetic force vanishes if **v** is either parallel or antiparallel to the direction of **B** (in these cases $\theta = 0$ or 180° and $\mathbf{v} \times \mathbf{B} = 0$), and (c) if **v** is at right angles to **B** ($\theta = 90°$), the deflecting force has its maximum value, given by Eq. 33-1, that is, $q_0 v B$.

This definition of **B** is similar in spirit, although more complex, than the definition of the electric field **E**, which we can cast into this form: *If we place a positive test charge q_0 at point P and if an (electric) force **F** acts on the stationary charge, an electric field **E** is present at P, where **E** is the vector satisfying the relation*

$$\mathbf{F} = q_0 \mathbf{E},$$

q_0 and **F** being measured quantities. In defining **E**, the only characteristic direction to appear is that of the electric force $\mathbf{F}_E$ which acts on the positive test body; the direction of **E** is taken to be that of $\mathbf{F}_E$. In defining **B**, *two* characteristic directions appear, those of **v** and of the magnetic force $\mathbf{F}_B$; they prove always to be at right angles. Although we can easily solve the above equation for **E**, we cannot solve Eq. 33-2 for **B**. Why not?

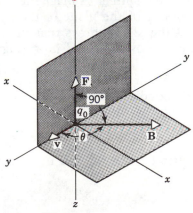

figure 33-3
Illustrating $\mathbf{F} = q_0 \mathbf{v} \times \mathbf{B}$ (Eq. 33-2). Test charge q_0 is fired through the origin with velocity **v**.

---

* You may wish to review Section 2-4, which deals with vector products.

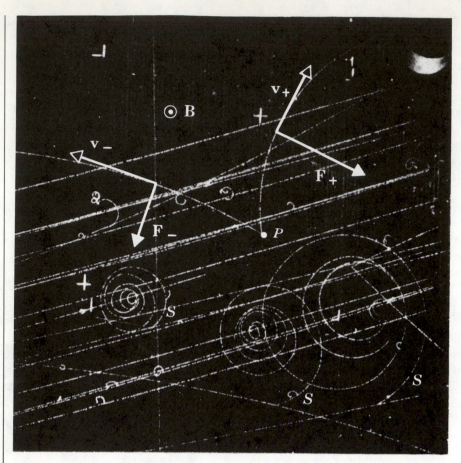

figure 33-4
A *bubble chamber* is a device for rendering visible, by means of small bubbles, the tracks of charged particles that pass through the chamber. The figure is a photograph taken with such a chamber immersed in a magnetic field **B** and exposed to radiations from a large cyclotron-like accelerator. The curved *V* at point *P* is formed by a positive and a negative electron, which deflect in opposite directions in the magnetic field. The spirals *S* are the tracks of three low-energy electrons. (Courtesy E. O. Lawrence Radiation Laboratory, University of California.)

In Fig. 33-4 a positive and a negative electron are created at point *P* in a bubble chamber. A magnetic field is perpendicular to the chamber, pointing out of the plane of the figure (symbol ⊙).* The relation $\mathbf{F} = q_0\mathbf{v} \times \mathbf{B}$ (Eq. 33-2) shows that the deflecting forces acting on the two particles are as indicated in the figure. These deflecting forces would make the tracks deflect as shown.

The unit of **B** that follows from Eq. 33-3 is the newton/(coulomb) (meter/second). This is given the SI name *tesla* (abbr. T) or weber/meter² (abbr. Wb/m²).** Recalling that a coulomb/second is an ampere, we have

$$1 \text{ tesla} = 1 \text{ weber/meter}^2 = 1 \text{ newton/(ampere} \cdot \text{meter)}.$$

An earlier unit for **B**, still in common use, is the *gauss;* the relationship is

$$1 \text{ tesla} = 1 \text{ weber/meter}^2 = 10^4 \text{ gauss.}†$$

The fact that the magnetic force is always at right angles to the direction of motion means that (for steady magnetic fields at least) the work done by this force on the particle is zero. For an element of the path of the particle of length $d\mathbf{l}$, this work dW is $\mathbf{F}_B \cdot d\mathbf{l}$; dW is zero because $\mathbf{F}_B$ and $d\mathbf{l}$ are always at right angles. Thus a static magnetic field cannot

---

* The symbol ⊗ indicates a vector into the page, the × being thought of as the tail of an arrow; the symbol ⊙ indicates a vector out of the page, the dot being thought of as the tip of an arrow.

** The weber is used to measure the magnetic flux, a concept we discuss in succeeding chapters. See also Sec. 33-1.

† See "Megagauss Physics" by C. M. Fowler, *Science,* April 1973 for a fascinating account of the properties of such enormous fields.

change the kinetic energy of a moving charge; it can only deflect it sideways.*

If a charged particle moves through a region in which both an electric field and a magnetic field are present, the resultant force is found by combining Eqs. 27-2 and 33-2, or

$$\mathbf{F} = q_0\mathbf{E} + q_0\mathbf{v} \times \mathbf{B}. \tag{33-4}$$

This is sometimes called the *Lorentz equation* in tribute to H. A. Lorentz who did so much to develop and clarify the concepts of the electric and magnetic fields.

EXAMPLE 1

A uniform magnetic field **B** points horizontally from south to north; its magnitude is 1.5 T. If a 5.0-MeV proton moves vertically downward through this field, what force will act on it?

The kinetic energy of the proton is

$$K = (5.0 \times 10^6 \text{ eV})(1.6 \times 10^{-19} \text{ J/eV}) = 8.0 \times 10^{-13} \text{ J}.$$

We can find its speed from the relation $K = \frac{1}{2}mv^2$, or

$$v = \sqrt{\frac{2K}{m}} = \sqrt{\frac{(2)(8.0 \times 10^{-13} \text{ J})}{1.7 \times 10^{-27} \text{ kg}}} = 3.1 \times 10^7 \text{ m/s}.$$

Equation 33-3 gives

$$F = qvB \sin \theta = (1.6 \times 10^{-19} \text{ C})(3.1 \times 10^7 \text{ m/s})(1.5 \text{ T})(\sin 90°)$$

$$= 7.4 \times 10^{-12} \text{ N}.$$

You can show that this force is about 4×10^{14} times greater than the weight of the proton.

The relation $\mathbf{F} = q\mathbf{v} \times \mathbf{B}$ shows that the *direction* of the deflecting force is to the east. If the particle had been negatively charged, the deflection would have been to the west. This is predicted automatically by Eq. 33-2 if we substitute $-e$ for q_0.

A current is an assembly of moving charges. Because a magnetic field exerts a sideways force on a moving charge, we expect that it will also exert a sideways force on a wire carrying a current. Figure 33-5 shows a length *l* of wire carrying a current *i* and placed in a magnetic field **B**. For simplicity we have oriented the wire so that the current density vector **j** is at right angles to **B**.

33-3
MAGNETIC FORCE ON A CURRENT

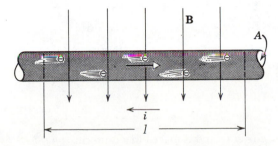

figure 33-5
A wire carrying a current *i* is placed at right angles to a magnetic field **B**. Only the drift velocity of the electrons, not their random motion, is suggested.

---

* Actually, $\mathbf{F}_B$ is at right angles to **v**, *even if* **B** varies with time. In this case, however, the theory of special relativity predicts that an *electric* field **E** will appear and act on the charged particle in such a way as to do work on it. See *Introduction to Special Relativity* by Robert Resnick, John Wiley & Sons, 1968 (Section 4.2) for details.

The current i in a metal wire is carried by the free (or conduction) electrons, n being the number of such electrons per unit volume of the wire. The magnitude of the average force on one such electron is given by Eq. 33-3, or, since $\theta = 90°$,

$$F' = q_0 vB \sin \theta = ev_d B$$

where v_d is the drift speed. From the relation $v_d = j/ne$ (Eq. 31-5),

$$F' = e \left(\frac{j}{ne} \right) B = \frac{jB}{n}.$$

The length l of the wire contains nAl free electrons, Al being the volume of the section of wire of cross section A that we are considering. The total force on the free electrons in the wire, and thus on the wire itself, is

$$F = (nAl)F' = nAl\,\frac{jB}{n}.$$

Since jA is the current i in the wire, we have

$$F = ilB. \tag{33-5}$$

The negative charges, which move to the right in the wire of Fig. 33-5, are equivalent to positive charges moving to the left, that is, in the direction of the current arrow. For such a positive charge the velocity $\mathbf{v}$ would point to the left and the force on the wire, given by Eq. 33-2 ($\mathbf{F} = q_0 \mathbf{v} \times \mathbf{B}$) points up, out of the page. This same conclusion follows if we consider the actual negative charge carriers for which $\mathbf{v}$ points to the right but q_0 has a negative sign. Thus by measuring the sideways magnetic force on a wire carrying a current and placed in a magnetic field, we cannot tell whether the current carriers are negative charges moving in a given direction or positive charges moving in the opposite direction.

Equation 33-5 holds only if the wire is at right angles to $\mathbf{B}$. We can express the more general situation in vector form as

$$\mathbf{F} = i\mathbf{l} \times \mathbf{B}, \tag{33-6a}$$

where $\mathbf{l}$ is a vector whose magnitude is the length of the wire and which points along the (straight) wire in the direction of the current. Equation 33-6a is equivalent to the relation $\mathbf{F} = q_0 \mathbf{v} \times \mathbf{B}$ (Eq. 33-2); either can be taken as a defining equation for $\mathbf{B}$. Note that the vector $\mathbf{l}$ in Fig. 33-5 points to the left and that the magnetic force $\mathbf{F}$ ($= i\mathbf{l} \times \mathbf{B}$) points up, out of the page. This agrees with the conclusion obtained by analyzing the forces that act on the individual charge carriers.

If we consider a differential element of a conductor of length $d\mathbf{l}$, we can find the force $d\mathbf{F}$ acting on it, by analogy with Eq. 33-6a, from

$$d\mathbf{F} = i\,d\mathbf{l} \times \mathbf{B}. \tag{33-6b}$$

By integrating this formula in an appropriate way we can find the force $\mathbf{F}$ on a conductor which is not straight.

A wire bent as shown in Fig. 33-6 carries a current i and is placed in a uniform magnetic field $\mathbf{B}$ that emerges from the plane of the figure. Calculate the force acting on the wire. The magnetic field is represented by lines of induction, shown emerging from the page. The dots show that the sense of $\mathbf{B}$ is up, out of the page.

EXAMPLE 2

figure 33-6
Example 2

723 MAGNETIC FORCE ON A CURRENT SEC. 33-3

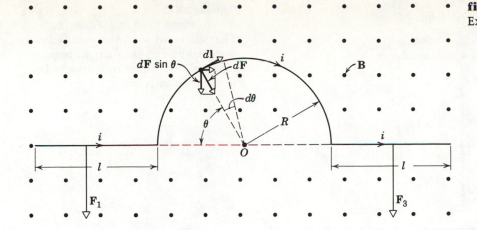

The force on each straight section, from Eq. 33-6a, has the magnitude

$$F_1 = F_3 = ilB$$

and points down as shown by the arrows in the figure.

A segment of wire of length dl on the arc has a force $d\mathbf{F}$ on it whose magnitude is

$$dF = iB\ dl = iB(R\ d\theta)$$

and whose direction is radially toward O, the center of the arc. Only the downward component of this force is effective, the horizontal component being canceled by an oppositely directed component associated with the corresponding arc segment on the other side of O. Thus the total force on the semicircle of wire about O points down and is

$$F_2 = \int_0^\pi dF \sin\theta = \int_0^\pi (iBR\ d\theta) \sin\theta = iBR \int_0^\pi \sin\theta\ d\theta = 2iBR.$$

The resultant force on the whole wire is

$$F = F_1 + F_2 + F_3 = 2ilB + 2iBR = 2iB(l + R).$$

Notice that this force is the same as that acting on a straight wire of length $2l + 2R$.

Figure 33-7 shows the arrangement used by Thomas, Driscoll, and Hipple in 1949 at the National Bureau of Standards to measure the magnetic field provided by a laboratory magnet such as that of Fig. 33-2. The rectangle is a coil of nine turns whose width a and length b are about 10 cm and 70 cm, respectively. The lower end of the coil is placed in the magnetic field $\mathbf{B}$ and the upper end is hung from the arm of a balance; $\mathbf{B}$ enters the plane of the figure at right angles.

An accurately known current i of about 0.10 A is set up in the coil in the direction shown, and weights are placed in the right-hand balance pan until the system is balanced. The magnetic force $\mathbf{F}$ ($= i\mathbf{l} \times \mathbf{B}$; see Eq. 33-6$a$) on the bottom leg of the coil points upward, as shown in the figure. Equation 33-5 also shows that the force on each wire at the bottom of the coil is iaB. Since there are nine wires, the total force on the bottom leg of the coil is $9iaB$. The forces on the vertical sides of the coil ($= i\mathbf{l} \times \mathbf{B}$) are sideways; because they are equal and opposite, they cancel and produce no effect.

After balancing the system, the experimenters reversed the direction of the current, which changed the sign of all the magnetic forces acting on the coil. In particular, $\mathbf{F}$ then pointed downward, which caused the rest point of the bal-

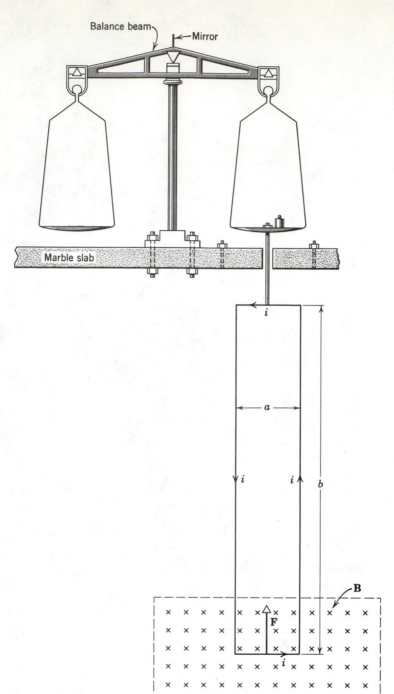

Balance beam
Mirror
Marble slab

figure 33-7
An apparatus used to measure **B**.
The zero point of the balance is
observed by means of a light beam
reflected from the mirror attached
to the balance beam.

ance to move. A mass m of about 8.78 g had to be added to the left balance pan
to restore the original rest point. The *change* in force when the current is reversed is $2F$, and this must equal the weight added to the left balance pan, or

$$mg = 2(9iaB) = 18iaB.$$

This gives

$$B = \frac{mg}{18ai} = \frac{(8.78 \times 10^{-3} \text{ kg})(9.80 \text{ m/s}^2)}{(18)(0.10 \text{ m})(0.10 \text{ A})} = 0.48 \text{ T} = 4800 \text{ gauss.}$$

The Bureau of Standards' workers made this measurement with much more
precision than these approximate numbers suggest. In one series of measurements, for example, they found a magnetic field of 4697.55 gauss; even greater
precision is possible today, by this and other methods.

Figure 33-8 shows a rectangular loop of wire of length a and width b placed in a uniform magnetic field **B**, with sides 1 and 3 always normal to the field direction. The normal nn' to the plane of the loop makes an angle θ with the direction of **B**.

33-4
TORQUE ON A CURRENT LOOP

Assume the current to be as shown in the figure. Wires must be provided to lead the current into the loop and out of it. If these wires are twisted tightly together, there will be no net magnetic force on the twisted pair because the currents in the two wires are in opposite directions. Thus the lead wires may be ignored. Also, some way of supporting the loop must be provided. Let us imagine it to be suspended from a long string attached to the loop at its center of mass. In this way the loop will be free to turn, through a small angle at least, about any axis through the center of mass. Alternatively, we can imagine the experiment to be carried out in an orbiting earth satellite (Skylab, say) in which the effective value of **g** is zero.

The net force on the loop is the resultant of the forces on the four sides of the loop. On side 2 the vector **l** points in the direction of the current and has the magnitude b. The angle between **l** and **B** for side 2 (see Fig. 33-8b) is $90° - \theta$. Thus the magnitude of the force on this side is

$$F_2 = ibB \sin (90° - \theta) = ibB \cos \theta.$$

From the relation $\mathbf{F} = i\mathbf{l} \times \mathbf{B}$ (Eq. 33-6a), we find the direction of $\mathbf{F}_2$ to be out of the plane of Fig. 33-8b. You can show that the force $\mathbf{F}_4$ on side 4 has the same magnitude as $\mathbf{F}_2$ but points in the opposite direction. Thus $\mathbf{F}_2$ and $\mathbf{F}_4$, taken together, have no effect on the motion of the loop. The net force they provide is zero, and, since they have the same line of action, the net torque due to these forces is also zero.

The common magnitude of $\mathbf{F}_1$ and $\mathbf{F}_3$ is iaB. These forces, too, are oppositely directed so that they do not tend to move the coil bodily. As Fig. 33-8b shows, however, they do *not* have the same line of action if the coil is in the position shown; there is a net torque, which tends to rotate the coil clockwise about the line xx'. The coil can be supported on a rigid axis that lies along xx', with no loss of its freedom of motion. This torque can be represented in Fig. 33-8b by a vector pointing into the figure at point x' or in Fig. 33-8a by a vector pointing along the xx' axis from right to left.

The magnitude of the torque τ' is found by calculating the torque caused by $\mathbf{F}_1$ about axis xx' and doubling it, for $\mathbf{F}_3$ exerts the same torque about this axis that $\mathbf{F}_1$ does. Thus

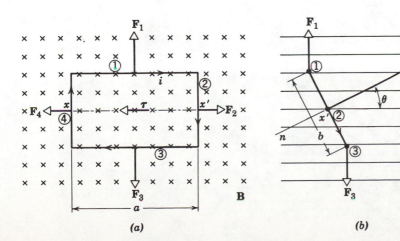

(a)   (b)

figure 33-8
A rectangular coil carrying a current i is placed in a uniform external magnetic field.

$$\tau' = 2(iaB)\left(\frac{b}{2}\right)(\sin\theta) = iabB\sin\theta.$$

This torque acts on every turn of the coil. If there are N turns, the torque on the entire coil is

$$\tau = N\tau' = NiabB\sin\theta = NiAB\sin\theta, \qquad (33\text{-}7)$$

in which A, the area of the coil, is substituted for ab.

This equation can be shown to hold for *all plane loops of area A, whether they are rectangular or not.* A torque on a current loop is the basic operating principle of the electric motor and of most electric meters used for measuring current or potential difference.

EXAMPLE 3

A galvanometer. Figure 33-9 shows the rudiments of a galvanometer, which is the basic operating mechanism of ammeters and voltmeters. The coil is 2.0 cm high and 1.0 cm wide; it has 250 turns and is mounted so that it can rotate about a vertical axis in a uniform *radial* magnetic field with $B = 2000$ gauss. A spring Sp provides a countertorque that cancels out the magnetic torque, resulting in a steady angular deflection ϕ corresponding to a given current i in the coil. If a current of 1.0×10^{-4} A produces an angular deflection of 30°, what is the *torsional constant* κ of the spring (see Eq. 15-21)?

Equating the magnetic torque to the torque caused by the spring (see Eq. 33-7) yields

$$\tau = NiAB\sin\theta = \kappa\phi$$

or

$$\kappa = \frac{NiAB\sin\theta}{\phi}$$

$$= \frac{(250)(1.0 \times 10^{-4}\ \text{A})(2.0 \times 10^{-4}\ \text{m}^2)(0.20\ \text{T})(\sin 90°)}{30°}$$

$$= 3.3 \times 10^{-8}\ \text{N}\cdot\text{m/degree}.$$

Note that the normal to the plane of the coil (that is, the pointer P) is always at right angles to the (radial) magnetic field so that $\theta = 90°$.

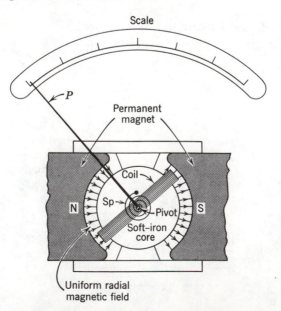

figure 33-9
Example 3. The elements of a galvanometer, showing the coil, the helical spring Sp, and pointer P.

A current loop orienting itself in an external magnetic field reminds us of the action of a compass needle in such a field. One face of the loop behaves like the north pole of the needle;* the other face behaves like the south pole. Compass needles, bar magnets, and current loops can all be regarded as *magnetic dipoles*. We show this here for the current loop, reasoning entirely by analogy with *electric* dipoles.

A structure is called an electric dipole if (a) when placed in an *external* electric field it experiences a torque given by Eq. 27-11,

$$\tau = \mathbf{p} \times \mathbf{E}, \tag{33-8}$$

where $\mathbf{p}$ is the electric dipole moment, and (b) it sets up a field of its own at distant points, described qualitatively by the lines of force of Fig. 29-10 and quantitatively by Eq. 29-11. These two requirements are not independent; if one is fulfilled, the other follows automatically.

The magnitude of the torque described by Eq. 33-8 is

$$\tau = pE \sin \theta, \tag{33-9}$$

where θ is the angle between $\mathbf{p}$ and $\mathbf{E}$. Let us compare this with Eq. 33-7, the expression for the torque on a current loop:

$$\tau = (NiA)B \sin \theta. \tag{33-7}$$

In each case the appropriate field (E or B) appears, as does a term $\sin \theta$. Comparison suggests that NiA in Eq. 33-7 can be taken as the *magnetic dipole moment* μ, corresponding to p in Eq. 33-9, or

$$\mu = NiA. \tag{33-10}$$

Equation 33-7 suggests that we write the torque on a current loop in vector form, in analogy with Eq. 33-8, or

$$\tau = \mu \times \mathbf{B}. \tag{33-11}$$

The magnetic dipole moment of the loop μ must be taken to lie along an axis perpendicular to the plane of the loop; its direction is given by the following rule: Let the fingers of the right hand curl around the loop in the direction of the current; the extended right thumb will then point in the direction of μ. If μ is defined by this rule and Eq. 33-10, check carefully that Eq. 33-11 correctly describes in every detail the torque acting on a current loop in an external field (see Fig. 33-8).

Since a torque acts on a current loop, or other magnetic dipole, when it is placed in an external magnetic field, it follows that work (positive or negative) must be done by an external agent to change the orientation of such a dipole. Thus a magnetic dipole has *potential energy* associated with its orientation in an external magnetic field. This energy may be taken to be zero for any arbitrary position of the dipole. By analogy with the assumption made for electric dipoles in Section 27-6, we assume that the magnetic energy U is zero when μ and $\mathbf{B}$ are at right angles, that is, when $\theta = 90°$. This choice of a zero-energy configuration for U is arbitrary because we are interested only in the *changes* in energy that occur when the dipole is rotated.

The magnetic potential energy in any position θ is defined as the work that an external agent must do to turn the dipole from its zero-energy position ($\theta = 90°$) to the given position θ. Thus

---

* The north pole of a compass needle is the end that points toward the geographic north.

$$U = \int_{90°}^{\theta} \tau \, d\theta = \int_{90°}^{\theta} NiAB \sin \theta \, d\theta = \mu B \int_{90°}^{\theta} \sin \theta \, d\theta = -\mu B \cos \theta,$$

in which Eq. 33-7 is used to substitute for τ. In vector symbolism we can write this relation as

$$U = -\boldsymbol{\mu} \cdot \mathbf{B}, \qquad (33\text{-}12)$$

which is in perfect correspondence with Eq. 27-13, the expression for the energy of an *electric* dipole in an external *electric* field,

$$U = -\mathbf{p} \cdot \mathbf{E}.$$

EXAMPLE 4

A circular coil of N turns has an effective radius a and carries a current i. How much work is required to turn it in an external magnetic field $\mathbf{B}$ from a position in which θ equals zero to one in which θ equals 180°? Assume that $N = 100$, $a = 5.0$ cm, $i = 0.10$ A, and $B = 1.5$ T.

The work required is the difference in energy between the two positions, or, from Eq. 33-12,

$$W = U_{\theta=180°} - U_{\theta=0} = (-\mu B \cos 180°) - (-\mu B \cos 0) = 2\mu B.$$

But $\mu = NiA$, so that

$$W = 2NiAB = 2Ni(\pi a^2) B$$

$$= (2)(100)(0.10 \text{ A})(\pi)(5 \times 10^{-2} \text{ m})^2(1.5 \text{ T}) = 0.24 \text{ J}.$$

33-5
THE HALL EFFECT

In 1879 E. H. Hall, at Harvard University, reported an experiment that gives the sign of the charge carriers in a conductor; see p. 676. Figure 33-10 shows a flat strip of copper carrying a current i in the direction shown. As usual, the direction of the current arrow, labeled i, is the direction in which the charge carriers would move *if* they were positive. The current arrow can represent either positive charges moving down (as in Fig. 33-10a) or negative charges moving up (as in Fig. 33-10b). The Hall effect can be used to decide between these two possibilities.*

A magnetic field $\mathbf{B}$ is set up at right angles to the strip by placing the strip between the polefaces of an electromagnet. This field exerts a deflecting force $\mathbf{F}$ on the strip (given by $i\mathbf{l} \times \mathbf{B}$), which points to the right in the figure. Since the sideways force on the strip is due to the sideways forces on the charge carriers (given by $q\mathbf{v} \times \mathbf{B}$), it follows that these carriers, whether they are positive or negative, will tend to drift toward the right in Fig. 33-10 as they drift along the strip, producing a *transverse Hall potential difference* V_{xy} between points such as x and y. The sign of the charge carriers is determined by the sign of this Hall poten-

---

* The connection between the eminent but self-effacing physicist H. A. Rowland and the Hall effect has often been glossed over. From the reference cited in the footnote on p. 717 we learn that Hall (Rowland's student at the Johns Hopkins University) used ". . . an experimental configuration devised by Rowland," and also, in Rowland's words (in 1894), ". . . I had already obtained the Hall effect on a small scale. . . ." Rowland's colleague, Joseph Ames, wrote ". . . There have been several striking cases where it might have seemed to an impartial observer that Rowland's name should have appeared on the title page."

The question of priority of discovery is complex. Before drawing conclusions, you should read carefully (a) the reference in the footnote on p. 717 and (b) Hall's paper, which can be found in "Source Book in Physics" by William Francis Magie, ed., McGraw-Hill Book Co., 1935, p. 541.

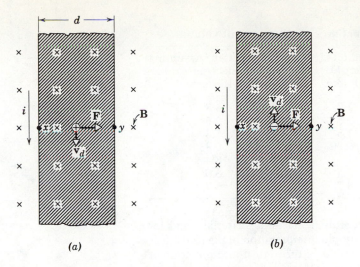

figure 33-10
A current i is set up in a copper strip placed in a magnetic field **B**, assuming (a) positive carriers and (b) negative carriers.

tial difference. If the carriers are positive, y will be at a higher potential than x; if they are negative, y will be at a lower potential than x. Experiment shows that in metals the charge carriers are negative.*

To analyze the Hall effect quantitatively, let us use the free-electron model of a metal, the same model used in Section 31-4 to help us understand why metals obey Ohm's law. The charge carriers can be assumed to move along the conductor with a certain constant drift speed v_d. The magnetic deflecting force that causes the moving charge carriers to drift toward the right edge of the strip is given by $q\mathbf{v}_d \times \mathbf{B}$ (see Eq. 33-2).

The charge carriers do not build up without limit on the right edge of the strip because the displacement of charge gives rise to a transverse *Hall electric field* $\mathbf{E}_H$, which acts, inside the conductor, to oppose the sideways drift of the carriers. This Hall electric field is another manifestation of the Hall potential difference and is related to it by

$$E_H = V_{xy}/d.$$

Eventually an equilibrium is reached in which the sideways magnetic deflecting force on the charge carriers is just canceled by the oppositely directed electric force $q\mathbf{E}_H$ caused by the Hall electric field, or

$$q\mathbf{E}_H + q\mathbf{v}_d \times \mathbf{B} = 0,$$

which we can write as

$$\mathbf{E}_H = -\mathbf{v}_d \times \mathbf{B}. \qquad (33\text{-}13)$$

This equation shows explicitly that if we measure $\mathbf{E}_H$ and $\mathbf{B}$, we can find $\mathbf{v}_d$ both in magnitude and direction; given the direction of $\mathbf{v}_d$, the sign of the charge carriers follows at once, as Fig. 33-10 shows.

The number of charge carriers per unit volume (n) can also be found from Hall effect measurements. If we write Eq. 33-13 in terms of magnitudes, for the case in which $\mathbf{v}_d$ and $\mathbf{B}$ are at right angles, we obtain $E_H = v_d B$. Combining this with Eq. 31-5 $(v_d = j/ne)$ leads to

$$E_H = \frac{j}{ne} B \quad \text{or} \quad n = \frac{jB}{eE_H}. \qquad (33\text{-}14)$$

The agreement between experiment and Eq. 33-14 is rather good for monovalent metals, as Table 33-1 shows.

---

* At the time of the Hall-Rowland experiments the electron was not yet discovered (see Section 33-8). Hall's analysis was based on a fluid model of electricity but the general conclusions remain unchanged.

Table 33-1
Number of conduction electrons per unit volume

Metal	Based on Hall Effect Data, $10^{22}/cm^3$	Calculated, Assuming One Electron/Atom, $10^{22}/cm^3$
Li	3.7	4.8
Na	2.5	2.6
K	1.5	1.3
Cs	0.80	0.85
Cu	11	8.4
Ag	7.4	6.0
Au	8.7	5.9

For nonmonovalent metals, for iron and similar magnetic materials, and for semiconductors such as germanium, the simple interpretation of the Hall effect in terms of the free-electron model is not valid. A theoretical interpretation of the Hall effect based on quantum physics gives a reasonable agreement with experiment in all cases.

EXAMPLE 5

A copper strip 2.0 cm wide and 1.0 mm thick is placed in a magnetic field with $B = 1.5$ T, as in Fig. 33-10. If a current of 200 A is set up in the strip, what Hall potential difference appears across the strip?

From Eq. 33-14,

$$E_H = \frac{jB}{ne};$$

but

$$E_H = \frac{V_{xy}}{d} \quad \text{and} \quad j = \frac{i}{A} = \frac{i}{dh},$$

where h is the thickness of the strip. Combining these equations gives

$$V_{xy} = \frac{iB}{neh} = \frac{(200 \text{ A})(1.5 \text{ T})}{(8.4 \times 10^{28}/m^3)(1.6 \times 10^{-19} \text{ C})(1.0 \times 10^{-3} \text{ } m)}$$

$$= 2.2 \times 10^{-5} \text{ V} = 22 \text{ } \mu\text{V}.$$

These potential differences, though quite measurable, are not large. See p. 678 for the calculation of n.

33-6
CIRCULATING CHARGES

Figure 33-11 shows a negatively charged particle introduced with velocity **v** into a uniform magnetic field **B**. We assume that **v** is at right angles to **B** and thus lies entirely in the plane of the figure. The relation $\mathbf{F} = q\mathbf{v} \times \mathbf{B}$ (Eq. 33-2) shows that the particle will experience a sideways deflecting force of magnitude qvB. This force will lie in the plane of the figure, which means that the particle cannot leave this plane.

This reminds us of a stone held by a rope and whirled in a horizontal circle on a smooth surface. Here, too, a force of constant magnitude, the tension in the rope, acts in a plane and at right angles to the velocity. The charged particle, like the stone, also moves with constant speed in a circular path. From Newton's second law we have

$$qvB = \frac{mv^2}{r} \quad \text{or} \quad r = \frac{mv}{qB}, \tag{33-15}$$

which gives the radius of the path. The three spirals in Fig. 33-4 show relatively low-energy electrons in a bubble chamber. The paths are not

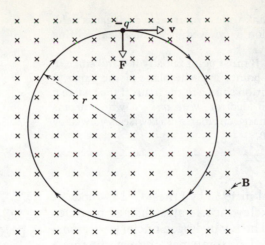

circles because the electrons lose energy by collisions in the chamber as they move.

The angular velocity ω is given by v/r or, from Eq. 33-15,

$$\omega = \frac{v}{r} = \frac{qB}{m}.$$

The frequency ν is given by

$$\nu = \frac{\omega}{2\pi} = \frac{qB}{2\pi m}. \qquad (33\text{-}16)$$

Note that ν does not depend on the speed of the particle. Fast particles move in large circles (Eq. 33-15) and slow ones in small circles, but all require the same time T (the *period*) to complete one revolution in the field.

The frequency ν is a characteristic frequency for the charged particle in the field and may be compared to the characteristic frequency of a swinging pendulum in the earth's gravitational field or to the characteristic frequency of an oscillating mass-spring system. It is sometimes called the *cyclotron frequency* of the particle in the field because particles circulate at this frequency in a cyclotron.

A 10-eV electron is circulating in a plane at right angles to a uniform magnetic field of 1.0×10^{-4} T (= 1.0 gauss).

(a) What is its orbit radius?

The velocity of an electron whose kinetic energy is K can be found from

$$v = \sqrt{\frac{2K}{m}} .$$

Verify that this yields 1.9×10^6 m/s for v. Then, from Eq. 33-15,

$$r = \frac{mv}{qB} = \frac{(9.1 \times 10^{-31} \text{ kg})(1.9 \times 10^6 \text{ m/s})}{(1.6 \times 10^{-19} \text{ C})(1.0 \times 10^{-4} \text{ T})} = 0.11 \text{ m} = 11 \text{ cm}.$$

(b) What is the cyclotron frequency? From Eq. 33-16,

$$\nu = \frac{qB}{2\pi m} = \frac{(1.6 \times 10^{-19} \text{ C})(1.0 \times 10^{-4} \text{ T})}{(2\pi)(9.1 \times 10^{-31} \text{ kg})} = 2.8 \times 10^6 \text{ Hz}.$$

(c) What is the period of revolution T?

$$T = \frac{1}{\nu} = \frac{1}{2.8 \times 10^6 \text{ Hz}} = 3.6 \times 10^{-7} \text{ s}.$$

EXAMPLE 6

Thus an electron requires 0.36 μs to make one revolution in a 1.0-gauss field. (*d*) What is the direction of circulation as viewed by an observer sighting along the field?

In Fig. 33-11 the magnetic force $q\mathbf{v} \times \mathbf{B}$ must point radially inward, since it provides the centripetal force. Since **B** points into the plane of the paper, **v** would have to point to the left at the position shown in the figure if the charge q were positive. However, the charge is an electron, with $q = -e$, which means that **v** must point to the right. Thus the charge circulates clockwise as viewed by an observer sighting in the direction of **B**.

The cyclotron, first made operational in 1932 by Ernest O. Lawrence at the University of California at Berkeley, accelerates charged particles such as hydrogen nuclei (protons) and heavy hydrogen nuclei (deuterons) to high energies so that they can be used in atom-smashing experiments.* Figure 33-12 shows a cyclotron formerly operated at the University of Pittsburgh. Although conventional cyclotrons of this type are no longer used, we discuss them for two reasons: (a) they provide an excellent framework within which to discuss the action of magnetic and electric fields on charged particles and (b) the cyclotron has led to the production of several generations of improved accelerators, the proton synchrotron being one. These later devices provide even more opportunities for studying the interactions of charged particles with magnetic and electric fields although that, of course, is not their primary purpose.

33-7
CYCLOTRONS AND SYNCHROTRONS

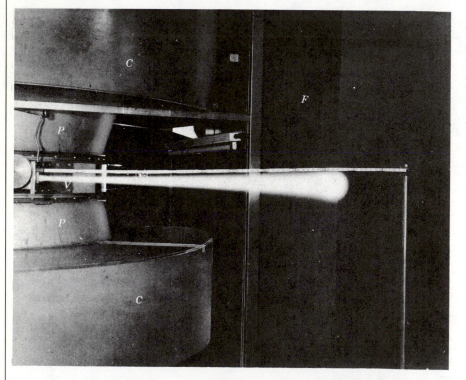

figure 33-12
The former University of Pittsburgh cyclotron. Note vacuum chamber *V*, magnet frame *F*, magnetic pole faces *P*, magnet coils *C*, and the deuteron beam emerging into the air of the laboratory through an aluminum foil "window." The rule is 6 ft. long. (Courtesy A. J. Allen.)

In an *ion source* at the center of the cyclotron molecules of deuterium (heavy hydrogen) are bombarded with electrons whose energy is high enough (say 100 eV) so that many positive ions are formed during the collisions. Many of these ions are free deuterons, which enter the cyclo-

---

* Lawrence received a Nobel prize in 1939 for this work.

tron proper through a small hole in the wall of the ion source and are available to be accelerated.

The cyclotron uses a modest potential difference for accelerating (say 10^5 V), but it requires the ion to pass through this potential difference a number of times. To reach 10 MeV with 10^5 V accelerating potential requires 100 passages. A magnetic field is used to bend the ions around so that they may pass again and again through the same accelerating potential.

Figure 33-13 is a top view of the part of the cyclotron that is inside the vacuum tank marked V in Fig. 33-12. The two D-shaped objects, called *dees*, are made of copper sheet and form part of an electric oscillator which establishes an accelerating potential difference across the gap between the dees. The direction of this potential difference is made to change sign some millions of times per second.

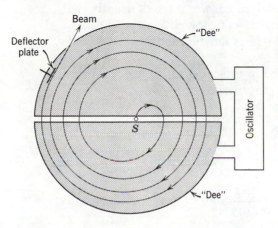

figure 33-13
The elements of a cyclotron showing the ion source S and the dees. The deflector plate, held at a suitable negative potential, deflects the particles out of the dee system.

The dees are immersed in a magnetic field $(B \cong 1.6$ T$)$ whose direction is out of the plane of Fig. 33-13. The field is set up by a large electromagnet, part of which is shown by F in Fig. 33-12. Finally, the space in which the ions move is evacuated to a pressure of about 10^{-6} mm-Hg. If this were not done, the ions would continually collide with air molecules.

Suppose that a deuteron, emerging from the ion source, finds the dee that it is facing to be negative; it will accelerate toward this dee and will enter it. Once inside, it is screened from electric fields by the metal walls of the dees. The magnetic field is not screened by the dees so that the ion bends in a circular path whose radius, which depends on the velocity, is given by Eq. 33-15, or

$$r = \frac{mv}{qB}.$$

After a time t_0 the ion emerges from the dee on the other side of the ion source. Let us assume that the accelerating potential has now changed sign. Thus the ion *again* faces a negative dee, is further accelerated, and again describes a semicircle, of somewhat larger radius (see Eq. 33-15), in the dee. *The time of passage through this dee, however, is still t_0.* This follows because the period of revolution T of an ion circulating in a magnetic field does not depend on the speed of the ion; see Eq. 33-16. This process goes on until the ion reaches the outer edge of one dee where it is pulled out of the system by a negatively charged deflector plate.

The key to the operation of the cyclotron is that the characteristic frequency ν at which the ion circulates in the field must be equal to the fixed frequency ν_0 of the electric oscillator, or

$$\nu = \nu_0.$$

The *resonance condition* says that if the energy of the circulating ion is to increase, energy must be fed to it at a frequency ν_0 that is equal to the natural frequency ν at which the ion circulates in the field. In the same way we feed energy to a swing by pushing it at a frequency equal to the natural frequency of oscillation of the swing.

From Eq. 33-16 ($\nu = qB/2\pi m$), we can rewrite the resonance equation as

$$\frac{qB}{2\pi m} = \nu_0. \tag{33-17}$$

Once we have selected an ion to be accelerated, q/m is fixed; usually the oscillator is designed to work at a single frequency ν_0. We then "tune" the cyclotron by varying B until Eq. 33-17 is satisfied and an accelerated beam appears.

The *energy* of the particles produced in the cyclotron depends on the radius R of the dees. From Eq. 33-15 ($r = mv/qB$), the velocity of a particle circulating at this radius is given by

$$v = \frac{qBR}{m}.$$

The kinetic energy is then

$$K = \tfrac{1}{2}mv^2 = \frac{q^2B^2R^2}{2m}. \tag{33-18}$$

EXAMPLE 7

The University of Pittsburgh cyclotron had an oscillator frequency of 12×10^6 Hz and a dee radius of 53 cm ($= 21$ in.). (a) What value of B is needed to accelerate deuterons?

From Eq. 33-17, $\nu_0 = qB/2\pi m$, so that

$$B = \frac{2\pi\nu_0 m}{q} = \frac{(2\pi)(12 \times 10^6 \text{ Hz})(3.3 \times 10^{-27} \text{ kg})}{1.6 \times 10^{-19} \text{ C}} = 1.6 \text{ T}.$$

Note that the deuteron has the same charge as the proton but (very closely) twice the mass. (b) What deuteron energy results?

From Eq. 33-18,

$$K = \frac{q^2B^2R^2}{2m} = \frac{(1.6 \times 10^{-19} \text{ C})^2(1.6 \text{ T})^2(0.53 \text{ m})^2}{(2)(3.3 \times 10^{-27} \text{ kg})}$$

$$= (2.8 \times 10^{-12} \text{ J}) \left(\frac{1 \text{ eV}}{1.6 \times 10^{-19} \text{ J}}\right) = 17 \text{ MeV}.$$

There are two reasons why the conventional cyclotron that we have described runs into difficulties at high energies. One deals with physics, the other with cost. Let us discuss each in turn.

(a) The cyclotron fails to operate at high energies because one of its assumptions, that the frequency of rotation of an ion circulating in a magnetic field is independent of its speed, is true only for speeds much less than that of light. As the particle speed increases, we must use the *relativistic mass m* in Eq. 33-16.

The relativistic mass increases with velocity (Eq. 8-20) so that at high enough speeds ν decreases with velocity. Thus the ions get out of step with the electric oscillator, and eventually the energy of the circulating ion stops increasing.

(b) The second difficulty associated with the acceleration of particles to high energies is that the size of the magnet that would be required to guide such particles in a circular path is very large. For a 30-GeV proton, for example, in a field of 1.5 T (= 15,000 gauss) the radius of curvature is 65 m. A magnet of the cyclotron type of this size (about 430 ft in diameter) would be prohibitively expensive. Incidentally, a 30-GeV proton has a speed equal to 0.99998 that of light.

Both the relativistic and the economic limitations have been removed by techniques that can be understood in terms of Eq. 33-17 ($2\pi\nu_0 m = qB$) in which m is now taken to be the relativistic mass, given by Eq. 8-20, or

$$ m = \frac{m_0}{\sqrt{1 - (v/c)^2}}, $$

v being the speed of the particle and c being that of light.

As the particle speed increases, the relativistic mass m also increases. To maintain the equality in Eq. 33-17, and thus insure resonance, one may decrease the oscillator frequency ν_0 as the particle (assumed to be a proton) accelerates in such a way that the product $\nu_0 m$ remains constant. Accelerators that use this technique have been called *synchrocyclotrons*.

To ameliorate the magnet cost limitation one can vary *both* B and ν_0 in a cyclic fashion in such a way that not only is Eq. 33-17 satisfied at all times but the particle orbit radius remains essentially constant during the acceleration process. This permits the use of an annular (or ring-shaped) magnet, rather than the conventional cyclotron type, at great saving in cost. With the *two* variables B and ν_0 at our disposal, it is possible to preserve *two* equalities during the acceleration process, one being Eq. 33-17 and the other being the relation

$$ v = \omega_0 R_o = (2\pi\nu_0)R_0 $$

in which R_0 is the desired (fixed) orbit radius. Accelerators that use this technique are called *synchrotrons*. Table 33-2 shows some characteristics of the accelerator built at the Brookhaven National Laboratory, embodying these principles.

Table 33-2
The Brookhaven Proton Synchrotron

Maximum proton energy	33 GeV
Mean orbit radius	128 m
Maximum orbit field	1.3 T
Injection energy	50 MeV
Pulse repetition rate	2.4 Hz
Beam aperture	18 cm × 8 cm
Total weight of magnets	4000 tons

Note that even the energy at which the protons are *injected* into this accelerator (50 MeV) far exceeds the capabilities of a conventional cyclotron.*

As of this writing (1977) the protron synchrotron producing the most energetic protons (500 GeV with 1000 GeV as a goal) is located in Batavia, Illinois.† The ring, of 954 separate magnets, is 6.3 km (= 4.1 miles) in circumference! The system for injecting protons into this ring is impressive in itself. The

---

* See, for example, "Introduction to Nuclear Physics" by Harald Enge, John Wiley & Sons, 1966, for a readable account with much more information about particle accelerators than we can present here.
† See "The Batavia Accelerator" by R. R. Wilson, *Scientific American*, February 1974.

protons are first accelerated to 750 keV by a transformer-rectifier arrangement. They are then directed into a 145 m-long linear accelerator from which they emerge with an energy of 200 MeV. The protons are then led into a "medium size" proton synchrotron (the "booster") from which they emerge with an energy of 80 GeV. Only then are they injected into the main accelerator ring.

In all of these processes magnetic and electric fields not only accelerate the protons, they direct them in desired directions so that they may be used for experiments; above all, they focus them so that a well-defined proton beam emerges after as much as 10^6 miles of travel through the accelerator complex. Although this is not its purpose, no better "laboratory" for demonstrating the action of magnetic and electric fields on charged particles has yet been devised. As an indication of its size and scope, we point out that its yearly budget for electric energy alone is several million (1976) dollars.

33-8 THE DISCOVERY OF THE ELECTRON

This crucial experiment, performed in 1897 by J. J. Thomson* in the Cavendish Laboratory in Cambridge, England, was a measurement of the ratio of charge e to mass m of the electron by observing its deflection in combined magnetic and electric fields. It amounted to the discovery of the electron as a fundamental particle and we discuss it here as another practical example of the action of magnetic and electric fields on charged particles.

In Fig. 33-14, which is a modernized version of Thomson's apparatus, electrons are emitted from hot filament F and accelerated by an applied potential difference V. They then enter a region in which they move at right angles to an electric field $\mathbf{E}$ and a magnetic field $\mathbf{B}$; $\mathbf{E}$ and $\mathbf{B}$ are right angles to each other. The beam is made visible as a spot of light when it strikes fluorescent screen S. The entire region in which the electrons move is highly evacuated so that collisions with air molecules will not occur.

The resultant force on a charged particle moving through an electric and a magnetic field is given by Eq. 33-4, or

$$\mathbf{F} = q_0\mathbf{E} + q_0\mathbf{v} \times \mathbf{B}.$$

Study of Fig. 33-14 shows that the electric field deflects the particle upward and the magnetic field deflects it downward. If these deflecting forces are to cancel (that is, if $\mathbf{F} = 0$), this equation, for this problem, reduces to

$$eE = evB$$

or
$$E = vB. \qquad (33\text{-}19)$$

Thus for a given electron speed v the condition for zero deflection can be satisfied by adjusting E or B.

Thomson's procedure was (a) to note the position of the undeflected beam spot, with $\mathbf{E}$ and $\mathbf{B}$ both equal to zero; (b) to apply a fixed electric field $\mathbf{E}$, measuring on the fluorescent screen the deflection so caused; and (c) to apply a magnetic field and adjust its value until the beam deflection is restored to zero.

---

* There is published evidence that the German physicist E. Wiechert discovered the electron some months before J. J. Thomson. In 1936 J. S. Townsend wrote ". . . the credit for having been the first to make [this] discovery is as much due to Weichert as the credit for having been the first to discover x-rays is due to Röntgen."

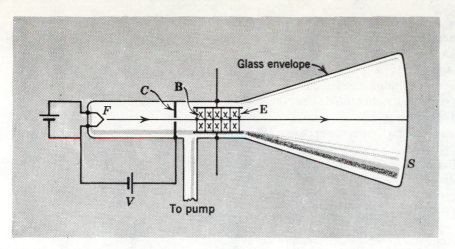

figure 33-14
Electrons from the heated filament F are accelerated by a potential difference V and pass through a hole in the screen C. After passing through a region in which perpendicular electric and magnetic fields are present, they strike the fluorescent screen S.

In Section 27-5 we saw that the deflection y of an electron in a purely electric field (step b), measured at the far edge of the deflecting plates, is given by Eq. 27-9, or, with small changes in notation,

$$y = \frac{eEl^2}{2mv^2},$$

where v is the electron speed and l is the length of the deflecting plates; y is not measurable directly, but it may be calculated from the measured displacement of the spot on the screen if the geometry of the apparatus is known. Thus y, E, and l are known; the ratio e/m and the velocity v are unknown. We cannot calculate e/m until we have found the velocity, which is the purpose of step c above.

If (step c) the electric force is set equal and opposite to the magnetic force, the net force is zero and we can write (Eq. 33-19)

$$v = \frac{E}{B}.$$

Substituting this equation into the equation for y and solving for the ratio e/m leads to

$$\frac{e}{m} = \frac{2yE}{B^2l^2}, \tag{33-20}$$

in which all the quantities on the right can be measured. Thomson's value for e/m was 1.7×10^{11} C/kg, in full agreement with the 1977 value of 1.758805×10^{11} C/kg.

questions

1. In a letter to the editor of *Sky and Telescope*, August 1976, Cicely M. Botley says, in part, "The [geomagnetic] pole in northern Canada is more correctly the *dip pole*, where a freely suspended magnetic needle is vertical. The *geomagnetic pole*, from which geomagnetic latitudes are measured, is in northwest Greenland." Discuss these two concepts.

2. Of the three vectors in the equation $\mathbf{F} = q\mathbf{v} \times \mathbf{B}$, which pairs are always at right angles? Which may have any angle between them?

3. Why do we not simply define the magnetic field **B** to point in the direction of the magnetic force that acts on the moving charge?

4. Imagine that you are sitting in a room with your back to one wall and that an electron beam, traveling horizontally from the back wall toward the front wall, is deflected to your right. What is the direction of the magnetic field that exists in the room?

5. If an electron is not deflected in passing through a certain region of space, can we be sure that there is no magnetic field in that region?

6. If a moving electron is deflected sideways in passing through a certain region of space, can we be sure that a magnetic field exists in that region?

7. A beam of protons is deflected sideways. Could this deflection be caused (a) by an electric field? (b) By a magnetic field? (c) If either could be responsible, how would you be able to tell which was present?

8. A conductor, even though it is carrying a current, has zero net charge. Why, then, does a magnetic field exert a force on it?

9. In Example 2 (see Fig. 33-6) we saw that the magnetic force was the same as if the semicircular arc had been replaced by a straight wire of length $2R$. Would this same conclusion hold if we replaced the semicircular arc by a curve of irregular shape? Give a specific example to prove your point, one way or the other.

10. Does Eq. 33-6a ($\mathbf{F} = i\mathbf{l} \times \mathbf{B}$) hold for a straight wire whose cross section varies irregularly along its length (a "lumpy" wire)?

11. A straight copper wire carrying a current i is immersed in a magnetic field $\mathbf{B}$, at right angles to it. We know that $\mathbf{B}$ exerts a sideways force on the free (or conduction) electrons. Does it do so on the bound electrons? After all, they are not at rest. Discuss.

12. In Section 33-3 we state that a magnetic field $\mathbf{B}$ exerts a sideways force on the conduction electrons in, say, a copper wire carrying a current i. We have tacitly assumed that this same force acts on the conductor itself. Are there some missing steps in this argument?

13. Equation 33-11 ($\boldsymbol{\tau} = \boldsymbol{\mu} \times \mathbf{B}$) shows that there is no torque on a current loop in an external magnetic field if the angle between the axis of the loop and the field is (a) 0° or (b) 180°. Discuss the nature of the equilibrium (that is, is it stable, neutral, or unstable?) for these two positions.

14. In Example 4 we showed that the work required to turn a current loop end-for-end in an external magnetic field is $2\mu B$. Does this hold no matter what the original orientation of the loop was?

15. Imagine that the room in which you are seated is filled with a uniform magnetic field with $\mathbf{B}$ pointing vertically upward. A circular loop of wire has its plane horizontal. For what direction of current in the loop, as viewed from above, will the loop be in stable equilibrium with respect to forces and torques of magnetic origin?

16. A rectangular current loop is in an arbitrary orientation in an external magnetic field. Is any work required to rotate the loop about an axis perpendicular to its plane?

17. You wish to modify a galvanometer (see Example 3) to make it into (a) an ammeter and (b) a voltmeter. What do you need to do in each case?

18. (a) In measuring Hall potential differences, why must we be careful that points x and y in Fig. 33-10 are exactly opposite each other? (b) If one of the contacts is movable, what procedure might we follow in adjusting it to make sure that the two points are properly located?

19. A uniform magnetic field fills a certain cubical region of space. Can an electron be fired into this cube from the outside in such a way that it will travel in a closed circular path inside the cube?

20. Imagine the room in which you are seated to be filled with a uniform magnetic field with $\mathbf{B}$ pointing vertically downward. At the center of the room two electrons are suddenly projected horizontally with the same speed but in opposite directions. (a) Discuss their motions. (b) Discuss their motions if one particle is an electron and one a positron, that is, a positively charged electron.

21. In Fig. 33-4 why are the low-energy electron tracks spirals? That is, why does the radius of curvature change in the constant magnetic field in which the chamber is immersed?

22. What are the primary functions of (a) the electric field and (b) the magnetic field in the cyclotron?

23. What central fact makes the operation of a conventional cyclotron possible? Ignore relativistic considerations.

24. For Thomson's *e/m* experiment to work properly (Section 33-8), is it essential that the electrons have a fairly constant speed?

25. The arrangement of crossed electric and magnetic fields shown in the central part of Fig. 33-14 is sometimes called a *velocity filter*. How can this name be justified?

SECTION 33-2

1. Particles 1, 2, and 3 follow the paths shown in Fig. 33-15 as they pass through the magnetic field there. What can one conclude about each particle?
   Answer: Particle 1 is positive, particle 2 is neutral, particle 3 is negative.

2. The electrons in the beam of a television tube have an energy of 12 keV. The tube is oriented so that the electrons move horizontally from south to north. The earth's magnetic field points down and has $B = 5.5 \times 10^{-5}$ T. (a) In what direction will the beam deflect? (b) What is the acceleration of a given electron? (c) How far will the beam deflect in moving 20 cm through the television tube?

3. An electron has a velocity given in m/s by $\mathbf{v} = 2.0 \times 10^6 \mathbf{i} + 3.0 \times 10^6 \mathbf{j}$. It enters a magnetic field given in T by $\mathbf{B} = 0.03\mathbf{i} - 0.15\mathbf{j}$. (a) Find the magnitude and direction of the force on the electron. (b) Repeat your calculation for a deuteron having the same velocity.
   Answer: (a) 6.2×10^{-14} **k**, N. (b) -6.2×10^{-14} **k**, N.

4. A beam of electrons whose kinetic energy is K emerges from a "window" at the end of an accelerator tube. There is a metal plate a distance d from this window and at right angles to the direction of the emerging beam. Show that we can stop the beam from hitting the plate if we apply a magnetic field B such that

$$B \geq \left(\frac{2mK}{e^2 d^2}\right)^{1/2},$$

in which m and e are the electron mass and charge. How should **B** be oriented?

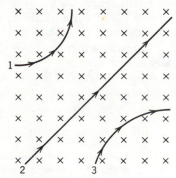

figure 33-15
Problem 1

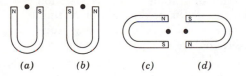

figure 33-16
Problem 5

SECTION 33-3

5. Figure 33-16 shows a magnet and a straight wire in which a current of electrons is flowing out of the page at right angles to it. In which case will there be a force on the wire that points toward the top of the page? *Answer:* (b).

6. A metal wire of mass m slides without friction on two horizontal rails spaced a distance d apart, as in Fig. 33-17. The track lies in a vertical uniform magnetic field **B**. A *constant current i* flows from generator G along one rail, across the wire, and back down the other rail. Find the velocity (speed and direction) of the wire as a function of time, assuming it to be at rest at $t = 0$.

7. A wire 1.0 m long carries a current of 10 A and makes an angle of 30° with a uniform magnetic field with $B = 1.5$ T. Calculate the magnitude and direction of the force on the wire.
   Answer: 7.5 N, perpendicular to both the wire and to **B**.

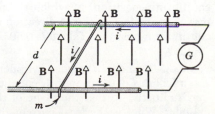

figure 33-17
Problem 6

8. A wire 50 cm long lying along the x-axis carries a current of 0.50 A in the positive x-direction. A magnetic field is present that is given in T by $\mathbf{B} = 0.0030\mathbf{j} + 0.010\mathbf{k}$. Find the components of the force on the wire.

9. A wire of 60 cm length and mass 10 g is suspended by a pair of flexible leads in a magnetic field of 0.40 T. What are the magnitude and direction of the current required to remove the tension in the supporting leads? See Fig. 33-18.   *Answer:* 0.41 A, from left to right.

10. An electron in a uniform magnetic field **B** has a velocity $\mathbf{v} = 4.0 \times 10^5\mathbf{i} + 7.1 \times 10^5\mathbf{j}$, in m/s. It experiences a force $\mathbf{F} = -2.7 \times 10^{-13}\mathbf{i} + 1.5 \times 10^{-13}\mathbf{j}$, in N. If $B_x = 0$, find the magnetic field.

11. Consider the possibility of a new design for an electric train. The engine is driven by the force due to the vertical component of the earth's magnetic field on a conducting axle. Current is passed down one rail, into a conducting wheel, through the axle, through another conducting wheel, and then back to the source via the other rail. (*a*) What current is needed to provide a modest 10,000 N force? Take the vertical component of **B** to be 10^{-5} T and the length of the axle to be 3.0 m. (*b*) How much power would be lost for each ohm of resistance in the rails? (*c*) Is such a train totally unrealistic or marginally unrealistic?
Answer: (*a*) 3.3×10^9 A. (*b*) 1.0×10^{17} W. (*c*) Totally unrealistic.

12. A U-shaped wire of mass m and length l is immersed with its two ends in mercury (Fig. 33-19). The wire is in a homogeneous magnetic field **B**. If a charge, that is, a current pulse $q = \int i\, dt$, is sent through the wire, the wire will jump up. Calculate, from the height h that the wire reaches, the size of the charge or current pulse, assuming that the time of the current pulse is very small in comparison with the time of flight. Make use of the fact that impulse of force equals $\int F\, dt$, which equals mv. (Hint: Try to relate $\int i\, dt$ to $\int F\, dt$.) Evaluate q for $B = 0.10$ T, $m = 10$ g, $l = 20$ cm, and $h = 3$ m.

13. Figure 33-20 shows a wire of arbitrary shape carrying a current i between points a and b. The wire lies in a plane at right angles to a uniform magnetic field **B**. Prove that the force on the wire is the same as that on a straight wire carrying a current i directly from a to b. (Hint: Replace the wire by a series of "steps" parallel and perpendicular to the straight line joining a and b.)

14. Figure 33-21 shows a wire ring of radius a at right angles to the general direction of a radially symmetric diverging magnetic field. The magnetic field at the ring is everywhere of the same magnitude B, and its direction at the ring is everywhere at an angle θ with a normal to the plane of the ring. The twisted lead wires have no effect on the problem. Find the magnitude and direction of the force the field exerts on the ring if the ring carries a current i as shown in the figure.

SECTION 33-4

15. An N-turn circular coil of radius R is suspended in a uniform magnetic field **B** that points vertically downward. The coil can rotate about a horizontal axis through its center. A mass m hangs by a string from the bottom of the coil. When a current i is put through the coil it eventually assumes an equilibrium position, in which the perpendicular to the plane of the coil makes an angle ϕ with the direction of **B**. Find ϕ and draw a sketch of this equilibrium position. Take $B = 0.50$ T, $R = 10$ cm, $N = 10$, $m = 500$ g, and $i = 1.0$ A.   *Answer:* $\phi = 72°$.

16. Figure 33-22 shows a wooden cylinder with a mass m of 0.25 kg, a radius R, and a length l of 0.1 m with N equal to ten turns of wire wrapped around it longitudinally, so that the plane of the wire loop contains the axis of the cylinder. What is the least current through the loop that will prevent the cylinder from rolling down an inclined plane whose surface is inclined at

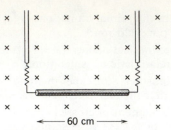
figure 33-18
Problem 9

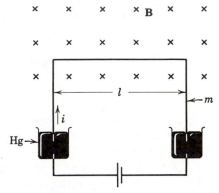
figure 33-19
Problem 12

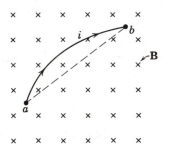
figure 33-20
Problem 13

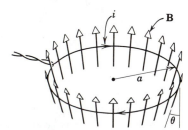
figure 33-21
Problem 14

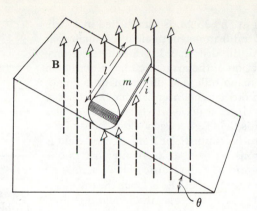

figure 33-22
Problem 16

an angle θ to the horizontal, in the presence of a vertical field of magnetic induction 0.5 T, if the plane of the windings is parallel to the inclined plane?

17. A certain galvanometer has a resistance of 75.3 Ω; its needle experiences a full-scale deflection when a current 1.62×10^{-3} A passes through its coil. (a) Determine the value of the auxiliary resistance required to convert the galvanometer into a voltmeter that reads 1.000 V at full-scale deflection. How is it to be connected? (b) Determine the value of the auxiliary resistance required to convert the galvanometer into an ammeter that reads 0.0500 A at full-scale deflection. How is it to be connected?
Answer: (a) 540 Ω, connected in series. (b) 2.52 Ω, connected in parallel.

18. A circular loop of wire having a radius of 8.0 cm carries a current of 0.20 A. A unit vector parallel to the dipole moment μ of the loop is given in A · m² by $0.60\mathbf{i} - 0.80\mathbf{j}$. If the loop is located in a magnetic field given in T by $\mathbf{B} = 0.25\mathbf{i} + 0.30\mathbf{k}$, find (a) the magnitude and direction of the torque on the loop and (b) the magnetic potential energy of the loop. Assume the same zero-energy configuration that we assumed in Section 33-4.

19. Figure 33-23 shows a rectangular twenty-turn loop of wire, 10 cm by 5.0 cm. It carries a current of 0.10 A and is hinged at one side. What torque (direction and magnitude) acts on the loop if it is mounted with its plane at an angle of 30° to the direction of a uniform magnetic field of 0.50 T?
Answer: 4.3×10^{-3} N·m. The torque vector is parallel to the long side of the coil and points down.

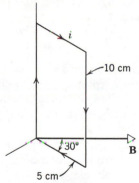

figure 33-23
Problem 19

20. Prove that the relation $\tau = NAiB \sin \theta$ holds for closed loops of arbitrary shape and not only for rectangular loops as in Fig. 33-8. (Hint: Replace the loop of arbitrary shape by an assembly of adjacent long, thin—approximately rectangular—loops which are equivalent to it as far as the distribution of current is concerned.)

21. A length l of wire carries a current i. Show that if the wire is formed into a circular coil, the maximum torque in a given magnetic field is developed when the coil has *one* turn only and the maximum torque has the value

$$\tau = \frac{1}{4\pi} l^2 iB.$$

SECTION 33-5

22. In a Hall effect experiment a current of 3.0 A lengthwise in a conductor 1.0 cm wide, 4.0 cm long, and 10^{-3} cm thick produced a transverse Hall voltage (across the width) of 1.0×10^{-5} V when a magnetic field of 1.5 T passed perpendicularly through the thin conductor. From these data, find (a) the drift velocity of the charge carriers and (b) the number of carriers per cubic centimeter. (c) Show on a diagram the polarity of the Hall voltage with a given current and magnetic field direction, assuming the charge carriers are (negative) electrons.

23. A current i, indicated by the crosses in Fig. 33-24, is established in a strip of copper of height h and width w. A uniform magnetic field **B** is applied at right angles to the strip. (a) Calculate the drift speed v_d for the electrons. (b) What are the magnitude and direction of the magnetic force **F** acting on the electrons? (c) What would the magnitude and direction of a homo-geneous electric field **E** have to be in order to counter-balance the effect of the magnetic field? (d) What is the voltage V necessary between two sides of the conductor in order to create this field **E**? Between which sides of the conductor would this voltage have to be applied? (e) If no electric field is applied *from the outside*, the electrons will be pushed somewhat to one side and therefore will give rise to a uniform Hall electric field $\mathbf{E}_H$ across the conductor until the forces of this electrostatic field $\mathbf{E}_H$ balance the magnetic forces encountered in part (b). What will be the magnitude and direction of the field $\mathbf{E}_H$? Assume that n, the number of conduction electrons per unit volume, is $1.1 \times 10^{29}/m^3$ and that $h = 0.020$ m, $w = 0.10$ cm, $i = 50$ A, and $B = 2.0$ T.
    Answer: (a) 1.4×10^{-4} m/s. (b) 4.5×10^{-23} N; down. (c) 2.8×10^{-4} V/m; down. (d) 5.7×10^{-6} V; top $+$, bottom $-$. (e) Same as (c).

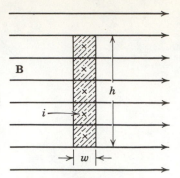

figure 33-24
Problem 23

24. (a) Show that the ratio of the Hall electric field E_H to the electric field E responsible for the current is

$$\frac{E_H}{E} = \frac{B}{ne\rho},$$

where ρ is the resistivity of the material.

(b) What is the angle between $\mathbf{E}_H$ and $\mathbf{E}$? (c) Evaluate this ratio for the con-ditions of Example 5.

SECTION 33-6
25. A proton, a deuteron, and an α-particle, accelerated through the same po-tential difference, enter a region of uniform magnetic field, moving at right angles to **B**. (a) Compare their kinetic energies. (b) If the radius of the pro-ton's circular path is 10 cm, what are the radii of the deuteron and the α-particle paths?   *Answer:* (a) $K_p = K_d = \frac{1}{2} K_\alpha$. (b) $R_d = R_\alpha = 14$ cm.

26. A proton, a deuteron, and an α-particle with the same kinetic energies enter a region of uniform magnetic field, moving at right angles to **B**. Compare the radii of their circular paths.

27. An electron is accelerated through 15,000 V and is then allowed to circulate at right angles to a uniform magnetic field with $B = 250$ gauss; 10^4 gauss $=$ 1 T. What is its path radius?   *Answer:* 1.7 cm.

28. In a nuclear experiment a 1.0-MeV proton moves in a uniform magnetic field in a circular path. What energy must (a) an alpha particle and (b) a deuteron have if they are to circulate in the same orbit?

29. (a) In a magnetic field with $B = 0.50$ T, for what path radius will an electron circulate at 0.10 the speed of light? (b) What will its kinetic energy be?
    Answer: (a) 0.34 mm. (b) 2.6 keV.

30. (a) What speed would a proton need to circle the earth at the equator, if the earth's magnetic field is everywhere horizontal there and directed along longitudinal lines? Take the magnitude of the earth's magnetic field to be 0.41×10^{-4} T at the equator. (b) Draw the velocity and magnetic field vectors corresponding to this situation.

31. What uniform magnetic field can be set up in space to permit a proton of speed 1.0×10^7 m/s to move in a circle the size of the earth's equator?
    Answer: 1.6×10^{-8} T.

32. An α-particle travels in a circular path of radius 0.45 m in a magnetic field with $B = 1.2$ T. Calculate (a) its speed, (b) its period of revolution, (c) its kinetic energy, and (d) the potential difference through which it would have to be accelerated to achieve this energy.

33. A neutral particle, viewed in a reference frame in which it is at rest, lies in a homogeneous magnetic field of magnitude B. At time $t = 0$ it decays into two charged particles each of mass m. (a) If the charge of one of the particles is $+q$, what is the charge of the other? The two particles move off in separate paths both of which lie in the plane perpendicular to **B**. (b) At a later time the particles collide. Express the time from decay until collision, t, in terms of m, B, and q. *Answer: (a) $-q$. (b) $t = \pi m/qB$.*

34. Singly ionized chlorine atoms of 35 u (= unified atomic mass units) and 37 u, traveling with speed 2.0×10^5 m/s, enter perpendicularly a uniform magnetic field of 0.50 T. After bending through 180° the atoms strike a photographic film. What is the separation distance between the two spots on the film? (1.00 u = 1.67×10^{-27} kg.)

35. Show that the radius of curvature of a charged particle moving at right angles to a magnetic field is proportional to its momentum.

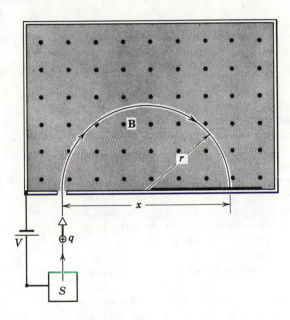

figure 33-25
Problem 36.

36. *Mass spectrometer.* Figure 33-25 shows an arrangement used by Dempster to measure the masses of ions. An ion of mass M and charge $+q$ is produced essentially at rest in source S, a chamber in which a gas discharge is taking place. The ion is accelerated by potential difference V and allowed to enter a magnetic field **B**. In the field it moves in a semicircle, striking a photographic plate at distance x from the entry slit is recorded. Show that the mass M is given by

$$M = \frac{B^2 q}{8V} x^2.$$

37. Two types of singly ionized atoms having the same charge q, and mass differing by a small amount ΔM are introduced into the mass spectrometer described in Problem 36. (a) Calculate the difference in mass in terms of V, q, M (of either), B, and the distance Δx between the spots on the photographic plate. (b) Calculate Δx for a beam of singly ionized chlorine atoms of masses 35 and 37 u if $V = 7.3 \times 10^3$ V and $B = 0.50$ T (1.00 u = 1.67×10^{-27} kg).

Answer: (a) $B\left(\dfrac{mq}{2\,V}\right)^{1/2} \Delta x$. (b) 8.2 mm.

figure 33-26
Problem 38.

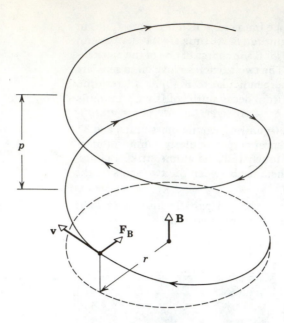

38. A 2-keV positron is projected into a uniform magnetic field **B** of 0.10 T with its velocity vector making an angle of 89° with **B**. (*a*) Convince yourself that the path will be a helix, its axis being the direction of **B**. Find (*b*) the period, (*c*) the pitch *p*, and (*d*) the radius *r* of the helix; see Fig. 33-26.

39. *Time-of-flight spectrometer.* S. A. Goudsmit has devised a method for measuring accurately the masses of heavy ions by timing their period of circulation in a known magnetic field. A singly charged ion of iodine makes 7.00 rev in a field of 4.50×10^{-2} T in about 1.29×10^{-3} s. What (approximately) is its mass in kilograms? Actually, the mass measurements are carried out to much greater accuracy than these approximate data suggest.
Answer: 2.11×10^{-25} kg, or about 127 proton masses.

40. *Zeeman effect.* In Bohr's theory of the hydrogen atom the electron can be thought of as moving in a circular orbit of radius *r* about the proton. Suppose that such an atom is placed in a magnetic field, with the plane of the orbit at right angles to **B**. (*a*) If the electron is circulating clockwise, as viewed by an observer sighting along **B**, will the angular frequency increase or decrease? (*b*) What if the electron is circulating counterclockwise? Assume that the orbit radius does not change. [Hint: The centripetal force is now partially electric ($\mathbf{F}_E$) and partially magnetic ($\mathbf{F}_B$) in origin.]

41. In Problem 40 show that the change in frequency of rotation caused by the magnetic field is given approximately by

$$\Delta\nu = \pm \frac{Be}{4\pi m}.$$

Such frequency shifts were observed by Zeeman in 1896. (Hint: Calculate the frequency of rotation without the magnetic field and also with it. Subtract, bearing in mind that because the effect of the magnetic field is very small, some—but not all—terms containing *B* can be set equal to zero with little error.)

42. (*a*) What is the cyclotron frequency of an electron with an energy of 100 eV in the earth's magnetic field of 1.0×10^{-4} T? (*b*) What is the radius of curvature of the path of this electron if its velocity is perpendicular to the magnetic field?

43. A 10-KeV electron is circulating in a plane at right angles to a uniform magnetic field. The orbit radius is 25 cm. Find (*a*) the magnetic field, (*b*) the cyclotron frequency, and (*c*) the period of the motion.
Answer: (*a*) 1.4×10^{-3} T. (*b*) 3.8×10^7 Hz. (*c*) 2.6×10^{-8} s.

SECTION 33-7

44. The cyclotron shown in Fig. 33-12 was normally adjusted to accelerate deuterons. (a) What energy of protons could it produce, using the same oscillator frequency as that used for deuterons? (b) What magnetic field would be required? (c) What energy of protons could be produced if the magnetic field was left at the value used for deuterons? (d) What oscillator frequency would then be required? (e) Answer the same questions for α-particles.

45. A deuteron in a large cyclotron is moving in a magnetic field with $B = 1.5$ T and an orbit radius of 2.0 m. Because of a grazing collision with a target, the deuteron breaks up, with a negligible loss of kinetic energy, into a proton and a neutron. Discuss the subsequent motions of each. Assume that the deuteron energy is shared equally by the proton and neutron at breakup.
    Answer: Neutron moves tangent to the original path. Proton moves in a circular orbit of 1.0-m radius.

46. In a certain cyclotron a proton moves in a circle of radius $r = 0.50$ m. The magnitude of the **B** field is 1.2 T. (a) What is the cyclotron frequency? (b) What is the kinetic energy of the proton?

47. Estimate the total path length traversed by a deuteron in the cyclotron shown in Fig. 33-12 during the acceleration process. Assume an accelerating potential between the dees of 80,000 V.   *Answer:* About 240 m.

SECTION 33-8

48. An electric field of 1500 V/m and a magnetic field of 0.40 T act on a moving electron to produce no force. (a) Calculate the minimum electron speed v. (b) Draw the vectors **E**, **B**, and **v**.

49. An electron is accelerated through a potential difference of 1000 V and directed into a region between two parallel plates separated by 0.02 m with a potential difference of 100 V between them. If the electron enters moving perpendicular to the electric field between the plates, what magnetic field is necessary perpendicular to both the electron path and the electric field so that the electron travels in a straight line?
    Answer: 2.7×10^{-4} T.

50. A 10-KeV electron moving horizontally enters a region of space in which there is a downward-directed electric field of magnitude 100 V/cm. (a) What are the magnitude and direction of the (smallest) magnetic field that will allow the electron to continue to move horizontally? Ignore the gravitational force, which is rather small. (b) Is it possible for a proton to pass through this combination of fields undeflected? If so, under what circumstances?

51. A positive point charge Q travels in a straight line with constant speed through an evacuated region in which there is a uniform electric field **E** and a uniform magnetic field **B**. (a) If **E** is directed vertically up and the charge travels horizontally from north to south with speed v, determine the least value of the magnitude of **B** and the corresponding direction of **B**. (b) Explain why **B** is not uniquely determined when **E** and **v** alone are given. (c) Suppose the charge is a proton which enters the region after having been accelerated through a potential difference of 3.10×10^5 V. If $E = 1.90 \times 10^5$ V/m, compute the value of B in part (a). (d) If in part (c) the electric field **E** is turned off, determine the radius r of the circle in which the proton now moves.
    Answer: (a) $B = E/v$; from east to west. (c) 2.47×10^{-2} T. (d) 3.26 m.

34
ampère's law

One class of problems involving magnetic fields, dealt with in Chapter 33, concerns the forces *exerted by* a magnetic field on a moving charge or on a current-carrying conductor and the torque exerted on a magnetic dipole (a bar magnet or a current loop, say). A second class of problems concerns the *production* of a magnetic field by a magnet, a current-carrying conductor, or by moving charges. This chapter deals with problems of this second class.

The discovery that currents produce magnetic effects was made by Hans Christian Oersted in 1820. Oersted made his discovery in connection with a classroom demonstration. In his paper entitled *The Action of Currents on Magnets*, Oersted wrote, translated from the Latin,

The first experiments on the subject which I undertook to illustrate were set on foot in the classes for electricity, galvanism, and magnetism, which were held by me in the winter just past.

Because of the importance of Oersted's discovery (a fundamental connection between electricity and magnetism) and especially because of its context in a teaching situation, the medal awarded annually by the American Association of Physics Teachers to a physics teacher especially noted for his or her impact on the teaching of physics is named after Oersted.

If we may deal for the moment with current-carrying wires as typical sources of magnetic fields and as typical objects on which magnetic fields may act, we may write in analogy with the argument of Section 27-1 for electric fields.

$$\text{current} \rightleftharpoons \text{field } (\mathbf{B}) \rightleftharpoons \text{current},$$

which suggests (a) that currents generate magnetic fields and (b) that magnetic fields exert forces on currents. We dealt with (b) in Section 33-3; we deal with (a) in this chapter.

Figure 34-1, which shows a wire surrounded by a number of small magnets, shows a modification of Oersted's experiment. If there is no current in the wire, all the magnets are aligned with the horizontal component of the earth's magnetic field. When a strong current is present, the magnets point so as to suggest that the magnetic field lines form closed circles around the wire. This view is strengthened by the experiment in Fig. 34-2, which shows iron filings on a horizontal glass plate, through the center of which a current-carrying conductor passes.

Today we write the quantitative relationship between current i and the magnetic field **B** as

$$\oint \mathbf{B} \cdot d\mathbf{l} = \mu_0 i, \tag{34-1}$$

which is known as *Ampère's law.* Ampère, being an advocate of the action-at-a-distance point of view, did not formulate his results in terms of fields; this was first done by Maxwell. Ampère's law, including an important extension of it made later by Maxwell, is one of the basic equations of electromagnetism (see Table 40-2).

We can gain an appreciation of the way Ampère's law developed historically by considering a hypothetical experiment which has, in fact, much in common with experiments that were actually carried out. The experiment consists of measuring **B** at various distances r from a long

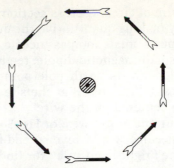

figure 34-1
An array of compass needles near a central wire carrying a strong current. The black ends of the compass needles are their north poles. The central dot shows the current emerging from the page. As usual, the direction of a current is taken as the direction of flow of positive charge.

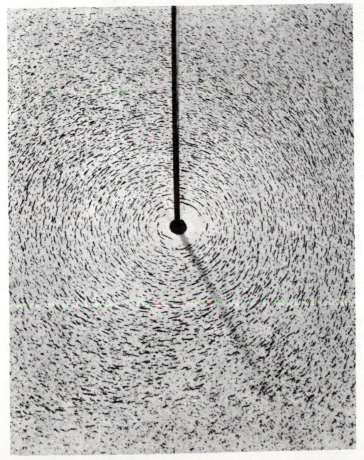

figure 34-2
Iron filings around a wire carrying a strong current. (Courtesy Physical Science Study Committee.)

straight wire of circular cross section and carrying a current i. This can be done by making quantitative the qualitative observation of Fig. 34-1.

Let us put a small compass needle a distance r from the wire. Such a needle, a small magnetic dipole, tends to line up with an external magnetic field, with its north pole pointing in the direction of **B**. Figure 34-1 makes it clear that **B** at the site of the dipole is tangent to a circle of radius r centered on the wire.

If the current in the wire of Fig. 34-1 is reversed in direction, all the compass needles would reverse end-for-end. This experimental result leads to the "right-hand rule" for finding the direction of **B** near a wire carrying a current i: *Grasp the wire with the right hand, the thumb pointing in the direction of the current. The fingers will curl around the wire in the direction of **B**.*

Let us now turn the dipole through an angle θ from its equilibrium position. To do this, we must exert an external torque just large enough to overcome the restoring torque τ that will act on the dipole. τ, θ, and **B** are related by Eq. 33-11 ($\boldsymbol{\tau} = \boldsymbol{\mu} \times \mathbf{B}$), which we can write in terms of magnitude as

$$\tau = \mu B \sin \theta \qquad (34\text{-}2)$$

in which μ is the magnitude of the magnetic moment of the dipole and θ is the angle between the vectors $\boldsymbol{\mu}$ and **B**. Even though we may not know the value of μ for the compass needle, we may take it to be a constant, independent of the position or orientation of the needle. Thus by measuring τ and θ in Eq. 34-2 we can obtain a *relative* measure of B for various distances r and for various currents i in the wire. We can describe the experimental results by the proportionality

$$B \propto \frac{i}{r}. \qquad (34\text{-}3)$$

We can convert this proportionality into an equality by inserting a proportionality constant. As in the case of Coulomb's law, and for similar reasons (see Section 26-4), we do not write this constant simply as, say, k but in a more complex form, namely $\mu_0/2\pi$, in which μ_0 is called the *permeability constant.** Equation 34-3 then becomes

$$B = \frac{\mu_0 i}{2\pi r}, \qquad (34\text{-}4)$$

which we choose to write in the form

$$(B)(2\pi r) = \mu_0 i. \qquad (34\text{-}5)$$

The left side of Eq. 34-5 is $\oint \mathbf{B} \cdot d\mathbf{l}$ for a path consisting of a circle of radius r centered on the wire. For all points on this circle **B** has the same (constant) magnitude B and $d\mathbf{l}$, which is always tangent to the path of integration, points in the same direction as **B**, as Fig. 34-3 shows. Thus

$$\oint \mathbf{B} \cdot d\mathbf{l} = \oint B \, dl = B \oint dl = (B)(2\pi r),$$

$\oint dl$ being simply the circumference of the circle. In this special case, therefore, we can write the experimentally observed connection between the field and the current as

$$\oint \mathbf{B} \cdot d\mathbf{l} = \mu_0 i, \qquad (34\text{-}1)$$

---

* μ_0 has no connection with the dipole moment μ that appears in Eq. 34-2.

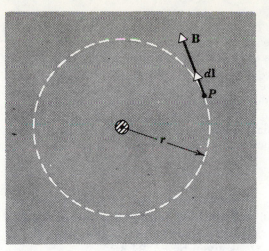

figure 34-3
A circular path of integration surrounding a wire. The central dot suggests a current i in the wire emerging from the page. Note that the angle between $\mathbf{B}$ and $d\mathbf{l}$ is zero so that $\mathbf{B} \cdot d\mathbf{l} = B\, dl$.

which is Ampère's law. A host of other experiments suggests that Eq. 34-1 is true in general for *any* magnetic field configuration, for *any* assembly of currents, and for *any* path of integration.*

In applying Ampère's law in the general case, we construct a *closed linear path* in the magnetic field as shown in Fig. 34-4. This path is divided into elements of length $d\mathbf{l}$, and for each element the quantity $\mathbf{B} \cdot d\mathbf{l}$ is evaluated. Recall that $\mathbf{B} \cdot d\mathbf{l}$ has the magnitude $B\, dl \cos\theta$ and can be interpreted as the product of dl and the component of $\mathbf{B}$ ($= B \cos\theta$) parallel to $d\mathbf{l}$. The integral is the sum of the quantities $\mathbf{B} \cdot d\mathbf{l}$ for all path elements in the complete loop; it is a *line integral* around a closed path. The term i on the right of Eq. 34-1 is the *net* current that passes through the area bounded by the closed path.

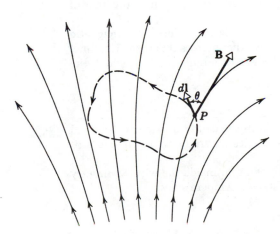

figure 34-4
A path of integration in a magnetic field.

The permeability constant in Ampère's law has an assigned value of

$$\mu_0 = 4\pi \times 10^{-7} \text{ tesla} \cdot \text{meter/ampere}.$$

Both this and the permittivity constant (ϵ_0) occur in electromagnetic formulas when SI units are used.

---

* We must modify Eq. 34-1 if a time-varying electric field or magnetic material are present within the path of integration. In this chapter we assume that if electric fields are present, they are constant in magnitude and direction and that no magnetic material is present.

You may wonder why ϵ_0 in Coulomb's law is a measured quantity, whereas μ_0 in Ampère's law is an assigned quantity. The answer is that the ampere, which is the SI unit for the current i in Ampère's law, is defined by a laboratory technique (the *current balance*) that involves forces exerted by magnetic fields and in which this same constant μ_0 appears. In effect, as we show in detail in Section 34-4, the size of the current that we agree to define as one ampere is adjusted so that μ_0 may have exactly the value assigned to it above. In Coulomb's law, on the other hand, the quantities **F**, q, and r are measured in ways in which the constant ϵ_0 plays no role. This constant must then take on the particular value that makes the left side of Coulomb's law equal to the right side; no arbitrary assignment is possible.

34-2
B NEAR A LONG WIRE

We have seen that the lines of **B** for a long straight cylindrical wire carrying a current i are concentric circles centered on the axis of the wire and that B at a distance r from this axis is given by Eq. 34-4:

$$B = \frac{\mu_0 i}{2\pi r}. \tag{34-4}$$

We may regard this as an experimental result consistent with, and readily derivable from, Ampère's law.

It is interesting to compare Eq. 34-4 with the expression for the electric field near a long line of charge, or

$$E = \frac{1}{2\pi\epsilon_0}\frac{\lambda}{r}. \tag{28-10}$$

In each case there are multiplying constants, namely $\mu_0/2\pi$ and $1/2\pi\epsilon_0$, and factors describing the device responsible for the field, namely i and λ. Finally, each field varies as $1/r$.

Equation 28-10 may be derived from Gauss's law by relating the electric field at a Gaussian surface to the net charge within this surface. The (surface) integral in Gauss's law is evaluated for a closed cylindrical surface to which the lines of **E** are everywhere perpendicular.

Equation 34-4 may be derived from Ampère's law by relating the magnetic field at a path of integration to the net current that pierces this path. The (line) integral in Ampère's law is evaluated for a closed circular path to which the lines of **B** are everywhere tangent.

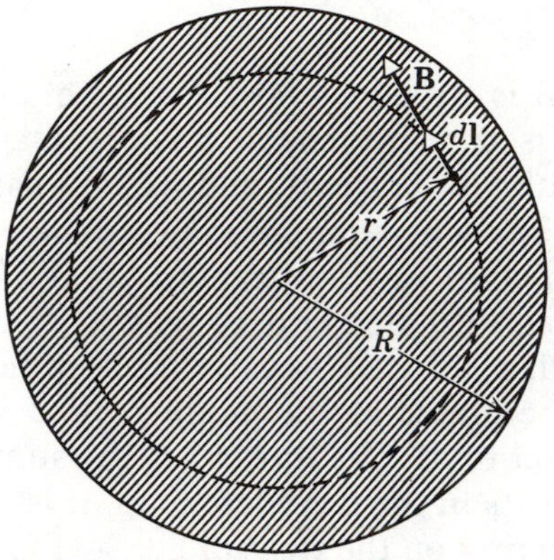

figure 34-5
Example 1. A circular path of integration inside a wire. A current i_0, distributed uniformly over the cross section of the wire, emerges from the page.

EXAMPLE 1

Derive an expression for **B** at a distance r from the center of a long cylindrical wire of radius R, where $r < R$. The wire carries a current i_0, distributed uniformly over the cross section of the wire.

Figure 34-5 shows a circular path of integration inside the wire. Symmetry suggests that **B** is tangent to the path as shown. Ampère's law,

$$\oint \mathbf{B} \cdot d\mathbf{l} = \mu_0 i,$$

gives

$$(B)(2\pi r) = \mu_0 i_0 \frac{\pi r^2}{\pi R^2},$$

since only the fraction of the current that passes through the path of integration is included in the factor i on the right. Solving for B and dropping the subscript on the current yields

$$B = \frac{\mu_0 i r}{2\pi R^2}.$$

At the surface of the wire $(r = R)$ this equation reduces to the same expression as that found by putting $r = R$ in Eq. 34-4 $(B = \mu_0 i/2\pi R)$.

EXAMPLE 2

Figure 34-6 shows a flat strip of copper of width a and negligible thickness carrying a current i. Find the magnetic field **B** at point P, at a distance R from the center of the strip, at right angles to the strip.

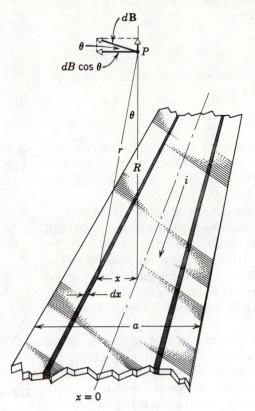

figure 34-6
Example 2. A flat strip of width a carries a current i.

Let us subdivide the strip into long infinitesimal filaments of width dx, each of which may be treated as a wire carrying a current di given by $i(dx/a)$. The field contribution dB at point P in Fig. 34-6 is given, for the element shown, by the differential form of Eq. 34-4, or

$$dB = \frac{\mu_0}{2\pi} \frac{di}{r} = \frac{\mu_0}{2\pi} \frac{i(dx/a)}{R \sec \theta}.$$

Note that the vector $d\mathbf{B}$ is at right angles to the line marked r.

Only the horizontal component of $d\mathbf{B}$, namely $dB \cos \theta$, is effective, the vertical component being canceled by the contribution associated with a symmetrically located filament on the other side of the origin. Thus B at point P is given by the (scalar) integral

$$B = \int dB \cos \theta = \int \frac{\mu_0 i(dx/a)}{2\pi R \sec \theta} \cos \theta$$

$$= \frac{\mu_0 i}{2\pi a R} \int \frac{dx}{\sec^2 \theta}.$$

The variables x and θ are not independent, being related by

$$x = R \tan \theta$$

or

$$dx = R \sec^2 \theta \, d\theta.$$

Bearing in mind that the limits on θ are $\pm \tan^{-1}(a/2R)$ and eliminating dx from this expression for B, we find

$$B = \frac{\mu_0 i}{2\pi a R} \int \frac{R \sec^2 \theta \, d\theta}{\sec^2 \theta}$$

$$= \frac{\mu_0 i}{2\pi a} \int_{-\tan^{-1} a/2R}^{+\tan^{-1} a/2R} d\theta = \frac{\mu_0 i}{\pi a} \tan^{-1} \frac{a}{2R}.$$

At points far from the strip, $a/2R$ is a small angle, for which $\tan^{-1} \alpha \cong \alpha$. Thus we have, as an approximate result,

$$B \cong \frac{\mu_0 i}{\pi a} \left(\frac{a}{2R} \right) = \frac{\mu_0}{2\pi} \frac{i}{R}.$$

This result is expected because at distant points the strip cannot be distinguished from a cylindrical wire (see Eq. 34-4).

Figure 34-7 shows the lines of **B** representing the field of **B** near a long straight wire. Note the increase in the spacing of the lines with increasing distance from the wire. This represents the $1/r$ decrease in B predicted by Eq. 34-4.

Figure 34-8 shows the resultant magnetic lines associated with a current in a wire that is oriented at right angles to a uniform *external* field $\mathbf{B}_e$ directed to the right. At any point the resultant magnetic field **B** will be the vector sum of $\mathbf{B}_e$ and $\mathbf{B}_i$, where $\mathbf{B}_i$ is the magnetic field set up by the current in the wire. The fields $\mathbf{B}_e$ and $\mathbf{B}_i$ tend to cancel above the wire and to reenforce each other below the wire. At point P in Fig. 34-8 $\mathbf{B}_e$ and $\mathbf{B}_i$ cancel exactly. Very near the wire the field is represented by circular lines and is essentially $\mathbf{B}_i$.

Michael Faraday, who originated the concept of magnetic lines, endowed them with more reality than they are currently given. He imagined that, like stretched rubber bands, they represent the site of mechanical forces. On this picture can we not visualize that the wire in Fig. 34-8 will be pushed up? Today we use lines of **B** largely for purposes of visualization. For quantitative calculations we use the field vectors, describing the force on the wire in Fig. 34-8, for example, from the relation $\mathbf{F} = i\mathbf{l} \times \mathbf{B}$.

In applying this relation to Fig. 34-8, we recall that **B** is always the *external field* in which the wire is immersed; that is, it is $\mathbf{B}_e$ and thus points to the right. Since **l** points out of the page, the magnetic force on the wire $(= i\mathbf{l} \times \mathbf{B}_e)$ does indeed point up. It is necessary to use only the external field in such calculations because the field set up by the current in the wire cannot exert a force on the wire, just as the gravitational

34-3
LINES OF **B**

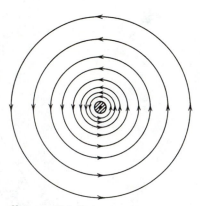

figure 34-7
Lines of **B** near a long, circularly cylindrical wire. A current i, suggested by the central dot, emerges from the page.

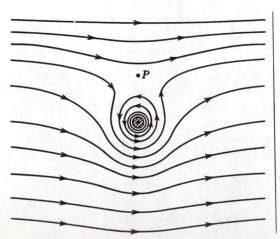

figure 34-8
Lines of **B** near a long current-carrying wire immersed in a uniform external field $\mathbf{B}_e$ that points to the right. The current i is emerging from the page. At point P, $\mathbf{B} = 0$.

field of the earth cannot exert a force on the earth itself but only on another body. In Fig. 34-7, for example, there is no magnetic force on the wire because no external magnetic field is present.

Faraday's idea of lines of **B** was instrumental in overthrowing the older action-at-a-distance theory of magnetic (and electric) attraction. Like many new ideas, it was not immediately accepted. In 1851, for example, Faraday wrote:

> I cannot refrain from again expressing my conviction of the truthfulness of the representation, which the idea of lines of force affords in regard to magnetic action. All the points that are experimentally established in regard to that action—i.e., all that is not hypothetical—appear to be well and truly represented by it.

On the other hand, four years later another well-known British scientist, Sir George Airy, wrote:

> I declare that I can hardly imagine anyone who practically and numerically knows this agreement [with the action-at-a-distance theory] to hesitate an instant in the choice between this simple and precise action, on the one hand, and anything so vague as lines of force, on the other hand.

Students who imagine that scientific pronouncements are absolute can do well to compare these two statements, each made by a distinguished contemporary. Modern examples are not lacking.

34-4
TWO PARALLEL CONDUCTORS

Figure 34-9 shows two long parallel wires separated by a distance d and carrying currents i_a and i_b. It is an experimental fact, noted by Ampère only one week after word of Oersted's experiments reached Paris, that two such conductors attract each other.

Some of Ampère's colleagues thought that in view of Oersted's experiment this attraction between two conductors was an obvious result and did not need to be proved. They reasoned that if wire a and wire b each exert forces on a compass needle they should exert forces on each other. This conclusion is wrong. When he heard it, Arago, a contemporary of Ampère, drew two iron keys from his pocket and replied, "Each of these keys attracts a magnet. Do you believe that they therefore also attract each other?"

Wire a in Fig. 34-9 will produce a magnetic field $\mathbf{B}_a$ at all nearby points. The magnitude of $\mathbf{B}_a$, due to the current i_a, at the site of the second wire is, from Eq. 34-4,

$$B_a = \frac{\mu_0 i_a}{2\pi d}.$$

The right-hand rule shows that the direction of $\mathbf{B}_a$ at wire b is down, as shown in the figure.

Wire b, which carries a current i_b, finds itself immersed in an *external magnetic field* $\mathbf{B}_a$. A length l of this wire will experience a sideways magnetic force (= $i\mathbf{l} \times \mathbf{B}$) whose magnitude is

$$F_b = i_b l B_a = \frac{\mu_0 l i_b i_a}{2\pi d}. \tag{34-6}$$

The vector rule of signs tells us that $\mathbf{F}_b$ lies in the plane of the wires and points to the left in Fig. 34-9.

We could have started with wire b, computed the magnetic field which it produces at the site of wire a, and then computed the force on wire a. The force on wire a would, for parallel currents, point to the right. The forces that the two wires exert on each other are equal and

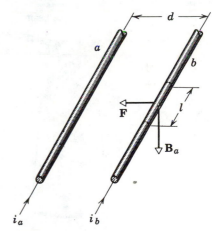

figure 34-9
Two parallel wires that carry parallel currents attract each other.

opposite, as they must be according to Newton's law of action and reaction. For antiparallel currents the two wires repel each other.

This discussion reminds us of our discussion of the electric field between two point charges in Section 27-1. There we saw that the charges act on each other through the intermediary of the electric field. The conductors in Fig. 34-9 act on each other, as we have said earlier, through the intermediary of the magnetic field **B.** We think in terms of

$$\text{current} \rightleftharpoons \text{field (B)} \rightleftharpoons \text{current}$$

and not, as in the action-at-a-distance point of view, in terms of

$$\text{current} \rightleftharpoons \text{current}.$$

The attraction between long parallel wires is used to define the ampere. Suppose that the wires are one meter apart ($d = 1.0$ m, exactly) and that the two currents are equal ($i_a = i_b = i$). If this common current is adjusted until, by measurement, the force of attraction per unit length between the wires is 2×10^{-7} N/m exactly, the current is defined to be one ampere. From Eq. 34-6,

$$\frac{F}{l} = \frac{\mu_0 i^2}{2\pi d} = \frac{(4\pi \times 10^{-7} \text{ T}\cdot\text{m/A})(1 \text{ A})^2}{(2\pi)(1 \text{ m})}$$

$$= 2 \times 10^{-7} \text{ N/m}$$

as expected.*

At the National Bureau of Standards primary measurements of current are made with a *current balance*. This consists of a carefully wound coil placed between two other coils, as in Fig. 34-10. The outer pair of coils is fastened to

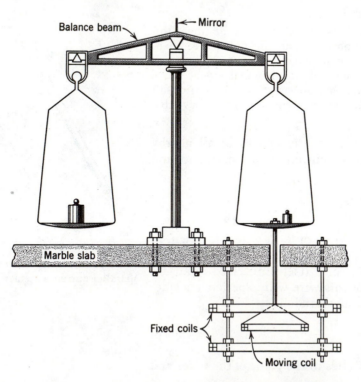

Balance beam — Mirror
Marble slab
Fixed coils
Moving coil

figure 34-10
A current balance.

---

* Note that μ_0 appears in this relation used to define the ampere. As stated on page 749, μ_0 is assigned the (arbitrary) value of $4\pi \times 10^{-7}$ tesla·meter/ampere, and the size of the current that we define as one ampere is adjusted to give the required force of attraction per unit length.

the table, and the inner one is hung from the arm of a balance. The coils are so connected that the current to be measured exists, as a common current, in all three of them.

The coils exert forces on one another—just as the parallel wires of Fig. 34-9 do—which can be measured by loading weights on the balance pan. The current is defined in terms of this measured force and the dimensions of the coils. The current balance is perfectly equivalent to the long parallel wires of Fig. 34-9 but is a much more practical arrangement. Current balance measurements are used primarily to standardize other, more convenient, secondary methods of measuring currents.

EXAMPLE 3

A long horizontal rigidly supported wire carries a current i_a of 100 A. Directly above it and parallel to it is a fine wire that carries a current i_b of 20 A and weighs 0.0050 lb/ft (= 0.073 N/m). How far above the lower wire should this second wire be strung if we hope to support it by magnetic repulsion?

To provide a repulsion, the two currents must point in opposite directions. For equilibrium, the magnetic force per unit length must equal the weight per unit length and must be oppositely directed. Solving Eq. 34-6 for d yields

$$d = \frac{\mu_0 i_a i_b}{2\pi(F/l)} = \frac{(4\pi \times 10^{-7}\ \text{T·m/A})(100\ \text{A})(20\ \text{A})}{(2\pi)(0.073\ \text{N/m})}$$

$$= 5.5 \times 10^{-3}\ \text{m} = 5.5\ \text{mm}.$$

We assume that the wire diameters are much smaller than their separation. This assumption is necessary because in deriving Eq. 34-6 we tacitly assumed that the magnetic field produced by one wire is uniform for all points within the second wire.

Is the equilibrium of the suspended wire stable or unstable against vertical displacements? This can be tested by displacing the wire vertically and examining how the forces on the wire change.

Suppose that the fine wire is suspended *below* the rigidly supported wire. How may it be made to "float"? Is the equilibrium against vertical displacements stable or unstable?

EXAMPLE 4

Two parallel wires a distance d apart carry equal currents i in opposite directions. Find the magnetic field for points between the wires and at a distance x from one wire.

Study of Fig. 34-11 shows that $\mathbf{B}_a$ due to the current i_a and $\mathbf{B}_b$ due to the current i_b point in the same direction at P. Each is given by Eq. 34-4 ($B = \mu_0 i/2\pi r$) so that

$$B = B_a + B_b = \frac{\mu_0 i}{2\pi}\left(\frac{1}{x} + \frac{1}{d-x}\right).$$

This relationship does not hold for points inside the wires because Eq. 34-4 is not valid there.

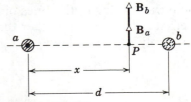

figure 34-11
Example 4

34-5
B *FOR A SOLENOID*

A *solenoid* is a long wire wound in a close-packed helix and carrying a current i. We assume that the helix is very long compared with its diameter. What is the nature of the field of $\mathbf{B}$ that is set up?

For points close to a single turn of the solenoid, the observer is not aware that the wire is bent in an arc. The wire behaves magnetically almost like a long straight wire, and the lines of $\mathbf{B}$ due to this single turn are almost concentric circles.

The solenoid field is the vector sum of the fields set up by all the

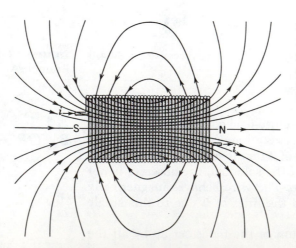

figure 34-12
A loosely wound solenoid.

turns that make up the solenoid. Figure 34-12, which shows a "solenoid" with widely spaced turns, suggests that the fields tend to cancel between the wires. It also suggests that, at points inside the solenoid and reasonably far from the wires, **B** is parallel to the solenoid axis. In the limiting case of adjacent square tightly packed wires, the solenoid becomes essentially a cylindrical current sheet and the requirements of symmetry then make the statement just given necessarily true. We assume that it is true in what follows.

For points such as P in Fig. 34-12 the field set up by the upper part of the solenoid turns (marked $\odot$) points to the left and tends to cancel the field set up by the lower part of the solenoid turns (marked $\otimes$), which points to the right. As the solenoid becomes more and more ideal, that is, as it approaches the configuration of an infinitely long cylindrical current sheet, the field **B** at outside points approaches zero. Taking the external field to be zero is not a bad assumption for a practical solenoid if its length is much greater than its diameter and if we consider only external points near the central region of the solenoid, that is, away from the ends. Figure 34-13 shows the lines of **B** for a real solenoid, which is far from ideal in that the length is not much greater than the diameter. Even here the spacing of the lines of **B** in the central plane shows that the external field is much weaker than the internal field.

Let us apply Ampère's law,

figure 34-13
A solenoid of finite length. The right end, from which lines of **B** emerge, behaves like the north pole of a compass needle. The left end behaves like the south pole.

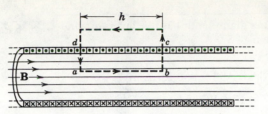

figure 34-14

A section of an ideal solenoid, made of adjacent square turns, equivalent to an infinitely long cylindrical current sheet.

$$\oint \mathbf{B} \cdot d\mathbf{l} = \mu_0 i,$$

to the rectangular path *abcd* in the ideal solenoid of Fig. 34-14. We write the integral $\oint \mathbf{B} \cdot d\mathbf{l}$ as the sum of four integrals, one for each path segment:

$$\oint \mathbf{B} \cdot d\mathbf{l} = \int_a^b \mathbf{B} \cdot d\mathbf{l} + \int_b^c \mathbf{B} \cdot d\mathbf{l} + \int_c^d \mathbf{B} \cdot d\mathbf{l} + \int_d^a \mathbf{B} \cdot d\mathbf{l}.$$

The first integral on the right is Bh, where B is the magnitude of $\mathbf{B}$ inside the solenoid and h is the arbitrary length of the path from a to b. Note that path ab, though parallel to the solenoid axis, need not coincide with it.

The second and fourth integrals are zero because for every element of these paths $\mathbf{B}$ is at right angles to the path. This makes $\mathbf{B} \cdot d\mathbf{l}$ zero and thus the integrals are zero. The third integral, which includes the part of the rectangle that lies outside the solenoid, is zero because we have taken $\mathbf{B}$ as zero for all external points for an ideal solenoid.

Thus $\oint \mathbf{B} \cdot d\mathbf{l}$ for the entire rectangular path has the value Bh. The net current i that passes through the area bounded by the path of integration is not the same as the current i_0 in the solenoid because the path of integration encloses more than one turn. Let n be the number of turns per unit length; then

$$i = i_0(nh).$$

Ampère's law then becomes

$$Bh = \mu_0 i_0 n h$$

or

$$B = \mu_0 i_0 n. \qquad (34\text{-}7)$$

Although we derived Eq. 34-7 for an infinitely long ideal solenoid, it holds quite well for actual solenoids for internal points near the center of the solenoid. It shows that B does not depend on the diameter or the length of the solenoid and that B is constant over the solenoid cross section. A solenoid is a practical way to set up a known uniform magnetic field for experimentation, just as a parallel-plate capacitor is a practical way to set up a known uniform electric field.

The solenoid provides a good context in which to discuss Φ_B, the *flux of the magnetic field* $\mathbf{B}$. We discussed the flux Φ_E of the electric field $\mathbf{E}$ in Section 28-3, restricting ourselves largely, for reasons having to do with Gauss's law, to the flux for closed surfaces. We did, however, discuss the flux Φ_E for open surfaces; see Fig. 28-1.

We can, in a similar way, define the flux Φ_B for a magnetic field $\mathbf{B}$, for either a closed or an open surface. In either case it is given by

$$\Phi_B = \int \mathbf{B} \cdot d\mathbf{S},$$

in strict analogy to the discussion of Φ_E in Section 28-3. The SI unit of $\mathbf{B}$, as we have seen, is the *tesla* (1 tesla = 1 weber/meter²) while that of Φ_B is simply the *weber* (abbr. Wb).

EXAMPLE 5

A solenoid is 1.0 m long and 3.0 cm in inner diameter. It has five layers of windings of 850 turns each and carries a current of 5.0 A.

(a) What is B at its center? From Eq. 34-7,

$$B = \mu_0 i_0 n = (4\pi \times 10^{-7} \text{ T·m/A})(5.0 \text{ A})(5 \times 850 \text{ turns/m})$$

$$= 2.7 \times 10^{-2} \text{ T} = 2.7 \times 10^{-2} \text{ Wb/m}^2.$$

We can use Eq. 34-7 even if the solenoid has more than one layer of windings because the diameter of the windings does not enter.

(b) What is the magnetic flux Φ_B for a cross section of the solenoid at its center? To the extent that **B** is constant, we can calculate the flux from

$$\Phi_B = \int \mathbf{B} \cdot d\mathbf{S} = BA,$$

where A is the effective cross-sectional area. Let us take A as the area of a circular disk whose diameter is the inner diameter of the windings (3.0 cm). The effective area can then be shown to be 7.1×10^{-4} m², and

$$\Phi_B = BA = (2.7 \times 10^{-2} \text{ Wb/m}^2)(7.1 \times 10^{-4} \text{ m}^2)$$

$$= 1.9 \times 10^{-5} \text{ Wb}.$$

EXAMPLE 6

A toroid. Figure 34-15 shows a toroid, which we may describe as a solenoid bent into the shape of a doughnut. Calculate **B** at interior points.

From symmetry the lines of **B** form concentric circles inside the toroid, as shown in the figure. Let us apply Ampère's law to the circular path of integration of radius r:

$$\oint \mathbf{B} \cdot d\mathbf{l} = \mu_0 i$$

or

$$(B)(2\pi r) = \mu_0 i_0 N,$$

where i_0 is the current in the toroid windings and N is the total number of turns. This gives

$$B = \frac{\mu_0}{2\pi} \frac{i_0 N}{r}.$$

In contrast to the solenoid, B is not constant over the cross section of a toroid. Show from Ampère's law that B equals zero for points outside an ideal toroid.

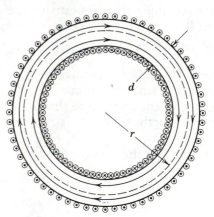

figure 34-15
Example 6. A toroid.

34-6
THE BIOT-SAVART LAW*

We can use Ampère's law to calculate magnetic fields only if the symmetry of the current distribution is high enough to permit the easy evaluation of the line integral $\oint \mathbf{B} \cdot d\mathbf{l}$. This requirement limits the usefulness of the law in practical problems. The law does not fail; it simply becomes difficult to apply in a useful way.

Similarly, in electrostatics, we use Gauss's law to calculate electric fields only if the symmetry of the charge distribution is high enough to permit the easy evaluation of the surface integral $\oint \mathbf{E} \cdot d\mathbf{S}$. We can, for example, use Gauss's law to find the electric field due to a long uni-

---

* See "Ampère as a Contemporary Physicist" by R. A. Tricker, *Contemporary Physics*, August 1962; and "Electromagnetism as a Second Order Effect. III: The Biot-Savart Law" by W. G. V. Rosser, *Contemporary Physics*, October 1961, for more information about Ampère's law and the law of Biot and Savart.

formly charged rod but we cannot apply it usefully to an electric dipole, for the symmetry is not high enough in this case.

To compute **E** at a given point for an *arbitrary* charge distribution, we divided the distribution into *charge elements dq* and (see Section 27-4) we used Coulomb's law to calculate the field contribution *d***E** due to each element at the point in question. We found the field **E** at that point by adding, that is, by integrating, the field contributions *d***E** for the entire distribution.

We now describe a similar procedure for computing **B** at any point due to an arbitrary current distribution. We divide the current distribution into *current elements* and, using the law of Biot and Savart (which we describe below), we calculate the field contribution *d***B** due to each current element at the point in question. We find the field **B** at that point by integrating the field contributions for the entire distribution.

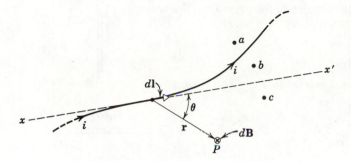

figure 34-16
The current element *d*l establishes a magnetic field contribution *d***B** at point *P*.

Figure 34-16 shows an arbitrary current distribution consisting of a current *i* in a curved wire. The figure also shows a typical current element; it is a length *d*l of the conductor carrying a current *i*. Its direction is that of the tangent to the conductor (dashed line). A current element cannot exist as an isolated entity because a way must be provided to lead the current into the element at one end and out of it at the other. Nevertheless, we can think of an actual circuit as made up of a large number of current elements placed end to end.

Let *P* be the point at which we want to know the magnetic field *d***B** associated with the current element. According to the Biot-Savart law, *d***B** is given in magnitude by

$$dB = \frac{\mu_0 i}{4\pi} \frac{dl \, \sin\theta}{r^2}, \qquad (34\text{-}8)$$

where **r** is a displacement vector from the element to *P* and θ is the angle between this vector and *d***l**. The direction of *d***B** is that of the vector *d***l** × **r**. In Fig. 34-16, for example, *d***B** at point *P* for the current element shown is directed into the page at right angles to the plane of the figure. Note that Eq. 34-8, being an inverse square law that describes the magnetic field due to a current element, may be viewed as the magnetic equivalent of Coulomb's law, which is an inverse square law that describes the electric field due to a charge element.

We may write the law of Biot and Savart in vector form as

$$d\mathbf{B} = \frac{\mu_0 i}{4\pi} \frac{d\mathbf{l} \times \mathbf{r}}{r^3}. \qquad (34\text{-}9)$$

This formulation reduces to that of Eq. 34-8 when we express it in terms of magnitudes; it also gives complete information about the direction of *d***B**, namely, that it is the same as the direction of the vector *d***l** × **r**.

The resultant field at P is found by integrating Eq. 34-9, or

$$\mathbf{B} = \int d\mathbf{B}, \qquad (34\text{-}10)$$

where the integral is a vector integral.

A long straight wire. We illustrate the law of Biot and Savart by applying it to find $\mathbf{B}$ due to a current i in a long straight wire. We discussed this problem at length in connection with Ampère's law in Section 34-1.

Figure 34-17, a side view of the wire, shows a typical current element $d\mathbf{x}$. The magnitude of the contribution $d\mathbf{B}$ of this element to the magnetic field at P is found from Eq. 34-8, or

$$dB = \frac{\mu_0 i}{4\pi} \frac{dx \sin \theta}{r^2}.$$

The directions of the contributions $d\mathbf{B}$ at point P for all elements are the same, namely, into the plane of the figure at right angles to the page. Thus the vector integral of Eq. 34-10 reduces to a scalar integral, or

$$B = \int dB = \frac{\mu_0 i}{4\pi} \int_{x=-\infty}^{x=+\infty} \frac{\sin \theta \, dx}{r^2}.$$

Now, x, θ, and r are not independent, being related (see Fig. 34-17) by

$$r = \sqrt{x^2 + R^2}$$

and

$$\sin \theta \; [= \sin (\pi - \theta)] = \frac{R}{\sqrt{x^2 + R^2}},$$

so that the expression for B becomes

$$B = \frac{\mu_0 i}{4\pi} \int_{-\infty}^{+\infty} \frac{R \, dx}{(x^2 + R^2)^{3/2}}$$

$$= \frac{\mu_0 i}{4\pi R} \left. \frac{x}{(x^2 + R^2)^{1/2}} \right|_{x=-\infty}^{x=+\infty}$$

$$= \frac{\mu_0}{2\pi} \frac{i}{R}.$$

This is the result that we arrived at earlier for this problem (see Eq. 34-4). The law of Biot and Savart will always yield results that are consistent with Ampère's law and with experiment.

This problem reminds us of its electrostatic equivalent. We derived an expression for $\mathbf{E}$ due to a long charged rod, using Gauss's law (Section 28-8); we also solved this problem by integration methods, using Coulomb's law (Section 27-4).

EXAMPLE 7

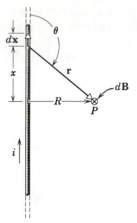

figure 34-17
Example 7

EXAMPLE 8

A circular current loop. Figure 34-18 shows a circular loop of radius R carrying a current i. Calculate $\mathbf{B}$ for points on the axis.

The vector $d\mathbf{l}$ for a current element at the top of the loop points perpendicularly out of the page. The angle θ between $d\mathbf{l}$ and $\mathbf{r}$ is $90°$, and the plane formed by $d\mathbf{l}$ and $\mathbf{r}$ is normal to the page. The vector $d\mathbf{B}$ for this element is at right angles to this plane and thus lies in the plane of the figure and at right angles to $\mathbf{r}$, as the figure shows.

Let us resolve $d\mathbf{B}$ into two components, one, $d\mathbf{B}_{\parallel}$, along the axis of the loop and another, $d\mathbf{B}_{\perp}$, at right angles to the axis. Only $d\mathbf{B}_{\parallel}$ contributes to the total magnetic field $\mathbf{B}$ at point P. This follows because the components $d\mathbf{B}_{\parallel}$ for all current elements lie on the axis and add directly; however, the components $d\mathbf{B}_{\perp}$ point in different directions perpendicular to the axis, and their result for the complete loop is zero, from symmetry. Thus

$$B = \int dB_{\parallel},$$

where the integral is a simple scalar integration over the current elements.

For the current element shown in Fig. 34-18 we have, from the Biot-Savart law (Eq. 34-8),

$$dB = \frac{\mu_0 i}{4\pi} \frac{dl \, \sin 90°}{r^2}.$$

We also have

$$dB_{\parallel} = dB \cos \alpha.$$

Combining gives

$$dB_{\parallel} = \frac{\mu_0 i \cos \alpha \, dl}{4\pi r^2}.$$

Figure 34-18 shows that r and α are not independent of each other. Let us express each in terms of a new variable x, the distance from the center of the loop to the point P. The relationships are

$$r = \sqrt{R^2 + x^2}$$

and

$$\cos \alpha = \frac{R}{r} = \frac{R}{\sqrt{R^2 + x^2}}.$$

Substituting these values into the expression for $dB_{\parallel}$ gives

$$dB_{\parallel} = \frac{\mu_0 i R}{4\pi (R^2 + x^2)^{3/2}} \, dl.$$

Note that i, R, and x have the same values for all current elements. Integrating this equation, noting that $\int dl$ is simply the circumference of the loop $(= 2\pi R)$, yields

$$B = \int dB_{\parallel} = \frac{\mu_0 i R}{4\pi (R^2 + x^2)^{3/2}} \int dl$$

$$= \frac{\mu_0 i R^2}{2(R^2 + x^2)^{3/2}}. \tag{34-11}$$

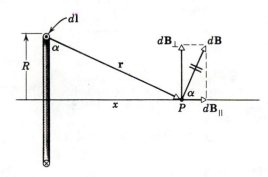

figure 34-18
Example 8. A ring of radius R carrying a current i.

If we put $x \gg R$ in Example 8 so that points close to the loop are not considered, Eq. 34-11 reduces to

$$B = \frac{\mu_0 i R^2}{2x^3}.$$

Recalling that πR^2 is the area A of the loop and considering loops with N turns, we can write this equation as

$$B = \frac{\mu_0}{2\pi} \frac{(NiA)}{x^3} = \frac{\mu_0}{2\pi} \frac{\mu}{x^3},$$

where μ is the *magnetic dipole moment* of the current loop. This reminds us of the result derived in Problem 25, Chapter 27 [$E = (1/2\pi\epsilon_0)(p/x^3)$], which is the formula for the *electric* field on the axis of an *electric* dipole.

Thus we have shown in two ways that we can regard a current loop as a magnetic dipole: It experiences a torque given by $\tau = \mu \times B$ when we place it in an *external* magnetic field (Eq. 33-11); it generates its own magnetic field given, for points on the axis, by the equation just developed.

Table 34-1 is a summary of the properties of electric and magnetic dipoles.

Table 34-1
Some dipole equations

Property	Dipole Type	Equation
Torque in an external field	electric	$\tau = p \times E$
	magnetic	$\tau = \mu \times B$
Energy in an external field	electric	$U = -p \cdot E$
	magnetic	$U = -\mu \cdot B$
Field at distant points along axis	electric	$E = \dfrac{1}{2\pi\epsilon_0} \dfrac{p}{x^3}$
	magnetic	$B = \dfrac{\mu_0}{2\pi} \dfrac{\mu}{x^3}$
Field at distant points along perpendicular bisector	electric	$E = \dfrac{1}{4\pi\epsilon_0} \dfrac{p}{x^3}$
	magnetic	$B = \dfrac{\mu_0}{4\pi} \dfrac{\mu}{x^3}$

EXAMPLE 9

In the Bohr model of the hydrogen atom the electron circulates around the nucleus in a path of radius 5.3×10^{-11} m at a frequency ν of 6.5×10^{15} Hz ($=$ rev/s). (a) What value of B is set up at the center of the orbit?

The current is the rate at which charge passes any point on the orbit and is given by

$$i = e\nu = (1.6 \times 10^{-19} \text{ C})(6.5 \times 10^{15} \text{ Hz}) = 1.0 \times 10^{-3} \text{ A}.$$

B at the center of the orbit is given by Eq. 34-11 with $x = 0$, or

$$B = \frac{\mu_0 i R^2}{2(R^2 + x^2)^{3/2}} = \frac{\mu_0 i}{2R}$$

$$= \frac{(4\pi \times 10^{-7} \text{ T} \cdot \text{m/A})(1.0 \times 10^{-3} \text{ A})}{(2)(5.3 \times 10^{-11} \text{ m})}$$

$$= 12 \text{ T}$$

(b) What is the equivalent magnetic dipole moment? From Eq. 33-10,

$$\mu = NiA = (1)(1.0 \times 10^{-3} \text{ A})(\pi)(5.3 \times 10^{-11} \text{ m})^2$$

$$= 8.8 \times 10^{-24} \text{ A} \cdot \text{m}^2.$$

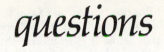

1. Can the path of integration around which we apply Ampère's law pass *through* a conductor?

2. Suppose we set up a path of integration around a cable that contains twelve wires with different currents (some in opposite directions) in each wire. How do we calculate i in Ampère's law in such a case?

3. Apply Ampère's law qualitatively to the three paths shown in Fig. 34-19.

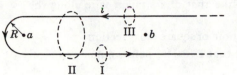

figure 34-19
Question 3,
Problem 27

4. Discuss and compare Gauss's law and Ampère's law.

5. (a) Must Ampère's law (Eq. 34-1) be applied always around a *closed* loop? Can we, for example, apply it to a semicircular arc? (b) Can we apply Ampère's law to a closed *surface?*

6. Give details of three ways in which you can measure the magnetic field **B** at a point P, a perpendicular distance r from a long straight wire carrying a constant current i. Base them on: (a) firing a particle of charge q through point P with velocity **v**, parallel, say, to the wire; (b) measuring the force per unit length exerted on a second wire, parallel to the first wire and carrying a current i'; (c) measuring the torque exerted on a small magnetic dipole located a perpendicular distance r from the wire.

7. A current is set up in a long copper pipe. Is there a magnetic field (a) inside and (b) outside the pipe?

8. Equation 34-4 ($B = \mu_0 i/2\pi r$) suggests that a strong magnetic field is set up at points near a long wire carrying a current. Since there is a current i and a magnetic field **B,** why is there not a force on the wire in accord with the equation $\mathbf{F} = i\mathbf{l} \times \mathbf{B}$?

9. A beam of 20-MeV protons emerges from a cyclotron-like device. Is a magnetic field associated with these particles?

10. Does it necessarily follow from symmetry arguments alone that the lines of **B** around a long straight wire carrying a current i must be concentric circles?

11. A long straight wire of radius R carries a steady current i. In what sense, if any, does the magnetic field generated by this current depend on R?

12. A long straight wire carries a constant current i. Does Ampère's law (Eq. 34-1) hold for (a) a path of integration that encloses the wire but is not circular, (b) a path of integration that does not enclose the wire, and (c) a path that encloses the wire but does not all lie in one plane? Discuss.

13. Two long straight wires pass near one another at right angles. If the wires are free to move, describe what happens when currents are sent through them.

14. Is **B** constant in magnitude for points that lie on a given magnetic field line?

15. In electronics, wires that carry equal but opposite currents are often twisted together to reduce their magnetic effect at distant points. Why is this effective?

16. Two long parallel conductors carry equal currents i in the same direction. Sketch roughly the resultant lines of **B** due to the action of *both* currents. Does your figure suggest an attraction between the wires (in the same sense that Fig. 34-8 suggests an upward force on the wire in that figure)?

17. In Fig. 34-8 explain the relation between the figure and Eq. 33-6a ($\mathbf{F} = i\mathbf{l} \times \mathbf{B}$).

18. Test the "floating" wire of Example 3 for equilibrium under *horizontal* displacements. Consider that the wire floats above the rigidly supported wire and also below it. Summarize the equilibrium situation for both wire positions and for both vertical and horizontal displacements.

19. A current is sent through a vertical spring from whose lower end a weight is hanging; what will happen?

20. Comment on this statement: "The magnetic field **B** outside a long solenoid cannot be zero, if only for the reason that the helical nature of the windings produces a field for external points like that of a straight wire along the solenoid axis."

21. Does Eq. 34-7 $(B = \mu_0 in)$ hold for a solenoid of square cross section?

22. In your own words convince yourself that $\mathbf{B} = 0$ for points outside an ideal solenoid, such as that of Fig. 34-14.

23. What is the direction of the magnetic fields at points a, b, and c in Fig. 34-16 set up by the particular current element shown?

24. In a circular loop of wire carrying a current i, is **B** uniform for all points within the loop?

25. Discuss analogies and differences between Coulomb's law and the Biot-Savart law.

26. Equation 34-9 gives the law of Biot and Savart in vector form. Write its electrostatic equivalent [that is, Eq. 27-6, or $dE = dq/(4\pi\epsilon_0 r^2)$] in vector form.

27. How might you measure the magnetic dipole moment of a compass needle?

28. What is the basis for saying that a current loop is a magnetic dipole?

29. A circular loop of wire lies on the floor of the room in which you are sitting and it carries a constant current i in a clockwise direction, as viewed from above. What is the direction of the magnetic dipole moment of this current loop?

30. As an exercise in vector representation, contrast and compare Fig. 18-12 which deals with fluid flow, with Fig. 34-7, which deals with the magnetic field. How strong an analogy can you make?

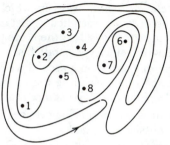

figure 34-20
Problem 1

problems

SECTION 34-1

1. Eight wires cut the page perpendicularly at the points shown in Fig. 34-20. A wire labeled with the integer k $(k = 1, 2, \ldots 8)$ bears the current ki_0. For those with odd k, the current flows up out of the page; for those with even k it flows down into the page. Evaluate $\oint \mathbf{B} \cdot \mathbf{dl}$ along the closed path shown in the direction indicated by the arrowhead.
   Answer: $-10\mu_0 i_0$ (Why the minus sign?).

2. Show that it is impossible for a uniform magnetic field **B** to drop abruptly to zero as one moves at right angles to it, as suggested by the horizontal arrow in Fig. 34-21 (see point a). In actual magnets fringing of the lines of **B** always occurs, which means that **B** approaches zero in a continuous and gradual way. (Hint: Apply Ampère's law to the rectangular path shown by the dashed lines.)

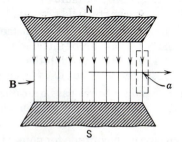

figure 34-21
Problem 2

SECTION 34-2

3. A #10 bare copper wire (0.10 in. in diameter) can carry a current of 50 A without overheating. For this current, what is B at the surface of the wire?
   Answer: 7.9×10^{-3} T.

4. A surveyor is using a compass 20 ft below a power line in which there is a steady current of 100 A. Will this interfere seriously with the compass reading? The horizontal component of the earth's magnetic field at the site is 0.20 gauss.

5. If a point charge of magnitude $+q$ and speed v is a distance d from the axis of a long straight wire carrying a current i and is traveling perpendicular to

the axis of the wire, what are the direction and magnitude of the force acting on it if the charge is moving (a) toward, or (b) away from the wire?

Answer: (a) $\dfrac{\mu_0}{2\pi}\dfrac{qvi}{r}$, antiparallel to *i*. (b) Same magnitude, parallel to *i*.

6. A long straight wire carries a current of 50 A. An electron, traveling at 1.0×10^7 m/s, is 5.0 cm from the wire. What force acts on the electron if the electron velocity is directed (a) toward the wire, (b) parallel to the wire, and (c) at right angles to the directions defined by (a) and (b)?

7. A long solid cylindrical copper wire of radius R carries a current *i* distributed uniformly over the cross section of the wire. Sketch roughly the magnitude of the magnetic field **B** as a function of the distance *r* from the axis of the wire for (a) $r < R$ and (b) $r > R$.

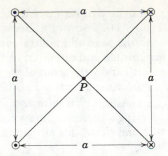

figure 34-22
Problem 8

8. Four long copper wires are parallel to each other, their cross section forming a square 20 cm on edge. A 20-A current is set up in each wire in the direction shown in Fig. 34-22. What are the magnitude and direction of **B** at the center of the square?

9. A long coaxial cable consists of two concentric conductors with the dimensions shown in Fig. 34-23. There are equal and opposite currents *i* in the conductors. (a) Find the magnetic field B at *r* within the inner conductor ($r < a$). (b) Find B between the two conductors ($a < r < b$). (c) Find B within the outer conductor ($b < r < c$). (d) Find B outside the cable ($r > c$).

Answer: (a) $\dfrac{\mu_0 ir}{2\pi a^2}$. (b) $\dfrac{\mu_0 i}{2\pi r}$. (c) $\dfrac{\mu_0 i}{2\pi r}\left(\dfrac{c^2 - r^2}{c^2 - b^2}\right)$. (d) Zero.

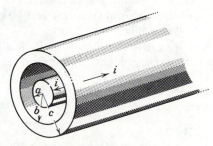

figure 34-23
Problem 9

10. Two long wires a distance *d* apart carry equal antiparallel currents *i*, as in Fig. 34-24. (a) Show that B at point P, which is equidistant from the wires, is given by

$$B = \frac{2\mu_0 i d}{\pi(4R^2 + d^2)}.$$

(b) In what direction does **B** point?

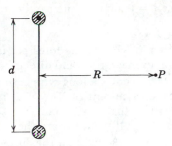

figure 34-24
Problem 10

11. Figure 34-25 shows a hollow cylindrical conductor of radii *a* and *b* which carries a current *i* uniformly spread over its cross section. (a) Show that the magnetic field B for points inside the body of the conductor (that is, $a < r < b$) is given by

$$B = \frac{\mu_0 i}{2\pi(b^2 - a^2)}\frac{r^2 - a^2}{r}.$$

Check this formula for the limiting case of $a = 0$. (b) Make a rough plot of the general behavior of $B(r)$ from $r = 0$ to $r \to \infty$.

figure 34-25
Problem 11

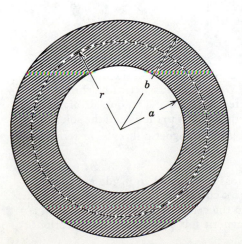

12. Two long straight wires a distance d (10 cm) apart each carry a current i (100 A). Figure 34-26 shows a cross section, with the wires running perpendicular to the page and point P lying on the perpendicular bisector of d. Find the magnitude and direction of the magnetic field at P when the current in the left-hand wire is out of the page and the current in the right-hand wire is (a) in the same direction and (b) in the opposite direction.

13. A conductor consists of an infinite number of adjacent wires, each infinitely long and carrying a current i. Show that the lines of **B** will be as represented in Fig. 34-27 and that B for all points in front of the infinite current sheet will be given by

$$B = \tfrac{1}{2}\mu_0 n i,$$

where n is the number of conductors per unit length. Derive both by direct application of Ampère's law and by considering the problem as a limiting case of Example 2.

14. A long straight conductor has a circular cross section of radius R and carries a current i. Inside the conductor there is a cylindrical hole of radius a whose axis is parallel to the conductor axis at a distance b from it. Use superposition ideas and obtain an expression for the magnetic field **B** inside the hole.

SECTION 34-3

15. A long wire carrying a current of 100 A is placed in a uniform external magnetic field of 50 gauss. The wire is at right angles to this external field. Locate the points at which the *resultant* magnetic field is zero.
Answer: $B = 0$ along a line parallel to the wire and 4.0 mm from it.

SECTION 34-4

16. In Problem 8 what is the force per unit length (N/m) acting on the lower left wire, in magnitude and direction?

17. Figure 34-28 shows a long wire carrying a current of 30 A. The rectangular loop carries a current of 20 A. Calculate the resultant force acting on the loop. Assume that $a = 1.0$ cm, $b = 8.0$ cm, and $l = 30$ cm.
Answer: 3.2×10^{-3} N, toward the wire.

18. Suppose, in Fig. 34-22, that the currents are all in the same direction. What is the force per unit length (N/m, magnitude and direction) on any one wire? In the analogous case of parallel motion of charged particles in a plasma, this is known as the pinch effect.

19. Two long parallel wires of negligible radius are a distance d apart. Let there be a current i in each wire (a) in the same direction and (b) in opposite directions. Letting r be the perpendicular distance from the center of one wire, find the magnitude B of the magnetic field in the region between the wires at points in the plane of the two wires.
Answer: (a) $\dfrac{\mu_0 i}{2\pi}\left[\dfrac{d-2r}{r(d-r)}\right]$. (b) $\dfrac{\mu_0 i}{2\pi}\left[\dfrac{d}{r(d-r)}\right]$.

SECTION 34-5

20. A 200-turn solenoid having a length of 25 cm and a diameter of 10 cm carries a current of 0.30 A. (a) What is the magnitude of the magnetic field **B** near the center of the solenoid? (b) What is the magnetic flux through an annular ring having an inside diameter of 2.0 cm and an outside diameter of 8.0 cm if the plane of the ring is perpendicular to the axis of the solenoid?

21. Express (a) the magnetic field **B** and (b) the magnetic flux Φ_B in terms of mass, length, time, and charge (M, L, T, Q).
Answer: (a) M/QT. (b) ML^2/QT.

22. A toroid having a 5.0 cm × 5.0 cm cross section and an inside radius of 15 cm has 500 turns of wire and carries a current of 0.80 A. (a) What is the magnitude of the magnetic field **B** in the center of the toroid (that is, at a radius of 17.5 cm)? (b) What is the magnetic flux through the cross section?

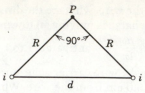

figure 34-26
Problem 12

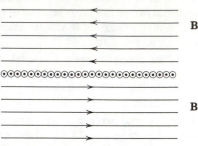

figure 34-27
Problem 13

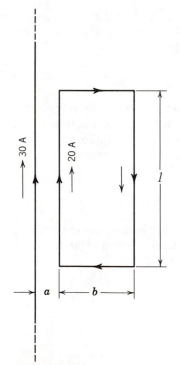

figure 34-28
Problem 17

23. A long straight wire of radius a carries a constant current i (a) Consider a concentric hypothetical circle of radius $2a$, its plane being at right angles to the wire. What magnetic flux Φ_B passes through this circle? (b) If we were to double the current i, what do you imagine would happen to this flux? *Answer:* (a) Zero. (b) Still zero; no lines of **B** pierce the circle in either case.

24. Derive the solenoid equation (Eq. 34-7) starting from the expression for the field on the axis of a circular loop (Example 8). (Hint: Subdivide the solenoid into a series of current loops of infinitesimal thickness and integrate.)

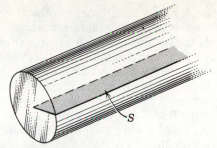

figure 34-29
Problem 25

25. A long copper wire carries a current of 10 A. Calculate the magnetic flux per meter of wire through a plane surface S inside the wire, as in Fig. 34-29. *Answer:* 1.0×10^{-6} Wb/m.

26. Two long, parallel #10 copper wires (diameter = 0.10 in.) carry currents of 10 A in opposite directions. (a) If their centers are 2.0 cm apart, calculate the flux per meter of wire length that exists in the space between the axes of the wires. (b) What fraction of the flux in (a) lies inside the wires? (c) Repeat the calculation of (a) for parallel currents.

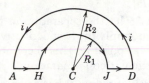

figure 34-30
Problem 28

SECTION 34-6

27. A long "hairpin" is formed by bending a piece of wire as shown in Fig. 34-19. If a 10-A current is set up, what are the direction and magnitude of **B** at point a? At point b? Take $R = 0.50$ cm. *Answer:* (a) 1.0×10^{-3} T, out of figure. (b) 8.0×10^{-4} T, out of figure.

28. Use the Biot-Savart law to calculate the magnetic field **B** at C, the common center of the semicircular arcs AD and HJ, of radii R_2 and R_1 respectively, forming part of the circuit $ADJHA$ carrying current i, as shown in Fig. 34-30.

29. Consider the circuit of Fig. 34-31. The curved segments are part of circles of radii a and b. The straight segments are along the radii. Find the magnetic field **B** at P, assuming a current i in the circuit.

Answer: $B = \dfrac{\mu_0 i \theta}{4\pi}\left(\dfrac{1}{b} - \dfrac{1}{a}\right)$, out of page.

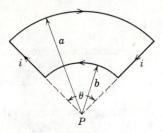

figure 34-31
Problem 29

30. The wire shown in Fig. 34-32 carries a current i. What is the magnetic field **B** at the center C of the semicircle arising from (a) each straight segment of length l, (b) the semicircular segment of radius R, and (c) the entire wire?

31. A straight conductor is split into identical semicircular turns as shown in Fig. 34-33. What is the magnetic field at the center C of the circular loop so formed? *Answer:* Zero.

32. A circular loop of radius 10 cm carries a current of 15 A. At its center is placed a second loop of radius 1.0 cm, having 50 turns and a current of 1.0 A. (a) What magnetic field **B** does the large loop set up at its center? (b) What torque acts on the small loop? Assume that the planes of the two loops are at right angles and that **B** provided by the large loop is essentially uniform throughout the volume occupied by the small loop.

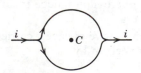

figure 34-32
Problem 30

33. You are given a closed circuit with radii a and b, as shown in Fig. 34-34, carrying a current i. (a) What are the magnitude and direction of **B** at point P? (b) Find the dipole moment of the circuit.

Answer: (a) $\dfrac{\mu_0 i}{4}\left(\dfrac{1}{a} + \dfrac{1}{b}\right)$, into the page. (b) $\dfrac{i\pi}{2}(a^2 + b^2)$, into the page.

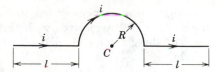

figure 34-33
Problem 31

figure 34-34
Problem 33

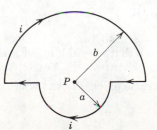

34. (a) A long wire is bent into the shape shown in Fig. 34-35, without cross-contact at P. The radius of the circular section is R. Determine the magnitude and direction of **B** at the center C of the circular portion when the current i flows as indicated. (b) The circular part of the wire is rotated without distortion about its (dashed) diameter perpendicular to the straight portion of the wire. The magnetic moment associated with the circular loop is now in the direction of the current in the straight part of the wire. Determine **B** at C in this case.

35. *Helmholtz coils.* Two 300-turn coils are arranged a distance apart equal to their radius, as in Fig. 34-36. For $R = 5.0$ cm and $i = 50$ A, plot B as a function of distance x along the common axis over the range $x = -5$ cm to $x = +5$ cm, taking $x = 0$ at point P. (Such coils provide an especially uniform field of B near point P.)

36. In Problem 35 let the separation of the coils be a variable z. Show that if $z = R$, then not only the first derivative (dB/dx) but also the second (d^2B/dx^2) of B is zero at point P. This accounts for the uniformity of B near point P for this particular coil separation.

37. A plastic disk of radius R has a charge q uniformly distributed over its surface. If the disk is rotated at an angular frequency ω about its axis, show that (a) the magnetic field at the center of the disk is

$$B = \frac{\mu_0 \omega q}{2\pi R}$$

and (b) the magnetic dipole moment of the disk is

$$\mu = \frac{\omega q R^2}{4}.$$

(Hint: The rotating disk is equivalent to an array of current loops; see Example 8.)

38. A messy loop of limp wire is placed on a table and anchored at points a and b as shown in Fig. 34-37. If a current i is now passed through the wire, will it try to form a circular loop or will it try to bunch up further?

39. A straight wire segment of length l carries a current i. (a) Show that the magnetic field **B** associated with this segment, at a distance R from the segment along a perpendicular bisector (see Fig. 34-38), is given in magnitude by

$$B = \frac{\mu_0 i}{2\pi R} \frac{l}{(l^2 + 4R^2)^{1/2}}.$$

(b) Does this expression reduce to an expected result as $l \to \infty$? *Answer:* (b) Yes.

40. A square loop of wire of edge a carries a current i. Show that the value of B at the center is given by

$$B = \frac{2\sqrt{2}\,\mu_0 i}{\pi a}.$$

41. A square loop of wire of edge a carries a current i. (a) Show that B for a point on the axis of the loop and a distance x from its center is given by

$$B = \frac{4\mu_0 i a^2}{\pi(4x^2 + a^2)(4x^2 + 2a^2)^{1/2}}.$$

(b) Does this reduce to the result of Problem 40 for $x = 0$? (c) Does the square loop behave like a dipole for points such that $x \gg a$? If so, what is its dipole moment? *Answer:* (b) Yes. (c) Yes; $\mu = ia^2$.

42. (a) Show that B at the center of a rectangle of length l and width d, carrying a current i, is given by

$$B = \frac{2\mu_0 i}{\pi} \frac{(l^2 + d^2)^{1/2}}{ld}.$$

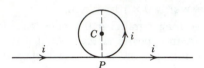

figure 34-35
Problem 34

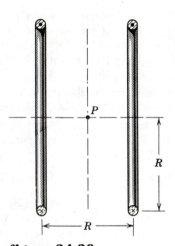

figure 34-36
Problems 35, 36

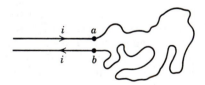

figure 34-37
Problem 38

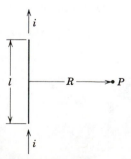

figure 34-38
Problem 39.

(b) What does B reduce to for $l \gg d$? Is this a result that you expect?

43. (a) A wire in the form of a regular polygon of n sides is just enclosed by a circle of radius a. If the current in this wire is i, show that the magnetic field **B** at the center of the circle is given in magnitude by

$$B = \frac{\mu_0 ni}{2\pi a} \tan{(\pi/n)}.$$

(b) Show that as $n \to \infty$ this result approaches that of a circular loop.

44. Calculate **B** approximately at point P in Fig. 34-39. Assume that $i = 10$ A and $a = 8.0$ cm.

45. You are given a length l of wire in which a current i may be established. The wire may be formed into a circle or a square. Which yields the larger value for B at the central point? See Problem 40. *Answer:* The square.

46. (a) A current i flows in a straight wire of length L in the direction shown in Fig. 34-40a. Starting from the Biot-Savart law, determine the resulting magnetic field (**B**$_P$, **B**$_Q$, **B**$_R$, **B**$_S$, respectively-direction and magnitude in each case) at each of the four points P, Q, R, S (all coplanar with the wire). (b) Using the results of part (a), compute the magnetic field **B** (magnitude and direction) resulting at the point T from the current flowing as indicated in the six-sided rectilinear closed loop shown in Fig. 34-40b. (Everything drawn is meant to lie in the same plane and all angles are 90°.)

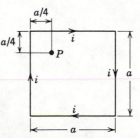

figure 34-39
Problem 44.

figure 34-40
Problem 46.

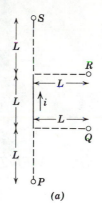

(a)

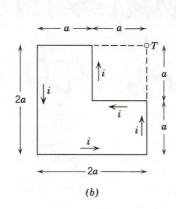

(b)

35
faraday's law of induction

For some physical laws it is hard to find experiments that lead in a direct and convincing way to the formulation of the law. Gauss's law, for example, emerged only slowly as the common factor with whose aid all electrostatic experiments could be interpreted and correlated. In Chapter 28 we found it best to state Gauss's law first and then to show that the underlying experiments were consistent with it.

Faraday's law of induction, which is one of the basic equations of electromagnetism (see Table 40-2), is different in that there are a number of simple experiments from which the law can be—and was—deduced directly. Such experiments were carried out by Michael Faraday in England in 1831 and by Joseph Henry in the United States at about the same time.

Faraday and Henry had several parallels in their lives. Both were apprentices at an early age. Faraday, at age 14, was apprenticed to a London bookbinder. He wrote, "There were plenty of books there and I read them." Henry, at age 13, was apprenticed to a watchmaker in Albany, New York.

In later years Faraday was appointed director of the Royal Institution in London, whose founding was due in large part to the American, Benjamin Thompson (Count Rumford). Henry, on the other hand, became secretary of the Smithsonian Institution in Washington, D.C., which was founded by an endowment from an Englishman, James Smithson.

Their greatest scientific overlap was that each discovered the law of electromagnetic induction, with which this chapter deals, independently and at about the same time. Even though Faraday published his results first, which gives him priority of discovery, the SI unit of inductance (see Chapter 36) is called the *henry* (abbr. H). On the other hand the SI unit of capacitance is, as we have seen, called the *farad* (abbr. F). In Section 38-1, in which we discuss oscil-

35-1
FARADAY'S EXPERIMENTS

lations in capacitative-inductive circuits, we shall see how appropriate it is to link the names of these two talented contemporaries in a single context.

Figure 35-1 shows the terminals of a coil connected to a galvanometer. Normally we would not expect this instrument to deflect because there seems to be no electromotive force in this circuit; but if we push a bar magnet toward the coil, with its north pole facing the coil, a remarkable thing happens. *While the magnet is moving*, the galvanometer deflects, showing that a current has been set up in the coil. If we hold the magnet stationary with respect to the coil, the galvanometer does not deflect. If we move the magnet *away* from the coil, the galvanometer again deflects, but in the opposite direction, which means that the current in the coil is in the opposite direction. If we use the south pole end of a magnet instead of the north pole end, the experiment works as described but the deflections are reversed. Further experimentation shows that *what matters is the relative motion of the magnet and the coil.* It makes no difference whether we move the magnet toward the coil or the coil toward the magnet.

The current that appears in this experiment is called an *induced current* and is said to be set up by an *induced electromotive force.* Note that there are no batteries anywhere in the circuit. Faraday was able to deduce from experiments like this the law that gives their magnitude and direction. Such emfs are very important in practice. The chances are good that the lights in the room in which you are reading this book are operated from an induced emf produced in a commercial electric generator.

In another experiment the apparatus of Fig. 35-2 is used. The coils are placed close together but at rest with respect to each other. When we close the switch S, thus setting up a steady current in the right-hand coil, the galvanometer deflects momentarily; when we open the switch, thus interrupting this current, the galvanometer again deflects momentarily, but in the opposite direction. No gross objects are moving in this experiment. In Faraday's words:

When the contact was made, there was a sudden and very slight effect at the galvanometer, and there was also a similar slight effect when the contact with the battery was broken. But whilst the voltaic current was continuing to pass through the one helix, no galvanometrical appearances nor any effect like induction upon the other helix could be perceived, although the active power of the battery was proved to be very great. . . .

Experiment shows that there will be an induced emf in the left coil of Fig. 35-2 whenever the current in the right coil is *changing.* It is the *rate at which the current is changing* and *not the size of the current* that is significant.

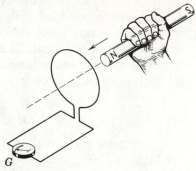

figure 35-1
Galvanometer G deflects while the magnet is moving with respect to the coil. Only their relative motion counts.

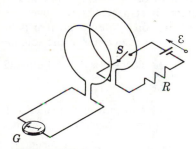

figure 35-2
Galvanometer G deflects momentarily when switch S is closed or opened. No motion is involved.

35-2
FARADAY'S LAW OF INDUCTION

Faraday had the insight to perceive that the change in the flux Φ_B for the left coil in the preceding experiments is the important common factor. This flux may be set up by a bar magnet or a current loop. Faraday's law of induction says that the induced emf ε in a circuit is equal (except for a minus sign) to the rate at which the flux through the circuit is changing. If the rate of change of flux is in webers/second, the emf ε will be in volts. In equation form

$$\varepsilon = -\frac{d\Phi_B}{dt}.$$

(35-1)

This is *Faraday's law of induction*. The minus sign is an indication of the *direction* of the induced emf, a matter we discuss in Section 35-3.*

If we apply Eq. 35-1 to a coil of N turns, an emf appears in every turn and these emfs are to be added. If the coil is so tightly wound that each turn can be said to occupy the same region of space, the flux through each turn will then be the same. The flux through each turn is also the same for (ideal) toroids and solenoids (see Section 34-5). The induced emf in all such devices is given by

$$\varepsilon = -N\frac{d\Phi_B}{dt} = -\frac{d(N\Phi_B)}{dt}, \qquad (35\text{-}2)$$

where $N\Phi_B$ measures the so-called *flux linkages* in the device.

Figures 35-1 and 35-2 suggest that there are at least two ways in which we can make the flux through a circuit change and thus induce an emf in that circuit. Speaking loosely, the coil that is connected to the galvanometer cannot tell in which of these experiments it is participating; it is aware only that the flux passing through its cross-sectional area is changing. The flux through a circuit can also be changed by changing its shape, that is, by squeezing or stretching it.

EXAMPLE 1

A long solenoid has 200 turns/cm and carries a current of 1.5 A; its diameter is 3.0 cm. At its center we place a 100-turn, close-packed coil of diameter 2.0 cm. This coil is arranged so that **B** at the center of the solenoid is parallel to its axis. The current in the solenoid is reduced to zero and then raised to 1.5 A in the other direction at a steady rate over a period of 0.050 s. What induced emf appears in the coil while the current is being changed?

The field B at the center of the solenoid is given by Eq. 34-7, or

$$B = \mu_0 ni = (4\pi \times 10^{-7}\,\text{T·m/A})(200 \times 10^2\,\text{turns/m})(1.5\,\text{A})$$

$$= 3.8 \times 10^{-2}\,\text{T}.$$

The area of the coil (not of the solenoid) is $3.1 \times 10^{-4}\,\text{m}^2$. The initial flux Φ_B through each turn of the coil is given by

$$\Phi_B = BA = (3.8 \times 10^{-2}\,\text{T})(3.1 \times 10^{-4}\,\text{m}^2) = 1.2 \times 10^{-5}\,\text{Wb}.$$

The flux goes from an initial value of 1.2×10^{-5} Wb to a final value of -1.2×10^{-5} Wb. The *change* in flux $\Delta\Phi_B$ for each turn of the coil during the 0.050-s period is thus twice the initial value. The induced emf is given by

$$\varepsilon = -\frac{N\Delta\Phi_B}{\Delta t} = -\frac{(100)(2 \times 1.2 \times 10^{-5}\,\text{Wb})}{0.050\,\text{s}} = -4.8 \times 10^{-2}\,\text{V} = -48\,\text{mV}.$$

The minus sign deals with the *direction* of the emf, as we explain below.

35-3
LENZ'S LAW

So far we have not specified the directions of the induced emfs. Although we can find these directions from a formal analysis of Faraday's law, we prefer to find them from the conservation-of-energy principle which, in this context, takes the form of Lenz's law, deduced by Heinrich Friedrich Lenz (1804–1865) in 1834: *The induced current will appear in such a direction that it opposes the change that produced it.* The minus sign in Faraday's law suggests this opposition. In mechanics

---

* Faraday, being untrained in mathematics, did not express his law of induction in equation form. In fact, in his three-volume "Experimental Researches in Electricity," a landmark work in the development of physics and chemistry, not a single equation appears!

the energy principle often allows us to draw conclusions about mechanical systems without analyzing them in detail. We use the same approach here.

Lenz's law refers to induced *currents*, which means that it applies only to closed conducting circuits. If the circuit is open, we can usually think in terms of what would happen if it *were* closed and in this way find the direction of the induced emf.

Consider the first of Faraday's experiments described in Section 35-1. Figure 35-3 shows the north pole of a magnet and a cross section of a nearby conducting loop. As we push the magnet toward the loop (or the loop toward the magnet) an induced current is set up in the loop. What is its direction?

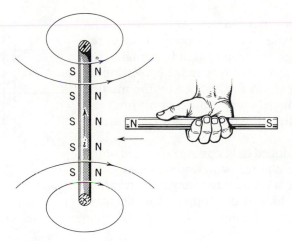

figure 35-3
If we move the magnet toward the loop, the induced current points as shown, setting up a magnetic field that opposes the motion of the magnet.

A current loop sets up a magnetic field at distant points like that of a magnetic dipole, one face of the loop being a north pole, the opposite face being a south pole. The north pole, as for bar magnets, is that face *from* which the lines of **B** emerge. If, as Lenz's law predicts, the loop in Fig. 35-3 is to oppose the motion of the magnet toward it, the face of the loop toward the magnet must become a north pole. The two north poles —one of the current loop and one of the magnet—will repel each other. The right-hand rule shows that for the magnetic field set up by the loop to emerge from the right face of the loop, the induced current must be as shown. The current will be counterclockwise as we sight along the magnet toward the loop.

When we push the magnet toward the loop (or the loop toward the magnet), an induced current appears. In terms of Lenz's law this pushing is the "change" that produces the induced current, and, according to this law, the induced current will oppose the "push." If we pull the magnet away from the coil, the induced current will oppose the "pull" by creating a *south* pole on the right-hand face of the loop of Fig. 35-3. To make the right-hand face a south pole, the current must be opposite to that shown in Fig. 35-3. Whether we pull or push the magnet, its motion will always be automatically opposed.

The agent that causes the magnet to move, either toward the coil or away from it, will always experience a resisting force and will thus be required to do work. From the conservation-of-energy principle this work done on the system must be exactly equal to the thermal energy produced in the coil, since these are the only two energy transfers that occur in the system. If we move the magnet more rapidly, we will have to do work at a faster rate and the rate of production of thermal energy will increase correspondingly. If we cut the loop and then perform the experiment, there will be no induced current, no thermal energy, no force on the magnet, and no work required to move it. There will still be an emf in the loop, but, like a battery connected to an open circuit, it will not set up a current.

If the current in Fig. 35-3 were in the *opposite* direction to that shown, the face of the loop toward the magnet would be a south pole, which would *pull* the bar magnet toward the loop. We would only need to push the magnet slightly to start the process and then the action would be self-perpetuating. The magnet would accelerate toward the loop, increasing its kinetic energy all the time. At the same time thermal energy would appear in the loop at a rate that would increase with time. This would indeed be a something-for-nothing situation! Needless to say, it does not occur.

Let us apply Lenz's law to Fig. 35-3 in a different way. Figure 35-4 shows the lines of **B** for the bar magnet.* On this point of view the "change" is the increase in Φ_B through the loop caused by bringing the magnet nearer. The induced current opposes this change by setting up a field that tends to oppose the increase in flux caused by the moving magnet. Thus the field due to the induced current must point from left to right through the plane of the coil, in agreement with our earlier conclusion.

It is not significant here that the induced field opposes the *field* of the magnet but rather that it opposes the *change*, which in this case is the *increase* in Φ_B through the loop. If we withdraw the magnet, we reduce Φ_B through the loop. The induced field will now oppose this decrease in Φ_B (that is, the change) by *reenforcing* the magnet field. In each case the induced field opposes the change that gives rise to it.

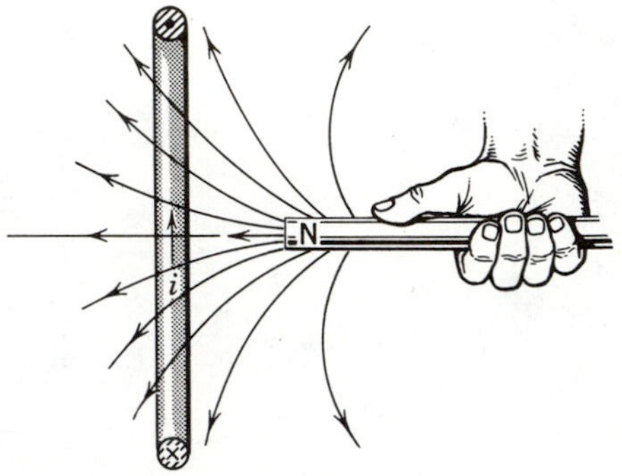

figure 35-4
If we move the magnet toward the loop, we increase Φ_B through the loop.

figure 35-5
A rectangular loop is pulled out of a magnetic field with velocity **v**.

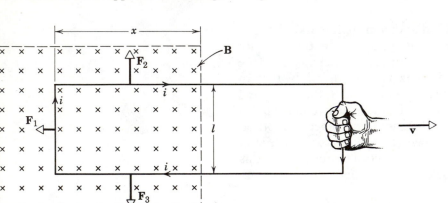

The example of Fig. 35-4, although easy to understand qualitatively, does not lend itself to quantitative calculations. Consider then Fig. 35-5, which shows a rectangular loop of wire of width l, one end of which is in a uniform field **B** pointing at right angles to the plane of the loop. This field **B** may be produced in the gap of a large electromagnet like that of Fig. 33-2. The dashed lines show the assumed limits of the magnetic field. The experiment consists in pulling the loop to the right at a constant speed v.

Note that the situation described by Fig. 35-5 does not differ in any

35-4
INDUCTION — A QUANTITATIVE STUDY

* There are two fields of **B** in this problem — one connected with the current loop and one with the bar magnet. You must always be certain which one is meant.

essential particular from that of Fig. 35-4. In each case a conducting loop and a magnet are in relative motion; in each case the flux of the field of the magnet through the loop is being caused to change with time.

The flux Φ_B enclosed by the loop in Fig. 35-5 is

$$\Phi_B = Blx,$$

where lx is the area of that part of the loop in which B is not zero. We find the emf ε, from Faraday's law, or

$$\varepsilon = -\frac{d\Phi_B}{dt} = -\frac{d}{dt}(Blx) = -Bl\frac{dx}{dt} = Blv, \qquad (35\text{-}3)$$

where we have set $-dx/dt$ equal to the speed v at which the loop is pulled out of the magnetic field. Note that the only dimension of the loop that enters into Eq. 35-3 is the length l of the left end conductor. As we shall see later, the induced emf in Fig. 35-5 may be regarded as localized here. An induced emf such as this, produced by pulling a conductor through a magnetic field (or conversely), is sometimes called a *motional emf.*

The emf Blv sets up a current in the loop given by

$$i = \frac{\varepsilon}{R} = \frac{Blv}{R}, \qquad (35\text{-}4)$$

where R is the loop resistance. From Lenz's law, this current (and thus ε) must be clockwise in Fig. 35-5; it opposes the "change" (the decrease in Φ_B) by setting up a field that is parallel to the external field within the loop.

The current in the loop will cause forces $\mathbf{F}_1$, $\mathbf{F}_2$, and $\mathbf{F}_3$ to act on the three conductors, in accord with Eq. 33-6a, or

$$\mathbf{F} = i\mathbf{l} \times \mathbf{B}. \qquad (35\text{-}5)$$

Because $\mathbf{F}_2$ and $\mathbf{F}_3$ are equal and opposite, they cancel each other; $\mathbf{F}_1$, which is the force that opposes our effort to move the loop, is given in magnitude from Eqs. 35-4 and 35-5 as

$$F_1 = ilB \sin 90° = \frac{B^2l^2v}{R}.$$

The agent that pulls the loop must do work at the steady rate of

$$P = F_1v = \frac{B^2l^2v^2}{R}. \qquad (35\text{-}6)$$

From the principle of the conservation of energy, thermal energy must appear in the resistor at this same rate. We introduced the conservation-of-energy principle into our derivation when we wrote down the expression for the current (Eq. 35-4); recall that the relation $i = \varepsilon/R$ for single-loop circuits is a direct consequence of this principle. Thus we should be able to write down the expression for the rate of thermal energy production in the loop with the expectation that we will obtain a result identical with Eq. 35-6. Recalling Eq. 35-4, we put

$$P_J = i^2R = \left(\frac{Blv}{R}\right)^2 R = \frac{B^2l^2v^2}{R},$$

which is indeed the expected result. This example provides a quantitative illustration of the conversion of mechanical energy (the work done by an external agent) into electrical energy (associated with the induced emf) and finally into thermal energy.

Figure 35-6 shows a side view of the coil in the field. In Fig. 35-6*a* the coil is stationary; in Fig. 35-6*b* we are moving it to the right; in Fig. 35-6*c* we are moving it to the left. The lines of **B** in these figures represent the *resultant field* produced by the vector addition of the field **B**$_0$ due to the magnet and the field **B**$_i$ due to the induced current, if any, in the coil. These lines suggest convincingly that the agent moving the coil always experiences an opposing force.

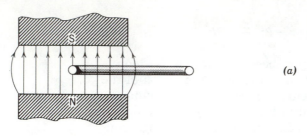

(*a*)

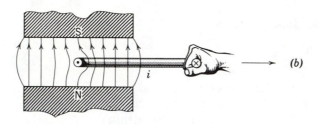

(*b*)

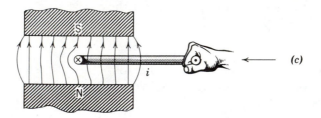

(*c*)

figure 35-6
Side view of a rectangular loop in a magnetic field showing the loop (*a*) at rest, (*b*) being pulled out, and (*c*) being pushed in. The configuration of the lines suggests an opposing force in both (*b*) and (*c*).

Figure 35-7 shows a rectangular loop of resistance *R*, width *l*, and length *a* being pulled at constant speed *v* through a region of thickness *d* in which a uniform magnetic field **B** is set up by a magnet.

(*a*) Plot the flux Φ_B through the loop as a function of the coil position *x*. Assume that $l = 4$ cm, $a = 10$ cm, $d = 15$ cm, $R = 16$ Ω, $B = 2.0$ T, and $v = 1.0$ m/s.

The flux Φ_B is zero when the loop is not in the field; it is *Bla* when the loop is entirely in the field; it is *Blx* when the loop is entering the field and $Bl[a - (x - d)]$ when the loop is leaving the field. These conclusions, which you should verify, are shown graphically in Fig. 35-8*a*.

(*b*) Plot the induced emf ε.

The induced emf ε is given by $\varepsilon = -d\Phi_B/dt$, which we can write as

$$\varepsilon = -\frac{d\Phi_B}{dt} = -\frac{d\Phi_B}{dx}\frac{dx}{dt} = -\frac{d\Phi_B}{dx} v,$$

where $d\Phi_B/dx$ is the slope of the curve of Fig. 35-8*a*. $\varepsilon(x)$ is plotted in Fig. 35-8*b*. Lenz's law, from the same type of reasoning as that used for Fig. 35-5, shows that when the coil is entering the field, the emf ε acts counterclockwise as seen from above. Note that there is no emf when the coil is entirely in the magnetic field because the flux Φ_B through the coil is not changing with time, as Fig. 35-8*a* shows.

(*c*) Plot the rate *P* of thermal energy production in the loop.

EXAMPLE 2

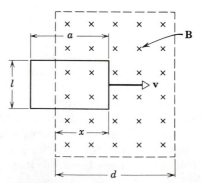

figure 35-7
Example 2. A rectangular loop is caused to move with a velocity **v** through a magnetic field. The position of the loop is measured by *x*, the distance between the effective left edge of the field **B** and the right end of the loop.

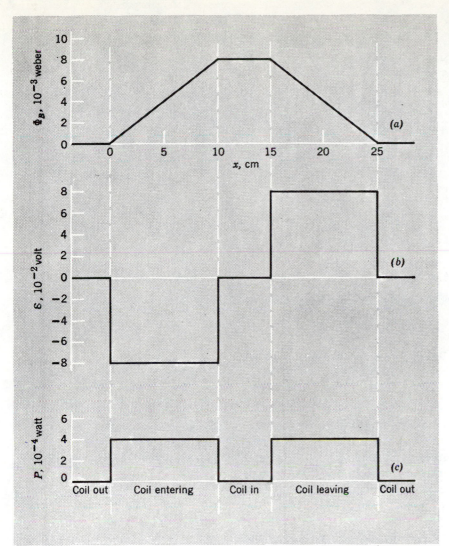

figure 35-8
Example 2. In practice the sharp corners would be rounded.

This is given by $P = \mathcal{E}^2/R$. It may be calculated by squaring the ordinate of the curve of Fig. 35-8b and dividing by R. The result is plotted in Fig. 35-8c.

If the fringing of the magnetic field, which cannot be avoided in practice (see Problem 34-2), is taken into account, the sharp bends and corners in Fig. 35-8 will be replaced by smooth curves. What changes would occur in the curves of Fig. 35-8 if the coil were open circuited?

EXAMPLE 3

A copper rod of length L rotates at angular frequency ω in a uniform magnetic field $\mathbf{B}$ as shown in Fig. 35-9. Find the emf $\mathcal{E}$ developed between the two ends of the rod. (We might measure this emf if we place a conducting rail along the dashed circle in the figure and connect a voltmeter between the rail and point O.

If a wire of length $d\mathbf{l}$ is moved at velocity $\mathbf{v}$ at right angles to a field $\mathbf{B}$, a motional emf $d\mathcal{E}$ will be developed (see Eq. 35-3) given by

$$d\mathcal{E} = Bv \, dl.$$

The rod of Fig. 35-9 may be divided into elements of length dl, the linear speed v of each element being ωl. Each element is perpendicular to $\mathbf{B}$ and is also moving in a direction at right angles to $\mathbf{B}$ so that, since the $d\mathcal{E}$'s are "in series",

$$\mathcal{E} = \int d\mathcal{E} = \int_0^L Bv \, dl = \int_0^L B(\omega l) \, dl = \tfrac{1}{2} B\omega L^2.$$

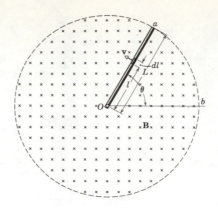

figure 35-9
Example 3

For a second approach, consider that at any instant the flux enclosed by the sector aOb in Fig. 35-9 is given by

$$\Phi_B = BA = B\left(\tfrac{1}{2}L^2\theta\right),$$

where $\tfrac{1}{2}L^2\theta$ can be shown to be the area of the sector. Differentiating gives

$$\frac{d\Phi_B}{dt} = \tfrac{1}{2}BL^2\frac{d\theta}{dt} = \tfrac{1}{2}B\omega L^2.$$

From Faraday's law, this is precisely the magnitude of ε, and agrees with the result just derived.

So far we have considered emfs induced by the relative motion of magnets and coils. In this section we assume that there is no physical motion of gross objects but that the magnetic field may vary with time. If we place a conducting loop in such a time-varying field, the flux through the loop will change and an induced emf will appear in the loop. This emf will set the charge carriers in motion, that is, it will induce a current.

From a microscopic point of view we can say, equally well, that the changing flux of **B** sets up an induced electric field **E** at various points around the loop. These induced electric fields are just as real as electric fields set up by static charges and will exert a force **F** on a test charge q_0 given by $\mathbf{F} = q_0\mathbf{E}$. Thus we can restate Faraday's law of induction in a loose but informative way as: *A changing magnetic field produces an electric field.*

To fix these ideas, consider Fig. 35-10, which shows a uniform magnetic field **B** at right angles to the plane of the page. We assume that **B** is increasing in magnitude at the same constant rate dB/dt at every point. This could be done by causing the current in the windings of the electromagnet that establishes the field to increase with time in the proper way.

The circle of arbitrary radius r shown in Fig. 35-10 encloses, at any instant, a flux Φ_B. Because this flux is changing with time, an induced emf given by $\varepsilon = -d\Phi_B/dt$ will appear around the loop. The electric fields **E** induced at various points of the loop must, from symmetry, be tangent to the loop. Thus the electric lines of force that are set up by the changing magnetic field are in this case concentric circles.

If we consider a test charge q_0 moving around the circle of Fig. 35-10, the work W done on it per revolution is, in terms of the definition of an emf, simply εq_0. From another point of view it is $(q_0E)(2\pi r)$, where q_0E is the force that acts on the charge and $2\pi r$ is the distance over

35-5
TIME-VARYING MAGNETIC FIELDS

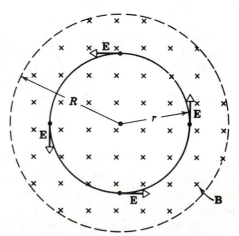

figure 35-10
The induced electric fields at four points produced by an increasing magnetic field. (The magnetic field cannot end abruptly at radius R but must approach zero gradually. This "fringing" does not change any of the arguments of this section.)

which the force acts. Setting the two expressions for W equal and canceling q_0 yields

$$\varepsilon = E2\pi r. \tag{35-7}$$

In a more general case than that of Fig. 35-10 we must write

$$\varepsilon = \oint \mathbf{E} \cdot d\mathbf{l}. \tag{35-8}$$

If this integral is evaluated for the conditions of Fig. 35-10, we obtain Eq. 35-7 at once. If Eq. 35-8 is combined with Eq. 35-1 ($\varepsilon = -d\Phi_B/dt$), we can write Faraday's law of induction as

$$\oint \mathbf{E} \cdot d\mathbf{l} = -\frac{d\Phi_B}{dt}, \tag{35-9}$$

which is the form in which this law is expressed in Table 40-2.

EXAMPLE 4

Let B in Fig. 35-10 be increasing at the rate dB/dt. Let R be the radius of the cylindrical region in which the magnetic field is assumed to exist. What is the magnitude of the electric field $\mathbf{E}$ at any radius r? Assume that $dB/dt = 0.10$ T/s and $R = 10$ cm.

(a) For $r < R$, the flux Φ_B through the loop is

$$\Phi_B = B(\pi r^2).$$

Substituting into Faraday's law (Eq. 35-9),

$$\oint \mathbf{E} \cdot d\mathbf{l} = -\frac{d\Phi_B}{dt}$$

yields

$$(E)(2\pi r) = -\frac{d\Phi_B}{dt} = -(\pi r^2)\frac{dB}{dt}.$$

Solving for E yields

$$E = -\tfrac{1}{2}r\frac{dB}{dt}.$$

The minus sign is retained to suggest that the induced electric field $\mathbf{E}$ acts to *oppose* the change of the magnetic field. Note that $E(r)$ depends on dB/dt and not on B. Substituting numerical values, assuming $r = 5.0$ cm, yields, for the magnitude of E,

$$E = \tfrac{1}{2}r\frac{dB}{dt} = (\tfrac{1}{2})(5 \times 10^{-2}\text{ m})(0.10\text{ T/s}) = 2.5 \times 10^{-3}\text{ V/m}.$$

(b) For $r > R$ the flux through the loop is

$$\Phi_B = \int \mathbf{B} \cdot d\mathbf{S} = B(\pi R^2).$$

This equation is true because $\mathbf{B} \cdot d\mathbf{S}$ is zero for those points of the loop that lie outside the effective boundary of the magnetic field.

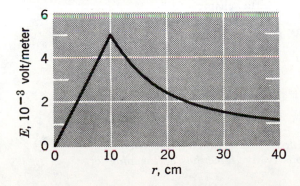

figure 35-11
Example 4. If the fringing of the field in Fig. 35-10 were to be taken into account, the result would be a rounding of the sharp cusp at $r = R$ ($= 10$ cm).

From Faraday's law (Eq. 35-9),

$$(E)(2\pi r) = -\frac{d\Phi_B}{dt} = -(\pi R^2)\frac{dB}{dt}.$$

Solving for E yields

$$E = -\frac{1}{2}\frac{R^2}{r}\frac{dB}{dt}.$$

These two expressions for $E(r)$ yield the same result, as they must, for $r = R$. Figure 35-11 is a plot of the magnitude of $\mathbf{E}(r)$ for the numerical values given.

In applying Lenz's law to Fig. 35-10, imagine that a circular conducting loop is placed concentrically in the field. Since Φ_B through this loop is increasing, the induced current in the loop will tend to oppose this "change" by setting up a magnetic field of its own that points up within the loop. Thus the induced current i must be counterclockwise, which means that the lines of the induced electric field $\mathbf{E}$, which is responsible for the current, must also be counterclockwise. If the magnetic field in Fig. 35-10 were *decreasing* with time, the induced current and the lines of force of the induced electric field $\mathbf{E}$ would be clockwise, again opposing the *change* in Φ_B.

Figure 35-12 shows four of many possible loops to which Faraday's law may be applied. For loops 1 and 2, the induced emf ε is the same because these loops lie entirely within the changing magnetic field and, having the same area, thus have the same value of $d\Phi_B/dt$. Note that even though the emf ε $(= \oint\mathbf{E}\cdot d\mathbf{l})$ is the same for these two loops, the distribution of electric fields $\mathbf{E}$ around the perimeter of each loop, as indicated by the electric lines of force, is different. For loop 3 the emf is less because Φ_B and $d\Phi_B/dt$ for this loop are less, and for loop 4 the induced emf is zero.

The induced electric fields that are set up by the induction process are not associated with charges but with a changing magnetic flux. Although both kinds of electric fields exert forces on charges, there is a difference between them. The simplest manifestation of this difference is that lines of $\mathbf{E}$ associated with a changing magnetic flux can form closed loops (see Fig. 35-12); lines of $\mathbf{E}$ associated with charges cannot form closed loops but can always be drawn to start on a positive charge and end on a negative charge.

Equation 29-5, which defined the potential difference between two points a and b, is

$$V_b - V_a = \frac{W_{ab}}{q_0} = -\int_a^b \mathbf{E}\cdot d\mathbf{l}.$$

We have insisted that if potential is to have any useful meaning this integral (and W_{ab}) must have the same value for every path connecting a and b. This proved to be true for every case examined in earlier chapters.

An interesting special case comes up if a and b are the same point. The path connecting them is now a closed loop; V_a must be identical with V_b and this equation reduces to

$$\oint\mathbf{E}\cdot d\mathbf{l} = 0. \tag{35-10}$$

However, when changing magnetic flux is present, $\oint\mathbf{E}\cdot d\mathbf{l}$ is precisely *not* zero but is, according to Faraday's law (see Eq. 35-9), $-d\Phi_B/dt$. Electric fields associated with stationary charges are *conservative*, but those associated with changing magnetic fields are *nonconservative*; see Section 8-2. Electric potential, which can be defined only for a conservative force, *has no meaning for electric fields produced by induction*.

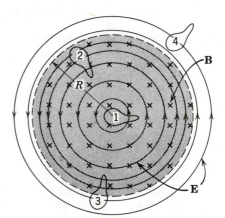

figure 35-12
Showing the circular lines of $\mathbf{E}$ from an increasing magnetic field. The four loops are imaginary paths around which an emf can be calculated.

The betatron is a device used to accelerate electrons to high speeds by allowing them to be acted upon by induced electric fields that are set up by a changing magnetic flux. It provides an excellent illustration of the "reality" of such induced fields and it is in this context that we discuss it. The energetic electrons can be used for fundamental research in physics or to produce penetrating X-rays which are useful in cancer therapy and in industry.

Figure 35-13 shows a 100-MeV betatron that was constructed at the General Electric Company. At this energy the electron speed is $0.999986c$, where c is the speed of light, so that relativistic mechanics must certainly be used in the analysis of its operation. Figure 35-14 shows a vertical cross section through the central part of the betatron to which the man in Fig. 35-13 is pointing.

The magnetic field in the betatron has several functions: (a) it guides the electrons in a circular path; (b) the changing magnetic field generates an electric field (Section 35-5) that accelerates the electrons in this path; (c) it keeps the radius of the orbit in which the electrons are moving essentially constant; (d) it introduces the electrons into the or-

figure 35-13
A 100-MeV betatron. M shows the magnet, C the magnetizing coils, and D the region in which the "doughnut" is located. (Courtesy General Electric Company.)

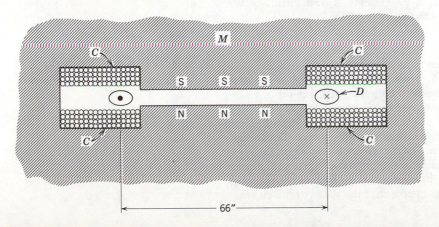

figure 35-14
Cross section of a betatron, showing magnet M, coils C, and "doughnut" D. Electrons emerge from the page at the left and enter it at the right. See Problem 33.

bit initially and removes them from the orbit after they have reached full energy; and finally (e) it provides a restoring force that resists any tendency for the electrons to leave their orbit, either vertically or radially. It is remarkable that it is possible to do all these things by proper shaping and control of the magnetic field.

The object marked D in Fig. 35-14 is an evacuated glass "doughnut" inside which the electrons travel. Their orbit is a circle at right angles to the plane of the figure. The electrons emerge from the plane at the left (·) and enter it at the right (×). In the General Electric machine the radius of the electron path is 33 in. The coils C and the 130-ton steel magnet shown in Fig. 35-13 provide the magnetic flux that passes through the plane of this orbit.

The current in coils C is made to alter periodically, 60 times/second to produce a changing flux through the orbit, shown in Fig. 35-15. Here Φ_B is taken as positive when **B** is pointing up, as in Fig. 35-14. If the electrons are to circulate in the direction shown, they must do so during the positive half-cycle, marked ac in Fig. 35-15. You should verify this (see Section 33-6). The electrons are accelerated by electric fields set up by the changing flux. The direction of these induced fields depends on the sign of $d\Phi_B/dt$ and must be chosen to accelerate, and not to decelerate, the electrons. Thus only half the positive half-cycle in Fig. 35-15 can be used for acceleration; it will prove to be ab.

The average value of $d\Phi_B/dt$ during the quarter-cycle ab is the slope of the dashed line, or

$$\frac{d\Phi_B}{dt} = \frac{1.8 \text{ Wb}}{4.2 \times 10^{-3} \text{ s}} = 430 \text{ V}.$$

From Faraday's law (Eq. 35-1), this is also the emf in volts. The electron will thus increase its energy by 430 eV every time it makes a trip around the orbit in the changing flux. If the electron gains only 430 eV of energy per revolution, it must make about 230,000 rev to gain its full 100 MeV. For an orbit radius of 33 in., this corresponds to a length of path of some 750 miles.

The betatron provides a good example of the fact that electric potential has no meaning for electric fields produced by induction. If a potential exists, it must be true that, as Eq. 35-10 shows, $\oint \mathbf{E} \cdot d\mathbf{l} = 0$ for any closed path. In the betatron, however, this integral, evaluated around the orbit, is precisely *not* zero but is, in our example, 430 volts. It must not be thought, of course, that the betatron violates the conservation-of-energy principle. The gain in kinetic energy of the circulating electron (430 eV/rev) must be supplied by an identifiable energy source. It comes, in fact, from the generator that energizes the magnet coils, thus providing the changing magnetic field. The energy is transmitted to the electron through the intermediary of this changing field.

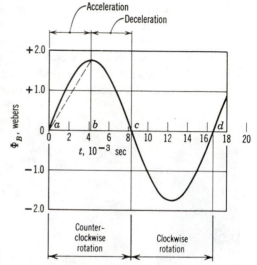

figure 35-15
The flux through the orbit of a betatron, during one cycle. Rotation of the electrons in the desired direction (counterclockwise as viewed from above in Fig. 35-14) is possible only during half-cycle ac.

EXAMPLE 5

In the betatron of Fig. 35-13, which is the "accelerating" quarter-cycle?

Let us assume that it is ab in Fig. 35-15, during which Φ_B through the orbit is *increasing*. If a conducting loop were placed to coincide with the orbit, an induced current would appear in the loop to oppose the tendency of Φ_B to increase. This means that a magnetic field would be set up that would oppose the field of the large magnet. Thus **E** would point outward at the right side of "doughnut" D in Fig. 35-14 and inward on the left side. The force $(-e\mathbf{E})$ acting on the electron is in the opposite direction to **E** because of the negative charge of the electron. Thus the tangential force acting on the electron is in the same direction as that at which it circulates in its orbit; this means that the speed of the electron will increase, as desired. You should go through this same analysis carefully, assuming (incorrectly, as it will turn out) that the accelerating half-cycle is *bc* in Fig. 35-15 rather than *ab*.

35-7
INDUCTION AND RELATIVE MOTION

Faraday's law, in the form $\varepsilon = -d\Phi_B/dt$, gives correctly the induced emf no matter whether the change in Φ_B is produced by moving a coil, moving a magnet, changing the strength of a magnetic field, changing the shape of a conducting loop, or in other ways. However, observers who are in relative motion with respect to each other, even though they would all agree on the numerical value of the emf, would give different microscopic descriptions of the induction process. In electromagnetic systems, as well as in mechanical systems, it is important that the state of motion of the observer with respect to his environment be made perfectly clear.

Figure 35-16 shows a closed loop which is caused to move at velocity **v** with respect to a magnet that provides a uniform field **B** in the region shown. We consider first an observer, identified as *S*, who is *at rest with respect to the magnet* used to establish the field **B**; see Fig. 35-16a. The induced emf in this case is called a *motional emf* because the conducting loop is moving with respect to this observer.

Consider a positive charge carrier at the center of the left end of the conducting loop. To observer *S*, this charge, constrained to move to the right along with the loop, is a charge *q* moving with a velocity **v** in the magnetic field **B** and as such it experiences a sideways magnetic deflecting force given by Eq. 33-2 ($\mathbf{F} = q\mathbf{v} \times \mathbf{B}$). This force causes the carriers to move upward along the conductor, so that they acquire a drift velocity v_d also, as shown in Fig. 35-16a.

The resultant equilibrium speed of the carrier is now **V**, which we find by adding **v** and v_d vectorially in Fig. 35-16a. The magnetic deflecting force $\mathbf{F}_m$ is, as always, at right angles to the resultant velocity **V** of the carrier and is given by

$$\mathbf{F}_m = q\mathbf{V} \times \mathbf{B}. \qquad (35\text{-}13)$$

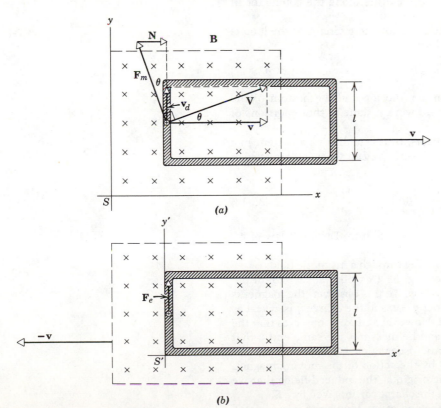

(a)

(b)

figure 35-16
A closed conducting loop is in relative motion with respect to a magnet. (*a*) An observer *S*, fixed with respect to the magnet that produces the field **B**, sees the loop moving to the right. He also sees (see text) a *magnetic* force (Eq. 33-2) equal to $F_m \cos \theta$ acting upward on the positive charge carriers. (*b*) An observer *S'*, fixed with respect to the loop, sees the magnet moving toward the left. He also sees an *electric* force (see text) acting upward on the positive charge carriers. In both figures there is, of course, an average internal collision force (not shown; see Sections 31-4 and 31-5) that keeps the charge carriers from accelerating.

Now $\mathbf{F}_m$ acting alone would tend to push the carriers through the left wall of the conductor. Because this does not happen the conductor wall must exert a normal force $\mathbf{N}$ on the carriers (see Fig. 35-16a) of magnitude such that $\mathbf{v}_d$ lies parallel to the axis of the wire; in other words, $\mathbf{N}$ exactly cancels the horizontal component of $\mathbf{F}_m$, leaving only the component $\mathbf{F}_m \cos \theta$ that lies along the direction of the conductor. This component of force on the carrier is also canceled out, this time by the average force $\overline{\mathbf{F}}_i$ associated with the internal collisions that the carrier experiences as it drifts with (constant) speed v_d through the wire.

The kinetic energy of the charge carrier as it drifts through the wire remains constant. This is consistent with the fact that the resultant force acting on the charge carrier ($= \mathbf{F}_m + \overline{\mathbf{F}}_i + \mathbf{N}$) is zero. The work done by $\mathbf{F}_m$ is zero because magnetic forces, acting at right angles to the velocity of a moving charge, can do no work on that charge. Thus the (negative) work done on the carrier by the average internal collision force $\mathbf{F}_i$ must be exactly canceled by the (positive) work done on the carrier by the force $\mathbf{N}$. In the last analysis $\mathbf{N}$ is exerted by the agent who pulls the loop through the magnetic field, and the mechanical energy expended by this agent appears as thermal energy in the loop, as we have seen in Section 35-4.

Let us then calculate the work dW done on the carrier in time dt by the force $\mathbf{N}$; it is

$$dW = N(v \, dt) \qquad (35\text{-}14)$$

in which $v \, dt$ is the distance that the loop (and the carrier) have moved to the right in Fig. 35-16a in time dt. We can write for N (see Eq. 35-13 and Fig. 35-16a).

$$N = F_m \sin \theta = (qVB)(v_d/V) = qBv_d. \qquad (35\text{-}15)$$

Substituting Eq. 35-15 into Eq. 35-14 yields

$$\begin{aligned} dW &= (qBv_d)(v \, dt) \\ &= (qBv)(v_d \, dt) = qBv \, dl \end{aligned} \qquad (35\text{-}16)$$

in which dl ($= v_d dt$) is the distance the carrier drifts along the conductor in time dt.

The work done on the carrier as it makes a complete circuit of the loop is found by integrating Eq. 35-16 around the loop and is

$$W = \oint dW = qBvl. \qquad (35\text{-}17)$$

This follows because work contributions for the top and the bottom of the loops are opposite in sign and cancel and no work is done in those portions of the loop that lie outside the magnetic field.

An agent that does work on charge carriers, thus establishing a current in a closed conducting loop, can be viewed as an emf. We can write, making use of Eq. 35-17,

$$\varepsilon = \frac{W}{q} = \frac{qBvl}{q} = Blv, \qquad (35\text{-}18)$$

which is, of course, the same result that we derived from Faraday's law of induction; see Eq. 35-3. Thus a motional emf is intimately connected with the sideways deflection of a charged particle moving through a magnetic field.

We now consider how the situation of Fig. 35-16 would appear to an observer S' who is *at rest with respect to the loop*. To this observer, the magnet is moving to the left in Fig. 35-16b with velocity $-\mathbf{v}$, and the charge q is at rest as far as its left-to-right motion is concerned. However S', like S, observes that the charge drifts clockwise around the loop and he measures the same emf ε that S measures. S' accounts for this, at the microscopic level, by postulating that an electric field $\mathbf{E}$ is induced in the loop by the action of the moving magnet. This induced field $\mathbf{E}$, which has the same origin as the induced fields that we discussed in Section 35-5, exerts a force on the charge carrier given by $q\mathbf{E}$.

The induced field $\mathbf{E}$, which exists in the end of the loop only, is associated with an emf ε and generates a current in the closed loop. Note that, in any closed loop in which there is a current, an internal electric field must exist at every point at which charges are moving. These electric fields, however, are set up by the emf, as in the case of a closed loop connected to a battery, and are *not* induced by the motion of the magnet. It is only this induced field **E** that we associate with the emf, through the relation (Eq. 35-8)

$$\varepsilon = \oint \mathbf{E} \cdot d\mathbf{l}$$

which reduced to

$$\varepsilon = El \qquad (35\text{-}19)$$

in this case. This is so because no *induced* electric field appears in the upper and lower bars (because of the nature of their motions), and none appears in the part of the loop outside the magnetic field.

The emfs given by Eqs. 35-18 and 35-19 must be identical because the relative motion of the loop and the magnet is identical in the two cases shown in Fig. 35-16. Equating these relations yields

$$El = Blv,$$

or

$$E = vB. \qquad (35\text{-}20a)$$

In Fig. 35-16*b* the vector **E** points upward along the axis of the left end of the conducting loop because this is the direction in which positive charges are observed to drift. The directions of **v** and **B** are clearly shown in this figure. We see, then, that Eq. 35-20*a* is consistent with the more general vector relation

$$\mathbf{E} = \mathbf{v} \times \mathbf{B}. \qquad (35\text{-}20b)$$

We have not proved Eq. 35-20*b* except for the special case of Fig. 35-16; nevertheless it proves to be true in general, that is, no matter what the angle between **v** and **B**.

We interpret Eq. 35-20*b* in the following way: Observer S fixed with respect to the magnet is aware only of a magnetic field. The force to him arises from the motion of the charges through **B**. Observer S' fixed on the charge carrier is aware of an electric field **E** also and attributes the force on the charge (at rest with respect to him initially) to the electric field. S says the force is of purely magnetic origin and S' says the force is of purely electric origin. From the point of view of S, the induced emf is given by $\oint (\mathbf{v} \times \mathbf{B}) \cdot d\mathbf{l}$. From the point of view of S', the same induced emf is given by $\oint \mathbf{E} \cdot d\mathbf{l}$, where **E** is the (induced) electric vector that he observes at points along the circuit.

For a third observer S'' who judges that both the magnet and the loop are moving, the force tending to move charges around the loop is neither purely electric nor purely magnetic, but a bit of each. In summary, in the equation

$$\mathbf{F}/q = \mathbf{E} + \mathbf{v} \times \mathbf{B},$$

different observers form different assessments of **E**, **B**, and **v** but, when these are combined, all observers form the same assessment of $\mathbf{F}/q$ and all obtain the same value for the induced emf in the loop (this depends only on the relative motion). That is, the total force is the same for all observers, but each observer forms a different estimate of the separate electric and magnetic forces contributing to the same total force.

The essential point is that what seems like a magnetic field to one observer may seem like a mixture of an electric field and a magnetic field to a second observer in a different inertial reference frame. Both observers would agree, however, on the gross measurable result, in the case of Fig. 35-16, the current in the loop. We are forced to conclude that magnetic and electric fields are *not* independent of each other and have no separate unique existence; they depend on the inertial frame.

Einstein started to think about relative motion at age 16 and published his famous paper on the theory of special relativity in 1905, when he was a 26-year-old patent examiner in the Swiss Patent Office in Berne. He was *not* led to this theory by early considerations of the nature of space and time but precisely by the problems raised in this section.

This is not only made clear by the title of his paper (translated from the German), "On the Electromagnetic Forces Acting on Moving Bodies," but is strongly reenforced by the opening lines of his paper, which are:

It is known that Maxwell's electrodynamics—as usually understood at the present time—when applied to moving bodies, leads to asymmetries which do not appear to be inherent in the phenomena. Take, for example, the reciprocal electrodynamic action of a magnet and a conductor. The observable phenomenon here depends only on the relative motion of the conductor and the magnet, whereas the customary view draws a sharp distinction between the two cases in which either the one or the other of these bodies is in motion. For if the magnet is in motion and the conductor at rest, there arises in the neighbourhood of the magnet an electric field with a certain definite energy, producing a current at the places where parts of the conductor are situated. But if the magnet is stationary and the conductor in motion, no electric field arises in the neighbourhood of the magnet. In the conductor, however, we find an electromotive force, to which in itself there is no corresponding energy, but which gives rise—assuming equality of relative motion in the two cases discussed—to electric currents of the same path and intensity as those produced by the electric forces in the former case.

If you want to pursue this matter further, please read Supplementary Topic V carefully and then refer to *Introduction to Special Relativity* (Chapter IV) by Robert Resnick, John Wiley & Sons (1968).

EXAMPLE 6

In Fig. 35-16 assume that $B = 2.0$ T, $l = 10$ cm, and $v = 1.0$ m/s. Calculate (a) the induced electric field observed by S' and (b) the emf induced in the loop.

(a) The electric field, which is apparent only to observer S', is associated with the moving magnet and is given in magnitude (see Eq. 35-20a) by

$$E = vB$$
$$= (1.0 \text{ m/s})(2.0 \text{ T})$$
$$= 2.0 \text{ V/m}.$$

(b) Observer S would calculate the induced (motional) emf from

$$\varepsilon = Blv$$
$$= (2.0 \text{ T})(1.0 \times 10^{-1} \text{m})(1.0 \text{ m/s})$$
$$= 0.20 \text{ V}.$$

Observer S' would not regard the emf as motional and would use the relationship

$$\varepsilon = El$$
$$= (2.0 \text{ V/m})(1.0 \times 10^{-1} \text{ m})$$
$$= 0.20 \text{ V}.$$

As must be the case, both observers agree as to the numerical value of the emf.

---

questions

1. In Figs. 35-1, 35-2, and 35-3, etc. we show, for simplicity, coils of only one turn. Explain the advantage of increasing the number of turns.
2. Are "induced" emfs and currents different in any way from emfs and currents provided by a battery connected to a conducting loop? Discuss.
3. Although we discussed these matters in earlier chapters, can you now explain more clearly in your own words the difference between a magnetic field **B** and the flux of a magnetic field Φ_B? Are they vectors or scalars? In what units may each be expressed? How are these units related? Are either or both (or neither) properties of a given point in space? Discuss fully.
4. A magnet is dropped from the ceiling along the axis of a copper loop lying flat on the floor. If the falling magnet is photographed with a time sequence

camera, what differences, if any, will be noted if (a) the loop is at room temperature and (b) the loop is packed in dry ice?

5. Two conducting loops face each other a distance d apart (Fig. 35-17). An observer sights along their common axis from left to right. If a clockwise current i is suddenly established in the larger loop, (a) what is the direction of the induced current in the smaller loop? (b) What is the direction of the force (if any) that acts on the smaller loop?

6. What is the direction of the induced emf in coil Y of Fig. 35-18 when (a) coil Y is moved toward coil X and (b) the current in coil X is decreased, without any change in the relative positions of the coils?

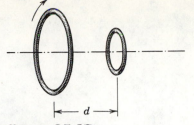

figure 35-17
Question 5

figure 35-18
Question 6

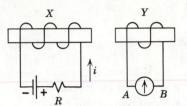

7. The north pole of a magnet is moved away from a metallic ring, as in Fig. 35-19. In the part of the ring farthest from the reader, which way does the current point?

8. *Eddy currents.* A sheet of copper is placed in a magnetic field as shown in Fig. 35-20. If we attempt to pull it out of the field or push it further in, an automatic resisting force appears. Explain its origin. (Hint: Currents, called eddy currents, are induced in the sheet in such a way as to oppose the motion.)

9. A current-carrying solenoid is moved toward a conducting loop as in Fig. 35-21. What is the direction of circulation of current in the loop as we sight toward it as shown?

10. If the resistance R in the left-hand circuit of Fig. 35-22 is increased, what is the direction of the induced current in the right-hand circuit?

figure 35-19
Question 7

figure 35-20
Question 8

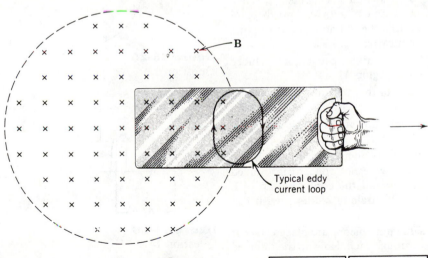

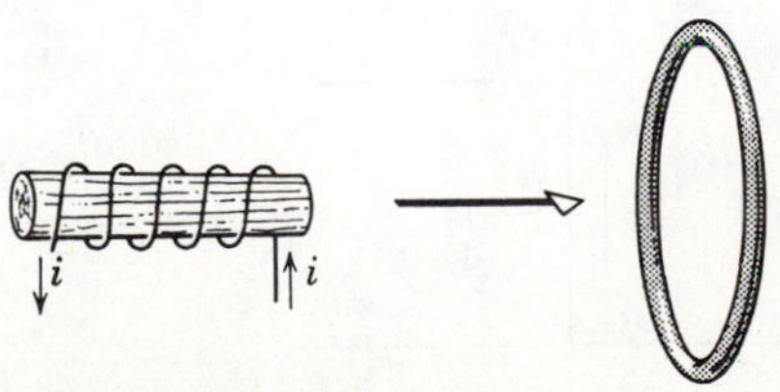

Typical eddy current loop

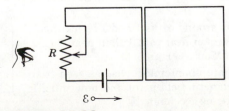

figure 35-21
Question 9

figure 35-22
Question 10

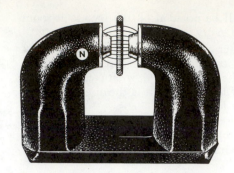

figure 35-23
Question 11

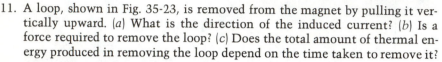

figure 35-24
Question 12

11. A loop, shown in Fig. 35-23, is removed from the magnet by pulling it vertically upward. (*a*) What is the direction of the induced current? (*b*) Is a force required to remove the loop? (*c*) Does the total amount of thermal energy produced in removing the loop depend on the time taken to remove it?

12. *Electromagnetic shielding.* Consider a conducting sheet lying in a plane perpendicular to a magnetic field **B**, as shown in Fig. 35-24. (*a*) If **B** suddenly changes, the full *change* in **B** is not immediately detected in region *P*. Explain. (*b*) If the resistivity of the sheet is zero, the *change* is not ever detected at *P*. Explain. (*c*) If **B** changes periodically at high frequency and the conductor is made of a material of low resistivity, the region near *P* is almost completely shielded from the *changes* in flux. Explain. (*d*) Is such a conductor useful as a shield from *static* magnetic fields? Explain.

13. *Magnetic damping.* A strip of copper is mounted as a pendulum about *O* in Fig. 35-25. It is free to swing through a magnetic field normal to the page. If the strip has slots cut in it as shown, it can swing freely through the field. If a strip without slots is substituted, the vibratory motion is strongly damped. Explain. (Hint: Use Lenz's law; consider the paths that the charge carriers in the strip must follow if they are to oppose the motion.)

14. What is the direction, if any, of the conventional current through resistor *R* in Fig. 35-26 (*a*) immediately after switch *S* is closed, (*b*) some time after switch *S* was closed, and (*c*) immediately after switch *S* is opened. (*d*) When switch *S* is held closed, which end of the long coil acts as a north pole? (*e*) How do the free charges in the coil containing *R* know about the flux within the long coil? What really gets them moving?

15. In Fig. 35-27 the movable wire is moved to the right, causing an induced current as shown. What is the direction of **B** in region *A*?

16. Account qualitatively for the configurations of the lines of **B** in Figs. 35-6 *b* and *c*.

17. How would Fig. 35-8 be changed if we took into account the necessary fringing of the magnetic field **B** in Fig. 35-7?

18. (*a*) In Fig. 35-10, need the circle of radius *r* be a conducting loop for **E** and ε to be present? (*b*) If the circle of radius *r* were not concentric (moved slightly to the left, say), would ε change? Would the configuration of **E** around the circle change? (*c*) For a concentric circle of radius *r*, with *r* > *R*, does an emf exist? Do electric fields exist?

19. A copper ring and a wooden ring of the same dimensions are placed so that there is the same changing magnetic flux through each. How do the induced electric fields in each ring compare?

20. In Fig. 35-12 how can the induced emfs around paths 1 and 2 be identical? The induced electric fields are much weaker near path 1 than near path 2, as the spacing of the lines of force shows. See also Fig. 35-11.

21. In a certain betatron the electrons rotate counterclockwise as seen from above. In what direction must the magnetic field point and how must it change with time while the electron is being accelerated?

22. Why can a betatron be used for acceleration only during one-quarter of a cycle?

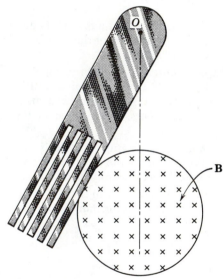

figure 35-25
Question 13

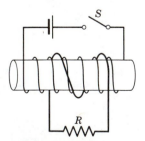

figure 35-26
Question 14

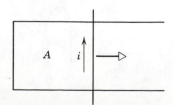

figure 35-27
Question 15

23. To make the electrons in a betatron orbit spiral outward, would it be necessary to increase or to decrease the central flux? Assume that **B** at the orbit remains essentially unchanged.

24. A cyclotron (see Section 33-7) is a so-called *resonance device.* Does a betatron depend on resonance? Discuss.

25. In Fig. 35-16a we can see that a force $(F_m \cos \theta)$ acts on the charge carriers in the left branch of the loop. However, if there is to be a continuous current in the loop, and there is, a force of some sort must act on charge carriers in the other three branches of the loop to maintain the same drift speed v_d in these branches. What is its source? (Hint: Consider that the left branch of the loop was the only conducting element, the other three being non-conducting. Would not positive charge pile up at the top of the left half and negative charge at the bottom?)

26. Show that one volt = one weber/second.

problems

SECTION 35-2

1. A uniform magnetic field **B** is normal to the plane of a circular ring 10 cm in diameter made of #10 copper wire (diameter = 0.10 in.). At what rate must B change with time if an induced current of 10 A is to appear in the ring?
Answer: 1.3 T/s.

2. You are given 50 cm of #18 copper wire (diameter = 0.040 in.). It is formed into a circular loop and placed at right angles to a uniform magnetic field that is increasing with time at the constant rate of 0.010 T/s. At what rate (in watts) is thermal energy generated in the loop?

3. A hundred turns of insulated copper wire are wrapped around an iron cylinder of cross-sectional area 0.001 m² and are connected to a resistor. The total resistance in the circuit is 10 Ω. If the longitudinal magnetic field in the iron changes from 1.0 T in one direction to 1.0 T in the opposite direction, how much charge flows through the circuit?   *Answer:* 2.0×10^{-2} C.

4. The current in the solenoid of Example 1 changes, not as in that Example, but according to $i = 3.0t + 1.0t^2$, where i is in amperes and t in seconds. (a) Plot quantitatively the induced emf in the coil from $t = 0$ to $t = 4.0$ s. (b) The resistance of the coil is 0.15 Ω. What is the instantaneous current in the coil at $t = 2.0$ s?

5. A small loop of area A is inside of, and has its axis in the same direction as, a long solenoid of n turns per unit length and current i. If $i = i_0 \sin \omega t$, find the emf ε in the loop.   *Answer:* $-\mu_0 n A i_0 \omega \cos \omega t$.

6. *Alternating current generator.* A rectangular loop of N turns and of length a and width b is rotated at a frequency ν in a uniform magnetic field **B,** as in Fig. 35-28. (a) Show that an induced emf given by

$$\varepsilon = 2\pi\nu NbaB \sin 2\pi\nu t = \varepsilon_0 \sin 2\pi\nu t$$

appears in the loop. This is the principle of the commercial alternating-current generator. (b) Design a loop that will produce an emf with $\varepsilon_0 = 150$ V when rotated at 60 rev/s in a magnetic field of 0.50 T.

7. In Fig. 35-29 a closed single turn with a copper coil resistance of 5.0 Ω is placed *outside* a solenoid like that of Example 1. If the current in the solenoid is changed as in that example, (a) what current appears in the loop while the solenoid current is being changed? (b) How do the free charges

figure 35-28
Problem 6

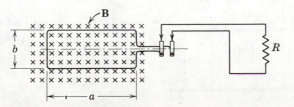

in the loop "get the message" from the solenoid that they should start moving (to establish a current)? After all, the magnetic flux is entirely confined to the interior of the solenoid.
Answer: (a) 2.1×10^{-4} A.

8. Figure 35-30 shows a copper rod moving on conducting rails with velocity **v** parallel to a long straight wire carrying a current i. Calculate the induced emf ε in the rod, assuming that $v = 5.0$ m/s, $i = 100$ A, $a = 1.0$ cm, and $b = 20$ cm.

9. A circular loop of wire 10 cm in diameter is placed with its normal making an angle of 30° with the direction of a uniform 0.50-T magnetic field. The loop is "wobbled" so that its normal rotates about the field direction at the constant rate of 100 rev/min; the angle between the normal and the field direction (= 30°) remains unchanged during this process. What emf appears in the loop?
Answer: Zero.

10. A stiff wire bent into a semicircle of radius R is rotated with a frequency ν in a uniform magnetic field **B**, as shown in Fig. 35-31. What are the amplitude and frequency of the induced emf and of the induced current when the internal resistance of the meter M is R_M and the remainder of the circuit has negligible resistance?

11. Figure 35-32 shows a copper rod moving with velocity **v** parallel to a long straight wire carrying a current i. Calculate the induced emf in the rod, assuming that $v = 5.0$ m/s, $i = 100$ A, $a = 1.0$ cm, and $b = 20$ cm.
Answer: 3.0×10^{-4} V.

12. A uniform magnetic field **B** is changing in magnitude at a constant rate dB/dt. You are given a mass m of copper which is to be drawn into a wire of radius r and formed into a circular loop of radius R. Show that the induced current in the loop does not depend on the size of the wire or of the loop and, assuming **B** perpendicular to the loop, is given by

$$i = \frac{m}{4\pi\rho\delta}\frac{dB}{dt},$$

where ρ is the resistivity and δ the density of copper.

13. A circular loop of radius r (10 cm) is placed in a uniform magnetic field **B** (0.80 T) normal to the plane of the loop. The radius of the loop begins shrinking at a constant rate dr/dt (80 cm/s). (a) What is the emf ε induced in the loop? (b) At what constant rate would the area have to shrink to induce this same emf?
Answer: (a) 0.40 V. (b) 0.50 m²/s.

14. Prove that the electric field **E** in a charged parallel-plate capacitor cannot drop abruptly to zero as one moves at right angles to it, as suggested by the arrow in Fig. 35-33 (see point a). In actual capacitors fringing of the lines of force always occurs, which means that **E** approaches zero in a continuous and gradual way; see Problem 34-2. (Hint: Apply Faraday's law to the rectangular path shown by the dashed lines.)

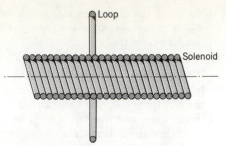

figure 35-29
Problem 7

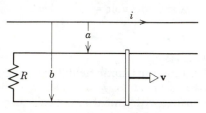

figure 35-30
Problem 8

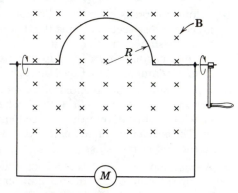

figure 35-31
Problem 10

figure 35-32
Problem 11

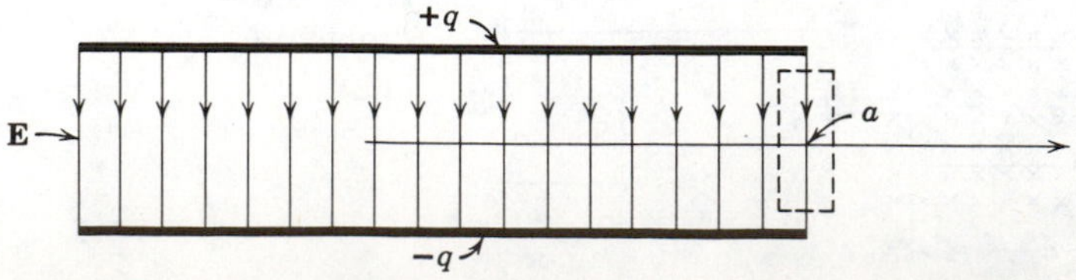

figure 35-33
Problem 14

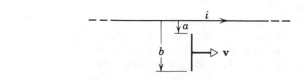

15. An electromagnetic "eddy current" brake consists of a disk of conductivity σ and thickness t rotating about an axis through its center with a magnetic field **B** applied perpendicular to the plane of the disk over a small area a^2 (see Fig. 35-34). If the area a^2 is at a distance r from the axis, find an approximate expression for the torque tending to slow down the disk at the instant its angular velocity equals ω. *Answer:* $\tau = B^2a^2r^2\,\omega\,\sigma\,t$.

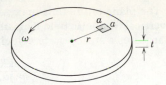

figure 35-34
Problem 15

SECTION 35-3

16. A small bar magnet is pulled rapidly through a conducting loop, along its axis. Sketch qualitatively (a) the induced current and (b) the rate of thermal energy production as a function of the position of the center of the magnet. Assume that the north pole of the magnet enters the loop first and that the magnet moves at constant speed. Plot the induced current as positive if it is clockwise as viewed along the path of the magnet.

17. A metal wire of mass m slides without friction on two rails spaced a distance d apart, as in Fig. 35-35. The track lies in a vertical uniform magnetic field **B**. (a) A *constant current i* flows from generator G along one rail, across the wire, and back down the other rail. Find the velocity (speed and direction) of the wire as a function of time, assuming it to be at rest at $t = 0$. (b) The generator is replaced by a battery with *constant emf ε*. The velocity of the wire now approaches a constant final value. What is this terminal speed? (c) What is the current in part (b) when the terminal speed has been reached? *Answer:* (a) $Bidt/m$, away from G. (b) ε/Bd. (c) Zero.

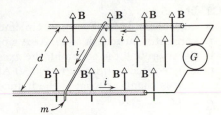

figure 35-35
Problem 17

18. In Fig. 35-36 the magnetic flux through the loop perpendicular to the plane of the coil and directed into the paper is varying according to the relation

$$\Phi_B = 6t^2 + 7t + 1,$$

where Φ_B is in milliwebers (1 milliweber $= 10^{-3}$ weber) and t is in seconds. (a) What is the magnitude of the emf induced in the loop when $t = 2.0$s? (b) What is the direction of the current through R?

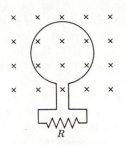

figure 35-36
Problem 18

SECTION 35-4

19. In the arrangement of Example 3 put B $= 1.2$ T and L $= 5.0$ cm. If $\varepsilon = 1.0$ V, what acceleration will a point at the end of the rotating rod experience? *Answer:* 2300 "g".

20. In Fig. 35-37 a conducting rod AB makes contact with the metal rails AD and BC which are 50 cm apart in a uniform magnetic field of 1.0 T perpendicular to the plane of the paper as shown. The total resistance of the circuit $ABCD$ is 0.4 Ω (assumed constant). (a) What is the magnitude and direction of the emf induced in the rod when it is moved to the left with a velocity of 8.0 m/s? (b) What force is required to keep the rod in motion? (c) Compare the rate at which mechanical work is done by the force **F** with the rate of development of thermal energy in the circuit.

21. In Fig. 35-38, $l = 2.0$ m and $v = 50$ cm/s. **B** is the earth's magnetic field, directed perpendicularly out of the page and having a magnitude 6.0×10^{-5} T at that place. The resistance of the circuit $ADCB$, assumed constant (explain how this may be achieved approximately), is $R = 1.2 \times 10^{-5}$ Ω. (a) What is the emf induced in the circuit? (b) What is the electric field in the wire AB? (c) What force does each electron in the wire experience due to the motion of the wire in the magnetic field? (d) What is the magnitude and direction of the current in the wire? (e) What force must an external agency exert in order to keep the wire moving with this constant velocity? (f) Compute the rate at which the external agency is doing work. (g) Compute the rate at which electrical energy is being converted into thermal energy. *Answer:* (a) 6.0×10^{-5} V. (b) 3.0×10^{-5} V/m. (c) 4.8×10^{-24} N. (d) 5.0 A. (e) 6.0×10^{-4} N. (f) 3.0×10^{-4} W. (g) 3.0×10^{-4} W.

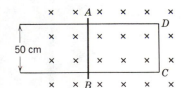

figure 35-37
Problem 20

22. A square wire of length l, mass m, and resistance R slides without friction down parallel conducting rails of negligible resistance, as in Fig. 35-39. The rails are connected to each other at the bottom by a resistanceless rail parallel to the wire, so that the wire and rails form a closed rectangular conduct-

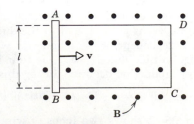

figure 35-38
Problem 21

ing loop. The plane of the rails makes an angle θ with the horizontal, and a uniform vertical magnetic field **B** exists throughout the region. (a) Show that the wire acquires a steady-state velocity of magnitude

$$v = \frac{mgR \sin \theta}{B^2 l^2 \cos^2 \theta}.$$

(b) Prove that this result is consistent with the conservation-of-energy principle. (c) What change, if any, would there be if **B** were directed down instead of up?

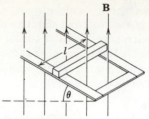

figure 35-39
Problem 22

SECTION 35-5

23. A long solenoid of radius r (2.5 cm) and turns per unit length n (100/cm) carries an initial current of i_0 (1.0 A). A single loop of wire of diameter D (10 cm) is placed around the solenoid, the axes coinciding. The current in the solenoid is reduced uniformly to i (0.50 A) over a period T (0.010 s). While the current is changing what is the induced emf ε in the surrounding loop?
   Answer: 1.2×10^{-3} V.

24. A circular coil of radius r (10 cm) is made of wire of resistance R (10 Ω). Perpendicular to the plane of the coil is a uniform magnetic field **B**. (a) At what constant rate must B increase so that there is a steady current i (0.010 A) in the circuit? (b) What power is dissipated in the resistor?

25. (a) For the arrangement of Fig. 34-28, what would be the current induced around the rectangular loop if the current in the wire decreased uniformly from 30 A to zero in 1.0 s? Assume no initial current in the loop and a resistance for the loop of 0.020 Ω. Take $a = 1.0$ cm, $b = 8.0$ cm, and $l = 30$ cm. (b) How much energy would be transferred to the loop in the 1.0-s interval?
   Answer: (a) 1.9×10^{-4} A. (b) 7.2×10^{-10} J.

26. The perpendicular field B through a one-turn circular loop of wire of negligible resistance changes with time as shown in Fig. 35-40. The loop is of radius r (10 cm) and is connected to a resistor R (10 Ω). (a) Plot the emf appearing across the resistor. (b) Plot the current i through the resistor R. (c) Plot the rate of thermal energy production in the resistor.

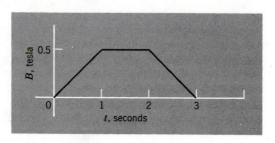

figure 35-40
Problem 26

27. A loop of wire of area A is connected to a resistance R. The loop is exposed to a time-varying **B** field (see Fig. 35-41). (a) Derive an expression for the net charge transferred through the resistor between $t = t_1$ and $t = t_2$. Show that your answer is proportional to the difference $\Phi_B(t_2) - \Phi_B(t_1)$, and is otherwise independent of the manner in which **B** is changing. (b) Suppose the change in flux, $\Phi_B(t_2) - \Phi_B(t_1)$ is zero. Does it then follow that no thermal energy production occurred during this time interval?

   Answer: (a) $Q = -\dfrac{1}{R}[\Phi_B(t_2) - \Phi_B(t_1)]$. (b) No.

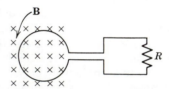

figure 35-41
Problem 27

28. Figure 35-42 shows two loops of wire having the same axis. The smaller loop is above the larger one, by a distance x which is large compared to the radius R of the larger loop. Hence, with current i flowing as indicated in the

larger loop, the consequent magnetic field is nearly constant throughout the plane area πr^2 bounded by the smaller loop. Suppose now that x is not constant but is changing at the constant rate $dx/dt = v$ (x increasing). (a) Determine the magnetic flux across the area bounded by the smaller loop as a function of x. (b) Compute the emf generated in the smaller loop at the instant when $x = NR$. (c) Determine the direction of the induced current flowing in the smaller loop if $v > 0$.

29. A wire is bent into three circular segments of radius r (10 cm) as shown in Fig. 35-43. Each segment is a quadrant of a circle, ab lying in the x-y plane, bc lying in the y-z plane, and ca lying in the z-x plane. (a) If a spatially uniform magnetic field $\mathbf{B}$ points in the x-direction, what is the magnitude of the emf ε developed in the wire when $\mathbf{B}$ increases at the rate of 3.0×10^{-3} T/s? (b) What is the direction of the current in the segment bc?
Answer: (a) 2.4×10^{-5} V. (b) From c to b.

30. Figure 35-44 shows a uniform magnetic field $\mathbf{B}$ confined to a cylindrical volume of radius R. $\mathbf{B}$ is decreasing in magnitude at a constant rate of 0.010 T/s. What is the instantaneous acceleration (direction and magnitude) experienced by an electron placed at a, at b, and at c? Assume $r = 5.0$ cm. (The necessary fringing of the field beyond R will not change your answer as long as there is axial symmetry about a perpendicular axis through b.)

31. A closed loop of wire consists of a pair of equal semicircles, radius 3.70 cm, lying in mutually perpendicular planes. The loop was formed by folding a circular loop along a diameter until the two halves became perpendicular. A uniform magnetic field $\mathbf{B}$ of magnitude 0.076 T is directed perpendicular to the fold diameter and makes equal angles (45°) with the planes of the semicircles as shown in Fig. 35-45a. (a) The magnetic field is reduced at a uniform rate to zero during a time interval 4.50×10^{-3} s. Determine the magnitude of the induced emf and the sense of the induced current in the loop during this interval. (b) How would the answers change if $\mathbf{B}$ is directed as shown in Fig. 35-45b, perpendicular to the direction first given for it but still perpendicular to the "fold-diameter?"
Answer: (a) 51×10^{-3} V. (b) $\varepsilon = 0$.

32. A uniform magnetic field $\mathbf{B}$ fills a cylindrical volume of radius R. A metal rod of length l is placed as shown in Fig. 35-46. If B is changing at the rate dB/dt, show that the emf that is produced by the changing magnetic field and that acts between the ends of the rod is given by

$$\varepsilon = \frac{dB}{dt} \frac{l}{2} \sqrt{R^2 - \left(\frac{l}{2}\right)^2} \; .$$

SECTION 35-6

33. Some measurements of the maximum magnetic field as a function of radius for the betatron described on page 781 are as follows:

r, cm	B, tesla	r, cm	B, tesla
0	0.400	81.2	0.409
10.2	0.950	83.7	0.400
68.2	0.950	88.9	0.381
73.2	0.528	91.4	0.372
75.2	0.451	93.5	0.360
77.3	0.428	95.5	0.340

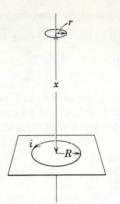

figure 35-42
Problem 28

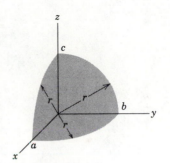

figure 35-43
Problem 29

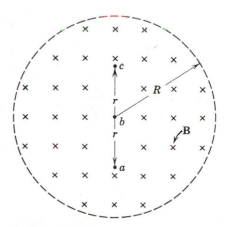

figure 35-44
Problem 30

figure 35-45
Problem 31

(a)

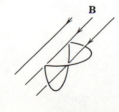

(b)

Show by graphical analysis that the relation $\overline{B} = 2B_R$ is satisfied at the orbit radius, $R = 84$ cm. (Hint: Note that $\overline{B} = \dfrac{1}{\pi R^2}\displaystyle\int_0^R B(r)(2\pi r)\,dr$ and evaluate the integral graphically.)

SECTION 35-7

34. (a) Estimate θ in Fig. 35-16a. Recall (see Section 31-1) that $v_d = 4 \times 10^{-2}$ cm/s in a typical case. Assume $v = 10$ cm/s. (b) It is clear that θ will be small. However, must we have $\theta \neq 0$ for the arguments presented in connection with this figure to be valid?

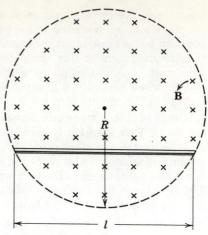

figure 35-46
Problem 32

36
inductance

If two coils are near each other, a current i in one coil will set up a flux Φ_B through the second coil. If this flux is changed by changing the current, an induced emf will appear in the second coil according to Faraday's law; see Fig. 35-2. However, two coils are not needed to show an inductive effect. An induced emf appears in a coil if the current *in that same coil* is changed. This is called *self-induction* and the electromotive force produced is called a *self-induced emf*. It obeys Faraday's law of induction just as other induced emfs do.

Consider first a "close-packed" coil, a toroid, or the central section of a long solenoid. In all three cases the flux Φ_B set up in each turn by a current i is essentially the same for every turn. Faraday's law for such coils (Eq. 35-2)

$$\varepsilon = -\frac{d(N\Phi_B)}{dt} \tag{36-1}$$

shows that the number of *flux linkages* $N\Phi_B$ (N being the number of turns) is the important characteristic quantity for induction. For a given coil, provided no magnetic materials such as iron are nearby, this quantity is proportional to the current i, or

$$N\Phi_B = Li, \tag{36-2}$$

in which L, the proportionality constant, is called the *inductance* of the device.

From Faraday's law (see Eq. 36-1) we can write the induced emf as

$$\varepsilon = -\frac{d(N\Phi_B)}{dt} = -L\frac{di}{dt}. \tag{36-3a}$$

Written in the form $$L = -\frac{\varepsilon}{di/dt},$$ (36-3b)

this relation may be taken as the defining equation for inductance for coils of all shapes and sizes, whether or not they are close-packed and whether or not iron or other magnetic material is nearby. It is analogous to the defining relation for capacitance, namely,

$$C = \frac{q}{V}.$$

If no iron or similar materials are nearby, L depends only on the geometry of the device. In an *inductor* (symbol ⌇⌇⌇) the presence of a *magnetic field* is the significant feature, corresponding to the presence of an *electric* field in a *capacitor*.

The SI unit of inductance, from Eq. 36-3b, is the volt · second/ampere. A special name, the *henry* (abbr. H), has been given to this combination of units, or

$$1 \text{ henry} = 1 \text{ volt · second/ampere}.$$

As we saw in Section 35-1, the SI unit of inductance is named after Joseph Henry (1797–1878), an American physicist and a contemporary of Faraday.

We can find the direction of a self-induced emf from Lenz's law; see Section 35-3. Suppose that a steady current i, produced by a battery, exists in a coil. Let us suddenly *decrease* the (battery) emf in the circuit. The current i will start to *decrease* at once. This decrease in current, in the language of Lenz's law, is the "change" which the self-induction must oppose. To oppose the falling current, the induced emf must point in the *same* direction as the current, as shown in Fig. 36-1a.

However, if we suddenly *increase* the (battery) emf, the current i will start to *increase* at once. This increase in current is now the "change" which self-induction must oppose. To oppose the rising current the induced emf must point in the *opposite* direction to the current, as Fig. 36-1b shows. In each case the self-induced emf acts to oppose the *change* in the current. The minus sign in Eq. 36-3 shows that ε and di/dt are opposite in sign, since L is always a positive quantity.

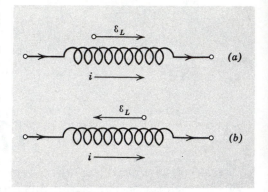

figure 36-1
In (a) the current i is made to *decrease* and in (b) it is made to *increase*. The self-induced emf ε_L opposes the *change* in each case.

36-2
CALCULATION OF INDUCTANCE

We were able to make a direct calculation of capacitance in terms of geometrical factors for a few special cases, such as the parallel-plate capacitor. In the same way, we can calculate the self-inductance L for a few special cases.

For a close-packed coil with no iron nearby, we have, from Eq. 36-2,

$$L = \frac{N\Phi_B}{i}.$$ (36-4)

Let us apply this equation to calculate L for a section of length l near the center of a long solenoid. The number of flux linkages in the length l of the solenoid is

$$N\Phi_B = (nl)(BA),$$

where n is the number of turns per unit length, B is the magnetic field inside the solenoid, and A is the cross-sectional area. From Eq. 34-7, B is given by

$$B = \mu_0 ni.$$

Combining these equations gives

$$N\Phi_B = \mu_0 n^2 liA.$$

Finally, the inductance, from Eq. 36-4, is

$$L = \frac{N\Phi_B}{i} = \mu_0 n^2 lA. \qquad (36\text{-}5)$$

The inductance of a length l of a solenoid is proportional to its volume (lA) and to the square of the number of turns per unit length. Note that it depends on geometrical factors only. The proportionality to n^2 is expected. If we double the number of turns per unit length, not only is the *total* number of turns N doubled but the flux *through each turn* Φ_B is also doubled, an over-all factor of four for the flux linkages $N\Phi_B$, hence also a factor of four for the inductance (Eq. 36-4).

EXAMPLE 1

Derive an expression for the inductance of a toroid of rectangular cross section as shown in Fig. 36-2. Evaluate for $N = 10^3$, $a = 5.0$ cm, $b = 10$ cm, and $h = 1.0$ cm.

The lines of **B** for the toroid are concentric circles. Applying Ampère's law,

$$\oint \mathbf{B} \cdot d\mathbf{l} = \mu_0 i,$$

to a circular path of radius r yields

$$(B)(2\pi r) = \mu_0 i_0 N,$$

where N is the number of turns and i_0 is the current in the toroid windings; recall that i in Ampère's law is the *total* current that passes through the path of integration. Solving for B yields

$$B = \frac{\mu_0 i_0 N}{2\pi r}.$$

The flux Φ_B for the cross section of the toroid is

$$\Phi_B = \int \mathbf{B} \cdot d\mathbf{S} = \int_a^b (B)(h\ dr) = \int_a^b \frac{\mu_0 i_0 N}{2\pi r} h\ dr$$

$$= \frac{\mu_0 i_0 Nh}{2\pi} \int_a^b \frac{dr}{r} = \frac{\mu_0 i_0 Nh}{2\pi} \ln \frac{b}{a},$$

where $h\ dr$ is the area of the elementary strip shown (dashed lines) in the figure.

The inductance follows from Eq. 36-4, or

$$L = \frac{N\Phi_B}{i_0} = \frac{\mu_0 N^2 h}{2\pi} \ln \frac{b}{a}.$$

Substituting numerical values yields

$$L = \frac{(4\pi \times 10^{-7}\ \text{Wb/A} \cdot \text{m})(10^3)^2(1.0 \times 10^{-2}\ \text{m})}{2\pi} \ln \frac{10 \times 10^{-2}\ \text{m}}{5 \times 10^{-2}\ \text{m}}$$

$$= 1.4 \times 10^{-3}\ \text{Wb/A} = 1.4\ \text{mH}.$$

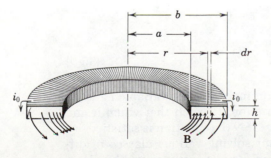

figure 36-2
Example 1. A cross section of a toroid, showing the current in the windings and the magnetic field.

In Section 32-8 we saw that if we suddenly introduce an emf ε, perhaps by using a battery, into a single loop circuit containing a resistor R and a capacitor C, the charge does not build up immediately to its final equilibrium value $(= C\varepsilon)$ but approaches it in an exponential fashion described by Eq. 32-15, or

$$q = C\varepsilon\ (1 - e^{-t/\tau_C}). \qquad (36\text{-}6)$$

The delay in the rise of the charge is described by the *capacitive time constant* τ_C, defined from

$$\tau_C = RC. \qquad (36\text{-}7)$$

If in this same circuit the battery emf ε is suddenly removed, the charge does not immediately fall to zero but approaches zero in an exponential fashion, described by Eq. 32-18b, or

$$q = C\varepsilon\ e^{-t/\tau_C}. \qquad (36\text{-}8)$$

The same time constant τ_C describes the fall of the charge as well as its rise.

An analogous delay in the rise (or fall) of the current occurs if we suddenly introduce an emf ε into (or remove it from) a single-loop circuit containing a resistor R and an inductor L. When the switch S in Fig. 36-3 is closed on a, for example, the current in the resistor starts to rise. If the inductor were not present, the current would rise rapidly to a steady value ε/R. Because of the inductor, however, a self-induced emf ε_L appears in the circuit and, from Lenz's law, this emf opposes the *rise* of current, which means that it opposes the battery emf ε in polarity. Thus the resistor responds to the difference between two emfs, a constant one ε due to the battery and a variable one ε_L $(= -L\ di/dt)$ due to self-induction. As long as this second emf is present, the current in the resistor will be less than ε/R.

As time goes on, the rate at which the current increases becomes less rapid and the self-induced emf ε_L, which is proportional to di/dt, becomes smaller. Thus a time delay is introduced, and the current in the circuit approaches the value ε/R asymptotically.

When the switch S in Fig. 36-3 is thrown to a, the circuit reduces to that of Fig. 36-4. Let us apply the loop theorem, starting at x in this figure and going clockwise around the loop. For the direction of current shown, x will be higher in potential than y, which means that we encounter a drop in potential of $-iR$ as we traverse the resistor. Point y is higher in potential than point z because, for an increasing current, the induced emf will oppose the *rise* of the current by pointing as shown. Thus as we traverse the inductor from y to z we encounter a drop in potential of $-L(di/dt)$. We encounter a rise in potential of $+\varepsilon$ in traversing the battery from z to x. The loop theorem thus gives

$$-iR - L\frac{di}{dt} + \varepsilon = 0$$

or

$$L\frac{di}{dt} + iR = \varepsilon. \qquad (36\text{-}9)$$

Equation 36-9 is a *differential equation* involving the variable i and its first derivative di/dt. We seek the function $i(t)$ such that when it and its first derivative are substituted in Eq. 36-9 the equation is satisfied.

Although there are formal rules for solving various classes of differential equations (and Eq. 36-9 can, in fact, be easily solved by direct

36-3
AN LR CIRCUIT

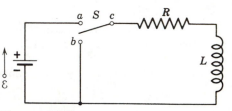

figure 36-3
An *LR* circuit.

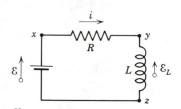

figure 36-4
The circuit of Fig. 36-3 just after switch S is closed on a.

integration, after rearrangement) we often find it simpler to guess at the solution, guided by physical reasoning and by previous experience. We can test any proposed solution by substituting it in the differential equation and seeing whether this equation reduces to an identity.

The solution to Eq. 36-9 is, we assert,

$$i = \frac{\varepsilon}{R} (1 - e^{-Rt/L}). \tag{36-10}$$

To test this solution by substitution, we find the derivative di/dt, which is

$$\frac{di}{dt} = \frac{\varepsilon}{L} e^{-Rt/L}. \tag{36-11}$$

Substituting i and di/dt into Eq. 36-9 leads to an identity, as you can easily verify. Thus Eq. 36-10 is a solution of Eq. 36-9. Figure 36-5 shows how the potential difference V_R across the resistor $(= iR$; see Eq. 36-10) and V_L across the inductor $(= L\, di/dt$; see Eq. 36-11) vary with time for particular values of ε, L, and R. Compare this figure carefully with the corresponding figure for an RC circuit (Fig. 32-11).

We can rewrite Eq. 36-10 as

$$i = \frac{\varepsilon}{R} (1 - e^{-t/\tau_L}), \tag{36-12}$$

in which τ_L, the *inductive time constant*, is given by

$$\tau_L = L/R. \tag{36-13}$$

Note the correspondence between Eqs. 36-6 and 36-12.

To show that the quantity τ_L $(= L/R)$ has the dimensions of time, we put

$$\frac{1 \text{ henry}}{\text{ohm}} = \frac{1 \text{ henry}}{\text{ohm}} \left(\frac{1 \text{ volt} \cdot \text{second}}{1 \text{ henry} \cdot \text{ampere}} \right) \left(\frac{1 \text{ ohm} \cdot \text{ampere}}{1 \text{ volt}} \right) = 1 \text{ second.}$$

The first quantity in parentheses is a conversion factor based on the defining equation for inductance $[L = -\varepsilon/(di/dt)$; see Eq. 36-3b]. The second conversion factor is based on the relation $V = iR$.

The physical significance of the time constant follows from Eq. 36-12. If we put $t = \tau_L = L/R$ in this equation, it reduces to

$$i = \frac{\varepsilon}{R} (1 - e^{-1}) = (1 - 0.37) \frac{\varepsilon}{R} = 0.63 \frac{\varepsilon}{R}.$$

Thus the time constant τ_L is that time at which the current in the circuit will reach a value within $1/e$ (about 37%) of its final equilibrium value (see Fig. 36-5).

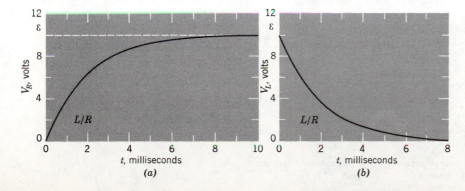

(a)   (b)

figure 36-5
If in Fig. 36-3 we assume that $R = 2000\ \Omega$, $L = 4$ H, and $\varepsilon = 10$ V, then (a) shows the variation of V_R with t during the current buildup after switch S is closed on a, and (b) the variation of V_L with t. The time constant is $L/R = 2.0 \times 10^{-3}$ s. Compare this figure carefully with Fig. 32-11, the corresponding figure for an RC circuit.

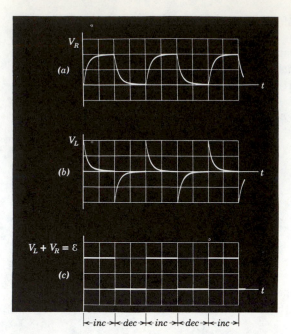

figure 36-6
Oscilloscope photograph showing
the variation with time of (a) the
potential drop V_R across the resistor,
(b) the potential drop V_L across the
inductor, and (c) the applied emf ε.
During the intervals marked *inc*
the current is increasing; during
those marked *dec* it is decreasing.
Compare with Fig. 32-13. (Courtesy
E. K. Hege, Rensselaer Polytechnic
Institute.)

If the switch S in Fig. 36-3, having been left in position *a* long enough
for the equilibrium current ε/R to be established, is thrown to *b**, the
effect is to remove the battery from the circuit. The differential equation
that governs the subsequent decay of the current in the circuit can be
found by putting $\varepsilon = 0$ in Eq. 36-9, or

$$L \frac{di}{dt} + iR = 0. \tag{36-14}$$

You can show by the test of substitution that the solution of this differential equation is

$$i = \frac{\varepsilon}{R} e^{-t/\tau_L}. \tag{36-15}$$

Just as for the *RC* circuit, the behavior of the circuit of Fig. 36-3 can be investigated experimentally, using a cathode-ray oscilloscope. If switch S in this figure
is thrown periodically between *a* and *b*, the applied (battery) emf alternates
between the values ε and zero. If the terminals of an oscilloscope are connected
across *b* and *c* in Fig. 36-3, the oscilloscope will display the waveform of this
applied emf on its screen, as in Fig. 36-6c.

If the terminals of the oscilloscope are connected across the resistor, the
waveform displayed (Fig. 36-6a) will be that of the current in the circuit, since
the potential drop across *R*, which determines the oscilloscope deflection, is
given by $V_R = iR$. During the intervals marked *inc* in Fig. 36-6, the current is
increasing and the waveform (see Eq. 36-12) is given by

$$V_R (= iR) = \varepsilon(1 - e^{-t/\tau_L}).$$

During the intervals marked *dec*, the current is decreasing and V_R (see Eq.
36-15) is given by

$$V_R (= iR) = \varepsilon e^{-t/\tau_L}.$$

Note that both the growth and the decay of the current are delayed.

If the oscilloscope terminals are connected across the inductor, the screen
will show a plot of the potential difference across it as a function of time
(Fig. 36-6b). While the current is increasing, the equation of the trace (see Eq.
36-11) should be

---

* The connection to *b* must be made before the connection to *a* is broken. A switch
which does this is called a "make before break" switch.

$$V_L \left(= L \frac{di}{dt}\right) = \varepsilon e^{-t/\tau_L}.$$

When the current is decreasing, V_L is given in terms of the time derivative of Eq. 36-15 and is

$$V_L \left(= L \frac{di}{dt}\right) = -\varepsilon e^{-t/\tau_L}.$$

Note that V_L is opposite in sign when the current is increasing (di/dt positive) and when it is decreasing (di/dt negative), as is true also for the induced emf $\varepsilon_L [=-L(di/dt)=-V_L]$.

Examination of Fig. 36-6 shows that at any instant the sum of curves a and b always yields curve c. This is an expected consequence of the loop theorem, as Eq. 36-9 shows.

EXAMPLE 2

A solenoid has an inductance of 50 H and a resistance of 30 Ω. If it is connected to a 100-V battery, how long will it take for the current to reach one-half its final equilibrium value?

The equilibrium value of the current is reached as $t \rightarrow \infty$; from Eq. 36-12 it is ε/R. If the current has half this value at a particular time t_0, this equation becomes

$$\frac{1}{2} \frac{\varepsilon}{R} = \frac{\varepsilon}{R} (1 - e^{-t_0/\tau_L}).$$

Solving for t_0 yields

$$t_0 = \tau_L \ln 2 = 0.69 \frac{L}{R}.$$

Using the values given, this reduces to

$$t_0 = 0.69\tau_L = 0.69 \left(\frac{50 \text{ H}}{30 \ \Omega}\right) = 1.2 \text{ s}.$$

36-4
ENERGY AND THE MAGNETIC FIELD

When we lift a stone we do work, which we can get back again by lowering the stone. It is convenient to think of the work done to lift the stone being stored temporarily in the gravitational field between the earth and the lifted stone, and being withdrawn from this field when we lower the stone.

When we pull two unlike charges apart we like to say that the work we do is stored in the electric field between the charges. We can get it back from the field by letting the charges move closer together again.

In the same way we can store energy in a magnetic field. For example, two long, rigid, parallel wires carrying current in the same direction attract each other and we must do work to pull them apart. We can get this stored energy back at any time by letting the wires move back to their original positions.

To derive a quantitative expression for the storage of energy in the magnetic field, consider Fig. 36-4, which shows a source of emf ε connected to a resistor R and an inductor L.

$$\varepsilon = iR + L \frac{di}{dt}, \tag{36-9}$$

is the differential equation that describes the growth of current in this circuit. We stress that this equation follows immediately from the loop theorem and that the loop theorem in turn is an expression of the principle of conservation of energy for single-loop circuits. If we multiply each side of Eq. 36-9 by i, we obtain

$$\varepsilon i = i^2 R + Li\frac{di}{dt}, \qquad (36\text{-}16)$$

which has the following physical interpretation in terms of work and energy:

1. If a charge dq passes through the seat of emf ε in Fig. 36-4 in time dt, the seat does work on it in amount $\varepsilon\,dq$. The *rate* of doing work is $(\varepsilon\,dq)/dt$, or εi. Thus the left term in Eq. 36-16 is the *rate at which the seat of emf delivers energy to the circuit.*

2. The second term in Eq. 36-16 is the *rate at which energy appears as thermal (Joule) energy in the resistor.*

3. Energy that does not appear as thermal (Joule) energy must, by our hypothesis, be stored in the magnetic field. Since Eq. 36-16 represents a statement of the conservation of energy for LR circuits, the last term must represent the *rate dU_B/dt at which energy is stored in the magnetic field*, or

$$\frac{dU_B}{dt} = Li\frac{di}{dt}. \qquad (36\text{-}17)$$

We can write this as $\qquad dU_B = Li\,di.$
Integrating yields

$$U_B = \int_0^{U_B} dU_B = \int_0^i Li\,di = \tfrac{1}{2}Li^2, \qquad (36\text{-}18)$$

which represents the total stored magnetic energy in an inductance L carrying a current i.

We can compare this relation with the expression for the energy associated with a capacitor C carrying a charge q, namely,

$$U_E = \frac{1}{2}\frac{q^2}{C}.$$

Here the energy is stored in an electric field. In each case the expression for the stored energy was derived by setting it equal to the work that must be done to set up the field.

EXAMPLE 3

A coil has an inductance of 5.0 H and a resistance of 20 Ω. If a 100-V emf is applied, what energy is stored in the magnetic field after the current has built up to its maximum value ε/R?

The maximum current is given by

$$i = \frac{\varepsilon}{R} = \frac{100\text{ V}}{20\ \Omega} = 5.0\text{ A}.$$

The stored energy is given by Eq. 36-18:

$$U_B = \tfrac{1}{2}Li^2 = \tfrac{1}{2}(5.0\text{ H})(5.0\text{ A})^2 = 63\text{ J}.$$

Note that the time constant for this coil ($= L/R$) is 0.25 s. After how many time constants will *half* of this equilibrium energy be stored in the field?

EXAMPLE 4

A 3.0 H inductor is placed in series with a 10-Ω resistor, an emf of 3.0 V being suddenly applied to the combination. At 0.30 s (which is one inductive time constant) after the contact is made, (a) what is the rate at which energy is being delivered by the battery?

The current is given by Eq. 36-12, or

$$i = \frac{\varepsilon}{R} (1 - e^{-t/\tau_L}),$$

which at $t = 0.30$ s $(= \tau_L)$ has the value

$$i = \left(\frac{3.0 \text{ V}}{10 \text{ }\Omega}\right) (1 - e^{-1}) = 0.189 \text{ A}.$$

The rate P_ε at which energy is delivered by the battery is

$$P_\varepsilon = \varepsilon i$$

$$= (3.0 \text{ V})(0.189 \text{ A})$$

$$= 0.567 \text{ W}.$$

(b) At what rate does energy appear as thermal (Joule) energy in the resistor? This is given by

$$P_J = i^2 R$$

$$= (0.189 \text{ A})^2 (10 \text{ }\Omega)$$

$$= 0.357 \text{ W}.$$

(c) At what rate P_B is energy being stored in the magnetic field?

This is given by the last term in Eq. 36-16, which requires that we know di/dt. Differentiating Eq. 36-12 yields

$$\frac{di}{dt} = \left(\frac{\varepsilon}{R}\right)\left(\frac{R}{L}\right) e^{-t/\tau_L}$$

$$= \frac{\varepsilon}{L} e^{-t/\tau_L}.$$

At $t = \tau_L$ we have

$$\frac{di}{dt} = \left(\frac{3.0 \text{ V}}{3.0 \text{ H}}\right) e^{-1} = 0.37 \text{ A/s}.$$

From Eq. 36-17, the desired rate is

$$P_B = \frac{dU_B}{dt} = Li \frac{di}{dt}$$

$$= (3.0 \text{ H})(0.189 \text{ A})(0.37 \text{ A/s})$$

$$= 0.210 \text{ W}.$$

Note that as required by the principle of conservation of energy (see Eq. 36-16)

$$P_\varepsilon = P_J + P_B,$$

or

$$0.567 \text{ W} = 0.357 \text{ W} + 0.210 \text{ W}$$

$$= 0.567 \text{ W}.$$

We now derive an expression for the *density* of energy u_B in a magnetic field. Consider a length l near the center of a very long solenoid; Al is the volume associated with this length. The stored energy must lie entirely within this volume because the magnetic field outside such a solenoid is essentially zero. Moreover, the stored energy must be uniformly distributed throughout the volume of the solenoid because the magnetic field is uniform everywhere inside.

36-5
ENERGY DENSITY AND THE MAGNETIC FIELD

Thus, we can write $\qquad u_B = \dfrac{U_B}{Al}$

or, since $\qquad U_B = \frac{1}{2}Li^2,$

we have $\qquad u_B = \dfrac{\frac{1}{2}Li^2}{Al}.$

To express this in terms of the magnetic field, we can substitute for L in this equation, using the relation $L = \mu_0 n^2 lA$ (Eq. 36-5). Also we can solve Eq. 34-7 $(B = \mu_0 in)$ for i and substitute in this equation. Doing so yields finally

$$u_B = \frac{1}{2}\frac{B^2}{\mu_0}. \qquad (36\text{-}19)$$

This equation gives the energy density stored at any point (in a vacuum or in a nonmagnetic substance) where the magnetic field is **B**. The equation is true for all magnetic field configurations, even though we derived it by considering a special case, the solenoid. Equation 36-19 is to be compared with Eq. 30-9,

$$u_E = \frac{1}{2}\epsilon_0 E^2, \qquad (36\text{-}20)$$

which gives the energy density (in a vacuum) at any point in an electric field. Note that both u_B and u_E are proportional to the square of the appropriate field quantity, B or E.

The solenoid plays a role with relationship to magnetic fields similar to the role the parallel-plate capacitor plays with respect to electric fields. In each case we have a simple device that can be used for setting up a uniform field throughout a well-defined region of space and for deducing, in a simple way, some properties of these fields.

EXAMPLE 5

A long _coaxial cable_ (Fig. 36-7) consists of two concentric cylinders with radii a and b. Its central conductor carries a steady current i, the outer conductor providing the return path. (a) Calculate the energy stored in the magnetic field for a length l of such a cable.

In the space between the two conductors Ampère's law,

$$\oint \mathbf{B} \cdot d\mathbf{l} = \mu_0 i,$$

leads to $\qquad (B)(2\pi r) = \mu_0 i$

or $\qquad B = \dfrac{\mu_0 i}{2\pi r}.$

Ampère's law shows further that the magnetic field is zero for points outside the outer conductor (why?). Magnetic fields exist _inside_ each of the conductors; although we can find their values from Ampère's law, we choose to ignore them, on the assumption that the cable dimensions are chosen so that most of the stored magnetic energy is in the space between the conductors.

The energy density for points between the conductors, from Eq. 36-19, is

$$u_B = \frac{1}{2\mu_0}B^2 = \frac{1}{2\mu_0}\left(\frac{\mu_0 i}{2\pi r}\right)^2 = \frac{\mu_0 i^2}{8\pi^2 r^2}.$$

Consider a volume element dV consisting of a cylindrical shell whose radii are r and $r + dr$ and whose length is l. The energy dU contained in it is

$$dU = u_B dV = \frac{\mu_0 i^2}{8\pi^2 r^2}(2\pi rl)(dr) = \frac{\mu_0 i^2 l}{4\pi}\frac{dr}{r}.$$

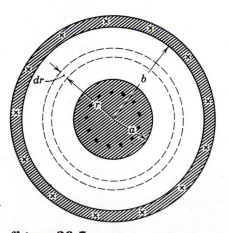

figure 36-7
Example 5. Cross section of a coaxial cable, showing steady equal but opposite currents in the central and outer conductors.

The total stored magnetic energy is found by integration, or

$$U = \int dU = \frac{\mu_0 i^2 l}{4\pi} \int_a^b \frac{dr}{r} = \frac{\mu_0 i^2 l}{4\pi} \ln \frac{b}{a},$$

which is the desired expression.

(b) What is the inductance of a length l of coaxial cable?

We can find the inductance L from Eq. 36-18 $(U = \frac{1}{2}Li^2)$, which leads to

$$L = \frac{2U}{i^2} = \frac{\mu_0 l}{2\pi} \ln \frac{b}{a}.$$

You should also derive this expression directly from the definition of inductance, using the procedures of Example 1.

EXAMPLE 6

Compare the energy required to set up, in a cube 10 cm on edge (a) a uniform electric field of 1.0×10^5 V/m and (b) a uniform magnetic field of 1.0 T $(= 10^4$ gauss). Both these fields would be judged reasonably large but they are readily available in the laboratory.

(a) In the electric case we have, where V_0 is the volume of the cube,

$$U_E = u_E V_0 = \tfrac{1}{2}\epsilon_0 E^2 V_0$$

$$= (0.5)(8.9 \times 10^{-12} \text{ C}^2/\text{N} \cdot \text{m}^2)(10^5 \text{ V/m})^2(0.1 \text{ m})^3$$

$$= 4.5 \times 10^{-5} \text{ J}.$$

(b) In the magnetic case, from Eq. 36-19, we have

$$U_B = u_B V_0 = \frac{B^2}{2\mu_0} V_o = \frac{(1.0 \text{ T})^2(0.1 \text{ m})^3}{(2)(4\pi \times 10^{-7} \text{ T} \cdot \text{m/A})}$$

$$= 400 \text{ J}.$$

In terms of fields normally available in the laboratory, much larger amounts of energy can be stored in a magnetic field than in an electric one, the ratio being about 10^7 in this example. Conversely, much more energy is required to set up a magnetic field of reasonable laboratory magnitude than is required to set up an electric field of similarly reasonable magnitude.

36-6
MUTUAL INDUCTANCE

In Section 35-2 we saw that if two coils are close together, as in Fig. 35-2, a steady current i in one coil will set up a magnetic flux Φ linking the other coil. If we change i with time, an emf ε given by Faraday's law (Eq. 35-2) appears in the second coil; we called this process *induction*. We could better have called it *mutual induction*, to suggest the mutual interaction of the two coils and to distinguish it from *self-induction*, in which only one coil is involved as described in the preceding sections of this chapter.

Let us look a little more quantitatively at mutual induction. Figure 36-8 shows two circular close-packed coils near each other and sharing a common axis. There is a current i_1 in coil 1, set up by an external circuit not shown. This current produces a magnetic field suggested by the lines of $\mathbf{B}_1$ in the figure. Coil 2 stands alone as a closed circuit, with no external connections to a battery; a magnetic flux Φ_{21} passes through (links) it.

We define the *mutual inductance* M_{21} of coil 2 with respect to coil 1 as

$$M_{21} = \frac{N_2\Phi_{21}}{i_1}. \tag{36-21a}$$

Compare this with Eq. 36-4 $(L = N\Phi/i)$, the definition of (self-) inductance. We can recast Eq. 36-21a as

$$M_{21}i_1 = N_2\Phi_{21}. \qquad (36\text{-}21b)$$

If, by external means, we cause i_1 to vary with time, we have

$$M_{21}\frac{di_1}{dt} = N_2\frac{d\Phi_{21}}{dt}. \qquad (36\text{-}22)$$

The right side of this equation, from Faraday's law (Eq. 35-2), is, apart from a change in sign, just the emf ε_2 appearing in coil 2 due to the changing current in coil 1, or

$$\varepsilon_2 = -M_{21}\frac{di_1}{dt} \qquad (36\text{-}23a)$$

which you should compare with Eq. 36-3a $(\varepsilon = -L\, di/dt)$ for self-inductance.

Let us now interchange the roles of coils 1 and 2 in Fig. 36-8. That is, we set up a current i_2 in coil 2, by external circuitry not shown, and this produces a magnetic flux Φ_{12} that links coil 1 (from which the external circuitry has now been removed). If we change i_2 with time, we have, by the same argument given above,

$$\varepsilon_1 = -M_{12}\frac{di_2}{dt}; \qquad (32\text{-}23b)$$

compare Eq. 36-23a.

Thus we see that the emf in *either* coil is proportional to the rate of change of current in the *other* coil. The proportionality constants M_{21} and M_{12} seem to be different, but we assert, without proof, that they are in fact the same so that no subscripts are needed. This conclusion is in no way obvious. Thus we have

$$M_{21} = M_{12} = M \qquad (32\text{-}24)$$

and we can rewrite Eqs. 32-23 as

$$\varepsilon_2 = -M\, di_1/dt \quad \text{and} \quad \varepsilon_1 = -M\, di_2/dt. \qquad (32\text{-}25)$$

The induction is indeed *mutual.* The SI unit for M (compare Eqs. 36-3 and 36-23) is the henry.

The calculation of M, like that of L, depends on the geometry of the system. The simplest case is that in which *all* of the flux from one coil links the other coil. Example 7 shows such a situation.

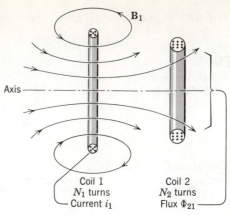

figure 36-8
The arrangement used to define the mutual inductance M_{21} of coil 2 with respect to coil 1.

In the toroid of Example 1 (see Fig. 36-2) let us relabel the turns in the winding shown (coil 1) as N_1 and the current in this winding as i_1. Now let us superimpose a second winding (coil 2) over the first, having N_2 turns. The coils are electrically insulated from each other. In terms of the geometrical factors of Example 1, what is the mutual inductance M for the two windings?

From Eq. 36-21a we have

$$M = \frac{N_2\Phi_{21}}{i_1}$$

in which Φ_{21} is here identical with Φ, the total common flux in coils 1 and 2 due to i_1. From Example 1 we see that, with the noted changes in notation,

$$\Phi = \frac{\mu_0 i_1 N_1 h}{2\pi} \ln\frac{b}{a},$$

EXAMPLE 7

so that

$$M = \frac{\mu_0 N_1 N_2 h}{2\pi} \ln \frac{b}{a}.$$

Note that if $N_1 = N_2$, the inductances L_1 and L_2 of the two windings are virtually the same (= L). A comparison with Example 1 shows that, in this case, $M = L$; see Problem 36 for a more general result.

---

questions

1. Under what conditions can we write Eq. 36-1 [$\varepsilon = d(N\Phi_B/dt)$] as $\varepsilon = N(d\Phi_B/dt)$? Can you think of a physical situation in which changes in N alone with time would produce an induced emf?

2. In a case of mutual induction, as in Fig. 35-2, is self-induction also present? Discuss.

3. Is the inductance per unit length for a solenoid near its center (a) the same as, (b) less than, or (c) greater than the inductance per unit length near its ends?

4. Two solenoids, A and B, have the same diameter and length and contain only one layer of copper windings, with adjacent turns touching, insulation thickness being negligible. Solenoid A contains many turns of fine wire and solenoid B contains fewer turns of heavier wire. (a) Which solenoid has the larger inductance? (b) Which solenoid has the larger inductive time constant?

5. If the flux passing through each turn of a coil is the same, the inductance of the coil may be computed from $L = N\Phi_B/i$ (Eq. 36-4). How might one compute L for a coil for which this assumption is not valid?

6. Show that the dimensions of the two expressions for L, $N\Phi_B/i$ (Eq. 36-4) and $\varepsilon/(di/dt)$ (Eq. 36-3b), are the same.

7. You are given a length l of copper wire. How would you arrange it to obtain the maximum self-inductance?

8. You want to wind a coil so that it has resistance R but essentially no inductance. How would you do it?

9. Does the time required for the current in a particular LR circuit to build up to any given fraction of its equilibrium value depend on the value of the applied constant emf?

10. A steady current is set up in a coil with a very large inductive time constant. When the current is interrupted with a switch, a heavy arc tends to appear at the switch blades. Explain. (Note: Interrupting currents in highly inductive circuits can be dangerous.)

11. In an LR circuit like that of Fig. 36-4, can the self-induced emf ever be larger than the battery emf?

12. In an LR circuit like that of Fig. 36-4, is the current in the resistance *always* the same as the current in the inductance?

13. In the circuit of Fig. 36-3 the self-induced emf is a maximum at the instant the switch is closed on a. How can this be since there is no current in the inductance at this instant?

14. The switch in Fig. 36-3, having been closed on a for a "long" time, is thrown to b. What happens to the energy initially stored in the inductor?

15. A coil has a (measured) inductance L and a (measured) resistance R. Is its inductive time constant given by Eq. 36-13? Bear in mind that we derived that equation (see Fig. 36-3) for a situation in which the inductive and resistive elements were physically separated. Discuss.

16. In Section 36-3 we show that Eq. 36-10 is a solution of Eq. 36-9. Can you be sure that it is the *only* solution?

17. If a current in a source of emf is in the direction of the emf, the energy of the source decreases; if a current is in a direction opposite to the emf (as in

charging a battery), the energy of the source increases. Do these statements apply to the inductor in Fig. 36-1a and 36-1b?

18. Can you make an argument based on the manipulation of bar magnets to suggest that energy may be stored in a magnetic field?

19. Does Eq. 36-18 ($U = \frac{1}{2}Li^2$) shed any light on the fact that (see Eq. 36-5) the inductance of a length l of a long solenoid is proportional to its volume?

20. Draw all the formal analogies that you can between a parallel-plate capacitor (for electric fields) and a long solenoid (for magnetic fields).

21. In a toroid is the energy density larger near the inner radius or near the outer radius?

22. Two coils are connected in series. Does their equivalent inductance depend on their geometrical relationship to each other?

23. You are given two similar flat circular coils of N turns each. For what geometry will their mutual inductance M be the greatest? For what geometry will it be the least? Assume that the coils are close together.

24. You are given two coils geometrically close together. Need they be electrically connected for them to exhibit mutual inductance? If they *are* electrically connected, can they still exhibit mutual inductance?

25. A flat circular coil is placed completely outside a long solenoid, near its center, the axes of the coil and the solenoid being parallel. Is there a mutual induction effect? Suppose the coil surrounds the solenoid? In each case justify your answer.

26. A circular coil of N turns surrounds a long solenoid. Is the mutual inductance greater when the coil is near the center of the solenoid or when it is near its end? Explain.

SECTION 36-1

1. The inductance of a close-packed coil of 400 turns is 8.0 mH. What is the magnetic flux through the coil when the current is 5.0×10^{-3} A?
Answer: 1.0×10^{-7} Wb.

2. Each item (a) coulomb·ohm·meter/weber, (b) volt·second, (c) coulomb·ampere/farad, (d) kilogram·volt·meter²/(henry·ampere)², (e) (henry/farad)$^{1/2}$ is equal to one of the items in the following list: meter, second, kilogram, dimensionless number, newton, joule, volt, ohm, watt, coulomb, ampere, weber, henry, farad. Give the equalities.

3. A 10-H inductor carries a steady current of 2.0 A. How can a 100-V self-induced emf be made to appear in the inductor?
Answer: Let the current change at 10 A/s.

SECTION 36-2

4. A solenoid is wound with a single layer of #10 copper wire (diameter, 0.10 in.). It is 4.0 cm in diameter and 2.0 m long. What is the inductance per unit length for the solenoid near its center? Assume that adjacent wires touch and that insulation thickness is negligible.

5. A long thin solenoid can be bent into a ring to form a toroid. Show that if the solenoid is long and thin enough, the equation for the inductance of a toroid (see Example 1) reduces to that for a solenoid of comparable length (Eq. 36-5).

6. *Inductors in Series* Two inductors L_1 and L_2 are connected in series and are separated by a large distance. (a) Show that the equivalent inductance L is $L_1 + L_2$. (b) Why must their separation be large?

7. Show that the self-inductance for a length l of a long wire associated with the flux *inside* the wire only is $\mu_0 l/8\pi$. Assume a uniform distribution of the current over the cross section of the wire.

problems

8. *Inductors in Parallel* Two inductors L_1 and L_2 are connected in parallel and separated by a large distance. (a) Show that the equivalent inductance L is given by

$$\frac{1}{L} = \frac{1}{L_1} + \frac{1}{L_2}.$$

(b) Why must their separation be large for this relationship to hold?

9. Two long parallel wires whose centers are a distance d apart carry equal currents in opposite directions. Show that, neglecting the flux within the wires themselves, the inductance of a length l of such a pair of wires is given by

$$L = \frac{\mu_0 l}{\pi} \ln \frac{d-a}{a},$$

where a is the wire radius. See Example 4, Chapter 34.

10. Calculate the self-inductance of two concentric hollow cylinders of radii a and b, and of length $l \gg a,b$. At one end the cylinders are connected by a flat conducting plate so that the current travels down in the inner cylinder and back in the outer. See Example 5 for hints.

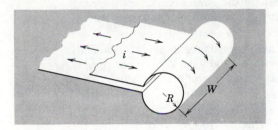

figure 36-9
Problem 11

11. A very wide copper strip of width W is bent into a piece of slender tubing of radius R with two plane extensions, as shown in Fig. 36-9. A current i flows through the strip, distributed uniformly over its width. In this way a "one-turn solenoid" has been formed. (a) Find the magnitude of the magnetic field **B** in the tubular part (far away from the edges). (Hint: Assume that the field outside this one-turn solenoid is negligibly small.) (b) Find the inductance of this one-turn solenoid, neglecting the two plane extensions. *Answer:* (a) $\mu_0 i/W$. (b) $\pi \mu_0 R^2/W$.

SECTION 36-3

12. The current in an LR circuit builds up to one-third of its steady-state value in 5.0 s. What is the inductive time constant?

13. How many "time constants" must we wait for the current in an LR circuit to build up to within 0.10 percent of its equilibrium value?   *Answer:* 6.9.

14. A 50-V potential difference is suddenly applied to a coil with $L = 50$ mH and $R = 180 \, \Omega$. At what rate is the current increasing after 0.001 s?

15. A wooden toroidal core with a square cross section has an inner radius of 10 cm and an outer radius of 12 cm. It is wound with one layer of #18 wire (diameter, 0.040 in.; "resistance", 160 ft/Ω). What are (a) the inductance and (b) the inductive time constant? Ignore the thickness of the insulation. *Answer:* (a) 2.8×10^{-4} H. (b) 2.7×10^{-4} s.

16. How long would it take for the potential difference across the resistor in an LR circuit ($L = 1.0$ H, $R = 1.0 \, \Omega$) to drop to 10% of its initial value?

17. A solenoid having an inductance of 6.0×10^{-6} H is connected in series to a $1.0 \times 10^3\text{-}\Omega$ resistor. (a) If a 10-V battery is switched across the pair, how long will it take for the current through the resistor to reach 80% of its final value? (b) What is the current through the resistor after one time constant? *Answer:* (a) 9.7×10^{-9} s. (b) 6.3×10^{-3} A.

18. The current in an *LR* circuit drops from 1.0 A at $t = 0$ to 0.010 A one second later. If *L* is 10 H, find the resistance *R* in the circuit.

19. In the circuit shown in Fig. 36-10, $\varepsilon = 10$ V, $R_1 = 5.0\ \Omega$, $R_2 = 10\ \Omega$, and $L = 5.0$ H. For the two separate conditions (I) switch *S* just closed and (II) switch *S* closed for a very long time, calculate (a) the current i_1 through R_1, (b) the current i_2 through R_2, (c) the current i through the switch, (d) the potential difference across R_2, (e) the potential difference across *L*, and (f) di_2/dt.
Answer: I. (a) 2.0 A. (b) Zero. (c) 2.0 A. (d) Zero. (e) 10 V. (f) 2.0 A/S II. (a) 2.0 A. (b) 1.0 A. (c) 3.0 A. (d) 10 V. (e) Zero. (f) Zero.

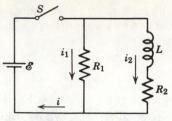

figure 36-10
Problem 19

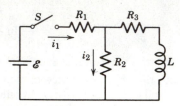

figure 36-11
Problem 20

20. In Fig. 36-11, $\varepsilon = 100$ V, $R_1 = 10\ \Omega$, $R_2 = 20\ \Omega$, $R_3 = 30\ \Omega$ and $L = 2.0$ H. Find the values of i_1 and i_2 (a) immediately after switch *S* is closed; (b) a long time later; (c) immediately after switch *S* is opened again; (d) a long time later.

21. Show that the inductive time constant τ_L can also be defined as the time that would be required for the current in an *LR* circuit to reach its equilibrium value *if it continued to increase at its initial rate*.

22. The switch *S* in Fig. 36-3 is thrown from *b* to *a*. After one inductive time constant show that (a) the total energy transformed to thermal energy in the resistor is $0.168\varepsilon^2\tau_L/R$ and that (b) the energy stored in the magnetic field is $0.200\varepsilon^2\tau_L/R$. (c) Show that the equilibrium energy stored in the magnetic field is $0.500\varepsilon^2\tau_L/R$.

23. A coil with an inductance of 2.0 H and a resistance of 10 Ω is suddenly connected to a resistanceless battery with $\varepsilon = 100$ V. (a) What is the equilibrium current? (b) How much energy is stored in the magnetic field when this current exists in the coil? *Answer:* (a) 10 A (b) 100 J.

24. A coil with an inductance of 2.0 H and a resistance of 10 Ω is suddenly connected to a resistanceless battery with $\varepsilon = 100$ V. At 0.10 s after the connection is made, what are the rates at which (a) energy is being stored in the magnetic field, (b) thermal energy is appearing, and (c) energy is being delivered by the battery?

25. A given coil is connected in series with a 10,000-Ω resistor. When a 50-V battery is applied to the two, the current reaches a value of 2.0 mA after 5.0 ms. (a) Find the inductance of the coil. (b) What is the energy stored in the inductance at this same moment? *Answer:* (a) 98 H. (b) 2.0×10^{-4} J.

26. A long wire carries a current *i* uniformly distributed over a cross section of the wire. Show that the magnetic energy per unit length stored *within* the wire equals $\mu_0 i^2/16\pi$. Note that it does not depend on the wire diameter.

27. The coaxial cable of Example 5 has $a = 1.0$ mm, $b = 4.0$ mm, and $c = 5.0$ mm (*c* is the radius of the outer surface of the outer conductor). It carries a current of 10 A in the inner conductor and an equal but oppositely directed return current in the outer conductor. Calculate and compare the stored magnetic energy per meter of cable length (a) within the central conductor, (b) in the space between the conductors, and (c) within the outer conductor.
Answer: (a) 2.5×10^{-6} J/m. (b) 14×10^{-6} J/m. (c) 0.80×10^{-6} J/m.

28. Prove that when switch *S* in Fig. 36-3 is thrown from *a* to *b* all the energy stored in the inductor appears as thermal energy in the resistor.

SECTION 36-5

29. What is the energy density in the magnetic field near the center of the solenoid of Problem 23, Chapter 35? *Answer:* 63 J/m³.

30. A circular loop of wire 5.0 cm in radius carries a current of 100 A. What is the energy density at the center of the loop?

31. A length of #10 copper wire carries a current of 10 A. Calculate (a) the magnetic energy density and (b) the electric energy density at the surface of the wire. The wire diameter is 0.10 in. and its resistance per unit length is 1.0 ohm/1000 ft.
    Answer: (a) 0.99 J/m³. (b) 4.8×10^{-15} J/m³.

32. (a) What is the magnetic energy density of the earth's magnetic field of 5.0×10^{-5} T? (b) Assuming this to be relatively constant over distances small compared with the earth's radius and neglecting the variations near the magnetic poles, how much energy would be stored in a shell between the the earth's surface and 16 km above the surface?

33. (a) Find an expression for the energy density as a function of the radius for the toroid of Example 1. (b) Integrating the energy density over the volume of the toroid, find the total energy stored in the field of the toroid; assume $i = 0.50$ A. (c) Using Eq. 36-18 evaluate the energy stored in the toroid directly from the inductance and compare with (b).

    Answer: (a) $\dfrac{\mu_0 i^2 N^2}{8\pi^2 r^2}$. (b) 1.8×10^{-4} J. (c) 1.8×10^{-4} J.

34. What must be the magnitude of a uniform electric field if it is to have the same energy density as that possessed by a 0.50-tesla magnetic field?

35. What is the magnetic energy density at the center of a circulating electron in the hydrogen atom (see Example 9, Chapter 34)?
    Answer: 5.7×10^7 J/m³.

SECTION 36-6

36. In Example 7 show that, if $N_1 \neq N_2$, the mutual inductance is given by

    $$M = \sqrt{L_1 L_2}.$$

    Do you suppose that this relation holds even if the situation is not like that of Example 7, that is, if it is not true that all the flux from one coil links all the turns of the other coil?

37. Two short cylindrical coils are connected in series; the coils are reasonably close together and share the same axis. (a) Show that the effective inductance of the combination is

    $$L = L_1 + L_2 \pm 2M.$$

    (b) What is the significance of the $\pm$ sign? Does it have anything to do with the relative sense (clockwise or counterclockwise) in which the coils are wound?

37
magnetic
properties
of matter

In electricity the *isolated charge q* is the simplest structure that can exist. If two such charges of opposite sign are placed near each other, they form an *electric dipole,* characterized by an electric dipole moment **p.** In magnetism isolated magnetic poles (usually called *magnetic monopoles*) which would correspond to isolated electric charges, apparently do not exist. The simplest magnetic structure is the *magnetic dipole,* characterized by a magnetic dipole moment **μ.** Table 34-1 summarizes some characteristics of electric and magnetic dipoles.

A current loop, a bar magnet, and a solenoid of finite length are examples of magnetic dipoles. We can identify their north poles (from which the lines of **B** emerge) by suspending them as compass needles and observing which end points north. We can find their magnetic dipole moments by placing the dipole in an external magnetic field **B,** measuring the torque τ that acts on it, and computing μ from Eq. 33-11, or

$$\tau = \mu \times \mathbf{B}. \tag{37-1}$$

Alternatively, we can measure **B** due to the dipole at a point along its axis a (large) distance r from its center and compute μ from the expression in Table 34-1, or

$$B = \frac{\mu_0}{2\pi} \frac{\mu}{r^3}. \tag{37-2}$$

Figure 37-1, which shows iron filings sprinkled on a sheet of paper under which there is a bar magnet, suggests that this dipole might be viewed as two "poles" separated by a distance d. However, all attempts

figure 37-1
A bar magnet is a magnetic dipole.
The iron filings suggest the pattern
of lines of force in Fig. 37-4a. See
Fig. 27-6b for the electrostatic
analog. (Courtesy Physical Science
Study Committee.)

to isolate these poles fail. If we break the magnet, as in Fig. 37-2, the
fragments prove to be dipoles and not isolated poles. Where one north
pole and one south pole existed there are now three of each. If we break
up a magnet into the electrons, protons, and neutrons that make up its
atoms, we will find that even these elementary particles are magnetic
dipoles. Figure 37-3 contrasts the electric and the magnetic characters
of the free electron.

All electrons have a characteristic *"spin" angular momentum* about
a certain axis, which has the value of

$$L_s = 0.5272943 \times 10^{-34} \text{ joule} \cdot \text{second}.$$

This is suggested by the vector $\mathbf{L}_s$ in Fig. 37-3b. Such a spinning charge
can be viewed classically as being made up of infinitesimal current

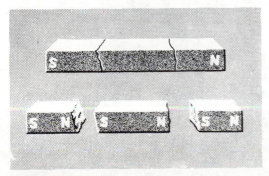

figure 37-2
If we break a bar magnet, each
fragment becomes a dipole. There
will always be equal numbers of
north and south poles, associated
in pairs.

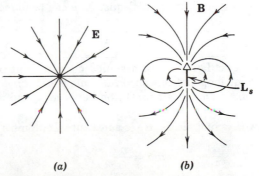

figure 37-3
(a) The lines of **E** and (b) the lines of
B for an electron. The magnetic
dipole moment of the electron, $\boldsymbol{\mu}_l$,
is directed opposite to the spin
angular momentum vector, $\mathbf{L}_s$.

loops. Each such loop is a tiny magnetic dipole, its moment being given by (see Eq. 33-10)

$$\mu = NiA, \tag{37-3}$$

where i is the equivalent current in each infinitesimal loop and A is the loop area. The number of turns, N, is unity for each loop. The magnetic dipole moment of the spinning charge can be found by integrating over the moments of the infinitesimal current loops that make it up; see Problem 8.

Although this model of the spinning electron is too mechanistic and is not in accord with modern quantum physics, it remains true that the magnetic dipole moments of elementary particles are closely connected with their intrinsic angular momenta, or spins. Those particles and nuclei whose spin angular momentum is zero (the α-particle, the pion, the O^{16} nucleus, etc.) have no magnetic dipole moment. We must distinguish the "intrinsic" or "spin" magnetic moment of the electron from any additional magnetic moment it may have because of its *orbital* motion in an atom; see Example 2.

EXAMPLE 1

Devise a method for measuring μ for a bar magnet.

(*a*) Place the magnet in a uniform external magnetic field **B**, with μ making an angle θ with **B**. The magnitude of the torque acting on the magnet (see Eq. 37-1) is given by

$$\tau = \mu B \sin \theta.$$

Clearly we can find μ if we measure τ, B, and θ.

(*b*) A second technique is to suspend the magnet from its center of mass and to allow it to oscillate about its stable equilibrium position in the external field **B**. For *small* oscillations, $\sin \theta$ can be replaced by θ and the equation just given becomes

$$\tau = -(\mu B)\theta = -\kappa\theta,$$

where κ is a constant. The minus sign has been inserted to show that τ is a *restoring torque*. Since τ is proportional to θ, the condition for simple angular harmonic motion is met. The frequency ν is given by the reciprocal of Eq. 15-24, or

$$\nu = \frac{1}{2\pi}\sqrt{\frac{\kappa}{I}} = \frac{1}{2\pi}\sqrt{\frac{\mu B}{I}},$$

in which I is the rotational inertia. With this equation μ can be found from the measured quantities ν, B, and I.

EXAMPLE 2

An electron in an atom circulating in an assumed circular orbit of radius r behaves like a tiny current loop and has an *orbital magnetic dipole moment* usually represented by μ_l.*,† Derive a connection between μ_l and the *orbital angular momentum* L_l.

Newton's second law ($F = ma$) yields, if we substitute Coulomb's law for F,

$$\frac{1}{4\pi\epsilon_0}\frac{e^2}{r^2} = ma = \frac{mv^2}{r}$$

or

$$v = \sqrt{\frac{e^2}{4\pi\epsilon_0 mr}}. \tag{37-4}$$

---

* This must not be confused with the magnetic dipole moment μ_s of the electron spin, which is also present.

† Although this model is too mechanistic and not in the spirit of modern quantum physics, it is nevertheless instructive to examine it.

The angular velocity ω is given by

$$\omega = \frac{v}{r} = \sqrt{\frac{e^2}{4\pi\epsilon_0 mr^3}}.$$

The current for the orbit is the rate at which charge passes any given point, or

$$i = ev = e\left(\frac{\omega}{2\pi}\right) = \sqrt{\frac{e^4}{16\pi^3\epsilon_0 mr^3}}.$$

The orbital dipole moment μ_l is given from Eq. 37-3 if we put $N = 1$ and $A = \pi r^2$, or

$$\mu_l = NiA = (1)\sqrt{\frac{e^4}{16\pi^3\epsilon_0 mr^3}}\,(\pi r^2) = \frac{e^2}{4}\sqrt{\frac{r}{\pi\epsilon_0 m}}. \qquad (37\text{-}5)$$

The orbital angular momentum L_l is

$$L_l = (mv)r.$$

Combining with Eq. 37-4 leads to

$$L_l = \sqrt{\frac{e^2 mr}{4\pi\epsilon_0}}.$$

Finally, eliminating r between this equation and Eq. 37-5 yields

$$\mu_l = L_l\left(\frac{e}{2m}\right),$$

which shows that the orbital magnetic moment of an electron is proportional to its orbital angular momentum. Convince yourself that the vectors $\boldsymbol{\mu}_l$ and $\mathbf{L}_l$ point in opposite directions.

For $r = 5.3 \times 10^{-11}$ m, which corresponds to hydrogen in its normal state, we have, from Eq. 37-5,

$$\mu_l = \frac{e^2}{4}\sqrt{\frac{r}{\pi\epsilon_0 m}}$$

$$= \frac{(1.6 \times 10^{-19}\text{ C})^2}{4}\sqrt{\frac{5.3 \times 10^{-11}\text{ m}}{(\pi)(8.9 \times 10^{-12}\text{ C}^2/\text{N}\cdot\text{m}^2)(9.1 \times 10^{-31}\text{ kg})}}$$

$$= 9.2 \times 10^{-24}\text{ A}\cdot\text{m}^2.$$

It is clear in Fig. 37-4 that, for the bar magnet and the solenoid, lines of **B** emerge from the top end and enter the bottom end. These are localized regions called the "north pole" (top end) and the "south pole" (bottom end). There are no sharply defined points so defined, as there are for the electrostatic dipole of Fig. 37-4c. Note that the lines of **B** and **E** in Fig. 37-4 are much alike (dipole field) at distances from the device that are much larger than its dimensions but that the behavior is much different for points very close to the device, including internal points.

The symmetry of nature has always been a guiding principle for physicists. For example, the existence of the (negative) electron suggested the existence of a positive electron, or positron, which was eventually discovered. In the same way the existence of a (positive) proton suggested that there might also be a negative proton and a large accelerator was built at the University of California (Berkeley) primarily to search for this particle; it was found. The discoveries of the positron and of the negative proton each were associated with Nobel prizes; see Appendix K.

With this motivation it is not surprising that physicists have long sought experimental evidence for the existence of magnetic monopoles.

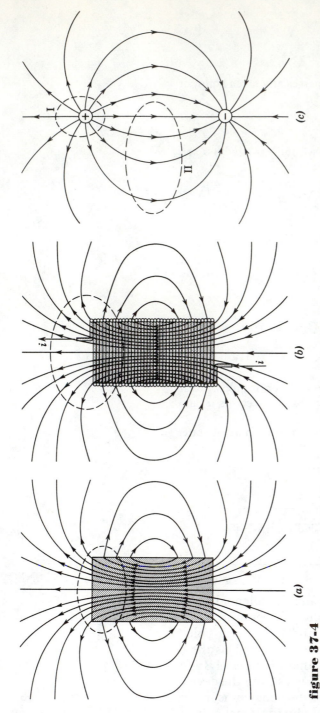

(a)

(b)

(c)

figure 37-4

Lines of **B** (*a*) for a bar magnet and (*b*) for a short solenoid. In each case the top of the figure is a north pole, the bottom being a south pole. (*c*) Lines of **E** for an electric dipole. The four dashed curves are intersections with the plane of the figure of closed Gaussian surfaces. Note that Φ_B equals zero for (*a*) or (*b*). Φ_E equals zero for surfaces like II in (*c*), which do not contain any charge, but Φ_E is not zero for surfaces like I. See Section 28-3.

Their absence, as we shall see in more detail in Chapter 40, represents a serious lack of symmetry between electricity and magnetism. Actually, magnetic monopoles were predicted to exist, on the basis of theory, by P. A. M. Dirac in 1931 and physicists have been searching for them constantly ever since.*

Gauss's law for magnetism, which is one of the basic equations of electromagnetism (see Table 40-2), is a formal way of stating a conclusion that seems to be forced on us by the facts of magnetism, namely, that *isolated magnetic poles do not seem to exist.* This equation asserts that the flux Φ_B through any *closed* Guassian surface must be zero, or

$$\Phi_B = \oint \mathbf{B} \cdot d\mathbf{S} = 0, \tag{37-6}$$

where the integral is to be taken over the entire closed surface. We contrast this with Gauss's law for electricity, which is

$$\epsilon_0 \oint \mathbf{E} \cdot d\mathbf{S} = q. \tag{37-7}$$

The fact that a zero appears at the right of Eq. 37-6, but not at the right of Eq. 37-7, means that in magnetism there seems to be no counterpart to the free charge q in electricity.

Figure 37-4a shows a Gaussian surface that encloses one end of a bar magnet. Note that the lines of **B** enter the surface inside the magnet and leave it outside the magnet. There is thus an inward (or negative) flux inside the magnet and an outward (or positive) flux outside it. The total flux for the whole surface is zero.

Figure 37-4b shows a similar surface for a solenoid of finite length which, like a bar magnet, is also a magnetic dipole. Here, too, Φ_B equals zero. Figures 37-4a and b show clearly that there are no "sources" of **B**; that is, there are no points from which lines of **B** emanate. Also, there are no "sinks" of **B**; that is, there are no points toward which **B** converges. In other words, *there are no free magnetic poles.*

Figure 37-4c shows a Gaussian surface (I) surrounding the positive end of an *electric* dipole. Here there *is* a net flux of the lines of **E**. There is a "source" of **E**; it is the charge q. If q is negative, we have a "sink" of **E** because the lines of **E** end on negative charges. For surfaces like surface II in Fig. 37-4c for which the charge inside is zero, the flux of **E** over the surface is also zero.

37-2 GAUSS'S LAW FOR MAGNETISM

37-3 THE MAGNETISM OF THE EARTH

The magnet with which we are most familiar is that on which we live, the earth. The supposition that the earth is a large magnet, with magnetic poles and a magnetic equator, was first made by Sir William Gilbert (1544–1603), physician to Queen Elizabeth I. Gilbert made a small spherical *terrella* ("little earth") out of a naturally occurring magnetic loadstone (literally "leading stone" or compass) and traced its lines of magnetism. In those days of navigation and exploration there was a natural interest in the compass and in the earth's magnetism.

* See "Quest for the Magnetic Monopole" by Richard A. Carrigan, Jr., *The Physics Teacher*, October 1975, for more information.

Figure 37-5*a* is an idealized sketch of the lines of **B** associated with the earth's magnetic field, both at and above its surface. To a first approximation we can represent this field by imagining a strong bar magnet located at the center of the earth, as in Fig. 37-5*b*. Note that the earth's magnetic axis and its rotational axis *RR* do not coincide, being separated by about 15°.

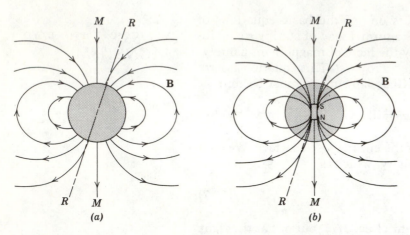

(a)   (b)

figure 37-5
(*a*) An idealized representation of the lines of **B** associated with the earth's magnetic field. *MM* is the earth's magnetic axis and *RR* is its rotational axis. (*b*) We can approximate the earth's external magnetic field by imagining that a srong bar magnet is located at the center of the earth.

The magnetic pole in the northern hemisphere is in arctic Canada. Note that it is a *south* magnetic pole because lines of **B** *converge* toward it, as in Fig. 37-4. There is a *north* magnetic pole in the southern hemisphere, in Antarctica; lines of **B** *emerge* from it.

There is, of course, no bar magnet buried at the center of the earth. The earth's magnetism must be related to the facts that the central core of the earth, whose radius is 55% of the earth's radius, is (*a*) liquid, (*b*) highly conducting, and (*c*) partakes of the earth's rotation. A dynamo effect, involving circulating currents in the earth's core whose mechanism is not yet fully understood, is almost certainly operating.*

Several other planets in our solar system, Mercury and Jupiter among them, also have magnetic fields. So do the sun and many other stars. There is also a magnetic field associated with our galaxy, that is, with the family of stars whose plane of symmetry is defined by the Milky Way. The galactic magnetic field is relatively weak (about 2 pT on the average) but its effects can be important because it extends over such large distances.

Two simple tools for exploring the earth's magnetic field are the compass needle and the dip needle, the latter being a gravitationally balanced magnetized needle whose axis of rotation is horizontal rather than vertical. In Tucson, Arizona, for example, the north pole of a compass needle (as of 1964) pointed about 13° east of geographic north. Such *declinations* must be known and taken into account when using the compass for navigation or direction finding. The horizontal component of the earth's magnetic field $\mathbf{B}_h$, to which the compass needle responds, is, at Tucson, 26 μT (=0.26 gauss).

Now let us orient the (horizontal) rotational axis of a dip needle at Tucson until it is at right angles to the horizontal component of **B**. Can you convince yourself that the needle will now point in the direction of **B**? It does, and we find that, at this site and at this date, the north end of the needle points downward toward the earth, the needle making an angle φ_i of about 59° (the *inclination*) with a horizontal plane containing

* See "The Earth's Magnetism" by S. K. Runcorn, *Scientific American*, September 1955.

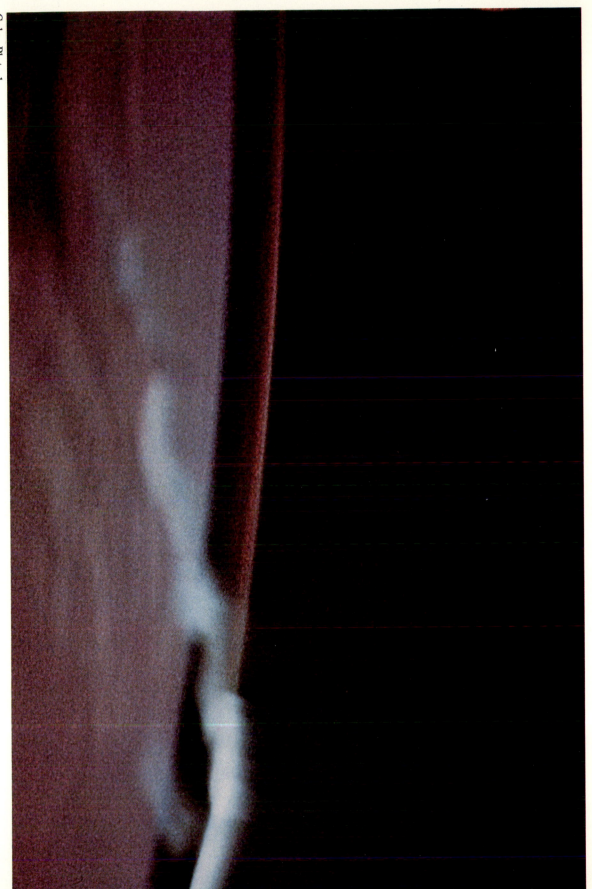

Color Plate 1

At the right we see an aurora viewed from Sky Lab 3, from somewhere over the Indian Ocean. The bright band extending to the left is the normal "airglow", a phenomenon not visible from earth to the naked eye. It represents the emission of radiation from atoms and molecules in the upper atmosphere that absorb energy from sunlight during the daytime. The full moon illuminates a largely clouded earth. A piece of Sky Lab barely shows at the upper left. (This photograph was taken by astronaut Owen K. Garriott by a hand-held camera in Sky Lab. We are indebted to him and to the National Aeronautics and Space Administration for permission to publish the photograph.)

its rotational axis. This shows that, as we expect from Fig. 37-5*a*, lines of **B** are *entering* the earth's surface at this point.

The earth's magnetic field is neither as regular nor as static as the idealized field in Fig. 37-5 suggests. Also, there are observable phenomena, going far beyond deflections of the compass needle, that would not occur if the earth had no magnetic field. Consider the following:

1. *Local variations.* The earth's magnetic field has important local variations, caused by differences in the magnetic properties of the rocks that make up the earth's crust and by the presence of concentrated magnetic ore bodies.

2. *Changes with time.* The mean magnetic declination and inclination vary measurably from year to year at any given location. For example, between the years 1600 and 1800 the measured magnetic declination at London varied in a continuous way from 11° east to 24° west. The north magnetic pole (as of 1948) was measured to be moving north west at about 8 km/y.

This wandering of the earth's magnetic axis, and thus the variation with time of local declinations and inclinations, has resulted in a new archeological specialty, *archeomagnetism,* by means of which dates may be assigned to ancient kilns, ovens, and hearths. The principle is based on the fact that most clays, from which such structures were made, contain small amounts of magnetic materials, whose orientation is frozen into position by heating during normal use. By comparing the present direction of the earth's magnetic field with that of the "frozen in" direction of the magnetization, an approximate archeological date may be established.

On a longer (geological) time scale there is evidence that the earth's magnetic axis has actually completely reversed in direction as many as nine times during the last 4×10^6 years.[*] The evidence rests on measurements of the (weak) magnetism frozen into rocks of known geological age at the time of their formation.

3. *Interactions with the solar wind.* The sun emits a steady stream of ionized hydrogen atoms (protons) and electrons, that sweeps through the solar system at supersonic speeds. This ever present "solar wind" interacts strongly in several ways with the earth's magnetic field.[†] (*a*) Occasional sharp increases in the intensity of the solar wind produce terrestrial *magnetic storms,* which seriously interfere with long-distance radio communication. (*b*) The protons and electrons of the solar wind, acted upon by forces given by Eq. 33-2 ($\mathbf{F} = q_0 \mathbf{v} \times \mathbf{B}$), spiral along the lines of **B,** moving back and forth between the magnetic north and south polar regions. These "trapped" electrons and protons form the so called *Van Allen radiation belts.* They were discovered by James A. Van Allen, of the State University of Iowa, in early satellite experiments.[‡] (*c*) The trapped solar wind particles, as they interact with the earth's atmosphere, produce the dazzling spectacle of the *aurora,* which is most prominent at about ±75° geomagnetic latitude; see Plate I, opposite.[**]

[*] "Reversals of the Earth's Magnetic Field" by Allan Cox, G. Brent Dalrymple, and Richard R. Doell, *Scientific American,* February 1967.
[†] "The Solar Wind" by E. N. Parker, *Scientific American,* April 1964.
[‡] See "Radiation Belts" by Brian J. O'Brien, *Scientific American,* May 1963. See also "The Magnetosphere" by J. A. Ratcliffe, *Contemporary Physics,* vol. 18, 1977.
[**] "Aurora and Airglow" by C. T. Elvery and Franklin A. Roach, *Scientific American,* September 1955.

EXAMPLE 3

From data given earlier in this section find (a) the vertical component B_v of the earth's magnetic field and (b) the resultant magnetic field **B** at Tucson.

Figure 37-6 shows the situation. We have

(a)
$$B_v = B_h \tan \varphi_i$$

$$= (26 \ \mu\text{T})(\tan 59°)$$

$$= 44 \ \mu\text{T} \ (= 0.44 \text{ gauss}),$$

and

(b)
$$B = \sqrt{B_h{}^2 + B_v{}^2}$$

$$= \sqrt{(26 \ \mu\text{T})^2 + (44 \ \mu\text{T})^2}$$

$$= 51 \ \mu\text{T} \ (= 0.51 \text{ gauss}).$$

Note that the magnetic declination at Tucson plays no role in this problem.

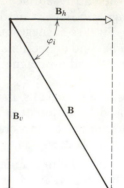

figure 37-6
Example 3.

37-4
PARAMAGNETISM

Magnetism as we know it in our daily experience is an important but special branch of the subject called *ferromagnetism*; we discuss this in Section 37-6. Here we discuss a weaker form of magnetism called *paramagnetism*.

For most atoms and ions, the magnetic effects of the electrons, including both their spins and orbital motions, exactly cancel so that the atom or ion is not magnetic. This is true for the rare gases such as neon and for ions such as Cu^+, which makes up ordinary copper.* For other atoms or ions the magnetic effects of the electrons do *not* cancel, so that the atom as a whole has a magnetic dipole moment $\boldsymbol{\mu}$. Examples are found among the transition elements, such as Mn^{++}; the rare earths, such as Gd^{+++}; and the actinide elements, such as U^{++++}.

If we place a sample of N atoms, each of which has a magnetic dipole moment $\boldsymbol{\mu}$, in a magnetic field, the elementary atomic dipoles tend to line up with the field. This tendency to align is called *paramagnetism*. For perfect alignment, the sample as a whole would have a magnetic dipole moment of $N\boldsymbol{\mu}$. However, the aligning process is seriously interfered with by thermal agitation effects. The importance of thermal agitation may be measured by comparing two energies: one $(= \frac{3}{2}kT)$ is the mean kinetic energy of translation of a gas atom at temperature T; the other $(=2\mu B)$ is the difference in energy between an atom lined up with the magnetic field and one pointing in the opposite direction. As Example 4 shows, the effect of the collisions at ordinary temperatures and fields is very great. The sample acquires a magnetic moment when placed in an external magnetic field, but this moment is usually very much smaller than the maximum possible moment $N\mu$.

EXAMPLE 4

A paramagnetic gas, whose atoms (see Example 2) have a magnetic dipole moment of about $10^{-23} \ \text{A} \cdot \text{m}^2$, is placed in an external magnetic field of magnitude 1.0 T $(= 10^4 \text{ gauss})$. At room temperature $(T = 300 \text{ K})$ calculate and compare U_T, the mean kinetic energy of translation $(= \frac{3}{2}kT)$, and U_B, the magnetic energy $(= 2\mu B)$:

*Cu^+ indicates a copper atom from which one electron has been removed; Al^{+++} indicates an aluminum atom from which three electrons have been removed, etc.

$$U_T = \tfrac{3}{2}kT = (\tfrac{3}{2})(1.38 \times 10^{-23} \text{ J/K})(300 \text{ K}) = 6 \times 10^{-21} \text{ J},$$

$$U_B = 2\mu B = (2)(10^{-23} \text{ A} \cdot \text{m}^2)(1.0 \text{ T}) = 2 \times 10^{-23} \text{ J}.$$

Because U_T equals $300 \, U_B$, we see that energy exchanges in collisions can interfere seriously with the alignment of the dipoles with the external field.

If we place a specimen of a paramagnetic substance in a nonuniform magnetic field, such as that near the pole of a strong magnet, it will be attracted toward the region of higher field, that is, toward the pole. We can understand this by drawing an analogy with the corresponding electric case of Fig. 37-7, which shows a dielectric specimen (a sphere) in a nonuniform electric field. The net electric force points to the right in the figure and is

$$F_e = q(E_0 + \Delta E) - q(E_0 - \Delta E) = q(2\Delta E),$$

which we can write as

$$F_e = \frac{(q \, \Delta x)}{\Delta x} 2\Delta E = p \left(\frac{2\Delta E}{\Delta x} \right) \cong p \left(\frac{dE}{dx} \right)_{\text{max}}$$

Here $p \, (= q \, \Delta x)$ is the induced electric dipole moment of the sphere. In the differential limit of a very small sphere $(2\Delta E/\Delta x)$ approaches $(dE/dx)_{\text{max}}$, the gradient of the electric field at the center of the sphere.

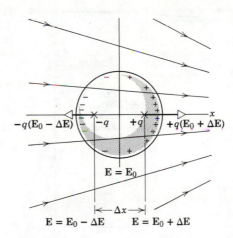

E = E_0

E = $E_0 - \Delta E$ E = $E_0 + \Delta E$

figure 37-7
A dielectric sphere in a nonuniform electric field. The effective induced charges are represented by the point charges $+q$ and $-q$

In the corresponding magnetic case we have, by analogy,

$$F_m = \mu \left(\frac{dB}{dx} \right)_{\text{max}} \tag{37-8}$$

Thus, by measuring the magnetic force F_m that acts on a small paramagnetic specimen when we place it in a nonuniform magnetic field whose field gradient $(dB/dx)_{\text{max}}$ is known, we can learn its magnetic dipole moment μ. The *magnetization* **M** of the specimen is defined as the magnetic moment per unit volume, or

$$\mathbf{M} = \frac{\mu}{V},$$

where V is the volume of the specimen. It is a vector because μ, the dipole moment of the specimen, is a vector.

In 1895 Pierre Curie (1859–1906) discovered experimentally that the magnetization M of a paramagnetic specimen is directly proportional to

B, the effective value of magnetic field in which the specimen is placed, and inversely proportional to the temperature, or

$$M = C \frac{B}{T}, \qquad (37\text{-}9)$$

in which C is a constant. This equation is known as *Curie's law*. The law is physically reasonable in that increasing B tends to align the elementary dipoles in the specimen, that is, to increase M, whereas increasing T tends to interfere with this alignment, that is, to decrease M. Curie's law is well verified experimentally, provided that the ratio B/T does not become too large.

M cannot increase without limit, as Curie's law implies, but must approach a value M_{max} $(= \mu N/V)$ corresponding to the complete alignment of the N dipoles contained in the volume V of the specimen. Figure 37-8 shows this saturation effect for a sample of $CrK(SO_4)_2 \cdot 12H_2O$. The chromium ions are responsible for all the paramagnetism of this salt, all the other elements being paramagnetically inert. To achieve 99.5% saturation, it is necessary to use applied magnetic fields as high as 50,000 gauss ($= 5.0$ T) and temperatures as low as 1.3 K. Note that for more readily achievable conditions, such as $B = 10,000$ gauss ($= 1.0$ T) and $T = 10$ K, the abscissa in Fig. 37-8 is only 1.0 so that Curie's law would appear to be well obeyed for this and for all lower values of B/T.

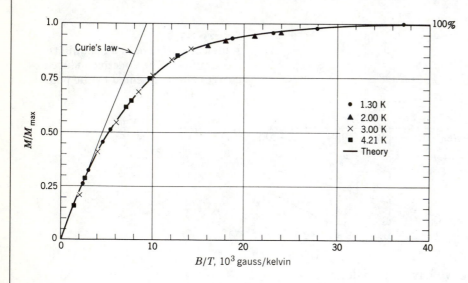

figure 37-8
The ratio M/M_{max} for a paramagnetic salt (chromium potassium alum) in various magnetic fields and at various temperatures. The curve through the experimental points is a theoretical curve calculated from modern quantum physics. (From measurements by W. E. Henry.)

The curve that passes through the experimental points in this figure is calculated from a theory based on modern quantum physics; it is in excellent agreement with experiment.

In 1846 Michael Faraday discovered that a specimen of bismuth brought near to the pole of a strong magnet is *repelled*. He called such substances *diamagnetic* (in contrast with paramagnetic specimens, which are *attracted*). Diamagnetism, present in all substances, is such a feeble effect that its presence is masked in substances made of atoms that have a net magnetic dipole moment, that is, in paramagnetic or ferromagnetic substances.

Figures 37-9a and b show an electron circulating in a diamagnetic atom at angular frequency ω_0 in an assumed circular orbit of radius r.

37-5
DIAMAGNETISM

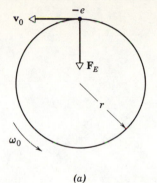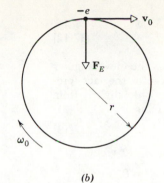

(a) *(b)*

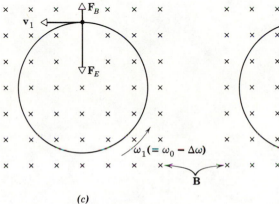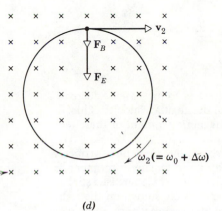

(c) *(d)*

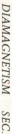

figure 37-9
(a) An electron circulating in an atom. (b) An electron circulating in the opposite direction. (c) A magnetic field is introduced, *decreasing* the linear speed of the electron in (a), that is, $v_1 < v_0$. (d) The magnetic field *increases* the linear speed of the electron in (b), that is, $v_2 > v_0$. See problem 15 for a different way of looking at diamagnetism. Our treatment of diamagnetism, though mechanistic and classical, nevertheless yields results in reasonable agreement with experiment.

Each electron is moving under the action of a centripetal force $\mathbf{F}_E$ of electrostatic origin where, from Newton's second law,

$$F_E = ma = m\omega_0^2 r. \qquad (37\text{-}10)$$

Each revolving electron has an orbital magnetic moment, but for the atom as a whole the orbits are randomly oriented so that there is no *net* magnetic effect. In Fig. 37-9a, for example, the magnetic dipole moment $\boldsymbol{\mu}_l$ points into the page; in Fig. 37-9b it points out and the suggested net effect for the two orbits shown is cancellation. This cancellation is also suggested at the left in Fig. 37-10.

If we apply an external field $\mathbf{B}$ as in Figs. 37-9c and d, an *additional* force, given by $-e(\mathbf{v} \times \mathbf{B})$, acts on the electron. This magnetic force acts always at right angles to the direction of motion; its magnitude is

$$F_B = evB = e(\omega r)B. \qquad (37\text{-}11)$$

Show that in Fig. 37-9c $\mathbf{F}_B$ and $\mathbf{F}_E$ point in opposite directions and that in Fig. 37-9d they point in the same direction. Note that since the centripetal force changes when we turn on the magnetic field (the radius can be shown to remain constant), the angular velocity must also change; thus ω in Eq. 37-11 differs from ω_0 in Eq. 37-10.

Applying Newton's second law to Figs. 37-9c and d, and allowing for both directions of circulation, yields for the *resultant* forces on the electrons

$$F_E \pm F_B = ma = m\omega^2 r.$$

Substituting Eqs. 37-10 and 37-11 into this equation yields

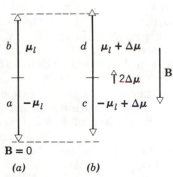

$\mathbf{B} = 0$

(a) *(b)*

figure 37-10
The magnetic moments of the two oppositely circulating electrons in an atom cancel when there is no external magnetic field, as in (a), but do not cancel when a field is applied, as in (b). Note that the resultant moment in (b) points in the *opposite* direction to $\mathbf{B}$. Compare carefully with Fig. 37-9.

$$m\omega_0^2 r \pm e\omega r B = m\omega^2 r$$

or
$$\omega^2 \mp \left(\frac{eB}{m}\right)\omega - \omega_0^2 = 0. \qquad (37\text{-}12)$$

We could solve this quadratic equation for ω, the new angular velocity. Rather than doing this, we take advantage of the fact (presented without proof; see Problem 14) that ω differs only slightly from ω_0, even in the strongest external magnetic fields. Thus

$$\omega = \omega_0 + \Delta\omega, \qquad (37\text{-}13)$$

where $\Delta\omega \ll \omega_0$. Substituting this equation into Eq. 37-12 yields

$$[\omega_0^2 + 2\omega_0\,\Delta\omega + (\Delta\omega)^2] \mp [\beta\omega_0 + \beta\Delta\omega] - \omega_0^2 = 0,$$

where β is a convenient abbreviation for eB/m. The two terms ω_0^2 cancel each other; the terms $(\Delta\omega)^2$ and $\beta\Delta\omega$ are small compared to the remaining terms and may be set equal to zero with only small error. This leads, as an excellent approximation, to

$$\Delta\omega \cong \mp \tfrac{1}{2}\beta = \pm\frac{eB}{2m} \qquad (37\text{-}14)$$

or, from Eq. 37-13,
$$\omega = \omega_0 \pm \frac{eB}{2m}.$$

Thus the effect of applying a magnetic field is to increase or decrease (depending on the direction of circulation) the angular velocity. This, in turn, increases or decreases the orbital magnetic moment of the circulating electron (see Example 2).

In Fig. 37-9c the angular velocity is reduced (because the centripetal force is reduced) so that the magnitude of the magnetic moment is reduced. In Fig. 37-9d, however, the angular velocity is increased so that the magnitude of $\boldsymbol{\mu}_l$ is increased. These effects are shown on the right in Fig. 37-10, where it will be noted that the two magnetic moments *no longer cancel.*

We see that if we apply a magnetic field **B** to a diamagnetic substance, a magnetic moment will be *induced* whose direction (out of the plane of Fig. 37-9) is *opposite* to **B**; see also Fig. 37-10. This is precisely the reverse of paramagnetism, in which the (*permanent*) magnetic dipoles tend to point in the *same* direction as the applied field.

We can now understand why a diamagnetic specimen is repelled when brought near to the pole of a strong magnet. If the pole is a north pole, there exists a nonuniform magnetic field **B** pointing away from the pole. If a sphere made of a diamagnetic material (bismuth, say) is brought near to this pole, the magnetization **M** that is induced in it points toward the pole, that is, *opposite to* **B.** Thus the side of the sphere closest to the magnet behaves like a north pole and is *repelled* by the nearby north pole of the magnet. For a paramagnetic sphere, the vector **M** points *along the direction of* **B** and the side of the sphere nearest to the magnet is a south pole, which is *attracted* to the north pole of the magnet.

EXAMPLE 5

Calculate the *change* in magnetic moment for a circulating electron, as described in Example 2, if a magnetic field **B** of 2.0 T (= 20,000 gauss) acts at right angles to the plane of the orbit.

We obtain μ from Eq. 37-3, or

$$\mu = NiA = (1)(e\nu)(\pi r^2) = (1)\left(\frac{e\omega}{2\pi}\right)(\pi r^2) = \tfrac{1}{2}er^2\omega.$$

The *change* in μ is

$$\Delta\mu = \tfrac{1}{2}er^2\,\Delta\omega$$

or, from Eq. 37-14,

$$\Delta\mu = \pm\,\tfrac{1}{2}er^2\left(\frac{eB}{2m}\right) = \pm\,\frac{e^2Br^2}{4m}.$$

Substituting numbers yields

$$\Delta\mu = \pm\,\frac{(1.6\times10^{-19}\text{ C})^2(2.0\text{ T})(5.3\times10^{-11}\text{ m})^2}{(4)(9.1\times10^{-31}\text{ kg})}$$

$$= \pm4.0\times10^{-29}\text{ A}\cdot\text{m}^2.$$

In Example 2 the moment μ_l was 9.2×10^{-24} A·m², so that the change induced by even a strong external magnetic field is rather small, the ratio $\Delta\mu/\mu_l$ being about 4×10^{-6}.

37-6
FERROMAGNETISM

For three elements (Fe, Co, and Ni) and for a variety of alloys of these and other elements, a special effect occurs which permits a specimen to achieve a high degree of magnetic alignment in spite of the randomizing tendency of the thermal motions of the atoms. In such materials, described as *ferromagnetic*, a special form of interaction called *exchange coupling* occurs between adjacent atoms, coupling their magnetic moments together in rigid parallelism.* If the temperature is raised above a certain critical value, called the *Curie temperature*, the exchange coupling suddenly disappears and the materials become simply paramagnetic. For iron the Curie temperature is 1043 K. Ferromagnetism is evidently a property not only of the individual atom or ion but also of the interaction of each atom or ion with its neighbors in the crystal lattice (see Fig. 21-5) of the solid.

Figure 37-11 shows a *magnetization curve* for a specimen of iron. To obtain such a curve, we form the specimen, assumed initially unmagnetized, into a ring and wind a toroidal coil around it, as in Fig. 37-12, to form a so-called *Rowland ring*.† When a current i is set up in the coil, *if the iron core is not present*, a magnetic field is set up within the toroid given by (see Eq. 34-7)

$$B_0 = \mu_0 ni, \tag{37-15}$$

where n is the number of turns per unit length for the toroid. Although this formula was derived for a long solenoid, we can apply it to a toroid

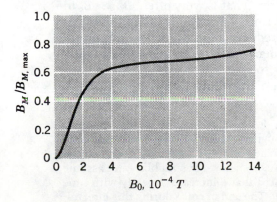

figure 37-11
A magnetization curve for iron.

---

*Exchange coupling, a purely quantum effect, cannot be "explained" in terms of classical physics.

† See Section 33-1 for further information about H. A. Rowland.

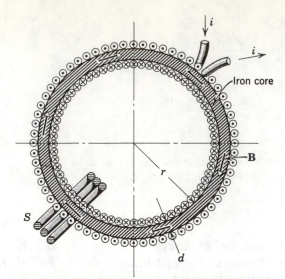

figure 37-12
A Rowland ring, showing a
secondary coil S.

if $d \ll r$ in Fig. 37-12. Because of the iron core, the actual value of **B** in the toroidal space will exceed **B**$_0$, by a large factor in many cases, since the elementary atomic dipoles in the core line up with the applied field **B**$_0$, thereby setting up their own magnetic field. Thus we can write

$$B = B_0 + B_M \qquad (37\text{-}16)$$

where B_M is the magnetic field contribution due to the core; it is proportional to the magnetization M of the core. Often $B_M \gg B_0$.

The field B_0 is proportional to the current in the toroid and we can calculate it readily, using Eq. 37-15; B can be measured in a way that we describe below. An experimental value for B_M can be derived from Eq. 37-16. It has a maximum value $B_{M,\text{max}}$ corresponding to complete alignment of the atomic dipoles in the iron. Thus we can plot, as in Fig. 37-11, the fractional degree of alignment $(= B_M/B_{M,\text{max}})$ as a function of B_0. For this specimen a value of 96.5% saturation is achieved at $B_0 = 0.13$ T $(= 1300$ gauss; this point is about 16 ft to the right of the origin in the figure); increasing B_0 to 1.0 T $(= 10,000$ gauss; about 120 ft to the right in Fig. 37-11) increases the fractional saturation only to 97.7%.

The use of iron in transformers, electromagnets, etc., greatly increases the strength of the magnetic field that can be generated by a given current in a given set of windings. That is, very often, $B_M \gg B_0$ in Eq. 37-16. However, the presence of iron also sets a limit to the maximum magnetic field that can be produced because of the saturation effect suggested in Fig. 37-11. To generate magnetic fields greater than this saturation limit, it is necessary to abandon the use of iron and to rely on the "brute force" application of very large (and often transient) currents.*

To measure B in the system of Fig. 37-12, let the current in the toroid windings be increased from zero to i. The flux through the secondary coil S will change by BA, where A is the area of the toroid. While the flux is changing, an induced emf will appear in coil S, according to Faraday's law. For simplicity, we assume that the current in the toroid is so adjusted that B increases linearly with time for an interval Δt, as shown in Fig. 37-13a. The emf in coil S during this interval, from Faraday's law,† will then be

---

* See "Megagauss Physics" by C. M. Fowler, *Science*, April 1973.
† We ignore the minus sign because we are concerned only with the magnitude of ε.

$$\varepsilon = N\frac{\Delta\Phi_B}{\Delta t} = \frac{NBA}{\Delta t},$$

where N is the number of turns in coil S. This emf will set up a current i_s in coil S given by

$$i_s = \frac{\varepsilon}{R} = \frac{NBA}{R\,\Delta t}$$

or

$$B = \frac{(i_s\,\Delta t)R}{NA} = \frac{qR}{NA},$$

in which R is the resistance of coil S and $i_s\,\Delta t$ is the charge q that passes through this coil during time Δt. If a so-called *ballistic galvanometer* is connected to S, its deflection will be a measure of the charge q. Thus it is possible to find B for any value of the current i in the toroid windings. A more detailed analysis shows that it is not necessary that the curve $B(t)$ in Fig. 37-13a be linear during the interval Δt.

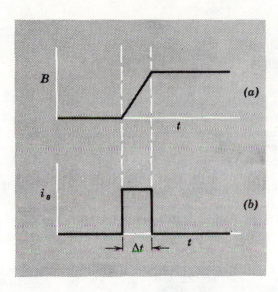

figure 37-13
(a) B for a Rowland ring as the current in the windings is increased from zero during an interval Δt.
(b) The corresponding induced current in the secondary coil. Both curves are idealized; in practice, the sharp corners would be rounded off.

Magnetization curves for ferromagnetic materials *do not retrace themselves* as we increase and then decrease the toroid current. Figure 37-14 shows the following operations with a Rowland ring: (1) starting with the iron unmagnetized (point a), increase the toroid current until $B_0 \ (= \mu_0 ni)$ has the value corresponding to point b; (2) reduce the current in the toroid winding back to zero (point c); (3) reverse the toroid current and increase it in magnitude until point d is reached; (4) reduce the current to zero again (point e); (5) reverse the current once more until point b is reached again. The lack of retraceability shown in Fig. 37-14 is called *hysteresis*. Note that at points c and e the iron core is magnetized, even though there is no current in the toroid windings; this is the familiar phenomenon of *permanent magnetism*.

We explained the magnetization curve for paramagnetism (see Fig. 37-8) in terms of the mutually opposing tendencies of alignment with the external field and of randomization because of thermal agitation. In ferromagnetism, however, we have assumed that adjacent atomic dipoles are locked in rigid parallelism. Why, then, does the magnetic moment of the specimen not reach its saturation value for very low — even zero — values of B_0? The current interpretation is to assume the existence within the specimen of *domains*, that is, of local regions

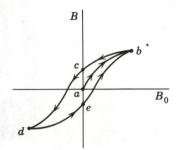

figure 37-14
A magnetization curve (ab) for a specimen of iron and an associated hysteresis loop ($ebcde$).

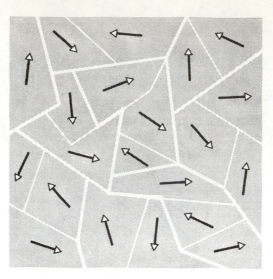

within which there is essentially perfect alignment. The domains themselves, however, as Fig. 37-15 suggests, are not parallel at moderately
low values of B_0.

Figure 37-16 shows some domain photographs, taken by sprinkling a
colloidal suspension of finely powdered iron oxide on a properly etched
single crystal of iron. The domain boundaries, which are thin regions in
which the alignment of the elementary dipoles changes from a certain
direction in one domain to an entirely different direction in the other,
are the sites of intense but highly localized and nonuniform magnetic
fields. The suspended colloidal particles are attracted to these regions.
Although the atomic dipoles in the individual domains are completely
aligned, the specimen as a whole may have a very small resultant magnetic moment. This is the state of affairs in an unmagnetized iron nail.

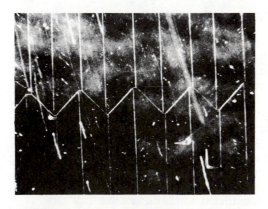

As we magnetize a piece of iron by placing it in an external magnetic field, two effects take place. One is a growth in size of the domains
that are favorably oriented at the expense of those that are not, as in Fig.
37-17. Second, the direction of orientation of the dipoles within a domain may swing around as a unit, becoming closer to the field direction.
Hysteresis comes about because the domain boundaries do not move
completely back to their original positions when the external field B_0
is removed.

Two other types of magnetism, closely related to ferromagnetism, are
antiferromagnetism and *ferrimagnetism* (note spelling). In antiferro-

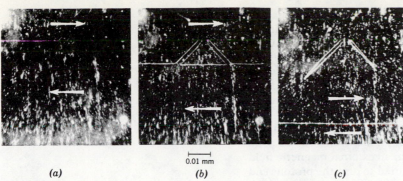

(a)   *(b)*   *(c)*

figure 37-17

(*a*) A boundary between two domains, with the magnetization in each domain shown by the white arrows. (*b*) If an external magnetic field pointing from left to right is imposed on the specimen, the upper domain will grow at the expense of the lower. The domain boundary will move down as the elementary dipoles reverse themselves. (*c*) The process continues. The boundary has moved across a region in which there is a crystal imperfection. (Courtesy H. J. Williams, Bell Telephone Laboratories.)

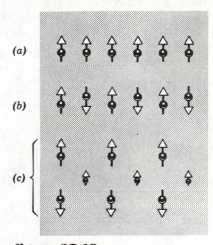

figure 37-18

Showing how elementary magnetic dipoles are oriented by the interatomic exchange coupling in (*a*) ferromagnetism, (*b*) antiferromagnetism, and (*c*) ferrimagnetism. The actual arrangements are, of course, three-dimensional.

magnetic substances, of which MnO_2 is an example, the exchange coupling to which we referred on page 825 serves to lock adjacent ions into rigid *antiparallelism* (see Fig. 37-18). Such materials exhibit very little gross external magnetism. However, if they are heated above a certain temperature, called the *Néel temperature*, the exchange coupling ceases to act and the material becomes paramagnetic. In ferrimagnetic substances, of which iron ferrite is an example, two different kinds of magnetic ions are present. In iron ferrite the two ions are Fe^{++} and Fe^{+++}. The exchange coupling locks the ions into a pattern like that of Fig. 37-18*c*, in which the external effects are intermediate between ferromagnetism and antiferromagnetism. Here, too, the exchange coupling disappears if the material is heated above a certain characteristic temperature.

37-7
NUCLEAR MAGNETISM

Many nuclei have magnetic dipoles, and the possibility arises that a specimen of matter may exhibit gross external magnetic effects associated with its nuclei. However, nuclear magnetic moments are several orders of magnitude smaller than those associated with the electronic motions in an atom or ion. The magnetic moment of an electron associated with its spin, for example, exceeds that of the proton (the nucleus of hydrogen) by a factor of 660.

Gross external effects for nuclear magnetism are smaller than the corresponding (ionic) paramagnetic effects by the *square* of ratios of this order of magnitude, because (*a*), *all else being equal*, the external magnetism is reduced by such a ratio, but (*b*) the very fact that the magnetic dipole moment of the nucleus is smaller means that (see Example 4) the thermal vibrations are proportionally (to a good approximation) more effective in reducing the degree of alignment of the elementary dipoles in an external magnetic field; thus all else is *not* equal and the ratio enters twice.

Techniques such as the Rowland ring (see Fig. 37-12) are far too insensitive to detect nuclear magnetism. We describe here a *nuclear resonance technique* by means of which nuclear magnetism can readily reveal itself. This method is also vastly useful for studying paramagnetism, ferromagnetism, antiferromagnetism, and ferrimagnetism, in all of which cases the magnetic effects are associated not with the nuclei but with the atomic electrons. The nuclear-resonance technique was developed in 1946 by E. M. Purcell and his co-workers at Harvard. Simultaneously and independently, F. Bloch and his co-workers at Stanford discovered a very similar method. For these achievements the two physicists shared a Nobel prize.

We focus our attention on the problem of measuring the magnitude μ of the magnetic moment of the proton. In principle, this can be done by placing a specimen containing protons in an external magnetic field **B** and by measuring the energy $(=2\mu B)$ required to turn the protons end for end. A rigorously correct description of the procedures cannot be given without using quantum physics. The description given, although based entirely on classical physics, nevertheless leads to the correct conclusions.

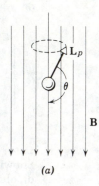
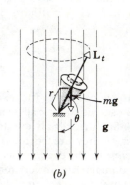

(a)   (b)

figure 37-19

(a) A spinning proton precessing in an external magnetic field and (b) a spinning top precessing in an external gravitational field. $\mathbf{L}_p$ and $\mathbf{L}_t$ are the two angular momentum vectors.

Figure 37-19a shows a spinning proton with its axis making an angle θ with a uniform external magnetic field **B**. Figure 37-19b shows a spinning top with its axis making an angle θ with a uniform external gravitational field **g**. In each case there is a torque that tends to align the axis of the spinning object with the field. For the proton (see Eq. 33-11) it is given by

$$\tau_p = \mu B \sin\theta. \tag{37-17a}$$

For the top it is given by $\quad \tau_t = mgr\sin\theta, \tag{37-17b}$

where r locates the center of mass of the top and m is its mass.

In Section 13-2, we saw that the spinning top precesses about a vertical axis with an angular frequency given by

$$\omega_t = \frac{mgr}{L_t}, \tag{37-18a}$$

in which L_t is the spin angular momentum of the top.

The proton, which has a quantized spin angular momentum L_p, will also precess about the direction of the (magnetic) field because of the action of the (magnetic) torque. You should derive the expression for the frequency of precession, being guided by the derivation of Section 13-2, but using the magnetic torque (see Eq. 37-17a) instead of the gravitational torque (see Eq. 37-17b). The relation is

$$\omega_p = \frac{\mu B}{L_p}. \tag{37-18b}$$

EXAMPLE 6

831 NUCLEAR MAGNETISM SEC. 37-7

What is the precession frequency of a proton in a magnetic field of 0.50 T (= 5000 gauss)?

The quantities μ and L_p in Eq. 37-18b are 1.4×10^{-26} A·m^2 and 0.53×10^{-34} J·s. This equation then yields

$$\nu_p = \frac{\omega_p}{2\pi} = \frac{\mu B}{2\pi L_p} = \frac{(1.4 \times 10^{-26}\ \text{A·m}^2)(0.50\ \text{T})}{(2\pi)(0.53 \times 10^{-34}\ \text{J·s})} = 2.1 \times 10^7\ \text{Hz}.$$

This frequency (=21 MHz) is in the radio-frequency range.

It is possible to change the energy of any system in periodic motion if we allow an external influence to act on it at the same frequency as that of its motion. This is the familiar *resonance* condition. As an "external influence" for the precessing proton, we use a small alternating magnetic field $\mathbf{B}_{\text{osc}}$ arranged to be at right angles to the steady field $\mathbf{B}$. This oscillating field combines vectorially with the steady field so that the *resultant* field rocks back and forth between the limits shown by the dashed lines in Fig. 37-20. Typical values for B and for the amplitude of $\mathbf{B}_{\text{osc}}$ are 5000 gauss and one gauss, respectively, so that the rocking angle α in the figure is quite small. If the angular frequency ω_0 of the oscillating field is chosen equal to the angular precession frequency ω_p of the proton, it turns out that the precessing proton can absorb energy. An increase in energy means an increase in θ in Fig. 37-19a.

The resonance condition

$$\omega_0 = \frac{\mu B}{L_p} \tag{37-19}$$

can be used to measure μ. We place the spinning proton in a known field $\mathbf{B}$, apply a "perturbing field" at right angles to it, and vary the angular frequency ω_0 of this perturbing field until resonance occurs. It is possible to tell when Eq. 37-19 is satisfied because, at resonance, many spinning protons will tend to turn end for end in the field, absorbing energy which can be detected by appropriate electronic techniques.

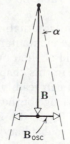

figure 37-20
In the nuclear magnetic resonance method a small oscillating magnetic field $\mathbf{B}_{\text{osc}}$ is placed at right angles to a steady field $\mathbf{B}$.

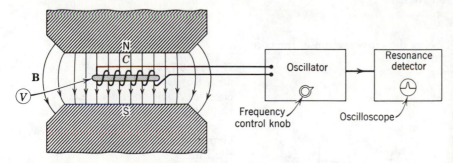

figure 37-21
An arrangement to observe nuclear resonance. The oscillating field is horizontal within the coil.

Figure 37-21 is a schematic diagram of an experimental arrangement. The protons, present as hydrogen nuclei in a small vial V of water, are immersed in a strong steady magnetic field caused by the electromagnet whose pole faces N and S are shown. A rapidly alternating current in the small coil C provides the (horizontal) weak, perturbing magnetic field $\mathbf{B}_{\text{osc}}$. This current is provided by a radio-frequency oscillator whose angular frequency ω_0 can be varied; an electronic "resonance detector," also connected to the oscillator, serves to indicate when energy is being drained from the oscillator and used to "flip the protons." In principle, the oscillator angular frequency ω_0 is varied until the resonance detector shows that Eq. 37-19 is satisfied (see Fig. 37-22). The magnetic moment μ can then be determined by measuring B and ω_0. Surprisingly enough, magnetic moments can be measured in this and similar ways to a much greater accuracy than we can measure μ for a bar magnet. For the proton we have

$$\mu_p = 1.410617 \times 10^{-26}\ \text{A·m}^2.$$

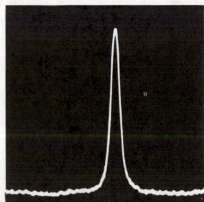

figure 37-22
An oscilloscope photograph of a proton resonance peak showing energy absorbed from the oscillator (vertical axis) versus oscillator frequency (horizontal axis). (From Bloembergen et al.

In Chapter 30 we saw that if a dielectric is placed in an electric field, polarization charges will appear on its surface. These surface charges, which find their origin in the elementary electric dipoles (permanent or induced) that make up the dielectric, set up a field of their own that modifies the original field. For the simple case discussed in Chapter 30—a dielectric slab in a parallel-plate capacitor—this complication can readily be handled in terms of the electric field vector **E** and some knowledge of the electric properties of the slab material, such as the dielectric constant. For more complex problems we asserted that it was useful to introduce two other (subsidiary) electric vectors, the *electric polarization* **P** and the *electric displacement* **D**. Table 30-2 shows some of the characteristics of these three vectors.

In magnetism we find a similar situation. If magnetic materials are placed in an external magnetic field, the elementary magnetic dipoles (permanent or induced) will act to set up a field of their own that will modify the original field. For the simple case discussed in this chapter—a Rowland ring with a ferromagnetic core—this complication can readily be handled in terms of the magnetic field vector **B** and some knowledge of the magnetic properties of the ring material, such as that provided by Fig. 37-11. For more complex problems we find it useful to introduce two other (subsidiary) magnetic vectors, the *magnetization* **M** and the *magnetic field strength* **H**. We do so largely so that the student who takes a second course in electromagnetism will have some familiarity with them.

Consider a Rowland ring carrying a current i_0 in its windings and designed so that its core, assumed to be iron, can be removed. The magnetic field **B**, measured by the methods of Section 37-6, will be much greater when the core is in place than when it is not, assuming that the current in the windings remains unchanged.

Physically we can understand the large value of B in the iron core in terms of the alignment of the elementary dipoles in the iron. A hypothetical slice out of the iron core, as in Fig. 37-23b, has a magnetic moment $d\mu$ equal to the vector sum of all of the elementary dipoles contained in it. We define our first subsidiary vector, the *magnetization* **M**, as the magnetic moment per unit volume of the core material. For the slice of Fig. 37-23b we have

$$d\mu = \mathbf{M}(A\ dl),$$

where $(A\ dl)$ is the volume of the slice, A being the cross section of the core.

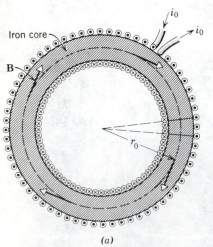

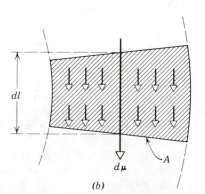

(a)   (b)

37-8
THREE MAGNETIC VECTORS

figure 37-23
(a) A Rowland ring with an iron core. (b) A slice of the core, showing its magnetic moment $d\mu$ caused by the alignment of the elementary magnetic dipoles in the iron.

When we discussed Ampère's law in Chapter 34, we assumed that no magnetic materials were present. If we apply this law, namely,

$$\oint \mathbf{B} \cdot d\mathbf{l} = \mu_0 i, \tag{37-20}$$

to the circular path of integration shown in Fig. 37-23a, we obtain

$$(B)(2\pi r_0) = \mu_0(N_0 i_0), \tag{37-21}$$

in which r_0 is the mean radius of the core, N_0 is the number of turns, and i_0 is the current in each turn. We see at once that Ampère's law, in the form of Eq. 37-20, *is not valid when magnetic materials are present.* Equation 37-21 predicts that, since the right side is the same whether or not the core is in place, the magnetic field B should also be the same, a prediction not in accord with experiment.

We can increase B in the absence of the iron core to the value that it has when the core is in place if we increase the current in the windings by an amount $i_{M,0}$. The magnetization of the iron core is thus *equivalent in its effect on* **B** to such a hypothetical current increase. We choose to modify Ampère's law by arbitrarily inserting a *magnetizing current* term i_M on the right, obtaining

$$\oint \mathbf{B} \cdot d\mathbf{l} = \mu_0(i + i_M). \tag{37-22}$$

If we give i_M a suitable value when the iron core is in place, it is clear that Ampère's law, in this new form, can remain valid. It remains to relate this (largely hypothetical) magnetizing current to something more physical, the magnetization **M**.*

Applying Eq. 37-22 to the iron ring of Fig. 27-23a yields

$$(B)(2\,\pi r_0) = \mu_0(N_0 i_0 + N_0 i_{0M,0}). \tag{37-23}$$

We can relate $i_{M,0}$ to the magnetization **M** if we recall (Eq. 33-10) that the magnetic moment of a magnetic dipole in the form of a current loop is given by

$$\mu = NiA,$$

where N is the number of turns in the loop, i is the loop current, and A is the loop area. Let us this equation to find what increase $i_{M,0}$ in current in the windings around the slice of Fig. 37-23b would produce a magnetic moment equivalent to that actually produced by the alignment of elementary dipoles in the slice. We have

$$M(A\ dl) = \left(N_0 \frac{dl}{2\pi r_0}\right)(i_{M,0})(A),$$

the quantity in the first parentheses on the right being the number of turns associated with the slice of thickness dl. This reduces to

$$N_0 i_{M,0} = M(2\pi r_0). \tag{37-24}$$

Substituting this into Eq. 37-23 yields

$$(B)(2\pi r_0) = \mu_0(N_0 i_0) + \mu_0(M)(2\pi r_0). \tag{37-25}$$

We now choose to generalize from the special case of the Rowland ring by writing Eq. 37-25 as

$$\oint \mathbf{B} \cdot d\mathbf{l} = \mu_0 i + \mu_0 \oint \mathbf{M} \cdot d\mathbf{l}$$

or

$$\oint \left(\frac{\mathbf{B} - \mu_0 \mathbf{M}}{\mu_0}\right) \cdot d\mathbf{l} = i.$$

The quantity $(\mathbf{B} - \mu_0 \mathbf{M})/\mu_0$ occurs so often in magnetic situations that we give it a special name, the *magnetic field strength* **H**, or

$$\mathbf{H} = \frac{\mathbf{B} - \mu_0 \mathbf{M}}{\mu_0}$$

which we write as $\qquad\qquad \mathbf{B} = \mu_0 \mathbf{H} + \mu_0 \mathbf{M}. \tag{37-26}$

---

* It is possible to give reality to the magnetizing current by viewing it as a real current that flows around the magnet at its surface, being the resultant macroscopic effect of all the microscopic current loops that constitute the atomic electron orbits. This *Amperian current* viewpoint, however, does not take the magnetization due to electron spin readily into account. Since we do not attempt to measure magnetizing currents experimentally, other than through their (postulated) magnetic effects, we prefer to view the magnetizing current as a convenient formalism.

Ampère's law can now be written in the simple form

$$\oint \mathbf{H} \cdot d\mathbf{l} = i, \tag{37-27}$$

which holds in the presence of magnetic materials and in which i is the *true current only*, that is, it does not include the magnetizing current. This reminds us that the electric displacement vector **D** permitted us to write Gauss's law for the case in which dielectric materials are present, in a form involving *free charges only*, that is, not polarization charges; see Table 30-2.

We state without proof (see Problems 23 and 24) that at a boundary between two media (1) the component of **H** tangential to the surface has the same value on each side of the surface* and (2) the component of **B** perpendicular to the surface has the same value on each side of the surface. These *boundary conditions* are of great value in solving complex problems.

To find H in our Rowland ring, let us apply Ampère's law in the generalized form of Eq. 37-27. We have

$$(H)(2\pi r_0) = N_0 i_0,$$

where i_0 is the (true) current in the windings. This gives

$$H = \left(\frac{N_0}{2\pi r_0}\right) i_0 = n i_0, \tag{37-28}$$

in which n is the number of turns per unit length. Since we have not introduced any information describing the core into Eq. 37-27, the value of H computed from Eq. 37-28 is independent of the core material.

B can be measured experimentally by the method of Section 37-6 and M can then be calculated from Eq. 37-26. Note in passing (see Eq. 37-15) that the abscissa B_0 in Fig. 37-11 is proportional to $H(= \mu_0 H)$, the ordinate being proportional to B. Curves such as this and that of Fig. 37-14 are called *B-H curves*.

Let us assume that we have made measurements of **H**, **B**, and **M** for a wide variety of magnetic materials, using either the technique described or an equivalent one. For *paramagnetic* and *diamagnetic* materials we would find, as an experimental result, that **B** is directly proportional to **H**, or

$$\mathbf{B} = \kappa_m \mu_0 \mathbf{H}, \tag{37-29}$$

in which κ_m, the *permeability* of the magnetic medium, is a constant for a given temperature and density of the material. Eliminating **B** between Eqs. 37-29 and 37-26 allows us to write

$$\mathbf{M} = (\kappa_m - 1)\mathbf{H}, \tag{37-30}$$

which is another expression of the linear or proportional character of paramagnetic and diamagnetic materials.

For a vacuum, in which there are no magnetic dipoles present to be aligned, the magnetization **M** must be zero. Putting $\mathbf{M} = 0$ in Eq. 37-26 leads to

$$\mathbf{B} = \mu_0 \mathbf{H} \qquad \text{(in vacuum)}. \tag{37-31}$$

Comparison with Eq. 37-29 shows that a vacuum must be described by $\kappa_m = 1$. Equation 37-30 verifies that the magnetization vanishes if we

---

* Assuming that there are no true currents at the surface, as there are in the Rowland ring in Fig. 37-23a, for example.

Table 37-1
Three magnetic vectors

Name	Symbol	Associated with	Boundary Condition
Magnetic induction*	**B**	All currents	Normal component continuous
Magnetic field strength	**H**	True currents only	Tangential component continuous†
Magnetization (magnetic dipole moment per unit volume)	**M**	Magnetization currents only	Vanishes in a vacuum

Defining equations for **B**	$\mathbf{F} = q\mathbf{v} \times \mathbf{B}$	Eq. 33-2
or	$= i\mathbf{l} \times \mathbf{B}$	Eq. 33-6a
General relation among the three vectors	$\mathbf{B} = \mu_0\mathbf{H} + \mu_0\mathbf{M}$	Eq. 37-26
Ampère's law when magnetic materials are present	$\oint \mathbf{H} \cdot d\mathbf{l} = i$ (i = true current only)	Eq. 37-27
Empirical relations for certain magnetic materials**	$\mathbf{B} = \kappa_m \mu_0 \mathbf{H}$	Eq. 37-29
	$\mathbf{M} = (\kappa_m - 1)\mathbf{H}$	Eq. 37-30

* We have usually called **B** simply "the magnetic field." Here we call it by its alternative name, "magnetic induction," to avoid confusion with "magnetic field strength," the name for **H**.
** For paramagnetic and diamagnetic materials only, if κ_m is to be independent of **H**.
† Assuming no true currents exist at the boundary.

put κ_m equal to unity. For paramagnetic materials κ_m is slightly greater than unity. For diamagnetic materials it is slightly less than unity; Eq. 37-30 shows that this requires **M** and **H** to be oppositely directed, a fact implicit in Section 37-5.

In ferromagnetic materials the relationship between **B** and **H** is far from linear, as Figs. 37-11 and 37-14 show. Experimentally, κ_m proves to be a function not only of the value of H but also, because of hysteresis, of the magnetic and thermal history of the specimen.*

An interesting special case of ferromagnetism is the permanent magnet, for which **H**, **M**, and **B** all have nonvanishing values inside the magnet even though there is no true current. Figure 37-24 shows typical lines of **B** and **H** associated with such a magnet. The lines of **B** may be drawn as continuous loops, the boundary condition (2), mentioned above, being satisfied where the lines enter and leave the magnet.

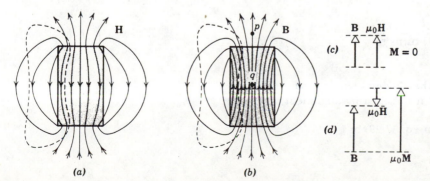

(a)   (b)

figure 37-24
(a) The lines of **H** and (b) the lines of **B** for a permanent magnet. Note that the lines of **H** change direction at the boundary. The closed dashed curves are paths of integration around which Ampère's law may be applied. The relation $\mathbf{B} = \mu_0\mathbf{H} + \mu_0\mathbf{M}$ is shown to be satisfied for (c) a particular outside point p and (d) a particular inside point q.

* In the dielectric case there are waxy materials, called *ferroelectrics*, for which the relationship between **D** and **E** is nonlinear, which exhibit hysteresis, and from which quasi-permanent electric dipoles (*electrets*) can be constructed. However, most commonly useful dielectric materials are linear, whereas most commonly useful magnetic materials are nonlinear.

Equation 37-22 shows that the vector **B** is associated with the *total* current, both true and magnetizing. In Fig. 37-24b $\oint \mathbf{B} \cdot d\mathbf{l}$ around any loop such as that shown by the dashed curve is not zero and must be associated with a hypothetical magnetizing current i_M, imagined to circulate around the magnet at its surface; actual or true currents (i) do not exist in this problem. Figure 37-24a shows that **H** reverses direction at the boundary. Since **H** (see Eq. 37-27) is associated with true currents only, we must have $\oint \mathbf{H} \cdot d\mathbf{l} = 0$ around any loop such as that shown by the dashed lines. The reversal of **H** at the boundary makes this possible. Note that **M** and **H** point in opposite directions within the magnet. Table 37-1 summarizes the properties of the three vectors **B**, **H**, and **M**.

EXAMPLE 7

In the Rowland ring the (true) current i_0 in the windings is 2.0 A and the number of turns per unit length (n) in the toroid is 10 turns/cm. B, measured by the technique of Section 37-6, is 1.0 T. Calculate (a) H, (b) M, and (c) the magnetizing current $i_{M,0}$ both when the core is in place and when it is removed. (d) For these particular operating conditions, what is κ_m?

(a) H is independent of the core material and may be found from Eq. 37-28:

$$H = ni$$

$$= (10 \text{ turns/cm})(2.0 \text{ A})$$

$$= 2.0 \times 10^3 \text{ A/m}.$$

(b) M is zero when the core is removed. With the core in place, we may solve Eq. 37-26 for **M**, obtaining for the magnitude of **M**,

$$M = \frac{B - \mu_0 H}{\mu_0}$$

$$= \frac{(1.0 \text{ T}) - (4\pi \times 10^{-7} \text{ T} \cdot \text{m/A})(2.0 \times 10^3 \text{ A/m})}{(4\pi \times 10^{-7} \text{ T} \cdot \text{m/A})}$$

$$= 7.9 \times 10^5 \text{ A/m}.$$

(c) The effective magnetizing current follows from Eq. 37-24:

$$i_{M,0} = M \left(\frac{2\pi r_0}{N_0} \right) = \frac{M}{n}$$

$$= \frac{7.9 \times 10^5 \text{ A/m}}{2.0 \times 10^3 \text{ turns/m}}$$

$$= 390 \text{ A}.$$

An *additional* current of this amount in the windings would produce the same value of B in the absence of a core as that obtained by alignment of the elementary dipoles, with the core in place.

(d) The permeability can be found from Eq. 37-29, or

$$\kappa_m = \frac{B}{\mu_0 H}$$

$$= \frac{1.0 \text{ T}}{(4\pi \times 10^{-7} \text{ T} \cdot \text{m/A})(2.0 \times 10^3 \text{ A/m})}$$

$$= 397.$$

We emphasize that this value of κ_m holds only for the special conditions of this experiment.

questions

1. Two iron bars are identical in appearance. One is a magnet and one is not. How can you tell them apart? You are not permitted to suspend either bar as a compass needle or to use any other apparatus.

2. Two iron bars always attract, no matter the combination in which their ends are brought near each other. Can you conclude that one of the bars must be unmagnetized?

3. The neutron, which has no charge, has a magnetic dipole moment. Is this possible on the basis of classical electromagnetism, or does this evidence alone indicate that classical electromagnetism has broken down?

4. Do all permanent magnets have to have identifiable north and south poles? Consider geometries other than the straight bar magnet.

5. If magnetic monopoles are shown to exist, is it conceivable that there might be a whole family of them, with different masses, magnetic pole strengths, intrinsic angular momenta, etc.? Compare Appendix F.

6. A certain short iron rod is found, by test, to have a north pole at each end. You sprinkle iron filings over the rod. Where (in the simplest case) will they cling? Make a rough sketch of what the lines of B must look like, both inside and outside the rod. See "A Three-Pole Bar Magnet?" by Jerry D. Wilson, *The Physics Teacher*, September 1976.

7. Consider these two situations. (*a*) A (hypothetical) magnetic monopole is pulled through a single-turn conducting loop along its axis, at a constant speed. (*b*) A short bar magnet (a magnetic dipole) is similarly pulled. Compare qualitatively the net amounts of charge transferred through any cross section of the loop during these two processes.

8. Cosmic rays are charged particles that strike our atmosphere from some external source. We find that more low-energy cosmic rays reach the earth near the north and south magnetic poles than at the (magnetic) equator. Why is this so?

9. How might the magnetic dipole moment of the earth be measured?

10. Give three reasons for believing that the flux Φ_B of the earth's magnetic field is greater through the boundaries of Alaska than through those of Texas.

11. You are a manufacturer of compasses. (*a*) Describe ways in which you might magnetize the needle. (*b*) The end of the needle that points north is usually painted a characteristic color. Without suspending the needle in the earth's field, how might you find out which end of the needle to paint? (*c*) Is the painted end a north or a south magnetic pole?

12. Can you think of a mechanism by which a magnetic storm, that is, a violent perturbation of the earth's magnetic field, can interfere with radio communication?

13. Convince yourself, in terms of the relation $\mathbf{F} = q_0\mathbf{v} \times \mathbf{B}$ (Eq. 33-2) that electrons and protons of the solar wind that are trapped in the earth's Van Allen radiation belts will indeed spiral around the lines of **B** and be reflected backward near the earth's north and south magnetic poles (the magnetic mirror effect). In what way do the motions of the trapped electrons and protons differ?

14. Aurorae are most frequently observed, not at the north and south magnetic poles, but at magnetic latitudes about 23° away from these poles (passing through Hudson Bay, for example, in the northern geomagnetic hemisphere). Can you think of any reason, however qualitative, why the auroral activity should not be strongest at the poles themselves?

15. Is the magnetization at saturation for a paramagnetic substance very much different from that for a saturated ferromagnetic substance of about the same size?

16. The magnetization induced in a given diamagnetic sphere by a given external magnetic field does not vary with temperature, in sharp contrast to the situation in paramagnetism. Is this understandable in terms of the description that we have given of the origin of diamagnetism?

17. Explain why a magnet attracts an unmagnetized iron object such as a nail.

18. Does any net force or torque act on (a) an unmagnetized iron bar or (b) a permanent bar magnet when placed in a uniform magnetic field?

19. A nail is placed at rest on a smooth table top near a strong magnet. It is released and attracted to the magnet. What is the source of the kinetic energy it has just before it strikes the magnet?

20. Compare the magnetization curves for a paramagnetic substance (see Fig. 37-8) and for a ferromagnetic substance (see Fig. 37-11). What would a similar curve for a diamagnetic substance look like? Do you think that it would show saturation effects in strong applied fields (say 10 T)?

21. Why do iron filings line up with a magnetic field, as in Fig. 37-1? After all, they are not intrinsically magnetized.

22. Distinguish between the precession frequency and the cyclotron frequency of a proton in a magnetic field.

23. In our discussion of nuclear magnetism we said that energy absorption occurs because the dipoles are turned end for end. However, a given dipole might initially be lined up either with the field or against it. In the first case there would be an *absorption* of energy, but in the second case there would be a *release* of energy, each amount being $2\mu B$. Why do we observe a *net* absorption? These two events would seem to cancel.

24. Discuss similarities and differences in Tables 30-2 and 37-1.

25. In what sense do a parallel plate capacitor filled with a dielectric and a Rowland ring (see Fig. 37-12) with an iron core show formal similarities as far as E and B (and their related vectors) are concerned? Discuss in terms of Tables 30-2 and 37-1.

26. A Rowland ring (see Fig. 37-12) carries a constant current. If we cut a small slot in the iron core, leaving an air gap, what changes occur in **B**, **H**, and **M** within the core?

27. You are given a bar magnet, with its north pole pointing up and its south pole pointing down. What are the directions of **B**, **H**, and **M** for points (a) inside the magnet near its center, (b) outside and just above the magnet, and (c) outside and just below the magnet?

problems

SECTION 37-1

1. The dipole moment of a small, single-turn, current loop is 2.0×10^{-4} A·m². What is the magnetic field (due to the dipole) on the axis of the dipole 8.0 cm away from the loop?   *Answer:* 7.8×10^{-8} T, pointing along the axis.

2. A simple bar magnet is suspended by a string, as shown in Fig. 37-25. If a uniform magnetic field **B** directed parallel to the ceiling is then established, sketch the resulting orientation of string and magnet.

3. Calculate (a) the electric field **E** and (b) the magnetic field **B** at a point 0.10 nm away from a proton, measured along its axis of spin. The magnetic moment of the proton is 1.4×10^{-26} A·m².
   Answer: (a) 1.4×10^{11} V/m. (b) 2.8×10^{-3} T.

4. Show that, classically, a spinning positive charge will have a spin magnetic moment that points in the same direction as its spin angular momentum.

5. An electron has a spin angular momentum L of 0.53×10^{-34} J·s and a magnetic moment μ of 9.3×10^{-24} A·m². Compare μ/L and e/m for the electron. *Answer:* They are equal.

6. A total charge q is distributed uniformly on a dielectric ring of radius r. If the ring is rotated about an axis perpendicular to its plane and through its center at an angular speed ω, find the magnitude and direction of its resulting magnetic moment.

7. Show, by sketching the magnetic field of a magnetic dipole, that (a) if the dipole moments of two nearby dipoles are parallel, they will not tend to remain that way and (b) if they are anti-parallel, they *will* tend to remain that way.

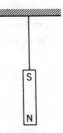

figure 37-25
Problem 2

In each case consider the torques on the second dipole in the field of the first.

8. Assume that the electron is a small sphere of radius R, its charge and mass being spread uniformly throughout its volume. Such an electron has a "spin" angular momentum L of 0.53×10^{-34} J·s and a magnetic moment μ of 9.3×10^{-24} A·m². Show that $e/m = 2\mu/L$. Is this prediction in agreement with experiment? (Hint: The spherical electron must be divided into infinitesimal current loops and an expression for the magnetic moment found by integration. This model of the electron is too mechanistic to be in the spirit of quantum physics.)

SECTION 37-3

9. From the data given in the text find (a) the average vertical component of the earth's magnetic field at Tucson in 1964 and (b) the average magnitude of **B**. *Answer:* (a) 44μT (=0.44 gauss). (b) 51μT (=0.51 gauss).

10. The earth has a magnetic dipole moment of 8.0×10^{22} A·m². (a) What current would have to be set up in a single turn of wire going around the earth at its magnetic equator if we wished to set up such a dipole? (b) Could such an arrangement be used to cancel out the earth's magnetism at points in space well above the earth's surface? (c) On the earth's surface?

SECTION 37-4

11. (a) What is the magnetic moment due to the orbital motion of an electron in an atom when the orbital angular momentum is one quantum unit $(= h/2\pi = 1.05 \times 10^{-34}$ J · s). (b) The intrinsic spin magnetic moment of an electron is 0.928×10^{-23} A·m². What is the difference in the magnetic potential energy U between the states in which the magnetic moment is aligned with and aligned in the opposite direction to an external magnetic field of 1.2 T? (c) What absolute temperature would be required so that the energy difference in (b) would equal the mean thermal energy $kT/2$? *Answer:* (a) 9.2×10^{-24} A · m². (b) 2.2×10^{-23} J. (c) 3.2 K.

12. At what temperature will the average thermal energy of a paramagnetic gas be equal to the magnetic energy in a field of 0.50 T (=5000 gauss) if the dipole moments of the atoms are about 10^{-23} A·m²?

SECTION 37-5

13. An electron travels in a circular orbit about a fixed positive point charge in the presence of a uniform magnetic field **B** directed normal to the plane of its motion. The electric force has precisely N times the magnitude of the magnetic force on the electron. (a) Determine the two possible angular speeds of the electron's motion. (b) Evaluate these speeds numerically if $B = 0.427$ T (=4.27×10^3 gauss) and $N = 100$.

Answer: (a) $(N \pm 1) \dfrac{eB}{m}$. (b) 7.43×10^{12} rad/s; 7.57×10^{12} rad/s.

14. Prove that $\Delta\omega \ll \omega_0$ in Eq. 37-13.

15. Can you give an explanation of diamagnetism based on Faraday's law of induction? In Figs. 37-9a and b, for example, what inductive effects can be expected as the magnetic field is built up from zero to the final value **B**?

SECTION 37-6

16. The dipole moment associated with an atom of iron in an iron bar is 1.8×10^{-23} A·m². Assume that all the atoms in the bar, which is 5.0 cm long and has a cross-sectional area of 1.0 cm², have their dipole moments aligned. (a) What is the dipole moment of the bar? (b) What torque must be exerted to hold this magnet at right angles to an external field of 1.5 T (=15,000 gauss)? The density of iron is 7.9 g/cm³.

17. A Rowland ring is formed of ferromagnetic material. It is circular in cross section, with an inner radius of 5.0 cm and an outer radius of 6.0 cm and is wound with 400 turns of wire. (a) What current must be set up in the windings to attain $B_0 = 2.0 \times 10^{-4}$ T in Fig. 37-11? (b) A secondary coil wound around the toroid has 50 turns and has a resistance of 8.0 Ω. If, for this value

of B_0, we have $B_M = 800 B_0$, how much charge moves through the secondary coil when the current in the toroid windings is turned on?

Answer: (a) 0.14 A. (b) 7.9×10^{-5} C.

18. *Dipole-dipole interaction.* The exchange coupling mentioned in Section 37-6 as being responsible for ferromagnetism is *not* the mutual magnetic interaction energy between two elementary magnetic dipoles. To show this (a) compute B a distance a ($=10$ nm) away from a dipole of moment μ ($= 1.8 \times 10^{-23}$ A·m²); (b) compute the energy ($= 2\mu B$) required to turn a second similar dipole end for end in this field. What do you conclude about the strength of this dipole-dipole interaction? Compare with the results of Example 4. (Note: For the same distance, the field in the median plane of a dipole is only half as large as on the axis; see Eq. 37-2.)

SECTION 37-7

19. Assume that the hydrogen nuclei (protons) in 1.0 g of water could all be aligned. What magnetic field B would be produced 5.0 cm from the sample, along its alignment axis?   *Answer:* 7.5×10^{-6} T.

20. It is possible to measure e/m for the electron by measuring (a) the cyclotron frequency ν_c of electrons in a given magnetic field and (b) the precession frequency ν_p of protons in the same field. Show that the relation is

$$\frac{e}{m} = \frac{\nu_c}{\nu_p} \frac{\mu_s}{L_s}.$$

SECTION 37-8

21. The magnetic energy density can be shown to be given in its most general form as

$$\mu_B = \tfrac{1}{2} \mathbf{B} \cdot \mathbf{H}.$$

Does this reduce to a familiar result for a vacuum?   *Answer:* Yes.

22. An iron magnet containing iron of relative permeability 5000 has a flux path 1.0 m long in the iron and an air gap 0.01 m long each with cross-sectional areas of 0.02 m². What current is necessary in a 500 turn coil wrapped around the iron to give a flux density in the air gap of 1.8 T?

23. *Boundary condition for* **H.** Prove that at the boundary between two media the tangential component of **H** has the same value on each side of the surface, assuming that there is no current at the surface. (Hint: Construct a closed rectangular loop, the two opposite longer sides being parallel to the surface, with one side in each medium. Use Ampère's law in the form that applies when magnetic materials are present.)

24. *Boundary condition for* **B.** Prove that at the boundary between two media the normal component of **B** has the same value on each side of the surface. (Hint: Construct a closed Gaussian surface shaped like a flat pillbox with one face in each medium and apply Gauss's law for magnetism.)

38
electromagnetic oscillations

The LC system in Fig. 38-1 (assumed resistanceless*) resembles the mass-spring system of Fig. 8-4 (assumed frictionless*) in that, among other things, each system has a characteristic frequency of oscillation. The analogy actually goes far beyond this, as we will see quantitatively in Section 38-3. Let us first, however, treat LC oscillations from a physical but semiquantitative point of view.

Assume that initially the capacitor C in Fig. 38-1a carries a charge q_m and the current i in the inductor is zero. At this instant the energy stored in the capacitor is given by Eq. 30-7, or

$$U_E = \frac{1}{2}\frac{q_m^2}{C}. \tag{38-1}$$

The energy stored in the inductor, given by

$$U_B = \tfrac{1}{2}Li^2, \tag{38-2}$$

is zero because the current is zero. The capacitor now starts to discharge through the inductor, positive charge carriers moving counterclockwise, as shown in Fig. 38-1b. This means that a current i, given by dq/dt and pointing down in the inductor, is established.

As q decreases, the energy stored in the electric field in the capacitor also decreases. This energy is transferred to the magnetic field that appears around the inductor because of the current i that is building up

---

* A common term to describe the closely related situations of Figs. 8-4 and 38-1 is "undamped." If friction is present (as in Fig. 15-19) or resistance (as in Fig. 38-3), we describe the situation as "damped."

there. Thus the electric field decreases, the magnetic field builds up, and energy is transferred from the former to the latter.

At a time corresponding to Fig. 38-1c, all the charge on the capacitor will have disappeared. The electric field in the capacitor will be zero, the energy stored there having been transferred entirely to the magnetic field of the inductor. According to Eq. 38-2, there must then be a current—and indeed one of maximum value—in the inductor. Note that even though q equals zero, the current (which is dq/dt) is *not* zero at this time.

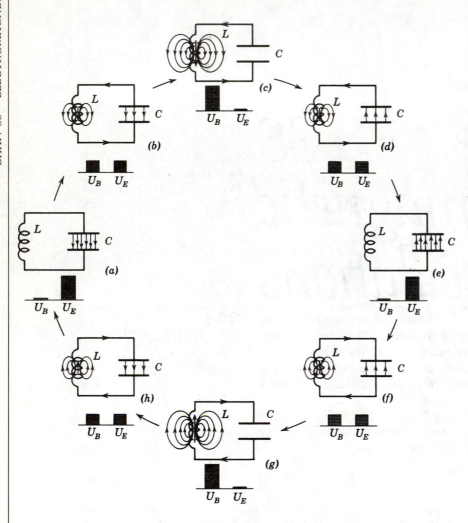

figure 38-1
Showing eight stages in a cycle of oscillation of a resistanceless LC circuit. The bar graphs below each figure show the stored magnetic and electric energy. The vertical arrows on the inductor axis show the current. Compare this figure in detail with Fig. 8-4, to which it exactly corresponds.

The large current in the inductor in Fig. 38-1c continues to transport positive charge from the top plate of the capacitor to the bottom plate, as shown in Fig. 38-1d; energy now flows from the inductor back to the capacitor as the electric field builds up again. Eventually, the energy will have been transferred completely back to the capacitor, as in Fig. 38-1e. The situation of Fig. 38-1e is like the initial situation, except that the capacitor is charged oppositely.

The capacitor will start to discharge again, the current now being clockwise, as in Fig. 38-1f. Reasoning as before, we see that the circuit eventually returns to its initial situation and that the process continues at a definite frequency ν (measured, say, in hertz, or cycles/second) to which corresponds a definite *angular* frequency ω (= $2\pi\nu$ and measured, say, in radians/second). Once started, such LC oscillations (in the ideal case described, in which the circuit contains no resistance) continue indefinitely, energy being shuttled back and forth between the electric field in the capacitor and the magnetic field in the inductor. Any configuration in Fig. 38-1 can be set up as an initial condition. The oscillations will then continue from that point, proceeding clockwise around the figure. Compare these oscillations carefully with those of the mass-spring system described in Fig. 8-4.

To measure the charge q as a function of time, we can measure the variable potential difference $V_C(t)$ that exists across capacitor C. The relation

$$V_C = \left(\frac{1}{C}\right) q$$

shows that V_C is proportional to q. To measure the current we can insert a small resistance R in the circuit and measure the potential difference across it. This is proportional to i through the relation

$$V_R = (R)i.$$

We assume here that R is so small that its effect on the behavior of the circuit is negligible. Both q and i, or more correctly V_C and V_R, which are proportional to them, can be displayed on a cathode-ray oscilloscope. This instrument can plot automatically on its screen graphs proportional to $q(t)$ and $i(t)$, as in Fig. 38-2.

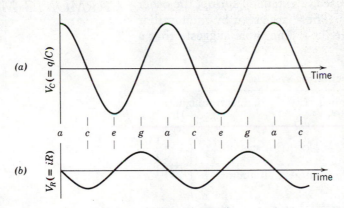

(a)

(b)

figure 38-2
A drawing of an oscilloscope screen showing potential differences proportional to (a) the charge and (b) the current, in the circuit of Fig. 38-1, as a function of time. The letters indicate corresponding phases of oscillation in that figure. Note that because $i = dq/dt$ the lower curve is proportional to the derivative of the upper. Can you verify this?

A 1.0-μF capacitor is charged to 50 V. The charging battery is then disconnected and a 10-mH coil is connected across the capacitor, so that LC oscillations occur. What is the maximum current in the coil? Assume that the circuit contains no resistance.

The maximum stored energy in the capacitor must equal the maximum stored energy in the inductor, from the conservation-of-energy principle. This leads, from Eqs. 38-1 and 38-2, to

$$\frac{1}{2}\frac{q_m^2}{C} = \tfrac{1}{2}Li_m^2,$$

where i_m is the *maximum* current and q_m is the *maximum* charge. Note that the maximum current and the maximum charge do not occur at the same time but one-fourth of a cycle apart; see Figs. 38-1 and 38-2. Solving for i_m and substituting CV_0 for q_m gives

$$i_m = V_0\sqrt{\frac{C}{L}} = (50\text{ V})\sqrt{\frac{1.0 \times 10^{-6}\text{ F}}{10 \times 10^{-3}\text{ H}}} = 0.50\text{ A}.$$

EXAMPLE 1

In an actual *LC* circuit the oscillations will not continue indefinitely because there is always some resistance present that will drain energy from the electric and magnetic fields and dissipate it as thermal (Joule) energy; the circuit will warm up. The oscillations, once started, will die away, as in Fig. 38-3. Compare this figure with Fig. 15-19, which shows the decay of the mechanical oscillations of a mass-spring system caused by frictional damping.

It is possible to have sustained electromagnetic oscillations if arrangements are made to supply, automatically and periodically (once a cycle, say), enough energy from an outside source to compensate for that dissipated as thermal energy.* We are reminded of a clock escapement, which is a device for feeding energy from a spring or a falling weight into an oscillating pendulum, thus compensating for frictional losses that would otherwise cause the oscillations to die away. *LC* oscillators, whose frequency may be varied between certain limits, are commercially available as packaged units over a wide range of frequencies, extending from low audio-frequencies (lower than 10 Hz) to microwave frequencies (higher than 10 GHz).

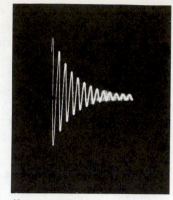

figure 38-3
A photograph of an oscilloscope trace showing how the oscillations in an *RCL* circuit die away because energy is dissipated as thermal energy in the resistor. The figure is a plot of the potential difference across the resistor as a function of time.

38-2
ANALOGY TO SIMPLE HARMONIC MOTION

Figure 8-4 shows that in a mass-spring system performing simple harmonic motion, as in an oscillating *LC* circuit, two kinds of energy occur. One is potential energy of the compressed or extended spring; the other is kinetic energy of the moving mass. These are given by the familiar formulas in the first column of Table 38-1. The table suggests that a

Table 38-1
Energy in oscillating systems

Mechanical (Fig. 8-4)		Electromagnetic (Fig. 38-1)	
spring	$U_P = \frac{1}{2}kx^2$	capacitor	$U_E = \frac{1}{2}\frac{q^2}{C}$
mass	$U_K = \frac{1}{2}mv^2$	inductor	$U_B = \frac{1}{2}Li^2$
	$v = dx/dt$		$i = dq/dt$

capacitor is in some formal way like a spring and an inductor is like a mass and that certain electromagnetic quantities "correspond" to certain mechanical ones, namely,

q corresponds to x,
i corresponds to v,
C corresponds to $1/k$,
L corresponds to m.

Comparison of Fig. 38-1, which shows the oscillations of a resistanceless *LC* circuit, with Fig. 8-4, which shows the oscillations in a frictionless mass-spring system, indicates how close the correspondence is. Note how v and i correspond in the two figures; also x and q. Note, too, how in each case the energy alternates between two forms, magnetic

---

* From another point of view it is possible to supply an electronically generated "negative resistance" to the circuit, large enough in magnitude to just cancel the actual resistance. See "Negative Resistor to Provide Self-Oscillation in RLC Circuits" by Edwin A. S. Lewis, *American Journal of Physics*, December 1976.

and electric for the LC system, and kinetic and potential for the mass-spring system.

In Section 15-3 we saw that the natural angular frequency of oscillation of a (frictionless) mass-spring system is

$$\omega = 2\pi\nu = \sqrt{\frac{k}{m}}.$$

The method of correspondences suggests that to find the natural frequency for a (resistanceless) LC circuit, k should be replaced by $1/C$ and m by L, obtaining

$$\omega = 2\pi\nu = \sqrt{\frac{1}{LC}}. \qquad (38\text{-}3)$$

This formula is indeed correct, as we show in the next section.

We now derive an expression for the frequency of oscillation of a (resistanceless) LC circuit, our derivation being based on the conservation-of-energy principle. The total energy U present at any instant in an oscillating LC circuit is given by

$$U = U_B + U_E = \tfrac{1}{2}Li^2 + \frac{1}{2}\frac{q^2}{C},$$

which expresses the fact that at any arbitrary time the energy is stored partly in the magnetic field in the inductor and partly in the electric field in the capacitor. If we assume the circuit resistance to be zero, there is no energy transfer to thermal energy and U remains constant with time, even though i and q vary. In more formal language, dU/dt must be zero. This leads to

$$\frac{dU}{dt} = \frac{d}{dt}\left(\tfrac{1}{2}Li^2 + \frac{1}{2}\frac{q^2}{C}\right) = Li\frac{di}{dt} + \frac{q}{C}\frac{dq}{dt} = 0. \qquad (38\text{-}4)$$

Now, q and i are not independent variables, being related by

$$i = \frac{dq}{dt}.$$

Differentiating yields $\qquad \dfrac{di}{dt} = \dfrac{d^2q}{dt^2}.$

Substituting these two expressions into Eq. 38-4 leads to

$$L\frac{d^2q}{dt^2} + \frac{1}{C}q = 0. \qquad (38\text{-}5)$$

This is the differential equation that describes the oscillations of a (resistanceless) LC circuit. To solve it, note that Eq. 38-5 is mathematically of *exactly* the same form as Eq. 15-6,

$$m\frac{d^2x}{dt^2} + kx = 0, \qquad (15\text{-}6)$$

which is the differential equation for the mass-spring oscillations. Fundamentally, it is by comparing these two equations that the correspondences on p. 844 arise.

The solution of Eq. 15-6 proved to be

$$x = A\cos(\omega t + \phi), \qquad (15\text{-}8)$$

where $A\ (= x_m)$ is the amplitude of the motion and ϕ is an arbitrary *phase angle.* Since q corresponds to x, we can write the solution of Eq. 38-5 as

$$q = q_m \cos(\omega t + \phi), \qquad (38\text{-}6)$$

where ω is the still unknown angular frequency of the electromagnetic oscillations.

We can test whether Eq. 38-6 is indeed a solution of Eq. 38-5 by substituting it and its second derivative in that equation. To find the second derivative, we write

$$\frac{dq}{dt} = i = -\omega q_m \sin(\omega t + \phi) \qquad (38\text{-}7a)$$

and

$$\frac{d^2q}{dt^2} = -\omega^2 q_m \cos(\omega t + \phi). \qquad (38\text{-}7b)$$

Substituting q and d^2q/dt^2 into Eq. 38-5 yields

$$-L\omega^2 q_m \cos(\omega t + \phi) + \frac{1}{C} q_m \cos(\omega t + \phi) = 0.$$

Canceling $q_m \cos(\omega t + \phi)$ and rearranging leads to

$$\omega = \sqrt{\frac{1}{LC}}.$$

Thus, if ω is given the constant value $1/\sqrt{LC}$, Eq. 38-6 is indeed a solution of Eq. 38-5. This expression for ω agrees with Eq. 38-3, which we arrived at by the method of correspondences.

The phase angle ϕ in Eq. 38-6 is determined by the conditions that prevail at $t = 0$. If the initial condition is as represented by Fig. 38-1a, then we put $\phi = 0$ in order that Eq. 38-6 may predict $q = q_m$ at $t = 0$. What initial condition in Fig. 38-1 is implied if we select $\phi = 90°$?

EXAMPLE 2

(a) In an oscillating LC circuit, what value of charge, expressed in terms of the maximum charge, is present on the capacitor when the energy is shared equally between the electric and the magnetic field? (b) How much time is required for this condition to arise, assuming the capacitor to be fully charged initially? Assume that $L = 10$ mH and $C = 1.0$ μF.

(a) The stored energy and the *maximum* stored energy in the capacitor are, respectively,

$$U_E = \frac{q^2}{2C} \qquad \text{and} \qquad U_{E,m} = \frac{q_m^2}{2C}.$$

Substituting $U_E = \tfrac{1}{2} U_{E,m}$ yields

$$\frac{q^2}{2C} = \frac{1}{2}\frac{q_m^2}{2C} \qquad \text{or} \qquad q = \frac{1}{\sqrt{2}} q_m.$$

(b) To find the time, we write, assuming $\phi = 0$ in Eq. 38-6,

$$q = q_m \cos \omega t = \frac{1}{\sqrt{2}} q_m,$$

which leads to

$$\omega t = \cos^{-1} \frac{1}{\sqrt{2}} = \frac{\pi}{4} \qquad \text{or} \qquad t = \frac{\pi}{4\omega}.$$

The angular frequency ω is found from Eq. 38-3, or

$$\omega = \sqrt{\frac{1}{LC}} = \sqrt{\frac{1}{(10 \times 10^{-3}\ \text{H})(1.0 \times 10^{-6}\ \text{F})}} = 1.0 \times 10^4\ \text{rad/s}.$$

The time t is then

$$t = \frac{\pi}{4\omega} = \frac{\pi}{(4)(1.0 \times 10^4 \text{ rad/s})} = 7.9 \times 10^{-5} \text{ s } (=79 \text{ } \mu\text{s}).$$

What is the frequency ν in Hz?

The stored electric energy in the LC circuit, using Eq. 38-6, is

$$U_E = \frac{1}{2}\frac{q^2}{C} = \frac{q_m^2}{2C}\cos^2(\omega t + \phi), \qquad (38\text{-}8)$$

and the magnetic energy, using Eq. 38-7a, is

$$U_B = \frac{1}{2}Li^2 = \frac{1}{2}L\omega^2 q_m^2 \sin^2(\omega t + \phi).$$

Substituting the expression for ω (see Eq. 38-3) into this last equation yields

$$U_B = \frac{q_m^2}{2C}\sin^2(\omega t + \phi). \qquad (38\text{-}9)$$

Figure 38-4 shows plots of $U_E(t)$ and $U_B(t)$ for the case of $\phi = 0$. Note that (a) the maximum values of U_E and U_B are the same $(= q_m^2/2C)$; (b) at any instant the sum of U_E and U_B is a constant $(= q_m^2/2C)$; and (c) when U_E has its maximum value, U_B is zero and conversely. This analysis supports the qualitative analysis of Section 38-1. Compare this discussion with that given in Section 15-4 for the energy transfers in a mass-spring system.

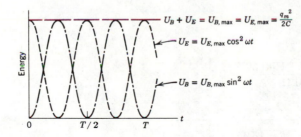

figure 38-4
The stored magnetic (—————) and electric (—————) energies in the circuit of Fig. 38-1. Note that their sum is a constant. $T (=1/\nu)$ is the period of the oscillation.

The RCL circuit. (a) Derive an expression for the quantity $q(t)$ for a single-loop circuit containing a resistance R as well as an inductance L and a capacitance C. (b) After how long a time will the charge oscillations decay to half-amplitude if $L = 10$ mH, $C = 1.0$ μF, and $R = 0.10$ Ω?

(a) If U is the total stored field energy, we have, as before,

$$U = \frac{1}{2}Li^2 + \frac{1}{2}\frac{q^2}{C}.$$

U is no longer constant but rather

$$\frac{dU}{dt} = -i^2 R,$$

the minus sign signifying that the stored energy U *decreases* with time, being converted to thermal (Joule) energy at the rate i^2R. Combining these two equations leads to

$$Li\frac{di}{dt} + \frac{q}{C}\frac{dq}{dt} = -i^2 R.$$

EXAMPLE 3

Substituting dq/dt for i and d^2q/dt^2 for di/dt leads finally to

$$L \frac{d^2q}{dt^2} + R \frac{dq}{dt} + \frac{1}{C} q = 0,$$

which is the differential equation that describes the damped LC oscillations. If we put $R = 0$, this equation reduces, as expected, to Eq. 38-5.

Compare this differential equation for damped LC oscillations with Eq. 15-37, or

$$m \frac{d^2x}{dt^2} + b \frac{dx}{dt} + kx = 0, \tag{15-37}$$

which describes damped mass-spring oscillations. Once again the equations are mathematically identical, the resistance R corresponding to the mechanical damping constant b and q corresponding to x.

The solution of the RCL circuit follows at once, by correspondence, from the solution of Eq. 15-37. It is (see Eqs. 15-38 and 15-39) for R *reasonably small*, and for an initial condition in which the capacitor has a maximum charge (that is, $\phi = 0$),

$$q = q_m e^{-Rt/2L} \cos \omega' t, \tag{38-10}$$

where

$$\omega' = \sqrt{\frac{1}{LC} - \left(\frac{R}{2L}\right)^2}. \tag{38-11}$$

Note that Eq. 38-10, which can be described as a cosine function with an exponentially decreasing amplitude, is the equation of the decay curve of Fig. 38-3. Note, too (see Eq. 38-11), that the presence of resistance reduces the oscillation frequency. These two equations reduce to familiar results as $R \to 0$.

(b) The oscillation amplitude will have decreased to half the initial amplitude when the amplitude factor $e^{-Rt/2L}$ in Eq. 38-10 has the value one-half, or

$$\tfrac{1}{2} = e^{-Rt/2L}.$$

At what time t will this occur? What is ω'?

We have

$$t = \frac{2L}{R} \ln 2 = \frac{(2)(10 \times 10^{-3} \text{ H})(0.69)}{0.10 \ \Omega} = 0.14 \text{ s}.$$

The angular frequency, from Eq. 38-11, is

$$\omega' = \sqrt{\frac{1}{(10 \times 10^{-3} \text{ H})(1.0 \times 10^{-6} \text{ F})} - \left(\frac{0.10 \ \Omega}{2 \times 10 \times 10^{-3} \text{ H}}\right)^2}$$

$$= \sqrt{10^8 \ (\text{rad/s})^2 - 25 \ (\text{rad/s})^2} = 1.0 \times 10^4 \text{ rad/s}.$$

Note that the second term is quite small, so that in this case, as in many practical cases, the resistance has a negligible effect on the frequency. Can you show that 0.14 s, the time at which the oscillations decrease to half-amplitude, corresponds to about 220 cycles of oscillation? The damping is much less severe than that shown in Fig. 38-3.

38-4
LUMPED AND DISTRIBUTED ELEMENTS

In this section we put electromagnetic oscillations aside for the moment and return to mechanical oscillating systems, so that we can develop the concepts of *lumped* and *distributed* elements. In Section 38-5 we will return to electromagnetic systems and make a similar comparison.

In the oscillating mass-spring system in Fig. 8-4 the two kinds of energy involved appear in quite separate parts of the system, the potential energy being associated with the spring and the kinetic energy with the moving mass. A closed organ pipe (see Fig. 20-6b) is a mechanical oscillating system in which the two kinds of energy are *not* separated in space. Kinetic energy, associated with moving elements of air, and potential energy, associated with compressions and rarifactions of elements of air, are both present throughout the volume of the pipe. We say that such an oscillating system has *distributed* rather than *lumped* (as in the mass-spring system) elements.

We see at once one difference between lumped and distributed mechanical oscillating systems. A lumped system has a single frequency of oscillation, given for the mass-spring system by Eq. 15-11; see Section 15-5 for other examples of lumped oscillating mechanical systems. On the other hand, distributed systems, such as the organ pipes in Fig. 20-6 and the vibrating string of Fig. 20-4, have a number of discrete oscillation frequencies (harmonics), the values for these two cases being given in Section 20-5.

Now let us examine in more detail, from the point of view of distributed elements, the behavior of a closed organ pipe, which we will idealize as an *acoustic resonant cavity.*

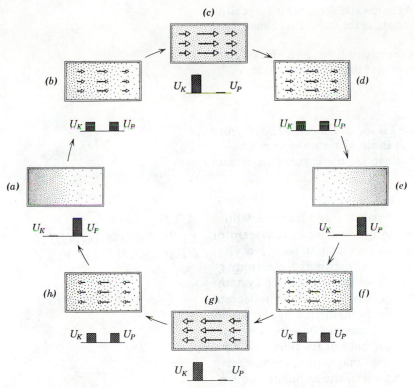

figure 38-5
Showing eight stages in a cycle of oscillation of a cylindrical acoustic resonant cavity, suggestive of a closed organ pipe. The bar graphs below each figure show the kinetic and potential energy. The arrows represent the directed velocities of small volume elements of the gas.

Figure 38-5, a series of "snapshots" taken one-eighth of a cycle apart, shows the pressure and velocity variations in the fundamental mode of a particular acoustic resonant cavity. There is a pressure node at the center and a pressure antinode at each end. There is a velocity* node at each end and a velocity antinode at the center. When the pressure variation is the greatest, the velocity is zero (Figs. 38-5a and e). When the pressure is uniform, the velocities have their maximum values (Figs. 38-5c, and g).

The energy in the acoustic resonator alternates between kinetic

---

*The velocity of interest here is the directed velocity $\mathbf{v}_{gas}$ of small volume elements of the gas which are, however, large enough to contain a great number of molecules. The thermal velocities of the molecules have no directional preference and are ignored.

energy associated with the moving gas and potential energy associated with the compression and rarefaction of the gas. In Figs. 38-5c and g the energy is all kinetic and in Figs. 38-5a and e it is all potential. In intermediate phases it is part kinetic and part potential.

The kinetic energy of a small mass Δm of the gas, which is moving parallel to the cylinder axis with a speed v_{gas}, is $\frac{1}{2}\Delta m v_{gas}^2$. The *kinetic energy density*, that is, the kinetic energy per unit volume, is

$$u_K = \frac{1}{2}\frac{\Delta m}{\Delta V}v_{gas}^2 = \frac{1}{2}\rho_0 v_{gas}^2,$$

in which ΔV is the volume of the gas element and ρ_0 is the mean density of the gas.

The potential energy per unit volume in the gas, that is, the *potential energy density*, associated with the compressions and rarefactions of the gas is given (we state) by

$$u_P = \frac{1}{2}B\left(\frac{\Delta\rho}{\rho_0}\right)^2.$$

Here B is the bulk modulus* of elasticity of the gas and $\Delta\rho/\rho_0$, which is positive for a compression and negative for a rarefaction, is the fractional change in gas density at any point.

The angular frequency of oscillation for the cavity of Fig. 38-5, in the fundamental (or lowest frequency) mode, shown in that figure, is found from

$$\omega_1 = 2\pi\nu = \frac{2\pi v}{\lambda} = \frac{\pi v}{l'}$$

where v is the speed of sound in the gas and l is the length of the cavity. From Eq. 20-1 we may write v as $\sqrt{B/\rho_0}$. Note that in the above we have put $\lambda = 2l$, corresponding to the fundamental mode. What are the angular frequencies ω_2, ω_3, etc., of the higher frequency modes?

38-5
ELECTROMAGNETIC RESONANT CAVITY

Consider now a second closed cylinder, of radius a and length l, whose walls are made of copper or some other good conductor. A system of oscillating electric and magnetic fields can be set up in such a cavity. Such an *electromagnetic resonant cavity* is a distributed electromagnetic oscillator, in contrast to an LC circuit, which is a lumped system. As for the acoustic resonator, many modes of oscillation with discrete frequencies are possible; we describe only the fundamental mode, which has the lowest frequency. The cavity oscillations can be set up by suitably connecting the cavity, through a small hole in its side wall, to a source of electromagnetic radiation such as a magnetron. If the cavity dimensions are of the order of a few centimeters, the resonant frequencies will be of the order of 10^{10} Hz. This is far higher than the acoustic frequencies in cavities of the same size, reflecting the fact that the speed of electromagnetic waves in free space ($= 3 \times 10^8$ m/s) is much greater than the speed of sound in air (about 350 m/s).

Figure 38-6, which is a series of "snapshots" taken one-eighth of a cycle apart, shows, by the horizontal lines, how the electric field **E** varies with time in the cavity. The electric lines of force originate on positive charges at one end of the cylinder and terminate on negative charges at the other end. As **E** changes with time, eventually reversing itself, currents flow from one end of the cylinder to the other on the inner cylinder wall. At any point in the cavity, energy is stored in the electric field in an amount per unit volume given by Eq. 30-9, or

$$u_E = \frac{1}{2}\epsilon_0 E^2. \tag{38-12}$$

* See p. 435.

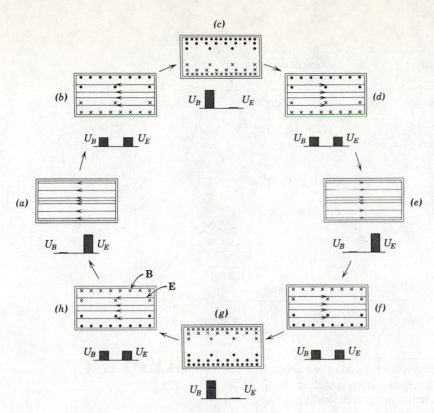

figure 38-6
Showing eight stages in a cycle of oscillation of a cylindrical electromagnetic resonant cavity. The bar graphs below each figure show the stored electric and magnetic energy. The dots and crosses represent circular lines of **B**; the horizontal lines represent **E**. Compare with Fig. 38-5; both figures are examples of distributed elements.

Figure 38-6 also shows, by the dots and crosses, how the magnetic field **B** varies with time. The magnetic lines form circles about the cylinder axis. Note that the magnetic field has a maximum value when the electric field is zero, and conversely. At any point in the cavity energy is stored in the magnetic field in an amount per unit volume given by Eq. 36-19, or

$$u_B = \frac{1}{2\mu_0} B^2. \tag{38-13}$$

Thus, as in the LC circuit, energy is shuttled back and forth between the electric and the magnetic fields. However, these fields no longer occupy completely separate regions of space. In Section 40-5 we clarify just *why* the electric and magnetic fields in Fig. 38-6 interact in the way that we have described.

We state without proof that the angular frequency of oscillation for the electromagnetic cavity in Fig. 38-6 is, in the fundamental mode shown in that figure,

$$\omega_1 = \frac{2.41c}{a},$$

in which a is the cavity radius and c is the speed of electromagnetic radiation in free space. We will see in Section 41-8 that c may be written in terms of electromagnetic quantities as $1/\sqrt{\mu_0\epsilon_0}$. As the field patterns in Fig. 38-6 suggest, the resonant frequency of oscillation of the cavity, for the mode of oscillation shown, depends only on the radius of the cavity and not on its length.

Table 38-2 summarizes some characteristics of the four oscillating systems that we have discussed so far. For lumped systems it gives expressions for the two kinds of energy involved and for the (single) oscillation frequency. For the distributed systems it gives expressions for the two kinds of energy density involved and for the oscillation frequency in the fundamental mode. You should study carefully all the correspondences, similarities, and differences that occur in the table.

Table 38-2
Four oscillating systems

	Mechanical Systems (frictionless)	Electromagnetic Systems (resistanceless)
Lumped systems	Mass + spring $U_K = \frac{1}{2}mv^2$ $U_P = \frac{1}{2}kx^2$ $\omega = \sqrt{\dfrac{k}{m}}$ (Fig. 8-4)	LC circuit $U_B = \frac{1}{2}Li^2$ $U_E = \frac{1}{2}(1/C)q^2$ $\omega = \sqrt{\dfrac{1}{LC}}$ (Fig. 38-1)
Distributed systems	Acoustic cavity $u_K = \frac{1}{2}\rho_0 v_{gas}^2$ $u_P = \frac{1}{2}B(\Delta\rho/\rho_0)^2$ $\omega_1 = \dfrac{\pi v}{l}; \quad v = \sqrt{\dfrac{B}{\rho_0}}$ (Fig. 38-5)	Electromagnetic cavity $u_B = \frac{1}{2}(1/\mu_0)B^2$ $u_E = \frac{1}{2}\epsilon_0 E^2$ $\omega_1 = \dfrac{2.41c}{a}; \quad c = \sqrt{\dfrac{1}{\epsilon_0\mu_0}}$ (Fig. 38-6)

EXAMPLE 4

In the cavity of Fig. 38-6, what is the relationship between the "average" value of E throughout the cavity, measured at the instant corresponding to Fig. 38-6a, to the "average" value of B, measured at the instant corresponding to Fig. 38-6c?

At the first instant the energy is all electric and at the second it is all magnetic. The total energy U, found by integrating the energy density over the volume of the cavity, must be the same at these two instants, or

$$U = \int u_{E,m}\, dV = \int u_{B,m}\, dV,$$

where dV is a volume element in the cavity and $u_{E,m}$ and $u_{B,m}$ are the *maximum* values of u_E and of u_B at the site of this volume element; these maximum values occur one-fourth of a cycle apart, as Fig. 38-6 shows. Substituting Eqs. 38-12 and 38-13 leads to

$$\int \frac{\epsilon_0 E_m^2}{2}\, dV = \int \frac{B_m^2}{2\mu_0}\, dV$$

or

$$\mu_0\epsilon_0 \int E_m^2\, dV = \int B_m^2\, dV.$$

The quantity $\int E_m^2\, dV$ can be written as $\overline{E_m^2}V$, where V is the cavity volume and $\overline{E_m^2}$ is the average value of E_m^2 throughout the cavity. Treating B_m in the same way leads to

$$\mu_0\epsilon_0 \overline{E_m^2} = \overline{B_m^2}$$

or, taking square roots, $\quad \sqrt{\overline{B_m^2}} = \sqrt{\mu_0\epsilon_0}\, \sqrt{\overline{E_m^2}}.$

We can represent $\sqrt{\overline{B_m^2}}$ by B_{rms}, a so-called "root-mean-square" value of B_m. In computing B_{rms}, note that the averaging is done throughout the volume of the cavity, at the instant corresponding to Fig. 38-6c. It is not a time average for a particular point in the cavity. Doing the same for E yields

$$E_{rms}/B_{rms} = \frac{1}{\sqrt{\mu_0\epsilon_0}}$$

$$= [(4\pi \times 10^{-7}\ \text{T·m/A})(8.9 \times 10^{-12}\ \text{C}^2/\text{N·m}^2)]^{-1/2}$$

$$= 3.00 \times 10^8\ \text{m/s}.$$

If this reminds you of the speed of light, it is no coincidence because that is exactly what it is; we will explore this in detail in Section 41-8.

If E_{rms} above equals 1.0×10^4 V/m, a reasonable value, then

$$B_{rms} = (1.0 \times 10^4 \text{ V/m})/(3.00 \times 10^8 \text{ m/s})$$

$$= 3.3 \times 10^{-5} \text{ T} = 0.33 \text{ gauss.}$$

What is the total stored energy in the cavity under these conditions, assuming the cavity to be 10 cm long and 3.0 cm in diameter?

Recall that in Example 6, Chapter 36, we showed the energy density for a magnetic field of "ordinary" laboratory magnitude (say, 1 T) to be enormously greater than that for an electric field of "ordinary" magnitude (say, 10^5 V/m). This fact is consistent with the present example.

questions

1. Why doesn't the *LC* circuit of Fig. 38-1 simply stop oscillating when the capacitor has been completely discharged?

2. How might you start an *LC* circuit into oscillation with its initial condition being represented by Fig. 38-1c? Devise a switching scheme to bring this about.

3. In an oscillating *LC* circuit, assumed resistanceless, what determines (a) the frequency and (b) the amplitude of the oscillations?

4. In connection with Figs. 38-1c and g, explain how there can be a current in the inductor even though there is no charge on the capacitor.

5. In Fig. 38-1 is it possible to have (a) an *LC* circuit without resistance, (b) an inductor without inherent capacitance, or (c) a capacitor without inherent inductance? Discuss the practical validity of the *LC* circuit of Fig. 38-1, in which each of the above possibilities is ignored. See "Self-resonant Effects in Coils and Capacitors: an Experiment" by Samuel Derman, *The Physics Teacher*, September 1976.

6. In Fig. 38-1, what changes are called for if the oscillations are to proceed *counterclockwise* around the figure?

7. In Fig. 38-1, what phase angles ϕ in Eq. 38-6 correspond to the eight circuit situations shown?

8. All practical *LC* circuits have to contain *some* resistance and thus are *RCL* circuits. However, one can buy a packaged audio oscillator in which the output maintains a constant amplitude indefinitely and does not decay, as in Fig. 38-3. How can this happen? (*Hint:* Consider the analogy to the pendulum clock, in which falling weights are involved.)

9. Can you see any physical reason for assuming that R is "small" in Eqs. 38-10 and 38-11? (*Hint:* Consider what might happen if the damping R were so large that Eq. 38-10 could not even go through one cycle of oscillation before q was reduced to essentially zero. Could this happen? If so, what do you imagine Fig. 38-3 would look like?)

10. Tabulate as many mechanical or electric systems as you can think of that possess a natural frequency, along with the formula for that frequency if given in the text.

11. Discuss the periodic flow of energy, if any, from point to point in an acoustic resonant cavity.

12. Can a given circuit element (a capacitor, say) behave like a "lumped" element under some circumstances and like a "distributed" element under others?

13. List as many (a) lumped and (b) distributed mechanical oscillating systems as you can.

14. Are oscillating systems (mechanical, say) *either* lumped *or* distributed? That is, is there no middle ground? (a) Consider a lumped system such as an

idealized mass-spring arrangement. How might you change it physically to make it more distributed? (b) Consider a distributed system such as a vibrating string. How might you change it physically to make it more lumped?

15. A coil has a measured inductance L. In a practical case it also has a capacitance C, adjacent windings behaving as "plates." It can be made to oscillate at a certain frequency without attaching it to an external capacitance. Is this a case of distributed elements? Do you suppose that it can oscillate at more than one frequency? Discuss.

16. A violin string is an oscillating mechanical system with distributed elements. Give some qualitative details as to why this is so. For example, where are the kinetic and potential energies to be found? (Compare Figs. 19-16 and 38-5.)

17. An air-filled acoustic resonant cavity and an electromagnetic resonant cavity of the same size have resonant frequencies that are in the ratio of 10^6 or so. Which has the higher frequency and why?

18. What constructional difficulties would you encounter if you tried to build an LC circuit of the type shown in Fig. 38-1 to oscillate (a) at 0.01 Hz, or (b) at 10^{10} Hz?

19. Electromagnetic cavities are often silver-plated on the inside. Why?

problems

SECTION 38-1

1. Find the capacitance of an LC circuit if the maximum charge on the capacitor is 1.0 μC and the total energy is 1.4×10^{-4} J.   *Answer:* 3.6×10^{-9} F.

2. A 1.5-mH inductor in an LC circuit stores a maximum energy of 1.0×10^{-5} J. What is the peak current?

3. In an oscillating LC circuit $L = 1.0$ mH, $C = 4.0$ μF and the maximum charge on C is 3.0 μC. Find the maximum current.   *Answer:* 47 mA.

4. An oscillating LC circuit consisting of a 1.0 nF (= 1.0 nanofarad = 1.0×10^{-9} F) capacitor and a 3.0-mH coil carries a peak voltage of 3.0 V. (a) What is the maximum charge on the capacitor? (b) What is the peak current through the circuit? (c) What is the maximum energy stored in the magnetic field of the coil?

5. In an oscillating LC circuit, (a) in terms of the maximum charge on the capacitor, what value of charge is present when the energy is shared equally between the electric and the magnetic fields? (b) What fraction of a period must elapse following the time the capacitor is fully charged for this condition to arise?   *Answer:* (a) $q = q_m/\sqrt{2}$. (b) $t = T/8$.

6. At some instant in an oscillating LC circuit, three-fourths of the total energy is stored in the magnetic field of the inductor. (a) In terms of the maximum charge on the capacitor, what is the charge on the capacitor at this instant? (b) In terms of the maximum current in the inductor, what is the current in the inductor at this instant?

SECTION 38-2

7. Given a 1.0-mH inductor, how would you make it oscillate at 1.0 MHz (= 1.0×10^6 Hz)?
Answer: Connect a 25-pF capacitor across it and use it as the resonant element in an oscillator.

8. You are given a 10-mH inductor and two capacitors, of 5.0- and 2.0-μF capacitance. (a) Can you find four resonant frequencies that can be obtained by connecting these elements in various ways? (b) Are there more than four such frequencies?

9. An LC circuit has an inductance $L = 3.0$ mH and a capacitance $C = 10$ μF. (a) Calculate the angular frequency ω of oscillation. (b) Find the period T of the oscillation. (c) At time $t = 0$ the capacitor is charged to 200 μC, and the current is zero. Sketch roughly the charge on the capacitor as a function of time.   *Answer:* (a) 5.8×10^3 rad/s. (b) 1.1×10^{-3} s.

10. How long will it take for an uncharged 4.0-pF capacitor in an LC circuit to charge if its final voltage is 1.0 mV and the maximum current is 50 mA?

11. An inductor is connected across a capacitor whose capacitance can be varied by turning a knob. We wish to make the frequency of the LC oscillations vary linearly with the angle of rotation of the knob, going from 2×10^5 Hz to 4×10^5 Hz as the knob turns through 180°. If $L = 1.0$ mH, plot C as a function of angle for the 180° rotation.
    Answer: 0°, 45°, 90°, 135°, and 180° correspond respectively to 6.4, 4.1, 2.8, 2.1, and 1.6×10^{-10} F.

12. A variable capacitor with a range from 10 to 365 pF is used with a coil to tune the input to a radio. (a) What ratio of maximum to minimum frequencies may be tuned with such a capacitor? (b) If this capacitor is to tune from 0.54 to 1.60 MHz, the ratio computed in (a) is too large. By adding a capacitor in parallel to the variable capacitor this range may be adjusted. How large should this capacitor be and what inductance should be chosen in order to tune the desired range of frequencies?

13. An oscillating LC circuit is designed to operate at a peak current i (30 mA). The inductance L (0.042 H) is fixed and the frequency is varied by changing C. (a) If the capacitor has a maximum peak voltage V_m (50 V), can the circuit safely operate at a frequency ν of 1.0 MHz? (b) What is the maximum safe operating frequency? (c) What is the minimum capacitance?
    Answer: (a) 1300 V; no. (b) 6300 Hz. (c) 1.8×10^{-8} F.

14. A 10-kg mass oscillates on a spring that, when extended 2.0 cm from equilibrium, has a restoring force of 5.0 N. (a) Find the capacitance of the analogous LC system with $L = 1.0 \times 10^{-3}$ H. (b) Would it be a simple matter to construct the analogous circuit?

15. Initially the 900-μF capacitor is charged to 100 V, and the 100-μF capacitor is uncharged in Fig. 38-7. (a) Describe in detail how one may charge the 100-μF capacitor to 300 V using S_1 and S_2 appropriately. (b) Describe in detail the mass + spring mechanical analogy of this problem.
    Answer: Let T_2 be the period of the inductor and 900 μF capacitor and T_1 the period of inductor and 100 μF capacitor. Then (a) close S_2, wait $T_2/4$; quickly close S_1, then open S_2; wait $T_1/4$ and then open S_1.

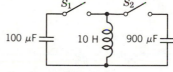

figure 38-7
Problem 15

SECTION 38-3

16. A circuit has $L = 10$ mH and $C = 1.0$ μF. How much resistance must be inserted in the circuit to reduce the (undamped) resonant frequency by 0.01%?

17. In an oscillating LC circuit $L = 3.0$ mH and $C = 2.7$ μF. At $t = 0$ the charge $q = 0$ and the current $i = 2.0$ A. (a) What is the maximum charge that will appear on the capacitor? (b) In terms of the period T of oscillation, how much time after $t = 0$ will elapse until the energy stored in the capacitor will be increasing at its greatest rate? (c) What is this greatest rate at which the energy stored in the capacitor increases?
    Answer: (a) 1.8×10^{-4} C. (b) $T/8$. (c) 67 W.

18. In a damped LC circuit, find the time required for the maximum energy present in the capacitor during one oscillation to fall to half the maximum energy present during the first oscillation. Assume $q = q_m$ at $t = 0$ (i.e., use Eq. 38-10).

19. Derive the differential equation for an LC circuit (Eq. 38-5) using the loop theorem.

20. Show that, for low damping, the current in a damped LC circuit is given approximately by

$$i = -q_m \omega' e^{-Rt/2L} \sin (\omega' t + \phi),$$

in which
$$\phi = \tan^{-1} \frac{R}{2L\omega'}.$$

Start from Eq. 38-10.

21. Suppose that in an oscillating *RCL* circuit the amplitude of the charge oscillations drops to one-half its initial value after *n* cycles. Show that the fractional reduction in the frequency of resonance, caused by the presence of the resistor, is given to a close approximation by

$$\frac{\omega - \omega'}{\omega} = \frac{0.0061}{n^2},$$

which is independent of *L*, *C*, or *R*. Apply to the decay curve of Fig. 38-3.

22. *"Q" for a circuit.* In the damped *LC* circuit of Example 3 show that the fraction of the energy lost per cycle of oscillation, $\Delta U/U$, is given to a close approximation by $2\pi R/\omega L$. The quantity $\omega L/R$ is often called the *"Q"* of the circuit (for "quality"). A "high-*Q*" circuit has low resistance and a low fractional energy loss per cycle $(= 2\pi/Q)$.

SECTION 38-5

23. What would be the dimensions of a cylindrical electromagnetic resonant cavity (like that described in the text) operating, in the fundamental mode, at 60 Hz, the frequency of household alternating current?
    Answer: Radius $= 1.9 \times 10^3$ km, independent of its length.

24. Sketch diagrams like those shown in Figure 38-6 showing a cycle of oscillation of a cylindrical electromagnetic resonant cavity operating, not in the fundamental mode, but in the first overtone.

39
alternating currents

Thus far we have discussed the response, that is, the generated currents, when emfs that vary with time in different ways are applied to circuits containing elements of resistance R, capacitance C, and inductance L, in various combinations.

In Chapter 32 we discussed the (steady) currents generated when we apply (steady) emfs to purely resistive networks. In Section 32-8 we discussed the response of a single-loop RC circuit when a "square wave" emf is applied, as in Fig. 32-13. In Section 36-3 we described the behavior of a single-loop LR network when a similar square wave is applied, as in Fig. 36-6. Finally, in Chapter 38, we analyzed the responses of single-loop LC and RCL circuits in which *no* emf is applied other than the transient emf needed to start the "free oscillations"; see Figs. 38-1, 38-2b, and 38-3.

This chapter deals with *alternating currents* set up in single-loop RCL circuits, such as that of Fig. 39-1, by an emf that varies with time as

$$\mathcal{E} = \mathcal{E}_m \sin \omega t, \tag{39-1}$$

$\omega (= 2\pi\nu$, where ν is measured in hertz) being a fixed angular frequency. An emf of this type might be established, for example, by an alternating current generator in a commercial power plant, where usually $\nu = 60$ Hz. (See Problem 6, Chapter 35.) The symbol for a source of alternating emf such as that described by Eq. 39-1 is —⊙— Such a device is called an alternating current generator or an *ac generator*.*

Alternating currents are important for two reasons: (1) On the practical side, modern technology and indeed the life-style in technologically

39-1
INTRODUCTION

---

* The term 'alternating current' is used loosely here: strictly this is an alternating voltage.

advanced nations, would be very different if emfs such as those given by Eq. 39-1, and the alternating currents to which they correspond, were not available. Power distribution systems, radio, television, satellite communication systems, computor systems, and so on, would be either much less effective or impossible without alternating emfs and the alternating currents generated by them. (2) On the theoretical side, if we know the response of *any RCL* circuit (no matter how many elements or loops are involved) to the emf of Eq. 39-1, we can find the response, that is, the currents generated, to *any* arbitrary emf, no matter how complicated its waveform. We rely here on the facts that we can write any complex waveform as the sum of separate sine (and cosine) terms in a Fourier series and that we can apply the principle of superposition. See Section 19-4 and Fig. 19-5 for an analogy.

A central aim of this chapter is to find the alternating current i in the circuit of Fig. 39-1 in terms of ε_m, ω, R, C, and L. Note that, for the conditions assumed in Fig. 39-1,* the current in all parts of the loop is the same (as for single-loop direct current circuits in Chapter 32) and we may safely assume that i is given by

$$i = i_m \sin(\omega t - \phi) \qquad (39\text{-}2)$$

in which ω is the angular frequency of the applied alternating emf of Eq. 39-1.

In Eq. 39-2 i_m is the current amplitude and ϕ is the phase angle between the alternating current of Eq. 39-2 and the alternating emf of Eq. 39-1. It is our job to express i_m and ϕ in Eq. 39-2 in terms of ε_m, ω, R, C, and L.†

To do so, let us break down the problem suggested by Fig. 39-1 into three separate problems, in which we consider R, C, and L separately and in turn.‡ We start with R.

1. A resistive circuit. Figure 39-2a shows a circuit containing a resistive element only, acted on by the alternating emf of Eq. 39-1. From the loop theorem and from the definition of resistance we can write

$$V_R = \varepsilon_m \sin \omega t \quad \text{(loop theorem)} \qquad (39\text{-}3a)$$
$$V_R = i_R R \quad \text{(definition of } R) \qquad (39\text{-}3b)$$

or

$$i_R = \left(\frac{\varepsilon_m}{R}\right) \sin \omega t. \qquad (39\text{-}3c)$$

Comparison of Eqs. 39-3a and 39-3c shows that the time-varying quantities V_R and i_R are *in phase*, that is, they reach their maximum values at the same time. As expected from Eq. 39-2 they also have the same angular frequency ω. We show both of these things in Fig. 39-2b, a plot of Eqs. 39-3a and 39-3c.

39-2
RCL ELEMENTS, CONSIDERED SEPARATELY

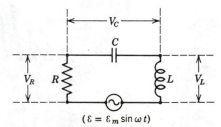

$(\varepsilon = \varepsilon_m \sin \omega t)$

figure 39-1
A single-loop *RCL* circuit contains an ac generator. V_R, V_C, and V_L are the time-varying potential differences across the resistor, the capacitor, and the inductor, respectively.

---

* The condition is, as Fig. 39-1 implies, that we can localize R, C, and L in separate, physically identifiable parts of the circuit. We discuss the basis of this assumption (that of *lumped impedances*) fully in Chapter 40. The lower the frequency the more justifiable this assumption becomes.

† As we shall see (Eq. 39-14), ϕ actually does not depend on ε_m.

‡ We could apply the loop theorem (see Section 32-2) to the circuit of Fig. 39-1 and solve the differential equation that results but this would involve us in mathematical difficulties beyond the scope of this book. We adopt an indirect but equally valid and more physically informative approach.

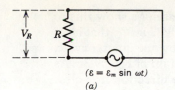

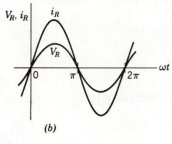

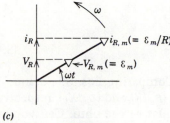

figure 39-2
(a) A single-loop resistive circuit containing an ac generator. (b) The current and the potential difference across the resistor are in phase. In Eq. 39-2, $\phi=0$. (c) A phasor diagram shows the same thing. The arrows on the vertical axis are instantaneous values.

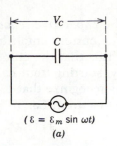

Figure 39-2c shows another way of looking at the situation. It is sometimes called a *phasor diagram*, in which the phasors, represented by the open arrows, rotate counterclockwise with an angular frequency ω about the origin. The phasors have these properties: (a) The length of the phasor is proportional to the *maximum* value of the alternating quantity involved, thus ε_m for V_R (Eq. 39-3a) and (ε_m/R) for i_R (Eq. 39-3c), and (b) the projection of the phasors on the vertical axis gives the *instantaneous* values of the alternating currents involved. Thus the arrows on the vertical axis represent the time-varying quantities V_R and i_R, as in Eqs. 39-3a and 3c, respectively. That V_R and i_R are in phase follows from the fact that their phasors lie along the same line in Fig. 39-2c.

Follow the rotation of the phasors in Fig. 39-2c and convince yourself that this phasor diagram completely and correctly describes Eqs. 39-3a and 3c.

2. *A capacitive circuit.* Figure 39-3a shows a circuit containing a capacitive element only, acted on by the alternating emf of Eq. 39-1. From the loop theorem and from the definition of capacitance we can write

$$V_C = \varepsilon_m \sin \omega t \quad \text{(loop theorem)} \tag{39-4a}$$

and

$$V_C = q/C \quad \text{(definition of } C\text{).} \tag{39-4b}$$

From these relations we have

$$q = \varepsilon_m C \sin \omega t$$

or

$$i_C\left(=\frac{dq}{dt}\right) = \omega C \varepsilon_m \cos \omega t. \tag{39-4c}$$

Comparison of Eqs. 39-4a and 39-4c shows that the time-varying

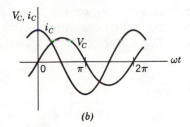

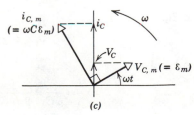

figure 39-3
(a) A single-loop capacitive circuit containing an ac generator. (b) The potential difference across the capacitor *lags* the current by a quarter-cycle. In Eq. 39-2, $\phi = -90°$. (c) A phasor diagram shows the same thing. The arrows on the vertical axis are instantaneous values.

quantities V_C and i_C are one-quarter cycle out of phase. We show this in Fig. 39-3b, a plot of Eqs. 39-4a and 39-4c. We see that V_C lags i_C, that is, as time goes on, V_C reaches its maximum *after* i_C does, by one-quarter cycle.

We show this with equal clarity in the phasor diagram of Fig. 39-3c. As the phasors rotate in the assumed counterclockwise direction it is clear that the $V_{C,m}$ phasor *lags* behind the $i_{C,m}$ phasor by one-quarter cycle.

The phase angle ϕ between V_C and i_C in Fig. 39-3 is $-90°$. To show this put $\phi = -90°$ in Eq. 39-2 and expand. We obtain

$$i = i_m \cos \omega t,$$

in agreement with Eq. 39-4c.

For reasons of symmetry of notation we choose to rewrite Eq. 39-4c as

$$i_C = \left(\frac{\varepsilon_m}{X_C}\right) \cos \omega t \qquad (39\text{-}5a)$$

in which we must have

$$X_C = \frac{1}{\omega C}. \qquad (39\text{-}5b)$$

We call X_C the *capacitive reactance*. Comparing Eqs. 39-3c and 39-5a we see that the current amplitudes are (ε_m/R) and (ε_m/X_C), respectively. We see from this that the unit of X_C must be the ohm. Can you prove this explicitly, starting from Eq. 39-5b?

Also, let us recognize that ε_m in Eq. 39-4a is the maximum value of V_C $(= V_{C,m})$. Thus, from Eq. 39-5a we can write

$$V_{C,m} = i_{C,m} X_C. \qquad (39\text{-}6a)$$

This suggests that when *any* alternating current, of amplitude i_m and angular frequency ω, exists in a capacitor the maximum potential difference across the capacitor (*regardless of the complexity of the circuit in which the capacitor is involved*) is given by

$$V_{C,m} = i_m X_C. \qquad (39\text{-}6b)$$

EXAMPLE 1

In Fig. 39-3a, let $C = 150\ \mu\text{F}$, $\nu = 60$ Hz, and $\varepsilon_m = 300$ V. Find (a) $V_{C,m}$, (b) X_C, and (c) $i_{C,m}$.

(a) From Eq. 39-4a,

$$V_{C,m} = 300\text{ V}.$$

(b) From Eq. 39-5b,.

$$X_C = \frac{1}{\omega C} = \frac{1}{2\pi\nu C}$$
$$= \frac{1}{(2\pi)(60\text{ Hz})(150 \times 10^{-6}\text{ F})}$$
$$= 18\ \Omega.$$

(c) From Eq. 39-6a,

$$i_{C,m} = \frac{V_{C,m}}{X_C}$$
$$= \frac{300\text{ V}}{18\ \Omega} = 17\text{ A}.$$

How would $i_{C,m}$ change if you doubled the frequency? Does this seem intuitively reasonable?

---

3. *An inductive circuit.* Figure 39-4a shows a circuit containing an inductive element only, acted on by the alternating emf of Eq. 39-1. From the loop theorem and the definition of inductance we can write

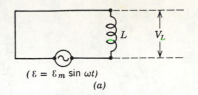

$$V_L = \mathcal{E}_m \sin \omega t \text{ (loop theorem)} \qquad (39\text{-}7a)$$

and

$$V_L = L(di/dt) \text{ (definition of } L). \qquad (39\text{-}7b)$$

From these relations we see that

$$di = (\mathcal{E}_m/L) \sin \omega t \, dt$$

or

$$i_L = \int di = - (\mathcal{E}_m/\omega L) \cos \omega t. \qquad (39\text{-}7c)$$

Comparison of Eqs. 39-7a and 7c shows that the time-varying quantities V_L and i_L are a quarter-cycle out of phase. We show this in Fig. 39-4b, a plot of Eqs. 39-7a and 39-7c. We see that V_L *leads* i_L, that is, as time goes on V_L reaches its maximum *before* i_L does, by a quarter-cycle.

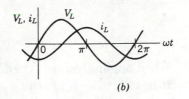

We show this with equal clarity in the phasor diagram of Fig. 39-4c. As the phasors rotate in the counterclockwise direction it is clear that the $V_{L,m}$ phasor precedes (that is, *leads*) the $i_{L,m}$ phasor by a quarter-cycle.

The phase angle ϕ between V_L and i_L in Fig. 39-4 is +90°. To show this put $\phi = +90°$ in Eq. 39-2 and expand. We obtain

$$i = -i_m \cos \omega t,$$

in agreement with Eq. 39-7c.

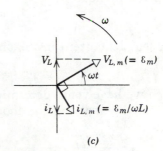

figure 39-4
(a) A single-loop inductive circuit containing an ac generator. (b) The potential difference across the inductor *leads* the current by a quarter-cycle. In Eq. 39-2, $\phi = +90°$. (c) A phasor diagram shows the same thing. The arrows on the vertical axis are instantaneous values.

Again, for reasons of compactness of notation we choose to rewrite Eq. 39-7c as

$$i_L = - (\mathcal{E}_m/X_L) \cos \omega t \qquad (39\text{-}8a)$$

in which we must have

$$X_L = \omega L. \qquad (39\text{-}8b)$$

We call X_L the *inductive reactance.* As for the capacitive reactance (Eq. 39-5) we see that the SI unit for X_L must also be the ohm. Can you prove this explicitly, starting from Eq. 39-8b?

Also, let us recognize that $\mathcal{E}_m$ in Eq. 39-7a is the maximum value of $V_L (= V_{L,m})$. Thus from Eq. 39-7a we can write

$$V_{L,m} = i_{L,m} X_L. \qquad (39\text{-}9a)$$

This suggests that when *any* alternating current of amplitude i_m and angular frequency ω exists in an inductor, the maximum potential difference across the inductor (*regardless of the complexity of the circuit in which the inductor is involved*) is given by

$$V_{L,m} = i_m X_L. \qquad (39\text{-}9b)$$

In Fig. 39-4 let $L = 60$ mH, $\nu = 60$ Hz, and $\mathcal{E}_m = 300$ V. Find (a) $V_{L,m}$, (b) X_L, and (c) $i_{L,m}$.

EXAMPLE 2

(a) From Eq. 39-7a,

$$V_{L,m} = \mathcal{E}_m = 300 \text{ V}.$$

(b) From Eq. 39-8b,

$$X_L = \omega L = 2\pi\nu L$$
$$= (2\pi)(60 \text{ Hz})(60 \times 10^{-3} \text{ H})$$
$$= 23 \ \Omega.$$

(c) From Eq. 39-9a,

$$i_{L,m} = \frac{V_{L,m}}{X_L}$$

$$= \frac{300 \text{ V}}{23 \text{ }\Omega} = 13 \text{ A}.$$

How would $i_{L,m}$ change if you doubled the frequency? Does your answer seem intuitively reasonable?

39-3
THE SINGLE-LOOP RCL CIRCUIT

Having finished our analysis of separate single-loop R, C, and L circuits, we now return to Fig. 39-1, in which all three elements are present. The emf is given by Eq. 39-1

$$\mathcal{E} = \mathcal{E}_m \sin \omega t \qquad (39\text{-}1)$$

and the (single) current in the circuit has the form shown in Eq. 39-2

$$i = i_m \sin (\omega t - \phi) \qquad (39\text{-}2)$$

in which we have yet to determine i_m and ϕ.

We start by applying the loop theorem to the circuit of Fig. 39-1, obtaining

$$\mathcal{E} = V_R + V_C + V_L \qquad (39\text{-}10)$$

These are all sinusoidally time-varying quantities, their maximum values being, in order, $\mathcal{E}_m$ (see Eq. 39-1), $V_{R,m}$ ($= i_m R$), $V_{C,m}$ ($= i_m X_C$; see Eq. 39-6b), and $V_{L,m}$ ($= i_m X_L$; see Eq. 39-9b).

Although Eq. 39-10 is true at any instant of time we cannot easily use it to find i_m and ϕ in Eq. 39-2 because of the phase differences that exist among the separate terms. We therefore turn to the phasor diagram of Fig. 39-5a, which shows the maximum values of i, V_R, V_C, and V_L. Check Section 39-2 to see that the phase differences are correct, that is, V_R is *in phase* with the current (Fig. 39-2), V_C *lags* the current by a quarter-cycle (Fig. 39-3), and V_L *leads* the current by a quarter-cycle (Fig. 39-4).

As in Section 39-2, the projections of the phasors on the vertical axis give the *instantaneous* values of the quantities involved. Thus the *algebraic* sum of V_R, V_C, and V_L on the vertical axis equals $\mathcal{E}$, as Eq. 39-10 requires.

On the other hand, we assert that the *vector* sum of the phasor amplitudes $V_{R,m}$, $V_{C,m}$, and $V_{L,m}$ yields a phasor whose amplitude is the $\mathcal{E}_m$ of Eq. 39-1. The projection of $\mathcal{E}_m$ on the vertical axis would, of course, be the time-varying $\mathcal{E}$ of Eq. 39-1, that is, it would be $V_R + V_C + V_L$ as Eq. 39-10 asserts. Note that, in vector operations, the (algebraic) sum of the projections of any number of vectors on a given straight line is equal to the projection on that line of the (vector) sum of those vectors.

We can find $\mathcal{E}_m$ from Fig. 39-5b, in which we have formed the phasor $V_{L,m} - V_{C,m}$. This is at right angles to $V_{R,m}$ and we have

$$\mathcal{E}_m = \sqrt{V_{R,m}^2 + (V_{L,m} - V_{C,m})^2}$$

$$= \sqrt{(i_m R)^2 + (i_m X_L - i_m X_C)^2}$$

$$= i_m \sqrt{R^2 + (X_L - X_C)^2} \qquad (39\text{-}11)$$

We call the quantity multiplying i_m the *impedance Z* of the circuit of Fig. 39-1. Thus we can write

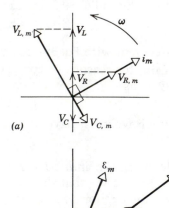

(a)

(b)

figure 39-5

(a) A phasor diagram corresponding to the circuit of Fig. 39-1. Note the phasor amplitudes and the instantaneous values shown on the vertical axis. (b) The same diagram, showing the relationship between the emf ξ and the current i of Eqs. 39-1 and 39-2, respectively. The phase angle ϕ of Eq. 39-2 is clearly shown.

$$i_m = \frac{\varepsilon_m}{Z}, \qquad (39\text{-}12)$$

which reminds us of the relation $i = \varepsilon/R$ (Eq. 32-2) for single-loop resistive networks with steady emfs. The SI unit of impedance is evidently the ohm.

We can write out Eq. 39-12 in full detail (see Eqs. 39-11, 39-5b, and 39-8b) as

$$i_m = \frac{\varepsilon_m}{\sqrt{R^2 + (\omega L - 1/\omega C)^2}}. \qquad (39\text{-}13)$$

Thus we have solved the first problem posed in the second paragraph of Section 39-2; we have expressed i_m in terms of ε_m, ω, R, C, and L. It remains to find the phase angle ϕ between i and ε; compare Eqs. 39-1 and 39-2.

We show the angle ϕ in Fig. 39-5b and we can write

$$\tan \phi = \frac{V_{L,m} - V_{C,m}}{V_{R,m}}$$

$$= \frac{i_m(X_L - X_C)}{i_m R}$$

$$= \frac{X_L - X_C}{R}. \qquad (39\text{-}14)$$

Thus we have solved the second problem posed in the second paragraph of Section 39-2; we have expressed ϕ in terms of ω, R, C, and L. As we said earlier, ϕ does not depend on ε_m. Increasing ε_m increases i_m (see Eq. 39-13) but it does not change ϕ; the *scale* of the operation is changed but not its *nature*.

We drew Fig. 39-5b arbitrarily with $X_L > X_C$; that is, we assumed the circuit of Fig. 39-1 to be more inductive than capacitive. For this assumption ε_m *leads* i_m (although not by so much as a quarter-cycle but less than $+90°$, as it did in the purely inductive circuit of Fig. 39-4). The phase angle ϕ in Eq. 39-14 (and thus in Eq. 39-2) is positive.

If, on the other hand, we had $X_C > X_L$, the circuit would be more capacitive than inductive and ε_m would *lag* i_m (although not by as much as a quarter-cycle, as it did in the purely capacitive circuit of Fig. 39-3). Consistent with this change from leading to lagging, the angle ϕ in Eq. 39-14 (and thus in Eq. 39-2) would automatically become negative.

EXAMPLE 3

In Fig. 39-1 let $R = 4.0\ \Omega$, $C = 150\ \mu\text{F}$, $L = 60\ \text{mH}$, $\nu = 60\ \text{Hz}$, and $\varepsilon_m = 300\ \text{V}$. Find (a) X_C, (b) X_L, (c) Z, (d) i_m, and (e) ϕ.

(a) $X_C = 18\ \Omega$ as in Example 1. Note that X_C depends only on C and ω (Eq. 39-5b) and *not* on the nature of the circuit in which C is an element.

(b) $X_L = 23\ \Omega$, as in Example 2. Note that X_L depends only on L and ω (Eq. 39-8b) and *not* on the nature of the circuit in which L is an element.

(c) From Eq. 39-11 we have

$$Z = \sqrt{R^2 + (X_L - X_C)^2}$$
$$= \sqrt{(4.0\ \Omega)^2 + (23\ \Omega - 18\ \Omega)^2}$$
$$= 6.4\ \Omega.$$

Note that this circuit is more inductive than capacitive, that is, $X_L > X_C$, as in Fig. 39-5.

(d) From Eq. 39-12 we have

$$i_m = \frac{\mathcal{E}_m}{Z} = \frac{300\ \text{V}}{6.4\ \Omega} = 47\ \text{A}.$$

(e) From Eq. 39-14 we have

$$\tan\phi = \frac{X_L - X_C}{R}$$
$$= \frac{23\ \Omega - 18\ \Omega}{4.0\ \Omega} = 1.25$$

or $\phi = +51°$. Because $X_L > X_C$, ϕ is positive and $\mathcal{E}_m$ *leads* i_m, as Fig. 39-5b suggests, but, as we expect, by less than $+90°$.

Power is an important practical matter. If you turn on the lights in a room, you are dealing directly with the subject matter of this section. We start by realizing that power dissipation in *RCL* circuits (Fig. 39-1 is an example on which we will focus), occurs *only* in the resistive element R; there is no mechanism for dissipating power in purely capacitive or purely inductive elements.

Let us therefore consider the purely resistive single-loop circuit of Fig. 39-2a, from two points of view. First let us replace the alternating emf by a *steady* emf of magnitude $\mathcal{E}_m$. The (steady) power dissipated in R (see Eq. 31-17) is given by

$$P = \frac{\mathcal{E}_m^2}{R} \quad \text{(steady emf)}. \tag{39-15}$$

If the alternating emf of Eq. 39-1 applies, we have, however

$$P(t) = \frac{(\mathcal{E}_m \sin \omega t)^2}{R} \quad \text{(alternating emf)}. \tag{39-16}$$

Our real concern is not so much for $P(t)$, which is sometimes zero, but for the average power $P(t)\ (= P_{av})$, or

$$P_{av} = \frac{\mathcal{E}_m^2}{R} \overline{(\sin \omega t)^2}. \tag{39-17}$$

Figure 39-6 shows (a) a plot of $\sin \omega t$ and (b) a plot of $(\sin \omega t)^2$. Note that $\overline{\sin \omega t} = 0$ (the positive parts just cancel the negative parts). However, $\overline{(\sin \omega t)^2} = \frac{1}{2}$ (the parts above the line "$\frac{1}{2}$" just cancel those below that line; there are no negative values. Thus,

$$P_{av} = \frac{1}{2}\frac{\mathcal{E}_m^2}{R} = \left(\frac{\mathcal{E}_m}{\sqrt{2}}\right)^2 \frac{1}{R}. \tag{39-18}$$

We choose to call $\mathcal{E}_m/\sqrt{2}$ the *root-mean-square* (abbr. rms) value of $\mathcal{E}\ (= \mathcal{E}_{rms})$. The notation is justified. First we *squared* $\mathcal{E}_m$, then we averaged it (or took the *mean* of it) over an integral number of cycles (the averaging factor, in this case of sinusoidal functions, being $\frac{1}{2}$), and

39-4
POWER IN ALTERNATING CURRENT CIRCUITS

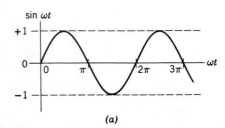

(a)

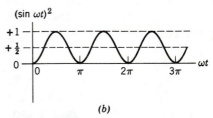

(b)

figure 39-6
(a) A plot of $\sin \omega t$, showing that its average value over an integral number of cycles is zero. The parts of the curve above the horizontal axis just cancel those below. (b) A plot of $(\sin \omega t)^2$, showing that its average value over an integral number of cycles is $\frac{1}{2}$. The parts of the curve above the horizontal line "$\frac{1}{2}$" just cancel those below it; the curve has no negative values.

finally we took the *square root.** We can thus write Eq. 39-18 as

$$P_{av} = \frac{\mathcal{E}_{rms}^2}{R}. \tag{39-19}$$

Equation 39-19 looks much like Eq. 31-17 and the message is that if we use rms quantities for $\mathcal{E}$, for V, and for i, the *average* power dissipation, which is all that usually matters, will be the same for alternating current circuits as for direct current circuits with a constant emf. Alternating current instruments, such as ammeters and voltmeters, are almost always too sluggish mechanically to follow the actual cycles of oscillation. They are deliberately calibrated to read $\mathcal{E}_{rms}$, V_{rms}, and i_{rms}. Thus, if you plug an alternating current voltmeter into a household electric outlet and it reads "120 V," the maximum value of the potential difference at the outlet is $\sqrt{2}$ (120 V) or 170 V. The sole reason for using rms values in alternating current circuits is to let us use the familiar direct current power relationships of Section 31-5. Thus, we summarize:

$$\mathcal{E}_{rms} = \mathcal{E}_m/\sqrt{2}, \quad V_{rms} = V_m/\sqrt{2}, \quad \text{and} \quad i_{rms} = i_m/\sqrt{2}. \tag{39-20}$$

Because the proportionality factor $(= 1/\sqrt{2})$ is the same in each case, we can write the important Eq. 39-13 as

$$i_{rms} = \frac{\mathcal{E}_{rms}}{\sqrt{R^2 + (\omega L - 1/\omega C)^2}} \tag{39-21}$$

and, indeed, this is the form that we almost always use.

EXAMPLE 4

(a) A steady emf $(\mathcal{E}_0 = 120 \text{ V})$ is applied to a single-loop resistive circuit with $R = 150 \ \Omega$. What is the power dissipation? (b) The steady emf $\mathcal{E}_0$ is replaced by an alternating emf $(\mathcal{E} = \mathcal{E}_m \sin \omega t$. If the average power is to remain unchanged, what must $\mathcal{E}_m$ be?

(a)
$$P = \frac{\mathcal{E}_0^2}{R} = \frac{(120 \text{ V})^2}{150 \ \Omega} = 96 \text{ W}.$$

(b) From Eq. 39-18 and the conditions of the problem,

$$P_{av} (= 96 \text{ W}) = \left(\frac{\mathcal{E}_m}{\sqrt{2}}\right)^2 \frac{1}{R}$$

or

$$\mathcal{E}_m = \sqrt{2P_{av}R}$$
$$= \sqrt{(2)(96 \text{ W})(150 \ \Omega)}$$
$$= 170 \text{ V}.$$

From Eq. 39-20 we have

$$\mathcal{E}_{rms} = \mathcal{E}_m/\sqrt{2} = (170 \text{ V})/\sqrt{2}$$
$$= 120 \text{ V},$$

just as we expect.

Now we turn from power considerations for the resistive circuit of Fig. 39-2 to those for the more general circuit of Fig. 39-1, in which all three elements, R, C, and L, appear. For the instantaneous power we can write (compare Eq. 31-15)

---

* We have encountered rms quantities before, namely, in the rms speeds of molecules (see Section 23-4; also, see p. 852).

$$P(t) = \mathcal{E}(t)\, i\,(t)$$

$$= [\mathcal{E}_m \sin \omega t]\,[i_m \sin (\omega t - \phi)]. \qquad (39\text{-}22)$$

If we expand the factor $\sin (\omega t - \phi)$ by a trignometric identity that is evident by comparison with Eq. 39-23, we obtain

$$P(t) = (\mathcal{E}_m\, i_m)(\sin \omega t)(\sin \omega t \cos \phi - \cos \omega t \sin \phi)$$

$$= \mathcal{E}_m\, i_m \sin^2 \omega t \cos \phi - \mathcal{E}_m\, i_m \sin \omega t \cos \omega t \sin \phi. \qquad (39\text{-}23)$$

If we now find $\overline{P(t)}$ $(= P_{av})$, which is our principle practical concern, we have

$$P_{av} = \tfrac{1}{2}\mathcal{E}_m\, i_m \cos \phi + 0. \qquad (39\text{-}24)$$

This comes about because, as we have seen, $\overline{\sin^2 \omega t} = \tfrac{1}{2}$ and we assert (see Problem 9) that $\overline{\sin \omega t \cos \omega t} = 0$. Because $\mathcal{E}_{rms} = \mathcal{E}_m/\sqrt{2}$ and $i_{rms} = i_m/\sqrt{2}$ (Eq. 39-20) we can write Eq. 39-24 as

$$P_{av} = \mathcal{E}_{rms}\, i_{rms} \cos \phi \qquad (39\text{-}25)$$

in which we call $\cos \phi$ the *power factor*. Figure 39-2, which shows a purely resistive load, is characterized by $\phi = 0$ (and thus $\cos \phi = 1$)so that Eq. 39-24 becomes

$$P_{av} = \mathcal{E}_{rms}\, i_{rms} \quad \text{(resistive load)}. \qquad (39\text{-}26)$$

Subscripts aside, this is just the relation that we would write for direct current circuits (compare Eq. 31-15).

EXAMPLE 5

Let us use the same parameters for Fig. 39-1 that we did in Example 3, namely, $R = 4.0\ \Omega$, $C = 150\mu F$, $L = 60$ mH, $\nu = 60$ Hz, and $\mathcal{E}_m = 300$ V. Find (a) $\mathcal{E}_{rms}$, (b) i_{rms}, (c) ϕ, (d) $\cos\phi$, and (e) P_{av}.

(a) $\mathcal{E}_{rms} = \mathcal{E}_m/\sqrt{2} = (300\ \text{V})/\sqrt{2} = 210$ V.

(b) $i_{rms} = i_m/\sqrt{2} = (47\ \text{A})/\sqrt{2} = 33$ A; see Example 3, part (d).

(c) $\phi = 51°$, as in Example 3, part (e).

(d) The power factor, $\cos\phi$, is $\cos 51°$ or 0.63.

(e) $P_{av} = \mathcal{E}_{rms}\, i_{rms} \cos \phi$

$$= (210\ \text{V})(33\ \text{A})(0.63)$$

$$= 4.4\ \text{kW}.$$

39-5
RESONANCE IN ALTERNATING CURRENT CIRCUITS

In this section we return to the circuit of Fig. 39-1 and consider the effect, as far as the current i_{rms} is concerned, of varying the angular frequency ω of the applied emf of Eq. 39-1, assuming that $\mathcal{E}_m$, R, C, and L remain fixed.

We have stressed the analogy between mechanical systems, such as mass-spring arrangements (Fig. 8-4) and electromagnetic systems (Fig. 38-1). With this in mind we are entitled to look upon the *RCL* circuit of Fig. 39-1 as possessing a "natural" frequency of oscillation, and to view the circuit as acted upon by an external influence, in this case the applied alternating emf, given by $\mathcal{E} = \mathcal{E}_m \sin \omega t$ (Eq. 39-1), in which ω is the angular frequency of the "driving force." We expect a maximum "response," defined here by the i_{rms} of Eq. 39-21, when the angular frequency ω of the driving force just equals the natural frequency of oscillation ω_0 for the free oscillations of the circuit. Let us see whether these analogies and expectations turn out to be true.

From Eq. 39-21 we see that the maximum value of i_{rms} occurs when $X_L = X_C$ and has the value

$$i_{rms,max} = \frac{\mathcal{E}_{rms}}{R}, \qquad (39\text{-}27)$$

that is, i_{rms} is limited only by the circuit resistance. If $R \to 0$, $i_{rms,max} \to \infty$. Putting $X_L = X_C$ yields

$$\omega L = \frac{1}{\omega C}$$

or

$$\omega = \frac{1}{\sqrt{LC}}. \qquad (39\text{-}28)$$

The quantity on the right (see Eq. 38-3) is just the natural angular frequency ω_0 of the circuit of Fig. 39-1, with $\mathcal{E}$ and R removed. That is:

the maximum value of i_{rms} occurs when the frequency ω of the driving force [the $\mathcal{E}(t)$ of Eq. 39-1], is exactly equal to the natural frequency ω_0 of the *undamped* ($R = 0$) circuit of Fig. 39-1.

This condition

$$\omega = \omega_0 \qquad (39\text{-}29)$$

is called *resonance*. We met it earlier in connection with the cyclotron (see Section 33-7) and in many equivalent mechanical situations.

Figure 39-7 is a plot of i_{rms} versus ω (see Eq. 39-21), for fixed values of $\mathcal{E}_m$, C, and L, but for three different values of R. Note how rapidly the sharpness of the resonance peak broadens as R increases.

Figure 39-7 suggests to us the common experience of tuning a radio set. What we do here when we turn the tuning knob is to adjust the natural frequency ω_0 of an internal circuit to the frequency ω of the signal transmitted by the station antenna, until Eq. 39-29 is satisfied. In a metropolitan area, where there are many incoming signals whose fre-

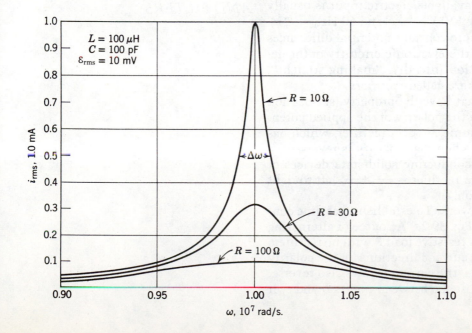

figure 39-7
Showing resonance in the *RCL* circuit of Fig. 39-1 for three different values of R. Along the horizontal axis we plot the (variable) angular frequency ω of the impressed emf. The arrow marked "$\Delta\omega$" on the curve for $R = 10\ \Omega$ is called the "full width of the curve at half maximum" or, more concisely, the *half width*.

quencies are not too far apart, the sharpness (or "quality"; see Problem 17) of tuning becomes very important.

Now let us consider the phenomenon of resonance from the point of view of the phasor diagram of Fig. 39-5b. If $X_L = X_C$, as resonance requires, the phasors $V_{L.m}$ $(=i_m X_L)$ and $V_{C.m}$ $(=i_m X_C)$ just cancel each other and the phase angle $\phi = 0$. The fact that $\phi = 0$ if $X_L = X_C$ is explicitly shown to be true by Eq. 39-14. If $\phi > 0$, as in Fig. 39-5b, we have $X_L > X_C$ and the circuit is predominantly inductive. On the other hand, if $\phi < 0$, we have $X_C > X_L$ and the circuit is predominantly capacitive.

EXAMPLE 6

In Fig. 39-7 (a) show that the angular frequency of the impressed emf at which resonance occurs is indeed 1.0×10^7 rad/s; (b) show also that the maximum value of i_{rms} at resonance, for $R = 10\ \Omega$, is 1.0 mA.

(a) At resonance $X_L = X_C$ or

$$\omega L = \frac{1}{\omega C}$$

which leads to

$$\omega = \frac{1}{\sqrt{LC}} = \frac{1}{\sqrt{(100\ \mu H)(100\ pF)}}$$

$$= 1.0 \times 10^7 \text{ rad/s.}$$

(b) At resonance Eq. 39-27 applies, or

$$i_{rms.max} = \frac{\mathcal{E}_{rms}}{R}$$

$$= \frac{10\text{ mV}}{10\ \Omega} = 1.0\text{ A.}$$

39-6

ALTERNATING CURRENT RECTIFIERS AND FILTERS

We often have available a source of alternating emf (given, say, by Eq. 39-1, $\mathcal{E} = \mathcal{E}_m \sin\omega t$) and we want to derive from it, by electronic means, a constant potential difference. For example, in television sets, sound reproducing systems, etc., the available electric input is usually an alternating emf, often described by 120 V $(= \mathcal{E}_{rms})$ and 60 Hz $(= \omega/2\pi)$. From this we need to derive one or more constant potential differences (50 V, 300 V, 1500 V, etc.) to operate the electronic circuitry of the device. This process is called *rectification* (literally, "making straight") and the devices that make it possible are called *rectifiers*.

Rectifiers are nonohmic devices that have the property (see Section 31-3) that their resistance depends on the polarity of the applied potential difference. In this section we assume an ideal rectifier, which has $R = 0$ for a given polarity and $R \to \infty$ when that polarity is reversed.

Physically, rectifiers may be semiconducting solid-state devices or vacuum tube diodes. The symbol for a rectifier is ⟶▷⟶, left to right being the direction of "easy conduction."

Figure 39-8 suggests how rectifiers work. To establish a baseline consider Fig. 39-8a, which is identical to Fig. 39-2a. A source of alternating emf $\mathcal{E}$ $(= \mathcal{E}_m \sin \omega t)$ is connected with a resistive load R with no rectifiers in the circuit. The arrows indicate the current directions for the polarity shown. When the polarity of $\mathcal{E}$ reverses, the current arrows also reverse.

If we connect a cathode ray oscilloscope between points b and c, it

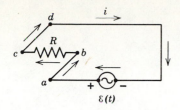

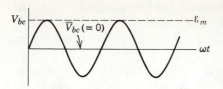

(a)

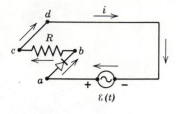

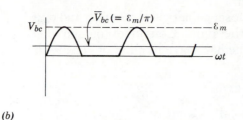

(b)

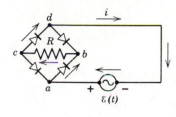

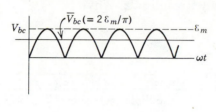

(c)

figure 39-8
An ac generator ε ($= \varepsilon_m \sin \omega t$) acts on the three circuits shown. The potential difference V_{bc} is displayed on the right. Note that $\overline{V_{bc}}$ increases from zero to ε_m/π to $2\varepsilon_m/\pi$.

displays the waveform shown on the right. Note that $\overline{V_{bc}} = 0$ in this case, the positive halves of the sine wave just cancelling the negative halves. No rectification occurs, which is not surprising because there are no rectifiers in the circuit.

If we connect a dc voltmeter* between points b and c in Fig. 39-8a, it would read zero because the coil assembly is too sluggish to follow the waveform shown in Fig. 39-8a. For a typical dc voltmeter, at 60 Hz, say, the pointer will oscillate rapidly with small amplitude about its zero position. Why?

In Fig. 39-8b we insert a rectifier in the path ab, the direction of "easy conduction" being from a to b. When the polarity reverses, the circuit is virtually open because the (ideal) rectifier has infinite resistance.

If we connect a cathode ray oscilloscope between points b and c, it displays the waveform shown on the right. It is clear that $\overline{V_{bc}} > 0$. In fact,

$$\overline{V_{bc}} = \frac{\varepsilon_m}{2\pi}\left\{\int_0^\pi \sin \omega t \, d(\omega t) + \int_\pi^{2\pi} 0 \, d(\omega t)\right\} = \frac{\varepsilon_m}{\pi}. \qquad (39\text{-}30)$$

The arrangement of Fig. 39-8b is called a *half-wave rectifier*. It does not yet produce a truly constant potential difference but it is a step in

---

* We can construct such a voltmeter by putting a large resistor R_0 in series with the coil of the sensitive galvanometer shown in Fig. 33-9. The large series resistor R_0 ($\gg R$ in Fig. 39-8a) is needed to insure that touching the voltmeter terminals to points b and c in Fig. 39-8a will not appreciably change the current distribution in that circuit.

that direction. If we connect a dc voltmeter between points b and c in Fig. 39-8b, it will give a definite reading. As for Fig. 39-8a there will be rapid small-amplitude oscillations of the pointer about its equilibrium position.

In Fig. 39-8c we introduce four rectifiers. For the polarity shown the current follows the path indicated by the arrows. When the polarity of the emf reverses, the current becomes counterclockwise and its path through the rectifier network is *dbca*. *Note that the direction of the current through R remains unchanged*; it is *always* from b to c, regardless of the polarity of the source of emf.

If we connect a cathode ray oscilloscope between points b and c, it displays the waveform shown on the right. Convince yourself that $\overline{V_{bc}}$ is just twice the value given for the half-wave rectifier of Fig. 39-8b, that is,

$$\overline{V_{bc}} = 2\ \varepsilon_m/\pi. \tag{39-31}$$

The arrangement of Fig. 39-8c is called a *full-wave rectifier*. If we connect a dc voltmeter between points b and c, it will deflect twice as far as it did in connection with Fig. 39-8b. Oscillations of the pointer will still occur but they will have smaller amplitude. Why?

In Fig. 39-8 the emf is $\varepsilon = \varepsilon_m \sin \omega t$, in which $\varepsilon_m = 300$ V and $R = 15\ \Omega$. What average power P_{av} is dissipated in the resistor for the three cases shown?

EXAMPLE 7

(a) For Fig. 39-8a we have, from Eqs. 39-18 and 39-19,

$$P_{av} = \frac{\varepsilon_m{}^2}{2R} \left(= \frac{\varepsilon_{rms}{}^2}{R} \right)$$

$$= \frac{(300\ V)^2}{(2)(15\ \Omega)} = 3.0\ kW.$$

(b) Because the circuit of Fig. 39-8b is nonconducting (open circuit) for half the time, P_{av} is just half the value shown in (a) above, or 1.5 kW.

(c) In Fig. 39-8c P_{av} has the same value (= 3.0 kW) as in (a) above. Note that P_{av} does not depend on the *direction* of i or V; both quantities occur as squares.

The waveform of Fig. 39-8c is *still* not a nontime-varying potential difference. We can, however, resolve it into a steady component and a series of sinusoidally alternating potential differences with various angular frequencies ω, amplitudes V, and relative phases ϕ. We often call these ac components *the ripple*. Qualitatively we can see that this division of the waveform of Fig. 39-8c is reasonable, just by inspection. Quantitatively we could (but won't) derive it by Fourier analysis of the waveform.

What we need to do now is to connect the potential difference between points b and c of the circuit of Fig. 39-8 c, that is, the waveform of that figure, through a *filter circuit*, which has these properties: (1) it passes the dc component from input to output with negligible reduction in value, and (2) it greatly decreases the amplitude of the time-varying components, that is, of the ripple.

Figure 39-9 shows a simple filter circuit. It contains an ideal inductor L (that is, it has no resistive or capacitive properties) and an ideal capacitor C (that is, it has no resistive or inductive properties). The input V_{in} to the filter may be steady or sinusoidally oscillating. To explore the behavior of the filter we will look at these two cases separately.

For V_{in} = a constant we see that,

$$V_{out} = V_{in} = \text{the same constant.}$$

Neither L nor C has any effect. In fact, L might be replaced by a straight wire (shorted out) and C removed from (clipped out of) the circuit, with no noticeable effect on V_{out}.

For an ac input, however, the situation is quite different. We assume from the beginning that both L and C are "large" so that $X_L (=\omega L) \gg X_C (=1/\omega C)$. If ω and C are large enough, $X_C \to 0$ and the capacitor acts as a virtual short circuit for ac components, even though it has no effect on the dc component.

Assume

$$V_{in} = V_{in,m} \sin \omega t. \tag{39-32}$$

From Eq. 39-2 we have

$$i = i_m \sin (\omega t - \phi). \tag{39-2}$$

From Eq. 39-13, with $R = 0$ and ε_m replaced $V_{in,m}$, we have

$$i = \frac{V_{in,m}}{X_L - X_C} \sin (\omega t - \phi). \tag{39-33}$$

Because we have already assumed that $X_L \gg X_C$ we can write this as

$$i \cong \frac{V_{in,m}}{X_L} \sin (\omega t - \phi). \tag{39-34}$$

To find the phase angle ϕ we turn to Eq. 39-14, or

$$\tan \phi = \frac{X_L - X_C}{R}. \tag{39-14}$$

With $X_L \gg X_C$ and $R \to 0$ we have $\tan \phi \to +\infty$ or $\phi = +90°$. Thus Eq. 39-33 becomes

$$i = -\frac{V_{in,m}}{X_L} \cos \omega t. \tag{39-35}$$

Note that a cosine function is just 90° out of phase with a sine function.

The output V_{out} is given by q/C so that we must find q for the capacitor. From Eq. 39-35 we have

$$q = \int i \,(dt) = \frac{1}{\omega} \int i \, d(\omega t)$$

$$= -\left(\frac{1}{\omega}\right)\left(\frac{V_{in,m}}{X_L}\right) \int \cos \omega t \, d(\omega t)$$

$$= -\left(\frac{1}{\omega}\right)\left(\frac{V_{in,m}}{X_L}\right) \sin \omega t. \tag{39-36}$$

Because $X_C = 1/\omega C$ we have

$$V_{out} = \frac{q}{C} = -(X_C)\left(\frac{V_{in,m}}{X_L}\right) \sin \omega t. \tag{39-37}$$

The ac attenuation α is the ratio of the amplitude of V_{out} (see Eq. 39-37) to that of V_{in} (see Eq. 39-32), or

$$\alpha = V_{out,m}/V_{in,m} = X_C/X_L. \tag{39-38}$$

Because we have chosen $X_L \gg X_C$ we see that the attenuation of the "ripple components" is large. Much more effective filters than those of Fig. 39-9 are possible, as amateur builders of sound reproducing equipment are well aware (see Problem 23).

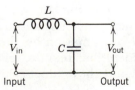

figure 39-9
A simple filter circuit designed to pass dc potential differences from input to output without appreciable attenuation but to attenuate time-varying components greatly. Convince yourself that if L, C, and ω all have "large" values, the filter element will be more effective in reducing the amplitudes of ac input components.

EXAMPLE 8

In Fig. 39-9 let $V_{in} = V_{in,m} \sin \omega t$, in which $V_{in,m} = 300$ V, and $\omega/2\pi = 60$ Hz. Let $L = 10$ H and $C = 300$ μF. Find $V_{out,m}$.

From Eq. 39-37, we have

$$V_{out,m} = (V_{in,m})(X_C)\left(\frac{1}{X_L}\right)$$

$$= (V_{in,m})\left(\frac{1}{\omega C}\right)\left(\frac{1}{\omega L}\right)$$

$$= \frac{V_{in,m}}{\omega^2 L C}$$

$$= \frac{(300 \text{ V})}{(2\pi \times 60 \text{ Hz})^2 (10 \text{ H})(300 \times 10^{-6} \text{ F})}$$

$$= 0.70 \text{ V}.$$

In this case the attenuation factor is

$$(V_{out,m}/V_{in,m}) = (0.70 \text{ V})/(300 \text{ V})$$

or 2.3×10^{-3}. Convince yourself that this is the ratio X_C/X_L, in which we have assumed $X_L \gg X_C$.

39-7
THE TRANSFORMER

In dc circuits the power dissipation in a resistive load is given by Eq. 31-15 ($P = iV$). This means that, for a given power requirement, we have our choice of a relatively large current i and a relatively small potential difference V or just the reverse, provided that their product remains constant. In the same way, for ac circuits the average power dissipation is given by Eq. 39-25 ($P_{av} = i_{rms} V_{rms}$) and we have the same choice as to the relative values of i_{rms} and V_{rms}.*

In electric power distribution systems it is clear that at both the generating end (the electric power plant) and the receiving end (the home or factory) it is desirable, both for reasons of safety and the efficient design of equipment, to deal with relatively low voltages. For example, no one wants an electric toaster or a child's electric train to operate at, say, 10 kV.

On the other hand, in the transmission of electric energy from the generating plant to the consumer, we want the *lowest* possible current (and thus the *largest* possible potential difference) so as to minimize the i^2R ohmic losses in the transmission line. $V_{rms} = 350$ kV is not uncommon. Thus there is a fundamental "mismatch" between the requirements for efficient transmission on the one hand and efficient and safe generation and consumption on the other. We need a device that can, as design considerations require, raise (or lower) the potential difference in a circuit, keeping the product iV essentially constant. The alternating current *transformer* of Fig. 39-10 is such a device. It has no direct current counterpart of equivalent simplicity, which is why dc distribution systems, strongly advocated by Edison, have now been essentially totally replaced by ac systems, strongly advocated by Tesla and others.

Within the last few decades the use of dc transmission over very long distances or for special situations such as underwater or underground transmission cables has undergone a revival. In the Soviet Union, for example, there is a 300-mile line from Volgograd to the Donbass operating at a dc potential difference of 800 kV. The power is generated as ac, stepped up to the transmission voltage, converted to dc, reconverted

---

* We assume a resistive load so that $\cos \phi = 1$ in Eq. 39-25.

to ac at the receiving point, and stepped down to lower voltages for transmission. The advantage of high-voltage dc transmission under these circumstances is that, for dc, the capacitive and inductive properties of the transmission line can be ignored. Note that transformers, the subject of this section, are still very much required.*

In Fig. 39-10 two coils are shown wound around a soft iron core. The *primary* winding, of N_1 turns, is connected with an alternating current generator whose emf $\varepsilon_1(t)$ is given by $\varepsilon_1 = \varepsilon_m \sin \omega t$. The *secondary* winding, of N_2 turns, is on open circuit as long as switch S is open, which we assume for the present. Thus there is no secondary current. We assume further that the resistances of the primary and secondary windings and also the magnetic "losses" in the iron core are negligible. Actually, well-designed, high-capacity transformers can have energy losses as low as one percent so that our assumption of an ideal transformer is not unreasonable.

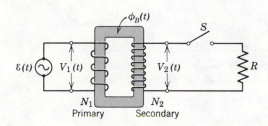

figure 39-10
An ideal transformer, showing two coils wound on the same soft iron core. An ac generator is connected with the primary winding. The secondary winding may be on open circuit (if switch S is open) or connected with a resistive load R (if switch S is closed).

For the above conditions the primary winding is a pure inductance; compare Fig. 39-4. Thus the (very small) primary current, called the *magnetizing current* $i_{mag}(t)$, lags the primary potential difference $V_1(t)$ by 90°; the power factor ($=\cos \phi$ in Eq. 39-25) is zero, and thus no power is delivered from the generator to the transformer.

However, the small alternating primary current $i_{mag}(t)$ induces an alternating magnetic flux $\Phi_B(t)$ in the iron core and we assume that all this flux links the turns of the secondary windings. From Faraday's law of induction (see Eq. 35-2) the emf *per turn* ε_T is the same for both the primary and secondary windings. Thus, on an rms basis, we can write

$$(\varepsilon_T)_{rms} = \left(-\frac{d\Phi_B}{dt}\right)_{rms} = \frac{V_{1,rms}}{N_1} = \frac{V_{2,rms}}{N_2}$$

or

$$V_{2,rms} = V_{1,rms}(N_2/N_1). \qquad (39\text{-}39)$$

If $N_2 > N_1$, we speak of a *step-up transformer*; if $N_2 < N_1$, we speak of a *step-down transformer*.

In all of the above we have assumed an open circuit secondary so that no power is transmitted through the transformer. If we now close switch S in Fig. 39-10, however, we have a more practical situation in which the secondary winding is connected with a resistive load R. In the general case the load would also contain inductive and capacitive elements but we confine ourselves to this special case.

Several things happen when we close switch S. (1) A current $i_2(t)$ appears in the secondary circuit, with a corresponding power dissipation $i_{2,rms}^2 R$ ($=V_{2,rms}^2/R$) in the resistive load. (2) This current induces its

* See L. O. Barthold and H. G. Pfeiffer, "High-Voltage Power Transmission," *Scientific American*, May 1964.

own alternating magnetic flux in the iron core and this flux induces (from Faraday's law and Lenz's law) an *opposing* emf in the primary windings. The two windings appear now as a fully coupled mutual inductance; see Section 36-6. (3) $V_1(t)$, however, cannot change in response to this opposing emf because it must *always* equal $\mathcal{E}(t)$ as provided by the generator; closing switch S cannot change this fact. (4) For this reason a *new* resultant current $i_1(t)$ must appear in the primary windings, its magnitude and phase angle being just that needed to cancel the opposing emf generated in the primary windings by $i_2(t)$. In particular the phase angle ϕ between $i_1(t)$ and $\mathcal{E}(t)$ for an ideal transformer must approach $0°$, so that the power factor in Eq. 39-25, $\cos \phi$, must approach unity.

All of the above is consistent with energy conservation. When we close switch S, power is dissipated in the resistive load. This requires that the generator provide an equal power to the (ideal) transformer, or, assuming $\phi = 0°$ ($\cos \phi = 1$),

$$\mathcal{E}_{\text{rms}} i_{1,\text{rms}} = V^2_{2,\text{rms}}/R. \qquad (39\text{-}40)$$

This relation expresses the fact that, for an ideal transformer with a resistive load, the power provided by the generator on the primary side just equals that dissipated in the resistive load on the secondary side.

EXAMPLE 9

A transformer on a utility pole operates at $V_{1,\text{rms}} = 8.0$ kV (see Fig. 39-10) on the primary side and supplies electric energy to a number of nearby houses at $V_{2,\text{rms}} = 120$ V.

(a) What is the turns ratio N_1/N_2? (b) If the average power consumption in the houses for a given time interval is 70 kW, what are the rms currents in the primary and secondary windings of the transformer? Assume an ideal transformer, a resistive load, and a power factor of unity. (c) What is the equivalent resistive load R in the secondary circuit?

(a) From Eq. 39-39,

$$N_1/N_2 = V_{1,\text{rms}}/V_{2,\text{rms}}$$

$$= 8.0 \text{ kV}/120 \text{ V}$$

$$= 67.$$

(b) From Eq. 39-25 we have, with $\cos \phi = 1$

$$i_{2,\text{rms}} = P_{\text{av}}/(V_{2,\text{rms}})(\cos \phi)$$

$$= (70 \text{ kW})/(120 \text{ V})(1)$$

$$= 580 \text{ A},$$

and also

$$i_{1,\text{rms}} = P_{\text{av}}/(V_{1,\text{rms}})(\cos \phi)$$

$$= (70 \text{ kW})/(8.0 \text{ kV})(1)$$

$$= 8.8 \text{ A}.$$

Note that, as required for an ideal transformer,

$$i_{1,\text{rms}} V_{1,\text{rms}} = i_{2,\text{rms}} V_{2,\text{rms}} = 70 \text{ kW}.$$

(c) Here we have

$$R = (V_{2,\text{rms}})^2/P_{\text{av}}$$

$$= (120 \text{ V})^2/70 \text{ kW}$$

$$= 0.21 \ \Omega.$$

Show that the same result follows by solving Eq. 39-40 for R.

questions

1. In the relation $\omega = 2\pi\nu$ we measure ω in radians/second and ν in hertz or cycles/seconds. The radian is a measure of angle. What connection do angles have with alternating currents?

2. Problem 35-6 suggests how an alternating emf such as that described by Eq. 39-1 can be generated. If the output of such an ac generator is connected to an *RCL* circuit such as that of Fig. 39-1, what is the ultimate source of the power dissipated in the circuit? In what part of the circuit does the power dissipation occur?

3. In the circuit of Fig. 39-1, why is it safe to assume that (a) the alternating current of Eq.39-2 has the same angular frequency $\mathcal{E}_m$ as the alternating emf of Eq. 39-1, and (b) that the phase angle ϕ in Eq. 39-2 does not vary with time? What would happen if either of these (true) statements were false?

4. Is there an analogy, however loose, between the facts that (a) the phase angle ϕ in Eq. 39-2 does not depend on the value of ω in Eq. 39-1, and (b) the oscillation frequency of a mass-spring system does not depend on the amplitude of the oscillation? Distinguish between the *nature* of a physical event and its *scale*.

5. How does a phasor differ from a vector? We know, for example, that emfs, potential differences, and currents are *not* vectors. How then can we justify constructions such as Fig. 39-5?

6. In the purely resistive circuit of Fig. 39-2 how does the maximum value of the alternating current i_m vary with the angular frequency of the applied emf?

7. Would any of the discussion of Section 39-2 be invalid if the phasor diagrams were to rotate in the clockwise direction, rather than the counterclockwise direction which we assumed?

8. Does it seem intuitively reasonable that the capacitive reactance $(= 1/\omega C)$ should vary inversely with the angular frequency, whereas the inductive reactance $(=\omega L)$ varies directly with this quantity?

9. During World War II, at a large research laboratory in this country, an alternating current generator was located a mile or so from the laboratory building which it served. A technician increased the speed of the generator to compensate for what he called "the loss of frequency along the transmission line" connecting the generator with the laboratory building. Comment.

10. Discuss in your own words what it means to say that a potential difference "leads" or "lags" an alternating current.

11. If, as we stated in Section 39-3, a given circuit is "more inductive than capacitive," that is, that $X_L > X_C$, does this mean, for a fixed angular frequency, that (a) L is relatively "large" and C is relatively "small," or (b) L and C are both relatively "large"? (c) For fixed values of L and C does this mean that ω is relatively "large" or relatively "small"?

12. Assume that in Fig. 39-1 we let $\omega \to 0$. Does Eq. 39-13 approach an expected value? Discuss.

13. How is it that in Fig. 39-2 the current phasor points in the direction of the voltage phasor but in Figs. 39-3, 39-4, and 39-5 the current phasors point in three different directions? The applied emf (Eq. 39-1) is the same in all four cases.

14. Consider this statement: "If $X_L > X_C$, then, regardless of the frequency, we must have $L > 1/C$." What is wrong with this statement?

15. Do Kirchoff's loop theorem (see Section 32-2) and Kirchoff's junction theorem (see Section 32-5) apply to multiloop ac circuits as well as to multiloop dc circuits?

16. In Example 5 what would be the effect on P_{av} if you increased (a) R, (b) C, and (c) L? How would ϕ in Eq. 39-25 change in these three cases?

17. Do commercial power station engineers like to have a low power factor (see Eq. 39-25) or a high one, or does it make any difference to them? Between what values can the power factor fluctuate? What determines the power factor; is it characteristic of the generator, of the transmission line,

of the circuit to which the transmission line is connected, or of some combination of these?

18. If you know the power factor (=cos ϕ in Eq. 39-25) for a given RCL circuit, can you tell whether or not the applied alternating emf is leading or lagging the current? If so, how? If not, why not?

19. If $R = 0$ in the circuit of Fig. 39-1, there can be no Joule internal energy dissipation in the circuit. However, an alternating emf and an alternating current are still present. Discuss the energy flow in the circuit under these conditions.

20. If you want to reduce your electric bill, do you hope for a small or a large power factor (=cos ϕ in Eq. 39-25) or does it make any difference? If it does, is there anything that you can do about it? Discuss.

21. If $R = 0$ in Fig. 39-1, how would the instantaneous current vary with time? What value would the phase angle ϕ have? What would be the average power dissipation?

22. In Eq. 39-25 is ϕ the phase angle between $\mathcal{E}(t)$ and $i(t)$ or between $\mathcal{E}_{rms}$ and i_{rms}? Discuss.

23. Resonance in RCL circuits, as judged by Eq. 39-29 and Fig. 39-7, occurs when the angular frequency of the alternating emf (the driving force) is exactly equal to the natural angular frequency of the (undamped) LC circuit. In Section 15-10, however, we saw that resonance for damped mass-spring systems, as judged by Eq. 15-41 and Fig. 15-20, occurs when the angular frequency of the driving force is close to, *but not exactly equal to*, the natural angular frequency of the (undamped) mass-spring system. Is there a failure of the principle of correspondence here? (*Hint:* Do the quantities displayed on the vertical axes of Figs. 15-20 and 39-7 truly "correspond"? See Problem 15.)

24. Convince yourself that filter circuits, such as those shown in Fig. 39-9 become, for fixed values of L and C, more effective as the angular frequency ω of V_{in} increases. What does "more effective" mean in this connection?

25. In Fig. 39-8c why does the current approaching junction c not divide equally between paths cd and ca? Both are potentially conducting paths.

26. Sketch roughly the waveforms of Fig. 39-8b,c if the rectifiers are not "ideal," that is, if the resistance in the forward direction, though small, is not actually zero, and if the resistance in the backward direction, though large, is not infinitely large.

27. Sketch an $i - V$ curve (see Figs. 31-4, 31-5, and 31-6) for the idealized rectifier assumed in Section 39-6. Include both polarities of the applied potential difference V (and of the corresponding current i) in your plot.

28. A doorbell transformer is designed for a primary rms input of 120 V and a secondary rms output of 6.0 V. What would happen if the primary and secondary connections were accidentally interchanged during installation? Would you have to wait for someone to push the doorbell to find out? Discuss.

29. You are given a transformer enclosed in a wooden box, its primary and secondary terminals being available at two opposite faces of the box. How could you find its turns ratio without opening the box?

problems

SECTION 39-1

1. A commercial alternating current generator is characterized by $\nu = 60$ Hz. What is the angular frequency ω and in what units is it expressed?
   Answer: $\omega = 2\pi\nu = 380$ rad/s.

SECTION 39-2

2. An 0.50-μF capacitor is connected, as shown in Fig. 39-3a, to an ac generator with $\mathcal{E}_m = 300$ V. What is the amplitude i_m of the resulting alternating current if the angular frequency ω is (a) 100 rad/s, and (b) 1000 rad/s?

3. A 45-mH inductor has a reactance X_L of 1300Ω. What must be (a) the applied angular frequency ω and (b) the applied frequency ν for this to be true? (c) If, as in Fig. 39-4a, an alternating emf with $\varepsilon_m = 300$ V is applied, what is the amplitude i_m of the alternating current that results? *Answer:* (a) 2.9×10^{-4} rad/s. (b) 4.6 kHz. (c) 0.23A.

4. A 1.5-μF capacitor has a reactance X_C of 12 Ω. What must be (a) the applied frequency ν and (b) the applied angular frequency ω for this to be true? (c) If, as in Fig. 39-3a, an alternating emf with $\varepsilon_m = 300$ V is applied, what is the amplitude i_m of the alternating current that results?

5. (a) At what frequency ν would a 6.0-mH inductor and a 10-μF capacitor have the same reactance? (b) What would this reactance be? (c) How would this frequency compare with the natural resonant frequency of free oscillations if the components were connected as a (resistanceless) LC oscillator, as in Fig. 38-1? *Answer:* (a)650 Hz. (b) 24 Ω. (c) They are equal.

SECTION 39-3

6. Recalculate Example 3 for $C = 20$ μF, all other given quantities remaining the same. Will ε_m lead or lag i_m in this new situation? Draw a rough diagram corresponding to Fig. 39-5.

7. Redraw (roughly) Figs. 39-5a and 5b for the cases of $X_C > X_L$ and $X_C = X_L$.

SECTION 39-4

8. We have seen that $\overline{\sin^2 \omega t} = \frac{1}{2}$ (average value). Find the average value of $\sin^2(\omega t + \phi)$ where ϕ is a (constant) phase angle?

9. Show that (see Eq. 39-23) $\overline{\sin \omega t \cos \omega t} = 0$ by using the trignometric identity $2 \sin \omega t \cos \omega t = \sin 2\omega t$. Also, plot $\sin \omega t \cos \omega t$ roughly and show graphically that its average value is zero.

10. The average value of $\varepsilon_m \sin \omega t$ over many cycles (see Fig. 39-6a) is zero. We have also seen that $\varepsilon_{rms} = \varepsilon_m/\sqrt{2}$. How does the average value of $\varepsilon_m \sin \omega t$, averaged over one-half of a cycle only, compare with ε_m and with ε_{rms}?

11. In an RCL circuit such as that of Fig. 39-1, assume $R = 5.0$ Ω, $L = 60$ mH, $\nu = 60$ Hz, and $\varepsilon_m = 300$ V. For what values of C would P_{av} for the circuit be (a) a maximum and (b) a minimum? (c) What are these maximum and minimum values and what are the corresponding phase angles and power factors? (d) If $\cos \phi$ in Eq. 39-25 is negative, it would seem that P_{av} in Eq. 39-25 would be negative. What can this mean? Can you find a value of C above that will give a negative value for $\cos \phi$? Discuss. *Answer:* (a) 120 μF. (b) Infinity. (c) 9000W, 420W; 0°, 78°; 1.0, 0.22.

SECTION 39-5

12. An RCL circuit, such as that of Fig. 39-1, has $R = 20$ Ω, $C = 20$ μF, and $L = 1.0$ H. (a) At what angular frequency ω of the ac generator will the circuit resonate with maximum response? (b) At what angular frequencies will the response be one-half the maximum value? We define "response" to be measured by the rms current in the circuit, as in Fig. 39-7.

13. A resistor-capacitor-inductor circuit R_1, C_1, L_1, connected as in Fig. 39-1, exhibits resonance at the same frequency as a second, separate, combination R_2, C_2, L_2. If the two combinations are connected in series in a single circuit, at what frequency would the combined circuit resonate?

Answer: $\omega = \sqrt{\dfrac{1}{L_1 C_1}} = \sqrt{\dfrac{1}{L_2 C_2}}$, independent of R_1 and R_2.

14. In Fig. 39-7 show that for frequencies higher than the resonant frequency the circuit is predominantly inductive and for frequencies lower than the resonant frequency it is predominantly capacitive. What does this statement mean? How do you interpret it in terms of Fig. 39-5b?

15. Show that the amplitude of the *charge* (not current) oscillations in an RCL circuit such as that of Fig. 39-1 is

$$q_m = \frac{\varepsilon_m}{\sqrt{(\omega^2 L - 1/C)^2 + (\omega R)^2}}.$$

(a) For what value of ω will q_m be a maximum? (b) Does this result shed any light on the comparison of Figs. 15-20 and 39-7, suggested in Question 23?

Answer: (a) $\omega = \sqrt{\dfrac{1}{LC} - \dfrac{R^2}{2L^2}}$.

16. Figure 39-11 shows an ac generator connected, through terminals a and b, to a "black box" containing an RCL circuit whose elements and arrangements we do not know. An alternating current, given by $i = i_m \sin(\omega t + \phi)$ appears in the lead-in wires. (a) What is the power factor? (b) Is the circuit in the box capacitive or inductive in nature? (c) Does the emf lag the current or lead it? (d) What average power P_{av} is delivered to the box by the source of emf if $\varepsilon_m = 750$ V and $i_m = 12$ A? (e) Why don't you need to know the angular frequency ω to answer the preceding question? (f) If you wanted the circuit to resonate, in the sense of Fig. 39-7, what is the nature of the circuit element that you would connect between terminals a and b? (g) At resonance, what values would ϕ and P_{av} have?

17. Show that the fractional half-width of the resonance curves of Fig. 39-7 is given, to a close approximation, by

$$\frac{\Delta\omega}{\omega} = \frac{\sqrt{3}R}{\omega L},$$

in which ω is the resonant frequency and $\Delta\omega$ is the width of the resonance peak at $i = \frac{1}{2}i_m$. Note (see Problem 38-22) that this expression may be written as $\sqrt{3}/Q$ which shows clearly that a "high-Q" circuit has a sharp resonance peak, that is, a small $\Delta\omega/\omega$.

SECTION 39-6

18. In Fig. 39-8 what are (a) $V_{bc,rms}$ and (b) P_{av} for the resistor R in each of the cases shown?

19. In Fig. 39-8, shade the areas, in all three cases, that cancel to form $\overline{V_{bc}}$.

20. For the full-wave rectified waveform of Fig. 39-8c, what is the fundamental (lowest) frequency ν_0 of the "ac ripple"? Assume that, for the ac generator, $\nu = \omega/2\pi = 60$ Hz.

21. In Fig. 39-12 in which $R_1 \gg R_2$ and both V_{in} and V_{out} are constant potential differences, show that the attenuation factor $V_{out}/V_{in} \cong R_2/R_1$. Compare with Fig. 39-9 and Eq. 39-38, in which the (ac) attenuation factor is (approximately) X_C/X_L. Discuss analogies and differences.

22. (a) Show that the attenuation factor $V_{out,m}/V_{in,m}$ (see Eq. 39-38 and Fig. 39-9) can be written, for $\omega \gg \omega_0$, as $(\omega_0/\omega)^2$. Here ω is the input angular frequency and ω_0 is the resonant angular frequency $(=\sqrt{1/LC})$ of the LC filter combination. (b) Show that $\omega \gg \omega_0$ corresponds to $X_L \gg X_C$. Is this reasonable?

23. Show by qualitative argument that the three-stage filter of Fig. 39-13 is more effective than the one-stage filter of Fig. 39-9. Sketch its dc equivalent, that is, replace L by R_1 and C by R_2, with $R_1 \gg R_2$. Derive the dc attenuation factor V_{out}/V_{in}, both of these quantities being dc potentials. Answer: $(R_2/R_1)^3$.

24. Show that, in low current, high-voltage applications (such as power supplies for television picture tubes) we can replace the inductor L of Fig. 39-9 by a "large" resistor R and still achieve substantial reduction of the ac component of V_{in} without too much reduction of the dc component.

SECTION 39-7

25. An ac transmission line transfers energy at the rate $P_{av} = 5.0$ MW from a generating plant to a factory. (a) What current i_{rms} is present in the line if the transmission voltage V_{rms} is 120 V? (b) If $V_{rms} = 80$ kV? (c) What is the ratio of the thermal (Joule) energy losses in the line for these two cases? Assume that the power factor $\cos \phi = 1$. Answer: (a) 42 kA, (b) 63A, (c) 4.5×10^5.

26. A transformer has 500 primary turns and 10 secondary turns. (a) If $V_{1,rms}$ for the primary is 120 V, what is $V_{2,rms}$ for the secondary, assumed on open

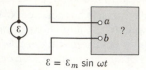

figure 39-11
Problem 16

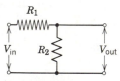

figure 39-12
Problem 21

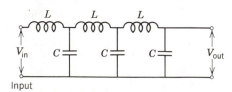

figure 39-13
Problem 23

circuit? (b) If the secondary now has a resistive load R of 15Ω, what are $i_{1,\text{rms}}$ and $i_{2,\text{rms}}$? Assume an ideal transformer with $\phi = 0$.

27. The output of a full-wave rectifier (see Fig. 39-8c) is fed into an (ideal) transformer with a step-up ratio of 2:1. Sketch roughly the waveform appearing at the secondary, assumed to be on open circuit.

28. In Fig. 39-10 compare these quantities for (a) switch S open, and (b) switch S closed: $\Phi_B(t)$, $\mathcal{E}(t)$, $V_1(t)$, $V_2(t)$, $V_{1,\text{rms}}$, $V_{2,\text{rms}}$, $i_1(t)$, $i_2(t)$, $i_{1,\text{rms}}$, and $i_{2,\text{rms}}$. Assume an ideal transformer, with $\phi = 0$.

29. In Fig. 39-10 show that $i_1(t)$ in the primary circuit remains unchanged if a resistance $R'[=R(N_1/N_2)^2]$ is connected directly across the generator, the transformer and the secondary circuit being removed. That is,

$$i_1(t) = \frac{\mathcal{E}(t)}{R'}.$$

In this sense we see that a transformer not only "transforms" potential differences and currents but also resistances. In the more general case, in which the secondary load in Fig. 39-10 contains capacitive and inductive elements as well as resistive, we say that a transformer transforms *impedances*. See Problem 31 for an example.

30. In Problem 7 of Chapter 32 we asserted (see Fig. 32-3a) that, for dc circuits, the power dissipated in an external resistance R is a maximum when $R = r$, where r is the internal resistance of the (dc) source of emf. In Fig. 39-14 show that, in the same way, the average power dissipation P_{av} in R is a maximum when $R = r$, in which r is the internal resistance of the ac generator. In the text we have tacitly assumed that $r = 0$; here we assume that $r \neq 0$.

31. *Impedance Matching.* We have seen in Problem 29 that a transformer can serve as a resistance (generally, impedance) transforming device. We saw also, in Problem 30 that (see Fig. 39-14) the transfer of power from an ac generator (internal resistance r) to a resistive load R is a maximum when $R = r$. Suppose that, in Fig. 39-14, $r = 1.0$ kΩ, $R = 10$ Ω, $\omega/2\pi = 60$ Hz, and $\mathcal{E}_{\text{rms}} = 120$ V. Design a transformer, to be interposed between the ac generator and the load, that will insure maximum power transfer to R. Assume an ideal transformer with $\phi = 0$. Such a technique is used, for example, when it is necessary to transfer power efficiently from a (high-impedance) audio amplifier to a (low-impedance) loudspeaker.
Answer: $N_1/N_2 = 10$.

figure 39-14
Problem 30

40
maxwell's equations

In classical mechanics and thermodynamics we sought to identify the smallest, most compact set of equations or laws that would define the subject as completely as possible. In mechanics we found this in Newton's three laws of motion (Sections 5-2, 5-4, and 5-5) and in the associated force laws, such as Newton's law of gravitation (Section 16-2). In thermodynamics we found it in the three laws described in Sections 21-2, 22-7, and 25-4.

In this chapter we seek to do the same thing for electromagnetism, proceeding in several steps. First we display, in Table 40-1, a *tentative* set of such equations. After studying this table, we will conclude from arguments of symmetry that these equations are not yet complete and that there may be (and indeed *is*) a missing term in one of them.

40-1
THE BASIC EQUATIONS OF ELECTROMAGNETISM

Table 40-1
Tentative* basic equations of electromagnetism

Symbol	Name	Equation	Text Reference
I	Gauss's law for electricity	$\epsilon_0 \oint \mathbf{E} \cdot d\mathbf{S} = q$	Eq. 28-6
II	Gauss's law for magnetism	$\oint \mathbf{B} \cdot d\mathbf{S} = 0$	Eq. 37-6
III	Faraday's law of induction	$\oint \mathbf{E} \cdot d\mathbf{l} = -\dfrac{d\Phi_B}{dt}$	Eq. 35-9
IV	Ampère's law	$\oint \mathbf{B} \cdot d\mathbf{l} = \mu_0 i$	Eq. 34-1

* "Tentative" suggests, as we shall see below, that Eq. IV is not yet complete and requires an additional term; see Table 40-2.

The missing term will prove to be no trifling correction but will round out the complete description of electromagnetism and, beyond this, will establish optics as an integral part of electromagnetism. In particular it will allow us to prove that the speed of light c in free space is related to purely electric and magnetic quantities by

$$c = \frac{1}{\sqrt{\epsilon_0 \mu_0}}. \qquad (40\text{-}1)$$

It will also lead us to the concept of the electromagnetic spectrum, which lies behind the experimental discovery of radio waves.

We have seen how the principle of symmetry permeates physics and how it has often led to new insights or discoveries. For example, (a) if body A attracts body B with a force **F**, then body B attracts body A with a force $-\mathbf{F}$ (it does), and (b) if there is a negative electron, there may well be a positive electron (there is), etc.

Let us examine Table 40-1 from this point of view. First we say that when we are dealing with symmetry considerations alone (that is, not making quantitative calculations) we can ignore ϵ_0 and μ_0. These constants result from our choice of unit systems and play no role in arguments of symmetry. There are in fact unit systems in which $\epsilon_0 = \mu_0 = 1$.

With this in mind we see that the left sides of the equations in Table 40-1 are completely symmetrical, in pairs. Equations I and II are surface integrals of **E** and **B**, respectively, over closed surfaces. Equations III and IV are line integrals of **E** and **B**, respectively, around closed loops.

On the right side of these equations, things are *not* symmetrical and, in fact, there are two kinds of asymmetries, which we shall discuss separately.

1. The first asymmetry, which is not really the concern of this chapter, deals with the apparent fact that although there are isolated centers of charge (electrons and protons, say) there seem not to be isolated centers of magnetism (magnetic monopoles; see Section 37-1). Thus we account for the "q" on the right of Eq. I and for the "0" on the right of Eq. II. In the same way the term $\mu_0 i (= \mu_0 dq/dt)$ appears on the right of Eq. IV but no similar term (a current of magnetic monopoles) appears on the right of Eq. III. The resolution of *this* asymmetry depends on the not as yet discovered magnetic monopole. Considerations of symmetry have motivated physicists to search for the magnetic monopole in great earnest and in many ways. It is as though nature were hinting and guiding physicists in their explorations.

2. The second asymmetry, with which this chapter deals, sticks out like a sore thumb. On the right side of Eq. III (Faraday's law of induction; see Eq. 35-9) we find the term $-d\Phi_B/dt$ and we interpret this law loosely by saying:

If you change a magnetic field $(d\Phi_B/dt)$, you produce an electric field $(\oint \mathbf{E} \cdot d\mathbf{l})$.

We learned this in Section 35-1 where we showed that if you shove a bar magnet through a closed conducting loop, you do indeed induce an electric field, and thus a current, in that loop.

From the principle of symmetry we are entitled to suspect that the symmetrical relation holds, that is:

If you change an electric field $(d\Phi_E/dt)$, you produce a magnetic field $(\oint \mathbf{B} \cdot d\mathbf{l})$.

This symmetry principle indeed meets the test of experiment and we discuss it fully in the next section. This supposition supplies us with the important "missing" term in Eq. IV in Table 40-1.

Here we discuss in detail the evidence for the supposition of the previous section, namely:

"A changing electric field induces a magnetic field."

Although we will be guided by considerations of symmetry alone, we will also point to direct experimental verification.

Figure 40-1a shows a uniform electric field **E** filling a cylindrical region of space. It might be produced by a circular parallel-plate capacitor, as suggested in Fig. 40-1b. We assume that E is increasing at a steady rate dE/dt, which means that charge must be supplied to the capacitor plates at a steady rate; to supply this charge requires a steady current i into the positive plate and an equal steady current i out of the negative plate.

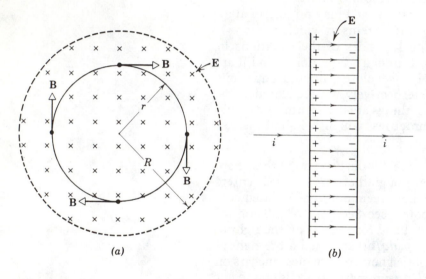

(a)   (b)

40-2
INDUCED MAGNETIC FIELDS

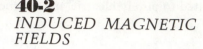

figure 40-1
(a) Showing the induced magnetic fields **B** at four points, produced by a changing electric field **E**. The electric field is increasing in magnitude. Compare Fig. 35-10.
(b) Such a changing electric field may be produced by charging a parallel-plate capacitor as shown.

Experiment shows (see Problem 4) that *a magnetic field is set up by this changing electric field*. Figure 40-1a shows **B** for four arbitrary points. Figure 40-1 suggests a beautiful example of the symmetry of nature. A changing *magnetic* field induces an *electric* field (Faraday's law); now we see that a changing *electric* field induces a *magnetic* field.

To describe this new effect quantitatively, we are guided by analogy with Faraday's law of induction,

$$\oint \mathbf{E} \cdot d\mathbf{l} = -\frac{d\Phi_B}{dt},$$ (40-2)

which asserts that an electric field (left term) is produced by a changing magnetic field (right term). For the symmetrical counterpart we might write*

$$\oint \mathbf{B} \cdot d\mathbf{l} = \mu_0 \epsilon_0 \frac{d\Phi_E}{dt}.$$ (40-3)

Equation 40-3 asserts that a magnetic field (left term) can be produced

---

* Our system of units requires that we insert the constants ϵ_0 and μ_0 in Eq. 40-3. In some unit systems they would not appear.

by a changing electric field (right term). Compare carefully Fig. 35-10, which illustrates the production of an electric field by a changing magnetic field, with Fig. 40-1a. In each case the appropriate flux Φ_B or Φ_E is *increasing*. However, experiment shows that the lines of **E** in Fig. 35-10 are *counterclockwise*, whereas those of **B** in Fig. 40-1a are *clockwise*. This difference requires that the minus sign of Eq. 40-2 be omitted from Eq. 40-3.

In Section 34-1 we saw that a magnetic field can also be set up by a current in a wire. We described this quantitatively by Ampère's law:

$$\oint \mathbf{B} \cdot d\mathbf{l} = \mu_0 i,$$

in which i is the conduction current passing through the loop around which the line integral is taken. Thus there are at least two ways of setting up a magnetic field: (a) by a changing electric field and (b) by a current. In general, both possibilities must be allowed for, or*

$$\oint \mathbf{B} \cdot d\mathbf{l} = \mu_0 \epsilon_0 \frac{d\Phi_E}{dt} + \mu_0 i. \qquad (40\text{-}4)$$

Maxwell is responsible for this important generalization of Ampère's law. It is a central and vital contribution, as we have pointed out earlier.

In Chapter 34 we assumed that no changing electric fields were present so that the term $d\Phi_E/dt$ in Eq. 40-4 was zero. In the discussion just given we assumed that there were no conduction currents in the space containing the electric field. Thus the term i in Eq. 40-4 is zero. We see now that each of these situations is a special case.

EXAMPLE 1

A parallel-plate capacitor with circular plates is being charged as in Fig. 40-1. (a) Derive an expression for the induced magnetic field at various radii r. Consider both $r < R$ and $r > R$.

From Eq. 40-3,

$$\oint \mathbf{B} \cdot d\mathbf{l} = \mu_0 \epsilon_0 \frac{d\Phi_E}{dt},$$

we can write, for $r \lessgtr R$,

$$(B)(2\pi r) = \mu_0 \epsilon_0 \frac{d}{dt} \left[(E)(\pi r^2) \right] = \mu_0 \epsilon_0 \pi r^2 \frac{dE}{dt}.$$

Solving for B yields $\qquad B = \tfrac{1}{2}\mu_0 \epsilon_0 r \dfrac{dE}{dt} \qquad (r \lessgtr R).$

For $r \gtrless R$, Eq. 40-3 yields

$$(B)(2\pi r) = \mu_0 \epsilon_0 \frac{d}{dt} \left[(E)(\pi R^2) \right] = \mu_0 \epsilon_0 \pi R^2 \left(\frac{dE}{dt} \right),$$

or $\qquad B = \dfrac{\mu_0 \epsilon_0 R^2}{2r} \dfrac{dE}{dt} \qquad (r \gtrless R).$

---

* Actually, there is a third way of setting up a magnetic field, by the use of magnetized bodies. In Section 37-8 we saw that this could be accounted for by inserting a *magnetizing current* term i_M on the right side of Ampère's law. This law would then read, in its full generality,

$$\oint \mathbf{B} \cdot d\mathbf{l} = \mu_0 \epsilon_0 \frac{d\Phi_E}{dt} + \mu_0 i + \mu_0 i_M.$$

In all that follows we assume that no magnetic materials are present so that $i_M = 0$.

(b) Find B at $r = R$ for $dE/dt = 10^{12}$ V/m · s and for $R = 5.0$ cm.

At $r = R$ the two equations for B reduce to the same expression, or

$$B = \tfrac{1}{2}\mu_0\epsilon_0 R \frac{dE}{dt}$$

$$= (\tfrac{1}{2})(4\pi \times 10^{-7} \text{ T} \cdot \text{m/A})(8.9 \times 10^{-12} \text{ C}^2/\text{N} \cdot \text{m}^2)(5.0 \times 10^{-2} \text{ m})(10^{12} \text{ V/m} \cdot \text{s})$$

$$= 2.8 \times 10^{-7} \text{ T} = 0.0028 \text{ gauss.}$$

This shows that the induced magnetic fields in this example are so small that they can scarcely be measured with simple apparatus, in sharp contrast to induced *electric* fields (Faraday's law), which can be demonstrated easily. This experimental difference is in part due to the fact that induced emfs can easily be multiplied by using a coil of many turns. No technique of comparable simplicity exists for magnetic fields. In experiments involving oscillations at very high frequencies dE/dt above can be very large, resulting in significantly larger values of the induced magnetic field.

Equation 40-4 shows that the term $\epsilon_0 \, d\Phi_E/dt$ has the dimensions of a current. Even though no motion of charge is involved, there are advantages in giving this term the name *displacement* current. Thus we can say that a magnetic field can be set up either by a conduction current i or by a displacement current i_d $(=\epsilon_0 d\Phi_E/dt)$, and we can rewrite Eq. 40-4 as†

$$\oint \mathbf{B} \cdot d\mathbf{l} = \mu_0(i_d + i). \tag{40-5}$$

40-3
DISPLACEMENT CURRENT

The concept of displacement current permits us to retain the notion that *current is continuous,* a principle established for steady conduction currents in Section 31-1. In Fig. 40-1b, for example, a current i enters the positive plate and leaves the negative plate. The *conduction* current is *not* continuous across the capacitor gap because no charge is transported across this gap. However, the displacement current i_d in the gap will prove to be exactly i, thus retaining the concept of the continuity of current.

To calculate the displacement current, recall (see Eq. 30-2) that E in the gap is given by

$$E = \frac{q}{\epsilon_0 A}.$$

Differentiating gives $\quad \dfrac{dE}{dt} = \dfrac{1}{\epsilon_0 A} \dfrac{dq}{dt} = \dfrac{1}{\epsilon_0 A} i.$

The displacement current i_d is by definition

$$i_d = \epsilon_0 \frac{d\Phi_E}{dt} = \epsilon_0 \frac{d(EA)}{dt} = \epsilon_0 A \frac{dE}{dt}.$$

Eliminating dE/dt between these two equations leads to

---

* The word "displacement" was introduced for historical reasons that need not concern us here.

† We may write this more generally, taking the presence of magnetic materials into account, as

$$\oint \mathbf{B} \cdot d\mathbf{l} = \mu_0(i_d + i + i_M).$$

See the footnote on p. 883.

$$i_d = (\epsilon_0 A)\left(\frac{1}{\epsilon_0 A}\, i\right) = i,$$

which shows that the displacement current in the gap is identical with the conduction current in the lead wires.

EXAMPLE 2

What is the displacement current for the situation of Example 1? From the definition of displacement current,

$$i_d = \epsilon_0 \frac{d\Phi_E}{dt} = \epsilon_0 \frac{d}{dt}\,[(E)(\pi R^2)] = \epsilon_0 \pi R^2 \frac{dE}{dt}$$

$$= (8.9 \times 10^{-12}\ \text{C}^2/\text{N} \cdot \text{m}^2)(\pi)(5.0 \times 10^{-2}\ \text{m})^2(10^{12}\ \text{V/m} \cdot \text{s})$$

$$= 0.070\ \text{A}.$$

Even though this displacement current is reasonably large, it produces only a small magnetic field (see Example 1) because it is spread out over a large area. In contrast, the *conduction* current i in the lead wires (also = 0.070 A, because $i = i_d$) can produce magnetic effects close to the (thin) wires that are easily detectable by a compass needle. The discussion of Section 34-2 shows that the magnetic effect is greatest at the surface of the wire. The capacitor of Fig. 40-1 behaves like a "conductor" of radius 5.0 cm, carrying a (displacement) current of 0.070 A. Its largest magnetic effect is at the capacitor edge, that is, at $r = 5.0$ cm, and is given by Eq. 34-4 ($B = \mu_0 i/2\pi r$).

40-4
MAXWELL'S EQUATIONS

Equation 40-4 completes our presentation of the basic equations of electromagnetism, called *Maxwell's equations.** They are summarized in Table 40-2, which rounds out the "tentative" Table 40-1 by supplying the "missing" term in Eq. IV of that table. All equations of physics that serve, as these do, to correlate experiments in a vast area and to predict new results have a certain beauty about them and can be appreciated, by those who understand them, on an aesthetic level. This is true for Newton's laws of motion, for the laws of thermodynamics, for the theory of relativity, and for the theories of quantum physics. As for Maxwell's equations, the physicist Ludwig Boltzmann (quoting a line from Goethe) wrote, "Was it a god who wrote these lines. . . ." In more recent times J. R. Pierce, in a book chapter entitled 'Maxwell's Wonderful Equations' says, "To anyone who is motivated by anything beyond the most narrowly practical, it is worth while to understand Maxwell's equations simply for the good of his soul." The scope of these equations is remarkable, including as it does the fundamental operating principles of all large-scale electromagnetic devices such as motors, synchrotrons, television, and microwave radar.†

We suggested in Section 40-1 that Maxwell's equations (as they appear in

---

* In the preceding sections we have leaned heavily on symmetry arguments in developing Maxwell's equations as displayed in Table 40-2. Actually Maxwell did not base his work on such arguments. The British physicist Oliver Heaviside (1850–1925) seems to have been the first to point out the symmetry between **E** and **B** that these equations display. See "Maxwell, Displacement Current, and Symmetry" by Alfred M. Bork, *American Journal of Physics*, November, 1963, for an interesting historical account.
† For a recent spectacular example of an application of Maxwell's equations, see "Electromagnetic Flight" by Henry H. Kolm and Richard D. Thornton, *Scientific American*, October 1973. The article states: "The future of high-speed ground transportation may well lie not with wheeled trains but with vehicles that 'fly' a foot or so above a guideway, lifted and propelled by electromagnetic forces."

Table 40-2
The basic equations of electromagnetism (Maxwell's equations) *

Number	Name	Equation	Describes		Crucial Experiment	General Text Reference
I	Gauss's law for electricity	$\epsilon_0 \oint \mathbf{E} \cdot d\mathbf{S} = q$	Charge and the electric field	1.	Like charges repel and unlike charges attract, as the inverse square of their separation.	Chapter 28
				1 .	A charge on an insulated conductor moves to its outer surface.	
II	Gauss's law for magnetism	$\oint \mathbf{B} \cdot d\mathbf{S} = 0$	The magnetic field	2.	It has thus far not been possible to verify the existence of a magnetic monopole.	Section 37-2
III	Faraday's law of induction	$\oint \mathbf{E} \cdot d\mathbf{l} = -\dfrac{d\Phi_B}{dt}$	The electrical effect of a changing magnetic field	4.	A bar magnet, thrust through a closed loop of wire, will set up a current in the loop.	Chapter 35
IV	Ampère's law (as extended by Maxwell)	$\oint \mathbf{B} \cdot d\mathbf{l}$ $= \mu_0 \left(\epsilon_0 \dfrac{d\Phi_E}{dt} + i \right)$	The magnetic effect of a changing electric field or of a current	3.	The speed of light can be calculated from purely electromagnetic measurements.	Section 41-8
				3'.	A current in a wire sets up a magnetic field near the wire.	Chapter 34

* Written on the assumption that no dielectric or magnetic material is present.

Table 40-2) bear the same relationship to electromagnetism that Newton's laws of motion do to mechanics. There is, however, an important difference. Einstein presented his special theory of relativity (see Supplementary Topic V) in 1905, roughly 200 years after Newton's laws appeared and about 40 years after Maxwell's equations. As it turns out, Newton's laws had to be drastically modified in cases in which the relative speeds approached that of light. However, *no changes whatever were required in Maxwell's equations.* In the language of Supplementary Topic V we say that "Maxwell's equations are invariant under a Lorentz transformation but Newton's laws of motion are not."

Now that we have identified Maxwell's equations as the basic equations of electromagnetism, we want to test them in several situations. In this section we address ourselves to the electromagnetic cavity oscillations of Section 38-5 (see Fig. 38-6). In Chapter 41 we will show that Maxwell's equations predict the existence of electromagnetic waves, that visible light is such a wave, and that the speed c of electromagnetic radiation (light) is indeed given by Eq. 40-1.

A completely formal treatment of the cavity oscillations of Fig. 38-6, which is beyond our scope here, would start from Maxwell's equations and would end with mathematical expressions for the variation of **B** and **E** with time and with position in the cavity for all modes of oscillation of the cavity. We confine ourselves to the fundamental mode only, shown in Fig. 38-6, for which we *postulated* the variations of **B** and **E** given in that figure; we will show that these postulated fields are completely consistent with Maxwell's equations.

Figure 40-2 presents two views of the cavity of Fig. 38-6*d*, in which both electric and magnetic fields are present. Study of Fig. 38-6 reveals that **B** is *decreasing* in magnitude and **E** is *increasing* at this phase of the cycle of oscillation. Let us apply Faraday's law,

$$\oint \mathbf{E} \cdot d\mathbf{l} = -\frac{d\Phi_B}{dt},$$

40-5
MAXWELL'S EQUATIONS AND CAVITY OSCILLATIONS

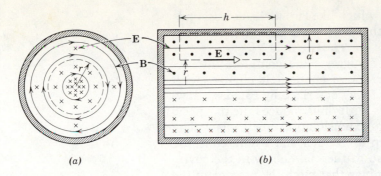

(a) (b)

to the rectangle of dimensions h and $a - r$. There is a definite flux Φ_B through the rectangular path in question, and this flux is decreasing with time because **B** is decreasing. The line integral above is

$$\oint \mathbf{E} \cdot d\mathbf{l} = hE(r),$$

in which $E(r)$ is the value of E at a radius r from the axis of the cavity.

Note that **E** equals zero for the upper leg of the integration path (which lies in the cavity wall) and that **E** and $d\mathbf{l}$ are at right angles on the two side legs. Combining these equations yields

$$E(r) = -\frac{1}{h}\frac{d\Phi_B}{dt}. \qquad (40\text{-}6)$$

Equation 40-6 shows that $E(r)$ depends on the rate at which Φ_B through the path shown is changing with time and that it has its maximum value when $d\Phi_B/dt$ is a maximum. This occurs when **B** is zero, that is, when **B** is changing its direction; recall that a sine or cosine is changing most rapidly, that is, it has the steepest tangent, at the instant it crosses the axis between positive and negative values. Thus the electric field pattern in the cavity will have its *maximum* value when the magnetic field is *zero* everywhere, consistent with Figs. 38-6a and 38-6e and with the concept of the interchange of energy between electric and magnetic fields. You can show, by applying Lenz's law, that the electric field in Fig. 40-2b indeed points to the *right*, as shown, if the magnetic field is *decreasing*.

Figure 40-2a shows an end view of the cavity; the electric lines of force are entering the page at right angles to the page and the magnetic lines form clockwise circles. Let us apply Ampère's law in the form

$$\oint \mathbf{B} \cdot d\mathbf{l} = \mu_0\epsilon_0 \frac{d\Phi_E}{dt} + \mu_0 i, \qquad (40\text{-}4)$$

to the circular path of radius r shown in the figure. No charge is transported through the ring so that the conduction current i in Eq. 40-4 is zero. The line integral on the left is $(B)(2\pi r)$ so that the equation reduces to

$$B(r) = \frac{\mu_0\epsilon_0}{2\pi r}\frac{d\Phi_E}{dt}. \qquad (40\text{-}7)$$

Equation 40-7 shows that the magnetic field $B(r)$ is proportional to the rate at which the electric flux Φ_E through the ring is changing with time. The field $B(r)$ has its maximum value when $d\Phi_E/dt$ is at its maximum; this occurs when $\mathbf{E} = 0$, that is, when **E** is reversing its direction. Thus we see that **B** has its *maximum* value when **E** is *zero* for all points in the cavity. This is consistent with Figs. 38-6c and 38-6g and with the concept of the interchange of energy between electric and magnetic forms. A comparison with Fig. 40-1a, which like Fig. 40-2a, corresponds to an increasing electric field, shows that the lines of **B** are indeed clockwise, as viewed along the direction of the electric field.

Comparison of Eqs. 40-6 and 40-7 suggests the complete interdependence of **B** and **E** in the cavity. As the magnetic field changes with time, it induces the electric field in a way described by Faraday's law. The electric field, which also changes with time, induces the magnetic field in a way described by Maxwell's extension of Ampère's law. The oscillations, once established, sustain each other and would continue indefinitely were it not for losses due to thermal energy in the cavity walls or leakage of energy from openings that might be present in the walls. In Chapter 41 we show that a similar interplay of **B** and **E** occurs not only in standing electromagnetic waves in cavities but also in traveling electromagnetic waves, such as radio waves or visible light.

In Fig. 40-2 analyze the currents (both conduction and displacement) that occur in the cavity (both in its conducting walls and within its volume). Show the relationship between these currents and the electric and magnetic fields and also show that, considering both conduction and displacement currents together, it is reasonable to conclude that current is continuous around closed loops.

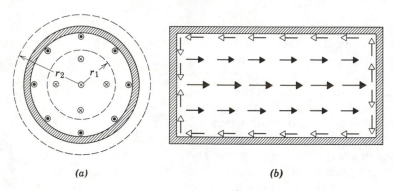

(a)   *(b)*

Figure 40-3 shows two views of the cavity, at an instant corresponding to that of Fig. 40-2. For simplicity, we do not show the **E** and **B** fields; the arrows represent currents. Because E is increasing in Figs. 40-2 and 40-3, the positive charge on the left end cap must be increasing. Thus there must be conduction currents in the walls pointing from right to left in Fig. 40-3b. These currents are also shown by the dots (representing the tips of arrows) near the cavity walls in Fig. 40-3a.

Bearing in mind that $\epsilon_0 \, d\Phi_E/dt$ is a displacement current, we can write Eq. 40-7 as

$$B(r) = \frac{\mu_0}{2\pi r}\left(\epsilon_0 \frac{d\Phi_E}{dt}\right) = \frac{\mu_0}{2\pi r}\, i_d.$$

This equation stresses that **B** in the cavity is associated with a displacement current; compare Eq. 34-4. Applying the right-hand rule in Fig. 40-2a shows that the displacement current i_d must be directed into the plane of the figure if it is to be associated with the clockwise lines of **B** that are present.

The displacement current is represented in Fig. 40-3b by the arrows that point to the right and in Fig. 40-3a by the crosses that represent arrows entering the page. Study of Fig. 40-3 shows that the current is continuous, being directed up the walls as a conduction current and then back down through the volume of the cavity as a displacement current. If we apply Ampère's law as extended by Maxwell,

EXAMPLE 3

figure 40-3
The cavity of Fig. 40-2, showing (a) the conduction current coming up the walls and displacement current going down the cavity volume and (b) the displacement current (black arrowheads) in the volume of the cavity and the conduction currents (white arrowheads) at the walls. The arrows in each case represent current densities. Note the continuity of current (conduction + displacement), that is, it is possible to form closed current loops.

$$\oint \mathbf{B} \cdot d\mathbf{l} = \mu_0(i_d + i), \qquad\qquad (40\text{-}8)$$

to the circular path of radius r_1 in Fig. 40-3a, we see that $\mathbf{B}$ at that path is due entirely to the displacement current, the conduction current i within the path being zero.

For the path of radius r_2, the net current enclosed is zero because the conduction current in the walls is exactly equal and opposite to the displacement current in the cavity volume. Since i equals i_d in magnitude, but is oppositely directed, it follows from Eq. 40-8 that B must be zero for all points outside the cavity, in agreement with observation.

questions

1. In your own words explain why Eq. III of Table 40-1 can be interpreted by saying, "A changing magnetic field can generate an electric field."

2. If (as is true) there are unit systems in which ϵ_0 and μ_0 do not appear, how can Eq. 40-1 be true?

3. If a uniform flux Φ_E through a plane circular ring decreases with time, is the induced magnetic field (as viewed along the direction of $\mathbf{E}$) clockwise or counterclockwise? Does it make any difference (a) whether the electric field is uniform or (b) whether the plane of the ring is perpendicular to the electric field?

4. Compare Tables 40-1 and 40-2. Is it enough to rely on the principle of symmetry alone or do we really need experimental verification for the "missing" term in Eq. IV?

5. Why is it so easy to show that "a changing magnetic field produces an electric field" but so hard to show in a simple way that "a changing electric field produces a magnetic field"?

6. In Fig. 40-1a consider a circle with $r > R$. How can a magnetic field be induced around this circle, as Example 1 shows? After all, there is no electric field at the location of this circle and $dE/dt = 0$ here.

7. In Fig. 40-1a, $\mathbf{E}$ is into the figure and is increasing in magnitude. Find the direction of $\mathbf{B}$ if (a) $\mathbf{E}$ is into the figure and decreasing, (b) $\mathbf{E}$ is out of the figure and increasing, (c) $\mathbf{E}$ is out of the figure and decreasing, and (d) $\mathbf{E}$ remains constant.

8. In Fig. 38-1c a displacement current is needed to maintain continuity of current in the capacitor. How can one exist, considering that there is no charge on the capacitor?

9. At what parts of the cycle will (a) the conduction current and (b) the displacement current in the cavity of Fig. 38-6 be zero?

10. In Figs. 40-1a and 40-1b what is the direction of the displacement current i_d? In this same figure, can you find a rule relating the directions of $\mathbf{B}$ and $\mathbf{E}$?

11. What advantages are there in calling the term $\epsilon_0 \, d\Phi_E/dt$ in Eq. IV, Table 40-2 a displacement *current?*

12. Why are the magnetic effects of conduction currents in wires so easy to detect but the magnetic effects of displacement currents in capacitors so hard to detect?

13. Discuss the time variation during one complete cycle of the charges that appear at various points on the inner walls of the oscillating electromagnetic cavity of Fig. 38-6.

14. Would you expect that the arrangement of the magnetic and electric fields in Fig. 40-2 is the only possible arrangement? If there are other arrangements, would you expect them to have higher or lower frequencies than those shown in Fig. 40-2?

15. In connection with Fig. 40-3, in what sense can the end caps be considered as capacitor plates? In what sense can the cylindrical walls be considered as an inductor? (*Note:* Figure 40-3 is clearly a case of distributed elements but there must be a smooth transition between distributed and lumped elements.)

16. (a) In Fig. 40-2a is it possible to apply Faraday's law usefully to the dashed circle? (b) Is it possible to apply Ampère's law usefully to the dashed rectangle of Fig. 40-2b? Discuss.

problems

SECTION 40-2

1. (a) Convince yourself that, for the conditions of Fig. 40-1, the induced magnetic field is greatest at the edge of the capacitor. (b) Make a rough, non-numerical, plot of $B(r)$ from $r = 0$ to $r \gg R$.

2. You are given a 1.0-μF parallel-plate capacitor. How would you establish an (instantaneous) displacement current of 1.0 A in the space between its plates?

3. Prove that the displacement current in a parallel-plate capacitor can be written as

$$i_d = C \frac{dV}{dt}.$$

4. In 1929 M. R. Van Cauwenerghe succeeded in measuring directly, for the first time, the displacement current i_d between the plates of a parallel-plate capacitor to which an alternating potential difference was applied, as suggested by Fig. 40-1. He used circular plates whose effective radius was 40 cm and whose capacitance was 1.0×10^{-10} F. The applied potential difference had a maximum value V_m of 174 kV at a frequency of 50 Hz. (a) What maximum displacement current is present between the plates? (See Problem 3.) (b) Why is the applied potential difference chosen to be as high as it is? [The delicacy of these measurements is such that they were only performed in a direct manner more than 60 years after Maxwell enunciated the concept of displacement current! The reference is *Journal de Physique*, No. 8, p. 303, (1929)].

SECTION 40-3

5. Figure 40-4 shows the plates P_1 and P_2 of a circular parallel-plate capacitor of radius R. They are connected as shown to long straight wires in which a constant conduction current i exists. A_1, A_2, and A_3 are hypothetical circles of radius r, two of them outside the capacitor and one between the plates.

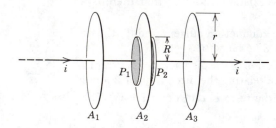

figure 40-4
Problem 5

Write expressions for the magnetic field B at the circumference of each of these circles.

Answer: $\frac{\mu_0 i}{2\pi r}$, in all cases.

6. In Example 1 show that the *displacement current density* j_d is given, for $r < R$, by

$$j_d = \epsilon_0 \frac{dE}{dt}.$$

7. A parallel-plate capacitor has square plates 1.00 m on a side as in Fig. 40-5. There is a charging current of 2.00 A flowing into (and out of) the capacitor. (a) What is the displacement current through the region between the plates?

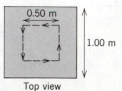

Edge view   Top view

figure 40-5
Problem 7

(*b*) What is dE/dt in this region? (*c*) What is the displacement current through the square dashed path between the plates? (*d*) What is $\int \mathbf{B} \cdot d\mathbf{l}$ around this square dashed path?
Answer: (*a*) 2.00 A. (*b*) 2.3×10^{11} V/m·s. (*c*) 0.50 A. (*d*) 6.3×10^{-7} T·m.

8. The capacitor in Fig. 40-6 consisting of two circular plates with area $A = 0.10$ m² is connected to a source of potential $\varepsilon = \varepsilon_m \sin \omega t$, where $\varepsilon_m = 200$ V and $\omega = 100$ rad/s. The maximum value of the displacement current is $i_d = 8.9 \times 10^{-6}$ A. Neglect fringing of the electric field at the edges of the plates. (*a*) What is the maximum value of the current i? (*b*) What is the maximum value of $d\Phi_E/dt$, where Φ_E is the electric flux through the region between the plates? (*c*) What is the separation d between the plates? (*d*) Find the maximum value of the magnitude of **B** between the plates at a distance $R = 0.10$ m from the center.

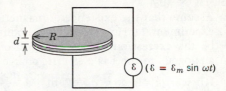

figure 40-6
Problem 8

9. In Example 1 how does the displacement current through a hypothetical concentric circular loop of radius r vary with r? Consider both (*a*) $r < R$ and (*b*) $r > R$.   *Answer:* (*a*) $\pi\epsilon_0 r^2(dE/dt)$. (*b*) $\pi\epsilon_0 R^2(dE/dt)$.

10. A parallel-plate capacitor with circular plates 20.0 cm in diameter is being charged as in Fig. 40-1*a*. The displacement current density throughout the region is uniform, into the paper in the diagram, and has a value of 20.0 A/m². (*a*) Calculate the magnetic field B at a distance $R = 5.0$ cm from the axis of symmetry of the region. (*b*) Calculate dE/dt in this region.

11. Identify the Maxwell equation that is equivalent to or includes: (*a*) Electric lines of force end only on electric charges. (*b*) The displacement current. (*c*) Under static conditions, there cannot be any charge inside a conductor. (*d*) A changing electric field must be accompanied by a magnetic field. (*e*) The net magnetic flux through a closed surface is always zero. (*f*) A changing magnetic field must be accompanied by an electric field. (*g*) Magnetic flux lines have no ends. (*h*) The net electric flux through a closed surface is proportional to the total charge inside. (*i*) An electric charge is always accompanied by an electric field. (*j*) There are no magnetic monopoles. (*k*) An electric current is always accompanied by a magnetic field. (*l*) Coulomb's law. (*m*) The electrostatic field is conservative.
Answer: See Table 40-2. (*a*) I. (*b*) IV. (*c*) I. (*d*) IV. (*e*) II. (*f*) III. (*g*) II. (*h*) I. (*i*) I. (*j*) II. (*k*) IV. (*l*) I. (*m*) III.

12. Collect and tabulate expressions for the following four quantities, considering both $r < R$ and $r > R$. Copy down the derivations side by side and study them as interesting applications of Maxwell's equations to problems having cylindrical symmetry. (*a*) $B(r)$ for a current i in a long wire of radius R (see Section 34-2). (*b*) $E(r)$ for a long uniform cylinder of charge of radius R (see Section 28-8; also Problem 27, Chapter 28). (*c*) $B(r)$ for a parallel-plate capacitor, with circular plates of radius R, in which E is changing at a constant rate (see Section 40-2). (*d*) $E(r)$ for a cylindrical region of radius R in which a uniform magnetic field B is changing at a constant rate (see Section 35-5).

13. A long cylindrical conducting rod with radius a is centered on the x-axis as shown in Fig. 40-7. A narrow saw cut is made in the rod at $x = b$. A conduction current i, increasing with time and given by $i = \alpha t$, flows toward the right in the rod; α is a (positive) proportionality constant. At $t = 0$ there is no charge on the cut faces near $x = b$. (*a*) Find the magnitude of the charge on these faces, as a function of time. (*b*) Use Eq. I in Table 40-2 to find E in the gap as a function of time. (*c*) Sketch the lines of **B** for $r < a$, where r is the distance from the x-axis. (*d*) Use Eq. IV in Table 40-2 to find $B(r)$ in the gap for $r < a$. (*e*) Compare the above answer with $B(r)$ in the *rod* for $r < a$.
Answer: (*a*) $\frac{1}{2}\alpha t^2$. (*b*) $\alpha t^2/2\pi\epsilon_0 a^2$. (*c*) As you sight along the axis of the rod from left to right in Fig. 40-7 the lines of **B** form clockwise circles, both in the rod and in the gap. (*d*) $\mu_0 \alpha t r/2\pi a^2$. (*e*) Same as (*d*).

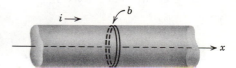

figure 40-7
Problem 13

14. Using the definitions of flux Φ, volume charge density ρ, and current density $\mathbf{j}$, write the four Maxwell equations of Table 40-2 in such a manner that all the fluxes, currents, and charges appear as volume or surface integrals.

SECTION 40-4

15. Assume that the existence of magnetic monopoles is firmly established by experiment. (a) How would you modify the equations of Table 40-2? Let q_m be the expression for the strength of the presumed magnetic monopole, analogous to the basic electric charge e. (b) What SI units would q_m have?

    Answer: (a) $\oint \mathbf{B} \cdot d\mathbf{S} = q_m$, $\oint \mathbf{E} \cdot d\mathbf{l} = -\dfrac{d\Phi_B}{dt} + \mu_0 \dfrac{-dqm}{dt}$; other equations are unchanged. (b) V $\cdot$ s.

16. *A self-consistency property of two of the Maxwell equations* (numbers III and IV in Table 40-2). Two adjacent closed paths $abcda$ and $efcbe$ share the common edge bc as shown in Fig. 40-8. (a) We may apply $\oint \mathbf{E} \cdot d\mathbf{l} = -d\Phi_B/dt$ to each of these closed paths separately. Show that from this alone, $\oint \mathbf{E} \cdot d\mathbf{l} = -d\Phi_B/dt$ is *automatically* satisfied for the composite closed path $abefcda$. (b) Repeat using $\dfrac{1}{\mu_0} \oint \mathbf{B} \cdot d\mathbf{l} = i + \epsilon_0 \dfrac{d\Phi_E}{dt}$.

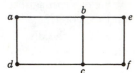

figure 40-8
Problem 16

17. *A self-consistency property of two of the Maxwell equations* (numbers I and II in Table 40-2). Two adjacent parallelepipeds share a common face as shown in Fig. 40-9. (a) We may apply $\epsilon_0 \oint \mathbf{E} \cdot d\mathbf{S} = q$ to each of the two closed surfaces separately. Show that, from this alone, it follows that $\epsilon_0 \oint \mathbf{E} \cdot d\mathbf{S} = q$ is *automatically* satisfied for the composite closed surface. (b) Repeat using $\oint \mathbf{B} \cdot d\mathbf{S} = 0$.

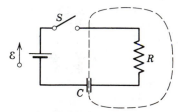

figure 40-9
Problem 17

SECTION 40-5

18. A cylindrical electromagnetic cavity 5.0 cm in diameter and 7.0 cm long is oscillating in the mode shown in Fig. 38-6. (a) Assume that, for points on the axis of the cavity, $E_m = 10^4$ V/m. For such axial points what is the maximum rate $(dE/dt)_m$ at which E changes? (b) Assume that the average value of $(dE/dt)_m$, for all points over a cross section of a cavity, is about one-half the value found above for axial points. On this assumption, what is the maximum value of B at the cylindrical surface of the cavity?

19. in microscopic terms the principle of continuity of current may be expressed as

$$\oint (\mathbf{j} + \mathbf{j}_d) \cdot d\mathbf{S} = 0,$$

in which $\mathbf{j}$ is the conduction current density and $\mathbf{j}_d$ is the displacement current density. The integral is to be taken over any closed surface; the equation essentially says that whatever current flows into the enclosed volume must also flow out. (a) Apply this equation to the surface shown by the dashed lines in Fig. 40-10 shortly after switch S is closed. (b) Apply it to various surfaces that may be drawn in the cavity of Fig. 40-3, including some that cut the cavity walls.

figure 40-10
Problem 19

41
electromagnetic waves

In Chapter 40 we stated without proof that Maxwell's equations predict the existence of electromagnetic waves whose speed in a vacuum is given by Eq. 40-1, or*

$$c = \frac{1}{\sqrt{\mu_0 \epsilon_0}}.$$ (40-1)

We have called c the "speed of light" but, because neither λ nor ν (recall that $c = \lambda\nu$) appears in Eq. 40-1, c is the speed of electromagnetic waves in general, no matter what their wavelengths (λ) or their frequencies (ν).

We can classify waves as standing waves or as traveling waves. In Section 40-5 (see Figs. 40-2 and 40-3) we studied *standing* electromagnetic waves in an electromagnetic resonant cavity. In this chapter we study *traveling* electromagnetic waves, either confined within the conducting walls of a transmission line (Sections 41-4, 5, 6) or in free space (Sections 41-3, 7, 8).

Figure 41-1 suggests the range of the electromagnetic spectrum as we now know it.† From Maxwell's equations we conclude that all these waves have the same nature and speed and that they differ only in frequency, and thus in wavelength. The names attached to the various regions of the spectrum are identified with the experimental techniques

41-1
INTRODUCTION

41-2
THE ELECTROMAGNETIC SPECTRUM

---

* See Section 41-8 and Supplementary Topic VI for a proof of this relationship.
† *Spectrum* is a Latin word that means "ghost" or "apparition." It was first used in this way by Isaac Newton in 1671 to describe the wavering, rainbowlike image formed on the wall of a darkened room when he held a prism in the path of a beam of sunlight entering through a small hole in his window shade.

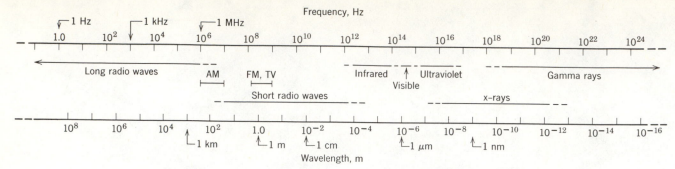

figure 41-1
The electromagnetic spectrum. Note that both scales are logarithmic. The terms μm and nm are *micrometer* and *nanometer,* respectively; both are accented on the first syllable.

for producing and detecting the waves in question. For the AM and the FM-TV bands the frequency ranges are also a matter of legal definition and are sharply defined.

There are no gaps in the spectrum. For example, we can produce electromagnetic waves with $\nu \cong 3 \times 10^{11}$ Hz, either by microwave techniques (microwave oscillators) or by infrared techniques (heated sources). Also, there are no realized upper or lower limits to the frequency or wavelength scales. As an example at one end of the spectrum, electromagnetic waves with $\nu \cong 10^{-2}$ Hz, which corresponds to a period of about 100 s and to a wavelength of about 5000 earth radii, have been detected at the earth's surface.*

It is hard to realize the extent to which we are bathed in electromagnetic waves. The sun is our predominant source, in the sense that its radiations define the environment to which we as a species have adapted.

Let us also consider earth-originated sources of electromagnetic radiation. We are criss-crossed by radio and TV signals. Microwaves from radar systems, from telephone relay systems, and so forth, may reach us. There are electromagnetic waves from light bulbs, from heated engine blocks in automobiles, from X-ray machines, from fireflies, from lightning flashes, from γ-radiation from the radioactive materials in the earth, and so on.

Many electromagnetic waves reach us from extraterrestrial sources and in fact essentially all that we know about the universe comes to us in this way. Electromagnetic waves from outside the earth in the visual range have, of course, been observed since the dawn of the human race. Three technological developments have greatly extended our horizons as far as the study of electromagnetic waves from space is concerned. They are:

The Telescope (see Section 44-6)

This instrument was first used for astronomical observations by Galileo in 1610. With it he (a) discovered the mountains and craters on the moon, which had previously been assumed to be a perfectly spherical body; (b) discovered that the Milky Way was composed of multitudes of individual stars; (c) discovered the four innermost satellites of Jupiter, revealing a model of the Copernican solar system; (d) observed the phases of the planet Venus, an important support for the Copernican theory of the solar system; (e) shed some light on the distinctive character of Saturn (his telescope could not resolve the rings of Saturn but at least he claimed it to differ from other planets) and (f) made a step forward in understanding sunspots. It would be hard to claim, for a single person and in so short a span of years, such major discoveries with astro-

* See "The Longest Electromagnetic Waves" by James R. Heirtzler, *Scientific American,* March 1962.

nomical telescopes. Ground-based optical astronomy reaches its present potential with the construction of the 6.0m (= 236 in.) reflecting telescope in the Soviet Union.*

In October, 1957 the USSR satellite Sputnik I (mass = 83 kg) orbited the earth. From this beginning has developed the major space effort with which we are all familiar. The UHURU satellite, launched by the United States in 1970, initiated the study of X-ray emmission from extraterrestrial objects.† Since that time many other satellites have studied electromagnetic waves from space in this and other regions of the electromagnetic spectrum.

Orbiting Satellites and Other Spacecraft

A Bell Laboratories engineer, Karl G. Jansky,‡ was investigating electromagnetic disturbances to transoceanic telephone traffic when he realized, in 1931, that a source of extraterrestrial signals at radio frequencies was present. This was the beginning of the science of radioastronomy,

The Development of Radioastronomy

figure 41-2
One of eight similar components of the 5-km radio telescope array at the Mullard Radio Astronomy Observatory near Cambridge, England. At a wavelength of 6 cm this array has a resolving power of 2 seconds of arc, comparable with that of large optical telescopes. See "Radio Astronomy and Cosmology" by John B. Irwin, *Sky and Telescope,* December 1976.

---

* See "The Soviet 6-meter Altazimuth Reflector" by Bazart K. Ioannisiani, *Sky and Telescope,* November 1977.
† See "Some Recent Advances in X-Ray Astronomy" by Alan P. Lightman, *Sky and Telescope,* October 1976. Satellites made it possible to study electromagnetic radiations from space which otherwise would be absorbed by the earth's atmosphere.
‡ The unit of energy flux in radioastronomy is called the *jansky* in his honor.

based on high performance radiotelescopes such as the 1000-ft diameter parabolic reflector at Arecibo, Puerto Rico. The radiotelescope array at the Mullard Radio Astronomy Observatory, near Cambridge, England, is another example. This 5-km array, which became operational in 1972, employs four fixed and four movable parabolic reflectors of the type shown in Fig. 41-2.

In this section we consider just four of the important discoveries about the nature of our universe made by studying electromagnetic waves from space. We restrict ourselves to the radio frequency region of the spectrum.

41-3
ELECTROMAGNETIC WAVES FROM SPACE

1. Electromagnetic radiation with $\lambda = 21.1$ cm ($\nu = 1420$ MHz) is emitted by neutral hydrogen atoms* that populate the spaces between the stars that comprise our galaxy.† The density of hydrogen atoms in interstellar space is only about 1 atom/cm³ and the mean time spent by a hydrogen atom capable of radiating before it does so is about 10^7 y. Nevertheless our galaxy is so huge that this electromagnetic radiation is readily detectable.

For hydrogen atoms at rest with respect to the terrestrial detector of their radiations the precise wavelength is 21.1061 cm ($\nu = 1420.406$ MHz). Our galaxy, however, is not a rigid structure, its outer regions move with respect to the inner core. Just as for sound, we can measure the radial velocities of light-emitting sources by the Doppler effect; see Sections 20-7 and 42-5. By observing these frequency shifts we can learn a lot about the structure and the internal motions of our galaxy. In recent years electromagnetic radiation from molecules in the interstellar regions of our galaxy has been detected. We may list formaldehyde (HCHO), ammonia (NH_3), and carbon monoxide (CO).

2. Quasi-stellar radio sources (*quasars*) were discovered in 1962. These are optical objects associated normally with large radioemissions. What makes them so interesting is that they have very large "red shifts" in their spectra which, with our present understanding of astrophysics, would lead us to believe that they are at enormous distances from us, well outside our own galaxy. On the other hand, if they are really so far away, they must have enormous energy outputs for which we cannot account. Their nature is under active study today.

3. The primeval fireball radiation.‡ Most astrophysicists today believe that the universe originated about 7×10^9 years ago in the explosion of a very highly concentrated matter-radiation complex. This is the so-called "Big Bang" theory of the origin of the universe. The debris from this explosion expanded outward uniformly in all directions (hence the expanding universe concept) and the matter gradually condensed to form galaxies, stars, and planets as we know them today.

In 1965 the Princeton University physicist Robert Dicke recognized that, if the remnants of the electromagnetic radiation from this primeval fireball, expected to be detectable by radiotelescopes, could be found, it would provide impressive evidence in support of the Big-Bang theory.

* See "Radio Emission from Interstellar Neutral Hydrogen," by R. D. Davies, *Contemporary Physics,* August 1961.
† Our galaxy is a disc-shaped cluster of stars about 10^5 light years in diameter. The Milky Way defines the plane of the disc. Our sun is located about halfway from the galactic center to the galactic edge. There are many other galaxies in our universe; every new high-performance telescope reveals more.
‡ See "The Primeval Fireball" by P. J. E. Peebles and D. T. Wilkinson, *Scientific American,* June 1976.

Unknown to Dicke two Bell Laboratory scientists, A. Penzias and R. Wilson, had actually discovered this electromagnetic radiation from space without realizing its cosmological significance. They made their discovery during engineering studies of communication satellite problems. Within a few years the reality of the primeval fireball radiation was well established.

This episode reminds us that Jansky, also of the Bell Laboratories, founded the science of radio astronomy in 1931. In each case a strictly engineering endeavor led to unexpected insights into our understanding of the universe.*

4. In 1968 a group of British radio astronomers at Cambridge University, headed by Antony Hewish, discovered the first of a series of objects that emit strong short bursts of electromagnetic radiation separated by the astronomically incredibly short time period of about one second. These objects were called *pulsars*.†

It is now generally agreed that pulsars are ordinary stars in which, during the process of shrinking under their own attractive gravitational forces, the electrons of the star combine with the protons, leaving a compact ball of neutrons. Typical *neutron stars* (see Problems 4-23 and 16-5) have these properties: (a) radii of about 10 km (the solar radius is about 7×10^5 km), (b) densities of about 10^{16} kg/m³ (the mean solar density is about 10^3 kg/m³), and (c) periods of rotation of about 10^{-2} s (the solar rotation period is about 10^6 s). The observed pulses seem to arise from a radio-emitting spot on the surface of the neutron star. The emitted electromagnetic rotation sweeps by a terrestrial observer as the star rotates; the analogy to the revolving beam from a lighthouse beacon is clear.

41-4
TRANSMISSION LINES

In this section we turn our attention from the passive observation of electromagnetic waves that reach us to the deliberate generation of such waves and their transmission from point to point by a *transmission line*. Figure 41-3 shows one type of line, a *coaxial cable*, its input end being connected to a switch *S*. For the time being we assume that the cable is infinitely long and that the cable elements have zero resistance. (The Atlantic cable, first laid successfully in 1866, is a coaxial cable, although it is far from resistanceless.)

When switch *S* is closed on *b*, the central and the outer conductors are at the same potential. If we then throw the switch to *a*, a potential difference *V* suddenly appears between these elements. This potential

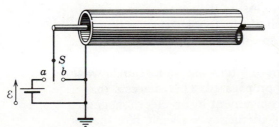

figure 41-3
An electromagnetic signal can be sent along the transmission line (coaxial cable) by throwing switch *S* from *b* to *a*.

* See "The Roots of Solid-State Research at Bell Labs" by Lillian Hartmann Hoddeson, *Physics Today*, March 1977, for a fascinating historical account of the interaction of science and technology in a particular industrial laboratory. In 1937 and 1956 Bell Laboratory scientists were awarded Nobel Prizes.
† See "Pulsars" by A. Hewish, *Scientific American*, October 1968, and "The Nature of Pulsars" by J. P. Ostriker, *Scientific American*, January 1971. Hewish received a shared Nobel Prize in 1974 for this discovery.

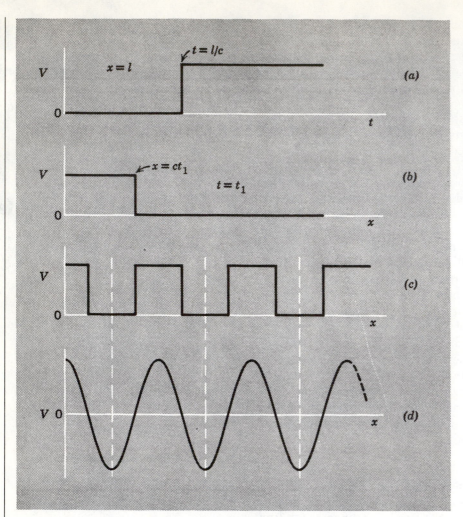

figure 41-4
(*a*) The variation with time of the potential difference between the conductors of a coaxial cable at a distance *l* from the input end. (*b*) An instantaneous "snapshot" of the pulse in the cable at a certain time t_1. (*c*) The waveform if switch *S* in Fig. 41-3 is periodically thrown between *a* and *b*. (*d*) The waveform if we replace switch *S* by an electromagnetic oscillator with a sinusoidal output.

difference does not appear instantaneously all along the line but is propagated with a finite speed *c* that will turn out to be exactly that of light, assuming a resistanceless line. Figure 41-4*a* shows that the potential difference between the conductors at a distance *l* along the line suddenly rises, at a time given by $t = l/c$, from zero to a value determined by the battery emf. We can also consider the variation of *V* with position *x* along the line at a given time t_1 after closing the switch. Figure 41-4*b* shows such an instantaneous "snapshot." It, too, suggests a traveling "wavefront" moving along the line at speed *c*. At $t = t_1$ the signal has not yet reached points where $x > ct_1$.

If we throw switch *S* periodically from *b* to *a* and back again, a wave disturbance like that of Fig. 41-4*c* is propagated. This suggests that if we replace the battery and switch arrangement by an electromagnetic oscillator with a sinusoidal output of frequency *ν* a wave like that of Fig. 41-4*d* will be propagated.

A traveling wave in a resistanceless transmission line will exhibit a wavelength λ given by

$$\lambda = \frac{c}{\nu}.$$

If the oscillator frequency is 60 Hz, the common commercial power frequency, the wavelength is 5×10^6 m, which is about 3000 miles. At this low frequency traveling waves are not apparent in any line of normal length. By the time the polarity of the oscillator has changed appreciably, the energy fed into the line at the oscillator end has been delivered to the load.

Frequencies in the radio or the microwave range are much higher and the wavelengths correspondingly smaller. Commercial television frequencies as established by the Federal Communications Commission range from 54 to 980 MHz. In terms of wavelength this is a range of 5.6 to 0.31 m. At these wavelengths the patterns of potential difference in the transmission lines used to send television signals across the country can be described aptly as traveling waves. Microwaves, used in radar systems and for communication purposes, have even smaller wavelengths, in the range of about 20 cm to about 0.5 mm.

These considerations suggest another way of viewing the difference between lumped and distributed circuit elements. A system is "distributed" if the wavelength is about the same size as, or less than, the dimensions of the system. If the wavelength is much larger than the dimensions of the system, we are dealing with lumped components. A transmission line 50 m long would be a lumped system for electromagnetic radiation at 60 Hz $(\lambda = 5 \times 10^6$ m$)$ but a distributed system at 100 MHz $(\lambda = 3$ m$)$. In a lumped system the circuit analysis is normally carried out in terms of lumped system parameters such as L, C, and R; in a distributed system the analysis is often carried out in terms of the fields that are set up and the charges and currents that are related to them.

EXAMPLE 1

A potential difference given by

$$V_0 = V_m \sin \omega t$$

is applied between the terminals of a long resistanceless transmission line; the frequency $\nu (= \omega / 2\pi)$ is 3×10^9 Hz $(=3$ GHz; see Table 1-2$)$. Write an equation for $V(t)$ at a point P which is 1.5 wavelengths down the line from the oscillator.

The general equation for a wave traveling in the x direction (see Eq. 19-10a) is

$$V = V_m \sin (\omega t - kx),$$

where $k (= 2\pi / \lambda)$ is the wave number. At $x = 0$ this gives correctly the time variation of the input terminal potential difference. At $x = 1.5\lambda$ we have

$$V_P = V_m \sin \left[\omega t - \left(\frac{2\pi}{\lambda} \right) (1.5\lambda) \right] = V_m \sin (\omega t - 3\pi)$$

$$= -V_m \sin \omega t.$$

Thus V_P is always equal in magnitude to V_0 but is opposite in sign. What is the wavelength in this example?

41-5
COAXIAL CABLE— FIELDS AND CURRENTS

Figures 41-5a and 41-5b are "snapshots" of the electric and magnetic field configurations in a coaxial cable. The electric field is radial and the magnetic field forms concentric lines about the central conductor. The entire pattern moves along the line, assumed resistanceless, at speed c.

The field patterns in this figure obey the *boundary condition* required for a line that is assumed to be resistanceless, namely, that **E** for all points on either conducting surface has no tangential component (see Example 7, Chapter 28). We can find the field patterns mathematically from Maxwell's equations by imposing this requirement. The configuration

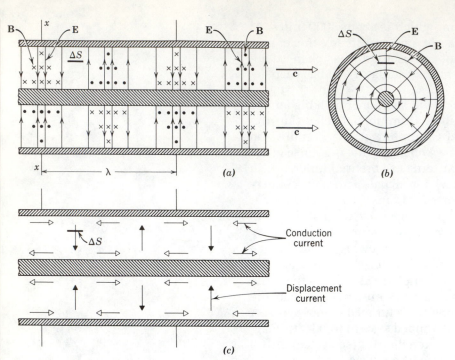

figure 41-5
(a) The electric and magnetic fields in a coaxial cable, showing a wave traveling to the right at speed c.
(b) A cross-sectional view at a plane through xx in (a); the wave is emerging from the page. (c) Conduction currents in the walls (open arrows) and displacement currents in the space between the conductors (filled arrows) associated with the wave in (a); the arrows in each case represent current density vectors.

shown is the simplest of many different wave patterns that can travel along the line. The coaxial cable, unlike the electromagnetic cavity of Fig. 38-6, is not a resonant device. The angular frequency ω of waves that travel along it can be varied continuously, as is the case for all traveling waves, such as transverse waves in a long stretched cord.

Figure 41-5c shows the currents in the cable at the instant corresponding to Figs. 41-5a and 41-5b. The arrows parallel to the cable axis represent conduction currents in the central and the outer conductors. The vertical arrows with filled heads represent displacement currents that exist in the space between the conductors. Note that the conduction current and the displacement current arrows form closed loops, preserving the concept of the continuity of current.

EXAMPLE 2

Verify that the displacement current represented in Fig. 41-5c is consistent with the pattern of **B** and **E** shown in Fig. 41-5a.

Consider a small surface element Δ**S** shown edge-on in Fig. 41-5; it is shown as viewed from above in Fig. 41-6. This hypothetical element is stationary with respect to the cable while the field configuration moves through it at speed c. Figure 41-6a shows the electric lines of force in and near this element. It is clear from symmetry that, at the instant shown, the net flux Φ_E through this area is zero. However, even though Φ_E is zero in magnitude, it is, at this instant, *changing at its most rapid rate,* since **E** at the element ΔS is at the very moment of reversing its direction as the wave moves through. Thus the displacement current, which is given by

$$i_d = \epsilon_0 \frac{d\Phi_E}{dt},$$

also has its maximum value.

Figure 41-6b shows the magnetic field in the vicinity of the element of area. Let us apply the generalized form of Ampère's law,

$$\oint \mathbf{B} \cdot d\mathbf{l} = \mu_0(i + i_d),$$

to the element. The conduction current i is zero since no charge is transported

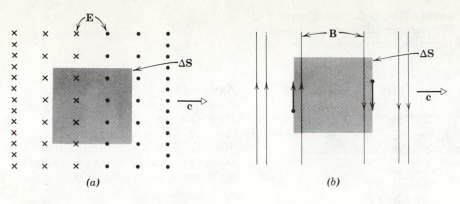

(a)　　　　　　　　　　　　(b)

figure 41-6
The area element S in Fig. 41-5 enlarged and viewed from above, showing the adjacent (a) electric and (b) magnetic fields.

through ΔS. The displacement current i_d is not zero, having, in fact, its maximum value. Thus, since the right side of this equation $(=\mu_0 i_d)$ does not vanish, the left side must not vanish. Study of Fig. 41-6b shows that $\oint \mathbf{B} \cdot d\mathbf{l}$ around the boundary of this square has, indeed, a nonzero value. Thus the field and displacement current configurations of Fig. 41-5a–c are consistent. We have not discussed the direction of i_d; that is, does it point into the plane of Fig. 41-6, as Fig. 41-5c asserts, or out of it? We leave this as a question. You may be guided by considering the direction of the displacement current in Figs. 40-1a and 40-1b.

EXAMPLE 3

Show that the conduction currents in Fig. 41-5c are appropriately related to the magnetic field pattern.

Figure 41-7 shows a cross section of the cable at a plane through xx in Fig. 41-5a. Let us apply Ampère's law,

$$\oint \mathbf{B} \cdot d\mathbf{l} = \mu_0(i + i_d),$$

to the ring of radius r. The displacement current through this ring is zero, this current being at right angles to the central conductor. The conduction current i in the central conductor, shown by the $\times$'s in the figure, does pass through this ring so that the equation becomes

$$(B)(2\pi r) = \mu_0 i,$$

or

$$B = \frac{\mu_0 i}{2\pi r}.$$

Note that $\mathbf{B}$ is related to the current in the central conductor by the usual right-hand rule and that the expression for B is that found earlier (Eq. 34-4) for a long straight wire carrying a steady current.

If we apply Ampère's law to the large ring of radius r', we must put the *net* conduction current equal to zero because the current in the outer conductor, shown by the dots, is equal and opposite to that in the inner conductor. This means that $\mathbf{B}$ must be zero for points outside the cable, in agreement with experiment.

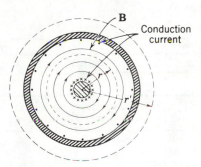

figure 41-7
Example 3. The coaxial cable of Fig. 41-5b, showing the conduction currents in the central and outer conductors. The wave is emerging from the page.

It is interesting to compare the electromagnetic oscillations in a typical *traveling* wave, such as that of Fig. 41-5, with those in a cavity resonator, such as that of Fig. 38-6. The latter oscillations are an electromagnetic *standing* wave. In a traveling wave $\mathbf{E}$ and $\mathbf{B}$ are in phase, which means that at a given position along the transmission line they reach their maxima at the same time. However, Fig. 38-6 shows that at a given position in a cavity resonator $\mathbf{E}$ and $\mathbf{B}$ reach their maxima one-fourth of a cycle apart; they are 90° out of phase.

A complete analogy exists in mechanical systems. In the acoustic resonator of Fig. 38-5 the time variations of pressure and velocity for the standing acoustic wave are also 90° out of phase, in exact correspondence to the electromagnetic cavity oscillations of Fig. 38-6. The acoustic analogy to a transmission line (Fig. 41-8) would be an infinitely long gas-filled tube, one end being connected to an acoustic oscillator such as a loud-speaker. The entire configuration of Fig. 41-8 moves to the right with speed v. The pressure variations, suggested by the dots, and the instantaneous velocities, suggested by the arrows, are *in phase*, just as are **E** and **B** in the coaxial cable of Fig. 41-5.

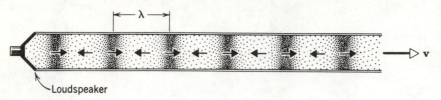

Loudspeaker

figure 41-8
An acoustic transmission line, showing a sound wave traveling to the right. The small arrows with filled heads show the directed drift velocities for small volume elements of the gas. Compare Fig. 41-5.

It is possible to send electromagnetic waves through a hollow metal pipe that has no central conductor. We assume that the inner walls of such a pipe, or *waveguide* as it is called, are resistanceless and that the cross section is rectangular.

Figure 41-9 shows a typical electric and magnetic field pattern. We imagine that a microwave oscillator is connected to the left end and sends electromagnetic energy down the guide. Figure 41-9(*i*) shows a side view of the guide and Fig. 41-9(*ii*), a top view; Fig. 41-9(*iii*) shows the cross section. As for the coaxial cable, the field patterns are such that **E** has no tangential component for any point on the inner surface of the guide. The fields **E** and **B** are in phase, again like the coaxial cable.

41-6
WAVEGUIDE

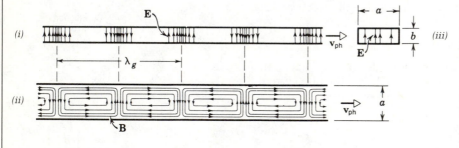

figure 41-9
A waveguide, showing (*i*) a side view of the lines of **E**, (*ii*) a top view of the lines of **B**, and (*iii*) a cross-sectional view of the lines of **E**. In (*iii*) the wave is emerging from the page. For simplicity the lines of **B** are not shown in (*i*) and (*iii*), nor are the lines of **E** shown in (*ii*).

As for all traveling waves, the angular frequency ω of electromagnetic waves traveling down a guide can be varied continuously. In a waveguide of given dimensions, however, there exists, for every mode of transmission, that is, for every pattern of **E** and **B,** a so-called *cutoff frequency* ω_0. A given guide will not transmit waves in a given mode if their frequency is below the cutoff value for that mode in that guide. The field patterns of Fig. 41-9 show the *dominant mode* for a rectangular guide; this is the mode with the lowest cutoff frequency. Given the frequency ω of electromagnetic waves to be transmitted, it is common practice to select a guide whose dimensions are such that ω is larger than the cutoff frequency ω_0 for the dominant mode but smaller than the cutoff frequencies of all other modes. Under these conditions the dominant mode of propagation is the only one possible.

In a (resistanceless) coaxial cable the wave patterns travel at speed c. In the acoustic transmission line of Fig. 41-8 (assumed "resistanceless") the waves also travel at a speed v, which is the same as the propagation speed in an infinite medium. In a waveguide, however, the speed is *not c*. In waveguides we must distinguish between (*a*) the *phase speed* $v_{\rm ph}$, which is the speed at which the wave patterns of Fig. 41-9 travel, and (*b*) the *group speed* $v_{\rm gr}$, which is the speed at which electromagnetic energy or information-carrying "signals" travel along the guide. These speeds, which are identical for electromagnetic waves in a coaxial cable and for acoustic waves in a tube, are different for waves in a waveguide.

The phase speed is not directly measurable. The wave pattern is a repetitive structure, and there is no way to distinguish one wave maximum from another. We can observe the waves entering one end of the guide and leaving at the other, but there is no way to identify a particular wave maximum so that we can time its passage down the guide. We can put a "signal" on the wave by increasing the power level of the oscillator for a short time. We can time this power pulse as it passes through the guide, but there is no guarantee that it travels at the same speed as the wave pattern and, indeed, it does not. The speed of such signals or markers is the speed at which *energy* is propagated, that is, the group speed.

From Maxwell's equations we can show that the phase speed and the group speed for the mode of Fig. 41-9 are

$$v_{ph} = \frac{c}{\sqrt{1 - \left(\frac{\lambda}{2a}\right)^2}} \qquad (41\text{-}1)$$

and

$$v_{gr} = c\sqrt{1 - \left(\frac{\lambda}{2a}\right)^2}, \qquad (41\text{-}2)$$

in which *a* is the width of the guide and λ the free-space wavelength. Note that as $a \to \infty$, which corresponds to free-space conditions, $v_{ph} = v_{gr} = c$.

The phase speed v_{ph} is *greater* than the velocity of light, the group speed v_{gr} being correspondingly less. In relativity theory we learn that no speed at which signals or energy travel can be faster than that of light. However, signals or energy *cannot* be transmitted down a guide at the phase speed v_{ph}; they travel with speed v_{gr} which is *always* less than *c* so there is no conflict with the theory of relativity.

The wavelength λ in Eqs. 41-1 and 41-2 is the wavelength that would be measured for the oscillations in free space, that is,

$$\lambda = \frac{c}{\nu}, \qquad (41\text{-}3)$$

where *c* is the speed in free space and *ν* is the frequency. For waves of a given frequency, the wavelength exhibited in a guide (λ_g) must differ from the free-space wavelength λ because the speed v_{ph} has changed. The so-called *guide wavelength* λ_g is given by

$$\lambda_g = \frac{v_{ph}}{\nu} = \frac{v_{ph}}{c/\lambda} = \lambda\,\frac{v_{ph}}{c}.$$

From Eq. 41-1 this yields

$$\lambda_g = \frac{\lambda}{\sqrt{1 - \left(\frac{\lambda}{2a}\right)^2}}. \qquad (41\text{-}4)$$

Thus the guide wavelength, which is the wavelength exhibited by the field patterns in Fig. 41-9, is larger than the free-space wavelength.

EXAMPLE 4

What must be the width *a* of a rectangular guide such that the energy of electromagnetic radiation whose free-space wavelength is 3.0 cm travels down the guide (*a*) at 95% of the speed of light? (*b*) At 50% of the speed of light?

From Eq. 41-2 we have

$$v_{gr} = 0.95c = c\sqrt{1 - \left(\frac{\lambda}{2a}\right)^2}.$$

Solving for *a* yields $a = 4.8$ cm; repeating for $v_{gr} = 0.50c$ yields $a = 1.7$ cm.

This illustrates the cutoff phenomenon described above. If $\lambda = 2a$, then $v_{gr} = 0$ and energy cannot travel down the guide. For the radiation considered in this example $\lambda = 3.0$ cm, so that the guide must have a width *a* of *at least* $\frac{1}{2} \times 3.0$ cm $= 1.5$ cm if it is to transmit this wave. The guide whose width we calculated in (*a*) above can transmit radiations whose free-space wavelength is 2×4.8 cm $= 9.6$ cm *or less*.

The acoustic transmission line of Fig. 41-8 cannot be infinitely long. Its far end may be sealed by a solid cap or left open, or it may have a flange, a horn, or some similar device mounted on it. If the far end is not sealed, energy will escape into the medium beyond. We call this *acoustic radiation*. In general, some energy will also be reflected back down the transmission line. If acoustic radiation is desirable, the designer's task is to fashion a termination (that is, an "acoustic antenna") for the transmission line such that the smallest possible fraction of the incident energy will be reflected back down the line. Such a termination might take the form of a flared horn. Acoustic radiation, of course, requires a medium such as air in order to be propagated.

An electromagnetic transmission line such as a coaxial cable or a waveguide can also be terminated in many ways, and energy can escape from the end of the line into the space beyond. In contrast to sound waves, a physical medium is not required. Thus electromagnetic energy can be radiated from the end of the transmission line, to form a traveling electromagnetic wave in free space.

Figure 41-10 shows an effective termination for a coaxial cable; it consists of two wires arranged as shown and is called an *electric dipole antenna*. The potential difference between the two conductors alternates sinusoidally as the wave reaches them, the effect being that of an electric dipole whose dipole moment **p** varies with time, both in magnitude and direction.

figure 41-10
An electric dipole antenna on the end of a coaxial cable.

figure 41-11
Showing the fields of **E** and **B** radiated away from an oscillating electric dipole. We show only the fields at distances from the dipole that are large compared with its dimensions. The electromagnetic wave sweeping through distant point P is a plane wave, moving in the x-direction.

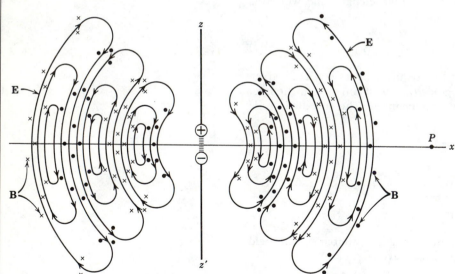

Figure 41-11 shows an electromagnetic wave generated by such an oscillating electric dipole. The figure is a section through a figure of revolution about the dipole axis zz' and the wave moves out in any direction from the dipole with speed c. The fields shown are those that exist at distances from the dipole which are large compared with the dipole dimensions. We call this the *radiation field* and it is the focus of our concern here. The fields close to the dipole (the *near field*) are more complex but they have no interest for us here because they die out rapidly and we want to study only the fields at large distances.

Figure 41-12 shows eight cyclical "snapshots" of the fields of **E** and **B** sweeping past an observer at point P in Fig. 41-11. We assume that P is so far distant from the oscillating dipole that the wave fronts passing

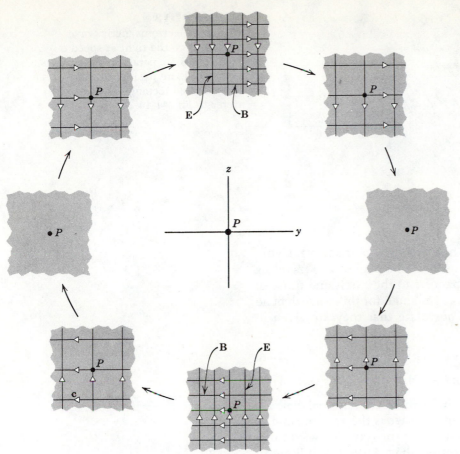

figure 41-12
Showing eight cyclical "snapshots" of the plane electromagnetic wave radiated from the oscillating dipole of Fig. 41-11 through point P. The direction of the wave (the x-direction) is out of the plane of the figure.

through and near P describe a plane wave; see Section 19-2. The speed c of the wave in free space is given by $c = \nu\lambda$, which we can write as

$$c = \frac{\omega}{k}, \qquad (41\text{-}5)$$

where ω, the angular frequency, and k, the wave number, are related to the frequency ν and the wavelength λ by

$$\omega = 2\pi\nu \quad \text{and} \quad k = 2\pi/\lambda.$$

41-8
TRAVELING WAVES AND MAXWELL'S EQUATIONS

In earlier sections we have postulated the existence of certain magnetic and electric field distributions, in resonant cavities, coaxial cables, and waveguides, and we have shown that these postulated distributions are consistent with Maxwell's equations, as are the distributions of conduction and displacement currents associated with the fields. If you pursue your studies of electromagnetism, you will learn how to derive mathematical expressions for $\mathbf{E}$ and $\mathbf{B}$ by subjecting Maxwell's equations to the boundary conditions appropriate to the problem at hand. In this section we continue our program by showing that the postulated patterns of $\mathbf{E}$ and $\mathbf{B}$ for a traveling electromagnetic wave are completely consistent with Maxwell's equations. In doing so, we will be able to show that the speed of such waves in free space is that of visible light and thus that visible light is itself an electromagnetic wave.

If the observer at P in Fig. 41-11 is, as we have assumed, at a considerable distance from the source, the *wavefronts* described by the electric and magnetic fields that reach him (see Fig. 41-12) will be

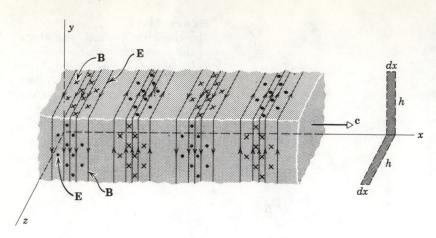

figure 41-13
A plane electromagnetic wave
traveling to the right at speed c.
Lines of **B** are parallel to the y axis;
those of **E** are parallel to the z axis.
The shaded rectangles on the right
refer to Fig. 41-14.

planes and the wave that moves past him will be a *plane wave*. Figure
41-13 shows a three-dimensional "snapshot" of a plane wave traveling
in the x direction. The lines of **E** are parallel to the z axis and those of
B are parallel to the y axis. The values of **B** and **E** for this wave depend
only on x and t (not on y or z). We postulate that they are given in
magnitude by

$$B = B_m \sin (kx - \omega t) \qquad (41\text{-}6)$$

and $\qquad\qquad E = E_m \sin (kx - \omega t). \qquad (41\text{-}7)$

Figure 41-14 shows two sections through the three-dimensional diagram
of Fig. 41-13. In Fig. 41-14a the plane of the page is the xz plane and in
Fig. 41-14b it is the xy plane. Note that, as for the traveling waves in a
coaxial cable (Fig. 41-5a) and in a waveguide (Fig. 41-9), **E** and **B** are in
phase, that is, at any point through which the wave is moving they
reach their maximum values at the same time.

The shaded rectangle of dimensions dx and h in Fig. 41-14a is fixed in
space. As the wave passes over it, the magnetic flux Φ_B through the rectangle will change, which will give rise to induced electric fields around

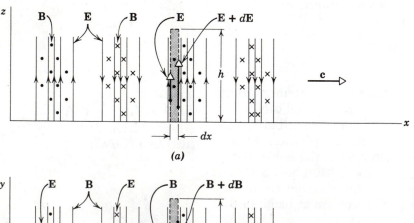

(a)

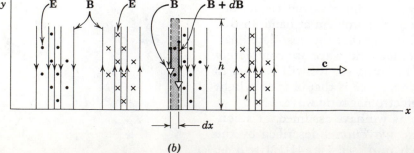

(b)

figure 41-14
The wave of Fig. 41-13 viewed (a) in
the xz plane and (b) in the xy plane.

the rectangle, according to Faraday's law of induction. These induced electric fields are, in fact, simply the electric component of the traveling wave.

Let us apply Lenz's law to this induction process. The flux Φ_B for the shaded rectangle of Fig. 41-14a is *decreasing* with time because the wave is moving through the rectangle to the right and a region of weaker magnetic fields is moving into the rectangle. The induced field will act to oppose this change, which means that if we imagine that the boundary of the rectangle is a conducting loop, a *counterclockwise* induced current would appear in it. This current would produce a field of **B** that, within the rectangle, would point out of the page, thus opposing the decrease in Φ_B. There is, of course, no conducting loop, but the net induced electric field **E** does indeed act counterclockwise around the rectangle because $E + dE$, the magnitude of **E** at the right edge of the rectangle, is greater than E, the magnitude of **E** at the left edge. Thus the electric field configuration is entirely consistent with the concept that it is induced by the changing magnetic field.

For a more detailed analysis let us apply Faraday's law of induction, or

$$\oint \mathbf{E} \cdot d\mathbf{l} = -\frac{d\Phi_B}{dt}, \qquad (41\text{-}8)$$

going counterclockwise around the shaded rectangle of Fig. 41-14a. There is no contribution to the integral from the top or bottom of the rectangle because **E** and $d\mathbf{l}$ are at right angles here. The integral then becomes

$$\oint \mathbf{E} \cdot d\mathbf{l} = [(E + dE)(h)] - [(E)(h)] = dE\,h.$$

The flux Φ_B for the rectangle is

$$\Phi_B = (B)(dx\,h),$$

where B is the magnitude of **B** at the rectangular strip and $dx\,h$ is the area of the strip. Differentiating gives

$$\frac{d\Phi_B}{dt} = h\,dx\,\frac{dB}{dt}.$$

From Eq. 41-8 we then have

$$dE\,h = -h\,dx\,\frac{dB}{dt},$$

or

$$\frac{dE}{dx} = -\frac{dB}{dt}. \qquad (41\text{-}9)$$

Actually, both B and E are functions of x and t; see Eqs. 41-6 and 41-7. In evaluating dE/dx, we assume that t is constant because Fig. 41-14a is an "instantaneous snapshot." Also, in evaluating dB/dt we assume that x is constant since what is required is the time rate of change of B at a particular place, the strip in Fig. 41-14a. The derivatives under these circumstances are called *partial derivatives*, and a somewhat different notation is used for them; see Example 2, p. 414 and p. 636. In this notation Eq. 41-9 becomes

$$\frac{\partial E}{\partial x} = -\frac{\partial B}{\partial t}. \qquad (41\text{-}10)$$

The minus sign in this equation is appropriate and necessary, for, although E is increasing with x at the site of the shaded rectangle in Fig. 41-14a, B is decreasing with t. Since $E(x,t)$ and $B(x,t)$ are known (see

Eqs. 41-6 and 41-7), Eq. 41-10 reduces to

$$kE_m \cos(kx - \omega t) = \omega B_m \cos(kx - \omega t),$$

or (see Eq. 41-5)

$$\frac{\omega}{k} = \frac{E_m}{B_m} = c. \tag{41-11a}$$

Thus the speed of the wave c is the ratio of the amplitudes of the electric and the magnetic components of the wave. From Eqs. 41-6 and 41-7 we see that the ratio of amplitudes is the same as the ratio of instantaneous values, or

$$E = cB. \tag{41-11b}$$

This important result will be useful in later sections.

We now turn our attention to Fig. 41-14b, in which the flux Φ_E for the shaded rectangle is decreasing with time as the wave moves through it. According to Equation IV, Table 40-2 (with $i = 0$, because there are no conduction currents in a traveling electromagnetic wave),

$$\oint \mathbf{B} \cdot d\mathbf{l} = \mu_0 \epsilon_0 \frac{d\Phi_E}{dt}, \tag{41-12}$$

this changing flux will induce a magnetic field at points around the periphery of the rectangle. This induced magnetic field is simply the magnetic component of the electromagnetic wave. Thus, as in the cavity resonator of Section 38-5, the electric and the magnetic components of the wave are intimately connected with each other, each depending on the time rate of change of the other.

Comparison of the shaded rectangles in Fig. 41-14 shows that for each the appropriate flux, Φ_B or Φ_E, is *decreasing* with time. However, if we proceed counterclockwise around the upper and lower shaded rectangles, we see that $\oint \mathbf{E} \cdot d\mathbf{l}$ is *positive*, whereas $\oint \mathbf{B} \cdot d\mathbf{l}$ is *negative*. This is as it should be. If you compare Figs. 35-10 and 40-1a, you will be reminded that although the fluxes Φ_B and Φ_E in those figures are changing with time in the same way (both are increasing) the lines of the induced fields, **E** and **B**, respectively, circulate in opposite directions.

The integral in Eq. 41-12, evaluated by proceeding counterclockwise around the shaded rectangle of Fig. 41-14b, is

$$\oint \mathbf{B} \cdot d\mathbf{l} = [-(B + dB)(h)] + [(B)(h)] = -h\,dB,$$

where B is the magnitude of **B** at the left edge of the strip and $B + dB$ is its magnitude at the right edge.

The flux Φ_E through the rectangle of Fig. 41-14b is

$$\Phi_E = (E)(h\,dx).$$

Differentiating gives

$$\frac{d\Phi_E}{dt} = h\,dx\,\frac{dE}{dt}.$$

Thus we can write Eq. 41-12 as

$$-h\,dB = \mu_0 \epsilon_0 \left(h\,dx\,\frac{dE}{dt} \right)$$

or, substituting partial derivatives,

$$-\frac{\partial B}{\partial x} = \mu_0 \epsilon_0 \frac{\partial E}{\partial t}. \tag{41-13}$$

Again, the minus sign in this equation is appropriate and necessary, for, although B is increasing with x at the site of the shaded rectangle in Fig. 41-14b, E is decreasing with t.

Combining this equation with Eqs. 41-6 and 41-7 yields

$$-kB_m \cos(kx - \omega t) = -\mu_0 \epsilon_0 \omega E_m \cos(kx - \omega t),$$

or (see Eq. 41-5)

$$\frac{E_m}{B_m} = \frac{k}{\mu_0 \epsilon_0 \omega} = \frac{1}{\mu_0 \epsilon_0 c}. \qquad (41\text{-}14)$$

Eliminating E_m/B_m between Eqs. 41-11a and 41-14 yields

$$c = \frac{1}{\sqrt{\mu_0 \epsilon_0}}. \qquad (41\text{-}15)$$

Substituting numerical values yields

$$c = \frac{1}{\sqrt{(4\pi \times 10^{-7}\ \text{T}\cdot\text{m/A})(8.9 \times 10^{-12}\ \text{C}^2/\text{N}\cdot\text{m}^2)}}$$

$$= 3.0 \times 10^8\ \text{m/s}, \qquad (41\text{-}16)$$

which is the speed of light in free space! This emergence of the speed of light from purely electromagnetic considerations is the crowning achievement of Maxwell's electromagnetic theory. Maxwell made this prediction before radio waves were known and before it was realized that light was electromagnetic in nature. His prediction led to the concept of the electromagnetic spectrum and to the discovery of radio waves by Heinrich Hertz in 1890. It made it possible to discuss optics as a branch of electromagnetism and to derive its fundamental laws from Maxwell's equations.

A conclusion as basic as Eq. 41-15 must be subject to rigorous experimental testing and indeed this has been done. Of the three quantities in this equation, one, μ_0, has an exact assigned value of $4\pi \times 10^{-7}$ T·m/A. The other two, c and ϵ_0, are measurable quantities. Our confidence in Maxwell's equations, bolstered by many successful predictions and agreements with experiment, is such that we now use the measured speed of light (2.99792458×10^8 m/s; see Appendix B) to determine ϵ_0, by way of Eq. 41-15. This gives a much better value than we could measure by the methods of Section 30-2.

Curiously enough Maxwell himself did not view the propagation of electromagnetic waves and electromagnetic phenomena in general, in anything like the terms suggested by, say, Fig. 41-13. Like all physicists of his day he believed firmly that space was permeated by a subtle substance called the *ether* and that electromagnetic phenomena could be accounted for in terms of rotating vortices in this aether.

It is a tribute to Maxwell's genius that, with such mechanical models in his mind, he was able to deduce the laws of electromagnetism that bear his name. These laws, as we have pointed out, not only required no change when Einstein's special theory of relativity came on the scene almost half-century later but, indeed, were strongly confirmed by that theory. Today we no longer believe in the aether concept; see Supplementary Topic V, "Special Relativity—A Summary of Conclusions."

41-9
THE POYNTING VECTOR

One of the important characteristics of an electromagnetic wave is that it can transport energy from point to point. As we show below, we can describe the rate of energy flow per unit area in a plane electromagnetic wave by a vector $\mathbf{S}$, called the *Poynting vector* after John Henry Poynting (1852–1914), who first pointed out its properties. We define $\mathbf{S}$ from

$$\mathbf{S} = \frac{1}{\mu_0} \mathbf{E} \times \mathbf{B}. \qquad (41\text{-}17)$$

In SI units **S** is expressed in watts/meter²; the direction of **S** gives the direction in which the energy moves. The vectors **E** and **B** refer to their instantaneous values at the point in question. If we apply Eq. 41-17 to the traveling plane electromagnetic wave of Fig. 41-13, it is clear that **E** × **B**, hence **S**, point in the direction of propagation. Note, too, that **S** points parallel to the axis for all points in the coaxial cable of Fig. 41-5.

We get meaningful results if we extend the Poynting vector concept to other electromagnetic situations involving either traveling or standing electromagnetic waves, as we will see in Examples 5 and 6. If we extend it to circuit situations involving steady or almost steady currents and lumped circuit elements, we are led to some interesting conclusions, which we explore in Problems 28, 31, and 32.

EXAMPLE 5

Analyze energy flow in the cavity of Fig. 38-6, using the Poynting vector.

Study of Fig. 41-15 shows that when the energy is all electric (Figs. 41-15a and 41-15e) it is concentrated along the axis, because this is the region in which **E** has its maximum value. When the energy is all magnetic (Figs. 41-15c and 41-15g), it is concentrated near the walls. Thus the energy surges back and forth periodically between the central region and the region near the walls. The figure

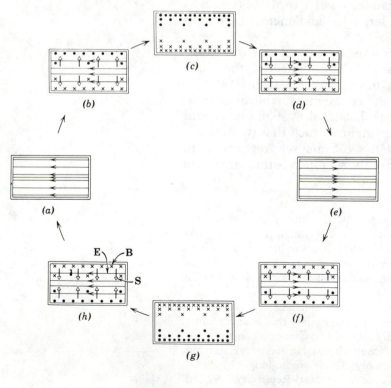

figure 41-15
Example 5. Energy surges back and forth periodically between the central region of the cavity and the region near the walls, as indicated by the Poynting vector **S** (vertical arrows).

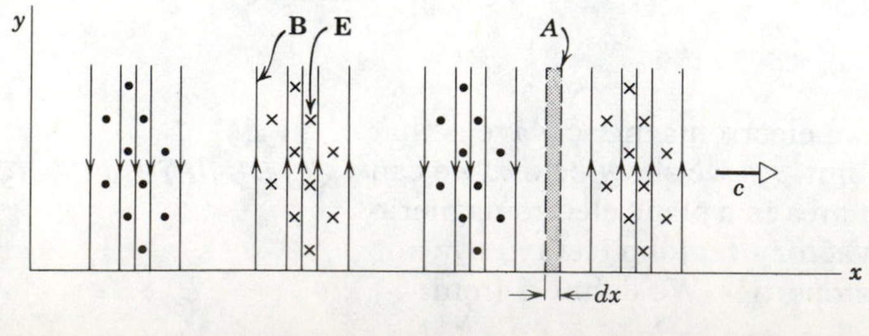

figure 41-16
A plane wave is traveling to the right at speed c; compare Fig. 41-14. The dashed rectangle in this figure represents a three-dimensional box, of area A and thickness dx, that extends at right angles to the plane of the figure.

shows, by the open arrows, the direction of **S** at various points in the cavity and at various times in the cycle. Note that **S** equals zero for Figs. 41-15*a, c, e,* and *g,* which is appropriate because at these instants of time the field configurations are momentarily stationary and energy is not flowing. A pendulum bob at the end of its swing forms a mechanical analogy. Verify from Eq. 41-17 that these arrows point in the correct directions.

We can use Figure 41-16 to derive the Poynting relation for the special case of a traveling plane electromagnetic wave. It shows a cross section of a traveling plane wave, along with a thin "box" of thickness dx and area A. The box, a mathematical construction, is fixed with respect to the axes while the wave moves through it. At any instant the energy stored in the box, from Eqs. 30-9 and 36-19, is

$$dU = dU_E + dU_B = (u_E + u_B)(A \ dx)$$

$$= \left(\tfrac{1}{2}\epsilon_0 E^2 + \frac{1}{2\mu_0} B^2\right)A \ dx, \qquad (41\text{-}18)$$

where $A \ dx$ is the volume of the box and E and B are the instantaneous values of the field vectors in the box.

Using Eq. 41-11*b* $(E = cB)$ to eliminate *one* of the E's in the first term in Eq. 41-18 and *one* of the B's in the second term leads to

$$dU = \left[\tfrac{1}{2}\epsilon_0 E(cB) + \frac{1}{2\mu_0} B\left(\frac{E}{c}\right)\right] A \ dx$$

$$= \frac{(\mu_0\epsilon_0 c^2 + 1)(EBA \ dx)}{2\mu_0 c}.$$

From Eq. 41-15, however, $\mu_0\epsilon_0 c^2 = 1$, so that

$$dU = \frac{EBA \ dx}{\mu_0 c}.$$

This energy dU will pass through the right face of the box in a time dt equal to dx/c. Thus the energy per unit area per unit time, which is S, is given by

$$S = \frac{dU}{dt \ A} = \frac{EBA \ dx}{\mu_0 c \ (dx/c) \ A} = \frac{1}{\mu_0} EB.$$

This is exactly the prediction of the more general relation Eq. 41-17 for a traveling plane wave.

This relation refers to values of S, E, and B at any instant of time. We are usually more interested in the *average* value of S, taken over one or more cycles of the wave. An observer making intensity measurements on a wave moving past him would measure this average value $\overline{S}$. We can easily show (see Example 6) that $\overline{S}$ is related to the *maximum* values of E and B by

$$\overline{S} = \frac{1}{2\mu_0} E_m B_m.$$

EXAMPLE 6

An observer is at a distance r from a point light source whose power output is P_0. Calculate the magnitudes of the electric and the magnetic fields. Assume that the source is monochromatic, that it radiates uniformly in all directions, and that at distant points it behaves like the traveling plane wave of Fig. 41-13.

The power that passes through a sphere of radius r is $(\overline{S})(4\pi r^2)$, where $\overline{S}$ is the *average* value of the Poynting vector at the surface of the sphere. This power must equal P_0, or

$$P_0 = \overline{S}4\pi r^2.$$

From the definition of **S** (Eq. 41-17), we have

$$\overline{S} = \overline{\left(\frac{1}{\mu_0} EB\right)}.$$

Using the relation $E = cB$ (Eq. 41-11b) to eliminate B leads to

$$\overline{S} = \frac{1}{\mu_0 c} \overline{E^2}.$$

The average value of E^2 over one cycle is $\frac{1}{2} E_m^2$, since E varies sinusoidally (see Eq. 41-7). This leads to

$$P_0 = \left(\frac{E_m^2}{2\mu_0 c}\right)(4\pi r^2),$$

or

$$E_m = \frac{1}{r}\sqrt{\frac{P_0 \mu_0 c}{2\pi}}.$$

For $P_0 = 10^3$ W and $r = 1.0$ m this yields

$$E_m = \frac{1}{(1.0 \text{ m})} \sqrt{\frac{(10^3 \text{ W})(4\pi \times 10^{-7} \text{ Wb/A·m})(3 \times 10^8 \text{ m/s})}{2\pi}}$$

$$= 240 \text{ V/m}.$$

The relationship $E_m = cB_m$ (Eq. 41-11a) leads to

$$B_m = \frac{E_m}{c} = \frac{240 \text{ V/m}}{3 \times 10^8 \text{ m/s}} = 8 \times 10^{-7} \text{ T}.$$

Note that E_m is appreciable as judged by ordinary laboratory standards but that B_m (= 0.008 gauss) is quite small.

---

questions

1. If you are asked what fraction of the electromagnetic spectrum lies in the visible range, how would you respond?

2. Project Seafarer was an ambitious program to construct an enormous antenna, buried underground on a site about 4000 square miles in area. Its purpose was to transmit signals to submarines while they were deeply submerged. If the effective wavelength was, say, about 10^4 earth radii, what would be the frequency and the period of the radiations emitted? Ordinarily electromagnetic radiations do not penetrate very far into conductors such as sea water. Can you think of any reason why such ELF (extremely low frequency) radiations should penetrate more effectively? Think of the limiting case of zero frequency. (Why not transmit signals at zero frequency?)

3. How would you characterize electromagnetic radiation that has a frequency of 10 kHz? 10^{20} Hz? or a wavelength of 500 nm? 10 km? 0.50 nm?

4. Electromagnetic waves reach us from the farthest depths of the universe. Do they tell us what the universe is like now, what it was like at some time in the past, or something in between? Discuss.

5. *The Dark Night Sky Paradox:** Perhaps the simplest astronomical observation that you can make is this: When the sun sets, the sky becomes dark. This is true and seems obvious but an argument can be made that it should not be so. Consider:

"Assuming an infinite universe, uniformly populated by stars more or less like our sun, we can say that a straight line projected from the observer in any direction will eventually hit a star. The *distances R* of most of these stars will be very great indeed so that the stars illuminate him only weakly, the illumination varying as $1/R^2$. On the other hand, the *number* of distant stars located within a spherical shell whose radii are R and $R + \Delta R$ increases as R^2 (assuming that ΔR is constant). Can you prove this last statement? These two effects seem to cancel precisely. Thus the night sky should be virtually infinitely bright, the observer being illuminated by an infinity of suns."

Can you see any flaw in this argument? Think of the finite speed of light, the large scale of the universe, the expanding universe concept, the finite

* Usually called *Olbers' paradox.*

lifetime of stars, and so on. See "The Dark Night Sky Paradox" by E. R. Harrison, *American Journal of Physics*, February 1977, for an excellent historical review and a lucid explanation.

6. When ordinary stars condense to form neutron stars (pulsars), why does the angular rotation rate become so large?

7. In the coaxial cable of Fig. 41-3, what are the directions of the conduction current (a) in the central conductor and (b) in the outer conductor, shortly after the switch is thrown to position a? Consider points that have been reached by the wavefront of Fig. 49-4b and those that have not.

8. What is the relation between the wavelength in the cable of Fig. 41-5 and that in free space for a coaxial cable?

9. What is the direction of the displacement current in Fig. 41-6? Give an argument to support your answer.

10. Compare a coaxial cable and a waveguide, used as a transmission line. Point out both similarities and differences.

11. Can traveling waves with a continuous range of wavelengths be sent down (a) a coaxial cable and (b) a waveguide? Can standing waves with a continuous range of wavelengths be set up in a resonant cavity? Develop mechanical or acoustical analogies to support your answers.

12. If a certain wavelength is larger than the cutoff wavelength for a guide in its dominant mode, can energy be sent down it in any other mode?

13. Explain why the term $\epsilon_0 \, d\Phi_E/dt$ is needed in Ampère's equation to understand the propagation of electromagnetic waves.

14. In the equation $c = 1/\sqrt{\mu_0 \epsilon_0}$ (Eq. 40-1), how can c always have the same value if μ_0 is *arbitrarily* assigned and ϵ_0 is measured?

15. Is it conceivable that electromagnetic theory might some day be able to predict the value of c (3×10^8 m/s), not in terms of μ_0 and ϵ_0, but directly and numerically without recourse to any measurements?

16. Figure 41-17 shows a magnetic dipole activated by an oscillating LC circuit. Discuss the nature of the traveling wave at a distant point P.

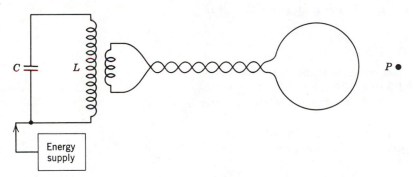

figure 41-17
Question 16

17. In a coaxial cable is the energy transported in the conductors, through the agency of the currents, or in the space between them, through the agency of the fields?

SECTION 41-2
1. In Fig. 41-1, showing the electromagnetic spectrum, the frequency ν and the wavelength λ are related by $c = \lambda\nu$. If the frequency intervals in the figure are evenly spaced, show that the wavelength intervals are also necessarily equally spaced. Bear in mind that both scales are logarithmic.

problems

SECTION 41-3
2. The earth's mean radius is 6.4×10^6 m and the mean earth-sun distance is 1.5×10^8 km. What fraction of the electromagnetic radiation emitted by the

sun is intercepted by the disc of the earth? Assume an inverse square law for the decrease of intensity, measured perhaps in watts/meter², with distance.

3. The intensity of direct solar radiation that was unabsorbed by the atmosphere on a lazy summer day is 100 W/m². How close would you have to stand to a 1.0-kW electric heater to feel the same intensity? Assume that the heater is 100% efficient and radiates equally in all directions.
Answer: 0.89 m.

4. Our closest stellar neighbor, α-Centauri, is 4.3 light years away. It has been suggested that TV programs from our planet (The Tonight Show, for example) have reached this star and may have been viewed by the hypothetical inhabitants of a hypothetical planet orbiting this star. The moon is 3.8×10^5 km from earth and perhaps its hypothetical inhabitants (who always hide when a space ship approaches) also watch these programs. From the inverse square law, what would be the ratio of the intensities of such signals, measured perhaps in W/m², for the moon and for α-Centauri?

SECTION 41-5

5. A coaxial cable is made of a center wire of radius a surrounded by a thin metal tube of radius b. The space between the conductors is evacuated. (*a*) Find the capacitance per unit length of this coaxial cable (*Hint:* Imagine equal but opposite charges to be on the wire and the tube). (*b*) Find the inductance per unit length of this coaxial cable (*Hint:* Imagine a current i flowing down the center wire and back along the tube.)
Answer: (*a*) $2\pi\epsilon_0/\ln b/a$. (*b*) $(\mu_0/2\pi) \ln b/a$.

6. Using Gauss's law, sketch the instantaneous charges that appear on the conductors of the coaxial cable of Fig. 41-5 and show that this pattern of charges is appropriately related to the conduction currents shown in Fig. 41-5c.

7. A resonant cavity is constructed by closing each end of the coaxial cable of Fig. 41-5 with a metal cap. The cavity contains three half-waves. Describe the patterns of **E** and **B** that occur, assuming the same mode of oscillation as that shown in Fig. 41-5. (Hint: Remember that **E** can have no tangential component at a conducting surface and that **B** and **E** must be 90° out of phase.)

SECTION 41-6

8. (*a*) For a rectangular guide of width 3.0 cm, what must the free-space wavelength of radiation be if it is to require 1.0 μs (= 10^{-6} s) for energy to traverse a 100-m length of guide? (*b*) What is the phase speed under these circumstances?

9. For a rectangular guide of width 3.0 cm, plot the phase speed, the group speed, and the guide wavelength as a function of the free-space wavelength. Assume the dominant mode.

10. Under what conditions will the guide wavelength in the guide of Fig. 41-9 be double the free-space wavelength?

11. (*a*) What guide wavelength does 10-cm radiation (free-space wavelength) exhibit in a rectangular guide whose width is 6.0 cm? Assume the dominant mode. (*b*) What is the cutoff wavelength for this guide?
Answer: (*a*) 18 cm. (*b*) 12 cm.

SECTION 41-7

12. (*a*) At a distance of 80 miles from a radio transmitter, how much later would you observe a wave emitted from the antenna? (*b*) If the radiation were a radio wave emitted from the sun? (*c*) If it were emitted from a stellar radio source 380 light years distant?

SECTION 41-8

13. An electromagnetic wave is traveling in the negative y-direction. At a particular position and time, the electric field is along the positive z-axis and has a magnitude of 100 V/m. What are the direction and magnitude of the magnetic field?   *Answer: 3.3×10^{-7} T, in the negative x-direction.*

14. The magnetic field equations for an electromagnetic wave in free space are $B_x = B \sin(ky + \omega t)$, $B_y = B_z = 0$. (a) What is the direction of propagation? (b) Write the electric field equations.

15. How does the displacement current density vary with space and time in a traveling plane electromagnetic wave?

16. (a) Starting with Eqs. 41-10 and 41-13, show that $E(x, t)$ satisfies the "wave equation"

$$\frac{\partial^2 E}{\partial x^2} = \mu_0 \epsilon_0 \frac{\partial^2 E}{\partial t^2}.$$

    (b) What "wave equation" does $B(x, t)$ satisfy?

17. Show that (a) through (d) below satisfy Eqs. 41-10 and 41-13. In each of these, A is a constant. (a) $E = Ac(x - ct)$, $B = A(x - ct)$. (b) $E = Ac(x + ct)^{15}$, $B = -A(x + ct)^{15}$. (c) $E = Ace^{(x-ct)}$, $B = Ae^{(x-ct)}$. (d) $E = Ac \ln(x + ct)$, $B = -A \ln(x + ct)$. (e) Generalize these examples to show that $E = Acf(x - ct)$, $B = Af(x - ct)$ is a solution where f is *any* function of $(x - ct)$. What is the corresponding situation for functions of $(x + ct)$?

18. Show that the directions of the fields **E** and **B** in Fig. 41-11 are consistent with the direction of propagation of the radiation.

19. Show that the directions of the fields E and B in the various "patches" in Fig. 41-12 are consistent with the direction of propagation of the radiation.

SECTION 41-9

20. Sunlight strikes the earth, outside its atmosphere, with an intensity of 2.0 cal/cm²-min. Calculate E_m and B_m for sunlight, assuming it to be a wave like that of Fig. 41-12.

21. A #10 copper wire (diameter, 0.10 in.; resistance per 1000 ft, 1.00 Ω) carries a current of 25 A. Calculate (a) **E**, (b) **B**, and (c) **S** for a point on the surface of the wire.
    Answer: (a) $\mathbf{E} = 8.2 \times 10^{-2}$ V/m, parallel to the wire. (b) $\mathbf{B} = 3.9 \times 10^{-3}$ T, tangent to the wire and perpendicular to its axis. (c) $\mathbf{S} = 260$ W/m², radially inward.

22. Prove that for any point in an electromagnetic wave such as that of Fig. 41-13 the density of energy stored in the electric field equals that stored in the magnetic field.

23. A plane radio wave has $E_m \cong 10^{-4}$ V/m. Calculate (a) B_m and (b) the intensity of the wave, as measured by $\overline{\mathbf{S}}$.
    Answer: (a) 3.3×10^{-13} T. (b) 1.3×10^{-11} W/m².

24. If a coaxial cable has resistance, energy must flow from the fields into the conducting surfaces to provide Joule heating. How must the electric lines of force of Fig. 41-5a be modified in this case? (*Hint:* The Poynting vector near the surface must have a component pointing toward the surface.)

25. Analyze the flow of energy in the waveguide of Fig. 41-9, using the Poynting vector.

26. A coaxial cable (inner radius a, outer radius b) is used as a transmission line between a battery ε and a resistor R, as shown in Fig. 41-18. (a) Calculate **E**, **B** for $a < r < b$. (b) Calculate the Poynting vector **S** for $a < r < b$. (c) By suitably integrating the Poynting vector, show that the total power flowing across the annular cross section $a < r < b$ is ε^2/R. Is this reasonable?

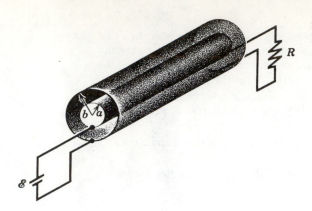

figure 41-18
Problem 26

(d) Show that the direction of **S** is always from the battery to the resistor, no matter which way the battery is connected.

27. An airplane flying at a distance of 10 km from a radio transmitter receives a signal of intensity 10 μW/m². Calculate (a) the (average) electric field at the airplane due to this signal; (b) the (average) magnetic field at the airplane; (c) the total power radiated by the transmitter, assuming the transmitter to radiate isotropically and the earth to be a perfect absorber.
Answer: (a) 6.1×10^{-2} V/m. (b) 2.1×10^{-10} T. (c) 13 kW.

28. A cube of edge a has its edges parallel to the x-, y-, and z-axes of a rectangular coordinate system. A uniform electric field **E** is parallel to the y-axis and a uniform magnetic field **B** is parallel to the x-axis. Calculate (a) the rate at which, according to the Poynting vector point of view, energy may be said to pass through each face of the cube and (b) the net rate at which the energy stored in the cube may be said to change.

29. Consider the possibility of standing waves:

$$E = E_m(\sin \omega t)(\sin kx)$$
$$B = B_m(\cos \omega t)(\cos kx).$$

(a) Show that these satisfy Eqs. 41-10 and 41-13 if E_m is suitably related to B_m and ω suitably related to k. What are these relationships? (b) Find the (instantaneous) Poynting vector. (c) Show that the time average power flow across any area is zero. (d) Describe the flow of energy in this problem.
Answer: (a) Eq. 41-11a must hold for standing waves as well as for traveling waves. (b) $S = (E_m^2/4\mu_0 c) \sin 2\omega t \sin 2kx$.

30. Figure 41-19 shows a long resistanceless transmission line, delivering power from a battery to a resistive load. A steady current i exists as shown. (a) Sketch qualitatively the electric and magnetic fields around the line, and (b) show that, according to the Poynting vector point of view, energy travels from the battery to the resistor through the space around the line and not through the line itself. (*Hint:* Each conductor in the line is an equipotential surface, since the line has been assumed to have no resistance.)

31. Figure 41-20 shows a cylindrical resistor of length l, radius a, and resistivity ρ, carrying a current i. (a) Show that the Poynting vector **S** at the surface of the resistor is everywhere directed normal to the surface, as shown. (b) Show that the rate P at which energy flows into the resistor through its cylindrical surface, calculated by integrating the Poynting vector over this surface, is equal to the rate at which thermal energy is produced; that is,

$$\int \mathbf{S} \cdot d\mathbf{A} = i^2 R,$$

where $d\mathbf{A}$ is an element of area of the cylindrical surface. This shows that, according to the Poynting vector point of view, the energy that appears in a

figure 41-19
Problem 30

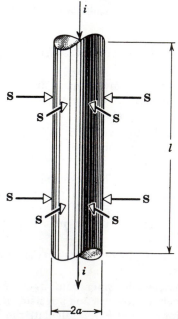

figure 41-20
Problem 31

resistor as thermal energy does not enter it through the connecting wires but through the space around the wires and the resistor. (*Hint:* **E** is parallel to the axis of the cylinder, in the direction of the current; **B** forms concentric circles around the cylinder, in a direction given by the right-hand rule.)

32. Figure 41-21 shows a parallel-plate capacitor being charged. (*a*) Show that the Poynting vector **S** points everywhere radially into the cylindrical volume. (*b*) Show that the rate P at which energy flows into this volume, calculated by integrating the Poynting vector over the cylindrical boundary of this volume, is equal to the rate at which the stored electrostatic energy increases; that is, that

$$\int \mathbf{S}\cdot d\mathbf{A} = Ad\,\frac{d}{dt}\left(\tfrac{1}{2}\epsilon_0 E^2\right),$$

where Ad is the volume of the capacitor and $\tfrac{1}{2}\epsilon_0 E^2$ is the energy density for all points within that volume. This analysis shows that, according to the Poynting vector point of view, the energy stored in a capacitor does not enter it through the wires but through the space around the wires and the plates. (*Hint:* To find **S** we must first find **B**, which is the magnetic field set up by the displacement current during the charging process; see Fig. 40-1. Ignore fringing of the lines of **E**.)

33. A long hollow nonconducting cylinder (radius R, length l) carries a uniform charge per unit area of σ on its surface. An externally applied torque causes the cylinder to rotate at constant acceleration $\omega(t) = \alpha t$ about the cylinder axis. (*a*) Find **B** within the cylinder (treat it as a solenoid). (*b*) Find **E** at the inner surface of the cylinder. (*c*) Find **S** at the inner surface of the cylinder. (*d*) Show that the flux of **S** entering the interior volume of the cylinder is equal to $\dfrac{d}{dt}\left(\dfrac{\pi R^2 l}{2\mu_0}\,B^2\right)$.

Answer: (*a*) $B = \sigma R \omega \mu_0$. (*b*) $E = \tfrac{1}{2}\mu_0 \sigma R^2 \alpha$. (*c*) $S = \tfrac{1}{2}\mu_0 \sigma^2 R^3 \alpha^2 t$.

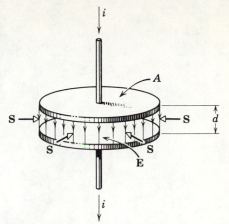

figure 41-21
Problem 32

42
the nature and propagation of light

This chapter connects what is true about the entire electromagnetic spectrum (see Fig. 41-1) to visible light, which is, of course, a part of that spectrum. In chapters that immediately follow we will focus on visible light. This is important to do because almost all of us, having adapted by genetic selection to the sun, have two electromagnetic receptors that function in this range and, indeed, define the range. These, of course, are our eyes. We must always remember that whatever we say about the visible part of the spectrum holds, at its foundations, for all other parts of the spectrum as well. The differences are largely in the nature of the means of production and detection in the various ranges.

Here we define "light" as radiation that can affect the eye. Figure 42-1, which shows the relative eye sensitivity of an assumed *standard*

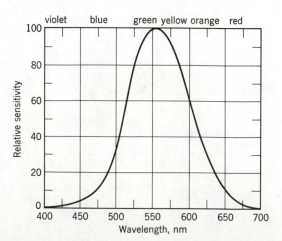

figure 42-1
The relative eye sensitivity of an assumed *standard observer* at different wavelengths for a specified level of illumination.

observer to radiations of various wavelengths, shows that the center of the visible region is about 5.55×10^{-7} m or 555 nm. Light of this wavelength produces the sensation of yellow-green.*

In optics we often use the micrometer (abbr. μm) the nanometer (abbr. nm), and the Ångstrom (abbr. Å) as units of wavelength. They are defined from

$$1 \text{ micrometer} = 10^{-6} \text{ meter}$$

$$1 \text{ nanometer} = 10^{-9} \text{ meter}$$

$$1 \text{ Ångstrom} = 10^{-10} \text{ meter}.$$

Thus we can say that the center of the visible region is 0.555 μm, 555 nm, or 5550 Å.

The limits of the visible spectrum are not well defined because the eye sensitivity curve approaches the axis asymptotically at both long and short wavelengths. If we take the limits, arbitrarily, as the wavelengths at which the eye sensitivity has dropped to 1% of its maximum value, these limits are about 430 and 690 nm, less than a factor of two in wavelength. The eye can detect radiation beyond these limits if it is intense enough. In many experiments in physics one can use photographic plates or light-sensitive electronic detectors in place of the human eye.

Energy is carried by electromagnetic waves from the sun to the earth or from an open fire to a hand placed nearby. We described the transport of energy by such a wave in free space in Section 41-9 by the Poynting vector **S**, or

$$\mathbf{S} = \frac{1}{\mu_0} \mathbf{E} \times \mathbf{B}, \qquad (42\text{-}1)$$

where **E** and **B** are the instantaneous values of the electric and magnetic field vectors.

Less familiar is the fact that electromagnetic waves may also transport linear momentum. In other words, it is possible to exert a pressure (a *radiation pressure*†) on an object by shining a light on it. Such forces must be small in relation to forces of our daily experience because we do not ordinarily notice them. We do not, after all, fall over backward when we raise a window shade in a dark room and let sunlight shine on us. Radiation pressure effects are, however, important in the life cycles of stars because of the incredibly high temperatures (2×10^7 K for our sun) that we associate with stellar interiors.

The first measurement of radiation pressure was made in 1901–1903 by Nichols and Hull at Dartmouth College and by Lebedev in Russia, about thirty years after the existence of such effects had been predicted theoretically by Maxwell.

42-2
ENERGY AND MOMENTUM

---

* See "The Retinex Theory of Color Vision" by Edwin H. Land, *Scientific American*, December 1977, and "Color and Perception: the Work of Edwin Land in the Light of Current Concepts" by M. H. Wilson and R. W. Brocklebank, *Contemporary Physics*, December 1961, for a fascinating discussion of the problems of perception and the distinction between color as a characteristic of light and color as a perceived property of objects.

† See "Radiation Pressure" by G. E. Henry, *Scientific American*, June 1957; see also "The Pressure of Laser Light" by Arthur Ashkin, *Scientific American*, February 1972.

Let a parallel beam of light fall on an object for a time t, the incident light being *entirely absorbed* by the object. If energy U is absorbed during this time, the momentum p delivered to the object is given, according to Maxwell's prediction, by

$$p = \frac{U}{c} \qquad \text{(total absorption)}, \qquad (42\text{-}2a)$$

where c is the speed of light. The direction of **p** is the direction of the incident beam. If the light energy U is *entirely reflected*, the momentum delivered will be twice that given above, or

$$p = \frac{2U}{c} \qquad \text{(total reflection)}. \qquad (42\text{-}2b)$$

In the same way, twice as much momentum is delivered to an object when a perfectly elastic tennis ball is bounced from it as when it is struck by a perfectly inelastic ball (a lump of putty, say) of the same mass and speed. If the light energy U is partly reflected and partly absorbed, the delivered momentum will lie between U/c and $2U/c$.

EXAMPLE 1

A parallel beam of light with an energy flux S of 10 W/cm² falls for 1 hr on a perfectly reflecting plane mirror of 1.0-cm² area. (*a*) What momentum is delivered to the mirror in this time and (*b*) what force acts on the mirror?

(*a*) The energy that is reflected from the mirror is

$$U = (10 \text{ W/cm}^2)(1.0 \text{ cm}^2)(3600 \text{ s}) = 3.6 \times 10^4 \text{ J}.$$

The momentum delivered after 1 hour's illumination is

$$p = \frac{2U}{c} = \frac{(2)(3.6 \times 10^4 \text{ J})}{3 \times 10^8 \text{ m/s}} = 2.4 \times 10^{-4} \text{ kg m/s}.$$

(*b*) From Newton's second law, the average force on the mirror is equal to the average rate at which momentum is delivered to the mirror, or

$$F = \frac{p}{t} = \frac{2.4 \times 10^{-4} \text{ kg m/s}}{3600 \text{ s}} = 6.7 \times 10^{-8} \text{ N}.$$

This is a small force.

Nichols and Hull, in 1903, measured radiation pressures and verified Eq. 42-2, using a torsion balance technique. They allowed light to fall on mirror M in Fig. 42-2; the radiation pressure caused the balance arm to turn through a measured angle θ, twisting the torsion fiber F. Assuming a suitable calibration for their torsion fiber, the experimenters could arrive at a numerical value for this pressure. Nichols and Hull measured the intensity of their light beam by allowing it to fall on a blackened metal disk of known absorptivity and by measuring the temperature rise of this disk. In a particular run these experimenters measured a radiation pressure of 7.01×10^{-6} N/m²; for their light beam, the value predicted, using Eq. 42-2, was 7.05×10^{-6} N/m², in excellent agreement. Assuming a mirror area of 1 cm², this represents a force on the mirror of only 7×10^{-10} N, about 100 times smaller than the force calculated in Example 1.

The success of the experiment of Nichols and Hull was the result in large part of the care they took to eliminate spurious deflecting effects caused by changes in the speed distribution of the molecules in the gas

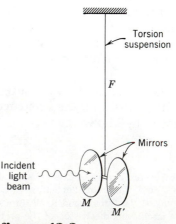

figure 42-2
Suggesting the experiment of Nichols and Hull, used to measure radiation pressure.

surrounding the mirror. These changes were brought about by the small rise in the temperature of the mirror as it absorbed light energy from the incident beam. This "radiometer effect" is responsible for the spinning action of the familiar toy radiometers when placed in a beam of sunlight. In a perfect vacuum such effects would not occur, but in the best vacuums available in 1903 radiometer effects were present and had to be taken specifically into account in the design of the experiment.

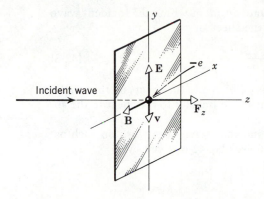

figure 42-3
An incident plane light wave falls on an electron in a thin resistive sheet. Instantaneous values of **E**, **B**, the electron velocity **v**, and the radiation force $\mathbf{F}_z$ are shown.

To demonstrate the transport of momentum from Maxwell's equations in a particular case, let a plane electromagnetic wave traveling in the z-direction fall on a large thin sheet of a material of high resistivity as in Fig. 42-3. A small part of the incident energy will be absorbed within the sheet, but most of it will be transmitted if the sheet is thin enough.*

The incident wave vectors **E** and **B** vary with time at the sheet as

$$\mathbf{E} = \mathbf{E}_m \sin \omega t \qquad (42\text{-}3)$$

and
$$\mathbf{B} = \mathbf{B}_m \sin \omega t \qquad (42\text{-}4)$$

where **E** is parallel to the $\pm y$ axis and **B** is parallel to the $\pm x$ axis.

In Section 31-4 we saw that the effect of a (constant) electric force ($= -eE$) on a conduction electron in a metal was to make it move with a (constant) drift speed v_d. The electron behaves as if it is immersed in a viscous fluid, the electric force acting on it being counterbalanced by a "viscous" force, which may be taken as proportional to the electron speed. Thus for a constant field E, after equilibrium is established,

$$eE = bv_d, \qquad (42\text{-}5)$$

where b is a resistive damping coefficient. The electron equilibrium speed, dropping the subscript d, is thus

$$v = \frac{eE}{b}. \qquad (42\text{-}6)$$

If the applied electric field varies with time and if the variation is slow enough, the electron speed can continually readjust itself to the changing value of E so that its speed continues to be given essentially by its equilibrium value (Eq. 42-6) at all times. These readjustments are more rapidly made the more viscous the medium, just as a stone falling in air reaches a constant equilibrium rate of descent only relatively slowly but one falling in a viscous oil does so quite rapidly. We assume that the sheet in Fig. 42-3 is so viscous, that is, that its resistivity is so high, that Eq. 42-6 remains true even for the rapid oscillations of E in the incident light beam.

As the electron vibrates parallel to the y-axis, it experiences a *second* force

---

* Some of the incident energy will also be reflected, but the reflected wave is of such low intensity that we can ignore it in the derivation that follows.

due to the *magnetic* component of the wave. This force $\mathbf{F}_z$ $(= -e\mathbf{v} \times \mathbf{B})$ points in the z-direction, being at right angles to the plane formed by $\mathbf{v}$ and $\mathbf{B}$, that is, the xy-plane. The instantaneous magnitude of $\mathbf{F}_z$ is given by

$$F_z = evB = \frac{e^2 EB}{b}.$$
(42-7)

$\mathbf{F}_z$ always points in the positive z-direction because $\mathbf{v}$ and $\mathbf{B}$ reverse their directions simultaneously; this force is, in fact, the mechanism by which the radiation pressure acts on the sheet of Fig. 42-3.

From Newton's second law, F_z is the rate dp_e/dt at which the incident wave delivers momentum to each electron in the sheet, or

$$\frac{dp_e}{dt} = \frac{e^2 EB}{b}.$$
(42-8)

Momentum is delivered at this rate to every electron in the sheet and thus to the sheet itself. It remains to relate the momentum transfer to the sheet to the absorption of energy within the sheet.

The electric field component of the incident wave does work on each oscillating electron at an instantaneous rate (see Eq. 42-6) given by

$$\frac{dU_e}{dt} = F_E v = (eE)\left(\frac{eE}{b}\right) = \frac{e^2 E^2}{b}.$$

Note that the magnetic force $\mathbf{F}_z$, always being at right angles to the velocity $\mathbf{v}$, does no work on the oscillating electron. Equation 41-11b shows that for a plane wave in free space B and E are related by

$$E = Bc.$$

Substituting above for *one* of the E's leads to

$$\frac{dU_e}{dt} = \frac{e^2 EBc}{b}.$$
(42-9)

This equation represents the rate, per electron, at which energy is absorbed from the incident wave.

Comparing Eqs. 42-8 and 42-9 shows that

$$\frac{dp_e}{dt} = \frac{1}{c}\frac{dU_e}{dt}.$$

Integrating yields

$$\int_0^t \frac{dp_e}{dt}\,dt = \frac{1}{c}\int_0^t \frac{dU_e}{dt}\,dt,$$

or

$$p_e = \frac{U_e}{c},$$
(42-10)

where p_e is the momentum delivered to a single electron in any given time t and U_e is the energy absorbed by that electron in the same time interval. Multiplying each side by the number of free electrons in the sheet leads to Eq. 42-2a.

Although we derived this relation (Eq. 42-10) for a particular kind of absorber, no characteristics of the absorber—for example, the resistive damping coefficient b—remain in the final expression. This is as it should be because Eq. 42-10 is a general property of radiation absorbed by *any* material.

42-3 THE SPEED OF LIGHT*

Light travels so fast that there is nothing in our daily experience to suggest that its speed is not infinite. It calls for considerable insight even to ask, "How fast does light travel?" Galileo asked himself this question and actually tried to answer it experimentally. His book, *Two New Sciences*, published in 1638, is written in the form of a conversation

* See "The Speed of Light" by J. H. Rush, *Scientific American*, August 1955.

among three persons called Salviati, Sagredo, and Simplicio. Here is part of what they say about the speed of light.

Simplicio: Everyday experience shows that the propagation of light is instantaneous; for when we see a piece of artillery fired, at a great distance, the flash reaches our eyes without lapse of time; but the sound reaches the ear only after a noticeable interval.

Sagredo: Well, Simplicio, the only thing I am able to infer from this familiar bit of experience is that sound, in reaching our ear, travels more slowly than light; it does not inform me whether the coming of the light is instantaneous or whether, although extremely rapid, it still occupies time. . . .

Salviati, who speaks with Galileo's voice, then describes a possible method (actually carried out) for measuring the speed of light. He and an assistant stand facing each other some distance apart, at night. Each carries a lantern which can be covered or uncovered at will. Galileo started the experiment by uncovering his lantern. When the light reached the assistant he uncovered his own lantern, whose light was then seen by Galileo. Galileo tried to measure the time between the instant at which he uncovered his own lantern and the instant at which the light from his assistant's lantern reached him. For a one-mile separation we now know that the round trip travel time would be only 11 μs. This is much less than human reaction times, so the method fails.

To measure a large velocity directly, we must either measure a small time interval or use a long base line. This situation suggests that astronomy, which deals with great distances, might be able to provide an experimental value for the speed of light; this proved to be true. Although it would be desirable to time the light from the sun as it travels to the earth, there is no way of knowing when the light that reaches us at any instant left the sun; we must use subtler astronomical methods.

Note, however, that microwave pulses are quite regularly reflected from the moon; this gives a 7.68×10^8-m base line (there and back) for timing purposes. The speed of light (and of microwaves) is so well known now from other experiments that we use these measurements to measure the lunar distance accurately. Microwave signals have also been reflected from Venus.

In 1675 Ole Roemer, a Danish astronomer working in Paris, made some observations of the moons of Jupiter (see Problem 20) from which a speed of light of 2×10^8 m/s may be deduced. About fifty years later James Bradley, an English astronomer, made some astronomical observations of an entirely different kind from which a value of 3.0×10^8 m/s may be deduced:

In 1849 Hippolyte Louis Fizeau (1819–1896), a French physicist, first measured the speed of light by a nonastronomical method, obtaining a value of 3.13×10^8 m/s. Figure 42-4 shows Fizeau's apparatus. Let us first ignore the toothed wheel. Light from source S is made to converge

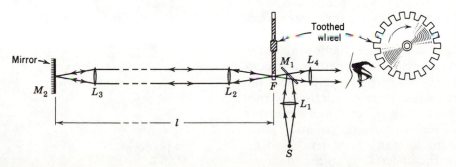

figure 42-4
Fizeau's apparatus for measuring the speed of light.

by lens L_1, is reflected from mirror M_1, and forms in space at F an image of the source. Mirror M_1 is a so-called "half-silvered mirror"; its reflecting coating is so thin that only half the light that falls on it is reflected, the other half being transmitted.

Light from the image at F enters lens L_2 and emerges as a parallel beam; after passing through lens L_3 it is reflected back along its original direction by mirror M_2. In Fizeau's experiment the distance l between M_2 and F was 8630 meters or 5.36 miles. When the light strikes mirror M_1 again, some will be transmitted, entering the eye of the observer through lens L_4.

The observer will see an image of the source formed by light that has traveled a distance $2l$ between the wheel and mirror M_2 and back again. To time the light beam a marker of some sort must be put on it. This is done by "chopping" it with a rapidly rotating toothed wheel. Suppose that during the round-trip travel time of $2l/c$ the wheel has turned just enough so that, when the light from a given "burst" returns to the wheel, point F is covered by a tooth. The light will hit the face of the tooth that is toward M_2 and will not reach the observer's eye.

If the speed of the wheel is exactly right, the observer will not see any of the bursts because each will be screened by a tooth. The observer measures c by increasing the angular speed ω of the wheel from zero until the image of source S disappears. Let θ be the angular distance from the center of a gap to the center of a tooth. The time needed for the wheel to rotate a distance θ is the round-trip travel time $2l/c$. In equation form,

$$\frac{\theta}{\omega} = \frac{2l}{c} \quad \text{or} \quad c = \frac{2\omega l}{\theta}. \tag{42-11}$$

This "chopped beam" technique, suitably modified, is used today to measure the speeds of neutrons and other particles.

EXAMPLE 2

The wheel used by Fizeau had 720 teeth. What is the smallest angular speed at which the image of the source will vanish?

The angle θ is 1/1440 rev; solving Eq. 42-11 for ω gives

$$\omega = \frac{c\theta}{2l} = \frac{(3.00 \times 10^8 \text{ m/s})(1/1440 \text{ rev})}{(2)(8630 \text{ m})} = 12.1 \text{ rev/s}.$$

The French physicist Foucault (1819–1868) greatly improved Fizeau's method by substituting a rotating mirror for the toothed wheel. The American physicist Albert A. Michelson (1852–1931) conducted an extensive series of measurements of c, extending over a fifty-year period, using this technique.

We must view the speed of light within the larger framework of the speed of electromagnetic radiation in general. It is a significant experimental confirmation of Maxwell's theory of electromagnetism that the speed in free space of waves in all parts of the electromagnetic spectrum has the same value c. Table 42-1 shows some selected measurements that have been made of the speed of electromagnetic radiation since Galileo's day. It stands as a monument to man's persistence and ingenuity. Note in the last column how the uncertainty in the measurement has improved through the years. Note also the international character of the effort and the variety of methods.

Table 42-1
The speed of electromagnetic radiation in free space (some selected measurements)*

Date	Experimenter	Country	Method	Speed (km/s)	Uncertainty (km/s)
1600(?)	Galileo	Italy	Lanterns and shutters	"If not instantaneous, it is extraordinarily rapid"	
1675	Roemer	France	Moons of Jupiter	200,000	
1729	Bradley	England	Aberration of starlight	304,000	
1849	Fizeau	France	Toothed wheel	313,300	
1862	Foucault	France	Rotating mirror	298,000	500
1876	Cornu	France	Toothed wheel	299,990	200
1880	Michelson	U.S.A.	Rotating mirror	299,910	50
1883	Newcomb	England	Rotating mirror	299,860	30
1906	Rosa and Dorsey	U.S.A.	Electromagnetic theory	299,781	10
1923	Mercier	France	Standing waves on wires	299,782	15
1926	Michelson	U.S.A.	Rotating mirror	299,796	4
1928	Karolus and Mittelstaedt	Germany	Kerr cell	299,778	10
1932	Michelson, Pease, and Pearson	U.S.A.	Rotating mirror	299,774	11
1941	Anderson	U.S.A.	Kerr cell	299,776	14
1950	Bergstrand	Sweden	Geodimeter	299,792.7	0.25
1950	Essen	England	Microwave cavity	299,792.5	3
1950	Bol and Hansen	U.S.A.	Microwave cavity	299,789.3	0.4
1951	Aslakson	U.S.A.	Shoran radar	299,794.2	1.9
1952	Rank, Ruth, and Ven der Sluis	U.S.A.	Molecular spectra	299,776	7
1952	Froome	England	Microwave interferometer	299,792.6	0.7
1954	Florman	U.S.A.	Microwave interferometer	299,795.1	1.9
1957	Bergstrand	Sweden	Geodimeter	299,792.85	0.16
1958	Froome	England	Microwave interferometer	299,792.50	0.10
1965	Kolibayev	U.S.S.R.	Geodimeter	299,792.6	0.06
1967	Grosse	West Germany	Geodimeter	299,792.5	0.05
1973	Evenson, Wells, Peterson, Danielson, and Day	U.S.A.	Laser techniques	299,792.4574	0.0012

* See "Some Recent Determinations of the Velocity of Light, III" by Joseph F. Mulligan, *American Journal of Physics*, October 1976. Here the 1974 measurements of Blayney et al. at the National Physical Laboratory in England are described. They are in very close agreement with the results of Evenson et al. (1973).

The task of arriving at a single "best" value for any physical quantity, such as c, from many independent measurements is usually difficult because it involves a careful evaluation of each measurement and a complex averaging process, which takes into account other physical quantities with which the quantity in question may be associated. In the case of c, however, the matter is straightforward. The latest (1973) compilers† of the "best" values of the physical constants state: "All past measurements of c have been rendered obsolete by Evenson et al.'s recent measurement . . ."; this measurement is the last entry in Table 42-1.

These workers, at the National Bureau of Standards in Boulder, Colorado, measured the frequency ν of a certain radiation emitted by a helium-neon laser by comparing it directly with the oscillation frequency of the cesium clock, which is used to define the second (see

† See footnote to Appendix B.

Appendix A). Then, using accurate measurements of the wavelength of this radiation made by several groups of workers, they computed c from the relation $c = \lambda \nu$ and deduced the value displayed in Appendix B, namely,

$$c = (299,792.4574 \pm 0.0012) \text{ km/s}.$$

The largest source of uncertainty in this measurement is that of the definition of the meter (needed to find the wavelength) in terms of radiations from the krypton-86 atom (see Appendix A).

It is now clear that the best measurements of c are *not* made by timing the passage of light over a measured distance, as by Fizeau in 1849 and by Michelson et al. in 1932 (see Table 42-1). They are made by measuring the frequency ν and the wavelength λ of the light and computing c from $c = \lambda \nu$. This holds true for either traveling waves, as in the measurements of Evenson and his collaborators, described above, or for standing waves, an example of which we now describe. These standing wave experiments will also remind us that c in free space has the same value throughout the entire electromagnetic spectrum and is not confined to visible light.

We describe here the "microwave cavity method" used by Essen in England and by Bol and Hansen in the U.S.A. It employs standing electromagnetic waves confined to a cavity rather than traveling waves in free space.

It is possible to convert a section of waveguide such as that of Fig. 41-9 into a resonant cavity by closing it with two metal caps; see Fig. 42-5. The pattern of oscillations in the cavity is closely related to that in the guide and exhibits the same "guide wavelength" λ_g. The guide wavelength is related to the cavity length l by

$$\lambda_g = \frac{2l}{n} \qquad n = 1, 2, 3, \cdots, \tag{42-12}$$

which is the same relationship used for acoustic waves in closed pipes; n (=3 for Fig. 42-5) gives the number of half-waves contained in the cavity.

The procedure is to measure λ_g for such a cavity, which has been tuned to resonance, and then, using Eq. 41-4,

$$\lambda_g = \frac{\lambda}{\sqrt{1 - \left(\frac{\lambda}{2a}\right)^2}}, \tag{42-13}$$

calculate the free-space wavelength λ. From the measured resonant frequency, the speed c can be found from $c = \lambda \nu$.

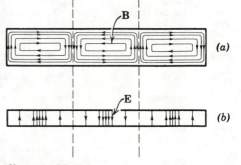

figure 42-5
A resonant cavity constructed from a section of waveguide; compare Fig. 41-9. For simplicity the lines of **E** are not shown in (a) and those of **B** are not shown in (b).

EXAMPLE 3

Essen of the National Physical Laboratory in England made a resonant cavity measurement of the speed of electromagnetic waves in 1950; see Table 42-1. His cavity was made of a circular waveguide rather than a rectangular one; it can be shown that, for the oscillation pattern used by him, the geometrical factor $2a$ in Eq. 42-13 must be replaced by $1.64062R$, where R is the guide radius. The cavity radius was 3.25876 cm; the cavity length was 15.64574 cm and it proved to resonate at 9.498300×10^9 Hz. At resonance it was determined that there were eight half-waves in the cavity. What value of c results?

From Eq. 42-12, computing only an approximate result,

$$\lambda_g = \frac{2l}{n} = \frac{(2)(15.6 \text{ cm})}{8} = 3.90 \text{ cm}.$$

Substituting into Eq. 42-13, suitably modified for a circular waveguide, yields

$$3.90 \text{ cm} = \frac{\lambda}{\sqrt{1 - \left(\frac{\lambda}{(1.64)(3.26 \text{ cm})}\right)^2}}.$$

Solving this equation for λ yields $\lambda = 3.15$ cm. Finally, we have

$$c = \lambda \nu = (3.15 \text{ cm})(9.50 \times 10^9 \text{ Hz}) = 2.99 \times 10^8 \text{ m/s}.$$

For practical reasons Essen analyzed his data in a more roundabout way than that given. His final result, based on many measurements under different conditions and carried to much greater accuracy than that illustrated in the above example, was 299,792.5 km/s with an uncertainty of 3.0 km/s.

When we say that the speed of sound in dry air at 0°C is 331.7 m/s, we imply a reference frame fixed with respect to the air mass. When we say that the speed of light in free space is 2.99792458×10^8 m/s, what reference frame is implied? It cannot be the medium through which the light wave travels because, in contrast to sound, no medium is required.

The concept of a wave requiring no medium was abhorrent to the physicists of the nineteenth century, influenced as they then were by a false analogy between light waves and sound waves or other purely mechanical disturbances. These physicists postulated the existence of an *ether*, which was a tenuous substance that filled all space and served as a medium of transmission for light. The ether was required to have a vanishingly small density to account for the fact that it could not be observed by any known means in an evacuated space.

The ether concept, although it proved useful for many years, did not survive the test of experiment. In particular, careful attempts to measure the speed of the earth through the ether always gave the result of zero.† Physicists were not willing to believe that the earth was permanently at rest in the ether and that all other bodies in the universe were in motion through it. Other hypotheses about the nature of the propagation of light also proved unsatisfactory for one reason or another.

Einstein in 1905 resolved the difficulty of understanding the propagation of light by making a bold postulate: If a number of observers are moving (at uniform velocity) with respect to each other and to a source of light and if each observer measures the speed of the light emerging from the source, *they will all obtain the same value.* This is the fundamental assumption of Einstein's special theory of relativity. It does away with the need for an ether by asserting that the speed of light is the same in *all* reference frames; none is singled out as fundamental. The theory of relativity, derived from this postulate, has been subject to many experimental tests, from which agreement with the predictions of theory has always emerged. These agreements, extending over half a century, lend strong support to Einstein's basic postulate about light propagation.

Figure 42-6 focuses specifically on the fundamental problem of light propagation. A source of light, at rest in reference frame S', emits a

42-4
MOVING SOURCES AND OBSERVERS*

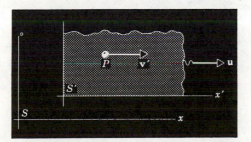

figure 42-6
Observers S and S', who are in relative motion, each observe a light pulse P. The pulse is emitted from a source, not shown, that is at rest in the S' frame of reference.

* For further information about the material of this and the next section see Supplementary Topic V.
† See Section 45-7, which describes the crucial experiment of Michelson and Morley.

light pulse P whose speed v' is measured by an observer at rest in this same frame. From the point of view of an observer in reference frame S, frame S' and its associated observer are moving in the positive x-direction at speed u. Question: What speed v would observer S measure for the light pulse P? Einstein's hypothesis asserts that *each* observer would measure the same speed c, or that

$$v = v' = c.$$

This hypothesis contradicts the classical law of addition of velocities (see Section 4-6), which asserts that

$$v = v' + u. \tag{42-14}$$

This law, which is so familiar that it seems (incorrectly) to be intuitively true, is in fact based on observations of gross moving objects in the world about us. Even the fastest of these—an earth satellite, say—is moving at a speed that is quite small compared to that of light. The body of experimental evidence that underlies Eq. 42-14 thus represents a severely restricted area of experience, namely, experiences in which $v' \ll c$ and $u \ll c$. If we assume that Eq. 42-14 holds for all particles regardless of speed, we are making a gross extrapolation. Einstein's theory of relativity predicts that this extrapolation is indeed not valid and that Eq. 42-14 is a limiting case of a more general relationship that holds for light pulses and for material particles, whatever their speed, or

$$v = \frac{v' + u}{1 + v'u/c^2}. \tag{42-15}$$

Equation 42-15 is quite indistinguishable from Eq. 42-14 at low speeds, that is, when $v' \ll c$ and $u \ll c$; see Example 4.

If we apply Eq. 42-15 to the case in which the moving object is a light pulse, and if we put $v' = c$, we obtain

$$v = \frac{c + u}{1 + cu/c^2} = c.$$

This is consistent, as it must be, with the fundamental assumption on which the derivation of Eq. 42-15 is based; it shows that *both* observers measure the same speed c for light. Equation 42-14 predicts (incorrectly) that the speed measured by S will be $c + u$. Figure 42-7 shows that the (correct) Eq. 42-15 and the (approximate) Eq. 42-14 cannot be distin-

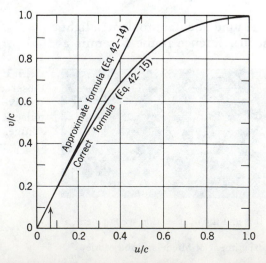

figure 42-7
The speed of a particle P, as seen by observer S in Fig. 42-6, for the special case of $v' = u$. All speeds are expressed as a ratio to c, the speed of light. The vertical arrow corresponds to about 0.25×10^8 m/s. ($=5 \times 10^7$ mi/h.)

EXAMPLE 4

Suppose that $v' = u = 1.0 \times 10^5$ m/s ($=0.0003c = 230,000$ mi/h). What error is made in using Eq. 42-14 rather than Eq. 42-15 to calculate v?

Equation 42-14 gives

$$v = v' + u = 1.0 \times 10^5 \text{ m/s} + 1.0 \times 10^5 \text{ m/s} = 2.0 \times 10^5 \text{ m/s}.$$

Equation 42-15 gives

$$v = \frac{v' + u}{1 + v'u/c^2}$$

$$= \frac{1.0 \times 10^5 \text{ m/s} + 1.0 \times 10^5 \text{ m/s}}{1 + \left(\dfrac{1.0 \times 10^5 \text{ m/s}}{3.0 \times 10^8 \text{ m/s}}\right)^2}$$

$$= \frac{2.0 \times 10^5 \text{ m/s}}{1.00000011}.$$

Even at 230,000 mi/hr the error in Eq. 42-14 is immeasurably small.

EXAMPLE 5

Two electrons are ejected in opposite directions from radioactive atoms in a sample of radioactive material. Let each electron have a speed, as measured by a laboratory observer, of $0.6c$ (this corresponds to a kinetic energy of 130 keV). What is the speed of one electron as seen from the other?

Equation 42-14 gives

$$v = v' + u = 0.6c + 0.6c = 1.2c.$$

Equation 42-15 gives

$$v = \frac{v' + u}{1 + v'u/c^2} = \frac{0.6c + 0.6c}{1 + (0.6c)^2/c^2} = 0.88c.$$

This example shows that for speeds that are comparable to c, Eqs. 42-14 and 42-15 yield rather different results. A wealth of indirect experimental evidence points to the latter result as being correct.

42-5
DOPPLER EFFECT

We have seen that the same speed is measured for light no matter what the relative speeds of the light source and the observer are. The measured frequency and wavelength will change, but always in such a way that their product, which is the velocity of light, remains constant. Such frequency shifts are called *Doppler shifts*, after Johann Doppler (1803–1853), who first predicted them.

In Section 20-7 we showed that if a source of *sound* is moving away from an observer at a speed u, the frequency heard by the observer (see Eq. 20-10, which we have rearranged and in which we have substituted u for v_s) is

$$v' = v \, \frac{1}{1 + u/v} \cdot \quad \begin{cases} 1. \text{ sound wave} \\ 2. \text{ observer fixed in medium} \\ 3. \text{ source receding from observer} \end{cases} \quad (42\text{-}16)$$

In this equation v is the frequency heard when the source is at rest and v is the speed of sound.

If the source is at rest in the transmitting medium but the observer is moving away from the source at speed u, the observed frequency (see Eq. 20-9b, in which u has been substituted for v_0) is

$$\nu' = \nu\left(1 - \frac{u}{v}\right) \cdot \quad \begin{cases} 1. \text{ sound wave} \\ 2. \text{ source fixed in medium} \\ 3. \text{ observer receding from source} \end{cases} \quad (42\text{-}17)$$

Even if the relative separation speeds u of the source and the observer are the same, the frequencies predicted by Eqs. 42-16 and 42-17 are different. This is not surprising, because a sound source moving through a medium in which the observer is at rest is physically different from an observer moving through that medium with the source at rest, as comparison of Figs. 20-10 and 20-11 shows.

We might be tempted to apply Eqs. 42-16 and 42-17 to light, substituting c, the speed of light, for v, the speed of sound. For light, as contrasted with sound, however, it has proved impossible to identify a medium of transmission relative to which the source and the observer are moving. This means that "source receding from observer" and "observer receding from source" are physically identical situations and must exhibit *exactly the same* Doppler frequency. As applied to light, either Eq. 42-16 or Eq. 42-17 or both must be incorrect. The Doppler frequency predicted by the theory of relativity is, in fact,

$$\nu' = \nu\,\frac{1 - u/c}{\sqrt{1 - (u/c)^2}} \cdot \quad \begin{cases} 1. \text{ light wave} \\ 2. \text{ source and observer} \\ \quad \text{separating} \end{cases} \quad (42\text{-}18)$$

In all three of the foregoing equations we obtain the appropriate relations for the source and the observer *approaching* each other if we replace u by $-u$.

Equations 42-16, 42-17, and 42-18 are not so different as they seem if the ratio u/c is small enough. This was made clear in Example 3, Chapter 20, for the first two of these equations. Let us expand Eqs. 42-16, 42-17, and 42-18 by the binomial theorem, as in the example referred to. The equations then become, substituting c for v,

$$\nu' = \nu\left[1 - \frac{u}{c} + \left(\frac{u}{c}\right)^2 + \cdots\right], \qquad (42\text{-}16a)$$

$$\nu' = \nu\left(1 - \frac{u}{c}\right), \qquad (42\text{-}17a)$$

and

$$\nu' = \nu\left[1 - \frac{u}{c} + \frac{1}{2}\left(\frac{u}{c}\right)^2 + \cdots\right]. \qquad (42\text{-}18a)$$

The ratio u/c for all available monochromatic light sources, even those of atomic dimensions, is small. This means that successive terms in these equations become small rapidly and, depending on the accuracy required, only a limited number of terms need be retained.

Under nearly all circumstances the differences among these three equations are not important. Nevertheless, it is of extreme interest to carry out at least one experiment precisely enough to serve as a test of Eq. 42-18a and thus, in part, of the theory of relativity.

H. E. Ives and G. R. Stilwell carried out such a precision experiment in 1938. They sent a beam of hydrogen atoms, generated in a gas discharge, down a tube at speed u, as in Fig. 42-8a. They could observe light emitted by these atoms in a direction opposite to **u** (atom 1, for example) using a mirror, and also in a direction parallel to **u** (atom 2, for example). With a precision spectrograph, they could photograph a particular characteristic spectrum line in this light, obtain-

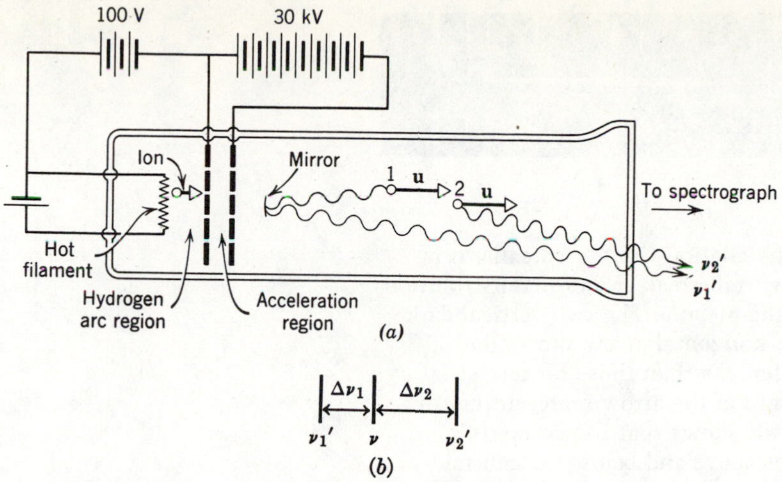

figure 42-8
The Ives-Stilwell experiment.

ing, on a frequency scale, the lines marked ν_1' and ν_2' in Fig. 42-8b. It is also possible to photograph, on the same photographic plate, a line corresponding to light emitted from *resting* atoms; such a line appears as ν in Fig. 42-9b. A fundamental measured quantity in this experiment is $\Delta\nu/\nu$, defined from

$$\frac{\Delta\nu}{\nu} = \frac{\Delta\nu_2 - \Delta\nu_1}{\nu},$$ (42-19)

(see Fig. 42-8b). It measures the extent to which the frequency of the light from resting atoms fails to lie halfway between the frequencies ν_1' and ν_2'. Table 42-2 shows that the measured results agree with the formula predicted by the theory of relativity (Eq. 42-18a) and not with the classical formula borrowed from the theory of sound propagation in a material medium (Eq. 42-16a).

Table 42-2
The Ives-Stilwell experiment*

$\dfrac{\Delta\nu}{\nu}$, 10^{-5}	Speed of moving atoms $(=u)$, 10^6 m/s			
	0.865	1.01	1.15	1.33
Theoretical value according to classical theory (Eq. 42-16a)	1.67	2.26	2.90	3.94
Theoretical value according to the theory of relativity (Eq. 42-18a)	0.835	1.13	1.45	1.97
Experimental value	0.762	1.1	1.42	1.9

* See Eq. 42-19; the table shows only part of the data taken by Ives and Stilwell.

Ives and Stilwell did not present their experimental results as evidence for the support of Einstein's theory of relativity but rather gave them an alternative theoretical explanation. Modern observers, looking not only at their excellent experiment but at the whole range of experimental evidence, now give the Ives-Stilwell experiment the interpretation we have described for it above.

The Doppler effect for light finds many applications in astronomy, where it is used to determine the speeds at which luminous heavenly bodies are moving toward us or receding from us. Such Doppler shifts measure only the radial or line-of-sight components of the relative velocity. All galaxies† for which such measurements have been made

† See "The Red-Shift" by Allen R. Sandage, *Scientific American*, September 1956.

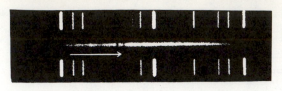

(a)　　　　　　　　　　　　*(b)*

figure 42-9

(*a*) The central spot is a nebula in the constellation Corona Borealis; it is 130,000,000 light years distant. (*b*) The central streak shows the distribution in wavelength of the light emitted from this nebula. The two vertical dark bands show the presence of calcium. The horizontal arrow shows that these calcium lines occur at longer wavelengths than those for terrestrial light sources containing calcium, the length of the arrow representing the wavelength shift. Measurement of this shift shows that the galaxy is receding from us at 13,400 mi/s. The lines above and below the central streak represent light from a terrestrial source, used to establish a wavelength scale. (Courtesy Mount Wilson and Mount Palomar Observatories.)

(Fig. 42-9) appear to be receding from us, the recession velocity being greater for the more distant galaxies; these observations are the basis of the expanding-universe concept.

EXAMPLE 6

Certain characteristic wavelengths in the light from a galaxy in the constellation Virgo are observed to be increased in wavelength, as compared with terrestrial sources, by about 0.4%. What is the radial speed of this galaxy with respect to the earth? Is it approaching or receding?

If λ is the wavelength for a terrestrial source, then

$$\lambda' = 1.004\lambda.$$

Since we must have $\lambda'\nu' = \lambda\nu = c$, we can write this as

$$\nu' = 0.996\nu.$$

This frequency shift is so small that, in calculating the source velocity, it makes no practical difference whether we use Eq. 42-16, 42-17, or 42-18. Using Eq. 42-17 we obtain

$$\nu' = 0.996\nu = \nu\left(1 - \frac{u}{c}\right).$$

Solving yields $u/c = 0.004$, or $u = (0.004)(3 \times 10^8 \text{ m/s}) = 1.2 \times 10^6$ m/s or 2.7×10^6 mi/h. The galaxy is *receding*; had u turned out to be negative, the galaxy would have been moving toward us.

questions

1. How might an eye-sensitivity curve like that of Fig. 42-1 be measured?
2. Why are danger signals in red, when the eye is most sensitive to yellow-green?
3. Comment on this definition of the limits of the spectrum of visible light given by a physiologist: "The limits of the visible spectrum occur when the eye is no better adapted than any other organ of the body to serve as a detector."
4. The human body can "detect" electromagnetic radiations in parts of the electromagnetic spectrum apart from the visual range. Give examples for the infrared, ultraviolet, X-ray, and gamma ray regions. Some of the "detection processes" are not very pleasant.

5. In connection with Fig. 42-1 (a) do you think it possible that the wavelength of maximum sensitivity could vary if the intensity of the light is changed? (b) What might the curve of Fig. 42-1 look like for a group of color blind people who could not, for example, distinguish red from green? (See "The Science of Color" by S. J. Edwards, *Physics Education*, June 1975.)

6. Suppose that human eyes were insensitive to visible light but were very sensitive to infrared light. What environmental changes would be needed if you were to (a) walk down a long corridor and (b) drive a car? Would the phenomenon of color exist? How would traffic lights have to be modified?

7. How can an object absorb light energy without absorbing momentum?

8. Name two historic experiments, in addition to the radiation pressure measurements of Nichols and Hull, in which a torsion balance was used. Both are described in this book, one in Part 1 and one in Part 2.

9. A searchlight sends out a parallel beam of light. Does the searchlight experience any force associated with the emission of light?

10. In Section 42-2 we stated that the force on the mirror in the radiation pressure experiment of Nichols and Hull (see Fig. 42-2) was about 7×10^{-10} N. Identify an object whose weight at the earth's surface is about this magnitude (Table 1-4 might help).

11. Discuss this (1972) statement by Lewis M. Branscomb, Director of the National Bureau of Standards. "Scientists have looked forward to the possibility of using one gauge—one 'yardstick so to speak'—not only for the three dimensions of space but for the fourth dimension of time as well. To interchange clocks and rules scientists must know the speed with which light travels. . . . With this demonstration that both the space (wavelength) and time (frequency) dimensions of a single light source can be measured with prodigious accuracy, this goal is now within our grasp." What does he mean? (See *Science News*, February 1972.)

12. As you recline in a beach chair in the sun why are you so very conscious of the thermal energy that is delivered to you, but totally unresponsive to the linear momentum delivered from the same source?

13. Some advocate that the speed of light, now known to such high accuracy, be proclaimed as a defined constant rather than a measured one. What are the implications of this in terms of the present definitions of the meter and the second (Appendix A)?

14. When a parallel beam of light falls on an object the momentum transfers are given by Eqs. 42-2. Do these equations still hold if the light source is moving rapidly toward or away from the object at, perhaps, a speed of 0.1 c?

15. How could Galileo test experimentally that reaction times were an overwhelming source of error in his attempt to measure the speed of light, described on p. 923?

16. It has been suggested that the velocity of light may change slightly in value as time goes on. Can you find any evidence for this in Table 42-1?

17. Can you think of any "every day" observation (that is, without experimental apparatus) to show that the speed of light is not infinite? Think of lightning flashes, possible discrepancies between the predicted time of sunrise and the observed time, radio communications between earth and astronauts in orbiting space ships, and so on? Discuss.

18. A friend asserts that Einstein's postulate (that the speed of light is not affected by the uniform motion of the source or the observer) must be discarded because it violates "common sense." How would you answer him?

19. Why is the rotating mirror method for measuring the speed of light better than the toothed wheel method of Fizeau? (See Fig. 42-4.)

20. In a vacuum, does the speed of light depend on (a) the wavelength, (b) the frequency, (c) the intensity, (d) the speed of the source, or (e) the speed of the observer?

21. Can a galaxy be so distant that its recession speed equals c? If so, how can we see the galaxy? That is, will its light ever reach us?

22. How do the Doppler effects for light and for sound differ? In what ways are they the same?

23. Gamma rays are electromagnetic radiation emitted from radioactive nuclei. In free space, do they travel with the same speed as visible light? Does their speed depend on the speed of the nucleus that emits them?

problems

SECTION 42-1

1. (*a*) At what wavelengths does the eye sensitivity have half its maximum value? (*b*) What are the frequency and the period of the light for which the eye is most sensitive? See Fig. 42-1.
Answer: (*a*) 510 and 610 nm. (*b*) 5.5×10^{14} Hz and 1.8×10^{-15} s.

SECTION 42-2

2. Radiation from the sun striking the earth has an intensity of 1400 W/m². Assuming that the earth behaves like a flat disk at right angles to the sun's rays and that all the incident energy is absorbed, calculate the force on the earth due to radiation pressure. Compare it with the force due to the sun's gravitational attraction.

3. What is the radiation pressure 1.0 m away from a small 500-W light bulb? Assume that the surface on which the pressure is exerted faces the bulb and is perfectly absorbing and that the bulb radiates uniformly in all directions.
Answer: 1.3×10^{-7} Pa.

4. Prove, for a plane wave at normal incidence on a plane surface, that the radiation pressure on the surface is equal to the energy density in the beam outside the surface. This relation holds no matter what fraction of the incident energy is reflected.

5. Prove, for a stream of bullets striking a plane surface at right angles, that the "pressure" is *twice* the (kinetic) energy density in the stream above the surface; assume that the bullets are completely "absorbed" by the surface. Contrast this with the behavior of light. (See preceding problem.)

6. Show that for complete absorption of a parallel beam of light, the radiation pressure on the absorbing object is given by $p = S/c$, where S is the magnitude of the Poynting vector and c is the speed of light in free space.

7. It has been proposed that a spaceship might be propelled in the solar system by radiation pressure, using a large sail made of aluminum foil. How large must the sail be if the radiation force is to be equal in magnitude to the sun's gravitational attraction? Assume that the mass of the *ship + sail* is 1500 kg, that the sail is perfectly reflecting, and that the sail is oriented at right angles to the sun's rays. The sun's mass is 1.97×10^{30} kg. (Incidentally, NASA had plans to build such a "space clipper" to intercept Halley's Comet on its next swing into the inner solar system in 1986. The basic design was a sail made of polymer film coated with an aluminum film a few atoms thick. The sail area would be about 100 acres. The space clipper ship would have been well instrumented but unmanned.)
Answer: 9.2×10^5 m² (3200 ft on edge).

8. A particle in the solar system is under the combined influence of the sun's gravitational attraction and the radiation force due to the sun's rays. Assume that the particle is a sphere of density 1.0 g/cm³ and that all of the incident light is absorbed. (*a*) Show that all particles with radius less than some critical radius, R_0, will be blown out of the solar system. (*b*) Calculate R_0. (*c*) Does R_0 depend on the distance from the earth to the sun? (See the appendices for the necessary constants.)

9. Show that $\epsilon_0 \mathbf{E} \times \mathbf{B}$ has the dimensions of $\dfrac{\text{momentum}}{\text{area-time}}$, whereas $\dfrac{1}{\mu_0} \mathbf{E} \times \mathbf{B}$ has

the dimensions of $\dfrac{\text{energy}}{\text{area-time}}$. (The vector $\epsilon_0 \mathbf{E} \times \mathbf{B}$ may be used for computing momentum flow in the same manner that $\mathbf{S} = \dfrac{1}{\mu_0} \mathbf{E} \times \mathbf{B}$ is used to compute energy flow.)

10. A small spaceship whose mass, with occupant, is 1500 kg is drifting in outer space, where no gravitational field exists. If it shines a searchlight, which radiates 10^4 W, into space, what speed would the ship attain in one day because of the reaction force associated with the momentum carried away by the light beam?

11. A plane electromagnetic wave propagating through a vacuum is described by the following electric and magnetic fields: $E_x = E_0 \sin (kz - \omega t)$, $B_y = \dfrac{E_0}{c} \sin (kz - \omega t)$, $E_y = E_z = B_x = B_z = 0$. (a) What is the direction of propagation of the wave? (b) Find the average power per unit area. (c) What is the rate at which momentum is delivered to a perfectly absorbing surface of area A that is normal to the direction of propagation?
    Answer: (a) +z-direction. (b) $E_0^2/2\mu_0 c$. (c) $\epsilon_0 E_0^2 A/2$.

12. A plane electromagnetic wave, with wavelength 3.0 m, travels in free space in the +x-direction with its electric vector **E**, of amplitude 300 V/m, directed along the y-axis. (a) What is the frequency ν of the wave? (b) What is the amplitude of the **B** field associated with the wave? (c) If $E = E_m \sin (kx - \omega t)$, what are the values of k and ω for this wave? (d) What is the time-averaged rate of energy flow per unit area associated with this wave? (e) If the wave fell upon a perfectly absorbing sheet of area A, what momentum would be delivered to the sheet per second and what is the radiation pressure exerted on the sheet?

SECTION 42-3

13. On page 926 the uncertainty in the measurement of the speed of light c is given as ±0.0012 km/s. In Appendix B it is given as ±0.004 parts per million. Show that these uncertainties are consistent.

14. For the value of the speed of light quoted on p. 926 (and in Appendix B) what is the uncertainty of the measurement, in feet/second? How many feet does light travel in one second?

15. Suppose that light is timed over a one-mile base line and its speed is measured to the accuracy quoted on p. 926. How large an error in the length of the base line could be tolerated, assuming other sources of error to be negligible.
    Answer: 3×10^{-4} in.

16. *Bradley's method for determining the speed of light.* Consider a star located on a line through the sun, drawn perpendicular to the plane of the earth's orbit about the sun. The star's distance is much greater than the radius of the earth's orbit. Show that, due to the finite speed of light, a telescope through which the star is seen must be tilted at an angle of 20.5" to the perpendicular, in the direction the earth is moving. This phenomenon, called *aberration, is* noticeable and was first explained by James Bradley in 1729. (Hint: draw an analogy to a person running through rain while holding an umbrella.)

17. Suppose that we were able to establish radio communication with the hypothetical inhabitants of a hypothetical planet orbiting our nearest star, α-Centauri, which is 4.2 light years from us. How long would it take to receive a reply to a message? Repeat for the Great Nebula in Andromeda, one of our closest extragalactic neighbors but 2×10^6 light years distant. What do these considerations lead you to conclude about the nature of our possible communication with extragalactic peoples?
    Answer: 8.4 y; 4×10^6 y.

18. In Table 42-1, which of the four determinations of c reported by Michelson and his collaborators best agrees with the currently accepted value of c as displayed in Appendix B? Take the uncertainties of measurement into ac-

count. You might make a graph showing, on a suitable scale, the three values and their uncertainties displayed as error bars.

19. The uncertainty of the distance to the moon, as measured by the reflection of radar waves from it, is about 0.8 km. Assuming that this uncertainty is associated only with the measurement of the elapsed time, what uncertainty in this time is implied? *Answer: 5.3 μs.*

20. Roemer's method for measuring the speed of light consisted in observing the apparent times of revolution of one of the moons of Jupiter. The true period of revolution is 42.5 hr. (*a*) Taking into account the finite speed of light, how would you expect the apparent time of revolution to alter as the earth moves in its orbit from point *x* to point *y* in Fig. 42-10? (*b*) What observations would be needed to compute the speed of light? Neglect the motion of Jupiter in its orbit. Figure 42-10 is not drawn to scale.

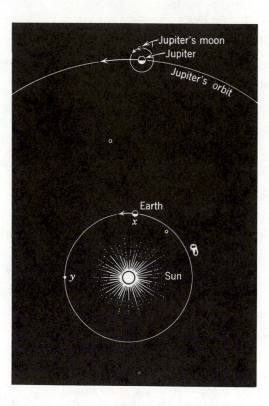

figure 42-10
Problem 20

SECTION 42-4

21. Assume that $v' = u$ in Eqs. 42-14 and 42-15. At what speed u would these expressions differ by 1.0%? *Answer: 0.1 c.*

SECTION 42-5

22. Show that a Doppler frequency shift in kHz can be converted to a radial velocity in km/s by multiplying by −0.211.

23. A rocketship is receding from the earth at a speed of 0.2c. A light in the rocketship appears blue to passengers on the ship. What color would it appear to be to an observer on the earth? (See Fig. 42-1.)
Answer: Yellow-orange.

24. The difference in wavelength between an incident microwave beam and one reflected from an approaching or receding car is used to determine automobile speeds on the highway. (*a*) Show that if *v* is the speed of the car and ν the frequency of the incident beam, the change of frequency is approximately $2v\nu/c$, where *c* is the speed of the electromagnetic radiation. (*b*) For microwaves of frequency 2450 MHz, what is the change of frequency per mi/hr of speed?

25. The period of rotation of the sun at its equator is 24.7 days; its radius is 7.0×10^8 m. What Doppler wavelength shifts are expected for characteristic wavelengths in the vicinity of 550 nm emitted from the edge of the sun's disk? *Answer:* 3.8×10^{-3} nm.

26. An earth satellite, transmitting on a frequency of 40 MHz (exactly), passes directly over a radio receiving station at an altitude of 250 miles and at a speed of 18,000 mi/hr. Plot the change in frequency, attributable to the Doppler effect, as a function of time, counting $t = 0$ as the instant the satellite is over the station. (*Hint:* The speed u in the Doppler formula is not the actual velocity of the satellite but its component in the direction of the station. Use the nonrelativistic formula (Eq. 42-16a) and neglect the curvature of the earth and of the satellite orbit.)

27. The "red shift" of radiation from a distant nebula consists of the light (H_γ), known to have a wavelength of 4340 Å when observed in the laboratory, appearing to have a wavelength of 6562 Å. (*a*) What is the speed of the nebula in the line of sight relative to the earth? (*b*) Is it approaching or receding? *Answer:* (*a*) 1.2×10^8 m/s. (*b*) Receding.

28. Show that, for slow speeds, the Doppler shift can be written in the approximate form

$$\frac{\Delta\lambda}{\lambda} = \frac{u}{c},$$

where $\Delta\lambda$ is the change in wavelength.

29. In the experiment of Ives and Stilwell the speed u of the hydrogen atoms in a particular run was 8.61×10^5 m/s. Calculate $\Delta\nu/\nu$, on the assumptions that (*a*) Eq. 42-18a is correct and (*b*) that Eq. 42-16a is correct; compare your results with those given in Table 42-2 for this speed. Retain the first three terms only in Eqs. 42-18a and 42-16a.
Answer: (*a*) $\Delta\nu/\nu = 0.825 \times 10^{-5}$. (*b*) $\Delta\nu/\nu = 1.65 \times 10^{-5}$.

43
reflection and refraction-plane waves and plane surfaces

So far we have considered the electromagnetic spectrum, including visible light, only in unimpeded free space. Here we examine it as reflected from plane surfaces such as glass or water and especially as it behaves when it passes through such transparent materials. As we shall see in the next chapter, without the consideration of these properties, such devices as cameras, telescopes, spectacles, microscopes, etc., could not be explained. As always, what we say in this chapter for visible light has its counterpart in other regions of the electromagnetic spectrum.

In Fig. 43-1a a light beam falling on a water surface is both reflected from the surface and bent (that is, *refracted*) as it enters the water.* We represent the incident beam in Fig. 43-1b by a single line, the *incident ray*, parallel to the direction of propagation. We assume the incident beam in Fig. 43-1b to be a *plane wave*, the wavefronts being normal to the incident ray. The reflected and refracted beams are also represented

43-1
REFLECTION AND REFRACTION

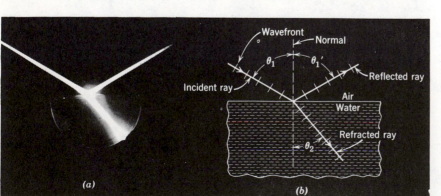

(a)   *(b)*

figure 43-1
(*a*) A photograph showing reflection and refraction at an air-water interface. (*b*) A representation using rays.

* A common and spectacular example of reflection and refraction is the rainbow. See "The Theory of the Rainbow" by H. Moysés Nussenzveig, *Scientific American*, April 1977.

by rays. The angles of *incidence* (θ_1), of *reflection* (θ_1'), and of *refraction* (θ_2) are measured between the normal to the surface and the appropriate ray, as shown in the figure.

By experiment, we find these laws governing reflection and refraction:

1. The reflected and the refracted rays lie in the plane formed by the incident ray and the normal to the surface at the point of incidence, that is, the plane of Fig. 43-1*b*.

2. For reflection:

$$\theta_1' = \theta_1. \qquad (43\text{-}1)$$

3. For refraction:

$$\frac{\sin \theta_1}{\sin \theta_2} = n_{21}, \qquad (43\text{-}2)$$

where n_{21} is a constant called the *index of refraction* of medium 2 with respect to medium 1. Table 43-1 shows the indices of refraction for some common substances with respect to a vacuum for a wavelength (sodium light) of 589 nm (= 5890 Å).

The index of refraction of one medium with respect to another generally varies with wavelength, as Fig. 43-2 shows. Because of this fact

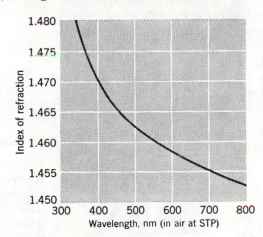

figure 43-2
The index of refraction of fused quartz with respect to a vacuum.

Table 43-1
Some indices of refraction*
[for $\lambda = 589$ nm (= 5890 Å)]

Medium	Index of Refraction
Water	1.33
Ethyl alcohol	1.36
Carbon bisulfide	1.63
Air (1 atm and 20°C)	1.0003
Methylene iodide	1.74
Fused quartz	1.46
Glass, crown	1.52
Glass, dense flint	1.66
Sodium chloride	1.53

* Measured with respect to a vacuum. The index with respect to air (except, of course, the index of air itself) will be negligibly different in most cases.

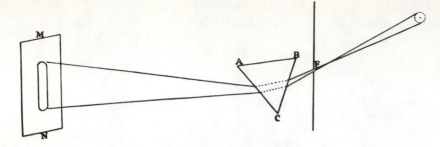

refraction, unlike reflection, can be used to analyze a beam of light into its component wavelengths. Figure 43-3, taken from Newton's *Opticks,* shows how Newton, using glass prism *ABC*, formed a spectrum of sunlight entering his window through a small hole at *F.*

The law of reflection was known to Euclid. That of refraction was discovered experimentally by Willebrod Snell (1591–1626) and deduced from the early corpuscular theory of light by René Descartes (1596–1650). The law of refraction is known as Snell's law or (in France) as Descartes' law.

We can derive the laws of reflection and refraction from Maxwell's equations, which means that these laws should hold for all regions of the electromagnetic spectrum. Figure 43-4*a* shows an experimental setup for investigating the reflection of microwaves from a large metal sheet. Figure 43-4*b* shows the reading of the detector as a function of the angular position of the mirror. The existence of a reflected beam at the proper angle confirms the law of reflection for microwaves. There is ample experimental evidence that Eqs. 43-1 and 43-2 correctly describe the behavior of reflected and refracted beams in all parts of the electromagnetic spectrum.

It is common knowledge that a polished steel surface will form a well-defined reflected beam if an incident beam falls on it, but a sheet of paper will reflect

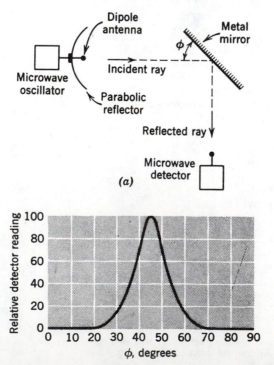

figure 43-4
(*a*) An apparatus to test the law of reflection for microwaves. (*b*) A reflected beam (for $\lambda \cong 10$ cm) appears for the expected orientation of the mirror.

light more or less in all directions (diffuse reflection). It is largely by diffuse re-flection that we see nonluminous objects around us. The difference between diffuse and specular (that is, mirrorlike) reflection is a matter of surface rough-ness; a reflected beam will be formed only if the average depth of the surface irregularities of the reflector is substantially less than the wavelength of the incident light. This criterion of surface roughness has different implications in different regions of the electromagnetic spectrum. The bottom of a cast-iron skillet, for example, is a good reflector for microwaves of wavelength 0.5 cm but is not a good reflector for visible light (that is, one cannot shave or put on makeup by it).

A second requirement for the existence of a reflected beam is that the trans-verse dimensions of the reflector must be substantially larger than the wave-length of the incident beam. If a beam of visible light falls on a polished metal disk the size of a dime, a reflected beam will be formed. However, if the same disk is placed in a beam of short radio waves with, say, $\lambda = 1.0$ m, radiation will be scattered in all directions from it, and no well-defined unidirectional beam will appear. We investigate this phenomenon of *diffraction* in Chapter 46. The requirements that surfaces be "smooth" and "large" also apply to the formation of refracted beams. If these two requirements are not met the description of re-flection and refraction in terms of rays, whose behavior is governed by Eqs. 43-1 and 43-2, is not valid.

EXAMPLE 1

Figure 43-5 shows an incident ray i striking a plane mirror MM' at angle of incidence θ. Trace this ray.

The reflected ray makes an angle θ with the normal at b and falls as an inci-dent ray on mirror $M'M''$. Its angle of incidence θ' on this mirror is $\pi/2 - \theta$. A second reflected ray r' makes an angle θ' with the normal erected at b'. Rays i and r' are antiparallel for any value of θ. To see this, note that

$$\phi = \pi - 2\theta' = \pi - 2\left(\frac{\pi}{2} - \theta\right) = 2\theta.$$

Two lines are parallel if their opposite interior angles for an intersecting line (ϕ and 2θ) are equal.

Repeat the problem if the angle between the mirrors is 120° rather than 90°.

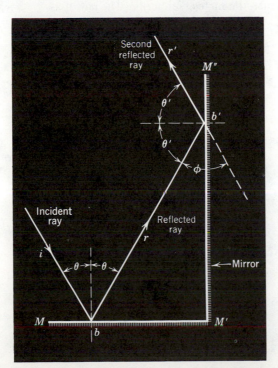

figure 43-5
Example 1.

EXAMPLE 2

An incident ray in air falls on the plane surface of a block of quartz and makes an angle of 30° with the normal. This beam contains two wavelengths, 400 and 500 nm. The indices of refraction for quartz with respect to air (n_{qa}) at these wavelengths are 1.4702 and 1.4624, respectively. What is the angle between the two refracted beams?

From Eq. 43-2 we have, for the 400-nm beam,

$$\sin\theta_1 = n_{qa}\sin\theta_2,$$

or

$$\sin 30° = (1.4702)\sin\theta_2,$$

which leads to

$$\theta_2 = 19.88°.$$

For the 500-nm beam we have

$$\sin 30° = (1.4624)\sin\theta_2',$$

or

$$\theta_2' = 19.99°.$$

The angle $\Delta\theta$ between the beams is 0.11°, the shorter wavelength component being bent through the larger angle, that is, having the smaller angle of refraction.

EXAMPLE 3

An incident ray falls on one face of a glass prism in air as in Fig. 43-6. The angle θ is so chosen that the emerging ray also makes an angle θ with the normal to the other face. Derive an expression for the index of refraction of the prism material with respect to air.

Note that $\angle abc = \alpha$, the two angles having their sides mutually perpendicular. Therefore,

$$\alpha = \tfrac{1}{2}\phi \text{ in which } \phi \text{ is the prism angle.} \tag{43-3}$$

The deviation angle ψ is the sum of the two opposite interior angles in triangle aed, or

$$\psi = 2(\theta - \alpha).$$

Substituting $\tfrac{1}{2}\phi$ for α and solving for θ yields

$$\theta = \tfrac{1}{2}(\psi + \phi). \tag{43-4}$$

At point a, θ is the angle of incidence and α the angle of refraction. The law of refraction (see Eq. 43-2) is

$$\sin\theta = n_{ga}\sin\alpha,$$

in which n_{ga} is the index of refraction of the glass with respect to air.

From Eqs. 43-3 and 43-4 this yields

$$\sin\frac{\psi + \phi}{2} = n_{ga}\sin\frac{\phi}{2}$$

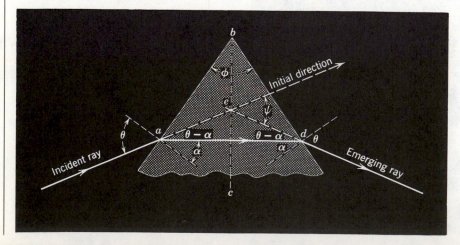

figure 43-6
Example 3.

or
$$n_{ga} = \frac{\sin \frac{1}{2}(\psi + \phi)}{\sin (\phi/2)}, \qquad (43\text{-}5)$$

which is the desired relation. This equation holds only for θ so chosen that the light ray passes symmetrically through the prism. For this condition the deviation angle ψ is a minimum; if θ is either increased or decreased, a larger deviation will be produced. ψ is called the *angle of minimum deviation.*

A theory of light would not be accepted if it were not able to predict the well-established laws of reflection and refraction. We could derive these laws from Maxwell's equations, but mathematical complexity prevents us from doing so here. Fortunately, we can derive these and several other laws of optics on the basis of a simpler but less comprehensive theory of light, put forward by the Dutch physicist Christian Huygens in 1678. This theory simply assumes that light is a wave rather than, say, a stream of particles. It says nothing about the nature of the wave and, in particular—since Maxwell's theory of electromagnetism appeared only after the lapse of a century—gives no hint of the electromagnetic character of light. Huygens did not know whether light was a transverse wave or a longitudinal one; he did not know the wavelengths of visible light; he had little knowledge of the speed of light. Nevertheless, his theory was a useful guide to experiment for many years and remains useful today for pedagogic and certain other practical purposes. We must not expect it to yield the same wealth of detailed information that Maxwell's more complete electromagnetic theory does.

Huygens' theory is based on a geometrical construction, called *Huygens' principle,* that allows us to tell where a given wavefront will be at any time in the future if we know its present position; it is: *All points on a wavefront can be considered as point sources for the production of spherical secondary wavelets. After a time t the new position of the wavefront will be the surface of tangency to these secondary wavelets.*

We illustrate this by a trivial example: Given a wavefront (ab in Fig. 43-7) in a plane wave in free space, where will the wavefront be a

43-2
HUYGENS' PRINCIPLE

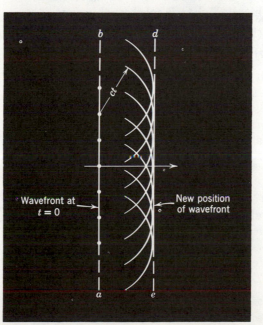

Wavefront at $t = 0$

New position of wavefront

figure 43-7
The propagation of a plane wave in free space is described by the Huygens construction. Note that the ray (horizontal arrow) representing the wave is perpendicular to the wavefronts.

time t later? Following Huygens' principle, we let several points on this plane (see dots) serve as centers for secondary spherical wavelets. In a time t the radius of these spherical waves is ct, where c is the speed of light in free space. We represent the plane of tangency to these spheres at time t by de. As we expect, it is parallel to plane ab and a perpendicular distance ct from it. Thus plane wavefronts are propagated as planes and with speed c. Note that the Huygens method involves a three-dimensional construction and that Fig. 43-7 is the intersection of this construction with the plane of the page.

We might expect that, contrary to observation, a wave should be radiated backward as well as forward from the dots in Fig. 43-7. This result is avoided by assuming that the intensity of the spherical wavelets is not uniform in all directions but varies continuously from a maximum in the forward direction to a minimum of zero in the back direction. This is suggested by the shading of the circular arcs in Fig. 43-7. Huygens' method can be applied quantitatively to *all* wave phenomena; see Problem 14. The method was put on a firm mathematical footing by Augustin Fresnel (1788–1827).

43-3
HUYGENS' PRINCIPLE AND THE LAW OF REFLECTION

Figure 43-8*a* shows three wavefronts in a plane wave falling on a plane mirror MM'. For convenience they are chosen to be one wavelength apart. Note that θ_1, the angle between the wavefronts and the mirror, is the same as the angle between the incident ray and the normal to the mirror. In other words, θ_1 is the *angle of incidence*. The three wavefronts are related to each other by the Huygens construction, as in Fig. 43-7.

In Fig. 43-8*b* a Huygens wavelet centered on point a will expand to include point l after a time λ/c. Light from point p in this same wavefront cannot move beyond the mirror but must expand upward as a spherical Huygens wavelet. Setting a compass to radius λ and swinging an arc about p provides a semicircle to which the reflected wavefront must be tangent. Since point l must lie on the new wavefront, this tangent must pass through l. Note that the angle θ_1' between the wavefront and the mirror is the same as the angle between the reflected ray and the normal to the mirror. In other words, θ_1' is the *angle of reflection*.

Consider right triangles alp and $a'lp$. They have side lp in common and side al $(=\lambda)$ is equal to side $a'p$. The two right triangles are thus congruent and we may conclude that

$$\theta_1 = \theta_1',$$

as required by the law of reflection. If you recall that the Huygens construction is three-dimensional and that the arcs shown represent segments of spherical surfaces, you will be able to convince yourself that the reflected ray lies in the plane formed by the incident ray and the normal to the mirror, that is, the plane of Fig. 43-8. This is also a requirement of the law of reflection; see p. 939.

43-4
HUYGENS' PRINCIPLE AND THE LAW OF REFRACTION

Figure 43-9 shows four stages in the refraction of three successive wavefronts in a plane wave falling on an interface between air (medium 1) and glass (medium 2). For convenience, we assume that the incident wavefronts are separated by λ_1, the wavelength as measured in medium 1. Let the speed of light in air be v_1 and that in glass be v_2. We assume that

$$v_2 < v_1. \tag{43-6}$$

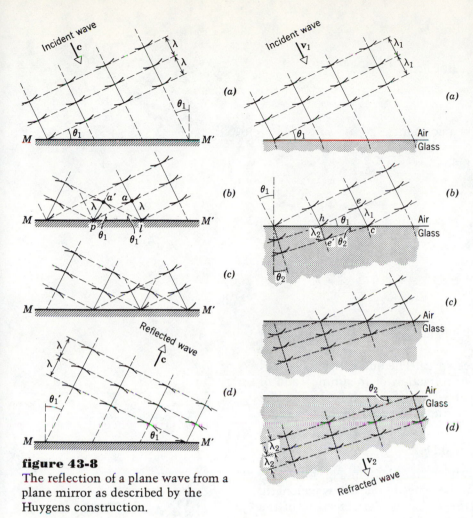

figure 43-9
The refraction of a plane wave at a plane interface as described by Huygens construction; we do not show the reflected wave, for simplicity. Note the change in wavelength on refraction.

figure 43-8
The reflection of a plane wave from a plane mirror as described by the Huygens construction.

This assumption about the speeds is vital to the derivation that follows. It was not possible to test it experimentally because of technical difficulties until 1850, when the assumption was shown by Foucault to be correct.

The wavefronts in Fig. 43-9a are related to each other by the Huygens construction of Fig. 43-7. As in Fig. 43-8, θ_1 is the angle of incidence. In Fig. 43-9b consider the time ($=\lambda_1/v_1$) during which a Huygens wavelet from point e moves to include point c. Light from point h, traveling through glass at a reduced speed (recall the assumption of Eq. 43-6) will move a shorter distance

$$\lambda_2 = \lambda_1 \frac{v_2}{v_1} \tag{43-7}$$

during this time. The refracted wavefront must be tangent to an arc of this radius centered on h. Since c lies on the new wavefront, the tangent must pass through this point, as shown. Note that θ_2, the angle between the refracted wavefront and the air-glass interface, is the same as the angle between the refracted ray and the normal to this interface. In other words, θ_2 is the *angle of refraction*. Note, too, that the wavelength in glass (λ_2) is less than the wavelength in air (λ_1).

For the right triangles hce and hce' we may write

$$\sin \theta_1 = \frac{\lambda_1}{hc} \qquad \text{(for } hce\text{)}$$

and

$$\sin \theta_2 = \frac{\lambda_2}{hc} \qquad \text{(for } hce'\text{)}.$$

Dividing and using Eq. 43-7 yields

$$\frac{\sin \theta_1}{\sin \theta_2} = \frac{\lambda_1}{\lambda_2} = \frac{v_1}{v_2} = \text{a constant.} \qquad (43\text{-}8)$$

The law of refraction, as stated in Eq. 43-2, is

$$\frac{\sin \theta_1}{\sin \theta_2} = n_{21}, \qquad (43\text{-}2)$$

so that n_{21} is now revealed as the ratio of the speeds of light in the two media, or

$$n_{21} = \frac{v_1}{v_2}. \qquad (43\text{-}9)$$

We may rewrite Eq. 43-8 as

$$\left(\frac{c}{v_1}\right) \sin \theta_1 = \left(\frac{c}{v_2}\right) \sin \theta_2, \qquad (43\text{-}10)$$

in which c is the speed of light in free space. The quantities (c/v_1) and (c/v_2) (see Eq. 43-9) are the indices of refraction of medium 1 and of medium 2, respectively, with respect to a vacuum. Introducing the symbols n_1 and n_2 for these quantities allows us to write the law of refraction as

$$n_1 \sin \theta_1 = n_2 \sin \theta_2. \qquad (43\text{-}11)$$

If we assume that the medium above the glass in Fig. 43-9 is a vacuum rather than air, the speed v_1 becomes c and the wavelength, called λ_1 in Fig. 43-9, assumes a value λ that is characteristic of the wave in free space. Equation 43-7 may thus be written

$$\lambda_2 = \lambda \frac{v_2}{c} = \frac{\lambda}{n_2} \qquad (43\text{-}12a)$$

or

$$\lambda_n = \frac{\lambda}{n}. \qquad (43\text{-}12b)$$

This shows specifically that the wavelength of light in a material medium is less than the wavelength of the same wave in a vacuum. Figure 43-9 shows clearly the difference in wavelength in the two media.

The application of Huygens' principle to refraction requires that if a light ray is bent toward the normal in passing from air to an optically dense medium then the speed of light in that optically dense medium (glass, say) must be *less* than that in air; see Eq. 43-6. This requirement holds for all wave theories of light. For the early particle theory of light put forward by Newton, refraction can be explained only if the speed of light in the medium in which light is bent toward the normal (the optically dense medium) is *greater* than that in air. The dense medium was thought to exert attractive forces on the light "corpuscles" as they neared the surface, speeding them up and changing their direction to cause them to make a smaller angle with the normal. Figure 43-10 shows a figure from a 1637 work of René Descartes, in which he makes an analogy between the refraction of light and the motion of a tennis ball on entering a medium in which it moves more slowly.

An experimental comparison of the speed of light in water and in air is decisive, therefore, between the wave and corpuscular theories of light. Such a measurement was first carried out by Foucault in 1850; he showed conclusively

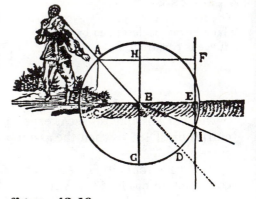

figure 43-10
According to the early (incorrect) corpuscular theory of light, ABI is the path of a ray on entering a medium in which its speed is less. The (correct) wave theory predicts that for the ray shown the speed in the lower medium (below BE) must be greater; from Descartes' La Dioptrique (1637).

that light travels more slowly in water than in air, thus ruling out the corpuscular theory of Newton.

43-5
TOTAL INTERNAL REFLECTION

Let light rays in an optically dense medium (glass, say) fall on a surface on the other side of which is a less optically dense medium (air, say); see Fig. 43-11. As the angle of incidence θ is increased, a situation is reached (see ray e) at which the refracted ray points along the surface, the angle of refraction being 90°. For angles of incidence larger than this *critical angle* θ_c no refracted ray exists, giving rise to a phenomenon called *total internal reflection*.

We find the critical angle by putting $\theta_2 = 90°$ in the law of refraction (see Eq. 43-11):

$$n_1 \sin \theta_c = n_2 \sin 90°,$$

or
$$\sin \theta_c = \frac{n_2}{n_1}. \tag{43-13}$$

For glass and air $\sin \theta_c = (1.00/1.50) = 0.667$, which yields $\theta_c = 41.8°$. Total internal reflection does not occur when light originates in the medium of lower index of refraction.

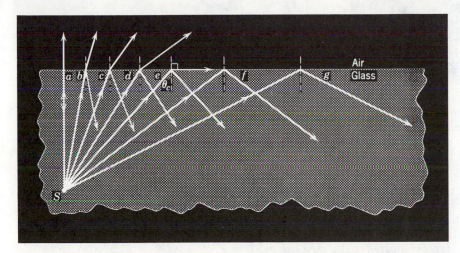

figure 43-11
Showing the total internal reflection of light from a source S; the *critical angle* is θ_e.

Light can be "piped" from one point to another with little loss by allowing it to enter one end of a rod of transparent plastic. The light will undergo total internal reflection at the boundary of the rod and will follow its contour, emerging at its far end. Images may be transferred from one location to another, using a bundle of fine glass fibers, each fiber transmitting a small fraction of the image.*

Fiber optics techniques make possible many useful optical devices for transmitting and transforming luminous images. Figure 43-12 shows a short fiber bundle constructed so that the fibers taper in diameter along its length. The wide end is shown placed over the letter S in the printed word "OPTICS." We see, with the aid of a mirror placed above the bundle, that a letter S, reduced in size, has been transmitted by total internal reflection through the bundle to its narrow end.

* See "Fiber Optics" by N. S. Kapany, *Scientific American*, November 1960.

figure 43-12
A bundle of tapered fibers (below) is placed over the letter *S*. Above, with the aid of a mirror, we see that the image, reduced in size, is transmitted to the top of the bundle by total internal reflection in the individual fibers. (Courtesy Dr. N. S. Kapany, Optics Technology, Inc.)

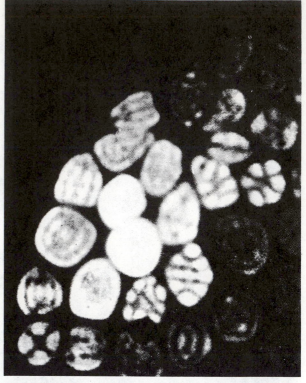

figure 43-13
A photomicrograph of light emerging from the end of a bundle of fibers. The fiber diameters approach the wavelength of light so that each fiber acts like an "optical waveguide." We have here convincing visual evidence that light is an electromagnetic wave. (Courtesy of Dr. N. S. Kapany, Optics Technology, Inc.)

Figure 43-13 is an enlarged view of a cross section of such a fiber bundle in which the diameters of the individual fibers are made so small that they are of the order of magnitude of the wavelength of light. This condition violates the spirit of our assumption of p. 941, namely, that the transverse dimensions of reflecting and refracting surfaces would be large compared to the wavelength of light. Consequently, a description of the reflection and refraction of light in terms of rays, as in Figs. 43-1*b* and 43-11, is not possible. Figure 43-13 is readily interpreted, however, on the basis of the electromagnetic wave theory of light and provides convincing pictorial supporting evidence for that theory. The fibers behave like waveguides (see Section 41-6), and the patterns of darkness and light represent the distribution of the **E** and **B** vectors for various modes of oscillation of the electromagnetic waves traveling down the "guides."

As of 1977 several experimental, field condition, tests of telephone communication* by optical fiber are underway. In a test in Chicago 24 hairlike fibers formed, with their protective sheath, a cable 0.50 in. in diameter and 1.5 miles long. The incoming voice signal is transformed into coded light pulses which are sent along the fiber at 4.47 pulses/second; the pulses are reconstituted at the far end to form a replica of the input signal. The design capability of the cable is 8064 simultaneous two-way conversations. The chemical purity requirements of the fiber materials and the design requirements of the electrical ⇌ optical input and output devices at the ends of the cable push solid-state and materials research technologies close to their present-day limits.

* See "Communication by Optical Fiber" by J. S. Cook, *Scientific American,* November 1973; "Telephoning by Light, I: The Breakthroughs" by John H. Douglas, *Science News,* July 1975; "Light-Wave Communications" by W. S. Boyle, *Scientific American,* August 1977.

EXAMPLE 4

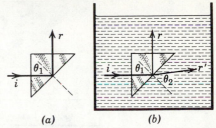

figure 43-14
Examples 4, 5

Figure 43-14*a* shows a triangular prism of glass, a ray incident normal to one face being totally reflected. If θ_1 is 45°, what can you conclude about the index of refraction *n* of the glass?

The angle θ_1 must be equal to or greater than the critical angle θ_c where θ_c is given by Eq. 43-13:

$$\sin \theta_c = \frac{n_2}{n_1} = \frac{1}{n},$$

in which, for all practical purposes, the index of refraction of air $(=n_2)$ is set equal to unity. Suppose that the index of refraction of the glass is such that total internal reflection just occurs, that is, that $\theta_c = 45°$. This would mean

$$n = \frac{1}{\sin 45°} = 1.41.$$

Thus the index of refraction of the glass must be *equal to or larger than* 1.41. If it were less, total internal reflection would not occur.

EXAMPLE 5

What happens if the prism in Example 4 (assume that $n = 1.50$) is immersed in water $(n = 1.33)$? See Fig. 43-14*b*.

The new critical angle, given by Eq. 43-13, is

$$\sin \theta_c = \frac{n_2}{n_1} = \frac{1.33}{1.50} = 0.887,$$

which corresponds to $\theta_c = 62.5°$. The actual angle of incidence $(=45°)$ is less than this so that we do *not* have total internal reflection.

There is a reflected ray, with an angle of reflection of 45°, as Fig. 43-14*b* shows. There is also a refracted ray, with an angle of refraction given by

$$n_1 \sin \theta_1 = n_2 \sin \theta_2$$

$$(1.50)(\sin 45°) = (1.33) \sin \theta_2,$$

which yields $\theta_2 = 52.9°$. Show that as $n_2 \rightarrow n_1$, $\theta_c \rightarrow 90°$.

Maxwell's equations permit us to calculate how the incident energy is divided between the reflected and the refracted beams. Figure 43-15 shows the theoretical prediction for (*a*) a light beam in air falling on a glass-air interface and (*b*) a light beam in glass falling on such an interface. Figure 43-15*a* shows that for angles of incidence up to about 50°, less than 10% of the light energy is reflected. At grazing incidence, however (that is, angles of incidence near 90°), the surface becomes an excellent reflector. We are all familiar with the high reflecting power of a wet road for light from automobile headlights that strikes near grazing incidence.

Figure 43-15*b* shows clearly that at a certain critical angle (41.8° in this case; see Eq. 43-13) *all* the light is reflected. For angles of incidence appreciably below this value, about 4% of the energy is reflected.

43-6
FERMAT'S PRINCIPLE

In 1650 Pierre Fermat discovered a remarkable principle which we can express in these terms: *A light ray traveling from one point to another will follow a path such that, compared with nearby paths, the time required is either a minimum or a maximum or will remain unchanged (that is, it will be stationary).*

We can readily derive the laws of reflection and refraction from this

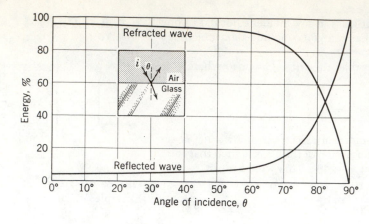

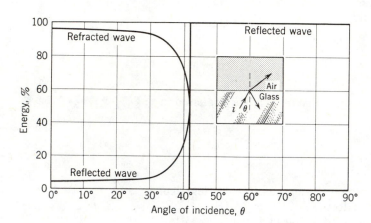

figure 43-15

(a) The percent of the energy reflected and refracted when an incident wave in air falls on glass (n = 1.50). The same for the incident wave in glass, showing total internal reflection.

principle. Figure 43-16 shows two fixed points A and B and a ray APB connecting them.* The total length l of this ray is

$$l = \sqrt{a^2 + x^2} + \sqrt{b^2 + (d - x)^2},$$

where x locates the point P at which the ray touches the mirror.

According to Fermat's principle, P will have a position such that the time of travel of the light must be a minimum (or a maximum or must remain unchanged). Expressed in another way, the total length l of the ray must be a minimum (or a maximum or must remain unchanged). In either case, the methods of the calculus require that dl/dx be zero. Taking this derivative yields

$$\frac{dl}{dx} = (\tfrac{1}{2})(a^2 + x^2)^{-1/2}(2x) + \tfrac{1}{2}[b^2 + (d - x)^2]^{-1/2}(2)(d - x)(-1) = 0,$$

which we can rewrite as

$$\frac{x}{\sqrt{a^2 + x^2}} = \frac{d - x}{\sqrt{b^2 + (d - x)^2}}.$$

Comparison with Fig. 43-16 shows that we can rewrite this as

$$\sin \theta_1 = \sin \theta_1',$$

or

$$\theta_1 = \theta_1',$$

which is the law of reflection.

To prove the law of refraction from Fermat's principle, consider Fig.

*We assume that ray APB lies in the plane of the figure; see Problem 26.

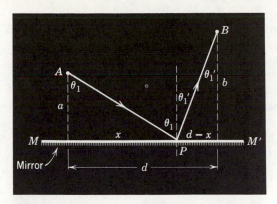

figure 43-16
A ray from A passes through B after reflection at P.

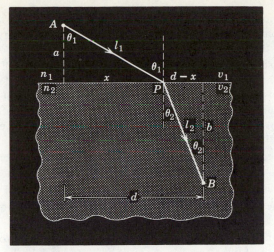

figure 43-17
A ray from A passes through B after refraction at P.

43-17, which shows two points A and B in two different media and a ray APB connecting them. The time t is given by

$$t = \frac{l_1}{v_1} + \frac{l_2}{v_2}.$$

Using the relation $n = c/v$ we can write this as

$$t = \frac{n_1 l_1 + n_2 l_2}{c} = \frac{l}{c}.$$

The quantity $l(= n_1 l_1 + n_2 l_2)$ is called the *optical path length* of the ray. Equation 43-12b $(\lambda_n = \lambda/n)$ shows that the optical path length is equal to the length that this same number of waves would have if the medium were a vacuum. We must not confuse the optical path length with the geometrical path length, which is $l_1 + l_2$.

Fermat's principle requires that l be a minimum (or a maximum or must remain unchanged) which, in turn, requires that x be so chosen that $dl/dx = 0$. The optical path length is

$$l = n_1 l_1 + n_2 l_2 = n_1 \sqrt{a^2 + x^2} + n_2 \sqrt{b^2 + (d - x)^2}.$$

Differentiating yields

$$\frac{dl}{dx} = n_1(\tfrac{1}{2})(a^2 + x^2)^{-1/2}(2x) + n_2(\tfrac{1}{2})[b^2 + (d - x)^2]^{-1/2}(2)(d - x)(-1) = 0,$$

which we can recast as

$$n_1 \frac{x}{\sqrt{a^2 + x^2}} = n_2 \frac{d - x}{\sqrt{b^2 + (d - x)^2}}.$$

Comparison with Fig. 43-17 shows that this, in turn, we can write as

$$n_1 \sin \theta_1 = n_2 \sin \theta_2,$$

which is the law of refraction.

In each of the examples of this section the time required, or, what is equivalent, the optical path length, proves to be a *minimum*.

questions

1. Describe what your immediate environment would be like if all objects were totally absorbing. Sitting in a chair in a room, could you see anything? If a person entered the room could you see her?

2. Would you expect sound waves to obey the laws of reflection and of refraction obeyed by light waves? Discuss the propagation of spherical and of cylindrical waves, using Huygens' principle. Does Huygens' principle apply to sound waves in air? If Huygens' principle predicts the laws of reflection and refraction, why is it necessary or desirable to view light as an electromagnetic wave, with all its attendant complexity?

3. A street light, viewed by reflection across a body of water in which there are ripples, appears very elongated. Explain.

4. The light beam in Fig. 43-1a is broadened on entering the water. Explain.

5. By what percent does the speed of blue light in fused quartz differ from that of red light?

6. Can (a) reflection phenomena or (b) refraction phenomena be used to determine the wavelength of light?

7. How can one determine the indices of refraction of the media in Table 43-1 relative to water, given the data in that table?

8. You are given a cube of glass. How can you find the speed of light (from a sodium light source) in this cube?

9. Describe and explain what a fish sees as he looks in various directions above his "horizon."

10. How did Foucault's measurement of the speed of light in water decide between the wave and the particle theories of light?

11. Why does a diamond "sparkle" more than a glass imitation cut to the same shape?

12. Is it plausible that the wavelength of light should change in passing from air into glass but that its frequency should not? Explain.

13. Light has (a) a wavelength, (b) a frequency, and (c) a speed. Which, if any, of these quantities remains unchanged when light passes from a vacuum into a slab of glass?

14. In reflection and refraction why do the reflected and refracted rays lie in the plane defined by the incident ray and the normal to the surface? Can you think of any exceptions?

15. What causes mirages? Does it have anything to do with the fact that the index of refraction of air is not constant but varies with its density? See "Mirages" by Alistair B. Fraser and William B. Mach, *Scientific American*, January, 1976.

16. Design a periscope, taking advantage of total internal reflection. What are the advantages compared with silvered mirrors?

17. What characteristics must a material have in order to serve as an efficient "light pipe"?

18. A certain toothbrush has a red plastic handle into which rows of nylon bristles are set. The tops of the bristles (but not their sides) appear red. Explain.

19. Discuss the formation of rainbows. See "The Theory of the Rainbow" by H. Moysés Nussenzveig, *Scientific American*, April 1977.

20. Why are optical fibers potentially more effective carriers of information than, say, microwaves or cables? Think of the frequencies involved.

21. What does "optical path length" mean? Can the optical path length ever be less than the geometrical path length? Ever greater?

22. A solution of copper sulphate appears blue when we view it through transmitted light. Does this mean that a copper sulphate solution absorbs blue light selectively? Discuss.

23. If you are an English literature major and interested in James Joyce, what would the letters *alp* and *hce* in Figures 43-8 and 43-9 mean to you?

problems

1. The wavelength of yellow sodium light in air is 5890 Å. (*a*) What is its frequency? (*b*) What is its wavelength in glass whose index of refraction is 1.52? (*c*) From the results of (*a*) and (*b*) find its speed in this glass. *Answer:* (*a*) 5.1×10^{14} Hz. (*b*) 3880 Å. (*c*) 1.98×10^8 m/s.

2. Prove that if a mirror is rotated through an angle α, the reflected beam is rotated through an angle 2α. Is this result reasonable for $\alpha = 45°$?

3. Ptolemy, who founded the city of Alexandria in Egypt toward the end of the first century A.D., gave the following measured values for the angle of incidence θ_1 and the angle of refraction θ_2 for a light beam passing from air to water:

θ_1	θ_2	θ_1	θ_2
10°	7°45′	50°	35°0′
20°	15°30′	60°	40°30′
30°	22°30′	70°	45°30′
40°	29°0′	80°	50°0′

Are these data consistent with Snell's law? If so, what index of refraction results? These data are interesting as the oldest recorded physical measurements. *Answer:* Yes; $n = 1.34$ vs. today's value of 1.33.

4. What is the speed in fused quartz of light of wavelength 550 nm (= 5500 Å)? (See Fig. 43-2.)

5. The speed of yellow sodium light in a certain liquid is measured to be 1.92×10^8 m/s. What is the index of refraction of this liquid, with respect to air, for sodium light? *Answer:* 1.56.

6. Suppose that the speed of light in air has been measured with an uncertainty of, say, 1 km/s. In calculating the speed in vacuum, suppose that it is not certain whether n for air is 1.00029 or 1.00030. (*a*) How much extra uncertainty is introduced into the calculated value for c? (*b*) Estimate how accurately n should be known for this purpose.

7. A bottom-weighted vertical pole extends 2.0 m above the bottom of a swimming pool and 0.5 m above the water. Sunlight is incident at 45°. What is the length of the shadow of the pole on the bottom of the pool? *Answer:* 1.4 m.

8. A 60° prism is made of fused quartz. A ray of light falls on one face, making an angle of 45° with the normal. Trace the ray through the prism graphically with some care, showing the paths traversed by rays representing (*a*) blue light, (*b*) yellow-green light, and (*c*) red light. (See Figs. 43-2 and 43-6.)

9. Prove that a ray of light incident on the surface of a sheet of plate glass of thickness t emerges from the opposite face parallel to its initial direction but displaced sideways, as in Fig. 43-18. Show that, for small angles of incidence θ, this displacement is given by

$$x = t\theta \frac{n-1}{n}$$

where n is the index of refraction and θ is measured in radians.

10. Assume that the index of refraction of the earth's atmosphere varies, with altitude only, from the value one at the "edge" of the atmosphere to some larger value at the surface of the earth. (*a*) Neglecting the earth's curvature, show that the apparent angle of a star from the zenith direction is independent of how the refractive index of the atmosphere varies with altitude and depends only on the value of n at the earth's surface. (*Hint:* Compare a uniform atmosphere with one consisting of layers of increasing refractive index.) (*b*) How does the earth's curvature affect the analysis?

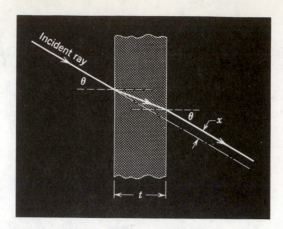

figure 43-18
Problem 9

11. Two perpendicular mirrors form the sides of a vessel filled with water, as shown in Fig. 43-19. A light ray is incident from above, normal to the water surface. (a) Show that the emerging ray is parallel to the incident ray. Assume that there are two reflections at the mirror surfaces. (b) Repeat the analysis for the case of oblique incidence, the ray lying in the plane of the figure. (c) Using three mirrors, state and prove the three-dimensional analog to this problem.

figure 43-19
Problem 11

12. Show that for a thin prism (ϕ small) and light not far from normal incidence (θ_1 small) the deviation angle is independent of the angle of incidence and is equal to $(n - 1)\,\phi$ (see Fig. 43-6).

13. In Fig. 43-6 show by graphical ray tracing, using a protractor, that if θ for the incident ray is *either* increased *or* decreased, the deviation angle ψ is increased. The symmetrical situation shown in this figure is called the *position of minimum deviation*.

SECTION 43-2

14. One end of a stick is dragged through water at a speed v which is greater than the speed u of water waves. Applying Huygens' construction to the water waves, show that a conical wavefront is set up and that its half-angle α is given by

$$\sin \alpha = u/v.$$

This is familiar as the bow wave of a ship or the shock wave caused by an object moving through air with a speed exceeding that of sound, as in Fig. 20-12.

15. When an electron moves through a medium at a speed exceeding the speed of light in that medium, it radiates electromagnetic energy (the Cerenkov effect, see Section 20-7). What minimum speed must an electron have in a liquid of refractive index 1.54 in order to radiate?
Answer: 1.9×10^8 m/s.

SECTION 43-5

16. A ray of light is incident normally on the face ab of a glass prism ($n = 1.52$), as shown in Fig. 43-20. (a) Assuming that the prism is immersed in air, find the largest value for the angle ϕ so that the ray is totally reflected at face ac. (b) Find ϕ if the prism is immersed in water.

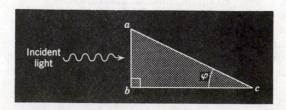

figure 43-20
Problem 16

17. A light ray falls on a square glass slab as in Fig. 43-21. What must the index of refraction of the glass be if total internal reflection occurs at the vertical face? *Answer: n > 1.22.*

18. A given monochromatic light ray, initially in air, strikes the 90° prism at P (see Fig. 43-22) and is refracted there and at Q to such an extent that it just grazes the right-hand prism surface after it emerges into air at Q. (a) Determine the index of refraction, relative to air, of the prism for this wavelength in terms of the angle of incidence θ_1 which gives rise to this situation. (b) Give a numerical upper bound for the index of refraction of the prism. (c) Show, by a ray diagram, what happens if the angle of incidence at P is slightly greater than θ_1, is slightly less than θ_1.

19. A glass prism with an apex angle of 60° has $n = 1.60$. (a) What is the smallest angle of incidence for which a ray can enter one face of the prism and emerge from the other? (b) What angle of incidence would be required for the ray to pass through the prism symmetrically, as in Fig. 43-6? *Answer: (a) 36°. (b) 53°.*

20. A point source is 80 cm below the surface of a body of water. Find the diameter of the largest circle at the surface through which light can emerge from the water.

21. A drop of liquid may be placed on a semicircular slab of glass as in Fig. 43-23. (a) Show how to determine the index of refraction of the liquid by observing total internal reflection. The index of refraction of the glass is unknown and must also be determined. Is the range of indices of refraction that can be measured in this way restricted in any sense? (b) In reality, how practical is this method?

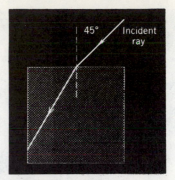

figure 43-21
Problem 17

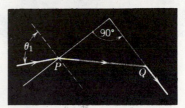

figure 43-22
Problem 18

figure 43-23
Problem 21

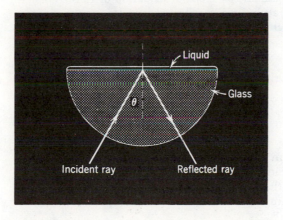

22. A point source of light is placed a distance h below the surface of a large deep lake. (a) Show that the fraction f of the light energy that escapes directly from the water surface is independent of h and is given by

$$f = \tfrac{1}{2}(1 - \sqrt{1 - 1/n^2})$$

where n is the index of refraction of water. (*Note:* Absorption within the water and reflection at the surface—except where it is total—have been neglected.) (b) Evaluate this fraction for $n = 1.33$.

23. Figure 43-24 shows a *constant-deviation prism*. Although made of one piece of glass, it is equivalent to two 30° 60° 90° prisms and one 45° 45° 90° prism. White light is incident in the direction i. θ_1 is changed by rotating the prism so that, in turn, light of any desired wavelength may be made to follow the path shown, emerging at r. Show that, if $\sin \theta_1 = \tfrac{1}{2}n$, then $\theta_2 = \theta_1$ and beams i and r are at right angles.

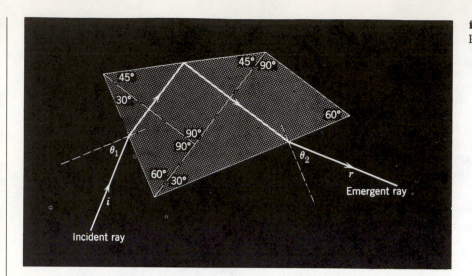

figure 43-24
Problem 23

24. A plane wave of white light traveling in fused quartz strikes a plane surface of the quartz, making an angle of incidence θ. Is it possible for the internally reflected beam to appear (a) bluish or (b) reddish? Roughly what value of θ must be used? (*Hint:* White light will appear bluish if wavelengths corresponding to red are removed from the spectrum.)

25. A glass cube has a small spot at its center. (a) What parts of the cube face must be covered to prevent the spot from being seen, no matter what the direction of viewing? (b) What fraction of the cube surface must be so covered? Assume a cube edge of 1.0 cm and an index of refraction of 1.50. (Neglect the subsequent behavior of an internally reflected ray.)
Answer: (a) Cover the center of each face with an opaque disk of radius 0.45 cm. (b) About 0.63.

SECTION 43-6

26. Using Fermat's principle, prove that the reflected ray, the incident ray, and the normal lie in one plane.

27. Prove that the optical path lengths for reflection and refraction in Figs. 43-16 and 43-17 are minimal when compared with other nearby paths connecting the same two points.

28. Light of free space wavelength 600 nm (= 6000 Å) travels 1.6×10^{-4} cm in a medium of index of refraction 1.5. Find (a) the optical path length, (b) the wavelength in the medium, and (c) the phase difference after moving that distance, with respect to light traveling the same distance in free space.

44

reflection and refraction-spherical waves and spherical surfaces

In Chapter 43 we described the reflection and refraction of plane waves at plane surfaces. In this chapter we consider the more general case of spherical waves falling on spherical reflecting and refracting surfaces. All of the results of Chapter 43 will emerge as special cases of the results of this chapter, because we can view a plane as a spherical surface with an infinite radius of curvature.

Both in Chapter 43 and in this chapter we make extensive use of *rays*. Although a ray is a convenient construction, it proves impossible to isolate one physically. Figure 44-1a shows schematically a plane wave of wavelength λ falling on a slit of width $a = 5\lambda$. We find that the light flares out into the geometrical shadow of the slit, a phenomenon called *diffraction*, which we will study in Chapter 46. Figures 44-1b ($a = 3\lambda$) and 44-1c ($a = \lambda$) show that diffraction becomes more pronounced as $a/\lambda \to 0$ and that attempts to isolate a single ray from the incident plane wave are futile.

Figure 44-2 shows water waves in a shallow ripple tank, produced by tapping the water surface periodically with the edge of a stick. We see that the plane wave so generated also flares out by diffraction when it encounters a gap in a barrier placed across it. Diffraction is characteristic of waves of all types. We can hear around corners, for example, because of the diffraction of sound waves.

The diffraction of waves at a slit (or at an obstacle such as a wire) is expected from Huygens' principle. Consider the portion of the wavefront that arrives at the position of the slit in Fig. 44-1. We can view every point on it as the site of an expanding spherical Huygens' wavelet. The "bending" of light into the region of the geometrical shadow is as-

44-1
GEOMETRICAL OPTICS AND WAVE OPTICS

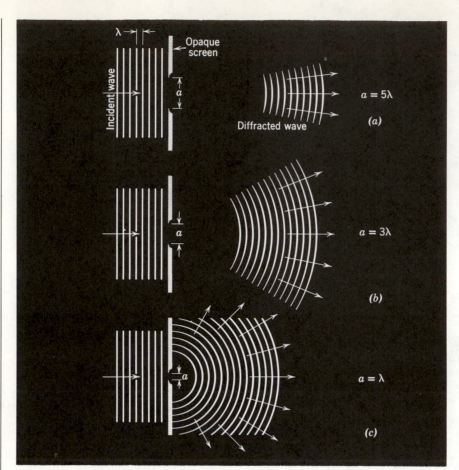

sociated with the blocking off of Huygens' wavelets from those parts of
the incident wavefront that lie behind the slit edges.

Figure 44-3 was made by allowing parallel light to fall on a slit placed
50 cm in front of a photographic plate. In Fig. 44-3*a* the slit width was
about 6 μm. The central band of light is much wider than this, showing that light has "flared out" into the geometric shadow of the slit.
In addition, many secondary maxima, omitted from Fig. 44-1 for simplicity, appear. Figure 44-3*b* shows what happens when the slit width
is reduced by a factor of two. The central maximum becomes wider, in
agreement with Fig. 44-1. Figure 44-3*c* shows the effect of reducing the
slit width by an additional factor of 7, to 0.4 μm. The central maximum
is now much wider and the secondary maxima, whose intensities relative to the central maximum have been deliberately overemphasized by
long exposure, are very evident.

Diffraction can be ignored if the ratio *a*/λ is large enough, *a* being a
measure of the smallest sideways dimension of the slit or obstacle. If
$a \gg \lambda$, light appears to travel in straight lines which we can represent by
rays that obey the laws of reflection (Eq. 43-1) and refraction (Eq. 43-2).
In Chapter 43 this condition, called *geometrical optics*, prevailed, the
lateral dimensions of all mirrors, prisms, etc., being much greater than
the wavelength. We also assume in this chapter that the conditions for
geometrical optics are *also* satisfied.

If the requirement for geometrical optics is not met, we cannot describe the behavior of light by rays but must take its wave nature specifically into account. This subject is called *wave optics*; it includes
geometrical optics as an important limiting case in much the same

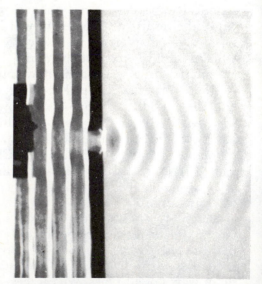

figure 44-2
Diffraction of water waves at a slit
in a ripple tank. Note that the slit
width is about the same size as the
wavelength. Compare with Fig.
44-1*c*. (Courtesy of Educational
Services Incorporated.)

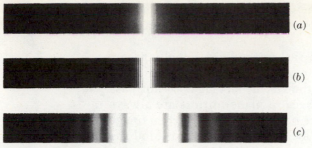

figure 44-3
(*a*) The intensity of light diffracted from a slit of width $a \cong 6$ μm and falling on a screen 50 cm beyond. (*b*) The slit width is reduced by a factor of two. (*c*) The slit width is further reduced by an additional factor of seven. Note that secondary maxima, made prominent in this case by deliberate overexposure, appear on either side of the central maximum. These secondary maxima have been omitted from Fig. 44-1 for simplicity.

sense that we can view an ideal gas as a limiting case of a real gas; both abstractions are exceedingly useful. We will treat wave optics in later chapters.

Figure 44-4 shows a point source of light O, the *object*, placed a distance o in front of a plane mirror. The light falls on the mirror as a spherical wave represented in the figure by rays emanating from O.* At the point at which each ray strikes the mirror we construct a reflected ray. If we

44-2
SPHERICAL WAVES—PLANE MIRROR

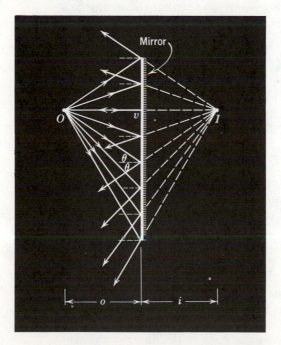

figure 44-4
A point object O forms a virtual image I in a plane mirror. The rays *appear* to emanate from I, but actually light does not pass through this point.

---

*In our discussion of reflection from mirrors in Chapter 43 (see Fig. 43-8) we assumed an incident *plane* wave; the incident rays are parallel to each other in that case. Here we have a *point* source and the rays striking the mirror are *diverging* from that point source.

extend the reflected rays backward, they intersect at a point *I* which is the same distance behind the mirror that the object *O* is in front of it; *I* is called the *image* of *O*.

Images may be *real* or *virtual*. In a real image light actually passes through the image point; in a virtual image the light *behaves* as though it diverges from the image point, although, in fact, it does not pass through this point; see Fig. 44-4. Images in plane mirrors are always virtual. We know from daily experience how "real" such a virtual image appears to be and how definite is its location in the space behind the mirror, even though this space may, in fact, be occupied by a brick wall.

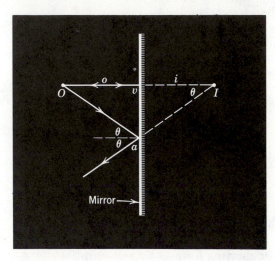

figure 44-5
Two rays from Fig. 44-4; ray *Oa* makes an arbitrary angle θ with the normal.

Figure 44-5 shows two rays from Fig. 44-4. One strikes the mirror at *v*, along a perpendicular line. The other strikes it at an arbitrary point *a*, making an angle of incidence θ with the normal at that point. Elementary geometry shows that the angles *aOv* and *aIv* are also equal to θ. Thus the right triangles *aOva* and *aIva* are congruent and

$$o = -i, \tag{44-1}$$

in which we arbitrarily introduce the minus sign to show that *I* and *O* are on opposite sides of the mirror. Equation 44-1 does not involve θ, which means that *all* rays striking the mirror pass through *I* when extended backward, as we have seen above. Beyond assuming that the mirror is truly plane and that the conditions for geometrical optics hold, we have made no approximations in deriving Eq. 44-1. A point object produces a point image in a plane mirror, with $o = -i$, no matter how large the angle θ in Fig. 44-5.

Because of the finite diameter of the pupil of the eye, only rays that lie fairly close together can enter the eye after reflection at a mirror. For the eye position shown in Fig. 44-6 only a small patch of the mirror near point *a* is effective in forming the image; the rest of the mirror may be covered up or removed. If we move our eye to another location, a different patch of the mirror will be effective; the location of the virtual image *I* will remain unchanged, however, as long as the object remains fixed.

If the object is an extended source such as the head of a person, a virtual image is also formed. From Eq. 44-1, every point of the source has an image point that lies an equal distance directly behind the plane of

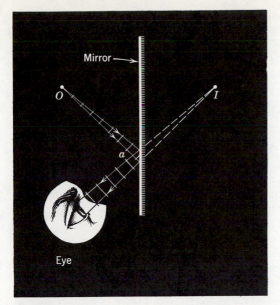

figure 44-6
A "pencil" of rays from O enters the eye after reflection at the plane mirror. Only a small portion of the mirror near a is effective. The small arcs represent portions of spherical wavefronts.

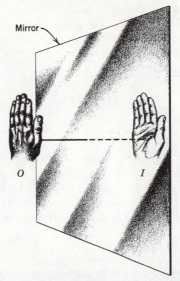

figure 44-7
A plane mirror reverses left and right. The object O is a left hand; the image I is a right hand. Try it in a mirror.

the mirror. Thus the image reproduces the object point by point. Most of us prove this every day by looking into a mirror.

Images in plane mirrors differ from objects in that left and right are interchanged. The image of a printed page is different from the page itself. Similarly, if a top is made to spin clockwise, the image, viewed in a vertical mirror, will seem to spin counterclockwise. Figure 44-7 shows an image of a left hand, constructed by using point-by-point application of Eq. 44-1; the image has the symmetry of a right hand.*

How tall must a vertical mirror be if a person 6 ft high is to be able to see her entire length? Assume that her eyes are 4 in. below the top of her head.

Figure 44-8 shows the paths followed by light rays leaving the top of the

EXAMPLE 1

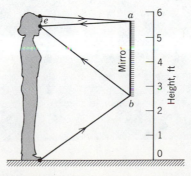

figure 44-8
Example 1. A person can view her full-length image in a mirror that is only half her height.

---

* See "The Overthrow of Parity" by Philip Morrison, *Scientific American*, April 1957, for a discussion of the distinction in nature between right and left.

woman's head and the tips of her toes. These rays, chosen so that they will enter the eye e after reflection, strike the vertical mirror at points a and b, respectively. The mirror need occupy only the region between these two points. Calculation shows that b is 2 ft, 10 in. and a is 5 ft, 10 in. above the floor. The length of the mirror is thus 3.0 ft, or half the height of the person. Note that this height is independent of the distance between the person and the mirror. Mirrors that extend below point b only show reflections of the floor between the person and the mirror.

EXAMPLE 2

Two plane mirrors are placed at right angles, and a point object O is located on the perpendicular bisector, as shown in Fig. 44-9a. Locate the images. Images I_1 and I_2 are formed in mirrors ab and cd, respectively. There is also a third image; it may be considered to be the image of I_1 in mirror cd or the image of I_2 in mirror ab. The three images and the object O lie on a circle whose center is on the line of intersection of the mirrors and whose plane is at right angles to that line.

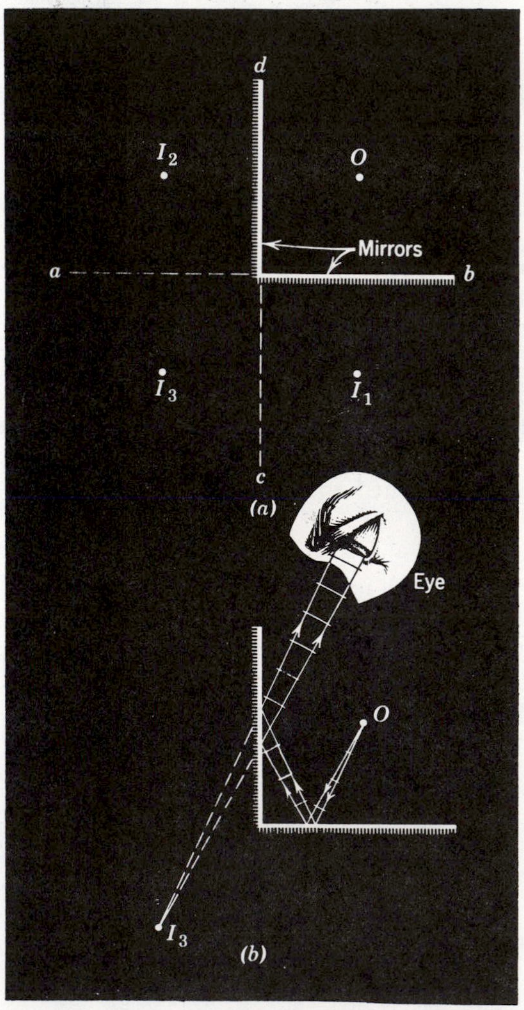

figure 44-9
Example 2. (a) Real object O has three virtual images. (b) A typical bundle of rays used to view I_3. The light originates from O.

In viewing I_3 the light entering the observer's eye is reflected *twice* after leaving the source. Figure 44-9b shows a typical bundle of rays. In viewing I_1 or I_2, the light is reflected only once, as in Fig. 44-6.

In Fig. 44-10 a spherical light wave from a point object O falls on a concave spherical mirror whose radius of curvature is r.* A line through O and the center of curvature C makes a convenient reference axis.

A ray from O that makes an arbitrary angle α with this axis intersects the axis at I after reflection from the mirror at a. A ray that leaves O along the axis will be reflected back along itself at v and will also pass through I. Thus, for these two rays at least, I is the image of O; it is a *real* image because light actually passes through I. Let us find the location of I.

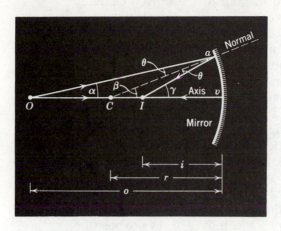

figure 44-10
Two rays from O converge after reflection in a spherical concave mirror, forming a real image at I.

A useful theorem is that the exterior angle of a triangle is equal to the sum of the two opposite interior angles. Applying this to triangles $OaCO$ and $OaIO$ in Fig. 44-10 yields

$$\beta = \alpha + \theta$$

and

$$\gamma = \alpha + 2\theta.$$

Eliminating θ between these equations leads to

$$\alpha + \gamma = 2\beta. \qquad (44\text{-}2)$$

In radian measure we can write angles α, β, and γ as

$$\alpha \cong \frac{av}{vO} = \frac{av}{o}$$

$$\beta = \frac{av}{vC} = \frac{av}{r} \qquad (44\text{-}3)$$

$$\gamma \cong \frac{av}{vI} = \frac{av}{i}.$$

Note that only the equation for β is exact, for the reason that the center of curvature of arc av is at C and not at O or I. However, the equations for α and for γ are approximately correct if these angles are sufficiently small. *In all that follows we assume that the rays diverging from the object make only a small angle α with the axis of the mirror.* We call such rays, which lie close to the mirror axis, *paraxial rays*. We did not find it necessary to make such an assumption for plane mirrors. Sub-

---

* A spherical shell, viewed from inside, is everywhere *concave*; viewed from outside it is everywhere *convex*. In this chapter we will always judge concave and convex from the point of view of an observer sighting along the direction of the incident light.

stituting these equations into Eq. 44-2 and canceling av yields

$$\frac{1}{o} + \frac{1}{i} = \frac{2}{r}, \qquad (44\text{-}4)$$

in which o is the *object distance* and i is the *image distance*. Both these distances are measured from the *vertex* of the mirror, which is the point v at which the axis intercepts the mirror.

Significantly, Eq. 44-4 does not contain α (or β, γ, or θ), so that it holds for all rays that strike the mirror provided that they are sufficiently paraxial. In an actual case the rays can be made as paraxial as one likes by putting a circular diaphragm in front of the mirror, centered about the vertex v; this will impose a certain maximum value of α.

As α in Fig. 44-10 is permitted to become larger, it will become less true that a point object will form a point image; the image will become extended and fuzzy. No sharp criterion for deciding whether a given ray is paraxial can be laid down. If the maximum permitted value of α is reduced, the rays will become more paraxial and the image will become sharper. Unfortunately, the image will also become fainter because less total light energy will be reflected from the mirror. A compromise must often be made between image brightness and image quality.

As for plane mirrors, the image (real or virtual) in a spherical mirror can be seen only if the eye is located so that light rays from the object can enter it after reflection. In Fig. 44-11 a bundle of light rays is shown entering the eye in position x; only the small patch of mirror near a is effective for this eye position. If the observer moves his eye to position y, the image will vanish for him because the mirror does not exist near point a'.

Although Eq. 44-4 was derived for the special case in which the object is located beyond the center of curvature, it is generally true, no matter where the object is located. It is also true for convex mirrors, as in Fig. 44-12.

In applying Eq. 44-4, we must be careful to follow a consistent convention of signs for o, i, and r. As the basis for the sign conventions to be used in this book, we start from this statement:

> In Fig. 44-10, in which light *diverges* from a *real* object, falls on a *concave* mirror, and *converges* after reflection to form a *real* image, the quantities o, i, and r in Eq. 44-4 and f in Eq. 44-5 are given positive numerical values.

We used Fig. 44-10 to derive Eq. 44-4 and you should associate them in your mind as an aid in getting the signs correct.

Let us fix our minds on the side of the mirror from which incident light comes. Because mirrors are opaque, the light, after reflection, must remain on this side, and if an image is formed here it will be a *real* image. Therefore, we call the side of the mirror from which the light comes the *R-side* (for *real image*). We call the back of the mirror the *V-side* (for *virtual image*), because images formed on this side of the mirror must be virtual, no light being actually present on this side.

In the indented statement above we associated real images with positive image distances. This suggests our first sign convention:

1. The image distance i is positive if the image (real) lies on the R-side of the mirror, as in Fig. 44-10; i is negative if the image (virtual) lies on the V-side, as in Fig. 44-12.

If the mirror in Fig. 44-10, which is concave as viewed from the direction of the incident light, is made convex, the rays will *diverge* after re-

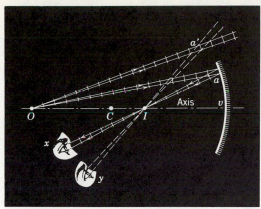

figure 44-11
The eye must be properly located to
see (real) image *I*. An observer at *x*
can see the image. One at *y* cannot.

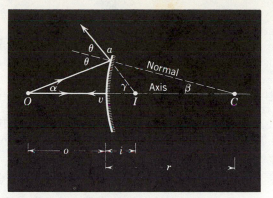

figure 44-12
Two rays from *O* diverge after
reflection in a spherical convex
mirror, forming a virtual image at *I*,
the point from which they appear to
originate. Compare Fig. 44-10.

flection and will form a *virtual* image, as Fig. 44-12 shows. Thus the
indented statement above suggests our second sign convention:

2. The radius of curvature *r* is positive if the center of curvature of the
mirror lies on the R-side, as in Fig. 44-10; *r* is negative if the center of
curvature lies on the V-side, as in Fig. 44-12.

You should not commit these sign conventions to memory but should
deduce them in each case from the basic statement on p. 964, using
Figs. 44-10, 12 as mnemonic aids.

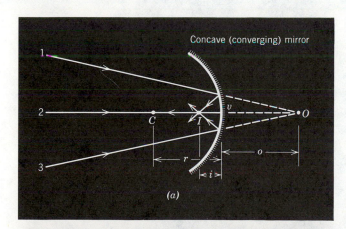

figure 44-13
(*a*) *Converging* rays (see 1, 2, and 3)
fall on a *concave* mirror. A *virtual*
object *O* produces a *real* image *I*;
note that no light passes through *O*
but it does pass through *I*. Also
note that *o* is negative but *i, r* (and
thus *f*), are positive. (*b*) Again
converging rays (see 1, 2, and 3) fall
on a *convex* mirror. A *virtual* object
O produces a *virtual* image *I*. Note
again that no light passes through
O or *I*. Here *o, i, r* (and thus *f*), are
all negative. Compare these figures
with Fig. 44-10, in which a *real*
object (*diverging* rays) produces a
real image on reflection from a
concave mirror. Here *o, i, r* (and
thus *f*), are all positive. See also
Fig. 44-12, in which a *real* object
(*diverging* rays) produces a *virtual*
image in a *convex* mirror. Here *o* is
positive but *i, r,* (and thus *f*), are
negative. These four figures show all
possibilities for arrangements of
virtual/real objects and virtual/real
images in mirrors. Compare them
carefully. Figure 44-23 shows a
similar comparison for thin lenses.

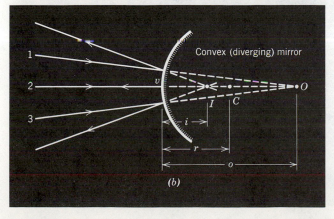

For all cases that we have considered so far, including plane mirrors, we have assumed that the incident light striking the mirror was *diverging* from a point source (the *object*) when it struck the mirror. In such cases, as we have seen, we have taken the object distance o in Eq. 44-4 as *positive*.

It is possible, by arrangements of mirrors and/or lenses, to make *converging* light fall on the mirror in question. In such cases we declare the *object* to be *virtual* and we assign a *negative* sign to the object distance o in Eq. 44-4. Figure 44-13 is an example. No matter what the source of light, O, which lies on the V-side, is a virtual object and o, the object distance, is taken as negative.

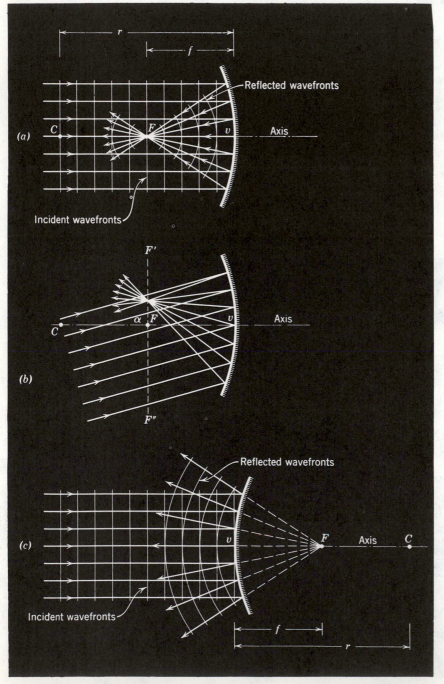

figure 44-14
(a) The focal point for a concave spherical mirror, showing both the rays and the wavefronts. *F* and *C* lie on the R-side, the focal point is real, and the focal length *f* of the mirror is positive (as is *r*). (b) The same, except that the incident light makes an angle α with the mirror axis; the rays are focused at a point in the *focal plane F′ F″*. (c) Same as (a) except that the mirror is convex; *F* and *C* lie on the V-side of the mirror. The focal point is virtual and the focal length *f* is negative (as is *r*).

EXAMPLE 3

967

SPHERICAL WAVES—SPHERICAL MIRROR   SEC. 44-3

A convex mirror has a radius of curvature of 20 cm. If a point source is placed 14 cm away from the mirror, as in Fig. 44-12, where is the image?

A rough graphical construction, applying the law of reflection at *a* in the figure, shows that the image will be on the V-side of the mirror and thus will be virtual. We may verify this quantitatively and analytically from Eq. 44-4, noting that *r* is negative here because the center of curvature of the mirror is on its V-side. We have

$$\frac{1}{o} + \frac{1}{i} = \frac{2}{r}$$

or

$$\frac{1}{+14 \text{ cm}} + \frac{1}{i} = \frac{2}{-20 \text{ cm}},$$

which yields $i = -5.8$ cm, in agreement with the graphical prediction. The negative sign for *i* reminds us that the image is on the V-side of the mirror and thus is virtual.

When *parallel* light falls on a mirror (Fig. 44-14), we call the image point (real or virtual) the *focal point F* of the mirror. The focal length *f* is the distance between *F* and the vertex. If we put $o \to \infty$ in Eq. 44-4, thus insuring parallel incident light, we have

$$i = \tfrac{1}{2}r = f.$$

Equation 44-4 can then be rewritten

$$\frac{1}{o} + \frac{1}{i} = \frac{1}{f}, \qquad (44\text{-}5)$$

where *f*, like *r*, is taken as positive for mirrors whose centers of curvature are on the R-side (that is, for *concave*, or *converging* mirrors; see Fig. 44-14a) and negative for those whose centers of curvature are on the V-side (that is, for *convex*, or *diverging* mirrors; see Fig. 44-13c). Figure 44-14b shows an incident plane wave that makes a small angle α with the mirror axis. The rays are focused at a point in the *focal plane* of the mirror. This is a plane at right angles to the mirror axis at the focal point.

We now consider objects that are not points. Figure 44-15 shows a candle in front of (a) a concave mirror and (b) a convex mirror. We choose

figure 44-15
The image of an extended object in (a) a concave mirror and (b) a convex mirror is located graphically. Any two of the three special rays shown are sufficient.

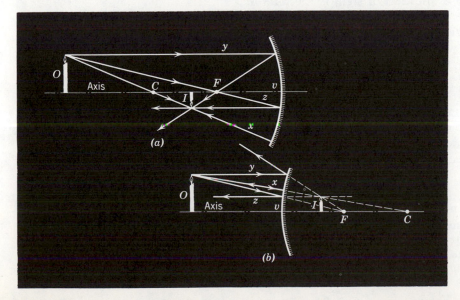

to draw the mirror axis through the foot of the candle and, of course, through the center of curvature. We can find the image of any off-axis point, such as the tip of the candle, graphically, using the following facts:

1. A ray that strikes the mirror after passing (either directly or upon being extended) through the center of curvature C returns along itself (ray x in Fig. 44-15). Such rays strike the mirror at right angles.

2. A ray that strikes the mirror parallel to its axis passes (or will pass when extended) through the focal point (ray y).

3. A ray that strikes the mirror after passing (either directly or upon being extended) through the focal point emerges parallel to the axis (ray z).

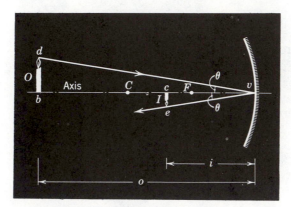

figure 44-16
A particular ray for the arrangement of Fig. 44-15, used to show that the *lateral magnification m* is given by $-i/o$.

Figure 44-16 shows a ray (dve) that originates on the tip of the object candle of Fig. 44-15a, is reflected from the mirror at point v, and passes through the tip of the image candle. The law of reflection demands that this ray make equal angles θ with the mirror axis as shown. For the two similar right triangles in the figure we can write

$$\frac{ce}{bd} = \frac{vc}{vb}.$$

The quantity on the left (apart from a question of sign) is the *lateral magnification m* of the mirror. Since we want to represent an *inverted* image by a *negative* magnification, we arbitrarily define m for this case as $-(ce/bd)$. Since $vc = i$ and $vb = o$, we have at once

$$m = -\frac{i}{o}. \tag{44-6}$$

This equation gives the magnification for spherical and plane mirrors under all circumstances. For a plane mirror, $o = -i$ and the predicted magnification is $+1$ which, in agreement with experience, indicates an *erect* image the same size as the object.

Images in spherical mirrors suffer from several "defects" that arise because the assumption of paraxial rays is never completely justified. In general, a point source will not produce a point image; see Problem 14. Apart from this, distortion arises because the magnification varies somewhat with distance from the mirror axis, Eq. 44-6 being strictly correct only for paraxial rays. Superimposed on these defects are *diffraction effects* which come about because the basic assumption of geometrical optics, that light travels in straight lines, must always be considered an approximation.

Figure 44-17 shows a point source O near a convex spherical *refracting* surface of radius of curvature r. The surface separates two media whose indices of refraction differ, that of the medium in which the incident light falls on the surface being n_1 and that on the other side of the surface being n_2.

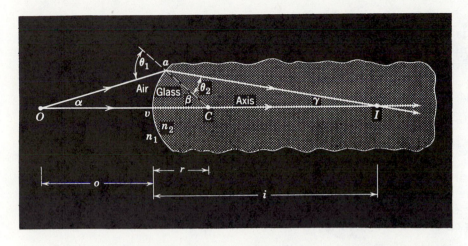

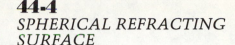

figure 44-17
Two rays from O converge after refraction at a spherical surface, forming a real image at I. Here o (diverging rays), i, and r are all positive; see Example 4.

From O we draw a line through the center of curvature C of the refracting surface, thus establishing a convenient axis which intercepts the surface at vertex v. From O we draw a ray that makes a small but arbitrary angle α with the axis and strikes the refracting surface at a, being refracted according to

$$n_1 \sin \theta_1 = n_2 \sin \theta_2.$$

The refracted ray intersects the axis at I. A ray from O that travels along the axis will not be bent on entering the surface and will also pass through I. Thus, for these two rays at least, I is the image of O.

As in the derivation of the mirror equation, we use the theorem that the exterior angle of a triangle is equal to the sum of the two opposite interior angles. Applying this to triangles $COaC$ and $ICaI$ yields

$$\theta_1 = \alpha + \beta \tag{44-7}$$

and

$$\beta = \theta_2 + \gamma. \tag{44-8}$$

As α is made small, angles β, γ, θ_1, and θ_2 in Fig. 44-17 also become small. We assume that α, hence all these angles, *are arbitrarily small*. We made this same paraxial ray assumption for spherical mirrors. Replacing the sines of the angles by the angles themselves—since the angles are required to be small—permits us to write the law of refraction as

$$n_1\theta_1 \cong n_2\theta_2. \tag{44-9}$$

Combining Eqs. 44-8 and 44-9 leads to

$$\beta = \frac{n_1}{n_2}\theta_1 + \gamma.$$

Eliminating θ_1 between this equation and Eq. 44-7 leads, after rearrangement, to

$$n_1\alpha + n_2\gamma = (n_2 - n_1)\beta. \tag{44-10}$$

In radian measure the angles α, β, and γ in Fig. 44-17 are

$$\alpha \cong \frac{av}{o}$$

$$\beta = \frac{av}{r} \qquad (44\text{-}11)$$

$$\gamma \cong \frac{av}{i}.$$

Only the second of these equations is exact. The other two are approximate because I and O are *not* the centers of circles of which av is an arc. However, for paraxial rays (α small enough) the inaccuracies in Eq. 44-11 can be made as small as desired.

Substituting Eqs. 44-11 into Eq. 44-10 leads readily to

$$\frac{n_1}{o} + \frac{n_2}{i} = \frac{n_2 - n_1}{r}. \qquad (44\text{-}12)$$

This equation holds whenever light is refracted from point objects at spherical surfaces, assuming only that the rays are paraxial. As with the mirror formula, care must be taken to use Eq. 44-12 with consistent signs for o, i, and r. Once again we establish our sign conventions by physical reasoning from a particular case, that of Fig. 44-17:

> In Fig. 44-17, in which light *diverges* from a *real* object, falls on a *convex* refracting surface, and *converges* after refraction to form a *real* image, the quantities o, i, and r in Eq. 44-12 have positive numerical values.

We used Fig. 44-17 to derive Eq. 44-12, and you should associate it in your mind as an aid in getting the signs correct. This basic statement is quite similar to that which we made for mirrors on p. 964.

We fix our attention on the side of the refracting surface from which the incident light falls on the surface. In contrast to mirrors, the light energy *passes through* a refracting surface to the other side, and if a real image is formed it must appear on the far side, which we call the R-side. The side from which the incident light comes is called the V-side because virtual images must appear here. Figure 44-18 suggests this important distinction between reflection and refraction.

In the indented statement above we associated real images with positive-image distances. Thus we are led to the sign convention:

1. The image distance i is positive if the image (real) is on the R-side of the refracting surface, as in Fig. 44-17; i is negative if the image (virtual) lies on the V-side, as in Fig. 44-19.

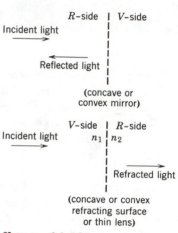

figure 44-18

Real images are formed on the same side as the incident light for mirrors but on the opposite side for refracting surfaces and thin lenses. This is so because the incident light is reflected back by mirrors but is transmitted through by refracting surfaces. Note that o is positive if rays *diverge* from a *real* object O.

figure 44-19

Two rays from O diverge after refraction at a spherical surface, forming a virtual image at I. Here o is positive (the rays from O are diverging) but i and r are negative.

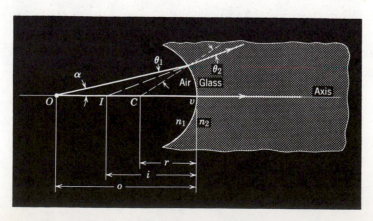

The refracting surface in Fig. 44-17 is convex. If it is made concave (still assuming that $n_2 > n_1$), the rays will *diverge* after refraction and form a *virtual* image, as Fig. 44-19 shows. Thus we are led to our second sign convention:

2. The radius of curvature r is positive if the center of curvature of the refracting surface lies on the R-side, as in Fig. 44-17; r is negative if the center of curvature is on the V-side, as in Fig. 44-19.

The sign conventions for refracting surfaces are the same as for mirrors (p. 964), the fundamental difference between the two situations being absorbed in the definitions of R-side and V-side in Fig. 44-18. This difference is easily remembered on physical grounds.

EXAMPLE 4

Locate the image for the geometry shown in Fig. 44-17, assuming the radius of curvature to be 10 cm, n_2 to be 2.0, and n_1 to be 1.0. Let the object be 20 cm to the left of v.

From Eq. 44-12,

$$\frac{n_1}{o} + \frac{n_2}{i} = \frac{n_2 - n_1}{r},$$

we have

$$\frac{1.0}{+20 \text{ cm}} + \frac{2.0}{i} = \frac{2.0 - 1.0}{+10 \text{ cm}}.$$

Note that r is positive because the center of curvature of the surface lies on the R-side. This relation yields $i = +40$ cm in agreement with the graphical construction. The light energy actually passes through I so that the image is real, as indicated by the positive sign for i.

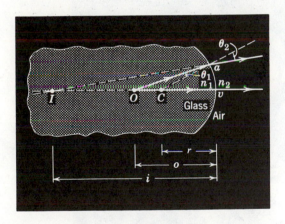

figure 44-20
Two rays from O appear to originate from I (virtual image) after refraction at a spherical surface.

EXAMPLE 5

An object is immersed in a medium with $n_1 = 2.0$, being 15 cm from the spherical surface whose radius of curvature is -10 cm, as in Fig. 44-20, r is negative because C lies on the V-side. Locate the image.

Figure 44-20 shows a ray traced through the surface by applying the law of refraction at point a. A second ray from O along the axis emerges undeflected at v. We find the image I by extending these two rays backward; it is virtual.

From Eq. 44-12,

$$\frac{n_1}{o} + \frac{n_2}{i} = \frac{n_2 - n_1}{r},$$

we have

$$\frac{2.0}{+15 \text{ cm}} + \frac{1.0}{i} = \frac{1.0 - 2.0}{-10 \text{ cm}},$$

which yields $i = -30$ cm, in agreement with Fig. 44-20 and with the sign conventions. Note that n_1 always refers to the medium on the side of the surface from which the light comes.

What is the relationship between i and o if the refracting surface is plane?   **EXAMPLE 6**

A plane surface has an infinite radius of curvature. Putting $r \to \infty$ in Eq. 44-12 leads to

$$i = -o\,\frac{n_2}{n_1}.$$

Figure 44-21 illustrates the situation graphically (a) for an object in air as seen from below water and (b) for an object in water with air above. This shows that a diver, looking upward at, say, an overhanging tree branch, will think it higher than it is by the factor 1.33/1.00. Similarly, an observer in air will think that objects on the bottom of a water tank are closer to the surface than they actually are, in the ratio 1.00/1.33. These considerations, being based on Eq. 44-12, hold

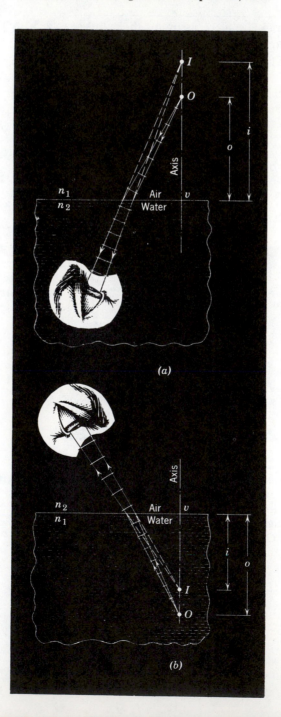

(a)

(b)

figure 44-21
Refraction at a plane surface at near-normal incidence, showing a pencil of rays and the corresponding wavefronts entering the pupil. (a) Source in air and (b) source in water.

only for paraxial rays, which means that the incident rays can make only a small angle with the normal; this angle has been exaggerated in the figure for clarity. Note again that n_1 is always identified with the medium that lies on the side of the surface containing the incident light.

In most refraction situations there is more than one refracting surface. This is true even for a spectacle lens, the light passing from air into glass and then from glass into air. In microscopes, telescopes, cameras,* etc., there are often many more than two surfaces.

Figure 44-22a shows a thick glass "lens" of length l whose surfaces are ground to radii r' and r''. A point object O' is placed near the left surface as shown. A ray leaving O' along the axis is not deflected on entering or leaving the lens because it falls on each surface along a normal.

A second ray leaving O', at an arbitrary angle α with the axis, strikes the surface at point a', is refracted, and strikes the second surface at point a''. The ray is again refracted and crosses the axis at I'', which, being the intersection of two rays from O'', is the image of point O', formed after refraction at two surfaces.

44-5
THIN LENSES

*For insight into the difficult task of designing real high-performance lenses see "The Photographic Lens" by William H. Price, *Scientific American,* August 1976.

figure 44-22
(a) Two rays from O' intersect at I'' (real image) after refraction at two spherical surfaces. (b) The first surface and (c) the second surface shown separately. The quantities α and n have been exaggerated for clarity.

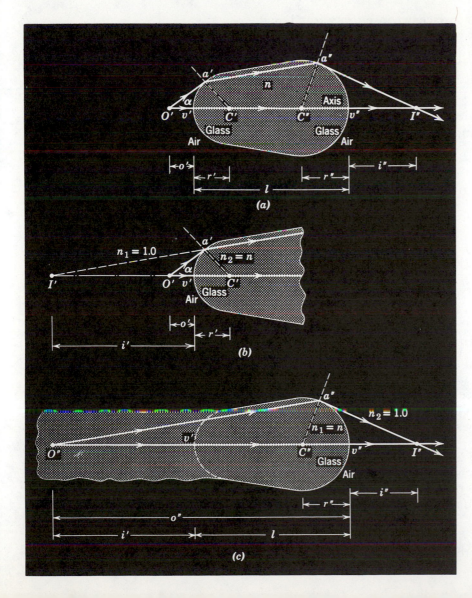

Figure 44-22*b* shows the first surface, which forms a virtual image of O' at I'. To locate I', we use Eq. 44-12,

$$\frac{n_1}{o} + \frac{n_2}{i} = \frac{n_2 - n_1}{r}.$$

Putting $n_1 = 1.0$ and $n_2 = n$ and bearing in mind that the image distance is negative (that is, $i = -i'$ in Fig. 44-22*b*), we obtain

$$\frac{1}{o'} - \frac{n}{i'} = \frac{n-1}{r'}. \tag{44-13}$$

In this equation i' will be a positive number because we have arbitrarily introduced the minus sign appropriate to a virtual image.

Figure 44-22*c* shows the second surface. Unless an observer at point a'' were aware of the existence of the first surface, he would think that the light striking that point originated at point I' in Fig. 44-22*b* and that the region to the left of the surface was filled with glass. Thus the (virtual) image I' formed by the first surface serves as a real object O'' for the second surface. The distance of this object from the second surface is

$$o'' = i' + l. \tag{44-14}$$

In applying Eq. 44-12 to the second surface, we insert $n_1 = n$ and $n_2 = 1.0$ because the object behaves as if it were imbedded in glass. If we use Eq. 44-14, Eq. 44-12 becomes

$$\frac{n}{i' + l} + \frac{1}{i''} = \frac{1-n}{r''}. \tag{44-15}$$

Let us now assume that the thickness l of the "lens" in Fig. 44-22 is so small that we can neglect it in comparison with other linear quantities in this figure (such as o', i', o'', i'', r', and r''). In all that follows we make this *thin-lens approximation*. Putting $l = 0$ in Eq. 44-15 leads to

$$\frac{n}{i'} + \frac{1}{i''} = -\frac{n-1}{r''}. \tag{44-16}$$

Adding Eqs. 44-13 and 44-16 leads to

$$\frac{1}{o'} + \frac{1}{i''} = (n-1)\left(\frac{1}{r'} - \frac{1}{r''}\right).$$

Finally, calling the original object distance simply o and the final image distance simply i leads to

$$\frac{1}{o} + \frac{1}{i} = (n-1)\left(\frac{1}{r'} - \frac{1}{r''}\right). \tag{44-17}$$

This equation holds only for paraxial rays and only if the lens is so thin that it essentially makes no difference from which surface of the lens the quantities o and i are measured. In Eq. 44-17 r' refers to the first surface struck by the light as it traverses the lens and r'' to the second surface.

The sign conventions for Eq. 44-17 are the same as those for mirrors and for single refracting surfaces. Because the lens is assumed to be thin, we refer to the R-side and the V-side of the lens itself (see Fig. 44-18) rather than those of its separate surfaces. The sign conventions are suggested by Fig. 44-23.

1. The image distance i is positive if the image (real) lies on the R-side

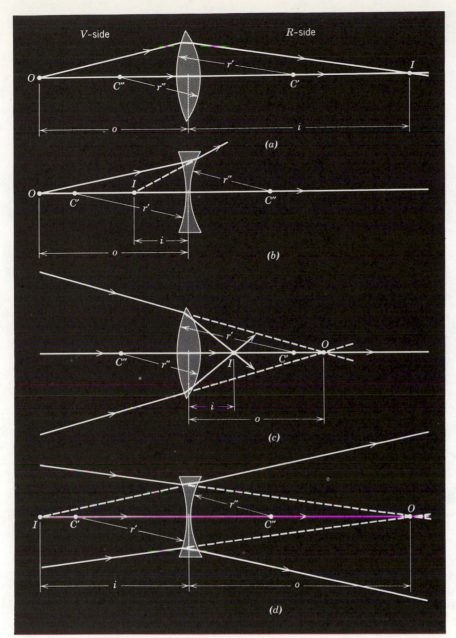

figure 44-23
(a) A real image and a real object. Both o and i are positive. (b) A virtual image and a real object. Note that o is positive but i is negative. (c) A real image and a virtual object. Note that o is negative but i is positive. (d) A virtual image and a virtual object. Both o and i are negative. Note that the sign of o is determined by whether the incident rays are diverging (first two cases) or converging (second two cases) when they strike the thin lens. For diverging incident rays the object is real; for converging rays it is virtual. The sign of i is determined by whether I is on the R-side of the lens (in which case I is real and i is positive, as in the first and third figures) or by whether I is on the V-side of the lens (in which case I is virtual and i is negative, as in the second and fourth figures).

of the lens, as in Figs. 44-23a and 44-23c. It is negative if the image (virtual) lies on the V-side of the lens, as in Figs. 44-23b and 44-23d.

2. The object distance o is positive if *diverging* rays fall on the lens, as in Figs. 44-23a and 44-23b; the object is real in such cases. The object distance o is negative if *converging* rays fall on the lens, as in Figs. 44-23c and 44-23d; the object is virtual in such cases.

3. The radii of curvature r' and r'' refer, respectively, to the first and to the second surfaces to be struck by the incident light. Each of these quantities is positive if the corresponding centers of curvature C' and C'' lie on the R-side of the lens; they are negative otherwise.

Figure 44-24a and 44-24c shows parallel light from a distant object falling on a thin lens. The image location is called the *second focal point*

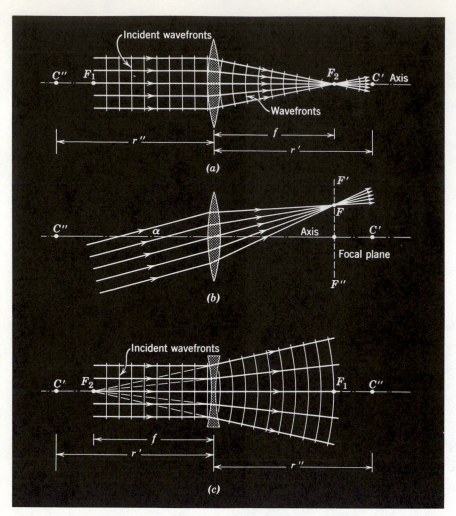

figure 44-24
(a) Parallel light passes through the second focal point F_2 of a converging lens. (b) The incident light makes an angle α with the lens axis, the rays being focused in the focal plane $F'F''$. (c) Parallel light, passing through a diverging lens, seems to originate at the second focal point F_2. C' and C'' are centers of curvature for the lens surfaces; F_1 is the first focal point.

F_2 of the lens. The distance from F_2 to the lens is called the *focal length f*. The *first focal point* for a thin lens (F_1 in figure) is the object position for which the image is at infinity. For thin lenses the first and second focal points are on opposite sides of the lens and are equidistant from it.

We can find the focal length from Eq. 44-17 by inserting $o \rightarrow \infty$ and $i = f$. This yields

$$\frac{1}{f} = (n - 1)\left(\frac{1}{r'} - \frac{1}{r''}\right).$$ (44-18)

This relation is called the *lens maker's equation* because it allows us to compute the focal length of a lens in terms of the radii of curvature and the index of refraction of the material. Combining Eqs. 44-17 and 44-18 allows us to write the thin-lens equation as

$$\frac{1}{o} + \frac{1}{i} = \frac{1}{f}.$$ (44-19)

Figure 44-24*b* shows parallel incident rays that make a small angle α with the lens axis; they are brought to a focus in the *focal plane $F'F''$*, as shown. This is a plane normal to the lens axis at the focal point.

In Fig. 44-24*a* we note that all rays in the figure contain the same number of wavelengths; in other words, they have the same *optical path*

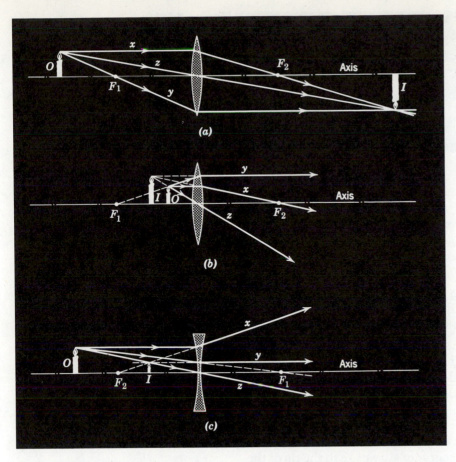

lengths; see Section 43-6. The optical path lengths are the same because the wavefronts are surfaces over which the wave disturbance has the same constant value and because all the rays shown pass through the same number of wavefronts.

EXAMPLE 7

The lenses of Fig. 44-24 have radii of curvature of magnitude 40 cm and are made of glass with $n = 1.65$. Compute their focal lengths.

Since C' lies on the R-side of the lens in Fig. 44-24a, r' is positive ($=+40$ cm). Since C'' lies on the V-side, r'' is negative ($=-40$ cm). Substituting in Eq. 44-18 yields

$$\frac{1}{f} = (n-1)\left(\frac{1}{r'} - \frac{1}{r''}\right) = (1.65 - 1)\left(\frac{1}{+40 \text{ cm}} - \frac{1}{-40 \text{ cm}}\right),$$

or
$$f = +31 \text{ cm.}$$

A positive focal length indicates that in agreement with Fig. 44-24a the focal point F_2 is on the R-side of the lens and parallel incident light converges after refraction to form a real image.

In Fig. 44-24c C' lies on the V-side of the lens so that r' is negative ($=-40$ cm). Since r'' is positive ($=+40$ cm), Eq. 44-18 yields

$$f = -31 \text{ cm.}$$

A negative focal length indicates that in agreement with Fig. 44-24c the focal point F_2 is on the V-side of the lens and incident light diverges after refraction to form a virtual image.

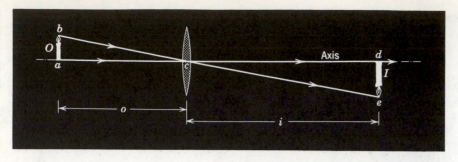

figure 44-26
Two rays for the situation of Fig. 44-25a.

We can locate the image of an extended object such as a candle (Fig. 44-25) graphically by using the following three facts:

1. A ray parallel to the axis and falling on the lens passes, either directly or when extended, through the second focal point F_2 (ray x in Fig. 44-25).

2. A ray falling on a lens after passing, either directly or when extended, through the first focal point F_1 will emerge from the lens parallel to the axis (ray y).

3. A ray falling on the lens at its center will pass through undeflected. There is no deflection because the lens, near its center, behaves like a thin piece of glass with parallel sides. The direction of the light rays is not changed and the sideways displacement can be neglected because the lens thickness has been assumed to be negligible (ray z).

Figure 44-26, which represents part of Fig. 44-25a, shows a ray passing from the tip of the object through the center of curvature to the tip of the image. For the similar triangles abc and dec we may write

$$\frac{de}{ab} = \frac{dc}{ac}.$$

The right side of this equation is i/o and the left side is $-m$, where m is the *lateral magnification*. The minus sign is required because we wish m to be negative for an inverted image. This yields

$$m = -\frac{i}{o}, \tag{44-20}$$

which holds for all types of thin lenses and for all object distances.

EXAMPLE 8

A converging thin lens has a focal length of +24 cm. An object is placed 9.0 cm from the lens as in Fig. 44-25b. Describe the image.

From Eq. 44-19,

$$\frac{1}{o} + \frac{1}{i} = \frac{1}{f},$$

we have

$$\frac{1}{+9.0 \text{ cm}} + \frac{1}{i} = \frac{1}{+24 \text{ cm}},$$

which yields $i = -14.4$ cm, in agreement with the figure. The minus sign means that the image is on the V-side of the lens and is thus virtual.

The lateral magnification is given by

$$m = -\frac{i}{o} = -\frac{-14.4 \text{ cm}}{+9.0 \text{ cm}} = +1.6,$$

again in agreement with the figure. The plus signifies an erect image.

Images formed by lenses suffer from defects similar to those discussed for mirrors on p. 968. There are effects connected with the failure of a point object to form a point image, with the variation of magnification with distance from the lens axis, and with diffraction. For lenses, but not for mirrors, there are also *chromatic aberrations* associated with the fact that the refracting properties of the lens vary with wavelength because the index of refraction of the lens material does. If a point object on the lens axis emits white light, the image, neglecting other lens defects, will be a series of colored points spread out along the axis. We have all seen the colored images produced by inexpensive lenses. A great deal of ingenious optical engineering goes into the design of lenses (more commonly lens systems) in which the various lens defects are minimized. The lens surfaces are normally not spherical nor are the lenses "thin."

Note that we can write the mirror formula (Eq. 44-5) and the (identical) thin lens formula (Eq. 44-19) in this form:

$$\frac{1}{o/|f|} + \frac{1}{i/|f|} = \pm 1 \qquad (44\text{-}21)$$

in which $|f|$, the absolute value of the focal length f, is always positive. On the right side of the equation we choose $+1$ for a converging lens or a concave mirror and -1 for a diverging lens or a convex mirror. See Problem 41.

Figure 44-27 is a graphical representation of Eq. 44-21, for both mirrors and lenses, Fig. 44-27*a* being for concave mirrors and converging lenses and Fig. 44-27*b* for convex mirrors and diverging lenses.

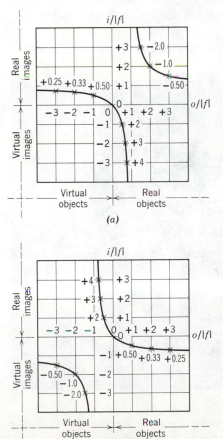

(a)

(b)

figure 44-27
(*a*) A representation of $i/|f|$ and $o/|f|$ for concave mirrors and converging lenses. Note that (lower left quadrant) a virtual object cannot produce a virtual image. The numbers near the *x*'s are the magnifications (see Eq. 44-20). A positive sign means an erect image, a negative sign an inverted image. Compare previous figures.

(*b*) A representation of $i/|f|$ and $o/|f|$ for convex mirrors and diverging lenses. Note that (upper right quadrant) a real object cannot produce a real image. Can you verify this? Once again, the numbers near the *x*'s are the magnifications (Eq. 44-20). See "Image Formation in Lenses & Mirrors, a Complete Representation" by Albert A. Bartlett, *The Physics Teacher*, May 1976.

The human eye is a remarkably effective organ* but its range can be extended in many ways by a host of optical instruments such as spectacles, simple magnifiers, motion picture projectors, cameras (including TV cameras), microscopes, telescopes, etc. In many cases these devices extend the scope of our vision beyond the visible range; satellite-borne infrared cameras and X-ray microscopes are examples.

In almost all cases of modern sophisticated optical instruments the mirror and thin lens formulas (Eqs. 44-5 and 44-19) hold only as an approximation. The rays may not be paraxial, as anyone who has used a camera knows; in astronomical telescopes, however, the rays are indeed paraxial. In typical laboratory microscopes the lens cannot be considered "thin" in the sense in which that term was defined in Section 44-5. In most optical instruments lenses are compound, that is, they are made of several components, cemented together, the interfaces rarely being exactly spherical. This is done to improve image quality and brightness and to relax greatly the dependence on paraxial rays.

In what follows we describe three optical instruments, assuming, for simplicity of illustration, that the thin lens formula applies.

a. The Simple Magnifier. The normal human eye can focus a sharp image of an object on the retina if the object O is located anywhere from infinity, the stars, say, to a certain point called the near point P_n, which we take to be about 25 cm from the eye. If you move the object within the near point the perceived retinal image becomes fuzzy. The location of the near point normally varies with age. We have all heard stories about people who claim not to need glasses but who read their newspapers at arm's length; their near points are receding! Find your own near point by moving this page closer to your eyes, considered separately, until you reach a position at which the image begins to become indistinct.

Figure 44-28a shows an object O placed at the near point P_n. The size

44-6
OPTICAL INSTRUMENTS

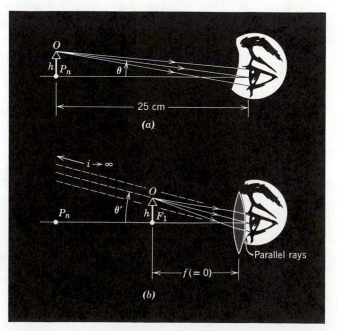

figure 44-28
(a) An object O of height h is placed at the near point P_n of the human eye. If moved any closer, it would fail to form a clear image on the retina. (b) A converging lens (that is, a simple magnifier) is placed close to the eye and the object O is moved from P_n to F_1. Not drawn to scale.

* Insect's eyes are even more versatile than human eyes. (Have you ever tried to catch a fly?) See "The Compound Eye of Insects" by G. Adrian Horridge, *Scientific American*, July 1977.

of the perceived image on the retina is measured by the angle θ. In Fig. 44-28b we insert a converging lens of focal length f just in front of the eye and move the object O to the first focal point F_1 of the lens. The eye now perceives an image at infinity, the angle of the image rays being θ', where $\theta' > \theta$. The *angular magnification* m_θ, not to be confused with the *lateral magnification* m given by Eq. 44-20, can be found from

$$m_\theta = \theta'/\theta$$

where
$$\theta \cong h/25 \text{ cm} \quad \text{and} \quad \theta' \cong h/f,$$

or

$$m_\theta \cong 25 \text{ cm}/f. \tag{44-22}$$

Note that, as expected, if $f = 25$ cm, then $m_\theta = 1$, that is, $\theta' = \theta$. Lens aberrations limit the angular magnifications for a simple converging lens to a few orders of magnitude. This is enough, however, for stamp collectors and for actors portraying Sherlock Holmes. More sophisticated magnifier designs have appreciably greater angular magnifications.

b. A Compound Microscope. Figure 44-29 shows a thin lens version of a compound microscope, used for viewing small objects that are very close to the instrument. The object O, of height h, is placed just outside the first focal point F_1 of the objective lens, whose focal length is f_{ob}. A real, inverted image I of height h' is formed by the objective, the lateral magnification being given by Eq. 44-20, or

$$m = \frac{h'}{h} = -\frac{s \tan \theta}{f_{ob} \tan \theta} = -\frac{s}{f_{ob}}. \tag{44-23}$$

As usual, the minus sign indicates an inverted image.

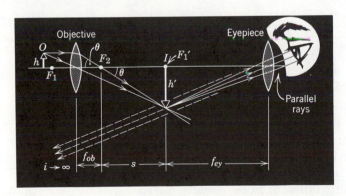

figure 44-29
A simplified version of a compound microscope, using "thin" lenses.

The distance s (sometimes called the tube length) is so chosen that the image I falls on the first focal point F_1' of the eyepiece, which then acts as a simple magnifier as in subsection a above. Parallel rays enter the eye and a final image I' forms at infinity. The final magnification M is given by the product of the linear magnification m for the objective lens, given by Eq. 44-23, and the angular magnification m_θ for the eyepiece, given by Eq. 44-22, or

$$M = m \times m_\theta = -\left(\frac{s}{f_{ob}}\right)\left(\frac{25 \text{ cm}}{f_{ey}}\right). \tag{44-24}$$

c. An Astronomical Telescope. Like microscopes, telescopes come in a large variety of forms. The form we describe here is the simple re-

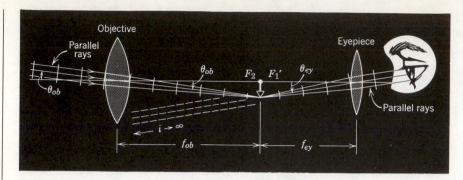

figure 44-30
A simplified version of an astronomical telescope, using "thin" lenses.

fracting telescope* that consists of an objective lens and an eyepiece, both represented in Fig. 44-30 by thin lenses, although in practice, as for microscopes, they will each be compound lens systems.

At first glance it may seem that the lens arrangements for telescopes (Fig. 44-30) and for microscopes (Fig. 44-29) are similar. However, telescopes are designed to view large objects, such as galaxies, stars, and planets, at large distances, whereas microscopes are designed for just the opposite purpose. Note also that in Fig. 44-30 the second focal point of the objective F_2 coincides with the first focal point of the eyepiece F_1', but in Fig. 44-29 these points are separated by a distance s (sometimes called the *tube length*).

In Fig. 44-30 parallel rays from a distant object strike the objective lens, making an angle θ_{ob} with the telescope axis and forming a real, inverted image at the common focal point F_2, F_1'. This image acts as an object for the eyepiece and a (still inverted) virtual image is formed at infinity. The rays defining the image make an angle θ_{ey} with the telescope axis.

The angular magnification m_θ of the telescope is given by θ_{ey}/θ_{ob}. For paraxial rays we can write $\theta_{ob} = h'/f_{ob}$ and $\theta_{ey} = -h'/f_{ey}$ or

$$m_\theta = -\frac{f_{ob}}{f_{ey}}. \tag{44-25}$$

Magnification is only one of the design factors of an astronomical telescope and is indeed easily achieved (How?). There is also *light gathering power* which determines how bright the image is. This is important when viewing faint objects such as distant galaxies and is accomplished by making the objective lens diameter as large as possible. There is *field of view*. An instrument designed for galactic observation (narrow field of view) must be quite different from one designed for the observation of meteors (wide field of view). Also, there are lens and mirror aberrations including *spherical aberration* (that is, lenses and mirrors with truly spherical surfaces do not form sharp images) and *chromatic aberration* (that is, for simple lenses the focal length varies with wavelength so that fuzzy images are formed, displaying unnatural colors). There is also *resolving power* which describes the ability of any optical instrument to distinguish between two objects (stars, say) whose

---

* For an example of the diversity of astronomical telescopes, see "The MMT Observatory on Mount Hopkins" by Nathaniel P. Carleton and Thomas E. Hoffman, *Sky and Telescope*, July 1976. "MMT" stands for Multiple Mirror Telescope.
* See also *Physics Today*, p. 18, April 1977 for a preview of a 2.4-meter (aperture) Space Telescope, planned to orbit the earth in 1983. The device is 46 ft long and will weigh 17,000 lb. It will be launched into orbit by a space shuttle, which will return every two or three years to make repairs and change or add instruments.

angular separation is small. We will discuss this fully in Section 46-5. This by no means exhausts the design parameters of astronomical telescopes. We could also make a similar listing for compound microscopes and, indeed, for any high performance optical instrument.

EXAMPLE 9

Figure 44-31 shows a simple astronomical telescope like that of Fig. 44-30 with the exceptions that (a) the incident parallel rays are parallel to the telescope axis, and (b) the incident rays fill the objective lens, whose diameter is d_{ob}, as, they normally would. Find the diameter d of the *emergent pencil*, as it is called, which contains all the information that is available for entering the eye.

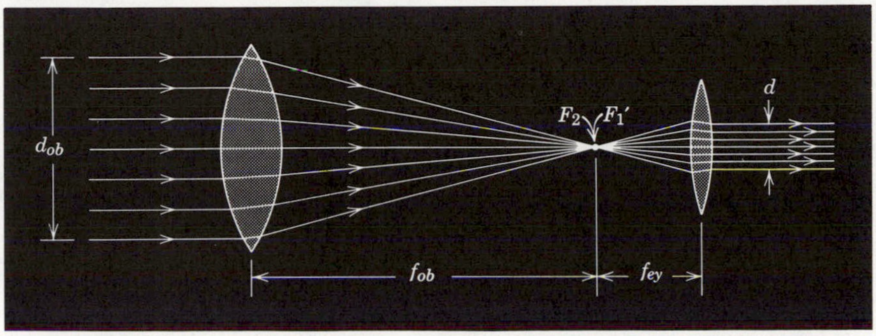

figure 44-31
The formula for d (Eq. 44-26) holds even if the incident rays are not parallel to the telescope axis, as in Fig. 44-30. The rays must be close to the axis, however, or most of the quantitative considerations of this section do not hold true.

In Fig. 44-31 we have, from similar triangles,

$$\frac{d_{ob}/2}{f_{ob}} = \frac{d/2}{f_{ey}},$$

or (see Eq. 44-25)

$$d = d_{ob}\left(\frac{f_{ey}}{f_{ob}}\right) = -\frac{d_{ob}}{m_\theta}. \tag{44-26}$$

Note that m_θ is inherently negative so that d is positive, as it must be.

We must compare d with d_p, the diameter of the pupil of the eye. The pupil diameter is not constant but varies with illumination, between the limits of about 2 mm (sunlight) to 9 mm (darkness). The ideal condition is $d = d_p$. For example, if $d > d_p$ some light does not enter the eye and is wasted.

Actually these considerations are modified in practice. For most major optical instruments we use photographic recording. Also, telescopes (and cameras) in orbiting satellites and space probes often send data to earth electronically, where it is processed by computor techniques into photographic form. The emergent pencil concept remains important however, not only for visual observation but for many lens systems design considerations.

---

questions

1. Can you think of a simple test or observation to prove that the law of reflection is the same for all wavelengths, under conditions in which geometrical optics prevails?

2. We all know that when we look into a mirror right and left are reversed. Our right hand will seem to be a left hand; if we part our hair on the left it will seem to be parted on the right, etc. Can you think of a system of mirrors that would let us see ourselves as others see us? If so draw it and prove your point by sketching some typical rays.

3. We have seen that a single reflection in a plane mirror reverses right and left. When we drive down a highway, for example, the letters on the highway signs are reversed as seen through the rear view mirror. And yet, as

seen through this same mirror, you still seem to be driving down the right lane. Why does the mirror reverse the signs and not the lanes? Or does it? Discuss.

4. A young women peers closely at her face in a high-quality plane mirror. She sees an image of her face in perfect focus. (Ignore the right-left reversal). Why the "perfect focus"? After all her nose is closer to the mirror than her earlobes are.

5. Note (Fig. 44-8) that any portion of the mirror that extends below *b* or above *a* is totally wasted if you wish to see yourself at full length in a mirror. If you have a "full length" mirror available convince yourself that this is true by taping a sheet of newspaper across the mirror at points *a* and *b*, leaving only the region between *a* and *b* exposed. Then, by stepping backward and forward, convince yourself that this theorem is true no matter what your distance from the mirror. Finally tape a sheet of paper *between* *a* and *b* and note what you see in the mirror.

6. If a mirror reverses right and left, why doesn't it reverse up and down?

7. Devise a system of plane mirrors that will let you see the back of your head. Trace the rays to prove your point.

8. If converging rays fall on a plane mirror, is the image virtual?

9. In many city buses a convex mirror is suspended over the door, in full view of the driver. Why not a plane or a concave mirror?

10. Dentists and dental hygienists use a small mirror with a long handle attached to examine your teeth. Is the mirror concave, convex, or plane, and why?

11. What approximations were made in deriving the mirror equation (Eq. 44-4):

$$\frac{1}{o} + \frac{1}{i} = \frac{2}{r}?$$

12. Under what conditions will a spherical mirror, which may be concave or convex, form (*a*) a real image, (*b*) an inverted image, and (*c*) an image smaller than the object?

13. Is it possible to see a perfect reflector? That is, if a perfectly reflecting concave spherical mirror is placed in a dark room and illuminated by, say, a luminous point object placed on its optical axis, could you see the surface of the mirror? Discuss.

14. Can a virtual image be projected onto a screen? Photographed? If you put a piece of paper at the site of a virtual object (assuming a high-intensity light beam) will it ignite after sufficient exposure? Consider these two questions for real images and real objects and note the differences, if any. Discuss.

15. The human eye has often been likened to a camera and, indeed, this comparison is valid. There is a difference however; the eye has no shutter (the eyelids do not serve this purpose). If you leave the shutter of a camera open and scan it across the horizon, the developed image on the film will be a blur. On the other hand, if you scan the horizon with your eye you will see every object distinctly. Have you any ideas to explain the difference between these two situations? See "Visual Motion Perception" by Gunnar Johansson, *Scientific American*, June 1975.

16. In connection with Question 15 consider this further difference between the eye and the camera. If you photograph, say, an elm tree, the image on the film can easily be overexposed beyond the point of recognition. On the other hand, if you simply look at an elm tree your perceived image remains constant for hours on end. Why this difference?

17. For the human eye we assert that the main focusing device is the cornea, that is, the curved outer surface of the eye and that the "lens" of the eye serves to make minor focussing adjustments. True or false? What function does the iris have? Can you think of advantages that the camera has over the human eye? There are some.

18. Does the apparent depth of an object below water depend on the angle of view of the observer in air? Explain and illustrate with ray diagrams.

19. An unsymmetrical thin lens forms an image of a point object on its axis. Is the image location changed if the lens is reversed?

20. Why has a lens two focal points and a mirror only one?

21. Under what conditions will a thin lens, which may be converging or diverging, form (a) a real image, (b) an inverted image, and (c) an image smaller than the object?

22. A skin diver wants to use an air-filled plastic bag as a converging lens for underwater use. Sketch a suitable cross section for the bag.

23. What approximations were made in deriving the thin lens equation (Eq. 44-19):

$$\frac{1}{o} + \frac{1}{i} = \frac{1}{f}?$$

24. Under what conditions will a thin lens have a lateral magnification (a) of -1 and (b) of $+1$?

25. How does the focal length of a glass lens for blue light compare with that for red light, assuming the lens is (a) diverging and (b) converging?

26. Does the focal length of a lens depend on the medium in which the lens is immersed? Is it possible for a given lens to act as a converging lens in one medium and a diverging lens in another medium?

27. Are the following statements true for a glass lens in air? (a) A lens that is thicker at the center than at the edges is a converging lens for parallel light. (b) A lens that is thicker at the edges than at the center is a diverging lens for parallel light. Explain and illustrate, using wavefronts.

28. Under what conditions would the lateral magnification ($m = -i/o$) for lenses and mirrors become infinite? Is there any practical significance to such a condition?

29. Light rays are reversible. Discuss the situation in terms of objects and images if all rays in Figs. 44-10, 44-14, 44-17, 44-19, 44-23, and 44-25 are reversed in direction.

30. In connection with Fig. 44-24a, we pointed out that all rays originating on the same wavefront in the incident wave have the same optical path length to the image point. Discuss this in connection with Fermat's principle (Section 43-6).

31. Is the focal length of a spherical mirror affected by the medium in which it is immersed? . . . of a thin lens . . . ? What's the difference?

32. What is the significance of the origin of coordinates in Figs. 44-27a and 44-27b?

33. In Fig. 44-23 what are the signs of o, i, r', and r'' in the four cases shown?

34. In Fig. 44-27a how do you interpret $o/|f| = +1$ and $i/|f| = +1$? Draw a ray diagram to illustrate these two situations, for thin converging lenses. In Fig. 44-27b answer the same question for $o/|f| = -1$ and $i/|f| = -1$, for thin diverging lenses.

35. Why is the magnification of a simple magnifier (see the derivation leading to Eq. 44-22) defined in terms of angles rather than image/object size?

36. Ordinary spectacles do not magnify but a simple magnifier does. What then, is the function of spectacles?

37. The "f-number" of a camera lens (see Problem 47) is its focal length divided by its aperture, that is, its effective diameter. Why is this useful to know in photography? How can the f-number of the lens by changed? How is exposure time related to f-number?

38. Does it matter whether (a) an astronomical telescope, (b) a compound microscope, (c) a simple magnifier, (d) a camera, including a TV camera, or (e) a projector, including a slide projector and a motion picture projector produce erect or inverted images? What about real or virtual images? Discuss each case.

39. Why does chromatic aberration occur in simple lenses but not in mirrors?

40. The unaided human eye produces a real but *inverted* image on the retina. (a) Why then don't we perceive objects such as people and trees as upside down? (b) We don't, of course, but suppose that we wore special glasses so that we did. If you then turned this book upside down, could you read this question with the same facility that you do now? Discuss.

41. In the movies, when the director wishes to portray a scene as viewed through a pair of binoculars, a mask with a horizontal figure-eight opening usually appears on the screen. What is wrong with this?

42. Why are all recent large astronomical telescopes of the reflecting rather than the refracting variety? Think of mechanical mounting problems for lenses and mirrors, the difficulties of shaping (that is, "figuring") the various optical surfaces involved, problems with small flaws in the optical glass blanks used to make lenses and mirrors, etc.

SECTION 44-2

1. A small object is 10 cm in front of a plane mirror. If you stand behind the object, 30 cm from the mirror, and look at its image, for what distance must you focus your eyes?   *Answer:* 40 cm.

2. Suppose you wished to photograph an object seen in a plane mirror. If the object is 5.0 m to your right and 1.0 m closer to the plane of the mirror than you, for what distance must you focus the lens of your camera?

3. A point object is 10 cm away from a mirror while the eye of an observer (pupil diameter 5.0 mm) is 20 cm away. Assuming both the eye and the point to be on the same line perpendicular to the surface, find the area of the mirror used in observing the reflection of the point.   *Answer:* 2.2 mm².

4. Two plane mirrors make an angle of 90° with each other. What is the largest number of images of an object placed between them that can be seen by a properly placed eye? The object need *not* lie on the mirror bisector.

5. Solve Example 2 if the angle between the mirrors is (a) 45°, (b) 60°, (c) 120°, the object always being placed on the bisector of the mirrors.
Answer: (a) 7, (b) 5, (c) 2.

6. How many images of himself can an observer see in a room whose ceiling and two adjacent walls are mirrors? Explain.

7. A small object O is placed one-third of the way between two parallel plane mirrors as in Fig. 44-32. Trace appropriate bundles of rays for viewing the four images that lie closest to the object.

8. In Fig. 44-6 you rotate the mirror 30° counterclockwise, leaving the point object O in place. Is the (virtual) image point displaced? If so, where is it? Can the eye still see the image without being moved? Sketch a figure showing the new situation, with O and the eye remaining unchanged in position.

9. You are peering into a peephole into an illuminated box. A plane mirror is on the interior wall facing you. A square wire framework is rotating slowly in a counterclockwise direction (as seen from above) about a vertical axis.

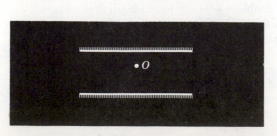

figure 44-32
Problem 7

The distance from the peephole to the mirror is 15 cm. (a) How far from the mirror should the vertical axis of rotation of the square be located so that its image in the mirror would appear at half size? (b) Compare the directions of rotation of the object and image. (c) Describe in your own words what you would see, assuming that you have not been told that the box contains a mirror.

10. Extend Fig. 44-9 to three dimensions by adding a mirror perpendicular to the common axis of the two mirrors shown. This forms a *corner reflector*, much used in optical, microwave, and other applications. It has the property that an incident ray is returned, after three reflections, along the same direction. Can you prove this?

SECTION 44-3

11. For clarity, the rays in figures like Fig. 44-10 are not drawn paraxial enough for Eq. 44-4 to hold with great accuracy. With a ruler, measure r and o in this figure and calculate, from Eq. 44-4, the predicted value of i. Compare this with the measured value of i.

12. Fill in this table, each column of which refers to a spherical mirror and a real object. Check your results by graphical analysis. Distances are in centimeters; if a number has no plus or minus sign in front of it, find the correct sign.

	a	b	c	d	e	f	g	h
Type	Concave						Convex	
f	20		+20			20		
r				−40			40	
i				−10			4	
o	+10	+10	+30	+60				+24
m		+1		−0.5		+0.10		0.50
Real image?		no						
Erect image?								no

Answer: For alternate vertical columns: (a) +, +40, −20, +2, no, yes. (c) Concave, +40, +60, −2, yes, no. (e) Convex, −20, +20, +0.5, no, yes. (g) −20, −, −, +5, no, yes.

13. A short linear object of length l lies on the axis of a spherical mirror, a distance o from the mirror. (a) Show that its image will have a length l' where

$$l' = l\left(\frac{f}{o-f}\right)^2.$$

(b) Show that the *longitudinal magnification* $m'(=l'/l)$ is equal to m^2 where m is the lateral magnification discussed in Section 44-3. (c) Is there any condition such that, neglecting all mirror defects, the image of a small cube would also be a cube?   *Answer:* (c) Yes; object at center of curvature.

14. Redraw Fig. 44-33 on a large sheet of paper and trace carefully the reflected rays, using the law of reflection. Is a point focus formed? Discuss.

15. A thin flat plate of partially reflecting glass is a distance b from a convex mirror. A point source of light S is placed a distance a in front of the plate (see Fig. 44-34) so that its image in the partially reflecting plate coincides with its image in the mirror. If $b = 7.5$ cm and the focal length of the mirror is $f = -30$ cm, find a and draw the ray diagram.   *Answer:* 23 cm.

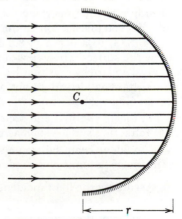

figure 44-33
Problem 14

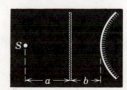

figure 44-34
Problem 15

16. Verify that Eq. 44-4 is consistent with the situations of Figs. 44-5, 10, 12 (see Example 3), 13, 14, and 15. In some cases only a qualitative answer is possible.

17. Modify Fig. 44-12 in such a way (possibly by adding a second mirror to the left of the one shown) that the object for Fig. 44-12 is virtual. Trace the rays.

SECTION 44-4

18. A penny lies on the bottom of a swimming pool 10 ft. deep. What is its apparent depth as viewed from above the water? The index of refraction of water is 1.33.

19. Fill in the following table, each column of which refers to a spherical surface separating two media with different indices of refraction. Distances are measured in centimeters. The object is real in all cases.

	a	b	c	d	e	f	g	h
n_1	1.0	1.0	1.0	1.0	1.5	1.5	1.5	1.5
n_2	1.5	1.5	1.5		1.0	1.0	1.0	
o	+10	+10		+20	+10		+70	+100
i		−13	+600	−20	−6	−7.5		+600
r	+30		+30	−20		−30	+30	−30
Real image?								

Draw a figure for each situation and construct the appropriate rays graphically. Assume a point object.

Answer: For alternate vertical columns: (a) −18, no. (c) +71, yes. (e) +30, no. (g) −26, no.

20. A layer of water ($n = 1.33$) 2.0 cm thick floats on carbon tetrachloride ($n = 1.46$) 4.0 cm thick. How far below the water surface, viewed at normal incidence, does the bottom of the tank seem to be?

21. As an example of the importance of the paraxial ray assumption, consider this problem. You place a coin at the bottom of a swimming pool filled with water ($n = 1.33$) to a depth of 8.0 ft. What is the apparent depth of the coin below the surface when viewed (a) at near normal incidence (that is, by paraxial rays) and (b) by rays that leave the coin making an angle of 30° with the normal (that is, definitely not paraxial rays)? What do you conclude? *Answer:* (a) 6.0 ft. (b) 5.6 ft.

22. Define and locate the first and second focal points (see p. 975) for a single spherical refracting surface such as that of Fig. 44-17.

23. A parallel incident beam falls on a solid glass sphere at normal incidence. Locate the image in terms of the index of refraction n and the sphere radius r.

Answer: Assuming the light incident from the left, $i = \dfrac{2-n}{2(n-1)} r$, to the right

of the right edge of the sphere if n < 2, as it is for glass.

SECTION 44-5

24. A double-convex lens is to be made of glass with an index of refraction of 1.50. One surface is to have twice the radius of curvature of the other and the focal length is to be 6.0 cm. What are the radii?

25. A lens is made of glass having an index of refraction of 1.5. One side of the lens is flat and the other convex with a radius of curvature of 20 cm. (a) Find the focal length of the lens. (b) If an object is placed 40 cm to the left of the

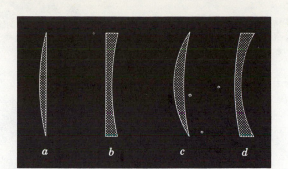

figure 44-35
Problem 26

lens, where will the image be located? (c) Would either of these answers change if we rotated the lens through 360°?

Answer: (a) +40 cm. (b) Image to the right at +∞. (c) No.

26. Using the lensmaker's equation (Eq. 44-18), decide which of the thin lenses in Fig. 44-35 are converging and which diverging for parallel incident light.

27. An object is placed at a center of curvature of a double-concave lens, both of whose radii of curvature have the same magnitude. (a) What are the signs of the two radii of curvature? (b) Find the location of the image in terms of the radius of curvature r and the index of refraction, n, of the glass. (c) Describe the nature of the image. (d) Verify your result with a ray diagram.

Answer: (a) r' is negative and r'' is positive; see Fig. 44-23b and 44-23c.
(b) $i = -\dfrac{r}{2n-1}$. (c) Virtual and erect.

28. Show that the focal length f' for a thin lens whose index of refraction is n and which is immersed in a medium, water, say, whose index of refraction is n' is given by

$$\frac{1}{f'} = \frac{n-n'}{n'}\left(\frac{1}{r'} - \frac{1}{r''}\right).$$

29. Reproduce Fig. 44-27a from first principles, that is, from Eq. 44-19. How do you know (a) that the lens is diverging or converging? (b) That the image is real or virtual? (c) That the object is real or virtual? (d) That the lateral magnification is > 1 or < 1?

30. Fill in this table, each column of which refers to a thin lens, to the extent possible. Check your results by graphical analysis. Distances are in centimeters; if a number (except in row n) has no plus sign or minus sign in front of it, find the correct sign.

	a	b	c	d	e	f	g	h	i
Type	converging								
f	10	+10	10	10					
r'					+30	−30	−30		
r''					−30	+30	−60		
i									
o	+20	+5	+5	+5	+10	+10	+10	+10	+10
n					1.5	1.5	1.5		
m			>1	<1				0.5	0.5
Real image?									yes
Erect image?								yes	

Draw a figure for each situation and construct the appropriate rays graphically. The object is real in all cases.

Answer: Alternate vertical columns (an X means that the quantity cannot be found from the data given): (*a*) +, X, X, +20, X, −1, yes, no. (*c*) Converging, +, X, X, −10, X, no, yes. (*e*) Converging, +30, −15, +1.5, no, yes. (*g*) Diverging, −120, −9.2, +0.92, no, yes. (*i*) Converging, +3.3, X, X, +5, X, −, no.

31. A luminous object and a screen are a fixed distance D apart. (*a*) Show that a converging lens of focal length f will form a real image on the screen for two positions that are separated by

$$d = \sqrt{D(D - 4f)}.$$

(*b*) Show that the ratio of the two image sizes for these two positions is

$$\left(\frac{D - d}{D + d}\right)^2.$$

32. A converging lens with a focal length of +20 cm is located 10 cm to the left of a diverging lens having a focal length of −15 cm. If a real object is located 40 cm to the left of the first lens, locate and describe completely the image formed.

33. Two thin lenses of focal length f_1 and f_2 are in contact. Show that they are equivalent to a single thin lens with a focal length given by

$$f = \frac{f_1 f_2}{f_1 + f_2}.$$

34. Show that the distance between a real object and its real image formed by a thin converging lens is always greater than or equal to four times the focal length of the lens.

35. The formula

$$\frac{1}{o} + \frac{1}{i} = \frac{1}{f}$$

is called the *Gaussian* form of the thin lens formula. Another form of this formula, the *Newtonian* form, is obtained by considering the distance x from the object to the first focal point and the distance x' from the second focal point to the image. Show that

$$xx' = f^2.$$

36. An erect object is placed a distance in front of a converging lens equal to twice the focal length f_1 of the lens. On the other side of the lens is a converging mirror of focal length f_2 separated from the lens by a distance $2(f_1 + f_2)$. (*a*) Find the location, nature and relative size of the final image. (*b*) Draw the appropriate ray diagram. See Fig. 44-36.

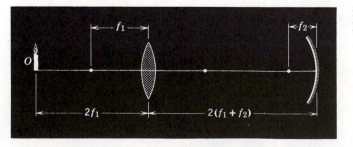

figure 44-36
Problem 36

37. Two thin lenses, one having $f = +12.0$ cm and the other having $f = −10.0$ cm, are separated by 7.0 cm. A small object is placed 43.5 cm from the center of

the lens system on the principal axis first on one side and next on the other side. Find the location of the final image in each case.

Answer: For object nearer to the converging lens the image is 8.5 cm to the opposite side of the center of the lens system. For an object nearer the diverging lens, the image is 63.5 cm to the opposite side of the center of the lens system.

38. (a) Show that a thin converging lens of focal length f followed by a thin diverging lens of focal length $-f$ will bring parallel light to a focus beyond the second lens provided that the separation of the lenses L satisfies $0 < L < f$. (b) Does this property change if the lenses are interchanged? (c) What happens when $L = 0$?

39. An object is placed 1.0 m in front of a converging lens, of focal length 0.5 m, which is 2.0 m in front of a plane mirror. (a) Where is the final image, measured from the lens, that would be seen by an eye looking toward the mirror through the lens? (b) Is the final image real or virtual? (c) Is the final image erect or inverted? (d) What is the lateral magnification?

Answer: (a) 0.60 m on the side of the lens away from the mirror. (b) Real. (c) Erect. (d) +0.20.

40. An object is 20 cm to the left of a lens with a focal length of +10 cm. A second lens of focal length +12.5 cm is 30 cm to the right of the first lens. (a) Using the image formed by the first lens as the object for the second, find the location and relative size of the final image. (b) Verify your conclusions by drawing the lens system to scale and constructing a ray diagram. (c) Describe the final image.

41. Show that Eq. 44-21 is correct.

42. An object is placed 1.0 m in front of a converging lens, of focal length 0.50 m, which is 2.0 m in front of a plane mirror. (a) Where is the final image, measured from the lens, that would be seen by an eye looking toward the mirror through the lens? (b) Is the final image real or virtual? (c) Is the final image erect or inverted? (d) What is the lateral magnification?

SECTION 44-6

43. In connection with Fig. 44-28b, (a) show that if the object O is moved from the first focal point F_1 toward the eye, the image moves in from infinity and the angle θ' (and thus the angular magnification m_θ) is increased. (b) If you continue this process, at what image location will m_θ have its maximum useable value? (c) Show that the maximum useable value of m_θ is $1 + (25\text{ cm})/f$. (d) Show that in this situation the angular magnification is equal to the linear magnification.

44. A microscope of the type shown in Fig. 44-29 has a focal length for the objective lens of 4.0 cm and for the eyepiece lens of 8.0 cm. The distance between the lenses is 25 cm. (a) What is the distance s in Fig. 44-29? (b) To reproduce the conditions of Fig. 44-29 how far beyond F_1 in that figure should the object be placed? (c) What is the lateral magnification m of the objective? (d) What is the angular magnification m_θ of the eyepiece? (e) What is the overall magnification M of the microscope?

45. *The eye—the basic optical instrument:* Figure 44-37a suggests a normal human eye. Parallel rays, entering a relaxed eye gazing at infinity, produce a real, inverted image on the retina. The eye thus acts as a converging lens. Most of the refraction occurs at the outer surface of the eye, the *cornea*. Assume a focal length f for the eye of 2.50 cm.

In Fig. 44-37b the object is moved in to a distance o (=40.0 cm) from the

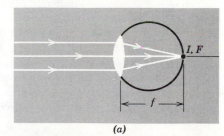

(a)

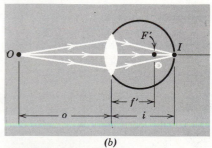

(b)

figure 44-37
Problem 45

eye. To form an image on the retina the effective focal length of the eye must be reduced to f'. This is done by the action of the ciliary muscles that change the shape of the lens and thus the effective focal length of the eye.

(a) Find f' from the above data. (b) Would the effective radii of curvature of the lens become larger or smaller in the transition from a to b in Fig. 44-37? (In the figure the structure of the eye is only roughly suggested and Fig. 44-37b is not to scale.) *Answer:* (a) 2.35 cm. (b) Smaller.

46. In an eye that is *farsighted* the eye focuses parallel rays so that the image would form behind the retina, as in Fig. 44-38a. In an eye that is *nearsighted* the image is formed in front of the retina, as in Fig. 44-38b. (a) How would you design a corrective lens for each eye defect? Make a ray diagram for each case. (b) If you need spectacles only for reading, are you nearsighted or farsighted? (c) What is the function of bifocal spectacles, in which the upper parts and lower parts have different focal lengths? (d) Some musicians in symphony orchestras (and others as well) wear *trifocals.* What occupational optical problem is involved here?

47. *The camera:* Figure 44-39 shows an idealized camera focused on an object at infinity. A real, inverted image I is formed on the film, the image distance i being equal to the (fixed) focal length f (=5.0 cm, say) of the lens system. In Fig. 44-39b the object O is closer to the camera, the object distance o being, say, 100 cm. To focus an image I on the film, we must extend the lens away from the camera (why?). (a) Find i' in Fig. 44-39b. (b) By how much must the lens be moved? Note that the camera differs from the eye (see Problem 45) in this respect. In the camera, f remains constant and the image distance i must be adjusted by moving the lens. For the eye the image distance i remains constant and the focal length f is adjusted by distorting the lens. Compare Figs. 44-37 and 44-39 carefully.
Answer: (a) 5.3 cm. (b) 3.0 mm.

48. *The reflecting telescope:* Isaac Newton, having convinced himself (erroneously as it turned out) that chromatic aberration was an inherent property of refracting telescopes, invented the reflecting telescope, shown schematically in Fig. 44-40. He presented his second model of this telescope, which has a magnifying power of 38, to the Royal Society, which still has it.

In Fig. 44-40 incident light falls, closely parallel to the telescope axis, on the objective mirror M. After reflection from small mirror M' (the figure is not to scale), the rays form a real, inverted image in the focal plane through F. This image is then viewed through an eyepiece.

(a) Show that the angular magnification m_θ is also given by Eq. 44-25, or

$$m_\theta = -f_{ob}/f_{ey}$$

where f_{ob} is the focal length of the objective mirror and f_{ey} that of the eyepiece. (b) The 200-in. mirror of the reflecting telescope at Mt. Palomar in California has a focal length of 16.8 m. Estimate the size of the object formed

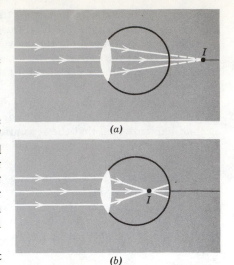

figure 44-38
Problem 46

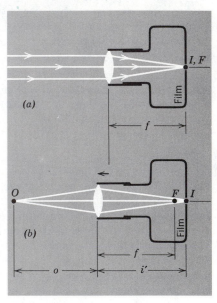

figure 44-39
Problem 47

figure 44-40
Problem 48

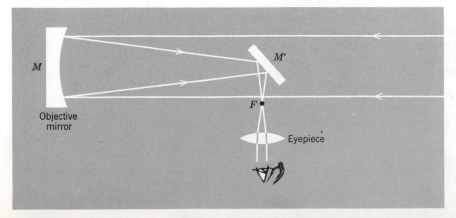

in the focal plane of this mirror when the object is a meter stick 2.0 km away. Assume parallel incident rays. (c) The mirror of a reflecting astronomical telescope has an effective radius of curvature ("effective" because such mirrors are ground to a parabolic rather than a spherical shape, to eliminate spherical aberration defects) of 10 m. To give an angular magnification of 200, what must be the focal length of the eyepiece?

45
interference

In Section 19-7 we saw that if two mechanical waves of the same frequency travel in approximately the same direction and have a phase difference that remains constant with time, they may combine so that their energy is not distributed uniformly in space but is a maximum at certain points and a minimum (possibly even zero) at others. The demonstration of such *interference* effects for light by Thomas Young in 1801 first established the wave theory of light on a firm experimental basis. Young was able to deduce the wavelength of light from his experiments, the first measurement of this important quantity.

Young allowed sunlight to fall on a pinhole S_0 punched in a screen A in Fig. 45-1. The emerging light spreads out by diffraction (see Section 44-1) and falls on pinholes S_1 and S_2 punched into screen B. Again diffraction occurs and two overlapping spherical waves expand into the space to the right of screen B.

The condition for geometric optics, namely, that $a \gg \lambda$ where a is the diameter of the pinholes, is definitely *not* met in this experiment. The pinholes do not cast geometrical shadows but act as sources of expanding Huygens' wavelets. We are dealing here (and in the three succeeding chapters) with *wave optics* rather than with geometrical optics.

Figure 45-2, taken from an 1803 paper of Young, shows the region between screens B and C. The blackening represents the minima of the wave disturbance; the white space between represents the maxima. If you hold the page with your eye close to the left edge and look at a grazing angle along the figure, you will see that along lines marked by x's there is cancellation of the wave; between them there is reinforcement. If we place a screen anywhere across the superimposed waves, we expect to find alternate bright and dark spots on it. Figure 45-3 shows a

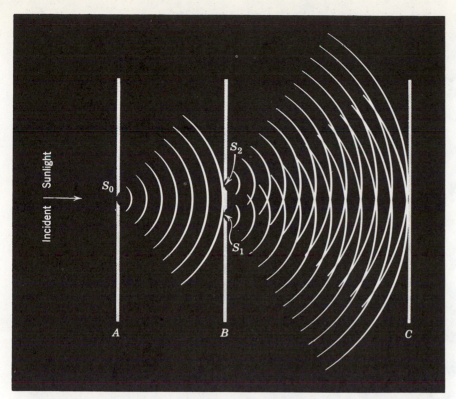

figure 45-1
Showing how Thomas Young produced an interference pattern by allowing diffracted waves from pinholes S_1 and S_2 to overlap on screen C.

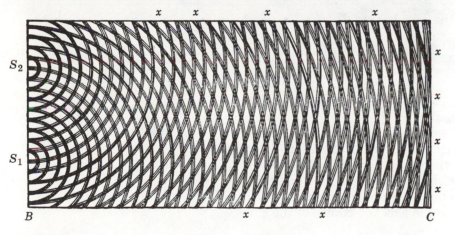

figure 45-2
Thomas Young's original drawing showing interference effects in overlapping waves. Place the eye near the left edge and sight at a grazing angle along the figure. (From Thomas Young, *Phil. Transactions*, 1803.) See also *Great Experiments in Physics*, p. 93, Morris H. Shamos, ed., Holt and Company, N.Y., 1959, for a readable annotated account of this experiment.

figure 45-3
Interference fringes for monochromatic light, made with an arrangement like that of Fig. 45-1, using long narrow slits rather than pinholes.

photograph of such *interference fringes;* in keeping with modern technique, long narrow slits rather than pinholes were used in preparing this figure.

Interference is not limited to light waves but is a characteristic of all wave phenomena. Figure 45-4, for example, shows the interference pattern of water waves in a shallow *ripple tank.* The waves are generated by two vibrators that tap the water surface in synchronism, producing two expanding spherical waves.

Let us analyze Young's experiment quantitatively, assuming that the incident light consists of a single wavelength only. In Fig. 45-5 P is an arbitrary point on the screen, a distance r_1 and r_2 from the narrow slits S_1 and S_2, respectively. Let us draw a line from S_2 to b in such a way that the lines PS_2 and Pb are equal. If d, the slit spacing, is much smaller than the distance D between the two screens (the ratio d/D in the figure has been exaggerated for clarity), S_2b is then almost perpendicular to both r_1 and r_2. This means that angle S_1S_2b is almost equal to angle PaO, both

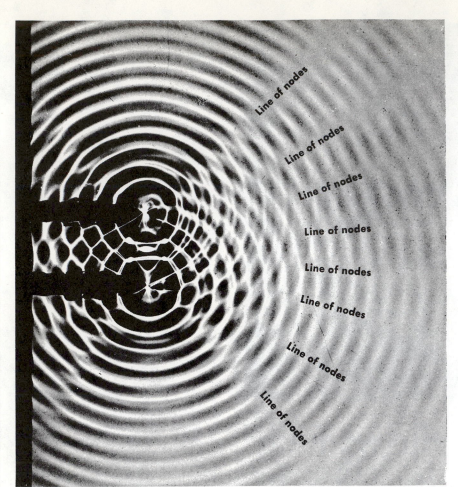

Line of nodes

Line of nodes

Line of nodes

Line of nodes

Line of nodes

Line of nodes

Line of nodes

Line of nodes

figure 45-4
The interference of water waves in a ripple tank. There is destructive interference along the lines marked "Line of nodes" and constructive interference between these lines. (Courtesy Physical Science Study Committee.)

angles being marked θ in the figure. This is equivalent to saying that the lines r_1 and r_2 may be taken as parallel.

We often put a lens in front of the two slits, as in Fig. 45-6, the screen C being in the focal plane of the lens. Under these conditions light focused at P must have struck the lens parallel to the line Px, drawn from P through the center of the (thin) lens. Under these conditions rays r_1 and r_2 are strictly parallel even though the requirement $D \gg d$ is not met. The lens L may in practice be the lens and cornea of the eye, screen C being the retina.

The two rays arriving at P in Figs. 45-5 or 45-6 from S_1 and S_2 are in phase at the source slits, both being derived from the same wavefront in the incident plane wave. Because the rays have different optical path lengths, they arrive at P with a phase difference. The number of wavelengths contained in S_1b, which is the path difference, determines the nature of the interference at P.

To have a *maximum* at P, S_1b $(=d \sin \theta)$ must contain an integral number of wavelengths, or

$$S_1b = m\lambda \qquad m = 0, 1, 2, \ldots,$$

which we can write as

$$d \sin \theta = m\lambda \qquad m = 0, 1, 2, \ldots \text{(maxima)}. \qquad (45\text{-}1)$$

Note that each maximum above O in Figs. 45-5 and 45-6 has a symmetrically located maximum below O. There is a central maximum described by $m = 0$.

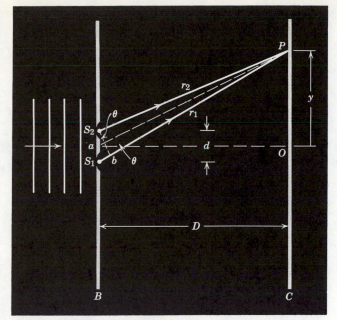

figure 45-5
Rays from S_1 and S_2 combine at P. The wave fronts of light falling on screen B have been taken as parallel. Actually, $D \gg d$, the figure being distorted for clarity. We represent the midpoint of the slits by a.

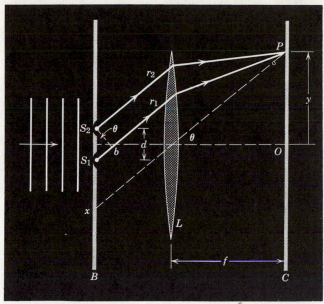

figure 45-6
A lens is normally used to produce interference fringes; compare with Fig. 45-5. The figure is again distorted for clarity in that $f \gg d$ in practice.

For a *minimum* at P, S_1b $(= d \sin \theta)$ must contain a half-integral number of wavelengths, or

$$d \sin \theta = (m + \tfrac{1}{2})\lambda \qquad m = 0, 1, 2, \ldots \text{(minima)}. \qquad (45\text{-}2)$$

If a lens is used as in Fig. 45-6, it may seem that a phase difference should develop between the rays beyond the plane represented by S_2b, the path lengths between this plane and P being clearly different. In Section 44-5, however, we saw that for such parallel rays focused by a lens the *optical path lengths* are identical. Two rays with the same optical path lengths contain the same number of wavelengths, so that no phase difference will result because of the light passing through the lens.

The double-slit arrangement in Fig. 45-5 is illuminated with light from a mercury vapor lamp so filtered that only the strong green line ($\lambda = 546$ nm or 5460 Å) is effective. The slits are 0.10 mm apart, and the screen on which the interference pattern appears is 20 cm away. What is the angular position of the first minimum? Of the tenth maximum?

At the first minimum we put $m = 0$ in Eq. 45-2, or

$$\sin \theta = \frac{(m + \tfrac{1}{2})\lambda}{d} = \frac{(\tfrac{1}{2})(546 \times 10^{-9} \text{ m})}{0.10 \times 10^{-3} \text{ m}} = 0.0027.$$

This value for $\sin \theta$ is so small that we can take it to be the value of θ, expressed in radians; expressed in degrees it is 0.16°.

At the tenth maximum (not counting the central maximum) we must put $m = 10$ in Eq. 45-1. Doing so and calculating as before leads to an angular position of 3.8°. For these conditions we see that the angular spread of the first dozen or so fringes is small.

EXAMPLE 1

In Example 1 what is the linear distance on screen C between adjacent maxima?   **EXAMPLE 2**

If θ is small enough, we can use the approximation

$$\sin \theta \cong \tan \theta \cong \theta.$$

From Fig. 45-5 we see that $\tan \theta = \dfrac{y}{D}.$

Substituting this into Eq. 45-1 for $\sin \theta$ leads to

$$y = m \frac{\lambda D}{d} \qquad m = 0, 1, 2, \ldots \text{(maxima)}.$$

The positions of any two adjacent maxima are given by

$$y_m = m \frac{\lambda D}{d}$$

and $y_{m+1} = (m + 1) \dfrac{\lambda D}{d}.$

We find their separation Δy by subtracting

$$\Delta y = y_{m+1} - y_m = \frac{\lambda D}{d}$$

$$= \frac{(546 \times 10^{-9} \text{ m})(20 \times 10^{-2} \text{ m})}{0.10 \times 10^{-3} \text{ m}} = 1.09 \text{ mm}.$$

As long as θ in Figs. 45-5 and 45-6 is small, the separation of the interference fringes is independent of m; that is, the fringes are evenly spaced. Note that if the incident light contains more than one wavelength the separate interference patterns, which will have different fringe spacings, will be superimposed.

We can use Eq. 45-1 to measure the wavelength of light; to quote Thomas Young:

From a comparison of various experiments, it appears that the breadth of the undulations [that is, the wavelength] constituting the extreme red light must be supposed to be, in air, about one 36 thousandth of an inch, and those of the extreme violet about one 60 thousandth; the mean of the whole spectrum, with respect to the intensity of light, being one 45 thousandth.

Young's value for the average effective wavelength present in sunlight (1/45,000 in.) can be written as 570 nm, which agrees rather well with the wavelength at which the eye sensitivity is a maximum, 555 nm (see Fig. 42-1). It must not be supposed that Young's work was received without criticism. One of his contemporaries, a firm believer in the corpuscular theory of light, wrote in part:

We wish to raise our feeble voice against innovations that can have no other effect than to check the progress of science, and renew all those wild phantoms of the imagination which Bacon and Newton put to flight from her temple. This paper contains nothing that deserves the name of either experiment or discovery.

Needless to say, posterity has decided in favor of Young.

Analysis of the derivation of Eqs. 45-1 and 45-2 shows that a fundamental requirement for the existence of well-defined interference fringes on screen C in Figs. 45-5, 6 is that the light waves that travel from S_1 and S_2 to any point P on this screen must have a sharply defined phase difference ϕ that remains constant with time. If this condition is satisfied, a stable, well-defined fringe pattern will appear. At certain points P, ϕ

45-2
COHERENCE

will be given, independent of time, by $n\pi$ where $n = 1, 3, 5, \ldots$ so that the resultant intensity will be strictly zero and will remain so throughout the time of observation. At other points ϕ will be given by $n\pi$ where $n = 0, 2, 4 \ldots$ and the resultant intensity will be a maximum. Under these conditions the two beams emerging from slits S_1 and S_2 are said to be completely *coherent*.

Let the source in Fig. 45-5 be removed and let slits S_1 and S_2 be replaced by two completely independent light sources, such as two fine incandescent wires placed side by side in a glass envelope. No interference fringes will appear on screen C but only a relatively uniform illumination. We can interpret this if we make the reasonable assumption that for completely independent light sources the phase difference between the two beams arriving at P will vary with time in a random way. At a certain instant conditions may be right for cancellation and a short time later (perhaps 10^{-8} s) they may be right for re-enforcement. This same random phase behavior holds for all points on screen C with the result that this screen is uniformly illuminated. The intensity at any point is equal to the sum of the intensities that each source S_1 and S_2 produces separately at that point. Under these conditions the two beams emerging from S_1 and S_2 are said to be completely *incoherent*.

Note that for completely *coherent* light beams one (1) combines the amplitudes vectorially, taking the (constant) phase difference properly into account, and then (2) squares this resultant amplitude to obtain a quantity proportional to the resultant intensity. For completely *incoherent* light beams, on the other hand, one (1) squares the individual amplitudes to obtain quantities proportional to the individual intensities, and then (2) adds the individual intensities to obtain the resultant intensity. This procedure is in agreement with the experimental fact that for completely independent light sources the resultant intensity at every point is always greater than the intensity produced at that point by either light source acting alone.

It remains to investigate under what experimental conditions coherent or incoherent beams may be produced and to give an explanation for coherence in terms of the mode of production of the radiation. Consider first a parallel beam of microwave radiation emerging from an antenna connected by a coaxial cable to an oscillator based on an electromagnetic resonant cavity. The cavity oscillations (see Section 38-5) are completely periodic with time and produce, at the antenna, a completely periodic variation of **E** and **B** with time. The radiated wave at large enough distances from the antenna is well represented by Fig. 41-13. Note that (1) the wave has essentially infinite extent in time, including both future times ($t > 0$, say) and past times ($t < 0$); see Fig. 45-7a. At any point, as the wave passes by, the wave disturbance (i.e., **E** or **B**) varies with time in a perfectly periodic way. (2) The wavefronts at points

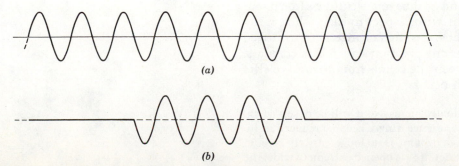

(a)

(b)

figure 45-7
(a) A section of an infinite wave and (b) a wavetrain.

far removed from the antenna are parallel planes of essentially infinite extent at right angles to the propagation direction. At any instant of time the wave disturbance varies with distance along the propagation direction in a perfectly periodic way.

Two beams generated from a single traveling wave like that of Fig. 41-13 will be completely coherent. One way to generate two such beams is to put an opaque screen containing two slits in the path of the beam. The waves emerging from the slits will always have a constant phase difference at any point in the region in which they overlap and interference fringes will be produced. Coherent radio beams can also be readily established, as can coherent elastic waves in solids, liquids and gases. The two prongs of the vibrating tapper in Fig. 45-4, for example, generate two coherent waves in the water of the ripple tank.

The technique of producing two beams from a single beam (and thus from a single source) tests specifically whether the wavefronts in the single parallel beam are truly planes, that is, whether all points in a plane at right angles to the direction of propagation have the same phase at any given instant. By dividing the beam in another way it is possible to test whether the beam is truly periodic over a large number of cycles of oscillation. This can be done, as we show in detail in Section 45-7, by inserting in the beam at 45° to it a thin sheet of material possessing the property that two beams are produced, one (which will be at right angles to the incident beam) by reflection and a second (which will be in the direction of the incident beam) by transmission. In the visible region such a sheet, called a *half-silvered mirror*, may be formed from a glass plate by depositing on it an appropriately thin film of silver. By appropriate use of mirrors (see Section 45-7) these two sub-beams can be recombined into a single beam traveling in a chosen direction. If the beams travel different distances before they are recombined, we are comparing in the combined beam a sample of the original beam with another sample an arbitrarily large number of cycles away. If the original beam is truly periodic in space and time, the two sub-beams will be completely coherent and interference fringes will be produced when they are recombined.

If we turn from microwave sources to common sources of visible light, such as incandescent wires or an electric discharge passing through a gas, we become aware of a fundamental difference. In both of these sources the fundamental light emission processes occur in individual atoms and these atoms do not act together in a cooperative (that is, *coherent*) way. The act of light emission by a single atom takes, in a typical case, about 10^{-8} s and the emitted light is properly described as a *wavetrain* (Fig. 45-7b) rather than as a wave (Fig. 45-7a). For emission times such as these the wavetrains are a few meters long.

Interference effects from ordinary light sources may be produced by putting a very narrow slit (S_0 in Fig. 45-1) directly in front of the source. This insures that the wavetrains that strike slits S_1 and S_2 in screen B in this figure originate from the same small region of the source. The diffracted beams emerging from S_1 and S_2 thus represent the same population of wavetrains and are coherent with respect to each other. If the phase of the light emitted from S_0 changes, this change is transmitted simultaneously to S_1 and S_2. Thus, at any point on screen C, a constant phase difference is maintained between the beams from these two slits and a stationary interference pattern occurs.

If the width of slit S_0 in Fig. 45-1 is gradually increased, it will be observed experimentally that the maxima of the interference fringes become reduced in intensity and that the intensity in the fringe minima is no longer strictly zero. In other words, the fringes become less distinct. If S_0 is opened extremely wide, the lowering of the maximum intensity and the raising of the minimum intensity will be such that the fringes disappear, leaving only a uniform illumination. Under these conditions we say that the beams from S_1 and S_2 pass continuously from a condition of complete coherence to one of complete incoherence. When not at either of these two limits, the beams are said to be partially coherent.

Partial coherence can also be demonstrated in two beams that are produced, as described on p. 1000, by inserting a "half-silvered mirror" in a beam at an angle of 45° to the direction of propagation. The two beams so produced, by reflection and transmission, traverse paths of different lengths before they are recombined. If the path difference is small compared with the average length of a wavetrain, the interference fringes will be sharply defined and will go essentially to zero at their minima. If the path difference is deliberately made longer, the fringes will become less distinct, and finally, when the path difference is larger than the average length of a wavetrain, the fringes will disappear altogether. Thus it is possible once again to progress smoothly in an experimental arrangement from complete coherence, through partial coherence, to complete incoherence.

The lack of coherence of the light from ordinary sources such as glowing wires is due to the fact that the emitting atoms do not act cooperatively (that is, *coherently*). Since 1960 it has proved possible to construct sources of visible light in which the atoms *do* act cooperatively and in which the emitted light is highly coherent. Such devices are called *lasers**; their light output is extremely monochromatic, intense, and highly collimated. The coherence of the emitted light can be demonstrated by placing a screen with two holes in it in the beam emerging from the laser. An interference pattern results, as Fig. 45-8 shows. These methods permit, for the first time, a degree of control of visible light approaching that possible for radio and for microwaves. The practical applications of lasers, which include the amplification of weak light signals, the use of light beams as highly efficient carriers of in-

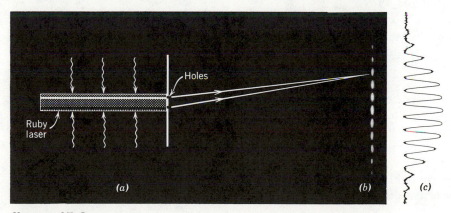

figure 45-8
A screen with two small holes is placed against the end of a laser. (*a*) The light passing through the holes forms an interference pattern on a strip of photographic film placed at (*b*). The fact that the pattern is formed shows that the light emitted from the laser is coherent across the beam cross section. At (*b*) is shown the image as formed on a photographic strip exposed in this manner. At (*c*) is an intensity plot of the film made by measuring the degree to which the film has been exposed. (Courtesy of D. F. Nelson and R. J. Collins, Bell Telephone Laboratories.)

---

* Laser, is a coined word, meaning "light amplification through stimulated emission of radiation." For an early report, see "Optical Masers" by Arthur L. Schawlow, *Scientific American*, June 1961.

formation from point to point,* and the production of high temperatures by intense local heating, are being actively exploited.†

45-3
INTENSITY IN YOUNG'S EXPERIMENT

Let us assume that the electric field components of the two waves in Fig. 45-5 vary with time at point P as

$$E_1 = E_0 \sin \omega t \tag{45-3}$$

and

$$E_2 = E_0 \sin (\omega t + \phi) \tag{45-4}$$

where $\omega \ (= 2\pi\nu)$ is the angular frequency of the waves and ϕ is the phase difference between them. Note that ϕ depends on the location of point P, which, in turn, for a fixed geometrical arrangement, is described by the angle θ in Figs. 45-5 and 45-6. We assume that the slits are so narrow that the diffracted light from each slit illuminates the central portion of the screen uniformly. This means that near the center of the screen E_0 is independent of the position of P, that is, of the value of θ.

The resultant wave disturbance‡ at P is found from

$$E = E_1 + E_2 \tag{45-5}$$

and (see Eq. 19-16) proves to be

$$E = E_\theta \sin (\omega t + \beta), \tag{45-6a}$$

where

$$\beta = \tfrac{1}{2}\phi \tag{45-6b}$$

and

$$E_\theta = 2E_0 \cos \beta = E_m \cos \beta. \tag{45-6c}$$

E_m, the maximum possible amplitude for E_θ, is equal to twice the amplitude of the combining waves $(= 2E_0)$, corresponding to complete reinforcement. You should verify Eq. 45-6 carefully. The amplitude E_θ of the resultant wave disturbance, which determines the intensity of the interference fringes, will turn out to depend strongly on the value of θ, that is, on the location of point P in Figs. 45-5 and 45-6.

In Section 19-6 we showed that the intensity of a wave I, measured perhaps in watts/meter², is proportional to the square of its amplitude. For the resultant wave then, ignoring the proportionality constant,

$$I_\theta \propto E_\theta^2. \tag{45-7}$$

This relationship seems reasonable if we recall (Eq. 30-9) that the energy density in an electric field is proportional to the *square* of the electric field strength. This is true for rapidly varying electric fields, such as those in a light wave, as well as for static fields.

The ratio of the intensities of two light waves is the ratio of the squares of the amplitudes of their electric fields. If I_θ is the intensity of the resultant wave at P, and I_0 is the intensity that a single wave acting alone would produce, then

$$\frac{I_\theta}{I_0} = \left(\frac{E_\theta}{E_0}\right)^2. \tag{45-8}$$

---

* See "Light-Wave Communications" by W. S. Boyle, *Scientific American*, August 1977.
† For an example of one of the many applications of lasers, see "Processing Materials with Lasers" by Edward N Breinan, Bernard H. Kear, and Conrad M. Banas, *Physics Today*, November 1976.
‡ The electric field **E** in the light wave rather than the magnetic field **B** is normally identified with the "wave disturbance" because the effects of **B** on the human eye and on various light detectors are exceedingly small. Note too that, although Eq. 45-5 should be a vector equation, in most cases of interest the **E** vectors in the two interfering waves are closely parallel so that an algebraic equation suffices.

Combining with Eq. 45-6c leads to

$$I_\theta = 4I_0 \cos^2 \beta = I_m \cos^2 \beta. \qquad (45\text{-}9)$$

Note that the intensity of the resultant wave at any point P varies from zero [for a point at which $\phi\,(=2\beta) = \pi$, say] to I_m, which is four times the intensity I_0 of each individual wave [for a point at which $\phi\,(=2\beta) = 0$, say]. Let us compute I_θ as a function of the angle θ in Figs. 45-5 or 45-6.

The phase difference ϕ in Eq. 45-4 is associated with the path difference $S_1 b$ in Fig. 45-5 or 45-6. If $S_1 b$ is $\frac{1}{2}\lambda$, ϕ will be π; if $S_1 b$ is λ, ϕ will be 2π, etc. This suggests that

$$\frac{\text{phase difference}}{2\pi} = \frac{\text{path difference}}{\lambda},$$

$$\phi = \frac{2\pi}{\lambda}\,(d\,\sin\,\theta),$$

or, finally, from Eq. 45-6b,

$$\beta = \tfrac{1}{2}\phi = \frac{\pi d}{\lambda}\,\sin\,\theta. \qquad (45\text{-}10)$$

We can substitute this expression for β into Eq. 45-9 for I_θ, yielding the latter quantity as a function of θ. For convenience we collect here the expressions for the amplitude and the intensity in double-slit interference.

[Eq. 45-6c]	$E_\theta = E_m \cos \beta$	interference	$(45\text{-}11a)$
[Eq. 45-9]	$I_\theta = I_m \cos^2 \beta$	from narrow	$(45\text{-}11b)$
		slits (that is,	
[Eq. 45-10]	$\beta\,(=\tfrac{1}{2}\phi) = \dfrac{\pi d}{\lambda}\,\sin\,\theta$	$a \ll \lambda)$	$(45\text{-}11c)$

To find the positions of the intensity maxima, we put

$$\beta = m\pi \qquad m = 0, 1, 2, \ldots$$

in Eq. 45-11b. From Eq. 45-11c this reduces to

$$d\,\sin\,\theta = m\lambda \qquad m = 0, 1, 2, \ldots \text{ (maxima)},$$

which is the equation derived in Section 45-1 (Eq. 45-1). To find the intensity minima we write

$$\frac{\pi d\,\sin\,\theta}{\lambda} = (m + \tfrac{1}{2})\pi \qquad m = 0, 1, 2, \ldots \text{ (minima)},$$

which reduces to the previously derived Eq. 45-2.

Figure 45-9 shows the intensity pattern for double-slit interference.

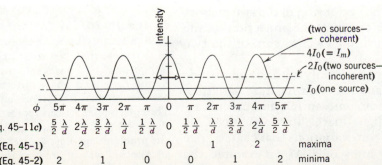

figure 45-9
The intensity pattern for double-slit interference. The heavy arrow in the central peak represents the half-width of the peak. This figure is constructed on the assumption that the two interfering waves each illuminate the central portion of the screen uniformly, that is, I_0 is independent of position as shown.

The horizontal solid line is I_0; this describes the (uniform) intensity pattern on the screen if one of the slits is covered up. If the two sources were incoherent, the intensity would be uniform over the screen and would be $2I_0$; see the horizontal dashed line in Fig. 45-9. For coherent sources we expect the energy to be merely redistributed over the screen, because energy is neither created nor destroyed by the interference process. Thus the *average* intensity in the interference pattern should be $2I_0$, as for incoherent sources. This follows at once if, in Eq. 45-11b, we substitute one-half for the cosine-squared term and if we recall that $I_m = 4I_0$. We have seen several times that the average value of the square of a sine or a cosine term over one or more half-cycles is one-half.

In Section 45-3 we combined two time-varying wave disturbances, namely,

$$E_1 = E_0 \sin \omega t \qquad (45\text{-}3)$$

and

$$E_2 = E_0 \sin (\omega t + \phi), \qquad (45\text{-}4)$$

which have the same angular frequency ω and amplitude E_0 but which have a phase difference ϕ between them. In this case the result (Eqs. 45-11a and 45-11c) is easily obtained algebraically.

In later chapters we will want to add larger numbers of wave disturbances, often an infinite number, with infinitesimal individual amplitudes. Since analytic methods become more difficult in such cases we describe a graphical method, illustrating it by rederiving Eq. 45-11a.

A sinusoidal wave disturbance such as that represented by Eq. 45-3 can be represented graphically, using a rotating vector. In Fig. 45-10a a vector of magnitude E_0 is allowed to rotate about the origin in a counterclockwise direction with an angular frequency ω. Following electrical engineering practice we call such a rotating vector a *phasor*; see Section 39-2. The alternating wave disturbance E_1 (Eq. 45-3) is represented by the projection of this phasor on the vertical axis.

A second wave disturbance E_2, which has the same amplitude E_0 but a phase difference ϕ with respect to E_1,

$$E_2 = E_0 \sin (\omega t + \phi), \qquad (45\text{-}4)$$

can be represented graphically (Fig. 45-10b) as the projection on the vertical axis of a second phasor of magnitude E_0 which makes an angle ϕ with the first phasor. As this figure shows, the sum E of E_1 and E_2 is the sum of the projections of the two phasors on the vertical axis. This is revealed more clearly if we redraw the phasors, as in Fig. 45-10c, placing the foot of one arrow at the head of the other, maintaining the proper phase difference, and letting the whole assembly rotate counterclockwise about the origin.

In Fig. 45-10c E can also be regarded as the projection on the vertical axis of a phasor of length E_θ, which is the vector sum of the two phasors of magnitude E_0. Note that the (algebraic) sum of the projections of the two phasors is equal to the projection of the (vector) sum of the two phasors.

In most problems in optics we are concerned only with the *amplitude* E_θ of the resultant wave disturbance and not with its time variation. This is because the eye and other common measuring instruments respond to the resultant intensity of the light (that is, to the square of the amplitude) and cannot detect the rapid time variations that characterize visible light. For sodium light ($\lambda = 589$ nm $= 5890$ Å), for example, the

45-4
ADDING WAVE DISTURBANCES

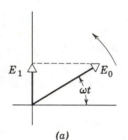

(a)

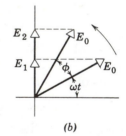

(b)

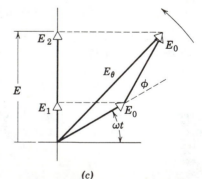

(c)

figure 45-10
(a) A time-varying wave disturbance E_1 is represented by a rotating vector or *phasor*. (b) Two wave disturbances E_1 and E_2, with a phase difference ϕ between them, are so represented. These two phasors can represent the two wave disturbances in the double-slit problem; see Eqs. 45-3 and 45-4. (c) Another way of drawing (b).

$\nu(=\omega/2\pi)$ is 5.1×10^{14} Hz. Often, then, we need not consider the rotation of the phasors but can confine our attention to finding the magnitude of the resultant phasor.

Figure 45-11a shows the phasors for double-slit interference at time $t = 0$; compare Fig. 45-10c. We see that

$$E_\theta = 2E_0 \cos \beta = E_m \cos \beta,$$

in which, from the theorem that the exterior angle of a triangle (ϕ) is equal to the sum of its two opposite interior angles $(\beta + \beta)$,

$$\beta = \tfrac{1}{2}\phi.$$

This is exactly the result arrived at earlier algebraically; compare Eqs. 45-11a and 45-11c.

In a more general case we might want to find the resultant of a number (>2) of sinusoidally varying wave disturbances. The general procedure is the following:

1. Construct a series of phasors representing the functions to be added. Draw them end to end, maintaining the proper phase relationships between adjacent phasors.

2. Construct the vector sum of this array. Its length gives the amplitude of the resultant. The angle between it and the first phasor is the phase of the resultant with respect to this first phasor. The projection of this phasor on the vertical axis gives the time variation of the resultant wave disturbance.

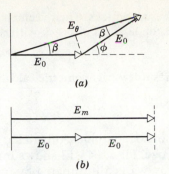

(a)

(b)

figure 45-11
(a) A construction to find the amplitude E_θ of two wave disturbances of amplitude E_0 and phase difference ϕ. (b) The maximum possible amplitude for these wave disturbances occurs for $\phi = 0$ and has the value $E_m = 2E_0$.

EXAMPLE 3

Find graphically the resultant $E(t)$ of the following wave disturbances:

$$E_1 = 10 \sin \omega t$$
$$E_2 = 10 \sin (\omega t + 15°)$$
$$E_3 = 10 \sin (\omega t + 30°)$$
$$E_4 = 10 \sin (\omega t + 45°).$$

Figure 45-12 in which E_0 equals 10, shows the assembly of four phasors that

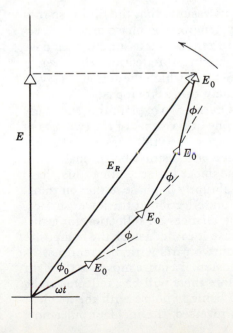

figure 45-12
Example 3. Four wave disturbances are added graphically, using the method of phasors.

represents these functions. Their vector sum, by graphical measurement, has an amplitude E_R of 38 and a phase ϕ_0 with respect to E_1 of 23°. In other words,

$$E(t) = E_1 + E_2 + E_3 + E_4 = 38 \sin{(\omega t + 23°)}.$$

Check this result by trigonometric calculation.

The colors of soap bubbles, oil slicks, and other thin films are the result of interference. Figure 45-13 shows interference effects in a thin vertical film of soapy water illuminated by monochromatic light.

Figure 45-14 shows a film of uniform thickness d and index of refraction n, the eye being focused on spot a. The film is illuminated by a broad source of monochromatic light S. There exists on this source a point P such that two rays, identified by the single and double arrows, respectively, can leave P and enter the eye as shown, after passing through point a. These two rays follow different paths in going from P to the eye, one being reflected from the upper surface of the film, the other from the lower surface. Whether point a appears bright or dark depends on the nature of the interference between the two waves that diverge from a. These waves are coherent because they both originate from the same point P on the light source.

If the eye looks at another part of the film, say a', the light that enters the eye must originate from a different point P' of the extended source, as suggested by the dashed lines in Fig. 45-14.

For near-normal incidence ($\theta \cong 0$ in Fig. 45-14) the geometrical path difference for the two rays from P will be close to $2d$. We might expect the resultant wave reflected from the film near a to be an interference maximum if the distance $2d$ is an integral number of wavelengths. This statement must be modified for two reasons.

First, the wavelength must refer to the wavelength of the light in the film λ_n and not to its wavelength in air λ; that is, we are concerned with optical path lengths rather than geometrical path lengths. The wavelengths λ and λ_n (see Eq. 43-12b) are related by

$$\lambda_n = \lambda/n. \tag{45-12}$$

To bring out the second point, let us assume that the film is so thin that $2d$ is very much less than a wavelength. The phase difference between the two waves would be close to zero on our assumption, and we would expect such a film to appear bright on reflection. However, it appears dark. This is clear from Fig. 45-13, in which the action of gravity produces a wedge-shaped film, extremely thin at its top edge. As drainage continues, the dark area increases in size. To explain this and many similar phenomena, we assume that one or the other of the two rays of Fig. 45-14 suffers an abrupt phase change of π ($=180°$) associated either with reflection at the air-film interface or transmission through it. As it turns out, the ray reflected from the upper surface suffers this phase change. The other ray is not changed abruptly in phase, either on transmission through the upper surface or on reflection at the lower surface.

In Section 19-9 we discussed phase changes on reflection for transverse waves in strings. To extend these ideas, consider the composite string of Fig. 45-15, which consists of two parts with different masses per unit length, stretched to a given tension. If a pulse moves to the right in Fig. 45-15a, approaching the junction, there will be a reflected and a transmitted pulse, the reflected pulse being *in phase* with the incident pulse. In Fig. 45-15b the situation is reversed, the incident pulse now

45-5
INTERFERENCE FROM THIN FILMS

figure 45-13
A soapy water film on a wire loop, viewed by reflected light. The black segment at the top is not a tear. It arises because the film, by drainage, is so thin here that there is destructive interference between the light reflected from its front surface and that reflected from its back surface. We shall see that these two waves differ in phase by 180°.

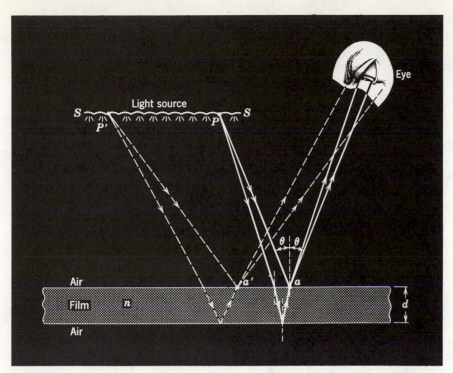

figure 45-14
Interference by reflection from a thin film, assuming an extended source S.

being in the lighter string. In this case the reflected pulse will differ in phase from the incident pulse by π (=180°). In each case the transmitted pulse will be in phase with the incident pulse.

Figure 45-15a suggests a light wave in glass, say, approaching a surface beyond which there is a less optically dense medium (one of lower index of refraction) such as air. Figure 45-15b suggests a light wave in air approaching glass. To sum up the optical situation, when reflection occurs from an interface beyond which the medium has a *lower* index of refraction, the reflected wave undergoes *no phase change*; when the medium beyond the interface has a *higher* index, there is a phase change of π.* The transmitted wave does not experience a change of phase in either case.

We are now able to take into account both factors that determine the nature of the interference, namely, differences in optical path length and phase changes on reflection. For the two rays of Fig. 45-14 to combine to give a *maximum* intensity, assuming normal incidence, we must have

$$2d = (m + \tfrac{1}{2})\lambda_n \qquad m = 0, 1, 2, \ldots.$$

The term $\tfrac{1}{2}\lambda_n$ is introduced because of the phase change on reflection, a phase change of 180° being equivalent to half a wavelength. Substituting λ/n for λ_n yields finally

$$2dn = (m + \tfrac{1}{2})\lambda \qquad m = 0, 1, 2, \ldots \text{ (maxima)}. \qquad (45\text{-}13)$$

The condition for a *minimum* intensity is

$$2dn = m\lambda \qquad m = 0, 1, 2, \ldots \text{ (minima)}. \qquad (45\text{-}14)$$

* These statements, which can be proved rigorously from Maxwell's equations (see also Section 45-6), must be modified for light falling on a less dense medium at an angle such that total internal reflection occurs. They must also be modified for reflection from metallic surfaces.

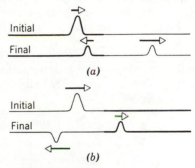

figure 45-15
Phase changes on reflection at a junction between stretched composite string. (a) Incident pulse in heavy string and (b) incident pulse in light string.

These equations hold when the index of refraction of the film is either greater or less than the indices of the media on *each* side of the film. Only in these cases will there be a relative phase change of 180° for reflections at the two surfaces. A water film in air and an air film in the space between two glass plates provide examples of cases to which Eqs. 45-13 and 45-14 apply. Example 5 provides a case in which they do not apply.

If the film thickness is not uniform, as in Fig. 45-13, where the film is wedge-shaped, constructive interference will occur in certain parts of the film and destructive interference will occur in others. Lines of maximum and of minimum intensity will appear—these are the interference fringes. They are called *fringes of constant thickness*, each fringe being the locus of points for which the film thickness d is a constant. If the film is illuminated with white light rather than monochromatic light, the light reflected from various parts of the film will be modified by the various constructive or destructive interferences that occur. This accounts for the brilliant colors of soap bubbles and oil slicks.

Only if the film is "thin," which implies that d is no more than a few wavelengths of light, will fringes of the type described, that is, fringes that appear localized on the film and associated with a variable film thickness, be possible. For very thick films (say $d \cong 1$ cm), the path difference between the two rays of Fig. 45-14 will be many wavelengths and the phase difference at a given point on the film will change rapidly as we move even a small distance away from a. For "thin" films, however, the phase difference at a also holds for reasonably nearby points; there is a characteristic "patch brightness" for any point on the film, as Fig. 45-13 shows. Interference fringes can be produced for thick films; they are not localized on the film but are at infinity. See Section 45-7.

EXAMPLE 4

A water film ($n = 1.33$) in air is 320 nm thick. If it is illuminated with white light at normal incidence, what color will it appear to be in reflected light?

By solving Eq. 45-13 for λ,

$$\lambda = \frac{2dn}{m + \frac{1}{2}} = \frac{(2)(320 \text{ nm})(1.33)}{m + \frac{1}{2}} = \frac{850 \text{ nm}}{m + \frac{1}{2}} \quad \text{(maxima)}.$$

From Eq. 45-14 the minima are given by

$$\lambda = \frac{850 \text{ nm}}{m} \quad \text{(minima)}.$$

Maxima and minima occur for the following wavelengths:

m	0 (max)	1 (min)	1 (max)	2 (min)	2 (max)
λ, nm	1700	850	570	425	340

Only the maximum corresponding to $m = 1$ lies in the visible region (see Fig. 42-1); light of this wavelength appears yellow-green. If white light is used to illuminate the film, the yellow-green component will be enhanced when viewed by reflection.

EXAMPLE 5

*Nonreflecting glass.** Lenses are often coated with thin films of transparent substances such as MgF$_2$ ($n = 1.38$) in order to reduce the reflection from the glass surface, using interference. How thick a coating is needed to produce a minimum reflection at the center of the visible spectrum (550 nm)?

---

* See "Optical Interference Coatings" by Philip Baumeister and Gerald Pineus, *Scientific American*, December 1970.

We assume that the light strikes the lens at near-normal incidence (θ is exaggerated for clarity in Fig. 45-16), and we seek destructive interference between rays r and r_1. Equation 45-14 does not apply because in this case a phase change of 180° is associated with *each* ray, for at *both* the upper and lower surfaces of the MgF$_2$ film the reflection is from a medium of greater index of refraction.

There is no net change in phase produced by the two reflections, which means that the optical path difference for destructive interference is $(m + \frac{1}{2})\lambda$ (compare Eq. 45-13), leading to

$$2dn = (m + \tfrac{1}{2})\lambda \qquad m = 0, 1, 2, \ldots \text{ (minima)}.$$

Solving for d and putting $m = 0$ yields

$$d = \frac{(m + \frac{1}{2})\lambda}{2n} = \frac{\lambda}{4n} = \frac{550 \text{ nm}}{(4)(1.38)} = 100 \text{ nm}.$$

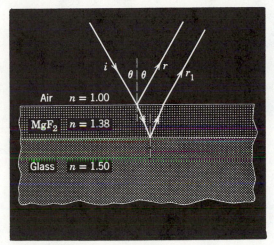

figure 45-16
Example 5. Unwanted reflections from glass can be reduced by coating the glass with a thin transparent film.

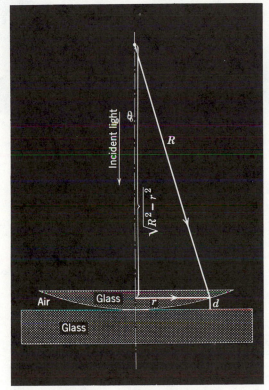

figure 45-17
Example 6. Apparatus for observing Newton's rings.

EXAMPLE 6

Newton's rings. Figure 45-17 shows a lens of radius of curvature R resting on an accurately plane glass plate and illuminated from above by light of wavelength λ. Figure 45-18 shows that circular interference fringes (Newton's rings) appear, associated with the variable thickness air film between the lens and the plate. Find the radii of the circular interference maxima.

Here it is the ray from the *bottom* of the (air) film rather than from the top that undergoes a phase change of 180°, for it is the one reflected from a medium of higher refractive index. The condition for a maximum remains unchanged (Eq. 45-13), however, and is

$$2d = (m + \tfrac{1}{2})\lambda \qquad m = 0, 1, 2, \ldots, \tag{45-15}$$

the index of refraction of the air film being assumed to be unity. From Fig. 45-17

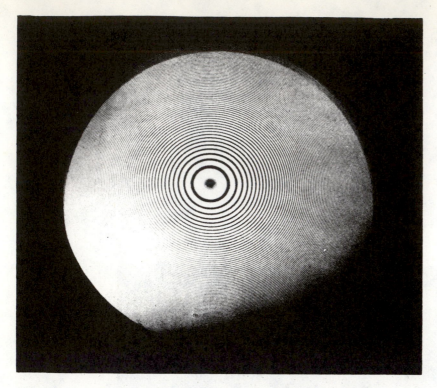

we can write

$$d = R - \sqrt{R^2 - r^2} = R - R\left[1 - \left(\frac{r}{R}\right)^2\right]^{1/2}$$

If $r/R \ll 1$, we can expand the square bracket by the binomial theorem, keeping only two terms, or

$$d = R - R\left[1 - \frac{1}{2}\left(\frac{r}{R}\right)^2 + \cdots\right] \cong \frac{r^2}{2R}.$$

Combining with Eq. 45-15 yields

$$r = \sqrt{(m + \tfrac{1}{2})\lambda R} \qquad m = 0, 1, 2, \ldots,$$

which gives the radii of the bright rings. If white light is used, each spectrum component will produce its own set of circular fringes, the sets all overlapping.

G. G. Stokes (1819–1903) used the principle of optical reversibility to investigate the reflection of light at an interface between two media. The principle states that if there is no absorption of light, a light ray that is reflected or refracted will retrace its original path if its direction is reversed. This reminds us that any mechanical system can run backward as well as forward, provided there is no absorption of energy because of friction, etc.

Figure 45-19a shows a wave of amplitude E reflected and refracted at a surface separating media 1 and 2, where $n_2 > n_1$. The amplitude of the reflected wave is $r_{12}E$, in which r_{12} is an *amplitude reflection coefficient*. The amplitude of the refracted wave is $t_{12}E$, where t_{12} is an *amplitude transmission coefficient*.

We consider only the possibility of phase changes of 0 or 180°. If $r_{12} = +0.5$, for example, we have a reduction in amplitude on reflection by one-half and no change in phase. For $r_{12} = -0.5$ we have a phase

45-6
OPTICAL REVERSIBILITY AND PHASE CHANGES ON REFLECTION

change of 180° because

$$E \sin (\omega t + 180°) = -E \sin \omega t.$$

Figure 45-19b suggests that if we reverse these two rays, they should combine to produce the original ray reversed in direction. Ray $r_{12}E$, identified by the single arrows in the figure, is reflected and refracted, producing the rays of amplitudes r_{12}^2E and $r_{12}t_{12}E$. Ray $t_{12}E$, identified by the triple arrows, is also reflected and refracted, producing the rays of amplitudes $t_{12}t_{21}E$ and $t_{12}r_{21}E$ as shown. Note that r_{12} describes a ray in medium 1 reflected from medium 2, and r_{21} describes a ray in medium 2 reflected from medium 1. Similarly, t_{12} describes a ray that passes from medium 1 to medium 2; t_{21} describes a ray that passes from medium 2 to medium 1.

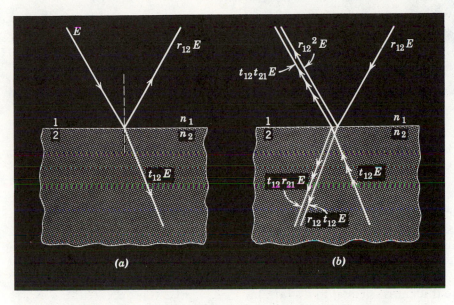

figure 45-19
(a) A ray is reflected and refracted at an air-glass interface. (b) The optically reversed situation; the two rays in the lower left must cancel.

The two rays in the upper left of Fig. 45-19b must be equivalent to the incident ray of Fig. 45-19a, reversed; the two rays in the lower left of Fig. 45-19b must cancel. This second requirement leads to

$$r_{12}t_{12}E + t_{12}r_{21}E = 0,$$

or
$$r_{12} = -r_{21}.$$

This result tells us that if we compare a wave reflected from medium 1 with one reflected from medium 2, they behave differently in that one or the other undergoes a phase change of 180°. We must rely on experiment or analysis by Maxwell's equations to show that, as we pointed out earlier, the ray reflected from the more optically dense medium is the one that experiences the phase change of 180°.

An *interferometer* is a device that can be used to measure lengths or changes in length with great accuracy by means of interference fringes. We describe the form originally built by A. A. Michelson (1852–1931) in 1881.

Consider light that leaves point P on extended source S (Fig. 45-20) and falls on half-silvered mirror M. This mirror has a silver coating just

45-7
MICHELSON'S INTERFEROMETER*

* See "Michelson: America's first Nobel Prize Winner in Science" by R. S. Shankland, *The Physics Teacher*, January 1977. See also "Michelson and his Interferometer" by R. S. Shankland, *Physics Today*, April 1974.

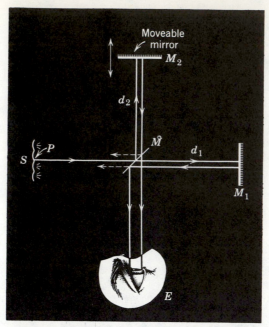

figure 45-20
Michelson's interferometer, showing the path of a particular ray originating at point P of an extended source S.

thick enough to transmit half the incident light and to reflect half; in the figure we have assumed that this mirror, for convenience, possesses negligible thickness. At M the light divides into two waves. One proceeds by transmission toward mirror M_1; the other proceeds by reflection toward M_2. The waves are reflected at each of these mirrors and are sent back along their directions of incidence, each wave eventually entering the eye E. Because the waves are coherent, being derived from the same point on the source, they will interfere.

If the mirrors M_1 and M_2 are exactly perpendicular to each other, the effect is that of light from an extended source S falling on a uniformly thick slab of air, between glass, whose thickness is equal to $d_2 - d_1$. Interference fringes appear, caused by small changes in the angle of incidence of the light from different points on the extended source as it strikes the equivalent air film. For *thick* films a path difference of one wavelength can be brought about by a very small change in the angle of incidence.

If M_2 is moved backward or forward, the effect is to change the thickness of the equivalent air film. Suppose that the center of the (circular) fringe pattern appears bright and that M_2 is moved just enough to cause the first bright circular fringe to move to the center of the pattern. The path of the light beam striking M_2 has been changed by one wavelength. This means (because the light passes twice through the equivalent air film) that the mirror must have moved one-half a wavelength.

The interferometer is used to measure changes in length by counting the number of interference fringes that pass the field of view as mirror M_2 is moved. Length measurements made in this way can be accurate if large numbers of fringes are counted.

Michelson measured the length of the standard meter, kept in Paris, in terms of the wavelength of certain monochromatic red light emitted from a light source containing cadmium. He showed that the standard meter was equivalent to 1,553,163.5 wavelengths of the red cadmium light. For this he received the Nobel prize in 1907.

Physicists have long speculated on the advantages of discarding the standard meter bar as the basic standard of length and of *defining* the meter in terms of the wavelength of some carefully chosen monochromatic radiation. This would make the primary length standard readily available in laboratories all over the world. It would improve the accuracy of length measurements, since one would no longer need to compare an unknown object with a standard object (the meter bar), using interferometer techniques, but could measure the unknown object *directly* and in an absolute sense, using these techniques. There is the additional advantage that if the standard meter bar were destroyed, it could never be replaced, whereas light sources and interferometers will (presumably) always be available.

In 1961 such an atomic standard of length was adopted by international agreement. Quoting from an article* describing the event:

The wavelength of the orange-red light of krypton-86 has replaced the platinum iridium bar as the world standard of length. Formerly the wavelength of this light was defined as a function of the length of the meter bar. Now the meter is defined as a multiple (1,650,763.73) of the wavelength of the light.

The light from krypton-86 was used in preference to that from other atomic sources because it produces sharper interference fringes in the interferometer over the long optical paths sometimes used in length measurement.

45-8
MICHELSON'S INTERFEROMETER AND LIGHT PROPAGATION†

In Section 42-4 we presented Einstein's hypothesis, now well verified, that in free space light is propagated with the same speed c no matter what the relative velocity of the source and the observer may be. We pointed out that this hypothesis contradicted the views of nineteenth-century physicists regarding wave propagation. It was difficult for these physicists, trained as they were in the classical physics of the time, to believe that a wave could be propagated without a medium. If such a medium could be established, the speed c of light would naturally be construed as the speed *with respect to that medium,* just as the speed of sound always refers to a medium such as air.

Although no medium for light propagation was obvious, the physicists postulated one, called the *ether*‡ and hypothesized that its properties were such that it was undetectable by ordinary means such as weighing.

In 1881 (24 years before Einstein's hypothesis) A. A. Michelson set himself the task of forcing the ether, assuming that it existed, to submit to direct physical verification. In particular, Michelson, later joined by E. W. Morley, tried to measure the speed u with which the earth moves through the ether. Michelson's interferometer was their instrument of choice for this now-famous Michelson-Morley experiment.

The earth together with the interferometer moving with velocity **u** through the ether is equivalent to the interferometer at rest with the ether streaming through it with velocity −**u,** as shown in Fig. 45-21. Consider a wave moving along the path MM_1M and one moving along MM_2M. The first corresponds classically to a man rowing a boat a distance d downstream and the same distance upstream; the second corresponds to rowing a boat a distance d across a stream and back.

---

* *Scientific American,* p. 75, December 1960.
† See Supplementary Topic V.
‡ More fully, the *luminiferous* (or light-carrying) *ether.*

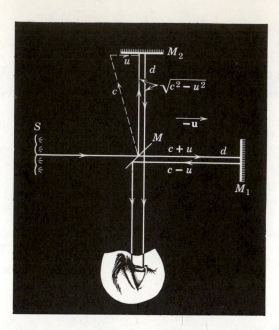

On the ether hypothesis the speed of light on the path MM_1 is $c + u$; on the return path M_1M it is $c - u$. The time required for the complete trip is

$$t_1 = \frac{d}{c + u} + \frac{d}{c - u} = d\frac{2c}{c^2 - u^2} = \frac{2d}{c}\frac{1}{1 - (u/c)^2}.$$

The speed of light, *on the ether hypothesis*, for path MM_2 is $\sqrt{c^2 - u^2}$, as Fig. 45-21 suggests. This same speed holds for the return path M_2M, so that the time required for this complete path is

$$t_2 = \frac{2d}{\sqrt{c^2 - u^2}} = \frac{2d}{c}\frac{1}{\sqrt{1 - (u/c)^2}}.$$

The difference of time for the two paths is

$$\Delta t = t_1 - t_2$$
$$= \frac{2d}{c}\left\{\left[1 - \left(\frac{u}{c}\right)^2\right]^{-1} - \left[1 - \left(\frac{u}{c}\right)^2\right]^{-1/2}\right\}$$

Assuming $u/c \ll 1$, we can expand the quantities in the square brackets by using the binomial theorem, retaining only the first two terms. This leads to

$$\Delta t = \frac{2d}{c}\left\{\left[1 + \left(\frac{u}{c}\right)^2 + \cdots\right] - \left[1 + \frac{1}{2}\left(\frac{u}{c}\right)^2 + \cdots\right]\right\}$$
$$= \frac{2d}{c}\left\{\frac{1}{2}\left(\frac{u}{c}\right)^2\right\} = \frac{du^2}{c^3}. \tag{45-16}$$

Now let the entire interferometer be rotated through 90°. This will interchange the roles of the two light paths, MM_1M now being the "cross-stream" path and MM_2M the "down- and up-stream" path. The time difference between the two waves entering the eye is also reversed; this changes the phase difference between the combining waves and alters the positions of the interference maxima. *The experiment consists of looking for a shift of the interference fringes as the apparatus is rotated.*

The *change* in time difference is $2\Delta t$, which corresponds to a fringe shift of $2\Delta t/T$ where $T(= \lambda/c)$ is the period of vibration of the light. The expected maximum shift in the number of fringes on a 90° rotation (see Eq. 45-16) is

$$\Delta N = \frac{2\Delta t}{T} = \frac{2\Delta tc}{\lambda} = \frac{2d}{\lambda}\left(\frac{u}{c}\right)^2 \qquad (45\text{-}17)$$

In the Michelson-Morley interferometer let $d = 11$ m (obtained by multiple reflection in the interferometer) and $\lambda = 5.9 \times 10^{-7}$ m. If u is assumed to be roughly the orbital speed of the earth, then $u/c \cong 10^{-4}$. The expected maximum fringe shift when the interferometer is rotated through 90° is then

$$\Delta N = \frac{2d}{\lambda}\left(\frac{u}{c}\right)^2 = \frac{(2)(11\text{ m})}{5.9 \times 10^{-7}\text{ m}}(10^{-4})^2 = 0.4.$$

Even though a shift of only about 0.4 of a fringe was expected, Michelson and Morley were confident that they could observe a shift of 0.01 fringe. *They found from their experiment, however, that there was no observable fringe shift!*

The analogy between a light wave in the supposed ether and a boat moving in water, which seemed so evident in 1881, is simply incorrect. The derivation based on this analogy is incorrect for light waves. When the analysis is carried through on Einstein's hypothesis, the observed negative result is clearly predicted, the speed of light being c for all paths. The motion of the earth around the sun and the rotation of the interferometer have, in Einstein's view, no effect whatever on the speed of the light waves in the interferometer.

It should be made clear that although Einstein's hypothesis is completely consistent with the negative result of the Michelson-Morley experiment, this experiment standing alone cannot serve as a proof for Einstein's hypothesis. Einstein said that no number of experiments, however large, could prove him right but that a single experiment could prove him wrong. Our present-day belief in Einstein's hypothesis rests on consistent agreement in a large number of experiments designed to test it. The "single experiment" that might prove Einstein wrong has never been found.

questions

1. Is Young's experiment an interference experiment or a diffraction experiment, or both?

2. Do interference effects occur for sound waves? Recall that sound is a longitudinal wave and that light is a transverse wave.

3. In Young's double-slit interference experiment, using a monochromatic laboratory light source, why is screen A in Fig. 45-1 necessary? What would happen if one gradually enlarged the hole in this screen?

4. What changes occur in the pattern of interference fringes if the apparatus of Fig. 45-5 is placed under water?

5. If interference between light waves of different frequencies is possible, one should observe light beats, just as one obtains sound beats from two sources of sound with slightly different frequencies. Discuss how one might experimentally look for this possibility.

6. Why are parallel slits preferable to the pinholes that Young used in demonstrating interference?

7. Is coherence important in reflection and refraction?

8. If your source of light is a laser beam, you do not need (see Fig. 45-8) the equivalent of screen A in Fig. 45-1. Why?

9. Defend this statement: Fig. 45-7a is a sine (or cosine) wave but Fig. 45-7b is not. Indeed you cannot assign a unique frequency to the curve of Fig. 45-7b. Why not? (Hint: think of Fourier analysis.)

10. Most of us are familiar with rotating or oscillating radar antennas which produce rotating or oscillating beams of microwave radiation. It is also possible to produce an oscillating beam of microwave radiation *without* any mechanical motion of the transmitting antenna. This is done by periodically changing the phase of the radiation as it emerges from various sections of the (long) transmitting antenna. Convince yourself that, by constructive interference from various parts of the fixed antenna, an oscillating microwave beam could indeed be so produced. A very large radar installation of this type has been installed by the United States in the Aleutian Islands to monitor USSR missile experiments.

11. Describe the pattern of light intensity on screen C in Fig. 45-5 if one slit is covered with a red filter and the other with a blue filter, the incident light being white.

12. Define carefully, and distinguish between, the angles θ and ϕ that appear in Eq. 45-10.

13. If one slit in Fig. 45-5 is covered, what change occurs in the intensity of light in the center of the screen?

14. In Young's double-slit experiment suppose that screen A in Fig. 45-1 contained *two* very narrow parallel slits instead of one. (*a*) Show that if the spacing between these slits is properly chosen, the interference fringes can be made to disappear. (*b*) Under these conditions, would you call the beams emerging from slits S_1 and S_2 in screen B coherent? They do not produce interference fringes. (*c*) Discuss what would happen to the interference fringes in the case of a single slit in screen A if the slit width were gradually increased.

15. What are the requirements for a maximum intensity when viewing a thin film by *transmitted* light?

16. In a Newton's rings experiment, is the central spot, as seen by reflection, dark or light? Explain.

17. Why must the film of Fig. 45-14 be "thin" for us to see an interference pattern of the type described?

18. Why do coated lenses (see Example 5) look purple by reflected light?

19. A person wets his eyeglasses to clean them. As the water evaporates he notices that for a short time the glasses become markedly more nonreflecting. Explain.

20. A lens is coated to reduce reflection, as in Example 5. What happens to the energy that had previously been reflected? Is it absorbed by the coating?

21. Very small changes in the angle of incidence do not change the interference conditions much for "thin" films but they do change them for "thick" films. Why?

22. In connection with the phase change on reflection at an interface between two transparent media, do you think that phase shifts other than 0 or π are possible? Do you think that phase shifts can be calculated rigorously from Maxwell's equations?

23. Consider the following objects that produce colors when exposed to sunlight: (1) soap bubbles, (2) rose petals, (3) the inner surface of an oyster shell (irridescence), (4) thin oil slicks, (5) non-reflecting coatings on camera lenses, and (6) peacock tail feathers. All but one of these are purely interference phenomena, no pigments being involved. Which one is it? Discuss.

24. Why does a film (soap bubble, oil slick, etc.) have to be "thin" to display interference effects? Or does it? How thin is "thin"? Discuss.

25. An *optical flat* is a slab of glass that has been ground flat to within a small fraction of a wavelength. How may it be used to test the flatness of a second slab of glass?

26. The directional characteristics of a certain radar antenna as a receiver of radiation are known. What can be said about its directional characteristics as a transmitter?

27. A person in a dark room, looking through a small window, can see a second person standing outside in bright sunlight. The second person cannot see the first person. Is this a failure of the principle of optical reversibility? Assume no absorption of light.

28. Why is it necessary to rotate the interferometer in the Michelson-Morley experiment?

29. How is the negative result of the Michelson-Morley experiment interpreted according to Einstein's theory of relativity?

30. An automobile directs its headlights onto the side of a barn. Why are interference fringes not produced?

31. In principle, could you construct an acoustical version of Michelson's interferometer and use it to measure wind velocities? Discuss some of its design difficulties. Are there simpler ways to measure wind velocities?

problems

SECTION 45-1

1. Design a double-slit arrangement that will produce interference fringes 1° apart on a distant screen. Assume sodium light ($\lambda = 589$ nm $= 5890$ Å). *Answer:* Slit separation must be 0.034 mm.

2. In a double-slit arrangement the distance between slits is 5.0 mm and the slits are 1.0 m from the screen. Two interference patterns can be seen on the screen, one due to light of 480 nm and the other 600 nm. What is the separation on the screen between the third-order interference fringes of the two different patterns?

3. Sodium light ($\lambda = 589$ nm) falls on a double slit of separation $d = 2.0$ mm. D in Fig. 45-5 is 4 cm. What percent error is made in locating the tenth bright fringe if it is *not* assumed that $D \gg d$? *Answer:* 0.03%

4. In a double-slit arrangement the slits are separated by a distance equal to 100 times the wavelength of the light passing through the slits. (a) What is the angular separation between the first and second maxima? (b) What is the linear distance between the first and second maxima if the screen is at a distance of 50 cm from the slits?

5. A double-slit arrangement produces interference fringes for sodium light ($\lambda = 589$ nm) that are 0.20° apart. For what wavelength would the angular separation be 10 percent greater? *Answer:* 650 nm

6. Sodium light ($\lambda = 589$ nm) falls on a double slit of separation $d = 0.20$ mm. A thin lens ($f = +1.0$ m) is placed near the slit as in Fig. 45-6. What is the linear fringe separation on a screen placed in the focal plane of the lens?

7. In the front of a lecture hall, a coherent beam of monochromatic light from a helium-neon laser ($\lambda = 640$ nm $= 6400$ Å) illuminates a double slit. From there it travels a distance d (20 m) to a mirror at the back of the hall, and returns the same distance to a screen. (a) In order that the distance between interference maxima be s (10 cm) what should be the distance between the two slits? (b) State briefly what you will see if the lecturer slips a thin sheet of cellophane over one of the slits. The optical path length through the cellophane is 2.5 wavelengths longer than the equivalent air path length. *Answer:* (a) 0.26 mm. (b) In place of the central maximum you get a minimum.

8. In Young's interference experiment in a large ripple tank (see Fig. 45-4) the coherent vibrating sources are placed 12.0 cm apart. The distance between

maxima 2.0 m away is 18.0 cm. If the speed of ripples is 25.0 cm/s, find the frequency of the vibrators.

9. A thin flake of mica ($n = 1.58$) is used to cover one slit of a double-slit arrangement. The central point on the screen is occupied by what used to be the seventh bright fringe. If $\lambda = 550$ nm, what is the thickness of the mica?

Answer: 6.4×10^{-3} mm

10. One slit of a double-slit arrangement is covered by a thin glass plate of refractive index 1.4, and the other by a thin glass plate of refractive index 1.7. The point on the screen where the central maximum fell before the glass plates were inserted is now occupied by what had been the fifth bright fringe before. Assume $\lambda = 480$ nm and that the plates have the same thickness t and find the value of t.

11. A double-slit arrangement produces interference fringes for sodium light ($\lambda = 589$ nm) that are 0.20° apart. What is the angular fringe separation if the entire arrangement is immersed in water?

Answer: 0.15°.

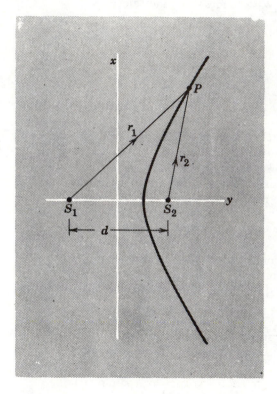

figure 45-22
Problem 12

12. Two point sources in Fig. 45-22 emit coherent waves. Show that curves, such as that given, over which the phase difference for rays r_1 and r_2 is a constant, are hyperbolas. (*Hint:* A constant phase difference implies a constant difference in length between r_1 and r_2.)

13. In Fig 45-23, the source emits monochromatic light of wavelength λ. S is a narrow slit in an otherwise opaque screen I. A plane mirror, whose surface includes the axis of the lens shown, is located a distance h below S. Screen II is at the focal plane of the lens. (a) Find the condition for maximum and minimum brightness of fringes on screen II in terms of the usual angle θ, the wavelength λ, and the distance h. (b) Do fringes appear only in region A (above the axis of the lens), only in region B (below the axis of the lens), or in both regions A and B? Explain. (*Hint:* Consider the image of S formed by the mirror.)

Answer: (a) 2 h sin θ = mλ (minimum); 2 h sin θ = (m + $\frac{1}{2}$)λ (maximum).
   (b) Only in region A.

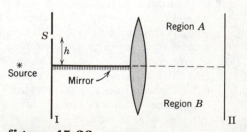

figure 45-23
Problem 13

14. Design, semi-quantitatively, a Young-type double slit interference experiment (see Fig. 45-1) for sound waves in air. Assume a frequency of 500 Hz and a speed of 330 m/s. Discuss some of the design parameters such as the nature of the source, the size of screen B, the width, height, and separation of the two slits, the "fringe" separation, the nature of the fringe detector, etc.

SECTION 45-3

15. Light of wavelength 600 nm is incident normally on a system of two parallel narrow slits separated by 0.60 mm. Sketch the intensity in the pattern observed on a distant screen as a function of angle θ, as in Fig. 45-5, for the range of values $0 \leqslant \theta \leqslant 0.0040$ radians.

16. Show that the half-width $\Delta\theta$ of the double-slit interference fringes (see arrow in Fig. 45-9) is given by

$$\Delta\theta = \frac{\lambda}{2d}$$

if θ is small enough so that $\sin\theta \cong \theta$.

17. One of the slits of a double-slit system is wider than the other, so that the amplitude of the light reaching the central part of the screen from one slit, acting alone, is twice that from the other slit, acting alone. Derive an expression for I_θ in terms of θ, corresponding to Eqs. 45-11b and 45-11c.

Answer: $I_\theta = \frac{1}{9} I_m \left[1 + 8 \cos^2\left(\frac{\pi d \sin\theta}{\lambda}\right)\right]$

18. S_1 and S_2 in Fig. 45-24 are effective point sources of radiation, excited by the same oscillator. They are coherent and in phase with each other. Placed 4.0 m apart, they emit equal amounts of power in the form of 1.0-m wavelength electromagnetic waves. (*a*) Find the positions of the first (that is, the nearest), the second, and the third maxima of the received signal, as the detector is moved out along Ox. (*b*) Is the intensity at the nearest minimum equal to zero? Justify your answer.

figure 45-24
Problem 18

SECTION 45-4

19. Find the sum of the following quantities (*a*) by the vector method and (*b*) analytically:

$$y_1 = 10 \sin \omega t$$

$$y_2 = 8 \sin (\omega t + 30°).$$

Answer: $y = 17 \sin (\omega t + 13°)$.

20. Add the following quantities graphically, using the vector method:

$$y_1 = 10 \sin \omega t$$

$$y_2 = 15 \sin (\omega t + 30°)$$

$$y_3 = 5 \sin (\omega t - 45°).$$

21. Consider the problem of determining the sum

$$A_1 \sin (\omega t + \phi_1) + A_2 \sin (\omega t + \phi_2) + \cdots + A_n \sin (\omega t + \phi_n)$$

from the phasor diagram.
(*a*) Show that the sum may always be written in the form

$$B \sin \omega t + C \cos \omega t.$$

(*b*) Show that $B^2 + C^2 \leq (A_1 + A_2 + \cdots + A_n)^2$.
(*c*) When does the equality sign in (*b*) hold?
Answer: (c) When all phase angles ϕ_i are equal.

SECTION 45-5

22. We wish to coat a flat slab of glass ($n = 1.50$) with a transparent material ($n = 1.25$) so that light of wavelength 600 nm (in vacuum) incident normally is not reflected. How can this be done?

23. A thin film 4.0×10^{-5} cm thick is illuminated by white light normal to its surface. Its index of refraction is 1.5. What wavelengths within the visible spectrum will be intensified in the reflected beam?
    Answer: 480 nm (blue).

24. White light reflected at perpendicular incidence from a soap bubble has, in the visible spectrum, a single interference maximum (at $\lambda = 600$ nm) and a single minimum at the violet end of the spectrum. If $n = 1.33$ for the film, calculate its thickness.

25. A plane wave of monochromatic light falls normally on a uniformly thin film of oil which covers a glass plate. The wavelength of the source can be varied continuously. Complete destructive interference of the reflected light is observed for wavelengths of 500 and 700 nm and for no other wavelengths in between. If the index of refraction of the oil is 1.30 and that of the glass is 1.50, find the thickness of the oil film.
    Answer: 670 nm.

26. A plane monochromatic light wave in air falls at normal incidence on a thin film of oil which covers a glass plate. The wavelength of the source may be varied continuously. Complete destructive interference in the reflected beam is observed for wavelengths of 500 and 700 nm and for no other wavelength in between. The index of refraction of glass is 1.50. Show that the index of refraction of the oil must be less than 1.50.

27. A thin film of acetone (refractive index 1.25) is floated on a thick glass plate (refractive index 1.50). Plane light waves of variable wavelengths are incident normal to the film. When one views the reflected wave it is noted that complete destructive interference occurs at 600 nm and constructive interference at 700 nm. Calculate the thickness of the acetone film.
    Answer: 840 nm.

28. White light reflected at perpendicular incidence from a soap film has, in the visible spectrum, an interference maximum at 600 nm (= 6000 Å) and a minimum at 450 nm (= 4500 Å), with no minimum in between. If $n = 1.33$ for the film, what is the film thickness, assumed uniform?

29. In Example 5 assume that there is zero reflection for light of wavelength 550 nm at normal incidence. Calculate the factor by which the reflection is diminished by the coating at 450 and at 650 nm.
    Answer: The intensity is diminished by 88% at 450 nm and by 94% at 650 nm.

30. A tanker has dumped a large quantity of kerosene ($n = 1.2$) into the Gulf of Mexico, creating a large slick on top of the water ($n = 1.3$). (a) If you are looking straight down from an airplane onto a region of the slick where its thickness is 460 nm, for which wavelength(s) of visible light is the reflection the greatest? (b) If you are scuba-diving directly under this same region of the slick, for which wavelengths of visible light is the transmitted intensity the strongest?

31. Light of wavelength 630 nm is incident normally on a thin wedge-shaped film of index of refraction 1.5. There are ten bright and nine dark fringes over the length of film. By how much does the film thickness change over this length?   *Answer:* 1.9×10^{-6} m ($= 1.9$ μm).

32. A broad source of light ($\lambda = 680$ nm) illuminates normally two glass plates 12 cm long that touch at one end and are separated by a wire 0.048 mm in diameter at the other (Fig. 45-25). How many bright fringes appear over the 12-cm distance?

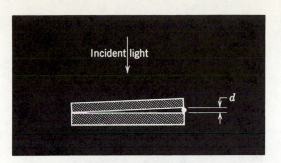

figure 45-25
Problem 32

33. An oil drop $(n = 1.20)$ floats on a water $(n = 1.33)$ surface and is observed from above by reflected light (see Fig. 45-26). (a) Will the outer (thinnest) regions of the drop correspond to a bright or a dark region? (b) Approximately how thick is the oil film where one observes the third blue region from the outside of the drop? (c) Why do the colors gradually disappear as the oil thickness becomes larger?   *Answer:* (a) Bright. (b) 594 nm.

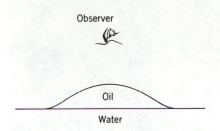

Observer

Oil

Water

figure 45-26
Problem 33

34. The diameter of the tenth bright ring in a Newton's rings apparatus changes from 1.40 to 1.27 cm as a liquid is introduced between the lens and the plate. Find the index of refraction of the liquid.

35. In a Newton's rings experiment the radius of curvature R of the lens is 5.0 m and its diameter is 2.0 cm. (a) How many rings are produced? (b) How many rings would be seen if the arrangement were immersed in water $(n = 1.33)$? Assume that $\lambda = 589$ nm.   *Answer:* (a) 34. (b) 46.

36. In the Newton's rings experiment, use the result of Example 6 to show that the difference in radius between adjacent rings is

$$\Delta r = r_{m+1} - r_m \cong \tfrac{1}{2} \sqrt{\frac{\lambda R}{m}}.$$

Assume $m \gg 1$ and use the binomial theorem. Is this result qualitatively consistent with Fig. 45-18?

37. In the Newton's rings experiment use the result of Example 6 and Problem 36 to show that the *area* between adjacent rings is given, for $m \gg 1$, by

$$A_m = \pi \lambda R \ (= \text{a constant}).$$

Is this result qualitatively consistent with Fig. 45-18?

SECTION 45-7

38. If mirror M_2 in Michelson's interferometer is moved through 0.233 mm, 792 fringes are counted. What is the wavelength of the light?

39. A thin film with $n = 1.40$ for light of wavelength 589 nm $(= 5890$ Å) is placed in one arm of a Michelson interferometer. If a shift of 7.0 fringes occurs, what is the film thickness?   *Answer:* 5200 nm.

40. A Michelson interferometer is used with a sodium discharge tube as a light source. The yellow sodium light consists of two wavelengths, 589.0 (=5890 Å) and 589.6 nm (=5896 Å). It is observed that the interference pattern disappears and reappears periodically as one moves mirror M_2 in Fig. 45-20. (a) Explain this effect. (b) Calculate the change in path difference between two successive reappearances of the interference pattern.

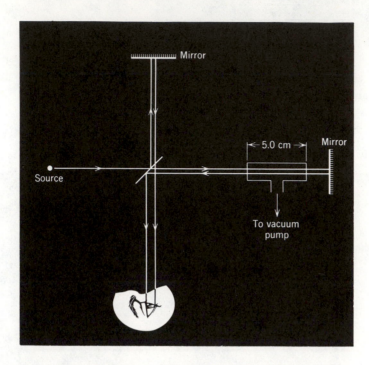

figure 45-27
Problem 41

41. An air-tight chamber 5.0 cm long with glass windows is placed in one arm of a Michelson interferometer as indicated in Fig. 45-27. Light of $\lambda = 5000$ Å is used. The air is slowly evacuated from the chamber using a vacuum pump. While the air is being removed, 60 fringes are observed to pass through the view. From these data, find the index of refraction of air at atmospheric pressure.   *Answer:* 1.0003

42. Write an expression for the intensity observed in Michelson's interferometer (Fig. 45-20) as a function of the position of the moveable mirror. Measure the position of the mirror from the point at which $d_1 = d_2$.

46
diffraction

Diffraction, which is illustrated in Fig. 44-3, is the bending of light around an obstacle such as the edge of a slit. We can see the diffraction of light by looking through a crack between two fingers at a distant light source such as a tubular neon sign or by looking at a street light through a cloth umbrella. Usually diffraction effects are small and we must look for them carefully. Also, most sources of light have an extended area so that a diffraction pattern produced by one point of the source will overlap that produced by another. Finally, common sources of light are not monochromatic. The patterns for the various wavelengths overlap and again the effect is less apparent.

Diffraction was discovered by Francesco Grimaldi (1618–1663), and the phenomenon was known both to Huygens and to Newton. Newton did not see in it any justification for a wave theory for light. Huygens, although he believed in a wave theory, did not believe in diffraction! He imagined his secondary wavelets to be effective only at the point of tangency to their common envelope, thus denying the possibility of diffraction (see Section 43-3). In his words:

And thus we see the reasons why light . . . proceeds only in straight lines in such a way that it does not illuminate any object except when the path from the source to the object is open along such a line.

Jean Augustin Fresnel (1788–1827) correctly applied Huygens' principle to explain diffraction. In these early days the light waves were believed to be mechanical waves in an all-pervading ether. We have seen (Section 41-8) how Maxwell showed that light waves were not mechanical in nature but electromagnetic. Einstein rounded out our

46-1
INTRODUCTION

modern view of light waves by eliminating the need to postulate an ether (see Section 42-4).

Figure 46-1 shows a general diffraction situation. Surface *A* is a wavefront that falls on *B*, which is an opaque screen containing an aperture of arbitrary shape; *C* is a diffusing screen that receives the light that passes through this aperture. We can calculate this pattern of light intensity on *C* by subdividing the wavefront into elementary areas *d***S**, each of which becomes a source of an expanding Huygens' wavelet. The light intensity at an arbitrary point *P* is found by superimposing the wave disturbances (that is, the **E** vectors) caused by the wavelets reaching *P* from all these elementary radiators.

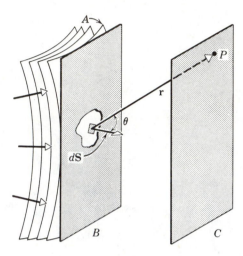

figure 46-1
Light is diffracted at the aperture in screen *B* and illuminates screen *C*. The intensity at *P* is found by dividing the wavefront at *B* into elementary radiators *d***S** and combining their effects at *P*.

The wave disturbances reaching *P* differ in amplitude and in phase because (*a*) the elementary radiators are at varying distances from *P*, (*b*) the light leaves the radiators at various angles to the normal to the wavefront (see Section 43-2), and (*c*) some radiators are blocked by screen *B*; others are not. Diffraction calculations—simple in principle— may become difficult in practice. The calculation must be repeated for every point on screen *C* at which we wish to know the light intensity. We followed exactly this program in calculating the double-slit intensity pattern in Section 45-3. The calculation there was simple because we assumed only two elementary radiators, the two narrow slits.

Figure 46-2*a* shows the general case of *Fresnel diffraction*, in which the light source and/or the screen on which the diffraction pattern is displayed are a finite distance from the diffracting aperture; the wavefronts that fall on the diffracting aperture in this case and that leave it to illuminate any point *P* of the diffusing screen are not planes; the corresponding rays are not parallel.

A simplification results if source *S* and screen *C* are moved to a large distance from the diffraction aperture, as in Fig. 46-2*b*. This limiting case is called *Fraunhofer diffraction*. The wavefronts arriving at the diffracting aperture from the distant source *S* are planes, and the rays associated with these wavefronts are parallel to each other. Fraunhofer conditions can be established in the laboratory by using two converging lenses, as in Fig. 46-2*c*. The first of these converts the diverging wave from the source into a plane wave. The second lens causes plane waves leaving the diffracting aperture to converge to point *P*. All rays that

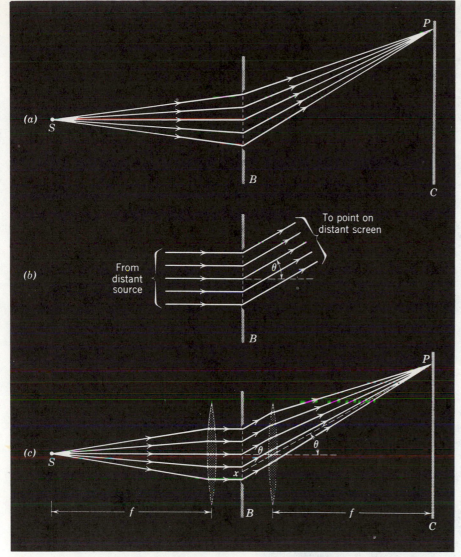

illuminate P will leave the diffracting aperture parallel to the dashed line Px drawn from P through the center of this second (thin) lens. We assumed Fraunhofer conditions for Young's double-slit experiment in Section 45-1 (see Fig. 45-5).

Although Fraunhofer diffraction is a limiting case of the more general Fresnel diffraction, it is an important limiting case and is easier to handle mathematically. This book deals only with Fraunhofer diffraction.

46-2
SINGLE SLIT

Figure 46-3 shows a plane wave falling at normal incidence on a long narrow slit of width a. Let us focus our attention on the central point P_0 of screen C. The parallel rays extending from the slit to P_0 all have the same optical (though not geometrical) path lengths, as we saw in Section 44-5. Since they are in phase at the plane of the slit, they will still be in phase at P_0, and the central point of the diffraction pattern that appears on screen C has a maximum intensity.

We now consider another point on the screen. Light rays which reach

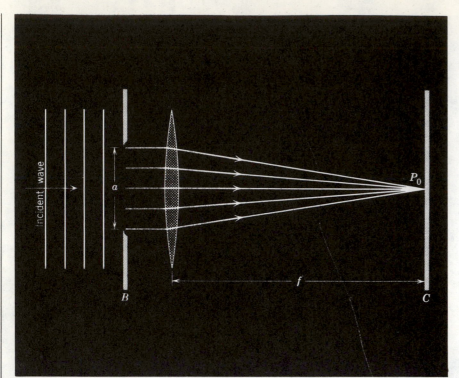

figure 46-3
Conditions at the central maximum of the diffraction pattern. The slit extends a distance above and below the figure, this distance being much greater than the slit width *a*.

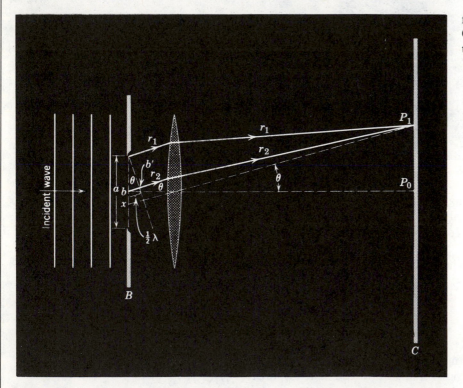

figure 46-4
Conditions at the first minimum of the diffraction pattern.

P_1 in Fig. 46-4 leave the slit at an angle θ as shown. (Note that the ray represented by the dashed line xP_1 is drawn to pass through the center of the lens and is thus undeflected; this ray determines θ.) Ray r_1 originates at the top of the slit and ray r_2 at its center. If θ is chosen so that the distance bb' in the figure is one-half a wavelength, r_1 and r_2 will be out of phase and will produce no effect at P_1.* In fact, every ray from the upper half of the slit will be canceled by a ray from the lower half, originating at a point $a/2$ below the first ray. The point P_1, the first minimum of the diffraction pattern, will have zero intensity (compare Fig. 44-3).

The condition shown in Fig. 46-4 is

$$\frac{a}{2} \sin \theta = \frac{\lambda}{2},$$

or

$$a \sin \theta = \lambda. \qquad (46\text{-}1)$$

As we stated earlier (see Fig. 44-1), the central maximum becomes wider as the slit is made narrower. If the slit width is as small as one wavelength $(a = \lambda)$, the first minimum occurs at $\theta = 90°$ ($\sin \theta = 1$ in Eq. 46-1), which implies that the central maximum fills the entire forward hemisphere. We assumed a condition approaching this in our discussion of Young's double-slit interference experiment in Section 45-1.

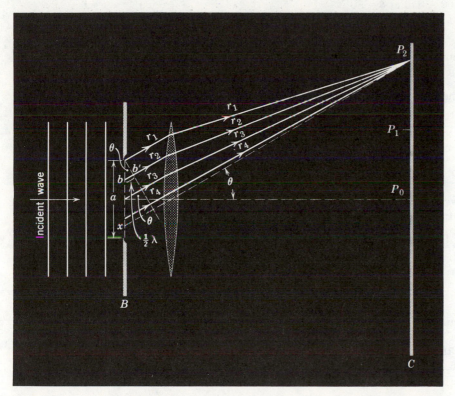

figure 46-5
Conditions at the second minimum of the diffraction pattern.

In Fig. 46-5 the slit is divided into four equal zones, with a ray leaving the top of each zone. Let θ be chosen so that the distance bb' is one-half a wavelength. Rays r_1 and r_2 will then cancel at P_2. Rays r_3 and r_4 will also be half a wavelength out of phase and will also cancel. Consider four other rays, emerging from the slit a given distance below the four rays above. The two rays below r_1 and r_2 will cancel uniquely, as will the two rays below r_3 and r_4. We can proceed across the entire slit and conclude again that no light reaches P_2; we have located a second point of zero intensity.

---

* Whatever phase relation exists between r_1 and r_2 at the plane represented by the sloping dashed line in Fig. 46-4 that passes through b' also exists at P_1, not being affected by the lens (see Section 44-5).

The condition described (see Fig. 46-5) requires that

$$\frac{a}{4} \sin \theta = \frac{\lambda}{2},$$

or

$$a \sin \theta = 2\lambda.$$

By extension, the general formula for the minima in the diffraction pattern on screen C is

$$a \sin \theta = m\lambda \qquad m = 1, 2, 3, \ldots \text{ (minima).} \qquad (46\text{-}2)$$

There is a maximum approximately halfway between each adjacent pair of minima. You should consider the simplification that has resulted in this analysis by examining Fraunhofer (Fig. 46-2c) rather than Fresnel conditions (Fig. 46-2a).

EXAMPLE 1

A slit of width a is illuminated by white light. For what value of a will the first minimum for red light ($\lambda = 650$ nm $= 6500$ Å) fall at $\theta = 30°$?

At the first minimum we put $m = 1$ in Eq. 46-2. Doing so and solving for a yields

$$a = \frac{m\lambda}{\sin \theta} = \frac{(1)(650 \text{ nm})}{\sin 30°} = 1300 \text{ nm.}$$

Note that the slit width must be twice the wavelength in this case.

EXAMPLE 2

In Example 1 what is the wavelength λ' of the light whose first diffraction maximum (not counting the central maximum) falls at $\theta = 30°$, thus coinciding with the first minimum for red light?

This maximum is about halfway between the first and second minima. We can find it without too much error by putting $m = 1.5$ in Eq. 46-2, or

$$a \sin \theta \cong 1.5\lambda'.$$

From Example 1, however, $\qquad a \sin \theta = \lambda.$

Dividing gives $\qquad \lambda' = \dfrac{\lambda}{1.5} = \dfrac{650 \text{ nm}}{1.5} = 430 \text{ nm.}$

Light of this color is violet. The second maximum for light of wavelength 430 nm will *always* coincide with the first minimum for light of wavelength 650 nm, no matter what the slit width. If the slit is relatively narrow, the angle θ at which this overlap occurs will be relatively large.

46-3
SINGLE SLIT— QUALITATIVE

In Section 46-2 we located the positions of the maxima and minima for a single-slit diffraction pattern. We now wish to find an expression for the intensity of the *entire* diffraction pattern as a function of the diffraction angle θ. We do so qualitatively in this section and quantitatively in the next section.

Figure 46-6 shows a slit of width a divided into N parallel strips of width Δx. Each strip acts as a radiator of Huygens' wavelets and produces a characteristic wave disturbance at point P, whose position on the screen, for a particular arrangement of apparatus, can be described by the angle θ.

If the strips are narrow enough—which we assume—all points on a given strip have essentially the same optical path length to P, and therefore all the light from the strip will have the same phase when it ar-

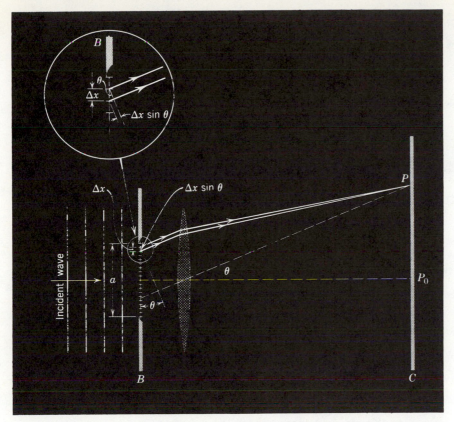

figure 46-6
We divide a slit of width a into N strips of width Δx. The insert shows conditions at the second strip more clearly. In the differential limit the slit is divided into an infinite number of strips (that is, $N \to \infty$) of differential width dx. For clarity in this and the following figure, we take $N = 18$.

rives at P. We may take the amplitudes ΔE_0 of the wave disturbances at P from the various strips may be taken as equal if θ in Fig. 46-6 is not too large.

We limit our considerations to points that lie in, or very close to, the plane of Fig. 46-6. It can be shown that this procedure is valid for a slit whose length is much greater than its width a. We made this same assumption tacitly both earlier in this chapter and in Chapter 45; see Figs. 45-5 and 46-3, for example.

The wave disturbances from adjacent strips have a constant phase difference $\Delta\phi$ between them at P given by

$$\frac{\text{phase difference}}{2\pi} = \frac{\text{path difference}}{\lambda},$$

or
$$\Delta\phi = \left(\frac{2\pi}{\lambda}\right)(\Delta x \sin \theta), \tag{46-3}$$

where $\Delta x \sin \theta$ is, as the figure insert shows, the path difference for rays originating at the top edges of adjacent strips. Thus, at P, N vectors with the same amplitude ΔE_0, the same frequency, and the same phase difference $\Delta\phi$ between adjacent members combine to produce a resultant disturbance. We ask, for various values of $\Delta\phi$ [that is, for various points P on the screen, corresponding to various values of θ (see Eq. 46-3)], what is the amplitude E_θ of the resultant wave disturbance? We find the answer by representing the individual wave disturbances ΔE_0 by phasors and calculating the resultant phasor amplitude, as described in Section 45-4.

At the center of the diffraction pattern θ equals zero, and the phase shift between adjacent strips (see Eq. 46-3) is also zero. As Fig. 46-7a shows, the phasor arrows in this case are laid end to end and the ampli-

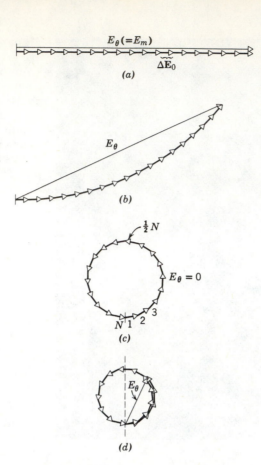

tude of the resultant has its maximum value E_m. This corresponds to the center of the central maximum.

As we move to a value of θ other than zero, $\Delta\phi$ assumes a definite nonzero value (again see Eq. 46-3), and the array of arrows is now as shown in Fig. 46-7*b*. The resultant amplitude E_θ is less than before. Note that the length of the "arc" of small arrows is the same for both figures and indeed for all figures of this series. As θ increases further, we reach a situation (Fig. 46-7*c*) in which the chain of arrows curls around through 360°, the tip of the last arrow touching the foot of the first arrow. This corresponds to $E_\theta = 0$, that is, to the first minimum. For this condition the ray from the top of the slit (1 in Fig. 46-7*c*) is 180° out of phase with the ray from the center of the slit ($\frac{1}{2}N$ in Fig. 46-7*c*). These phase relations are consistent with Fig. 46-4, which also represents the first minimum.

As θ increases further, the phase shift continues to increase, and the chain of arrows coils around through an angular distance greater than 360°, as in Fig. 46-7*d*, which corresponds to the first maximum beyond the central maximum. This maximum is much smaller than the central maximum. In making this comparison, recall that the arrows marked E_θ in Fig. 46-7 correspond to the *amplitudes* of the wave disturbance and not to the *intensity*. The amplitudes must be squared to obtain the corresponding relative intensities (see Eq. 45-7).

The "arc" of small arrows in Fig. 46-8 shows the phasors representing, in amplitude and phase, the wave disturbances that reach an arbitrary point P on the screen of Fig. 46-6, corresponding to a particular angle θ. The resultant amplitude at P is E_θ. If we divide the slit of Fig. 46-6 into

46-4
SINGLE SLIT—
QUANTITATIVE

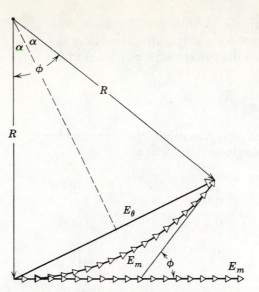

infinitesimal strips of width dx, the arc of arrows in Fig. 46-8 approaches the arc of a circle, its radius R being indicated in that figure. The length of the arc is E_m, the amplitude at the center of the diffraction pattern, for at the center of the pattern the wave disturbances are all in phase and this "arc" becomes a straight line as in Fig. 46-7a.

The angle ϕ in the lower part of Fig. 46-8 is revealed as the difference in phase between the infinitesimal vectors at the left and right ends of the arc E_m. This means that ϕ is the phase difference between rays from the top and the bottom of the slit of Fig. 46-6. From geometry we see that ϕ is also the angle between the two radii marked R in Fig. 46-8. From this figure we can write

$$E_\theta = 2R \sin \frac{\phi}{2}.$$

In radian measure ϕ, from the figure, is

$$\phi = \frac{E_m}{R}.$$

Combining yields

$$E_\theta = \frac{E_m}{\phi/2} \sin \frac{\phi}{2},$$

or

$$E_\theta = E_m \frac{\sin \alpha}{\alpha}, \qquad (46\text{-}4)$$

in which

$$\alpha = \frac{\phi}{2}. \qquad (46\text{-}5)$$

From Fig. 46-6, recalling that ϕ is the phase difference between rays from the top and the bottom of the slit and that the path difference for these rays is $a \sin \theta$, we have

$$\frac{\text{phase difference}}{2\pi} = \frac{\text{path difference}}{\lambda},$$

or

$$\phi = \left(\frac{2\pi}{\lambda}\right)(a \sin \theta).$$

Combining with Eq. 46-5 yields

$$\alpha = \frac{\phi}{2} = \frac{\pi a}{\lambda} \sin \theta. \qquad (46\text{-}6)$$

Equation 46-4, taken together with the definition of Eq. 46-6, gives the amplitude of the wave disturbance for a single-slit diffraction pattern at any angle θ. The intensity I_θ for the pattern is proportional to the square of the amplitude, or

$$I_\theta = I_m \left(\frac{\sin \alpha}{\alpha}\right)^2. \tag{46-7}$$

For convenience we display together, and renumber, the formulas for the amplitude and the intensity in single-slit diffraction.

[Eq. 46-4] $E_\theta = E_m \dfrac{\sin \alpha}{\alpha}$ (46-8a)

single-

[Eq. 46-7] $I_\theta = I_m \left(\dfrac{\sin \alpha}{\alpha}\right)^2$ slit   (46-8b)

diffraction

[Eq. 46-6] $\alpha \left(= \tfrac{1}{2}\phi\right) = \dfrac{\pi a}{\lambda} \sin \theta$ (46-8c)

Figure 46-9 shows plots of I_θ for several values of the ratio a/λ. Note that the pattern becomes narrower as we increase a/λ; compare this figure with Figs. 44-1 and 44-3.

Minima occur in Eq. 46-8b when

$$\alpha = m\pi \qquad m = 1, 2, 3, \ldots. \tag{46-9}$$

Combining with Eq. 46-8c leads to

$$a \sin \theta = m\lambda \qquad m = 1, 2, 3, \ldots \text{ (minima)},$$

which is the result derived in the preceding section (Eq. 46-2). In that section, however, we derived *only* this result, obtaining no quantitative information about the intensity of the diffraction pattern at places in which it was not zero. Here (Eqs. 46-8) we have complete intensity information.

Intensities of the secondary diffraction maxima. Calculate, approximately, the relative intensities of the secondary maxima in the single-slit Fraunhofer diffraction pattern.

The secondary maxima lie approximately halfway between the minima and are found (compare Eq. 46-9) from

$$\alpha \cong \left(m + \tfrac{1}{2}\right)\pi \qquad m = 1, 2, 3, \ldots.$$

Substituting into Eq. 46-8b yields

$$I_\theta = I_m \left[\frac{\sin \left(m + \tfrac{1}{2}\right)\pi}{\left(m + \tfrac{1}{2}\right)\pi}\right]^2,$$

which reduces to $\dfrac{I_\theta}{I_m} = \dfrac{1}{\left(m + \tfrac{1}{2}\right)^2 \pi^2}.$

This yields, for $m = 1, 2, 3, \ldots$, $I_\theta/I_m = 0.045, 0.016, 0.0083$, etc. The successive maxima decrease rapidly in intensity.

EXAMPLE 3

Width of the central diffraction maximum. Derive the *half-width* $\Delta\theta$ of the central maximum in a single-slit Fraunhofer diffraction (see Fig. 46-9b). The half-width is the angle between the two points in the pattern where the intensity is one-half that at the center of the pattern.

EXAMPLE 4

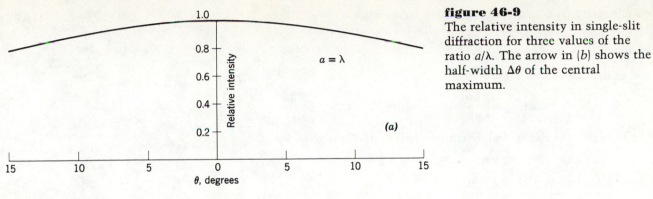

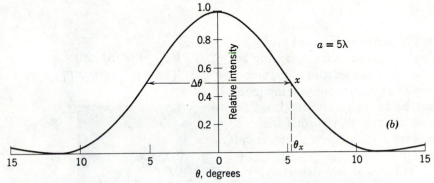

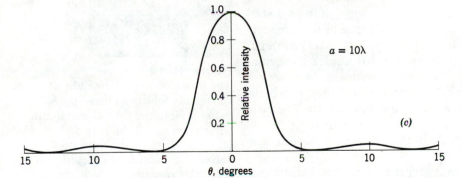

figure 46-9
The relative intensity in single-slit diffraction for three values of the ratio a/λ. The arrow in (b) shows the half-width $\Delta\theta$ of the central maximum.

Point *x* in Fig. 46-9*b* is so chosen that $I_\theta = \frac{1}{2}I_m$, or, from Eq. 46-8*b*,

$$\frac{1}{2} = \left(\frac{\sin \alpha_x}{\alpha_x}\right)^2.$$

This equation cannot be solved analytically for α_x. We can solve it graphically, as accurately as we wish, by plotting the quantity $(\sin \alpha_x/\alpha_x)^2$ as ordinate versus α_x as abscissa and noting the value of α_x at which the curve intersects the line "one half" on the ordinate scale (see Problem 11). However, if only an approximate answer is desired, it is often quicker to use trial-and-error methods.

We know that α equals π at the first minimum; we guess that α_x is perhaps $\pi/2$ (= 90° = 1.57 rad). Trying this in Eq. 46-8*b* yields

$$\frac{I_\theta}{I_m} = \left[\frac{\sin (\pi/2)}{\pi/2}\right]^2 = 0.4.$$

This intensity ratio is *less* than 0.5, so that α_x must be *less* than 90°. After a few more trials we find easily enough that

$$\alpha_x = 1.40 \text{ rad} = 80°$$

does yield a ratio close to the correct value of 0.5.

We now use Eq. 46-8c to find the corresponding angle θ:

$$\alpha_x = \frac{\pi a}{\lambda} \sin \theta_x = 1.40,$$

or, noting that $a/\lambda = 5$ for Fig. 46-9b,

$$\sin \theta_x = \frac{1.40\lambda}{\pi a} = \frac{1.40}{5\pi} = 0.0892.$$

The half-width $\Delta\theta$ of the central maximum (see Fig. 46-9b) is given by

$$\Delta\theta = 2\theta_x = 2 \sin^{-1} 0.0892 = 2 \times 5.1° = 10.2°,$$

which is in agreement with the figure.

Diffraction will occur when a wavefront is partially blocked off by an opaque object such as a metal disk or an opaque screen containing an aperture. Here we consider diffraction at a circular aperture of diameter d, the aperture constituting the boundary of a circular converging lens.

Our previous treatment of lenses was based on geometrical optics, diffraction being specifically assumed not to occur. A rigorous analysis would be based from the beginning on wave optics, since geometrical optics is always an approximation, although often a very good one. Diffraction phenomena would emerge in a natural way from such a wave-optical analysis.

Figure 46-10 shows the image of a distant point source of light (a star) formed on a photographic film placed in the focal plane of the converging lens of a telescope.* It is not a point, as the (approximate) geometrical optics treatment suggests, but a circular disk surrounded by several progressively fainter secondary rings. Comparison with Fig. 44-3c leaves little doubt that we are dealing with a diffraction phenomenon in which, however, the aperture is a circle rather than a long narrow slit. The ratio d/λ, where d is the diameter of the lens (or of a circular aperture placed in front of the lens), determines the scale of the diffraction pattern, just as the ratio a/λ does for a slit.

Analysis shows that the first minimum for the diffraction pattern of a circular aperture of diameter d, assuming Fraunhofer conditions, is given by

$$\sin \theta = 1.22 \frac{\lambda}{d}. \qquad (46\text{-}10)$$

This is to be compared with Eq. 46-1, or

$$\sin \theta = \frac{\lambda}{a},$$

which locates the first minimum for a long narrow slit of width a. The factor 1.22 emerges from the mathematical analysis when we integrate over the elementary radiators into which the circular aperture may be divided.

In actual lenses the image of a distant point object will be somewhat larger than that shown in Fig. 46-10 and may not have radial symmetry. This is caused by the various lens "defects" mentioned in Section 44-5. However, even if all of these defects could be eliminated by suitable shaping of the lens surfaces or by intro-

46-5
DIFFRACTION AT A CIRCULAR APERTURE

figure 46-10
The image of a star formed by a converging lens is a diffraction pattern. Note the central maximum, sometimes called the Airy disk (after Sir George Airy, who first solved the problem of diffraction at a circular aperture in 1835), and the circular secondary maximum. Other secondary maxima occur at larger radii but are too faint to be seen.

---

*Diffraction phenomena also occur in microscopes and other optical instruments. They are, in fact, characteristic of the entire electromagnetic spectrum.

ducing correcting lenses, the diffraction pattern of Fig. 46-10 would remain. It is an inherent property of the lens aperture and of the wavelength of light used.

The fact that lens images are diffraction patterns is important when we wish to distinguish two distant point objects whose angular separation is small. Figure 46-11 shows the visual appearances and the corresponding intensity patterns for two distant point objects (stars, say) with small angular separations and approximately equal central intensities. In (a) the objects are not resolved; that is, they cannot be distinguished from a single point object. In (b) they are barely resolved and in (c) they are fully resolved.*

In Fig. 46-11b the angular separation of the two point sources is such that the maximum of the diffraction pattern of one source falls on the first minimum of the diffraction pattern of the other. This is called *Rayleigh's criterion.* This criterion, though useful, is arbitrary; other criteria for deciding when two objects are resolved are sometimes used. From Eq. 46-10, two objects that are barely resolvable by Rayleigh's criterion must have an angular separation θ_R of

$$\theta_R = \sin^{-1}\frac{1.22\lambda}{d}.$$

Since the angles involved are rather small, we can replace $\sin\theta_R$ by θ_R, or

$$\theta_R = 1.22\frac{\lambda}{d}, \tag{46-11}$$

in which θ_R is expressed in radians. If the angular separation θ between the objects is greater than θ_R, we can resolve the two objects; if it is less, we cannot. The angle θ_R is the smallest angular separation for which resolution is possible, using Raleigh's criterion.

EXAMPLE 5

A converging lens 3.0 cm in diameter has a focal length f of 20 cm. (a) What angular separation must two distant point objects have to satisfy Rayleigh's criterion? Assume that $\lambda = 550$ nm.

From Eq. 46-11,

$$\theta_R = 1.22\frac{\lambda}{d} = \frac{(1.22)(5.5 \times 10^{-7}\text{ m})}{3.0 \times 10^{-2}\text{ m}} = 2.2 \times 10^{-5}\text{ rad.}$$

(b) How far apart are the centers of the diffraction patterns in the focal plane of the lens? The linear separation is

$$x = f\theta = (20\text{ cm})(2.2 \times 10^{-5}\text{ rad}) = 4400\text{ nm.}$$

This is 8.0 wavelengths of the light employed.

When we wish to use a lens to resolve objects of small angular separation, it is desirable to make the central disk of the diffraction pattern as small as possible. This can be done (see Eq. 46-11) by increasing the lens diameter or by using a shorter wavelength. One reason for constructing large telescopes is to produce *sharper* images so that we can examine celestial objects in finer detail. The images are also *brighter*, not only because the energy is concentrated into a smaller diffraction

*Diffraction effects set the so-called "bottom line" on resolution. However, other effects combine to hamper resolution. For atmospheric effects on the observation of stars see "Toward the Diffraction Limit" by Dietrick E. Thomsen, *Science News*, August 1975.

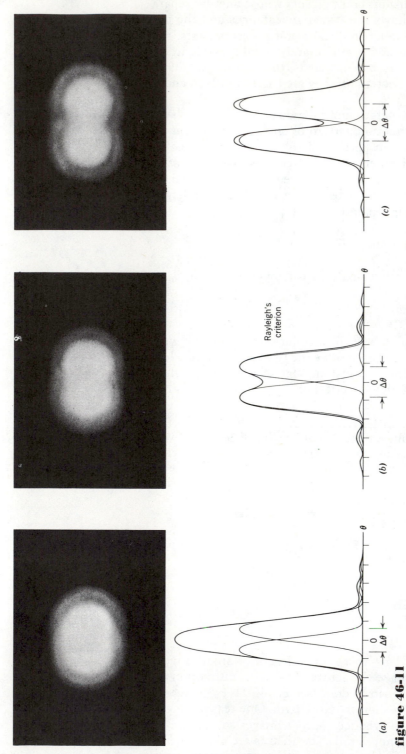

figure 46-11

The images of two distant point objects (simulated stars) are formed by a converging lens whose diameter (= 10 cm) is 200,000 times the effective wavelength (= 500 nm). Sketches of the images as they appear in the focal plane of the telescope objective lens are shown with the corresponding intensity plots below them. (*a*) The angular separation of the objects (see vertical ticks) is so small that the images are not resolved. (*b*) The objects are further apart and the images meet Rayleigh's criterion for resolution. (*c*) The objects are still farther apart and the images are well resolved.

disk but because the larger lens collects more light. Thus fainter objects, for example, distant galaxies, can be seen.

To reduce diffraction effects in *microscopes* we often use ultraviolet light, which, because of its shorter wavelength, permits finer detail to be examined than would be possible for the same microscope operated with visible light. We shall see in Chapter 50 that beams of electrons behave like waves under some circumstances. In the *electron microscope* such beams may have an effective wavelength of 4×10^{-3} nm, of the order of 10^5 times shorter than visible light ($\lambda \cong 500$ nm or 5000 Å). This permits the detailed examination of tiny objects like viruses. If a virus were examined with an optical microscope, its structure would be hopelessly concealed by diffraction.

46-6
DOUBLE SLIT INTERFERENCE AND DIFFRACTION COMBINED

In Young's double-slit experiment (Section 45-1) we assumed that the slits were arbitrarily narrow (that is, $a \ll \lambda$), which means that the central part of the diffusing screen was uniformly illuminated by the diffracted waves from each slit. When such waves interfere, they produce fringes of uniform intensity, as in Fig. 45-9. This idealized situation cannot occur with actual slits because the condition $a \ll \lambda$ cannot usually be met. Waves from the two actual slits combining at different points of the screen will have intensities that are *not* uniform but are governed by the diffraction pattern of a single slit. The effect of relaxing the assumption that $a \ll \lambda$ in Young's experiment is to leave the fringes relatively unchanged in location but to alter their intensities.

The interference pattern for infinitesimally narrow slits is given by Eq. 45-11*b* and 45-11*c*, or, with a small change in nomenclature,

$$I_{\theta,\text{int}} = I_{m,\text{int}} \cos^2 \beta, \tag{46-12}$$

where
$$\beta = \frac{\pi d}{\lambda} \sin \theta \tag{46-13}$$

in which d is the distance between the center-lines of the slits.

The intensity for the diffracted wave from either slit is given by Eqs. 46-8*b* and 46-8*c*, or, with a small change in nomenclature,

$$I_{\theta,\text{dif}} = I_{m,\text{dif}} \left(\frac{\sin \alpha}{\alpha}\right)^2, \tag{46-14}$$

where
$$\alpha = \frac{\pi a}{\lambda} \sin \theta. \tag{46-15}$$

We find the combined effect by regarding $I_{m,\text{int}}$ in Eq. 46-12 as a variable amplitude, given in fact by $I_{\theta,\text{dif}}$ of Eq. 46-14. This assumption, for the combined pattern, leads to

$$I_\theta = I_m (\cos \beta)^2 \left(\frac{\sin \alpha}{\alpha}\right)^2, \tag{46-16}$$

in which we have dropped all subscripts referring separately to interference and diffraction.

Let us express this result in words. At any point on the screen the available light intensity from each slit, considered separately, is given by the diffraction pattern of that slit (Eq. 46-14). The diffraction patterns for the two slits, again considered separately, coincide because parallel rays in Fraunhofer diffraction are focused at the same spot (see Fig. 46-5). Because the two diffracted waves are coherent, they will interfere.

The effect of interference is to redistribute the available energy over

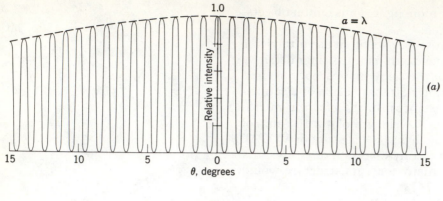

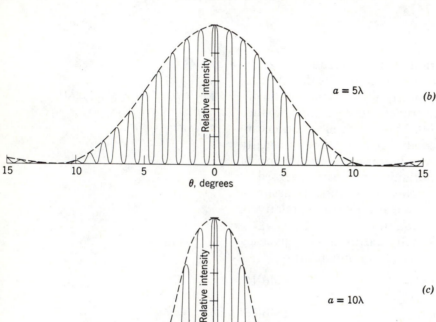

figure 46-12
Interference fringes for a double slit with slit separation $d = 50\lambda$. Three different slit widths, described by $a/\lambda = 1, 5,$ and 10 are shown.

the screen, producing a set of fringes. In Section 45-1, where we assumed $a \ll \lambda$, the available energy was virtually the same at all points on the screen so that the interference fringes had virtually the same intensities (see Fig. 45-9). If we relax the assumption $a \ll \lambda$, the available energy is *not* uniform over the screen but is given by the diffraction pattern of a slit of width a. In this case the interference fringes will have intensities that are determined by the intensity of the diffraction pattern at the location of a particular fringe. Equation 46-16 is the mathematical expression of this argument.

Figure 46-12 is a plot of Eq. 46-16 for $d = 50\lambda$ and for three values of a/λ. It shows clearly that for narrow slits ($a = \lambda$) the fringes are nearly uniform in intensity. As the slits are widened, the intensities of the fringes are markedly modulated by the "diffraction factor" in Eq. 46-16, that is, by the factor $(\sin \alpha/\alpha)^2$.

Equation 46-16 shows that the fringe envelopes of Fig. 46-12 are precisely the single-slit diffraction patterns of Fig. 46-9. This is especially clear in Fig. 46-13, which shows, for the curve of Fig. 46-12b, (a) the "interference factor" in Eq. 46-16 (that is, the factor $\cos^2 \beta$), (b) the "diffraction factor" $(\sin \alpha/\alpha)^2$, and (c) their product.

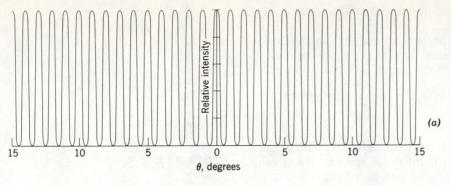

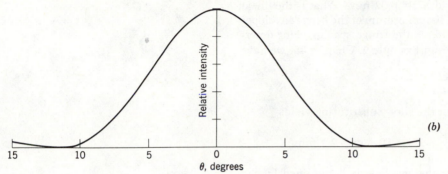

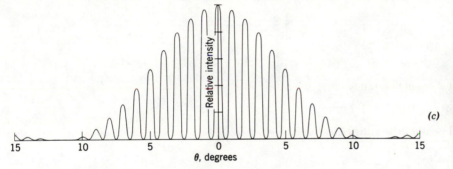

figure 46-13
(a) The "interference factor" and (b) the "diffraction factor" in Eq. 46-16 and (c) their product; compare Fig. 46-12b.

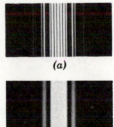

figure 46-14
(a) Interference fringes for a double-slit system in which the slit width is *not* negligible in comparison with the wavelength. The fringes are modulated in intensity by the diffraction pattern of a single slit. (b) If one of the slits is covered up, the interference fringes disappear and we see the single slit diffraction pattern. (Courtesy G. H. Carragan, Rensselaer Polytechnic Institute.)

If we put $a = 0$ in Eq. 46-16, then (see Eq. 46-15) $\alpha = 0$ and $\sin \alpha/\alpha \underset{\alpha \to 0}{=} \alpha/\alpha = 1$. Thus this equation reduces, as it must, to the intensity equation for a pair of vanishingly narrow slits (Eq. 46-12). If we put $d = 0$ in Eq. 46-16, the two slits coalesce into a single slit of width a, as Fig. 46-17 shows; $d = 0$ implies $\beta = 0$ (see Eq. 46-13) and $\cos^2 \beta = 1$. Thus Eq. 46-16 reduces, as it must, to the diffraction equation for a single slit (Eq. 46-14).

Figure 46-14 shows some actual double-slit interference photographs. The uniformly spaced interference fringes and their intensity modulation by the diffraction pattern of a single slit are clear. If one slit is covered up, as in Fig. 46-14b, the interference fringes disappear and we see the diffraction pattern of a single slit.

Starting from the curve of Fig. 46-12b, what is the effect of (a) increasing the slit width, (b) increasing the slit separation, and (c) increasing the wavelength?

(a) If we increase the slit width a, the envelope of the fringe pattern changes so that its central peak is sharper (compare Fig. 46-12c). The fringe spacing, which depends on d/λ, does not change.

EXAMPLE 6

(b) If we increase d, the fringes become closer together, the envelope of the pattern remaining unchanged.

(c) If we increase λ, the envelope becomes broader and the fringes move further apart. Increasing λ is equivalent to decreasing both of the ratios a/λ and d/λ. The general relationship of the envelope to the fringes, which depends only on d/a, does not change with wavelength.

EXAMPLE 7

In double-slit Fraunhofer diffraction what is the fringe spacing on a screen 50 cm away from the slits if they are illuminated with blue light ($\lambda = 480$ nm = 4800 Å), if $d = 0.10$ mm, and if the slit width $a = 0.02$ mm? What is the linear distance from the central maximum to the first minimum of the fringe envelope?

The intensity pattern is given by Eq. 46-16, the fringe spacing being determined by the interference factor $\cos^2 \beta$. From Example 2, Chapter 45, we have

$$\Delta y = \frac{\lambda D}{d},$$

where D is the distance of the screen from the slits. Substituting yields

$$\Delta y = \frac{(480 \times 10^{-9} \text{ m})(50 \times 10^{-2} \text{ m})}{0.10 \times 10^{-3} \text{ m}} = 2.4 \times 10^{-3} \text{ m} = 2.4 \text{ mm}.$$

The distance to the first minimum of the envelope is determined by the diffraction factor $(\sin \alpha/\alpha)^2$ in Eq. 46-16. The first minimum in this factor occurs for $\alpha = \pi$.

From Eq. 46-15,

$$\sin \theta = \frac{\alpha \lambda}{\pi a} = \frac{\lambda}{a} = \frac{480 \times 10^{-9} \text{ m}}{0.02 \times 10^{-3} \text{ m}} = 0.024.$$

This is so small that we can assume that $\theta \cong \sin \theta \cong \tan \theta$, or

$$y = D \tan \theta \cong D \sin \theta = (50 \text{ cm})(0.024) = 1.2 \text{ cm}.$$

There are about ten fringes in the central peak of the fringe envelope.

EXAMPLE 8

What requirements must be met for the central maximum of the envelope of the double-slit Fraunhofer pattern to contain exactly eleven fringes?

The required condition will be met if the sixth minimum of the interference factor $(\cos^2 \beta)$ in Eq. 46-16 coincides with the first minimum of the diffraction factor $(\sin \alpha/\alpha)^2$.

The sixth minimum of the interference factor occurs when

$$\beta = \tfrac{11}{2}\pi$$

in Eq. 46-12.

The first minimum in the diffraction term occurs for

$$\alpha = \pi.$$

Dividing (see Eqs. 46-13 and 46-15) yields

$$\frac{\beta}{\alpha} = \frac{d}{a} = \frac{11}{2}.$$

This condition depends only on the slit geometry and not on the wavelength. For long waves the pattern will be broader than for short waves, but there will always be eleven fringes in the central peak of the envelope.

The double-slit problem as illustrated in Fig. 46-12 combines interference and diffraction in an intimate way. At root both are superposition effects and depend on adding wave disturbances at a given point, taking phase differences properly into account. If the waves to be combined originate from a *finite* (and usually small) number of elementary coherent radiators, as in Young's double-slit experiment, we call the effect *interference*. If the waves to be combined originate by subdividing a wave into *infinitesimal* coherent radiators, as in our treatment of a single slit (Fig. 46-6), we call the effect *diffraction*. This distinction between interference and diffraction is convenient and useful. However, it should not cause us to lose sight of the fact that both are superposition effects and that often both are present simultaneously, as in Young's experiment.*

questions

1. Why is the diffraction of sound waves more evident in daily experience than that of light waves?

2. Why do radio waves diffract around buildings, although light waves do not?

3. A person holds a single narrow vertical slit in front of the pupil of his eye and looks at a distant light source in the form of a long heated filament. Is the diffraction pattern that he sees a Fresnel or a Fraunhofer pattern?

4. Do diffraction effects occur for virtual images as well as for real images? Explain.

5. Do diffraction effects occur for images formed by (*a*) plane mirrors and (*b*) spherical mirrors? Explain.

6. Comment on this statement: "Diffraction occurs in all regions of the electromagnetic spectrum." Consider the X-ray region and the micro-wave region for example, and give your arguments for believing it to be true or false.

7. Figure 46-1 shows the general case of diffraction from an aperture of arbitrary shape in an otherwise opaque screen. Discuss in general terms the inverse case, following the arguments of Section 46-1, in which diffraction occurs around an opaque *object*, such as a key or a paper clip, no screen B being present.

8. Distinguish between Fresnel and Fraunhofer diffraction. Do different physical principles underlie them? If so, what are they? If the same broad principle underlies them, what is it?

9. Most of us have noticed that the reception quality on a car radio deteriorates when we drive under a high-voltage transmission line. Does this have anything to do with the current in the line; with diffraction effects; with interference effects? Discuss.

10. *Fresnel's Bright Spot:* If a coin or a ball bearing (or even a bowling ball) is suspended between a point source of light and a photographic film a bright spot, called the Fresnel Bright Spot appears at the center of the geometrical shadow; see Fig. 46-15. Diffraction rings appear, both within the shadow and out of it. One might think that the center of the geometrical shadow, being most shielded from the light source, would be dark but just the reverse is true. Can you see the qualitative possibility that this experimental result can be consistent with Fresnel's diffraction theory?

*In this chapter we have not discussed *holography*, a technique in which a *three-dimensional* image of an object (rather than a two-dimensional image as in ordinary photography) can be produced. Diffraction plays a major role. To follow this up see "Holography, 1948–1971" by Dennis Gabor, *Science*, July 1972. Gabor won the Nobel Prize in 1971 for this discovery (made in 1948) and this article is an adaptation of his lecture on receiving the prize. See also "An Elementary Introduction to Practical Holography," by Alan G. Porter and S. George, *American Journal of Physics*, November 1975.

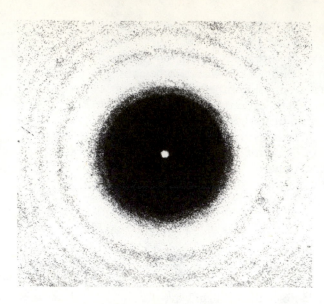

figure 46-15
Question 10
Fresnel's Bright Spot.

(Fresnel, who was trained as a civil engineer, submitted an essay on the diffraction of light to the French Academy of Sciences in response to a prize competition. He was then 30 years old. Siméon Poisson, a member of the prize committee and a distinguished mathematical physicist, was a confirmed believer in Newton's particle theory of light and totally disbelieved in the wave theory of Fresnel and others. He pointed out (as Fresnel had not) that Fresnel's theory predicted the existence of the bright spot that now bears Fresnel's name. Laboratory confirmation quickly followed. Although Fresnel won the prize, Poisson remained unconvinced of the wave theory of light and at his death, 22 years later, he still held to Newton's ideas).

11. A loud-speaker horn has a rectangular aperture 4 ft high and 1 ft wide. Will the pattern of sound intensity be broader in the horizontal plane or in the vertical?

12. We have claimed (correctly) that Maxwell's equations predict all the classical optical phenomena. Yet in Chapter 45 (Interference) and in this chapter (Diffraction) there is no mention of Maxwell's equations. Is there an inconsistency here? Where is the impact of Maxwell's equations felt? Discuss.

13. A radar antenna is designed to give accurate measurements of the height of an aircraft but only reasonably good measurements of its direction in a horizontal plane. Must the height-to-width ratio of the radar antenna be less than, equal to, or greater than unity?

14. In a single-slit Fraunhofer diffraction, what is the effect of increasing (a) the wavelength and (b) the slit width?

15. Sunlight falls on a single slit of width 10^3 nm. Describe qualitatively what the resulting diffraction pattern looks like.

16. In Fig. 46-5 rays r_1 and r_3 are in phase; so are r_2 and r_4. Why isn't there a *maximum* intensity at P_2 rather than a minimum?

17. Describe what happens to a Fraunhofer single-slit diffraction pattern if the whole apparatus is immersed in water.

18. When we speak of diffraction by a single "slit" we imply that the width of the slit must be much less than its height. Suppose that, in fact, the height was equal to twice the width. Make a rough guess at what the diffraction pattern would look like? An exact solution is of course not asked for.

19. In Fig. 46-3 we stated, correctly, that the optical path lengths from the slit to point P_0 are all the same. Why? Also, why are the optical path lengths of the rays in Fig. 46-5 from the slit to point P_2 *not* the same? What is the difference between these two figures?

20. In Fig. 46-7d why is E_θ, which represents the first maximum beyond the central maximum, not vertical? (Hint: Consider the effects of a slight winding or unwinding of the coil of vectors in this figure.)

21. Distinguish clearly between θ, α, and ϕ in Eq. 46-8c.

22. If we were to redo our analysis of the properties of lenses in Section 44-5 by the methods of geometrical optics but *without* restricting our considerations to paraxial rays and to "thin" lenses, would diffraction phenomena, such as that of Fig. 46-10, emerge from the analysis? Discuss.

23. Give at least two reasons why the usefulness of large reflecting telescopes increases as we increase the mirror diameter.

24. We have seen that diffraction limits the resolving power of optical telescopes (see Fig. 46-11). Does it also do so for radio telescopes, such as that of Fig. 41-2?

25. Distinguish carefully between interference and diffraction in Young's double-slit experiment.

26. In what way are interference and diffraction similar? In what way are they different?

27. In double-slit interference patterns such as that of Fig. 46-14a we said that the interference fringes were modulated in intensity by the diffraction pattern of a single slit. Could we reverse this statement and say that the diffraction pattern of a single slit is intensity-modulated by the interference fringes? Discuss.

SECTION 46-1

problems

1. *Babinet's principle:* A monochromatic beam of parallel light is incident on a "collimating" hole of diameter $x \gg \lambda$. Point P lies in the geometrical shadow region on a distance screen, as shown in Fig. 46-16a. Two obstacles, shown in Fig. 46-16b, are placed in turn over the collimating hole. A is an opaque circle with a hole in it and B is the "photographic negative" of A. Using superposition concepts, show that the intensity at P is identical for each of the two diffracting objects A and B.*

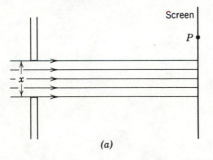

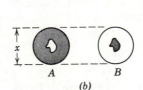

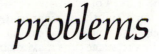

figure 46-16
Problem 1

SECTION 46-2

2. If the distance between the first and fifth minima of a single slit pattern is 0.35 mm with the screen 40 cm away from the slit and using light having a wavelength of 550 nm, what is the width of the slit?

3. In a single-slit diffraction pattern the distance between the first minimum on the right and the first minimum on the left is 5.2 mm. The screen on which the pattern is displayed is 80 cm from the slit and the wavelength is 5460 Å. Calculate the slit width. *Answer:* 0.17 mm.

4. For a single slit the first minimum occurs at $\theta = 90°$, thus filling the forward hemisphere beyond the slit with light. What must be the ratio of the slit width to the wavelength be for this to take place?

5. A slit 1.0 mm wide is illuminated by light of wavelength 589 nm = 5890 Å. We see a diffraction pattern on a screen 3.0 m away. What is the distance

---

* In this connection, it can be shown that the diffraction pattern of a wire is that of a slit of equal width. See "Measuring the Diameter of a Hair by Diffraction" by S. M. Curry and A. L. Schawlow, *American Journal of Physics,* May 1974.

between the first two diffraction minima on either side of the central diffraction maximum?   *Answer:* 1.8 mm.

6. A plane wave ($\lambda = 590$ nm) falls on a slit with $a = 0.40$ mm. A converging lens ($f = +70$ cm) is placed behind the slit and focuses the light on a screen. What is the linear distance on the screen from the center of the pattern to (*a*) the first minimum and (*b*) the second minimum?

SECTION 46-4

7. What is the half-width of a diffracted beam for a slit whose width is (*a*) 1, (*b*) 5, and (*c*) 10 wavelengths?   *Answer:* (*a*) 53°, (*b*) 10°, (*c*) 5.1°.

8. A single slit is illuminated by light whose wavelengths are λ_a and λ_b, so chosen that the first diffraction minimum of λ_a coincides with the second minimum of λ_b. (*a*) What relationship exists between the two wavelengths? (*b*) Do any other minima in the two patterns coincide?

9. In Fig. 46-7*d*, calculate the angle E_θ makes with the vertical, assuming the slit to be divided into infinitesimal strips of width dx. See Question 20. *Answer:* 12°.

10. (*a*) Show that the values of α at which intensity maxima for single-slit diffraction occur can be found exactly by differentiating Eq. 46-8*b* with respect to α and equating to zero, obtaining the condition

$$\tan \alpha = \alpha.$$

(*b*) Find the values of α satisfying this relation by plotting graphically the curve $y = \tan \alpha$ and the straight line $y = \alpha$ and finding their intersections; alternatively, use a pocket calculator. (*c*) Find the (nonintegral) values of m corresponding to successive maxima in the single-slit pattern. Note that the secondary maxima do not lie exactly halfway between minima.

11. In Example 4 solve the transcendental equation

$$\frac{1}{2} = \left(\frac{\sin \alpha_x}{\alpha_x}\right)^2$$

graphically for α_x, to an accuracy of three significant figures.
Answer: 79.7°.

12. If you double the width of a single slit the intensity of the central maximum of the diffraction pattern increases by a factor of four, even though the energy passing through the slit only doubles. Can you explain this quantitatively?

SECTION 46-5

13. The two headlights of an approaching automobile are 140 cm apart. At what maximum distance will the eye resolve them? Assume a pupil diameter of 5.0 mm and $\lambda = 550$ nm. Assume also that diffraction effects alone limit the resolution.   *Answer:* 10 km (= 6.3 mi).

14. An astronaut in a satellite claims he can just barely resolve two point sources on the earth, 160 km below him. What is their separation, assuming ideal conditions? Take $\lambda = 550$ nm, and the pupil diameter to be 5.0 mm.

15. How closely may two small objects be located if they are to be resolved when viewed through the telescope of a transit having a 3.0-cm objective lens if the transit is 370 m from the objects? Take the wavelength to be 550 nm.   *Answer:* 8.3 mm.

16. (*a*) How small is the angular separation of two stars if their images are barely resolved by the Thaw refracting telescope at the Allegheny Observatory in Pittsburgh? The lens diameter is 30 in. and its focal length is 46 ft. Assume $\lambda = 550$ nm. (*b*) Find the distance between these barely resolved stars if each of them is 10 light years distant from the earth. (*c*) For the image of a single star in this telescope, find the diameter of the first dark ring in the diffraction pattern, as measured on a photographic plate placed at the focal plane.

Assume that the star image structure is associated entirely with diffraction at the lens aperture and not with (small) lens "errors," etc.

17. The wall of a large room is covered with acoustic tile in which small holes are drilled 5.0 mm from center to center. How far can a person be from such a tile and still distinguish the individual holes, assuming ideal conditions? Assume the diameter of the pupil to be 4.0 mm and λ to be 550 nm. *Answer:* 37 m.

18. Find the separation of two points on the moon's surface that can just be resolved by the 200-in. telescope at Mount Palomar, assuming that this distance is determined by diffraction effects. The distance from the earth to the moon is 3.8×10^5 km. Assume $\lambda = 550$ nm ($= 5500$ Å).

19. Under ideal conditions, estimate the linear separation of two objects on the planet Mars which can just be resolved by an observer on earth (a) using the naked eye, and (b) using the 200-in. Mt. Palomar telescope. Use the following data: distance to Mars $= 8.0 \times 10^7$ km; diameter of pupil $= 5.0$ mm; wavelength of light $= 550$ nm. *Answer:* (a) 1.1×10^4 km. (b) 11 km.

20. A "spy in the sky" satellite orbitting at, say, 100 mi above the earth's surface, has a lens with a focal length of 8.0 ft. Its resolving power for objects on the ground is 1.2 ft, that is, it could easily detect an automobile. What is the effective lens diameter, determined by diffraction considerations alone? Assume $\lambda = 550$ nm. Even more effective satellites are reported to be in operation today. See "Reconnaissance and Arms Control" by Ted Greenwood, *Scientific American* February 1973.

21. (a) A circular diaphragm 0.60 m in diameter oscillates at a frequency of 25 kHz in an underwater source of sound for submarine detection. Far from the source the sound intensity is distributed as a Fraunhofer diffraction pattern for a circular hole whose diameter equals that of the diaphragm. Take the speed of sound in water to be 1450 m/s and find the angle between the normal to the diaphragm and the direction of the first minimum. (b) Repeat for a source having an (audible) frequency of 1.0 kHz. *Answer:* (a) 6.7°. (b) Because $1.22\lambda > d$, there is no answer for 1.0 kHz; explain.

22. It can be shown that, except for $\theta = 0$, a circular obstacle produces the same diffraction pattern as a circular hole of the same diameter. Furthermore, if there are many such obstacles, water droplets, say, located randomly, then the interference effects vanish leaving only the diffraction associated with a single obstacle. (a) Explain why one sees a "ring" around the moon on a foggy night. Usually the ring is reddish in color: explain. (b) Calculate the size of the water droplet in the air if the ring around the moon appears to have a diameter 1.5 times that of the moon. (c) At what distance from the moon might a bluish ring be seen? Sometimes the rings are white: explain. (d) The color arrangement is opposite to that in a rainbow: why should this be so?

23. Suppose that, as in Example 8, the envelope of the central peak contains eleven fringes. How many fringes lie between the first and second minima of the envelope? *Answer:* 5.

24. For $d = 2a$ in Fig. 46-17, how many interference fringes lie in the central diffraction envelope?

25. If we put $d = a$ in Fig. 46-17, the two slits coalesce into a single slit of width $2a$. Show that Eq. 46-16 reduces to the diffraction pattern for such a slit.

26. Construct qualitative vector diagrams like those of Fig. 46-7 for the double slit interference pattern. For simplicity consider $d = 2a$ (see Fig. 46-17). Can you interpret the main features of the intensity pattern this way?

27. (a) How many complete fringes appear between the first minima of the fringe envelope to either side of the central maximum for a double slit pattern if $\lambda = 550$ nm, $d = 0.15$ mm and $a = 0.030$ mm? (b) What is the ratio of the intensity of the third fringe to the side of the center to that of the central fringe? *Answer:* (a) 9. (b) 0.25.

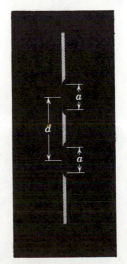

figure 46-17
Problems 24, 25, and 26

28. (*a*) Design a double-slit system in which the fourth fringe, not counting the central maximum, is missing. (*b*) What other fringes, if any, are also missing?

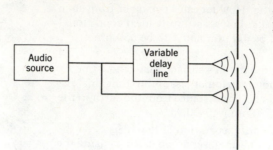

figure 46-18
Problem 29

29. An acoustic double-slit system (slit separation *d*, slit width *a*) is driven by two loudspeakers as shown in Fig. 46-18. By use of a variable delay line, the phase of one of the speakers may be varied. Describe in detail what changes occur in the intensity pattern at large distances as this phase difference is varied from zero to 2π. Take both the interference and diffraction effects into account.

47
gratings
and spectra

In connection with Young's experiment (Sections 45-1 and 45-3) we discussed the interference of two coherent waves formed by diffraction at two elementary radiators (pinholes or slits). In our first treatment we assumed that the slit width was much less than the wavelength, so that light diffracted from each slit illuminated the observation screen essentially uniformly. Later, in Section 46-6, we took the slit width into account and showed that the intensity pattern of the interference fringes is modulated by a "diffraction factor" $(\sin\alpha/\alpha)^2$ (see Eq. 46-16).

Here we extend our treatment to cases in which the number N of radiators or diffracting centers is larger—and usually much larger—than two. We consider two situations:

1. An array of N parallel equidistant slits, called a *diffraction grating*.
2. A three-dimensional array of periodically arranged radiators—the atoms in a crystalline solid such as NaCl, for example. In this case the average spacing between the elementary radiators is so small that interference effects must be sought at wavelengths much smaller than those of visible light. We speak of *X-ray diffraction*.

In each case we distinguish carefully between the diffracting properties of a single radiator (slit or atom) and the interference of the waves diffracted, coherently, from the assembly of radiators.

A logical extension of Young's double-slit interference experiment is to increase the number of slits from two to a larger number N. An arrangement like that of Fig. 47-1, usually involving many more slits (as many as 10^4 slits/cm is not uncommon), is called a *diffraction grating*. As for

47-1
INTRODUCTION

47-2
MULTIPLE SLITS

1047

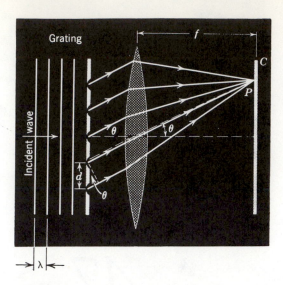

figure 47-1
An idealized diffraction grating containing five slits. The slit width a is shown for convenience to be considerably smaller than λ, although this condition is not realized in practice. The figure is also distorted in that f is much greater than d in practice.

a double slit, the intensity pattern that results when monochromatic light of wavelength λ falls on a grating consists of a series of interference fringes. The *angular separations* of these fringes are determined by the ratio λ/d, where d is the spacing between the centers of adjacent slits. The relative *intensities* of these fringes are determined by the diffraction pattern of a single grating slit, which depends on the ratio λ/a, where a is the slit width.

Figure 47-2, which compares the intensity patterns for $N = 2$ and $N = 5$, shows clearly that the "interference" fringes are modulated in intensity by a "diffraction" envelope, as in Fig. 46-14. Figure 47-3 presents a theoretical calculation of the intensity patterns for three fringes near the centers of the patterns of Fig. 47-2. These two figures show that increasing N (a) does not change the spacing between the (principal) interference fringe maxima, provided d and λ remain unchanged, (b) sharpens the (principal) maxima, and (c) introduces small secondary maxima between the principal maxima. Three such secondaries are present (but not readily visible) between each pair of adjacent principal maxima in Fig. 47-2b.

A principal maximum in Fig. 47-1 will occur when the path difference between rays from adjacent slits ($= d \sin \theta$) is given by

$$d \sin \theta = m\lambda \qquad m = 0, 1, 2, \ldots \qquad \text{(principal maxima)}, \quad (47\text{-}1)$$

where m is called the *order number.* This equation is identical with Eq. 45-1, which locates the intensity maxima for a double slit. The *loca-*

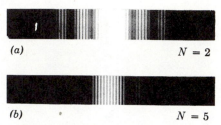

figure 47-2
Intensity patterns for "gratings" with (a) $N = 2$ and (b) $N = 5$ for the same value of d and λ. Note how the intensities of the fringes are modulated by a diffraction envelope as in Fig. 46-14; thus the assumption $a \ll \lambda$ is not realized in these actual "gratings." For $N = 5$ three very faint secondary maxima, not visible in this photograph, appear between each pair of adjacent primary maxima.

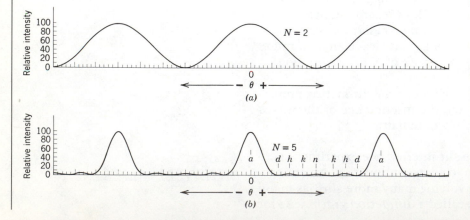

figure 47-3
Calculated intensity patterns for (a) a two-slit and (b) a five-slit grating for the same value of d and λ. This figure shows the sharpening of the principal maxima and the appearance of faint secondary maxima for $N > 2$. The letters on the five-slit pattern refer to Fig. 47-5. Three fringes only, centered around $\theta = 0$ in Fig. 47-1, are shown; compare Fig. 47-2. We may thus assume essentially equal intensities for each of the principal maxima shown.

tions of the (principal) maxima are thus determined only by the ratio λ/d and are independent of N. As for the double slit, the ratio a/λ determines the relative *intensities* of the principal maxima but does not alter their locations appreciably.

The sharpening of the principal maxima as N is increased can be understood by a graphical argument, using phasors. Figures 47-4a and 47-4b show conditions at any of the principal maxima for a two-slit and a nine-slit grating. The small arrows represent the amplitudes of the wave disturbances arriving at the screen at the position of each principal maximum. For simplicity we consider the central principal maximum only, for which $m = 0$, and thus $\theta = 0$, in Eq. 47-1.

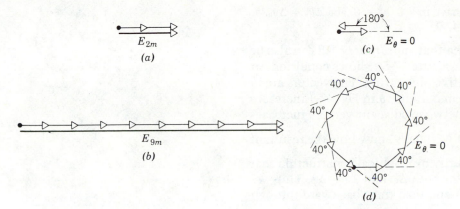

figure 47-4
Drawings (*a*) and (*b*) show conditions at the central principal maximum for a two-slit and a nine-slit grating, respectively. Drawings (*c*) and (*d*) show conditions at the minimum of zero intensity that lies on either side of this central principal maximum. In going from (*a*) to (*c*) the phase shift between waves from adjacent slits changes by 180° ($\Delta\phi = 2\pi/2$); in going from (*b*) to (*d*) it changes by 40° ($\Delta\phi = 2\pi/9$).

Consider the angle $\Delta\theta_0$ corresponding to the position of zero intensity that lies on either side of the central principal maximum. Figures 47-4c and 47-4d show the phasors at this point. The phase difference between waves from adjacent slits, which is zero at the central principal maximum, must increase by an amount $\Delta\phi$ chosen so that the array of phasors just closes on itself, yielding zero resultant intensity. For $N = 2$, $\Delta\phi = 2\pi/2$ (=180°); for $N = 9$, $\Delta\phi = 2\pi/9$ (=40°). In the general case it is given by

$$\Delta\phi = \frac{2\pi}{N}.$$

This increase in phase difference for adjacent waves corresponds to an increase in the path difference Δl given by

$$\frac{\text{phase difference}}{2\pi} = \frac{\text{path difference}}{\lambda},$$

or

$$\Delta l = \left(\frac{\lambda}{2\pi}\right) \Delta\phi = \left(\frac{\lambda}{2\pi}\right)\left(\frac{2\pi}{N}\right) = \frac{\lambda}{N}.$$

From Fig. 47-1, however, the path difference Δl at the first minimum is also given by $d \sin \Delta\theta_0$, so that we can write

$$d \sin \Delta\theta_0 = \frac{\lambda}{N},$$

or

$$\sin \Delta\theta_0 = \frac{\lambda}{Nd}.$$

Since $N \gg 1$ for actual gratings, $\sin \Delta\theta_0$, will ordinarily be quite small (that is, the lines will be sharp), and we may replace it by $\Delta\theta_0$, expressed in radians, to good approximation, or

$$\Delta\theta_0 = \frac{\lambda}{Nd} \qquad \text{(central principal maximum).} \qquad (47\text{-}2)$$

This equation shows specifically that if we increase N for a given λ and d, $\Delta\theta_0$ will decrease, which means that the central principal maximum becomes sharper.

We state without proof,* and for later use, that for principal maxima other than the central one (that is, for $m \neq 0$) the angular distance between the position θ_m of the principal maximum of order m and the minimum that lies on either side is given by

$$\Delta\theta_m = \frac{\lambda}{Nd\,\cos\theta_m} \qquad \text{(any principal maximum).} \qquad (47\text{-}3)$$

For the central principal maximum we have $m = 0$, $\theta_m = 0$, and $\Delta\theta_m = \Delta\theta_0$, so that Eq. 47-3 reduces, as it must, to Eq. 47-2.

The origin of the secondary maxima that appear for $N > 2$ can also be understood using the phasor method. Figure 47-5a shows conditions at the central principal maximum for a five-slit grating. The vectors are in phase. As we depart from the central maximum, θ in Fig. 47-1 increases from zero and the phase difference between adjacent vectors increases from zero to $\Delta\phi = \frac{2\pi}{\lambda} d \sin\theta$. Successive figures show how the resultant wave amplitude E_θ varies with $\Delta\phi$. Verify by graphical construction that a given figure represents conditions for both $\Delta\phi$ and $2\pi - \Delta\phi$. Thus we start at $\Delta\phi = 0$, proceed to $\Delta\phi = 180°$, and then trace backward through the sequence, following the phase differences shown in parentheses, until we reach $\Delta\phi = 360°$. This sequence corresponds to traversing the

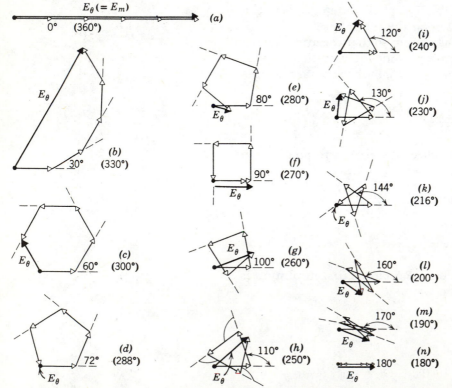

figure 47-5
The figures taken in sequence from (a) to (n) and then from (n) to (a) show conditions as the intensity pattern of a five-slit grating is traversed from the central principal maximum to an adjacent principal maximum. Phase differences between waves from adjacent slits are shown directly or, when going from (n) to (a), in parentheses. Principal maxima occur at (a), secondary maxima at, or near, (h) and (n), and points of zero intensity at (d) and (k). Compare Fig. 47-3b.

* See Problem 10.

intensity pattern from the central principal maximum to an adjacent one. Figure 47-5, which should be compared with Fig. 47-3b, shows that for $N = 5$ there are three secondary maxima, corresponding to $\Delta\phi = 110°$, 180°, and 250°. Make a similar analysis for $N = 3$ and show that only one secondary maximum occurs. In actual gratings, which commonly contain 10,000 to 50,000 "slits," the secondary maxima lie so close to the principal maxima or are so reduced in intensity that they cannot be distinguished from them experimentally.

The *grating spacing d* for a typical grating that contains 12,000 "slits" distributed over a 1-in. width is 2.54 cm/12,000, or 2100 nm. Gratings are often used to measure wavelengths and to study the structure and intensity of spectrum lines. Few devices have contributed more to our knowledge of modern physics.

47-3 DIFFRACTION GRATINGS

Gratings are made by ruling equally spaced parallel grooves on a glass or a metal* plate, using a diamond cutting point whose motion is automatically controlled by an elaborate ruling engine. Once such a master grating has been prepared, replicas can be formed by pouring a collodion solution on the grating, allowing it to harden, and stripping it off. The stripped collodion, fastened to a flat piece of glass or other backing, forms a good grating.

Figure 47-6 shows a cross section of a common type of grating ruled on glass. In the rudimentary grating of Fig. 47-1 open slits were separated by opaque strips; the *amplitude* of the wave disturbance varied in a periodic way as light passed through the grating, dropping to zero on the opaque strips. The grating of Fig. 47-6 is transparent everywhere, so that there is little periodic change in amplitude as the grating is crossed. The effect of the rulings is to change the *optical thickness* of the grating in a periodic way, rays traversing the grating between the rulings (b in Fig. 47-6) containing more wavelengths than rays traversing the grating in the center of the rulings (a in Fig. 47-6). This results in a periodic change of *phase* as light crosses the grating at right angles to the rulings. Reflection gratings also depend for their operation on a periodic change in phase of the reflected wave as one crosses the grating, the change in amplitude under these conditions being negligible. The principal maxima for *phase gratings*, assuming that the incident light falls on the grating at right angles, is given by the same formula derived earlier for idealized amplitude or slit gratings, namely,

$$d \sin \theta = m\lambda \qquad m = 0, 1, 2 \ldots,$$

where d is the distance between the rulings and the integer m is called the *order* of the particular principal maximum. Essentially all gratings used in the visible spectrum, whether of the transmission type, as in Fig. 47-6, or the reflection type, are phase gratings.

Figure 47-7 shows a simple grating spectroscope, used for viewing the spectrum of a light source, assumed to emit a number of discrete wavelengths, or *spectrum lines*. The light from source S is focused by lens L_1 on a slit S_1 placed in the focal plane of lens L_2. The parallel light emerging from collimator C falls on grating G. Parallel rays associated with a particular interference maximum occurring at angle θ fall on lens L_3, being brought to a focus in plane F-F'. The image formed in this

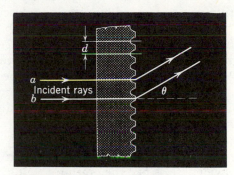

figure 47-6
An enlarged cross section of a diffraction grating ruled on glass. Such gratings, in which the phase of the emerging wave changes as one crosses the grating, are called *phase gratings.*

---

* Gratings ruled on metal are called *reflection gratings* because the interference effects are viewed in reflected rather than in transmitted light. For an insight into the highly complex technique of preparing gratings, see "Ruling Engines," by A. G. Ingolls, *Scientific American*, June, 1952, and "Diffraction Gratings," by E. W. Palmer and J. F. Verrill, *Contemporary Physics*, May, 1968.

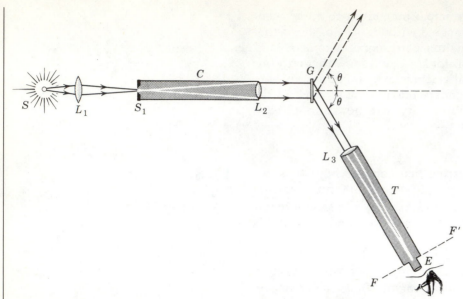

figure 47-7
A simple type of grating spectroscope
used to analyze the wavelengths of
the light emitted by source *S*.

plane is examined, using a magnifying lens arrangement *E*, called an
eyepiece. A symmetrical interference pattern is formed on the other side
of the central position, as shown by the dotted lines. The entire spectrum can be viewed by rotating telescope *T* through various angles. Instruments used for scientific research or in industry are more complex
than the simple arrangement of Fig. 47-7. They invariably employ photographic or photoelectric recording and are called *spectrographs.* Figure
49-12 shows a small portion of the spectrum of iron, produced by examining the light produced in an arc struck between iron electrodes, using
a research type spectrograph with photographic recording. Each line in
the figure represents a different wavelength that is emitted from the
source.

Grating instruments can be used to make absolute measurements of
wavelength, since the grating spacing *d* in Eq. 47-1 can be measured accurately with a traveling microscope. Several spectra are normally produced in such instruments, corresponding to $m = \pm 1, \pm 2$, etc., in Eq.
47-1 (see Fig. 47-8). This may cause some confusion if the spectra overlap. Further, this multiplicity of spectra reduces the recorded intensity
of any given spectrum line because the available energy is divided
among a number of spectra.

This disadvantage of the grating instrument can be overcome by shaping the
profile of the grating grooves so that a large fraction of the light is thrown into a
particular order on a particular side (for a given wavelength). This technique,
called *blazing,* so alters the diffracting properties of the individual grooves (by
controlling their profiles) that the light of wavelength λ diffracted by a single
groove has a sharp peak of maximum intensity at a selected angle $\theta\ (\neq 0)$.

Gratings can, of course, separate wavelengths that are distributed
continuously and not as sharp spectral lines. The emission from a red
hot poker is an example. At the Flandrau Planetarium at the University of Arizona a "live" (m = 1) spectrum* of the visible radiation

---

* See "Eight Feet of the Solar Spectrum," O. Richard Norton, *Sky and Telescope,* September 1977. Even here sharp spectral lines are present. They appear as *absorption*
lines (rather than *emission* lines) as the sun's radiation passes through its atmosphere.
These are the so-called (dark) Fraunhofer lines. The element helium (from a Greek word
meaning the sun) was discovered from an analysis of these lines. This is the only one of
the elements not first discovered on earth.

from the sun is displayed to visitors; it is 8 ft long and about 5.5 in. high.

Light can also be analyzed into its component wavelengths if the grating in Fig. 47-7 is replaced by a prism. In a *prism spectrograph* each wavelength in the incident beam is deflected through a definite angle θ, determined by the index of refraction of the prism material for that wavelength. Curves such as Fig. 43-2, which gives the index of refraction of fused quartz as a function of wavelength, show that the shorter the wavelength, the larger the angle of deflection θ. Such curves vary from substance to substance and must be found by measurement. Prism instruments are not adequate for accurate *absolute* measurements of wavelength because the index of refraction of the prism material at the wavelength in question is usually not known precisely enough. Both prism and grating instruments make accurate *comparisons* of wavelength, using a suitable comparison spectrum such as that shown in Fig. 49-12, in which careful absolute determinations have been made of the wavelengths of the spectrum lines. The prism instrument has an advantage over a grating instrument in that its light energy is concentrated into a single spectrum so that brighter lines may be produced.

A grating with 8000 rulings/in. is illuminated with white light at perpendicular incidence. Describe the diffraction pattern. Assume that the wavelength of the light extends from 4000 to 7000 Å (= 400 to 700 nm).

EXAMPLE 1

The grating spacing d is 2.54 cm/8000, or 3170 nm. The central or zero-order maximum corresponds to $m = 0$ in Eq. 47-1. All wavelengths present in the incident light are superimposed at $\theta = 0$, as Fig. 47-8 shows.

The first-order diffraction pattern corresponds to $m = 1$ in Eq. 47-1. The 400 nm line occurs at an angle given by

$$\theta = \sin^{-1}\frac{m\lambda}{d} = \sin^{-1}\frac{(1)(400 \text{ nm})}{3170 \text{ nm}} = \sin^{-1} 0.126 = 7.3°.$$

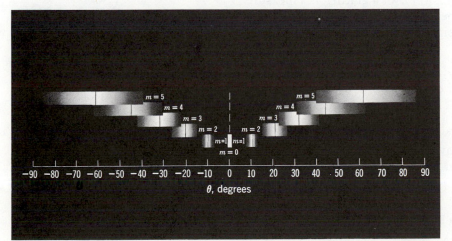

figure 47-8
Example 1. The spectrum of white light as viewed in a grating instrument like that of Fig. 47-7. The different orders, identified by the order number m, are shown separated vertically for clarity. As actually viewed, they would not be so displaced. The central line in each order corresponds to $\lambda = 550$ nm (= 5500 Å).

In the same way the angle for the 7000-A line is found to be 12.8°, and the entire pattern of Fig. 47-8 can be calculated. Note that the *first-order spectrum* ($m = 1$) is isolated but that the second-, third-, and fourth-order spectra overlap.

EXAMPLE 2

A diffraction grating has 10^4 rulings uniformly spaced over 1 in. It is illuminated at normal incidence by yellow light from a sodium vapor lamp. This light contains two closely spaced lines (the well-known *sodium doublet*) of wavelengths 5890.0 and 5895.9 Å (= 589.00 and 589.59 nm). (*a*) At what angle will the first-order maximum occur for the first of these wavelengths?

The grating spacing d is 10^{-4} in., or 2540 nm. The first-order maximum corresponds to $m = 1$ in Eq. 47-1. We thus have

$$\theta = \sin^{-1}\frac{m\lambda}{d} = \sin^{-1}\frac{(1)(589 \text{ nm})}{2540 \text{ nm}} = \sin^{-1}0.232 = 13.3°.$$

(*b*) What is the angular separation between the first-order maxima for these lines?

The straightforward way to find this separation is to repeat this calculation for $\lambda = 589.59$ nm and to subtract the two angles. A difficulty, which can best be appreciated by carrying out the calculation, is that we must carry a large number of significant figures to obtain a meaningful value for the difference between the angles. To calculate the difference in angular positions *directly*, let us write down Eq. 47-1, solved for $\sin \theta$, and differentiate it, treating θ and λ as variables:

$$\sin \theta = \frac{m\lambda}{d}$$

$$\cos \theta \, d\theta = \frac{m}{d} \, d\lambda.$$

If the wavelengths are close enough together, as in this case, $d\lambda$ can be replaced by $\Delta\lambda$, the actual wavelength difference; $d\theta$ then becomes $\Delta\theta$, the quantity we seek. This gives

$$\Delta\theta = \frac{m \, \Delta\lambda}{d \cos \theta} = \frac{(1)(0.59 \text{ nm})}{(2540 \text{ nm})(\cos 13.3°)} = 2.4 \times 10^{-4} \text{ rad} = 0.014°.$$

Note that although the wavelengths involve five significant figures, our calculation, done this way, involves only two or three, with consequent reduction in numerical manipulation.

The quantity $d\theta/d\lambda$, called the *dispersion D* of a grating, is a measure of the angular separation produced between two incident monochromatic waves whose wavelengths differ by a small wavelength interval. From this example we see that

$$D = \frac{d\theta}{d\lambda} = \frac{m}{d \cos \theta}. \qquad (47\text{-}4)$$

To distinguish light waves whose wavelengths are close together, the maxima of these wavelengths formed by the grating should be as narrow as possible. Expressed otherwise, the grating should have a high *resolving power R*, defined from

47-4
RESOLVING POWER OF A GRATING

$$R = \frac{\lambda}{\Delta\lambda}. \qquad (47\text{-}5)$$

Here λ is the mean wavelength of two spectrum lines that can barely be recognized as separate and $\Delta\lambda$ is the wavelength difference between them. The smaller $\Delta\lambda$ is, the closer the lines can be and still be resolved; hence the greater the resolving power R of the grating. It is to achieve a high resolving power that gratings with many rulings are constructed.

The resolving power of a grating is usually determined by the same consideration (that is, the Rayleigh criterion) that we used in Section 46-5 to determine the resolving power of a lens. If two principal maxima are to be barely resolved, they must, according to this criterion, have an angular separation $\Delta\theta$ such that the maximum of one line coincides with the first minimum of the other; see Fig. 46-11. If we apply this criterion, we can show that

$$R = Nm, \qquad (47\text{-}6)$$

where N is the total number of rulings in the grating and m is the order. As expected, the resolving power is zero for the central principal maximum ($m = 0$), all wavelengths being undeflected in this order.

Let us derive Eq. 47-6. The angular separation between two principal maxima whose wavelengths differ by $\Delta\lambda$ is found from Eq. 47-4, which we recast as

$$\Delta\theta = \frac{m\,\Delta\lambda}{d\,\cos\theta}. \qquad (47\text{-}4)$$

The Rayleigh criterion (Section 46-5) requires that this be equal to the angular separation between a principal maximum and its adjacent minimum. This is given from Eq. 47-3, dropping the subscript m in $\cos\theta_m$, as

$$\Delta\theta_m = \frac{\lambda}{Nd\,\cos\theta}. \qquad (47\text{-}3)$$

Equating Eqs. 47-4 and 47-3 leads to

$$R\,(=\lambda/\Delta\lambda) = Nm,$$

which is the desired relation.

EXAMPLE 3

In Example 2 how many rulings must a grating have if it is barely to resolve the sodium doublet in the third order?

From Eq. 47-5 the required resolving power is

$$R = \frac{\lambda}{\Delta\lambda} = \frac{589\text{ nm}}{(589.59 - 589.00)\text{nm}} = 1000.$$

From Eq. 47-6 the number of rulings needed is

$$N = \frac{R}{m} = \frac{1000}{3} = 330.$$

This is a modest requirement.

The resolving power of a grating must not be confused with its dispersion. Table 47-1 shows the characteristics of three gratings, each illuminated with light of $\lambda = 589$ nm ($= 5890$ Å), the diffracted light being viewed in the first order ($m = 1$ in Eq. 47-1).

Table 47-1
Some characteristics of three gratings
$(\lambda = 589 \text{ nm}, m = 1)$

Grating	N	d, nm	θ	R	D 10^{-2} degrees/nm
A	10,000	2540	13.3°	10,000	2.32
B	20,000	2540	13.3°	20,000	2.32
C	10,000	1370	25.5°	10,000	4.64

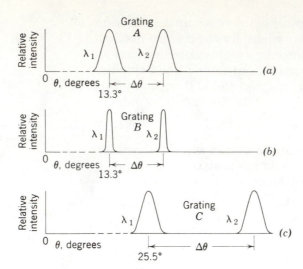

figure 47-9
The intensity patterns for light of wavelengths λ_1 and λ_2 near 589 nm (= 5890 Å), incident on the gratings of Table 47-1. Grating B has the highest resolving power and grating C the highest dispersion.

You should verify that the values of D and R given in the table can be calculated from Eqs. 47-4 and 47-6, respectively.

For the conditions of use noted in Table 47-1, gratings A and B have the same *dispersion* and A and C have the same *resolving power*. Figure 47-9 shows the intensity patterns that would be produced by these gratings for two incident waves of wavelengths λ_1 and λ_2, in the vicinity of $\lambda = 589$ nm. Grating B, which has high resolving power, has narrow intensity maxima and is inherently capable of distinguishing lines that are much closer together in wavelength than those of Fig. 47-9. Grating C, which has high dispersion, produces twice the angular separation between rays λ_1 and λ_2 that grating B does.

EXAMPLE 4

The grating of Example 1 has 8000 lines illuminated by light from a mercury vapor discharge. (*a*) What is the expected dispersion, in the third order, in the vicinity of the intense green line ($\lambda = 546$ nm $= 5460$ Å)?

Noting that $d = 3170$ nm, we have, from Eq. 47-1,

$$\theta = \sin^{-1}\frac{m\lambda}{d} = \sin^{-1}\frac{(3)(546 \text{ nm})}{3170 \text{ nm}} = \sin^{-1} 0.517 = 31.1°.$$

From Eq. 47-4 we have

$$D = \frac{m}{d \cos\theta} = \frac{3}{(3170 \text{ nm})(\cos 31.1°)} = 1.1 \times 10^{-3} \text{ rad/nm} = 6.3 \times 10^{-2} \text{ deg/nm}.$$

(*b*) What is the expected resolving power of this grating in the fifth order? Equation 47-6 gives

$$R = Nm = (8000)(5) = 40,000.$$

Thus near $\lambda = 546$ nm a wavelength difference $\Delta\lambda$ given by Eq. 47-5, or

$$\Delta\lambda = \frac{\lambda}{R} = \frac{546 \text{ nm}}{40,000} = 0.014 \text{ nm},$$

can be distinguished.

47-5
X-RAY DIFFRACTION

Figure 47-10 shows how X-rays are produced when electrons from a heated filament F are accelerated by a potential difference V and strike a metal target T. X-rays are electromagnetic radiation with wavelengths of the order of 0.1 nm (or 1 Å). This value is to be compared with 550 nm

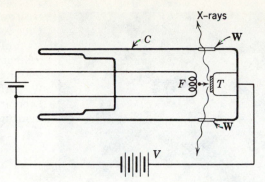

figure 47-10
X-rays are generated when electrons from heated filament F, accelerated by potential difference V, are brought to rest on striking metallic target T. W is a "window"—transparent to X-rays—in the evacuated metal container C.

(or 5500 Å) for the center of the visible spectrum. For such small wavelengths a standard optical diffraction grating, as normally employed, cannot be used. For $\lambda = 0.10$ nm and $d = 3000$ nm, for example, Eq. 47-1 shows that the first-order maximum occurs at

$$\theta = \sin^{-1}\frac{m\lambda}{d} = \sin^{-1}\frac{(1)(0.10\text{ nm})}{3 \times 10^3\text{ nm}} = \sin^{-1}0.33 \times 10^{-4} = 0.002°.$$

This is too close to the central maximum to be practical. A grating with $d \cong \lambda$ is desirable, but, because X-ray wavelengths are about equal to atomic diameters, such gratings cannot be constructed mechanically.

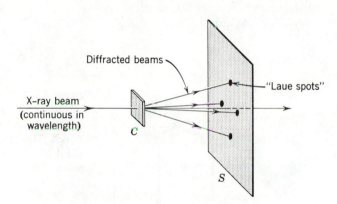

Diffracted beams

X-ray beam
(continuous in
wavelength)

"Laue spots"

C

S

figure 47-11
A nonmonochromatic beam of X-rays falls on a crystal C, which may be NaCl. Strong diffracted beams appear in certain directions, forming a so-called Laue pattern on a photographic film S.

In 1912 it occurred to physicist Max von Laue that a crystalline solid, consisting as it does of a regular array of atoms, might form a natural three-dimensional "diffraction grating" for X-rays. Figure 47-11 shows that if a collimated beam of X-rays, continuously distributed in wavelength, is allowed to fall on a crystal, such as sodium chloride, intense beams corresponding to constructive interference from the many diffracting centers of which the crystal is made up appear in certain sharply defined directions. If these beams fall on a photographic film, they form an assembly of "Laue spots." Figure 47-12, which is an actual example of these spots, shows that the hypothesis of Laue is indeed correct. The atomic arrangements in the crystal can be deduced from a careful study of the positions and intensities of the Laue spots* in much the same way that we might deduce the structure of an optical grating (that is, the detailed profile of its slits) by a study of the positions and intensities of the lines in the interference pattern.

* Other experimental arrangements have supplanted the Laue technique to a considerable extent today; the principle remains unchanged, however (see Question 21).

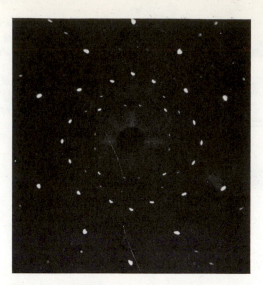

figure 47-12
Laue X-ray diffraction pattern from sodium chloride. A crystal of ordinary table salt was used in making this plate. (Courtesy of W. Arrington and J. L. Katz, X-ray Laboratory, Rensselaer Polytechnic Institute.)

Figure 47-13 shows how sodium and chlorine atoms (strictly, Na^+ and Cl^- ions) are stacked to form a crystal of sodium chloride. This pattern, which has *cubic* symmetry, is one of the many atomic arrangements exhibited by solids. The model represents the *unit cell* for sodium chloride. This is the smallest unit from which the crystal may be built up by repetition in three dimensions. You should verify that no smaller assembly of atoms possesses this property. For sodium chloride the length of the cube edge of the unit cell is 0.562737 nm.

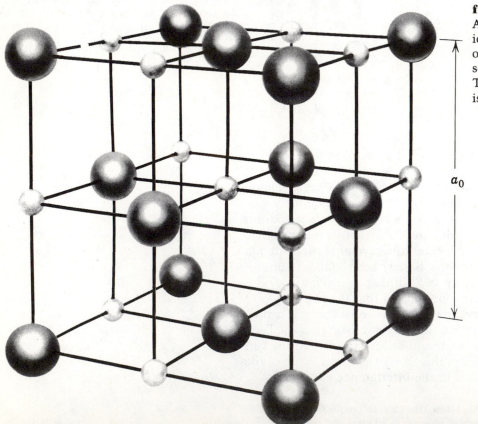

figure 47-13
A model showing how Na^+ and Cl^- ions are stacked to form a unit cell of NaCl. The small spheres represent sodium ions, the large ones chlorine. The edge a_0 of the (cubical) unit cell is 0.562737 nm.

Each unit cell in sodium chloride has four sodium ions and four chlorine ions associated with it. In Fig. 47-13 the sodium ion in the center belongs entirely to the cell shown. Each of the other twelve sodium ions shown is shared with three adjacent unit cells so that each contributes one-fourth of an ion to the cell under consideration. The total number of sodium ions is then $1 + \frac{1}{4}(12) = 4$. By similar reasoning you can show that although there are fourteen chlorine ions in Fig. 47-13, only four are associated with the unit cell shown.

The unit cell is the fundamental repetitive diffracting unit in the crystal, corresponding to the slit (and its adjacent opaque strip) in the optical diffraction grating of Fig. 47-1. Figure 47-14a shows a particular plane in a sodium chloride crystal. If each unit cell intersected by this plane is represented by a small cube, Fig. 47-14b results. You may imagine each of these figures extended indefinitely in three dimensions.

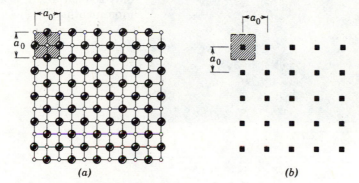

(a) (b)

figure 47-14
(a) A section through a crystal of sodium chloride, showing the sodium and chlorine ions. (b) The corresponding unit cells in this section, each cell being represented by a small black square.

Let us treat each small cube in Fig. 47-14b as an elementary diffracting center, corresponding to a slit in an optical grating. The *directions* (but not the intensities) of all the diffracted X-ray beams that can emerge from a sodium chloride crystal (for a given X-ray wavelength and a given orientation of the incident beam) are determined by the geometry of this three-dimensional lattice of diffracting centers. In exactly the same way the *directions* (but not the intensities) of all the diffracted beams that can emerge from a particular optical grating (for a given wavelength and orientation of the incident beam) are determined only by the geometry of the grating, that is, by the grating spacing d. Representing the unit cell by what is essentially a point, as in Fig. 47-14b, corresponds to representing the slits in a diffraction grating by lines, as we did in discussing Young's experiment in Section 45-1.

The *intensities* of the lines from an optical diffraction grating are determined by the diffracting characteristics of a single slit, as Fig. 46-14 shows. In the idealized case of Fig. 47-1 these characteristics depend on the slit width a. In practical optical gratings these characteristics depend on the detailed shape of the profile of the grating rulings.

In exactly the same way the *intensities* of the diffracted beams emerging from a crystal depend on the diffracting characteristics of the unit cell.* Fundamentally the X-rays are diffracted by electrons, diffraction by nuclei being negligible in most cases. Thus the diffracting characteristics of a unit cell depend on how the electrons are distributed

---

* For some directions in which a beam might be expected to emerge, from interference considerations, no beam will be found because the diffracting characteristics of the unit cell are such that no energy is diffracted in that direction. Similarly, in optical gratings some lines, permitted by interference considerations, may not appear if their predicted positions coincide with a null in the single-slit diffraction pattern (see Fig. 46-12).

(a)

(b)

figure 47-15
(a) A photograph of an oscilloscope screen arranged to display projected electron density contours for phthalocyanine ($C_{32}H_{18}N_8$). Such plots, constructed electronically from X-ray diffraction data by an analog computer, provide a vivid picture of the structure of molecules. (Courtesy of Ray Pepinsky.) (b) A structural representation of the molecule phthalocyanine. The student should make a detailed comparison with (a), locating the various atoms identified in (b). Note that hydrogen atoms, which contain only a single electron, are not prominent in (a).

throughout the volume of the cell. By studying the *directions* of diffracted X-ray beams, we can learn the basic symmetry of the crystal. By also studying the *intensities* we can learn how electrons are distributed in the unit cell. Figure 47-15 shows the average density of electrons projected onto a particular plane through a unit cell of a crystal of phthalocyanine. This remarkable figure suggests something of the full power of X-ray methods for studying the structure of solids.

Bragg's law predicts the conditions under which diffracted X-ray beams from a crystal are possible. In deriving it, we ignore the structure of the unit cell, which is related only to the intensities of these beams. The dashed sloping lines in Fig. 47-16a represent the intersection with the plane of the figure of an arbitrary set of planes passing through the ele-

47-6
BRAGG'S LAW

mentary diffracting centers. The perpendicular distance between adjacent planes is d. Many other such families of planes, with different *interplanar spacings*, can be defined.

Figure 47-16b shows a plane wave that lies in the plane of the figure falling on one member of the family of planes defined in Fig. 47-16a, the incident rays making an angle θ with the plane.* Consider a family of diffracted rays lying in the plane of Fig. 47-16b and making an angle β with the plane containing the elementary diffracting centers. The diffracted rays will combine to produce maximum intensity if the path difference between adjacent rays is an integral number of wavelengths or

$$ae - bd = h(\cos \beta - \cos \theta) = l\lambda \qquad l = 0, 1, 2, \ldots . \qquad (47\text{-}7)$$

For $l = 0$ this leads to $\qquad\qquad \beta = \theta,$

and the plane of atoms acts like a mirror for the incident wave, no matter what the value of θ.

For other values of l, β does not equal θ, but the diffracted beam can always be regarded as being "reflected" from a *different* set of planes than that shown in Fig. 47-16a with a different interplanar spacing d. Since we wish to describe each diffracted beam as a "reflection" from a particular set of planes and since we are dealing in the present argument only with the particular set of planes shown in Fig. 47-16a, we ignore all values of l other than $l = 0$ in Eq. 47-7. It can also be shown that a plane of diffracting centers acts like a mirror (that is, $\beta = \theta$) whether or not the incident wave lies in the plane of Fig. 47-16b.

Figure 47-16c shows an incident wave striking the *family* of planes, a single member of which was considered in Fig. 47-16b. For a single plane, mirrorlike "reflection" occurs for *any* value of θ, as we have seen. To have a constructive interference in the beam diffracted from the entire family of planes in the direction θ, the rays from the separate planes must reinforce each other. This means that the path difference for rays from adjacent planes (*abc* in Fig. 47-16c) must be an integral number of wavelengths or

$$2d \sin \theta = m\lambda \qquad m = 1, 2, 3, \ldots . \qquad (47\text{-}8)$$

This relation is called *Bragg's law* after W. L. Bragg who first derived it. The quantity d in this equation (the interplanar spacing) is the perpendicular distance between the planes. For the planes of Fig. 47-16a analysis shows that d is related to the unit cell dimension a_0 by

$$d = \frac{a_0}{\sqrt{5}}. \qquad (47\text{-}9)$$

If an incident *monochromatic* X-ray beam falls at an *arbitrary* angle θ on a particular set of atomic planes, a diffracted beam will *not* result because Eq. 47-8 will not, in general, be satisfied. If the incident X-rays are *continuous* in wavelength, diffracted beams will result when wavelengths given by

$$\lambda = \frac{2d \sin \theta}{m} \qquad m = 1, 2, 3, \ldots$$

are present in the incident beam (see Eq. 47-8).

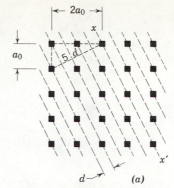

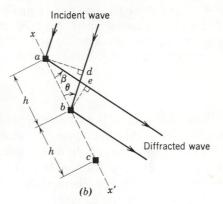

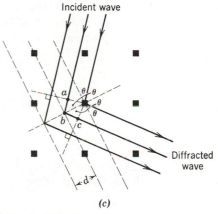

figure 47-16
(*a*) A section through the NaCl unit cell lattice of Fig. 45-14b. The dashed sloping lines represent an arbitrary family of planes, with interplanar spacing d. (*b*) An incident wave falls, at grazing angle θ, on one of the planes, *xx'*, shown in (*a*). (*c*) An incident wave falls on the entire family of planes shown in (*a*). A strong diffracted wave is formed.

---

* In X-ray diffraction it is customary to specify the direction of a wave by giving the angle between the ray and the plane (the *glancing angle*) rather than the angle between the ray and the normal.

EXAMPLE 5

At what angles must an X-ray beam with $\lambda = 0.110$ nm fall on the family of planes represented in Fig. 47-16c if a diffracted beam is to exist? Assume the material to be sodium chloride.

The interplanar spacing d for these planes is given by Eq. 47-9 or

$$d = \frac{a_0}{\sqrt{5}} = \frac{0.563 \text{ nm}}{2.24} = 0.252 \text{ nm}.$$

Equation 47-8 gives

$$\sin \theta = \frac{m\lambda}{2d} = \frac{(m)(0.110 \text{ nm})}{(2)(0.252 \text{ nm})} = 0.218 \ m.$$

Diffracted beams are possible at $\theta = 12.6°$ $(m = 1)$, $\theta = 25.9°$ $(m = 2)$, $\theta = 40.9°$ $(m = 3)$, and $\theta = 60.7°$ $(m = 4)$. Higher-order beams cannot exist because they require $\sin \theta$ to exceed unity. Actually, the odd-order beams $(m = 1, 3)$ prove to have zero intensity because the unit cell in cubic crystals such as NaCl has diffracting properties such that the intensity of the light scattered in these orders is zero.

X-ray diffraction is a powerful tool for studying the arrangements of atoms in crystals.* To do so quantitatively requires that the wavelength of the X-rays be known. In one of several approaches to this problem the unit cell dimension for NaCl† is determined by a method that does not involve X-rays. X-ray diffraction measurements on NaCl can then be used to determine the wavelength of the X-ray beam which in turn can be used to determine the structures of solids other than NaCl.

If ρ is the measured density of NaCl, we have, for the unit cell of Fig. 47-13,‡ recalling that each unit cell contains four NaCl "molecules,"

$$\rho = \frac{m}{V} = \frac{4m_{NaCl}}{a_0^3}.$$

Here m_{NaCl}, the mass of a NaCl molecule, is given by

$$m_{NaCl} = \frac{M}{N_0},$$

where M is the molecular weight of NaCl and N_0 is Avogadro's number. Combining these two equations and solving for a_0 yields

$$a_0 = \left(\frac{4M}{N_0\rho}\right)^{1/3}$$

which permits us to calculate a_0. Once a_0 is known, the wavelengths of monochromatic X-ray beams can be found, using Bragg's law (Eq. 47-8).

---

questions

1. Discuss this statement: "A diffraction grating can just as well be called an interference grating."
2. How would the spectrum of an enclosed source that is formed by a diffraction grating on a screen change (if at all) when the source, grating and screen are all submerged in water?

---

* See "X-ray Crystallography" by Sir Lawrence Bragg, *Scientific American*, July 1968.
† In practice, calcite ($CaCO_3$) proves to be more useful as a standard crystal for a number of technical reasons.
‡ This relation cannot be written down unless it is known that the structure of NaCl is cubic. This can be determined, however, by inspection of the symmetry of the spots in Fig. 47-12; the wavelength of the X-rays need not be known.

3. Assume that the limits of the visible spectrum are 4300 and 6800 Å. Is it possible to design a grating, assuming that the incident light falls normally on it, such that the first-order spectrum *barely overlaps* the second-order spectrum?

4. In a grating spectrograph, several lines having different wavelengths and formed in different orders might appear near a certain angle. How could you distinguish between their orders?

5. For the simple spectroscope of Fig. 47-7, show (a) that θ increases with λ for a grating and (b) that θ decreases with λ for a prism.

6. You are given a photograph of a spectrum on which the angular positions and the wavelengths of the spectrum lines are marked. (a) How can you tell whether the spectrum was taken with a prism or a grating instrument? (b) What information could you gather about either the prism or the grating from studying such a spectrum?

7. Explain in your own words why increasing the number of slits, N, in a diffraction grating sharpens the maxima. Why does decreasing the wavelength do so? Why does increasing the slit spacing, d, do so?

8. (a) What is a "phase grating"? (b) Can you make a diffraction grating out of parallel rows of fine wires strung closely together?

9. How much information can you discover about the structure of a diffraction grating by analyzing carefully the spectrum it forms of a monochromatic light source. Let $= 589$ nm, for an example.

10. (a) Why does a diffraction grating have closely spaced rulings? (b) Why does it have a large number of rulings?

11. Two nearly equal wavelengths are incident on a grating of N slits and are just not resolvable. However, they become resolved if the number of slits is increased. Formulas aside, is the explanation of this that: (a) more light can get through the grating? (b) the principal maxima become more intense and hence resolvable? (c) the diffraction pattern is spread more and hence the wavelengths become resolved? (d) there are a larger number of orders? or (e) the principal maxima become narrower and hence resolvable?

12. The relation $R = Nm$ suggests that the resolving power of a given grating can be made as large as desired by choosing an arbitrarily high order of diffraction. Discuss.

13. Show that at a given wavelength and a given angle of diffraction the resolving power of a grating depends only on its width $W\ (= Nd)$.

14. According to Eq. 47-3 the principal maxima become wider (that is, $\Delta\theta_m$ increases) the higher the order m (that is, the larger θ_m becomes). According to Eq. 47-6 the resolving power becomes greater the higher the order m. Explain this apparent paradox.

15. How would you *measure* (a) the dispersion D and (b) the resolving power R for either a prism or a grating spectrograph.

16. Is the pattern of Fig. 47-12 more properly described as a diffraction pattern or as an interference pattern?

17. For a given family of planes in a crystal, can the wavelength of incident X-rays be (a) too large or (b) too small to form a diffracted beam?

18. Why cannot a simple cube of edge $a_0/2$ in Fig. 47-13 be used as a unit cell for sodium chloride?

19. In some respects Bragg reflection *differs from* plane grating diffraction. Of the following statements which one is *true for Bragg reflection but not true for grating diffraction?* (a) two different wavelengths may be superposed. (b) radiation of a given wavelength may be sent in more than one direction. (c) long waves are deviated more than short waves. (d) there is only one grating spacing. (e) diffraction maxima of a given wavelength occur only for particular angles of incidence.

20. If a parallel beam of X-rays of wavelength λ is allowed to fall on a randomly oriented crystal of any material, generally no intense diffracted beams will

occur. Such beams appear if (a) the X-ray beam consists of a continuous distribution of wavelengths rather than a single wavelength or (b) the specimen is not a single crystal but a finely divided powder. Explain.

21. In Fig. 47-17(a) we show schematically the Debye-Scherrer experimental arrangement and in Fig. 47-17(b) a corresponding x-ray diffraction pattern. (a) Keeping in mind that the Laue method uses a large single crystal and an x-ray beam continuously distributed in wavelength, explain the origin of the spots. (Hint: each spot corresponds to the direction of scattering from a family of planes.)

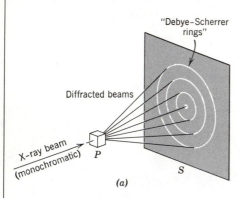
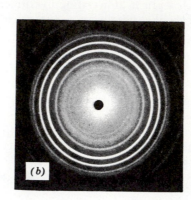

figure 47-17
Question 21 (a) and (b)
(a) A monochromatic beam of X-rays falls on an aggregate P of very small crystals oriented at random (called a microcyrstalline powder). Strong diffracted beams form rings surrounding a central spot, the so-called Debye-Scherrer rings, on a photographic film S.
(b) Debye-Scherrer X-ray diffraction pattern from zirconium oxide crystals. (From U. Fano and L. Fano, *Basic Physics of Atoms and Molecules*, John Wiley's Sons, 1959)

(b) Keeping in mind that the Debye-Scherrer method uses a large number of small single crystals randomly oriented and a monochromatic beam of x-rays, explain the origin of the rings. (Hint: because the small crystals are randomly oriented, all possible angles of incidence are obtained.)

SECTION 47-2

problems

1. Show that in a grating with alternately transparent and opaque strips of *equal* width, all the even orders (except $m = 0$) are absent.

2. The central intensity maximum formed by a grating, along with its subsidiary secondary maxima, can be viewed as the diffraction pattern of a single "slit" whose width is that of the entire grating. Treating the grating as a single wide slit, assuming that $m = 0$, and using the methods of Section 46-4, show that Eq. 47-2 can be derived.

3. A diffraction grating is made up of slits of width 300 nm with a 900 nm separation between their centers. The grating is illuminated by monochromatic plane waves, $\lambda = 600$ nm, the angle of incidence being zero. (a) How many diffracted lines are there? (b) What is the angular width of the spectral lines observed, if the grating has 1000 slits? *Angular width* is defined to be the angle between the two positions of zero intensity on either side of the maximum.   *Answer:* (a) 3, corresponding to $m = 0, \pm 1$. (b) 1.8×10^{-3} rad.

4. Assume that light is incident on a grating at an angle ψ as shown in Fig. 47-18. Show that the condition for a diffraction maximum is

$$d(\sin \psi + \sin \theta) = m\lambda \qquad m = 0, 1, 2, \dots.$$

Only the special case $\psi = 0$ has been treated in this chapter (compare Eq. 47-1).

5. A transmission grating with $d = 1.50 \times 10^{-4}$ cm is illuminated at various angles of incidence by light of wavelength 600 nm. Plot as a function of angle of incidence (0 to 90°) the angular deviation of the first-order diffracted beam from the incident direction.

6. Derive this expression for the intensity pattern for a three-slit "grating":

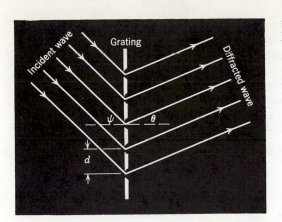

figure 47-18
Problem 4

$$I_\theta = \tfrac{1}{9} I_m (1 + 4 \cos \phi + 4 \cos^2 \phi),$$

where
$$\phi = \frac{2\pi d \sin \theta}{\lambda}.$$

Assume that $a \ll \lambda$ and be guided by the derivation of the corresponding double-slit formula (Eq. 45-9).

7. (a) Using the result of Problem 6, show that the half-width of the fringes for a three-slit diffraction pattern, assuming θ small enough so that $\sin \theta \cong \theta$, is

$$\Delta\theta \cong \frac{\lambda}{3.2d}.$$

(b) Compare this with the expression derived for the two-slit pattern in Problem 16, Chapter 45. (c) Do these results support the conclusion that for a fixed slit spacing the interference maxima become sharper as the number of slits is increased?

8. (a) Using the result of Problem 6, show that a three-slit "grating" has only one secondary maximum. Find (b) its location and (c) its relative intensity.

9. A three-slit grating has separation d between adjacent slits. If the middle slit is covered up, will the half-width of the intensity maxima become broader or narrower? See Problem 7.
Answer: Narrower by the factor 0.78.

10. Derive Eq. 47-3, that is, the expression for $\Delta\theta_m$, the angular distance between a principal maximum of order m and either adjacent minimum.

11. A diffraction grating has a large number N of slits, each of width d. Let I_{max} denote the intensity at some principal maximum, and let I_k denote the intensity of the kth adjacent secondary maximum. (a) If $k \ll N$, show from the phasor diagram that, approximately, $I_k/I_{max} = 1/(k + \tfrac{1}{2})^2 \pi^2$. (Compare this with the single-slit formula.) (b) For those secondary maxima which lie roughly midway between two adjacent principal maxima, show that roughly $I_k/I_{max} = 1/N^2$. (c) Consider the central principal maximum and those adjacent secondary maxima for which $k \ll N$. Show that this part of the diffraction pattern quantitatively resembles that for one single slit of width Nd.

SECTION 47-3

12. A diffraction grating 2.0 cm wide has 6000 rulings. At what angles will maximum-intensity beams occur if the incident radiation has a wavelength of 589 nm?

13. With light from a gaseous discharge tube incident normally on a grating with a distance 1.73×10^{-4} cm between adjacent slit centers, a green line appears with sharp maxima at measured transmission angles $\theta = \pm 17.6°$, $37.3°$, $-37.1°$, $65.2°$, and $-65.0°$. Compute the wavelength of the green line that best fits the data. *Answer:* 524 nm.

14. A diffraction grating has 5000 rulings/in., and a strong diffracted beam is noted at $\theta = 30°$. (a) What are the possible wavelengths of the incident light? (b) What colors are they (see Figure 42-1)?

15. A grating has 3150 rulings/cm. For what wavelengths in the visible spectrum can fifth-order diffraction be observed?
   Answer: All wavelengths shorter than 635 nm.

16. Given a grating with 4000 lines/cm, how many orders of the entire visible spectrum (400–700 nm) can be produced?

17. A diffraction grating 3.00 cm wide produces a deviation of 30.0° in the second order with light of wavelength 600 nm. What is the total number of lines on the grating?   *Answer:* 12,500.

18. Light of wavelength 600 nm is incident normally on a diffraction grating. Two *adjacent* principal maxima occur at $\sin \theta = 0.2$ and $\sin \theta = 0.3$, respectively. The fourth order is a missing order. (a) What is the separation between adjacent slits? (b) What is the smallest possible individual slit width? (c) Name *all* orders actually appearing on the screen with the values chosen in (a) and (b).

19. Assume that the limits of the visible spectrum are arbitrarily chosen as 430 and 680 nm ($= 4300$ and .6800 Å). Design a grating that will spread the first-order spectrum through an angular range of 20°.
   Answer: 11,000 lines/cm.

20. Light containing a mixture of two wavelengths, 500 nm and 600 nm, is incident normally on a diffraction grating. It is desired (1) that the first and second principal maxima for each wavelength appear at $\theta \le 30°$, (2) that the dispersion be as high as possible, *and* (3) that the third order for 600 nm be a missing order. (a) What is the separation between adjacent slits? (b) What is the smallest possible individual slit width? (c) Name *all* orders for 600 nm that actually appear on the screen with the values chosen in (a) and (b).

21. A narrow beam of monochromatic light strikes a grating at normal incidence and produces sharp maxima at the following angles from the normal: 6°40′, 13°30′, 20°20′, 35°40′. No other maxima appear at any angle between 0° and 35°40′. The separation between adjacent slit centers in the grating is 5.04×10^{-4} cm. (a) Compute the wavelength of the light used. (b) Make the most complete *quantitative* statement that can be inferred from the above data concerning the width of each slit.
   Answer: (a) 586 nm. (b) The slit width lies between 1.01 and 1.68 μm.

22. A grating designed for use in the infrared region of the electromagnetic spectrum is "blazed" to concentrate all its intensity in the first order ($m = 1$) for $\lambda = 8,000$ nm. If visible light (400 nm $< \lambda <$ 700 nm) were allowed to fall on this grating, what visual appearance would the diffracted beams present?

23. Show that the dispersion of a grating can be written as

$$D = \frac{\tan \theta}{\lambda}.$$

24. A grating has 3500 rulings/cm and is illuminated at normal incidence by white light. A spectrum is formed on a screen 30 cm from the grating. If a 1.0-cm square hole is cut in the screen, its inner edge being 5.0 cm from the central maximum, what range of wavelengths passes through the hole?

25. Two spectral lines have wavelengths λ and $\lambda + \Delta\lambda$, respectively, where $\Delta\lambda \ll \lambda$. Show that their angular separation $\Delta\theta$ in a grating spectrometer is given approximately by $\Delta\theta = \Delta\lambda/\sqrt{(d/m)^2 - \lambda^2}$, where d is the separation of adjacent slit centers and m is the order at which lines are observed. Notice that the angular separation is greater in the higher orders.

26. An optical grating with a spacing $d = 15,000$ Å is used to analyze soft X-rays of wavelength $\lambda = 5.0$ Å. The angle of incidence θ is $90° - \gamma$, where γ is a *small* angle. The first-order maximum is found at an angle $\theta = 90° - 2\beta$. Find the value of β.

SECTION 47-4

27. A grating has 6000 rulings/cm and is 6.0 cm wide. (a) What is the smallest wavelength interval that can be resolved in the third order at $\lambda = 500$ nm? (b) How many higher orders can be seen? Assume normal incidence of light on the grating throughout.   *Answer:* (a) 4.6×10^{-3} nm. (b) None.

28. A grating has 40,000 rulings spread over 3.0 in. (a) What is its expected dispersion D for sodium light ($\lambda = 589$ nm $= 5890$ Å) in the first three orders? (b) What is its resolving power in these orders?

29. A source containing a mixture of hydrogen and deuterium atoms emits a red doublet at $\lambda = 656.3$ nm ($= 6563$ Å) whose separation is 0.18 nm ($= 1.8$ Å). Find the minimum number of lines needed in a diffraction grating which can resolve these lines in the first order.   *Answer:* 3650.

30. In a particular grating the sodium doublet (see Example 2) is viewed in third order at 10° to the normal and is barely resolved. Find (a) the grating spacing and (b) the total width of the rulings.

31. A diffraction grating has a resolving power $R = \lambda/\Delta\lambda = Nm$. (a) Show that the corresponding frequency range $\Delta\nu$ that can just be resolved is given by $\Delta\nu = c/Nm\lambda$. (b) From Fig. 47-1, show that the "times of flight" of the two extreme rays differ by an amount $\Delta t = Nd \sin\theta/c$. (c) Show that $(\Delta\nu)(\Delta t) = 1$, this relation being independent of the various grating parameters. Assume $N \gg 1$.

32. A diffraction grating is made up of slits of width 300 nm with a 900 nm separation between centers. The grating is illuminated by monochromatic plane waves, $\lambda = 600$ nm, the angle of incidence being zero. (a) How many diffracted lines are there? (b) What is the angular width of the spectral lines observed, if the grating has 1000 slits? (c) How is the angular width of the spectral lines related to the resolving power of the grating?

SECTION 47-5

33. Consider an infinite two-dimensional square lattice as in Fig. 47-14b. One interplanar spacing is obviously a_0 itself. (a) Calculate the next five smaller interplanar spacings by sketching figures similar to Figure 47-16a. (b) Show that your answers obey the general formula

$$d = a_0/\sqrt{h^2 + k^2}$$

where h, k are both relatively prime integers (no common factor other than unity).

SECTION 47-6

34. Monochromatic high energy x-rays are incident on a crystal. If first order reflection is observed at Bragg angle 3.4°, at what angle would second order reflection be expected?

35. Monochromatic X-rays are incident on a NaCl crystal whose lattice spacing is 0.3 nm. When the beam is rotated 60° from the normal, first-order Bragg reflection is observed. What is the wavelength of the x-rays?
Answer: 0.30 nm.

36. In comparing the wavelengths of two monochromatic X-ray lines, it is noted that line A gives a first-order reflection maximum at a glancing angle of 30° to the smooth face of a crystal. Line B, known to have a wavelength of 0.097 nm, gives a third-order reflection maximum at an angle of 60° from the same face of the same crystal. Find the wavelength of line A.

37. Monochromatic X-rays ($\lambda = 1.25$ Å) fall on a crystal of sodium chloride, making an angle of 45° with the reference line shown in Fig. 47-19. The planes shown are those of Fig. 47-16a, for which $d = 2.52$ Å. Through what angles must the crystal be turned to give a diffracted beam associated with the planes shown? Assume that the crystal is turned about an axis that is perpendicular to the plane of the page. Ignore the possibility (see Problem 40) that some of these beams may be of zero intensity.
Answer: 30.6°, 15.3°, (clockwise); 3.1°, 37.8°, (counterclockwise).

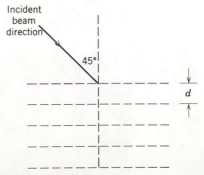

figure 47-19
Problems 37, 38

38. Assume that the incident X-ray beam in Fig. 47-19 is not monochromatic but contains wavelengths in a band from 0.095 to 0.130 nm. Will diffracted beams, associated with the planes shown, occur? Assume $d_0 = 0.275$ nm.

39. Prove that it is impossible to determine both wavelength of radiation and spacing of Bragg reflecting planes in a crystal by measuring the angles for Bragg reflection in several orders.

40. *Missing orders in X-ray diffraction.* In Example 5 the $m = 1$ beam, permitted by interference considerations, has zero intensity because of the diffracting properties of the unit cell for this geometry of beams and crystal. Prove this. (*Hint:* Show that the "reflection" from an atomic plane through the top of a layer of unit cells is canceled by a "reflection" from a plane through the middle of this layer of cells. All odd-order beams prove to have zero intensity.)

48
polarization

Light, like all electromagnetic radiation, is predicted by electromagnetic theory to be a *transverse wave*, the directions of the vibrating electric and magnetic vectors being at right angles to the direction of propagation instead of parallel to it, as in a longitudinal wave. The transverse waves of Figs. 48-1 and 41-13 have the additional characteristic that they are *plane-polarized.* This means that the vibrations of the **E** vector are parallel to each other for all points in the wave. At any such point the vibrating **E** vector and the direction of propagation form a plane, called the *plane of vibration;* in a plane-polarized wave all such planes are parallel.

The transverse nature of light waves cannot be deduced from the interference or diffraction experiments so far described because longitudinal waves such as sound waves also show these effects. Thomas Young in 1817 provided an experimental basis for believing that light waves are transverse. Two of his contemporaries, Dominique-François Arago (1786–1853) and Augustin Jean Fresnel (1788–1827), were able, by allowing a light beam to fall on a crystal of calcite, to produce two separate

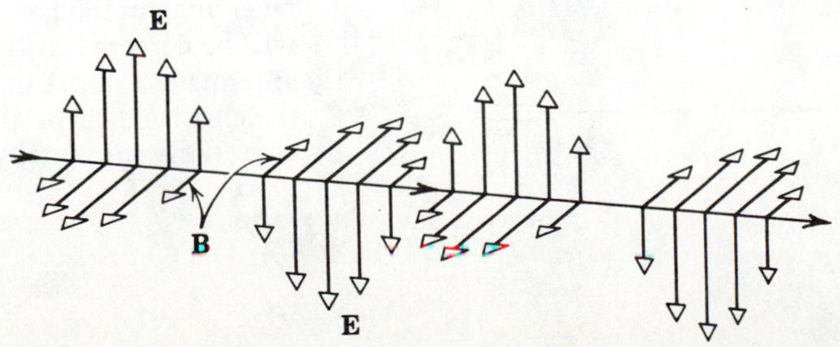

figure 48-1
An instantaneous "snapshot" of a plane-polarized wave showing the vectors **E** and **B** along a particular ray. The wave is moving to the right with speed *c*. The plane containing the vibrating **E** vector and the direction of propagation is a *plane of vibration.*

beams (see Section 48-4). Astonishingly, these beams, although coherent, produced no interference fringes but only a uniform illumination. Young deduced from this that light must be a transverse wave and that the planes of vibration in the two beams must be at right angles to each other. Wave disturbances that act at right angles to each other cannot show interference effects; you are asked to prove this in Problem 1. Young's words to Arago were these:

I have been reflecting on the possibility of giving an imperfect explanation of the affection of light which constitutes polarization without departing from the genuine doctrine of undulations. It is a principle in this theory that all undulations are simply propagated through homogeneous mediums in concentric spherical surfaces like the undulations of sound, consisting simply in the direct and retrograde motions of the particles in the direction of the radius with their concomitant condensation and rarefactions [that is, longitudinal waves]. And yet, it is possible to explain in this theory a transverse vibration, propagated also in the direction of the radius, and with equal velocity, the motions of the particles being in a certain constant direction with respect to that radius; and this is a *polarization*.

Note how Young presents the possibility of a transverse vibration as a novel idea, light having been generally — but incorrectly — assumed to be a longitudinal vibration.

In a plane-polarized transverse wave it is necessary to specify two directions, that of the wave disturbance (**E**, say) and that of propagation. In a longitudinal wave these directions are identical. In plane-polarized transverse waves, but not in longitudinal waves, we may thus expect a lack of symmetry about the direction of propagation. Electromagnetic waves in the radio and microwave range exhibit this lack of symmetry readily. Such a wave, generated by the surging of charge up and down in the dipole that forms the transmitting antenna of Fig. 48-2, has (at large distances from the dipole and at right angles to it) an electric field vector parallel to the dipole axis. When this plane-polarized wave falls on a second dipole connected to a microwave detector, the alternating electric component of the wave will cause electrons to surge back and forth in the receiving antenna, producing a reading on the detector. If we turn the receiving antenna through 90° about the direction of propagation, the detector reading drops to zero. In this orientation the electric field vector is not able to cause charge to move along the dipole axis because it points at right angles to this axis. We can reproduce the experiment of Fig. 48-2 by turning the receiving antenna of a television set (assumed an electric dipole type) through 90° about an axis that points toward the transmitting station.

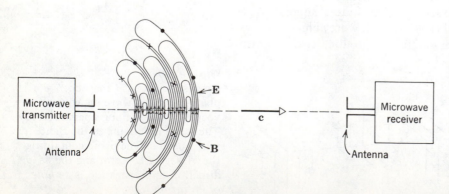

figure 48-2
The vectors **E** in the transmitted wave are parallel to the axis of the receiving antenna so that the wave will be detected. If the receiving antenna is rotated through 90° about the direction of propagation, essentially no signal will be detected.

Common sources of visible light differ from radio and microwave sources in that the elementary radiators, that is, the atoms and molecules, act independently. The light propagated in a given direction consists of independent wavetrains whose planes of vibration are randomly oriented about the direction of propagation, as in Fig. 48-3b. Such light, though still transverse, is *unpolarized.* The random orientation of the planes of vibration produces symmetry about the propagation direction, which, on casual study, conceals the true transverse nature of the waves. To study this transverse nature, we must find a way to unsort the different planes of vibration.

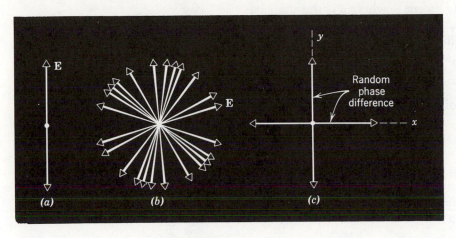

figure 48-3
(a) A plane-polarized transverse wave moving toward the reader, showing only the electric vector. (b) An unpolarized transverse wave viewed as a random superposition of many plane-polarized wavetrains. (c) A second, completely equivalent, description of an unpolarized transverse wave; here the unpolarized wave is viewed as two plane-polarized waves with a random phase difference. The orientation of the x and y axes about the propagation direction is completely arbitrary.

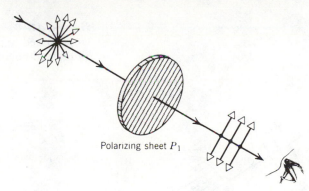

Polarizing sheet P_1

figure 48-4
A polarizing sheet produces plane-polarized light from unpolarized light. The parallel lines, which are not actually visible on the sheet, suggest the characteristic polarizing direction of the sheet.

Figure 48-4 shows unpolarized light falling on a sheet of commercial polarizing material called *Polaroid.** There exists in the sheet a certain characteristic polarizing direction, shown by the parallel lines. The sheet will transmit only those wave-train components whose electric vectors vibrate parallel to this direction and will absorb those that vibrate at right angles to this direction. The emerging light will be plane-polarized. This polarizing direction is established during the manufacturing process by embedding certain long-chain molecules in a flexible plastic sheet and then stretching the sheet so that the molecules are aligned parallel to each other. Polarizing sheets more than 2 ft wide and 100 ft long may be produced.

In Fig. 48-5 the polarizing sheet or *polarizer* lies in the plane of the

48-2
POLARIZING SHEETS

* There are other ways of producing polarized light without using this well-known commerical product. We mention some of them below. Also see "The Amateur Scientist" by Jearl Walker, *Scientific American*, December 1977 for ways of making polarizing sheets, quarter-wave and half-wave plates and doing various experiments with them.

page and the direction of propagation is into the page. The arrow **E** shows the plane of vibration of a randomly selected wavetrain falling on the sheet. Two vector components, E_x (of magnitude $E \sin \theta$) and E_y (of magnitude $E \cos \theta$), can replace **E**, one parallel to the polarizing direction and one at right angles to it. Only the former will be transmitted; the latter is absorbed within the sheet.

Let us place a second polarizing sheet P_2 (usually called, when so used, an *analyzer*) as in Fig. 48-6. If P_2 is rotated about the direction of propagation, there are two positions, 180° apart, at which the transmitted light intensity is almost zero; these are the positions in which the polarizing directions of P_1 and P_2 are at right angles.

If the amplitude of the plane-polarized light falling on P_2 is E_m, the amplitude of the light that emerges is $E_m \cos \theta$, where θ is the angle between the polarizing directions of P_1 and P_2. Recalling that the intensity of the light beam is proportional to the square of the amplitude, we see that the transmitted intensity I varies with θ according to

$$I = I_m \cos^2 \theta, \qquad (48\text{-}1)$$

in which I_m is the maximum value of the transmitted intensity. It occurs when the polarizing directions of P_1 and P_2 are parallel, that is, when $\theta = 0$ or 180°. Figure 48-7a, in which two overlapping polarizing sheets are in the parallel position ($\theta = 0$ or 180° in Eq. 48-1) shows that the light transmitted through the region of overlap has its maximum value. In Fig. 48-7b one or the other of the sheets has been rotated through 90° so that θ in Eq. 48-1 has the value 90 or 270°; the light transmitted through the region of overlap is now a minimum.

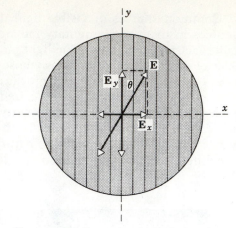

figure 48-5
A wavetrain **E** is equivalent to two component wavetrains $\mathbf{E}_y$ and $\mathbf{E}_x$. Only $\mathbf{E}_y$ is transmitted by the polarizer.

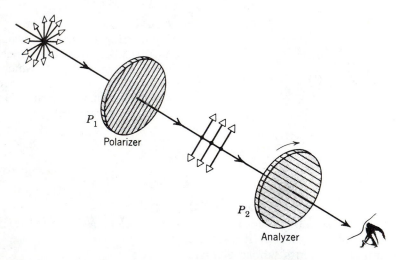

figure 48-6
Unpolarized light is not transmitted by crossed polarizing sheets.

P_1
Polarizer

P_2
Analyzer

Equation 48-1, called the law of Malus, was discovered by Étienne Louis Malus (1775–1812) experimentally in 1809, using polarizing techniques other than those so far described (see Section 48-3). Equation 48-1 describes precisely the lack of symmetry about the propagation direction that must be exhibited by plane-polarized transverse waves. Longitudinal waves could not possibly show such effects. Interestingly enough the human eye, under certain conditions, can detect polarized light.*

---

* The so-called *Haidinger's brushes;* the interested student is referred to *Concepts of Classical Optics* by John Strong, W. H. Freeman & Co., 1958.

figure 48-7
Two square sheets of Polaroid are laid over a book. In (a) the axes of polarization of the two sheets are parallel and light passes through both sheets. In (b) one sheet has been rotated 90° and no light passes through. The book is opened to an illustration of the Luxembourg Palace in Paris. Malus discovered the phenomenon of polarization by reflection while looking at sunlight reflected off the palace windows through a calcite crystal.

EXAMPLE 1

Two polarizing sheets have their polarizing directions parallel so that the intensity I_m of the transmitted light is a maximum. Through what angle must either sheet be turned if the intensity is to drop by one-half?

From Eq. 48-1, since $I = \frac{1}{2}I_m$, we have

$$\tfrac{1}{2}I_m = I_m \cos^2 \theta$$

or

$$\theta = \cos^{-1} \pm \frac{1}{\sqrt{2}} = \pm 45°, \pm 135°.$$

The same effect is obtained no matter which sheet is rotated or in which direction.

Historically polarization studies were made to investigate the nature of light. Today we reverse the procedure and deduce something about the nature of an object from the polarization state of the light emitted by, or scattered from, that object. It has been possible to deduce, from studies of the polarization of light reflected from them, that the grains of cosmic dust present in our galaxy have been oriented in the weak galactic magnetic field ($\sim 5 \times 10^{-10}$ T) so that their long dimension is parallel to this field. Polarization studies have shown that Saturn's rings consist of ice crystals. The size and shape of virus particles can be determined by the polarization of ultraviolet light scattered from them. Much useful information about the structure of atoms and nuclei is gained from polarization studies of their emitted radiations in all parts of the electromagnetic spectrum. Thus we have a useful research technique for structures ranging in size from a galaxy ($\sim 10^{+?}$ m) to a nucleus ($\sim 10^{-15}$ m). Polarized light also has many practical applications in industry and in engineering science.

Malus discovered in 1809 that light can be partially or completely polarized by reflection. Anyone who has watched the sun's reflection in water, while wearing a pair of sunglasses made of polarizing sheet, has probably noticed the effect. It is necessary only to tilt the head from side to side, thus rotating the polarizing sheets, to observe that the intensity of the reflected sunlight passes through a minimum.

48-3
POLARIZATION BY REFLECTION

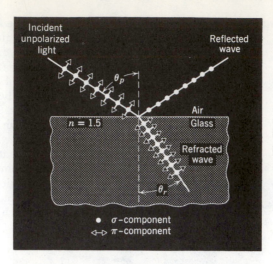

figure 48-8
For a particular angle of incidence θ_p, the reflected light is completely polarized, as shown. The transmitted light is partially polarized.

Figure 48-8 shows an unpolarized beam falling on a glass surface. The **E** vector for each wavetrain in the beam can be resolved into two components, one perpendicular to the plane of incidence—which is the plane of Fig. 48-8—and one lying in this plane. The first component, represented by the dots, is called the *σ-component*, from the German *senkrecht*, meaning perpendicular. The second component, represented by the arrows, is called the *π-component* (for *parallel*). On the average, for completely unpolarized incident light, these two components are of equal amplitude.

Experimentally, for glass or other dielectric materials, there is a particular angle of incidence, called the *polarizing angle* θ_p, at which the reflection coefficient for the π-component is zero. This means that the beam reflected from the glass, although of low intensity, is plane-polarized, with its plane of vibration at right angles to the plane of incidence. This polarization of the reflected beam can easily be verified by analyzing it with a polarizing sheet.

The π-component at the polarizing angle is entirely refracted; the

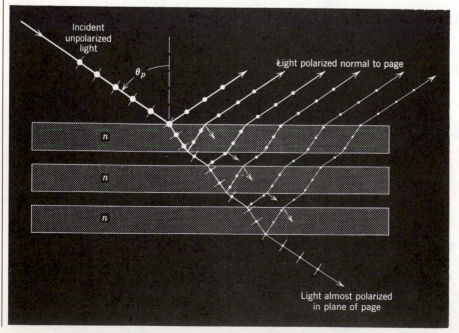

figure 48-9
Polarization of light by a stack of glass plates. Unpolarized light is incident on a stack of glass plates at Brewster's angle θ_p. (Polarization in the plane of the page is shown by the short lines and polarization normal to the page by the dots.) All light reflected out of the original ray is polarized normal to the page. After passing through several reflecting interfaces, the transmitted light no longer contains any appreciable component polarized normal to the page.

σ-component is only partially refracted. Thus the transmitted beam, which is of high intensity, is only partially polarized. By using a stack of glass plates rather than a single plate, we obtain reflections from successive surfaces and we can increase the intensity of the emerging reflected (σ-component) beam (see Fig. 48-9). By the same token, the σ-components are progressively removed from the transmitted beam, making it more completely π-polarized.

At the polarizing angle it is found experimentally that the reflected and the refracted beams are at right angles, or (Fig. 48-8)

$$\theta_p + \theta_r = 90°.$$

From Snell's law, $\quad\quad n_1 \sin \theta_p = n_2 \sin \theta_r.$

Combining these equations leads to

$$n_1 \sin \theta_p = n_2 \sin (90° - \theta_p) = n_2 \cos \theta_p$$

or $\quad\quad\quad \tan \theta_p = \dfrac{n_2}{n_1},$ $\quad\quad\quad\quad$ (48-2)

where the incident ray is in medium 1 and the refracted ray in medium 2. This can be written as

$$\tan \theta_p = n, \quad\quad\quad\quad (48\text{-}3)$$

where $n \ (= n_2/n_1)$ is the index of refraction of medium 2 with respect to medium 1. Equation 48-3 is known as *Brewster's law* after Sir David Brewster (1781–1868), who deduced it empirically in 1812. It is possible to prove this law rigorously from Maxwell's equations (see also Question 6).

EXAMPLE 2

We wish to use a plate of glass ($n = 1.50$) as a polarizer. What is the polarizing angle? What is the angle of refraction?

From Eq. 48-3,

$$\theta_p = \tan^{-1} 1.50 = 56.3°.$$

The angle of refraction follows from Snell's law:

$$(1) \ \sin \theta_p = n \sin \theta_r$$

or $\quad\quad\quad \sin \theta_r = \dfrac{\sin 56.3°}{1.50} = 0.555 \quad\quad \theta_r = 33.7°.$

48-4
DOUBLE REFRACTION

In earlier chapters we assumed that the speed of light, and thus the index of refraction, is independent of the direction of propagation in the medium and of the state of polarization of the light. Liquids, amorphous solids such as glass, and crystalline solids having cubic symmetry normally show this behavior and are said to be *optically isotropic*. Many other crystalline solids are optically *anisotropic* (that is, not isotropic).*

Solids may be anisotropic in many properties. Mica cleaves readily in

---

* Many transparent amorphous solids such as glasses and plastics become optically anisotropic when they are mechanically stressed. This fact is useful in engineering design studies in that strains in gears, bridge structures, etc., can be studied quantitatively by building plastic models, stressing them appropriately, and examining the optical anisotropy that results, using polarization techniques. The interested reader should consult "Photoelasticity," a chapter by H. T. Jessop in Vol. 6 of the *Encyclopedia of Physics*, edited by H. Flugge (1958), Springer Verlag, Berlin.

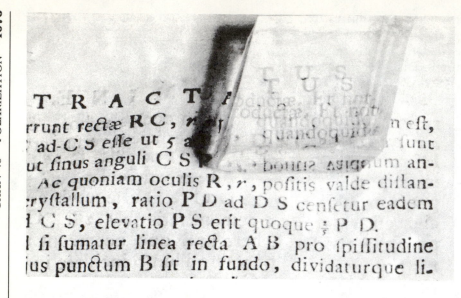

figure 48-10
Two images, one polarized 90°
relative to the other, are formed by a
calcite crystal. The book on which
the crystal is lying is Huygens' *Opera
Reliqua*, wherein the phenomenon
of double refraction is discussed.

one plane only; a cube of crystalline graphite does not have the same
electric resistance between all pairs of opposite faces; a cube of crystalline nickel magnetizes more readily in certain directions than in others,
etc. If a solid is a mixture of a large number of tiny crystallites, it may
appear to be isotropic because of the random orientations of the crystallites. Powdered mica, for example, compacted to a solid mass with a
binder, does not exhibit the cleavage properties that characterize the
crystallites making it up.

Figure 48-10, in which a polished crystal of calcite ($CaCO_3$) is laid
over some printed letters, shows the optical anisotropy of this material;
the image appears double. Figure 48-11 shows a beam of unpolarized
light falling on a calcite crystal at right angles to one of its faces. The

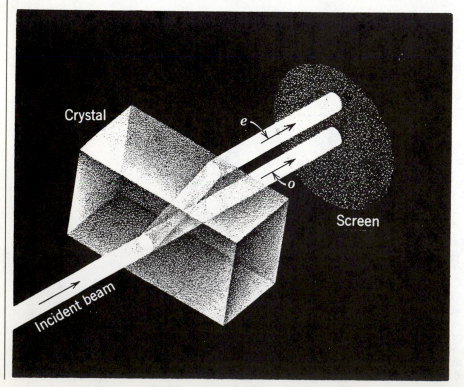

figure 48-11
A beam of unpolarized light falling
on a calcite crystal is split into two
beams which are polarized at right
angles to each other.

single beam splits into two at the crystal surface. The "double-bending" of a beam transmitted through calcite, exhibited in Figs. 48-10 and 48-11, is called *double refraction.*

If the two emerging beams in Fig. 48-11 are analyzed with a polarizing sheet, they are found to be plane-polarized with their planes of vibration at right angles to each other, a fact discovered by Huygens in 1678. Huygens used a second calcite crystal to investigate the polarization states of the beams labeled *o* and *e* in the figure.

If experiments are carried out at various angles of incidence, one of the beams in Fig. 48-11 (represented by the *ordinary ray*, or *o*-ray) will be found to obey Snell's law of refraction at the crystal surface, just like a ray passing from one isotropic medium into another. The second beam (represented by the *extraordinary ray*, or *e*-ray) will not. In Fig. 48-11, for example, the angle of incidence for the incident light is zero but the angle of refraction of the *e*-ray, contrary to the prediction of Snell's law, is not. In general, the *e*-ray does not even lie in the plane of incidence.

This difference between the waves represented by the *o*- and *e*-rays with respect to Snell's law can be explained in these terms:

1. The *o*-wave travels in the crystal with the same speed v_o in all directions. In other words, the crystal has, for this wave, a single index of refraction n_o, just like an isotropic solid.

2. The *e*-wave travels in the crystal with a speed that varies with direction from v_o to a larger value (for calcite) v_e. In other words, the index of refraction, defined as c/v, varies with direction from n_o to a smaller value (for calcite) n_e.

The quantities n_o and n_e are called the *principal indices of refraction* for the crystal. Problem 12 suggests how to measure them. Table 48-1 shows these indices for six doubly refracting crystals. For three of them the *e*-wave is slower; for the other three it is faster than the *o*-wave. Some doubly refracting crystals (mica, topaz, etc.) are more complex optically than calcite and require *three* principal indices of refraction for a complete description of their optical properties. Crystals whose basic crystal structure is cubic (see Fig. 47-13) are optically isotropic, requiring only *one* index of refraction.

Table 48-1
Principal indices of refraction of several doubly refracting crystals (for sodium light, $\lambda = 589$ nm $= 5890$ Å)

Crystal	Formula	n_o	n_e	$n_e - n_o$
Ice	H_2O	1.309	1.313	+0.004
Quartz	SiO_2	1.541	1.553	+0.012
Wurzite	ZnS	2.356	2.378	+0.022
Calcite	$CaCO_3$	1.658	1.486	0.172
Dolomite	$CaO \cdot MgO \cdot 2CO_2$	1.681	1.500	−0.181
Siderite	$FeO \cdot CO_2$	1.875	1.635	−0.240

The behavior for the speeds of the two waves traveling in calcite is summarized by Fig. 48-12, which shows two wave surfaces spreading out from an imaginary point light source S imbedded in the crystal. The *o*-wave surface is a sphere, as we would expect if the medium were isotropic. The *e*-wave surface is an ellipsoid of revolution about a characteristic direction in the crystal called the *optic axis*. The two wave

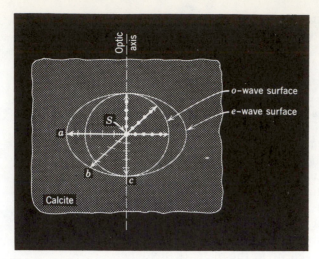

figure 48-12
Huygens' wave surfaces generated by a point source S imbedded in calcite. The polarization states for three o-rays and three e-rays are shown by the dots and lines, respectively. Note that in general (ray Sb) the bars representing the polarization direction are not perpendicular to the e-rays.

surfaces represent light having two different polarization states. If we consider for the present only rays lying in the plane of Fig. 48-12, then (a) the plane of polarization for the o-rays is perpendicular to the figure, as suggested by the dots, and (b) that for the e-rays coincides with the plane of the figure, as suggested by the short lines. We describe the polarization states more fully at the end of this section.

Figure 48-13, which shows a typical calcite crystal that may be obtained by cleavage from a naturally occurring crystal, shows how to locate the optic axis. The edges of calcite crystals may have any lengths but the angles at which the edges intersect always have one or another of two values, 78° 05' or 101° 55'. The optic axis is found by erecting a line at either of the two corners where three obtuse angles meet (the "blunt" corners), making equal angles with the crystal edges. *Any line in the crystal parallel to this line is also an optic axis.*

We can use Huygens' principle to study the propagation of light waves in doubly refracting crystals. Figure 48-14a shows the special case in which unpolarized light falls at normal incidence on a calcite slab cut from a crystal in such a way that the optic axis is normal to the surface. Consider a wavefront that, at time $t = 0$, coincides with the crystal surface. Following Huygens, we may let any point on this surface serve as a radiating center for a double set of Huygens' wavelets, such as those in Fig. 48-12. The plane of tangency to these wavelets represents the new position of this wavefront at a later time t. The incident beam in Fig. 48-14a is propagated through the crystal without deviation at speed v_o. The beam emerging from the slab will have the same polarization character as the incident beam. The calcite slab, in these special circumstances only, behaves like an isotropic material, and no distinction can be made between the o- and the e-waves.

Figure 48-14b shows two views of another special case, namely, unpolarized incident light falling at right angles on a slab cut so that the optic axis is parallel to its surface. In this case too the incident beam is propagated without deviation. However, we can now identify o- and e-waves that travel through the crystal with different speeds, v_o and v_e, respectively. These waves are polarized at right angles to each other.

Some doubly refracting crystals have the interesting property, called *dichroism*, in which one of the polarization components is strongly absorbed within the crystal, the other being transmitted with little loss. Dichroism, illustrated in Fig. 48-15, is the basic operating principle of the commercial Polaroid sheet. The

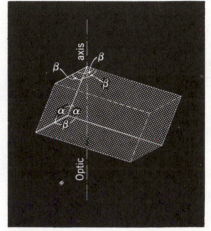

figure 48-13
A calcite crystal; α is 78° 05'; β is 101° 55'.

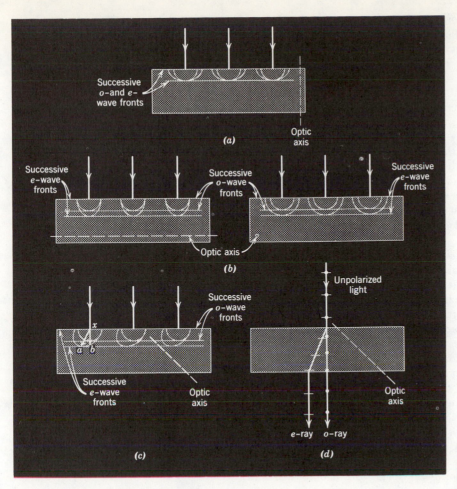

figure 48-14
Unpolarized light falls at normal incidence on slabs cut from a calcite crystal. The Huygens' wavelets are appropriate sections of the figure of revolution about the optic axis represented by Fig. 48-13. (a) No double refraction or speed difference occurs. (b) No double refraction occurs but there is a speed difference. (c) Both double refraction and a speed difference occur. (d) Same as (c) but showing the polarization states and the emerging rays.

many small crystallites, imbedded in a plastic sheet with their optic axes parallel, have a polarizing action equivalent to that of a single large crystal slab.

Figure 48-14c shows unpolarized light falling at normal incidence on a calcite slab cut so that its optic axis makes an arbitrary angle with the crystal surface. Two spatially separated beams are produced, as in Fig. 48-11. They travel through the crystal at different speeds, that for the o-wave being v_o and that for the e-wave being intermediate between v_o and v_e. Note that ray xa represents the shortest *optical* path for the transfer of light energy from point x to the e-wavefront. Energy transferred along any other ray, in particular along ray xb, would have a longer transit time, a consequence of the fact that the speed of e-waves varies with direction.* Figure 48-14d represents the same case as Fig. 48-14c. It shows the rays emerging from the slab, as in Fig. 48-11, and makes clear that the emerging beams are polarized at right angles to each other, that is, they are *cross-polarized*.

We now seek to understand, in terms of the atomic structure of optically anisotropic crystals, how cross-polarized light waves with different speeds can exist. Light is propagated through a crystal by the action of the vibrating **E** vectors of the wave on the electrons in the crystal. These electrons, which experience electrostatic restoring forces if they are moved from their equilibrium positions, are set into forced periodic oscillation about these positions and pass along the transverse wave disturbance that constitutes the light wave. The strength of the

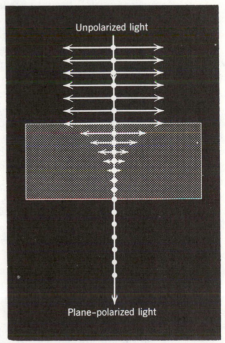

figure 48-15
Showing the absorption of one polarization component inside a dichroic crystal of the type used in Polaroid sheets.

* The reader who has not previously read Section 43-6 on Fermat's principle may wish to do so now.

restoring forces may be measured by a force constant k, as for the simple harmonic oscillator discussed in Chapter 15 (see Eq. 15-4).

In optically isotropic materials the force constant k is the same for all directions of displacement of the electrons from their equilibrium positions. In doubly refracting crystals, however, k varies with direction. For electron displacements that lie in a plane at right angles to the optic axis k has the constant value k_o, no matter how the displacement is oriented in this plane. For displacements parallel to the optic axis, k has the larger value (for calcite) k_e.* Note carefully that the speed of a wave in a crystal is determined by the direction in which the **E** vectors vibrate and *not* by the direction of propagation. It is the transverse **E**-vector vibrations that call the restoring forces into play and thus determine the wave speed. Note too that the stronger the restoring force, that is, the larger k, the faster the wave. For waves traveling along a stretched cord, for example, the restoring force for the transverse displacements is determined by the tension F in the cord. Equation 19-12 shows that an increase in F means an increase in the wave speed v.

Figure 48-16, a long weighted "tire chain" supported at its upper end, provides a one-dimensional mechanical analogy for double refraction. It applies specifically to o- and e-waves traveling at right angles to the optic axis, as in Fig. 48-14b. If the supporting block is oscillated, as in Fig. 48-16a, a transverse wave travels along the chain with a certain speed. If the block is oscillated lengthwise, as in Fig. 48-16b, another transverse wave is also propagated. The restoring force for the second wave is greater than for the first, the chain being more rigid in the plane of Fig. 48-16b than in the plane of Fig. 48-16a. Thus the second wave travels along the chain with a greater speed.

In the language of optics we would say that the speed of a transverse wave in the chain depends on the orientation of the plane of vibration of the wave. If we oscillate the top of the chain in a random way, the wave disturbance at a point along the chain can be described as the sum of two waves, polarized at right angles and traveling with different speeds. This corresponds exactly to the optical situation of Fig. 48-14b.

For waves traveling parallel to the optic axis, as in Fig. 48-14a, or for waves in optically isotropic materials, the appropriate mechanical analogy is a single weighted hanging chain. Here there is only one speed of propagation, no matter how the upper end is oscillated. The restoring forces are the same for all orientations of the plane of polarization of waves traveling along such a chain.

These considerations allow us to understand more clearly the polarization states of the light represented by the double-wave surface of Fig. 48-12. For the (spherical) o-wave surface, the **E**-vector vibrations must be everywhere at right angles to the optic axis. If this is so, the same force constant k_o will always hold, and the o-waves will travel with the same speed in all directions. More specifically, if we draw a ray in Fig. 48-12 from S to the o-wave surface, considered three-dimensionally (that is, as a sphere), the **E**-vector vibrations will always be at right angles to the plane defined by this ray and the optic axis. Thus these vibrations will always be at right angles to the optic axis.

For the (ellipsoidal) e-wave surface, the **E**-vector vibrations in general have a component parallel to the optic axis. For rays such as Sa in Fig. 48-12 or for the e-rays of Fig. 48-14b, the vibrations are completely parallel to this axis. Thus a relatively strong force constant (in calcite) k_e is operative, and the wave speed v_e will be relatively high. For e-rays, such as Sb in Fig. 48-12, the parallel component of the **E**-vector vibrations is less than 100%, so that the corresponding wave speed will be less than v_e. For ray Sc, in Fig. 48-12, the parallel component is zero, and the distinction between o- and e-rays disappears.

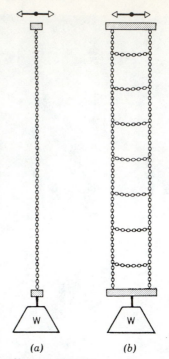

figure 48-16
Two views of a one-dimensional mechanical model for double refraction.

---

* For doubly refracting crystals with $n_e > n_o$ (see Table 48-1) k for displacements parallel to the optic axis is *smaller* than for those at right angles to it. Also, for crystals with three principal indices of refraction, there will be three principal force constants. Such crystals have two optic axes and are called *biaxial*. The crystals listed in Table 48-1 have only a single optic axis and are called *uniaxial*.

Let plane-polarized light of angular frequency ω $(=2\pi\nu)$ fall at normal incidence on a slab of calcite cut so that the optic axis is parallel to the face of the slab, as in Fig. 48-17. The two waves that emerge will be plane-polarized at right angles to each other, and, if the incident plane of vibration is at 45° to the optic axis, they will have equal amplitudes. Since the waves travel through the crystal at different speeds, there will be a phase difference ϕ between them when they emerge from the crystal. If the crystal thickness is chosen so that (for a given frequency of light) $\phi = 90°$, the slab is called a *quarter-wave plate*. The emerging light is said to be *circularly polarized*.

In Section 15-7 we saw that the two emerging plane-polarized waves just described (vibrating at right angles with a 90° phase difference) can be represented as the projections on two perpendicular axes of a vector rotating with angular frequency ω about the propagation direction. These two descriptions of circularly polarized light are completely equivalent. Figure 48-18 clarifies the relationship between these two descriptions.

48-5
CIRCULAR POLARIZATION

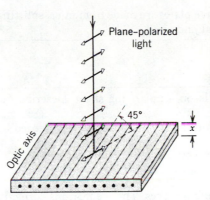

figure 48-17
Plane-polarized light falls on a doubly refracting slab of thickness x cut with its optic axis parallel to the surface. The plane of vibration of the incident light is oriented to make an angle of 45° with the optic axis.

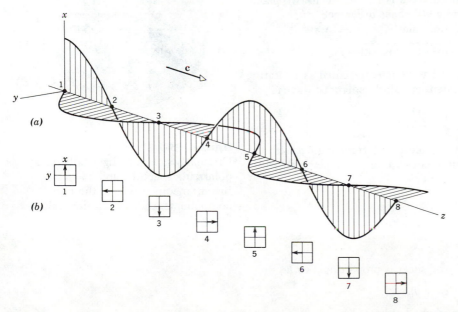

figure 48-18
(a) Two plane-polarized waves of equal amplitude and at right angles to each other are moving in the z direction. They differ in phase by 90°; where one wave has maximum values, the other is zero. (b) Views of the resultant amplitude of the approaching wave as seen by observers located at the positions shown on the z axis. Note that the resultant vector rotates clockwise with time.

A quartz quarter-wave plate is to be used with sodium light ($\lambda = 589$ nm $= 5890$ Å). What must its thickness be?

Two waves travel through the slab at speeds corresponding to the two principal indices of refraction given in Table 48-1 ($n_e = 1.553$ and $n_o = 1.541$). If the

EXAMPLE 3

crystal thickness is x, the number of wavelengths of the first wave contained in the crystal is

$$N_e = \frac{x}{\lambda_e} = \frac{xn_e}{\lambda},$$

where λ_e is the wavelength of the e-wave in the crystal and λ is the wavelength in air. For the second wave the number of wavelengths is

$$N_o = \frac{x}{\lambda_o} = \frac{xn_o}{\lambda},$$

where λ_o is the wavelength of the o-wave in the crystal. The difference $N_e - N_o$ must be one-fourth, or

$$\frac{1}{4} = \frac{x}{\lambda} (n_e - n_o).$$

This equation yields

$$x = \frac{\lambda}{4(n_e - n_o)} = \frac{589 \text{ nm}}{(4)(1.553 - 1.541)} = 0.012 \text{ mm}.$$

This plate is rather thin; most quarter-wave plates are made from mica, splitting the sheet to the correct thickness by trial and error.

EXAMPLE 4

A beam of circularly polarized light falls on a polarizing sheet. Describe the emerging beam.

The circularly polarized light, as it enters the sheet, can be represented by

$$E_x = E_m \sin \omega t$$

and

$$E_y = E_m \cos \omega t,$$

where x and y represent arbitrary perpendicular axes. These equations correctly represent the fact that a circularly polarized wave is equivalent to two plane-polarized waves with equal amplitude and a 90° phase difference.

The resultant amplitude in the incident circularly polarized wave is

$$E_{cp} = \sqrt{E_x{}^2 + E_y{}^2} = \sqrt{E_m{}^2 (\sin^2 \omega t + \cos^2 \omega t)} = E_m,$$

an expected result if the circularly polarized wave is represented as a rotating vector. The resultant intensity in the incident circularly polarized wave is proportional to $E_m{}^2$, or

$$I_{cp} \propto E_m{}^2. \tag{48-4}$$

Let the polarizing direction of the sheet make an arbitrary angle θ with the x axis as shown in Fig. 48-19. The instantaneous value of the plane-polarized wave transmitted by the sheet is

$$E = E_y \sin \theta + E_x \cos \theta$$

$$= E_m \cos \omega t \sin \theta + E_m \sin \omega t \cos \theta$$

$$= E_m \sin (\omega t + \theta).$$

The intensity of the wave transmitted by the sheet is proportional to E^2, or

$$I \propto E_m{}^2 \sin^2 (\omega t + \theta).$$

The eye and other measuring instruments respond only to the average intensity $\bar{I}$, which is found by replacing $\sin^2 (\omega t + \theta)$ by its average value over one or more cycles ($= \frac{1}{2}$), or

$$\bar{I} \propto \tfrac{1}{2} E_m{}^2.$$

Comparison with Eq. 48-4 shows that inserting the polarizing sheet reduces the

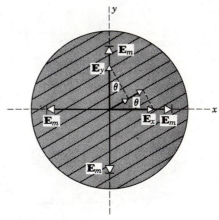

figure 48-19
Circularly polarized light falls on a polarizing sheet. $\mathbf{E}_x$ and $\mathbf{E}_y$ are instantaneous values of the two components, their maximum values being $\mathbf{E}_m$.

intensity by one-half. The orientation of the sheet makes no difference, since θ does not appear in this equation; this is to be expected if circularly polarized light is represented by a rotating vector, all azimuths about the propagation direction being equivalent. Inserting a polarizing sheet in an *unpolarized* beam has just the same effect, so that a simple polarizing sheet cannot be used to distinguish between unpolarized and circularly polarized light.

EXAMPLE 5

A beam of light is thought to be circularly polarized. How may this be verified?

Insert a quarter-wave plate. If the beam is circularly polarized, the two components will have a phase difference of 90° between them. The quarter-wave plate will introduce a further phase difference of ±90° so that the emerging light will have a phase difference of either zero or 180°. In either case the light will now be *plane-polarized* and can be made to suffer complete extinction by rotating a polarizer in its path.

Does the quarter-wave plate have to be oriented in any particular way to carry out this test?

EXAMPLE 6

A plane-polarized light wave of amplitude E_0 falls on a calcite quarter-wave plate with its plane of vibration at 45° to the optic axis of the plate, which is taken as the y axis; see Fig. 48-20. The emerging light will be circularly polarized. In what direction will the rotating electric vector appear to rotate? The direction of propagation is out of the page.

The wave component whose vibrations are parallel to the optic axis (the e-wave) can be represented as it emerges from the plate as

$$E_y = (E_0 \cos 45°) \sin \omega t = \frac{1}{\sqrt{2}} E_0 \sin \omega t = E_m \sin \omega t.$$

The wave component whose vibrations are at right angles to the optic axis (the o-wave) can be represented as

$$E_x = (E_0 \sin 45°) \sin (\omega t - 90°) = -\frac{1}{\sqrt{2}} E_0 \cos \omega t = -E_m \cos \omega t,$$

the 90° phase shift representing the action of the quarter-wave plate. Note that E_x reaches its maximum value one-fourth of a cycle *later* than E_y does, for, in calcite, wave E_x (the o-wave) travels *more slowly* than wave E_y (the e-wave).

To decide the direction of rotation, let us locate the tip of the rotating electric vector at two instants of time, (a) $t = 0$ and (b) a short time t_1 later chosen so that ωt_1 is a small angle. At $t = 0$ the coordinates of the tip of the rotating vector (see Fig. 48-20a) are

$$E_y = 0 \quad \text{and} \quad E_x = -E_m.$$

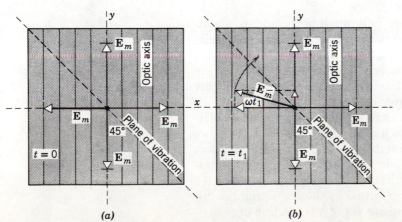

(a)   (b)

figure 48-20
Plane-polarized light falls from behind on a quarter-wave plate oriented so that the light emerging from the page is circularly polarized. In this case the electric vector $\mathbf{E}_m$ rotates clockwise as seen by an observer facing the light source.

At $t = t_1$ these coordinates become, approximately,

$$E_y = E_m \sin \omega t_1 \cong E_m(\omega t_1)$$

$$E_x = -E_m \cos \omega t_1 \cong -E_m.$$

Figure 48-20b shows that the vector which represents the emerging circular polarized light is rotating clockwise; by convention such light is called *right-circularly polarized*, the observer always being considered to face the light source.

If the plane of vibration of the incident light in Fig. 48-20 is rotated through $\pm 90°$, the emerging light will be *left-circularly polarized*.

48-6
ANGULAR MOMENTUM OF LIGHT

That light waves can deliver *linear momentum* to an absorbing screen or to a mirror is in accord with classical electromagnetism, with quantum physics, and with experiment. The facts of circular polarization suggest that light so polarized might also have *angular* momentum associated with it. This is indeed the case; once again the prediction is in accord with classical electromagnetism and with quantum physics. Experimental proof was provided in 1936 by Richard A. Beth, who showed that when circularly polarized light is produced in a doubly refracting slab, the slab experiences a reaction torque.

The angular momentum carried by light plays a vital role in understanding the emission of light from atoms and of γ-rays from nuclei. If light carries away angular momentum as it leaves the atom, the angular momentum of the residual atom must change by exactly the amount carried away; otherwise the angular momentum of the isolated system *atom plus light* will not be conserved.

Classical and quantum theory both predict that if a beam of circularly polarized light is completely absorbed by an object on which it falls, an angular momentum given by

$$L = \frac{U}{\omega} \tag{48-5}$$

is transferred to the object, where U is the amount of absorbed energy and ω the angular frequency of the light. You should verify that the dimensions in Eq. 48-5 are consistent.

48-7
SCATTERING OF LIGHT

A light wave, falling on a transparent solid, causes the electrons in the solid to oscillate periodically in response to the time-varying electric vector of the incident wave. The wave that travels through the medium is the resultant of the incident wave and of the radiations from the oscillating electrons. The resultant wave has a maximum intensity in the direction of the incident beam, falling off rapidly on either side. The lack of sideways scattering, which would be essentially complete in a large "perfect" crystal, comes about because the oscillating charges in the medium act cooperatively or coherently.

When light passes through a gas, we find much more sideways scattering. The oscillating electrons in this case, being separated by relatively large distances and not being bound together in a rigid structure, act independently rather than cooperatively. Thus the rigid cancellation of wave disturbances that are not in the forward direction is less likely to occur; there is more sideways scattering.

Light scattered sideways from a gas can be wholly or partially polarized, even though the incident light is unpolarized. Figure 48-21 shows an unpolarized beam moving upward on the page and striking a gas atom at *a*. The electrons at *a* will oscillate in response to the electric components of the incident wave, their motion being equivalent to two oscillating dipoles whose axes are represented by the arrow and the dot at *a*. An oscillating dipole does not radiate along its own line of action.

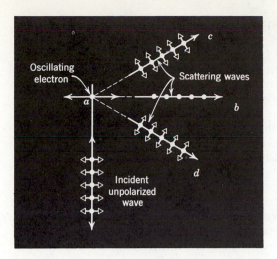

figure 48-21
Light is polarized either partially (c and d) or completely (b) by scattering from a gas molecule at a.

Thus an observer at b would receive no radiation from the dipole represented by the arrow at a. The radiation reaching him would come entirely from the dipole represented by the dot at a; thus this radiation would be plane-polarized, the plane of vibration passing through the line ab and being normal to the page.

Observers at c and d would detect partially polarized light, since the dipole represented by the arrow at a would radiate somewhat in these directions. Observers viewing the transmitted or the back-scattered light would not detect any polarization effects because both dipoles at a would radiate equally in these two directions.

A familiar example is the scattering of sunlight by the molecules of the earth's atmosphere. If the atmosphere were not present, the sky would appear black except when we looked directly at the sun. This has been verified by measurements made above the atmosphere in manned and unmanned space vehicles. We can easily check with a polarizer that the light from the cloudless sky is at least partially polarized. This fact is used in polar exploration in the so-called *solar compass.* In this device we establish direction by noting the nature of the polarization of the scattered sunlight. As is well known, magnetic compasses are not useful in these regions. It has been learned* that bees orient themselves in their flights between their hive and the pollen sources by means of polarization of the light from the sky; bees' eyes contain built-in polarization-sensing devices.

It still remains to be explained why the light scattered from the sky is predominantly blue and why the light received directly from the sun—particularly at sunset when the length of the atmosphere that it must traverse is greatest—is red. The cross section of an atom or molecule for light scattering depends on the wavelength, blue light being scattered more effectively than red light. Since the blue light is largely scattered, the transmitted light will have the color of normal sunlight with the blues largely removed; it is therefore more reddish in appearance.

The fact that the scattering cross section for blue light is higher than that for red light can be made reasonable. An electron in an atom or molecule is bound there by strong restoring forces. It has a definite natural frequency, like a small mass suspended in space by an assembly of springs. The natural frequency for

* See "Polarized Light and Animal Navigation" by Talbot H. Watermann in *Scientific American*, July 1955; *Bees: Their Vision, Chemical Sense, and Language* by K. von Frisch, Cornell University Press, 1950; and "Polarized-Light Navigation by Insects" by Rüdiger Wehner, *Scientific American*, July 1976.

electrons in atoms and molecules is usually in a region corresponding to violet or ultraviolet light.

When light is allowed to fall on such bound electrons, it sets up forced oscillations at the frequency of the incident light beam. In mechanical resonant systems it is possible to "drive" the system most effectively if we impress on it an external force whose frequency is as close as possible to that of the natural resonant frequency. In the case of light the blue is closer to the natural resonant frequency of the bound electron than is the red light. Therefore, we would expect the blue light to be more effective in causing the electron to oscillate, and thus it will be more effectively scattered.

When X-rays were discovered in 1898, there was much speculation whether they were waves or particles. In 1906 they were established as transverse waves by Charles Glover Barkla (1877–1944) by means of a polarization experiment.

When the unpolarized X-rays strike scattering block S_1 in Fig. 48-22, they set the electrons into oscillatory motion. The considerations of the preceding section require that the X-rays scattered toward the second block be plane-polarized as shown in the figure. Let this wave be scattered from the second scattering block, and let us examine the radiation scattered from it by rotating a detector D in a plane at right angles to the line joining the blocks. The electrons will oscillate parallel to each other, and the positions of maximum and zero intensity will be as shown. A plot of detector reading as a function of the angle ϕ supports the hypothesis that X-rays are transverse waves. If the X-rays were a stream of particles or a longitudinal wave, these effects could by no means be so readily understood. Thus Barkla's important experiment established that X-rays are a part of the electromagnetic spectrum.

48-8
DOUBLE SCATTERING

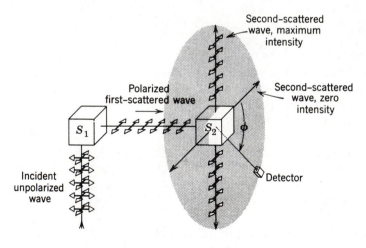

figure 48-22
A double-scattering experiment used by Barkla to show that X-rays are transverse waves.

In later studies you will learn that beams of particles such as electrons, protons, and pions can be viewed as waves. Scattering (including double scattering) techniques are often used to investigate the polarization characteristics of such beams.*

questions

1. It is said that light from ordinary sources is unpolarized. Can you think of any common sources that emit polarized light?
2. Why do sunglasses made of polarizing materials have a marked advantage over those that simply depend on absorption effects?
3. Unpolarized light falls on two polarizing sheets so oriented that no light is

---

*See "Polarized Accelerator Targets" by Gilbert Shapiro, *Scientific American*, July, 1976.

transmitted. If a third polarizing sheet is placed between them, can light be transmitted?

4. 3-D movies were very popular in the early 1950's. Viewers wore polarizing glasses and a polarizing sheet was placed in front of each of the *two* projectors needed. Explain how the system worked. Can you suggest any problems that may have led to the early abandonment of the system?

5. A wire grid, consisting of an array of wires arranged parallel to one another, can polarize an incident unpolarized beam of electromagnetic waves that pass through it. Explain the facts that (a) the diameter of the wires and the spacing between them must be much less than the incident wavelength to obtain effective polarization and (b) the transmitted component is the one whose electric vector oscillates in a direction perpendicular to the wires.

6. Brewster's law determines the polarizing angle on reflection from a dielectric. A plausible interpretation for zero reflection of the π-component at that angle is that the charges in the dielectric are caused to oscillate parallel to the reflected ray by this component and produce no radiation in this direction. Explain this and comment on the plausibility.

7. Can polarization by reflection occur if the light is incident on the interface from the side with the higher index of refraction (glass to air, for example)?

8. Devise a way to identify the polarizing direction of a sheet of Polaroid.

9. Is the optic axis of a doubly refracting crystal simply a line or a direction in space? Has it a direction sense, like an arrow? What about the characteristic direction of a polarizing sheet?

10. If ice is doubly refracting (see Table 48-1), why don't we see two images of objects viewed through an ice cube?

11. Is it possible to produce interference effects between the o-beam and the e-beam, which are separated by the calcite crystal from the incident unpolarized beam in Fig. 48-11, by recombining them? Explain your answer.

12. From Table 48-1, would you expect a quarter-wave plate made from calcite to be thicker than one made from quartz?

13. Does the *e*-wave in doubly refracting crystals always travel at a speed given by c/n_e?

14. In Fig. 48-14a and 48-14b describe qualitatively what happens if the incident beam falls on the crystal with an angle of incidence that is not zero. Assume in each case that the incident beam remains in the plane of the figure.

15. Devise a way to identify the direction of the optic axis in a quarter-wave plate.

16. If plane-polarized light falls on a quarter-wave plate with its plane of vibration making an angle of (a) 0° or (b) 90° with the axis of the plate, describe the transmitted light. (c) If this angle is arbitrarily chosen, the transmitted light is called *elliptically polarized*; describe such light.

17. You are given an object which may be (a) a disk of grey glass, (b) a polarizing sheet, (c) a quarter-wave plate, or (d) a half-wave plate (see Problem 14). How could you identify it?

18. Can a plane-polarized light beam be represented as a sum of two circularly polarized light beams of opposite rotation? What effect has changing the phase of one of the circular components on the resultant beam?

19. How can a right-circularly polarized light beam be transformed into a left-circularly polarized beam?

20. Could (a) a radar beam and (b) a sound wave in air be circularly polarized?

21. A beam of light is said to be unpolarized, plane-polarized, or circularly polarized. How could you choose among them experimentally?

22. A parallel beam of light is absorbed by an object placed in its path. Under what circumstances will (a) linear momentum and (b) angular momentum be transferred to the object?

23. When observing a clear sky through a polarizing sheet, one finds that the intensity varies by a factor of two on rotating the sheet. This does not happen when one views a cloud through the sheet. Can you devise an explanation?

24. In 1949 it was discovered that light from distant stars in our galaxy is slightly plane-polarized, with the preferred plane of vibration being parallel to the plane of our galaxy. This is probably due to non-isotropic scattering of the starlight by elongated and slightly aligned interstellar grains. If the grains are oriented with their long axis parallel to the interstellar magnetic field lines, as discussed in the text, and they absorb and radiate electromagnetic waves like the oscillating electrons in a radio antenna, how must the magnetic field be oriented with respect to the galactic plane?

problems

SECTION 48-1

1. Prove that two plane-polarized light waves of equal amplitude, their planes of vibration being at right angles to each other, cannot produce interference effects. (*Hint:* Prove that the intensity of the resultant light wave, averaged over one or more cycles of oscillation, is the same no matter what phase difference exists between the two waves.)

SECTION 48-2

2. Unpolarized light falls on two polarizing sheets placed one on top of the other. What must be the angle between the characteristic directions of the sheets if the intensity of the transmitted light is (*a*) one-third the maximum intensity of the transmitted beam or (*b*) one-third the intensity of the incident beam? Assume that the polarizing sheet is ideal, that is, that it reduces the intensity of unpolarized light by exactly 50%.

3. A beam of plane polarized light strikes two polarizing sheets. The characteristic direction of the first sheet is at an angle θ with respect to the incident beam, the characteristic direction of the second sheet being at 90° with respect to the incident beam. Find the angle θ for a transmitted beam intensity that is 1/10 the incident beam intensity.   *Answer:* 20° or 70°.

4. An unpolarized beam of light is incident on a group of four polarizing sheets which are lined up so that the characteristic direction of each is rotated by 30° clockwise with respect to the preceding sheet. What fraction of the incident intensity is transmitted?

5. A beam of light is a mixture of plane polarized light and randomly polarized light. When it is sent through a Polaroid sheet, it is found that the transmitted intensity can be varied by a factor of five depending on the orientation of the Polaroid. Find the relative intensities of these two components of the incident beam.
   Answer: Plane polarized $\frac{2}{3}$, randomly polarized $\frac{1}{3}$.

6. Partially polarized light (a mixture of unpolarized and plane-polarized beams) can be represented as two plane-polarized beams of unequal intensities, I along the *x*-axis and i along the *y*-axis say, with a random phase difference. The degree of polarization is defined as $p = (I - i)/(I + i)$.

   (*a*) Suppose a beam of partially polarized light passes through a Polaroid sheet with its characteristic direction at an angle θ with the *x*-axis; show that the transmitted intensity I_t is

   $$I_t = I \frac{1 + p \cos 2\theta}{1 + p}.$$

   (*b*) Does this reduce to expected results for $p = 1$ and $p = 0$?

7. It is desired to rotate the plane of polarization of a beam of plane-polarized light by 90°. (*a*) How might this be done using only Polaroid sheets? (*b*) How many sheets are required in order that the total intensity loss is less than 5%? Assume that each Polaroid sheet is ideal.   *Answer:* 48.

SECTION 48-3

8. (a) At what angle of incidence will the light reflected from water be completely polarized? (b) Does this angle depend on the wavelength of the light?

9. Calculate the range of polarizing angles for white light incident on fused quartz. Assume that the wavelength limits are 400 and 700 nm (4000 and 7000 Å) and use the dispersion curve of Fig. 43-2.
   Answer: 55°30′ to 55°46′.

SECTION 48-4

10. Plane-polarized light of wavelength 525 nm strikes, at normal incidence, a wurzite crystal, cut with its faces parallel to the optic axis. What is the smallest possible thickness of the crystal if the emergent o and e-rays combine to form plane-polarized light? See Table 48-1.

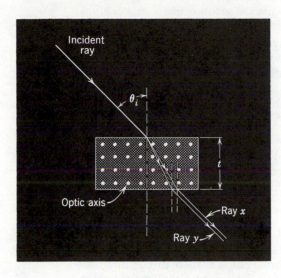

figure 48-23
Problem 11

11. A narrow beam of unpolarized light falls on a calcite crystal cut with its optic axis as shown in Fig. 48-23. (a) For $t = 1.0$ cm and for $\theta_i = 45°$, calculate the perpendicular distance between the two emerging rays x and y. (b) Which is the o-ray and which the e-ray? (c) What are the states of polarization of the emerging rays? (d) Describe what happens if a polarizer is placed in the incident beam and rotated. (*Hint:* Inside the crystal the E-vector vibrations for one ray are always perpendicular to the optic axis and for the other ray they are always parallel. The two rays are described by the indices n_o and n_e; *in this plane* each ray obeys Snell's law.)
   Answer: (a) 0.55 mm. (b) x is the e-ray, y is the o-ray. (c) E in ray y lies in the plane of the figure; E in ray x lies at right angles to the plane of the figure. (d) Every 90° one or the other beam, alternately, will be extinguished.

12. A prism is cut from calcite so that the optic axis is parallel to the prism edge as shown in Fig. 48-24. Describe how such a prism might be used to measure the two principal indices of refraction for calcite. (*Hint:* See hint in Problem 11; see also Example 3, Chapter 43.)

13. How thick must a sheet of mica be if it is to form a quarter-wave plate for yellow light ($\lambda = 589$ nm $= 5890$ Å)? Mica cleaves in such a way that the appropriate indices of refraction, for transmission at right angles to the cleavage plane, are 1.6049 and 1.6117.   *Answer:* 0.022 mm.

SECTION 48-5

14. What would be the action of a *half-wave plate* (that is, a plate twice as thick as a quarter-wave plate) on (a) plane-polarized light (assume the plane of vibration to be at 45° to the optic axis of the plate), (b) circularly polarized light, and (c) unpolarized light.

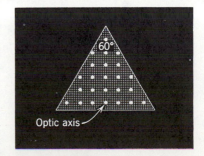

figure 48-24
Problem 12

15. Describe the state of polarization represented by these sets of equations:

$$(a)\ E_x = E \sin (kz - \omega t)$$
$$E_y = E \cos (kz - \omega t),$$
$$(b)\ E_x = E \cos (kz - \omega t)$$
$$E_y = E \cos \left(kz - \omega t + \frac{\pi}{4} \right),$$
$$(c)\ E_x = E \sin (kz - \omega t)$$
$$E_y = -E \sin (kz - \omega t).$$

Answer: Assuming that a right-handed coordinate system is used, (a) circular, counterclockwise as seen facing the scource; (b) elliptical, counterclockwise as seen facing the scource, major axis of ellipse along $y = x$; (c) plane, along $y = -x$.

16. A sheet of Polaroid and a quarter-wave plate are glued together in such a way that, if the combination is placed with face A against a shiny coin, the face of the coin can be seen when illuminated with light of appropriate wavelength. When the combination is placed with face A away from the coin, the coin cannot be seen. (a) Which component is on face A and (b) what is the relative orientation of the components?

SECTION 48-6

17. Show that in a parallel beam of circularly polarized light the angular momentum per unit volume L_r is given by

$$L_r = \frac{P}{\omega c},$$

where P is the power per unit area of the beam. Start from Eq. 48-5.

18. Assume that a parallel beam of circularly polarized light whose intensity is 100 W is absorbed by an object, (a) At what rate is angular momentum transferred to the object? (b) If the object is a flat disk of diameter 5.0 mm and mass 1.0×10^{-2} g, after how long a time (assuming it is free to rotate about its axis) would it attain an angular speed of 1.0 rev/s? Assume a wavelength of 500 nm.

49
light and
quantum physics

We have studied many properties of light, including its propagation, reflection, refraction, diffraction, and interference. This chapter deals in part with the production of light and with the way that such studies led, in 1900, to the birth of quantum physics.

The most common light sources are heated solids and gases through which an electric discharge is passing. The tungsten filament of an incandescent lamp and the familiar neon sign are examples in each category. By analyzing the light from a source with a spectrometer, we can learn how strongly it radiates at various wavelengths. Figure 49-1, which is typical of spectra for heated solids, shows the results of such measurements for a heated tungsten ribbon at 2000 K.

The ordinate $\mathcal{R}_\lambda$ in Fig. 49-1 is called the *spectral radiancy*, defined so that the quantity $\mathcal{R}_\lambda \, d\lambda$ is the rate at which energy is radiated per unit area of surface for wavelengths lying in the interval λ to $\lambda + d\lambda$. Typical units for $\mathcal{R}_\lambda$ are $W/cm^2 \cdot \mu m$; the corresponding units of $\mathcal{R}_\lambda \, d\lambda$ are W/cm^2. In measuring $\mathcal{R}_\lambda$, all radiation emerging into the forward hemisphere is included.

Sometimes we wish to discuss the total radiated energy without regard to its wavelength. An appropriate quantity here is the *radiancy $\mathcal{R}$*, defined as the rate per unit surface area at which energy is radiated into the forward hemisphere, appropriate units being W/cm^2. We can find it by integrating the radiation present in all wavelength intervals:

$$\mathcal{R} = \int_0^\infty \mathcal{R}_\lambda \, d\lambda. \qquad (49\text{-}1)$$

The radiancy $\mathcal{R}$ can be interpreted as the area under the plot of $\mathcal{R}_\lambda$ against λ. In Fig. 49-1 this area—and thus $\mathcal{R}$—is 23.5 W/cm^2. Note the

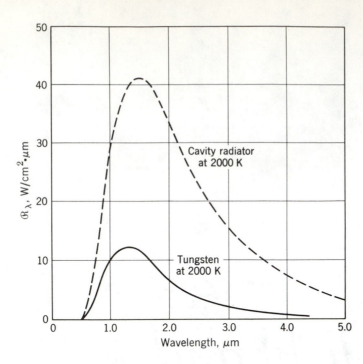

figure 49-1
The spectral radiancy of tungsten at 2000 K. The dashed curve refers to a cavity radiator at the same temperature. One micrometer $(=1 \ \mu m) = 10^{-6} \ m = 10^{4} \ Å = 10^{3} \ nm.$

formal similarity between such curves and the Maxwell speed distribution curve of Section 24-2.

For every material there exists a family of spectral radiancy curves like that of Fig. 49-1, one curve for every temperature. If such families of curves are compared, no obvious regularities stand out. A quantitative understanding in terms of a basic theory presents serious difficulties. Fortunately, it is possible to work with an idealized heated solid, called a *cavity radiator.* Its light-emitting properties prove to be independent of any particular material and to vary in a simple way with temperature. In much the same way it proved convenient earlier to deal with an ideal gas rather than to analyze the properties of the infinite variety of real gases. The cavity radiator is the *ideal solid* as far as its light-emitting characteristics are concerned. We shall describe in the next two sections how the theoretical study of cavity radiation in 1900 by the German physicist Max Planck (1858–1947) laid the foundations of modern quantum physics.

49-2
CAVITY RADIATORS

Let us construct a cavity in each of three metal blocks through the walls of which a small hole is drilled. Let the blocks be made of any suitable materials; for example, tungsten, tantalum, and molybdenum. Let each block be raised to the same uniform temperature (say 2000 K) as determined by a suitable thermometer. Finally, let us observe the blocks by their emitted light in a dark room. Measurements of $\mathscr{R}$ and $\mathscr{R}_\lambda$ show the following:

1. The radiation from the cavity interior is always more intense than the radiation from the outside wall. Comparison of the two curves in Fig. 49-1 makes this clear for tungsten. For the three materials given,

at 2000 K the ratio of the radiancy for the outside surface to that for the cavity is 0.259 (tungsten), 0.212 (molybdenum), and 0.232 (tantalum).

2. At a given temperature the radiancy of the hole is *identical for all three radiators,* in spite of the fact that the radiancies of the outer surfaces are different. At 2000 K the cavity radiancy (that is, the hole radiancy) is 90.0 W/cm².

3. In contrast to the radiancy of the outer surfaces, the cavity radiancy $\mathcal{R}_c$ varies with temperature in a simple way, namely, as

$$\mathcal{R}_c = \sigma T^4, \tag{49-2}$$

where σ is a universal constant (the Stefan-Boltzmann constant) whose measured value is 5.67×10^{-8} W/m² · K⁴. The radiancy of the outer surfaces varies with temperature in a more complicated way and is different for different materials. We often write it as

$$\mathcal{R} = e\mathcal{R}_c = e\sigma T^4, \tag{49-3}$$

where e, the *emissivity,* depends on the material and the temperature.

4. $\mathcal{R}_\lambda$ for the cavity radiation varies with temperature in the way shown in Fig. 49-2. These curves depend only on the temperature and are quite independent of the material and of the shape and size of the cavity.

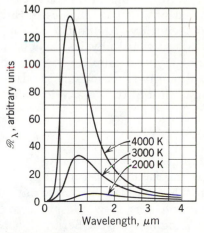

figure 49-2
The spectral radiancy for cavity radiation at three different temperatures.

Figure 49-3 shows an actual cavity, consisting of a hollow thin-walled cylinder of tungsten heated by sending an electric current through it. The cylinder is mounted in an evacuated glass bulb, and a tiny hole is drilled through the cylinder wall. You can see from the photograph that the radiancy of the cavity interior is greater than that of the cavity walls.

We can deduce many of the facts just given about cavity radiation from Fig. 49-4, which shows two cavities made of different materials, of arbitrary shapes, but with the same wall temperature T. Radiation, described by $\mathcal{R}_A$, goes from cavity A to cavity B and radiation described by $\mathcal{R}_B$ moves in the opposite direction. If these two rates of energy transfer are not equal, one end of the composite block will start to heat up and the other end will start to cool down, which is a violation of the second law of thermodynamics. (Why?) Thus we must have

$$\mathcal{R}_A = \mathcal{R}_B = \mathcal{R}_c, \tag{49-4}$$

where $\mathcal{R}_c$ describes the total radiation for *all* cavities.

Not only the total radiation but also the distribution of radiant energy with wavelength must be the same for each cavity in Fig. 49-4. We can show this by placing a filter between the two cavity openings, so chosen that it permits only a selected narrow band of wavelengths to pass. Applying the same argument, we can show that we must have

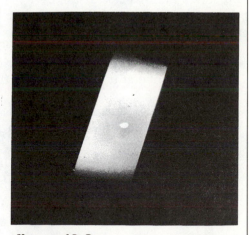

figure 49-3
An incandescent tungsten tube with a small hole drilled in its wall. The radiation emerging from the hole is cavity radiation.

figure 49-4
Two radiant cavities initially at the same temperature are placed together as shown.

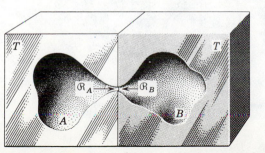

$$\mathcal{R}_{\lambda A} = \mathcal{R}_{\lambda B} = \mathcal{R}_{\lambda c}, \qquad (49\text{-}5)$$

where $\mathcal{R}_{\lambda c}$ is a spectral radiancy characteristic of all cavities.

A theoretical explanation for the cavity radiation was the outstanding unsolved problem in physics during the years before the turn of the present century. A number of capable physicists advanced theories based on classical physics, which, however, had only limited success. Figure 49-5, for example, shows the theory of Wien; the fit to the experimental points is reasonably good, but definitely not within the experimental error of the data. Wien's formula is

$$\mathcal{R}_{\lambda} = \frac{c_1}{\lambda^5} \frac{1}{e^{c_2/\lambda T}},$$

where c_1 and c_2 are constants that must be determined empirically by fitting the theoretical formula to the experimental data.

In 1900 Max Planck pointed out that if Wien's formula were modified in a simple way, it would prove to fit the data precisely. Planck's formula, announced to the Berlin Physical Society on October 19, 1900, was

$$\mathcal{R}_{\lambda} = \frac{c_1}{\lambda^5} \frac{1}{e^{c_2/\lambda T} - 1}. \qquad (49\text{-}6)$$

This formula, though interesting and important, was still empirical at that stage and did not constitute a theory.

Planck sought such a theory in terms of a detailed model of the atomic processes taking place at the cavity walls. He assumed that the atoms that make up these walls behave like tiny electromagnetic oscillators, each with a characteristic frequency of oscillation. The oscillators emit electromagnetic energy into the cavity and absorb electromagnetic energy from it. Thus it should be possible to deduce the characteristics of the cavity radiation from those of the oscillators with which it is in equilibrium.

Planck was led to make two radical assumptions about the atomic oscillators. As eventually formulated, these assumptions are the following:

1. An oscillator cannot have *any* energy but only energies given by

$$E = nh\nu, \qquad (49\text{-}7)$$

49-3
PLANCK'S RADIATION FORMULA

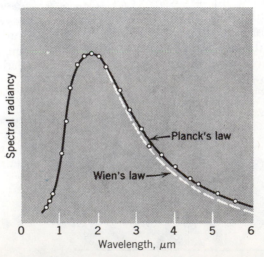

figure 49-5
The circles show the experimental spectral radiancy data of Coblentz for cavity radiation. The theoretical formulas of Wien and Planck are also shown, Planck's providing an excellent fit to the data.

where ν is the oscillator frequency, h is a constant (now called *Planck's constant*), and n is a number (now called a *quantum number*) that can take on only integral values. Equation 49-7 asserts that the oscillator energy is *quantized*. Later developments show that the correct formula for a harmonic oscillator is $E = (n + \frac{1}{2})h\nu$. This change makes no difference to Planck's conclusions, however.

2. The oscillators do not radiate energy continuously, but only in "jumps," or *quanta*. These quanta of energy are emitted when an oscillator changes from one to another of its quantized energy states. Thus, if n changes by one unit, Eq. 49-7 shows that an amount of energy given by

$$\Delta E = \Delta n h \nu = h \nu \qquad (49\text{-}8)$$

is radiated. As long as an oscillator remains in one of its quantized states (or *stationary states* as they are called), it neither emits nor absorbs energy.

These assumptions were radical ones and, indeed, Planck himself resisted accepting them for many years. In his words, "My futile attempts to fit the elementary quantum of action [that is, the quantity h] somehow into the classical theory continued for a number of years, and they cost me a great deal of effort."

Consider the application of Planck's hypotheses to a large-scale oscillator such as a mass-spring system or an *LC* circuit. It would be a stoutly defended common belief that oscillations in such systems could take place with *any* value of total energy and not with only certain discrete values. In the decay of such oscillations (by friction in the mass-spring system or by resistance and radiation in the *LC* circuit), it would seem that the mechanical or electromagnetic energy would decrease in a perfectly continuous way and not by "jumps." There is no basis in everyday experience, however, to dismiss Planck's assumptions as violations of "common sense," for Planck's constant proves to have a very small value, namely,

$$h = 6.626 \times 10^{-34} \text{ joule} \cdot \text{second}.$$

The following example makes this clear.

EXAMPLE 1

A mass-spring system has a mass $m = 1.0$ kg and a spring constant $k = 20$ N/m and is oscillating with an amplitude of 1.0 cm. (*a*) If its energy is quantized according to Eq. 49-7, what is the quantum number n?

From Eq. 15-11 the frequency is

$$\nu = \frac{1}{2\pi} \sqrt{\frac{k}{m}} = \frac{1}{2\pi} \sqrt{\frac{20 \text{ N/m}}{1.0 \text{ kg}}} = 0.71 \text{ Hz}.$$

From Eq. 7-8 the mechanical energy is

$$E = \tfrac{1}{2} k x_{max}^2 = \tfrac{1}{2}(20 \text{ N/m})(1.0 \times 10^{-2} \text{ m})^2 = 1.0 \times 10^{-3} \text{ J}.$$

From Eq. 49-7 the quantum number is

$$n = \frac{E}{h\nu} = \frac{1.0 \times 10^{-3} \text{ J}}{(6.6 \times 10^{-34} \text{ J} \cdot \text{s})(0.71 \text{ Hz})} = 2.1 \times 10^{30}.$$

(*b*) If n changes by unity, what fractional change in energy occurs?

The fractional change in energy is given by dividing Eq. 49-8 by Eq. 49-7, or

$$\frac{\Delta E}{E} = \frac{h\nu}{nh\nu} = \frac{1}{n} = \sim 10^{-30}.$$

Thus for large-scale oscillators the quantum numbers are enormous and the quantized nature of the energy of the oscillations will not be apparent. Similarly, in large-scale experiments we are not aware of the discrete nature of mass and the quantized nature of charge, that is, of the existence of atoms and electrons.

On the basis of his two assumptions, Planck was able to derive his radiation law (Eq. 49-6) entirely from theory. His theoretical expressions for the hitherto empirical constants c_1 and c_2 were

$$c_1 = 2\pi c^2 h \qquad \text{and} \qquad c_2 = \frac{hc}{k},$$

where k is Boltzmann's constant (see Section 23-5) and c is the speed of light. By inserting the experimental values for c_1 and c_2, Planck was able to derive the values of both h and k. Planck described his theory to the Berlin Physical Society on December 14, 1900. Quantum physics started on that date. Planck's ideas soon received re-enforcement from Einstein, who, in 1905, applied the concepts of energy quantization to a new area of physics, the photoelectric effect.

Before discussing this effect, it is important to realize that although Planck had quantized the energies of the oscillators in the cavity walls he still treated the radiation within the cavity as an electromagnetic wave. Einstein's analysis of the photoelectric effect first pointed out the inadequacy of the wave picture of light in certain situations.

Figure 49-6 shows an apparatus used to study the photoelectric effect. Monochromatic light, falling on metal plate *A*, will liberate *photoelectrons*, which can be detected as a current if they are attracted to metal cup *B* by means of a potential difference *V* applied between *A* and *B*. Galvanometer *G* serves to measure this *photoelectric current*.

49-4
PHOTOELECTRIC EFFECT

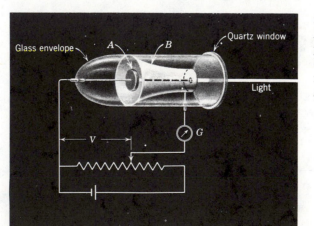

figure 49-6
An apparatus used to study the photoelectric effect. *V* can be varied continuously and can also be reversed in sign by a switching arrangement not shown.

Figure 49-7 (curve *a*) is a plot of the photoelectric current in an apparatus like that of Fig. 49-6, as a function of the potential difference *V*. If *V* is large enough, the photoelectric current reaches a certain limiting value at which all photoelectrons ejected from plate *A* are collected by cup *B*.

If we reverse *V* in sign, the photoelectric current does not immediately drop to zero, which proves that the electrons are emitted from *A*

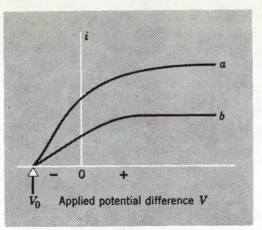

figure 49-7
Some data taken with the apparatus of Fig. 49-6. The applied potential difference V is called positive when the cup B in Fig. 49-6 is positive with respect to the photoelectric surface A. In curve b the incident light intensity has been reduced to one-half that of curve a.

with a finite velocity. Some will reach cup B in spite of the fact that the electric field opposes their motion. However, if this reversed potential difference is made large enough, a value V_0 (the *stopping potential*) is reached at which the photoelectric current does drop to zero. This potential difference V_0, multiplied by the electron charge, measures the kinetic energy K_{max} of the fastest ejected photoelectron. In other words,

$$K_{max} = eV_0. \tag{49-9}$$

Here K_{max} turns out to be independent of the intensity of the light as shown by curve b in Fig. 49-7, in which the light intensity has been reduced to one-half.

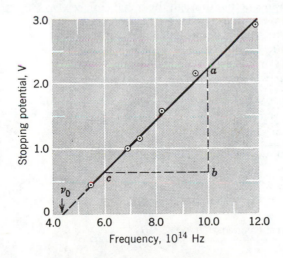

figure 49-8
A plot of Millikan's measurements of the stopping potential at various frequencies for sodium. The cutoff frequency ν_0 is 4.39×10^{14} Hz.

Figure 49-8 shows the stopping potential V_0 as a function of the frequency of the incident light for sodium. Note that there is a definite cutoff frequency ν_0, below which no photoelectric effect occurs. These data were taken by R. A. Millikan (1868–1953), whose painstaking work on the photoelectric effect won him the Nobel prize in 1923. Because the photoelectric effect is largely a surface phenomenon, it is necessary to avoid oxide films, grease, or other surface contaminants. Millikan devised a technique to cut shavings from the metal surface under vacuum conditions, a "machine shop *in vacuo*" as he called it.

Three major features of the photoelectric effect cannot be explained in terms of the wave theory of light:

1. Wave theory suggests that the kinetic energy of the photoelectrons should increase as the light beam is made more intense. However, Fig. 49-7 shows that K_{max} $(=eV_0)$ is independent of the light intensity; this has been tested over a range of intensities of 10^7.

2. According to the wave theory, the photoelectric effect should occur for any frequency of the light, provided only that the light is intense enough. However, Fig. 49-8 shows that there exists, for each surface, a characteristic *cutoff frequency* ν_0. For frequencies less than this, the photoelectric effect disappears, no matter how intense the illumination.

3. If the energy of the photoelectrons is "soaked up" from the incident wave by the metal plate, it is not likely that the "effective target area" for an electron in the metal is much more than a few atomic diameters. Thus, if the light is feeble enough, there should be a measurable time lag (see Example 2) between the impinging of the light on the surface and the ejection of the photoelectron. During this interval the electron should be "soaking up" energy from the wave until it has accumulated enough energy to escape. However, no detectable time lag has ever been measured. This disagreement is particularly striking when the photoelectric substance is a gas; under these circumstances the energy of the emitted photoelectron must certainly be "soaked out of the wave" by a single atom.

EXAMPLE 2

A metal plate is placed 5 m from a monochromatic light source whose power output is 10^{-3} W. Consider that a given ejected photoelectron may collect its energy from a circular area of the plate as large as ten atomic diameters (10^{-9} m) in radius. The energy required to remove an electron through the metal surface is about 5.0 eV. Assuming light to be a wave, how long would it take for such a "target" to soak up this much energy from such a light source?

The target area is $\pi\,(10^{-9}$ m$)^2$ or 3×10^{-18} m^2; the area of a 5-m sphere centered on the light source is $4\pi\,(5$ m$)^2 \cong 300$ m^2. Thus, if the light source radiates uniformly in all directions, the rate P at which energy falls on the target is given by

$$P = (10^{-3}\text{ W})\left(\frac{3 \times 10^{-18}\text{ m}^2}{300\text{ m}^2}\right) = 10^{-23}\text{ J/s}.$$

Assuming that all this power is absorbed, we may calculate the time required from

$$t = \left(\frac{5\text{ eV}}{10^{-23}\text{ J/s}}\right)\left(\frac{1.6 \times 10^{-19}\text{ J}}{1\text{ eV}}\right) \cong 20\text{ hours}.$$

However, no detectable time lag has been measured.

49-5
EINSTEIN'S PHOTON THEORY

Einstein succeeded in explaining the photoelectric effect by making a remarkable assumption, namely, that the energy in a light beam travels through space in concentrated bundles, later called *photons*.* The

---

* The word *photon* was coined by an American physical chemist, G. N. Lewis, in 1926. He wrote, "I therefore take the liberty of proposing for this hypothetical new atom . . . the name *photon*." When Max Planck proposed Einstein for membership in the Berlin Physical Society, he deprecated Einstein's 1905 paper on the photon theory, saying that such a great man should be excused a little fault. Planck, you will recall, was the (almost reluctant) discoverer of quantization. Einstein received the Nobel prize in 1921, not for the theory of relativity but for his concept of photons.

energy E of a single photon (see Eq. 49-8) is given by

$$E = h\nu. \tag{49-10}$$

Recall that Planck believed that light, although emitted from its source discontinuously, travels through space as an electromagnetic wave. Einstein's hypothesis suggests that light traveling through space behaves not like a wave at all but like a particle. Millikan, whose experiments verified Einstein's ideas in every detail, spoke of Einstein's "bold, not to say reckless, hypothesis."

If we apply Einstein's photon concept to the photoelectric effect, we obtain

$$h\nu = E_0 + K_{max} \tag{49-11}$$

where $h\nu$ is the energy of the photon. Equation 49-11 says that a photon carries an energy $h\nu$ into the surface. Part of this energy (E_0) is used in causing the electron to pass through the metal surface. The excess energy $(h\nu - E_0)$ is given to the electron in the form of kinetic energy; if the electron does not lose energy by internal collisions as it escapes from the metal, it will exhibit it all as kinetic energy after it emerges. Thus K_{max} represents the maximum kinetic energy that the photoelectron can have outside the surface; in nearly all cases it will have less energy than this because of internal losses.

Consider how Einstein's photon hypothesis meets the three objections raised against the wave-theory interpretation of the photoelectric effect. As for objection 1 (the lack of dependence of K_{max} on the intensity of illumination), there is complete agreement of the photon theory with experiment. If we double the light intensity, we double the number of photons and thus double the photoelectric current; we do not change the energy $(=h\nu)$ of the individual photons or the nature of the individual photoelectric processes described by Eq. 49-11.

Objection 2 (the existence of a cutoff frequency) follows from Eq. 49-11. If K_{max} equals zero, we have

$$h\nu_0 = E_0,$$

which asserts that the photon has just enough energy to eject the photoelectrons and none extra to appear as kinetic energy. This quantity E_0 is called the *work function* of the substance. If ν is reduced below ν_0, the individual photons, no matter how many of them there are (that is, no matter how intense the illumination), will not have enough energy to eject photoelectrons.

Objection 3 (the absence of a time lag) follows from the photon theory because the required energy is supplied in a concentrated bundle. It is not spread uniformly over a large area, as in the wave theory.

Although the photon hypothesis certainly fits the facts of photoelectricity, it seems to be in direct conflict with the wave theory of light which, as we have seen in earlier chapters, has been verified in many experiments. Our modern view of the nature of light is that it has a dual character, behaving like a wave under some circumstances and like a particle, or photon, under others. We discuss the wave-particle duality at length in Chapter 50. Meanwhile, let us continue our studies of the firm experimental foundation on which the photon concept rests.

Let us rewrite Einstein's photoelectric equation (Eq. 49-11) by substituting eV_0 for K_{max} (see Eq. 49-9). This yields, after rearrangement,

$$V_0 = \frac{h}{e}\nu - \frac{E_0}{e}. \tag{49-12}$$

Thus Einstein's theory predicts a linear relationship between V_0 and ν, in complete agreement with experiment; see Fig. 49-8. The slope of the experimental curve in this figure should be h/e, or

$$\frac{h}{e} = \frac{ab}{bc} = \frac{2.20\ \text{V} - 0.65\ \text{V}}{(10 \times 10^{14} - 6.0 \times 10^{14})\ \text{Hz}} = 3.9 \times 10^{-15}\ \text{V} \cdot \text{s}.$$

We can find h by multiplying this ratio by the electron charge e,

$$h = (3.9 \times 10^{-15}\ \text{V} \cdot \text{s})(1.6 \times 10^{-19}\ \text{C}) = 6.2 \times 10^{-34}\ \text{J} \cdot \text{s}.$$

From a more careful analysis of these and other data, including data taken with lithium surfaces, Millikan found the value $h = 6.57 \times 10^{-34}$ J·s with an accuracy of about 0.5%. This agreement with the value of h derived from Planck's radiation formula is a striking confirmation of Einstein's photon concept.*

EXAMPLE 3

Deduce the work function that follows from Fig. 49-8.

The intersection of the straight line in Fig. 49-8 with the horizontal axis is the cutoff frequency ν_0. Substituting these values yields

$$E_0 = h\nu_0 = (6.63 \times 10^{-34}\ \text{J} \cdot \text{s})(4.39 \times 10^{14}\ \text{Hz})$$

$$= 2.92 \times 10^{-19}\ \text{J} = 1.82\ \text{eV}.$$

49-6
THE COMPTON EFFECT

Compelling confirmation of the concept of the photon as a concentrated bundle of energy was provided in 1923 by A. H. Compton (1892–1962) who earned a Nobel prize for this work in 1927.† Compton allowed a beam of X-rays of sharply defined wavelength λ to fall on a graphite block, as in Fig. 49-9, and he measured, for various angles of scattering, the intensity of the scattered X-rays as a function of their wavelength. Figure 49-10 shows his experimental results. We see that although the incident beam consists essentially of a single wavelength λ the scattered X-rays have intensity peaks at *two* wavelengths; one of them is the same as the incident wavelength, the other, λ', being larger by an amount $\Delta\lambda$. This so-called *Compton shift* $\Delta\lambda$ varies with the angle at which the scattered X-rays are observed.

The presence of a scattered wave of wavelength λ' cannot be understood if the incident X-rays are regarded as an electromagnetic wave like that of Fig. 41-13. On this picture the incident wave of frequency ν causes electrons in the scattering block to oscillate at that same frequency. These oscillating electrons, like charges surging back and forth in a small radio transmitting antenna, radiate electromagnetic waves that again have this same frequency ν. Thus, on the wave picture

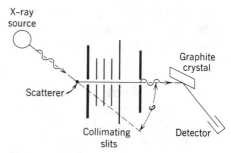

figure 49-9
Compton's experimental arrangement. Monochromatic X-rays of wavelength λ fall on a graphite scatterer. The distribution of intensity with wavelength is measured for X-rays scattered at any selected angle φ. The scattered wavelengths are measured by observing Bragg reflections from a crystal; see Eq. 47-8. Their intensities are measured by a detector, such as an ionization chamber.

---

* In an actual experiment, we would have to take into account any difference in the contact potential that might exist between the metal of plate A and the (different) metal of cup B in Fig. 49-6. See, for example, *Experiments in Modern Physics* by A. C. Melissinos, Academic Press, New York, 1966 for an exact treatment of such a situation.

† For an historical account of Compton's researches read "The Scattering of X Rays as Particles," by A. H. Compton, *American Journal of Physics*, December 1961. In this article Compton recalled that the distinguished Indian physicist, Sir C. V. Raman, who won a Nobel prize in 1930, said to him: "Compton, you are a good debator but the truth is not in you." This, and the earlier comment by Planck on Einstein, suggests the reluctance with which new ideas are accepted even by brilliant scientists, such as Planck and Raman. Compton (and Einstein) turned out to be right.

the scattered wave should have the same frequency ν and the same wavelength λ as the incident wave.

Compton* was able to explain his experimental results by postulating that the incoming X-ray beam was not a wave but an assembly of photons of energy E $(=h\nu)$ and that these photons experienced billiard-ball-like collisions with the free electrons in the scattering block. The "recoil" photons emerging from the block constitute, on this view, the scattered radiation. Since the incident photon transfers some of its energy to the electron with which it collides, the scattered photon must have a lower energy E'; it must therefore have a lower frequency $\nu'(=E'/h)$, which implies a larger wavelength $\lambda'(=c/\nu')$. This point of view accounts, at least qualitatively, for the wavelength shift $\Delta\lambda$. Notice how different this particle model of X-ray scattering is from that based on the wave picture. Now let us analyze a single photon-electron collision quantitatively.

Figure 49-11 shows a collision between a photon and an electron, the electron assumed to be initially at rest and essentially free, that is, not bound to the atoms of the scatterer. Let us apply the law of conservation of energy to this collision. Since the recoil electrons may have a speed v that is comparable with that of light, we must use the relativistic expression for the kinetic energy of the electron. From Eqs. 49-10 and 8-21 we may write

$$h\nu = h\nu' + (m - m_0)c^2,$$

in which the second term on the right is the relativistic expression for the kinetic energy of the recoiling electron, m being the relativistic mass and m_0 the rest mass of that particle. Substituting c/λ for ν (and c/λ' for ν') and using Eq. 8-20 to eliminate the relativistic mass m leads us to

$$\frac{hc}{\lambda} = \frac{hc}{\lambda'} + m_0 c^2 \left(\frac{1}{\sqrt{1 - (v/c)^2}} - 1 \right). \qquad (49\text{-}13)$$

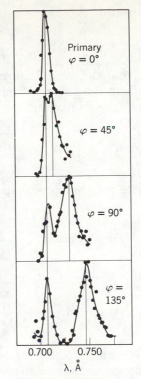

figure 49-10

Compton's experimental results. The solid vertical line on the left corresponds to the wavelength λ, that on the right to λ'. Results are shown for four different angles of scattering φ. Note that the Compton shift $\Delta\lambda$ for $\varphi = 90°$ is $h/m_0c = 0.0243$ Å.

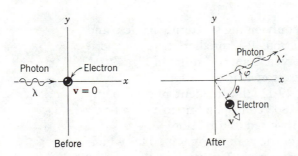

figure 49-11

A photon of wavelength λ is incident on an electron at rest. On collision, the photon is scattered at an angle φ with increased wavelength λ', whereas the electron moves off with speed v in direction θ.

Now let us apply the (vector) law of conservation of linear momentum to the collision of Fig. 49-11. We first need an expression for the momentum of a photon. In Section 42-2 we saw that if an object completely absorbs an energy U from a parallel light beam that falls on it, the light beam, according to the wave theory of light, will simultaneously transfer to the object a linear momentum given by U/c. On the photon picture we imagine this momentum to be carried along by the individual photons, each photon transporting linear momentum in amount $p = h\nu/c$, where $h\nu$ is the photon energy. Thus, if we substitute

* P. W. Debye simultaneously and independently offered the same interpretation.

$$p = \frac{E}{c} = \frac{h\nu}{c} = \frac{h}{\lambda}. \qquad (49\text{-}14)$$

This conclusion, that the momentum of a photon is given by h/λ, may also be deduced from the theory of relativity.

For the electron, the relativistic expression for the linear momentum is given by Eq. 9-13, or

$$\mathbf{p}_e = \frac{m_0 \mathbf{v}}{\sqrt{1 - (v/c)^2}}.$$

We can then write for the conservation of the x-component of linear momentum

$$\frac{h}{\lambda} = \frac{h}{\lambda'} \cos \varphi + \frac{m_0 v}{\sqrt{1 - (v/c)^2}} \cos \theta \qquad (49\text{-}15)$$

and for the y-component

$$0 = \frac{h}{\lambda'} \sin \varphi - \frac{m_0 v}{\sqrt{1 - (v/c)^2}} \sin \theta. \qquad (49\text{-}16)$$

Our immediate aim is to find $\Delta\lambda \; (=\lambda' - \lambda)$, the wavelength shift of the scattered photons, so that we may compare it with the experimental results of Fig. 49-10. Compton's experiment did not involve observations of the recoil electron in the scattering block. Of the five collision variables $(\lambda, \lambda', v, \varphi, \text{ and } \theta)$ that appear in the three equations (49-13, 49-15, and 49-16) we may eliminate two. We chose to eliminate v and θ, which deal only with the electron, thereby reducing the three equations to a single relation among the variables.

Carrying out the necessary algebraic steps (see Problem 31) leads to this simple result:

$$\Delta\lambda \; (=\lambda' - \lambda) = \frac{h}{m_0 c}(1 - \cos \varphi). \qquad (49\text{-}17)$$

Thus the Compton shift $\Delta\lambda$ depends only on the scattering angle φ and *not* on the initial wavelength λ. Equation 49-17 predicts within experimental error the experimentally observed Compton shifts of Fig. 49-10. Note from the equation that $\Delta\lambda$ varies from zero (for $\varphi = 0$, corresponding to a "grazing" collision in Fig. 49-11, the incident photon being scarcely deflected) to $2h/m_0 c$ (for $\varphi = 180°$, corresponding to a "head-on" collision, the incident photon being reversed in direction).

It remains to explain the presence of the peak in Fig. 49-10 for which the wavelength does *not* change on scattering. This peak can be understood as resulting from a collision between a photon and electrons bound in an ionic core in the scattering block. During photon collisions the bound electrons behave like the free electrons that we considered in Fig. 49-11, with the exception that their effective mass is much greater. This is because the ionic core as a whole recoils during the collision. The effective mass M for a carbon scatterer is approximately the mass of a carbon nucleus. Since this nucleus contains six protons and six neutrons, we have approximately, $M = 12 \times 1840 m_0 = 22,000 m_0$. If we replace m_0 by M in Eq. 49-17, we see that the Compton shift for collisions with tightly bound electrons is immeasurably small.

As in the cavity radiation problem (see Eq. 49-7) and the photoelectric effect (see Eq. 49-11), Planck's constant h is centrally involved in the

Compton effect. The quantity h is the central constant of quantum physics. In a universe in which $h = 0$ there would be no quantum physics and classical physics would be valid in the subatomic domain. In particular, as Eq. 49-17 shows, there would be no Compton effect (that is, $\Delta\lambda = 0$) in such a universe.

EXAMPLE 4

X-rays with $\lambda = 1.00$ Å ($=0.10$ nm) are scattered from a carbon block. The scattered radiation is viewed at 90° to the incident beam. (a) What is the Compton shift $\Delta\lambda$? (b) What kinetic energy is imparted to the recoiling electron?

(a) Putting $\varphi = 90°$ in Eq. 49-17, we have, for the Compton shift,

$$\Delta\lambda = \frac{h}{m_0 c}(1 - \cos\varphi)$$

$$= \frac{6.63 \times 10^{-34} \text{ J·s}}{(9.11 \times 10^{-31} \text{ kg})(3.00 \times 10^8 \text{ m/s})}(1 - \cos 90°)$$

$$= 2.43 \times 10^{-12} \text{ m} = 0.0243 \text{ Å} = 2.43 \times 10^{-3} \text{ nm}.$$

(b) If we put K for the kinetic energy of the electron, we can write Eq. 49-13 as

$$\frac{hc}{\lambda} = \frac{hc}{\lambda'} + K.$$

Since $\lambda' = \lambda + \Delta\lambda$, we obtain

$$\frac{hc}{\lambda} = \frac{hc}{\lambda + \Delta\lambda} + K,$$

which reduces to

$$K = \frac{hc\,\Delta\lambda}{\lambda(\lambda + \Delta\lambda)}$$

$$= \frac{(6.63 \times 10^{-34} \text{ J·s})(3.00 \times 10^8 \text{ m/s})(2.43 \times 10^{-12} \text{ m})}{(1.00 \times 10^{-10} \text{ m})(1.00 + 0.024) \times 10^{-10} \text{ m}}$$

$$= 4.73 \times 10^{-17} \text{ J} = 295 \text{ eV}.$$

You may show that the initial photon energy E in this case ($=h\nu = hc/\lambda$) is 12,400 eV so that the photon lost about 2.3% of its energy in this collision. A photon whose energy was ten times as large ($= 124,000$ eV) can be shown to lose 23% of its energy in a similar collision. This follows from the fact that $\Delta\lambda$ does not depend on the initial wavelength. Hence more energetic X-rays, which have smaller wavelengths, will experience a larger percent increase in wavelength and thus a larger percent loss in energy.

49-7
LINE SPECTRA

We have seen how Planck successfully explained the nature of the radiation from heated solid objects of which the cavity radiator formed the prototype. Such radiations form *continuous spectra* and are contrasted with *line spectra* such as those of Fig. 49-12, which show the radiation emitted from iron ions and atoms in an electric arc struck between iron electrodes. We shall see that Planck's quantization ideas, suitably ex-

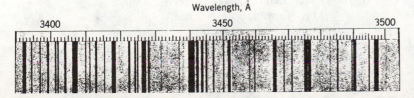

Wavelength, Å

3400   3450   3500

figure 49-12
A small portion of the spectrum of iron, in the region of 3400 to 3500 Å ($=340$ to 350 nm).

tended, lead to an understanding of line spectra also. The prototype for the study of line spectra is that of atomic hydrogen; being the simplest atom it has the simplest spectrum.

Line spectra are common in all parts of the electromagnetic spectrum. Figure 49-13 shows a spectrum of the γ-rays ($\lambda \cong 10^{-12}$ m) emitted from a particular radioactive nucleus, an isotope of mercury. Figure 49-14 shows a spectrum of X-rays ($\lambda \cong 10^{-10}$ m) emitted from a molybdenum target when struck by a 35-KeV electron beam. The sharp emission lines are superimposed on a continuous background.

Figure 49-15 shows a spectrum associated with the molecule HCl. It occurs in the infrared, with $\lambda \cong 10^{-6}$ m. This is an *absorption* spectrum rather than an emission spectrum, as in Fig. 49-12. Experiment shows that isolated atoms and molecules absorb radiation, as well as emit it, at discrete wavelengths.

Figure 49-16 shows a portion of the absorption spectrum of ammonia (NH$_3$) in the microwave region ($\lambda \cong 10^{-2}$ m). Finally, Fig. 49-17 shows how radiation in the radio-frequency region ($\lambda \cong 43$ m) is absorbed by hydrogen molecules placed in a magnetic field.

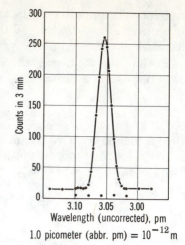

1.0 picometer (abbr. pm) = 10^{-12} m

figure 49-13
A wavelength plot for a gamma ray emitted by the nucleus Hg^{198}. (From data by DuMond and co-workers.)

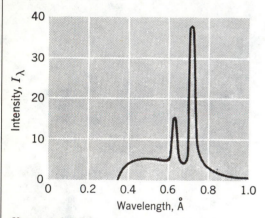

figure 49-14
X-rays from a molybdenum target struck by 35-keV electrons. Note the two sharp lines rising above a broad continuous base. The wavelength of the most intense line is 7.1×10^{-2} nm or 0.71 Å. (From data by Ulrey.)

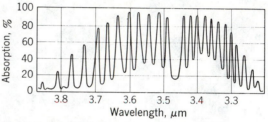

figure 49-15
An absorption spectrum of the HCl molecule near $\lambda = 3.5 \times 10^{-6}$ m = 3.5 μm. (From data by E. S. Imes.)

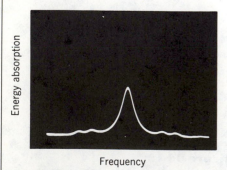

figure 49-16
An oscilloscope trace showing one strong line and four weak lines in the absorption spectrum of ammonia at microwave frequencies.

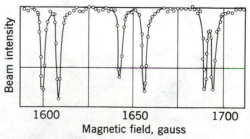

figure 49-17
A portion of the absorption spectrum of the protons in molecular hydrogen at $\lambda \cong 43$ m. In this technique the frequency is left fixed and the sample is placed in a magnetic field, which is varied to scan the spectrum. (From data by Kellogg, Rabi, and Zacharias.)

49-8
*ATOMIC MODELS—THE
BOHR HYDROGEN
ATOM*

1105

ATOMIC MODELS—THE BOHR HYDROGEN ATOM   SEC. 49-8

The attempts of physicists to explain observable phenomena in terms of theoretical models of the physical world which can be given mathematical expression is nowhere better illustrated than in the development of models of the atom. In this case, key evidence leading finally to the wave-mechanical atom was the line spectrum of hydrogen.

In 1815 Prout (1785–1850) suggested that the elements were made up of hydrogen, using as evidence the fact that the atomic weights of many elements are nearly integer multiples of that of hydrogen. With the discovery of the electron by J. J. Thomson (1856–1940) in 1897 the level of sophistication increased greatly. Thomson proposed the "plum pudding" model where the positive charge of the atom was thought to be spread out through the whole atom (a sphere of radius about 10^{-10} m) with the electrons located here and there like plums in the pudding. Then in 1911 Ernest Rutherford (1871–1937) showed the inconsistency between the α-particle scattering experiments of Geiger and Marsden and Thomson's atom and proposed the nuclear model of the atom which is the basis of present theories. Here the positive charge is confined to the nucleus, a very small sphere of radius about 10^{-14} m. The electrons circulate about the nucleus in a volume of the same order of magnitude as Thomson's sphere.

Investigation of the hydrogen spectrum led Niels Bohr (1885–1962) to postulate that the circular orbits of the electrons were quantized, that is, that their angular momentum could have only integral multiples of a basic value. We shall present this Bohr atom in some detail here. The Bohr atom, while deficient in several details, illustrates the ideas of quantization within the simpler mathematical framework of classical physics. Before proceeding, however, we should point out that the wavemechanical atom subsequently replaced the Bohr atom, as we will point out in Sections 50-3 and 50-4. Furthermore, models of the nucleus, while retaining the basic assumptions of Rutherford, have been highly refined and now assume the presence of subnuclear particles which themselves move within and make up the nucleus.

We might wish to associate the frequency of an emitted spectrum line, such as from hydrogen (Fig. 49-18), with the frequency of an electron revolving in an orbit inside the atom. Classical electromagnetism predicts that charges will radiate energy when they are accelerated. In this way electromagnetic waves are emitted from a radio transmitting antenna in which electrons are caused to surge back and forth. This radiation represents a loss of energy for the moving electrons which, in a radio antenna, is compensated for by supplying energy from an oscillator. In an isolated atom, however, no energy is supplied from external sources. We would expect the frequency of the electron, and thus that of the emitted radiation, to change continuously as the energy drains away. This prediction of classical theory cannot be reconciled with the existence of sharp spectrum lines. Thus classical physics cannot explain the hydrogen, or any other, spectrum.

Bohr circumvented this difficulty by assuming that, like Planck's oscillators, the hydrogen atom exists in certain stationary states in which it does not radiate. Radiation occurs only when the atom makes a transition from one state, with energy E_k, to a state with lower energy E_j. In equation form

$$h\nu = E_k - E_j, \tag{49-18}$$

where $h\nu$ is the quantum of energy carried away by the photon that is emitted from the atom during the transition.

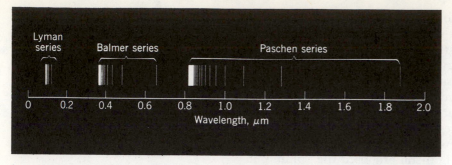

figure 49-18
The spectrum of hydrogen. It consists of a number of series of lines, three of which are shown. Within each series the spectrum lines follow a regular pattern, approaching a so-called *series limit* at the short-wave end of the series.

In order to learn the allowed frequencies predicted by Eq. 49-18, we need to know the energies of the various stationary states in which a hydrogen atom can exist. This calculation was first carried out by Bohr on the basis of a specific model for the hydrogen atom put forward by him. Bohr's model was highly successful for hydrogen and had a tremendous influence on the further development of the subject; it is now regarded as an important preliminary stage in the development of a more complete theory of quantum physics.

Let us assume that the electron in the hydrogen atom moves in a circular orbit of radius r centered on its nucleus. We assume that the nucleus, which is a single proton, is so massive that the center of mass of the system is essentially at the position of the proton. Let us calculate the energy E of such an atom.

Writing Newton's second law for the motion of the electron, we have (using Coulomb's law)

$$F = ma,$$

or

$$\frac{e^2}{4\pi\epsilon_0 r^2} = m\,\frac{v^2}{r}.$$

This allows us to calculate the kinetic energy of the electron, which is

$$K = \tfrac{1}{2}mv^2 = \frac{e^2}{8\pi\epsilon_0 r}. \tag{49-19}$$

The potential energy U of the proton-electron system is given by

$$U = V(-e) = -\frac{e^2}{4\pi\epsilon_0 r}, \tag{49-20}$$

where V $(=e/4\pi\epsilon_0 r)$ is the potential of the proton at the radius of the electron.

The total energy E of the system is

$$E = K + U = -\frac{e^2}{8\pi\epsilon_0 r}. \tag{49-21}$$

Since the orbit radius can apparently take on any value, so can the energy E. The problem of quantizing E reduces to that of quantizing r.

Every property of the orbit is fixed if the radius is given. Equations 49-19, 49-20, and 49-21 show this specifically for the energies K, U, and E. From Eq. 49-19 we can show that the linear speed v for the electron is also given in terms of r by

$$v = \sqrt{\frac{e^2}{4\pi\epsilon_0 mr}}. \tag{49-22}$$

The rotational frequency ν_0 follows at once from

$$\nu_0 = \frac{v}{2\pi r} = \sqrt{\frac{e^2}{16\pi^3\epsilon_0 mr^3}}. \tag{49-23}$$

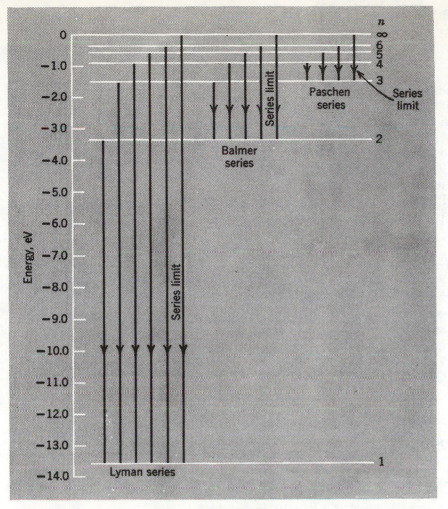

figure 49-19
An energy level diagram for hydrogen showing the quantum number n for each level and some of the transitions that appear in the spectrum. An infinite number of levels is crowded in between the levels marked $n = 6$ and $n = \infty$. Compare this figure carefully with Fig. 49-18.

The linear momentum p follows from Eq. 49-22:

$$p = mv = \sqrt{\frac{me^2}{4\pi\epsilon_0 r}}. \tag{49-24}$$

The angular momentum L is given by

$$L = pr = \sqrt{\frac{me^2 r}{4\pi\epsilon_0}}. \tag{49-25}$$

Thus if r is known, the orbit parameters K, U, E, v, ν_o, p, and L are also known. If any one of these quantities is quantized, all of them must be.

At this stage Bohr had no rules to guide him and so made (after some indirect reasoning which we do not reproduce) a bold hypothesis, namely, that the necessary quantization of the orbit parameters shows up most simply when applied to the angular momentum and that, specifically, L can take on only values given by

$$L = n\frac{h}{2\pi} \qquad n = 1, 2, 3, \ldots . \tag{49-26}$$

Planck's constant appears again in a fundamental way; the integer n is a *quantum number*.

Combining Eqs. 49-25 and 49-26 leads to

$$r = n^2 \frac{h^2\epsilon_0}{\pi me^2} \qquad n = 1, 2, 3, \ldots, \tag{49-27}$$

which tells how r is quantized. Substituting Eq. 49-27 into Eq. 49-21 produces

$$E = -\frac{me^4}{8\epsilon_0^2 h^2} \frac{1}{n^2} \qquad n = 1, 2, 3, \ldots, \qquad (49\text{-}28)$$

which gives directly the energy values of the allowed stationary states.

Figure 49-19 shows the energies of the stationary states and their associated quantum numbers. Equation 49-27 shows that the orbit radius increases as n^2. The upper level in Fig. 49-19, marked $n = \infty$, corresponds to a state in which the electron is completely removed from the atom (that is, $E = 0$ and $r = \infty$). Figure 49-19 also shows some of the quantum jumps that take place between the different stationary states.

Combining Eqs. 49-18 and 49-28 allows us to write a completely theoretical formula for the frequencies of the lines in the hydrogen spectrum. It is

$$\nu = \frac{me^4}{8\epsilon_0^2 h^3} \left(\frac{1}{j^2} - \frac{1}{k^2} \right) \qquad (49\text{-}29)$$

in which j and k are integers describing, respectively, the lower and the upper stationary states. The corresponding wavelengths can easily be found from $\lambda = c/\nu$. Table 49-1 shows some wavelengths so calculated; it should be compared carefully with Figs. 49-18 and 49-19.

Table 49-1
The hydrogen spectrum
(Some selected lines)

Name of series	Quantum Number		Wavelength, Å
	j (lower state)	k (upper state)	
Lyman	1	2	1216
	1	3	1026
	1	4	970
	1	5	949
	1	6	940
	1	∞	912
Balmer	2	3	6563
	2	4	4861
	2	5	4341
	2	6	4102
	2	7	3970
	2	∞	3650
Paschen	3	4	18751
	3	5	12818
	3	6	10938
	3	7	10050
	3	8	9546
	3	∞	8220

EXAMPLE 5

Calculate the binding energy of the hydrogen atom (the energy binding the electron to the nucleus) from Eq. 49-28.

The binding energy is numerically equal to the energy of the lowest state in Fig. 49-19. The largest negative value of E in Eq. 49-28 is found for $n = 1$. This yields

$$E = -\frac{me^4}{8\epsilon_0^2 h^2}$$

$$= -\frac{(9.11 \times 10^{-31} \text{ kg})(1.60 \times 10^{-19} \text{ C})^4}{(8)(8.85 \times 10^{-12} \text{ C}^2/\text{N} \cdot \text{m}^2)^2 (6.63 \times 10^{-34} \text{ J} \cdot \text{s})^2}$$

$$= -2.17 \times 10^{-18} \text{ J} = -13.6 \text{ eV},$$

which agrees with the experimentally observed binding energy for hydrogen.

In principle any negatively charged particle could replace the electron in the hydrogen (or other) atom, still be bound to the nucleus, and orbit the nucleus in a manner similar to the electron. Such a newly formed atom is called an *exotic atom*.* In practice, it is difficult to build exotic atoms, for many of the substitute negative particles are unstable or, when captured by a nucleus, form unstable atoms. Nevertheless, such atoms have been made and studied.

The Bohr model also applies to an exotic hydrogenlike atom, and Eq. 49-27 shows that the radii are smaller the larger the mass of the orbiting particle. A muon (see Appendix F), for example, will orbit much closer to the nucleus than an electron and its behavior will be strongly affected by the distribution of charge within the nucleus. Studies of muonic atoms (see Problem 49-46) give us information, therefore, about how the positive charges (protons) are distributed within nuclei.

In general, the motivation to investigate exotic atoms is either to learn about the size and detailed structure of the nucleus or to learn about properties (such as mass) of the negative particles themselves, which are more easily determined when they are in a bound system than when they are moving about freely.

49-9
THE CORRESPONDENCE PRINCIPLE

Although all theories in physics have limitations, they usually do not break down abruptly but in a continuous way, yielding results that agree less and less well with experiment. Thus the predictions of Newtonian mechanics become less and less accurate as the speed is made to approach that of light. A similar relationship must exist between quantum physics and classical physics; it remains to find the circumstances under which the latter theory is revealed as a special case of the former.

The radius of the lowest energy state in hydrogen (the so-called *ground state*) is found by putting $n = 1$ in Eq. 49-27; it turns out to be 5.3×10^{-11} m. If $n = 10,000$, however, the radius is $(10,000)^2$ times as large or 5.3 mm. This "atom" is so large that we suspect that its behavior should be accurately described by classical physics. Let us test this by computing the frequency of the emitted light on the basis of both classical and quantum assumptions. These calculations should differ at small quantum numbers but should agree at large quantum numbers. The fact that quantum physics reduces to classical physics at large quantum numbers is called the *correspondence principle*. This principle, credited to Niels Bohr, was very useful during the years in which quantum physics was being developed. Bohr, in fact, based his theory of the hydrogen atom on correspondence principle arguments.

Classically, the frequency of the light emitted from an atom is equal to ν_0, its frequency of revolution† in its orbit. We can express this in terms of a quantum number n by combining Eqs. 49-23 and 49-27 to

---

* See "Exotic Atoms" by Clyde E. Wiegand in *Scientific American*, November 1972.
† Integral multiples of the frequency also exist but may be ignored without affecting the present argument.

obtain

$$\nu_0 = \frac{me^4}{8\epsilon_0^2 h^3} \frac{2}{n^3}. \tag{49-30}$$

Quantum physics predicts that the frequency ν of the emitted light is given by Eq. 49-29. Considering a transition between an orbit with quantum number $k = n$ and one with $j = n - 1$ leads to

$$\nu = \frac{me^4}{8\epsilon_0^2 h^3} \left[\frac{1}{(n-1)^2} - \frac{1}{n^2} \right]$$

$$= \frac{me^4}{8\epsilon_0^2 h^3} \left[\frac{2n-1}{(n-1)^2 n^2} \right]. \tag{49-31}$$

As $n \to \infty$ the expression in the square brackets above approaches $2/n^3$ so that $\nu \to \nu_0$ as $n \to \infty$. Table 49-2 illustrates this example of the correspondence principle.

Table 49-2
The correspondence principle as applied to the hydrogen atom

Quantum Number, n	Frequency of Revolution in Orbit (Eq. 49-30) Hz	Frequency of Transition to Next Lowest State (Eq. 49-31) Hz	Difference, %
2	8.20×10^{14}	24.6×10^{14}	67
5	5.26×10^{13}	7.38×10^{13}	29
10	6.57×10^{12}	7.72×10^{12}	14
50	5.25×10^{10}	5.42×10^{10}	3
100	6.578×10^9	6.677×10^9	1.5
1,000	6.5779×10^6	6.5878×10^6	0.15
10,000	6.5779×10^3	6.5789×10^3	0.015
25,000	4.2099×10^2	4.2102×10^2	0.007

questions

1. "Pockets" formed by the coals in a coal fire seem brighter than the coals themselves. Is the temperature in such pockets appreciably higher than the surface temperature of an exposed glowing coal?

2. The relation $\mathscr{R} = \sigma T^4$ (Eq. 49-2) is exact for true cavities and holds for all temperatures. Why don't we use this relation as the basis of a definition of temperature at, say, 100°C?

3. Do all incandescent solids obey the fourth-power law of temperature, as Eq. 49-3 seems to suggest?

4. A hole in the wall of a cavity radiator is sometimes called a *black body*. Why?

5. It is stated that if we look into a cavity whose walls are maintained at a constant temperature, no details of the interior are visible. Does this seem reasonable?

6. A piece of metal glows with a bright red color at 1100 K. At this same temperature, however, a piece of quartz does not glow at all. Explain. (*Hint:* Quartz is transparent to visible light.)

7. Are there quantized quantities in classical physics? If so, give examples. Is energy quantized in classical physics?

8. Does it make sense to speak of charge quantization in physics? How, if at all, is this different from energy quantization?

9. Show that Planck's constant has the dimensions of angular momentum.

Does this necessarily suggest that angular momentum is a quantized quantity?

10. For quantum effects to be "everyday" phenomena in our lives, what order of magnitude value would h need to have? (See G. Gamow, *Mr. Tompkins in Wonderland* (Cambridge University Press, Cambridge, 1957), for a delightful popularization of a world in which the physical constants c, G, and h make themselves obvious.)

11. In Fig. 49-7, why doesn't the photoelectric current rise vertically to its maximum (saturation) value when the applied potential difference is slightly more positive than V_0?

12. In the photoelectric effect, why does the existence of a cutoff frequency speak in favor of the photon theory and against the wave theory?

13. Why are photoelectric measurements so sensitive to the nature of the photoelectric surface?

14. Why is it that even for incident radiation that is monochromatic the photoelectrons are emitted with a spread of velocities?

15. How can a *photon* energy be given by $E = h\nu$ (Eq. 49-10) when the very presence of the frequency ν in the formula implies that light is a *wave*?

16. Does Einstein's theory of photoelectricity, in which light is postulated to be a photon, invalidate Young's interference experiment?

17. Explain the statement that one's eyes could not detect faint starlight if light were not corpuscular.

18. Assume that the emission of photons from a source of radiation is random in direction. Would you expect the intensity (or energy density) to vary inversely as the square of the distance from the source in the photon theory as it does in the wave theory?

19. What is the direction of a Compton scattered electron with maximum kinetic energy compared with the direction of the incident monochromatic photon beam?

20. Why, in the Compton scattering picture (Fig. 49-11), would you expect $\Delta\lambda$ to be independent of the materials of which the scatterer is composed?

21. Why don't we observe a Compton effect with visible light?

22. Light from distant stars is Compton scattered many times by free electrons in outer space before reaching us. This shifts the light toward the red. How can this shift be distinguished from the Doppler red shift due to the motion of receding stars?

23. Why was the Balmer series, rather than the Lyman or Paschen series, the first to be detected and analyzed in the hydrogen spectrum?

24. Only a relatively small number of Balmer lines can be observed from laboratory discharge tubes, whereas a large number are observed in stellar spectra. Explain this in terms of the small density, high temperature, and large volume of gases in stellar atmospheres.

25. In Bohr's theory for the hydrogen atom orbits, what is the implication of the fact that the potential energy is negative and is greater in magnitude than the kinetic energy?

26. (a) Can a hydrogen atom absorb a photon whose energy exceeds its binding energy (13.6 eV)? (b) What minimum energy must a photon have to initiate the photoelectric effect in hydrogen gas?

27. On emitting a photon, the hydrogen atom recoils to conserve momentum. Explain the fact that the energy of the emitted photon is less than the energy difference between the energy levels involved in the emission process.

28. Would you expect to observe all the lines of atomic hydrogen if such a gas were excited by electrons of energy 13.6 eV? Explain.

29. How would you estimate the temperature of hydrogen gas at which atomic collisions cause significant ionization of the atoms?

30. List and discuss the assumptions made by Planck in connection with the cavity radiation problem, by Einstein in connection with the photoelectric effect, and by Bohr in connection with the hydrogen atom problem.

31. Describe several methods that can be used to determine experimentally the value of Planck's constant h.

32. Discuss Example 1 in terms of the correspondence principle.

33. According to classical mechanics, an electron moving in an orbit should be able to do so with any angular momentum whatever. According to Bohr's theory of the hydrogen atom, however, the angular momentum is quantized according to $L = nh/2\pi$. Reconcile these two statements, using the correspondence principle.

34. The correspondence principle is expressed by taking n to infinity or by taking h to zero. Explain the relation between these two techniques for finding the classical limit.

35. Can you use the device of letting $h \to 0$ to obtain classical results from quantum results in the case of the photoelectric effect? Explain.

36. (a) Newton's light corpuscles were assumed to behave according to the laws of Newtonian mechanics. Is the photon concept a return to this idea of a light corpuscle? (b) The ether was invented as a medium in which light waves undulate. Does the photon concept eliminate the need for an ether?

problems

SECTION 49-2

1. Figure 49-1 compares the spectral radiancy of tungsten at 2000 K with that of a cavity radiator. If the radiancy $\mathscr{R}$—the area under the curve for tungsten—is 23.3 W/cm², verify that the emissivity of tungsten at 2000 K is 0.259.

2. What is the power radiated from a nichrome wire 1.0 m long and having a diameter of 0.060 in. at a temperature of 800°C if the emissivity of nichrome is 0.92?

3. The power rating of a light bulb tells how much electrical power is supplied. (a) An incandescent bulb of power rating $P = 100$ W has a tungsten filament of diameter $d = 0.40$ mm and length $l = 30$ cm. The emissivity of tungsten is 0.26. At what temperature does the filament operate? (b) A fluorescent tube rated at 40 W gives as much visible light as a 100-W incandescent bulb. Explain why.   *Answer:* (a) 1470 K.

4. The average rate of solar radiation incident per unit area on the earth is 0.485 cal/cm² · min (or 335 W/m²). (a) Explain the consistency of this number with the solar constant (the solar energy falling per unit time at normal incidence on unit area of the earth's surface) whose value is 1.94 cal/cm² · min (or 1340 W/m²). (b) Consider the earth to behave like a cavity radiating energy into space at this same rate. What surface temperature would the earth have under these circumstances?

5. (a) Assuming the surface temperature of the sun to be 5700 K, use the Stefan-Boltzmann law to determine the rest mass lost per second to radiation by the sun. Take the sun's diameter to be 1.4×10^9 m. (b) What fraction of the sun's rest mass is lost each year by electromagnetic radiation? Take the sun's rest mass to be 2.0×10^{30} kg.
Answer: (a) 4.1×10^9 kg. (b) 6.5×10^{-14}.

6. An oven with an inside temperature $T = 227$°C is in a room having a temperature $T_r = 27$°C. There is a small opening of area 5.0 cm² in one side of the oven. How much net power is transferred from the oven to the room? (*Hint:* Consider both oven and room as cavities.)

SECTION 49-3

7. The wavelength λ_{max} at which the spectral radiancy has its maximum value per unit wavelength for a particular temperature T is given by the

Wien displacement law, $\lambda_{max} T =$ constant. At what wavelength does a cavity radiator at 6000 K radiate most per unit wavelength? The experimentally determined value of Wien's constant is 2.898×10^{-3} m·K.
Answer: 4800 Å = 480 nm.

8. Show that Wien's law is a special case of Planck's law (Eq. 49-6) for short wavelengths or low temperatures.

9. Prove the Stefan-Boltzmann law by showing directly from Planck's formula that

$$\int_0^\infty \mathscr{R}_\lambda \, d\lambda = AT^4$$

where A is a constant.

10. A cavity whose walls are held at 4000 K has a circular aperture 5.0 mm in diameter. (*a*) At what rate does energy in the visible range (defined to extend from 0.40 to 0.70 μm) escape from this hole? (*b*) What fraction of the total radiation escaping from the cavity does this represent? Solve either analytically or graphically.

SECTION 49-4

11. In Example 2 suppose that the "target" is a single gas atom of 1.0Å (= 0.10 nm) radius and that the intensity of the light source is reduced to 1.0×10^{-5} W. If the binding energy of the most loosely bound electron in the atom is 2.0 eV, what time lag for the photoelectric effect is expected on the basis of the wave theory of light? *Answer:* 10 years.

SECTION 49-5

12. Show that the energy E of a photon (in eV) is related to the wavelength λ (in nm) by

$$E = 1.24 \times 10^3/\lambda.$$

13. Solar radiation falls on the earth at a rate of 2.0 cal/cm²·min. How many photons/cm²·min is this, assuming an average wavelength of 550 nm?
Answer: 2.3×10^{19}.

14. A 100-W sodium vapor lamp radiates uniformly in all directions. (*a*) At what distance from the lamp will the average density of photons be 10/cm³? (*b*) What is the average density of photons 2.0 m from the lamp? Assume the light to be monochromatic, with $\lambda = 589$ nm (= 5890 Å).

15. An atom absorbs a photon having a wavelength of 375 nm and immediately emits another photon having a wavelength of 580 nm. How much energy was absorbed by the atom in this process? Ease the computation by using the result of Problem 12. *Answer:* 1.17 eV.

16. What are (*a*) the frequency, (*b*) the wavelength, and (*c*) the momentum of a photon whose energy equals the rest energy of the electron?

17. The energy required to remove an electron from sodium is 2.3 eV. Does sodium show a photoelectric effect for orange light, with $\lambda = 680$ nm (= 6800 Å)? *Answer:* No.

18. You wish to pick a substance for a photocell operable with visible light. Which of the following will do (work function in parentheses): tantalum (4.2 eV); tungsten (4.5 eV); aluminum (4.2 eV); barium (2.5 eV); lithium (2.3 eV)?

19. Incident photons strike a sodium surface having a work function $E_0 = 2.2$ eV causing photoelectric emission. When a stopping potential $V_0 = 5.0$ V is imposed, there is no photocurrent. What is the wavelength of the incident photons? *Answer:* 172 nm.

20. Find the maximum kinetic energy of photoelectrons if the work function of the material is 2.0×10^{-19} J and the frequency of the radiation is 3.0×10^{15} Hz.

21. (*a*) If the work function for a metal is 1.8 eV, what would be the stopping

potential for light having a wavelength of 4000 Å? (b) What would be the maximum velocity of the emitted photoelectrons at the metal's surface? *Answer:* (a) 1.3 V. (b) 6.8×10^5 m/s.

22. Light of a wavelength 200 nm falls on an aluminum surface. In aluminum 4.2 eV are required to remove an electron. What is the kinetic energy of (a) the fastest and (b) the slowest emitted photoelectrons? (c) What is the stopping potential? (d) What is the cutoff wavelength for aluminum?

23. The work function for a clean lithium surface is 2.3 eV. Make a rough plot of the stopping potential V_o versus the frequency of incident light for such a surface.

24. Show, by analyzing a collision between a photon and a free electron (using relativistic mechanics), that it is impossible for a photon to give all of its energy to the free electron. In other words, the photoelectric effect cannot occur for completely free electrons; the electrons must be bound in a solid or in an atom.

SECTION 49-6

25. Photons of wavelength 0.024 Å are incident on free electrons. (a) Find the wavelength of a photon which is scattered 30° from the incident direction. (b) Do the same if the scattering angle is 120°.
Answer: (a) 0.027 Å. (b) 0.060 Å.

26. What is the maximum kinetic energy of the Compton-scattered electrons from a sheet of copper struck by a monochromatic photon beam in which the incident photons each have a wavelength of 1.4×10^{-3} nm? (See Example 4.)

27. An X-ray photon of wavelength $\lambda = 0.10$ Å strikes an electron head on ($\varphi = 180°$). Determine (a) the change in wavelength of the photon, (b) the change in energy of the photon, and (c) the final kinetic energy of the electron.   *Answer:* (a) 0.049 Å. (b) −41 keV. (c) 41 keV.

28. A 0.20-nm photon falling on a carbon block is scattered by a Compton collision and its frequency is shifted by 0.010%. (a) Through what angle is the photon scattered? (b) How much energy does the electron which scattered the photon gain? [Note that for any wave motion, $\Delta\nu = -\left(\dfrac{c}{\lambda^2}\right)\Delta\lambda$.]

29. Calculate the percent change in photon energy for a Compton collision with φ in Fig. 49-11 equal to 90° for radiation in (a) the microwave range, with $\lambda = 3.0$ cm, (b) the visible range, with $\lambda = 5000$ Å = 500 nm, (c) the X-ray range, with $\lambda = 1.0$ Å = 0.10 nm, and (d) the gamma-ray range, the energy of the gamma-ray photons being 1.0 MeV. What are your conclusions about the importance of the Compton effect in these various regions of the electromagnetic spectrum, judged solely by the criterion of energy loss in a single Compton encounter?
Answer: (a) 8.1×10^{-9}%. (b) 4.9×10^{-4}%. (c) 2.4%. (d) 68%.

30. A photon "hits" an electron "head-on" and recoils backward directly along the line of incidence. If the electron moves off at a speed βc, where $\beta \ll 1$ (=10^{-3}, for example), show that the ratio of the final electron kinetic energy to the initial photon energy is just β. (*Hint:* Set up the problem as a non-relativistic Compton "collision.")

31. Carry out the necessary algebra to eliminate v and θ from Eqs. 49-13, 49-15, and 49-16 to obtain the Compton shift relation (Eq. 49-17).

SECTION 49-8

32. A line in the X-ray spectrum of gold consists of photons all having nearly the same wavelength 0.185 Å. If the energy in these photons comes from the atoms jumping from one specific energy level, −13.7 KeV, to another lower one, what is the second energy?

33. Light of wavelength 4863 Å = 486.3 nm is emitted by a hydrogen atom. (a) What transition of the hydrogen atom is responsible for this radiation? (b) To what series does this radiation belong?
Answer: (a) $n = 4$ to $n = 2$. (b) The Balmer series.

34. Show on an energy-level diagram for hydrogen, the quantum numbers corresponding to a transition in which the wavelength of the emitted photon is 1216 Å = 121.6 nm.

35. What are (a) the energy, (b) momentum, and (c) wavelength of the photon that is emitted when a hydrogen atom undergoes a transition from the state $n = 3$ to $n = 1$? *Answer:* (a) 12 eV. (b) 6.5×10^{-27} kg·m/s. (c) 1030 Å.

36. (a) Using Bohr's formula for the frequencies of the lines of the hydrogen spectrum, calculate the three longest wavelengths of the Balmer series. (b) Between what wavelength limits does the Balmer series lie?

37. In the ground state of the hydrogen atom, according to Bohr's theory, what are (a) the quantum number, (b) the orbit radius, (c) the angular momentum, (d) the linear momentum, (e) the angular velocity, (f) the linear speed, (g) the force on the electron, (h) the acceleration of the electron, (i) the kinetic energy, (j) the potential energy, and (k) the total energy?
Answer: (a) 1. (b) 5.3×10^{-11} m. (c) 1.1×10^{-34} J·s. (d) 2.0×10^{-24} kg·m/s. (e) 4.1×10^{16} rad/s. (f) 2.2×10^6 m/s. (g) 8.2×10^{-8} N. (h) 9.0×10^{22} m/s². (i) +13.6 eV. (j) −27.2 eV. (k) −13.6 eV.

38. How do the quantities (b) to (k) in Problem 37 vary with the quantum number?

39. How much energy is required to remove an electron from a hydrogen atom in the state with $n = 8$? *Answer:* 0.21 eV.

40. A hydrogen atom is excited from a state with $n = 1$ to one with $n = 4$. (a) Calculate the energy that must be absorbed by the atom. (b) Calculate and display on an energy-level diagram the different photon energies that may be emitted if the atom returns to the $n = 1$ state. (c) Calculate the recoil speed of the hydrogen atom, assumed initially at rest, if it makes the transition from $n = 4$ to $n = 1$ in a single quantum jump.

41. A hydrogen atom in a state having a *binding energy* (the energy required to remove an electron) of 0.85 eV makes a transition to a state with an *excitation energy* (the difference in energy between the state and the ground state) of 10.2 eV. (a) Find the energy of the emitted photon. (b) Show this transition on an energy-level diagram for hydrogen, labeling the appropriate quantum numbers.
Answer: (a) 2.6 eV. (b) $n = 4$ to $n = 2$.

42. A neutron, with kinetic energy of 6.0 eV, collides with a resting hydrogen atom in its ground state. Show that this collision must be elastic (that is, kinetic energy must be conserved).

43. From the energy-level diagram for hydrogen, explain the observation that the frequency of the second Lyman-series line is the sum of the frequencies of the first Lyman-series line and the first Balmer-series line. This is an example of the empirically discovered *Ritz combination principle*. Use the diagram to find some other valid combination.

44. Using Bohr's theory, calculate the energy required to remove the electron from the ground state of singly ionized helium, that is, helium with one electron removed.

45. Use Bohr's theory to compare the spectrum of singly ionized helium with the spectrum of hydrogen.
Answer: $\lambda_{He} = \frac{1}{4}\lambda_H$, for corresponding spectral lines.

46. *Muonic atoms.* Apply Bohr's theory to a muonic atom, which consists of a nucleus of charge Ze with a negative muon (an elementary particle with a charge of −e and a mass m that is 207 times as large as the electron mass) circulating about it. Calculate (a) the radius of the first Bohr orbit, (b) the ionization energy, and (c) the wavelength of the most energetic photon that can be emitted. Assume that the muon is circulating about a hydrogen nucleus (Z = 1).

47. *Positronium.* Apply Bohr's theory to the positronium atom. This consists of a positive and a negative electron revolving around their center of mass, which lies halfway between them. (a) What relationship exists between this

spectrum and the hydrogen spectrum? (*b*) What is the radius of the ground state orbit? (*Hint:* It will be necessary to analyze this problem from first principles because this "atom" has no nucleus; both particles revolve about a point halfway between them.)
Answer: (*a*) Corresponding positronium wavelengths are longer by a factor of two. (*b*) Radius to center of mass is equal to the corresponding radius for hydrogen.

48. Perhaps an atom could be formed by an electron and a neutron binding together by gravitational forces. Calculate the ground state radius of an electron in such an atom by using a Bohr-type model in which the Coulomb attractive electrical force is replaced by the attractive gravitational force.

49. A diatomic gas molecule consists of two atoms of mass m separated by a fixed distance d rotating about an axis as indicated in Fig. 49-20. Assuming that its angular momentum is quantized as in the Bohr atom, determine (*a*) the possible angular velocities, and (*b*) the possible quantized rotational energies. (*c*) Show these on an energy-level diagram.
Answer: (*a*) $nh/\pi md^2$. (*b*) $n^2h^2/4\pi^2md^2$.

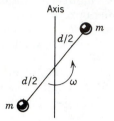

figure 49-20
Problem 49

SECTION 49-9

50. (*a*) If the angular momentum of the earth due to its motion around the sun were quantized according to Bohr's relation $L = nh/2\pi$, what would the quantum number be? (*b*) Could such quantization be detected if it existed?

51. In Table 49-2 show that the quantity in the last column is given by

$$\frac{100(\nu - \nu_0)}{\nu} \cong \frac{150}{n}$$

for large quantum numbers.

52. If an electron is revolving in an orbit at frequency ν_0, classical electromagnetism predicts that it will radiate energy not only at this frequency but also at $2\nu_0$, $3\nu_0$, $4\nu_0$, and so on. Show that this is also predicted by Bohr's theory of the hydrogen atom in the limiting case of large quantum numbers.

50
waves and particles

In 1924 Louis de Broglie of France reasoned that (*a*) nature is strikingly symmetrical in many ways; (*b*) our observable universe is composed entirely of light and matter; (*c*) if light has a dual, wave-particle nature, perhaps matter has also. Since matter was then regarded as being composed of particles, de Broglie's reasoning suggested that one should search for a wavelike behavior for matter.

De Broglie's suggestion might not have received serious attention had he not predicted what the expected wavelength of the so-called matter waves would be. We recall that about 1680 Huygens put forward a wave theory of light that did not receive general acceptance, not only because it seemed to contradict Newton's particle — or corpuscular — theory but also because Huygens was not able to state what the wavelength of the light was. When Thomas Young rectified this defect in 1800, the wave theory of light started on its way to acceptance.

De Broglie assumed that the wavelength of the predicted matter waves was given by the same relationship that held for light, namely, Eq. 49-14, or

$$\lambda = \frac{h}{p},$$ (50-1)

which connects the wavelength of a light wave with the momentum of the associated photons. The dual nature of light shows up strikingly in this equation and also in Eq. 49-10 ($E = h\nu$). Each equation contains within its structure both a wave concept (ν and λ) and a particle concept

50-1
MATTER WAVES

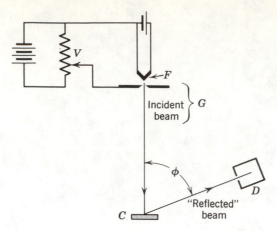

(E and p). De Broglie predicted that the wavelength of matter waves would also be given by Eq. 50-1, where p would now be the momentum of the particle of matter.

EXAMPLE 1

What wavelength is predicted by Eq. 50-1 for a beam of electrons whose kinetic energy is 100 eV?

We can find the speed of the electrons from $K = \frac{1}{2}mv^2$, or

$$v = \sqrt{\frac{2K}{m}} = \sqrt{\frac{(2)(100 \text{ eV})(1.6 \times 10^{-19} \text{ J/eV})}{9.1 \times 10^{-31} \text{ kg}}}$$

$$= 5.9 \times 10^6 \text{ m/s}.$$

The momentum follows from

$$p = mv = (9.1 \times 10^{-31} \text{ kg})(5.9 \times 10^6 \text{ m/s}) = 5.4 \times 10^{-24} \text{ kg} \cdot \text{m/s}.$$

The wavelength (called the *de Broglie wavelength*) is found from Eq. 50-1 or

$$\lambda = \frac{h}{p} = \frac{6.6 \times 10^{-34} \text{ J} \cdot \text{s}}{5.4 \times 10^{-24} \text{ kg} \cdot \text{m/s}} = 1.2 \text{ Å } (=0.12 \text{ nm}).$$

This is the same order of magnitude as the size of an atom or the spacing between adjacent planes of atoms in a solid.

In 1926 Elsasser pointed out that the wave nature of matter might be tested in the same way that the wave nature of X-rays was first tested, namely, by allowing a beam of electrons of the appropriate energy to fall on a crystalline solid. The atoms of the crystal serve as a three-dimensional array of diffracting centers for the electron "wave"; we should look for strong diffracted peaks in certain characteristic directions, just as for X-ray diffraction.

This idea was tested by C. J. Davisson and L. H. Germer in this country and by G. P. Thomson in Scotland.* Figure 50-1 shows the apparatus of Davisson and Germer. Electrons from a heated filament are accelerated by an adjustable potential difference V and emerge from the "electron gun" G with kinetic energy eV. This electron beam is allowed to

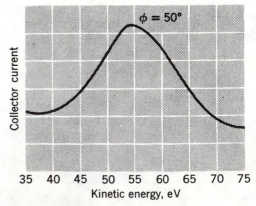

figure 50-2
The collector current in detector D in Fig. 50-1 as a function of the kinetic energy of the incident electrons, showing a diffraction maximum. The angle ϕ in Fig. 50-1 is adjusted to 50°. If an appreciably smaller or larger value is used, the diffraction maximum disappears.

* For an historical account of Thomson's researches see "Early Work in Electron Diffraction" by Sir George Thomson, *Scientific American,* December 1961. Interestingly, G. P. Thomson, who demonstrated that electrons are wave-like, is the son of J. J. Thomson who demonstrated (see Section 33-8) that they are particle-like. Both father and son received Nobel prizes, 31 years apart.

fall at normal incidence on a single crystal of nickel at C. Detector D is set at a particular angle ϕ and readings of the intensity of the "reflected" beam are taken at various values of the accelerating potential V. Figure 50-2 shows that a strong beam occurs at $\phi = 50°$ for $V = 54$ volts.

All such strong "reflected" beams can be accounted for by assuming that the electron beam has a wavelength, given by $\lambda = h/p$, and that "Bragg reflections" occur from certain families of atomic planes precisely as described for X-rays in Section 47-6.

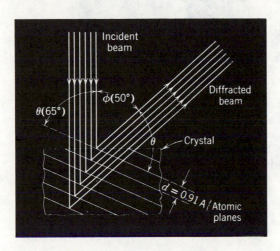

figure 50-3
The strong diffracted beam at $\phi = 50°$ and $V = 54$ volts arises from wavelike "reflection" from the family of atomic planes shown, for $d = 0.91$ Å. The Bragg angle θ is 65° For simplicity, refraction of the wave as it enters and leaves the surface is ignored.

Figure 50-3 shows such a Bragg reflection, obeying the Bragg relationship

$$m\lambda = 2d \sin \theta \qquad m = 1, 2, 3, \ldots. \qquad (50\text{-}2)$$

For the conditions of Fig. 50-3 the effective interplanar spacing d can be shown by X-ray analysis to be 0.91 Å. Since ϕ equals 50°, it follows that θ equals $90° - \frac{1}{2} \times 50°$ or 65°. The wavelength to be calculated from Eq. 50-2, if we assume $m = 1$, is

$$\lambda = 2d \sin \theta = 2(0.91 \text{ Å})(\sin 65°) = 1.65 \text{ Å} = 0.165 \text{ nm}.$$

The wavelength calculated from the de Broglie relationship $\lambda = h/p$ is, for 54-eV electrons (see Example 1), 1.64 Å = 0.164 nm. This excellent agreement, combined with much similar evidence*, is a convincing argument for believing that electrons are wavelike in some circumstances.

Not only electrons but all other particles, charged or uncharged, show wavelike characteristics. Beams of slow neutrons from nuclear reactors are routinely used to investigate the atomic structure of solids.† See Fig. 50-4, which suggests the wave-like character of (a) electrons and (b) neutrons.

The evidence for the existence of matter waves with wavelengths given by Eq. 50-1 is strong indeed. Nevertheless, the evidence that matter is composed of particles remains equally strong; see Fig. 10-12. Thus, for matter as for light, we must face up to the existence of a dual character; matter behaves in some circumstances like a particle and in others like a wave.

---

* An interferometer operating with electron beams has been designed, and diffraction patterns have been obtained from electron beams passing through a multiple slit system. See "An Experiment on Electron Interference" by O. Donati, G. F. Missiroli, and G. Possi, *American Journal of Physics*, May 1973, and "Electron Diffraction at Multiple Slits" by Clauss Jönsson, *American Journal of Physics*, January 1974.

† An apparatus that exhibits the interference of neutron waves has been constructed, as well. See "Neutrons as Waves" by Dietrick E. Thomsen in *Science News*, April 24, 1976.

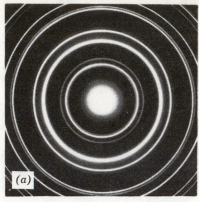 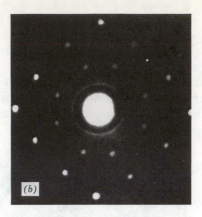

figure 50-4

In Figure 50-4a we show a Debye-Scherrer pattern of *electron* diffraction by gold crystals (compare with the X-ray diffraction pattern of Fig. 47-17) and in Fig. 50-4b we show a Laue pattern of diffraction of *neutrons* from a nuclear reactor by a single sodium chloride crystal (compare with the X-ray diffraction pattern of Fig. 47-12). (Figure 50-4a is from U. Fano and L. Fano, *Basic Physics of Atoms and Molecules*, John Wiley and Sons, 1959, and Fig. 50-4b is from Blackwood et al., *Outline of Atomic Physics*, John Wiley and Sons, 1955.)

The motion of electrons in beams is not bounded or limited in the beam direction. We can make an analogy to a sound wave in a long gas-filled tube, a wave traveling down a long string, or an electromagnetic wave in a long waveguide. All four cases can be described by appropriate traveling waves and, significantly, waves of *any* wavelength (within a certain range) can be propagated.

Let these last three waves be bounded by imposing physical restrictions. For the sound wave this corresponds to inserting end walls on a section of the long gas-filled pipe, thus forming an acoustic resonant cavity (Fig. 38-5). For the waves in the string it corresponds to removing a finite section of string and clamping it at each end, as a violin string (Fig. 20-4). For the electromagnetic wave it corresponds to inserting end caps on a finite length of waveguide, thus forming an electromagnetic resonant cavity (Fig. 38-6).

Two important changes occur: (a) the motions are now represented by standing rather than traveling waves and (b) only certain wavelengths (or frequencies) can now exist. This *quantization* of the wavelength is a direct result of confining the wave to a finite region. We expect that if electrons are limited in their motions by being localized in an atom, that (a) the electron motion can be represented by a standing matter wave, and (b) the electron motion will become quantized, that is, its energy can take on only certain discrete values.

De Broglie was able to derive the Bohr quantization condition for angular momentum by applying proper boundary conditions to matter waves in the hydrogen atom. Figure 50-5 suggests an instantaneous "snapshot" of a standing matter wave associated with an orbit of radius r. The de Broglie wavelength ($\lambda = h/p$) has been chosen so that the orbit of radius r contains an integral number n of the matter waves, or

50-2
ATOMIC STRUCTURE AND STANDING WAVES

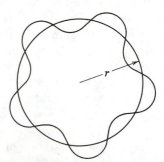

figure 50-5

Showing how an electron wave can be adjusted in wavelength to fit an integral number of times around the circumference of a Bohr orbit of radius r. This concept, like the Bohr orbit concept, is now regarded as oversimplified.

$$\frac{2\pi r}{\lambda} = \frac{2\pi r}{(h/p)} = n \qquad n = 1, 2, 3, \ldots.$$

This leads at once to

$$L = pr = n\frac{h}{2\pi} \qquad n = 1, 2, 3, \ldots .$$

which (see Section 49-8) is the Bohr quantization condition for L.

50-3
WAVE MECHANICS

The idea that the stationary states in atoms correspond to standing matter waves was taken up by Erwin Schrödinger in 1926 and used by him as the foundation of *wave mechanics,* one of several equivalent formulations of quantum physics.

An important quantity in wave mechanics is the *wave function* Ψ, which measures the "wave disturbance" of matter waves. For waves on strings the "wave disturbance" may be measured by a transverse displacement y; for sound waves it may be measured by a pressure variation p; for electromagnetic waves it may be measured by the electric field vector $\mathbf{E}$.

We make the physical meaning of the wave disturbance Ψ clear in Section 50-4. Meanwhile, let us study the wave function $\Psi(x)$ for a simple, one-dimensional problem, that of the possible motions of a particle of mass m confined between rigid walls of separation l as in Fig. 50-6b. The wave function can be obtained by analogy with a known mechanical problem, that of the natural modes of vibration of a string of length l, clamped at each end as in Fig. 50-6a.

In the vibrating string the boundary conditions require that nodes exist at each end. This means that the wavelength λ must be chosen so that

$$l = n\frac{\lambda}{2} \qquad n = 1, 2, 3, \ldots ,$$

or that the wavelength λ is "quantized" by the requirement that

$$\lambda = \frac{2l}{n} \qquad n = 1, 2, 3, \ldots ; \tag{50-3}$$

The wave disturbance for the string is represented by a standing wave whose spatial dependence is $A \sin kx$ (see Section 19-9), where A is a constant and $k(= 2\pi/\lambda)$ is the wave number. Since λ is quantized, k must be also, or

$$k = \frac{2\pi}{\lambda} = \frac{n\pi}{l} \qquad n = 1, 2, 3, \ldots ,$$

which leads to

$$y = A \sin\frac{n\pi x}{l} \qquad n = 1, 2, 3, \ldots . \tag{50-4}$$

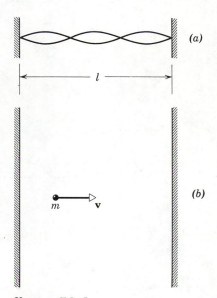

figure 50-6
(a) A stretched string of length l clamped between rigid supports. (b) A particle of mass m and velocity v confined to move between rigid walls a distance l apart.

Inspection of Eq. 50-4 shows that no matter what value of n we select, nodes exist at $x = 0$ and at $x = l$, as required by the boundary conditions. Figure 50-7 shows plots of this equation (the spatial dependence of the standing wave) for the modes of vibration of the string corresponding to $n = 1, 2,$ and 3.

Consider now the particle confined between rigid walls. Since we assume the walls to be perfectly rigid, the particle cannot penetrate them so that Ψ, which represents the particle motion in some way not

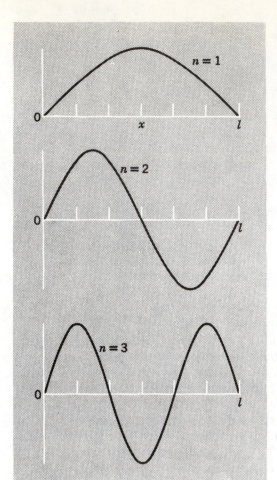

figure 50-7
Three quantized modes of vibration for the string of Fig. 50-6a. The figure also represents three of the quantized wave functions for the particle of Fig. 50-6b. The ordinate is a displacement amplitude in the first case and a wave-function amplitude in the second.

yet clearly specified (see Section 50-4), must vanish at $x = 0$ and $x = l$. The allowed wavelengths of the matter waves must be given by Eq. 50-3, or

$$\lambda = \frac{2l}{n}.$$

Replacing λ by h/p (see Eq. 50-1) leads to

$$p = \frac{nh}{2l}, \qquad (50\text{-}5)$$

which shows that the linear momentum of the particle is quantized. The momentum $p(= mv)$ is related to the energy E (which is entirely kinetic and is equal to $\frac{1}{2}mv^2$) by

$$p = \sqrt{2mE}. \qquad (50\text{-}6)$$

Combining Eqs. 50-5 and 50-6 leads to the quantization condition for E, or

$$E = n^2 \frac{h^2}{8ml^2} \qquad n = 1, 2, 3, \ldots \qquad (50\text{-}7)$$

The particle cannot have any energy, as we would expect classically, but only energies given by Eq. 50-7.

We describe the matter wave in strict analogy with Eq. 50-4, by

$$\Psi = A \sin\frac{n\pi x}{l} \qquad n = 1, 2, 3, \ldots. \qquad (50\text{-}8)$$

Figure 50-7 can serve equally well to show how the amplitude of the standing matter waves for the states of motion corresponding to $n = 1, 2,$ and 3 varies throughout the box. We see clearly in this problem how the act of localizing or bounding a particle leads to energy quantization.

EXAMPLE 2

Consider an electron ($m = 9.1 \times 10^{-31}$ kg) confined by electrical forces to move between two rigid "walls" separated by 1.0×10^{-9} m, which is about five atomic diameters. Find the quantized energy values for the three lowest stationary states.

From Eq. 50-7, for $n = 1$, we have

$$E = n^2 \frac{h^2}{8ml^2} = (1)^2 \frac{(6.6 \times 10^{-34} \text{ J} \cdot \text{s})^2}{(8)(9.1 \times 10^{-31} \text{ kg})(1.0 \times 10^{-9} \text{ m})^2}$$

$$= 6.0 \times 10^{-20} \text{ J} = 0.38 \text{ eV}.$$

The energies for the next two states ($n = 2$ and $n = 3$) are $2^2 \times 0.38$ eV $= 1.5$ eV and $3^2 \times 0.38$ eV $= 3.4$ eV.

EXAMPLE 3

Consider a grain of dust ($m = 1.0 \ \mu\text{g} = 1.0 \times 10^{-9}$ kg) confined to move between two rigid walls separated by 0.1 mm ($= 10^{-4}$ m). Its speed is only 10^{-6} m/s, so that it requires 100 s to cross the gap. What quantum number describes this motion?

The energy is

$$E \ (= K) = \tfrac{1}{2}mv^2 = \tfrac{1}{2}(10^{-9} \text{ kg})(10^{-6} \text{ m/s})^2$$

$$= 5 \times 10^{-22} \text{ J}.$$

Solving Eq. 50-7 for n yields

$$n = \sqrt{8mE}\,\frac{l}{h} = \sqrt{(8)(10^{-9} \text{ kg})(5 \times 10^{-22} \text{ J})} \left(\frac{10^{-4} \text{ m}}{6.6 \times 10^{-34} \text{ J} \cdot \text{s}}\right)$$

$$= 3 \times 10^{14}.$$

Even in these extreme conditions the quantized nature of the motion would never be apparent; we cannot distinguish experimentally between $n = 3 \times 10^{14}$ and $n = 3 \times 10^{14} + 1$. Classical physics, which fails completely for the problem of Example 2, works extremely well for this problem.

50-4
MEANING OF Ψ

Max Born first suggested that the quantity Ψ^2 at any particular point is a measure of the probability that the particle will be near that point. More exactly, if a volume element dV is constructed at that point, the probability that the particle will be found in the volume element at a given instant is $\Psi^2 \, dV$. This interpretation of Ψ provides a statistical connection between the wave and the associated particle; it tells us where the particle is likely to be, not where it is.

For the particle confined between rigid walls, the probability that the particle will lie between two planes that are distance x and $x + dx$ from one wall (see Fig. 50-8) is given by

$$\Psi^2 \, dx = A^2 \sin^2 \frac{n\pi x}{l} \, dx.$$

Thus the *probability density* is

$$\Psi^2 = A^2 \sin^2 \frac{n \pi x}{l} \qquad n = 1, 2, 3, \ldots . \qquad (50\text{-}9)$$

Figure 50-8 shows Ψ^2 for the three stationary states corresponding to $n = 1$, 2, and 3. Note that for $n = 1$ the particle is more likely to be near the center than the ends. This is in sharp contradiction to the results of classical physics, according to which the particle has the same probability of being located anywhere between the walls, as shown by the horizontal line in Fig. 50-8.

The problem of a particle confined between rigid walls has little real application in physics. We would prefer to illustrate the wave mechanics of Schrödinger by applying it to a more experimentally realizable situation, such as the hydrogen atom. Only mathematical complexity prevents us from doing this. We state without proof that when this problem is solved by wave mechanics, the motion of the electron in the ground state of the atom, defined by putting $n = 1$ in Eq. 49-28, is described by the following wave function:

$$\Psi = \sqrt{\frac{1}{\pi a^3}} \, e^{-r/a} \qquad (50\text{-}10)$$

where

$$a = \frac{h^2 \epsilon_0}{\pi m e^2}.$$

Putting $n = 1$ in Eq. 49-27 shows that a is the radius of the ground-state orbit in Bohr's theory. This special interpretation has little meaning in wave mechanics; a is taken here merely as a convenient unit of length when dealing with atomic problems, having the value $0.529 \text{ Å} = 0.0529$ nm.

Consider two hypothetical spherical shells centered on the nucleus of a hydrogen atom with radii r and $r + dr$. What is the probability $P(r)$ that the electron will lie between these shells, as a function of r?

This probability is $\Psi^2 \, dV$, where dV is the volume between the shells, or $4\pi r^2 \, dr$. Thus

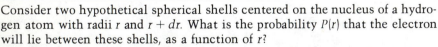

$$\Psi^2 \, dV = \left(\sqrt{\frac{1}{\pi a^3}} \, e^{-r/a} \right)^2 (4\pi r^2 \, dr) = P(r) \, dr$$

or

$$P(r) = \frac{4r^2}{a^3} \, e^{-2r/a}.$$

Figure 50-9 shows a plot of this function. Note that the most probable location for the electron corresponds to the first Bohr radius. Thus in wave mechanics we do not say that the electron in the $n = 1$ state in hydrogen goes around the nucleus in a circular orbit of 0.529 Å radius, but only that the electron is more likely to be found at this distance from the nucleus than at any other distance, either larger or smaller.

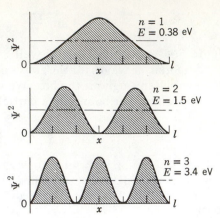

figure 50-8
The "probability density" for three states of motion of the particle of Fig. 50-6*b*, along with the corresponding quantized energies for the conditions of Example 2. The horizontal lines show the predictions of classical mechanics, in which the probability function is constant for all positions of the particle.

EXAMPLE 4

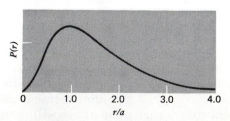

figure 50-9
The probability function for the ground state of the hydrogen atom, as calculated from wave mechanics. The separation between the nucleus and the electron is r; a is the radius of the first Bohr orbit (0.529×10^{-10} m), used here merely as a convenient unit of distance.

Some physicists believe that only those quantities that can be measured have any real meaning in physics. If we could focus a "super" microscope on an electron in an atom and see it moving around in an orbit, we would declare that such orbits have meaning. However, we shall show that it is fundamentally impossible to make such an observation—even with the most ideal instruments that could conceivably be constructed. Therefore, one might say that such orbits have no physical meaning.

We observe the moon traveling around the earth by means of the sunlight that it reflects in our direction. Now light transfers linear momentum to an object from which it is reflected. In principle, this reflected light would disturb the course of the moon in its orbit, although a little thought shows that this disturbing effect is negligible in this case.

For electrons the situation is quite different. Here, too, we can hope to "see" the electron only if we reflect light, or another particle, from it. In this case the recoil that the electron experiences when the light (photon) bounces from it completely alters the electron's motion in a way that cannot be avoided or even corrected for.

It is not surprising that the probability curve of Fig. 50-9 is the most detailed information that we can hope to obtain, by measurement, about the distribution of negative charge in the hydrogen atom. If orbits such as those envisaged by Bohr existed, they would be broken up completely in our attempts to verify their existence. Under these circumstances, we prefer to say that it is the probability function, and not the orbits, that represents physical reality.

The fact that we can't describe the motions of electrons in a classical way finds expression in the *uncertainty principle,* enunciated by Werner Heisenberg in 1927.* To formulate this principle, consider a beam of monoenergetic electrons of speed v_0 moving from left to right in Fig. 50-10. Let us set ourselves the task of measuring the position of a particular electron in the vertical (y) direction and also its velocity component v_y in this direction. If we succeed in carrying out these measurements with unlimited accuracy, we can then claim to have established the position and motion of the electron (or one component of it at least) with precision. However, we shall see that it is impossible to make these two measurements simultaneously with unlimited accuracy.

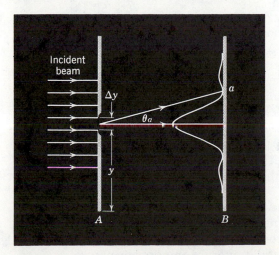

figure 50-10
An incident beam of electrons is diffracted at the slit in screen A, forming a typical diffraction pattern on screen B. If the slit is made narrower, the pattern becomes wider.

* See "The Uncertainty Principle" by G. Gamow in *Scientific American,* January 1958.

To measure y we block the beam with an absorbing screen A in which we put a slit of width Δy. If an electron gets through the slit, its vertical position must be known to this accuracy. By making the slit narrower, we can improve the accuracy of this vertical position measurement as much as we wish.

Since the electron behaves like a wave, it will undergo diffraction at the slit, and a photographic plate placed at B in Fig. 50-10 will reveal a typical diffraction pattern. The existence of this diffraction pattern means that there is an uncertainty in the values of v_y possessed by the electrons emerging from the slit. Let v_{ya} be the value of v_y that corresponds to an electron landing at the first minimum on the screen, marked by point a and described by a characteristic angle θ_a. We take v_{ya} as a rough measure of the uncertainty Δv_y in v_y for electrons emerging from the slit.

The first minimum in the diffraction pattern is given by Eq. 46-2, or

$$\sin \theta_a = \frac{\lambda}{\Delta y}.$$

If we assume that θ_a is small enough, we can write this equation as

$$\theta_a \cong \frac{\lambda}{\Delta y}. \tag{50-11}$$

To reach point a, v_{ya} $(= \Delta v_y)$ must be such that

$$\theta_a \cong \frac{\Delta v_y}{v_0}. \tag{50-12}$$

Combining Eqs. 50-11 and 50-12 leads to

$$\frac{\Delta v_y}{v_0} = \frac{\lambda}{\Delta y},$$

which we rewrite as $\qquad \Delta v_y \, \Delta y \cong \lambda v_0. \tag{50-13}$

Now λ, the wavelength of the electron beam, is given by h/p or h/mv_0; substituting this into Eq. 50-13 yields

$$\Delta v_y \, \Delta y \cong \frac{h v_0}{m v_0}.$$

We rewrite this as $\qquad \Delta p_y \, \Delta y \cong h. \tag{50-14}$

In Eq. 50-14 Δp_y $(= m \Delta v_y)$ is the uncertainty in our knowledge of the vertical momentum of the electrons; Δy is the uncertainty in our knowledge of their vertical position. The equation tells us that, since the product of these uncertainties is a constant, we cannot measure p_y and y simultaneously with unlimited accuracy.

If we want to improve our measurement of y (that is, if we want to reduce Δy), we use a finer slit. However, this will produce a wider diffraction pattern (see Eq. 50-11). A wider pattern means that our knowledge of the vertical momentum component of the electron has deteriorated, or, in other words, Δp_y has increased; this is exactly what Eq. 50-14 predicts.

The limits on measurement imposed by Eq. 50-14 have nothing to do with the crudity of our measuring instruments. We are permitted to postulate the existence of the finest conceivable measuring equipment. Equation 50-14 represents a fundamental limitation, imposed by nature.

Equation 50-14 is a derivation, for a special case, of a general principle known as the *uncertainty principle*. As applied to position and momentum measurements, it asserts that

$$\Delta p_x \, \Delta x \gtrsim h$$

$$\Delta p_y \, \Delta y \gtrsim h$$

$$\Delta p_z \, \Delta z \gtrsim h. \qquad (50\text{-}15)$$

Thus no component of the motion of an electron, free or bound, can be described with unlimited precision.

Planck's constant h probably appears nowhere that has more deep-seated significance than in Eq. 50-15. If this product had been zero instead of h, the classical ideas about particles and orbits would be correct; it would then be possible to measure both momentum and position with unlimited precision. The fact that h appears means that the classical ideas are wrong; the magnitude of h tells us under what circumstances these classical ideas must be replaced by quantum ideas. Gamow* has speculated, in an interesting and readable fantasy, what our world would be like if the constant h were much larger than it is, so that nonclassical ideas would be apparent to our sense perceptions.

EXAMPLE 5

An electron has a speed of 300 m/s, accurate to 0.010%. With what fundamental accuracy can we locate the position of this electron?

The electron momentum is

$$p = mv = (9.1 \times 10^{-31} \text{ kg})(300 \text{ m/s}) = 2.7 \times 10^{-28} \text{ kg}\cdot\text{m/s}.$$

The uncertainty in momentum is given to be 0.010% of this, or

$$\Delta p = (0.00010)(2.7 \times 10^{-28} \text{ kg}\cdot\text{m/s}) = 2.7 \times 10^{-32} \text{ kg}\cdot\text{m/s}.$$

The minimum uncertainty in position, from Eq. 50-15, is

$$\Delta x = \frac{h}{\Delta p} = \frac{6.6 \times 10^{-34} \text{ J}\cdot\text{s}}{2.7 \times 10^{-32} \text{ kg}\cdot\text{m/s}}$$

$$= 2.4 \text{ cm}.$$

If the electron momentum has really been determined by measurement to have the accuracy stated, there is no hope whatever that its position can be known to any better accuracy than that stated, namely, about 1 in. The concept of the electron as a tiny dot is not very valid under these circumstances.

EXAMPLE 6

A bullet has a speed of 300 m/s, accurate to 0.010%. With what fundamental accuracy can we locate its position? Its mass is 50 g (=0.050 kg).

This example is the same as Example 5 in every respect save the mass of the particle involved. The momentum is

$$p = mv = (0.05 \text{ kg})(300 \text{ m/s}) = 15 \text{ kg}\cdot\text{m/s}$$

and

$$\Delta p = (0.00010)(15 \text{ kg}\cdot\text{m/s}) = 1.5 \times 10^{-3} \text{ kg}\cdot\text{m/s}.$$

Equation 50-15 yields

$$\Delta x = \frac{6.6 \times 10^{-34} \text{ J}\cdot\text{s}}{1.5 \times 10^{-3} \text{ kg}\cdot\text{m/s}} = 4.4 \times 10^{-13} \text{ m}.$$

This is so far beyond the possibility of measurement (a nucleus is only about 10^{-15} m in diameter) that we can assert that for heavy objects like bullets the uncertainty principle sets no limits whatever on our measuring procedures. Once again the correspondence principle shows us how quantum physics reduces to classical physics under the appropriate circumstances.

* *Mr. Tompkins in Wonderland*, New York: Macmillan, 1940.

The uncertainty relation shows us why it is possible for both light and matter to have a dual, wave-particle, nature. It is because these two views, so obviously opposite to each other, can never be brought face to face in the same experimental situation. If we invent an experiment that forces the electron to reveal its wave character strongly, its particle character will always be inherently fuzzy. If we change the experiment to bring out the particle character more strongly, the wave character necessarily becomes fuzzy. Matter and light are like coins that can be made to display either face at will but not both simultaneously. Niels Bohr first pointed out in his principle of complementarity how the ideas of wave and of particle complement rather than contradict each other.

questions

1. How can the wavelength of an electron be given by $\lambda = h/p$ when the very presence of the momentum p in this formula implies that the electron is a particle?

2. How could Davisson and Germer be sure that the "54-eV" peak of Fig. 50-2 was a first-order diffraction peak, that is, that $m = 1$ in Eq. 50-2?

3. In a repetition of Thomson's experiment for measuring e/m for the electron (see Section 33-8), a beam of electrons is collimated by passage through a slit. Why is the beamlike character of the emerging electrons not destroyed by diffraction of the electron wave at this slit? (See Problem 9.)

4. Why is the wave nature of matter not more apparent to our daily observations?

5. Considering the wave behavior of electrons, we should expect to be able to construct an electron microscope. This, indeed, has been done. (a) Compare the lens system of a light microscope with that of an electron microscope. (b) What advantages might an electron microscope have over a light microscope? (For recent developments, see "The Scanning Electron Microscope" by T. E. Everhart and T. L. Hayes, *Scientific American*, January 1972; "A High-Resolution Scanning Electron Microscope" by Albert V. Crewe, *Scientific American*, April 1971.)

6. How can electron diffraction be used to study properties of the surface of a solid? (See, in this connection, "Electron Microscopy of Atoms in Crystals" by John M. Cowley and Sumio Iijima, *Physics Today*, March 1977.)

7. How would an electron interferometer work? . . . a neutron interferometer? (See footnotes, p. 1119).

8. Is an electron a particle? Is it a wave? Explain.

9. Considering electrons and photons to be particles, how are they different from each other?

10. Can the de Broglie wavelength of a particle be smaller than a linear dimension of the particle? . . . ; Larger? Is there any relation necessarily between such quantities?

11. If, in the de Broglie formula $\lambda = h/mv$, we let $m \to \infty$, do we get the classical result for particles of matter?

12. Apply the correspondence principle to the problem of a particle confined between rigid walls, showing that those features which seem "strange" (that is, the quantization of energy and the nonuniformity of the probability functions of Fig. 50-8) become undetectable experimentally at large quantum numbers.

13. In the $n = 1$ mode, for a particle confined between rigid walls, what is the probability that the particle will be found in a small volume element at the surface of either wall?

14. A standing wave can be viewed as the superposition of two traveling waves.

Can you apply this to the problem of a particle confined between rigid walls, giving an interpretation in terms of the motion of the electron?

15. Discuss similarities and differences between a matter wave and an electromagnetic wave.

16. What is the physical significance of the wave function Ψ?

17. How can the predictions of wave mechanics be so exact if the only information we have about the positions of the electrons is statistical?

18. (a) Give examples of how the process of measurement disturbs the system being measured. (b) Can the disturbances be taken into account ahead of time by suitable calculations?

19. Why does the concept of Bohr orbits violate the uncertainty principle?

20. Make up some numerical examples to show the difficulty of getting the uncertainty principle to reveal itself during experiments with an object whose mass is about 1 g.

21. Figure 50-8 shows that for $n = 3$ the probability function for a particle confined between rigid walls is zero at two points between the walls. How can the particle ever move across these positions? (*Hint:* Consider the implications of the uncertainty principle.)

22. We can state the uncertainty principle in terms of angular quantities (compare Eq. 50-15) as

$$\Delta L \; \Delta \phi \gtrsim h$$

where ΔL is the uncertainty in the *angular* momentum and $\Delta \phi$ the uncertainty in the *angular* position. For electrons in atoms the angular momentum has definite quantized values, with no uncertainty whatever. What can we conclude about the uncertainty in the angular position and about the validity of the orbit concept?

problems

SECTION 50-1

1. A bullet of mass 40 g travels at 1000 m/s. (a) What wavelength can we associate with it? (b) Why does the wave nature of the bullet not reveal itself through diffraction effects? *Answer:* (a) $\lambda = 1.7 \times 10^{-35}$ m.

2. If the de Broglie wavelength of a proton is 1.0×10^{-13} m, (a) what is the speed of the proton and (b) through what electric potential would the proton have to be accelerated to acquire this speed?

3. Sodium ions are accelerated through a potential difference of 300 V. (a) What is the momentum acquired by the ions? (b) What is their de Broglie wavelength? *Answer:* (a) 1.9×10^{-21} kg·m/s. (b) 3.5×10^{-13} m.

4. What wavelength do we associate with a beam of neutrons whose energy is 0.025 eV?

5. An electron and a photon each have a wavelength of 2.0 Å = 0.20 nm. What are their (a) momenta and (b) energies?
Answer: (a) 3.3×10^{-24} kg·m/s for each. (b) 38 eV for the electron and 6200 eV for the photon.

6. The 50-GeV electron accelerator at Stanford (1 GeV = 10^9 eV) provides an electron beam of small wavelength, suitable for probing the fine details of nuclear structure by scattering experiments. What will this wavelength be and how does it compare with the size of an average nucleus? (At these energies it is sufficient to use the extreme relativistic relationship between momentum and energy, namely, $p = E/c$. This is the same relationship used for light (Eq. 42-2a) and is justified whenever the kinetic energy of a particle is much greater than its rest energy $m_0 c^2$, as in this case.)

7. The highest achievable resolving power of a microscope is limited only by the wavelength used; that is, the smallest detail that can be separated is about equal to the wavelength. Suppose one wishes to "see" inside an atom. Assuming the atom to have a diameter of 1.0 Å = 0.10 nm, this means that

we wish to resolve detail of separation about 0.10 Å = 0.010 nm. (a) If an electron microscope is used, what minimum energy of electrons is needed? (b) If a light microscope is used, what minimum energy of photons is needed? (c) Which microscope seems more practical for this purpose?
Answer: (a) 1.5×10^4 eV. (b) 1.2×10^5 eV.

8. What accelerating voltage would be required for electrons in an electron microscope to obtain the same ultimate resolving power as that which could be obtained from a gamma-ray microscope using 0.20 MeV-gamma rays?

9. In a repetition of Thomson's experiment for measuring e/m for the electron, a beam of 1.0×10^4-eV electrons is collimated by passage through a slit of width 0.50 mm. Why is the beamlike character of the emergent electrons not destroyed by diffraction of the electron wave at this slit?
Answer: The de Broglie wavelength (0.12 Å) is much smaller than the slit width.

10. In the experiment of Davisson and Germer (a) at what angles would the second- and third-order diffracted beams corresponding to a strong maximum of Fig. 50-2 occur, provided they are present, and (b) at what angle would the first-order diffracted beam occur if the accelerating potential were changed from 54 to 60 V? The experimenter is free to rotate the crystal.

11. A neutron crystal spectrometer utilizes crystal planes of spacing $d =$ 0.7323 Å in a beryllium crystal. What must be the Bragg angle θ so that only neutrons of energy $K = 4.0$ eV are reflected? Consider only first-order reflections.
Answer: 5.6°.

12. Make a plot of de Broglie wavelength against kinetic energy for (a) electrons and (b) protons. Restrict the range of energy values to those in which classical mechanics applies reasonably well. A convenient criterion is that the maximum kinetic energy on each plot only be about 5% of the rest energy $m_0 c^2$ for the particular particle.

13. What is the wavelength of a hydrogen atom moving with a velocity corresponding to the mean kinetic energy for thermal equilibrium at 20°C?
Answer: 1.5 Å = 0.15 nm.

SECTION 50-2

14. According to the correspondence principle, as $n \to \infty$ we expect classical results in the Bohr atom. Hence, the de Broglie wavelength associated with the electron (a quantum result) should get smaller compared with the radius of the orbit as n increases. Indeed, we expect that $\lambda/r \to 0$ as $n \to \infty$. Is this the case?

SECTION 50-3

15. (a) What is the separation in energy between the lowest two energy levels for a container 20 cm on a side containing argon atoms? (b) How does this compare with the thermal energy of the argon atoms at 300 K? (c) At what temperature does the thermal energy equal the spacing between these two energy levels?
Answer: (a) 3.9×10^{-22} eV. (b) The thermal energy is about 10^{20} times as great. (c) 3.0×10^{-18} K.

16. (a) Find, approximately, the smallest allowed energy of an electron confined to an atomic nucleus (diameter about 1.4×10^{-14} m)? (b) Compare this with the several MeV of energy binding protons and neutrons inside the nucleus; on this basis should we expect to find electrons inside nuclei?

SECTION 50-4

17. If an electron moves from a state represented by $n = 3$ in Fig. 50-8 to one represented by $n = 2$ in that figure, emitting electromagnetic radiation in the process, what are (a) the energy of the single emitted photon and (b) the corresponding wavelength?
Answer: (a) 1.9 eV. (b) 6600 Å.

18. A particle is confined between rigid walls separated by a distance l. What is the probability that it will be found within a distance $l/3$ from one wall

(a) for $n = 1$, (b) for $n = 2$, (c) for $n = 3$, and (d) under the assumption of classical physics?

19. In the ground state of the hydrogen atom show that the probability P_r that the electron lies within a sphere of radius r is given by

$$P_r = 1 - e^{-2r/a}\left(\frac{2r^2}{a^2} + \frac{2r}{a} + 1\right).$$

Does this yield expected values for (a) $r = 0$ and (b) $r = \infty$? (c) State clearly the difference in meaning between this expression and that given in Section 50-4.   *Answer:* (a) Zero. (b) 1; both as expected.

20. In the ground state of the hydrogen atom, what is the probability that the electron will lie within a sphere whose radius is that of the first Bohr orbit?

SECTION 50-5

21. A microscope using photons is employed to locate an electron in an atom to within a distance of 0.10 Å = 0.010 nm. What is the uncertainty in the momentum of the electron located in this way?
Answer: 6.6×10^{-23} kg·m/s.

22. The uncertainty in the position of an electron is given as about 0.050 nm, which is the radius of the first Bohr orbit in hydrogen. What is the uncertainty in the linear momentum of the electron?

23. Show that if the uncertainty in the location of a particle is equal to its de Broglie wavelength, the uncertainty in its velocity is equal to its velocity.

24. In Example 2, the electron's energy was determined exactly by the size of the box. How do you reconcile this with the fact that the uncertainty in the location of the electron cannot exceed 1.0×10^{-9} m if the uncertainty principle is to be obeyed?

25. From the uncertainty principle $\Delta p_x \, \Delta x \geq h$, show that, if L is the angular momentum component along a line perpendicular to the x-direction and φ is the azimuthal angle about this line (see Fig. 50-11), then

$$\Delta L \Delta \varphi \geq h;$$

(see Question 22).

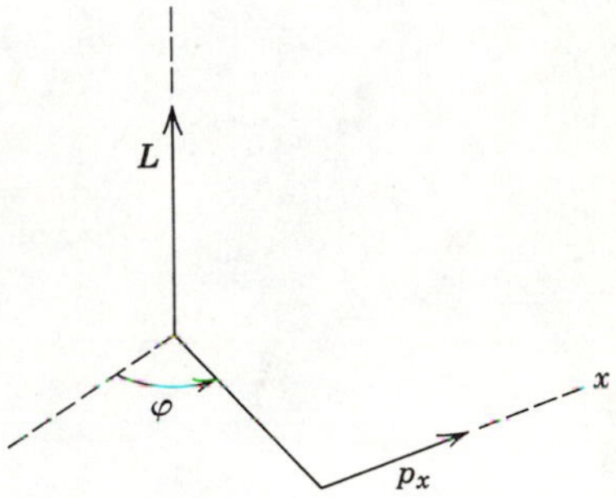

figure 50-11
Problem 25

supplementary topics

In Section 11-6 we discussed the relations between the linear and angular kinematic variables for a particle moving in a plane but confined to move in a circle about an axis at right angles to the plane. Such a particle might be any particle in a rigid body rotating about a fixed axis. Here we relax the restriction and allow the particle to move freely in the plane. A planet moving in an elliptical orbit about the sun is an example.

We start from Eq. 11-11, $\mathbf{r} = \mathbf{u}_r r$, in which, however, we now take *both* $\mathbf{u}_r$ and r to be variables; the particle is no longer confined to a circle of constant radius. We find the velocity by differentiation, or

$$\mathbf{v} = \frac{d\mathbf{r}}{dt} = \mathbf{u}_r \frac{dr}{dt} + r \frac{d\mathbf{u}_r}{dt}.$$

Equation 11-13 shows us that $d\mathbf{u}_r/dt = \mathbf{u}_\theta \omega$. Thus we can write

$$\mathbf{v} = \mathbf{u}_r \frac{dr}{dt} + \mathbf{u}_\theta \omega r, \tag{I-1}$$

which shows that $\mathbf{v}$ has two components, a radial component $v_r = dr/dt$ and a component at right angles, $v_\theta = \omega r$. If we hold r constant, then $dr/dt = 0$ and Eq. I-1 reduces to Eq. 11.14*a*, as it must.

To find the acceleration we differentiate Eq. I-1, remembering that *all five* quantities on the right are variables. We obtain

$$\mathbf{a} = \frac{d\mathbf{v}}{dt} = \mathbf{u}_r \frac{d^2 r}{dt^2} + \frac{dr}{dt} \frac{d\mathbf{u}_r}{dt} + (\mathbf{u}_\theta)\left(\omega \frac{dr}{dt} + r \frac{d\omega}{dt}\right) + (\omega r)\left(\frac{d\mathbf{u}_\theta}{dt}\right).$$

Now $d\mathbf{u}_r/dt = \mathbf{u}_\theta \omega$, $d\mathbf{u}_\theta/dt = -\mathbf{u}_r \omega$ (see Eq. 11-16), and $d\omega/dt = \alpha$. Substituting and rearranging leads us finally to

$$\mathbf{a} = \mathbf{u}_r\left(\frac{d^2 r}{dt^2} - \omega^2 r\right) + \mathbf{u}_\theta\left(\alpha r + 2\omega \frac{dr}{dt}\right). \tag{I-2}$$

$$= \mathbf{u}_r(a_r - \omega^2 r) + \mathbf{u}_\theta(\alpha r + 2\omega v_r).$$

Once again, if $r =$ a constant, then $dr/dt = d^2r/dt^2 = 0$ and Eq. I-2 reduces to Eq. 11-17, which we derived especially for this case.

The two new terms in Eq. I-2, $\mathbf{u}_r d^2r/dt^2$ and $\mathbf{u}_\theta 2\omega\, dr/dt$, need a little explanation. We can understand the first of these terms by imagining that the particle moving in the plane is *not* rotating about the axis. If we put $\omega = \alpha = 0$ in Eq. I-2 this equation reduces to

$$\mathbf{a} = \mathbf{u}_r \frac{d^2r}{dt^2} = \mathbf{u}_r a_r,$$

which is just the familiar acceleration of a particle moving along a straight line. Hence this term in Eq. I-2 gives the radial acceleration due to the change in the *magnitude* of $\mathbf{r}$, the other radial acceleration term arising from the changing *direction* of $\mathbf{r}$ as the particle rotates.

There are also two θ-directed acceleration terms. The first one, $\mathbf{u}_\theta \alpha r$, arises simply from the angular acceleration α of a particle in circular motion ($r =$ constant) and is the tangential acceleration of Section 11-5. To understand the second term, $\mathbf{u}_\theta 2\omega\, dr/dt$, consider a man walking outward along a radial line painted on the floor of a merry-go-round. The merry-go-round is rotating with constant angular velocity ω so that its angular acceleration α is now zero. If the man were simply to stand still on the merry-go-round, ($d^2r/dt^2 = dr/dt = 0$, and $r =$ constant) his acceleration, as seen by an observer in a reference frame on the ground (see Eq. I-2), would be simply the familiar centripetal acceleration $-\mathbf{u}_r \omega^2 r$, directed radially inward. If he walks outward, however, $dr/dt \neq 0$ and then Eq. I-2 predicts that the ground observer would also measure a θ-directed acceleration given by $\mathbf{u}_\theta 2\omega v_r$, where $v_r = dr/dt$. This is called a *Coriolis acceleration*. It arises from the fact that even though the angular velocity of the man is constant his speed increases as r increases. Let us convince ourselves that this effect really exists.*

Figure I-1a shows the walking man (point P) as he appears to the ground observer at times t and $t + \Delta t$. We show at time t his radially directed velocity $\mathbf{v}_r(= \mathbf{u}_r\, dr/dt)$ and also a θ-directed velocity caused by the rotation of the merry-go-round and given by $\mathbf{v}_\theta(= \mathbf{u}_\theta \omega r)$. At a time Δt later each of these velocities has changed. The radial velocity has changed in direction, although its magnitude remains dr/dt. The θ-directed velocity has not only changed direction (we have learned to account for this as a centripetal acceleration), but, because the man has moved outward to a point at which the floor is moving faster, its *magnitude* has also changed, from ωr to $\omega(r + \Delta r)$.

Figure I-1b shows the change in velocity caused by the change in direction of the radial line along which the man is walking. If $\Delta\theta$ in the triangle shown is small enough, we have

$$\Delta v_r = v_r \, \Delta\theta.$$

Dividing by Δt and letting Δt approach zero yields

$$a' = \frac{dv_r}{dt} = v_r \frac{d\theta}{dt} = v_r\omega.$$

This is just half the term $2\omega v_r$ in Eq. I-2. However, we have considered only the change in the *radial* velocity; there is also a change in the *tangential* velocity.

The change in tangential velocity, caused by the fact that the man is moving radially outward, is

$$\Delta v_\theta = \omega(r + \Delta r) - \omega r = \omega\Delta r.$$

Dividing by Δt and letting Δt approach zero yields

$$a'' = \frac{dv_\theta}{dt} = \omega \frac{dr}{dt} = \omega v_r.$$

Now both a' and a'' are magnitudes of vectors that point in the same direction,

* See "The Coriolis Effect," James E. McDonald, *Scientific American*, May 1952 and also "The Case of the Coriolis Force," Malcolm Correll, *The Physics Teacher*, January 1976.

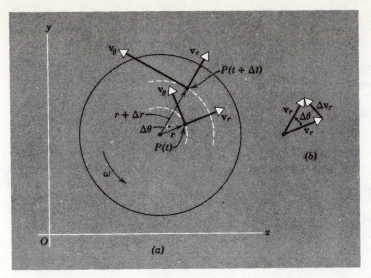

figure 1-1
(a) A merry-go-round, rotating about a fixed axis, is observed by an observer in inertial reference frame x, y. A man walks along a radial line at constant speed v. In a time interval Δt this line, as seen by the ground observer, sweeps through an angle $\Delta\theta$ and the man moves between the positions shown. His r- and θ-directed velocities are shown for each position. *(b)* Showing the change Δv_r in the walking man's r-directed velocity. Note that, as $\Delta t \to 0$, Δv_r points in the θ-direction at P.

namely the direction of increasing θ at point $P(t)$. The total acceleration in this direction is then

$$a' + a'' = v_r\omega + \omega v_r = 2\omega v_r,$$

which is just what we set out to prove.

If there is indeed an acceleration in the θ-direction in Fig. I-1, there must be a force in this direction. For a man walking outward along a radial line on a rotating merry-go-round this force can only be provided by the friction between his feet and the floor.

We remember that we can interpret classical mechanics most simply if we always view events from an inertial frame. If we do so, we can always associate accelerations with forces exerted by bodies that we can point to in the environment. We can still apply classical mechanics, however, if we select a noninertial reference frame, such as a rotating frame. The small penalty that we must pay is that we must introduce inertial forces, that is, forces that we cannot associate with objects in the environment and which cannot be detected by an observer in an inertial frame. In Section 6-4 we saw that centrifugal force is such an inertial force.

Consider an observer on the rotating merry-go-round watching a man walk along a radial line at a constant speed $v_r = dr/dt$. He would say that the man is in equilibrium because he has no acceleration. Yet the floor is exerting a (very real) frictional force on the soles of the man's feet. This force has one component $(-\mathbf{u}_r F_r)$ that points radially inward and one $(\mathbf{u}_\theta F_\theta)$ that points in the θ-direction, that is, in the direction of rotation.

From the point of view of the ground observer these forces are understandable and, indeed, quite necessary. F_r is associated with the centripetal acceleration $\omega^2 r$ and F_θ with the Coriolis acceleration $2\omega v_r$. The observer on the merry-go-round does not see either of these accelerations however; to him the walking man is in equilibrium. How can this be, in view of the frictional forces that act on the soles of the walking man's shoes? The man himself is well aware of these forces; if he did not lean to compensate for their turning effect, they would knock him off his feet!

The observer on the merry-go-round saves the situation by declaring that two inertial forces act on the walking man, just canceling the (real) frictional forces. One of these inertial forces, called the *centrifugal force*, has magnitude F_r and acts radially *outward*. The other, called the *Coriolis force*, has magnitude F_θ and acts in the negative θ-direction, that is, *opposite* to the direction of rotation. By introducing these forces, which seem quite "real" to him although he cannot point to any body in the environment that is causing them, the observer in the rotating (noninertial) reference frame can apply classical mechanics in the usual way. The ground observer, who is in an inertial frame, cannot detect these inertial forces. Indeed there is no need for them—and no room for them—in his applications of classical mechanics.

Equations I-1 and I-2 are general kinematical descriptions for the motion of a particle in two dimensions. An obvious extension, which we will not attempt here, is to derive corresponding descriptions for motion in three dimensions; this will require us to introduce a third unit vector to define the third dimension.*

Some vectors called *axial vectors*, such as **ω**, **α**, **τ**, and **l**, differ in a rather important way from other vectors called *polar vectors*, of which **r**, **v**, **a**, **F**, and **p** are examples. Although we shall not need to take this difference into account in this book, it may prove to be instructive and interesting to examine briefly what the difference is.

Consider a typical polar vector such as **r.** If a student leaves his dormitory and goes to a classroom, his displacement vector **r** points *from* the dormitory *to* the classroom; there is no question as to our choice of direction. This direction is both "physical" and "natural." Similar remarks apply to the other typical polar vectors listed, namely, **v, a, F,** and **p.**

If a student sees a wheel rotating about a fixed axis, he can assign an angular velocity **ω** to the wheel and can give direction to **ω** by the right-hand rule (see Section 11-4). This direction, however, *is a convention only,* based on this arbitrary rule. A left-hand rule would have given the opposite direction. The things that are "physical" and "natural" about the wheel are the axis of rotation and the sense of rotation, that is, is it going clockwise or counterclockwise as the student looks at it from a particular end of the axis? Whether **ω** is chosen to point in one way or the other along the axis does not really matter as long as we are consistent. The same remarks apply to the angular acceleration **α** and to the other axial vectors listed, namely, **τ**(= **r** × **F**) and **l**(= **r** × **p**). It is for this reason that we sometimes find it more comfortable to say "torque *around* an axis" than "torque *along* an axis" although they mean the same thing. All vectors defined as the vector product of two *polar* vectors are axial vectors because they all depend for their direction assignment on the (arbitrary) right-hand rule.

We have stressed that the laws of physics remain the same no matter how we change the inertial reference frame in which they are expressed. In Section 2-5 we discussed this for translations and rotations of the reference frame and noted that laws expressed in vector form remained unchanged (that is, *invariant*) under such transformations. We also noted that something special may occur when we change the reference frame in another way, namely, by substituting a left-handed frame for a right-handed one. There is an easy way to make such a transformation: Build a right-handed frame and look at its image in a mirror; it will be converted to a left-handed frame (see Fig. II-1) because of the well-known property of a mirror to reverse right and left.

Figure II-1*a* shows the vector displacement of a student from his dormitory to each of three classrooms. In the mirror each displacement is *still* from the dormitory *D* to a classroom *C*. In Fig. II-1*b*, however, we show a rotating wheel in three orientations. If we establish the directions of **ω** for both the wheels and their mirror images by the right-hand rule, we see that the image vectors are reversed in comparison to the corresponding image vectors in Fig. II-1*a* (toward the origin rather than away from the origin). Polar vectors and axial vectors behave differently when we transform reference frames by mirror reflection! This behavior of axial vectors under mirror reflection is not hard to understand. If we imagine ourselves physically applying the right-hand rule to a real rotating wheel, in the mirror, we shall *seem* to be applying a left-hand rule because the image of our right hand is our left hand. A left-hand rule, of course, will give us the opposite direction for **ω.**

Hence an axial vector is a vector whose sense of direction depends on the

---

SUPPLEMENTARY TOPIC II
POLAR VECTORS AND AXIAL VECTORS

---

* See, for example, *Mechanics*, Section 3-5, by Keith R. Symon, Addison-Wesley Publishing Co., 3d ed., 1971.

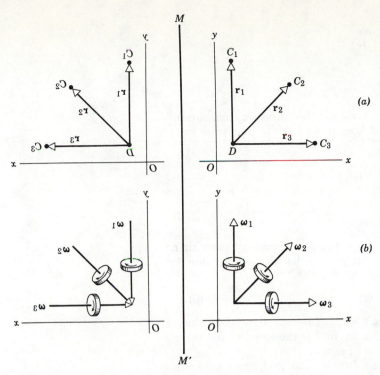

figure II-1

(a) Polar vectors, showing, on the right, the displacements r_1, r_2, and r_3 between a dormitory D and three classrooms C_1, C_2, and C_3. On the left we have the mirror images of D, C_1, C_2, and C_3, along with the corresponding displacements. (b) Axial vectors, showing, on the right, the angular velocities ω_1, ω_2, and ω_3 of three wheels rotating as shown. On the left we have the mirror images of these wheels, along with the angular velocities assigned using the usual right hand rule.

handedness of the reference frame. It is sometimes called a *pseudovector*. A polar vector is a vector that has a direction independent of the reference frame. We mention these facts (1) to stress the arbitrary character of the direction assigned to axial vectors and (2) to stress the importance of testing experiments and physical laws for invariance under translation, rotation, and mirror reflection of the inertial reference frame. In Section 2-5 we referred briefly to some experiments that were *not* invariant under a reflection transformation. This fact, which constituted a violation under certain circumstances of a law of physics previously thought to be well founded (the law of *conservation of parity*), has posed some challenging problems and is leading us to an understanding of the physical world at a deeper level.*

Figure III-1 shows a section of a long string which is under tension F. The string has been pulled transversely in the y-direction so that a displacement wave travels along the string in the x-direction. We consider a differential element of the string dx and apply Newton's second law of motion to it in order to find how the wave moves along the string.

Let μ be the mass per unit length of the string, so that the mass of element dx is $\mu\,dx$. The net force in the y-direction acting on this element is

$$F \sin \theta_{x+dx} - F \sin \theta_x.$$

We consider only small transverse displacements of the string, so that the restoring force will vary linearly with displacement and the principle of superposition will hold (see Section 19-4). This means that θ in Fig. III-1 will be small, so that we may replace $\sin \theta$ by $\tan \theta$. Now $\tan \theta$ is simply the slope of the string, that is, it equals $\partial y/\partial x$. We must use partial derivatives because the transverse displacement y depends not only on x but also on t. The net force in the y-direction is then

$$F\left(\frac{\partial y}{\partial x}\right)_{x+dx} - F\left(\frac{\partial y}{\partial x}\right)_{x},$$

*See "The Overthrow of Parity," by Philip Morrison, *Scientific American*, April 1957.

SUPPLEMENTARY TOPIC III
THE WAVE EQUATION FOR A STRETCHED STRING

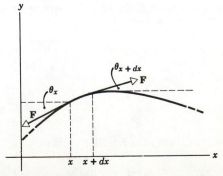

figure III-1

which we may write as

$$F \frac{\partial}{\partial x}\left(\frac{\partial y}{\partial x}\right) dx$$

or

$$F \frac{\partial^2 y}{\partial x^2} dx.$$

The mass of the element of the string is $\mu \, dx$ and its transverse acceleration is simply $\partial^2 y / \partial t^2$. Hence, Newton's second law, applied to the transverse motion of the string, is

$$F \frac{\partial^2 y}{\partial x^2} dx = (\mu \, dx) \frac{\partial^2 y}{\partial t^2}$$

or

$$\frac{\partial^2 y}{\partial x^2} = \frac{\mu}{F} \frac{\partial^2 y}{\partial t^2}. \qquad \text{(III-1)}$$

Equation III-1, called the *wave equation*, is the differential equation that describes wave propagation in a string of mass per unit length μ and tension F.

To prove this we show that Eqs. 19-2 and 19-3

$$y = f(x \pm vt), \qquad \text{(III-2)}$$

which is the general equation representing a wave of any shape traveling along x, is a solution of Eq. III-1. Recall that v in Eq. III-2 is the speed of the wave disturbance and f is any reasonable function of $(x \pm vt)$.

Let us see whether Eq. III-2 is indeed a solution of Eq. III-1 by substituting the former equation into the latter. To do so we note that the two second partial derivatives of y are

$$\frac{\partial^2 y}{\partial x^2} = f'' \qquad \text{and} \qquad \frac{\partial^2 y}{\partial t^2} = v^2 f''$$

in which f'' is the second derivative of the function f of Eq. III-2 with respect to $(x \pm vt)$. Substitution of these derivatives into Eq. III-1 yields

$$f'' = \frac{\mu}{F} v^2 f'',$$

which we may write as (see Eq. 19-12)

$$v = \sqrt{\frac{F}{\mu}}. \qquad \text{(III-3)}$$

Thus we conclude that Eq. III-2 is indeed a solution of the partial differential equation Eq. III-1 if the speed of the wave disturbance described by this equation is given by Eq. III-3.

In particular, let us check that Eq. 19-10

$$y = y_m \sin (kx \pm \omega t) \qquad \text{(19-10)}$$

is a solution of Eq. III-1. We know that it must be because Eq. 19-10 is simply a special case of the general relation Eq. III-2, which we have just shown to be a solution. Even so it is instructive to test this important specific function of $(x \pm vt)$ by substitution into Eq. III-1.

The second derivatives of Eq. 19-10 are

$$\frac{\partial^2 y}{\partial x^2} = -k^2 y_m \sin (kx \pm \omega t)$$

and

$$\frac{\partial^2 y}{\partial t^2} = -\omega^2 y_m \sin (kx \pm \omega t).$$

Substitution into Eq. III-1 yields

$$-k^2 y_m \sin (kx \pm \omega t) = \left(\frac{\mu}{F}\right)[-\omega^2 y_m \sin (kx \pm \omega t)]$$

or
$$\frac{\omega}{k} = \sqrt{\frac{F}{\mu}}.$$

Since $\omega/k = v$ (see Eq. 19-11), this relation is identical with Eq. III-3, and Eq. 19-10, as we expect, is indeed a solution of Eq. III-1.

SUPPLEMENTARY TOPIC IV
DERIVATION OF MAXWELL'S SPEED DISTRIBUTION LAW

Boltzmann, in 1876, derived the Maxwell speed distribution law (see Eq. 24-2) from this line of argument: Let a uniform gravitational field g act on an ideal gas maintained at a fixed temperature T. The number of molecules per unit volume n_v will then decrease with altitude z according to the law of atmospheres (see Example 1, Chapter 17). From what we know about the statistical-mechanical interpretation of temperature, however, the speed distribution law —whose form we assume that we do not yet know—must remain the same at all altitudes because it depends only on the temperature. However, this law determines the rate at which molecules move vertically in the atmosphere at any altitude and must thus be intimately related to the decrease of n_v with z. By exploring this relationship in detail we can, in fact, deduce the speed distribution law.

The weight of gas per unit area between the levels z and $z + dz$ in Fig. IV-1 is $n_v mg\, dz$ in which m is the mass of a single molecule. For equilibrium, this weight per unit area must equal the difference in pressure between z and $z + dz$, or

$$n_v mg\, dz = -dp \qquad \text{(IV-1)}$$

in which we have inserted a minus sign because p decreases as z increases.

We can write the equation of state of an ideal gas, $pV = nRT$, as

$$p = n_v kT \qquad \text{(IV-2)}$$

because $n = n_v V/N_0$, where $N_0 (= R/k)$ is Avogadro's number, the number of molecules per mole, and k is Boltzmann's constant. Combining Eqs. IV-1 and IV-2 yields

$$\frac{dp}{p} = \frac{dn_v}{n_v} = -\frac{mg}{kT}\, dz.$$

For a constant temperature, we can integrate this relation to yield

$$n_v = \text{constant } e^{-mgz/kT} \qquad \text{(IV-3)}$$

which, in view of Eq. IV-2, agrees with the result of Example 1, Chapter 17.

We can find the change in n_v as we go from z to $z + dz$ by differentiating Eq. IV-3, or

$$dn_v = -\text{constant } e^{-mgz/kT}\, dz. \qquad \text{(IV-4)}$$

We associate this decrease in n_v over the interval dz with the fact that, at $z = 0$ (which can be any level we choose) there are some upward-directed molecules —we call them "special molecules" temporarily for convenience—whose vertical velocity components lie in a particular range v_z to $v_z + dv_z$ such that (neglecting collisions; see below) they can rise as high as z but not as high as $z + dz$. Such molecules pass upward through the level z, reverse their direction and pass downward again, as Fig. IV-1 shows. At this point we see more clearly the relationship between Eq. IV-3 and the speed distribution law. Molecules that pass through the interval dz (from above or below) or molecules that never reach the interval cannot contribute to the decrease dn_v of Eq. IV-4.

The rate per unit area at which "special molecules" leave level $z = 0$ (and arrive at level z) is $v_z n_v(v_z)\, dv_z$. Here $n_v(v_z)\, dv_z$ is the number of molecules per unit volume whose vertical velocity components lie between v_z and $v_z + dv_z$.

Now the rate per unit area at which the "special molecules" arrive at level z, but not as high as level $z + dz$, is proportional to the magnitude of the density difference dn_v between z and $z + dz$, or, from Eq. IV-4,

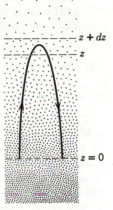

figure IV-1

$$v_z n_r(v_z) \, dv_z = \text{constant } e^{-mgz/kT} \, dz, \tag{IV-5}$$

in which the constant is independent of z. Equation IV-5, which requires that the change dn_v be accounted for by the "special molecules" is, in fact, the defining equation for $n_v(v_z)$.

From conservation of energy the special molecules have the property that*

$$\tfrac{1}{2}mv_z^2 = mgz$$

or

$$mv_z \, dv_z = mg \, dz.$$

We use these two relations to eliminate z and dz from Eq. IV-5, obtaining, as you should verify,

$$n_v(v_z) \, dv_z = \text{constant } e^{-mv_z^2/2kT} \, dv_z \tag{IV-6a}$$

in which $n_v(v_z) \, dv_z$ is the number of molecules per unit volume whose vertical velocity components lie between v_z and $v_z + dv_z$. Note that Eq. IV-6a does not contain g or z. The gravitational field of Fig. IV-1, introduced to allow us to calculate the speed distribution, has served its purpose. We may apply Eq. IV-6a to a gas for which $g = 0$ or in which gravitational effects are negligible. In such a case the vertical direction, which we have identified as the z-direction, no longer has any special meaning. That is, the speed distribution for one component of velocity should be the same for another component of velocity since there is no special or preferred direction in a gas in equilibrium free of external forces. Thus we can write

$$n_v(v_x) \, dv_x = \text{constant } e^{-mv_x^2/2kT} \, dv_x \tag{IV-6b}$$

and

$$n_v(v_y) \, dv_y = \text{constant } e^{-mv_y^2/2kT} \, dv_y, \tag{IV-6c}$$

for the other two velocity components.

We now seek to find Maxwell's speed distribution (Eq. 24-2); it is expressed in terms of the speed v, rather than in terms of the separate components v_x, v_y, and v_z. We are not concerned here with the direction of $\mathbf{v}$, because we assume it to be completely random. We can represent any velocity $\mathbf{v}$ as a vector extending from the origin in Fig. IV-2; the projections of the vector in the x-, y-, and z-directions are v_x, v_y, and v_z, respectively. We commonly say that the axes of Fig. IV-2 define a "velocity space," which has many formal similarities to ordinary (or coordinate) space, in which the axes are x, y, and z.

We also show in Fig. IV-2 a small "volume" element, whose sides are dv_x, dv_y, and dv_z; we say that this element has a volume $dv_x \, dv_y \, dv_z$ in velocity space. A point in this element corresponds to a particle whose velocity components lie between v_x and $v_x + dv_x$; v_y and $v_y + dv_y$; and v_z and $v_z + dv_z$. We can regard $n_v(v_z)$ in Eq. IV-6a as giving the *probability* that a given molecule will have a velocity component in the specified range v_z to $v_z + dv_z$, with similar interpretations for $n_v(v_x)$ and $n_v(v_y)$. The probability that a given molecule will have *all three* of its velocity components in the specified ranges, that is, the probability that the tip of the velocity vector $\mathbf{v}$ will lie inside the volume element of Fig. IV-2, is the *product* of the three (independent) probabilities given in Eq. IV-6, or

$$\text{constant } e^{-mv_x^2/2kT}e^{-mv_y^2/2kT}e^{-mv_z^2/2kT} \, dv_x \, dv_y \, dv_z$$

which, since

$$v^2 = v_x^2 + v_y^2 + v_z^2,$$

we may write as

$$\text{constant } e^{-mv^2/2kT}(dv_x \, dv_y \, dv_z). \tag{IV-7}$$

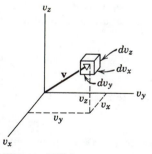

figure IV-2

* If we consider collisions, this result is still true *on the average* for the many molecules that start at $z = 0$ with a given value of v_z and move to the interval z to $z + dz$, having $v_z = 0$ there, even though such molecules would follow very erratic paths because of the collisions.

The quantity in parentheses above is a volume element in velocity space. Since in Maxwell's speed distribution law we are not concerned with the direction of molecular velocities but only with their speeds, it is more convenient to substitute a different volume element for the above, namely, one corresponding to all molecules whose speeds lie between v and $v + dv$, *regardless of direction*. This volume element is not a "cube" but is the space between two concentric spheres, one of radius v and one of radius $v + dv$. The volume of this element in velocity space is $(4\pi v^2)(dv)$. Substituting this for the quantity enclosed in parentheses in Eq. IV-7 yields for the number of molecules per unit volume whose speeds lie between v and $v + dv$,

$$n_v(v) \, dv = \text{constant } e^{-mv^2/2kT}(4\pi v^2 \, dv)$$

or

$$n_v(v) = Cv^2 e^{-mv^2/2kT}$$

in which C is a constant. If we sum up over all possible speeds, we simply obtain the total number of molecules per unit volume, regardless of speed. Hence, we can find C by requiring that

$$\int_0^\infty n_v(v) \, dv = n_v,$$

where n_v is the total number of particles per unit volume, regardless of speed. Guided by the methods of Example 3 (Chapter 24), you should show that

$$C = 4\pi n_v(m/2\pi kT)^{3/2}$$

so that

$$n_v(v) = 4\pi n_v(m/2\pi kT)^{3/2}v^2 e^{-mv^2/2kT}. \tag{IV-8}$$

Let us consider a finite number N of molecules contained in a box of volume V. If we multiply each side of the above equation by V, we can replace $n_v V$ on the right by N and $n_v(v)V$ on the left by $N(v)$, which gives us Eq. 24-2.

Here we simply display in one place some conclusions drawn from the special theory of relativity (hereafter, SR) proposed by Einstein in 1905. We omit all proofs and make only a modest attempt to make the conclusions acceptable in terms of "common sense."

SUPPLEMENTARY TOPIC V
*SPECIAL RELATIVITY— A SUMMARY OF CONCLUSIONS**

V-1
Introduction

V-2
The Postulates (RR, Section 1.9)

Einstein based his theory on two postulates and *all* of the conclusions of SR derive from them.

a. The First Postulate. From the time of Galileo it was known that the laws of mechanics were the same in all inertial frames (see Fig. V-1 and p. 66). This means that all inertial observers having relative motion, even though they may measure different values for the velocities, momenta, etc., of the particles involved in a given experiment (a game of pool, perhaps), would nevertheless agree on the laws of mechanics involved (conservation of linear momentum, etc.) and on the outcome of the experiment (who won).

Einstein took the bold step of extending this invariance principle to *all* of physics, not only mechanics, including especially electromagnetism. Einstein's first postulate is:

---

*For a fuller treatment, geared to the level of this book, see *Introduction to Special Relativity*, Robert Resnick, John Wiley and Sons, Inc., New York, 1968. References to this work will be in the style RR, p. 187; RR, Section 1.9, etc.

The laws of physics are the same in all inertial frames. No preferred inertial frame exists.

b. The Second Postulate. In pre-SR days a vexing question was this: Given that the speed of light c is 2.988×10^8 m/s, with respect to what is this speed measured? For sound waves in air the answer is simple—it is with respect to the medium (air) through which the sound wave travels. Light, however, travels through a vacuum. Even so, is there a tenuous medium (the luminiferous, or light carrying, ether) that plays the same role for light that air does for sound? If such an ether exists, can we detect it? Alternatively, should c be measured with respect to the source that emits the light?

All attempts to make experimental verifications along these lines failed completely (see Section 45-8* and RR, Sections 1.5 through 1.8). Einstein made a second bold postulate.

The speed of light is the same in all inertial frames.

Note that no ether is needed or involved. This second postulate means, for example, that if you consider three light sources (a) one at rest with respect to you, (b) one moving toward you at speed $0.9\,c$, say, and (c) one moving away from you at speed $0.9\,c$, you would measure the *same* speed of light from all three sources.

This second postulate has been tested directly (see RR, p. 34) using as a moving "light" source π^0 mesons generated in a proton synchrotron at speeds of $0.99975\,c$. These mesons disintegrate by emitting γ-rays which, like light, are electromagnetic in character and travel with the same speed. The speed measured for the radiation emitted by these fast moving sources was, within experimental error, just c, as Einstein's second postulate predicts.

V-3
Special Relativity and Newtonian Mechanics (RR, Section 2.8)

Many of the conclusions of SR simply don't seem reasonable on the basis of everyday experience. Even Einstein's second postulate seems to violate common sense. If you catch a pitched baseball thrown by a pitcher (a) at rest with respect to you, (b) moving toward you (in an automobile, say) at 30 mi/h and, (c) moving away from you at this same speed, you expect a different baseball speed in each case with respect to you. But if you extend this experience to a source (the pitcher) emitting light (photons), you would contradict Einstein's second postulate. And yet experiment shows that light does have the same speed in each case, in support of Einstein's postulate.

The solution to this dilemma comes about when we realize that the basis of our "common sense" experience is very limited indeed. It is restricted to speeds v such that $v \ll c$, where c is the speed of light. For example, the speed of a satellite in earth orbit may be about 8000 m/s, which seems fast to us, but in terms of the speed of light (3.0×10^8 m/s) it is only $0.000027\,c$. We simply have no personal experience in regions of high relative velocity.

As an example, to accelerate an average person (to say nothing of a spaceship) to $0.90\,c$ would require no less than 13 percent of this country's 1971 total energy consumption. However, the particles of physics (electrons, mesons, protons, etc.) can readily be accelerated to high speeds. Electrons emerging from the two-mile long linear accelerator at Stanford University have speeds of $0.999\,c$, for example. In the arena of particle physics SR is absolutely necessary for the solution of mechanical problems.

It turns out that in nature there is a certain finite speed that cannot be exceeded and which we call the limiting speed. This limiting speed is the speed

*This book is published in a combined volume (Chapters 1–50) and separate volumes (Part I, Chapters 1–25; Part II, Chapters 26–50). Whether cited references are accessible depends on which of these three volumes you are reading.

of light, c, the greatest speed with which signals can be transmitted. Classical physics assumes that signals can be transmitted with infinite speed, but nature contradicts that assumption, and it really does seem fanciful that such a signal could exist. Experiment confirms c as the limiting speed, so that in a sense the speed of light plays the role in relativity that infinity does in classical physics. It is then not difficult to understand—in fact, it becomes very plausible —that the finite speed of the source of light cannot affect the measured value of the speed of an emitted signal already having the limiting value.

The world in which we live and develop our sense perceptions is a world of Newtonian mechanics, in which $v \ll c$. Newtonian mechanics is revealed as a special case of SR for the limit of low speeds. Indeed, a test of SR is to allow $c \to \infty$ (in which case $v \ll c$ always holds true) and see that the corresponding formulas of Newtonian mechanics emerge.

Newtonian mechanics, although a special case, is an all-important one. It describes the essential motions of our solar system, the tides, our space ventures, the behavior of baseballs and pinball machines, etc. It works beautifully in the vastly important realm $v \ll c$. But it breaks down at speeds approaching that of light.

Few theories have been subject to more rigorous experimental tests than SR. Not the least among them is the fact that particle accelerators work. They are designed using SR at the level of engineering and technology. An accelerator designed on the basis of Newtonian mechanics simply would not work. Nuclear reactors and, alas, nuclear bombs, are further proof that SR really works.

Einstein once said that no number of experiments could prove him right but a single experiment could prove him wrong. To date this single experiment has not been found.

The basic observation made in SR (or in Newtonian mechanics for that matter) is this. Consider observers to be in different inertial frames, S and S' (Fig. V-1). The corresponding axes of S and S' are parallel, the x-x' axes being common, and S' moves to the right with speed v as seen by S; the two origins coincide at $t = t' = 0$. Each observer, S and S', records the same event, which might be the detonation of a tiny flashbulb, and assigns space and time coordinates to the event, namely, x, y, z, t and x', y', z', t'. What are the relations between these two sets of numbers written down in the observers' notebooks?

Before SR the accepted relations were

$$x' = x - vt \qquad y' = y$$
$$t' = t \qquad z' = z, \tag{V-1}$$

called the *Galilean transformation equations* (RR, Section 1.2). Though impressively correct in the important region $v \ll c$ they nevertheless fail as $v \to c$.

The corresponding equations used in SR, called the *Lorentz transformation equations*, are (RR, Table 2-1)

$$x' = \frac{x - vt}{\sqrt{1 - (v/c)^2}} \qquad y' = y$$
$$t' = \frac{t - (v/c^2)x}{\sqrt{1 - (v/c)^2}} \qquad z' = z \tag{V-2}$$

Note certain things about these equations. (a) Space and time coordinates are thoroughly intertwined. In particular, time is not the same for each observer; t' depends on x as well as on t. (b) If we let $c \to \infty$, the Lorentz equations reduce to the Galilean equations, as promised! Finally, (c) We must have $v < c$ or else the quantities x' and t' become indeterminate ($v = c$) or imaginary ($v > c$). The speed of light is an upper limit for the speeds of material objects.

The Lorentz equations, like everything else in SR, can be derived from Einstein's two postulates (RR, Section 2.2).

V-4
The Transformation Equations
(RR, Section 2.2)

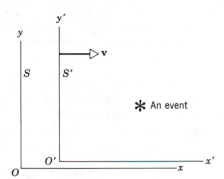

figure V-1
Two inertial frames with parallel axes, the $x - x'$ axis being common. S' moves to the right with speed v as seen by S. At $t = t' = O$ the two origins, O and O', coincide.

Let S' observe two events that occur at the same place in his reference frame. They might be two successive positions of the hand of a clock located at a fixed position, x'. Let S' measure the time interval $\Delta t'$ between these events. S, for whom the clock appears to be moving, observes the same two events and measures a different time interval Δt, which is given by

$$\Delta t = \frac{\Delta t'}{\sqrt{1 - (v/c)^2}}. \qquad (V\text{-}3)$$

This fact, that $\Delta t > \Delta t'$, is called *time dilation*, and we often verbalize it as "moving clocks run slow." Observer S records a longer time interval than shown to have transpired on the moving clock.

Equation V-3 has been tested experimentally and found to be correct. In one test the "moving clocks" were fast particles called pions ($\pi \pm$). Pions are radioactive, and their rate of radioactive decay is a measure of their time-keeping ability. See RR, Example 3, p. 75.

Now let us consider a rod, parallel to the $x - x'$ axes, to be at rest in the S' frame. S' will measure a length $\Delta x'$ for it. S, however, measuring the same rod, which is moving with respect to him, would find a length Δx, which is given by

$$\Delta x = \sqrt{1 - (v/c)^2}\, \Delta x' \qquad (V\text{-}4)$$

This fact, that $\Delta x < \Delta x'$, is called *length contraction*.

The length contraction has been verified in the design of, say, the linear electron accelerator at Stanford University. At an exit speed of $v = 0.999975c$, each meter of the accelerating tube seems like 7.1 mm to an observer moving with the electron. If these length contraction considerations had not been taken into account the machine simply would not work.

The simplest way to understand these results—the time dilation and the length contraction—is to note that one observer, S', is at rest with respect to what he is measuring (clock or rod) whereas for the other observer, S, the objects are in motion. Relativity therefore asserts that *motion affects measurement*. If we had interchanged frames, letting the clock and rod be at rest in S, for example, we would have found the observers again disagreeing on the measured values, but now we would have $\Delta x' < \Delta x$ and $\Delta t' > \Delta t$. So the results are reciprocal, neither observer being "absolutely" right or wrong.

What both observers *will* agree on however, is the *rest length* of a given rod (they will both measure the rod to have the same length when the rod is at rest with respect to their measuring instruments) and the *proper time interval* of a given clock (they will both measure the successive positions of the hand of the clock to have taken the same elapsed time when the clock is at rest with respect to their measuring instruments).

That motion should affect measurement is not a strange idea, even in classical physics. For example, the measured frequency of sound or of light depends on the motion of the source with respect to the observer; we call this the Doppler effect and everyone is familiar with it. And, in mechanics the measured values of the speed, the momentum, or kinetic energy of moving particles are different for observers on the ground than those on a moving train. However, in classical physics measurements of space intervals and time intervals *are* absolute whereas in SR such measurements are relative to the observer. Not only does experiment contradict classical physics but only by adopting the relativity of space and time do we arrive at the invariance (the absoluteness) of all of the laws of physics for all observers. Surely, giving up the absoluteness of the laws of physics (would they then be laws?), as classical notions of time and length require, would leave us with an arbitrary and complex world. By comparison, relativity is absolute and simple.

Let S observe a particle moving with speed u' parallel to the x'-axis. What speed u would S measure? From the Galilean transformation equations (Eq. V-1) we can easily show that

$$u = u' + v \qquad (V\text{-}5)$$

This relation, which seems to most of us to be intuitively obvious, is, alas, not

V-5
Time Dilation and Length Contraction (RR, Sections 2.3 and 2.4)

V-6
Relativistic Addition of Velocities and the Doppler Effect (Sections 4.6, 6.5, 42.4, and 42.5; RR, Sections 2.6 and 2.7)

true (except for the very important special case of $v \ll c$). The Lorentz transformation equations lead us to

$$u = \frac{u' + v}{1 + (u'v/c^2)}. \tag{V-6}$$

As we expect, for $c \rightarrow \infty$, Eq. V-6 reduces to Eq. V-5. Prove that if $u' < c$ and $v < c$, then it must always be true that $u < c$. There is no way to generate speeds $\geq c$ by compounding velocities.

Using the relativistic velocity addition result (Eq. V-6), we can deduce the Doppler effect for light. In relativity theory there is no difference between the two cases, which are different in classical theory (namely, source at rest — observer moving and observer at rest — source moving); only the relative motion v of source and observer counts. This fact and the result

$$\nu = \nu'\sqrt{\frac{c \pm v}{c \mp v}} \tag{V-7}$$

are in agreement with experiment. Here, ν' is the frequency of the source at rest in S' and ν is the frequency observed in frame S with respect to which the source moves at speed v; the upper signs refer to source and observer moving *toward* one another and the lower signs refer to source and observer moving *away* from one another. Equation V-7 is called the *longitudinal* Doppler effect, and v refers to the relative velocity of source and observer along the line connecting them.

There is in relativity, however, an effect not predicted by classical physics, namely a *transverse* Doppler effect; that is, if the relative velocity v is at right angles to the line connecting source and observer, we find

$$\nu = \nu'\sqrt{1 - v^2/c^2}. \tag{V-8}$$

This result, confirmed by experiment, can be interpreted simply as a time dilation, moving clocks appearing to run slow.

We have seen that time and length measurements are functions of velocity v. Should mass be any different? SR tells us that the *relativistic mass m* of a particle moving at speed v with respect to the observer is

$$m = \frac{m_0}{\sqrt{1 - (v/c)^2}} \tag{V-9}$$

in which m_0 is the rest mass, that is, the mass measured when the particle is at rest ($v = 0$) with respect to the observer.

It is m and not m_0 that must be taken into account when designing magnets to bend charged particles in arcs of circles. By these techniques, Eq. V-9 has been thoroughly tested. Incidentally, the ratio m/m_0 for electrons emerging from the Stanford University linear accelerator at $K = 30$ GeV is the order of 60,000.

To preserve the law of conservation of linear momentum in SR, we redefine the momentum of a particle of rest mass m_0 and speed v as,

$$p = mv = \frac{m_0 v}{\sqrt{1 - (v/c)^2}}. $$

As a result of the considerations above, in SR the kinetic energy of a particle is no longer given by $\frac{1}{2} m_0 v^2$ but by

$$K = mc^2 - m_0 c^2$$

$$= m_0 c^2 \left(\frac{1}{\sqrt{1 - (v/c)^2}} - 1 \right). \tag{V-10}$$

Can you show that $K \rightarrow \frac{1}{2} m_0 v^2$ as $c \rightarrow \infty$?

V-7
Mass, Momentum, and Kinetic Energy (Sections 8.9 and 9.3; RR, Sections 3.3 and 3.5)

The best known result of SR is the so-called mass-energy equivalence. That is, the conservation of total energy is equivalent to the conservation of relativistic mass. Mass and energy are equivalent; they form a single invariant that we can call mass-energy. The relation

$$E = mc^2 \qquad\qquad (V\text{-}11)$$

expresses the fact that mass-energy can be expressed in energy (E) units or equivalently in mass $(m = E/c^2)$ units. In fact, it has become common practice to refer to masses in terms of electron volts, such as saying that the rest mass of an electron is 0.51 MeV, for convenience in energy calculations. Likewise entities of zero rest mass, such as photons, may be assigned an effective mass equivalent to their energy. We associate mass with each of the various forms of energy.

In classical physics we had two separate conservation principles: (1) the conservation of (classical) mass, as in chemical reactions, and (2) the conservation of energy. In relativity, these merge into one conservation principle, that of the conservation of mass-energy. The two classical laws may be viewed as special cases that would be expected to agree with experiment only if energy transfers into or out of the system are so small compared with the system's rest mass that the corresponding fractional change in rest mass of the system is too small to be measured.

For example, the rest mass of a hydrogen atom is 1.00797 u (= 938.8 MeV). If enough energy (13.58 eV) is added to ionize hydrogen, that is, to break it up into its constituent parts, a proton and an electron, the fractional change in rest mass of the system is

$$\frac{13.58\ \text{eV}}{938.8 \times 10^6\ \text{eV}} = 1.45 \times 10^{-8}$$

or 1.45×10^{-6} percent, too small to measure. However, for a nucleus such as the deuteron, whose rest mass is 2.01360 u (= 1876.4 MeV), one must add an energy of 2.22 MeV to break it up into its constituent parts, a proton and a neutron. The fractional change in rest mass of the system is

$$\frac{2.22\ \text{MeV}}{1876.4\ \text{MeV}} = 1.18 \times 10^{-3}$$

or 0.12 percent, which is readily measureable. This is characteristic of the fractional rest-mass changes in nuclear reactions, so that one must use the relativistic law of conservation of mass-energy to get agreement between theory and experiment in nuclear reactions. The classical (rest) mass is *not* conserved, but total energy (mass-energy) is.

In Chapter 41 we sought to make the existence of electromagnetic waves plausible by showing that such waves are consistent with Maxwell's equations as expressed in Table 40-2. Here we seek to start from Maxwell's equations and derive from them a differential equation whose solutions will describe electromagnetic waves. We will show directly that the speed c of such waves is given by Eq. 40-1, or $c = 1/\sqrt{\epsilon_0\mu_0}$.

We followed a similar program in Supplementary Topic III for mechanical waves on a stretched string. Starting from Newton's laws of motion we derived a differential equation (Eq. III-1) whose solutions (Eq. III-2) described such waves. We showed further that the speed v of these waves is given by Eq. III-3, or $v = \sqrt{F/\mu}$.

In Table 40-2 we wrote Maxwell's equations as

$$\epsilon_0 \oint \mathbf{E} \cdot d\mathbf{S} = q, \qquad\qquad (VI\text{-}1)$$

* Supplementary Topics I to V appear in Part I.

V-8
The Equivalence of Mass and Energy (Section 8.9; RR, Section 3.6)

SUPPLEMENTARY TOPIC VI*
THE DIFFERENTIAL FORM OF MAXWELL'S EQUATIONS AND THE ELECTROMAGNETIC WAVE EQUATION

VI-1
Introduction

$$\oint \mathbf{B} \cdot d\mathbf{S} = 0, \tag{VI-2}$$

$$\oint \mathbf{B} \cdot d\mathbf{l} = \mu_0(i + \epsilon_0 \, d\Phi_E/dt), \tag{VI-3}$$

and

$$\oint \mathbf{E} \cdot d\mathbf{l} = -d\Phi_B/dt. \tag{VI-4}$$

These equations are said to be written in integral form. The field variables $\mathbf{E}$ and $\mathbf{B}$, which are usually the unknown quantities, appear in the integrands. Only in a few symmetric cases (see Sections 28-8 and 34-2, for example) can we "factor them out." In more general problems we cannot do so.

The situation is somewhat analogous to computing the density ρ of a body if we know its mass m and volume τ. In general these are related by the integral equation

$$m = \int \rho \, d\tau.$$

Only if ρ is a constant over all parts of the volume can we factor it out and write $\rho = m/\tau$.

To carry out our program it is desirable to recast Maxwell's equations in the form of equalities that apply at each *point* in space rather than as integrals that apply to various *regions* of space. In other words, we wish to convert Maxwell's equations from the integral form of Eqs. VI-1 to 4 into *differential form*. We will then be able to relate $\mathbf{E}$ and $\mathbf{B}$ at a point to the charge density and current density at that point.

VI-2
The Operator ∇

To transform Maxwell's equations into differential form we must deepen our understanding of vector methods and, in particular, become familiar with the vector operator ∇.

In Section 29-7 we saw how to obtain the components of the (vector) electrostatic field $\mathbf{E}$ at any point from the (scalar) potential function $V(x,y,z)$ by partial differentiation. Thus,

$$E_x = -\frac{\partial V}{\partial x}, \quad E_y = -\frac{\partial V}{\partial y}, \quad \text{and} \quad E_z = -\frac{\partial V}{\partial z}$$

so that the electrostatic field

$$\mathbf{E} = \mathbf{i}E_x + \mathbf{j}E_y + \mathbf{k}E_z$$

can be written as

$$\mathbf{E} = -\left(\mathbf{i}\frac{\partial V}{\partial x} + \mathbf{j}\frac{\partial V}{\partial y} + \mathbf{k}\frac{\partial V}{\partial z}\right). \tag{VI-5}$$

We can write Eq. VI-5 in compact vector notation as

$$\mathbf{E} = -\nabla V,$$

where ∇ ("del") is a vector operator defined as

$$\nabla - \mathbf{i}\frac{\partial}{\partial x} + \mathbf{j}\frac{\partial}{\partial y} + \mathbf{k}\frac{\partial}{\partial z}. \tag{VI-6}$$

This operator is useful in dealing with scalar and vector fields (see Sections 16-8 and 18-7 for examples of such fields). Given any scalar field ψ we may form a vector field, called the gradient of ψ and written as grad ψ or $\nabla\psi$, simply by applying the operator ∇ to ψ. Given a vector field $\mathbf{U} = U_x\mathbf{i} + U_y\mathbf{j} + U_z\mathbf{k}$ we may apply the operator ∇ to it in two different ways. One way is to take the dot product of ∇ and $\mathbf{U}$, yielding the scalar field called the divergence of $\mathbf{U}$ and written as div $\mathbf{U}$ or $\nabla \cdot \mathbf{U}$. The other way is to take the cross product of ∇ and $\mathbf{U}$, yielding the vector field called the curl of $\mathbf{U}$ and written curl $\mathbf{U}$ or $\nabla \times \mathbf{U}$. These operations may be summarized as

$$\text{grad } \psi \equiv \nabla\psi = \mathbf{i}\,\frac{\partial\psi}{\partial x} + \mathbf{j}\,\frac{\partial\psi}{\partial y} + \mathbf{k}\,\frac{\partial\psi}{\partial z}, \tag{VI-7}$$

$$\text{div } \mathbf{U} \equiv \nabla\cdot\mathbf{U} = \frac{\partial U_x}{\partial x} + \frac{\partial U_y}{\partial y} + \frac{\partial U_z}{\partial z}, \tag{VI-8}$$

$$\text{curl } \mathbf{U} \equiv \nabla\times\mathbf{U} = \mathbf{i}\left(\frac{\partial U_z}{\partial y} - \frac{\partial U_y}{\partial z}\right)$$

$$+\,\mathbf{j}\left(\frac{\partial U_x}{\partial z} - \frac{\partial U_z}{\partial x}\right)$$

$$+\,\mathbf{k}\left(\frac{\partial U_y}{\partial x} - \frac{\partial U_x}{\partial y}\right). \tag{VI-9}$$

Note that grad ψ and curl $\mathbf{U}$ are vectors, whereas div $\mathbf{U}$ is a scalar. You can gain some familiarity with these operations by the following exercises: (1) prove that curl (grad ψ) = 0 and (2) prove that div (curl $\mathbf{U}$) = 0.

Another frequently occurring operator is ∇^2 ("del squared"). It is simply $\nabla\cdot\nabla$, or, as the student can show from Eq. VI-6,

$$\nabla^2 = \nabla\cdot\nabla = \frac{\partial^2}{\partial x^2} + \frac{\partial^2}{\partial y^2} + \frac{\partial^2}{\partial z^2}.$$

When we apply ∇^2 to a scalar field ψ, we obtain

$$\nabla^2\psi = \frac{\partial^2\psi}{\partial x^2} + \frac{\partial^2\psi}{\partial y^2} + \frac{\partial^2\psi}{\partial z^2}. \tag{VI-10}$$

For a vector field $\mathbf{U}$, the operation $\nabla^2\mathbf{U}$ is

$$\nabla^2\mathbf{U} = \mathbf{i}\left(\frac{\partial^2}{\partial x^2} + \frac{\partial^2}{\partial y^2} + \frac{\partial^2}{\partial z^2}\right)U_x + \mathbf{j}\left(\frac{\partial^2}{\partial x^2} + \frac{\partial^2}{\partial y^2} + \frac{\partial^2}{\partial z^2}\right)U_y + \mathbf{k}\left(\frac{\partial^2}{\partial x^2} + \frac{\partial^2}{\partial y^2} + \frac{\partial^2}{\partial z^2}\right)U_z \tag{VI-11}$$

As an exercise, show that curl (curl $\mathbf{U}$) = $-\nabla^2\mathbf{U}$ + grad (div $\mathbf{U}$).

In this section we show how to cast the first two of Maxwell's equations (Eqs. VI-1, 2) into differential form. Let us apply Eq. VI-1 to a differential volume element shaped like a rectangular parallelepiped and containing a point P at (and near) which an electric field exists (see Fig. VI-1a). Point P is located at x, y, z in the reference frame of Fig. VI-1b and the edges of the parallelepiped have lengths dx, dy, and dz.

We can write the surface area vector for the rear face of the parallelepiped as $d\mathbf{S} = -\mathbf{i}\,dy\,dz$. The minus sign enters because $d\mathbf{S}$ is defined to point in the direction of the *outward* normal, which is defined by $-\mathbf{i}$. For the front face we have $d\mathbf{S} = +\mathbf{i}\,dy\,dz$.

If the electric field at the rear face is $\mathbf{E}$, that at the front face, which is a distance dx away from the rear face, is $\mathbf{E} + (\partial\mathbf{E}/\partial x)\,dx$, the latter term representing the change in $\mathbf{E}$ associated with the change dx in x.

The flux through the entire surface of the parallelepiped is $\oint \mathbf{E}\cdot d\mathbf{S}$ and the contribution to this flux due to these two faces alone is

$$(\mathbf{E})\cdot(-\mathbf{i}\,dy\,dz) + \left(\mathbf{E} + \frac{\partial\mathbf{E}}{\partial x}\,dx\right)\cdot(+\mathbf{i}\,dy\,dz)$$

$$= dx\,dy\,dz\left(\frac{\partial\mathbf{E}}{\partial x}\cdot\mathbf{i}\right) = dx\,dy\,dz\,\frac{\partial}{\partial x}(\mathbf{E}\cdot\mathbf{i})$$

$$= dx\,dy\,dz\,\frac{\partial E_x}{\partial x}.$$

With similar contributions from the other four faces the total electric flux becomes

$$\oint \mathbf{E}\cdot d\mathbf{S} = dx\,dy\,dz\left(\frac{\partial E_x}{\partial x} + \frac{\partial E_y}{\partial y} + \frac{\partial E_z}{\partial z}\right).$$

VI-3
Maxwell's Equations in Differential Form—I

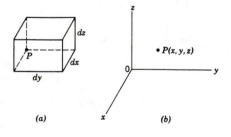

(a)   *(b)*

From Eq. VI-8 we may write this as

$$\oint \mathbf{E} \cdot d\mathbf{S} = dx\,dy\,dz\,\text{div }\mathbf{E}. \tag{VI-12}$$

Now the right-hand side of Eq. VI-1, which gives the charge enclosed by the surface, may be written in general as $q = \int \rho\,d\tau$ and, in particular, for the differential volume element at P, as

$$q = \rho\,dx\,dy\,dz, \tag{VI-13}$$

where ρ is the charge per unit volume at P. Substituting Eqs. VI-12 and 13 into Eq. VI-1 and canceling the common factor $dx\,dy\,dz$, we have finally

$$\epsilon_0\,\text{div }\mathbf{E} = \rho, \tag{VI-14}$$

which is Maxwell's first equation (Eq. VI-1) in differential form.

Using the same technique, we can express Maxwell's second equation (Eq. VI-2) in differential form as

$$\text{div }\mathbf{B} = 0. \tag{VI-15}$$

We now seek to transform Maxwell's third and fourth equations (Eqs. VI-3, 4) into differential form. We start by applying Eq. VI-3 to a differential surface element of rectangular shape at a point P in some region of a magnetic field, as shown in Fig. VI-2a. The point P is located at x, y, z in the reference frame of Fig. VI-2b and the sides of the rectangle, which is parallel to the x-y plane, have lengths dx and dy. Going around the path, as shown by the arrows, we have

$$\oint \mathbf{B} \cdot d\mathbf{l} = \mathbf{B} \cdot (-\mathbf{j}\,dy) \qquad \text{(rear side)}$$

$$+ \mathbf{B} \cdot (+\mathbf{i}\,dx) \qquad \text{(left side)}$$

$$+ \left(\mathbf{B} + \frac{\partial \mathbf{B}}{\partial x}\,dx\right) \cdot (+\mathbf{j}\,dy) \qquad \text{(front side)}$$

$$+ \left(\mathbf{B} + \frac{\partial \mathbf{B}}{\partial y}\,dy\right) \cdot (-\mathbf{i}\,dx), \qquad \text{(right side)}$$

where $\mathbf{B}$ is the magnetic field at P.

Collecting terms we obtain

$$\oint \mathbf{B} \cdot d\mathbf{l} = dx\,dy\left(\frac{\partial \mathbf{B}}{\partial x}\cdot\mathbf{j} - \frac{\partial \mathbf{B}}{\partial y}\cdot\mathbf{i}\right)$$

$$= dx\,dy\left[\frac{\partial}{\partial x}(\mathbf{B}\cdot\mathbf{j}) - \frac{\partial}{\partial y}(\mathbf{B}\cdot\mathbf{i})\right]$$

$$= dx\,dy\left(\frac{\partial B_y}{\partial x} - \frac{\partial B_x}{\partial y}\right). \tag{VI-16}$$

Now in the right-hand side of Eq. VI-3 i is the current enclosed by the path and $d\Phi_E/dt$ is the change in electric flux through the enclosed surface. Hence, if $\mathbf{J}$ is taken to represent the current density and $d\mathbf{S}(= \mathbf{k}\,dx\,dy)$ the surface area vector, we can write

$$i = \mathbf{J} \cdot d\mathbf{S} = \mathbf{J} \cdot (\mathbf{k}\,dx\,dy) = dx\,dy\,J_z \tag{VI-17}$$

and

$$\frac{d\Phi_E}{dt} = \frac{\partial \mathbf{E}}{\partial t} \cdot d\mathbf{S} = \frac{\partial \mathbf{E}}{\partial t} \cdot (\mathbf{k}\,dx\,dy)$$

or

$$\frac{d\Phi_E}{dt} = \frac{\partial E_z}{\partial t}\,dx\,dy. \tag{VI-18}$$

Substituting Eqs. VI-16, 17, and 18 into Eq. VI-3 and canceling the common factor $dx\,dy$, we get

$$\frac{1}{\mu_0}\left(\frac{\partial B_y}{\partial x} - \frac{\partial B_x}{\partial y}\right) = J_z + \epsilon_0\frac{\partial E_z}{\partial t}. \tag{VI-19}$$

VI-4

Maxwell's Equations in Differential Form — II

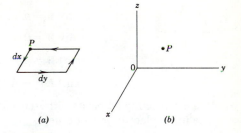

(a)   (b)

We could have proceeded exactly as above if we had started with a rectangle parallel to the *y-z* plane or one parallel to the *z-x* plane. Each rectangle would have given us a different component of an arbitrarily oriented differential surface at *P*. Equation VI-19 is obviously the *z*-component equation corresponding to Eq. VI-3. If we multiply it by **k** and add to it the two similar vector equations, which may be obtained by cyclically permuting *x, y, z* and **i, j, k**, corresponding to the *x*-component and *y*-component equations, we obtain

$$\text{curl } \mathbf{B} = \mu_0(\mathbf{J} + \epsilon_0 \, \partial\mathbf{E}/\partial t), \qquad (\text{VI-20})$$

which is the third Maxwell equation in differential form.

Similarly, starting with Eq. VI-4, we may show that

$$\text{curl } \mathbf{E} = -\partial\mathbf{B}/\partial t, \qquad (\text{VI-21})$$

which is the fourth Maxwell equation in differential form.

We have derived four differential equations (see VI-22 to VI-25 below) from the four integral equations (VI-1 to VI-4). It can be shown that the integral equations can be derived from the differential equations, that is, the two sets of equations are *equivalent*.

We have now obtained from their integral form the four basic equations of electromagnetism, Maxwell's equations, in differential form. Corresponding to the integral equations, Eqs. VI-1, 2, 3, and 4 respectively, we have

VI-5
The Wave Equation

$$\epsilon_0 \text{ div } \mathbf{E} = \rho, \qquad (\text{VI-22})$$

$$\text{div } \mathbf{B} = 0, \qquad (\text{VI-23})$$

$$\text{curl } \mathbf{B} = \mu_0(\mathbf{J} + \epsilon_0 \, \partial\mathbf{E}/\partial t), \qquad (\text{VI-24})$$

$$\text{curl } \mathbf{E} = -\partial\mathbf{B}/\partial t, \qquad (\text{VI-25})$$

which are four coupled partial differential equations. They apply at each point of space in an electromagnetic field.

We will now derive the wave equation for electromagnetic waves in free space. In free space, the charge density ρ and the current density **J** are zero, so that the Maxwell equations there become

$$\text{div } \mathbf{E} = 0,$$

$$\text{div } \mathbf{B} = 0,$$

$$\text{curl } \mathbf{B} = \mu_0\epsilon_0 \, \partial\mathbf{E}/\partial t,$$

and
$$\text{curl } \mathbf{E} = -\partial\mathbf{B}/\partial t.$$

Let us take the curl of the equation for curl **E**; we obtain

$$\text{curl curl } \mathbf{E} = -\text{curl } \frac{\partial\mathbf{B}}{\partial t} = -\frac{\partial}{\partial t} \text{curl } \mathbf{B}.$$

But, from above, curl $\mathbf{B} = \mu_0\epsilon_0(\partial\mathbf{E}/\partial t)$, so that

$$\text{curl curl } \mathbf{E} = -\mu_0\epsilon_0 \frac{\partial^2\mathbf{E}}{\partial t^2}. \qquad (\text{VI-26})$$

From the exercise in VI-2, we know that curl curl $\mathbf{E} = -\nabla^2 E + \text{grad div } \mathbf{E}$ and from above that div $\mathbf{E} = 0$. Thus,

$$\text{curl curl } \mathbf{E} = -\nabla^2\mathbf{E}. \qquad (\text{VI-27})$$

Combining Eqs. VI-26 and VI-27, we obtain finally

$$\nabla^2\mathbf{E} = \mu_0\epsilon_0 \frac{\partial^2\mathbf{E}}{\partial t^2}. \qquad (\text{VI-28}a)$$

Proceeding as above, show that **B** satisfies the same equation, or

$$\nabla^2\mathbf{B} = \mu_0\epsilon_0 \frac{\partial^2\mathbf{B}}{\partial t^2}. \qquad (\text{VI-28}b)$$

Equations VI-28 are the equations of electromagnetic wave motion. Being vector equations, they are equivalent to six scalar equations, one for each component of **E** and of **B**.

There are many solutions of Eqs. VI-28, corresponding to different kinds of electromagnetic waves—plane, spherical, and cylindrical waves being three examples. Let us consider a solution in which two components of **E** and two of **B** vanish, that is, in which

$$E_x = E_z = 0 \quad \text{and} \quad B_x = B_y = 0.$$

Equations VI-28 are satisfied for these assumptions. For the nonvanishing components, E_y and B_z, Eqs. VI-28 reduce to (see Eq. VI-11)

$$\frac{\partial^2 E_y}{\partial x^2} + \frac{\partial^2 E_y}{\partial y^2} + \frac{\partial^2 E_y}{\partial z^2} = \mu_0 \epsilon_0 \frac{\partial^2 E_y}{\partial t^2} \tag{VI-29a}$$

and

$$\frac{\partial^2 B_z}{\partial x^2} + \frac{\partial^2 B_z}{\partial y^2} + \frac{\partial^2 B_z}{\partial z^2} = \mu_0 \epsilon_0 \frac{\partial^2 B_z}{\partial t^2}. \tag{VI-29b}$$

If we make the additional assumption that E_y and B_z are functions of x and t only, the simplified wave equation that results, namely

$$\frac{\partial^2 E_y}{\partial x^2} = \mu_0 \epsilon_0 \frac{\partial^2 E_y}{\partial t^2} \quad \text{and} \quad \frac{\partial^2 B_z}{\partial x^2} = \mu_0 \epsilon_0 \frac{\partial^2 B_z}{\partial t^2}, \tag{VI-29c}$$

is similar to Eq. III-1 for the vibrating string.

A solution to these equations, as you may verify by substitution, is

$$E_y = E_m \sin (kx - \omega t) \tag{VI-30a}$$

and

$$B_z = B_m \sin (kx - \omega t). \tag{VI-30b}$$

We interpret Eqs. VI-30 as an electromagnetic wave traveling in the positive x-direction, as in Fig. 41-13, with a speed $c = \omega/k$. Show that substituting Eq. VI-30a into Eq. VI-29a (or Eq. VI-30b into Eq. VI-29b) yields

$$c = \omega/k = 1/\sqrt{\epsilon_0 \mu_0},$$

which (see Eq. 40-1) gives the speed of electromagnetic waves in free space.

appendices

SI base units[a]

Quantity	Name	Symbol	Definition
length	meter	m	" . . . the length equal to 1,650,763.73 wavelengths in vacuum of the radiation corresponding to the transition between the levels $2p_{10}$ and $5d_5$ of the krypton-86 atom." (1960)
mass	kilogram	kg	" . . . this prototype [a certain platinum-iridium cylinder] shall henceforth be considered to be the unit of mass." (1889)
time	second	s	" . . . the duration of 9,192,-631,770 periods of the radiation corresponding to the transition between the two hyperfine levels of the ground state of the cesium-133 atom." (1967)

* Adapted from "The International System of Units (SI)," National Bureau of Standards Special Publication 330, 1972 edition.

[a] The definitions of these base units were adopted by the General Conference of Weights and Measures, an international body, on the dates shown. In this book we will not use the candela.

SI base units (Continued)

Quantity	Name	Symbol	Definition
electric current	ampere	A	" . . . that constant current which, if maintained in two straight parallel conductors of infinite length, of negligible circular cross section, and placed 1 meter apart in vacuum, would produce between these conductors a force equal to 2×10^{-7} newton per meter of length." (1946)
thermodynamic temperature	kelvin	K	" . . . the fraction 1/273.16 of the thermodynamic temperature of the triple point of water." (1967)
amount of substance	mole	mol	" . . . the amount of substance of a system which contains as many elementary entities as there are atoms in 0.012 kilogram of carbon-12." (1971)
luminous intensity	candela	cd	" . . . the luminous intensity, in the perpendicular direction, of a surface of 1/600,000 square meter of a blackbody at the temperature of freezing platinum under a pressure of 101,325 newton per square meter." (1967)

Some SI derived units with special names

	SI unit			
Quantity	Name	Symbol	Expression in terms of other units	Expression in terms of SI base units
frequency	hertz	Hz		s^{-1}
force	newton	N		$m \cdot kg/s^2$
pressure	pascal	Pa	N/m^2	$kg/m \cdot s^2$
energy, work, quantity of heat	joule	J	$N \cdot m$	$kg \cdot m^2/s^2$
power, radiant flux	watt	W	J/s	$kg \cdot m^2/s^3$
quantity of electricity, electric charge	coulomb	C		$A \cdot s$
electric potential, potential difference, electromotive force	volt	V	W/A	$kg \cdot m^2/A \cdot s^3$
capacitance	farad	F	C/V	$A^2 \cdot s^4/kg \cdot m^2$
electric resistance	ohm	Ω	V/A	$kg \cdot m^2/A^2 \cdot s^3$
conductance	siemens	S	A/V	$A^2 \cdot s^3/kg \cdot m^2$
magnetic flux	weber	Wb	$V \cdot s$	$kg \cdot m^2/A \cdot s^2$
magnetic field	tesla	T	Wb/m^2	$kg/A \cdot s^2$
inductance	henry	H	Wb/A	$kg \cdot m^2/A^2 \cdot s^2$

Some symbols for units of physical quantities

SI Symbols		Symbols other than SI that are Commonly Used	
Name	Abbreviation	Name	Abbreviation
ampere	A	angstrom	Å
candela	cd	British thermal unit	Btu
coulomb	C	calorie	cal
farad	F	day	d
henry	H	degree	°
hertz	Hz	dyne	dyn
joule	J	electron volt	eV
kelvin	K	foot	ft
kilogram	kg	gauss	G
meter	m	gram	g
mole	mol	horsepower	hp
newton	N	hour	h
ohm	Ω	inch	in.
pascal	Pa	mile	mi
radian	rad	minute (of arc)	'
second	s	minute (of time)	min
siemens	S	pound	lb
steradian	sr	revolution	rev
tesla	T	second (of arc)	"
volt	V	standard atmosphere	atm
watt	W	unified atomic mass unit	u
weber	Wb	year	yr

Over the years many hundreds of measurements of fundamental physical quantities, alone and in combination, have been made by hundreds of scientists in many countries. These measurements have different precisions and, most important, they are interdependent. For example, the direct measurements of e, e/m, h/e, etc., are obviously interrelated. Sorting out the best values of e, m, h, etc., from a large mass of overlapping data is not simple.†

For most problems in this book three significant figures will do, and the computational (rounded) values may be used.

APPENDIX B
SOME FUNDAMENTAL CONSTANTS OF PHYSICS*

Constant	Symbol	Computational value	Best (1973) Value Value[a]	Uncertainty[b]
Speed of light in a vacuum	c	3.00×10^8 m/s	2.99792458	0.004
Elementary charge	e	1.60×10^{-19} C	1.6021892	2.9
Electron rest mass	m_e	9.11×10^{-31} kg	9.109534	5.1
Permittivity constant	ϵ_0	8.85×10^{-12} F/m	8.854187818	0.008
Permeability constant	μ_0	1.26×10^{-6} H/m	4π (exactly)	—
Electron charge to mass ratio	e/m_e	1.76×10^{11} C/kg	1.7588047	2.8
Proton rest mass	m_p	1.67×10^{-27} kg	1.6726485	5.1
Ratio of proton mass to electron mass	m_p/m_e	1840	1836.15152	0.38
Neutron rest mass	m_n	1.68×10^{-27} kg	1.6749543	5.1
Muon rest mass	m_μ	1.88×10^{-28} kg	1.883566	5.6
Planck constant	h	6.63×10^{-34} J·s	6.626176	5.4
Electron Compton wavelength	λ_C	2.43×10^{-12} m	2.4263089	1.6
Molar gas constant	R	8.31 J/mol·K	8.31441	31
Avogadro constant	N_A	6.02×10^{23}/mol	6.022045	5.1
Boltzmann constant	k	1.38×10^{-23} J/K	1.380662	32
Molar volume of ideal gas at STP[c]	V_m	2.24×10^{-2} m³/mol	2.241383	31
Faraday constant	F	9.65×10^4 C/mol	9.648456	2.8
Stefan–Boltzmann constant	σ	5.67×10^{-8} W/m²·K⁴	5.67032	125
Rydberg constant	R	1.10×10^7/m	1.097373177	0.075
Gravitational constant	G	6.67×10^{-11} m³/s²·kg	6.6720	615
Bohr radius	a_0	5.29×10^{-11} m	5.2917706	0.82
Electron magnetic moment	μ_e	9.28×10^{-24} J/T	9.284832	3.9
Proton magnetic moment	μ_p	1.41×10^{-26} J/T	1.4106171	3.9
Bohr magneton	μ_B	9.27×10^{-24} J/T	9.274078	3.9
Nuclear magneton	μ_N	5.05×10^{-27} J/T	5.050824	3.9

[a] Same unit and power of ten as the computational value.
[b] Parts per million.
[c] STP-standard temperature and pressure=0° C and 1.0 atm.

* The values in this table were selected from a longer list developed by E. Richard Cohen and B. N. Taylor, *Journal of Physical and Chemical Reference Data*, vol. 2, no. 4 (1973).
† See "A Pilgrim's Progress in Search of the Fundamental Constants," by J. W. M. Du Mond, *Physics Today*, October 1965, and "The Fundamental Physical Constants" by Taylor, Langenberg, and Parker, *Scientific American*, October, 1970.

The sun

Mass	1.99×10^{30} kg
Radius	6.96×10^5 km
Mean density	1,410 kg/m³
Surface gravity	274 m/s²
Surface temperature	6000 K
Total radiation rate	3.92×10^{26} W

APPENDIX C
SOLAR, TERRESTRIAL, AND LUNAR DATA

The earth

Mass	5.98×10^{24} kg
Equatorial radius	6.378×10^6 m
	3963 mi
Polar radius	6.357×10^6 m
	3950 mi
Radius of a sphere of the same volume	6.37×10^6 m
	3600 mi
Mean density	5522 kg/m³
Acceleration of gravity[a]	9.80665 m/s²
	32.1740 ft/s²
Mean orbital speed	29,770 m/s
	18.50 mi/s
Angular speed	7.29×10^{-5} rad/s
Solar constant[b]	1340 W/m²
Magnetic field (at Washington, D.C.)	5.7×10^{-5} T
Magnetic dipole moment	8.1×10^{22} A·m²
Standard atmosphere	1.013×10^5 Pa
	14.70 lb/in.²
	760.0 mm-Hg
Density of dry air at STP[c]	1.29 kg/m³
Speed of sound in dry air at STP	331.4 m/s
	1089 ft/s
	742.5 mi/h

[a] This value, adopted by the General Committee on Weights and Measures in 1901, approximates the value at 45° latitude at sea level.
[b] This is the rate per unit area at which solar energy falls, at normal incidence, just outside the earth's atmosphere.
[c] STP = standard temperature and pressure = 0°C and 1 atm.

The moon

Mass	7.36×10^{22} kg
Radius	1738 km
Mean density	3340 kg/m³
Surface gravity	1.67 m/s²
Mean earth-moon distance	3.80×10^5 km

APPENDIX D
THE SOLAR SYSTEM*

	MERCURY	VENUS	EARTH	MARS	JUPITER	SATURN	URANUS	NEPTUNE	PLUTO
Maximum distance from sun (10^6 km)	69.7	109	152.1	249.1	815.7	1,507	3,004	4,537	7,375
Minimum distance from sun (10^6 km)	45.9	107.4	147.1	206.7	740.9	1,347	2,735	4,456	4,425
Mean distance from sun (10^6 km)	57.9	108.2	149.6	227.9	778.3	1,427	2,869.6	4,496.6	5,900
Mean distance from sun (astronomical units)	.387	.723	1	1.524	5.203	9.539	19.18	30.06	39.44
Period of revolution	88 d	224.7 d	365.26 d	687 d	11.86 y	29.46 y	84.01 y	164.8 y	247.7 y
Rotation period	59 d	−243 d retrograde	23 h 56 min 4 s	24 h 37 min 23 s	9 h 50 min 30 s	10 h 14 min	−11 h retrograde	16 h	6 d 9 h
Orbital velocity (km/s)	47.9	35	29.8	24.1	13.1	9.6	6.8	5.4	4.7
Inclination of axis	<28°	3°	23°27′	23°59′	3°05′	26°44′	82°5′	28°48′	?
Inclination of orbit to ecliptic	7°	3.4°	0°	1.9°	1.3°	2.5°	.8°	1.8°	17.2°
Eccentricity of orbit	.206	.007	.017	.093	.048	.056	.047	.009	.25
Equatorial diameter (km)	4,880	12,104	12,756	6,787	142,800	120,000	51,800	49,500	6,000 (?)
Mass (earth = 1)	.055	.815	1	.108	317.9	95.2	14.6	17.2	.1 (?)
Volume (earth = 1)	.06	.88	1	.15	1,316	755	67	57	.1 (?)
Density (water = 1)	5.4	5.2	5.5	4.0	1.3	.7	1.2	1.7	?
Oblateness	0	0	.003	.009	.06	.1	.06	.02	?
Atmosphere (main components)	none	carbon dioxide	nitrogen, oxygen	carbon dioxide, argon	hydrogen, helium	hydrogen, helium	hydrogen, helium, methane	hydrogen, helium, methane	none detected
Mean temperature at visible surface (degrees Celsius) S = solid, C = clouds	350(S) d −170(S) night	−33 (C) 480 (S)	22 (S)	−23 (S)	−150 (C)	−180 (C)	−210 (C)	−220 (C)	−230(?)
Atmospheric pressure at surface (millibars)	10^{-9}	90,000	1,000	6	?	?	?	?	?
Surface gravity (earth = 1)	.37	.88	1	.38	2.64	1.15	1.17	1.18	?
Mean apparent diameter of sun as seen from planet	1°22′40″	44′15″	31′59″	21′	6′09″	3′22″	1′41″	1′04″	49″
Known satellites	0	0	1	2	13	10	5	2	0

* Reprinted by permission from "The Solar System," Carl Sagan, *Scientific American*, September 1975.

APPENDIX E
PERIODIC TABLE OF THE ELEMENTS

Key:
- atomic number
- atomic mass

1
H
1.0079

PERIODS

Main table

Period	IA	IIA	IIIB	IVB	VB	VIB	VIIB	VIII	VIII	VIII	IB	IIB	IIIA	IVA	VA	VIA	VIIA	VIIIA (O) NOBLE GASES
1	1 **H** 1.00797																	2 **He** 4.00260
2	3 **Li** 6.941	4 **Be** 9.01218											5 **B** 10.81	6 **C** 12.01115	7 **N** 14.0067	8 **O** 15.9994	9 **F** 18.99840	10 **Ne** 20.179
3	11 **Na** 22.98977	12 **Mg** 24.305											13 **Al** 26.98154	14 **Si** 28.086†	15 **P** 30.97376	16 **S** 32.06	17 **Cl** 35.453	18 **Ar** 39.948
4	19 **K** 39.098	20 **Ca** 40.08	21 **Sc** 44.9559	22 **Ti** 47.90	23 **V** 50.9414	24 **Cr** 51.996	25 **Mn** 54.9380	26 **Fe** 55.847	27 **Co** 58.9332	28 **Ni** 58.71	29 **Cu** 63.546	30 **Zn** 65.38	31 **Ga** 69.72	32 **Ge** 72.59	33 **As** 74.9216	34 **Se** 78.96	35 **Br** 79.904	36 **Kr** 83.80
5	37 **Rb** 85.4678	38 **Sr** 87.62	39 **Y** 88.9059	40 **Zr** 91.22	41 **Nb** 92.9064	42 **Mo** 95.94	43 **Tc** 98.9062	44 **Ru** 101.07	45 **Rh** 102.9055	46 **Pd** 106.4	47 **Ag** 107.868	48 **Cd** 112.40	49 **In** 114.82	50 **Sn** 118.69	51 **Sb** 121.75	52 **Te** 127.60	53 **I** 126.9045	54 **Xe** 131.30
6	55 **Cs** 132.9054	56 **Ba** 137.34	57 ***La** 138.9055	72 **Hf** 178.49	73 **Ta** 180.9479	74 **W** 183.85	75 **Re** 186.2	76 **Os** 190.2	77 **Ir** 192.22	78 **Pt** 195.09	79 **Au** 196.9665	80 **Hg** 200.59	81 **Tl** 204.37	82 **Pb** 207.19	83 **Bi** 208.9804	84 **Po** (210)	85 **At** (210)	86 **Rn** (222)
7	87 **Fr** (223)	88 **Ra** 226.0254	89 **†Ac** (227)	104 (261)	105 (260)	106 (263)												

* Lanthanides

58 **Ce** 140.12	59 **Pr** 140.9077	60 **Nd** 144.24	61 **Pm** (147)	62 **Sm** 150.4	63 **Eu** 151.96	64 **Gd** 157.25	65 **Tb** 158.9254	66 **Dy** 162.50	67 **Ho** 164.9304	68 **Er** 167.26	69 **Tm** 168.9342	70 **Yb** 173.04	71 **Lu** 174.97

† Actinides

90 **Th** 232.0381	91 **Pa** 231.0359	92 **U** 238.029	93 **Np** 237.0482	94 **Pu** (244)	95 **Am** (243)	96 **Cm** (247)	97 **Bk** (247)	98 **Cf** (251)	99 **Es** (254)	100 **Fm** (257)	101 **Md** (258)	102 **No** (255)	103 **Lr** (256)

Family name			Symbol		Spin	Charge, e	Strangeness	Rest mass, MeV	Mean life, seconds	Typical decay mode
			Particle	Antiparticle						
—		Photon	γ	γ	1	0	0	0	Stable	—
L E P T O N S		Electron	e^-	$\overline{e^-}$	$\frac{1}{2}$	∓ 1	0	0.5110	Stable	—
		Muon	μ^+	$\overline{\mu^+}$	$\frac{1}{2}$	± 1	0	105.7	2.197×10^{-6}	$e + \nu + \overline{\nu}$
		Electron's neutrino	ν_e	$\overline{\nu_e}$	$\frac{1}{2}$	0	0	0	Stable	—
		Muon's neutrino	ν_μ	$\overline{\nu_\mu}$	$\frac{1}{2}$	0	0	0	Stable	—
H A D R O N S	M E S O N S	Pion	π^+ π^0	$\overline{\pi^+}$ $\overline{\pi^0}$	0 0	± 1 0	0 0	139.6 135.0	2.603×10^{-8} 8.28×10^{-17}	$\mu + \nu$ $\gamma + \gamma$
		K-meson	K^+	$\overline{K^+}$	0	± 1	± 1	493.7	1.237×10^{-8}	$\mu + \nu$
			K^0	$\overline{K^0}$	0	0	± 1	497.7	$\begin{cases} 8.930 \times 10^{-11} \\ 5.181 \times 10^{-8} \end{cases}$	$\pi^+ + \pi^-$ $\pi^0 + \pi^0 + \pi^0$
		Eta-meson	η^0	η^0	0	0	0	548.8	?	$\gamma + \gamma$
	B A R Y O N S	N U C L E O N — Proton	p	$\overline{p}$	$\frac{1}{2}$	± 1	0	938.3	Stable	—
		Neutron	n	$\overline{n}$	$\frac{1}{2}$	0	0	939.6	918	$p + e^- + \nu$
		Lambda particle	Λ^0	$\overline{\Lambda^0}$	$\frac{1}{2}$	0	∓ 1	1116	2.578×10^{-10}	$p + \pi^-$
		Sigma particle	Σ^+	$\overline{\Sigma^+}$	$\frac{1}{2}$	$+1$	∓ 1	1189	8.00×10^{-11}	$p + \pi^0$
			Σ^0	$\overline{\Sigma^0}$	$\frac{1}{2}$	0	∓ 1	1192	$< 1.0 \times 10^{-14}$	$\Lambda^0 + \gamma$
			Σ^-	$\overline{\Sigma^-}$	$\frac{1}{2}$	-1	∓ 1	1197	1.482×10^{-10}	$n + \pi^-$
		Xi particle	Ξ^0	$\overline{\Xi^0}$	$\frac{1}{2}$	0	∓ 2	1315	2.96×10^{-10}	$\Lambda^0 + \pi^0$
			Ξ^-	$\overline{\Xi^-}$	$\frac{1}{2}$	∓ 1	∓ 2	1321	1.652×10^{-10}	$\Lambda^0 + \pi^-$
		Omega particle	Ω^-	$\overline{\Omega^-}$	$\frac{3}{2}$	∓ 1	∓ 3	1672	1.3×10^{-10}	$\Xi^0 + \pi^-$

* See (1) "Review of Particle Properties," *Reviews of Modern Physics*, vol. 48, no. 2, Part II, April (1976).
(2) "Quarks with Color and Flavor," by Sheldon Lee Glashow, *Scientific American*, October (1975).
(3) "The New Elementary Particles and Charm," by Lewis Ryder, *Physics Education*, January (1976) for fuller information. Particle physics is one of the sharp cutting edges of contemporary physics.

Conversion factors may be read off directly from the tables. For example, 1 degree $= 2.778 \times 10^{-3}$ revolutions, so $16.7° = 16.7 \times 2.778 \times 10^{-3}$ rev. The SI quantities are capitalized. The prefix "ab" refers to electromagnetic units (emu); "stat" refers to electrostatic units (esu). Adapted in part from G. Shortley and D. Williams, *Elements of Physics*, Prentice-Hall, Englewood Cliffs, N.J., 1965.

Plane angle

	°	′	″	RADIAN	rev
1 degree =	1	60	3600	1.745×10^{-2}	2.778×10^{-3}
1 minute =	1.667×10^{-2}	1	60	2.909×10^{-4}	4.630×10^{-5}
1 second =	2.778×10^{-4}	1.667×10^{-2}	1	4.848×10^{-6}	7.716×10^{-7}
1 RADIAN =	57.30	3438	2.063×10^{5}	1	0.1592
1 revolution =	360	2.16×10^{4}	1.296×10^{6}	6.283	1

Solid angle

1 sphere = 4π steradians = 12.57 steradians

Length

	cm	METER	km	in.	ft	mi
1 centimeter =	1	10^{-2}	10^{-5}	0.3937	3.281×10^{-2}	6.214×10^{-6}
1 METER =	100	1	10^{-3}	39.3	3.281	6.214×10^{-4}
1 kilometer =	10^{5}	1000	1	3.937×10^{4}	3281	0.6214
1 inch =	2.540	2.540×10^{-2}	2.540×10^{-5}	1	8.333×10^{-2}	1.578×10^{-5}
1 foot =	30.48	0.3048	3.048×10^{-4}	12	1	1.894×10^{-4}
1 mile =	1.609×10^{5}	1609	1.609	6.336×10^{4}	5280	1

1 angstrom = 10^{-10} m
1 nautical mile = 1852 m = 1.151 miles = 6076 ft
1 light-year = 9.4600×10^{12} km
1 parsec = 3.084×10^{13} km
1 fathom = 6 ft
1 yard = 3 ft
1 rod = 16.5 ft
1 mil = 10^{-3} in.

Area

	METER²	cm²	ft²	in.²	circ mil
1 SQUARE METER =	1	10^{4}	10.76	1550	1.974×10^{9}
1 square centimeter =	10^{-4}	1	1.076×10^{-3}	0.1550	1.974×10^{5}
1 square foot =	9.290×10^{-2}	929.0	1	144	1.833×10^{8}
1 square inch =	6.452×10^{-4}	6.452	6.944×10^{-3}	1	1.273×10^{6}
1 circular mil =	5.067×10^{-10}	5.067×10^{-6}	5.454×10^{-9}	7.854×10^{-7}	1

1 square mile = 2.788×10^{8} ft² = 640 acres   1 acre = 43,600 ft²
1 barn = 10^{-28} m²

Volume

	METER³	cm³	li	ft³	in.³
1 CUBIC METER =	1	10^6	1000	35.31	6.102×10^4
1 cubic centimeter =	10^{-6}	1	1.000×10^{-3}	3.531×10^{-5}	6.102×10^{-2}
1 liter =	1.000×10^{-3}	1000	1	3.531×10^{-2}	61.02
1 cubic foot =	2.832×10^{-2}	2.832×10^4	28.32	1	1728
1 cubic inch =	1.639×10^{-5}	16.39	1.639×10^{-2}	5.787×10^{-4}	1

1 U. S. fluid gallon = 4 U. S. fluid quarts = 8 U. S. pints = 128 U. S. fluid ounces = 231 in.³
1 British imperial gallon = 277.4 in.³   1 liter = 10^{-3} m³.

Mass

Quantities in the shaded areas are not mass units but are often used as such. When we write, for example, 1 kg "=" 2.205 lb this means that a kilogram is a *mass* that *weighs* 2.205 pounds under standard condition of gravity (g = 9.80665 m/s²).

	gm	KG	slug	u	oz	lb	ton
1 gram =	1	0.001	6.852×10^{-5}	6.024×10^{23}	3.527×10^{-2}	2.205×10^{-3}	1.102×10^{-6}
1 KILOGRAM =	1000	1	6.852×10^{-2}	6.024×10^{26}	35.27	2.205	1.102×10^{-3}
1 slug =	1.459×10^4	14.59	1	8.789×10^{27}	514.8	32.17	1.609×10^{-2}
1 u =	1.660×10^{-24}	1.660×10^{-27}	1.137×10^{-28}	1	5.855×10^{-26}	3.660×10^{-27}	1.829×10^{-30}
1 ounce =	28.35	2.835×10^{-2}	1.943×10^{-3}	1.708×10^{25}	1	6.250×10^{-2}	3.125×10^{-5}
1 pound =	453.6	0.4536	3.108×10^{-2}	2.732×10^{26}	16	1	0.0005
1 ton =	9.072×10^5	907.2	62.16	5.465×10^{29}	3.2×10^4	2000	1

Density

Quantities in the shaded areas are weight densities and, as such, are dimensionally different from mass densities. See note for mass table.

	slug/ft³	KG/METER³	g/cm³	lb/ft³	lb/in.³
1 slug per ft³ =	1	515.4	0.5154	32.17	1.862×10^{-2}
1 KILOGRAM per METER³ =	1.940×10^{-3}	1	0.001	6.243×10^{-2}	3.613×10^{-5}
1 gram per cm³ =	1.940	1000	1	62.43	3.613×10^{-2}
1 pound per ft³ =	3.108×10^{-2}	16.02	1.602×10^{-2}	1	5.787×10^{-4}
1 pound per in.³ =	53.71	2.768×10^4	27.68	1728	1

Time

	yr	d	h	min	SECOND
1 year =	1	365.2	8.766×10^3	5.259×10^5	3.156×10^7
1 day =	2.738×10^{-3}	1	24	1440	8.640×10^4
1 hour =	1.141×10^{-4}	4.167×10^{-2}	1	60	3600
1 minute =	1.901×10^{-6}	6.944×10^{-4}	1.667×10^{-2}	1	60
1 SECOND =	3.169×10^{-8}	1.157×10^{-5}	2.778×10^{-4}	1.667×10^{-2}	1

Speed

	ft/s	km/h	METER/SECOND	mi/h	cm/s	knot
1 foot per second =	1	1.097	0.3048	0.6818	30.48	0.5925
1 kilometer per hour =	0.9113	1	0.2778	0.6214	27.78	0.5400
1 METER per SECOND =	3.281	3.6	1	2.237	100	1.944
1 mile per hour =	1.467	1.609	0.4470	1	44.70	0.8689
1 centimeter per second =	3.281×10^{-2}	3.6×10^{-2}	0.01	2.237×10^{-2}	1	1.944×10^{-2}
1 knot =	1.688	1.852	0.5144	1.151	51.44	1

1 knot = 1 nautical mi/h   1 mi/min = 88.00 ft/s = 60.00 mi/h

Force

Quantities in the shaded areas are not force units but are often used as such. For instance, if we write 1 gram-force "=" 980.7 dynes, we mean that a gram-*mass* experiences a *force* of 980.7 dynes under standard conditions of gravity ($g = 9.80665$ m/s^2).

	dyne	NEWTON	lb	pdl	gf	kgf
1 dyne =	1	10^{-5}	2.248×10^{-6}	7.233×10^{-5}	1.020×10^{-3}	1.020×10^{-6}
1 NEWTON =	10^5	1	0.2248	7.233	102.0	0.1020
1 pound =	4.448×10^5	4.448	1	32.17	453.6	0.4536
1 poundal =	1.383×10^4	0.1383	3.108×10^{-2}	1	14.10	1.410×10^{-2}
1 gram-force =	980.7	9.807×10^{-3}	2.205×10^{-3}	7.093×10^{-2}	1	0.001
1 kilogram-force =	9.807×10^5	9.807	2.205	70.93	1000	1

Pressure

	atm	dyne/cm^2	inch of water	cm-Hg	PASCAL	lb/in.^{2}	lb/ft^2
1 atmosphere =	1	1.013×10^6	406.8	76	1.013×10^5	14.70	2116
1 dyne per cm^2 =	9.869×10^{-7}	1	4.015×10^{-4}	7.501×10^{-5}	0.1	1.450×10^{-5}	2.089×10^{-3}
1 inch of water[a] at 4° C =	2.458×10^{-3}	2491	1	0.1868	249.1	3.613×10^{-2}	5.202
1 centimeter of mercury[a] at 0° C =	1.316×10^{-2}	1.333×10^4	5.353	1	1333	0.1934	27.85
1 PASCAL =	9.869×10^{-6}	10	4.015×10^{-3}	7.501×10^{-4}	1	1.450×10^{-4}	2.089×10^{-2}
1 pound per in.^{2} =	6.805×10^{-2}	6.895×10^4	27.68	5.171	6.895×10^3	1	144
1 pound per ft^2 =	4.725×10^{-4}	478.8	0.1922	3.591×10^{-2}	47.88	6.944×10^{-3}	1

[a] Where the acceleration of gravity has the standard value 9.80665 m/s^2.

1 bar = 10^6 dyne/cm^2 = 0.1 MPa   1 millibar = 10^3 dyne/cm^2 = 10^2 Pa

Energy, work, heat

Quantities in the shaded areas are not properly energy units but are included for convenience. They arise from the relativistic mass-energy equivalence formula $E = mc^2$ and represent the energy released if a kilogram or unified atomic mass unit (u) is completely converted to energy.

	Btu	erg	ft·lb	hp·h	JOULE	cal	kW·h	eV	MeV	kg	u
1 British thermal unit =	1	1.055×10^{10}	777.9	3.929×10^{-4}	1055	252.0	2.930×10^{-4}	6.585×10^{21}	6.585×10^{15}	1.174×10^{-14}	7.074×10^{12}
1 erg =	9.481×10^{-11}	1	7.376×10^{-8}	3.725×10^{-14}	10^{-7}	2.389×10^{-8}	2.778×10^{-14}	6.242×10^{11}	6.242×10^{5}	1.113×10^{-24}	670.5
1 foot-pound =	1.285×10^{-3}	1.356×10^{7}	1	5.051×10^{-7}	1.356	0.3239	3.766×10^{-7}	8.464×10^{18}	8.464×10^{12}	1.509×10^{-17}	9.092×10^{9}
1 horsepower-hour =	2545	2.685×10^{13}	1.980×10^{6}	1	2.685×10^{6}	6.414×10^{5}	0.7457	1.676×10^{25}	1.676×10^{19}	2.988×10^{-11}	1.800×10^{16}
1 JOULE =	9.481×10^{-4}	10^{7}	0.7376	3.725×10^{-7}	1	0.2389	2.778×10^{-7}	6.242×10^{18}	6.242×10^{12}	1.113×10^{-17}	6.705×10^{9}
1 calorie =	3.968×10^{-3}	4.186×10^{7}	3.087	1.559×10^{-6}	4.186	1	1.163×10^{-6}	2.613×10^{19}	2.613×10^{13}	4.659×10^{-17}	2.807×10^{10}
1 kilowatt-hour =	3413	3.6×10^{13}	2.655×10^{6}	1.341	3.6×10^{6}	8.601×10^{5}	1	2.247×10^{25}	2.247×10^{19}	4.007×10^{-11}	2.414×10^{16}
1 electron volt =	1.519×10^{-22}	1.602×10^{-12}	1.182×10^{-19}	5.967×10^{-26}	1.602×10^{-19}	3.827×10^{-20}	4.450×10^{-26}	1	10^{-6}	1.783×10^{-36}	1.074×10^{-9}
1 million electron volts =	1.519×10^{-16}	1.602×10^{-6}	1.182×10^{-13}	5.967×10^{-20}	1.602×10^{-13}	3.827×10^{-14}	4.450×10^{-20}	10^{6}	1	1.783×10^{-30}	1.074×10^{-3}
1 kilogram =	8.521×10^{13}	8.987×10^{23}	6.629×10^{16}	3.348×10^{10}	8.987×10^{16}	2.147×10^{16}	2.497×10^{10}	5.610×10^{35}	5.610×10^{29}	1	6.025×10^{26}
1 unified atomic mass unit =	1.415×10^{-13}	1.492×10^{-3}	1.100×10^{-10}	5.558×10^{-17}	1.492×10^{-10}	3.564×10^{-11}	4.145×10^{-17}	9.31×10^{8}	931.0	1.660×10^{-27}	1

Power

	Btu/h	ft·lb/s	hp	cal/s	kW	WATT
1 British thermal unit per hour =	1	0.2161	3.929×10^{-4}	7.000×10^{-2}	2.930×10^{-4}	0.2930
1 foot-pound per second =	4.628	1	1.818×10^{-3}	0.3239	1.356×10^{-3}	1.356
1 horsepower =	2545	550	1	178.2	0.7457	745.7
1 calorie per second =	14.29	3.087	5.613×10^{-3}	1	4.186×10^{-3}	4.186
1 kilowatt =	3413	737.6	1.341	238.9	1	1000
1 WATT =	3.413	0.7376	1.341×10^{-3}	0.2389	0.001	1

Charge

	abcoul	A·h	COULOMB	statcoul
1 abcoulomb =	1	2.778×10^{-3}	10	2.998×10^{10}
1 ampere-hour =	360	1	3600	1.079×10^{13}
1 COULOMB =	0.1	2.778×10^{-4}	1	2.998×10^{9}
1 statcoulomb =	3.336×10^{-11}	9.266×10^{-14}	3.336×10^{-10}	1

1 electronic charge = 1.602×10^{-19} coulomb

Current

	abamp	AMPERE	statamp
1 abampere =	1	10	2.998×10^{10}
1 AMPERE =	0.1	1	2.998×10^9
1 statampere =	3.336×10^{-11}	3.336×10^{-10}	1

Potential, electromotive force

	abvolt	VOLT	statvolt
1 abvolt =	1	10^{-8}	3.336×10^{-11}
1 VOLT =	10^8	1	3.336×10^{-3}
1 statvolt =	2.998×10^{10}	299.8	1

Resistance

	abohm	OHM	statohm
1 abohm =	1	10^{-9}	1.113×10^{-21}
1 OHM =	10^9	1	1.113×10^{-12}
1 statohm =	8.987×10^{20}	8.987×10^{11}	1

Capacitance

	abf	FARAD	μF	statf
1 abfarad =	1	10^9	10^{15}	8.987×10^{20}
1 FARAD =	10^{-9}	1	10^6	8.987×10^{11}
1 microfarad =	10^{-15}	10^{-6}	1	8.987×10^5
1 statfarad =	1.113×10^{-21}	1.113×10^{-12}	1.113×10^{-6}	1

Inductance

	abhenry	HENRY	μH	mH	stathenry
1 abhenry =	1	10^{-9}	0.001	10^{-6}	1.113×10^{-21}
1 HENRY =	10^9	1	10^6	1000	1.113×10^{-12}
1 microhenry =	1000	10^{-6}	1	0.001	1.113×10^{-18}
1 millihenry =	10^6	0.001	1000	1	1.113×10^{-15}
1 stathenry =	8.987×10^{20}	8.987×10^{11}	8.987×10^{17}	8.987×10^{14}	1

Magnetic flux

	maxwell	WEBER
1 maxwell =	1	10^{-8}
1 WEBER =	10^8	1

Magnetic field

	gauss	TESLA	milligauss
1 gauss =	1	10^{-4}	1000
1 TESLA =	10^4	1	10^7
1 milligauss =	0.001	10^{-7}	1

1 tesla = 1 weber/meter^2

Mathematical Signs and Symbols

$=$ equals
$\cong$ equals approximately
$\neq$ is not equal to
$\equiv$ is identical to, is defined as
$>$ is greater than ($\gg$ is much greater than)
$<$ is less than ($\ll$ is much less than)
$\geq$ is more than or equal to (or, is no less than)
$\leq$ is less than or equal to (or, is no more than)
$\pm$ plus or minus ($\sqrt{4} = \pm 2$)
$\propto$ is proportional to (Hooke's law: $F \propto x$, or $F = -kx$)
Σ the sum of
$\bar{x}$ the average value of x

The Greek Alphabet

Alpha	A	α		Nu	N	ν
Beta	B	β		Xi	Ξ	ξ
Gamma	Γ	γ		Omicron	O	o
Delta	Δ	δ		Pi	Π	π
Epsilon	E	ϵ		Rho	P	ρ
Zeta	Z	ζ		Sigma	Σ	σ
Eta	H	η		Tau	T	τ
Theta	Θ	θ		Upsilon	Y	υ
Iota	I	ι		Phi	Φ	ϕ, φ
Kappa	K	κ		Chi	X	χ
Lambda	Λ	λ		Psi	Ψ	ψ
Mu	M	μ		Omega	Ω	ω

Geometry

Circle of radius r: circumference $= 2\pi r$; area $= \pi r^2$.
Sphere of radius r: area $= 4\pi r^2$; volume $= \frac{4}{3}\pi r^3$.
Right circular cylinder of radius r and height h: area $= 2\pi r^2 + 2\pi rh$;
volume $= \pi r^2 h$.

Quadratic Formula

If $ax^2 + bx + c = 0$, then $x = \dfrac{-b \pm \sqrt{b^2 - 4ac}}{2a}$.

Trigonometric Functions of Angle θ

$$\sin \theta = \frac{y}{r} \qquad \cos \theta = \frac{x}{r}$$

$$\tan \theta = \frac{y}{x} \qquad \cot \theta = \frac{x}{y}$$

$$\sec \theta = \frac{r}{x} \qquad \csc \theta = \frac{r}{y}$$

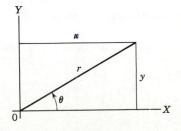

Pythagorean Theorem

$$x^2 + y^2 = r^2$$

Trigonometric Identities

$$\sin^2\theta + \cos^2\theta = 1 \qquad \sec^2\theta - \tan^2\theta = 1 \qquad \csc^2\theta - \cot^2\theta = 1$$

$$\sin 2\theta = 2\sin\theta\cos\theta$$

$$\cos 2\theta = \cos^2\theta - \sin^2\theta = 2\cos^2\theta - 1 = 1 - 2\sin^2\theta$$

$$\sin\theta = \frac{e^{i\theta} - e^{-i\theta}}{2i} \qquad \cos\theta = \frac{e^{i\theta} + e^{-i\theta}}{2}$$

$$e^{\pm i\theta} = \cos\theta \pm i\sin\theta$$

$$\sin(\alpha \pm \beta) = \sin\alpha\cos\beta \pm \cos\alpha\sin\beta$$

$$\cos(\alpha \pm \beta) = \cos\alpha\cos\beta \mp \sin\alpha\sin\beta$$

$$\tan(\alpha \pm \beta) = \frac{\tan\alpha \pm \tan\beta}{1 \mp \tan\alpha\tan\beta}$$

$$\sin\alpha \pm \sin\beta = 2\sin\tfrac{1}{2}(\alpha \pm \beta)\cos\tfrac{1}{2}(\alpha \mp \beta)$$

Taylor's Series

$$f(x_0 + x) = f(x_0) + f'(x_0)x + f''(x_0)\frac{x^2}{2!} + f'''(x_0)\frac{x^3}{3!} + \cdots$$

Binomial Expansion

$$(1 + x)^n = 1 + \frac{nx}{1!} + \frac{n(n-1)}{2!}x^2 + \cdots$$

Exponential Expansion

$$e^x = 1 + x + \frac{x^2}{2!} + \frac{x^3}{3!} + \cdots$$

Logarithmic Expansion

$$\ln(1 + x) = x - \tfrac{1}{2}x^2 + \tfrac{1}{3}x^3 - \cdots$$

Trigonometric Expansions (θ in radians)

$$\sin\theta = \theta - \frac{\theta^3}{3!} + \frac{\theta^5}{5!} - \cdots$$

$$\cos\theta = 1 - \frac{\theta^2}{2!} + \frac{\theta^4}{4!} - \cdots$$

Derivatives and Indefinite Integrals

In what follows, the letters u and v stand for any functions of x, and a and m are constants. To each of the integrals should be added an arbitrary constant of integration. The *Handbook of Chemistry and Physics* (Chemical Rubber Publishing Co.) gives a more extensive tabulation.

1. $\dfrac{dx}{dx} = 1$

2. $\dfrac{d}{dx}(au) = a\dfrac{du}{dx}$

3. $\dfrac{d}{dx}(u + v) = \dfrac{du}{dx} + \dfrac{dv}{dx}$

4. $\dfrac{d}{dx}x^m = mx^{m-1}$

1. $\int dx = x$

2. $\int au\,dx = a\int u\,dx$

3. $\int(u + v)\,dx = \int u\,dx + \int v\,dx$

4. $\int x^m\,dx = \dfrac{x^{m+1}}{m+1} \qquad (m \neq -1)$

5. $\dfrac{d}{dx} \ln x = \dfrac{1}{x}$

5. $\displaystyle\int \dfrac{dx}{x} = \ln |x|$

6. $\dfrac{d}{dx} (uv) = u \dfrac{dv}{dx} + v \dfrac{du}{dx}$

6. $\displaystyle\int u \dfrac{dv}{dx} dx = uv - \int v \dfrac{du}{dx} dx$

7. $\dfrac{d}{dx} e^x = e^x$

7. $\displaystyle\int e^x dx = e^x$

8. $\dfrac{d}{dx} \sin x = \cos x$

8. $\int \sin x\, dx = - \cos x$

9. $\dfrac{d}{dx} \cos x = - \sin x$

9. $\int \sin x\, dx = \sin x$

10. $\dfrac{d}{dx} \tan x = \sec^2 x$

10. $\int \tan x\, dx = \ln |\sec x|$

11. $\dfrac{d}{dx} \cot x = - \csc^2 x$

11. $\int \cot x\, dx = \ln |\sin x|$

12. $\dfrac{d}{dx} \sec x = \tan x \sec x$

12. $\int \sec x\, dx = \ln |\sec x + \tan x|$

13. $\dfrac{d}{dx} \csc x = - \cot x \csc x$

13. $\int \csc x\, dx = \ln |\csc x - \cot x|$

14. $\dfrac{d}{dx} \arctan x = \dfrac{1}{1 + x^2}$

14. $\displaystyle\int \dfrac{dx}{1 + x^2} = \arctan x$

15. $\dfrac{d}{dx} \arcsin x = \dfrac{1}{\sqrt{1 - x^2}}$

15. $\displaystyle\int \dfrac{dx}{\sqrt{1 - x^2}} = \arcsin x$

16. $\dfrac{d}{dx} \operatorname{arcsec} x = \dfrac{1}{x\sqrt{x^2 - 1}}$

16. $\displaystyle\int \dfrac{dx}{x\sqrt{x^2 - 1}} = \operatorname{arcsec} x$

Vector Products

Let $\mathbf{i}, \mathbf{j}, \mathbf{k}$ be unit vectors in the x, y, z directions. Then

$$\mathbf{i} \cdot \mathbf{i} = \mathbf{j} \cdot \mathbf{j} = \mathbf{k} \cdot \mathbf{k} = 1, \qquad \mathbf{i} \cdot \mathbf{j} = \mathbf{j} \cdot \mathbf{k} = \mathbf{k} \cdot \mathbf{i} = 0,$$

$$\mathbf{i} \times \mathbf{i} = \mathbf{j} \times \mathbf{j} = \mathbf{k} \times \mathbf{k} = 0,$$

$$\mathbf{i} \times \mathbf{j} = \mathbf{k}, \qquad \mathbf{j} \times \mathbf{k} = \mathbf{i}, \qquad \mathbf{k} \times \mathbf{i} = \mathbf{j}.$$

Any vector $\mathbf{a}$ with components a_x, a_y, a_z along the x, y, z axes can be written

$$\mathbf{a} = a_x\mathbf{i} + a_y\mathbf{j} + a_z\mathbf{k}.$$

Let $\mathbf{a}, \mathbf{b}, \mathbf{c}$ be arbitrary vectors with magnitudes a, b, c. Then

$$\mathbf{a} \times (\mathbf{b} + \mathbf{c}) = \mathbf{a} \times \mathbf{b} + \mathbf{a} \times \mathbf{c}$$

$$(s\mathbf{a}) \times \mathbf{b} = \mathbf{a} \times (s\mathbf{b}) = s(\mathbf{a} \times \mathbf{b}) \qquad (s - \text{a scalar}).$$

Let θ be the smaller of the two angles between $\mathbf{a}$ and $\mathbf{b}$. Then

$$\mathbf{a} \cdot \mathbf{b} = \mathbf{b} \cdot \mathbf{a} = a_x b_x + a_y b_y + a_z b_z = ab \cos \theta$$

$$\mathbf{a} \times \mathbf{b} = -\mathbf{b} \times \mathbf{a} = \begin{vmatrix} \mathbf{i} & \mathbf{j} & \mathbf{k} \\ a_x & a_y & a_z \\ b_x & b_y & b_z \end{vmatrix} = (a_y b_z - b_y a_z)\mathbf{i} + (a_z b_x - b_z a_x)\mathbf{j} + (a_x b_y - b_x a_y)\mathbf{k}$$

$$|\mathbf{a} \times \mathbf{b}| = ab \sin \theta$$

$$\mathbf{a} \cdot (\mathbf{b} \times \mathbf{c}) = \mathbf{b} \cdot (\mathbf{c} \times \mathbf{a}) = \mathbf{c} \cdot (\mathbf{a} \times \mathbf{b})$$

$$\mathbf{a} \times (\mathbf{b} \times \mathbf{c}) = (\mathbf{a} \cdot \mathbf{c})\mathbf{b} - (\mathbf{a} \cdot \mathbf{b})\mathbf{c}$$

Degrees	Radians	Sine	Tangent	Cotangent	Cosine		
0	0	0	0	∞	1.0000	1.5708	**90**
1	.0175	.0175	.0175	57.290	.9998	1.5533	89
2	.0349	.0349	.0349	28.636	.9994	1.5359	88
3	.0524	.0523	.0524	19.081	.9986	1.5184	87
4	.0698	.0698	.0699	14.301	.9976	1.5010	86
5	.0873	.0872	.0875	11.430	.9962	1.4835	**85**
6	.1047	.1045	.1051	9.5144	.9945	1.4661	84
7	.1222	.1219	.1228	8.1443	.9925	1.4486	83
8	.1396	.1392	.1405	7.1154	.9903	1.4312	82
9	.1571	.1564	.1584	6.3138	.9877	1.4137	81
10	.1745	.1736	.1763	5.6713	.9848	1.3963	**80**
11	.1920	.1908	.1944	5.1446	.9816	1.3788	79
12	.2094	.2079	.2126	4.7046	.9781	1.3614	78
13	.2269	.2250	.2309	4.3315	.9744	1.3439	77
14	.2443	.2419	.2493	4.0108	.9703	1.3265	76
15	.2618	.2588	.2679	3.7321	.9659	1.3090	**75**
16	.2793	.2756	.2867	3.4874	.9613	1.2915	74
17	.2967	.2924	.3057	3.2709	.9563	1.2741	73
18	.3142	.3090	.3249	3.0777	.9511	1.2566	72
19	.3316	.3256	.3443	2.9042	.9455	1.2392	71
20	.3491	.3420	.3640	2.7475	.9397	1.2217	**70**
21	.3665	.3584	.3839	2.6051	.9336	1.2043	69
22	.3840	.3746	.4040	2.4751	.9272	1.1868	68
23	.4014	.3907	.4245	2.3559	.9205	1.1694	67
24	.4189	.4067	.4452	2.2460	.9135	1.1519	66
25	.4363	.4226	.4663	2.1445	.9063	1.1345	**65**
26	.4538	.4384	.4877	2.0503	.8988	1.1170	64
27	.4712	.4540	.5095	1.9626	.8910	1.0996	63
28	.4887	.4695	.5317	1.8807	.8829	1.0821	62
29	.5061	.4848	.5543	1.8040	.8746	1.0647	61
30	.5236	.5000	.5774	1.7321	.8660	1.0472	**60**
31	.5411	.5150	.6009	1.6643	.8572	1.0297	59
32	.5585	.5299	.6249	1.6003	.8480	1.0123	58
33	.5760	.5446	.6494	1.5399	.8387	.9948	57
34	.5934	.5592	.6745	1.4826	.8290	.9774	56
35	.6109	.5736	.7002	1.4281	.8192	.9599	**55**
36	.6283	.5878	.7265	1.3764	.8090	.9425	54
37	.6458	.6018	.7536	1.3270	.7986	.9250	53
38	.6632	.6157	.7813	1.2799	.7880	.9076	52
39	.6807	.6293	.8098	1.2349	.7771	.8901	51
40	.6981	.6428	.8391	1.1918	.7660	.8727	**50**
41	.7156	.6561	.8693	1.1504	.7547	.8552	49
42	.7330	.6691	.9004	1.1106	.7431	.8378	48
43	.7505	.6820	.9325	1.0724	.7314	.8203	47
44	.7679	.6947	.9657	1.0355	.7193	.8029	46
45	.7854	.7071	1.0000	1.0000	.7071	.7854	**45**
	Cosine	Cotangent	Tangent	Sine		Radians	Degrees

1901	Wilhelm Konrad Röntgen	1845–1923	for the discovery of the remark-able rays subsequently named after him
1902	Hendrik Antoon Lorentz Pieter Zeeman	1853–1928 1865–1943	for their researches into the influence of magnetism upon radiation phenomena
1903	Antoine Henri Becquerel	1852–1908	for his discovery of spontaneous radioactivity

* See *Nobel Lectures, Physics,* 1901–1970, Elsevier Publishing Company, for the Nobel presentations, lectures and biographies. The attributions are, almost without exception, quotations from the Nobel Prize citations.

	Pierre Curie Marie Sklowdowska-Curie	1859–1906 1867–1934	for their joint researches on the radiation phenomena discovered by Professor Henri Becquerel
1904	Lord Rayleigh (John William Strutt)	1842–1919	for his investigations of the densities of the most important gases and for his discovery of argon
1905	Philipp Eduard Anton von Lenard	1862–1947	for his work on cathode rays
1906	Joseph John Thomson	1856–1940	for his theoretical and experimental investigations on the conduction of electricity by gases
1907	Albert Abraham Michelson	1852–1931	for his optical precision instruments and metrological investigations carried out with their aid
1908	Gabriel Lippmann	1845–1921	for his method of reproducing colors photographically based on the phenomena of interference
1909	Guglielmo Marconi Carl Ferdinand Braun	1874–1937 1850–1918	for their contributions to the development of wireless telegraphy
1910	Johannes Diderik van der Waals	1837–1923	for his work on the equation of state for gases and liquids
1911	Wilhelm Wien	1864–1928	for his discoveries regarding the laws governing the radiation of heat
1912	Nils Gustaf Dalén	1869–1937	for his invention of automatic regulators for use in conjunction with gas accumulators for illuminating lighthouses and buoys
1913	Heike Kamerlingh Onnes	1853–1926	for his investigations of the properties of matter at low temperatures which led, *inter alia*, to the production of liquid helium
1914	Max von Laue	1879–1960	for his discovery of the diffraction of Röntgen rays by crystals
1915	William Henry Bragg William Lawrence Bragg	1862–1942 1890–1971	for their services in the analysis of crystal structure by means of Röntgen rays
1917	Charles Glover Barkla	1877–1944	for his discovery of the characteristic Röntgen radiation of the elements
1918	Max Planck	1858–1947	for his discovery of energy quanta
1919	Johannes Stark	1874–1957	for his discovery of the Doppler effect in canal rays and the splitting of spectral lines in electric fields
1920	Charles-Édouard Guillaume	1861–1938	for the service he has rendered to precision measurements in Physics by his discovery of anomalies in nickel steel alloys
1921	Albert Einstein	1879–1955	for his services to Theoretical Physics, and especially for his discovery of the law of the photoelectric effect
1922	Niels Bohr	1885–1962	for the investigation of the structure of atoms, and of the radiation emanating from them
1923	Robert Andrews Millikan	1868–1953	for his work on the elementary charge of electricity and on the photoelectric effect
1924	Karl Manne Georg Siegbahn	1886–1954	for his discoveries and research in the field of x-ray spectroscopy
1925	James Franck Gustav Hertz	1882–1964 1887–1975	for their discovery of the laws governing the impact of an electron upon an atom

1926	Jean Baptiste Perrin	1870–1942	for his work on the discontinuous structure of matter, and especially for his discovery of sedimentation equillibrium
1927	Arthur Holly Compton	1892–1962	for his discovery of the effect named after him
	Charles Thomson Rees Wilson	1869–1959	for his method of making the paths of electrically charged particles visible by condensation of vapor
1928	Owen Willans Richardson	1879–1959	for his work on the thermionic phenomenon and especially for the discovery of the law named after him
1929	Prince Louis-Victor de Broglie	1892–	for his discovery of the wave nature of electrons
1930	Sir Chandrasekhara Venkata Raman	1888–1970	for his work on the scattering of light and for the discovery of the effect named after him
1932	Werner Heisenberg	1901–1976	for the creation of quantum mechanics, the application of which has, among other things, led to the discovery of the allotropic forms of hydrogen
1933	Erwin Schrödinger Paul Adrien Maurice Dirac	1887–1961 1902–	for the discovery of new productive forms of atomic theory
1935	James Chadwick	1891–1974	for his discovery of the neutron
1936	Victor Franz Hess	1883–1964	for his discovery of cosmic radiation
	Carl David Anderson	1905–	for his discovery of the positron
1937	Clinton Joseph Davisson George Paget Thomson	1881–1958 1892–1975	for their experimental discovery of the diffraction of electrons by crystals
1938	Enrico Fermi	1901–1954	for his demonstrations of the existence of new radioactive elements produced by neutron irradiation, and for his related discovery of nuclear reactions brought about by slow neutrons
1939	Ernest Orlando Lawrence	1901–1958	for the invention and development of the cyclotron and for results obtained with it, especially with regard to artificial radioactive elements
1943	Otto Stern	1888–1969	for his contribution to the development of the molecular ray method and his discovery of the magnetic moment of the proton
1944	Isidor Isaac Rabi	1898–	for his resonance method for recording the magnetic properties of atomic nuclei
1945	Wolfgang Pauli	1900–1958	for the discovery of the Exclusion Principle, also called the Pauli Principle
1946	Percy Williams Bridgman	1882–1961	for the invention of an apparatus to produce extremely high pressures, and for the discoveries he made therewith in the field of high-pressure physics
1947	Sir Edward Victor Appleton	1892–1965	for his investigations of the physics of the upper atmosphere, especially for the discovery of the so-called Appleton layer
1948	Patrick Maynard Stuart Blackett	1897–1974	for his development of the Wilson cloud chamber method, and his discoveries therewith in the fields of nuclear physics and cosmic radiation

1949	Hideki Yukawa	1907–	for his prediction of the existence of mesons on the basis of theoretical work on nuclear forces
1950	Cecil Frank Powell	1903–1969	for his development of the photographic method of studying nuclear processes and his discoveries regarding mesons made with this method
1951	Sir John Douglas Cockcroft Ernest Thomas Sinton Walton	1897–1967 1903–	for their pioneer work on the transmutation of atomic nuclei by artificially accelerated atomic particles
1952	Felix Bloch Edward Mills Purcell	1905– 1912–	for their development of new methods for nuclear magnetic precision methods and discoveries in connection therewith
1953	Frits Zernike	1888–1966	for his demonstration of the phase-contrast method, especially for his invention of the phase-contrast microscope
1954	Max Born	1882–1970	for his fundamental research in quantum mechanics, especially for his statistical interpretation of the wave function
	Walther Bothe	1891–1957	for the coincidence method and his discoveries made therewith
1955	Willis Eugene Lamb	1913–	for his discoveries concerning the fine structure of the hydrogen spectrum
	Polykarp Kusch	1911–	for his precision determination of the magnetic moment of the electron
1956	William Shockley John Bardeen Walter Houser Brattain	1910– 1908– 1902–	for their researches on semiconductors and their discovery of the transistor effect
1957	Chen Ning Yang Tsung Dao Lee	1922– 1926–	for their penetrating investigation of the so-called parity laws which has led important discoveries regarding the elementary particles
1958	Pavel Aleksejevič Čerenkov Il' ja Michajlovič Frank Igor' Evgen'evič Tamm	1904– 1908– 1895–1971	for the discovery and the interpretation of the Čerenkov effect
1959	Emilio Gino Segrè Owen Chamberlain	1905– 1920–	for their discovery of the antiproton
1960	Donald Arthur Glaser	1926–	for the invention of the bubble chamber
1961	Robert Hofstadter	1915–	for his pioneering studies of electron scattering in atomic nuclei and for his thereby achieved discoveries concerning the structure of the nucleons
	Rudolf Ludwig Mössbauer	1929–	for his researches concerning the resonances absorption of γ-radiation and his discovery in this connection of the effect which bears his name
1962	Lev Davidovič Landau	1908–	for his pioneering theories of condensed matter, especially liquid helium
1963	Eugene P. Wigner	1902–	for his contributions to the theory of the atomic nucleus and the elementary particles, particularly through the discovery and application of fundamental symmetry principles
	Maria Goeppert Mayer J. Hans D. Jensen	1906–1972 1907–1973	for their discoveries concerning nuclear shell structure

1964	Charles H. Townes Nikolai G. Basov Alexander M. Prochorov	1915– 1922– 1916–	for fundamental work in the field of quantum electronics which has led to the construction of oscillators and amplifiers based on the maser-laser principle
1965	Sin-Itiro Tomonaga Julian Schwinger Richard P. Feynman	1906– 1918– 1918–	for their fundamental work in quantum electrodynamics, with deep-ploughing consequences for the physics of elementary particles
1966	Alfred Kastler	1902–	for the discovery and development of optical methods for studying Hertzian resonance in atoms
1967	Hans Albrecht Bethe	1906–	for his contributions to the theory of nuclear reactions, especially his discoveries concerning the energy production in stars
1968	Luis W. Alvarez	1911–	for his decisive contribution to elementary particle physics, in particular the discovery of a large number of resonance states, made possible through his development of the technique of using hydrogen bubble chamber and data analysis
1969	Murray Gell-Mann	1929–	for his contributions and discoveries concerning the classification of elementary particles and their interactions
1970	Hannes Alvén	1908–	for fundamental work and discoveries in magneto-hydrodynamics with fruitful applications in different parts of plasma physics
	Louis Néel	1904–	for fundamental work and discoveries concerning antiferromagnetism and ferrimagnetism which have led to important applications in solid state physics
1971	Dennis Gabor	1900–	for his discovery of the principles of holography
1972	John Bardeen Leon N. Cooper J. Robert Schrieffer	1908– 1930– 1931–	for their development of a theory of superconductivity
1973	Leo Esaki	1925–	for his discovery of tunneling in semiconductors
	Ivar Giaever	1929–	for his discovery of tunneling in superconductors
	Brian D. Josephson	1940–	for his theoretical prediction of the properties of a super-current through a tunnel barrier
1974	Antony Hewish	1924–	for the discovery of pulsars
	Sir Martin Ryle	1918–	for his pioneering work in radioastronomy
1975	Aage Bohr Ben Mottelson James Rainwater	1922– 1926– 1917–	for the discovery of the connection between collective motion and particle motion and the development of the theory of the structure of the atomic nucleus based on this connection
1976	Burton Richter Samuel Chao Chung Ting	1931– 1936–	for their (independent) discovery of an important fundamental particle.
1977	Philip Warren Anderson Nevill Francis Mott John Hasbrouck Van Vleck	1923– 1905– 1899–	for their fundamental theoretical investigations of the electronic structure of magnetic and disordered solids

Much of the literature of physics is written, and continues to be written, in the Gaussian system of units. In electromagnetism many equations have slightly different forms depending on whether it is intended, as in this book, that mks variables be used or that Gaussian variables be used. Equations in this book can be cast in Gaussian form by replacing the symbols listed below under "SI" by those listed under "Gaussian." For example, Eq. 37-26,

$$\mathbf{B} = \mu_0(\mathbf{H} + \mathbf{M})$$

becomes

$$\frac{\mathbf{B}}{c} = \left(\frac{4\pi}{c^2}\right)\left(\frac{c}{4\pi}\mathbf{H} + c\mathbf{M}\right)$$

or

$$\mathbf{B} = \mathbf{H} + 4\pi\mathbf{M}$$

in Gaussian form. Symbols used in this book that are not listed below remain unchanged. The quantity c is the speed of light.

APPENDIX L
THE GAUSSIAN SYSTEM OF UNITS

Quantity	SI	Gaussian
Permittivity constant	ϵ_0	$1/4\pi$
Permeability constant	μ_0	$4\pi/c^2$
Electric displacement	$\mathbf{D}$	$\mathbf{D}/4\pi$
Magnetic induction	$\mathbf{B}$	$\mathbf{B}/c$
Magnetic flux	Φ_B	Φ_B/c
Magnetic field strength	$\mathbf{H}$	$c\mathbf{H}/4\pi$
Magnetization	$\mathbf{M}$	$c\mathbf{M}$
Magnetic dipole moment	μ	$c\mu$

In addition to casting the equations in the proper form it is of course necessary to use a consistent set of units in those equations. Below we list some equivalent quantities in SI and Gaussian units. This table can be used to transform units from one system to the other.

Quantity	Symbol	SI	Gaussian system
Length	l	1 m	10^2 cm
Mass	m	1 kg	10^3 g
Time	t	1 s	1 s
Force	$\mathbf{F}$	1 N	10^5 dynes
Work or Energy	W, E	1 J	10^7 ergs
Power	P	1 W	10^7 ergs/s
Charge	q	1 C	3×10^9 statcoul
Current	i	1 A	3×10^9 statamp
Electric field strength	$\mathbf{E}$	1 V/m	$\frac{1}{3} \times 10^{-4}$ statvolt/cm
Electric potential	V	1 V	$\frac{1}{300}$ statvolt
Electric polarization	$\mathbf{P}$	1 C/m²	3×10^5 statcoul/cm²
Electric displacement	$\mathbf{D}$	1 C/m²	$12\pi \times 10^5$ statvolt/cm
Resistance	R	1 Ω	$\frac{1}{9} \times 10^{-11}$ s·cm^{-1}
Capacitance	C	1 F	9×10^{11} cm
Magnetic flux	Φ_B	1 Wb	10^8 maxwells
Magnetic induction	$\mathbf{B}$	1 T	10^4 gauss
Magnetic field strength	$\mathbf{H}$	1 A-turn/m	$4\pi \times 10^{-3}$ oersted
Magnetization	$\mathbf{M}$	1 Wb/m²	$1/4\pi \times 10^4$ gauss
Inductance	I	1 H	$\frac{1}{9} \times 10^{-11}$

All factors of 3 in the above table, apart from exponents, should be replaced by 2.99792458 for accurate work; this arises from the numerical value of the speed of light. For example the SI unit of capacitance

(= 1 farad) is actually $8.987551787 \times 10^{11}$ cm rather than 9 (= 3^2) $\times 10^{11}$ cm as listed above. This example also shows that not only units but also the dimensions of physical quantities may differ between the two systems.

Consult *Classical Electromagnetism*, second edition, by J. D. Jackson (John Wiley and Sons, 1975) for a fuller treatment of units and dimensions.

index

SOME USEFUL NUMBERS

$\sqrt{2} = 1.414$ $\quad$ $\sqrt{3} = 1.732$ $\quad$ $\sqrt{10} = 3.162$ $\quad$ $\pi = 3.142$

$\pi^2 = 9.870$ $\quad$ $\sqrt{\pi} = 1.773$ $\quad$ $\log \pi = 0.4971$ $\quad$ $4\pi = 12.57$

$e = 2.718$ $\quad$ $1/e = 0.3679$ $\quad$ $\log e = 0.4343$ $\quad$ $\ln 2 = 0.6932$

$\sin 30° = \cos 60° = 0.5000$ $\qquad$ $\cot 30° = \tan 60° = 1.7321$

$\cos 30° = \sin 60° = 0.8660$ $\qquad$ $\sin 45° = \cos 45° = 0.7071$

$\tan 30° = \cot 60° = 0.5774$ $\qquad$ $\tan 45° = \cot 45° = 1.0000$

Change of Base

$\log x = \ln x/\ln 10 = 0.4343 \ln x$

$\ln x = \log x/\log e = 2.303 \log x$